2192

LYNDON STATE COLLEGE

Dangerous Properties of Industrial Materials

THIRD EDITION

Dangerous Properties of Industrial Materials

N. IRVING SAX

Director, Radiological Sciences Laboratory, New York State Health Department, Albany, N.Y.

ASSISTED BY: MILTON S. DUNN / BENJAMIN FEINER / JOSEPH J. FITZGERALD / LEONARD J. GOLDWATER

JOHN H. HARLEY / DWIGHT F. METZLER / BERNARD L. OSER / ALEXANDER RIHM / MEREDITH THOMPSON

VNR Van Nostrand Reinhold Company
New York   Cincinnati   Toronto   London   Melbourne

To My Father
Louis Sax

Van Nostrand Reinhold Company Regional Offices:
New York Cincinnati Chicago Millbrae Dallas

Van Nostrand Reinhold Company International Offices:
London Toronto Melbourne

Manufactured in the United States of America

Published by Van Nostrand Reinhold Company
450 West 33rd Street, New York, N.Y. 10001

15 14 13 12 11 10 9 8 7 6 5 4 3

PREFACE TO THIRD EDITION

The purpose of this work is to provide a single source for quick, up-to-date, concise, hazard-analysis information about more than 12,000 common industrial and laboratory materials. The information is set forth in Section 12, General Chemicals. This section, which comprises the body of the book, is designed to expedite the retrieval of hazard information by categorizing the data into:

1 General information about substances listed, such as synonyms, description, formula, and the physical constants.

2 Hazard analyses, which include a toxic hazard rating or toxicology paragraph; a fire hazard rating; an explosion hazard rating, and a disaster hazard rating, to give some idea of the hazards produced when quantities of a material become involved in disasters such as fire, explosion or flood.

3 Countermeasures, or what may be done to mitigate the effects of using a given material, for instance, shipping regulations, storage and handling, first aid measures for exposed individuals, fire-fighting measures, ventilation controls, and personnel protection.

The concise material in Section 12 is all keyed to the textual material in Sections 1-11, which include a number of discussions of ways and means of bringing about the protective measures suggested in Section 12. There is detailed information about the meaning of the terminology used in Section 12.

In recognition of the great level of public interest now focused upon exploitation of the environment to the detriment of public health, Section 4 has been included. It deals with air and water pollution problems, as well as the burgeoning problem of solid waste disposal. The problem of environmental pollution by radioactive materials is covered extensively in Sections 5 and 8. The problem of food additives is taken up in detail in Section 10. All of this has been keyed into the hazard analyses of Section 12.

In order to fit all this material into one book of reasonable size, it has been necessary to forego repetitive explanations and stereotyped instructions, warnings or reasons for a given recommendation. As a result, a vast number of cross references were required; these have been used with care and should not be overlooked by the user of the book, as they are essential in obtaining a complete picture of a given material.

The physical size of the book has been reduced by using smaller type in order to make it easier to handle.

N. Irving Sax

Albany, New York
August, 1968

ACKNOWLEDGMENTS

The author wishes to acknowledge with thanks the invaluable assistance of his wife, Pauline Sax, in all phases of the compilation of the manuscript. He wishes also to acknowledge the assistance of N. Nyman in posting data and checking constants, as well as the help and suggestions of many others.

CONTENTS

SECTION 1

TOXICOLOGY

Leonard J. Goldwater, M. D.
Professor of Occupational Medicine
Columbia University
New York, N. Y.

This section has been prepared for the purpose of providing a certain amount of background for those without special training in industrial toxicology who may be called upon to understand and interpret data which are largely medical in nature.

Toxicity Defined

Toxicity is the ability of a chemical molecule or compound to produce injury once it reaches a susceptible site in or on the body. Toxicity hazard is the probability that injury may be caused by the manner in which the substance is used.

Acute. This term will be used in the medical sense to mean "of short duration." As applied to materials which are inhaled or absorbed through the skin, it refers to a single exposure of a duration measured in seconds, minutes or hours. As applied to materials which are ingested, it refers generally to a single quantity or dose.

Chronic. This term will be used in contrast to "acute" and means "of long duration." As applied to materials which are inhaled or absorbed through the skin, it refers to prolonged or repeated exposures of a duration measured in days, months or years. As applied to materials which are ingested, it refers to repeated doses over a period of days, months, or years. The term "chronic" will not refer to severity of symptoms but will carry the implication of exposures or doses which would be relatively harmless unless extended or repeated over long periods of time (days, months or years). In this book the term "chronic" includes exposures which might also be called "subacute," i.e., somewhere between "acute" and "chronic."

Local. This term refers to the site of action of an agent and means that the action takes place at the point or area of contact. The site may be skin, mucous membranes of the eyes, nose, mouth, throat, or anywhere along the respiratory or gastrointestinal system. Absorption does not necessarily occur.

Systemic. This term refers to a site of action other than the point of contact and presupposes that absorption has taken place. It is possible, however, for toxic agents to be absorbed through a channel (skin, lungs or intestinal canal) and produce later manifestations on one of those channels which are not a result of the original direct contact. Thus it is possible for some agents to produce harmful effects on a single organ or tissue as a result of both "local" and "systemic" actions.

Absorption. A material is said to have been absorbed only when it has gained entry into the blood stream and consequently may be carried to all parts of the body. Absorption requires that a substance pass through the skin, a mucous membrane, or the air sacs (alveoli) of the lungs. It may also be produced by means of a needle (subcutaneous, intravenous, etc.) but this is not of importance in industrial toxicology.

Toxicity Ratings

An explanation of the toxicity ratings is given in the following paragraphs:

U = **Unknown.** This designation is given to substances which fall into one of the following categories:

(a) No toxicity information could be found in the literature and none was known to the authors.

(b) Limited information based on animal experiments was available but in the opinion of the authors this information could not be applied to human exposures. In some cases this information is mentioned so that the reader may know that some experimental work has been done.

(c) Published toxicity data were felt by the authors to be of questionable validity.

0 = No Toxicity. This designation is given to materials which fall into one of the following categories:

(a) Materials which cause no harm under any conditions of use.

(b) Materials which produce toxic effects on humans only under the most unusual conditions or by overwhelming dosage.

1 = Slight Toxicity:

(a) Acute Local. Materials which on single exposures lasting seconds, minutes or hours cause only slight effects on the skin or mucous membranes regardless of the extent of the exposure.

(b) Acute Systemic. Materials which can be absorbed into the body by inhalation, ingestion or through the skin and which produce only slight effects following single exposures lasting seconds, minutes, or hours, or following ingestion of a single dose, regardless of the quantity absorbed or the extent of exposure.

(c) Chronic Local. Materials which on continuous or repeated exposures extending over periods of days, months, or years cause only slight harm to the skin or mucous membranes. The extent of exposure may be great or small.

(d) Chronic Systemic. Materials which can be absorbed into the body by inhalation, ingestion or through the skin and which produce only slight effects following continuous or repeated exposures extending over days, months or years. The extent of the exposure may be great or small.

In general, those substances classified as having "slight toxicity" produce changes in the human body which are readily reversible and which will disappear following termination of exposure, either with or without medical treatment.

2 = Moderate Toxicity:

(a) Acute Local. Materials which on single exposure lasting seconds, minutes or hours cause moderate effects on the skin or mucous membranes. These effects may be the result of intense exposure for a matter of seconds or moderate exposure for a matter of hours.

(b) Acute Systemic. Materials which can be absorbed into the body by inhalation, ingestion or through the skin and which produce moderate effects following single exposures lasting seconds, minutes or hours, or following ingestion of a single dose.

(c) Chronic Local. Materials which on continuous or repeated exposures extending over periods of days, months or years cause moderate harm to the skin or mucous membranes.

(d) Chronic Systemic. Materials which can be absorbed into the body by inhalation, ingestion or through the skin and which produce moderate effects following continuous or repeated exposures extending over periods of days, months or years.

Those substances classified as having "moderate toxicity" may produce irreversible as well as reversible changes in the human body. These changes are not of such severity as to threaten life or produce serious permanent physical impairment.

3 = Severe Toxicity:

(a) Acute Local. Materials which on single exposures lasting seconds or minutes cause injury to skin or mucous membranes of sufficient severity to threaten life or to cause permanent physical impairment or disfigurement.

(b) Acute Systemic. Materials which can be absorbed into the body by inhalation, ingestion or through the skin and which can cause injury of sufficient severity to threaten life following a single exposure lasting seconds, minutes or hours, or following ingestion of a single dose.

(c) Chronic Local. Materials which on continuous or repeated exposures extending over periods of days, months or years can cause injury to skin or mucous membranes of sufficient severity to threaten life or to cause permanent impairment, disfigurement or irreversible change.

(d) Chronic Systemic. Materials which can be absorbed into the body by inhalation, ingestion or through the skin and which can cause death or serious physical impairment following continuous or repeated exposures to small amounts extending over periods of days, months or years.

Toxicology Defined

In simple terms, toxicology may be defined as the study of the action of poisons on the living organism. Industrial toxicology is concerned with the human organism and consequently lies within the broad field of medicine. Since medicine cannot be considered an exact science in the same sense as chemistry, physics or mathematics, *toxicologic phenomena cannot always be predicted with accuracy or explained on the basis of physical or chemical laws.* It is this unpredictability which frequently reduces conclusions and decisions to opinion rather than fact.

Poison Defined

In attempting to define the term "poison," a major consideration is that of *quantity* or *dosage*. A poison may be defined as a substance which causes harm to living tissues when applied in relatively small doses. It is not always easy to make a clear-cut distinction between poisonous and nonpoisonous substances. (See further discussion on p. 7).

Effective Dosage

Generally speaking, industrial toxicology is concerned with the effects of substances which gain entry into some part of the human body. Almost any substance can cause harm when applied directly to the skin. Among the factors which are concerned in effective dosage, the most important are:

(1) Quantity or concentration of the material.

(2) Duration of exposure.

(3) State of dispersion (size of particle or physical state, e.g., dust, fume, gas, etc.).

(4) Affinity for human tissue.

(5) Solubility in human tissue fluids.

(6) Sensitivity of human tissues or organs.

Obviously there are possibilities for wide variations in any of these factors.

Methods of Expressing Effective Dosage

I. Threshold Limit Values. TLV (formerly maximum allowable concentration or MAC). In the USA, threshold limit values (TL or TLV) set by the American Conference of Governmental Industrial Hygienists have received wide acceptance. According to the ACGIH these values "represent conditions under which it is believed that nearly all workers may be repeatedly exposed day after day, without adverse effect." Most of the TLV's refer to time-weighted average concentrations for a normal work day, but some are levels which should not be exceeded at any time.

Closely related to TLV's are the so-called acceptable concentration standards promulgated by the American Standards Association. According to the ASA, these standards are designed to prevent "(1) undesirable changes in body structure or biochemistry; (2) undesirable functional reactions that may have no discernible effects on health; (3) irritation or other adverse sensory effects."

For gases and vapors the TL value is usually expressed in parts per million (ppm), that is, parts of the gas or vapor per million parts of air.

For fumes and mists, and for some dusts, the TL is usually given as milligrams per cubic meter (mg/m^3) or per 10 cubic meters of air.

For some dusts, particularly those containing silica, the TL is usually expressed as millions of particles per cubic foot of air (MPPCF).

Literal application of TL values is dangerous for the following reasons:

(a) A great majority of all published TL values are based either on speculation, opinion or limited experimentation on rats, mice, guinea pigs or other laboratory animals. In very few instances have the values been established firmly on a basis of examinations of human subjects correlated with adequate environmental observations.

(b) Concentrations of toxic or harmful materials in any working environment rarely remain constant throughout a work day. The occurrence of "surges" is well known.

(c) Industrial exposures frequently involve mixtures rather than single compounds. Very little is known about the effects of mixtures.

(d) Individuals vary tremendously in their sensitivity or susceptibility to toxic substances. The factors controlling this variability are not well understood. It should not be assumed that conditions which are safe for some individuals will be safe for all.

(e) There may be a tendency to consider one substance half as toxic as another if the TL is twice as great.

(f) A single TL value is usually given for substances which occur in the form of salts or compounds of different solubility or in different physical states (e.g., lead, mercury).

(g) TL values may be written into legal instruments (laws, codes) and thus be used for purposes for which they are not intended.

If the above limitations are understood and accepted, the published TL values can be used to great practical advantage. Their principal usefulness is in connection with the design of ventilating systems (see Section 2). Ventilation engineers must have a concrete figure to serve as the basis for the design of ventilating equipment. It should not be assumed, however, that attainment of concentrations at or even below the published TL values will necessarily prevent all cases of occupational poisoning; nor should it be assumed that concentrations which exceed the given limits will necessarily result in cases of poisoning. The concept known as Haber's Law, which involves the product of concentration and time ($C \times T = K$), expresses an index of the degree of toxic effect. This too, may be misleading, since the relative effectiveness of large doses over a short period of time may bear little relationship to the effects of smaller doses over a longer time period.

II. Minimal Lethal Dose and LD50 Test. In

experimental toxicology it is common practice to determine the quantity of poison per unit of body weight of an experimental animal which will have a fatal effect. (A scale commonly used is milligrams of poison per kilogram of body weight.) That amount per unit of body weight which will cause even one fatality in a group of experimental animals is known as the minimal lethal dose (MLD). A more commonly used figure in experimental industrial toxicology is the amount which will kill one-half of a group of experimental animals. This is known as the LD50 test (Lethal Dose 50 percent) representing 50 percent fatalities. For application of the LD50 test, see p. 8 Testing for Toxicity.

III. "Range Finding" Test. This approach to determining and expressing the degree of toxicity of chemicals used in industry has been developed primarily by H. F. Smyth, Jr., and his collaborators. Its greatest usefulness is in testing new compounds for which no toxicological information exists. The basis of the test is a comparison of the potency of an unknown compound with that of a more familiar material. This is possible since there are a number of chemicals for which fairly extensive toxicological data are already available. By this technique a certain amount of valuable information can be obtained within a space of about three weeks.

IV. Hazard Rating. This term is used in this book to indicate whether a material has high, moderate, or slight toxicity hazard, or none at all. It is obviously a somewhat crude method but it will serve as a rough guide to the risk involved in exposure to various chemicals until further information can be obtained.

The hazard ratings of industrial chemicals are based on an interpretation of all available information, particularly TL, LD50, and Range Finding data, as well as human experience.

Toxicity by Analogy

Because of the paucity of toxicological information on many chemical compounds used in industry there frequently is a tendency to assume that compounds which are closely related chemically will have similar toxic properties. While this may be true for a limited number of substances it is by no means universally true.

As mentioned elsewhere, many chemical compounds when absorbed by the body undergo a series of changes (detoxication processes) before they are excreted. The intermediate products depend largely on the chemical structure of the original material, and minor differences in structure may result in totally different intermediate and end products. This principle is well illustrated in the case of benzene and toluene; these are closely related chemically, but their metabolism is dissimilar and their degree of toxicity is also very different. "Toxicity by analogy" can be very dangerous and misleading.

Classes of Toxic Substances

Toxic or harmful substances encountered in industry may be classified in various ways. A simple and useful classification is given below, together with definitions adopted by the American Standards Association.

Dusts. Solid particles generated by handling, crushing, grinding, rapid impact, detonation and decrepitation of organic or inorganic materials such as rocks, ore, metal, coal, wood, grain, etc. Dusts do not tend to flocculate except under electrostatic forces; they do not diffuse in air, but settle under the influence of gravity.

Fumes. Solid particles generated by condensation from the gaseous state, generally after volatilization from molten metals, etc., and often accompanied by a chemical reaction such as oxidation. Fumes flocculate and sometimes coalesce.

Mists. Suspended liquid droplets generated by condensation from the gaseous to the liquid state or by breaking up a liquid into a dispersed state, such as by splashing, foaming and atomizing.

Vapors. The gaseous form of substances which are normally in the solid or liquid state and which can be changed to these states by either increasing the pressure or decreasing the temperature alone. Vapors diffuse.

Gases. Normally formless fluids which occupy the space of enclosure and which can be changed to the liquid or solid state only by the combined effect of increased pressure and decreased temperature. Gases diffuse.

This classification does not include the obvious categories of solids and liquids which may be harmful, nor does it encompass physical agents. The latter, strictly speaking, cannot be considered "substances." Living agents, such as bacteria, molds and other parasites comprise another group of "substances" that would appear in a comprehensive classification of industrial health hazards.

Routes of Absorption

In the physiologic sense, a material is said to have been absorbed only when it has gained entry into the blood stream and consequently is carried to all parts of the body. Something which is swallowed and which is later excreted more or less unchanged in the feces has not necessarily been absorbed, even though it may have remained within the gastrointestinal tract for hours or even days. The industrial toxicologist is concerned primarily with three routes of absorption or portals of entry through which materials may enter the blood stream: the skin, the gastrointestinal tract and the lungs.

Absorption through the Skin. Before the introduction of modern methods in the treatment of syphilis, a part of the standard therapy consisted of mercury treatment. Its effectiveness depended on the fact that certain forms of mercury can be absorbed through the intact skin. It is now recognized that skin absorption may be a significant factor in occupational mercury poisoning as well as in a number of other industrial diseases. In the case of metals other than mercury, however, entry through the skin is relatively unimportant except for some organometallic compounds such as lead tetraethyl.

Skin absorption attains its greatest importance in connection with the organic solvents. It is generally recognized that significant quantities of these compounds may enter the body through the skin either as a result of direct accidental contamination or indirectly when the material has been spilled on the clothing. An additional source of exposure is found in the fairly common practice of using industrial solvents for removing grease and dirt from the hands and arms, in other words, for washing purposes. This procedure, incidentally, is a fruitful source of dermatitis.

Gastrointestinal Absorption. The mere fact that something has been taken into the mouth and swallowed does not necessarily mean that it will be absorbed. Naturally, the less soluble the material is, the less is the likelihood of absorption. In the past it has been common practice to attribute certain cases of occupational poisoning to uncleanly habits on the part of the victim, particularly failure to wash the hands before eating. There is no doubt that some toxic materials used industrially can be absorbed through the intestinal tract but it is now generally believed that with certain notable exceptions this portal of entry is of minor importance. One outstanding exception is the case of the radium dial painters who followed the practice of "pointing" their brushes between their lips, thus ingesting lethal quantities of radioactive material. Accidental swallowing of harmful amounts of poisonous compounds in single large doses has also been known to occur. In general it can be said that intestinal absorption of industrial poisons is of minor importance and that the "dirty hands" theory of poisoning has been pretty well discredited.

Absorption through the Lungs. The inhalation of contaminated air is by far the most important means by which occupational poisons gain entry into the body. It seems safe to estimate that at least 90 percent of all industrial poisoning (exclusive of dermatitis) can be attributed to absorption through the lungs. Harmful substances may be suspended in the air in the form of dust, fume, mist or vapor, and may be mixed with the respired air in the case of true gases. Since an individual under conditions of moderate exertion will breathe about 10 cubic meters of air in the course of an ordinary 8-hour working day it is readily understood that any poisonous material present in the respired air offers a serious threat.

Fortunately all foreign matter which is inhaled is not necessarily absorbed into the blood. A certain amount, particularly that which is in a very finely divided state, will be immediately exhaled. Another portion of respired particulate matter is trapped by the mucus which lines the air passages and is subsequently brought up in the sputum. (In this connection it might be mentioned that some of the sputum may be consciously or unconsciously swallowed, thus affording an opportunity for intestinal absorption.) Other particles are taken up by "scavenger cells" following which they may enter the blood stream or be deposited in various tissues or organs. True gases will pass directly from the lungs into the blood in the same manner as the oxygen in inspired air. Because of the fact that a great majority of the known industrial poisons may at some time be present as atmospheric contaminants and thus constitute a potential threat to health, programs directed toward the prevention of occupational poisoning generally place major emphasis on ventilation to reduce the hazard (see Section 2).

Storage and Excretion

Some toxic substances can be retained or stored in the body for indefinite periods of time, being excreted slowly over periods of months or years. Lead, for example, is stored primarily in the bones and mercury principally in the kidneys. Smaller amounts may be stored in other organs or tissues. Particulate matter when inhaled can be phagocytosed and remain in regional lymph nodes where it may have little effect as in the case of coal dust, or may produce pathological changes as in the case of silica and beryllium.

The excretion of toxic agents takes place through the same channels as does absorption, namely, lungs, intestines and skin, but the kidneys (urine) are the main excretory organs for many substances. Sweat, saliva and other body fluids may participate to a small extent in the excretory process. Gases and volatile vapors are commonly excreted in the lungs and breath. This can be used as a measure of earlier absorption.

Many organic compounds are not excreted unchanged, but pass through what is known as biotransformation. The processes by which this occurs are also "detoxication mechanisms." The resulting new compounds, or metabolites, can be found in the urine and are used as evidence of absorption of the parent substance (see Table 9, p. 22.

Individual Susceptibility

The term "individual susceptibility" has long been used to express the well-known fact that under conditions of like exposure to potentially harmful substances there is usually a marked variability in the manner in which individuals will respond. Some may show no evidence of intoxication whatsoever; others may show signs of mild poisoning, while still others may become severely or even fatally poisoned. Comparatively little is known about the factors which are responsible for this variability. It is believed that differences in the anatomical structure of the nose may be concerned with different degrees of efficiency in filtering out harmful dusts in the inspired air. Previous infections of the lungs, particularly tuberculosis, are known to enhance susceptibility to silicosis. Most industrial toxicologists believe that obesity is an important predisposing factor among persons who are subject to occupational exposure to organic solvents and related compounds. Age and sex are also believed to play a part and previous illnesses may be significant.

Other possible factors relating to individual susceptibility are even less understood than those just mentioned. It has been suggested that different rates of working speed, resulting in variations in respiratory rate, in depth of respiration and in pulse rate may play a part. The action of the cilia, those tiny hairs present in the cells which line the air passages, may have some importance. The permeability of the lungs may influence absorption and the efficiency of the kidneys may govern the rate at which toxic materials are excreted, but the underlying nature of these possible variations is not known. Since the liver plays a major role in the detoxification and excretion of harmful substances, subnormal functioning of that organ may lead to increased susceptibility.

There is considerable literature purporting to show that nutritional factors may have something to do with susceptibility to occupational poisoning. Most of the published material is rather unscientific and unconvincing, but a few reports strongly suggest that there actually exists a relationship between the nature of the diet and susceptibility to poisoning. There is as yet no substantial evidence that the addition of vitamin concentrates, milk or special foods have any protective value, but when diets are deficient in some of the essential nutritional elements it appears that poisoning is more likely to occur. There is considerable evidence that indulgence in ethyl alcohol will significantly increase the possibility of occupational poisoning occurring, particularly from organic solvents.

Acute and Chronic Effects

Industrial toxicology is generally concerned with the effects of low grade sub-lethal exposures which are continued over a period of months or years. It is, of course, true that toxicological problems are not infrequently presented as a result of accidents which create sudden massive exposures to overwhelming concentrations of toxic compounds. The acute poisoning which results may cause unconsciousness, shock or collapse, severe inflammation of the lungs or even sudden death. An understanding of the nature of the action of the offending agent may be of great value in the treatment of acute poisoning but in some instances the only application of toxicological knowledge will be in establishing the cause of death.

The detection of minute amounts of toxic agents in the atmosphere and in body fluids (blood and urine) and the recognition of the effects of exposure to small quantities of poisons are among the principal jobs of the industrial toxicologist. The manifestations of chronic poisoning are often so subtle that the keenest judgment is required in order to detect and interpret them. The most refined techniques of analytical chemistry and of clinical pathology are called into play, involving studies of the working environment and of exposed individuals.

In order to demonstrate that chronic industrial poisoning has taken place or is a possibility it must be shown that an offending agent is present in significant concentrations, that it has been absorbed, and that it has produced in the exposed subject disturbances compatible with poisoning by the suspected substance. Significant concentrations are ordinarily expressed in terms of threshold limits (TL). Absorption of a substance may be proved by demonstrating its presence in the blood or urine in concentrations above those found in non-exposed persons, or by finding certain metabolic products in the excreta. To prove that disturbances have occurred in an exposed subject may require the application of all the diagnostic procedures used in medicine, including a medical history, physical examination, blood counts, urinalysis, x-ray studies and other measures.

A few of the more widely used industrial chemicals, notably lead and benzene, will produce changes in the blood in the very early stages of poisoning. Other chemicals, particularly chlorinated hydrocarbons, give no such early evidence of their action. Heavy metals such as mercury and lead produce their chronic harmful effects through what is known as "cumulative action." This means that over a period of time the material which is absorbed is only partially excreted and that increasing amounts accumulate in the body. Eventually the quantity becomes great enough to cause physiologic disturbances. Volatile compounds do not accumulate in the body but probably produce their chronic toxic effects by causing a series of small insults to one or more of the vital organs.

Site of Action of Poisons

Brief mention has already been made of the fact that different poisons act on different parts of the body. Many substances can produce a local or direct action upon the skin. The fumes and mists arising from strong acids, some of the war gases and many other chemicals have a direct irritating effect on the eyes, nose, throat and lower air passages. If they reach the lungs they may set up a severe inflammatory reaction called chemical pneumonitis. These local effects are of greatest importance in connection with acute poisoning. More important to the industrial toxicologist are the so-called systemic effects.

Systemic or indirect effects occur when a toxic substance has been absorbed into the blood stream and distributed throughout the body. Some materials such as arsenic, when absorbed in toxic amounts, may cause disturbances in several parts of the body: blood, nervous system, liver, kidneys and skin. Benzene, on the other hand, appears to affect only one organ, namely, the blood-forming bone marrow. Carbon monoxide causes asphyxia by preventing the hemoglobin of the blood from carrying out its normal function of transporting oxygen from the lungs to the tissues of the body. Although oxygen starvation occurs equally in all parts of the body, brain tissue is most sensitive; consequently the earliest manifestations are those due to damage to the brain. An understanding of what organ or organs can be damaged, and the nature and manifestations of the damage caused

by various compounds, is among the more important functions of the industrial toxicologist.

At the cellular level, toxic agents may act on the cell surface or within the cell, depending on "receptors" or binding sites. A familiar example is the affinity of arsenic and mercury for sulfhydryl (S—H) groups in biological material.

Absorption and Poisoning

As mentioned above, with the exception of external irritants, toxic substances generally must be absorbed into the body and distributed through the body by means of the blood stream in order for poisoning to occur. In other words, poisoning ordinarily does not occur without absorption. On the other hand, absorption does not necessarily or always result in poisoning. The human body is provided with an elaborate system of protective mechanisms and is able to tolerate to an amazing degree the presence of many toxic materials. Some foreign materials are excreted unchanged through the urine and feces. Toxic gases, following absorption, may be given off through the lungs. Some chemical compounds go through processes of metabolism and are excreted in an altered form. Some of these processes are known as detoxication (or detoxification) mechanisms. In some instances the intermediate products in a detoxication process may be more toxic than the original substance, e.g., formic acid and formaldehyde from methyl alcohol.

Since absorption must precede poisoning, the question often arises as to where the dividing line between absorption and poisoning is to be drawn. An answer to this question frequently entails considerable difficulty. There is no doubt that when absorption reaches a point where it causes impairment of health, poisoning has occurred. Impaired health manifests itself by the presence of altered structure, altered function, altered chemistry, or a combination of these. These impairments, in turn, result in abnormal symptoms, abnormal physical or laboratory findings, or combinations of these.

When absorption has produced both abnormal symptoms and abnormal objective findings there is no question that poisoning has occurred. In the opinion of the writer, absorption which produces objective evidence of altered structure or function should also be called poisoning, even though there may be no abnormal subjective symptoms. When subjective symptoms constitute the only basis for distinguishing between absorption and poisoning, the distinction becomes a matter of medical opinion requiring individual evaluation.

Causal Relationship and Competent Producing Cause

The industrial toxicologist frequently becomes involved in medicolegal problems since actual or suspected cases of occupational disease often result in workmen's compensation claims or negligence suits. A successful legal action on the part of the claimant or plaintiff depends upon his ability to demonstrate, usually through medical or other expert testimony, that occupational exposure harmed his health.

A competent producing cause is one which conceivably *could* have produced the harmful effect. It involves possibility.

Causal relationship exists when a competent producing cause actually *did* produce the harmful effect. It involves probability.

Medicolegal cases are usually adjudicated on the basis of opinion because of the fact that medicine is not an exact science. It has been said, and truthfully, that in medicine anything can happen. Decisions, then, must be made on the basis of the most probable explanation of a set of circumstances. Medical opinion, to be convincing, must be based on actual facts or observations, but the same set of facts or observations may be subject to more than one interpretation. Hence the importance of opinion.

Workmen's compensation laws are usually written or interpreted in such a way that in cases of doubt (sometimes reasonable doubt) a decision is rendered in favor of the claimant. Socially this is probably correct, at least in theory. This practice, however, tends to place the burden of proof on the defendant rather than on the claimant. Competent producing cause is all too often considered to be synonymous with causal relationship. Often it devolves upon a defendant to attempt to prove that a competent producing cause was in fact not the actual cause of an illness. Obviously this may entail considerable difficulty. It is not sufficient for the defendant merely to deny the existence of causal relationship. Successful defense requires an opinion (diagnosis) other than that of an occupational disease which will provide a more convincing explanation of the observed facts. Often this requires the highest degree of diagnostic acumen as well as the most astute legal procedure.

Toxicity Testing

Legislation designed to protect consumers against accidental poisoning prescribes a number of standard procedures for testing toxicity. The most frequently used tests are:

(1) Acute oral LD_{50} (see above pp. 3, 4).
(2) Acute dermal toxicity. This test gives the effects of skin absorption following a single application.
(3) Acute or primary irritation. This concerns the immediate effects of a chemical on skin, eyes or mucous membranes.
(4) Acute inhalation.
(5) Subacute and chronic feeding or ingestion.
(6) Subacute and chronic skin absorption.
(7) Subacute and chronic inhalation.
(8) Skin sensitization.

Details on performance of most of these tests are given in a publication entitled "Principles and Procedures for Evaluating the Toxicity of Household Substances," Publication 1138 of the National Science Foundation—National Research Council, Washington, D.C., 1964 and in the text of the Federal Hazardous Substances Labeling Act (Public Law 86-613). The latter also contains instructions on labeling requirements.

Principles of Prevention

Any effective program for the prevention of occupational poisoning depends on teamwork. Key members of the preventive team are the industrial physician with his knowledge of toxicology, the industrial hygiene engineer with his understanding of control measures, and the chemist who provides basic analytical data. The industrial nurse, because of her intimate contact with employed personnel, often provides the first clue to the existence of a potentially dangerous situation. Safety men and supervisors also play an important part, while a well-informed working force, instructed to recognize danger signs, constitutes an additional safeguard.

It was pointed out above that an understanding of the basic principles of industrial toxicology is fundamental in any program of prevention. One must know what materials cause poisoning and the relative toxicity of various compounds and groups of compounds. Another essential is an understanding of how toxic substances are absorbed, and the relative importance of the various routes of absorption. Realization of the fact that inhalation is the principal mode of entry means that preventive measures will be directed mainly toward reducing atmospheric contamination. For protection against chemicals that can be absorbed through the skin, suitable impervious gloves and work clothing can be provided.

Although very little is known about "individual susceptibility," the few facts which are available may be helpful in selectively placing certain persons in jobs where their age, sex, state of nutrition or previous illnesses will not constitute any extra threat to their well-being.

Thorough familiarity with the so-called threshold levels, their interpretation and applications is essential for the engineer who is called upon to design ventilation systems. All who are concerned with a preventive program must know that months or years of exposure may elapse before toxic manifestations appear, and the physician in particular must be prepared to detect and recognize the earliest evidences of poisoning. Failure to apply these few fundamentals may be costly not only in terms of lost production and lost income but also in terms of health and even of life.

Preventive procedures are often spoken of as falling into two major categories: (1) medical control and (2) engineering control.

Medical Control. This term is used to describe those procedures which are applied to the employed person. It includes:

(a) Preplacement physical examinations. One purpose of these examinations is to protect workers with known susceptibility against any potentially harmful exposure. An individual found to have healed pulmonary tuberculosis, for example, should not be placed in a job entailing exposure to silica dust nor should one with a former liver disease be exposed to chlorinated hydrocarbons.

(b) Periodic examinations. A major purpose of these is to detect any existing evidence of poisoning at an early stage when corrective measures can be expected to result in complete recovery. Correction may call for improved industrial hygiene practices, for temporary or permanent change of job assignment or both of these.

(c) Education. The purpose of this is to inform workers and supervisory personnel of the nature of any potentially harmful materials with which they may come in contact. An informed working force may be expected to accept and observe recommended precautionary measures.

(d) Personal protective devices. With few exceptions, personal protective devices should be relied upon only when the application of engineering measures offers insurmountable difficulties. Protective clothing is usually worn to prevent injuries and so will not be discussed here.

Protective masks have been widely used, or rather misused, to reduce the inhalation of potentially harmful dusts, fumes, vapors, mists and gases. In general it can be said that reliance on masks is justified only when exposures are of short duration (minutes) and low frequency (10–20 times daily).

Protective masks or respirators obviously should be properly selected for the purpose for which they are intended. To be effective they must be individually fitted to the user. In the filter or canister types the filtering units must be replaced regularly and with adequate frequency. A program of regular cleaning, repairing and replacing of worn-out parts is highly desirable. Those who are obliged to wear respirators or masks should be fully instructed as to proper use and maintenance. Unless these rules are followed it cannot be said that masks and respirators are being properly used and it might be preferable not to use them at all (see Section 3).

Engineering Control (see also Section 2). In this category are included those procedures which are applied to the working environment rather than to the individual. The most important engineering control measures are:

(a) Substitution of a less toxic in place of a more toxic material, when this is possible technologically. Common examples of this approach are the use of certain ketones in place of benzene and the use of steel shot in place of silica sand in sand-blasting.

(b) Enclosure of a process. This has its widest application in the chemical industries where frequently it is possible and practicable to design totally enclosed systems for carrying out the manufacture or processing of chemical compounds.

(c) Segregation. This may be accomplished by confining a potentially dangerous process to a segregated or enclosed area to prevent contamination of adjacent work spaces. In some situations segregation can be accomplished by locating a process in an open shed or even completely out of doors.

(d) Ventilation. This is perhaps the most important engineering control measure. Ventilation may be general or local. General ventilation consists in rapid dilution of contaminated air with fresh air usually by fans located in windows or overhead in work areas. Fans or blowers may operate by bringing fresh air into a space, thus forcing contaminated air out through natural openings such as doors and windows, or by drawing off contaminated air thus creating a partial vacuum which is filled by the entry of fresh air. Unpleasant drafts of air, particularly near open doors and windows, are sometimes caused when natural ventilation is produced in large volume.

Local ventilation usually consists in providing air suction close to the point where potentially harmful dusts, fumes, vapors, mists or gases are generated. This permits removal of the contaminants with relatively small quantities of air and prevents contamination of adjacent work spaces. Collection and disposal of contaminants removed by local exhaust ventilation sometimes presents a major engineering problem. In some processes, especially those involving the use of volatile chemicals, it is common practice to install a recovery system as part of the ventilating equipment. This may result in savings sufficient to defray the cost of installing and operating the ventilating system.

(e) Wetting. The use of water to limit the dispersal of atmospheric contaminants finds its chief application in the control of dust. This procedure is widely used in rock-drilling and it is also useful when sweeping is done in a dusty workroom.

(f) Neutralization or inactivation of chemical compounds is sometimes useful in connection with local exhaust ventilation and in cleaning up contaminated areas.

(g) General housekeeping procedures, while perhaps the simplest of all control measures, are none the less extremely important and valuable. Regular cleanup schedules, particularly where dust is a problem, are essential in any control program.

In addition to the specific procedures just enumerated, it is often important to conduct regular appraisals of the working environment by means of dust counts, air analyses and similar tests, thereby checking on the effectiveness of the preventive measures.

FIRST AID

Emergency Treatment of Acute Poisoning

Acute poisoning may be the result of entry into the body of large or concentrated doses of a poison through

(a) Breathing (inhalation),
(b) Swallowing (ingestion),
(c) Skin absorption,
(d) Injection (hypodermic or intravenous entry).

It is obvious that the route of entry will influence the type of emergency treatment.

In every case of acute poisoning, summon medical assistance immediately. The names and telephone numbers of one or more on-call physicians, the nearest hospital and ambulance service should be posted near appropriate telephones.

If the police department, fire department or utility company maintains an emergency service, its telephone numbers should also be posted.

Every industrial establishment, no matter how small, should have at least one person on duty at all times who is trained and designated to take charge in the event of an emergency due to poisoning. This individual should be conversant with the emergency handling of the particular situations that may arise.

Improper first aid may be more harmful than none at all.

Although prompt action is always important, there are relatively few situations in which a delay of seconds or minutes will have a significant bearing on the outcome.

When possible, a sample of the suspected poison, or the container from which it came, should be preserved for the guidance of the treating physician, the police or the medical examiner (coroner).

General Procedures

(A) **Inhalation:**
(1) Remove victim from contaminated area. Rescuers should be properly protected or provided with life lines.
(2) Keep victim warm (not hot) and quiet. Lying flat is usually the best position.
(3) If breathing has stopped, give artificial respiration.
(4) Administer oxygen, if it is available.
(5) Keep breathing passages open. Examine mouth for false teeth and chewing gum and remove them if present.

(B) **Ingestion:**
(1) Attempt to empty the stomach by causing vomiting or use of an emetic. This should be done even if a period of several hours has passed since the poison was swallowed.
Exceptions: Corrosive chemicals such as strong acids or caustic alkalies, victim having convulsions, victim unconscious.
(2) Dilute the poison by administering fluids in any of the following forms:
(a) Plain tap water: 3–4 glasses.
(b) Soapy water: 2–3 glasses.
(c) Table salt in warm water: one tablespoon to an ordinary 8 ounce tumbler.
(d) Milk: 3–4 glasses.
If these fluids are vomited, which is desirable, the doses may be repeated several times.
(3) Give the victim a "universal antidote." A mixture of powdered burnt toast (charcoal), strong tea and milk of magnesia will absorb and neutralize many poisons. (One piece of toast and 4 tablespoons of milk of magnesia in a cup of strong tea.)

(C) **Skin Contact:**
(1) Dilute the contaminating substance with large amounts of water. This is best done in a shower, but may also be done with a hose, buckets or other means. The water should be lukewarm if possible.
(2) Remove contaminated clothing. Those

assisting the victim should protect their own skin with gloves, if available.

(3) Chemical burns of the eye should be treated with large amounts of water or with a weak solution of bicarbonate of soda (a level teaspoonful of bicarbonate to 1 quart of warm, clean water).

(D) Injected Poisons:

(1) Absorption may be delayed by the application of a tourniquet above the point of injection if this is in the arm or leg. The tourniquet should not be so tight that it impairs the flow of arterial blood.

(2) Excretion can be hastened by administering large doses of water or other fluid.

(3) Attempt to determine the nature of the poison so that the proper antidote can be given.

LABORATORY PROCEDURES

Application and Interpretation

It is axiomatic that accurate diagnosis is basic to the proper handling of any medical problem. Diagnosis is rarely, if ever, established by any single observation or test. A total picture, as determined through the history, the physical examination and diagnostic tests, must be obtained and interpreted. Frequently there is a tendency on the part of persons who have not had medical training to place undue emphasis on the importance of laboratory procedures. This is understandable since deviations from the normal which can be expressed in terms of numbers may be more comprehensible than data obtained by the taking of a history or performing a physical examination. The important point is that ordinarily a reliable diagnosis can be arrived at only by securing and interpreting all available information.

Laboratory procedures in industrial toxicology can be divided into two major classes: (1) those applied to human beings and (2) those applied to the environment. The present discussion is concerned only with the former.

Basic Principles

(1) Specificity of Laboratory Tests. The human organism is limited in the manner in which it can react to harmful influences from outside or morbid processes originating within the body. In other words, *similar or identical changes may be produced by widely differing causes.* A familiar example of this is the presence of stippled red blood cells which can be found in lead poisoning but which also occur in a number of blood disorders such as anemias and leukemias. There are few, if any, laboratory tests which can be said to be pathognomonic, that is, so selective and specific that a positive result in and of itself alone is sufficient to determine a diagnosis.

(2) Normal Values. Those who have been trained in the physical sciences are accustomed to think of physical constants in terms of fixed or absolute values. Such things as freezing and melting point, vapor pressure and many other characteristics of materials have a single fixed value when measured under the same conditions. The situation in the biological sciences is not so simple. It can be stated categorically that *in the quantitative study of biological materials (blood, urine, etc.) there is no such thing as a single, fixed normal value.* There is, however, a normal range. Failure to understand this principle has been responsible for many erroneous conclusions. Unless one knows what is normal, it is impossible to know what is abnormal.

(3) Trends and Mass Data. Very often a single, isolated laboratory test will be inconclusive as to whether it indicates a normal or abnormal state. Serial observations made over a period of time and showing a trend in a certain direction may be highly significant even though none of the individual tests yields results falling outside the normal range. For example, a series of blood counts starting in the upper part of the normal range and gradually but steadily falling to the lower normal limit may mean that poisoning has occurred, in spite of the fact that the lowest count still falls within the normal range.

Another effective means of detecting subtle changes is through the use of mass data. Using the example of blood counts again, it may be found that the average of the values for a group of workers having a common exposure is different from that of a group of similar individuals who have had no such exposure. Although this approach may not reveal which individuals in the group have been affected, it will show that excessive concentrations of a noxious agent are present and bring out the necessity for corrective measures. Trends can be used in the interpretation of mass data as well as in following the status of individuals.

(4) Control Observations. Although standard values for the normal ranges encountered in the performance of laboratory tests have been published in various textbooks, it is not always entirely safe to use these standards in evaluating occupational exposure to toxic compounds. This principle is universally recognized in biological experimentation, and "control" observations are made to obviate erroneous conclusions. This means that a group to be tested or treated is compared with a similar or identical group under

conditions involving a single variable. For industrial toxicology the application of this principle means that if a group of workers is to be tested for possible toxic manifestations, the same tests should also be performed on a group of workers similar in every respect, but not exposed to any potentially toxic agent. Such observations may entail some difficulty, but they are important particularly in interpreting mass data.

(5) The Importance of Simple Tests. Some people have a tendency to believe that because a diagnostic test is complicated and involves the use of expensive equipment it necessarily yields information that is of greater significance than that obtained by simpler tests. This is by no means always the case. It is true, of course, that some of the most valuable diagnostic procedures used in industrial toxicology, as well as in ordinary medical diagnosis, are fairly complicated. This should not result in the fact being overlooked that equally important information can be obtained through very simple tests. Not infrequently the simple tests are ignored completely, resulting in diagnostic errors.

New laboratory tests are constantly being developed. Some of these prove to have lasting value, others enjoy a brief vogue and then fall into disuse. A test of proven worth should not be discarded capriciously in favor of a newer one until the latter has conclusively demonstrated its superiority.

(6) Laboratory Errors. The performance of diagnostic tests, as well as their interpretation, involves human factors. Anyone who has had any experience with laboratory work is well aware of the many possibilities for error. Some of the more important are listed below.

(a) Errors in Collection. Many diagnostic procedures require the most scrupulous care in the collection of the sample to be tested or analyzed. This is particularly true of specimens to be analyzed quantitatively for heavy metals. The slightest deviation from the approved technique may vitiate the results of the test. The same principle applies to many other laboratory procedures.

(b) Accidental Contamination. A specimen may be properly collected but may become contaminated after collection and before analysis.

(c) Deterioration. It is always desirable and often essential that a sample be analyzed as soon as possible after collection. Physical, chemical or morphological changes may occur if there is undue delay between the time of collection and the time of analysis. There are methods of preservation which can and should be applied when delay cannot be avoided.

(d) Technical Errors. There is always a possibility, even under the best circumstances, and even when the competence of the technician is unquestionable, for technical errors to creep into laboratory procedures. These can be reduced to a minimum if duplicate samples are analyzed and if "blanks" are run to test the purity of the reagents used.

(e) Errors in Calculation. Some tests require the application of mathematical formulas in the calculation of the results. This introduces a possible source of error.

(f) Mixed Specimens. It is rare, but by no means unknown in busy laboratories, for mistakes to occur in the identification of specimens. This obviously could result in serious error.

(g) Errors in Reporting and Recording. In addition to the erroneous reporting that would result from mixed specimens, there are other possibilities for mistakes. Slips of the pen may occur either when the results of a test are entered on the original report or when these results are transcribed to another document. Illegible writing may be a source of confusion or error.

(h) Wilful Error. The possibility of error as a result of slovenliness or even of outright dishonesty, while rare, should not be overlooked.

The laboratory can be a valuable adjunct to, but should never be a substitute for, medical judgment. When the two appear to be in conflict, it is always wise to check the possibility of error in the laboratory.

The techniques for performing laboratory tests can be found in a number of excellent textbooks. In the discussion which follows no attempt is made to describe techniques.

Common Laboratory Tests and Their Significance

An attempt will be made in this section to present, in tabular form, some of the most commonly used laboratory tests together with ranges of normal values. The tests to be considered here are those which are ordinarily performed in any clinical laboratory and which require relatively simple equipment and techniques. The more specialized and complicated procedures will be summarized elsewhere.

The Urine. Laboratory examination of the urine is one of the simplest and at the same time one of the most valuable diagnostic aids. A routine urinalysis should be included as part of the examination of any case of suspected poisoning. Periodic testing of the urine for specific changes is often useful in periodic examinations to detect early evidence of poisoning. The periodic examination need not always embrace a complete urinalysis.

TABLE 1. Normal Values of Urine

Test	Range of Normal Values	Significance
1. Color	Pale straw to deep amber	Low specific gravity usually associated with pale color, high specific gravity with a deeper color.
2. Turbidity	Usually clear, if specimen is freshly voided	Turbidity not necessarily an indication of abnormal status.
3. Acidity	pH 4.8–7.5	Fresh urine is usually slightly acid. On standing, urine may become alkaline due to decomposition with formation of ammonia.
4. Specific gravity	1.001–1.030	Specific gravity depends on fluid intake. In certain kidney diseases it may become fixed at 1.010.
5. Sugar (glucose or reducing bodies)	None	A meal high in carbohydrate may give transient sugar in urine. Its presence does not necessarily mean diabetes.
6. Albumin (protein)	None—by ordinary methods; 2–8 mg/100 ml by quantitative method	Albumin in the urine usually denotes the presence of kidney disease. Occasionally albumin appears in the urine of normal persons following long standing (postural albuminuria).
7. Casts Red blood cells Leukocytes and epithelial cells	0–9,000 in 12 hr 0–1,500,000 in 12 hr 32,000–4,000,000 in 24 hr	In routine examinations of urine a few casts and blood cells may be found in normal specimens.
8. Porphyrins	Trace—qualitative; 0.001–0.010 mg/100 ml —quantitative	May be increased in exposure to lead.
9. Concentration	Sp. gr. above 1.020	A rough, but useful measure of kidney function.
10. Dilution	Sp. gr. 1.002	Same.
11. Dye excretion	15 min 30–50% 30 min 15–25% 60 min 10–15% 120 min 3–10% Total of 70–80% in 2 hr	Values given are based on intravenous injection of phenolsulfonphthalein (PSP). Low dye excretion means poor kidney function.

Interpretation of Urine Tests. Abnormalities in the urine may be caused by diseases of the kidneys or of other parts of the urinary tract: ureters, urinary bladder, urethra and accessory glandular structures. Only when the kidneys themselves are involved will there be disturbances in function. Casts originate only in the kidneys; other formed elements (blood cells, epithelial cells) may enter the urine from other parts of the urinary system. Disturbances in kidney function may not be present unless there has been fairly extensive damage to kidney tissue.

A number of diseases of nonoccupational nature can produce abnormal findings in the urine. Those occupational exposures which may produce urinary changes are listed in Table 2.

Disturbances in the Blood Due to Occupational Poisoning. In this section attention will be limited to those components of the blood included in Table 3. The tests involved can be and usually are performed in any clinical laboratory. Some toxic materials affect several of the blood elements while others produce changes in a single component. Table 5 represents an at-

TABLE 2. Occupational Poisons Which May Produce Abnormalities
Detectable by Ordinary Urinalysis

Substance	Findings	Significance
Aniline dye inter-mediates	Dark color, blood cells	Suggests presence of cancer of bladder
Arsenic	Albumin, blood cells	Due to damage to the kidneys
Benzol	Red blood cells	Severe poisoning with bleeding into urinary tract
Cadmium	Proteinuria	Due to kidney injury
Carbon disulfide	Albumin	Kidney damage
Carbon tetra-chloride	Albumin, casts, blood cells, impaired function	Due to kidney damage
Chlordane		Reported as causing kidney injury
Chlorinated di-phenyls		Reported as causing kidney injury
Chlorinated naphthalenes		Reported as causing kidney injury
Chlorobenzenes	Albumin, red blood cells, dark color	Kidney damage
Chloroform		Reported as causing kidney injury
Cobalt		Questionable
Cresol	Albumin, red blood cells	Due to kidney irritation
DDT	Albumin	Kidney degeneration
Di(p-aminophenyl) methane		Reported as causing kidney injury
Dichloroethyl ether		Kidney damage produced in experimental animals
Dimethyl formamide		Reported as causing kidney injury
Dimethyl sulfate		Kidney damage produced in experimental animals
Dioxane	Red blood cells, disturbed function	Hemorrhagic nephritis
Ethylene chloro-hydrin		Kidney damage produced in experimental animals
Ethylene dichloride		Kidney damage produced in experimental animals
Glycols	Red blood cells, impaired function	Has occured following ingestion
Lead	Increases porphyrins; Albumin	Questionable
Mercury	Albumin, impaired function	Due to kidney injury

TABLE 2.   Occupational Poisons Which May Produce Abnormalities Detectable by Ordinary Urinalysis (continued)

Substance	Findings	Significance
Methyl bromide	Albumin	Kidney damage
Methyl chloride	Albumin	Kidney damage
Methylene chloride	Albumin	Kidney damage
Methyl formate		Kidney injury has been produced in experimental animals
Naphthols	Albumin, blood cells	Due to irritation of kidneys
Naphthylamines		See aniline dye intermediates
Nitrobenzol	Albumin, blood cells	Kidney irritation; dark color due to various pigments
Oxalic acid	Albumin	Occurs in severe poisoning

TABLE 3.   Normal Blood Values

Test	Normal Range	Comments

Red blood cells
  Males
  Females
	
	4.5–6.0 million/cmm
	4.0–5.0 million/cmm
		Values for U.S.A.

Hemoglobin
  Males
  Females
	
	14–18 g/100 cc
	12–15 g/100 cc
		Values given in percent are meaningless unless the gravimetric equivalent of 100% is stated.

White blood cells
  Total
  Differential
    Band forms
    Segmented
    Lymphocytes
    Monocytes
    Eosinophiles
    Basophiles
	
	5,000–10,000/cmm
	
	0– 5%
	35–70%
	20–60%
	2–10%
	0– 5%
	0– 2%
		Total white cell counts fluctuate widely from hour to hour. Differential counts remain fairly static.

Platelets	200,000–500,000/cmm	
Reticulocytes	0.1–0.5%	High values indicate active blood regeneration

Erthrocyte sedimentation rate (ESR)	Wintrobe method
	  Males 0–9 mm in 1 hr
	  Females 0–20 mm in 1 hr
	Westergren method
	  Less than 20 mm in 1 hr
		Values differ in males and females

Hematocrit	40–48% cell volume	Affected by dehydration
Mean corpuscular volume (MCV)	80–94 cu. microns	Average size of red blood cells
Mean corpuscular hemoglobin (MCH)	27–32 micromicrograms	Average hemoglobin content per cell

TABLE 3.   Normal Blood Values (continued)

Test	Normal Range	Comments
Mean corpuscular hemoglobin concentration (MCC)	32–38%	Average concentration of hemoglobin in red blood cells
Red cell fragility	Hemolysis starts in solution of 0.42% NaCl, complete in 0.32% NaCl	
Bleeding time	Less than 3 min	Capillary bleeding
Coagulation time	6–10 min	Venous blood
Prothrombin time	12–15 sec	
Appearance of red blood cells	Uniform size and staining, appear circular	Parasites such as those of malaria may be found in stained blood smears.

tempt to summarize the hematologic abnormalities that may be caused by occupational exposure to toxic compounds.

The interpretation of laboratory tests on the blood should be governed by the basic principles which apply to all laboratory diagnostic procedures (see p. 11). In particular, the importance of normal or control observations should be stressed. Furthermore, there is nothing specific about blood examinations which permits a conclusion that if abnormalities are found in a worker who has been exposed to a known hemotoxic agent the change is necessarily due to the poison. At best, the evidence is only presumptive and forms only a part of the total picture.

Since complete laboratory examination of the blood is somewhat time consuming it is often impracticable to perform complete studies on large groups of workers. As a routine procedure, determination of the hemoglobin and examination of a stained blood smear may be useful. The combination of these two tests will reveal or give a clue to the existence of most abnormalities of the blood. Single tests or partial examinations are often all that are required in periodic examinations to detect early poisoning. In cases where differential diagnosis is involved, more complete blood studies are usually required.

Liver Function Studies. As mentioned elsewhere, disturbances of liver function detectable by laboratory methods may not be present until there has been a substantial degree of damage to the liver. This means that liver function tests are of very limited value in detecting early poisoning and that normal results in liver function studies do not necessarily mean that the liver is intact.

A variety of causes may produce abnormalities in liver function and it is often difficult to determine the causative factor. In studying liver function it is customary to perform a battery of tests. Various combinations of positive and negative results are helpful in arriving at an opinion or conclusion as to the true nature of the disease process.

Some of the tests which are frequently used in studying liver function have been included in Table 4, p. 17. Some of the more specific tests are summarized in Table 6. Interpretation should be guided by the general principles set forth on p. 16.

Table 7 is a list of substances commonly used in industry which may produce injury to the liver and consequently may, in severe poisoning, result in abnormal liver function. This list of substances is given without further comment since available pertinent information is somewhat limited. For many of the substances listed, the evidence is confined to observations on experimental laboratory animals.

Miscellaneous Tests of General Importance. *Serological Tests for Syphilis.* Testing the blood for evidence of past or present syphilitic infection is commonly practiced as part of a preplacement or routine periodic physical examination. The oldest and best known of the blood serological tests is the Wassermann reaction. In recent years there have been developed a number of modifications of the Wassermann test, such as the Kolmer, Kahn, Kline, Mazzini, Hinton, Eagle and V.D.R.L. tests. All of these are applied for the same purpose, i.e., to detect syphilis.

In the interpretation of serological tests for

TABLE 4. Normal Blood Chemistry Values

Test	Normal Range	Comments
Blood urea	20–35 mg/100 cc blood	Increased in kidney disease
Urea nitrogen	9–17 mg/100 cc blood	
Non Protein	25–40 mg/100 cc blood	Increased in kidney disease
Nitrogen (NPN)		
Total protein	6.0–8.0 g/100 cc blood	
(L.F.)		
Serum albumin	3.5–5.5 mg/100 cc blood	
(L.F.)		
Serum globulin	2.0–3.0 mg/100 cc blood	
Albumin-globulin	1.5:1 to 2.5:1	
ratio (A/G)(L.F.)		
Uric acid	2–7 mg/100 cc blood	Females slightly lower
Total cholesterol	150–250 mg/100 cc	Increased in biliary obstruction, decreased in diseases of liver
Cholesterol	60–75% of total cholesterol	Decreased in liver disease
esters (L.F.)		
Serum bilirubin (L.F.)		See Table 6
Alkaline phosphatase	1.0–4.0 Bodansky Units/100 cc	High values in biliary obstruction.
(L.F.)	of serum	Little change in intrinsic liver disease
Acid phosphatase	0.5–2.0 Gutman Units/100 cc	Not related to liver disease
	of serum	
Blood sugar	80–120 mg/100 cc blood	
Icterus index (L.F.)	2–8 Units	Frequently used as a measure of jaundice, but unreliable because of interfering substances
Calcium	8.5–11.5 mg/100 cc serum	4.5–5.7 meg/liter
Chlorides as NaCl	350–400 mg/100 cc plasma. 100–110 meq/liter	
Creatinine	1–2 mg/100 cc blood	
Oxygen capacity	18–24 cc/100 cc blood	Arterial blood
CO_2 combining power	45–70 vol/100 cc 21–30 meq/liter	
Cholinesterase	R.B.C. 0.67–0.86 pH units/hr Plasma 0.70–0.97 pH units/hr	Michel Method. Decreased in organic phosphate poisoning.
Potassium	16–22 mg/100 cc serum 4.0–6.0 meq/liter	
Sodium	315–340 mg/100 cc serum 136–145 meq/liter	
Phosphorus, inorganic	3–4 mg/100 cc serum	1.5–2.8 meq/liter
Magnesium	1.5–2.4 meq/liter	

Tests marked (L. F.) are used in liver function studies.

TABLE 5.   Occupational Poisons Which May Produce Abnormalities
Detectable by Studies of the Blood

Substance	Findings	Comments
Acrylonitrile	Anemia, leukocytosis	Reported but not definitely established
Allyl isopropyl acetyl carbamide	Thrombopenia	
Aniline	Anemia, stippled red cells, leukocytosis	Findings differ in acute and chronic exposure
Antimony	Leukopenia	Effects may resemble those of arsenic
Arsenic	Anemia, leukopenia	Industrial poisoning may be caused by arsine
Benzene	Decrease in all formed elements	Death may result from depression of the bone marrow
Carbon disulfide	Anemia, leukocytosis, immature W.B.C.	Conflicting reports in medical literature
Carbon tetrachloride	Anemia, leukopenia	Questionable
Cobalt	Increased R.B.C.	Animal experiments only
DDT	Anemia, leukopenia	Rarely occurs
Ethyl silicate	Anemia, leukocytosis	Animal experiments only
Ethylene glycol monomethyl	Decrease in all formed elements, increased percentage of immature W.B.C.	Based on human observations
Ethylene oxide		Questionable
Fluorides	Anemia	
Lead	Mild anemia, stippling of R.B.C.	Well established
Manganese	Leukopenia	Questionable
Mercury	High hemoglobin	Reported but unconfirmed
Methyl chloride	Decrease in formed elements	Animal experiments
Nitrobenzenes (nitrophenols)	Reduced R.B.C. with signs of regeneration	Due to blood destruction
Nitrous fumes	Decreased W.B.C.	Reported but unconfirmed
Phenylhydrazine	Anemia with signs of regeneration	Due to blood destruction
Radium	Decrease in all formed elements	Well established in humans
Selenium	Anemia	Animal experiments
Tetrachlorethane	Signs of blood destruction	Not fully established in humans
Thallium	Increased W.B.C. and eosinophiles	Reported but unconfirmed

TABLE 5.   Occupational Poisons Which May Produce Abnormalities
Detectable by Studies of the Blood (continued)

Substance	Findings	Comments
Thorium	Decrease in all formed elements	Due to radioactivity
Toluene	Anemia, leukopenia	Mild when compared with benzene
Toluene diamine	Anemia	
Toluidine	Anemia	Questionable
Trichlorethylene	Anemia	Reported but unconfirmed
Trinitrotoluene	Anemia, leukopenia	Well established
Uranium	Anemia, leukopenia	Due to radioactivity
Vanadium	Anemia	Questionable
Xylene	Anemia, leukopenia	Questionable

TABLE 6.   Liver Function Tests
(Other than those given in Table 4)

Test	Normal Range	Comments
Bromsulfone-phthalein test (BSP)	Less than 5% retention after 45 min	Disease of liver tissue, biliary obstruction or circulatory obstruction may cause greater retention
Cephalin flocculation	0–1+ flocculation in 48 hr	Increased flocculation found in disease of liver tissue, associated with abnormalities of serum proteins
Serum bilirubin	Total 0.2–1.0 mg % Direct 0.1–0.7 mg % Indirect 0.1–0.3 mg %	
Thymol turbidity	0–4 Maclagan units	Significance same as cephalin flocculation
Urine urobilinogen	0.5–2.0 mg/24 hr or dilution of 1:4–1:30	May be abnormal in biliary obstruction or in disease of liver tissue

syphilis there are a few principles which are particularly important:

(1) Conclusions should not be drawn or therapy instituted as a result of a single positive report. Before action is taken, the test should be repeated and perhaps one or more of the modified forms of the test done as well.

(2) A positive test, in and of itself, does not necessarily mean that there has been a syphilitic infection. Positive reactions can be caused by other diseases (e.g., malaria) and furthermore there occasionally occur false positives.

(3) A negative test does not necessarily rule out the possibility of syphilis. It takes time for the blood to develop the antibodies which produce positive reactions so that in the very early stages of the disease a positive test may not be obtained. In such cases a dark field examination of material obtained directly from a suspected lesion may reveal the presence of the spirochaetes which cause syphilis. False negative reactions are rare, but can occur.

(4) A positive reaction may be found in persons who have had syphilis even though they have received adequate treatment and to all intents and purposes have been cured. This con-

TABLE 7.   Occupational Poisons Which May Produce
Abnormalities in Liver Function

Acrylonitrile	Diphenyl
Antimony	Ethylene chlorohydrin
Arsenic*	Ethylene dichloride
Beryllium*	Methyl bromide
Cadmium*	Methyl chloride*
Carbon disulfide*	Methyl formate
Carbon tetrachloride*	Methylene chloride
Chlordane	Nitrobenzol*
Chlorinated diphenyls*	Phenol*
Chlorinated naphthalenes*	Phenylhydrazine
Chloroform	Phosphorus*
Cobalt	Tetrachlorethane
Cycloparaffins	Trichlorethylene
(cyclopropane, -butane, etc.)	Trifluorochloroethylene
DDT	Trinitrotoluene*
Diethylene dioxide* (dioxane)	(TNT)
Dimethylformate	Uranium
Dinitrophenol*	

*Substances marked with an asterisk have been shown to produce liver
injury.

dition is known as being "Wassermann fast." A
positive test, then, does not necessarily mean the
presence of active disease.

(5) The various serological tests have a relatively high degree of specificity but are, nonethe-
less, subject to all of the possible errors and pit-
falls of any laboratory procedure (see p. 12).

Other Serological Tests. Tests performed on
the blood serum may provide valuable diagnostic evidence in a number of infectious diseases. Some of these may be occupational in
origin, such as brucellosis (undulant fever),
tularemia, Weil's Disease (spirochaetal jaundice), psittacosis, glanders, and others.
Interpretation of the results of serological
tests applied to these infections should be in accordance with the general principles which are
stated on p. 11.

Bacteriologic Examinations. This category
includes direct examination of stained or un-
stained material such as sputum, pus or urine
sediment and the study of cultures of suspected
specimens having the same general origin. These
tests are usually applied as part of a general diagnostic study. Yeasts, molds, and fungi may be
studied by methods similar to those applied to
bacteria.

Infectious diseases of occupational origin, in
addition to those mentioned previously, include
anthrax, tetanus, rabies, erysipeloid, blastomy-
cosis, actinomycosis, moniliosis, coccidiodomy-
cosis, aspergillosis and cryptococcal infection.

Stool Examination. Fresh or stained specimens of the stools (feces) are often examined to
detect the presence of intestinal parasites such as
worms or amoebae. A search is made either for
the parasite itself or its eggs. Another common
test applied to the stools is that for blood, to secure evidence of cancer or other lesions of the
gastrointestinal tract.

Interpretation of stool examinations should be
guided by the general principles of all laboratory diagnostic procedures (see p. 11).

Electrocardiogram (EKG). This test has found
wide application in the diagnosis of heart
disease. The reading or interpretation of an
electrocardiogram requires a high degree of skill.
There is nothing magical about the instrument
which produces the electrocardiogram, the elec-
trocardiograph. All it can do is to give a record
of the manner in which a nerve impulse travels
through the heart muscle. Disturbances of the
conduction system and abnormalities of the
heart muscle may produce characteristic abnormalities of the electrocardiogram. A normal
tracing does not necessarily mean that there is
nothing wrong with the heart, and an abnormal
tracing does not necessarily mean that heart
disease is present.

Basal Metabolic Rate (BMR). This is a
method for measuring the rate at which oxygen
is consumed by the body. Its most frequent
application is in the study of suspected or actual
diseases of the thyroid gland. Many conditions
other than thyroid disease can produce abnormalities of the BMR.

Roentgenologic (X-ray) Examinations. The
use of the x-ray has wide application as a diag-

nostic tool. It is important to realize that all the x-ray can do diagnostically is to provide a record of the density of the structures through which the rays pass. Dense objects obstruct the passage of the beam, resulting in less darkening of the film. The darker areas on an x-ray film are those through which the rays pass most readily. The usefulness of the x-ray can be enhanced by the use of radio-opaque contrast media, such as the familiar barium mixtures used in the roentgen examination of the intestinal tract.

The basic principles of the interpretation of laboratory tests apply to the interpretation of data obtained by roentgenologic examination.

Other Tests. There are many other tests used in diagnosis. For details the reader is referred to recent textbooks on laboratory methods.

Special Laboratory Tests. In the previous pages a summary has been given of the more important laboratory procedures which are used in medical diagnosis. These tests are used in studying nonoccupational as well as occupational illnesses. There are, in addition, a number of tests which are utilized only in the diagnosis of diseases caused by noxious substances. Exposure to such substances occurs principally as an occupational hazard although occasionally there is a nonoccupational origin. Examples of the latter would be the inhalation of toxic gases produced by faulty heating or refrigerating equipment or the ingestion of poisonous materials in contaminated food and beverages. Occupational poisoning may occur in the practice of hobbies or through work in one's own home and is not necessarily the result of adverse conditions of employment.

There are two main types of tests used to detect the absorption of potentially harmful substances.

(1) Detection of the material itself in body fluids, particularly in urine and blood. This is used principally in connection with metallic poisons.

(2) Detection of the products of metabolism (metabolites) of a toxic substance. This is used principally in connection with toxic organic compounds.

In addition to these two main types there are several of lesser importance such as:

(3) Examination of expired air for the detection of toxic gases.

(4) Examination of sputum for dust particles or asbestosis bodies.

(5) Stool examination for heavy metals.

(6) Biopsy examination of living tissue.

(7) Post mortem examination (autopsy) and chemical analysis of organs.

The general principles of interpretation (see p. 16) apply to the special tests as well as to all others. Of particular importance is the avoidance of contamination of specimens. Many of the analytical procedures used in determining toxic substances in blood and urine are so sensitive that the slightest error in the technique of collecting and handling the specimen can vitiate the results of the analysis.

"Normal" Content of Metals in Urine and Blood. A number of metals which can cause occupational poisoning may be found in the urine or blood of persons who have had no occupational exposure to these substances. Lead, arsenic and mercury are common examples of

TABLE 8.　Materials for Which Analysis of Urine
and Blood Is Useful

Substance	Normal Range	Comments
Arsenic, Urine	0–0.85 mg/liter	Arsenical medication or a fish diet can result in arsenic appearing in urine and blood
Fluoride, Urine	0.2–0.4 mg/liter	Fluorine content of drink-
Blood	0.01 mg/liter	ing water is important
Lead, Urine	0.01–0.08 mg/liter	Average 0.03 mg/liter
Blood	0.01–0.05 mg/100 cc	Average 0.03 mg/100 cc
Mercury,* Urine	0–0.020 mg/liter	
Blood	0–0.005 mg/100 cc	
Selenium, Urine	0–0.10 mg/liter	Depends on soil content of selenium
Carbon monoxide, Blood	0–10%	CO measured as carboxy-hemoglobin; influenced by smoking

*Tentative values.

TABLE 9. Common Urine Tests for the Presence of Metabolites

Metabolite	Agent	Normal Range	Comments
Sulfates	Aniline Benzene Phenol Xylidine	Inorganic sulfate 85–90% Organic sulfate 10–15%	Can be influenced by diet Can be influenced by diet
Hippuric acid	Toluene Ethylbenzene	0.6–0.7 g in 24 hr	Also present in other benzene homologs which are oxidized to benzoic acid
Glucuronates	Aniline Benzene Phenol Terpenes Xylidine	0.7–1.4 g in 24 hr	
Trichloracetic acid	Trichlorethylene		Trichloroethanol also produced
Formic acid	Methyl alcohol	30–120 mg in 24 hr	
Thiocyanate	Nitriles	0–14 mg in 24 hr	Influenced by smoking
2,6-Dinitro-4-amino-toluene	TNT	None	
p-Amino-phenol	Aniline	None	
DDA	DDT	None	Acetic acid metabolite of DDT

this. Extensive studies on lead have shown that the so-called normal content may be as high as 0.08 mg/liter in urine and 0.05 mg/100 cc in blood. Reliable quantitative data of a similar nature is somewhat limited for other metals.

In interpreting the results of analyses of blood and urine for heavy metals a few points are particularly important.

(1) The presence of a toxic metal indicates absorption but not necessarily poisoning.

(2) The metal could have gained access to the body from nonoccupational as well as occupational sources.

(3) It is important to know the amount or concentration of toxic metal, not merely whether it is present or absent.

(4) Since little is known about correlations between blood and urine values for heavy metals and between these values and clinical manifestations it is good practice to study both urine and blood for maximum information.

(5) There are no fixed values for heavy metals in blood or urine which definitely establish a diagnosis of poisoning. The findings merely enter into the total picture and require interpretation by a physician skilled in diagnosis.

Pulmonary Function Tests

The interpretation of tests of pulmonary (lung) function require an understanding of a number of terms which are common usage.

Volumes. There are four primary volumes which do not overlap.

(1) Tidal Volume, or the depth of breathing, is the volume of gas inspired or expired during each respiratory cycle.

(2) Inspiratory Reserve Volume, (formerly complemental or complementary air minus tidal volume) is the maximal amount of gas that can be inspired from the end-inspiratory position.

(3) Expiratory Reserve Volume, (formerly reserve or supplemental air) is the maximal volume of gas that can be expired from the end-expiratory level.

(4) Residual Volume, (formerly residual capacity or residual air) is the volume of gas remaining in the lungs at the end of a maximal expiration.

Capacities. There are four capacities, each of which includes two or more of the primary volumes.

(1) Total Lung Capacity, (formerly total lung volume) is the amount of gas contained in the lung at the end of a maximal inspiration.

(2) Vital Capacity, is the maximal volume of gas that can be expelled from the lungs by forceful effort following a maximal inspiration.

(3) Inspiratory Capacity, (formerly complemental or complementary air) is the maximal volume of gas that can be inspired from the resting expiratory level.

(4) Functional Residual Capacity, (formerly

functional residual air, equilibrium capacity, or mid-capacity), is the volume of gas remaining in the lungs at the resting expiratory level. The resting end-expiratory position is used here as a base line because it varies less than the end-inspiratory position.

Respiratory Dead Space. Anatomical dead space is the internal volume of the conducting airway from the nose and mouth down to the alveoli. It is called "dead" space because no rapid exchange of O_2 and CO_2 occurs there between gas and blood.

Physiological dead space includes the anatomical dead space and two additional volumes: (1) the volume of inspired gas ventilating alveoli which have no pulmonary capillary blood flow and (2) the volume of inspired gas ventilating some alveoli in excess of that required to arterialize the pulmonary capillary blood flowing around them. It is approximately equal to cubic centimeters per pound of body weight.

Frequency is the rate of breathing.

Alveolar ventilation per minute is equal to (tidal volume – dead space) times frequency.

Minute volume equals tidal volume times frequency.

"Alveolar capillary block" is a decrease in the capacity of oxygen to diffuse from the alveolar gas into the pulmonary capillary blood.

Mechanical Properties of the Lung. Compliance is a measure of the distensibility of the lungs and thorax, the volume change per unit pressure change (liters per centimeter of H_2O).

Resistance is made up of the resistance of the tracheo-bronchial tree to the flow of air as well as the viscous resistance of the tissues as they are being deformed. It is equal to pressure difference per liter per second.

Maximal breathing capacity is the greatest volume of gas that can be breathed per minute.

Timed vital capacity is the timed segment of a rapidly exhaled vital capacity, usually the volume exhaled in 1, 2, or 3 seconds as the percentage of that vital capacity.

Oxygen capacity is the amount of oxygen contained in blood that is fully oxygenated.

Arterial oxygen saturation is the percentage of

$$\frac{\text{ml } O_2 \text{ actually combined with hemoglobin}}{\text{max. ml } O_2 \text{ capable of combining with hemoglobin}}$$

Normal Ranges of Measurements Used in Evaluating Pulmonary Function:

Vital Capacity
 75 to 80% of total lung capacity.
  Young adult male   25 ml/cm of ht.
  Young adult female 20 ml/cm of ht.
Residual Volume
 20–55% of total lung capacity.
Functional Residual Capacity
 40–50% of total lung capacity.
Physiologic Dead Space
 ml/lb body weight.
Minute Volume (basal)
 2.5–5.0 liters/min/m^2
Walking Ventilation
 (180 ft/min) = 8–11 liters/min/m^2
Timed Vital Capacity

	Lower Limit of Normal	Average Normal
1st second	75% of V.C.	83% of V.C.
2nd second	90% of V.C.	94% of V.C.
3rd second		97% of V.C.

Maximal Breathing Capacity
  Young adult male   150 liters/min
  Young adult female 100 liters/min
Arterial blood oxygen saturation
 97% ± 1%
Arterial blood,
 partial pressure oxygen 95 ± 5 mm Hg
Arterial blood,
 partial pressure CO_2 40 ± 2 mm Hg
Oxygen Capacity of Hemoglobin =
 20.6 ml O_2/100 ml whole blood

Application of Pulmonary Function Tests. In occupational medicine, the principal application of pulmonary function tests is found in studying individuals who are suffering from fibrotic diseases of the lungs due to inhalation of harmful dust. When groups of these tests are applied it is possible to determine, to a certain extent, the degree of functional impairment. Attempts have been made to correlate pulmonary function tests with x-ray findings, but the results have generally not been satisfactory. Abnormalities of the heart or blood vessels may affect the results of pulmonary function tests to a marked degree and consequently these tests cannot be applied to estimating the degree of lung injury when cardiovascular disease coexists.

SECTION 2

INDUSTRIAL AIR CONTAMINANT CONTROL

Benjamin Feiner

Head, Environmental Control Unit, Engineering Section
Division of Industrial Hygiene
New York State Labor Department
New York, N.Y.

When materials are used industrially so that air contaminants are created, generated, or released in concentrations which may injure the health of workers, the usual method of providing protection is by means of ventilation, usually local exhaust. There are, however, other methods of protection which should be investigated before consideration of ventilation control.

CONTROL BY METHODS OTHER THAN VENTILATION

The basic principle of industrial hygiene is to prevent dangerous materials from coming into contact with workers.

While ventilation control is the most widely used method of achieving this protection, a number of other, simpler procedures are available. These can often effect considerable operational economies as well as reduce the cost of ventilation or eliminate its need entirely, and are frequently the most practical method of providing such protection.

Design into Plant or Process

It is becoming increasingly common for plant and design engineers to consult with the industrial hygiene engineer at the design stage of a new plant or process. Consideration of industrial hygiene control principles at this point can eliminate or simplify the costly exhaust ventilation which would otherwise be necessary. Automation and automatic operations which require few if any workers are examples of this principle. Other examples include the design of continuous, enclosed chemical processes instead of batch processes, underground trenches for ventilation ductwork and grouping of hazardous operations to localize control, such as central shakeout stations, etc.

Process Change

A simple process change can often not only reduce contaminant dispersion but improve production efficiency. The following are examples of decreasing contaminant dispersal or otherwise reducing exposure: metal joining by welding or crimping instead of soldering; a change in temperature, speed or pressure of a chemical reaction; automatic electrostatic paint spraying instead of manual compressed air paint spraying; mechanical continuous hopper charging instead of manual batch charging.

A change in the physical condition or container specifications of raw materials received by a plant for further processing may be salutary. Thus, use of pelletized or briquetted materials that are ordinarily dusty, such as carbon powder, may drastically reduce atmospheric dust contamination at several steps in a process. Batch charging of slightly wetted materials or in paper bags rather than in a dry bulk state may eliminate or reduce the need for control in storage bins and batch mixers.

Substitution

An often effective and usually inexpensive method of control is the substitution of nontoxic or less toxic materials for highly toxic ones. The classic examples of substitution as a control measure include replacing of mercury used in the processing of fur into hatter's felt with non-mercurial compounds; replacement of white lead in paint pigments by zinc, barium or titanium oxides; the use of mixtures of paraffin hydrocarbons instead of benzene (benzol) in the rubber products industries; and the almost complete disappearance of the hazards of carbon tetrachloride from industry by substitution of much less toxic chlorinated hydrocarbons. We may also cite the use of steel shot instead of sand for abrasive blasting, synthetic rather than sandstone grinding wheels, and nonsilica parting compounds in foundry molding operations.

Frequently, such substitution carries with it a bonus in the form of an operational improvement. For example, the replacement of benzene by toluene in the manufacture of self-sealing aircraft gasoline tanks during World War II resulted in elimination of the troublesome problem of too rapid drying of the benzene between the various steps of the process.

Substitution often requires a good selling job to overcome the reluctance of production engineers to change the status quo. Fortunately, it can usually be demonstrated that substitution can result in an overall increase in economy.

Isolation in Time or Space

Many operations which do not readily lend themselves to ventilation control because of their nature or extent may generate contaminants in such quantities as to permeate an entire workroom or building and expose all the workers to a hazard, although only a few of them are actually engaged in the operations. In such instances, an attempt should be made to perform the operation so that only those workers immediately concerned with it need be within its influence.

One such method is isolation in time or space. In a foundry without central shakeout this operation may be performed after the regular shift has gone for the day. The three or four shakeout workers can be provided with suitable respirators for the one or two hours during which they are exposed to the silica dust. Blasting in mines at the end of or between shifts and housekeeping procedures or plant painting at night are other examples of work that can be scheduled so as to minimize the number of workers exposed to a hazard.

An operation can also be isolated in space. The ordinary furnace is a prime example of such isolation. Complete enclosure of a sand blast operation with an air line respirator for the worker within the enclosure is another instance. Some operations require complete enclosure and remote control so that nobody is exposed, as in many processes involving nuclear radiation.

Enclosing a dangerous operation or locating one or more dangerous operations together in a separate room or building not only sharply reduces the number of workers exposed but greatly simplifies the necessary control procedures. Plating tanks, lead melting pots, paint dipping operations and similar processes, when located in a separate room and grouped together, can usually be provided with efficient and relatively inexpensive local exhaust systems. Where continuous supervision of an operation by a worker is not necessary, only general ventilation may be required to prevent escape of contaminants into the main workroom. If necessary, the exposed worker can in such a case be provided with a respirator for use during his brief periods of exposure.

Segregation of Personnel

An opposite approach, where contaminant-producing operations must be carried out over a large area, is to segregate the worker from the operation. The crane operator in a large foundry, bulk material storage building, or cement clinker shed can be provided with a completely enclosed cab ventilated under positive pressure to keep contaminants out. In automatic stone crushing, grinding and conveying processes, where only periodic or emergency attendance is required by an operator, small ventilated rooms strategically located within the large workroom can be occupied by the workers during the major part of the work day. If tempered air is supplied to these rooms, the necessity of heating a very large building can frequently also be avoided.

Local Suppression of Contaminants

Mechanical or physical prevention or lessening of contaminant release can reduce or eliminate the need for ventilation control. The wetting of dust with water or other liquids is one of the oldest methods of control and may be very effective if properly used. Wet drilling in mines and quarries; water sprays at blasting, crushing, conveying operations and foundry shakeout; wet grinding and machining—all these can reduce dust dispersion if properly applied. The application of water must be designed to blanket the dust source completely. The particles must be thoroughly wetted by means of high pressure sprays, wetting agents, deluge sprays or other procedures as indicated. Means must be provided for containment or continuous removal and disposal of the wetted dust. Baffles can be used around an operation releasing dust at high speed. Considerable mist and vapor suppression can be accomplished by flotation of plastic chips, plastic foam or nontoxic, nonmiscible liquids with low vapor pressure on the top of a large, high vapor pressure liquid surface which is releasing a contaminating vapor.

Housekeeping

The conventional need for and advantages of good housekeeping are familiar to everyone. Less commonly appreciated is the role of good industrial housekeeping as a positive factor in contaminant control. Dust or other contaminants which fall or settle onto the floor, workbenches, machines, walls, rafters and ledges may become air-borne again by the action of ambient air currents, drafts, vibration, and normal plant activity. Constant good housekeeping by vacuum cleaning or wet washing is necessary to remove these materials continually and to prevent their becoming an additional source of air contamination. This is frequently true of gross materials as well as dusts. Lead pot dross accumulating on the floor may be reduced to airborne lead particles by the abrading action of being stepped on continuously.

In many industries, air recontamination resulting from inadequate housekeeping may be as significant a source of contaminant as the operations themselves. Lead storage battery manufacture, high-silica stone crushing and screening, and mercury thermometer manufacture are examples of the need for proper housekeeping practice as a means of contaminant control.

Design is an important factor. Light-colored walls, good illumination, smooth, impervious, coved floors, and careful workspace design simplify and increase the incentive for proper housekeeping.

Personal Respiratory Protection

This will be discussed in Section 3.

CONTROL BY VENTILATION

The design of exhaust ventilation usually depends on several factors, such as the physical state of the contaminant, i.e., dust, fume, smoke, mist, gas or vapor, the manner in which it is generated, the velocity and direction of release to the atmosphere, and its relative toxicity.

Section 1 dealt with the methods by which relative toxicity is determined and discussed two methods for describing it quantitatively. The concept of maximum allowable concentration (MAC), now more commonly described as threshold limit values (TLV), expresses toxicity in numerical values. The units may be in terms of parts of the contaminant per million parts of air (ppm) for gases and vapors, milligrams of contaminant per cubic meter of air (mg/m^3) for solids and mists, and millions of respirable particles per cubic foot of air (mppcf) for mineral dusts. These values, when used with understanding, give a semiquantitative index of concentrations which a normal worker can safely tolerate for an eight hour daily exposure for an indefinite length of time at the rate of 5 days per week. The manner in which these values have been derived, their limitations, and the precautions with which they must be used have already been described in Section 1. These considerations are stressed here because, as we shall see later, the TLV data are sometimes used in ventilation design calculations. They will also influence selection of ventilation type and must be carefully interpreted and never used as absolute values. A list of TLV's is published annually by the American Conference of Governmental Industrial Hygienists (AGGIH).

The other method of delineating relative toxicity described in Section 1 is by the more general concept of slightly toxic, moderately toxic, and highly toxic. Wherever toxicity information is available on the materials listed in Section 12 they have been classified into categories of slightly, moderately, or highly toxic. Substances of unknown toxicity should usually be treated as highly toxic, until evidence to the contrary becomes available. The substances for which numerical TLV's have been established may also be classified as slightly, moderately, and highly toxic in accordance with Table 1.

TABLE 1. Threshold Limit Values

Toxicity	ppm	mg/m^3	mppcf
Slight	over 500	over 0.5	50
Moderate	101–500	0.101–0.5	20
High	0–100	0–0.1	5

A reliable determination of the health hazard created by an industrial process and the need for control requires careful evaluation by a trained industrial hygienist. Chemical analyses of air for contaminants are usually necessary. The concentration of contaminant thus measured, its toxicity rating, and an evaluation of such pertinent factors as rate of work, length of work day, time of year, housekeeping, and working conditions will determine the need for and extent of ventilation control.

In general, it may be stated that an operation or process releasing highly toxic air contaminants *almost always* requires control; an operation releasing moderately toxic contaminants *usually* requires control; one releasing slightly toxic contaminants *occasionally* requires control.

In all cases, one of the most important factors in determining the need for control will be the toxicity and concentration of contaminants created—because the more toxic the contaminant, the lower is the atmospheric concentration which can be tolerated without injury and without the need for control.

The ventilation design data which appear later in this section have been related to the toxicity ratings in Section 12. Ventilation control is applicable to all air-borne substances which are shown in Section 12 as having a potential systemic toxicity upon entering the body by inhalation. In addition, those substances which, when air-borne, can cause local injury to the mucous membranes or skin are also amenable to control by ventilation and will be so treated.

The basic ventilation methods are *local exhaust* ventilation and *dilution* (general) ventilation.

Local Exhaust Ventilation

In local exhaust ventilation, the air-borne contaminant is removed or captured from the environment at or as close as possible to the source.

It is the method of choice wherever contaminants are released from a sharply defined and narrowly circumscribed operation. Its basic advantages are that contaminant control can be positive and efficient, exhaust air volumes are relatively low, and the cost of make-up air is kept at a minimum. Its disadvantages are high initial cost of installation, elaborate system of hoods and piping frequently required which use valuable plant space, and relatively high power requirements.

The basic principles and elements of local exhaust must be understood before design can be attempted. These fall into definite categories.

General. Air will flow from one point to another when there exists a difference in "head" or pressure between the two points. In nature, such flow is random, as the flow of air from a high pressure weather system to a low pressure one.

A local exhaust system is a mechanical device for creating a pressure differential between two points so as to direct air flow. At one end of a system the fan creates pressure in a hood or orifice lower than the air just outside by means of piping connecting the hood and the fan. The air which flows into the hood because of the pressure difference travels to the fan and is then conveyed from the fan by means of discharge piping to a desired location.

This series of events performs a number of functions. The hood, which is located around,

at, or near a source of atmospheric contamination, prevents the escape of the contaminant, captures it, or acts as a curtain between the contaminant and the workroom. The piping through which the air is moved by the fan serves both to pull air through the hood and to convey away the contaminant captured by the hood. The contaminant-laden air reaching the fan may be discharged by the fan to the outdoors untreated, to the outdoors after being cleaned of contaminant by means of an air cleaner, or in certain instances, back to the workroom after being cleaned (see discussion of Recirculation, p. 30).

Hoods. The function of an exhaust system is to protect the worker from exposure to potentially hazardous contaminants created in the workroom. As such, the heart of the system is the hood or orifice. Design of a system begins with the hood, which is usually a compromise between the ideal and the practical. Hood design is governed by a consideration of the physical state of the contaminant to be controlled as well as such details as temperature of process, velocity and direction of contaminant trajectory, nature of operation, degree and closeness of attention required by the operation, and extent of control needed. In simple terms, the designer will first mentally enclose the entire source of contaminant and then make such openings or changes in it as are dictated by the operation. He will depend upon air flow and velocity to make up for the openings or design compromises.

Hood types include complete enclosure, partial enclosure, shaped enclosure, lateral rear or overhang, canopy, and free hanging. The *complete enclosure* is typified by the abrasive blasting room where the worker is inside the room but is completely protected by an air supplied helmet. The function of the enclosure is to protect not the operator but other workers in the vicinity. The standard spray booth is a *partial enclosure* where the worker sprays objects inside the enclosure through an open front or from within the enclosure. Here, the flow of air through the frontal opening acts as a curtain between the point of spray or contaminant release and the workroom. It provides a capture velocity between the worker and source.

An example of a *shaped enclosure* is the ordinary grinding-wheel hood which assumes the contours of the wheel with a minimum sized working opening through which, ideally, a velocity high enough to capture the fine particles carried around the wheel is maintained. In such hoods, the pipe take-off is set tangential to the point of operation so that the heavy particles projected with high velocity travel directly into the pipe. Inner baffles are often used to decrease particle velocity.

Open surface tanks are usually provided with *lateral slots*, *vertical* or *partially overhanging* rear hoods, or rarely, *overhead canopy* hoods with one or more sides. When properly designed, these hoods rely on a number of factors to provide good control. There are, for instance, a capturing effect, a control effect produced by bringing a mass of air down to the tank and into the hood, and a dilution effect, whereby that portion of the contaminant which is not captured is diluted to safe levels.

A common *free-hanging* hood is the welding fume hood consisting of a flexible pipe with a flanged orifice which can easily be located near the point of weld. Here the contaminant is captured as it is released and carried into the hood before it becomes dispersed and thus difficult to capture.

The design of hood shapes is governed by two inherent characteristics of air entering an opening under suction: (1) air will enter the opening from all directions, and (2) in general the velocity of this air at distances from the opening follows an inverse square law, with some exceptions to be noted later. Thus, if we assume an ideal point source of suction, air will travel to the point equally from all directions, and the equal velocity contour at any distance from the point will have the shape of a sphere whose radius equals the distance. Round, square, or essentially round or square openings are practical counterparts of the ideal point source of suction. They will have essentially spherical velocity contours, and the velocity at a distance from the opening will largely follow the inverse square law. Suitable baffling or flanging of the hood opening can limit the directions from which air will flow into the hood and thus create air ve-

locities at a selected point higher than with an unflanged hood.

Similarly, one can postulate an ideal line source of suction. Flow into the line, which will come from all directions perpendicular to the line, will have a cylindrical contour. The slot is the practical counterpart of the ideal line source. Experimental data indicate that flow into a slot will be such that for practical purposes velocity at a distance from the slot will vary inversely as the distance rather than as the square of the distance. Here also, flanging can increase capture efficiency.

Rectangular hoods with large width-to-length ratios are essentially similar to round or square openings. As the hood shape changes from square to increasingly narrow rectangular, the characteristics of air flow will vary gradually from those of square to those of slot hoods. Since the characteristics of air flow into these transition hoods are complex, it may be arbitrarily assumed that rectangular hoods with width-to-length ratios higher than 0.3 are square hoods, and hoods with ratios of 0.3 and less are slot hoods.

These considerations are utilized in the derivation of hood equations and in the design of hoods to the end that the hoods provide a capture velocity at the point of contaminant release with the least possible air flow. Typical capture velocities are shown in Table 2A.

Piping. Piping in an exhaust system serves the dual function of acting as a link between the fan and hood, so that the desired air quantity is drawn into the hood, and of conveying the contaminant-laden air to a desired location. Pipe construction, including elbows, fittings, tapers, and joints, requires due consideration for aerodynamic principles so that resistance to flow is kept at a minimum. Material and gauge of piping will depend on the length and diameter of

TABLE 2A.   Range of Capture Velocities

Condition of Dispersion of Contaminant	Examples	Capture Velocity (fpm)
Released with no velocity into quiet air	Evaporation from tanks; degreasing, etc.	50–100
Released at low velocity into moderately still air	Spray booths; intermittent container filling; low speed conveyor transfers; welding; plating; pickling	100–200
Active generation into zone of rapid air motion	Spray painting in shallow booths; barrel filling; conveyor loading; crushers	200–500
Released at high initial velocity into zone of very rapid air motion	Grinding; abrasive blasting; tumbling	500–2000

the pipe, the pressure of the system, and the abrasive or corrosive nature of the contaminant being conveyed. Design pipe velocities, and therefore diameters, are a function of the minimum required transport velocity for a particular contaminant. Since a truly air-borne contaminant is part of the air stream, there is no minimum pipe velocity requirement for such contaminants and pipe diameters may be selected on the basis of compromise. Low velocity (large diameter) pipe calls for less fan horsepower but more metal and plant space. High velocity (small diameter) pipe requires less metal and space but more horsepower. For air-borne contaminants, (gases, vapors, and most fumes and mists) a pipe velocity of 2000 fpm has been found to be optimum. For particulate matter, pipe velocities required for transport will range from 2500–5000 fpm, with usual velocities in the 3000–4000 fpm range, depending on the density, size, and shape of the contaminant particles. Typical transport velocities are shown in Table 2B.

A very simple, but useful equation of flow, $Q = AV$, is basic to exhaust system calculations. Q is rate of flow in cubic feet per minute (cfm) of air flowing through a certain area (A) in square feet, usually a hood opening or pipe, at a velocity (V) in linear feet per minute (fpm). If any two terms are known or specified, the third can be obtained by calculation. In this manner, hood or slot face area, pipe diameter, and pipe or hood air flow or velocity can be determined.

TABLE 2B. Average Transport
Velocities for Dust

Material	Minimum Transport Velocity (fpm)
Very fine, light dusts	2500
Fine, dry dusts and powders	3000
Average industrial dusts	3500
Coarse dusts	4000–4500
Heavy or moist dust loading	4500 and up

Fans. Numerous fan types are available to the exhaust system designer, but as a rule the individual exhaust system limits the type of fan which may be selected. The propeller fan, usually with a cast metal blade, is used principally for spray booths and similar applications for movement of large volumes of air at relatively low pressures. The axial (vane or tube) fan is similar, but it can generate higher pressures and is used where higher pressures must be

maintained and where an in-line fan is advantageous.

There are three principal types of centrifugal fans. Each has its own application, although there is some overlapping of characteristics. The forward curve (commonly known as "squirrel cage" or "sirocco") fan is a relatively high volume, quiet fan with low space requirements; it is used principally to move air at low pressure with some applications for fume or mist handling. The radial (paddle wheel) fan is a pressure exhauster used for dust-handling systems. The backward curved fan is an efficient, non-overloading fan used to move clean air or air containing gases and vapors at moderate pressures, although it can be used for dust systems when located on the clean side of a dust collector.

Selection of a fan is governed by the material to be handled, need for flexibility, location, space limitations, noise generated, and necessary efficiency. Although selection is usually done from the manufacturer's certified multi-rating tables, the fan laws, which describe the relationship between cfm, pressure, horsepower, speed, and size in fan systems can be used. This permits selection of a fan, evaluation of the adequacy of an existing fan for a new system, and determination of the effect of a change in conditions in an existing system.

Air Cleaning. Air cleaning, when required, is an integral part of an exhaust system and must be designed to be compatible with all the other components of the system. The nature and amount of contaminants and required degree of air cleaning will dictate the design and size of air cleaner to be used. This in turn will influence the type and horsepower of the fan to be used. (See Air Cleaning, p. 45).

Make-up Air. One of the most important and least understood considerations in exhaust system design is the necessity to supply tempered make-up air to the workroom to replace the air removed by the exhaust system. All too frequently this need is overlooked and natural infiltration is relied on. In some instances, natural infiltration may be acceptable, particularly where the exhaust volumes are low and plant heating capacity is adequate. However, this method usually produces adverse effects, particularly in a new "tight" building which will not permit enough make-up air to enter by infiltration. This will increase the resistance to be overcome by the exhaust fan and thus reduce the exhaust volume and the degree of control provided by the exhaust hoods. Other adverse effects include negation of the action of natural draft stacks with introduction of contaminants into the workroom; and creation of cold drafts, particularly in undesirable locations. Such drafts can cause workers to shut off exhaust fans and can result in inefficient or inadequate heating. High velocity cross drafts may interfere with the operation of exhaust hoods and can reintroduce settled contaminants into the workroom air.

A properly designed make-up air system should supply tempered air approximately equal in volume to the air exhausted. A slight excess of supply over exhaust is usually desirable for most efficient operation. In some cases, an excess of exhaust is required to prevent contaminants from entering adjacent workrooms. Make-up air can be provided by means of a central-heated supply system, with filtered intakes drawing air from uncontaminated outside areas, and ductwork terminating in suitable grilles or registers. This technique is preferable since the supply air can be directed to desired locations to achieve such incidental benefits as spot cooling or heating and dilution of air-borne contaminants near their point of origin. However, make-up can also be provided by means of fresh air supply unit heaters properly located in windows, exterior walls or on the roof. These systems cannot be carefully designed, but they are flexible and eliminate the need for expensive ductwork. In some instances, untempered make-up air can be brought directly to the vicinity of an exhausted operation when there are no workers in that vicinity. It is frequently possible to design supply systems to provide comfort ventilation at the living zone of the plant (below the 8–10 foot level) and thus eliminate the need for heating air in unoccupied zones in high ceilinged workrooms. Where possible, the supply air should be zoned from clean to increasingly contaminated areas to achieve a dilution effect (see Dilution Ventilation, p. 38).

Recirculation. A discussion of make-up air must inevitably lead to a consideration of recirculation. This is the process whereby air discharged by an exhaust system is completely or partially returned to the workroom after suitable cleaning, obviating the need for costly supply air systems, or making possible much smaller systems. Recirculation has always been attractive to management because of the savings in both capital costs of installing a tempered supply system and heating fuel costs. Recent developments have accelerated the trend to install recirculation. Many plants are now air conditioned in summer as well as heated in winter. The new sophisticated technologies frequently require the maintenance of workroom air to rigid temperature, humidity, and dust concentration specifications. Continuous discharge of

this expensively conditioned air imposes a severe economic burden. Furthermore, the increasing emphasis on air pollution control is resulting in the requirement of air cleaning devices where none was needed before or the use of expensive, high efficiency air cleaners where cheaper collectors were previously acceptable. Obviously, it would be highly desirable to minimize the financial impact of costly air cleaning by recirculating the cleaned air. Indeed, under conditions where recirculation is permissible, it may be economically feasible to provide high efficiency air cleaning with recirculation in cases where air pollution requirements would permit low efficiency cleaners with discharge to the outdoors. Lastly, there are cases where recirculation is mandatory when exhaust systems are so located that it is virtually impossible to run a discharge pipe to the outdoors.

Most states and other governmental agencies having applicable jurisdiction require prior approval for recirculation. Recirculation from systems handling highly toxic contaminants is almost never granted because very few cleaning devices are efficient enough to afford the required high degree of air cleaning. Even where such air cleaners are available, recirculation of highly toxic contaminants is contraindicated because of the severe potential hazard created by improper operation or maintenance of air cleaners or by a sudden massive collector failure. Systems handling moderately toxic contaminants may be recirculated if compelling reasons exist, if favorable conditions are present, and if proper safeguards are provided. However, recirculation from systems handling contaminants of little or no toxicity is becoming increasingly acceptable.

There are a number of mandatory or desirable conditions for recirculation which must be considered.

(1) High efficiency air cleaning must be provided so that the concentration of contaminant in the returned air is a fraction of the threshold limit value (TLV). Usually it should be not greater than 20 percent of the TLV and lower if possible. It should be noted in this connection that rated cleaning efficiency of collectors, which is expressed in percent by weight of contaminant removed from the air stream, may be misleading when choosing a collector for a recirculating system since many TLV's are expressed in terms of parts per million parts of air (by volume) or millions of particles of dust per cubic foot of air. Thus a collector with even 99 percent efficiency by weight may be inadequate and a number of such collectors in series may be required.

(2) Wherever possible recirculation should be

limited to a large workroom with good ventilation so that odors or contaminants in the returned air are effectively diluted.

(3) Year-round, 100 percent recirculation should usually be avoided for large systems or systems handling odors or moderately toxic contaminants. It is good practice to provide the discharge from the air cleaner with a Y connection, one arm of which is ducted to the outdoors and the other back to the workroom. A motorized damper installed in the Y will direct the air outdoors during those times of the year when the workroom air is not tempered and return it to the workroom at other times. This has the advantage of providing additional comfort or general ventilation when recirculation is not needed. Frequently, the damper motor is thermostatically controlled so that air is automatically returned to the workroom only when the outside temperature is below a predetermined level during the heating season and above a certain level during the cooling season. Ability to discharge to the outdoors is also a useful safety factor in the event of sudden air cleaner failure. (A pressure actuated alarm device should be used to signal such failure.) In certain cases it is desirable to limit the amount and duration of recirculation as much as possible, and yet take advantage of its desirable features to the fullest. This can be accomplished by automatic temperature actuated modulating dampers set for complete recirculation during peak heating or cooling times, no recirculation when neither heating or cooling is required, and varying percentages of recirculation, depending on the temperature, at all other times.

(4) Recirculation should not create drafts on workers or excessive "high velocity air noise" which is usually in the more injurious high frequency range and may require mufflers or other acoustic control devices.

System Testing. Since exhaust system design involves empirical as well as theoretical considerations, it is necessary that all systems be tested after installation to ensure their effectiveness. Such tests are made to ensure that the design exhaust air quantity is being maintained into the hoods and that this design air quantity is actually providing the necessary control. This latter point is particularly important where toxic materials are to be controlled.

A swinging vane anemometer (velometer) will suffice to determine the air velocity into a simple hood with a clearly defined opening. Usually velocity measurements by means of a Pitot tube in the duct behind an exhaust hood are necessary. Other physical measurements which are made to evaluate an exhaust system will include static pressure or throat suction at all hoods,

static pressure at both sides of the fan and collector, fan rpm, and brake horsepower.

Determination of the degree of control provided by a design air quantity can usually only be made by chemical air analyses for the contaminant at the hood and in the worker's breathing zone. Chemical analyses of air are almost always required also for testing a dilution ventilation system. Such tests are also frequently made at the inlet and outlet of a collector to determine its efficiency.

Exhaust System Maintenance. Proper maintenance is vital because systems will decrease in efficiency for a number of reasons. An inefficient exhaust system may be worse than none at all since it provides a false sense of security. Obvious maintenance procedures include routine inspection to replace broken or corroded hoods, ducts, fan blades, and collector elements, as well as routine cleaning of hoods, ducts, and collectors. Constant surveillance to warn against gradual diminution of air flow is the critical element in maintenance.

Fortunately, such surveillance is relatively simple. After a system has been designed, installed, and tested, the hood suction and the fan inlet suction measured when the system is new and performing at maximum efficiency are accurate indices of the proper quantity of air flow into each hood and the system as a whole. (Systems with cloth dust arresters should be checked in this manner after several weeks of use.) A pressure tap at the fan inlet pipe and at each hood, connected to a suitable pressure indicating device, will reveal when the air flow decreases significantly. The pressure created by proper air flows can be posted at each gauge which should be designed to give a visual or audible indication of air flow diminution. In some instances, automatic devices are used to shut down operations if the air flow drops too much.

In summary, a local exhaust system is a mechanical air handling and cleaning system with a number of interrelated components. The inadequacy of any one of these components because of improper design or the failure of any one of these components because of poor maintenance will result in a loss of efficiency and the possible introduction of a health hazard.

Specific Hood Design Considerations

Local exhaust ventilation comprises a more or less elaborately engineered system containing a number of interrelated components. In usual practice, design of such systems is an engineering procedure involving the utilization of aerodynamic principles, mechanical engineering fundamentals, knowledge of the behavior of particles in an air stream, as well as industrial toxicology.

It does not fall within the scope of this book to treat of these matters thoroughly or in detail, nor do space limitations permit an exhaustive discussion. Complete treatments are available in numerous published works, a number of which have been listed at the end of this section. When possible, exhaust systems should be designed by an industrial hygiene engineer or the standard books in the field should be consulted as a complement to the design features to be described below.

However, industrial hygiene engineering is a specialized profession, with very few private consultants available to industry. There is a need, in work of this kind, for a simplified approach to local exhaust hood and dilution ventilation design which can be used by the plant or production engineer. Management will not very often have industrial hygiene engineering personnel; there will frequently be the need for emergency installation of exhaust ventilation; and there may not always be the time or personnel for careful study and application of the engineering data in the standard works.

Therefore a simplified approach is offered to meet this pressing need. Required cfm or fpm for basic hood types under average conditions are indicated and modifying factors for unusual or varying conditions are given. The values used are derived in the main from a consensus of published data tempered by many years of experience. It is felt that intelligent application of the material to follow will result in sufficient data for adequate control. No specific information on design and selection of piping, fans and collectors will be given, since a wealth of information is available in the various sources of information listed at the end of this section.

The hoods to be described are basic and can be applied, as shown in the figures, to many operations. They can be modified, adapted or even combined in many ways, limited only by the designer's imagination, to meet a particular situation.

As has been previously noted, the hood is the heart of the local exhaust system and dictates the design of all its other elements. Several considerations bear repeating, i.e., the hood should enclose the operation as much as possible or be as close to it as possible; it should permit air passage with a minimum of extraneous restrictions and obstructions; baffles and flanges should be provided to concentrate flow of air in the desired areas; and the hood cfm should provide a velocity high enough to capture or contain the containment (see Table 2A).

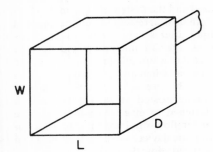

$$\text{cfm} = (L)(W)(V)(a)$$

where

L and W = open face dimensions (feet),
     V = required control velocity at the face (fpm),
     D = depth from face to rear or line of baffles,
     a = a temperature correction factor.

V^a Values for Different Depths $(D)^b$

Toxicity	Shallow	Average	Deep
Slight	125	75	50
Moderate	150	100	75
High	200	150	125

Rate of Evolution and Temperature Factor (a)

Temperature (°F)	Rate	a
Up to 150	Low	1.0
151–300	Moderate	1.25
301–600	High	1.50
601 and up	—	2.0

[a] For extremely toxic and some radioactive substances, V may have to be 200 or 250 for average depths and increased proportionally for shallow hoods. Where a contaminant is released within the hood at a high velocity, particularly toward the opening, the selected V should also be increased.

[b] The booth should completely contain the operation and allow working room. Depth to the rear or the line of baffles should be as large as possible. Expressing depth as a function of W or L (whichever is the larger), depths from 65–85% of W or L may be considered average. Depths up to 65% may be considered shallow, and depths above 85% may be considered deep.

Figure 1. Enclosing Hood or Booth.

Special attention must be given to insure uniformity of flow into the hood face or slot. For a simple hood or slot, a gradually tapered transformation from the face to the duct inlet will suffice to provide uniform face velocities. Long hoods can have several transformations. Where a slot is part of a manifold, internal vanes can provide uniform air velocity into the slot. In most cases, uniformity of distribution is designed by providing a restriction at the face, slot, or within the hood so that the velocity (and therefore resistance to air flow) in the restriction will be much higher than in the hood face. When this is done, air flow through the restriction will distribute itself uniformly through the restriction, and consequently through the hood opening.

In certain instances, a slot acts as both the restriction and the hood. A simple illustration of this principle is the spray booth line of baffles, where the velocity into the spaces between the baffle plates is much higher than the velocity in the section of the booth behind the baffles. Here, air velocity through the baffle plates, and into the booth face will be uniform. Other examples of providing for distribution include oversized low velocity manifolds behind narrow, high velocity slots and plenums with perforated plates.

Theoretically, the air flow required to control a contaminant using a specific hood depends in most cases only on such physical considerations as temperature, method of contaminant generation, velocity and trajectory of propulsion of contaminant, and distance of hood face to point of contaminant release. However hood cfm design data must show increasing air flow requirements with increasing toxicity of contaminant. This is so because very few exhaust hoods are 100 percent effective in controlling all the contaminant released at an operation, and control for the more toxic contaminants which have much lower TLV's must be greater than for the less toxic substances.

The basic hood types include free-hanging plain (square or round) openings, free-hanging slot openings, enclosure with frontal opening, lateral exhaust at a tank or table top, down-

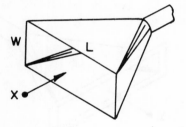
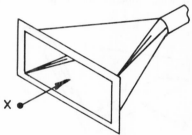

$$cfm = (V)(10X^2 + A)^a$$

where V is required velocity (fpm) at distance X (feet) from the hood face of area A (square feet).

V Values for Different Rates of Contaminant Release[b]

Toxicity	No Upward Velocity or Low Rate of Evolution	Moderate Upward Velocity or Moderate Rate	Active Generation or High Rate	Violent[c] Generation
Slight	50	75	125	500
Moderate	100	150	250	1000
High	150	225	400	1500

For flanged hoods or hoods on table tops the cfm may be decreased by 25%.

[a]This formula may be used for openings with W/L ratios above 0.3 as well as square or round openings. This type of hood should not be used where significant cross drafts may be present unless side shields are provided.

[b]Where reliable capture data are available for a particular operation and contaminant, these data should be used in place of the V values shown above.

[c]This type of hood is not recommended for violently generated contaminants.

Figure 2.   Free Hanging Plain Openings.

draft and canopy hoods. A considerable volume of data on hood design and cfm requirements for specific operations is available (see references), and the discussion to follow must of necessity be general. The design data for each type of hood are shown in Figures 1 through 6.

Enclosing Hood (Fig. 1). This is the method of choice whenever operations permit its use. It is the simplest to build, usually does not interfere with operations, and can provide the most efficient control with minimum air flow rates. The enclosure principle finds many applications in local exhaust ventilation, including spray booths, melting pots, cabinet blasting, tunnel drying, belt transfer, mulling and screening. Most of the specially shaped hoods for such equipment as woodworking machines and buffers are adaptations of the enclosure principle, taking advantage of the characteristics of each machine.

Free-hanging Plain Openings (Fig. 2). These are essentially round or square openings or rectangular openings with width-to-length ratios above 0.3. They are used where the contaminant is derived from a point source or a very small area and where operations do not permit en-closure or encumbrance by hoods. They may be permanently mounted in position or may be attached to a flexible pipe for adjustment as operations require. Applications will include arc welding, soldering, tail pipe exhaust, and high toxicity high precision milling and machining. The latter use requires very careful design and operation.

The basic formula in Figure 2 is derived from Dalla Valle. The formula indicates that the cfm value needed to maintain a required velocity varies as the square of the distance of the hood face from the contaminant source. For this reason, the hood must be kept as close as possible to the contaminant source. If close placement is not possible, this type of hood may become prohibitively expensive in terms of required cfm. Flanging the hood greatly increases its efficiency and should be used whenever possible.

Free-hanging Slot Hood (Fig. 3). This is a rectangular hood similar to the free-hanging round or square hood which is used in similar circumstances where the zone of contaminant generation requires a long narrow source of suction. The hood may be considered a slot when the ratio of width to length is 0.3 or less.

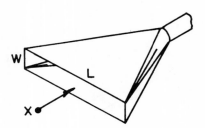 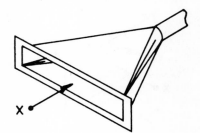

$$cfm = (3.7)(V)(X)(L)^a$$

where V is the required velocity (fpm) at distance X (feet) from the face of a slot L feet long.

V Values for Different Rates of Contaminant Release[b]

Toxicity	No Upward Velocity or Low Rate of Evolution	Moderate Upward Velcity or Moderate Rate	Active Generation or High Rate	Violent[c] Generation
Slight	50	75	125	500
Moderate	100	150	250	1000
High	150	225	400	1500

For flanged hoods or hoods on table tops the cfm may be decreased by 25%.

[a] This formula may be used for openings with W/L ratios of 0.3 or less. This type of hood should not be used where significant cross drafts are present unless side shields are provided.

[b] Where reliable capture data are available for a particular operation and contaminant, these data should be used in place of the V values shown above.

[c] This type of hood is not recommended for violently generated contaminants.

Figure 3. Free-hanging Slot Opening.

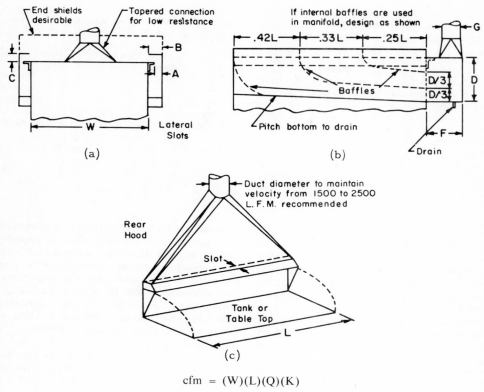

(a)   (b)

(c)

$$cfm = (W)(L)(Q)(K)$$

where

Q = cfm per square foot of tank or table surface,

W and L = open surface of tank or table top dimensions (feet),

K = a correction factor for different width-length (W/L) ratios.

Q Values for Lateral Exhaust at Tanks or Tables Where W/L Is 0.5[a]

Toxicity	Flash Point (°F)	Low Rate of Evolution	Moderate Rate of Evolution	High Rate of Evolution
Slight	Over 200	50–75	75–100	100–150
Moderate	100–200	75–100	100–150	125–175
High	Below 100	100–150	125–175	150–200

[a] The lower figure in each pair is for tanks or tables against a wall or baffle; the higher figure for free standing tanks or tables (the hood shown in Figure 4(c) also serves as a baffle). See also notes on p. 36.

Correction Factors (K) for Different W/L Ratios

W/L	K	W/L	K
0.1	0.70	0.6	1.05
0.2	0.80	0.7	1.10
0.3	0.90	0.8	1.20
0.4	0.95	0.9	1.30
0.5	1.0	1.0	1.40

Figure 4. Lateral Ventilation.

Notes to Figure 4:

(1) (a) and (b) show end and side views of a tank with lateral slots. Balanced flow through slot C can be achieved by a high velocity inner slot (A), large manifold cross-section area (B × D), or internal baffle vanes as in (b).

(2) Rear hood shown in (c) may be used when the front or ends of the tank must be unencumbered. Balance can be secured by 60° transformation pieces (one for each 5 feet of hood length) or high velocity inner slot.

(3) When slots on two opposite sides are provided, W is halved for the purpose of determing the correction factor. Side shields should be provided when cross drafts are present.

(4) Rate of evolution depends on temperature, gassing, evaporation rate, and agitation. At normal room temperature, quiescent operations will have a low rate.

(5) Slots can normally be sized for velocities from 1500–2000 fpm. Rear hood faces can normally be sized for velocities from 200–600 fpm.

(6) Flash points of flammable liquids are also a factor in this type of ventilation design. Both the toxicity and the flash point of such a liquid should be considered and the higher required Q value used.

(7) Vapor phase degreasers with condensing coils constitute a special case. Poorly operated or improperly designed degreasers will have a moderate or high rate of evolution. Good degreaser operation will have a low rate of evolution.

The formula developed by the late Professor L. Silverman, of Harvard University School of Public Health, indicates that required cfm varies as the distance from the contaminant, and close placement, while important, is not as critical as with the round or square hood. Again, flanging increases efficiency.

Lateral Tank or Table Ventilation (Fig. 4). A modification of the free-hanging slot which finds widespread use in local exhaust ventilation is lateral ventilation at tanks and tables, where operations make enclosing hoods impractical. This design utilizes slots along one or preferably two sides of the tank or table, although rear hoods along one side of the tank or table may also be used. Wherever possible, the slot or hood should be located along the long side or sides.

A basic ventilation design has evolved for this type of exhaust in which required ventilation is expressed in terms of cfm per square foot of tank or table top surface. As previously noted in the general discussion of local exhaust ventilation, this method of control has three features —a capture effect, a mass control effect, and a dilution effect (see discussion of Dilution later in this section) for the uncaptured contaminants.

These hoods can be utilized in an operation where the contaminants are released at or immediately above a flat surface. Tank or table top operations, plating and all associated tank operations, degreasing, paint dipping, rubber cementing, and air drying are among the operations amenable to this design. The tank or table width-to-length ratio is an important factor in determining cfm since it reflects the distance over which the air flow must provide control and should be as small as possible. Figure 4 shows basic rates for tanks with a width-length (W/L) ratio of 0.5 with modifying factors for other ratios and for varying conditions.

Downdraft Hoods (Fig. 5). These consist of a grille top table or platform above a hopper or plenum through which air is exhausted. They are used where the operation must be unencumbered on all sides or where it is desirable to have a downward flow of air past the source of contaminant. Examples of operations where these hoods may be used, include welding, soldering, metallizing, shakeout, portable tool chipping and grinding, spray painting, and solvent drying. Their efficiency is adversely affected by cross drafts and thermal updrafts, and they are usually used where no other hood type is feasible.

Canopy Hoods (Fig. 6). The familiar canopy hood should never be used where the operations require the worker to place his head under the hood since he will thus be exposed to the contaminant-laden air. Nor is its use advised in locations where cross drafts are prevalent. Where only infrequent tending of the operation is needed, the duct to which the hood is attached can be provided with a telescoping joint so that the hood can be lowered very close to the tank or table to overcome the effect of cross drafts.

Canopy hoods are often used at hot processes to take advantage of the thermal updraft. A frequently overlooked feature of such use is the secondary air induced into the area under the hood by the heated air rising into it. The fan must be selected to accommodate this secondary air as well as the design air quantity. Otherwise, air will spill out of the top of the hood. Design of canopy hoods for hot processes is fully discussed in Hemeon, "Plant and Process Ventilation."[4]

If carefully designed, canopy hoods can be used in a natural draft system sometimes more successfully than in a fan system. Of critical importance in such use are the shape and size of the hood, adequate duct diameter to move the air at

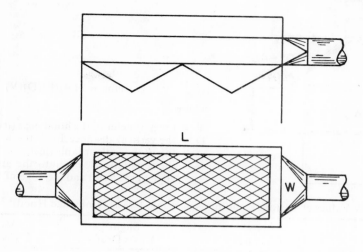

$$cfm = (W)(L)(V)(a)$$

where

V = required velocity through grille (based on gross open area),[a]

a = a temperature correction factor.

V Values at Various Conditions of Room Cross Drafts and Rates of Evolution[b]

	Rate of Evolution or Cross Drafts			
Toxicity	Low Rate or 0–35 fpm Draft Velocity	Moderate Rate or 35–75 fpm Draft Velocity	High Rate or 75–150 fpm Draft Velocity	Strong[c] 150–300 fpm Draft Velocity
Slight	100	125	250	350
Moderate	125	150	300	400
High	150	200	350	500

Temperature Correction Factor (a)

Temperature of Operations (°F)	a
Up to 100	1.0
100 to 150	1.25
150 to 300	1.50
Above 300	2.00

[a] For a required velocity above the grille use the formula in Figure 2.

[b] Baffles or side shields on one or more sides should be used wherever possible. These will permit selection of a V value from a lower cross draft category.

[c] This type of hood is not recommended for highly toxic contaminants, very hot processes, or, when unshielded, in locations with strong cross drafts.

Figure 5. Downdraft Hoods.

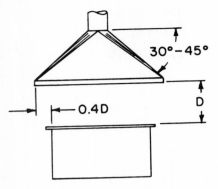

$$cfm = (1.4)(P)(D)(V)$$

where

P = the perimeter of the hood base (feet),
D = the vertical distance from the hood base to the top of the tank or table (feet),
V = the required velocity into the area between the edge of the hood and the edge of the table (fpm).

V Values at Various Conditions of Room Drafts (Unbaffled Hood)[a,b]

Toxicity	No Cross Drafts	Slight Cross Drafts	Moderate Cross Drafts
Slight	75	125	175
Moderate	125	175	225
High	175	225	275

[a] See Figure 5 for definition of cross draft categories.
[b] See text for a discussion of canopies for hot processes.

Effect of Baffles on Required V Values[c]

Baffles on two opposite sides—decrease V by 25
Baffles on two adjacent sides—decrease V by 50
Baffles on three sides —see Figure 1, enclosing hoods

[c] V cannot be less than 50 fpm.

Note:
Unbaffled canopy hoods should not be used in strong drafts. Hoods with baffles on two adjacent sides may be used in strong drafts, selecting V values from the table above under moderate cross drafts. In all cases where baffles are provided at canopy hoods, they should be perpendicular to the prevailing cross draft direction and on the lee side.

Figure 6. Canopy Hoods.

low velocities, sufficient length of vertical pipe to create a thermal head and, most important, a positive make-up air supply. In workrooms without adequate make-up air, during winter conditions air flow through natural draft systems may reverse and flood the workroom with contaminants. Wherever possible, canopy hoods should be provided with one or more sides.

Dilution Ventilation

Dilution ventilation consists of general ventilation of a workroom so designed that the contaminants released into the atmosphere are continuously diluted by the introduction of uncontaminated air to levels to which a worker can be safely exposed for eight hours a day. It is usually applied to the control of contaminants released over such a large area or in such a manner that local exhaust ventilation is impossible, impractical, or prohibitively expensive. It should almost never be used where local exhaust ventilation is feasible or where highly toxic air contaminants are involved. It is most successfully used to control the vapors evaporated from liquids such as solvents or thinners. Operations at which dilution ventilation may be used include dip or roller coating and air drying, use of volatile liquids in open surface tank operations, and cementing.

Dilution ventilation must be designed and tailored to meet a particular situation. Design factors will include rate of contaminant release as well as physical constants of the contaminant, toxicity, and workroom conditions. The random placement of one or two exhaust fans in the wall will usually be inadequate. Similarly, the

$$\text{Dilution (cfm)} = \frac{\text{Prevailing Concentration (ppm)} \times \text{Infiltration (cfm)} \times K}{\text{TLV (ppm)}} \tag{1}$$

obsolete concept of air changes per hour cannot be used because it cannot be related to contaminant generation rate. Another obvious drawback to the air change method of ventilation design is that it is a function of room volume. For example, an operation involving volatile solvents may evaporate one pint of acetone per minute and general room ventilation at twelve air changes per hour might be specified. In a workroom 100 × 100 × 20 feet, this is equivalent to 40,000 cfm. Identical conditions may exist in a workroom 20 × 30 × 10 feet. Twelve air changes an hour will require 1200 cfm. The former cfm would be excessive; the latter, much too low for adequate control.

The principal disadvantage of dilution ventilation lies in the high exhaust rates usually required which result in the need for heating and tempering large volumes of make-up air for efficient operation. The principal advantages

tions, the required dilution ventilation will be found by the following formula:

where TLV = the threshold limit value,
K = a factor of safety (see Table 3).

However, these data are difficult to obtain and are subject to error, and the cfm obtained cannot be extrapolated to a future change in conditions which may require lower or higher ventilation rates.

A more rational method is to use data on rate of contaminant release to determine rate of exhaust necessary to continuously dilute the contaminant to safe levels, where

sp. gr. = specific gravity of parent liquid (Section 12, d:),
M.W. = molecular weight of parent liquid (Section 12 mol. wt.)
TLV = threshold limit value of the vapor in ppm, (Section 12)
K = factor of safety (see Table 3).

The formula is:

$$\text{cfm} = \frac{(403)(\text{sp. gr.})(10^6)(\text{pints of liquid evaporated/min})(K)}{(\text{M.W.})(\text{TLV})} \tag{2}$$

its simplicity, low original cost, and low power requirements.

$$\text{cfm} = \frac{(387)(10^6)(\text{lb of liquid evaporated/min})(K)}{(\text{M.W.})(\text{TLV})} \tag{3}$$

Where the evolution rate is expressed in pounds per minute, the formula becomes:

Dilution ventilation is used primarily to control vapors and gases which mix readily and uniformly with the workroom air and whose rate of evolution is relatively constant and can be readily determined. It is rarely applied to the control of particulate matter such as dust, fumes, and mists since these do not mix uniformly and their rate of evolution is usually impossible to ascertain. In some instances, empirical data for fume or mist control by dilution are available. One such case is black iron or mild steel welding where cfm dilution rates have been set up on the basis of number of welders in a workroom and size of welding rod used.

Calculation of Dilution Ventilation Rates. One method of calculating cfm dilution rates involves a knowledge of prevailing conditions during normal operations. If tests can be made to determine the actual concentration of contaminant and the rate of natural infiltration of air into the workroom during existing condi-

Equation (2) above is only applicable to volatile organic liquids. Dilution ventilation for gases can also be designed, but determination of the rate of generation is much more difficult than with liquids. If the rate can be determined with reasonable accuracy, it should be expressed in terms of pounds per minute, and equation (3) should be used. Table 3 describes a simplified method of calculating dilution ventilation for substances for which TLV's have not been established using the general formulas and safety factor data shown.

Rate of Contaminant Generation. The amount of liquid evaporated per minute must be determined as accurately as possible. Several methods for such determinations are available. Usually process control data are the most reliable sources. For example, in a coating operation, the amount of solids applied to the product can be determined, and this information, coupled with production rate and coating material composition data, can be used to calculate pints of volatile material released per minute. Often, operating and supervisory personnel can supply fairly reliable information on rate of use of volatile materials. Purchasing department data over a sufficiently large period of time can also be utilized to determine the average amount of volatile material used per month. This can be translated into pints evaporated per working minute. Where an operation is performed for only part of a day, the time during which the vapor is being released is used to determine the rate of contaminant generation.

Effect of Distribution and Make-up Air. One of the most important factors in the design of dilution ventilation, and the one most frequently overlooked, is the need to supply sufficient and properly directed make-up air to replace the air exhausted. Ideally, the make-up air will be positively introduced and distributed in such a manner that the source of contaminant is situated between the worker and the exhaust outlet, the exhaust is located as close as possible to the source of contamination, all the air will pass through the zone of contamination and the contaminant will be diluted to the design levels as soon as it is generated. Under such conditions, the K factor in the formulas could be 1, since no factor of safety would be required. However, ideal distribution conditions can never be assumed and actual conditions will vary considerably.

Average conditions may be assumed to be those where the toxicity of the contaminant is moderate, the exhaust fan or fans are placed reasonably close to the operation releasing vapors, make-up is by infiltration through doors, windows and walls so that a reasonable amount of dilution occurs as the contaminant is generated, and the worker is not too near or in the zone of concentrated contaminant release. Under such circumstances, the average K factor may be taken as 4 so that the average concentration in the workroom is $\frac{1}{4}$ of the TLV and the

concentration in any part of the workroom will probably not be higher than the TLV.

However, average distribution conditions are probably as elusive as ideal conditions and the need arises for a more rational basis of selecting a factor of safety to reflect actual conditions, whether poor, average, good, or excellent. Poor distribution, such as may be caused by short circuiting the make-up air, inadequacy of infiltration as a source of make-up air, location of the worker directly downstream from or in the evolving vapors, will require a higher average factor of safety than 4. A lower factor may be used with good distribution, which will usually consist of a positive mechanical make-up air

K Factors at Different Distribution Conditions
(See Figure 7)

Toxicity	Poor	Average	Good	Excellent
Slight	7	4	3	2
Moderate	8	5	4	3
High	11	8	7	6

supply system with properly placed supply and exhaust fans. Excellent distribution will be provided by properly located exhaust fans with a mechanical supply system directing tempered air in the optimum manner by means of diffusers, ductwork, or a perforated plenum. This will make possible a further reduction in safety factor. The influence of distribution on safety factors is shown in Table 3. Typical distribution conditions are shown in Figure 7.

Effect of Toxicity. The TLV or relative toxicity of the substance being controlled enters into the design directly since it is part of the calculation formula. Where the TLV (in ppm) of a substance has been promulgated by a recognized authority such as the ACGIH or the American Standards Association, it is recommended that dilution ventilation rates be calculated using

TABLE 3.
Dilution cfm Formula

Slightly toxic substances	cfm =	$\dfrac{(1 \times 10^6)(\text{sp. gr.}^a)(\text{pints evaporated/min})(K)}{\text{M.W.}^b}$
Moderately toxic substances	cfm =	$\dfrac{(2 \times 10^6)(\text{sp. gr.})(\text{pints evaporated/min})(K)}{\text{M.W.}}$
Highly toxic substances	cfm =	$\dfrac{(8 \times 10^6)(\text{sp. gr.})(\text{pints evaporated/min})(K)}{\text{M.W.}}$

[a] sp. gr. = specific gravity of liquid (Section 12, d.).
[b] M.W. = molecular weight of liquid (Section 12, mol. wt.).

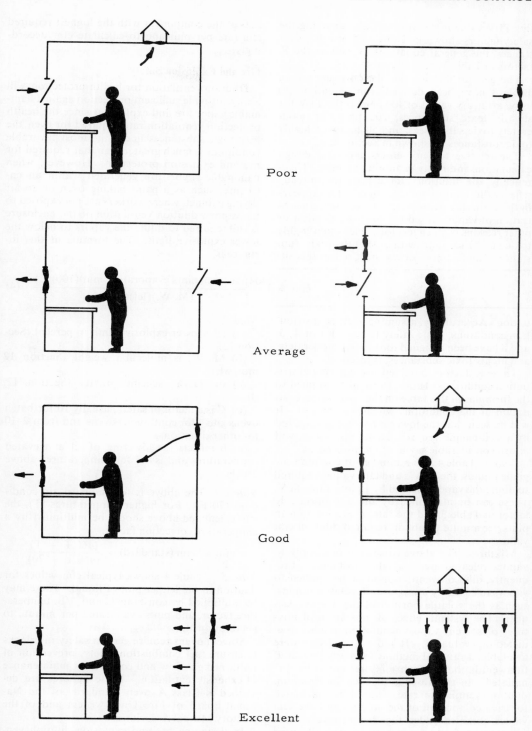

Poor

Average

Good

Excellent

Figure 7. Typical Distribution Conditions. (*Courtesy, Industrial Ventilation Manual, Seventh ed., American Conference of Governmental Industrial Hygienists*)

this TLV value in equation (2) and applying the safety factors shown in Table 3. For other substances, the general equations as well as the K factors in Table 3 should be used.

The general dilution ventilation equations in Table 3 have been derived from equation (2) using arbitrary values of 400 ppm as the TLV for slightly toxic substances, 200 ppm for moderately toxic substances, and 50 ppm for highly toxic substances, as shown in Section 12.

The relative toxicity enters into the design calculations indirectly in another manner. Theoretically the dilution cfm calculation already reflects the relative toxicity since TLV appears in the formula. However, a local high concentration of vapor caused by poor distribution or other reasons is much less tolerable with a highly toxic substance than with a slightly toxic substance. The factor of safety must reflect this situation. Accordingly, assuming average distribution conditions, basic safety factors (K) of 4, 5, and 8 have been assigned to slightly, moderately, and highly toxic substances, respectively.

These K factors modified for different distribution conditions (Table 3) should be applied to the formulas in this table on the basis of an evaluation of the conditions which will prevail. It will be seen that the lower air volumes required by a well-engineered exhaust-supply system will afford considerable savings in heating costs.

Note: Table 4 lists a number of common organic liquids, the TLV's and cfm values required to dilute the vapor released by 1 pint to the TLV. These can be multiplied by the appropriate K factor in Table 3 and the rate of evaporation in pints per minute to obtain required dilution cfm values.

Mixtures. The above considerations apply to vapors released by a single substance. Frequently, dilution ventilation must be applied to vapors released by a mixture of volatile liquids, such as the volatile portion of the average lacquer. The systemic effect of inhalation of mixtures of solvent vapors cannot be readily predicted or evaluated. If more knowledge were available, a rational method for designing dilution ventilation for such mixtures could be formulated. In the absence of such information, dilution ventilation rates should be calculated for each component of the mixture and the cfm values obtained added together to give a total cfm rate. A method which is occasionally used to afford greater safety is to assume that the entire volume of mixture being evaporated consists of the component with the highest required cfm rate per pint and to calculate cfm accordingly.

Fire and Explosion Safety

Dilution ventilation for health protection will always provide sufficient protection against flammable vapor fire and explosions since the health protection ventilation rate required for even the least toxic substances under the most favorable conditions is much greater than that required for fire and explosion protection. However, when flammable vapors are generated within an enclosure, such as a paint baking oven or an air drying cabinet, where workers are not exposed to the vapors, dilution ventilation of the enclosure is still required to dilute the vapors to below the lower explosive limit. The formula in this instance is:

$$cfm = \frac{(403)(sp.\ gr.)(pints\ evaporated/min)(100)(C)}{(M.\ W.)(lel)(B)} \quad (4)$$

Note 1:
(a) lel is lower explosive limit in percent (Section 12).
(b) M. W. is molecular weight (Section 12 mol. wt.).
(c) sp. gr. is specific gravity (Section 12, d).
(d) C is a factor of safety, usually 10 for batch ovens and 4 for continuous ovens and from 2–10 for other situations.
(e) B reflects the lowering of lel at elevated temperatures and is 0.7 for temperatures above 250°F.

Note 2: The above is for standard air conditions (70°F). For higher temperatures (T), the cfm calculated above should be multiplied by a temperature correction factor.

$$cfm = cfm\ (standard) \times \frac{(460°F + T)}{(460°F + 70)}$$

Note 3: Table 4 shows typical cfm values for dilution to the lel per pint of liquid. These may be multiplied by constants C and 1/B, temperature factor, and pints evaporated per minute to get dilution cfm.

Solvent ovens require certain safety interlocks to ensure fuel combustion safety, prevention of unburned gas flow and continuous maintenance of exhaust ventilation. These features are described in Class A oven standards of the National Board of Fire Underwriters and of the Factory Mutual Associates.

In designing fire and explosion dilution ventilation for mixtures, it is common practice to regard the entire mixture as consisting of the

TABLE 4. Properties of Common Flammable and Toxic Solvents[a]

Name	M.W.	sp.gr.	Flash Point (°F)	lel (%/vol.)	Boiling Point (°F)	Evaporation Rate	TLV (ppm)	cu ft air/pint solvent required for dilution — lel	cu ft air/pint solvent required for dilution — TLV[b]
Acetone	58	0.79	0	2.2	134	Fast	1000	210 × C	5,500 × K
n-Amyl acetate	130	0.88	77	1.1	300	Slow	100	244 × C	27,200 × K
Benzol (benzene)	78	0.88	12	1.4	176	Fast	25	290 × C	Not recommended
n-Butanol (butyl alcohol)	74	0.81	100	1.7	243	Slow	100	254 × C	44,100 × K
n-Butyl acetate	116	0.88	72	1.7	260	Medium	200	177 × C	15,300 × K
Butyl "Cellosolve"	118	0.90	141	—	340	Nil	50	—	Not recommended
Carbon tetrachloride	154	1.60	None	None	170	Fast	10	Nonflammable	Not recommended
"Cellosolve"	90	0.93	104	2.6	275	Slow	200	156 × C	20,800 × K
"Cellosolve" acetate	132	0.98	124	1.7	313	Slow	100	168 × C	20,970 × K
Chloroform	119	1.48	None	None	142	Fast	50	Nonflammable	Not recommended
1-2 Dichloroethylene	97	1.29	43	9.7	141	Fast	200	50 × C	26,900 × K
Ethyl acetate	88	0.90	24	2.2	171	Fast	400	184 × C	10,300 × K
Ethyl alcohol (ethanol)	46	0.79	55	3.4	173	Fast	1000	193 × C	6,900 × K
Ethyl ether	74	0.71	−49	1.9	95	Fast	400	205 × C	9,630 × K
Ethylene dichloride (1,2-dichloroethane)	99	1.26	56	6.2	181	Fast	50	77 × C	Not recommended
Methyl acetate	74	0.92	14	4.1	140	Fast	200	118 × C	25,000 × K
Methyl alcohol (methanol)	32	0.79	52	6.0	147	Fast	200	156 × C	49,100 × K
Methyl isobutyl ketone	100	0.80	73	—	244	Medium	100	—	32,300 × K
Methyl "Cellosolve"	76	0.97	105	—	255	Nil	25	—	Not recommended
Methyl "Cellosolve" acetate	118	1.00	132	—	289	Nil	25	—	Not recommended
Methyl chloroform (1,1,1-trichloroethane)	133	1.33	None	None	165	Fast	350	Nonflammable	11,520 × K
Methyl ethyl ketone (2-butanone)	72	0.81	30	1.8	176	Fast	200	250 × C	22,750 × K
Methyl propyl ketone	86	0.82	60	1.6	216	Medium	200	236 × C	19,000 × K
Methylene chloride (dichloromethane)	85	1.34	None	None	104	Fast	500	Nonflammable	12,750 × K
Monochlorobenzene	113	1.11	85	1.8	270	Medium	75	218 × C	5,290 × K
Naphtha (petroleum)									
Low boiling, 140–206°F	—	0.70	50	0.9	—	Fast	500	344 × C	7,000 × K
Medium boiling, 196–250°F	—	0.73	50	to	—	Fast	500	313 × C	6,000 × K
High boiling, 217–288°F	—	0.76	100	1.3	—	Medium	500	282 × C	5,400 × K
Safety solvent, 300–400°F	—	0.80	100–110	1.3	—	Slow	500	250 × C	4,600 × K
Perchloroethylene (tetrachloroethylene)	166	1.62	None	None	249	Medium	100	Nonflammable	39,600 × K
Isopropyl alcohol	60	0.79	53	2.5	181	Fast	400	206 × C	13,200 × K
Isopropyl acetate	102	0.89	40	1.8	194	Fast	200	192 × C	17,500 × K
Toluol (toluene)	92	0.87	40	1.3	232	Medium	200	295 × C	19,000 × K
Trichloroethylene	131	1.47	None	None	189	Fast	100	Nonflammable	45,000 × K
Xylol (xylene)	106	0.88	63	1.0	291	Medium	100	332 × C	33,000 × K

[a] Courtesy, Division of Industrial Hygiene, New York State Department of Labor.
[b] 1967 Threshhold Limit Values, American Conference of Governmental Industrial Hygienists.

component requiring the largest amount of dilution air per pint and to calculate the air quantity on that basis (see also Flammable Atmosphere Ovens, p. 47).

CONTROL OF SPECIFIC INDUSTRIAL PROCESSES

Vapor Phase Degreasers

I. Ventilation Design. Properly designed, operated, and located degreasers do not normally require local exhaust ventilation if adequate, draft-free general ventilation is available to the area. However, the usual degreasing operation is such that local exhaust ventilation is required to control the vapors released.

Design data will be found in Figure 4. The most commonly used degreasing solvents—trichlorethylene and perchlorethylene—fall into the high toxicity class (Hazard Rating 3, Section 12). The "Freons" which have recently been introduced for degreasing are mostly of low toxicity (Hazard Rating 1, Section 12).

The rate of solvent vapor evolution requires individual subjective evaluation. Any conditions of design, operation, or location which cause excessive "dragout" of solvent vapors will create a high rate of evolution. Average conditions will result in a low or moderate rate. Where operations are such that the work is not dry upon removal from the tank, secondary ventilation as in Figures 4(c) or 5 may be required to control the concentration of vapors evaporated from the work subsequent to its removal from the tank.

The most commonly used hood for degreasing is the "lateral slot hood" shown in Figures 4(a) and 4(b). The outer slot, C, should be sized for velocities between 400 and 600 fpm to minimize loss of solvent by dragout of vapors.

II. Equipment Design:

(a) Thermostatic control is required to prevent the temperature of the liquid from rising in excess of 20°F above its boiling point, in the vapor phase to prevent the rise of the vapor level above the condenser, and in the cooling water to maintain condensing temperatures between room temperature and 110°F.

(b) The size of the tank should be such that the horizontal cross-section area of the work basket or loading rack is less than two-thirds of the horizontal cross-section area of the tank.

(c) The height of the tank should be such that during normal operations the distance between the top of the vapor zone and the top of the tank is at least 15 inches and not less than two-thirds of the tank width.

(d) The combustion chambers of gas or oil heated tanks shall be enclosed except for necessary openings and independently vented to the outdoors by means of a corrosion-proof pipe. Natural draft pipes shall have draft diverters.

(e) Tight-fitting tank covers shall be provided and kept closed whenever the degreaser is idle. The covers shall not interfere with the exhaust hood when in an open position.

III. Location:

(a) Wherever possible, the room in which a degreaser is located shall have a volume of at least 20,000 cubic feet. Degreasers located in smaller rooms shall have local exhaust ventilation, even if not otherwise deemed necessary.

(b) Ample clearance shall be provided on all sides.

(c) Degreasers should not be located near pits in the floor, corner pockets or dead air spaces. Pits containing large degreasers should be large enough for a worker to enter conveniently for cleanup and maintenance, and should be provided with mechanical ventilation.

(d) Degreasers should not be located near other exhaust systems in strong cross drafts or in areas with ambient air velocities in excess of 50 fpm. When such a location is unavoidable, provide a shield or baffle upstream.

(e) Do not locate near open flames, electric heaters, welding torches, or other sources of ignition or excessively hot surfaces.

IV. Operation:

(a) The operator should be taught to follow the manufacturer's instructions.

(b) If possible provide a mechanical loading hoist with a maximum vertical speed of 11 fpm. If not, the work should be introduced and removed as slowly as possible and hooks or stops should be provided to permit convenient holding of the work in the boiling liquid and vapor zone without the necessity of the worker bending over the tank.

(c) Use only the designated solvent containing suitable inhibitors.

(d) Remove adhering aluminum dust or chips from parts before degreasing.

(e) For small parts, use only metal racks, wire, or perforated metal baskets. Place the work so as to facilitate draining and to prevent cupping or retention of liquid. Do not load beyond design capacity.

(f) Hold the work in the vapor zone until drainage and condensation of liquid on the work stops and the work is completely dry. Hold above the vapor zone briefly before removal from the tank.

(g) When removal of wet work is unavoidable, provide a ventilated drying station as described above in Ventilation Design.

(h) When spray nozzles are used, always operate with the nozzle below the vapor level.

V. Maintenance and Cleaning:

(a) Inspect all piping, valves, pumps, thermostats, etc., periodically in accordance with the manufacturer's instructions.

(b) Clean tanks and pits frequently.

(1) Allow tank to reach room temperature, drain contents completely, open all cleanout doors, shut off and lock all power switches, and ventilate thoroughly.

(2) Provide the worker with suitable protective clothing, air line respirator and, if it is necessary for him to enter the tank or pit, a safety harness. A trained worker should be stationed nearby to render immediate assistance in the event of an emergency. All ventilation should be operated continuously during the cleaning process.

(3) Store collected sludge and residue in a covered container until it can be disposed of safely.

VI. Air Contaminant Emissions:

(a) The nature of vapor degreasing is such that the concentration of contaminant discharged by the exhaust system is rarely sufficiently high to constitute an air pollution problem. In addition, the high cost of most degreasing solvents acts as an economic self-limiting factor. A degreasing exhaust system will create excessive air contamination only when operating conditions are very poor. Steps should be taken to improve operations, not only to avoid the installation of expensive air cleaning equipment but also to prevent the loss of costly solvent.

(b) Potential solvent emissions of a system can be estimated from its operational losses which consist of liquid solvent entrapped in the sludge at the bottom of the tank, evaporation to the workroom, and vapors captured by the exhaust hood and discharged to the outdoors. If we assume that 25 percent of the losses are accounted for by the first two factors, 75 percent of losses will be emitted as environmental contaminants.

VII. Air Cleaning. The only practical air cleaning methods for vapors released by degreasing are adsorption and condensation. Combustion of trichlorethylene is not practical because of the high temperature required and the hydrochloric acid produced which is more objectionable than the original contaminant (see Industrial Air Cleaning—Gases and Vapors).

Spray Coating

I. Design Data (see also Fig. 1):

(a) Both worker and operator inside of booth:

W = Width of widest work plus 6 feet

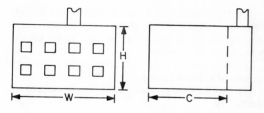

Figure 8.

H = Height of tallest work plus 3 feet
C = Depth of deepest work plus at least 3 foot clearance between work and baffle line and between work and face of booth
V = 100–200 lfm through booth cross-section
cfm = W × H × V

(b) Work inside booth—operator in front of booth:

W = width of widest work to be sprayed plus 12 inches
H = Height of tallest work to be sprayed plus 12 inches
C = Depth of deepest work plus 1 foot rear and 1.5 foot front clearance. Depth shall be at least equal to three-quarters of the height or width, which ever is greater.
V = 100–200 lfm through face area and all other openings
cfm = W × H minus any approved face obstructions down from booth roof or up from floor × V

(c) 100 fpm minimum velocity; 150–200 fpm is required velocity for shallow booths, spraying toxic material (Hazard Rating 3, Section 12), presence of strong cross drafts, unavoidable rebound because of size and shape of work.

(d) Baffles—open area between baffle plates from 25–50 percent of cross-sectional booth area. Distance between line of baffles and bottom of booth from 3–5 times the distance between baffles and the sides and top of the booth.

(e) Where worker is outside of the booth, reduce face area by baffling the frontal opening as much as possible.

II. Fan and Piping:

(a) Cast, non-ferrous metal bladed propellor fan, belt-driven, motor outside of duct or air stream, belt covered, all moving parts electrically grounded. For systems with high resistance, use tube axial or centrifugal (backward curved or radial blade) fans. Forward curved centrifugal fans (squirrel cage or sirocco) *should not be used*.

(b) Locate fan as close as possible to the point of discharge to the atmosphere.

(c) Select fan and pipe diameter for approximately 2000 fpm (except where long runs dictate smaller diameters and higher velocities).

(d) Pipe seams and joints should be of air-tight construction. Airtight construction mandatory for pipes under pressure inside of the workroom or building.

(e) Interlocking of fan with spray gun by means of solenoid desirable.

(f) No dampers, blast gates, bird screens in ductwork.

III. Construction and Installation:

(a) Gauge of metal from 18–22 depending on size. Standard airtight construction.

(b) Inner booth walls of galvanized sheet steel or equivalent fireproof, smooth, hard, impervious construction. Floor of non-sparking, nonflammable material. No pipes, cabinets, beams or other encumbrances in booth.

(c) Windows, if used, shall be of 1/4 inch safety glass or wired glass, maximum dimensions, 18 × 54 inches.

(d) Locate booth away from strong drafts and at least 20 feet from open flames, sparks, welding, or other sources of ignition.

(e) Ground all metal parts, paint containers, paint hose, etc.

(f) Sprinklers to be protected against paint deposition by a film of light grease or light weight paper bags. Provide fire extinguisher.

(g) No storage or combustible construction within 3 feet on all sides.

(h) For heavy, continuous spraying in booths without water wash or paint filters, provide flanged pipe joints for convenient cleaning. Flanges every fourth joint recommended in all cases.

IV. Electrical:

(a) Fan motors not to be located in any hood, booth or duct. Fan belts to be effectively grounded and completely enclosed.

(b) Wiring within a booth and within 10 feet of booth opening to be in sealed metal conduit with explosion-proof fittings.

(c) Lamps are not permitted in a booth when they may be subject to accumulations of flammable residues.

(d) Lamps within a booth, when permitted, and within 10 feet of booth opening to be totally enclosed, protected against breakage by location or guards, switchless, and provided with sockets with non-metallic shells.

(e) When such lamps are behind transparent panels in the walls or ceiling of a booth, the panels shall comply with Item No. IIIc, shall effectively isolate the interior of the booth from the lamps, and shall be separated from the lamps by an air space.

(f) Switches, controls, and all other electrical equipment within a booth or within 10 feet of the booth opening shall be explosion proof.

V. Operations:

(a) Locate the work as deep in the booth as possible. Position work and operate gun so that spray is directed at work at an angle upstream toward the rear of the booth. Spray gun to be held at optimum distance from work so that proper coverage is achieved with minimum overspray and rebound.

(b) Use turntable when possible. Turntable to be rotatable 360° while supporting the largest piece of work.

(c) When in the booth do not spray toward the front or perpendicular to the air stream. If such spraying is unavoidable, and in all cases where toxic materials are sprayed inside the booth, use approved respirator.

(d) Alternate spraying of materials, the combination of which may result in spontaneous ignition, should be avoided. Where it must be done, clean booth, fan, and piping before each change of material.

(e) Separate booths must be provided for exclusive spraying of hydrogen peroxide, perchlorates, and other strong oxidizers.

VI. Special Types and Applications:

(a) Downdraft (see also Fig. 5): Average conditions; 100 fpm minimum velocity into gross downdraft area; other conditions, see Fig. 5; avoid cross drafts; provide side shields if possible; grates to be non-ferrous metal; projected area of work to no greater than 67 percent of gross area of grate.

(b) Conveyorized booth: Area of side conveyor openings as small as possible; openings as far from front as possible; side opening area to be included with face area for cfm calculations. For automatic painting, interlock so that paint flow ceases if conveyor stops or if fire starts.

(c) Electrostatic spraying:

(1) Location: Electrostatic apparatus and devices used in connection with paint spraying, dipping, or other coating operations shall have transformers and power packs located outside spray booths or enclosures and areas subject to paint deposit.

(2) Supports: Electrodes and electrostatic atomizing heads shall be rigidly supported and permanently installed and shall be effectively insulated from ground.

(3) Clearance: Space of at least 2 times the sparking distance shall be maintained between articles being finished and electrodes, electrostatic atomizing heads, or conductors.

(4) Articles on conveyors shall be so arranged and supported as to maintain the required clearances.

(5) Electrostatic apparatus shall be equipped with automatic controls arranged to stop the entire operation if the ventilation of the

spraying or drying area falls below a velocity of 100 fpm, if an article on the conveyor projects into the required clearance, or if the conveyor stops.

(6) Guarding and isolation of process: The electrical field and all parts of the equipment carrying high potential shall be located, guarded, and fenced off to provide safe isolation of the process. Guards and fences shall be of conducting material and shall be grounded.

(7) Signs: A suitable sign stating the sparking distance shall be posted conspicuously near the assembly. Signs designating the process zone as dangerous by reason of fire and accident hazard shall be posted.

(8) Insulators: Insulators shall meet test standards for the potential used and shall be maintained clean and dry.

(d) Airless spraying (hot, steam, hydrostatic): Locate heaters outside of booth; interlock heater with exhaust fan; provide suitable hose, valves and fittings for high pressure applications; locate equipment, including paint hose, so that in the event of a rupture flammable material will not be discharged into an ignition source.

(e) Multiple booth systems: With a single fan, balanced air flow should be provided by system design and not with blast gates. If individual fans are used, all fan motors to be interlocked so that all must operate together.

VII. Maintenance:

(a) Clean booth, fan, and piping frequently. Solvents, if used, must be nonflammable and nontoxic (Hazard Rating 1)

(b) Interior walls of booth can be coated with nonflammable strippable or water soluble compound.

(c) Use non-ferrous cleaning tools only. If deposits are dry, wet them before cleaning.

(d) Wet and clean booths, fans and piping before repair, dismantling, or alteration.

(e) Check fan belts and pulleys periodically.

(f) Store cleaning rags, residues, loaded filters, etc., in covered containers, preferably under water, and dispose of daily.

VIII. Air Contaminant Emissions:

(a) Minimize overspray by proper spraying techniques.

(b) Minimize overspray by electrostatic or airless spraying.

(c) Potential emissions will vary widely with operations. They can be estimated as follows:

Overspray = Paint used minus paint adhering to work; (if latter cannot be determined, assume gallon of paint to cover 300 square feet of coated surface.)

Potential emissions (before air cleaning, if any) = Overspray minus paint adhering to booth walls, floor, fan, ductwork, etc. (If latter

cannot be determined, assume that the potential emissions are equal to 100 percent of the volatile component of the overspray and 50 percent of the nonvolatiles)

IX. Air Cleaning:

(a) Water wash: Interlock water pump with fan; maintain water filter, pump, and nozzles in clean condition; dispose of slurry in wet condition.

(b) Dry paint filters: Do *not* use for materials or combinations of materials which are susceptible to spontaneous ignition; provide pressure actuated interlock between filters and spray gun so that the latter cannot operate when the booth ventilation velocity falls below the design amount (replace filters at this point); provide access doors and sprinklers in plenum behind the filters.

(c) Baffles: Air distributing baffles provide an incidental air cleaning effect by impingement; cleaning can be improved by countercurrent flanges on baffle plates and by using triple row of staggered, flanged baffles.

Cleaning Efficiency

	Volatiles (%)	Nonvolatiles (%)
Water wash	0–10	80–90
Filters	0	90
Baffles	0	Indeterminate

Flammable Atmosphere Ovens

The flammable atmosphere oven is one of the most prevalent pieces of equipment found in industry. It is used to evaporate flammable liquids from objects coated or impregnated with a variety of materials as well as for any process requiring evaporation of volatile flammable liquids. Two hazards must be controlled: fire and explosion may result from any source of ignition in the presence of flammable concentrations of vapor, and escape of toxic vapors into the workroom may create a health hazard.

Common Industrial Ovens. There are two main types of oven: the batch type in which the work is loaded, heated for a period of time, and removed; and the conveyor type through which the work passes continuously, entering freshly coated through one opening, and emerging dry through a second opening, or the same opening.

Ovens are usually arranged with either a recirculating system in which a fixed portion of the heated vapor-laden air is continuously recirculated through the oven and the balance exhausted to the outside, or non-recirculating with continuous exhaust of air from the oven. In direct fired ovens, the source of heat may be in contact with the air within the oven. In indirect ovens, an intermediate heat exchanger is used to convey heat from the source to the oven atmosphere. Gas, oil, steam, or electricity are the heat sources. Indirect fired gas or oil ovens require independent venting of the combustion chamber to remove the products of combustion. In all other cases, the ventilation aspects of oven design are concerned only with removal of vapors, which in the case of direct fired gas or oil heated ovens are intimately mingled with the products of combustion.

Ventilation. Oven ventilation has two functions: to provide safety by diluting the evolving vapor concentration within all parts of the oven to well below the lel (see Fire and Safety Dilution Ventilation this section and also Section 6). This ventilation is particularly important at the peak evaporation rates which occur during the early part of a batch cycle or near the entrance end of a continuous oven. Also, the ventilation must provide protection for the worker by maintaining a controlling velocity into all operating openings of the oven to prevent escape of vapor into the workroom. Ventilation rates are calculated on the basis of both criteria, and the higher rate is used. Economical oven design will minimize both the amount of solvent introduced into the oven as well as the area of the oven opening (the term "solvent" denotes all volatile flammable liquids).

(a) Safety Ventilation:

(1) Determination of Solvent Quantity Evaporated into the Oven. Before ventilation design can be undertaken, the amount and nature of the solvents evaporated into the oven by the drying process must be known, since safety ventilation rates will depend on these factors.

The amount and composition of solvent present in the coating or impregnating material must be determined, taking into account the amount of thinner added to the original material. This information can usually be obtained from the supplier. In the average dipping process, the coating material may contain up to 50 percent by volume of solvent. In spray coating, the amount may be as high as 85 percent. In fabric coating, the material will usually contain less solvent. A typical baking paint or enamel will contain petroleum naphtha; some formulations contain toluene or xylene as well. Some processes require a coating material containing mixtures of the alphatic and aromatic hydrocarbons, various esters, ketones, and alcohols. Thus, fairly accurate information on the ingredients present is important.

A simple and widely used method of estimating the quantity of solvent evaporated is to utilize plant records of coating material used per

specified period of time and number of oven batches or hours of oven operation in the same period. If the percent solvent is known, a simple calculation will yield volume of solvent per batch or per hour of oven operation. Another method of determining solvent use involves weighing the solids deposited per unit of surface area and calculating back to the original coating material. The increased accuracy of this method is usually not sufficient to warrant the trouble involved unless the information is also desired for operational purposes.

The difference between the solvent in the coating material as applied and that on the work entering the oven may arise from a number of conditions, as follows:

(i) Overspray: This will depend to a large extent on the spraying technique and the size and shape of the work being sprayed. With good technique, the overspray for large flat surfaces may be as low as 10 percent, i.e., 10 percent of the material leaving the spray gun does not reach the work. For small or open surfaced objects, or with poor technique, the overspray may reach 40–80 percent.

(ii) Dipping: In dipping, especially with low viscosity substances, there will be some material dripping from the object as it is raised from the tank. Some of this will drip back on to the drainboard or floor before reaching the oven. Evaporation from the tank is also a factor.

(iii) Evaporation: There is an inevitable evaporative loss of solvent between the time the article is coated and the time it reaches the oven. This will depend on the coating process, the ambient room air temperature and velocity, distance from oven, speed of conveyor travel, amount of incidental or operational preliminary air drying, thickness of coat, evaporation rate, and other factors.

These pre-oven solvent evaporation losses are so difficult to estimate with accuracy that they are frequently ignored. Where they may be large enough to drastically affect the required oven heat capacity and ventilation rates, they can be considered, but extreme precaution must be observed because to overestimate the pre-oven solvent losses in the design of oven ventilation may introduce a serious fire and explosion hazard.

(2) Ventilation Rates. The internal oven ventilation rate required to dilute the solvent vapor evaporated within the oven can be calculated from the following formula based on Avogadro's number at 1 atmosphere and 70°F.

$$\text{cfm (at oven conditions)} = \frac{(403)(\text{sp. gr.})(\text{pints/min})(100)\dfrac{(T + 460)}{(530)}}{(\text{M.W.})(\text{lel})(B)} \ (C)$$

sp. gr. is the specific gravity of the solvents.

Pints/min is pints of solvent evaporated into the oven per minute. For batch ovens this is determined by dividing the amount of solvent evaporated per batch by the time of the batch. For continuous ovens, it is the amount of solvent entering the oven per minute.

T is the oven temperature in °F. The T factor in the equation reflects the volume of expanded air at the elevated temperature of the oven.

C is a factor of safety designed to maintain at least one-fourth of lel at all times. For continuous ovens it is 4. For batch ovens, where the average evaporation rate in pints per minute will not reflect the higher and usually indeterminate peak evaporation rate, the C factor is taken at 10 to ensure maintenance of one-fourth of the lel at the peak rate.

M.W. is the molecular weight of the solvent.

lel is the lower explosive limit in percent of the solvent vapor at standard conditions.

B is a factor which reflects the decrease in lel at elevated temperatures. It may be ignored for temperatures up to 250°F. For the usual oven temperatures between 250 and 500°F it is assumed to equal 0.7.

The formula refers to single substances. Since the lel of a particular mixture of solvent vapors can usually not be calculated accurately and is difficult to determine experimentally, it is normal practice to regard a mixture as consisting entirely of that component requiring the largest amount of dilution air per pint and to determine the air quantity on that basis (see Table 4 for constants and dilution air per pint data for a number of common solvents).

(b) Health Hazard Ventilation. A ventilation rate of from 50–100 cfm/ft^2 of oven opening will be adequate to prevent vapors from escaping into the workroom. The total cfm thus obtained, multiplied by the temperature correction factor, will be the hygienic ventilation rate. The required ventilation will then be this rate or the safety ventilation rate, whichever is greater.

In most batch ovens the latter will be greater. However, in many continuous ovens, the entrance and exit openings may be so large that the hygienic ventilation rate will be greater and may result in a cfm so excessive as to make it prohibitively expensive to maintain the required oven temperature. To overcome this, "air seals" may be used.

In the simplest and safest form of the air seal, the ends of the oven are provided with independently ventilated vestibules or canopy hoods to control the contaminated air which may spill out of the oven which is ventilated at the safety ventilation rate. Alternate designs make use of one fan to provide both oven safety ventilation and an "air curtain" at the oven opening by cycling part of the air handled by the fan into a vestibule and back to the oven. The latter technique has the operational advantage of preheating the work and the air entering the oven. However, unless carefully designed, it will not provide adequate control of vapors escaping from the oven and, more important, the internal oven exhaust feature of the system may not provide adequate dilution ventilation in all parts of the oven.

Equipment Design Features:

(1) Location of make-up air inlets, recirculating vents and exhaust outlets to ensure continuous ventilation in all parts of the oven, especially near the flame or heating element.

(2) Oven heating system (except steam) to be located at least 20 feet from operations and/or storage areas involving flammable materials (Section 7). Floor surface of oven to be of incombustible material and to extend at least 1 foot beyond oven outline.

(3) Oven to be constructed of noncombustible material throughout. Expansion joints to be provided in oven framing. Oven interiors to have smooth surfaces.

(4) Explosion relief vents (1 square foot area per 15 cubic feet of oven volume) to be provided, preferably in top. Inner panel edges of the vents to be maintained free of condensate.

(5) Discharge ducts not to be connected to other ventilating systems, chimney, or flue. Ducts to be of noncombustible material and free of obstructions and screens.

(6) Two inch clearance or 3/4 inch insulation to be provided for ducts passing through combustible walls, floors, and roofs. Duct external surface temperature not to exceed 160°F at these points.

(7) Products of combustion from indirect fired ovens to be exhausted or vented to the outside, preferably by means of an independent system.

Safeguards and Interlocks:

(1) Exhaust fan to be interlocked with heat supply by means of air flow switch so that heat supply is cut off unless fan is maintaining required air flow. Switch to be installed preferably on suction side of fan.

(2) Recirculating fans to be interlocked with heat supply by means of air flow switch so that heat supply is cut off unless fans are operating. Rotational or electrical interlock recommended as additional safety measure.

(3) Conveyors to be interlocked with oven exhaust and recirculating fans so that conveyors cannot operate unless fans are on.

(4) Switch for fans to include time delay relay to secure proper preventilation. If necessary, oven doors to have limit switch interlock to prevent time delay relay from operating unless oven doors are wide open.

(5) Air supply or exhaust volume control dampers to be properly adjusted and permanently fixed to maintain proper air supply. Means to be provided to prevent full closure at any time.

(6) Excess temperature limit switch with manual reset to be provided in order to cut off heat supply when temperature exceeds safe limit.

(7) Pressure switch with manual reset (gas and oil ovens) to be installed in fuel line to shut off fuel supply when pressure is insufficient.

(8) Combustion (flame) failure safeguards to be provided to shut off fuel supply in the event of flame failure (gas and oil ovens).

Operation. Ovens must be designed to meet a particular set of conditions. Many fires and explosions occur as a result of changing these conditions without making the necessary changes in the oven ventilation and safety features. The worst conditions to be encountered in a particular oven operation must be used as a design basis, and a suitable placard or warning notice placed on the oven outlining these conditions.

Any change in these condition will cause a change in the amount or rate of solvent evaporated into the oven.

In batch ovens, particularly in production shops, the maximum amount of solvent per batch, once it has been determined, will not materially change unless there is a change in product or process. If the oven has been designed for the maximum and is so operated, no problems will arise.

In continuous ovens, however, a number of conditions such as conveyor or web speed, doctor blade adjustment, composition of coating material, etc., can be readily and frequently changed during the operation. The general principle of designing for maximum conditions is thus particularly important.

Continuous ovens, particularly those used in conjunction with cloth coating and roller coating operations, lend themselves to a design feature which has operational as well as safety advantages.

It is possible to install continuous vapor concentration indicators and controls calibrated in terms of percent of lel. The sampling probes are located near the entrance of the oven at the zone of highest vapor concentrations. It is common practice to set the control to actuate an alarm or turn the heat off when the concentration at this point exceeds 40 percent of the lel. This is usually equivalent to an average concentration below 25 percent of the lel.

In many instances, these indicators are being used as an additional safeguard since the conveyor speed can be adjusted and maintained at such a rate that 40 percent of the lel is not reached. However, it must be borne in mind that these devices are very sensitive and may be inaccurate. They must be carefully calibrated and adjusted, zeroed with fresh air every day, serviced frequently, and always used as a check on the ventilation and not as a sole means of determining safe operations (coating materials containing silicones and other substances may poison the indicator and render it useless).

Any of the following changes in design and operating conditions may make the design safety ventilation inadequate and require a recalculation of the adequacy of the ventilation (conversely, a change may introduce conditions which will permit a decrease in ventilation with a resultant saving in heat or time):

(1) Elevation in temperature.

(2) Increase in amount of coated work in a batch, or per unit time.

(3) Increase in thickness of coating.

(4) Increase in conveyor speed.

(5) Change in size or surface area of work.

(6) Change in composition or nature of solvent or coating material.

(7) Change in physical shape of the work. A coated object with a high surface area to mass ratio may result in a much higher peak evaporation rate and higher average rate of evaporation than one with a low surface area to mass ratio.

(8) Change in physical nature of work. Solvent evaporation from a coated impervious surface will be much greater initially and come to completion in a much shorter time than from a coated or impregnated porous surface.

Maintenance. The interior of ovens and ductwork must periodically be cleaned to remove dripped paint, deposits and condensates. Non-ferrous scraping tools must be used, and if necessary, ductwork should be taken apart and scraped, steamed or, in unusual circumstances, burned clean in a safe location.

All controls, probes, switches, etc., must be examined and cleaned. Explosion vents must be carefully cleaned and loosened to prevent locking by polymerization of condensed fume at the joints.

Controls, switches, interlocks, conveyors, etc., must periodically be checked to ensure proper operation.

Air Contaminant Emissions. The solvent evaporation rates which are used to determine required ventilation will also be the potential emission rates and need only be converted to pounds per hour.

Air Cleaning. Ovens and their ventilation features are carefully designed to ensure operational efficiency and safety. If air cleaning is required, extreme caution must be taken to ensure that the air cleaner does not interfere with the oven or introduce excessive back-pressure to reduce the air flow and introduce an explosion hazard. Similarly, the air cleaner itself may constitute a fire or explosion hazard unless properly designed, located, and operated. Additional interlocks and alarms for the operation of the air cleaner may be advisable.

The usual air cleaning methods consist of adsorption on activated charcoal, incineration with after-burners, or condensation. Local ordinances may prohibit one or more of the above techniques because of the fire hazard.

Air Cleaning Efficiency. Equipment can usually be designed to provide any desired efficiency. However, the usual criteria may not always be valid for oven emissions. Depending on the binders, drying oils, or plasticizers used in the coating formulations, fumes may be created in very low concentrations which will nevertheless have a high pollution potential because of their odor and visibility. In such cases, cleaning efficiency higher than otherwise needed may be required.

Plating, Metal Finishing and Cleaning, and Allied Processes

Table 5 lists the more common plating-room and allied open surface tank operations. The usual contaminants released to the atmosphere, their hazard rating, usual rates of evolution, and recommended hood types are given (see Hoods in this section). Special operations may involve contaminants and rates differing from those shown, and the required ventilation in such cases would have to be modified upward or downward accordingly.

Since most of these operations require relatively unencumbered access to the surface of the tank, the hood of choice is usually one of those shown in Figure 4. For certain operations with a very high rate of evolution, such as boiling, caustic etching of aluminum, or where contaminant continues to be released from the work as it is raised from the tank, as in nitric acid bright dipping of copper, hoods shown in Figures 1 or 4(c) will provide better control.

Since virtually all the contaminants listed are highly soluble in water, air cleaning, when required, can readily be accomplished by a simple wet scrubber. For contaminants in the form of mists, as in chromic acid plating, special mist collectors requiring little or no water are available.

See also Industrial Air Cleaning, p. 51, and hazard ratings, Section 12.

Miscellaneous Operations

Table 8 lists some of the more common industrial processes for which exhaust ventilation is usually required. Hood and air cleaning data will be found elsewhere in this section.

INDUSTRIAL AIR POLLUTION CONTROL

Increasing concern over the proliferating problem of environmental pollution decreases the options available to management in carrying out its responsibility for the control of industrially produced air contaminants.

Air pollution control rules and ambient air quality standards for the environment are becoming more stringent and now the design of industrial exhaust systems must always include consideration of immediate or possible future need for suitable air cleaning equipment to control emissions of air contaminants.

The atmosphere is a natural resource and can no longer be considered a sink into which unlimited quantities of air-borne contaminants may be discharged without deleterious effects upon the public health. Careful plant siting remote from settled areas can no longer be relied upon as a totally adequate substitute for expensive air cleaning devices and procedures since many such locations have become populated, often as a result of the plant's presence.

Many industrial exhaust systems which now discharge into the atmosphere, without treatment, gases, fumes and mists from such operations as plating, paint dipping and die casting will ultimately be required to include air cleaning. Also, in many places where dust collectors have long been required to control particulate emissions, the collection efficiency will have to be greater than now acceptable. Specific emission and air quality standards are being written into air pollution control rules and the air cleaner selected will have to meet these standards.

However, the economic burden of providing high efficiency air cleaning can be mitigated. As discussed previously under Recirculation, exhaust systems with air cleaners handling contaminants of little or no toxicity (Hazard Rating 0 or 1, Section 12) can frequently be designed to return the cleaned air to the workroom. This eliminates the need for a costly tempered make-up air system to replace the air discharged to the outdoors and results in a considerable saving in heating, air conditioning, and expensive ductwork installation.

Material collected by an air cleaning system can have economic value as in cement manufac-

TABLE 5. Plating, Metal Finishing and Cleaning, and Allied Processes

Operation	Contaminant	Toxic or Flammable Rating	Rate of Evolution	Hood Type	Comments
Chrome plating	Chromic acid mist	High	High	4	Hydrogen and oxygen released— possible explosion hazard with foam blanket
Cyanide plating					
(a) Brass-bronze	Cyanide and alkali mist	Slight	Low	4	
(b) Zinc	Cyanide and alkali mist	Slight	Low	4	
(c) Copper (other than conventional)	Cyanide and alkali mist	Slight	Low	4	
(d) Gold, silver, cadmium, conventional copper	Cyanide and alkali mist	Slight	Nil–Low	4	Local exhaust not usually necessary
Strike (copper and silver)	Cyanide mist	Slight	Moderate	4	
Acid plating					
(a) Copper	Sulfuric acid mist	Slight	Nil–Low	4	Local exhaust not usually necessary
(b) Nickel (hydrofluoric acid)	Hydrogen fluoride gas	High	Low–moderate	4	Magnesium base metal
(c) Tin (halide bath)	Halide mists	Slight	Moderate	4	
(d) Zinc (chloride bath)	Chloride mists	Slight	Low	4	
(e) Nickel (sulfate and chloride)	—	—	—	—	Local exhaust not usually necessary, otherwise see Zinc (chloride bath) above
(f) Tin and zinc (sulfate)	—	—	—	—	
Fluoborate plating					
(a) Lead	Fluoborates, hydrogen fluoride gas	High	Low	4	
(b) Lead-tin, nickel, tin cadmium, copper, zinc	Fluoborate mist	Moderate	Low–moderate	4	
Electroless copper	Formaldehyde gas	High	High	4	
Anodizing					
(a) Chromic acid	Chromic acid mist	High	High	4	
(b) Sulfuric acid	Sulfuric acid mist	Slight	Moderate–high	4	
Electropolish					
(a) Brass, bronze, copper	Acid mists, arsine	High	Low	4	High toxic rating because of arsine possibility
(b) All other	Acid mists, arsine	High	Moderate	4	High toxic rating because of arsine possibility
Etching					
(a) Aluminum	Alkali mists	Slight	High	4	
(b) Copper	Hydrogen chloride gas	High	Moderate	4	
Surface treatment					
(a) Parkerize, bonderize	Steam, inorganic salts	Nil or high	High	1,4	Hood No. 1 preferable, high toxic rating if

(b) Alkaline oxidizing	Alkali mists	Slight	High	1,4	"Black Magic," "Ebanol," etc. Hood No. 1 or 4(c) recommended
(c) Stainless steel descaling	Nitric, sulfuric, hydrochloric acids	High	Moderate–high	1,4	
Pickling					
(a) Aluminum-nitric acid	Oxides of nitrogen (gas)	High	Moderate	1,4	Hood No. 1 or 4(c) recommended
(b) Aluminum-chromic acid	Chromic acid mists	High	Low	1,4	
(c) Aluminum-sodium hydroxide	Alkali mists	Slight	Moderate–high	1,4	
(d) Copper-sulfuric acid	Sulfuric acid mists	Slight	Low–moderate	1,4	
(e) Iron, steel, nickel (hydrochloric, sulfuric acid)	Hydrogen chloride gas and sulfuric acid mist	High	Moderate–high	1,4	
(f) Stainless steel-nitric acid	Oxides of nitrogen (gas)	High	Moderate	1,4	Passivation, immunization, hood No. 1 or 4(c) recommended
Acid dipping					
(a) Bright dip-copper, brass, bronze, aluminum	Oxides of nitrogen, sulfuric acid mist	High	Moderate–high	1,4	Hood No. 1 or 4(c) recommended
(b) Zinc-hydrochloric acid	Hydrogen chloride gas	High	Low	1,4	
Stripping agents					
(a) Hydrochloric acid	Hydrogen chloride gas	High	Low–moderate	1,4	These items refer to the stripping of a plated coating from a base metal. The usual stripping agents and rates of evolution are shown. For mixtures of the agents, the component with the more stringent requirement is the basis of design.
(b) Sulfuric acid	Sulfuric acid mists	Slight	Low–moderate	1,4	
(c) Nitric acid	Oxides of nitrogen	High	Moderate–high	1,4	
(d) Hydrofluoric acid	Hydrogen fluoride gas	High	Low–moderate	1,4	
(e) Chromic acid	Chromic acid mists	High	Low–moderate	1,4	
(f) Sodium hydroxide	Alkali mists	Slight	Low–moderate	1,4	
(g) Ammonium hydroxide	Ammonia gas	High	Low–moderate	1,4	
Metal cleaning					
(a) Vapor degreasing	See Degreasing (this section)				
(b) Alkaline cleaning	Alkali mist	Slight	Moderate–high	1,4	The lower rate of evolution is for cold baths, the higher for heated or air agitated baths
(c) Emulsion cleaning (petroleum and coal tar solvents)	Solvent vapors	Slight–moderate	Low–high	1,4	
(d) Emulsion cleaning (chlorinated hydrocarbons)	Chlorinated hydrocarbon vapor	Moderate–high	Moderate–high	1,4	

TABLE 6. Miscellaneous Operations

Operation	Contaminant	Toxic or Flammable Rating	Rate of Evolution	Hood Type	Comments	Type of Air Cleaner (see Industrial Air Cleaners)
Dipping and drying						
(a) Paint and enamel	Vapors of petroleum naphtha, and possibly xylene	Slight–moderate	Low–moderate	1,4	Hood No. 1 or 4(c) recommended	
(1) Dipping		Slight–moderate	Low–moderate	1,4(c)	See also Dilution Ventilation, this section	
(2) Draining		Slight–moderate	Low–moderate	1,4(c)		
(3) Air drying						
(b) Lacquer	Vapors of naphtha, esters, aromatics	High	High	1,4	Hood No. 1 or 4(c) recommended	
(1) Dipping		High	High	1,4(c)	See also Dilution Ventilation, this section	
(2) Draining		High	High	1,4(c)		
(3) Air drying						
Flow coating						
(a) Doctor blade	Varies	Slight–high	Moderate–high	1,3,6	Hood No. 3 may consist of the entrance to a drying tunnel. Hood No. 6 as close as possible to operation.	
(b) Deluge	Varies	Slight–high	Moderate–high	1,4(c)	Hood No. 1 to enclose the operation as much as possible. Ventilation must also take into account explosive concentrations. See also Dilution Ventilation, Fire Safety.	
Spraying	See Spraying					
Forced drying and baking	See Ovens					
Melting	Metal fume, gases	Slight–high	High	1,6	Hood No. 6 as close as possible to pot. Products of combustion to be vented to hood or independently.	Cloth filters, electrostatic precipitator, high efficiency scrubbers
Toxic metal machining	Metal dust	High	Moderate–high	1,2	Hood No. 2 to be shaped to suit operation, if possible, 4 in. maximum from tool.	Cloth filters, electrostatic precipitator, absolute filter
Welding (unconfined)						
(a) Iron and mild steel	Iron oxide fume	Slight	High	1,2	Also, dilution ventilation at 1000–2000 cfm per welder	Not usually needed
(b) All other welding (fluoride coated rod, toxic metal, inert gas shielded, etc.)	Fumes, oxides of nitrogen, ozone	High	High	1,2		If needed, as in melting

Operation	Contaminant			Hood	Remarks	Air cleaning
Welding (confined)	—	—	—	—	Air line respirator for welder and dilution ventilation. If other workers exposed, local exhaust also.	Not usually needed
Soldering	Lead and cadmium and flux fume	High	Low–moderate	1,2,3,4(c),5	Not usually needed	Not usually needed
Flame cutting	Metal fume, oxides of nitrogen, ozone	High	High	1,5	As in melting	As in melting
Portable grinding, chipping, snagging	Metal, abrasive and possibly silica dust	Slight–high	High	1,2,5	High efficiency cyclone, cloth filter	High efficiency cyclone, cloth filter
Garages (a) Local tailpipe exhaust	Carbon monoxide, aldehydes, etc.	High	High	Special	Hood or open pipe end to fit over tailpipe. *CFM per Branch* Vehicles up to 200 hp—100 cfm Vehicles above 200 hp—200 cfm Diesels—400 cfm	Not usually needed
(b) General ventilation	Carbon monoxide, aldehydes, etc.	High	Varies		5000 cfm per running auto 10,000 cfm per running truck 100 cfm/hp per running diesel	
Lab fume hoods (a) Routine (b) Radioactive or toxic	Varies Varies	Moderate–high High	Low Low	1 (special) 1 (special)	Smooth inner surface and coved corners desirable, special baffle	Absolute filter, special disposal of used filters
(c) explosive	Perchloric acid, etc.	High	Low	1 (special)	Smooth inner surface, coved corners, special baffle, drain, provisions for pipe and hood wash down	Not usually needed or as above
Epoxy mixing, potting, etc.	Vapors of amines, styrene, etc.	High	Low–moderate	1,2	Amine vapors can cause contact dermatitis. Hood No. 1 preferable. Ventilation must also maintain required flow at operator's hands.	Not usually needed

ture where it can be reintroduced into the process; in agricultural lime manufacture where it can be sold as a high quality product; in jewelry buffing where it can be reclaimed for its intrinsic value; or, in the case of collected wood dust, where it can be utilized as fuel.

Since the cost and size of air cleaners are almost directly proportional to the quantity of air to be cleaned and the concentration of contaminants, both of these parameters should be minimized. Careful design of hoods (q.v.) can minimize the amount of exhaust air required to provide adequate in-plant control and reduce the amount of contaminant introduced into the system. Low velocity take-offs will diminish the amount of contaminant collected by the exhaust pipe. Hoods can also act as inertial settling chambers to keep heavier particles from getting into the air stream.

Contaminant generation can in many cases be reduced by the selection of appropriate equipment, a change in operations, or improved working technique. Correct paint spraying procedures limit overspray and, concomitantly, air contaminants. Electrostatic or airless spraying minimizes overspray; and paint dipping or flow coating, instead of spraying, will reduce airborne particulates to zero. Flotation of plastic chips or surfactant foam on the surface of plating baths will act as a mist suppressant and frequently eliminate the need for a collector. A properly designed and operated degreaser (see Vapor Phase Degreasers) will require no air cleaning and perhaps no exhaust system at all. Hydroblast which eliminates particulate generation can replace dry abrasive blasting.

Many more examples can be cited but they would merely underline the principle, i.e., a system must be designed not as an array of individual elements, each by itself functioning as efficiently as possible, but rather as an interrelated unit with each element functioning efficiently not only by itself but as part of an integrated whole.

The cost of air cleaning can also be minimized by providing separate exhaust systems for groups of machines or operations generating similar contaminants. Thus it is not advisable to combine into a single exhaust system operations which generate particulates requiring air cleaning and processes which generate gases or vapors. The latter may not require air cleaning or may require more expensive air cleaners. Using separate systems, especially if only one will need an air cleaner, will result in lower air cleaning costs. Similar considerations apply to such installations as woodworking shops where sanders create dust requiring high efficiency air

cleaning, and planers and jointers create chip for which less efficient cleaners are adequate.

Another approach for particulates is to provide two different collectors in series. The primary collector, usually an inexpensive, low pressure cyclone, removes most of the large particles. The remaining large particles and all the fine particles are collected by a high efficiency cleaner which can be much smaller and less frequently overloaded than if no primary collector were used.

Control by Atmospheric Dispersion and Dilution

Although this method of control is encountering increasing disfavor and opposition, its use will continue in certain applications. The use of a stack to dilute and disperse contaminants is one of the oldest and until recently most widely used techniques for control of air pollution. In theory, a contaminated air stream emitted from a stack will be diluted to acceptable levels by the time it reaches ground level. Most industrial applications of this technique consist merely of an exhaust pipe discharging 6 or 10 feet above the roof of the building. The use of a standard weather cap is contraindicated since it will direct the contaminated air downward. Preferably the means of weather protection should permit discharge of the air directly upward, adding to the effective stack height. The usual pragmatic measure of efficiency of this method is the number of complaints elicited from residents in the environs. However, ground level concentrations can be measured.

A number of factors influence the efficiency of the atmospheric dispersion and dilution technique. These include local topography, micrometeorology, height, distribution and configuration of buildings and other structures in the vicinity, all of the weather variables, rate and nature of contaminant emissions, height and cross-sectional area of stack, and stack gas temperature and flow rate.

A number of empirical equations have been derived, utilizing the above factors, for the prediction of ground level concentrations at various distances and cardinal points from the source. These can be used to design stacks against a desired ground level concentration. However, they yield only approximations and are highly dependent on accurate but not readily accessible long term meteorological data at the source.

Present day control standards severely limit dispersion and dilution as an acceptable technique. It cannot be used in locations which have stable atmospheres or frequent temperature inversions since turbulent mixing is required. Also and more seriously, air pollution is becoming so ubiquitous that the concept of acceptable ground level concentrations at a certain distance from a specific emitter is rapidly losing validity. Many separate sources may each be contributing an "acceptable" ground level concentration to the same point, making the total effect intolerable.

As a result of this developing problem, the concept of ambient air quality standards has emerged. Accordingly, discharge from a particular emitter will be evaluated not only in terms of its contaminant concentration but for its effect on ambient air quality and impact on the environment as well (ambient air quality is discussed at greater length in Section 4).

Accordingly, industrial air pollution control is coming more and more to mean industrial air cleaning.

AIR CLEANING

It is the function of this section not to supply technical design information but rather to discuss general principles and design features as a guide to the selection of suitable air cleaning equipment. It is seldom feasible or desirable for the plant to build its own equipment since all types of equipment are now commercially available. Manufacturers of such equipment can furnish valuable guidance and also guarantee specific performance standards to meet the requirements of the law.

Air Cleaning Design Factors

(1) Cleaning Efficiency. The most important factor is required air cleaning efficiency, and this is dictated by existing or probable future air pollution control rules of the agency having applicable jurisdiction. Efficiency is usually expressed as percent by weight of contaminant removed from the air stream by the air cleaner. The criteria for determining required efficiency are usually expressed in terms of the nature of the contaminant and potential emission rate (pounds per hour) before air cleaning. Contaminants with high air pollution potential (toxicity, nuisance or odor factor, property damage) require higher cleaning efficiency. High potential emission rates also require higher efficiencies.

As previously indicated, high efficiency cleaning can usually be most economically provided by a two stage cleaning cycle, with a lower efficiency collector preceding the higher efficiency one. The latter can then be much smaller than if only one high efficiency collector was provided. The overall cost of such a two cycle system is usually less than a single high efficiency collector. However, a two cycle system may be contraindicated by the lack of space since the primary collector will usually be large.

Although a low efficiency collector may be acceptable at the time of the original installation, it is usually advisable to plan for the possible need for higher efficiency cleaning at a later date and initially provide excess fan and motor capacity and reserve space for a second collector.

(2) Air Stream Characteristics:

(a) Temperature: High temperature air filters may require special filtering cloth media and special construction. Consideration should be given to the possible economic advantage of precooling the air before it enters the collector, permitting the use of standard equipment.

(b) Steam or moisture: Excessive moisture content of the air to be filtered may result in condensation, packing of the contaminants, and plugging of filter fabric or other components of a dry collector.

(3) Contaminant Characteristics:

(a) Corrosive or oxidizing contaminants may require neutralizing wet cleaners or corrosion resistant construction.

(b) Wax impregnating, buffing, rubber manufacture and other processes involving waxy or greasy contaminants can quickly plug cloth arrestors and some dynamic air cleaners and may require special equipment.

(c) Abrasive materials require wet or specially designed collectors to prevent excessive wear.

(d) Particle size and shape and size distribution will dictate the efficiency of the collector with the smallest particles usually the determining factor.

(e) Combustibility of the contaminant may require explosion-proof construction, venting, and fire protection.

(4) Collected Material Disposal:

(a) Dry collectors may be continuously or intermittently unloaded through special valves or gates to conveyors or containers. Manual unloading should be directly into bags or containers. In all cases, the unloading should be conducted in such a manner that a secondary air pollution problem does not arise and defeat the purpose of the air cleaner. When dust bins are not too large, they may be lined with heavy-duty plastic bags which can be tied before removal.

(b) Wet collectors can be continuously discharged into a sewer, settling basin, etc., or can periodically be unloaded in the form of a slurry. Wherever possible, wastes should be clarified or treated before discharge. Water pollution regulations may circumscribe the manner in which wet waste may be disposed. In many areas, such requirements limit the use of wet collectors or make necessary costly treatment.

Air Cleaning Devices

Dust Collectors. (See Figure 9) *Cloth Filters.* One of the most efficient and widely used media for removing dry particulates from an air stream is the cloth filter, also known as the dust arrester, or bag filter. Practically any fibrous, woven or felted material can be used as the filtering agent.

The usual design takes the form of tubes, screens, envelopes, or mats. Dust laden air enters one side of the fabric and emerges through the other side as cleaned air to be discharged out of doors or, in some cases, recirculated back to the workroom. In most commercial designs, the fabric is mounted in a suitable framework enclosed in a housing. Air enters from below, usually via a plenum large enough to permit settling of the larger particles by gravity before they reach the filter zone. A manual or motorized shaking mechanism is usually provided to permit periodic shaking of the fabric to dislodge adhering aggregates. Mat filters can either be shaken, replaced or cleaned by other means. The dust falls into a hopper for subsequent removal.

The cleaning process is not simple filtration since the pores of the medium are usually much larger than the particles to be collected. When new, much dust will pass through the filter until a bed of deposited dust is built up on the fabric. The process is a complicated one and not clearly understood. It probably involves impingement of the dust particles on the fibers as well as deposition under the influence of settling, Brownian motion, and static electricity created by the flowing air. The dust mat will rapidly build up on the medium, and it is this mat, rather than the fabric, which acts as the filtering agent. A permanent dust base will be created within the pores which will not be dislodged by shaking, so that cleaning efficiency remains high.

The most commonly used fabric medium for normal applications is cotton or wool sateen or felt. Operating temperature is usually the determining factor in the selection of fabric. Cotton may be used for temperatures up to 180°F. Wool is acceptable for temperatures of 200°F, and synthetics such as acrylics are available for temperatures up to 350°F and higher. Glass-fiber and asbestos fabrics have been used for applications up to 650°F, but the fragility of the fibers results in failure after a number of shaking cycles.

The size of the filter, i.e., the area of filter medium through which a given stream of air passes, will affect the resistance to the air flow and therefore the required fan horsepower. In designing a cloth filter, the optimum size will usually be a compromise between available space, initial cost, and required horsepower. In some instances, the nature of the pollutants being exhausted or their toxicity will be the governing factor. The usual practice is to select a filtering velocity of 3 linear fpm (cfm to cloth area ratio of 3 to 1).

Normal resistance of the filter to flow ranges from 2 to 6 inches of water, the usual resistance for a filter with a ratio of 3 to 1 being about 3 inches. Ratios higher than 3 to 1 will make possible smaller units, but the resultant high resistance will require more horsepower and the lower filtering area will necessitate more frequent shaking. A less recognized consequence is greater variation in cfm developed by the fan when smaller units are used since cfm will decrease as the resistance increases. With systems involving the control of highly toxic substances, where minimum flow rates at the hood must be exceeded at all times, lower cloth ratios are always used.

An important feature of the design of filters is the shaking mechanism and cycle. The fabric must be cleansed of adhering dust sufficiently often and thoroughly so that the rate of exhaust does not fall below the required minimum. The shaking mechanism must vigorously rap, shake and flex the fabric to dislodge the dust which may adhere very firmly. The shaking cycle must, as indicated, be frequent enough to ensure continuous maintenance of the design cfm. For many normal applications, shaking can be performed periodically at normal process shutdowns such as lunch time or at the end of the day. An automatic interlock can be provided which will operate the shaking mechanism for a predetermined period each time the fan motor is turned off.

When abnormally heavy dust loading exists, it may be necessary to shut down the process *during* a shift in order to permit shaking. However, it is usually not advisable and frequently not safe to shut down operations involving the control of highly toxic, flammable, or explosive contaminants. Similarly, it may not be feasible to shut down operations performed on a 24 hour a day basis. In such situations, continuous cleaning must be provided.

One method of providing continuous cleaning is to use a three compartment cleaner, each compartment being independent of the others and provided with a separate shaking mechanism. In this design, each compartment is successively and automatically removed from air cleaning service for shaking. Although only two-thirds of the cloth area is available at any one time for filtration, the cloth ratio can be the same as for a conventional arrester since the

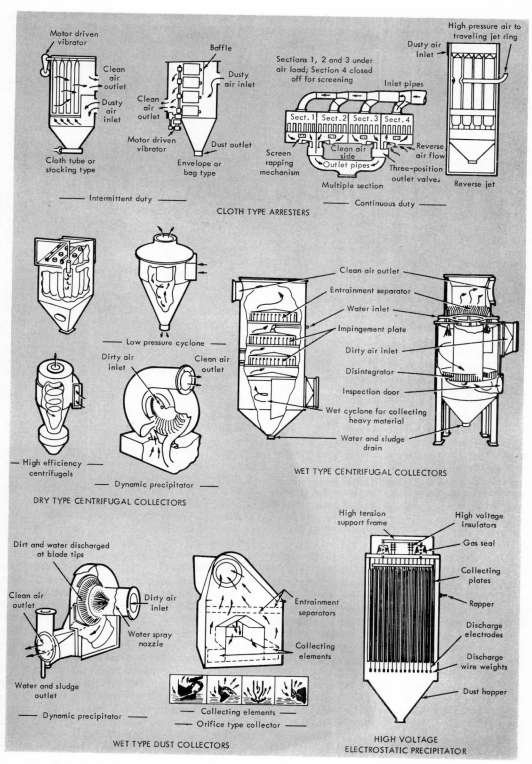

Figure 9. Dust Collectors. (*Courtesy Heating, Piping and Air Conditioning, January 1968*).

disadvantage of utilizing only two-thirds of the cloth is overcome by the more frequent cleaning.

A modification of the standard cloth arrester was developed a number of years ago and is now available commercially as a reverse jet arrester. It may consist of the standard cloth tubes or a single large diameter bag, usually of wool felt. The bag or tubes are circumscribed by a pipe ring which travels up and down the surface continuously. Jets of compressed air are continuously blown through the fabric, countercurrent to the flow of cleaned air out of the cloth, through a slot or orifices. The material caked on the cloth is continuously dislodged from the surface and, since it is agglomerated, is not reentrained in the air stream but falls to the collecting hopper. Continuous cleaning prevents excessive build up of resistance, making possible much smaller units with cloth ratios from 5–10 times greater than with standard arresters. The resistance, after initial build-up, is essentially constant, resulting in essentially constant cfm at the hoods and making possible more efficient fan selection and operation.

There are many special types of filters, including dry paint filters, mats or panels of treated paper or cloth, and dry or oiled wire mesh, which have applications for light dust loading or special conditions. The "absolute" filter is a special deep bed fibrous filter with extremely high efficiency for dust and fume particulates as fine as 0.3 μ. These are particularly useful for extremely toxic materials such as beryllium dust, radioactive particulates, and virulent biologicals. Their special advantage lies not only in high efficiency but also in the fact that, when loaded, the complete filter unit can be discarded, usually by companies specializing in hazardous wastes disposal. Disposal of hazardous dusts collected by conventional filters can create a secondary air pollution problem as well as an industrial hygiene problem.

Filters have wide application and can be used for almost all types of dusts and fumes in the fine particle range with efficiencies up to 99 percent. Use of less efficient, low pressure pre-cleaners will reduce the required size of the more expensive arrester and will increase its useful life by removing the larger more abrasive particles and by making possible shorter and less frequent cleaning cycles.

Dynamic Precipitation:

(1) Cyclones. The cyclone is basically a cylinder set on top of a cone. The dust laden air enters the top of the cylinder tangentially and travels radially, contiguous to its inside surface. The change in direction of travel of the dust particles, resulting from centrifugal forces, causes them to move to the inner wall of the cylinder in a zone of lower air velocity. Here they are precipitated from the air stream and fall to a dust hopper at the bottom of the cone by the action of gravity. The cleaned air loses velocity as it travels downward in a helical motion until it reverses direction and is discharged axially out the top of the cylinder through a concentric outlet pipe.

Efficiency of collection is a function of entrance and radial velocities within the cylinder and is therefore directly proportional to pressure drop or resistance. Length of cone is also a factor since it increases the time available for the dust to separate from the air stream. The standard low pressure cyclone is useful for relatively coarse dusts under conditions of moderate to heavy loading. Efficiency will be about 75–80 percent for particles down to 40 μ. Resistance will be about 1–3 inches of water.

Since radial velocity directly affects efficiency, for a given air volume efficiency will increase with decrease in cylinder diameter. Modifications of the dimensions of the standard cyclone which have been developed empirically over the years have resulted in the so-called high pressure or high efficiency cyclone. Performance can be further improved by using a number of very small cyclones in parallel instead of a single unit. Multiple units with efficiencies up to 90 percent for particles down to 10 μ can thus be obtained with pressure drops ranging from 2–10 inches.

Principal advantages of the cyclone include its simplicity, lack of moving parts, and constant air flow.

(2) Dry Dynamic Precipitator. This collector uses a specially shaped impeller to precipitate dust from the air stream by centrifugal and dynamic forces. The dust travels along the impeller blade surfaces and is projected into a separate dust chamber in the impeller housing from which it is transported through an opening to a hopper below. Collection efficiency and resistance are similar to the multiple cyclone units described above. The principal advantage of this equipment is that the impeller serves as both the air mover and the dust collector, resulting in a lower initial cost and space requirement than for a separate fan and cyclone with similar characteristics. However, it cannot be used for large air quantities and is not suitable for fibrous or adhesive dusts or for very large particles.

(3) Louver Collectors. These consist of a series of louvers or plates set at an angle to air stream. Separation results from the centrifugal forces created by the rapid and frequent change in direction of air flow as the air passes over the louver surfaces. Decrease in the spaces between the louver surfaces results in an increase

in efficiency but also increases the possibility of plugging by deposition of dust in the spaces. Characteristics and advantages are similar to those of the cyclone, but the louver collector is not as adaptable for large volumes of air because of its space requirements. The usual resistance will be between 0.3 and 1.0 inches of water.

(4) Inertial Separators. These are large settling chambers in which dust settles from the air stream by gravity. The pressure drop is very low, and the settling chamber will have some limited usefulness where space is available and the contaminant consists principally of large, heavy particles. Chambers as long as possible with cross-sections as large as possible will result in low particle velocity with maximum possible distance of travel to permit the particles to reach terminal settling velocity before the air is discharged outdoors.

The settling chamber principle has effectively been used in the low velocity plenum main where the "cleaned air" is either discharged directly to the outdoors or into a suitable collector. The latter can usually be quite small because of the ultimately very light dust loading. A bonus advantage of such equipment is that branch pipes can be added, removed, or relocated at will without affecting the dynamic balance of the system.

Inertial separators can also be provided with suitably arranged baffle plates so that there is a continuous change of direction of air flow. The resultant lowering of particle velocity plus an impingement factor increases the rate of settling. Such an arrangement is particularly suited for removal of mists from an air stream.

Wet Collectors

The basic principle of the wet collector is to provide intimate contact between the collecting liquid and the contaminant in the air to be cleaned. When the contaminant is particulate, the wetting of the particle increases its mass and it becomes relatively easy to remove from the air stream. Wet collection is also suitable for gases or vapors which are either soluble in, or react with, the collecting medium. The intimate mixture of the air and liquid provides a large interface area between the two media so that solution or reaction of the contaminant is enhanced. There are many designs of wet collectors with a wide range of pressure drops and efficiencies. As in all cleaners, the former varies directly with the latter.

The principal advantages of wet collectors are the maintenance of a constant air flow, applicability to corrosive or hot air streams, suitability for both particulate and non-particulate contaminants, and the absence of a secondary air contamination during disposal of collected contaminants.

Among the disadvantages are the creation or required use of corrosive liquids in the unit (increasing construction and maintenance costs), elaborate installations frequently required, the need for a reliable and plentiful water supply, and disposal problems which may require expensive pretreatment. Suitable protection against freezing must be provided if located outdoors.

Efficiency for particle collection can be as high as 98 percent. With proper design and operation, essentially 100 percent efficiency can be obtained for gases and vapors.

The basic wet collectors include the following:

(1) Spray Chamber. This is perhaps the simplest wet collector. It consists of sprays, usually in combination with scrubber plates, through which the contaminated air flows. Efficiency will be governed by the pressure, number, and location of the spray nozzles. The standard dry cyclone can also be converted to a wet centrifugal collector by suitably placed nozzles within the cyclone to provide both centrifugal and wetting action, with the latter enhancing the former.

Water may be distributed by nozzles, an elevated reservoir, or the air stream itself. Pressure drop will vary from 2-6 inches. Water consumption is usually from 3-5 gal/1000 cfm of air.

(2) Wet Dynamic. The dry dynamic precipitator previously described can be modified by the addition of spray nozzles at the inlet designed to maintain a film of water on the impeller surfaces. This increases the efficiency of collection retention of the particles on the impeller blade surface, as well as the disposal of the material. Water consumption will be from $\frac{1}{2}$-1 gal/1000 cfm of air.

(3) Packed Towers. These are used extensively by the chemical industry for process purposes, but they can also function as air cleaners. Essentially they consist of shells of steel, ceramic, plastic or other suitable material packed with such materials as irregularly shaped ceramic saddles or rings, or granular materials such as sand, coke and gravel. Fibrous packing such as glass or steel wool may also be used. Water, at the usual rate of about 10 gal/1000 cfm of air is introduced at the top and contaminated air at the bottom. Efficiency depends on period of contact between the air and the wetted packing so that air velocity through the unit should be as low as possible and the surface area and depth of packing as great as possible. Pressure drops may vary widely depending on design but the usual range is from 1.5-3.5 inches. The simplicity of design of these towers

minimizes maintenance problems and makes them particularly adaptable for hot or corrosive air streams. Heavy dust loading will however require cleaning or replacement of the packing.

(4) Wet Orifice Collectors. In these, the passage of the contaminated air at a high velocity through an orifice or baffled opening located above a water reservoir creates a water curtain with considerable turbulence. The contaminants are in this manner thoroughly wetted and retained in the water reservoir. The lack of ledges, obstructions or moving parts makes the design features of orifice collectors fairly simple. This, plus the presence of water, makes orifice collectors particularly adaptable for explosive, pyrophoric, sticky or fibrous contaminants. The pressure drop varies from 2.5–6 inches and higher. Facilities must be provided for separating entrained moisture from the effluent before discharge. Maintenance of a constant water level in the reservoir is critical for the creation of the proper water curtain.

(5) High Efficiency Wet Collectors. As the size of the water droplets created by a wet collector approaches the size of the smallest particles to be collected, the probability of contact, and therefore collection increases. For this reason, the fog filter and the venturi scrubber are among the most efficient of wet collectors since their action results in the dispersion of very fine water particles.

Fog filters consist of many small, high pressure nozzles located in centrifugal tower collectors. Water pressures of 250–600 psi at flow rates of 5–10 gal/1000 cfm of air are used. The very fine orifices at high pressures require special nozzle construction. The water may have to be pre-filtered to prevent plugging, and for this same reason recirculation of the water is not practical.

In the venturi scrubber, the contaminated air is passed through a venturi throat at very high velocity. Water, which is simultaneously fed into the throat, is atomized by the air into a very fine fog. The intimate mixing of the contaminant and water in the turbulence created by the venturi results in rapid wetting and collection. Pressure losses may be as high as 12–16 inches, but efficiencies may run as high as 99 percent in the submicron particle size range.

Electrostatic Precipitators. The basic principle involves the creation of a strong unidirectional electrostatic field between two electrodes. The dust particles in the contaminated air passing between the electrodes are electrically charged and attracted toward an oppositely charged collecting electrode where they are deposited.

There are two general types of electrostatic precipitators:

(1) The high potential, single stage precipitator in which ionization and collection are simultaneous throughout the unit. These are used for the collection of particulates (dust, fume, mist) in industrial exhaust systems where loading can be high and will generate from 50,000–75,000 volts.

(2) The low potential two-stage precipitator is divided into a pre-ionizing section followed by a non-ionizing collection section. It is used for general air cleaning or industrial air cleaning where the concentration of particulates is low. The potential ranges from 12,000–15,000 volts.

The design of these units will depend on air flow rate, characteristics of the particles and the air stream, and required efficiency. Although efficiencies approaching 100 percent can be obtained, it is not usually practical to do so since the cost rises very sharply as this figure is approached. Such factors as local air pollution laws, available space, reclaim value of collected material, etc., will affect the design.

The major advantages of the electrostatic precipitator include constant air flow rate and low pressure drop. Among its disadvantages are high initial and operating cost, unsuitability for flammable or explosive atmospheres, frequent necessity to condition the entering air stream, and the need for primary separators when high dust loading (more than 25 grains/cu ft of air) may be encountered.

Gas and Vapor Collectors. (See Figures 10 and 11) As indicated above, since atmospheric control of gases and vapors by dispersion and dilution is becoming less acceptable, air cleaners for these contaminants, which were formerly seldom required, will in many cases ultimately become mandatory. Therefore, the problem of design of air cleaning for gases and vapors is rapidly attracting much more attention than in the past, particularly for relatively small installations.

The nature of gases (the term gases will be used to signify both gases and vapors) is such that their removal from an air stream is limited to four basic methods: absorption by solution and/or reaction in a suitable liquid medium, adsorption on the surface of a solid material, combustion to yield innocuous products, and condensation to the liquid state. Proper design can provide any desired degree of efficiency up to virtually 100 percent. Resistance to air flow may be as high as 20–30 inches of static pressure.

(1) Absorption. This is normally accomplished by any of the wet collectors previously described for particulates. (Dynamic scrubbers, however, serve no useful purpose in gas collec-

TABLE 7. Approximate Characteristics of Dust and Mist Collection Equipment

Equipment Type	Purchase Cost[b] ($/cu ft/min)	Smallest Particle Collected (μ)[c]	Pressure Drop (in. H_2O)	Power Used[d] (kW/1000 cu ft/min)	Remarks
Settling chambers					
(1) Simple	0.1	40	0.1–0.5	0.1	Large, low pressure drop, precleaner
(2) Multiple tray	0.2–0.6	10	0.1–0.5	0.1	Difficult to clean, warpage problem
Inertial separators					
(1) Baffle chamber	0.1	20	0.5–1.5	0.1–0.5	Power plants, rotary kilns, acid mists
(2) Orifice impaction	0.1–0.3	2	1–3	0.2–0.6	Acid mists
(3) Louver type	0.1–0.3	10	0.3–1	0.1–0.2	Fly ash, abrasion problem
(4) Gas reversal	0.1	40	0.1–0.4	0.1	Precleaner
(5) Rotating impeller	0.2–0.6	5	—	0.5–2	Compact
Cyclones					
(1) Single	0.1–0.2	15	0.5–3	0.1–0.6	Simple, inexpensive, most widely used
(2) Multiple	0.3–0.6	5	2–10	0.5–2	Abrasion and plugging problems
Filters					
(1) Tubular	0.3–2	<0.1	2–6	0.5–1.5	High efficiency, temperature and humidity limits
(2) Reverse jet	0.7–1.2	<0.1	2–6	0.7–1.5	More compact, constant flow
(3) Envelope	0.3–2	<0.1	2–6	0.5–1.5	Limited capacity, constant flow possible
Electrical precipitators					
(1) One-stage	0.6–3	<0.1	0.1–0.5	0.2–0.6	High efficiency, heavy duty, expensive
(2) Two-stage	0.2–0.6	<0.1	0.1–0.3	0.2–0.4	Compact, air-conditioning service
Scrubbers					
(1) Spray tower	0.1–0.2	10	0.1–0.5	0.1–0.2	Common, low water use
(2) Jet	0.4–1	2	—	2–10	Pressure gain, high velocity liquid jet
(3) Venturi	0.4–1.2	1	10–15	2–10	High velocity gas stream
(4) Cyclonic	0.3–1	5	2–8	0.6–2	Modified dry collector
(5) Inertial	0.4–1	2	2–15	0.8–8	Abrasion problem
(6) Packed	0.3–0.6	5	0.5–10	0.6–2	Channeling problem
(7) Rotating impeller	0.4–1.2	2	—	2–10	Abrasion problem

[a] From "Dust Collector Review," David G. Stephan, Air Pollution Engineering Research, U.S. Department of Health, Education and Welfare, Public Health Service, Cincinnati, Ohio, 1960.
[b] Steel construction, not installed, includes necessary auxiliaries, 1960 prices.
[c] With 90–95% efficiency by weight.
[d] Includes pressure loss, water pumping, electrical energy.

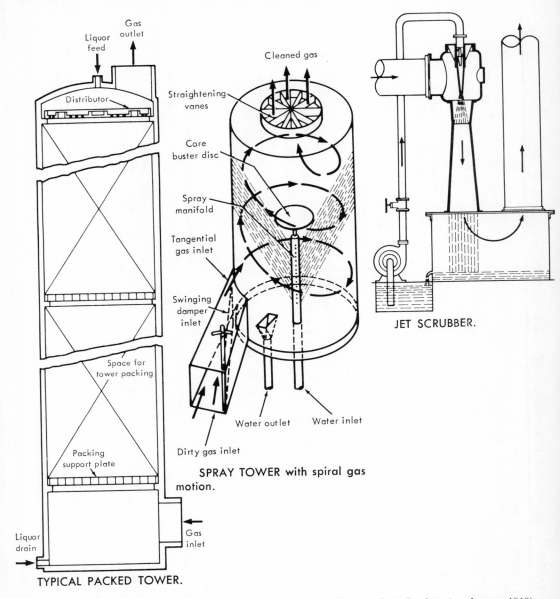

Gas outlet

Liquor feed

Distributor

Cleaned gas

Straightening vanes

Core buster disc

Spray manifold

Tangential gas inlet

Swinging damper inlet

Space for tower packing

Packing support plate

Water outlet   Water inlet

Dirty gas inlet

SPRAY TOWER with spiral gas motion.

JET SCRUBBER.

Liquor drain

Gas inlet

TYPICAL PACKED TOWER.

Figure 10. Wet Gas and Vapor Collectors. (*Courtesy Heating, Piping and Air Conditioning, January 1968*).

tion except when the dynamic features of the design also increase the contact between the contaminant and the scrubbing medium.) Since gaseous molecules will not be physically entrained and retained by a liquid, the medium must be one in which the contaminant is highly soluble or with which it will react. In many instances water alone will be satisfactory. In other cases, caustic soda or other chemicals or reactants must be added to the water. The scrubber should be designed to provide an in-

timate mixture of contaminated air and absorbing liquid for sufficient duration and magnitude of contact so that maximum solution or reaction takes place. The medium is usually completely or partially recirculated, particularly when water is in short supply, when disposal of liquid wastes is a problem, or when chemical solutions are used as the medium. Arrangements must be made for continuous or periodic replenishment to prevent the medium from becoming saturated or vitiated with a resultant decrease in absorbing

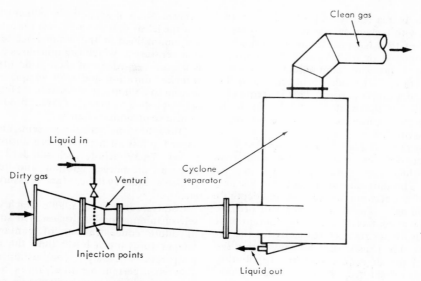

VENTURI TYPE SCRUBBER.

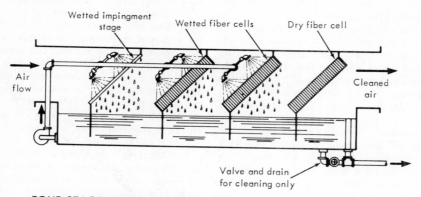

FOUR-STAGE CAPILLARY WASHER

Figure 10. (*Cont'd*)

efficiency. Mist entrainment devices are usually necessary in the discharge stack.

If the contaminant is collected in a viable condition, it can frequently be recovered for sale or reintroduction into the process. Otherwise, disposal of waste liquid must be to a sewer, stream, ground, or sump, with due consideration for local or state rules. Frequently, dilution, clarification, or chemical treatment is required before disposal. The necessity for costly treatment of waste liquid or the unavailability of a reliable water supply may contraindicate the use of wet collectors.

(2) Adsorption. Adsorption of gases on the surface of a solid is possible because of the existence of available molecular binding energy on the surface of the adsorbent. Maximum adsorption of gases from an air stream will occur when there is a high concentration of gases, a large adsorbing surface (finely divided material) absence of interfering substances, low temperature, and favorable characteristics of the gas molecule with respect to the molecular structure of the adsorbent (shape, size, and polarity of the respective molecules). All other factors being equal, efficiency and resistance to air flow will bear a direct relationship to fineness and shape of the adsorbing particles, and thickness of the bed of adsorbent.

The usual adsorbents consist of activated charcoal or silica gel, although a number of substances can be used. The efficiency can frequently be increased by impregnation of the adsorbent with a chemical agent which converts the pollutant to a harmless or much more adsorbable material or with a catalyst which oxidizes or decomposes the contaminant to a more desirable state. The adsorbent can also be coated on to an inert carrier. This is frequently done when the adsorbent may be an expensive material which cannot be reactivated for reuse after it is saturated.

The adsorbent is placed in a perforated canister, flat bed, or other suitable container through which the contaminated air can be directed. The units are arranged in series and parallel so an optimum surface area and depth of bed is obtained. The design basis will provide for maximum contact between the contaminant and the adsorbent, sufficient capacity for the desired service life, resistance to air flow within the capacity of the air mover, uniform distribution over the adsorbent bed with avoidance of channeling, pretreatment of the air if necessary to remove particulates and other interfering substances, and provision for renewing or replacing the adsorbent periodically.

The collected material may be removed from the adsorbent by suitable means (this also regenerates the adsorbent for further use); the saturated adsorbent may be discarded (highly toxic or radioactive contaminants may require special handling); or the contaminant may be oxidized or combusted on the surface of the adsorbent.

(3) Combustion. When a gas or vapor can be oxidized to an innocuous substance (as with most hydrocarbons), combustion is frequently used as a method for removing such contaminants from the air stream. In some instances, where suitable liquid or solid collecting media are not available or practical for a particular contaminant, combustion may be the only feasible method. Design of a system for maximum combustion requires bringing the oxygen into intimate contact with the gas molecules at adequate temperatures for a sufficient duration. This will require a combination of high temperature and turbulent mixing in a properly designed combustion chamber.

When the contaminant is present in the air in concentrations between the lel (lower explosive limit) and the uel (upper explosive limit), provision of a source of ignition will initiate combustion which will then be self-sustaining. In those rare instances where the contaminant is present in concentrations above the uel, provision of additional dilution air to bring the level to below the uel will make this process possible. However, in most industrial applications, the concentration of gases in the effluent will be below the lel so that the air stream must be heated and maintained at the autogenous temperature (temperature at which organic gases and vapors will burn regardless of their concentration) for sufficient duration and with adequate turbulent mixing to complete combustion. This is usually accomplished by means of direct fired burners in a furnace or similar chamber.

Direct fired incineration presents an explosion hazard unless all necessary precautions are followed. These include such standard safety features as pre-purge, ignition and temperature controls, limit switches, fuel pressure controls, etc.

Catalysts are now available which will initiate self-sustaining combustion on their surface, even of dilute concentrations of contaminant, at temperatures much lower than the autogenous temperature. Catalytic combustion is not suitable when certain metallic vapors or other inorganics which may poison the catalyst are present in the air stream. Pre-filtration may also be necessary to remove particulates from the air.

(4) Condensation. Vapors can be removed from effluent by condensation to the liquid state by standard techniques or refrigeration. The liquid is collected in suitable containers for subsequent disposal. This technique is quite expensive and is usually economically feasible only for substances with a relatively high boiling point, where low air volumes are to be treated, where the concentration of the contaminant in the air is relatively high, or where the collected material can be reused or sold.

(5) Masking. Masking is a control method of very limited usefulness. Low concentrations of contaminants which are sufficiently odorous to be objectionable but which are otherwise innocuous may be treated by injection of masking agents into the air stream before discharge to the outdoors. These agents, which should not create a secondary air pollution problem, are designed to mask the objectionable odor or neutralize it.

INFORMATION SOURCES

Governmental Industrial Hygiene Agencies

There are many governmental industrial hygiene agencies which are available for consultation and advice on problems of industrial ventilation. Most of them offer data sheets and informational brochures for the asking. A majority of the state and local agencies have promulgated codes and standards which must be followed when installing an industrial ventilation system.

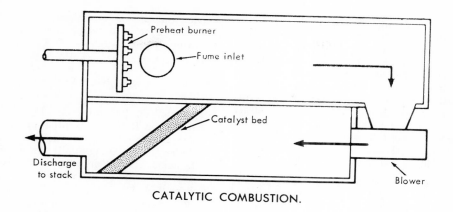

CATALYTIC COMBUSTION.

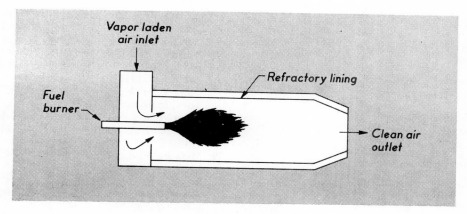

INCINERATION.

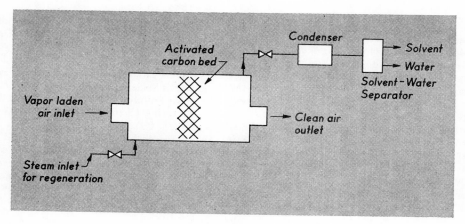

ADSORPTION.

Figure 11. Gas and Vapor Cleaning. (*Courtesy Heating, Piping and Air Conditioning, December 1959*).

States. Nearly all the states and Puerto Rico have industrial hygiene or occupational health divisions within the state health departments. In New York, Massachusetts, and Illinois, the industrial hygiene division is within the state labor department. California, Ohio, Oregon, and Washington have such divisions in both the health and labor departments, with the larger one located in the health department.

Counties. The following counties have industrial hygiene services of varying magnitudes, all located in the county health department:

California	Alameda, Contra Costa, Los Angeles, Orange, San Bernadino, San Diego, San Mateo, Santa Clara, Stanislaus
Georgia	Fulton
Kansas	Wichita, Sedgwick
Kentucky	Louisville, Jefferson
Pennsylvania	Allegheny
Texas	Harris

Cities. The following cities have industrial hygiene services:

California	Albany, Berkeley, Long Beach, Los Angeles, Pasadena, San Bernardino, San Jose, Vernon
Colorado	Denver City
Maryland	Baltimore
Michigan	Detroit
Minnesota	Minneapolis
Missouri	St. Louis
New Mexico	Albuquerque
Ohio	Cincinnati, Cleveland
Pennsylvania	Philadelphia
Texas	Dallas, Houston
Wisconsin	Milwaukee

Federal. The Division of Occupational Health, U.S. Public Health Service, Department of Health, Education and Welfare, 4th and C Sts., N.W., Washington, D.C., is engaged in research, laboratory, and field activities in the field of industrial hygiene and ventilation. Many of its publications in this field are available.

Other federal agencies which publish in this field include the Department of Labor, Bureau of Mines, Atomic Energy Commission, and the Defense Department.

Other Countries. In most of the countries of the world, industrial ventilation activities will be carried out primarily within the Ministry of Labor. In some cases they are performed within the Ministry of Health.

BIBLIOGRAPHY

1. "Industrial Ventilation," 9th edition, A Manual on Recommended Practice, Committee on Industrial Ventilation, P.O. Box 453, Lansing, Michigan, 1967.
2. Alden, J. S., "Design of Industrial Exhaust Systems," New York, The Industrial Press, 1959.
3. Brandt, A. D., "Industrial Health Engineering," New York, John Wiley & Sons, 1947.
4. Hemeon, W. C. L., "Plant and Process Ventilation," New York, Industrial Press, 1955.
5. Drinker, P., and Hatch, T., "Industrial Dust," New York, McGraw-Hill Book Co., 1954.
6. Patty, F. A., editor, "Industrial Hygiene and Toxicology," Vol. 1, New York, Interscience Publishers, 1958.
7. Dalla Valle, J. M., "Exhaust Hoods," New York, Industrial Press, 1952.
8. Dalla Valle, J. M., "The Industrial Environment and Its Control," New York, Pitman, 1948.
9. McCabe, L. C., "Air Pollution," New York, McGraw-Hill Book Co., 1952.
10. Mallette, F. S., "Problems and Control of Air Pollution," New York, Reinhold Publishing Corp., 1955.
11. Magill, P. L., "Air Pollution Handbook," New York, McGraw-Hill Book Co., 1956.
12. "Handbook on Air Cleaning," U.S. Atomic Energy Commission, Washington, D.C., 1952.
13. "Heating, Ventilating and Air Conditioning Guide," Am. Soc. of Heating, Air Cond. Engineers, New York, 1958.
14. "Engineering Manual for Control of In-Plant Environment," American Foundrymen's Society, Des Plaines, Ill., 1956.
15. "Air Pollution Abatement Manual," Manufacturing Chemists' Association, Washington, D.C.
16. Stern, Arthur C., editor, "Air Pollution," Vol. II, New York, Academic Press, 1962.
17. "Handbook of Industrial Loss Prevention," Factory Mutual Engineering Division, New York, McGraw-Hill Book Co., 1959.

RESPIRATORY PROTECTION AND PERSONAL HYGIENE

Benjamin Feiner

Head, Environmental Control Unit, Engineering Section
Division of Industrial Hygiene
New York State Labor Department
New York, N.Y.

RESPIRATORY PROTECTION

Principles

Personal respiratory protection is one of the classical methods of control in industrial hygiene. However, the importance of this technique is often misunderstood. All the other methods of control, when properly designed and applied, can provide adequate continuous protection of a worker against harmful contaminants under normal working conditions. Personal respiratory protection, however, finds its chief usefulness as an emergency or short term means of protection. It should be used as a primary protective device for normal operations only when no other method of control is possible. At times, it may be useful as an adjunct to exhaust ventilation or other control measures.

Respirators are thus emergency devices. They are needed when it is necessary to enter a highly contaminated atmosphere for a short time for rescue or emergency repair work; as a means of escape from a suddenly highly contaminated atmosphere; for periodic, short term inspection, maintenance or repair of equipment located in a contaminated atmosphere; and for normal operations in conjunction with other control measures where the contaminant is so toxic that a single control measure such as ventilation cannot safely be relied on.

A respirator must be designed and selected for the particular environment in which it is to be used. The type of contaminant, its probable maximum concentration, the possibility of oxygen deficiency, the useful life of the respirator, the escape routes available—all these and other factors must be considered in selecting a respirator for emergency use, for periodic use, or for stand-by purposes. Where these factors are not known with certainty, the device providing the widest spectrum of protection must be used.

Respirators must fit well, if possible without discomfort, should permit breathing without undue effort, should not interfere with vision, and should permit complete freedom of movement where danger may otherwise result. Respirators should be cleaned after each use and sterilized frequently, assigned to individual workers, and stored in dust free cabinets readily available for use; where necessary, they should be available both in the workroom for escape and just outside the workroom in a safe atmosphere for rescue. Careful records should be kept of length of time in use, the remaining oxygen or air supply or condition of purifying element, and they should be renewed when necessary.

Atmospheric Hazards

The type and degree of hazard to be encountered will govern the kind of respiratory protective device to be used. Several classifications may be described to assist in proper selection of a device:

(1) Oxygen deficiency,
(2) Gases or vapors immediately dangerous to life,
(3) Gases or vapors not immediately dangerous to life,

(4) Particulates,

(5) Combination of particulates and gases or vapors immediately dangerous to life,

(6) Combination of particulates and gases or vapors not immediately dangerous to life.

Where the degree of hazard is not known, it must be assumed to be immediately dangerous to life.

Oxygen Deficiency. Normal air contains approximately 21 percent of oxygen by volume. This oxygen content may be reduced by such factors as dilution or displacement of oxygen by other gases and loss of oxygen by reaction with other substances or by absorption in certain materials. These conditions are usually found in such confined spaces as storage bins or silos, tanks, sewers, wells, mines, and ships' holds, and may also exist in burning rooms and in closed areas where natural oxidation of materials can occur.

Atmospheres containing 16 percent or less of oxygen may cause serious injury or death to people breathing them, depending on the actual concentration, length of exposure, and physical activity of the exposed persons. In such atmospheres, the respiratory protective device must be of a type which supplies fresh air or oxygen to the wearer. If the atmosphere is such that the wearer cannot safely escape, without respiratory protection, from the remotest location he is likely to be in to an uncontaminated location, the device chosen must be a self-contained type.

Gas or Vapors. Gases and vapors may be classified as toxic or inert. Toxic gases and vapors may cause injury or death depending on the concentration present. Inert gases can displace oxygen and are dangerous only under the conditions described above.

Where gases are so toxic that even in low concentrations they may be immediately dangerous to life, the respiratory protective device must be chosen to provide positive, reliable control for the conditions to be encountered.

Particulates. Particulates may assume the form of dispersions of solids, such as dusts, fumes, and smokes; liquids, such as mists and fogs; and combinations of the two, such as mists created by sprays of suspensions of dusts and paints. With rare exceptions, such as the organic phosphorus insecticides and, possibly, massive concentrations of the more highly toxic metals, particulate contaminants are not immediately dangerous to life. Mechanical filtration of inspired air is the usual protection method used.

Particulates may be classified as toxic, pneumoconiosis producing, and non-specific or nuisance. The design of a respirator depends on the type of dust to be encountered. Toxic particulates, such as lead, fluoride, and phosphorus dusts, enter the blood stream directly from the lungs and cause systemic poisoning. Pneumoconiosis-producing dusts (silica, asbestos) remain in the lungs and cause localized fibrotic diseases. Nuisance dusts (flour, wool, wood) may do either, but usually do not produce local or systemic effects. However, they may be irritating or allergenic and, in massive concentrations, may cause debility by their physical presence in the lungs.

Combination of Particulates and Gases. Special respiratory devices designed to protect against mixtures of more than one type of contaminant are available. These should be selected to provide protection against the maximum expected concentration of each contaminant. Where one of the contaminants may be immediately dangerous to life, this should dictate the type of protection used.

Types of Respirators

There are two principal types of respiratory protective devices, i.e., oxygen or air supply and air purifying. There are, in turn, several classifications within each type. Table 1 lists all the common types and describes their principal limitations and necessary precautions to be observed when they are used.

(A) Supplied Air or Oxygen Respirators:

(1) Self-contained. These are completely independent of the atmosphere surrounding them. They are entirely self-contained and can be worn by the user, may be used in any atmosphere, and permit complete freedom of movement. All self-contained units are designed for specific hours of use. There are three principal types, as follows:

(a) Recirculating Compressed Oxygen. Compressed oxygen, in a cylinder carried by the user, is made available for breathing to a breathing bag through a suitable tube connected to a mouthpiece or facepiece by means of check and pressure regulating valves. Exhaled breath passes through another check valve to a canister where carbon dioxide is absorbed and the breath is returned to the breathing bag through a cooler. In the event of any failure of this system, it can be bypassed by means of a manually operated valve which admits oxygen directly to the breathing bag.

(b) Demand Compressed Air or Oxygen. Compressed oxygen or air, contained in a cylinder worn by the user, is supplied through pressure-reducing valves to the facepiece only when the wearer inhales, in quantities governed

TABLE 1.   Respiratory Protective Equipment, Applications, Limitations, Precautions

Atmosphere	Recommended Type of Respirator	Applications and Limitations	Precautions
All particulates, gases, vapors, oxygen deficiency	*Self-contained* Recirculating compressed oxygen Demand compressed air or oxygen Self-generating oxygen	Use in any atmosphere, allows freedom of movement, allows worker to leave atmosphere by any route; limited time of use, careful training required for proper use.	Wearer should be in good physical condition, thoroughly trained; assure plentiful supply of air or oxygen in tank, check for proper and tight fit, use with life line, leave at once if an odor is detected, do not remove until out into respirable air.
	Supply Air Hose mask with blower Hose mask without blower Air line respirator	Unlimited time of use, use in any atmosphere (except air line not to be used in oxygen deficient or immediately dangerous atmosphere), not to be used where worker cannot escape unharmed without protection, must exit by entrance route, 150 ft maximum from exit (75 ft hose mask without blower), limits freedom of movement.	Place inlet in respirable air location, adjust fit and air lines properly, test before entering dangerous atmosphere, use life line, protect air line or hose from sharp edges or falling objects, leave at once if air flow is interrupted, do not remove until in respirable air (air line respirators must have a clean supply of air free from dust, oil and carbon monoxide).
Particulates alone	*Mechanical Filter* Special filter respirators	Allows freedom of movement, not to be used in excessively dusty atmospheres, in oxygen deficient atmospheres, or in atmospheres containing gases or vapors, not to be used for abrasive blasting, relatively difficult to breathe, limited time of use.	Use clean filter and change when plugged, ensure good fit and good operating conditions, leave at once if difficulty in breathing increases significantly.
Gases and vapors alone	*Chemical Absorbers* Universal gas mask Special canister gas mask Special cartridge respirator	Allows freedom of movement, do not use in atmospheres deficient in oxygen or containing excessive contaminants (above 2% with gas masks, above 1000 ppm with cartridge type), used for limited time and specific contaminant only (cartridge respirators not to be used in atmospheres immediately dangerous to life), relatively difficult to breathe, limited time of use.	Adjust properly, insure good tight fit, check operating condition, always use fresh canister or cartridge at start of use if possible, enter atmosphere cautiously, whenever odor is detected leave at once. Leave at once also if difficulty in breathing increases significantly.
Combination of particulates and gases and vapors	*Chemical-Mechanical Filters* Gas mask with filter Filter respirator with chemical cartridge	See mechanical filters and chemical absorbers.	See mechanical filters and chemical absorbers.

by his breathing. There is no recirculation and the exhaled breath is directed to the atmosphere.

(c) Self-generating Oxygen (Recirculating). In this device, the exhaled breath is directed to a chemical canister which simultaneously releases oxygen by the action of the moisture in the breath on the chemicals and absorbs carbon dioxide from the breath. The breath containing the released oxygen enters the breathing bag for inhalation and the exhaled breath repeats the cycle.

(2) Hose Type. These respirators supply air

to the wearer from an uncontaminated source and are therefore independent of the workroom atmosphere.

(a) Hose Mask (with or without Blower). A large diameter hose connected to a motor or hand-operated air mover located in a clean atmosphere supplies air to a full facepiece worn by the user. The hose diameter is sufficiently large so that the wearer can inhale clean air even if the blower is not operating. This type of equipment is frequently used without a blower. In this event, the inlet end is anchored in a respirable atmosphere location and may be provided with a screen to filter out coarse dust.

(b) Air-line Respirators. These may consist of a full or half-mask facepiece or a head-covering helmet or hood to which air is supplied, by means of suitable reducing valves, from a source of compressed air, usually an air compressor, although a compressed air tank may be used. They may be of the continuous flow type or the demand type which is governed by the wearer's breathing. The latter is used only with a facepiece. The abrasive blasting respirator is a special type of air line respirator with a hood designed to be resistant against abrasion and to protect the head, face and neck of the wearer against abrasive particles.

(B) **Air-purifying Respirators.** In these, gaseous contaminants are removed from otherwise respirable air by absorption or chemical reaction and particulate contaminants are removed by mechanical filtration. They cannot be used in an oxygen deficient atmosphere and must not be used in atmospheres containing contaminants in concentrations higher than those for which they were designed.

(1) Chemical Respirators. The inspired air is drawn over suitable chemicals, where gaseous contaminants are removed before inhalation. Breath is exhaled to the atmosphere. The respirator is usually designed for a particular contaminant but may also be designed for combinations of contaminants.

(a) Gas Masks. These consist of a full facepiece attached to a canister containing suitable contaminant-removing chemicals, and may be used for emergency purposes under certain conditions. The Universal Gas-Mask Canister protects against a number of contaminants including carbon monoxide.

(b) Chemical Cartridge Respirators. These consist of a half-mask facepiece attached to one or more cartridges containing suitable air purifying chemicals. They are for non-emergency use only and are usually designed for a single gas or vapor or single classes of gas or vapor.

(c) Self-rescue Respirators. These are similar to the chemical cartridge type. They are not used routinely but carried on the person for use to escape to a safe atmosphere in an emergency or catastrophic situation.

(2) Mechanical Filters. These are similar to the chemical cartridge respirators except that the purifying chemicals in the cartridge are replaced by filters, usually a felt pad. Filter respirators are designed to remove a specific single contaminant or class of particulate contaminant but are also available for several different particulate contaminants. Certain dusts, such as mercury compounds, may have a vapor pressure so high that when caught on a filter, inspired air passing over them may introduce toxic vapors into the air being breathed. Respirators for these may require a chemical cartridge in series with the filter.

(3) Combination Respirators for Gases and Particulates. These are combined respirators, in one unit, for simultaneous protection against gas, vapor and particulates. They consist of gas mask canisters or chemical cartridges with mechanical filters in series, so that inspired air passes first through the filter and then over the chemical granules.

Respirator Testing and Approval

The U. S. Bureau of Mines has established a testing section where respiratory devices are inspected and tested. It has prepared a series of Approval Schedules which set forth minimum requirements for respirators. The Bureau also tests commercial respiratory protective devices and issues approvals to those which meet its standards. These approvals are for a particular design of device for a specific use. Such devices bear a Bureau of Mines Approval Number and contain a legend describing the conditions for which the approval was granted. Lists of such approved respiratory protective devices are published periodically by the Bureau of Mines. Many unapproved devices are available commercially and extravagant claims are frequently made for them. Maximum safety demands that only U. S. Bureau of Mines approved equipment be used and only for the purposes for which designed.

Selection of Respirators

Tables 2A and 2B give a basis for selection of a respiratory protective device according to the degree of toxicity and expected concentration of the contaminant against which protection is desired for emergency or short term use. There

TABLE 2A. Selection of Respirators for Emergency or Short Term Use on the Basis of Hazard and Expected Concentration

Toxicity	Expected Concentrations of Gases or Vapors			
	Two to five times TLV or up to 1000 ppm	Five to ten times TLV or 1000–5000 ppm	Above ten times TLV or 5000–20,000 ppm	Oxygen deficiency, emergency or above 20,000 ppm
Low	No respirator, or chemical cartridge needed	Canister gas mask	Canister gas mask or hose type respirator	Self-contained air or oxygen
Moderate	Chemical cartridge	Canister gas mask or hose type respirator	Hose type or self-contained air or oxygen	Self-contained air or oxygen
High	Canister gas mask	Hose type respirator	Self-contained air or oxygen	Self-contained air or oxygen

NOTES:

(1) TLV refers to the threshold limit values for a number of substances published by the American Conference of Governmental Industrial Hygienists. (See Section 12).

(2) See Sections 1 and 2 for a discussion of toxicity ratings and their relation to TLV values.

(3) When unavoidable conditions necessitate using respirators for longer periods (above 1 hr), use equipment in a higher protective category than shown above.

are, however, a number of other factors to be considered.

Type of Contaminant. In addition to the toxicity and expected concentration of the contaminant, both the physical form and identity of the contaminant must be known so that the proper respirator can be selected.

Period of Required Protection. The maximum time to be spent in the contaminated air must be known, so that a device which will function properly for that period can be selected. Self-contained or air purifying respirators have limited periods of use. Hose type respirators have relatively unlimited usefulness.

Availability of Safe Atmosphere. Use of hose type respirators in contaminated areas is limited to a distance equal to the maximum amount of hose which can be used as well as the need to enter and leave the area by the same route. Other conditions will usually necessitate self-contained apparatus or, in less severe situations, air purifying respirators.

Activity of the Wearer. Certain types of activity, such as traveling over a large area, climbing, manipulating equipment, etc., preclude the use of hose type respirators. Hard physical labor not only increases the discomfort of some types of respiratory equipment but also increases the breathing rate three- or fourfold that of rest conditions. This will result in a much more rapid depletion of the air or oxygen

supply in a self-contained respiratory device and likewise much more rapid overwhelming of a canister, cartridge or filter.

Ease of Use. All users of respiratory devices require careful instruction in proper selection, fitting and operation of the devices. Some are much more complicated than others, and this may be a governing factor in the selection of one device over another, assuming equal protection is provided by each one. All users or potential users of devices should be required to familiarize themselves completely with their principles, use, operation and emergency procedures.

Care of Respirators

Maintenance. Wherever possible, care of respiratory devices should be the full-time function of a specially trained operator. Careful records should be kept of time of use of each device so that air or oxygen cylinders are recharged when necessary and canisters, cartridges and filters are replaced before exhaustion or plugging. Worn or broken parts should quickly be repaired or replaced.

Inspection. All devices, whether for routine or emergency use, should be inspected periodically to ensure that they will operate properly when needed. Rubber parts should be checked for deterioration, metal parts for corrosion, and plastic or glass parts for cracks or breaks.

TABLE 2B. Selection of Respirators for Emergency or Short Term Use on the Basis of Hazard and Expected Concentration

Toxicity	Expected Concentrations of Particulate Matter (Dusts, Fumes and Mists)			
	Two to five Times TLV	Five to Twenty Times TLV	Above Twenty Times TLV	Oxygen Deficient, Emergency, Highly corrosive
Low	Respirator not usually needed	Filter	Filter or hose type respirator	Where exposure is to extremely corrosive dusts or to dusts in an oxygen deficient atmosphere, a self-contained air or oxygen respirator must be used.
Moderate or high (toxicity no greater than lead)	Filter	Filter or hose type respirator	Hose type or self-contained air or oxygen	
Extremely high (toxicity greater than lead)	Filter or hose type respirator	Hose type respirator	Self-contained air or oxygen	

NOTES:
(1) TLV refers to the threshold limit values for a number of substances published by the American Conference of Governmental Industrial Hygienists. (See Section 12).
(2) See Sections 1 and 2 for a discussion of toxicity ratings and their relation to TLV values.
(3) Expected concentration of particulate matter have been shown only as multiples of the threshold limit values. Where these values are not available, the following concentrations may be used as a guide:

Mineral Dusts		Other Dusts, Fumes and Mists
2–5 (TLV)	Up to 50 MPPCF*	Up to 0.5 mg/m³
5–20 (TLV)	50–1000 MPPCF*	0.5–10 mg/m³
Above 20 (TLV)	Above 1000 MPPCF*	Above 10 mg/m³

*Millions of particles of dust per cubic foot of air

(4) When unavoidable conditions necessitate using respirators for longer periods (above 1 hr), use equipment in a higher protective category than shown above.

Cleaning. Equipment, whether assigned permanently to individual wearers or available to a number of workers, should be cleaned, sterilized and dried after each use to prevent the spread of infection and to ensure continuing efficient operation. Recommendations of the manufacturer of the device for proper cleaning and sterilization methods should be followed.

Storage. Respirators should be stored in clean compartments protected against humidity, extremes of temperature, and sunlight. Non-emergency respirators may be centrally stored. Equipment for emergency use should be stored in suitable compartments within work areas where the emergency may occur, for escape purposes, as well as outside the potentially dangerous areas for use in entering the area for rescue or repair purposes. Replacement canisters, cartridges and filters should be similarly stored.

PERSONAL HYGIENE

Personal hygiene is another tool available for the protection of the worker against the harmful features of an industrial environment. Its principal function is the prevention of industrial dermatosis caused by contact with harmful substances but, in some instances, it is also an important element in the prevention of systemic poisoning by ingestion or inhalation of toxic materials.

It has been estimated that 90 percent of industrial dermatitis cases are caused by chemical agents; 80 percent of these cases involve such primary irritants as alkalies, organic and inorganic acids, corrosive salts, petroleum oils and solvents, tars and pitches, and a host of other substances. These produce contact dermatitis through direct local action. About 20 percent of the cases are caused by such sensitizing agents as dyes and dye intermediates, amines, natural and synthetic resins, and waxes and bichromates. In these, allergic reactions may set in so that each subsequent exposure may lead to a dermatitis which is hard to cure, prolonged and recurrent. In some cases prolonged contact with some substances may cause skin cancer (see Section 9).

Control measures usually involve a multiphasic preventive and corrective approach which will include substitution of less irritating substances where possible, redesign of operations or change or work habits to prevent contact, provision of enclosures, splash guards or shields as a physical barrier against contact, proper washing facilities, work clothing and storage facilities, protective clothing, and barrier creams. Medical control is essential to treat breaks, cuts, or abrasions of the skin and to detect dermatitis in the earliest possible stages.

One outstanding example of substitution is the considerable work that has been done to formulate cutting oils which minimize the occurrence of dermatitis. Automatic machining operations and automatic degreasing and paint dipping are instances of virtual elimination of a contact dermatitis hazard by a process change. Hot caustic cleaning tanks can be provided with shields to prevent spattering of hot caustic solution onto the skins of workers.

Proper washing facilities include the provision of a suitable number of conveniently located sinks with hot water so that washing may be performed as frequently as possible. At the very least, washing should be performed before coffee breaks or rest periods, before lunch, and at the end of the shift. More frequent washing may be necessary if highly hazardous contaminants are handled, if contact is frequent, or if heavy soil occurs rapidly. Solvents should never be used. Skin cleansers should not be of the poor quality, harsh alkaline type, but the mild industrial skin cleansers which are available in a variety of forms. Where abrasives in the cleanser are necessary to remove heavy soil, they should be such natural organic substances as corn meal or wood flour rather than fine sand or pumice which may add to the dermatitis problem.

A good industrial skin cleanser will fulfill a number of requirements:

(1) It should quickly and efficiently remove heavy industrial soil.

(2) It should not remove the natural fats and oils from the skin.

(3) It should not abrade the skin or act as a skin sensitizer.

(4) It should be pleasant to use.

(5) In powder form, it should flow easily through dispensers and not clog the plumbing.

(6) It should not deteriorate or become insect infested.

Showers should be provided in all cases where possible, and their use should be mandatory where high hazard chemicals are employed. Their provision and use are strongly indicated where whole-body exposure to solvents or oils is possible, where exposure exists to toxic agents which may be absorbed through the skin, and where such extremely toxic dusts as beryllium are present. Emergency deluge showers should be strategically located throughout the plant when accidental whole-body exposure to highly corrosive and other similarly dangerous chemicals may occur. Eye baths should be similarly located.

Clean work clothing, changed frequently, is highly desirable in all work situations. Where known dermatitis-causing chemicals are present, such clothing should be changed daily and in many cases undergarments should similarly be provided and changed. Clothing lockers should be provided and, where possible, should be equipped with double compartments for separation of street and work clothes. Where special skin hazards or extremely toxic substances are present it is especially desirable to have a double set of locker rooms separated by showers. On coming to work, the worker strips in the first locker room and passes through the shower room to the second locker room where clean work clothes await him. At the end of a shift, the worker strips in the second locker room, deposits his soiled work clothing in a hamper, showers, and proceeds to the first locker to put on his street clothes. (An increasingly common, and highly desirable, practice is to furnish the second locker room with automatic washing machines into which the worker places the soiled clothing as soon as he removes them. These are then washed and machine dried by a cleanup man and are ready for the next day's use.

In general, it is advisable for management to provide laundry service for work clothing. The additional cost will be slight and will be warranted by the decrease in dermatitis cases. Where work clothing may be heavily contaminated with highly toxic materials, it should be kept apart from other work clothing and given special laundry service, either in the plant or by a commercial laundry alerted to the hazard to prevent exposure of laundry workers to the contaminant. Where highly toxic materials such as beryllium are involved, the daily change of work clothing should be mandatory, not only to protect the worker but to safeguard members of his family from exposure to the toxic dusts he might otherwise carry home.

Where chemicals can penetrate or saturate ordinary work clothing, resulting in continuous skin exposure, impervious aprons, trousers, jackets, and shoes are necessary. Goggles, face shields, and impervious gloves also have a useful protective function.

Barrier creams, when intelligently used, can serve the dual purpose of providing some protection for the skin and making it easier to wash up. They are usually mandatory when operations do not permit wearing gloves. They can be applied to the face, neck, and hands, and are often the only available means of protecting these areas. Protective ointments should be non-irritating and non-sensitizing, should offer protection against the irritating agent, should be easily applied, stay on during work, and be easily removable at the end of a shift.

Barrier creams come in a number of types.

(1) Vanishing cream. This fills the pores of the skin and facilitates removal of dirt from the skin at the end of a shift.

(2) "Invisible glove" creams. These leave a film of resin or wax on the skin which acts as a physical barrier between the irritant and the skin. The water soluble type gives some protection against solvents and other organic compounds and is easily washed off. The water insoluble type protects against water soluble irritants. A special cleanser must be used for its removal.

(3) Fat-containing ointments. These coat the skin with a layer of fat which provides a greater degree of protection against water soluble material than the above type.

(4) Detoxifying creams. These contain a chemical which will neutralize some materials. Mild acids to protect against alkalies and soap or weak hydroxides to protect against acids are examples of this type.

(5) Inert creams. Creams containing inert powders such as talc can provide more protection than fatty creams against allergens and physical irritants such as glass fibers.

The level of the personal hygiene program which should be instituted in any industrial situation cannot be related directly to the potential severity of chemicals, which have been rated as "low," "moderate," and "high" in Section 12. In general, where exposure is to a highly hazardous substance, the most rigorous personal hygiene should be insisted upon—wash, showers, daily clothing changes, laundry provisions, and all the other features described above.

However, other factors may influence the scope of the personal hygiene program, including the length of exposure, area of skin exposed, and action of the contaminant on the skin. In some instances, exposure to a contaminant with a low hazard rating may require personal hygiene procedures fully as extensive as those for a high hazard substance. In the design of new plants, it is highly desirable, and usually more economical, to provide complete washing, shower, and locker facilities. In all cases, the personal hygiene program should be as extensive as is economically feasible and consistent with the degree of hazard.

REFERENCES

1. Bulletin 226—Respiratory Protective Equipment, Bureau of Labor Standards, U. S. Department of Labor, Washington, D. C., 1961.
2. American Standard Safety Code for Head, Eye and Respiratory Protection, Z2.1-1959, American Standards Association, 10 E. 40 St., New York 16, N. Y.
3. List of Respiratory Protective Devices Approved by the Bureau of Mines, Bureau of Mines, U. S. Department of the Interior, Washington, D. C.
4. Schwartz, L., Tulipan, L., and Peck, S. M., "Occupational Diseases of the Skin," 3rd edition, Philadelphia, Lea and Febiger, 1957.

SECTION 4

CONTROL OF ENVIRONMENTAL POLLUTION

PART ONE

MANAGING ENVIRONMENTAL QUALITY

Dwight F. Metzler
Deputy Commissioner, New York State Health Department
Albany, New York

Public concern is increasing over the growing population and accelerating conversion of resources, which factors cause damaging and often poorly understood interactions with the environment. Until we understand the environment sufficiently to anticipate reactions likely to occur as a result of our activities we may find that our "solutions" are only partial or that they actually worsen the problem.

For instance, many communities solved the nuisance of overflowing cesspools and pollution of ground water by providing sewerage systems. The effluent from such systems has in turn frequently created serious stream pollution problems.

Man can no longer live as a despot, considering only his wants without also considering the consequences. To understand the consequences, he must study the natural laws that govern the physical environment and the interactions between air, and land and water.

The first smog episodes in some United States cities occurred within the last 5–10 years. Usable water shortages are increasing in severity because of pollution and the effect of large urban areas on rainfall and runoff. In some places these shortages are at the crisis stage, and have spawned huge remedial programs for air pollution control, water quality management and solid waste disposal.

Before discussing air, soil and water pollutants, or suitable controls, consider the interrelationships of these classes of pollutants.

Environmental pollution[1] has been defined as "the unfavorable alteration of our surroundings wholly or largely as a by-product of man's actions, through direct or indirect effects of changes in energy patterns, radiation levels, chemical and physical constitution and abundances of organisms. These changes may affect man directly, or through his supplies of water and of agricultural and other biological products, his physical objects or possessions, or his opportunities for recreation and appreciation of nature."

The wastes from agriculture, industry, human metabolism and combustion disperse via the air and water and often affect distant populations as regards man's health, joy of living, working and playing. Animal populations (most public concern is for fish) such as birds, insects and lower life forms are also adversely affected.

The changes in quality of water as a natural resource due to the addition of nutrients from waste effluents, the environmental effects of the global increase in CO_2, the long-term effects of increased lead content of ocean waters and the problems associated with rapid planetary distribution of radioactive wastes from nuclear explosions are a few examples of the pervasiveness of today's pollutants. Their control is important, especially in the developed portions of the globe; for example, uncontrolled pollution can feed upon itself because, as it reaches a point where certain microorganisms die, the overall effect may be greatly increased since some microorganisms combat degradation of the environment and serve as a major natural mechanism for reducing water, soil and air pollution.

These paragraphs are directed primarily at pollutants which affect man, and those which kill microorganisms, alter plant life, or impair the quality of food also affect man, though indirectly.

77

Thus, Na in irrigation water, B from industrial wastes and Cu, As or Pb from insecticides and fungicides all impair the ability of soil to produce. Some toxic compounds such as chlorinated hydrocarbons (see Section 12) are absorbed by plants and thus enter the food chain.

Synergism

Some toxic materials exhibit increased toxicity when absorbed into an organism in combination with some other agent or compound. This is known as potentiation or synergism; for example, epidemiologic evidence points to an incidence of chronic bronchitis in areas polluted with SO_2, such as New York City, which is higher than that expected from SO_2 alone. Such studies and animal experiments show that the toxicity of some air pollutants can be greatly changed when they act in combination with other pollutants. For instance, cigarette smoking appears to have a greater impact on public health than might be expected. The co-effects of other air pollutants have yet to be evaluated.

This effect seems negligible in water pollutants, especially when they pass through the gastrointestinal tract. Possibly dilution from water and food, the fact that the gastrointestinal tract is a tougher organ, and the fact that the stomach is acid and the duodenum alkaline, all act to reduce the physiological impact of water pollutants. Note: This may not be true for persons on long-term medication.

WATER POLLUTION

Pollution—Types and Sources

Water is said to be polluted when its quality is degraded by sewage, industrial wastes or natural seepage to the point where its beneficial use is unreasonably affected. While the beneficial use of water usually refers to agricultural, domestic or industrial water supply, the definition is changing to include aesthetics as well as an environment for aquatic life; for example, both federal and international agencies deplore the appearance of wastes discharged at the canyon wall at Niagara Falls, even though no actual use is affected. This is not an exception. Most states now emphasize appearance in their stream quality standards and programs for pollution abatement.

The four main types of pollution are: biological, chemical, physical and radioactive.

All living (biological) matter affects water use and quality. Viruses, such as infectious hepatitis virus, occur in sewage, which must be treated to reduce their numbers to a point where the danger of infection to humans or animals is minimum.

Pathogenic bacteria, of which the Salmonellae are the best known, are found in domestic sewage. Other bacteria, the simplest form of plant life, are essential to the sewage purification process since they metabolize nutrients to help stabilize the organic matter content. Under anaerobic conditions, bacteria produce odors, and attack and destroy other living things. Fungi function like bacteria. Algae, in reasonable amounts, aid the purification process and add oxygen to the water, but in water enriched by nitrogen and phosphorus (entrophication), they respond with excessive growth, which prevents penetration by sunlight, depletes oxygen and causes taste and odor problems.

The industrial expansion of the last two decades has added greatly to the environmental burden of chemical contaminants. To naturally occurring chlorides, nitrates and sulfates are being added acids, alkalies, arsenic compounds, aziridines, barium, compounds of boron and cadmium, carbon, chloroform extractables, carbamates, compounds of cesium, chlorinated hydrocarbons, chlorides, chromium and copper compounds, cyanides, dyes, fluorides, greases, hydrocarbons, chromium and copper compounds, cyanides, dyes, fluorides, greases, hydrocarbons, hydroxides, hydrogen sulfide, compounds of iron, nickel, industrial plastics (epoxy, diepoxy and episulfide), lead, manganese, nitrosamines, oils, organic chemicals, organic thiosulfate, oxidizing agents, phenols, phosphates, reducing agents, compounds of potassium, zinc, selenium, strontium, and sulfur (Table 1).

The principal physical pollutants are temperature and turbidity. Sudden temperature changes adversely affect fish, and elevated temperatures accelerate the depletion of oxygen by increasing the rate at which bacteria and other plants metabolize the nutrients. Concern over the effect of raising temperatures is growing, and it appears that all states will soon limit the permissible temperature rise of a waterway to 5–7°F. above natural. More stringent requirements are probable for coastal and estuarine waters. Land runoff is the major cause of turbidity in water, although many industrial processes also contribute. A reduction in turbidity is not always beneficial; for example, a threefold reduction in turbidity of the Missouri River by dams above Omaha, Nebraska, contributed to a major increase in taste and odor problems at downstream waterworks because the clarified water stimulated the growth of algae and other plants which cause taste and odor problems.

For a discussion of radioactive wastes, see Section 8.

TABLE 1. Partial List of Pollutants[3] Which May Enter the Sewer System of a
Municipality Such as Rochester, New York

Acetic acid	Detergents	Resins
Boric acid, borates	Diesel oil	Salt
Chromic acid, chromates	Ferric oxide	Shellac
Hydrochloric acid	"Freon" solvent	Silicones
Nitric acid	Gold	Silver
Sulfuric acid	Glutamates	Silver nitrate
Albumen	Gum arabic	Sodium bisulfate
Alcohol	Inorganic phosphates	Sodium cyanide
Amines	Lead and its compounds	Sodium hydroxide
Anionic surfactants	Methyl chloride	Sodium hypochlorite
Arsenic	Nickel and its compounds	Sodium nitrate
Butyl alcohol	Non-ionic surfactants	Sodium silicate
Cadmium	"Perchloron"	Sodium silicofluoride
"Carbitol" solvent	Petrolatum	Sodium sulfate
Casein	Pine oil	Stoddard solvent
Copper compounds	Potash	Tantalum
Cyanides	Potassium hydroxide	"Triacetate"
Caustic etching and	Potassium chromate	Trichloroethylene
wetting compounds		Zinc and its compounds
		Zirconium oxide

The definitions of some water pollution terms are given below:[2]

Biochemical Oxygen (BOD): The quantity of oxygen utilized in the biochemical oxidation of organic matter in a specified time and at a specified temperature. It is not related to the oxygen requirements in chemical combustion, but is determined entirely by the availability of the material as a biological food and by the amount of oxygen utilized by the microorganisms during oxidation.

Biodegradable Organics. Organic compounds which can be broken down by microorganisms to form stable simple compounds, such as carbon dioxide and water.

Pathogenic Bacteria. Bacteria which can cause disease.

Coliform Bacteria—Coliform Group: A group of bacteria, predominantly inhabitants of the intestines of man but also found on vegetables, including all aerobic and facultative anaerobic gram-negative, non-spore-forming bacilli that ferment lactose with gas formation. This group includes five tribes of which the very great majority are Eschericheae. The eschericheae tribe comprises three genera and ten species, of which Escherichia coli and aerobacter aerogenes are dominant. The Escherichia coli are normal inhabitants of the intestine of man and all vertebrates whereas aerobacter aerogenes normally are found on grain and plants, and only to a varying degree in the intestine of man and animals. Formerly referred to as B. Coli, Coli group, Coli-Aerogenes Group.

Influent: Sewage, water, or other liquid, raw or partly treated, flowing into a reservoir, basin, or treatment plant, or part thereof.

Effluent: (1) A liquid which flows out of a containing space. (2) Sewage, water, or other liquid, partially or completely treated, or in its natural state, flowing out of a reservoir, basin, or treatment plant, or part thereof.

Digestion: The anaerobic decomposition of organic matter, resulting in partial gasification, liquefaction, and mineralization.

Trickling Filter: A filter consisting of an artificial bed of coarse material, such as broken stone, clinkers, slate, slats, or brush, over which sewage is distributed in drops, films, or spray, from troughs, drippers, moving distributors, or fixed nozzles, and through which it trickles to the underdrains, allowing for the formation of zoogleal slimes which clarify and oxidize the sewage.

Population Equivalent: (1) The population calculated to contribute a given amount of biochemical oxygen demand (BOD) per day. A common base is 0.167 pound of 5-day BOD per capita per day. (2) For an industrial waste, the estimated number of people contributing sewage equal in effect to a unit volume of the waste or to some other unit involved in producing or manufacturing a particular commodity.

Settleable Solids: Suspended solids which will settle in quiescent water, sewage, or other liquid in a reasonable period. A reasonable period is commonly, though arbitrarily, taken as 2 hours. Also called settling solids.

Suspended Solids: (1) The quantity of material retained when a sample of water, sewage, or other liquid is filtered through an asbestos mat in a Gooch crucible. (2) Solids that either float on the surface of, or are in suspension in, water, sewage, or other liquids, and which are largely removable by laboratory filtering. See Matter, Suspended.

Total Solids: All the solids in water, sewage, or other liquids; it includes the suspended solids (largely removable by filter paper) and the filterable solids (those which pass through filter paper).

Sludge, Activated: Sludge floc produced in raw or settled sewage by the growth of zoogleal bacteria and other organisms in the presence of dissolved oxygen, and accumulated in sufficient concentration by returning floc previously formed.

COD—Chemical Oxygen Demand: A measure of the oxygen equivalent of that portion of the organic matter in a sample that is susceptible to oxidation by a strong chemical oxidant.

Lagoon or Oxidation Pond: An aerobic treatment device. It may have anaerobic zones and/or facultative zones. The stabilization of the organic matter is brought about by bacteria, and in the case of aerated lagoons, the oxygen is supplied by mechanical or diffused aeration.

Organic Nitrogen: All nitrogen present in organic compounds may be considered organic nitrogen. Most of the organic nitrogen present in domestic sewage is in the form of proteins or their degradation products.

Municipal Sewage

Municipal sewage refers to mixed wastes from domestic sources with small to major addition of wastes from commerce and industry. The volume of such waste approximates 70 percent of the water usage and is expressed in terms of gallons per capita per day. About 125 million people are served by public sewers in the United States. Approximately 10 percent of this sewage is discharge without treatment, and an additional 8 percent is inadequately treated.

The 5-day BOD and suspended solids content of raw sewage varies from 125–350 mg/liter. The grease content ranges from 5–120 mg/liter with an average of 40. Both untreated and treated sewage contain nutrients, especially N,K and P which fertilize the receiving waters. The P content has steadily increased in the last decade, primarily from detergents used in homes and commerce. Investigations around Lake Erie revealed that about 80 percent of the P which enters the lake comes from municipal sewage. Concentrations of 8–25 mg/liter of PO_4 are common in raw sewage. When the receiving waters are a lake, pond or other confined body of water, the importance of this contribution of P can better be understood by calculating it in pounds per day or tons per year. The total for Lake Erie is estimated at 24,000 a year. Total N in raw sewage varies from 10–55 mg/liter, and K from 10–20 mg/liter.

Industrial Wastes

The nature of industrial waste is as complex as the great variety of products which contribute to the highest standard of living the world has ever known. Constant changes occur in the wastes from industry, which introduces at least 10,000 new products annually. Because of the great variability possible, each industrial waste requires individual evaluation to determine its impact upon receiving waters. Wherever possible, the concept of population equivalents is also used to measure the industrial waste load. While no better parameter has been developed, the variability of industrial wastes makes this less meaningful than for domestic sewage; for example, the population equivalent of spent sulfite liquor in the state of Washington in 1960 was 12 million as compared to its census of 2.9 million people. Nationwide data for industrial wastes either receiving or needing treatment are not available. Connecticut and Kansas were reported to be the first states which completed an inventory of all their industrial wastes. South Carolina has completed, perhaps, the most comprehensive survey of its industrial waste pollution in a project supported by both industry and the South Carolina Department of Health. While the urban population of South Carolina is 1.1 million persons, the wastes from just three classes of industry—textile, pulp and paper, and food processing—yield a population equivalent of 4.6 million people. It has been estimated that the pollution load from industrial wastes is equal to that from municipal wastes, but these estimates have no sound basis. Industrial and municipal wastes are not comparable, and without a survey of each state, estimates are no more than educated guesses. Efforts have been made to compile a national inventory by the Conference of State Sanitary Engineers since 1961, and the United States Public Health Service initiated a study in 1963.

(For a toxic hazard rating of the components of Table 1, see Section 12). In addition, wastes which are hauled by scavenger companies to a sewage treatment plant and put into the system also contain industrial components.

Industrial officials and their advisors should note the change that has occurred in federal policy, as well as the policy of some states, regarding the wastes from industry. All communities in the United States are eligible for federal grants which can total 30–55 percent of the cost of a municipal waste treatment works. To meet the definition of a municipal facility, the treatment works must be publicly owned. It may, however, treat the wastes of a single industry.

If appropriations back up the promise of the legislation, the federal subsidy will discourage the construction of waste treatment facilities by industry and encourage industry to seek public acceptance of its wastes. Where states have construction grants programs the advantage will be even greater. In New York State, for example, a public body is eligible to receive as much as 85 percent assistance with the costs of construction plus 33 percent of the costs of maintaining and operating a treatment plant.

LAND DRAINAGE

Land drainage is the main cause of stream pollution. The natural leaching of chloride and sulfate deposits degrade large volumes of water in the Southwest. The rapid trend toward more efficient mass feeding of farm animals, poultry, hogs and cattle, instead of on scattered individual farms, makes this source one of the most serious land drainage problems of the next decade. Conservative estimates place the amount of excrement from farm animals at ten times that of the human population. Because of the large volume of material, the difficulty of returning it to productive agricultural land and the cost of other disposal measures, much of it leaches or is washed into watercourses.

Of the many pesticides and herbicides in use, the chlorinated hydrocarbons, such as aldrin, chlordane, dieldrin, DDT, endrin, heptachlor, methoxychlor and toxaphene, are the most persistent and of greatest concern (see Section 12). While less than 5 percent of the land surface of the continental United States is treated annually, the applications are concentrated in areas which are the most productive; for instance, as much as 400 pounds of DDT may be applied to an acre of land.

In New York State, chlorinated hydrocarbon pesticides are widely dispersed and are carried

TABLE 2. Summary of Chlorinated Hydrocarbon Pesticides in Surface Waters[4]

1964–1966

| Compound | Date Collected | Micrograms per Liter | | Remarks |
		Maximum	Minimum	
DDD	8/ 6/64 9/27/65	.083	.006	Maximum value found in Niagara River; no trace since 1964
DDE	9/27/65 10/11/65	.011	.011	Pesticide detected once in Seneca River and Lake Champlain
TDE	8/15/66	.021	.021	Pesticide detected once in Peconic River
Tedion	8/10/64	0.11	0.11	Pesticide detected once in Niagara River
Dieldrin	9/20/65 9/27/65	0.088	0.0004	Maximum value found in Mohawk River; no trace in 1966
Endrin	8/11/64	0.11	0.11	Pesticide detected in Oswego River
Toxaphene	8/10/64	0.75	0.75	Pesticide detected in Mohawk River; no trace since 1964
Aldrin	8/10/64 9/27/65	0.26	0.011	Maximum value found in Hudson River; no trace since 1964
Lindane	10/11/65	0.015	0.015	Pesticide detected once in Lake Champlain

TABLE 3

Constituent	Range	Mean	Pounds/Year/Acre Urban Runoff	Raw Sewage
Turbidity	30–1000 units	170 units	—	—
pH	5.3–8.7	7.5	—	—
Cl	3–35 mg/liter	12 mg/liter	—	—
Suspended solids	5–1200 mg/liter	210 mg/liter	730	540
COD	20–610 mg/liter	99 mg/liter	240	960
BOD	2–84 mg/liter	19 mg/liter	33	540
PO_4	—	—	2.5	27
Total N	—	—	8.9	81

away by surface waters. Dieldrin is the most prevalent, with an unusual apparent persistence after an agricultural growing season through the following spring runoff.

Accidental massive releases of insecticides have caused fish kills in limited areas, but there is no evidence that normal levels are toxic to mammals or fish. It is, nevertheless, true that chlorinated hydrocarbons are present in the surface waters used by some New York public water supplies. A listing of concentration ranges in New York State streams, rivers and lakes follows (Table 2).

Salinity from irrigation return flows is a major source of pollution in 17 western states, i.e., the interior valleys of California, the Great Basin, the Colorado and Rio Grande basins, and parts of the Arkansas and the Red River basins. In the case of the Colorado and Rio Grande rivers, the concern is international.

The magnitude and importance of storm-produced urban runoff is often underestimated or ignored. Its characteristics vary and must be determined for each city. In some communities,

the adverse effects increase with the intensity of the storm and the quality of the runoff becomes worse as the storm progresses. Investigators in other cities report that stormwater runoff quality improves with time. No pattern is apparent.

Measurements have shown that 25–40 percent of the annual production of suspended solids may be caused by stormwater flushing. The percentage of bacteria similarly flushed into waterways is about the same. One-third of the total BOD and suspended solids about Buffalo, New York, is flushed untreated from its sewers, though this represents a sewage overflow of only 2–3 percent.[5] Table 3 lists the constituent con-

TABLE 4

Indicator Organism	Counts Exceeded in Designated Percent of Samples		
	90%	50%	10%
Coliforms	2900	58,000	460,000
Fecal coli	500	10,900	76,000
Fecal strep	4900	20,500	110,000

TABLE 5

City and Year of Study	BOD (mg/liter)	COD (mg/liter)	Suspended Solids (mg/liter)	Total Solids (mg/liter)	Coliforms/100 ml
Detroit, 1949 (range)	96–234	—	—	310–914	25,000–930,000
Detroit, 1960	—	—	102 and 203	—	2,300–430,000
Pretoria, S. Africa	34	28	—	—	230,000
Stockholm, 1945–48 (median/high)	17/80	188/3100	—	300/3000	4000/200,000
Leningrad, USSR, 1948–50	36	—	14,541	—	—
Moscow, USSR, 1936 (range)	186/285	—	1000/3500	—	—
Oxney, U.K.	Up to 100	—	Up to 2045	—	—

centrations from the study, and Table 4 lists the range of bacterial counts.

Some urban runoff characteristics from widely scattered cities are listed in Table 5.

In a 1963 Detroit study,[6] a sample was obtained from the Trenton Channel during an exceptionally severe storm (>4 inches of rain). Approximately one-fourth of the river's flow passes through this Channel. The total coliform count was 1,600,000/100 ml, approaching that of raw sewage.

Applying the runoff rates and loadings from Tables 3, 4 and 5, the weights of pollutants from stormwater runoff are:

ppb (parts per billion) for drinking water, primarily because at higher levels unpleasant tastes and odors occur. The reported CCE of some natural waters are given in Table 7.

Table 8 shows the percentage of time that CCE concentrations are <100 ppb for five major rivers.

The potential carcinogenic and mutagenic effects of the organic nitrosamines have been pointed out by European investigators based on the theory that the nitrosamine group attaches to critical proteins and converts them into mutated forms. Industrial plastics, chemical sterilants and missile fuels are examples of such

TABLE 6

City	Area (square miles)	Population Density (capita/acre)	Urban Runoff Load per Year (tons)				
			S.S.	BOD	COD	P	N
New York	315	39	74,000	3300	24,000	250	900
Buffalo	39.4	21	9,000	412	3,000	31	100
Detroit	139.6	19	33,000	1500	11,000	110	400
Cincinnati	77.3	10	18,000	820	6,000	60	440

HEALTH EFFECTS

The chemical change which man has wrought in his environment during the past two decades is so great that it can only be likened to a revolution, and much of this environmental change relates to chemicals in water. To ignore the implication of these changes would be as unwise as to create nonexistent dangers and hazards. The recovery from water supplies of chlorinated hydrocarbons, aromatic ethers, nitriles, o-nitrochlorobenzene and other toxic compounds emphasizes the need to quantitate those present in waste.

The Robert A. Taft Sanitary Engineering Center developed the carbon-chloroform extract (CCE) method to measure a group of unknown organic chemical pollutants originating, for the most part, in industry. Some of these pollutants are carcinogenic when concentrated in the carbon column. Stockinger[7] notes that while substances with carcinogenic potential can be found in water courses, none of them "pose a substantial threat to health" separately, or in combination, when passed through the gastrointestinal tract. However, they do add to the exposure burden from all sources, including tobacco smoke. Also the toxicity hazard of these substances to man differs when they pollute the air and are inhaled.

The USPHS recommends a level of CCE <200

chemicals; however, these substances are unstable and it is doubtful that they will persist long enough in water to be a health hazard. Gentry[8] presents epidemiologic evidence that

TABLE 7.   Average CCE Levels in Raw Water

Source	Average Levels (ppb)
Ohio River	100–360
Columbia River	24
Lake Superior	5
Colorado River	24
Lake Erie	69
Mississippi River	89
Missouri River	169
Genesee River	190
Hudson River (Poughkeepsie)	262
St. Lawrence River (Massena)	46

TABLE 8

	Percent
Ohio River	50
Mississippi	72
Missouri	90
Colorado	92
Columbia	96

chemical mutagens may be more effective in combination than might be expected from the individual components. The teratogenic potential of some materials was brought to broad public attention by the drug, thalidomide. Laboratory experiments exposing mice and chickens to "Diazinon," "Phaltan," "Captan," carbany, 5-fluorouracil and 6-mercaptopyridine[9] showed increased malformations of newborn; however, drinking water levels of these materials are now so low that their significance is minor.

Manganese compounds are common water pollutants which may have some relation to Parkinsonism, but their intake from normal food far exceeds the amount which might at present be derived from water.

Stockinger concludes that until present levels of water pollution are exceeded, "no significant human health effects may be anticipated to arise from water pollutants per se." There is no direct or substantial body of evidence that synergized or potentiated effects from chemical water pollutants are, in themselves, menacing health; at present levels, toxicologic analysis indicates that, at worst, water pollutants are contributing relatively little to the body burden of environmental pollutants generally.

"Interpretation is still difficult for the conditions of greatest concern—teratogenesis, mutagenesis, and carcinogenis—because even if human threshold doses were known, the effect on dosage of cofactors, promoters, accelerators and potentiators is not. But with substances productive of conditions of this type that interfere with genetic nuclear material, we are on particularly uncertain ground; a convincing case can be made for a 'no threshold' dose-effect for substances that interfere with nuclear material, either by competitive inhibition from structural similarity or by free radical action. The optimistic estimate is that present water pollutant levels may be contributing to, but not originating, serious long-term health effects."[7]

TREATMENT METHODS

The objective of treatment of sewage and industrial waste is the prevention of degradation of watercourses and rivers, ponds and lakes, coastal waters and estuaries. Most of the basic treatment methods have been known and practiced for many years, leading to the statement that nothing new has happened to sewage treatment in half a century, but the same might be said of the automobile. There have been refinements, increased automation, improved design and more effective operation. In addition, the requirements for high degrees of treatment and increased possibilities of reuse have generated new processes and applications for the tertiary treatment of wastes.

Sewage treatment can incorporate various combinations of processes, and the great variety of wastes, combined with the differing nature and use of the receiving waters, requires individual analysis and design for each location. Municipalities and other public bodies usually select a competent engineering firm to plan, design and supervise construction of the works. A few design the works with their own engineering staff.

Industries, on the other hand, often contract with engineer-contractors or with equipment manufacturers and contractors to build their waste treatment facilities. This difference in approach sometimes contributes to problems between industry and the regulatory agencies because of the lack of engineering plans which can serve as a basis for approval and the issuance of a permit. Industrial officials will avoid much difficulty by insisting upon detailed engineering plans and approval by the regulatory agency, regardless of the arrangement for engineering services.

Treatment methods may be divided into five categories for most circumstances:

(1) Removal of settleable and floating material;

(2) Removal or stabilization of organic matter which is in suspension or solution;

(3) Disinfection;

(4) Sludge treatment, digestion and removal;

(5) Tertiary treatment, providing for the removal of nutrients and trace contaminants.

Removal of settleable and floating material, when accompanied by sludge digestion is usually called *primary* treatment. The process includes the manual or mechanical screening of coarse suspended and floating material which can be shredded by comminuters for return to the sewage and removal by subsequent process. Grit is manually or mechanically removed by controlling the velocity of sewage flow which allows the grit to settle but keeps the organic matter in suspension. This is necessary to prevent damage to equipment and accumulations in the sludge digesters.

Greases, oils and other floating material are removed by skimming, usually as a part of primary settling tanks. Primary sedimentation removes the mineral and organic matter which will settle. It is then removed as sludge from the bottom of the tanks for disposal, dewatering or digestion. Coagulants may be added to increase settling efficiency. A typical flow diagram is shown in Figure 1.

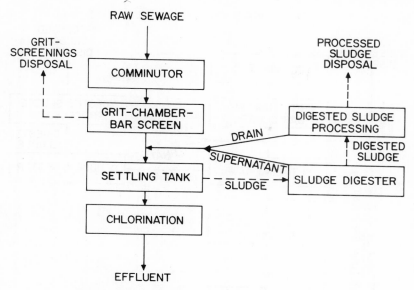

Figure 1. Typical flow diagram for primary treatment.

Secondary treatment usually refers to the biologic process in which soluble organic compounds are converted into bacterial cells and inorganic elements which, in turn, provide food for algae, protozoa, rotifers and crustaceans. Various processes are used but the most successful supply a continuous flow of food and oxygen to the bacteria. Trickling filters, activated sludge and lagoons are the most frequently used processes. Numerous variations of each process are used. Modifications of standard activated sludge include step-aeration, contact-stabilization, extended aeration and complete mixing. Some advantages of the activated sludge process include lower capital cost, a somewhat higher degree of purification and compatability with nutrient removal processes. However, it is more sensitive to upset and has a higher operating cost than filters or lagoons.

Primary treatment removes about 30–40 percent BOD and up to 60 percent suspended solids. *Secondary* treatment removes 80–95 percent BOD and up to 98 percent of the suspended solids.

Figure 2 shows a typical flow diagram.

Tertiary or chemical processes and filtration are becoming increasingly important in the treatment of wastes, both because of a greater variety and types to treat and because of the requirements for treatment in excess of secondary. Some of these have been developed for water desalination and are in the initial stages of being applied to sewage and industrial wastes.

These processes include coagulation and adsorption, which have been basic tools in water treatment programs for many years. Dialysis and electrodialysis employ membranes to recover the product and reduce the dissolved solids. Cation and anion exchange resins are increasing in use as they are developed to meet many different requirements for the reduction of dissolved solids. Oxidation-reduction reactions are effective in the reduction of some metallic ions using such materials as Cl_2, $FeSO_4$ and SO_2.

Sand filters or a combination of sand and graded crushed anthracite are used increasingly where the degree of purification must exceed that of secondary treatment.

Where higher degrees of treatment are required, consideration should be given to reuse of the water. In these cases, the water which is released is likely of better quality than the water in the natural environment. With the tools at hand to return it to its original quality, the question naturally occurs, "Why not reuse it?"

POLLUTION CONTROL THROUGH PLANT CHANGES

What can industry do by in-plant changes to control pollution? In the following paragraphs philosophy and procedures are presented, and examples are cited.

Comprehensive Industrial Waste Survey

The first step in developing an industrial waste water pollution abatement program is to define

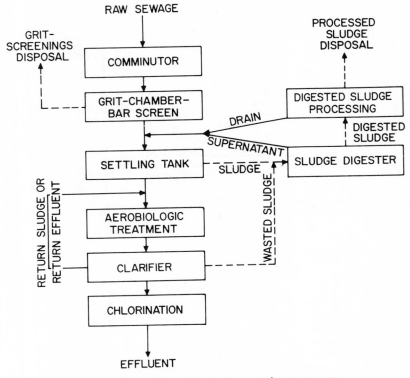

Figure 2. Typical flow diagram for secondary treatment.

the problem. This means determination of the character and volume of both the combined flow and the individual wastes within the plant. Comprehensive industrial waste surveys serve this purpose. Data obtained from such studies not only locate major waste sources but also provide information on which to base process modification and/or by-product recovery. Consideration must always be given to the effect of specific wastes on subsequent water uses. Maximum benefits accrue from a coordinated stream and industrial waste survey.

Water Use and Reuse

A strong tendency exists to use more water per unit of production when an abundance of fresh, cheap water is available. Invariably, the quantity of liquid wastes is a function of the quantity of fresh water used. When fresh water use is curtailed, a net reduction in waste may be anticipated. Common methods of reducing fresh water use include: the reduction of pressure on fresh water lines, the judicious control of valves, the installation of automatic shut-off valves on hose lines, the reuse of process water instead of fresh water, and the installation of water meters in each department.

Reuse of process water has for years been a favorite method of abating water pollution. Attention was first directed toward water reuse systems in paper mills, and these are now incorporated into the design of new mills. Savings result from recovery of chemicals and heat, and frequently from reduced pumping costs. It is true that complications in operation may result from the influence of recovered chemicals on the quantity of new chemicals to be added, from foam production, and from slime formation, but these are not insurmountable problems.

There are numerous ways of reducing waste volume in the textile industry. Strong rinse waters from dye operations may be reused to make up new baths. Weak rinses may be recycled through in-plant water treatment units. Counter-flow systems may be employed where rinse waters are circulated in the opposite direction to movement of the cloth.

The beet sugar industry has made good progress in reusing pulp screen water and pulp press water. These process waters, formerly wasted, are now returned to the diffuser water supply. Barometric condenser water can be pumped over cooling towers and recycled. Flume water can be screened, settled, and reused with bleed-off being limited to that drawn off with sludge from the clarifier.

There appears to be a trend to treat and reuse roll-cooling water in steel mills. Following conventional scale pits, the cooling water is put through a clarifier for removal of settleable solids and oil. The clarified waste water is then filtered through high-rate sand filters and returned to the mill water supply. This system of treatment and reuse has been confined, in the past, to water-scarce areas such as the Kaiser Steel Mill at Fontana, California. It is now being applied as a water pollution control measure by steel mills operating on the Great Lakes near Chicago and Buffalo.

The wet process industries reuse process water to varying degrees. The foregoing examples are typical. Reuse of process water, either directly or following treatment, constitutes a positive means of abating water pollution.

Good Housekeeping Practice

Waste prevention is one of the most important considerations in water pollution control. Loss from leakage can be minimized by care in assembly of equipment and by proper maintenance. Overflow can be prevented by the use of float-operated or electronic level controls that either shut off the flow or sound an alarm indicating that the tank is nearly full. Spillage is generally the result of careless handling, improper design or poor equipment layout.

The metal-finishing industry has done much to improve in-plant waste control. "Dragout" is reduced by allowing time for parts to drain either over the plating tank or into a separate dry-reclaim tank. Synchronized fog or spray rinse nozzles are installed at the exit of each plating tank and over the plating tank itself, allowing concentrated rinse to drain back into the tank. Drip pans are placed between tanks in such a way that solution draining from the work returns to the tank from which the work emerged. Air blow-off is occasionally used. The first rinse immediately after the plating tank is now a nonflowing or stagnant solution which is fed back to the plating tank as part of the make-up. Multiple countercurrent rinsing conserves water and permits recovery of some plating chemicals.

Production, refining, and use of petroleum products present many opportunities to benefit from in-plant control. Oil-in-water emulsions, phenols, mercaptans, etc., are most effectively treated at the source. Spent soluble oil coolants, common to machine shops, are normally not discharged to sewer systems. Dilution of oily wastes complicates treatment and should be avoided.

The dairy industry has made phenomenal progress in reducing milk losses. Twenty years ago usual milk losses amounted to as much as 2 percent of the total milk received at the plant. Today a milk plant cannot operate economically if whole milk losses amount to more than one-half of 1 percent.

The term "good housekeeping" implies an attempt to keep solid wastes out of plant sewers. Water contact with solids invariably solublizes some materials, thus adding to the concentration of the plant effluent. However, solids from the operating floor can be collected by squeegee and removed instead of being flushed into the sewer. In some instances, almost half the raw waste tonnage received from fruit and vegetable canneries must eventually be handled as solid waste.

Substitution of Chemicals

The total waste load from some manufacturing operations can be reduced by substituting chemicals having a low BOD for those with a higher BOD. The textile industry has used this approach with some success. Examples of such substitutions include:

(1) Carboxymethylcellulose for starch in slashing;

(2) Mineral acid for acetic acid;

(3) Synthetic detergents for soap;

(4) Polyvinyl alcohol and polystyrenes for starch in finishing.

Dimethylamine sulfate (DMA) is being used by the leather tanning industry to replace sulfides and sulfhydrates in unhairing systems. Proteolytic enzymes are also gaining favor for loosening the hair. These new materials eliminate the use of lime in the unhairing operation with a corresponding reduction of solids in the tannery waste effluent.

In the development of new chemical processes, it is not uncommon to withhold approval until the process wastes have been characterized with respect to treatment and effects on subsequent uses of the receiving waters.

Segregation of Wastes

Wastes from a manufacturing plant are segregated to facilitate treatment and to reduce the size of the treatment works. New factories will have at least three drainage systems, namely, process wastes, cooling water and storm drain, and sanitary wastes. Highly concentrated wastes or wastes requiring special treatment may be removed from the general process drain. Some wastes are rendered innocuous by addition of alkali, others require acid, and occasionally an acid waste can be used to neutralize an alkaline waste in which case equalization may prove beneficial.

Process Modifications

Conversion of calcium-base sulfite pulp mills to magnesia-base in recent years constitutes a major change in both process and equipment, with effluent improvements as one of the objectives. The magnesia-base sulfite process possess other advantages, including:

(1) A wider variety of wood species can be pulped.

(2) Magnesia-base pulp gives higher yield and greater strength.

(3) Full chemical and semichemical pulp can be produced with a single recovery process.

(4) Heat and chemical recovery are highly efficient. An 80–90 percent recovery of the magnesia-base may be expected in this process. The 1962 production of magnesia-base sulfite pulp in the United States was estimated at 790,000 tons out of a total sulfite capacity of 3,400,000 tons, or 23 percent.

Summary

The foregoing illustrations should interest industrial officials confronted with liquid waste abatement requirements. Internal studies may lead to profitable by-product recovery, but these opportunities have been largely exploited. Research is expected to develop economically feasible recovery or treatment processes for some as yet unsolved industrial waste problems. Coincident with this continued research, organizations administering anti-stream pollution laws must become familiar with the intricacies of industrial waste problems. They should recognize that losses sustained in liquid wastes frequently are not apparent to those who are primarily interested in production of an established product.

THE ROLE OF GOVERNMENT

An understanding of the relative roles of various levels in government and trends in government policy is essential for those who deal with problems of environmental pollution. The full impacts of air and water pollution were recognized long before the actual mechanisms were understood. As these had their roots in the protection of health, local and state programs were organized in health departments. In urban areas, many of the action and service programs are carried out by local government. The local health departments are often charged with monitoring air quality and enforcing local air pollution ordinances. They may maintain some surveillance of the quality of public water supply and have programs for supervising sewage disposal in new housing developments with individual disposal systems or small community-type treatment works. They usually provide consultation to the local collection agency on refuse collection and disposal and, in a few cases, actually manage the collection and disposal. Local departments of public works, departments of water supply, and departments of sewers and sewage treatment more often supply the direct services. Using these services is generally considered superior to providing a private water system, sewerage system or solid waste disposal. The conditions under which certain types of industrial wastes may be accepted vary with the type and capacity of the local works; however, the tendency for some uniform rules for accepting liquid or solid industrial wastes is growing.

In New York state, for example, all local communities must follow a careful program of surveillance of the industrial wastes they receive if they are to be eligible for state aid.

Comprehensive programs for water and air pollution are carried out at the state level. Table 9 lists the state agencies which have the primary responsibility for water pollution control, air sanitation and solid waste disposal.

The first federal legislation dealing with air pollution became law in 1963 and additional laws were enacted in 1965 (P.L. 89–272) and 1966 (P.L. 89–675).

The specific federal acts, as they deal with local and state units of government and with industry, will be discussed later. Certain trends have developed in federal and state legislation which will probably accelerate during the next 10 years. An understanding of the governmental posture and of these trends is essential to public officials and to industry managers. A movement has started toward national monitoring of air, soil and water to identify both specific problems and general trends. There is a major gap in the technology for accomplishing this, but it will gradually be corrected as more money becomes available for research and development. The government will play a more important role in encouraging research and development of specific machinery and of products which are essential to the reduction of air, soil and water pollution. As in the space program, the federal government will stimulate research and development with grants, contracts and developmental work in government laboratories.

The government will use its authority to require changes in products, i.e., the threatened legislation in 1965 to force manufacturers of detergents to find materials which were biodegradable. This approach has also been followed with respect to automobile safety and air pollution

TABLE 9. Pollution Agencies

State	Water Pollution Agencies	Air Pollution Agencies	Agencies Responsible for Solid Wastes Planning
ALABAMA	Water Improvement Commission Montgomery, Ala. Division of Game and Fish Department of Conservation Montgomery, Alabama	Department of Public Health Bureau of Environmental Health Room 328 State Office Building Montgomery, Ala. 36104	
ALASKA	Branch of Sanitation and Engineering Department of Health and Welfare Fourth and Main Juneau, Alaska	Dept. of Health and Welfare, Division of Public Health Pouch H Juneau, Alaska 99801	State Department of Health and Welfare Alaska Office Building P. O. Box 3-200 Juneau, Alaska 99801
ARIZONA	Bureau of Sanitation State Department of Health Phoenix, Ariz.	Department of Health Div. of Environmental Health State Office Building Phoenix, Ariz. 85007	Arizona State Department of Health Division of Environmental Health Phoenix, Ariz. 85007
ARKANSAS	Water Pollution Control Commission 921 West Markham St. Little Rock, Ark. Game and Fish Commission Game and Fish Commission Building Little Rock, Ark.	Pollution Control Commission 1100 Harrington Ave. Little Rock, Ark. 72202	Arkansas Pollution Control Commission 1100 Harrington Ave. Little Rock, Ark. 72202
CALIFORNIA	Department of Water Resources P. O. Box 388 Sacramento, Calif. Division of Environmental Sanitation 2151 Berkeley Way Berkeley, Calif. 94704 Water Pollution Control Board Room 316 1227 O St. Sacramento, Calif. Department of Fish and Game 722 Capitol Ave. Sacramento, Calif.	Bureau of Air Sanitation Dept. of Public Health 2151 Berkeley Way Berkeley, Calif. 94704	State Department of Public Health 2151 Berkeley Way Berkeley, Calif. 97404
COLORADO	Division of Sanitation Department of Public Health State Office Building Denver, Colo.	Department of Public Health Division of Occupational and Radiological Health 4210 East 11th Ave. Denver, Colo. 80220	State Department of Public Health 4210 East 11th Ave. Denver, Colo. 80220

TABLE 9. Pollution Agencies (Continued)

State	Water Pollution Agencies	Air Pollution Agencies	Agencies Responsible for Solid Wastes Planning
COLORADO (Cont'd)	Department of Game and Fish 1530 Sherman St. Denver, Colo.		
CONNECTICUT	Water Resources Commission Department of Agriculture, Conservation and Natural Resources Room 317 State Office Building Hartford, Conn.	Department of Health 79 Elm St. Hartford, Conn. 06115	State Department of Health 79 Elm St. Hartford, Conn. 06115
DELAWARE	Water Pollution Commission Dover, Del. Board of Game and Fish Commissioners North St. Dover, Del.	Water and Air Resources Commission Federal and D Streets Dover, Del. 19901	State Board of Health State Health Building Dover, Del. 19901
DISTRICT OF COLUMBIA		Air Pollution Division Bureau of Public Health Engineering Department of Public Health Judiciary Building Sixth and Indiana Ave., N. W. Washington, D. C. 20001	Director of Sanitary Engineering District Building Washington, D. C.
FLORIDA	Bureau of Sanitary Engineering Board of Health P. O. Box 210 Jacksonville, Fla.	Bureau of Sanitary Engineering State Board of Health P. O. Box 210 Jacksonville, Fla. 32201	Bureau of Sanitary Engineering Florida State Board of Health P. O. Box 210 Jacksonville, Fla. 32201
GEORGIA	Water Quality Division Environmental Health Branch Department of Public Health Atlanta, Ga. Game and Fish Commission State Capitol Atlanta, Ga.	Industrial Hygiene Service Department of Public Health 47 Trinity Ave., S. W. Atlanta, Ga. 30334	Environmental Health Branch Department of Public Health 47 Trinity Ave., S. W. Atlanta, Ga. 30334

HAWAII	Bureau of Sanitary Engineering Environmental Health Division Department of Health P. O. Box 3378 Honolulu, Hawaii	Department of Health Environmental Health Division Health Engineering Branch Air Sanitation Section P. O. Box 3378 Kinau Hale Honolulu, Hawaii 96801	Hawaii State Department of Health Environmental Health Division P. O. Box 3378 Honolulu, Hawaii 96801
IDAHO	Department of Fish and Game 518 Front Street Boise, Idaho	Department of Health Air Pollution Control Commission Statehouse Boise, Idaho 83701	Idaho Department of Health Statehouse Boise, Idaho 83701
ILLINOIS	Sanitary Water Board Springfield, Ill.	Air Pollution Control Board 616 State Office Building 400 South Spring St. Springfield, Ill. 62706	Illinois Department of Public Health State Office Building 400 South Spring St. Springfield, Ill. 62706
	Division of Fisheries Department of Conservation 102 State Office Building Springfield, Ill.		
INDIANA	Division of Sanitary Engineering Bureau of Environmental Sanitation Indiana State Board of Health 1330 West Michigan Street Indianapolis, Ind.	Air Pollution Control Board 1330 West Michigan St. Indianapolis, Ind. 46207	
	Stream Pollution Control Board 1330 West Michigan Street Indianapolis, Ind.		
	Division of Fish and Game Department of Conservation Indianapolis, Ind.		
IOWA	Division of Public Health Engineering State Department of Health State House Des Moines, Ia.	State Department of Health State Office Building Des Moines, Ia. 50319	Environmental Hygiene and Engineering Services State Department of Health State Office Building Des Moines, Ia. 50319
	Conservation Commission East 7th and Court Ave. Des Moines, Ia.		
KANSAS	Environmental Health Services Department of Health 510 State Office Building Topeka, Kans.	Department of Health Environmental Health Services Industrial, Radiation and Air Hygiene Program	Kansas State Department of Health State Office Building Topeka Ave. at 10th Topeka, Kans. 66612

TABLE 9.   Pollution Agencies (Continued)

State	Water Pollution Agencies	Air Pollution Agencies	Agencies Responsible for Solid Wastes Planning
KANSAS (Cont'd)	Forestry, Fish and Game Commission Box 581 Pratt, Kans.	State Office Building 10th and Harrison Streets Topeka, Kans. 66612	
KENTUCKY	Division of Public Health Engineering Department of Health Frankfort, Ky. Water Pollution Control Commission Department of Health Frankfort, Ky. Department of Fish and Wildlife Resources Frankfort, Ky.	State Department of Health Division of Environmental Health Air Pollution Program 275 East Main Street Frankfort, Ky. 40601	Kentucky State Department of Health 275 East Main Street Frankfort, Ky. 40601
LOUISIANA	Division of Public Health Engineering Department of Health State Office Building New Orleans, La. Stream Control Commission P. O. Box 9055 University Station Baton Rouge, La. Department of Wildlife and Fisheries New Orleans, La.	Air Control Commission c/o Division of Public Health Engineering State Board of Health P. O. Box 60630 New Orleans, La. 70160	State Board of Health Civic Center P. O. Box 60630 New Orleans, La. 70160
MAINE	Water Improvement Commission State House Augusta, Me. 04330	Division of Sanitary Engineering Department of Health and Welfare State House Augusta, Me. 04330	Maine Department of Health and Welfare Bureau of Health State House Augusta, Me. 04330
MARYLAND	Division of Sanitary Engineering 301 West Preston St. Baltimore, Md. 21201 Water Pollution Control Commission State Office Building Annapolis, Md.	Division of Air Quality Bureau of Resources Protection Department of Health Room 409 State Office Building 301 West Preston St. Baltimore, Md. 21201	State Department of Health State Office Building 301 West Preston St. Baltimore, Md. 21201

MASSACHUSETTS	Division of Sanitary Engineering Massachusetts Department Public Health 511 State House Boston, Mass. 02133	Division of Sanitary Engineering Department of Public Health State House Boston, Mass. 02133	Massachusetts Department of Public Health 546 State House Boston, Mass. 02133
MICHIGAN	Water Resources Commission 200 Mill St., Station B Lansing, Mich.	Department of Public Health Division of Occupational Health 3500 North Logan St. Lansing, Mich. 48914	Michigan Department of Health 3500 North Logan St. Lansing, Mich. 48914
MINNESOTA	Division of Environmental Health Department of Health University Campus Minneapolis, Minn. 55440 Water Pollution Control Commission Minneapolis, Minn.	Department of Health Division of Environmental Health, Section of Radiation and Occupational Health University Campus Minneapolis, Minn. 55440	Division of Environmental Health State Department of Health University Campus Minneapolis, Minn. 55440
MISSISSIPPI	Division of Sanitary Engineering Board of Health Jackson, Miss. Game and Fish Commission Jackson, Miss.	Air and Water Pollution Control Commission P. O. Box 827 Jackson, Miss. 39205	No state agency
MISSOURI	Water Pollution Board 112 West High P. O. Box 154 Jefferson City, Mo. Conservation Commission Farm Bureau Building Jefferson City, Mo.	Air Conservation Commission Jefferson City, Mo. 65101	Missouri Department of Public Health and Welfare State Office Building 221 West High St. Jefferson City, Mo. 65102
MONTANA	Division of Environmental Sanitation Montana State Board of Health Helena, Mont. Department of Fish and Game Mitchell Building Helena, Mont.	State Board of Health Disease Control Division State Laboratory Building Helena, Mont. 59601	State Board of Health Cogswell Building Helena, Mont. 59601
NEBRASKA	Division of Sanitation Nebraska Department of Health Box 4757 State House Station Lincoln, Nebr. 68509	No state agency	State Department of Health State House Station Box 4757 Lincoln, Nebr. 68509

94

TABLE 9. Pollution Agencies (Continued)

State	Water Pollution Agencies	Air Pollution Agencies	Agencies Responsible for Solid Wastes Planning
NEBRASKA (Cont'd)	Water Pollution Control Council Box 4757 State House Station Lincoln, Nebr. 68509	Bureau of Environmental Health Division of Health 790 Sutro St. Reno, Nev. 89502	Bureau of Environmental Health 790 Sutro St. Reno, Nev. 89502
NEVADA	Bureau of Environmental Sanitation Division of Public Health Engineering 755 Ryland St. Reno, Nev. Fish and Game Commission 51 Grove St. Reno, Nev.		
NEW HAMPSHIRE	Water Supply and Pollution Control Commission 61 South Spring St. Concord, N. H. 03301 Fish and Game Department 34 Bridge St. Concord, N. H.	State Department of Health and Welfare 61 South Spring St. Concord, N. H. 03301	Division of Public Health State Department of Health and Welfare State Health Building 61 South Spring St. Concord, N. H. 03301
NEW JERSEY	Division of Fish and Game Department of Conservation and Economic Development 230 West State St. Trenton, N. J. Division of Shell Fisheries Department of Conservation and Economic Development Millville, N. J. Bureau of Public Health Engineering Division of Environmental Health New Jersey State Department of Health State House Trenton, N. J.	State Department of Health Division of Environmental Health Air Sanitation Program P. O. Box 1540 Trenton, N. J. 08625	State Commissioner of Health P. O. Box 1540 Trenton, N. J. 08625
NEW MEXICO	Environmental Sanitation Services Department of Public Health 408 Galisteo St. Santa Fe, N. Mex. 87501	Division of Occupational Health- Air Pollution Office of Environmental Factors	New Mexico Department of Public Health 408 Galisteo St. Santa Fe, N. Mex. 87501

NEW MEXICO (Cont'd)	Game and Fish Department State Capitol Santa Fe, N. Mex. Oil Conservation Commission	Department of Public Health 408 Galisteo St. Santa Fe, N. Mex. 87501	
NEW YORK	Division of Pure Waters Department of Health 84 Holland Ave. Albany, N. Y. 12208 Division of Fish and Game New York State Conservation Department State Campus Albany, N. Y.	Division of Air Pollution Control State Department of Health 84 Holland Ave. Albany, N. Y. 12208	Division of General Engineering and Radiological Health New York State Department of Health 84 Holland Ave. Albany, N. Y. 12208
NORTH CAROLINA	Department of Water and Air Resources P. O. Box 9392 Raleigh, N. C.	State Board of Health Raleigh, N. C. 27602	Sanitary Engineering Division North Carolina State Board of Health Raleigh, N. C. 27602
NORTH DAKOTA	Environmental Sanitation Services State Department of Health State Capitol Building Bismarck, N. D. Game and Fish Department Bismarck, N. D.	State Department of Health Environmental Health and Engineering Services State Capitol Bismarck, N. Dak. 58501	Division of General Sanitation State Department of Health Capitol Building Bismarck, N. Dak. 58501
OHIO	Division of Sanitary Engineering Ohio Department of Health 101 North High Street Columbus, Ohio Water Pollution Control Board 306 Ohio Departments Building Columbus, Ohio Division of Wildlife Ohio Department of Natural Resources 1500 Dublin Rd. Columbus, Ohio	Department of Health Division of Engineering P. O. Box 118 Columbus, Ohio 43216	Ohio Department of Health 450 East Town Columbus, Ohio 43215

TABLE 9. Pollution Agencies (Continued)

State	Water Pollution Agencies	Air Pollution Agencies	Agencies Responsible for Solid Wastes Planning
OKLAHOMA	Water Resources Board P. O. Box 3324 State Capitol Station Oklahoma City, Okla. Division of Sanitary Engineering Oklahoma State Department of Health Oklahoma City, Okla. Department of Wildlife Conservation Room 118 State Capitol Bldg. Oklahoma City, Okla. The Corporation Commission	State Department of Health Division of Sanitary Engineering, Occupational and Radiological Health Section 3400 North Eastern Oklahoma City, Okla. 73105	State Department of Health 3400 North Eastern Oklahoma City, Okla. 73105
OREGON	Division of Sanitation and Engineering Oregon State Board of Health Salem, Ore. Sanitary Authority 968 Portland State Office Building Portland, Ore. Fish Commission 307 Portland State Office Building Portland, Ore.	Air Quality Control State Sanitary Authority State Board of Health 1400 Southwest Fifth Ave. Portland, Ore. 97201	Vector Control Program State Board of Health 1400 South West Fifth Ave. Portland, Ore. 97201
PENNSYLVANIA	Division of Sanitary Engineering Health & Welfare Building Harrisburg, Pa. 17120 Fish Commission 41 South Office Building Harrisburg, Pa.	Department of Health Division of Air Pollution Control P. O. Box 90 Harrisburg, Pa. 17120	Pennsylvania Department of Health State Capitol Health and Welfare Building Harrisburg, Pa. 17120
RHODE ISLAND	Division of Harbors and Rivers Department of Public Works 216 State Office Building Providence, R. I. Division of Sanitary Engineering 335 State Office Building Providence, R. I. Division of Fish and Game Veteran's Memorial Building	No state agency	Rhode Island Department of Health State Office Building Providence, R. I. 02903

SOUTH CAROLINA

Division of Sanitary Engineering
State Board of Health
Hampton State Office Building
Columbia, S. C.

Water Pollution Control Authority
Hampton State Office Building
Columbia, S. C.

Pollution Control Authority
Room 137
J. Marion Sims Building
2600 Bull St.
Columbia, S. C. 29201

Environmental Sanitation
Division of Local Health Services
State Board of Health
Columbia, S. C. 29201

SOUTH DAKOTA

Division of Sanitary Engineering
South Dakota Department of Health
Pierre, S. D.

Committee on Water Pollution
South Dakota Department of Health
Pierre, S. D.

Game, Fish and Parks Department
State Office Building
Pierre, S. D.

Occupational and Radiological
    Health Section
Division of Sanitary Engineering
State Department of Health
Pierre, S. Dak. 57501

South Dakota State Planning
    Commission
State Capitol
Pierre, S. Dak. 57501

TENNESSEE

Stream Pollution Control Board
226 Cordell Hull Building
Nashville, Tenn. 37219

Game and Fish Commission
226 Cordell Hull Building
Nashville, Tenn. 37219

Division of Industrial Service
Department of Public Health
727 Cordell Hull Building
Sixth Ave., North
Nashville, Tenn. 37219

Tennessee Department of Public Health
Division of Sanitary Engineering
Cordell Hull Building
Sixth Ave., North
Nashville, Tenn. 37219

TEXAS

Division of Water Pollution Control
Texas State Department of Health
1100 West 49th St.
Austin, Tex. 78756

Game and Fish Commission
Austin, Tex.

Air Control Board
1100 West 49th Street
Austin, Tex. 78756

State Department of Health
1100 West 49th St.
Austin, Tex. 78756

UTAH

Bureau of Sanitation
State Department of Health
Salt Lake City, Utah

Water Pollution Control Board
45 Fort Douglas Boulevard
Salt Lake City, Utah 84113

State Department of Health
44 Medical Dr.
Salt Lake City, Utah 84113

State Department of Health
45 Fort Douglas Boulevard
Salt Lake City, Utah 84113

TABLE 9. Pollution Agencies (Continued)

State	Water Pollution Agencies	Air Pollution Agencies	Agencies Responsible for Solid Wastes Planning
VERMONT	Bureau of Environmental Sanitation State Department of Health 115 Colchester Ave. Burlington, Vt. 05402 Water Conservation Board State Office Building Montpelier, Vt. Fish and Game Commission Montpelier, Vt. 05602	Industrial Hygiene Division Department of Health P. O. Box 333 32 Spaulding St. Barre, Vt. 05641	Bureau of Environmental Sanitation Division of Sanitary Engineering Vermont State Department of Health 115 Colchester Ave. Burlington, Vt. 05402
VIRGINIA	Water Pollution Control 415 West Franklin Street P. O. Box 5285 Richmond, Va.	Division of Engineering Bureau of Industrial Hygyiene State Department of Health Richmond, Va. 23219	Virginia State Department of Health Bank and Governor Streets Richmond, Va. 23219
WASHINGTON	Division of Engineering and Sanitation State Department of Health Seattle, Wash. Pollution Control Commission 224 Old Capitol Building Olympia, Wash. Department of Fisheries 4015 – 20th Avenue West Seattle, Wash.	Air Sanitation and Radiation Control Section Division of Engineering and Sanitation State Department of Health 1510 Smith Tower Seattle, Wash. 98104	State Department of Health Public Health Building Olympia, Wash. 98502
WEST VIRGINIA	Division of Sanitary Engineering Health Department State Capitol Charleston, W. Va. Water Resources Commission 1709 Washington St. East Charleston, W. Va.	Air Pollution Control Commission 4108 MacCorkle Ave., Southeast Charleston, W. Va. 25304	West Virginia State Department of Health State Office Building No. 1 1800 Washington St., East Charleston, W. Va. 25305
WISCONSIN	Section of Sanitary Engineering State Board of Health 453 State Office Building Madison, Wis. Committee on Water Pollution Room 453, State Office Building 1 West Wilson St. Madison, Wis. 53701	Air Pollution Control Division State Board of Health State Office Building 1 West Wilson St. P. O. Box 309 Madison, Wis. 53701	State Board of Health 1 West Wilson St. Madison, Wis. 53702

WISCONSIN (Cont'd)	Conservation Commission P. O. Box 450 Madison, Wis.	
WYOMING	Division of Environmental Sanitation State Department of Public Health State Office Building Cheyenne, Wyo. 82001	Division of Industrial Hygiene Department of Public Health State Office Building Cheyenne, Wyo. 82001
	Game and Fish Commission Box 378 Cheyenne, Wyo.	State Department of Public Health State Capitol Building Cheyenne, Wyo. 82001
GUAM		Director of Public Health and Welfare Government of Guam P. O. Box 2876 Agana, Guam 96910
PUERTO RICO		Secretary of Health Puerto Rico Department of Health Ponce de Leon Ave. San Juan, P. R. 00908
AMERICAN SAMOA		American Samoa Department of Public Works Pago Pago, American Samoa
VIRGIN ISLANDS		Bureau of Environmental Sanitation Virgin Islands Department of Health Public Health Service P. O. Box 1442 Charlotte Amalie, V. I. 00801

from automobiles. If the solid waste problem continues to mount, it may result in forbidding the manufacture of packaging materials which are resistant to decay.

The trend toward placing responsibility upon manufacturers for the toxicity of their products is accelerating. The time may be near at hand when each manufacturer will be required to present a toxicity hazard analysis for any new chemicals or materials which he plans to market.

The 1965 amendments to the Federal Water Pollution Control Act required that standards be set for all interstate waters in the country. This is but the beginning of a trend which will continue, and more attention will be given to the effects of combinations of various pollutants.

A few years ago, a water or air polluter could move to another state if he was required to abate his pollution. This is no longer an option, for the federally approved standard for water quality and increased federal demands for uniform programs among the states will eliminate any havens of pollution. Congress will increase its demands that departments, such as Housing and Urban Development, Transportation, Interior, and Health, Education and Welfare, require some uniformity of objective as conditions for making federal grants available. Concern with other environmental factors such as noise, crowding, hazards at work, and radiation will increase.

Both industry and municipal officials must appreciate the increasing trend toward management of problems on a regional or shared basis. The costs of sewage treatment works for each community and each industry are unnecessarily high, often with regard to initial cost and always in respect to maintenance and operation. Regional systems of water management, such as the Delaware River Basin Commission, and water quality management, such as the Ohio River Sanitation Commission, have considerable appeal. Water resources planning on a multi-state regional basis is encouraged by the Water Resources Planning Act of 1965 and by other federal legislation and policies.

In summary, the governmental posture will be one of increased stimulation of research and development, requirement of more responsibility upon the manufacturers of new products, greatly improved performance by local and state pollution abatement agencies, and insistence upon the regional solution of many problems regardless of county, city or state boundaries. Both state and federal laws recognize that the primary responsibility for water and air pollution control rests at the state level. This control is usually achieved by requiring the presentation of engineering plans before issuance of a permit and periodic surveillance of the system to evaluate its performance. The authority for beneficial water use also rests with the state, so it is natural that it should also have the responsibility of reducing its use for waste disposal.

Some states have carried out effective water and/or air pollution abatement programs. Citizens of those which have not have turned to the federal government for relief. This trend started with enactment of the first federal water pollution law in 1948. As public concern increased, federal legislation accelerated, with major amendments in 1956, 1961, 1965 and 1966.

Each of the amendments has strengthened the federal role in enforcement. The Secretary of Interior may, on his own motion in certain cases, and upon the complaint of state officials in others, call an enforcement conference. The states and the federal government each appoint conferees. They hear testimony for the Secretary and recommend a course of action to him. If remedial action does not result, the Secretary may call a hearing at which a determination is made. If the board finds pollution occurring and unsatisfactory progress to abate it, it recommends appropriate action to the Secretary. If necessary, the U. S. Attorney General enforces the recommendations. Experience has demonstrated this to be a highly effective procedure. Industries or municipalities faced by such action will be wise to get the facts and proceed expeditiously to abate their pollution.

ORGANIZATIONAL APPROACHES TO MANAGING THE PROBLEM

A comprehensive water quality management program should be developed in a framework of state health and state water resource policy. This requires determination of the needs in sufficient detail to provide a basis for making state policy. An adequate legal base is necessary if the policies are to be implemented, as is assignment of responsibility to a single state agency, adequately staffed. In addition, decisions must be made as to the balance which will be required between state assistance, public education and enforcement.

No two states are alike in their physical problems or in the organizational structures available to solve the problems. Each state must tailor its program to the particular needs and resources of its communities, commensurate with state objectives and policies. Individual feeferences are important. A program which is successful in California will probably not be applicable to Maine, and a program tailored to

New York will not be successful in Wyoming. The variation in hydrologic features, the amount of local self-government allowed cities and towns, the relation of industry to local governments and the adequacy of state financing laws are a few examples of the differences which will affect a state water quality control program. The uses of water, the financial capability, sociological traditions and public concern all need to be considered in developing a comprehensive state water pollution program. In considering such a program, attention should be given to seven factors:

(1) Water quality surveillance;
(2) The establishment of stream standards;
(3) Comprehensive planning;
(4) Issuance of permits;
(5) Enforcement;
(6) Construction grants;
(7) Evaluation of operation and maintenance with possible assistance.

Water Quality Surveillance

Water quality management requires an understanding of the system to be managed. This means that basic data on the quantity and quality of water must be available. The federal government has long maintained stations to measure the flow of water in important streams throughout the country, and the United States Public Health Service established a nationwide network, which has been continued by the Department of Interior, to measure the quality of water in interstate streams. A much more extensive network of stations to measure quantity and quality is necessary to implement a water quality program. These data need to be analyzed and published yearly.

New York maintains about seventy stations where samples are collected about once a month, transported to the laboratory and analyzed. This makes available a representative picture of water quality in every section of the state. Other states have similar programs.

Instrumentation for the continuous monitoring of surface waters is in an early stage of development. Until recently, very few companies offered a completely automated system to measure multiple water quality parameters and to transmit the data to a central system for storage, manipulation and retrieval. Sensing probes currently available represent adaptations of laboratory sensors housed in a protective structure for field use where minimum attention can be provided; therefore, users of automatic monitoring equipment have had to maintain a routine program of maintenance, calibration and standardizatton to minimize malfunction and to obtain reliable data.

Two basic types of instrumentation can be considered for water monitoring. One type, the electrochemical system, uses specialized electrodes which relate concentrations of parameters to voltages produced; the other type applies colorimetric analytical procedures to samples to determine concentration of parameters. Great strides have been made in the automation of this type of analysis although units are not presently available for commercial purchase. A demonstration model to measure 12 parameters of water quality has been developed and is being field tested. The components, it measures are: sulfate, ammonia, phenol, fluoride, chemical oxygen demand, nitrite, nitrate, iron, phosphate, methylene blue alkyl sulfonate, bicarbonate and carbonate alkalinity. A major advantage of the wet chemistry system is the wide range of analyses it can perform compared to the probe-type system; however, the cost of the equipment, replenishment and storage of chemicals and the size of the unit represent disadvantages.

Probe devices are simpler and more compact, are commercially available, and have been used by water pollution control agencies for a number of years. Parameters normally measured include pH, temperature, dissolved oxygen, dissolved chloride, specific conductance and oxidation-reduction potential. Electrodes for measuring other parameters are being developed; however, the type of instrumentation and the water quality measurements made at a specific site will depend on the objective of the monitoring program.

The unit cost of data collection by continuous monitoring is small per parameter measured; however, the voluminous data must be logically handled to avoid being inundated by information. One monitor, measuring 6 parameters an hour, provides about 52,500 items of data annually. Provisions for automatic data processing will enable the collector to make maximum use of the monitoring system and the information provided. The data can be handled as follows:

(1) Editing, validating and storage for subsequent analysis and retrieval;
(2) Statistical evaluation to determine average, maximum and minimum ranges, standard error and confidence limits;
(3) Correlation studies to determine the relationship of parameters;
(4) Analysis of parameters with respect to time to determine trends, cyclical changes, fluctuations and durations.

Statistical analysis will determine the optimum sampling frequency and which parameters to measure. Automated monitoring systems can:

(1) Determine compliance with stream standards,

(2) Detect abnormal changes in water quality,

(3) Forecast water quality conditions and the effects on related water usages,

(4) Measure long-term water quality trends,

(5) Provide water quality data for special studies.

Common parameters of water quality which can be routinely measured with some reliability normally include temperature, pH, dissolved oxygen and specific conductance. Chloride analyses are made in situations where industrial wastes or natural conditions make such data desirable. Oxidation-reduction potential measurements provide useful data on waters with low dissolved oxygen.

While the potential of automation for monitoring surface waters is unlimited, the technology and state of the art are still primitive. The potential user should determine the reliability of the sensing devices he intends to purchase, the problem of data handling, and the maintenance and operation of all equipment associated with the system.

Stream Standards

Public Law 89-234 was enacted by the Congress in October, 1965. Each state was required to establish water quality standards to protect the public health and welfare and enhance the quality of water. The law made the classification of waters and the adoption of stream standards a necessary part of each state program. National policy is revealed in the word "enhance." Stream quality standards are for the purpose of providing a base to improve the quality of water, not to legalize pollution. The standards should give attention to existing water uses and anticipated future uses, including aesthetic values. All surface water should be free of materials which will settle to form objectionable deposits, floating debris, oil, scum, substances producing objectionable color, odor, taste or turbidity, materials which are toxic or which produce undesirable physiological responses in human, fish and other animal life, and substances in concentrations which have an adverse effect upon aquatic life.

The suitability of water for swimming, water skiing and other direct contact should be decided upon the basis of a sanitary survey combined with a measure of the levels of fecal coliform count.

Streams and lakes which are used as a source of raw water supply should have CCE—a measure of organic pollution—at or below 0.2 mg/liter. Because conventional water treatment plants cannot adequately remove pesticides, some authorities suggest that criteria be set for the following pesticides: aldrin, DDT, dieldrin, endrin, heptachlor, heptachlor epoxide, lindane, methoxychlor, toxaphene, 2,4-D, 2,4,5-T, 2,4,5-TP, the organo-phosphates and carbamates. Boron and uranium compounds (see Section 12) should also be included. The Public Health Service drinking water standards can be used as a guide to define toxic substances.

The levels for the coliform and fecal coliform groups become particularly important for public water supplies which are treated only by chlorination, and a level of 50 coliforms/100 ml is usually regarded as the maximum for chlorination alone. Less agreement exists with respect to the coliform count for bathing waters. Most state standards require an average below 1000–2400/100 ml.

The establishment of stream standards also requires a decision on the part of the regulatory agency as to the quality and quantity of wastes which can be discharged and still meet these standards. In effect, it requires the agency to adopt effluent standards. This amounts to a kind of land use zoning and represents a drastic, though necessary, step forward in controlling pollution. Table 10 summarizes the standards for New York, and Table 11 gives a summary of the Ohio River Sanitation Commission (Orsanco) standards.

Comprehensive Planning

A state's investment for water supply and pollution abatement constitutes most of all the money spent for water resource development. For this reason, decisions in planning for local improvements should be made within the framework of local needs and overall state water policy. Ideally, comprehensive studies for pollution abatement should be undertaken only after comprehensive water studies are made of the river basins of the state. This has not occurred in a single state; in fact, planning for water pollution abatement usually precedes and, in fact, develops support for comprehensive water resources planning.

Local communities should be encouraged to develop master plans for water and waste water facilities in order to assure the maximum benefits and economies from the program. These studies help to develop the framework for expanding utility systems and for promoting multi-municipal planning and joint-venture cooperation. A

TABLE 10. New York State Classes and Standards for Fresh Surface Waters

STANDARDS OF QUALITY

Clean and Best Use	Dissolved Oxygen (mg/liter)	Monthly Coliform Bacteria Median (per 100 ml[a])	pH	Toxic Wastes, Deleterious Substances, Colored Wastes, Heated Liquids, and Taste and Odor Producing Substances	Floating Solids, Settleable Solids, Oil, Sludge Deposits
AA—Source of unfiltered public water supply and any other usage	5.0 minimum (trout) 4.0 minimum (non-trout)	Not to exceed 50; 20% of samples not to exceed 240	6.5 – 8.5	None in sufficient amounts or at such temperatures as to be injurious to fish life or make the waters unsafe or unsuitable	None attributable to sewage, industrial wastes or other wastes
A—Source of filtered public water supply and any other usage	Same as class AA	Not to exceed 5000; 20% of samples not to exceed 20,000	6.5 – 8.5	None in sufficient amounts or at such temperature as to be injurious to fish life or make the waters unsafe or unsuitable[b]	None which are readily visible and attributable to sewage, industrial wastes or other wastes
B—Bathing and any other usages except as a source of public water supply	Same as class AA	Not to exceed 2400; 20% of samples not to exceed 5000	6.5 – 8.5	None in sufficient amounts or at such temperatures as to be injurious to fish life or make the waters unsafe or unsuitable	None which are readily visible and attributable to sewage, industrial wastes or other wastes
C—Fishing and any other usages except public water supply and bathing	Same as class AA	Same as class B	6.5 – 8.5	None in sufficient amounts or at such temperatures as to be injurious to fish life or impairs the waters for any other best usage	None which are readily visible and attributable to sewage, industrial wastes or other wastes
D—Natural drainage, agriculture and industrial water supply	3.0 minimum	Same as class B	6.0 – 9.5	None in sufficient amounts or at such temperatures as to prevent fish survival or impair the waters for agricultural purposes or any other best usage	None which are readily visible and attributable to sewage, industrial wastes or other wastes

[a]Waste effluents discharging into public water supply and recreation waters must be effectively disinfected.
[b]Phenolic compounds cannot exceed .005 mg/liter; no odor producing substances causing threshold odor number to exceed 8.

multiplicity of waste water plants on a single watershed is not desirable since the staffing requirements for numerous small plants become burdensome and the operating costs are greater.

New York has maintained such a program for both water supply and waste water collection and treatment. The purposes of this program are to encourage comprehensive planning for the immediate and future needs, to promote intermunicipal cooperation, to develop the engineering and economic facts which will lead to the construction of the needed facilities and to implement the modernization of outmoded systems as changing requirements evolve. Officials are encouraged to undertake these studies on a county basis, since in New York state the county provides a unit of sufficient size to demonstrate the economies and yet small enough to obtain the necessary intermunicipal cooperation.

These studies will require updating from time to time, but they provide a master plan against which subsequent improvements are measured. Once a plan is accepted, future plans for waste treatment works should generally conform to it unless something occurs which justifies a change.

In summary, the comprehensive study proposes a general conceptual plan of a waste water utility, unified as to boundaries, engineering economics, organization and management. It will concentrate responsibility for economical de-

TABLE 11. Orsanco Stream Quality Criteria and Minimum Conditions

Water Use	Monthly Arithmetic Coliform Average (per 100 ml)	Dissolved Oxygen (mg/liter)	pH	Solids and Sludge Deposits[a]	Toxic and Deleterious Substances, Colored, Taste and Odor Producing Substances, and Heated Liquids[a]
Public water supply	Not to exceed 5000; 20% of samples not to exceed 5000; 5% of samples not to exceed 20,000.			Dissolved solids: maximum monthly arithmetic average 500; maximum value of 750 mg/liter.	Gross beta activity (in absence of strontium 90 and alpha emitters) not to exceed 1000 pCi/liter at any time. Chemical constituents: See footnote b.
Industrial water supply		Maximum daily average of 2.0; absolute minimum of 1.0.	5.0–9.0	Dissolved solids: maximum monthly arithmetic average 750; maximum value of 1000 mg/liter.	Maximum temperature of 95°F at any time.
Aquatic life		Maximum of 5.0 during 16 hr of any 24-hr period; absolute minimum of 3.0 at any time.	Minimum: 5.0; maximum: 9.0; daily average between 6.5 and 8.5.		Maximum temperature of 93°F at any time during the months of May through November; maximum of 73°F at any time from December through April. Ionic substances cannot exceed 0.1, the 48-hr median tolerance limit; other limiting concentrations permitted at discretion of regulatory agency.
Recreation	Not to exceed 1000; 20% of samples not to exceed 1000; maximum of 2400 on any day.				
Agricultural of stock watering					

[a] All waters at all times and at all places must be free from sludge deposits; from unsightly or deleterious floating debris, oil, scum and other floating materials; from nuisances due to color, odor or other conditions; from toxic or harmful discharges attributable to municipal, industrial or other discharges.
[b] Maximum permissible concentrations (mg/liter): arsenic 0.05, barium 1.0, cadmium 0.01, chromium (hexavalent) 0.05, cyanide 0.2, fluoride 2.0, lead 0.05, selenium 0.01, silver 0.05.

sign, sound planning, orderly construction and effective performance of the facilities in one or two governing bodies rather than dispersing it over a number of such groups.

Issuance of Permits

An effective state program requires that it control the sources of pollution through the issuance of permits. A permit should be required for each existing or new outlet, or to construct, operate or use a waste water disposal system.

The issuance of permits is based upon the rights of states to control their waters and to protect the public health. The release of wastes into waters represents a use of these waters equal to actually removing water from the stream for agricultural, industrial or municipal purposes. Most states also require approval for the extension of sewer systems. This is such a burdensome requirement, in the large states at least, that the difficulty of administering it exceeds its value. In the past and now, permits for sewer extensions were used as incentives to encourage the building of treatment works; however, the advent of a truly national program, combined with stream quality standards and adequate enforcement measures, seems to make this requirement less necessary.

The legal base for permits should be sufficiently broad to include all industrial wastes, including drainage from such agricultural enterprises as duck farms and livestock feed lots, and each permit should be issued on the basis of engineering plans. A mechanism needs to be established for the revocation of permits because of non-compliance, usually after appropriate investigation and public hearing.

Enforcement Measures

A vigorous enforcement program is essential to any state water pollution control program. Successful implementation of such a program requires: (1) meaningful and streamlined laws, (2) organization and personnel; and (3) systematic identification and evaluation of pollution problems.

Pollution control laws should provide procedures for speedy adjudication of cases—first, on the administrative level and, then, on the court level—while retaining the concept of "due process" to which the respondent polluter is entitled. Organization should reflect the centralized control necessary to advance a statewide program by reposing the authority in a single administrative unit. Specialists, such as scientists, engineers, attorneys and technicians are needed to do the job. All pollutional situations should be identified, evaluated and categorized

in terms of location, type and order of importance.

The philosophy of enforcement is that it is not an end in itself but only a means to achieve a goal—the control and abatement of water pollution. The basic purpose of enforcement is to bridge the gap betweeen achievement of pollution abatement as a general ideal and its application to specific municipal and industrial pollutional situations. Thus, a legal framework is established within which municipal and industrial corporations can operate to provide necessary treatment facilities. Sometimes this framework is necessary for management to justify to its constituents (either voters or stockholders) the expenditure of large sums of money.

An enforcement unit is necessary for the everyday operation, management and coordination of an enforcement program. The principal responsibilities of this unit would be to:

(1) Set up priorities for enforcement action;
(2) Provide liaison between the field offices and the legal staff;
(3) Coordinate enforcement with other water pollution control components, such as plan review, construction grants, and operation and maintenance;
(4) Review and evaluate case reports and abatement timetables;
(5) Establish and maintain record-keeping systems, including follow-up on timetables;
(6) Implement policy through legal action.

A field organization, adequately staffed, is necessary for the initial identification and evaluation of pollutional situations and for the preparation of case reports. Field personnel initially develop and negotiate abatement timetables which are subsequently incorporated into administrative or court orders. The dates in these timetables must be followed up by field personnel to determine compliance with orders.

Legal services are provided by attorneys working directly in the enforcement unit. The duties of the legal staff are to arrange hearing schedules, prepare cases for trial, and prepare notices, complaints, stipulations, orders and all other legal documents relating to the program. A brief review of New York's experience follows to illustrate these general comments.

Enactment of Chapter 180 of the Laws of 1965 as part of the New York Pure Waters legislation marked the beginning of the enforcement phase of the program. This legislation amended the basic water pollution control law by streamlining and consolidating enforcement procedures.

An enforcement section in the Bureau of Water Quality Management, Division of Pure

Waters, is responsible for the management and coordination of the enforcement program.

The Department's five regional offices are responsible for providing field information on pollutional situations and incorporating it in case reports. Regional and/or local health office personnel, conservation specialists and other witnesses testify at hearings. Abatement timetables are initially developed and negotiated by regional office personnel. Follow-up on such timetables, once incorporated in the Commissioner's Order, is the responsibility of the section in conjunction with regional personnel.

Legal services are provided by the Department's Office of Counsel. The staff consists of five attorneys working full time on a water pollution control.

Over 90 percent of the Commissioner's Orders issued have been stipulated to by the respondent polluters.

Obviously, an enforcement program does not terminate with the issuance of a Commissioner's Order. Systematic and vigorous follow-up to determine compliance with the terms of orders is necessary to sustain and consummate the program. The large number of orders and the abatement-step due dates contained in these orders necessitate the use of automatic data processing for an effective follow-up system. Serious or flagrant violations, when identified, are referred via the Office of Counsel to the Attorney General's Office, with a recommendation that court action be initiated. At the end of 1967, twelve such referrals had been made. Court action, when undertaken, has been effective. Continuous and vigorous pressure has to be applied in all phases of the program in order to complete the statewide clean-up by 1973.

Construction Grants

The building of municipal sewage treatment works can be accelerated by a state program of financial incentives through construction grants. A state program can supplement the federal construction grants activities program, with the state legislation "geared" to the federal legislation. Construction grants for municipal sewage treatment works were first authorized in 1956, with the most recent amendments being in the Clean Water Restoration Act of 1966. Financial assistance is not given for street sewers, laterals, and trunks, but interceptor lines, pumping stations, force mains, sewage treatment plants and plant outfall lines are eligible.

The Federal Clean Water Restoration Act of 1966 provides that states are eligible for a basic 30 percent grant. The grant is increased to 40 percent of the project cost if the state matches

25 percent of the cost of all projects for which federal grants are given in each fiscal year. The maximum federal grant eligibility increases to 50 percent if state water quality standards have been established for the receiving waters and, for interstate waters, if the standards have been approved by the federal government. In addition, 10 percent of the grant amount is added to the actual federal share if the project is certified by a state, metropolitan, or regional planning agency as conforming with the metropolitan or regional plan.

States with construction grants programs are eligible for either 50 or 55 percent federal participation. In these cases, additions of the state portion lowers the local contribution to not more than 25 percent and in some cases to 15 percent.

If, as in New York State, the state has the basic 30 percent state grant and also provides a pre-financing grant guaranteeing up to 30 percent of the inadequate federal grant, then municipalities may be guaranteed a minimum of 60 percent of project cost (after completion of construction), and would be eligible for the maximum state and federal participation of 80 or 85 percent.

A state construction grants office needs to be organized similarly to the regional construction grants office of the Federal Water Pollution Control Administration, with both an administrative section and an engineering section. A state grants office without comprehensive planning functions and without design review responsibilities would consist of approximately one-half engineers and one-half administrative and fiscal personnel, with the numbers dependent upon the size of the program. The large force of administrative personnel eliminates the necessity for engineers to perform administrative functions such as checking performance and payment bonds, insurance, computation of payments, real estate clearance certificates, application review, and the many other administrative chores deriving from a grant program.

Maintenance and Operation

The key to a successful pollution abatement program lies in the effective maintenance and operation of the treatment works. Numerous studies have shown that this is the weakest part of the total program. The states have recognized this, and some have maintained programs of treatment plant evaluation for half a century.

The recent adoption of stream quality standards by the states will focus more attention upon the efficient performance of waste treatment works. Local ordinances are needed to control the types and quantities of wastes which can be released into public sewers. Oils, toxic materials and trash are examples of wastes which should be kept from the sewer. A visit to the bar screens at some public sewage treatment plants will reveal that industries use the sanitary sewers for the disposal of solid wastes more properly disposed of by other means.

The increased demand for quality performance will place more emphasis upon the need for qualified operators. It will require more laboratory surveillance and better record keeping. Some states offer incentives of financial assistance to communities which meet certain minimum standards. Most of these aid programs are only partially successful. During the first two years of the program in New York, only half of the eligible communities applied. In some cases at least, local officials recognizing the inadequacy of their plant operation chose to ignore the opportunity to receive state aid rather than risk a searching look at their performance.

Regional approaches to water management have increased in popularity, and recent federal legislation has given additional impetus to solving problems on the basis of the large areas incorporated in major river basins. An understanding of the influence which these agencies will exert on the use of water is important to industry leaders and local officials.

The Water Resources Planning Act of 1965 (P. L. 89-90) authorizes the formation of river basin commissions to carry out comprehensive multi-purpose water and related land resource planning and to coordinate the efforts of federal, interstate and state agencies. The supporters of this legislation anticipated nine or ten such commissions to cover the United States, so that very large areas and diverse interests are included in each. These commissions are being established and will have a major effect upon water development policies, including those related to water quality management.

Water pollution control on a regional basis is not new. In 1935, New York, New Jersey and Connecticut negotiated a compact, known as the Interstate Sanitation Commission, to abate pollution in the New York City harbor and in the adjacent waterways. The Ohio River Sanitation Commission represented a truly pioneering effort when, in 1948, eight states and the federal government signed a compact to abate pollution of the Ohio River and its major tributaries. Two other compacts were negotiated in 1941—the New England Interstate Water Pollution Control Commission and the Interstate Commission on the Potomac River.

The Delaware River Basin compact represents the latest cooperative effort at water resources planning and management, including water quality control. It involves New York, New Jersey, Pennsylvania and Delaware, as well as the federal government. The four governors are commissioners, and the federal representative is the Secretary of Interior. The compact gives the commission authority to plan and require new projects to conform to the plan, to construct projects and to allocate water among the states. The personal involvement of the governors in the decision making, and the broad authority, makes this compact unique. Only time will tell whether it will live up to its promises but, in the meantime, federal policy is forcing the special purpose compacts such as the Ohio River Sanitation Commission to consider broadening their mission.

While the actual direction of regional approaches to solving water pollution problems is not certain, future efforts will be more multi-purpose oriented. They will combine waste treatment with the controlled release from reservoirs. They will encourage uniformity in standards of quality and eliminate havens for polluters through monitoring and enforcement.

BIBLIOGRAPHY

1. "Restoring the Quality of our Environment." Report of the Environmental Pollution Panel, President's Science Advisory Committee, The White House, Nov. 1965.
2. "Glossary—Water and Sewage Control Engineering," under the joint sponsorship of APHA, ASCE, AWWA, and Federation of Sewage Works Associations.
3. Taken from preliminary copy of "Report on Sewage Treatment Facilities for City of Rochester, New York," as prepared by Black and Veatch, Consulting Engineers, Kansas City, Missouri, 1967.
4. New York State Water Quality Surveillance Network.
5. Camp, Thomas R., verbal report, ASCE Symposium, April 1962, on storm sewage overflows.
6. Burm, R. S., "The Bacteriological Effect of Combined Sewer Overflows on the Detroit River," WPCF Journal 39, 3, 410 (1967).
   Palmer, C. L., "Feasibility of Combined Sewer Systems," WPCF Journal 35, 2, 162 (1963).
   Weibel, S. R., Anderson, R. J., and Woodward, R. L., "Urban Land Runoff as a Factor in Stream Pollution," WPCF Journal 36, 7, 914 (1964).
   Camp, T. R., "The Problem of Separation in

Planning Sewer Systems," WPCF Journal 36, 12, 1959 (1966).

7. Stockinger, H. H., "Man's Environment in the 21st Century," Dept. of Environmental Sciences and Engineering, Publ. No. 105, School of Public Health, University of North Carolina, July 1965.

8. Wilson, J. G., and Balassa, J. J., "Teratogenic Interactions of Chemical Agents in the Rat," Jour. Pharm. Exptl. Therap. 144, 429 (1964).

9. Gentry, J. T., "Current Epidemiological Findings in Congenital Malformations," University of North Carolina Seminar, Dec. 1964.

PART TWO

AIR POLLUTION

Alexander Rihm, S.M.
Assistant Commissioner
New York State Health Department
Albany, N. Y.

The average man can live weeks without food, days without water, but only minutes without air. He can purify his food and water supply before use, but he must breathe as it is the air on which his life depends.

The air surrounding this earth is a limited resource which man now realizes must be used and guarded carefully. The growth of our population in the past has in part been determined by the ability of the environment to support it. Solutions to the problems of food supply, water supply, waste disposal, transportation and housing have increased the potential capacity of the urban environment to support life. Future progress undoubtedly will depend in part on man's ability to solve the problem of air pollution.

This is not a new problem since it is often due to uncontrollable natural sources as well as the activities of human beings. It is a dynamic problem, changing constantly as man changes his ways of doing things.

Air pollution is also an "area-unique" problem, for type, quantity and nature of contaminants discharged to the atmosphere of one area rarely resemble those of another, anymore than do the topography, meterology, or other factors of metropolitan New York resemble those of greater Chicago, Los Angeles County, greater London, Moscow, Tokyo or thousands of other places throughout the world.

The first thorough consideration of the problem can be traced back to 1661. John Evelyn, in his pamphlet "Fumifugium," advocated the use of better fuels as a means of reducing air pollution and more comprehensive urban planning to avoid effects on health, property and vegetation by consideration of such factors as weather, topography and proper source location.

Air pollution became a widespread problem with the industrial revolution and associated technological and economic changes, the population explosion, and increased population concentration in urban centers. In 1920, about half the U.S. population lived in urban areas. By 1965, two-thirds of the population was urban, and it is estimated that by the year 2000, 85 percent of the people will be living on less than 10 percent of the land. Great segments of the United States and other countries will become gigantic urban tracts and consequently will present an interrelated complex of air pollution problems.

The seriousness of the situation is rapidly increasing and the difficulties of applying effective controls are becoming progressively .greater. Nevertheless, air pollution can and must be controlled by abating atmospheric pollution at its source (through proper design, maintenance and operation of equipment, and strict enforcement of regulatory ordinances) and by constant cooperation of all involved, namely the public, commerce, industry and government.

DEFINITIONS

A few of the frequently used terms in air pollution control work follow:

(a) Air Contaminant. A dust, fume, gas,

mist, smoke, vapor, odor, or other foreign matter, or any combination thereof, not present in "normal" air. An air contaminant can be natural as well as man-made.

(b) *Air Contamination Source.* Any source at, from, or by reason of which any air contaminant is emitted into the atmosphere.

(c) *Air Contamination.* Presence in the outdoor atmosphere of one or more air contaminants which contribute, or are likely to contribute, to a condition of air pollution.

(d) *Air Pollution.* The presence in the outdoor atmosphere of one or more air contaminants injurious to human, plant or animal life, or to property; or which unreasonably interfere with the comfortable enjoyment of life and property.

The fact that air may be contaminated without being polluted is particularly important to the use of air as a natural resource. In essence, air pollution exists only when one or more contaminants are present and persist so as to create (1) an adverse effect on human, plant or animal life, or (2) an adverse social or economic effect. Air pollution control is the means taken to abate the pollution of the air and to improve and eliminate adverse or harmful conditions.

(e) *Dust.* Solid particles of material.

(f) *Fly Ash.* Gas-borne particulates arising from combustion of fuels and consisting essentially of fused ash and/or unburned material.

(g) *Fumes.* Solid particles, formed by condensation from the gaseous state, suspended in the atmosphere.

(h) *Mist.* Liquid droplets suspended in the atmosphere.

(i) *Smoke.* Fine, gas-borne carbonaceous particles resulting from incomplete combustion and in sufficient concentration to be visible.

The most all-inclusive definition of smoke is that it consists of any observable emission from a chimney, stack, duct or open fire or from the burning of fuel or refuse. Smokes are actually complex colloidal systems formed by the incomplete combustion of carbonaceous matter. Smoke particles are usually less than $0.3–0.5\mu$ in diameter.

The Ringelmann chart was introduced into the United States in 1897, and along with such substitutes as the Micro-Ringelmann chart and the smoke scope it has been used extensively in air pollution control since the early 1900's. It is a means by which the reduction in light transmission through a smoke plume can be estimated and, until about 1950, was about the only tool available to control agencies for estimating discharges from individual sources.

(j) *Vapor.* The gaseous state of substances which are normally solid or liquid.

(k) *Odor.* Molecules of a substance which stimulates the olfactory organ.

(l) *Gas.* A formless state of matter.

Gas molecules, because of Brownian movement, do not agglomerate and tend to diffuse throughout any available volume. A packet of gas, however, will rise or fall, depending on its density relative to the surrounding air, but once dispersed it shows no tendency to reform the packet.

(m) *Aerosol.* Colloidal suspension in air of particles or droplets, possible because of their small size.

Aerosols disappear from the atmosphere by evaporation, precipitation, diffusion, or settling. Larger particles tend to settle; small ones disperse by diffusion. An aerosol's stability is chiefly affected by Brownian oscillations and gravity settling. Droplets collide and adhere or coalesce to form larger droplets, while particles form loose aggregates which are roughly spherical or filamentous in shape. The greater the concentration, the greater the coagulation rate.

The terms "smoke," "fog," "haze," "dust," "fly ash," "fumes" and "mist" are used to describe particular types of aerosols, depending on the size, shape and characteristic behavior of the dispersed particles.

(n) *Smog.* A combination of smoke and fog.

The term "smog" is being broadened to include all conditions of high atmospheric pollution caused by a complex mixture of gases and suspended matter, characterized by a marked reduction in visibility and by certain physiological responses in human beings such as eye, nose and throat irritations. It can also damage vegetation.

The term "smaze" designates a combination of smoke and haze. It, too, reduces visibility.

(o) *Air Cleaning Installation.* Any process or equipment which removes, reduces or renders less noxious air contaminants discharged into the atmosphere.

SOURCES OF CONTAMINATION

The sources of air contamination can be classified in a number of different ways. In this chapter, they will be classified in three categories as follows: combustion sources, process sources and natural sources.

Combustion

Combustible material is basically hydrocarbon. In ideal combustion, reactions with oxygen produce CO_2 and H_2O.

$$C_xH_y + (x + y/4)O_2 \rightarrow xCO_2 + y/2H_2O$$

Insufficient air results in unburned or partially burned hydrocarbon products which may be simpler organic compounds such as acids and aldehydes. Fly ash, smoke and CO are also emitted. Complete combustion requires sufficient O_2, turbulent mixing and a high enough temperature for an adequate period of time and is usually not attained. To promote more complete burning, 50–100 percent excess air is often used, but if this cools the flame the amount of unburned or partially burned material is increased. On the other hand, reliance upon combustion at elevated temperatures creates increasing quantities of oxides of nitrogen (NO and NO_2) from the reaction of atmospheric N_2 with O_2.

A second source of air contaminants is the impurities such as sulfur in the combustible hydrocarbons, e.g., which when oxidized produces SO_2 and SO_3.

Other mineral impurities in most hydrocarbons are left behind as ash particulates along with unburned carbon and can be carried into the air by the draft of the rising gases. Fluorides, which damage plants, are important in this category.

Both anthracite and bituminous coal have long been used for domestic and industrial heating and power production. Even in well-operated furnaces particulates and sulfur oxides are produced. More contaminants are created from bituminous coal since it usually contains more non-combustible material. It requires more excess air to burn properly which produces greater air velocities than for anthracite and increases the amount of solids entrained in the stack gases.

Much of the domestic heating requirement is met by distillate fuel oils such as kerosene and No. 2 which burn cleaner than coal but emit SO_2 and organic acids. Residual fuel oils or the heavier fractions contain more impurities than distillate oils and thus create larger amounts of air contaminants.

Of all heating fuels, natural gas burns cleanest giving rise to almost no solids but proportionally more oxides of nitrogen caused by higher combustion temperatures. Its unavailability in many areas is a drawback.

For comparison, Table 1 shows contaminant emissions per 10^6 Btu's of heat delivered for various fuels.

Much of the air contaminants in a community's atmosphere comes from the burning gasoline and diesel fuel in motor vehicles. Incomplete combustion accounts for unburned hydrocarbons, CO and aldehydes, additives account for lead, nickel and phosphorus compounds, and the high temperatures account for oxides of nitrogen. Diesel fuel, used for larger engines, emits more particulates and odorous compounds than gasoline.

Where no adequate refuse collection is provided, large quantities of refuse are burned in backyard burners. This situation is compounded in autumn by the tremendous amount of leaf burning practiced in many communities. Both backyard trash and leaf burning yield large amounts of organic and particulate pollution, often accompanied by odors.

Refuse is usually collected and reduced by burning in municipal incinerators, by deposition in sanitary landfills, or by burning in open dumps. When combustion is good, municipal incineration tends to produce less organic contaminants than other types of waste burning. Landfill causes little air pollution but some odors. Open burning dumps are like backyard burning, but the pollution is confined to fewer locations.

The relative pollution importance of combustion sources varies with the climate of a region. For instance, in cold climates, fuel for heating may be most important. In warmer climates, motor vehicles emissions may be most important. The social and economic development of a region may also influence this; e.g., on the East Coast of the United States large quantities of residual oil and bituminous coal are burned,

TABLE 1. Contaminant Emissions (in pounds) per 10^6 Btu of Fuel

Contaminant	Bituminous Coal	Anthracite Coal	Residual Oil	Distillate Oil	Natural Gas
Solids	3.68	0.60	0.074	0.071	0.010
SO_2	2.76	0.99	1.554	0.604	0.001
NO_2	0.74	0.79	0.888	0.497	0.200
Hydrocarbons	0.74	0.12	0.037	0.036	0.040
Organic acids	1.10	0.20	0.111	0.107	0.060
Aldehydes	0.07	0.04	0.007	0.014	0.010
NH_3	0.07	0.08	0.015	0.007	0.020

whereas on the West Coast, gaseous and liquid fuels predominate. Mass transportation facilities can also effect the relative importance of pollutants; e.g., many commuters in the New York metropolitan area daily utilize mass transportation to and from work, whereas in some other urban areas the automobile is used almost exclusively.

The potential annual emissions in any area resulting from fuel burning in any area can be estimated readily if the amount of fuel used is known.

TABLE 2. Annual Fuel Use in New York Metropolitan Area[a]

Type	Amount burned
Coal	
Bituminous	1.1×10^7 tons
Anthracite	2.1×10^6 tons
Oil	
Distillate	3.7×10^9 gal
Residual	4.9×10^9 gal
Natural gas	4.1×10^{11} cu ft

[a]Seventeen counties—eight in New York State, nine in New Jersey. Land area 3493 sq. miles. Total population 15,392,000 (January 1, 1965, estimated).

In addition, 3.4×10^6 tons of refuse and garbage are burned yearly in the municipal incinerators, 1.6×10^6 tons in domestic, commercial or industrial incinerators, and 3.2×10^5 tons of refuse in the open. It is calculated that this yields about 1.6×10^6 tons of sulfurous pollutants each year.

Almost 5.3×10^6 tons of CO (about 95% from motor vehilces) and 7×10^6 pounds of lead are emitted annually.

Eric P. Grant estimated that in 1965, prior to inception of controls, over 2×10^7 gal/day of gasoline were being used in California and about 10% was being emitted unburned into the atmosphere.

The density of motor vehicles and their mode of operation in a given area is important. Thus, the number of motor vehicles registered in New York City is less than in Los Angeles County, but the area of Los Angeles County is about 2×10^3 square miles, while that of New York City is 3×10^2 square miles, so that the density of motor vehicles, particularly in Manhattan and Brooklyn, is greater than that of Los Angeles and the amount of contaminants being emitted per square mile is greater. Although there are many complaints of air pollution from the emissions of diesel-powered buses and trucks, the actual pollution from these sources is not large, for diesels are only a small portion of the total number of motor vehicles.

Process

Process sources of air contaminants are divided into operational categories: crushing and grinding, drying and baking, evaporation, chemical reaction and nuclear reaction. Many processes use more than one and quite often they are combined.

(1) Crushing and grinding includes all processes where air contaminants are created from larger particles by physical means such as polishing, abrading, pulverizing and chipping. The contaminants in this case are always particulates, e.g., as in pulverizing operations in cement mills and metal grinding in machine shops.

(2) Drying and baking operations in which a liquid is removed from a solid material. Some solid may be lifted into the atmosphere by the air stream or in handling operations. Heat is generally applied and the liquid either evaporates or is separated by filtration as in a cement kiln or a detergent powder spray drier.

(3) Evaporation is conversion of material to the vapor state from the liquid or directly (sublimation) from the solid state. Vapors, mists and fumes, as previously defined, are created in this manner. Occasionally, process turbulence is so high that solid particles are carried away from the process without undergoing a change of state. Common evaporation examples are iron oxide fumes created in an open hearth furnace and volatilization losses in the petroleum industry. Large losses formerly occurred in gasoline engines where raw gaseous fuel was forced around the piston rings to the road draft tube and to the atmosphere. In new vehicles, the crankcase emission-control system prevents this by returning the unburned gasoline to the induction system.

(4) Chemical reactions include the oxidation or reduction of compounds in a process. A vast variety of chemical reaction processes release products and/or intermediates to the ambient air.

(5) Nuclear reactions include any changes which occur in the nucleus of an atom resulting from radioactive decay, nuclear fission or fusion, or bombardment with neutrons or charged particles. A common natural atmospheric contaminant is radon resulting from the radioactive decay of radium in the soil. A common artificial contaminant is strontium 90 resulting from fission in the atmosphere of uranium during weapons testing (see Sections 5, 8 and 12).

Examples of contaminants commonly found in the atmosphere are given in Table 3.

TABLE 3.　Classification of Air Contaminants

Classifications	Examples
(1) Aerosols	
(a) Solid or liquid particulates	Carbon, fly ash, metal oxides, fumes, mists, dusts,
(b) Pollens	Ragweed Pollen
(2) Gases and vapors	
(a) Sulfur compounds	SO_2, SO_3, H_2S, mercaptans
(b) Nitrogen compounds	NO, NH_3, NO_2
(c) Halogens and their compounds	HF, HCl, F_2, Cl_2
(d) Organic compounds	Hydrocarbons, aldehydes, organic acids
(3) Radioactive substances	Radioactive gases and aerosols
(4) Smoke	Carbon particles
(5) Odors	
(6) Others (in particular instances, sizeable emissions of special and significant contaminants may not fit clearly into one of the above categories or may require classification in more than one category.	

Any of these contaminants may be used in an individual community as a contamination index. However, since the nature of pollution varies, no contaminant can be a universal index. For instance SO_2, CO and smoke shade (an indication of suspended material in the atmosphere) have been used as indices of total contamination in the New York metropolitan area. Such a combination undoubtedly could not be used as an index of total contamination in the Los Angeles atmosphere.

Natural

Natural sources, often overlooked in evaluating an area's air pollution problem, are important because in some areas (generally rural) they contribute more than artificial ones. Even in urban areas these sources constitute a significant "background." Examples are volcanic ash and fumes, tree pollen, turpenes, dried leaf dust, dried insect parts, dust from cultivated and uncultivated land and from roads, ozone (as a result of lightning), and salt particles from the ocean.

POLLUTION INFLUENCING FACTORS

Many factors influence the nature and extent of a community's air pollution problem. Some factors are seasonal, i.e., related to heating or leaf burning; some are related to economic cycles, such as diminished pollutant emissions during periods of low industrial production;

others are dependent on technological efficiency, illustrated by higher emissions from improperly operated or equipped individual motor vehicles, buses or trucks. But the 3 major factors are: total population and population density, topography, and weather.

(a) Population. Pollution is a product of people from heating, waste disposal, transportation and industrial activity. The more concentrated the population and its activities, the greater is the resulting air pollution.

(b) Topography. This is important because it influences the movement of pollution from the source to the environment. It helps determine whether a specific region may be subject to intense localized air pollution, chronic air pollution, or both by affecting wind speed, wind direction and the temperature profile through air drainage and radiation. There is generally a greater diurnal temperature variation in valleys than in surrounding hills. Cold air tends to drain away from the hills and settle in the valleys, especially during the night. Heating and cooling of a valley may occur more rapidly on one side than on the other, creating local eddy currents. Fumigations may result when high concentrations of contaminants in the upper air are brought down to the ground surface due to local air movement created by topographic influence.

Los Angeles is typical of unfavorable topography. It lies in a basin surrounded by mountains tall enough to inhibit horizontal air movement when inversions occur. Thus, contaminant concentrations increase until the inversion layer rises above the mountains.

Topography is sometimes a favorable factor, e.g., the flat terrain surrounding the New York metropolitan area permits the huge amount of air pollutants generated in the city to be blown out, diluted and dispersed over the Atlantic Ocean. Rough terrain causes mechanical and convective turbulence; smooth terrain does not.

(c) Weather. Unequal solar heating of the earth's surface in different latitudes causes the movement of large homogenous air masses which change the weather along their route and are themselves modified by passage over the earth's surface.

Low and high pressure areas accompany the movement of air masses. Low pressure areas are characterized by moderate wind speeds, small diurnal temperature changes, widespread cloudiness, and rain. High pressure areas are usually accompanied by light winds, clear skies, and large diurnal temperature changes.

The air within a high pressure area is dense in relation to the surrounding air and tends to flow downward and outward. In a low pressure area, the air is lighter and flows upward and outward. On a macroscale this creates an air flow pattern between adjacent high and low pressure areas. There is generally a good dispersion of air contaminants when an area experiences low pressure and poor during high pressure periods.

On a micro-scale, daytime solar heating of the earth and nighttime radiation cooling cause heating and cooling of layers of the atmosphere adjacent to the earth's surface. Thus after dark the low lying air is cooler and tends to remain in place forming a temperature inversion which will persist unless modified by strong winds, cloudiness and frontal movements.

Air will decrease in temperature and density as it expands; this decrease in temperature with increase in altitude is known as the lapse rate. The dry adiabatic lapse rate is $5.4°F/1000$ ft ($1°C/100$ m). When this lapse rate is modified by solar warming of the earth and the actual temperature drop with height exceeds the normal lapse rate, the atmosphere is said to be unstable and the rate of diffusion of contaminants is good. The greatest turbulence usually occurs on hot, sunny summer days.

When the actual drop in temperature is less than the adiabatic lapse rate, a stable atmospheric condition exists. When temperature increases with height or warm air overlies cold, denser air, an extremely stable or inversion condition prevails. In a stable atmosphere, a volume of air displaced vertically tends to return to its starting point. Under these conditions, contaminants remain concentrated in a fairly compact volume.

Almost all air pollution incidents have occurred as the result of a stagnating high pressure air mass accompanied by severe nocturnal inversions. This occurs frequently over the eastern United States, particularly in the fall of the year during "Indian Summer" when skies are blue and fronts are almost stationary. This is when heavy concentrations of contaminants are noted close to the ground surface.

Cloud cover affects the stability of the air by reflecting or absorbing incoming short wave solar radiation, or by reflecting back to earth any outgoing long wave terrestrial radiation (heat energy). The first tends to keep the maximum temperature lower and the second keeps the minimum temperature higher beneath the cloud layer.

Rain, snow and other forms of precipitation act to "wash out" contaminants from the atmosphere, but this process is generally inefficient.

HEALTH ASPECTS

Gathering proof that air pollution causes or contributes to disease is complicated by a number of factors. For instance, the nature and extent of pollution of outdoor air constantly changes. The human, except in industry, is rarely exposed to one contaminant at a time. Contaminants may combine to produce an effect different than either would produce alone. The evidence that air pollution contributes to disease is mounting rapidly.

The health effects of air pollution fall into three categories: (a) the highly dramatic incident usually associated with a fairly clearly defined source or source complex and associated with adverse meteorological conditions, e.g., the Meuse Valley in Belgium in 1930, Donora, Pennsylvania in 1948, and Poza Rica, Mexico in 1950; (b) generalized or chronic air pollution conditions to which both industrial and domestic sources contribute air contamination, e.g., the fogs of London and the smogs of Los Angeles and New York; (c) the more specific relationships between air pollution and respiratory disease, heart disease, allergies, and lung cancer.

Specific Incidents

(1) Meuse Valley. From December 1–5, 1930, the Meuse Valley in Belgium was blanketed by a heavy cover of smoke and fog. Many people suffered respiratory tract irritation and 60 died. In this highly industrialized area, the high level of air pollution resulted from an intense weather inversion that persisted for 4 days.

(2) Donora, Pennsylvania. In 1948, public health officials in the United States were shocked into a rude awakening. On October 26, in Donora, the weather was foggy and raw, and the wind velocity was low. This condition of atmospheric stability continued through October 30. In the valley of the Monongahela River, the Donora episode caused illness to 43 percent of the population, and 20 died (17 of them on the third day) in a community of less than 15,000. The principal symptoms were first of the respiratory tract and second of the gastrointestinal tract. The severity of illness appeared to be independent of sex, race, occupation, length of residence in the area or physical exertion at the time of the incident. A direct relationship was noted between symptom severity and age. More than 60 percent of those over 60 reported some symptoms due to the smog; half of these were affected severely. The fatalities ranged from 52–84 years of age, with an average age of 65 years.

An impressive fact, learned from a survey conducted 10 years later, was that the mortality rate for the group reportedly ill during the smog was greater for the 10-year period than for the group not affected.

(3) Poza Rica, Mexico. This was the site of an accidental release of H_2S on November 24, 1950, which killed 22 people and sickened more than 200. Unburned H_2S surged through a clogged burner and escaped. An intense weather inversion occurred about 4 a.m. so that the escaping poison was trapped in the valley in which the town is situated. Some 320 persons were affected and 22 were killed. The Poza Rica incident differs from those of Donora and the Meuse Valley in that H_2S specifically caused the illness and deaths, while in the other incidents no specific cause has been established.

Generalized Incidents

(1) London. In 1952, during a fog and temperature inversion from December 5 through 9, an unusually large number of deaths occurred and many were made ill. Deaths between 3500 and 4000 in excess of normal were reported. The mortality rate remained elevated for a number of weeks and 8000 additional deaths may have occurred because of this incident, many from respiratory illnesses. The deaths from pneumonia were four times higher, and from bronchitis nine times higher, than normal. While the increase in mortality rate affected all age groups, those over 45 were affected the most. As to the very young, newborn death rate was doubled, and that for infants aged 1–12 months more than doubled. In fact, those least able to bear additional stress were affected most severely.

In retrospect, other episodes of somewhat less severity occurred from December 9–11, 1873, January 26–29, 1880, December 28–30, 1892, November 26 to December 1, 1948, and again in 1956, 1958, 1960 and 1962.

Such air pollution incidents are not limited to London. In 1928, in February, Manchester had 13 days of smog and the death rate rose to 34.4 per thousand by the end of the month, as compared with the annual rate of 13.2 per thousand. Episodes are recorded for Liverpool, Birmingham, Leeds, Glasgow and other cities in Great Britain.

(2) Los Angeles. The Los Angeles basin has received the most notoriety in the recent past as an area of high and repeated air pollution. However, it was noted as far back as 1542 that smoke tended to remain in the basin. It is now known that the meteorological and topographical features of the Los Angeles area are conducive to the retention of air contaminants for relatively long periods. About 60 days per year, Los Angeles is subject to smogs which cause reduced visibility and nose, throat and eye irritation. These smogs also result in considerable plant damage.

It has been noted that similar conditions, but to a lesser extent, occur in the San Francisco Bay area and in other regions of the West Coast. The Los Angeles County Air Pollution Control District, organized in 1947, has conducted an intensive campaign to reduce emissions to the atmosphere. Until recently, however, there was no technology for controlling one of the most vexing sources in the Los Angeles basin, namely, automobile emissions. The hydrocarbons that discharge to the atmosphere from these mobile sources undergo a complex reaction in the presence of sunlight, to produce a photochemical smog containing ozone and peroxyacetyl nitrate (PAN).

(3) New York City. From November 15–24, 1953, under a high pressure air mass, there was a pronounced increase in the concentration of air pollutants, particularly from November 18–22. An analysis of deaths in New York city for the 5 years before and after this incident showed some 165 deaths above average for the period, with most occurring in 3 age groups: 65 and over, 45–64, and under 1 year. In all probability, additional deaths occured elsewhere in the area covered by this stagnant air mass but there are no data available. From October 14–19, 1957, there was a similar occurrence. The area experienced another stagnating high pressure area with a period of intense air pollution from October 15–17. This incident occurred during an influenza epidemic. Normally the number of deaths per day in New York city during October averages around 200. During this epidemic the average rose to 277. On October 16, the day of the highest air pollution in this incident, the number of deaths rose to 360, the average for the 3-day period October 15–17 being 330. No excess deaths were reported in 1962 in New York City under meteorological conditions similar to those in 1953. Smoke shade in 1962 was reported to be lower; SO_2 higher.

For 3 days beginning November 23, 1966, the East Coast experienced another intensive pollution episode. It was estimated by Dr. Leonard Greenburg that between 152 and 168 excess deaths occurred in New York City alone. Though the incident lasted only 3 days, its effect, as measured by excess deaths, lasted an additional 4 days.

Long periods of high air pollution occur in New York relatively infrequently, but when they do occur they have a marked adverse effect on the health of the residents. Metropolitan New York has intense inversions lasting 36 hours or more, only about one-tenth as often as Los Angeles.

(4) New Orleans. On October 26, 1955, during a smog incident the number of asthma attacks increased sharply; two persons over 50 years of age died and some 350 persons went to hospitals for treatment. Some relationship seems to exist between the attacks and winds from the south and west southwest, since these incidents have occurred on more than one occasion.

Air Pollution and Disease

A direct relationship has been established between high air pollution and eye, nose and throat irritation, reduction in visibility, and reduction in the amount of visible and ultraviolet light. Indisputable increases in mortality have occurred during high pollution incidents, as already described. But the relationship has not been as well established between air pollution and certain disease categories such as: (1) heart disease, (2) lung cancer, (3) respiratory allergies, (4) respiratory infections, and (5) chronic respiratory disease. Available data are often obscured by differences in disease nomenclature, instrumentation, statistical methods, and insufficient documentation. Chronic bronchitis in Great Britain, for example, does not generally denote the same disease entity as in the US. Furthermore, personal air pollution (smoking of cigarettes) complicates analysis, as do occupational exposures and socioeconomic status. From both experimental and epidemiological studies there is every reason to believe that some individuals are more sensitive to air pollution than others.

(1) Heart Disease. Sulfates in areas of high population density are found in concentrations known to make breathing difficult. Ozone, by scarring the linings of the respiratory tract, can also make breathing difficult. Such effects can lead to overburdening of the heart and it is not surprising that deaths from heart disease almost tripled during the London smog disaster of 1952.

Carbon monoxide in relatively low concentrations interferes with the transport of oxygen by hemoglobin. It has been estimated that continued exposure to 30 ppm of CO would tie up about 5 percent of the body's circulating hemoglobin, making it unavailable for transport of oxygen. While most people would be unaffected by a small change in the oxygen-carrying capacity of the blood, a few individuals with borderline involvement with diseases of

heart, lungs or blood vessels may be seriously affected. Almost all CO in urban atmospheres is generated by motor vehicles. The role of the cigarette in CO exposure, however, is an important factor.

Nitrogen dioxide can also scar lung tissue, placing added burden on the heart.

(2) Lung Cancer. Carcinogenic substances are produced during the combustion of any organic matter—even simple substances like natural gas. While the role of general air pollution in increasing the risk of lung cancer is considered less than the role of cigarette smoking, nevertheless, the highest incidence of lung cancer usually occurs in areas of high atmospheric pollution. It has been estimated that in the Liverpool area of England, 50 percent of the lung cancer deaths result from smoking and 35 percent from air pollution.

Data substantiate the conclusion that the excess of urban lung cancer deaths over rural lung cancer deaths is due to air pollution. It has been shown that British immigrants to New Zealand and the Union of South Africa have a higher lung cancer death rate than native born New Zealanders and South Africans, even though the lung cancer death rate of these immigrants is less than that found in England.

The lung cancer death rate is higher for males than for females, indicating that perhaps other factors such as occupational exposures are involved. On the other hand, ozonized gasoline containing benzo-(a)-pyrene has been demonstrated to cause lung cancer in mice previously infected with virus influenza in about the same male to female ratio experienced in the human population.

(3) Respiratory Allergies. About 12 million Americans are reported to suffer from allergic symptoms such as asthma and hay fever resulting in 33 million bed-disability day annually. This is primarily due to air-borne pollens, especially of ragweed which is an important natural pollutant ignored by most air pollution agencies.

Asthma has been linked to man-made pollution, although such illness is reported due to air contaminants when no pollens are present in the atmosphere. Such illnesses are probably due to other allergic reactions and may occur even when concentrations of such air contaminants are very low. Asthma has been linked to man-made air pollution in Nashville, Tennessee; Pasadena, California; and New Orleans, Louisiana, to mention a few places. An unusual type of asthma, referred to as "Yokohama Asthma," was a major cause of illness among the U.S. Armed Forces personnel in the Tokyo-Yokohama region of Japan. This resulted in an increased percentage of permanent impairment of pulmonary function, disability, and death among those affected. In these latter cases, the exact pollutants causing the allergic reaction have yet to be determined.

(4) Respiratory Infections. The incidence of "colds" and respiratory disease in general has been related to air pollution levels as indicated by suspended particulates, dustfall, and sulfate concentrations. During the 1962 smog episode in New York City, upper respiratory infections increased significantly among inmates of homes for the aged.

The London fogs are representative of the effect of air pollution on respiratory diseases. The average number of deaths from pneumonia was 31, and from bronchitis, 51 for the 4 weeks ending November 29, 1952 (the weeks immediately preceding the December 5–10, 1952 air pollution incident). For the week ending December 13, 1952 (the week including most of the days of the fog and the days immediately following) there were 160 deaths from pneumonia

and 704 from bronchitis, a five- and fourteen-fold increase, respectively.

(5) Chronic Respiratory Disease. Studies of chronic disease rates in Great Britain in relation to air contaminant levels have demonstrated that there is an impressive relationship between morbidity and mortality rates from chronic bronchitis and air pollution. In Great Britain, chronic bronchitis is the third leading cause of death and the leading cause of disability. It is often found combined with bronchial asthma and emphysema. In the United States, the evidence is not as clear-cut. The rates for chronic bronchitis are, in general, lower than in Great Britain. In this country, the presence of bronchitis is often ignored and the presence of emphysema is noted exclusively or considered a more important diagnosis. It has been shown, however, that patients suffering from chronic bronchitis do have more severe symptoms during periods of high air pollution.

Dr. Warren Winkelstein, reporting on a study made in Buffalo, New York, has shown that

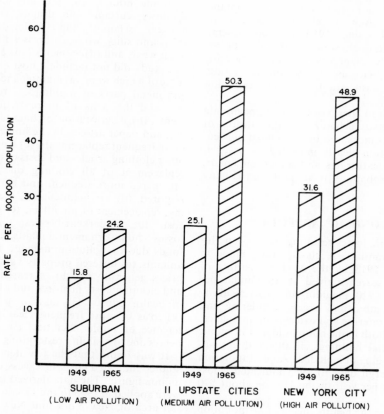

Figure 1. Pneumonia Death Rate by Level of Air Pollution, 1949 and 1965.

chronic respiratory disease death rates are related to air pollution levels as indicated by suspended particulates. One example is shown in Table 4.

TABLE 4.   Average Annual Death Rates per 100,000 Population from Chronic Respiratory Disease according to Economic and Air Pollution Levels, and Age: White Men 50–69 years, Buffalo and Environs, 1959–61

Economic Level	Air Pollution Level				Suspended Particulates (μg/m^3/24 hr)
	< 80	80–100	100–135	> 135	Total
1 (low)	—	0[a]	126	188	133
2	64	75	96	105	84
3	—	65	51	103[a]	64
4	35	47	114	—	52
5 (high)	42	63	0[a]	—	50

[a] Rate based on less than 5 deaths.

Short-term impaired lung function among school children in polluted areas of Japan has been noted during periods of heavy air contamination. The seven- to eightfold increase of deaths in the US since 1950 from emphysema is alarming, although some of this can be attributed to better diagnosis. The urban rate for emphysema is twice as high as the rural rate, indicating an association with air pollution levels.

A study made in New York state of pneumonia death rates versus urbanization and degree of air pollution as measured by suspended particulate concentration clearly indicated an association between death rates and an urban factor, which in all probability is air pollution (Figure 1).

ECONOMIC ASPECTS

The economic losses resulting from air pollution are substantial. A classic study made in Pittsburgh in 1913 by the Mellon Institute of Industrial Research arrived at an annual per capita figure of $20, including the costs resulting from poor combustion, excess laundering and dry cleaning, interior and exterior maintenance, and lighting, both in the household and in public and commercial buildings; health and agricultural effects were not included. An extrapolation of the Pittsburgh data to 1965 indicated a loss of $12 billion per year in the US, or about $70 a person.

Even though the health effects of air pollution can be demonstrated, individuals tend to think that this is something which happens to someone else. The public might be more highly motivated to control air pollution if it was convinced that its everyday living costs would thereby be reduced.

Estimates indicate that air pollution in the US can be controlled to acceptable levels by spending about $17 per person annually. If such estimates are accurate, we can save about $9 billion each year by spending $3 billion.

The $70 per person per year cost of air pollution is admittedly a rough estimate and a crude average. This rough estimate may be far too low and the crude average hides some startlingly high figures.

Irving Michelson has reported on a study made in 1960 in the upper Ohio River Valley of several cities with high air pollution levels. The extra costs involved in living in high air pollution areas were determined by comparison of costs for the same items in a nearby community having only average urban air pollution levels. The study covered only the extra costs in maintaining: (a) exterior walls and windows of frame houses, (b) the interior walls and windows, curtains and drapes, rugs and upholstery, (c) laundry and cleaning costs of men's shirts and suits, women's dresses and children's outer wear, and (d) women's hair and face care. The study did not include a host of other items, some of which were: (a) maintenance of masonry and metal parts of buildings, screens, gutters, etc., and the costs of more frequent replacements, (b) deterioration of linens, leather, rubber, and paper articles in the home and costs of more frequent replacements, (c) laundering of all other clothing articles and costs of more frequent replacement of all clothing due to corrosive attack and more frequent and harsher cleaning required, (d) car maintenance, including washing, replacement of air filters, (e) extra lighting costs due to darkened skies, (f) agricultural losses—home ornamental plants, higher costs of foods due to pollution damage to crops and animals, (g) lowered property values, (h) higher prices generally due to increased maintenance and cleaning costs in stores and offices, (i) costs of health effects, (j) loss of esthetic values, (k) costs of more frequent cleaning and maintenance of public facilities, (l) transportation delays due to visibility restrictions.

It was found that the few items in the cities studied indicated costs in excess of the $70 per annum figure. The data derived from this study were applied in 1965 to the 17-county metropolitan area of New York and New Jersey. This application showed that the cost per person per year in this area was approximately $200. Naturally, the average per capita cost varied considerably from one part of the area to another, depending on the air pollution levels, with a range from about $30 to over $350 per person per year.

Other economic losses not even included in the foregoing discussion are highly significant.

(a) Poor Combustion: The losses attributable to improper combustion are almost beyond calculation. Fuel is consumed to obtain heat and power. If part of this fuel is not burned or is incompletely burned it produces lesser amounts of heat and thus is wholly or partially wasted. In addition, contaminants are emitted to the atmosphere which result in other economic damage. Such losses can be very expensive. Improvement in the efficiency of fuel consumption by power plants has reduced the amount of coal required to produce 1 kilowatt-hour of electricity from 1.17 pounds in 1951 to a little less than 0.7 pound in 1967. Such losses are serious, not only to large consumers but to the householder as well. As much as 25 percent of the fuel bought for heating his home may be lost.

Perhaps the greatest fuel waste results from the operation of motor vehicles which can approach 10 percent.

(b) Plant Damage: Losses resulting from damage to vegetation and crops are very severe in certain regions with high air pollution. Some crops are more susceptible to damage than others. Spinach, endive and mustard are more readily damaged by Los Angeles smog than celery, cabbage and cauliflower. Some air contaminants are more likely to cause plant damage than others. Tomatoes and orchids are adversely affected by minute amounts of ethylene, an air pollutant from automobile exhaust. Fluorine compounds may also damage plants; thus gladioli are affected by fluoride concentration on the order of parts per billion.

A wide range of plant life from lichens to trees has fared badly in our urban centers. The failure of lichens to grown in cities has been attributed to air pollution.

(c) Corrosion: Corrosion is the result of slow chemical and electrochemical reactions between a metal and its environment. The oxidation or rusting of iron is the most common example, but there are many others.

Corrosive damage is due to certain agents which may be grouped as: (1) oxidants, including O_2 and O_3, nitrogen oxides, HNO_3 and organic peroxides; (2) acidic materials, including SO_2, H_2SO_4, H_2S, HCl, CO_2, H_2CO_3 and tar acids; (3) salts, including $(NH_y)_2SO_4$ and other ammonium salts and NaCl, and (4) alkalies. The potential for corrosion of the latter two groups should not be minimized. A common cause of corrosion is the use of salt in the winter for melting and removing snow. Most metals except Al and Zn are resistant to attack by alkalies.

(d) Deterioration of Physical Structures: The term "corrosion" in the strict sense is limited to action on metals, but equally important is deterioration of building materials such as stone and masonry, textiles, protective coatings, rubber and leather.

Stone building materials may be placed into two groups, namely, resistant and nonresistant to the destructive action of air pollutants. The acids in the air, such as H_2SO_4, can be very destructive. Sulfuric acid attacks carbonate-containing stone such as limestone and converts the $CaCO_3$ to $CaSO_4$ which is soluble in rain water, causing pitting of the stone. Encrustations may be formed in the building materials because of crystallization of soluble salts on evaporation of water. These eventually break away from the stone and leave pitted surfaces. Porosity of the stone is caused by analogous reactions. Masonry is subject to similar attack.

Among the little known damages caused by air pollution are the irrevocable losses sustained by art objects. It is not often possible to make adequate restitution. Once an art object is damaged, its value is lost forever. A classic example is the ancient Egyptian obelisk which has stood in Central Park for 90 years. The boundaries of the large quartz grains in this granite obelisk have been eaten away. In the winter, water penetrates this network of cracks and upon freezing, flakes off pieces of granite. Cleopatra's Needle has deteriorated more since it was brought to New York in 1881 than during its previous 3000 years in Egypt.

Lead-containing paints can be damaged by H_2S which turns the paints black. Deposition of particles on wooden and masonry structures requires frequent cleanings. Rubber cracks because of exposure to O_3 and other oxidants in the air.

There are other serious losses attributable to air pollution which are not readily apparent. For instance, SO_2 dissolved by cooling water from the air may damage air conditioning systems.

Fluorosis of animals has become fairly well defined. While fluorine toxicosis is usually produced by a less direct process than inhalation, the chemical agent is airborne. The fluoride-containing effluents from industries may contaminate the forage crops which are subsequently ingested by animals.

SOCIAL ASPECTS

It is difficult to separate social from economic aspects of air pollution. Certainly, people take pride in living in a clean community. Conversely, communities or parts of communities which have heavily contaminated air are usually run down and ill kept. The loss of property values associated with such decline is certainly an economic effect.

Michelson's data strongly suggest that many persons would have spent more on keeping clean if they felt they could afford it. This was noted particularly among the lower income families. It is impossible to attach a dollar value to this cost to human dignity.

No one likes to live in a community where odors constantly or frequently are present. Even odors which are pleasant to the average individual, such as those associated with coffee roasting or with the making of chocolate, become objectionable in a community exposed to them constantly. The condition is worse when due to basically objectionable odors such as H_2S or mercaptans. Here again there may be an association with economic effects, since the commerce and trade of a community where odors are present are bound to suffer.

Society is beginning to demand no less than an absolutely clean stack. People object to the discharge of smoke or fly ash from chimneys or the discharge of smoke from jet aircraft, even when such discharging may contribute little to the total atmospheric burden.

From another viewpoint, many social practices must be altered if air pollution control is to be achieved. The burning of leaves in the fall of the year, for instance, formerly considered a delightful social pastime, contributes huge quantities of contaminants to the atmosphere. Likewise, the burning in an open dump of solid wastes collected from the community contributes its share to the atmospheric burden.

AIR QUALITY MANAGEMENT

Air quality management denotes a reasonable use of the air resource to dispose of society's wastes. Thus use must be in a manner compatible with, but to an extent no greater than, the capability of the ambient air to tolerate the use without undue detriment to man and his environment. Air pollution measures instituted for the preservation of pristine air for preservation reasons alone are a waste of a valuable resource, particularly when these measures would not produce a significant reduction in the hazard to man or to those elements in the environment for which man has a real concern.

Utilizing this concept, good air quality management programs—not just air pollution control programs—are indicated.

Air quality management involves a number of steps which fit quite neatly into three categories: (1) the establishment of air quality standards, (2) surveillance and monitoring to determine if the standards are being exceeded, and (3) the application of air pollution controls, when and where indicated, for the benefit of society.

The ambient air over a region represents a resource capability which varies from place to place in its volume, climate, meteorological patterns, geographical location, and places of origin. These variations determine the relative worth to a region of its air and the uses to which the air can be put under a balanced air resource management program. The ambient air over a region is therefore a resource which varies in value from region to region, just as do mineral, water, climatological and other resources. It is a resource which a region has a right to guard, protect, and use competitively with other regions to its advantage, as it does with other resources. Control officials, industry, the general public, the news media, and elected officials have a responsibility to acquire knowledge about the state of the art of air pollution control, the real significance of air contamination and the benefits versus the risks of air pollution controlled or uncontrolled, before making potentially unreasonable demands upon each other.

AIR QUALITY STANDARDS

Air quality standards, objectives or goals must be determined, first, on the basis of health requirements, and, *second*, on the basis of a reasonable use of the air resource in accordance with the economic and social needs of an area.

The establishment of such standards is no easy task, for a number of reasons: (a) There is not universal agreement even for specific air contaminants as to what concentration will cause adverse health effects in a community. (b) Because two or more contaminants present in the community's atmosphere may have a greater physiological effect than any one contaminant acting alone, the picture becomes even more clouded. Data are not now available (and may never be) upon which to establish precise levels at which health effects occur. (c) Man is a complex organism. The point at which an adverse physiological effect begins varies with the individual. Is the standard to be set for the most susceptible individual or group of individuals? Admittedly, standards must be arbitrary to some extent, based on available data and best judgment. But, is this not true of any standard? If a standard is set, there is no reason why this must remain unchanged forever. A standard can be revised as more evidence becomes available. Maximum allowable concentrations for industrial exposures have been in use for many years and are constantly undergoing change as new data are developed.

When it comes to establishing standards based on social and economic aspects, the problem becomes even more complex. If the standards are too stringent in an industrial community, for instance, it may place the industries in that area in an uncompetitive position, forcing them to close, with a consequent loss of jobs. On the other hand, if the standard is too lenient, in say a recreational area, this too can have an adverse effect on the economy, resulting from a loss of tourist and vacation business.

The fact remains, however, that a standard or goal must be established before a control program can be administered in a practical and reasonable manner.

SAMPLING AND ANALYSIS

Atmospheric sampling is performed to determine: (1) the nature and amount of pollution in a region, (2) the geographical distribution of contaminants in the atmosphere, both in terms of nature and concentrations, (3) how serious the pollution problem is, (4) whether or not established air quality standards are being achieved, (5) what priority should be placed on the control of various problems, and (6) the relationship between adverse effects and the concentrations of various contaminants. In the last instance, air sampling operations are usually conducted as part of a special study.

Stack or effluent sampling is performed: (1) to determine whether a given emitter is complying with the emissions limitations established by official agencies and (2) to provide routine data for process operating guidance.

Some of the basic problems involved in the collection of particulate matter from the outside air are: (a) efficient separation of materials in quantities sufficient for weighing, or chemical or other analysis, and (b) accurate measurement of

sampled air volume. The volume of air sampled must be measured with a precision equal that required of the results. In stack sampling, the contaminant concentration in the effluent stream is usually greater, but the effluent volume being sampled may be hard to measure accurately.

The collection and analysis of gaseous samples also presents problems and may involve expensive equipment. It is sometimes difficult to measure extremely low concentrations of gases found in the outdoor atmosphere, particularly when other substances are present which interfere with the analysis.

Equipment which continuously samples and analyzes a gas stream is more expensive, but reduces the man-hours required for sampling. Continuous recording equipment has advantages because it records momentary changes in pollutant intensity. It also records peak concentrations which might otherwise escape attention and which may be important in evaluating effects of air contamination.

Atmospheric Sampling

It is virtually impossible to locate a single sampling station which will produce results representative of a region's atmosphere. Atmospheric sampling is expensive in terms of equipment and personnel time. To be of value, the data must be analyzed, and this should be considered before any sampling network is set up. Because of the many factors involved, any sampling network represents, at best, a compromise between the money available and the data needed.

Some of the important factors which must be considered in establishing a sampling network and choosing sites for samplers are:

(1) Size of region;
(2) Horizontal distribution (grid or radial distribution of sampling sites, etc.);
(3) Meteorology (upwind and downwind locations from specific sources, etc.);
(4) Vertical distribution of equipment (street level, use of helicopters);
(5) Topography (hills, valleys);
(6) Adjacent obstructions (trees and buildings may influence collections);
(7) Availability of reliable power;
(8) Accessibility of site (snow, locked doors on weekends, etc.);
(9) Personnel available to conduct sampling program and to analyze samples;
(10) Cooperation of other agencies and individuals;
(11) Amount and type of equipment available;

(12) Expected duration of sampling program (people may tolerate short-term operation of noisy equipment but object on a long-term basis);

(13) Problems associated with sampler operation (noise, frequency of needed calibration);

(14) Public sentiment;

(15) Space requirements;

(16) Protection of equipment and site (elements, vandalism, theft);

(17) Possible hazards involved (electric motors in explosive atmosphere, fire, solution spillage, storage of cylinders of gases under high pressure).

Micro-meteorological conditions also influence sampling results. Pollutant concentrations at gound level depend to a great extent on wind velocity and direction and on atmospheric stability. It is relatively easy to measure wind direction and velocity, air turbulence and air temperature, but it is extremely difficult to measure vertical temperature profiles.

It is not possible to dicuss in detail within the bounds of this chapter all of the sampling equipment in use today. Only a brief description will be given of the more common instruments.

The methods used for sampling and analysis of the outdoor air may be placed into six major groups. These are methods to determine: (a) suspended particulate matter, (b) settled particulate matter, (c) gaseous contaminants, (d) reduction in visibility, (e) odor, (f) radionuclides.

(a) Suspended Particulate Matter. Various methods have been used for the collection of particles suspended in the atmosphere. Most depend upon filtration for the collection of a sample, but some employ electrostatic precipitation and others use thermal precipitation. The period of sampling may vary from a relatively short time to 24 hours or more. The volume of air sampled in such procedures also varies over a wide range—from a fraction of a cubic meter to over 100,000 cubic feet of air. Various types of filters used to collect the sample may be classified as: (1) accordion-pleated circular filters, (2) matted filters, (3) filter papers corresponding to Whatman No. 4, (4) fast flow filter papers, (5) fiber-glass filter webs, and (6) membrane filters.

(1) High Volume Air Samplers. An electrically driven centrifugal fan draws air to be sampled through a filter paper, most commonly a glass-fiber filter web, at a rate of about 60 cfm. The filters are weighed before and after exposure so that the suspended particulate

matter can be measured in micrograms per cubic meter of air. The usual sampling period is 24 hours, or one calendar day. A limited number of chemical analyses can be made on the collected material, such as benzene solubles, nitrates and sulfates. However, chemical analysis of particulates collected on filters must be approached with a great deal of caution because the filter itself may contain substances similar to those being analyzed for.

(2) Soiling. The degree of blackness of a spot or trace produced by filtering a known volume of air through a known area of filter paper is known as soiling. The instrument usually used for this measurement is the AISI (American Iron and Steel Institute) sampler.

The blackness or opacity of the stain produced can be measured either manually or by means of more sophisticated photometric and recording equipment. It can be measured either by reflectance or by reduction in light transmission. Light reflectance is usually expressed as RUDS (reflectance units of dirt shade). These are arbitrary units in which the reflectance of clean filter paper is set at 100 on the photovolt reflectance meter. In this system, a dirt shade of zero is "absolutely" clean and a dirt shade of 100 is "absolutely" black. Values above 100 are theoretically possible, but have not been found in practice.

Light transmission is usually expressed in terms of COH units (an abbreviation for coefficient of haze). A COH unit is defined as that quantity of light scattering solids producing an optical density of 0.01 when measured by light transmission. Subjectively, these units imply that 1 COH represents a brilliantly clear day, whereas 10 COH represent a day of very low visibility. Because of variation in density of filter papers and a need for frequent zeroing, there is a trend to the use of RUDS rather than COH units.

(3) Dust Count. A known value of air can be drawn through a membrane filter. The filter is then placed on a slide, a drop of cedar oil is added, and the number of particles present is counted; either an oil immersion objective is used, or the slide is placed on a microprojector and a count is made with the aid of this instrument. Dust counts obtained by this counting procedure are approximately ten times larger than those obtained by counting impinger samples, an instrument commonly used in sampling industrial atmospheres.

(4) Adhesive Paper. Papers mounted on a cylinder are sometimes used to obtain information relative to the directional origin of con-

taminants. The cylinder is mounted vertically and the direction is noted on the paper.

(5) Nuclei Counters. Condensation nuclei counters are available which give a fairly continuous record of the number of condensation nuclei present in the atmosphere. In this instrument, a sample of air is first passed through a humidifying chamber where its relative humidity is raised to 100 percent. Next, the sample is drawn into a cloud chamber and expanded adiabatically. The result of this rapid cooling is a supersaturation of the air (relative humidity about 400 percent). Condensation of the water vapor in the sample occurs on the nuclei present and the resulting droplets will grow large enough to scatter light. The amount of light scattered is proportional to the number of water droplets in the sample and is measured by a photomultiplier tube.

Concentrations of condensation nuclei can be obtained which range from $10^1 - 10^7$ particles/cc. The smallest nuclei counted can be 1^{-3} μ radius.

Once the cycle is complete and the count recorded, the instrument automatically repeats the operation to produce a series of condensation nuclei counts.

(b) Settled Particulate Matter:

(1) Sootfall. The measurement of sootfall or dustfall is perhaps the most widely used procedure for measuring air pollution and has been in use for about 50 years. Open-topped, wide-mouthed, flat-bottomed, straight-wall jars are the preferred collection vessels, but generally commercially available 1-gallon jars, having a mouth diameter of about 4 inches, are used. Many variations of the dustfall jar have been developed but have not found wide usage.

The jar is used either dry or filled to one-half its volume with water, to which is added in the winter time antifreeze such as isopropyl alcohol to prevent freezing, and in the summertime, a fungicide. The jar is usually exposed for a period of 1 month, after which it is covered and replaced by another container and returned to the laboratory. The settleable particulate matter collected is reported as milligrams per square centimeter per 30 days. The collected material can be analyzed for insoluble solids, soluble solids, insoluble ash and soluble ash, and numerous chemical elements and radicles.

(2) Pollen Count. The Durham pollen sampler is another instrument in wide use. A greased microscopic slide is mounted upon a stand protected with a weatherproof top. The sampling period generally is 24 hours, and the pollen count which is obtained microscopically is reported as grains per square centimeter. The slides upon which the pollen is collected are usually coated with a sticky material.

(c) Gaseous Contaminants. The principal gaseous inorganic air contaminants are: SO_2, SO_3, H_2S, CO, CO_2, NO_x, NH_3 and O_3 (oxidants). In special instances, HCl, HF and HCN can be important. The aerosols fomed by the salts of these compounds can be included in this group. Chief among these are H_2SO_4 and some of its salts like $(NH_4)_2SO_4$, $CaSO_4$, $NaCl$, particularly in marine atmosphere and where rock salt is used for snow removal, and some carbonates. In limited instances, fluorides, phosphates, and cyanides may be important air contaminants (Section 12).

The organic gases result chiefly from improper and incomplete combustion of fuels. Evaporation and volatilization of liquid fuels also contribute. Each time a tank full of gasoline is added to the fuel tank of a motor vehicle or a storage tank, a tank of gasoline vapor is expelled into the atmosphere. There is also some synthesis of organic pollutants in most fuel combustion.

The principal organic air pollutants are saturated and unsaturated hydrocarbons such as are present in motor vehicle exhaust and in gasoline and other liquid fuel vapors; oxygenated organic compounds, particularly aldehydes but also organic acids and alcohols formed in the incomplete combustion of fuel; phenols; polynuclear hydrocarbons, some of which have been shown to be carcinogens. Among the organic contaminants that have been isolated from air are: methane, hexane, and related aliphatic hydrocarbons; ethylene, acetylene and related unsaturated aliphatic hydrocarbons; benzene and other aromatic hydrocarbons; the aldehydes, formaldehyde, acetaldehyde, and acrolein; phenol; and among the polycyclic hydrocarbons, benzo-(a)-pyrene.

The range of instrument complexity is much more pronounced in the gas sampling field than it is in particulate sampling. Instruments range from the simple to the bizarre.

(1) Lead Peroxide Candle. Like the dustfall jar, this method for sampling sulfur compounds in the atmosphere has been in use for many years. A glass tube wrapped with gauze impregnated with PbO_2 paste is exposed to the atmosphere. The sulfur compounds, principally SO_2, in the atmosphere react with the PbO_2 to form $PbSO_4$. Integrated monthly samples are usually collected. Results of subsequent chemical analysis are expressed as milligrams of sulfate ion per square centimeter of peroxide surface per 30 days. Direct correlation between

SO_2 in the atmosphere as measured by other methods has been attempted, with questionable validity. While PbO_2 candle results are indicative of a community's total pollution, they provide a poor measure of the SO_2 content of the atmosphere. Since a result is obtained from a sample collected over an extended period of time it gives no indication of peak concentrations.

(2) Detector Tubes. Detector tubes are used to obtain instantaneous information on the concentration of specific contaminants in the atmosphere. A measured amount of air is drawn through a prepared tube, usually containing a colorimetric chemical reagent which absorbs and reacts with a specific contaminant. The resulting color change is measured against a standard calibration chart. The color indicates both the presence and the concentration of a specific contaminant.

(3) Impregnated Tape Sampler. This sampler has commonly been used for H_2S, but it can be used for sampling for other contaminants as well. It is essentially the same unit used to measure soiling. It is modified by impregnating the filter tape with lead acetate (when measuring for H_2S). The air is prefiltered before being passed through the filter tape. The H_2S in the air sampled reacts with the lead acetate to form black lead sulfide. The optical density of the exposed spots are proportional to the H_2S concentrations. The flow rate is approximately 0.25–1 cfm.

(4) Impingers. Air to be sampled is drawn into absorbing solutions contained in giant impingers, usually at a flow rate of 1 1/2-2 liters/min. In sampling for SO_2, an absorbing solution of sodium tetrachlormercurate is used. Saltzman reagent is used for NO_2 sampling. Alkaline KI is used for oxidants. Analysis of the color change of the absorbing solution yields the respective gas concentrations in parts per million. The sampling period is usually 24 hours.

(5) Sulfur Dioxide Samplers. A number of instruments have been developed for measuring the SO_2 concentration in the atmosphere. Almost all of these have limitations because of other substances usually present in the atmosphere which interfere with the analytical procedure.

(I) Hydrogen Peroxide Analyzer (Thomas Autometer): Air is bubbled through an absorber at a constant rate. The absorbent is distilled water containing a small amount of H_2O_2 for converting SO_2 to H_2SO_4. A set of electrodes operates a conductivity monitor to indicate the SO_2 concentration in the air sampled. The concentration is based on the change in conductivity of a previously standardized solution. Any substances, such as strong acidic or basic materials, which will change the conductivity of the solution will interfere with the method.

(II) Electroconductivity Analyzer (Davis): The air sample is drawn through a chamber containing distilled water where the SO_2 ionizes. Measurement is based upon the differential conductivity before and after ionization. Any contaminant that will ionize will interfere with the method.

(III) Potentiometric Analyzer (Titrilog): The air sampled passes through a solution which absorbs the SO_2. A continuous reaction takes place in the solution between the absorbed SO_2 and electrolytically generated Br_2. The bromine generating current, which is continuously recorded, varies directly with the amount of SO_2 present in the sampled air. Reducing and oxidizing substances present in the air interfere, with resulting low or high values.

(6) Continuous Wet Chemistry Methods (Technicon Autoanalyzer). Metered air is pumped through a gas absorption system. Absorbed contaminants are mixed with reagents and color indicator. The solution is conducted to a colorimeter, where the color developed is compared with a reference standard. The voltage produced is recorded continuously. By changing reagents and reprogramming the flexible tubing in the solution pump, the instrument can be used for different determinations. The instrument is commonly used for measuring SO_2, NO, NO_2, aldehydes (aliphatic) and oxidants.

(7) Infrared Analyzers. Air to be sampled is admitted to a sampling cell and the contaminant absorbs some of the infrared energy transmitted through the cell. (Different reference cells are used for different contaminants.) The remaining infrared energy is transmitted to a director divided into two chambers. The difference in the detected infrared energy between the sample cell and a reference cell is directly proportional to the concentration of the contaminant in the sample. The instrument is commonly used for measuring CO and hydrocarbons.

(8) Hydrogen Flame Ionization. The air sample being analyzed for total hydrocarbon content is drawn continuously into the flame of burning hydrogen, causing ionization of carbon atoms in the sample which produces changes in current flow proportional to the carbon content of the sampled air. An electrometer circuit provides the driving voltage for a meter or potentiometric strip-chart recorder. Measurement signals are a function of the instantaneous

quantities of carbon atoms present. A disadvantage of the method lies in its inability to distinquish between various hydrocarbon compounds, and in this respect it does not produce a true measure of the concentration of hydrocarbon compounds present in the air.

(9) Coulometric Ozone Sampler. (Mast). The air to be sampled passes through an electrolytic cell where it is exposed to a buffered solution of KI. The ozone present releases hydrogen and iodine, which combine to form HI. This compound releases current-producing electrons. The current produced is directly proportional to the ozone concentration. A recorder continuously records the current as ozone concentration.

(d) Reduction in Visibility. The most frequent means for measuring reduction in visibility is selecting numerous objects at various distances from an observer and noting which objects can and cannot be seen at the various distances under varying atmospheric conditions.

An attempt to become more scientific led to the development of the use of long-path light beams and photoelectric cells. Recently, it has been suggested that lasers be used for this purpose, and this approach shows promise.

(e) Odors. Odors are common and a major part of complaints received by regulatory agencies. There are two principal categories of odor analysis; namely, (1) subjective or organoleptic methods and (2) objective methods.

Nearly all subjective method of odor evaluation are performed by dilution; i.e. a known volume of air to be tested is diluted with a known volume of deodorized air, and the volume of purified air required to dilute the sample to a concentration at which it can no longer be detected by smell is determined. Syringes and tubes are employed for the dilution steps.

Objective methods of odor evaluation depend on the isolation of the odorants by physico-chemical techniques such as adsorption on activated carbon, with subsequent elution and identification of the separated substances, or on general or specific chemical methods. Reduction of $KMnO_4$ by an odorous gas stream is a general method. A specific chemical method would determine the amount of a specific odorant or malodorant, such as formaldehyde in a gas stream.

(f) Radionuclide Sampling and Analysis. Radionuclide sampling and analysis comprise three principal types of collection: (1) dustfall collection, in which a sample is collected for a period of one month, (2) 24-hour samples collected on sticky paper, and (3) high volume membrane filter collections.

The sampling is carried out in a manner identical in each case to collecting non-radioactive material from the atmosphere (Section 5).

Stack Sampling

Because many regulatory agencies impose restrictions on the quantity of particulates and gases that may be exhausted to the air, air pollution sampling and analysis comprise the sampling and analysis not only of the contaminants in the outdoor air but also of materials exhausted to the atmosphere through a stack. Generally, dust and fly ash emissions from stacks are determined by passing samples of dust-laden gas in a stack through filters or thimbles of known weight, and at the same time measuring the volume of gas sampled. In collecting dust samples it is particularly important that the samples be collected isokinetically, i.e. at the same velocity as the gases passing through the stack. If this is not done, the kinetic energy of the particles will influence the amount of dust collected and the sample will not be truly representative. At the end of the sampling period, the thimbles or filter papers are removed from their special holders, dried and reweighed. The gain in weight of the filter paper or the thimbles divided by the volume of the gases sampled, corrected to stack conditions, or to STP (O°C and 760 mm), give the dust loading of a stack in units of weight per unit volume, such as grains per standard cubic foot. If an abatement device is used and it is desired to determine the efficiency of the control equipment, then simultaneous samples of the effluent gas stream are taken from the inlet and outlet ports of the control equipment.

The methods used for the analysis of pollutants in stack gases are generally modifications of the methods used for the determination of such pollutants in the general atmosphere or in industrial hygiene analyses.

A stack sampling train can be either simple or complex, depending on the contaminant or contaminants for which the sampling is being performed. Some of the particular components which may be included in such a train are: (1) coarse filter or cyclone, (2) fine filter, (3) thermometers, (4) various absorbing solutions for gases, (5) adsorbents, (6) freeze-out trains, (7) moisture traps, (8) flow control valve, (9) flow meter, (10) vacuum pump.

Perhaps the most commonly used piece of equipment for stack sampling is the Orsat apparatus for determing the composition of flue gases. It is commonly used for measuring the efficiency of combustion in boilers as well as incinerators.

The principal constituents of flue gas (CO_2, CO and O_2) can be measured in the Orsat apparatus by passing a sample of gas successively into three solutions, each having a high absorptive capacity for one of the constituent gases.

CO_2 is absorbed in a solution of KOH (caustic potash). O_2 is absorbed in a solution of pyrogallic acid. Ammonical solution of CuCl absorbs the CO.

While not technically a stack sampling technique, since the sampling is not done at the stack, the most frequent measurement of stack emissions is that made of smoke density. The degree of blackness of the smoke issuing from a stack is reported in terms of the Ringelmann scale. The measurements fall into two classifications: (a) the unaided visual measurements such as those made directly by the use of the Ringelmann chart, the Micro-Ringelmann chart, the Public Health Service photo strip, and similar aids (in these cases the density or shade of the smoke is compared with the chart visually), and (b) instrumental visual methods employing the smoke scope, Umbrascope, Bacharach smoke tester, and the von Brand smoke recorder.

All these measurements are technically accurate for measuring shades of gray only, Ringelmann No. 1 being equivalent to a reduction of 20 percent of light transmission through the plume; Ringelmann No. 2 being equivalent to 40 percent reduction of light transmission through the plume, and so on.

In the last few years, the equivalent opacity measurement has come into fairly widespread use. Observers are trained through the use of equipment which generates white "smoke" which has equivalent light-attenuating power to gray smoke. Many ordinances now are being written in terms of equivalent opacity.

Various types of equipment have been developed and are in use to aid operators in the control of combustion. The most common of these are the photoelectric cell installed in the stack and the closed circuit TV camera.

Tracer Detection

Tracers too are not, in the strict sense of the word, a stack sampling procedure. But this method can be used to give some indication of the contribution to the total atmospheric burden of a single source located within a source complex. The tracers used are: (a) fluorescent materials, (b) dyes, (c) radioactive substances, (d) characteristic substances, and (e) odors.

Fluorescent Tracers. The presence of fluorescent particles in ordinary air is rare. Consequently, such materials can be employed to locate point sources of pollution and to estimate their contribution to the total observed pollution. Methods have been developed by means of which fluorescent particles are injected into a stack gas stream at the same velocity at which the stack gas moves. Using a blower type aerosol generator, 1 gram of material will produce an aerosol cloud containing up to 6×10^{10} discretely identifiable particles. The particles are generally collected by filtration in the area being tested, and the number of fluorescent particles on the filters are counted microscopically. The principal tracer materials used are zinc sulfide, a mixture of zinc and cadmium sulfides and zinc silicate. Cadmium sulfide is a toxic material (see section 12).

Dye Tracers. The use of dye tracers such as uranine has the advantage that the dye in the air sample may be dissolved and the depth of color of the dye determined by photometric means. This is a much simpler procedure than counting under a microscope, with special arrangements for fluorescent counting.

Radioactive Tracers. Some non-radioactive compounds, such as antimony oxide, which are not a common component of the air have been used as a dispersal tracer. The field samples collected are made radioactive by exposure in an atomic pile, and the amount of tracer substance is determined by measuring the induced radioactivity.

The use of short-lived radioactive isotopes has also been considered, but the wisdom of distributing material such as this directly into the air is doubtful.

Characteristic Substances. This method of tracing a point source of pollution is particularly valuable when used in conjunction with investigations of sources producing or using a characteristic material. The catching and identification of such a specific material, as for instance a dye, drug, dye intermediate, or other intermediate, can be used to locate the source, provided there is only one producer of that type of material in the neighborhood. In an analogous manner, metals such as zinc, copper or mercury can be used as tracers for a smelter or refinery.

Photomicrographs of typical emissions have been used with some success in making comparison with photomicrographs of materials collected from the atmosphere and thus identifying the sample with its source.

Odor Tracers. If a pollution source emits an odor, the odorous substance may also be used as a tracer in a manner similar to the use of characteristic materials.

For methods of control, see Section 2.

BIBLIOGRAPHY

1. McCabe, L. C., Editor, "Air Pollution," New York, McGraw-Hill Book Co., 1951.
2. Faith, W. L., "Air Pollution Control," New York, John Wiley & Sons, 1959.
3. Magill, P. L., Holden, F. R., and Ackley, C., Editors, "Air Pollution Handbook," New York, McGraw-Hill Book Co., 1956.
4. Mallette, F. S., Editor, "Problems and Control of Air Pollution," New York, Reinhold Publishing Corp., 1955.
5. "Air Pollution Abatement Manual," Manufacturing Chemists' Association, Washington, D. C., 1952.
6. Jacobs, M. B., "The Chemical Analysis of Air Pollutants," New York, Interscience Publishers, 1960.
7. Jacobs, M. B., "The Analytical Chemistry of Industrial Poisons, Hazards, and Solvents," 2nd edition, New York, Interscience Publishers, 1949.
8. "Laboratory Methods," Air Pollution Control District, County of Los Angeles, 1958.
9. ASTM Standards, "Methods of Atmospheric Sampling and Analysis," in Part 23, Philadelphia, 1966.
10. Stern, A. C., Editor, "Air Pollution," New York, Academic Press, 1962, 2 vols.
11. Strauss, W., Editor, "Industrial Gas Cleaning," Vol. 8, New York, Pergamon Press, 1966.
12. Restoring the Quality of Our Environment, Report of the Environmental Pollution Panel, President's Science Advisory Committee, The White House, November 1965.
13. Gilpin, Alan, "Control of Air Pollution," London, Butterworths, 1963.
14. White, H. J., "Industrial Electrostatic Precipitation," Reading, Mass., Addison-Wesley Publishing Co., 1963.
15. "North American Combustion Handbook," 1st edition, 2nd printing, Cleveland, Ohio, North American Manufacturing Co., 1957.
16. World Health Organization, "Air Pollution," New York, Columbia University Press, 1961.
17. Proceedings National Conference on Air Pollution, November 1958, Public Health Service Publication No. 654, Govt. Printing Office, 1959.
18. Proceedings National Conference on Air Pollution, December 1962, Public Health Service Publication No. 1022, Govt. Printing Office, 1963.
19. Proceedings National Conference on Air Pollution, December 1966, Public Health Service Publication No. 1649, Govt. Printing Office, 1967.

PART THREE

THE PROBLEM OF SOLID WASTES

Meredith Thompson, D.Eng.
Assistant Commissioner
New York State Health Department
Albany, N. Y.

The solid wastes problem consists of the storage, collection, transportation and disposal of unwanted wastes from residential, commercial, industrial and agricultural sources. Such wastes include not only solids but also the liquid component of industrial and institutional wastes which for various reasons are not handled by liquid waste treatment facilities.

Solid wastes include garbage, rubbish and trash, collectively called refuse. Garbage is the wasted material resulting from the distribution, preparation and serving of food in homes and restaurants. Rubbish consists of dry waste such as paper, bottles, cans, ashes and similar materials. Trash is a term for the remaining community wastes such as leaves, lawn rakings, refrigerators, furniture, building-demolition materials, diseased trees and industrial wastes. Also included are special industrial wastes such as toxic liquid and solid industrial wastes, radioactive wastes, pesticides, farm wastes from animals and poultry, junked autos, incinerator residues, and sewage treatment sludges.

The variety of materials which contribute to the unwanted waste problem is a measure of the complexity of its composition and disposition which has escalated with the growth and technological advance of the country. The developing problems are caused by the quantity and nature of the wastes and such factors as technology, geography, economics, government and, to a large extent, public attitudes.

No useful data are available on specific types of wastes accumulated in different segments of the community. For 1967 the per capita daily rate of accumulation is estimated at 4.5 pounds, or an astounding 250 billion pounds of municipal solid wastes annually! Collection and disposal costs are estimated at 3 billion dollars a year. By 1980 the per capita output is estimated to be 5.5 lb/day, exclusive of industrial, agricultural and demolition wastes, or an estimated total waste of about 8 pounds per capita per day.

The 1965 President's Science Advisory Committee on Restoring the Quality of Our Environment reported that "solid wastes of industrial origin comprise another category of major magnitude. The interactions of industrial material flow are complex and industry contributes both to the total waste problem and to the conservation of useful raw material. Significant amounts of waste are salvaged and recycled. Industrial scrap, an obvious example of the magnitude being generated, is calculated at 12–15 million tons annually. The gross tonnage of non-ferrous scrap is not available, but a clue to its magnitude is that in 1963, recovery from it yielded 974,000 tons of Cu, 493,470 tons of Pb and 268,250 tons of Zn."

The paper industry generates approximately 30 million tons of waste paper and paper products. Depending on economic conditions, up to 10 million tons of waste paper may be salvaged and recycled to the industry.

In some parts of the country large volumes of waste develop from the mining of solid fuels, metals and non-metallic minerals in the form of rock, mill-tailings and smelter slag.

Collection, Treatment and Disposal Methods

These methods vary because the wastes emanate from residential, commercial and agricultural sources. The methods adopted also depend upon whether an urban or rural area is to be served.

Municipal collection services may be provided to all premises, including residential, commercial and industrial, or they may be limited for reasons of economy. Industrial refuse is generally excluded because of its wide variation in type and quantity. Standard collection equipment is the enclosed compactor type vehicle. Compaction provides units with a density of 400–500 lb/yd^3 of rated capacity.

Bulk storage containers used for commercial and industrial establishments can be hauled directly to the disposal area or emptied into large compactor type trucks.

As distance to a disposal site increases, the cost of transportation becomes a more important factor in site determination and method of disposal. An intermediate step in refuse collection and transportation is the transfer station. Tractor-trailer units which may be compactor units are used to transport refuse from a transfer station to a distant but economical disposal site.

The transfer station may consist of only a ramp to an elevated platform from which the collection trucks dump directly into the trailers.

Disposal Methods. The disposal of varied wastes is an ever-increasing problem which must be solved. As times change, new methods must be developed.

For instance, hog feeding was a method of garbage disposal which has nearly disappeared because it requires separation of garbage from refuse. The householder generally will not effectively separate and store his garbage and refuse. Where hog feeding is permitted, state law usually requires pasteurization of garbage.

Residential and commercial use of garbage grinders, which dispose of garbage to a municipal or private sewer system, is expanding greatly. However, the volume of waste handled by this method is small compared to the total volume. Garbage grinders can reduce the refuse collection problem, but this household convenience must be weighed against the economics of collection and the increased load on the sewers and sewage treatment plants.

Incineration is not a method of disposal; it merely reduces the total volume of wastes by about 80 percent. The residue still needs a final land or water disposal site. Incinerator units are available for use by private residences, apartment houses, commercial and industrial establishments, and even municipalities. Residential and apartment house incinerators are generally unable to meet ambient air quality standards and therefore will be used less in the future unless more efficient units can be developed. Municipal incineration becomes a method of choice in many urban areas as less land becomes available for sanitary land-fill and no better methods are developed. Generally, little effort is made to reduce the source volume of waste materials going to incinerators, although this is probably the most expensive method of refuse reduction. Most municipal incinerators are unable to meet the ambient air quality standards promulgated by federal and state governments. The cost of air pollution control equipment for incinerators will in some instances double the total cost of construction. Control equipment includes electrostatic precipitators, sedimentation chambers, water sprays and centrifugal separators.

Two new incinerators for New York city able to handle about 7000 tons of waste with a residue of 1500 tons will cost $50 million each.

Disposal of unwanted wastes is accomplished in many instances by deposition on so-called dumps. Dumps are areas where any individual, industrial, or community wastes may be dumped at any time, with little or no control. The open dump is still very much in existence although contrary to the laws of many states. Dumps are unsightly, breed rats and flies, are a source of smoke and odor which contribute to air, surface and ground water pollution, and generally create nuisance conditions.

The logical improvement to control dumps is the sanitary landfill method of disposal, which is the engineered process of controlled compaction and cover. A true sanitary landfill operation is insect-, rodent-, nuisance- and odor-free refuse disposal. The sanitary landfill is not the solution to all unwanted waste problems, but when competently planned and efficiently operated, it causes little difficulty and the land can be reclaimed for use as parking lots and recreation areas. Sanitary landfills for large municipalities require large land areas, normally about 1 acre a year for every 8 feet of fill per 10,000 population.

Composting is a volume reduction method for handling refuse. It produces final products which must be disposed of by sale, incineration or sanitary landfill. It is a process of biodegradation under controlled conditions of ventilation, temperature and moisture. Extensive handling and processing equipment is required to sort and separate metals, glass, and other non-degradable materials. Grinding, shredding, screening, rasping and tumbling equipment is required to reduce the aggregate size. Composting is undertaken in closed vessels or in open-air windrows under aerobic conditions. Composting in closed vessels takes approximately 5 days, compared with 8 weeks in open windrows. Composting has not been successfully demonstrated in the United States because of inability to profitably dispose of the soil-conditioning end product. As a matter of fact, in most instances it has been difficult even to give away the end product. Apparently commercial fertilizers are more acceptable.

While the United States is studying "composting" at the municipal level, many foreign countries which utilized the method for many years are dropping it. Japan has closed all its composting plants in favor of incineration.

Interrelationships of Unwanted Wastes. Some methods of disposal of unwanted wastes are actually reduction methods. Sound engineering guidance is needed to prevent the development of new environmental problems from the solution of a solid wastes problem. For instance, in determining the location of a sanitary landfill site there are more important problems to be considered than isolation. The site must have geological and hydrological characteristics such as to minimize the possibility of surface and ground water pollution. Seepage from toxic residential, commercial and industrial liquid and solid wastes deposited in the landfill can pollute ground water used for private and municipal water supplies. The California Water Pollution Control Board found that seepage through incinerator ash will leach alkalies and salts from the fill. The Board calculated that an acre of ash 12 inches thick produces 1.5 tons of sodium and potassium, 0.9 ton of chloride and 0.24 ton of sulfate.

The pollution of surface or ground water may be detrimental to drinking water supplies and to recreational areas and waters used for industrial purposes.

Municipal incinerators can add to air pollution unless properly designed and sited. Incinerators not only produce a residue which must be disposed of, but can create new environmental problems. Consideration must be given to the consequence of each proposed action. If a rule is promulgated which prohibits apartment house incinerators in order to control air pollution, the problem of municipal collection and disposal of these wastes remains. Probably neither collection nor disposal facilities would be able to handle the increased load.

Newer Methods of Unwanted Wastes Disposal. An extension of the sanitary landfill technique is the utilization and reclamation of abandoned strip mines, salt caverns, quarries and the like. Philadelphia and New York city propose via transfer stations and rail transportation to dispose of wastes economically at such sites. Needed equipment is being developed which can greatly compress wastes. The Japanese compress garbage at about 3000 lb/in^2 into hollow metal cubes which are welded together for use as a building material.

The D and J Press Company, Inc., of North Tonawanda, N. Y. has developed a 70-foot-long refuse disposal machine. Refuse is dumped from packer trucks into a hopper, which evacuates into a press box that forces the material into an extrusion chamber under 350 tons shear. The material is then fed to the bottom of an 8-foot trench, dug by the machine, under 150 tons of pressure. The excavation dirt from the trenches is fed back by conveyor to cover the refuse where it is tamped under 4000 foot-pounds of pressure.

Britain and Sweden have installed pneumatic tubes in large apartment complexes to carry household refuse to central incinerators a mile away.

Composting is being seriously considered as an economical method of refuse reduction. The USPHS in its Solid Wastes Program has a 7-year TVA-operated research study at Johnson City, Tennessee, which involves adding raw sewage sludge to compost to increase its fertilizer value. One question to be resolved is the period of time that pathogens in the sludge will survive.

The National Waste Conversion Corporation of New York is developing a method for adding plant foods to ground garbage from which metallic particles have been removed. The ground garbage is pulped and fed to decks of aerobic digesters. The pulp is moved continuously from deck to deck by mechanical rakes. The compost is dried, compacted and granulated to farm fertilizer grade. This plant food provides slow releasing nitrogen for sustained product growth, and the process takes approximately 72 hours per cycle.

New methods of packaging foods and non-returnable bottles and cans have added to the waste disposal problem. In an attempt to resolve this phase of the problem, Dow Chemical is trying to develop a biodegradable bottle. Also under study by federal agencies and private companies are edible food wrappings. Researchers are trying to develop a sprayed-on package wrapper which either would be part of the food or could be washed off before cooking or eating.

Health Effects

It might be helpful at this time to define environmental health so that a basis for desired action can be determined. WHO defines environmental health as the control of "those factors in man's physical environment which exercise or may exercise a deleterious effect on his physical, mental or social well-being."

A study of the health implications of solid wastes was completed in 1967 by Dr. Thrift G. Hanks of the Life Systems Division of Aerojet-General Corporation under a contract with the USPHS. This study indicated that "hypothetically, solid wastes can produce undesirable effects by biological, chemical, physical, mechanical or psychological means. For example, human pathogens in feces provide a biological threat, industrial wastes create chemical hazards, flammable materials provide a physical hazard of fires or explosions, and broken glass and other sharp-edged wastes create mechanical hazards. These hazards, plus unsightliness, costs of waste disposal, special interest and jurisdictional disputes, threats to property and other factors, provide a basis for potential psychological and behavioral disturbances."

Solid Wastes Administration

The organization and administration required to handle the total unwanted wastes problem efficiently for a community can vary considerably. Many experts indicate that pollutional problems must be attacked on a "problem shed" basis. In other words, air pollution problems are handled by an organization which covers the air pollution area concerned. Similar reasoning is given for the control of water and land pollution. Since the various sheds do not normally cover identical areas, separate and distinct "shed" control agencies are recommended. This increase in control agencies over and beyond normal local, state and federal agencies can compound the administrative and enforcement problems.

Continued effective, efficient and economical solid wastes disposal depends on present methods and on future problems and new methods of collection, transportation and disposal. For instance, further research is required on closed waste systems which might be developed for the individual home, apartment or community.

Comprehensive studies by an experienced and competent engineering firm are necessary for communities within an economic service area to be served individually and collectively by the most efficient collection, transportation and disposal system. Comprehensive studies should be developed for no less an area than a county,

unless there are specific circumstances which would dictate the study of a smaller area. The completed study of a county should not be considered the end result; adjoining county studies should be evaluated to determine if a better solution to an individual county waste management problem would be a multi-county or regional plan. Even the development of a regional plan should not be assumed to be the best solution. An engineering consideration of all the county and regional comprehensive studies might indicate a solution which was state-wide in nature such as a rail transportation system served by transfer stations.

Comprehensive studies must identify the present and future problems and provide alternative solutions. These studies must include the volume and character of wastes from all sources already in or coming into the study area. It is necessary not only to determine and plan for the total volume of wastes being produced within the study area, but more important is the recognition by all concerned of the need to reduce the volume of waste where it is generated. Consequently industry has a responsibility to study its routine operations and determine how processes can be revamped to reduce the quantity of unwanted (and costly!) wastes.

The comprehensive study must cover present and proposed collection routes within each community and the proposed methods and costs of disposal at various time intervals. Comprehensive studies do not develop final engineering plans for construction purposes; these are developed from economic projects in the study. The comprehensive study must also cover unwanted wastes problems and also the effects of any recommended courses of action upon air and water pollution control activities.

The full report should be produced in several steps. One should include all the data and statistics gathered in connection with the study together with a detailed narrative. A second narrative step should be developed for the needs of governmental and related agencies. The third and probably most important part is public education on the magnitude and importance of this problem.

In summary, the problems of collection, transportation, treatment and disposal of solid wastes from residential, commercial industrial and agricultural sources are closely related to the public health and welfare and as such merit a high priority in the budget of a community. The ultimate disposal of such wastes in the most economical manner is often hampered by the psychological fact that no one wants his neighbors' wastes in his own backyard, town or even county. This reaction has developed through years of unsatisfactory experience due to poor design, location and operation of incinerators and dumps.

The flow of solid wastes is increasing in quantity and complexity and we have to cope with it. Some of the "musts" for each community are:

(a) Develop a detailed engineering study and report on a multi-municipal or county-wide basis, and by a competent consulting engineer, covering all related factors including geology, hydrology and topography for the area being studied.

(b) Educate the people to the fact that adequately designed, located and operated facilities can be a community asset.

(c) Employ qualified and competent personnel to operate waste treatment and disposal facilities.

SECTION 5

RADIATION HAZARDS

John H. Harley, Ph.D.
Director, Health and Safety Laboratory
U. S. Atomic Energy Commission
New York, N. Y.

The field of atomic energy has undergone a very considerable change since about 1960. Prior to that time the major applications of radioactive material were to operations being carried out in government laboratories or under government contract. Medical applications were largely limited to the use of radium. Since 1960 the applications of radioactive materials have increased tremendously, and by 1965 the industry might be considered to have come of age, with economic nuclear power, the development of isotopic power sources and numerous tracer applications.

The rapid growth of the industry means that a sizeable fraction of the population now has a possible exposure to radiation and that control of radioactive materials must be exercised on a large scale. The industry is subject to regulations, but these apply to the overall performance of radiation protection and not to the means of controlling exposure. This section will discuss the principles and describe techniques of radiation protection.

Radiation protection is the prevention of illness or injury from overexposure to x-rays and nuclear radiation. Nuclear radiation and x-rays are invisible, and except in the case of acute overexposure such as might occur in an atomic explosion, the effects do not appear for some time. In the case of radium poisoning, for example, delayed effects are still being discovered as a result of exposures incurred in the 1920's. At the present time, there is no cure for the various effects produced, so all our efforts must be concentrated on prevention.

To operate a plant or a laboratory safely where x-rays or isotopes are used, it is necessary to have a general understanding of the nature of radiation, its measurement and the possible effects of overexposure. On the basis of this knowledge, methods of exposure evaluation and control may be set up.

When we consider radiation exposure, we must think of that radiation which is added to the normal background radiation to which man has been exposed for hundreds of generations. This background radiation from the naturally occurring radioisotopes and from cosmic rays is the basic level of exposure and cannot be reduced.

The theoretical ideal of radiation control would be maintenance of exposure at background level, but this is often not possible. Economic considerations become increasingly important as the reduction in exposure is continued to lower and lower levels. On the other hand, while the maximum permissible levels given are considered safe based on past experience, designing for exposure at just these levels is not necessarily a guarantee of good radiation protection. Frequently, by careful planning, exposures can be reduced considerably below the permissible level without extensive changes in facilities or without additional costs. In such cases, these reductions should be made. If further reduction of radiation levels requires large changes in equipment or would be very expensive, there must be a compromise between the desired and the practical levels.

The experience of radiologists, radiographers and isotope users has provided a reasonable basis for adopting a program for personnel protection. This experience has resulted in a series of maximum permissible levels for exposure to external radiation, for inhalation or ingestion of isotopes, and for laboratory or plant waste products disposal. The accepted levels contain a reasonable factor of safety. However, needless exposure to any potential hazard is foolish and every reasonable effort should be made to minimize such exposure.

Fortunately, exposures can be measured and can be controlled so that permissible levels are not exceeded, and thus preventable damage from radiation can be avoided.

A radiation protection program must consist of two parts—evaluation and control. Evaluation consists of the measurements required to estimate possible radiation exposures and the necessary interpretation of these measurements. Control consists of taking measures required to reduce radiation exposure. A good control program minimizes necessary exposures and eliminates needless exposures. A continuing evaluation program will indicate the effectiveness of the controls and will give a continuous record of all exposures. This section will emphasize these two facets of a realistic radiation protection program.

REGULATIONS

Radiation exposure is almost always subject to some form of regulation. These regulations usually cover the purchase, possession, use and disposal of radioactive material and the use of x-ray equipment.

The federal government is responsible for most of the radioactive material although it is relinquishing this authority to the individual states as the states request this authority and show that they are competent to handle radiation problems. Some natural radioactive materials and all x-ray equipment are not covered by federal regulation.

The federal government has classified radioactive materials into three groups; source materials including uranium and thorium, special nuclear materials including plutonium and uranium that is enriched in the ^{233}U or ^{235}U isotopes, and by-product material including any other radioactive material resulting from the production or utilization of special nuclear material. The federal regulations have been set forth in the Code of Federal Regulations, Title 10, Chapter I. The parts of this chapter of interest here are indicated below:*

*Copies of the AEC Rules and Regulations applying to work with radioactive materials are available on subscription from the Superintendent of Documents, Government Printing Office in Washington, D. C.

Part 20. Standards for Protection against Radiation
Part 30. Licensing of By-Product Material
Part 31. Radiation Safety Requirements for Radiographic Operations
Part 34. Licenses for Radiography and Radiation Safety Requirements for Radiographic Operations
Part 40. Licensing of Source Material
Part 50. Licensing of Production and Utilization Facilities
Part 70. Special Nuclear Material

The federal government or the states issue specific licenses for purchase, possession, use or disposal of source material, special nuclear material or by-product material. The licensee must show that a suitable radiation protection program is set up before obtaining the license and he is subject to inspection for compliance with the terms of the license. Violations of the radiation protection requirements subject the licensee to possible revocation of the license.

Certain small quantities of various radioisotopes are available without license. The quantities are considered to be harmless under any conditions of usage and are available for experimental work. These exempt quantities will be discussed later and are listed in Table 7.

While the Code of Federal Regulations applies specifically to licensees, the same standards are applied to operations under government contracts. Both groups are subject to inspection but the basic responsibility for compliance with regulations and standards lies with the operator. Thus, he must have an adequate radiation protection program.

In most states, x-rays and naturally occurring radioactive materials are included under a code of either health or labor regulations. These codes of radiation protection standards usually apply to medical installations as well as those of an industrial nature. In general, there has been an attempt to standardize codes developed by the states to follow the federal regulations where they are applicable.

NATURE OF RADIATION

Radiation is a form of energy, and as such, can be put to use for a variety of purposes. As with other forms of energy, it can be dangerous when uncontrolled. To control radiation intelligently, it is necessary to understand its nature and then proceed to the practical aspects of measurement, evaluation and control.

In the broad sense, radiation includes light, radio waves, cosmic rays and many other forms, but our immediate interest is limited to nuclear radiation and x-rays. These two forms represent controllable industrial or laboratory hazards that are utilized for the very properties that make them dangerous.

The fundamental properties of radiation are mass, electrical charge and energy. While all the types of radiation that we are interested in have energies within a rather narrow range, they differ markedly in the properties of mass and charge. The radiations all originate within the atom, and their similarities and differences are characteristic of their origin.

Atomic Structure

The scientists' picture of the structure of matter has changed radically in the last century. Knowledge has progressed from the concept of the atom as an indivisible unit of a chemical element, through the planetary atom consisting of a dense central nucleus surrounded by electrons moving in orbits around it, to the modern quantum theory. The latter theory is invaluable in theoretical physics for describing the behavior of atoms, but it has almost completely eliminated any simple physical picture of atomic structure. For the moment, however, we may consider the atom as a positively charged nucleus surrounded by a negatively charged electron cloud.

For many years, every atom of each chemical element was considered to have a particular atomic mass, known as its atomic weight, which was characteristic of that element. In practice, the weights of individual atoms were not known, and the atomic weights used represented the average of a large number of atoms. These atomic weights were used in calculating the proportions of various substances entering chemical reactions and were considered to be among the most basic physical properties of atoms.

In studying atomic structure however, it became apparent that the average atomic weights did not have as great a regularity as might have been expected. Actually the irregular nature of these atomic weights stems from the fact that most of the elements consist of mixtures of atoms having different atomic weights. The range of these differences is small, being only one to a few atomic mass units. The individual atoms of an element having a particular atomic weight make up an isotope of the element. The naturally occurring mixture of the various isotopes produces the average atomic weight, which is the one used for chemical calculations. It is possible to have more than one form of a particular isotope. The two forms differ in their energy content, and one is usually indicated as a

metastable state. The most general term applied to the isotopes of all elements in all forms is nuclide, and radionuclide is a proper term for the radioactive form of any element having a particular energy state and atomic weight. Common usage, however, has accepted the term radioisotope as an exact equivalent and it will be used here.

We can describe the properties of the atoms, their components and their radiations in physical units; the mass in atomic mass units (amu) which are $1/16$ the mass of the most abundant isotope of oxygen, the charge in units of the charge on a single electron, and the energy in terms of the energy acquired by an electron when it is accelerated by a potential of 1 volt. This latter quantity, the electron-volt (eV) is so small that we commonly speak of thousands (keV) or millions (MeV) of electron volts. The absolute energy of an atom or its components is of little interest, but the energy changes involved in the production of radiation are of primary interest to us.

Within the nucleus are protons with a mass of 1 amu and a positive charge of 1 unit, and neutrons with a mass of 1 amu and no electrical charge. Outside the nucleus are electrons (extra-nuclear electrons) with a mass of about 1/1800 amu and a negative charge of one unit. The characteristics of an atom are due to the number of protons and neutrons which make it up. Since the atom is electrically neutral, the number of positively charged protons in the nucleus and negatively charged electrons outside the nucleus must be equal.

The number of protons in the nucleus determines the particular chemical species or element that the atom belongs to, while its mass is determined by the number of protons and neutrons in the nucleus. We may completely describe any atom in terms of its mass and its elemental species.

The different chemical species are designated by the element symbol or by the atomic number (the number of protons in the nucleus). In the scientific shorthand used, $^{12}_{6}C$ describes an atom of carbon with an atomic number of 6 and a mass of 12. A different atomic number indicates a different element, but isotopes of the same element can have different masses. Our example, carbon, exists as three isotopes of mass 12, 13, and 14. These would be represented as $^{12}_{6}C$, $^{13}_{6}C$, $^{14}_{6}C$, respectively.

Actually, since the element symbol also designates the atomic number, the more usual terminology is ^{12}C, ^{13}C, ^{14}C (or C-12, C-13 and C-14).

The chemical nature of an element is almost independent of the particular isotope, so that in a chemical reaction all of the isotopes follow the same reaction pattern. Even the isotopes of hydrogen, where the ratios of masses are 1,2 and 3, show only slight differences in chemical reactivity. In cases of heavier elements where the relative differences in mass between isotopes are smaller, variations in chemical behavior can usually not be observed. If it is necessary to separate the isotopes of an element, physical properties highly dependent on mass must be used. Some of those which have proved practicable are the deflection of ions in a magnetic field, variations in diffusion rate of gases, and thermal diffusion. Isotopes separated in this way can hardly be distinguished chemically but may be distinguished by instruments in which measurement is based upon mass, for example, the mass spectrograph or the optical spectrograph.

Over 300 stable isotopes of the elements are now known. The total number of isotopes has been increased by the production of artificially radioactive isotopes by particle accelerators and by atomic fission. Up to the year 1965, over 1100 unstable isotopes had been identified.

A large fraction of the naturally existing isotopes are stable, but some of the natural isotopes and most of the man-made isotopes disintegrate spontaneously. The process of disintegration (also called radioactive decay) produces nuclear radiation, and the emitting isotopes (nuclides) are called radioisotopes (radionuclides). The most common emissions are alpha, beta and gamma radiation. Each type will be discussed, together with the nuclear transformation which accompanies it.

Nuclear Radiation

The radiations with which we are concerned are x-rays and the alpha, beta and gamma emissions from radioactive isotopes. There are installations where exposure to neutrons or accelerator-produced particles is possible. These should also be under competent safety supervision, but such a discussion is outside the scope of this section.

X-rays and gamma radiation are electromagnetic; i.e., their properties are similar to ultraviolet rays or visible light, only they are much more penetrating. Beta and alpha radiations are particulate; they are extremely small particles moving at high velocity.

All types of radiation share the property of losing energy by absorption in passing through matter. In any given case the degree of absorption depends upon the type of radiation, but all types are absorbed to some extent. Also, the process of absorption in all cases results in ionization, i.e., the removal of an extranuclear electron from an atom of the absorbing material. It is this process of ionization that both produces damage in tissue and allows us to design instruments for detection and measurement.

Since the properties of the various radiations determine not only the protective measures needed but also the methods of measurement, a brief general description of the different radiations will be given at this point.

Alpha particles are emitted by many of the heavy artificial and naturally radioactive elements. The alpha particle is the same as the nucleus of the helium atom, with a mass of 4 amu and a positive charge of 2. A typical alpha decay could be expressed,

$$^{226}_{90}Ra \rightarrow {}^{4}_{2}He + {}^{222}_{88}Rn$$

where radium is transformed to radon by alpha emission. The relatively high mass of the alpha particle means that for a given energy, the velocity is relatively low. The exact relationship is shown by the kinetic energy relationship for mass and velocity,

$$E = \frac{1}{2}mv^2$$

where

E = kinetic energy,
m = mass,
v = velocity.

The heavy, slow-moving, highly charged alpha particle has a great opportunity to interact with an absorber. Thus, on the average it dissipates its energy in a short path length (e.g., most alpha particles are completely absorbed by a few centimeters of air, or less than 0.005 mm of aluminum). When all its kinetic energy has been lost, the alpha particle will take up two electrons from its surroundings and become an electrically netural atom of helium. Most of the energies of alphas given off by different emitters lie in the range of 4–8 MeV but all the particles emitted from a single isotope will have the same energy.

Beta emission is a property of both heavy and light radioactive elements. The beta particle is an electron that possesses kinetic energy because of the speed with which it is emitted from the nucleus. In representing a beta emission, the change in mass of the atom is negligible, but the loss of the unit charge causes a change in atomic number; e.g.,

$$^{14}_{6}C \rightarrow e + {}^{14}_{7}N$$

The energies of the beta particles given off by a single emitting isotope cover a range from 0 MeV to a maximum value characteristic of the isotope. Their average energy is approximately one-third of the maximum value. The maximum energies of different isotopes lie in the range of 0–4 MeV. The velocities of the more energetic betas approach the speed of light; they may be computed from the kinetic energy formula. The range of beta particles in air may be more than a meter for energetic betas, and such particles will penetrate several millimeters of aluminum or plastic.

Gamma emission is a secondary process following rapidly after certain alpha or beta decays. The emission of the latter particles may leave the nucleus in an unstable state, and the excess energy is released in the form of gamma radiation. Since gamma radiation has neither mass nor charge, there is no change in the atomic number or atomic weight of the emitting isotope.

The energies of gamma rays are mostly in the range of 0–2 MeV, and the radiation from a single isotope will be made up of one or more groups of rays each having a single value of energy. Gamma rays can be very penetrating. Absorbing material does not completely stop gamma rays, but only reduces their intensity. For example, the gamma radiation from a radium source is reduced to 50 percent of its incident value by 0.5 inch of lead. One inch would pass 25 percent; 1.5 inches, 12.5 percent and so on. Therefore, we speak of the half-thickness of a given absorber material for a particular gamma emitter rather than its range.

Most of the known radioisotopes emit alpha or beta particles, sometimes accompanied by gamma rays. Three other less common processes—positron emission, electron capture, and internal conversion—should however be considered.

The *positron* is similar to a beta particle but has a positive charge. A typical reaction would be

$$^{30}_{15}P \rightarrow e^+ + {}^{30}_{14}Si$$

All the effects of beta radiation are present, and in addition, 0.51 MeV gamma rays are always emitted. These result when the positron loses its kinetic energy by absorption and combines with a normal electron. The two particles are annihilated and their mass is completely transformed into two gamma photons with an energy of 0.51 MeV.

The reaction for *electron capture* is similar to positron emission, for example,

$$^{64}_{29}Cu + e \rightarrow {}^{64}_{28}Ni$$

The electron is one of the extranuclear electrons from the original atom. Most commonly it is one of the inner electrons and as the inner electron is replaced by an outer electron, x-rays are

emitted. Whenever the nucleus is left in an unstable state after capture, gamma rays are also emitted.

When a nucleus emits gamma radiation, it is possible that some of the gamma rays may be absorbed by one of the extranuclear electrons in the atom. If this occurs, the electron may be ejected from the atom with considerable kinetic energy. This process is known as *internal conversion* and may be exemplified by

$$^{60}_{27}Co \rightarrow {}^{60}_{27}Co^+ + e^-$$

The electron emission leaves a charged cobalt ion which must combine with another electron to resume electrical neutrality. (Note that since the electron involved is extranuclear, there is no change in atomic number).

Internal conversion electrons may be distinguished from beta particles in that they have a single energy rather than a range from zero to a maximum. Theoretically, they could appear with any gamma emission. Possibly they do, but where the gamma ray energy is greater than 0.5 MeV, the probability of internal conversion is usually very low. With lower energy gammas, the fraction of gamma rays causing internal conversion may become quite large.

All the disintegration or decay processes described are the result of energy changes, which eventually end up in the production of a stable nucleus. The excess energy of the unstable nucleus is released by one or more of these processes according to characteristic rates. These decay rates are discussed later in this section.

X-radiation

The production of x-rays is not a spontaneous process. Also, in contrast to nuclear radiation, it is not the nucleus that is involved but the extranuclear electrons. These electrons may exist in many different energy states or levels. By the addition of energy, one of the electrons may be raised to a higher energy level. If this level is unstable, the electron will return to a lower energy level, and the atom will emit the energy in the form of radiation. Depending on the quantity of energy involved, the radiation may appear as visible light, ultraviolet radiation, or x-rays. X-rays have the highest energy, and are consequently the most penetrating and present the greatest hazard, although any of the others may cause damage to the skin.

The excitation energy for the production of x-rays is derived from the acceleration of a stream of electrons by a high electric potential. A schematic diagram of an x-ray tube which converts electrical energy into x-ray energy is

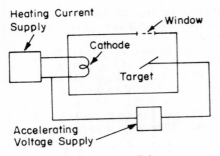

Figure 1. X-ray Tube.

shown in Figure 1. A stream of electrons emitted by the heated cathode is accelerated and strikes a target. Most of these electrons have their energy dissipated as heat in the target, but a fraction of them excite an extranuclear electron in an atom of the target material and produce x-rays.

Formerly, x-rays were considered to have lower energy than gamma rays, but the advent of multi-million volt x-ray installations has brought about an overlapping of their energy ranges. The true distinction lies not in the energy but rather in the origin. Gamma radiation comes spontaneously from nuclear transformations, while x-rays are produced by excitation of extra-nuclear electrons. Like gamma radiation, x-rays have no mass or charge, only energy. The energies of the x-radiation produced by a tube depend on the target material and the acceleration voltage. Any target gives a continuous band of x-rays with peak intensities at energies characteristic of the target material.

Their maximum energy cannot be greater than the energy of the accelerating potential, and it is customary to describe x-rays in terms of the voltage used for acceleration, for example 75 or 200 keV. Actually, only a portion of the x-radiation produced would have energies of 75 or 200 keV and a considerable quantity of softer (lower energy) x-rays are always emitted. The intensity of an x-ray source is usually expressed not in terms of a number of x-rays of a particular energy produced but in terms of the electric current flowing in the anode or target circuit.

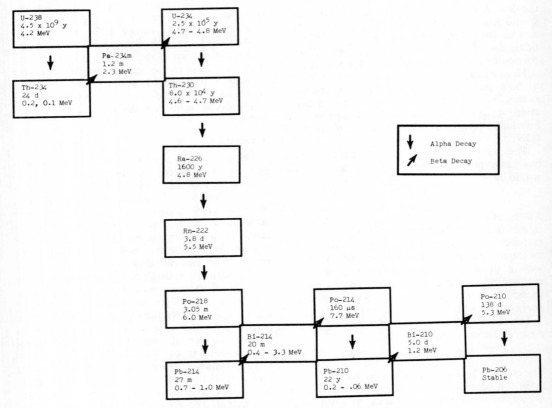

Figure 2. Principal Decay Scheme of the Uranium Series.

The description of x-radiation in terms of the instrument characteristics of voltage, current and target material may be roughly translated into the more fundamental physical units as will be shown later. Hazards and protective measures in this system will be discussed later.

The Natural Radioisotopes

The majority of the naturally occurring radioisotopes are members of three series of elements. These series all have a long-lived radioisotope as the first member, pass through a sequence of radioactive disintegrations involving alpha, beta, and gamma emission, and end up as a stable isotope of lead. The principal decay schemes for the three natural series are shown in Figures 2, 3 and 4.

Other important naturally-occurring radioisotopes are ^{3}H, ^{14}C, and ^{40}K. These undergo simple decay yielding stable isotopes. ^{3}H and ^{14}C are not long-lived on the scale of uranium and thorium. They are always present however because they are continually being produced in the atmosphere by cosmic ray bombardment.

Artificial Radioisotopes

Many man-made isotopes are now available. They include not only new isotopes of the natural elements, but entirely new elements as well. The new elements are those having an atomic number greater than that of uranium-92 and are called the transuranic elements. The most familiar one is plutonium, produced by neutron irradiation of $^{238}_{92}$U in a reactor.

The neutron bombardment of ^{238}U produces another radioactive series, the neptunium series which decays in a fashion similar to the natural series described above.

The nuclear reactor is designed to sustain a controlled fission process, where a fissionable element is split into two lighter elements. For example:

$$^{235}_{92}U + ^{1}_{0}n \rightarrow ^{95}_{38}Sr + ^{139}_{54}Xe + 2^{1}_{0}n$$

The reaction produces about 200 MeV of energy as well as 1–3 neutrons. These neutrons in turn cause fission in more atoms of the fissionable material (fuel). This chain reaction starts

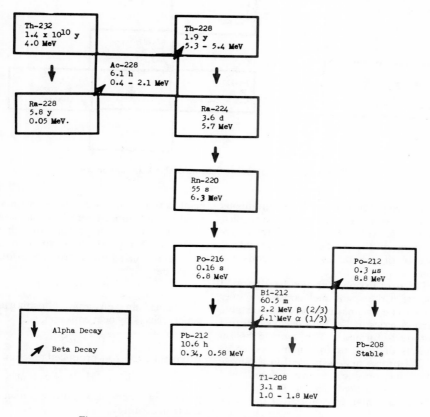

Figure 3. Principal Decay Scheme of the Thorium Series.

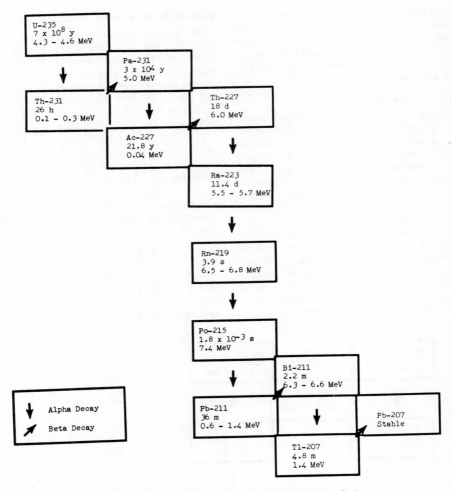

Figure 4. Principal Decay Scheme of the Actinium Series.

either by injecting neutrons or by spontaneous fission of a few atoms of fuel and is controlled by absorbing neutrons in excess of those required to maintain the reaction. The reactor is thus a source of a high flux of neutrons for irradiation.

The products of fission that may be separated chemically from the spent fuel consist of over 30 elements. The original isotopes are unstable and undergo a series of 1–6 beta decays before a stable isotope is formed. The average fission series consists of three or four members, so as many as 100 isotopes may be present in a sample of mixed fission products. The different elements may be separated chemically, but this process does not separate isotopes of the same element.

The major sources today of individual arti-ficial radioisotopes are neutron irradiation (in a reactor) or bombardment in an accelerator such as a cyclotron with protons or other charged particles. A typical transformation obtainable in a reactor, using the notation previously given, would be

$$^{31}_{15}P + ^{1}_{0}n \rightarrow ^{32}_{15}P$$

producing an unstable phosphorus isotope which decays by beta emission,

$$^{32}_{15}P \rightarrow ^{32}_{16}S + e$$

A typical cyclotron-produced reaction would be

$$^{7}_{3}Li + ^{1}_{1}H \rightarrow ^{7}_{4}Be + ^{1}_{0}n$$

producing a beryllium isotope which decays by electron capture

$$^{7}_{4}Be + e \rightarrow ^{7}_{3}Li$$

These processes or irradiation and bombardment as well as the chemical separation of specific elements from fission products and from natural series produce radioisotopes of practically every element. The use of these isotopes in the laboratory, in medicine and in industry has brought many benefits, and while most of them are hazardous materials, they can be handled safely.

The radioisotopes covered by federal regulations and their radiation properties age given under each element in Section 12.

Radioactive Decay

Radioactive isotopes may be characterized by the rate at which they disintegrate. All isotopes follow the same law of disintegration but the rates differ. This law states that a fixed fraction of the number of atoms present disintegrates in a unit time. The fraction, which differs for different isotopes, is known as the disintegration constant and is a fundamental constant for each radioisotope. Another way of expressing the same thought is that for each isotope there is a period of time during which half of the atoms initially present will disintegrate. This time is defined as the half-life of the radioisotope.* Thus if a given number of atoms is present at a particular instant, one-half of them will have decayed after 1 half-life, three fourths after 2 half-lives, and so on. Mathematically the number of atoms remaining at any given time may be expressed

$$N = N_0 e^{-\lambda t}$$

where

N_0 = the original number of atoms,
λ = the disintegration constant,
t = the time

(t and λ must be expressed in consistent units).

The half-life relationship may be arrived at from the above equation as

$$T_{0.5} = 0.693/\lambda$$

This simple relationship between half-life and disintegration constant is quite useful since λ is the basic unit used by the physicist, while $T_{0.5}$ is the common unit. The actual half-lives of the known isotopes range from less than 10^{-6}

*All values for half-life and energy in this section and in Section 12 were selected as follows:
Mass Numbers to 228:
Nuclear Data Group, Nuclear Data Sheets, 1959–1965, Academic Press, 1966, New York.
  Mass Numbers greater than 228:
Reviews of Modern Physics "Table of Isotopes," Strominger, D., *et al.*, (April 1958).

second to more than 10^{17} seconds. The half-life is constant since the rate of radioactive decay is independent of physical variables such as temperature, pressure or concentration.

Radioactive Series

As indicated in Figures 2, 3 and 4, the parent of a natural radioactive series undergoes a series of disintegrations before reaching a stable form. Shorter series frequently occur in the fission products where the so-called fission product chains usually have four or five members. The importance of this stepwise disintegration is that the radiation effect may be greater than from the parent alone or the chemical differences arising as decay proceeds along the chain may be of metabolic significance. These factors cannot be discussed here, but it is necessary to keep in mind that such decay series exist and that they may require consideration in hazard evaluation.

Probably the most significant chains are those following ^{222}Rn and ^{220}Rn. Radon is a gaseous element produced by the decay of radium. As a gas it may emanate from the solid material containing the radium and be distributed in the air of the workroom or the atmosphere. When the radon decays, the daughter products are solid and these, in turn, decay to additional solid members of the chain. Chemically, the principal elements are polonium, lead and bismuth. On formation, the atoms generally attach themselves to small aerosol particles and their hazard is considered to be by inhalation. The shorter-lived daughters are always readily measurable on air filters while ^{210}Pb and ^{210}Po are found in all vegetation, in foods and in man. They are distributed by a process of natural fallout that carries them to the soil and to plant surfaces.

Environmental Radiation.

Radiation in the environment comes from both natural and man-made sources. The natural external radiation includes gamma radiation from potassium, the members of the thorium and uranium series present in the ground, and cosmic radiation from outer space. Internally, the body contains small amounts of radium, uranium and the daughter products of radon, as well as potassium, ^{14}C and minor radioactive elements. The man-made contribution is chiefly radioactive fallout from nuclear weapons tests. The contributions from nuclear accidents or from waste disposal into the environment have so far not been a measurable contribution to worldwide exposure.

If we consider the possible radiation hazards from a single source such as a reactor, we must differentiate sharply between normal operations and accidents. In the course of normal operation, the releases should be minimal and depending on the type of reactor, would consist of tritium, ^{41}A, ^{85}Kr and possibly traces of fission products, particularly volatile radioisotopes of intermediate half-life such as ^{131}I. These are normally monitored directly at the source, the point of highest concentration. Reactor accidents have an extremely small probability, and reactors are designed so that the maximum expected accident will not release significant amounts of radioactivity into the environment. If an accident were to occur, the expected release would probably be the more volatile fission products such as the gases and radio-iodine.

While accidents are extremely unlikely, it is necessary to guard against the gradual build-up of man-made radioactivity in the environment from chronic low-level releases. This could eventually result in problems comparable to those presently encountered in air and water pollution by non-radioactive materials.

Absorption of Radiation

The single property shared by all forms of radiation is energy. Any radiation is a form of energy, and it is the dissipation of this energy in matter that causes biological damage or that allows us to detect and measure radiation. X-rays and gamma rays are electromagnetic energy; the alpha and beta particles possess kinetic energy. While these two types of energy are quite different, the effect they produce in matter is much the same.

In passing through matter, nuclear and x-radiations lose their energy principally by causing ionization in the absorbing material. In this process, neutral atoms of the absorber are ionized, yielding an electron and a positively charged ion. The electrons may acquire sufficient kinetic energy so that they cause further ionization in the absorber. The original ion-electron pairs are spoken of as primary ionization, and those produced by the electrons as secondary ionization. Each ion pair requires the same amount of energy for production; the excess energy imparted in any ionizing event appears as kinetic energy of the electron which then causes secondary ionization. Thus a certain amount of energy when dissipated in an absorber will produce a certain number of ion pairs. This is true whether the energy is introduced as x-rays or as nuclear radiation.

The average energy required to produce an ion pair in air is 32.5eV for either primary or secondary ionization. The total energy of any particle or ray will be expended in producing these ion pairs, so that the number of ion pairs resulting from any radiation may be calculated. When the ions and electrons recombine, the ionization energy will appear in the absorber as heat. In fact, one method of determining the total energy produced by strong sources is by measurement with a calorimeter, an instrument designed to determine the amount of heat produced in a process.

Particulate radiation, alpha and beta, has a finite range in an absorber and this range is dependent on the energy of the particle. One method of determining the approximate energy of alpha or beta radiation is to determine the amount of absorber which just absorbs the radiation completely.

Electromagnetic radiation, x-ray and gamma, does not have a finite range but is reduced in intensity as described earlier. The half-thickness or amount of absorber required to reduce the radiation intensity to one-half is a characteristic of electromagnetic radiation dependent on the energy of the rays. Determination of the half-thickness may be used to measure the energy of x- and gamma radiation. A further description of absorption will be found on p. 175.

RADIATION UNITS

In relating the measured values of isotope radiation to the predicted or actual biological effects produced by this radiation, two distinct types of units have arisen. The first, based on the number of radioactive disintegrations per unit time, is a measure of the quantity of the isotope present. This is, in effect, a unit of radiation flux whether measured at the source or at a considerable distance. The second type of unit is one of radiation dose, measured in terms of the quantity of radiation that is absorbed, and is a more direct measure of the possible biological effect.

The relationship between flux and dose units for radioisotopes is not simple and cannot be readily calculated with any degree of accuracy. Such a relationship can be determined empirically for an individual isotope but is only valid under conditions which are identical with those used in determining the relationship.

A similar relationship of flux and dose exists for x-rays. This section will define and explain the various systems of units in common usage for radiation intensity, energy and dose.

Units of Radiation Intensity

The units of radiation flux differ for x-ray and isotope radiation. However, the basic principle is that radiation flux is the measure of the strength of the source. In the case of x-rays, this can only be expressed in terms of the amount of energy dissipated by the x-ray tube. The common units are milliamperes of plate current flowing in the tube. This energy includes both that emitted as x-radiation and that dissipated as heat within the tube. However, if the operating voltage, plate current and the efficiency of the x-ray tube are known, the radiation flux may be calculated; i.e., the product of voltage and current expresses the wattage dissipated, and the product of this wattage and efficiency would give the wattage of x-rays produced.

The energy of the individual x-rays is determined by the target material in the x-ray tube and the operating voltage. Flux, which is a measure of the number of emitted rays, must include the other factors mentioned. However, the number of x-rays emitted by a source is not a common characteristic for describing flux.

In the case of isotopes, the source flux (activity) means the number of radioactive disintegrations occurring in the source per unit time. As in the case of x-rays, this is the total of the radiation emitted and dissipated within the source itself. The actual emission can only be determined by measurement and the correction factor obtained would include both the geometrical effects and the absorption of the radiation within the source.

The practical units for radioisotopes are disintegrations per minute (dpm) or per second (dps). More intense sources may be expressed in terms of the curie, which is equivalent to 2.2×10^{12} dpm. Smaller units such as the millicurie (mCi), microcurie (μCi) and picocurie (pCi) are 10^{-3}, 10^{-6} and 10^{-12} curie, respectively. Again, it should be emphasized that these units express only the source activity or the quantity of radioisotope present and not the radiation dose which this quantity would produce.

Units of Energy

The energy of any radiation is expressed in terms of electron volts. An electron volt is the energy which would be acquired by an electron accelerated by an electrical potential of 1 volt. In terms of the more common units of physics, the electron volt is 1.60×10^{-12} erg. This unit is so small that energies are usually given in thousands of electron volts (keV) or millions of electron volts (MeV). The introduction of high-powered accelerators has also brought billion electron volts (BeV) into the physicist's vocabulary.

The usual range of energies for the common radiations are tabulated below:

Radiation	Energy
X-rays	0–3 MeV
Alpha Particles	4–8 MeV
Beta particles	0–4 MeV
Gamma rays	0–2 MeV

These ranges include most of the radiations from the common isotopes and x-ray equipment, but higher energies of radiation do exist.

The energies shown are for single particles or rays. The total energy rate of emission of a source would be the product of the individual energy and the number of particles or rays per unit time. This product would give a different measure of source activity (radiation flux), as for example million electron volts per minute.

Units of Radiation Dose

Radiation dose is a measure of the amount of energy that is absorbed by the material being irradiated. *This is the only quantity which can be related to biological effect.* The instruments and measurements for health protection therefore should be designed in terms of radiation dose. The basic quantity is the amount of energy liberated in a unit mass of material and the unit is the roentgen (R). This is a purely arbitrary unit which has been set up as the quantity of x- or gamma radiation such that the associated ionization per cubic centimeter of air at standard conditions will produce 1 electrostatic unit (esu) of electricity of either sign. A more useful definition is that the roentgen is the quantity of x- or gamma radiation which results in the absorption of 83.4 ergs per gram of air.

This unit was set up at a time when only the dosage from x- or gamma radiation was considered to be important. The equivalent absorption for tissue was found to be approximately the same as that for air. When it became necessary to consider the dosage from beta and alpha particles the unit of roentgen equivalent physical (rep) was developed. This unit is no longer recommended but may still appear in the literature.

In addition, the biological effects of a given quantity of absorbed energy were supposed to be higher for alpha particles, neutrons and protons than for the others. The corresponding factor for alpha radiation was set at 10 (originally 20) and a new unit was devised to incorporate this factor. For alpha radiation in particular the roentgen equivalent man (rem) is defined as the quantity of radiation which when absorbed

TABLE 1. Dose Rates Due to External and Internal
Irradiation from Natural Sources in "Normal" Regions

Source of Irradiation	Dose Rate (mrad/y) Gonad	Bone Marrow
External irradiation		
Cosmic rays (including neutrons)	29	29
Terrestrial radiation (including air)	50	50
Internal Irradiation		
^{40}K	20	15
^{14}C	0.7	1.6
Other nuclides	1	1
Total	101	96

in tissue produces an effect equivalent to the absorption of one roentgen of x- or gamma radiation. The factor of 10 is called the relative biological effectiveness (RBE) and is used as a multiplying factor. For example, if measurement in air of alpha radiation showed one roentgen, the possible biological effect would be more closely approximated if the result were given as 10 rem.

As a further attempt to simplify dose units, the International Commission on Radiation Units (ICRU) in 1953 adopted the rad as the unit for absorbed dose. This unit is defined as the equivalent of 100 ergs per gram absorbed in the material of interest. Since this quantity does not include any factor for RBE, it must be specified which type of radiation is involved. For soft tissues, the numerical value for rads is sufficiently close to the numerical value of roentgens for protective purposes.

By applying these and various later terms, it was hoped to develop a system for expressing dosages which would be additive when exposure to several types of radiation had occurred. There is still hope that a simple compatible system of dose units will be agreed on in the future.

Two distinctions must be kept in mind. The first is that measurements are ordinarily made in terms of dose rate, i.e., roentgens (rems, rads) per hour, while chronic biological effects are assessed in terms of integrated dose, i.e., total roentgens (rems, rads). The second distinction is that the roentgen or rad is defined in terms of energy absorption per unit weight or volume. Thus, a total body dose of 100 mr would indicate a much greater total quantity of radiation absorbed than a 100 mr dose to one hand or arm. No description of dose is complete unless it tells the amount of material or portion of the body involved.

TABLE 2. Diagnostic X-ray Exposures

14 × 17 in. chest plate	0.05 r
Photofluoroscopic chest	0.1–1.2 r
Extremities	0.25–1.0 r
Skull	1.3 r
Abdomen	1.3 r
Gastrointestinal series	0.65 r/plate
Lumbar, spine lateral	5.7 r
Pregnancy lateral	9.0 r
Fluoroscopy	0.28 r/sec

The instruments and measurements for health protection should be designed in terms of radiation dose. This is rather complex in execution since it is difficult to produce a detector which absorbs radiation in the same way as air or tissue. A dose measurement made with a simple instrument therefore must be carefully interpreted to yield valid estimates of dose to man.

While the units of intensity and energy of radiation are readily understood, the units of dose are somewhat more difficult to put into perspective. For this reason, a comparison will be made here of various doses.

The permissible levels for occupational radiation exposure which have been adopted are based on the value of a maximum of 5 rem/y to the whole body.* With our present knowledge, it is believed that this level can be absorbed by a man for a working lifetime without the appearance of any sign of body damage. An idea of the magnitude of the permissible level can

*Permissible levels are covered in detail on p. 148.

TABLE 3.   Expected Effects of Acute Whole-body Radiation Doses

Acute dose (r)	Probable Effect
0–50	No obvious effect, except possibly minor blood changes.
80–120	Vomiting and nausea for about 1 day in 5–10% of exposed personnel. Fatigue but no serious disability.
130–170	Vomiting and nausea for about 1 day, followed by other symptoms of radiation sickness in about 25% of personnel. No deaths anticipated.
180–220	Vomiting and nausea for about 1 day, followed by other symptoms of radiation sickness in about 50% of personnel. No deaths anticipated.
270–330	Vomiting and nausea in nearly all personnel on first day, followed by other symptoms of radiation sickness. About 20% deaths within 2–6 weeks after exposure; survivors convalescent for about 3 months.
400–500	Vomiting and nausea in all personnel on first day, followed by other symptoms of radiation sickness. About 50% deaths within 1 month; survivors convalescent for about 6 months.
550–750	Vomiting and nausea in all personnel within 4 hours from exposure, followed by other symptoms of radiation sickness. Up to 100% deaths; few survivors convalescent for about 6 months.
1000	Vomiting and nausea in all personnel with 1–2 hours. Probably no survivors from radiation sickness.
5000	Incapacitation almost immediately. All personnel will be fatalities within 1 week.

be obtained by a study of Tables 1, 2 and 3 which show various radiation dosages and effects.

Table 1, taken from the report of the U. N. Scientific Committee on the Effects of Atomic Radiation[1], cites the normal radiation levels from natural sources, to which everyone is exposed. Table 2, from a report given at a Health Physics Insurance Seminar in March, 1951,[2] shows the radiation exposure received in various diagnostic x-rays. Table 3, from "The Effects of Nuclear Weapons,"[3] shows the effects that can be expected from various large instantaneous doses of gamma radiation.

EFFECTS OF RADIATION

The effects of radiation are largely based on the ionization produced when the energy of the radiation is absorbed in matter. While the physical and even some of the biological effects are understood, many of the biological mechanisms and the ultimate radiation damage in humans are not.

Physical Effects

Since the same physical processes are involved in the absorption of radiation, whether the absorber is air, metal or tissue, we must first consider the general action of radiation when it is absorbed in any material.

The primary effect of the absorption of radiation in matter is that of ionization. Atoms of the absorber are given sufficient energy so that they ionize; i.e., an electron is removed from the atoms leaving a positively charged residue or ion. This ionization takes place in any substance which absorbs radiation. Very roughly, if the amount of absorber in the path of the radiation is expressed in terms of the weight per unit area, the degree of absorption is independent of the material.

For both scientific and practical purposes the common units are grams or milligrams per square centimeter of absorber. This quantity is obtained by multiplying the thickness of absorber, expressed in centimeters, by the density, expressed in grams per cubic centimeter. This product will be in units of grams per square centimeter. For example, if the beta radiation from a source would be completely absorbed by 1100 mg/cm^2 of aluminum it would also be absorbed by 1100 mg/cm^2 of air. However, the absorber thickness would be only about $1/4$ inch for the aluminum and over 30 feet for air.

The result of ionization is merely a conversion of the radiation energy into another form of energy within the absorber. This energy may still be released in such a way as to cause other effects within the absorber, and it is these secondary effects which are of the greatest importance in radiation protection work.

The primary effects of ionization and the distribution of this ionization over various path lengths in different absorbers have been mentioned previously. The different types of radiation also show different degrees of absorption and these differences also are biologically significant. Alpha particles are heavy, slow moving, and expend their energy in a relatively short path. They are, therefore, spoken of as showing high specific ionization; i.e., a large number of ions are formed per unit length of path in the absorber. Gamma and x-radiations, on the other hand, require a great thickness of absorber for complete absorption. Gamma rays and x-rays have a low specific ionization; i.e., the ionization is spread out over the long path required for complete absorption. Beta particles are intermediate in their specific ionization.

Biological Effects

The biological effects of radiation are considered here only in sufficient detail to be of assistance in problems of radiation protection. Some of the information is also required for an understanding of the concepts that have gone into the formulation of permissible levels.

X-rays or gamma rays, because of their penetrating nature may dissipate only a fraction of their energy in passing through the body. This is particularly true of high energy rays. The energy dissipated is, of course, the dose delivered to the body or portion of the body.

Radioisotopes, in contrast, may present a further hazard when the material is taken into the body and irradiates the tissues or organs internally. The most serious effects from this standpoint are produced by the alpha emitters such as radium, uranium and plutonium. They are particularly marked because alpha emitters outside the body expend their energy either in the clothing or in the dead cells of the epidermis and the emitters cannot penetrate to living cells. Once they are taken into the body, this same property of short range and high specific ionization increases their relative effect considerably. Alpha emitters in a small section of tissue will irradiate that small section very heavily.

Beta emitters can be both an internal and an external hazard. The range of most external beta radiation is great enough that the outer tissues, at least, will be penetrated. The most common external effects have been radiation burns and malignancies of the skin. Internally, they may produce a considerable effect. Their specific ionization is high although not as great as that for alpha radiation.

The preceding paragraphs have emphasized the ionization effects, particularly specific ionization. Many secondary effects can be caused by the ionization process. It may disrupt molecules, it may destroy body cells or the energy may merely appear in final form as heat released within the absorber. Depending on the location of the absorbing atom within the molecule, the ionization may or may not disrupt the molecule. If this molecule is in a critical place within the cell, the cell, its function or its ability to reproduce itself may be destroyed. Many of these processes are reversible; that is, damage caused by molecule disruption or cell destruction can be reversed by the usual reparative mechanism of the body. This is confirmed by experimental data which show that a fixed total dose spread out over a period of weeks produces a smaller effect than the same dose delivered in a few minutes.

However, in the case of a large acute dose or continued chronic overexposure, there is the possibility that nonreversible damage will occur.

Another type of cell change which is possible is that the regulative functions of a tissue may be destroyed. In this case a carcinoma (cancer) may be produced. Although the mechanism is not fully understood, there is direct evidence that continued insult to a tissue may produce this result. The high rates of leukemia among radiologists, bone cancer among radium dial painters, and lung cancer among miners of the Czechoslovakian and German uranium mines all point to radiation as the causative agent. This irreversible damage in chronic radiation exposure was apparently cumulative and the cumulative effects led to the illnesses.

Internal Emitters. The biological effects of radiation from radioisotopes in the body are complicated by several factors. In any determination of radiation effects, whether on working populations or in animal experiments, the following factors must be considered: (1) the location of specific isotopes in the body and (2) the relative sensitivity of different tissues to radiation.

The general effects of external radiation have been previously described but there are certain modifications in the consideration of radiation from internal sources. The first is that different elements tend to localize in different organs of the body, e.g., calcium or strontium in bone, iron in the red blood cells and iodine in the thyroid. This is true for any material which is metabolized following either inhalation or ingestion. Of course, many difficultly soluble substances will remain in the lungs for long periods after inhalation. This means that the total amount of such a radioactive material is not distributing its dose uniformly but rather is concentrating the effect in a relatively small fraction of the body.

Most of the heavy metals tend to be deposited in the bone structure. After deposition, there is usually a continuous excretion of the isotope which gradually reduces the amount present. The excretion rate of such materials has been considered to follow much the same pattern as the radioactive decay of an isotope. The time required by the body to eliminate one-half the total quantity it contains is thus referred to as "biological half-life." Most of the experimental data on excretion seem to fit a power function rather than an exponential function, but the concept of biological half-life is still used in deriving permissible levels.

Such body deposits may depend on many physiological factors both in the process of deposition and of excretion. For many years a high calcium diet was recommended for radium workers, as it was supposed that a large excess of calcium entering the body would reduce the amount of radium deposition. Actually, the relative radium deposition is a function of the ratio of radium to calcium in the blood stream. Unless the calcium level of the blood is maintained at a very high value there will still be deposition of radium. The increase in the blood calcium required to cut the radium deposition by even a factor of 3 would be impossible to attain.

Besides the bone structure, common sites of deposition are the lungs and lymph nodes for inhaled particles, and specific organs for certain isotopes, such as the thyroid for iodine and the spleen for iron.

A second consideration is that certain organs or tissues are more radiosensitive than others. The membranes lining the bronchi are supposedly quite sensitive to radiation and this is the primary site of many lung cancers attributed to inhaled radioactive material. The spleen is also sensitive to radiation and relatively small doses have produced more irreversible damage in that organ than in other parts of the body. The organ most likely to be damaged because of the combined effects of concentration and radiosensitivity is known as the critical organ for a particular isotope. In general, any cell in the process of division (mitosis) is radiosensitive and for that reason a person is more sensitive to radiation during his growing period than as an adult.

Radiation Injury. The effects of radiation are non-specific; i.e., other agents or diseases can cause the same damage. For example, it is impossible to distinguish between radiation-induced anemia and normally incident anemia. Other possible effects such as lung cancer, leukemia and bone cancer present similar difficulties.

In any case, where the effects of radiation are being studied, conclusions can only be drawn on the basis of incidence of a particular type of damage above that normally occurring in a comparable population. If tabulations are made of incidence in a particular group, such as chemical operators exposed to radiation in a process plant, the radiation effects can only be evaluated by a statistically valid comparison with a similar group exposed to the same chemical hazards but not the radiation.

Different people exhibit differing degrees of sensitivity to toxic agents. This is also true of radiation effects. Many of the early injuries to radium dial painters occurred with relatively small amounts of radium deposited in the body. Other workers with many times as much radium deposited in their bodies showed no injury even after many years.

The direct effects of ionizing radiation with which we are concerned are those on the cells or tissues exposed to the radiation. The five principal damaging effects are:

(a) Superficial injuries such as skin damage or erythema;

(b) General effects on the body, particularly the blood-forming organs, and nonspecific shortening of life span;

(c) Induction of cancer;

(d) Miscellaneous effects such as cataracts or impaired fertility;

(e) Genetic effects.

All of the effects mentioned are possible with excessive exposure to ionizing radiation. The first four, the somatic effects, were the basis for setting permissible dose levels for occupational exposure.

Not only the total dose, but the rate at which it is delivered, will influence the nature of the effects. For example, about 250 roentgens would be necessary to give a superficial radiation burn from x-rays in a single dose to the skin. The lifetime dose permitted under present standards is about 250 roentgens to the whole body spread over 50 years, and this should produce no measurable effects.

All of the experience resulting in human injury has been gained from large doses of radiation. The effects of low doses of radiation are still not clear. Many scientists believe that all toxic materials including radiation require a certain minimum threshold dose before effects can appear. This is probably true of somatic effects, i.e., injury to the individual receiving the dose, although this is not conceded by all. In the case of genetic effects, it is considered that a linear relationship exists—a very small dose of radiation will produce a very small effect. It is not possible to demonstrate this linear hypothesis since the expected effects could only be shown on a statistical basis by using large populations. The hypothesis cannot be disproved, and it is presently accepted as a conservative assumption in evaluating radiation hazards. This has led to the general concept of comparative risk as a basis of deciding the permissible amounts of radiation to large populations.

The evaluation of possible genetic effects is still controversial. However, genetic data caused some radiation protection advisory groups to recommend lower permissible levels when larger population groups are involved.

PERMISSIBLE LEVELS

A maximum permissible level (mpl) is a recommendation on exposure to a toxic material or a hazardous condition. In chemical exposures, it is clear that there is a threshold below which no damage occurs. In the case of radiation, this is not fully accepted, and the permissible levels of exposure are considered as levels which constitute an acceptable risk.

The initial experience with internal radionuclides came from the radium dial painters who tipped their brushes with their tongues. The experience led to the selection of the permissible body burden, 0.1 μCi of radium, and this in turn was used to develop comparable standards for other radionuclides in the body, particularly alpha emitters. In the same way, the early experience of radiologists led to the acceptance of permissible levels for x-rays. The first permissible levels used were in terms of the erythema dose—the dose causing reddening of the skin. This obviously did not allow for the delayed affects of radiation and the permissible level has been reduced considerably since this early value.

An mpl can be set up in two ways: one is on the basis of the largest quantity of radiation or radioactive material which is known to have caused no damage to the exposed individual; the other is the smallest amount that is known to have caused damage to exposed individuals. Fixing permissible levels could be done best if both the maximum and minimum figures were known. Our experience has been that the maximum value is usually obtained from known exposures of working populations, while the minimum values are obtained from animal experimentation.

Of course, in the cases of x-rays and ingested radium, the minimum values also come from human experiences. The weakness in evaluating human exposure is that the hazard is not recognized until some clinical evidence of damage appears, and at that time, satisfactory records of exposure are often not available. Animal experimentation is unsatisfactory because there must be a final extrapolation from animal experience to an estimate of what might be expected in humans. However, in spite of the failings of the available methods of determining mpl's, the values which have been set can serve as a standard for operation.

The approach to setting maximum levels in the case of radiation and radioisotopes has been more unified than the approach to setting levels for other toxic materials. The large proportion of work with radioactive isotopes in the period of development either was a direct government function or was under private contract with government agencies. This has meant that a centralized group has been studying the common problems appearing in all fields where radiation is a possible hazard. In addition, recognition by scientists that such a hazard existed has resulted in considerable group and community activity in evaluating hazards.

The reductions in permissible levels that have occurred since the concept first arose have been based on a change in the type of damage that was considered to be limiting. The reductions in 1958 were largely based on a change from considerations of somatic damage to the individual to possible genetic damage to the population at large. This also involved the knowledge that an increasingly larger proportion of the population as a whole was receiving some exposure to radiation either occupationally or from other man-made sources.

ICRP

The International Commission on Radiological Protection (ICRP) has actually considered three groups. The first, those occupationally exposed, is the smallest and from genetic considerations is allowed the highest radiation dose (5 rem/yr). The next smallest group has the permissible level set at one-tenth of the occupational level. This special group includes those not occupationally exposed but having access to areas of higher than background radiation, those living in the vicinity of atomic energy installations and others having casual exposure. The third group, which includes the rest of the population, is much larger, and its permissible level is set at one-thirtieth of the occupational level (5 rem in 30 years). The rationale for the three groups is that the total genetic exposure to the population is a function of both the exposure of each group and the number in the group.

Based on the considerations previously described, a set of permissible levels of direct radiation exposure, of amounts in the body (body burden) and of permissible concentrations of radioactive materials in air and water has been established.[4]

The detailed data on distribution of materials in the body and permissible levels for various organs may be found in the report of the ICRP.[4]

In contrast to previous sets of permissible levels, limits are not given for weekly exposure. The period for summing exposures is generally 13 weeks for radiation workers, with a maximum for the year as well, and one year for the population.

The maximum permissible levels set by the ICRP and other groups are for 40 hours exposure weekly over a working lifetime or for 168 hours a week exposure over a lifetime for the general population. This must be remembered in comparing single acute exposures with the ICRP values. These acute exposures should be compared with some integrated value over some reasonable period of time such as 13 weeks or even a calendar year. This, of course, is for cases where the exposure will not be repeated.

NCRP

The major national authority in the field at present is the National Commission on Radiation Protection and Measurements (NCRP). This committee was organized in its present form in 1946 and, in cooperation with other committees and the ICRP, has set the present national working standards for radiation protection. The Committee has been organized into subcommittees of specialists in different fields, and these subcommittees have published several handbooks on different types of radiation hazards under the auspices of the National Bureau of Standards. The current handbooks are listed in the bibliography and additional ones will no doubt appear from time to time, both on new subjects and on modifications of the current material.

In general, the NCRP had adopted the same general levels[5] as the ICRP since there is considerable overlapping of membership. While the recommendations of these two groups do not have the force of law in themselves, they have been adopted in drafting both federal and state codes on radiation protection.

FRC

Since so many agencies of the U. S. Government are concerned with radiation and radiation protection, the Federal Radiation Council (FRC) was formed in 1959, to provide a federal policy on human radiation exposure. The FRC relies quite heavily on the recommendations of the NCRP and ICRP in its deliberations. It has avoided the use of the terms "maximum permis-

TABLE 4

Type of Exposure	Condition	Dose[a] (rem)
Radiation worker:		
(a) Whole body, head and trunk, active blood forming organs, gonads, or lens of eye	Accumulated dose 13 weeks	5 times number of years beyond age 18 3
(b) Skin of whole body and thyroid	Year 13 weeks	30 10
(c) Hands and forearms, feet and ankles	Year 13 weeks	75 25
(d) Bone	Body burden	0.1 μg of ^{226}Ra or its biological equivalent
(e) Other organs	Year 13 weeks	15 5
Population:		
(a) Individual[b]	Year	0.5 (whole body)
(b) Average[b]	30 years	5 (gonads)

[a] Minor variations here from certain other recommendations are not considered significant in light of present uncertainties.

[b] The distinction between individual and average values is necessary since under some conditions only measurements of average values will be possible. The assumption is made here that the individual values will not exceed 3 times the average.

sible level" and "maximum permissible concentration" and has adopted the following:[6]

(1) "Radiation Protection Guide (RPG) is the radiation dose which should not be exceeded without careful consideration of the reasons for doing so; every effort should be made to encourage the maintenance of radiation doses as far below this guide as practicable."

(2) "Radioactivity Concentration Guide (RCG) is the concentration of radioactivity in the environment which is determined to result in whole body or organ doses equal to the Radiation Protection Guide."

These guides will certainly be expected to provide the target levels for operation of government facilities and licensees. Table 4 gives the RPG values for occupational exposures and for the general population.

Only a few RCG values have been issued, and the permissible concentrations for air and water given in Table 5 are those shown in Part 20 of the Code of Federal Regulations, Chapter I. These are reprinted from the Federal Register as revised to December 1965.

The exposures considered by the NCRP and ICRP are theoretically under control; i.e., they are received under industrial conditions and there is some control of the source. The FRC has also attempted to develop guides for uncontrolled situations such as those arising from nuclear weapons tests or nuclear accidents. The principle invoked is the balancing of risk against benefit or the comparative risk of receiving a radiation dose as against the possible consequences of actions taken to avoid the dose.

It must be stressed that the permissible levels set by the ICRP, NCRP and other groups are for continuous exposure for a working lifetime at the occupational levels or continuous exposure for a lifetime at the non-occupational levels. Measurements in the working environment or of the surroundings will probably show peak values, and some of these may exceed the recommended levels. This is not necessarily serious. It does show that a problem exists, but if the levels are merely a few times the permissible, corrections can be instituted in an orderly manner and not on a crash basis. If the levels are hundreds of times the permissible, then immediate action is desirable but it is still not expected that any individual would suffer radiation damage as long as the exposure is continued for only a day or two. This warning on the use of maximum permissible levels may seem unnecessary but experience has shown that this is a major source of misunderstanding.

It should also be clearly understood that the maximum permissible levels have been set up for radiation protection purposes. They do follow the latest proven scientific information as closely as possible; however, the various constants and metabolic factors that have proved to be quite acceptable in radiological health calculations are not satisfactory for more exact scientific purposes.

Many of the assumptions made in developing maximum permissible levels are intentionally conservative. This does not mean that the operator can be allowed leeway in his operations to expose workers to several times the permissible levels on a continuing basis. It does mean, however, that the various calculations and levels set up for protection purposes are not adequate for attempting to predict actual radiation damage to individuals.

TABLE 5.   Concentrations in Air and Water above Natural Background
Table I: In restricted areas (occupational)
Table II: Outside restricted areas

Element-Z	Isotope[a]	Table I		Table II	
		Column 1 Air (μCi/ml)	Column 2 Water (μCi/ml)	Column 1 Air (μCi/ml)	Column 2 Water (μCi/ml)
Actinium-89	^{227}Ac S	2×10^{-12}	6×10^{-5}	8×10^{-14}	2×10^{-6}
	I	3×10^{-11}	9×10^{-3}	9×10^{-13}	3×10^{-4}
	^{228}Ac S	8×10^{-8}	3×10^{-3}	3×10^{-9}	9×10^{-5}
	I	2×10^{-8}	3×10^{-3}	6×10^{-10}	9×10^{-5}
Americium-95	^{241}Am S	6×10^{-12}	1×10^{-4}	2×10^{-13}	4×10^{-6}
	I	1×10^{-10}	8×10^{-4}	4×10^{-12}	2×10^{-5}
	^{242m}Am S	6×10^{-12}	1×10^{-4}	2×10^{-13}	4×10^{-6}
	I	3×10^{-10}	3×10^{-3}	9×10^{-12}	9×10^{-5}

See footnotes at end of table.

TABLE 5 *(Continued)*

Element-Z	Isotope[a]		Table I Column 1 Air (μCi/ml)	Table I Column 2 Water (μCi/ml)	Table II Column 1 Air (μCi/ml)	Table II Column 2 Water (μCi/ml)
Americium-95 continued	^{242}Am	S	4×10^{-8}	4×10^{-3}	1×10^{-9}	1×10^{-4}
		I	5×10^{-8}	4×10^{-3}	2×10^{-9}	1×10^{-4}
	^{243}Am	S	6×10^{-12}	1×10^{-4}	2×10^{-13}	4×10^{-6}
		I	1×10^{-10}	8×10^{-4}	4×10^{-12}	3×10^{-5}
	^{244}Am	S	4×10^{-6}	1×10^{-1}	1×10^{-7}	5×10^{-3}
		I	2×10^{-5}	1×10^{-1}	8×10^{-7}	5×10^{-3}
Antimony-51	^{122}Sb	S	2×10^{-7}	8×10^{-4}	6×10^{-9}	3×10^{-5}
		I	1×10^{-7}	8×10^{-4}	5×10^{-9}	3×10^{-5}
	^{124}Sb	S	2×10^{-7}	7×10^{-4}	5×10^{-9}	2×10^{-5}
		I	2×10^{-8}	7×10^{-4}	7×10^{-10}	2×10^{-5}
	^{125}Sb	S	5×10^{-7}	3×10^{-3}	2×10^{-8}	1×10^{-4}
		I	3×10^{-8}	3×10^{-3}	9×10^{-10}	1×10^{-4}
Argon-18	^{37}A	Sub[b]	6×10^{-3}	—	1×10^{-4}	—
	^{41}A	Sub	2×10^{-6}	—	4×10^{-8}	—
Arsenic-33	^{73}As	S	2×10^{-6}	1×10^{-2}	7×10^{-8}	5×10^{-4}
		I	4×10^{-7}	1×10^{-2}	1×10^{-8}	5×10^{-4}
	^{74}As	S	3×10^{-7}	2×10^{-3}	1×10^{-8}	5×10^{-5}
		I	1×10^{-7}	2×10^{-3}	4×10^{-9}	5×10^{-5}
	^{76}As	S	1×10^{-7}	6×10^{-4}	4×10^{-9}	2×10^{-5}
		I	1×10^{-7}	6×10^{-4}	3×10^{-9}	2×10^{-5}
	^{77}As	S	5×10^{-7}	2×10^{-3}	2×10^{-8}	8×10^{-5}
		I	4×10^{-7}	2×10^{-3}	1×10^{-8}	8×10^{-5}
Astatine-85	^{211}At	S	7×10^{-9}	5×10^{-5}	2×10^{-10}	2×10^{-6}
		I	3×10^{-8}	2×10^{-3}	1×10^{-9}	7×10^{-5}
Barium-56	^{131}Ba	S	1×10^{-6}	5×10^{-3}	4×10^{-8}	2×10^{-4}
		I	4×10^{-7}	5×10^{-3}	1×10^{-8}	2×10^{-4}
	^{140}Ba	S	1×10^{-7}	8×10^{-4}	4×10^{-9}	3×10^{-5}
		I	4×10^{-8}	7×10^{-4}	1×10^{-9}	2×10^{-5}
Berkelium-97	^{249}Bk	S	9×10^{-10}	2×10^{-2}	3×10^{-11}	6×10^{-4}
		I	1×10^{-7}	2×10^{-2}	4×10^{-9}	6×10^{-4}
	^{250}Bk	S	1×10^{-7}	6×10^{-3}	5×10^{-9}	2×10^{-4}
		I	1×10^{-6}	6×10^{-3}	4×10^{-8}	2×10^{-4}
Beryllium-4	^{7}Be	S	6×10^{-6}	5×10^{-2}	2×10^{-7}	2×10^{-3}
		I	1×10^{-6}	5×10^{-2}	4×10^{-8}	2×10^{-3}
Bismuth-83	^{206}Bi	S	2×10^{-7}	1×10^{-3}	6×10^{-9}	4×10^{-5}
		I	1×10^{-7}	1×10^{-3}	5×10^{-9}	4×10^{-5}
	^{207}Bi	S	2×10^{-7}	2×10^{-3}	6×10^{-9}	6×10^{-5}
		I	1×10^{-8}	2×10^{-3}	5×10^{-10}	6×10^{-5}
	^{210}Bi	S	6×10^{-9}	1×10^{-3}	2×10^{-10}	4×10^{-5}
		I	6×10^{-9}	1×10^{-3}	2×10^{-10}	4×10^{-5}
	^{212}Bi	S	1×10^{-7}	1×10^{-2}	3×10^{-9}	4×10^{-4}
		I	2×10^{-7}	1×10^{-2}	7×10^{-9}	4×10^{-4}
Bromine-35	^{82}Br	S	1×10^{-6}	8×10^{-3}	4×10^{-8}	3×10^{-4}
		I	2×10^{-7}	1×10^{-3}	6×10^{-9}	4×10^{-5}
Cadmium-48	^{109}Cd	S	5×10^{-8}	5×10^{-3}	2×10^{-9}	2×10^{-4}
		I	7×10^{-8}	5×10^{-3}	3×10^{-9}	2×10^{-4}
	^{115m}Cd	S	4×10^{-8}	7×10^{-4}	1×10^{-9}	3×10^{-5}
		I	4×10^{-8}	7×10^{-4}	1×10^{-9}	3×10^{-5}
	^{115}Cd	S	2×10^{-7}	1×10^{-3}	8×10^{-9}	3×10^{-5}
		I	2×10^{-7}	1×10^{-3}	6×10^{-9}	4×10^{-5}

See footnotes at end of table.

TABLE 5 (*Continued*)

Element-Z	Isotope[a]	Table I Column 1 Air (μCi/ml)	Table I Column 2 Water (μCi/ml)	Table II Column 1 Air (μCi/ml)	Table II Column 2 Water (μCi/ml)
Calcium-20	^{45}Ca S	3×10^{-8}	3×10^{-4}	1×10^{-9}	9×10^{-6}
	I	1×10^{-7}	5×10^{-3}	4×10^{-9}	2×10^{-4}
	^{47}Ca S	2×10^{-7}	1×10^{-3}	6×10^{-9}	5×10^{-5}
	I	2×10^{-7}	1×10^{-3}	6×10^{-9}	3×10^{-5}
Californium-98	^{249}Cf S	2×10^{-12}	1×10^{-4}	5×10^{-14}	4×10^{-6}
	I	1×10^{-10}	7×10^{-4}	3×10^{-12}	2×10^{-5}
	^{250}Cf S	5×10^{-12}	4×10^{-4}	2×10^{-13}	1×10^{-5}
	I	1×10^{-10}	7×10^{-4}	3×10^{-12}	3×10^{-5}
	^{251}Cf S	2×10^{-12}	1×10^{-4}	6×10^{-14}	4×10^{-6}
	I	1×10^{-10}	8×10^{-4}	3×10^{-12}	3×10^{-5}
	^{252}Cf S	2×10^{-11}	7×10^{-4}	7×10^{-13}	2×10^{-5}
	I	1×10^{-10}	7×10^{-4}	4×10^{-12}	2×10^{-5}
Californium-98	^{253}Cf S	8×10^{-10}	4×10^{-3}	3×10^{-11}	1×10^{-4}
	I	8×10^{-10}	4×10^{-3}	3×10^{-11}	1×10^{-4}
	^{254}Cf S	5×10^{-12}	4×10^{-6}	2×10^{-13}	1×10^{-7}
	I	5×10^{-12}	4×10^{-6}	2×10^{-13}	1×10^{-7}
Carbon-6	^{14}C S	4×10^{-6}	2×10^{-2}	1×10^{-7}	8×10^{-4}
	(CO_2) Sub	5×10^{-5}	—	1×10^{-6}	—
Cerium-58	^{141}Ce S	4×10^{-7}	3×10^{-3}	2×10^{-8}	9×10^{-5}
	I	2×10^{-7}	3×10^{-3}	5×10^{-9}	9×10^{-5}
	^{143}Ce S	3×10^{-7}	1×10^{-3}	9×10^{-9}	4×10^{-5}
	I	2×10^{-7}	1×10^{-3}	7×10^{-9}	4×10^{-5}
	^{144}Ce S	1×10^{-8}	3×10^{-4}	3×10^{-10}	1×10^{-5}
	I	6×10^{-9}	3×10^{-4}	2×10^{-10}	1×10^{-5}
Cesium-55	^{131}Cs S	1×10^{-5}	7×10^{-2}	4×10^{-7}	2×10^{-3}
	I	3×10^{-6}	3×10^{-2}	1×10^{-7}	9×10^{-4}
	^{134m}Cs S	4×10^{-5}	2×10^{-1}	1×10^{-6}	6×10^{-3}
	I	6×10^{-6}	3×10^{-2}	2×10^{-7}	1×10^{-3}
	^{134}Cs S	4×10^{-8}	3×10^{-4}	1×10^{-9}	9×10^{-6}
	I	1×10^{-8}	1×10^{-3}	4×10^{-10}	4×10^{-5}
	^{135}Cs S	5×10^{-7}	3×10^{-3}	2×10^{-8}	1×10^{-4}
	I	9×10^{-8}	7×10^{-3}	3×10^{-9}	2×10^{-4}
	^{136}Cs S	4×10^{-7}	2×10^{-3}	1×10^{-8}	9×10^{-5}
	I	2×10^{-7}	2×10^{-3}	6×10^{-9}	6×10^{-5}
	^{137}Cs S	6×10^{-8}	4×10^{-4}	2×10^{-9}	2×10^{-5}
	I	1×10^{-8}	1×10^{-3}	5×10^{-10}	4×10^{-5}
Chlorine-17	^{36}Cl S	4×10^{-7}	2×10^{-3}	1×10^{-8}	8×10^{-5}
	I	2×10^{-8}	2×10^{-3}	8×10^{-10}	6×10^{-5}
	^{38}Cl S	3×10^{-6}	1×10^{-2}	9×10^{-8}	4×10^{-4}
	I	2×10^{-6}	1×10^{-2}	7×10^{-8}	4×10^{-4}
Chromium-24	^{51}Cr S	1×10^{-5}	5×10^{-2}	4×10^{-7}	2×10^{-3}
	I	2×10^{-6}	5×10^{-2}	8×10^{-8}	2×10^{-3}
Cobalt-27	^{57}Co S	3×10^{-6}	2×10^{-2}	1×10^{-7}	5×10^{-4}
	I	2×10^{-7}	1×10^{-2}	6×10^{-9}	4×10^{-4}
	^{58m}Co S	2×10^{-5}	8×10^{-2}	6×10^{-7}	3×10^{-3}
	I	9×10^{-6}	6×10^{-2}	3×10^{-7}	2×10^{-3}
	^{58}Co S	8×10^{-7}	4×10^{-3}	3×10^{-8}	1×10^{-4}
	I	5×10^{-8}	3×10^{-3}	2×10^{-9}	9×10^{-5}
	^{60}Co S	3×10^{-7}	1×10^{-3}	1×10^{-8}	5×10^{-5}
	I	9×10^{-9}	1×10^{-3}	3×10^{-10}	3×10^{-5}

See footnotes at end of table.

TABLE 5 (*Continued*)

Element-Z	Isotope[a]	Table I Column 1 Air (μCi/ml)	Table I Column 2 Water (μCi/ml)	Table II Column 1 Air (μCi/ml)	Table II Column 2 Water (μCi/ml)
Copper-29	^{64}Cu S	2×10^{-6}	1×10^{-2}	7×10^{-8}	3×10^{-4}
	I	1×10^{-6}	6×10^{-3}	4×10^{-8}	2×10^{-4}
Curium-96	^{242}Cm S	1×10^{-10}	7×10^{-4}	4×10^{-12}	2×10^{-5}
	I	2×10^{-10}	7×10^{-4}	6×10^{-12}	3×10^{-5}
	^{243}Cm S	6×10^{-12}	1×10^{-4}	2×10^{-13}	5×10^{-6}
	I	1×10^{-10}	7×10^{-4}	3×10^{-12}	2×10^{-5}
	$^{244\alpha}$Cm S	9×10^{-12}	2×10^{-4}	3×10^{-13}	7×10^{-6}
	I	1×10^{-10}	8×10^{-4}	3×10^{-12}	3×10^{-5}
	^{245}Cm S	5×10^{-12}	1×10^{-4}	2×10^{-13}	4×10^{-6}
	I	1×10^{-10}	8×10^{-4}	4×10^{-12}	3×10^{-5}
	^{246}Cm S	5×10^{-12}	1×10^{-4}	2×10^{-13}	4×10^{-6}
	I	1×10^{-10}	8×10^{-4}	4×10^{-12}	3×10^{-5}
	^{247}Cm S	5×10^{-12}	1×10^{-4}	2×10^{-13}	4×10^{-6}
	I	1×10^{-10}	6×10^{-4}	4×10^{-12}	2×10^{-5}
	^{248}Cm S	6×10^{-13}	1×10^{-5}	2×10^{-14}	4×10^{-7}
	I	1×10^{-11}	4×10^{-5}	4×10^{-13}	1×10^{-6}
	^{249}Cm S	1×10^{-5}	6×10^{-2}	4×10^{-7}	2×10^{-3}
	I	1×10^{-5}	6×10^{-2}	4×10^{-7}	2×10^{-3}
Dysprosium-66	^{165}Dy S	3×10^{-6}	1×10^{-2}	9×10^{-8}	4×10^{-4}
	I	2×10^{-6}	1×10^{-2}	7×10^{-8}	4×10^{-4}
	^{166}Dy S	2×10^{-7}	1×10^{-3}	8×10^{-9}	4×10^{-5}
	I	2×10^{-7}	1×10^{-3}	7×10^{-9}	4×10^{-5}
Einsteinium-99	^{253}Es S	8×10^{-10}	7×10^{-4}	3×10^{-11}	2×10^{-5}
	I	6×10^{-10}	7×10^{-4}	2×10^{-11}	2×10^{-5}
	^{254m}Es S	5×10^{-9}	5×10^{-4}	2×10^{-10}	2×10^{-5}
	I	6×10^{-9}	5×10^{-4}	2×10^{-10}	2×10^{-5}
	^{254}Es S	2×10^{-11}	4×10^{-4}	6×10^{-13}	1×10^{-5}
	I	1×10^{-10}	4×10^{-4}	4×10^{-12}	1×10^{-5}
	^{255}Es S	5×10^{-10}	8×10^{-4}	2×10^{-11}	3×10^{-5}
	I	4×10^{-10}	8×10^{-4}	1×10^{-11}	3×10^{-5}
Erbium-68	^{169}Er S	6×10^{-7}	3×10^{-3}	2×10^{-8}	9×10^{-5}
	I	4×10^{-7}	3×10^{-3}	1×10^{-8}	9×10^{-5}
	^{171}Er S	7×10^{-7}	3×10^{-3}	2×10^{-8}	1×10^{-4}
	I	6×10^{-7}	3×10^{-3}	2×10^{-8}	1×10^{-4}
Europium-63	^{152}Eu S (T/2 = 9.2 hr)	4×10^{-7}	2×10^{-3}	1×10^{-8}	6×10^{-5}
	I	3×10^{-7}	2×10^{-3}	1×10^{-8}	6×10^{-5}
	^{152}Eu S (T/2 = 13 yr)	1×10^{-8}	2×10^{-3}	4×10^{-10}	8×10^{-5}
	I	2×10^{-8}	2×10^{-3}	6×10^{-10}	8×10^{-5}
	^{154}Eu S	4×10^{-9}	6×10^{-4}	1×10^{-10}	2×10^{-5}
	I	7×10^{-9}	6×10^{-4}	2×10^{-10}	2×10^{-5}
	^{155}Eu S	9×10^{-8}	6×10^{-3}	3×10^{-9}	2×10^{-4}
	I	7×10^{-8}	6×10^{-3}	3×10^{-9}	2×10^{-4}
Fermium-100	^{254}Fm S	6×10^{-8}	4×10^{-3}	2×10^{-9}	1×10^{-4}
	I	7×10^{-8}	4×10^{-3}	2×10^{-9}	1×10^{-4}
	^{255}Fm S	2×10^{-8}	1×10^{-3}	6×10^{-10}	3×10^{-5}
	I	1×10^{-8}	1×10^{-3}	4×10^{-10}	3×10^{-5}
	^{256}Fm S	3×10^{-9}	3×10^{-5}	1×10^{-10}	9×10^{-7}
	I	2×10^{-9}	3×10^{-5}	6×10^{-11}	9×10^{-7}
Fluorine-9	^{18}F S	5×10^{-6}	2×10^{-2}	2×10^{-7}	8×10^{-4}
	I	3×10^{-6}	1×10^{-2}	9×10^{-8}	5×10^{-4}

See footnotes at end of table.

TABLE 5 (*Continued*)

Element-Z	Isotope[a]		Table I		Table II	
			Column 1 Air (μCi/ml)	Column 2 Water (μCi/ml)	Column 1 Air (μCi/ml)	Column 2 Water (μCi/ml)
Gadolinium-64	^{153}Gd	S	2×10^{-7}	6×10^{-3}	8×10^{-9}	2×10^{-4}
		I	9×10^{-8}	6×10^{-3}	3×10^{-9}	2×10^{-4}
	^{159}Gd	S	5×10^{-7}	2×10^{-3}	2×10^{-8}	8×10^{-5}
		I	4×10^{-7}	2×10^{-3}	1×10^{-8}	8×10^{-5}
Gallium-31	^{72}Ga	S	2×10^{-7}	1×10^{-3}	8×10^{-9}	4×10^{-5}
		I	2×10^{-7}	1×10^{-3}	6×10^{-9}	4×10^{-5}
Germanium-32	^{71}Ge	S	1×10^{-5}	5×10^{-2}	4×10^{-7}	2×10^{-3}
		I	6×10^{-6}	5×10^{-2}	2×10^{-7}	2×10^{-3}
Gold-79	^{196}Au	S	1×10^{-6}	5×10^{-3}	4×10^{-8}	2×10^{-4}
		I	6×10^{-7}	4×10^{-3}	2×10^{-8}	1×10^{-4}
	^{198}Au	S	3×10^{-7}	2×10^{-3}	1×10^{-8}	5×10^{-5}
		I	2×10^{-7}	1×10^{-3}	8×10^{-9}	5×10^{-5}
	^{199}Au	S	1×10^{-6}	5×10^{-3}	4×10^{-8}	2×10^{-4}
		I	8×10^{-7}	4×10^{-3}	3×10^{-8}	2×10^{-4}
Hafnium-72	^{181}Hf	S	4×10^{-8}	2×10^{-3}	1×10^{-9}	7×10^{-5}
		I	7×10^{-8}	2×10^{-3}	3×10^{-9}	7×10^{-5}
Holmium-67	^{166}Ho	S	2×10^{-7}	9×10^{-4}	7×10^{-9}	3×10^{-5}
		I	2×10^{-7}	9×10^{-4}	6×10^{-9}	3×10^{-5}
Hydrogen-1	^{3}H	S	5×10^{-6}	1×10^{-1}	2×10^{-7}	3×10^{-3}
		I	5×10^{-6}	1×10^{-1}	2×10^{-7}	3×10^{-3}
		Sub	2×10^{-3}	—	4×10^{-5}	—
Indium-49	^{113m}In	S	8×10^{-6}	4×10^{-2}	3×10^{-7}	1×10^{-3}
		I	7×10^{-6}	4×10^{-2}	2×10^{-7}	1×10^{-3}
	^{114m}In	S	1×10^{-7}	5×10^{-4}	4×10^{-9}	2×10^{-5}
		I	2×10^{-8}	5×10^{-4}	7×10^{-10}	2×10^{-5}
	^{115m}In	S	2×10^{-6}	1×10^{-2}	8×10^{-8}	4×10^{-4}
		I	2×10^{-6}	1×10^{-2}	6×10^{-8}	4×10^{-4}
	^{115}In	S	2×10^{-7}	3×10^{-3}	9×10^{-9}	9×10^{-5}
		I	3×10^{-8}	3×10^{-3}	1×10^{-9}	9×10^{-5}
Iodine-53	^{125}I	S	5×10^{-9}	4×10^{-5}	8×10^{-11}	2×10^{-7}
		I	2×10^{-7}	6×10^{-3}	6×10^{-9}	2×10^{-4}
	^{126}I	S	8×10^{-9}	5×10^{-5}	9×10^{-11}	3×10^{-7}
		I	3×10^{-7}	3×10^{-3}	1×10^{-8}	9×10^{-5}
	^{129}I	S	2×10^{-9}	1×10^{-5}	2×10^{-11}	6×10^{-8}
		I	7×10^{-8}	6×10^{-3}	2×10^{-9}	2×10^{-4}
	^{131}I	S	9×10^{-9}	6×10^{-5}	1×10^{-10}	3×10^{-7}
		I	3×10^{-7}	2×10^{-3}	1×10^{-8}	6×10^{-5}
	^{132}I	S	2×10^{-7}	2×10^{-3}	3×10^{-9}	8×10^{-6}
		I	9×10^{-7}	5×10^{-3}	3×10^{-8}	2×10^{-4}
	^{133}I	S	3×10^{-8}	2×10^{-4}	4×10^{-10}	1×10^{-6}
		I	2×10^{-7}	1×10^{-3}	7×10^{-9}	4×10^{-5}
	^{134}I	S	5×10^{-7}	4×10^{-3}	6×10^{-9}	2×10^{-5}
		I	3×10^{-6}	2×10^{-2}	1×10^{-7}	6×10^{-4}
	^{135}I	S	1×10^{-7}	7×10^{-4}	1×10^{-9}	4×10^{-6}
		I	4×10^{-7}	2×10^{-3}	1×10^{-8}	7×10^{-5}
Iridium-77	^{190}Ir	S	1×10^{-6}	6×10^{-3}	4×10^{-8}	2×10^{-4}
		I	4×10^{-7}	5×10^{-3}	1×10^{-8}	2×10^{-4}
	^{192}Ir	S	1×10^{-7}	1×10^{-3}	4×10^{-9}	4×10^{-5}
		I	3×10^{-8}	1×10^{-3}	9×10^{-10}	4×10^{-5}

See footnotes at end of table.

TABLE 5 (*Continued*)

Element-Z	Isotope[a]	Table I Column 1 Air (μCi/ml)	Table I Column 2 Water (μCi/ml)	Table II Column 1 Air (μCi/ml)	Table II Column 2 Water (μCi/ml)
Iridium-77 continued	^{194}Ir S	2×10^{-7}	1×10^{-3}	8×10^{-9}	3×10^{-5}
	I	2×10^{-7}	9×10^{-4}	5×10^{-9}	3×10^{-5}
Iron-26	^{55}Fe S	9×10^{-7}	2×10^{-2}	3×10^{-8}	8×10^{-4}
	I	1×10^{-6}	7×10^{-2}	3×10^{-8}	2×10^{-3}
	^{59}Fe S	1×10^{-7}	2×10^{-3}	5×10^{-9}	6×10^{-5}
	I	5×10^{-9}	2×10^{-3}	2×10^{-9}	5×10^{-5}
Krypton-36[b]	^{85m}Kr Sub	6×10^{-6}	—	1×10^{-7}	—
	^{85}Kr Sub	1×10^{-5}	—	3×10^{-7}	—
	^{87}Kr Sub	1×10^{-6}	—	2×10^{-8}	—
Krypton-36	^{88}Kr Sub	1×10^{-6}	—	2×10^{-8}	—
Lanthanum-57	^{140}La S	2×10^{-7}	7×10^{-4}	5×10^{-9}	2×10^{-5}
	I	1×10^{-7}	7×10^{-4}	4×10^{-9}	2×10^{-5}
Lead-82	^{203}Pb S	3×10^{-6}	1×10^{-2}	9×10^{-8}	4×10^{-4}
	I	2×10^{-6}	1×10^{-2}	6×10^{-8}	4×10^{-4}
	^{210}Pb S	1×10^{-10}	4×10^{-6}	4×10^{-12}	1×10^{-7}
	I	2×10^{-10}	5×10^{-3}	8×10^{-12}	2×10^{-4}
	^{212}Pb S	2×10^{-8}	6×10^{-4}	6×10^{-10}	2×10^{-5}
	I	2×10^{-8}	5×10^{-4}	7×10^{-10}	2×10^{-5}
Lutetium-71	^{177}Lu S	6×10^{-7}	3×10^{-3}	2×10^{-8}	1×10^{-4}
	I	5×10^{-7}	3×10^{-3}	2×10^{-8}	1×10^{-4}
Manganese-25	^{52}Mn S	2×10^{-7}	1×10^{-3}	7×10^{-9}	3×10^{-5}
	I	1×10^{-7}	9×10^{-4}	5×10^{-9}	3×10^{-5}
	^{54}Mn S	4×10^{-7}	4×10^{-3}	1×10^{-9}	1×10^{-4}
	I	4×10^{-8}	3×10^{-3}	1×10^{-9}	1×10^{-4}
	^{56}Mn S	8×10^{-7}	4×10^{-3}	3×10^{-8}	1×10^{-4}
	I	5×10^{-7}	3×10^{-3}	2×10^{-8}	1×10^{-4}
Mercury-80	^{197m}Hg S	7×10^{-7}	6×10^{-3}	3×10^{-8}	2×10^{-4}
	I	8×10^{-7}	5×10^{-3}	3×10^{-8}	2×10^{-4}
	^{197}Hg S	1×10^{-6}	9×10^{-3}	4×10^{-8}	3×10^{-4}
	I	3×10^{-6}	1×10^{-2}	9×10^{-8}	5×10^{-4}
	^{203}Hg S	7×10^{-8}	5×10^{-4}	2×10^{-9}	2×10^{-5}
	I	1×10^{-7}	3×10^{-3}	4×10^{-9}	1×10^{-4}
Molybdenum-42	^{99}Mo S	7×10^{-7}	5×10^{-3}	3×10^{-8}	2×10^{-4}
	I	2×10^{-7}	1×10^{-3}	7×10^{-9}	4×10^{-5}
Neodymium-60	^{144}Nd S	8×10^{-11}	2×10^{-3}	3×10^{-12}	7×10^{-5}
	I	3×10^{-10}	2×10^{-3}	1×10^{-11}	8×10^{-5}
	^{147}Nd S	4×10^{-7}	2×10^{-3}	1×10^{-8}	6×10^{-5}
	I	2×10^{-7}	2×10^{-3}	8×10^{-9}	6×10^{-5}
	^{149}Nd S	2×10^{-6}	8×10^{-3}	6×10^{-8}	3×10^{-4}
	I	1×10^{-6}	8×10^{-3}	5×10^{-8}	3×10^{-4}
Neptunium-93	^{237}Np S	4×10^{-12}	9×10^{-5}	1×10^{-13}	3×10^{-6}
	I	1×10^{-10}	9×10^{-4}	4×10^{-12}	3×10^{-5}
	^{239}Np S	8×10^{-7}	4×10^{-3}	3×10^{-8}	1×10^{-4}
	I	7×10^{-7}	4×10^{-3}	2×10^{-8}	1×10^{-4}
Nickel-28	^{59}Ni S	5×10^{-7}	6×10^{-3}	2×10^{-8}	2×10^{-4}
	I	8×10^{-7}	6×10^{-2}	3×10^{-8}	2×10^{-3}
	^{63}Ni S	6×10^{-8}	8×10^{-4}	2×10^{-9}	3×10^{-5}
	I	3×10^{-7}	2×10^{-2}	1×10^{-8}	7×10^{-4}
	^{65}Ni S	9×10^{-7}	4×10^{-3}	3×10^{-8}	1×10^{-4}
	I	5×10^{-7}	3×10^{-3}	2×10^{-8}	1×10^{-4}

See footnotes at end of table.

TABLE 5 (*Continued*)

Element-Z	Isotope[a]	Table I		Table II	
		Column 1 Air (μCi/ml)	Column 2 Water (μCi/ml)	Column 1 Air (μCi/ml)	Column 2 Water (μCi/ml)
Niobium-41	^{93m}Nb S	1×10^{-7}	1×10^{-2}	4×10^{-9}	4×10^{-4}
(Columbium)	I	2×10^{-7}	1×10^{-2}	5×10^{-9}	4×10^{-4}
	^{95}Nb S	5×10^{-7}	3×10^{-3}	2×10^{-8}	1×10^{-4}
	I	1×10^{-7}	3×10^{-3}	3×10^{-9}	1×10^{-4}
	^{97}Nb S	6×10^{-6}	3×10^{-2}	2×10^{-7}	9×10^{-4}
	I	5×10^{-6}	3×10^{-2}	2×10^{-7}	9×10^{-4}
Osmium-76	^{185}Os S	5×10^{-7}	2×10^{-3}	2×10^{-8}	7×10^{-5}
	I	5×10^{-8}	2×10^{-3}	2×10^{-9}	7×10^{-5}
	^{191m}Os S	2×10^{-5}	7×10^{-2}	6×10^{-7}	3×10^{-3}
	I	9×10^{-6}	7×10^{-2}	3×10^{-7}	2×10^{-3}
	^{191}Os S	1×10^{-6}	5×10^{-3}	4×10^{-8}	2×10^{-4}
	I	4×10^{-7}	5×10^{-3}	1×10^{-8}	2×10^{-4}
	^{193}Os S	4×10^{-7}	2×10^{-3}	1×10^{-8}	6×10^{-5}
	I	3×10^{-7}	2×10^{-3}	9×10^{-9}	5×10^{-5}
Palladium-46	^{103}Pd S	1×10^{-6}	1×10^{-2}	5×10^{-8}	3×10^{-4}
	I	7×10^{-7}	8×10^{-3}	3×10^{-8}	3×10^{-4}
	^{109}Pd S	6×10^{-7}	3×10^{-3}	2×10^{-8}	9×10^{-5}
	I	4×10^{-7}	2×10^{-3}	1×10^{-8}	7×10^{-5}
Phosphorus-15	^{32}P S	7×10^{-8}	5×10^{-4}	2×10^{-9}	2×10^{-5}
	I	8×10^{-8}	7×10^{-4}	3×10^{-9}	2×10^{-5}
Platinum-78	^{191}Pt S	8×10^{-7}	4×10^{-3}	3×10^{-8}	1×10^{-4}
	I	6×10^{-7}	3×10^{-3}	2×10^{-8}	1×10^{-4}
	^{193m}Pt S	7×10^{-6}	3×10^{-2}	2×10^{-7}	1×10^{-3}
	I	5×10^{-6}	3×10^{-2}	2×10^{-7}	1×10^{-3}
	^{197m}Pt S	6×10^{-6}	3×10^{-2}	2×10^{-7}	1×10^{-3}
	I	5×10^{-6}	3×10^{-2}	2×10^{-7}	9×10^{-4}
	^{197}Pt S	8×10^{-7}	4×10^{-3}	3×10^{-8}	1×10^{-4}
	I	6×10^{-7}	3×10^{-3}	2×10^{-8}	1×10^{-4}
Plutonium-94	^{238}Pu S	2×10^{-12}	1×10^{-4}	7×10^{-14}	5×10^{-6}
	I	3×10^{-11}	8×10^{-4}	1×10^{-12}	3×10^{-5}
	^{239}Pu S	2×10^{-12}	1×10^{-4}	6×10^{-14}	5×10^{-6}
	I	4×10^{-11}	8×10^{-4}	1×10^{-12}	3×10^{-5}
Plutonium-94	^{240}Pu S	2×10^{-12}	1×10^{-4}	6×10^{-14}	5×10^{-6}
	I	4×10^{-11}	8×10^{-4}	1×10^{-12}	3×10^{-5}
	^{241}Pu S	9×10^{-11}	7×10^{-3}	3×10^{-12}	2×10^{-4}
	I	4×10^{-8}	4×10^{-2}	1×10^{-9}	1×10^{-3}
	^{242}Pu S	2×10^{-12}	1×10^{-4}	6×10^{-14}	5×10^{-6}
	I	4×10^{-11}	9×10^{-4}	1×10^{-12}	3×10^{-5}
	^{243}Pu S	2×10^{-6}	1×10^{-2}	6×10^{-8}	3×10^{-4}
	I	2×10^{-6}	1×10^{-2}	8×10^{-8}	3×10^{-4}
	^{244}Pu S	2×10^{-12}	1×10^{-4}	6×10^{-14}	4×10^{-6}
	I	3×10^{-11}	3×10^{-4}	1×10^{-12}	1×10^{-5}
Polonium-84	^{210}Po S	5×10^{-10}	2×10^{-5}	2×10^{-11}	7×10^{-7}
	I	2×10^{-10}	8×10^{-4}	7×10^{-12}	3×10^{-5}
Potassium-19	^{42}K S	2×10^{-6}	9×10^{-3}	7×10^{-8}	3×10^{-4}
	I	1×10^{-7}	6×10^{-4}	4×10^{-9}	2×10^{-5}
Praseodymium-59	^{142}Pr S	2×10^{-7}	9×10^{-4}	7×10^{-9}	3×10^{-5}
	I	2×10^{-7}	9×10^{-4}	5×10^{-9}	3×10^{-5}
	^{143}Pr S	3×10^{-7}	1×10^{-3}	1×10^{-8}	5×10^{-5}
	I	2×10^{-7}	1×10^{-3}	6×10^{-9}	5×10^{-5}

See footnotes at end of table.

TABLE 5 (*Continued*)

Element-Z	Isotope[a]	Table I Column 1 Air (μCi/ml)	Table I Column 2 Water (μCi/ml)	Table II Column 1 Air (μCi/ml)	Table II Column 2 Water (μCi/ml)
Promethium-61	^{147}Pm S	6×10^{-8}	6×10^{-3}	2×10^{-9}	2×10^{-4}
	I	1×10^{-7}	6×10^{-3}	3×10^{-9}	2×10^{-4}
	^{149}Pm S	3×10^{-7}	1×10^{-3}	1×10^{-8}	4×10^{-5}
	I	2×10^{-7}	1×10^{-3}	8×10^{-9}	4×10^{-5}
Protoactinium-91	^{230}Pa S	2×10^{-9}	7×10^{-3}	6×10^{-11}	2×10^{-4}
	I	8×10^{-10}	7×10^{-3}	3×10^{-11}	2×10^{-4}
	^{231}Pa S	1×10^{-12}	3×10^{-5}	4×10^{-14}	9×10^{-7}
	I	1×10^{-10}	8×10^{-4}	4×10^{-12}	2×10^{-5}
	^{233}Pa S	6×10^{-7}	4×10^{-3}	2×10^{-8}	1×10^{-4}
	I	2×10^{-7}	3×10^{-3}	6×10^{-9}	1×10^{-4}
Radium-88	^{223}Ra S	2×10^{-9}	2×10^{-5}	6×10^{-11}	7×10^{-7}
	I	2×10^{-10}	1×10^{-4}	8×10^{-12}	4×10^{-6}
	^{224}Ra S	5×10^{-9}	7×10^{-5}	2×10^{-10}	2×10^{-6}
	I	7×10^{-10}	2×10^{-4}	2×10^{-11}	5×10^{-6}
	^{226}Ra S	3×10^{-11}	4×10^{-7}	3×10^{-12}	3×10^{-8}
	I	5×10^{-11}	9×10^{-4}	2×10^{-12}	3×10^{-5}
	^{228}Ra S	7×10^{-11}	8×10^{-7}	2×10^{-12}	3×10^{-8}
	I	4×10^{-11}	7×10^{-4}	1×10^{-12}	3×10^{-5}
Radon-86	^{220}Rn S	3×10^{-7}	—	1×10^{-8}	—
	I	—	—	—	—
	^{222}Rn S	1×10^{-7}	—	3×10^{-9}	—
Rhenium-75	^{183}Re S	3×10^{-6}	2×10^{-2}	9×10^{-8}	6×10^{-4}
	I	2×10^{-7}	8×10^{-3}	5×10^{-9}	3×10^{-4}
	^{186}Re S	6×10^{-7}	3×10^{-3}	2×10^{-8}	9×10^{-5}
	I	2×10^{-7}	1×10^{-3}	8×10^{-9}	5×10^{-5}
	^{187}Re S	9×10^{-6}	7×10^{-2}	3×10^{-7}	3×10^{-3}
	I	5×10^{-7}	4×10^{-2}	2×10^{-8}	2×10^{-3}
	^{188}Re S	4×10^{-7}	2×10^{-3}	1×10^{-8}	6×10^{-5}
	I	2×10^{-7}	9×10^{-4}	6×10^{-9}	3×10^{-5}
Rhodium-45	^{103m}Rh S	8×10^{-5}	4×10^{-1}	3×10^{-6}	1×10^{-2}
	I	6×10^{-5}	3×10^{-1}	2×10^{-6}	1×10^{-2}
	^{105}Rh S	8×10^{-7}	4×10^{-3}	3×10^{-8}	1×10^{-4}
	I	5×10^{-7}	3×10^{-3}	2×10^{-8}	1×10^{-4}
Rubidium-37	^{86}Rb S	3×10^{-7}	2×10^{-3}	1×10^{-8}	7×10^{-5}
	I	7×10^{-8}	7×10^{-4}	2×10^{-9}	2×10^{-5}
	^{87}Rb S	5×10^{-7}	3×10^{-3}	2×10^{-8}	1×10^{-4}
	I	7×10^{-8}	5×10^{-3}	2×10^{-9}	2×10^{-4}
Ruthenium-44	^{97}Ru S	2×10^{-6}	1×10^{-2}	8×10^{-8}	4×10^{-4}
	I	2×10^{-6}	1×10^{-2}	6×10^{-8}	3×10^{-4}
	^{103}Ru S	5×10^{-7}	2×10^{-3}	2×10^{-8}	8×10^{-5}
	I	8×10^{-8}	2×10^{-3}	3×10^{-9}	8×10^{-5}
	^{105}Ru S	7×10^{-7}	3×10^{-3}	2×10^{-8}	1×10^{-4}
	I	5×10^{-7}	3×10^{-3}	2×10^{-8}	1×10^{-4}
	^{106}Ru S	8×10^{-8}	4×10^{-4}	3×10^{-9}	1×10^{-5}
	I	6×10^{-9}	3×10^{-4}	2×10^{-10}	1×10^{-5}
Samarium-62	^{147}Sm S	7×10^{-11}	2×10^{-3}	2×10^{-12}	6×10^{-5}
	I	3×10^{-10}	2×10^{-3}	9×10^{-12}	7×10^{-5}
	^{151}Sm S	6×10^{-8}	1×10^{-2}	2×10^{-9}	4×10^{-4}
	I	1×10^{-7}	1×10^{-2}	5×10^{-9}	4×10^{-4}
	^{153}Sm S	5×10^{-7}	2×10^{-3}	2×10^{-8}	8×10^{-5}
	I	4×10^{-7}	2×10^{-3}	1×10^{-8}	8×10^{-5}

See footnotes at end of table.

TABLE 5 (*Continued*)

Element-Z	Isotope[a]	Table I		Table II	
		Column 1 Air (μCi/ml)	Column 2 Water (μCi/ml)	Column 1 Air (μCi/ml)	Column 2 Water (μCi/ml)
Scandium-21	^{46}Sc S	2×10^{-7}	1×10^{-3}	8×10^{-9}	4×10^{-5}
	I	2×10^{-8}	1×10^{-3}	8×10^{-10}	4×10^{-5}
	^{47}Sc S	6×10^{-7}	3×10^{-3}	2×10^{-8}	9×10^{-5}
	I	5×10^{-7}	3×10^{-3}	2×10^{-8}	9×10^{-5}
	^{48}Sc S	2×10^{-7}	8×10^{-4}	6×10^{-9}	3×10^{-5}
	I	1×10^{-7}	8×10^{-4}	5×10^{-9}	3×10^{-5}
Selenium-34	^{75}Se S	1×10^{-6}	9×10^{-3}	4×10^{-8}	3×10^{-4}
	I	1×10^{-7}	8×10^{-3}	4×10^{-9}	3×10^{-4}
Silicon-14	^{31}Si S	6×10^{-6}	3×10^{-2}	2×10^{-7}	9×10^{-4}
	I	1×10^{-6}	6×10^{-3}	3×10^{-8}	2×10^{-4}
Silver-47	^{105}Ag S	6×10^{-7}	3×10^{-3}	2×10^{-8}	1×10^{-4}
	I	8×10^{-8}	3×10^{-3}	3×10^{-9}	1×10^{-4}
	^{110m}Ag S	2×10^{-7}	9×10^{-4}	7×10^{-9}	3×10^{-5}
	I	1×10^{-8}	9×10^{-4}	3×10^{-10}	3×10^{-5}
	^{111}Ag S	3×10^{-7}	1×10^{-3}	1×10^{-8}	4×10^{-5}
	I	2×10^{-7}	1×10^{-3}	8×10^{-9}	4×10^{-5}
Sodium-11	^{22}Na S	2×10^{-7}	1×10^{-3}	6×10^{-9}	4×10^{-5}
	I	9×10^{-9}	9×10^{-4}	3×10^{-10}	3×10^{-5}
	^{24}Na S	1×10^{-6}	6×10^{-3}	4×10^{-8}	2×10^{-4}
	I	1×10^{-7}	8×10^{-4}	5×10^{-9}	3×10^{-5}
Strontium-38	^{85m}Sr S	4×10^{-5}	2×10^{-1}	1×10^{-6}	7×10^{-3}
	I	3×10^{-5}	2×10^{-1}	1×10^{-6}	7×10^{-3}
	^{85}Sr S	2×10^{-7}	3×10^{-3}	8×10^{-9}	1×10^{-4}
	I	1×10^{-7}	5×10^{-3}	4×10^{-9}	2×10^{-4}
	^{89}Sr S	3×10^{-8}	3×10^{-4}	3×10^{-10}	3×10^{-6}
	I	4×10^{-8}	8×10^{-4}	1×10^{-9}	3×10^{-5}
	^{90}Sr S	1×10^{-9}	1×10^{-5}	3×10^{-11}	3×10^{-7}
	I	5×10^{-9}	1×10^{-3}	2×10^{-10}	4×10^{-5}
	^{91}Sr S	4×10^{-7}	2×10^{-3}	2×10^{-8}	7×10^{-5}
	I	3×10^{-7}	1×10^{-3}	9×10^{-9}	5×10^{-5}
	^{92}Sr S	4×10^{-7}	2×10^{-3}	2×10^{-8}	7×10^{-5}
	I	3×10^{-7}	2×10^{-3}	1×10^{-8}	6×10^{-5}
Sulfur-16	^{35}S S	3×10^{-7}	2×10^{-3}	9×10^{-9}	6×10^{-5}
	I	3×10^{-7}	8×10^{-3}	9×10^{-9}	3×10^{-4}
Tantalum-73	^{182}Ta S	4×10^{-8}	1×10^{-3}	1×10^{-9}	4×10^{-5}
	I	2×10^{-8}	1×10^{-3}	7×10^{-10}	4×10^{-5}
Technetium-43	^{96m}Tc S	8×10^{-5}	4×10^{-1}	3×10^{-6}	1×10^{-2}
	I	3×10^{-5}	3×10^{-1}	1×10^{-6}	1×10^{-2}
	^{96}Tc S	6×10^{-7}	3×10^{-3}	2×10^{-8}	1×10^{-4}
	I	2×10^{-7}	1×10^{-3}	8×10^{-9}	5×10^{-5}
	^{97m}Tc S	2×10^{-6}	1×10^{-2}	8×10^{-8}	4×10^{-4}
	I	2×10^{-7}	5×10^{-3}	5×10^{-9}	2×10^{-4}
	^{97}Tc S	1×10^{-5}	5×10^{-2}	4×10^{-7}	2×10^{-3}
	I	3×10^{-7}	2×10^{-2}	1×10^{-8}	8×10^{-4}
	^{99m}Tc S	4×10^{-5}	2×10^{-1}	1×10^{-6}	6×10^{-3}
	I	1×10^{-5}	8×10^{-2}	5×10^{-7}	3×10^{-3}
	^{99}Tc S	2×10^{-6}	1×10^{-2}	7×10^{-8}	3×10^{-4}
	I	6×10^{-8}	5×10^{-3}	2×10^{-9}	2×10^{-4}
Tellurium-52	^{125m}Te S	4×10^{-7}	5×10^{-3}	1×10^{-8}	2×10^{-4}
	I	1×10^{-7}	3×10^{-3}	4×10^{-9}	1×10^{-4}

See footnotes at end of table.

TABLE 5 (*Continued*)

Element-Z	Isotope[a]	Table I Column 1 Air (μCi/ml)	Table I Column 2 Water (μCi/ml)	Table II Column 1 Air (μCi/ml)	Table II Column 2 Water (μCi/ml)
Tellurium-52 continued	^{127m}Te S	1×10^{-7}	2×10^{-3}	5×10^{-9}	6×10^{-5}
	I	4×10^{-8}	2×10^{-3}	1×10^{-9}	5×10^{-5}
	^{127}Te S	2×10^{-6}	8×10^{-3}	6×10^{-8}	3×10^{-4}
	I	9×10^{-7}	5×10^{-3}	3×10^{-8}	2×10^{-4}
	^{129m}Te S	8×10^{-8}	1×10^{-3}	3×10^{-9}	3×10^{-5}
	I	3×10^{-8}	6×10^{-4}	1×10^{-9}	2×10^{-5}
	^{129}Te S	5×10^{-6}	2×10^{-2}	2×10^{-7}	8×10^{-4}
	I	4×10^{-6}	2×10^{-2}	1×10^{-7}	8×10^{-4}
	^{131m}Te S	4×10^{-7}	2×10^{-3}	1×10^{-8}	6×10^{-5}
	I	2×10^{-7}	1×10^{-3}	6×10^{-9}	4×10^{-5}
	^{132}Te S	2×10^{-7}	9×10^{-4}	7×10^{-9}	3×10^{-5}
	I	1×10^{-7}	6×10^{-4}	4×10^{-9}	2×10^{-5}
Terbium-65	^{160}Tb S	1×10^{-7}	1×10^{-3}	3×10^{-9}	4×10^{-5}
	I	3×10^{-8}	1×10^{-3}	1×10^{-9}	4×10^{-5}
Thallium-81	^{200}Tl S	3×10^{-6}	1×10^{-2}	9×10^{-8}	4×10^{-4}
	I	1×10^{-6}	7×10^{-3}	4×10^{-8}	2×10^{-4}
	^{201}Tl S	2×10^{-6}	9×10^{-3}	7×10^{-8}	3×10^{-4}
	I	9×10^{-7}	5×10^{-3}	3×10^{-8}	2×10^{-4}
	^{202}Tl S	8×10^{-7}	4×10^{-3}	3×10^{-8}	1×10^{-4}
	I	2×10^{-7}	2×10^{-3}	8×10^{-9}	7×10^{-5}
	^{204}Tl S	6×10^{-7}	3×10^{-3}	2×10^{-8}	1×10^{-4}
	I	3×10^{-8}	2×10^{-3}	9×10^{-10}	6×10^{-5}
Thorium-90	^{228}Th S	9×10^{-12}	2×10^{-4}	3×10^{-13}	7×10^{-6}
	I	6×10^{-12}	4×10^{-4}	2×10^{-13}	10^{-5}
	^{230}Th S	2×10^{-12}	5×10^{-5}	8×10^{-14}	2×10^{-5}
	I	10^{-11}	9×10^{-4}	3×10^{-13}	3×10^{-5}
	^{232}Th S	3×10^{-11}	5×10^{-5}	10^{-12}	2×10^{-6}
	I	3×10^{-11}	10^{-3}	10^{-12}	4×10^{-5}
	Th (natural) S	3×10^{-11}	3×10^{-5}	10^{-12}	10^{-6}
	I	3×10^{-11}	3×10^{-4}	10^{-12}	10^{-5}
	^{234}Th S	6×10^{-8}	5×10^{-4}	2×10^{-9}	2×10^{-5}
	I	3×10^{-8}	5×10^{-4}	10^{-9}	2×10^{-5}
Thulium-69	^{170}Tm S	4×10^{-8}	1×10^{-3}	1×10^{-9}	5×10^{-5}
	I	3×10^{-8}	1×10^{-3}	1×10^{-9}	5×10^{-5}
	^{171}Tm S	1×10^{-7}	1×10^{-3}	4×10^{-9}	5×10^{-5}
	I	2×10^{-7}	1×10^{-2}	8×10^{-9}	5×10^{-4}
Tin-50	^{113}Sn S	4×10^{-7}	2×10^{-3}	1×10^{-8}	9×10^{-5}
	I	5×10^{-8}	2×10^{-3}	2×10^{-9}	8×10^{-5}
	^{125}Sn S	1×10^{-7}	5×10^{-4}	4×10^{-9}	2×10^{-5}
	I	8×10^{-8}	5×10^{-4}	3×10^{-9}	2×10^{-5}
Tungsten-74 (Wolfram)	^{181}W S	2×10^{-6}	1×10^{-2}	8×10^{-8}	4×10^{-4}
	I	1×10^{-7}	1×10^{-2}	4×10^{-9}	3×10^{-4}
	^{185}W S	8×10^{-7}	4×10^{-3}	3×10^{-8}	1×10^{-4}
	I	1×10^{-7}	3×10^{-3}	1×10^{-9}	1×10^{-4}
	^{187}W S	4×10^{-7}	2×10^{-3}	2×10^{-8}	7×10^{-5}
	I	3×10^{-7}	2×10^{-3}	1×10^{-8}	6×10^{-5}
Uranium-92	^{230}U S	3×10^{-10}	1×10^{-4}	1×10^{-11}	5×10^{-6}
	I	1×10^{-10}	1×10^{-4}	4×10^{-12}	5×10^{-6}
	^{232}U S	1×10^{-10}	8×10^{-4}	3×10^{-12}	3×10^{-5}
	I	3×10^{-11}	8×10^{-4}	9×10^{-13}	3×10^{-5}

See footnotes at end of table.

TABLE 5 (*Continued*)

Element-Z	Isotope[a]	Table I Column 1 Air (μCi/ml)	Table I Column 2 Water (μCi/ml)	Table II Column 1 Air (μCi/ml)	Table II Column 2 Water (μCi/ml)
Uranium-92 continued	^{233}U S	5×10^{-10}	9×10^{-4}	2×10^{-11}	3×10^{-5}
	I	1×10^{-10}	9×10^{-4}	4×10^{-12}	3×10^{-5}
	^{234}U S	6×10^{-10}	9×10^{-4}	2×10^{-11}	3×10^{-5}
	I	1×10^{-10}	9×10^{-4}	4×10^{-12}	3×10^{-5}
	^{235}U S	5×10^{-10}	8×10^{-4}	2×10^{-11}	3×10^{-5}
	I	1×10^{-10}	8×10^{-4}	4×10^{-12}	3×10^{-5}
	^{236}U S	6×10^{-10}	1×10^{-3}	2×10^{-11}	3×10^{-5}
	I	1×10^{-10}	1×10^{-3}	4×10^{-12}	3×10^{-5}
	^{238}U S	7×10^{-11}	1×10^{-3}	3×10^{-12}	4×10^{-5}
	I	1×10^{-10}	1×10^{-3}	5×10^{-12}	4×10^{-5}
	^{240}U S	2×10^{-7}	1×10^{-3}	8×10^{-9}	3×10^{-5}
	I	2×10^{-7}	1×10^{-3}	6×10^{-9}	3×10^{-5}
	U (natural) S	7×10^{-11}	5×10^{-4}	3×10^{-12}	2×10^{-5}
	I	6×10^{-11}	5×10^{-4}	2×10^{-12}	2×10^{-5}
Vanadium-23	^{48}V S	2×10^{-7}	9×10^{-4}	6×10^{-9}	3×10^{-5}
	I	6×10^{-8}	8×10^{-4}	2×10^{-9}	3×10^{-5}
Xenon-54	^{131m}Xe Sub	2×10^{-5}	—	4×10^{-7}	—
	^{133}Xe Sub	1×10^{-5}	—	3×10^{-7}	—
	^{133m}Xe Sub	1×10^{-6}	—	3×10^{-7}	—
	^{135}Xe Sub	4×10^{-6}	—	1×10^{-7}	—
Ytterbium-70	^{175}Yb S	7×10^{-7}	3×10^{-3}	2×10^{-8}	1×10^{-4}
	I	6×10^{-7}	3×10^{-3}	2×10^{-8}	1×10^{-4}
Yttrium-39	^{90}Y S	1×10^{-7}	6×10^{-4}	4×10^{-9}	2×10^{-5}
	I	1×10^{-7}	6×10^{-4}	3×10^{-9}	2×10^{-5}
	^{91m}Y S	2×10^{-5}	1×10^{-1}	8×10^{-7}	3×10^{-3}
	I	2×10^{-5}	1×10^{-1}	6×10^{-7}	3×10^{-3}
	^{91}Y S	4×10^{-8}	8×10^{-4}	1×10^{-9}	3×10^{-5}
	I	3×10^{-8}	8×10^{-4}	1×10^{-9}	3×10^{-5}
	^{92}Y S	4×10^{-7}	2×10^{-3}	1×10^{-8}	6×10^{-5}
	I	3×10^{-7}	2×10^{-3}	1×10^{-8}	6×10^{-5}
	^{93}Y S	2×10^{-7}	8×10^{-4}	6×10^{-9}	3×10^{-5}
	I	1×10^{-7}	8×10^{-4}	5×10^{-9}	3×10^{-5}
Zinc-30	^{65}Zn S	1×10^{-7}	3×10^{-3}	4×10^{-9}	1×10^{-4}
	I	6×10^{-8}	5×10^{-3}	2×10^{-9}	2×10^{-4}
	^{69m}Zn S	4×10^{-7}	2×10^{-3}	1×10^{-8}	7×10^{-5}
	I	3×10^{-7}	2×10^{-3}	1×10^{-8}	6×10^{-5}
	^{69}Zn S	7×10^{-6}	5×10^{-2}	2×10^{-7}	2×10^{-3}
	I	9×10^{-6}	5×10^{-2}	3×10^{-7}	2×10^{-3}
Zirconium-40	^{93}Zr S	1×10^{-7}	2×10^{-2}	4×10^{-9}	8×10^{-4}
	I	3×10^{-7}	2×10^{-2}	1×10^{-8}	8×10^{-4}
	^{95}Zr S	1×10^{-7}	2×10^{-3}	4×10^{-9}	6×10^{-5}
	I	3×10^{-8}	2×10^{-3}	1×10^{-9}	6×10^{-5}
	^{97}Zr S	1×10^{-7}	5×10^{-4}	4×10^{-9}	2×10^{-5}
	I	9×10^{-8}	5×10^{-4}	3×10^{-9}	2×10^{-5}
Any single radionu-clide not listed above with decay mode other than alpha emission or	Sub	1×10^{-6}	—	3×10^{-8}	—

See footnotes at end of table.

TABLE 5 (*Continued*)

Element-Z	Isotope[a]	Table I		Table II	
		Column 1 Air (μCi/ml)	Column 2 Water (μCi/ml)	Column 1 Air (μCi/ml)	Column 2 Water (μCi/ml)
spontaneous fission and with radioactive half-life less than 2 hr					
Any single radionuclide not listed above with decay mode other than alpha emission or spontaneous fission and with radioactive half-life greater than 2 hr		3×10^{-9}	9×10^{-5}	1×10^{-10}	3×10^{-6}
Any single radionuclide not listed above, which decays by alpha emission or spontaneous fission		6×10^{-13}	4×10^{-7}	2×10^{-14}	3×10^{-8}

Element and Isotope

If it is known that ^{90}Sr, ^{125}I, ^{126}I, ^{129}I, ^{131}I, (^{133}I, Table II only), ^{210}Pb, ^{210}Po, ^{211}At, ^{223}Ra, ^{224}Ra, ^{226}Ra, ^{227}Ac, ^{228}Ra, ^{230}Th, ^{231}Pa, ^{232}Th, Th (natural), ^{248}Cm, ^{254}Cf, and ^{256}Fm are not present			9×10^{-5}		3×10^{-6}
If it is known that ^{90}Sr, ^{125}I, ^{126}I, ^{129}I, (^{131}I, ^{133}I, Table II only), ^{210}Pb, ^{210}Po, ^{223}Ra, ^{226}Ra, ^{228}Ra, ^{231}Pa, Th (natural), ^{248}Cm, ^{254}Cf, and ^{256}Fm are not present			6×10^{-5}		2×10^{-6}
If it is known that ^{90}Sr, ^{129}I, (^{125}I, ^{126}I, ^{131}I, Table II only), ^{210}Pb, ^{226}Ra, ^{228}Ra, ^{248}Cm, and ^{254}Cf are not present			2×10^{-5}		6×10^{-7}
If it is known that (^{129}I, Table II only), ^{226}Ra, and ^{228}Ra are not present			3×10^{-6}		1×10^{-7}
If it is known that alpha emitters and ^{90}Sr, ^{129}I, ^{210}Pb, ^{227}Ac, ^{228}Ra, ^{230}Pa, ^{241}Pu, and ^{249}Bk are not present			3×10^{-9}		1×10^{-10}
If it is known that alpha emitters and ^{210}Pb, ^{227}Ac, ^{228}Ra, and ^{241}Pu are not present			3×10^{-10}		1×10^{-11}
If it is known that alpha emitters and ^{227}Ac are not present			3×10^{-11}		1×10^{-12}
If it is known that ^{227}Ac, ^{230}Th, ^{231}Pa, ^{238}Pu, ^{239}Pu, ^{240}Pu, ^{242}Pu, ^{244}Pu, ^{248}Cm, ^{249}Cf, and ^{251}Cf are not present			3×10^{-12}		1×10^{-13}

See footnotes on p. 162.

[a] Soluble (S); Insoluble (I).

[b] "Sub" means that values given are for submersion in a semispherical infinite cloud of air-borne material.

NOTE: In any case where there is a mixture in air or water of more than one radionuclide, the limiting values should be determined as follows:

(1) If the identity and concentration of each radionuclide in the mixture are known, the limiting values should be derived as follows: Determine, for each radionuclide in the mixture, the ratio between the quantity present in the mixture and the limit otherwise established for the specific radionuclide when not in a mixture. The sum of such ratios for all the radionuclides in the mixture may not exceed "1" (i.e., "unity").

EXAMPLE: If radionuclides A, B, and C are present in concentrations C_A, C_B, and C_C, and if the applicable MPC's, are MPC_A, and MPC_B, and MPC_C respectively, then the concentrations shall be limited so that the following relationship exists:

$$\frac{C_A}{MPC_A} + \frac{C_B}{MPC_B} + \frac{C_C}{MPC_C} \leq 1$$

(2) If either the identity or the concentration of any radionuclide in the mixture is not known, the limiting values shall be:

(a) For purposes of Table I, Col. 1—6 × 10^{-13}
(b) For purposes of Table I, Col. 2—4 × 10^{-7}
(c) For purposes of Table II, Col. 1—2 × 10^{-14}
(d) For purposes of Table II, Col. 2—3 × 10^{-8}

(3) If any of the conditions specified below are met, the corresponding values specified below may be used in lieu of those specified in paragraph (2) above.

(a) If the identity of each radionuclide in the mixture is known but the concentration of one or more of the radionuclides in the mixture is not known, the concentration limit for the mixture is the limit specified for the radionuclide in the mixture having the lowest concentration limit; or

(b) If the identity of each radionuclide in the mixture is not known, but it is known that certain radionuclides specified are not present in the mixture, the concentration limit for the mixture is the lowest concentration limit specified for any radionuclide which is not known to be absent from the mixture; or

(4) If the mixture of radionuclides consists of uranium and its daughter products in ore dust prior to chemical processing of the uranium ore, the values specified below may be used in lieu of those determined in accordance with paragraph (1) above or those specified in paragraphs (2) and (3) above.

(a) For purposes of Table I, Col. 1—1 × 10^{-10} μCi/ml gross alpha activity; or 2.5 × 10^{-11} μCi/ml natural uranium; or 75 μg/m^3 of air natural uranium.

(b) For purposes of Table II, Col. 1—3 × 10^{-12} μCi/ml gross alpha activity; or 8 × 10^{-13} μCi/ml natural uranium; or 3 μg/m^3 of air natural uranium.

(5) For purposes of this note, a radionuclide may be considered as not present in a mixture if (a) the ratio of the concentration of that radionuclide in the mixture (C_A) to the concentration limit for that radionuclide specified (MPC_A) does not exceed $1/10$, (i.e., $C_A/MPC_A \leq 1/10$) and (b) the sum of such ratios for all the radionuclides considered as not present in the mixture does not exceed $1/4$, i.e.,

$$\frac{C_A}{MPC_A} + \frac{C_B}{MPC_B} + \cdots \leq 1/4)$$

RADIATION MEASUREMENT

The measurement of radiation for the purposes of protection requires both the proper instruments and the proper techniques of using them. This section must necessarily concentrate on the second of these requirements.

There are two major types of instruments for measurement of radiation and radioactivity. These are the field survey instrument used for measurements on the spot and the laboratory instruments which require that a sample be brought back into the laboratory. The latter, of course, are usually more accurate, but field equipment is adequate for the accuracy required in radiation surveys.

Laboratory and field survey equipment for the

measurement of radiation are generally based on two types of detectors. In one, the ionization produced in a gas is collected to give an electrical pulse current which can be measured. These gas ionization instruments include the ionization chamber, the proportional counter, and the Geiger counter. The second type is based on conversion of radiation energy into a light pulse by a scintillator. The light pulse is then reconverted to an electrical pulse or current by means of a photomultiplier tube. Various scintillating materials, such as zinc sulfide for alpha particles and sodium iodide for gamma rays have been widely applied. Instruments based on solid-state detectors will undoubtedly be the choice for many purposes in about 1970, but are presently still under development. It is beyond the scope of this section to go into the theory and operation of counters, but references are appended in which such information is available.

A valuable tool for determination of the body burden of a gamma emitting isotope is the whole-body counter. This instrument usually consists of a large sodium iodide scintillation detector in a shielded room with a multichannel analyzer as auxiliary equipment. Such a spectrometer system allows qualitative and quantitative estimates of nuclides in the body. This is important since a normal man contains about 150 grams of potassium which is naturally radioactive and can mask small amounts of other radioisotopes if qualitative identification is not carried out.

For body burdens of gamma emitters approaching the mpl's, a simpler, more portable shield can be used. Several of these have been reported in the literature and their use does allow measurements to be made without the costly shielded room.

While details of instruments are not given in this section, it is desirable to point out certain precautions that must be observed in their operation.

(1) Instruments must be maintained in good repair and calibrated at intervals to assure the desired accuracy.

(2) The instrument used must be selected so that it is sensitive to the radiation to be measured. For example, the standard Geiger survey meter does not respond well to alpha or weak beta radiation.

(3) As a corollary, the instrument calibrations must be carried out with isotopes having the same or similar energy to those being measured.

(4) Personnel handling or using radiation instruments must have an understanding of their characteristics and of the possible interpretation of readings obtained.

Survey instruments should be calibrated in terms of the particular radiation which will be present in the actual operations. However, it is still standard practice to calibrate such instruments in terms of the gamma radiation from a radium or ^{60}Co source. This same source may also be used for film calibration. Radium and cobalt sources are prepared in terms of rhm units, i.e., roentgens per hour at a distance of 1 meter. Other dose rates may be obtained by varying the distance between the source and the instrument.

Table 6 gives the rhm values[7] for 1 curie of various isotopic sources, illustrating the possible differences between sources.

Radium calibration sources in themselves pre-

TABLE 6.   Calculated Gamma Radiation Levels[a] for 1 Curie of Some Radioisotopes

Isotope	Half-life	Roentgens per Hour at 1 Foot	Roentgens per Hour at 1 Meter
^{22}Na	2.6 yr	14.1	1.31
^{24}Na	15 hr	20.6	1.92
^{52}Mn	5.7 days	20.8	1.93
^{54}Mn	314 days	5.22	0.485
^{59}Fe	45 days	7.0	0.651
^{58}Co	71 days	6.02	0.560
^{60}Co	5.3 yr	14.3	1.3
^{64}Cu	12.8 hr	4.52	0.42
^{65}Zn	245 days	3.22	0.30
^{76}As	26.5 hr	5.81	0.54
^{82}Br	36 hr	16.1	1.50
^{128}I	25 min	0.194	0.018
^{130}I	12.5 hr	13.5	1.25
^{131}I	8.1 days	2.54	0.236
^{137}Cs	30 yr	3.4	0.32
^{170}Tm	127 days	0.04	0.004
^{192}Ir	74 days	5.4	0.5
^{198}Au	2.7 days	2.5	0.23
Radium in equilibrium (0.5 mm Pt filtration)		9.04	0.84

[a]For an unshielded point source except as noted. The values shown have been modified from the original.

sent an additional hazard above that of their gamma radiation. The powdered radium salts are enclosed in platinum or a glass and platinum capsule. If the capsule should be ruptured, a considerable quantity of alpha emitting material may be released. Cobalt sources, on the other hand, are solid metallic material and are not subject to rupture. Their only disadvantage is the change in their activity with time—the ^{60}Co decaying with a half-life of 5.3 years.

Personnel Monitoring

Licensees under the Atomic Energy Commission are required to provide appropriate personnel monitoring equipment under the following circumstances:

(1) Any individual entering a restricted area who is likely to receive in excess of 25 percent of the permissible dose in any calendar quarter.

(2) Any individual under age 18 who is likely to receive an excess of 5 percent of the permissible dose in any calendar quarter.

(3) Any individual who enters a high radiation area which is defined as an area where a major portion of the body could receive in excess of 100 mrem in any one hour.

Photographic film is still the accepted standard for evaluation of personnel exposure to ·low radiation levels. The film most commonly used is x-ray film because of its high sensitivity. The ordinary dental film packet originally manufactured for dental x-ray work was very satisfactory. Special monitoring films are now readily available in convenient package form. Naturally, the degree of blackening of the film must be interpreted by means of a microphotometer or densitometer in terms of radiation dose. This is done by calibrating the film with known amounts of radiation.

The film packet can be merely pinned to the work clothes, or if only penetrating radiation is encountered, it can be carried in the pocket. However, the common method of wearing is to enclose the packet in a holder (film badge) for convenience. This film badge allows identification of the film with the individual, it holds the film in the proper place with respect to the body and by being worn outside the clothing it can be made more responsive to less penetrating radiation such as beta particles. It also can be designed to distinguish between different energies or types of radiation. Many badge designs are currently available and all of them have advantages for the particular installation using them. Unless an installation employs a large number of workers with possible radiation exposure, the expense of setting up and operating a film badge service, including the calibrating and processing of films, is very high. Fortunately, several reliable organizations offer a film badge service for smaller installations at a reasonable cost.

The photographic film is not a perfect dosimeter, and considerable effort has been devoted to the development of glass dosimeters and thermoluminescent dosimeters for personnel monitoring. The sensitivity of the latter is somewhat better than film, and it is expected that thermoluminescent dosimetry (TLD) will find increasingly wide use. In principle, calcium fluoride or lithium fluoride store energy when exposed to ionizing radiation. When the phosphor is heated to several hundred degrees centigrade, this energy is released in the form of visible light and can be measured with a photometer.

The film badge and thermoluminescent dosimeters are limited by the requirement that they be processed before a measure of radiation exposure can be obtained. This is perfectly satisfactory where the personnel monitoring is done only as a matter of record and there is no danger of high level exposures. If such exposures are possible, a pocket ionization chamber (dosimeter) in addition to the film is desirable. Pocket chambers are available in various ranges of total radiation dose and are made either as direct reading dosimeters with their own internal scale or as simple chambers which must be inserted in a reader to obtain the dose measurement. All of these operate by charging the chamber before use. Exposure to radiation causes ionization within the chamber and allows a portion of this charge to leak off. After wearing, the residual charge is measured and the amount of leakage related to the radiation exposure. The internal scale or the reader scale is calibrated directly in terms of dose received. Dosimeters can be read daily or hourly to follow the exposure level if the circumstances require. However, they are chiefly sensitive to gamma or x-radiation, since the walls of the chamber will effectively cut out alpha and most beta particles.

The recommended monitoring period, e.g., the time of wearing a film badge, is usually one month in the case of predictable low level exposure. When high or extremely variable exposure is possible, daily readings of dosimeters are recommended. If the operator can show that the maximum exposure is less than one-fourth of the permissible occupational level, individual personnel monitoring is not required.

Radiation Surveys

The general area survey made with a portable survey meter is most useful for locating sources of radiation and evaluating possible hazard, but is also valuable for the immediate determination of the effectiveness of corrective measures or of cleanup procedures. When used for exposure evaluation, particular care should be taken to make measurements with the detector probe in the locations occupied by the operator's body, as well as measurements throughout the entire working area.

Area surveys are carried out with a portable dose rate survey meter sensitive to the radiation present. General area surveys for beta and gamma radiation are best performed by holding the detector away from the body at waist level and tracing a systematic path through the area, noting any rise in radiation level and marking these spots either directly with chalk or on a diagram. Radiation levels above the permissible level require cleaning or supplementary shielding and re-surveying until the level is reduced to acceptable limits. Levels markedly above background but below the permissible level may be attacked in a more leisurely manner.

Surface contamination from alpha or beta emitters can also be surveyed by moving the detector slowly over a systematic path covering the floor, walls, and working surfaces. Areas indicating contamination well above background may be checked with a wipe test as described in the discussion on surface contamination.

Permissible radiation levels are set up based on the dose received above that due to the radiation that is normally present. This background radiation is variable and must be considered when surveying work areas. Background level can best be measured with the actual instrument used for the survey in a spot well away from the work area. Readings on survey instruments ranging from 0.01–0.02 mr/hr are quite common and the background may be even higher in geographical areas where large deposits of radioactive elements are found.

Any instrument used for area surveys should be sensitive to the least penetrating radiation present. Survey meters are worthless unless they are maintained in proper condition and are calibrated frequently enough to assure the accuracy of the readings. Small "button" sources are available from the instrument manufacturers for frequent rough checks of operating conditions, but a calibration of the instrument on all scales should be made three or four times a year. Inaccurate or uncalibrated instruments can be dangerous if they give low readings in the presence of high radiation fields.

On the other hand, incorrect high readings may cause considerable expense in trying to remedy a situation which does not actually exist. Calibration services are available from instrument manufacturers and from consulting laboratories in the field of radiation protection. If several instruments are in use, the cost of calibrating sources and their shielding will be negligible compared with the convenience of being able to run calibrations without having the instruments out of service for extended periods.

Surface Contamination

The contamination of working surfaces (or other portions of the working area) with radioactive material is referred to as surface contamination. If the isotopes are beta or gamma emitters, the contamination is a source of external radiation. The magnitude of this radiation can best be evaluated with survey equipment. Alpha emitters are not an external radiation hazard, but alpha contamination that later becomes air-borne or is transported to the body

of the individual can offer an internal hazard. This is true to a lesser extent for beta emitters.

Surface contamination is evaluated purely in terms of the amount present on the surface. The critical factor for alpha emitters is the amount of material which can become air-borne and available for inhalation or ingestion. Several methods of measurement have been devised, none of which is completely satisfactory. The most widely accepted method is to wipe a fixed area with a piece of filter paper and measure the activity taken up by the paper. The results are expressed in terms of disintegrations per minute per hundred square centimeters of similar units. Since the correlation between the amount that can be wiped off with a paper and the amount that can become air-borne or can be picked up on the hands is not good, this cannot be considered very meaningful. Low but positive results from a wipe test should not lead to ignoring a removable surface contamination, as it is always desirable to keep working areas as free as possible from radioactivity. However, the cleaning of such surfaces need not be performed on an emergency basis.

Air-borne Radioactivity

Almost all operations involving radioisotopes may result in some of the activity becoming air-borne. This may arise from the dusting of dry materials, spraying from solutions, or the liberation of radioactive gaseous products. It is common practice in both laboratory and plant processes to carry out all operations in enclosed hoods or with good general ventilation.

The high toxicity of small amounts of some radioisotopes when taken into the body make the collection and evaluation of air-borne radioactivity a critical part of most radiation protection systems. Solid or liquid materials can be determined by drawing known quantities of contaminated air through a filter and measuring the activity collected on the filter. This can be an extremely rapid and accurate method of analysis.

Two types of hazard evaluation are possible. One is the continuous sampling of air-borne material in the general work area for an overall picture of the probable exposure. The other is the sampling of air in the operator's breathing zone during the time that the actual process which may give rise to air-borne activity is being carried out.

Gaseous isotopes, of course, will not be collected on a filter. (The natural radon isotopes can be estimated, however, by the collection of their solid daughter products on a filter.) Other radioactive gases must be absorbed in a gas washer or an impinger and the resulting solution treated to convert the activity to a form suitable for counting. Even this will not work with some of the inert gases such as krypton and argon and more elaborate methods are needed for their measurement. However, their production and measurement are very specialized problems only encountered in reactor operation and fuel reprocessing.

Bioassay

Measurement of air-borne dust or other activity is difficult on a continuous basis, particularly if it is desirable to monitor the exposure of each individual. Some evaluation is possible by measurements on the individual himself (bioassay). For example, analysis for the amount of an isotope excreted in the urine of an individual may furnish an indication of relative exposure. While the conversion of the concentration of isotope in the urine to the body burden or the concentration of that isotope in the air breathed is not always possible, the excretion does constitute a measure of the relative intake and can be valuable as a secondary means of exposure evaluation. No permissible levels for urine excretion have been set up, since in addition to the usual variables found in biological systems, others such as diet and fluid intake can cause marked fluctuations in the urinary concentration of an isotope. Urine analyses, therefore, are only valuable when a continuing sequence is maintained. Methods of analysis have been recommended by a joint panel of the World Health Organization, the International Atomic Energy Agency and the Food and Agriculture Organization.[8]

One specific bioassay method is accepted for radium exposure evaluation. That is the measurement of the radon concentration in the expired air. Radon is one of the rare gases and an immediate daughter of radium. A portion of the radon formed by radium decay in the body is transported to the lungs and appears in the exhaled breath. The value of 10^{-12} curie of radon/liter has been set as an indication of 0.1 μg of radium in the body. This latter value is the maximum permissible amount based on the experience of the radium dial painters. The actual measurement of breath radon content requires specialized apparatus.

Bioassay samples are taken at various intervals depending on the degree of exposure. Slightly exposed personnel may be sampled yearly while those engaged in actual processing may be sampled quarterly. In a few cases where there is exposure to highly toxic substances such as plutonium, urine analysis may be performed more frequently.

Total Exposure

The overall evaluation of plant personnel exposure to radiation must include all possible factors. For example, if there is total body external irradiation, radiation from inhaled or ingested isotopes and surface irradiation of the hands when handling isotopes, it is necessary that the sum of these exposures does not exceed the maximum permissible level. It is not sufficient that each individual factor does not exceed the level set when it is considered to be the only exposure. Such a summation can be a complicated calculation. However, a reasonable approximation can be made, relatively simply, by taking the percentage of permissible level represented by each exposure and summing them up. If they are less than 100 percent, the total exposure should be acceptable.

Overexposures are handled according to their severity. Where a weekly dose is excessive, it should be possible to maintain the 13-week summation below the permissible level by reducing subsequent exposure, or if necessary by removing the individual entirely from possible exposure. Any exposure greater than that which can be treated in this way is serious enough so that the USAEC regulations as well as any applicabe state or local codes will require notification* of such an occurrence, the reason for it, and procedures instituted to prevent recurrence. Whenever overexposure occurs, the source must be located by a thorough survey. The source must be removed and adequate safeguards set up to prevent recurrence. It should be noted here, however, that no system of rules and safeguards can completely protect careless or ignorant individuals. Therefore, considerable stress must be placed on training the worker in safe practices.

Waste Monitoring

Since a radiation protection program includes prevention of exposure of those outside the plant as well as plant personnel, it is necessary to monitor plant effluents and wastes before they can be disposed of through the usual channels. The levels of activity permitted in waste materials are quite low and direct measurement with survey instruments is not usually accurate enough to be adequate. In most cases,

*Requirements for notification of the AEC are detailed in Part 20 of Title 10, Chapter I, of the Code of Federal Regulations.

only a very few isotopes can possible be present. Thus samples of the waste material can be put through relatively simple chemical separations to isolate a particular isotope from the bulk of the material and to prepare it in a form suitable for counting. A few simple cases may be handled by merely concentrating the isotope, for example, by evaporation of waste solutions or by ashing of combustible solid wastes. Where the isotope is a gamma emitter, these concentration procedures may be perfectly adequate, but for beta emitters, particularly those of low energy, the sample bulk may absorb a large fraction of the emitted particles. Then it is necessary to resort to some form of chemical separation. The chemical methods required for such analyses are described in the literature. The sampling process is frequently much more complex, particularly where the wastes are not homogeneous. For this reason, many installations where only small amounts of wastes are accumulated do not bother with sampling or analysis, but treat all such material as contaminated and dispose of it by appropriate methods.

RESPONSIBILITY

The specific responsibilities of the licensee handling isotopes (by-product material), source materials or radio-graphic sources are set forth in the Code of Federal Regulations, and similar responsibilities apply to government contractors. Medical x-ray operation is usually covered in state and local codes. All regulations place the responsibility for the control of radiation exposure squarely on the management of the installation. They must design facilities and plan operations so that overexposures do not occur.

The responsibility of the supervisor of work with x-ray equipment or with radioisotopes is perhaps even greater than that of a supervisor in charge of dangerous chemical operations. The immediate effects of most chemical accidents are fully appreciated, and the chemical agent can usually be neutralized or destroyed to prevent further exposure. This is not true for radiation or radiation emitters, and the lack of immediate effect makes supervisory control more difficult.

In considering the handling of reasonably long-lived isotopes, it must be realized that responsibility involves being certain not only that the workers who handle the material are not overexposed, but also that none of the material escapes to endanger others.

All instruction designed to minimize the radiation exposure of plant workers should originate with the technically qualified man in charge. Most of the attention must be directed to giving the operating supervisors a complete picture of the nature of the hazards, the ways in which the resulting exposures can be minimized, and the value of the monitoring and control procedures to be used. Depending on the size of the organization and the quality of the personnel, this same instruction may be given to all workers or else the complete responsibility may be given to the operating supervisor. In any case, the workers should know that they are being exposed to radiation, the steps being taken to protect them from overexposure, and the safety regulations which are made necessary by the nature of radiation. It is sometimes difficult to teach safe practices without causing an excessive alarm in some people, but the fear of alarm cannot be used to justify withholding such information. For example, a licensee for radiography must describe an adequate training program of:

    (a) Initial training,

    (b) Periodic training,

    (c) On-the-job training,

    (d) Means to be used by the licensee to determine the radiographer's assistant's knowledge of, understanding of, and ability to comply with, the AEC regulations and licensing requirements and the operating and emergency procedures of the licensee.

The paragraphs below will explain the responsibility of supervisory personnel to the various groups under their control, as well as to those outside the plant. Management, operational supervisors and technical personnel must all play a part in minimizing radiation exposure.

Workers

The most immediate responsibility of management is to the workers directly concerned with radiation. Responsible persons must see that the exposure of the workers is continuously evaluated, that records of their exposure are maintained, and that continued overexposure is not allowed. With a group of non-technical personnel this responsibility can best be delegated to the operation supervisor. From his other duties, he is familiar with the operations and personnel and is able to issue direct orders. It is not necessary that he be completely trained in the technical aspects of radiation protection, but rather that he understand the measurements made and the meaning of the numbers obtained. In addition, there must be someone available to give technical advice on radiation protection problems. While a large organization may have its own radiation safety officer or similarly qualified person, the smaller organization may have to depend on consultants in the field. In either case, the direct day-to-day supervision can best be handled through the operational supervisor or a radiation safety supervisor. In technical groups such as laboratories or hospitals, there is still need for someone to have charge of records and to perform the duties of an operational supervisor. This person must have the authority to issue direct orders concerning exposure. Almost all of the recorded cases of radiation overexposure have been to technical personnel who knew the hazard and ignored the possible consequence. Management is still responsible for the actions of these employees. To protect the workers and themselves, they must set up an adequate supervisory system and follow the records to see that the program is being carried out. Those working directly with radiation will have the highest potential exposure and must be given the most attention in a radiation protection program. This must include exposure evaluation, exposure records and job training to minimize hazards.

All other employees may be split into two classifications: those who enter the work area for relatively short periods, such as maintenance workers, and those who do not enter the area. In an operation involving low radiation levels, exposure can be considered negligible if that of the direct workers is well below the maximum permissible levels. In any case, casual visits to the work area should be minimized in length and cut to a very small number by proper planning. If the radiation level in the work area is higher, it becomes necessary to include anyone entering the work area in the complete control program of personnel monitoring and exposure records. In a large organization this can become a matter of considerable expense, but again, by careful planning, the number of these casual visitors can be kept to a minimum. The casual worker in the operating area has the added disadvantage that he is not fully acquainted with the operations or the hazards involved and may do something which would expose him to more radiation than someone familiar with the operations. The supervisor for the area is responsible for the safety of these part-time workers and must either supervise their activities or instruct them in the safety procedures of the area. Employees outside the work area must receive only a minimum exposure and this is the responsibility of the technically qualified person in charge. Generally, minimizing such exposure is accomplished in the original design of the installation, but periodic checks of the operation of control measures must be made. Personnel monitoring should not be required, as any design which would allow a measurable exposure outside the work area is entirely unsatisfactory, and no operations should be carried out in such an area.

Neighborhood

Minimizing the hazards of exposure to persons in the neighborhood of the plant is the responsibility of the qualified technical man in charge. Direct radiation should not be a hazard; the problems are rather those of radioisotope waste disposal. Periodic checks of contamination of the neighborhood must be made, and if the level of activity handled in the installation increases, continuous monitoring may be required. The permissible levels of exposure for plant personnel are not satisfactory for those living or working in the neighborhood. All plant workers must realize the hazards involved in their jobs and, in accepting a job, must also accept the hazards. This is true in any field and is not unique for those engaged in work with radiation. People outside the plant are not aware of the hazards and have not entered into any agreement with the company. Therefore, every effort should be made to see that their possible exposure is far below that allowed for plant personnel. The limitation of exposure for those in the neighborhood has been set at one-tenth of that allowed for occupational exposure.

The responsibility to those even more remote from the work area is one of reducing the possible buildup of the worldwide radiation level through careless handling of wastes. Although procedures for waste handling may originate with the technically qualified man in charge, responsibility for carrying out these procedures must rest directly on the operating supervisor.

RADIATION CONTROL

A radiation protection program has two facets —the continuous evaluation of exposure and the reduction of exposure by any applicable control measure. The fact that all exposures are maintained below maximum permissible levels is an indication that the control procedures are working, but since any unnecessary exposure is foolish, the possibility of maintaining values as low as possible should be continuously kept in mind.

A major factor in control is the proper training of operating personnel. It is part of supervisory responsibility that every worker know what he is working with, what the hazards are, and what measures are being taken to secure his safety. He must be trained in safe techniques and know what to do in case of an accident. And he must be made to realize that observance of safety rules and personnel monitoring requirements are just as much a part of the job as the actual operations performed.

At the present time, the use of most radioisotopes requires licensing through the Atomic Energy Commission or the states. This means that a certain degree of control is exercised in the allocation of isotopes. Applications are reviewed, and applicants are required to describe the purpose for which the isotopes are to be used and the facilities and equipment available for radiation protection. The applicant agrees to keep records of the receipt, use, and distribution of the isotopes, and further agrees not to use them for purposes other than those described or even to redistribute the materials to others without authorization.

Licensees and contractors of the Atomic Energy Commission and other users of radiation sources are subject to various types of inspection. It has been pointed out, however, that these inspections are not intended to control radiation, but merely to determine compliance with regulations. Responsibility for control rests with the operator.

There are many laboratories working with microcurie amounts of various radioisotopes which do not require any special handling or special equipment. The federal government has issued a list of exempt quantities of various radioisotopes. This list is reproduced in Table 7.[9]

The quantities are exempted based on the concept that this amount of the given isotope cannot cause any damage under any conditions of use or misuse. On this basis, these quantities indicate whether or not the laboratory should take any precautions in handling.

The measurement and evaluation of exposure have been discussed previously; the following discussion will introduce the various methods of reducing radiation exposure.

Reducing Exposure

The three methods of reducing radiation exposure are: (1) increasing shielding, (2) decreasing the time of exposure and (3) increasing the distance between worker and source. In the usual x-ray or isotope installation the shielding is part of the original design, but modifications of equipment or changes in operating technique may require shielding changes. Such shielding is subject to the same criteria as that in the original design and will be covered later. The

TABLE 7. Quantities of Radioisotopes That Are Exempt from License Requirements

Material	Microcuries	Material	Microcuries
^{105}Ag	1	^{103}Pd and ^{103}Rh	50
^{111}Ag	10	^{109}Pd	10
^{76}As, ^{77}As	10	^{147}Pm	10
^{198}Au	10	^{210}Po	0.1
^{199}Au	10	^{143}Pr	10
^{140}Ba and ^{140}La	1	^{239}Pu	1
^{7}Be	50	^{226}Ra	0.1
^{14}C-14	50	^{86}Rb	10
^{45}Ca-45	10	^{186}Re	10
^{109}Cd and ^{109}Ag	10	^{105}Rh	10
^{144}Ce and ^{144}Pr	1	^{106}Ru and ^{106}Rh	1
^{36}Cl	1	^{35}S	50
^{60}Co	1	^{124}Sb	1
^{51}Cr	50	^{46}Sc	1
^{137}Cs and ^{137}Ba	1	^{153}Sm	10
^{64}Cu	50	^{113}Sn	10
^{154}Eu	1	^{89}Sr	1
^{18}F	50	^{90}Sr and ^{90}Y	0.1
^{55}Fe	50	^{182}Ta	10
^{59}Fe	1	^{96}Tc	1
^{72}Ga	10	^{99}Tc	1
^{71}Ge	50	^{127}Te	10
^{3}H (HTO or T$_2$O)	250	^{129}Te	1
^{131}I	10	Th (natural)	50
^{114}In	1	^{204}Tl	50
^{192}Ir	10	Tritium (see ^{3}H)	250
^{42}K	10	U (natural)	50
^{140}La	10	^{233}U	1
^{52}Mn	1	^{234}U, ^{235}U	50
^{56}Mn	50	^{48}V	1
^{99}Mo	10	^{185}W	1
^{22}Na	10	^{90}Y	1
^{95}Nb	10	^{91}Y	1
^{59}Ni	1	^{65}Zn	10
^{63}Ni	1	Unidentified radioactive materials or any of the above in unknown mixtures	0.1
^{32}P	10		

actual physical setup may vary, however, if only temporary shielding is required. Temporary shielding is set up with interlocking lead bricks or steel plates of the proper thickness. Such shields can be readily assembled or stored when no longer needed. If these temporary shielding materials are made easily available in the work area they will be used more often.

Reducing the time of exposure is not just a matter of calculating the permissible time in a certain area which will keep the exposure just at the maximum permissible level. Exposure time should be reduced to the minimum consistent with sound economical operation. This can be facilitated by careful planning of procedures. In a practical sense this means that isotopes should be kept in shielded containers when not in actual use, x-ray workers should not enter any area where measurable scattering from x-ray equipment occurs except when absolutely necessary, and all those not directly concerned with an immediate operation should be excluded from the irradiated area. Again, these steps are not taken merely to maintain exposures below maximum permissible level but to keep overall exposure to a minimum.

One method of reducing exposure time which may be applicable in some instances is rotation of personnel. The operators or technicians may be trained to interchange jobs, and the higher exposure operations rotated through the largest possible group.

The factor of distance is probably the most valuable means of controlling radiation exposure. This has been emphasized indirectly in the preceding paragraph in that those not directly concerned with the operation should be excluded from the work area and also that operating personnel should stay away from radiation sources when their presence is not required.

Radiation intensity decreases according to the inverse square law; i.e., doubling the distance drops the exposure rate to $1/4$, tripling the distance, to $1/9$, and so on. This rapid fall off with distance can be of greatest assistance when properly used. It is certainly the cheapest barrier to overexposure.

Actual work areas where the possible exposure approaches permissible levels should be posted with signs indicating the level of radioactivity, and whenever possible, temporary barriers should be set up to keep those not working directly with the radiation source at a safe distance.

Federal regulations require posting of radiation areas (greater than 5 mrem/hr), high radiation areas (greater than 100 mrem/hr) and air-borne radioactivity areas (greater than 25 percent of the maximum permissible air concentration).

When handling isotopes, various personnel protection equipment may be required. Depending on the activity level and the physical state of the active material, coveralls, gloves or respirators may be required. This will be discussed further in connection with decontamination. In cases where heavy beta exposure is possible, it is sometimes advisable that the workers wear heavy-lensed goggles to reduce eye exposure.

Decontamination

Isotope spills may occur, and prompt decontamination is necessary to minimize exposure. This is not only required for the reduction of direct radiation but to prevent spilled material from becoming air-borne or being picked up by contact. The supervisor and preferably the workers as well should know the proper means of cleanup for any possible isotope spill in the operations which they are performing. When an accident occurs the operator should first move to a safe distance and call for assistance. The greatest difficulty with spills in the past has been that the workers, scientifically trained or not, try to clean up by themselves and spread contamination throughout the laboratory or plant.

The first operation in decontamination is to check the personnel to see if their clothing or bodies have become contaminated. This should be done by the supervisor or technically qualified person. If the workers are not contaminated, they may proceed with the area cleanup.

If it is found that the workers are contaminated, their clothing should be removed and their bodies checked for radioactivity. Any active areas should be washed with plenty of water, then soap and water, and finally scrubbed with a soft brush. Decontaminating agents are not always satisfactory for application on skin. Two agents recommended in the National Bureau of Standards Handbook 48 are titanium dioxide paste or a saturated solution of potassium permanganate followed by a 5 percent sodium bisulfite solution rinse.[10]

The following detailed directions for these decontamination procedures are taken from NBS Handbook No. 48.

(1) Apply a liberal portion of titanium dioxide paste to the hands. Work this paste over the affected surface and adjacent areas of the skin for at least 2 minutes. Use water, sparingly, only to keep the paste moist. Rinse with warm water, and follow by thorough washing with soap, brush, and water. Be sure that no paste is allowed to remain around the nails. Monitor. Repeat the entire process if necessary. It should be noted that the condition of the titanium dioxide paste is very important. In order to be effective, the paste must be prepared by mixing

precipitated titanium dioxide (a very thick slurry, never permitted to dry) with a small amount of lanolin.

(2) Part A: Mix equal volumes of a saturated solution of potassium permanganate and 0.2N sulfuric acid. Pour this over the wet hands, rubbing the entire surface and using a hand brush for not more than 2 minutes. Note: This application will remove a layer of skin if allowed to remain in contact with the hands too long; consequently, the time stated here should not be exceeded for any single application. Be sure that all areas are thoroughly covered. Rinse with warm water and proceed as follows:

Part B: Apply a freshly prepared 5 percent solution of sodium bisulfite ($NaHSO_3$) in the same manner as above, using a hand brush and tepid water for not more than 2 minutes. Wash with soap and water, and rinse thoroughly. The above procedure may be repeated several times as long as the permanganate solution is not applied for more than 2 minutes during any one washing. Applications to parts of the body other than the hands may be facilitated by the use of swabs steeped in the solutions. Lanolin or hand cream should be applied after washing. Note: A hand decontamination kit should be maintained in each washroom associated with laboratories in which work is done with radioisotopes.

If a significant* quantity of radioactive material has been ingested, the treatment should follow that given for general chemical poisoning such as dosing with an emetic. Introduction of significant quantities of radioisotopes into a puncture wound or cut requires immediate action. The injury should be washed with large amounts of water and treatment similar to that for snake bite given immediately afterward. The treatments in this paragraph should be under the direction of a physician.

The problems of skin contamination, ingestion and entry through wounds have been mentioned. However, it must be remembered that these are treatments indicated where possibly toxic quantities of material are present. If small quantities of isotopes are being handled, there is no point in using drastic means of first aid when an accident occurs. It is the responsibility of the operational supervisor to know whether an acci-

TABLE 8.   Decontamination Methods

Surface	Method	Advantages	Disadvantages
Paint	Water	Most practical method of gross decontamination from a distance. Contamination reduced by approximately 50%.	Protection needed from contaminated spray. Runoff must be controlled. Water under high pressure should not be used on a surface covered with contaminated dust.
	Steam (with detergent if available)	Most practical method for decontaminating large horizontal, vertical, and overhead surfaces. Contamination reduced by approximately 90%.	Same as for water.
	Soapless detergents	Where effective, reduces activity to safe level in 1 or 2 applications.	Mild action.
	Complexing agents:[a] oxalates, carbonates, citrates[b]	Holds contamination in solution. Contamination on unweathered surfaces reduced by approximately 75% in 4 min. Easily stored, nontoxic, noncorrosive.	Requires application from 5–30 min for effectiveness. Has little penetrating power; hence of small value on weathered surfaces.
	Organic solvents	Quick dissolving action makes solvents useful on vertical and overhead surfaces.	Toxic and flammable. Requires good ventilation and fire precautions.
	Caustics[c]	Minimum contact with contaminated surface. Contamination reduced almost 100%.	Applicable only on horizontal surfaces. Personnel hazard. Not to be used on aluminum or magnesium.

*Several times the permissible body burden.

TABLE 8. Decontamination Methods (*Continued*)

Surface	Method	Advantages	Disadvantages
	Abrasion (wet sand-blasting)	Complete removal of surface and contamination. Feasible for large-scale operations.	Contaminated sand spread over large area. Method too harsh for many surfaces.
Metal	Water	Contamination reduced by approximately 50%.	Same as for painted surfaces.
	Detergents	Removal of oil or grease films.	Same as for painted surfaces.
	Organic solvents	Stripping of grease	Same as for painted surfaces.
	Complexing agents:[a] oxalates, carbonates, citrates[b]	Holds contamination in solution.	Difficult to keep in place on any but horizontal surfaces. Limited value on weathered porous surfaces.
	Inorganic acids	Fast, complete decontamination.	Good ventilation required; acid fumes toxic to personnel. Possibility of excessive corrosion. Acid mixture cannot be safely heated.
	Acid mixtures	Action of weak acid. Reduces contamination of unweathered surfaces.	Same as for inorganic acids.
	Abrasion (buffers, grinders)	Useful for detailed cleaning.	Follow-up procedure required to pick up powdered contamination.
	Abrasion (wet sand blasting)	Same as for painted surfaces.	Same as for painted surfaces.
Concrete	Abrasion (vacuum blasting)	Direct removal of contaminated dust.	Contamination of equipment.
	Vacuum cleaning	Same as for vacuum blasting on concrete.	Same as for vacuum blasting on concrete.
	Flame cleaning	Only method of trapping contamination on surface.	Slow and painstaking. Fire and air-borne radiation hazard is great.
Brick	Same as for concrete	Same as for concrete.	Same as for concrete.
Asphalt	Abrasion	No direct contact with surface; contamination may be reduced to safe level.	Residual contamination fixed into asphalt. If road is subject to further contamination, may require re-covering.
Wood	Flame cleaning	Same as for flame cleaning on concrete.	Same as for flame cleaning on concrete.

[a]Another complexing agent would be a solution of 1% ethylenediaminetetraacetic acid ("Versene," "Sequestrene") in water.

[b]Oxalates, carbonates or citrates may be the ammonium or sodium salts.

[c]Caustics should preferably be trisodium phosphate or sodium sesquicarbonate rather than hydroxides since they are less hazardous.

dent can be a source of hazard and what procedures should be followed in the event of an accident.

The reason for immediate action in area decontamination is to prevent the spread of radioactive material. The return of the area to operating condition is usually not a critical factor and the cleanup can proceed in an orderly, planned fashion. All cleaning should be from the outer part of the area working in toward the center of contamination. The operation should be monitored for personnel overexposure and to check that contamination is not being spread. In most cases a preliminary washing with water and detergent will remove the majority of the isotope. After that, smaller areas may be cleaned up with decontaminating agents suitable for the isotope and the surface to be cleaned. Table 8 lists decontamination methods recommended by the Federal Civil Defense Agency.[11]

The maximum level of beta and gamma radiation which is frequently recommended is 1.0 mrad/hr for large surfaces. While this may seem to be a low level it is not impossible to attain, and maintenance of contamination below this value will keep the general level of the work area within satisfactory limits.

The decontamination procedure should be followed with survey instruments and the working area should not be reopened for operations without the approval of the qualified technical person in charge. If the quantity of isotope involved in the accident is at the toxic level, the cleanup crew should wear gloves and other protective clothing. Any possible air-borne contamination makes necessary the use of respirators. When the cleanup procedure is completed, the bodies of the crew should be monitored, after removal of their protective clothing, and any radioactive material on the body removed as described above.

Ventilation

Ventilation is needed only when dealing with isotopes that may become air-borne. This can happen much more readily than is commonly imagined, since transfer of dry solids, boiling of liquids or other normal processes frequently give rise to air-borne contamination. Such operations should be carried out in well-ventilated areas only, and monitoring of the level of air-borne activity must be a part of the protection program. This is true only if the total amount of isotope handled is possibly toxic; it is not required for lower levels of activity.

General ventilation requirements for the handling of toxic materials are covered in Section 2. The general requirements apply as well to handling radioactive materials. The only additional factor to consider is that hoods for use with isotopes emitting penetrating radiation may also require shielding. This shielding may disturb the air flow characteristics of the hood, and measurements of air flow should be made before any operations are carried out to insure adequate ventilation. Almost all operations involving small amounts of isotopes can be carried out in a totally enclosed hood or dry box. The dry box is set up so that the worker performs all operations through gloved ports. The ventilation intake is at the ends of the box, and the outlet at the rear or top is connected to suitable ducts. Such a totally enclosed system offers a great margin of safety while a small amount of practice allows normal operations to be carried out with facility.

Personnel Contamination

The hazard from radioisotopes taken internally has been described earlier in this section. Actually, effective prevention of isotope ingestion depends on the personal habits of the worker. The common preliminary to ingestion is transfer of activity from contaminated surfaces or objects to the mouth by way of contaminated hands, food, etc., when eating or smoking. It is an absolute requirement for laboratories handling toxic amounts of unsealed isotopes that edible materials should not be allowed into the work area. The general spread of contamination by hands to personal clothing and personal items such as pocketbooks can only be stopped by frequent hand cleaning, particularly before leaving the work area. This cannot readily be a supervised process, but whenever dangerous quantities of isotopes are handled, the workers must be explicitly instructed in the requirements of personal cleanliness and good housekeeping.

The requirement for protective clothing in isotope handling depends on the level of activity handled. At any level of activity, it is desirable that street clothing should not be worn in the plant or laboratory. Nor are laboratory coats good protection for prevention of contamination. Actually, when levels above tracer quantities are handled in such a way that contamination can occur, a change room should be provided with separate lockers for street and work clothing and with adequate shower and scrubbing facilities. (At this same level, the laundering of work clothing should be done at the site. Contaminated clothing should not be sent to commercial laundries.)

Rubber or plastic gloves are a very good means of preventing the spread of radioactivity by the hands. Some delicate manual operations may require surgical gloves, but whenever possible cloth-lined heavy gloves should be worn. These should never be removed from the work area until ready for disposal. More specialized equipment such as rubbers, overshoes or shoe covers are rarely required, unless serious floor contamination has occurred.

Reevaluation

A radiation control program is almost useless unless it is continuously reevaluated. Many changes can take place without being apparent. Shields may develop cracks or other openings, spills of radioisotopes may cause contamination in an unexpected part of the work area, ventilation blocks may occur, or some change in operating technique may present a new source of radiation. Therefore, a reevaluation of the control system must be made periodically. A portion of this is done through the personnel monitoring program, but surveys with portable instruments can be of great value in detecting sources of radiation before they appear in the results of personnel monitoring. This concept is true of many types of health protection. No control program can be considered completely static and free from the possibility of breakdown.

DESIGN OF INSTALLATIONS

The design of the installation where radiation exposure is possible is probably the most important single factor in maintaining low exposure levels. The addition of radiation control measures to an operating laboratory or plant is always more expensive and usually less effective than when control measures are built into the original design. The various factors of responsibility must be included when designing a work area, and a complete knowledge of the processes to be carried out in the area must be available to the designer. Redesign may be necessary if the quantity of isotopes or the x-ray energy is increased and this will add very considerably to the cost. Suitable protection for the foreseeable future should be part of the initial plan.

The overall design must be viewed from the standpoint of exposure reduction in isotope and x-ray installations. Adequate shielding and electrical safeguards are the primary considerations for x-rays. In isotope laboratories such additional factors as storage space, materials of construction and ventilation are also important. In both cases every effort should be made to use distance as a method of exposure reduction.

No working areas should be laid out near a high level radiation source. Partitioning may be designed to act not only as shielding but as a physical barrier to prevent passage of workers through or near areas of high radiation exposure. Wherever possible, separate buildings or wings should be adapted for isotope or x-ray operations.

The primary attention is given to the working area, called a "restricted area" in federal regulations and a "controlled area" by the NCRP. The NCRP definition is "a defined area in which the occupational exposure of personnel to radiation or to radioactive materials is under the supervision of an individual in charge of radiation protection. (This implies that a 'controlled area' is one that requires control of access, occupancy, and working conditions for radiation protection purposes.)"

Installations where radiation is a hazard may be classed according to whether the source of radiation is electrical, such as x-ray equipment, or radioisotopes. Under the classification of x-ray operations we have medical units for diagnosis and therapy and industrial units for testing of products as well as laboratory equipment for x-ray diffraction and other scientific measurements. The isotope field may be divided into laboratory installations using isotopes for experimental purposes and industrial operations such as radiography. The large-scale utilization of low priced isotopes, such as mixed fission products, for sterilization of foods or drugs or similar processes is still in the developmental stages.

One factor in laboratory design which lies outside the field of protection is the requirement that radiation measuring equipment be installed in areas which have low radiation background and which are not easily subject to contamination. Counters and other equipment are efficient only when the ratio of background to the desired activity is maintained at a low level. This usually requires isolation or shielding of the measuring area beyond the requirements for eliminating health hazards.

Medical X-ray Installations

The medical x-ray technician and radiologist have given less consideration in the past to their safety precautions than anyone else with hazardous exposure. Most of our knowledge leading to the setting up of permissible levels came from overexposures incurred in the medical use of x-rays. It is also probable that many patients received excessive exposures in the course of diagnosis. X-ray therapy is always a calculated risk, where the possible damage from the high x-ray dose is balanced against the possible effects of the disease. Here again it is probable that the hazard was not given full weight in decisions to use extensive x-ray therapy.

Medical x-ray facilities do not come under federal regulations but may come under state and local codes. Both the ICRP[12] and the NCRP[13] have issued recommendations on protective measures, and the references should be consulted for details. This section will deal only with the general features of design.

The basic criterion for x-ray operation is that the radiologist or technician not receive more than the maximum permissible dose. Reduction of dose rate can best be accomplished by proper design of the installation. The two sources of exposure are the useful beam from the x-ray tube and scattered radiation from the patient and the four walls and ceiling of the room. It is not possible to shield the useful beam, although 1 mm of aluminum should be used as a minimum filter to reduce the low energy x-rays which are of little value in radiography yet present considerable biological hazard. The useful beam, of course, must also be prevented from leaving the room and exposing those outside the x-ray installation. The requirements for lead, iron, and concrete shielding vary for beams of different energies and intensities as does the distance between source and shield.

As was previously discussed, x-rays and gamma rays cannot be completely blocked, but only reduced in intensity by shielding. Thickness of shielding is expressed as half value (HVL) or tenth value (TVL) layers, the thickness required to reduce the radiation intensity to one-half or one-tenth of the initial value. Figure 5 shows the HVL's in lead and in concrete for radiation of various energies. Figure 6 shows the number of HVL's required for various degrees of reduction.

The usual description of an x-ray source is in terms of the applied voltage and the target current. These parameters may be converted approximately to dose units by the graph shown in Fig. 7.[12] This figure gives the roentgens per minute per milliampere at various high voltages with various filters built into the x-ray equipment.

Figures 5, 6 and 7 may be used to calculate whether existing shielding is adequate or for approximate design work. From the characteristics of the source, Figure 7 may be used to obtain the roentgens per milliampere-minute at 1 meter. This may be converted to the required distance by the inverse square law to give the dose rate at the point of interest. Figure 6 will then give the number of HVL's to reduce this to the per-

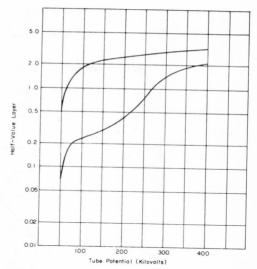

Figure 5. Approximate HVL's for X-rays at High Filtration for Various Tube Potentials: (a) Centimeters of Concrete (Upper Curve), (b) Millimeters of Lead (Lower Curve).

missible level, and Figure 5 will give the HVL layers.

For example, a wall between the radiation area and an adjoining office is to be shielded. The source is a 150 kV machine with 2 mm Al filtration and operates at 5 mA, 16 min/day (average), 5 days a week. The wall is 3 meters from the source.

From Figure 7, the dose rate is about 2.1 R/mA-min, so

$$(2.1)(5)(16)(5) = 840 \text{ R/week}$$
$$\text{at 1 meter}$$

At 3 meters this would be

$$840/9 = 93\text{R/week}.$$

This would require a reduction of

$$93/0.01 = 9300$$

or 13 HVL's (Figure 6). From Figure 5, the HVL for 150 kV x-rays in lead is 0.3 mm and the total shield would be 3.9 mm to attain 0.01 R/week.

The actual dose delivered by an x-ray machine, of course, depends on the time of use as well as the instrument characteristics. For the purpose of computation the weekly workload (W) is computed. This is taken as the number of milliampere-minutes per week and may be averaged over a long period. Since not all areas being shielded have the same occupancy or are exposed to the beam for the same time, the oc-

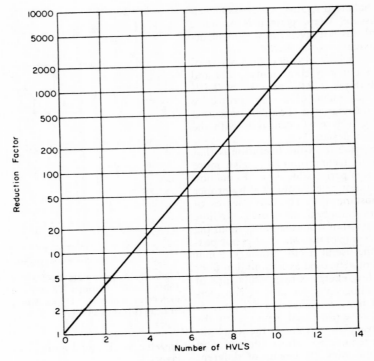

Figure 6. Reduction Factor as a Function of Number of HVL's.

TABLE 9.　Occupancy Factors

Full occupancy (T = 1)	Control space, offices, corridors and waiting space large enough to hold desks, darkrooms, workrooms and shops, nurse stations, rest and lounge rooms routinely used by occupationally exposed personnel, living quarters, children's play areas, occupied space in adjoining buildings.
Partial occupancy (T = 1/4)	Corridors too narrow for desks, utility rooms, rest and lounge rooms not used routinely by occupationally exposed personnel, wards and patients' rooms, elevators using operators, unattended parking lots.
Occasional occupancy (T = 1/16)	Closets too small for future occupancy, toilets not used routinely by occupationally exposed personnel, stairways, automatic elevators, sidewalks, streets.

Use Factors

Full use (U = 1)	Floors of radiation rooms except dental installations, doors, wall and ceiling areas of radiation rooms routinely exposed to the useful beam.
Partial use (U = 1/4)	Doors and wall areas of radiation rooms not routinely exposed to the useful beam, floors of dental installations.
Occasional use (U = 1/16)	Ceiling areas of radiation rooms not routinely exposed to the useful beam.

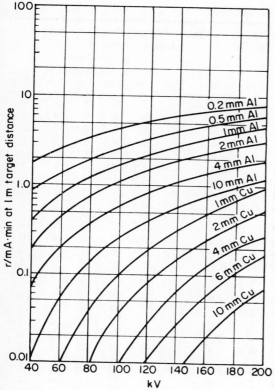

r/mA·min at 1 m target distance

0.2 mm Al
0.5 mm Al
1mm Al
2mm Al
4 mm Al
10 mm Al
1mm Cu
2 mm Cu
4 mm Cu
6 mm Cu
10 mm Cu

kV

Figure 7. Approximate Radiation Output in Roentgens per Milliampere-minute at a Distance of 1 Meter, Measured in Air, of the Primary Radiation from a Tube with Tungsten Target and Total Filtrations from 0.2 mm Al to 10 mm Cu at 40 – 200 kV Constant Potential.

cupancy factor (T) and use factor (U) are applied to compensate. These have been defined by the ICRP[12] and are shown in Table 9.

From the product (WUT) the requirements for a primary protective barrier may be determined from Table 10. It should be noted that if the shielded area is not part of the radiation facility, an additional TVL is required to reduce the dose to the permissible level.

The previous example could be handled by the table as follows,

$$WUT = (5)(16)(5)(1/4)(1) = 100$$

From Table 10 with a 150 kV source and WUT = 100 we find the required shield to be 1.3 mm for 0.1 R/week and 1.3 + 0.9 = 2.2 mm for 0.01 R/week. This is if the wall is not routinely exposed to the useful beam. If it is, WUT would be 400 and the required shielding would be about 1.9 + 0.9 = 2.8, which is in reasonable agreement with the 3.3 found previously.

It is apparent that extending this barrier to the floor, ceiling and all walls would be extremely expensive, both for the shielding material itself and for suitable building construction to support such a weight of shielding. Therefore, it is more economical mechanically to limit the direction in which the beam can be pointed. If this is done, a primary barrier is required to cover only the field which can be intercepted by the useful beam.

The primary barriers described for shielding do not completely cover the protection requirements. Since the tube housing gives off some radiation and since some low energy x-rays are

TABLE 10. Primary Protective Barrier Requirements for 100 mR/week

The tabulated values give the shielding required to reduce the exposure dose to 100 mR in a week, the value assumed for design purposes in controlled areas. To compute the shielding required outside controlled areas it is necessary to add half a TVL to reduce the weekly exposure dose to about 30 mR and one TVL to reduce it to 10 mR.

Tube Voltage, Constant Potential (kV)	WUT (mA-min/week)	Primary Protective Barrier Requirements (mm of lead) at Target Distance of					Primary Protective Barrier Requirements (cm of 2.35 g/cm³ concrete) at Target Distance of				
		TVL	1 m	2 m	3 m	5 m	10 m	TVL	1 m	2 m	3 m	5 m	10 m
	10,000		0.7	0.6	0.5	0.4	0.3		7	6	5	4	3
	3,000		0.6	0.5	0.4	0.3	0.2		6	5	4	3	2
50	1,000	0.2	0.5	0.4	0.3	0.2	0.1	2	5	4	3	2	1
	300		0.4	0.3	0.2	0.2	0.1		4	3	2	2	1
	100		0.3	0.2	0.2	0.1	0.1		3	2	2	1	1
	10,000		1.6	1.3	1.1	0.9	0.7		14	12	10	8	6

TABLE 10 (*Continued*)

Tube Voltage, Constant Potential (kV)	WUT (mA-min/ week)	Primary Protective Barrier Requirements (mm of lead) at Target Distance of						Primary Protective Barrier Requirements (cm of 2.35 g / cm³ concrete) at Target Distance of					
		TVL	1 m	2 m	3 m	5 m	10 m	TVL	1 m	2 m	3 m	5 m	10 m
70	3,000	0.5	1.4	1.1	0.9	0.7	0.5	4	12	10	8	6	4.5
	1,000		1.1	0.9	0.7	0.5	0.4		10	8	6	4.5	3.5
	300		0.9	0.7	0.5	0.4	0.2		8	6	4.5	3.5	2
	100		0.7	0.5	0.4	0.3	0.1		6	4.5	3.5	2.5	1
85	10,000	0.8	2.7	2.2	1.9	1.5	1.1	6.5	23	19	16	13	9.5
	3,000		2.3	1.8	1.5	1.2	0.8		19.5	15.5	13	11	7
	1,000		1.8	1.4	1.1	0.9	0.6		15.5	12.5	9.5	8	5
	300		1.4	1.1	0.8	0.6	0.4		12.5	9.5	7	5	3.5
	100		1.1	0.8	0.6	0.4	0.2		9.5	7	5	3.5	2
100	10,000	0.85	3.3	2.8	2.5	2.1	1.6	7	26.5	22	20	17	13
	3,000		2.9	2.4	2.0	1.7	1.2		23	19	16	14	10
	1,000		2.5	2.0	1.6	1.3	0.8		20	16	13	10.5	6.5
	300		2.0	1.5	1.2	0.9	0.5		16	12	10	7.5	4
	100		1.6	1.1	0.8	0.6	0.3		13	9	6.5	5	2.5
125	10,000	0.9	3.7	3.2	2.8	2.5	1.9	7	30	26	24	21	16.5
	3,000		3.3	2.7	2.4	2.0	1.5		27	23	20	17	13
	1,000		2.8	2.3	1.9	1.6	1.0		24	19	16.5	14	9
	300		2.4	1.8	1.5	1.1	0.7		20	16	13	10	6
	100		1.9	1.4	1.1	0.8	0.4		16.5	12	10	7	3
150	10,000	0.9	3.9	3.4	3.1	2.7	2.1	8	33	29	26	23	18
	3,000		3.5	2.9	2.6	2.2	1.6		30	25	22	19	14
	1,000		3.0	2.5	2.2	1.7	1.2		25.5	21	19	14.5	10.5
	300		2.6	2.1	1.7	1.3	0.8		22	18	15	11	7
	100		2.2	1.6	1.3	0.9	0.5		19	14	12	8	4
200	30,000	2	8	6.5	6	5	4	9	49	42	39	34	30
	10,000		7	5.5	5	4.2	3.3		44	37	34	30	25
	3,000		6	4.5	4	3.3	2.5		39	32	29	25	21
	1,000		5	3.8	3.3	2.7	1.8		34	27	25	22	16
	300		4	3.0	2.5	1.9	1.2		30	23	21	17	12
	100		3.3	2.4	1.9	1.3	0.9		25	20	17	13	8
250	30,000	3	13.5	12	10.5	9	7.5	10	55	49	45	41	35
	10,000		12	10.5	9	7.5	6		50	45	40	35	30
	3,000		10.5	8.5	7.5	6	4.5		45	39	35	30	25
	1,000		9	7	6	5.5	3.5		40	34	30	26	20
	300		7.5	5.5	4.5	3.5	2.5		35	28	25	20	15
	100		6	4.5	3.5	2.5	1.5		30	25	20	15	10
300	30,000	6	24	20	18	15.5	12	10	58	51	48	44	38
	10,000		21	17	15	12.5	9.5		53	46	43	39	33
	3,000		18	14	12	10	7		48	41	38	33	28
	1,000		15	11.5	10	7.5	5		43	36	33	29	23
	300		12	9	7.5	5.5	3.5		38	32	29	24	18
	100		9.5	7	5.5	4	2.5		33	28	24	19	15

produced by scattering from the subject, additional protection is required.

All walls out of the direct beam and the floor and ceiling must have suitable secondary bar-

riers to absorb scattered radiation plus leakage radiation through the tube shield if the area on the other side of the shield is inhabited. These secondary barriers are computed on the basis that the scattered radiation is 0.1 percent of the primary beam at a distance of 1 meter from the scatterer for x-rays up to 500 kV.

Secondary barriers based on the assumptions described are shown in Table 11. This table is adequate for determining whether barriers installed are probably adequate for the particular application. In designing a new installation it may be desirable for economic reasons to follow the more detailed calculations given in the original references.[12,13]

All of the computations are intended to yield adequate barriers for the conditions described. It must be pointed out, however, that the final test is a measurement of the performance of these barriers under working conditions and that the criterion is the protection of the operators.

The operator of medical x-ray equipment might possibly receive an excessive dose from scattered radiation. This may be reduced somewhat by wearing protective clothing, but this method is certainly not reliable. It is much bet-

TABLE 11.   Secondary Protective Barrier Requirements for 100 mR/week

The tabulated values give the shielding required to reduce the exposure dose to 100 mR in a week, the value assumed for design purposes in controlled areas. To compute the shielding required outside controlled areas it is necessary to add half a TVL to reduce the weekly exposure dose to about 30 mR and one TVL to reduce it to 10 mR (for values of TVL, see Table 10).

Tube Voltage, Constant Potential (kV)	WUT (mA-min/ week)	Secondary Protective Barrier Requirements (mm of lead) at Target Distance of					Secondary Protective Barrier Requirements (cm of 2.35 g/ cm^3 concrete) at Target Distance of				
		1 m	2 m	3 m	5 m	10 m	1 m	2 m	3 m	5 m	10 m
50[a]	10,000	0.35	0.25	0.2	0.1	0	3.5	2.5	2.0	1	0
	3,000	0.25	0.15	0.1	0.1	0	2.5	1.5	1	1	0
	1,000	0.2	0.1	0.1	0	0	2	1	1	0	0
	300	0.1	0	0	0	0	1	0	0	0	0
	100	0	0	0	0	0	0	0	0	0	0
70[a]	10,000	0.9	0.7	0.5	0.3	0.1	7	5.5	4	2.5	1
	3,000	0.7	0.5	0.3	0.1	0	5.5	4	2.5	1	0
	1,000	0.5	0.3	0.1	0	0	4	2.5	1	0	0
	300	0.3	0.1	0	0	0	2.5	1	0	0	0
	100	0.1	0	0	0	0	1	0	0	0	0
85[a]	10,000	1.4	1.0	0.8	0.4	0.2	12	8	6.5	4	2
	3,000	1.1	0.7	0.4	0.2	0	9	6	4	2	0
	1,000	0.8	0.4	0.2	0	0	6.5	4	2	0	0
	300	0.4	0.2	0	0	0	4	2	0	0	0
	100	0.2	0	0	0	0	2	0	0	0	0
100[a]	10,000	1.6	1.1	0.9	0.5	0.2	13	10	7	4	2
	3,000	1.2	0.8	0.5	0.2	0	10	7	4	2	0
	1,000	0.9	0.4	0.2	0	0	7	4	2	0	0
	300	0.5	0.2	0	0	0	4	2	0	0	0
	100	0.2	0	0	0	0	2	0	0	0	0
125[a]	10,000	1.8	1.4	1.0	0.5	0.2	14.5	11	8	4	2
	3,000	1.4	0.9	0.5	0.2	0	11	7.5	4	2	0
	1,000	1.0	0.5	0.2	0	0	7.5	4	2	0	0
	300	0.5	0.2	0	0	0	4	2	0	0	0
	100	0.2	0	0	0	0	2	0	0	0	0
150[a]	10,000	1.9	1.5	1.0	0.6	0.2	15	11	8	5	2
	3,000	1.5	0.9	0.6	0.2	0	11	7.5	5	2	0
	1,000	1.0	0.6	0.2	0	0	8	5	2	0	0
	300	0.6	0.2	0	0	0	5	2	0	0	0
	100	0.2	0	0	0	0	2	0	0	0	0

TABLE 11 (*Continued*)

Tube Voltage, Constant Potential (kV)	WUT (mA-min/ week)	Secondary Protective Barrier Requirements (mm of lead) at Target Distance of					Secondary Protective Barrier Requirements (cm of 2.35 g/ cm^3 concrete) at Target Distance of				
		1 m	2 m	3 m	5 m	10 m	1 m	2 m	3 m	5 m	10 m
200[b]	30,000	5.1	3.9	3.2	2.6	1.6	36	29	25	22	15
	10,000	4.1	3.0	2.4	1.8	0.8	31	24	20	16	9
	3,000	3.2	2.2	1.6	0.9	0.3	26	19	15	10	5
	1,000	2.4	1.4	0.9	0.3	0	21	14	10	5	0
	300	1.6	0.7	0.3	0	0	15	8	5	0	0
250[b]	30,000	7.5	6.0	5.0	4.0	2.5	37	31	27	23	17
	10,000	6.2	4.8	3.8	2.7	1.2	32	26	22	18	10
	3,000	5.0	3.6	2.6	1.4	0.5	27	21	17	11	5
	1,000	3.8	2.4	1.4	0.5	0	22	16	11	5	0
	300	2.5	1.0	0.5	0	0	17	9	5	0	0
300[b]	30,000	14.5	11	9.5	7.5	5	40	32	28	25	19
	10,000	12	8.5	7	5	3	34	27	24	19	14
	3,000	9.5	6	5	3	1	28	22	19	14	8
	1,000	7	4	3	1	0	24	16	14	8	0
	300	5	2.5	1	0	0	19	12	8	0	0

The figures in the table allow for both scattered and leakage radiation. The scattered radiation at 1 m was assumed to be 0.1% of the incident beam. To compute the leakage radiation it was assumed that the maximum ratings were 180 mA-min in 1 hr and 900 mA-min in 1 hr for the diagnostic-type protective tube housing (a) and the therapeutic-type protective tube housing (b) respectively.

ter to include a shielded booth for the operator in the original design. A proper design would require that the operator be in the booth in order for the x-ray equipment to operate. Thus, after placement of the patient, he would enter the booth and begin the exposure. While automatic timers are desirable, they may fail and the presence of an operator at all times will insure that the patient does not get an excessive dose. The patient may be observed through lead glass or heavy-liquid filled windows.

The installation described is one step away from the so-called totally protected system. Such a system includes electrical interlocks so that it is impossible for the x-ray unit to be operated unless everyone is out of the x-ray room with the exception of the patient. Such an installation is most desirable where heavy doses are administered for therapeutic purposes. The less rigid design, requiring only that the operator be at the control panel behind a shield before radiation can begin, is suitable for diagnostic x-ray.

Since the computations described have all determined the thickness of lead in inches or in millimeters, and lead is sold in units of pounds per square foot, a comparison of these units is shown in Figure 8.[14]

Industrial X-ray Installations

Industrial radiography requirements have brought about the development of higher and higher voltage x-ray equipment. The need for seeking flaws in large castings as well as similar non-destructive tests requires extremely penetrating, high energy radiation. In addition, radiographic equipment is in operation a larger fraction of the time than ordinary medical equipment. Therefore, even more care is required in industrial work than in medical work. One advantage of industrial design is that the direct beam can usually be absolutely fixed in a vertical downward position on the lowest floor of the building and the need for primary shielding can be eliminated.

There is no excuse for any industrial installation not using a totally protected system. The operator is dealing with inanimate objects which can be set immovably in place and the exposure must be controlled to produce a usable film. Therefore, the operator should be in a booth where he can control the total exposure without being exposed to either primary or secondary radiation. If the objects to be radiographed are changed by other workers, there must also be an interlock and signal system so

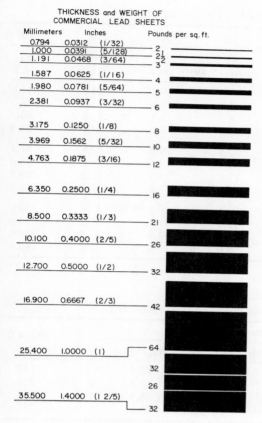

THICKNESS and WEIGHT OF
COMMERCIAL LEAD SHEETS

Millimeters	Inches		Pounds per sq. ft.
0.794	0.0312	(1/32)	2
1.000	0.0391	(5/128)	2½
1.191	0.0468	(3/64)	3
1.587	0.0625	(1/16)	4
1.980	0.0781	(5/64)	5
2.381	0.0937	(3/32)	6
3.175	0.1250	(1/8)	8
3.969	0.1562	(5/32)	10
4.763	0.1875	(3/16)	12
6.350	0.2500	(1/4)	16
8.500	0.3333	(1/3)	21
10.100	0.4000	(2/5)	26
12.700	0.5000	(1/2)	32
16.900	0.6667	(2/3)	42
25.400	1.0000	(1)	64
			32
			26
35.500	1.4000	(1 2/5)	32

Figure 8.

that even a careless x-ray operator cannot start the exposure unless everyone is clear of the area.

The high voltage industrial installations usually rely on concrete shielding rather than lead. The high mechanical strength and simple construction of concrete barriers more than compensates for the greater thickness required in comparison to lead. The space requirement can be reduced somewhat with heavy concrete, and these compositions will be described later. There has been some question of the efficiency of concrete shielding for broad x-ray beams, and concrete-lead sandwiches have been used to advantage. No design data are available.

Gamma-ray Radiography

Isotopic sources for industrial or medical radiography are more closely allied to x-ray equipment than to isotopes as used for laboratory purposes. The source must be a gamma emitter of high intensity and usually of high energy. When installed, it is protected by heavy shielding to give a useful beam very close in properties to the useful beam from x-ray equipment and is completely encapsulated to prevent escape of the isotope. All of the design criteria for x-ray installations would apply to the use of gamma sources for radiography.

The common isotopes in radiography include ^{226}Ra, ^{60}Co and ^{137}Cs. In addition, others such as ^{192}Ir, ^{153}Eu and ^{170}Tm are being used. Radiographic sources are similar to x-ray equipment in their shielding requirements. The rhm factors for curie quantities of various isotopes have been given in Table 6, and these give the basic unshielded dose rate required for shield design. As for x-rays, the design dose rates are 0.1 R/week in the working area and 0.01 R/week in uncontrolled areas.

The advantages of concrete for shielding are very marked when designing for energetic gamma emitters. Table 12 shows the barrier characteristics of plain and heavy concrete for gamma shielding.[15] The composition of the concretes is given in Table 13. Baryte aggregate, which is frequently mentioned for shielding, yields a concrete very similar in its properties to the magnetite concrete shown in Table 12.

As an example, a 5 curie ^{137}Cs source (0.66 MeV) is to be shielded from the work area. From Table 6 the unshielded dose rate is 1.6 R/hr at 1 meter or 64 R per 40 hour week. The required reduction factor is 640. From Table 12, assuming 1 MeV and a reduction of 1000, a shield of 24.5 inches would be needed. This is conservative since both the energy and required reduction are less.

More exact values may be obtained for shielding of ^{226}Ra, ^{60}Co, and ^{137}Cs from Figures 9, 10 and 11. These are taken from NBS Handbook 73 which also has charts for iron and lead shielding plus requirements for secondary shielding of leakage and scattered radiation.[16]

In the above example, Figure 11 would give a value of 20 inches of concrete for a ^{137}Cs source with 1.6 rhm. This is a better value than the one above, but applies only to standard concrete.

For reduction of radiation exposure to nonoccupational levels, an additional TVL should be added. This is about 6.2 inches for ^{137}Cs and 8.6 inches for ^{60}Co.

If the use and occupancy factors from Table 9 are applicable, they may be used to reduce the required shielding. They should be multiplied in to get a corrected rhm value for the source.

Figures 9, 10 and 11 also indicate the value of distance in reducing exposure and shielding requirements. For the ^{137}Cs source noted, the shield would only need to be 10 inches thick if

TABLE 12. Gamma-ray Shielding with Concrete

Reduction Factor		2	10	10^2	10^3	10^6
Energy of γ Rays	Type of Concrete	ST[a] (in.)	ST (in.)	ST (in.)	ST (in.)	ST (in.)
1 MeV	Plain	3.2	9.3	17.1	24.5	45.6
	Magnetite	2.0	5.8	10.6	15.3	28.3
	Limonite	2.8	8.2	15.1	21.6	40.3
	Limonite + iron	1.7	5.0	9.2	13.1	24.3
3 MeV	Plain	5.3	15.9	29.8	42.9	81.1
	Magnetite	3.1	9.4	17.5	25.4	48.0
	Limonite	4.7	13.6	25.5	36.9	69.8
	Limonite + iron	2.6	7.9	15.0	21.5	40.9
5 MeV	Plain	6.8	20.9	39.8	57.9	110.0
	Magnetite	3.5	11.1	21.4	31.3	59.6
	Limonite	5.4	16.9	32.4	47.3	89.9
	Limonite + iron	2.8	9.0	17.4	25.6	48.8

[a] Shield thickness.

TABLE 13. Composition of Some Plain and Heavy Concretes

Type	Composition (% by weight)			lb/cu ft
	Water	Cement	Aggregate	
Ordinary	8.4	17.7	73.9	147
Magnetite	5.5	11	83.5	236
Limonite	10.5	21	68.5	164
Limonite + iron	4.7	9.5	85.8	275

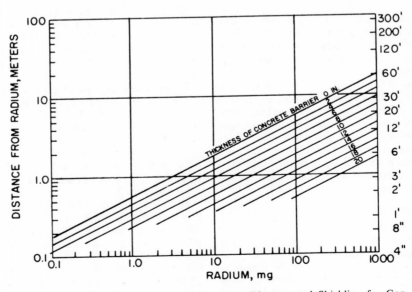

Figure 9. Relation between Amount of Radium, Distance and Shielding for Controlled Areas.

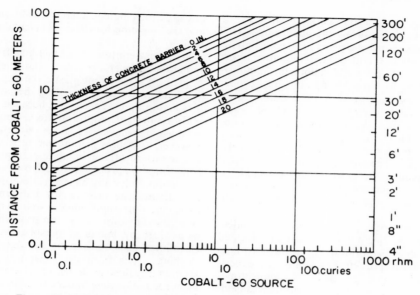

Figure 10. Relation between rhm, Distance and Shielding for Controlled Areas.

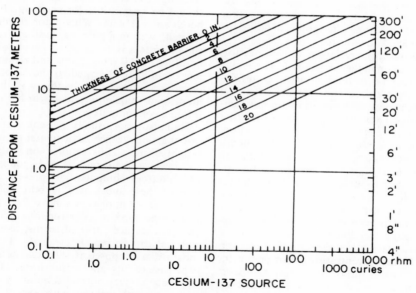

Figure 11. Relation between rhm, Distance and Shielding for Controlled Areas.

the area to be shielded were 6 meters away rather than one. Distance is always the cheapest way of reducing exposure.

Isotope Installations

Operations with isotopes that are not encapsulated still may require shielding in the same way as the radiographic sources described

above. They do offer additional hazards, however, over those found with encapsulated sources.

First, materials may become air-borne, so the external radiation from an isotopic source may be negligible compared with the hazard from its inhalation or ingestion.

Secondly, appreciable amounts of the radioisotope may be left on working surfaces or equipment surfaces during the operations. This material in turn may be generally spread about by later work. Good design, therefore, must include not only proper shielding for minimizing exposure to direct radiation but proper ventilation, proper methods of handling, containers where necessary, and means for ready decontamination of working surfaces and equipment. Shielding from gamma radiation may follow the design outlined for x-radiation of the same energy. The actual intensity is known from the quantity of isotope available, and the shielding requirements may be obtained by calculation. Again, the rhm values of Table 6 will be useful.

Laboratory operations with isotopes, unless they are of a routine nature, are usually shielded by interlocking lead bricks. This is the most flexible system for providing shielding in experimental work. The quantities of radioisotopes involved are usually such that primary shielding alone is sufficient. Of course, this cannot be assumed, but should be checked with survey instruments just as any operation would be checked. Also, it must not be assumed that bench tops or table tops offer good shielding for radiation directed downward.

Industrial tracer operations with isotopes follow those common in the laboratory. The major differences are that ordinarily larger quantities of isotopes are required and that the personnel with possible exposure to radiation are not technically trained. Therefore, it is more critical that they be given proper instruction and protection. Such operations are so varied that only generalized statements may be made. The installation must be designed around the process to be carried out. Protective measures can be made less expensively if included in the original design. This is usually not possible as it is more likely that equipment previously used for normal operations would be modified for the new process.

The basic criterion is the prevention of overexposure. If the concentration of isotope in the material is known, the possible dose rate from the total amount of isotope can be calculated. The safe procedure is to design any required shielding so that if all the isotope were concentrated at any point in the system, the shielding would be sufficient to maintain the dose rate outside the shield below the maximum permissible level. This may not be possible with extremely large quantities of radioisotopes. Where the equipment is extensive it may not be necessary, particularly when dealing with solutions which may be readily homogenized. In such a case the amount of isotope possibly present in any area may be considered as a point source and the shielding for the area calculated on this basis.

With beta emitters, the shielding from direct radiation is less of a problem, and thick ($\frac{1}{2}$ inch or greater) transparent plastic barriers will give sufficient protection. This simplifies any laboratory manipulative operations, because the source is always in sight. Alpha emitters should require no shielding whatsoever. Both of these emitters, however, as well as gamma emitters, do require consideration of the inhalation, ingestion and contamination problems in design. The problem of ventilation for any hazard is described in Section 2, and the criteria for handling various toxic materials apply equally well to radioactive isotopes. The direct ingestion problem can only be avoided by habits of personal caution and cleanliness and wearing of suitable protective clothing. It is from the viewpoint of decontaminability that the design of laboratory equipment for radioactive materials differs from that for other toxic substances.

An isotope laboratory should have special cleaning services. The laboratory personnel should be responsible for cleaning all benches, hoods and cabinets. When floor cleaning is performed, wet mopping or vacuuming through suitable filters is required; sweeping should not be allowed. All pails, mops and cleaning equipment should be assigned to the laboratory and never used elsewhere. In this way they can be monitored frequently and discarded or cleaned if found contaminated.

Radioisotopes usually present a greater hazard than any other toxic material on a pure weight basis. Thus smaller amounts of these substances present real hazards, and the design of laboratory equipment to prevent accumulation of these substances is both difficult and costly. Two approaches have been proposed; one is the type of construction involving impervious surfaces, free of seams, using a material that can be readily cleaned. The material most often proposed is stainless steel, which is expensive to buy and to fabricate. The second approach utilizes normal laboratory equipment and covers all surfaces with expendable materials that can be stripped off and discarded. Both methods have their merits and will be discussed in some detail.

One fallacy in the initial concept of stainless steel or other "impervious" surfaces is the belief that they are truly impervious. This has been shown to be false. Stainless steel after one vigorous cleaning is found to deteriorate in that more and more material may be absorbed or adsorbed and retained on the surface. Successive cleanings have been found to become more difficult and to require more vigorous methods of decontamination. Glass always shows high adsorptive characteristics for traces of most elements, and even the drastic methods of cleaning chemical glassware do not always remove all of the adsorbed material. On the other hand, "impervious" surfaces can be fabricated to be free of cracks, seams and crevices so that large quantities of an isotope cannot become permanently lodged in the structure.

The removable expendable covering for laboratory surfaces is probably less expensive than the impervious construction. In a typical operation, the bench surfaces may be divided into areas with stainless or plastic trays. These in turn are covered with an absorbent paper with a waterproof backing. In case of spills the paper is discarded as radioactive waste and the trays taken up and cleaned as vigorously as required. Floors are made of rubber or asphalt tile which can be removed and replaced in small areas at a time when or if they become contaminated. Several strippable plastic coatings have been developed for covering vertical surfaces subject to contamination. These resemble paints in texture and application except that they do not form a firm bond on the surface to which they are applied. If they become contaminated, they can be stripped off and replaced.

The replaceable material concept can be carried one step further. Movable work benches of the type used in machine shops can be fitted to chemical procedures by covering with glass, linoleum or "Transite." The coverings can then be discarded if they become excessively contaminated, and even the benches are not expensive to replace. In any case, separation of work tables and cabinets is desirable for preventing the spread of contamination.

The major objections to the replaceable type of construction is that large quantities of radioactive wastes are accumulated and that the removal and disposal of contaminated sections may expose the workers to more radiation or radioactive material than a cleaning procedure carried out on an impervious surface. However, such an installation is cheaper, particularly when existing laboratory facilities are being converted to isotope operations, and procedures now exist to allow decontamination to be accomplished safely.

Both methods of construction require periodic monitoring to determine when decontamination is required. As in every other phase of radiation protection, there is no need for assumptions that everything is operating properly. Measurements can be made and interpreted and there is no excuse for the statement that any exposure "shouldn't have happened."

Figure 12 shows the properties of various construction materials with regard to decontamina-

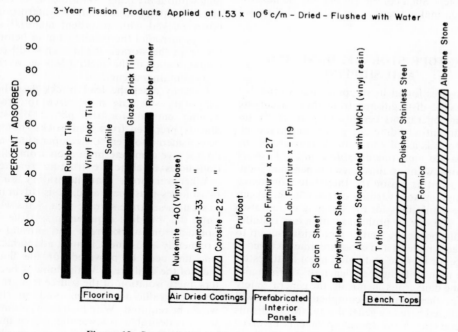

Figure 12. Susceptibility to Fission Product Contamination.

tion.[17] While these data were obtained a number of years ago and information is not available for all nuclides, conditions, and materials, such a compilation does indicate a possible choice of materials for design purposes.

The choice of technique, impervious or removable, in setting up a laboratory is governed by cost since either method can be utilized to maintain exposure at levels below the maximum permissible values.

Waste Disposal

The preceding paragraphs have discussed the aspects of three factors—shielding, decontaminability and ventilation—with respect to design. A fourth factor which must be considered in isotope installations is that of liquid waste disposal. This will be covered later but it must also be considered in the design of an installation.

Sinks and drains for carrying off highly radioactive solutions cannot be fed directly into the normal channels. Such installations require two entirely separate waste systems: one for inactive material going directly to the regular sewer system and one for radioactive material which is led into tanks for holdup or processing prior to disposal. In extreme cases, these waste lines may also require shielding to prevent overexposure to workers in the area. In any case they should be designed to minimize concentration of isotopes in the lines. Such concentrations can be hazardous when making repairs on high level disposal systems.

ISOTOPE STORAGE, HANDLING AND SHIPPING

Exposures found in shipping and storage of isotopes are distinguished from those occurring in actual operations because they generally involve an entirely different group of personnel. This group is usually untrained, unaware of the hazards and must operate solely on the basis of directions, without allowance for personal judgment. For this reason the Interstate Commerce Commission and various carriers have set up definite regulations for the shipping of radioisotopes. The storage problems within any installation must be met by similar regulations set up to fit the activity level and exposure time involved.

Since the storage and handling of isotopes includes many of the problems of storing and handling dangerous chemicals, the general safety rules for these must be complied with and will be discussed here in a general way. More details will be given on the specific problems involved in storing and handling radioisotopes.

Storage

Storage facilities are dictated by the type of emitter and the quantity being stored. Laboratory quantities of radioisotopes may not even require a separate storage area but are usually kept away from counting areas to prevent an increased counter background from stray radiation or contamination. Alpha or beta emitters are usually quite well shielded by the actual bottle or other container and additional shielding is best provided by a plastic container having walls that are about a centimeter thick. These containers can be more useful if they are designed to be liquid tight and thus to retain liquids or solutions if the inner container becomes broken. Additional protection is provided by filling with vermiculite or other absorbing material. Loose fitting heavy caps provide sufficient shielding and ease of handling. Laboratory quantities of gamma emitters are best stored in a glass or plastic inner container held within a solid lead block of suitable thickness. Such storage blocks are commercially available or may be readily fabricated. The block should have a heavy, loose-fitting cap to provide sufficient shielding from the top and to allow ready removal. Radioisotope solutions, whether in tracer quantities or large amounts, should preferably be contained in an unbreakable plastic bottle. Polyethylene bottles resist the action of most chemicals and can stand accidental dropping without breaking. With the larger amounts of radioactivity, a secondary waterproof container packed with absorbent material should always surround the primary bottle before it is placed in the storage shield. This will prevent contamination of the shield if leakage of the primary container occurs.

In many cases the lead bricks described for temporary shielding may serve to protect a gamma emitter for the period of storage. Shields, prepared from lead bricks should only be considered for temporary operations as they do not give complete protection from all directions unless an elaborate structure is erected.

Whenever larger than laboratory quantities of isotopes must be stored, it is advantageous to set up a separate storage area. This will minimize the shielding requirements, if the area is isolated from work areas and normal traffic. Frequently a basement is most adaptable, as no shielding need be provided for the floor and most of the wall area. With alpha or beta emitters, mere isolation of the source from the work areas or traffic is sufficient and no shielding would be required. With high level gamma emitters it is desirable to lower the source into steel tanks or cylinders imbedded in concrete beneath the cellar or basement floor. The cylinder or tank acts as a container and does provide some shielding from radiation coming through the walls at an angle which would allow it to reach a populated area. The most important shielding is on the top which is the direct radiation field. A lead or steel cap for the tank or cylinder can be fitted with a ring bolt and raised or lowered by means of a chain hoist. Ventilation is required, particularly if radium or thorium sources are kept in the storage room.

Any open industrial storage installation should meet the requirement that no exposure above the permissible level is possible with all sources in their storage containers and capped. While this requirement may seem excessive, since personnel ordinarily do not operate within the storage area, it is desirable because the workers involved are frequently untrained. For the very few installations using multicurie quantities of isotopes, a time and distance requirement may be set up and higher dose rates allowed. The storage area in such an installation should be locked and should only be opened by a responsible supervisor. When entering such an area the handling crew should be under the control of the responsible supervisor who should remain outside the doorway to the storage area. The doors, although locked from the outside, should be capable of being readily opened from the inside to prevent anyone from locking himself into the room. The general instructions for layout of a completely protected radiation source, given in pp. 186–187 should apply to the storage area for multicurie lots of isotopes.

Handling

As in all operations, the handling of radioisotopes must be set up to prevent excessive exposure to personnel and in fact to prevent any unnecessary exposure. The operational techniques, whether for the plant or laboratory, must be carefully planned to prevent direct radiation exposure and inhalation or ingestion of radioactive materials. After planning any operation, it is advisable to make one or more test runs of the complete procedure using inactive materials. It will be found that this will reduce exposure markedly by familiarizing personnel with the techniques so that they may be performed more rapidly. In addition, it is frequently found that such tests will show how process improvements may be made. The advantage is even more marked where unfamiliar apparatus is required or where there is a necessity for working around or over shields where normal body movement is not possible.

During usual handling operations on either a laboratory or plant scale, the greatest exposure is likely to occur during transfer of the radioisotopes. Transfer is involved in preparation of solutions, weighing or measuring quantities into new containers or loading reaction facilities. Regardless of the type of emitter or its physical state, the operations are best carried out in a well-ventilated hood or glove box. This is most important with finely divided solids or other materials which may become air-borne. The hood interior should be readily decontaminated to remove the normal spillage which occurs during transfer operations. This is discussed in the section on decontamination (pp. 170–173).

The shielding of actual operations depends entirely on the type of emission produced by the isotope. Alpha emitters require no shielding, beta emitters are best shielded with transparent plastic, while strong gamma sources may require lead or steel. If the operation requires heavy shielding, remote control of the apparatus must also generally be provided. This would not necessarily be a complete mechanical robot system but might merely be an extension of all controls outside the shield. Remote control equipment, such as tongs or solution transfer devices are commercially available which allow handling at a distance. The usual beta emitter operations would take place behind a plastic shield with cutouts for arms so that the operator can reach inside to carry on the process. Where high activity is present or if possible hand contamination is a problem, the operator may wear long gauntlet gloves where the cuffs are sealed to the shield. Such complete units, called either glove boxes or dry boxes, are available commercially and are very convenient for carrying out small scale laboratory or pilot plant operations.

Shipping

Radioisotope shipments are controlled by regulations of the carrier, the Interstate Commerce Commission, the Civil Aeronautics Board, the Post Office Department and the U. S. Coast Guard. These regulations are designed both to prevent any excessive exposure to shipping personnel and to prevent damage to materials such as photographic film which are sensitive to radiation.

Since the limiting factor of most regulations is the dose rate at the surface of the package, it has become customary to increase the size of the outer packing container rather than to increase the weight of shielding. This is usually more effective if packing and shipping costs are considered. In other words, a 1 inch cube source might have to be contained in a 4 inch cube of lead for adequate shielding. The same reduction in surface dose rate (for high energy gamma) could be obtained by placing the original source in a corrugated carton 1 foot in each dimension. The latter method would be much less expensive. However, the shipping carton must be strong enough to survive handling and the source must be fixed in the center of the carton so that it cannot shift to one of the carton faces.

The shipping regulations applying to the transportation of radioactive materials are included in Section 11.

RADIOISOTOPE WASTE

All operations involving radioisotopes produce radioactive material which is no longer suitable for use. Such material must be handled properly and this is a major responsibility of those in charge of the operations. The actual methods of management depend on the activity level, the type of waste and the radiochemical properties of the isotopes. The two major methods of management are dilution for disposal and containment, and these will be discussed in relation to the considerations mentioned and the quantity and type of inert waste material associated with the radioisotope.

As the total quantity of isotopes being handled in the world increases, more and more emphasis is being placed on containment, and this method should always be given primary consideration. The adoption of a conservative attitude in waste management has several advantages. First, little is actually known about the process of concentration whereby plant and animal life may collect appreciable amounts of an isotope. Second, poor public relations may result from the exposure of large off-site populations to any measurable radioactivity. Third, local public health officials may yield to public appeal and set further restrictions on disposal if levels near the maximum permissible are found.

Most of the federal regulations are based on the performance of the radiation protection system. In the case of disposal of radioactive waste, they also require prior approval of the method to be used. Systems involving incineration or disposal at sea are likely to receive particularly close scrutiny. In essence, approval is required for any disposal except that of small amounts to the sanitary sewers.

Activity Level

For the purposes of this discussion, a low activity level for gases and liquids will be defined as one where the concentration can be reduced below the maximum permissible level by reasonable dilution. Other considerations may dictate

the use of a containment procedure, but this definition will assist in waste disposal planning. High activity levels are those that cannot be treated by moderate dilution. Where large quantities of such wastes must be stored, it becomes necessary to reduce the bulk of material by concentration. Even though this process raises the relative activity of the waste to a much higher level, the storage becomes much less of a space problem.

Radiochemical Properties

The important properties of waste for disposal are the type and energy of the radiation and the half-life and relative biological hazard of the isotope.

The probable fate of the waste will regulate the relative amounts of these materials that can be disposed of safely through normal channels. Alpha emitters are all heavy metals and tend to be deposited in the bone. Thus they must be kept out of food or drinking water and must be prevented from becoming air-borne. However, they may be stored in large quantities without hazard, so they are not ordinarily released into the normal disposal channels. Long-lived beta emitters such as ^{90}Sr are subject to the same considerations. Storage of gamma emitters requires isolation or even shielding to prevent exposure to possible external radiation hazard.

Half-life is an important consideration, as short-lived emitters may be economically stored until their activity has decayed to a negligible value. Also, the biological hazard from the short-lived materials is considerably less than for long-lived emitters. As the half-life becomes longer the cost of storing large quantities for the required period of years becomes appreciable. Safe methods are therefore sought for low cost disposal of these materials by burial.

Waste Management

As mentioned earlier, it is possible to dispose of radioactive wastes after suitable dilution in some cases. In others it is necessary to contain them and to store the material. Dilution is the process of mixing the waste with sufficient inert material to reduce the concentration of activity below the permissible levels. An example would be the flushing of microcurie amounts of an isotope into the regular sewer system. Containment is the storage of waste under conditions that keep it isolated to prevent direct radiation exposure.

The three forms of waste material produced in isotope operations are solid, liquid, and air-borne. In most contaminated solid materials, the activity per pound of waste is relatively low.

Such material presents a considerable storage problem and efforts are usually made to reduce its bulk by incineration or baling. Liquid wastes are also bulky and must be concentrated by evaporation or the activity must be removed by some process such as precipitation, flocculation or ion exchange. Air-borne wastes may involve either radioactive gases as in reactor effluents, or air-borne dusts such as hood exhausts. The containment and storage of waste gases is impossible, and reduction in contamination must be made by collecting the radioactive material by absorption, scrubbing or removal of particles by filtration. Contaminated gases must be diluted to satisfactory levels before release. Present regulations require measurement of concentrations at the point of release even though the stack effluent may be diluted manyfold before reaching an inhabited area.

Segregation of high and low level wastes was found to be a valuable step at the Knolls Atomic Power Laboratory of the General Electric Company which has described how slightly contaminated wastes may be baled with a lower installation and operating cost than incineration.[18] The gain in volume reduction is only 4 to 1 for incineration over baling for combustible materials, and even less for mixed solid wastes.

The incineration process also suffers from the requirement that considerable care must be taken to remove radioactivity from the flue gases of the incinerator. Entrainment of solid material by the hot gases can carry appreciable amounts of material out the stack, thereby making some form of stack gas cleaning necessary. As noted above, it is necessary to obtain approval before incineration may be used.

These two procedures for reducing the bulk of solid waste are the only ones in current use. Many installations with small quantities of waste, of course, find it possible to handle their material without bulk reduction.

Liquid wastes are ordinarily solutions or suspensions of radioactive materials in a large bulk of solvent. A simple method of bulk reduction is evaporation, but the amounts of heat required for dilute solutions may make this an expensive operation in spite of its simplicity. As in the case of incineration, the exit gases or vapors from the evaporator must be cleaned before discharged to the atmosphere. The amount of liquid entrained in vapor coming off a boiling solution can represent several percent of the total liquid volume. This can only be removed at considerable cost as any filters in the system must be heated to prevent excessive condensation.

An isotope in solution or suspension can fre-quently be concentrated by precipitation with specific materials or by general collection with a flocculating agent. The alum process used for water purification will scavenge many different isotopes and yield a resulting liquid which may be suitable for disposal through the normal sewer system. Other methods such as the precipitation of calcium phosphate have given effective scavenging from aqueous solutions. Any precipitation or flocculation process requires filtration to remove the concentrated activity and the carrying material. Sedimentation processes are not sufficiently efficient, and a large volume of liquid waste requires a sand bed or other large filter. Small volumes may be treated with standard laboratory equipment. Setting aside a small area for such processes will reduce possible contamination of other laboratory areas.

Considerable research has been done on the removal of isotopes from liquids by ion exchange. Both the ion exchange resins and the montmorillonite clays have been tested as scavenging agents. These have the advantage for small installations that they can be run efficiently on an intermittent basis. When their capacity has been reached, as indicated by activity passing through the system, the resin or clay can be treated as a compact solid waste for disposal. The clays have the advantage that high temperature baking gives a massive product which is readily handled.

Any intermittent handling process for liquid wastes requires that adequate storage tank capacity be available to contain the waste accumulated in the period between disposals. Naturally, with short-lived isotopes a system of hold-back tanks may be the only requirement for waste disposal. The liquids are stored until a test shows that the activity level is below the limit permissible for disposal into the normal sewer system. Long-lived isotopes may merely be held until processing facilities are ready to accept the waste.

Since the usual laboratory problem encountered is the disposal of small amounts of waste, the requirements of Part 20 are paraphrased here.

No licensee shall discharge licensed material into a sanitary sewerage system unless:

(a) It is readily soluble or dispersible in water;

(b) The quantity of any licensed or other radioactive material released into the system by the licensee in any one day does not exceed the larger of:

(1) The quantity which, if diluted by the average daily quantity of sewage released into the sewer by the licensee, will result in an aver-

age concentration equal to the limits specified in Table 5 (Table I, column 2) or

(2) Ten times the quantity of such material specified in Table 7.

(c) The quantity of any licensed or other radioactive material released in any one month, if diluted by the average monthly quantity of water released by the licensee, will not result in an average concentration exceeding the limits specified in Table 5 (Table I, column 2).

(d) The gross quantity of licensed and other radioactive material released into the sewerage system by the licensee does not exceed 1 curie/yr.

Excreta from individuals undergoing medical diagnosis or therapy with radioactive material shall be exempt from any limitations.

Air-borne wastes may be concentrated by filtration of particulate matter or by absorption or scrubbing of radioactive gases. These processes yield solid or liquid material as a concentrate, and the wastes are disposed of as described above.

For the small isotope user, there are waste disposal services in several large cities which have been authorized to collect, store and dispose of isotopes.

Equipment which has become contaminated cannot be disposed of simply by dumping, as it may appear in a critical location later as second hand equipment. Frequently the cheapest method of disposal is to decontaminate the article sufficiently so that it can be returned to the normal channels of commerce. For example, this has been done successfully with scrap steel contaminated with uranium.

MEDICAL PROGRAMS AND RECORDS

As radioisotope and x-ray utilization in industry increases it becomes much more important that adequate knowledge of the exposure of everyone employed in these fields is maintained. The employer must protect himself and his employees having the possibility of significant radiation exposure with a complete medical protection program and must maintain complete medical records and radiation exposure records on each employee. These will be invaluable if any question of radiation damage occurs and in addition will furnish future information for the scientific evaluation of long-term exposure to low radiation levels.

The medical program will be most valuable in maintaining the general health of workers. The clinical symptoms of radiation damage occur only with a considerable overexposure.

Therefore, the responsibility for prevention of radiation damage rests entirely on the personnel monitoring and control systems.

The World Health Organization in its technical report No. 196 has outlined the requirements for medical supervision in radiation work.[19] This is a clear description of the program required, including the role of the physician.

At the present time there is only a limited number of medical tests available for radiation protection. Most exposure information is still obtained from personnel and area monitoring. Any radiation protection program is a failure if clinical evidence of radiation damage appears. Such a statement is not always true for other toxic materials, as clinical evidence sometimes may appear and a course of treatment may be indicated which can restore the worker to complete health. In case of radiation, the first clinical signs appear after considerable non-reversible damage has occurred. Thus, medical tests are not as much a part of a protection program as they are a confirmation that some acute overexposure has occurred.

A complete blood count was often recommended as a check on radiation exposure, but is not in favor with medical men interested in radiation protection. The most marked change is ordinarily seen in the leucocyte count and more specifically in the lymphocytes. The normal variation in various cell counts obscures the small variation caused by low level exposure. When a positive indication is given by blood count it means that the subject has received definite overexposure approaching the acute stage. A systematic following of each individual could possibly give trends indicating chronic low level exposure, but such data have not been available on industrial or laboratory populations in the past.

Some work is going on testing the chromosome changes that occur as a result of relatively low doses of radiation, perhaps less than 25 rads. This involves measuring chromosome abnormalities in certain types of blood cells after suitable culture. While this has not been proved diagnostically, it may be helpful in evaluating the dose received in accidents.

One of the most valuable functions of a medical staff in a radiation protection program is in the selection of employees. Those candidates showing symptoms of normal disease which may also be attributable later to radiation should not go into a job requiring radiation exposure. This will obviously protect the employer and, less obviously, the candidate, as radiation may aggravate an existing condition. A typical example would be those showing signs of anemia.

The examination for selection of employees should be as complete as the periodic examination. In addition, it is desirable to obtain the entire past radiation history of the candidate, particularly if he has incurred industrial radiation exposure. An exposure history should not be grounds for turning down the candidate, but a knowledge of previous exposure is desirable and should be included in the employee's record.

If the medical examiner has some idea of the operations, he may be more useful in selecting candidates. Certain positions require specific mental characteristics as well as physical features, particularly where failure of the individual may endanger others. This has been a familiar industrial problem for many years and is not peculiar to jobs with a possible radiation exposure.

In particular, the medical examiner should be familiar with the details of metabolism and the permissible levels of any isotopes handled in the plant or laboratory under his control. This is required not for the treatment of overexposure but rather for the recognition of the hazards which may exist.

Complete annual medical examinations of all employees with significant exposures are desirable. They should include a full chest x-ray and blood count as well as the normal examination. In addition, the condition of the skin, nails, and eyes should be noted in the examination. It should be remembered that while the blood count is not a reliable indication of chronic overexposure, it is valuable as an aid in diagnosing acute overexposure. Additional examinations should be performed whenever the employee exposure record indicates an acute overexposure of 25 R or greater.

A complete set of records should be maintained for each employee. This should include the results of each medical examination, history of illnesses, exposure data and the results of any special tests such as urinary excretion of isotopes or breath radon measurements. Such records should be maintained in an individual file for each employee and should be reviewed at each periodic physical examination.

The exact form of medical records is subject to individual preference. Monitoring records for AEC licensees should be kept on Form AEC-5 or in a form containing all of the information required by that form. This includes records of personnel monitoring with personnel dosimeters, the results of radiation surveys with instruments and any disposal of radioactive materials to the environment. The regulations contain requirements for notification of over-exposures and excessive levels and concentrations. In addition, licensees are required to notify employees of their radiation exposure on an annual basis.

It is also worthwhile for the person in charge of radiation protection to review the cumulative records of personnel monitoring at least semi-annually. This will indicate which employees and which operations are showing the greatest exposure. Such knowledge may enable him to correct minor difficulties and reduce exposure appreciably. This again is part of the concept that maintenance of exposure below the permissible levels is not a sign of a perfect protection program and that efforts should be directed toward reducing all exposure to the minimum.

BIBLIOGRAPHY

1. U. N. Committee on the Effects of Atomic Radiation Report, New York, 1966.
2. Health Physics Insurance Seminar, U. S. Atomic Energy Commission Report TID-388, March 1951.
3. "The Effects of Nuclear Weapons," U. S. Government Printing Office, Washington, 1962.
4. "Permissible Dose for Internal Radiation," ICRP Publication 2, New York, Pergamon Press, 1959.
5. "Maximum Permissible Body Burdens and Maximum Permissible Concentrations of Radionuclides in Air and in Water for Occupational Experience," National Bureau of Standards Handbook 69, U. S. Government Printing Office, Washington, 1959.
6. "Background Material for the Development of Radiation Protection Standards," Federal Radiation Council, Report No. 1, U. S. Government Printing Office, Washington, 1960.
7. "Nuclear Science Series, Preliminary Report," National Research Council Publication 251, 1951.
8. "Methods of Radiochemical Analysis," World Health Organization, Geneva, 1966.
9. "Code of Federal Regulations," Title 10, Chapter I, Part 20.
10. "Control and Removal of Radioactive Contamination in Laboratories," National Bureau of Standards Handbook 48, U. S. Government Printing Office, Washington, 1951.
11. "Radiological Decontamination in Civil Defense," Federal Civil Defense Agency, TM-11-6.
12. "Protection against X-rays up to Energies of 3 MeV and Beta- and Gamma-Rays from Sealed Sources," International Commission

on Radiological Protection, ICRP Publication 3, New York, 1960.

13. "X-ray Protection," National Bureau of Standards Handbook 60, U. S. Government Printing Office, Washington, 1955.

14. "Lead Handbook for the Chemical Process Industries," American Smelting and Refining Co., 1954.

15. Callan, Edwin J., J. Am. Concrete Inst., 17 (September 1953).

16. "Protection Against Radiations from Sealed Gamma Sources, National Bureau of Standards Handbook 73, U. S. Government Printing Office, Washington, 1960.

17. "Laboratory Design for Handling Radioactive Materials," American Institute of Architects and U. S. Atomic Energy Commission, U. S. AEC Report NP-3873, May 1952.

18. Larson, R. E., and Simon, R. H., "Solid Waste Disposal at the Knolls Atomic Power Laboratory," U. S. Atomic Energy Commission Report KAPL-936, June 1953.

19. "Medical Supervision in Radiation Work," World Health Organization, Technical Report Series No. 196, Geneva, 1960.

GENERAL REFERENCES

"Radiation Protection Criteria and Standards. Hearings before the Joint Committee on Atomic Energy," U. S. Government Printing Office, Washington, 1960.

Glasstone, "Sourcebook on Atomic Energy," Princeton, N. J., D. Van Nostrand Co., 1950.

Friedlander, Kennedy and Miller, "Nuclear and Radiochemistry," 2nd edition, New York, John Wiley & Sons, 1964.

Evans, "The Atomic Nucleus," New York, McGraw-Hill Book Co., 1955.

Blatz, editor, "Radiation Hygiene Handbook," New York, McGraw-Hill Book Co., 1959.

"Los Alamos Handbook of Radiation Monitoring," 3rd edition, U. S. Atomic Energy Commission Report LASL-1839, November 1958.

Price, "Nuclear Radiation Detection," 2nd edition, New York, McGraw-Hill Book Co., 1964.

"Industrial Waste Disposal. Hearings before the Joint Committee on Atomic Energy," January 28–February 3, 1959, U. S. Government Printing Office, Washington, 1959.

INTERNATIONAL COMMISSION ON RADIOLOGICAL PROTECTION

1 "Recommendations of the International Commission on Radiological Protection" (adopted September 9, 1958)

2 "Report of Committee II on Permissible Dose for Internal Radiation" (1959)

3 "Report of Committee III on Protection against X-Rays up to Energies of 3 MeV and Beta- and Gamma-Rays from Sealed Sources" (1960)

4 "Report of Committee IV (1953–1959) on Protection against Electromagnetic Radiation above 3 MeV and Electrons, Neutrons and Protons" (adopted 1962, with revisions adopted in 1963)

5 "Report of Committee V on the Handling and Disposal of Radioactive Materials in Hospitals and Medical Research Establishments" (1964).

6 "Recommendations of the International Commission on Radiological Protection" (as amended 1959 and revised 1962)

7 "Principles of Environmental Monitoring Related to the Handling of Radioactive Materials" (adopted September 13, 1965)

8 "The Evaluation of Risks from Radiation"

9 "Recommendations of the International Commission on Radiological Protection" (adopted September 17, 1965)

NATIONAL BUREAU OF STANDARDS HANDBOOKS

23 "Radium Protection"

27 "Safe Handling of Radioactive Luminous Compound"

42 "Safe Handling of Radioactive Isotopes"

48 "Control and Removal of Radioactive Contamination in Laboratories"

49 "Recommendations for Waste Disposal of Phosphorus-32 and Iodine-131 for Medical Users"

51 "Radiological Monitoring Methods and Instruments"

53 "Recommendations for the Disposal of Carbon-14 Wastes"

55 "Protection against Betatron-Synchrotron Radiations up to 100 Million Electron Volts"

57 "Photographic Dosimetry of X- and Gamma Rays"

58 "Radioactive-Waste Disposal in the Ocean"

59 "Permissible Dose from External Sources of Ionizing Radiation"

60 "X-Ray Protection"

61 "Regulation of Radiation Exposure by Legislative Means"

62 "Report of the International Commission on Radiological Units and Measurements (ICRU)"

63 "Protection Against Neutron Radiation up to 30 Million Electron Volts"

64 "Design of Free-Air Ionization Chambers"
65 "Safe Handling of Bodies Containing Radioactive Isotopes"
66 "Safe Design and Use of Industrial Beta-Ray Sources"
69 "Maximum Permissible Body Burdens and Maximum Permissible Concentrations of Radionuclides in Air and in Water for Occupational Exposure"
72 "Measurement of Neutron Flux and Spectra for Physical and Biological Applications"
73 "Protection against Radiations from Sealed Gamma Sources"
76 "Medical X-Ray Protection up to Three Million Volts"
78 "Report of International Commission on Radiological Units and Measurements (ICRU)"
80 "A Manual of Radioactivity Procedures"
84 "Radiation Quantities and Units, International Commission on Radiological Units and Measurements (ICRU) Report 10a, 1962"
84 "Errata to Accompany Radiation Quantities and Units, International Commission on Radiological Units and Measurements (ICRU)"
85 "Physical Aspects of Irradiation, Recommendations of International Commission on Radiological Units and Measurements (ICRU)"
86 "Radioactivity Recommendations of International Commission on Radiological Units and Measurements (ICRU)"
87 "Clinical Dosimetry, Recommendations of International Commission on Radiological Units and Measurements (ICRU) Report 10d"
87 "Insert to Accompany Clinical Dosimetry"
88 "Radiobiological Dosimetry, Recommendations of International Commission on Radiological Units and Measurements (ICRU) Report 10e 1962"
89 "Methods of Evaluating Radiological Equipment and Materials, Recommendation of International Commission on Radiological Units and Measurements (ICRU)"
92 "Safe Handling of Radioactive Materials, Recommendation of National Committee on Radiation Protection (NCRP Report 30)"

93 "Safety Standards for Non-Medical X-Ray and Sealed Gamma-Ray Sources"
97 "Shielding for High-Energy Electron Accelerator Installations, Recommendations of National Committee on Radiation Protection and Measurements (NCRP Report 31)"

Available from Superintendent of Documents, Government Printing Office, Washington, D. C. 20402.

INTERNATIONAL ATOMIC ENERGY AGENCY SAFETY SERIES

1 "Safe Handling of Radioisotopes"
2 "Safe Handling of Radioisotopes—Health Physics Addendum"
3 "Safe Handling of Radioisotopes—Medical Addendum"
4 "Safe Operation of Critical Assemblies and Research Reactors"
5 "Radioactive Waste Disposal into the Sea"
6 "Regulations for the Safe Transport of Radioactive Materials (1964 Revised Edition)"
7 "Regulations for the Safe Transport of Radioactive Materials: Notes on Certain Aspects of the Regulations"
8 "The Use of Film Badges for Personnel Monitoring"
9 "Basic Safety Standards for Radiation Protection"
10 "Disposal of Radioactive Wastes into Fresh Water"
11 "Methods of Surveying and Monitoring Marine Radioactivity"
12 "The Management of Radioactive Waste Produced by Radioisotope Users"
13 "The Provision of Radiological Protection Services"
14 "The Basic Requirements for Personnel Monitoring"
15 "Radioactive Waste Disposal into the Ground"
16 "Manual on Environmental Monitoring in Normal Operation"
17 "Techniques for Controlling Air Pollution from the Operation of Nuclear Facilities"
18 "Environmental Monitoring in Emergency Situations"

These are for sale by the National Agency for International Publications, Inc., 317 East 34th Street, New York, New York 10016.

SECTION 6

INDUSTRIAL FIRE PROTECTION

N. Irving Sax
Director, Radiological Sciences Laboratory
New York State Health Department
Albany, New York
Adjunct Professor, Division of Bioenvironmental Engineering
School of Engineering, Rensselaer Polytechnic Institute
Troy, New York
Consultant on Pollution Control

WHAT IS FIRE?

Fire, combustion, or burning requires three things: (1) a *fuel* (any oxidizable material), (2) *oxygen* (usually air), and (3) a certain *temperature* (heat). Fire is the chemical union of oxygen with fuel, accompanied by evolution of thermal energy, indicated by incandescence or flame. If any one of these three constituents is not present in the proper proportions or degree, no fire will occur. If a fire exists and even one of them is sufficiently altered, the fire will go out. Therefore, in its simplest form, all fire control or extinguishment reduces to a manipulation of these three essential constituents.

Classification of Fires

Class A Fires. These are fires in *ordinary combustible materials* where the quenching and cooling effects of quantities of water or solutions containing large percentages of water are of first importance. Ordinary combustible materials tend to produce glowing embers after burning, and these must be quenched to prevent rekindling.

Class B Fires. These are fires in *flammable liquids* (oils, gasoline, solvents, etc.), where a blanketing or smothering effect is essential to put the fire out. This effect keeps oxygen away from the fuel, and can be obtained with carbon dioxide, dry chemical (essentially sodium bicarbonate), foam, or a vaporizing-liquid type of extinguishing agent. Water is most effective when used as a fine spray or mist.

Class C Fires. These are fires in *electrical equipment*, where the use of a nonconducting extinguishing agent is essential. Water spray, carbon dioxide, dry chemical, or vaporizing liquid is satisfactory.

Metal and Gas Fires. Such fires have not yet been classified. These require special agents and techniques. For a discussion of extinguishing agents and techniques for these classes of fire See p. 206.

Definition of Terms

The meanings of fire control terms as used in this book are as follows:

Flash Point (flash p). This is the lowest temperature at which a liquid will give off enough flammable vapor at or near its surface such that in intimate mixture with air and a spark or flame it ignites. The flash point of liquids is usually determined by the Standard Method of Test for Flash Point with the Tag Closed Cup Tester (ASTM D56-52, available from the American Society for Testing Materials, 1916 Race St., Philadelphia, Pa.). This method is also the standard of the American Standards Association (ASA Z11.24-1952, available from the American Standards Association, 70 East 45th St., New York, N.Y.). The Interstate Commerce Commission uses the Tag Open Cup (TOC) Tester giving results 5–10°F higher (less flammable). Other methods frequently used are Cleveland Open Cup (COC) and Pensky-Martens (PM). The closed cup flash point value is usually several degrees lower (more flammable) than the open cup, as the test in the former case is made on a saturated vapor-air mixture, whereas in the latter case the vapor has free access to air and thus is slightly less concentrated. For this reason, open cup values more nearly simulate actual conditions (see pp. 196–198).

Fire Point (fire pt.). This is the lowest temperature at which a mixture of air and vapor continue to burn in an open container when ignited. It is usually above the flash point. Where the flash point is available, only it is given; if it is not, the fire point may be given. It is at least as significant as the flash point as an indication of the fire hazard of a material.

Autoignition Temperature (autoign. temp.). This is the temperature at which a material (solid, liquid, or gas) will self-ignite and sustain combustion in the absence of a spark or flame (ASTM Designation D286-36). This value is influenced by the size, shape and material of the heated surface, the rate of heating (in the case of a solid), and other factors.

Vapor Density (vap. d.). This value expresses the ratio of the density of a vapor to the density of air. The vapors of most flammable liquids are heavier than air, thus they can readily flow into low areas, excavations and similar localities. Hence, ventilating outlets in a plant should be located near ground level. For combustible gases and vapors which are lighter than air, ventilating outlets should be near the ceiling.

Melting Point (mp). This is the temperature at which the solid and liquid forms of a substance exist in equilibrium. This value indicates at what temperature flammable materials that are solid at room temperature may become flammable liquids.

Boiling Point (bp). This is the temperature at which a continuous flow of vapor bubbles occurs in a liquid being heated in an open container. The boiling point may be taken as an indication of the volatility of a material. Thus, in the case of a flammable liquid, boiling point can be a direct measure of the hazard involved in its use.

Formula. In the event of a lack of information regarding a material, its formula can give a clue to its fire hazard. For instance, all materials composed solely of carbon and hydrogen are combustible and in some degree flammable. If they are liquids with a low boiling point they can be assumed to be fire hazards.

Underwriter's Laboratories Classification (ulc). This is a standard classification for grading the relative fire hazard of flammable liquids against the following standards:

Ether class	100
Gasoline class	90–100
Ethyl alcohol class	60–70
Kerosene class	30–40
Paraffin oil class	10–20

Where this value is known it is an excellent measure of the relative hazard of a flammable liquid. Unfortunately, it is available in only a few instances.

Susceptibility to Spontaneous Heating. Many materials combine with atmospheric oxygen at ordinary temperatures and liberate heat. If the heat is evolved faster than it is dissipated due to poor housekeeping, a fire can start, particularly in the presence of easily ignited waste, etc. ["Factory Mutual Modified Mackey Method," Industrial and Engineering Chemistry (March 1927)].

Explosive Range or Flammability Limits. These values expressed in percent by volume of fuel vapor in air are the ranges of concentration over which a particular vapor or gas mixture with air will burn when ignited. If a mixture within its explosive range of concentrations is ignited, flame propagation will occur. This range will be indicated by lel for lower explosive limit or uel for upper explosive limit. The values given, unless otherwise indicated, are for normal conditions of temperature and pressure.

FIRE PROTECTION

The two main aspects of fire protection are *prevention* and *loss limitation*.

Prevention

Fire prevention is an inseparable requirement of fire safety. Since, in order for a fire to start, all three necessary constituents—fuel, oxygen, heat—must be represented, effective fire prevention simply boils down to manipulation of these constituents to the extent that a fire cannot start. For instance, where a flammable material such as acetone is used out in an open work shop, two of the three needed constituents are immediately present, i.e., the fuel and a supply of oxygen. Now the only thing lacking to start a fire is heat. Thus by referring to acetone in Section 12 it is found that the flash point is 14°F, which means that at any temperature above 14°F, acetone can evolve enough vapor to form a flammable mixture with air which will catch fire if exposed to a spark, flame or other source of ignition. Thus, strictly from the standpoint of fire prevention in an installation using acetone, the following avenues are open:

(1) The working (ambient) temperature must be kept below 14°F, or

(2) The supply of atmospheric oxygen must be cut off, or

(3) Sources of ignition, such as flames, glowing cigars and cigarettes, sparks, etc., must be eliminated from the area, or

(4) The area must be ventilated so that even though the acetone gives off enough vapor to form a flammable mixture with air, the vapor will be drawn out of the area by means of fume exhaust equipment as rapidly as it is evolved, thus preventing the build-up of dangerous concentrations of vapors.

Naturally, since conditions (1) and (2) above are relatively difficult to attain on an industrial scale, conditions (3) and (4) are the ones most likely to be used.

Furthermore, although total removal of any one of the necessary conditions for a fire will absolutely prevent its occurrence, such stringent restrictions on industrial operations are seldom economically feasible. Industrial materials are, however, studied with a view to ascertaining just how much leeway there is, so that a compromise between absolute fire prevention and economy of operation may be reached. It is for this reason that, while we know how to prevent fires, they still do start, and why *loss limitation* is such an important part of industrial fire protection.

Below is some discussion of the three essentials of fire.

Oxygen. Although under certain unusual circumstances it is possible to produce combustion-like chemical reactions with materials such as chlorine or sulfur, it is safe to say that nearly all combustion requires the presence of oxygen. Also the higher the concentration of oxygen in an atmosphere, the more rapidly will burning proceed. Industrially it is difficult to manipulate the oxygen concentration in a working area, particularly since a concentration of oxygen far enough below normal to keep fires from starting would also be too low to support human life.

When industry has found it necessary to work with materials so sensitive to oxygen that they would catch fire at ordinary temperatures merely upon being exposed to air, it has found it possible to isolate such materials from air, either in a vacuum chamber or in a chamber filled with an inert atmosphere, such as argon, helium, or nitrogen. In Section 12 the materials which require such isolation are so noted.

Heat. As a necessary component of fire, this is often manipulated to render an industrial set-up safe from fire. The most difficult aspect of controlling the heat component of a fire is the easily overlooked fact that to *start* a fire it is often necessary to heat to a sufficient degree only a *very small quantity of fuel and oxygen mixture.* Then, since fires are by definition exothermic, the very small fire started by a tiny heat source supplies to its surroundings more heat than it absorbs, thus enabling it to ignite more fuel and oxygen mixture, and so on, until very quickly there is more heat available than is needed to propagate a large fire. The heat may be provided by various sources of ignition, such as high environmental (ambient) temperatures, hot surfaces, mechanical friction, sparks, or open flame.

Fuel. The third aspect of fire prevention will be discussed in detail in the following section.

Sources of Ignition

The following are the chief sources of ignition and suggestions for reducing the hazard due to them:

(1) Open Flames. At or near a flammable-liquid installation it is necessary to check for such sources as burners, matches, lamps, welding torches, lighting torches, lanterns, small furnaces, and the possibility of broken gas or oil lines becoming flaming torches. Ample isolation may often be obtained by means of partitions. In this respect the partition should be substantial enough to contain the fire while the sprinklers or other fire-fighting apparatus put it out. Fire-resistant construction (brick or concrete walls) is generally recommended.

It is important to confine the flammable liquid while it is in use. Safety cans should be used for transporting small quantities of flammable liquids about a working area, as well as for storage at a bench. Wherever possible, closed systems should be used to prevent the spread of fumes, etc. In the event of a fire, it is imperative to prevent spreading of the fire. Hence all tanks should have trapped overflow drains leading to a safe place. Dikes must be used to contain the overflow of burning liquid; otherwise fires could easily spread over large areas, trapping personnel and causing great damage. The principle behind this form of protection is to contain the fire at all costs. Installations of flammable liquids in upper stories should be made only in such a fashion that burning liquid will be prevented from flowing down stairwells, pipe openings, cracks in walls, etc., by means of waterproof floors, dikes, overflow pipes, etc.

(2). Electrical Sources (electric power supply and generating equipment, heating equipment, and lighting equipment). The following precautions for good maintenance are suggested. A complete listing cannot be given here. The provisions of the National Electrical Code are the recognized standard and these should be carefully observed in installing electrical equipment in hazardous locations.

(a) Use special wiring and conduit.

(b) Use explosion-proof motors, particularly if located at ground level or in pits or low places.

(c) Use only specially engineered heating units, keeping in mind the autoignition temperature of the material in use (hot water or steam heating units are much to be preferred).

(d) Controls for motors, thermal cut-outs, switches, relays, transformer contactors, etc., which are liable to spark or heat up should not be installed in flammable liquid storage areas. Use only explosion-proof, push-button control switches within such an area.

(e) In dangerous atmospheres and for storage, only vapor-tight globes with electric lamps may be used. In well-ventilated areas, ordinary lamps will do. Fixed lamp installations are to be preferred to extension cords. Also, approved safety flashlights are preferred to portable lamps.

(f) Do not install fuses or circuit breakers in hazardous locations except in explosion-proof cases.

(g) Motor frames, control boxes, conduits, etc., should all be grounded in accordance with the general requirements for installation of electric power as outlined in the National Electrical Code.

(3) **Overheating** (excessive temperatures at points requiring heat). Such processes should be kept out of combustible buildings and closely supervised. The use of automatic temperature controls and high temperature limit switches is recommended, although supervision is still important.

(4) **Hot Surfaces.** The incomplete immersion of hot metal in quenching baths, the contact of flammable vapors and hot combustion chambers, hot dryers, ovens, boilers, ducts and steam lines all are frequent causes of flammable vapor fires. Care should be taken that material whose auto-ignition point is lower than the temperature sometimes reached by operating equipment be kept at a safe distance from such equipment. This equipment should be carefully supervised and maintained to prevent accidental overheating, etc.

(5) **Spontaneous Ignition.** Many fires are caused by spontaneous heating of materials, accelerated by external heat from processes such as dryers, ovens, ducts, impregnating or steam lines adjacent to piles of waste materials. Sometimes the accumulated heat in a closed, unventilated warehouse will be sufficient to accelerate oxidation to the point of an actual fire. Wherever flammable liquids are handled, particularly those which are known to be liable to spon-

taneous heating, it is important to pay particular attention to housekeeping and ventilation. Fires are almost sure to follow neglect of these matters. All equipment and buildings should be kept free of deposits and accumulations of wiping rags, waste materials, oil mops, etc.

(6) **Sparks, etc.** Sparks from mechanical tools and equipment, hot ashes from smoking, unprotected extension lights, boilers and furnaces, backfire from gasoline engines, are all potential causes of fire. Smoking should be prohibited in areas where flammable liquids are stored or are used in the open. All equipment in such areas should be maintained in first class condition. Wherever possible, spark-proof or non-sparking tools and materials should be used.

(7) **Static Electricity.** This is due to electrical impulses generated on the surface of a material by friction, such as calendering, printing, and the like. Many fires are caused in the rubber and paper industries by this means. Most of these occur during the months when humidity is relatively low, and artificial heat is used. Maintaining a relative humidity of from 40–50 percent in rooms where flammable liquids are used will greatly reduce the chance of static sparks. Electrical grounding, static discharge devices, etc., should be mandatory, and all flammable liquid tanks, piping and equipment should be so interconnected and grounded that the chances for static sparks are minimized. In all this type of equipment, belts should be eliminated and direct or chain drives used wherever possible. If belt drives must be used, the belt speed should be kept below 150 ft/min, or a special belt dressing should be used which will reduce the possibility of the formation of a static spark.

(8) **Friction.** Many fires are caused by mechanical friction, i.e., from fan impellers rubbing on casings, poorly lubricated fan bearings, grinding processes and machining, etc. Fans and other equipment should be frequently inspected and maintained in the best possible condition. Other processes known to generate a good deal of heat due to friction should be well separated from locations where flammable liquids are stored or used.

It is vitally necessary that a complete program for the handling, transfer and use of flammable liquids be set up and maintained. This program should start when the process is initially under construction. Where flammable liquids are called for in the original write-up of a process, the first question is to determine whether the flammable material can be replaced by a non-flammable one. If the question of cost arises it should be remembered that to the possibly low cost of the flammable liquid should be added the cost of special protection needed to use it safely as well as its effect upon the insurance rate. It may well be that in the final analysis, the cost of a flammable material is not as favorable as it seemed at first. However, there are many flammable materials in constant use for which there are no substitutes, but even they can be safely handled if proper precautions are taken.

(9) **Fuel.** Combustion takes place most readily between oxygen and a fuel in its vapor or other finely divided state. Solids are most easily ignited when reduced to powders or vaporized by the application of heat, but except in a few cases the temperatures required for the vaporization of solids are well above normal ambient temperatures.

Liquids present a different case. Some liquids will give off dangerous quantities of flammable vapors well below normal room temperature (the vapor pressure of a liquid is a measure of this effect); others do so at points only slightly above room temperature, and still others at much higher temperatures. It is apparent that the temperature at which a liquid evolves vapors which can form flammable mixtures with air is a measure of its hazard potential. This is indicated by the flash point.

So indicative of fire hazard is the flash point of a liquid that the Interstate Commerce Commission rates any liquid whose flash point is 80°F or below as a *high* fire hazard, the theory being that 80°F represents the upper limit of normal or "room" temperatures; any liquid which will flash at or below this point is dangerous. A flash point of from 80–350°F indicates a *moderate* fire hazard; above 350°F the fire hazard is considered slight. The national Fire Protection Association rates as a *high* hazard a liquid whose flash point is less than 20°F; *moderate* from 20–70°F, and *slight* from 70–200°F. Only liquids having a flash point less than 200°F are generally called flammable by the National Fire Protection Association.

In this book we use the ICC classification for liquids of flash point at 80°F or less as being *dangerous* fire hazards; we use the range from about 80–225°F to indicate *moderate* fire hazard; flash points in excess of about 225°F are rated as *slightly* hazardous. It should be understood, however, that practically all organic materials will burn if exposed to sufficiently high temperatures. The ratings given above are merely an indication of the risk involved in handling or storing them.

It is important to isolate a potential fire hazard. Thus it is necessary to use closed and vented tanks to hold flammable liquids. In this way the possibilities of igniting the tank of liquid are greatly reduced, as is the chance of materials at a distance from such a tank becoming involved in fires. It is also important that flammable materials be housed in fire-resistant structures because burning liquids can generate great heat and often set fire to the buildings in which they are burning.

A vital point in flammable liquids safety is the prevention of the accumulation of explosive concentrations of vapors in closed off areas (see Section 2 for a full discussion of ventilation control for toxic vapors). Wherever either moderately or highly flammable liquids are used or stored, ventilation is a very important consideration.

The amount needed, whether natural or mechanical (fans and blowers) depends upon the materials and the conditions involved. No dependence should be placed upon the odor of the material as a warning, because some flammable vapors are heavy and tend to settle and because smell is deceptive. The safe procedure is continual testing with an explosion or flammable vapor indicator.

Besides flammable liquid fires, the results of which can be *somewhat* mitigated by effective loss limitation techniques, there are two more types of disaster, protection from which is nearly entirely dependent upon prevention.

(10) **Dust Explosions.** Practically any combustible, when in the form of dust and mixed with air in the proper proportion, will burn so rapidly as to cause a severe explosion if ignited by heat, a spark or flame. Ignorance of this fact has led to many serious disasters. Grain, flour, coal dust, and metal powders all constitute hazards in this regard. Explosions have been known to occur in plants handling fertilizers, wood dust, powdered milk, soap powder, paper dust, cocoa, spices, cork, sulfur, heard rubber dust, leather dust, and many other products. For the prevention of dust explosions good housekeeping is of the utmost importance. All equipment must be dust-tight and kept so. Explosion vents should lead outdoors to a safe location, and the vent ducts themselves should be strong enough to withstand the force of the explosion. Vacuum cleaning is superior to sweeping. The use of compressed air to blow dust off equipment and thus create dust clouds should be *forbidden*!

Ledges, exposed piping, beams, etc., in the ceiling should be kept free from accumulation of dust. Where a dusty operation is to be installed in a location where there is piping and projections overhead, it is often erroneously considered satisfactory to install a smooth ceiling below the piping and other projections. This does not eliminate the hazard and may intensify it; unless the ceiling is extremely well designed and installed, dust will penetrate it and settle not only on the piping but on the upper side of the ceiling itself. Then a shock may be sufficient to fill the entire false space with a combustible dust cloud which a spark may set off. If piping cannot be relocated or eliminated, it would be better to leave it exposed and provide for a regular cleaning program.

It was for reasons such as the above that a starch dust plant recently constructed in the southwest, where weather conditions are moderate the year round, has been built entirely of open construction so that there is no confinement of the force of any explosion and the constant flow of air through the plant provides little opportunity for dust layers to build up.

As in the case of flammable liquid fires and explosions, the control of dust explosions is based upon prevention of ignition and secondarily limitation of damage in the event ignition does occur.

To prevent ignition, open flames, smoking, and cutting or welding are prohibited until the area is made dust free. Electrical wiring should be of the type suitable for a dusty atmosphere and static electricity, too, must be eliminated. Highly dangerous materials of this sort are handled most satisfactorily in enclosed systems in which suitable inert gases are introduced into the system to replace the air normally present. This precaution is particularly applicable to the field of powder metallurgy. The kind of inert gas used must be chosen on the basis of its suitability for the operation in question.

(11) Salt-bath Explosions. The third type of disaster in which after-the-fact protection is much less important than prevention is the molten salt-bath explosion. There have been serious disasters involving such baths, because personnel involved on both the management and the operating level failed to appreciate the potential hazards of the situation. Due to mechanical failure or human failure, or a combination of both, molten salt baths have been allowed to explode. The hazards of molten salt baths may be summarized as follows:

(1) Violent generation of steam due to water introduced as "carry-over" on a piece of work from a preliminary cleansing or quenching bath, condensation on overhead service piping, leaky roofs and operation of automatic sprinklers, also contact with liquid foods placed on ledges near the baths for "warming-up" by workmen.

(2) Sudden and explosive expansion of air occluded in blow-holes of castings and that trapped in tubes, closed piping, or hollow metal

work when immersed in molten baths without pre-warming.

(3) Violent and uncontrollable chemical reactions between nitrate baths and carbonaceous materials such as oils, soot, graphite, and cyanide carry-over from adjacent carburizing baths.

(4) Vigorous and explosive reaction between overheated nitrate baths and aluminum alloys.

(5) Explosive reaction between normally heated nitrate baths and carelessly introduced magnesium alloys.

(6) Thermit-like reaction between aluminum alloy articles lost in bath and the iron oxide sludge blanketing and insulating the bottom of bath container.

(7) Structural failure of bath container while in operation under conditions tending to lower the normal durability; reaction between metal of bath container and nitrate due to localized overheating.

(8) Failure of temperature controls, with consequent overheating of nitrate bath.

(9) Storage and handling of bulk supply of sodium nitrate, and careless disposal and storage of waste nitrate without regard to active properties of the salt.

(10) Accidental or uninstructed setting of temperature control above safe operating limits.

The precautions for safe operation of molten salt baths are summarized as follows:

(a) Guard against the introduction of any extraneous matter.

(b) Protect completely from overheating by automatic control and temperature readings taken at regular intervals.

(c) Isolate the operation as far as practicable.

(d) Instruct all personnel thoroughly and completely in regular and emergency procedures.

From the foregoing it can be clearly seen that the handling of flammable liquids, flammable dust or molten salt baths are the three chief operations in which prevention is the most important phase of fire protection. In each of these cases whatever action is taken after the act and whatever physical protection is provided can only furnish some degree of mitigation of loss and it is often insufficient to prevent a large scale disaster.

Loss Limitation

The other aspect of a realistic fire protection program is limitation of loss due to a fire which includes a provision for the *prompt discovery* and equally *prompt extinguishment* of the fire. It is certain that everyone has at one time or another wondered why a particularly destructive fire was allowed to happen, when supposedly a great deal of effort is constantly being devoted to the prevention of such fires.

It may be that too often the prevention aspect of fire protection has been the total or nearly total effort at protection, with the result that when even a small fire starts it has a good chance to become a calamity. However, even the utmost vigilance would have been of no avail in guarding against some of the fires which are on record, and once the fire has started its cause is immaterial. The cause of the *loss* is much more important, and the facts which determine *that* are the physical conditions and those measures which have or have not been taken to limit the extension of the fire.

One of the means of preventing the extension of fire is to segregate hazardous processes and storage into separate buildings. But even where hazardous processes are not involved, the concentration of too much value in one fire area must be guarded against. This is best accomplished by the erection of separate buildings, adequately spaced, which in turn presents problems of maintenance and operation. Suppose a major plant has an operation involving flammable liquids. Such processes as spray painting or dipping use tremendous quantities of flammable solvent; extraction processes and the manufacture of products of which flammable solvents are major constituents are typical examples of processes which require subdivision. The flammable liquid operations and all its appurtenances must be physically separated from the rest of the plant for maximum safety. Care should be taken that the separate buildings are no larger than production efficiency demands; in other words, only as much should be put under one roof as is necessary to be in one building. Where space or production requirements preclude separate buildings, the area in which flammable liquids are handled must be physically separated from the rest of the plant by approved fire walls or, where these are not practicable, by water-curtain type sprinklers.

Subdivision of one large risk into smaller fire areas may also be accomplished by means of fire walls which stop the spread of fire from one area to another. To accomplish this, the wall must be carried through the roof and either go through the side walls for a distance of at least 36 inches or turn back on both ends for a distance of several feet to provide a barrier around which the fire cannot travel.

Many otherwise sound fire walls have failed because holes were made in the walls to permit the passage of pipes, conduit, etc., and then never properly closed. Every hole in a fire wall must be sealed at the time such work is being done. The weakest point in a fire wall is the fire door provided to permit access from one section to another. At any given time a high percentage of such doors are found upon inspection to be useless as fire barriers. They must be test operated regularly. The chief deficiencies are missing fusible links, damage to the doors by materials handling equipment, which damage would prevent their operation, or blocking by material left in the doorway so that the door cannot close.

Proper maintenance includes regular inspection, physical guarding to prevent damage, the painting of "keep clear" lines on the floor, and a constant program of education. Wherever practicable, such doors should be closed at night to insure that they will be closed in the event of fire. Where the use or occupancy of the building has changed and fire doors are not needed any longer, the openings should be bricked up to the same thickness as the original walls. Even a pair of fire doors, one on each side of the opening, is less resistant to the passage of fire than the fire wall in which they are installed. In normal operation, because of the aisle space leading to the opening, the heat on the door should be less than elsewhere in the building. If the doors are closed and contents are piled against the doors, fire may be transmitted from one side of the wall to the other. In all cases, the doors provided should be of the type approved for the opening in the wall. In many cases, doors approved for use only on vertical enclosures such as stairways are installed in fire walls and will not serve their intended purpose in the event of fire. If openings are necessary in fire walls for the passage of conveyors and no type of door installation is practical, then the openings should be specially protected by hooded automatic sprinkler heads directly over the opening on each side of the wall. A fire wall should be thought of as a dam which any small leak can cause to fail.

The spread of fire from floor to floor in a building is prevented by the proper enclosure of vertical openings such as stairways, elevators, shafts, and process openings through the floor. The question of stairways deserves particular comment because in so many cases, self-closing stairway doors are found to be wedged open to permit easy passage from floor to floor. Such examples of poor management entirely negate the cost of closing the stairway off by providing the doors in the first place.

One of the primary reasons for enclosing stairways is to permit the passage without injury of personnel from upper floors to the street level past the floor which is on fire. If the stairway doors are wedged open this may be impossible. The stairways can immediately become choked with hot air, smoke and gases from the stairwell.

Provision of fusible link arrangements to close such doors is not very satisfactory because fumes and smoke will pass through without operating the link. In at least one laboratory the problem of stairway doors being wedged open versus the desirability of having them closed in a hurry has been solved by providing for electric latches. These latches hold all the doors open, but connected to the fire alarm system is a relay which causes all the electric locks to release when a fire alarm is sounded, thus closing the doors.

A further loss-limiting device which is useful where flammable dusts and vapors are used is the explosion vent. It is important to install explosion vents in areas where flammable liquids or dusts are used because of the possibility of great damage due to explosive ignition of such mixtures and air. Therefore, on a practical basis, properly designed explosion vents are a suitable safeguard, as they reduce the chances of destruction indoors by allowing the force of the explosion to be transmitted outdoors.

In order to fully relieve the pressures produced by explosions in vapor and air mixtures, a vent area as large as 1 square foot for every 10 cubic feet of room volume would be necessary. However, it is unlikely that more than a fraction of the total volume of a room will at any time be within the explosive range. Therefore for a small room with a floor area of about 200 square feet, the venting area should be at least 1 square foot for each 30 cubic feet of room volume. For larger areas, this proportion may not be obtainable, but in no case should the vent area be less than 1 square foot for each 50 cubic feet of volume.

Approved explosion venting windows are available. Also, sky-lights, roof hatches, or light windows hinged at the top and carefully installed to swing outward under even slight pressure can be useful. Under some conditions, doors equipped with releasing latches may be utilized as vents.

Furthermore, where the conservation of heat in a plant is important, and the walls of the building are otherwise of strong construction, a section of exterior wall may be built of light wood, or hollow tile or some other material which is relatively weak compared to the rest of the building so that in case of an explosion, these sections will give first.

It is important that snow or ice be kept from collecting on explosion vents so that they can operate freely in case of an explosion.

FIRE DISCOVERY AND TURNING IN THE ALARM

With the exception of such occurrences as flammable liquid explosions, dust explosions and molten salt bath explosions, most fires are quite small at first and can be readily extinguished if discovered early. Although no one would dispute the truth of this statement, it is amazing how many organizations completely ignore the fact and make no provision for the prompt discovery and sounding of an alarm, in the event of a fire.

The greatest single cause of large fires is delayed alarms. This does not necessarily mean delayed discovery; the files are replete with examples of people who discover fires and take entirely the wrong action because they had not been properly trained.

The detection of fire is too often left to chance. If it is left to the chance passerby it can be assumed that no alarm will be turned in until flames are actually leaping from the front of the building, at which point a heavy fire loss has already taken place. The best solution to this problem is an automatic fire alarm system connected to an office manned by responsible personnel on an around the clock basis. In larger installations this may be the plant's own security office; in smaller installations, the office of a private company specializing in this service. In small cities or rural locations this service can often terminate at the police station or fire house. Such an alarm system, *if properly installed and maintained*, assures that fires will be discovered and reported promptly in their incipient stages. Automatic sprinkler systems with provision for an alarm can serve the same purpose.

Most automatic fire alarm systems are one of two general types; one measures the rate at which the temperature of the air rises; the other sounds an alarm when the temperature reaches a certain predetermined point. Each has its merits for a particular installation, and some types of systems combine both features. Any such system should bear the approval of Factory Mutual or Underwriters Laboratories.

Under some circumstances, such as the storage of furs or documents, or as in electrical equipment rooms, a slow-burning, smoldering type of fire may be expected. In such a case, neither the fixed temperature nor the temperature-rate-of-rise device may function rapidly enough to transmit an alarm before serious damage has been done. A photoelectric cell device might be used here to trigger the alarm when traces of smoke pass through the light beam. Such a system also starts all necessary action, such as shutting down fans, operating dampers and transmitting the alarm. This type of protection is a "must" in the modern windowless factory which depends for its ventilation entirely upon a mechanical air-conditioning system into which the introduction of smoke at any point can cause it to be transmitted throughout the area, with consequent panic and possible loss of life.

The most ancient and still most common form of fire protection is the watchman. He is an extremely important employee. In many cases the entire physical property of an organization together with its business future is entrusted to his care over half the time. However, he is rarely chosen with these responsibilities in mind. Just as rarely is an effort made to see to it that the watchman is schooled to react automatically to take proper action if the emergency against which he was employed does occur. Too often the position is used for the semi-retirement of the superannuated employee.

Even when the watchman is hired outright, the pay scale is often so low as to attract only the completely unfit. There are scores of instances of watchmen of valuable properties who cannot write or read English and in truth can hardly speak it, who do not know how to use the dial telephone, who do not know how to turn in a fire alarm—in short, who do more to present a hazard to a property than to protect it.

There are three considerations in watchman protection which cannot be overlooked: (a) selection of personnel, (b) training, and (c) supervision. The watchman should be neither too young nor too old. If too young he may lack judgment and in addition is very likely to be bored with the job. If too old he may lack the physical stamina which the job requires. He must be thoroughly and carefully trained in the proper action to be taken in the event of any emergency in the plant.

Particular stress should be laid upon obtaining assistance for the watchman before attempt-to fight any fire or deal with any other dangerous emergency. Many concerns use a "watchman daily report form," which also contains an illustrated paragraph of instruction on some phase of the watchman's duties. If the watchman's service is to be performed effectively, regular routes must be laid out and means provided to determine that the watchman is or has been in a required location at the required time. The watchman's clock is the most familiar example

of such a device, but it has a serious limitation —no check can be made until the next day.

There is no instantaneous assurance that the rounds are being made in accordance with instructions, or in the case of a single watchman in the plant, there is no provision for investigation should the watchman be injured, become ill, or leave his post for any other reason.

The best type of protection is that provided by electric watchman-reporting-stations; most often they are incorporated in fire alarm boxes whereby the watchman sends a signal to the supervisory office and indicates his location at the same time. If the watchman does not report at each location within the time specified, an immediate investigation can be made. In a large plant, this service would probably terminate at the plant guard office; in the smaller plant it would be provided by a company specializing in fire protection.

Even when the plant is in operation, the question of sounding the alarm must not be left to chance. The fire protection organization composed of plant personnel should provide for a specific person (with a number of alternates) to turn in the alarm, and emphasis should be laid at all times on the matter of doing this promptly.

No fire department objects to being called to trivial fires. Over 95 percent of the responses of all fire departments, large and small, are to fires which could be classed as trivial. A high proportion of the other 5 percent could have been kept in the trivial class had the fire department been called immediately when the fire was discovered. *Delayed alarms cause fire disasters.*

EXTINGUISHMENT OF FIRE

To extinguish a fire it is necessary to remove or sufficiently lower the concentration of only one of the three requirements mentioned previously.

Normally, the removal of fuel is not considered as a practical fire-extinguishing method; however, in such cases as multiple flammable liquid tanks which are interconnected, it is often entirely practicable to pump the flammable contents of a tank which is burning at its surface to another tank, and thus extinguish the fire by leaving the burning tank nearly empty. In the main, however, fire extinguishment operates by either cooling the burning material below its ignition temperature or cutting off its supply of oxygen. Water, foam, carbon dioxide gas, bicarbonate of soda (dry chemical), and halogenated hydrocarbons comprise the most widely used fire-extinguishing media. Burning metals require special extinguishing media and techniques.

The chief problem is to determine which medium to use and in what manner. For all the media there are one or more methods of application varying both in scale and in type of operation, i.e., manual or automatic, for single spot delivery or general coverage.

Water

The oldest, cheapest, and most commonly used fire-extinguishing agent is *water*. As an example of the use of bulk or "coarse" water, an extremely effective fire extinguisher for small blazes is a pail of water. However, it is inadvisable to use bulk water on a fire involving oils, gasoline, paints or solvents; since they may be lighter than water, they will tend to be physically dispersed by the impact of a stream. As a result, the burning material will be scattered about the area and spread the fire. Moreover, no blanketing action is possible, as the oil or solvent may float on the water. Water should never be used to extinguish fires of metals which may react violently with it to cause an explosion.

Special fire pails are purchased for protective purposes; they have rounded bottoms which make them unsuitable for general service. The use of 55 gallon used oil drums full of water equipped with two or three fire pails is practical for yard storage where the material is not piled too high, as well as for the early stages of construction projects before permanent water supplies are installed. Calcium chloride can be added to the water to prevent freezing during cold weather.

Water is usually most effective when applied in the form of fine droplets or spray. This has a blanketing action, and avoids the difficulty of impact scattering of materials lighter than water. Proper fire extinguishers should always be at hand in areas where flammable solids or liquids are used or stored. Modern sprinkler systems are the most efficient means of supplying such a spray over a wide area.

Fire Extinguishers

A standard $2\frac{1}{2}$ gallon soda-acid fire extinguisher provides $2\frac{1}{2}$ gallons of water. When the extinguisher is inverted, a small bottle of sulfuric acid is overturned into the water in which bicarbonate of soda has been dissolved. The pressure of carbon dioxide gas so formed expels the water. Such an extinguisher requires annual recharge and must be protected from freezing. The soda-acid extinguisher is rapidly being supplanted by a type in which the carbon dioxide gas is provided by a small high pressure cylinder

which can be punctured when the extinguisher is put into operation. The water is then driven out by the gas. This type of extinguisher can be winterized by the addition of antifreeze chemicals if desired; in addition, some manufacturers provide an antifreeze additive which also increases the extinguishing effectiveness of the unit. The maintenance required is an annual check on the weight of the carbon dioxide gas cylinder.

Automatic Sprinkler Systems

A properly designed, installed, and maintained automatic sprinkler system is by far the most efficient fire protection device as yet produced. It is no exaggeration to state that American industry could not exist as it does today without automatic sprinklers. Statistics kept over a 50 year period by the National Fire Protection Association covering some 70,000 fires show that sprinklers extinguished or held in check 96 percent of the fires in sprinklered buildings. Of the failures, about 1 percent were due to water being shut off on the sprinkler system or the water supply being defective. Another 1 percent were due to the fact that the system could not get into operation fast enough because the fire started in an unprotected portion of the building and had gained tremendous headway by the time it entered the protected section. Obstructions to distribution of the water from the sprinklers which in effect rendered at least portions of the building unsprinklered also caused some failures.

Objection is sometimes made to installing automatic sprinklers for fear that premature operation would cause excessive water damage. The possibility of a sprinkler head misoperating from defects in manufacture is negligible—only one chance in 8,400,000 per year. Premature operation from other causes can be easily prevented: from overheating and corrosion by using special sprinkler heads; from mechanical damage by installing wire guards; from freezing by providing antifreeze solutions or by use of dry pipe systems.

A hose stream delivers the equivalent of a dozen sprinklers, whereas five heads or less extinguish 73 percent of all reported fires where sprinklers operate. Sprinkler piping is tested for a pressure of at least 200 lb/in.^{2} before the system is accepted. No such tests apply to ordinary water piping, and the possibility of damage from a burst water pipe is therefore much greater in the case of ordinary domestic service. In addition, no sprinkler system is complete without an alarm device which notifies responsible authorities when any water flows in the system—a feature entirely lacking in domestic and process piping.

An automatic sprinkler system is essentially an arrangement of pipes throughout the protected area provided at suitable intervals with sprinkler heads equipped with fusible (wood's metal) links to release the water when the temperature at the head reaches the desired point.

Wet type automatic sprinkler systems consist of such an array of piping together with an alarm check valve which prevents back flow from the system to the water supply and provides a means whereby a local central station alarm can be received in the event that there is a flow of water in the system due to a fire or accident.

In a low building where water pressure and volume are good, the water supply may come solely from the street main. In other cases, it may be necessary to provide an elevated supply of water so that the water will be automatically available under gravity pressure. A fire department connection is provided so that the fire department can pump into the system and increase the volume of water available in the event of a serious fire. The maintenance is simple but the utmost care must be taken to make sure that there is water in the system at all times.

The equipment is designed to facilitate inspection. The outside valve leading from the street main is almost invariably a post indicator valve, i.e., a valve brought up above the surface of the ground with an indicator as part of the valve, which shows whether the valve is open or shut. Interior valves are of the outside stem and yoke type (OS & Y); i.e., the valve stem protrudes through the valve yoke and is visible for its full length when the valve is open. If the stem cannot be seen the valve is closed.

Seals and padlock arrangements are often used to prevent unauthorized tampering with sprinkler valves. A tag system including notification of responsible authority should be used in connection with routine work to make sure that the valves are reopened when the work is completed. If extensive alterations are necessary, they should be carefully planned so that the maximum portion of the system can be plugged off and so be kept in operation until the alterations and additions are ready to be tied back into the main system. Every effort should be made to put the system back in service to at least give partial coverage during nights and over weekends during the period of repair. There are many instances of fires getting headway in plants which had an excellent sprinkler system which was out of service for repairs at the critical moment.

The ultimate in protection of sprinkler valves

against tampering whether accidental or malicious is an electrical alarm device, attached to the valve, which will transmit a signal to a central headquarters when the valve is closed.

When alteration or new construction is planned for a sprinklered building, a part of the alteration cost must be provision for the necessary changes, extensions or additions to the sprinkler system. An unsprinklered mezzanine constructed within the plant or an unsprinklered shed constructed outside can become so heavily involved in a fire that by the time the fire reaches the sprinklered area, the system is overpowered.

In areas subject to freezing, a *dry pipe* automatic sprinkler system may be installed. In such a system, air under pressure in the piping keeps the water from entering the unheated area through the dry pipe valve where a differential type of valve seat enables a relatively low air pressure to hold back a much higher pressure of water. Such a system requires a dependable supply of compressed air to make up any losses in air pressure due to air leakage in the system. If the air pressure fails, the system "goes wet," and an alarm will be received. In non-freezing weather the system may be allowed to remain wet for a short period of time before it is drained and the dry pipe system restored. A word of caution, however; once the system has "gone wet" no further alarm will be received from a flow of water due to the operation of a sprinkler head. During freezing weather, of course, immediate action must be taken to drain the system and restore the valve to prevent freezing. Large dry pipe systems are provided with exhaust valves to vent the air in the system quickly and thus speed up the water discharge.

When settling of a building occurs, the pitch of the piping must be carefully checked to see that all sections of the system will drain back to the valve and that no pockets are created where water may remain and cause the pipe to burst during freezing weather. A dry pipe system necessarily operates more slowly than a wet pipe system. In the event that the situation changes and a permanent heating system is installed in the building, the sprinkler system should be converted to a wet pipe system. Another practice that should be guarded against is the "pumping up" of the air pressure on the sprinkler system to an excessively high point to prevent the accidental tripping of the dry pipe valve because of leakage of air. By filling this system with many times the normal volume of air, a situation is created where in any emergency it may take several minutes for air to exhaust itself from the system and finally permit the flow of water from a sprinkler head. The proper way to correct air leakage is to inspect the piping, to locate and eliminate the air leaks.

Modern warehousing may present difficult new problems, such as the high piling of stock now permitted by means of mechanical handling equipment; the palletizing of material providing for horizontal and vertical spaces through the pile; and the storage of such materials as rubber tires.

Most sprinkler heads now in use project the water upward in a very fine spray; the spray patterns from the various sprinkler heads join just below the ceiling to provide a barrier against the transmission of heat to the building proper. It is highly effective as a heat absorber, which is after all the primary purpose for using water in fire extinguishment. These new heads are by no means a cure-all, particularly for an inherently unsatisfactory situation, but they will definitely provide a higher degree of protection than has heretofore been possible in the type of occupancies discussed above.

Fire Hoses

Fire hose lines fall into two classes: interior and exterior. Interior fire hoses are generally $1\frac{1}{2}$ inches inside diameter with a $\frac{1}{2}$ inch open nozzle. The water supply may come from a take-off of the sprinkler system, from domestic or process water pipes large enough to deliver a sufficient quantity of water at the required pressure or, in multi-story buildings, from a standpipe system.

In multi-story buildings the standpipe system is provided because it may be used by the fire department as well. No fire hose could stand the pressure necessary to overcome the friction and gravity loss involved in delivering a stream of water to a fire at the top of a skyscraper. In addition, the time consumed for such an operation would be intolerable. Such systems generally get their primary supply of water from roof tanks or, in very high buildings, from a series of tanks provided at various floor levels throughout the building. Outlets of $2\frac{1}{2}$ inches diameter are provided for fire department use. Where the head pressure from an elevated tank would cause excessive pressure at the valve, reducing valves are installed.

Easily broken sealing devices are provided so that the fire department with its trained personnel can quickly obtain the full flow from the standpipe if desired. Unlined linen hose, air pressure tested at the factory, is usually used. This hose should never be wet except during an actual fire, since without proper facilities for drying, it will invariably mildew and deteriorate. Another common cause of rot in such hose is

drip from valves which are not properly seated or are cracked open slightly because of tampering with the valve. Hose set in wall cabinets is much less likely to have this trouble than that exposed on open racks. When the linen hose is first used it leaks a little until its pores swell and it becomes reasonably watertight. For its purpose, however, it is superior to cotton rubber-lined hose because of the smaller space required for storage and the small amount of maintenance required as long as the hose is kept dry.

For outside use, $2\frac{1}{2}$ inch cotton rubber-lined hose is standard. Hose should be double jacketed to be able to stand maximum wear. Because of the minimum amount of attention usually accorded hose in the average industrial plant, mildew-proof hose should be purchased. Its higher initial cost will be more than offset by the longer maintenance-free service to be expected from it. Such hose is provided either mounted on wheeled carts which will carry about 500 feet together with nozzles, wrenches, etc., as a compact unit, or in wooden structures or hose houses built around the fire hydrant to provide protection for several hundred feet of the hose attached to the hydrant and about 300 more feet in reserve in the hose house. The hose house is also used for the storage of other emergency equipment. Building and equipping individual hose houses has become increasingly expensive.

When the type of hazard demands heavy-duty outside fire protection and the organization of a thoroughly trained brigade, serious consideration should be given to the provision of fewer strategically located hose reels and the elimination of hose houses. For the cost of several hose houses a light weight truck can be purchased and readily adapted as a hose and equipment carrier, thus providing greater flexibility and ease of maintenance at lower cost.

Foam

Another important type of fire-extinguishing material is *foam*. Most foam installations are small enough to be portable and very effective on fires covering a relatively small area. There are a few installations, however, which have installed permanent piping to bring foam to bear on relatively large areas of acute fire hazard. Foam, sometimes referred to as alcohol foam, is particularly important in the case of liquids which are lighter than water and which would float on the surface of the water and spread out, thereby carrying the fire to points beyond which it originated. Chemically this foam is often produced by bubbling carbon dioxide or some other gas through a liquid containing a foam producing chemical. The hard resilient bubbles of foam thus formed coat a burning surface with what is essentially the gas used to bubble through the liquid in the first place, and if this gas is carbon dioxide it effectively smothers the fire by preventing access of oxygen.

Other Agents

Still another technique for fire extinguishment is the use of carbon dioxide, dry chemical, and halogenated hydrocarbons such as carbon tetrachloride or methyl bromide. Here a portable extinguisher of carbon dioxide or dry chemical type is applicable to relatively small fires. Large scale protection would require permanently piped carbon dioxide systems so that enough gas can be brought to bear on the burning mass to smother it effectively. As to the use of halogenated hydrocarbons, it must be remembered that when the most commonly used members of this group are heated to high temperatures they decompose and evolve highly toxic decomposition products. In the case of carbon tetrachloride, one of the decomposition products is the well-known poison, phosgene. However, in the case of a fire where the ventilation is relatively good, it may be advisable to put up with the comparatively small amount of phosgene formed in order to put out the fire quickly and effectively. An ideal example of where *not* to use halogenated hydrocarbon types of fire extinguishers is in public conveyances where many people might be trapped, such as a subway train without adequate ventilation. The use of halogenated hydrocarbon fire extinguishers should be restricted to relatively small fires in well-ventilated areas.

Gas Fires

The extinguishment of gas fires may be carried out with carbon dioxide, dry chemical, or in some cases with water supply. But it must be borne in mind that if the gas fire is put out and the gas supply is not shut off, the gas will continue to pour out, ultimately filling a large room or building with a flammable or even explosive mixture which upon ignition could cause more damage than the original burning gas. The best method of handling gas fires is to stop the flow of the gas. If this cannot be done at once, it is often wise to simply spray the surrounding area and surfaces with water to keep them cool and thus prevent their ignition until the supply of gas is exhausted or a valve controlling the flow can be closed.

Burning Metals

Fires caused by burning metals are very difficult to extinguish and cannot be handled in the ordinary manner. For instance, to spray water upon burning metal might cause an explosion which would spatter flaming particles of metal to great distances. Usually the best way to attack burning metal is with specially formulated dry type fire-extinguishing agents. For instance, when one is planning to use a metal in a form in which it might readily become ignited, it is wise to discuss the situation with the manufacturer or supplier of this metal and obtain from him explicit instructions for the storage, handling, and fire extinguishment of the metal. This is particularly applicable to the use of sodium, potassium, lithium, zirconium, uranium, thorium, and magnesium. For instance, it has been found that ordinary sand, even when dry, is a very poor material for extinguishing metal fires; it may react with the hot metal and add more heat to an already intense fire. Often salt (sodium chloride), sodium bicarbonate, graphite, magnesium carbonate, magnesium oxide, or mixtures of all of these materials have been found effective; in every case the supplier of this material will know how it must be handled. Water should never be applied to burning metals.

INTANGIBLE LOSSES DUE TO FIRE

While ordinary fire insurance may pay the dollar losses incurred in a fire, and insurance coverage may also be provided for the loss of profit during the shutdown period, what protection is there against the loss of market and the fact that customers unable to be supplied may turn to a competitor?

When it seems that all the normal precautions have been taken and that there is a generally good level of protection throughout a plant, it is necessary to consider the entire operation and determine if there are bottlenecks where a relatively minor loss could result in a severe one by a consequent shutdown in production. This may be within the organization or elsewhere in the production chain. Power supply, for instance, is an extremely critical item. Inquiry should be made of the local utility to determine just how dependable it is. Should the local utility have a serious breakdown in its own facilities, arrangements should be made for possible connections to a utility grid in order to continue an uninterrupted flow of power.

Many areas experience periodic shortages of natural gas, during which industrial organizations must conserve its use in order that more may be available for home heating. If such fuel is used, arrangements should be made for the emergency use of liquefied petroleum gas. Safety should be provided for in advance; if precautions are improvised at the last minute, unnecessary risks may be taken. As to utility arrangements, transformers and other substation equipment should be adequately guarded against damage or destruction from fire in the plant.

Another very important source of serious loss results from the destruction of records. The old iron safe in the treasurer's office does not represent adequate protection for the vital records it contains, should there be a destructive fire. The Underwriters Laboratories and the Safe Manufacturers Association test and label safes and cabinets for their fire resistance. Very often we find that while some documents easily recognized as vital are well protected, those considered to be just ordinary "business records" are housed in sheet metal filing cabinets which are no more protection than wood in the event of a severe fire. The loss of Accounts Receivable records can be disastrous, and other records, even though not directly concerned with the collection of payments and accounts, may require reconstruction in the event of loss. Also, an accurate proof of loss is necessary for the prompt and proper payment of insurance claims. If the records supporting such a proof of loss have been destroyed, it may be necessary to resort to litigation or accept a lesser figure because of lack of proof.

Engineering drawings of the plant and its equipment may be almost priceless, especially when they relate to underground or otherwise concealed installations.

Provision of fire-sale record cabinets is in itself not enough; precaution must be taken to close the files and record containers in the event of a fire. Normally, in case of fire employees would immediately evacuate the area without any attention to this important detail, which would render the entire installed protection useless during the working day. The production chain should further be carefully surveyed for vital bottlenecks which might be subjected to fire loss due to the type of operation or to their exposure to other fire-producing potentials. The protection provided should be thoroughly considered to determine whether or not it is adequate under the circumstances. If adequate protection for one reason or another is unduly costly, a study should be undertaken to determine where else the work could be carried on if the operation were wiped out. In some cases it may be cheaper to assume the risk, because in the event of a fire the work could be contracted out to other organizations in the vicinity with similar facilities. Or it may be found that such an arrangement would be impractical and that the loss of the spray painting and detearing steps, for example, would completely shut down production; continuity of production might then best be protected by rearranging the area and subdividing it with fire walls so that only a portion of the facilities would be destroyed at any time.

Having examined the production bottlenecks in a plant and determined the best way with which each one must be dealt to prevent it from shutting down production, it is necessary to examine the situation with respect to suppliers. It is quite a shock to receive word that the XYZ Corporation has burned to the ground, and then realize that it was the sole supplier of an important subassembly of one's production and that there is only a single week's inventory on hand. When subcontracts for any portion of production are undertaken, it is necessary to consider the subcontractor's plant in the same light as one's own. There is no reason why this matter cannot be brought out during the contract negotiations, but even prior to this time, a discreet survey of the situation can be made in connection with the ordinary inspection of his facilities. If the situation looks dangerous, the least precaution that should be taken is to divide up the subcontract and allow a portion of it to another contractor, thus spreading the risk. Or one might advance his delivery schedule to provide himself with a larger inventory.

Then there is the question of storage. Many industrial organizations produce the year round at a relatively steady rate to meet the requirements of a seasonal market. This requires the utilization of storage space often not available at one's own plant. And in too many cases storage space is leased with little or no inquiry into the fire safety of the situation, with the result that months of production effort can go up in smoke and a market may be lost forever. Here again, the basic principle of subdivision should be employed to the fullest extent practicable, even though this may involve some inconvenience and expense.

For bibliography, see p. 225.

SECTION 7

STORAGE AND HANDLING OF HAZARDOUS MATERIALS

N. Irving Sax

Director, Radiological Sciences Laboratory
New York State Health Department
Albany, New York
Adjunct Professor, Division of Bioenvironmental Engineering
School of Engineering, Rensselaer Polytechnic Institute
Troy, New York
Consultant on Pollution Control

Due to the wide range of materials which have to be stored in factories, laboratories and warehouses, good storage practice has become very important and is steadily becoming more so. Economically, it is probably not feasible to store each item of inventory in an environment which is ideally safe. It is therefore necessary to group items for storage in such a way that the best use is made of available space. To effect a compromise between perfect safety and the competitive demands of an industrial economy, a source of information is necessary. The description of each material listed in Section 12, for instance, contains as much of this information as was judged to be reliable and useful for the purposes of this book.

The disastrous effects of neglecting the physical and chemical properties of stored materials are fires, explosions, emission of toxic gases, vapors, dusts and radiation, and various combinations of these effects.

In order to make it possible to discuss most of the materials with which we are concerned, we shall group the many thousands of individual items into categories as follows:

(1) **Explosives.** This classification includes materials which under certain conditions of temperature, shock or chemical action can decompose rapidly to evolve either large volumes of gas or so much heat that the surrounding air is forced to expand very rapidly; in either case, an explosion results. These materials are particularly dangerous when involved in shock- or heat-producing disaster conditions. Appendix 1 to this section discusses in detail the storage, handling, and disposal of explosives.

(2) **Flammable Materials.** Practically all combustion takes place between oxygen and a fuel in its vapor or some other finely divided state. Thus, in order to start a solid burning it is necessary to heat at least a portion of it to a point where enough vapor is evolved to catch fire. Since at a given temperature liquids are generally at a higher vapor pressure than solids, flammable liquids are usually very easy to ignite. Flammable dusts are about as easy to ignite as vapors or gases.

A further, quantitative, difference exists between propagation rates of solid, liquid, and gas fires. Fires of solids propagate slowly, fires of liquids propagate relatively rapidly, and gas, vapor, or dust fires propagate so rapidly they often seem to explode.

Section 12 lists, in alphabetical order, thousands of materials with varying degrees of flammability. In each case where sufficiently reliable data are available the degree of fire hazard of the material is indicated.

Materials which ignite easily under normal industrial conditions are considered to be dangerous fire hazards, e.g., pyrophoric substances such as finely divided metals, hydrides of boron, and phosphorus, liquids with flash points of 80°F or lower, and flammable gases. Such materials must be stored in places that are cool enough to prevent accidental ignition in the event that vapors of the fuel mix with the air. It is important to provide adequate ventilation in the storage space, so that normal leakage of such vapors from containers will be diluted enough to prevent a spark from igniting them. A further precaution is to locate the storage area well away from areas of fire hazard, for example, where torch cutting of metals is performed, etc. Highly flammable materials must be kept apart from powerful oxidizing agents, materials which are susceptible to spontaneous heating, explosives, or materials which react with air or moisture to evolve heat.

Ample fire-fighting equipment, either automatic or manual, must be readily available for emergency use. The storage area must be posted to prevent smoking or striking of matches. Bare filament heaters or other sources of ignition must be kept away. The area must be electrically grounded and periodically inspected or equipped with automatic smoke or fire detection equipment (see Section 6).

All combustible materials are dangerous in a fire and/or explosion, both because heat will cause them to catch fire and add to the smoke and fume hazard, and because small explosions have been known to scatter clouds of flammable dusts which in turn have caused still greater explosions than the original one which gave rise to them (see Appendix 3 to this section).

(3) Oxidizing Agents. Oxidizing agents are sources of oxygen, one of the necessary components of a fire. Normal air with its 20 percent content of oxygen is the primary source. However, many other materials are so constituted chemically that they can supply oxygen to a reaction, even in the absence of air. Some of these oxygen suppliers require heat before they will yield oxygen; others evolve significant amounts of it at room temperature. When containers of oxidizing materials are damaged, for example, by the forces unleashed during a disaster, their contents can mix with the contents of other damaged containers and possibly start fires and explosions. This is a strong argument in favor of *separate* storage over *mixed* storage. The following classes of compounds are noted for their ability to supply oxygen: organic and inorganic peroxides, oxides, permanganates, perrhenates, chlorates, perchlorates, persulfates, organic and inorganic nitrites, organic and inorganic nitrates, nitrates, iodates, periodates, bromates, perselenates, perbromates, chromates, dichromates, ozone, and perborates.

In Section 12 each material which is an oxidizing agent is so rated. It is important to know this when planning storage.

In general, it is unsafe to store oxidizers close to liquids of low flash point. In fact, even slightly flammable materials should be isolated from oxidizers. For instance, ordinary glycerol and potassium permanganate, when simply blended at room temperature for a few minutes, react violently to produce a hot fire.

As a general rule, it is wise to keep all flammables away from an area where oxidizing agents are stored. This storage area should be kept cool and ventilated, and should be fireproof. Normal fire fighting equipment is less useful here, since the blanketing or smothering effect of fire extinguishers is less effective because the oxidizers supply their own oxygen. With these materials it is best to keep fuel away.

(4) Water Sensitive Fire and Explosion Hazards. These are materials which react with water, steam or water solutions to evolve heat, flammable gases, or explosive gases. Examples of such materials are: lithium, sodium, potassium, calcium, rubidium, cesium, alloys and amalgams of the above, hydrides, nitrides, sulfides, carbides, borides, silicides, tellurides, selenides, arsenides, phosphides, acid anhydrides, and concentrated acids or alkalies.

Of the above list, the first eight are materials which react exothermally with moisture and evolve hydrogen gas (see Section 12); the next nine (nitrides through phosphides) react rapidly with moisture to evolve volatile, flammable, sometimes spontaneously flammable and/or explosive hydrides. Acid anhydrides and concentrated acids or alkalies react with moisture to evolve heat. Such materials must be stored in well-ventilated, cool, *dry* areas. Because many of these materials are also flammable, it is essential that NO automatic sprinkler system be used in a storage area which houses them; in fact, such an area should have no water coming into it at all. Heating may be electrical or with hot, dry air. The building must be waterproof, located on high ground, and separated from other storage. This building should conform to that required for storage of hydrogen.

Particular attention must be paid to the following: pocketing of light gases under the roof, introduction of sources of ignition, periodic inspection, automatic detection and alarm due to dangerous concentrations of flammable gases.

Since many disasters cause wetting of extensive areas either by flooding (if that is the nature of the disaster), by breaking of water pipes and sprinkler systems, or possibly by damaging the warehouse so that rain can get in, it is important that the items listed above receive special storage consideration for protection against disasters (see Appendix 3 to this section).

(5) Fire and Explosion Hazards of Acid and

Acid Fume Sensitive Materials. These are materials which react with acid and acid fumes to evolve heat, hydrogen, and flammable and/or explosive gases. Examples are: lithium through phosphides in the foregoing list, concentrated alkalies, metals (including structural alloys), arsenic, selenium, tellurium, and cyanides. Therefore it follows that acids should not be stored in close proximity to these materials. If, however, such storage is contemplated, the following precautions are called for: the area must be kept cool; it must be ventilated and periodically inspected; sources of ignition must be kept away; construction as for hydrogen storage is required.

In fact, keeping in mind that acid and/or acid fumes can attack structural alloys and evolve hydrogen, acid storage might well be in ventilated wooden sheds. Or if metal is used in construction, it should be painted or otherwise rendered immune to attack by acid. If hydrogen can be evolved, the building should be so constructed that possible hydrogen pockets are eliminated.

Disaster conditions of high temperatures, much vibration and/or flooding can bring acids or acid fumes into contact with lithium, sodium, potassium, calcium, and rubidium; unless a good deal of thought has gone into the problem of storing such materials far enough apart to avoid this, fires and explosions may result (see Appendix 3 to this section).

(6) Compressed Gases. Tanks of compressed gases should be stored upright and chained or otherwise securely attached to some substantial support to minimize the chance of falling over and breaking or straining the valve or other part of the tank. The tank storage area should be kept cool, out of the direct rays of the sun, away from hot pipes, in a ventilated area where care has been taken in construction so that pocketing of gases which may escape from their containers will be kept to a minimum.

The building which houses the tanks should be fireproof. It should provide some means (such as a sprinkler system) of keeping the tanks cool in case of external or internal fire. The reason for concern about overheating is that this will raise the pressure of the gas in the tank to where the safety disc or a valve might rupture, thus releasing a significant volume of, say, hydrogen or acetylene, which might then catch fire and possibly explode.

While it is true that today the handling of compressed gas tanks is relatively safe, due in great part to engineering improvements and standardization of components, certain precautions are still advisable. Aside from upright,

cool storage, care must be taken to keep from damaging tanks in handling; valves must be operated carefully and kept in good condition. Avoid drastic changes in temperature; do not hammer valve cocks; keep valve cover on whenever possible; do not interchange reduction gauges from one gas to another without checking for compatibility; discourage tampering with tanks in any way; put tank storage in charge of personnel competent to handle it. For further information on compressed gas tank storage consult the Compressed Gas Association of New York City, your gas supplier, or your local fire department.

Disaster conditions including high ambient temperatures due to nearby fires, shock and vibration due to nearby explosions, etc., are particularly trying for compressed gas storage. Great care should be exercised in the storage of large quantities of compressed gases, particularly in areas of high population density (see Appendix 3 to this section).

(7) Toxic Hazards. These are materials which under either normal conditions or disaster conditions, or both, can be dangerous to living things around them. Thus carbon tetrachloride, for example, if stored in a poorly ventilated place under standard conditions can evolve enough vapor to render the storage area toxic. Under disaster conditions of high temperature, if carbon tetrachloride is decomposed it can form significant quantities of the highly toxic phosgene.

Since it is nearly impossible to seal containers perfectly, it is always to be expected that some of whatever volatile materials are stored will escape into the atmosphere of the storage area. Likewise, air, atmospheric moisture, and carbon dioxide will come into contact with the contents of imperfectly sealed containers. Some initially well-sealed containers will build up enough internal pressure to break a seal or even burst the sealed unit.

Materials which are toxic because of their radioactivity are also included in this category. Such materials are dangerous if allowed to become air-borne, or to be ingested or inhaled into the body. Thus under normal storage conditions, small amounts of radioactive materials may escape their containers and contaminate the atmosphere; disaster conditions such as high temperatures, severe shocks, floods or combinations of these can burst containers and volatilize or scatter by the force of an explosion or spread about by means of flood waters much larger quantities of radioactive materials. Furthermore, it must be borne in mind that sources of radiation enclosed in lead shielding can become dangerous in a disaster because disaster conditions can be such as to melt the shield or volatilize a radioactive material (see also Sections 5 and 8, and Appendix 3 to this section).

In general, materials which are toxic as stored or which can decompose into toxic components due to contact with heat, moisture, acids or acid fumes, should be stored in a cool, well-ventilated place, out of the direct rays of the sun, away from areas of high fire hazard, and should be periodically inspected and monitored. Incompatible materials should be isolated from each other.

(8) Corrosive Materials. Corrosive materials include acids, acid anhydrides, and alkalies. Such materials often destroy their containers and get into the atmosphere of a storage area; some are volatile, others react violently with moisture. Acid fumes react to evolve toxic fumes with sulfides, sulfites, cyanides, arsenides, tellurides, phosphides, borides, silicides, carbides, fluorides, selenides; they liberate hydrogen upon contact with metals and hydrides. Alkalies may liberate hydrogen upon contact with aluminum, etc. Acid mists or fumes corrode structural materials and equipment and are toxic to personnel. Such materials should be kept cool, but well above their freezing points. Acetic acid, for instance, can freeze in an unheated room and, due to differential expansion, crack a glass container. There should be sufficient ventilation to prevent accumulation of fumes; there must also be regular inspection. Containers of corrosive materials should be carefully handled, kept closed, and labeled. All exposed metal in the vicinity of such storage should be painted and checked for weakening by corrosion. Corrosive materials should be isolated from materials noted above (cyanides, sulfides, etc.), reaction with which can produce highly toxic fumes.

It should be added that strong acids and alkalies will cause serious burns and eye damage to personnel, and hence adequate protection in the form of gloves, aprons, goggles, etc., should be worn when handling them into and out of storage areas (see Section 3).

Some examples of materials which warrant special care in storage are: alkali metals, calcium, acid anhydrides, concentrated acids, concentrated alkalies, arsenic, arsenic compounds, beryllium, beryllium compounds, borides, phosphides, phosphorus, silicides, cyanides, nitrides, nitrates, nitrites, sulfur, sulfides, tellurium, tellurium compounds, selenium, selenium compounds, carbides, halogenated hydrocarbons, mercury, mercury compounds, lead, lead compounds, cadmium, cadmium compounds, etc.

Storage of significant quantities of any of the above materials should be brought to the attention of the safety director of the plant owning the material, the local fire department, and the local Civil Defense headquarters.

The term "significant quantity" is a difficult one to define. In an unventilated room even 1 pound is significant, while in well ventilated storage areas it may take as much as 100–1000 pounds to be significant, particularly to a safety director and a fire department. Ton lots and over are of concern to Civil Defense authorities, as well as to fire departments and safety directors. In Section 12 of this book, the disaster hazard potential of many materials is given. Often indicated are the most toxic capabilities of the materials and the disaster conditions which might release them. Naturally, the decomposition of complex materials yields a number of products, and it is beyond the scope of this book to list them all. However, when at least one toxic material is known to be formed, it is listed. Often the product listed is not the one formed in the greatest abundance. An effort has been made to list at least one product of toxicological significance.

Appendix 3 to this section discusses in some detail the problem of planning in advance for the time when a combination of circumstances approximating a disaster might occur in an industrial installation.

A disaster can occur in a plant from without, due to flood, fire, bombing or earthquake, or from within, due to fire or explosion.

Whatever the cause, the effects are much the same and the information in Appendix 3 will be found to apply.

Handling of Toxic and Corrosive Materials

When loading, transporting, or unloading such materials, HANDLE CONTAINERS CAREFULLY! Damaged drums may leak and, in an enclosed space may reach dangerous concentrations of vapors or dusts.

IF A LEAKY CONTAINER IS DISCOVERED, put on protective clothing (see below). After thoroughly airing out the space, enter and turn the leaking container on its side or end to stop the leak. Block or rope off the contaminated area and post with DANGER signs. Then report the leak immediately to your safety department or to the supplier. Wire or phone and report the following information:

(1) What is leaking? (See drum label.)
(2) Where is the leaky container now located?
(3) Where can you be reached for instructions? (Give phone number or address for telegram.)
(4) Has anything or anybody been contami-

nated by the spilled material? (If any persons are contaminated, call a local doctor.)

Finally, be sure that the affected area will be properly guarded in your absence by a responsible person until necessary instructions are received.

Before you open a container or handle a toxic or corrosive material in any way, PUT ON COMPLETE PROTECTIVE CLOTHING. Because your *eyes* could be harmed, you should wear goggles (or a plastic face shield). Since you could be poisoned by breathing vapors or getting the poison into your *mouth*, you should wear an effective respirator. Since many poisons can be absorbed through the *skin*, you should wear a clean cap, clean rubber gloves (rubber can absorb dangerous amounts of poison after prolonged use), rubber boots, and coveralls which enclose your neck.

Drum Handling. HANDLE DRUMS CAREFULLY! Don't cause leaks. Store and use drums in a cool, well-ventilated area. Do not store drums near steam pipes, boilers, or other sources of heat.

To Open Drums. Set the drum with the bung end up under a ventilating hood. Unscrew the bung slowly to gradually release any internal pressure. Then fit a valve in place of the top bung. Be sure it is tight. Now tip the drum on its side with the side bung up and fit another valve in place. Be sure that both connections are leak-proof. Support the drum in its cradle with one valve uppermost.

To Empty Drum. Connect the upper valve to an open air vent pipe. Connect the other valve to a closed system storage or processing unit. Watch for leaky connections and correct immediately if found. IF LIQUID SPILLS, absorb spillage with dry clay or sawdust. Bury or burn these sweepings. (Avoid the smoke. It may carry poisonous vapors.) Scrub the spill area with soda ash and soap. Flush this thoroughly and repeat the operation as often as necessary to destroy or remove as much of the poison as possible. *Don't track spilled poisons around.*

To Discard Empty Drums Safely. NEVER USE CONTAMINATED DRUMS FOR ANYTHING ELSE no matter how clean they seem to be. Empty drums must be decontaminated and disposed of as follows: (1) Thoroughly scrub out the open drum with hot water and soda ash or 5 percent caustic solution. (2) Flush and wash again until as much as possible of the poison has been removed. (3) To be sure drums are *not* reused, perforate sides, top and bottom before discarding. (4) For greater safety, burn out the perforated drum (avoid the smoke) before finally dumping it in a safe area.

Rain water may rinse traces of poison out of a drum onto soil or into drainage streams where the poison could eventually reach animals, fish or even humans. Take no chances of this in drum disposal. Perform all operations carefully and with thoroughness.

Appendix 1

EXPLOSIVES

Explosives are shock sensitive to varying degrees (see Explosives, High, Section 12, for exact data). Explosives manufacturers as well as other organizations have issued numerous recommendations as to safe and effective storage. The Institute of Makers of Explosives has published a pamphlet, "Standard Storage Magazines Recommended by the Institute of Makers of Explosives" and has also issued the "American Table of Distances, Specifying Distances to Be Maintained between Magazines for Explosives and Inhabited Buildings, Public Railways, and Public Highways." In addition, the Institute has issued a pamphlet entitled "Suggested State Law Compiled by Institute of Makers of Explosives," which has been the foundation of the law of at least some states with respect to explosives.

Those in close contact with explosives are of the opinion that conditions of storage largely determine not only the safety but also the efficiency that may be expected when explosives are used. Extremes of heat or cold, of air dryness, or of moisture in storage, the roughness or carefulness with which handled, the length of time stored, the length of time out of their original container before used—these and many other considerations have a vital influence on the behavior of a commercial explosive. The manufacturers issue pamphlets in connection with their products, and many excellent suggestions are made with respect to the safe and efficient transportation and storage as well as use of explosives; a particularly informative pamphlet entitled "Safety in the Handling and Use of Explosives" was issued by the Institute of Makers of Explosives under the auspices of representative explosives manufacturing organizations of the United States.

The Bureau of Mines of the U. S. Department of Interior has issued much information about the testing, transportation, storage and use of explosives in the mining and allied industries, but more especially with respect to coal mining. The information has been published in numerous pamphlets (Bulletins, Technical Papers, Reports of Investigations, and Information Circulars), many of which are now out of print. Other publications on explosives can be obtained from the Superintendent of Documents, Washington, D. C.; Reports of Investigations can be obtained from the Bureau of Mines, U. S. Department of Interior, Washington, D. C. Safety in the use of explosives is treated in publications of the National Safety Council, the American Standards Association, as well as in various other technical journals.

Location and Construction of Magazines

Magazines should be situated far enough away from other buildings or vital structures so that in case of an explosion in the magazine the least possible damage will be done to buildings or persons in the surrounding area. Some states and cities have laws or regulations that specify this distance. Where such laws do not exist, it is recommended that, insofar as feasible, the magazine location comply with the American Table of Distance (Table 1).

The mandatory regulations as to the location of surface magazines under the Federal Explosives Act state (as of June 1, 1944):

Section 24(c) (4): All storage magazines must be properly constructed, safely located, and securely locked.

Section 25(b)(2)— Location of box-type magazines: Box type magazines, when located outside a building, shall be securely anchored. No magazine shall be placed in a building containing oil, grease, gasoline, waste paper, or other highly flammable materials nor shall a magazine be placed less than 20 feet from a stove or furnace or open fire or flame, or less than 5 feet from other sources of external heat.

In the administration of the Federal Explosives Act, it is also recommended that surface magazines be detached from other buildings in conformity with the American Table of Distances and that they not be nearer than 200 feet from any power plant, hoist house, mill, dam, or other vital structure or to any mine shaft, tunnel, or slope opening at the surface. In other words, the magazine should be at least 200 feet from any mine structure or opening, and the explosives and detonator magazines should be at least 100 feet apart if not barricaded, or 50 feet if barricaded. Barricading is defined in Table 1.

TABLE 1. American Table of Distances for Storage of Explosives
(as Revised and Approved by The Institute of Makers of Explosives September 30, 1955)

Explosives		Distances When Storage Is Barricaded (ft)			
Pounds over	Pounds Not over	Inhabited Buildings	Passenger Railways	Public Highways	Separation of Magazines
2	5	70	30	30	6
5	10	90	35	35	8
10	20	110	45	45	10
20	30	125	50	50	11
30	40	140	55	55	12
40	50	150	60	60	14
50	75	170	70	70	15
75	100	190	75	75	16
100	125	200	80	80	18
125	150	215	85	85	19
150	200	235	95	95	21
200	250	255	105	105	23
250	300	270	110	110	24
300	400	295	120	120	27
400	500	320	130	130	29
500	600	340	135	135	31
600	700	355	145	145	32
700	800	375	150	150	33
800	900	390	155	155	35
900	1,000	400	160	160	36
1,000	1,200	425	170	165	39
1,200	1,400	450	180	170	41
1,400	1,600	470	190	175	43

TABLE 1 (Continued)

Explosives		Distances When Storage Is Barricaded (ft)			
Pounds over	Pounds Not over	Inhabited Buildings	Passenger Railways	Public Highways	Separation of Magazines
1,600	1,800	490	195	180	44
1,800	2,000	505	205	185	45
2,000	2,500	545	220	190	49
2,500	3,000	580	235	195	52
3,000	4,000	635	255	210	58
4,000	5,000	685	275	225	61
5,000	6,000	730	295	235	65
6,000	7,000	770	310	245	68
7,000	8,000	800	320	250	72
8,000	9,000	835	335	255	75
9,000	10,000	865	345	260	78
10,000	12,000	875	370	270	82
12,000	14,000	885	390	275	87
14,000	16,000	900	405	280	90
16,000	18,000	940	420	285	94
18,000	20,000	975	435	290	98
20,000	25,000	1055	470	315	105
25,000	30,000	1130	500	340	112
30,000	35,000	1205	525	360	119
35,000	40,000	1275	550	380	124
40,000	45,000	1340	570	400	129
45,000	50,000	1400	590	420	135
50,000	55,000	1460	610	440	140
55,000	60,000	1515	630	455	145
60,000	65,000	1565	645	470	150
65,000	70,000	1610	660	485	155
70,000	75,000	1655	675	500	160
75,000	80,000	1695	690	510	165
80,000	85,000	1730	705	520	170
85,000	90,000	1760	720	530	175
90,000	95,000	1790	730	540	180
95,000	100,000	1815	745	545	185
100,000	110,000	1835	770	550	195
110,000	120,000	1855	790	555	205
120,000	130,000	1875	810	560	215
130,000	140,000	1890	835	565	225
140,000	150,000	1900	850	570	235
150,000	160,000	1935	870	580	245
160,000	170,000	1965	890	590	255
170,000	180,000	1990	905	600	265
180,000	190,000	2010	920	605	275
190,000	200,000	2030	935	610	285
200,000	210,000	2055	955	620	295
210,000	230,000	2100	980	635	315
230,000	250,000	2155	1010	650	335
250,000	275,000	2215	1040	670	360
275,000	300,000	2275	1075	690	385

NOTES FOR TABLE 1:

(1) "Explosives" means any chemical compound, mixture, or device, the primary or common purpose of which is to function by explosion, i.e., with substantially instantaneous release of gas and heat, unless such compound, mixture, or device is otherwise specifically classified by the Interstate Commerce Commission.

(2) "Magazine" means any building or structure, other than an explosives manufacturing building, used for the permanent storage of explosives.

(3) "Natural barricade" means natural features of the ground, such as hills or timber of sufficient density that the surrounding exposures which require protection cannot be seen from the magazine when the trees are bare of leaves.

(4) "Artificial barricade" means an artificial mound or revetted wall of earth of a minimum thickness of 3 ft.

(5) "Barricaded" means that a building containing explosives is effectually screened from a magazine, building, railway, or highway, either by a natural barricade, or by an artificial barricade of such height that a straight line from the top of any sidewall of the building containing explosives to the eave line of any magazine, or building, or to a point 12 ft above the center of a railway or highway, will pass through such intervening natural or artificial barricade.

(6) When a building containing explosives is not barricaded, the distances shown in the table should be doubled.

(7) "Inhabited building" means a building regularly occupied in whole or in part as a habitation for human beings, or any church, schoolhouse, railroad station, store, or other structure where people are accustomed to assemble, except any building or structure occupied in connection with the manufacture, transportation, storage, or use of explosives.

(8) "Railway" means any steam, electric, or other railroad or railway which carries passengers for hire.

(9) "Highway" means any public street or public road.

(10) When two or more storage magazines are located on the same property, each magazine must comply with the minimum distances specified from inhabited buildings, railways, and highways, and in addition, they should be separated from each other by not less than the distances shown for "Separation of Magazines," except that the quantity of explosives contained in cap magazines shall govern in regard to the spacing of said cap magazines from magazines containing other explosives. If any two or more magazines are separated from each other by less than the specified "Separation of Magazines" distances, then such two or more magazines, as a group, must be considered as one magazine, and the total quantity of explosives stored in such group must be treated as if stored in a single magazine located on the site of any magazine of the group, and must comply with the minimum of distances specified from other magazines, inhabited buildings, railways, and highways.

(11) The Institute of Makers of Explosives does not approve the permanent storage of more than 300,000 lb of commercial explosives in one magazine or in a group of magazines which is considered as one magazine.

(12) This table applies only to the manufacture and permanent storage of commercial explosives. It is not applicable to transportation of explosives, or any handling or temporary storage necessary or incident thereto. It is not intended to apply to bombs, projectiles, or other heavily encased explosives.

(13) All types of blasting caps in strengths through No. 8 cap should be rated at $1\frac{1}{2}$ lb of explosives per 1000 caps. For strengths higher than No. 8 cap, consult the manufacturer.

The magazine site should be selected with the thought of safety and protection as well as economy of operation. The topography of the ground should be carefully considered to take full advantage of protection offered by hills, rock ledges, soil banks, dense woods, etc. A site on sandy soil is preferable to one on rocky ground, though rock ledges can afford considerable protection. Wooded regions are highly desirable, but precautions must be taken against brush or other fires.

The site should have a good drainage to prevent dampness or moisture from penetrating the magazine. Dugouts and tunnels in the sides of hills offer good protection for the storage of explosives, but are not recommended unless they can be well ventilated and well drained.

In brief, the magazine should be located so that in the event of an explosion there would not be serious damage to life and property. It should also be located so as to prevent damage to or deterioration of explosives because of dampness or unduly high or low temperatures; although the magazine should be so isolated by natural or other barriers as to minimize loss of life or property in the event of an explosion, it should not be so far away that its contents could be stolen by mischievous persons.

An artificial barricade should consist of a constructed mound of earth or sand fill with a minimum width of 3 feet at the top. One or more of the sides should be supported by a concrete, timber or masonry wall.

Barricades. The location and construction of magazines and the use of barricades to minimize destruction of life and property should the contents of the magazine be detonated are closely related. Magazines should be protected by barricades, either natural or artificial, regardless of their distance from other buildings. The distances given in Table 1 should be used in choosing a site and should be doubled where the magazine is not screened or protected by natural or artificial barriers or barricades. An efficient barricade will limit or control the forces of an accidental explosion by diverting them upward and by shortening the trajectories of flying fragments. In some cases, it is not practical or even possible to separate magazines the required distances, but the danger can be greatly minimized by building efficient barricades.

The choice of a barrier will be governed largely by the material that is readily available or the ease with which it is procured. A simple barricade can be erected by scraping or hauling dirt and piling it close to the magazine. A concrete or timbered wall can be erected at least 4 feet from the magazine, and the dirt may be piled against the wall; the width at the top of the pile or bank should not be less than 3 feet. An excavation made in the side of a hill for a magazine, with the excavated dirt built up in front, makes an efficient barrier. When the structures to be protected are on the same plane as the magazine, all sides should be barricaded essentially to or above the height of the magazine.

Foundations. Magazines should be so constructed as to prevent the stored explosives from absorbing dampness and to allow free circulation of air beneath the floor if it is a surface structure. Foundations should preferably be of brick, stone, concrete, hollow tile or concrete blocks filled with concrete, or a combination of these. The height of the foundation above the ground at any point should be at least 1 foot. Foundations of magazines used for temporary storage may be wood posts, but the space between the ground and the floor should be enclosed and the enclosure covered with galvanized or sheet iron. In fact, the space under the floor of building type magazines of any kind should be enclosed to prevent entrance of persons, animals, sparks, or firebrands. Screened vents should be left in the foundation walls for ventilation.

Ventilation. To be properly stored, explosives should be kept in a magazine which is dry and of uniform temperature; adequate ventilation is probably the best means of attaining such a condition. By means of vents in the foundation, walls and roof, air can be made to circulate freely in the magazine.

More consideration would be given to magazine ventilation if it were realized that the two primary causes for the deterioration of explosives are moisture and excessive fluctuation in temperature. Unless air circulates freely in the magazine, the atmosphere may become excessively hot, cold, or humid. Long exposure to such atmospheric conditions has essentially the same effect upon some explosives as dampness, with the result that there may be misfires or incomplete detonation or burning. With some types of explosives, unfavorable atmospheric conditions may cause the nitroglycerin to separate from the other ingredients causing nitroglycerin "leaks" which make the explosives more sensitive to shock or frictional impact.

Materials of Construction. The field representatives of the Bureau of Mines engaged in administering the Federal Explosives Act find magazines of various types and construction. Many of them are substantially built of concrete, stone, or hard-burnt brick; more frequently, however, magazines are built flimsily, as for instance, a light framework of wood covered with corrugated iron, with little or no attempt to make them bulletproof, theftproof, fireproof, or resistant to heat, cold, or moisture.

Stone, concrete, or hard brick magazines are generally thought to be undesirable because of the possibility of flying fragments in case of an explosion. The material used in the construction of a magazine should disintegrate and shatter easily if an explosion were to occur within the magazine, yet construction should be substantial enough to prevent theft.

The mandatory regulations relating to construction of surface explosives storage magazines are given in the Federal Explosives Act.

In addition, those in charge of the administration of the Act offer the following recommendations:

Recommendations as to Surface Magazines

(1) Surface magazines should be well ventilated.

(2) Natural light or light by portable electric storage lamps or by protected electrical systems, or flood lights from the outside of the magazine should be provided.

(3) Floors should be made of wood or other non-sparking material and have no metal exposed.

(4) Magazines should be grounded if constructed of steel or covered with sheet iron.

Precautions against Theft

Explosives must be protected against theft by storing them in magazines constructed and locked as required by regulations. Additional protection may be secured in appropriate cases by using:

(1) Mortise locks wherever practicable. When padlocks are used they should be protected by steel hoods or shields placed around the locks to prevent tempering. The hood or shield may be made from pipe or of steel large enough in diam-

eter to enclose and protect the lock and hasp and at least 4 inches in length, securely fastened to the door by welding or by lugs that are riveted or bolted or fastened by some other secure method.

(2) Fences.

(3) Floodlights.

(4) Gas bombs.

(5) Alarm systems, such as the "electric eye," etc.

(6) Additional guards or inspection by watch-men.

Precautions against Fires

In order to prevent fires the following precautions should be taken:

(1) The magazine should be kept clean.

(2) Wastepaper, sawdust, used empty boxes and containers, and other combustible material should not be left to accumulate in or around any magazine.

(3) All surrounding area for 25 feet, and preferably 50 feet in all directions, should be kept free of rubbish, dry grass, or other material of a combustible nature, and where feasible the area should be covered with a material to prevent the growth of grass, weeds, and brush.

(4) Smoking, open lights, or other flame, or carrying of matches and smoker's articles into or around any magazine should be prohibited.

(5) Black powder and high explosives should be stored in separate magazines.

General Rules

The following rules, although not directly applicable to magazine construction, do apply to the safe operation of explosives and detonator magazines.

No detonators, tools, or other materials should be stored in a magazine containing explosives.

No high—or low—explosives or blasting device heaters should be stored in a detonator magazine.

Lighting in the magazine should be natural or by permissible lights; if a magazine is electrically lighted, the lamps should be of the vaporproof type, the switch should be outside the building, and the wiring should be in conduit.

Only tools made of wood or other nonmetallic material should be used in opening cases of explosives.

Explosives should be stored so that the oldest stock is used first.

Cases of explosives should be stored topside up, in other words, so that cartridges are lying flat. However, they should be turned at regular intervals as this will help to prevent their deterioration.

Preferably, cases of explosives should not be piled in stacks more than 6 feet high.

Cases containing explosives should always be lifted and set down carefully; never slide one over on another or drop from one level to another or otherwise mishandle them.

Only authorized persons should be allowed in the magazine. One person should be made responsible for the operation and be held responsible and accountable for the contents of the magazine.

Safety with Explosives

Safe use of explosives is the result of planning and doing the thing right. The user must remember that he is dealing with a powerful force, and that various devices and methods have been prepared to enable him to direct this force. He should realize that this force, if misdirected, may seriously and permanently injure him and his fellow-workers.

The following list of "Don'ts" has been prepared after many years of experience, and furnishes a comprehensive guide to safe practices. Nothing in this book is to be construed as superseding federal, state, corporation, or municipal laws, ordinances, or regulations.

DON'TS (Adopted by The Institute of Makers of Explosives, May 8, 1951). For the purposes of this list of DON'TS, the terms contained therein shall be as follows:

The term "explosives" shall signify any or all of the following: dynamite, black blasting powder, pellet powder, blasting caps, and electric blasting caps.

The term "elastic blasting cap" shall signify any or all of the following: instantaneous electric blasting caps, delay electric blasting caps, and delay electric igniters with blasting caps attached.

(1) DON'T purchase, possess, store, transport, handle, or use explosives except in strict accordance with organizational, local, state, and federal regulations.

(2) DON'T store explosives anywhere except in a magazine which is clean, dry, well ventilated, properly located, substantially constructed, and securely locked.

(3) DON'T allow persons under 18 years of age to handle or use explosives, or to be present where explosives are being handled or used.

(4) DON'T leave explosives lying around where children can get them.

(5) DON'T allow leaves, grass, brush, or debris to accumulate within 25 feet of an explosives magazine.

(6) DON'T smoke or have matches, open lights, or other fire or flame, in or near an explosives magazine, or have them nearby while handling or loading explosives.

(7) DON'T shoot into explosives with any firearm, or allow shooting in the vicinity of an explosives magazine.

(8) DON'T store any metallic tools or implements in an explosives magazine.

(9) DON'T drop, throw, or slide packages of explosives or handle them roughly in any manner.

(10) DON'T open kegs or cases of explosives in a magazine.

(11) DON'T open kegs or wooden cases of explosives with metallic tools. Use a wooden wedge and wooden, rubber, or fiber mallet. Metallic slitters may be used for opening fiberboard cases, provided that the metallic slitter does not come in contact with the metallic fasteners of the case.

(12) DON'T store or leave packages of explosives which have been opened without replacing the cover.

(13) DON'T use empty explosives cases for kindling.

(14) DON'T permit any paper product used in the packing of explosives to leave your possession. Accumulations of fiberboard cases, paper case liners, cartons, or cartridge paper should be destroyed by burning after they have been carefully examined to make sure that they are empty.

(15) DON'T use explosives that are obviously deteriorated.

(16) DON'T attempt to reclaim or use fuse, blasting caps, electric blasting caps, or any other explosives that have been water soaked, even if they have dried out. Consult the manufacturer.

(17) DON'T carry explosives in pockets of clothing.

(18) DON'T make up primers of explosives in a magazine or near excessive quantities of explosives.

(19) DON'T force cartridges of any explosives into a bore hole or past any obstruction in a bore hole.

(20) DON'T allow explosives, or drilled holes while being loaded with explosives, to be exposed to sparks from steam shovels, locomotives, or any other source.

(21) DON'T spring a bore hole near another hole loaded with explosives.

(22) DON'T load a sprung bore hole with another charge of explosives until it has cooled sufficiently.

(23) DON'T tamp with metallic bars or tools. Use only a wooden stick with no exposed metal parts.

(24) DON'T use combustible material for stemming.

(25) DON'T allow near the danger area of a blast any persons not essential to the blasting operations.

(26) DON'T fire a blast until all surplus explosives are in a safe place, all persons and vehicles are at a safe distance or under sufficient cover, and until adequate warning has been given.

(27) DON'T return to the face until the smoke and fumes from the blast have been dissipated by adequate ventilation.

(28) DON'T attempt to investigate a misfire too soon. Follow all applicable rules and regulations, or, if no rules are in effect, wait at least an hour.

(29) DON'T drill, bore, or pick out a charge of explosives that has misfired. Misfires should be handled only by a competent and experienced man.

(30) DON'T abandon any explosives. Dispose of or destroy them in strict accordance with the methods recommended by the manufacturer.

(31) DON'T store cases of dynamite so that the cartridges stand on end.

(32) DON'T leave dynamite, black blasting powder, or pellet powder in a field or any place where livestock can get at them.

(33) DON'T take surplus quantities of permissible dynamite, black blasting powder or pellet powder into a mine at any one time. These explosives deteriorate rapidly in a damp atmosphere.

(34) DON'T use black blasting powder or pellet powder with permissible explosives or dynamite, nor dynamite with permissible explosives, in the same bore hole in a coal mine.

(35) DON'T tamp pellet powder in a bore hole hard enough to crush the pellets, because of danger of premature explosion.

(36) DON'T store blasting caps or electric blasting caps in the same box, container, or magazine with other explosives.

(37) DON'T leave blasting caps or electric blasting caps exposed to the direct rays of the sun.

(38) DON'T insert a wire, a nail, or any other implement into the open end of a blasting cap to remove it from a box.

(39) DON'T strike, tamper with, or attempt to remove or investigate the contents of a blasting cap or an electric blasting cap.

(40) DON'T try to pull the wires out of an electric blasting cap.

(41) DON'T connect blasting caps or electric blasting caps to Primacord except by methods recommended by the manufacturer.

(42) DON'T attempt to fire a circuit of electric blasting caps except by an adequate quantity of delivered current.

(43) DON'T use in the same circuit electric blasting caps made by more than one manufacturer.

(44) DON'T handle explosives during the approach or progress of an electrical storm. All persons should retire to a place of safety.

(45) DON'T make electrical connections without first making sure that the ends of the wires are bright and clean.

(46) DON'T allow electrical connections to come in contact with other connections, bare wire, rails, pipes, and ground, or other possible sources of current or paths of leakage.

(47) DON'T leave electric wires or cables of any kind near electric blasting caps or bore holes charged with explosives except at the time of, and for the purpose of, firing the blast.

(48) DON'T use electric blasting caps in very wet work unless they have adequate water resistance and suitably insulated leg wires.

(49) DON'T use any means other than a blasting galvanometer containing a silver chloride cell for testing electric blasting caps, singly or when connected in a circuit.

(50) DON'T use damaged leading or connecting wire in blasting circuits.

(51) DON'T use Duplex leading wire except for single shot firing.

(52) DON'T tamper with or change the circuit of a blasting machine in any way for any purpose.

(53) DON'T spare force or energy in operating a blasting machine.

(54) DON'T store fuse or fuse lighters in a wet or damp place, or near oil, gasoline, kerosene, distillates, or similar solvents.

(55) DON'T store fuse near radiators, steam pipes, boilers, or stoves.

(56) DON'T handle fuse carelessly in cold weather. If possible, it should be warmed slightly before using to avoid cracking the waterproof coat.

(57) DON'T use short fuse. Cut fuse long enough to extend beyond the collar of the hole and to allow time to retire safely from the blast. Never use less than 2 feet.

(58) DON'T cut fuse until you are ready to insert it into a blasting cap. Cut off an inch or two to insure a dry end.

(59) DON'T cut fuse on a slant. Cut it square across with a clean, sharp blade. Seat the fuse lightly against the cap charge and avoid twisting after it is in place.

(60) DON'T crimp blasting caps to fuse with a knife or with the teeth. Use a standard cap crimper and make sure that the cap is securely fastened to the fuse.

(61) DON'T use fuse and blasting caps in wet work without having a thoroughly waterproof joint between the fuse and cap.

(62) DON'T kink fuse in making up primers or in tamping a charge.

(63) DON'T hold the primer cartridge in the hand when lighting fuse.

(64) DON'T light fuse in any bore hole until the holes contain sufficient stemming to protect explosives from sparks from the end spit of fuse or a flying match head.

(65) DON'T try to light fuse with burning paper, other flammable refuse, or improvised torches.

(66) DON'T light fuse near blasting caps or any explosives, other than those being used in the blast.

Destruction of Explosives*

Explosives to be destroyed may be fresh material from damaged packages, or material that has deteriorated either from natural aging or from improper storage to the point where it is unfit for use.

Deteriorated explosives may be more dangerous to handle than explosives in good condition. When there is any question about the safety of the undertaking, a representative of the manufacturer of the particular lot of explosives should be consulted, or a request for assistance should be made to an authorized representative of the Bureau of Mines or to some one else known to have had the necessary experience. This is especially true if large quantities of explosives must be destroyed.

Most explosives, except detonators, are best destroyed by burning. The hazard of an explosion is always present, even under the most favorable conditions, so it is of prime importance to select a site where no damage will be done, either to persons or property, if the explosives detonate. This means a safe distance from any structure, railroad, or highway and from any place where one or more persons may even accidentally be exposed to danger, including that from flying fragments.

During the destruction of any type of explosives the possibility of preignition should be prevented by eliminating smoking and open lights.

Only one type of explosive should be destroyed at a time, and the utmost care should be taken to see that no detonators are accidentally included in explosives to be destroyed by burning.

*U.S. Bureau of Mines Circular 7335.

High explosives should never be burned in cases or in deep piles. Dynamites, especially permissible gelatins, become increasingly sensitive when overheated before ignition. Quantities of dynamite to be burned should not exceed 100 pounds of regular dynamite or 10 pounds of permissible gelatin. Local conditions may limit destruction to much smaller amounts, but when more than these maximum quantities must be destroyed, a new space should be selected for each lot, as it is not safe to place explosives on ground heated by the preceding burning.

No attempt should be made to return to the site as long as any flame or smoke can be observed.

As soon as all dynamite has been burned, it is believed to be good practice to plow the ground, as the residue may contain salts said to be attractive to livestock, which if eaten can produce serious results.

Dynamite. When properly stored, dynamite should remain in good condition for a long time, but it may, and usually does deteriorate rapidly if improperly stored or handled. The most common signs of deterioration are discoloration, leakiness, hardness or excessive softness, or the formation of crystals on the outside of the wrapper. Frequently a combination of two or more of these signs can be noted.

Many persons believe that the crystals are nitroglycerin and that they are especially dangerous; while these are not facts, because the crystals actually are salts that have exuded through the wrapper, whereas nitroglycerin is an oily liquid at normal temperatures, their presence on the outside of the wrapper or on the container shows that the dynamite has deteriorated to some degree.

Care should be exercised in handling deteriorated explosives, whether loose or in containers. Most persons experience undesirable effects, especially headaches of varying degrees of severity, by absorption of nitroglycerin through the skin when handling leaky or loose dynamite, and some persons are so sensitive as to have headaches after working over loose dynamite for only a short time, even without touching it. Therefore, if leaky or loose dynamite must be handled, gloves should be worn and then burned as often as they become impregnated with nitroglycerin.

Some dynamites are rather difficult to ignite, especially when wet, so it is best to prepare a bed of dry, combustible material, such as excelsior, wood shavings, or sawdust; to maintain combustion it is sometimes necessary before igniting the pile or bed to pour a little kerosene over the dynamite and the fuel bed. The area of the bed should be such that the dynamite can be arranged in a single layer if sticks or part sticks are being destroyed, or arranged not to exceed 2 inches in thickness if loose dynamite is to be burned. The bed should be long and narrow rather than square or circular.

It is often recommended that each stick of dynamite be slit and the loose material scattered on the fuel bed, but considering the extra hazard to the operator, it seems preferable to deposit whole sticks of the more common sizes on the fuel bed as carefully as possible without slitting. If the cartridges are of large diameter, such as those often used in quarry blasting, the loose material should be spread; "free running" (loose) dynamite may be so spread, but never in a thickness exceeding 2 inches.

When the bed has been formed and the dynamite deposited on it, a train of paper or similar readily ignitable material should be laid to it, preferably on the downwind side, and the explosives ignited thus. The train should be long enough to permit the operator to reach a safe place.

Dynamite should ordinarily burn quietly, with a bluish flame. If solid pieces remain, as sometimes happens, especially if the dynamite is wet, it is dangerous to poke about the debris or attempt to handle the pieces for reburning until it is certain that they are cool. The containers should be burned separately.

Detonators. Blasting caps, electric blasting caps and delay electric blasting caps which have so deteriorated from age or improper storage that they are unfit for use should be destroyed. These devices should also be destroyed if they have ever been under water as, for example, during a flood, regardless of whether or not they have been subsequently dried out. In some cases, the shells of caps that have been wet and then dried will show signs of corrosion. Such caps may be very dangerous to handle, and it is recommended that they not be disturbed until a representative of the manufacturer has had an opportunity to pass on them. The method most generally used for destroying detonators is to explode them under some confinement as described below. *Detonators should not be thrown into small bodies of water* such as rivers, creeks, ponds, or wells.

If possible, it is advisable to explode ordinary (fuse) blasting caps in the original container with the cover removed. Otherwise, they should be prepared for blasting as follows: They should be placed in a small box or bag; a hole should be dug in the ground, preferably in dry sand, at least 1 foot deep; the container should be placed in the bottom of the hole and primed with one cartridge of dynamite and a good electric blasting cap or ordinary cap and fuse; the caps and the primed cartridge should be carefully covered with paper and then with dry sand or fine dirt and fired from a safe distance. It is recommended that not more than 100 caps be destroyed at a time and that the ground around the shots be thoroughly examined after the shot to make certain that no unexploded caps remain. The same hole should not be used for successive shots.

To destroy electric blasting caps or delay electric blasting caps, it is necessary first to cut the wires off about 1 inch from the top of the cap, preferably with a pair of tin snips. No attempt should be made to cut the wires from more than one cap at a time. Not more than 100 caps should be placed in a box or paper bag, primed with a cartridge of dynamite and a good electric blasting cap, buried under paper and sand or dirt, and exploded as described above. The same precautions mentioned above should be observed.

Blasting caps should never be destroyed by placing them in a hole which is to be shot, especially by dropping them into well drill holes. Many bad accidents have occurred in this way.

Electric Squibs and Delay Electric Squibs. These devices should be destroyed by the same procedure as used for electric blasting caps.

Safety Fuse. This material may be disposed of very satisfactorily by burning in a bonfire.

Primacord. The preferred method of destroying Primacord is by burning. It should not be burned on the spool but should be strung out in parallel lines one-half inch or more apart on paper or dry straw.

Removal of Nitroglycerin from Magazine Floors. Contaminated floors should be scrubbed well with a stiff broom, hard brush or mop, using an ample volume of a solution in the proportion of $1\frac{1}{2}$ quarts of water, $3\frac{1}{2}$ quarts of denatured alcohol, 1 quart of acetone, and 1 pound of sodium sulfide (60 percent commercial). The liquid should be used freely to decompose the nitroglycerin thoroughly. If the magazine floor is covered with "Ruberoid" or any material impervious to nitroglycerin, this portion of the floor should be swept thoroughly with dry sawdust and the sweepings taken to a safe distance from the magazine and destroyed by burning.

It was formerly considered good practice to cover explosives magazine floors with rubber or similar material, but the best practice now is generally believed to be a tight, smooth floor, free from exposed nails or other metal fasten-

ings, so as to eliminate holes and ragged edges that sooner or later result from excessive wear. Floor coverings often hide floor stains and, when worn, render proper cleaning difficult or impossible unless the whole covering is removed. With the tight wood floor now considered best, cleaning is more certain and, if necessary, parts of the floor can be carefully removed and replaced.

Conclusion

Manufacturers frequently employ other methods of destroying explosives and blasting supplies than those described above, but when employed these should be used only under direction of the manufacturer's representative.

As indicated previously, the information given here is not for the purpose of encouraging persons unfamiliar with the proper methods of destroying explosives to undertake such work unassisted. It is realized however that expert advice and assistance are not always available and it is believed that destruction of unwanted explosives can be accomplished with less hazard by following the foregoing suggestions (see also Explosives; Explosives, High; Explosives, Low in Sections 6 and 12).

Appendix 2

TANKS FOR STORAGE OF FLAMMABLE LIQUIDS

The contents of storage tanks have played a part in some very serious fires. There are three types of tank storage (a) underground; (b) above ground, outside building; (c) above ground, inside building.

Underground Tanks

Plans for underground storage tanks designed for flammable liquids should be discussed with local building groups, national regulatory groups, and an insurance carrier. The following is a check list of items to be considered:

(a) The optimum size and position of the buried tank, e.g., whether it shall be vertical or horizontal.

(b) Location of the tank with respect to the terrain in the general area which is to be considered, the location of buildings, cellars, pits, etc., in the same area. Also the possibility of corrosion is an important factor.

(c) Adequate anchorage for the tank, keeping in mind the possibility of flood, heavy rainfall, or other events and what effect these might have on the installation.

(d) Sufficient covering for the tank; this will

improve the safety of the installation and help with the problem of anchorage.

(e) In selecting a location in which to bury a tank, cognizance must be taken of the possibilities for corrosion, to prolong the life of the tank. It is important to determine if the region chosen is reasonably free of corrosive effluents from nearby plants, corrosive cinder-fill, or possibly corrosive ground water. If corrosion is considered likely, tanks should be painted with at least one coat of red lead in linseed oil primer and then one coat of asphalt or coal tar base paint. Other formulas may be equally effective.

(f) It must be remembered that the equivalence of a below ground location may be obtained with a partially buried tank. In such a situation the exposed part of the tank may be covered with a dirt fill and a concrete cap as well as a concrete retaining wall for anchorage.

Above Ground Tanks

Above ground tanks can be a source of extreme danger, if not planned carefully. For instance, rupture of an above ground tank or a leak in such a tank at a point below the liquid level may very easily lead to a serious fire. Therefore, such tanks should be located on ground sloping away from main buildings and plant utility installations. Protection against fire spread is provided on level ground by tank spacing, proper drainage facilities and adequate dikes. In hilly terrain provision must be made for safe drainage past installations at lower levels. The following considerations must be kept in mind when planning a flammable-liquid storage tank above ground:

(a) The possibility of damage to nearby buildings or tanks.

(b) The amount of flammable material contained in the unit under consideration.

(c) The values involved in the unit under consideration and those adjoining.

(d) The burning and ignition characteristics of the materials involved.

(e) The provision of adequate room for fire-fighting operations and access for fire-fighting equipment.

(f) Dike capacity of such an installation: In the ideal case the dike capacity should equal the total tank capacity inside the dike area. Where the flammable liquid involved is such as may be subject to boilover, i.e., fuel oil, etc., the diking capacity must be 1.5 times the total liquid capacity of the tanks in the dike area.

(g) Adequate drainage arrangements must be made to prevent the accumulation of water in dike areas.

Tank Construction

The following points must be considered in making sure that tank construction is adequate:

(a) In the case of horizontal steel tanks, it should have an Underwriters' label indicating that it meets definite construction standards.

(b) All horizontal tanks must be given a hydrostatic strength and leakage test before installation is complete.

(c) In the case of above ground tanks, they must be constructed upon fire-resisting supports such as brick or reinforced concrete.

(d) Vertical steel tanks must be constructed in accordance with the American Petroleum Institute's specification for standard tanks.

(e) A leakage test is recommended for vertical tanks.

(f) Provision must be made for adequate venting facilities and flame arrestors.

Tank Connections and Fittings

(a) Connections to horizontal tanks

(b) The use of steel flanges or wrought iron couplings to make pipe connections

(c) Provision for manholes

(d) Provision for shut-off valves

(e) The use of heating equipment

(f) The use of electrical equipment with the attendant necessities for prevention of sparks, i.e., grounding, electrical bonding, etc.

Fire Protection

The protection required for tanks containing flammable liquids is determined by the tank size, type, location, exposure to or from buildings or other tanks, value of contents, flash point of the liquid stored, and the probability of the interruption to production by loss of tank contents.

Conditioning and Use of Old Tanks

Failure to take proper precautions during the cleaning and repair of used tanks can result in serious explosion hazards. To avoid such hazards the following precautions should be observed.

(a) Tanks which have contained flammable liquids should be purged with steam or filled with water and allowed to drain, or combinations of these methods may be employed.

(b) Use a portable flammable vapor indicator to determine that vapor concentrations inside such tanks have been reduced to safe levels.

(c) Remove all remaining scale and sludge from inside such tanks with non-ferrous scrapers. The use of detergents to clean such tanks, while helpful in actually removing the residues, should be undertaken with care, because for example, caustic solutions promote the decomposition of nitrocellulose residues, and the reaction thus started will move with explosive violence under the proper conditions.

(d) Take the additional precaution of filling the tank with water or an inert gas such as carbon dioxide, if cutting or welding torches must be used on it.

(e) Proper supervision of the workers and sufficient ventilation is essential inside a tank.

(f) While such reconditioning or repair is in progress, it is wise to move any readily flammable materials from the neighborhood.

Appendix 3

INDUSTRIAL DISASTER CONTROL

The function of disaster control in industry is taking on increased importance. Such factors as industrial plant decentralization and the advent of newer and more powerful weapons of war have made it necessary for a plant to be as self-sufficient as possible, particularly during times of emergency. All that disaster planning involves is an extension of routine fire and emergency programs to cope with disaster.

The first requirement of a disaster control program is common sense. No highly specialized knowledge or backlog of experience is necessary. The essence of disaster planning is thorough analysis of the existing plant facilities, with emphasis on emergency facilities, and observance of how they may be affected by a disaster. Unless a plant is in the zone of total destruction of an atom bomb, or some similar destructive agency, an adequate disaster control plan is important. Although it is possible to protect against an atomic detonation, the protection necessary is so elaborate and so costly that it may not be economically feasible.

Model Disaster Control Program

Make a complete survey of existing functions and facilities, including possible emergency facilities. It is necessary to determine what is done under routine conditions and what is to be done when these conditions are disrupted, e.g., during the emergencies created by fires, hurricanes, earthquakes, release of toxic materials, bombs, etc.

Determine the key production points in the plant—the departments and processes which, if interrupted, would result in a complete production stoppage. Such key locations should have the best possible protection. A plan, to be successful, must be sufficiently broad in scope to be adaptable to any emergency. Specific operational details, of course, need not be included, since these will depend upon the actual nature of the emergency. The following factors, however, must be taken into consideration: executive authority and duties; handling the alarm; emergency headquarters; operations of the police force; operations of maintenance force; operations of fire department; operations of medical department; operation of public information and industrial relations group; operation of monitoring group for toxic inhalants and radioactivity.

Executive Authority and Duties

A person responsible for coordinating the operation of all emergency groups must be designated in advance to serve as plant emergency director. Provision must be made for succession of command. There must be no confusion as to who is in charge. The plant manager is usually a good choice during office hours, and the general shift foreman for nights and holidays. It is recognized that someone with authority must be available during emergencies.

The plant emergency director shall coordinate the efforts of the emergency groups which shall include: determination of seriousness of emergency condition; sounding general alarm throughout plant; directing operation of all plant forces from emergency headquarters; promptly summoning any outside aid which may be deemed necessary. (In calling outside assistance, the plant emergency director may secure excellent advice from the heads of his own forces, e.g., the plant fire chief might be consulted on the necessity of summoning outside fire assistance). Aid of various types should be promptly summoned. If necessary, it can be turned back if not needed upon arrival. Delay can be disastrous; inform the press as to requirements of public interest in an attempt to avoid promoting misinformation; attempt to maintain a log of events to serve as a record.

Handling the Alarm

Upon discovering an emergency, the individual employee shall:

For a fire or explosion, pull nearest fire alarm box, or telephone fire department. (The importance of indoctrinating all personnel in advance as to how to transmit a fire alarm *cannot be overemphasized*.)

For other emergencies, telephone for ambulance, police patrol, emergency repair, medical, safety or decontamination aid as may be necessary.

For notification, the plant emergency director should be notified immediately after completing the call for assistance. The supervisor shall, immediately upon learning of an emergency in his area, proceed to the scene and see that the appropriate notifications as required above have been made. He will direct the forces at hand in initiating appropriate action and then report to the plant emergency director. An emergency procedure should be established for the telephone switchboard operators to screen telephone calls, restricting the use of lines to official calls.

When an emergency is reported, a general alarm should be sounded alerting the entire plant; the method of sounding the alarm depends on the equipment available at the plant. In such an eventuality, all personnel except those initiating on-the-spot action or assigned to some specific function, should evacuate their respective buildings, closing all doors but not locking them, and assemble in front of their buildings to await further orders. All radioactive, highly toxic or corrosive materials should be locked up.

Emergency Headquarters

Primary and secondary emergency plant headquarters from which operations will be directed should be established in advance. The communications room in the police or guard headquarters is usually a good location for the primary headquarters. A secondary headquarters is necessary in the event that the primary location is involved in a disaster; if both the primary and secondary locations are knocked out, an alternate will have to be set up on the spot by the executive authority. Depending upon the plant, the secondary or alternate headquarters might be established at the fire headquarters, plant management office, safety director's office, etc., or in the field. A portable sign reading "Emergency Headquarters" should be provided at the appropriate location. (Red lettering on a white background is recommended.) Red electric lanterns should be used at night in emergencies to designate the headquarters. Both primary and secondary headquarters should be equipped with: telephone facilities, both interplant and outside; radio (if available); map showing fire mains, fire alarm boxes, fire hydrants, general piping layouts and similar pertinent data; telephone numbers (plant and home) and home addresses of key personnel, together with information as to the time and route required for each person to report to the plant in response to an emergency call; same information as above for all other personnel; specific information as to how to contact and/or obtain the following services: fire department, police department, medical services,

ambulances, public utilities, and other public services; the means of contacting nearby military establishments if assistance is needed, e.g., air transportation.

Operation of Police Force

The chief of the police force is responsible for maintaining the material and data required at primary and secondary emergency headquarters. In case of emergency, the senior police officer on the premises shall: maintain liaison with the plant emergency director; alert guards on duty; maintain order at all scenes of emergency; assist plant emergency director in placing calls for outside assistance and recalling off-duty personnel; escort emergency vehicles and personnel returning to the plant to appropriate locations; if equipped with radio, dispatch radio patrol cars to locations determined by the plant emergency director in order to maintain communications, particularly in the event of telephone failure (if both telephone and radio fail, police will have to handle communications within plant and with outside agencies by messenger); stand by to evacuate entire plant if necessary; if the emergency requires outside personnel, the police should guide them to the emergency headquarters where the director will direct their admittance and disposition.

Operation of Maintenance Staff

The maintenance staff initiates emergency action if large scale plant emergency groups are not maintained; e.g., in the event a plant does not have a full-time fire department, the maintenance personnel organized and trained as a fire brigade should immediately combat a fire and attempt to prevent it from spreading pending arrival of the outside fire department.

Maintenance personnel should be organized by the chief of maintenance into appropriate squads to handle electrical repairs, emergency lighting, rescue, maintenance of power and water supply, salvage operations to minimize water damage, transportation, etc. The senior maintenance man should maintain liaison with the plant emergency director and direct the operations of his various divisions. Maintenance personnel should be kept informed of hazardous materials in the plant and how to cope with them.

Operation of Fire Department

The cooperation between the plant and municipal fire departments should be outlined in detail on a mutual aid agreement plan. Regardless of how an alarm is received, the fire department immediately responds in accordance with the prearranged plan. Upon arrival at the scene, the senior fire officer present is in charge of all fire department apparatus and personnel. He maintains liaison with the plant emergency director and with the monitors as concerns radiological and toxic chemicals safety.

An extra supply of rubber boots, clothing, helmets, and fire hose is desirable where regular equipment may become contaminated.

Fire department personnel should be kept informed of all hazardous materials in a plant and how to cope with same.

Operation of Medical Department

The chief of the medical department shall be responsible for first aid training and for the procurement, maintenance, and availability of appropriate medical supplies.

The senior medical man present at the time of an emergency shall: maintain liaison with plant emergency director; care for injured; transport casualties to municipal hospitals as may be necessary; arrange for appropriate ambulances and for other vehicles to serve as ambulances in cases of catastrophe; when casualties are sent to outside hospitals, maintain liaison with the hospital personnel as to the nature of injuries and whether or not radioactivity is a factor; coordinate efforts of Red Cross units which may be called in.

Depending upon the outside hospital facilities available, the chief of the medical department may want to have mimeographed copies of medical data available relative to contamination control for distribution to municipal hospital medical personnel in times of emergencies. Immediately after a disaster the medical division, in cooperation with industrial hygiene personnel, shall evaluate the degree of severity of exposures. Special clinical and laboratory examinations of blood and urine should then be performed on all individuals who may have received significant exposure. Follow-up examinations will have to be made at the discretion of the medical department.

Operation of Public Information and Industrial Relations Group

Public information releases should generally be prepared by the plant emergency director or some other authoritative individual designated in advance. It is recommended that these be in writing, if time permits, to prevent misinterpretation. A list of editors and radio newsmen should be prepared well in advance, so that these individuals can be contacted as quickly as possible and informed that a written release statement is being prepared with authoritative information relative to the emergency.

The senior industrial relations representative, the safety director, and the senior insurance department representative shall maintain liaison with the plant emergency director. The insurance representative shall, in addition: obtain names, addresses, and severity of injury to all casualties; when ordered by plant emergency director, notify the families of the casualties as to the nature of injury and location of injured; prepare necessary reports for insurance carrier.

Operations of Monitoring Group for Toxic Inhalants and Radioactivity

When an emergency occurs this group shall assemble at its base and immediately contact the plant emergency director, who shall instruct the monitoring group leader as to the areas to be monitored and the contaminants or radiations to be measured. While maintaining constant contact with the plant emergency director both for direction and to report monitoring results, this group has the following responsibilities:

In accordance with a prearranged plan to have personnel evacuated from contaminated or high radiation level areas; to assist in the control of exposures to radiation and toxic materials by firemen or other emergency workers; to monitor possible contaminated personnel in order to control the spreading of contamination as well as to be able to advise personal decontamination procedures; to help in decontamination of any contaminated areas; to make checks upon possible environmental contamination, e.g., fall-out, radiation, contamination of water, air or food, livestock, houses, etc. *Since hastily given numbers are often misunderstood, it is a prime responsibility of a monitor to see to it that his data go only to the plant emergency director to use as he sees fit.*

While the foregoing discussion does not provide a complete plan for controlling plant disasters, it is hoped that it will serve as a guide and provide stimulus for thought and planning relative to continuity of production and safeguarding of personnel during actual emergencies.

BIBLIOGRAPHY

Hartmann, Irving, "The Explosibility of Titanium, Zirconium, Thorium, Uranium and Their Hydrides," Bureau of Mines Report No. 3202, NYO-1562.

Bararas, G. D., et al., "Zirconium Metal Powder Precautionary Handling Suggestions," J. Am. Chem. Soc. 70, 877(1948).

Birchall, James D., "The Classification of Fire Hazards and Extinction Methods," London, Ernest Benn Ltd., 1954.

"Fire and Explosion Hazards of Thermal Insecticidal Fogging," National Board Fire Underwriters Research Report No. 9.

"Safety in the Mining Industry," BM 481, Bureau of Mines, Washington, D.C.

"Investigations on the Explosibility of Ammonium Nitrate," Bureau of Mines Report of Investigations 4994, August 1953.

"Potential Hazards in Molten Salt Baths for Heat Treatment of Metals," National Board Fire Underwriters Research Report No. 2, 1954.

"Fire Hazards and Safeguards for Metal Working Industries," Technical Survey No. 2, National Board of Fire Underwriters, 1954.

"Flammable Liquid Pumping Equipment," Factory Mutual Bulletin of Loss Prevention No. 13.24, Associated Factory Mutual Fire Insurance Companies, December 1949.

"Explosives Drivers' Handbook," OP 2239, Department of the Navy, Bureau of Ordnance, March 29, 1956.

"Storage, Handling and Use of Flammable Liquids," National Board Fire Underwriters No. 30, July 1956.

"Dangerous Chemicals Code," 1951 Edition, Bureau of Fire Prevention, City of Los Angeles Fire Department, Parker and Co., Los Angeles, Calif.

Marshall Sittig, "Sodium, Its Manufacture, Properties, Uses," New York, Reinhold Publishing Corp., 1956.

Robinson, Clark Shove, "Explosions, Their Anatomy and Destructiveness," New York, McGraw-Hill Book Co., 1944.

"Blasters' Handbook," 12th edition, E.I. du Pont de Nemours and Co., Inc., Wilmington, Delaware.

"Handbook of Chemistry and Physics," 47th edition, Cleveland, Chemical Rubber Publishing Co.

"Storage Tanks for Flammable Liquids," Factory Mutual Bulletin of Loss Prevention No. 13.23, Associated Factory Mutual Fire Insurance Companies.

"Flammable Liquid Safeguards," Factory Mutual Bulletin of Loss Prevention No. 13.20, Associated Factory Mutual Fire Insurance Companies, December 1940.

"Ventilation and Operation of Open Surface Tanks," American Standard Safety Code No. Z9.1-1951, American Standards Association, Inc., New York.

Babcock, Chester I., "Ammonium Nitrate—Behavior in Fires, NFPA Quarterly (January 1960).

"Standard for Magnesium," NFPA No. 48 (June 1959).

"Guide for Zirconium," NFPA No. 482M (June 1959).

"Standard for Titanium," NFPA No. 481 (June 1959).

Guise, A. B., "The Chemical Aspects of Fire Extinguishment," NFPA Quarterly (April 1960).

Nickerson, M. H., "Palletized Storage Fire Problems," NFPA Quarterly (July 1957).

Betz, G. M., "Organic Peroxides Storage and Handling," NFPA Quarterly (July 1962).

"Handling Cryogenic Fluids," NFPA Quarterly (July 1960).

"Fire Protection Guide on Hazardous Materials," 1st edition, NFPA, 1966.

Le Vine, R. Y., "Electrical Equipment in Chemical Atmospheres," NFPA Quarterly (April 1964).

Bahme, C. W., "Our Protection for Chemicals," NFPA International, 1961.

"National Fire Codes," NFPA International, 60 Batterymarch Street, Boston, Mass., Vols. 1 thru 10 for 1966–1967.

RADIOLOGICAL ENVIRONMENTAL POLLUTION CONTROL

Joseph J. Fitzgerald

*President and Technical Director of Cambridge Nuclear and
Sanders Nuclear Corporations*

and

N. Irving Sax

*Director, Radiological Sciences Laboratory
New York State Health Department
Albany, New York*

*Adjunct Professor, Division of Bioenvironmental Engineering
School of Engineering, Rensselaer Polytechnic Institute
Troy, New York
Consultant on Pollution Control*

The air we breathe, the water we drink and the food we eat have always contained radioactive materials. Man has lived and society has flourished in this background of activity. The discovery of the fission process and its subsequent harnessing via the nuclear reactor for power and research purposes as well as weapons have made available an additional large quantity of artificially produced radioactive materials and the potential for producing enormously larger amounts.

It is universally agreed that the interaction of penetrating radiation and matter is such that in the case of living tissue, particularly in as complex an organism as man, the effects of exposure are overwhelmingly destructive. Furthermore, the amount of destruction appears to be proportional to the dose or quantity of radiant energy absorbed.

While man appears willing to accept the possibly deleterious effects from exposure to natural activities (of course, he has little practical alternative), he is usually violently opposed to exposure, particularly involuntary exposure, to artificially produced radioactivity. In fact, the violence of opposition is often out of all proportion to present day concepts of "dose-damage" relationships.

It is, therefore, necessary from the standpoint of public health and public relations that our large and rapidly increasing nuclear technology be safety-engineered to hold the resulting environmental pollution to a practical minimum. This requires continuous or continual surveillance over radioactive materials discharged to the environment and the application of effective countermeasures to assure compliance with "maximum permissible concentrations." Facility inspections by competent personnel should be made on a regular schedule. An early warning system should be developed to notify a qualified public health agency of a possible emergency in the event of a malfunction so that intensive surveillance of the facility and population at risk might be instituted in time to intercept and test for contaminated food and water supplies which could if necessary be held up or diverted until safe foods are substituted.

SOURCES OF POLLUTION

The major sources of radiological environmental pollution are: (1) nuclear detonations, (2) nuclear fuels fabrication, (3) nuclear reactor operations, (4) nuclear fuels reprocessing, and (5) isotope production and uses.

By agreement, nuclear detonations in the United States and the USSR are presently limited to underground use and devices of relatively low yield. Non-agreement nations, however, have in recent years added somewhat to the air-borne and upper atmosphere inventory of fission products. The future use of nuclear explosives to shape harbors and to build canals (Plowshare) will undoubtedly be continued only if the release of activity can be adequately controlled. Recent experience has verified that the potential population dose from underground detonations is considerably less than from above ground tests of equal yield.

The particles of radioactive debris injected into the atmosphere from a nuclear explosion and which eventually fall to earth are called fallout. In a nuclear explosion, a great deal of heat is generated in a relatively small amount of material and in a very short time. Temperatures in the immediate vicinity of the bomb may reach millions of degrees, and a fireball is produced which contains very hot gases, melted and vaporized fission products, and environmental materials. As the fireball cools and expands upon its journey into the troposphere and stratosphere, the vaporized materials cool into particulates, and depending upon particle size and dispersion, this material and all the other particulates sucked up into the fireball begin to fall back to earth. The time for half of the debris to fall to earth is called the "half-residence" time. In the equatorial latitudes this is now considered to vary from approximately 1 year for debris injected into the lower levels to 5 years or more for debris injected into higher levels. During this half residence time the activity is dispersed in a band that circles the earth between 20 and 60° North or South latitude (depending on the point of release). Although this dispersion reduces the ground level concentration, the number of persons exposed is increased. Megaton bombs propel most of their radioactive debris into the stratosphere (the higher levels), whereas low yield explosions deposit most of their debris in the troposphere, as shown in Figure 1.

The radionuclides of major public health concern from environmental pollution due to fallout are ^{90}Sr, ^{137}Cs, and ^{131}I because fission weapon detonations yield relatively large quantities of these radionuclides, they are readily assimilated into the biosphere upon falling to earth, and each of them has a short, direct pathway to man. As a single example, consider fallout upon grazing lands, ingestion by cows with forage and subsequent rapid excretion into milk which quickly reaches and is ingested by people. Significant amounts of these nuclides can also find their way to man via fresh vegetables, grains and grain products, and even meat (Figures 2, 3, 4 and 5).

PATTERNS OF GLOBAL FALLOUT MOVEMENT

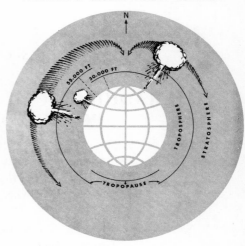

Figure 1. Patterns of Global Fallout Movement. The Arrows Represent the Ultimate Movement of Debris that Occurs Only After Constant Mixing by Random Eddies, and not as Simple One-way Circulation. (*Courtesy U. S. Atomic Energy Commission*)

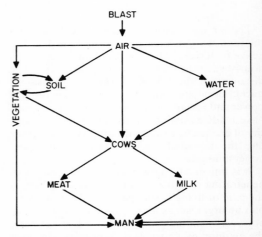

Figure 2. Pathways to Man of Radioactive Fallout from Nuclear Detonations.

^{131}I activity can yield a high dose to the thyroids of the milk drinking population for a short period of time after the detonation. On a long term basis, the dose from ingestion of foods contaminated with ^{90}Sr is of most concern. While ^{137}Cs is as long lived and as abundant as ^{90}Sr, its biological effect is substantially less because once ingested and absorbed, its biological half-life or body retention time is much shorter than

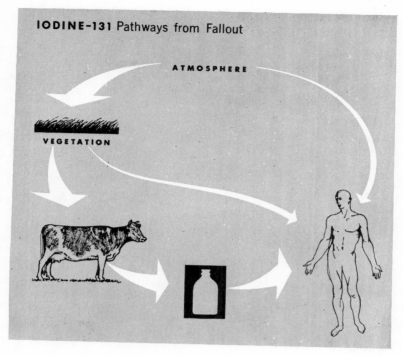

Figure 3. (*Courtesy U. S. Atomic Energy Commission*)

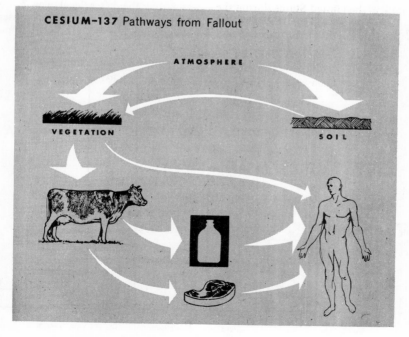

Figure 4. (*Courtesy U. S. Atomic Energy Commission*)

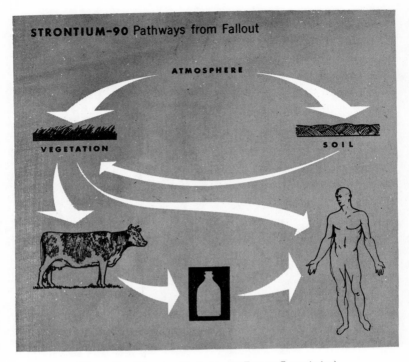

Figure 5. (*Courtesy U. S. Atomic Energy Commission*)

⁹⁰Sr. Some average activity levels (in picocuries per liter) for milk in the United States during the last quarter of 1966 were 11.7 of ^{90}Sr, 0 of ^{131}I and 20 of ^{137}Cs. In Canada the levels were 14.1 for ^{90}Sr and 35 for ^{137}Cs. In Chile, the ^{90}Sr level was 2 and the ^{137}Cs level was 7. More than 1000 picocuries/liter of ^{131}I was observed in the United States in 1962. Table 1 compares the in-

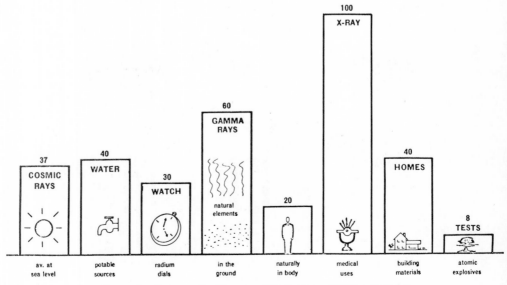

Figure 6. Comparative Sources of Radiation—Milliroentgen/yr.

TABLE 1.   Daily Intake of
Various Radionuclides during 1964

Radionuclides	Picocuries per Day	Source
^{40}K	2500	Natural
^{137}Cs	90	Nuclear tests
^{90}Sr	25	Nuclear tests
^{144}Ce	14	Nuclear tests
^{210}Pb	4	Natural
^{226}Ra	2	Natural
^{239}Pu	0.1	Nuclear tests

take of activity from natural sources and 1964 nuclear test sources, and a comparison of calculated annual doses from natural versus artificial sources is presented in Figure 6.

The Plowshare Program includes research, development and experimental testing of peaceful uses for nuclear explosives such as nuclear excavation technology which may bring us such boons as the creation of a sea level canal across the Central American Isthmus and improved management of natural resources, i.e., increasing gas well productivity and controlling subterranean water movement.

As indicated in Figure 7 a nuclear explosion (A or G) melts, vaporizes and fractures the adjacent rock, and sends out (B or H) a shock wave from the expanding cavity which is a result of the pressure of hot vapor. If the explosive is buried to the proper depth, the cavity (C) grows preferentially toward the surface (D,E). The explosion lifts most of the rock and dirt (F), some falling back inside, the rest outside the crater. If, however, the explosive is buried very deeply, then there is no crater, and as the cavity begins to cool (I), in most types of rock, fractured rock begins to fall back into it. This collapse continues upward leaving a column or chimney (J) of broken rock.

The chief public health consideration is leakage of radioactive materials from the blast area (see Figure 8). This hazard is controllable; in a soil medium which turns to slag, the particulate radioactive material is concentrated in the melted rock; cratering types of explosions do release some radioactivity to the surface and thus into the lower atmosphere, resulting in local fallout and some environmental contamination. Some idea of potential population exposure from this may be obtained from Figure 9. Note the rapid reduction in hazard with time due mainly to improved explosives design and emplacement techniques.

Nuclear Fuels

At this juncture it is convenient to discuss the environmental pollution problems associated with the production of nuclear fuels, the use of nuclear fuels in power and research reactors, and the reprocessing of spent nuclear fuel elements (Figure 10).

The major uranium ore mining areas in the United States are in northwestern New Mexico, central Wyoming, and the Colorado-Utah border region. At present, there are more than 20 mills that extract uranium from the ore. The milled material is refined to purity standards more characteristic of the pharmaceutical industry than of heavy chemical manufacture. The refined material is enriched in content of ^{235}U which has a high thermal neutron cross section for fission. Three gaseous diffusion plants in the United States separate ^{235}U from ^{238}U and are located at Oak Ridge, Tennessee, Paducah, Kentucky, and Portsmouth, Ohio.

After refining, the fuel may be converted from the hexafluoride of the separation process to a suitable form for fabrication such as uranium oxide pellets which are loaded into thin walled, cladded tubes of stainless steel or an alloy of zirconium.

Fission reactor operations produce amounts of mixed fission products proportional to the reactor energy generated. This accumulates in the fuel elements.[1] In reprocessing spent fuel elements, the accumulated mixed fission products are freed and separated from the unfissioned uranium and plutonium. Ultimately, the uraand plutonium are separated, and the enrichment process is repeated (Figure 11).[2] In the separation process, large quantities of byproduct radioactive waste accumulates and is usually stored in underground tanks.[3-5] However, this so-called waste contains substantial amounts of ^{90}Sr, ^{144}Ce, and ^{147}Pm, all of which qualify as potential heat sources for radioisotopic power generation. A fourth radioisotope ^{137}Cs is also present in large amounts and is a potential source of energy for process radiation purposes. Thus, today's waste material may be useful tomorrow.

In the future, more trans-plutonium elements such as ^{241}Am, ^{242}Cm, ^{244}Cm, and ^{252}Cf will be produced. Many uses will be found for these isotopes as heat, power, and irradiation sources for industry and medicine as discussed below.

The environmental pollution problems which emerge from the above technology are all concerned with radioactive contamination of the air and water and subsequent spread into the biosphere (Figure 11).

In the case of fuels production and fabrication technology, the public health problems are concerned with the environmental effects of disposal of by-products from the concentration of

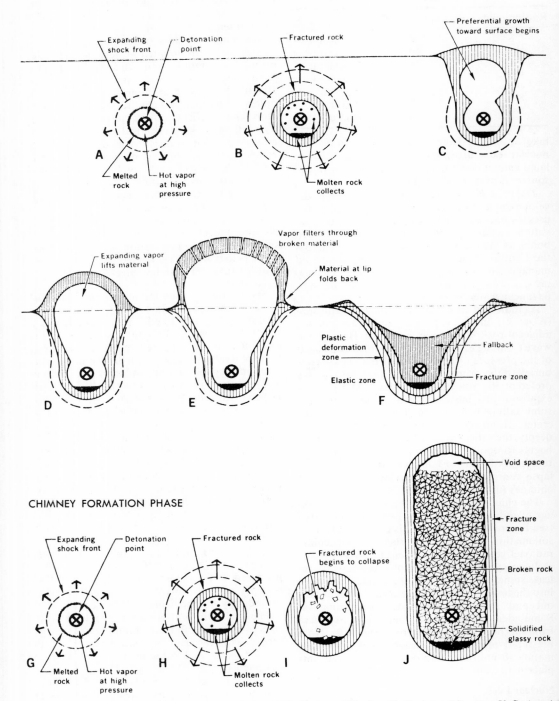

Figure 7. Crater and Chimney Formation Phases of Underground Nuclear Explosions. (*Courtesy U. S. Atomic Energy Commission*)

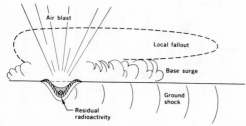

Several aspects of a nuclear cratering explosion, shown here, could be hazardous to man if not properly controlled.

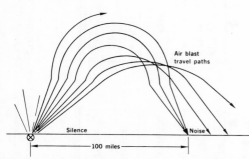

The air blast resulting from the release of energy to the atmosphere during a cratering explosion may be reflected back to earth many miles from the point of detonation.

Figure 8. Environmental Effects from Nuclear Cratering Explosions. Several Aspects of a Nuclear Cratering Explosion Shown Here, Could be Hazardous to Man if not Properly Controlled. (*Courtesy U. S. Atomic Energy Commission*)

natural radioactivities such as ores of uranium and thorium.

The reprocessing waste disposal problem was studied in great detail by USPHS[15] as it applied to potential contamination of a river in New Mexico by a refining operation. It appears from this study that the problems arise from piling up of the thousands of tons of uranium and thorium depleted ores close to but outside of the processing plants. This treated waste, in re-establishing its broken equilibria with respect to radioactive daughter products, can become a source of radio-toxic nuclides which by means of natural weathering find their way into the biosphere and then to surrounding populations.

This is not now a difficult control problem. It can be controlled by containment of wastes. Adequate surveillance techniques exist and are relatively easy to apply to check upon the efficiency of control measures.

Nuclear Power Reactors

Now let us consider the public health implications of the nuclear power reactor operations. Actually the demonstration by Enrico Fermi in Chicago at 3:20 p.m. on December 2, 1942 that

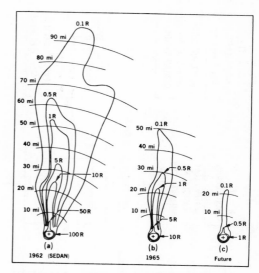

Figure 9. Computed Radiation Exposures from Underground Nuclear Explosions. The radioactivity escaping from a nuclear excavation depends both on the total amount produced by the explosion and the fraction that escapes into the atmosphere. The diagram on the left shows the pattern of the fallout that was observed in 1962 from the 100 kiloton SEDAN experiment. The center pattern indicates the fallout that might have been expected if SEDAN had been conducted with 1965 technology. Explosives development and improvements in emplacement techniques are expected to reduce the radioactivity released from nuclear excavations to that shown in the right-hand drawing. The decrease from the left-hand to the right-hand pattern is about 100-fold. The amount of radioactivity released is relatively independent of the size of the explosion. These fallout patterns, shown in terms of infinite dose, indicate the dose of external gamma radiation a person living outdoors for a lifetime might receive at various distances from the excavation. For comparison, the average external gamma dose a person in the United States receives from natural sources of radiation is about 0.1 r/yr. (*Courtesy U. S. Atomic Energy Commission*)

it was possible to control the enormous outpouring of energy from nuclear fission preceded the first nuclear detonation, named Trinity, at Alamogordo, New Mexico in 1945. Nuclear detonations compared to reactors are *uncontrolled* releases of fission energy much as the burning of a stream of H_2 in air is a controlled energy release versus the explosion of a volume of H_2 in air which is an uncontrolled release of energy. Per gram of H_2 burned or exploded, the same energy is produced. The difference lies in that the controlled burning, like the controlled release of nuclear energy, is sustained and rate

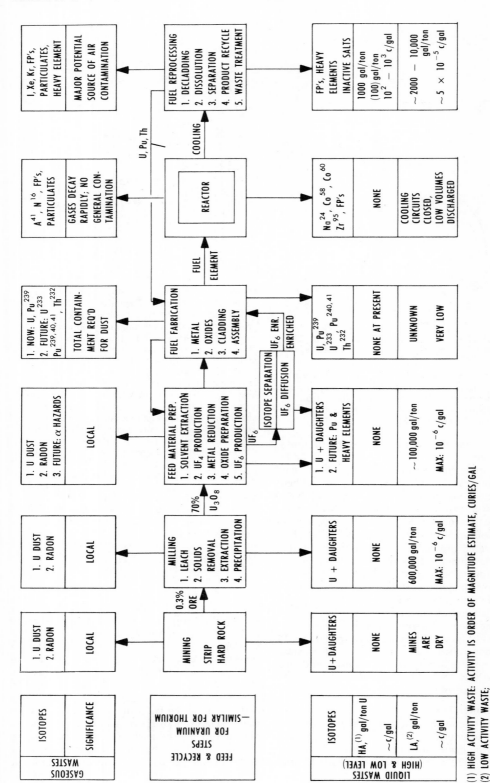

Figure 10. Schematic Flow Chart for the Origin of Radioactive Waste. (From Industrial Radioactive Waste Disposal Hearings Before the Special Committee on Radiation of the Joint Committee on Atomic Energy, Congress of the United States, 86th Congress, January 28 to February 3, 1959.

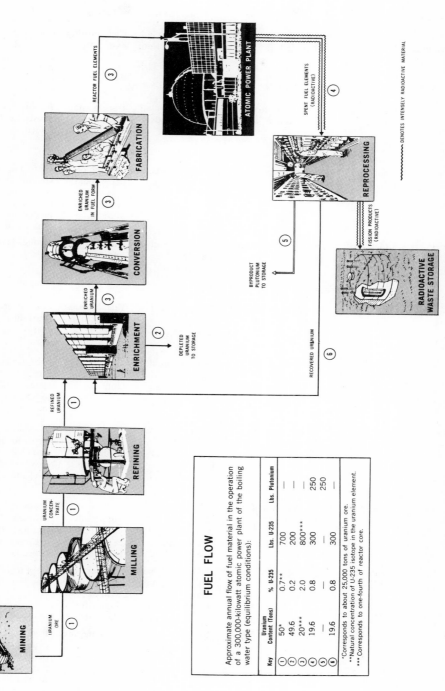

Figure 11. Steps in the Supply of Atomic Fuel. (*Courtesy U. S. Atomic Energy Commission*)

MINING

MILLING

REFINING

ENRICHMENT

CONVERSION

FABRICATION

ATOMIC POWER PLANT

REPROCESSING

RADIOACTIVE WASTE STORAGE

URANIUM ORE

URANIUM CONCEN-TRATE

REFINED URANIUM

ENRICHED URANIUM

ENRICHED URANIUM IN FUEL FORM

REACTOR FUEL ELEMENTS

SPENT FUEL ELEMENTS (RADIOACTIVE)

DEPLETED URANIUM TO STORAGE

BYPRODUCT PLUTONIUM TO STORAGE

RECOVERED URANIUM

FISSION PRODUCTS (RADIOACTIVE)

DENOTES INTENSELY RADIOACTIVE MATERIAL

FUEL FLOW

Approximate annual flow of fuel material in the operation of a 300,000-kilowatt atomic power plant of the boiling water type (equilibrium conditions):

Key	Uranium Content (Tons)	% U-235	Lbs. U-235	Lbs. Plutonium
(1)	50*	0.7**	700	—
(2)	49.6	0.2	200	—
(3)	20***	2.0	800***	—
(4)	19.6	0.8	300	250
(5)	—	—	—	250
(6)	19.6	0.8	300	—

*Corresponds to about 25,000 tons of uranium ore.
**Natural concentration of U-235 isotope in the uranium element.
*** Corresponds to one-fourth of reactor core.

TABLE 2. Behavior of Fission Products

Reactor	Date of Incident	Reactor Type	Reactor Fuel	Type of Incident	Cause of Incident
ORNL graphite reactor	1947–1948	Graphite moderated, air cooled	Aluminum-clad natural uranium slugs	Fuel burning and melting	Cladding failures resulted in oxidation of uranium, and, in some cases, the uranium oxide plugged the channel and caused fuel to melt
NRX	December 1952	D_2O moderated, water cooled	Aluminum-clad uranium rods	Fuel melting and U-H_2O reaction	Reduced coolant flow caused light water to be boiled out
NRX	July 1955	D_2O moderated, water cooled	Aluminum-clad uranium rods	Melting of experimental plutonium-aluminum alloy fuel rod	Water entered sheath and stream pressure lifted sheath out of contact with casting can; alloy melted and penetrated sheaths and reactor wall
GI	October 1956	Graphite moderated, air cooled	Aluminum-clad natural uranium rods	Fuel burning and melting	Reduced coolant flow
Windscale No. 1	October 1957	Graphite moderated, air cooled	Magnox-clad natural uranium rods	Fuel burning and melting	Local overheating in reactor during Wigner energy release
EL-3	April 1958	D_2O cooled and moderated	Aluminum-clad, hollow, uranium-2% Mo alloy	Fuel melting and oxidation	Vibration caused break in poorly supported fuel element, which dropped to bottom of tank and melted
NRU	May 1958	D_2O moderated and water cooled	Aluminum-clad uranium plates	Fuel burning following explosive failure of waterlogged fuel element	Waterlogged element burst, lodged in core, and broke during removal efforts; part of element burned
OMRE	October 1958	Organic cooled and moderated	Stainless steel-clad, UO_2-stainless steel dispersion plates	Melting of aluminum cladding of experimental fuel element	Aluminum fins on an experimental fuel assembly strained out particles in organic coolant and reduced flow
BORAX-IV	March 1959	Boiling water	UO_2-ThO_2 pellets, aluminum clad lead bonded	Deliberate operation with defective fuel elements	22 of 69 fuel elements had cladding defects
SRE	July 1959	Sodium cooled, graphite moderated	Stainless steel-clad uranium rods bonded with Na and K	Fuel melting	Reduced coolant flow caused by decomposition of "Tetralin" that leaked into system resulted in formation of iron-uranium eutectic and cladding failure

in Reported Reactor Fuel Failure Incidents

Extent of Contamination	Major Fission Products Released	Amounts of Fission Products Released (Ci)	Comments
UO_2 particles and accompanying fission products contaminated a large part of the restricted area adjacent to the stack	Mixed nonvolatile fission products in UO_2 particles found on ground; volatile fission products dispersed to atmosphere through stack	Not reported; can be calculated from irradiation times, estimated flux, and slug weight	Showed need of stack filters; particles found ranged from about 90–400 μ
Reactor building badly contaminated; active cloud carried detectable activity 1/4 mile; low level river contamination	Volatile and gaseous	10,000; long-lived fission products in coolant water; high intensity cloud of air-borne short-lived fission products	Non-typical operating conditions existed at time of accident due to tests with reactor
Reactor vessel contaminated with plutonium	Apparently not reported; can be assumed that mixed fission products, excluding xenon and krypton, accompanied the plutonium	Not reported	Only reported melt-down of a plutonium fuel element in an operating reactor, except for the experimental reactor Clementine
Blocked fuel channel held most of fission products; activity below maximum permissible limit outside station	Iodine and rare gases	20–50 (estimated)	Reactor remained at full power for about 20 min after trouble started
Widespread ^{131}I contamination of milk supply of large area	^{131}I, ^{132}Te, ^{137}Cs, ^{89}Sr, ^{90}Sr, ^{106}Ru, ^{144}Ce	^{131}I, 30,000; ^{132}Te, 12,000; ^{137}Cs, 600; ^{89}Sr, 80; ^{106}Ru, 80; ^{144}Ce, 80; ^{90}Sr, 2	Only reactor incident to date that caused significant damage to public
D_2O coolant and helium cover gas	Xenon and iodine isotopes	^{133}Xe, 100 Ci in helium cover gas; ^{131}I, ^{133}I, ^{131}Xe, ^{133}Xe, and ^{135}Xe, in D_2O, amounts not reported	1.6 kg of uranium (estimated) oxidized; showed that only rare gases transferred to cover gas
Reactor building badly contaminated; detectable contamination in about 100 acres of adjacent land	Mixed fission products found in surface contamination and stack filters; less ruthenium and $^{140}Ba + {}^{140}La$ than expected	Total amounts not measured; surface contamination up to 2.5 r/hr; air, 100 yards from building 1 hr after burning, 200,000 dis/min-m^3	This is the most serious incident reported involving a waterlogged fuel element
Mixed fission products in coolant; xenon in cover gas	Xe, Kr, Rb, I, Ba, La, and Te	1000 (total)	Showed need of more efficient filtration of coolant; coolant was purified and reused
Reactor water and air around pool	Fission gases, ^{138}Xe and ^{88}Kr	^{138}Xe, 4 Ci/min; ^{83}Kr, 0.7 Ci/min	This cannot be considered an accident
Sodium coolant only	$^{95}Zr + {}^{95}Nb$, ^{141}Ce, ^{137}Cs, $^{140}Ba + {}^{140}La$, ^{103}Ru, ^{131}I, ^{134}Cs	$^{95}Zr + {}^{95}Nb$, 306; ^{141}Ce, 131; $^{140}Ba + {}^{140}La$, 36 each; ^{137}Cs, 27.7; ^{103}Ru, 20.9; ^{131}I, 16.3; ^{134}Cs, 0.4	Since all fission products except rare gases remained in coolant, the fraction of fission products released could be accurately determined (0.3% of fuel element activity)

TABLE 2. (*Continued*)

Reactor	Date of Incident	Reactor Type	Reactor Fuel	Type of Incident	Cause of Incident
WTR	April 1960	Water cooled and moderated	Uranium-aluminum alloy in aluminum-clad plates	Fuel melting	Deliberate operation with one-third normal coolant flow or defective fuel or both
SL-1	January 1961	Boiling water	Uranium-aluminum alloy in aluminum-clad plates	Fuel melting and high pressure steam generation	Withdrawal of control rod beyond specified limit
ETR	December 1961	Water cooled and moderated	Uranium-aluminum alloy in aluminum-clad plates	Fuel melting	Reduced coolant flow (35% of normal) caused by "Lucite" sight glass
MTR	November 1962	Water cooled, beryllium moderated	Aluminum-clad uranium-aluminum alloy plates	Fuel melting	Gasket material from roof of seal tank restricted coolant flow
ORR	July 1963	Water cooled and moderated	Uranium-aluminum alloy in aluminum-clad plates	Fuel melting	Reduced coolant flow caused by neoprene gasket

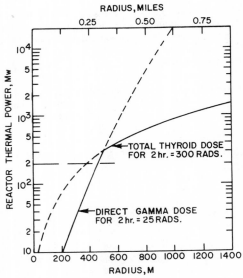

Figure 12. Exclusion Area—Radius Determination.

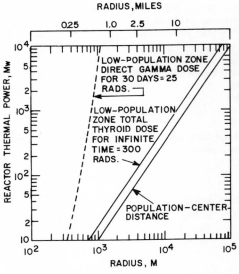

Figure 13. Population—Radius Determination.

TABLE 2. (*Continued*)

Extent of Contamination	Major Fission Products Released	Amounts of Fission Products Released (Ci)	Comments
Primary coolant and air in vicinity	Fission gases, xenon, and krypton	5000, mixed fission products to coolant, 261, airborne xenon and krypton just after accident; 800, total	A normal fuel element would probably not have failed
Building badly contaminated; extensive low level deposition of ^{131}I	^{131}I outside building; mixed fission products inside	^{131}I, 84 outside building; no report on mixed fission products	Reactor not operating at time of incident; only reactor incident involving fatalities
Primary coolant and atmosphere	Mixed fission products to coolant water; fission gases and daughters (particles) to stack	6.0, fission gases; 0.4, particulate; immediate release through stack; 42, mixed fission products to water; continuing release to water and atmosphere from unclad ^{235}U	Showed need of more precautions in order to keep extraneous materials out of core
Mixed fission products in coolant water, some airborne activity in building and at area monitors	^{131}I, ^{133}I, ^{135}I, ^{140}Ba, and ^{91}Sr found in coolant water	Release as factor increase in activity over normal; total, 13–16; ^{131}I and ^{133}I, 62; ^{135}I, 126; ^{140}Ba, 25; ^{91}Sr, 36	Loss of 0.7 g of ^{235}U from one fuel plate (see also ETR and ORR incidents)
Core and pool water; air in reactor building and some surface contamination from fission-gas daughters	^{133}I, ^{134}I, ^{135}I, ^{138}Cs, ^{88}Kr, ^{135}Xe, and ^{138}Xe found in coolant water; ^{88}Kr, ^{88}Rb, ^{138}Xe, and ^{138}Cs air-borne	1000–3000, in coolant water	See ETR, above; short operating time after restart (~ 30 min) resulted in nonequilibrium fission; product distribution

limited and thus is enormously more useful to mankind. However, not only is the same amount of energy available, but the same fission products or combustion products accumulate from the reaction.

A nuclear reactor is a system for the controlled production of nuclear energy from the fissioning of uranium atoms. A nuclear reactor is not a nuclear bomb. A bomb is designed to be an effective explosive. A nuclear reactor is of its essence designed to be safe. The first controlled fission reaction (or the birth of controlled reactor operations) was achieved in 1942.

In a nuclear detonation in the atmosphere particularly, the very large amount of energy, largely heat, is produced so suddenly that the bomb container is utterly destroyed, even vaporized, so that all of the energy and fission products are suddenly injected into a point in the atmosphere free to dissipate as they will. In a nuclear reactor, however, the rate of fissioning is carefully controlled so that the heat produced may be continuously carried away to do useful work while the fission products, both radioactive and stable, are kept behind for later use or disposal.

The reason it is possible to run a nuclear reactor with much less environmental pollution per unit of useful energy produced than a coal or oil fired energy producer, is that the energy producing reaction (fission) can be induced in a canned fuel by irradiating it with neutrons, a penetrating subatomic particle (see Section 5). Thus the fuel for a nuclear reactor can be sealed in a can and fission induced in such a manner that all that comes out of the can is heat or radiant energy. The heat is carried away by a circulating coolant which, because it comes into such intimate contact with the cans of fissioning fuel, is also irradiated by neutrons to become radioactive.

A perfectly engineered and operating nuclear reactor poses no public health problems to its community.

However, since we can only approach perfection, the history of reactor operation is not completely free from incidents of concern to the public health (see Table 2). Although the record of reactor-caused public health injury is outstandingly good in comparison with the power generation field as a whole, the continuing development of improved controls, reactor data processing systems, and automatic computer controlled reactor operating systems promises increased safety improvements.

To summarize (Table 3), public health problems from imperfect engineering or operation of nuclear reactors can come from leakage into the environment of either fission products out of the sealed containers or activated coolant from its closed system.

The potential health hazards of reactor malfunction can be radically reduced by careful siting. Since nuclear reactors are rapidly replacing conventional power generating equipment and since the demand for new power is itself rising rapidly, siting becomes a very important consideration.

Criteria have been set forth in the Federal Register 10-CFR-Part 100 to guide selection of a suitable reactor site.

The three important considerations in site selection are availibility of an exclusion area, a low population area, and a suitable population center distance. These parameters divide the area surrounding the reactor into three regions:

(1) *Exclusion Area*—The region immediately surrounding the facility. It must be under full control of the reactor licensee to allow rapid and effective evacuation of persons from the site in the event of an accident. The limiting dose to a person at the boundary of this area for a period of 2 hours following an incident is 25 rem to the whole body or 300 rem to the thyroid from radioactive iodine.

(2) *Low Population Area*—The region surrounding the exclusion area; it may contain residents. The total number and density of population should be such that evasive or protective measures can be effectually taken in the case of an accident. The total dose to an individual who remains at the outer boundary should not exceed 25 rem to the whole body or 300 rem to the thyroid.

(3) *Population Center Distance*—This is measured from the reactor to the nearest boundary of a populated center, i.e. one with more than about 25,000 residents. It should be at least four-thirds the distance from the reactor to the outer boundary of the low population area.

Simplified, general, and specific equations for calculating exclusion area and low population zone distances are given in Table 4.

To illustrate, assume the power level is 125 MW then the estimated low population zone distance for the meteorological conditions and leakage given (see Table 4) is

$$X_0 = 0.1 \times (125)^{2/3} = 2.5 \text{ miles}$$

Figures 12 and 13 show the exclusion area, low population zone, and population center

distances that are required for an appropriate site under 10-CFR-100 at a given reactor power level.[9] Comparisons of calculated site distances and the actual site distances (approved prior to the development of the site criteria) are shown in Table 5. This indicates that the distances permitted in earlier reactor site determinations are in general agreement with those based on the newer guidelines.

In practice, reactor sites are frequently chosen by the potential licensee from economic and other practical considerations rather than purely from the criteria set forth in 10-CFR-100. It then becomes necessary to evaluate and attempt to justify the site compared to the guides set forth in 10-CFR-100.

It should be noted that keeping the hazards from a reactor malfunction to a low level does not depend on siting alone but on the efficacy of the built-in mechanical and electronic protective features and a well-worked-out schedule of design and operations. Take containment for example. Containment vessels can be built to withstand major accidents without releasing to the environs more than 0.05–1 percent of the building air each 24 hours.[10] A summary of some of the specified versus measured leak rates from existing containment buildings is given in Table 6. Each facility requires individual treatment of its containment problems.

To reduce the possibility of releasing fission products to the environs, one should assess the degree of containment required; provide a method to measure the leak rate from the container, protect the integrity of the containment buildings from missiles generated by an accident, include an automatic or manual shutdown ventilation system, establish an effective warning and alarm system, consider partial containment, use water and foam sprays and stand-by filters within the building, and provide an isolation system for all containment penetrations which could provide a path for the release of fission products to the environment.

The reactor site selection policies of some other countries are: In Britain, a mathematical site rating plan is used, based on the product of population density and the square of air concentrations over distances of 1–12 miles from the reactor. For distances less than 1 mile, special considerations apply and exposure control action is taken to prevent the general population from experiencing exposures greater than 25 rem to the thyroid, 15 rem to the bone, and 25 rem to the whole body.

In Germany, environmental hazard analyses have been made in which reliance was placed on the integrity of the outer containment building. The limiting exposures were 25 rem to the whole

TABLE 3.   Potential Reactor Incidents and Methods of Prevention

Types of Incidents	Methods of Prevention
Directly Induced Nuclear Incidents	
Reactivity Incidents	
Start up	(1) Limit rod rates. (2) Use of adequate multiple channel instrumentation. (3) Provide period and/or flux level scram. (4) Avoid positive temperature coefficient. (5) Insert a large neutron source in the reactor.
Cold coolant	(1) Install temperature difference interlocks to prevent an inadvertent injection of cold coolant into the reactor core. (2) Design reactor such that sufficient volume for coolant mixing is provided to limit the rate at which reactivity can be introduced. (3) Provide means to keep the coolant in the isolated loop warm. (4) Make available a flux scram.
Rod jump	(1) Design such that maximum shock does not cause rod to jump and increase the reactivity by an appreciable amount.
Foreign material introduction	(1) Design reactor in manner that virtually prevents the introduction of foreign material or detects presence before a significant incident can be incurred.
Shock damage	(1) Design against a significant shock.
Indirect Nuclear Incident	
Mechanically induced incident	
Loss of coolant	(1) Design such that coolant loop rupture does not empty the reactor core of coolant. (2) Use an emergency injection system to replace coolant. (3) Design system with adequate safety margin. (4) Test pressure vessel and loops to detect any possible source of loss of coolant.
Loss of coolant flow	(1) Use multiple pumps and separate power supplies which are considered reliable. (2) Provide rapid means to detect loss of coolant flow and to scram in an adequate period. (3) Assure adequate thermal safety margin in design.
Control rod ejection	(1) Complete casualty analyses and strengthen areas where control rod ejection may be probable. (2) Protect control drive mechanism by an adequate shield. (3) Provide assurance that a single casualty cannot eject all safety rods.
Chemically induced incident	
Metal-water reaction	
Production of gases and explosion	

body and to critical organs within a period of 50 years.

In Canada, there is no general code of regulations. The location of a reactor is selected by the correlation of all safety aspects of its design and operation.

In Japan, the siting of reactors is based on containment and the exclusion area principle. Seismologic considerations constitute an additional problem which must be worked into the design of reactor structure and piping.

During 1966, nuclear reactor power plants became competitive with conventional power plants, and from then on the vast majority of new power plants were and will be nuclear. Figure 14 illustrates as of May 1966 the existing and planned atomic power plants for the United States. It is now obvious that both the number of planned future plants and the average power level per plant are increasing annually. Therefore, the quantity of radioactive waste materials generated, although already great, will be enormously greater in the future. Thus, while the radiological safety record has been excellent

TABLE 4.   Exclusion and Low Population Zone Formulas

Conditions	General Equations Distance (m)	Specific Equations Distance (m)	Distance (miles)
Exclusion area (2 hr exposure)			
Thyroid (300 rads)	$x_0 = \left[\dfrac{2.79 \times 10^8 P_0 \lambda_1}{v\pi C_y C_z}\right]^{\frac{1}{2-n}}$	$x_0 = 5.56 P_0^{2/3}$	$x_0 = 0.0035 P_0^{2/3}$
Whole body (25 rads)	$x_0 = \left[\dfrac{1.21 \times 10^7 P_0 \lambda_1}{v\pi C_y C_z}\right]^{\frac{1}{2-n}}$	$x_0 = 1.36 P_0^{2/3}$	$x_0 = 0.00085 P_0^{2/3}$
Low population zone (continuous exposure)			
Thyroid (300 rads)	$x_0 = \left[\dfrac{1.55 \times 10^{10} P_0 \lambda_1}{v\pi C_y C_z}\right]^{\frac{1}{2-n}}$	$x_0 = 160 P_0^{2/3}$	$x_0 = 0.1 P_0^{2/3}$
Whole body (25 rads)	$x_0 = \left[\dfrac{1.01 \times 10^8 P_0 \lambda_1}{v\pi C_y C_z}\right]$	$x_0 = 11.1 P_0^{2/3}$	$x_0 = 0.007 P_0^{2/3}$

P = Reactor power level in megawatts (thermal)
λ_1 = Leak rate from the building = 1.16×10^{-8} sec^{-1}
n = Stability parameter = 0.5
v = Cloud velocity = 1 m/sec
C_y = Diffusion coefficients in the y direction = 0.4 $m^{n/2}$
C_z = Diffusion coefficients in the z direction = 0.07 $m^{n/2}$
h = Height above ground = 0 m

TABLE 5.   Calculated and Actual Site Distances for Selected Reactors

Reactor	Thermal Power Level (MW)	Exclusion Area Radius (miles) Calculated	Actual	Low Population Zone Distance, Calculated (miles)	Population Center Distance (miles) Calculated	Actual
Dresden	630	0.50	0.50	7.4	9.9	14.0
Consolidated Edison	585	0.48	0.30	7.0	9.4	17.0
Yankee	485	0.42	0.50	6.3	8.4	21.0
PRDC	300	0.31	0.75	4.5	6.1	7.5
PWR	270	0.31	0.40	4.1	5.6	7.5
Consumers	240	0.30	0.50	3.9	5.2	135.0
Hallam	240	0.30	0.25	3.9	5.2	17.0
Pathfinder	203	0.29	0.50	3.4	4.6	3.5
PG&E	202	0.29	0.25	3.4	4.6	3.0
Philadelphia Electric	115	0.26	0.57	2.4	3.2	21.0
NASA	60	0.22	0.50	1.6	2.1	3.0
CVTR	60	0.22	0.50	1.6	2.1	25.0
Elk River	58	0.22	0.23	1.5	2.0	20.0
VBWR	50	0.21	0.40	1.4	1.9	15.0
Piqua	48	0.21	0.14	1.4	1.8	27.0

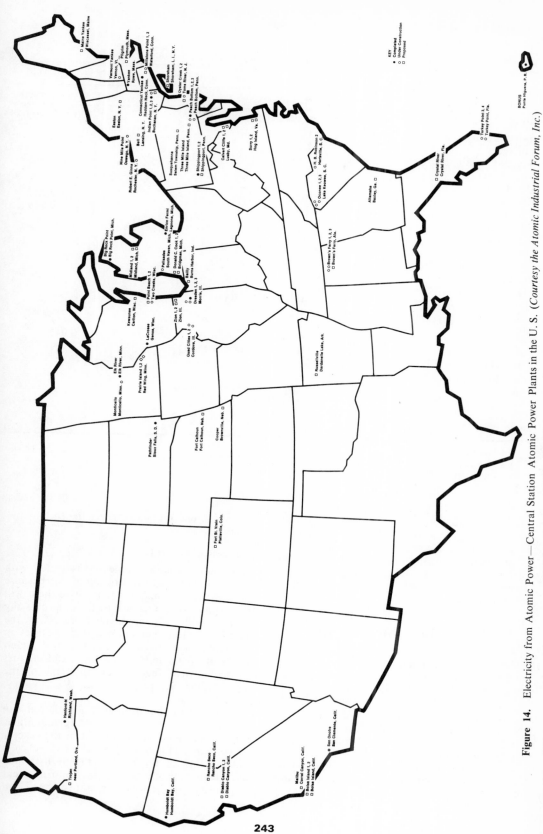

Figure 14. Electricity from Atomic Power—Central Station Atomic Power Plants in the U. S. *(Courtesy the Atomic Industrial Forum, Inc.)*

TABLE 6.   Specified and Measured Leak Rates from Nuclear Reactors

Short Name of Plant	Leakage Rate at Design Pressure (%/day)[a]	
	Specified	Observed
PWR	0.1	0.15 ± 0.075
Indian Point	0.1	
Yankee	0.1	
N.S. Savannah	0.2	
Saxton	0.2	
SPWR	0.5	
VBWR	1.0	<0.02
Dresden	0.5	0.016
Elk River	0.1	
Pathfinder	[0.2]	
Humboldt Bay	0.5 (suppression chamber) [0.05 dry-well vessel]	
Bonus	0.2	
Big Rock Point	0.5	
Hallam	320 (building) [0.001 (reactor cavity)]	
Enrico Fermi	0.06	
ECCR	~ 2 (based on rupture of one contaminated in-pile loop)	0.047[b]
FWCND	0.25	
Peach Bottom	[0.2]	
Piqua	0.2%/day/psig (maximum: 5%/day at 5 psig)	
CVTR	0.1	
HWCTR	1.0 (acceptable)[c]	0.58

[a] Nuclear Safety 2, No. 1.
[b] ANS Transaction 7, No. 1, 185 (June 1964).
[c] Nuclear Safety 3, No. 4.

up to now, much more care is needed since the activity available for dispersion in the environs is rapidly increasing as are the number of points of production and the population at risk. We can continue an excellent safety record if we adhere to reactor safeguards and radiation protection standards.[1,3]

PROTECTION AND CONTROL

To provide, to the best of our present state of knowledge, radiological protection for populations from the operation of nuclear facilities, environmental contamination must be kept below the "maximum permissible levels."[16]

Although nuclear operations are so designed, they are nevertheless carefully monitored and controlled by electronic instrumentation to make sure that all is operating as planned.

Furthermore, to insure against undetected releases or miscalculations of the impact of normal operational emissions upon the environs, a program of environmental surveillance is maintained.

Reprocessing Hazards

The public health problems associated with reprocessing nuclear fuel are important because as nuclear reactor operations increase there will be a proportional increase in fuel reprocessing activities with a consequent increase in the hazard. Nuclear fuel reprocessing facilities in the United States are located at Hanford, Idaho Falls, and Savannah River facilities, and the first privately owned fuel processing plant of the Nuclear Fuel Services, Inc., at West Valley, New York.

Compared to the amount a population can tolerate in its environment, very large quantities of fission products are produced in the reactor fuel container (Figure 15) by the time even a small fraction of the nuclear fuel is fissioned. These fission products absorb neutrons that would otherwise be used for more fission, and

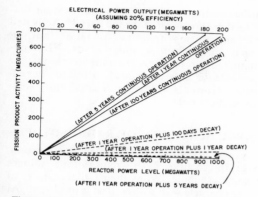

Figure 15. Production of Fission Products in a Nuclear Reactor.

it soon becomes economically necessary to separate them in order to reuse the uranium and plutonium and to maintain efficient power production.

The spent fuel cans or slugs are removed from the reactor and usually stored (cooled) in a pool of water for months before processing. The cooling period permits much of the short-lived activity to decay before the actual reprocessing.

Several techniques are used for separation; they must be operable in "hot laboratories" and remotely through shields several feet thick:

(1) Precipation,
(2) Ion exchange,
(3) Solvent extraction,
(4) Volatilization,
(5) High temperature pyrometallurgy.

In (1), (2) and (3) above the spent uranium fuel and its container of aluminum or stainless steel is chemically converted to an aqueous solution of one of its salts. Then aqueous precipitation, ion exchange, or solvent extraction can be applied. However, ion exchange methods have been more or less relegated to laboratory scale processes. Solvent extraction is much used by operators in Great Britain and Canada.

Great decontamination factors for the separation of uranium from its fission products can be achieved by volatilization which involves first reacting fluorides with the spent fuel. Then, since fuel fluorides are much more volatile than fission product fluorides, a separation can be effected.

Pyrometallurgical processes involve melting the spent fuel in the presence of a limited amount of oxygen such that the high neutron cross section lanthanide oxides and the highly volatile radiogases such as ^{133}Xe and Cs_2O are separated from the uranium.

POTENTIAL ENVIRONMENTAL PROBLEMS

Once the fuel jacket is dissolved, the radiogases ^{85}Kr and ^{133}Xe and volatile radioiodine may be released from the system. The quantity of radioiodine in the spent fuel will vary considerably with the cooling period prior to jacket dissolution. The longest half-life iodine isotope of any consequence is ^{131}I, which has a half-life of only 8 days.

Those separation processes wherein the uranium and plutonium are separated from the fission products by $KMnO_4$ can liberate the very volatile RuO_4. In the later purification stages of reprocessing a variety of fission products will be seen in the effluent.

From a public health protection viewpoint the stack effluent must be monitored at the point of discharge where concentrations must not exceed the maximum permissible levels of intake to the body. Short term radiological environmental hazards may be governed by the ^{131}I deposition in vegetation. Long term contamination problems may be set by longer half-life materials such as the ^{90}Sr released to the environs. Intermediate environmental contamination problems may be caused by the ^{106}Ru released by accident or as a result of a faulty air-cleaning system. The radiogases ^{85}Kr and ^{133}Xe may expose persons in the vicinity of the stack.

Deep bed activated charcoal filters may be effective in the partial retention of ^{85}Kr and ^{133}Xe. ^{133}Xe has a short half-life so it can be retained long enough for substantial decay before it is released from the stack. The longer half-life of ^{85}Kr will not permit this. ^{131}I particulates may be collected effectively by filtration via activated charcoal filters or caustic scrubbers. For iodine vapors, caustic scrubbers and copper mesh filters have been successful. Radioactive iodine can deposit on surfaces of the building from which it is released and on particles in the air stream. If the particle concentration in the air stream is high, it may initially be collected with a high efficiency on fiber filter units. If the air stream is at an elevated temperature, iodine may sublime off the particles upon which it was deposited and be released as vapor to the environs. RuO_4, too, can "plate out" on surfaces if released to the atmosphere.

The degree to which air-borne material is diluted by and dispersed from a stack ultimately determines its potential contamination hazard. Some useful contamination minimizing devices are: dilution of the small quantities of contaminated air from the separation processing vessels with large amounts of clean office and laboratory air; filtration of the contaminated air prior to dilution, to remove much of its activity; a high linear velocity of the effluent gases from the stack so that the effluent continues to rise before it begins its dispersion, thus approximating a taller stack. The location of the stack is also a very important factor in the effective dilution and dispersion of the effluent.[3] Further control of the stack effluent contamination is achieved by constant filtering before discharge and constant monitoring of the results. Continuous assay of specific activities in the effluent downstream of the filter units can be achieved with warning and alarm signals set to indicate when action levels of radioactivity occur. Such signals can be made to activate a ventilation shutdown valve to prevent further releases. Such precautions are necessary because, for example, the loss of caustic solution or change in temperature of the silver nitrate solution of a scrubber may substantially re-reduce its ability to collect and retain ^{131}I. Ruptures in high efficiency filter units which may occur during or prior to operations could quickly result in the release of substantial quantities of activity to the environment, and if this is not picked up at once and corrective action taken, some population exposure can result.

To summarize, good engineering and siting are necessary. Operational controls at the plant are necessary also. However, for the protection of the public, a careful system of environmental surveillance about the plant must be maintained by a qualified radiation protection and control organization or a public health agency.

ESTIMATED GROWTH IN OPERATIONS

The growth of the reprocessing of spent fuel elements industry will parallel the growth of nuclear reactor operations. It is projected that by 1980, nuclear electrical power capacity will approach 10^5 MW. Since 1 MW (electrical) may produce 2×10^6 curies of fission product waste, we face a production of perhaps 10^{12} curies of waste in that year. Experience at Hanford, Idaho Falls, Savannah River and the commercial site in New York, as well as the conclusions drawn from earlier studies of the separation stack effluents by Fitzgerald[17] at the Knolls Atomic Power Laboratory, indicate that even the processing of spent nuclear fuels can be carried out safely if enough thought is given to the problems discussed above.

The only problem which becomes more serious with time is ^{85}Kr. We presently can control it only by dilution with fresh air. If the present rate of growth of the demand for fission power continues to the year 2000, ^{85}Kr may be the en-

vironmental pollutant whose concentration limits further expansion of the industry.[18] Activated charcoal can slow up its rate of release but not stop it. It has been possible to store ^{85}Kr contaminated air in large compression tanks for later release at a controlled rate into the environs. However, this is not a satisfactory solution to the problem.

LARGE SCALE USE OF RADIOISOTOPES

As our inventory of fission products grows, large (megacurie, MCi) radioisotope sources become available. When their radiant energy of decay interacts with matter, heat is produced. At an easily attained conversion efficiency of 10 percent to electric power, such sources become kilowatt level generators which makes them useful to supply electrical power for space, marine or terrestrial applications. Properly encapsulated MCi radioisotope sources have important applications for process radiation in industry and for diagnostic, therapeutic and even prosthetic uses in the healing arts.

The public health implications relate to leaks of activity from the shields or containers during normal operation or malfunction, leading to exposure of operating personnel as well as the public via contamination of the environment. Due to the many thousands, often millions, of curies involved, even a small leak could cause a widespread contamination problem. Therefore, safe handling procedures must be enforced during production, shipping and use of these sources. also at the point of use, procedures to test the sources periodically for leaks and to monitor the environs so that the unnoticed escape of any activity may be quickly detected are necessary.

Several metallic containers are often used and found to be suitable for containment of a large isotope source. The containers must be fabricated of highly corrosion resistant and long term chemically compatible metals, and are considered acceptable only after they have satisfactorily passed a series of severe qualification tests against fire, thermal shock, vibration, pressure, weld porosity, and impaction. The selection of qualification tests depends upon the use for which the source was designed.

Use of Isotopes in Space as
Power and Heat Sources

The feasibility of efficiently converting radioactive decay energy into power has led to the development of compact, reliable, long-lived, nuclear power supplies for use where conventional sources of power are less than satisfactory. Such power sources are called SNAP (Systems for Nuclear Auxiliary Power) devices and are already capable of producing up to kilowatts (and soon 10 kw power levels will be achieved) of electrical power. SNAP units are presently used to power space satellites, remote automatic weather stations such as those near the poles, navigational buoys, and underwater electronic equipment as well as a wide variety of more mundane equipment.

In space exploration, radioisotope sources may be used to produce heat. For example, ^{210}Po or ^{170}Tm sources can be used to heat the liquid hydrogen propellant of an isotopic thruster which can produce a low thrust for a long time by expelling the propellant through a nozzle (see Figure 16) as in the POODLE propulsion stage of a Titan missile where, to provide sufficient thrust to explore deep space, a power level of 5×10^3 thermal watts or 1.5×10^5 curies of ^{210}Po or 2.1×10^6 curies of ^{170}Tm are required.

Space vehicles containing large radioisotope sources may become public health problems consequent to malfunction during the transportation, launch, operation in space, and reentry phases as illustrated in Figure 17. Such malfunctions may easily involve the release of enough radioactivity to the environment to constitute a severe contamination problem. Thus if a 10^5 curie source of ^{90}Sr (a qualified SNAP fuel) were to be injected into the atmosphere via reentry burnup of the vehicle, it could have the same ^{90}Sr atmospheric contamination effect as detonation of a 500 kiloton atomic bomb.

REENTRY FROM ORBIT
BURNUPS AND DISPERSAL AT HIGH ALTITUDE
PREORBITAL ABORT
ASCENSION ISLAND
LAUNCH ABORT
ATLANTIC OCEAN
WATER IMMERSION
CAPE KENNEDY
LAUNCH PAD ABORTS
TRANSPORTATION PHASE

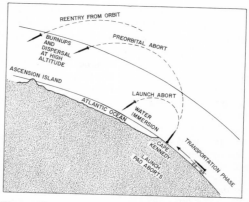

Figure 17. Possible Accidents for Space Applications of Radioisotope Sources.

The source capsule within the space vehicle

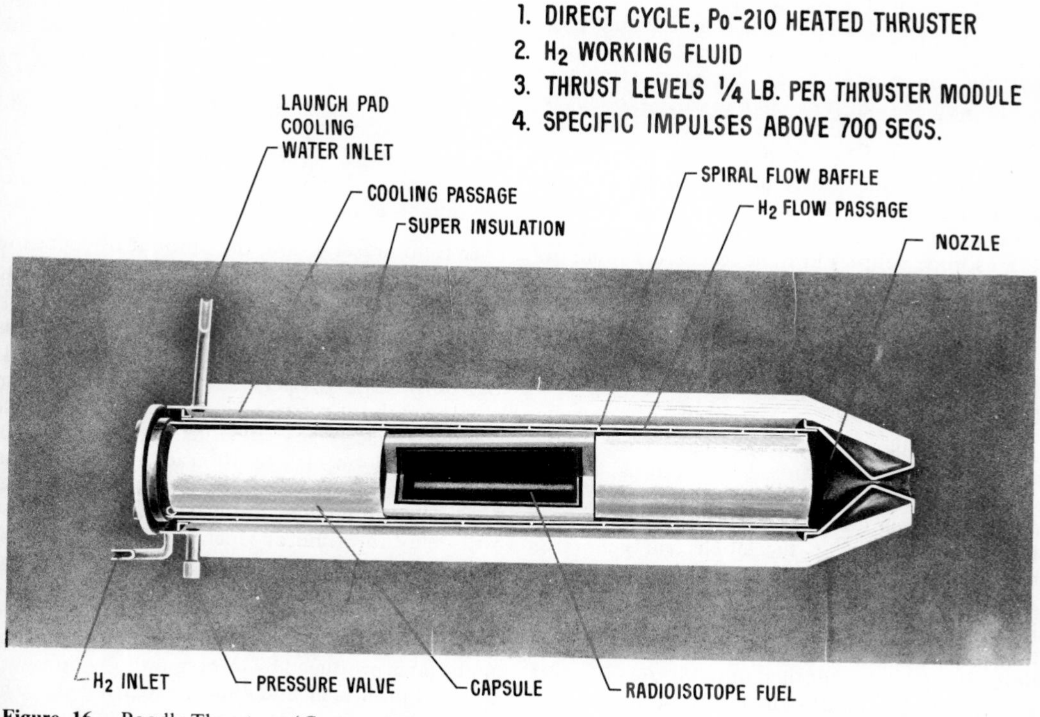

1. DIRECT CYCLE, Po-210 HEATED THRUSTER
2. H2 WORKING FLUID
3. THRUST LEVELS 1/4 LB. PER THRUSTER MODULE
4. SPECIFIC IMPULSES ABOVE 700 SECS.
LAUNCH PAD COOLING WATER INLET
COOLING PASSAGE
SUPER INSULATION
SPIRAL FLOW BAFFLE
H2 FLOW PASSAGE
NOZZLE
H2 INLET
PRESSURE VALVE
CAPSULE
RADIOISOTOPE FUEL

Figure 16. Poodle Thruster. (*Courtesy TRW Space Technology Laboratories, Thompson Ramo Wooldridge Inc.*)

should be designed to maintain its integrity against fire, explosion, and impaction. It should be fueled with the shortest-lived isotopes compatible with its mission so that an accidental dispersion of the isotopes into the upper atmosphere would pose that much less serious a problem, added to which the residence time of the debris in space would allow much of the activity to decay before it fell to earth. Thus, it would be safer also to select a source material with relatively short physical half-life compared to its half-residence time in the upper atmosphere.

Isotope Sources for Use in the Ocean as Power and Heat

There are many potential underwater applications of radioisotope sources. These include

$$A = \frac{(\text{dose in rads})(\text{process rate in lb/hr})}{(\text{fraction of energy absorbed})(D \text{ in rad lb/hr curie})}$$

where

$$D = 1.17 \times 10^4 \ \frac{\text{rad-lb}}{\text{hr-curies}} \quad \text{for total absorption of energy in the irradiated material}$$

sound generation, instrument electrical power supply, propulsion for small submersibles, electrical and thermal power for undersea platforms and small isotope heaters such as the one proposed by Sanders Nuclear (Figure 18). Note that the power requirement for even a navigational light system, i.e., 300 mW (e) to 30 W (e) or a 10 W (e) ^{90}Sr powered thermoelectric generator (SNAP-7A) with a design life of 10 years takes in the source strength range of approximately 10^3–10^5 curies.

Public health considerations of the use of radioisotope sources in the ocean are similar to those for space with the additional problem of maintaining source integrity against the corrosive properties of ocean water.

Isotopes Sources for Use on Land for Power and Heat

SNAP units are expected to have wide application on land in seismological stations, microwave relay stations, unmanned tracking sites, automatic weather stations, lighthouses, etc. For example, a ^{90}Sr thermoelectric generator became operational in 1961 on the uninhabited Axel-Heiberg Island 700 miles from the North Pole and provided continuous electric power to a United States Weather Bureau Automatic Station for 2 years. Another unit, SNAP-7C, was installed on Minna Bluff in 1962 to provide electricity for an unattended weather station 700 miles from the South Pole.

To ensure public safety, the SNAP unit should be designed to maintain its integrity of containment in case of accident. Environmental analyses must be made around the location of the unit to provide adequate safeguards against contamination problems as a result of any malfunction.

Use of Isotopes for Process Radiation

Large isotope sources are used as irradiators for processing chemicals, preserving food, sterilizing drugs and hospital supplies, and changing the properties of materials, etc. Such sources can take up to 10^6 or more curies of radioisotope such as ^{60}Co. For ^{60}Co sources, the relationship between activity of source, dose delivered per unit time, amount of material irradiated, and the absorption factor is given below;

To illustrate: Assume a dose of 10^5 rads is required for 24,000 pounds of material per day with an approximately 50 percent absorption factor of the total gamma energy being absorbed in the irradiated material; then the activity of the source should be

$$A = \frac{(10^5 \text{ rad})(1000 \text{ lb/hr})}{(0.5)\left(1.17 \times 10^4 \ \dfrac{\text{rad-lb}}{\text{hr-curie}}\right)}$$

$$= 1.7 \times 10^4 \text{ curies of } ^{60}\text{Co}$$

In food preservation, the amount of radiation delivered depends upon the food itself and the result desired. If the goal is to prolong the shelf life or storage time, a dose of 10^5 to 5×10^5 rads may be sufficient. This is pasteurization dose. If the purpose is to sterilize food for long term storage without refrigeration, the required dose is approximately 5×10^6 rads. At low doses in the range of 4×10^3 to 10^4 rads, radiation can prevent sprouting of potatoes and onions. Recent work on bananas indicates that a low dose will delay the time of ripening. The resultant savings in spoilage losses through shelf life extension has major economic significance.

In the plastics industry, radiation processed wood-plastic materials have been developed. A material is produced by impregnating wood with a liquid monomer and then irradiating it. The radiation polymerizes the monomer and yields a solid wood-plastic composite which exhibits im-

plasma for localization of tumors (particularly of the brain). There are also many therapeutic uses for large radioisotope sources, particularly in cancer therapy. A teletherapy unit may contain as much as 10^4 curies of either ^{60}Co or ^{137}Cs. Recently, an artificial implantable cardiac pacemaker as well as an artificial implantable heart were proposed and are now in a stage of development using SNAP power supplies.

SNAP chemical battery powered pacemakers are presently used at the rate of more than 10^4 per year. A nuclear powered pacemaker may contain more than several curies of an alpha emitter or hundreds of curies of weak beta emitter. Isotope powered hearts will require substantially more energy and consequently larger quantities of isotopes.

Under a National Heart Institute contract, Fitzgerald showed that an isotope powered artificial heart is feasible. The implementation of this program could result in the prolongation of hundreds of thousands of lives each year. Fitzgerald's work[18] showed that a large isotope source could be safely implanted in the body without the introduction of a significant dose to the patient. The source can be safely contained in the body without significant hazard to the person or the environment.

RADIOACTIVE WASTE DISPOSAL

Radioactive waste is generated by nuclear operations and cannot be disposed of by conventional means because radioactivity is not biologically degraded and may be long lived. Special methods are required to protect populations against exposure.

The greatest source of radioactive wastes is the nuclear fuel cycle—mining, milling, fabrication, irradiation and subsequent processing. Wastes are also produced by neutron activation of non-fuel materials in and around reactors. Additional wastes are produced from the radioisotopes used in industries, laboratories, and hospitals. Wastes can be solid, liquid, or gaseous. *Solid wastes* include fission products, activated reactor materials, contaminated filter units, fabrics, wood and metals as well as contaminated absorbent papers, rags, hypodermic needles or glassware used in working with radioactive materials. *Liquid waste* may include fission products, scrubber solutions from air-cleaning systems, reactor cooling water, fuel reprocessing liquids, and washing or rinsing wastes associated with the handling of radioactive materials. *Gaseous or aerosol waste* material includes radioactive effluent from separation processes, reactor stacks, glove boxes,

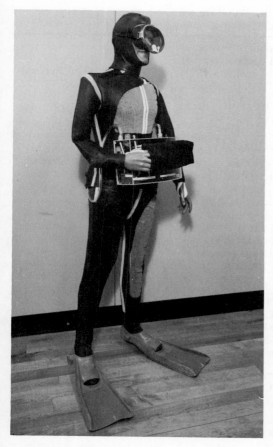

Figure 18. Sanders Nuclear Heated Undergarment for Diving Suit. (*Courtesy Sanders Nuclear Corporation, a jointly owned corporation of Sanders Associates, Inc. and Cambridge Nuclear Corporation*)

provements in hardness, compression strength, moisture resistance and toughness over wood and yet retains the inherent natural beauty of wood.

For public health purposes, routine environmental surveys must be performed to assure adequate protection and control from any leakage of activity from the source.

Use of Isotopes in Medicine

Radioisotopes are widely used in medicine. In addition to the universal use of tracers for biomedical research, radioisotopes have made possible many advances in medical diagnosis. Dynamic *in vivo* tracer diagnostic tests of organ functions have been developed for thyroid, heart, liver, kidney, bone marrow, and spleen. Tracer tests are used to find the total volume of

TABLE 7. Typical Quantities of Radioactive Waste from Establishments Operated for the United States Atomic Energy Commission[a]

Site	Type of Waste	Amount of Waste and Radioactive Content	Method of Disposal
Hanford (1959)	Liquid	3×10^3 Ci/day beta 1200 Ci/day ^{51}Cr 70 Ci/day ^{65}Zn 0.2 Ci/day ^{90}Sr	Pipeline to river
(1944–58)	Liquid	1.5×10^7 m^3 2.5×10^6 Ci	Seepage into ground
(1956–58)	Gaseous	1 Ci/day ^{131}I	Discharge from stacks
Idaho (1954–58)	Liquid	1700 Ci/yr beta about 100 Ci/yr of half-life greater than 100 days	Discharge to ground water
(1956–58)	Solid	7×10^3 yd^3/yr 10^4 Ci/yr	Burial in soil
(1956–58)	Gaseous	10^5 Ci/yr beta, mainly very short half-life, and noble gases	Discharge from stacks
Oak Ridge National Laboratory (1954–57)	Liquid	10^6 m^3/yr 250 Ci/yr beta 50 Ci/yr ^{90}Sr	Discharge to local stream
Oak Ridge Gaseous	Liquid	3000 m^3/yr 0.02 Ci/yr α (U)	Discharge to stream and in settling basin
Diffusion Facility	Gaseous	0.25 Ci/yr α (U)	Discharge from stacks
Brook-haven (1957–58)	Liquid	5×10^5 m^3/yr 0.1 Ci/yr beta	Discharge to stream
	Solid	10^3 Ci/yr	Dump in drums on bed of Atlantic Ocean
	Gaseous	700 Ci/hr ^{41}A	Discharge from stacks

[a]This table taken from "Disposal of Radioactive Wastes into Fresh Water," International Atomic Energy Agency Safety Series No. 10, Vienna, 1963.

fume hood exhausts, or incinerators. Typical quantities of waste generated by large nuclear facilities in the United States are listed in Table 7.

Radioactive wastes vary widely in concentration of radioactivity and can be classified with regard to potential hazard as low, intermediate, and high. The ranges of activity are presented in Table 8.

There are three general concepts of radioactive waste management:

(1) Concentrate and Contain: High level radioactive wastes can be concentrated, even solidified, and stored underground in large permanently controlled concrete encased steel tanks.

(2) Dilute and Disperse: Low level liquid and

TABLE 8. Radioactive Waste Levels, Concentrations and Sources

Level	Gaseous[a]	Liquid	Solid	Possible Sources of Waste	Amount Generated per Year
		Phase			
High	$>10^{-7}\mu Ci(\alpha)/cc$ $>10^{-4}\mu Ci(\beta,\gamma)/cc$	>1 Ci/liter $>10^3\mu Ci/cc$	$>10^3$ Ci/ft^3	separations processing stack, fuel reprocessing, waste liquid, waste solution	10^6 gal
Intermediate	10^{-13}–$10^{-7}\mu Ci(\alpha)/cc$[b] 10^{-10}–$10^{-4}\mu Ci(\beta,\gamma)/cc$[b]	10^{-3}–1 Ci/liter 10^0–$10^3\mu Ci/cc$	10–10^3 Ci/ft^3	Fragments of spent fuel elements, expended resin	
Low	$<10^{-13}\mu Ci(\alpha)/cc$[c] $<10^{-10}\mu Ci(\beta,\gamma)/cc$	$<10^{-3}$ Ci/liter $<10^0\mu Ci/cc$	<10 Ci/ft^3	Reactor stack effluent, reactor coolant water, research laboratories, hospital and industry	8×10^9 gal 2×10^6 ft^3

[a]Refers to concentration levels in the environs.
[b]Could be reduced to acceptable inhalation levels by the use of fine filters (DF = 1000) and the use of dilution in the environs by passage through a high stack (addition reduction by a factor of 1000).
[c]Does not require filtration or dilution.

gaseous waste may be diluted by large volumes of water or air and dispersed in the environment.

(3) Detain and Decay: Intermediate level radioactive wastes may be held up in retention tanks to permit decay of shorter-lived components before release to the environment as low level waste. This method has found its widest application in a disposal method of burying radioactive waste in selected soils. The slow movement of the activity through the soil and via ground water provides for considerable decay time before the activity reaches the food chain. Figure 19 illustrates methods of radioactive waste disposal.

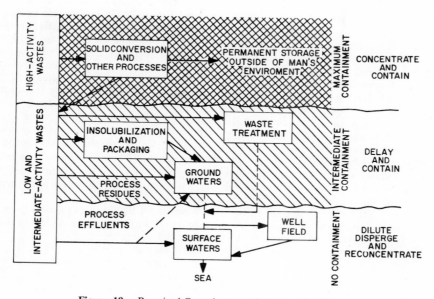

Figure 19. Required Containment of Radioactive Wastes.

Disposal in the Sea

Certain types and amounts of solid radioactive waste may be dumped into the sea. Such wastes packaged in suitable containers, usually steel drums lined with concrete, have been so disposed at designated locations in the Atlantic and Pacific Oceans since 1946. These locations are off the continental shelves where the water is more than 1000 fathoms deep. Sea disposal for solid waste, however, is usually more expensive than ground disposal and presently over 95 percent of low level solid wastes are buried in the ground.

Liquid radioactive wastes may be discharged through a pipe which extends into the sea. In Great Britain, studies were made and experiments performed to estimate limiting values for continual discharge into the Irish Sea without endangering the public health or marine life. After years of experience including careful surveillance of the shore, the sea bed and edible marine products, a discharge level of 1000 curies a day seemed safe provided that no more than 8000 curies of ^{106}Ru and 2800 curies of ^{90}Sr were released over a period of 28 days. More recent data indicates that it might be possible to release up to 10^5 curies each month.

At the Third United Nations International Conference on the Peaceful Uses of Atomic Energy held in Geneva (1964) the attitudes towards sea disposal of radioactive waste were varied and controversial. As summarized by Belter, Ferguson and Culler[11] for Great Britain, their experience in sea disposal of both liquid and solid wastes has been quite satisfactory; for the USSR, the authors reiterated earlier warnings of the "grave dangers" of sea disposal; for the French, oceanographic studies in the Mediterranean Sea indicated that it would be technically feasible to dispose of their packaged low level solid wastes from a health and safety viewpoint (they have not done so because of public reaction); for the United States, it is felt that sea disposal is as safe and acceptable as land burial for packaged low level solid wastes.

Recommended waste disposal practices in the ocean are set forth in the NBS Handbook 58 which includes discussions on the public relations problem, accidental hazards and the finality of ocean disposal, characteristics of the ocean, the fate of radioactive materials introduced into the ocean, considerations for the selection of sites of disposal, and recommendations for the transportation of radioactive materials and for the containers for ocean disposal.

At the present time, approximately nine firms and several government agencies are qualified by AEC for disposal in the sea of waste generated from their operations. These agencies include the National Institutes of Health, the U.S. Naval Radiological Defense Laboratory, and the U.S. Fish and Wildlife Service. The licenses provide for disposal of packaged low level radioactive waste in sea at least 1000 fathoms deep. The waste must be packaged in shielded containers so that they will not easily be damaged or broken. The package must be of sufficient density (10 lb/gal) to go to the ocean bottom, must be appropriately labeled, and must conform to applicable shipping regulations (Section 11).

Disposal into the Atmosphere

Air-borne radioactive waste can present significant problems unless an effective air-cleaning system is maintained before it is allowed to disperse into the atmosphere. High efficiency air filters must be used to collect particulates, and caustic scrubbers, packed towers or activated charcoal filters are used to collect iodine and ruthenium that may be volatilized and pass through the particulate filters. In general, the non-reactive gases, xenon and krypton pass through these units but are diluted with large volumes of air from office and service areas and outside before entering the environs. The filter units and the contaminated caustic solutions must then be disposed of as solid and liquid waste.

Air-borne radioactive materials from fume hoods, dry boxes, and hot cells are generally drawn through a series of filters and scrubbers before being diluted and dispersed to the environs. Figure 20 illustrates a method of gaseous waste disposal that has been effectively used at ORNL. To reduce the dose rate problem from the release of radiogases, high stacks are used and the outlets are sometimes tapered to increase the velocity of the air stream at the point of discharge, thereby increasing the dilution factors. In air-cooled reactors, large volumes of dilution air may flow through. The Oak Ridge Graphite Reactor discharges approximately 500 curies of ^{41}A a day through a 200-foot stack with a flow rate of 120,000 ft^3 min. The inlet air is prefiltered to remove the particulates that may otherwise be activated in their pass through the reactor.

When short-lived gaseous radioisotopes are to be released through a stack, it may be possible to collect the daughter products in high efficiency filters if they are particulate. In such cases, the passage time through the stack and filters is adjusted to give adequate time for the gaseous parents to decay to the particulate daughter product.

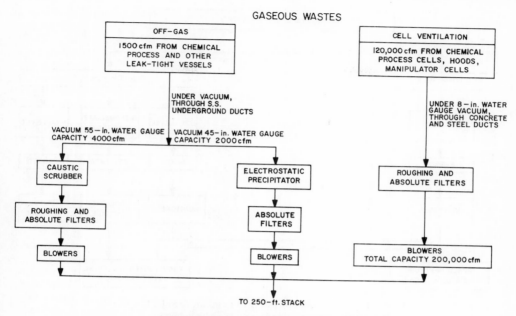

Figure 20. Processing of Gaseous Wastes.

When large quantities of non-reactive radiogases are generated, a method of storing under compression has been used. Then at a more desirable time and after some decay, the longer-lived radiogases may be released under control into the environs so that permissible levels are not exceeded.

Disposal in the Ground

In certain instances, solid and liquid radioactive wastes may be disposed in the ground. High level wastes are permanently kept in large underground tanks. Intermediate level liquid wastes are sometimes temporarily stored in retention tanks before discharge into water systems or the ground. Figures 21 and 22 present schematic diagrams for disposal of solid and liquid wastes into the land and waterways, respectively. In general, the burial of high level waste on land has been less expensive and more widely used.

Many solid radioactive wastes can be compressed or burned in order to concentrate them. Concentration by incineration can reduce the bulk of combustible waste by a factor of approximately 20. Compression can reduce the volume by a factor of 6 or 7. However, the equipment is significantly less expensive than that for incineration, and the operational cost of compression is usually smaller.

Open-field burning of low level combustible waste has been studied by Harris and Weinstein.[12] This method of concentration eliminates the need for a costly incinerator or the expense of shipping a large quantity of low level waste to a burial ground. However, it can result in contamination of the environs by the uncontrolled release of radioactive materials not detected prior to incineration. Caution is recommended in the use of this mode of waste management.

High level waste involves long term storage,[13] and the tank, as shown in Figure 23, must be made of non-corrodible material. A secondary container (a saucer) should be provided to contain the release of radioactive material until corrective action is taken. Since high level waste evolves much heat, effluent gases and aerosols must be decontaminated, and a sensitive monitoring system must be included in the design of the system. The storage tank should be located in an area where the release of activity from the tank would not result in an environmental problem.

Fixation of high level waste in glasses[14] appears at this time to be a solution to the "ultimate" disposal problem. General agreement was found at the last Geneva Conference among the larger nations (France, England, Russia, and the United States) that storage of liquids in underground tanks was only an interim solution to the problem and that fixation in solids held the most long term promise.

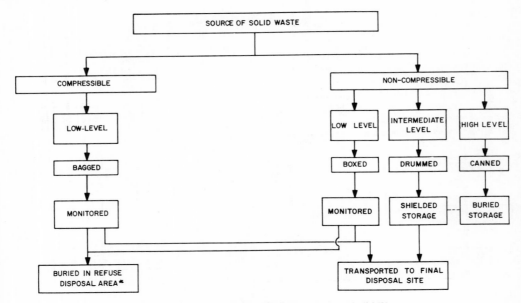

Figure 21. Proposed Processing of Laboratory Solid Waste.
*This material is segregated and regulated nonhazardous waste.

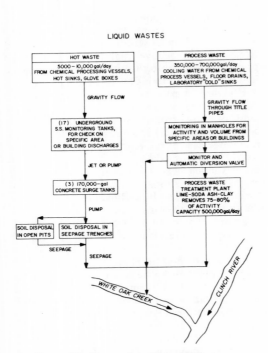

Figure 22. Processing of Liquid Wastes.

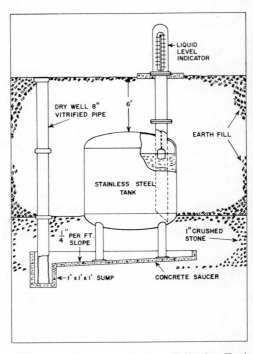

Figure 23. Dry Well and Waste Collection Tank Installation.

It is now proposed by the US, the UK and France that high-level liquid waste will be fixed in glass and the glasses will be encased in a stainless steel can. The US proposes to store in a dry underground location, such as a salt mine; the UK has suggested storage in an air cooled vault; the French have not yet indicated a choice, the USSR is considering burying uncanned glass blocks in pits in the earth, because on the basis of an achieveable leach rate of 10^{-8}g/cm^2/day, glass containing 10^3 curies/liter can be safely stored in a covered pit and even higher concentrations may also be stored safely, if one takes into account the absorption capacity of the soil for fission products. It should be noted here that earlier tests on the leach rate of fission products from the BNL phosphate glass yielded a rate of approximately 10^{-5} curie/cm^2/day. Highly porous clay balls have been developed by LASL and COORS Porcelain Company to soak up radioactive liquid waste and upon heat treatment, the waste is solidified into a ball.

The most recent and most promising method of direct intermediate-level liquid waste disposal is the method of hydraulic fracturing as shown in Figure 24. A hydrofracturing pilot plant was built at Oak Ridge in 1964 after several years of study using small amounts of radioactive cesium and cobalt. The method involves drilling a well and injecting a radioactive waste-cement-clay mixture under high pressure into an impermeable formation such as shale by hydraulic fracturing.

Specific details on burial conditions are given for land disposal; an AEC licensee in accordance with Federal Regulations (10-CFR-Part 20, Section 20.304) may dispose of licensed material by burials in the soil provided that:

(1) "The total amount of licensed or other radioactive materials buried at any one location and time does not exceed, at the time of burial. 1000 times the amount specified in Appendix C of this part."

(2) "Burial is at a minimum depth of 4 feet.

(3) "Successive burials are separated by distances of at least 6 feet.

(4) "Not more than twelve burials are made in any year."

Alpha Activity Buried in Soil. Solid waste contaminated with alpha activity has been buried in holes approximately 15 feet in diameter and 15 feet in depth. After the waste material is placed in the hole, it is covered with a foot of earth. Then 8 inches of concrete is poured on top of the earth, and 2 or more feet of earth is placed on top of the concrete.

Beta-Gamma Activity Buried in Soil. Solid wastes that are contaminated by beta-gamma activity may be buried by conventional sanitary landfill methods in trenches about 10 feet wide and 15 feet deep. The wastes are put into the trenches and then covered by 3 or more feet of earth. The higher level wastes require more shielding to reduce the dose rate at the surface to a very low level. All burials must be recorded with respect to quantity and type of radioactive waste.

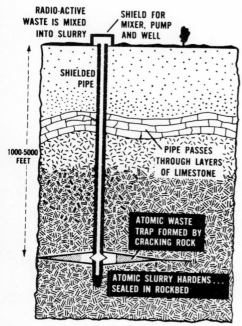

Figure 24. Waste Disposal by Hydraulic Fracturing. (*Atomics, May–June 1965, p. 13*)

BIBLIOGRAPHY

1. Sax, N. Irving, "Dangerous Properties of Industrial Materials," 2nd edition, New York, Reinhold Publishing Corp., 1963.
2. Hogerton, John F., "Atomic Fuel," U.S. Atomic Energy Commission, Division of Technical Information, 1964.
3. Fitzgerald, Joseph J., "Applied Radiation Protection and Control," to be published by U.S. Atomic Energy Commission, Division of Technical Information.
4. Mawson, C. A., "Management of Radioactive Waste," Princeton, N.J., D. Van Nostrand Co., 1965.
5. Straub, Conrad, "Low-Level Radioactive Waste," U.S. Government Printing Office, 1964.

6. Industrial Radioactive Waste Disposal, Hearings Before the Special Committee on Radiation of the Joint Committee on Atomic Energy, Congress of the United States, 86th Congress, January 28–February 3, 1959.

7. "Operational Accidents and Radiation Exposure Experience within the USAEC—1943–1964," Division of Operational Safety, 1965.

8. Nuclear Safety Quarterly Technical Progress Report 5, No. 4 (Summer 1964).

9. Nuclear Safety Quarterly Technical Progress Report 4, No. 1 (September 1962).

10. ANS Transaction 7, No. 1 (June 1964).

11. Waste Management: Technological Advances and Attitude of Safety, ANS Nuclear News (October 1964). P. 94. D. E. Ferguson, ORNL; F. L. Culler, Jr., ORNL; W. G. Belter, USAEC.

12. Harris, W. B., and Weinstein, M. S., "Open-field Burning of Low-level Radioactive Contaminated Combustible Wastes," AEC Report TID-7512.

13. Fox, Charles H., "Radioactive Wastes," Atomic Energy Commission, Division of Technical Information, 1965.

14. "Radioactive Wastes, How and Where They Go," 18, No. 3 (May–June 1965).

15. San Juan Basin Physiological Research Project, Research Branch, Div. of Radiological Health, Bureau of State Services, PHS-HEW, 1962.

16. 10 CFR-20 Federal Register.

17. Fitzgerald, Joseph J., "Evaluation of KAPL Separations Process Plant Effluent," Jan. 5, 1954.

18. Fitzgerald, Joseph J., Nuclear Section on Report to NIH concerning Feasibility of Artifical Heart, 1966.

ALLERGIC DISEASE IN INDUSTRY

Milton S. Dunn, M.S., M.D.
Former Medical Director, Rensselaer Division
General Aniline and Film Corporation
Rensselaer, New York

This section is included primarily to acquaint the nonmedical reader with some representative industrial medical problems which until now have in many cases been obscurely defined. It has been found that the concept of allergy explains many of these problems. What heretofore have been considered to be purely toxicological reactions have in some instances, after careful evaluation bearing allergic reaction in mind, turned out to be allergic phenomena.

Allergy or, more accurately, allergic disease is presently estimated to affect the lives of from 10–30 percent of the population. Generally, an allergic disease is not fatal. It may, however, be disabling. In industry, the problem of lost time due to on-the-job contact with allergens is a perplexing one and is complicated by off-the-job contacts over which the industrial hygienist has little control.

The unfortunate individual who cannot eat eggs, drink milk, or eat fish, etc., without suffering symptoms such as wheezing, coughing, sneezing, running eyes, or even swelling of the skin is familiar to all. This allergic individual has his counterpart in industry where he cannot touch various chemicals or inhale various dusts, etc., without suffering from allergic disease.

Mechanism of Allergic Reactions

The mechanism by which an allergic reaction takes place is rather complicated; the simplified explanation below may permit a general understanding of the problem.

Only an allergic person is able to produce within his blood stream substances called *reagins.* These are the body's response to the presence of *allergens;* reagins are in fact, antibodies and the ability to produce them is an in-

*Presently Area Medical Consultant to the New York State Department of Social Services.

herited characteristic which differentiates the allergic from the non-allergic individual. Reagins are specifics; i.e., if cotton allergen enters the body of one allergic to cotton, his body generates a specific reagin against cotton. These reagins circulate by way of the blood stream and fasten themselves to tissues of specific organs. Frequently, the tissues involved are "covering tissues" such as the lining of the nose, the lining of the outside of the eyeball, the linings of the lungs, the linings of sinuses or the skin itself. It therefore happens that we can have the cotton allergen and its reagin both attached to a single tissue cell. When the allergen and the reagin *unite* as happens in the tissues of an allergic individual, the cell forming the third member of the combination produces a substance that resembles histamine. Since it cannot be exactly identified as histamine, it is called the "H" substance or histamine-like substance. To sum up, we now have a situation in which the cotton allergen unites with the cotton reagin which in turn has been attached to a tissue cell of the specific organ of, let us say, the skin, and the cell in the skin elaborates the "H" substance. This "H" substance is then carried by the blood stream or by the lymphatic system to the lining walls of the tiny arterial blood vessels. When this "H" substance reacts with the lining of the tiny arteries, it causes the walls of these arteries to become less tense, less of a barrier against the outward flow of the fluid contained in them; as a result, this fluid moves out of the blood stream into the tissues and into the spaces in between the tissues. It waterlogs them, interferes with circulation, and causes the tissue to become swollen. This is called *edema.* This swelling can be localized as hives, which look like mosquito bites, or it can take place over relatively large areas (angioedemas), e.g., the whole hand may swell, or the whole foot, or the lips or eyelids.

The "H" substance in some cases also attacks smooth muscle—the kind of muscle found around the air passages in the lungs, around the stomach and the bowel, and in various other parts of the body. The "H" substance seems to irritate these smooth muscles and causes them to contract. This is called *spasm.* When these smooth muscles contract they compress the passage around which they are situated. If, for example, an air passage in the lung is normally 1 cm in diameter, then spasm of the smooth muscles around it reduces this to perhaps half a centimeter or less. The amount of narrowing depends upon the amount of the "H" substance which has been liberated. It also depends upon many other factors which will be discussed below. The mechanism of the allergic reaction then, goes like this: (1) an *allergen* enters the body; (2) its *reagin* is produced; (3) both of these attach themselves to *cells* which (4) liberate the *"H" substance;* (5) the "H" substance then causes lining tissues to *swell* or smooth muscles to go into *spasms;* (6) the allergic person becomes aware of symptoms.

The following illustration is a simple representation of this allergen-reagin-cell-"H" substance effect. The allergen is represented as a key. The reagin is represented as a lock whose keyhole is fitted only by that specific key. The cell is represented as a rectangular box in which there is an "H" representing the "H" substance. The smooth muscle and wall of the small arteries are also represented diagrammatically. Now it can be observed that when the key fits exactly into the lock and it in turn is united with a cell, the "H" substance is released. This "H" substance then reacts with the wall of the tiny artery or the smooth muscle which in turn leads to the allergic manifestation.

Antihistamines. In the above illustration, the "H" substance or histamine-like material is represented as being the trigger that sets off the symptoms. Based on this idea the "antihistaminic" drugs have been produced and are now enjoying some favor for the treatment of allergic disease. This is an ill-founded approach because the antihistaminics relieve only the symptoms. *They do not relieve the disease.* In order to relieve the disease they would have to prevent the reaction between the allergen and the reagin, which produces the "H" substance. All that the antihistaminics can do is prevent the "H" substance from reacting with the small blood vessels or smooth muscle which results in the

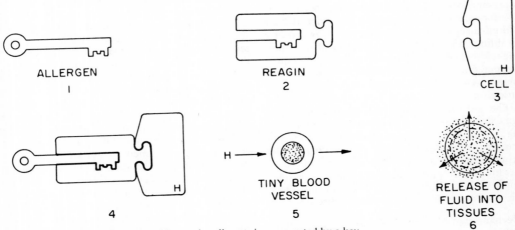

(1) An allergen enters the body. Above, the allergen is represented by a key.
(2) The allergen's specific reagin is produced. The reagin is represented as a lock whose keyhole is fitted only by that specific key.
(3) When the allergen and the reagin unite as happens in the tissues of an allergic individual, the cell forming the third member of the combination produces a substance that resembles histamine. This cell containing the "H" substance is represented above as a rectangular box containing an "H."
(4) When the key fits exactly into the lock and it in turn is united with a cell, the "H" substance is released.
*(5) The "H" substance reacts with the lining of the tiny arteries, and causes the walls of these arteries to become less tense, less of a barrier against the outward flow of the fluid contained in them.
(6) The fluid moves out of the blood stream into the tissues and into the spaces in between the tissues, causes the tissues to become swollen, and leads to the allergic manifestation.

*The "H" substance in some cases attacks smooth muscle and causes the muscle to go into spasms.

Figure 1. Simplified Allergen Reagin Mechanism.

allergic manifestation. Now, once the body reacts to an allergen by manufacturing a reagin to the specific allergen, an equilibrium is established. The more the body is in contact with that allergen, the more reagin it is going to produce. It follows, therefore, that after exposure to a massive dose of allergen, so much reagin is produced that a great deal of "H" substance is formed and a tremendous amount of antihistaminic would be necessary to merely relieve the symptoms.

Antihistaminic drugs have beneficial actions, but they may also have *toxic* side reactions. It may become necessary to take so much antihistaminic to relieve the symptoms that the patient may suffer more from the toxic reaction to the antihistaminic drug than he would have suffered from the allergic symptoms. The antihistaminics may "stop the smoke from coming out of the chimney but they don't put out the fire in the furnace." Another general disadvantage of the antihistaminic drugs is their *sedative* effect. Thus an individual who is under the influence of these drugs may become sleepy or sluggish to the extent that his coordination is affected, and he may become a hazard to himself and those about him in the discharge of his normal duties at work or even by driving his car. It is important that the amount of antihistaminic which he takes is carefully controlled by a physician. He should not decide the dosage for himself.

Furthermore, by depending upon the antihistaminic drug to control the allergic symptoms the patient and even his supervisor are prone to neglect the *cause* of the trouble. Only by searching out and identifying the exciting agent can proper measures be taken either to remove it from contact with the worker or to protect him from it. Palliative measures are therefore justified only if at the same time a concerted effort is being made to deal with the causal agent.

General Causes of Allergic Reactions

Any substance which produces an allergic reaction is called an *allergen*. Almost any substance can be allergenic ("allergy producing") to an allergically predisposed person. The paragraphs below describe some of the main causes of these curious reactions.

(1) An allergen affects only the susceptible individual because an allergic person exhibits an abnormal response to the particular allergen. Usually the substance is otherwise harmless; in fact, some usually wholesome foods—eggs, wheat, milk, oranges—may be specific causes of allergic disease.

(2) Allergic susceptibility is hereditary. If one parent is allergic or of an allergic family, about 30 percent of the children will show some allergic symptom. If both parents are allergic or of allergic families, it is quite certain that about 90 percent, perhaps all, of the children of that union will be allergic. There is today no known way to prevent the passage of these inherited factors from parent to offspring. An individual is born an allergic, will live his life an allergic and will pass that allergic predisposition on to his children in about the ratio noted above.

(3) The allergic reaction is, in general, a *quantitative* one; i.e., *slight* contact with an allergen produces *light* symptoms, contact with a large amount of the same allergen produces major symptoms. An appreciation of this is important in order to understand why at times the worker may show very slight symptoms and at other times, when the allergen reaches him in greater quantity, his symptoms may be severe.

(4) Another peculiarity of the allergic disease is that the *allergy is to a specific substance*. Thus, if one is allergic to animals, he may not be allergic to all animals; or if to furs, not to all furs; or all dairy products, or all citrus fruits, or all chemicals. This specificity can usually permit a worker sensitized to one type of reagin to work with a substitute. We say "usually," because sometimes the individual becomes sensitive to the new so rapidly that the symptoms seem to persist.

Specificity solves many industrial problems. The idea is to discover the offending substance (allergen) and, if possible, exchange it for a harmless one.

It usually requires *long contact* with an allergen to become allergic to that substance. Rarely is an individual born allergic to a specific substance. He develops the allergy because he is predisposed, and contact with an allergen does the rest. This is frequently seen in workers who are in close contact with animals. For instance, few are ordinarily allergic to the fur or dander of the rat, but in pharmacological laboratories where white rats are used as test animals, allergically predisposed workers may develop allergic reactions to these animals. Contact with white mice, white rats, guinea pigs and rabbits may result in severe bronchial asthma.

Cross Sensitization

An allergic victim can become *cross sensitized;* i.e., having become allergic to one substance, he finds himself allergic to related substances. The connection is not always easy to discover. A typical case of cross sensitization was that of a young man who, two months before, had had a very severe throat infection. His physician had given him an injection of penicillin which promptly cured it. However, this man was of an allergic predisposition, and about 10 days after the injection of penicillin, he began to have itching and swelling of the skin in various places. His hands and feet in particular were so swollen that he was unable to wear shoes or to work. After about two weeks, these symptoms subsided and he returned to work. Several months later he presented himself to his physician with the same marked swelling of the hands and feet, eyelids, and lips that he had exhibited after his illness. Investigation showed that the first episode of swelling was a reaction to penicillin and that the second episode of swelling was a reaction to a substance which is botanically closely related to it. It is known that the drug penicillin is manufactured from a mold which belongs to the same plant family as the mushroom and the young man's second attack was undoubtedly a reaction to some "spent spawn" with which he had covered his grounds in order to grow a good lawn and which he had bought by the truck load from a nearby mushroom grower.

Also, a person sensitive to a particular substance in industry may have the same symptoms at home, because at home he may be in contact with exactly the same substance. Thus he may become sensitive to the wool at a felt mill and from this develop, let us say, asthmatic and nasal symptoms while at work. Then while at home he may breathe the dust from wool filled comforters or old woolen blankets, which also can cause the allergic reaction. This extra-occupational sensitization can be puzzling. After all, which is the primary cause of the allergy, the wool at the factory or the wool at home? The answer is that if the worker can show that he had no disability until his occupation brought him into contact with the wool at a particular mill, and that only after that contact did he begin to show symptoms at home, it would be fair to assume that the primary allergization took place at the mill.

Summation of Effect

An allergic disability is often the result of a combination of causes; it must be understood as a *summation of effect*. For example, in order for one to become aware that his breathing is uncomfortable, his vital capacity (ability to breathe air in and out) must be reduced below normal. Let us assume that this awareness requires a reduction to 75 percent of normal. Now all the factors which cause this reduction of 25 percent must be accounted for. Allow 5 percent of the reduction for normal fatigue, another 10 percent for an infection, and another 5 percent for a toxic state. This adds up to a 20 percent reduction in breathing which may not be enough

to cause an awareness of breathing discomfort. Then if superimposed upon this situation is an allergic reaction which further reduces the vital capacity, say, 5 percent, it alone would be blamed for causing the awareness of breathing discomfort. It is important to realize that *all* the factors *taken together* which reduce his vital capacity are the "cause" of his sensation. Any combination of contributing causes may bring about the result if they add up to a sufficient reduction in vital capacity.

We frequently hear that asthma is due to "nerves"; some say infection; some say the allergic reaction. Actually, all three factors could be the answer. If it happens that the infection is the last factor needed to reach the 25 percent reduction or awareness point, it is easy to assume that the infection alone is the cause. This is no more the truth than that it is "nerves" alone. Actually all the factors taken together are the final cause of the disability.

It is the problem of the safety supervisor to remove as many of the factors as he can. He may be able to remove allergenic dust by proper ventilation; he may be able to keep the plant warmer and more comfortable, thereby reducing the liability to respiratory infection; he may help the worker resolve his emotional tensions, either by suggesting medical aid or discussing the work problem if that is a cause. Many approaches suggest themselves to the supervisor who is aware of the factors involved in the summation of effects which produces the asthma.

General Symptoms of Allergic Disease

Allergic symptoms of industrial origin are the same as those due to any other origin such as household dust. The problem in industry is to determine whether symptoms exhibited by a worker are caused by some material at work or are due to some contact outside of work. Table 1 lists easily recognizable symptoms in general order of frequency.

Complications of Allergic Diseases

Allergic diseases are generally not fatal, but they may be disabling, and a cause of lost time and inefficiency in production. Also, there may be many complicating disabilities as a result of a chronic allergy. One of these is sinusitis. Thus in an individual with a chronically obstructed nose due to inhaling substances to which he is allergic, the membranes of the nose are more liable to infection which, if it occurs, causes a blockage of the sinuses so that the infection within them is unable to drain out. The result is infected, blocked sinuses which may be so severe that the worker is disabled.

TABLE 1.   General Symptoms of Allergic Disease

Organ Affected	Description of Symptoms
Skin	A rash, usually of the eczematous type, seen as reddened patches with some oozing, crusting, scaling, and swelling. These patches may even at times run together to form larger areas sometimes enough to cover the whole face, neck, hands, arms, trunk, chest, etc.
Respiratory	Cough, wheezing, shortness of breath, or even at times complete or almost complete obstruction to breathing in the lungs. This obstructive type of breathing may be seen as bouts of *asthma* causing a blueness of the lips and fingernails. Sometimes such severe obstruction may occur, as to result in fainting from lack of oxygen. This obstruction is caused by spasm of the smaller bronchial tubes, swelling of the inside lining of the tubes and production of large amounts of thick and sticky mucus. Such an obstruction can in time produce emphysema, which is characterized by chronic bronchitis, stretching of the air sacs in the lungs, and a barrel shaped chest. This condition may cripple the victim permanently. The barrel shaped chest is the most easily recognizable sign and will be discussed further in these paragraphs.
Eye	Watering, often with severe itching and swelling of the lids. The eyes may be so badly affected that reading, working, or carrying on usual duties is greatly impaired.
Nasal system	The nasal passages are stuffed up and the nose runs as if the individual had a heavy cold. Blockage of the nose may result in ear disease. The ears may become congested, and this congestion at times may lead to infection in the middle ear (the so-called water in the ear feeling). Bouts of sneezing are characteristic of nasal allergy, and the victim may sneeze fifteen or twenty times in succession.
Sinuses	Nasal membranes that are constantly congested by an allergy cannot properly throw off infection. Thus a worker with a chronic nasal allergy is especially subject to upper respiratory infections and colds. Besides, the infection is very likely to spread to the paranasal sinuses and to persist until the allergic congestion is relieved.

The point is that the infection which actually causes the sinusitis is not occupational, but the groundwork was laid by the allergic disability of the nose, and this may have been occupational.

Asthmatic bronchitis is another complicating disease and one frequently found among workers subject to allergic lung disease. If a worker has a chronic cough and produces mucus in his lungs because he is in contact with something to which he is allergic, and if superimposed upon this he develops an infection in the lungs, he may very easily develop a chronic bronchitis of the asthmatic or wheezing type which is called asthmatic bronchitis. This can become deep seated and last for years even though the infection is treated thoroughly. The only real solution to the problem is to control the allergic reaction. Only then will the asthmatic bronchitis be controlled. Both the infection and the allergy must be treated.

If a patient suffers from bonchial asthma for a relatively long time, the chest takes on a characteristic *barrel shape*. His lungs are constantly in a state of expansion. It would seem as though he could take in a tremendous amount of air and be able to breathe very well, whereas actually the chest and lungs can inflate and deflate only slightly. This state in which the lung is in a constantly expanded position and the elasticity of the lung tissue has been lost is called *emphysema*. However, emphysema is a relatively late complication of chronic asthma and is sometimes seen in workers who have been exposed to the prolonged onslaught of inhalant substances to which they are allergic.

Futhermore, chronic allergic lung disease can in time damage the *heart*, although such conditions are rare. The blood circulation through the lung can be so hindered that the right side of the heart (which pumps the blood through the lung) is chronically overworked and thus becomes enlarged and inefficient. If this inefficiency continues, left-sided heart distention may result and this in turn may result in congestive heart failure.

A *skin infection* can very easily be superimposed upon areas which have been scratched, rubbed or otherwise irritated due to an allergenic contact. The infection again, is not an allergic disease but is superimposed upon the allergic disease which is the basis of the trouble.

If the *eyes* are constantly irritated and the patient rubs them frequently, he rubs germs into the edges of the eyelids and *styes* may result. The sty is a relatively frequent complication of allergic irritation of the eyelids, as is conjunctivitis, which is actually an infection superimposed upon the allergic irritation of the outer lining of the eyeball.

Suggested Materials for Emergency Treatment

A physician called to an emergency room to treat the victim of a very severe allergic reaction should have available various drugs, medications and materials. Time may be precious. One of the more important materials is an oxygen breathing apparatus. It is also suggested that hypodermic syringes and needles be kept available and in a sterile condition so that if the physician needs them, they will be quickly at hand. Table 2 indicates the most useful materials which should be kept available in the emergency room. None is expensive, and they can save lives when at hand to be used by the physician in charge. It should be stressed that the dosage, technique of administration, and frequency of administration, as well as the decision as to when to use any of the procedures mentioned for treatment, should be *left solely to the discretion of a physician.*

Factors Influencing Allergic Disease

Psychosomatic. Psychosomatic influences or simply the influence of emotion on the severity of allergic disease must be taken into considera-

TABLE 2.   Suggested Materials for Emergency Treatment

Drug or Apparatus	Form and/or Concentration	Purpose
Oxygen breathing apparatus	100% medical oxygen	To supply oxygen to the asthmatic patient who is suffering from oxygen lack.
Hypodermic syringes (sterile)	1, 2, 10, and 20 cc, disposable	To be available to physician if needed in an emergency.
Hypodermic needles (sterile)	20 gauge 1 and 1½ in., 25 gauge ³/₈ in., disposable	To be available to physician if needed in an emergency.
Epinephrine	1–1000 aqueous, 30 cc rubber capped vial	Injected subcutaneously for rapid relief of asthma and hives.
Epinephrine	1–100 in oil, 2 cc glass vial	Injected intramuscularly for prolonged relief of asthma and hives.
Aminophyllin	3¾ grains, 2 cc glass vial	Injected intramuscularly for relief of asthma.
Aminophyllin	7½ grains, 20 cc glass vial	Injected intravenously, slowly, for relief of asthma.
Aminophyllin	1½ grain plain tablets or capsules	To be taken by mouth for sustained relief of asthma after the emergency is treated.
Aminophyllin	3 grain enteric coated tablets	To be taken by mouth for sustained relief of asthma after the emergency is treated. These tablets have a coating which dissolves off in about 4 hr and then releases the medication—good for bedtime dosage.
"Demerol"	30 cc rubber capped vial	Injected intramuscularly for the relief of severe apprehension in patient with asthma.
Ephedrine sulfate	³/₈ grain tablets	To be taken by mouth for treatment after asthma emergency is treated.
Phenobarbital	¼ grain tablets	To be taken by mouth for sustained relief of apprehension after the emergency is treated in asthmatic patient.

TABLE 2 (*Continued*)

Drug or Apparatus	Form and/or Concentration	Purpose
Antihistaminic	Injectable	Injected for severe hives in emergency.
Cortisone	25 mg tablets	To be taken by mouth for severe allergic disease.
Cortisone	Injectable	To be injected intramuscularly for severe allergic disease.
Hydrocortisone	20 mg tablets	To be taken by mouth for severe allergic disease.
ACTH	Injectable	To be injected intramuscularly for severe allergic disease.
Calcium salts	Glass vials	To be injected either intramuscularly or intravenously for emergency treatment of severe hives.
Potassium iodide	5 grain enteric coated tablets	To be taken by mouth as a starting treatment after acute asthma emergency is treated, used to liquefy tenacious pulmonary mucus. The enteric coated tablet is much less irritating to the stomach than is the saturated solution.
Sodium iodide	10%, 10 cc glass vials	Administered intravenously for the liquefaction of pulmonary mucus in asthma.
Intravenous infusion set-up	Sterile and ready to use 1000 cc flasks of water, glucose, saline, etc.	Given by intravenous route for emergency treatment of asthma. Also used as vehicle to administer aminophyllin, NaI, etc., by intravenous route.
"Isuprel" or "Aludrine"	Sublingual tablets	To be kept under the tongue until completely dissolved acting in same way as epinephrine. This method is very useful.
Nebulizer	"Vaponefrin" or De Vilbis #40	May be used as a rapid means of bringing relief in asthma. May use "Vaponefrin," "Aerolone," "Isuprel," or epinephrine 1–100.
"Alevaire"	Solution	May be used in nebulizer for purpose of thinning out pulmonary secretions in severe asthma or asthmatic bronchitis.

tion, but may be overrated. This is only one of all the factors which result in the disability. To be sure, if an allergic disability results from any type of contact, it is frequently made more severe by a psychosomatic or emotional upset. This type of upset cannot be said to cause the original allergic disease; it does, however, make a patient more acutely aware of his disability. It is not the root of the allergic disease, but it may be the trigger which sets it off. Also, it often happens that an emotional upset may aggravate the allergic disease and the allergic disease in turn may cause greater emotional upset. This cycle is sometimes very difficult to interrupt. Suffice it here to say that a worker suffering from allergic disease often needs help to remove the cause of his emotional disturbance if such exists.

Fatigue and Toxic States. Fatigue is another real factor in the aggravation of the allergic reaction. If a worker is extremely tired, all types of allergic disability tend to become much more severe. Toxic states due to infection, as well as the use of tobacco, drugs, or alcohol, aggravate the preexisting allergic disease.

Sex. The sex factor is important also. For one thing, women usually keep themselves cleaner while at work than men do, and therefore, though women may fuss a good deal about minor skin disabilities they have far fewer major skin disabilities than men. Women know that a clean skin prevents a great deal of trouble. Allergic sensitivity is affected also by menstruation, menopause, and pregnancy. This is reasonable since these functions cause changes in the body chemistry. For instance, during menopause the skin is sometimes more reactive to allergens, and pregnancy often makes women more sensitive to allergenic materials. They can then develop severe wheezing and coughing symptoms from materials which normally would not affect them.

Age. The age of the patient is very important. The skin of the young or middle aged adult is considered normal. It is soft and supple and contains a moderate amount of moisture. Such skin can better withstand the effects of irritating substances than that of an aged person whose skin is relatively dry and thin. Also, the defatting influence of soaps, alkalies, and solvents is much more marked on dry, thin skin than on moist, normal skin.

Extra-occupational Contacts. Another factor is the effect of extra-occupational allergic disease which may be aggravated by contacts on the job. The cumulative effect can disable a worker, e.g., in an individual with hay fever who during the time of year when he is exposed to pollen of the hay fever producing plant is also in contact with an allergen at his work. Even if his hay fever symptoms are mild, if he comes in contact with even small amounts of allergens such as wool, cotton dust, animal or vegetable fiber, these may be enough added burden on the allergic process to make him markedly uncomfortable. This is a summation of effect. Any evaluation of industrial allergic disability must take both factors into account.

Weather. Weather changes in humidity and temperature of the air can aggravate allergic disease. If a worker has a nasal disability which causes his nose to be constantly wet and stuffed, he will find it worse on damp, humid days. If the weather is clear and dry, mucus in the nose tends to dry out more quickly and he breathes more easily. The same can be said in general for disabilities of the lung. In some cases, however, increased humidity of the air helps in that it tends to thin the secretions in the nose and lungs so they can be more easily removed by coughing or blowing of the nose. It can generally be said that if the respiratory mucus is thin and watery, a humid atmosphere will tend to make symptoms worse, whereas if the secretions are thick and sticky, a humid atmosphere would tend to soften and loosen them and thus to make symptoms less severe. Changes in weather from stormy to clear or clear to stormy also affect the patient with a respiratory allergy. It may be noted, for instance, that just preceding a violent summer storm some respiratory patients will feel very uncomfortable, and that just preceding a sudden drop in temperature of the order of 20–25°F, many types of allergic respiratory disease are markedly affected.

Common Allergic Diseases

Hay Fever. One of the most common allergic diseases is hay fever which is caused by the pollen of various plants. These pollens are given off by the flowering plant and are air-borne in the form of particles so small that they can be carried great distances or into homes and factories.

Mold Allergy. Another equally common but not too frequently recognized allergic symptom is what seems to be hay fever, but actually is allergy to the spores of mold. Moldy foods, such as bread, are familiar to all. The mold is formed by the mold spore which is suspended in the air and falls onto the surface of the bread. It takes root and grows into a mold plant, which we recognize as a greenish or purplish or sometimes varicolored velvety growth on the surface of the bread. This growth gives off tremendous numbers of mold spores which are air-borne. This is its method of reproduction. Suspended in the air, the spores are breathed in and, when they come in contact with the membranes of the nose or lungs, can produce allergic reactions. Some industries are especially liable to the mold hazard. Any type of installation which is damp is more likely to have mold growing on the surface of the walls, floors, and furniture than is one which is kept dry, light and well aired. Laundries, agricultural establishments, fruit and vegetable packing plants have particularly high concentrations of mold spores suspended in their atmospheres. If a worker is allergic to the mold spores in the air of the factory, he may develop allergic symptoms which could then be considered occupationally caused.

Contact Dermatitis. The most important aspect of contact dermatitis from the standpoint of a safety engineer is that the material which causes the rash does so primarily on the *exposed areas* of the body. Thus, if a worker handles leather goods his hands may become involved with disease. If a worker wears a protective mask over the face while spraying lacquer, he may develop a dermatitis outside of and around the protected area but not within the limits of the mask. This is an important point to understand for it frequently is a clue to the cause of trouble. Many such cases have been seen, as for instance, the worker with a rash on only the right hand who touched a plastic control knob with the right hand; the case of the woman with a skin rash in the shape of a triangle extending from the bridge of her nose over each cheek and across her chin, due to the rubber of her filter mask touching that part of the face.

In some instances the allergens might only cause the dermatitis when they become dissolved

TABLE 3.   Some Familiar Causes of Contact Dermatitis

(A) *Plant Products*

(1) Grasses

Bermuda grass	Orchard grass	Rye grass
Blue grass	Philodendron	Sugar cane
Johnson grass	Red top	Timothy grass
Jute	Rice	Tobacco
Morning glory	Rice dust	Wheat

(2) Flowers

Angelica	Gaillardia	Primrose
Bleeding heart	Geranium	Primula
Chrysanthemum	Helenium	Pyrethrum
Clover	Hops	Rain flower
Cotton	Marigold	Tansy
Cotton seed	Orris root	Tulip bulbs
Daffodil		

(3) Weeds

Bell heather	Mayweed	Ragweed, dwarf
Burweed marsh elder	Milfoil	Ragweed, giant
Camomile	Plantain	Ragweed, western
Clematis	Poison ivy	Sagebrush
Clover	Poison oak	Sweet vernal
Cocklebur	Poison sumac	Thistle
Lambs quarter	Primrose	

(4) Trees and Lumber Products

Acacia	Elm	Pine
African boxwood	Eucalyptus	Poplar
Alder	Hickory	Prune
Aroeira	Japanese hardwood	Redwood
Ash	Laurel	Resins
Australian dogwood	Lemon wood	Rosewood
Beech	Macassar wood	Rungus
Birch	Magnolia	Satinwood
Bitterwood	Mahjong	Sawdust
Borneo rosewood	Mahogany	Silver spruce
Brazilian walnut	Mangowood	Spruce
Cedar	Maple	Sycamore
Chestnut	Marsh elder	Teakwood
Cocobolo	Mesquite	Turpentine
Cocowood	Oak	Virginia creeper
Creosote bush	Olive wood	Yew
Ebony	Partridge wood	

(5) Nuts

Almond	Cocoanut	Peanut
Bilhawanol	Filbert	Pecan
Brazil	Hazelnut	Walnut, black
Cashew	Hickory	Walnut, English
Chestnut		

(6) Spices

Allspice	Ginger	Peppermint
Aniseed	Horseradish	Pimento
Caraway	Mace	Poppyseed
Cinnamon	Mustard	Sage
Clove	Nutmeg	Thyme
Coriander	Paprika	Vanilla
Fennel	Pepper, black	Wintergreen

TABLE 3 (*Continued*)

(A) *Plant Products (Cont'd)*

   (7) Fruits and Vegetables

Artichoke	Garlic	Parsnip
Asparagus	Grapefruit	Potato
Carrot	Grapes	Prickley pear
Celery	Irish potatoes	Radish
Cinnamon	Lemon	Spinach
Corn	Lime	Tomato
Corn starch	Mint	Turnip greens
Cow parsnip	Mustard	Vanilla
Figs	Orange	Watercress

(B) *Cosmetics*[a]

Anhydrotics	Face packs	Polish removers
Artificial lashes	Hair dressing	Rouges
Bleaching creams	Hair dyes	Sachet powders
Body powders	Hair lotion	Scalp creams
Cleansing tissues	Hair tonics	Scented soaps
Cold creams	Hand lotions	Shampoos
Deodorants	Lipstick	Toilet waters
Depilatories	Massage creams	Tooth paste
Eyebrow pencil	Nail polish	Vanishing creams
Eyelash ointments	Orris root	Volatile oils
Eye shadow	Perfumes	Wave sets
Face powders	Pigments	Wrinkle removers

(C) *Animal Products*

Beeswax	Egg white	Horse hair
Cat dander	Feathers	Lanolin
Cat hair	Furs	Leather
Chicken blood	Glue	Rattlesnake venom
Camel hair	Goat dander	Sheep dander
Cod liver oil	Goat hair	Sheep wool
Dog dander	Hog intestine	Silk
Dog hair	Horse dander	Wool
Egg		

(D) *Drugs*

Acetone	Dental amalgam	"Novocaine"
Adhesive plaster	Dental plates	"Nupercaine"
Almond oil	Derris root	Oil of Mirbane
Alum	Dextrins	Opium
Ammoniated mercury	Emetine	Orris root
Arnica	Ephedrine	p-Chlor-m-cresol
Arsphenamine	Etherial oils	p-Nitraniline
Atropine	Ethylaminobenzoate	Phenol
Balsam of Peru	Formalin	Phenylenediamine
Belladonna	Glycerine	Phenylhydrazine
Bergamot, oil of	Green soap	Physostigmanine
Bitter almond	Hair dyes	Picric acid
Bitter orange	Hexylresorcinol	"Procaine"
Boric acid	Icthyol	Pyrogallic acid
"Butesin picrate"	Ink	Quinine
Butyn	Insect powder	Quinones
Camphor	Iodine	Resorcin
Cantharides	Iodoform	Salicylic acid
Cassia	Lacquer	Salicylates
Celloidin	Lavender	Scarlet red

TABLE 3 (*Continued*)

(D) *Drugs* (*Cont'd*)

Chloral hydrate	Lanolin	Silver nitrate
Chrysarabin	Linseed oil	Strychnine
Citronella, oil of	"Lysol"	Sulfonamides
Clove, oil of	"Mentholatum"	Sulfur
Codeine	"Mercurochrome"	Tannic acid
Colocynth	Mercury bichloride	Tar
Cresol	"Merthiolate"	"Xeroform"
Croton oil	Morphine	Zinc oxide
Crude coal tar	Mustard	

(E) *Chemicals*

Acetone		
Alizarin	Explosives	Photographic developers
Aluminum salts	Formalin	Picric acid
Aniline dyes	Gasoline	Picrates
Arsenic salts	Germacides	Plastics
Auto polishes	Gold	Polishes
"Bakelite"	Hexamethylenetetramine	Pyridine
Bergamot, oil of	Hexanitrodiphenylamine	Pyrogallic acid
Bichromates	Hexylresorcinol	Pyrogallol
Bleaching powders	Inks	Resins
Bluings	Lacquers	Rubber
Boric acid	Lead acetate	Salicylates
"Butesin picrate"	Mercury	Silver nitrate
Butyn	Mercurial compounds	Soaps
Chromium	*m*-Phenylenediamine	Stains
Chrome salts	Nickel	Sulfur monochloride
Citronella, oil of	Nickel compounds	Sulfur
Cobalt	Nitrobenzene	Tannates
Cobalt salts	*p*-Nitraniline red	Tannic acid
Copper sulfate	*p*-Phenylenediamine	Tars
Dimethyl sulfate	*p*-Dichlorbenzene	Waxes
Dyes	Persulfates	Zinc
Ethyl gasoline	Petroleum products	Zinc compounds
	Phenol	

[a]Most of the contact dermatoses resulting from cosmetics occur among the users of the cosmetics rather than among the workers who produce them.

in sweat. A dust which is suspended in the air and permeates the clothing might only cause a dermatitis of the armpits or the groin because only here was there sufficient moisture from perspiration to dissolve the dust and carry it through the clothing and into the skin.

Friction from clothing also may be a factor in rubbing the allergen into the skin at the point of contact. Thus contact dermatitis may develop more easily under the waist band or under the collar or cuffs.

The chemical causes of occupational dermatoses can be divided into two main categories: (1) *primary irritants* and (2) *sensitizers*. Only the sensitizers are allergy producing. A *primary irritant* is a chemical substance that causes dermatitis by its direct action on the skin at the point of contact. It may dissolve or emulsify the fat in the skin. It may precipitate the proteins, such as in a tanning action, or it may cause a chemical oxidation or reduction of the outer skin. It may act to dehydrate, it may dissolve the keratin or outer layer, or it may cause a keratogenic or thickening action. Primary irritation produces a skin inflammation like a burn and will occur in almost all skins that touch the substance. A common example of this type of substance is kerosene. It will produce in the skins of many persons a reaction that varies from the diffuse redness of a first degree sunburn to the blistering of a second degree burn. Susceptibility is greatest in thin skinned blondes, but almost anyone will react to it to some degree. It is obvious then that the patch testing with such a substance at full strength is useless, since it will elicit a positive reaction from almost everyone.

The *sensitizers* do not necessarily cause demonstrable skin changes during the first contact, but after a contact of several days or more, a further contact may cause the skin to break out in a dermatitis. This is an allergic response to a substance in one who has developed sensitivity to it. The substances capable of eliciting this response are legion and include many contacted outside the working environment, such as cosmetics, clothing dyes, etc. It is probable that no one is now allergic to a substance that he has yet not contacted. It appears that initial contact with the allergenic material "sensitizes" that individual to later contact with the same substance. Thus typical sensitization dermatitis does not appear the first time an individual contacts the material. Some substances have a high sensitizing index; i.e., they will induce sensitivity in a considerable fraction of people exposed. Others have a low sensitizing index and only rarely will they produce an allergic dermatitis. Between these two is a wide range of substances that can cause sensitization dermatitis. The reaction within the skin in an allergic dermatitis varies in degree. It follows a common pattern of reddening of the skin, formation of vesicles or tiny blisters, oozing or weeping when these vesicles break, swelling, formation of scales, and, usually, itching. If the sensitization skin reaction is mild, the patient at the discretion of the physician may be permitted to continue the exposure contact in hopes of becoming "hardened," i.e., developing a relative degree of immunity, but this must be done with caution and only under continuing medical supervision.

Oil folliculitis, which is *not* an allergic reaction but is often confused with allergic disease, is an inflammation and infection localized at the hair follicle and caused by oil. Severe cases are often called "oil boils." This skin affliction results from continued contact of the hairy skin with oil so that bacteria are carried down into the follicles and infection occurs. Dark, oily skinned persons are more likely to develop this condition than are the fair skinned. Of utmost importance here is personal hygiene; the most effective prevention lies in adequate washing facilities.

Another kind of occupational skin trouble is defatting. This is not strictly a "dermatitis" since there is no inflammation. Almost any solvent can cause this if it dissolves out the oils and fat of the skin. This results in dryness, cracking, and scaliness, usually without inflammation. However, when the cracks deepen and widen to become fissures, local inflammation may develop. Treatment consists of avoiding contact with the defatting agent, using superfatted soaps, and rubbing lanolin into the skin after washing to keep the skin soft and pliable. Frequently the source of this condition is a solvent used to remove oil or stain from the hands. In such a case, the use of a protective hand cream will permit easy subsequent removal of the oil or stain with soap and water and thus avoid the necessity for defatting solvent. Gloves also may prevent the need for such solvent cleansing.

Dermatophytosis is commonly called "athlete's foot." This fungus type disease can cause a skin eruption of the hands as well. The dermatophytid, or "id," reaction can be easily confused with contact dermatitis. In many cases the reaction on the hands due to an "athlete's foot" infection opens the way for further difficulty due to contact with chemicals or other substances acting as sensitizers.

Criteria for Occupational Dermatoses

(A) An occupational dermatosis is one in which the role of an occupational causal factor has (at some previous time) been established beyond reasonable doubt.

(B) The person has been working in contact with an agent known to be capable of producing similar changes in the skin.

(C) The time relationship between exposure to the agent and the onset of the dermatosis is correct for that particular agent and that particular abnormality of the skin.

(D) The site of the onset of the cutaneous disease and the site of maximum involvement are consistent with the site of maximum exposure.

(E) The lesions present are consistent with those known to have followed the reputed exposure or trauma.

(F) The person is employed in an occupation in which similar cases have previously occurred.

(G) Some of the person's fellow workers using the same agent are known by the examiner to have similar manifestations due to the same cause.

(H) So far as the examiner can ascertain, there has been no exposure outside of occupation that can be implicated.

(I) If the diagnosis is dermatitis, the following items are important:

(1) Attacks after exposure to an agent followed by improvement and clearing after cessation of exposure constitute most convincing evidence of the occupational factor as a cause.

(2) The results of patch tests performed and interpreted by competent medical personnel corroborate the history and examination in the majority of cases.

These criteria with the exception of a few words are copied verbatim from "Industrial Dermatoses, A Report by the Committee on Industrial Dermatoses of the Section on Dermatology and Syphilology of the American Medical Association," J.A.M.A., 118, 613 (1942). It is not implied, nor is it likely that *all* of these criteria will be met in any individual case.

The Patch Test

Patch tests can be dangerous to the patient, are nearly always hard to interpret but are useful in diagnosis. Patch tests should be administered by no one but a physician or a competent assistant under the supervision of a physician. A person upon whom a patch test is placed sometimes develops severe local or possibly general reactions to this minute amount of substance which has been placed on his skin. Therefore, the concentration of the substance used in the test is important. The actual substances used are important. Some substances are so toxic that no matter how they are treated, they should not be used for patch testing.

Another type of testing is the intracutaneous test and the scratch test. *Scratch* tests are fraught with similar dangers and call for similar warning: again, only the *physician* is competent to do this type of testing and not non-medical personnel. The *ophthalmic* test is another one sometimes used to detect sensitivity to inhaled substances. This test, however, is quite unpredictable and is not to be recommended unless the physician feels that there is no other way to determine sensitivity.

The techniques for performing all of these tests are described in standard textbooks of dermatology.

TABLE 4.   Occupations and Likely Sensitizers

Key to Abbreviations
R = Respiratory Symptoms—Asthma and Nasal Allergy
S = Skin Symptoms—Eczema and Hives
C = Contact Dermatitis

Occupation	Symptoms	Sensitizers—Allergens
Agricultural	R, S, C	Animal, plant and mold products. Insecticides, chemicals, drugs, solvents, detergents, etc.
Airplane workers	C	"Dope" solvents, paints, cutting oils, woods, glues, caustic cleansers, electroplating compounds
Asphalt and pitch workers	C	Pitch, tars, and petroleum products
Autotyper	C	Bichromate form of chromium
Bakers and millers	R, S, C	Flours, conditioners, bleaches, flavoring extracts, vegetable dyes, sugars, eggs, chocolate, mold, etc.
Barbers	C	Cosmetics, depilatories, hair dyes, hair preparations containing mercury, quinine, resorcin, and sulfur
Brewery workers	R, S, C	Barley, grains, hops, malt, yeast, mold, glues, paper products, inks, detergents, etc.
Bricklayers, masons	C	Cements, paints and stains
Bronzers	C	Arsenic and metallic salts
Burnishers	C	Mercurials
Butchers	R, S, C (hog itch)	Animal hairs and danders, intestinal contents, molds, detergents, insecticides
Cabinet makers and carpenters	R, S, C	Sawdusts, resins, glues, lacquers, solvents, varnish, thinners
Candy makers	R, S, C	Nuts, spices, flavorings, fruits, sugars, dyes, vegetable colorings
Canners of fruits and vegetables	C	Fruits, vegetables, juices, cleansing alkalies
Cattlemen	R, S, C	Cottonseed, feeds, grains, hair and danders, molds, pollens, detergents, cleansers, etc.

TABLE 4 (*Continued*)

Occupation	Symptoms	Sensitizers—Allergens
Chemists	C	Irritants and allergenic substances in common use in the laboratory
Circus workers	R, S, C	Animal hairs and danders, disinfecting, deodorizing and cleansing chemicals, dyes, fabrics, foods, grains, pollens, molds, sawdust
Compositors	C	Benzene and chromium compounds
Cooks	R, S, C	Flours, fruits, vegetables, flavorings, vegetable dyes, spices, soaps, detergents, insecticides, mold
Dentists and dental technicians	C	Antibiotics such as streptomycin, penicillin, etc., "Iodoform," "Novocaine," various caine drugs, mercury amalgam
Dye workers	C	Dyes of all sorts and intermediates
Electroplaters	C	Chromium compounds, cyanides
Electrotypers	C	Copper compounds
Explosive manufacturers	C	Ammonium nitrate, DNT, mercury fulminate, "Hexite," picric acid, picrates, tetryl, TNT
Exterminators	R, S, C	Arsenic compounds, fluorides, formalin, mercury compounds, pyrethrum, cereal grains, DDT, "Lethane," p-dichlorobenzene, rotenone
Felt workers	R, S, C	Bichromates, hair and dander, mercury salts, sizings
Foresters	R, S, C	Insecticide sprays, "poisonous" shrubs, pollens, sawdusts
Fruit and vegetable handlers	R, S, C	Dyes, fruit colorings, fruits, vegetables, glues, insecticides, mold, oils, paper, waxes, woods
Fur cleaners	R, S, C	Cleansers, dusts, dyes, fabrics, hair, danders, solvents
Furniture workers	R, S, C	Dusts, fabrics, glue, lacquers, metals, mold, paint, plastics, sawdusts, solvents, woods, kapok
Furriers	R, S, C	Aniline dye, arsenic compounds, p-amidophenol, p-phenylenediamine
Garage workers	C	Chromium compounds, copper salts, detergents, gasolines, oils, rubber, soaps, solvents
Gardeners	R, S, C	Arsenic compounds, fertilizers, insecticides, lime dust, mold, plants, pollens, and substances from trees and plants
Grocers	R, S, C	Sugar, flours, flavorings, spices, fruits, vegetables, burlap, glue
Hair dressers and beauticians	C	Ammonium thioglycolate, hair dyes, wave set solutions, cosmetics, perfumes, resins, solvents
Insecticide workers	R, S, C	Insecticides, pyrethrum, DDT
Interior decorators	R, S, C	House dusts, dyes, colorings, paints, lacquers, resins, solvents, fabrics
Iron and steel workers	C	Chromium compounds, lime, "pickling" chemicals, nitrates, cyanides, "case hardeners," lubricants, quenching oils, mold makers' sand
Jewelers	C	Chromium salts, cyanides, nickel salts
Leather workers	C	Amido-azo-toluene hydrochloride, arsenic, sodium sulfide, bismark brown, dyes, lime, nigrosene, vegetable and chemical tanning agents, bichromates, sumac, tannins
Machinists	C	Solvents, oils, greases, cutting oils (especially insoluble)

TABLE 4 (*Continued*)

Occupation	Symptoms	Sensitizers—Allergens
Milliners	R, S, C	Dyes, arsenic, feathers, wool, felts, hair, dusts, glues, solvents, fabrics
Nurses	C	Mercury bichloride, caine drugs, formalin, lye, detergents, soaps, medicated oils and alcohols, streptomycin, penicillin, antiseptics
Painters	C	Aniline dyes, linseed oil, paint removers, pigments, dyes, arsenic compounds, solvents, thinners, tar distillates, turpentine, varnish removers, zinc chloride, bichromates
Paper makers	C	Dyes, glue, resins, wood pulp, mucilage, bleaching solutions, digesting solutions
Pharmacists	C	Chemicals, drugs, dyes, coloring, glue, solvents, oils, waxes, etc.
Photoengravers and lithographers	C	Bichromates, photodevelopers, solvents, inks, dyes
Photographers	C	Bichromates in blueprint developers, hydroquinones, metol, mercury salts, silver and platinum salts, pyrogallol, sulfides, p-phenylenediamine
Photogravure workers	C	p-Nitranilin red
Photostat workers	C	Chromates
Piano workers	C	Aniline dyes, arsenic in felts
Plywood workers	C	Casein, glues (especially those made from the synthetic resins), woods and wood dusts
Polishers of metals	C	Bichromates, oxalic acid, turpentine
Preserve packers	C	Fruits and vegetables, glues, inks, colorings
Printers	C	Inks, soaps, solvents, dyes, oils
Rayon manufacturers	C	Bleaches, carbon bisulfide, coning oils, wood pulp treated with sodium hydroxide
Resin manufacturers	C	Formaldehyde, hexamethylenetetramine, monomers of polymerized resins, phenol, urea
Restaurant workers	C	Alkali cleansers, fruits, vegetables, juices, flavorings, extracts, spices, colorings, flours, soaps, mold, insecticides
Road workers	R, S, C	Asphalt, tar, "poison" plants, pollens of plants in the vicinity
Soap manufacturers	C	Aromatics added to soaps, soap, detergents, volatile oils, colorings
Solderers	C	Zinc chloride, chromium compounds
Varnishers and lacquerers	C	Resins, solvents, thinners, dyes, pigments, lacquers

Allergenic Potential

It has been noted that allergenic materials are a cause of allergic disease. However, this should be further qualified. The allergenicity of all substances is not the same. Some materials, such as poison ivy, buckwheat, p-phenylenediamine, cotton-seed, castor bean dust, chocolate, fish glue, wintergreen, etc., are very powerful allergens. These are said to have a *high allergenic potential*. It is found that extremely small amounts of these allergens acting over a relatively short time are able to produce severe symptoms in the allergic person. Conversely, allergins in large amounts which act over a long period of time producing only slight symptoms are said to have a low allergenic potential. Examples of low potential materials are boric acid, rice dust, and tea. A material which in moderate amounts acting over a moderate period of time causes only moderate symptoms is said to have a moderate allergenic potential. Most substances are in this class.

Table 5 contains a list of substances which are among the more frequent causes of allergic difficulty. They are arranged whenever possible according to a general classification, such as fruits, furs, waxes, etc. This classification does not mean that all the materials in this class have the same ability to cause allergic troubles, but it will allow for a more rapid survey of the contactants. The column "Danger of Sensitization by Testing" has been included because, even though it is not a frequent phenomenon, it is an extremely important one and should be avoided if possible.

Pre-employment Evaluation

History. In the pre-employment physical examination stress should be laid on the history of allergic disease in the person and in his family. Here the physician may prevent the development of many chronically allergic patients. If a patient has hay fever or allergic eczema or is known to show hives or angioedema under specific circumstances, it is better that he be placed in a job as free from allergens as possible because he is much more likely to develop allergic symptoms when in an atmosphere of allergenic substances than if he were a non-allergic individual.

Patch Test. The pre-employment patch test is not recommended to discover which workers may be allergic. Such a test may sometimes cause sensitization in an individual who previously was not sensitive to a particular substance. Another disadvantage is that a worker who is rejected because of a positive patch test may find that he has legal claim against the company, since after the test he had a skin eruption which he did not have before.

Protective Measures

Protective measures for personnel against the effects of allergens are precisely the same as those described elsewhere (see Section 2).

However, to the extent that allergenic sensitivity is often a personal thing in that *individual* allergies are so common, it is often much more practical to consider some of the individual protective measures described below.

TABLE 5. Some Common Substances Causing Allergy

Key to Abbreviations

Symptoms	Danger of Sensitization	Allergenic Potential
C = Contact dermatitis	Negl = Negligible	Mod = Moderate
R = Respiratory: nasal, pulmonary	Slt = Slight	Mkd = Marked
	Mod = Moderate	Slt = Slight
S = Skin: hives, eczema	Var = Variable	Var = Variable
	Mkd = Marked	P.I. = Primary irritant

Type of Test	Vehicle	Type of Action
P = Patch test	Alc = Alcohol	Cont = Contact
I.C. = Intracutaneous (or scratch test)	Aq = Aqueous	Inh = Inhalation
	Acet = Acetone	
	Petr = Petrolatum	

The concentration of the test material for scratch or intracutaneous testing is left to the attending physician.

Substance	Symptoms	Type of Test	Concentration or % for Testing	Vehicle	Danger of Sensitization by Testing	Allergenic Potential	Action: Contact, Inhalation
Adhesive plaster	C	P	As is	—	Negl	Low-Mod	Cont
Ammonium fluoride	C	P	1%	Aq	Slt	Mod	Cont
Barley	R, S	I.C.	—	—	Negl	Mod	Inh
Buckwheat	R, S	I.C.	—	—	Negl	Mkd	Inh
"Butesin" picrate	C	P	1%	Alc	Negl	Mod	Cont
Almond oil	C	P	100%	—	Negl	Mod	Cont
Alum	C	P	10%	Aq	Negl	Low	Cont
Ammoniated mercury	C	P	5%	Petr	Negl	Mod	Cont

TABLE 5 (*Continued*)

Substance	Symptoms	Type of Test	Concentration or % for Testing	Vehicle	Danger of Sensitization by Testing	Allergenic Potential	Action: Contact, Inhalation
Auto polishes	C	P	10%	Aq	Slt to Mod	Mod	Cont
Bleaching powders	C	P	10%	Aq	Slt to Mod	Mod	Cont
Bluing	C	P	As is	—	Negl	Low	Cont
Boric acid	C	P	50%	Petr	Negl	Low	Cont
Chrysanthemum	C	P	50–	Aq	Slt	Mod	Cont
	R, S	I.C.	100%		Slt	Mkd	Inh
	C	P	As is	—	Negl	Mod	Cont
Casein	R, S	I.C.			Negl	Mod	Inh
Castor Bean (dust)	R, S	I.C.	(Great Caution Advised)			Very Mkd	Inh
Cereal	C	P	As is	—	Negl	Slt to Mod	Cont
grains	R, S	I.C.			Negl	Mod	Inh
Cocoa	C	P	100%	—	Negl	Mkd	Cont
	R, S	I.C.			Negl	Mkd	Inh
Copper sulfate	C	P	1–5%	Aq	Negl	Mod	Cont
Corn	C	P	As is	—	Negl	Mod	Cont
	R, S	I.C.			Negl	Mod	Inh
Cosmetics	C	P	100%	—	Negl	Mod	Cont
	R, S	I.C.			Negl	Mkd	Inh
Cottonseed	R, S	I.C.	(Use caution)		Negl	Very Mkd	Inh
Crayons	C	P	50–100%	Petr	Negl	Low	Cont
Daffodil	C	P	Leaf or bulb	—	Negl	Low	Cont
Deodorants	C	P	50–100%	Aq	Negl	Low	Cont
Egg white	C	P	50–100%	Aq	Negl	Mod	Cont
	R, S	I.C.			Negl	Mkd	Inh
Feathers	C	P	As is	—	Negl	Mod	Cont
	R, S	I.C.			Negl	Mod	Inh
Fish	C	P	Var?	Aq	Negl	Mod-Mkd	Cont
	R, S	I.C.	(Use caution)		Negl	Mod-Mkd	Inh
Formaldehyde	C	P	2%	Aq	Negl	Mod	Cont
Flaxseed	C	P	As is	—	Negl	Mod	Cont
	R, S	I.C.	(Use caution)		Negl	Mkd	Inh
Flavoring oils	C	P	2%	Alc	Negl	Slt-Mod	Cont
Floor wax	C	P	10%	Petr	Var	Var	Cont
Flour	C	P	As is	—	Negl	Var	Cont
	R, S	I.C.	(Use caution)		Negl	Var	Inh
Flowers	C	P	As is	—	Var	Var	Cont
	R, S	I.C.			Var	Var	Inh
Fruits	C	P	50–100%	Aq	Negl but some slt	Slt to Mkd	Cont
Furs	C	P	As is	—	Negl	Mod	Cont
	R, S	I.C.			Negl	Mod	Inh
Glue	C	P	10–50%	Aq	Negl	Mkd	Cont
	R, S		(Dangerous—Use great caution)			Mkd	Inh
Glycerine	C	P	As is	—	Negl	Slt	Cont
Green soap	C	P	1–5%	Aq	Negl	Slt	Cont
Hair—all kinds	C	P	As is	Aq	Negl	Mod	Cont
	R, S	I.C.			Negl	Mod	Inh
Hair dyes	C	P	10–100%	Aq	Var and undetermined	Slt to Mkd	Cont

TABLE 5 (*Continued*)

Substance	Symptoms	Type of Test	Concentration or % for Testing	Vehicle	Danger of Sensitization by Testing	Allergenic Potential	Action: Contact, Inhalation
House dust	C	P	As is	—	Negl	Slt	Cont
	R, S	I.C.			Negl	Mod	Inh
Hops	C	P	As is	—	Negl	Mod	Cont
	R, S	I.C.			Negl	Low	Inh
Inks	C	P	10–100%	Aq	Negl-Slt	Slt-Mod	Cont
	R, S	I.C.			Negl	Slt-Mod	Inh
Iodoform	C	P	25%	Petr	Negl	Slt	Cont
Juniper oil	C	P	1%	Alc	Negl	Slt	Cont
Kapok	R, S	I.C.	—	—	Negl	Mod	Inh
Karaya gum	R, S	I.C.	—	—	Negl	Mod	Inh
Kerosene	C	P	25–50%	Petr	Negl	P.I.	Cont
Lacquers	C	P	100%	—	Negl-Mod	Mod-Mkd	Cont
Lanolin	C	P	100%	—	Negl	Mod	Cont
Latex	C	P	100%	—	Negl	Mod	Cont
Leathers	C	P	As is	—	Negl	Slt	Cont
Lysol	C	P	1%	Aq	Negl	Mod	Cont
Mercury bichloride	C	P	0.1%	Aq	Negl	Mod	Cont
Metals	C	P	As is	—	Negl	Slt-Mod	Cont
Metal alloys	C	P	As is	—	Negl	Slt-Mod	Cont
Methyl salicylate	C	P	2%	Petr	Mkd	Mod	Cont
Mold spores	R, S	I.C.	—	—	Negl	Mod-Mkd	Inh
Mucilage	C	P	100%	—	Negl	Mod	Cont
Mustard oil	C	P	1%	Alc	Negl	Mod	Cont
Nickel sulfate	C	P	5%	Aq	Negl	Mod	Cont
Novocaine	C	P	1%	Aq	Negl	Mod	Cont
Nylon	C	P	As is	—	Negl	Slt	Cont
	R, S	I.C.			Negl	Slt	Inh
Oats	C	P	As is	—	Negl	Slt	Cont
	R, S	I.C.			Negl	Slt	Inh
Orange oil	C	P	1%	Alc	Negl	Mod	Cont
Orris root	C	P	As is	—	Negl	Slt	Cont
	R, S	I.C.			Negl	Mod	Inh
Peanut	R, S	I.C.	—	—	Negl	Mkd	Inh
Penicillin	C	P	Var	Aq	Slt	Mkd	Cont
	R, S	I.C.	Var		Slt	Mkd	Inh
Perfume oils	C	P	1%	Petr	Negl	Mod	Cont
Petroleum	C	P	10–20%	Petr	Negl	Slt	Cont
Phenol	C	P	2%	Aq	Negl	Slt	Cont
p-Phenylenediamine	C	P	2%	Petr	Mod	Mkd	Cont
Photographic developers	C	P	5%	Aq	Negl	Mod	Cont
Picric acid	C	P	1–5%	Aq	Negl	Mod	Cont
Plants	C	P	As is	—	Slt-Mkd	Slt-Mkd	Cont
Poison ivy	C	P	0.01%	Acet	Very Mkd	Very Mkd	Cont
Poison oak	C	P	0.01%	Acet	Very Mkd	Very Mkd	Cont
Poison sumac	C	P	0.01%	Acet	Very Mkd	Very Mkd	Cont
Powder, cleansing	C	P	As is	—	Negl	Slt	Cont
Potassium dichromate	C	P	0.5%	Aq	Negl	Mod	Cont

TABLE 5 (*Continued*)

Substance	Symptoms	Type of Test	Concentration or % for Testing	Vehicle	Danger of Sensitization by Testing	Allergenic Potential	Action: Contact. Inhalation
Primrose	C	P	25%	Aq	Very Mkd	Mkd	Cont
Pyrethrum	C	P	20%	Alc	Negl	Mkd	Cont
	R, S	I.C.			Negl	Mkd	Inh
Ragweed leaf	C	P	As is	—	Negl	Mod	Cont
Ragweed pollen	C	P	1–5%	Petr	Negl	Mod	Cont
	R, S	I.C.			Negl	Mkd	Inh
Resins	C	P	50–100%	Alc or Petr	Negl to Mkd	Mod to Mkd	Cont
Rubber, natural	C	P	As is	—	Negl	Mod	Cont
Rubber, synthetic	C	P	As is	—	Negl	Mod	Cont
Sawdust	C	P	As is	—	Negl	Low	Cont
	R, S	I.C.			Negl	Low	Inh
Shellacs	C	P	50–100%	Alc	Negl to Slt	Mod	Cont
Silk	R, S	I.C.	Test Dil	—	Negl	Mod	Inh
Silver nitrate	C	P	5%	Aq	Negl	Slt	Cont
Soaps	C	P	1–3%	Aq	Negl	Slt–Mkd	Cont
Sulfonamides	C	P	5–50%	Petr	Mod	Mkd	Cont
Sulfur, ppt.	C	P	5%	Petr	Negl	Mod	Cont
Tannic acid	C	P	1%	Aq	Negl	Mod	Cont
Tars	C	P	As is	Do not cover	Negl	Mod	Cont
Tobacco	C	P	As is	—	Negl	Mod	Cont
	R, S	I.C.			Negl	Mod	Inh
Tulip bulb	C	P	As is	—	Negl	Low	Cont
Turpentine	C	P	25–50%	Petr	Negl	Slt	Cont
Varnishes	C	P	25–100%	Alc or Petr	Negl to Slt	Slt to Mod	Cont
Vegetables	C	P	As is	—	Negl	Slt–Mod	Cont
Waxes	C	P	25–50%	Petr	Negl	Slt–Mod	Cont
Wheat	C	P	As is	—	Negl	Mod	Cont
	R, S	I.C.			Negl	Mod	Cont
Wintergreen, oil	C	P	0.5%	Alc	Very Mkd	Mkd	Cont
Wood and pulp	C	P	As is	—	Negl	Low	Cont
	R, S	I.C.			Negl	Low	Inh
Wools	C	P	As is	—	Negl	Mod	Cont
	R, S	I.C.			Negl	Mod	Inh

Protective Ointments. Protective ointment, even though it is not the best method of keeping allergens from reaching the skin, is usually the most practical—more so than masks and gloves. The protective ointment may be easily washed off with soap and water, and the washing not only removes the protectant but the irritant as well. Many types of protective ointments are available to prevent skin irritations, but if they are to do the job, they must have certain properties. They should be non-irritating and non-sensitizing. They should really act as a barrier to the irritant. They should be of such consistency as to apply easily. They should stay on the skin while the worker is exposed but yet should be easily removed by soap and water after work. In recent years, for instance, the silicone containing protective ointments and creams have been used successfully.

Respirators. Mechanical filter respirators can be used to prevent the inhalation of various dusts, fumes, fogs, or mists. The mechanical trapping of particles suspended in air that is breathed can effect a tremendous amount of relief on the part of the user of such a mask. Since various filter respirators are available to industry, the proper type of filter should be ascertained for each individual substance before it is recommended to the worker. Mechanical filter respirators are not very comfortable, it is true, but a worker exposed to large amounts of irritating suspended substances should be advised to wear one.

Protective Clothing. Protective clothing, if it is properly designed, is of great value in the prevention of occupational dermatitis. Closely woven fabrics of cotton are more or less impervious to dusts, but this type of clothing should be laundered at the end of each day. Rubber or rubberized cloth, of course, affords better protection, but under certain industrial conditions this material deteriorates rapidly. This may be overcome by the use of a synthetic resin film, which can be manufactured into all sorts of protective covering such as aprons, hoods, sleeves, and coveralls. The synthetic resin films can be nonflammable, comparatively cheap, and easy to keep clean with soap and water. If gloves are used, they should extend well up onto the arm to prevent irritation due to entry of the material at the cuff. Aprons should reach well up to the neck and extend below the knees. Cleanliness is one of the most important measures in the prevention of industrial dermatitis.

Even with control of all the factors, such as protection by masks, proper clothing, ventilation or respirators, a worker may be so extremely sensitive to the irritant that even the minute amounts that still get through to him may be enough to produce his disability. In such a case it is necessary to transfer him to another job.

Inoculation Treatment. To make a worker less sensitive to inhaled allergenic materials such as wool dust, the furs or dander of various animals, pollens, and the dusts from bedding or upholstered furniture, relatively effective inoculation treatment is available. The spacing, the amounts, and the concentrations of injections used should be left completely to the discretion of the attending physician. There are a number of substances which must not be administered by the inoculation method. Some of them are so unpredictable in their reactions that to subject the patient to such a series of desensitizations is dangerous.

Legal Considerations

The laws of various states differ as to just what constitutes an "allergic disease" in industry. It is recommended that before a case decision is made, the law be carefully investigated. Some states require that trauma be present at the time that the material or substance in question was first introduced into the skin before a "resultant allergic dermatitis" can be legally declared. In other states, this trauma is not required. There are various interpretations as to specific cause and aggravation, and in most states the law has been very carefully worded to include all of the usual circumstances. It is only recommended that those substances which seem likely to cause trouble be carefully watched, and either avoided or handled in such a way that a minimum of disability results from contact.

All the previous discussions are worded, whenever possible, in such a way as to be easily understood by non-medical readers. While this policy makes inevitable some inaccuracy in interpretation and use of terms, it has not been carried so far as to misinform the reader.

SECTION 10

FOOD ADDITIVES

Bernard L. Oser, Ph.D.
Food and Drug Research Laboratories, Inc.
Maspeth, New York

All foods are composed of chemical substances. The major ones are the nutrients, carbohydrates, proteins, lipids, vitamins and "minerals." In addition, many organic and inorganic substances, present in lesser amounts, contribute to the color, texture, taste, aroma and other qualities of food. Natural constituents may be present in their unchanged, natural form, but some are modified by reactions during production or processing. Added substances may likewise be present in both unchanged or chemically modified forms.

A general definition of a food additive is "a substance or mixture of substances, other than a basic food stuff, which is present in a food as a result of any aspect of production, processing, storage, or packaging."[1] Accidental contaminants are not included in this definition, which differs significantly from that contained in the Food, Drug, and Cosmetic Act,[2] namely,

"The term 'food additive' means any substance the intended use of which results or may reasonably be expected to result, directly or indirectly, in its becoming a component or otherwise affecting the characteristics of any food (including any substance intended for use in producing, manufacturing, packing, processing, preparing, treating, packaging, transporting, or holding food; and including any source of radiation intended for any such use), if such substance is not generally recognized, among experts qualified by scientific training and experience to evaluate its safety, as having been adequately shown through scientific procedures (or, in the case of a substance used in food prior to January 1, 1958, through either scientific procedures or experience based on common use in food) to be safe under the conditions of its intended use; · · · ".

In the implementation of the food additives law the term "safe" means that there is reasonable certainty, among qualified scientists, that no harm will result from its use. It is not required to prove beyond any possible doubt that the additive is safe under all conceivable circumstances of use or misuse.

The statute forbids the issuance of a food additive regulation if a fair evaluation of the data fails to establish that the proposed use of the additive will be safe. However, if an additive "is found to induce cancer when ingested by man or animal," it cannot be considered safe. Several potential food additives have been denied continued use by reason of this principle, e.g., the flavoring substances, oil of calamus and oil of sassafras (and its major constituent safrole), the food colors, butter yellow (dimethylamino-azobenzene) and FD & C Yellows 3 and 4 (in the latter cases because of the possible presence of traces of the bladder carcinogen 2-naphthylamine).

Nevertheless, it must be recognized that certain substances known to be carcinogenic in experimental animals are naturally present in trace amounts in various foods. Oft-cited examples are selenium, present in grains grown on seleniferous soil, but present in minute traces accompanying sulfur in foods, and aflatoxin, a carcinogenic metabolite of various species of molds, particularly of the *Aspergillus* genus. For administrative purposes, tolerances for these carcinogenic substances are set below minimal detectable levels as determined by analytical procedures of acceptable sensitivity. This realistic approach has been adopted on a purely pragmatic basis, but it does not assure the complete absence of the compounds in question.

Food additives may be classified in various ways. *Direct additives* (see above) are those intentionally added to foods to achieve some desirable purpose or effect for which they are usually expected to remain as components of the foods. *Indirect additives* are non-intentional in the sense that they may remain in the food only because they are unavoidable as a result of their use in some stage of production, processing, or packaging.

Food additives may also be classified according to their functions. For example, among direct food additives there are:

(1) Acids, alkalies, buffers, and neutralizing agents;
(2) Preservatives and antioxidants;
(3) Bleaching and maturing agents;
(4) Emulsifying, stabilizing, or thickening agents;
(5) Flavoring materials;
(6) Coloring agents;
(7) Nutritional supplements.

Indirect food additives generally fall into one or the other of the following functional categories:

(1) Pesticides used in the production of raw agricultural commodities but which remain as residues in processed foods;
(2) Veterinary drugs and growth stimulants remaining as residues in meat, milk, or eggs;
(3) Components of packaging materials which migrate into foods in contact with the surface of their containers,
(4) Lubricants used on the surfaces of processing equipment.

Under the terms of Section 409 of the Federal Food, Drug, and Cosmetic Act, food additives either must be covered by a regulation prescribing the conditions under which they may be safely used or must be exempt from such a regulation. Hence under conditions of proper (i.e., safe) use, these substances can hardly be regarded as "dangerous" chemicals. Even under normal handling conditions in food manufacturing establishments, they are rarely considered to be dangerous.

In the pages which follow, direct food additives have been listed in accordance with classifications under the Federal regulations, viz.:

    *Subpart A.   Definitions and procedural and interpretative regulations.

---

*The "Subpart" designations are those of Part 121 of the Code of Federal Regulations.

Subpart B.   Substances exempt from the requirements of tolerances on the ground that they are generally recognized as safe ("GRAS"; see Table 1).

Subpart C.   Food additives permitted in feed and drinking water of animals or for the treatment of food-producing animals (Table 2).

Subpart D.   Food additives permitted in food for human consumption (Table 3).

Subpart E.   Substances employed in the manufacture of food packaging materials, whose use has been exempt under the prior sanction provisions of the Act.

Subpart F.   Food additives resulting from contact with containers or equipment, or otherwise affecting food.

Certain common food ingredients such as salt, pepper, sugar, vinegar, baking powder, and sodium glutamate are regarded as foods rather than as food additives. Chemicals used as food additives are required to be of appropriate "food grade" and must be used in accordance with good manufacturing practice. The latter is defined to restrict the quantity employed to the minimum necessary to accomplish the intended physical or other technical effect in the food, and which quantity is reduced to the extent reasonably possible. The Food Protection Committee has published the Food Chemicals Codex,[3] a collection of monographs about 500 food chemicals describing their chemical and physical properties. These are regarded as standard specifications for "food grade" chemicals by the Food and Drug Administration.

Substances listed under Subpart B are classified in Section 121.101 of the Code of Federal Regulations[4] according to their functions. Subparagraphs (h) and (i) of this Section list substances which may migrate into food from paper or cotton materials used in food packaging.

Subpart C includes antibiotics, hormones, and other drugs added to animal feed in the form of concentrates or premixes for the purpose of stimulating growth, or preventing or treating diseases (such as mastitis, coccidiosis, respiratory infections, or parasitic infestations). It also includes adjuvants used in the manufacture of feeds. It should be noted that animal feeds are "foods" as defined in the Food, Drug, and Cosmetic Act.

Additives used to impart color to food are covered by the Color Additives Amendments of the Food, Drug, and Cosmetic Act. A color additive in food is deemed to be unsafe unless it is used in accordance with regulations permitting its use or exempting it from such regulations, and unless it is from a certified batch of the color, if the regulation so prescribes (Table 4).

Whereas it is not a "substance" in the usual sense of the word, radiation used in the production, processing, and handling of food comes under the scope of the Food Additives Amendment. The permitted sources of radiation and their respective regulations are as follows:

High dose gamma radiation	121.3002
Low dose gamma radiation	121.3003
High dose electron beam radiation	121.3004
High dose X-radiation	121.3005
Ultraviolet radiation	121.3006
Low dose electron beam radiation	121.3007

Each of these types of radiation has certain specific applications and no others are permitted except by amendment of the Regulations.

Part 121 of the Code of Federal Regulations referred to above[4] lists by functional categories substances permitted under regulations or exempt therefrom. Additions or amendments to regulations are published in the Federal Register. The Food Protection Committee of the National Academy of Sciences—National Research Council has published a monograph entitled "Chemicals Used in Food Processing"[5] which includes, in addition to the name and chemical formula of each substance, the foods in which it is used and the approximate levels of use.

A considerable number of chemicals appear on several lists of food additives because they have different functional applications.

Since meat and poultry products are under the jurisdiction of the U. S. Department of Agriculture rather than the Food and Drug Administration, food additives permitted in or on these products appear on separate lists. However, virtually all substances permitted by the U. S. Department of Agriculture under the Meat Inspection Act and the Poultry Products Inspection Act are included in the regulations under Section 409 (the Food Additives Amendment) of the Federal Food, Drug, and Cosmetic Act. The reason for the separate listing is that under the definition of food additives in the Food, Drug, and Cosmetic Act, substances are exempt if they are "used in accordance with a sanction or approval granted prior to the enactment of the Amendement, under the Meat Inspection Act, as amended and extended, or under the Poultry Products Inspection Act."

Included in the accompanying tables are the numbers of the sections of the Regulations describing each additive and its permitted uses. In the last column of Tables 2 and 3 are shown the major or representative uses of each substance or class of substances, as shown in the Regulations, except where the designation of the class or substance indicates its function.

It will be recalled that in some cases the substances mentioned are not used per se but are present as reaction products or are expressed in terms of the compound or element named in the tolerances (e.g., bromide or arsenic). It will be apparent from the tables that many substances perform either similar functions in different types of food or a variety of functions in the same class of foods.

Hazards associated with the use of pesticides or drugs are discussed in Section 12 as are those involved in the handling of chemical agents, such as acids, bases, solvents, etc.

Precautions for the safe handling of chemicals, as described in the labeling, are so generally observed in food manufacturing operations that accidents are extremely rare. A very few substances have been disallowed as food additives because of their potentially hazardous nature. Among these were dulcin, an artificial sweetening agent; safrole and coumarin, components of natural flavoring substances; acetoaminofluorene, a potent pesticide; and several azo dyes which were once permitted as certified food coloring agents. However, it was animal tests for toxicity or carcinogenicity, rather than adverse human experience, that led to the prohibition of continued use of these substances.

Among the substances sanctioned prior to enactment of the Food Additives Amendment to the Federal Food, Drug, and Cosmetic Act were a number of classes of chemicals used in the manufacture of food packaging materials. These are listed under Subpart E of the Food Additive Regulations in the following functional groups: antioxidants, antimycotics, driers, drying oils (as components of finished resins), release agents, stabilizers, and a miscellaneous group of substances used in the manufacture of paper and paperboard products. The use of these substances is predicated on their being of good commercial grade, suitable for association with food, and on good manufacturing practice (Table 5).

Food packaging materials are fabricated of metal, paper, plastics, rubber, cotton and wood. In certain cases, such as "tin" cans, the surfaces in contact with the food are coated with resins or lacquers of both natural and synthetic origin. They, like the synthetic plastic coatings, films, sheets, etc., may also contain plasticizers, moisture repellants, antioxidants, pigments, etc., not to mention the low molecular weight fractions (monomers, dimers, etc.) of polymeric resins. Many of these are capable of migrating to varying degrees into foods in direct contact with the surfaces of containers. In this connection, mention should also be made of sizings on cotton bags, preservatives in wood and paper (residues from pulp), and the components of adhesives employed in the manufacture of paper cartons.

The food packaging materials per se present no chemical hazard worthy of mention in this context. However, their constituents come within the scope of the food additives law because under conditions of use they may become indirect additives. Such hazards as the chemicals themselves may pose through direct contact would occur only in the plants where the plastics, adhesives, or other packaging materials are manufactured.

Many food additives are included among the items listed in Subpart F which covers substances which may migrate into food from containers or equipment. Because of the extremely remote possibility that they present any hazard as food additives, these items will not be listed here. They include such categories as:

Nylon resins	121.2502
Slimicides	121.2505
Materials used in resinous and polymeric coatings	121.2514
Chelating agents	121.2515
Defoaming agents	121.2557
Adhesives	121.2520

Lubricants	121.2531
Textiles and textile fibers	121.2535
Resin-bonded filters	121.2536
Sanitizing solutions	121.2547
Rubber articles intended for repeated use	121.2562
Antioxidants and/or stabilizers for polymers	121.2566
Resinous and polymeric coatings	121.2569
Components of paper and paperboard	121.2571

BIBLIOGRAPHY

1. Principles and Procedures for Evaluating the Safety of Food Additives, Food Protection Committee, National Academy of Sciences-National Research Council, Publication 750 (1959).
2. Federal Food, Drug, and Cosmetic Act, United States Code Title 21, U. S. Government Printing Office, Washington, D.C., May 1966.
3. Food Chemicals Codex First Edition, National Academy of Sciences-National Research Council, Publication 1406, Washington, D.C. (1966).
4. Code of Federal Regulations, Title 21 Part 121, U. S. Government Printing Office, Washington, D.C., January 1, 1967.
5. Chemicals Used in Food Processing, National Academy of Sciences-National Research Council, Publication 1274, Washington, D.C. (1965).

TABLE 1.   Substances That Are Generally Recognized as Safe for Their Intended Use

Part A

Product	Tolerance	Limitations or Restrictions
(1) Anticaking Agents		
Aluminum calcium silicate	2%	In table salt
Calcium silicate	5%	In baking powder
Calcium silicate	2%	In table salt
Magnesium silicate	do	do
Sodium aluminosilicate (sodium silicoaluminate).	do	
Sodium calcium aluminosilicate, hydrated (sodium calcium silicoaluminate).	do	
Tricalcium silicate	do	do
(2) Chemical Preservatives		
Ascorbic acid		
Ascorbyl palmitate		
Benzoic acid	0.1%	
Butylated hydroxyanisole	Total content of antioxidants not over 0.02% of fat or oil content, including essential (volatile) oil content of food	
Butylated hydroxytoluene	do	
Calcium ascorbate		
Calcium propionate		
Calcium sorbate		
Caprylic acid		In cheese wraps
Dilauryl thiodipropionate	Total content of antioxidants not over 0.02% of fat or oil content, including essential (volatile) oil content of the food	
Erythorbic acid		
Gum guaiac	0.1% (equivalent antioxidant activity 0.01%)	In edible fats or oils
Methylparaben (methyl-*p*-hydroxybenzoate)	0.1%	
Nordihydroguaiaretic acid	Total content of antioxidants not over 0.02% of fat or oil content including essential (volatile) oil content of the food	
Potassium bisulfite		Not in meats or in food recognized as source of vitamin B_1
Potassium m-bisulfite		do
Potassium sorbate		
Propionic acid		
Propyl gallate	Total content of antioxidants not over 0.02% of fat or oil content, including essential (volatile) oil content of the food	
Propylparaben (propyl *p*-hydroxybenzoate)	0.1%	
Sodium ascorbate		
Sodium benzoate	0.1%	
Sodium bisulfite		do
Sodium m-bisulfite		do

TABLE 1. Part A (*Continued*)

Product	Tolerance	Limitations or Restrictions
Sodium propionate		
Sodium sorbate		
Sodium sulfite		do
Sorbic acid		
Stannous chloride	0.0015% calculated as tin	
Sulfur dioxide		do
Thiodipropionic acid	Total content of antioxidants not over 0.02% of fat or oil content, including essential (volatile) oil content of the food	
Tocopherols		

(3) Emulsifying Agents

Product	Tolerance	Limitations or Restrictions
Cholic acid	0.1%	Dried egg whites
Desoxycholic Acid	do	do
Diacetyl tartaric acid esters of mono- and diglycerides of edible fats or oils, or edible fat-forming fatty acids		
Glycocholic acid	0.1%	do
Mono- and diglycerides of edible fats or oils, or edible fat-forming acids.		
Monosodium phosphate derivatives of mono- and diglycerides of edible fats or oils, or edible fat-forming fatty acids.		
Propylene glycol		
Ox bile extract	0.1%	Dried egg whites
Taurocholic acid (or its sodium salt)	do	do

(4) Nonnutritive Sweeteners

Ammonium saccharin
Calcium cyclamate (calcium cyclohexyl sulfamate)
Calcium saccharin
Magnesium cyclamate (magnesium cyclohexyl sulfamate)
Potassium cyclamate (potassium cyclohexyl sulfamate)
Saccharin
Sodium cyclamate (sodium cyclohexyl sulfamate)
Sodium saccharin

(5) Nutrients and/or Dietary Supplements[a]

Alanine (L- and DL-forms)
Arginine (L- and DL-forms)
Ascorbic acid
Aspartic acid (L- and DL-forms)
Biotin
Calcium carbonate

TABLE 1.   Part A (*Continued*)

Product	Tolerance	Limitations or Restrictions
Calcium citrate		
Calcium glycerophosphate		
Calcium oxide		
Calcium pantothenate		
Calcium phosphate (mono-, di-, tribasic)		
Calcium pyrophosphate		
Calcium sulfate		
Carotene		
Choline bitartrate		
Choline chloride		
Copper gluconate	0.005%	
Cuprous iodide	0.01%	In table salt as a source of dietary iodine
Cysteine (L-form)		
Cystine (L- and DL-forms)		
Ferric phosphate		
Ferric pyrophosphate		
Ferric sodium pyrophosphate		
Ferrous gluconate		
Ferrous lactate		
Ferrous sulfate		
Glycine (aminoacetic acid)		In animal feeds
Histidine (L- and DL-forms)		
Inositol		
Iron, reduced		
Isoleucine (L- and DL-forms)		
Leucine (L- and DL-forms)		
Linoleic acid (prepared from edible fats and oils and free from chick edema factor)		
Lysine (L- and DL-forms)		
Magnesium oxide		
Magnesium phosphate (di-, tribasic)		
Magnesium sulfate		
Manganese chloride		
Manganese citrate		
Manganese gluconate		
Manganese glycerophosphate		
Manganese hypophosphite		
Manganese sulfate		
Manganous oxide		
Mannitol	5%	In special dietary foods
Methionine		Animal feeds
Methionine hydroxy analog and its calcium salts		do
Niacin		
Niacinamide		
D-Pantothenyl alcohol		
Phenylalanine (L- and DL-forms)		
Potassium chloride		
Potassium glycerophosphate		

TABLE 1. Part A (*Continued*)

Product	Tolerance	Limitations or Restrictions
Potassium iodide	0.01%	In table salt as a source of dietary iodine
Proline (L- and DL-forms)		
Pyridoxine hydrochloride		
Riboflavin		
Riboflavin-5-phosphate		
Serine (L- and DL-forms)		
Sodium pantothenate		
Sodium phosphate (mono-, di-, tribasic)		
Sorbitol	7%	In foods for special dietary use
Thiamine hydrochloride		
Thiamine mononitrate		
Threonine (L- and DL-forms)		
Tocopherols		
α-Tocopherol acetate		
Tryptophan (L- and DL-forms)		
Tyrosine (L- and DL-forms)		
Valine (L- and DL-forms)		
Vitamin A		
Vitamin A acetate		
Vitamin A palmitate		
Vitamin B_{12}		
Vitamin D_2		
Vitamin D_3		
Zinc sulfate		
Zinc gluconate		
Zinc chloride		
Zinc oxide		
Zinc stearate (prepared from stearic acid free from chick edema factor)		

(6) Sequestrants

Product	Tolerance	Limitations or Restrictions
Calcium acetate		
Calcium chloride		
Calcium citrate		
Calcium diacetate		
Calcium gluconate		
Calcium hexa m-phosphate		
Calcium phosphate, monobasic		
Calcium phytate		
Citric acid		
Dipotassium phosphate		
Disodium phosphate		
Isopropyl citrate	0.02%	
Monoisopropyl citrate		
Potassium citrate		
Sodium acid phosphate		
Sodium citrate		
Sodium diacetate		
Sodium gluconate		
Sodium hexa m-phosphate		
Sodium m-phosphate		

TABLE 1. Part A (*Continued*)

Product	Tolerance	Limitations or Restrictions
Sodium phosphate (mono-, di-, tribasic)		
Sodium potassium tartrate		
Sodium pyrophosphate		
Sodium pyrophosphate, tetra		
Sodium tartrate		
Sodium thiosulfate	0.1%	In salt
Sodium tripolyphosphate		
Stearyl citrate	0.15%	
Tartaric acid		

(7) Stabilizers

Acacia (gum arabic)
Agar-agar
Ammonium alginate
Calcium alginate
Carob bean gum (locust bean gum)
Chondrus extract (carrageenin)
Ghatti gum
Guar gum
Potassium alginate
Sodium alginate
Sterculla gum (karaya gum)
Tragacanth (gum tragacanth)

(8) Miscellaneous and/or General Purpose Food Additives

Product	Tolerance	Limitations or Restrictions
Acetic acid		
Adipic acid		Buffer and neutralizing agent
Aluminum ammonium sulfate		
Aluminum potassium sulfate		
Aluminum sodium sulfate		
Aluminum sulfate		
Ammonium bicarbonate		
Ammonium carbonate		
Ammonium hydroxide		
Ammonium phosphate (mono- and dibasic)		
Ammonium sulfate		
Beeswax (yellow wax)		
Beeswax, bleached (white wax)		
Bentonite		
Butane		
Caffeine	0.02%	In cola type beverages
Calcium carbonate		
Calcium chloride		
Calcium citrate		
Calcium gluconate		
Calcium hydroxide		
Calcium lactate		
Calcium oxide		
Calcium phosphate (mono-, di-, tribasic)		

TABLE 1. Part A (*Continued*)

Product	Tolerance	Limitations or Restrictions
Caramel		
Carbon dioxide		
Carnauba wax		
Citric acid		
Dextrans (of average molecular weight below 100,000)		
Ethyl formate	0.0015%	As fumigant for cashew nuts
Glutamic acid		Salt substitute
Glutamic acid hydrochloride		do
Glycerin		
Glyceryl monostearate		
Helium		
Hydrochloric acid		Buffer and neutralizing agent
Hydrogen peroxide		Bleaching agent
Lactic acid		
Lecithin		
Magnesium carbonate		
Magnesium hydroxide		
Magnesium oxide		
Magnesium stearate		As migratory substance from packaging materials when used as a stabilizer
Malic acid		
Methylcellulose (U.S.P. methylcellulose, except that the methoxy content shall not be less than 27.5% and not more than 31.5% on a dry-weight basis)		
Monoammonium glutamate		
Monopotassium glutamate		
Nitrogen		
Nitrous oxide		Propellant for certain dairy and vegetable-fat toppings in pressurized containers
Papain		
Phosphoric acid		
Potassium acid tartrate		
Potassium bicarbonate		
Potassium carbonate		
Potassium citrate		
Potassium hydroxide		
Potassium sulfate		
Propane		
Propylene glycol		
Rennet (rennin)		
Silica aerogel (finely powdered microcellular silica foam having a minimum silica content of 89.5%)		Component of anti-foaming agent
Sodium acetate		
Sodium acid pyrophosphate		
Sodium aluminum phosphate		
Sodium bicarbonate		

TABLE 1. Part A (*Continued*)

Product	Tolerance	Limitations or Restrictions
Sodium carbonate		
Sodium citrate		
Sodium carboxymethylcellu- lose (the sodium salt of car- boxymethylcellulose not less than 99.5% on a dry-weight basis, with maximum substi- tution of 0.95 carboxymethyl groups per anhydroglucose unit, and with a minimum viscosity of 25 centipoises for 25% by weight aqueous solu- tion at 25°C)		
Sodium caseinate		
Sodium citrate		
Sodium hydroxide		
Sodium pectinate		
Sodium phosphate (mono-, di-, tribasic)		
Sodium potassium tartrate		
Sodium sesquicarbonate		
Sodium tripolyphosphate		
Succinic acid		
Sulfuric acid		
Tartaric acid		
Triacetin (glyceryl triacetate)		
Triethyl citrate	0.25%	Dried egg whites

[a]Amino acids listed may be free, hydrochloride salt, hydrated, or anhydrous form, where applicable.
[b]For the purpose of this list, no attempt has been made to designate those sequestrants that may also func- tion as chemical preservatives.

TABLE 1

Part B. Spices, Seasonings, Essential Oils, Oleoresins, and Natural Extractives[c]

(1) Spices and Other Natural Seasonings and Flavoring (Leaves, Roots, Barks, Berries, etc.)

Common Name	Botanical Name of Plant Source
Alfalfa herb and seed	*Medicago sativa* L.
Allspice	*Pimenta officinalis* Lindl.
Ambrette seed	*Hibiscus abelmoschus* L.
Angelica	*Angelica archangelica* L. or other spp. of *Angelica*
Angelica root	do
Angelica seed	do
Angostura (cusparia bark)	*Galipea officinalis* Hancock.
Anise	*Pimpinella anisum* L.
Anise, star	*Illicium verum* Hook. f.
Balm (lemon balm)	*Melissa officinalis* L.
Basil, bush	*Ocimum minimum* L.
Basil, sweet	*Ocimum basilicum* L.
Bay	*Laurus nobilis* L.
Calendula	*Calendula officinalis* L.
Camomile (chamomile), English or Roman	*Anthemis nobilis* L.
Camomile (chamomile), German or Hungarian	*Matricaria chamomilla* L.
Capers	*Capparis spinosa* L.
Capsicum	*Capsicum frutescens* L. or *Capsicum annuum* L.
Caraway	*Carum carvi* L.
Caraway, black (black cumin)	*Nigella sativa* L.
Cardamom (cardamon)	*Elettaria cardamomum* Maton.
Cassia, Chinese	*Cinnamomum cassia* Blume.
Cassia, Padang or Batavia	*Cinnamomum burmanni* Blume.
Cassia, Saigon	*Cinnamomum loureiril* Nees.
Cayenne pepper	*Capsicum frutescens* L. or *Capsicum annuum* L.
Celery seed	*Apium graveolens* L.
Chervil	*Anthriscus cerefolium* (L.) Hoffm.
Chives	*Allium schoenoprasum* L.
Cinnamon, Ceylon	*Cinnamomum zeylanicum* Nees.
Cinnamon, Chinese	*Cinnamomum cassia* Blume.
Cinnamon, Saigon	*Cinnamomum loureirii* Nees.
Clary (clary sage)	*Salvia sclarea* L.
Clover	*Trifolium* spp.
Cloves	*Eugenia caryophyllata* Thunb.
Coriander	*Coriandrum sativum* L.
Cumin (cummin)	*Cuminum cyminum* L.
Cumin, black (black caraway)	*Nigella sativa* L.
Dill	*Anethum graveolens* L.
Elder flowers	*Sambucus canadensis* L.
Fennel, common	*Foeniculum vulgare* Mill.
Fennel, sweet (finocchio, Florence fennel)	*Foeniculum vulgare* Mill. var. dulce (DO.) Alef.
Fenugreek	*Trigonella foenum-graecum* L.
Galanga (galangal)	*Alpinia officinarum* Hance.
Garlic	*Allium sativum* L.
Geranium	*Pelargonium* spp.
Ginger	*Zingiber officinale* Rosc.
Glycyrrhiza	*Glycyrrhiza glabra* L. and other spp. of *Glycyr-rhiza.*
Grains of paradise	*Amomum melegueta* Rosc.
Horehound (hoarhound)	*Marrubium vulgare* L.

[c]To this list and that under Part C below should be added the natural and synthetic flavoring substances declared safe for use under "omnibus" regulations (Sections 121.1163 and 121.1164).

TABLE 1.   Part B *(Continued)*

Common Name	*Botanical Name of Plant Source*
Horseradish	*Armoracia lapathifolia* Gilib.
Hyssop	*Hyssopus officinalis* L.
Lavender	*Lavandula officinalis* Chaix.
Licorice	*Glycyrrhiza glabra* L. and other spp. of *Glycyrrhiza.*
Linden flowers	*Tilia* spp.
Mace	*Myristica fragrans* Houtt.
Marigold, pot	*Calendula officinalis* L.
Marjoram, pot	*Majorana onites* (L.) Benth.
Marjoram, sweet	*Majorana hortensis* Moench.
Mustard, black or brown	*Brassica nigra* (L.) Koch.
Mustard, brown	*Brassica juncea* (L.) Coss.
Mustard, white or yellow	*Brassica hirta* Moench.
Nutmeg	*Myristica fragrans* Houtt.
Oregano (oreganum, Mexican oregano, Mexican sage, origan)	*Lippia* spp.
Paprika	*Capsicum annuum* L.
Parsley	*Petroselinum crispum* (Mill.) Mansf.
Pepper, black	*Piper nigrum* L.
Pepper, cayenne	*Capsicum frutescens* L. or *Capsicum annuum* L.
Pepper, red	do.
Pepper, white	*Piper nigrum* L.
Peppermint	*Mentha piperita* L.
Poppy seed	*Papaver somniferum* L.
Pot marigold	*Calendula officinalis* L.
Pot marjoram	*Majorana onites* (L.) Benth.
Rosemary	*Rosmarinus officinalis* L.
Rue	*Ruta graveolens* L.
Saffron	*Crocus sativus* L.
Sage	*Salvia officinalis* L.
Sage, Greek	*Salvia triloba* L.
Savory, summer	*Satureia hortensis* L. (*Satureja*).
Savory, winter	*Satureia montana* L. (*Satureja*).
Sesame	*Sesamum indicum* L.
Spearmint	*Mentha spicata* L.
Star anise	*Illicium verum* Hook. f.
Tarragon	*Artemisia dracunculus* L.
Thyme	*Thymus vulgaris* L.
Thyme, wild or creeping	*Thymus serpyllum* L.
Turmeric	*Curcuma longa* L.
Vanilla	*Vanilla planifolia* Andr. or *Vanilla tahitensis* J. W. Moore
Zedoary	*Curcuma zedoaria* Rosc.

(2) Essential Oils, Oleoresins (Solvent Free), and Natural Extractives (Including Distillates)

Common Name	*Botanical Name of Plant Source*
Alfalfa	*Medicago sativa* L.
Allspice	*Pimenta officinalis* Lindl.
Almond, bitter (free from prussic acid)	*Prunus amygdalus* Batsch, *Prunus armeniaca* L., or *Prunus persica* (L.) Batsch.
Ambrette (seed)	*Hibiscus moschatus* Moench.
Angelica root	*Angelica archangelica* L.
Angelica seed	do.
Angelica stem	do.
Angostura (cusparia bark)	*Galipea officinalis* Hancock.

TABLE 1.   Part B (*Continued*)

Common Name	Botanical Name of Plant Source
Anise	*Pimpinella anisum* L.
Asafetida	*Ferula assa-foetida* L. and related spp. of *Ferula*.
Balm (lemon balm)	*Melissa officinalis* L.
Balsam of Peru	*Myroxylon pereirae* Klotzsch.
Basil	*Ocimum basilicum* L.
Bay leaves	*Laurus nobilis* L.
Bay (myrcia oil)	*Pimenta racemosa* (Mill.) J. W. Moore.
Bergamot (bergamot orange)	*Citrus aurantium* L. subsp. bergamia Wright et Arn.
Bitter almond (free from prussic acid)	*Prunus amygdalus* Batsch, *Prunus armeniaca* L. or *Prunus persica* (L.) Batsch.
Bois de rose	*Aniba rosaeodora* Ducke.
Cacao	*Theobroma cacao* L.
Camomile (chamomile) flowers, Hungarian	*Matricaria chamomilla* L.
Camomile (chamomile) flowers, Roman or English	*Anthemis nobilis* L.
Cananga	*Cananga odorata* Hook. f. and Thoms.
Capsicum	*Capsicum frutescens* L. and *Capsicum annuum* L.
Caraway	*Carum carvi* L.
Cardamon seed (cardamon)	*Elettaria cardamomum* Maton.
Carob bean	*Ceratonia siliqua* L.
Carrot	*Daucus carota* L.
Cascarilla bark	*Croton eluteria* Benn.
Cassia bark, Chinese	*Cinnamomum cassia* Blume.
Cassia bark, Padang or Batavia	*Cinnamomum burmanni* Blume.
Cassia bark, Saigon	*Cinnamomum loureirii* Nees.
Celery seed	*Aplum graveolens* L.
Cherry, wild, bark	*Prunus serotina* Ehrh.
Chervil	*Anthriscus cerefolium* (L.) Hoffm.
Chicory	*Cichorium intybus* L.
Cinnamon bark, Ceylon	*Cinnamomum zeylanicum* Nees.
Cinnamon bark, Chinese	*Cinnamomum cassia* Blume.
Cinnamon bark, Saigon	*Cinnamomum loureirii* Nees.
Cinnamon leaf, Ceylon	*Cinnamomum zeylanicum* Nees.
Cinnamon leaf, Chinese	*Cinnamomum cassia* Blume.
Cinnamon leaf, Saigon	*Cinnamomum loureirii* Nees.
Citronella	*Cymbopogon nardus* Rendle.
Citrus peels	*Citrus* spp.
Clary (clary sage)	*Salvia sclarea* L.
Clove bud	*Eugenia caryophyllata* Thunb.
Clove leaf	do.
Clove stem	do.
Clover	*Trifolium* spp.
Coca (decocainized)	*Erythroxylum coca* Lam. and other spp. of *Erythroxylium*.
Coffee	*Coffea* spp.
Cola nut	*Cola acuminata* Schott and Endl., and other spp. of *Cola*.
Coriander	*Coriandrum sativum* L.
Corn silk	*Zea mays* L.
Cumin (cummin)	*Cuminum cyminum* L.
Curacao orange peel (orange, bitter peel)	*Citrus aurantium* L.
Cusparia bark	*Galipea officinalis* Hancock.
Dandelion	*Taraxacum officinale* Weber and *T. laevigatum* DC.

TABLE 1.   Part B (*Continued*)

Common Name	Botanical Name of Plant Source
Dandelion root	do.
Dill	*Anethum graveolens* L.
Dog grass (quackgrass, triticum)	*Agropyron repens* (L.) Beauv.
Elder flowers	*Sambucus canadensis* L. and *S. nigra* L.
Estragole (esdragol, esdragon, tarragon)	*Artemisia dracunculus* L.
Estragon (tarragon)	do.
Fennel, sweet	*Foeniculum vulgare* Mill.
Fenugreek	*Trigonella foenum-graecum* L.
Galanga (galangal)	*Alpinia officinarum* Hance.
Garlic	*Allium sativum* L.
Geranium	*Pelargonium* spp.
Geranium, East Indian	*Cymbopogon martini* Stapf.
Geranium, rose	*Pelargonium graveolens* L'Her.
Ginger	*Zingiber officinale* Rosc.
Glycyrrhiza	*Glycyrrhiza glabra* L. and other spp. of *Glycyrrhiza*.
Glycyrrhizin, ammoniated	do.
Grapefruit	*Citrus paradisi* Macf.
Guava	*Psidium* spp.
Hickory bark	*Carya* spp.
Horehound (hoarhound)	*Marrubium vulgare* L.
Hops	*Humulus lupulus* L.
Horsemint	*Monarda punctata* L.
Hyssop	*Hyssopus officinalis* L.
Immortelle	*Helichrysum augustifolium* DC.
Jasmine	*Jasminum officinale* L. and other spp. of *Jasminum*.
Juniper (berries)	*Juniperus communis* L.
Kola nut	*Cola acuminata* Schott and Endl., and other spp. of *Cola*.
Laurel berries	*Laurus nobilis* L.
Laurel leaves	*Laurus* spp.
Lavender	*Lavandula officinalis* Chaix.
Lavender, spike	*Lavandula latifolia* Vill.
Lavandin	Hybrids between *Lavandula officinalis* Chaix and *Lavandula latifolin* Vill.
Lemon	*Citrus limon* (L.) Burm. f.
Lemon balm (see balm)	
Lemon grass	*Cymbopogon citratus* DC. and *Cymbopogon flexuosus* Stapf.
Lemon peel	*Citrus limon* (L.) Burm. f.
Licorice	*Glycyrrhiza glabra* L. and other spp. of *Glycyrrhiza*.
Lime	*Citrus aurantifolia* Swingle.
Linden flowers	*Tilia* spp.
Locust bean	*Ceratonia siliqua* L.
Lupulin	*Humulus lupulus* L.
Mace	*Myristica fragrans* Houtt.
Malt (extract)	*Hordeum vulgare* L., or other grains.
Mandarin	*Citrus reticulata* Blanco.
Marjoram, sweet	*Majorana hortensis* Moench.
Maté	*Ilex paraguariensis* St. Hil.
Melissa (see balm)	
Menthol	*Mentha* spp.
Menthyl acetate	do.

TABLE 1. Part B (*Continued*)

Common Name	Botanical Name of Plant Source
Molasses (extract)	*Saccharum officinarum* L.
Mustard	*Brassica* spp.
Naringin	*Citrus paradisi* Macf.
Neroli, bigarade	*Citrus aurantium* L.
Nutmeg	*Myristica fragrans* Houtt.
Onion	*Allium cepa* L.
Orange, bitter, flowers	*Citrus aurantium* L.
Orange, bitter, peel	do.
Orange leaf	*Citrus sinensis* (L.) Osbeck.
Orange, sweet	do.
Orange, sweet, flowers	do.
Orange, sweet, peel	do.
Origanum	*Origanum* spp.
Palmarosa	*Cymbopogon martini* Stapf.
Paprika	*Capsicum annuum* L.
Parsley	*Petroselinum crispum* (Mill.) Mansf.
Pepper, black	*Piper nigrum* L.
Pepper, white	do.
Peppermint	*Mentha piperita* L.
Peruvian balsam	*Myroxylon pereirae* Klotzsch.
Petitgrain	*Citrus aurantium* L.
Petitgrain lemon	*Citrus limon* (L.) Burm. f.
Petitgrain mandarin or tangerine	*Citrus reticulata* Blanco.
Pimenta	*Pimenta officinalis* Lindl.
Pimenta leaf	do.
Pipsissewa leaves	*Chimaphila umbellata* Nutt.
Pomegranate	*Punica granatum* L.
Prickly ash bark	*Xanthoxylum* (or *Zanthoxylum*) *Americanum* Mill. or *Xanthoxylum clava-herculis* L.
Rose absolute	*Rosa alba* L., *Rosa centifolia* L., *Rosa damascena* Mill., *Rosa gallica* L., and vars. of these spp.
Rose (otto of roses, attar of roses)	do.
Rose buds	do.
Rose flowers	do.
Rose fruit (hips)	do.
Rose geranium	*Pelargonium graveolens* L'Her.
Rose leaves	*Rosa* spp.
Rosemary	*Rosmarinus officinalis* L.
Rue	*Ruta graveolens* L.
Saffron	*Crocus sativus* L.
Sage	*Salvia officinalis* L.
Sage, Greek	*Salvia triloba* L.
Sage, Spanish	*Salvia lavandulaefolia* Vahl.
St. John's bread	*Ceratonia siliqua* L.
Savory, summer	*Satureia hortensis* L.
Savory, winter	*Satureia montana* L.
Schinus molle	*Schinus molle* L.
Sloe berries (blackthorn berries)	*Prunus spinosa* L.
Spearmint	*Mentha spicata* L.
Spike lavender	*Lavandula latifolia* Vill.
Tamarind	*Tamarindus indica* L.
Tangerine	*Citrus reticulata* Blanco.
Tannic acid	Nutgalls of *Quercus infectoria* Oliver and related spp. of *Quercus*. Also in many other plants.
Tarragon	*Artemisia dracunculus* L.

TABLE 1.　Part B *(Continued)*

Common Name	*Botanical Name of Plant Source*
Tea	*Thea sinensis* L.
Thyme	*Thymus vulgaris* L. and *Thymus zygis* var. *gracilis* Boiss.
Thyme, white	do.
Thyme, wild or creeping	*Thymus serpyllum* L.
Triticum (see dog grass)	
Tuberose	*Polianthes tuberosa* L.
Turmeric	*Curcuma longa* L.
Vanilla	*Vanilla planifolia* Andr. or *Vanilla tahitensis* J. W. Moore.
Violet flowers	*Viola odorata* L.
Violet leaves	do.
Violet leaves absolute	do.
Wild cherry bark	*Prunus serotina* Ehrh.
Ylang-ylang	*Cananga odorata* Hook. f. and Thoms.
Zedoary bark	*Curcuma zedoaria* Rosc.

(3) Natural Substances Used in Conjunction with Spices and Other Natural Seasonings and Flavorings

Common Name	*Botanical Name of Plant Source*
Algae, brown (kelp)	*Laminaria* spp. and *Nereocystis* spp.
Algae, red	*Porphyra* spp. and *Rhodymenia palmata* (L.) Grev.
Dulse	*Rhodymenia palmata* (L.)

(4) Natural Extractives (Solvent Free) Used in Conjunction with Spices, Seasonings, and Flavorings

Common Name	*Botanical Name of Plant Source*
Algae, brown	*Laminaria* spp. and *Nereocystis* spp.
Algae, red	*Porphyra* spp. and *Rhodymenia palmata* (L.) Grev.
Apricot kernel (persic oil)	*Prunus armeniaca* L.
Dulse	*Rhodymenia palmata* (L.) Grev.
Kelp (see algae, brown).	
Peach kernel (persic oil)	*Prunus persica* Sieb. et Zucc.
Peanut stearine	*Arachis hypogaea* L.
Persic oil (see apricot kernel and peach kernel)	
Quince seed	*Cydonia oblonga* Miller.

(5) Miscellaneous

Common Name	*Derivation*
Ambergris	*Physeter macrocephalus* L.
Castoreum	*Castor fiber* L. and *C. canadensis* Kuhl.
Civet (zibeth, zibet, zibetum)	Civet cats, *Viverra civetta* Schreber and *Viverra Zibetha* Schreber.
Cognac oil, white and green	Ethyl oenanthate, so-called.
Musk (Tonquin musk)	Musk deer, *Moschus moschiferus* L.

TABLE 1

Part C.   Trace Minerals Added to Animal Feeds as Nutritional Dietary Supplements
at Levels Consistent with Good Feeding Practice[c]

Element	Source Compounds	Element	Source Compounds
Cobalt	Cobalt acetate	Iron	Iron ammonium citrate
	Cobalt carbonate		Iron carbonate
	Cobalt chloride		Iron chloride
	Cobalt oxide		Iron gluconate
	Cobalt sulfate		Iron oxide
Copper	Copper carbonate		Iron phosphate
	Copper chloride		Iron pyrophosphate
	Copper gluconate		Iron sulfate
	Copper hydroxide		Reduced iron
	Copper o-phosphate	Manganese	Manganese acetate
	Copper oxide		Manganese carbonate
	Copper pyrophosphate		Manganese citrate (soluble)
	Copper sulfate		Manganese chloride
Iodine	Calcium iodate		Manganese gluconate
	Calcium iodobehenate		Manganese o-phosphate
	Cuprous iodide		Manganese phosphate (dibasic)
	3,5-Diiodosalicylic acid		Manganese sulfate
	Ethylenediamine dihydroiodide		Manganous oxide
	Potassium iodate	Zinc	Zinc acetate
	Potassium iodide		Zinc carbonate
	Sodium iodate		Zinc chloride
	Sodium iodide		Zinc oxide
	Thymol iodide		Zinc sulfate

[c]All substances listed may be in anhydrous or hydrated form.

TABLE 1.

Part D.   Synthetic Flavoring Substances and Adjuvants That are Generally
Recognized as Safe for Their Intended Use[d]

Acetaldehyde (ethanal)

Acetoin (acetyl methylcarbinol)

Aconitic acid (equisetic acid, citridic acid, achilleic acid)

Anethole (p-propenyl anisole)

Benzaldehyde (benzoic aldehyde)

Brominated vegetable oils

n-Butyric acid (butanoic acid)

d- or l-Carvone (carvol)

Cinnamaldehyde (cinnamic aldehyde)

Citral (2,6-dimethyloctadien-2,6-al-8, geranial, neral)

Decanal (n-decylaldehyde, capraldehyde, capric aldehyde, caprinaldehyde, aldehyde C-10)

Diacetyl (2,3-butanedione)

Ethyl acetate

Ethyl butyrate

3-Methyl-3-phenylglycidic acid ethyl ester (ethyl methylphenylglycidate, so-called strawberry aldehyde, C-16 aldehyde).

Ethyl vanillin

Eugenol

Geraniol (3,7-dimethyl-2,6 and 3,6-octadien-1-ol)

Geranyl acetate (geraniol acetate)

Glycerol (glyceryl) tributyrate (tributyrin, butyrin)

Limonene (d-, l-, and dl-)

Linalool (linalol, 3,7-dimethyl-1,6-octadien-3-ol)

Linalyl acetate (bergamol)

l-Malic acid

Methyl anthranilate (methyl-2-aminobenzoate)

Piperonal (3,4-methylenedioxy-benzaldehyde, heliotropin)

Vanillin

[d]See footnote p. 288

TABLE 1

Part E.   Substances Migrating to Food from Paper and Paperboard Products
Used in Food Packaging That Are Generally Recognized as Safe for Their Intended Use,
Within the Meaning of Section 409 of the Act

Acetic acid
Alum (double sulfate of aluminum and ammonium potassium, or sodium)
Aluminum hydroxide
Aluminum oleate
Aluminum palmitate
Ammonium chloride
Ammonium hydroxide
Calcium chloride
Calcium hydroxide (lime)
Calcium sulfate
Casein
Cellulose acetate
Clay (kaolin)
Copper sulfate
Cornstarch
Corn sugar (sirup)
Dextrin
Diatomaceous earth filler
Ethyl cellulose
Ethyl vanillin
Ferric sulfate
Ferrous sulfate
Formic acid or sodium salt
Glycerin
Guar gum
Invert sugar
Iron, reduced
Locust bean gum (carob bean gum)
Magnesium carbonate
Magnesium chloride
Magnesium hydroxide
Magnesium sulfate
Methyl and ethyl acrylate
Mono- and diglycerides from glycerolysis of edible fats and oils

Oleic acid
Oxides of iron
Potassium sorbate
Propionic acid
Propylene glycol
Silicon dioxides
Pulps from wood, straw, bagasse, or other natural sources
Soap (sodium oleate, sodium palmitate)
Sodium aluminate
Sodium carbonate
Sodium chloride
Sodium hexa m-phosphate
Sodium hydrosulfite
Sodium hydroxide
Sodium phosphoaluminate
Sodium silicate
Sodium sorbate
Sodium sulfate
Sodium thiosulfate (additive in salt)
Sodium tripolyphosphate
Sorbitol
Soy protein, isolated
Sulfamic acid
Sulfuric acid
Starch, acid modified
Starch, pregelatinized
Starch, unmodified
Sucrose
Talc
Urea
Vanillin
Zinc hydrosulfite
Zinc sulfate

TABLE 1

Part F.   Substances Migrating to Food from Cotton and Cotton Fabrics
Used in Dry Food Packaging

Acacia (gum arabic)	Sodium bicarbonate
Acetic acid	Sodium carbonate
Beef tallow	Sodium chloride
Calcium chloride	Sodium hydroxide
Carboxymethylcellulose	Sodium sulfate
Coconut oil, refined	Sodium silicate
Corn dextrin	Sodium tripolyphosphate
Cornstarch	Sorbose
Fish oil (hydrogenated)	Soybean oil (hydrogenated)
Gelatin	Stearic acid
Guar gum	Talc
Hydrogen peroxide	Tall oil
Japan wax	Tallow (hydrogenated)
Lard	Tallow flakes
Lard oil	Tapioca starch
Lecithin (vegetable)	Tartaric acid
Locust bean gum (carob bean gum)	Tetrasodium pyrophosphate
Oleic acid	Urea
Peanut oil	Wheat starch
Potato starch	Zinc chloride
Sodium acetate	

TABLE 2.　Food Additives Permitted in Feed and Drinking Water of Animals or for the Treatment of Food-producing Animals

Substance	Section	Principal Use
Aklomide (2-chloro-4-nitrobenzamide)	121.269	Drug
Aluminum phosphide	121.281	Fumigant
Ammoniated rice hulls	121.285	Nutrient
Amprolium	121.210	Drug
Antibiotics for growth promotion and feed efficiency	121.225	Drugs
Arsanilic acid	121.253	Drug
Bacitracin	121.232	Drug
Bacitracin methylene disalicylate	121.252	Drug
Bithionol	121.238	Drug
Buquinolate (ethyl-4-hydroxy-6,7-diisobutoxy-3-quinoline carboxylate)	121.291	Drug
Calcium silicate	121.250	Anticaking agent
Carbophenothion	121.274	Pesticide
Choline xanthate	121.231	Nutrient
Chlortetracycline	121.208	Drug
Glutamic acid fermentation product	121.206	Protein
DDT	121.226	Pesticide
Demeton	121.221	Pesticide
Diethylcarbamazine	121.214	Drug
Diethylstilbestrol	121.241	Drug
Diammonium phosphate	121.265	Nutrient
Dienestrol diacetate	121.266	Drug
Dihydrostreptomycin sulfate	121.268	Drug
Dimetridazole	121.258	Drug
3,5-Dinitrobenzamide	121.263	Drug
Dioxathion	121.204	Pesticide
Disodium EDTA	121.271	Chelating agent
Diuron	121.218	Pesticide
Estradiol benzoate	121.245	Drug
Estradiol monopalmitate	121.257	Drug
Ethion (O, O, O', O'-tetraethyl-S, S'-methylenebisphosphorodithioate)	121.211	Pesticide
Ethoxyquin in animal feeds	121.202	Antioxidant
Ethoxyquin in certain dehydrated forage crops	121.201	Antioxidant
Ethromycin thiocyanate	121.292	Drug
Ethyl cellulose	121.230	Filler
Food additives for use in milk-producing animals	121.249	Drugs
Furazolidone	121.255	Drug
Griseofulvin	121.267	Drug
Hemicellulose extract	121.275	Nutrient
Hexachlorophene	121.278	Drug
Hygromycin B	121.213	Drug
Bromides, inorganic	121.270	Fumigation
Iron ammonium citrate	121.282	Nutrient
Iron-choline citrate complex	121.247	Nutrient
Lignin sulfonates	121.234	Pelleting aid
Malathion	121.228	Pesticide
Medroxyprogesterone acetate	121.276	Drug
Menadione dimethylpyrimidinol bisulfite	121.286	Nutrient
Methiotriazamine	121.239	Drug
Methyl esters of higher fatty acids	121.224	Fat
Mineral oil	121.246	Lubricant, etc.
Nihydrazone	121.237	Drug
3-Nitro-4-hydroxyphenylarsonic acid	121.262	Drug
Nitrodan (3-methyl-5-[(p-nitrophenyl)azo]-rhodanine)	121.288	Drug
Nitrofurazone	121.248	Drug

TABLE 2. (*Continued*)

Substance	Section	Principal Use
Novobiocin	121.212	Drug
Nystatin	121.220	Drug
O,O-Diethyl S-2-(ethylthio)ethyl phosphorodithioate	121.215	Pesticide
O,O-Dimethyl S-4-oxo-1,2,3-benzotriazin-3-(4H)-ylmethyl phosphorodithioate	121.240	Pesticide
Petroleum hydrocarbons, odorless light	121.277	Insecticide formulation
Oxytetracycline	121.251	Drug
Petrolatum	121.261	Lubricant, etc.
Phenothiazine	121.279	Drug
Piperonyl butoxide	121.289	Pesticide
Polyethylene glycols	121.259	Vehicle
Polyoxyethylene glycol (400) mono- and dioleates	121.203	Emulsifier
Polysorbate 60 [polyoxyethylene (20) sorbitan monostearate]	121.236	Emulsifier
Polysorbate 80	121.235	Emulsifier
Procaine penicillin	121.256	Drug
Progesterone	121.243	Drug
Promazine hydrochloride	121.219	Drug
Pyrethrins	121.290	Pesticide
Pyrophyllite	121.273	Anticaking agent
Reserpine	121.205	Drug
Ronnel	121.209	Pesticide
Silicon dioxide	121.229	Anticaking agent
Sodium arsanilate	121.254	Drug
Sodium 2,2-dichloropropionate	121.216	Pesticide
Sodium nitrite	121.223	Preservative
Sorbitan monostearate	121.272	Emulsifier
Sulfaethoxypyridazine	121.280	Drug
Sulfanitran [acetyl-(p-nitrophenyl)-sulfanilamide]	121.264	Drug
Synthetic isoparaffinic petroleum hydrocarbons	121.287	Insecticide formulation
TDE (DDD)	121.227	Pesticide
Testosterone propionate	121.244	Drug
Thiabendazole	121.260	Drug
Tylosin	121.217	Drug
Verxite	121.222	Bulking agent
Yellow prussiate of soda	121.284	Anticaking agent
Zinc bacitracin	121.233	Drug
Zoalene	121.207	Drug

TABLE 3.   Food Additives Permitted in Food for Human Consumption

Substance	Section	Principal Use
Acetone	121.1042	Solvent
Acetone peroxides	121.1023	Dough conditioner
Acetylated monoglycerides	121.1018	Emulsifier
Acetyl-(p-nitrophenyl)-sulfanilamide	121.1169	Drug residue
Acrylamide-acrylic acid resin	121.1092	Clarifying agent
Aluminum nicotinate	121.1141	Nutrient
Aluminum phosphide	121.1178	Fumigant
Amprolium	121.1022	Drug residue
Arabinogalactan	121.1174	Emulsifier
Arsenic	121.1138	Drug residue
Azodicarbonamide	121.1085	Flour maturing agent
Butylated hydroxyanisole (BHA)	121.1035	Antioxidant
Butylated hydroxytoluene (BHT)	121.1034	Antioxidant
Bacitracin, zinc bacitracin, manganese bacitracin, bacitracin methylene disalicylate	121.1005	Drug
Bacterial catalase	121.1170	Enzyme
Bithionol	121.1107	Drug
Boiler water additives	121.1088	
Buquinolate (ethyl-4-hydroxy-6,7-diisobutoxy-3-quinoline carboxylate)	121.1002	Drug
1,3-Butylene glycol	121.1176	Solvent
Calcium disodium EDTA	121.1017	Chelating agent
Calcium lactobionate	121.1162	Firming agent
Calcium lignosulfonate	121.1102	Dispersing agent
Calcium pantothenate, calcium chloride double salt	121.1037	Nutrient
Calcium silicate	121.1135	Anticaking agent
Calcium stearyl-2-lactylate	121.1047	Dough conditioner
Captan	121.1061	Pesticide
Carrageenan	121.1066	Thickener
Carrageenan with polysorbate 80	121.1193	Thickener
Castor oil	121.1028	Release agent
Chemicals used for the control of microorganisms in cane sugar mills	121.1155	Sanitation
Chemicals used in washing fruits and vegetables	121.1091	Sanitation
Chewing gum base	121.1059	
Chlorobutanol	121.1131	Sanitation
Chlorhexidine	121.1175	Drug
2-Chloro-4-nitrobenzamide	121.1177	Drug
Chloropentafluoroethane	121.1181	Propellant
Chlortetracycline	121.1014	Drug
Coatings on fresh citrus fruit	121.1179	
Combustion product gas	121.1060	Air displacement
Coumarone-indene resin	121.1050	Coating
DDT	121.1093	Pesticide
Defoaming agents	121.1099	
Dehydroacetic acid	121.1089	Preservative
Dienestrol diacetate	121.1173	Drug
Diethyl pyrocarbonate	121.1117	Preservative
Dihydrostreptomycin	121.1051	Drug
Dimetridazole	121.1167	Drug
3,5-Dinitrobenzamide	121.1168	Drug
Dioctyl sodium sulfosuccinate	121.1137	Processing aid
Disodium EDTA	121.1056	Chelating agent
Disodium guanylate	121.1109	Flavor enhancer

TABLE 3. (*Continued*)

Substance	Section	Principal Use
Disodium inosinate	121.1090	Flavor enhancer
Dried yeasts	121.1125	Nutrient
D-Pantothenamide	121.1123	Nutrient
Erythromycin	121.1143	Drug
Estradiol benzoate	121.1128	Drug
Estradiol monopalmitate	121.1150	Drug
Ethion	121.1126	Pesticide
Ethopabate	121.1106	Drug
Ethoxyquin	121.1001	Antioxidant
Ethyl cellulose	121.1087	Filler
Ethyl formate	121.1096	Fumigant
Ethylenediamine	121.1184	
Ethylene dichloride	121.1040	Solvent
Ethylene oxide polymer	121.1161	Foam stabilizer
Fatty acids	121.1070	Lubricant, etc.
Food starch, modified	121.1031	Thickener
Fumaric acid and salts of fumaric acid	121.1130	Acidulant
Fumigants for grain mill machinery	121.1133	
Fumigants for processed grains used in production of fermented malt beverages	121.1152	
Furaltadone	121.1094	Drug
Furazolidone	121.1145	Drug
Furcelleran	121.1068	Thickener
Gibberellic acid and its potassium salt	121.1010	Malting
Glycerol ester of wood rosin	121.1084	Suspending agent
Glyceryl-lacto esters of fatty acids	121.1004	Emulsifiers
n-Heptyl p-hydroxybenzoate	121.1186	Preservative
Hexachlorophene	121.1188	Drug
Hexane residues	121.1045	Solvent
Hop extract, modified	121.1082	Flavor
Hydrogen cyanide	121.1072	Pesticide
Hydroxylated lecithin	121.1027	Emulsifier
Hydroxypropyl cellulose	121.1160	Thickener
Hydroxypropyl methylcellulose	121.1021	Emulsifier
Hygromycin B	121.1024	Drug
Bromides, inorganic	121.1020	Fumigant
Ion-exchange membranes	121.1180	Deacidifier
Ion-exchange resins	121.1148	
Iron ammonium citrate	121.1190	Nutrient
Iron-choline citrate complex	121.1100	Nutrient
Isopropyl alcohol	121.1043	Solvent
Kelp	121.1149	Nutrient
Lactylic esters of fatty acids	121.1048	Emulsifiers
Malathion	121.1172	Pesticide
Maleic hydrazide	121.1006	Pesticide
Mannitol	121.1115	Bodying agent
Medroxyprogesterone acetate	121.1187	Drug
Metaldehyde	121.1052	Pesticide
Methacrylic acid-divinylbenzene copolymer	121.1136	Vehicle
Methiotriazamine	121.1108	Drug
Methyl chloride	121.1083	Propellant
Methyl ethyl cellulose	121.1112	Thickener
Methyl formate	121.1062	Fumigant
Methyl glucoside-coconut oil ester	121.1151	Processing aid
Methylene chloride	121.1039	Solvent

TABLE 3. (*Continued*)

Substance	Section	Principal Use
Methylparaben	121.1158	Preservative
Modified hop extract	121.1082	Flavor
Modified polyacrylamide resin	121.1192	Clarifying agent
Monoglyceride citrate	121.1036	Stabilizer
Morpholine	121.1105	In coatings
Natural flavoring substances and natural substances used in conjunction with flavors	121.1163	
Neomycin	121.1104	Drug
Neomycin, polymyxin, hydrocortisone acetate, hydrocortisone sodium succinate in milk from dairy cows	121.1003	Drugs
Nicotinamide-ascorbic acid complex	121.1095	Nutrient
Nihydrozone	121.1103	Drug
Novobiocin	121.1033	Drug
Nystatin	121.1055	Drug
Octafluorocyclobutane	121.1065	Propellant
Petroleum hydrocarbons, odorless light	121.1182	Processing aid
Oleandomycin	121.1011	Drug
Oxystearin	121.1016	Release agent
Oxytetracycline	121.1046	Drug
Paraformaldehyde	121.1079	Pesticide
Partially defatted, cooked cottonseed flour or similar products derived from cottonseed intended for human consumption	121.1019	Food
Penicillin	121.1026	Drug
Petrolatum	121.1166	Lubricant
Petroleum wax	121.1156	Lubricant
Phenothiazine	121.1189	Drug
Piperonyl butoxide	121.1074	Pesticide
Polyacrylamide	121.1119	In gelatin capsules
Polyethylene glycol 400	121.1185	Emulsifier
Polyethylene glycol 6000	121.1057	Emulsifier
Polyethylene glycol, minimum molecular weight 1300	121.1121	Emulsifier
Polyglycerol esters of fatty acids	121.1120	Emulsifier
Polyoxyethylene (20) sorbitan tristearate	121.1008	Emulsifier
Polysorbate 60 [polyoxyethylene (20) sorbitan monostearate]	121.1030	Emulsifier
Polysorbate 80	121.1009	Emulsifier
Polyvinylpolypyrrolidone	121.1110	Clarifying agent
Potassium bromate	121.1194	Nutrient
Potassium iodide	121.1073	Nutrient
Potassium nitrate	121.1132	Curing agent
Prednisolone	121.1147	Drug
Prednisone	121.1157	Drug
Progesterone	121.1127	Drug
Promazine hydrochloride	121.1054	Drug
Propylene glycol alginate	121.1015	Thickener
Propylparaben	121.1159	Preservative
Propylene glycol mono- and diesters of fats and fatty acids	121.1113	Emulsifier
Propylene oxide	121.1076	Fumigant
Pteroylglutamic acid	121.1134	Nutrient
Pyrethrins	121.1075	Pesticide
Quinine	121.1081	Drug
Reserpine	121.1007	Drug
Rhizopus oryzae	121.1165	Enzyme
Rice bran wax	121.1098	Coating

TABLE 3. (*Continued*)

Substance	Section	Principal Use
Ronnel	121.1078	Pesticide
Safrole-free extract of sassafras	121.1097	Flavor
Salicylic acid	121.1140	Drug
Salts of carrageenan	121.1067	Thickener
Salts of fatty acids	121.1071	Anticaking agents
Salts of furcelleran	121.1069	Thickeners
Silicon dioxide	121.1058	Adsorbent
Sodium lauryl sulfate	121.1012	Emulsifier
Sodium methyl sulfate	121.1171	In pectin
Sodium nitrate	121.1063	Curing agent
Sodium nitrite	121.1064	Curing agent
Sodium stearyl fumarate	121.1183	Dough conditioner
Sorbitol	121.1053	Bodying agent
Sorbitan monostearate	121.1029	Emulsifier
Sperm oil, hydrogenated	121.1101	Lubricant
Stearyl monoglyceridyl citrate	121.1080	Emulsifier
Streptomycin	121.1025	Drug
Succinylated monoglycerides	121.1195	Emulsifier
Succistearin stearoyl propylene glycol hydrogen succinate)	121.1197	Emulsifier
Sugar beet extract flavor base	121.1086	Flavor
Sulfaethoxypyridazine	121.1144	Drug
Sulfamethazine	121.1124	Drug
Synthetic flavoring substances and adjuvants	121.1164	
Synthetic glycerin produced by the hydrogenolysis of carbohydrates	121.1111	Miscellaneous
Synthetic isoparaffinic petroleum hydrocarbons	121.1154	In coatings
Terpene resin	121.1077	Moisture barrier
Testosterone propionate	121.1129	Drug
Tetradifon	121.1038	Pesticide
THBP	121.1116	Antioxidant
Thiabendazole	121.1153	Drug
Toxaphene	121.1196	Pesticide
Trichloroethylene	121.1041	Solvent
Tylosin	121.1049	Drug
White mineral oil	121.1146	Lubricant
Xylitol	121.1114	Bodying agent
Yellow prussiate of soda	121.1032	Anticaking agent
Zoalene	121.1013	Drug

TABLE 4.   (a) Color Additives for Food and (b) Permitted Diluents

(a) Color Additive	Restrictions
Algae meal	
Annatto extract	
β-Apo-8'-carotenal	
Caramel	
β-Carotene	
Cottonseed flour	Toasted, partially defatted, cooked
Fruit juice	
Grape skin extract	
Paprika (and oleoresin)	
Saffron	
Tagetes, meal and extract	
Titanium oxide	
Turmeric (and oleoresin)	
Ultramarine blue	
Vegetable juice	
Citrus Red No. 2	Subject to certification
Orange B	Subject to certification
FD & C Yellow No. 5	Subject to certification
FD & C Green No. 3	Provisional until March 31, 1968
	Subject to certification
FD & C Yellow No. 6	Provisional until March 31, 1968
	Subject to certification
FD & C Red No. 2	Provisional until March 31, 1968
	Subject to certification
FD & C Red No. 3	Provisional until March 31, 1968
	Subject to certification
FD & C Blue No. 1	Provisional until March 31, 1968
	Subject to certification
FD & C Blue No. 2	Provisional until March 31, 1968
	Subject to certification
FD & C Violet No. 1	Provisional until March 31, 1968
	Subject to certification
FD & C Red No. 4	Provisional until March 31, 1968 in maraschino cherries; subject to certification
Lakes of FD & C color additives	See individual color additives

(b) Diluents	Restrictions
Castor oil	Not more than 500 ppm
(1) In Marking Inks for Gum and Confectionery	
n-Butyl alcohol	No residue in finished confectionery or gum
Cetyl alcohol	No residue in finished confectionery or gum
Cyclohexane	No residue in finished confectionery or gum
Ethyl cellulose	
Ethylene glycol monoethyl ether	No residue in finished confectionery or gum
Isobutyl alcohol	No residue in finished confectionery or gum
Isopropyl alcohol	No residue in finished confectionery or gum
Polyoxyethylene sorbitan monooleate (polysorbate 80)	
Polyvinyl acetate	
Polyvinylpyrrolidone	
Rosin and rosin derivatives	
Shellac, purified	

TABLE 4. (*Continued*)

(b) Diluents	Restrictions
(2) In Marking Inks for Fruit and Vegetables	
Acetone	No residue
Alcohol, SDA-3A	No residue
Benzoin	
Copal, Manila	
Ethyl acetate	No residue
Ethyl cellulose	
Methylene chloride	No residue
Polyvinylpyrrolidone	
Rosin and rosin derivatives	
Silicon dioxide	Not more than 2% of ink solids
Terpene resins, natural	
Terpene resins, synthetic	
(3) In Mixtures for Coloring Shell Eggs	
Alcohol, SDA-23A	No penetration
Damar gum (resin)	No penetration
Diethylene glycol distearate	No penetration
Dioctyl sodium sulfosuccinate	No penetration
Ethyl cellulose	No penetration
Ethylene glycol distearate	No penetration
Japan wax	No penetration
Limed rosin	No penetration
Naphtha	No penetration
Pentaerythritol ester of fumaric acid-rosin adduct	No penetration
Polyethylene glycol 6000	No penetration
Polyvinyl alcohol	No penetration
Rosin and rosin derivatives	No penetration

TABLE 5.   Substances Employed in the Manufacture of Food-packaging Materials for Which Prior Sanctions Have Been Granted

(a) Antioxidants (limit of addition to food, 0.005%)
   Butylated hydroxyanisole
   Butylated hydroxytoluene
   Dilauryl thiodipropionate
   Distearyl thiodipropionate
   Gum guaiac
   Nordihydroguaiaretic acid
   Propyl gallate
   Thiodipropionic acid
   2,4,5-Trihydroxybutyrophenone

(b) Antimycotics
   Calcium propionate
   Methylparaben (methyl *p*-hydroxybenzoate)
   Propylparaben (propyl *p*-hydroxybenzoate)
   Sodium benzoate
   Sodium propionate
   Sorbic acid

(c) Driers
   Cobalt caprylate
   Cobalt linoleate
   Cobalt naphthenate
   Cobalt tallate
   Iron caprylate
   Iron linoleate
   Iron naphthenate
   Iron tallate
   Manganese caprylate
   Manganese linoleate
   Manganese naphthenate
   Manganese tallate

(d) Drying Oils (as components of finished resins)
   Chinawood oil (tung oil)
   Dehydrated castor oil
   Linseed oil
   Tall oil

(e) Plasticizers
   Acetyl tributyl citrate
   Acetyl triethyl citrate
   p-tert-Butylphenyl salicylate
   Butyl stearate
   Butylphthalyl butyl glycolate
   Dibutyl sebacate
   Di(2-ethylhexyl) phthalate (for foods of high water content only)
   Diethyl phthalate
   Diisobutyl adipate
   Diisooctyl phthalate (for foods of high water content only)
   Diphenyl-2-ethylhexyl phosphate
   Epoxidized soybean oil (iodine number maximum 6; and oxirane oxygen, minimum, 6.0%)
   Ethylphthalyl ethyl glycolate
   Glycerol monooleate
   Monoisopropyl citrate
   Mono, di-, and tristearyl citrate
   Triacetin (glycerol triacetate)
   Triethyl citrate
   3-(2-Xenoyl)-1,2-epoxypropane

(f) Release Agents
   Dimethylpolysiloxane (substantially free from hydrolyzable chloride and alkoxy groups, no more than 18% loss in weight after heating 4 hr at 200°C; viscosity 300 centistokes, 600 centistokes at 25°C, specific gravity 0.96–0.97 at 25°C, refractive index 1.400–1.404 at 25°C)
   Linoleamide (linoleic acid amide)
   Oleamide (oleic acid amide)
   Palmitamide (palmitic acid amide)
   Polyethylene glycol 400
   Polyethylene glycol 1500
   Polyethylene glycol 4000
   Stearamide (stearic acid amide)

(g) Stabilizers
   Aluminum mono-, di-, and tristearate
   Ammonium citrate
   Ammonium potassium hydrogen phosphate
   Calcium acetate
   Calcium carbonate
   Calcium glycerophosphate
   Calcium phosphate
   Calcium hydrogen phosphate
   Calcium oleate
   Calcium ricinoleate
   Calcium stearate
   Disodium hydrogen phosphate
   Magnesium glycerophosphate
   Magnesium stearate
   Magnesium phosphate
   Magnesium hydrogen phosphate
   Mono-, di-, and trisodium citrate
   Mono-, di-, and tripotassium citrate
   Potassium oleate
   Potassium stearate
   Sodium pyrophosphate
   Sodium stearate
   Sodium tetrapyrophosphate
   Stannous stearate (not to exceed 50 ppm tin as a migrant in finished food)
   Zinc o-phosphate (not to exceed 50 ppm zinc as a migrant in finished food)
   Zinc resinate (not to exceed 50 ppm zinc as a migrant in finished food)

TABLE 5. (*Continued*)

(h) *Substances Employed in the Manufacture of Paper and Paperboard Products Used in Food Packaging*

- Aliphatic polyoxyethylene ethers
- 1-Alkyl $(C_6$-$C_{18})$-amino-3-aminopropane monoacetate
- Borax or boric acid for use in adhesives, sizes, and coatings
- Butadiene-styrene copolymer
- Chromium complex of perfluoro-octane sulfonyl glycine for use on paper and paperboard which is waxed
- Disodium cyanodithioimidocarbamate with ethylenediamine and potassium *N*-methyl dithiocarbamate and/or sodium 2-mercaptobenzothiazole (slimicides)
- Ethyl acrylate and methyl methacrylate copolymers of itaconic acid or methacrylic acid for use only on paper and paperboard which is waxed
- Hexamethylenetetramine as a setting agent for protein, including casein
- 1-(2-Hydroxyethyl)-1-(4-chlorobutyl)-2-alkyl $(C_6$-$C_{17})$ imidazolinium chloride
- Itaconic acid (polymerized)
- Melamine formaldehyde polymer
- Methyl acrylate (polymerized)
- Methyl ethers or mono-, di-, and tri-propylene glycol
- Myristo chromic chloride complex
- Nitrocellulose
- Polyethylene glycol 400
- Polyvinyl acetate
- Potassium pentachlorophenate as a slime control agent
- Potassium trichlorophenate as a slime control agent
- Resins from high and low viscosity polyvinyl alcohol for fatty foods only
- Rubber hydrochloride
- Sodium pentachlorophenate as a slime control agent
- Sodium trichlorophenate as a slime control agent
- Stearato-chromic chloride complex
- Titanium dioxide
- Urea-formaldehyde polymer
- Vinylidine chlorides (polymerized)

SECTION 11

SHIPPING REGULATIONS

This section is included in this book for the following reasons:

(1) By clearly indicating that certain materials present handling and shipping problems, it is hoped to awaken in more people who handle and ship or receive potentially hazardous materials an awareness of the shippers' responsibility for the materials shipped.

(2) By stating for each material chosen (a) a hazard classification, (b) required or suggested labels, (c) acceptable shipping quantities and warnings to indicate clearly which materials have thus far (1967) been deemed worthy of regulation, and (d) which of the regulating agencies apply restrictions and what the restrictions are, we hope to prevent some accident due to ignorance of basic shipping and storage practice.

The information contained in the Shipping Regulations entries is taken from the following sources:

Agent T. C. George's
Tariff No. 19, publishing
Interstate Commerce Commission
Regulations for Transportation of
  Explosives and Other Dangerous Articles
  by Land and Water in Rail Freight Service
  and by Motor Vehicle (Highway) and Water
  including
  Specifications for Shipping Containers
  Issued 8/5/66—Effective 9/5/66
  Available from: T. C. George, Agent
    63 Vesy Street
    New York, N.Y. 10007

IATA regulations
relating to
the carriage of
restricted articles
by air
Eleventh Edition (original English)
effective 1st October 1967 until further notice
Issued by Authority of the Traffic Director

International Air Transport Association
1155 Mansfield Street, Montreal 2, Quebec, Canada

Guide to
Precautionary Labeling of
Hazardous Chemicals (Manual L-1 6th Edition 1961)
Published by
Manufacturing Chemists' Association, Inc.
1825 Connecticut Avenue, N.W.
Washington 9, D.C.

Regulations Prescribed by the Commandant of the Coast Guard of the United States Coast Guard, Treasury Department published, as required by law as Code of Federal Regulations 46, parts 146–149. Available from the Government Printing Office.

It is urgently recommended that individuals responsible for shipping, receiving, and storing of potentially hazardous materials make themselves familiar with the publications referred to above and apply such information, as well as that outlined in this book, to their responsibilities.

In Section 12, which follows, the readers' attention is directed to the format which was designed to assist the shipper of potentially hazardous materials by

(a) the cross-referencing of common and technical names of materials for easy identification

(b) the listing of physical properties

(c) the listing of toxic hazard ratings or toxicology of many materials useful in the event of accident

(d) the listing of fire and explosion hazard ratings useful in the event of accident as well as fire fighting suggestions

(e) the listing of known shipping regulations with each entry. These may require a word of explanation:

(1) The entry ICC, under "Shipping Regulations" in Section 12, indicates that the material is covered by one of Agent T. C. George's Tariffs.

The following excerpt material from Tariff No. 19 is the basis for the entries immediately following the letters ICC: in Section 12.

PART 72—COMMODITY LIST OF EXPLOSIVES AND OTHER DANGEROUS ARTICLES CONTAINING THE SHIPPING NAME OR DESCRIPTION OF ALL ARTICLES SUBJECT TO PARTS 71-79

Sec.
72.1 Proper shipping name.
72.2 Articles not described.
72.3 Labels required and prohibited articles.
72.4 Explanation of signs and abbreviations.
72.5 List of explosives and other dangerous articles.

§ 72.1 Proper shipping name. (a) The proper shipping name which must be used and shown on outside shipping containers appears in Roman type (not italics). The abbreviations N. O. I. and N. O. I. B. N. may be used in lieu of the abbreviation n. o. s. where it appears in the list of explosives and other dangerous articles.

§ 72.2 Articles not described. (a) For an article not described by name shown in § 72.5 of this part, when such article is classified as dangerous by §§ 73.115, 73.150, 73.151, 73.240, 73.300, 73.326, 73.343, 73.381, or 73.391, the article must be prepared and offered for shipment in compliance with the regulations for the group within which it is properly classified.

§ 72.3 Labels required and prohibited articles. (a) Section 72.5 of this part also shows the kind of label when required on shipments of explosives and other dangerous articles and the articles which are prohibited for transportation.

§ 72.4 Explanation of signs and abbreviations. (a) An asterisk indicates that articles may or may not be classed as flammable liquids, flammable solids, oxidizing materials, compressed gases, poisons, or corrosive liquids by Parts 71-79. If so classed, such articles are subject to the regulations prescribed for articles within these definitions.

F. L.—Flammable liquid.
F. S.—Flammable solid.
Oxy. M.—Oxidizing material.
Cor. L.—Corrosive liquid.
Nonf. G.—Nonflammable compressed gas.
F. G.—Flammable compressed gas.
Pois. A.—Poison gas or liquid, class A.
Pois. B.—Poisonous liquid or solid, class B.
Pois. C.—Tear gas, class C.
Pois. D.—Radioactive materials, class D.
Expl. A.—Class A explosives.
Expl. B.—Class B explosives.
Expl. C.—Class C explosives.
Not accepted—Means not to be offered or accepted for transportation.
Forbidden—Means prohibited by law.
∅ Indicates that articles may be transported as rail baggage.
\# Required for rail express shipments only.
\#\# Required for rail express and water shipments only.
N. O. S.—Means not otherwise specified.
N. O. I.—Means not otherwise indexed.
N. O. I. B. N.—Means not otherwise indexed by name.

Note 1: Where the word "INFLAMMABLE" is now painted, stencilled, or otherwise permanently marked on tank cars, cargo tank motor vehicles, portable tanks, or other containers, it may be so continued until such tanks or other containers are repainted, restencilled, or both, and at such times shall be replaced with the word "FLAMMABLE" unless otherwise ordered by the Commission.

Next is listed a word description (pp. 310–311) of the label required by ICC for shipment of the entry:

§ 73.402 **Labeling dangerous articles.** (a) Each package containing any dangerous article as defined by Parts 71–79 must be conspicuously labeled by the shipper as follows, except as otherwise provided:

(1) **"Red label"** as described in § 73.405 on containers of flammable liquids, except when exempted from the regulations by § 73.118. If flammable liquid is also a class A poison or a radioactive material poison D, the "poison gas" label or "Radioactive materials" label must also be applied to the package.

(2) **"Yellow label"** as described in § 73.406 on containers of flammable solids and oxidizing materials, except when exempted from the regulations by §§ 73.153 and 73.182. If flammable solid or oxidizing material is also a class A poison or a radioactive material poison D, the "poison gas" label or "radioactive materials" label must also be applied to the package.

(3) **"White label"** as described in § 73.407 (a) (1), (2) and (3) on containers of acids, alkaline caustic liquids or corrosive liquids, except when exempted from regulations by § 73.244. If the acid, alkaline caustic liquid or corrosive liquid is also a class A poison or a radioactive material poison D, the "poison gas" label or "radioactive materials" label must also be applied to the package.

(4) **"Red label"** as described in § 73.408 (a) (1) on containers of flammable compressed gases, except when exempted from the regulations by § 73.302. If the flammable compressed gas is also a class A poison or a radioactive material poison D, the "poison gas" label or "radioactive materials" label must also be applied to the package.

(5) **"Green label"** as described in § 73.408 (a) (2) on containers of nonflammable compressed gases, except when exempted from the regulations by § 73.302. If the nonflammable compressed gas is also a class A poison or a radioactive material poison D, the "poison gas" label or "radioactive materials" label must also be applied to the package.

(6) **"Poison gas"** label as described in § 73.409 (a) (1) on containers of class A poisons.

(7) **"Poison"** label as described in § 73.409 (a) (2) on containers of class B poison liquids or solids, except when exempted from the regulations by § 73.345 and § 73.364. If the class B poison liquid or solid is also a radioactive material poison D, the "radio-active materials" label must also be applied to the package,

(8) **"Radioactive Materials"**, label as described in § 73.414 (a) or (c) on containers of Groups I, II and IV radioactive materials, except when exempted by § 73.392.

(9) **"Radioactive Materials"**, label as described in § 73.414 (b) or (d) on containers of class D poisons, Group III, except when exempted by § 73.392.

(10) **"Radioactive Materials"**, label as described in § 73.414 (e) on bundles, boxes, barrels or crates of magnesium-thorium alloys, and on packages of uranium, normal or depleted, in solid form as provided for by § 73.392 (e) and (f) respectively.

(11) **"Tear gas"** label as described in § 73.409 (a) (3) on containers of poisons, class C.

(12) **"Bung label"** as described in § 73.119 (i) of this part on metal barrels or drums containing flammable liquids with vapor pressure exceeding 16 pounds per square inch absolute.

(13) **"Empty label"** as described in § 73.413 of this part must be applied to containers which have been emptied and on which the old label has not been removed. obliterated, or destroyed. It must be so placed on the container as to completely cover the old label.

(14) Labels authorized for shipments of explosives and other dangerous articles by air, as shown in §§ 73.405 (b), 73.406 (b), 73.407 (b), 73.408 (b), 73.409 (b), 73.410 (b), 73.411 (b), 73.412 (b), and 73.414 (c), may be used in lieu of labels otherwise prescribed for surface transportation to or from airport.

(b) Labels when applied to packages offered for transportation by rail express, rail baggage, or other forms of transportation for which a certified shipping order, bill of lading, or other shipping paper is not required, must show shipper's name in printing, stamping, or writing underneath the certificate printed thereon.

(1) Labels authorized for shipments of explosives and other dangerous articles by air are shown in §§ 73.405 (b), 73.406 (b), 73.407 (b), 73.408 (b), 73.409 (b), 73.410 (b), 73.411 (b), 73.412 (b), and 73.414 (c). Shipments so labeled must be tendered with a signed certificate, in duplicate, reading as follows (one signed copy shall accompany each shipment and the other signed copy shall be retained by the original carrier):

This is to certify that the contents of this package are properly described by name and are packed and marked and are in proper condition for transportation according to regulations prescribed by the Interstate Commerce Commission and the Administrator of the Federal Aviation Agency. (For shipments on passenger-carrying aircraft the following must be added to certificate: This shipment is within the limitations prescribed for passenger-carrying aircraft.)

(c) Labels and marking name of contents are not required on carload or truckload quantities of dangerous articles, except class A, class C, or class D poisons, by rail freight, rail express, or highway, when such shipments are unloaded by the consignee or his duly authorized agent from the car or motor vehicle in which originally loaded.

(1) Carload or truckload shipments of class A, class B, class C, or class D poisons offered for transportation by, for, or to the Departments of the Army, Navy, and Air Force of the United States Government are exempt from labeling requirements when shipments are loaded or unloaded by the shipper or his duly authorized agent and such shipments are accompanied by qualified personnel supplied with equipment to repair leaks or other container failure which will permit escape of contents.

(d) Except on class A, class C, or class D poisons, labels are not required on less-than-carload shipments by motor vehicle by public highway when the articles are readily identifiable by reason of type of container or when the container is plainly marked to indicate its contents and;

(1) When the shipment is transported from origin to destination without transfer between vehicles and;

(2) When the shipper or its employees are in direct control or perform the loading, transporting and unloading.

(e) When it is known that subsequent shipments of these packages may be made by consignees in less-than-carload or less-than-truckload quantities, or in carload or truckload quantities to a point where they may be handled by other than the original consignee, the original shipper should attach labels to the packages as would be required for less-than-carload or less-than-truckload shipments.

§ 73.403 **Labels for mixed packing.** (a) Use red label only when red and other labels are prescribed except when poison gas label or radioactive materials label are prescribed then both the red label and the posion gas label or red label and radioactive materials label must be used.

(b) Use White acid (alkaline caustic liquid or corrosive liquid) label only when white acid (alkaline caustic liquid or corrosive liquid) and yellow or poison labels are prescribed or poison labels (class B) are prescribed, except when poison gas label or radioactive materials label are prescribed then both the white acid label and the poison gas label or white acid and radioactive materials label must be used.

(c) Use yellow label only when yellow and poison labels are prescribed except when poison gas label or radioactive materials label are prescribed then both the yellow label and the poison gas label or the yellow label and the radioactive materials label must be used.

§ 73.404 **Labels.** (a) Shippers must furnish and attach the labels prescribed for their packages. Labels should be applied to that part of the package bearing consignee's name and address.

(b) Labels must not be applied to packages containing articles which are not subject to Parts 71-79 or are exempted therefrom.

(c) Shippers must not use labels which by their size, shape, and color, may readily be confused with the standard caution labels prescribed in this part.

(d) Labels must conform to standards as to size, printing and color, and samples will be furnished, on request, by the Bureau of Explosives.

(e) A combination diamond-shaped label-tag of proper size and color, bearing on one side the shipping information and on the reverse side the wording prescribed in this part, will be permitted.

(f) As certification of compliance with regulations is also required by other Government agencies, and to avoid multiplicity of certifications, there may be added to the certificate on labels "and the Commandant of the Coast Guard," or "and the Administrator of the Federal Aviation Agency," or "and the Post Office Department," as is necessary.

(g) The carrier's name and stationery form number, or the shipper's name and address, may be printed on the labels, in type not larger than 10 point, if placed within the black-line border and in the upper or lower corner of the diamond.

(h) Labels remaining on hand and which were authorized by regulations in effect on December 31, 1950, may be used until present stocks are exhausted.

The final entry under ICC is descriptive of the volume, weight, or in the case of radio-active materials, the number of Curies in the package or the level of radiation per-mitted to be emitted from the package as a maximum acceptable for shipment in one outside container by rail express.

For example, the ICC entry in Section 12 for Benzene is:

ICC: flammable liquid, Red label, 10 gallons

A list of ICC regulations governing ship-ments of explosives and other dangerous articles appears on pages 318–362.

(2) the entry IATA, under "Shipping Regu-lations" in Section 12, indicate that the material is covered by International Air Transport Association regulations for re-stricted articles.

Thus, as listed in Section 12 under "Ship-ping Regulations," restricted articles will be indicated by the letters IATA: B, C, D, where B provides the class of material such as Combustible liquid, Corrosive liquid, explosive, flammable compressed gas, flammable liquid, flammable solid, non-flammable compressed gas, not other-wise specified (n.o.s.), other restricted articles Class A, other restricted articles Class B, other restricted articles Class C, oxidizing material, poisonous article Class A, poisonous article Class B, poisonous article Class C, radioactive materials.

C gives the name of the label when re-quired to be attached to each package.

D shows whether or not the article may be carried in a passenger aircraft, the maximum net quantity per package based on the metric system, and the type of pack-aging and marking required. If the article is shown as "not acceptable," it cannot be carried.

E shows whether or not the article may be carried in a cargo aircraft, the max-imum net quantity per package based on the metric system, and the type of packag-ing and marking required. If the article is listed "not acceptable," it cannot be car-ried.

Note: the quantity limitations of D and E apply only to the amount contained in one package, not in the shipment or aircraft.

For example, under Benzene in Section 12 is listed IATA: B, C, D, E where

B = Flammable liquid,

C = Red label

D = 1 liter (passenger)

E = 40 liters (cargo)

Some general information on the transport of dangerous material by air is contained in the following article, reprinted by the author's per-mission. Before transporting such materials by air, however, consult the IATA Regulations, Eleventh Edition.

TRANSPORT OF DANGEROUS GOODS BY AIR*

No part of the air transport business is more important than safety. And no part requires more attention to detail by airlines, regulatory agencies and the shipping public. But before reviewing developments relating to the carriage of commodities with dangerous characteristics (generally referred to as restricted articles) which include radioactive materials, a brief outline of IATA's activity in air freight appears desirable.

Part of IATA's commercial activity is con-cerned with the standardization of procedures, forms, handling agreements and many other factors that facilitate the rapid despatch of air freight consignments. Today's air shippers have a complete worldwide network of air freight services at their disposal. They can make the simplest or the most complicated of bookings over as many airlines' routes as necessary, at a standard rate, and still be assured of speedy delivery.

The airlines offer much more than a series of freight rates and space aboard aircraft. Safety, simplicity, speed, reliability and savings are the most important features. Airline services cover pre-shipment activity, such as information on packing, documentary requirements and sched-ules, actual carriage by air, interline transfers, handling at destination and delivery from air-port to consignee, all part of a door-to-door service.

The smooth functioning of the vast world-wide network depends on a large degree of standardization and cooperation. Indeed, the fact that more than 125 countries and territories are served by an intricate pattern of air routes, makes standardization vital. Present agreement on forms and procedures, governing rules and clear determination of responsibilities is exten-sive. High on the safety and standardization list is the "IATA Regulations Relating to the Car-riage of Restricted Articles by Air" and there can be no compromise with these safety regu-lations.

*A. Groenewege, International Air Transport Association.

It is of the utmost importance that any person shipping or accepting air freight consignments be fully familiar with the specific packaging, labelling and handling requirements provided for in the IATA Restricted Articles Regulations. As such, an immediate and permanent responsibility rests with the shipping public, agents, air freight forwarders, and airline personnel for the proper and strict application of these regulations.

Development of IATA Restricted Articles Regulations

Before 1950, very few countries permitted the carriage of restricted articles by air at all (the term "restricted articles" includes a wide variety of articles which have dangerous characteristics that make it necessary to control, restrict, or forbid acceptance for air shipment). It was apparent, however, that in order to meet commercial needs, at least small quantities of restricted articles should be permitted in air transportation under well defined conditions. Moreover, it was important that these commodities should be able to be carried in a manner acceptable to surface transport throughout the nations of the world.

As a beginning, the "IATA Conditions of Carriage for Goods" listed acceptability of restricted articles under several general headings. However, such a method of presentation made the proper classification of a particular article very difficult. Because of this, IATA considered that a single list in alphabetical sequence would be more convenient to handle and that it would reduce to a minimum the possibility of error in actual application. To produce such a list, the IATA Restricted Articles Working Group was established in 1950. Its members are technical, traffic handling and safety experts and they act on behalf of the airline industry as a whole and not just for the particular benefit of their own companies.

Pursuant to the authority of the IATA Traffic Conferences, the present task assigned to the IATA Group can best be summarized as follows:

(i)   to make it possible to carry, subject to the paramount needs of safety, restricted articles under specific conditions and without advance arrangements, whenever possible;

(ii) to take clearly into account the characteristics peculiar to air transport, such as pressures at high altitude and overall loading and storage requirements;

(iii) to review constantly the IATA Restricted Articles Regulations in the light of new

technical developments and changing requirements of the industry and air transport.

As a first step, the IATA Group studied all available regulations for the carriage of dangerous goods by different means of transport. In this process, consideration was also given to the Interstate Commerce Commission Regulations which had been in force for many years in the United States and which had been accepted by the U.S. Civil Aeronautics Board for air transportation in 1942. The 1942 rules covered only explosives, but were later expanded to include corrosives, flammable and poisonous articles, gases and oxidising materials. In 1949, direct reference was made to the ICC Regulations for the definition of classes of dangerous goods and for packaging, marking and labelling requirements.

Careful examination proved that the ICC Regulations could be used as a basis for the development of the IATA Restricted Articles Regulations for international transport. However, the IATA Group had the very difficult task of fitting into the ICC Regulations, all other governmental regulations, so that the final version of the IATA Restricted Articles Regulations would be acceptable to all interested Governments.

Another problem in preparing the IATA Restricted Articles Regulations was to make them as complete as possible without filling volumes. As a general principle, regulatory authorities have found it necessary to restrict the listing of articles to those which are in regular movement. For that reason, IATA has attempted to include those commodities which are listed by the principal governmental regulations and are moving regularly in air transport. The magnitude of the various practical and technical problems involved can best be illustrated by the fact that surface transport has not yet been able to produce a set of international regulations for worldwide application.

As a basic principle, the IATA Group also adopted the ruling that carriers and approving Governments may be more restrictive than the IATA Restricted Articles Regulations, but *not* less restrictive. If a Member airline decides not to carry certain restricted articles, an appropriate exception is included in the IATA Restricted Articles Regulations. Basically, carrier exceptions are not filed for reasons of safety, but rather to avoid internal traffic handling difficulties.

Only a few carrier and governmental exceptions, of limited scope, are listed in the existing

IATA Restricted Articles Regulations. Obviously, governmental exceptions could place a serious commercial hardship on the airlines operating services to, from, or over territories of such countries. Realizing the many difficulties arising from the existence of such governmental exceptions, and the tremendous amount of work and research done by the IATA Group for the last fifteen years, most Governments have accepted the IATA Restricted Articles Regulations unconditionally.

After six years of work by the IATA Group, the first worldwide regulations governing the carriage of restricted articles became effective January 1, 1956, on the services of all IATA Member airlines. As a result, it is now possible to move shipments of restricted articles freely over the services of the majority of the airlines of the world, without delays for repacking and relabelling, and under regulations which are readily understood by all personnel involved. This is particularly important in the case of industrial and medical isotopes and other radioactive substances whose effective life is limited.

The IATA Restricted Articles Regulations are binding upon the 99 IATA Member airlines in scheduled and unscheduled operations by virtue of IATA Traffic Conference Resolution 618. In addition, 64 non-IATA carriers, participating in the IATA Interline Cargo Handling Agreement, are applying these regulations. Prior to the implementation of airline regulations for the carriage of restricted articles, the IATA Restricted Articles Regulations were approved by interested Governments. Meanwhile, a number of Governments have incorporated the IATA Restricted Articles Regulations into their respective aeronautical regulations; thus making them binding upon all aircraft operating services under the registry of those countries.

It is important to note that the offering of articles in violation of the IATA Restricted Articles Regulations may be a breach of law, and subject to legal penalties. It is prohibited to offer any package or container for air transport that will cause dangerous evolution of heat or gas, or produce corrosive substances under conditions normally incident to transport.

The IATA Restricted Articles Regulations (15,000 copies are distributed throughout the world) list some 2,000 commodities requiring special packing and handling and include compressed gases, corrosive liquids, explosives, flammable liquids, flammable solids, magnetized materials, oxidizing materials, poisons, polymerizable materials, radioactive materials, noxious or irritating substances, articles liable to damage aircraft structures and articles possessing other inherent characteristics which make them unsuitable for air carriage unless properly prepared for shipment. The carriage of certain other specifically listed goods is strictly forbidden. For those commodities that can be carried, the regulations prescribe special packing requirements, handling methods, storage and labelling, and specify the maximum net quantity permitted per package for both passenger and cargo aircraft.

The IATA Restricted Articles Regulations also require restricted articles acceptable for air transport to show the proper shipping name and to be accompanied by any necessary instructions for safe handling during transport. They must be packed, marked and labelled in accordance with the specific IATA safety provisions. To facilitate handling, eight special labels are available and shippers are required to attach the appropriate label to each package. The airlines will not accept restricted articles unless the shipper or his authorized agent certifies that the shipper or his authorized agent certifies that the contents of a consignment are properly described by name, and meet all the conditions specified in the IATA Restricted Articles Regulations.

It is also important to note that IATA maintains a close liaison with the International Civil Aviation Organization (ICAO) the United Nations Committee of Experts on the Transport of Dangerous Goods, the International Atomic Energy Agency (IAEA) and the Atomic Energy Authorities of the main isotope producing countries, and many other governmental and technical organizations concerned with the transport of dangerous goods. IATA fully supports the important work undertaken by the United Nations Committee in order to bring about the greatest possible degree of uniformity in the carriage of dangerous goods by all means of transport. This uniformity is particularly important when two different means of transport have to be used.

IATA hopes that all countries with important international trade in chemical substances and other "restricted articles" will be able to implement in coming years the framework of the United Nations recommendations, as may be amended, for all means of transport. However, it is important to note that amendments to the IATA Restricted Articles Regulations are subject to prior approval by the Governments concerned. Because of the importance of maintaining worldwide uniformity and a single set of regulations as far as air transport is concerned, amendments of principle with respect to classification, labelling and packing, cannot be implemented unless all interested Governments, including the United States of America, have agreed to amend their present regulations.

Packaging Requirements for Restricted Articles

The type of packaging used for the carriage of restricted articles is one of the most important considerations. In general, restricted articles can be carried safely in transport, provided the packaging is adequate and the net quantity per package is restricted. However, some commodities, such as certain types of explosives, are too dangerous to ship by air. On the other hand, many restricted articles can be carried safely if the quantity per package is small and the article is properly packed. It should also be noted that some commodities are not acceptable for shipment in passenger aircraft, but only on all-cargo aircraft.

Only relatively small quantities of restricted articles are permitted on passenger-carrying aircraft. In general, much larger quantities of restricted articles can be carried per package on all-cargo aircraft. To carry these larger quantities safely, special minimum packaging requirements have been developed for worldwide application, and these have to be met in all circumstances. The various containers must also comply with specific construction requirements and minimum standards. Those carried by cargo aircraft are subject to performance tests.

In 1959, the IATA Group initiated a comprehensive study for the development of packaging specifications for the carriage of larger quantities of restricted articles on all-cargo aircraft. As a result, detailed packaging provisions and container standards were incorporated in the IATA Restricted Articles Regulations and became effective April 1, 1962. This packaging programme covered the following main points:

(i)   general container requirements applicable to all containers used in air transport;

(ii) a range of general type of inside containers made of earthenware, glass, hard rubber, gutta percha, wax, plastic, metal, fibre and paper. Where necessary, these inside containers were related to performance tests;

(iii) a range of general types of outside or single containers in varying capacities of up to 220 litres (50 Imp. gallons/55 U.S. gallons) for liquids, and 140 kilograms (300 pounds) for solids, covering steel barrels and drums (including removable head, single trip, re-usable and lined types), aluminum drums, wooden boxes and kegs, fibre and plywood drums and fibreboard boxes, all related to performance tests;

(iv) a range of some 90 Packaging Notes covering over 800 articles in Section IV of the IATA Restricted Articles Regulations. These articles are related to various acceptable combinations or ranges of the containers, mentioned in paragraphs (ii) and (iii) above, or in some cases to specialized containers and packaging.

At present, the United Nations Committee is studying the possible adoption of the basic principles developed by IATA with respect to the important question of packaging specifications and container standards, including the use of performance tests. The experience with the IATA packaging requirements has shown that these are entirely satisfactory; and there is every reason to believe that they would continue to meet both commercial needs and the overall safety requirements of the IATA Member airlines in future. Therefore, it is hoped that the United Nations Committee will see its way clear to adopt the IATA packaging principles, which have been widely recognized as the only practical approach to achieve worldwide harmonization in this complex field.

Development of IATA Radioactive Materials Regulations

Almost ten years ago, IATA recognized the importance and need of regulating the acceptance of radioactive materials, particularly in the light of the increasing use of air transport for the carriage of a wide range of radioactive isotopes for commercial, medical, research and other peaceful purposes. The original IATA requirements for radioactive materials, which form one of the classes of restricted articles covered by the IATA Restricted Articles Regulations, were closely based on the ICC Regulations. These regulations became effective June 1, 1958 and provided for shipment of larger sources activity of radioactive materials than heretofore permitted.

At the same time, new and detailed packaging provisions were introduced for these larger quantities of radioactive materials, in accordance with Packaging Note 44 of the Third Edition of the IATA Regulations. (In the Tenth Edition of the IATA Regulations, effective 1st April, 1965, Packaging Notes 700–703 inclusive cover the carriage of radioactive materials). One of the requirements was that the container should be of such mechanical strength and design so as to minimize the likelihood of breakage or leakage in a severe accident or fire. Shippers and producers were also required to supply the office of the accepting airline with a certificate in duplicate (original to accompany the consignment), issued and signed by the competent Government Authority in the country of origin testifying that the container complied with all the requirements specified. This certificate was in addition to the normal Shipper's Certification required for the carriage of restricted articles.

At present, radioactive materials are defined in the IATA Restricted Articles Regulations as any material or combination of materials which spontaneously emits ionizing radiation. For the purpose of classification, radioactive materials are divided into the following three groups according to the type of radiation emitted at any time during transport:

(i) Group I radioactive materials are those materials which emit any gamma radiation either alone or with electrically charged particles or corpuscules (alpha, beta, etc.).

(ii) Group II radioactive materials are those materials, which emit neutrons and either or both of the types of radiation characteristic of Group I radioactive materials.

(iii) Group III radioactive materials are those materials which emit electrically charged corpuscular rays only (alpha, beta, etc.) or any other that is so shielded that the gamma radiation at the surface of the package does not exceed 10 milliroentgens per 24 hours at any time during transport.

IATA also followed with a great deal of interest the work done by the International Atomic Energy Agency (IAEA) with a view to developing a set of regulations for the safe transport of radioactive materials by all means of transport. The expanding use of radioactive materials for peaceful purposes and the many technical problems posed in international transport because of the hazards associated with the movement of these materials, had clearly demonstrated the need for uniform regulations. When the original provisions contained in IAEA Safety Series No. 6 were examined in the light of comments received from all interested parties, IATA was invited to participate in the discussions of the IAEA Panel.

It was the general view of the IATA Group, that the amended provisions of the IAEA Regulations represented a very considerable improvement over existing conditions, and that the efforts of the IAEA Panel were of great importance to the transport industry. Furthermore, IATA supported whole-heartedly the working programme established by IAEA for the development of container specifications and performance tests as this represented another important step forward in regulating the safe transport of radioactive materials.

After careful examination of the various technical and practical aspects involved, the IATA Group agreed in April 1963, that the basic principles adopted by IAEA provided an acceptable frame-work for the development of more specific regulations for the carriage of radioactive materials by air. This was with the understanding, however, that there might be a need to incorporate additional requirements or limitations so as to reflect the characteristics peculiar to air transport and to ensure safety to the maximum extent possible.

Subsequently, the IATA Group examined in detail the amended IAEA Regulations as developed by the IAEA Panel, with a view to formulating detailed provisions for the carriage of radioactive materials by air. In the last two years, a great deal of work has been done by the IATA Group and a number of joint meetings were held with representatives of IAEA and the Atomic Energy Authorities of the main isotope producing countries in order to discuss matters of mutual interest.

At present, the final provisions of the IATA Regulations on Radioactive Materials, embodying the IAEA principles, are being drafted. It should be noted that, based on practical and technical considerations involved, some additional restrictions have been imposed compared with the IAEA Regulations. For example, large radioactive sources will be accepted only on all-cargo aircraft and not on passenger aircraft. Similarly, bulk loads, i.e. loose radioactive materials, and pyrophoric radioactive materials will not be permitted in the air transport even if the package design has the approval of the competent authority.

The actual date of implementation of the new IATA Regulations on Radioactive Materials by Member airlines is largely dependent upon the legislative acceptance by IAEA Member States with respect to the amended IAEA Regulations, in particular the main isotope producing countries. In any event, it seems inevitable that there will be an interim period when the IATA Restricted Articles Regulations are not compatible with the governing national regulations and, therefore, two sets of regulations have to be published. This would mean that radioactive materials may be accepted for air shipment if they comply with either of these two sets of regulations as required by the countries of origin, transit and destination.

In the meantime, further work has to be done by the IATA Group in devising an acceptable format for the integration of the new radioactive materials regulations and in the most convenient manner, in the Eleventh Edition of the IATA Restricted Articles Regulations. This Edition is most likely to become effective towards the end of 1966.

In conclusion, I wish to emphasize that IATA is anxious to cooperate with all interested parties in pursuit of the important objectives of safety and uniform regulations in the transport of dangerous goods by all means of transport throughout the world. Clearly, only through close cooperation and teamwork can these objectives be realized.

---

(3) The entry MCA warning label indicates that the material listed in Section 12 has been selected by the Manufacturing Chemists' Association (MCA) as worthy of a special label to warn users, shippers, etc. of the hazards of the material. The Sixth Edition, 1961 of the Guide to Precautionary Labeling of Hazardous Chemicals, published by Manufacturing Chemists' Association, Inc., is the source of these entries.

(4) Coast Guard listing of restricted articles closely follows ICC. Reference is made to CFR 46, parts 146–149, for the regulations concerning Explosives or Other Dangerous Articles on Board Vessels.

§ 72.5 List of explosives and other dangerous articles. (a) For explanation of signs and abbreviations see § 72.4 of this part.

LIST OF EXPLOSIVES AND OTHER DANGEROUS ARTICLES

Article	Classed as	Exemptions and Packing (see sec.)	Label required if not exempt	Maximum quantity in 1 outside container by rail express
Acetaldehyde (ethyl aldehyde)	F. L.	73.118, 73.119	Red	10 gallons.
Acetone	F. L.	73.118, 73.119	Red	10 gallons.
Acetone cyanhydrin	Pois. B.	73.345, 73.346	Poison	55 gallons.
*Acetone oils	F. L.	73.118, 73.119	Red	10 gallons.
Acetonitrile	F. L.	73.118, 73.119	Red	10 gallons.
Acetyl benzoyl peroxide, solid	Not accepted			Not accepted.
Acetyl benzoyl peroxide, solution	Oxy. M.	No exemption, 73.222	Yellow	1 quart.
Acetyl chloride	Cor. L.	73.244, 73.247	White	1 gallon.
Acetyl peroxide, solid	Not accepted			Not accepted.
Acetyl peroxide, solution	Oxy. M.	73.153 (b), 73.222	Yellow	1 quart.
Acetylene	F. G.	73.306, 73.303	Red Gas	300 pounds.
Acid carboys, empty	See § 73.29 (c)			
Acids, liquid, n. o. s.	Cor. L.	73.244, 73.245	White	5 pints.
Acid picric. See *Picric acid.				
*Acid sludge. See *Sludge acid.				
Acrolein, inhibited	F. L.	No exemption, 73.122	Red	1 quart.
Acrylonitrile	F. L.	73.118, 73.119	Red	10 gallons.
Actuating cartridges, explosive, fire extinguisher or valve	Expl. C.	73.114		150 pounds.
*Adhesives, n.o.s. See Cement, liquid, n.o.s.				
Aeroplane flares. See Special fireworks.				
Aerosol products. See Compressed gases, n.o.s.				
Air, compressed	Nonf. G.	73.306, 73.302	Green	300 pounds.
Aircraft rocket engines (commercial)	F. S.	No exemption, 73.238	Yellow	550 pounds.
Aircraft rocket engine igniters (commercial)	F. S.	No exemption, 73.238	Yellow	25 pounds.
*Alcohol, n. o. s.	F. L.	73.118, 73.125	Red	10 gallons.
Alcohol, allyl	Pois. B.	73.345, 73.346	Poison	55 gallons.
Aldrin	Pois. B.	73.364, 73.376	Poison	200 pounds.
*Aldrin, cast solid	See § 73.376 (b)			
*Aldrin mixtures, liquid, with 60 percent or less aldrin	See § 73.361 (b)			
Aldrin mixtures, liquid, with more than 60 percent aldrin	Pois. B.	73.345, 73.361	Poison	55 gallons.
Aldrin mixtures, dry, with more than 65 percent aldrin	Pois. B.	73.364, 73.376	Poison	200 pounds.

Article	Classification	Regulation sections	Label	Exemption quantity
*Aldrin mixtures, dry, with 65 percent or less aldrin.	See § 73.376 (b).			
*Alkaline caustic liquids, n. o. s.	Cor. L.	73.244, 73.249	White	10 gallons.
Alkaline corrosive battery fluid.	Cor. L.	73.244, 73.249, 73.257	White	10 gallons.
Alkaline corrosive battery fluid with storage battery.	Cor. L.	No exemption, 73.258	White	400 pounds.
Alkaline corrosive liquids, n. o. s.	Cor. L.	73.244, 73.249	White	10 gallons.
Allyl alcohol. See Alcohol, allyl.				
Alkyl aluminum halides. See Pyroforic liquids, n.o.s.				
Allyl bromide.	F. L.	73.118, 73.119	Red	10 gallons.
Allyl chlorocarbonate. See Allyl chloroformate.				
Allyl chloroformate.	Cor. L.	No exemption, 73.288	White	5 pints.
Allyl trichlorosilane.	Cor. L.	No exemption, 73.280	White	10 gallons.
Aluminum alkyls. See Pyroforic liquids, n.o.s.				
Aluminum dross.	See § 73.173			
*Aluminum liquid (or paint). See *Paint, enamel, lacquer, stain, shellac, varnish, etc.				
Aluminum nitrate.	Oxy. M.	73.153, 73.182	Yellow	100 pounds.
Aluminum triethyl.	F. L.	No exemption, 73.134	Red	2 ounces.
Aluminum trimethyl.	F. L.	No exemption, 73.134	Red	2 ounces.
Amatol. See High explosives.				
Ammonia, anhydrous. See Anhydrous ammonia.				
Ammonium arsenate, solid.	Pois. B.	73.364, 73.365	Poison	200 pounds.
Ammonium bichromate (ammonium dichromate).	F. S.	73.153, 73.154, 73.235	Yellow	100 pounds.
Ammonium nitrate.	Oxy. M.	73.153, 73.182	Yellow	100 pounds.
Ammonium nitrate (organic coating).	Oxy. M.	73.153, 73.182	Yellow	100 pounds.
*Ammonium nitrate—phosphate.	Oxy. M.	73.153, 73.182	Yellow	100 pounds.
Ammonium nitrate—carbonate mixtures.	Oxy. M.	73.153, 73.182	Yellow	100 pounds.
*Ammonium nitrate mixed fertilizer.	Oxy. M.	73.153, 73.182	Yellow	100 pounds.
Ammonium nitrate fertilizer, containing 90 percent or more ammonium nitrate with no organic coating.	Oxy. M.	73.153, 73.182	Yellow	100 pounds.
Ammonium perchlorate.	Oxy. M.	73.153, 73.154, 73.239 (a).	Yellow	100 pounds.
Ammonium permanganate.	Oxy. M.	73.153, 73.154	Yellow	100 pounds.
Ammonium picrate. See High explosives.				
Ammonium picrate, wet (not to exceed 16 ounces).	See § 73.192			
Ammunition, chemical (containing class A poisons, liquids, or gases). See Chemical ammunition.				
Ammunition, chemical (containing class B poisons, liquids, or gases). See Chemical ammunition.				
Ammunition, chemical (containing class C poisons, liquids, or solids). See Chemical ammunition.				

* See § 72.4.

LIST OF EXPLOSIVES AND OTHER DANGEROUS ARTICLES—*Continued*

Article	Classed as	Exemptions and Packing (see sec.)	Label required if not exempt	Maximum quantity in 1 outside container by rail express
Ammunition, chemical, explosive	See § 73.59			
Ammunition material	See § 73.55			
Ammunition non-explosive	See § 73.55			
Ammunition for cannon with empty projectiles	Expl. B	No exemption, 73.89		Not accepted.
Ammunition for cannon with explosive projectiles	Expl. A	No exemption, 73.54		Not accepted.
Ammunition for cannon with gas projectiles	Expl. A	No exemption, 73.54		Not accepted.
Ammunition for cannon with illuminating projectiles	Expl. A	No exemption, 73.54		Not accepted.
Ammunition for cannon with incendiary projectiles	Expl. A	No exemption, 73.54		Not accepted.
Ammunition for cannon with inert-loaded projectiles	Expl. B	No exemption, 73.89		Not accepted.
Ammunition for cannon with smoke projectiles	Expl. A	No exemption, 73.54		Not accepted.
Ammunition for cannon with solid projectiles	Expl. B	No exemption, 73.89		Not accepted.
Ammunition for cannon without projectiles	Expl. B	No exemption, 73.89		Not accepted.
Ammunition, rocket. See Rocket ammunition.				
Ammunition, small-arms. See Small-arms ammunition.				
Ammunition for small-arms with explosive projectiles	Expl. A	No exemption, 73.58		Not accepted.
Ammunition for small arms with incendiary projectiles	Expl. A	No exemption, 73.58		Not accepted.
*Amyl acetate	F. L.	73.118, 73.119	Red	10 gallons.
Amyl chloride	F. L.	73.118, 73.119	Red	10 gallons.
Amyl mercaptan	F. L.	No exemption, 73.141	Red	10 gallons.
Amyl nitrite	F. L.	73.118, 73.119	Red	10 gallons.
Amyl trichlorosilane	Cor. L.	No exemption, 73.280	White	10 gallons.
⊘Anhydrous ammonia	Nonf. G.	73.306, 73.304, 73.314, 73.315	Green	300 pounds.
Anhydrous hydrazine. See Hydrazine, anhydrous.				
Anhydrous hydrofluoric acid. See Hydrofluoric acid, anhydrous.				
Aniline oil drums, empty	See § 73.347 (d)			
Aniline oil, liquid	Pois. B.	No exemption, 73.347	Poison	55 gallons.
Anisoyl chloride	Cor. L.	73.244, 73.279	White	1 quart.
*Anti-freeze compounds, liquid	F. L.	73.118, 73.119	Red	10 gallons.

*Anti-freeze preparations, proprietary, liquid	F. L.	73.118, 73.119.	Red.	10 gallons.
Antimony pentachloride	Cor. L.	73.244, 73.247.	White.	1 quart.
Antimony pentachloride solution	Cor. L.	73.244, 73.245.	White.	5 pints.
Antimony pentafluoride	Cor. L.	No exemption, 73.246.	White.	25 pounds.
*Apparatus. See Refrigerating machines, comp. gas or flammable liquid.				
*Aqua ammonia solution containing anhydrous ammonia	Nonf. G.	73.306, 73.304, 73.314, 73.315.	Green.	300 pounds.
Argon	Nonf. G.	73.306, 73.302, 73.314.	Green.	300 pounds.
Argon, pressurized liquid. See Lead arsenate.	Nonf. G.	No exemption, 73.304.	Green.	300 pounds.
Arsenate of lead. See Lead arsenate.				
Arsenic acid, liquid	Pois. B.	73.345, 73.348.	Poison.	55 gallons.
Arsenic acid, solid	Pois. B.	73.364, 73.366.	Poison.	200 pounds.
Arsenic bromide, solid	Pois. B.	73.364, 73.365.	Poison.	200 pounds.
Arsenic chloride (arsenous), liquid	Pois. B.	73.345, 73.346.	Poison.	55 gallons.
Arsenic iodide, solid	Pois. B.	73.364, 73.365.	Poison.	200 pounds.
Arsenic pentoxide, solid	Pois. B.	73.364, 73.366.	Poison.	200 pounds.
Arsenic solid	Pois. B.	73.364, 73.365.	Poison.	200 pounds.
Arsenic sulfide (powder), solid	Pois. B.	73.364, 73.365.	Poison.	200 pounds.
Arsenic trichloride, liquid	Pois. B.	73.345, 73.346.	Poison.	55 gallons.
Arsenic trioxide, solid (arsenic, white, solid, arsenous acid, solid)	Pois. B.	73.364, 73.366, 73.368.	Poison.	200 pounds.
Arsenic, white, solid	Pois. B.	73.364, 73.366.	Poison.	200 pounds.
Arsenical compounds or mixtures, n.o.s., liquid	Pois. B.	73.345, 73.346.	Poison.	55 gallons.
Arsenical compounds or mixtures, n.o.s., solid	Pois. B.	73.364, 73.367.	Poison.	200 pounds.
Arsenical dip, liquid (sheep dip)	Pois. B.	73.345, 73.346.	Poison.	55 gallons.
Arsenical dust	Pois. B.	73.364, 73.368.	Poison.	200 pounds.
Arsenical flue dust	Pois. B.	73.364, 73.368.	Poison.	200 pounds.
Arsenous acid, solid	Pois. B.	73.364, 73.365.	Poison.	200 pounds.
Arsenous and mercuric iodide solution, liquid	Pois. B.	73.345, 73.346.	Poison.	55 gallons.
*Asphalt, cut-back. See *Road asphalt, or tar, liquid.				
Automobiles, motorcycles, tractors or other self-propelled vehicles.	See §§ 73.120, 73.306.			
Automobiles, motorcycles, tractors or other self-propelled vehicles, engines or other mechanical apparatus, with charged electric storage batteries, wet.	See § 73.250.			
1-Aziridinyl phosphine oxide (tris). See Tris-(1-aziridinyl) phosphine oxide.				
Bags, nitrate of soda, empty and unwashed	F. S.	No exemption, 73.155 (a).	Yellow.	25 pounds.
Barium azide—50 percent or more water wet.	F. S.	No exemption, 73.239.	Yellow.	1 pound.

* See § 72.4.　　∅ See § 72.4.

LIST OF EXPLOSIVES AND OTHER DANGEROUS ARTICLES—*Continued*

Article	Classed as	Exemptions and Packing (see sec.)	Label required if not exempt	Maximum quantity in 1 outside container by rail express
Barium chlorate.	Oxy. M.	73.153, 73.163	Yellow	100 pounds.
Barium chlorate, wet.	Oxy. M.	73.153, 73.163 (a) (6)	Yellow	200 pounds.
Barium cyanide, solid.	Pois. B.	73.370	Poison	200 pounds.
Barium nitrate.	Oxy. M.	73.153, 73.182	Yellow	100 pounds.
Barium perchlorate.	Oxy. M.	73.153, 73.154	Yellow	100 pounds.
Barium permanganate.	Oxy. M.	73.153, 73.154	Yellow	100 pounds.
Barium peroxide (*binoxide, dioxide*)	Oxy. M.	73.153, 73.156	Yellow	100 pounds.
Barrels, empty. *See* Drums, empty.				
Batteries, dry.	Not regulated			
Batteries, electric storage, wet.	Cor. L.	73.260	White	600 pounds.
Batteries, electric storage, wet, with automobiles, auto parts, engines or other mechanical apparatus.	Cor. L.	73.250, 73.260	White	No limit.
Batteries, electric storage, wet, with containers of corrosive battery fluid.	Cor. L.	No exemption, 73.258.	White	2 gallons.
Battery charger with electrolyte (acid) or battery fluid. See Electrolyte (acid) or Alkaline corrosive battery fluid.	See § 73.259			
Benzene (benzol).	F. L.	73.118, 73.119	Red	10 gallons.
Benzine.	F. L.	73.118, 73.119	Red	10 gallons.
Benzol (benzene).	F. L.	73.118, 73.119	Red	10 gallons.
Benzoyl chloride.	Cor. L.	73.244, 73.247	White	1 quart.
Benzoyl peroxide.	Oxy. M.	No exemption, 73.157, 73.158.	Yellow	25 pounds.
Benzyl bromide (bromotoluene, alpha).	Cor. L.	No exemption, 73.281.	White	5 pints.
Benzyl chloride.	Cor. L.	73.244, 73.295.	White	1 quart.
Benzyl chlorocarbonate. *See* Benzyl chloroformate.				
Benzyl chloroformate.	Cor. L.	No exemption, 73.288.	White	5 pints.
Beryllium compounds, n. o. s.	Pois. B.	73.364, 73.365.	Poison	200 pounds.
Black blasting powder. *See* Black powder.				
Black pellet powder. *See* Black powder.				
Black powder.	Expl. A.	No exemption, 73.60.		Not accepted.
Black powder igniters with empty cartridge bags.	Expl. C.	No exemption, 73.106.		150 pounds.

Article	Classification	Exemption	Label	Quantity
Black rifle powder. *See* Black powder.				
Blasting caps—1,000 or less	Expl. C	No exemption, 73.103		See § 73.86.
Blasting caps—more than 1,000	Expl. A	No exemption, 73.66		Not accepted.
Blasting caps with metal clad mild detonating fuse—1,000 or less	Expl. C	No exemption, 73.103		See § 73.86.
Blasting caps with metal clad mild detonating fuse—more than 1,000	Expl. A	No exemption, 73.66 (c), 73.67		Not accepted.
Blasting caps with safety fuse—1,000 or less	Expl. C	No exemption, 73.103		See § 73.86.
Blasting caps with safety fuse—more than 1,000	Expl. A	No exemption, 73.66 (c), 73.67		Not accepted.
Blasting caps, electric. *See* Electric blasting caps.				
Blasting gelatin. *See* High explosives.				
Blasting powder. *See* Black powder.				
*Boiler compound, liquid	Cor. L	73.244, 73.249	White	10 gallons.
Bombs, explosive. *See* Explosive bomb.				
Bombs, explosive, with gas, smoke, or incendiary material. *See* Explosive bomb.				
Bombs, explosive, incendiary. *See* Explosive bombs.				
Bombs, fireworks. *See* Special fireworks.				
Bombs, gas, smoke or incendiary, non-explosive. *See* Chemical ammunition.				
Bombs, incendiary or smoke without bursting charges. *See* Special fireworks.				
Bombs, practice, with electric primers or electric squabs				
Bombs, sand-loaded or empty	See § 73.55			
Boosters (explosive)	Expl. A	No exemption, 73.69		Not accepted.
Bordeaux arsenites, liquid	Pois. B	73.345, 73.346	Poison	55 gallons.
Bordeaux arsenites, solid	Pois. B	73.364, 73.365	Poison	200 pounds.
Boron trichloride	Cor. L	No exemption, 73.251	White	1 quart.
Boron trifluoride	Nonf. G	73.306, 73 302	Green	300 pounds.
Bottles, *acid or other corrosive liquids, empty*	See § 73.29			
*Box toe gum	F. L	73.118, 73.119	Red	10 gallons.
Boxes, *reused*	See § 73.28			
Bromacetone, liquid	Pois. A	No exemption, 73.329 (a)	Poison Gas	Not accepted.
Brombenzyl cyanide, liquid	Pois. C	No exemption, 73.382	Tear Gas	20 pounds.
Bromine	Cor. L	No exemption, 73.252	White	1 quart.
Bromine pentafluoride	Cor. L	No exemption, 73.284	White	100 pounds.
Bromine trifluoride	Cor. L	No exemption, 73.283	White	100 pounds.
Bromotoluene, alpha. *See* Benzyl bromide.				
*Bronze liquid (or paint). *See* *Paint, enamel, lacquer, stain, shellac, varnish, etc.				

* See § 72.4.

LIST OF EXPLOSIVES AND OTHER DANGEROUS ARTICLES—*Continued*

Article	Classed as	Exemptions and Packing (see sec.)	Label required if not exempt	Maximum quantity in 1 outside container by rail express
Brucine, solid (dimethoxy strychnine)	Pois. B	73.364, 73.365	Poison	200 pounds.
Burnt cotton (not repicked)	F. S.	No exemption, 73.159	Yellow	Not accepted.
Burnt fiber	F. S.	No exemption, 73.169	Yellow	Not accepted.
Bursters (explosive)	Expl. A	No exemption, 73.69		Not accepted.
Butadiene, inhibited	F. G.	73.306, 73.304, 73.314, 73.315	Red Gas	300 pounds.
Butane. See Liquefied petroleum gas.				
Butyl acetate	F. L.	73.118, 73.119	Red	10 gallons.
Butyl alcohol. See Alcohol, n. o. s.				
**Butyl mercaptan	F. L.	No exemption, 73.141	Red	10 gallons.
Butyl trichlorosilane	Cor. L.	No exemption, 73.280	White	10 gallons.
Butyraldehyde	F. L.	73.118, 73.119	Red	10 gallons.
Cacodylic acid, solid (dimethylarsenic)	Pois. B	73.364, 73.365	Poison	200 pounds.
Calcium arsenate, solid	Pois. B	73.364, 73.367, 73.368	Poison	200 pounds.
Calcium arsenite, solid	Pois. B	73.364, 73.365	Poison	200 pounds.
Calcium chlorate	Oxy. M	73.153, 73.163	Yellow	100 pounds.
Calcium chlorite	Oxy. M	No exemption, 73.160	Yellow	100 pounds.
Calcium cyanide. See Cyanide of calcium.				
Calcium hypochlorite compounds, dry, containing more than 39 percent available chlorine	Oxy. M	73.153, 73.217	Yellow	100 pounds.
Calcium, metallic	F. S.	73.153, 73.154	Yellow	100 pounds.
Calcium, metallic, crystalline	F. S.	No exemption, 73.231	Yellow	25 pounds.
Calcium nitrate	Oxy. M	73.153, 73.182	Yellow	100 pounds.
Calcium peroxide	Oxy. M	73.153, 73.156	Yellow	100 pounds.
Calcium permanganate	Oxy. M	73.153, 73.154	Yellow	100 pounds.
Calcium phosphide	F. S.	No exemption, 73.161	Yellow	25 pounds.
Calcium resinate	F. S.	No exemption, 73.166	Yellow	125 pounds.
Calcium resinate, fused	F. S.	No exemption, 73.166	Yellow	125 pounds.
Cannon primers	Expl. C	No exemption, 73.107	Yellow	150 pounds.
Caps, blasting. See Blasting caps.				
Caps, toy. See Toy caps.				
Caprylyl peroxide solution	Oxy. M	73.153 (b), 73.221	Yellow	1 quart.
Carbolic acid, fused solid. See Carbolic acid (phenol), solid.				
Carbolic acid (phenol), liquid, (liquid tar acid containing over 50 percent benzo-phenol)	Pois. B	73.345, 73.349	Poison	55 gallons.

Article	Label	Reference	Label color	Maximum quantity
Carbolic acid (phenol), solid	Pois. B.	73.369	Poison	250 pounds.
Carbon bisulfide (disulfide)	F. L.	No exemption, 73.121	Red	Not accepted.
Carbon dioxide gas, liquefied ("mining device")	Nonf. G.	73.304 (a) (2) Table Note 4	Green	6 pounds.
Ø Carbon dioxide, liquefied	Nonf. G.	73.306, 73.304, 73.314, 73.315	Green	300 pounds.
Carbon dioxide—nitrous oxide mixture	Nonf. G.	73.306, 73.304	Green	300 pounds.
Carbon dioxide—oxygen mixture	F. G.	73.306, 73.304	Green	300 pounds.
Carbon monoxide	F. L.	73.306, 73.304	Red Gas	150 pounds.
*Carbon remover, liquid	F. L.	73.118, 73.119	Red	10 gallons.
Carbonyl chloride. See Phosgene.				
Carboys, acid, empty. See § 73.29 (c).	See § 73.29 (c).			
Carboys, empty. See Acid carboys, empty.				
Cartridge bags, empty, with black powder igniters.	Expl. C.	No exemption, 73.106		150 pounds.
Cartridge cases, empty, primed.	Expl. C.	No exemption, 73.107		150 pounds.
Case oil. See Gasoline, *Naphtha.				
Casinghead gasoline. See Gasoline.				
Casks, empty. See Drums, empty.				
Caustic potash, liquid.	Cor. L.	73.244, 73.249	White	10 gallons.
Caustic soda, liquid.	Cor. L.	73.244, 73.249	White	10 gallons.
*Cement, adhesive, n.o.s. See Cement, liquid, n.o.s.				
*Cement, leather.	F. L.	73.118, 73.119	Red	12 gallons.
*Cement, linoleum, tile, wallboard, or container, liquid.	F. L.	73.118, 73.119, 73.132	Red	15 gallons.
*Cement, liquid, n.o.s.	F. L.	73.118, 73.119, 73.132	Red	15 gallons.
*Cement, pyroxylin.	F. L.	73.118, 73.119	Red	15 gallons.
*Cement, roofing, liquid.	F. L.	73.118, 73.132	Red	12 gallons.
*Cement, rubber.	F. L.	73.118, 73.132	Red	15 gallons.
Cesium-137	Pois. D.	73.392, 73.393	Radioactive materials Red	300 curies. See §73.393(L).
*Charcoal, activated.	F. S.	73.162	Yellow #	200 pounds.
Charcoal briquettes.	F. S.	73.162	Yellow #	200 pounds.
Charcoal, shell.	F. S.	73.162	Yellow #	200 pounds.
Charcoal, wood, ground, crushed, granulated or pulverized.	F. S.	73.162	Yellow	200 pounds.
Charcoal, wood, lump.	F. S.	73.162	Yellow	100 pounds.
Charcoal screenings, made from "pinon," wood.	F. S.	73.162	Yellow	200 pounds.
Charcoal, wood screenings other than "Pinon" wood screenings.	F. S.	No exemption, 73.162	Yellow #	Not accepted.
Charcoal screenings, wet.	Not accepted.			Not accepted.
Charcoal, wet.	Not accepted.			Not accepted.

* See § 72.4.   Ø See § 72.4.   # See § 72.4.

LIST OF EXPLOSIVES AND OTHER DANGEROUS ARTICLES—Continued

Article	Classed as	Exemptions and Packing (see sec.)	Label required if not exempt	Maximum quantity in 1 outside container by rail express
Charged oil well jet perforating guns (*total explosive contents in guns exceeding 20 pounds per motor vehicle*)	Expl. A	No exemption, 73.53 (u), 73.80		Not accepted.
Charged oil well jet perforating guns (*total explosive contents in guns not exceeding 20 pounds per motor vehicle*)	Expl. C	No exemption 73.53 (u), 73.110		Not accepted.
*Chemicals, n. o. s. *See* *Drugs, chemicals, medicines or cosmetics, n. o. s.				
Chemical ammunition (containing class A poisons, liquids, or gases)	Pois. A	No exemption, 73.330	Poison gas	Not accepted.
Chemical ammunition (containing class B poisons, liquids, or gases)	Pois. B	73.345, 73.350	Poison	55 gallons.
Chemical ammunition (containing class C poisons, liquids, or solids)	Pois. C	No exemption, 73.383	Tear Gas	20 pounds.
Chemical ammunition, explosive	See § 73.59			
*Chemical kits	See § 73.286			
Chloracetophenone, gas, liquid, or solid	Pois. C	No exemption, 73.382	Tear Gas	75 pounds.
Chloracetyl chloride	Cor. L	No exemption, 73.253	White	1 quart.
*Chlorate and borate mixtures	Oxy. M	73.153, 73.229	Yellow	100 pounds.
*Chlorate and magnesium chloride mixture	Oxy. M	73.153, 73.229	Yellow	100 pounds.
Chlorates, n. o. s.	Oxy. M	73.153, 73.163	Yellow	100 pounds.
Chlorates, n. o. s., wet	Oxy. M	73.153, 73.163 (a) (6)	Yellow	200 pounds.
Chlorae explosives, dry. See High explosives.				
Chlorate of potash	Oxy. M	73.153, 73.163	Yellow	100 pounds.
Chlorate of soda	Oxy. M	73.153, 73.163	Yellow	100 pounds.
Chlorate powders. See High explosives.				
Chloride of phosphorus. *See* Phosphorus trichloride.				
Chloride of sulfur. *See* Sulfur chloride.				
⊘Chlorine	Nonf. G	73.306, 73.304, 73.314, 73.315	Green	150 pounds.
Chlorine dioxide hydrate, frozen	Oxy. M	No exemption, 73.237	Yellow	Not accepted.
Chlorine trifluoride	Cor. L	No exemption, 73.285	White	100 pounds.

Chlorobenzoyl peroxide (para)	Oxy. M	No exemption, 73.157, 73.158	Yellow	25 pounds.
Chlorodinitrobenzol. See Dinitrochlorbenzol, solid.				
4-Chloro-o-toluidine hydrochloride	Pois. B	No exemption, 73.362	Poison	1 quart.
Chlorosulfonic acid	Cor. L	73.244, 73.254	White	1 quart.
Chlorosulfonic acid-sulfur trioxide mixture	Cor. L	73.244, 73.254	White	1 quart.
Chlorpicrin and nonflammable, nonliquefied compressed gas mixtures				Not accepted.
Chlorpicrin, liquid	Pois. A	No exemption, 73.329 (c)	Poison Gas	24 pounds.
Chlorpicrin, absorbed	Pois. B	No exemption, 73.357	Poison	75 pounds.
Chlorpicrin mixtures *(containing no compressed gas or poisonous liquid, class A)*	Pois. B	No exemption, 73.357	Poison	75 pounds.
Chlorpicrin and methyl chloride mixtures	Pois. A	No exemption, 73.329 (b)	Poison Gas	Not accepted.
Chromic acid	Oxy. M	73.153, 73.164	Yellow	100 pounds.
Chromic acid solution	Cor. L	73.244, 73.245, 73.287	White	1 gallon.
Chromic anhydride. See Chromic acid.				
Chromium trioxide. See Chromic acid.				
Chromyl chloride	Cor. L	No exemption, 73.247	White	1 gallon.
Cigar and cigarette lighter fluid	F. L.	73.118, 73.119	Red	10 gallons.
Cigarette lighters charged with fuel	See § 73.21 (d)			
Cigarette loads	Expl. C	No exemption, 73.111		150 pounds.
*Cleaning fluid or liquid	F. L.	73.118, 73.119	Red	10 gallons.
Cloud gas cylinders. See Chemical ammunition.	Not accepted			Not accepted.
Coal briquettes, hot.	Not accepted			Not accepted.
Coal gas. See Hydrocarbon gas, non-liquefied.				
Coal, ground bituminous, sea coal, coal facings, etc.	See § 73.165			
*Coal tar distillate	F. L.	73.118, 73.119	Red	10 gallons.
*Coal tar light oil	F. L.	73.118, 73.119	Red	10 gallons.
*Coal tar naphtha	F. L.	73.118, 73.119	Red	10 gallons.
*Coal tar oil	F. L.	73.118, 73.119	Red	10 gallons.
*Coating solution	F. L.	73.118, 73.132	Red	15 gallons.
*Cobalt-60	Pois. D	73.392, 73.393	Radioactive materials, red	300 curies. See § 73.393 (L).
Cobalt resinate, precipitated	F. S.	No exemption, 73.166	Yellow	125 pounds.
Cocculus, solid *(fish berry)*	Pois. B	73.364, 73.365	Poison	200 pounds.
Coke, hot.	Not accepted			Not accepted.
Collodion	F. L.	73.118, 73.119	Red	10 gallons.
Collodion cotton, wet. *See Wet nitrocellulose.*				
Cologne spirits *(alcohol)*	F. L.	73.118, 73.119	Red	10 gallons.
Colored fire. See Common fireworks.				
Columbian spirits *(wood alcohol)*	F. L.	73.118, 73.119	Red	10 gallons.
Combination fuzes.	Expl. C	No exemption, 73.105		150 pounds.

* See § 72.4.   Ø See § 72.4.

LIST OF EXPLOSIVES AND OTHER DANGEROUS ARTICLES—*Continued*

Article	Classed as	Exemptions and Packing (see sec.)	Label required if not exempt	Maximum quantity in 1 outside container by rail express
Combination primers	Expl. C.	No exemption, 73.107		150 pounds.
Commercial shaped charges. See High explosives.	See § 73.65 (h)			200 pounds.
Common fireworks	Expl. C.	No exemption.73.100(r),73.108		1 quart.
*Compounds, cleaning, liquid	Cor. L.	73.244, 73.245.	White	10 pints.
*Compounds, cleaning, liquid (containing hydrochloric (muriatic) acid)	Cor. L. / F. L.	73.244, 73.263 / 73.118, 73.119	White / Red	10 gallons. / 10 pints.
*Compounds, cleaning, liquid	Cor. L.	73.244, 73.256	White	55 gallons.
Compounds, cleaning, liquid (*Containing hydrofluoric acid*)	F. L.	73.118, 73.128	Red	1 gallon.
*Compounds, enamel	Cor. L.	73.244, 73.245.	White	55 gallons.
*Compounds, iron or steel rust preventing or removing	F. L.	73.118, 73.128.	Red	1 gallon.
*Compounds, lacquer, paint, or varnish, etc., removing, reducing or thinning, liquid	Cor. L.	73.244, 73.245.	White	55 gallons.
*Compounds, lacquer, paint, or varnish removing, liquid	F. L.	73.118, 73.129.	Red	10 gallons.
*Compounds, polishing, liquid	F. L.	73.118, 73.119.	Red	55 gallons.
*Compounds, tree or weed killing, liquid	Pois. B.	73.345, 73.346.	Poison	100 pounds.
*Compounds, tree or weed killing, liquid	Oxy. M.	73.153, 73.154, 73.229.	Yellow	10 gallons.
Compounds, tree or weed killing, solid	F. L.	73.118, 73.119.	Red	1 quart.
*Compounds, type-cleaning, liquid	Cor. L.	73.244, 73.245.	White	10 gallons.
*Compounds, vulcanizing, liquid	F. L.	73.118, 73.119.	Red	
*Compounds, vulcanizing, liquid	F. L.		Red	
Compressed gases, n.o.s.	Nonf. G.	73.306, 73.305, 73.304, 73.302	Green	300 pounds.
Compressed gases, n.o.s.	F. G.	73.306, 73.305, 73.304, 73.302	Red Gas	300 pounds.
*Containers, empty. See Acid carboys, empty; bottles empty, drums empty, cylinders empty.	See § 73.29.			
Containers, reused	See § 73.28.			
Copper acetoarsenite, solid (*emerald green, imperial green, Kings green, moss green, meadow green, mitis green, parrot green, Vienna green*)	Pois. B.	73.364, 73.367.	Poison	200 pounds.
Copper arsenite, solid (*Scheele's green, cupric green, copper orthoarsenite, Swedish green*)	Pois. B.	73.364, 73.365.	Poison	200 pounds.

Article	Classification	Exemptions	Label	Maximum quantity in one outside container
Copper cyanide	See § 73.370			
Cordeau detonant fuse	Expl. C.	No exemption, 73.104		300 pounds.
Corrosive battery fluid. *See* Electrolyte (acid), or Alkaline corrosive battery fluid.				
Corrosive liquid, n.o.s.	Cor. L.	73.244, 73.245	White.	5 pints.
Cotton, burnt. *See* Burnt cotton.				
Cotton waste, oily *with more than 5 percent of animal or vegetable oil.*				
Crotonaldehyde	F. S.	No exemption, 73.167	Yellow	Not accepted.
*Crude nitrogen fertilizer solution	F. L.	73.118, 73.119.	Red.	10 gallons.
*Crude oil, petroleum	Nonf. G.	73.306, 73.304, 73.314	Green.	300 pounds.
Cumene hydroperoxide	F. L.	73.118, 73.119	Red.	10 gallons.
Cupriethylene-diamine solution	Oxy. M.	73.153 (b), 73.224	Yellow.	1 quart.
Cyanide of calcium or cyanide of calcium mixtures, solid	Cor. L.	73.244, 73.249.	White.	1 gallon.
Cyanides of copper, zinc, lead, and silver.	Pois. B.	73.370 (c) and (d)	Poison.	200 pounds.
Cyanides or cyanide mixtures, dry.	See § 73.370.	See § 73.370.		
Cyanide of potassium, liquid.	Pois. B.	73.364, 73.370.	Poison.	200 pounds.
Cyanide of potassium, solid.	Pois. B.	73.345, 73.352.	Poison.	55 gallons.
Cyanide of sodium, liquid.	Pois. B.	73.370	Poison.	200 pounds.
Cyanide of sodium, solid.	Pois. B.	73.345, 73.352.	Poison.	55 gallons.
Cyanogen bromide.	Pois. B.	73.370	Poison.	200 pounds.
Cyanogen chloride containing less than 0.9 percent water.	Pois. A.	No exemption, 73.379.	Poison Gas.	25 pounds.
Cyanogen gas.	Pois. A.	No exemption, 73.328.	Poison Gas.	Not accepted.
Cyclohexane.	F. L.	No exemption, 73.328.	Red.	Not accepted.
Cyclohexanone peroxide not over 50 percent concentration.	Oxy. M.	73.118, 73.119.	Yellow.	10 gallons.
Cyclohexanone peroxide over 50 percent concentration but not exceeding 85 percent concentration.	Oxy. M.	73.153, 73.154.	Yellow.	25 pounds.
Cyclohexenyl trichlorosilane.	Cor. L.	No exemption, 73.157, 73.158.	White.	25 pounds.
Cyclohexyltrichlorosilane.	Cor. L.	No exemption, 73.280.	White.	10 gallons.
Cyclopentane.	F. L.	No exemption, 73.280.	Red.	10 gallons.
Cyclopentane, methyl.	F. L.	73.118, 73.119.	Red.	10 gallons.
Cyclopropane.	F. G.	73.118, 73.119. 73.306, 73.304.	Red Gas.	300 pounds.
Cyclotrimethylenetrinitramine, desensitized. *See* High Explosives.				
Cyclotrimethylenetrinitramine, wet with not less than 10 percent of water. *See* High explosives.				
Cylinders, empty.	See § 73.29			

* See § 72.4.

LIST OF EXPLOSIVES AND OTHER DANGEROUS ARTICLES—Continued

Article	Classed as	Exemptions and Packing (see sec.)	Label required if not exempt	Maximum quantity in 1 outside container by rail express
Decaborane	F. S.	No exemption, 73.236	Yellow	25 pounds.
Delay electric igniters	Expl. C.	No exemption, 73.106		150 pounds.
Denatured alcohol. *See* Alcohol, n. o. s.				
Depth bombs. See Explosive bomb.				
Detonating fuzes, class A explosives	Expl. A.	No exemption, 73.69		Not accepted.
Detonating fuzes, class A explosives, radioactive	Expl. A.	No exemption, 73.69		Not accepted.
Detonating fuzes, class C explosives	Expl. C.	No exemption, 73.113		150 pounds.
Detonating primers	Expl. A.	No exemption, 73.68		Not accepted.
Diazodinitrophenol. See Initiating explosive.				
Dichlorethylene	F. L.	73.118, 73.119	Red	10 gallons.
Dichlorodifluoromethane	Nonf. G.	73.306, 73.304, 73.314, 73.315	Green	300 pounds.
∅Dichlorodifluoro-methane-dichlorotetrafluoroethane mixture	Nonf. G.	73.306, 73.304, 73.314, 73.315	Green	300 pounds.
∅Dichlorodifluoromethane-monochlorodifluoromethane mixture	Nonf. G.	73.306, 73.304, 73.314	Green	300 pounds.
∅Dichlorodifluoromethane-trichloromonofluoromethane-monochlorodifluoromethane mixture	Nonf. G.	73.306, 73.304, 73.314	Green	300 pounds.
∅Dichlorodifluoromethane-trichlorotrifluoroethane mixture	Nonf. G.	73.306, 73.304, 73.314	Green	300 pounds.
*Dichlorodifluoromethane-monofluorotrichloromethane mixture	Nonf. G.	73.306, 73.304, 73.314, 73.315	Green	300 pounds.
Dichlorofluoromethane and difluoroethane mixture (*constant boiling mixture*)	Nonf. G.	73.306, 73.304, 73.314, 73.315	Green	300 pounds.
Dichloroisocyanuric acid, dry, containing more than 39% available chlorine	Oxy. M.	73.153, 73.217	Yellow	100 pounds.
Dicumyl peroxide, solid	Oxy. M.	73.153, 73.154	Yellow	25 pounds.
Dicumyl peroxide, 50% solution	Oxy. M.	73.153 (b), 73.224	Yellow	1 quart.
Diethyl aluminum chloride	F. L.	No exemption, 73.134	Red	2 ounces.
Diethyl dichlorosilane	Cor. L.	No exemption, 73.280	White	10 gallons.
Diethylamine	F. L.	73.118, 73.119	Red	10 gallons.
Diethylene glycol dinitrate	See § 73.51 (d)			
Difluoroethane	F. G.	73.306, 73.304, 73.314, 73.315	Red Gas	300 pounds.
Difluoromonochloroethane	F. G.	73.306, 73.304, 73.314	Red Gas	300 pounds.
Difluorophosphoric acid, anhydrous	Cor. L.	No exemption, 73.275	White	1 gallon.
*Di iso octyl acid phosphate	Cor. L.	73.244, 73.296	White	1 quart.

Article	Classification	Regulation	Label	Maximum quantity
Diisopropylbenzene hydroperoxide	Oxy. M.	73.153 (b), 73.224	Yellow	1 quart.
Dimethylamine, anhydrous	F.G.	73.306, 73.304, 73.314, 73.315	Red Gas	300 pounds.
Dimethylamine, aqueous solution	F.L.	73.118, 73.119	Red	10 gallons.
Dimethyldichlorosilane	F.L.	No exemption, 73.135	Red	10 gallons.
Dimethyl ether	F.G.	73.306, 73.304, 73.314	Red Gas	300 pounds.
Dimethylhexane dihydroperoxide	Oxy. M.	No exemption, 73.157, 73.158	Yellow	25 pounds.
Dimethylhydrazine, unsymmetrical	F.L.	No exemption, 73.145	Red	5 pints.
Dimethyl sulfate	Cor. L.	No exemption, 73.255	White	1 quart.
Dimethyl sulfide	F.L.	73.118, 73.119	Red	10 gallons.
Dinitrobenzol, solid	Pois. B.	73.364, 73.371	Poison	200 pounds.
Dinitrobenzol, liquid	Pois. B.	73.345, 73.346	Poison	55 gallons.
Dinitrochlorbenzol, solid (dinitrochlorobenzene, chlorodinitrobenzol)				
Dinitrophenol solutions	Pois. B.	73.364, 73.365	Poison	200 pounds.
Diphenyl dichlorosilane	Pois. B.	73.345, 73.362A	Poison	65 pounds.
Diphenylaminechlorarsine, gas, liquid, or solid	Cor. L.	No exemption, 73.280	White	10 gallons.
Diphenylchlorarsine, solid	Pois. C.	No exemption, 73.382	Tear Gas	75 pounds.
Diphosgene. *See* Phosgene.	Pois. C.	No exemption, 73.382	Tear Gas	20 pounds.
Dispersant gas, n.o.s.	See § 73.314 (c) Table Note 13, 73.315 (a) Table Note 9.			
*Distillate.	F.L.	73.118, 73.119	Red	10 gallons.
*Dodecyltrichlorosilane.	Cor. L.	No exemption, 73.280	White	10 gallons.
*Dressing, leather.	F.L.	73.118, 73.119	Red	10 gallons.
*Driers, paint, varnish, enamel, etc. *See* *Paint driers, liquid.				
Drill cartridges.	See § 73.55			
*Drugs, chemicals, medicines or cosmetics, n. o. s.	F.L.	73.118, 73.119	Red	10 gallons.
*Drugs, chemicals, medicines or cosmetics, n. o. s.	F.S.	73.153, 73.154	Yellow	100 pounds.
*Drugs, chemicals, medicines or cosmetics, n. o. s.	Oxy. M.	73.153, 73.154	Yellow	100 pounds.
*Drugs, chemicals, medicines or cosmetics, n. o. s.	Cor. L.	73.244, 73.245	White	1 quart.
*Drugs, chemicals, medicines or cosmetics, n. o. s. (liquid)	Pois. B.	73.345, 73.346	Poison	55 gallons.
*Drugs, chemicals, medicines or cosmetics, n. o. s. (solid)	Pois. B.	73.364, 73.365	Poison	200 pounds.
Drums, empty.	See § 73.29			
Dummy cartridges.	See § 73.55			
Dusts, by-product, poisonous. *See* Arsenical dust.				
*Dynamite. *See* High explosives.				
Electric blasting caps—more than 1,000	Expl. A.	No exemption, 73.66		Not accepted.

* See § 72.4.   ⊘ See § 72.4.

LIST OF EXPLOSIVES AND OTHER DANGEROUS ARTICLES—*Continued*

Article	Classed as	Exemptions and Packing (see sec.)	Label required if not exempt	Maximum quantity in 1 outside container by rail express
Electric blasting caps—1,000 or less	Expl. C	No exemption, 73.103		See § 73.86.
Electric squibs	Expl. C	No exemption, 73.106		150 pounds.
Electric storage batteries, wet. *See* Batteries, electric storage, wet.				
Electrolyte (acid), battery fluid	Cor. L	73.244, 73.257	White	5 gallons.
Electrolyte (acid) or alkaline corrosive battery fluid packed with storage batteries	Cor. L	No exemption, 73.258	White	2 gallons.
Electrolyte (acid) or alkaline corrosive battery fluid packed with battery charger, radio current supply device, or electronic equipment and actuating devices	Cor. L	No exemption, 73.259	White	6 quarts.
Empty cartridge bags with black powder igniters	Expl. C	No exemption, 73.106		150 pounds.
Empty cartridge cases, primed	Expl. C	No exemption, 73.107		150 pounds.
*Enamel. *See* *Paint, enamel, lacquer, stain, shellac, varnish, etc.				
Engines, internal combustion	See § 73.120			
Engine starting fluid	F. G.	No exemption, 73.304	Red Gas	60 pounds.
*Eradicators, paint or grease, liquid	F. L.	73.118, 73.119	Red	10 gallons.
Etching acid liquid, n.o.s.	Cor. L	No exemption, 73.299	White	10 pounds.
Ethane	F. G.	73.306, 73.304	Red Gas	300 pounds.
Ether	F. L.	73.118, 73.119	Red	10 gallons.
Ether, ethyl (*sulfuric*). *See* Ether.				
Ethyl acetate	F. L.	73.118, 73.119	Red	10 gallons.
Ethyl alcohol. *See* Alcohol, n. o. s.				
Ethyl aldehyde. *See* Acetaldehyde.				
Ethyl aluminum dichloride	F. L.	No exemption, 73.134	Red	2 ounces.
Ethyl aluminum sesquichloride	F. L.	No exemption, 73.134	Red	2 ounces.
Ethyl chloride	F. L.	No exemption, 73.123	Red	300 pounds in cylinders, 15 pounds in other containers.
Ethyl chlorocarbonate. *See* Ethyl chloroformate.				
Ethyl chloroformate	Cor. L	No exemption, 73.288	White	5 pints.
Ethyldichloroarsine	Pois. A	No exemption, 73.328	Poison Gas	Not accepted.
Ethyl dichlorosilane	F. L.	No exemption, 73.135	Red	10 gallons.

⊘Ethylene	F. G.	73.306, 73.304	Red Gas	300 pounds.
Ethylene dichloride	F. L.	73.118, 73.119	Red	10 gallons.
Ethylene imine, inhibited	F. L.	No exemption, 73.139	Red	5 pints.
Ethylene oxide	F. L.	No exemption, 73.124	Red	300 pounds in cylinders, 15 pounds in other containers.
Ethyl formate	F. L.	73.118, 73.119	Red	10 gallons.
Ethyl mercaptan	F. L.	No exemption, 73.141	Red	10 gallons.
Ethyl methyl ether	F. L.	73.118, 73.119	Red	10 gallons.
Ethyl methyl ketone	F. L.	73.118, 73.119	Red	10 gallons.
Ethyl nitrate (nitric ether)	F. L.	73.118, 73.119	Red	10 gallons.
Ethyl nitrite (nitrous ether)	F. L.	73.118, 73.119	Red	10 gallons.
Ethyl phenyl dichlorosilane	Cor. L.	No exemption, 73.280	White	10 gallons.
Ethyl trichlorosilane	F. L.	No exemption, 73.135	Red	10 gallons.
Explosive auto alarms	Expl. C.	No exemption, 73.111		150 pounds.
Explosive bomb	Expl. A.	No exemption, 73.56		Not accepted.
Explosive cable cutters	Expl. C.	No exemption, 73.102		150 pounds.
Explosives, class A	See §73.53			...
Explosives, class B	See §73.88			...
Explosives, class C	See §73.100			...
Explosive compositions	Expl. A or B	No exemption, 73.53, 73.61 to 73.93		10 pounds.
Explosive mine	Expl. A.	No exemption, 73.56		Not accepted.
Explosive power device, class B	Expl. B.	No exemption, 73.94	Red#	150 pounds.
Explosive power device, class C	Expl. C.	No exemption, 73.102		150 pounds.
Explosive projectile	Expl. A.	No exemption, 73.56		Not accepted.
Explosive release device	Expl. C.	No exemption, 73.102		150 pounds.
Explosive rivets	Expl. C.	No exemption, 73.100 (q)		150 pounds.
Explosive samples for laboratory examination	See §73.86			...
*Explosive torpedo	Expl. A.	No exemption, 73.56		Not accepted.
*Extracts, liquid, flavoring	F. L.	73.118, 73.119	Red	10 gallons.
Fabrics or fibers with *animal or vegetabel oil. See* Fibers or fabrics with *animal or vegetable oil.*				
Felt waste, wet. *See* Waste wool, wet.				
Ferric arsenate, solid	Pois. B.	73.364, 73.365	Poison	200 pounds.
Ferric arsenite, solid	Pois. B.	73.364, 73.365	Poison	200 pounds.
Ferrous arsenate (*iron arsenate*), solid	Pois. B.	73.364, 73.365	Poison	200 pounds.
*Fertilizer ammoniating solution *containing free ammonia*	Nonf. G.	73.306, 73.304, 73.314	Green	300 pounds.
Fertilizer, tankage. *See* Garbage, tankage.				

LIST OF EXPLOSIVES AND OTHER DANGEROUS ARTICLES—Continued

Article	Classed as	Exemptions and Packing (see sec.)	Label required if not exempt	Maximum quantity in 1 outside container by rail express
Fiber, burnt.	F. S.	No exemption, 73.169.	Yellow.	Not accepted.
Fibers or fabrics, with animal or vegetable oil.	F. S.	No exemption, 73.170.	Yellow.	Not accepted.
Film, motion-picture and toy pieces of. See Motion-picture film, toy and motion-picture film scrap.				
Firecrackers. See Common fireworks or Special fireworks.				
Firecracker salutes. See Common fireworks or Special fireworks.				
Fire extinguisher charges.	Cor. L.	73.261.	White.	1 gallon.
Fire extinguisher charges containing not to exceed 50 grains of propellant explosives per unit.	See § 73.88 (f) Note 1.			
Fire extinguishers.	Nonf. G.	73.306.	Green.	300 pounds.
Fireworks, common.	Expl. C.	No exemption, 73.100 (r). 73.108.		200 pounds.
Fireworks, exhibition display pieces. See Special fireworks.				
Fireworks, special.	Expl. B.	No exemption, 73.88 (d), 73.91.	Special Fireworks ##.	200 pounds.
Fish meal. See Fish scrap or fish meal.				
Fish scrap or fish meal containing less than 6 percent or more than 12 percent moisture, n.o.s.	F. S.	No exemption, 73.171.	Yellow.	Not accepted.
Fissile radioactive materials, n.o.s.	Pois. D.	73.392, 73.393.	Radioactive materials, red.	See §73.393(g) and (m).
*Flame retardant compound, liquid.	Cor. L.	73.244, 73.291.	White.	10 gallons.
Flammable liquids, n. o. s.	F. L.	73.118, 73.119.	Red.	10 gallons.
Flammable solids, n. o. s.	F. S.	73.153, 73.154.	Yellow.	25 pounds.
Flares. See Common fireworks.				
Flares, aeroplane. See Special fireworks.				
Flash cartridges. See Special fireworks and Low explosives.				
Flash crackers. See Common fireworks or Special fireworks.				

Article	Classification	Exemption	Label	Quantity
Flash powder. See Special fireworks and Low explosives.				
Flash sheets. See Special fireworks and Low explosives.				
Flexible linear shaped charges, metal clad	Expl. C.	No exemption, 73.104.		300 pounds.
Flue dust, poisonous	Pois. B.	73.364, 73.368.	Poison.	200 pounds.
Fluorine	F. G.	73.306, 73.302.	Red Gas.	6 pounds.
Fluosulfonic acid	Cor. L.	No exemption, 73.274.	White.	10 pints.
Formic acid	Cor. L.	73.244, 73.245, 73.289.	White.	5 gallons.
Formic acid, solution	Cor. L.	73.244, 73.245, 73.289.	White.	5 gallons.
Fulminate of mercury, dry	Forbidden explosive.			Forbidden explosive.
Fulminate of mercury, wet. See Initiating explosive.				
*Fumigants	See § 73.152 (a) Note 1.			
*Furniture polish. See *Polishes, metal, stove, furniture, and wood, liquid.				
*Furniture or wood stains, liquid. See *Paint, enamel, lacquer, stain, shellac, varnish, etc.				
Fuse igniters	Expl. C.	No exemption, 73.106.		150 pounds.
Fuse, instantaneous. See Instantaneous fuse.				
Fuse lighters	Expl. C.	No exemption, 73.106.		150 pounds.
Fuse, mild detonating, metal clad. See Mild detonating fuse, metal clad.				
Fuse, safety. See Safety fuse.				
Fusees. See Railway or highway fusees.				
Fuzes, combination	Expl. C	No exemption, 73.105.		150 pounds.
Fuzes, detonating, class A explosives	Expl. A	No exemption, 73.69.		Not accepted.
Fuzes, detonating, class A explosives, radioactive	Expl. A	No exemption, 73.69.		Not accepted.
Fuzes, detonating, class C explosives	Expl. C	No exemption, 73.113.		150 pounds.
Fuzes, percussion	Expl. C	No exemption, 73.105.		150 pounds.
Fuzes, time	Expl. C	No exemption, 73.105.		150 pounds.
Fuzes, tracer	Expl. C	No exemption, 73.105.		150 pounds.
Garbage tankage containing less than 8 percent of moisture	F. S. See § 73.29	No exemption, 73.209.	Yellow.	Not accepted.
Gas cylinders, empty.				
*Gas drips, hydrocarbon	F. L.	73.118, 73.119.	Red.	10 gallons.
Gas identification sets	Pois. A. and C.	No exemption, 73.331.	Poison Gas.	See § 73.331.
Gas mine. See Explosive mine.	See § 73.56 (d)			
Gasoline	F. L.	73.118, 73.119.	Red.	10 gallons.
Gelatine dynamite. See High explosives.				

* See § 72.4.   ## See § 72.4.

LIST OF EXPLOSIVES AND OTHER DANGEROUS ARTICLES—*Continued*

Article	Classed as	Exemptions and Packing (see sec.)	Label required if not exempt	Maximum quantity in 1 outside container by rail express
Gold-198	Pois. D	73.392, 73.393	Radioactive materials, red	300 curies. See §73.393 (L).
*Gold paint. *See* *Paint, enamel, lacquer, stain, shellac, varnish, etc.*				
Grenades, empty, primed	Expl. C	No exemption, 73.107		150 pounds.
Grenades, hand. *See* Hand grenades.				
Grenades, hand or rifle, explosive, with gas, smoke or incendiary material. *See* Hand or Rifle grenades.				
Grenades, hand or rifle, with gas, smoke or incendiary material but without bursting charges	See §§ 73.88 (d), 73.108, 73.330, 73.350			
Grenades, police. *See* Police grenades, poison gas, class A.				
Grenades, rifle. *See* Rifle grenades.				
Grenades, tear gas. *See* Tear gas grenades.				
Guanidine nitrate	Oxy. M	73.153, 73.182	Yellow	100 pounds.
Guanyl nitrosamino guanylidene hydrazine. *See* Initiating explosive.				
Guanyl nitrosamino guanyl tetrazene. *See* Initiating explosive.				
Guided missiles, with warheads. *See* Rocket ammunition with explosive, illuminating, gas, incendiary or smoke projectile.				
Guided missiles without warheads. *See* Rocket motors, class A explosives or Rocket motors, class B explosives.				
Guncotton. *See* High explosives.				
Hafnium metal, dry, chemically produced, finer than 20 mesh particle size	F. S	No exemption, 73.214	Yellow	75 pounds.
Hafnium metal, dry, mechanically produced, finer than 270 mesh particle size	F. S	No exemption, 73.214	Yellow	75 pounds.
Hafnium metal, wet, chemically produced, finer than 20 mesh particle size	F. S	No exemption, 73.214	Yellow	150 pounds.

Article	Classification	Sections	Label	Maximum quantity
Hafnium metal, wet, mechanically produced, finer than 270 mesh particle size	F. S.	No exemption, 73.214	Yellow	150 pounds.
Hair, wet	F. S.	No exemption, 73.172	Yellow	Not accepted.
Hand grenades	Expl. A.	No exemption, 73.56.		Not accepted.
Hand signal devices	Expl. C.	No exemption, 73.108.		200 pounds.
Heaters for refrigerator cars, liquid fuel type.	F. L.	73.146.		......
Helium	Nonf. G.	73.306, 73.302, 73.314	Green	300 pounds.
Helium—oxygen mixture.	Nonf. G.	73.306, 73.302.	Green	300 pounds.
Heptane.	F. L.	73.118, 73.119.	Red	10 gallons.
Hexadecyltrichlorosilane.	Cor. L.	No exemption, 73.280.	White	10 gallons.
Hexaethyl tetraphosphate and compressed gas mixture.				Not accepted.
Hexaethyl tetraphosphate, liquid.	Pois. A.	No exemption, 73.334.	Poison Gas	1 quart.
Hexaethyl tetraphosphate mixture, dry.	Pois. B.	No exemption, 73.358.	Poison	200 pounds.
Hexaethyl tetraphosphate mixture, liquid.	Pois. B.	73.377.	Poison	1 quart.
Hexafluorophosphoric acid.	Cor. L.	73.359.	Poison	1 gallon.
Hexafluoropropylene.	Nonf. G.	No exemption, 73.275.	White	300 pounds.
Hexamethylene diamine solution.	Cor. L.	73.306, 73.304, 73.314, 73.315	Green	10 gallons.
Hexane.	F. L.	73.244, 73.249, 73.292.	White	10 gallons.
Hexyl trichlorosilane.	Cor. L.	73.118, 73.119.	Red	10 gallons.
High explosives.	Expl. A.	No exemption, 73.280.	White	See § 73.86.
High explosives, liquid.	Expl. A.	No exemption, 73.61 to 73.87.		Not accepted.
Highway fusees.	Expl. C.	No exemption, 73.62.		200 pounds.
High wines. See Alcohol or alcohol, n. o. s.				
Hydraulic accumulators (pressurized with non-flammable, nonliquefied compressed gas).	See § 73.306 (e) (2)			......
Hydrazine, anhydrous.	Cor. L.	No exemption, 73.276.	White	5 pints.
Hydrazine solution containing 50 percent or less of water.	Cor. L.	No exemption, 73.245.	White	5 pints.
Hydriodic acid.	Cor. L.	73.244, 73.245.	White	1 gallon.
Hydrobromic acid.	Cor. L.	73.244, 73.262.	White	1 gallon.
Hydrobromic acid, anhydrous. See Hydrogen bromide.				
Hydrocarbon gas, liquefied.	F. G.	73.306, 73.304, 73.314	Red Gas	300 pounds.
Hydrocarbon gas, nonliquefied.	F. G.	73.306, 73.302.	Red Gas	300 pounds.
Hydrochloric (muriatic) acid.	Cor. L.	73.244, 73.263	White	10 pints.
Hydrochloric acid, anhydrous. See Hydrogen chloride.				
Hydrochloric acid mixtures.	Cor. L.	73.244, 73.263.	White	10 pints.
Hydrochloric acid solution, inhibited.	Cor. L.	73.244, 73.263.	White	10 pints.
Hydrocyanic acid (prussic), liquid.	Pois. A.	No exemption, 73.332.	Poison Gas	Not accepted.
Hydrocyanic acid (prussic), unstabilized.	Not accepted	No exemption, 73.332.	Poison Gas	Not accepted.
Hydrocyanic acid solutions.	Pois. B.	No exemption, 73.351.	Poison	25 pounds.

* See § 72.4.

LIST OF EXPLOSIVES AND OTHER DANGEROUS ARTICLES—*Continued*

Article	Classed as	Exemptions and Packing (see sec.)	Label required if not exempt	Maximum quantity in 1 outside container by rail express
Hydrofluoric acid	Cor. L	73.244, 73.264 (a)	White	10 pints.
Hydrofluoric acid, anhydrous	Cor. L	No exemption, 73.264 (b)	White	110 pounds.
Hydrofluoric and sulfuric acids, mixtures	Cor. L	No exemption, 73.290	White	10 pints.
Hydrofluosilicic acid	Cor. L	73.244, 73.265	White	10 pints.
⊘Hydrogen	F. G.	70.306, 73.302, 73.314	Red Gas	300 pounds.
Hydrogen bromide	Nonf. G.	73.306, 73.304	Green	300 pounds.
Hydrogen chloride	Nonf. G.	73.306, 73.304	Green	300 pounds.
Hydrogen, liquefied	F. G.	No exemption, 73.316	Red Gas	Not accepted.
Hydrogen peroxide (hydrogen dioxide) solution in water containing over 8 percent hydrogen peroxide by weight	Cor. L	73.244, 73.266	White	1 gallon.
Hydrogen sulfide	F. G.	73.306, 73.304, 73.314	Red Gas	300 pounds.
Hypochlorite solutions containing more than 7 percent available chlorine by weight	Cor. L	73.277	White	4 gallons.
Igniter cord	Expl. C	No exemption, 73.100 (s)		150 pounds.
Igniter fuse-metal clad	Expl. C	No exemption, 73.106		150 pounds.
Igniters	Expl. C	No exemption, 73.106		150 pounds.
Igniters, jet thrust (*jato*), class A explosives	Expl. A	No exemption, 73.79	None.	Not accepted.
Igniters, jet thrust (*jato*), class B explosives	Expl. B	No exemption, 73.92	Red #	550 pounds.
Igniters, rocket motor, class A explosives	Expl. A	No exemption, 73.79		Not accepted.
Igniters, rocket motor, class B explosives	Expl. B	No exemption, 73.92	Red #	550 pounds.
Illuminating projectiles fuzed or not fuzed, with expelling charges. See Special fireworks.				
Inflammable liquids, n. o. s. See Flammable liquids, n. o. s.				
Inflammable solids, n. o. s. See Flammable solids, n. o. s.				
Initiating explosive	Expl. A	No exemption, 73.70 to 73.78, incl.		Not accepted.
Diazodinitrophenol		No exemption, 73.70		
Fulminate of mercury		No exemption, 73.71		

Guanyl nitrosamino guanylidene hydrazine		No exemption, 73.72		
Lead azide, dextrinated type only		No exemption, 73.73		
Lead mononitroresorsinate		No exemption, 73.70		
Lead styphnate (lead trinitroresorcinate)		No exemption, 73.74		
Nitro mannite		No exemption, 73.75		
Nitrosoguanidine		No exemption, 73.76		
Pentaerythrite tetranitrate		No exemption, 73.77		
Tetrazene (guanyl nitrosamino guanyl tetrazene)		No exemption, 73.78		
*Ink	F. L.	73.118, 73.144	Red	10 gallons.
*Insecticide, dry	Pois. B	73.364, 73.365	Poison	200 pounds.
*Insecticide, liquid	Pois. B	73.345, 73.346	Poison	55 gallons.
Insecticide, liquefied gas	Nonf. G	73.306, 73.304	Green	300 pounds.
*Insecticide, liquid (vermin exterminator)	F. L.	73.118, 73.119	Red	10 gallons.
Instantaneous fuse	Expl. C	No exemption, 73.100 (m)	Red	150 pounds.
Iodine monochloride	Cor. L	No exemption, 73.293	White	1 quart.
Iridium-192	Pois. D	73.392, 73.393	Radioactive materials, red	300 curies. See §73.393 (L).
Iron mass, spent	F. S.	No exemption, 73.174	Yellow	Not accepted.
Iron sponge not properly oxidized	F. S.	No exemption, 73.174	Yellow	Not accepted.
Iron sponge, spent	F. S.	No exemption, 73.174	Yellow	Not accepted.
Isobutane. See Liquefied petroleum gas.				
Isobutylene. See Liquefied petroleum gas.				
Isooctane	F. L.	73.118, 73.119	Red	10 gallons.
Isopentane	F. L.	73.118, 73.119	Red	10 gallons.
Isoprene	F. L.	73.118, 73.119	Red	10 gallons.
*Isopropanol	F. L.	73.118, 73.119	Red	10 gallons.
Isopropyl acetate	F. L.	73.118, 73.119	Red	10 gallons.
Isopropyl mercaptan	F. L.	73.118, 73.119	Red	10 gallons.
Isopropyl percarbonate, stabilized	Cor. L	No exemption, 73.141	White	Not accepted.
Isopropyl percarbonate, unstabilized	F. S.	No exemption, 73.218, 73.282	Yellow	Not accepted.
Jet thrust igniters. See Igniters, jet thrust.				
Jet thrust unit (jato) class A explosives	Expl. A	No exemption, 73.79	Red	Not accepted.
Jet thrust unit (jato) class B explosives	Expl. B	No exemption, 73.92	Yellow	550 pounds.
Kegs, reused	See § 73.28		Red #	
*Lacquer. See *Paint, enamel, lacquer, stain, shellac, varnish, etc.				
*Lacquer base, liquid. See *Paint, enamel, lacquer, stain, shellac, varnish, etc.				
*Lacquer base or lacquer chips, dry	F. S.	73.153, 73.175	Yellow	100 pounds.

* See § 72.4.   # See § 72.4.

∅ See § 72.4.

LIST OF EXPLOSIVES AND OTHER DANGEROUS ARTICLES—*Continued*

Article	Classed as	Exemptions and Packing (see sec.)	Label required if not exempt	Maximum quantity in 1 outside container by rail express
*Lacquer, base or lacquer chips, plastic (*wet with alcohol or solvent*)*	F. L.	73.118, 73.127	Red	25 pounds.
*Lacquer removing, reducing, and thinning compounds. See *Compounds, lacquer, paint, or varnish removing, reducing, or thinning, liquid.*				
Lauroyl peroxide	Oxy. M.	73.153 (b), 73.157, 73.158	Yellow	25 pounds.
Lead arsenate, solid	Pois. B.	73.364, 73.367	Poison	200 pounds.
Lead arsenite, solid	Pois. B.	73.364, 73.365	Poison	200 pounds.
*Lead azide. See *Initiating explosive.*				
Lead cyanide	See § 73.370			
*Lead mononitroresorcinate. See *Initiating explosive.*				
Lead nitrate	Oxy. M.	73.153, 73.182	Yellow	100 pounds.
*Lead styphnate (lead trinitroresorcinate. See *Initiating explosive.*				
*Leather bleach	F. L.	73.118, 73.119	Red	10 gallons.
*Leather dressing	F. L.	73.118, 73.119	Red	10 gallons.
Lewisite	Pois. A.	No exemption, 73.328	Poison Gas	Not accepted.
*Lighter fluid. See *Cigar and cigarette lighter fluid.*				
⊘Liquefied carbon dioxide. See *Carbon dioxide, liquefied.*				
Liquefied carbon dioxide gas (mining device). See *carbon dioxide gas, liquefied (mining device).*				
Liquefied hydrocarbon gas	F. G.	73.306, 73.304, 73.314	Red Gas	300 pounds.
Liquefied nonflammable gases *charged with nitrogen, carbon dioxide, or air*	Nonf. G.	73.306, 73.304	Green	300 pounds.
⊘Liquefied petroleum gas	F. G.	73.306, 73.304, 73.314, 73.315	Red Gas	300 pounds.
*Liquids other than those classified as flammable, corrosive, or poisonous charged with nitrogen, carbon dioxide or air. See *Compressed gases, n. o. s.*				
Lithium aluminum hydride	F. S.	No exemption, 73.206	Yellow	25 pounds.
Lithium aluminum hydride, ethereal	F. L.	No exemption, 73.137	Red	1 quart.

Article	Class	Reference	Label	Maximum quantity
Lithium amide, powdered.	F. S.	73.153, 73.168.	Yellow	100 pounds.
Lithium ferro silicon.	F. S.	No exemption, 73.206.	Yellow	25 pounds.
Lithium hydride in solid forms	F. S.	No exemption, 73.206.	Yellow	25 pounds.
Lithium hypochlorite compounds, dry, containing more than 39 percent available chlorine.	F. S.	No exemption, 73.206.	Yellow	100 pounds.
Lithium metal.	Oxy. M.	73.153, 73.217.	Yellow	100 pounds.
Lithium metal, in cartridges.	F. S.	No exemption, 73.206.	Yellow	25 pounds.
Lithium peroxide.	See § 73.206			
Lithium silicon.	Oxy. M.	73.153 (a), 73.154.	Yellow	100 pounds.
London purple, solid.	F. S.	No exemption, 73.206.	Yellow	25 pounds.
Low explosives.	Pois. B.	73.364, 73.365.	Poison	200 pounds.
Low blasting explosives. *See* Low explosives.	Expl. A.	No exemption, 73.60.		Not accepted.
Machines or apparatus.	See § 73.130, 73.306.			
Magnesium arsenate, solid.	Pois. B.	73.364, 73.367.	Poison	200 pounds.
Magnesium dross.	See § 73.173			
*Magnesium, metallic, powdered, pellets, turnings, or ribbon.	F. S.	73.153, 73.220.	Yellow	100 pounds.
Magnesium nitrate.	Oxy. M.	73.153, 73.182.	Yellow	100 pounds.
Magnesium perchlorate.	Oxy. M.	73.153, 73.154.	Yellow	100 pounds.
Magnesium peroxide, solid.	Oxy. M.	73.153, 73.154.	Yellow	100 pounds.
Magnesium scrap (borings, clippings, shavings, sheets, turnings or scalpings).	F. S.	73.153, 73.220.	Yellow	100 pounds.
Magnesium-thorium alloys in formed shapes (*not powdered, and which shall contain not more than 4 percent nominal thorium 232*).	Pois. D	73.392 (e).	Radioactive materials, red	See §73.393(L).
Matches, block. See Matches, strike-anywhere.				
Matches, book, card, or strike-on-box, *with other articles*	See § 73.176 (g)			
Matches, book, card, or strike-on-box.	See § 73.176 (g)			
Matches, strike-anywhere.	F. S.	No exemption, 73.176.	Yellow	60 pounds.
*Medicines, n. o. s. *See* *Drugs, chemicals, medicines, or cosmetics, n. o. s.				
Memtetrahydro phthalic anhydride.	Cor. L.	No exemption, 73.298.	White	1 quart.
*Mercaptan mixtures, aliphatic.	F. L.	No exemption, 73.141.	Red	10 gallons.
*Mercurial liquid, n. o. s.	Pois. B.	73.345, 73.346.	Poison	55 gallons.
Mercuric acetate.	Pois. B.	73.364, 73.365.	Poison	200 pounds.
Mercuric-ammonium chloride, solid.	Pois. B.	73.364, 73.365.	Poison	200 pounds.
Mercuric bonzoate, solid.	Pois. B.	73.364, 73.365.	Poison	200 pounds.

* See § 72.4.   ∅ See § 72.4.

LIST OF EXPLOSIVES AND OTHER DANGEROUS ARTICLES—*Continued*

Article	Classed as	Exemptions and Packing (see sec.)	Label required if not exempt	Maximum quantity in 1 outside container by rail express
Mercuric bromide, solid.	Pois. B.	73.364, 73.365.	Poison.	200 pounds.
Mercuric chloride. *See* Mercury bichloride, solid.				
Mercuric cyanide, solid.	Pois. B.	73.370.	Poison.	200 pounds.
Mercuric iodide, solid.	Pois. B.	73.364, 73.365.	Poison.	200 pounds.
Mercuric iodide solution.	Pois. B.	73.345, 73.346.	Poison.	55 gallons.
Mercuric oleate, solid.	Pois. B.	73.364, 73.365.	Poison.	200 pounds.
Mercuric oxide (red). solid.	Pois. B.	73.364, 73.365.	Poison.	200 pounds.
Mercuric oxide (yellow), solid.	Pois. B.	73.364, 73.365.	Poison.	200 pounds.
Mercuric oxycyanide, solid.	Pois. B.	73.364, 73.365, 73.370.	Poison.	200 pounds.
Mercuric-potassium cyanide, solid.	Pois. B.	73.364, 73.365.	Poison.	200 pounds.
Mercuric-potassium iodide, solid.	Pois. B.	73.364, 73.365.	Poison.	200 pounds.
Mercuric salicylate, solid.	Pois. B.	73.364, 73.365.	Poison.	200 pounds.
Mercuric subsulfate, solid.	Pois. B.	73.364, 73.365.	Poison.	200 pounds.
Mercuric sulfate, solid.	Pois. B.	73.364, 73.365.	Poison.	200 pounds.
Mercuric sulfo cyanate, solid (mercuric thio-cyanate).	Pois. B.	73.364, 73.365.	Poison.	200 pounds.
Mercurol (*mercury nucleate*), solid.	Pois. B.	73.364, 73.365.	Poison.	200 pounds.
Mercurous bromide, solid.	Pois. B.	73.364, 73.365.	Poison.	200 pounds.
Mercurous gluconate, solid.	Pois. B.	73.364, 73.365.	Poison.	200 pounds.
Mercurous iodide, solid.	Pois. B.	73.364, 73.365.	Poison.	200 pounds.
Mercurous nitrate, solid.	Pois. B.	73.364, 73.365.	Poison.	200 pounds.
Mercurous oxide, black, solid.	Pois. B.	73.364, 73.365.	Poison.	200 pounds.
Mercurous sulfate, solid.	Pois. B.	73.364, 73.365.	Poison.	200 pounds.
Mercury acetate, solid.	Pois. B.	73.364, 73.365.	Poison.	200 pounds.
Mercury bichloride, solid (mercuric chloride)	Pois. B.	73.364, 73.372.	Poison.	200 pounds.
Mercury bisulfate, solid.	Pois. B.	73.364, 73.365.	Poison.	200 pounds.
Mercury compounds, n. o. s. (solid).	Pois. B.	73.364, 73.365.	Poison.	200 pounds.
Mercury cyanide, solid.	Pois. B.	73.370.	Poison.	200 pounds.
Mercury fulminate. See Initiating explosive.				
Metal kegs, reused. *See* *Polishes, metal, stove, furniture and wood, liquid.	See § 73.28.			
*Metal polish.				

Article		Exemption / reference	Label	Maximum quantity in one outside container
Metallic sodium or potassium. *See* Sodium or potassium, metallic.				
Methanol (methyl alcohol). *See* Wood alcohol.				
Methane.	F. G.	73.306, 73.301.	Red Gas.	300 pounds.
Methyl acetate.	F. L.	73.118, 73.119.	Red.	10 gallons.
Methyl acetone.	F. L.	73.118, 73.119.	Red.	10 gallons.
Methyl alcohol (methanol). *See* Wood alcohol.				
Methyl acetylene—15% to 20% propadiene mixture.	F. G.	73.306, 73.304, 73.314.	Red Gas.	300 pounds.
Methyl aluminum sesquibromide.	F. L.	No exemption, 73.134.	Red.	2 ounces.
Methyl aluminum sesquichloride.	F. L.	No exemption, 73.134.	Red.	2 ounces.
Methyl bromide and chlorpicrin mixture, liquid.	Pois. B.	No exemption, 73.353.	Poison.	55 gallons.
Methyl bromide and ethylene dibromide mixture, liquid.	Pois. B.	No exemption, 73.353.	Poison.	55 gallons.
Methyl bromide and nonflammable, nonliquefied compressed gas mixtures, liquid.	Pois. B.	No exemption, 73.353.	Poison.	300 pounds.
Methyl bromide, liquid (*bromomethane*).	Pois. B.	No exemption, 73.353.	Poison.	55 gallons.
Ø Methyl chloride.	F. G.	73.306, 73.304, 73.314, 73.315.	Red Gas.	300 pounds.
Methyl chloride-methylene chloride mixture.	Cor. L.	73.306, 73.304, 73.314.	Red Gas.	300 pounds.
Methyl chloroformate.	F. L.	No exemption, 73.288.	White.	5 pints.
Methylchloromethyl ether, anhydrous.	Pois. A.	No exemption, 73.143.	Red.	No accepted.
Methyldichlorarsine.	F. L.	No exemption, 73.328.	Poison Gas.	Not accepted.
Methyl dichlorosilane.	F. L.	No exemption, 73.136.	Red.	10 gallons.
Methyl ethyl ketone.	F. L.	73.118, 73.119.	Red.	10 gallons.
Methyl formate.	F. L.	73.118, 73.119.	Red.	10 gallons.
Methyl hydrate. *See* Alcohol or alcohol, n. o. s.				
Methylhydrazine.	F. L.	No exemption, 73.145.	Red.	5 pints.
Methyl iso-propenyl ketone, inhibited.	F. L.	73.118, 73.119.	Red.	10 gallons.
Methyl magnesium bromide in ethyl ether in concentrations not over 40 percent.	F. L.	No exemption, 73.149.	Red.	6 quarts.
Methyl mercaptan.	F. G.	73.306, 73.304, 73.314, 73.315.	Red Gas.	300 pounds.
Methyl methacrylate monomer.	F. L.	73.118, 73.119.	Red.	10 gallons.
Methyl parathion, liquid.	Pois. B.	No exemption, 73.358.	Poison.	1 quart.
Methyl parathion mixture, dry.	Pois. B.	73.377.	Poison.	200 pounds.
Methyl parathion mixture, liquid.	Pois. B.	73.359.	Poison.	1 quart.
Methyltrichlorosilane.	F. L.	No exemption, 73.135.	Red.	10 gallons.
Methyl vinyl ketone, inhibited.	F. L.	73.147.	Red.	10 gallons.
Mild detonating fuse, metal clad.	Expl. C.	No exemption, 73.104.	Red.	300 pounds.
Mine rescue equipment.	See § 73.306 (a)(2) and 76.703(d)			
	See § 73.55			
Mines, empty.				

* See § 72.4.   Ø See § 72.4.

LIST OF EXPLOSIVES AND OTHER DANGEROUS ARTICLES—Continued

Article	Classed as	Exemptions and Packing (see sec.)	Label required if not exempt	Maximum quantity in 1 outside container by rail express
Mines, explosive, with gas material. See Explosive mine.	See § 73.56 (d)			
Mixed acid. See Nitrating (mixed) acid.				
Mixtures of hydrofluoric and sulfuric acids. See Hydrofluoric and sulfuric acids, mixtures.				
Mixtures or solutions of liquefied nonflammable gases and liquids other than those classified as flammable, corrosive, or poisonous charged with nitrogen, carbon dioxide, or air. See Compressed gases, n. o. s.				
Monobromotrifluoromethane	Nonf. G.	73.306, 73.304, 73.314	Green	300 pounds.
Monochloracetone, stabilized	Pois. C.	No exemption, 73.384	Tear Gas	5 gallons.
Monochloracetone (unstabilized)	Not accepted			Not accepted.
Monochloroacetic acid, liquid	Cor. L.	73.244, 73.294	White	1 quart.
Monochlorodifluoromethane	Nonf. G.	73.306, 73.301, 73.314, 73.315	Green	300 pounds.
Monochloroethylene. *See Vinyl chloride.*				
Monochloropentafluoroethane	Nonf. G.	73.306, 73.304	Green	300 pounds.
Monochlorotetrafluoroethane	Nonf. G.	73.306, 73.304, 73.314	Green	300 pounds.
Monochlorotrifluoromethane	Nonf. G.	73.306, 73.304	Green	300 pounds.
Monoethylamine	F. L.	73.118, 73.148	Red	10 gallons.
Monofluorophosphoric acid, anhydrous	Cor. L.	No exemption, 73.275.	White	1 gallon.
Monomethylamine, anhydrous	F. G.	73.306, 73.304, 73.314, 73.315	Red Gas	300 pounds.
Monomethylamine, aqueous solution	F. L.	73.118, 73.119	Red	10 gallons.
*Mortar stain, liquid	F. L.	73.118, 73.128	Red	55 gallons.
⊘Motion-picture film, nitrocellulose base, including mixed shipments with nonflammable film	F. S.	No exemption, 73.177	Yellow	200 pounds.
Motion-picture film, old and wornout (nitrocellulose)	F. S.	No exemption, 73.178	Yellow	200 pounds.
Motion-picture film, old and wornout (slow-burning)	See § 73.181 (a) (1)			
⊘Motion-picture outfits, toy	See § 73.181 (a) (4)			

Article				
⊘Motion-picture film (*processed, positive or negative, nitrocellulose*)	F. S.	No exemption, 73.177	Yellow	200 pounds.
⊘Motion-picture film (*processed, positive or negative*)	See § 73.181 (a) (1)			
⊘Motion-picture film (*processed, positive or negative, slow-burning*)				
Motion-picture film scrap (*nitrocellulose*), samples of	F. S.	No exemption, 73.196	Yellow	25 pounds.
Motion-picture film scrap (*nitrocellulose*), *other than samples*	F. S.	No exemption, 73.195	Yellow	Not accepted.
Motion-picture film scrap (*slow-burning*)	See § 73.181 (a) (2)			
⊘Motion-picture film, toy (*nitrocellulose*)	F. S.	No exemption, 73.179	Yellow	200 pounds.
⊘Motion-picture film, toy (*slow-burning*)	See § 73.181 (a) (1)			
⊘Motion-picture film, toy pieces (*nitrocellulose*)	See § 73.181 (a) (3)			
Motion-picture film, unexposed (*nitrocellulose base*)	F. S.	73.180	Yellow##	250 pounds.
Motion-picture film, unexposed (*slow-burning*)	See § 73.181 (a) (1)			
Motorcycles	See § 73.120			
Motor fuel antiknock compound	Pois. B.	No exemption, 73.354	Poison	55 gallons.
*Motor fuel, n. o. s.	F. L.	73.118, 73.119	Red	10 gallons.
Motors, internal combustion. See § 73.120				
Muriatic acid. *See* Hydrochloric acid.				
Mustard gas	Pois. A.	No exemption, 73.328	Poison Gas	Not accepted.
*Naphtha	F. L.	73.118, 73.119	Red	10 gallons.
*Naphtha distillate	F. L.	73.118, 73.119	Red	10 gallons.
*Naphtha, petroleum. *See* *Petroleum naphtha.				
*Naphtha, solvent	F. L.	73.118, 73.119	Red	10 gallons.
Natural gasoline. See Gasoline.				
Neohexane	F. L.	73.118, 73.119	Red	10 gallons.
Neon gas	Nonf. G.	73.306, 73.302	Green	300 pounds.
New explosives or explosive devices	See §§ 73.51 (q), 73.86, and 74.502 (a) (8)			
Nickel carbonyl	F. L.	No exemption, 73.126	Red	Not accepted.
*Nickel catalyst, finely divided, activated or spent	F. S.	No exemption, 73.233	Yellow	100 pounds.
Nickel cyanide, solid	Pois. B.	73.370	Poison	200 pounds.
Nicotine hydrochloride	Pois. B.	73.345, 73.346	Poison	55 gallons.
Nicotine, liquid	Pois. B.	73.345, 73.346	Poison	55 gallons.

* See § 72.4.   ## See § 72.4.   ⊘ See § 72.4.

LIST OF EXPLOSIVES AND OTHER DANGEROUS ARTICLES—*Continued*

Article	Classed as	Exemptions and Packing (see sec.)	Label required if not exempt	Maximum quantity in 1 outside container by rail express
Nicotine salicylate	Pois. B	73.364, 73.365	Poison	200 pounds.
Nicotine sulfate, liquid	Pois. B	73.345, 73.346	Poison	55 gallons.
Nicotine sulfate, solid	Pois. B	73.364, 73.365	Poison	200 pounds.
Nicotine tartrate	Pois. B	73.364, 73.365	Poison	200 pounds.
Nitrate of aluminum. *See Aluminum nitrate.*				
Nitrate of ammonia. *See Ammonium nitrate.*				
Nitrate of ammonia explosives. *See High explosives.*				
Nitrate of ammonia fertilizer. *See Ammonium nitrate fertilizer, containing 90 percent or more ammonium nitrate with no organic coating.*				
Nitrate of barium. *See Barium nitrate.*				
Nitrate of lead. *See Lead nitrate.*				
Nitrate of potash. *See Potassium nitrate.*				
Nitrate of soda. *See Sodium nitrate.*				
Nitrate of soda and potash	Oxy. M	73.153, 73.182	Yellow	100 pounds.
Nitrate of soda bags, empty, unwashed. *See Bags, nitrate of soda, empty and unwashed.*				
Nitrate of strontia. *See Strontium nitrate.*				
Nitrates, n. o. s	Oxy. M	73.153, 73.182	Yellow	100 pounds.
Nitrating (mixed) acid	Cor. L	No exemption, 73.267	White	2½ pints.
Nitric acid	Cor. L	No exemption, 73.268	White	5 pints.
Nitric ether. *See Ethyl nitrate.*				
Nitric oxide	Pois. A	No exemption, 73.337	Poison Gas	Not accepted.
Nitrite of soda. *See Sodium nitrite.*				
Nitrobenzol, liquid (*oil of mirbane*)	Pois. B	73.345, 73.346	Poison	55 gallons.
Nitro carbo nitrate	Oxy. M	73.153, 73.182	Yellow	100 pounds.
Nitrocellulose, colloided, granular, flake or block, wet with alcohol or solvent. See Wet nitrocellulose colloided, granular or flake—20 percent alcohol or solvent; or block—25 percent alcohol.				
Nitrocellulose, colloided, granular, or flake, wet with 20 percent water. See Wet nitrocellulose, colloided, granular or flake—20 percent water.				

Article	Class	Reference	Label	Quantity
Nitrocellulose (collodion cotton), wet with water. See Wet nitrocellulose—20 percent water.				
Nitrocellulose (collodion cotton), wet with alcohol or solvent. See Wet nitrocellulose—30 percent alcohol or solvent.				
Nitrocellulose, dry. See High explosives.				
Nitrocellulose flakes, wet with alcohol or solvent. See Wet nitrocellulose flakes—20 percent alcohol or solvent.				
Nitrochlorbenzene, ortho, liquid.	Pois. B	73.345, 73.346	Poison	55 gallons.
Nitrochlorbenzene, meta or para, solid.	Pois. B	73.364, 73.374	Poison	200 pounds.
Nitrogen.	Nonf. G	73.306, 73.302, 73.314	Green	300 pounds.
Nitrogen dioxide, liquid.	Pois. A	No exemption, 73.336	Poison Gas	Not accepted.
*Nitrogen fertilizer solution.	Nonf. G	73.306, 73.304, 73.314	Green	300 pounds.
Nitrogen peroxide, liquid.	Pois. A	No exemption, 73.336	Poison Gas	Not accepted.
Nitrogen, pressurized liquid.	Nonf. G	No exemption, 73.304	Green	300 pounds.
Nitrogen tetroxide, liquid.	Pois. A	No exemption, 73.336	Poison Gas	Not accepted.
Nitrogen tetroxide-nitric oxide mixtures containing up to 33.2 percent weight nitric oxide.	Pois. A	No exemption, 73.338	Poison Gas	Not accepted.
Nitroglycerin, liquid.	See § 73.51 (d).			
Nitroglycerin liquid, desensitized. See High Explosive, liquid.				
Nitroglycerin, spirits of. See Spirits of nitroglycerin.				
Nitrohydrochloric acid.	Cor. L	No exemption, 73.278	White	5 pints.
Nitrohydrochloric acid, diluted.	Cor. L	No exemption, 73.278	White	5 pints.
Nitro manite. See Initiating explosive.				
Nitroguanidine, dry. See High explosives.				
Nitroguanidine, wet with water. See Wet nitroguanidine—20 percent water.				
Nitrosoguanidine. See Initiating explosive.				
Nitrostarch, dry. See High explosives.				
Nitrostarch, wet with alcohol or solvent. See Wet nitrostarch—30 percent alcohol or solvent.				
Nitrostarch, wet with water. See Wet nitrostarch—20 percent water.				
Nitrosylchloride.	Nonf. G	73.306, 73.304, 73.314	Green	300 pounds.
Nitrourea. See High explosives.				
⊘Nitrous oxide.	Nonf. G	73.306, 73.304, 73.315	Green	300 pounds.
Nitroxylol.	Pois. B	73.345, 73.346	Poison	55 gallons.
Nonliquefied gases. See Compressed gases, n. o. s.				

* See § 72.4.   ⊘ See § 72.4.

LIST OF EXPLOSIVES AND OTHER DANGEROUS ARTICLES—*Continued*

Article	Classed as	Exemptions and Packing (see sec.)	Label required if not exempt	Maximum quantity in 1 outside container by rail express
Nonliquefied hydrocarbon gas	F. G.	73.306, 73.302	Red Gas	300 pounds.
Nonyl trichlorosilane	Cor. L.	No exemption, 73.280	White	10 gallons.
Octadecyltrichlorosilane	Cor. L.	No exemption, 73.280	White	10 gallons.
Octyl trichlorosilane	Cor. L.	No exemption, 73.280	White	10 gallons.
*Oil, *described as* oil, oil, oil, n. o. s., petroleum oil, or petroleum oil, n. o. s.	F. L.	73.118, 73.119	Red	10 gallons.
Oil of mirbane. See Nitrobenzol, liquid.				
Oil of vitriol. See Sulfuric acid.				
Oil well cartridges	Expl. C.	No exemption, 73.112		150 pounds.
Oleum	Cor. L.	73.244, 73.272	White	10 pints.
*Organic phosphate compound, liquid, n. o. s.	Pois. B.	No exemption, 73.358	Poison	1 quart.
*Organic phosphate compound mixtures, liquid n. o. s.	Pois. B.	73.359	Poison	1 quart.
*Organic phosphate compound, mixtures, dry, n. o. s.	Pois. B.	73.377	Poison	200 pounds.
*Organic phosphates, n. o. s. mixed with compressed gas	Pois. A.	No exemption, 73.334	Poison Gas	Not accepted.
Ortho-nitroaniline	Pois. B.	73.364, 73.373	Poison	200 pounds.
Oxide, spent. See Spent oxide.				
Oxidizing material, n. o. s.	Oxy. M. *See* § 73.152 (a) Note 1	73.153, 73.154	Yellow	25 pounds.
Oxidizing materials with other articles—fumigants.				
⊘Oxygen	Nonf. G.	73.306, 73.302, 73.314	Green	300 pounds.
Oxygen, pressurized liquid	Nonf. G.	No exemption, 73.304	Green	300 pounds.
*Paint driers, liquid	F. L.	73.118, 73.128	Red	55 gallons.
*Paint, enamel, lacquer, stain, shellac, varnish, aluminum, bronze, gold, wood filler, liquid, and lacquer base liquid.	F. L.	73.118, 73.128	Red	55 gallons.
*Paint, lacquer and varnish removing, reducing or thinning compounds. See *Compounds, lac-				

quer, paint or varnish, reducing or thinning liquid, etc.

Paper cap ammunition for toy pistols. *See* Toy caps.

Paper caps. *See* Toy caps.

Item	Class	Sections	Label	Quantity
Paper stock, wet. *See* Waste paper, wet.	F. S.	No exemption, 73.185	Yellow	Not accepted.
Paper waste, wet. *See* Waste paper, wet.				
Para chlorobenzoyl peroxide. *See* Chlorobenzoyl peroxide, (para).				
Paramenthane hydroperoxide	Oxy. M.	73.153 (b), 73.224	Yellow	1 quart.
Paranitraniline (paranitroaniline), solid	Pois. B.	73.364, 73.373	Poison	200 pounds.
Parathion and compressed gas mixture	Pois. A.	No exemption, 73.334	Poison Gas	Not accepted.
Parathion, liquid	Pois. B.	No exemption, 73.358	Poison	1 quart.
Parathion, mixture, dry	Pois. B.	73.377	Poison	200 pounds.
Parathion mixture, liquid	Pois. B.	73.359	Poison	1 quart.
Paris green, solid	Pois. B.	73.364, 73.367	Poison	200 pounds.
Pentaborane	F. L.	No exemption, 73.138	Red	Not accepted.
Pentane	F. L.	73.118, 73.119	Red	10 gallons.
Pentane, methyl	F. L.	73.118, 73.119	Red	10 gallons.
Pentaerythrite tetranitrate. *See* Initiating explosive.				
Pentolite, dry. *See* High explosives.				
Peracetic acid	Oxy. M.	73.223	Yellow	5 pints.
Perchlorate of ammonia. *See* Ammonium perchlorate.				
Perchlorate of potash. *See* Potassium perchlorate.				
Perchlorates, n. o. s.	Oxy. M.	73.153, 73.154	Yellow	100 pounds.
Perchloric acid *in excess of 72 percent*	See § 73.269			
Perchloric acid *not in excess of 72 percent*	Cor. L.	73.244, 73.269	White	5 pints.
Perchloro-methyl-mercaptan	Pois. B.	73.345, 73.360	Poison	10 pounds.
Percussion caps	Expl. C.	No exemption, 73.107		150 pounds.
Percussion fuzes	Expl. C.	No exemption, 73.105		150 pounds.
Permanganate of potash	Oxy. M.	73.153, 73.154, 73.194	Yellow	100 pounds.
Permanganate of soda	Oxy. M.	73.153, 73.154	Yellow	100 pounds.
Permanganates, n. o. s.	Oxy. M.	73.153, 73.154	Yellow	100 pounds.
Peroxide of sodium. *See* Sodium peroxide.				
Peroxides, organic, liquids or solutions, n. o. s.	F. L.	No exemption, 73.119 (m)	Red	1 quart.
Peroxides, organic, liquids, n. o. s.	Oxy. M.	73.153, 73.221	Yellow	1 quart.
Peroxides, organic, solution, liquid, n. o. s.	Oxy. M.	73.153, 73.221	Yellow	1 quart.
Petroleum, crude. *See* Crude oil.				
Petroleum distillate	F. L.	73.118, 73.119	Red	10 gallons.
Petroleum ether	F. L.	73.118, 73.119	Red	10 gallons.

* See § 72.4.   Ø See § 72.4.

LIST OF EXPLOSIVES AND OTHER DANGEROUS ARTICLES—Continued

Article	Classed as	Exemptions and Packing (see sec.)	Label required if not exempt	Maximum quantity in 1 outside container by rail express
Petroleum gas, liquefied. *See* Liquefied petroleum gas.				
*Petroleum naphtha	F. L.	73.118, 73.119	Red	10 gallons.
Phenol. *See* Carbolic acid.				
Phenylcarbylamine chloride	Pois. A	No exemption, 73.328	Poison Gas	Not accepted.
Phenyldichlorarsine, liquid	Pois. B	No exemption, 73.355	Poison	30 gallons.
Phenyl trichlorosilane	Cor. L	No exemption, 73.280	White	10 gallons.
Phosgene (*diphosgene*)	Pois. A	No exemption, 73.333	Poison Gas	Not accepted.
Phosphoric anhydride	F. S.	No exemption, 73.188	Yellow	100 pounds.
Phosphorus, amorphous, red	F. S.	No exemption, 73.189	Yellow	11 pounds.
Phosphorus oxybromide	Cor. L	No exemption, 73.271	White	1 quart.
Phosphorus oxychloride	Cor. L	No exemption, 73.271	White	1 quart.
Phosphorus pentachloride	F. S.	No exemption, 73.191	Yellow	5 pounds.
Phosphorus sesquisulfide	F. S.	No exemption, 73.225	Yellow	11 pounds.
Phosphorus tribromide	Cor. L	No exemption, 73.270	White	1 quart.
Phosphorus trichloride	Cor. L	No exemption, 73.271	White	1 quart.
Phosphorus, white or yellow, dry	F. S.	No exemption, 73.190 (d) and (e)	Yellow	Not accepted.
Phosphorus, white or yellow, in water	F. S.	No exemption, 73.190 (b) and (c)	Yellow	25 pounds.
Photographic film scrap, X-ray film scrap. *See* Special fireworks or low explosives.	F. S.	No exemption, 73.195	Yellow	See § 73.196.
Photographic flash powder. See Special fireworks or low explosives.				
Picrates, dry. See High explosives.				
Picrate of ammonia. See High explosives.				
Picric acid, dry. See High explosives.				
Picric acid, wet, not exceeding 16 ounces	See § 73.192			
Picric acid, wet, with not less than 10 percent water, over 25 pounds. See High explosives.				
Picric acid, wet, with not less than 10 percent water, in excess of 16 ounces but not exceeding 25 pounds	F. S.	No exemption, 73.193	Yellow	25 pounds.
Pinwheels. See Common fireworks.				
*Plastic solvent, n. o. s.	F. L.	73.118, 73.119	Red	10 gallons.
Poisonous liquid or gas, n. o. s.	Pois. A	No exemption, 73.328	Poison Gas	Not accepted.
Poisonous liquids, n. o. s.	Pois. B	No exemption, 73.345, 73.346	Poison	55 gallons.
Poisonous liquids, n. o. s.	Pois. C	No exemption, 73.382	Tear Gas	75 pounds.

Poisonous solids, n. o. s.	Pois. B	73.364, 73.365	Poison	200 pounds.
Poisonous solids, n. o. s.	Pois. C	No exemption, 73.382	Tear Gas	75 pounds.
*Police grenades, poison gas, class A	Pois. A	No exemption, 73.335	Poison Gas	Not accepted.
*Polishes, metal, stove, furniture and wood, liquid	F. L.	73.118, 73.129	Red	55 gallons.
*Polymerizable materials	See § 73.21 (b)			
Potash, caustic, solution. See Caustic potash, liquid				
Potassium arsenate, solid	Pois. B	73.364, 73.365	Poison	200 pounds.
Potassium arsenite, solid	Pois. B	73.364, 73.365	Poison	200 pounds.
Potassium bromate (potash chlorate). See Chlorate of potash.	Oxy. M	73.153, 73.154	Yellow	100 pounds.
Potassium cyanide. See Cyanide of potassium.				
Potassium dichloroisocyanurate, dry, containing more than 39% available chlorine.	Oxy. M	73.153, 73.217	Yellow	100 pounds.
Potassium hydroxide solution. See Caustic potash, liquid.				
Potassium, metallic	F. S.	No exemption, 73.206	Yellow	25 pounds.
Potassium, metallic liquid alloy	F. S.	No exemption, 73.202	Yellow	1 pound.
Potassium nitrate	Oxy. M	73.153, 73.182	Yellow	100 pounds.
Potassium nitrate mixed (fused) with sodium nitrite	Oxy. M	73.153, 73.183	Yellow	100 pounds.
Potassium nitrite	Oxy. M	73.153, 73.154	Yellow	100 pounds.
Potassium perchlorate	Oxy. M	73.153, 73.219	Yellow	100 pounds.
Potassium permanganate. See Permanganate of potash.				
Potassium peroxide	Oxy. M	No exemption, 73.154	Yellow	100 pounds.
Potassium sulfide	F. S.	73.153, 73.207	Yellow	300 pounds.
Potato spray (arsenical), liquid. See *Insecticide, liquid.				
Pressurized products. See Compressed gases, n.o.s.				
Primers. See Cannon, combination, or small-arm primers.				
Primers, detonating. See Detonating primers.				
Projectiles, explosive. See Explosive projectile.				
Projectiles, gas, nonexplosive. See Chemical Ammunition, class A, B, or C.				
Projectiles, gas, smoke, or incendiary, with burster or booster with or without detonating fuze. See Explosive projectile.				
Projectiles, illuminating, incendiary or smoke with expelling charge but without bursting charge. See Special fireworks.				
Projectiles, sand-loaded, empty, or solid	See § 73.55.			

* See § 72.4.

LIST OF EXPLOSIVES AND OTHER DANGEROUS ARTICLES—*Continued*

Article	Classed as	Exemptions and Packing (see sec.)	Label required if not exempt	Maximum quantity in 1 outside container by rail express
Propane. See Liquefied petroleum gas.				
Propellant explosives (liquid), class A.	Expl. A.	No exemption, 73.64 (d).	Red #.	See § 73.86.
Propellant explosives (liquid), class B	Expl. B.	No exemption, 73.93.	Red #.	10 pounds.
Propellant explosives (solid), class B.	Expl. B.	No exemption, 73.93.	Red #.	10 pounds.
Propellant explosives in water (*smokeless powder for cannon or small arms*).	Expl. B.	No exemption, 73.93.		Not accepted.
Propellant explosives in water unstable, condemned or deteriorated (*smokeless powder for cannon or small arms*).	Expl. B.	No exemption, 73.93.		Not accepted.
Propellant explosives (solid), class B, and small-arms primers. See Propellant explosives, class B.				
Propylene imine, inhibited.	F. L.	No exemption, 73.139.	Red.	5 pints.
Propylene oxide.	F. L.	73.118, 73.119.	Red.	10 gallons.
Propyl alcohol. See Alcohol, n. o. s.				
Propyl mercaptan.	F. L.	No exemption, 73.141.	Red.	10 gallons.
Propyl trichlorosilane.	Cor. L.	No exemption, 73.280.	White.	10 gallons.
Propylene. See Liquefied petroleum gas.				
Prussic acid. See Hydrocyanic acid.				
*Pyridine.	F. L.	73.118, 73.119.	Red.	10 gallons.
Pyroforic liquids, n.o.s.	F. L.	No exemption, 73.134.	Red.	70 pounds.
Pyro sulfuryl chloride.	Cor. L.	73.244, 73.247.	White.	1 quart.
Pyroxylin cement. See Cement, pyroxylin.				
*Pyroxylin plastic scrap.	F. S.	No exemption, 73.195, 73.196.	Yellow.	See § 73.196.
Pyroxylin plastics, rods, sheets, rolls, tubes.	F. S.	73.197.	Yellow #.	350 pounds.
*Pyroxylin solution.	F. L.	73.118, 73.119.	Red.	10 gallons.
*Pyroxylin solvent, n. o. s.	F. L.	73.118, 73.119.	Red.	10 gallons.
Radio current supply device with electrolyte (acid) or battery fluid.	See § 73.259.			
Radioactive materials, n.o.s.	Pois. D.	73.392, 73.393.	Radioactive materials, blue or red	See §73.393(f) and (L).

Article	Classification	Exemption reference	Label	Maximum quantity
Rags, oily	F. S.	No exemption, 73.199	Yellow	Not accepted.
Rags, wet	F. S.	No exemption, 73.200	Yellow	Not accepted.
Railway fusees	Expl. C.	No exemption, 73.108		200 pounds.
Railway torpedoes	Expl. B.	No exemption, 73.91	Special Fireworks##	200 pounds.
*Reducing compounds, paint, varnish, lacquer, etc. See *Compounds, lacquer, paint, or varnish, etc., removing, reducing or thinning, liquid.				
Refrigerant gas, n. o. s.	See §§ 73.314 (c) table Note 13, 73.315 (a) table Note 9.			
Refrigerating machines	See § 73.130, 73.306.			
*Removing compounds, paint, varnish, lacquer, etc. See *Compounds, lacquer, paint, or varnish, etc., removing, reducing or thinning, liquid.				
*Resin solution (resin compound, liquid)	F. L.	73.118, 73.119	Red	55 gallons.
Resinate of cobalt, precipitated. See Cobalt resinate, precipitated.				
Rifle grenades	Expl. A.	No exemption, 73.56		Not accepted.
Rifle powder. See Propellant explosives, class A, Propellent explosives, class B, or Black powder.				
*Road asphalt or tar, liquid	F. L.	73.118, 73.131	Red	10 gallons.
Rocket bodies, with electric primers or electric squibs	See § 73.55.			
Rocket ammunition with empty projectiles	Expl. B.	No exemption, 73.90		Not accepted.
Rocket ammunition with explosive projectiles	Expl. A.	No exemption, 73.57		Not accepted.
Rocket ammunition with illuminating projectiles	Expl. A.	No exemption, 73.57		Not accepted.
Rocket ammunition with gas projectiles	Expl. A.	No exemption, 73.57		Not accepted.
Rocket ammunition with incendiary projectiles	Expl. A.	No exemption, 73.57		Not accepted.
Rocket ammunition with inert-loaded projectiles	Expl. B.	No exemption, 73.90		Not accepted.
Rocket ammunition with smoke projectiles	Expl. A.	No exemption, 73.57		Not accepted.
Rocket ammunition with solid projectiles	Expl. B.	No exemption, 73.90		Not accepted.
Rocket engines (liquid), class B explosives	Expl. B.	No exemption, 73.95		Not accepted.
Rocket fireworks. See Common fireworks.				
Rocket heads. See Explosive projectiles.				
Rocket motors, class A explosives	Expl. A.	No exemption, 73.79		Not accepted.
Rocket motors, class B explosives	Expl. B.	No exemption, 73.92	Red #	550 pounds.
Roman candles. See Common fireworks.				
*Rough ammoniate tankages	F. S.	No exemption, 73.210	Yellow	Not accepted.

* See § 72.4.   # See § 72.4.   ## See § 72.4.

LIST OF EXPLOSIVES AND OTHER DANGEROUS ARTICLES—Continued

Article	Classed as	Exemptions and Packing (see sec.)	Label required if not exempt	Maximum quantity in 1 outside container by rail express
*Rubber cement. See *Cement, rubber.				
*Rubber scrap or rubber buffings	F. S.	73.153, 73.201	Yellow	10 pounds.
*Rubber, shoddy, regenerated rubber, or reclaimed rubber	F. S.	73.153, 73.201	Yellow	10 pounds.
*Rum, denatured	F. L.	73.118, 73.119	Red	10 gallons.
Safety fuse	See § 73.100 (o)			
Safety squibs	Expl. C.	No exemption, 73.106		150 pounds.
Saltpeter. See Potassium nitrate.				
Saltpeter, Chile. See Sodium nitrate.				
Salutes. See Common fireworks and Special fireworks.				
Samples of explosives	See § 73.86			
Samples, New explosives	See § 73.86			
Samples of explosives and explosive articles	See § 73.86			
Scheele's green. See Copper arsenite.				
Self-lighting cigarettes	See § 73.21 (d)			
Self-propelled vehicles	See § 73.120, 73.257, 73.306			
Shaped charges, commercial. See High explosives.	See § 73.65 (h)			
*Shellac. See *Paint, enamel, lacquer, stain, shellac, varnish, etc.				
Shellac, liquid	F. L.	73.118, 73.128	Red	55 gallons.
Shells, fireworks. See Common fireworks or Special fireworks.				
Ship distress signals. See Special fireworks.				
Signal flares	Expl. C.	No exemption, 73.108	White	200 pounds.
Silicon chloride (tetrachloride)	Cor. L.	73.244, 73.247		1 gallon.
Silicon tetrafluoride	Nonf. G.	73.306, 73.302	Green	300 pounds.
Silver cyanide	See § 73.370	73.153, 73.182		
Silver nitrate	Oxy. M.	No exemption, 73.248	Yellow	100 pounds.
*Sludge acid	Cor. L.	No exemption, 73.101	White	1 quart.
Small-arms ammunition	Expl. C.	No exemption, 73.101		150 pounds.
Small-arms ammunition, tear gas cartridges	Expl. C.	No exemption, 73.101	Tear gas	150 pounds.
Small-arms primers	Expl. C.	No exemption, 73.107		150 pounds.

Smoke candles. *See Chemical ammunition, class B or C.*	Expl. C	No exemption, 73.108		200 pounds.
Smoke generators. *See Chemical ammunition, class B or C.*				
Smoke grenades.	Expl. C	No exemption, 73.108		200 pounds.
Smoke pots.	Expl. C	No exemption, 73.108		200 pounds.
Smoke signals.	Expl. C	No exemption, 73.108		200 pounds.
Smokeless powder for cannon or small arms. *See Propellant explosives, class A, or propellant explosives, class B.*				
Smokeless powder for small arms (100 pounds or less).	See §73.88 (f), §73.197a			
Smoke projectiles with bursting charges. See Explosive projectile.				
Smoke projectiles with expelling charge but without bursting charge. See Special fireworks.				
Soda amatol. See High explosives.				
Soda, caustic solution. See Caustic soda, liquid.				
Sodium aluminum hydride.	F. S.	No exemption, 73.206	Yellow	25 pounds.
Sodium aluminate, liquid.	Cor. L.	73.244, 73.249	White	10 gallons.
Sodium amide.	F. S.	No exemption, 73.206	Yellow	25 pounds.
Sodium arsenate, solid.	Pois. B.	73.364, 73.365, 73.368	Poison	200 pounds.
Sodium arsenite (solution), liquid.	Pois. B.	73.345, 73.346	Poison	55 gallons.
Sodium azide.	Pois. B.	73.364, 73.375	Poison	100 pounds.
Sodium bromate.	Oxy. M.	73.153, 73.154	Yellow	100 pounds.
Sodium cacodylate, solid (*sodium dimethyl arsenate*).	Pois. B.	73.364, 73.365	Poison	200 pounds.
Sodium chlorate (*soda chlorate*).	Oxy. M.	73.153, 73.163	Yellow	100 pounds.
Sodium chlorite.	Oxy. M.	No exemption, 73.160	Yellow	100 pounds.
Sodium chlorite solution (not exceeding 42 percent sodium chlorite).	Cor. L.	73.244, 73.263	White	4 gallons.
Sodium cyanide. *See Cyanide of sodium.*				
Sodium dichloroisocyanurate, dry, containing more than 39% available chlorine.	Oxy. M.	73.153, 73.217	Yellow	100 pounds.
Sodium hydride.	F. S.	No exemption, 73.198	Yellow	25 pounds.
Sodium hydrosulfite.	F. S.	73.153, 73.204	Yellow	100 pounds.
Sodium hydroxide solution. *See Caustic soda, liquid.*				
Sodium, metallic.	F. S.	No exemption, 73.206	Yellow	25 pounds.
Sodium, metallic, dispersion in organic solvent.	F. S.	No exemption, 73.230	Yellow	10 pounds.
Sodium, metallic liquid alloy.	F. S.	No exemption, 73.202	Yellow	1 pound.
Sodium methylate, dry.	F. L.	73.153, 73.154	Yellow	100 pounds.
*Sodium methylate, alcohol mixture.	F. L.	73.118, 73.119	Red	10 gallons.
Sodium nitrate.	Oxy. M.	73.153, 73.182	Yellow	100 pounds.
Sodium nitrite.	Oxy. M.	73.153, 73.154, 73.234	Yellow	100 pounds.

* See § 72.4.

LIST OF EXPLOSIVES AND OTHER DANGEROUS ARTICLES—Continued

Article	Classed as	Exemptions and Packing (see sec.)	Label required if not exempt	Maximum quantity in 1 outside container by rail express
Sodium nitrite mixed (fused) with potassium nitrate.	Oxy. M.	73.153, 73.183.	Yellow.	100 pounds.
Sodium nitrite mixtures (sodium nitrate, sodium nitrite, and potassium nitrate).	Oxy. M.	73.153, 73.234.	Yellow.	100 pounds.
Sodium permanganate.	Oxy. M.	73.153, 73.154.	Yellow.	100 pounds.
Sodium peroxide.	Oxy. M.	No exemption, 73.187.	Yellow.	100 pounds.
Sodium picramate wet with 20 percent of water.	F. S.	No exemption, 73.205.	Yellow.	25 pounds.
Sodium potassium alloys.	F. S.	No exemption, 73.206.	Yellow.	25 pounds.
*Sodium sulfide.	F. L.	73.153, 73.207.	Yellow.	300 pounds.
Solvents, n. o. s.	F. L.	73.118, 73.119.	Red.	15 Gallons.
*Sparklers. See Common fireworks.				
Special fireworks.	Expl. B.	No exemption, 73.88 (d), 73.91	Special Fireworks ##.	200 pounds.
Spent iron mass. See Iron mass, spent.				
Spent iron sponge. See Iron sponge, spent.				
Spent mixed acid.	Cor. L.	No exemption, 73.248.	White.	1 quart.
Spent oxide.	F. S.	No exemption, 73.174.	Yellow.	Not accepted.
Spent sulfuric acid.	Cor. L.	No exemption, 73.248.	White.	1 quart.
Spirits of nitroglycerin.	F. L.	No exemption, 73.133.	Red.	6 quarts.
Spirits of nitroglycerin *not exceeding 1 percent nitroglycerin by weight.*	F. L.	73.118, 73.133.	Red.	6 quarts.
Sporting powder. See Black powder or Propellant explosives, class B explosives.				
Spray starting fluid. See Engine starting fluid.				
Spreader cartridges. See Special fireworks.				
Squibs, electric or safety. See Electric squibs or Safety squibs.				
*Stain. See *Paint, enamel, lacquer, stain, shellac, varnish, etc.				
Starter cartridges, jet engine, class B explosives.	Expl. B.	No exemption, 73.92.	Red #.	200 pounds.
Starter cartridges, jet engine, class C explosives.	Expl. C.	No exemption, 73.102.		150 pounds.
Storage batteries, wet. See Batteries, electric storage, wet.				
Stores, ship or vessels.	See § 71.9.			

Article	Classification	See §	Label	Quantity limit
*Stove polish. See *Polishes, metal, stove, furniture and wood, liquid.				
Strike-anywhere matches. See Matches, strike-anywhere.				
Strike-on-box matches. See Matches, strike-on-box.				
Strontium arsenite, solid	Pois. B	73.364, 73.365	Poison	200 pounds.
Strontium chlorate	Oxy. M	73.153, 73.163	Yellow	100 pounds.
Strontium chlorate, wet	Oxy. M	73.153, 73.163 (a) (6)	Yellow	200 pounds.
Strontium nitrate	Oxy. M	73.153, 73.182	Yellow	100 pounds.
Strontium peroxide	Oxy. M	73.153 (a), 73.154	Yellow	100 pounds.
Strychnine and salts thereof, solid	Pois. B	73.364, 73.365	Poison	200 pounds.
Styphnate of lead. See Initiating explosive.				
Succinic acid peroxide	Oxy. M	73.153 (b), 73.157, 73.158	Yellow	25 pounds.
Sulfide of sodium. See Sodium sulfide.				
Sulfide of potassium. See Potassium sulfide.				
Sulfur chloride (mono and di)	Cor. L	No exemption, 73.247	White	1 gallon.
ØSulfur dioxide	Nonf. G	73.306, 73.304, 73.314, 73.315	Green	300 pounds.
Sulfur hexafluoride	Nonf. G	73.306, 73.304	Green	300 pounds.
Sulfur trioxide, stabilized	Cor. L	73.244, 73.273	White	1 gallon.
Sulfuric acid (oil of vitriol)	Cor. L	73.244, 73.272	White	10 pints.
Sulfuric acid, fuming (oleum) (Nordhausen). See Sulfuric acid.				
Sulfuryl chloride	Cor. L	73.244, 73.247	White	1 quart.
Sulfuryl fluoride	Nonf. G	73.306, 73.304, 73.314	Green	300 pounds.
Supplementary charges (explosive)	Expl. A	No exemption, 73.69		Not accepted
Tankage. See Garbage tankage.				
*Tankage fertilizers.	F. S.	No exemption, 73.209	Yellow	Not accepted.
*Tankages, rough ammoniate.	F. S.	No exemption, 73.210	Yellow	Not accepted.
Tank car, containing residual phosphorus and filled with water or inert gas.	See § 73.232 (d)			
Tank cars, empty (last contents poisons, class A)	See § 74.562 (d) and (e)			
Tank cars, gas (must not contain gases that combine chemically)	See § 73.301 (a)			
Tank truck, empty.	See § 73.29			
Tanks, empty.	See § 73.29			
*Tar, liquid.	F. L.	73.118, 73.131	Red.	10 gallons.
Tear gas candles.	Pois. C	No exemption, 73.385	Tear Gas	75 pounds.
Tear gas cartridges. See Small-arms ammunition tear gas cartridges.				
Tear gas grenades.	Pois. C	No exemption, 73.385	Tear Gas	75 pounds.
Tear gas material, liquid or solid, n. o. s.	Pois. C	No exemption, 73.382	Tear Gas	75 pounds.

* See § 72.4.   # See § 72.4.   # See § 72.4.   ## See § 72.4.   Ø See § 72.4.

LIST OF EXPLOSIVES AND OTHER DANGEROUS ARTICLES—*Continued*

Article	Classed as	Exemptions and Packing (see sec.)	Label required if not exempt	Maximum quantity in 1 outside container by rail express
*Tertiary alcohol. *See* Alcohol, n. o. s.				
Tertiary butylisopropyl benzene hydroperoxide	Oxy. M.	73.153 (b), 73.224	Yellow	1 quart.
Tetraethyl dithio pyrophosphate and compressed gas mixture	Pois. A.	No exemption, 73.334	Poison Gas	Not accepted.
Tetraethyl dithio pyrophosphate mixture, liquid	Pois. B.	No exemption, 73.358	Poison	1 quart.
Tetraethyl dithio pyrophosphate mixture, dry	Pois. B.	73.377	Poison	200 pounds.
Tetraethyl dithio pyrophosphate mixture, liquid	Pois. B.	73.359	Poison	1 quart.
Tetraethyl lead, liquid	Pois. B.	No exemption, 73.354	Poison	55 gallons.
Tetraethyl pyrophosphate and compressed gas mixture	Pois. A.	No exemption, 73.334	Poison Gas	Not accepted.
Tetraethyl pyrophosphate, liquid	Pois. B.	No exemption, 73.358	Poison	1 quart.
Tetraethyl pyrophosphate mixture, dry	Pois. B.	73.377	Poison	200 pounds.
Tetraethyl pyrophosphate mixture, liquid	Pois. B.	73.359	Poison	1 quart.
Tetrafluoroethylene, inhibited	F. G.	73.306, 73.304	Red Gas	300 pounds.
Tetranitromethane	Oxy. M.	No exemption, 73.203	Yellow	25 pounds.
Tetrazene (guanyl nitrosamino guanyl tetrazene). *See* Initiating explosive.				
Tetryl. *See* High explosives.				
Textile waste, wet	F. S.	No exemption, 73.211	Yellow	Not accepted.
Thallium salts, solid	Pois. B.	73.364, 73.365	Poison	200 pounds.
Thallium sulfate, solid	Pois. B.	73.364, 73.365	Poison	200 pounds.
*Thinning compounds, paint, varnish, lacquer, etc. *See* *Paint, enamel, lacquer, stain, shellac, varnish, etc.				
Thiocarbonyl-chloride. *See* Thiophosgene.				
Thionyl chloride	Cor. L.	No exemption, 73.247	White	1 gallon.
Thiophosgene (*Thiocarbonyl-chloride*)	Pois. B.	No exemption, 73.356	Poison	1 gallon.
Thiophosphoryl chloride	Cor. L.	No exemption, 73.271	White	1 quart.
Thorium metal, powdered	F. S.	73.226	Yellow	25 pounds.
Time fuzes	Expl. C.	No exemption, 73.105	White	150 pounds.
Tin tetrachloride, anhydrous	Cor. L.	73.244, 73.247	White	1 quart.
Titanium metal powder, dry	F. S.	No exemption, 73.208	Yellow	75 pounds.
Titanium metal powder, wet, with not less than 20 percent water	F. S.	No exemption, 73.208	Yellow	150 pounds.

Titanium sulfate solution containing not more than 45 percent sulfuric acid	Cor. L.	73.244, 73.297	White	1 gallon.
Titanium tetrachloride	Cor. L.	73.244, 73.247	White	10 gallons.
Toluol ((toluene))	F.L.	73.118, 73.119	Red	10 gallons.
Torches. See Common fireworks.				
Torpedoes, cap. See Special fireworks.				
Torpedoes, empty	See § 73.55			
Torpedoes, explosives. See Explosive torpedoes.				
Torpedoes, railway or track. See Railway torpedoes.				
Toy propellant devices	Expl. C.	No exemption, 73.111		150 pounds.
Toy smoke devices	Expl. C.	No exemption, 73.111		150 pounds.
Toy torpedoes. See Special fireworks.				
Toy caps	Expl. C.	No exemption, 73.100 (p), 73.109		150 pounds.
Tracers	Expl. C.	No exemption, 73.105		150 pounds.
Tracer fuzes	Expl. C.	No exemption, 73.105		150 pounds.
Tractors	See § 73.120			
Trailer or truck body with refrigerating or heating equipment on flat cars	See §§ 73.120 (c), 73.306			
Trichloroisocyanuric acid, dry, containing more than 39% available chlorine	Oxy. M.	73.153, 73.217	Yellow	100 pounds.
Trichlorosilane	F.L.	No exemption, 73.136	Red	10 gallons.
Trick matches	Expl. C.	No exemption, 73.111		150 pounds.
Trick noise makers, explosive	Expl. C.	No exemption, 73.111		150 pounds.
Trifluorochloroethylene	F.G.	73.306, 73.304, 73.314	Red Gas	300 pounds.
Trimethylamine, anhydrous	F.G.	73.306, 73.304, 73.314, 73.315	Red Gas	300 pounds.
Trimethylamine, aqueous solution	F.L.	73.118, 73.119	Red	10 gallons.
Trimethylchlorosilane	F.L.	No exemption, 73.135	Red	10 gallons.
Trinitrobenzene. See High explosives.				
Trinitrobenzene, wet (not to exceed 16 ounces)	F.S.	73.212		16 ounces.
Trinitrobenzoic acid, dry. See High explosives.				
Trinitrobenzoic acid, wet, not exceeding 16 ounces.	See § 73.192			
Trinitrobenzoic acid, wet with not less than 10 percent water, in excess of 16 ounces but not exceeding 25 pounds	F.S.	No exemption, 73.193	Yellow	25 pounds.
Trinitrobenzoic acid, wet, with not less than 10 percent water, over 25 pounds. See High explosives.				
Trinitroresorcinol. See High explosives.				
Trinitrotoluene. See High explosives.				

* See § 72.4.

LIST OF EXPLOSIVES AND OTHER DANGEROUS ARTICLES—Continued

Article	Classed as	Exemptions and Packing (see sec.)	Label required if not exempt	Maximum quantity in 1 outside container by rail express
Trinitrotoluene, wet (*not to exceed 16 ounces*)	F. S.	73.212	White	16 ounces.
Tris-(1-aziridinyl) phosphine oxide	Cor. L.	73.244, 73.299(a)	Red	1 gallon.
*Turpentine substitutes	F. L.	73.118, 73.119	Red	10 gallons.
Uranium, normal or depleted, in solid metal form (*not borings, chips, or pieces*)	Pois. D.	73.392(f)	Radioactive materials, red	See §73.393(L).
Urea nitrate wet with not less than 10 percent of water, over 25 pounds. *See* High explosives.				
Urea nitrate wet with not less than 10 percent of water, in excess of 16 ounces but not exceeding 25 pounds	F. S.	No exemption, 73.193	Yellow	25 pounds.
Urea nitrate, dry. *See* High explosives.				
Urea nitrate, wet with not less than 10 percent of water, not exceeding 16 ounces	See § 73.192	73.153 (b), 73.227	Yellow	16 ounces.
*Urea peroxide	Oxy. M.			25 pounds.
*Varnish. *See* *Paint, enamel, lacquer, stain, shellac, varnish, etc.				
*Varnish driers. *See* *Paint driers, liquid.				
*Varnish remover or reducer. *See* *Compounds, lacquer, paint, or varnish removing, reducing, or thinning, liquid.				
*Varnish thinning compounds. *See* *Compounds, lacquer, paint, or varnish removing, reducing, or thinning, liquid.				
Very signal cartridges	Expl. C.	No exemption, 73.108	Red	200 pounds.
Vinyl acetate	F. L.	73.118, 73.119	Red Gas	10 gallons.
Vinyl chloride	F. G.	73.306, 73.304, 73.314, 73.315	Red	300 pounds.
Vinylidene chloride, inhibited	F. L.	73.118, 73.119	Red	10 gallons.
Vinyl fluoride inhibited	F. G.	73.306, 73.304	Red Gas	300 pounds.
Vinyl methyl ether, inhibited	F. G.	73.306, 73.304, 73.314	Red Gas	20 pounds.
Vinyl trichlorosilane	F. L.	No exemption, 73.135	Red	10 gallons.

Article		Sections	Label	Maximum quantity in one outside container
War heads. See Explosive projectiles.				
Waste paper, wet.	F. S.	No exemption, 73.186.	Yellow	Not accepted.
Waste textile, wet.	F. S.	No exemption, 73.211.	Yellow	Not accepted.
Waste wool, wet.	F. S.	No exemption, 73.213.	Yellow	Not accepted.
*Water treatment compound, liquid.	Cor. L.	73.244, 73.249.	White	10 gallons.
*Weed killing compounds, liquid. See *Compounds, tree or weed killing, liquid.				
Wet hair. See Hair wet.				
Wet nitrocellulose, colloided, granular or flake—20 percent alcohol or solvent; or block—25 percent alcohol.	F. L.	73.118, 73.127	Red	25 pounds.
Wet nitrocellulose, colloided, granular or flake—20 percent water.	F. S.	73.153, 73.184	Yellow	100 pounds.
Wet nitrocellulose—30 percent alcohol or solvent.	F. L.	73.118, 73.127	Red	25 pounds.
Wet nitrocellulose—20 percent water.	F. S.	73.153, 73.184	Yellow	100 pounds.
Wet nitrocellulose flakes—20 percent alcohol or solvent.	F. L.	73.118, 73.127	Red	25 pounds.
Wet nitroguanidine—20 percent water.	F. S.	73.153, 73.184	Yellow	100 pounds.
Wet nitrostarch—20 percent water.		73.153, 73.184	Yellow	100 pounds.
Wet nitrostarch—30 percent alcohol or solvent.		73.118, 73.127	Red	25 pounds.
Wet paper stock. See Paper stock, wet.				
Wet rags. See Rags, wet.				
Wet textile waste. See Waste textile, wet.				
Wet waste paper. See Waste paper, wet.				
Wet waste wool. See Waste wool, wet.				
Wood alcohol (methanol, methyl alcohol).	F. L.	73.118, 73.125.	Red	10 gallons.
*Wood filler. See *Paint, enamel, lacquer, stain, shellac, varnish, etc.				
*Wood polish. See *Polishes, metal, stove, furniture and wood. liquid.				
*Wood stain, liquid. See *Paint, enamel, lacquer, stain, shellac, varnish, etc.				
Wool waste, wet. See Waste wool, wet.				
X-ray film (nitrocellulose base).	F. S.	No exemption, 73.177	Yellow	200 pounds.
X-ray film (slow-burning).	See § 73.181 (a) (1)			
X-ray film scrap (nitrocellulose base), samples of.	F. S.	No exemption, 73.196.	Yellow	25 pounds.
X-ray film scrap (nitrocellulose base), other than samples.	F. S.	No exemption, 73.195.	Yellow	Not accepted.

* See § 72.4.

LIST OF EXPLOSIVES AND OTHER DANGEROUS ARTICLES—*Concluded*

Article	Classed as	Exemptions and Packing (see sec.)	Label required if not exempt	Maximum quantity in 1 outside container by rail express
X-ray film scrap (*slow-burning*)	See § 73.181 (a) (2)	73.180	Yellow ##	250 pounds.
X-ray film, unexposed (*nitrocellulose base*)	F. S.	73.118, 73.119	Red	10 gallons.
Xylol (Xylene)	F. L.	No exemption, 73.382	Tear Gas	75 pounds.
Xylyl bromide	Pois. C.			
Zinc ammonium nitrite	Oxy. M.	No exemption, 73.228	Yellow	100 pounds.
Zinc arsenate	Pois. B.	73.364, 73.365	Poison	200 pounds.
Zinc arsenite, solid	Pois. B.	73.364, 73.365	Poison	200 pounds.
Zinc chlorate	Oxy. M.	73.153, 73.163	Yellow	100 pounds.
Zinc cyanide.... *See* Pyroforic liquids, n. o. s.	See § 73.370			
Zinc ethyl. *See* Nitrates, n. o. s.				
Zinc nitrate.	Oxy. M.	73.153, 73.154	Yellow	100 pounds.
Zinc permanganate	Oxy. M.	73.153, 73.154	Yellow	100 pounds.
Zinc peroxide	F. S.	No exemption, 73.214	Yellow	75 pounds.
Zirconium metal, dry, mechanically produced, finer than 270 mesh particle size	F. S.	No exemption, 73.214	Yellow	75 pounds.
Zirconium metal, dry, chemically produced, finer than 20 mesh particle size	F. S.	No exemption, 73.214	Yellow	150 pounds.
Zirconium metal, wet, mechanically produced, finer than 270 mesh particle size	F. S.	No exemption, 73.214	Yellow	150 pounds.
Zirconium metal, wet, chemically produced, finer than 20 mesh particle size	F. L.	No exemption, 73.140	Red	5 pounds.
Zirconium, metallic, solutions or mixtures thereof, liquid	Oxy. M.	No exemption, 73.216	Yellow	25 pounds.
Zirconium picramate, wet with 20 percent of water.	F. S.	73.153, 73.220	Yellow	100 pounds.
Zirconium scrap (borings, clippings, shavings, sheets, or turnings)				

* See § 72.4.   ## See § 72.4.

SECTION 12

GENERAL CHEMICALS

ABBREVIATIONS

ACGIH	American Conference of Government and Industrial Hygienists	MCA	Manufacturing Chemists Association
anh.	anhydrous	mCi	millicuries
ASA	American Standards Association	MeV	millions of electron volts
ASTM	American Society for Testing Materials	mg	milligrams
		ml	milliliters
atm.	atmospheres	mm	millimeters of mercury
autoign. temp.	autoignition temperature	mol wt	molecular weight
BeV	billions of electron volts	mp	melting point
bp	boiling point	mppcf	millions of particles per cubic foot
°C	degrees Centigrade	μCi	microcuries
cc	cubic centimeters	μg	micrograms
C.C.	closed cup	n-	normal
cfm	cubic feet per minute	nCi	nanocuries
CG	Coast Guard	o-	ortho
Ci	curies	O.C.	open cup
CNS	central nervous system	p-	para
C.O.C.	Cleveland open cup	pCi	picocuries
cpm	counts per minute	ppm	parts per million
d	density	sec-	secondary
decomp.	decomposes or decomposition	spont. htg.	spontaneous heating
dpm	disintegrations per minute	sym-	symmetrical
eV	electron volts	T.C.C.	Tag closed cup
°F	degrees Fahrenheit	tert-	tertiary
fCi	femtocuries	TLV	Threshold Limit Values
flash p	flash point	T.O.C.	Tag open cup
fp	freezing point	U	unknown
fpm	feet per minute	uel	upper explosive limit
g	grams	ulc	Underwriters Laboratory Classification
G.I.	gastrointestinal		
gpm	gallons per minute	uns-	unsymmetrical
IATA	International Air Transport Association	vap. d.	vapor density
		vap. press.	vapor pressure
ICC	Interstate Commerce Commission	1°	primary
		α	alpha
°K	degrees Kelvin	β	beta
keV	thousands of electron volts	γ	gamma
kg	kilograms	*Radiologic half life abbreviations*	
km	kilometers	d	days
lel	lower explosive limit	h	hours
lpm	liters per minute	m	minutes
m-	meta	s	seconds
m³	cubic meters	y	years

A

"666". See benzene hexachloride.

"1068". See chlordane.

AAMX. See acetoacet-*m*-xylidide.

ABALYN. See methyl abietate.

ABIETIC ACID
General Information
Synonyms: Abietinic acid, sylvic acid.
Description: Yellow powder.
Formula: $C_{20}H_{30}O_2$.
Constants: Mol wt: 302.4, mp: 172–175°C.
Hazard Analysis
Toxic Hazard Rating:
  Acute Local: Irritant 2; Ingestion 2.
  Acute Systemic: Ingestion 1.
  Chronic Local: Irritant 2.
  Chronic Systemic: U.
Fire Hazard: Slight; can react with oxidizing materials (Section 6).
Explosion Hazard: Slight, if dust is exposed to flame or sparks.
Countermeasures
Personnel Protection: Section 3.
Personal Hygiene: Section 3.
Storage and Handling: Section 7.

ABIETIC ACID, ETHYL ESTER. See ethyl abietate.

ABIETINIC ACID. See abietic acid.

ABITOL. See hydroabietyl alcohol.

"A" BLASTING POWDER. See explosives, low.

ABRIN
General Information
Synonym: Toxalbumin.
Description: Yellowish-white powder.
Hazard Analysis
Toxic Hazard Rating:
  Acute Local: 0
  Acute Systemic: Ingestion 3; Inhalation 3; Skin Absorption 1.
  Chronic Local: U.
  Chronic Systemic: U.
Caution: Exceedingly toxic in eye and nose. Smallest particle in a wound may be fatal.
Fire Hazard: Slight (Section 6).
Disaster Hazard: Dangerous, when heated to decomposition, emits highly toxic fumes.
Countermeasures
Ventilation Control: Section 2.
Personal Hygiene: Section 3.
Storage and Handling: Section 7.

ABSINTHIUM
General Information
Synonym: Wormwood.
Description: Dried leaves and flowering tops of Artemisia absinthium L.
Hazard Analysis
Toxic Hazard Rating:
  Acute Local: Allergen 2.

  Acute Systemic: Ingestion 3.
  Chronic Local: Allergen 2.
  Chronic Systemic: Ingestion 2.
Caution: Habitual users develop "absinthism," with tremors, vertigo, vomiting and hallucinations. Local contact may cause contact dermatitis (Section 9).
Fire Hazard: Slight.
Countermeasures
Ventilation Control: Section 2.
Personal Hygiene: Section 3.
Storage and Handling: Section 7.

ACACIA GUM
General Information
Synonym: Gum arabic.
Description: Spheres or "tears."
Constants: Mol wt: 240,000, d: 1.35–1.49.
Hazard Analysis
Toxic Hazard Rating:
  Acute Local: Allergen 1.
  Acute Systemic: 0.
  Chronic Local: Allergen 1.
  Chronic Systemic: 0.
Caution: Local contact may cause contact dermatitis. Inhalation as dust may cause respiratory symptions such as asthma, watery nose and eyes, cough and wheezing. Hives, eczema and angioedema may also result from inhalation or ingestion (Section 9).
Note: Used as a stabilizer food additive and is a migrating substance from packaging materials (Section 10).
Fire Hazard: Slight (Section 6).
Countermeasures
Personal Hygiene: Section 3.

ACACIA WOOD
Hazard Analysis
Toxic Hazard Rating:
  Acute Local: Allergen 1; Inhalation 1.
  Acute Systemic: 0.
  Chronic Local: Allergen 1; Inhalation 1.
  Chronic Systemic: 0.
Toxicology: Toxicity by inhalation is that of any wood dust. Local contact may cause contact dermatitis (Section 9).
Fire Hazard: Slight, can react with oxidizing materials (Section 6).
Countermeasures
Storage and Handling: Section 7.

ACANTHITE. See silver sulfide.

ACCELLERENE. See p-nitroso dimethyl aniline.

ACENAPHTHENE
General Information
Synonym: Naphthalene ethylene.
Description: White, elongated crystals.
Formula: $C_{10}H_6(CH_2)_2$.
Constants: Mol wt: 154.2, mp: 95°C, bp: 277.5°C, d: 1.024 at 99°/4°C, vap. press.: 10 mm at 131.2°C, vap. d.: 5.32.
Hazard Analysis
Toxicity: Details unknown. Irritating to skin and mucous membrane. Mild emetic.
Caution: May cause vomiting if swallowed in large quantities.

Fire Hazard: Slight; reacts with oxidizing materials (Section 6).

Countermeasures

Storage and Handling: Section 7.

ACERDOL. See calcium permanganate and manganese compounds.

ACETAL

General Information

Synonyms: 1, 1-Diethoxyethane; diethylacetal, acetaldehyde diethylacetal and ethylidene diethyl ether.

Description: Colorless, volatile liquid, agreeable odor; nutty aftertaste.

Formula: $CH_3CH(OC_2H_5)_2$.

Constants: Mol wt: 118.17, bp: 102.0°C, flash p: −5°F (C.C.), lel = 1.65%, uel = 10.4%, d: 0.831, autoign temp: 446°F, vap. press.: 10 mm at 8.0°C, vap. d.: 4.08, mp: −100°C.

Hazard Aanalysis

Toxic Hazard Rating:

Acute Local: Inhalation 1.

Acute Systemic: Ingestion 2; Inhalation 2.

Chronic Local: U.

Chronic Systemic: Ingestion 1; Inhalation 1.

Toxicology: No cases of industrial intoxication are known. It is narcotic and more toxic than paraldehyde. Conclusions are based upon animal experiments. High concentrations in air cause narcosis.

Fire Hazard: Dangerous, when exposed to heat or flame; can react vigorously with oxidizing materials.

Spontaneous Heating: No.

Explosion Hazard: Moderate, when exposed to flame. Old samples have been known to explode upon heating.

Disaster Hazard: Dangerous, due to fire and explosion hazards.

Countermeasures

Ventilation Control: Section 2.

To Fight Fire: Carbon dioxide, alcohol foam, dry chemical or carbon tetrachloride (Section 6).

Personal Hygiene: Section 3.

Storage and Handling: Section 7.

Shipping Regulations: Section 11.

IATA: Flammable liquid, red label, 1 liter (passenger), 40 liters (cargo).

ACETALDEHYDE

General Information

Synonyms: Acetic aldehyde; ethyl aldehyde.

Description: Colorless, fuming liquid; pungent, fruity odor.

Formula: CH_3CHO.

Constants: Mol wt: 44.05, mp: −123.5°C, bp: 20.8°C, lel = 4.0%, uel = 57%, flash p: −36°F (C. C.), d: 0.7827 at 20°/20°C, autoign. temp.: 365°F, vap. d.: 1.52.

Hazard Analysis

Toxic Hazard Rating:

Acute Local: Irritant 3; İngestion 3; Inhalation 3.

Acute Systemic: Ingestion 2; Inhalation 2.

Chronic Local: Irritant 1.

Chronic Systemic: Ingestion 2; Inhalation 2.

TLV: ACGIH; 200 parts per million in air; 360 milligrams per cubic meter of air.

Toxicology:A local irritant. It has a general narcotic action on the central nervous system. (Section 1). A synthetic flavoring substance and adjuvant. See section 10. A common air contaminant (Section 4).

Fire Hazard: Dangerous, when exposed to heat or flame; can react vigorously with oxidizing materials (Section 6).

Spontaneous Heating: No.

Explosion Hazard: Severe when exposed to flame.

Disaster Hazard: Highly dangerous due to fire and explosion hazard.

Countermeasures

Ventilation Control: Section 2.

Personnel Protection: Section 3.

Personal Hygiene: Section 3.

First Aid: Section 1.

Storage and Handling: Section 7.

To Fight Fire: Carbon dioxide, dry chemical, alcohol foam or carbon tetrachloride (Section 6).

Shipping Regulations: Section 11.

I.C.C.: Flammable liquid; red label, 10 gallons.

Coast Guard: Inflammable liquid; red label.

MCA warning label.

IATA: Flammable liquid, red label, 1 liter (passenger), 40 liters (cargo).

m-ACETALDEHYDE. See metaldehyde.

p-ACETALDEHYDE. See paraldehyde.

ACETALDEHYDE AMMONIA. See aldehyde ammonia.

ACETALDEHYDE CYANOHYDRIN. See lactonitrile.

ACETALDEHYDE DIETHYLACETAL. See acetal.

ACETALDEHYDE SODIUM BISULFITE

General Information

Description: White crystals decomposed by acids, soluble in water, insoluble in alcohol.

Formula: $CH_3CH(OH)OSO_2Na \cdot \frac{1}{2}H_2O$.

Constant: Mol wt: 157.

Hazard Analysis

Toxic Hazard Rating: U. Limited animal experiments suggest low toxicity and irritation.

Disaster Hazard Rating: Dangerous. When heated to decomposition it emits highly toxic fumes.

ACETALDOL. See aldol.

ACETAMIDE

General Information

Synonym: Acetic acid amine.

Description: Colorless, mobile liquid; mousy odor.

Formula: CH_3CONH_2.

Constants: Molecular weight: 59.07, melting point: 81°C, boiling point: 222°C, density: 1.159 at 20°/4°C, vap. press.: 1 mm at 65.0°C.

Hazard Analysis

Toxicity: Details unknown. Limited animal experimentation shows low toxicity.

Disaster Hazard: Dangerous; see cyanides.

Countermeasures

Personal Hygiene: Section 3.

Storage and Handling: Section 7.

ACETAMIDINE HYDROCHLORIDE

General Information

Synonym: Ethanamidine hydrochloride, α-amino-α-imino-ethane.

Description: Long, somewhat deliquescent prisms when crystallized from ethanol.

Formula: $C_2H_6N_2 \cdot HCl$.

Constants: Mol wt: 94.6, mp: 174°C.

Soluble in water and alcohols.

Hazard Analysis

Toxic Hazard Rating:

TOXIC HAZARD RATING CODE (For detailed discussion, see Section 1.)

0 NONE: (a) No harm under any conditions; (b) Harmful only under unusual conditions or overwhelming dosage.

1 SLIGHT: Causes readily reversible changes which disappear after end of exposure.

2 MODERATE: May involve both irreversible and reversible changes; not severe enough to cause death or permanent injury.

3 HIGH: May cause death or permanent injury after very short exposure to small quantities.

U UNKNOWN: No information on humans considered valid by authors.

Acute Local: Irritant 2, Ingestion 2, Inhalation 2.
Acute Systemic: U.
Chronic Local: U.
Chronic Systemic: U.
Disaster Hazard: Dangerous; see chlorides.
Countermeasures
Personal Hygiene: Section 3.
Storage and Handling: Section 7.

p-ACETAMIDO BENZALDEHYDE
General Information
Formula: $C_9H_9O_2N$.
Constant: Mol wt = 163.
Hazard Analysis
Toxic Hazard Rating: U. Limited animal experiments suggest low toxicity and irritation.

2-ACETAMIDOFLUORENE
General Information
Description: Powder.
Formula: $C_6H_4CH_2C_6H_3NHCOCH_3$.
Constant: Mol wt: 223.3.
Hazard Analysis
Toxicity: Details unknown; an insecticide.
Disaster Hazard: Dangerous; see cyanides.
Countermeasures
Storage and Handling: Section 7.

ACETANILIDE
General Information
Synonyms: N-Phenylacetamide, antifebrin.
Description: White, shining crystalline scales.
Formula: $CH_3CONHC_6H_5$.
Constants: Mol wt: 135.16, mp: 113.5°C, bp: 303.8°C, flash p: 345°F (O. C.), d: 1.2105 at 4°/4°C, autoign. temp.: 1015°F, vap. press.: 1 mm at 114.0°C, vap. d.: 4.65.
Hazard Analysis
Toxic Hazard Rating:
Acute Local: Allergen 1.
Acute Systemic: Ingestion 2; Inhalation 1.
Chronic Local: Allergen 1.
Chronic Systemic: Ingestion 2; Inhalation 2.
Toxicology: In experimental poisoning a variety of effects have been reported, depending on the animal or tissue used for testing. Acute poisoning in humans may occur if several grams are taken by mouth. Cyanosis is a prominent finding in both acute and chronic poisoning. Habitual use of acetanilide for relief of headache has resulted in damage to the blood forming organs, with anemia and cyanosis. The latter is due to the presence of methemoglobin. Sulfhemoglobin has also been demonstrated in such cases. Habitual use may cause serious injury to the blood (Section 1). It may also cause contact dermatitis and inhalation or ingestion may cause an eczematous eruption of the skin (Section 9).
Fire Hazard: Slight, when exposed to heat or flame (Section 6).
Spontaneous Heating: No.
To Fight Fire: Water, carbon dioxide, dry chemical or carbon tetrachloride (Section 6).
Disaster Hazard: Dangerous; see aniline.
Countermeasures
Ventilation Control: Section 2.
Personnel Protection: Section 3.
Storage and Handling: Section 7.

ACETATE RAYON
Hazard Analysis
Caution: Skin rashes from rayon garments are usually due to dyes, conditioners or other chemicals added to the rayon (Section 9).
Fire Hazard: Slight, when exposed to heat or flame; can react with oxidizing materials (Section 6).
Countermeasures
Storage and Handling: Section 7.

ACETIC ACID
General Information
Synonyms: Methane carboxylic acid; vinegar acid; ethanoic acid.
Description: Clear, colorless liquid; pungent odor.
Formula: CH_3COOH.
Constants: Mol wt: 60.05, mp: 16.7°C, bp: 118.1°C, flash p: 109°F (C.C.), lel = 5.4%, uel = 16.0% at 212°F, d: 1.049 at 20°/4°C, autoign. temp.: 800°F, vap. press.: 11.4 mm at 20°C, vap. d.: 2.07.
Hazard Analysis
Toxic Hazard Rating:
Acute Local: Irritant 3; Ingestion 3; Inhalation 2.
Acute Systemic: U.
Chronic Local: Irritant 2.
Chronic Systemic: U.
TLV: ACGIH; 10 ppm in air; 25 milligrams per cubic meter of air.
Toxicology: Caustic, irritating, can cause burns, lachrymation and conjunctivitis. It attacks the skin easily and can cause dermatitis and ulcers. Inhalation causes irritation of mucous membranes. In case of contact with skin, eyes or clothing, immediately flush skin or eyes with plenty of water and remove all contaminated clothing. If swallowed give magnesia, chalk or whiting in water (Section 1).
Note: It is a miscellaneous and/or general purpose food additive which can migrate to food from packaging materials. A common air contaminant (Section 4).
Fire Hazard: Moderate, when exposed to heat or flame; can react vigorously with oxidizing materials (Section 6).
Spontaneous Heating: No.
Caution: Particularly dangerous in contact with chromic acid, sodium peroxide, nitric acid.
Explosion Hazard: Moderate, when exposed to flame.
Disaster Hazard: Dangerous; when heated to decomposition, emits toxic fumes.
Countermeasures
Ventilation Control: Section 2.
To Fight Fire: Carbon dioxide, dry chemical, alcohol foam or carbon tetrachloride (Section 6).
Personnel Protection: Section 3.
Storage and Handling: Section 7.
Shipping Regulations: Section 11.
Coast Guard: Combustible liquid.
MCA warning label.
IATA: Corrosive liquid, white label, 1 liter (passenger), 40 liters (cargo).

ACETIC ACID 80%; 56–70%; 28%. See acetic acid.

ACETIC ACID AMINE. See acetamide.

ACETIC ACID SECONDARY BUTYL ESTER. See sec.-butylacetate.

ACETIC ACID DIMETHYLAMIDE. See N, N-dimethyl acetamide.

ACETIC ACID, GLACIAL. See acetic acid.

ACETIC ALDEHYDE. See acetaldehyde.

ACETIC ANHYDRIDE
General Information
Synonyms: Acetyl oxide; acetic oxide; ethanoic anhydride.
Description: Colorless, very mobile, strongly refractive liquid; very strong acidic odor.
Formula: $(CH_3CO)_2O$.
Constants: Mol wt: 102.09, mp: −73.1°C, bp: 140°C, flash p: 129°F (C. C.), d: 1.082 at 20°/4°C, lel = 2.7%, uel = 10.1%, autoign. temp.: 734°F, vap. press.: 10 mm at 36.0°C, vap. d.: 3.52.
Hazard Analysis
Toxic Hazard Rating:
Acute Local: Irritant 3; Ingestion 3; Inhalation 2.

Acute Systemic: U.
Chronic Local: Irritant 2.
Chronic Systemic: U.

TLV: ACGIH (Accepted); 5 parts per million in air; 21 milligrams per cubic meter of air.

Toxicology: An irritant, corrosive on contact with tissue, particularly the eyes and upper respiratory tract. Systemic effects can be avoided by heeding the warning of its presence—coughing and a burning sensation in nose and throat. Ingestion causes a burning pain in the stomach followed by nausea and vomiting. If not removed from skin at once it will cause a reddening which may be followed by wrinkling, whitening and peeling. Continued contact with skin can cause dermatitis (Section 9). It is especially hazardous to the eyes, with delayed action causing eye burns.

Fire Hazard: Moderate, when exposed to heat or flame.

Spontaneous Heating: No.

Explosion Hazard: Moderate, when exposed to flame.

Disaster Hazard: Dangerous; when heated to decomposition, it emits toxic fumes; can react vigorously with oxidizing materials; will react violently on contact with water or steam.

Countermeasures

Ventilation Control: Section 2.

To Fight Fire: Carbon dioxide, dry chemical, alcohol foam or carbon tetrachloride (Section 6).

Personnel Protection: Section 3.

Storage and Handling: Section 7.

Shipping Regulations: Section 11.
Coast Guard: Combustible liquid.
MCA warning label.
IATA: Corrosive liquid, white label, 1 liter (passenger), 5 liters (cargo).

ACETIC ETHER. See ethyl acetate.

ACETIC OXIDE. See acetic anhydride.

ACETIN. See glyceryl monoacetate.

ACETOACETANILIDE

General Information

Synonyms: β-Ketobutyranilide; α-acetyl acetanalide.

Description: White crystalline solid.

Formula: $CH_3COCH_2CONHC_6H_5$.

Constants: Mol wt: 177.2, mp: 85°C, bp: decomposes, flash p: 365°F (C. O. C.), d: 1.260 at 20°C, vap. press.: 0.01, mm at 20°C.

Hazard Analysis

Toxicity: Details unknown; a weak allergen (Section 9). See also acetanilide.

Fire Hazard: Slight, when exposed to heat or flame.

Disaster Hazard: Dangerous; see aniline and cyanides.

Countermeasures

Storage and Handling: Section 7.

To Fight Fire: Water, alcohol foam, carbon dioxide, dry chemical or carbon tetrachloride (Section 6).

ACETOACET-o-ANISIDIDE

General Information

Description: Crystals.

Formula: $CH_3COCH_2CONHC_6H_4OCH_3$.

Constants: Mol wt: 207.22, mp: 86.6°C, flash p: 325°F (O. C.), d: 1.132 at 86.6°/20°C, vap. d: =. 7.0.

Hazard Analysis

Toxicity: Unknown.

Fire Hazard: Slight, when exposed to heat or flame; can react with oxidizing materials.

Countermeasures

To Fight Fire: Water, foam, carbon dioxide, dry chemicals or carbon tetrachloride (Section 6).

ACETOACET-o-CHLORANILIDE

General Information

Description: Crystals.

Formula: $CH_3COCH_2CONHC_6H_4Cl$.

Constants: Mol wt: 211.65, mp: 107°C, bp: decomposes, flash p: 350°F (C. O. C.), d: 1.438 at 20°C, vap. press.: 0.01 mm at 20°C, vap. d.: 7.31.

Hazard Analysis

Toxicity: See acetanilide.

Fire Hazard: Slight, when exposed to heat or flame.

Disaster Hazard: Dangerous; see aniline, and cyanides; can react vigorously with oxidizing materials.

Countermeasures

Ventilation Control: Section 2.

To Fight Fire: Water, foam, carbon dioxide, dry chemical or carbon tetrachloride (Section 6).

Personal Hygiene: Section 3.

Storage and Handling: Section 7.

ACETOACET-p-CHLORANILIDE

General Information

Description: Crystals.

Formula: $CH_3COCH_2CONHC_6H_4Cl$.

Constants: Mol wt: 211.65, mp: 133°C, bp: decomposes, flash p: 320°F (O. C.), d: 1.348 at 20°C, vap. press.: <0.01 mm at 20°C, vap. d.: 7.31.

Hazard Analysis

Toxicity: See acetanilide.

Fire Hazard: Slight, when exposed to heat or flame.

Disaster Hazard: Dangerous; see aniline, phosgene and cyanides; can react vigorously with oxidizing materials.

Countermeasures

Ventilation Control: Section 2.

To Fight Fire: Water, foam, carbon dioxide, dry chemical, or carbon tetrachloride (Section 6).

Personal Hygiene: Section 3.

Storage and Handling: Section 7.

ACETOACETIC ACID

General Information

Synonym: Acetyl acetic acid.

Description: Colorless syrup.

Formula: CH_3COCH_2COOH.

Constants: Mol wt: 102.1, bp: <100°C (Decomp.).

Hazard Analysis

Toxicity: Details unknown. A strong acid, hence an irritant.

Fire Hazard: Slight; when heated, emits acrid fumes; reacts with oxidizing materials (Section 6).

Countermeasures

Storage and Handling: Section 7.

ACETOACETIC ESTER. See ethyl acetoacetate.

ACETOACET-p-PHENETIDIDE

General Information

Description: Crystals.

Formula: $CH_3COCH_2CONHC_6H_4OC_2H_5$.

Constants: Mol wt: 221.25, mp: 108.5°C, bp: decomposes, flash p: 325°F (O. C.), d: 1.220 at 20°C, vap. press.: 0.02 mm at 20°C, vap d.: 7.63.

Hazard Analysis

Toxicity: See acetanilide.

Fire Hazard: Slight, when exposed to heat or flame; can react with oxidizing materials.

TOXIC HAZARD RATING CODE (For detailed discussion, see Section 1.)

0 NONE: (a) No harm under any conditions; (b) Harmful only under unusual conditions or overwhelming dosage.

1 SLIGHT: Causes readily reversible changes which disappear after end of exposure.

2 MODERATE: May involve both irreversible and reversible changes; not severe enough to cause death or permanent injury.

3 HIGH: May cause death or permanent injury after very short exposure to small quantities.

U UNKNOWN: No information on humans considered valid by authors.

Countermeasures
To Fight Fire: Water, foam, carbon dioxide, dry chemical or carbon tetrachloride (Section 6).
Ventilation Control: Section 2.
Personal Hygiene: Section 3.
Storage and Handling: Section 7.

ACETOACET-o-TOLUIDIDE
General Information
Description: Crystals.
Formula: $CH_3COCH_2CONHC_6H_4CH_3$.
Constants: Mol wt: 191.22, mp: 106°C, bp: decomposes, d: 1.300 at 20°C, vap. press.: 0.01 mm at 20°C, flash point 320°F (C. O. C.).
For details see Aceto-p-phenetidide.

ACETOACET-m-XYLIDIDE
General Information
Synonyms: AAMX.
Description: White, to light yellow crystalline solid soluble in water to 0.5% at 25°C.
Formula: $CH_3COCH_2CONHC_6H_3(CH_3)_2$.
Constants: Mol wt: 205; mp: 89–90°C; d: 1.238, flash p: 340°F (OC).
Hazard Analysis
Toxic Hazard Rating: U.
Fire Hazard: Combustible.
Countermeasures
To Fight Fire: Alcohol foam (Section 6).
Storage and Handling: Section 7.

ACETOGUANAMINE
General Information
Description: Crystals.
Formula: $CH_3(C_3N_3)(NH_2)_2$.
Constants: Mol wt: 125.15, mp: 270°C (decomp.), d: 1.44.
Hazard Analysis
Toxicity: Details unknown. See amines.
Fire Hazard: Slight, when exposed to heat or flame; can react with oxidizing materials (Section 6).
Countermeasures
Storage and Handling: Section 7.

ACETOIN
General Information
Synonyms: Acetyl methyl carbinal, 3-hydroxy-2-butanone, dimethylketol.
Description: Slightly yellow liquid or crystalline solid. Soluble in alcohol, miscible with water.
Formula: $CH_3COCHOHCH_3$.
Constants: Mol wt: 88, d: 1.016, bp: 140–148°C, mp: 15°C.
Hazard Analysis
Toxic Hazard Rating: U. A synthetic flavoring substance and adjuvant (see Section 10).

ACETOL
General Information
Synonyms: Acetylsalicylic acid; aspirin.
Description: White crystals.
Formula: $CH_3COOC_6H_4COOH$.
Constants: Mol wt: 180.2, mp: 135°C, d: 1.35.
Hazard Analysis
Toxic Hazard Rating:
Acute Local: Allergen 1; Ingestion 2.
Acute Systemic: Ingestion 1; Inhalation 1.
Chronic Local: Allergen 1.
Chronic Systemic: U.
Caution: Ten grams may be fatal. Contact, inhalation or ingestion can cause asthma, sneezing, irritations and watering of eyes and nose as well as hives and eczema (Section 9).
Fire Hazard: Slight, when exposed to heat or flame (Section 6).
Countermeasures
Ventilation Control: Section 2.

Personal Hygiene: Section 3.
Storage and Handling: Section 7.

ACETONANYL. See 2,2,4-trimethyl-1,2,-dihydroquinoline.

ACETONE
General Information
Synonyms: Dimethyl ketone; ketone propane; propanone.
Description: Colorless liquid; fragrant mintlike odor.
Formula: CH_3COCH_3.
Constants: Mol wt: 58.08, mp: −94.6°C, bp: 56.48°C, ulc = 90, flash p: 0°F (C. C.), lel = 2.6%, uel = 12.8%, d: 0.7972 at 15°C, autoign. temp.: 1000°F, vap. press.: 400 mm at 39.5°C, vap. d.: 2.00.
Hazard Analysis
Toxic Hazard Rating:
Acute Local: Irritant 1; Ingestion 2; Inhalation 2.
Acute Systemic: Ingestion 2; Inhalation 2; Skin Absorption 2.
Chronic Local: Irritant 1.
Chronic Systemic: Ingestion 1; Inhalation 1; Skin Absorption 1.
TLV: ACGIH; 1000 parts per million in air; 2400 miligrams per cubic meter of air.
Toxicology: Acetone is narcotic in high concentrations. In industry, no injurious effects from its use have been reported, other than the occurrence of skin irritations resulting from its de-fatting action or headache from prolonged inhalation. A food additive permitted in food for human consumption (Section 10). A common air contaminant. See Section 4.
Fire Hazard: Dangerous, when exposed to heat or flame.
Explosion Hazard: Moderate, when vapor is exposed to flame.
Disaster Hazard: Dangerous, due to fire and explosion hazard; can react vigorously with oxidizing materials.
Countermeasures
Ventilation Control: Section 2.
To Fight Fire: Carbon dioxide, dry chemical, alcohol foam or carbon tetrachloride (Section 6).
Personnel Protection: Section 3.
Storage and Handling: Section 7.
Shipping Regulations: Section 11.
I. C. C.: Flammable liquid; red label, 10 gallons.
Coast Guard Classification: Inflammable liquid; red label.
MCA warning label.
IATA: Flammable liquid, red label, 1 liter (passenger), 40 liters (cargo).

ACETONE CHLOROFORM. See chlorobutanol.

ACETONE CYANHYDRIN
General Information
Synonym: Oxyisobutyric nitrile.
Description: Liquid.
Formula: CH_3COCH_3HCN.
Constants: Mol wt: 85.10, mp: −20°C, bp: 82°C at 23 mm, d: 0.932 at 19°C, autoign. temp.: 1270°F, flash p: 165°F, vap. d.: 2.93.
Hazard Analysis
Toxic Hazard Rating:
Acute Local: Irritant 1.
Acute Systemic: Ingestion 3; Inhalation 3; Skin Absorption 2.
Chronic Local: 0.
Chronic Systemic: U.
Caution: This material readily decomposes to hydrogen cyanide and acetone. It must be kept slightly acidified. It should not be stored for long periods and it must be kept cool. See also hydrocyanic acid and acetone.
Fire Hazard: Slight, when exposed to heat or flame.
Disaster Hazard: Dangerous; see cyanides.
Countermeasures
Ventilation Control: Section 2.
To Fight Fire: Carbon dioxide, dry chemical, alcohol foam or carbon tetrachloride (Section 6).

Personnel Protection: Section 3.
Storage and Handling: Section 7.
Shipping Regulations: Section 11.
  I. C. C.: Poison B; poison label; 55 gallons.
  Coast Guard Classification: Poison B; poison label.
  MCA warning label.
  IATA: Poison B, poison label, 1 liter (passenger), 220
    liters (cargo).

ACETONEDICARBOXYLIC ACID
General Information
Synonym: β-Ketoglutaric acid.
Description: White crystals.
Formula: $C_5H_6O_5$.
Constants: Mol wt: 146.10, mp: 135°C (decomp).
Hazard Analysis
Toxicity: Unknown.
Fire Hazard: Slight; when heated, it emits acrid fumes.
Countermeasures
Storage and Handling: Section 7.

ACETONE DICHLORIDE. See 2,2-dichloropropane.

ACETONE OIL
General Information
Description: (a) Standard: light, lemon-yellow. (b) Refined:
  almost water white. (c) Heavy: dark, orange-yellow.
Constants: bp: (a) 75-160°C, (b) 75–105°C, (c) 80–225°C,
  d: (a) 0.826–0.830, (b) 0.812, (c) 0.885–0.865.
Hazard Analysis
Toxicity: Unknown.
Fire Hazard: Dangerous; when exposed to heat or flame.
Explosion Hazard: Moderate, when exposed to flame.
Disaster Hazard: Dangerous; can react vigorously with
  oxidizing materials.
Countermeasures
Storage and Handling: Section 7.
To Fight Fire: Carbon dioxide, dry chemical, or carbon
  tetrachloride (Section 6).
Shipping Regulations: Section 11.
  I. C. C.: Flammable liquid; red label, 10 gallons.
  Coast Guard Classification: Inflammable liquid; red label.
  MCA warning label.
  IATA: Flammable liquid, red label, 1 liter (passenger),
    40 liters (cargo).

ACETONE PEROXIDE
General Information
Description: Liquid.
Hazard Analysis
Toxicity: Details unknown. See also peroxides, organic.
Fire Hazard: Moderate, by spontaneous chemical reaction;
  can react vigorously with reducing materials (Sec-
  tion 6).
Explosion Hazard: Moderate, when shocked.
Countermeasures
Storage and Handling: Section 7.

ACETONE SEMICARBAZONE
General Information
Formula: $(CH_3)_2CNNHCONH_2$.
Constant: Mol wt: 115.1.
Hazard Analysis
Toxicity: Details unknown; an insecticide.
Disaster Hazard: Moderately dangerous; when heated to
  decomposition, emits toxic fumes.
Countermeasures
Storage and Handling: Section 7. .

ACETONITRILE. See methylcyanide.

ACETONYL ACETONE
General Information
Synonyms: Hexanedione-2,5; 1,2-diacetyl-ethane.
Description:Colorless liquid.
Formula: $CH_3COCH_2CH_2COCH_3$.
Constants: Mol wt: 114.14, mp: −9°C, bp: 192.4°C, flash
  p: 174° F (C. C.), d: 0.970 at 20°/4°C, autoign. temp.:
  920° F, vap. d.: 3.94.
Hazard Analysis
Toxic Hazard Rating:
  Acute Local: Irritant 2; Ingestion 1; Inhalation 1.
  Acute Systemic: U.
  Chronic Local: U.
  Chronic Systemic: U.
Caution: Can irritate the eyes.
Fire Hazard: Moderate, when exposed to heat or flame; can
  react with oxidizing materials.
Spontaneous Heating: No.
Countermeasures
To Fight Fire: Carbon dioxide, dry chemical, alcohol foam
  or carbon tetrachloride (Section 6).
Ventilation Control: Section 2.
Personnel Protection: Section 3.
Storage and Handling: Section 7.

3-(αACETONYLBENZYL)-4-HYDROXYCOUMARIN.
  See warfarin.

3-α-ACETONYLFURFURYL)-4-HYDROXYCOUMARIN
General Information
Description: White Powder.
Hazard Analysis
Toxic Hazard Rating:
  Acute Local: U.
  Acute Systemic: Ingestion 3; Inhalation 3.
  Chronic Local: U.
  Chronic Systemic: Ingestion 3; Inhalation 3.
See also Warfarin.
Caution: This rodenticide is almost always used mixed with
  bait preparations unpalatable to humans and resembles
  warfarin in action. However, in case of accidental in-
  gestion, induce vomiting until fluid is clear. Administer
  vitamin K (oral or intravenous) in large doses. Call a
  physician immediately.
Countermeasures
Ventilation Control: Section 2.
Personnel Protection: Section 3.
First Aid: Section 1.
Storage and Handling: Section 7.

p-ACETOPHENETIDE. See acetophenetidine.

ACETOPHENETIDINE
General Information
Synonyms: p-Acetophenetide; phenacetin.
Description: Glistening, crystalline powder.
Formula: $CH_3CONHC_6H_4OC_2H_5$.
Constants: Mol wt: 179.21, mp: 138°C.
Hazard Analysis
Toxic Hazard Rating:
  Acute Local: Ingestion 2; Inhalation 1.
  Acute Systemic: Ingestion 2, Inhalation 1.
  Chronic Local: U.
  Chronic Systemic: Ingestion 2; Inhalation 1.
Toxicology: Although it is considered less toxic than aceta-
  nilide, the symptoms of poisoning from it are nearly
  identical: weakness, dizziness, depression, collapse and

TOXIC HAZARD RATING CODE *(For detailed discussion, see Section 1.)*

0 NONE: (a) No harm under any conditions; (b) Harmful only under unusual conditions or overwhelming dosage.

1 SLIGHT: Causes readily reversible changes which disappear after end
  of exposure.

2 MODERATE: May involve both irreversible and reversible changes;
  not severe enough to cause death or permanent injury.

3 HIGH: May cause death or permanent injury after very short exposure
  to small quantities.

U UNKNOWN: No information on humans considered valid by authors.

very evident cyanosis. Chronic effects consist of loss of weight, insomnia, shortness of breath, weakness and often anemia.
Disaster Hazard: Moderately dangerous; when heated to decomposition it emits toxic fumes.
Countermeasures
Ventilation Control: Section 2.
Personal Hygiene: Section 3.
Storage and Handling: Section 7.

ACETOPHENONE. See phenyl methyl ketone.

ACETOPHENONE OXIME
General Information
Formula: $C_6H_3O(NOH)CH_3$.
Constant: Mol wt: 139.2.
Hazard Analysis
Toxicity: Details unknown; an insecticide.
Disaster Hazard: Dangerous; see cyanides.
Countermeasures
Storage and Handling: Section 7.

p-ACETOTOLUIDIDE
General Information
Synonym: p-Tolylacetamide.
Description: Crystals.
Formula: $CH_3C_6H_4NHCOCH_3$.
Constants: Mol wt: 149.1, bp: 307°C, flash p: 335°F (C. C.), d: 1.212, vap. d.: 5.14.
Hazard Analysis
Toxicity: Details unknown. See acetophenetidine and acetanilide.
Fire Hazard: Slight, when exposed to heat or flame; can react with oxidizing materials.
Countermeasures
To Fight Fire: Water, foam, carbon dioxide, dry chemical, or carbon tetrachloride (Section 6).
Storage and Handling: Section 7.

ACETOXYISOSUCCINODINITRILE
Hazard Analysis
Toxic Hazard Rating: U. Limited animal experiments suggest high toxicity. See also nitriles.
Disaster Hazard: Dangerous. When heated to decomposition emits highly toxic fumes of nitriles.
Countermeasures
Storage and Handling: Section 7.

ACETOXY MERCURI BENZO THIOPHENE
Hazard Analysis
Toxicity: Highly toxic. See mercury compounds, organic.
Disaster Hazard: Dangerous; see mercury and oxides of sulfur.
Countermeasures
Storage and Handling: Section 7.

ACETOZONE. See acetyl benzoyl peroxide.

ACETULAN
General Information
Description: Pale yellow liquid.
Formula: Acetylated lanolin fractions.
Constants: D: 0.867.
Hazard Analysis
Toxicity: Limited information indicates virtually no toxicity.
Fire Hazard: Slight when exposed to heat or flame. Can react with oxidizers.
Countermeasures
To Fight Fire: Foam, carbon dioxide, dry chemical (Section 6).
Storage and Handling: Section 7.

α-ACETYLACETANILIDE. See acetoacetanilide.

ACETYL ACETIC ACID. See acetoacetic acid.

ACETYLACETONATE OF CHROMIUM. See chromium 2,4-pentanedione derivative.

ACETYLACETONATE OF COPPER. See copper 2,4-pentanedione derivative.

ACETYLACETONE. See pentanedione-2,4.

ACETYLAMINOAZOTOLUENE
Hazard Analysis
Toxic Hazard Rating: A possible carcinogen.

p-ACETYLAMINOBENZALDEHYDE THIOSEMICARBAZONE. See "thiosemicarbazone."

2-ACETYLAMINOFLUORENE
General Information
Synonym: N-2-Fluorenylacetamide.
Description: Crystals, insoluble in H_2O; soluble in glycols, alcohols, and fat solvents.
Constant: mp: 194°C.
Hazard Analysis
Toxic Hazard Rating: U. A potent carcinogen.

ACETYL BENZENE. See phenyl methyl ketone.

ACETYL BENZOYL ACONINE. See aconitine.

ACETYL BENZOYL PEROXIDE
General Information
Synonym: Acetozone.
Description: White crystals.
Formula: $C_6H_5COOOCOCH_3$.
Constants: Mol wt: 180.2, mp: 36–37°C, bp: 130°C at 19 mm.
Hazard Analysis
Toxic Hazard Rating:
  Acute Local: Irritant 3; Ingestion 2.
  Acute Systemic: U.
  Chronic Local: Irritant 2; Ingestion 1; Inhalation 2.
  Chronic Systemic: U.
Caution: This material is a powerful oxidizing agent which is corrosive to skin and mucous membranes.
Fire Hazard: Moderate, by spontaneous chemical reaction.
Explosion Hazard: Moderate, when shocked or exposed to heat.
Disaster Hazard: Dangerous; shock or heat will cause detonation with evolution of toxic fumes; will react with water or steam to produce heat; can react vigorously with reducing materials.
Countermeasures
Ventilation Control: Section 2.
To Fight Fire: Carbon dioxide or dry chemical (Section 6).
Personnel Protection: Section 3.
Storage and Handling: Section 7.
Shipping Regulations: Section 11.
  I.C.C. Classification: Not accepted.
  Coast Guard: Not Permitted.
IATA: Oxidizing material, not acceptable (passenger and cargo).

ACETYL BENZOYL PEROXIDE, SOLUTION. See also acetyl benzoyl peroxide.
Countermeasures
Shipping Regulations: Section 11.
  I.C.C.: Oxidizing material; yellow label, 1 quart.
  Coast Guard: Oxidizing material; yellow label.
  IATA: Oxidizing material, not acceptable (passenger and cargo); (containing more than 40% by weight of peroxide).

ACETYL BENZYL PEROXIDE, DRY
General Information
Description: Powder.
Hazard Analysis
See acetyl benzoyl peroxide and peroxides, organic.
Toxicity: Details unknown.
Fire Hazard: Moderate, by spontaneous chemical reaction (Section 6).
Caution: A powerful oxidizing agent.

Explosion Hazard: Moderate, when shocked or exposed to heat.

Disaster Hazard: Dangerous; shock will cause detonation with evolution of toxic fumes; will react with water and steam to produce heat; can react vigorously with reducing materials.

Countermeasures

Storage and Handling: Section 7.

IATA: Oxidizing material, yellow label, not acceptable (passenger), 1 liter (cargo); (containing not more than 40% by weight of peroxide in a non-volatile solvent).

ACETYL BENZYL PEROXIDE, WET
Hazard Analysis

Toxicity: Details unknown. See also peroxides, organic.

Fire Hazard: Moderate, by spontaneous chemical reaction (Section 6).

Disaster Hazard: Moderately dangerous; when heated to decomposition, it emits toxic fumes; can react vigorously with reducing materials.

Countermeasures

Storage and Handling: Section 7.

ACETYL BROMIDE
General Information

Synonym: Ethanoyl bromide.

Description: Colorless fuming liquid; turns yellow in air.

Formula: CH_3COBr.

Constants: Mol wt: 122.96, mp: $-96.5°C$, bp: $76.7°C$, d: 1.52 at $9.5°/4°C$.

Hazard Analysis

Toxic Hazard Rating:

Acute Local: Irritant 3; Ingestion 3; Inhalation 3.

Acute Systemic: U.

Chronic Local: U.

Chronic Systemic: U.

Caution: This material readily hydrolyzes. See also hydrobromic acid, acetic acid.

Explosion Hazard: Slight, by spontaneous chemical reaction; decomposes violently upon contact with moisture.

Disaster Hazard: Dangerous; when heated to decomposition, it emits highly corrosive and toxic fumes of carbonyl bromide and bromine; will react with water or steam to produce heat.

Countermeasures

Ventilation Control: Section 2.

To Fight Fire: Dry chemical, carbon dioxide, or carbon tetrachloride (Section 6).

Personnel Protection: Section 3.

Storage and Handling: Section 7.

Shipping Regulations: Section 11.

IATA: Corrosive liquid, white label, 1 liter (passenger), 5 liters (cargo).

α-ACETYLBUTYROLACTONE
General Information

Synonym: α-acetyl-γ-hydroxybutyric acid-γ-lactone.

Description: Liquid, fruity odor. Soluble in water.

Formula: $C_6H_8O_3$.

Constants: Mol wt: 128.1, d: 1.187 at 20/20.

Hazard Analysis

Toxic Hazard Rating:

Acute Local: Irritant 2; Ingestion 2; Inhalation 2.

Acute Systemic: U.

Chronic Local: U.

Chronic Systemic: U.

Remarks: Avoid prolonged contact with skin or any contact with eyes or mucous membranes.

Fire Hazard: Slightly hazardous when exposed to heat or fire. Can react with oxidizers.

Countermeasures

To Fight Fire: Water, foam, carbon dioxide, dry chemical, or carbon tetrachloride (Section 6).

Storage and Handling: Section 7.

Ventilation Control: Section 2.

Personnel Protection: Section 3.

ACETYL CHLORIDE
General Information

Synonym: Ethanoyl chloride.

Description: Colorless, fuming liquid.

Formula: CH_3COCl.

Constants: Mol wt: 78.50, mp: $-112°C$, bp: $51-52°C$, flash p: $40°F$ (C.C.), Autoign. temp.: $734°F$, d: 1.1051 at $20°/4°C$, vap. d.: 2.70.

Hazard Analysis

Toxic Hazard Rating:

Acute Local: Irritant 3; Ingestion 3; Inhalation 3.

Acute Systemic: U.

Chronic Local: U.

Chronic Systemic: U.

Caution: It readily hydrolyzes to form hydrogen chloride and acetic acid. See also hydrochloric acid, acetic acid.

Fire Hazard: Dangerous, when exposed to heat or flame; reacts violently with water.

Spontaneous Heating: No.

Explosion Hazard: Slight, by spontaneous chemical reaction.

Disaster Hazard: Dangerous; when heated to decomposition, emits highly toxic fumes of phosgene; will react with water or steam to produce heat and toxic or corrosive fumes.

Countermeasures

Ventilation Control: Section 2.

To Fight Fire: Carbon dioxide, or dry chemical (Section 6).

Personnel Protection: Section 3.

Storage and Handling: Section 7.

Shipping Regulations: Section 11.

I.C.C.: Corrosive liquid; white label, 1 gallon.

Coast Guard Classification: Corrosive liquid; white label.

MCA warning label.

IATA: Corrosive liquid, white label, 1 liter (passenger), 5 liters (cargo).

ACETYLENE
General Information

Synonyms: Ethyne; ethine.

Description: Colorless gas; garlic-like odor.

Formula: $HC \equiv CH$.

Constants: Mol wt: 26.04, bp: $-84.0°C$ (sublimes), lel: 2.5%, uel: 80%, mp: $-81.8°C$, flash p: $0°F$ (C.C.), d: 1.173 g/liter at $0°C$, autoign. temp.: $571°F$, vap. press.: 40 atm at $16.8°C$, vap d: 0.91.

Hazard Analysis

Toxic Hazard Rating:

Acute Local: 0.

Acute Systemic: Inhalation 2.

Chronic Local: 0.

Chronic Systemic: Inhalation 1.

Caution: When mixed with oxygen, in proportions of 40 per cent or more acetylene acts as a narcotic and has been used in anesthesia. Acetylene acts as a simple asphyxiant by diluting the oxygen in the air to a level which will not support life. However, the presence of impurities in commercial acetylene may result in the production of symptoms before an asphyxiant concentration is reached. Dizziness, headache, mild gastric symptoms, and in high concentrations, semi-asphyxia and brief loss of consciousness have all been reported. In general industrial practice, however, acetylene does not constitute a serious hazard. See argon for discussion of simple asphyxiants.

Fire Hazard: Dangerous, when exposed to heat or flame. (Section 6).

Spontaneous Heating: No.

Explosion Hazard: Moderate, when exposed to heat or flame or by spontaneous chemical reaction. At high pressures and even moderate temperatures, acetylene has been known to decompose explosively. Forms explosive compounds with copper and silver (see acetylides).

Disaster Hazard: Dangerous; when ignited it burns with an intensely hot flame; can react vigorously with oxidizing materials.

Countermeasures

Ventilation Control: Section 2.

To Fight Fire: Carbon dioxide, water spray, or dry chemical. Stop Flow of gas.

Storage and Handling: Section 7.

Shipping Regulations: Section 11.

   I.C.C.: Flammable gas; red gas label, 300 pounds.

   Coast Guard Classification: Inflammable gas; red gas label.

   IATA: Flammable gas, red label, not acceptable (passenger), 140 kilograms (cargo).

ACETYLENE CHLORIDE
General Information

Synonym: Chloroethyne.

Description: Gas.

Formula: CHCCl.

Constants: Mol wt: 60.5, bp: -32 to $-30°C$, d: 2.0.

Hazard Analysis

Toxicity: Details unknown. Probably has anesthetic properties if inhaled. See also chlorinated hydrocarbons, aliphatic.

Fire Hazard: Dangerous, by spontaneous chemical reaction (Section 6).

Spontaneous Heating: Spontaneously flammable in air.

Explosion Hazard: Severe, when shocked or exposed to heat.

Disaster Hazard: Dangerous; shock will explode it; when heated to decomposition, emits highly toxic fumes of phosgene; can react vigorously with oxidizing materials.

Countermeasures

Storage and Handling: Section 7.

ACETYLENE DICHLORIDE. See trans-dichloroethylene.

ACETYLENE DIUREINE
General Information

Synonym: Glycol uril.

Description: White, needle-like crystals.

Formula: NHCONHCNCHNHCONH.

Constants: Mol wt: 142.12, d: 1.599.

Hazard Analysis

Toxic Hazard Rating: U.

Fire Hazard: Slight; (Section 6).

Countermeasures

Storage and Handling: Section 7.

ACETYLENE TETRABROMIDE
General Information

Synonym: Tetrabromoethane.

Description: Colorless to yellow liquid.

Formula: $CHBr_2CHBr_2$.

Constants: Mol wt: 345.70, bp: 151°C at 54 mm, fp: $-1°C$, d: 2.9638 at 20°/4°C.

Hazard Analysis

Toxic Hazard Rating:

   Acute Local: Irritant 2; Ingestion 2; Inhalation 2.

   Acute Systemic: Inhalation 2; Skin Absorption 1.

   Chronic Local: Irritant 1.

   Chronic Systemic: Inhalation 1.

Data based on animal experiments. High concentrations are irritant and narcotic.

TLV: ACGIH (recommended); 1 part per million in air; 14 milligrams per cubic meter of air.

Disaster Hazard: Dangerous; when heated, it emits highly toxic fumes of carbonyl bromide.

Countermeasures

Ventilation Control: Section 2.

Personnel Protection: Section 3.

Storage and Handling: Section 7.

ACETYLENE TETRACHLORIDE
General Information

Synonyms: Tetrachloroethane-1, 1, 2, 2.

Description: Heavy, colorless, mobile liquid; chloroform-like odor.

Formula: $CHCl_2CHCl_2$.

Constants: Mol wt: 167.86, mp: $-43.8°C$, bp: 146.3°C, d: 1.600 at 20°/4°C.

Hazard Analysis

Toxic Hazard Rating:

   Acute Local: Irritant 2; Ingestion 3; Inhalation 3.

   Acute Systemic: Ingestion 3; Inhalation 3; Skin Absorption 3.

   Chronic Local: U.

   Chronic Systemic: Ingestion 3; Inhalation 3; Skin Absorption 3.

TLV: ACGIH (recommended); 5 parts per million in air; 34 milligrams per cubic meter of air. Absorbed via skin.

Toxicology: This is generally considered the most toxic of the common chlorinated hydrocarbons. It has a fairly strong irritant action on the mucous membranes of the eyes and upper respiratory tract; a concentration of 3 ppm produces a detectable odor. There is thus an initial warning effect. Its narcotic action is stronger than that of chloroform, but because of its low volatility, narcosis is less severe and much less common in industrial poisoning than in the case of other chlorinated hydrocarbons. The toxic action of this material is chiefly on the liver, where it produces acute yellow atrophy and cirrhosis. Fatty degeneration of the kidneys and heart, hemorrhage into the lungs and serous membranes, and edema of the brain have also been found in fatal cases. Some reports indicate a toxic action on the central nervous system, with changes in the brain and in the peripheral nerves. The effect on the blood is one of hemolysis, with appearance of young cells in the circulation and a monocytosis. Due to its solvent action on the natural skin oils, dermatitis is not uncommon (Section 9).

   The initial symptoms resulting from exposure to the vapor are lacrimation, salivation and irritation of the nose and throat. Continued exposure to high concentrations results in restlessness, dizziness, nausea and vomiting and narcosis. The latter, however, is rare in industry. More commonly, the exposure is less severe, and the complaints are vague and referable to the digestive and nervous systems. The patient's complaints gradually progress to serious illness, with development of a toxic jaundice, liver tenderness, etc., and possibly albuminuria and edema. With serious liver damage the jaundice increases and toxic symptoms appear, with somnolence, delirium, convulsions and coma usually preceding death.

   This material is considered to be a very severe industrial hazard and its use has been restricted or even forbidden in certain countries.

Explosion Hazard: Moderate, in contact with sodium or potassium. When heated in contact with solid potassium hydroxide a spontaneously flammable gas is evolved. Any water causes appreciable hydrolysis even at room temperature and both hydrolysis and oxidation become comparatively rapid above 110°C.

Disaster Hazard: Dangerous; when heated, it emits highly toxic decomposition products.

Countermeasures
Ventilation Control: Section 2.
Personnel Protection: Section 3.
First Aid: Section 1.
Storage and Handling: Section 7.
Shipping Regulations: Section 11.
   IATA: Poison B, poison label, 1 liter (passenger), 40 liters (cargo).

N-**ACETYL ETHANOLAMINE.** See hydroxyethylacetamide.

ACETYL GLYCOLIC ACID ETHYL ESTER
General Information
Description: Liquid.
Formula: $CH_3COOCH_2COOC_2H_5$.
Constants: Mol wt: 146, bp: 184–189°C, flash p: 349°F, d: 1.094.
Hazard Analysis
Toxic Hazard Rating:
   Acute Local: Ingestion 2; Inhalation 2.
   Acute Systemic: U.
   Chronic Local: U.
   Chronic Systemic: U.
Fire Hazard: Slight, when exposed to heat or flame; can react vigorously with oxidizing materials (Section 6).
Countermeasures
Ventilation Control: Section 2.
Personal Hygiene: Section 3.
Storage and Handling: Section 7.

ACETYLHYDRAZINE
Hazard Analysis
Toxic Hazard Rating: U. Limited animal experiments suggest toxicity, possible liver damage from chronic exposure. Large doses cause haemolysis. See also phenyl hydrazine.

ACETYL HYDROPEROXIDE. See peracetic acid (40% solution).

ACETYLIDES
General Information
Synonym: Carbides.
Hazard Analysis
Toxicity: See individual compounds.
Fire Hazard: Unknown.
Explosion Hazard: Severe, when shocked or exposed to heat. Acetylides are very sensitive to shock, friction and heat. They explode readily and are one of the few commercial explosives which contain no oxygen or nitrogen. Their explosion produces no gas but simply is an effect of the large amount of heat, instantaneously produced. Acetylides are used for detonating compositions, or in combination with lead azide in detonating rivets where the acetylides reduce the flash point of the more insensitive azides. They are in a class with the fulminates and the azides as primary detonants.
   Because these materials are so sensitive to shock and temperature, they must be handled with extreme care. They must be kept cool, and if they are to be stored, should be kept wet. (See fulminates for suggested precautions in storage and handling of acetylides). Metal powders, such as finely divided copper or silver should not be stored or kept with acetylene or acetylides, because it is possible for them to react with these metal powders to form very sensitive acetylides which, while they are not dangerous in themselves, can cause enough

of a flash to ignite a possibly explosive mixture of gases and thus cause an explosion in a warehouse or storage area (For destruction of acetylides, see Section 7). Examples of commercially used acetylides are silver acetylide and copper acetylide.
Countermeasures
Storage and Handling: Section 7.

ACETYL IODIDE
General Information
Synonym: Ethanoyl iodide.
Description: Brown, transparent, fuming liquid.
Formula: CH_3COI.
Constants: Mol wt: 169.96, bp: 104–106°C, d: 2.067.
Hazard Analysis
Toxic Hazard Rating:
   Acute Local: Irritant 3; Ingestion 3; Inhalation 3.
   Acute Systemic: U.
   Chronic Local: U.
   Chronic Systemic: U.
Disaster Hazard: Dangerous, see iodides; will react with water or steam to produce toxic or corrosive fumes.
Countermeasures
Ventilation Control: Section 2.
Personnel Protection: Section 3.
Storage and Handling: Section 7.
Shipping Regulations: Section 11.
   IATA: Corrosive liquid, white label, 1 liter (passenger), 5 liters (cargo).

ACETYL KETENE. See diketene.

ACETYL METHYL CARBINOL. See Acetoin.

3-ACETYL-6-METHYL-1, 2-PYRAN-2, 4(3H)-DIONE. See dehydroacetic acid.

N-**ACETYL MORPHOLINE**
General Information
Description: Liquid.
Formula: $CH_3CONCH_2CH_2OCH_2CH_2$.
Constants: Mol wt: 129.16, mp:14°C, bp: decomposes, flash p: 235°F (O.C.), d: 1.1164, vap. press.: 0.02 mm at 20°C, vap d: 4.46.
Hazard Analysis
Toxicity: Details unknown. Limited animal experiments suggest low toxicity. See morpholine.
Fire Hazard: Slight, when exposed to heat or flame; can react vigorously with oxidizing materials.
Countermeasures
To Fight Fire: Water, alcohol, foam, carbon dioxide, dry chemical or carbon tetrachloride (Section 6).
Storage and Handling: Section 7.

ACETYL (p-NITROPHENYL) SULFANILAMIDE
General Information
Synonym: Sulfaniltran.
Formula: $C_{14}H_{13}N_3O_5S$.
Constants: Mol wt: 335, mp: 260–261°C.
Hazard Analysis
Toxic Hazard Rating: U. A food additive permitted in the feed and drinking water of animals and/or for the treatment of food producing animals (Section 10).
Disaster Hazard: Dangerous, when heated to decomposition can give off highly toxic fumes.

ACETYL OXIDE. See acetic anhydride.

TOXIC HAZARD RATING CODE *(For detailed discussion, see Section 1.)*

0 NONE: (a) No harm under any conditions; (b) Harmful only under unusual conditions or overwhelming dosage.

1 SLIGHT: Causes readily reversible changes which disappear after end of exposure.

2 MODERATE: May involve both irreversible and reversible changes; not severe enough to cause death or permanent injury.

3 HIGH: May cause death or permanent injury after very short exposure to small quantities.

U UNKNOWN: No information on humans considered valid by authors.

ACETYL PEROXIDE
General Information
Synonym: Ethanoyl peroxide, diacetyl peroxide.
Description: Solid or colorless crystals.
Formula: $(CH_3CO)_2O_2$.
Constants: Mol wt: 118.1, mp: 30°C, bp: 63°C at 21 mm, d: 1.18.
Hazard Analysis
Toxic Hazard Rating:
  Acute Local: Irritant 2; Ingestion 2; Inhalation 2.
  Acute Systemic: U.
  Chronic Local: U.
  Chronic Systemic: U.
Fire Hazard: Dangerous, by spontaneous chemical reaction. A powerful oxidizing agent; can cause ignition of organic materials on contact.
Explosion Hazard: Severe, when shocked or exposed to heat. It may explode spontaneously possibly when more than 24 hours old; it should be used up as soon as prepared.
Disaster Hazard: Highly dangerous; shock will explode it; it will react with water or steam to produce heat; can react vigorously with reducing materials; can emit toxic fumes on contact with acid or acid fumes.
Countermeasures
Ventilation Control: Section 2.
To Fight Fire: Carbon dioxide, dry chemical, or carbon tetrachloride (Section 6).
Personnel Protection: Section 3.
Storage and Handling: Section 7. Must be kept below 27°C and not warmed over 30°C. Do not add to hot materials. Do not add accelerator to this material. Store in original container with vented cap. Avoid bodily contact. See peroxides, organic, also section 7. This material is nearly always stored and handled as a 25% solution in an inert solvent. See also acetyl peroxide 25% solution in dimethyl phthalate.
Shipping Regulations: Section 11.
  I.C.C.: Not accepted.
  Coast Guard Classification: Not permitted.
  IATA: Oxidizing material, not acceptable (passenger and cargo).

ACETYL PEROXIDE 25% SOLUTION (IN DIMETHYL PHTHALATE)
General Information
Synonym: Diacetyl peroxide solution.
Description: Crystal clear liquid.
Constants: Mp: −7°C, flash p: 113°F (O.C.), d: 1.18 at 20°C.
Hazard Analysis
Toxic Hazard Rating:
  Acute Local: Irritant 2; Ingestion 2; Inhalation 2.
  Acute Systemic: U.
  Chronic Local: U.
  Chronic Systemic: U.
See also peroxides, organic.
Fire Hazard: Moderate when exposed to heat or flame, or by spontaneous chemical reaction. An oxidizing agent.
Disaster Hazard: Moderately dangerous; when heated to decomposition, emits toxic fumes; can react vigorously with oxidizing materials.
Countermeasures
Ventilation Control: Section 2.
Personnel Protection: Section 3.
To Fight Fire: Foam, carbon dioxide, or carbon tetrachloride (Section 6).
Storage and Handling: Section 7.
Shipping Regulations: Section 11.
  I.C.C.: Oxidizing material; yellow label, 1 quart.
  Coast Guard Classification: Oxidizing material; yellow label.

ACETYL PHENOL. See phenyl acetate.
  IATA: Oxidizing material, yellow label, not acceptable (passenger), 1 liter (cargo).

ACETYL-p-PHENYLENEDIAMINE. See p-aminoacetanilide.

1-ACETYL-2-PHENYLHYDRAZINE
General Information
Formula: $CH_3CONHNHC_6H_5$.
Constant: Mol wt: 150.2.
Hazard Analysis
Toxicity: See phenylhydrazine.
Disaster Hazard: Moderately dangerous; when heated to decomposition, it may emit toxic fumes.
Countermeasures
Storage and Handling: Section 7.

ACETYLSALICYLIC ACID. See acetol.

ACETYL TRIBUTYL CITRATE
General Information
Description: Odorless powder.
Formula: $C_5H_7O_2(COOC_4H_9)_3$.
Constants: Mol wt: 402.5, bp: 172–174°C at 1 mm, flash p: 400°F (C.O.C.), d: 1.048 at 25°/25°C.
Hazard Analysis
Toxicity Hazard Rating:
  Acute Local: Irritant 1; Ingestion 1; Inhalation 1.
  Acute Systemic: U.
  Chronic Local: U.
  Chronic Systemic: U.
Disaster Hazard: Slight; when heated, it emits acrid fumes (Section 6).
Countermeasures
Personnel Protection: Section 3.
Storage and Handling: Section 7.

ACETYL TRIETHYL CITRATE
General Information
Constant: Flash p: 370°F (C.O.C.).
Hazard Analysis
See acetyl tributyl citrate.

ACETYL TRI-2-ETHYL HEXYLCITRATE
General Information
Description: Liquid.
Formula: $CH_3CO_2C(CO_2C_8H_{17})CH_2(CO_2C_8H_{17})$-$CH_2(CO_2C_8H_{17})$.
Constants: Mol wt: 570.8, bp: 225°C at 1 mm, flash p: 430°F (C.O.C.), d: 0.983 at 25°/25°C.
Hazard Analysis
Toxicity: Details unknown; probably low.
Fire Hazard: Slight, when exposed to heat or flame; can react with oxidizing materials (Section 6).
Countermeasures
Storage and Handling: Section 7.

ACID BARIUM OXALATE. See barium binoxalate.

ACID BUTYL PHOSPHATE
General Information
Synonym: N-Butyl acid phosphate, n-butyl phosphoric acid.
Description: Water white liquid; soluble in alcohol, acetone, and toluene; insoluble in water, petroleum and naphtha.
Constants: d: 1.120–1.125 at 25/40°C, flash p: 230°F (C.O.C.).
Hazard Analysis
Fire Hazard: Slight when exposed to heat or flame.
Disaster Hazard: Dangerous, when heated to decomposition it emits highly toxic fume.
Countermeasures
Shipping Regulations: Section 11.
  IATA: Other restricted articles, class B, no label required, no limit (passenger and cargo).
Storage and Handling: Section 7.

ACID CALCIUM PHOSPHATE. See calcium phosphate, monobasic.

ACID CARBOYS, EMPTY
Hazard Analysis
Warning: These containers may contain concentrated vapors or even some liquid acid remaining from their original contents. Therefore, they can give rise to all the hazards of their original contents.
Countermeasures
Shipping Regulations: See Section 11.
 Coast Guard Classification: Hazardous article.
 IATA: No limit (passenger and cargo).

ACID ETHYL SULFATE. See ethyl sulfuric acid.

ACID LIQUIDS, N.O.S. See corrosive liquids.
Countermeasures
Shipping Regulations: Section 11.
 I.C.C.: Corrosive liquid; white label, 5 pints.
 Coast Guard Classification: Corrosive liquid; white label.
 IATA: Corrosive liquid, white label, 1 liter (passenger), 2½ liters (cargo).

ACID, SPENT, or ACID, SLUDGE. See nitric acid, sulfuric acid.
Countermeasures
Shipping Regulations: Section 11.
 Coast Guard Classification: Corrosive liquid; white label.

ACONITE
General Information
Synonyms: Monkshood; aconitum; wolfsbane.
Description: The dried tuberous root of Aconitum Napellus, composed of several alkaloids, the chief one being aconitine.
Hazard Analysis
Toxic Hazard Rating:
 Acute Local: Ingestion 3; Inhalation 3.
 Acute Systemic: Ingestion 3; Inhalation 3; Skin Absorption 3.
 Chronic Local: U.
 Chronic Systemic: Ingestion 2; Inhalation 2; Skin Absorption 2.
Toxicology: The poisonous alkaloid can be absorbed through the skin sufficiently to cause death. Usually causes pupils to dilate, but may cause them to be contracted. A lethal dose causes diminution in force and frequency of pulse, a cold and clammy skin and a tingling and numbness of the mouth, face and throat. Somewhat larger doses cause burning of the throat and stomach, increased salivation, nausea, retching and vomiting, grinding of the teeth and extension of the numbness and tingling to other parts of the body. There is difficulty in swallowing and speaking and pain in eyes and head. The fatal period is from 8 minutes to 3 or 4 hours.
Fire Hazard: Slight, when exposed to heat or flame (Section 6).
Disaster Hazard: Dangerous; when heated to decomposition, it emits highly toxic fumes.
Countermeasures
Ventilation Control: Section 2.
Personnel Protection: Section 3.
First Aid: Section 1.
Storage and Handling: Section 7.
Treatment and Antidotes: Wash stomach with tannic acid or powdered charcoal. Heart stimulants, such as strong coffee or caffeine. Artificial respiration and oxygen if necessary. Keep patient warm. Call a physician.

ACONITIC ACID

General Information
Synonym: 1,2,3-Propenetricarboxylic acid.
Description: White crystalline powder.
Formula: $HOOCCHC(COOH)CH_2COOH$.
Constants: Mol wt: 174.11, mp: 195°C (decomp.).
Hazard Analysis
Toxicity: Unknown. A synthetic flavoring substance and adjuvant. (Section 10).
Fire Hazard: Slight; (Section 6).
Countermeasures
Storage and Handling: Section 7.

ACONITINE
General Information
Synonym: Acetyl benzoyl aconine.
Description: White crystalline alkaloid; feeble bitter taste.
Formula: $C_{34}H_{49}NO_{11}$.
Constants: Mol wt: 647.74, mp: 188–197.8°C.
Hazard Analysis
Toxic Hazard Rating:
 Acute Local: Ingestion 3; Inhalation 3.
 Acute Systemic: Ingestion 3; Inhalation 3; Skin Absorption 3.
 Chronic Local: U.
 Chronic Systemic: Ingestion 2; Inhalation 2; Skin Absorption 2.
Toxicology: This intensely poisonous alkaloid can be absorbed through the skin to cause death. Ingestion in small quantities may affect the eyes and cause blindness. Its toxicity is about equal to that of aconite.
Disaster Hazard: Dangerous; when heated to decomposition, it emits highly toxic fumes.
Countermeasures
Ventilation Control: Section 2.
Personnel Protection: Section 3.
First Aid: Section 1.
Storage and Handling: Section 7.

ACONITINE DIACETYL. See aconitine.

ACONITINE HYDROBROMIDE. See aconitine.

ACONITINE HYDROCHLORIDE. See aconitine.

ACONITINE NITRATE
General Information
Description: Crystals.
Formula: $C_{34}H_{49}NO_{11}HNO_3 \cdot 5H_2O$.
Constant: Mol wt: 800.84.
Hazard Analysis
Toxicity: See aconitine.
Fire Hazard: Moderate, by spontaneous chemical reaction; an oxidizing agent (Section 6).
Disaster Hazard: Dangerous; when heated, it emits highly toxic fumes of oxides of nitrogen; can react vigorously with reducing materials.
Countermeasures
Ventilation Control: Section 2.
Personnel Protection: Section 3.
Storage and Handling: Section 7.

ACONITINE SULFATE. See aconitine.

ACONITUM. See aconite.

ACONITUM FEROX
General Information
Synonyms: Bish; visha; Indian aconite.

TOXIC HAZARD RATING CODE (For detailed discussion, see Section 1.)

0 NONE: (a) No harm under any conditions; (b) Harmful only under unusual conditions or overwhelming dosage.

1 SLIGHT: Causes readily reversible changes which disappear after end of exposure.

2 MODERATE: May involve both irreversible and reversible changes; not severe enough to cause death or permanent injury.

3 HIGH: May cause death or permanent injury after very short exposure to small quantities.

U UNKNOWN: No information on humans considered valid by authors.

Hazard Analysis

Toxicity: Most powerful poison of the aconites. See also aconite.

Disaster Hazard: Dangerous; when heated to decomposition, it emits highly toxic fumes.

Countermeasures

Ventilation Control: Section 2.

Personnel Protection: Section 3.

First Aid: Section 1.

Storage and Handling: Section 7.

ACRIDINE

General Information

Description: Small colorless needles.

Formula: $C_{13}H_9N$.

Constants: Mol wt: 179.21, mp: 110.5°C, bp: 346°C, d: 1.1005 at 19.7°/4°C, vap. press.: 1 mm at 129.4°C.

Hazard Analysis

Toxic Hazard Rating:

Acute Local: Irritant 2; Ingestion 2; Inhalation 2.

Acute Systemic: U.

Chronic Local: U.

Chronic Systemic: U.

Toxicology: Occasional injuries stem from its industrial use either as a solid or as a vapor. Is strongly irritating to the skin and mucous surfaces. Upon inhalation it causes sneezing, itching or even violent burning of the skin; sometimes even inflammatory swelling. It is regarded as the effective irritant in tar and creosote or pitch, etc., which can sensitize the skin to light (Section 9).

Disaster Hazard: Moderately dangerous; when heated to decomposition, emits toxic fumes.

Countermeasures

Ventilation Control: Section 2.

Personnel Protection: Section 3.

Storage and Handling: Section 7.

ACROLEIC ACID. See acrylic acid.

ACROLEIN

General Information

Synonyms: Propenal; acrylic aldehyde; allyl aldehyde; acraldehyde.

Description: Colorless or yellowish liquid; disagreeable choking odor.

Formula: CH_2CHCHO.

Constants: Mol wt: 56.06, mp: −87.7°C, bp: 52.5°C, flash p: <0°F, d: 0.841 at 20°/4°C, autoign temp: unstable (532°F), lel: 2.8%, uel: 31%, vap d: 1.94.

Hazard Analysis

Toxic Hazard Rating:

Acute Local: Irritant 3; Allergen 1; Ingestion 3; Inhalation 3.

Acute Systemic: Ingestion 3; Inhalation 3; Skin Absorption 2.

Chronic Local: Irritant 3; Allergen 1.

Chronic Systemic: U.

Toxicology: Due to its extreme lachrymatory effect it serves as its own warning agent. Industry records one fatality ascribed to exposure to it formed by the heat of welding in an enclosed space. It affects particularly the membranes of the eyes and respiratory tract. It is a weak sensitizer; inhalation may cause asthmatic reaction (Section 9).

TLV: ACGIH (recommended); 0.1 part per million; 0.25 milligram per cubic meter.

Fire Hazard: Dangerous, when exposed to heat or flame.

Spontaneous Heating: No.

Explosion Hazard: Unknown.

Disaster Hazard: Dangerous; when heated to decomposition, emits highly toxic fumes; can react vigorously with oxidizing materials.

Countermeasures

Ventilation Control: Section 2.

Personnel Protection: Section 3.

To Fight Fire: Carbon dioxide, dry chemical or alcohol foam (Section 6).

Storage and Handling: Section 7.

Shipping Regulations: Section 11.

I.C.C.: Flammable liquid; red label, 1 quart.

Coast Guard Classification: Inflammable liquid; red label.

IATA: Flammable liquid, red label, not acceptable (passenger), 1 liter (cargo).

MCA warning label.

ACROLEIN DIMER

General Information

Synonym: 2-formyl-3,4-dihydro-2H-Pyran.

Description: Liquid; soluble in H_2O.

Formula: $OCH:CHCH_2CH_2CHCHCHO$.

Constants: Mol wt: 112, d: 1.0775 (20°C), bp: 151.3°C, fp: −100°C, flash p: 118°F (O.C.).

Hazard Analysis

Toxic Hazard Rating: U.

Fire Hazard: Moderately flammable when exposed to heat, flame, or powerful oxidizing agent.

Countermeasures

To Fight Fire: Alcohol foam.

Storage and Handling: Section 7.

ACRYLAMIDE

General Information

Description: White crystalline solid.

Formula: $CH_2CHCONH_2$.

Constants: Mol wt: 71.08, mp: 84.5 ± 0.3°C, bp: 125°C at 25 mm, d: 1.122 at 30°C, vap press.: 1.6 mm at 84.5°C, vap d: 2.45.

Hazard Analysis

Toxic Hazard Rating:

Acute Local: Irritant 3; Allergen 3: Ingestion 3; Inhalation 3.

Acute Systemic: Ingestion 3; Inhalation 2; Skin Absorption 2.

Chronic Local: Irritant 2; Allergen 2.

Chronic Systemic: Inhalation 2, Skin Absorption 2.

Toxicology: This material is dangerous because it can be absorbed through the unbroken skin. From animal experiments it would seem that its effect is toxic upon the central nervous system. Adult rats fed an average of 30 mg/kg of this material for 14 days were all partially paralyzed and had reduced their food consumption by 50%.

Disaster Hazard: Moderate when heated to decomposition, emits acrid fumes.

Countermeasures

Ventilation Control: Section 2.

Personnel Protection: Section 3.

Storage and Handling: Section 7.

ACRYLAMIDE—ACRYLIC ACID RESIN

General Information

Description: Produced by the polymerization of acrylamide with partial hydrolysis; or by copolymerization of acrylamide and acrylic acid.

Hazard Analysis

Toxic Hazard Rating: U. A food additive permitted in food for human consumption (Section 10).

ACRYLIC ACID

General Information

Synonyms: Propene acid; acroleic acid.

Description: Liquid; acrid odor.

Formula: C_2H_3COOH.

Constants: Mol wt: 72.06, mp: 14°C, bp: 141°C, d: 1.062, vap press.: 10 mm at 39.0°C, flash p: 130°F (OC), vap d: 2.5.

Hazard Analysis

Toxic Hazard Rating:

Acute Local: Irritant 3; Ingestion 3; Inhalation 3.

Acute Systemic: U.

Chronic Local: U.

Chronic Systemic: U.

Fire Hazard: Slight, when exposed to heat or flame (Section 6).

Explosion Hazard: Slight, by spontaneous chemical reaction; it polymerizes violently.

Disaster Hazard: Dangerous; when heated to decomposition, emits highly toxic fumes; it can react with oxidizing materials.

Countermeasures

Ventilation Control: Section 2.

Personnel Protection: Section 3.

Storage and Handling: Section 7.

ACRYLIC ACID METHYL ESTER. See methyl acrylate.

ACRYLIC ALDEHYDE. See acrolein.

ACRYLONITRILE
General Information

Synonyms: Propene nitrile; vinyl cyanide.

Description: Colorless, mobile liquid; mild odor.

Formula: CH_2CHCN.

Constants: Mol wt: 53.06, mp: $-82°C$, bp: $77.3°C$, fp: $-83°C$, flash p: $30°F$ (T.C.C.), lel: 3.1%, uel: 17%, d: 0.797 at $20°/4°C$, autoign temp: $898°F$, vap press.: 100 mm at $22.8°C$, vap d: 1.83.

Hazard Analysis

Toxicity: High. See cyanides and nitriles.

TLV: ACGIH (recommended): 20 parts per million in air; 43 milligrams per cubic meter of air. Absorbed via skin.

Toxicology: Acrylonitrile closely resembles hydrocyanic acid in its toxic action. By inhibiting the respiratory enzymes of tissue, it renders the tissue cells incapable of oxygen absorption. Poisoning is acute; there is little evidence of cumulative action on repeated exposure.

Exposure to low concentrations is followed by flushing of the face and increased salivation; further exposure results in irritation of the eyes, photophobia, irritation of the nose, deepened respiration, and, if exposure continues, shallow respiration, nausea, vomiting, weakness, an oppressive feeling in the chest, and occasionally headache and diarrhea are other complaints. Several cases of mild jaundice accompanied by mild anemia and leucocytosis have been reported. Urinalysis is generally negative, except for an increase in bile pigment. Serum and bile thiocyanates are raised. See also hydrocyanic acid.

Fire Hazard: Dangerous, when exposed to heat or flame.

Explosion Hazard: Moderate, when exposed to flame.

Disaster Hazard: Dangerous; see cyanides; can react vigorously with oxidizing materials.

Countermeasures

Ventilation Control: Section 2.

To Fight Fire: Carbon dioxide, dry chemical, carbon tetrachloride or alcohol foam.

Storage and Handling: Section 7.

Shipping Regulations: Section 11.

I.C.C.: Flammable liquid; red label, 10 gallons.

Coast Guard Classification: Inflammable liquid; red label.

MCA warning label.

IATA: Flammable liquid, red label, 1 liter (passenger), 40 liters (cargo).

2-ACRYLOXYETHYL DIMETHYLSULFONIUM METHYL SULFATE
Hazard Analysis

Toxic Hazard Rating: U. Limited animal experiments suggest moderate toxicity.

Disaster Hazard: Dangerous; when heated to decomposition emits highly toxic fumes.

Countermeasures

Storage and Handling: Section 7.

ACTIDIONE. See cycloheximide.

ACTINIC RADIATION
Hazard Analysis

Caution: Outdoor workers such as fishermen, sailors, soldiers and farmers show a high incidence of skin cancer. The commonest acute manifestation of actinic radiation effects on skin is sunburn. Section 5.

ACTINIUM
General Information

Description: Brownish-red granular powder.

Formula: Ac.

Constants: Mp: $1800°C$, at wt: 227.

Hazard Analysis

Radiation Hazard: Section 5. For permissible levels, see Table , p. .

Artificial isotope ^{227}Ac (Actinium Series), half life 21.8 d. Decays to radioactive ^{227}Th by emitting beta particles of 0.044 MeV. ^{227}Th in turn decays to ^{223}Ra by emitting alpha particles of 5.7–6.0 MeV and gamma rays of 0.03–0.3 MeV.

Natural isotope ^{228}Ac (Mesothorium-2, Thorium Series), half life 6.1h. Decays to radioactive ^{228}Th by emitting beta particles of 0.45–2.1 MeV. Also emits gamma rays of 0.06–1.64 MeV.

Countermeasures

Ventilation Control: Section 2.

Personnel Protection: Section 3.

ACTIVATED CARBON
General Information

Synonym: Charcoal, activated.

Description: Black amorphous mass.

Formula: C.

Constants: Mp: $>3500°C$, bp: $4000°C$, d: 3.51, at wt: 12.01.

Hazard Analysis

Toxic Hazard Rating:

Acute Local: Irritant 1; Allergen 1.

Acute Systemic: 0.

Chronic Local: Irritant 1; Allergen 1.

Chronic Systemic: 0.

Caution: Pure carbon is nontoxic. However, activated carbon, usually made from organic matter, is frequently found associated with small amounts of irritating and possibly toxic impurities. Avoid wetting and subsequent drying in storage. Store with ventilation. See also coal tar.

Fire Hazard: Slight, when exposed to heat or flame.

Spontaneous Heating: Yes.

Explosion Hazard: Slight, when dust is exposed to flame. (See dust explosions).

Countermeasures

Personal Hygiene: Section 3.

To Fight Fire: Water (Section 6).

Shipping Regulations: Section 11.

I.C.C.: Flammable solid; yellow label, 200 pounds.

Coast Guard Classification: Inflammable solid; yellow label.

IATA: Flammable solid, yellow label, 12 kilograms (passenger), 95 kilograms (cargo).

ADAMSITE. See diphenylamine chloroarsine.

TOXIC HAZARD RATING CODE *(For detailed discussion, see Section 1.)*

0 NONE: (a) No harm under any conditions; (b) Harmful only under unusual conditions or overwhelming dosage.

1 SLIGHT: Causes readily reversible changes which disappear after end of exposure.

2 MODERATE: May involve both irreversible and reversible changes; not severe enough to cause death or permanent injury.

3 HIGH: May cause death or permanent injury after very short exposure to small quantities.

U UNKNOWN: No information on humans considered valid by authors.

ADHESIVE PLASTER
General Information
Synonym: Adhesive tape.
Hazard Analysis
Toxic Hazard Rating:
  Acute Local: Allergen 1.
  Acute Systemic: 0.
  Chronic Local: Irritant 1; Allergen 1.
  Chronic Systemic: 0.
Caution: The reaction may be of allergic nature, manifested as itching and burning of the skin with redness and blister formation (Section 9).
Fire Hazard: Slight; when heated, emits acrid fumes (Section 6).
Countermeasures
Storage and Handling: Section 7.

ADHESIVE TAPE. See adhesive plaster.

ADHESIVES N.O.S. See cement, liquid, n.o.s.

ADIPIC ACID
General Information
Synonyms: Hexane dioic acid; 1,4-butane dicarboxylic acid.
Description: Fine white crystals or powder.
Formula: $COOH(CH_2)_4COOH$.
Constants: Mol wt: 146.14, mp: 152°C, bp: 337.5°C, flash p: 385°F (C.C.), d: 1.360 at 25°/4°C, vap press.: 1 mm at 159.5°C, vap d: 5.04.
Hazard Analysis
Toxicity: Details unknown; Limited animal experiments suggest low toxicity. A general purpose food additive (Section 10).
Fire Hazard: Slight, when exposed to heat or flame; can react with oxidizing materials.
Countermeasures
To Fight Fire: Water, foam, carbon dioxide, dry chemical or carbon tetrachloride (Section 6).
Storage and Handling: Section 7.

ADIPONITRILE
General Information
Synonyms: 1,4-Dicyanobutane; tetramethylene cyanide.
Description: Water-white liquid; practically odorless.
Formula: $CN(CH_2)_4CN$.
Constants: Mol wt: 108.14, mp: 2.3°C, bp: 295°C, flash p: 199.4°F (O.C.), d: 0.965 at 20°/4°C, vap d: 3.73.
Hazard Analysis
Toxic Hazard Rating:
  Acute Local: U.
  Acute Systemic: Ingestion 3; Inhalation 3; Skin Absorption 3.
  Chronic Local: U.
  Chronic Systemic: Ingestion 2; Inhalation 2.
Toxicology: It is toxic since the nitrile group will behave as a cyanide when ingested or absorbed in the body by contact. It produces disturbances of the respiration and circulation, irritation of the stomach and intestines, and loss of weight. Its low vapor pressure at room temperature makes exposure to harmful concentrations of its vapor unlikely if handled with reasonable care in well ventilated areas. Combustion products of adiponitrile may contain hydrocyanic acid. Accordingly, fires involving this material may be hazardous from a toxicological standpoint. See also hydrocyanic acid and nitriles.
Fire Hazard: Moderate, when exposed to heat or flame.
Disaster Hazard: Dangerous; when heated to decomposition, it emits highly toxic fumes; can react with oxidizing materials.
Countermeasures
Ventilation Control: Section 2.
To Fight Fire: Foam, carbon dioxide, dry chemical or carbon tetrachloride (Section 6).
Personnel Protection: Section 3.
Storage and Handling: Section 7.

"ADRENALINE"
General Information
Synonym: Epinephrine.
Description: Light brown or nearly white crystals.
Formula: $C_6H_3(OH)_2CHOHCH_2NHCH_3$.
Constants: Mol wt: 183.3, mp: 211–212°C.
Hazard Analysis
Toxic Hazard Rating:
  Acute Local: Allergen 1.
  Acute Systemic: Ingestion 0; Inhalation 2; Skin Absorption 3.
  Chronic Local: Allergen 1.
  Chronic Systemic: Ingestion 1; Inhalation 1; Skin Absorption 1.
Toxicology: May cause pallor, tremor, anxiety, nervousness, rapid, forceful pulse, rise in blood pressure and temperature, rapid breathing, dilation of the pupils. Allergic skin reactions have been known to result from contact causing contact dermatitis (Section 9).
Fire Hazard: Slight; when heated. (Section 6).
Countermeasures
Ventilation Control: Section 2.
Personnel Protection: Section 3.
First Aid: Section 1.
Treatment and Antidotes: Guanine or atropine may be helpful. Usually the symptoms are of short duration and clear up spontaneously.
Storage and Handling: Section 7.

AEROSOL IB. See diisobutyl sodium sulfosuccinate.

AGALMATOLITE. See pyrophyllite.

AGAR. See agar-agar.

AGAR-AGAR
General Information
Synonyms: Agar; gelose; Japanese, Bengal, Ceylon, or Chinese isinglass or gelatin; macassar gum.
Description: Unground—in thin, translucent membranous pieces; ground—pale, buff powder; soluble in boiling water; insoluble in cold water and organic solvents.
Hazard Analysis
Toxic Hazard Rating: U. A stabilizer food additive (Section 10).

AGE. See allyl glycidyl ether.

AGERITE WHITE. See p-benzyl oxyphenol.

AGGLUTININ. See ricin.

AIR (LIQUID AND COMPRESSED)
General Information
Description: Bluish, mobile liquid.
Formula: $O_2 + N_2$.
Constants: Bp: −189°C (liquid), flash p: none, autoign temp: none.
Hazard Analysis
Caution: Liquid air exists at so low a temperature that contact with it will destroy tissue. Personnel employed in areas of compressed air may develop caisson disease (the bends, the chokes) if decompression is too rapid.
Explosion Hazard: Moderate when containers under pressure are shocked or exposed to heat or flame. Flammable materials which have been in contact with liquid air may explode very easily. Ordinary oxidation is greatly accelerated in compressed air (Section 7).
Disaster Hazard: Moderately dangerous; can react vigorously with reducing materials.
Countermeasures
Storage and Handling: Section 7.
Shipping Regulations: See air, compressed.

AIR, COMPRESSED
Countermeasures
Shipping Regulations: Section 11.

I.C.C.: Nonflammable gas; green label, 300 pounds.

Coast Guard Classification: Nonflammable gas; green gas label.

IATA: Nonflammable gas, green label, 70 kilograms (passenger), 140 kilograms (cargo).

AIR, LIQUID, NON-PRESSURIZED
Countermeasures

Shipping Regulations: Section. 11.

IATA: Other restricted articles, class C, no label required, not acceptable (passenger and cargo).

AIR, LIQUID, LOW PRESSURE
Countermeasures

Shipping Regulations: Section 11.

IATA: Other restricted articles, class C, no label required, not acceptable (passenger); 140 kilograms (160 liters) (cargo).

AIR, LIQUID, PRESSURIZED
Countermeasures

Shipping Regulations: Section 11.

IATA: Nonflammable gas, green label, not acceptable (passenger), 140 kilograms (cargo).

AIRPLANE FLARES. See explosives, low.

AKLOMIDE. See 2-chloro-4-nitrobenzamide.

ALABANDITE. See manganous sulfide.

ALAMOSITE. See lead m-silicate.

ALANAP. See N-1-naphthyl phthalamic acid.

ALANINE
General Information

Synonyms: α-Alanine; α-amino propionic acid; 2-amino propanoic acid.

Description: A naturally occurring nonessential amino acid; colorless crystals; soluble in H_2O; slightly soluble in alcohol; insoluble in ether; occurs in DL, L, and D forms; the data here refer to the L and DL forms.

Formula: $CH_3CH(NH_2)COOH$.

Constants: Mol wt: 89, mp: (DL) 295°C (decomposes), (L) 297°C (decomposes).

Hazard Analysis

Toxic Hazard Rating: U. A nutrient and dietary supplement food additive.

α-ALANINE. See alanine.

ALCOHOL, DENATURED
General Information

Synonym: Denatured spirits.

Description: Liquid.

Composition: Alcohol and denaturants.

Hazard Analysis

Toxicity: This will depend upon the alcohol in question (generally ethyl alcohol) plus the denaturants used. Methyl alcohol is a common denaturant.

Fire Hazard: Dangerous; can react vigorously with oxidizing materials (Section 6).

Explosion Hazard: Moderate. See ethyl alcohol.

Countermeasures

Ventilation Control: Section 2.

Storage and Handling: Section 7.

Shipping Regulations: Section 11.

I.C.C. Classification: See alcohol or alcohol, N.O.S.

Coast Guard Classification: Inflammable liquid; red label.

ALCOHOLS, N.O.S. (See specific compound).
General Information

Description: A generic term applied to a series of compounds, the simplest of which has the general formula $C_nH_{2n+1}OH$.

Hazard Analysis

Toxicity: No general statement can be made as to the toxicity of alcohols, due to wide differences in toxic effects. When the term "alcohol" is used alone it usually applies to ethyl alcohol (C_2H_5OH), which is relatively nontoxic when compared with methyl and some other alcohols.

Fire Hazard: Dangerous; when exposed to heat or flame. (Section 6).

Countermeasures

Storage and Handling: Section 7.

Shipping Regulations: Section 11.

I.C.C.: Flammable liquid; red label, 10 gallons.

Coast Guard Classification: Inflammable liquid; red label.

IATA: Flammable liquid, red label, 1 liter (passenger), 40 liters (cargo).

ALCOHOL, TERTIARY. (See specific compound).
Countermeasures

Shipping Regulations: Section 11.

I.C.C. Classification: See alcohol or alcohol, N.O.S.

Coast Guard Classification: Inflammable liquid; red label.

ALDEHYDE AMMONIA
General Information

Synonyms: Acetaldehyde ammonia; 1-amino-ethanol.

Description: White crystalline solid.

Formula: $CH_3CH(NH_2)OH$.

Constants: Mol wt: 61.08, bp: 100°C, mp: 97°C.

Hazard Analysis

Toxic Hazard Rating:

Acute Local: Irritant 2; Ingestion 2; Inhalation 2.

Acute Systemic: U.

Chronic Local: U.

Chronic Systemic: Ingestion 2; Inhalation 2.

Fire Hazard: Moderate when exposed to heat or flame; readily decomposes into acetaldehyde and ammonia when heated (Section 6).

Explosion Hazard: Moderate, when exposed to heat or flame.

Explosive Range: See ammonia.

Disaster Hazard: Moderately dangerous; when heated to decomposition emits toxic fumes; can react with oxidizing materials.

Countermeasures

Ventilation Control: Section 2.

Personnel Protection: Section 3.

Storage and Handling: Section 7.

Shipping Regulations: Section 11.

IATA: Other restricted articles, class A, no label required, no limit (passenger and cargo).

ALDEHYDE C-8. See caprylaldehyde.

ALDEHYDE C-16. See ethyl methyl phenyl glycidate.

ALDEHYDES. (See also specific compounds).
Hazard Analysis

Toxic Hazard Rating:

Acute Local: Irritant 3; Ingestion 3; Inhalation 3.

Acute Systemic: Ingestion 2; Inhalation 2; Skin Absorption 1.

Chronic Local: U.

Chronic Systemic: Ingestion 1; Inhalation 1.

Toxicology: All the aldehydes possess anaesthetic proper-

TOXIC HAZARD RATING CODE (For detailed discussion, see Section 1.)

0 NONE: (a) No harm under any conditions; (b) Harmful only under unusual conditions or overwhelming dosage.

1 SLIGHT: Causes readily reversible changes which disappear after end of exposure.

2 MODERATE: May involve both irreversible and reversible changes; not severe enough to cause death or permanent injury.

3 HIGH: May cause death or permanent injury after very short exposure to small quantities.

U UNKNOWN: No information on humans considered valid by authors.

ties, but this is obscured by their highly irritant action on the eyes and mucous membranes of the respiratory tract. The lower aldehydes, very soluble in water, act chiefly on the eyes and tissues of the upper respiratory tract. The higher aldehydes, less soluble in water, tend to penetrate more deeply into the respiratory system and may affect the lungs.

ALDEHYDES, LIQUID N.O.S.
Shipping Regulations: Section 11.
  IATA: Flammable liquid, red label, 1 liter (passenger), 40 liters (cargo).

ALDEHYDINE. See 5-ethyl-2-methyl pyridine.

ALDOL
General Information
Synonyms: Acetaldol; 3-butanolal; oxybutyric aldehyde.
Description: Clear, white to yellow water; syrupy liquid.
Formula: $CH_3CH(OH) CH_2CHO$.
Constants: Mol wt: 72.10, bp: 182°C, flash p: 150°F. (O.C.), d: 1.11, autoign temp: 530°F, vap d: 3.04.
Hazard Analysis
Toxic Hazard Rating:
  Acute Local: Irritant 2; Ingestion 2; Inhalation 2.
  Acute Systemic: U.
  Chronic Local: U.
  Chronic Systemic: U.
Fire Hazard: Moderate, when exposed to heat or flame; it decomposes into crotonaldehyde and water when heated (see crotonaldehyde).
Spontaneous Heating: No.
Disaster Hazard: Dangerous; when heated to decomposition, it emits highly toxic fumes of crotonaldehyde; can react with oxidizing materials.
Countermeasures
Ventilation Control: Section 2.
To Fight Fire: Water, alcohol foam, carbon dioxide, dry chemical or carbon tetrachloride (Section 6).
Personnel Protection: Section 3.
Storage and Handling: Section 7.

ALDRIN
General Information
Synonyms: 1,2,3,4,10,10-hexachloro-1,4,4a,5,8,8a-hexahydro-1,4,5,8-dimethanonaphthalene; octalene; compound 118.
Description: Crystals.
Formula: $C_{12}H_8Cl_6$.
Constants: Mol wt: 365, mp: not less than 90°C; up to 100–101°C.
Hazard Analysis
Toxic Hazard Rating:
  Acute Local: U.
  Acute Systemic: Ingestion 3; Inhalation 3; Skin Absorption 3.
  Chronic Local: U.
  Chronic Systemic: Ingestion 3; Inhalation 3; Skin Absorption 3.
TLV: ACGIH (accepted); 0.25 milligrams per cubic meter of air. Absorbed via skin.
Toxicology: Ingestion, inhalation or absorption of this material into the body causes irritability, convulsions and depression in from 1 to 4 hours. Continued exposure to it causes liver damage.
Disaster Hazard: Dangerous; see chlorides.
Countermeasures
Ventilation Control: Section 2.
Personnel Protection: Section 3.
First Aid: Section 1.
Storage and Handling: Section 7.
Shipping Regulations: Section 11.
  Coast Guard Classification: Poison B; poison label.
  ICC: Poison B, poison label, 200 pounds.
  IATA: Poison B, poison label, 25 kilograms (passenger), 95 kilograms (cargo).

ALDRIN, CAST SOLID
Countermeasures
Shipping Regulations: Section 11.
  ICC: Section 73.376(b).
  IATA: Other restricted articles, class A, no label required, no limit (passenger and cargo).

ALDRIN MIXTURES, DRY, WITH >65% ALDRIN
Shipping Regulations: Section 11.
  ICC: Poison B, poison label, 200 pounds.
  IATA: Poison B, poison label, 25 kilograms (passenger), 95 kilograms (cargo).

ALDRIN MIXTURES, DRY, WITH 65% ALDRIN OR LESS
Shipping Regulations: Section 11.
  ICC: Section 73.376(b).
  IATA: Other restricted articles, class A, no label required, no limit (passenger and cargo).

ALDRIN MIXTURES, LIQUID, WITH MORE THAN 60% ALDRIN
Shipping Regulations: Section 11.
  ICC: Poison B, poison label, 55 gallons.
  IATA: Poison B, poison label, 1 liter (passenger), 220 liters (cargo).

ALDRIN MIXTURES, LIQUID, WITH 60% OR LESS ALDRIN
Countermeasures
Shipping Regulations: Section 11.
  ICC: Section 73.361(b).
  IATA: Other restricted articles, class A, no label required, no limit (passenger and cargo).

ALFALFA MEAL
Hazard Analysis
Toxicology: A mild sensitizer, which when inhaled may cause asthma, running nose, sneezing, coughing and tearing eyes (Section 1). Contact with skin may cause contact dermatitis (Section 9).
Fire Hazard: Moderate, when exposed to heat or flame; by spontaneous chemical reaction (Section 6).
Spontaneous Heating: Yes.
Caution: Avoid moisture content extremes. Fires may smolder for 72 hours before becoming noticeable.
Countermeasures
Storage and Handling: Section 7.

ALIPHATIC AMINES. See fatty amines.

ALIZARIN
General Information
Synonyms: 1,2-Dihydroxyanthraquinone.
Description: Orange-red crystals.
Formula: $C_6H_4(CO)_2C_6H_2(OH)_2$.
Constants: Mol wt: 240.2, bp: 430°C (sublimable), mp: 289°C.
Hazard Analysis
Toxic Hazard Rating:
  Acute Local: Allergen 1.
  Acute Systemic: U.
  Chronic Local: Allergen 1.
  Chronic Systemic: U.
Fire Hazard: Slight, when exposed to heat or flame; can react with oxidizing materials (Section 6).
Countermeasures
Personal Hygiene: Section 3.
Storage and Handling: Section 7.

ALIZARIN 778
Hazard Analysis
Toxic Hazard Rating:
  Acute Local: Allergen 1.
  Acute Systemic: U.
  Chronic Local: U.
  Chronic Systemic: U.

Fire Hazard: Slight; (Section 6).
Countermeasures
Personal Hygiene: Section 3.
Storage and Handling: Section 7.

ALIZARIN RED 1034
Hazard Analysis
Toxic Hazard Rating:
Acute Local: Allergen 1.
Acute Systemic: U.
Chronic Local: Allergen 1.
Chronic Systemic: U.
Fire Hazard: Slight; (Section 6).
Countermeasures
Personal Hygiene: Section 3.
Storage and Handling: Section 7.

ALIZARIN SULFATE. See alizarin red 1034.

ALKALIES. (See also specific compounds).
General Information
Description: A term loosely applied to the hydroxides and
carbonates of the alkali metals, alkaline earth metals as
well as the bicarbonate and hydroxide of ammonium.
They can neutralize acids, change the color of indicators
and impart a soapy taste and feel to aqueous solutions.
Hazard Analysis
Toxicity: Variable.
Toxicology: The alkalies, as a group, constitute one of the
commonest causes of occupational dermatitis. They act
on the skin as primary irritants. Alkaline solutions
soften and dissolve the keratin layer, and the skin becomes white, soggy, wrinkled and macerated. Repeated
exposure frequently results in the development of
chronic eczematous skin conditions (Section 9). The
stronger caustics may produce chemical burns which
are often deep and slow in healing. Systemically, the
only alkali presenting any hazard is ammonia.

ALKALIES, N.O.S. See sodium hydroxide.

ALKALINE CAUSTIC LIQUID, N.O.S.
General Information
Synonym: Caustic alkali solution.
Hazard Analysis
Toxic Hazard Rating:
Acute Local: Irritant 3; Ingestion 3; Inhalation 2.
Acute Systemic: U.
Chronic Local: Irritant 2.
Chronic Systemic: U.
Countermeasures
Ventilation Control: Section 2.
Personnel Protection: Section 3.
Shipping Regulations: Section 11.
I.C.C.: Corrosive liquid; white label, 10 gallons.
Coast Guard Classification: Corrosive liquid; white label.
IATA: Corrosive liquid, white label, 1 liter (passenger),
20 liters (cargo).

ALKALINE CORROSIVE BATTERY FLUID. See alka-
line caustic liquid, N.O.S.

ALKALINE CORROSIVE LIQUID, N.O.S. See alkaline
caustic liquids, N.O.S.

ALKALOIDS. See alkaloid salts.

ALKALOID SALTS
General Information
Synonym: Alkaloids.

Hazard Analysis
Toxicology: Practically all of the alkaloid salts are poison-
ous. Some of them also cause allergic symptoms in hu-
mans, such as contact dermatitis or asthma. See specific
alkaloid salt (Section 9).
Disaster Hazard: Dangerous; when heated to decomposi-
tion they emit highly toxic fumes.
Countermeasures
Ventilation Control: Section 2.
Personnel Protection: Section 3.
First Aid: Section 1.
Storage and Handling: Section 7.
Shipping Regulations: Section 11.
IATA (liquid): Poison B, poison label, 1 liter (passenger),
220 liters (cargo).
IATA (solid): Poison B, poison label, 25 kilograms (pas-
senger), 95 kilograms (cargo).

ALKANESULFONIC ACID, MIXED
General Information
Description: Liquid.
Constants: Mp: −40° C, bp: 120° C, d: 1.38.
Hazard Analysis
Toxicity: Details unknown; probably moderately toxic; when
heated to decomposition, it emits highly toxic fumes of
oxides of sulfur.
Countermeasures
Storage and Handling: Section 7.
Shipping Regulations: Section 11.
IATA: Corrosive liquid, white label, 1 liter (passenger), 5
liters (cargo).

ALKARGEN. See cacodylic acid.

ALKRON. See parathion.

ALKYL ALUMINUM HALIDES. See pyrophoric liquids,
N.O.S.

ALKYL ARYL POLYETHYLENE GLYCOL ETHER
General Information
Description: Liquid.
Constants: Bp: 150° C, flash p: 440° F, d: 1.06.
Hazard Analysis
Toxicity: Unknown; see glycols.
Fire Hazard: Slight, when exposed to heat or flame; can
react with oxidizing materials.
Countermeasures
To Fight Fire: Water, foam, carbon dioxide, dry chemical or
carbon tetrachloride (Section 6).
Storage and Handling: Section 7.

ALKYL DIMETHYL BENZYL AMMONIUM
CHLORIDE
General Information
Synonym: BTC.
Description: Clear, mobile liquid.
Formula: $C_6H_5CH_2N(CH_3)_2RCl$.
Constants: Mol wt: 365, d: 0.9884 at 20° C (50%).
Hazard Analysis
Toxic Hazard Rating:
Acute Local: Irritant 1; Ingestion 1.
Acute Systemic: Ingestion 1; Inhalation 1.
Chronic Local: U.
Chronic Systemic: U.
Fire Hazard: Slight; (Section 6).
Countermeasures
Personal Hygiene: Section 3.
Storage and Handling: Section 7.

TOXIC HAZARD RATING CODE (For detailed discussion, see Section 1.)

0 NONE: (a) No harm under any conditions; (b) Harmful only under unusual conditions or overwhelming dosage.

1 SLIGHT: Causes readily reversible changes which disappear after end
of exposure.

2 MODERATE: May involve both irreversible and reversible changes;
not severe enough to cause death or permanent injury.

3 HIGH: May cause death or permanent injury after very short exposure
to small quantities.

U UNKNOWN: No information on humans considered valid by authors.

ALKYL DISULFIDES
General Information
Synonym: Methyl, ethyl, n-propyl, n-butyl, isobutyl and isoamyl disulfides.
Hazard Analysis
Toxicity: Limited information, based mainly on animal experiments, suggests that these compounds are dangerous and may cause hemolytic anemia. They may also produce allergic dermatitis.

ALKYL MERCURIACETATES. See mercury compounds, organic.

ALKYL PHENYL POLYETHYLENE GLYCOL ETHER
General Information
Description: Liquid.
Constants: Mp: $-5°$C, d: 1.0643 at $20°/20°$C, autoign. temp.: $590°$C.
Hazard Analysis
Toxic Hazard Rating:
Acute Local: Irritant 1.
Acute Systemic: Ingestion 2; Inhalation 1; Skin Absorption 1.
Chronic Local: Irritant 1.
Chronic Systemic: Ingestion 1; Inhalation 1; Skin Absorption 1.
Toxicology: Water solutions of less than 1% of this material have irritating properties comparable to soap. Before this material is used in foods or drug preparations a complete physiological evaluation of it in its intended use should be made. See also glycols.
Fire Hazard: Slight, when exposed to heat or flame; can react with oxidizing materials.
Countermeasures
To Fight Fire: Water, foam, carbon dioxide, dry chemical or carbon tetrachloride (Section 6).
Ventilation Control: Section 2.
Personnel Protection: Section 3.
Storage and Handling: Section 7.

ALLENE
General Information
Synonyms: Dimethylene methane; propadiene.
Description: Colorless, unstable, flammable gas.
Formula: $H_2C:C:CH_2$.
Constants: Mol wt: 40.06, d: 1.787, mp: $-146°$C, bp: $-32°$C.
Hazard Analysis
Toxic Hazard Rating: U.
Fire Hazard: Dangerous, when exposed to heat, flame or powerful oxidizers.
Explosion Hazard: Moderate, when exposed to flame.
Countermeasures
Ventilation Control: Section 2.
To Fight Fire: Section 6.
Storage and Handling: Section 7.
Shipping Regulations: Section 11.
IATA: Flammable gas, red label, not acceptable (passenger), 140 liters (cargo).

ALLETHRIN
General Information
Synonyms: dl-2-allyl-4-hydroxy-3-methyl-2-cyclopenten-1-one ester of cis and trans-dl-chrysanthemummonocarboxylic acid.
Description: Viscous liquid.
Formula: $C_{19}H_{26}O_3$.
Constant: Mol wt: 302.4.
Hazard Analysis
Toxic Hazard Rating:
Acute Local: Allergen 1.
Acute Systemic: Ingestion 1; Inhalation 1.
Chronic Local: Allergen 1.
Chronic Systemic: Ingestion 1; Inhalation 1.
Toxicology: An insecticide. It can cause liver and kidney

damage by all routes of entry into the body. Lung congestion may occur due to exposure. Local contact may cause contact dermatitis. Inhalation may cause asthma, coughing, wheezing, running nose and eyes (Section 9).
Fire Hazard: Slight; (Section 6).
Countermeasures
Personal Hygiene: Section 3.
Storage and Handling: Section 7.
Shipping Regulations: Section 11.
IATA: Other restricted articles, class A, no label required, no limit (passenger and cargo).

ALLICIN
General Information
Description: Colorless oily liquid; sharp garlic odor.
Formula: $C_6H_{10}OS_2$.
Constants: Mol wt: 162.3, d: 1.112 at $20°$C.
Hazard Analysis
Toxic Hazard Rating:
Acute Local: Irritant 1; Ingestion 1; Inhalation 1.
Acute Systemic: U.
Chronic Local: U.
Chronic Systemic: U.
Disaster Hazard: Slightly dangerous; when heated to decomposition it emits toxic fumes.
Countermeasures
Ventilation Control: Section 2.
Personal Hygiene: Section 3.
Storage and Handling: Section 7.

ALLOXANTIN
General Information
Synonym: Uroxin.
Description: Crystalline powder; on exposure to air turns red; yellow at $225°$C.
Formula: $C_8H_6N_4O_8 \cdot 2H_2O$.
Constant: Mol wt: 322.19.
Hazard Analysis
Toxic Hazard Rating:
Acute Local: U.
Acute Systemic: Ingestion 2.
Chronic Local: U.
Chronic Systemic: Ingestion 2. Causes disturbed carbohydrate metabolism leading to diabetes.
Disaster Hazard: Moderately dangerous; when heated to decomposition, it emits toxic fumes.
Countermeasures
Personal Hygiene: Section 3.
Storage and Handling: section 7.

ALLSPICE
General Information
Synonym: Pimenta.
Composition: Eugenol, etc.
Hazard Analysis
Toxicology: A weak sensitizer which may cause dermatitis on local contact (Section 9).
Fire Hazard: Slight; (Section 6).
Storage and Handling: Section 7.

ALLTOX. See toxaphene.

ALLYL ACETATE
General Information
Synonym: 2-Propenyl methanoate.
Description: Liquid.
Formula: $CH_3CO_2CH_2CHCH_2$.
Constants: Mol wt: 100.1, vap d: 3.45, d: 0.928, bp: $104°$C
Hazard Analysis
Toxic Hazard Rating:
Acute Local: Irritant 3, Ingestion 3; Inhalation 3.
Acute Systemic: Ingestion 2, Skin Absorption 2.
Chronic Local: U.
Chronic Systemic: U.
Toxicology: Based upon animal experiments.

Fire Hazard: Slight; can react with oxidizing materials (Section 6).
Countermeasures
Storage and Handling: Section 7.

ALLYL ACETONE
General Information
Synonym: 5-hexen-2-one.
Description: Colorless liquid.
Formula: $CH_2CHCHCH_2COCH_3$.
Constants: Mol wt: 98.1, bp: 129.5°C, d: 0.841 at 20°/20°C, vap d: 3.39.
Hazard Analysis
Toxicity: Details unknown. Allyl compounds are generally toxic.
Fire Hazard: Moderate, when exposed to heat or flame; can react with oxidizing materials.
Countermeasures
To Fight Fire: Foam, carbon dioxide, dry chemical or carbon tetrachloride (Section 6).
Storage and Handling: Section 7.

ALLYL ALCOHOL
General Information
Synonym: 2-propen-1-ol.
Description: Limpid liquid; pungent odor. Soluble in water, alcohol and ether.
Formula: CH_2CHCH_2OH.
Constants: Mol wt: 58.08, mp: −129°C, bp: 96–97°c, lel: 2.5%, uel: 18%, flash p: 70°F (C.C.), d: 0.854 at 20°/4°C, autoign temp: 713°F, vap press.: 10 mm at 10.5°C, vap d: 2.00.
TLV: ACGIH (accepted) 2 parts per million; 5 milligrams per cubic meter of air. Absorbed via skin.
Hazard Analysis
Toxic Hazard Rating:
Acute Local: Irritant 3; Inhalation 3.
Acute Systemic: Ingestion 2; Inhalation 2; Skin Absorption 2.
Chronic Local: Irritant 3; Inhalation 3.
Chronic Systemic: Ingestion 2; Inhalation 2; Skin Absorption 2.
Toxicology: Animal experiments have resulted in marked irritation of skin, eyes, and respiratory tract as well as damage to kidneys and liver. Human exposure causes irritation of eyes, mucous membranes and skin. Systemic poisoning is possible but has not been reported.
Fire Hazard: Dangerous, when exposed to heat or flame.
Spontaneous Heating: No.
Explosion Hazard : Moderate, when exposed to flame.
Disaster Hazard: Dangerous; when heated, it emits toxic fumes; can react vigorously with oxidizing materials.
Countermeasures
Ventilation Control: Section 2.
To Fight Fire: Carbon dioxide, alcohol foam, dry chemical or carbon tetrachloride (Section 6).
Personnel Protection: Section 7.
Storage and Handling: Section 7.
Shipping Regulations: Section 11.
I.C.C.: Poison B; poison label, 55 gallons.
Coast Guard Classification: Poison B; poison label.
MCA warning label.
IATA: Poson B, poison label, 1 liter (passenger), 220 liters (cargo).

ALLYL ALDEHYDE. See acrolein.

ALLYL AMINE
General Information
Synonym: 2-propenylamine.
Description: Colorless liquid.
Formula: $CH_2CHCH_2NH_2$.
Constants: Mol wt: 57.09, bp: 53.2°C, d: 0.761 at 20°/4°C, flash p: −20°F, autoign temp: 705°F, vap d: 2.00. lel: 2.2%, uel: 22%.
Hazard Analysis
Toxic Hazard Rating:
Acute Local: Irritant 1.
Acute Systemic: Ingestion 3; Inhalation 3; Skin Absorption 2.
Chronic Local: U.
Chronic Systemic: U.
Toxicology: In animal experiments, inhalation produced irritation of nose and mouth, congestion of the eyes, irregular respiration, cyanosis, excitement, convulsions, coma and death. Extraordinary precautions against fumes is advised.
Fire Hazard: Moderate, when exposed to heat or flame.
Disaster Hazard: Dangerous; when heated to decomposition, emits toxic fumes, can react with oxidizing materials.
Countermeasures
Ventilation Control: Section 2.
To Fight Fire: Alcohol foam, carbon dioxide, dry chemical or carbon tetrachloride (Section 6).
Personnel Protection: Section 3.
Storage and Handling: Section 7.

ALLYL AMINE, ANHYDROUS. See allyl amine.

ALLYL BROMIDE
General Information
Description: Colorless liquid.
Formula: CH_2CHCH_2Br.
Constants: Mol wt: 121.0, mp: −119.4°C, bp: 71.3°C, flash p: 30°F, d: 1.3980 at 20°/4°C, autoign. temp.: 563°F, vap. d.: 4.17, lel: 4.4%, uel: 7.3%.
Hazard Analysis
Toxicity Hazard Rating:
Acute Local: Irritant 2; Ingestion 2; Inhalation 2.
Acute Systemic: Ingestion 3; Inhalation 3; Skin Absorption 2.
Chronic Local: U.
Chronic Systemic: U.
Fire Hazard: Dangerous, when exposed to heat or flame; can react vigorously with oxidizing materials.
Disaster Hazard: Dangerous; see bromides.
Countermeasures
Ventilation Control: Section 2.
To Fight Fire: Water, alcohol foam, carbon dioxide, dry chemical or carbon tetrachloride (Section 6).
Personnel Protection: Section 3.
First Aid: Section 1.
Storage and Handling: Section 7.
Shipping Regulations: Section 11.
I.C.C.: Flammable liquid; red label, 10 gallons.
Coast Guard Classification: Inflammable liquid; red label.
IATA: Flammable liquid, red label, 1 liter (passenger), 40 liters (cargo).

ALLYL CHLORIDE
General Information
Synonym: 3-Chloropropene.
Description: Colorless liquid.

TOXIC HAZARD RATING CODE (For detailed discussion, see Section 1.)

0 NONE: (a) No harm under any conditions; (b) Harmful only under unusual conditions or overwhelming dosage.

1 SLIGHT: Causes readily reversible changes which disappear after end of exposure.

2 MODERATE: May involve both irreversible and reversible changes; not severe enough to cause death or permanent injury.

3 HIGH: May cause death or permanent injury after very short exposure to small quantities.

U UNKNOWN: No information on humans considered valid by authors.

Formula: CH_2CHCH_2Cl.

Constants: Mol wt: 76.53, mp: $-136.4°C$, bp: $44.6°C$, flash p: $-25°F$, lel: 3.3%, uel: 11.2%, d: 0.938 at $20°/4°C$, autoign temp: $737°F$, vap d: 2.64.

Hazard Analysis

Toxic Hazard Rating:

Acute Local: Irritant 3; Ingestion 3; Inhalation 3.

Acute Systemic: Ingestion 2; Inhalation 2; Skin Absorption 2.

Chronic Local: Irritant 2; Ingestion 2; Inhalation 2.

Chronic Systemic: Ingestion 2; Inhalation 2; Skin Absorption 2.

TLV: ACGIH (recommended) 1 part per million in air; 3 milligrams per cubic meter of air.

Toxicology: The vapors of allyl chloride are quite irritating to the eyes, nose and throat and contact of the liquid with the skin, in addition to local vasoconstriction and numbness, may lead to rapid absorption and distribution through the body. If remedial measures are not taken promptly, such contact may result in burns and internal injuries. Inhalation may cause headache, dizziness, and in high concentration, loss of consciousness; however, even in low concentration, its odor and irritating effects furnish warning of its presence in most cases. Concentrations of the vapors high enough to cause serious effects, including damage to the lungs, especially on repeated exposure, may not be intolerable. Consequently, the warning characteristics should never be disregarded. In general, precautions should be taken at all times to avoid spillage and accumulation of noticeable concentrations of the vapors in the atmosphere. Acute exposure in experimental animals has resulted in marked inflammation of lungs, irritation of skin and swelling of the kidneys. Chronically exposed animals have shown degenerative changes in the liver and kidneys. Reported human exposures have been principally cases of irritation of eyes, skin and respiratory tract, sometimes accompanied by aches and pains in the bones. Liver and kidney injury is possible although not reported in the literature.

Storage and Handling: Keep away from heat, radiators and sunlight. Dry allyl chloride can be stored in steel for long periods without appreciable corrosion. Experiments show that cast iron, 18-8 stainless steel, monel "Inconel," nickel, "Hastelloy AB," stoneware and red brass are corroded less than 0.020 inch per year at storage temperatures. Aluminum, however, is corroded more than 0.05 inch per year by dry allyl chloride under similar conditions. When allyl chloride contains more than a trace of water, corrosion of carbon steel becomes appreciable and localized attack is frequently observed. This is due to the hydrogen chloride evolved during hydrolysis. Keep the liquid off the skin and clothes. Where danger of spillage upon the body may exist, workers should wear rubber gloves, goggles, aprons, and/or other protective clothing to avoid the consequences of spilling or splashing allyl chloride. Antisplash face masks are also recommended. Gloves made of materials which can be wet through, such as untreated canvas, are useless. Maintain good ventilation. Work in a fume hood or with closed system if possible; otherwise, use enough ventilation so that the odor of allyl chloride does not persist. If it should be necessary to enter an area in which the odor of allyl chloride is at all noticeable, use a gas mask equipped with an "organic vapor" canister. Do not disregard the warning odor or eye irritation of allyl chloride. See also Section 7.

Fire Hazard: Highly dangerous, when exposed to heat or flame. (Section 6).

Explosion Hazard: Moderate, when exposed to flame; it can react vigorously with oxidizing materials.

Disaster Hazard: Dangerous; see chlorides (Section 7).

Countermeasures

Ventilation Control: Section 2.

To Fight Fire: Carbon dioxide, alcohol foam, dry chemical, carbon tetrachloride or water spray (Section 6).

Personnel Protection: Section 3.

First Aid: See "Treatment and Antidotes" above. (See also Section 1).

Shipping Regulations: Section 11.

IATA: Flammable liquid, red label, 1 liter (passenger), 40 liters (cargo).

MCA warning label.

ALLYL CHLOROCARBONATE

General Information

Synonym: Allyl chloroformate.

Description: Liquid.

Formula: CH_2CHCH_2OCOCl.

Constants: Mol wt: 120.5, bp: $106-114°C$, flash p: $88°F$ (C.C.), d: 1.14, vap d: 4.2.

Hazard Analysis

Toxic Hazard Rating:

Acute Local: Irritant 3; Ingestion 3; Inhalation 3.

Acute Systemic: U.

Chronic Local: U.

Chronic Systemic: U.

Fire Hazard: Moderate, when exposed to heat or flame. (Section 6).

Disaster Hazard: Dangerous; when heated to decomposition, emits highly toxic fumes of phosgene; it can react with oxidizing materials.

Countermeasures

Ventilation Control: Section 2.

To Fight Fire: Water, alcohol foam, carbon dioxide, dry chemical or carbon tetrachloride (Section 6).

Personnel Protection: Section 3.

Storage and Handling: Section 7.

Shipping Regulations: Section 11.

I.C.C.: Corrosive liquid; white label, 5 pints.

Coast Guard Classification: Corrosive liquid; white label.

IATA: Corrosive liquid, white label, not acceptable (passenger), $2\frac{1}{2}$ liters (cargo).

ALLYL CHLOROFORMATE. See allyl chlorocarbonate.

ALLYL CHOROPHENYL CARBONATE

Hazard Analysis

Toxicity: Details unknown; herbicide. Allyl compounds are generally toxic.

Disaster Hazard: Dangerous; when heated to decomposition, it emits highly toxic fumes of phosgene.

Countermeasures

Storage and Handling: Section 7.

ALLYL CYANIDE

General Information

Synonym: Vinylacetonitrile.

Description: Colorless liquid.

Formula: CH_2CHCH_2CN.

Constants: Mol wt: 67.1, bp: $116-119°C$, d: 0.8318 at $20°/4°C$.

Hazard Analysis

Toxicity: Highly toxic. See cyanides and nitriles.

Disaster Hazard: Dangerous; emits highly toxic fumes when heated to decomposition or on contact with acids or acid fumes.

Countermeasures

Personnel Protection: Section 3.

Storage and Handling: Section 7.

ALLYL DIGLYCOL CARBONATE

General Information

Description: Liquid.

Constants: Bp: $162°C$, flash p: $378°F$, d: 1.14.

Hazard Analysis

Toxicity: Details unknown. The allyl compounds are generally toxic. (Section 1,).

Fire Hazard: Slight, when exposed to heat or flame; can react with oxidizing materials.

Countermeasures
To Fight Fire: Water, foam, carbon dioxide, dry chemical or carbon tetrachloride (Section 6).
Storage and Handling: Section 7.

ALLYL DISULFIDE. See allyl sulfide.

ALLYLENE
General Information
Synonym: Propyne, methyl acetylene.
Description: Gas.
Formula: CH_3CCH.
Constants: Mol wt: 40.06, mp: $-104.7°C$, lel: 1.7%, bp: $-23.3°C$, vap press.: 3876 mm Hg at 20°C, d: 1.787 g/l at 0°C, vap d: 1.38.
Hazard Analysis
Toxic Hazard Rating:
  Acute Local: 0.
  Acute Systemic: Inhalation 2.
  Chronic Local: 0.
  Chronic Systemic: U.
Toxicology: This compound is a simple anesthetic and in high concentrations is an asphyxiant.
TLV: ACGIH (recommended) 1000 parts per million of air; 1650 milligrams per cubic meter of air.
Fire Hazard: Dangerous, when exposed to heat or flame. (Section 6).
Explosion Hazard: Moderate, when exposed to flame.
Disaster Hazard: Moderately dangerous; can react vigorously with oxidizing materials.
Countermeasures
Ventilation Control: Section 2.
To Fight Fire: Stop flow of gas. Carbon dioxide, dry chemical or water spray (Section 6).
Storage and Handling: Section 7.
Shipping Regulations: Section 11.
  IATA: Flammable gas, red label, not acceptable (passenger), 140 kilograms (cargo).

ALLYL 3,4-EPOXY-6-METHYL CYCLOHEXANE CARBOXYLATE
Hazard Analysis
Toxic Hazard Rating:
  Acute Local: Irritant 1.
  Acute Systemic: Ingestion 2; Inhalation 1; Skin Absorption 2.
  Chronic Local: U.
  Chronic Systemic: U.
Data based on limited animal experiments.

ALLYL 9,10-EPOXY STEARATE
Hazard Analysis
Toxic Hazard Rating:
  Acute Local: Irritant 1.
  Acute Systemic: Ingestion 2; Inhalation 1; Skin Absorption 1.
  Chronic Local: U.
  Chronic Systemic: U.
Based on limited animal experiments.

ALLYL ESTER THIOCYANIC ACID. See allyl sulfocyanide.

ALLYL ETHER
General Information
Synonym: Diallyl ether.
Description: Liquid.
Formula: $(CH_2CHCH_2)_2O$.

Constants: Mol wt: 98.1, bp: 94.3°C, d: 0.805, vap d: 3.38, flash p: 20°F (O.C.).
Hazard Analysis
Toxicity: Details unknown. Animal experiments show high toxicity, as is the case with most allyl compounds. (Section 1).
Fire Hazard: Moderately dangerous; it can react with oxidizing materials. (Section 6).
Countermeasures
Storage and Handling: Section 7.
To Fight Fire: Alcohol foam (Section 6).

ALLYL FLUORIDE
General Information
Synonym: 3-Fluoropropene.
Description: Colorless gas.
Formula: CH_2CHCH_2F.
Constants: Mol wt: 60.07, bp: $-10°C$.
Hazard Analysis
Toxic Hazard Rating:
  Acute Local: Irritant 3; Ingestion 3; Inhalation 3.
  Acute Systemic: Ingestion 3; Inhalation 3.
  Chronic Local: Ingestion 3.
  Chronic Systemic: Ingestion 2; Inhalation 2.
Disaster Hazard: Dangerous; when heated to decomposition, it emits highly toxic fumes of fluorides; will react with water or steam to produce toxic and corrosive fumes.
Countermeasures
Ventilation Control: Section 2.
Personnel Protection: Section 3.
First Aid: Section 1.
Storage and Handling: Section 7.

ALLYL FORMATE
General Information
Synonym: 2-Propenyl methanoate.
Description: Liquid, slightly water soluble. Soluble in organic solvents.
Formula: $HCOOCH_2CHCH_2$.
Constants: Mol wt: 86.1, d: 0.948 at 18/4°C, bp: 83°C.
Hazard Analysis
Toxicity: Details unknown. Highly toxic to experimental animals with damage to liver. (Section 1).

ALLYL GLYCERYL ETHER
General Information
Synonym: 1-Allyloxy-2,3-epoxypropane.
Formula: $CH_2CHCH_2OCH_2CHOCH_2$.
Constants: Mol wt: 114.14, bp: 153.9°C, fp: $-100°C$ (forms glass), flash p: 135°F (O.C.), d: 0.9698 at 20°/4°C, vap press.: 21.59 mm at 60°C, vap d: 3.94.
Hazard Analysis
Toxic Hazard Rating:
  Acute Local: Irritant 2; Ingestion 2; Inhalation 2.
  Acute Systemic: Inhalation 2.
  Chronic Local: U.
  Chronic Systemic: U.
Fire Hazard: Moderate, when exposed to heat or flame; can react with oxidizing materials.
Countermeasures
To Fight Fire: Foam, carbon dioxide, dry chemical or carbon tetrachloride (Section 6).
Ventilation Control: Section 2.
Personnel Protection: Section 3.
Storage and Handling: Section 7.

TOXIC HAZARD RATING CODE *(For detailed discussion, see Section 1.)*

0 NONE: (a) No harm under any conditions; (b) Harmful only under unusual conditions or overwhelming dosage.

1 SLIGHT: Causes readily reversible changes which disappear after end of exposure.

2 MODERATE: May involve both irreversible and reversible changes; not severe enough to cause death or permanent injury.

3 HIGH: May cause death or permanent injury after very short exposure to small quantities.

U UNKNOWN: No information on humans considered valid by authors.

ALLYL GLYCIDYL ETHER
General Information
Synonym: AGE.
Hazard Analysis
Toxic Hazard Rating:
Acute Local: Irritant 3.
Acute Systemic: Ingestion 1; Inhalation 2; Skin Absorption 1.
Chronic Local: Allergen 2.
Chronic Systemic: Inhalation 2.
Toxicology: Can cause CNS depression and pulmonory edema. Data based on animal and human experience. See also epoxy resins.
TLV: ACGIH (recommended) 10 parts per million of air; 45 milligrams per cubic meter of air.

ALLYL HOMOLOG OF CINERIN I. See divinyl pyrethrin I.

ALLYLIDENE DIACETATE
General Information
Description: Liquid.
Formula: $CH_2CHCH(OCOCH_3)_2$.
Constants: Mol wt: 158.15, mp: $-36.6°C$, bp: $107°C$ at 50 mm, flash p: $180°F$ (O.C.), d: 1.0749 at $20°/20°c$, vap d: 5.46.
Hazard Analysis
Toxicity: Unknown. Limited animal experiments suggest low toxicity and irritation.
Fire Hazard: Moderate, when exposed to heat or flame; can react with oxidizing materials.
Countermeasures
To Fight Fire: Water may be used to blanket the fire. Foam, carbon dioxide, dry chemical or carbon tetrachloride (Section 6).
Storage and Handling: Section 7.

ALLYL IODIDE
General Information
Synonym: 3-Iodo propene.
Description: Yellow liquid.
Formula: CH_2CHCH_2I.
Constants: Mol wt: 168.0, mp: $-99.3°C$, bp: $103.1°C$, d: 1.825 at $20°/4°C$, vap d: 5.8.
Hazard Analysis
Toxic Hazard Rating:
Acute Local: Irritant 3; Ingestion 3; Inhalation 3.
Acute Systemic: Ingestion 3; Inhalation 3; Skin Absorption 2.
Chronic Local: U.
Chronic Systemic: U.
Fire Hazard: Moderate, when exposed to heat or flame (Section 1).
Disaster Hazard: Dangerous; when heated to decomposition, it emits highly toxic fumes of iodine and iodides; can react with oxidizing materials.
Countermeasures
Ventilation Control: Section 2.
Personnel Protection: Section 3.
To Fight Fire: Water, foam, carbon dioxide, dry chemical or carbon tetrachloride (Section 6).
Storage and Handling: Section 7.

ALLYL ISOPROPYL ACETYL CARBAMIDE
Hazard Analysis
Toxicology: Details unknown. Reported as causing purpura due to depression of blood platelets (Section 1).

ALLYL ISOTHIOCYANATE
General Information
Synonym: Allyl mustard oil.
Description: Colorless to pale yellow liquid; irritating odor.
Formula: CH_2CHCH_2NCS.
Constants: Mol wt: 99.15, mp: $-80°C$, bp: $150.7°C$, flash p: $115°F$, d: 1.013-1.016 at $25°/25°C$, vap press: 10 mm at $38.3°C$, vap d: 3.41.

Hazard Analysis
Toxic Hazard Rating:
Acute Local: Irritant 2; Ingestion 2; Inhalation 2.
Acute Systemic: U.
Chronic Local: Irritant 1.
Chronic Systemic: U.
Toxicology: A fumigant. Mild sensitizer and allergen. Local contact may cause contact dermatitis. Inhalation may cause asthma, watery eyes and nose and sneezing (Section 9).
Fire Hazard: Slight, when exposed to heat or flame (Section 6).
Disaster Hazard: Dangerous; when heated to decomposition, or on contact with acid or acid fumes, it emits highly toxic fumes of cyanides; can react with oxidizing materials.
Countermeasures
Ventilation Control: Section 2.
To Fight Fire: Foam, carbon dioxide, dry chemical or carbon tetrachloride (Section 6).
Personnel Protection: Section 3.
Storage and Handling: Section 7.
Shipping Regulations: Section 11.
IATA: Poison A, not acceptable (passenger and cargo).

ALLYL MALEATE. See diallyl maleate.

4-ALLYL-2-METHOXY PHENOL. See eugenol.

ALLYL-3,4-METHYLENE DIOXYBENZENE. See safrole.

ALLYL MUSTARD OIL. See allyl isothiocyanate.

1-ALLYLOXY-2,3-EPOXYPROPANE. See allyl glyceryl ether.

ALLYL PROPENYL. See 1,4-hexadiene.

ALLYL PROPYL DISULFIDE
General Information
Synonym: onion oil.
Description: A liquid.
Formula: $C_3H_5S_2C_3H_7$.
Constant: Mol wt: 148.16.
Hazard Analysis
Toxic Hazard Rating:
Acute Local: Irritant 3; Ingestion 3; Inhalation 3.
Acute Systemic: Ingestion 3; Inhalation 3.
Chronic Local: U.
Chronic Systemic: U.
TLV: ACGIH (recommended) 2 parts per million of air; 12 milligrams per cubic meter of air.
Toxicology: This material is particularly irritating to the eyes and respiratory passages. (Section 1).
Fire Hazard: Moderate (Section 6).
Disaster Hazard: Dangerous; when heated to decomposition, it emits highly toxic fumes of oxides of sulfur; can react with oxidizing materials.
Countermeasures
Ventilation Control: Section 2.
Personnel Protection: Section 3.
To Fight Fire: Foam, carbon dioxide, dry chemical or carbon tetrachloride (Section 6).
Storage and Handling: Section 7.

ALLYL SUCCINIC ANHYDRIDE
Hazard Analysis
Toxic Hazard Rating: Limited Animal experiments suggest high toxicity and irritation.

ALLYL SULFIDE
General Information
Synonym: Allyl disulfide; diallyl disulfide; garlic oil.
Description: A colorless liquid; garlic odor.
Formula: $(CH_2CHCH_2)S_2$.

Constants: Mol wt: 114.20, mp: $-83°C$, bp: $139°C$, d: 0.888, vap d: 3.90.

Hazard Analysis

Toxic Hazard Rating:

Acute Local: Irritant 3; Ingestion 3; Inhalation 3.

Acute Systemic: Ingestion 3; Inhalation 3.

Chronic Local: U.

Chronic Systemic: U.

Disaster Hazard: Dangerous; see sulfur compounds (Section 7).

Countermeasures

Ventilation Control:Section 2.

Personnel Protection: Section 3.

Storage and Handling: Section 7.

ALLYL SULFOCARBAMIDE. See allyl thiourea.

ALLYL SULFOCYANIDE

General Information

Synonym: Allyl ester thiocyanic acid.

Description: An oil.

Formula: CH_2CHCH_2CNS.

Constants: Mol wt: 99.2, bp: $161°C$, d: 1.056 at $15°C$.

Hazard Analysis

Toxicity: Details unknown. The allyl compounds are generally toxic.

Disaster Hazard: Dangerous; see cyanides and sulfur compounds.

Countermeasures

Personal Hygiene: Section 3.

Storage and Handling: Section 7.

ALLYL THIOCARBAMIDE. See allyl thiourea.

ALLYL THIOUREA

General Information

Synonym: Allyl sulfocarbamide; allyl thiocarbamide.

Description: White crystalline solid; slight garlic odor; bitter taste.

Formula: $C_3H_5NHCSNH_2$.

Constants: Mol wt: 116.1, mp: $72-74°C$, d: 1.22.

Hazard Analysis

Toxic Hazard Rating:

Acute Local: U.

Acute Systemic: U.

Chronic Local: Allergen 3.

Chronic Systemic: U.

Toxicology: Contact eczema due to sensitization in humans has been reported.

Disaster Hazard: Dangerous; see sulfur compounds.

Countermeasures

Ventilation Control: Section 2.

Personnel Protection: Section 3.

Storage and Handling: Section 7.

ALLYL TRICHLORIDE. See trichloropropene.

ALLYL TRICHLOROSILANE

General Information

Description: Colorless liquid; pungent, irritating odor.

Formula: $CH_2CHCH_2SiCl_3$.

Constant: Mol wt: 175.5, bp: $117.5°C$, d: 1.217 ($27°C$), flash p: $95°F$ (C.O.C.).

Hazard Analysis

Toxic Hazard Rating: U. See also silanes.

Fire Hazard: Moderate.

Disaster Hazard: Dangerous. See chlorides.

Countermeasures

Storage and Handling: Section 7.

Shipping Regulations:

ICC: Corrosive liquid, white label, 10 gallons.

IATA: Corrosive liquid, white label, not acceptable (passenger), 40 liters (cargo).

ALLYL TRISULFIDE

General Information

Synonym: Diallyl trisulfide.

Description: Liquid.

Formula: $(C_3H_5)_2S_3$.

Constants: Mol wt: 178.3, bp: $140°C$, d: 1.085 at $15°C$.

Hazard Analysis

Toxicity: Details unknown. The allyl compounds are generally toxic.

Disaster Hazard: Dangerous; see sulfur compounds.

Countermeasures

Storage and Handling: Section 7.

ALLYL VINYL ETHER

General Information

Synonym: Vinyl allyl ether.

Description: Very slightly soluble in H_2O.

Formula: $CH_2:CHOCH_2CH_2O(CH_2)_3CH_3$.

Constants: Mol wt: 144, d: 0.8, bp: $153°F$, flash p: $<68°F$ (O.C.).

Hazard Analysis

Toxic Hazard Rating: U. Limited animal experiments suggest high toxicity and low irritation.

Fire Hazard: Dangerous. See ethers.

Countermeasures

To Fight Fire: Water may be ineffective (Section 6).

Storage and Handling: Section 7.

ALMOND OIL

General Information

Synonyms: Almond oil expressed; almond oil sweet.

Description: Fixed, non-drying oil; oily liquid.

Composition: Oleic, linoleic, myristic, palmitic acids.

Constant: D: 0.910-0.915 at $25°/25°C$.

Hazard Analysis

Toxic Hazard Rating:

Acute Local: Allergen 1.

Acute Systemic: 0.

Chronic Local: Allergen 1.

Chronic Systemic: 0.

Toxicology: A weak sensitizer. Contact dermatitis may result from local contact (Section 9).

Fire Hazard: Slight, when exposed to heat or flame (Secion 6).

Countermeasures

Personal Hygiene: Section 3.

Storage and Handling: Section 7.

ALMOND OIL, BITTER

General Information

Description: Colorless oil, which turns to yellow; bitter almond odor.

Compositions: Chief known constituents are benzaldehyde, hydrocyanic acid, benzaldehyde cyanhydrin.

Constants: Bp: $179°C$, d: 1.045-1.070 at $15°C$.

Hazard Analysis

Toxic Hazard Rating:

Acute Local: Irritant 1; Allergen 1.

Acute Systemic: U.

Chronic Local: Allergen 1.

Chronic Systemic: U.

Toxicology: This material can be quite toxic if it has not

TOXIC HAZARD RATING CODE (For detailed discussion, see Section 1.)

0 NONE: (a) No harm under any conditions; (b) Harmful only under unusual conditions or overwhelming dosage.

1 SLIGHT: Causes readily reversible changes which disappear after end of exposure.

2 MODERATE: May involve both irreversible and reversible changes; not severe enough to cause death or permanent injury.

3 HIGH: May cause death or permanent injury after very short exposure to small quantities.

U UNKNOWN: No information on humans considered valid by authors.

been separated from its hydrogen cyanide. Weak sensitizer; may cause a contact dermatitis (Section 9).
Fire Hazard: Slight, when exposed to heat or flame (Section 6).
Disaster Hazard: Dangerous. See cyanides.
Countermeasures
Personal Hygiene: Section 3.
Storage and Handling: Section 7.

ALMOND OIL EXPRESSED. See almond oil.

ALMOND OIL SWEET. See almond oil.

ALODAN. See 1,2,3,4,7,7-hexachloro-5,6-bis-(chloromethyl)-2-norbornene.

ALPEROX C. See lauroyl peroxide.

ALPHA RAYS
General Information
Description: Particulate radiation emitted by certain radioactive isotopes. Alpha rays consist of heavy charged particles (helium nuclei) moving at high velocity. See Section 5.
Hazard Analysis
Radiation Hazard: See individual isotopes.

ALPHASOL IB. See diisobutyl sodium sulfosuccinate.

ALROSEPT MBC. See 1-tridecyl-2-benzyl-2-hydroxyimidazolium chloride.

ALROSEPT MM. See 1-tridecyl-2-methyl-2-hydroxyethylimidazolium chloride.

ALTAITE. See lead telluride.

ALUM
General Information
Synonym: Potassium aluminum sulfate.
Description: Colorless crystals.
Formula: $KAl(SO_4)_2 \cdot 12H_2O$.
Constants: Mol wt: 474.39, mp: 92.5°C, d: 1.725.
Hazard Analysis
Toxic Hazard Rating:
  Acute Local: Irritant 1; Allergen 1; Ingestion 1; Inhalation 1.
  Acute Systemic: 0.
  Chronic Local: Irritant 1; Allergen 1.
  Chronic Systemic: 0.
Toxicology: A general purpose food additive, it may migrate to food from packaging materials (Section 10). A weak sensitizer. Local contact may cause contact dermatitis (Section 9).
Countermeasures
Ventilation Control: Section 2.
Personal Hygiene: Section 3.

ALUM, AMMONIA. See Aluminum Ammonium Sulfate.

ALUMINA
General Information
Synonym: Aluminum oxide.
Description: White powder.
Formula: Al_2O_3.
Constants: Mol wt: 101.94, mp: 2050°C, bp: 2977°C, d.: 3.5–4.0, vap. press.: 1 mm at 2158°C.
Hazard Analysis
Toxic Hazard Rating:
  Acute Local: Inhalation 1.
  Acute Systemic: 0.
  Chronic Local: Inhalation 2.
  Chronic Systemic: 0.
Toxicology: There has been some record of lung damage due to the inhalation of finely divided aluminum oxide particles. However, this effect, known as Shaver's disease, is complicated by the presence in the inhaled air of such things as silica and oxides of iron.

TLV: ACGIH (recommended); 50 million particles per cubic foot.
Countermeasures
Ventilation Control: Section 2.

ALUMINA TRIHYDRATE. See Aluminum Hydroxide.

ALUMINUM
General Information
Description: A silvery ductile metal.
Formula: Al.
Constants: At wt: 26.97, mp: 660°C, bp: 2056°C, d: 2.702, vap. press.: 1 mm at 1284°C.
Hazard Analysis
Toxic Hazard Rating:
  Acute Local: 0.
  Acute Systemic: 0.
  Chronic Local: Inhalation 1.
  Chronic Systemic: 0.
TLV: ACGIH (recommended); 50 million particles per cubic foot of air.
Toxicology: Aluminum is not generally regarded as an industrial poison. Inhalation of finely divided aluminum powder has been reported as a cause of pulmonary fibrosis.
Fire Hazard of Dust: Moderate, when exposed to heat or flame or by chemical reaction.
Spontaneous Heating: No.
Explosion Hazard of Dust: Moderate, when exposed to heat or flame.
Countermeasures
Ventilation Control: Section 2.
To Fight Fire: Special mixtures of dry chemical (Section 6).

ALUMINUM ACETATE
General Information
Description: Amorphous white powder.
Formula: $Al(C_2H_3O_2)_3$.
Constants: Mol wt: 204.1, mp: decomposes.
Hazard Analysis
Toxic Hazard Rating:
  Acute Local: Irritant 1.
  Acute Systemic: 0.
  Chronic Local: Irritant 1.
  Chronic Systemic: 0.
Toxicology: Weak sensitizer. Local contact may cause contact dermatitis (Section 9).
Countermeasures
Personal Hygiene: Section 3.

ALUMINUM ACETOACETONATE. See aluminum acetate.

ALUMINUM ALKYLS. See pyrophoric liquids.

ALUMINUM AMMONIUM SULFATE
General Information
Synonym: Alum, ammonia; ammonium alum.
Description: Colorless crystals; odorless; soluble in H_2O, glycerine; insoluble in alcohol.
Formula: $Al_2(SO_4)_3(NH_4)_2SO_4 \cdot 24 H_2O$.
Constants: Mol wt: 906, d: 1,645, mp: 94.5°C, bp: loses $20H_2O$ at 120°C.
Hazard Analysis
Toxic Hazard Rating:
  Acute Local: Irritant 1; Ingestion 2.
  Acute Systemic: U.
  Chronic Local: Irritant 1; Inhalation 1.
  Chronic Systemic: U.
Toxicology: A mild astringent used as a general purpose food additive.
Disaster Hazard: Dangerous. See Sulfates.

ALUMINUM o-ARSENATE
General Information
Description: White powder.

Formula: $AlAsO_4 \cdot 8H_2O$.
Constants: Mol wt: 310, d: 3.011.
Hazard Analysis and Countermeasures
See arsenic compounds.

ALUMINUM ARSENIDE
General Information
Description: A solid.
Formula: AlAs.
Constant: Mol wt: 101.9.
Hazard Analysis and Countermeasures
See arsenic compounds and arsine.

ALUMINUM BENZOATE
General Information
Description: Crystalline powder; very slightly soluble in H_2O.
Formula: $Al(C_7H_5O_2)_3$.
Constant: Mol Wt: 390.3.
Hazard Analysis
Toxic Hazard Rating: U. Limited data indicate low toxicity. See also aluminum.

ALUMINUM BORIDE
General Information
Description: Powder.
Hazard Analysis
See Aluminum, also boron compounds.

ALUMINUM BOROFORMATE
General Information
Description: White lustrous scales; freely soluble in H_2O, and alcohol.
Composition: ~33% Al_2O_3, 20% H_3BO_3, 15% formic acid, 32% H_2O.
Hazard Analysis
See aluminum, also boron compounds.

ALUMINUM BOROHYDRIDE
General Information
Description: Liquid.
Formula: AlB_3H_{12}.
Constants: Mol wt: 71.53, bp: 44.5°C, mp: −64.5°C, vap. press. 400 mm at 28.1°C.
Hazard Analysis
Toxicity: Details unknown. See also hydrides and boron compounds (Section 1).
Fire Hazard: Dangerous, by spontaneous chemical reaction; ignites spontaneously in air (Section 6).
Explosion Hazard: Moderate; by spontaneous chemical reaction.
Caution: Evolves hydrogen on contact with water.
Disaster Hazard: Moderately dangerous; will react with water or steam to produce heat, hydrogen or toxic fumes; can react vigorously with oxidizing materials; can emit toxic fumes on contact with acid or acid fumes.
Countermeasures
Personal Hygiene: Section 3.
Storage and Handling: Section 7.
To Fight Fire: Carbon dioxide, dry chemical or carbon tetrachloride (Section 6).

ALUMINUM BOROTANNATE
General Information
Synonym: cutal.
Description: Light brown powder; insoluble in H_2O; soluble in dilute tartaric acid solution.
Hazard Analysis and Countermeasures
See aluminum, boron and tannic acid.

ALUMINUM BROMATE
General Information
Description: Crystals.
Formula: $Al(BrO_3)_3 \cdot 9H_2O$.
Constants: Mol wt: 572.9, mp: 62.3°C, bp: decomposes.
Hazard Analysis
Toxicity: Details unknown. See also bromates.
Fire Hazard: See bromates (Section 6).
Disaster Hazard: See bromates.
Countermeasures
Storage and Handling: Section 7.

ALUMINUM BROMIDE
General Information
Description: White to yellowish-red lumps.
Formula: $AlBr_3$.
Constants: Mol wt: 226.73, mp: 97.5°C, bp: 263.3°C at 748 mm, d: 3.0, vap. press.: 1 mm at 81.3°C.
Hazard Analysis
Toxicity: See bromides.
Disaster Hazard: See bromides.
Countermeasures
Storage and Handling: Section 7.
Shipping Regulations: Section 11.
    IATA (anhydrous): Other restricted articles, class B, no label required, not acceptable (passenger), 12 kilograms (cargo).

ALUMINUM CALCIUM SILICATE
Hazard Analysis
Toxic Hazard Rating: U.
An anti-caking agent food additive (Section 10).

ALUMINUM CARBIDE
General Information
Description: Greenish-gray, pulverized mass.
Formula: Al_4C_3.
Constants: Mol wt: 143.91, mp: stable to 1400°C, bp: decomposes at high temp., d: 2.36.
Hazard Analysis
Toxic Hazard Rating: U. Dust can cause pulmonary irritation.
See acetylides.
Countermeasures
Shipping Regulations: Section 11.
    IATA: Flammable solid, yellow label, 12 kilograms (passenger), 45 kilograms (cargo).

ALUMINUM CHLORATE
General Information
Description: Colorless, deliquescent crystals.
Formula: $Al(ClO_3)_3 \cdot 6H_2O$.
Constants: Mol wt: 385.4, mp: decomposes.
Hazard Analysis
Toxicity: Details unknown. See chlorates.
Fire Hazard: Moderate; by spontaneous chemical reaction; a powerful oxidizer; may ignite upon contact with combustibles (section 6).
Explosion Hazard: Moderate, when shocked, exposed to heat or by spontaneous chemical reaction with reducing agents.
Caution: When contaminated may become sensitized.
Disaster Hazard: Dangerous; shock or heat will explode it; See Chlorides.
Countermeasures
Storage and Handling: Section 7.

TOXIC HAZARD RATING CODE (For detailed discussion, see Section 1.)

0 NONE: (a) No harm under any conditions; (b) Harmful only under unusual conditions or overwhelming dosage.

1 SLIGHT: Causes readily reversible changes which disappear after end of exposure.

2 MODERATE: May involve both irreversible and reversible changes; not severe enough to cause death or permanent injury.

3 HIGH: May cause death or permanent injury after very short exposure to small quantities.

U UNKNOWN: No information on humans considered valid by authors.

ALUMINUM CHLORIDE
General Information
Description: Yellowish-white granular crystals.
Formula: $AlCl_3$.
Constants: Mol wt: 133.35, mp: 192.4°C, bp: 180.2°C (sublimes), d: 2.44, vap. press.: 1 mm at 100.0°C.
Hazard Analysis
Toxic Hazard Rating:
Acute Local: Irritant 3; Ingestion 3; Inhalation 3.
Acute Systemic: U.
Chronic Local: U.
Chronic Systemic: U.
Toxicity: See chlorides and hydrochloric acid.
Disaster Hazard: Dangerous; See hydrochloric acid; will react with water or steam to produce heat, toxic or corrosive fumes.
Countermeasures
Storage and Handling: Section 7.
Shipping Regulations: Section 11.
IATA (Anhydrous): Other restricted articles,,class B, no label required, not acceptable (passenger), 12 kilograms (cargo).

ALUMINUM COMPOUNDS
Most aluminum compounds have little or no toxicity. Exceptions to this rule are found in such substances as aluminum arsenide or aluminum silicofluoride in which the toxic properties are due to the radical which is joined to the aluminum atom. See also aluminum and alumina.

ALUMINUM CHLOROHYDROXIDE COMPLEX. See chlorhydrol.

ALUMINUM DIBORIDE
General Information
Description: Hexagonal crystals.
Formula: AlB_2.
Constants: Mol wt: 48.6, d: 3.19
Hazard Analysis
Toxicity: Details unknown. See also boron compounds.
Fire Hazard: See boron hydrides.
Explosion Hazard: See boron hydrides.
Disaster Hazard: Dangerous; will react with water or steam acid or acid fumes, it can emit toxic fumes.
Countermeasures
Storage and Handling: Section 7.

ALUMINUM DICHROMATE
General Information
Description: Solid.
Formula: $Al_2(Cr_2O_7)_3$.
Constant: Mol wt: 702
Hazard Analysis
Toxicity: See chromium compounds.
Fire Hazard: See chromates.
Disaster Hazard: Slight; reacts with reducing materials.
Countermeasures
Storage and Handling: Section 7.

ALUMINUM DIETHYL MONOCHLORIDE. See Diethyl aluminum chloride.

ALUMINUM DROSS. See aluminum.
Countermeasures
Shipping Regulations: Section 11.
I.C.C.: Section 73.173.
Coast Guard Classification: Inflammable solid.
IATA (dry): Not restricted (passenger and cargo).
(wet): Flammable solid, not acceptable (passenger and cargo).

ALUMINUM DUST
General Information
Synonym: Aluminum powder.
Description: Silvery metallic powder.
Formula: Al.
Constants: Mol wt: 26.97, mp: 660°C, d: 2.708.

Hazard Analysis
Toxic Hazard Rating:
Acute Local: Irritant 1; Inhalation 1.
Acute Systemic: 0.
Chronic Local: Inhalation 1.
Chronic Systemic: 0.
Toxicology: See aluminum.
Fire Hazard: Moderate, when exposed to heat or flame or by chemical reaction with oxidizing agents. (Section 6).
Explosion Hazard: Moderate, when exposed to flame.
Countermeasures
Ventilation Control: Section 2.
To Fight Fire: Special mixtures of dry chemicals (Section 6).
Personal Hygiene: Section 3.

ALUMINUM ETHOXIDE. See aluminum ethylate.

ALUMINUM ETHYLATE
General Information
Synonym: Aluminum ethoxide.
Description: Liquid; decomposed by H_2O.
Formula: $Al(OC_2H_5)_3$.
Constants: Mol wt: 162.15, bp: 200°C (6–8 mm) 175–180°C (3 mm), mp: 140°C.
Hazard Analysis
Toxic Hazard Rating:
Acute Local: Irritant 3; Inhalation 3.
Acute Systemic: U.
Chronic Local: Irritant 3; Inhalation 3.
Chronic Systemic: U.
See also organo metals.

ALUMINUM FERROCYANIDE
General Information
Description: Brown powder.
Formula: $Al_4[Fe(CN)_6]_3 \cdot 17H_2O$.
Hazard Analysis
Toxicity: See ferrocyanides.
Disaster Hazard: Dangerous, see ferrocyanides.
Countermeasures
Storage and Handling: Section 7.

ALUMINUM FERROSILICON
Shipping Regulations: Section 11.
IATA: Flammable solid, yellow label, 12 kilograms (passenger), 45 kilograms (cargo).

ALUMINUM FLUORIDE
General Information
Description: Colorless crystal.
Formula: AlF_3.
Constants: Mol wt: 84, mp: 1040°C, d: 3.07, vap. press: 1 mm at 1238°C, bp: 1537°C.
Hazard Analysis and Countermeasures
See fluorides.

ALUMINUM FLUOSILICATE
General Information
Synonym: Aluminum silicofluoride.
Description: White powder; slowly soluble in cold H_2O readily soluble in hot H_2O.
Formula: $Al_2(SiF_6)_3$.
Constant: Mol wt: 480.23.
Hazard Analysis and Countermeasures
See fluorides.

ALUMINUM FORMATE
General Information
Description: White powder.
Formula: $Al(CHO_2)_3$.
Constant: Mol wt: 162.0.
Hazard Analysis
Toxicity: Details unknown. See also formic acid. (Section 1).
Disaster Hazard: Slight; when heated it emits acrid fumes (Section 6).
Countermeasures
Storage and Handling: Section 7.

ALUMINUM FORMOACETATE
General Information
Description: White powder.
Formula: $Al(OH)(OOCH)(OOCCH_3)$.
Constant: Mol wt: 148.1.
Hazard Analysis
Toxicity: Slight.
Disaster Hazard: Slight.
Countermeasures
Storage and Handling: Section 7.

ALUMINUM HYDRATE. See aluminum hydroxide.

ALUMINUM HYDRIDE
General Information
Description: Powder.
Formula: AlH_3.
Hazard Analysis
Toxicity: Hydrides of some metals such as AsH_3 are extremely toxic; little is known about AlH_3. See also hydrides. (Section 1).
Fire Hazard: Moderate, by chemical reaction with water or oxidizing agents (Section 6).
Caution: It evolves hydrogen upon contact with moisture.
Explosion Hazard: Severe, by chemical reaction wherein hydrogen gas is produced.
Disaster Hazard: Moderately dangerous; will react with water or steam to produce heat and hydrogen; reacts with oxidizing materials; on contact with acid or acid fumes, it can emit toxic fumes.
Countermeasures
Storage and Handling: Section 7.
Shipping Regulations: Section 11.
  IATA: Flammable solid, yellow label, not acceptable (passenger), 12 kilograms (cargo).

ALUMINUM HYDROXIDE
General Information
Synonym: Alumina trihydrate; aluminum hydrate; hydrated alumina; hydrated aluminum oxide.
Description: White crystalline powder, balls, or granules; insoluble in H_2O; soluble in mineral acids and caustic soda.
Formula: $Al_2O_3 \cdot 3H_2O$ or $Al(OH)_3$.
Constants: Mol wt: (1) 156, (2) 78, d:2.42.
Hazard Analysis
Toxic Hazard Rating: U.
A substance migrating to food from packaging materials (Section 10).

ALUMINUM IODIDE
General Information
Description: White to brown plates.
Formula: AlI_3
Constants: Mol wt: 407.7, mp: 191°C, bp: 385.5°C, d: 3.98 at 25°C, vap. press.: 1 mm at 178.0°C (sublimes).
Hazard Analysis
Toxicity: See iodides.
Disaster Hazard: Dangerous; see iodides.
Countermeasures
Storage and Handling: Section 7.

ALUMINUM LACTATE
General Information
Description: White, yellowish powder.
Formula: $Al(C_3H_5O_3)_3$.
Constant: Mol wt: 294.2.

Hazard Analysis
Toxicity: See aluminum compounds.
Fire Hazard: Slight; (Section 6).
Countermeasures
Storage and Handling: Section 7.

ALUMINUM LIQUID (or PAINT)
General Information
Description: Silvery liquid.
Formula: Al in volatile vehicle.
Constant: Flash point 100°F or lower.
Hazard Analysis
Toxicity: See aluminum and vehicle involved.
Fire Hazard: Dangerous, when exposed to heat or flame.
Explosion Hazard: Moderate.
Disaster Hazard: Moderately dangerous; when heated to decomposition emits toxic fumes; can react vigorously with oxidizing materials.
Countermeasures
Storage and Handling: Section 7.
To Fight Fire: Water, foam, carbon dioxide, dry chemical or carbon tetrachloride (Section 6).
Shipping Regulations: Section 11.
  I.C.C.: Flammable liquid; red label, 55 gallons.
  Coast Guard Classification: Inflammable liquid; red label.
  IATA: Flammable liquid, red label, 1 liter (passenger), 220 liters (cargo).

ALUMINUM, METALLIC, POWDER
Countermeasures
Shipping Regulations: Section 11.
  IATA: Flammable solid, yellow label, 12 kilograms (passenger), 45 kilograms (cargo).

ALUMINUM METHYL
General Information
Synonym: Trimethyl aluminum.
Description: Colorless liquid.
Formula: $Al(CH_3)_3$.
Constants: Mol wt: 72.07, bp: 130°C, mp: 0°C.
Hazard Analysis
Toxicity: Related alkyl aluminum compounds show strong irritant properties and high toxicity.
Fire Hazard: Dangerous, by spontaneous chemical reaction with air and reacts violently with water (Section 6).
Explosion Hazard: Moderate, by chemical reaction with air; explodes on contact with water.
Disaster Hazard: Dangerous; when heated to decomposition, it emits toxic fumes, will explode on contact with moisture; react vigorously with oxidizing materials.
Countermeasures
To Fight Fire: Do not use water, foam, or halogenated extinguishing agents (Section 6).
Storage and Handling: Section 7.
  ICC: Flammable liquid, red label, 2 ounces.

ALUMINUM MONOPALMITATE. See Aluminum palmitate.

ALUMINUM NICOTINATE
Hazard Analysis
Toxic Hazard Rating: U.
A food additive permitted in food for human consumption Section 10).

ALUMINUM NITRATE
General Information
Description: White crystals.
Formula: $Al(NO_3)_3 \cdot 9H_2O$.

TOXIC HAZARD RATING CODE *(For detailed discussion, see Section 1.)*

0 NONE: (a) No harm under any conditions; (b) Harmful only under unusual conditions or overwhelming dosage.

1 SLIGHT: Causes readily reversible changes which disappear after end of exposure.

2 MODERATE: May involve both irreversible and reversible changes; not severe enough to cause death or permanent injury.

3 HIGH: May cause death or permanent injury after very short exposure to small quantities.

U UNKNOWN: No information on humans considered valid by authors.

Constants: Mol wt: 375.14, bp: decomposes at 150°C, mp: 70°C.

Hazard Analysis

Toxicity: See nitrates.

Fire Hazard: See nitrates.

Disaster Hazard: See nitrates.

Countermeasures

Storage and Handling: Section 7.

Shipping Regulations: Section 11.

   I.C.C.: Oxidizing material; yellow label, 100 pounds.

   Coast Guard Classification: Oxidizing material; yellow label.

   IATA: Oxidizing material, yellow label, 12 kilograms (passenger), 45 kilograms (cargo).

ALUMINUM NITRIDE
General Information

Description: White or colorless crystal.

Formula: AlN.

Constants: Mol wt: 41, mp: > 2200°C, bp: sublimes at 2000°C, d: 3.26.

Hazard Analysis

Toxicity: See nitrides and ammonia.

Disaster Hazard: Slight; will react with water or steam to produce toxic or corrosive fumes.

Countermeasures

Storage and Handling: Section 7.

ALUMINUM OLEATE
General Information

Description: Yellowish white viscous mass insoluble in H_2O; soluble in alcohol, benzene, ether, oil, and turpentine.

Formula: $Al(C_{18}H_{33}O_2)_3$.

Constant: Mol wt: 870.

Hazard Analysis

Toxic Hazard Rating: U. A substance migrating to food from packaging materials (Section 10).

ALUMINUM OXIDE. See alumina.

ALUMINUM OXALATE
General Information

Description: White powder.

Formula: $Al_2(C_2O_4)_3 \cdot 4H_2O$.

Constant: Mol wt: 390.1.

Hazard Analysis

Toxicity: Highly toxic. See oxalates.

ALUMINUM PAINT. See aluminum liquid or paint.

ALUMINUM PALMITATE
General Information

Synonym: Aluminum monopalmitate.

Description: White powder; insoluble in alcohol and acetone.

Formula: $Al(OH)_2(C_{16}H_{31}O_2)$.

Constant: Mol wt: 356.

Hazard Analysis

Toxic Hazard Rating: U. a substance migrating to food from packaging materials. See Section 10.

ALUMINUM 1-PHENOL-4-SULFONATE
General Information

Description: Reddish-white powder.

Formula: $Al(C_6H_5O_4S)_3$.

Constant: Mol wt: 546.5.

Hazard Analysis

Toxic Hazard Rating:

   Acute Local: Irritant 2; Allergen 1; Ingestion 2; Inhalation 1.

   Acute Systemic: U.

   Chronic Local: Allergen 1.

   Chronic Systemic: U.

Disaster Hazard: Dangerous; see phenol and sulfur compounds.

Countermeasures

Ventilation Control: Section 2.

Personnel Protection: Section 3.

Storage and Handling: Section 7.

ALUMINUM PHENOXIDE
General Information

Description: Gray-white powder or crystalline mass.

Formula: $Al(C_6H_5O)_3$.

Constants: Mol wt: 306.3, mp: 265°C (decomp.), d: 1.23.

Hazard Analysis

Toxic Hazard Rating:

   Acute Local: Irritant 2; Allergen 1; Ingestion 2; Inhalation 2.

   Acute Systemic: U.

   Chronic Local: Irritant 2; Allergen 1.

   Chronic Systemic: U.

Disaster Hazard: Dangerous; see phenol.

Countermeasures

Ventilation Control: Section 2.

Personnel Protection: Section 3.

Storage and Handling: Section 7.

ALUMINUM PHOSPHIDE
General Information

Description: Dark gray or dark yellow crystals.

Formula: AlP.

Constants: Mol wt: 57.96, d: 2.85 (25°/4°C).

Hazard Analysis

Toxicity: An insecticide and a fumigant; releases phosphine; see phosphine.

   A food additive permitted in the food and drinking water of animals and/or for the treatment of food producing animals. (Section 10).

Disaster Hazard: Dangerous; See Phosphides.

Countermeasures

Shipping Regulations: Section 11.

   IATA: Flammable; solid yellow label, 12 kilograms (passengers), 12 kilograms (cargo).

ALUMINUM PICRATE
General Information

Description: A solid.

Formula: $Al[(NO_2)_3C_6H_2O]_3$.

Constant: Mol wt: 771.3.

Hazard Analysis

Toxic Hazard Rating:

   Acute Local: Allergen 2; Ingestion 3; Inhalation 3.

   Acute Systemic: Ingestion 3; Inhalation 3.

   Chronic Local: Irritant 2; Allergen 2.

   Chronic Systemic: Ingestion 3; Inhalation 3.

Fire Hazard: Dangerous, by chemical reaction with reducing materials; a powerful oxidizer (Section 6).

Explosion Hazard: Severe, when shocked or exposed to heat. See also explosives, high.

Disaster Hazard: Highly dangerous; shock will explode it; when heated to decomposition, it emits highly toxic fumes of oxides of nitrogen and explodes; can react vigorously with reducing materials.

Countermeasures

Ventilation Control: Section 2.

Personnel Protection: Section 3.

Storage and Handling: Section 7.

ALUMINUM POWDER. See aluminum dust.

ALUMINUM PROPOXIDE
General Information

Description: White powder.

Formula: $Al(C_3H_7O)_3$.

Constants: Mol wt: 204.2, mp: 106°C, bp: 248°C, d: 1.058 at 20°/0°C.

Hazard Analysis

Toxicity: Details unknown. See also aluminum compounds.

Fire Hazard: Slight; a combustible material (Section 6).

Countermeasures

Storage and Handling: Section 7.

ALUMINUM RESINATE
General Information

Description: Brown mass.

Formula: $Al(C_{44}H_{63}O_5)_3$.
Constant: Mol wt: 2042.8.
Hazard Analysis
Toxicity: Unknown. Possibly an irritant.
Fire Hazard: Slight, when exposed to heat or flame (Section 6).
Countermeasures
Storage and Handling: Section 7.
Shipping Regulations: Section 11.
  IATA: Flammable solid, yellow label, 12 kilograms (passenger), 60 kilograms (cargo).

ALUMINUM RICINOLEATE
General Information
Description: Plastic solid, yellow-brown to dark brown.
Formula: $Al[CO_2(CH_2)_7CHCHCH_2CHOH(CH_2)_5-CH_3]_3$
Constants: Mol wt: 912, mp: 95°C.
Hazard Analysis
Toxicity: See aluminum compounds and castor oil.
Fire Hazard: Slight; (Section 6).
Countermeasures
Storage and Handling: Section 7.

ALUMINUM SILICON
Countermeasures
Shipping Regulations: Section 11.
  IATA (in ingots): Not restricted (passenger and cargo).
  IATA (powder): Flammable solid, yellow label, 12 kilograms (passenger and cargo).

ALUMINUM SODIUM SULFATE
General Information
Synonym: Sodium aluminum sulfate.
Description: Colorless crystals.
Formula: $NaAl(SO_4)_2 \cdot 12H_2O$.
Constants: Mol wt: 458.29, mp: 61°C, d: 1.675.
Hazard Analysis
Toxic Hazard Rating:
  Acute Local: Irritant 1; Allergen 1; Ingestion 1; Inhalation 1.
  Acute Systemic: 0.
  Chronic Local: Irritant 1; Allergen 1.
  Chronic Systemic: 0.
Toxicology: A weak sensitizer. A general purpose food additive.
  Local contact may cause contact dermatitis (Section 9).
Personal Hygiene: Section 3.

ALUMINUM SULFATE, DRY
General Information
Synonym: Cake alum.
Description: White powder.
Formula: $Al_2(SO_4)_3$.
Constants: Mol wt: 342.1, mp: decomposes at 770°C, d: 2.71.
Hazard Analysis
Toxicity: This material hydrolyzes readily to form some sulfuric acid which acts as a tissue irritant, particularly to the lungs. See sulfuric acid.
A general purpose food additive.

ALUMINUM SULFIDE
General Information
Description: Yellowish-gray lumps; odor of hydrogen sulfide.
Formula: Al_2S_3.
Constants: Mol wt: 150.12, mp: 1100°C, bp: subl. at 1550°C, d: 2.02.
Hazard Analysis
Toxicity: Highly toxic. See sulfides.

Fire Hazard: See sulfides.
Explosion Hazard: See sulfides.
Disaster Hazard: See sulfides.
Countermeasures
Storage and Handling: Section 7.

ALUMINUM SULFOCARBOLATE. See aluminum 1-phenol-4-sulfonate.

ALUMINUM TARTRATE
General Information
Description: White, odorless granules; slowly soluble in cold H_2O; readily soluble in hot H_2O; soluble in ammonia.
Formula: $Al_2(C_4H_4O_6)_3$.
Constant: Mol wt: 498.16.
Hazard Analysis
See aluminum and tartaric acid.

ALUMINUM THALLIUM SULFATE
General Information
Description: Cubic, octagonal, colorless crystals.
Formula: $AlTl(SO_4)_2 \cdot 12H_2O$.
Constants: Mol Wt: 639.7, mp: 91°C, d: 2.32.
Hazard Analysis
Toxicity: Highly toxic. See thallium compounds.
Disaster Hazard: See thallium compounds and sulfates.

ALUMINUM TRIETHYL. See triethyl aluminum.

ALUMINUM TRIMETHYL. See aluminum methyl.

ALUMINUM TRIPROPYL
General Information
Synonym: Tripropyl aluminum.
Description: Liquid.
Formula: $Al(C_3H_7)_3$.
Constant: Mol wt: 156.24.
Hazard Analysis
Toxic Hazard Rating: U. Related alkyl aluminum compounds are highly toxic. See also diisobutyl aluminum chloride.
Fire Hazard: Dangerous, by spontaneous chemical reaction with air; can react vigorously with oxidizing materials (Section 6).
Explosion Hazard: It hydrolyzes to evolve flammable vapor.
Countermeasures
Do not use water, foam, or halogenated extinguishing agents (Section 6).
Storage and Handling: Section 7.
Personal Hygiene: Section 3.

"ALUNDUM." See alumina.

"ALVAR"
General Information
Description: A solid polyvinyl acetal resin.
Hazard Analysis
Toxic Hazard Rating:
  Acute Local: Allergen 1.
  Acute Systemic: U.
  Chronic Local: Allergen 1.
  Chronic Systemic: U.
Toxicology: Local contact may cause a contact dermatitis (Section 9).
Fire Hazard: Slight, when exposed to heat or flame; can react with oxidizing materials (Section 6).
Countermeasures
Personal Hygiene: Section 3.
Storage and Handling: Section 7.

TOXIC HAZARD RATING CODE (For detailed discussion, see Section 1.)

0 NONE: (a) No harm under any conditions; (b) Harmful only under unusual conditions or overwhelming dosage.

1 SLIGHT: Causes readily reversible changes which disappear after end of exposure.

2 MODERATE: May involve both irreversible and reversible changes; not severe enough to cause death or permanent injury.

3 HIGH: May cause death or permanent injury after very short exposure to small quantities.

U UNKNOWN: No information on humans considered valid by authors.

ALYPIN
General Information
Description: White crystalline powder.
Formula: $C_{16}H_{26}O_2N_2$.
Constant: Mol wt: 278.4.
Hazard Analysis
Toxic Hazard Rating:
Acute Local: Allergen 1; Ingestion 1; Inhalation 1.
Acute Systemic: U.
Chronic Local: Allergen 1.
Chronic Systemic: U.
Caution: Poisonings have occurred from small doses but this is very unusual.
Disaster Hazard: Moderately dangerous; when heated to decomposition, it emits toxic fumes.
Countermeasures
Ventilation Control: Section 2.
Personal Hygiene: Section 3.
Storage and Handling: Section 7.

AMATOL
General Information
Description: A high explosive.
Composition: NH_4NO_3: 80%; T.N.T.: 20%.
Constant: D: 1.47.
Hazard Analysis
Toxic Hazard Rating:
Acute Local: Irritant 2; Allergen 2; Ingestion 2; Inhalation 2.
Acute Systemic: Ingestion 2; Inhalation 2; Skin Absorption 2.
Chronic Local: Allergen 2.
Chronic Systemic: Ingestion 3; Inhalation 3.
Caution: Local contact may cause contact dermatitis (Section 9).
Fire Hazard: High, as result of spontaneous chemical reaction. A powerful oxidizing mixture (Section 6).
Explosion Hazard: High, due to shock, spontaneous chemical reaction or exposure to flame (Section 7).
Disaster Hazard: Highly dangerous; emits highly toxic fumes; can react vigorously with oxidizing materials.
Countermeasures
Ventilation Control: Section 2.
Personal Protection: Section 3.
Storage and Handling: Section 7.
Shipping Regulations: Section 11.
IATA: Explosive, not acceptable (passenger and cargo).

AMERICIUM
General Information
Description: A metal.
Formula: Am.
Constant: At wt 241.
Hazard Analysis
Toxicity: Americium is radioactive and derives its toxicity from this fact.
Radiation Hazard: Section 5. For permissible levels, see Table 5, p. 150.
Artificial isotope ^{241}Am (Neptunium Series), half life 460 y. Decays to radioactive ^{237}Np by emitting alpha particles of 5.4 MeV.
Artificial isotope ^{242m}Am half life 16 hr. Decays to radioactive ^{242}Pu by electron capture (20%). Also decays to radioactive ^{242}Cm by emitting beta particles of 0.62, 0.67 MeV. Also emits weak gamma rays and X-rays.
Artificial isotope ^{242}Am, half life 100 y. Decays to radioactive ^{242}Pu by electron capture (10%). Also decays to radioactive ^{242}Cm by emitting beta particles of 0.59 MeV. Also emits weak gamma rays and X-rays.
Artificial isotope ^{244}Am, half life 26 m. Decays to radioactive ^{244}Cm by emitting beta particles of 1.5 mMeV.
Fire Hazard: Unknown.

AMBER OIL
General Information
Synonym: Oil of amber.

Description: Pale yellow to brown oil, volatile.
Composition: Phenols and terpenes.
Constant: Density 0.85–0.92.
Hazard Analysis
Toxic Hazard Rating:
Acute Local: Irritant 1; Allergen 1; Ingestion 1; Inhalation 1.
Acute Systemic: U.
Chronic Local: Allergen 1.
Chronic Systemic: U.
Caution: A weak sensitizer. Local contact may cause contact dermatitis (Section 9).
Fire Hazard: Slight, when exposed to heat, it emits acrid fumes (Section 6).
Countermeasures
Ventilation Control: Section 2.
Personal Hygiene: Section 3.
Storage and Handling: Section 7.

AMIDES
General Information
Description: Organic compounds containing the structural group $-CONH_2$, and closely related to the organic acids with the grouping COOH. Common examples are: acetamide, CH_3CONH_2, and urea, $CO(NH_2)_2$.
Hazard Analysis
Toxicity: Most of the saturated amides have low toxicity, but the unsaturated and N-substituted amides are frequently irritants, and may be absorbed via the skin. In animal experiments the latter two classes have caused injury to the liver, kidney and brain.

AMIDOL
General Information
Synonym: Diaminophenol hydrochloride.
Description: Grayish-white crystals.
Formula: $C_6H_8N_2O \cdot 2HCl$.
Constant: Mol wt: 197.07.
Hazard Analysis
Toxic Hazard Rating:
Acute Local: Irritant 1; Allergen 1.
Acute Systemic: U.
Chronic Local: Allergen 1.
Chronic Systemic: U.
Disaster Hazard: dangerous; see chlorides.
Countermeasures
Personal Hygiene: Section 3.
Storage and Handling: Section 7.

AMIDOSULFONIC ACID. See sulfamic acid.

"AMINE 220"
General Information
Description: Liquid.
Formula: $C_{17}H_{33}CNC_2H_4NC_2H_4OH$.
Constants: Mol wt: 350, bp: 235° C at 1 mm, flash p: 465° F (O.C.), d: 0.9300 at 20°/20° C, vap d: 12.1.
Hazard Analysis
Toxicity: Details unknown.
Fire Hazard: Slight, when exposed to heat or flame; can react with oxidizing materials.
Countermeasures
To Fight Fire: Foam, carbon dioxide, dry chemical or carbon tetrachloride (Section 6).
Storage and Handling: Section 7.

AMINES. See also specific compounds.
General Information
A large group of organic compounds, containing nitrogen and considered as being derived from ammonia (NH_3) by replacement of one or more H atoms by an organic radical.
Hazard Analysis
Toxicity: Variable; some are highly toxic, others only slightly. Many are skin irritants and some are sensitizers. See also fatty amines. Amines are common air contaminants (Section 4).

p-AMINOACETANILIDE
General Information
Synonym: Acetyl-p-phenylenediamine.
Description: White to reddish crystals.
Formula: $NH_2C_6H_4NHCOCH_3$.
Constant: Mol wt: 150.2.
Hazard Analysis
Toxicity: Details unknown. See also p-phenylene diamine.
Disaster Hazard: Moderately dangerous; when heated to decomposition it emits toxic fumes.
Countermeasures
Storage and Handling: Section 7.

AMINOACETIC ACID. See glycine.

o-AMINOACETOPHENONE
General Information
Synonym: o-Aminoacetylbenzene.
Description: Yellow, oily liquid.
Formula: $CH_3COC_6H_4NH_2$.
Constants: Mol wt: 135.2, bp: 250–252°C (slight decomp.).
Hazard Analysis
Toxicity: Unknown. See Amines.
Fire Hazard: Slightly dangerous; (Section 6).
Countermeasures
Storage and Handling: Section 7.

o-AMINOACETYL BENZENE. See o-aminoacetophenone.

α-AMINOANTHROQUINONE
General Information
Synonym: 1-Amino anthroquinone.
Description: Ruby red crystals; insoluble in water; soluble in alcohol, benzene, chloroform, ether, glacial acetic acid, hydrochloric acid.
Formula: $C_6H_4(CO_2)C_6H_3NH_2$(tri-cyclic).
Constants: Mol wt: 223.22, mp: 253°C, bp: sublines.
Hazard Analysis
Toxic Hazard Rating: U.
Toxicity: Has produced anemia and degenerative changes in liver and kidneys of experimental animals.

p-AMINOAZOBENZENE
General Information
Synonym: p-Phenylazoaniline.
Description: Yellow crystals.
Formula: $C_6H_5NNC_6H_4NH_2$.
Constants: Mol wt: 197.2, bp: > 360°C, mp: 126°C.
Hazard Analysis
Toxicity: Details unknown; see aniline.
Fire Hazard: Unknown.
Disaster Hazard: Moderately dangerous; when heated to decomposition, it emits toxic fumes.
Countermeasures
Storage and Handling: Section 7.

o-AMINOAZOTOLUENE
General Information
Synonym: 2-Amino-5-azotoluene; solvent yellow 3; tol-uazotoluidine.
Description: Reddish brown to yellow crystals; soluble in alcohol, ether, oils, and fats; slightly soluble in water.
Formula: $CH_3C_6H_4N_2C_6H_3NH_2CH_3$.
Constants: Mol wt: 225.3, mp: 99–117°C.
Hazard Analysis
Toxicity: Details unknown. Has produced cancer of the liver experimentally in rats and mice.
Fire Hazard: Unknown.

Disaster Hazard: Moderately dangerous; when heated to decomposition, it emits toxic fumes.
Countermeasures
Storage and Handling: Section 7.

AMINOAZOTOLUENE HYDROCHLORIDE
General Information
Synonym: 2-Amino-5-azotoluene hydrochloride.
Description: Crystals.
Formula: $C_{14}H_{14}N_3 \cdot HCl$.
Constant: Mol wt: 260.8.
Hazard Analysis
Toxicity: Details unknown. See also o-aminoazotoluene.
Disaster Hazard: dangerous; See chlorides.
Countermeasures
Storage and Handling: Section 7.

AMINOBENZENE. See aniline.

o-AMINOBENZENE SULFONIC ACID. See sulfanilic acid.

2-AMINOBENZENETHIOL
General Information
Synonym: 2-Amino phenyl mercaptan.
Description: Liquid.
Formula: $NH_2C_6H_4SH$.
Constants: Mol wt: 125.2, mp: 23°C, bp: 227.2°C, flash p: 175°F, d: 1.168, vap d: 4.3.
Hazard Analysis
Toxicity: Details unknown. See also mercaptans.
Fire Hazard: Moderate, when exposed to heat of flame.
Disaster Hazard: Dangerous; see sulfur compounds. Can react with oxidizing materials.
Countermeasures
Personal Hygiene: Section 3.
To Fight Fire: Water, foam, carbon dioxide, dry chemical or carbon tetrachloride (Section 6).
Storage and Handling: Section 7.

o-AMINOBENZOIC ACID. See anthranilic acid.

p-AMINOBENZOIC ACID
General Information
Synonyms: PABA, Paraminol.
Description: Yellowish to red crystals.
Formula: $NH_2C_6H_4COOH$.
Constants: Mol wt: 137.1, mp: 187°C.
Hazard Analysis
Toxic Hazard Rating:
  Acute Local: 0.
  Acute Systemic: Ingestion 2.
  Chronic Local: U.
  Chronic Systemic: Ingestion 1.
Toxicology: Large doses by mouth can cause nausea, vomiting, skin rash, methemoglobinemia and possibly toxic hepatitis.
Fire Hazard: Slight (Section 6).
Countermeasures
Personal Hygiene: Section 3.
Storage and Handling: Section 7.

1-AMINOBENZOTHIAZOLE
General Information
Description: Crystals.
Formula: $\dot{C}_6H_4SCNH_2N$.
Constant: Mol wt: 150.2.
Hazard Analysis
Toxicity: Details unknown, but probably toxic.

TOXIC HAZARD RATING CODE (For detailed discussion, see Section 1.)

0 NONE: (a) No harm under any conditions; (b) Harmful only under unusual conditions or overwhelming dosage.

1 SLIGHT: Causes readily reversible changes which disappear after end of exposure.

2 MODERATE: May involve both irreversible and reversible changes; not severe enough to cause death or permanent injury.

3 HIGH: May cause death or permanent injury after very short exposure to small quantities.

U UNKNOWN: No information on humans considered valid by authors.

Disaster Hazard: Dangerous, when heated to decomposition, it emits highly toxic fumes.
Countermeasures
Personal Hygiene: Section 3.
Storage and Handling: Section 7.

m-AMINOBENZOTRIFLUORIDE
General Information
Description: Colorless liquid with aniline-like odor.
Formula: $H_2NC_6H_4CF_3$.
Constants: Mol wt: 161.13, mp: 3°C, bp: 189°C, d: 1.303 at 15.5°/15.5°C, vap d: 5.56.
Hazard Analysis
Toxicity: Details unknown, but probably toxic. See fluorides.
Disaster Hazard: Dangerous. See fluorides.
Countermeasures
Personal Hygiene: Section 3.
Storage and Handling: Section 7.

p-AMINOBENZOYL-γ-DI-n-BUTYLAMINOPROPANOL SULFATE. See butyn.

p-AMINOBIPHENYL. See p-aminodiphenyl.

1-AMINOBUTANE. See butylamine.

2-AMINOBUTANE. See sec-butylamine.

2-AMINO-1-BUTANOL
General Information
Description: Water-white liquid.
Formula: $CH_3CH_2CHNH_2CH_2OH$.
Constants: Mol wt: 89.14, mp: $-2°C$, bp: 178°C,
flash point: 165°F (O.C.), d: 0.944 at 20°/20°C,
vap d: 3.06.
Hazard Analysis
Toxicity: Details unknown. Probably has low toxicity. See also alcohols and 3-amino-propanol.
Fire Hazard: Moderate, when exposed to heat or flame; can react with oxidizing materials.
Spontaneous Heating: No.
Countermeasures
To Fight Fire: Water spray, alcohol foam, carbon dioxide, dry chemical or carbon tetrachloride (Section 6).
Storage and Handling: Section 7.

1-AMINO-4-tert-BUTYL BENZENE. See p-tert-butyl aniline.

AMINOCYCLOHEXANE. See cyclohexyl amine.

p-AMINODIMETHYL ANILINE. See dimethyl-p-phenylene diamine.

AMINODIMETHYL BENZENE. See xylidene.

2-AMINO-4,6-DIMETHYLPYRIDINE
General Information
Description: Crystals.
Formula: $C_7H_{10}N_2$.
Constants: Mol wt: 122.17, bp: 235.3°C, mp: 68.5°C.
Hazard Analysis
Toxicity: Unknown. See also pyridine.
Disaster Hazard: Slight; on decomposition it emits toxic fumes.
Countermeasures
Storage and Handling: Section 7.

2-AMINO-4-6-DINITROPHENOL. See picramic acid.

p-AMINODIPHENYL
General Information
Synonym: Xenylamine; p-biphenylamine.
Description: Colorless crystals.
Formula: $C_6H_5C_6H_4NH_2$.
Constants: Mol wt: 169.2, mp: 53°C, bp: 302°C, d: 1.160 at 20°/20°C.

Hazard Analysis
Toxic Hazard Rating:
Acute Local: Irritant 1; Ingestion 1; Inhalation 1.
Acute Systemic: Ingestion 3; Inhalation 3; skin absorption 3.
Chronic Local: Irritant 1.
Chronic Systemic: Ingestion 3; Inhalation 3; skin absorption 3.
Toxicology: Has caused bladder cancer in humans and experimental animals. Effects resemble those of benzidine. See benzidine. (Section 1).
Disaster Hazard: Moderate. When strongly heated, emits toxic fumes.
Countermeasures
Ventilation Control: Section 2.
Personal Hygiene: Section 3.
Storage and Handling: Section 7.

AMINODITHIOFORMIC ACID. See dithiocarbamic acid.

AMINOETHANE. See ethylamine.

2-AMINOETHANESULFONIC ACID
General Information
Synonym: Taurine.
Description: Large rods; soluble in water; insoluble in alcohol.
Formula: $NH_2CH_2CH_2SO_3H$.
Constant: Mol wt: 125.14, decomposes at 300°C.
Hazard Analysis
Toxic Hazard Rating: U. Limited animal experiments suggest low toxicity and irritation.
Disaster Hazard: Dangerous. When heated to decomposition emits highly toxic fumes of oxides of sulfur.
Countermeasures
Storage and Handling: Section 7.

2-AMINOETHANOL. See monoethanolamine.

1-[(2-AMINOETHYL)AMINO]-2-PROPANOL
General Information
Synonym: N-(2-Hydroxy propyl) ethylene diamine.
Hazard Analysis
Toxic Hazard Rating: U. An epoxy resin curing agent. Probably a skin irritant and sensitizer. See also ethylenediamine.

o-AMINOETHYLBENZENE. See o-ethylaniline.

2-AMINO ETHYL ETHANOL AMINE
General Information
Synonym: (2-Hydroxyethyl) ethylenediamine.
Description: Colorless liquid—
Formula: $NH_2CH_2CH_2NHCH_2CH_2OH$.
Constants: Mol wt: 104.16, bp: 243.7°C, flash p: 275°F, d: 1.0304 at 20°/20°C, autoign. temp.: 695°F, vap. press.: <0.01 mm at 20°C, vap d: 3.59.
Hazard Analysis
Toxicity: Details unknown. Experiments indicate low toxicity
Fire Hazard: Moderate, when exposed to heat or flame; can react with oxidizing materials.
Countermeasures
To Fight Fire: Water, alcohol foam, carbon dioxide, dry chemical or carbon tetrachloride (Section 6).
Storage and Handling: Section 7.

1-AMINOETHYL-2-HEPTADECYL GLYOXALIDINE
Hazard Analysis
Toxicity: Details unknown; a plant fungicide.
Disaster Hazard: Slight; when exposed to heat, it emits acrid fumes.
Countermeasures
Storage and Handling: Section 7.

N-AMINO ETHYL MORPHOLINE
General Information
Description: Liquid.
Formula: $C_2H_4OC_2H_4NC_2H_4NH_2$.
Constants: Mol wt: 130.2, mp: 25.6°C, bp: 204.2°C, flash p: 347°F (O.C.), d: 0.9915 at 20°/20°C, vap d: 4.49.
Hazard Analysis
Toxicity: Details unknown. Experiments indicate low toxicity for rats and high toxicity for guinea pigs. (Section 1).
Fire Hazard: Moderate when exposed to heat or flame; can react with oxidizing materials.
Countermeasures
To Fight Fire: Alcohol foam, carbon dioxide, dry chemical or carbon tetrachloride (Section 6).
Storage and Handling: Section 7.

AMINOETHYL PIPERAZINE
General Information
Description: Light colored liquid.
Formula: $H_2NCH_2CH_2N(CH_2)_4NH$.
Constants: Mol wt: 129.2, d: 0.9852 at 20/20°C, mp: −19°C, bp: 220.4°C, flash p: 200°F (O.C.), vap d: 4.4.
Hazard Analysis
Toxicology: Details unknown. Limited animal experiments suggest moderate toxicity and irritation. See also piperazine.
To Fight Fire: Alcohol foam.

2-AMINO-2-ETHYL-1,3-PROPANEDIOL
General Information
Description: Colorless liquid or solid.
Formula: $CH_2OHC(C_2H_5)NH_2CH_2OH$.
Constants: Mol wt: 119.16, mp: 38°C, bp: 152°C at 10 mm, d: 1.099 at 20°/20°C, vap d: 4.11.
Hazard Analysis
Toxicity: Unknown. Probably has low toxicity.
Fire Hazard: Slight, when exposed to heat or flame; can react with oxidizing materials (Section 6).
Countermeasures
Storage and Handling: Section 7.

2-AMINOETHYL SULFAMATE
Hazard Analysis
Toxicity: Details unknown. Limited animal experiments suggest low toxicity.
Disaster Hazard: Dangerous. See Sulfamates.

4-AMINOFOLIC ACID. See aminopterin.

α-AMINOGLUTARIC ACID. See glutamic acid.

AMINO-4-GUANIDO VALERIC ACID. See argenine.

1-AMINO HEPTANE. See heptyl amine.

7-AMINOHEPTANOIC ACID, ISOPROPYL ESTER
Hazard Analysis
Toxic Hazard Rating: U. Limited animal experiments suggest moderate toxicity and irritation.

p-AMINOHYDROGENATED CARDANOL. See 4-amino-3-pentadecyl phenol.

1-AMINO-5-HYDROXYBENZOTHIAZOLE
General Information
Description: Crystals.
Formula: $(NH_2)(OH)(C_6H_3)SCN$.
Constant: Mol wt: 166.2.
Hazard Analysis
Toxicity: Details unknown; probably toxic.

Fire Hazard: Unknown.
Disaster Hazard: Dangerous, see sulfur compounds.
Countermeasures
Personal Hygiene: Section 3.

α-AMINO-β-HYDROXYBUTYRIC ACID. See threonine.

α-AMINO-β-p-HYDROXYPHENYL PROPIONIC ACID. See tyrosine.

α-AMINO-β-HYDROXYPROPIONIC ACID. See serine.

α-AMINO-β-IMIDOAZOLEPROPIONIC ACID. See histidine.

1,2-AMINO-3-INDOLEPROPIONIC ACID. See tryptophane.

2-AMINOISOBUTANE. See tert-butylamine.

α-AMINOISOCAPROIC ACID. See leucine.

α-AMINOISOVALERIC ACID. See valine.

AMINOMETHANE. See monomethylamine.

1-AMINO-2-METHOXY-4-NITROBENZENE
General Information
Synonyms: 4-Nitro-o-anisidine; 5-nitro-2-aminanisole; 2-methoxy-4-nitroaniline.
Formula: $C_6H_3(NO_2)(OCH_3)(NH_2)$.
Hazard Analysis
Toxicity: Details unknown. See also nitrobenzene.
Fire Hazard: See nitrates. (Section 6).
Disaster Hazard: See nitrates (Section 6).
Countermeasures
Storage and Handling: Section 7.

2-AMINO-4-METHYLPENTANE. See mono-sec-hexyl-amine.

2-AMINO-3-METHYL PENTANOIC ACID. See isoleucine.

1-AMINO-2-METHYLPROPANE. See isobutyl amine.

2-AMINO-2-METHYL-1,3-PROPANEDIOL
General Information
Description: Clear liquid.
Formula: $CH_2OHC(CH_3)NH_2CH_2OH$.
Constants: Mol wt: 105.14, mp: 110°C, bp: 151°C at 10 mm vap d: 3.63.
Hazard Analysis
Toxicity: Unknown. Probably has low toxicity.
Fire Hazard: Slight, when exposed to heat or flame; can react with oxidizing materials (Section 6).
Countermeasures
Storage and Handling: Section 7.

2-AMINO-2-METHYL-1-PROPANOL
General Information
Synonyms: AMP, isobutanolamine.
Description: Colorless liquid.
Formula: $CH_3CH_2NH_2CCH_2OH$.
Constants: Mol wt: 89.14, mp: 30–31°C, bp: 165°C, flash p: 153°F (Tag. O.C.), d: 0.934 at 20°/20°C, vap d: 3.04.
Hazard Analysis
Toxicity: Details unknown. Probably not highly toxic. See also alcohols and 3-amino propanol.
Fire Hazard: Moderate, when exposed to heat or flame, can react with oxidizing materials.

TOXIC HAZARD RATING CODE *(For detailed discussion, see Section 1.)*

0 NONE: (a) No harm under any conditions; (b) Harmful only under unusual conditions or overwhelming dosage.

SLIGHT: Causes readily reversible changes which disappear after end of exposure.

2 MODERATE: May involve both irreversible and reversible changes; not severe enough to cause death or permanent injury.
3 HIGH: May cause death or permanent injury after very short exposure to small quantities.
U UNKNOWN: No information on humans considered valid by authors.

Countermeasures
To Fight Fire: Alcohol foam, carbon dioxide, dry chemical
or carbon tetrachloride (Section 6).
Storage and Handling: Section 7.

2-AMINO-3-METHYLPYRIDINE
General Information
Description: Liquid.
Formula: $NC_5H_3(CH_3)NH_2$.
Constants: Mol wt: 108.08, bp: 221.1°C, mp: 33.3°C, vap d:
3.73.
Hazard Analysis
Toxicity: Details unknown. See also pyridine.
Fire Hazard: Slight; (Section 6).
Countermeasures
Storage and Handling: Section 7.

2-AMINO-4-METHYLPYRIDINE
General Information
Description: Crystals.
Formula: $NC_5H_3(CH_3)NH_2$.
Constants: Mol wt: 108.08, mp: 99°C, bp: 230.9°C,
vap. d.: 3.73.
Hazard Analysis
Toxicity: Details unknown. See also pyridine.
Fire Hazard: Slight; can react with oxidizing materials (sec-
tion 6).
Countermeasures
Storage and Handling: Section 7.
Countermeasures
Storage and Handling: Section 7.

2-AMINO-5-METHYLPYRIDINE
General Information
Description: Crystals.
Formula: $NC_5H_3(CH_3)NH_2$.
Constants: Mol wt: 108.18, mp: 76.6°C, bp: 227.1°C, vap
d: 3.73.
Hazard Analysis
Toxicity: Details unknown. See also pyridine.
Fire Hazard: Slight; on decomposition it emits toxic fumes;
can react with oxidizing materials (Section 6).
Countermeasures
Storage and Handling: Section 7.

2-AMINO-6-METHYLPYRIDINE
General Information
Description: Crystals.
Formula: $NC_5H_3(CH_3)NH_2$.
Constants: Mol wt: 108.08, mp: 43.7°C, bp: 214.4°C,
vap d: 3.73.
Hazard Analysis
Toxicity: Details unknown. See also pyridine,
Fire Hazard: Slight; can react with oxidizing materials.
Countermeasures
Storage and Handling: Section 7.

2-AMINO-4-(METHYLTHIO) BUTYRIC ACID. See
methionine.

α-AMINO-β-METHYLVALERIC ACID. See isoleucine.

α-AMINO-γ-METHYL VALERIC ACID. See leucine.

3-AMINO-2-NAPHTHOIC ACID
General Information
Synonym: 3-Aminonaphthalene-2-carboxyllic acid; 3-amino-
iso naphthoic acid.
Description: Yellow scales from dilute alcohol; soluble in
alcohol, ether.
Formula: $C_{11}H_9NO_2$.
Constants: Mol wt: 187.19, mp: 214°C.
Hazard Analysis
Toxic Hazard Rating: U. Limited animal experiments
suggest low toxicity and irritation.

1-AMINO-4-NITROBENZENE. See p-nitroaniline.

1-AMINO NONANE. See n-nonylamine.

4-AMINO-3-PENTADECYL PHENOL
General Information
Synonym: p-Amino hydrogenated cardanol.
Description: Tan-colored, crystalline solid.
Formula: $C_{15}H_{31}C_6H_3(NH_2)OH$.
Constants: Mol wt: 319.5, mp: 99–101°C, bp: 225–230°C
at 1 mm.
Hazard Analysis
Toxicity: Unknown.
Fire Hazard: Slight, when exposed to heat or flame; can
react with oxidizing materials (Section 6).
Countermeasures
Storage and Handling: Section 7.

1-AMINOPENTANE. See amylamine.

2-AMINOPENTANE. See sec-amylamine.

2-AMINOPHENETOLE. See o-phenetidine.

4-AMINOPHENETOLE. See p-phenetidine.

p-AMINOPHENOL
General Information
Synonyms: Rodinol; ursal P; p-hydroxy aniline.
Description: Colorless crystals. Slightly soluble in water,
alcohol and ether. Insoluble in chloroform.
Formula: $NH_2C_6H_4OH$.
Constant: Mol wt: 109.1, mp: 189.6–190.2°C, bp: 284°C
(decomp.).
Hazard Analysis
Toxic Hazard Rating:
Acute Local: Irritant 1; Allergen 1; Ingestion 1; Inhala-
tion 1.
Acute Systemic: Ingestion 2.
Chronic Local: Allergen 2.
Chronic Systemic: Ingestion 2.
Caution: This material resembles p-phenylene diamine. It
has reportedly been the cause of contact dermatitis and
bronchial asthma; can cause methemoglobinemia with
cyanosis (Section 9).
Disaster Hazard: Moderately dangerous; when heated to
decomposition, it emits toxic fumes.
Countermeasures
Ventilation Control: Section 2.
Personnel Protection: Section 3.
Storage and Handling: Section 7.

2-AMINOPHENOL ACETIC ACID, ISOAMYL ESTER
General Information
Formula: $C_6H_3(OH)NH_2CH_2COOC_5H_{11}$.
Constant: Mol wt: 237.
Hazard Analysis
Toxic Hazard Rating:
Acute Local: Allergin 3.
Acute Systemic: U.
Chronic Local: Allergin 3.
Chronic Systemic: U.
Human contact eczema has been reported.

AMINO PHENYL ARSINE ACID. See arsanilic acid.

AMINOPHENYL BORIC ACID
General Information
Description: White crystals. Slightly soluble in water.
Formula: $(NH_2C_6H_4)B(OH)_2$.
Constant: Mol wt: 137.0.
Hazard Analysis
Toxicology: Details unknown. See boron compounds.

p-AMINOPHENYL CADMIUM DILACTATE
Hazard Analysis
Toxicity: High. See cadmium compounds.
Disaster Hazard: Dangerous; when heated to decomposi-
tion, it emits highly toxic fumes.

Countermeasures
Personnel Protection: Section 3.

2-AMINOPHENYL MERCAPTAN. See 2-aminobenzene-thiol.

AMINO PHENYL MERCURIC ACETATE
General Information
Description: Colorless crystals. Insoluble in water.
Formula: $O_2Hg(C_6H_4)(NH_2)(C_2H_3)$.
Constants: Mol wt: 351.8, mp: 167°C.
Hazard Analysis
Toxicology: Highly toxic. See mercury compounds, organic.
Disaster Hazard: Dangerous. See mercury compounds, organic.

p-AMINOPHENYL MERCURIC LACTATE
Hazard Analysis
Toxicity: Highly toxic. See mercury compounds, organic.
Disaster Hazard: See mercury compounds, organic.
Countermeasures
Storage and Handling: Section 7.

m-AMINOPHENYL METHYL CARBINOL
General Information
Description: Solid; soluble in water.
Formula: $NH_2C_6H_4CH(OH)CH_3$.
Constants: Mol wt: 137, d: 1.12, bp: 217.3°C at 100 mm, fp: 66.4°C, flash p: 315°F (O.C.).
Hazard Analysis
Toxic Hazard Rating: U.
Fire Hazard: Slight.
Countermeasures
Storage and Handling: Section 7.
To Fight Fire: Alcohol foam (Section 6).

α-AMINO-β-PHENYLPROPIONIC ACID. See phenylalanine.

1-AMINOPROPANE. See n-propyl amine.

2-AMINOPROPANE. See isopropylamine.

2-AMINOPROPANOIC ACID. See alanine.

1-AMINO-2-PROPANOL
General Information
Synonym: MIPA, 2-hydroxypropylamine; isopropanolamine.
Description: Liquid; slight ammonia odor; soluble in water.
Formula: $CH_3CH(OH)CH_2NH_3$.
Constants: Mol wt: 85, d: 0.969, mp: 1.4°C, flash p: 171°F, vap d: 2.6.
Hazard Analysis
Toxic Hazard Rating: U.
Fire Hazard: Moderate.
Countermeasures
To Fight Fire: Alcohol foam (Section 6).
Storage and Handling: Section 7.

3-AMINOPROPANOL
General Information
Description: Colorless liquid.
Formula: $H_2NCH_2CH_2CH_2OH$.
Constants: Mol wt: 75.11, bp: 168°C at 500 mm, flash p: >175°F (T.O.C.), fp: 12.4°C, d: 0.9786 at 30°C, vap. press.: 2.1 mm at 60°C, vap. d.: 2.59.
Hazard Analysis
Toxic Hazard Rating:
  Acute Local: Irritant 2; Ingestion 2; Inhalation 1.
  Acute Systemic: Ingestion 2; Skin Absorption 1.

Chronic Local: U.
Chronic Systemic: U.
See also amines.
Caution: A moderately strong base which, from animal experiments, would seem to be a non-specific irritant.
Fire Hazard: Moderate, when exposed to heat or flame; can react with oxidizing materials.
Countermeasures
To Fight Fire: Foam, carbon dioxide, dry chemical or carbon tetrachloride (Section 6).
Ventilation Control: Section 2.
Personal Hygiene: Section 3.
Storage and Handling: Section 7.

α-AMINOPROPIONIC ACID. See alanine.

β-AMINOPROPIONITRILE. See 3-aminopropionitrile.

3-AMINOPROPIONITRILE
General Information
Synonym: β-Aminopropionitrile.
Description: Liquid with amine odor.
Formula: $H_2NCH_2CH_2CN$.
Constants: Mol wt: 70.1, bp: 185°C.
Hazard Analysis
Toxicology: Nitriles generally have cyanide-like effects. See hydrocyanic acid and cyanides.
Fire Hazard: See cyanides. Section 6.
Explosion Hazard: See cyanides.
Disaster Hazard: See cyanides.
Countermeasures
Ventilation Control: See cyanides. Section 2.
Personal Hygiene: See cyanides.
Storage and Handling: See cyanides.

p-AMINOPROPIOPHENONE
General Information
Synonym: PAPP.
Description: Yellow needlelike crystals. Soluble in water and alcohol.
Formula: $C_9H_{11}NO$.
Constants: Mol wt: 149.2, mp: 140°C.
Hazard Analysis
Toxic Hazard Rating:
  Acute Local: U.
  Acute Systemic: Ingestion 2.
  Chronic Local: U.
  Chronic Systemic: Ingestion 1.
Toxicology: Large doses can cause cyanosis.
Disaster Hazard: When strongly heated can emit highly toxic fumes, dangerous.

n-(3-AMINOPROPYL)CYCLOHEXYLAMINE
General Information
Description: Soluble in water.
Formula: $C_6H_{11}NHC_3H_6NH_2$.
Constants: Mol wt: 156, d: 0.9, vap d: 5.4, flash p: 175°F (O.C.).
Hazard Analysis
Toxic Hazard Rating: U.
Fire Hazard: Moderate.
Countermeasures
To Fight Fire: Alcohol foam (Section 6).
Storage and Handling: Section 7.

N-AMINOPROPYL MORPHOLINE
General Information
Description: Liquid.
Formula: $\overline{C_2H_4OC_2H_4N}C_3H_5NH_2$.

TOXIC HAZARD RATING CODE *(For detailed discussion, see Section 1.)*

0 NONE: (a) No harm under any conditions; (b) Harmful only under unusual conditions or overwhelming dosage.

1 SLIGHT: Causes readily reversible changes which disappear after end of exposure.

2 MODERATE: May involve both irreversible and reversible changes; not severe enough to cause death or permanent injury.

3 HIGH: May cause death or permanent injury after very short exposure to small quantities.

U UNKNOWN: No information on humans considered valid by authors.

Constants: Mol wt: 144.21, mp: $-15°C$, bp: $224.7°C$, flash p: $220°F$ (O.C.), d: 0.9872 at $20°/20°C$, vap. press.: 0.06 mm at $20°C$, vap d: 4.97.

Hazard Analysis

Toxicity: Details unknown. Experiments show low toxicity for rats and moderate toxicity for rabbits.

Fire Hazard: Moderate, when exposed to heat or flame; can react with oxidizing materials.

Countermeasures

To Fight Fire: Alcohol foam, carbon dioxide, dry chemical or carbon tetrachloride (Section 6).

Storage and Handling: Section 7.

AMINOPTERIN

General Information

Synonym: 4-Aminofolic acid.

Description: Yellow needles. Soluble in sodium hydroxide solution.

Formula: $C_{19}H_{20}N_8O_5 \cdot 2H_2O$.

Constants: Mol wt: 476.5.

Hazard Analysis

Toxic Hazard Rating:

Acute Local: U.

Acute Systemic: Ingestion 2.

Chronic Local: U.

Chronic Systemic: Ingestion 2.

Toxicology: An anti-metabolite sometimes used as a rodenticide. Large doses can produce bone marrow depression.

Disaster Hazard: Dangerous. When strongly heated can emit highly toxic fumes.

2-AMINOPYRIDINE

General Information

Synonym: α-Aminopyridine.

Description: White powder or crystals.

Formula: $C_5H_6N_2$.

Constants: Mol wt: 94.11, mp: $58.1°C$, bp: $210.6°C$.

Hazard Analysis

Toxic Hazard Rating:

Acute Local: U.

Acute Systemic: Ingestion 3; Inhalation 3.

Chronic Local: U.

Chronic Systemic: U.

Toxicology: A convulsant poison with effects resembling strychnine. In humans the symptoms are headache, weakness, collapse and epileptiform convulsions (Section 1).

TLV: ACGIH (recommended) 0.5 part per million; 2 milligrams per cubic meter of air.

Disaster Hazard: Dangerous; when heated to decomposition, it emits highly toxic fumes.

Countermeasures

Ventilation Control: Section 2.

Personal Hygiene: Section 3.

First Aid: Section 1.

Storage and Handling: Section 7.

AMINOPYRIDINO TRIBROMO GOLD

General Information

Description: Black crystals. Soluble in water.

Formula: $(NH_2)(C_5H_4N)AuBr_3$.

Constants: Mol wt: 531.1, mp: $106°C$ (decomp).

Hazard Analysis

Toxicology: Details unknown. See gold compounds.

Disaster Hazard: Dangerous. See Bromides.

AMINOPYRINE

General Information

Synonym: Dimethylaminoantipyrine.

Description: Colorless leaflets, somewhat soluble in water.

Formula: $C_3ON_3(CH_3)_4C_6H_5$.

Constants: Mol wt: 231.29, mp: $107-109°C$.

Hazard Analysis

Toxic Hazard Rating:

Acute Local: Allergen 1.

Acute Systemic: Ingestion 1.

Chronic Local: Allergen 1.

Chronic Systemic: Ingestion 1.

Toxicology: In sensitive individuals bone marrow depression can occur resulting particularly in leucopenia. Also, can cause drug rash.

Disaster Hazard: Moderately dangerous; when heated to decomposition, it emits toxic fumes.

Countermeasures

Ventilation Control: Section 2.

Personal Hygiene: Section 3.

Storage and Handling: Section 7.

p-AMINOSALICYLIC ACID

General Information

Synonym: Deapasil, PAS.

Description: Minute crystals (from alcohol). Soluble in dilute nitric acid and dilute sodium hydroxide.

Formula: $C_7H_7NO_3$.

Constants: Mol wt: 153.13, mp: $150°C$.

Hazard Analysis

Toxic Hazard Rating:

Acute Local: Allergen 1.

Acute Systemic: Ingestion 1.

Chronic Local: Allergen 1.

Chronic Systemic: Ingestion 1.

Toxicology: Intolerance or sensitivity may result in vomiting, diarrhea, skin rash and disturbed blood formation. In severe cases there is disturbed electrolyte balance, jaundice and anaphylactoid reactions.

Fire Hazard: Slight; (Section 6).

Countermeasures

Personal Hygiene: Section 3.

Storage and Handling: Section 7.

AMINOSUCCINIC ACID. See aspartic acid.

2-AMINO-5-SULFANILYLTHIAZOLE

General Information

Synonyms: Promizole, thiazosulfone.

Description: Fine needles from alcohol. Soluble in water.

Formula: $C_9H_9N_3O_2S_2$.

Constant: Mol wt: 255.32, mp: $220°C$.

Hazard Analysis

Toxic Hazard Rating:

Acute Local: U.

Acute Systemic: Ingestion 1; Inhalation 1.

Chronic Local: U.

Chronic Systemic: Ingestion 1, Inhalation 1.

Caution: Can cause a transient anemia. Over a period of time thyroid enlargement may occur. Large doses can cause hemolysis, hematuria, cyanosis, gastrointestinal and central nervous system disturbances.

Disaster Hazard: Dangerous; see sulfur compounds.

Countermeasures

Ventilation Control: Section 2.

Personal Hygiene: Section 3.

Storage and Handling: Section 7.

2-AMINOTHIAZOLE

General Information

Description: Light brown crystals.

Formula: $C_3H_2NSNH_2$.

Constants: Mol wt: 100.14, mp: $90°C$, bp: decomposes.

Hazard Analysis

Toxic Hazard Rating:

Acute Local: U.

Acute Systemic: Ingestion 1, Inhalation 1.

Chronic Local: U.

Chronic Systemic: U.

Caution: Belived to cause nausea, headache and general weakness after exposure to from 3-100 mg per cubic meter of air.

Disaster Hazard: Dangerous; See sulfur compounds.

Countermeasures
Ventilation Control: Section 2.
Personal Hygiene: Section 3.
Storage and Handling: Section 7.

α-AMINO-β-THIOLPROPIONIC ACID. See cysteine.

AMINOTOLUENE. See benzylamine.

3-AMINO-1H-1,2,4-TRIAZOLE
General Information
Synonym: Amizol.
Description: Crystalline. Soluble in water, alcohol and chloroform.
Formula: $C_2H_4N_4$.
Constants: Mol wt: 84.1, mp: 159°C.
Hazard Analysis
Toxicity: Limited animal experiments indicate relatively low toxicity but a possible carcinogen.
Disaster Hazard: Dangerous. When strongly heated it emits highly toxic fumes.

AMIODOXYL BENZOATE
General Information
Synonyms: Arthrytin, oxoate.
Description: White odorless, slightly bitter, crystalline powder.
Formula: $C_7H_8INO_4$.
Constant: Mol wt: 297.06.
Hazard Analysis
Toxic Hazard Rating:
  Acute Local: U.
  Acute Systemic: Ingestion 2; Inhalation 2.
  Chronic Local: U.
  Chronic Systemic: Ingestion 1; Inhalation 1.
Disaster Hazard: Moderately dangerous; when heated to decomposition, it emits toxic fumes.
Countermeasures
Personnel Protection: Section 3.
Storage and Handling: Section 7.

AMIZOL. See 3-amino-1H-1, 2, 4-triazole.

AMMATE. See ammonium sulfamate.

AMMONAL
Hazard Analysis
Toxicity: Details unknown, but probably toxic.
Fire Hazard: A powerful oxidizer.
Explosion Hazard: High, when shocked or exposed to heat.
Disaster Hazard: Highly dangerous; it can explode when shocked or heated; when heated to decomposition it emits toxic fumes; can react vigorously with reducing material.
Countermeasures
Storage and Handling: Section 7.

AMMONIA-d₃
General Information
Synonym: Trideuterio ammonia.
Description: Gas.
Formula: ND_3.
Constants: Mol wt: 20.05, mp: −74°C, bp: −33.4°C.
Hazard Analysis
Toxicity: See ammonia, anhydrous.
Fire Hazard: Moderate, when exposed to heat or flame (Section 6).
Explosion Hazard: Moderate, when exposed to flame (Section 7).

Disaster Hazard: Moderately dangerous; reacts rigorously with oxidizing materials.
Countermeasures
Storage and Handling: Section 7.

AMMONIA, ANHYDROUS
General Information
Synonym: Ammonia gas.
Description: Colorless gas, extremely pungent odor, liquefied by compression.
Formula: NH_3.
Constants: Mol wt: 17.03, mp: −77.7°C, bp: −33.35°C, lel: 16%, uel: 25%, d: 0.771 g/liter at 0°C, 0.817 g/ml at −79°C, autoign temp: 1204°F, vap press: 10 atm at 25.7°C, vap d: 0.6.
Hazard Analysis
Toxic Hazard Rating:
  Acute Local: Irritant 3; Ingestion 3; Inhalation 3.
  Acute Systemic: U.
  Chronic Local: Irritant 1.
  Chronic Systemic: U.
TLV: ACGIH (recommended) 50 parts per million in air; 35 milligrams per cubic meter of air.
Toxicology: Irritating to eyes and mucous membranes of respiratory tract. Signs and symptoms of exposure are irritation of the eyes, conjunctivitis, swelling of the eyelids, irritation of the nose and throat, coughing, dyspnoea and vomiting. Irritation of the skin may be experienced, especially if it is moist. Corneal ulcers have been reported following splashing of ammonia water in the eye (Section 1). A common air contaminant (Section 4).
Fire Hazard: Moderate, when exposed to heat or flame.
Spontaneous Heating: No. Requires high concentrations in air before it catches fire.
Explosion Hazard: Moderate, when exposed to flame. Forms explosive compounds in contact with silver or mercury (Section 7).
Disaster Hazard: Moderately dangerous; exposed to heat, it emits toxic fumes.
Countermeasures
Ventilation Control: Section 2.
To Fight Fire: Stop flow of gas, carbon dioxide, dry chemical or water spray (Section 6).
Personnel Protection: Section 3.
Personal Hygiene: Section 3.
First Aid: Section 1.
Storage and Handling: Section 7.
Shipping Regulations: Section 11.
  I.C.C.: Nonflammable gas; green label, 300 pounds.
  Coast Guard Classification: Noninflammable gas; green gas label.
  MCA warning label.
  IATA: Nonflammable gas, green label, not acceptable (passenger), 140 kilograms (cargo).

AMMONIA AQUA. See ammonium hydroxide.

AMMONIACAL COPPER CARBONATE
Hazard Analysis
Toxic Hazard Rating: U. A fungicide.

AMMONIA GAS. See ammonia anhydrous.

AMMONIUM ACID CARBONATE. See Ammonium bicarbonate.

AMMONIUM ACID PHOSPHATE. See ammonium phosphate, monobasic.

TOXIC HAZARD RATING CODE *(For detailed discussion, see Section 1.)*

0 NONE: (a) No harm under any conditions; (b) Harmful only under unusual conditions or overwhelming dosage.

1 SLIGHT: Causes readily reversible changes which disappear after end of exposure.

2 MODERATE: May involve both irreversible and reversible changes; not severe enough to cause death or permanent injury.

3 HIGH: May cause death or permanent injury after very short exposure to small quantities.

U UNKNOWN: No information on humans considered valid by authors.

AMMONIUM ALGINATE
General Information
Synonym: Ammonium polymannurate.
Description: Filamentous, grainy, granular, or powdered, colorless or slightly yellow; slowly soluble in water; insoluble in alcohol.
Formula: $(C_6H_7O_6 \cdot NH_4)_n$.
Constant: Mol wt: 32,000–250,000.
Hazard Analysis
Toxic Hazard Rating: U. A stabilizer food additive (Section 10).

AMMONIUM ALUM. See aluminum ammonium sulfate.

AMMONIA SOLUTIONS, containing free ammonia (NH_3), pressurized
Countermeasures
Shipping Regulations: Section 11.
  IATA: Nonflammable gas, green label, not acceptable (passenger), 140 kilograms (cargo).

AMMONIA SOLUTIONS, containing less than 10% ammonia (NH_3)
Shipping Regulations: Section 11.
  IATA: Not restricted (passenger and cargo).

AMMONIA SOLUTIONS, containing 10% or more ammonia (NH_3), non-pressurized
Shipping Regulations: Section 11.
  IATA: Other restricted articles, class A, 10 liters (passenger and cargo).

AMMONIUM ARSENATE
General Information
Description: White powder or crystal.
Formula: $(NH_4)_3AsO_4 \cdot 3H_2O$.
Constants: Mol wt: 247.1, mp: decomp. to yield NH_3.
Hazard Analysis
Toxicity: See arsenic compounds.
Disaster Hazard: See arsenic compounds.
Countermeasures
Storage and Handling: Section 7.
Shipping Regulations: Section 11.
  I.C.C.: Poison B; poison label, 200 pounds.
  Coast Guard Classification: Poison B; poison label.
  IATA: Poison B, poison label, 25 kilograms (passenger), 95 kilograms (cargo).

AMMONIUM m-ARSENITE
General Information
Description: White powder.
Formula: NH_4AsO_2.
Constant: Mol wt:125.
Hazard Analysis
Toxicity: See arsenic compounds.
Countermeasures
Storage and Handling: Section 7.

AMMONIUM AZIDE
General Information
Description: Colorless plates.
Formula: NH_4N_3.
Constants: Mol wt: 60.1, mp: 160°C, bp: 133.8°C, d: 1.346, vap. press.: 1 mm at 59.2°C (sublimes).
Hazard Analysis
Toxic Hazard Rating:
  Acute Local: U.
  Acute Systemic: Ingestion 3; Inhalation 3.
  Chronic Local: U.
  Chronic Systemic: U.
Fire Hazard: Moderate (Section 6).
Explosion Hazard: Moderate, when heated (Section 7).
Countermeasures
Ventilation Control: Section 2.
Personal Hygiene: Section 3.
First Aid: Section 1.
Storage and Handling: Section 7.

AMMONIUM BENZENE SULFONATE
General Information
Description: Crystals.
Formula: $NH_4C_6H_5SO_3$.
Constants: Mol wt: 175.2, mp: 271–275°C (decomposes), d:1.342.
Hazard Analysis
Toxicity: Details unknown, but probably toxic.
Disaster Hazard: Dangerous; see sulfonates.
Countermeasures
Storage and Handling: Section 7.

AMMONIUM BIBORATE
General Information
Synonym: Ammonium tetraborate.
Description: Colorless, tetragonal crystals.
Formula: $(NH_4)_2B_4O_7 \cdot 4H_2O$.
Constants: Mol wt: 263.4, mp: decomposes.
Hazard Analysis
Toxicity: Details unknown. A herbicide. See also boron compounds.

AMMONIUM BICAMPHORATE
General Information
Synonym: Acid ammonium camphorate; ammonium camphorate.
Description: Crystalline powder; freely soluble in water.
Formula: $NH_4HC_{10}H_{14}O_4 \cdot 3H_2O$.
Constant: Mol wt: 271.31.
Hazard Analysis
See camphor.

AMMONIUM BICARBONATE
General Information
Synonym: Ammonium acid carbonate; ammonium hydrogen carbonate.
Description: White crystals; soluble in water; insoluble in alcohol.
Formula: NH_4HCO_3.
Constants: Mol wt: 79, d: 1.586, mp decomposes at 36–60°C.
Hazard Analysis
Toxic Hazard Rating: U. A General purpose food additive (Section 10).

AMMONIUM BICHROMATE
General Information
Synonym: Ammonium dichromate.
Description: Yellow needles.
Formula: $(NH_4)_2Cr_2O_7$.
Constants: Mol wt: 252.10, mp: decomposes before it melts., d: 2.15 at 25°C.
Hazard Analysis
Toxic Hazard Rating:
  Acute Local: Irritant 3, Allergen 1, Ingestion 3, Inhalation 3.
  Acute Systemic: Ingestion 2.
  Chronic Local: Irritant 3, Allergen 1, Inhalation 3.
  Chronic Systemic: U.
Toxicology: If swallowed it causes prompt vomiting, but if retained, may lead to kidney injury and ulceration of stomach. Chrome ulcers or sores of skin are well known as is perforation of the nasal septum from chrome salts. Prolonged inhalation of dust can cause asthmatic symptoms.
Fire Hazard: Moderate; reacts with reducing agents.
Caution: An oxidizer.
Countermeasures
Storage and Handling: Section 7.
Shipping Regulations: Section 11.
  I.C.C.: Flammable solid; yellow label, 100 pounds.
  Coast Guard Classification: Inflammable solid; yellow label.
  IATA: Flammable solid, yellow label, 12 kilograms (passenger), 45 kilograms (cargo).

AMMONIUM BIFLUORIDE
General Information
Synonym: Ammonium hydrogen fluoride.
Description: White crystals.
Formula: NH_4FHF.
Constants: Mol wt: 57.05, d: 1.21 at $12°C/12°C$ (liquid).
Hazard Analysis
Toxicity: See fluorides.
Disaster Hazard: See fluorides.
Countermeasures
Storage and Handling: Section 7.
  IATA: Other restricted articles, class B, no label required,
    no limit (passenger), no limit (cargo).

AMMONIUM BIMALATE
General Information
Synonym: Acid ammonium malate.
Description: Crystals; soluble in 3 parts water; slightly
    soluble in alcohol.
Formula: $NH_4HC_2H_4O_5$.
Constant: Mol wt: 151.02, d: 1.51, mp: $161°C$.
Hazard Analysis
Toxic Hazard Rating:
  Acute Local: Irritant 2; Ingestion 2; Inhalation 2.
  Acute Systemic: U.
  Chronic Local: U.
  Chronic Systemic: U.

AMMONIUM BINOXALATE
General Information
Description: Colorless crystals.
Formula: $NH_4HC_2O_4 \cdot H_2O$.
Constants: Mol wt: 125.08, mp: decomposes, d: 1.556.
Hazard Analysis
Toxicity: See oxalates.

AMMONIUM BIPHOSPHATE. See ammonium
    phosphate, monobasic.

AMMONIUM BISULFATE. See ammonium hydrogen
    sulfate.

AMMONIUM BISULFITE
General Information
Synonym: Ammonium hydrogen sulfite.
Description: White crystals.
Formula: NH_4HSO_3.
Constants: Mol wt: 99.1, mp: decomposes.
Hazard Analysis
Toxicity: See bisulfites.
Disaster Hazard: See bisulfites.
Countermeasures
Storage and Handling: Section 7.

AMMONIUM BITARTRATE
General Information
Synonym: Acid ammonium tartrate.
Description: White crystals; soluble in water, acids and
    alkalies; insoluble in alcohol.
Formula: $(NH_4)HC_4H_4O_6$.
Constant: Mol wt: 167, d: 1.636.
Hazard Analysis
See tartaric acid.

AMMONIUM BORATE. See ammonium biborate.

AMMONIUM BOROFLUORIDE
Hazard Analysis
Toxicity: A strong irritant. See also fluorides.
Disaster Hazard: Dangerous. See fluorides.

AMMONIUM BROMATE
General Information
Description: Colorless crystals.
Formula: NH_4BrO_3.
Constants: Mol wt: 145.96, mp: explodes.
Hazard Analysis
Toxicity: See bromates.
Fire Hazard: See bromates.
Explosion Hazard: Severe.
Disaster Hazard: Dangerous. See Bromates.
Countermeasures
Storage and Handling: Section 7.

AMMONIUM BROMIDE
General Information
Description: Colorless, cubic, slightly hygroscopic crystals.
Formula: NH_4Br.
Constants: Mol wt: 98.0, mp: sublimes at $542°C$, bp:
    $396.0°C$, d: 2.429, vap. press.: 1 mm at $198.3°C$.
Hazard Analysis
Toxicity: See bromides.
Disaster Hazard: Dangerous. See bromides.
Countermeasures
Storage and Handling: Section 7.

AMMONIUM BROMOPLATINATE
General Information
Description: Red-brown, cubic crystals.
Formula: $(NH_4)_2PtBr_6$.
Constants: Mol wt: 710.8, mp: $145°C$ (decomposes),
    d: 4.625.
Hazard Analysis
Toxicity: Details unknown. See also bromides and platinum
    compounds.
Countermeasures
Storage and Handling: Section 7.

AMMONIUM BROMOSELENATE
General Information
Description: Red octagonal crystals.
Formula: $(NH_4)SeBr_6$.
Constants: Mol wt: 594.5, d: 3.326.
Hazard Analysis
Toxicity: See selenium compounds and bromides.
Disaster Hazard: Dangerous. See selenium compounds and
    Bromides.
Countermeasures
Storage and Handling: Section 7.

AMMONIUM BROMOSTANNATE
General Information
Description: Colorless crystals.
Formula: $(NH_4)_2SnBr_6$.
Constants: Mol wt: 634.3, mp: decomposes, d: 3.50.
Hazard Analysis
Toxicity: Details unknown. See also bromides and tin com-
    pounds.
Disaster Hazard: Dangerous; see bromides.
Countermeasures
Storage and Handling: Section 7.

AMMONIUM CADMIUM BROMIDE. See cadmium am-
    monium bromide.

AMMONIUM CADMIUM CHLORIDE
General Information
Description: Crystals.
Formula: $4NH_4Cl \cdot CdCl_2$.
Constants: Mol wt: 397.3, d: 2.01.

TOXIC HAZARD RATING CODE (For detailed discussion, see Section 1.)

0 NONE: (a) No harm under any conditions; (b) Harmful only under unusual conditions or overwhelming dosage.

1 SLIGHT: Causes readily reversible changes which disappear after end
  of exposure.

2 MODERATE: May involve both irreversible and reversible changes;
  not severe enough to cause death or permanent injury.

3 HIGH: May cause death or permanent injury after very short exposure
  to small quantities.

U UNKNOWN: No information on humans considered valid by authors.

Hazard Analysis
Toxicity: See cadmium compounds.
Disaster Hazard: See cadmium compounds and chlorides.
Countermeasures
Storage and Handling: Section 7.

AMMONIUM CALCIUM ARSENATE
General Information
Description: Colorless crystals.
Formula: $NH_4CaAsO_4 \cdot 6H_2O$.
Constants: Mol wt: 305.1, mp: 140°C (decomposes), d: 1.905.
Hazard Analysis
Toxicity: See arsenic compounds.
Disaster Hazard: See arsenic compounds.
Countermeasures
Storage and Handling: Section 7.

AMMONIUM CARBAMATE
General Information
Description: White, crystalline rhombic powder; soluble in water and alcohol; ammonia odor.
Formula: $NH_4CO_2NH_2$.
Constant: Mol wt: 78, sublimes at 60°C.
Hazard Analysis
Toxicity: Decomposes in air, releasing ammonia. See ammonia, anhydrous.
Disaster Hazard: Dangerous. See Carbamates.

AMMONIUM CARBAZOTATE. See ammonium picrate.

AMMONIUM CARBONATE
General Information
Synonym: Crystal ammonia.
Description: Colorless, crystalline plates.
Formula: $(NH_4)_2CO_3$.
Constant: Mol Wt: 96.09.
Hazard Analysis
Toxic Hazard Rating:
  Acute Local: Irritant 1; Ingestion 1; Inhalation 1.
  Acute Systemic: U.
  Chronic Local: Irritant 1.
  Chronic Systemic: U.
A general purpose food additive (Section 10).
Countermeasures
Ventilation Control: Section 2.
Personal Hygiene: Section 3.

AMMONIUM CHLORATE
General Information
Description: White crystal or mass.
Formula: NH_4ClO_3.
Constant: Mol wt: 101.6.
Hazard Analysis
Toxicity: Details unknown. See also chlorates.
Fire Hazard: Moderate fire hazard due to spontaneous chemical reaction with reducing agents. A powerful oxidizer. When contaminated by combustibles it may ignite (Section 6).
Explosion Hazard: High, due to shock, chemical reaction or exposure to heat (Section 7).
Caution: When contaminated it is very sensitive. It can be detonated.
Disaster Hazard: Highly dangerous; can explode when shocked or exposed to heat; when heated to decomposition, it emits highly toxic fumes; can react vigorously with reducing material.
Countermeasures
Storage and Handling: Section 7.
Shipping Regulations: Section 11.
  IATA: Oxidizing material, not acceptable (passenger and cargo).

AMMONIUM CHLORIDE
General Information
Synonym: Sal ammonia.
Description: White crystals.

Formula: NH_4Cl.
Constants: Mol wt: 53.50, mp: 520°C, bp: 337.8°C, d: 1.520, vap. press.: 1 mm at 160.4°C (sublimes).
Hazard Analysis
Toxic Hazard Rating:
  Acute Local: Irritant 1; Ingestion 1; Inhalation 1.
  Acute Systemic: U.
  Chronic Local: Irritant 1.
  Chronic Systemic: Ingestion 1.
Toxicology: A substance migrating to food from packaging materials. Large doses cause nausea, vomiting and acidosis.
Countermeasures
Ventilation Control: Section 2.
Personal Hygiene: Section 3.

AMMONIUM CHLOROAURATE
General Information
Description: Yellow crystals.
Formula: NH_4AuCl_4.
Constant: Mol wt: 357.1.
Hazard Analysis
Toxicity: See gold compounds.

AMMONIUM CHLOROGALLATE
General Information
Description: White crystals.
Formula: NH_4GaCl_4.
Constants: Mol wt: 229.6, mp: 275°C.
Hazard Analysis and Countermeasures
See gallium compounds and Chlorides.

AMMONIUM CHLOROIRIDATE
General Information
Description: Red-black crystals.
Formula: $(NH_4)_2IrCl_3$.
Constants: Mol wt: 441.9, mp: decomposes, d: 2.856.
Hazard Analysis and Countermeasures
See iridium compounds and chlorides.

AMMONIUM CHLOROIRIDITE
General Information
Description: Greenish-brown solid.
Formula: $(NH_4)_3IrCl_3 \cdot 1\frac{1}{2}H_2O$.
Constant: Mol wt: 487.
Hazard Analysis and Countermeasures
See iridium compounds and chlorides.

AMMONIUM CHLOROOSMATE
General Information
Description: Crystals.
Formula: $(NH_4)_2OsCl_6$.
Constants: Mol wt: 439, d: 2.93.
Hazard Analysis and Countermeasures
See osmium compounds and chlorides.

AMMONIUM CHLOROPALLADATE
General Information
Description: Red-brown crystals.
Formula: $(NH_4)_2PdCl_6$.
Constants: Mol wt: 355.5, mp: decomposes, d: 2.418.
Hazard Analysis and Countermeasures
See palladium compounds and chlorides.

AMMONIUM CHLOROPALLADITE
General Information
Description: Olive-green crystals.
Formula: $(NH_4)_2PdCl_4$.
Constants: Mol wt: 284.6, mp: decomposes, d: 2.17.
Hazard Analysis and Countermeasures
See palladium compounds and chlorides.

AMMONIUM CHLOROPLATINATE
General Information
Synonym: Platinic ammonium chloride.
Description: Cubic, yellow crystals.
Formula: $(NH_4)_2PtCl_6$.

Constants: Mol wt: 444.05, mp: decomposes, d: 3.065.
Hazard Analysis and Counter measures
See platinum compounds and chlorides.

AMMONIUM CHLOROPLATINITE
General Information
Synonym: Platinous ammonium chloride.
Description: Red crystals.
Formula: $(NH_4)_2PtCl_4$.
Constants: Mol wt: 373.1, mp: decomposes, d: 2.936.
Hazard Analysis and Countermeasures
See platinum compounds and chlorides.

AMMONIUM CHLOROPLUMBATE
General Information
Description: Yellow crystals.
Formula: $(NH_4)_2PbCl_6$.
Constants: Mol wt: 456, mp: 120°C (decomposes), d: 2.925.
Hazard Analysis and Countermeasures
See lead compounds and chlorides.

AMMONIUM CHLOROSTANNATE
General Information
Description: White crystals.
Formula: $(NH_4)_2SnCl_6$.
Constants: Mol wt: 367.5, mp: decomposes, d: 2.4.
Hazard Analysis and Countermeasures
See tin compounds and chlorides.

AMMONIUM CHROMATE
General Information
Description: Yellow, crystalline material.
Formula: $(NH_4)_2CrO_4$.
Constants: Mol wt: 152.1, mp: decomposes, d: 1.866.
Hazard Analysis
Toxicity: See chromium compounds.
Fire Hazard: See chromates.
Explosion Hazard: Slight, when shocked or heated (Section 7).
Disaster Hazard: Moderately dangerous; shock or heat will
    expode it; can react with reducing material.
Countermeasures
Storage and Handling: Section 7.

AMMONIUM CHROMIC SULFATE
General Information
Description: Green or violet crystals.
Formula: $NH_4Cr(SO_4)_2 \cdot 12H_2O$.
Constants: Mol wt: 478.4, mp: 94°C (−9H₂O at 100°C),
    d: 1.720.
Hazard Analysis
Toxicity: See chromium compounds.
Disaster Hazard: See sulfates.
Countermeasures
Storage and Handling: Section 7.

AMMONIUM COBALTOUS o-PHOSPHATE
General Information
Description: Violet crystals or powder.
Formula: $NH_4CoPO_4 \cdot H_2O$.
Constant: Mol wt: 190.
Hazard Analysis
Toxicity: See cobalt compounds.
Countermeasures
Ventilation Control: Section 2.
Personal Hygiene: Section 3.

AMMONIUM COBALTOUS SULFATE
General Information
Description: Red crystals.

Formula: $(NH_4)_2SO_4 \cdot CoSO_4 \cdot 6H_2O$.
Constants: Mol wt: 395.3, d: 1.902.
Hazard Analysis and Countermeasures
See cobalt compounds and sulfates.
Countermeasures
Ventilation Control: Section 2.
Personal Hygiene: Section 3.

AMMONIUM COPPER ARSENITE
Hazard Analysis and Countermeasures
See arsenic compounds.

AMMONIUM CUPRIC CHLORIDE
General Information
Description: Blue crystals.
Formula: $2NH_4Cl \cdot CuCl_2 \cdot 2H_2O$.
Constants: Mol wt: 277.5, mp: decomposes at 110°C,
    d: 1.993.
Hazard Analysis and Countermeasures
See copper compounds and chlorides.

AMMONIUM CUPRIC SULFATE
General Information
Description: Blue, violet crystals.
Formula: $(NH_4)_2SO_4 \cdot CuSO_4$.
Constant: Mol wt: 291.8.
Hazard Analysis and Countermeasures
See copper compounds and sulfates.

AMMONIUM CUPROUS IODIDE
General Information
Description: Crystals.
Formula: $NH_4I \cdot CuI \cdot H_2O$
Constant: Mol wt: 353.4.
Hazard Analysis and Countermeasures
See copper compounds and iodides.

AMMONIUM CYANATE
General Information
Description: White crystals.
Formula: NH_4OCN.
Constants: Mol wt: 60.1, mp: decomposes at 60°C.
Hazard Analysis and Countermeasures
See cyanates.

AMMONIUM CYANIDE
General Information
Description: Solid, white powder or crystal.
Formula: NH_4CN.
Constants: Mol wt: 44.1, mp: 36°C, bp: 31.7°C (sublimes),
    d: 1.002 g/liter at 100°C, vap. press.: 400 mm at 20.5°C.
Hazard Analysis
Toxicity: See cyanides.
Fire Hazard: Moderate, when exposed to heat or flame. At
    95° a flammable gas is evolved (Section 6).
Explosion Hazard: Moderate, when heated. At 95°F a gas is
    liberated. See also hydrogen cyanide.
Disaster Hazard: See cyanides.
Countermeasures
Storage and Handling: Section 7.

AMMONIUM CYANOAURATE
General Information
Description: Colorless plates.
Formula: $NH_4Au(CN)_4 \cdot H_2O$.
Constants: Mol Wt: 337.3, mp: decomposes at 200°C.
Hazard Analysis and Countermeasures
See cyanides and gold compounds.

TOXIC HAZARD RATING CODE (For detailed discussion, see Section 1.)

.0   NONE: (a) No harm under any conditions; (b) Harmful only under unusual conditions or overwhelming dosage.

1   SLIGHT: Causes readily reversible changes which disappear after end of exposure.

2   MODERATE: May involve both irreversible and reversible changes; not severe enough to cause death or permanent injury.

3   HIGH: May cause death or permanent injury after very short exposure to small quantities.

U   UNKNOWN: No information on humans considered valid by authors.

AMMONIUM CYANOAURITE
General Information
Description: Colorless crystals.
Formula: $NH_4Au(CN)_2$.
Constants: Mol wt: 267.3, mp: decomposes at 100°C.
Hazard Analysis and Countermeasures
See cyanides and gold compounds.

AMMONIUM CYANOPLATINITE
General Information
Description: Yellow crystals.
Formula: $(NH_4)_2Pt(CN)_4 \cdot H_2O$.
Constant: Mol wt: 353.4.
Hazard Analysis and Countermeasures
See cyanides and platinum compounds.

AMMONIUM DECABORATE. See ammonium penta-borate.

AMMONIUM DICHROMATE. See ammonium bichro-mate.

AMMONIUM DICYANOGUANIDINE
General Information
Description: Crystals.
Formula: $NH_4[NH_2CN(NCN)CN]$.
Constants: Mol wt: 126.2, mp: $> 300°C$.
Hazard Analysis and Countermeasures
See cyanides.

AMMONIUM DIFLUOPHOSPHATE
General Information
Description: Colorless crystals.
Formula: $NH_4PO_2F_2$.
Constants: Mol wt: 119, mp: 213°C.
Hazard Analysis and Countermeasures
See fluorides and phosphates.

AMMONIUM DIHYDROGEN ARSENATE
General Information
Description: Colorless crystals.
Formula: $NH_4H_2AsO_4$.
Constants: Mol wt: 159, mp: decomposes to lose NH_3, d: 2.311.
Hazard Analysis and Countermeasures
See arsenic compounds.

AMMONIUM DIHYDROGEN PHOSPHATE. See am-monium phosphate, monobasic.

AMMONIUM DINITRO-O-SEC-BUTYL PHENATE
Hazard Analysis
Toxic Hazard Rating: U. A herbicide. See also phenol com-pounds.
Disaster Hazard: Dangerous. When heated to decomposi-tion emits highly toxic fumes.
Countermeasures
Storage and Handling: Section 7.

AMMONIUM DINITRO-o-CRESOL
General Information
Description: Crystals.
Formula: $C_7H_9O_5N_3$.
Constant: Mol wt: 215.2.
Hazard Analysis
Toxic Hazard Rating:
Acute Local: Irritant 3; Ingestion 3.
Acute Systemic: U.
Chronic Local: Irritant 3.
Chronic Systemic: Ingestion 2; Inhalation 2; Skin Absorp-tion 2.
Fire and Explosion Hazard: See nitrates.
Disaster Hazard: Dangerous; see nitrates.
Countermeasures
Ventilation Control: Section 2.
Personnel Protection: Section 3.
First Aid: Section 1.

AMMONIUM DINITRO-o-CRESOLATE
Hazard Analysis
Toxic Hazard Rating: U. A herbicide. See also cresols.
Fire Hazard: Dangerous. A powerful oxidizer.
Explosion Hazard: U.
Disaster Hazard: Dangerous. A powerful oxidizer. When heated to decomposition emits highly toxic fumes.
Countermeasures
Storage and Handling: Section 7.
Shipping Regulations: Section 11.
IATA: Flammable solid, yellow label, not acceptable (pas-senger), 12 kilograms (cargo).

AMMONIUM DITHIOCARBAMATE
General Information
Synonym: Ammonium sulfocarbamate.
Description: Yellow, lustrous, almost odorless crystals when fresh; soluble in water; decomposes in air.
Formula: NH_2CSSNH_4.
Constant: Mol wt: 110.19.
Hazard Analysis
Toxic Hazard Rating: U. See ammonium sulfamate.

AMMONIUM EMBELATE
General Information
Description: Grayish-violet powder.
Formula: $(NH_4)_2C_{18}H_{26}O_4$.
Constant: Mol wt: 342.47.
Hazard Analysis
Toxic Hazard Rating:
Acute Local: Irritant 1; Ingestion 1; Inhalation 1.
Acute Systemic: U.
Chronic Local: U; Chronic Systemic: U.
Toxicology: If inhaled, may cause violent and prolonged sneezing.
Countermeasures
Ventilation Control: Section 2.
Personal Hygiene: Section 3.
Storage and Handling: Section 7.

AMMONIUM ETHYL SULFATE
General Information
Synonym: Ammonium sulfethylate.
Description: Hygroscopic crystals; freely soluble in water and alcohol.
Formula: $NH_4C_2H_5SO_4$.
Constant: Mol wt: 143.17, mp: 99°C.
Hazard Analysis
Toxic Hazard Rating: U. Probably an irritant.
Disaster Hazard: Dangerous. See sulfates.

AMMONIUM FERRICYANIDE
General Information
Description: Red crystals.
Formula: $(NH_4)_3Fe(CN)_6$.
Constants: Mol wt: 266.1, mp: decomposes.
Hazard Analysis and Countermeasures
See ferricyanides.

AMMONIUM FERROCYANIDE
General Information
Description: Yellow to blue crystals.
Formula: $(NH_4)_4Fe(CN)_6 \cdot 3H_2O$.
Constants: Mol wt: 338.2, mp: decomposes.
Hazard Analysis and Countermeasures
See ferrocyanides.

AMMONIUM FLUOANTIMONITE
General Information
Description: Colorless crystals.
Formula: $(NH_4)_2SbF_5$.
Constants: Mol wt: 252.8, mp: sublimes with decomposition.
Hazard Analysis and Countermeasures
See fluorides and antimony compounds.

AMMONIUM FLUOBORATE
General Information
Description: Crystals.

Formula: NH_4BF_4.
Constants: Mol wt: 104.9, mp: sublimes, d: 1.851.
Hazard Analysis and Countermeasures
See fluorides and boron compounds.

AMMONIUM FLUOGALLATE
General Information
Description: White crystals.
Formula: $(NH_4)_3GaF_6$.
Constants: Mol wt: 237.8, bp: decomposes to yield GaF_3.
Hazard Analysis and Countermeasures
See fluorides and gallium compounds.

AMMONIUM FLUOGERMANATE
General Information
Description: Colorless crystals.
Formula: $(NH_4)_2GeF_6$.
Constants: Mol wt: 222.1, d: 2.564 at 25°/25°C.
Hazard Analysis and Countermeasures
See fluorides and germanium compounds.

AMMONIUM FLUORIDE
General Information
Description: White crystals.
Formula: NH_4F.
Constants: Mol wt: 37.04, mp: sublimes, d: 1.315.
Hazard Analysis and Countermeasures
See fluorides.
Shipping Regulations: Section 11.
  IATA: Other restricted articles, class B, 12 kilograms (passenger), 45 kilograms (cargo).

AMMONIUM FLUOSILICATE
General Information
Synonyms: Cryptohalite; ammonium silico-fluoride.
Description: Colorless crystals.
Formula: $(NH_4)_2SiF_6$.
Constants: Mol wt: 178.1, mp: sublimes, d: 2.01.
Hazard Analysis and Countermeasures
See fluosilicates and fluorides.

AMMONIUM FLUOSULFONATE
General Information
Description: Colorless needles.
Formula: NH_4SO_3F.
Constants: Mol wt: 117.1, mp: 244.7°C.
Hazard Analysis and Countermeasures
See fluosulfonates.

AMMONIUM FLUOTITANATE
General Information
Description: Crystals.
Formula: $(NH_4)_2TiF_6$.
Constants: Mol wt: 198, mp: decomposes.
Hazard Analysis and Countermeasures
See fluorides.

AMMONIUM FLUOZIRCONATE
General Information
Description: Crystals.
Formula: $(NH_4)_2ZrF_6$.
Constants: Mol wt: 241.3, d: 1.154.
Hazard Analysis and Countermeasures
See fluorides.

AMMONIUM FORMATE
General Information
Description: White, deliquescent crystals.
Formula: NH_4COOH.

Constants: Mol wt: 63.1, mp: 116°C, bp: decomposes at 180°C, d: 1.266.
Hazard Analysis
Toxicity: Details unknown. See also esters and formic acid.

AMMONIUM GLUTAMATE. See monoammonium glutamate.

AMMONIUM HEXAFLUOPHOSPHATE
General Information
Synonym: Ammonium phosphorus hexafluoride.
Description: Colorless crystals.
Formula: NH_4PF_6.
Constants: Mol wt: 163, mp: decomposes, d: 2.180.
Hazard Analysis and Countermeasures
See fluorides and phosphates.

AMMONIUM HEXAFLUOPHOSPHATE FLUORIDE
General Information
Description: White crystals or powder.
Formula: $NH_4PF_6NH_4F$.
Constants: Mol wt: 200.06, mp: no melting, sublimes at about 140°C.
Hazard Analysis and Countermeasures
See fluorides and ammonium hexafluophosphate.

AMMONIUM HYDRATE. See ammonium hydroxide.

AMMONIUM HYDROGEN CARBONATE. See ammonium bicarbonate.

AMMONIUM HYDROGEN FLUORIDE. See ammonium bifluoride.

AMMONIUM HYDROGEN SELENATE
General Information
Description: Crystals.
Formula: NH_4HSeO_4.
Constants: Mol wt: 162.0, mp: decomposes, d: 2.162.
Hazard Analysis and Countermeasures
See selenium compounds.

AMMONIUM HYDROGEN SULFATE
General Information
Description: White, rhombic crystals; soluble in water, insoluble in acetone.
Formula: NH_4HSO_4.
Constants: Mol wt: 115.11, mp: 146.9°C, d: 1.78.
Hazard Analysis
Toxic Hazard Rating: See sulfuric acid.
Disaster Hazard: Dangerous, when heated to decomposition it emits highly toxic fumes and sulfuric acid and sulfur oxides.
Countermeasures
Storage and Handling: Section 7.
Shipping Regulations: Section 11.
  IATA: Other restricted articles class B, no label, 12 kilograms (passenger), 45 kilograms (cargo).

AMMONIUM HYDROGEN SULFIDE. See ammonium sulfhydrate.

AMMONIUM HYDROGEN SULFITE. See ammonium bisulfite.

AMMONIUM HYDROSULFIDE. See ammonium sulfhydrate.

AMMONIUM HYDROXIDE
General Information
Synonyms: Aqua ammonium; water of ammonia; aqua ammonia; ammonium hydrate.

TOXIC HAZARD RATING CODE *(For detailed discussion, see Section 1.)*

0 NONE: (a) No harm under any conditions; (b) Harmful only under unusual conditions or overwhelming dosage.

1 SLIGHT: Causes readily reversible changes which disappear after end of exposure.

2 MODERATE: May involve both irreversible and reversible changes; not severe enough to cause death or permanent injury.

3 HIGH: May cause death or permanent injury after very short exposure to small quantities.

U UNKNOWN: No information on humans considered valid by authors

Description: Colorless liquid.
Formula: NH_4OH.
Constants: Mol wt: 35.05, mp: $-77°C$.
Hazard Analysis
Toxic Hazard Rating:
Acute Local: Irritant 2; Ingestion 3; Inhalation 2.
Acute Systemic: U.
Chronic Local: Irritant 2.
Chronic Systemic: U.
A general purpose food additive which migrates to food from packaging materials (Section 10).
Fire Hazard: Slight; when heated, it emits toxic fumes; can react with oxidizing materials (Section 6).
Disaster Hazard: Dangerous; emits irritating fumes and liquid can inflict burns. Use with adequate ventilation.
Countermeasures
Ventilation Control: Section 2.
Personnel Protection: Section 3.
First Aid: Section 1.
Storage and Handling: Section 7.
MCA warning label.

AMMONIUM IODIDE
General Information
Description: Colorless, hygroscopic crystals.
Formula: NH_4I.
Constants: Mol wt: 145, mp: sublimes at 551°C, bp: 404.9°C, d: 2.514, vap. press.: 1 mm at 210.9°C (sublimes).
Hazard Analysis and Countermeasures
See iodides.

AMMONIUM MAGNESIUM ARSENATE
General Information
Description: Colorless crystals.
Formula: $NH_4MgAsO_4 \cdot 6H_2O$.
Constants: Mol wt: 289.4, mp: decomposes, 1.932 at 15°C.
Hazard Analysis and Countermeasures
See arsenic compounds.

AMMONIUM MAGNESIUM CHROMATE
General Information
Description: Yellow crystals.
Formula: $(NH_4)_2CrO_4 \cdot MgCrO_4 \cdot 6H_2O$.
Constants: Mol wt: 400.5, mp: decomposes, d: 1.84.
Hazard Analysis
Toxicity: See chromium compounds.
Fire Hazard: Moderate, as a result of chemical reaction with reducing agents (Section 6).
Caution: An oxidizer.
Disaster Hazard: Moderately dangerous; when heated, it can explode.
Countermeasures
Storage and Handling: Section 7.

AMMONIUM MOLYBDATE
General Information
Description: Colorless or slightly greenish or yellowish crystals.
Formula: $(NH_4)_6Mo_7O_{24} \cdot 4H_2O$.
Constants: Mol wt: 1236.0, mp: decomposes, d: 2.27.
Hazard Analysis
Toxic Hazard Rating:
Acute Local: Irritant 2, ingestion 2, Inhalation 2.
Acute Systemic: Ingestion 2.
Chronic Local: Irritant 1.
Chronic Systemic: Ingestion 1.
Toxicology: No cases of human poisoning have been reported. Animal experiments indicate relatively low systemic toxicity but moderately severe local irritation of skin, eyes and mucous membranes. Large doses have produced kidney damage in experimental animals. See molybdenum compounds.

AMMONIUM MOLYBDOTELLURATE
General Information
Description: Colorless crystals.

Formula: $(NH_4)_6(TeMo_6O_{24}) \cdot 7H_2O$.
Constants: Mol wt: 1321.7, mp: 550°C (decomp.), bp: decomposes, d: 2.78.
Hazard Analysis and Countermeasures
See tellurium compounds.

AMMONIUM MONOHYDROGEN ARSENATE (SOLID)
General Information
Description: White crystals or powder.
Formula: $(NH_4)_2HAsO_4$.
Constants: Mol wt: 176, mp: decomposes, d: 1.989.
Hazard Analysis and Countermeasures
See arsenic compounds.

AMMONIUM MONOSULFIDE. See ammonium sulfide.

AMMONIUM NICKEL CHLORIDE
General Information
Description: Green crystals.
Formula: $NH_4Cl \cdot NiCl_2 \cdot 6H_2O$.
Constants: Mol wt: 291.2, d: 1.645.
Hazard Analysis and Countermeasures
See nickel compounds and chlorides.

AMMONIUM NICKEL SULFATE
General Information
Synonym: Double nickel salt.
Description: Black to green crystals.
Formula: $(NH_4)_2SO_4 \cdot NiSO_4 \cdot 6H_2O$.
Constants: Mol wt: 395, d: 1.923.
Hazard Analysis and Countermeasures
See nickel compounds and sulfates.

AMMONIUM NITRATE
General Information
Description: Colorless crystals.
Formula: NH_4NO_3.
Constants: Mol wt: 80.05, mp: 169.6°C, bp: decomposes at 210°C, d: 1.725 at 25°C.
Hazard Analysis
Toxic Hazard Rating:
Acute Local: Irritant 1; Allergen 1; Inhalation 1.
Acute Systemic: U.
Chronic Local: Irritant 1; Allergen 1; Inhalation 1.
Chronic Systemic: U.
Toxicology: There have been reports of faintness and low blood pressure in workers exposed. These symptoms could be due to nitrites present as impurities. See also nitrates.
Fire Hazard: See nitrates.
To Fight Fire: Use water in large amounts. It is important that the mass of material be kept cool and that burning be extinguished promptly. Ventilate well (Section 6).
Explosion Hazard: May explode under confinement and high temperatures. Explosions have occurred in ships' holds, etc. There have been warehouse fires that did not detonate (Section 7). See also nitrates.
Caution: This material explodes more readily if contaminated, must be kept cool and unconfined. See also explosives, high.
Disaster Hazard: Dangerous; heat and confinement may explode it; when heated to decomposition it emits highly toxic fumes of oxides of nitrogen; can react vigorously with reducing materials.
Countermeasures
Personal Hygiene: Section 3.
Storage and Handling: Section 7.
Shipping Regulations: Section 11.
I.C.C.: Oxidizing material; yellow label, 100 pounds.
Coast Guard Classification: Oxidizing material; yellow label.

AMMONIUM NITRATE, FERTILIZER. See ammonium nitrate.
Countermeasures
Shipping Regulations: Section 11.

I.C.C.: Oxidizing material; yellow label, 100 pounds.
IATA: Oxidizing material, yellow label, 12 kilograms (passenger), 45 kilograms (cargo).

AMMONIUM NITRATE, MIXED FERTILIZER
Countermeasures
Shipping Regulations: Section 11.
  ICC: Oxidizing material, yellow label, 100 pounds.
  IATA: Oxidizing material, yellow label, 12 kilograms (passenger), 45 kilograms (cargo).

AMMONIUM NITRATE, WITH CARBONATES OR PHOSPHATES
Countermeasures
Shipping Regulations: Section 11.
  IATA: Oxidizing material, yellow label, 12 kilograms (passenger), 45 kilograms (cargo).

AMMONIUM NITRATE, WITH OR WITHOUT ORGANIC COATING
Countermeasures
Shipping Regulations: Section 11.
  IATA: Oxidizing material, yellow label, 12 kilograms (passenger), 45 kilograms (cargo).

AMMONIUM NITRITE
General Information
Description: White to yellow crystals.
Formula: NH_4NO_2.
Constants: Mol wt: 64.1, mp: decomposes, flash p: 158°F, vap. d: 1.69.
Hazard Analysis
Toxicity: See nitrites.
Fire Hazard: See nitrites.
Explosion Hazard: Severe, when shocked or exposed to heat (Section 7).
Disaster Hazard: See nitrites.
Countermeasures
Storage and Handling: Section 7.

AMMONIUM OXALATE
General Information
Description: Colorless crystals.
Formula: $(NH_4)_2C_2O_4 \cdot H_2O$.
Constants: Mol wt: 142.12, mp: decomposes, d: 1.50.
Hazard Analysis
Toxicity: See oxalates.

AMMONIUM PENTABORATE
General Information
Synonym: Ammonium decaborate.
Description: White solid.
Formula: $NH_4B_5O_8 \cdot 4H_2O$.
Constant: Mol wt: 272.20.
Hazard Analysis
Toxicity: See boron compounds.

AMMONIUM PERCHLORATE
General Information
Description: White crystals.
Formula: NH_4ClO_4.
Constants: Mol wt: 117.50, mp: decomposes, d: 1.95.
Hazard Analysis
Toxicity: See perchlorates.
Fire Hazard: Moderate; when exposed to heat or flame or by spontaneous chemical reaction with reducing materials (Section 6).
Caution: A very powerful oxidizer. Ignites violently with combustibles.

Explosion Hazard: Severe, when shocked, exposed to heat or by spontaneous chemical reaction. Explodes at high temperatures. When contaminated by reducing materials it can become a very sensitive high explosive (Section 7). See also perchlorates.
Disaster Hazard: See perchlorates and explosives, high.
Countermeasures
Storage and Handling: Section 7.
Shipping Regulations: Section 11.
  I.C.C.: Oxidizing material; yellow label, 100 pounds.
  Coast Guard Classification: Oxidizing material; yellow label.
  IATA: Oxidizing material, yellow label, 12 kilograms (passenger), 45 kilograms (cargo).

AMMONIUM PERCHROMATE. See ammonium peroxychromate.

AMMONIUM m-PERIODATE
General Information
Description: Colorless crystals.
Formula: NH_4IO_4.
Constants: Mol wt: 209, mp: explodes, d: 3.056.
Hazard Analysis
Toxicity: See iodates.
Fire Hazard: See iodates.
Explosion Hazard: See iodates.
Disaster Hazard: See iodates.
Countermeasures
Storage and Handling: Section 7.

AMMONIUM PERMANGANATE
General Information
Description: Crystalline solid.
Formula: NH_4MnO_4.
Constants: Mol wt: 137.0, mp: explodes, d: 2.208.
Hazard Analysis
Toxicity: See manganese compounds.
Fire Hazard: Moderate, by chemical reaction with reducing agents (Section 6).
Caution: A powerful oxidizer.
Explosion Hazard: Moderate, when shocked or exposed to heat. Can be exploded by percussion (Section 7).
Disaster Hazard: Moderately dangerous, shock and heat will explode it; when heated to decomposition, it emits toxic fumes; can react with reducing material.
Countermeasures
Storage and Handling: Section 7.
Shipping Regulations: Section 11.
  I.C.C.: Oxidizing material, yellow label, 100 pounds.
  Coast Guard Classification: Oxidizing material; yellow label.
  IATA: Oxidizing material, not acceptable (passenger and cargo).

AMMONIUM PEROXYBORATE
General Information
Description: White crystals.
Formula: $NH_4BO_3 \cdot \frac{1}{2}H_2O$.
Constants: Mol wt: 85.9, mp: decomposes.
Hazard Analysis
Toxicity: See boron compounds.
Fire Hazard: Slight, by chemical reaction with reducing agents (Section 6).
Caution: An oxidizer.
Countermeasures
Storage and Handling: Section 7.

TOXIC HAZARD RATING CODE (For detailed discussion, see Section 1.)

0 NONE: (a) No harm under any conditions; (b) Harmful only under unusual conditions or overwhelming dosage.

1 SLIGHT: Causes readily reversible changes which disappear after end of exposure.

2 MODERATE: May involve both irreversible and reversible changes; not severe enough to cause death or permanent injury.

3 HIGH: May cause death or permanent injury after very short exposure to small quantities.

U UNKNOWN: No information on humans considered valid by authors.

AMMONIUM PEROXYCHROMATE
General Information
Description: Red-brown crystals.
Formula. $(NH_4)_3CrO_8$.
Constants: Mol wt: 234.1, mp: decomposes at 40°C, bp: explodes at 50°C.
Hazard Analysis
Toxicity: See chromium compounds.
Fire Hazard: Moderate, by chemical reaction with reducing agents (Section 6).
Caution: A powerful oxidizer.
Explosion Hazard: Moderate, when heated (Section 7).
Disaster Hazard: Moderately dangerous; when heated to decomposition it emits toxic fumes and may explode.
Countermeasures
Storage and Handling: Section 7.

AMMONIUM PEROXYDISULFATE. See ammonium persulfate.

AMMONIUM PERRHENATE
General Information
Description: White plates.
Formula: NH_4ReO_4.
Constants: Mol wt: 268.4, mp: decomposes, bp: decomposes.
Hazard Analysis
Toxicity: Details unknown.
Fire Hazard: Moderate, by chemical reaction with reducing agents (Section 6).

AMMONIUM PERSULFATE
General Information
Synonym: Ammonium peroxydisulfate.
Description: White crystals.
Formula: $(NH_4)_2S_2O_8$.
Constants: Mol wt: 228.20, mp: decomposes at 120°C, d: 1.982.
Hazard Analysis
Toxic Hazard Rating:
Acute Local: Irritant 1; Ingestion 1; Inhalation 1.
Acute Systemic: U.
Chronic Local: Irritant 1.
Chronic Systemic: U.
Fire Hazard: Moderate, by chemical reaction with reducing agents (Section 6).
Caution: A powerful oxidizer.
Explosion Hazard: Slight; oxygen released quietly in a fire, probably at a low temperature.
Disaster Hazard: Dangerous. See sulfates. Can react vigorously with reducing agents.
Countermeasures
Ventilation Control: Section 2.
Personal Hygiene: Section 3.
Storage and Handling: Section 7.

AMMONIUM PHOSPHATE, DIBASIC
General Information
Synonyms: Ammonium phosphate, secondary; diammonium hydrogen phosphate; diammonium phosphate; DAP.
Description: White crystals or powder; soluble in water; insoluble in alcohol.
Formula: $(NH_4)_2HPO_4$.
Constants: Mol wt: 132, d: 1.619.
Hazard Analysis
Toxic Hazard Rating: U. A general purpose food additive which migrates to food from packaging materials (Section 10).
Disaster Hazard: See phosphates.

AMMONIUM PHOSPHATE, MONOBASIC
General Information
Synonyms: Ammonium acid phosphate; ammonium biphosphate; ammonium dihydrogen phosphate; ammonium phosphate, primary.

Description: Brilliant white crystals or powder, moderately soluble in water.
Formula: $NH_4H_2PO_4$.
Constants: Mol wt: 115, d: 1.803.
Hazard Analysis
Toxic Hazard Rating: U. A general purpose food additive (Section 10).
Disaster Hazard: Dangerous. See phosphates.

AMMONIUM PHOSPHATE, PRIMARY. See ammonium phosphate monobasic.

AMMONIUM PHOSPHATE, SECONDARY. See ammonium phosphate, dibasic.

AMMONIUM PHOSPHORUS HEXAFLUORIDE. See ammonium hexafluophosphate.

AMMONIUM PICRATE
General Information
Synonyms: Ammonium carbazotate; ammonium picronitrate.
Description: Yellow crystals.
Formula: $NH_4C_6H_2N_3O_7$.
Constants: Mol wt: 246.14, mp: decomposes, bp: explodes at 423°C, d: 1.719.
Hazard Analysis
Toxic Hazard Rating:
Acute Local: Irritant 2; Allergen 2; Ingestion 3; Inhalation 3.
Acute Systemic: Ingestion 3; Inhalation 3.
Chronic Local: Allergen 2.
Chronic Systemic: Ingestion 3; Inhalation 3; Skin Absorption 3.
Fire Hazard: Moderate, by spontaneous chemical reaction (Section 6).
Caution: A powerful oxidizer.
Explosion Hazard: Severe when shocked or exposed to heat or flame (Section 7). See also explosives, high.
Disaster Hazard: Highly dangerous; will explode when shocked; when heated to decomposition, it emits highly toxic fumes of oxides of nitrogen, etc.; can react vigorously with reducing materials.
Countermeasures
Ventilation Control: Section 2.
Personnel Protection: Section 3.
First Aid: Section 1.
Storage and Handling: Section 7.
Shipping Regulations: Section 11.
I.C.C.: High explosive.
Coast Guard Classification: High explosive.
IATA: Explosive, not acceptable (passenger and cargo).

AMMONIUM PICRATE, WET
General Information
Synonym: Ammonium carbazotate.
Hazard Analysis
Toxicity: See ammonium picrate.
Fire Hazard: Moderate, by chemical reaction with reducing agents (Section 6).
Caution: An oxidizer.
Explosion Hazard: Moderate, when heated (Section 7).
Disaster Hazard: Dangerous; when heated to decomposition, it emits toxic fumes of oxides of nitrogen, etc., and explodes.
Countermeasures
Storage and Handling: Section 7.
Shipping Regulations: Section 11.
I.C.C.: (Not to exceed 16 oz.) See Section 11.
Coast Guard Classification: (Not less than 10% water, not to exceed 16 oz.) Inflammable solid.
IATA (less than 10% water): Explosive, not acceptable (passenger and cargo).
IATA (not less than 10% water): Flammable solid, yellow label, ½ kilogram (passenger and cargo).

AMMONIUM PICRONITRATE. See ammonium picrate.

AMMONIUM POLYMANNURATE. See ammonium alginate.

AMMONIUM POLYSULFIDE. See ammonium sulfide.

AMMONIUM SACCHARIN
General Information
Description: White crystals or a white crystalline powder; freely soluble in water.
Formula: $C_7H_8N_2O_3S$.
Constant: Mol wt: 200.
Hazard Analysis
Toxic Hazard Rating: U. A nonnutritive sweetener food additive (Section 10).

AMMONIUM SELENATE
General Information
Description: Colorless crystals.
Formula: $(NH_4)_2SeO_4$.
Constants: Mol wt: 179.04, mp: decomposes, d: 2.194.
Hazard Analysis and Countermeasures
See selenium compounds.

AMMONIUM SELENIDE
General Information
Description: Brown crystals.
Formula: $(NH_4)_2Se$.
Constants: Mol wt: 115.0, bp: decomposes.
Hazard Analysis
Toxicity: See selenium compounds.
Fire Hazard: See hydrogen selenide.
Disaster Hazard: Dangerous; when heated to decomposition, or on contact with acid or acid fumes, it emits highly toxic fumes of selenium and it will react with water or steam to produce toxic and flammable vapors.
Countermeasures
Storage and Handling: Section 7.

AMMONIUM SELENITE
General Information
Description: Colorless or slightly reddish crystals.
Formula: $(NH_4)_2SeO_3 \cdot H_2O$.
Constant: Mol wt: 167.05.
Hazard Analysis and Countermeasures
See selenium compounds.

AMMONIUM SILICOFLUORIDE. See ammonium fluosilicate.

AMMONIUM SULFAMATE
General Information
Synonym: Ammate.
Description: Deliquescent crystalline material (white crystalline solid).
Formula: $NH_4OSO_2NH_2$.
Constants: Mol wt: 114.1, bp: 160°C (decomposes), mp: 125°C.
Hazard Analysis
Toxicology: A herbicide. Limited animal experiments indicate low toxicity. Ingestion may cause gastric irritation.
TLV: ACGIH (recommended), 15 milligrams per cubic meter of air.
Explosion Hazard: Slight, when exposed to heat or by spontaneous chemical reaction (hydrolysis); in a hot acid solution this material can undergo spontaneous hydrolysis liberating much heat.

Disaster Hazard: Dangerous; see sulfonates.
Countermeasures
Storage and Handling: Section 7.

AMMONIUM SULFATE
General Information
Description: Brownish-gray to white crystals.
Formula: $(NH_4)_2SO_4$.
Constants: Mol wt: 132.09, mp: decomposes, d: 1.77.
Hazard Analysis
Toxic Hazard Rating:
    Acute Local: Irritant 1; Ingestion 1; Inhalation 1.
    Acute Systemic: U.
    Chronic Local: U.
    Chronic Systemic: U.
A general purpose food additive (Section 10).
    Disaster Hazard: Dangerous. See sulfates.

AMMONIUM SULFHYDRATE
General Information
Synonyms: Ammonium hydrosulfide; ammonium hydrogen sulfide.
Description: Powder or crystals.
Formula: NH_4HS.
Constants: Mol wt: 51.11, mp: 118°C (decomposes), bp: 120°C (sublimes), vap. press. 400 mm at 21.8°C.
Hazard Analysis
Toxic Hazard Rating:
    Acute Local: Irritant 3; Ingestion 3; Inhalation 3.
    Acute Systemic: Ingestion 3; Inhalation 3; Skin Absorption 3.
    Chronic Local: U.
    Chronic Systemic: U.
Toxicity: Highly irritating. Penetrates skin readily.
Fire Hazard: See sulfides.
Disaster Hazard: See sulfides.
Explosion Hazard: See sulfides.
Countermeasures
Storage and Handling: Section 7.
Shipping Regulations: Section 11.
    IATA: Other restricted articles, class A, no label, 40 liters (passenger), 220 liters (cargo).

AMMONIUM SULFIDE
General Information
Synonym: Ammonium polysulfide.
Description: Yellow, hygroscopic crystals.
Formula: $(NH_4)_2S$.
Constants: Mol wt: 68.2, mp: decomposes.
Hazard Analysis
Toxicity: See sulfides. Evolves hydrogen sulfide on contact with acid or acid fumes. Fatal poisoning has been reported from use in hair waving lotions.
Fire Hazard: See sulfides.
Explosion Hazard: See sulfides.
Disaster Hazard: See sulfides.
Countermeasures
Storage and Handling: Section 7.
Shipping Regulations: Section 11.
    IATA (Solution): Other restricted articles, class A, no label 40 liters (passenger), 220 liters (cargo).

AMMONIUM SULFITE
General Information
Description: Colorless crystals.
Formula: $(NH_4)_2SO_3 \cdot H_2O$.
Constants: Mol wt: 134.16, mp: decomposes, bp: sublimes at 150°C, d: 1.41 at 25°C.

TOXIC HAZARD RATING CODE (For detailed discussion, see Section I.)

0 NONE: (a) No harm under any conditions; (b) Harmful only under unusual conditions or overwhelming dosage.

1 SLIGHT: Causes readily reversible changes which disappear after end of exposure.

2 MODERATE: May involve both irreversible and reversible changes; not severe enough to cause death or permanent injury.

3 HIGH: May cause death or permanent injury after very short exposure to small quantities.

U UNKNOWN: No information on humans considered valid by authors.

Hazard Analysis and Countermeasures
See sulfites.

AMMONIUM SULFOCYANATE. See ammonium thiocyanate.

AMMONIUM TELLURATE
General Information
Description: White powder.
Formula: $(NH_4)_2TeO_4$.
Constants: Mol wt: 227.7, mp: decomposes, d: 3.01 at 25° C.
Hazard Analysis and Countermeasures
See tellurium compounds.

AMMONIUM TETRABORATE. See ammonium biborate.

AMMONIUM TETRACHLOROZINCATE
General Information
Description: White, thin, shiny platelets. Hygroscopic and water soluble.
Formula: $ZnCl_2 \cdot 2NH_4Cl$.
Constants: Mol wt: 243.3, mp: 150° C (approx.), d: 1.879.
Hazard Analysis
Toxic Hazard Rating:
  Acute Local: Irritant 2; Ingestion 2; Inhalation 2.
  Acute Systemic: U.
  Chronic Local: Irritant 1.
  Chronic Systemic: U.
Toxicology: Effects are those of components: zinc chloride and ammonium chloride, both of which are irritants and are described under appropriate headings.
Disaster Hazard: Dangerous, See chlorides.

AMMONIUM THALLIUM CHLORIDE
General Information
Description: Colorless crystals.
Formula: $3NH_4Cl \cdot TlCl_3 \cdot 2H_2O$.
Constants: Mol wt: 507.3, d: 2.39.
Hazard Analysis and Countermeasures
See thallium compounds and chlorides.

AMMONIUM THIOANTIMONATE
General Information
Description: Yellow crystals.
Formula: $(NH_4)_3SbS_4 \cdot 4H_2O$.
Constants: Mol wt: 376.2, mp: decomposes.
Hazard Analysis
Toxicity: See antimony compounds.
Disaster Hazard: Dangerous; when heated to decomposition or on contact with acid or acid fumes, it emits highly toxic fumes of oxides of sulfur and antimony.
Countermeasures
Storage and Handling: Section 7.

AMMONIUM THIOCYANATE
General Information
Synonym: Ammonium sulfocyanate.
Description: Colorless solid.
Formula: NH_4SCN.
Constants: Mol wt: 76.1, mp: 149.6° C, bp: decomposes at 170° C, d: 1.305.
Hazard Analysis
Toxicity: A herbicide. See thiocyanates.
Disaster Hazard: See thiocyanates.
Countermeasures
Storage and Handling: Section 7.

AMMONIUM THIOGLYCOLATE
General Information
Description: Colorless liquid, strong skunklike odor.
Formula: $HSCH_2COONH_4$.
Constant: Mol wt: 109.1.
Hazard Analysis
Toxic Hazard Rating:
  Acute Local: Irritant 1; Allergen 2; Ingestion 2; Inhalation 2.

Acute Systemic: Ingestion 3; Inhalation 3.
Chronic Local: Allergen 2.
Chronic Systemic: Inhalation 3.
Caution: Emits hydrogen sulfide. Upon local contact it can cause a contact dermatitis (Section 9).
Disaster Hazard: Dangerous; when heated to decomposition or on contact with acid or acid fumes, it emits highly toxic fumes of sulfides.
Countermeasures
Ventilation Control: Section 2.
Personal Hygiene: Section 3.
First Aid: Section 1.
Storage and Handling: Section 7.

AMMONIUM THIOSULFATE
General Information
Description: White, monoclinic crystals.
Formula: $(NH_4)_2S_2O_3$.
Constant: Mol wt: 148.2.
Hazard Analysis
Toxicity: See thiosulfates.
Disaster Hazard: Dangerous; see sulfates.
Countermeasures
Storage and Handling: Section 7.

AMMONIUM TRICHLOROACETATE
General Information
Description: Colorless crystals.
Formula: $NH_4O_2CCCl_3$.
Constant: Mol wt: 180.6.
Hazard Analysis
Toxic Hazard Rating:
  Acute Local: Irritant 2; Ingestion 3; Inhalation 2.
  Acute Systemic: U.
  Chronic Local: Irritant 1.
  Chronic Systemic: Ingestion 1; Inhalation 1.
  A herbicide.
Disaster Hazard: Dangerous; when heated to decomposition or on contact with acid or acid fumes, it emits toxic fumes; will react with water or steam to produce toxic or corrosive fumes.
Countermeasures
Ventilation Control: Section 2.
First Aid: Section 1.
Storage and Handling: Section 7.

AMMONIUM TRIIODIDE
General Information
Description: Brown crystals.
Formula: NH_4I_3.
Constants: Mol wt: 398.8, d: 3.749.
Hazard Analysis and Countermeasures
See iodides.

AMMONIUM URANIUM FLUORIDE. See uranium ammonium fluoride.

AMMONIUM URANYL CARBONATE
General Information
Description: Yellow crystals.
Formula: $2(NH_4)_2CO_3 \cdot UO_2CO_3 \cdot 2H_2O$.
Constants: Mol wt: 558.3, mp: decomposes at 100° C, d: 2.773.
Hazard Analysis
Toxicity: See uranium compounds, soluble.
Radiation Hazard: See uranium.

AMMONIUM URANYL PENTAFLUORIDE
General Information
Description: Crystals.
Formula: $(NH_4)_3UO_2F_5$.
Constants: Mol wt: 419.2, mp: sublimes, d: 3.186.
Hazard Analysis
Toxicity: See uranium compounds, soluble and fluorides.
Radiation Hazard: See uranium.
Disaster Hazard: Dangerous; see fluorides.

Countermeasures
Storage and Handling: Section 7.

AMMONIUM m-VANADATE
General Information
Description: Colorless to yellow crystals.
Formula: NH_4VO_3.
Constants: Mol wt: 117, mp: decomposes, d: 2.326.
Hazard Analysis
Toxicity: See vanadium compounds.

AMMONIUM ZINC SULFATE
General Information
Description: White crystals.
Formula: $(NH_4)_2SO_4 \cdot ZnSO_4 \cdot 6H_2O$.
Constants: Mol wt: 401.7, mp: decomposes, d: 1.931.
Hazard Analysis and Countermeasures
See zinc compounds and sulfates.

AMMONIUM ZIRCONYL CARBONATE
General Information
Formula: $(NH_4)_3ZrOH(CO_3)_2 \cdot 2H_2O$.
Constant: Mol wt: 302.
Hazard Analysis
Toxic Hazard Rating: U. See zirconium compounds.

AMMONOBASIC MERCURIC CHLORIDE. See mercury compounds, inorganic.

AMMUNITION FOR CANNON
Countermeasures
Shipping Regulations: Section 11.
  I.C.C. (with empty projectiles): Explosive B, not accepted.
    (with explosive projectiles): Explosive A, not accepted.
    (with gas projectiles): Explosive A, not accepted.
    (with illuminating projectiles): Explosive A, not accepted.
    (with incendiary projectiles): Explosive A, not accepted.
    (with inert loaded projectiles): Explosive B, not accepted.
    (with smoke projectiles): Explosive A, not accepted.
    (with solid projectiles): Explosive B, not accepted.
    (without projectiles): Explosive B, not accepted.
  IATA: Explosives, not acceptable (passenger and cargo).

AMMUNITION, OR AMMUNITION MATERIAL, NON EXPLOSIVE
Shipping Regulations: Section 11.
  IATA: Not restricted (passenger and cargo).

AMOBARBITAL
General Information
Synonyms: 5-Ethyl-5-isoamylbarbituric acid; amytal.
Description: Slightly bitter crystals.
Formula: $C_{11}H_{18}N_2O_3$.
Constant: Mol wt: 226.27.
Hazard Analysis
Toxic Hazard Rating:
  Acute Local: Allergen 1; Ingestion 2.
  Acute Systemic: Ingestion 2.
  Chronic Local: Allergen 1.
  Chronic Systemic: Ingestion 1; Inhalation 1.
Caution: Continued use of this material may lead to habituation.
Countermeasures
Personal Hygiene: Section 3.

Storage and Handling: Section 7.

AMP. See 2-amino-2 methyl-1-propanol.

AMPHETAMINE. See "benzedrine."

AMPROLIUM
General Information
 Synonym: 1-[(4-amino-2-propyl-5-pyrimidyl) methyl] -2-picolinium chloride, hydrochloride.
Hazard Analysis
Toxic Hazard Rating: U. A food additive permitted in the feed and drinking water of animals and/or for the treatment of food producing animals. A food additive permitted in food for human consumption (Section 10).
Disaster Hazard: Dangerous. See chlorides.

"AMSCO" NAPHTHAS AND SOLVENTS
General Information
Description: Liquid petroleum hydrocarbons.
Composition: 40-90% aromatic hydrocarbons.
Hazard Analysis
Toxicity: Variable. See specific petroleum hydrocarbon.
Fire Hazard: Variable.
Explosion Hazard: Variable.

AMYGDALIC ACID. See mandelic acid.

AMYL ACETATE
General Information
Synonyms: Amyl acetic ether; pear oil.
Description: Colorless liquid, pear or banana-like odor.
Formula: $CH_3COO(CH_2)_4CH_3$.
Constants: Mol wt: 130.18, mp: $-78.5°C$, bp: $148°C$ at 737 mm, ulc: 55-60, lel: 1.1%, uel: 7.5%, flash p: $77°F$ (C.C.), d: 0.879 at $20°/20°C$, autoign. temp.: $714°F$, vap. d.: 4.5.
Hazard Analysis
Toxic Hazard Rating:
  Acute Local: Irritant 1, Ingestion 2; Inhalation 2.
  Acute Systemic: Ingestion 2; Inhalation 2.
  Chronic Local: Irritant 1.
  Chronic Systemic: U.
TLV: ACGIH (recommended); 100 parts per million in air; 525 mg per cubic meter of air.
Toxicology: Chronic toxicity is of a very low order. When inhaled in high concentrations, amyl acetate is irritating to the mucous membranes; it also possesses a narcotic effect, and from animal experiments it appears to be more toxic than butyl acetate. A concentration of 1,000 ppm, breathed for half an hour, has caused headache, fatigue, oppression in the chest, and irritation of the eyes and mucous membranes of the nose and throat, with excessive salivation. 5,000 ppm produces deep narcosis in cats in 30 minutes.
  Symptoms are burning of the eyes, lacrimation, headache, irritation and dryness of the throat, "dopiness," fatigue, and occasionally vague nervousness. See also esters, amly alcohol, acetic acid.
Fire Hazard: Dangerous, when exposed to heat or flame; when heated, it emits acrid fumes; can react with oxidizing materials.
Spontaneous Heating: No.
Explosion Hazard: Moderate, when exposed to flame (Section 7).
Countermeasures
Ventilation Control: Section 2.
To Fight Fire: Alcohol foam, carbon dioxide, dry chemical or carbon tetrachloride (Section 6).

TOXIC HAZARD RATING CODE *(For detailed discussion, see Section 1.)*

0 NONE: (a) No harm under any conditions; (b) Harmful only under unusual conditions or overwhelming dosage.

1 SLIGHT: Causes readily reversible changes which disappear after end of exposure.

2 MODERATE: May involve both reversible and irreversible changes; not severe enough to cause death or permanent injury.

3 HIGH: May cause death or permanent injury after very short exposure to small quantities.

U UNKNOWN: No information on humans considered valid by authors.

Personal Hygiene: Section 3.
Storage and Handling: Section 7.
Shipping Regulations: Section 11.
  I.C.C.: Flammable liquid; red label, 10 gallons.
  Coast Guard Classification: Inflammable liquid; red label.
  IATA: Flammable liquid, red label, 1 liter (passenger), 40 liter (cargo).

sec-AMYL ACETATE
General Information
Description: Colorless liquid.
Formula: $CH_3CO_2C_5H_{11}$.
Constants: Mol wt: 130.1, bp: 120°C, flash p: 89°F (C.C.), d: 0.862–0.866 at 20°/20°C, vap. d.: 4.48.
Hazard Analysis
Toxicity: Probably toxic. See also esters, amyl alcohol, acetic acid, amyl acetate.
TLV: ACGIH (recommended) 125 parts per million, 650 milligrams per cubic meter of air.
Fire Hazard: Dangerous, when exposed to heat or flame; can react with oxidizing materials.
Countermeasures
To Fight Fire: Alcohol foam, carbon dioxide, dry chemical or carbon tetrachloride (Section 6).
Storage and Handling: Section 7.

AMYL ACETATE, ISO-. See isoamyl acetate.

AMYL ACETIC ETHER. See amyl acetate.

AMYL ALCOHOL
General Information
Synonyms: 1-Pentanol; primary amyl alcohol.
Description: Clear liquid.
Formula: $CH_3CH_2CH_2CH_2CH_2OH$.
Constants: Mol wt: 88.1, mp:−79°C, bp: 137.8°C, flash p: 91°F (C.C.), d: 0.8168 at 20°/20°C, ulc: 40, lel: 1.2%, uel = 10% at 212°F, autoign. temp: 572°F, vap. press.: 1 mm at 13.6°C, 10 mm at 44.9°C, vap. d.: 3.04.
Hazard Analysis
Toxic Hazard Rating:
  Acute Local: Irritant 2, Inhalation 2.
  Acute Systemic: Ingestion 3; Inhalation 3.
  Chronic Local: Irritant 2.
  Chronic Systemic: Ingestion 3; inhalation 2; Skin Absorption 2.
Toxicology: According to animal experiments, the amyl alcohols are about four times as toxic as ethyl alcohol. However, because of their low volatility and their low solubility in the body fluids, they are absorbed slowly and only to a small extent. There is little definite evidence that their use in industry has resulted in poisoning. The vapor may, however, be irritant to the eyes and the upper respiratory tract. See also alcohols. Ingestion can cause headache, nausea, vomiting, delirium and methemoglobin formation.
Fire Hazard: Moderate, when exposed to heat or flame; can react with oxidizing materials.
Spontaneous Heating: No.
Explosion Hazard: Moderate, when exposed to flame (Section 7).
Countermeasures
Ventilation Control: Section 2.
To Fight Fire: Alcohol foam, carbon dioxide, dry chemical or carbon tetrachloride (Section 6).
Personal Hygiene: Section 3.
First Aid: Section 1.
Storage and Handling: Section 7.
Shipping Regulations: Section 11.
  Coast Guard Classification: Combustible liquid.

m-AMYL ALCOHOL. See 1-pentanol.

sec-n-AMYL ALCOHOL
General Information
Synonym: Methyl propyl carbinol, 3-pentanol, diethyl carbinol.
Description: Colorless liquid.
Formula: $CH_3CH_2CH_2CHOHCH_3$.
Constants: Mol wt: 88.2, bp: 116°C, flash p: 94°F (C.C.), ulc: 40–45, uel = 9.0%, fp: −50°C, d: 0.8169 at 20°/20°C, autoign. temp.: 650°F to 725°F, vap. d.: 3.04.
Hazard Analysis
Toxic Hazard Rating:
  Acute Local: Irritant 1; Ingestion 3; Inhalation 3.
  Acute Systemic: Ingestion 3; Inhalation 3.
  Chronic Local: U.
  Chronic Systemic: Ingestion 2; Inhalation 2; Skin Absorption 2.
Fire Hazard: Moderate, when exposed to heat or flame; can react with oxidizing materials.
Countermeasures
To Fight Fire: Alcohol foam, carbon dioxide, dry chemical or carbon tetrachloride (Section 6).
Ventilation Control: Section 2.
Personnel Protection: Section 3.
Storage and Handling: Section 7.

AMYL ALCOHOL, COMMERCIAL. See isoamyl alcohol.

AMYL ALCOHOL, PRIMARY, ACTIVE. See 2-methyl--1-butanol.

tert-AMYL ALCOHOL, REFINED
General Information
Synonyms: Dimethyl ethyl carbinol; 2-methyl-2-butanol.
Description: Colorless liquid.
Formula: $CH_3CH_2C(CH_3)OHCH_3$.
Constants: Mol wt: 88.15, mp: −11.9°C, bp: 101.8°C, flash p: 67°F (C.C.), d: 0.809, autoign. temp.: 819°F, vap. press.: 10 mm at 17.2°C, vap. d.: 3.03.
Hazard Analysis
Toxic Hazard Rating:
  Acute Local: Irritant 1; Ingestion 2; Inhalation 2.
  Acute Systemic: Ingestion 2; Inhalation 2.
  Chronic Local: U.
  Chronic Systemic: Ingestion 2; Inhalation 2; Skin Absorption 2.
Fire Hazard: Dangerous, when exposed to heat or flame; can react with oxidizing materials.
Countermeasures
To Fight Fire: Alcohol foam, carbon dioxide, dry chemical or carbon tetrachloride (Section 6).
Ventilation Control: Section 2.
Personnel Protection: Section 3.
Storage and Handling: Section 7.

AMYL ALCOHOL, SYNTHETIC. See pentasol.

AMYLAMINE
General Information
Synonym: 1-Amino pentane.
Description: Water-white liquid.
Formula: $CH_3(CH_2)_4NH_2$.
Constants: Mol wt: 87.16, mp: −55°C, bp: 104°C, flash p: 45°F (O.C.), d: 0.7614 at 20°/4°C, vap. d.: 3.01.
Hazard Analysis
Toxic Hazard Rating:
  Acute Local: Irritant 3; Inhalation 3.
  Acute Systemic: U.
  Chronic Local: U.
  Chronic Systemic: U.
  A strong irritant. See also amines.
Fire Hazard: Dangerous, when exposed to heat or flame; can react with oxidizing materials.

Countermeasures

To Fight Fire: Carbon dioxide, dry chemical or carbon tetra-chloride (Section 6).

Storage and Handling: Section 7.

Shipping Regulations: Section 11.

IATA: Flammable liquid, red label, 1 liter (passenger), 40 liters (cargo).

sec-AMYLAMINE

General Information

Synonyms: 2-Aminopentane; methyl propylcarbinylamine.

Formula: $CH_3(CH_2)_2CH(CH_3)NH_2$.

Constants: Mol wt: 87.16, d: 0.7, vap. d.: 3.0, bp: 198°F, flash p: 20°F.

Hazard Analysis

Toxic Hazard Rating: U. See also amines.

Fire Hazard: Dangerous, when exposed to heat or flame.

Countermeasures

To Fight Fire: Alcohol foam (Section 6).

Storage and Handling: Section 7.

AMYLAMINE LAURATE

General Information

Description: Liquid.

Constants: Flash p: 150°F, d: 0.88.

Hazard Analysis

Toxicity: Unknown. See amines.

Fire Hazard: Moderate, when exposed to heat or flame; can react with oxidizing materials.

Countermeasures

To Fight Fire: Water, carbon dioxide, dry chemical or carbon tetrachloride (Section 6).

Storage and Handling: Section 7.

AMYLAMINE OLEATE

General Information

Description: Liquid.

Constants: Flash p: 220°F, d: 0.89.

Hazard Analysis

Toxicity: Unknown. See amines.

Fire Hazard: Moderate, when exposed to heat or flame; can react with oxidizing materials.

Countermeasures

To Fight Fire: Water, carbon dioxide, dry chemical or carbon tetrachloride (Section 6).

Storage and Handling: Section 7.

AMYLAMINE STEARATE

General Information

Description: Liquid.

Constants: Flash p: 160°F, d: 0.88.

Hazard Analysis

Toxicity: Unknown. See amines.

Fire Hazard: Moderate, when exposed to heat or flame; can react with oxidizing materials.

Countermeasures

To Fight Fire: Water, carbon dioxide, dry chemical or carbon tetrachloride (Section 6).

Storage and Handling: Section 7.

p-tert-AMYLANILINE

General Information

Formula: $(C_2H_5)(CH_2)_2CC_6H_4NH_2$.

Constants: Mol wt: 161, d: 0.9, bp: 498–504°F, flash p: 215°F.

Hazard Analysis

Toxic Hazard Rate: U. See also aniline.

Fire Hazard: Moderate, when exposed to heat or flame.

AMYL AZIDE

General Information

Formula: $C_5H_{11}N_3$.

Constant: Mol wt: 113.

Hazard Analysis

Toxic Hazard Rating:

Acute Local: U.

Acute Systemic: Ingestion 2.

Chronic Local: U.

Chronic Systemic: U.

Toxicity: Causes fall in blood pressure.

AMYL BENZENE

General Information

Synonym: n-Pentyl benzene.

Description: Liquid.

Formula: $C_6H_5C_5H_{11}$.

Constants: Mol wt: 148.24, bp: 202.2°C, mp: −79.3°C, flash p: 150°F (O.C.), d: 0.8627 at 20°/4°C, vap. d.: 5.11.

Hazard Analysis

Toxicity: Details unknown. Probably irritating to mucous membrane and narcotic in high concentrations.

Fire Hazard: Moderate, when exposed to heat or flame; reacts with oxidizing materials.

Countermeasures

To Fight Fire: Foam, carbon dioxide, dry chemical or carbon tetrachloride (Section 6).

Storage and Handling: Section 7.

AMYL BENZOATE

General Information

Description: Clear liquid.

Formula: $C_6H_5COOC_5H_{11}$.

Constants: Mol wt: 192.3, bp: 260°C, flash p: 230°F, d: 0.98.

Hazard Analysis

Toxicity: Details unknown. See also esters, amyl alcohol, benzoic acid.

Fire Hazard: Moderate, when exposed to heat or flame; reacts with oxidizing materials.

Countermeasures

To Fight Fire: Carbon dioxide, dry chemical or carbon tetrachloride (Section 6).

Storage and Handling: Section 7.

N-AMYL BENZYLCYCLOHEXYLAMINE

General Information

Description: Liquid.

Formula: $C_{17}H_{27}N$.

Constant: Mol wt: 245.4.

Hazard Analysis

Toxic Hazard Rating:

Acute Local: Irritant 1; Ingestion 1; Inhalation 2.

Acute Systemic: Ingestion 2; Inhalation 2.

Chronic Local: U.

Chronic Systemic: U.

Fire Hazard: A moderately flammable liquid; can react with oxidizing materials.

Countermeasures

Ventilation Control: Section 2.

Personal Hygiene: Section 3.

Storage and Handling: Section 7.

AMYL BENZYL ETHER

General Information

Synonyms: Benzyl isoamyl ether; "gardenia oxide."

Description: Liquid.

TOXIC HAZARD RATING CODE (For detailed discussion, see Section 1.)

0 NONE: (a) No harm under any conditions; (b) Harmful only under unusual conditions or overwhelming dosage.

1 SLIGHT: Causes readily reversible changes which disappear after end of exposure.

2 MODERATE: May involve both irreversible and reversible changes; not severe enough to cause death or permanent injury.

3 HIGH: May cause death or permanent injury after very short exposure to small quantities.

U UNKNOWN: No information on humans considered valid by authors.

Formula: $C_5H_{11}OCH_2C_6H_5$.
Constants: Mol wt: 178.3, bp: 235°C, d: 0.965 at 15.5°/15.5°C.

Hazard Analysis
Toxicity: Details unknown. See also ethers.
Fire Hazard: Moderate, when exposed to heat or flame, can react with oxidizing materials.

Countermeasures
To Fight Fire: Foam, carbon dioxide, dry chemical, or carbon tetrachloride (Section 6).
Storage and Handling: Section 7.

AMYLBIPHENYL
General Information
Description: Liquid.
Formula: $C_5H_{11}C_6H_4C_6H_5$.
Constants: Mol wt: 224.3, mp: −60°C, bp: 305–337°C, flash p: 300°F, d: 0.958 at 20°/20°C, vap. d.: 7.73.

Hazard Analysis
Toxicity: Unknown.
Fire Hazard: Slight when exposed to heat or flame.
Disaster Hazard: Moderately dangerous; when heated to decomposition, it emits toxic fumes; can react with oxidizing materials.

Countermeasures
Storage and Handling: Section 7.
To Fight Fire: Foam, carbon dioxide, dry chemical or carbon tetrachloride (Section 6).

AMYL BORIC ACID
General Information
Description: Colorless crystals; water soluble.
Formula: $(C_5H_{11})B(OH)_2$.
Constants: Mol wt: 116.0, mp: 94°C (decomp).

Hazard Analysis
Toxicity: See boron compounds.

AMYL BROMIDE
General Information
Description: Colorless liquid.
Formula: $CH_3(CH_2)_4Br$.
Constants: Mol wt: 151.1, bp: 128–130°C, flash p: 90°F, fp: < −30°C, d: 1.211 at 25°/25°C.

Hazard Analysis
Toxicity: Details unknown. Probably a local irritant and narcotic in high concentrations. See also chlorinated hydrocarbons, aliphatic.
Fire Hazard: Moderate, when exposed to heat or flame.
Disaster Hazard: Dangerous; See bromides. Can react with oxidizing material.

Countermeasures
Storage and Handling: Section 7.
To Fight Fire: Water, foam, carbon dioxide, dry chemical or carbon tetrachloride (Section 6).

tert-AMYL CARBAMATE
General Information
Description: Crystals, camphor odor.
Formula: $CH_3CH_2CH_2CH_2CH_2CO_2NH_2$.
Constants: Mol wt: 131.2, mp: 83–86°C.

Hazard Analysis
Toxicity: Unknown.
Fire Hazard: Slight; when heated. (Section 6).

Countermeasures
Storage and Handling: Section 7.

AMYL CARBINOL. See n-hexyl alcohol.

AMYL CHLORIDE
General Information
Synonym: 1-chloropentane.
Description: Water-white, sweet odor.
Formula: $CH_3(CH_2)_3CH_2Cl$.
Constants: Mol wt: 106.60, mp: −99°C, bp: 108.2°C, flash p: 54°F (O.C.), d: 0.885 at 20°/4°C, autoign. temp.: 650°F, vap. d.: 3.67, lel: 1.4%, uel: 8.6%.

Hazard Analysis
Toxicity: Details unknown. See also chlorinated hydrocarbons, aliphatic.
Fire Hazard: Dangerous, when exposed to heat or flame.
Spontaneous Heating: No.
Explosion Hazard: Moderate (Section 7).
Disaster Hazard: Dangerous; when heated to decomposition, it emits highly toxic fumes of phosgene; can react with oxidizing materials.

Countermeasures
Storage and Handling: Section 7.
To Fight Fire: Foam, carbon dioxide, dry chemical or carbon tetrachloride (Section 6).
Shipping Regulations: Section 11.
　I.C.C.: Flammable liquid; red label, 10 gallons.
　Coast Guard Classification: Inflammable liquid; red label.
　IATA: Flammable liquid, red label, 1 liter (passenger), 40 liters (cargo).

tert-AMYL CHLORIDE
General Information
Description: Liquid.
Formula: $CH_3CH_2CCl(CH_3)CH_3$.
Constants: Mol wt: 106.6, bp: 87°C, d: 1.407, autoign. temp.: 649°F, vap. d.: 3.67, lel: 1.5%; uel: 7.4%.

Hazard Analysis
Toxicity: Details unknown. See also chlorinated hydrocarbons, aliphatic.
Fire Hazard: Moderate, when exposed to heat or flame.
Disaster Hazard: Dangerous; see chlorides can react with oxidizing materials.

Countermeasures
Storage and Handling: Section 7.
To Fight Fire: Water may used as a blanket. Foam, carbon dioxide, dry chemical or carbon tetrachloride (Section 6).

AMYL CHLORIDE, MIXED
General Information
Description: Straw to deep-purple colored.
Formula: $C_5H_{11}Cl$.
Constants: Mol wt: 106.6, bp: 85–109°C, flash p: 38°F (O.C.), d: 0.88 at 20°C. vap. d.: 3.67.

Hazard Analysis
Toxicity: Details unknown. See chlorinated hydrocarbons, aliphatic.
Fire Hazard: Dangerous, when exposed to heat or flame.
Disaster Hazard: Dangerous, see chlorides, can react vigorously with oxidizing materials.

Countermeasures
Storage and Handling: Section 7.
To Fight Fire: Foam, carbon dioxide, dry chemical or carbon tetrachloride (Section 6).

AMYL CHLORONAPHTHALENE
General Information
Description: Liquid.
Formula: $C_5H_{11}ClC_{10}H_6$.
Constants: Mol wt: 232.8, bp: 241°C, flash p: 295°F, d: 1.07.

Hazard Analysis
Toxicity: Details unknown. See also chlorinated naphthalene.
Fire Hazard: Slight, when exposed to heat or flame.
Disaster Hazard: Dangerous; see chlorides. Can react with oxidizing materials.

Countermeasures
Storage and Handling: Section 7.
To Fight Fire: Water, foam, carbon dioxide, dry chemical or carbon tetrachloride (Section 6).

p-tert-AMYL-o-CRESOL
General Information
Formula: $C_5H_{11}C_6H_3OHCH_3$.
Constants: Mol wt: 178.3, bp: 258°C, flash p: 240°F, d: 0.97.

Hazard Analysis
Toxic Hazard Rating:
  Acute Local: Irritant 3; Ingestion 3; Inhalation 3.
  Acute Systemic: U.
  Chronic Local: U.
  Chronic Systemic: U.
Toxicology: See also cresol.
Fire Hazard: Moderate, when exposed to heat or flame (Section 6).
Disaster Hazard: Dangerous; when heated, it emits highly toxic fumes; can react vigorously with oxidizing materials.

Countermeasures
Ventilation Control: Section 2.
Personal Hygiene: Section 3.
First Aid: Section 1.
Storage and Handling: Section 7.

4-tert-AMYL CYCLOHEXANOL
General Information
Description: Liquid.
Formula: $C_5H_{11}C_6H_4OH$.
Constants: Mol wt: 164.3, bp: 245°C, flash p: 212°F, d: 0.91.
Hazard Analysis
Toxicity: Details unknown. See also cyclohexanol.
Fire Hazard: Moderate, when exposed to heat or flame; can react with oxidizing materials (Section 6).
Countermeasures
To Fight Fire: Foam, carbon dioxide, dry chemical or carbon tetrachloride (Section 6).
Storage and Handling: Section 7.

AMYL DICHLOROSILANE
General Information
Description: Liquid.
Formula: $C_5H_{11}SiHCl_2$.
Constants: Mol wt: 171.2, bp: 235°C.
Hazard Analysis and Countermeasures
See chlorosilanes.

α-n-AMYLENE
General Information
Synonym: Propylethylene, methyl butene, 1-pentene.
Description: Liquid, highly disagreeable odor.
Formula: $CH_3(CH_2)_2CHCH_2$.
Constants: Mol wt: 70.13, mp: −124°C, bp: 36.7°C, lel: 1.6%, uel: 8.7%, flash p: 0°F (O.C.), d: 0.666, autoign. temp.: 523°F, vap. d.: 2.42.
Hazard Analysis
Toxic Hazard Rating:
  Acute Local: Ingestion 2; Inhalation 2.
  Acute Systemic: Ingestion 2; Inhalation 2; Skin Absorption 1.
  Chronic Local: U.
  Chronic Systemic: U.
Toxicology: Narcotic in high concentrations. A simple asphyxiant.
TLV: ACGIH (recommended) 1000 parts per million in air; 2950 milligrams per cubic meter of air.
Fire Hazard: Dangerous, when exposed to heat or flame; when heated to decomposition it emits toxic fumes; can react with oxidizing materials. (Section 6).
Spontaneous Heating: No.
Explosion Hazard: Moderate, when exposed to flame (Section 7).
Countermeasures
To Fight Fire: Foam, carbon dioxide, dry chemical or carbon tetrachloride (Section 6).
Ventilation Control: Section 2.

Shipping Regulations: Section 11.
  IATA: Flammable liquid, red label, 1 liter (passenger), 40 liters (cargo).

AMYLENE HYDRATE. See tert-amyl alcohol.

AMYLENES, MIXED
General Information
Description: Water-white liquid.
Formula: C_5H_{10}.
Constant: Mol wt: 70.58, bp: 32.2°C, flash p: above 0°F, d: 0.66 at 20°C.
Hazard Analysis
Toxic Hazard Rating:
  Acute Local: Irritant 2; Ingestion 2; Inhalation 2.
  Acute Systemic: U.
  Chronic Local: U.
  Chronic Systemic: U.
Fire Hazard: Dangerous, when exposed to heat or flame; reacts with oxidizing materials (Section 6).
Countermeasures
To Fight Fire: Foam, carbon dioxide, dry chemical or carbon tetrachloride (Section 6).
Ventilation Control: Section 2.
Personnel Protection: Section 3.
Storage and Handling: Section 7.

AMYL ETHER
General Information
Synonyms: Amyl oxide; diamyl ether.
Description: Liquid.
Formula: $[CH_3(CH_2)_3CH_2]_2O$.
Constants: Mol wt: 158.3, mp: −69.3°C, bp: 338°F, flash p: 135°F (O.C.), d: 0.774 at 20°/4°C, vap. d.: 5.46, autoign. temp.: 340°F.
Hazard Analysis
Toxicity: Details unknown. See also ethers.
Fire Hazard: Moderate, when exposed to heat or flame; reacts with oxidizing materials. See ethers.
Countermeasures
To Fight Fire: Foam, carbon dioxide, dry chemical or carbon tetrachloride (Section 6).
Storage and Handling: Section 7.

AMYL FORMATE
General Information
Description: Clear liquid.
Formula: $HCOO(CH_2)_4CH_3$.
Constants: Mol wt: 116.2, mp: −73.5°C, bp: 130.4°C, flash p: 80°F, d: 0.893 at 15°/4°C.
Hazard Analysis
Toxic Hazard Rating:
  Acute Local: Irritant 2; Ingestion 2; Inhalation 2.
  Acute Systemic: U.
  Chronic Local: U.
  Chronic Systemic: U.
Fire Hazard: Dangerous, when exposed to heat or flame; it can react vigorously with oxidizing materials.
Countermeasures
To Fight Fire: Foam, carbon dioxide, dry chemical or carbon tetrachloride (Section 6)
Ventilation Control: Section 2.
Personnel Protection: Section 3.
Storage and Handling: Section 7.
Shipping Regulations: Section 11.
  IATA: Flammable liquid, red label, 1 liter (passenger), 40 liters (cargo).

AMYL HYDRIDE. See n-pentane.

TOXIC HAZARD RATING CODE *(For detailed discussion, see Section 1.)*

.0 NONE: (a) No harm under any conditions; (b) Harmful only under unusual conditions or overwhelming dosage.

1 SLIGHT: Causes readily reversible changes which disappear after end of exposure.

2 MODERATE: May involve both irreversible and reversible changes; not severe enough to cause death or permanent injury.

3 HIGH: May cause death or permanent injury after very short exposure to small quantities.

U UNKNOWN: No information on humans considered valid by authors.

AMYL LACTATE
General Information
Description: Colorless liquid.
Formula: $CH_3CH(OH)COO(CH_2)_4CH_3$.
Constants: Mol wt: 160.2, bp: 210°C, flash p: 175°F, d: 0.960 at 20°C.
Hazard Analysis
Toxic Hazard Rating:
 Acute Local: Irritant 1; Ingestion 1; Inhalation 1.
 Acute Systemic: U.
 Chronic Local: U.
 Chronic Systemic: U.
See also esters.
Fire Hazard: Moderate, when exposed to heat or flame; can react with oxidizing materials.
Countermeasures
To Fight Fire: Foam, carbon dioxide, dry chemical or carbon tetrachloride (Section 6).
Personal Hygiene: Section 3.
Storage and Handling: Section 7.

AMYL LAURATE
General Information
Description: Liquid.
Formula: $C_5H_{11}O_2C(CH_2)_{10}CH_3$.
Constants: Mol wt: 270.44, bp: 290°C, flash p: 300°F, d: 0.86.
Hazard Analysis
Toxicity: Unknown. See also esters.
Caution: May de-fat skin and lead to dermatitis (Section 9).
Fire Hazard: Slight, when exposed to heat or flame; can react with oxidizing materials.
Countermeasures
To Fight Fire: Foam, carbon dioxide, dry chemical or carbon tetrachloride (Section 6).
Storage and Handling: Section 7.

AMYL MERCAPTAN
General Information
Synonym: 1-Pentanethiol.
Description: Water-white to yellow liquid.
Formula: $CH_3(CH_2)_4SH$.
Constants: Mol wt: 104.21, bp: 126.64°C, flash p: 65°F, d: 0.842 at 20°C, vap. press. 13.8 mm at 25°C, vap. d.: 3.59.
Hazard Analysis
Toxic Hazard Rating:
 Acute Local: Irritant 1; Allergen 1; Ingestion 1; Inhalation 1.
 Acute Systemic: U.
 Chronic Local: Allergen 1.
 Chronic Systemic: U.
Toxicology: See also mercaptans.
Caution: A weak sensitizer. Local contact may cause contact dermatitis (Section 9).
Fire Hazard: Dangerous, when exposed to heat or flame.
Disaster Hazard: Dangerous; see mercaptans; reacts with oxidizing materials.
Countermeasures
Ventilation Control: Section 2.
Storage and Handling: Section 7.
To Fight Fire: Foam, carbon dioxide, dry chemical or carbon tetrachloride (Section 6).
Shipping Regulations: Section 11.
 ICC: Flammable liquid, red label, 10 gallons.
 Coast Guard Classification: Inflammable liquid; red label.
 IATA: Flammable liquid, red label, 1 liter (passenger), 40 liters (cargo).

AMYL MERCAPTANS (mixed)
General Information
Formula: $C_5H_{11}SH$; $CH_3CH_2CH(CH_3)CH_2OH$.
Constants: d: 0.8; bp: 176–251°F, flash p: 65°F (O.C.).
Hazard Analysis
Toxic Hazard Rating: U.

Fire Hazard: Dangerous if exposed to heat or flame.
Disaster Hazard: Dangerous. See mercaptans.
Countermeasures
Storage and Handling: Section 7.

AMYL METHYL ALCOHOL
General Information
Description: Liquid.
Formula: $CH_3(CH_2)_2CH(CH_3)CH_2OH$.
Constants: Mol wt: 102.2, bp: 130°C, flash p: 114°F (C.C.), d: 0.804, vap. d.: 3.52.
Hazard Analysis
Toxicity: Details unknown. See also alcohols.
Fire Hazard: Moderate, when exposed to heat or flame; can react with oxidizing materials.
Countermeasures
To Fight Fire: Carbon dioxide, dry chemical or carbon tetrachloride (Section 6).
Storage and Handling: Section 7.

AMYL METHYL CARBINOL. See 2-heptanol.

AMYL METHYL KETONE
General Information
Description: Liquid.
Formula: $CH_3CO(CH_2)_4CH_3$.
Constants: Mol wt: 114.2, bp: 150°C, flash p: 120°F (O.C.), d: 0.817, autoign. temp.: 991°F, vap. d.: 3.94.
Hazard Analysis
Toxicity: Details unknown. See also ketones.
Fire Hazard: Moderate, when exposed to heat or flame; can react with oxidizing materials.
Countermeasures
To Fight Fire: Foam, carbon dioxide, dry chemical or carbon tetrachloride (Section 6).
Storage and Handling: Section 7.

AMYL NAPHTHALENE
General Information
Description: Liquid.
Formula: $C_{10}H_7C_5H_{11}$.
Constants: Mol wt: 198.3, mp: −30°C, bp: 288°C, flash p: 255°F (O.C.), d: 0.973, vap. d.: 6.86.
Hazard Analysis
Toxicity: Details unknown. See also naphthalene.
Fire Hazard: Slight, when exposed to heat or flame; can react with oxidizing materials.
Countermeasures
To Fight Fire: Foam, carbon dioxide, dry chemical or carbon tetrachloride (Section 6).
Storage and Handling: Section 7.

AMYL β-NAPHTHOL
General Information
Description: Liquid.
Formula: $C_5H_{11}C_{10}H_7O$.
Constants: Mol wt: 214.3, bp: 308°C, flash p: 325°F, d: 0.91.
Hazard Analysis
Toxicity: Details unknown. See also β-naphthol.
Fire Hazard: Slight, when exposed to heat or flame.
Disaster Hazard: Moderately dangerous; when heated to decomposition it emits toxic fumes; can react with oxidizing materials.
Countermeasures
Storage and Handling: Section 7.
To Fight Fire: Foam, carbon dioxide, dry chemical or carbon tetrachloride (Section 6).

AMYL NITRATE (MIXED ISOMERS)
General Information
Description: Liquid.
Formula: $C_5H_{11}NO_3$.
Constants: Mol wt: 133.15, bp: 145°C, flash p: 118°F, d: 0.99.

Hazard Analysis
Toxic Hazard Rating:
  Acute Local: Irritant 2; Ingestion 2; Inhalation 2.
  Acute Systemic: U.
  Chronic Local: U.
  Chronic Systemic: U.
Fire Hazard: Moderate, when exposed to heat or flame or by spontaneous chemical reaction.
Caution: An oxidizing agent.
Explosion Hazard: See nitrates.
Disaster Hazard: See nitrates.
Countermeasures
Ventilation Control: Section 2.
Personnel Protection: Section 3.
Storage and Handling: Section 7.
To Fight Fire: Foam, carbon dioxide, dry chemical or carbon tetrachloride (Section 6).
Shipping Regulations: Section 11.
  Coast Guard Classification: Inflammable liquid; red label.

AMYL NITRITE
General Information
Synonym: Isoamyl nitrite.
Description: Clear yellowish liquid, peculiar ethereal fruity odor and pungent aromatic taste.
Formula: $CH_3(CH_2)_4NO_2$.
Constants: Mol wt: 117.15, bp: 96–99°C, d: 0.8528 at 20°/4°C, autoign. temp.: 408°F, vap. d.: 4.0.
Hazard Analysis
Toxic Hazard Rating:
  Acute Local: Ingestion 2; Inhalation 2.
  Acute Systemic: Ingestion 2; Inhalation 2.
  Chronic Local: U.
  Chronic Systemic: Ingestion 1; Inhalation 1; Skin Absorption 1.
Toxicology: Causes flushing of skin, rapid pulse, headache and fall in blood pressure.
Fire Hazard: Moderate, when exposed to heat or flame or by spontaneous chemical reaction (Section 6).
Caution: An oxidizing material.
Disaster Hazard: Dangerous; see nitrates; can react with oxidizing or reducing materials.
Countermeasures
To Fight Fire: alcohol foam.
Ventilation Control: Section 2.
Personal Hygiene: Section 3.
Storage and Handling: Section 7.
Shipping Regulations: Section 11.
  I.C.C.: Flammable liquid; red label, 10 gallons.
  IATA: Flammable liquid, red label, 1 liter (passenger), 40 liters (cargo).

AMYL OCTYLAMINE
General Information
Descripton: Liquid.
Formula: $C_5H_{11}NHC_8H_{17}$.
Constant: Mol wt: 199.3.
Hazard Analysis
Toxicity: Details unknown. An insecticide. See also amines.
Fire Hazard: A moderately flammable material; can react with oxidizing materials (Section 6).
Countermeasures
Storage and Handling: Section 7.

AMYLOFORM
General Information
Description: Soft white powder. A formaldehyde-carbohydrate condensation product.

Hazard Analysis
Toxic Hazard Rating:
  Acute Local: Allergen 1.
  Acute Systemic: U.
  Chronic Local: Allergen 1.
  Chronic Systemic: U.
Caution: Local contact may cause contact dermatitis (Section 9).
Fire Hazard: Slight, when exposed to heat or flame; it can react with oxidizing materials.
Countermeasures
Personal Hygiene: Section 3.
Storage and Handling: Section 7.

AMYL OLEATE
General Information
Description: Liquid.
Formula: $CH_3(CH_2)_7CH:CH(CH_2)_7COOC_5H_{11}$.
Constants: Mol wt: 352.6, bp: 200°C, flash p: 366°F, d: 0.86.
Hazard Analysis
Toxic Hazard Rating:
  Acute Local: Irritant 2; Ingestion 2; Inhalation 2.
  Acute Systemic: U.
  Chronic Local: U.
  Chronic Systemic: U.
Fire Hazard: Slight, when exposed to heat or flame; can react with oxidizing materials.
Countermeasures
To Fight Fire: Foam, carbon dioxide, dry chemical or carbon tetrachloride (Section 6).
Ventilation Control: Section 2.
Personnel Protection: Section 3.
Storage and Handling: Section 7.

AMYL OXALATE
General Information
Formula: $(COOC_5H_{11})_2$.
Constants: Mol wt: 230, d: 1.0-, bp: 464–523°F.
Hazard Analysis
Toxic Hazard Rating: U. See esters.
Fire Hazard: Moderate, when exposed to heat or flame.
Countermeasures
Storage and Handling: Section 7.
To Fight Fire: Water, foam (Section 6).

AMYL OXIDE. See amyl ether.

o-AMYL PHENOL
General Information
Description: Liquid.
Formula: $C_5H_{11}C_6H_4OH$.
Constants: Mol wt: 164.24, bp: 342°C, flash p: 219°F (O.C.), d: 0.96–0.97, vap. d 5.66.
Hazard Analysis
Toxic Hazard Rating:
  Acute Local: Irritant 3; Ingestion 3.
  Acute Systemic: Ingestion 3.
  Chronic Local: Irritant 2.
  Chronic Systemic: U.
Fire Hazard: Slight, when exposed to heat or flame.
Disaster Hazard: Dangerous; when heated to decomposition, it emits toxic fumes; can react vigorously with oxidizing materials. (Section 7).
Countermeasures
Ventilation Control: Section 2.
Personnel Protection: Section 3.
First Aid: Section 1.

TOXIC HAZARD RATING CODE (For detailed discussion, see Section 1.)

0 NONE: (a) No harm under any conditions; (b) Harmful only under unusual conditions or overwhelming dosage.

1 SLIGHT: Causes readily reversible changes which disappear after end of exposure.

2 MODERATE: May involve both irreversible and reversible changes; not severe enough to cause death or permanent injury.
3 HIGH: May cause death or permanent injury after very short exposure to small quantities.
U UNKNOWN: No information on humans considered valid by authors.

To Fight Fire: Foam, carbon dioxide, dry chemical or carbon tetrachloride (Section 6).
Storage and Handling: Section 7.

o-sec-AMYL PHENOL
General Information
Description: Clear, straw colored liquid; very slightly soluble in water; soluble in oil and organic solvents.
Formula: $C_5H_{11}C_6H_4OH$.
Constants: Mol wt: 164.24, d (30/30°C): 0.955–0.971, initial bp: > 235°C, final bp: < 250.0°C, flash p: 200°F.
Hazard Analysis
Toxic Hazard Rate: U. See esters.
Fire Hazard: Moderate, when exposed to heat or flame.
Countermeasures
Storage and Handling: Section 7.
To Fight Fire: Foam, fog, dry chemical (Section 6).

o-tert-AMYL PHENOL
General Information
Description: Pale yellow liquid; slightly soluble in water; soluble in oil and organic solvents.
Formula: $(CH_3)_2(C_2H_5)CC_6H_4OH$.
Constants: Mol wt: 164.24, d: 0.96–0.97 (30°C), initial bp: not <233°C, final bp: not >245°C, flash p: 219°F.
Hazard Analysis
Toxic Hazard Rating: U. See esters.
Fire Hazard: Moderate, when exposed to heat or flame.
Countermeasures
To Fight Fire: Water, foam (Section 6).
Storage and Handling: Section 7.

p-sec-AMYL PHENOL
General Information
Formula: $C_5H_{11}C_6H_4OH$.
Constants: Mol wt: 164.24, d: <1.0, bp: 482–516°F, flash p: 270°F.
Hazard Analysis
Toxic Hazard Rating: U. See esters.
Fire Hazard: Slight when exposed to heat or flame.
Countermeasures
To Fight Fire: Water, foam (Section 6).
Storage and Handling: Section 7.

p-tert-AMYL PHENOL
General Information
Synonym: p-(alpha, alpha-dimethyl propyl) phenol, pentaphen.
Description: Colorless needles.
Formula: $CH_3CH_2C(CH_3)_2C_6H_4OH$.
Constants: Mol wt: 164.2, bp: 250°C, mp: 92–93°C, flash p: 232°F (O.C.).
Hazard Analysis
Toxicity: Details unknown. See also phenol.
Fire Hazard: A slightly flammable material (Section 6).
Disaster Hazard: Moderately dangerous; when heated to decomposition, it emits toxic fumes; it can react with oxidizing materials.
Countermeasures
To Fight Fire: Water, foam (Section 6).
Storage and Handling: Section 7.

p-tert-AMYL PHENOXYL ETHANOL
General Information
Description: Liquid.
Formula: $C_5H_{11}OC_6H_4CH_2CH_2OH$.
Constants: Mol wt: 212.3, bp: 291°C, flash p: 282°F, d: 1.01.
Hazard Analysis
Toxicity: Details unknown. See also alcohols.
Fire Hazard: Moderate, when exposed to heat or flame.
Disaster Hazard: Moderately dangerous; when heated to decomposition, it emits toxic fumes; can react with oxidizing materials.

Countermeasures
Storage and Handling: Section 7.
To Fight Fire: Foam, carbon dioxide, dry chemical or carbon tetrachloride (Section 6).

p-tert-AMYL PHENOXY ETHYL LAURATE
General Information
Description: Liquid.
Formula: $C_5H_{11}OC_6H_4CH_2CH_2OOC(CH_2)_{10}CH_3$.
Constants: Mol wt: 390.6, bp: 240°C, flash p: 410°F, d: 0.94.
Hazard Analysis
Toxicity: Unknown. See also esters.
Fire Hazard: Slight, when exposed to heat or flame, it can react with oxidizing materials.
Countermeasures
To Fight Fire: Foam, carbon dioxide, dry chemical or carbon tetrachloride (Section 6).
Storage and Handling: Section 7.

p-tert-AMYL PHENYL ACETATE
General Information
Description: Liquid.
Formula: $C_5H_{11}OC_6H_4CH_3COO$.
Constants: Mol wt: 222.3, bp: 253°C, flash p: 240°F, d: 0.99.
Hazard Analysis
Toxic Hazard Rating:
 Acute Local: Irritant 2; Ingestion 2.
 Acute Systemic: U.
 Chronic Local: U.
 Chronic Systemic: U.
Fire Hazard: Moderate, when exposed to heat or flame; (Section 6).
Countermeasures
To Fight Fire: Water, foam (Section 6).
Personnel Protection: Section 3.
Storage and Handling: Section 7.

p-tert-AMYL PHENYL-n-AMYL ETHER
General Information
Description: Liquid.
Formula: $C_5H_{11}C_6H_4OHOC_5H_{11}$
Constants: Mol wt: 251.4, bp: 285°C, flash p: 260°F, d: 0.9.
Hazard Analysis
Toxicity: Details unknown. See also ethers.
Fire Hazard: Moderate, when exposed to heat or flame, it can react with oxidizing materials.
Explosion Hazard: Details unknown. See also ethers.
Countermeasures
Storage and Handling: Section 7.
To Fight Fire: Foam, carbon dioxide, dry chemical, carbon tetrachloride (Section 6).

p-tert-AMYL PHENYL BUTYL ETHER
General Information
Formula: $C_5H_{11}C_6H_4OC_4H_9$.
Constants: Mol wt: 220, d: 0.9, bp: 507–511°F, flash p: 275°F.
Hazard Analysis
Fire Hazard: Slight, when exposed to heat or flame.
Countermeasures
To Fight Fire: Water, foam (Section 6).

AMYL PHENYL ETHER
General Information
Description: Liquid.
Formula: $C_5H_{11}OC_6H_5OH$.
Constants: Mol wt: 181.3, bp: 214°C, flash p: 185°F, d: 0.92, vap d: 5.7.
Hazard Analysis
Toxicity: Details unknown. See also ethers.
Fire Hazard: Moderate, when exposed to heat or flame, it can react with oxidizing materials.

Countermeasures
To Fight Fire: Foam, carbon dioxide, dry chemical or carbon tetrachloride (Section 6).
Storage and Handling: Section 7.

AMYL PHENYL METHYL ETHER
General Information
Description: Liquid.
Formula: $C_5H_{11}C_6H_4OHOCH_3$.
Constants: Mol wt: 195.3, bp: 240°C, flash p: 210°F, d: 0.94.
Hazard Analysis
Toxicity: Details unknown. See also ethers.
Fire Hazard: Moderate, when exposed to heat or flame, can react with oxidizing materials.
Countermeasures
To Fight Fire: Foam, carbon dioxide, dry chemical or carbon tetrachloride (Section 6).
Storage and Handling: Section 7.

AMYL PHTHALATE. See di-n-amyl phthalate.

AMYL PROPIONATE
General Information
Synonym: Pentyl propanoate.
Description: Stable, colorless, apple-like odor, liquid.
Formula: $CH_3CH_2COO(CH_2)_4CH_3$.
Constants: Mol wt: 144.21, mp: −73.1°C, bp: 106.2°C, flash p: 106°F (O.C.), d: 0.8761 at 15°/4°C, vap. press.: 10 mm at 46.3°C, vap. d.: 5.0, autoign. temp.: 712°F.
Hazard Analysis
Toxic Hazard Rating:
  Acute Local: Irritant 2; Ingestion 2; Inhalation 2.
  Acute Systemic: Ingestion 2; Inhalation 2; Skin Absorption 2.
  Chronic Local: U.
  Chronic Systemic: U.
Fire Hazard: Moderate, when exposed to heat or flame, can react with oxidizing materials.
Countermeasures
To Fight Fire: Foam, carbon dioxide, dry chemical or carbon tetrachloride (Section 6).
Ventilation Control: Section 2.
Personnel Protection: Section 3.
Storage and Handling: Section 7.

AMYL PYRIDINE
Hazard Analysis
Toxic Hazard Rating: U. Limited animal experiments suggest moderate toxicity. See also pyridine.

AMYL SALICYLATE
General Information
Description: Liquid.
Formula: $HOC_6H_4COOC_5H_{11}$.
Constants: Mol wt: 208.24, bp: 266.5°C, flash p: 270°F (C.C.), d: 1.065, vap. d.: 7.17.
Hazard Analysis
Toxic Hazard Rating:
  Acute Local: Irritant 2; Ingestion 2.
  Acute Systemic: U.
  Chronic Local: U.
  Chronic Systemic: U.
Caution: Mild sensitizer. Local contact may cause contact dermatitis (Section 9).
Fire Hazard: Moderate, when exposed to heat or flame, it can react with oxidizing materials.

Countermeasures
To Fight Fire: Foam, carbon dioxide, dry chemical or carbon tetrachloride (Section 6).
Personnel Protection: Section 3.
Storage and Handling: Section 7.

AMYL SILICATE
General Information
Description: Liquid.
Formula: $Si(OC_5H_9)_4$.
Constants: Mol wt: 368.6, mp: 148°C at 3 mm.
Hazard Analysis
Toxic Hazard Rating:
  Acute Local: Irritant 2; Ingestion 2.
  Acute Systemic: U.
  Chronic Local: U.
  Chronic Systemic: U.
Fire Hazard: Slight (Section 6).
Countermeasures
Personnel Protection: Section 3.
Storage and Handling: Section 7.

AMYL STEARATE
General Information
Description: Liquid.
Formula: $CH_3(CH_2)_{16}COOC_5H_{11}$.
Constants: Mol wt: 354.60, bp: 359.5°C, flash p: 368°F (O.C.), d: 0.860.
Hazard Analysis
Toxic Hazard Rating:
  Acute Local: Irritant 2; Ingestion 2.
  Acute Systemic: U.
  Chronic Local: U.
  Chronic Systemic: U.
Fire Hazard: Slight, when exposed to heat or flame; can react with oxidizing materials.
Spontaneous Heating: No.
Countermeasures
To Fight Fire: Foam, carbon dioxide, dry chemical or carbon tetrachloride (Section 6).
Personnel Protection: Section 3.
Storage and Handling: Section 7.

AMYL SULFIDE
General Information
Synonym: Diamyl sulfide.
Description: Liquid.
Formula: $(C_5H_{11})_2S$.
Constants: Mol wt: 174.3, bp: 170–180°C, flash p: 185°F (O.C.), d: 0.85–0.91.
Hazard Analysis
Toxic Hazard Rating:
  Acute Local: Irritant 2; Ingestion 2; Inhalation 2.
  Acute Systemic: Ingestion 2; Inhalation 2; Skin Absorption 2.
  Chronic Local: U.
  Chronic Systemic: U.
Fire Hazard: Moderate, when exposed to heat or flame.
Disaster Hazard: Dangerous; See sulfides; can react vigorously with oxidizing materials.
Countermeasures
Ventilation Control: Section 2.
Personnel Protection: Section 3.
To Fight Fire: Foam, carbon dioxide, dry chemical or carbon tetrachloride (Section 6).

AMYL SULFIDES (MIXED)
General Information
Formula: $C_5H_{11}S$.

TOXIC HAZARD RATING CODE *(For detailed discussion, see Section 1.)*

.0 NONE: (a) No harm under any conditions; (b) Harmful only under unusual conditions or overwhelming dosage.

1 SLIGHT: Causes readily reversible changes which disappear after end of exposure.

2 MODERATE: May involve both irreversible and reversible changes; not severe enough to cause death or permanent injury.

3 HIGH: May cause death or permanent injury after very short exposure to small quantities.

U UNKNOWN: No information on humans considered valid by authors.

Constants: Mol wt: 103, d: 0.9, bp: 338–356° F, flash p: 185° F (O.C.).

Hazard Analysis
Toxic Hazard Rate: U.
Fire Hazard: Moderate, when exposed to heat or flame.
Disaster Hazard: Dangerous. See sulfur compounds.

Countermeasures
Storage and Handling: Section 7.
To Fight Fire: Foam, fog, dry chemical (Section 6).

AMYL TETRATHIO-o-STANNATE
General Information
Description: Solid.
Formula: $(C_5H_{11}S)_4Sn$.
Constant: Mol wt: 531.5.

Hazard Analysis
Toxicology: Details unknown. See tin compounds.
Disaster Hazard: Dangerous; see sulfur compounds.

tert-AMYL TETRATHIO-o-STANNATE
General Information
Description: Solid.
Formula: $[CH_3CH_2C(CH_3)_2S]_4Sn$.
Constants: Mol wt: 531.5, mp: 44°C.

Hazard Analysis
Toxicology: Details unknown. See tin compounds.
Disaster Hazard: Dangerous; see sulfur compounds.

AMYL TOLUENE
General Information
Description: Liquid.
Formula: $C_5H_{11}C_6H_4CH_3$.
Constants: Mol wt: 162.25, bp: 205–210°C, flash p: 180°F (O.C.), d: 0.87, vap. d.: 5.6.

Hazard Analysis
Toxicity: Details unknown. See also toluene.
Fire Hazard: Moderate, when exposed to heat or flame; can react with oxidizing materials.

Countermeasures
To Fight Fire: Foam, carbon dioxide, dry chemical or carbon tetrachloride (Section 6).
Storage and Handling: Section 7.

AMYL TOLYL ETHER
General Information
Description: Liquid.
Formula: $C_5H_{11}OCH_2C_6H_5$.
Constants: Mol wt: 178.3, bp: 227°C, flash p: 195°F, d: 0.91.

Hazard Analysis
Toxicity: Details unknown. See also ethers.
Fire Hazard: Moderate, when exposed to heat or flame; can react with oxidizing materials.

Countermeasures
To Fight Fire: Foam, carbon dioxide, dry chemical or carbon tetrachloride (Section 6).
Storage and Handling: Section 7.

AMYL TRICHLOROSILANE
General Information
Description: Liquid.
Formula: $C_5H_{11}SiCl_3$.
Constant: Mol wt: 205.7.

Hazard Analysis
Toxicity: See chlorosilanes.
Disaster Hazard: See chlorosilanes.

Countermeasures
Storage and Handling: Section 7.
Shipping Regulations: Section 11.
I.C.C.: Corrosive liquid; white label, 10 gallons.
Coast Guard Classification: Corrosive liquid; white label.
IATA: Corrosive liquid, white label, not acceptable (passenger), 40 liters (cargo).

AMYL TRIETHOXYSILANE
General Information
Description: Liquid.
Formula: $C_5H_{11}Si(OC_2H_5)_3$.
Constants: Mol wt: 234.4, bp: 198°C, d: 0.889 at 20°C.

Hazard Analysis
Toxic Hazard Rating:
Acute Local: Irritant 1; Ingestion 1; Inhalation 1.
Acute Systemic: U.
Chronic Local: U.
Chronic Systemic: U.
Fire Hazard: A moderately flammable material; reacts with oxidizing materials (Section 6).
Personal Hygiene: Section 3.
Storage and Handling: Section 7.

AMYL TRIPHENYL GERMANIUM
General Information
Description: Colorless crystals. Insoluble in water. Soluble in organic solvents.
Formula: $Ge(C_5H_{11})(C_6H_5)_3$.
Constants: Mol wt: 375.0, mp: 43°C.

Hazard Analysis
Toxicity: Details unknown. See germanium compounds.

AMYL XYLYL ETHER
General Information
Description: Liquid.
Formula: $C_5H_{11}OC_6H_3(CH_3)_2$.
Constants: Mol wt: 192.29, bp: 250–260°C, flash p: 205°F (O.C.), d: 0.907.

Hazard Analysis
Toxicity: Details unknown. See also ethers.
Fire Hazard: Moderate, when exposed to heat or flame; it can react with oxidizing materials.

Countermeasures
To Fight Fire: Foam, carbon dioxide, dry chemical or carbon tetrachloride (Section 6).
Storage and Handling: Section 7.

AMYTAL. See amobarbital.

ANABASINE
General Information
Synonyms: Neonicotine; nicotine isomer.
Description: Colorless, oily liquid.
Formula: $C_{10}H_{14}N_2$.
Constants: Mol wt: 162.2, bp: 105°C at 2 mm Hg, d: 1.048 at 20°/20°C, fp: 9°C; miscible with water.
Toxic Hazard Rating:
Acute Local: U.
Acute Systemic: Ingestion 3; Inhalation 3; Skin Absorption 3.
Chronic Local: U.
Chronic Systemic: Ingestion 2; Inhalation 2; Skin Absorption 2.
Toxicology: Causes increased salivation, mental confusion, dizziness, disturbed hearing and vision, nausea, vomiting. In severe cases there is unconsciousness with convulsions.
Disaster Hazard: Dangerous; when heated to decomposition, it emits highly toxic fumes of nitrogen oxides and cyanides.

Countermeasures
Ventilation Control: Section 2.
Personnel Protection: Section 3.
First Aid: Section 1.
Storage and Handling: Section 7.

ANABASINE SULFATE
General Information
Synonym: Neonicotine sulfate.
Description: Colorless, oily liquid.
Formula: $(C_{10}H_{14}N_2)_2H_2SO_4$.
Constant: Mol wt: 422.5.

Hazard Analysis
Toxic Hazard Rating: See anabasine.
Toxicology: See anabasine.
Disaster Hazard: Dangerous; when heated to decomposition, it emits highly toxic fumes.

Countermeasures
Ventilation Control: Section 2.
Personal Protection: Section 3.
First Aid: Section 1.
Storage and Handling: Section 7.

ANASADOL. See salicylamide.

△ -ANDROSTEN-17(γ)-ol-3-one. See testosterone.

ANESTHESIA ETHER. See ethyl ether.

ANESTHESIN
General Information
Synonym: Ethyl-p-aminobenzoate.
Description: Crystals.
Formula: $C_9H_{11}NO_2$.
Constants: Mol wt: 165.2, mp: 88–90°C.
Hazard Analysis
Toxicity: Used medicinally and has low systemic toxicity when taken by mouth. See also esters, ethyl alcohol and p-amino benzoic acid.
Caution: A mild sensitizer. Local contact may cause contact dermatitis (Section 9).
Disaster Hazard: Moderately dangerous; when heated to decomposition, it emits toxic fumes.
Countermeasures
Personal Hygiene: Section 3.
Storage and Handling: Section 7.

ANETHOLE
General Information
Synonyms: p-propenyl anisole; anise camphor; p-methoxy-propenyl benzene.
Description: White crystals, anise oil odor; soluble in 8 volumes of 80% alcohol; almost immiscible with water.
Formula: $CH_3CH:CHC_6H_4OCH_3$.
Constants: Mol wt: 148, d: 0.983–0.987, mp: 22–23°C.
Hazard Analysis
Toxic Hazard Rate: U. A synthetic flavoring substance and adjuvant (Section 10).

ANGLISITE. See lead sulfate.

ANHALONIDINE
General Information
Description: White crystals.
Formula: $C_{12}H_{17}NO_3$.
Constant: Mol wt: 223.3.
Hazard Analysis
Toxic Hazard Rating:
  Acute Local: U.
  Acute Systemic: Ingestion 3; Inhalation 3.
  Chronic Local: U.
  Chronic Systemic: U.
Disaster Hazard: Dangerous; when heated to decomposition, it emits highly toxic fumes.
Countermeasures
Ventilation Control: Section 2.
Personal Hygiene: Section 3.
First Aid: Section 1.
Storage and Handling: Section 7.

ANILINE
General Information
Synonyms: Phenylamine; aminobenzene; aniline oil.
Description: Colorless, oily liquid.
Formula: $C_6H_5NH_2$.
Constants: Mol wt: 93.12, bp: 184.4°C, lel: 1.3%, ulc: 20–25, flash p: 158°F (C.C.), fp: −6.2°C, d: 1.02 at 20°/4°C, autoign. temp.: 1418°F, vap. press.: 1 mm at 34.8°C, vap. d: 3.22.

Hazard Analysis
Toxic Hazard Rating:
  Acute Local: Allergen 2.
  Acute Systemic: Ingestion 3; Inhalation 3; Skin Absorption 3.
  Chronic Local: Allergen 2.
  Chronic Systemic: Ingestion 3; Inhalation 3; Skin Absorption 3.
TLV: ACGIH (recommended); 5 parts per million in air; 19 milligrams per cubic meter of air. Absorbed via the skin.
Toxicology: The most important action of aniline on the body is the formation of methemoglobin, with the resulting anoxemia and depression of the central nervous system. Some investigators believe that aniline may also have a direct toxic action, resulting in a fall in blood pressure and cardiac arrhythmia. In acute exposures, which usually result from spilling the liquid on the skin and clothes, but which may also follow the inhalation of the vapor given off when aniline is heated, the signs are of methemoglobinaemia and anoxemia. In less acute exposure which has been prolonged over some weeks or months, there is usually hemolysis of the red blood cells, followed by stimulation of the bone marrow and attempts at regeneration. The red cells may show stippling; immature cells may be present. The white blood cells usually show little change either in number or morphology. The liver may be affected, with production of jaundice. The urine is frequently dark brown or wine colored, and may contain hemoglobin, hematoporphyrin, and in some cases, excretion products of aniline, such as p-aminophenol. Long continued employment in the manufacture of aniline dyes has been associated with the development of papillomatous growths of the bladder, some of which became malignant. Aniline itself has not been proven to be a carcinogen, but the intermediates benzedine and naphthylamines have been incriminated. See α-1 and β-naphthylamine. Note: A common air contaminant (Section 4).
Caution: Mild sensitizer. Local contact may cause contact dermatitis (Section 9).
Fire Hazard: Moderate, when exposed to heat or flame.
Spontaneous Heating: No.
Disaster Hazard: Dangerous; when heated to decomposition, it emits highly toxic fumes; can react vigorously with oxidizing materials.

Countermeasures
Ventilation Control: Section 2.
To Fight Fire: Alcohol foam, carbon dioxide, dry chemical or carbon tetrachloride (Section 6).
Personnel Protection: Section 3.
First Aid: Section 1.
Storage and Handling: Section 7.
Shipping Regulations: Section 11.
  I.C.C. Classification: Poison B; poison label.
  Coast Guard Classification: Poison B; poison label.

TOXIC HAZARD RATING CODE *(For detailed discussion, see Section 1.)*

0 NONE: (a) No harm under any conditions; (b) Harmful only under unusual conditions or overwhelming dosage.

1 SLIGHT: Causes readily reversible changes which disappear after end of exposure.

2 MODERATE: May involve both irreversible and reversible changes; not severe enough to cause death or permanent injury.

3 HIGH: May cause death or permanent injury after very short exposure to small quantities.

U UNKNOWN: No information on humans considered valid by authors.

IATA: Poison B, poison label, 25 kilograms (passenger), 95 kilograms (cargo).

ANILINE ACETATE
General Information
Description: Colorless liquid.
Formula: $C_6H_5NH_2CH_3COOH$.
Constants: Mol wt: 153.2, d: 1.071.
Hazard Analysis and Countermeasures
See aniline.

ANILINE ANTIMONYL TARTRATE
General Information
Synonym: Antimonyl aniline tartrate.
Description: White crystals.
Formula: $C_6H_5NH_2H[(SbO)C_4H_4O_6]$.
Constant: Mol wt: 380.
Hazard Analysis and Countermeasures
See aniline and antimony compounds.

ANILINE BLACK 870
General Information
Description: A dye.
Hazard Analysis
Toxicity: An allergen.
Caution: Mild sensitizer. Local contact may cause contact dermatitis (Section 9).
Disaster Hazard: Moderately dangerous; when heated to decomposition, it emits toxic fumes.

ANILINE BRILLIANT GREEN
Genteal Information
Description: A dye.
Hazard Analysis
Toxicity: An allergen.
Caution: Mild sensitizer. Local contact may cause contact dermatitis (Section 9).
Disaster Hazard: Moderately dangerous; when heated to decomposition, it emits toxic fumes.
Countermeasures
Storage and Handling: Section 7.

ANILINE CAMPHORATE
General Information
Description: Yellow crystals.
Formula: $(C_6H_5NH_2)_2C_{10}H_{16}O_4$.
Constant: Mol wt: 386.5.
Hazard Analysis and Countermeasures
See aniline.

ANILINE CHLORIDE. See aniline hydrochloride.

ANILINE DYES
Hazard Analysis
Toxicity: Variable. The finished dyes are generally very much less toxic than many of the intermediates occurring or used in the manufacture of the dyes. Some of the aniline dyes cause local irritative effects to the eyes, mucous membranes and skin; the basic dyes are believed to be more irritating than the acid dyes. Allergic responses to aniline dyes have been known to occur (Section 9). See also specific compounds.
Disaster Hazard: Moderately dangerous; when heated to decomposition, they emit toxic fumes.
Countermeasures
Storage and Handling: Section 7.

ANILINE FLUORIDE
General Information
Synonym: Aniline hydrofluoride.
Description: Crystalline powder.
Formula: $C_6H_5NH_2 \cdot HF$.
Constant: Mol wt: 113.1.
Hazard Analysis and Countermeasures
See aniline and fluorides.

ANILINE GREEN. See malachite green.

ANILINE HYDROBROMIDE
General Information
Synonym: Aniline bromide.
Description: White to reddish crystals.
Formula: $C_6H_5NH_2 \cdot HBr$.
Constants: Mol wt: 174.0, mp: 286°C.
Hazard Analysis and Countermeasures
See aniline and bromides.

ANILINE HYDROCHLORIDE
General Information
Synonym: Aniline chloride.
Description: Crystals.
Formula: $C_6H_5NH_2 \cdot HCl$.
Constants: Mol wt: 129.59, mp: 198°C, bp: 245°C, flash p: 380°F (O.C.), d: 1.22, vap. d.: 4.46.
Hazard Analysis
Toxicity: See aniline.
Fire Hazard: Slight, when exposed to heat or flame.
Spontaneous Heating: No.
Disaster Hazard: Dangerous; when heated to decomposition or on contact with acid or acid fumes, it emits highly toxic fumes of aniline and chlorine compounds; can react vigorously with oxidizing materials.
Countermeasures
Storage and Handling: Section 7.
To Fight Fire: Water, carbon dioxide, dry chemical or carbon tetrachloride (Section 6).
Shipping Regulations: Section 11.
IATA: Poison B, poison label, 25 kilograms (passenger), 95 kilograms (cargo).

ANILINE HYDROFLUORIDE. See aniline fluoride.

ANILINE OIL. See aniline.

ANILINE OIL DRUMS, EMPTY
Hazard Analysis
Toxicity: See aniline.
Fire Hazard: Slight, when exposed to heat or flame. Such drums may fill with vapors and under the proper conditions ignite.
Disaster Hazard: Dangerous, depending on the number involved; when heated, they emit highly toxic fumes of aniline.
Countermeasures
Storage and Handling: Section 7.
Shipping Regulations: Section 11.
Coast Guard Classification: Hazardous.

ANILINE OIL, LIQUID
Countermeasures
Shipping Regulations:
I.C.C.: Poison B, poison label, 55 gallons.
MCA warning label.

ANILINE SODIUM SULFONATE
General Information
Synonyms: Metanilic acid, sodium salt.
Description: Tan flakes.
Formula: $C_6H_4(NH_2)(NaSO_3)$.
Constant: Mol wt: 195.
Hazard Analysis
Toxic Hazard Rating:
Acute Local: Allergen 2.
Acute Systemic: Ingestion 3; Inhalation 3.
Chronic Local: Allergen 2.
Chronic Systemic: Ingestion 3; Inhalation 3.
Caution: A mild sensitizer. Local contact may cause contact dermatitis (Section 9). See also aniline and sulfonates.
Disaster Hazard: Dangerous; when heated to decomposition or on contact with acid or acid fumes, it emits highly toxic fumes of aniline and oxides of sulfur.
Countermeasures
Ventilation Control: Section 2.

Personal Hygiene: Section 3.
First Aid: Section 1.
Storage and Handling: Section 7.

ANILINE TRIBROMIDE. See 2,4,6-Tribromoaniline.

ANILINOBENZENE. See diphenylamine.

2-ANILINOETHANOL
General Information
Synonyms: β-Anilinoethanol ethoxyaniline; β-hydroxy-ethylaniline.
Formula: $C_6H_5NHCH_2CH_2OH$.
Constants: Mol wt: 137, d: 1.1, bp: 547°F, flash p: 305°F (O.C.).
Hazard Analysis
Toxic Hazard Rating: U.
Fire Hazard: Slight, when exposed to heat or flame.
Countermeasures
To Fighe Fire: Water, foam (Section 6).
Storage and Handling: Section 7.

β-ANILINOETHANOL ETHOXYANILINE. See 2-anilino ethanol.

ANILINOPHENOL. See p-hydroxy diphenylamine.

ANILITE
A high explosive mixture composed of liquid NO_2 and carbon disulfide or gasoline. It is extremely sensitive to shock. See Section 7.

o-ANISALDEHYDE. See o-methoxy benzaldehyde.

ANISE CAMPHOR. See anethole.

ANISE SEED OIL
General Information
Description: Colorless or pale yellow liquid.
Composition: 80–90% anethol, methyl chavicol, anisaldehyde.
Constant: d: 0.978–0.988 at 25°/25°C.
Hazard Analysis
Caution: A weak sensitizer. Local contact may cause contact dermatitis (Section 9).
Fire Hazard: Slight (Section 6).
Countermeasures
Storage and Handling: Section 7.

ANISIC ACID
General Information
Synonym: p-Methoxybenzoic acid.
Description: Needlelike crystals. Soluble in alcohol, chloroform, ether.
Formula: $CH_3OC_6H_4COOH$.
Constants: Mol wt: 152.1, bp: 150°C, d: 1.385, mp: 184°C.
Hazard Analysis
Toxicity: Unknown. Probably has low toxicity. See also anisole and benzoic acid.
Fire Hazard: Slight, when exposed to heat or flame; it can react with oxidizing materials (Section 6).
Countermeasures
Storage and Handling: Section 7.

o-ANISIC ALDEHYDE. See o-methoxy benzaldehyde.

o-ANISIDINE
General Information
Synonym: o-Methoxyaniline.
Description: Liquid.

Formula: $CH_3OC_6H_4NH_2$.
Constants: Mol wt: 123.15, mp: 5.2°C, bp: 225°C, d: 1.108 at 26°C.
Hazard Analysis
Toxic Hazard Rating:
Acute Local: Irritant 2; Allergen 1; Ingestion 2.
Acute Systemic: Ingestion 3; Inhalation 3; Skin Absorption 3.
Chronic Local: Irritant 2; Allergen 1.
Chronic Systemic: Ingestion 3; Inhalation 3; Skin Absorption 3.
TLV: ACGIH (recommended) 0.5 milligram per cubic meter; absorbed via the skin.
Caution: Local contact may cause a contact dermatitis (Section 9). See also aniline.
Disaster Hazard: Dangerous; when heated to decomposition, it emits toxic fumes (Section 6).
Countermeasures
Ventilation Control: Section 2.
Personnel Protection: Section 3.
First Aid: Section 1.

p-ANISIDINE
General Information
Synonym: p-Methoxyaniline.
Description: Liquid.
Formula: $CH_3OC_6H_4NH_2$.
Constants: Mol wt: 123.15, mp: 59°C, bp: 240°C, d: 1.071 at 55°/4°C, vap. d.: 4.28.
Hazard Analysis
Toxic Hazard Rating:
Acute Local: Irritant 2; Allergen 1; Ingestion 2.
Acute Systemic: Ingestion 3; Inhalation 3; Skin Absorption 3.
Chronic Local: Irritant 2; Allergen 1.
Chronic Systemic: Ingestion 3; Inhalation 3; Skin Absorption 3.
TLV: ACGIH (recommended) 0.5 milligram per cubic meter; absorbed via the skin.
Caution: Mild sensitizer. Local contact may cause a contact dermatitis (Section 9). See also aniline.
Disaster Hazard: Dangerous; when heated to decomposition, it evolves toxic fumes.
Countermeasures
Ventilation Control: Section 2.
Personnel Protection: Section 3.
First Aid: Section 1.

ANISOLE
General Information
Synonym: Phenyl methyl ether.
Description: Mobile liquid, clear straw color.
Formula: $C_6H_5OCH_3$.
Constants: Mol wt: 108.13, mp: −37.3°C, bp: 153.8°C, flash p: 125°F (C.O.C.), d: 0.996 at 18°/4°C, vap. press.: 10 mm at 42.2°C, vap. d.: 3.72.
Hazard Analysis
Toxic Hazard Rating:
Acute Local: U.
Acute Systemic: Ingestion 2; Inhalation 3; Skin Absorption 2.
Chronic Local: U.
Chronic Systemic: U.
Fire Hazard: Moderate, when exposed to heat or flame; can react with oxidizing materials (Section 6).
Countermeasures
To Fight Fire: Foam, carbon dioxide, dry chemical or carbon tetrachloride (Section 6).

TOXIC HAZARD RATING CODE (For detailed discussion, see Section 1.)

0 NONE: (a) No harm under any conditions; (b) Harmful only under unusual conditions or overwhelming dosage.

1 SLIGHT: Causes readily reversible changes which disappear after end of exposure.

2 MODERATE: May involve both irreversible and reversible changes; not severe enough to cause death or permanent injury.

3 HIGH: May cause death or permanent injury after very short exposure to small quantities.

U UNKNOWN: No information on humans considered valid by authors.

Ventilation Control: Section 2.
Personnel Protection: Section 3.
Storage and Handling: Section 7.

ANISOYL CHLORIDE
General Information
Synonym: p-Anisyl chloride.
Description: Needle-like crystals. Insoluble in water. Soluble in ether and acetone.
Formula: $CH_3OC_6H_4COCl$.
Constants: Mol wt: 170.6, mp: 22°C, bp: 262°–263°C (slight decomp.).
Hazard Analysis
Toxic Hazard Rating:
  Acute Local: Irritant 3; Ingestion 3; Inhalation 3.
  Acute Systemic: U.
  Chronic Local: U.
  Chronic Systemic: U.
Caution: Can readily evolve hydrochloric acid by hydrolysis.
Disaster Hazard: Dangerous; see chlorides.
Explosion Hazard: Can explode spontaneously at room temperature.
Countermeasures
Ventilation Control: Section 2.
Personnel Proection: Section 3.
First Aid: Section 1.
Storage and Handling: Section 7.
Shipping Regulations: Section 11.
  I.C.C.: Corrosive liquid; white label, 1 quart.
  Coast Guard Classification: Corrosive liquid; white label.
  IATA: Corrosive liquid, white label, 1 quart (passenger and cargo).

o-ANISYL BORIC ACID
General Information
Synonym: o-Methoxy phenyl boric acid.
Description: White crystals.
Formula: $CH_3OC_6H_4B(OH)_2$.
Constant: Mol wt: 152.0.
Hazard Analysis
Toxicity: See boron compounds.

m-ANISYL BORIC ACID
General Information
Synonym: m-Methoxy phenyl boric acid.
Description: White crystals.
Formula: $CH_3OC_6H_4B(OH)_2$.
Constant: Mol wt: 152.0.
Hazard Analysis
Toxicity: See boron compounds.

p-ANISYL BORIC ACID
General Information
Synonym: p-Methoxy phenyl boric acid.
Description: White crystals.
Formula: $CH_3OC_6H_4B(OH)_2$.
Constant: Mol wt: 152.0.
Hazard Analysis
Toxicity: See boron compounds.

p-ANISYL CHLORIDE. See anisoyl chloride.

ANOGON. See mercury compounds and iodine.

ANSOL M. See alcohol, denatured.

ANTHION. See potassium persulfate.

ANTHANTHRENE
Hazard Analysis
A polycyclic hydrocarbon found in air pollution studies.

ANTHRACENE
General Information
Synonyms: p-Naphthalene; green oil.
Description: Yellow crystals with blue fluorescence.

Formula: $C_6H_4(CH)_2C_6H_4$.
Constants: Mol wt: 178.22, mp: 217°C, bp: 345°C, lel: 0.6%, flash p:250°F (C.C.), d: 1.25 at 27°/4°C, autoign. temp.: 881°F, vap. press.: 1 mm at 145.0°C (sublimes), vap. d.: 6.15.
Hazard Analysis
Toxic Hazard Rating:
  Acute Local: Irritant 1; Allergen 1; Ingestion 1.
  Acute Systemic: U.
  Chronic Local: Irritant 3; Allergen 1.
  Chronic Systemic: U.
Caution: Although it is thought that the pure product is nontoxic, the product as used in industry is thought to be carcinogenic because of contained impurities.
Fire Hazard: Moderate, when exposed to heat or flame; reacts with oxidizing materials.
Spontaneous Heating: No.
Explosion Hazard: Moderate, when exposed to flame.
Countermeasures
Personal Hygiene: Section 3.
To Fight Fire: Water, foam, carbon dioxide, dry chemical or carbon tetrachloride (Section 6).
Storage and Handling: Section 7.

ANTHRACITE PARTICLES
General Information
Synonym: Coal dust.
Description: Black powder or dust.
Hazard Analysis
Toxic Hazard Rating:
  Acute Local: Inhalation 1.
  Acute Systemic: 0.
  Chronic Local: Inhalation 2.
  Chronic Systemic: 0.
TLV: ACGIH (recommended): 50 million particles per cubic foot of air.
Fire Hazard: Moderate, when exposed to heat or flame, or by chemical reaction with oxidizers (Section 6).
Explosion Hazard: Slight, when exposed to flame (Section 7).
Countermeasures
Ventilation Control: Section 2.

ANTHRALIN
General Information
Synonym: 1,8-Dihydroxy anthranol.
Description: Yellow, crystals, insoluble in water, soluble in chloroform, acetone and benzene.
Formula: $C_{14}H_{10}O_3$.
Constants: Mol wt: 226.2, mp: 176–181°C.
Hazard Analysis
Toxic Hazard Rating:
  Acute Local: Irritant 2.
  Acute Systemic: Ingestion 2, Inhalation 2; Skin Absorption 2.
  Chronic Local: Irritant 2.
  Chronic Systemic: Ingestion 2, Inhalation 2, Skin Absorption 2.
Toxicology: Locally it can cause folliculitis of skin. Absorption may result in kidney injury and intestinal disturbances.
Fire Hazard: Slight; when heated.
Countermeasures
Personal Hygiene: Section 3.
Storage and Handling: Section 7.

ANTHRANILIC ACID
General Information
Synonym: o-Amino benzoic acid.
Description: Needle-like crystals.
Formula: $C_6H_4(NH_2)COOH$.
Constants: Mol wt: 137.1, mp: 145°C, bp: sublimes.
Hazard Analysis
Toxicity: Unkown. See p-aminobenzoic acid.
Fire Hazard: Slight.

Countermeasures
Storage and Handling: Section 7.

ANTHRAQUINONE
General Information
Description: Powder.
Formula: $C_6H_4(CO)_2C_6H_4$.
Constants: Mol wt: 208.20, mp: 286°C, bp: 379.9°C, flash p: 365°F (C.C.), d: 1.438, vap. press.: 1 mm at 190.0°C, vap. d.: 7.16.
Hazard Analysis
Toxic Hazard Rating:
  Acute Local: Irritant 1; Allergen 1; Ingestion 1.
  Acute Systemic: U.
  Chronic Local: Allergen 1.
  Chronic Systemic: U.
Toxicology: Weak sensitizer. Local contact may cause contact dermatitis (Section 9).
Fire Hazard: Slight, when exposed to heat or flame.
Spontaneous Heating: No.
Countermeasures
To Fight Fire: Water, foam, carbon dioxide, dry chemical or carbon tetrachloride (Section 6).
Personal Hygiene: Section 3.
Storage ahd Handling: Section 7.

ANTHRAQUINONE BLUE S.R. 1089
General Information
Description: A dye.
Hazard Analysis
Toxicity: Weak sensitizer. Local contact may cause contact dermatitis (Section 9).

ANTHRAROBIN
General Information
Synonym: 3,4-Dihydroxy anthranol.
Description: Yellow-brown powder.
Formula: $C_{14}H_{10}O_3$.
Constants: Mol wt: 226.22, mp: 208°C.
Hazard Analysis
Toxicity: Mild allergen. Local contact may cause contact dermatitis (Section 9).
Fire Hazard: Slight; (Section 6).
Countermeasures
Storage and Handling: Section 7.

ANTIFEBRIN. See acetanilide.

ANTIFREEZE COMPOUND, LIQUID
See also ethylene glycol and methyl alcohol.
See specific components.
Shipping Regulations: Section 11.
  I.C.C.: Flammable liquid; red label, 10 gallons
  Coast Guard Classification: Combustible liquid; red label.
  Coast Guard Classification: Inflammable liquid; red label.
  IATA: Flammable liquid, red label, 1 liter (passenger), 40 liters (cargo).

ANTIFREEZE PREPARATIONS, LIQUID
See specific components.
Shipping Regulations: Section 11.
  I.C.C.: Flammable liquid; red label, 10 gallons.
  Coast Guard Classification: Combustible material; red label.
  Coast Guard Classification: Inflammable liquid; red label.
  IATA: Flammable liquid, red label, 1 liter (passenger), 40 liters (cargo).

ANTIMONIC CHLORIDE. See antimony pentachloride.

ANTIMONOUS BROMIDE. See antimony tribromide.

ANTIMONOUS CHLORIDE. See antimony trichloride.

ANTIMONY
General Information
Synonyms: Antimony regulus; stibium.
Description: Gray metal.
Formula: Sb.
Constants: At wt: 121.76, mp: ·630°C, bp: 1380°C, d: 6.684 at 25°C, vap. press.: 1 mm at 886°C.
Hazard Analysis
Toxicity: Highly toxic. See antimony compounds.
TLV: ACGIH (accepted); 0.5 milligrams per cubic meter of air.
Radiation Hazard: Section 5. For permissible levels, see Table 5, p. 150.
    Artificial isotope ^{122}Sb, half life 2.8 d. Decays to stable ^{122}Te by emitting beta particle of 1.40 (63%), 1.97 (30%) MeV. Also emits gamma rays of 0.56, 0.69, 1.26 MeV.
    Artificial isotope ^{124}Sb, half life 60 d. Decays to stable ^{124}Te by emitting beta particles of 0.05–2.31 MeV. Also emits gamma rays of 0.60–2.09 MeV.
    Artificial isotope ^{125}Sb, half life 2.7 y. Decays to stable ^{125}Te by emitting beta particles of 0.09–0.61 MeV. Also emits gamma rays of 0.04–0.67 MeV.
Fire Hazard: Moderate, in the form of dust or vapor, when exposed to heat or flame (Section 6). See also powdered metals.
Explosion Hazard: Moderate, in the form of dust when exposed to flame (Section 7). See also powdered metals.
Disaster Hazard: Moderately dangerous; when heated or on contact with acid, it emits toxic fumes.

ANTIMONY ARSENATE
General Information
Description: Heavy, white powder.
Composition: $Sb_2O_3 + 20\% H_4AsO_7$.
Hazard Analysis and Countermeasures
See arsenic and antimony compounds.

ANTIMONY ARSENITE
General Information
Description: Fine white powder.
Compositions: $Sb_2O_3 + As_2O_3$.
Hazard Analysis and Countermeasures
See antimony and arsenic compounds.

ANTIMONY BROMIDE. See antimony tribromide.

ANTIMONY CHLORIDE. See antimony pentachloride or antimony trichloride.

ANTIMONY COMPOUNDS
Hazard Analysis
Toxic Hazard Rating:
  Acute Local: Ingestion 3.
  Acute Systemic: Ingestion 3; Inhalation 3.
  Chronic Local: Irritant 2.
  Chronic Systemic: Ingestion 3; Inhalation 3.
TLV: ACGIH (recommended); 0.5 mg per cubic meter of air.
Toxicology: Because of the association with lead and arsenic in industry, it is often difficult to assess the toxicity of antimony and its compounds. Animals exposed to fumes of antimony oxide have developed pneumonitis, fatty degeneration of the liver, a decreased leucocyte count affecting in particular the polymorphonuclears, and damage to the heart muscle. In humans, complaints re-

---

TOXIC HAZARD RATING CODE (For detailed discussion, see Section 1.)

0 NONE: (a) No harm under any conditions; (b) Harmful only under un-usual conditions or overwhelming dosage.

1 SLIGHT: Causes readily reversible changes which disappear after end of exposure.

2 MODERATE: May involve both irreversible and reversible changes; not severe enough to cause death or permanent injury.

3 HIGH: May cause death or permanent injury after very short exposure to small quantities.

U UNKNOWN: No information on humans considered valid by authors.

ferable to the nervous system have been reported. In assessing human cases, however, the possibility of lead or arsenical poisoning must always be borne in mind. Locally antimony compounds are irritant to the skin and mucous membranes.

Signs and symptons may include irritation and eczematous eruption of the skin, inflammation of the mucous membranes of the nose and throat, metallic taste and stomatitis, gastrointestinal upset, with vomiting and diarrhea, and various nervous complaints, such as irritability, sleeplessness, fatigue, dizziness and muscular and neuralgic pains. See also specific compounds.

Countermeasures
Ventilation Control: Section 2.
Personnel Protection: Section 3.
First Air: Section 1.
Storage and Handling: Section 7.

ANTIMONY DIOXYSULFATE
General Information
Description: White powder.
Formula: $Sb_2O_2SO_4$.
Constants: Mol wt: 371.6, d: 4.89.
Hazard Analysis and Countermeasures
See antimony compounds and sulfates.

ANTIMONY ETHOXIDE
General Information
Synonym: Triethyl antimonite.
Description: Colorless liquid. Decomposes upon contact with moisture.
Formula: $Sb(C_2H_5O)_3$.
Constants: Mol wt: 256.9, d: 1.524 at 17°C, bp: 95°C at 11 mm Hg.
Hazard Analysis
Toxicity: Highly toxic. See antimony compounds.
Disaster Hazard: See antimony compounds and ethers.

ANTIMONY FLUORIDE. See antimony pentafluoride or antimony trifluoride.

ANTIMONY HYDRIDE
General Information
Synonyms: Stibine; hydrogen antimonide.
Description: Colorless gas.
Formula: SbH_3.
Constants: Mol wt: 124.78, mp: −88°C, bp: −17°C, d: 5.30 g/liter at 0°C.
Hazard Analysis
Toxic Hazard Rating:
  Acute Local: Irritant 0; Inhalation 2.
  Acute Systemic: Inhalation 3.
  Chronic Local: U.
  Chronic Systemic: U.
TLV: ACGIH (recommended): 0.1 parts per million in air; 0.5 milligrams per cubic meter of air.
Fire Hazard: Moderate, when exposed to flame (Section 6).
Explosion Hazard: U.
Disaster Hazard: Dangerous; when heated to decomposition, it emits highly toxic fumes; can react vigorously with oxidizing materials.
Countermeasures
Ventilation Control: Section 2.
Storage and Handling: Section 7.

ANTIMONY IODIDE. See antimony tri-iodide or antimony pentaiodide.

ANTIMONY LACTATE
General Information
Description: Tan-colored mass; soluble in water.
Formula: $Sb(C_3H_5O_3)_3$.
Constant: Mol wt: 300.
Hazard Analysis
Toxicity: See Antimony Compounds.
Disaster Hazard: See antimony.

Countermeasures
Shipping Regulations: Section 11.
  IATA (solid): Other restricted articles, class A no label, no limit (passenger), no limit (cargo).

ANTIMONYL ANILINE TARTRATE. See aniline antimonyl tartrate.

ANTIMONYL PYROGALLOL
General Information
Formula: $HOC_6H_3O(SbOH)O$.
Constant: Mol wt: 262.9.
Hazard Analysis
Toxicity: An allergen. See also antimony compounds.
Disaster Hazard: See antimony.
Countemeasures
Storage aħd Handling: Section 7.

ANTIMONY α-MERCAPTOACETAMIDE
General Information
Synonym: Antimony thioglycolamide.
Description: White crystals.
Formula: $Sb(C_2H_4NOS)_3$.
Constants: Mol wt: 392.1, mp: 139°C.
Hazard Analysis
Toxicity: See antimony compounds.
Disaster Hazard: Dangerous; See antimony and oxides of sulfur.
Countermeasures
Storage and Handling: Section 7.

ANTIMONY OXIDES. See antimony trioxide, antimony tetraoxide and antimony pentaoxide.

ANTIMONY OXYCHLORIDE
General Information
Synonyms: Basic antimony chloride; powder of algraoth.
Description: White, amorphous powder.
Formula: $SbOCl$.
Constants: Mol wt: 173.22, mp: decomposes at 170°C.
Hazard Analysis & Countermeasures
See antimony compounds & chlorides.

ANTIMONY PENTACHLORIDE
General Information
Synonyms: Antimonic chloride; antimony perchloride.
Description: Reddish-yellow, oily liquid, offensive odor.
Formula: $SbCl_5$.
Constants: Mol wt: 299.05 mp: 2.8°C, bp: 140°C, d: (lq) 2.336, vap. press.: 1 mm at 22.7°C.
Hazard Analysis & Countermeasures
See antimony compounds and hydrochloric acid.
Shipping Regulations: Section 11.
  I.C.C.: Corrosive liquid; white label, 1 quart.
  Coast Guard Classification: Corrosive liquid; white label.
  IATA: Corrosive liquid, white label, 1 liter (passenger and cargo).
  IATA (Solution): Corrosive liquid, white label, 1 liter (passenger), 2½ liters (cargo).

ANTIMONY PENTAFLUORIDE
General Information
Synonym: Antimony fluoride.
Description: Oily, colorless liquid.
Formula: SbF_5.
Constants: Mol wt: 216.76, mp: 71°C, bp: 149.5°C, d: (lq) 2.99 at 23°C.
Hazard Analysis & Countermeasures
See fluorides and antimony compounds.
Shipping Regulations: Section II.
  I.C.C.: Corrosive liquid; white label, 25 pounds.
  Coast Guard Classification: Corrosive liquid; white label.
  IATA: Corrosive liquid; white label, not acceptable (passenger), 12 kilogram (cargo).

ANTIMONY PENTAIODIDE
General Information
Synonym: Antimony iodide.
Description: Brown solid.
Formula: SbI_5.
Constants: Mol wt: 756.4, mp: 79°C, bp: 400.6°C.
Hazard Analysis & Countermeasures
See antimony compounds and iodides.
Ventilation Congrol: Section 2.
Personnel Protection: Section 3.
First Aid: Section 1.
Storage and Handling: Section 7.

ANTIMONY PENTAOXIDE
General Information
Synonym: Antimony pentoxide.
Description: Yellow powder.
Formula: Sb_2O_5.
Constants: Mol wt: 323.5, mp: 380°C (loses an oxygen molecule), bp: 930°C (loses 2 oxygen molecules), d: 3.78.
Hazard Analysis
Toxicity: See antimony compounds.
Fire Hazard: Moderate, by chemical reaction with reducing agents.
Caution: An oxidizer.
Disaster Hazard: Dangerous; See antimony.
Countermeasures
Storage and Handling: Section 7.

ANTIMONY PENTASULFIDE
General Information
Synonym: Antimony sulfide.
Description: Orange-yellow powder.
Formula: Sb_2S_5.
Constants: Mol wt: 403.82, mp: decomposes, d: 4.120.
Hazard Analysis
Toxicity: See antimony compounds and sulfides.
Fire Hazard: Moderate, when exposed to heat or by chemical reaction with powerful oxidizer.
Explosion Hazard: Moderate; when shocked or by spontaneous chemical reaction in contact with powerful oxidizers (Section 7).
Disaster Hazard: Dangerous; when heated to decomposition or on contact with acid or acid fumes, it emits highly toxic fumes of oxides of sulfur and antimony; it will react with water or steam to produce toxic and flammable vapors; it can react vigorously with oxidizing materials.
Countermeasures
Ventilation Control: Section 2.
To Fight Fire: Water (Section 6).
First Aid: Section 1.
Storage and Handling: Section 7.

ANTIMONY PENTOXIDE. See antimony pentaoxide.

ANTIMONY PERCHLORIDE. See antimony pentachloride.

ANTIMONY POTASSIUM OXALATE
General Information
Synonym: Potassium-antimony oxalate.
Description: White, crystalline powder.
Formula: $K_3Sb(C_2O_4)_3$.
Constant: Mol wt: 503.1.
Hazard Analysis
Toxicity: See antimony compounds and oxalates.

Disaster Hazard: Dangerous; See antimony.
Countermeasures
Ventilation Control: Section 2.
Personnel Protection: Section 3.
First Aid: Section 1.
Storage and Handling: Section 7.

ANTIMONY POTASSIUM TARTRATE
General Information
Synonyms: Tartar emetic; potassium antimony tartrate.
Description: Colorless crystals; white powder.
Formula: $KSbC_4H_4O_7 \cdot \frac{1}{2}H_2O$.
Constants: Mol wt: 333.94, mp: $-\frac{1}{2}H_2O$ at 100°C, d: 2.607.
Hazard Analysis
Toxicity: See antimony compounds.
Caution: This compound is used medicinally but the therapeutic dose is close to the toxic dose. It can cause cough, metallic taste, salivation, nausea and diarrhea. Skin rash may also occur. Large doses can cause severe damage to the liver.
Disaster Hazard; See antimony compounds.
Countermeasures
Storage and Handling: Section 7.
Shipping Regulations: Section 11.
IATA (solid): Other restricted articles, class A, no label, no limit (passenger), no limit (cargo).

ANTIMONY REGULUS. See antimony.

ANTIMONY SALT
General Information
Synonym: de Haens salt.
Description: White crystals.
Composition: Mixture of antimony trifluoride and sodium fluoride or ammonium sulfate.
Hazzard Analysis & Countermeasures
See antimony compounds and fluorides.

ANTIMONY SELENIDE. See antimony triselenide.

ANTIMONY SULFATE
General Information
Synonym: Antimony trisulfate.
Description: White powder.
Formula: $Sb_2(SO_4)_3$.
Constants: Mol wt: 531.70, mp: decomposes, d: 3.625 at 4°C.
Hazard Analysis & Countermeasures
See antimony compounds & Sulfate.

ANTIMONY SULFIDE. See antimony pentasulfide or antimony trisulflde.
Shipping Regulations: Section 11.
IATA (Solid): Other restricted articles, class A, no label, no limit (passenger), no limit (cargo).

ANTIMONY, SULFURATED
General Information
Synonym: Kermes mineral.
Description: Reddish-brown, odorless, tasteless powder.
Composition: 50-60% Sb.
Hazard Analysis & Countermeasures
See antimony compounds and sulfides.

ANTIMONY D-TARTRATE
General Information
Description: White crystals.
Formula: $Sb_2(C_4H_4O_6)_3 \cdot 6H_2O$.

TOXIC HAZARD RATING CODE *(For detailed discussion, see Section 1.)*

0 NONE: (a) No harm under any conditions; (b) Harmful only under unusual conditions or overwhelming dosage.

1 SLIGHT: Causes readily reversible changes which disappear after end of exposure.

2 MODERATE: May involve both irreversible and reversible changes; not severe enough to cause death or permanent injury.

3 HIGH: May cause death or permanent injury after very short exposure to small quantities.

U UNKNOWN: No information on humans considered valid by authors.

Constant: Mol wt: 795.8.
Hazard Analysis & Countermeasures
See antimony compounds.

ANTIMONY TELLURIDE. See antimony tritelluride.

ANTIMONY TETROXIDE
General Information
Synonym: Antimony oxide.
Description: White powder.
Formula: Sb_2O_4.
Constants: Mol wt: 307.5, mp: loses O_2 at 920°C, d: 4.07.
Hazard Analysis & Countermeasures
See antimony compounds.

ANTIMONY THIOGLYCOLAMIDE. See antimony α-mercaptoacetamide.

ANTIMONY TRIBROMIDE
General Information
Synonym: Antimonous bromide.
Description: Yellow, deliquescent, crystalline mass.
Formula: $SbBr_3$.
Constants: Mol wt: 361.51, mp: 96.6°C, bp: 275°C, d: 4.148, vap. press.: 1 mm at 93.9°C.
Hazard Analysis & Countermeasures
See antimony compounds and bromides.

ANTIMONY TRICHLORIDE
General Information
Synonyms: Antimonous chloride; butter of antimony.
Description: Colorless, transparent crystalline mass.
Formula: $SbCl_3$.
Constants: Mol wt: 228.13, mp: 73.4°C, bp: 219.0°C, d: 3.140 at 25°C, vap. press.: 1 mm at 49.2°C (sublimes).
Hazard Analysis
Toxic Hazard Rating:
Acute Local: Irritant 3; Ingestion 3; Inhalation 3.
Acute Systemic: Ingestion 3.
Chronic Local: Irritant 3; Inhalation 3.
Chronic Systemic: Ingestion 3; Inhalation 3.
Toxicology: Reacts vigorously with moisture generating heat and hydrogen chloride gas which is highly irritating and can cause pulmonary edema when antimony trichloride is inhaled into the lungs. Systemic effects can be caused by the antimony component. See antimony compounds.
Disaster Hazard: See antimony and hydrochloric acid.
Countermeasures
Ventilation Control: Section 2.
Personnel Protection: Section 3.
First Aid: Section 1.
Storage and Handling: Section 7.
Shipping regulations: Section 11.
IATA: Corrosive liquid, white label, 1 liter (passenger), 2½ liters (cargo).

ANTIMONY TRIETHYL
General Information
Synonym: Triethylstibine.
Description: Liquid. Water insoluble.
Formula: $Sb(C_2H_5)_3$.
Constant: Mol wt: 209.0, d: 1.324 at 16°C, mp: < −29°C, bp: 159.5°C.
Hazard Analysis
Toxicity: Alkyl metallics of this type are often highly toxic and corrosive. See also antimony compounds.
Fire Hazard: Dangerous, by chemical reaction. Spontaneously flammable in air (Section 6).
Disaster Hazard: Dangerous; when heated to decomposition it burns and emits highly toxic fumes of antimony; it can react vigorously with oxidizing materials.
Countermeasures
Storage and Handling: Section 7.

ANTIMONY TRIFLUORIDE
General Information
Synonym: Antimony fluoride.
Description: Octagonal crystals.
Formula: SbF_3.
Constants: Mol wt: 178.8, mp: 292°C, bp: sublimes, d: 4.379 at 20.9°C.
Hazard Analysis & Countermeasures
See antimony compounds and fluorides.

ANTIMONY TRIIODIDE
General Information
Synonym: Antimony iodide.
Description: Red to yellow crystals.
Formula: SbI_3.
Constants: Mol wt: 502.5, mp: 167°C, bp: 401°C, d: 4.768 at 22°C, vap. press.: 1 mm at 163.6°C.
Hazard Analysis & Countermeasures
See antimony compounds and iodides.

ANTIMONY TRIMETHYL
General Information
Synonym: Trimethyl stibine.
Description: Liquid. Slightly soluble in water.
Formula: $Sb(CH_3)_3$.
Constants: Mol wt: 166.9, d: 1.523 at 15°C, bp: 80.6°C.
Hazard Analysis
Toxicity: See antimony triethyl and antimony compounds.
Fire Hazard: Dangerous, by chemical reaction. Spontaneously flammable in air (Section 6).
Disaster Hazard: Dangerous; when heated to decomposition it emits highly toxic fumes of antimony; it can react vigorously with oxidizing materials.
Countermeasures
Storage and Handling: Section 7.

ANTIMONY TRIOXIDE
General Information
Description: White, odorless, tasteless, crystalline powder.
Formula: Sb_2O_3.
Constants: Mol wt: 291.52, mp: 656°C, bp: 1550°C (sublimes), d: 5.2, vap. press.: 1 mm at 574°C.
Hazard Analysis & Countermeasures
See antimony compounds.

ANTIMONY TRISELENIDE
General Information
Description: Gray powder.
Formula: Sb_2Se_3.
Constant: Mol wt: 480.40.
Hazard Analysis
Toxicity: See antimony and selenium compounds.
Disaster Hazard: Dangerous; See antimony and selenium; can react vigorously with oxidizing materials.
Countermeasures
Storage and Handling: Section 7.

ANTIMONY TRISULFIDE
General Information
Synonym: Antimony sulfide.
Description: Red to black crystals.
Formula: SbS_3.
Constants: Mol wt: 339.7, mp: 550°C, d: 4.64.
Hazard Analysis
Toxicity: See antimony compounds and sulfides.
Fire Hazard: Moderate, by spontaneous chemical reaction in contact with strong oxidizers (Section 6).
Explosion Hazard: Moderate, by spontaneous chemical reaction in contact with chlorates, perchlorates (Section 7).
Disaster Hazard: Dangerous; when heated to decomposition or on contact with acid or acid fumes, it emits highly toxic fumes of oxides of sulfur and antimony; it will react with water or steam to produce toxic and flam-

mable vapors; it can react vigorously with oxidizing materials.

Countermeasures

Storage and Handing: Section 7.

ANTIMONY TRISULFATE. See antimony sulfate.

ANTIMONY TRITELLURIDE
General Information

Synonym: Antimony telluride.

Description: Gray powder.

Formula: Sb_2Te_3.

Constants: Mol wt: 626.4, mp: 629°C.

Hazard Analysis

Toxicity: See antimony and tellurium compounds.

Fire Hazard: Moderate, by chemical reaction in contact with strong oxidizers (Section 6).

Explosion Hazard: Moderate, by chemical reaction in contact with chlorates and perchlorates (Section 7).

Disaster Hazard: Dangerous when heated to decomposition or on contact with acid or acid fumes, it emits highly toxic fumes of antimony and tellurium; it will react with water or steam to produce toxic and flammable vapors; can react vigorously with oxidizing materials.

Countermeasures

Storage and Handling: Section 7.

ANTIMONY YELLOW. See lead antimonate.

ANTIPYRINE
General Information

Synonym: Phenozone.

Description: Fine white, crystalline powder.

Formula: $C_{11}H_{12}N_2O$.

Constants: Mol wt: 188.23, mp: 113°C, bp: 319°C at 174 mm, d: 1.19.

Hazard Analysis

Toxic Hazard Rating:

Acute Local: U.

Acute Systemic: Ingestion 2; Inhalation 1.

Chronic Local: U.

Chronic Systemic: Ingestion 2; Inhalation 2.

Disaster Hazard: Moderately dangerous; when heated to decomposition, it emits toxic fumes.

Countermeasures

Ventilation Control: Section 2.

Personnel Protection: Section 3.

Storage and Handling: Section 7.

ANTISEPTIC OIL
General Information

Synonym: Hydroquinone in oil.

Description: Liquid.

Hazard Analysis

Toxicity: A weak sensitizer. Local contact may cause contact dermatitis (Section 9).

Fire Hazard: A combustible material. (Section 6).

Countermeasures

Storage and Handling: Section 7.

ANTOXYLIC ACID. See arsanilic acid.

ANTU. See naphthylthiourea.

APACHE COAL POWDERS. See explosives, high.

APHRODINE. See yohimbine.

APOATROPINE
General Information

Synonym: Atropamine.

Description: Prismatic crystals, soluble in alcohol, ether, chloroform, benzene, carbon tetrachloride. Insoluble in water.

Formula: $C_{17}H_{21}NO_2$.

Constants: Mol wt: 271.35, mp: 62°C.

Hazard Analysis

Toxic Hazard Rating:

Acute Local: 0.

Acute Systemic: Ingestion 3; Inhalation 3.

Chronic Local: 0.

Chronic Systemic: U.

Toxicology: A poisonous alkaloid which causes a marked conjunctivitis. Large doses cause death by respiratory failure. See also atropine.

Ventilation Control: Section 2.

Personal Hygiene: Section 3.

First Aid: Section 1.

Storage and Handling: Section 7.

APOATROPINE HYDROCHLORIDE. See atropine.

APOATROPINE SULFATE. See atropine.

APOCODEINE
General Information

Description: White crystalline solid.

Formula: $C_{18}H_{19}NO_2$.

Constants: Mol wt: 281.34, mp: 100–110°C (decomposes).

Hazard Analysis

Toxic Hazard Rating:

Acute Local: Allergen 1.

Acute Systemic: Ingestion 3; Inhalation 3.

Chronic Local: Allergen 1.

Chronic Systemic: U.

Toxicology: A poisonous alkaloid. See also codeine. Weak sensitizer. Local contact may cause contact dermatitis (Section 9).

Disaster Hazard: Dangerous; when heated to decomposition, it emits highly toxic fumes.

Countermeasures

Ventilation Control: Section 2.

Personal Hygiene: Section 3.

Fire Aid: Section 1.

Storage and Handling: Section 7.

APOCODEINE HYDROCHLORIDE. See apocodeine.

APOMORPHINE
General Information

Description: White crystalline alkaloid.

Formula: $C_{17}H_{17}NO_2$.

Constants: Mol wt: 267.32, mp: 170°C (decomposes).

Hazard Analysis

Toxic Hazard Rating:

Acute Local: Allergen 1.

Acute Systemic: Ingestion 3; Inhalation 3.

Chronic Local: Allergen 1.

Chronic Systemic: U.

Caution: Small doses cause depression while large doses produce excitement, convulsions and death. A powerful emetic. A weak sensitizer. Local contact may cause contact dermatitis (Section 9). Mild symptons require no treatment, but stimulants such as coffee or caffeine may be given. For severe cases, chloral hydrate, chloroform or ether. Call a physician.

TOXIC HAZARD RATING CODE (*For detailed discussion, see Section 1.*)

0 NONE: (a) No harm under any conditions; (b) Harmful only under unusual conditions or overwhelming dosage.

1 SLIGHT: Causes readily reversible changes which disappear after end of exposure.

2 MODERATE: May involve both irreversible and reversible changes; not severe enough to cause death or permanent injury.

3 HIGH: May cause death or permanent injury after very short exposure to small quantities.

U UNKNOWN: No information on humans considered valid by authors.

Disaster Hazard: Dangerous; when heated to decomposition, it emits highly toxic fumes.

Countermeasures
Ventilation Control: Section 2.
Personal Hygiene: Section 3.
First Aid: Section 1.
Storage and Handling: Section 7.

APOMORPHINE HYDROCHLORIDE. See apomorphine.

APPLE ACID. See malic acid.

AQUA AMMONIUM. See ammonium hydroxide.

AQUA FORTIS. See nitric acid.

AQUAMARINE. See beryl.

AQUAPHOR
Hazard Analysis
Toxicity: An allergen. Weak sensitizer. Local contact may cause contact dermatitis (Section 9).

AQUA REGIA
General Information
Synonym: Nitro hydrochloric acid.
Description: Fuming yellow, corrosive, suffocating volatile liquid.
Composition: (U.S.P.): 18 cc HNO_3 plus 82 cc HCl.
Hazard Analysis
Toxic Hazard Rating:
    Acute Local: Irritant 3; Ingestion 3; Inhalation 3.
    Acute Systemic: U.
    Chronic Local: Irritant 2.
    Chronic Systemic: U.
Fire Hazard: Moderate by chemical reaction with easily oxidized materials (Section 6).
Caution: A powerful oxidizer.
Disaster Hazard: Dangerous; when heated to decomposition, it emits highly toxic fumes of nitrosyl chloride; can react vigorously with reducing materials.
Countermeasures
Ventilation Control: Section 2.
Personnel Protection: Section 3.
First Aid: Section 1.
Storage and Handling: Section 7.

ARABINOGALACTIN
General Information
Description: A water soluble polysaccharide extracted from the timber of western larch trees. Dry light tan colored powder.
Constant: Mol wt: ≈ 72,000–92,000.
Hazard Analysis
Toxic Hazard Rating:U. A food additive permitted in food for human consumption.

ARAGONITE. See calcium carbonate.

ARALKONIUM CHLORIDE
General Information
Synonym: Dynaltone.
Formula: $C_{21}H_{36}Cl_3N$.
Constant: Mol wt: 408.9.
Hazard Analysis
Toxic Hazard Rating:
    Acute Local: Irritant 2.
    Acute Systemic: U.
    Chronic Local: Irritant 2.
    Chronic Systemic: Ingestion 1.
Toxicology: A quaternary ammonium compound. See alrosept.
Disaster Hazard: Dangerous. See chlorides.

ARAMITE. See sulfurous acid-2-(p-tert-butyl phenoxy)-1-methyl ethyl 2-chloroethyl ester.

ARATHANE. See crontonic acid-2,4-dinitrol-6 (1-methyl heptyl) phenyl ester.

ARECA NUT
General Information
Synonym: Betel.
Description: Brown, mottled with fawn color.
Hazard Analysis
Toxic Hazard Rating:
    Acute Local: 0.
    Acute Systemic: Ingestion 1.
    Chronic Local: 0.
    Chronic Systemic: U.
Fire Hazard: Slight; when heated (Section 6).
Countermeasures
Personal Hygiene: Section 3.
Storage and Handling: Section 7.

ARECOLINE BASE
General Information
Synonym: Methyl 1,2,5,6-tetrahydro-1-methyl nicotinate.
Description: Oily liquid.
Formula: $C_8H_{13}NO_2$.
Constants: Mol wt: 155.2, bp: 209°C.
Hazard Analysis
Toxic Hazard Rating:
    Acute Local: Irritant 2; Ingestion 2.
    Acute Systemic: Ingestion 3; Inhalation 3; Skin Absorption 3.
    Chronic Local: U.
    Chronic Systemic: U.
Fire Hazard: Combustible liquid (Section 6).
Disaster Hazard: Dangerous; when heated to decomposition it omits highly toxic fumes; it can react with oxidizing materials.
Countermeasures
Ventilation Control: Section 2.
Personnel Protection: Section 3.
First Aid: Section 1.
Storage and Handling: Section 7.

ARESKAP. See monobutyl phenyl phenol soldium sulfate.

ARESKET. See monobutyl diphenyl sodium monosulfonate.

ARGENTUM. See silver.

ARGININE
General Information
Synonyms: Guanidine amino valeric acid; amino-4-guanido-valeric acid.
Description: An essential amino acid for rats; occurs naturally in the L (+) form. Data here refer to L and DL forms. Prisms from water.
Formula: $NHC(NH_2) NH(CH_2)_3CH(NH_2)COOH$.
Constant: Mol wt: 174.
Hazard Analysis
Toxic HazardRating: U. A dietary supplement food additive (Section 10).

ARGON
General Information
Description: Colorless, inert gas.
Formula: A.
Constants: At wt: 39.94, mp: $-189.2°C$, bp: $-185.7°C$, d: 1.784 g/liter at $0°C$, 1.40 at $-186°C$ 1.65 at $-233°C$.
Hazard Analysis
Toxic Hazard Rating:
    Acute Local: 0.
    Acute Systemic: Inhalation 1.
    Chronic Local: U.
    Chronic Systemic: U.
Toxicology: Classified as a simple asphyxiant gas. Gases of this type have no specific toxic effect, but they act by excluding oxygen from the lungs. The effect of simple

asphyxiant gases is proportional to the extent to which they diminish the amount (partial pressure) of oxygen in the air that is breathed. The oxygen may be diminished to two-thirds of its normal percentage in air before appreciable symptoms develop and this in turn requires the presence of a simple asphyxiant in a concentration of 33 percent in the mixture of air and gas. When the simple asphyxiant reaches a concentration of 50 percent, marked symptons can be produced. A concentration of 75 percent is fatal in a matter of minutes.

The first symptons produced by simple asphyxiant gases, such as argon, are rapid respirations and air hunger. Mental alertness is diminished and muscular coordination is impaired. Later, judgment becomes faulty and all sensations are depressed. Emotional instability often results and fatigue occurs rapidly. As the asphyxia progresses, there may be nausea and vomiting, prostration and loss of consciousness, and finally, convulsions, deep coma and death.

Radiation Hazard: Section 5. For permissible levels, see Table 5, p. 150.

Artificial isotope ^{37}A, half life 34 d. Decays to stable ^{37}Cl by electron capture. Emits X-rays.

Artificial isotope ^{41}A, half life 1.8 h. Decays to stable ^{41}K by emitting beta particles of 1.2 MeV. Also emits gamma rays of 1.29 MeV. ^{41}A is a product of neutron bombardment of ^{40}A, the most abundant natural isotope, in air-cooled nuclear reactors.

Caution: Although ^{41}A has a short half-life, it may present a slight hazard as an effluent from nuclear reactors where it is formed by neutron bombardment of ^{40}A the most abundant natural isotope.

Countermeasures
Ventilation Control: Section 2.
Storage and Handling: Section 7.
Shipping Regulations: Section 11.
   I.C.C.: Nonflammable gas; green label, 300 pounds.
   Coast Guard Classification: Noninflammable gas; green gas label.
   IATA (gaseous): Nonflammable gas, green label, 70 kilograms (passenger), 140 kilograms (cargo).
      (liquid, non-pressurized); other restricted articles, class C, no label required, 50 liters (passenger and cargo).
      (liquid, low pressure): Other restricted articles, class C, no label required, not acceptable (passenger), 140 kilograms (cargo).
      (liquid, pressurized): Nonflammable gas, green label, not acceptable (passenger), 140 kilograms (cargo).

ARGYROL
General Information
Synonym: Mild silver protein.
Description: Brown to black crystals.
Composition: 19–23% Silver.
Hazard Analysis
Toxic Hazard Rating:
   Acute Local: Allergen 1.
   Acute Systemic: 0.
   Chronic Local: Allergen 1.
   Chronic Systemic: Ingestion 1.
Caution: Continued application can cause argyria. See also silver compounds. Local contact may cause contact dermatitis (Section 9).
Countermeasures
Personal Hygiene: Section 3.

ARICYL. See disodium acetoarsenate.

ARICYL ACID
General Information
Synonym: Arsonoacetic acid.
Description: Crystals.
Formula: $C_2H_5AsO_5$.
Constant: Mol wt: 184.1.
Hazard Analysis and Countermeasures
See arsenic compounds.

ARNICA
General Information
Synonyms: Wolfsbane; mountain tobacco.
Description: An alcoholic infusion.
Hazard Analysis
Toxic Hazard Rating:
   Acute Local: Irritant 2; Allergen 1; Ingestion 2; Inhalation 2.
   Acute Systemic: Ingestion 3; Skin Absorption 3.
   Chronic Local: Allergen 1.
   Chronic Systemic: U.
Caution: Overdose can be fatal. It can cause gastroenteritis, nervous disturbances and collapse. Local contact may cause contact dermatitis (Section 9).
Fire Hazard: Slight, when exposed to heat or flame; can react with oxidizing materials (Section 6).
Countermeasures
Ventilation Control: Section 2.
Personnel Protection: Section 3.
First Aid: Section 1.
Storage and Handling: Section 7.

AROCHLOR
General Information
Synonym: Chlorinated diphenyls.
Description: Colorless mobile oil.
Constant: d: 1.182–1.192 at 25/15.5°C; flash p: 286–302°F (COC), autoign. temp.: 349°F.
Hazard Analysis
See chlorinated diphenyls.

AROCHLOR 1232
General Information
Synonym: Chlorinated diphenyls.
Description: Colorless mobile oil.
Constants: D: 1.270–1.280 at 25/15.5°C, flash p: 305°–310°F (COC), autoign. temp.: 460°F.
Hazard Analysis
See chlorinated diphenyls.

AROCHLOR 1242
General Information
Synonym: Chlorinated diphenyls.
Description: Colorless mobile oil.
Constants: D: 1.381–1.392 at 25/15.5°C, flash p: 348–356°F, autoign. temp.: none; vap. press.: 10 mm Hg at 170°C.
Hazard Analysis
See chlorinated diphenyls.

AROCHLOR 1248
General Information
Synonym: Chlorinated diphenyls.
Description: Mobile oil.
Constants: D: 1.405–1.415 at 65°/15.5°C, flash p: 379–384°F (COC), autoign. temp.: none; vap. press.: 50 mm Hg at 225°C.

TOXIC HAZARD RATING CODE (For detailed discussion, see Section 1.)

0 NONE: (a) No harm under any conditions; (b) Harmful only under unusual conditions or overwhelming dosage.

1 SLIGHT: Causes readily reversible changes which disappear after end of exposure.

2 MODERATE: May involve both irreversible and reversible changes; not severe enough to cause death or permanent injury.

3 HIGH: May cause death or permanent injury after very short exposure to small quantities.

U UNKNOWN: No information on humans considered valid by authors.

Hazard Analysis
See chlorinated diphenyls.

AROCHLOR 1254, 1260, 1262, 4465, 2565, 5460, 1268
General Information
Synonym: Chlorinated diphenyls.
Description: Light yellow viscous oil.
Constants: D: 1.495–1.505 at 65°/15.5°C, flash p: none.
Hazard Analysis
See chlorinated diphenyls.

AROMATIC SPIRITS OF AMMONIA
General Information
Synonym: Spirit of Hartshorn.
Description: Colorless liquid, suffocating odor of ammonia.
Composition: 10% by weight of NH_3 in alcohol.
Hazard Analysis
Toxic Hazard Rating:
  Acute Local: Ingestion 1; Inhalation 1.
  Acute Systemic: Ingestion 1; Inhalation 1.
  Chronic Local: 0.
  Chronic Systemic: U.
Fire Hazard: Dangerous, in its usual form (solution of NH_3 in alcohol) (Section 6).
Explosion Hazard: Moderate, in its usual form (Section 7).
Disaster Hazard: Moderately dangerous; when heated it emits toxic fumes of ammonia; it can react with oxidizing materials.
Countermeasures
Ventilation Ccontrol: Section 2.
Personal Hygiene: Section 3.
Storage and Handling: Section 7.

ARSACETIN (SODIUM SALT)
General Information
Synonym: Sodium acetylarsanilate.
Description: White crystalline powder, odorless, tasteless.
Formula: $C_8H_9O_4NaNAs + 4H_2O$.
Constant: Mol wt: 353.
Hazard Analysis and Countermeasures
See arsenic compounds.

ARSANILIC ACID
General Information
Synonyms: Antoxylic acid; aminophenylarsine acid.
Description: White, crystalline powder.
Formula: $NH_2C_6H_4AsO(OH)_2$.
Constants: Mol wt: 217.0, mp:232°C.
Hazard Analysis
Toxicity: A grasshopper bait; a food additive permitted in the feed and drinking water of animals and/or for the treatment of food producing animals. See arsenic compounds and aniline.
Fire Hazard: Moderate. Decomposed by heat to yield flammable vapors.
Disaster Hazard: Dangerous; when heated to decomposition or on contact with acid or acid fumes, it emits highly toxic fumes of aniline and arsenic.
Countermeasures
Ventilation Control: Section 2.
Personnel Protection: Section 3.
First Aid: Section 1.
Storage and Handling: Section 7.

ARSANILIN ACID. See arsanilic acid.

ARSENATE OF IRON, FERRIC. See ferric arsenate.

ARSENATE OF IRON, FERROUS. See ferrous arsenate.

ARSENATES, N.O.S., LIQUID
Shipping Regulations: Section 11.
  IATA: Poison B, poison label, 1 liter (passenger), 220 liters (cargo).

ARSENATES, N.O.S., SOLID
Shipping Regulations: Section 11.

  IATA: Poison B, poison label, 25 kilograms (passenger), 95 kilograms (cargo).

ARSENIC. (See also arsenic vapor.)
General Information
Description: Silvery, brittle, crystalline metal.
Formula: As_4.
Constants: Mol wt: 299.64, mp: 814°C at 36 atm, bp: sublimes at 615°C, d: black crystals 5.724 at 14°C; black amor: 4.7, vap. press.: 1 mm at 372°C (sublimes).
Hazard Analysis
Toxicity: Used as a herbicede and an insecticide. Highly toxic. Used as a food additive in food for human consumption. See arsenic compounds.
TLV: ACGIH (recommended): 0.5 milligrams per cubic meter of air.
Radiation Hazard: Section 5. For permissible levels, see Table 5, p. 150.
  Artificial isotope [73]As, half life 76 d. Decays to stable [73]Ge by electron capture. Emits gamma rays of 0.05 MeV and X-rays.
  Artificial isotope [74]As, half life 18 d. Decays to stable [74]Ge by emitting positrons of 0.91 (26%), 1.51 (4%) MeV. Also decays to stable [74]Se by emitting beta particles of 0.72 (14%), 1.36 (18%) MeV. Also emits gamma rays of 0.60, 0.64 MeV and others.
  Artificial isotope [76]As, half life 26.5 h. Decays to stable [76]Se by emitting beta particles of 2.41 (31%), 2.97 (56%) MeV. Also emits gamma rays of 0.56 MeV and others.
  Artificial isotope [77]As, half life 39 h. Decays to stable [77]Se by emitting beta particles of 0.68 MeV.
Fire Hazard: Moderate in the form of dust when exposed to heat or flame or by chemical reaction with powerful oxidizers (Section 6).
Explosion Hazard: Slight in the form of dust when exposed to flame (Section 7).
Disaster Hazard; Dangerous; when heated or on contact with acid or acid fumes, it emits highly toxic fumes; it can react vigorously on contact with oxidizing materials.
Countermeasures
Ventilation Control: Section 2.
Personnel Protection: Section 3.
First Aid: Section 1.
Storage and Handling: Section 11.
  I.C.C. Classification: Poison B; poison label.
  Coast Guard Classification: Poison B; poison label.

m-ARSENIC ACID
General Information
Description: White crystals.
Formula: $HAsO_3$.
Constants: Mol wt: 123.9, mp: decomposes.
Hazard Analysis and Countermeasures
See arsenic compounds.

o-ARSENIC ACID
General Information
Synonyms: True arsenic acid.
Description: White, translucent crystals.
Formula: $H_3AsO_4 \cdot \frac{1}{2}H_2O$.
Constants: Mol wt: 150.9, mp: 35.5°C, bp: $-H_2O$ at 160°C, d: 2.0–2.5.
Hazard Analysis and Countermeasures
See arsenic compounds.
Shipping Regulations: Section 11.
  I.C.C.: Poison B; poison label, 200 pounds.
  Coast Guard Classification: Poison B; poison label.

ARSENIC ACID, LIQUID. See arsenic acid.
Shipping Regulations: Section 11.
  I.C.C.: Poison B; poison label, 55 gallons.
  Coast Guard Classification: Poison B; poison label.
  IATA: Poison B, poison label, 1 liter (passenger), 220 liters (cargo).

ARSENIC ACID, SOLID
Shipping Regulations: Section 11.
  IATA: Poison B, poison label, 25 kilograms (passenger), 95 kilograms (cargo).

ARSENICAL BABBITT
General Information
Description: A bearing metal.
Composition: Up to 3% As.
Hazard Analysis and Countermeasures
See arsenic compounds.

ARSENICAL COMPOUNDS OR MIXTURES, N.O.S. LIQUID
Hazard Analysis
Toxicity: See arsenic compounds.
Countermeasures
Shipping Regulations: Section 11.
  I.C.C.: Poison B; poison label, 55 gallons.
  Coast Guard Classification: Poison B; poison label.
  IATA: Poison B, poison label, 1 liter (passenger), 220 liters (cargo).

ARSENICAL DIP, LIQUID
General Information
Synonym: Sheep dip.
Hazard Analysis
Toxicity: See arsenic compounds.
Disaster Hazard: See arsenic compounds.
Countermeasures
Shipping Regulations: Section 11.
  I.C.C.: Poison B; poison label, 55 gallons.
  Coast Guard Classification: Poison B; poison label.
  IATA: Poison B, poison label, 1 liter (passenger), 220 liters (cargo).

ARSENICAL DUST. See arsenic.
Countermeasures
Shipping Regulations: Section 11.
  I.C.C.: Poison B; poison label, 200 pounds.
  Coast Guard Classification: Poison B; poison label.
  IATA: Poison B, poison label, 25 kilograms (passenger), 95 kilograms (cargo).

ARSENICAL FLUE DUST. See arsenic.
Countermeasures
Shipping Regulations: Section 11.
  I.C.C.: Poison B; poison label, 200 pounds.
  Coast Guard Classification: Poison B; poison label.
  IATA: Poison B, poison label, 25 kilograms (passenger), 95 kilograms (cargo).

ARSENICAL MIXTURES OR COMPOUNDS, N.O.S. SOLID. See arsenic compounds.
Countermeasures
Shipping Regulations: Section 11.
  I.C.C.: Poison B; poison label, 200 pounds.
  Coast Guard Classification: Poison B; poison label.
  IATA: Poison B, poison label, 25 kilograms (passenger), 95 kilograms (cargo).

ARSENIC BISULFIDE
General Information
Synonym: Arsenic sulfide; realgar.
Description: Red-brown crystals.
Formula: As_2S_2.
Constants: Mol wt: 214, bp: 565°C, mp: β = 307°C, d: α = 3.506 at 19°C, β = 3.254 at 19°C.

Hazard Analysis
Toxicity: See arsenic compounds and sulfides.
Fire Hazard: Moderate, in the form of dust when exposed to heat or flame (Section 6).
Explosion Hazard: Slight, when intimately mixed with powerful oxidizers (Section 7).
Disaster Hazard: Dangerous; see oxides of sulfur and arsenic; it will react with water or steam to produce toxic and flammable vapors; it can react vigorously with oxidizing materials.
Countermeasures
Ventilation Control: Section 2.
Personnel Protection: Section 3.
First Aid: Section 1.
Storage and Handling: Section 7.

ARSENIC BROMIDE
General Information
Synonyms: Arsenic tribromide; arsenous bromide.
Description: Yellowish-white crystals.
Formula: $AsBr_3$.
Constants: Mol wt: 314.7, mp: 32.8°C, bp: 220.0°C, d: 3.54 at 25°C, vap. press.: 1 mm at 41.8°C.
Hazard Analysis
Toxicity: See arsenic compounds and Bromides.
Countermeasures
Ventilation Control: Section 2.
Personnel Protection: Section 3.
First Aid: Section 1.
Storage and Handling: Section 7.
Shipping Regulations: Section 11.
  I.C.C.: Poison B; poison label, 200 pounds.
  Coast Guard Classification: Poison B; Poison label.
  IATA: Poison B, poison label, 25 kilograms (passenger), 95 kilograms (cargo).

ARSENIC CHLORIDE. See arsenic pentachloride.
Shipping Regulations: Section 11.
  I.C.C.: (Arsenous, liq.) poison B; poison label, 55 gallons.
  Coast Guard Classification: (Arsenous, liq.) poison B; poison label.

ARSENIC COMPOUNDS. See also specific compound.
Hazard Analysis
Toxic Hazard Rating:
  Acute Local: Irritant 2; Allergen 2; Ingestion 3.
  Acute Systemic: Ingestion 3; Inhalation 3.
  Chronic Local: Irritant 2, Allergen 2.
  Chronic Systemic: Ingestion 3, Inhalation 3.
Toxicology: Used as insecticides. Poisoning from arsenic compounds may be acute or chronic. Acute poisoning usually results from swallowing arsenic compounds; chronic poisoning from either swallowing or inhalation. Acute allergic reactions to arsenic compounds used in medical therapy have been fairly common. The type and severity of reaction depending upon the compound of arsenic.

Acute arsenic poisoning (from ingestion) results in marked irritation of the stomach and intestines with nausea, vomiting and diarrhea. In severe cases the vomitus and stools are bloody and the patient goes into collapse and shock with weak, rapid pulse, cold sweats, coma and death.

Chronic arsenic poisoning, whether through ingestion or inhalation, may manifest itself in many different ways. There may be disturbances of the digestive system such as loss of appetite, cramps, nausea, constipa-

---

TOXIC HAZARD RATING CODE *(For detailed discussion, see Section 1.)*

0 NONE: (a) No harm under any conditions; (b) Harmful only under unusual conditions or overwhelming dosage.

1 SLIGHT: Causes readily reversible changes which disappear after end of exposure.

2 MODERATE: May involve both irreversible and reversible changes; not severe enough to cause death or permanent injury.

3 HIGH: May cause death or permanent injury after very short exposure to small quantities.

U UNKNOWN: No information on humans considered valid by authors.

tion or diarrhea. Liver damage may occur, resulting in jaundice. Disturbances of the blood, kidneys and nervous system are not infrequent. Arsenic can cause a variety of skin abnormalities including itching, pigmentation and even cancerous changes. A characteristic of arsenic poisoning is the great variety of symptoms that can be produced.

In treating acute poisoning from ingestion the contents of the intestinal tract should be evacuated promptly by gastric lavage and saline cathartics. BAL (dimercaptol) is highly effective in both acute and chronic poisoning (Section 1). Note: Arsenic compounds are common air contaminants (Section 4).

Disaster Hazard: Dangerous; when heated to decomposition or on contact with acids or acid fumes, they emit highly toxic fumes of arsenic.

Countermeasures
Ventilation Control: Section 2.
Personnel Protection: Section 3.
First Aid: Section 1.

ARSENIC COPPER. See copper arsenide.

ARSENIC DIETHYL
General Information
Synonyms: Ethyl cacodyl; tetraethyl diarsine.
Descriptions: Liquid or oil.
Formula: $[As(C_2H_5)_2]_2$.
Constants: Mol wt: 266.2, bp: 185–190°C, d: about 1.
Hazard Analysis
Toxicity: See arsenic compounds.
Fire Hazard: Dangerous, by spontaneous chemical reaction (Section 6). A spontaneously flammable liquid.
Disaster Hazard: Dangerous; see arsenic; it can react vigorously with oxidizing materials.
Countermeasures
Ventilation Control: Section 2.
Personnel Protection: Section 3.
First Aid: Section 1.
Storage and Handling: Section 7.

ARSENIC DIIODIDE
General Information
Synonym: Arsenic iodide.
Description: Red crystals.
Formula: AsI_2.
Constants: Mol wt: 328.8, mp: decomposes at 136°C.
Hazard Analysis and Countermeasures
See arsenic compounds and iodides.

ARSENIC DIMETHYL
General Information
Synonym: Tetramethyl diarsyl.
Description: Colorless to yellow oily liquid.
Formula: $[As^1(CH_3)_2]_2$.
Constants: Mol wt: 210.0, mp: −6°C, bp: 186°C, d: 1.15.
Hazard Analysis
Toxicity: See arsenic compounds.
Fire Hazard: A moderately flammable liquid (Section 6).
Disaster Hazard: Dangerous; see arsenic.
Countermeasures
Ventilation Control: Section 2.
Personnel Protection: Section 3.
First Aid: Section 1.
Storage and Handling: Section 7.

ARSENIC DISULFIDE. See arsenic bisulfide.

ARSENIC FLUORIDE. See arsenic pentafluoride or arsenic trifluoride.

ARSENIC HEMISELENIDE
General Information
Description: Black crystals with metallic luster.
Formula: As_2Se.
Constant: Mol wt: 228.78.

Hazard Analysis
Toxicity: See arsenic compounds and selenium compounds.
Disaster Hazard: Dangerous; see arsenic and selenium; it can react vigorously with oxidizing materials.
Countermeasures
Ventilation Control: Section 2.
Personnel Protection: Section 3.
First Aid: Section 1.
Storage and Handling: Section 7.

ARSENIC HYDRIDE. See arsine.

ARSENIC IODIDE. See arsenic diiodide.
Countermeasures
Shipping Regulations: Section 11.
I.C.C.: Poison B; poison label, 200 pounds.
Coast Guard Classification: Poison B; poison label.
IATA: Poison B, poison label, 25 kilograms (passenger), 95 kilograms (cargo).

ARSENIC OXIDES. See arsenic trioxide or arsenic pentoxide.

ARSENIC OXYCHLORIDE
General Information
Description: Brown crystals.
Formula: AsOCl.
Constants: Mol wt: 126.4, bp: decomposes.
Hazard Analysis and Countermeasures
See arsenic compounds and chlorides.

ARSENIC PENTACHLORIDE
General Information
Description: Colorless liquid.
Formula: $AsCl_5$.
Constants: Mol wt: 252.2, mp: −40°C (approx.)
Hazard Analysis and Countermeasures
See arsenic compounds and HCl.

ARSENIC PENTAFLUORIDE
General Information
Synonym: Arsenic fluoride.
Description: Colorless gas.
Formula: AsF_5.
Constants: Mol wt: 169.9, mp: −80°C, bp: −53°C, d: 7.71 g/liter.
Hazard Analysis and Countermeasures
See arsenic compounds and fluorides.

ARSENIC PENTASELENIDE
General Information
Description: Black, brittle solid with a metallic luster.
Formula: As_2Se_5.
Constant: Mol wt: 544.62.
Hazard Analysis and Countermeasures
See arsenic and selenium compounds.

ARSENIC PENTASULFIDE
General Information
Description: Brownish-yellow, glassy amorphous, highly refractive mass.
Formula: As_2S_5.
Constant: Mol wt: 310.12.
Hazard Analysis
Toxicity: See arsenic compounds and sulfides.
Fire Hazard: A flammable material (Section 6).
Explosion Hazard: See sulfides.
Disaster Hazard: Dangerous; see oxides of sulfur and of arsenic; it will react with steam or water to produce toxic and corrosive fumes; it can react vigorously with oxidizing materisls.
Countermeasures
Ventilation Control: Section 2.
Personnel Protection: Section 3.
First Aid: Section 1.
Storage and Handling: Section 7.

ARSENIC PENTOXIDE
General Information
Synonym: Arsenic oxide.
Description: White, amorphous solid. Deliquescent.
Formula: As_2O_5.
Constants: Mol wt: 229.8, mp: 315°C (decomposes),
d: 4.086.
Hazard Analysis
Toxicity: See arsenic compounds.
Disaster Hazard: See arsenic compounds.
Countermeasures
Ventilation Control: Section 2.
Personnel Protection: Section 3.
First Aid: Section 1.
Shipping Regulations: Section 11.
 I.C.C.: Poison B; poison label, 200 pounds.
 Coast Guard Classification: Poison B; poison label.
 IATA: Poison B, poison label, 25 kilograms (passenger),
 95 kilograms (cargo).

ARSENIC PHOSPHIDE
General Information
Description: Brown to red powder.
Formula: AsP.
Constants: Mol wt: 105.9, mp: sublimes with decomposition.
Hazard Analysis
Toxicity: See arsenic compounds and phosphides.
Fire Hazard: Moderate, by spontaneous chemical reaction.
 Phosphine is liberated upon contact with moisture
 (Section 6).
Explosion Hazard: Details unknown.
Disaster Hazard: Dangerous; see oxides of phosphorus and
 of arsenic; it will react with water or steam to produce
 toxic and flammable vapors, it can react vigorously with
 oxidizing materials.
Countermeasures
Ventilation Control: Section 2.
Personnel Protection: Section 3.
First Aid: Section 1.
Storage and Handling: Section 7.

ARSENIC, SOLID
Shipping Regulations: Section 11.
 IATA: Poison B, poison label, 25 kilograms (passenger),
 95 kilograms (cargo).
 ICC: Poison B, poison label, 200 pounds.

ARSENIC SULFIDE (POWDER). See arsenic tri-
 sulfide, arsenic bisulfide and arsenic pentasulfide.
Countermeasures
Shipping Regulations: Section 11.
 I.C.C.: Poison B; poison label, 200 pounds.
 Coast Guard Classification: Poison B: poison label.
 IATA: Poison B, poison label, 25 kilograms (passenger),
 95 kilograms (cargo).

ARSENIC TRIBROMIDE. See arsenic bromide.

ARSENIC TRICHLORIDE
General Information
Description: Clear, almost colorless to pale yellow cor-
 rosive oily liquid, or needle-like crystals.
Formula: $AsCl_3$.
Constants: Mol wt: 181.3, mp: -18°C, bp: 130.2°C,
 d: (liquid) 2.163 at 14°/4°C, vap. press.: 10 mm at
 23.5°C, vap. d: 6.25.
Hazard Analysis and Countermeasures
See arsenic compounds and hydrochloric acid.

Shipping Regulations: Section 11.
 I.C.C.: Poison B; poison label, 55 gallons.
 Coast Guard Classification: Poison B; poison label.
 MCA warning label.
 IATA: Poison B, poison label, 1 liter (passenger;, 220
 liters (cargo).

ARSENIC TRIETHYL. See triethyl arsenic.

ARSENIC TRIFLUORIDE
General Information
Synonym: Arsenic fluoride.
Description: Oily liquid.
Formula: AsF_3.
Constants: Mol wt: 131.9, mp: -5.9°C, bp: 63°C at
 752 mm, d: (liquid) 2.666, vap. press.: 100 mm at
 13.2°C; 400 mm at 41.5°C.
Hazard Analysis and Countermeasures
See arsenic compounds and fluorides.

ARSENIC TRIMETHYL. See trimethyl arsenic.

ARSENIC TRIOXIDE
General Information
Synonyms: White arsenic.
Description: White, odorless, tasteless, amorphous powder.
Formula: As_2O_3.
Constants: Mol wt: 197.8, mp: 193°C (sublimes),
 d: (arsenalite) 3.865 at 25°C; (claudedite) 4.15;
 (amorphous) 4.09.
Hazard Analysis
Toxicity: A rodenticide. See arsenic compounds.
Disaster Hazard: See arsenic compounds.
Countermeasures
Ventilation Control: Section 2.
Personnel Protection: Section 3.
First Aid: Section 1.
Storage and Handling: Section 7.
Shipping Regulations: Section 11.
 I.C.C.: Poison B; poison label, 200 pounds.
 Coast Guard Classification: Poison B; poison label.
 IATA: Poison B, poison label, 25 kilograms (passenger),
 95 kilograms (cargo).

ARSENIC TRIPHENYL. See triphenyl arsenic.

ARSENIC TRISELENIDE
General Information
Synonym: Arsenious selenide.
Description: Brown crystals.
Formula: As_2Se_3.
Constants: Mol wt: 386.7, mp: 360°C, d: 4.75.
Hazard Analysis and Countermeasures
See arsenic compounds and selenium compounds.

ARSENIC TRISULFIDE
General Information
Synonym: Arsenic sulfide; orpiment.
Description: Yellow or red crystals.
Formula: As_2S_3.
Constants: Mol wt: 246.0, mp: 300°C, bp: 707°C, d: 3.43.
Hazard Analysis
Toxicity: See arsenic compounds and sulfides.
Fire Hazard: See sulfides.
Disaster Hazard: Dangerous; when heated to decomposi-
 tion or on contact with acid or acid fumes, it emits
 highly toxic fumes of sulfur and of arsenic; it will react
 with water or steam to produce toxic and flammable

TOXIC HAZARD RATING CODE (For detailed discussion, see Section 1.)

0 NONE: (a) No harm under any conditions; (b) Harmful only under unusual conditions or overwhelming dosage.

1 SLIGHT: Causes readily reversible changes which disappear after end
of exposure.

2 MODERATE: May involve both irreversible and reversible changes;
not severe enough to cause death or permanent injury.

3 HIGH: May cause death or permanent injury after very short exposure
to small quantities.

U UNKNOWN: No information on humans considered valid by authors.

vapors; it can react vigorously on contact with oxidizing materials.

Countermeasures
Ventilation Control: Section 2.
Personnel Protection: Section 3.
First Aid: Section 1.
Storage and Handling: Section 7.

ARSENIC VAPOR
General Information
Description: A vapor.
Formula: As.
Constant: At wt: 75.
Hazard Analysis
Toxicity: See arsenic compounds.
TLV: ACGIH (recommended); 0.5 milligrams per cubic meter of air.
Fire Hazard: Moderate, by chemical reaction with oxidizers (Section 6).
Disaster Hazard: See arsenic.
Countermeasures
Ventilation Control: Section 2.
Personnel Protection: Section 3.
First Aid: Section 1.

ARSENIOUS SELENIDE. See arsenic triselenide.

ARSENITES, n.o.s., liquid.
Shipping Regulations: Section 11.
   IATA: Poison B, poison label, 1 liter (passenger), 220 liters (cargo).

ARSENITES, n.o.s., solid
Shipping Regulations: Section 11.
   IATA: Poison B, poison label, 25 kilograms (passenger), 95 kilograms (cargo).

ARSENIURETTED HYDROGEN. See arsine.

ARSENOACETIC ACID
General Information
Description: Yellow crystals. Insoluble in water.
Formula: $(AsCH_2COOH)_2$.
Constants: Mol wt: 267.9, mp: $> 260°C$.
Hazard Analysis and Countermeasures
See arsenic compounds.

ARSENOBENZENE
General Information
Description: White crystals. Not soluble in water. Soluble in benzene.
Formula: $C_6H_5As_2C_6H_5$.
Constants: Mol wt: 304.0, mp: 212°C.
Hazard Analysis and Countermeasures
See arsenic compounds.

ARSENOUS ACID, SOLID. See arsenic trioxide.
Countermeasures
Shipping Regulations: Section 11.
   I.C.C.: Poison B; poison label, 200 pounds.
   Coast Guard Classification: Poison B; poison label.

ARSENOUS BROMIDE. See arsenic bromide.

ARSENOUS AND MERCURIC IODIDE SOLUTION, LIQUID
General Information
Description: Liquid.
Hazard Analysis
Toxicity: See arsenic compounds and mercury compounds.
Disaster Hazard: See arsenic and mercury compounds.
Countermeasures
Ventilation Control: Section 2.
Personnel Protection: Section 3.
First Aid: Section 1.
Shipping Regulations: Section 11.
   I.C.C.: Poison B; poison label, 55 gallons.

Coast Guard Classification: Poison B; poison label.
IATA: Poison B, poison label, 1 liter (passenger), 220 liters (cargo).

ARSINE
General Information
Synonyms: Arsenic hydride; arseniuretted hydrogen.
Description: Colorless gas, mild garlic odor.
Formula: AsH_3.
Constants: Mol wt: 77.93, mp: $-116.3°C$, bp: $-55°C$, d: 3.484 g/liter, vap. d.: 2.66.
Hazard Analysis
Toxic Hazard Rating:
   Acute Local: Inhalation 3.
   Acute Systemic: Inhalation 3.
   Chronic Local: U.
   Chronic Systemic: Inhalation 3.
TLV: ACGIH (recommended); 0.05 parts per million in air. 0.2 milligrams per cubic meter of air.
Toxicology: The toxicity of arsine is due to its hemolytic action. On entering the blood stream it combines with the hemoglobin of the red blood cells; gradually the arsenic in this hemoglobin-arsenic complex is oxidized, and the oxidation process is accompanied by hemolysis of the cell. The resulting anemia is responsible for the production of many of the symptoms accompanying arsine poisoning; other symptoms result from the hemolysis itself, and occur during the excretion of the hemoglobin. Hemoglobin and its degradation products are commonly found in the urine. Less commonly whole blood may be passed. Occasionally, the renal tubules may be plugged by debris, with resultant suppression of urine. Jaundice, which may be severe, is a common result of the hemolysis. Frequently there is edema of the lungs, which may be accompanied by cyanosis. Kidney damage is common in patients surviving acute effects of the gas.

Signs of poisoning usually develop within several hours of exposure. Headache, dizziness, nausea and vomiting, epigastric pain and weakness occur early, followed by tea-colored urine, or bloody urine in the more severe cases. Some time later, albumen and casts may appear in the urine, or, in serious cases, there may be suppression of urine. Jaundice and tenderness over the liver may appear about the same time. Blood examination shows an anemia which may be marked. In fatal cases, the patient may develop delirium, followed by coma and death. During the acute stage of poisoning and for some weeks after, arsenic may be demonstrated in the urine. See also arsenic and arsenic compounds.
Fire Hazard: Moderate, when exposed to flame (Section 6).
Explosion Hazard: Moderate, when exposed to flame (Section 7).
Disaster Hazard: Dangerous; when heated to decomposition, it emits highly toxic fumes of arsenic; can react vigorously with oxidizing materials.
Countermeasures
Ventilation Control: Section 2.
Storage and Handling: Section 7.
Shipping Regulations: Section 11.
   IATA: Poison A, no label, not acceptable (passenger), not acceptable (cargo).

ARSONOACETIC ACID. See aricyl acid.

ARSPHENAMINE
General Information
Synonym: 3-Diamino-4-dihydroxy-1-arsenobenzene hydrochloride.
Description: Light yellow, hygroscopic powder.
Formula: $C_{12}H_{12}As_2N_2O_2 \cdot 2HCl \cdot 2H_2O$.
Constant: Mol wt: 475.0.
Hazard Analysis & Countermeasures
See arsenic compounds.

ARSYSODILA. See sodium cacodylate.

ARTHRYTIN. See amiodoxyl benzoate.

ARTIFICIAL ALMOND OIL. See benzaldehyde.

ARTIFICIAL GUM. See dextrin.

ASBESTOS DUST. See asbestos particles.

ASBESTOS PARTICLES
General Information
Synonym: Asbestos dust.
Toxic Hazard Rating:
  Acute Local: Irritant 1; Inhalation 2.
  Acute Systemic: 0.
  Chronic Local: Inhalation 3.
  Chronic Systemic: U.
TLV: ACGIH (recommended); 5 million particles per cubic
  foot of air.
Hazard Analysis
Toxicology: The essential lesion produced by asbestos dust
  is a diffuse fibrosis which probably begins as a "collar"
  about the terminal bronchioles. Usually, at least 4 to 7
  years of exposure are required before a serious degree
  of fibrosis results. There is apparently less predisposition to tuberculosis than is the case with silicosis.
  Prolonged inhalation can cause cancer of the lung,
  pleura and peritoneum.
    Clinically, the most striking sign is shortness of
  breath of gradually increasing intensity, often associated with a dry cough. In the early stages physical
  signs are absent or slight; in the later stages rales may
  be heard, and in long-standing cases there is frequently
  clubbing of the fingers. In early stages of the disease the
  chest x-ray reveals a groundglass or granular change,
  chiefly in the lower lung fields; as the condition progresses the heart outline becomes "shaggy" and irregular patches of mottled shadowing may be seen.
  "Asbestos bodies" may be found in the sputum.
    At autopsy, the pleurae are thickened and adherent
  and thick subpleural fibrous plaques are often present.
  Where the disease is far advanced there are usually
  large areas of fibrosis, with emphysematous changes in
  the apices and bases. The alveolar walls are thickened,
  and the characteristic "asbestos bodies" are found.
Note: Common air contaminants (Section 4).
Countermeasures
Ventilation Control: Section 2.
Personal Hygiene: Section 3.

ASCARIDOLE
General Information
Synonym: Ascarisin.
  Description: Liquid.
Formula: $C_{10}H_{16}O_2$.
Constants: Mol wt: 168.2, bp: 115°C at 15 mm, d: 1.011 at
  13°/15°C.
Hazard Analysis
Toxicity: Details unknown. See oil of chenopodium. See
  also peroxides, organic.
Fire Hazard: Moderate, by spontaneous chemical reaction
  (Section 6).
Caution: An oxidizer.
Explosion Hazard: Explodes at 250°C.
Disaster Hazard: Dangerous; when heated, it emits toxic
  fumes and may explode; it reacts with reducing ma-
  terials.

Countermeasures
Storage and Handling: Section 7.

ASCARISIN. See ascaridole.

ASCORBIC ACID
General Information
Synonyms: L-ascorbic acid; vitamin C.
Description: White crystals; soluble in water; slightly sol-
  uble in alcohol; insoluble in ether, chloroform,
  benzene, petroleum ether, oils, and fats.
Formula: $OCOCOH:COHCHCHOHCH_2OH$.
Constants: Mol wt: 176, mp: 192°C.
Hazard Analysis
Toxic Hazard Rating: A chemical preservative food addi-
  tive; also a dietary supplement food additive (Section 10).

ASCORBYL PALMITATE
General Information
Description: A white or yellowish white powder; citrus
  like odor; soluble in alcohol, animal and vegetable
  oils; slightly soluble in water.
Formula: $C_{22}H_{38}O_7$.
Constants: Mol wt: 414, mp: 116–117°C.
Hazard Analysis
Toxic Hazard Rating: U. A chemical preservative food addi-
  tive (Section 10).

ASPARAGIC ACID. See aspartic acid.

ASPARAGINIC ACID. See aspartic acid.

ASPARTIC ACID
General Information
Synonym: Asparaginic acid; asparagic acid, aminosuccinic
  acid.
Description: A naturally occurring non-essential amino acid.
  The common form is L (+)—aspartic acid. The data
  below refer to both L and DL forms. Colorless crystals;
  soluble in water; insoluble in alcohol and ether.
Formula: $COOHCH_2CH(NH_2)COOH$.
Constants: Mol wt: 133, mp (DL)–278–280°C (decomposes),
  d: 1.663 (12/12°C), mp (L) 251°C.
Hazard Analysis
Toxic Hazard Rating: U. A dietary supplement food additive
  (Section 10).

ASPERIN. See acetol.

ASPHALT
General Information
Synonyms: Bitumen; petroleum pitch.
Description: Black or dark brown mass.
Constants: Bp: <470°C, flash p: 400+°F(C.C.), d: 0.95-1.1,
  autoign. temp.: 905°F.
Hazard Analysis
Toxic Hazard Rating:
  Acute Local: Irritant 2.
  Acute Systemic: U.
  Chronic Local: Irritant 2.
  Chronic Systemic: U.
Fire Hazard: Slight, when exposed to heat or flame.
Spontaneous Heating: No.
Countermeasures
To Fight Fire: Foam, carbon dioxide, dry chemical or
  carbon tetrachloride (Section 6).
Personnel Protection: Section 3.
Storage and Handling: Section 7.
Shipping Regulations: Section 11.
  Coast Guard Classification: Hazardous article.

TOXIC HAZARD RATING CODE *(For detailed discussion, see Section 1.)*

0 NONE: (a) No harm under any conditions; (b) Harmful only under unusual conditions or overwhelming dosage.

1 SLIGHT: Causes readily reversible changes which disappear after end
  of exposure.

2 MODERATE: May involve both irreversible and reversible changes;
  not severe enough to cause death or permanent injury.

3 HIGH: May cause death or permanent injury after very short exposure
  to small quantities.

U UNKNOWN: No information on humans considered valid by authors.

ASPHALT, CUTBACK
General Information
Description: A liquid petroleum product; solubility of residue from distillation in carbon tetrachloride = 99.5%.
Constant: Flash p: <50°F.
Hazard Analysis
Toxic Hazard Rating: U.
Fire Hazard: Dangerous, when exposed to heat or flame.
Countermeasures
To Fight Fire: Water, foam (Section 6).
Storage and Handling: Section 7.
Shipping Regulations:
  ICC: Flammable liquid, red label, 10 gallons.
  IATA: Flammable liquid, red label, 1 liter (passenger), 40 liters (cargo).

ASPHALT, LIQUID MEDIUM CURING
General Information
Constants: Flash p: 100°F(O.C.) (minimum) grades MC-0 and MC-1; 150°F(O.C.) (minimum) grades MC-2 through MC-5.
Hazard Analysis
Toxic Hazard Rating: U.
Fire Hazard: Moderate, when exposed to heat or flame.
Countermeasures
To Fight Fire: Water, foam (Section 6).
Storage and Handling: Section 7.

ASPHALT, LIQUID RAPID CURING
General Information
Constant: Flash p: 80°F (O.C.) minimum (grades RC-0 through RC-5).
Hazard Analysis
Toxic Hazard Rating: U.
Fire Hazard: Moderate when exposed to heat or flame.
Countermeasures
To Fight Fire: Water, foam (Section 6).
Storage and Handling: Section 7.

ASPHALT, LIQUID—SLOW CURING
General Information
Constants: Flash p: 150°F+ (O.C.) grades SC-0 plus SC-1; 175°F+ (O.C.) grades SC-2; 200°F+ (O.C.) grades SC-3; 225°F+ (O.C.) grades SC-4; 250°F+ (O.C.) grades SC-5.
Hazard Analysis
Toxic Hazard Rating: U.
Fire Hazard: Moderate, when exposed to heat or flame (grades SC-0 → SC-4); slight (grade SC-5).
Countermeasures
To Fight Fire: Water, foam (Section 6).
Storage and Handling: Section 7.

ASPIDIUM
General Information
Synonym: Male fern.
Hazard Analysis
Toxic Hazard Rating:
  Acute Local: Irritant 1; Allergen 1.
  Acute Systemic: Ingestion 2.
  Chronic Local: Allergen 1.
  Chronic Systemic: U.
Fire Hazard: Slight; (Section 6).
Countermeasures
Personal Hygiene: Section 3.
Storage and Handling: Section 7.

ASPIRIN. See acetol.

ASTATINE
General Information
Description: A member of the halogen family of elements (group VII). Occurs naturally in very small quantities.
Formula: At.
Constant: At wt: (210).

Hazard Analysis
Radiation Hazard: Section 5. For permissible levels, see Table 5, p. 150.
Artificial isotope ^{211}At, half life 7.2 h. Decays to radioactive ^{211}Po by electron capture (59%). Also decays to radioactive ^{207}Bi by emitting alpha particles of 5.86 MeV (41%). Also emits gamma rays of 0.67 MeV and x-rays.

ATABRINE DIHYDROCHLORIDE. See "atabrine" hydrochloride.

"ATABRINE" HYDROCHLORIDE
General Information
Synonym: Quinacrine hydrochloride.
Description: Bright yellow crystals.
Formula: $C_{23}H_{30}ClN_3O \cdot 2HCl \cdot 2H_2O$.
Constants: Mol wt: 508.9, mp: decomposes 248–250°C.
Hazard Analysis
Toxic Hazard Rating:
  Acute Local: Allergen 1.
  Acute Systemic: Ingestion 2.
  Chronic Local: Allergen 1.
  Chronic Systemic: Ingestion 2.
Disaster Hazard: Dangerous; see chlorides.

ATOXYL. See sodium arsanilate.

ATRAZINE. See 2-chloro-4-ethylamino-6-isopropylamino-5-triazine.

ATROPAMINE. See apoatropine.

ATROPIC ACID. See α-phenyl acrylic acid.

ATROPINE
General Information
Synonym: Daturine.
Description: Colorless, crystalline alkaloid.
Formula: $C_{17}H_{23}NO_3$.
Constants: Mol wt: 289.4, mp: 115.5°C; sublimes at 118°C.
Hazard Analysis
Toxic Hazard Rating:
  Acute Local: Allergen 1.
  Acute Systemic: Ingestion 3; Inhalation 3.
  Chronic Local: Allergen 1.
  Chronic Systemic: Ingestion 2.
Fire Hazard: Slight; (Section 6).
Countermeasures
Ventilation Control: Section 2.
Personal Hygiene: Section 3.
First Aid: Section 1.
Storage and Handling: Section 7.

ATROPINE METHYL BROMIDE. See atropine.

ATROPINE METHYL NITRATE. See atropine.

ATROPINE SULFATE. See atropine.

AURIC BROMIDE
General Information
Description: Gray powder; crystals brown.
Formula: $AuBr_3$.
Constants: Mol wt: 436.95, mp: $-Br_2$ at 160°C.
Hazard Analysis and Countermeasures
See gold compounds and bromides.

AURIC CHLORIDE
General Information
Synonym: Gold chloride.
Description: Claret red crystals.
Formula: $AuCl_3$.
Constants: Mol wt: 303.57, mp: 254°C decomposes, bp: sublimes 265°C, d: 3.9.
Hazard Analysis and Countermeasures
See gold compounds & chlorides.

AURIC CYANIDE. See cyanoauric acid.

AURIC HYDROGEN NITRATE
General Information
Synonym: Gold nitrate.
Description: Yellow crystals.
Formula: $AuH(NO_3)_4 \cdot 3H_2O$.
Constants: Mol wt: 500.29, mp: 72°C decomposes, d: 2.84.
Hazard Analysis and Countermeasures
See gold compounds and nitrates.

AURIC IODIDE
General Information
Synonym: Gold iodide.
Description: Dark green crystals.
Formula: AuI_3.
Constant: Mol wt: 577.96.
Hazard Analysis and Countermeasures
See gold compounds and iodides.

AURIC OXIDE
General Information
Synonym: Gold oxide.
Description: Brown black powder.
Formula: Au_2O_3.
Constants: Mol wt: 442.40, mp: $-O_2$ at 160°C, bp: $-3O_2$ at 250°C.
Hazard Analysis
Toxicity: See gold compounds.

AURIC SULFIDE
General Information
Synonym: Gold sulfide.
Description: Brown black powder.
Formula: Au_2S_3.
Constants: Mol wt: 490.60, mp: decomposes at 197°C, d: 8.754.
Hazard Analysis and Countermeasures
See gold compounds and sulfides.

AUROUS BROMIDE
General Information
Description: Yellowish-gray mass or crystalline powder.
Formula: AuBr.
Constants: Mol wt: 277.12, mp: decomposes 115°C, d: 7.9.
Hazard Analysis and Countermeasures
See gold compounds and bromides.

AUROUS CHLORIDE
General Information
Description: Yellow crystals.
Formula: AuCl.
Constants: Mol wt: 232.66, mp: decomposes to $AuCl_3$ at 170°C, bp: decomposes 289.5°C.
Hazard Analysis and Countermeasures
See gold compounds and chlorides.

AUROUS CYANIDE
General Information
Synonym: Gold cyanide.
Description: Light yellow, crystalline powder.
Formula: AuCN.
Constants: Mol wt: 223.22, mp: decomposes, d: 7.12.
Hazard Analysis and Countermeasures
See cyanides and gold compounds.

AUROUS IODIDE
General Information
Synonym: Gold iodide.

Description: Greenish-yellow powder.
Formula: AuI.
Constants: Mol wt: 324.12, mp: decomposes 120°C, d: 8.25.
Hazard Analysis and Countermeasures
See gold compounds and iodides.

AUROUS OXIDE
General Information
Description: Gray-violet crystals.
Formula: Au_2O.
Constants: Mol wt: 410.4, mp: $-O_2$ at 205°C, d: 3.6.
Hazard Analysis
Toxicity: See gold compounds.

AUROUS SULFIDE
General Information
Description: Brown black powder.
Formula: Au_2S.
Constants: Mol wt: 426.47, mp: decomposes 240°C.
Hazard Analysis and Countermeasures
See gold compounds and sulfides.

AUSTEN RED DIAMONDS. See explosives, high.

AUSTINITE. See calcium zinc arsenate monohydrate.

AUTARITE. See calcium iodate.

AUTO ALARMS, EXPLOSIVE
Shipping Regulations: Section 11.
  IATA: Explosive, explosive label, 25 kilograms (passenger), 70 kilograms (cargo).

AUTUNITE
General Information
Synonyms: Calco uranite; lime uranite.
Description: Bright yellow mineral of pearly luster.
Formula: $Ca(UO_2)_2P_2O_8 \cdot 8H_2O$.
Hazard Analysis
Toxicity: See uranium compounds, insoluble.
Radiation Hazard: See uranium.

AVERTIN. See tribromoethanol.

AVOGADRITE. See potassium fluoborate.

AZ. See azodicarbonamide.

AZELAIC ACID
General Information
Synonym: Nonanedioic acid.
Description: Powder.
Formula: $HOOC(CH_2)_7COOH$.
Constants: Mol wt: 188.22, mp: 106.5°C, bp: 356.5°C (decomposes), d: 1.038 at 110°C, vap. press.: 1 mm at 178.3°C.
Hazard Analysis
Toxic Hazard Rating: U. Closely related to glutaric acid and adipic acid. Probably has low toxicity.
Fire Hazard: Slight, when exposed to heat or flame; it can react with oxidizing materials (Section 6).
Countermeasures
Storage and Handling: Section 7.

AZIDES (See also specific compound.)
Hazard Analysis
Toxicity: Variable. Many azides cause fall in blood pressure and some inhibit enzyme action, thus resembling nitrites and cyanides.
Fire Hazard: Unknown.

TOXIC HAZARD RATING CODE (For detailed discussion, see Section 1.)

0 NONE: (a) No harm under any conditions; (b) Harmful only under unusual conditions or overwhelming dosage.

1 SLIGHT: Causes readily reversible changes which disappear after end of exposure.

2 MODERATE: May involve both irreversible and reversible changes; not severe enough to cause death or permanent injury.

3 HIGH: May cause death or permanent injury after very short exposure to small quantities.

U UNKNOWN: No information on humans considered valid by authors.

Explosion Hazard: Severe when shocked or exposed to heat (Section 7).

An azide is a compound of hydrogen or a metal and the monovalent -N₃ radical. The azides as a group are one of the few commercially produced explosives that contain no oxygen. Hydrogen azide or azoic acid and its sodium salt are soluble in water. All of its salts and the acid are unstable and decompose explosively; although lead azide, which is one of the most important azides, is not very sensitive and is used as a detonating agent, in which respect it is more efficient, weight for weight than fulminate. Azides can be used wherever fulminate may be used; in detonators and priming compositions as well. Their only drawbacks are their high ignition temperature (detonators) and relative insensitiveness to friction (priming compositions). However, by compounding them with other materials, this can be overcome. Mercury azide, however, is more sensitive than lead azide; in fact it is more sensitive then mercury fulminate. When packed in bulk, azide should contain not less than 20 percent of water and in this wet condition should be placed in bags of four-ounce or heavier duck. Each bag contains approximately 25 pounds, dry weight, of lead azide. In each bag and over the azide is placed a cap of the same cloth and of the diameter of the bag, and the bag is tied securely. Five of these bags are placed in a larger bag of four-ounce or heavier duck and this large bag is tied securely. The large bag should contain not more than 150 pounds, dry weight of lead azide. This bag is placed in the center of a container complying with ICC specification 5 or 5B (pertaining to metal barrels or drums), 17H (pertaining to metal drums, single trip) or 10B (pertaining to wooden barrels or kegs). The container is lined with a heavy close fitting bag of jute or other suitable bag material of equal strength. The large duck bag is completely surrounded within the jute bag by not less than 3 inches of well-packed sawdust, saturated with water and the jute bag is closed by securely sewing to prevent escape of sawdust. The outer container, drum or barrel is water-tight. When considered necessary to prevent freezing in shipment, or storage, an alternative method of packing is used. The lead azide is wet with not less than 20 percent of its weight of a solution of denatured ethyl alcohol in water containing not less than 33 percent by weight of ethyl alcohol. The 25-pound duck bags are placed in a strong waterproof bag instead of a duck bag and the drum or barrel is filled with sawdust saturated with the same 33 percent by weight solution of ethyl alcohol. Marking should be in accordance with U. S. Standard Specification for marking shipment no. 100-2. In addition, each barrel or cask is marked plainly "INITIATING EXPLOSIVE—DANGEROUS DO NOT STORE OR LOAD WITH ANY HIGH EXPLOSIVE." For further information concerning surface storage and the destruction of excess azide explosives, see Section 7.

Lead and mercury azides are the best examples of the azide type of explosive. Further uses of azide are in loading of fuse detonators and in the manufacture of priming compositions. Azides are used as substitutes for mercury fulminate. On account of its high temperature of ignition, lead azide is not easily ignited by the "spit" of a safety fuse. However, the addition of lead styphnate for instance, which is more readily ignited, overcomes the difficulty. It is also used in explosive rivets with the addition of silver acetylide or tetracine to lower its ignition temperature.

Disaster Hazard: Dangerous; shock and heat will explode it and when heated to decomposition, it emits highly toxic fumes.

Countermeasures
Storage and Handling: Section 7.

AZIMETHYLENE. See diazomethane.

AZIMINOBENZENE. See 1,2,3-benzotriazole.

AZINPHOS METHYL
Hazard Analysis
Toxicity: Highly toxic.
TLV: ACGIH (recommended) 0.2 milligram per cubic meter of air. May be absorbed via the skin.

1-AZIRIDINYL PHOSPHINE OXIDE (TRIS). See tris-(1-aziridinyl) phosphine oxide solution.

p-AZOANILINE. See p-diamino azobenzene.

AZOBENZENE
General Information
Synonyms: Azobenzol; azobenzide.
Description: Solid, orange-red crystals.
Formula: $C_6H_5NNC_6H_5$.
Constants: Mol wt: 182.2, mp: 68°C, d: 1.203 at 20°/4°C, vap. press.: 1 mm at 103.5°C.
Hazard Analysis
Toxic Hazard Rating:
  Acute Local: U.
  Acute Systemic: Ingestion 2; Inhalation 2.
  Chronic Local: U.
  Chronic Systemic: Ingestion 1.
An insecticide.
Toxicology: Experimentally has produced liver injury in rats.
Disaster Hazard: Dangerous. When strongly heated, it emits highly toxic fumes.

AZOBENZIDE. See azobenzene.

AZOBENZOL. See azobenzene.

1,1'-AZOBIS FORMAMIDE. See azodicarbonamide.

2,2'-AZOBISISOBUTYRONITRILE
General Information
Description: White powder; soluble in many organic solvents and in vinyl monomers; insoluble in water.
Formula: $(CH_3)_2C(CN)NNC(CN)(CH_3)_2$.
Constants: Mol wt: 172, mp: 105°C (decomposes).
Hazard Analysis and Countermeasures
See nitriles.

AZOCHLORAMIDE
General Information
Synonym: Chloroazodin.
Description: Bright-yellow crystals.
Formula: $C_2H_4Cl_2N_6$.
Constants: Mol wt: 183.0, mp: explodes with decomposition at 155°C.
Hazard Analysis
Toxic Hazard Rating:
  Acute Local: U.
  Acute Systemic: U.
  Chronic Local: Allergen 1.
  Chronic Systemic: U.
Explosion Hazard: Moderate, when exposed to heat (Section 7).
Explosive Range: At 155°C.
Disaster Hazard: Moderately dangerous; when heated to decomposition; it emits toxic fumes and may explode.
Countermeasures
Personal Hygiene: Section 3.
Storage and Handling: Section 7.

AZODICARBONAMIDE
General Information
Synonym: Az; 1,1'-azobisformamide.
Description: Yellow powder; insoluble in common solvents; soluble in dimethyl sulfoxide.
Hazard Analysis
Toxic Hazard Rating: A food additive permitted in food for human consumption (Section 10).

AZOIMIDE. See hydrazoic acid.

AZOTIC ACID. See nitric acid.

AZOXYBENZENE
General Information
Synonym: Azoxybenzide.
Description: Pale yellow crystals.
Formula: $C_{12}H_{10}N_2O$.
Constants: Mol wt: 198.2, mp: 36°C, d: 1.159 at 26°4°C.

Hazard Analysis
Toxicity: Details unknown; an insecticide. Limited animal experiments show moderate toxicity.
Fire Hazard: Slightly dangerous; on decomposition it emits toxic fumes (Section 6).
Countermeasures
Storage and Handling: Section 7.

AZOXYBENZIDE. See azoxybenzene.

TOXIC HAZARD RATING CODE (For detailed discussion, see Section 1.)

0 NONE: (a) No harm under any conditions; (b) Harmful only under un-usual conditions or overwhelming dosage.

1 SLIGHT: Causes readily reversible changes which disappear after end of exposure.

2 MODERATE: May involve both irreversible and reversible changes; not severe enough to cause death or permanent injury.

3 HIGH: May cause death or permanent injury after very short exposure to small quantities.

U UNKNOWN: No information on humans considered valid by authors.

B

BACOTRACIN
General Information
Description: An antibiotic. White to pale buff, hygroscopic powder; odorless or slight odor; freely soluble in water; soluble in alcohol, methanol and glacial acetic acid; insoluble in acetone, chloroform, and ether.
Hazard Analysis
Toxic Hazard Rating: U. A food additive permitted in feed and drinking water of animals and/or for the treatment of food producing animals. Also, a food additive permitted in food for human consumption (Section 10).

BACHITACIN METHYLENE DISALICYLATE
General Information
Description: White to gray-brown prowder; slight unpleasant odor; soluble in water, pyridine, ethanol; less soluble in acetone, ether, chloroform, benzene.
Hazard Analysis
Toxic Hazard Rating: U. A food additive permitted in feed and drinking water of animals and/or for the treatment of food producing animals. Also, a food additive permitted in food for human consumption (Section 10).

BACTERIAL CATALASE
General Information
Description: Derived from "Micrococcus lysodeikticus" by a pure culture fermentation process.
Hazard Analysis
Toxic Hazard Rating: U. A food additive permitted in food for human consumption (Section 10).

BADDELYTE. See zirconium dioxide.

BAGASSE DUST
Hazard Analysis
Toxic Hazard Rating:
  Acute Local: Irritant 2; Inhalation 2.
  Acute Systemic: 0.
  Chronic Local: Inhalation 2.
  Chronic Systemic: Inhalation 2.
Caution: A nuisance dust; inhalation can cause bronchial asthma, sneezing, rhinorrhea, pneumonitis etc. (Section 9).
Fire Hazard: Moderate, when exposed to heat or flame (Section 6).
Explosion Hazard: Moderate, when exposed to flame. See dust explosions.
Countermeasures
Ventilation Control: Section 2.
Personnel Protection: Section 3.
Storage and Handling: Section 7.

BAGS, NITRATE OF SODA, EMPTY AND UNWASHED. See also sodium nitrate.
Shipping Regulations: Section 11.
  I.C.C.: Flammable solid; yellow label, 25 pounds.
  Coast Guard Classification: Inflammable solid; yellow label.
  IATA: Flammable solid, yellow label, not acceptable (passenger), 12 kilograms (cargo).

"BAKELITE"
General Information
Description: Phenol-formaldehyde condensation-type plastic.

Hazard Analysis
Toxic Hazard Rating:
  Acute Local: Irritant 1; Inhalation 1.
  Acute Systemic: 0.
  Chronic Local: Allergen 1.
  Chronic Systemic: 0.
Disaster Hazard: Dangerous; when heated to decomposition it emits toxic fumes.
Countermeasures
Ventilation Control: Section 2.
Personal Hygiene: Section 3.
Storage and Handling: Section 7.

BAKING POWDER
Hazard Analysis
Toxic Hazard Rating:
  Acute Local: Irritant 1; Allergen 1.
  Acute Systemic: 0.
  Chronic Local: Allergen 1.
  Chronic Systemic: 0.

BAKING SODA. See sodium bicarbonate.

BAL. See 2,3-dimercapto-1-propanol.

BALATA
General Information
Description: Dried juice of the bully tree, mimusops balata. Resembles gutta percha.
Hazard Analysis
Toxic Hazard Rating:
  Acute Local: Irritant 1; Allergen 1.
  Acute Systemic: U.
  Chronic Local: Allergen 1.
  Chronic Systemic: U.
Fire Hazard: Slight, when exposed to heat or flame (Section 6).
Countermeasures
Personal Hygiene: Section 3.
Storage and Handling: Section 7.

BALM OF GILEAD. See canada balsam.

BALSAM COPAIBA. See copaiba.

BALSAM OF PERU
General Information
Synonym: Peruvian balsam.
Description: Dark brown viscid liquid; vanilla odor.
Hazard Analysis
Toxic Hazard Rating:
  Acute Local: Allergen 1.
  Acute Systemic: U.
  Chronic Local: Allergen 1.
  Chronic Systemic: U.
Fire Hazard: Slight; when heated (Section 6).
Countermeasures
Personal Hygiene: Section 3.
Storage and Handling: Section 7.

BANANA OIL. See iso amyl acetate.

BANANA PEEL OIL
Hazard Analysis
Toxic Hazard Rating:
  Acute Local: Allergen 1.

445

Acute Systemic: U.
Chronic Local: Allergen 1.
Chronic Systemic: U.
Fire Hazard: Slight (Section 6).
Countermeasures
Personal Hygiene: Section 3.
Storage and Handling: Section 7.

BARBITURATES
General Information
Synonyms: Derivatives of barbituric acid, i.e., barbital; barbitone; barbital sodium.
Hazard Analysis
Toxic Hazard Rating:
Acute Local: Allergen 1.
Acute Systemic: Ingestion 2; Inhalation 1.
Chronic Local: Allergen 1.
Chronic Systemic: Ingestion 1.
Toxicology: Taken internally in large doses they cause marked depression (sometimes preceded by excitation), prolonged coma and death. Allergic skin reactions may occur from contact (Section 9). May be habit-forming.
Fire Hazard: Slight; when heated. (Section 6).
Countermeasures
Ventilation Control: Section 2.
Personal Hygiene: Section 3.
Treatment and Antidotes: Induce vomiting or wash out stomach. Keep patient warm. Stimulants and artificial respiration in severe cases. Call a physician.
Storage and Handling: Section 7.

BARBITURIC ACID
General Information
Synonym: Malonylurea.
Description: Crystals or white to yellow-white powder.
Formula: $C_4H_4O_3N_2$.
Constants: Mol wt: 128.1, mp: 245°C, bp: 260°C (decomp.).
Hazard Analysis
Toxic Hazard Rating:
Acute Local: Irritant 3, Allergen 1; Ingestion 3; Inhalation 3.
Acute Systemic: U.
Chronic Local: Allergen 1.
Chronic Systemic: U.
Toxicology: A corrosive and irritating material. Has no hypnotic properties.
Fire Hazard: Slight (Section 6).
Countermeasures
Ventilation Control: Section 2.
Personnel Protection: Section 3.
First Aid: Section 1.
Storage and Handling: Section 7.

BARITE. See barium sulfate.

BARIUM
General Information
Description: Silver-white, slightly lustrous, somewhat malleable metal.
Formula: Ba.
Constants: At wt: 137.36, mp: 850°C, bp: 1140°C, d: 3.5 at 20°C, vap. press.: 10 mm at 1049°C.
Hazard Analysis
Toxicity: See barium compounds (soluble).
Radiation Hazard: Section 5. For permissible levels, see Table 5, p. 150.
Artificial isotope ^{131}Ba, half life 12 d. Decays to

radioactive ^{131}Cs by electron capture. Emits gamma rays of 0.09–1.05 MeV and X-rays.
Artificial isotope ^{140}Ba, half life 12.8 d. Decays to radioactive ^{140}La by emitting beta particles of 0.83–2.20 MeV. Also emits gamma rays of 0.03–0.54 MeV. ^{140}Ba exists in radioactive equilibrium with its daughter and permissible levels include the daughter activity.
Fire Hazard: Moderate in form of dust when exposed to heat or flame or by chemical reaction (Section 6).
Countermeasures
Ventilation Control: Section 2.
Personal Hygiene: Section 3.
Handling and Shipping: Sections 5 and 11.
Shipping Regulations:
IATA: Flammable solid, yellow label, 12 kilograms (passenger), 45 kilograms (cargo).

BARIUM ACETATE
General Information
Description: White crystals.
Formula: $Ba(C_2H_3O_2)_2 \cdot H_2O$.
Constants: Mol wt: 273.46, mp: – H_2O at 150°C, d: 2.19.
Hazard Analysis and Countermeasures
See barium compounds (soluble).

BARIUM ALLOYS, NON-PYROPHORIC
Shipping Regulations: Section 11.
IATA: Flammable solid, yellow label, 12 kilograms (passenger), 45 kilograms (cargo).

BARIUM ALLOYS, PYROPHORIC
Shipping Regulations: Section 11.
IATA: Flammable solid, not acceptable (passenger and cargo).

BARIUM AMIDE
General Information
Description: Gray-white crystals.
Formula: $Ba(NH_2)_2$.
Constants: Mol wt: 169.41, mp: 280°C.
Hazard Analysis and Countermeasures
See barium compounds (Soluble).

BARIUM AMYL SULFATE
General Information
Description: White crystals, fatty feel.
Formula: $Ba(C_5H_{11}SO_4)_2 \cdot H_2O$.
Constant: Mol wt: 507.80.
Hazard Analysis and countermeasures
See barium compounds (soluble) and sulfates.

BARIUM ANTIMONYL TARTRATE
General Information
Description: Crystals.
Formula: $BaSb_2(C_4H_4O_6)_4$.
Constant: Mol wt: 973.2.
Hazard Analysis and Countermeasures
Toxicity: See antimony and barium compounds.

BARIUM o-ARSENATE
General Information
Description: Black crystals.
Formula: $Ba_3(AsO_4)_2$.
Constant: Mol wt: 689.90.
Hazard Analysis and countermeasures
See arsenic compounds and barium compounds (soluble).

BARIUM ARSENIDE
General Information
Description: Brown crystals.

TOXIC HAZARD RATING CODE (For detailed discussion, see Section 1.)

0 NONE: (a) No harm under any conditions; (b) Harmful only under unusual conditions or overwhelming dosage.

1 SLIGHT: Causes readily reversible changes which disappear after end of exposure.

2 MODERATE: May involve both irreversible and reversible changes; not severe enough to cause death or permanent injury.

3 HIGH: May cause death or permanent injury after very short exposure to small quantities.

U UNKNOWN: No information on humans considered valid by authors.

Formula: Ba_3As_2.
Constants: Mol wt: 561.90, d: 4.1 at 15°C.
Hazard Analysis and Countermeasures
See arsenic compounds and barium compounds (soluble).

BARIUM AZIDE
General Information
Description: Monoclinic prisms.
Formula: $Ba(N_3)_2$.
Constants: Mol wt: 221.41, mp: $-N_2$ at about 120°C, bp: explodes, d: 2.936.
Hazard Analysis
Toxicity: See barium compounds (soluble) and azides.
Explosion Hazard: Moderate; when shocked or exposed to heat. (Section 7).
Disaster Hazard: Dangerous; shock and heat will explode it. (Section 7).
Countermeasures
Ventilation Control: Section 2.
Personal Hygiene: Section 3.
Storage and Handling: Section 7.
Shipping Regulations: Section 11.
   I.C.C. (50% or more water wet): Flammable solid, yellow label, 1 pound.
   IATA (dry or wet with less than 50% water): Flammable solid, not acceptable (passenger and cargo).
   IATA (wet with 50% or more water wet): Flammable solid, yellow label, not acceptable (passenger), ½ kilogram (cargo).

BARIUM BENZENE SULFONATE
General Information
Description: White nacreous leaflets.
Formula: $Ba(C_6H_5 \cdot SO_3)_2 \cdot H_2O$.
Constant: Mol wt: 469.71.
Hazard Analysis and Countermeasures
See barium compounds (soluble) and sulfonates.

BARIUM BENZOATE
General Information
Description: White nacreous leaflets.
Formula: $Ba(C_7H_5O_2)_2 \cdot 2H_2O$.
Constants: Mol wt: 415.61, mp: $-2H_2O$ at 100°C.
Hazard Analysis and Countermeasures
See barium compounds (soluble).

BARIUM BICHROMATE. See barium dichromate.

BARIUM BINOXALATE
General Information
Synonym: Acid barium oxalate.
Description: Crystals.
Formula: $Ba(HC_2O_4)_2 \cdot 2H_2O$.
Constant: Mol wt: 351.45.
Hazard Analysis and Countermeasures
See barium compounds (soluble) and oxalates.

BARIUM BINOXIDE. See barium peroxide.

BARIUM BOROTUNGSTATE. See barium borowolframate.

BARIUM BOROWOLFRAMATE
General Information
Synonym: Barium borotungstate.
Description: Crystals; effloresce rapidly in air.
Formula: $2BaO \cdot B_2O_3 \cdot 9WO_3 \cdot 18H_2O$.
Constant: Mol wt: 2787.93.
Hazard Analysis and Countermeasures
See barium compounds (soluble).

BARIUM BROMATE
General Information
Description: White crystals or crystalline powder.
Formula: $Ba(BrO_3)_2 \cdot H_2O$.
Constants: Mol wt: 411.21, mp: decomp. 260°C, d: 3.99 at 18°C.

Hazard Analysis
Toxicity: See barium compounds (soluble).
Fire Hazard: Moderate by chemical reaction with easily oxidized materials (Section 6).
Disaster Hazard: Dangerous; see bromides; can react with reducing materials.
Countermeasures
Ventilation Control: Section 2.
Personal Hygiene: Section 3.
Storage and Handling: Section 7.

BARIUM BROMIDE
General Information
Description: Colorless crystals.
Formula: $BaBr_2$.
Constants: Mol wt: 297.19, mp: 847°C, d: 4.781 at 24°C.
Hazard Analysis and Countermeasures
Toxicity: See barium compounds (soluble) and bromides.

BARIUM BROMIDE FLUORIDE
General Information
Description: Platelets.
Formula: $BaBr_2 \cdot BaF_2$.
Constants: Mol wt: 472.55, d: 4.96 at 18°C.
Hazard Analysis and Countermeasures
See fluorides and barium compounds (soluble) and bromides.

BARIUM BROMOPLATINATE
General Information
Description: Monoclinic crystals.
Formula: $BaPtBr_6 \cdot 10H_2O$.
Constants: Mol wt: 992.25, d: 3.71.
Hazard Analysis and Countermeasures
Toxicity: See barium compounds (soluble) and bromides.

BARIUM BUTYRATE
General Information
Description: Crystals.
Formula: $Ba(C_4H_7O_2)_2 \cdot 2H_2O$.
Constant: Mol wt: 347.59.
Hazard Analysis and Countermeasures
See barium compounds (soluble).

BARIUM CARBIDE
General Information
Description: Gray crystals.
Formula: BaC_2.
Constants: Mol wt: 161.4, d: 3.75.
Hazard Analysis
Toxicity: See barium compounds (soluble).
Fire Hazard: Moderate, by chemical reaction with moisture to form acetylene.
Explosion Hazard: Moderate; it evolves acetylene upon contact with moisture.
Disaster Hazard: Dangerous; it will react with water or steam to produce flammable vapors.
Countermeasures
Ventilation Control: Section 2.
To Fight Fire: Carbon dioxide, dry chemical or carbon tetrachloride (Section 6).
Storage and Handling: Section 7.

BARIUM CARBONATE
General Information
Description: White powder.
Formula: $BaCO_3$.
Constants: Mol wt: 197.37, mp: 1740 at 90 atm, bp: decomposes, d: 4.43.
Hazard Analysis and Countermeasures
See barium compounds (soluble).

BARIUM CHLORATE
General Information
Description: Colorless prisms or white powder.
Formula: $Ba(ClO_3)_2 \cdot H_2O$.
Constants: Mol wt: 322.29, mp: anhydrous 414°C, d: 3.18.

Hazard Analysis
Toxicity: See barium compounds (soluble).
Fire Hazard: See chlorates.
Explosion Hazard: See chlorates.
Disaster Hazard: See chlorates.
Countermeasures
Storage and Handling: Section 7.
Shipping Regulations: Section 11.
  I.C.C.: Oxidizing material; yellow label, 100 pounds.
  Coast Guard Classification: Oxidizing material; yellow label.
  IATA: Oxidizing material, yellow label, 12 kilograms (passenger), 45 kilograms (cargo).

BARIUM CHLORATE, WET. See barium chlorate.
Shipping Regulations: Section 11.
  I.C.C.: Oxidizing material, yellow label, 200 pounds.
  IATA: Oxidizing material, yellow label, 12 kilograms (passenger), 95 kilograms (cargo).

BARIUM CHLORIDE
General Information
Description: Colorless flat crystals.
Formula: $BaCl_2$.
Constants: Mol wt: 208.27, mp: transition at 925°C to cubic crystals, bp: 1560°C, d: 3.856 at 24°C.
Hazard Analysis
Toxicity: See barium compounds (soluble).
Countermeasures
Shipping Regulations: Section 11.
  MCA warning label.

BARIUM CHLOROPLATINATE
General Information
Description: Rhombic orange-yellow crystals.
Formula: $BaPtCl_6 \cdot 6H_2O$.
Constants: Mol wt: 653.43, mp: $-5H_2O$ at 70°C, d: 2.868.
Hazard Analysis and Countermeasures
See barium compounds (soluble) and chlorides.

BARIUM CHLOROPLATINITE
General Information
Description: Crystals.
Formula: $BaPtCl_4 \cdot 3H_2O$.
Constants: Mol wt: 528.47, d: 2.868.
Hazard Analysis and Countermeasures
See barium compounds (soluble) and chlorides.

BARIUM CHROMATE
General Information
Description: Heavy yellow crystalline powder.
Formula: $BaCrO_4$.
Constants: Mol wt: 253.37, d: 4.498 at 15°C.
Hazard Analysis
Toxicity: See barium compounds (soluble) and chromium compounds.
Fire Hazard: See chromates (Section 6).
Disaster Hazard: Dangerous; reacts vigorously with reducing materials.
Countermeasures
Ventilation Control: Section 2.
Personnel Protection: Section 3.
First Aid: Section 1.
Storage and Handling: Section 7.

BARIUM CITRATE
General Information
Description: White powder.
Formula: $Ba_3(C_6H_5O_7)_2 \cdot 7H_2O$.

Constant: Mol wt: 916.39.
Hazard Analysis and Countermeasures
See barium compounds (soluble).

BARIUM COMPOUNDS (SOLUBLE)
Hazard Analysis
Toxic Hazard Rating:
  Acute Local: Irritant 1; Ingestion 1; Inhalation 1.
  Acute Systemic: Ingestion 2; Inhalation 2.
  Chronic Local: Irritant 1.
  Chronic Systemic: Ingestion 1; Inhalation 1.
TLV: ACGIH (recommended): 0.5 milligrams per cubic meter of air.
Toxicology: The soluble barium salts, such as the chloride and sulfide, are poisonous when taken by mouth. The insoluble sulfate used in radiography is nonpoisonous. Few cases of industrial systemic poisoning have been reported, but one investigator describes a fatal case of poisoning attributed to barium oxide, the symptoms being severe abdominal pain with vomiting, dyspnoea, rapid pulse, paralysis of the arm and leg, and eventually cyanosis and death. The same investigator produced paralysis in animals with barium oxide and carbonate. The usual result of exposure to the sulfide, oxide and carbonate is irritation of the eyes, nose and throat, and of the skin, producing dermatitis (Section 9). The salts mentioned are somewhat caustic.
Countermeasures
Ventilation Control: Section 2.
Personal Hygiene: Section 3.
First Aid: Section 1.
Label Required: MCA warning label on barium chloride, barium fluosilicate and barium nitrate.
Shipping Regulations: Section 11.
  IATA (acid or water soluble): Poison B, poison label, 25 kilograms (passenger), 95 kilograms (cargo).

BARIUM CYANIDE
General Information
Description: White crystalline powder.
Formula: $Ba(CN)_2$.
Constant: Mol wt: 189.40.
Hazard Analysis and Countermeasures
See also cyanides and barium compounds (soluble).
Shipping Regulations: Section 11.
  I.C.C.: Poison B; poison label, 200 pounds.
  Coast Guard Classification: Poison B; poison label.
  IATA: Poison B, poison label, 12 kilograms (passenger), 95 kilograms (cargo).

BARIUM CYANOPLATINITE
General Information
Description: (a) Monoclinic yellow crystals; (b) rhombic green crystals.
Formula: $BaPt(CN)_4 \cdot 4H_2O$.
Constants: Mol wt: 508.73; mp: $-2H_2O$ at 100°C; d: (a) 2.076; (b) 2.085.
Hazard Analysis and Countermeasures
See barium compounds (soluble) and cyanides.

BARIUM DICHROMATE
General Information
Synonym: Barium bichromate.
Description: Brownish-red crystalline masses.
Formula: $BaCr_2O_7$.
Constant: Mol wt: 353.38.
Hazard Analysis
Toxicity: See barium compounds (soluble) and chromium compounds.

TOXIC HAZARD RATING CODE *(For detailed discussion, see Section 1.)*

0 NONE: (a) No harm under any conditions; (b) Harmful only under unusual conditions or overwhelming dosage.

1 SLIGHT: Causes readily reversible changes which disappear after end of exposure.

2 MODERATE: May involve both irreversible and reversible changes; not severe enough to cause death or permanent injury.

3 HIGH: May cause death or permanent injury after very short exposure to small quantities.

U UNKNOWN: No information on humans considered valid by authors.

Fire Hazard: Moderate by chemical reaction with easily oxidized materials; can react vigorously with reducing materials (Section 6).
Caution: A powerful oxidizer.
Countermeasures
Storage and Handling: Section 7.

BARIUM DIOXIDE. See barium peroxide.

BARIUM DIPHENYLAMINE SULFONATE
General Information
Description: Crystals.
Formula: $Ba(C_6H_5 \cdot NH \cdot C_6H_4SO_3)_2$.
Constant: Mol wt: 633.91.
Hazard Analysis
Toxicity: See barium compounds (soluble).
Disaster Hazard: Dangerous; see sulfonates.
Countermeasures
Storage and Handling: Section 7.

BARIUM DI-o-PHOSPHATE
General Information
Description: Rhombic white crystals.
Formula: $BaHPO_4$.
Constants: Mol wt: 233.35; d: 4.165 at 15°C.
Hazard Analysis and Countermeasures
See barium compounds (soluble) and phosphates.

BARIUM DITHIONATE
General Information
Synonym: Barium hyposulfate.
Description: Rhombic or monoclinic colorless crystals.
Formula: $BaS_2O_6 \cdot 2H_2O$.
Constants: Mol wt: 333.52; mp: decomposes; d: 4.536 at 13.5°C.
Hazard Analysis
Toxicity: See barium compounds (soluble).
Disaster Hazard: Dangerous; see sulfates.
Countermeasures
Ventilation Control: Section 2.
Storage and Handling: Section 7.

BARIUM DIURANATE. See uranium barium oxide.

BARIUM ETHYLSULFATE
General Information
Synonym: Barium sulfovinate.
Description: Colorless crystals; white lustrous leaf.
Formula: $Ba(C_2H_5SO_4)_2 \cdot 2H_2O$.
Constant: Mol wt: 423.63.
Hazard Analysis
Toxicity: See barium compounds (soluble).
Disaster Hazard: Dangerous; see sulfates.
Countermeasures
Ventilation Control: Section 2.
Personal Hygiene: Section 3.
Storage and Handling: Section 7.

BARIUM FERROCYANIDE
General Information
Description: Yellow crystals.
Formula: $Ba_2FE(CN)_6 \cdot 6H_2O$.
Constant: Mol wt: 594.77.
Hazard Analysis and Countermeasures
See barium compounds (soluble) and ferrocyanides.

BARIUM FLUOGALLATE
General Information
Description: White crystals.
Formula: $Ba_3(GaF_6)_2 \cdot H_2O$.
Constants: Mol wt: 797.54; mp: $-\frac{1}{2}H_2O$ at 110°C; $-\frac{1}{2}H_2O$ at 230°C; d: 4.06.
Hazard Analysis and Countermeasures
See barium compounds (soluble) and fluogallates.

BARIUM FLUORIDE
General Information
Description: White powder.
Formula: BaF_2.
Constants: Mol wt: 175.36; mp: 1280°C; bp: 2137°C, d: 4.83.
Hazard Analysis and Countermeasures
See fluorides and barium compounds (soluble).

BARIUM FLUORIDE CHLORIDE
General Information
Description: Tetragonal crystals.
Formula: $BaCl_2 \cdot BaF_2$.
Constants: Mol wt: 383.63, mp: 1008°C, d: 4.51 at 18°C.
Hazard Analysis and Countermeasures
See barium compounds (soluble) fluorides and chlorides.

BARIUM FLUORIDE IODIDE
General Information
Description: Plates.
Formula: $BaF_2 \cdot BaI_2$.
Constants: Mol wt: 566.56, d: 5.21 at 18°C.
Hazard Analysis and Countermeasures
See barium compounds (soluble) fluorides and iodides.

BARIUM FLUOSILICATE
General Information
Synonym: Barium silicofluoride.
Description: White Crystalline powder.
Formula: $BaSiF_6$.
Constants: Mol wt: 279.42, d: 4.29 at 21°C.
Hazard Analysis
Toxicity: An insecticide. See barium compounds (soluble) and fluosilicates.
Disaster Hazard: See fluosilicates.
Countermeasures
Storage and Handling: Section 7.
Shipping Regulations: Section 11.
MCA warning label.

BARIUM FORMATE
General Information
Description: Crystals.
Formula: $Ba(COOH)_2$.
Constants: Mol wt: 227.40, d: 3.21.
Hazard Analysis and Countermeasures
See barium compounds (soluble).

BARIUM d-GLUCONATE
General Information
Description: Prisms or rhombic leaflets.
Formula: $Ba(C_6H_{11}O_7)_2 \cdot 3H_2O$.
Constants: Mol wt: 581.71, mp: $-3H_2O$ at 100°C; 120°C decomposes.
Hazard Analysis and Countermeasures
See barium compounds (soluble).

BARIUM HEXABORIDE
General Information
Description: Cubic black metallic crystals.
Formula: BaB_6.
Constants: Mol wt: 202.28, mp: 2270°C, d: 4.36 at 15°C.
Hazard Analysis and Countermeasures
See barium compounds (soluble) and boron.

BARIUM HYDRATE. See barium hydroxide.

BARIUM HYDRIDE
General Information
Description: Gray crystals or lumps.
Formula: BaH_2.
Constants: Mol wt: 139.38, mp: decomposes 675°C, bp: 1400°C, d: 4.21 at 0°C.
Hazard Analysis and Countermeasures
See barium compounds (soluble) and hydrides.

BARIUM HYDROGEN PYROTELLURATE
General Information
Description: Voluminous precipitate; yellow when hot; white when cold.
Formula: $Ba(HTe_2O_7)_2 \cdot H_2O$.
Constant: Mol wt: 891.83.
Hazard Analysis and Countermeasures
See barium compounds (soluble) and tellurium compounds.

BARIUM HYDROSULFIDE
General Information
Description: Rhombic yellow crystals.
Formula: $Ba(HS)_2 \cdot 4H_2O$.
Constants: Mol wt: 275.57, mp: decomposes 50°C.
Hazard Analysis and Countermeasures
See barium compounds (soluble) and sulfides.

BARIUM HYDROXIDE
General Information
Synonyms: Barium hydrate; caustic baryta.
Description: White powder.
Formula: $Ba(OH)_2 \cdot 8H_2O$.
Constants: Mol wt: 315.51, mp: 78°C, bp: $-8H_2O$ at 780°C, d: 2.18 at 16°C; anhy. 4.50.
Hazard Analysis and Countermeasures
See barium compounds (soluble).

BARIUM HYPOCHLORITE
General Information
Description: Colorless crystals.
Formula: $Ba(ClO)_2 \cdot 2H_2O$.
Constants: Mol wt: 276.31, mp: decomposes.
Hazard Analysis and Countermeasures
See barium compounds (soluble) and hypochlorites.

BARIUM HYPOPHOSPHATE
General Information
Description: Needle-like crystals.
Formula: $BaPO_3$.
Constant: Mol wt: 216.34.
Hazard Analysis and Countermeasures
See barium compounds (soluble) and hypophosphates.

BARIUM HYPOPHOSPHITE
General Information
Description: Crystalline powder.
Formula: $Ba(H_2PO_2)_2 \cdot H_2O$.
Constants: Mol wt: 285.38, mp: decomposes, d: 2.90 at 17°C.
Hazard Analysis and Countermeasures
See barium compounds (soluble) and hypophosphites.

BARIUM HYPOSULFATE. See barium dithionate.

BARIUM HYPOSULFITE. See barium thiosulfate.

BARIUM IODATE
General Information
Description: White crystalline powder.
Formula: $Ba(IO_3)_2$.
Constants: Mol wt: 487.20, mp: decomposes, d: 4.998.
Hazard Analysis and Countermeasures
See barium compounds (soluble) and iodates.

BARIUM IODIDE
General Information
Description: Colorless crystals.
Formula: BaI_2.
Constants: Mol wt: 391.19, mp: 740°C, d: 4.917.
Hazard Analysis and Countermeasures
See barium compounds (soluble) and iodides.

BARIUM LACTATE
General Information
Description: White powder.
Formula: $Ba(CH_3 \cdot CH \cdot OH \cdot COO)_2$.
Constant: Mol wt: 315.50.
Hazard Analysis and Countermeasures
See barium compounds (soluble).

BARIUM LAURATE
General Information
Description: White, leaflet crystals.
Formula: $Ba(C_{12}H_{23}O_2)_2$.
Constants: Mol wt: 535.97, mp: 260°C.
Hazard Analysis and Countermeasures
See barium compounds (soluble).

BARIUM MALATE
General Information
Description: White powder.
Formula: $BaC_4H_4O_5$.
Constant: Mol wt: 269.43.
Hazard Analysis and Countermeasures
See barium compounds (soluble).

BARIUM MALONATE
General Information
Description: White powder.
Formula: $BaC_3H_2O_4 \cdot H_2O$.
Constant: Mol wt: 257.42.
Hazard Analysis and Countermeasures
See barium compounds (soluble).

BARIUM MANGANATE
General Information
Synonyms: Manganese green; Cassels green.
Description: Emerald green powder.
Formula: $BaMnO_4$.
Constants: Mol wt: 256.29, d: 4.85.
Hazard Analysis and Countermeasures
See barium compounds (soluble) and manganese compounds.

BARIUM MERCURY BROMIDE. See mercuric barium bromide.

BARIUM METHYLSULFATE
General Information
Description: Efflorescent crystals.
Formula: $Ba(CH_3SO_4)_2 \cdot 2H_2O$.
Constant: Mol wt: 395.59.
Hazard Analysis and Countermeasures
See barium compounds (soluble) and sulfates.

BARIUM MOLYBDATE
General Information
Description: White powder; tetragonal crystals.
Formula: $BaMoO_4$.
Constants: Mol wt: 297.31, d: 4.65–4.97.
Hazard Analysis and Countermeasures
See barium compounds (soluble).

BARIUM MONO-HYDROGEN o-ARSENATE
General Information
Description: Rhombic or monoclinic colorless crystals.
Formula: $BaHAsO_4 \cdot H_2O$.
Constants: Mol wt: 295.29, mp: $-H_2O$ at 150°C, d: 3.93 at 15°C.
Hazard Analysis and Countermeasures
See barium compounds (soluble) and arsenic compounds.

TOXIC HAZARD RATING CODE (For detailed discussion, see Section 1.)

0 NONE: (a) No harm under any conditions; (b) Harmful only under unusual conditions or overwhelming dosage.

1 SLIGHT: Causes readily reversible changes which disappear after end of exposure.

2 MODERATE: May involve both irreversible and reversible changes; not severe enough to cause death or permanent injury.

3 HIGH: May cause death or permanent injury after very short exposure to small quantities.

U UNKNOWN: No information on humans considered valid by authors.

BARIUM MONO o-PHOSPHATE
General Information
Description: Triclinic crystals.
Formula: $BaH_4(PO_4)_2$.
Constants: Mol wt: 331.35, d: 2.9 at 4°C.
Hazard Analysis and Countermeasures
See barium compounds (soluble) and phosphates.

BARIUM MONOSULFIDE
General Information
Synonym: Barium sulfide.
Description: Cubic colorless crystals.
Formula: BaS.
Constants: Mol wt: 169.43, d: 4.25 at 15°C.
Hazard Analysis
Toxicity: See barium compounds (soluble) and sulfides.
Fire Hazard: Moderate, by spontaneous chemical reaction;
  air, moisture or acid fumes may cause it to ignite.
Explosion Hazard: See sulfides.
Disaster Hazard: See sulfides.
Countermeasures
To Fight Fire: Carbon dioxide, dry chemical or carbon
  tetrachloride (Section 6).
Storage and Handling: Section 7.

BARIUM MONOXIDE. See barium oxide.

BARIUM MYRISTATE
General Information
Description: White crystalline powder.
Formula: $Ba(C_{14}H_{27}O_2)_2$.
Constant: Mol wt: 592.07.
Hazard Analysis and Countermeasures
See barium compounds (soluble).

BARIUM NITRATE
General Information
Synonym: Nitrobarite.
Description: Lustrous white crystals.
Formula: $Ba(NO_3)_2$.
Constants: Mol wt: 261.38, mp: 592°C, bp: decomposes,
  d: 3.24 at 23°C.
Hazard Analysis and Countermeasures
See barium compounds (soluble) and nitrates.
Shipping Regulations: Section 11.
  I.C.C.: Oxidizing material; yellow label, 100 pounds.
  Coast Guard Classification: Oxidizing material.
  MCA warning label.
  IATA: Oxidizing material, yellow label, 12 kilograms
  (passenger), 45 kilograms (cargo).

BARIUM NITRIDE
General Information
Description: Colorless crystals.
Formula: Ba_3N_2.
Constants: Mol wt: 440.10, mp: 1000°C.
Hazard Analysis
Toxicity: See barium compounds (Soluble).
Fire Hazard: Moderate, by spontaneous chemical reaction
  with water to liberate flammable vapor. See also am-
  monia.
Explosion Hazard: Slight, by spontaneous chemical reaction
  to liberate vapor which can form explosive mixtures
  with air. See also ammonia.
Disaster Hazard: See nitrides.
Countermeasures
Storage and Handling: Section 7.

BARIUM NITRITE
General Information
Description: White crystals.
Formula: $Ba(NO_2)_2$.
Constants: Mol wt: 229.38, mp: decomposes 217°C, d: 3.23
  at 23°C.
Hazard Analysis and Countermeasures
See barium compounds (soluble) and nitrites.

BARIUM OLEATE
General Information
Description: Yellowish-white granular masses.
Formula: $Ba(C_{18}H_{33}O_2)_2$.
Constant: Mol wt: 700.25.
Hazard Analysis and Countermeasures
See barium compounds (soluble).

BARIUM OXALATE
General Information
Description: White crystalline powder.
Formula: BaC_2O_4.
Constants: Mol wt: 225.38, d: 2.658.
Hazard Analysis and Countermeasures
See barium compounds (soluble) and oxalates.

BARIUM OXIDE
General Information
Synonyms: Barium monoxide; barium protoxide.
Description: White to yellowish-white powder.
Formula: BaO.
Constants: Mol wt: 153.36, mp: 1923°C, bp: 2000°C
  (approx.), d: 5.72.
Hazard Analysis
Toxicity: See barium compounds (soluble).
Fire Hazard: Slight by spontaneous chemical reaction;
  produces heat on contact with water or steam (Sec-
  tion 6).
Countermeasures
Storage and Handling: Section 7.
Shipping Regulations: Section 11.
  IATA: Other restricted articles, class B, no label re-
  quired, 12 kilograms (passenger), 45 kilograms (cargo).

BARIUM PALMITATE
General Information
Description: White crystalline powder.
Formula: $Ba(C_{16}H_{31}O_2)_2$.
Constants: Mol wt: 648.19, mp: decomposes.
Hazard Analysis
Toxicity: See barium compounds (soluble).

BARIUM PERCHLORATE
General Information
Description: Colorless crystals.
Formula: $Ba(ClO_4)_2 \cdot 3H_2O$.
Constants: Mol wt: 390.4, mp: decomposes at 400°C, d:
  2.74.
Hazard Analysis and Countermeasures
See barium compounds (soluble) and perchlorates.
Shipping Regulations: Section 11.
  I.C.C.: Oxidizing material; yellow label, 100 pounds.
  Coast Guard Classification: Oxidizing material; yellow
  label.
  IATA: Oxidizing material, yellow label, 12 kilograms
  (passenger), 45 kilograms (cargo).

BARIUM PERMANGANATE
General Information
Description: Brownish-violet crystals.
Formula: $Ba(MnO_4)_2$.
Constant: Mol wt: 375.22.
Hazard Analysis and Countermeasures
See barium compounds (soluble) and manganese compounds.
Shipping Regulations: Section 11.
  I.C.C.: Oxidizing material; yellow label, 100 pounds.
  Coast Guard Classification: Oxidizing material; yellow
  label.
  IATA: Oxidizing material, yellow label, 12 kilograms
  (passenger), 45 kilograms (cargo).

BARIUM PEROXIDE
General Information
Synonyms: Barium binoxide; barium dioxide.
Description: Grayish-white powder.
Formula: BaO_2.

Constants: Mol wt: 169.36, mp: 450°, bp: $-O_2$ at 800°C,
d: 4.96.
Hazard Analysis and Countermeasures
See barium compounds (soluble) and peroxides, inorganic.
Shipping Regulations: Section 11.
   I.C.C.: Oxidizing material; yellow label, 100 pounds.
   Coast Guard Classification: Oxidizing material; yellow
      label.
   IATA: Oxidizing material, yellow label, 12 kilograms
      (passenger), 45 kilograms (cargo).

BARIUM PEROXYDISULFATE
General Information
Description: Monoclinic white crystals.
Formula: $BaS_2O_8 \cdot 4H_2O$.
Constants: Mol wt: 401.57, mp: decomposes.
Hazard Analysis
Toxicity: See barium compounds (soluble).

BARIUM PHENOLSULFONATE
General Information
Synonyms: Barium sulfophenylate; barium sulfocarbolate.
Description: White odorless powder.
Formula: $C_{12}H_{12}BaO_9S_2$.
Constant: Mol wt: 501.71.
Hazard Analysis and Countermeasures
See barium compounds (soluble) and phenolsulfonates.

BARIUM PHOSPHATE, DIBASIC
General Information
Synonym: Secondary barium phosphate.
Description: Crystalline powder.
Formula: $BaHPO_4$.
Constant: Mol wt: 233.35.
Hazard Analysis and Countermeasures
See barium compounds (soluble) and phosphates.

BARIUM PHOSPHITE
General Information
Description: Crystalline powder.
Formula: $BaHPO_3$.
Constant: Mol wt: 217.35.
Hazard Analysis and Countermeasures
See barium compounds (soluble) and phosphites.

BARIUM PROPIONATE
General Information
Description: White powder.
Formula: $C_6H_{10}BaO_4 \cdot H_2O$.
Constants: Mol wt: 301.52, mp: 300°C decomposes.
Hazard Analysis
Toxicity: See barium compounds (soluble).

BARIUM PROTOXIDE. See barium oxide.

BARIUM PYROPHOSPHATE
General Information
Description: Rhombic white crystals.
Formula: $Ba_2P_2O_7$.
Constants: Mol wt: 448.68, d: 3.9 at 20°C.
Hazard Analysis and Countermeasures
See barium compounds (soluble) and phosphates.

BARIUM PYROVANADATE
General Information
Description: White crystals.
Formula: $Ba_2V_2O_7$.
Constants: Mol wt: 488.62, mp: 863°C.
Hazard Analysis and Countermeasures
See barium compounds (soluble) and vanadium compounds.

BARIUM RICINOLEATE
General Information
Description: Fine white powder.
Formula: $Ba[CO_2(CH_2)_7CHCHCH_2CHOH(CH_2)_5-CH_3]_2$.
Constants: Mol wt: 727, mp: 116°C, d: 1.21 at 25°/25°.
Hazard Analysis and Countermeasures
See barium compounds (soluble).

BARIUM SALICYLATE
General Information
Description: White needles.
Formula: $Ba(C_7H_9O_3)_2 \cdot H_2O$.
Constant: Mol wt: 429.60.
Hazard Analysis
Toxicity: An allergen. See also barium compounds (soluble).

BARIUM SELENATE
General Information
Description: Orthorhombic crystals.
Formula: $BaSeO_4$.
Constants: Mol wt: 280.32, mp: decomposes, d: 4.75.
Hazard Analysis and Countermeasures
See barium compounds (soluble) and selenium compounds.

BARIUM m-SILICATE
General Information
Description: Rhombic, colorless crystals.
Formula: $BaSiO_3$.
Constants: Mol wt: 213.42, mp: 1604°C, d: 4.399.
Hazard Analysis
Toxicity: See barium compounds (soluble).

BARIUM SILICOFLUORIDE. See barium fluosilicate.

BARIUM STEARATE
General Information
Description: White powder.
Formula: $Ba(C_{18}H_{35}O_2)_2$.
Constant: Mol wt: 709.4.
Hazard Analysis
Toxicity: See barium compounds (soluble).

BARIUM SUCCINATE
General Information
Description: Crystalline powder.
Formula: $BaC_4H_4O_4$.
Constant: Mol wt: 253.43.
Hazard Analysis
Toxicity: See barium compounds (soluble).

BARIUM SULFATE
General Information
Synonym: Blanc fixe, barite.
Description: White, heavy, odorless powder. Not soluble in
   water or dilute acids.
Formula: $BaSO_4$.
Constants: Mol wt: 233.4, d: 4.50 at 15°C, mp: 1580°C.
Hazard Analysis
Toxicity: 0. It is a non-toxic insoluble salt used as an opaque
   medium in radiography. Soluble impurities can lead to
   toxic reactions. See barium compounds (soluble).

BARIUM SULFHYDRATE
General Information
Description: Yellow crystals.
Formula: $Ba(SH)_2$.
Constant: Mol wt: 203.51.
Hazard Analysis and Countermeasures
See barium compounds (soluble) and sulfides.

TOXIC HAZARD RATING CODE (For detailed discussion, see Section 1.)

0 NONE: (a) No harm under any conditions; (b) Harmful only under unusual conditions or overwhelming dosage.

1 SLIGHT: Causes readily reversible changes which disappear after end
   of exposure.

2 MODERATE: May involve both irreversible and reversible changes;
   not severe enough to cause death or permanent injury.

3 HIGH: May cause death or permanent injury after very short exposure
   to small quantities.

U UNKNOWN: No information on humans considered valid by authors.

BARIUM SULFIDE. See barium monosulfide.

BARIUM SULFITE
General Information
Description: White powder.
Formula: $BaSO_3$.
Constant: Mol wt: 217.42.
Hazard Analysis and Countermeasures
See barium compounds (soluble) and sulfites.

BARIUM SULFOCARBOLATE. See barium phenol-sulfonate.

BARIUM SULFOCYANATE
General Information
Synonym: Barium sulfocyanide.
Description: White crystals.
Formula: $Ba(SCN)_2 \cdot 2H_2O$.
Constant: Mol wt: 289.55.
Hazard Analysis and Countermeasures
See also barium compounds (soluble) and thiocyanates.

BARIUM SULFOCYANIDE. See barium sulfocyanate.

BARIUM SULFOPHENYLATE. See barium phenolsulfonate.

BARIUM SULFOVINATE. See barium ethylsulfate.

BARIUM TARTRATE
General Information
Description: White granular powder.
Formula: $BaC_4H_4O_6 \cdot H_2O$.
Constants: Mol wt: 303.45, d: 2.980 at 20.8°C.
Hazard Analysis
Toxicity: See barium compounds (soluble).

BARIUM TELLURATE
General Information
Description: White crystals.
Formula: $BaTeO_4 \cdot 3H_2O$.
Constants: Mol wt: 383.02, mp: Decomp. $> 200°C$, d: 4.2.
Hazard Analysis and Countermeasures
See barium compounds (soluble) and tellurium compounds.

BARIUM TETRASULFIDE
General Information
Description: Rhombic red or yellow crystals or powder.
Formula: BaS_4.
Constants: Mol wt: 265.63, mp: decomposes 200°C, d: 2.988.
Hazard Analysis and Countermeasures
See barium compounds and sulfides.

BARIUM THIOCYANATE. See barium sulfocyanate.

BARIUM THIOSULFATE
General Information
Synonym: Barium hyposulfite.
Description: Crystalline powder.
Formula: BaS_2O_3.
Constants: Mol wt: 249.49, mp: decomposes.
Hazard Analysis and Countermeasures
See barium compounds (soluble) and sulfates.

BARIUM TITANATE
General Information
Synonym: Barium m-titanate.
Description: Light gray-buff powder (also exists in five crystal modifications); insoluble in water and alkalies; slightly soluble in dilute acids; soluble in concentrated sulfuric acid and hydrosulfuric acid.
Formula: $BaTiO_3$.
Constants: Mol wt: 233.26, mp: 3010°F, d: 5.95.
Hazard Analysis
Toxic Hazard Rate: U. Animal experiments show low toxicity. See also titanium compounds.

BARIUM o-TRIPHOSPHATE
General Information
Description: Cubic white crystals.
Formula: $Ba_3(PO_4)_2$.
Constants: Mol wt: 602.04, d: 4.1 at 16°C.
Hazard Analysis and Countermeasures
See barium compounds (soluble) and phosphates.

BARIUM TRISULFIDE
General Information
Description: Yellow green crystals.
Formula: BaS_3.
Constant: Mol wt: 233.54.
Hazard Analysis and Countermeasures
See barium compounds (soluble) and sulfides.

BARIUM TUNGSTATE
General Information
Description: Tetragonal, colorless crystals.
Formula: $BaWO_4$.
Constants: Mol wt: 385.28, d: 5.04.
Hazard Analysis
Toxicity: See barium compounds (soluble).
Countermeasures
Ventilation Control: Section 2.

BARIUM m-TUNGSTATE
General Information
Description: Rhombic crystals.
Formula: $Ba_3(H_2W_{12}O_{40}) \cdot 27H_2O$.
Constants: Mol wt: 3747.57, d: 4.30.
Hazard Analysis
Toxicity: See barium compounds (soluble).
Countermeasures
Ventilation Control: Section 2.

BARIUM ZIRCONATE
General Information
Description: Light gray-buff powder; insoluble in water and alkalies; slightly soluble in acid.
Formula: $BaZrO_3$.
Constants: Mol wt: 276, d: 5.52, mp: 4550°F.
Hazard Analysis
Toxic Hazard Rate:
  Acute Local: Inhalation 1.
  Acute Systemic: Ingestion 1.
  Chronic Local: Inhalation 2.
  Chronic Systemic: U.
Animal experiments show low oral toxicity and interstitial pneumonitis on inhalation. See also zirconium compounds.

BARLEY OIL
Hazard Analysis
Toxic Hazard Rating:
  Acute Local: Allergen 1.
  Acute Systemic: U.
  Chronic Local: Allergen 1.
  Chronic Systemic: U.
Fire Hazard: Slight; when heated.
Countermeasures
Personal Hygiene: Section 3.
Storage and Handling: Section 7.

BAS F. See potassium ammonium nitrate.

BASIC ANTIMONY CHLORIDE. See antimony oxychloride.

BASIC COPPER ZEOLITES
Hazard Analysis
Toxicity: See copper compounds.

BATTERIES, ELECTRIC STORAGE, WET
General Information
Description: Battery electrolytes which contain either sulfuric acid or sodium hydroxide.

Countermeasures
Shipping Regulations: Section 11.
  I.C.C.: Corrosive liquid; white label, 600 pounds.
  Coast Cuard Classification: Corrosive liquid; white label.
  IATA: Corrosive liquid, white label, not acceptable (passenger), no limit (cargo).

BATTERIES, ELECTRIC STORAGE, WET, WITH AUTOMOBILES OR AUTO PARTS. See sulfuric acid and/or sodium hydroxide, whichever electrolyte is present.
Shipping Regulations: Section 11.
  I.C.C.: Corrosive liquid; white label, no limit.
  Coast Guard Classification: Corrosive liquid; white label.
  IATA: Corrosive liquid, white label, not acceptable (passenger), 230 kilograms (gross) (cargo).

BATTERIES ELECTRIC STORAGE, WET, WITH OTHER EQUIPMENT
  IATA: Corrosive liquid, white label, not acceptable (passenger), 230 kilograms (gross) (cargo).

BATTERIES, ELECTRIC STORAGE, WET, WITH CONTAINERS OF CORROSIVE BATTERY FLUID. See sulfuric acid and/or sodium hydroxide, whichever electrolyte is present.
Shipping Regulations: Section 11.
  I.C.C.: Corrosive liquid; white label, 2 gallons.
  Coast Guard Classification: Corrosive liquid; white label.

BATTERIES, ELECTRIC STORAGE, WET, INSTALLED IN VEHICLES
Countermeasures
Shipping Regulations: Section 11.
  IATA: No limit (passenger and cargo).

BATTERIES, ELECTRIC STORAGE, WET, NON-SPILLABLE TYPE
Countermeasures
Shipping Regulations: Section 11.
  IATA: Corrosive liquids, white label, no limit (passenger and cargo).

BATTERIES, ELECTRIC STORAGE, EMPTY, WITH CONTAINERS OF CORROSIVE BATTERY FLUID
Countermeasures
Shipping Regulations: Section 11.
  IATA: Corrosive liquid, white label, not acceptable (passenger), 2½ liters (cargo).

BATTERY ACID. See sulfuric acid.

BATTERY CHARGER WITH ELECTROLYTE (ACID) OR ALKALINE CORROSIVE LIQUID. See sulfuric acid and/or sodium hydroxide, whichever is present.
Shipping Regulations: Section 11.
  I.C.C. Classification: Corrosive liquid; white label.
  Coast Guard Classification: Corrosive liquid; white label.
  IATA: Corrosive liquid, white label, not acceptable (passenger), 2½ liters (cargo).

BAUXITE. See alumina.

BAYBERRY OIL
Hazard Analysis
Toxic Hazard Rating:
  Acute Local: Allergen 1.
  Acute Systemic: U.
  Chronic Local: Allergen 1.

Chronic Systemic: U.
Fire Hazard: Slight; when heated (Section 6).
Countermeasures
Personal Hygiene: Section 3.
Storage and Handling: Section 7.

BAYTEX. See O,o-Dimethyl o-[4-(Methylthio)-m-Tolyl]-Phosphorothioate.

p-tert-BBA. See p-tert-butylbenzoic acid.

BBC. See bromobenzylcyanide.

B BLASTING POWDER. See explosives, low.

BEEF FAT OIL
Hazard Analysis
Toxic Hazard Rating:
  Acute Local: Allergen 1.
  Acute Systemic: U.
  Chronic Local: Allergen 1.
  Chronic Systemic: U.
Fire Hazard: Slight; when heated.
Countermeasures
Personal Hygiene: Section 3.
Storage and Handling: Section 7.

BEESWAX
General Information
Synonym: Yellow beeswax.
Description: Yellow to brownish-yellow, soft to brittle wax.
Constants: Mp: 62–65° C, d: 0.95–0.96.
Hazard Analysis
Toxic Hazard Rating:
  Acute Local: Allergen 1.
  Acute Systemic: U.
  Chronic Local: Allergen 1.
  Chronic Systemic: U.
A general purpose food additive. (Section 10).
Fire Hazard: Slight; when heated. (Section 6).
Countermeasures
Personal Hygiene: Section 3.
Storage and Handling: Section 7.

BEESWAX, WHITE
General Information
Synonym: White wax; cera alba.
Description: Beeswax bleached by sunlight or oxidizing agents. Yellowish white solid, translucent in thin layer. Other properties are those of beeswax.
Hazard Analysis
Toxic Hazard Rating: U. A general purpose food additive (Section 10).

"BEETLE"
General Information
Description: Urea-formaldehyde condensation product.
Hazard Analysis
Toxic Hazard Rating:
  Acute Local: Allergen 1; Inhalation 1.
  Acute Systemic: 0.
  Chronic Local: Irritant 1; Allergen 1; Inhalation 1.
  Chronic Systemic: U.
  Caution: Inhalation of dust may cause allergic response or irritation of lungs (Section 9).
Fire Hazard: Slight; (Section 6).
Countermeasures
Personal Hygiene: Section 3.
Storage and Handling: Section 7.

TOXIC HAZARD RATING CODE (For detailed discussion, see Section 1.)

0 NONE: (a) No harm under any conditions; (b) Harmful only under unusual conditions or overwhelming dosage.

1 SLIGHT: Causes readily reversible changes which disappear after end of exposure.

2 MODERATE: May involve both irreversible and reversible changes; not severe enough to cause death or permanent injury.

3 HIGH: May cause death or permanent injury after very short exposure to small quantities.

U UNKNOWN: No information on humans considered valid by authors.

BELLADONNA
General Information
Synonym: Deadly nightshade, from which the alkaloids, atropine and belladonnine are derived.
Hazard Analysis
Toxic Hazard Rating:
Acute Local: 0.
Acute Systemic: Ingestion 3; Inhalation 3.
Chronic Local: 0.
Chronic Systemic: Ingestion 2; Inhalation 2.
Toxicology: A poison. See also hyoscyamine and atropine. Local contact may cause a contact dermatitis (Section 9).
Countermeasures
Ventilation Control: Section 2.
Personal Hygiene: Section 3.
First Aid: Section 1.
Storage and Handling: Section 7.

BENTONITE
General Information
Synonym: Colloidal clay.
Description: A clay containing appreciable amounts of the clay mineral montmorillonite. Light yellow or green, cream, pink, gray, to black; plastic; insoluble in water and common organic solvents.
Hazard Analysis
Toxic Hazard Rating: U. A general purpose food additive (Section 10).

BENZAHEX. See benzene hexachloride.

BENZAL CHLORIDE
General Information
Synonyms: Benzyl dichloride; benzylidine chloride.
Description: Very refractive liquid.
Formula: $C_6H_5CHCl_2$.
Constants: Mol wt: 161.03, mp: $-16°C$, bp: $214°C$, d: 1.29.
Hazard Analysis
Toxicity: Details unknown. A strong irritant and lachrymator. High concentrations cause CNS depression.
Disaster Hazard: Dangerous; see chlorides.
Countermeasures
Ventilation Control: Section 2.
Personal Hygiene: Section 3.
Storage and Handling: Section 7.
Shipping Regulations: Section 11.
IATA: Poison B, poison label, 1 liter (passenger), 220 liters (cargo).

BENZALDEHYDE
General Information
Synonyms: Benzoic aldehyde; artificial almond oil.
Description: Refractive liquid.
Formula: C_6H_5CHO.
Constants: Mol wt: 106.12, mp: $-26°C$, bp: $179.0°C$, flash p: $148°F$, d: 1.050 at $15°/4°C$, autoign. temp.: $377°F$, vap. press.: 1 mm at $26.2°C$, vap. d.: 3.65.
Hazard Analysis
Toxic Hazard Rating:
Acute Local: Allergen 1.
Acute Systemic: Ingestion 2, Skin Absorption 1.
Chronic Local: Allergen 1.
Chronic Systemic: Ingestion 1.
A synthetic flavoring substance and adjuvant (section 10).
Caution: Acts as a feeble local anesthetic. See aldehydes. Local contact may cause contact dermatitis (Section 9). Causes CNS depression in small doses and convulsions in larger doses.
Fire Hazard: Slight, when exposed to heat or flame; reacts with oxidizing materials.
Spontaneous Heating: No.
Countermeasures
To Fight Fire: Foam, carbon dioxide, dry chemical or carbon tetrachloride (Section 6).

Personal Hygiene: Section 3.
Storage and Handling: Section 7.

BENZALDEHYDE CYANHYDRIN
General Information
Synonym: Mandelonitrile.
Description: Yellow viscous liquid.
Formula: $C_6H_5CH(OH)CN$.
Constants: Mol wt: 133.14, mp: $-10°C$, bp: $170°C$ decomposes, d: 1.124.
Hazard Analysis
Toxicity: Details unknown, but probably highly toxic. See also cyanides and nitriles.
Disaster Hazard: See cyanides.
Countermeasures
Storage and Handling: Section 7.

BENZALKONIUM CHLORIDE
General Information
Synonym: Zephiran chloride.
Description: White or yellowish-white powder. Aromatic odor. Very bitter taste.
Formula: Alkyl dimethyl benzylammonium chlorides.
Hazard Analysis
Toxic Hazard Rating:
Acute Local: Irritant 1; Ingestion 2.
Acute Systemic: Ingestion 2.
Chronic Local: Irritant 1.
Chronic Systemic: U.
Disaster Hazard: Dangerous. See chlorides.

BENZALMALONONITRILE
General Information
Description: Crystals.
Formula: $C_6H_5CH_2CH(CN)_2$.
Constant: Mol wt: 156.2.
Hazard Analysis
Toxicity: See cyanides.
Disaster Hazard: See cyanides.
COUNTERMEASURES
Storage and Handling: Section 7.

BENZANTHRONE
General Information
Description: Pale yellow needles.
Formula: $C_{17}H_{10}O$.
Constants: Mol wt: 230.25, mp: $174°C$, vap. press.: 1 mm at $225.0°C$.
Hazard Aanlysis
Toxicity: Unknown.
Fire Hazard; Slight; when heated. (Section 6).
Countermeasures
Storage and Handling: Section 7.

2-BENZAZINE. See isoquinoline.

3,4-BENZCHRYSENE. See picene.

"BENZEDRINE"
General Information
Synonym: Amphetamine.
Description: Liquid.
Formula: $C_9H_{13}N$.
Constants: Mol wt: 135.20, bp: $200°C$, flash p: $80°F$ (O.C.), d: 0.931, vap. d.: 4.65.
Hazard Analysis
Toxic Hazard Rating:
Acute Local: 0.
Acute Systemic: Ingestion 2; Inhalation 2.
Chronic Local: 0.
Chronic Systemic: Ingestion 1; Inhalation 1.
Caution: A stimulant. Overdoses cause hyperactivity, restlessness, insomnia, rapid pulse, rise in blood pressure, dilated pupils, dryness of the throat.
Fire Hazard: Dangerous; when exposed to heat or flame; can react with oxidizing materials.

Disaster Hazard: Dangerous, upon exposure to heat or flame.

Countermeasures

Ventilation Control: Section 2.

Personal Hygiene: Section 3.

Treatment and Antidotes: Evacuation of stomach if taken by mouth, sedatives. Call a physician.

To Fight Fire: Carbon dioxide or dry chemical or carbon tetrachloride (Section 6).

Storage and Handling: Section 7.

BENZENE

General Information

Synonyms: Benzol; phenyl hydride; coal naphtha.

Description: Clear colorless liquid.

Formula: C_6H_6.

Constants: Mol wt: 78.11, mp: 5.51°C, bp: 80.093°- 80.094°C, flash p: 12°F (C.C.), d: 0.8794 at 20°C, autoign. temp.: 1044°F, lel: 1.3%, uel: 7.1%, vap. press.: 100 mm at 26.1°C, vap. d.: 2.77.

Hazard Analysis

Toxic Hazard Rating:

Acute Local: Irritant 2; Ingestion 1; Inhalation 1.

Acute Systemic: Ingestion 2, Inhalation 2; Skin Absorption 2.

Chronic Local: 0.

Chronic Systemic: Ingestion 3; Inhalation 3; Skin Absorption 3.

TLV: ACGIH (recommended); 25 parts per million in air; 80 milligrams per cubic meter of air. May be absorbed via the skin.

Toxicology: Poisoning occurs most commonly through inhalation of the vapor, though benzene can penetrate the skin, and thus contribute to poisoning.

Locally, benzene has a comparatively strong irritating effect producing erythema and burning, and in more severe cases, edema and even blistering (Section 9). Exposure to high concentrations of the vapor (3,000 ppm or higher) results from accidents such as failure of equipment or spillage. Such exposure, while rare in industry, may result in acute poisoning, characterized by the narcotic action of benzene on the central nervous system. The anesthetic action of benzene is similar to that of other anesthetic gases, consisting of a preliminary stage of excitation followed by depression and, if exposure is continued, death through respiratory failure.

The chronic, rather than the acute form of benzene poisoning is important in industry; it has a toxic action on the blood-forming tissues. There is no specific blood picture occurring in cases of chronic benzol poisoning.

The bone marrow may be hypoplastic, normal, or hyperplastic, the changes being reflected in the peripheral blood. Anemia, leucopenia, macrocytosis, reticulocytosis, thrombocytopenia, high color index, and prolonged bleeding time may be present. Cases of myeloid leukemia have been reported. For the supervision of the worker, repeated blood examinations are necessary, including hemoglobin determinations, white and red cell counts and differential smears. Where a worker shows a progressive drop in either red or white cells, or where the white count remains below 5,000 per cu. mm., or the red count below 4.0 million per cu. mm., on two successive monthly examinations, he should be immediately removed from exposure. With this method of supervision of the worker, no permanent damage will result to the blood-forming system.

Following absorption of benzene, elimination is chiefly through the lungs, when fresh air is breathed. The portion that is absorbed, is oxidized, and the oxidation products are combined with sulfuric and glycuronic acids and eliminated in the urine. This may be used as a diagnostic sign. Benzene has a definite cumulative action, and exposure to relatively high concentrations are not serious from the point of view of causing damage to the blood-forming system, provided that the exposure is not repeated. On the other hand, daily exposure to concentrations of 100 ppm or less will usually cause damage if continued over a protracted period of time.

In acute poisoning, the worker becomes confused and dizzy, complains of tightening of the leg muscles and of pressure over the forehead, then passes into a stage of excitement. If allowed to remain in exposure, he quickly becomes stupefied and lapses into coma. In nonfatal cases, recovery is usually complete and no permanent disability occurs.

In chronic poisoning the onset is slow, with the symptoms vague; fatigue, headache, dizziness, nausea and loss of appetite, loss of weight, and weakness are common complaints in early cases. Later, pallor, nosebleeds, bleeding gums, menorrhagia, petechiae and purpura may develop. There is great individual variation in the signs and symptoms of chronic benzene poisoning. Note: Benzene is a common air contaminant.

Fire Hazard: Dangerous; when exposed to heat or flame; can react vigorously with oxidizing materials.

Spontaneous Heating: No.

Explosion Hazard: Moderate, when its vapors are exposed to flame. Use with adequate ventilation.

Disaster Hazard: Dangerous; highly flammable.

Countermeasures

Ventilation Control: Section 2.

Personnel Protection: Section 3.

To Fight Fire: Foam, carbon dioxide, dry chemical or carbon tetrachloride (Section 6).

First Aid: Section 1.

Storage and Handling: Section 7.

Shipping Regulations: Section 11.

I.C.C.: Flammable liquid; red label, 10 gallons.

Coast Guard Classification: Inflammable liquid; red label.

MCA warning label.

IATA: Flammable liquid, red label, 1 liter (passenger), 40 liters (cargo).

BENZENE ARSONIC ACID. See phenyl arsonic acid.

BENZENE CARBONYL CHLORIDE. See benzoyl chloride.

1, 3-BENZENDIOL. See resorcinol.

BENZENE DIAZOANILIDE. See β-diazoamidobenzol.

BENZENE DIAZONIUM CHLORIDE. See diazobenzene chloride.

BENZENE DIAZONIUM CHROMATE. See diazobenzol chromate.

BENZENE DIAZONIUM NITRATE. See diazonbenzene nitrate.

1, 3-BENZENE DICARBONITRILE. See o-dicyanobenzene.

BENZENE DICARBOXYLIC ACID. See phthalic acid.

1, 2-BENZENE DIOL. See pyrocatechol.

1, 4-BENZENE DIOL. See hydroquinone.

TOXIC HAZARD RATING CODE (*For detailed discussion, see Section 1.*)

0 NONE: (a) No harm under any conditions; (b) Harmful only under unusual conditions or overwhelming dosage.

1 SLIGHT: Causes readily reversible changes which disappear after end of exposure.

2 MODERATE: May involve both irreversible and reversible changes; not severe enough to cause death or permanent injury.

3 HIGH: May cause death or permanent injury after very short exposure to small quantities.

U UNKNOWN: No information on humans considered valid by authors.

BENZENE-1, 3-DIPHENYL. See m-terphenyl.

BENZENE-1, 4-DIPHENYL. See p-terphenyl.

m-BENZENEDISULFONIC ACID
General Information
Synonym: MBDSA.
Description: Gray crystalline hygroscopic powder.
Formula: $C_6H_6O_6S_2$.
Constant: Mol wt: 238.23.
Hazard Analysis
Toxic Hazard Rating:
  Acute Local: Irritant 3; Ingestion 3; Inhalation 2.
  Acute Systemic: U.
  Chronic Local: Irritant 2.
  Chronic Systemic: U.
Caution: In solutions it forms an extremely corrosive liquid.
Disaster Hazard: Dangerous; See sulfonates.
Countermeasures
Personnel Protection: Section 3.
First Aid: Section 1.
Storage and Handling: Section 7.

BENZENE HEXACHLORIDE. See hexachlorocyclohexane.

BENZENE PHOSPHONIC ACID
General Information
Description: Colorless crystals.
Formula: $C_6H_5PO(OH)_2$.
Constants: Mol wt: 158.1, mp: 165°C, d: 1.475.
Toxicity: Unknown.
Countermeasures
Storage and Handling: Section 7.

BENZENE PHOSPHORUS DICHLORIDE
General Information
Synonym: Phenyl dichlorophosphine.
Description: Colorless liquid while in air.
Formula: $C_6H_5PCl_2$.
Constants: Mol wt: 179, mp: −55°C, bp: 224.6°C, d: 1.319 at 20°/20°C, vap. d.: 6.2.
Hazard Analysis
Toxic Hazard Rating:
  Acute Local: Irritant 2; Ingestion 2; Inhalation 2.
  Acute Systemic: U.
  Chronic Local: U.
  Chronic Systemic: U.
Disaster Hazard: See hydrochloric acid.
Countermeasures
Ventilation Control: Section 2.
Personnel Protection: Section 3.
Storage and Handling: Section 7.

BENZENE PHOSPHORUS OXYDICHLORIDE
General Information
Synonym: Phenyl dichlorophosphine oxide.
Description: Colorless liquid; faint fruity odor.
Formula: $C_6H_5POCl_2$.
Contrants: Mol wt: 195, bp: 258°C, d: 1.375 at 20°/20°C, vap. d.: 6.7.
Hazard Analysis
Toxicity: Unknown.
Disaster Hazard: See chlorides and phosphates.
Countermeasures
Storage and Handling: Section 7.

BENZENE PHOSPHORUS THIODICHLORIDE
General Information
Synonym: Phenyl dichlorophosphine sulfide.
Description: A colorless liquid which fumes a little in air.
Formula: $C_6H_5PSCl_2$.
Constants: Mol wt: 211, bp: 205°C at 130 mm, d: 1.376 at 20°/13°C.
Hazard Analysis
Toxicity: Unknown.
Disaster Hazard: Dangerous; see chlorides, phosphates and

sulfates; will react with water or steam to product toxic and flammable vapors.
Countermeasures
Storage and Handling: Section 7.

o-NITRO-BENZENE SULFENYL CHLORIDE
General Information
Description: Yellow crystals.
Formula: $O_2NC_6H_4SCl$.
Constants: Mol wt: 189.6, bp: decomposes at 170°C (explosively), mp: 73–74°C.
Hazard Analysis
Explosion Hazard: Moderate, when heated.
Toxicity: Unknown.
Disaster Hazard: Dangerous; see sulfates and chlorides; can react vigorously with oxidizing materials.
Countermeasures
Storage and Handling: Section 7.

BENZENE SULFONIC ACID
General Information
Description: Deliquescent plates or tablets.
Formula: $C_6H_5SO_3H:1\frac{1}{2}H_2O$.
Constants: Mol wt: 185.2, mp: 43–44°C.
Hazard Analysis
Toxic Hazard Rating:
  Acute Local: Irritant 3; Ingestion 3; Inhalation 3.
  Acute Systemic: U.
  Chronic Local: U.
  Chronic Systemic: U.
Disaster Hazard: Dangerous. See sulfonates.
Countermeasures
Ventilation Control: Section 2.
Personnel Protection: Section 3.
First Aid: Section 1.
Storage and Handling: Section 7.

BENZENE SULFONYL FLUORIDE
General Information
Description: Clear liquid.
Formula: $C_6H_5SO_2F$.
Constants: Mol wt: 160, bp: 209°C, fp: −5°C, flash p: 196°F, d: 1.329, vap. press.: 8 mm at 80°C, vap. d.: 5.52.
Hazard Analysis
Toxicity: It appears to be highly toxic by vapor inhalation. Possibly slightly irritating to skin (Section 9).
Fire Hazard: Moderate, when exposed to heat or flame.
Disaster Hazard: Dangerous; see fluorides, chlorides and sulfates; can react vigorously with oxidizing materials.
Countermeasures
To Fight Fire: Water, foam, carbon dioxide, dry chemical or carbon tetrachloride (Section 6).
Storage and Handling: Section 7.

1, 2, 4, 5-BENZENE TETRACARBOXYLIC ACID. See pyromellitic acid.

BENZENE TETRACHLORIDE. See 1, 2, 4, 5-tetrachlorobenzene.

BENZENE THIOL. See phenyl mercaptan.

1, 2, 3-BENZENE TRIOL. See pyrogallol.

BENZETHONIUM CHLORIDE
General Information
Synonym: Hyamine 1622.
Description: Thin hexagonal plates, very soluble in water, alcohol, acetone.
Formula: $C_{27}H_{42}ClNO_2 \cdot H_2O$.
Constants: Mol wt: 466.1, mp: 164–166°C.
Hazard Analysis
Toxicity: Moderate. Estimated fatal dose is 1–3 grams. Ingestion may cause nausea, vomiting, collapse, convulsions or coma.
Disaster Hazard: Dangerous. See chlorides.

BENZHYDROL
General Information
Synonyms: Benzohydrol; diphenyl carbinol.
Description: Needle-like crystals.
Formula: $(C_6H_5)_2CHOH$.
Constants: Mol wt: 184.2, mp: 68–69°C, bp: 301.0°C at 50 mm, vap. press.: 1 mm at 110.0°C.
Hazard Analysis
Toxicity: Details unknown; an insecticide.
Fire Hazard: Slight; when heated.
Countermeasures
Storage and Handling: Section 7.

BENZIDINE
General Information
Synonyms: Benzidine base; p-diaminodiphenyl.
Description: Grayish, yellow crystalline powder; white or slightly reddish crystals, powders or leaf.
Formula: $NH_2C_6H_4C_6H_4NH_2$.
Constants: Mol wt: 184.23, mp: 127.5–128.7°C at 740 mm, bp: 401.7°C, d: 1.250 at 20°/4°C.
Hazard Analysis
Toxic Hazard Rating:
Acute Local: U.
Acute Systemic: Ingestion 3; Inhalation 3; Skin Absorption 3.
Chronic Local: U.
Chronic Systemic: Ingestion 3; Inhalation 3; Skin Absorption 3.
Toxicology: Can cause damage to blood including hemolysis and bone marrow depression. On ingestion causes nausea and vomiting which may be followed by liver and kidney damage. Is suspected of causing bladder tumors. Any exposure is considered extremely hazardous.
Disaster Hazard: Dangerous; when heated to decomposition, emits highly toxic fumes.
Countermeasures
Ventilation Control: Section 2.
Personal Hygiene: Section 3.
First Aid: Section 1.
Treatment and Antidotes: In case of contact, immediately wash skin with plenty of soap and water and flush eyes with plenty of water for at least 15 minutes. Secure medical attention at once.
Storage and Handling: Section 7.
Shipping Regulations: Section 11.
MCA warning label.
IATA: Poison B, poison label, 25 kilograms (passenger), 95 kilograms (cargo).

BENZIDINE BASE. See benzidine.

BENZIL
General Information
Synonym: Dibenzoyl.
Description: Yellow crystals.
Formula: $C_6H_5COCOC_6H_5$.
Constants: Mol wt: 210.2, mp: 95°C, bp: 346–348°C, d: 1.23 at 15°/4°C, vap. press.: 1 mm at 128.4°C.
Hazard Analysis
Toxicity: Details unknown; an insecticide.
Fire Hazard: Slight; when heated.
Countermeasures
Storage and Handling: Section 7.

BENZILIC ACID
General Information
Synonym: Diphenylglycolic acid.
Description: Light tan powder.

Formula: $C_6H_5C(OH)(COOH)C_6H_5$.
Constants: Mol wt: 228.24, mp: 146–149°C, bp: 180°C decomposes.
Hazard Analysis
Toxicity: Unknown.
Fire Hazard: Slight; when exposed to heat or flame; can react with oxidizing materials (Section 6).
Countermeasures
Storage and Handling: Section 7.

BENZIMIDAZOLE
General Information
Synonym: N', N'-Methenyl-o-phenylene diamine.
Formula: $C_7H_6N_2$.
Description: Tabular crystals soluble in alcohol, sparingly soluble in water.
Constants: Mol wt: 118.1, mp: 170.5°C, bp: > 360°C.
Hazard Analysis
Toxicity: Unknown. Limited animal experimentation suggests a moderate degree of toxicity.
Disaster Hazard: Dangerous; when heated to decomposition, it emits highly toxic fumes.

BENZINE. See petroleum spirits and naphtha, VM and P.
Shipping Regulations: Section 11.
ICC: Flammable liquid, red label, 10 gallons.
IATA: Flammable liquid, red label, 1 liter (passenger), 40 liters (cargo).

BENZINOFORM. See carbon tetrachloride.

BENZOATE OF SODA. See sodium benzoate.

BENZOCAINE. See anesthesin.

BENZOFLEX. See ethylene glycol dibenzoate.

BENZOGUANAMINE
General Information
Description: Crystals.
Formula: $C_6H_5(C_3N_3)(NH_2)_2$.
Constants: Mol wt: 187.2, mp: 227°C, d: 1.4.
Hazard Analysis
Toxicity: Unknown.
Fire Hazard: Slight.
Countermeasures
Storage and Handling: Section 7.

BENZOHYDROL. See benzhydrol.

BENZOIC ACID
General Information
Synonym: Phenylformic acid.
Description: White powder.
Formula: C_6H_5COOH.
Constants: Mol wt: 122.12, mp: 121.7°C, bp: 249°C, flash p: 250°F (C.C.), d: 1.316, autoign. temp.: 1065°F, vap. press.: 1 mm at 96.0°C (sublimes), vap. d.: 4.21.
Hazard Analysis
Toxic Hazard Rating:
Acute Local: Irritant 1; Ingestion 1; Inhalation 1.
Acute Systemic: Ingestion 1.
Chronic Local: 0.
Chronic Systemic: Ingestion 1.
A chemical preservative food additive (Section 10).
Fire Hazard: Slight, when exposed to heat or flame; can react with oxidizing materials.
Spontaneous Heating: No.
Countermeasures
To Fight Fire: Water, carbon dioxide, dry chemical or carbon tetrachloride (Section 6).

TOXIC HAZARD RATING CODE (For detailed discussion, see Section 1.)

0 NONE: (a) No harm under any conditions; (b) Harmful only under unusual conditions or overwhelming dosage.

1 SLIGHT: Causes readily reversible changes which disappear after end of exposure.

2 MODERATE: May involve both irreversible and reversible changes; not severe enough to cause death or permanent injury.

3 HIGH: May cause death or permanent injury after very short exposure to small quantities.

U UNKNOWN: No information on humans considered valid by authors.

Ventilation Control: Section 2.
Personal Hygiene: Section 3.
Storage and Handling: Section 7.

BENZOIC ACID ANHYDRIDE. See benzoic anhydride.

BENZOIC ACID α-METHYLBENZYL ESTER. See α-methylbenzyl benzoic acid.

BENZOIC ALDEHYDE. See benzaldehyde.

BENZOIC ANHYDRIDE
General Information
Synonym: Benzoic acid anhydride.
Description: Crystals.
Formula: $(C_6H_5CO)_2O$.
Constants: Mol wt: 226.2, mp: 42°C, bp: 360°C, d: 1.1989 at 15°/4°C, vap. press.: 1 mm at 143.8°C.
Hazard Analysis
Toxic Hazard Rating:
  Acute Local: Irritant 1; Allergen 1; Ingestion 1; Inhalation 1.
  Acute Systemic: U.
  Chronic Local: Allergen 1.
  Chronic Systemic: U.
Fire Hazard: Slight; when heated; will react with water or steam to produce heat (Section 6).
Countermeasures
Ventilation Control: Section 2.
Personal Hygiene: Section 3.
Storage and Handling: Section 7.

BENZOIC ETHER. See ethyl benzoate.

BENZOIN
General Information
Synonyms: Phenyl α-hydroxybenzyl ketone; phenylbenzoyl carbinol.
Description: Pale yellow crystals.
Formula: $C_{14}H_{12}O_2$.
Constants: Mol wt: 212.24, mp: 132°C, bp: 343°C, d: 1.310 at 20°/4°C, vap. press.: 1 mm at 135.6°C.
Hazard Analysis
Toxic Hazard Rating:
  Acute Local: Irritant 1; Allergen 1, Ingestion 1.
  Acute Systemic: U.
  Chronic Local: Allergen 1.
  Chronic Systemic: U.
Fire Hazard: Slight; when heated (section 6).
Countermeasures
Ventilation Control: Section 2.
Personal Hygiene: Section 3.
Storage and Handling: Section 7.

BENZOL. See benzene.
  Shipping Regulations: Section 11.
  I.C.C.: Flammable liquid; red label, 10 gallons.
  Coast Guard Classification: Inflammable liquid; red label.
  MCA warning label.

160° BENZOL. See naphtha (coal tar).

BENZOL DILUENT
General Information
Constant: Flash p: −25°F, autoign. temp. = 450°F (these values will vary depending on the manufacturer).
Hazard Analysis
Toxic Hazard Rating: U.
Fire Hazard: Dangerous, when exposed to heat or flame.
Countermeasures
To Fight Fire: Water, foam (Section 6).
Storage and Handling: Section 7.

BENZONITRILE
General Information
Synonym: Phenyl cyanide.
Description: Transparent, colorless oil; almond-like odor.

Formula: C_6H_5CN.
Constants: Mol wt: 135.2, d: 1.246 at 20°/4°C, bp: 228°C d: 1.0102 at 15°/15°C, vap. press.: 1 mm at 28.2°C, flash p: none discernible.
Hazard Analysis and Countermeasures
See nitriles.

BENZOPHENONE
General Information
Synonyms: Phenyl ketone; diphenyl ketone.
Description: Rhombic white crystals; persistent rose-like odor.
Formula: $C_6H_5COC_6H_5$.
Constants: Mol wt: 182.21, mp: α = 49°C; β = 26°C; γ = 47°C, bp: 305.4°C, d: α form = 1.0976 at 50°/50°C; β form = 1.108 at 23°/4°C, vap. press.: 1 mm at 108.2°C.
Hazard Analysis
Toxicity: Details unknown. See also ketones.
Fire Hazard: Slight; when heated. Can react with oxidizing materials (Section 6).
Countermeasures
Storage and Handling: Section 7.

BENZOPYRENE. See 3,4-benzpyrene.

1,2-BENZOPYRONE. See coumarin.

BENZOQUINONE. See quinone.

BENZOSULFINIDE. See saccharin.

BENZOTHIAZOLE
General Information
Description: Liquid, odor of quinoline, slightly water soluble.
Formula: C_7H_5NS.
Constants: Mol wt: 135.2, d: 1.246 at 20°/4°C, bp: 228°C at 765 mm Hg.
Hazard Analysis
Toxic Hazard Rating: U.
Toxicity: No data available on human toxicity. LD_{50} for mice has been found to be 100 mg/kg intravenously. This suggests high toxicity.
Disaster Hazard: Dangerous; see sulfides, cyanides.

2-BENZOTHIAZYL DISULFIDE
General Information
Description: Cream to light yellow powder.
Formula: $(C_6H_4SNCS)_2$.
Constants: Mol wt: 332.5, mp: 175°C, d: 1.5.
Hazard Analysis
Toxicity: Unknown.
Disaster Hazard: See sulfur compounds.
Countermeasures
Storage and Handling: Section 7.

1,2,3-BENZOTRIAZOLE
General Information
Synonym: Aziminobenzene.
Description: Needle-like crystals.
Formula: C_6H_4NHNN.
Constants: Mol wt: 119.1, mp: 100°C, bp: 204°C at 15 mm.
Hazard Analysis
Toxicity: Unknown. Limited animal experiments indicate moderate toxicity.
Disaster Hazard: Moderately dangerous; when heated to decomposition, emits toxic fumes.
Countermeasures
Storage and Handling: Section 7.

BENZOTRICHLORIDE
General Information
Synonym: α-Trichlorotoluene, phenylchloroform.
Description: Clear, colorless to yellowish liquid; penetrating odor.
Formula: $C_6H_5CCl_3$.

Constants: Mol wt: 195.46, mp: $-22°C$, bp: $214°C$, d: 1.38 at $15.5°/15.5°C$, vap. d.: 6.77.

Hazard Analysis

Toxic Hazard Rating:

Acute Local: Irritant 2; Ingestion 2; Inhalation 2.

Acute Systemic: Ingestion 2; Inhalation 2.

Chronic Local: U.

Chronic Systemic: U.

Toxicology: Can cause local irritation of skin and mucous membranes. Large doses can cause CNS depression.

Disaster Hazard: See hydrochloric acid.

Countermeasures

Ventilation Control: Section 2.

Personnel Protection: Section 3.

Storage and Handling: Section 7.

BENZOTRIFLUORIDE

General Information

Synonym: Trifluoromethylbenzene.

Description: Water-white liquid; aromatic odor.

Formula: $C_6H_5CF_3$.

Constants: Mol wt: 146.1, mp: $-28.5°C$, bp: $101°C$, flash p: $54°F$ (C.C.), d: 1.197 at $15.5°/15.5°C$, vap. d.: 5.04, vap. press.: 11 mm Hg at $0°C$.

Hazard Analysis

Toxic Hazard Rating:

Acute Local: U.

Acute Systemic: Ingestion 2; Inhalation 2.

Chronic Local: U.

Chronic Systemic: U.

Toxicity: In animal experiments has caused CNS depression. Slightly toxic by ingestion based on animal experiments.

Fire Hazard : Dangerous; when exposed to heat or flame.

Explosion Hazard: U.

Disaster Hazard: Dangerous; see fluorides; it can react vigorously with oxidizing materials.

Countermeasures

Storage and Handling: Section 7.

To Fight Fire: Water, foam, carbon dioxide, dry chemical or carbon tetrachloride (Section 6).

trans-β-BENZOYL ACRYLIC ACID. See β-benzoyl acrylic acid.

β-BENZOYL ACRYLIC ACID

General Information

Synonym: trans-β-Benzoyl acrylic acid.

Description: Straw-yellow crystals.

Formula: $C_6H_5COCHCHCOOH$.

Constants: Mol wt: 176.2, mp: $99°C$.

Hazard Analysis

Toxicity: Unknown. See also acrylic acid.

Fire Hazard: Slight; when heated.

Countermeasures

Storage ahd Handling: Section 7.

BENZOYL AMINO METHOXY CHLOR ANTHRAQUINONE

Hazard Analysis

Toxic Hazard Rating:

Acute Local: Allergen 1.

Acute Systemic: U.

Chronic Local: Allergen 1.

Chronic Systemic: U.

Disaster Hazard: Dangerous; see chlorides.

Countermeasures

Personal Hygiene: Section 3.

Storage and Handling: Section 7.

BENZOYL CHLORIDE

General Information

Synonym: Benzene carbonyl chloride.

Description: Colorless, fuming, pungent liquid, decomposes in water.

Formula: C_6H_5COCl.

Constants: Mol wt: 140.5, mp: $-0.5°C$, bp: $197°C$, flash p: $162°F$ (C.C.), d: 1.2187 at $15°/15°C$, vap. press.: 1 mm at $32.1°C$, vap. d.: 4.88.

Hazard Analysis

Toxic Hazard Rating:

Acute Local: Irritant 2; Ingestion 2; Inhalation 2.

Acute Systemic: U.

Chronic Local: Irritant 1.

Chronic Systemic: U.

Fire Hazard: Slight, when exposed to heat or flame.

Disaster Hazard: Dangerous; See chlorides. Will react with water or steam to produce heat and toxic and corrosive fumes; can react vigorously with oxidizing materials.

Countermeasures

Ventilation Control: Section 2.

Personnel Protection: Section 3.

To Fight Fire: Foam, carbon dioxide, dry chemical or carbon tetrachloride (Section 6).

Storage and Handling: Section 7.

Shipping Regulations: Section 11.

I.C.C.: Corrosive liquid; white label, 1 quart.

Coast Guard Classification: Corrosive liquid; white label.

MCA warning label.

IATA: Corrosive liquid, white label, 1 liter (passenger and cargo).

BENZOYL HYDROPEROXIDE. See perbenzoic acid.

BENZOYL PEROXIDE, DRY

General Information

Synonym: Lucidol.

Description: White, granular, tasteless, odorless. Not soluble in water. Soluble in benzene, acetone, chloroform.

Formula: $(C_6H_5CO)_2O_2$.

Constants: Mol wt: 242.22, mp: $103-105°C$ (decomp.), bp: decomposes explosively, autoign. Temp.: $176°F$.

Hazard Analysis

Toxic Hazard Rating:

Acute Local: Irritant 2; Ingestion 2; Inhalation 2.

Acute Systemic: U.

Chronic Local: Irritant 2.

Chronic Systemic: U.

Fire Hazard: Moderate, by spontaneous chemical reaction in contact with reducing agents, a powerful oxidizer.

Explosion Hazard: Moderate, when heated to above melting point or when overheated under confinement. Reacts violently in contact with various organic or inorganic acids, alcohols, amines, metallic naphthenates as well as with polymerization accelerators, i.e. dimethylaniline.

Disaster Hazard: Dangerous. May explode spontaneously.

Countermeasures

Personnel Protection: Section 3.

To Fight Fire: Water spray, foam (Section 6).

Storage and Handling: All precautions must be taken to guard against fire and explosion hazards. Keep in a cool place, out of the direct rays of the sun, away from sparks, open flames and other sources of heat; away from shock, rough handling, friction from grinding, etc. Isolated storage is required away from possible contact with acids, alcohols, ethers or other reducing agents or polymerization catalysts such as dimethyl-

---

TOXIC HAZARD RATING CODE (For detailed discussion, see Section 1.)

0 NONE: (a) No harm under any conditions; (b) Harmful only under unusual conditions or overwhelming dosage.

1 SLIGHT: Causes readily reversible changes which disappear after end of exposure.

2 MODERATE: May involve both irreversible and reversible changes; not severe enough to cause death or permanent injury.

3 HIGH: May cause death or permanent injury after very short exposure to small quantities.

U UNKNOWN: No information on humans considered valid by authors.

aniline. Complete instructions on storage and handling available from manufacturer. See also Section 7.

Shipping Regulations: Section 11.
Freight and Express: Yellow label.
Rail Express: Maximum weight; 25 lbs. per shipping case.
Parcel Post: Prohibited.
ICC: Oxidizing material, yellow label, 25 pounds.
IATA: Oxidizing material, yellow label, not acceptable (passenger), 12 kilograms (cargo).

BENZOYL PEROXIDE, WET
General Information
Description: A paste or wetted granular material containing at least 30% water.
Constant: Autoign. temp.: 176°F.
Hazard Analysis
Toxic Hazard Rating: See benzoyl peroxide, dry.
Toxicity: See benzoyl peroxide, dry.
Fire Hazard: Moderate by chemical reaction with reducing agents; a powerful oxidizer.
Disaster Hazard: See benzoyl peroxide, dry.
Note: Mixed with a large surplus of water (i.e., 30%) this material is relatively safe. It is most dangerous when it contains very low percentages of water (1% or less).
Countermeasures
Ventilation Control: Section 2.
Personnel Protection: Section 3.
To Fight Fire: Water, foam or spray (Section 6).
Storage and Handling: Care must be taken to prevent drying out of wet material. See benzoyl peroxide, dry. Section 7.
Shipping Regulations: Section 11.
With not less than 30% water.
Coast Guard Classification: Inflammable solid; yellow label.

3, 4-BENZYPYRENE
General Information
Synonym: Benzopyrene.
Description: Yellowish crystals. Insoluble in water. Soluble in benzene, toluene, xylene.
Formula: $C_{20}H_{12}$.
Constants: Mol wt: 252.3, mp: 179°C, bp: 312°C at 10 mm Hg.
Hazard Analysis
Toxic Hazard Rating: U.
Toxicology: Has been incriminated as a carcinogen. A common air contaminant (Section 4).

2,3-BENZPYROL. See indole.

BENZYL ACETATE
General Information
Description: Liquid.
Formula: $CH_3COOCH_2C_6H_5$.
Constants: Mol wt: 150.17, mp: −51.5°C, bp: 213.5°C, flash p: 216°F (C.C.), d: 1.06, autoign. temp.: 862°F, vap. press.: 1 mm at 45.0°C, vap. d.: 5.1.
Hazard Analysis
Toxic Hazard Rating:
Acute Local: Irritant 2; Ingestion 2; Inhalation 2.
Acute Systemic: U.
Chronic Local: Irritant 1.
Chronic Systemic: U.
See also esters.
Fire Hazard: Slight, when exposed to heat or flame; can react with oxidizing materials.
Spontaneous Heating: No.
Countermeasures
Ventilation Control: Section 2.
Personnel Protection: Section 3.
To Fight Fire: Foam, carbon dioxide or carbon tetrachloride (Section 6).
Storage and Handling: Section 7.
To Fight Fire: Water, foam, alcohol foam (Section 6).

BENZYL ALCOHOL
General Information
Synonyms: Hydroxytoluene; phenyl carbinol.
Description: Water-white liquid; faint aromatic odor.
Formula: $C_6H_5CH_2OH$.
Constants: Mol wt: 108.13, mp: −15.3°C, bp: 205.7°C, flash p: 213°F (C.C.), d: 1.050 at 15°/15°C, autoign. temp.: 817°F, vap. press.: 1 mm at 58.0°C, vap. d.: 3.72.
Hazard Analysis
Toxic Hazard Rating:
Acute Local: Irritant 1; Allergen 1; Ingestion 1; Inhalation 1.
Acute Systemic: Ingestion 1; Inhalation 1; Skin Absorption 1.
Chronic Local: Irritant 1; Allergen 1.
Chronic Systemic: U.
Toxicology: In addition to local irritation of skin and mucous membranes it can cause headaches, vertigo, nausea, vomiting and diarrhea.
Fire Hazard: Slight, when exposed to heat or flame; can react with oxidizing materials.
Spontaneous Heating: No.
Countermeasures
To Fight Fire: Foam, carbon dioxide, dry chemical or carbon tetrachloride. Section 6.
Ventilation Control: Section 2.
Personal Hygiene: Section 3.
Storage and Handling: Section 7.

BENZYLAMINE
General Information
Synonym: Aminotoluene.
Description: Strongly alkaline liquid, miscible with water, alcohol and ether.
Formula: $C_6H_5 \cdot CH_2 \cdot NH_2$.
Constants: Mol wt: 107.2, d: 0.983 at 19°/4°C, bp: 185°C.
Hazard Analysis
Toxic Hazard Rating:
Acute Local: Irritant 3; Ingestion 3; Inhalation 3.
Acute Systemic: Unknown.
Chronic Local: Irritant 2; Inhalation 2.
Chronic Systemic: Unknown.
See also amines.
Disaster Hazard: Moderately dangerous. When heated to decomposition it emits toxic fumes.

BENZYL BENZENE. See diphenyl methane.

BENZYL BENZOATE
General Information
Description: Liquid.
Formula: $C_6H_5COOCH_2C_6H_5$.
Constants: Mol wt: 212.2, mp: 21°C, bp: 324°C, flash p: 298°F (C.C.), d: 1.114, vap. d.: 7.3, autoign. temp.: 898°F.
Hazard Analysis
Toxic Hazard Rating:
Acute Local: Allergen 1; Ingestion 1; Inhalation 1.
Acute Systemic: U.
Chronic Local: Allergen 1.
Chronic Systemic: U.
Fire Hazard: Slight, when exposed to heat or flame; can react with oxidizing materials.
Spontaneous Heating: No.
Countermeasures
To Fight Fire: Water, foam, carbon dioxide, dry chemical or carbon tetrachloride (Section 6).
Ventilation Control: Section 2.
Personal Hygiene: Section 3.
Storage and Handling: Section 7.

BENZYL BROMIDE
General Information
Synonym: α-Bromotoluene.

Description: Clear refractive liquid; pleasant odor. Lachrymator. Insoluble in water.
Formula: $C_6H_5CH_2Br$.
Constants: Mol wt: 171.04, mp: $-4.0°C$, bp: $198°C$, d: 1.438 at $22°C/0°C$, vap. d.: 5.8.
Hazard Analysis
Toxic Hazard Rating:
Acute Local: Irritant 3; Ingestion 3; Inhalation 3.
Acute Systemic: Ingestion 3; Inhalation 3; Skin Absorption 3.
Chronic Local: U.
Chronic Systemic: U.
Toxicology: Intensely irritating to skin, eyes and mucous membranes. Large doses cause CNS depression.
Disaster Hazard: Dangerous; See bromides.
Countermeasures
Ventilation Control: Section 2.
Personnel Protection: Section 3.
First Aid: Section 1.
Storage and Handling: Section 7.
Shipping Regulations: Section 11.
I.C.C.: Corrosive liquid; white label, 5 pints.
Coast Guard Classification: Corrosive liquid; white label.
IATA: Corrosive liquid, white label, not acceptable (passenger), $2\frac{1}{2}$ liters (cargo).

BENZYL CARBINOL. See phenethyl alcohol.

BENZYL "CELLOSOLVE"
General Information
Synonym: Ethylene glycol monobenzyl ether.
Description: Water-white liquid; faint rose-like odor.
Formula: $C_6H_5CH_2OCH_2CH_2OH$.
Constants: Mol wt: 152.19, mp: $-75°C$, bp: $256°C$, flash p: $265°F$ (O.C.), d: 1.068, autoign. temp.: $665°F$, vap. d.: 5.25.
Hazard Analysis
Toxic Hazard Rating:
Acute Local: Irritant 1; Ingestion 2; Inhalation 1.
Acute Systemic: Ingestion 2; Inhalation 1.
Chronic Local: Irritant 1.
Chronic Systemic: U.
See also glycols.
Fire Hazard: Slight, when exposed to heat or flame; can react with oxidizing materials.
Spontaneous Heating: No.
Countermeasures
To Fight Fire: Carbon dioxide, dry chemical or carbon tetrachloride (Section 6).
Ventilation Control: Section 2.
Personal Hygiene: Section 3.
Storage and Handling: Section 7.

BENZYL CHLORIDE
General Information
Synonym: α-Chlorotoluene.
Description: Colorless liquid; very refractive. Irritating, unpleasant odor.
Formula: $C_6H_5CH_2Cl$.
Constants: Mol wt: 126.58, mp: $-43°C$, bp: $179°C$, lel: 1.1%, flash p: $153°F$, d: 1.1026 at $18°/4°C$, autoign. temp. $1085°F$, vap. d.: 4.36.
Hazard Analysis
Toxic Hazard Rating:
Acute Local: Irritant 3; Ingestion 3; Inhalation 3.
Acute Systemic: Ingestion 3; Inhalation 3; Skin Absorption 3.

Chronic Local: U.
Chronic Systemic: U.
Toxicology: Intensely irritating. See benzyl bromide.
TLV: ACGIH (recommended) 1 part per million of air; 5 milligrams per cubic meter of air.
Fire Hazard: Moderate, when exposed to heat or flame.
Spontaneous Heating: No.
Explosion Hazard: Moderate, when exposed to flame. The decomposition can reach explosive violence in presence of metals such as iron.
Disaster Hazard: Dangerous; see chlorides. Will react with water or steam to produce toxic and corrosive fumes; can react vigorously with oxidizing materials.
Countermeasures
Ventilation Control: Section 2.
Personnel Protection: Section 3.
To Fight Fire: Foam, carbon dioxide, dry chemical or carbon tetrachloride (Section 6).
Storage and Handling: Section 7.
Shipping Regulations: Section 11.
I.C.C.: Corrosive liquid; white label, 1 quart.
MCA warning label.
IATA: Corrosive liquid, white label, 1 liter (passenger and cargo).

BENZYL CHLOROCARBONATE. See benzyl chloroformate.

BENZYL CHLOROFORMATE
General Information
Synonym: Benzyl chlorocarbonate.
Description: Colorless to pale yellow liquid; odor of phosgene.
Formula: $ClCOOC_6H_5CH_2$.
Constant: Mol wt: 170.6.
Hazard Analysis
Toxic Hazard Rating:
Acute Local: Irritant 2; Ingestion 2; Inhalation 2.
Acute Systemic: U.
Chronic Local: U.
Chronic Systemic: U.
Disaster Hazard: Dangerous; see chlorides; it will react with water or steam to produce heat, toxic and corrosive fumes.
Countermeasures
Ventilation Control: Section 2.
Personnel Protection: Section 3.
Storage and Handling: Section 7.
Shipping Regulations: Section 11.
I.C.C.: Corrosive liquid; white label, 5 pints.
Coast Guard Classification: Corrosive liquid; white label.
IATA: Corrosive liquid, white label, not acceptable (passenger), $2\frac{1}{2}$ liters (cargo).

o-BENZYL-p-CHLOROPHENOL
General Information
Synonym: Santophen.
Description: Nearly colorless flakes.
Formula: $C_{13}H_{11}OCl$.
Constants: Mol wt: 218.6, mp: $49°C$, bp: $175°C$ at 5 mm, d: 1.2 at $55°/25°C$.
Hazard Analysis
Toxic Hazard Rating:
Acute Local: Irritant 2; Ingestion 2; Inhalation 2.
Acute Systemic: U.
Chronic Local: U.
Chronic Systemic: U.

TOXIC HAZARD RATING CODE *(For detailed discussion, see Section 1.)*

0 NONE: (a) No harm under any conditions; (b) Harmful only under unusual conditions or overwhelming dosage.

1 SLIGHT: Causes readily reversible changes which disappear after end of exposure.

2 MODERATE: May involve both irreversible and reversible changes; not severe enough to cause death or permanent injury.

3 HIGH: May cause death or permanent injury after very short exposure to small quantities.

U UNKNOWN: No information on humans considered valid by authors.

Disaster Hazard: Dangerous; see chlorides.
Countermeasures
Ventilation Control: Section 2.
Personnel Protection: Section 3.
Storage and Handling: Section 7.

BENZYL CINNAMATE
General Information
Synonym: Cinnamein.
Description: White crystals; aromatic odor.
Formula: $C_9H_7O_2C_7H_7$.
Constants: Mol wt: 238.27, mp: 39°C, bp: 350.0°C, vap.
    press.: 1 mm at 173.8°C.
Hazard Analysis
Toxic Hazard Rating:
   Acute Local: Irritant 1; Allergen 1; Ingestion 1; Inhala-
     tion 1.
   Acute Systemic: U.
   Chronic Local: Allergen 1.
   Chronic Systemic: U.
Fire Hazard: Slight; when heated.
Countermeasures
Personal Hygiene: Section 3.
Storage and Handling: Section 7.

BENZYL CYANIDE. See phenyl acetonitrile.

BENZYL CYANIDE, LIQUID
Countermeasures
Shipping Regulations: Section 11.
   IATA: Poison B, poison label, 1 liter (passenger), 220
   liters (cargo).

BENZYL DICHLORIDE. See benzal chloride.

BENZYL DIETHYLAMINE
General Information
Description: Liquid.
Formula: $C_6H_5CH_2N(C_2H_5)_2$.
Constants: Mol wt: 163.26, bp: 207–216°C, flash p: 170°F
   (O.C.), d: 0.89. vap. d.: 5.6.
Hazard Analysis
Toxicity: Details unknown. See also amines.
Fire Hazard: Moderate, when exposed to heat or flame; can
   react with oxidizing materials.
Spontaneous Heating: No.
Countermeasures
To Fight Fire: Foam, carbon dioxide, dry chemical or
   carbon tetrachloride (Section 6).
Storage and Handling: Section 7.

BENZYL ETHER. See dibenzyl ether.

BENZYL ETHYL ETHER
General Information
Description: Oily liquid; aromatic odor; insoluble in water;
   miscible with alcohol and ether.
Formula: $C_6H_5CH_2OC_2H_5$.
Constants: Mol wt: 136.19, d: 0.949, bp: 186°C.
Hazard Analysis
Toxic Hazard Rating: Probably narcotic in high concentra-
   tions. May also be an irritant.

BENZYL FORMATE
Hazard Analysis
Toxic Hazard Rating: U. Probably narcotic in high con-
   centrations.

BENZYLIDINE CHLORIDE. See benzal chloride.

BENZYL IODIDE
General Information
Synonym: α-Iodotoluene.
Description: Colorless crystals.
Formula: $C_6H_5CH_2I$.
Constants: Mol wt: 218.05, mp: 24°C, bp: 93°C at
   10 mm, d: 1.733 at 25°/4°C.

Hazard Analysis
Toxic Hazard Rating:
   Acute Local: Irritant 2; Ingestion 2; Inhalation 2.
   Acute Systemic: U.
   Chronic Local: U.
   Chronic Systemic: U.
Disaster Hazard: Dangerous;See iodides.
Countermeasures
Ventilation Control: Section 2.
Personnel Protection: Section 3.
Storage and Handling: Section 7.

BENZYL ISOAMYL ETHER. See amyl benzyl ether.

BENZYL MERCAPTAN
General Information
Synonyms: α-toluene thiol; phenyl methane thiol.
Description: A water-white, mobile liquid; strong odor.
Formula: C_7H_8S.
Constants: Mol wt: 124.2, bp: 194.8°C, flash p: 158°F
   (C.C.), d: 1.058 at 20°C, vap. d.: 4.28.
Hazard Analysis
Toxicity: Details unknown. See also mercaptans and
   benzene.
Fire Hazard: Moderate, when exposed to heat or flame.
Disaster Hazard: Dangerous; when heated to decomposition
   and on contact with acid or acid fumes, emits highly
   toxic fumes; can react vigorously with oxidizing ma-
   terials.
Countermeasures
Ventilation Control: Section 2.
To Fight Fire: Foam, carbon dioxide, dry chemical or
   carbon tetrachloride (Section 6).
Storage and Handling: Section 7.

BENZYL METHYL ETHER
General Information
Synonym: Methyl benzyl ether.
Description: Colorless liquid; insoluble in water; soluble in
   alcohol, ether.
Formula: $C_6H_5CH_2OCH_3$.
Constants: Mol wt: 122.16, d:0.987, bp: 174°C.
Hazard Analysis
Toxic Hazard Rating: Probably has narcotic properties
   typical of ether.

BENZYL MUSTARD OIL. See benzyl thiocyanate.

BENZYL NICOTINIUM CHLORIDE
General Information
Description: Crystals.
Formula: $C_{10}H_{11}N_2Cl(C_6H_5CH_2Cl)$.
Constant: Mol wt: 288.8.
Hazard Analysis
Toxic Hazard Rating:
   Acute Local: Irritant 2: Allergen 1; Ingestion 3; In-
    halation 2.
   Acute Systemic: Ingestion 3; Inhalation 3; Skin Ab-
    sorption 3.
   Chronic Local: Irritant 2; Allergen 1.
   Chronic Systemic: Ingestion 3; Inhalation 3; Skin Ab-
    sorption 3.
Disaster Hazard: Dangerous; See chlorides.
Countermeasures
Ventilation Control: Section 2.
Personnel Protection: Section 3.
First Aid: Section 1.
Storage and Handling: Section 7.

p-BENZYL OXYPHENOL. See hydroquinone mono-
   benzyl ether.

o-BENZYL PHENOL
General Information
Synonym: (2-Hydroxydiphenyl) methane.
Description: Crystals or liquid.

Formula: $C_6H_5CH_2C_6H_4OH$.
Constants: Mol wt: 184.2, mp: 20°C, bp: 154–156°C at 10 mm.
Hazard Analysis
Toxicity: Details unknown. See also phenol.
Disaster Hazard: Dangerous; when heated to decomposition, emits highly toxic fumes.
Countermeasures
Storage and Handling: Section 7.

p-BENZYL PHENOL
General Information
Synonym: (4-Hydroxydiphenyl) methane.
Description: Crystals; colorless to slightly pink powder; faintly pleasant phenolic odor.
Formula: $C_6H_5CH_2C_6H_4OH$.
Constants: Mol wt: 184.23, mp: 84°C, bp: 313°C, d: 1.10 at 25°/25°C.
Hazard Analysis
Toxicity: Details unknown. See also phenol.
Disaster Hazard: Dangerous; when heated to decomposition, emits highly toxic fumes.
Countermeasures
Storage and Handling: Section 7.

BENZYL PYRIDINE
General Information
Formula: $C_6H_5CH_2C_5H_4N$.
Constant: Mol wt: 169.2.
Hazard Analysis
Toxicity: Details unknown. See also pyridine.
Fire Hazard: A moderately flammable material (Section 6).
Disaster Hazard: Dangerous; when heated to decomposition, emits highly toxic fumes; can react vigorously with oxidizing materials.
Countermeasures
Storage and Handling: Section 7.

BENZYL SALICYLATE
General Information
Description: Thick liquid; pleasant odor.
Formula: $HOC_6H_4CO_2CH_2C_6H_5$.
Constants: Mol wt: 228.2, bp: 208°C at 26 mm, d: 1.175 at 20°C.
Hazard Analysis
Toxicity: Details unknown. See also benzyl alcohol and salicylic acid.
Fire Hazard: Slight when exposed to heat or flame; can react with oxidizing materials (Section 6).
Countermeasures
Storage and Handling: Section 7.

BENZYL THIOCYANATE
General Information
Synonym: Benzyl mustard oil.
Description: Orange-red crystalline solid.
Formula: $C_6H_5CH_2CNS$.
Constants: Mol wt: 149.2, mp: 41°C, bp: 230°C, d: 1.125.
Hazard Analysis
Toxicity: Details unknown. See also thiocyanates.
Fire Hazard: Moderately flammable (Section 6).

BENZYLTRIMETHYL AMMONIUM CHLORIDE
General Information
Description: Thick liquid.
Formula: $C_6H_5CH_2N(CH_3)_3Cl$.

Constants: Mol wt: 185.7, bp: >135°C (decomposes), fp: < −50°C (61% sol.), d: 1.07 at 20°/20° (61% sol.).
Hazard Analysis
Toxicity: Unknown.
Fire Hazard: Slight, when exposed to heat or flame; can react with oxidizing materials (Section 6).
Countermeasures
Storage and Handling: Section 7.

N-BENZYL-N,N,N-TRIMETHYL AMMONIUM HEXAFLUOPHOSPHATE
General Information
Description: Crystals.
Formula: $C_6H_5CH_2N(CH_3)_3PF_6$.
Constants: Mol wt: 295, mp: 1.52°C.
Hazard Analysis
Toxicity: Unknown. See also fluorides and phosphates.
Disaster Hazard: Dangerous; when heated to decomposition, or on contact with acid or acid fumes, emits highly toxic fumes.
Countermeasures
Storage and Handling: Section 7.

BENZYL TRIMETHYL AMMONIUM HYDROXIDE
General Information
Description: Solid.
Formula: $C_6H_5CH_2N(CH_3)_3OH$.
Constant: Mol wt: 150.3.
Hazard Analysis
Toxic Hazard Rating:
Acute Local: Irritant 2; Ingestion 2.
Acute Systemic: Ingestion 3.
Chronic Local: Irritant 2.
Chronic Systemic: U.
Fire Hazard: Moderate, when exposed to heat; can react with oxidizing materials; products of heat decomposition are flammable (Section 6).
Countermeasures
Personnel Protection: Section 3.
First Aid: Section 1.
Storage and Handling: Section 7.

BENZYL TRIPHENYL GERMANIUM
General Information
Description: Colorless crystals; insoluble in water. Soluble in organic solvents.
Formula: $(CH_2C_6H_5)(C_6H_5)_3Ge$.
Constants: Mol wt: 395.0, mp: 83.0°C.
Hazard Analysis
Toxicity: Details unknown. See germanium compounds.

BERBERINE
General Information
Description: White to yellow crystals.
Formula: $C_{20}H_{17}NO_4 \cdot 6H_2O$.
Constants: Mol wt: 443.44, mp: anh. 145°C.
Hazard Analysis
Toxic Hazard Rating:
Acute Local: U.
Acute Systemic: Ingestion 3; Inhalation 3.
Chronic Local: U.
Chronic Systemic: U.
Toxicology: An alkaloid poison. In toxic doses it lowers the temperature, increases peristalsis, and causes death by central paralysis.
Disaster Hazard: Dangerous; when heated to decomposition, emits highly toxic fumes.

TOXIC HAZARD RATING CODE *(For detailed discussion, see Section 1.)*

0 NONE: (a) No harm under any conditions; (b) Harmful only under unusual conditions or overwhelming dosage.

1 SLIGHT: Causes readily reversible changes which disappear after end of exposure.

2 MODERATE: May involve both irreversible and reversible changes; not severe enough to cause death or permanent injury.

3 HIGH: May cause death or permanent injury after very short exposure to small quantities.

U UNKNOWN: No information on humans considered valid by authors.

Countermeasures
Ventilation Control: Section 2.
Personal Hygiene: Section 3.
Storage and Handling: Section 7.
> Should carry a poison label. Should never be ingested without the advice of a physician. Should not be handled excessively, since it may be absorbed through the skin and have a toxic effect upon the body.

BERBERINE COMPOUNDS. See berberine.

BERGAMOT OIL
General Information
Description: Yellow-green liquid; agreeable odor.
Composition: 1-Linalyl acetate, 1-linalool, d-limonene, dipentene, bergaptene.
Constant: D: 0.875–0.880 at 25°/25°C.
Hazard Analysis
Toxic Hazard Rating:
  Acute Local: Allergen 1.
  Acute Systemic: U.
  Chronic Local: Allergen 1.
  Chronic Systemic: U.
Fire Hazard: Slight (Section 6).
Countermeasures
Personal Hygiene: Section 3.
Storage and Handling: Section 7.

BERKELIUM
General Information
Description: A synthetic radioactive element.
Formula: Bk.
Constant: At. wt: 247.
Hazard Analysis
Radiation Hazard: Section 5. For permissible levels, see Table 5, p. 150.
> Artificial isotope ^{249}Bk, half life about 300 d. Decays to radioactive ^{249}Cf by emitting beta particles of 0.1 MeV.
> Artificial isotope ^{250}Bk, half life 3 h. Decays to radioactive ^{250}Cf by emitting beta particles of 0.9, 1.9 MeV. Also emits gamma rays of 0.9 MeV.

BERLIN GREEN. See ferric ferricyanide.

BERYL
General Information
Description: Green, blue, yellow or white crystals.
Formula: $3BeO \cdot Al_2O_3 \cdot 6SiO_2$.
Constant: D: 2.63–2.91.
Hazard Analysis
Toxicity: Details unknown. Probably has little toxicity of itself as it has not been known to cause pneumoconiosis. See also beryllium.
Countermeasures
Ventilation Control: Section 2.

BERYLLIUM
General Information
Synonym: Glucinum.
Description: A grayish-white, hard light metal.
Formula: Be.
Constants: At wt: 9.013, mp: 1278°C, bp: 2970°C, d: 1.85.
Hazard Analysis
Toxic Hazard Rating:
  Acute Local: Inhalation 3.
  Acute Systemic: Inhalation 3.
  Chronic Local: Irritant 1; Inhalation 3.
  Chronic Systemic: Ingestion 3; Inhalation 3.
Radiation Hazard: Section 5. For permissible levels, see Table 5, p. 150.
> Artificial and natural isotope ^{7}Be, half life 53 d. Decays to stable ^{7}Li by electron capture. Emits gamma rays of 0.48 MeV and X-rays.
> Artificial isotope ^{10}Be, half life about 2.7×10^6 y.

Decays to stable ^{10}B by emitting beta particles of 0.56 MeV.
TLV: ACGIH (recommended) 0.002 mg cubic meter of air.
Toxicology: Beryllium is a common air contaminant (Section 4).
Fire Hazard: Moderate, in form of dust or powder; or when exposed to flame by spontaneous chemical reaction (Section 6).
Explosion Hazard: Slight, in the form of powder or dust.
Countermeasures
Ventilation Control: Section 2.
Personal Hygiene: Section 3.
First Aid: Section 1.
Storage and Handling: Section 7.
Shipping Regulations: Section 11.
> I.C.C. Classification: Metal powder, poison B; poison label.
> Coast Guard Classification: Metal powder, poison B; poison label.

BERYLLIUM ACETATE
General Information
Description: Plates.
Formula: $Be(C_2H_3O_2)_2$.
Constants: Mol wt: 127.10, mp: decomposes 300°C.
Hazard Analysis
Toxicity: See beryllium compounds.

BERYLLIUM ACETYLACETONATE
General Information
Description: Crystals. Practically insoluble in water. Freely soluble in alcohol, acetone, ether benzene, carbon disulfide and other organic solvents.
Formula: $C_{10}H_{11}BeO_4$
Constants: Mol wt: 207.23, mp: 108°C, bp: 270°C.
Hazard Analysis
See beryllium compounds.

BERYLLIUM ALGINATE
Toxicity: See beryllium compounds.

BERYLLIUM ALUMINATE
General Information
Synonym: Chrysoberyl.
Description: Rhombic crystals.
Formula: $BeAl_2O_4$.
Constants: Mol wt: 126.95, d: 3.76.
Hazard Analysis
Toxicity: See beryl.
Countermeasures
Ventilation Control: Section 2.

BERYLLIUM ALUMINUM SILICATE
General Information
Synonym: Euclase.
Description: Monoclinic crystals.
Formula: $Be_2Al_2(SiO_4)_2(OH)_2$.
Constants: Mol wt: 290.12, d: 3.1.
Hazard Analysis
Toxicity: See beryl.
Countermeasures
Ventilation Control: Section 2.

BERYLLIUM BENZENESULFONATE
General Information
Description: Monoclinic crystals.
Formula: $Be(C_6H_5O_3S)_2$.
Constant: Mol wt: 323.35.
Hazard Analysis
Toxicity: See beryllium compounds.
Disaster Hazard: Dangerous; when heated to decomposition, emits highly toxic fumes of beryllium oxide and oxides of sulfur.
Countermeasures
Storage and Handling: Section 7.

BERYLLIUM o-BORATE, BASIC
General Information
Synonym: Hambergite.
Description: Rhombic crystals.
Formula: $Be_2(OH)BO_3$.
Constants: Mol wt: 93.85, d: 2.35.
Hazard Analysis
Toxicity: See beryllium compounds.

BERYLLIUM BOROHYDRIDE
General Information
Description: Solid.
Formula: $Be(BH_4)_2$.
Constants: Mol wt: 38.72, mp: 123°C, sublimes at 91°C, vap. press.: 10 mm at 28.1°C.
Hazard Analysis and Countermeasures
See beryllium compounds, boron compounds and hydrides.

BERYLLIUM BROMIDE
General Information
Description: White deliquescent needles.
Formula: $BeBr_2$.
Constants: Mol wt: 168.85, mp: 490°C; sublimes at 474°C, bp: 520°C, d: 3.465 at 25°C.
Hazard Analysis and Countermeasures
See beryllium compounds and bromides.

BERYLLIUM BUTYRATE, BASIC
General Information
Description: Crystals.
Formula: $Be_4O(C_4H_7O_2)_6$.
Constants: Mol wt: 574.6, bp: 239 at 19 mm.
Hazard Analysis
Toxicity: See beryllium compounds.

BERYLLIUM CARBIDE
General Information
Description: Hexagonal yellow crystals.
Formula: Be_2C.
Constants: Mol wt: 30.0, mp: >2100°C decomposes, d: 1.90 at 15°C.
Hazard Analysis
Toxicity: See beryllium compounds.
Fire Hazard: Details unknown. See also carbides.
Explosion Hazard: See carbides.
Disaster Hazard: Dangerous; will react with water or steam to produce flammable vapors and highly toxic fumes of beryllium oxides.
Countermeasures
Storage and Handling: Section 7.

BERYLLIUM CARBONATE, BASIC
General Information
Description: White powder.
Formula: $BeCO_3 + Be(OH)_2$.
Constant: Mol wt: 112.05.
Hazard Analysis
Toxicity: See beryllium compounds.

BERYLLIUM CHLORIDE
General Information
Description: Colorless deliquescent needles.
Formula: $BeCl_2$.
Constants: Mol wt: 79.93, mp: 440°C, bp: 520°C, d: 1.899 at 25°C, vap. press.: 1 mm at 291°C (sublimes).
Hazard Analysis
Toxicity: See beryllium compounds.
Shipping Regulations: Section 11.

IATA: Poison B, poison label, 25 kilograms (passenger), 95 kilograms (cargo).

BERYLLIUM COMPOUNDS
Hazard Analysis
Toxic Hazard Rating:
Acute Local: Irritant 3; Inhalation 3.
Acute Systemic: Inhalation 3.
Chronic Local: Irritant 1; Allergin 1; Inhalation 3.
Chronic Systemic: Inhalation 3.
Toxicology: The extraction of beryllium from its ore is attended by exposure to acid salts of the metal, particularly the fluoride (BeF_2), the ammonium fluoride, and the sulfate ($BeSO_4$), and also to beryllium oxide (BeO), and hydroxide [$Be(OH)_2$]. Exposure to the oxide also occurs in the casting of beryllium alloys and in operations with beryllia ceramics. In the manufacture of fluorescent powders, lamps and sign tubes there may be exposure to beryllium carbonate and to more complex salts, such as zinc manganese beryllium silicate. Even alloys of low beryllium content have been shown to be dangerous.

Beryllium compounds can enter the body through inhalation of the dusts and fumes and they may act locally on the skin. Exposure to beryllium compounds encountered in the extraction of the metal or its oxide from the ore, particularly the halide salts, has been attended, in certain individuals, by the development of dermatitis of an edematous and papulovesicular type (Section 9), chronic skin ulcers, rhinitis, nasopharyngitis, epistaxis, bronchitis and in severe cases, by the development of an acute pneumonitis, with cough, scanty sputum, low-grade fever, rales, dyspnea and substernal pain. Radiographs show diffuse haziness throughout both lungs, followed by the appearance of soft, ill-defined opacities. The condition usually occurs while the worker is exposed, sometimes within one or two months of starting work, and recovery occurs within two months, as a rule, though radiographic changes sometimes persist for longer periods. Certain investigators have reported occasional failure of complete resolution, followed by fibrosis. In severe cases of pneumonitis the patient may die. Necropsies have revealed diffuse pulmonary edema, hemorrhagic extravasation, large numbers of plasma cells and a relative absence of polymorphonuclear infiltration. On the basis of experimental work with animals, certain investigators are of the opinion that the acute upper and lower respiratory effects are due chiefly to the acid radical present in the dust or fume, but this view has little support.

A delayed form of lung disease, characterized by the occurrence of granulomatous areas in the lung tissue, has been reported in workers manufacturing fluorescent powders, lamps and sign tubes, casting beryllium master alloys, and in the production of beryllium from beryl ore. Symptoms can start during exposure, but they might be delayed up to 5 years or more after leaving work. The commonest symptoms are coughing, shortness of breath, loss of appetite, loss of weight, and fatigue. Rales are usually present in the bases and axillae, and the red cell count is frequently elevated. Cyanosis is common and the pulse and respiratory rates are often increased. Radiographically, three stages of the disease are described: (1) a diffuse, uniform granular shadowing extending throughout both lung fields; (2) a diffuse reticular pattern on the granular background;

TOXIC HAZARD RATING CODE (*For detailed discussion, see Section 1.*)

0 NONE: (a) No harm under any conditions; (b) Harmful only under unusual conditions or overwhelming dosage.

1 SLIGHT: Causes readily reversible changes which disappear after end of exposure.

2 MODERATE: May involve both irreversible and reversible changes; not severe enough to cause death or permanent injury.

3 HIGH: May cause death or permanent injury after very short exposure to small quantities.

U UNKNOWN: No information on humans considered valid by authors.

(3) the appearance of distinct nodules scattered through the lungs, with some enlargement and blurring of the hilar shadows. The intensity of the shadowing is usually greater in the middle third of the lung fields. The prognosis is poor. Clinical improvement may occur gradually over a period of several years, but there appears to be little tendency for the radiographic shadowing to clear. In certain cases, the disease has progressed gradually for some months or years, with death resulting from respiratory and cardiac failure. In several instances necropsies have shown the presence of a diffuse fibrosis with coarse strands of hyalinized collagen between the aleveoli and, in some places, replacing them. The hyalinized areas contained granulomatous foci, the alveolar walls are thickened and fibrosed, the blood vessels being engorged and dilated. In some cases the hilar lymph nodes show granulomatous change and fibrosis. Granulomatous change has also been noted in the liver and hyaline fibrosis in the spleen. Two cases of delayed lung disease not coming to autopsy have presented papular lesions on the dorsum of the hands; on the biopsy these showed "sacroid-like" lesions with central necrosis.

Several cases have been reported in which localized granulomatous lesions developed following penetrating wounds caused by splinters of glass from broken fluorescent light tubes. Several weeks or months following the accident, swellings were noted in the injured areas and excision revealed granulomatous tumors, which in one case was shown to contain beryllium. Several cases of beryllium granuloma have been reported in persons residing near processing plants and in families of beryllium workers.

There is no specific treatment, but temporary remissions have been produced by ACTH and cortisone.

Countermeasures
Ventilation Control: Section 2.
Personal Hygiene: Section 3.
First Aid: Section 1.
Shipping Regulations: Section 11.
  I.C.C. (N.O.S): Poison B, poison label, 200 pounds.
  IATA (N.O.S.): Poison B, poison label, 25 kilograms (passenger), 95 kilograms cargo.

BERYLLIUM-COPPER ALLOY
General Information
Description: A metallic alloy.
Formula: Be_xCu_y.
Hazard Analysis
Toxicity: Cases of berylliosis have been reported from exposure to so called low beryllium alloys. See beryllium compounds.

BERYLLIUM ETHYL. See diethyl beryllium.

BERYLLIUM ETHYLENE DIAMINE CHLORIDE
General Information
Description: Crystals, practically insoluble in water.
Formula: $[Be(NH_2CH_2CH_2NH_2)_2]$.
Constant: Mol wt: 200.13.
Hazard Analysis
See beryllium compounds.

BERYLLIUM FLUORIDE
General Information
Description: Amorphous, colorless mass.
Formula: BeF_2.
Constants: Mol wt: 47.02, mp: 800°C, d: 1.986 at 25°C.
Hazard Analysis
Toxicity: See beryllium compounds and fluorides.

BERYLLIUM FORMATE
General Information
Description: Crystals.
Formula: $[Be(COOH)_2]_x$.

Constants: Mol wt: $(99.1)_x$, mp: decomposes at 150°C.
Hazard Analysis
Toxicity: See beryllium compounds.
Disaster Hazard: Dangerous; when heated to decomposition, it emits highly toxic fumes of beryllium oxide.
Countermeasures
Storage and Handling: Section 7.

BERYLLIUM HYDRIDE
General Information
Description: White solid. Reacts with water, dilute acids, methanol.
Formula: BeH_2.
Constant: Mol wt: 11.0.
Hazard Analysis
Toxicity: See beryllium compounds and hydrides.
Fire Hazard: This material liberates hydrogen rapidly when heated to 220°C. Also it reacts with water, dilute acids and methanol to liberate hydrogen. See hydrogen and hydrides.
Countermeasures
Storage and Handling: See hydrides (Section 7).

BERYLLIUM HYDROXIDE
General Information
Description: Amorphous powder or crystals.
Formula: $Be(OH)_2$.
Constants: Mol wt: 43.04, mp: decomposes at 138°C, d: 1.909 (cr).
Hazard Analysis
Toxicity: See beryllium compounds.
Shipping Regulations: Section 11.
  IATA: Poison B, poison label, 25 kilograms (passenger), 95 kilograms (cargo).

BERYLLIUM IODIDE
General Information
Description: Colorless needles.
Formula: BeI_2.
Constants: Mol wt: 262.85, mp: 480°C, bp: 488°C, d: 4.325 at 25°C, vap. press.: 1 mm at 283°C.
Hazard Analysis
Toxicity: See beryllium compounds.

BERYLLIUM METAL POWDER, OR FLAKE
Shipping Regulations: Section 11.
  IATA: Poison B, poison label, 25 kilograms (passenger), 95 kilograms (cargo).

BERYLLIUM NITRATE
General Information
Description: White, yellowish crystals; deliquescent.
Formula: $Be(NO_3)_2 \cdot 3H_2O$.
Constants: Mol wt: 187.08, mp: 60°C, bp: decomposes 100–200°C.
Hazard Analysis and Countermeasures
See beryllium compounds and nitrates.
Shipping Regulations: Section 11.
  IATA: Oxidizing material, yellow label, 12 kilograms (passenger), 45 kilograms (cargo).

BERYLLIUM NITRIDE
General Information
Description: Cubic colorless crystals.
Formula: Be_3N_2.
Constants: Mol wt: 55.06, mp: 2200 ± 100°C, bp: decomposes 2240°C.
Hazard Analysis and Countermeasures
See beryllium compounds and nitrides.

BERYLLIUM OXALATE
General Information
Description: Rhombic crystals.
Formula: $BeC_2O_4 \cdot 3H_2O$.
Constants: Mol wt: 151.08, mp: $-2H_2O$ at 100°C; $-3H_2O$ at 220°C, bp: decomposes at 350°C.

Hazard Analysis
Toxicity: See beryllium compounds and oxalates.

BERYLLIUM OXIDE
General Information
Synonym: Bromellete.
Description: White or amorphous powder.
Formula: BeO.
Constants: Mol wt: 25.0, mp: $2530 \pm 30°$ C, bp: $3900°$ C
(approx.), d: 3.025.
Hazard Analysis
Toxicity: See beryllium compounds.
Shipping Regulations: Section 11.
  IATA: Poison B, poison label, 25 kilograms (passenger),
  95 kilograms (cargo).

BERYLLIUM 2,4-PENTANEDIONE DERIVATIVE
General Information
Description: Monoclinic white crystals.
Formula: $Be(C_5H_7O_2)_2$.
Constants: Mol wt: 207.23, mp: $108°$ C, bp: $270°$ C,
d: 1.168 at $4°$ C.
Hazard Analysis
Toxicity: See beryllium compounds.

BERYLLIUM PERCHLORATE
General Information
Description: Very hygroscopic crystals; solubility in
water: 148.6 g/100 ml.
Formula: $Be(ClO_4)_2 \cdot 4H_2O$.
Constant: Mol wt: 279.49.
Hazard Analysis and Countermeasures
See beryllium compounds and perchlorates.

BERYLLIUM m-PHOSPHATE
General Information
Description: Crystals.
Formula: $Be(PO_3)_2$.
Constant: Mol wt: 167.0.
Hazard Analysis and Countermeasures
See beryllium compounds and phosphates.

BERYLLIUM o-PHOSPHATE
General Information
Description: Crystals.
Formula: $Be_3(PO_4)_2 \cdot 3H_2O$.
Constants: Mol wt: 271.05, mp: $-H_2O$ at $100°$ C.
Hazard Analysis and Countermeasures
See beryllium compounds and phosphates.

BERYLLIUM POTASSIUM FLUORIDE
General Information
Synonym: Glucinum potassium fluoride.
Description: White crystal masses.
Formula: $GlF_2 \cdot (KF)_2$; $BeF_2 \cdot (KF)_2$.
Constant: Mol wt: 163.30.
Hazard Analysis and Countermeasures
See beryllium compounds and fluorides.

BERYLLIUM POTASSIUM SULFATE
General Information
Description: Colorless, brilliant crystals; sparingly soluble
in water, soluble in concentrated K_2SO_4 solution;
insoluble in alcohol.
Formula: $K_2Be(SO_4)_2 \cdot 2H_2O$.
Constant: Mol wt: 315.37.
Hazard Analysis and Countermeasures
See beryllium compounds and sulfates.

BERYLLIUM PROPIONATE ACETATE, BASIC
General Information
Description: Crystals.
Formula: $Be_4O(C_2H_3O_2)_3 - (C_3H_5O_2)_3$.
Constants: Mol wt: 448.39, mp: $127°$ C, bp: $330°$ C.
Hazard Analysis
Toxicity: See beryllium compounds.

BERYLLIUM SELENATE
General Information
Description: Orthorhombic crystals; freely soluble in water.
Formula: $BeSeO_4 \cdot 4H_2O$.
Constants: Mol wt: 224.04, d: 2.03.
Hazard Analysis and Countermeasures
See beryllium compounds and selenium.

BERYLLIUM SODIUM-FLUORIDE
General Information
Synonym: Glucinum sodium fluoride.
Description: White crystalline mass.
Formula: $BeF_2 \cdot (NaF)_2$.
Constant: Mol wt: 131.10.
Hazard Analysis and Countermeasures
See beryllium compounds and fluorides.

BERYLLIUM STEARATE
General Information
Description: White waxy crystals.
Formula: $Be(C_{18}H_{35}O_2)_2$.
Constants: Mol wt: 575.94, mp: $45°$ C.
Hazard Analysis
Toxicity: See beryllium compounds.

BERYLLIUM SULFATE
General Information
Description: Crystals.
Formula: $BeSO_4$.
Constants: Mol wt: 105.08, mp: $550–600°$ C (decomposes),
d: 2.443.
Hazard Analysis and Countermeasures
See beryllium compounds and sulfates.
Shipping Regulations: Section 11.
  IATA: Poison B, poison label, 25 kilograms (passenger),
  95 kilograms (cargo).

BERYLLIUM SULFIDE
General Information
Description: Solid crystalline mass.
Formula: BeS.
Constants: Mol wt: 41.08, d: 2.36.
Hazard Analysis and Countermeasures
See beryllium compounds and sulfides.

BERYLLIUM ZINC SILICATE
General Information
Description: Crystalline solid.
Formula: $BeZn(SiO_4)$.
Constant: Mol wt: 166.5.
Hazard Analysis
Toxicity: See beryllium compounds.

BETA RAYS
General Information
Description: Particulate radiation emitted by certain
radioactive isotopes. Beta rays consist of electrons
moving at high velocity.
Hazard Analysis
Radiation Hazard: See Section 5 for complete discussion.

TOXIC HAZARD RATING CODE (For detailed discussion, see Section 1.)

0 NONE: (a) No harm under any conditions; (b) Harmful only under unusual conditions or overwhelming dosage.

1 SLIGHT: Causes readily reversible changes which disappear after end
of exposure.

2 MODERATE: May involve both irreversible and reversible changes;
not severe enough to cause death or permanent injury.

3 HIGH: May cause death or permanent injury after very short exposure
to small quantities.

U UNKNOWN: No information on humans considered valid by authors.

BETEL. See areca nut.

BFE. See bromotrifluoroethylene.

BGE. See n-butyl glycidyl ether.

BHA. See butylated hydroxyanisole.

BHC. See benzene hexachloride.

BHT. See di-tert-butyl-p-cresol.

BICALCIUM PHOSPHATE. See calcium phosphate, dibasic.

BICHROMATE OF SODA. See sodium dichromate.

cis-BICYCLO(2,2,1,)-5-HEPTANE-2,3-DICARBOXYLIC ACID DIMETHYL ESTER. See dimethyl carbate.

BICYCLOHEXYL
General Information
Synonym: Dicyclohexyl.
Description: Colorless oil; pleasant odor.
Formula: $C_{12}H_{22}$.
Constants: Mol wt: 166.3, mp: 2°C, bp: 240°C, flash p: 165°F, d: 0.883 at 25°/15.6°C, autoign. temp.: 471°F, vap. d.: 5.73, lel: 0.7% at 212°, uel: 5.1% at 302°.
Hazard Analysis
Toxicity: Unknown. See also cyclohexanol.
Fire Hazard: Slight, when exposed to heat or flame; can react with oxidizing materials.
Countermeasures
Storage and Handling: Section 7.
To Fight Fire: Alcohol foam, foam, carbon dioxide, dry chemical or carbon tetrachloride (Section 6).
Countermeasures
Storage and Handling: Section 7.
To Fight Fire: Alcohol foam (Section 6).

BICYCLOHEPTADIENE DIBROMIDE
General Information
Formula: $(C_7H_9Br)_2$.
Constant: Mol wt: 346.
Hazard Analysis
Toxic Hazard Rating:
  Acute Local: Irritant 3; Allergen 3; Inhalation 3.
  Acute Systemic: Ingestion 2; Inhalation 3.
  Chronic Local: Allergen 3.
  Chronic Systemic: Ingestion 3; Inhalation 3.
Toxicity: Human exposure has resulted in severe dermatitis, asthma, and injury to blood forming organs.
Disaster Hazard: Dangerous. See bromides.

BIDRIN. See 3-hydroxy-n,n-dimethyl-cis-crotonamide-dimethyl phosphate.

BI-(ETHYL MERCURIC) PERTHIOCYANATE
Hazard Analysis and Countermeasures
See mercury compounds, organic, and thiocyanates.

BIFORMYL. See glyoxal.

BIG REDS. See explosives, high.

BIMETHYL. See ethane.

BIOTIN
General Information
Synonyms: Vitamine H; 2'-keto-3,4-imidazolido-2-tetrahydrothiophene-n-valeric acid.
Description: Biotin is frequently referred to as a member of the Vitamin B complex. White crystals; soluble in water and alcohol; insoluble in petroleum ether and chloroform.
Formula: $C_{10}H_{16}N_2O_3S$.
Constants: Mol wt: 244, mp: 230–232°C.

Hazard Analysis
Toxic Hazard Rating: U. A nutrient and/or dietary supplement food additive (Section 10).

BIPHENYL. See diphenyl.

p-BIPHENYLAMINE. See p-amino diphenyl.

2,4'-BIPHENYLDIAMINE
General Information
Synonyms: Diphenyline; 2,4'-diphenyldiamine; o,p'-dianiline; 2,4'-diaminodiphenyl.
Description: Needles; very slightly soluble in alcohol and ether.
Formula: $(C_6H_4)_2(NH_2)_2$.
Constants: Mol wt: 184.23, mp: 45°C, bp: 363°C.
Hazard Analysis
See diphenylamine.

o-BIPHENYLENE METHANE. See fluorene.

BIPHENYL MERCURY
General Information
Description: Crystals, not easily dissolved in ordinary solvents.
Formula: $Hg(C_6H_5C_6H_4)_2$.
Constants: Mol wt: 507.0, mp: 216°C.
Hazard Analysis & Countermeasures
See mercury compounds, organic.

2-BIPHENYLYLDIPHENYL PHOSPHATE
General Information
Description: Clear mobile liquid.
Formula: $C_{24}H_{19}O_4P$.
Constants: Mol wt: 402.37, mp: <0°C, bp: 250–285°C at 5 mm, flash p: 437°F, d: 1.2 at 60°C, vap. d.: 13.8.
Hazard Analysis
Toxicity: Unknown.
Fire Hazard: Slight, when exposed to heat or flame.
Disaster Hazard: Dangerous; see phosphates; can react with oxidizing materials.
Countermeasures
Storage and Handling: Section 7.
To Fight Fire: Water, foam, carbon dioxide, dry chemical or carbon tetrachloride (Section 6).

1-(4-BIPHENYLYLOXY)-2-PROPANOL. See propylene glycol-4-biphenylyl ether.

BIRCH TAR OIL
General Information
Description: Brown liquid; leather-like odor.
Constant: D: 0.886–0.950.
Hazard Analysis
Toxic Hazard Rating:
  Acute Local: Irritant 2; Allergen 1; Ingestion 2.
  Acute Systemic: U.
  Chronic Local: Allergen 1.
  Chronic Systemic: U.
Fire Hazard: Slight, when exposed to heat or flame; can react with oxidizing materials (Section 6).
Countermeasures
Personnel Protection: Section 3.
Storage and Handling: Section 7.

BIRCH WOOD DUST
General Information
Description: A typical wood dust.
Hazard Analysis
Toxic Hazard Rating:
  Acute Local: Allergen 1; Inhalation 1.
  Acute Systemic: 0.
  Chronic Local: Allergen 1.
  Chronic Systemic: Inhalation 1.
Fire Hazard: Moderate, when exposed to heat or flame; can react with oxidizing materials (Section 6).

BIS-ACETYL ACETONE GERMANIUM DI-BROMIDE
General Information
Description: Colorless, tiny crystals.
Formula: $[CH(CCH_3O)_2]_2GeBr_2$.
Constants: Mol wt: 430.7, mp: 226°C.
Hazard Analysis
Toxicity: Details unknown. See germanium compounds.
Disaster Hazard: Dangerous. See bromides.

BIS-ACETYL ACETONE GERMANIUM DI-CHLORIDE
General Information
Description: Colorless crystals.
Formula: $GeCl_2 \cdot [CH(CCH_3O)_2]_2$.
Constants: Mol wt: 341.7, mp: 240°C (decomp.).
Hazard Analysis
Toxicity: Details unknown. See germanium compounds.
Disaster Hazard: Dangerous; see chloride fumes.

BIS(o-AMINO PHENYL DISULFIDE)
Hazard Analysis
Toxicity: Details unknown; an insecticide.
Disaster Hazard: Dangerous; see sulfides.
Countermeasures
Storage and Handling: Section 7.

BIS-(3-AMINO PROPYL)AMINE
General Information
Formula: $NH(C_3H_6NH_2)_2$.
Constant: Mol wt: 131.
Hazard Analysis
Toxic Hazard Rating: U. Limited animal experiments suggest high toxicity and irritation. See also amines.

N,N-BIS-(3-AMINOPROPYL)METHYL AMINE
General Information
Description: Liquid; completely miscible in water.
Formula: $CH_3N(C_3H_6NH_2)_2$.
Constants: Mol wt: 145, d: 0.9307 (20/20°C), bp: 240.6°C (760 mm), fp: −29.6°C, flash p: 220°F.
Hazard Analysis
Toxic Hazard Rating: U. Limited animal experiments suggest moderate toxicity and irritation. See also amines.
Fire Hazard: Moderate, when exposed to heat or flame.
Countermeasures
Storage and Handling: Section 7.
To Fight Fire: Foam, fog, dry chemical (Section 6).

1,3-BIS(2-BENZOTHIAZOLYL MERCAPTOMETHYL) UREA. See di-(2-benzothiazylthiomethyl)urea.

2,2-BIS(p-BROMOPHENYL)-1,1,1-TRI-CHLOROETHANE
General Information
Formula: $C_{14}H_9Br_2Cl_3$.
Constant: Mol wt: 443.43.
Hazard Analysis
Toxicity: Details unknown. The bromine analog of DDT. See DDT.
Disaster Hazard: Dangerous; see chlorides.

BIS(2-BUTYL OCTYL) SODIUM SULFOSUCCINATE
General Information
Description: White waxy pellets.
Formula: $(C_{12}H_{25}O_2C)_2CH_2CHSO_3Na$.
Constants: Mol wt: 556, mp: >200°C, d: 1.0 at 25°C.
Hazard Analysis
Toxicity: Unknown.
Disaster Hazard: Dangerous; see sulfonates.

Countermeasures
Storage and Handling: Section 7.

2,2-BIS(tert-BUTYLPEROXY)BUTANE
General Information
Synonym: 22-PB.
Description: Liquid.
Formula: $(CH_3)_3COOC(CH_2)_3OOC(CH_3)_3$.
Constants: Mol wt: 220.3, mp: −10° to −26°C, flash p: 84°F (O.C.), d: 0.8655 at 20°/4°C, vap. d.: 7.59.
Hazard Analysis
Toxicity: Unknown.
Fire Hazard: Dangerous; it burns when exposed to heat or flame or by spontaneous chemical reaction with reducing agents.
Caution: This is an oxidizer as well as a flammable material.
Countermeasures
To Fight Fire: Foam, carbon dioxide, dry chemical or carbon tetrachloride (Section 6).
Storage and Handling: Section 7.

BIS(2-p-tert-BUTYLPHENOXYETHYL) SODIUM SULFOSUCCINATE
General Information
Description: White powder.
Formula: $(C_{12}H_{17}OCO_2)_2CH_2SCHO_3Na$.
Constants: Mol wt: 540, mp: 130–140°C, d: 1.0 at 25°C.
Hazard Analysis
Toxicity: Unknown.
Disaster Hazard: Dangerous; see sulfonates.
Countermeasures
Storage and Handling: Section 7.

BIS(p-tert-BUTYL PHENYL) PHENYL PHOSPHATE
General Information
Description: Clear viscous liquid.
Formula: $C_{26}H_{31}O_4P$.
Constants: Mol wt: 438.49, bp: 260–275°C at 5 mm, fp: <0°C, flash p: 482°F, d: 1.11 at 25°/25°C, vap. d.: 15.1.
Hazard Analysis
Toxicity: Unknown.
Fire Hazard: Slight, when exposed to heat or flame.
Disaster Hazard: Dangerous; see phosphates; can react with oxidizing materials.
Countermeasures
To Fight Fire: Water, foam, carbon dioxide, dry chemical or carbon tetrachloride (Section 6).
Ventilation Control: Section 2.
Storage and Handling: Section 7.

BIS-β-CHLOROETHYL ETHER. See 2,2′-dichloroethyl ether.

BIS(5-CHLORO-2-HYDROXYPHENYL) METHANE
General Information
Formula: $(C_6H_3Cl(OH)_2CH_2)$.
Constant: Mol wt: 158.6.
Hazard Analysis
Toxicity: Details unknown; possibly similar to DDT. See DDT.
Disaster Hazard: Dangerous; see chlorides.
Countermeasures
Storage and Handling: Section 7.

3,3-BISCHLOROMETHYL OXYCYCLOBUTANE
Hazard Analysis
Toxic Hazard Rating:
Acute Local: Irritant 2; Inhalation 2.

TOXIC HAZARD RATING CODE (For detailed discussion, see Section 1.)

0 NONE: (a) No harm under any conditions; (b) Harmful only under unusual conditions or overwhelming dosage.

1 SLIGHT: Causes readily reversible changes which disappear after end of exposure.

2 MODERATE: May involve both irreversible and reversible changes; not severe enough to cause death or permanent injury.

3 HIGH: May cause death or permanent injury after very short exposure to small quantities.

U UNKNOWN: No information on humans considered valid by authors.

Acute Systemic: Inhalation 3.
Chronic Local: U.
Chronic Systemic: U.
Animal experiments show irritant and narcotic effects.
Disaster Hazard: Dangerous. See chlorides.

BIS(p-CHLOROPHENOXY METHANE)
General Information
Synonyms: Di(4-chlorophenoxy) methane; neotran.
Description: White crystals.
Formula: $C_{13}H_{10}Cl_2O_2$.
Constants: Mol wt: 269.1, mp: 70°C, bp: 189–194°C at 6 mm Hg.
Hazard Analysis
Toxic Hazard Rating:
Acute Local: U.
Acute Systemic: Ingestion 2.
Chronic Local: U.
Chronic Systemic: U.
Toxicology: In experimental animals, large doses have produced liver damage. No specific toxicologic data available for humans. Low toxicity in laboratory animals. An insecticide, possibly similar to DDT.
Disaster Hazard: Dangerous; see chlorides
Countermeasures
Storage and Handling: Section 7.

2,2-BIS-(p-CHLOROPHENYL)-1,1-DICHLOROETHANE. See DDD.

1,1-BIS(p-CHLOROPHENYL)-2-NITROBUTANE. See dilan.

1,1-BIS(p-CHLOROPHENYL)-2-NITROPROPANE. See dilan.

BIS-(o-CHLOROPHENYL) PHENYL PHOSPHATE
General Information
Description: Clear, pale straw colored mobile liquid.
Formula: $C_{18}H_{13}O_4Cl_2P$.
Constants: Mol wt: 395.18, mp: <0°C, bp: 255–275°C at 5 mm, flash p: >437°F, d: 1.34 at 25°/25°C, vap. d.: 13.6.
Hazard Analysis
Toxicity: Unknown.
Fire Hazard: Slight, when exposed to heat or flame.
Disaster Hazard: Dangerous; see chlorides and phosphates, can react with oxidizing materials.
Countermeasures
Ventilation Control: Section 2.
To Fight Fire: Water, foam, carbon dioxide, dry chemical or carbon tetrachloride (Section 6).
Storage and Handling: Section 7.

2,2-BIS-(p-CHLOROPHENYL)-1,1,1-TRICHLORO-ETHANE. See DDT.

1,1-BIS(p-CHLOROPHENYL)-2,2,2-TRICHLORO-ETHANOL
General Information
Synonym: Kelthane.
Hazard Analysis
Toxicity: Details unknown. Limited animal experiments suggest moderate toxicity. See also DDT.
Disaster Hazard: Dangerous; see chlorides.

BIS(DIETHYLTHIOCARBAMYL) DISULFIDE
General Information
Synonym: Tetraethylthiuram disulfide.
Description: Powder.
Formula: $C_{10}H_{20}N_2S_4$.
Constant: Mol wt: 296.5.
Hazard Analysis
Toxicity: Details unknown, but toxic when accompanied by ingestion of alcohol. See also bis(dimethyl thiocarbamyl) disulfide.

Disaster Hazard: Dangerous; see oxides of sulfur and nitrogen.
Countermeasures
Storage and Handling: Section 7.

BIS(1,3-DIMETHYLBUTYL) AMINE
General Information
Description: Liquid.
Formula: $[CH_3CH(CH_3)CH_2CH(CH_3)]_2NH$.
Constants: Mol wt: 185.2, bp: 190–200°C, d: 0.775 at 20°/20°C, vap. d.: 6.38.
Hazard Analysis
Toxicity: Details unknown. See also amines.
Fire Hazard: Slight; (Section 6).
Countermeasures
Storage and Handling: Section 7.

BIS-DIMETHYL GLYOXIME COBALTOCHLORIDE
General Information
Description: Pale green crystals soluble in water.
Formula: $Co(C_8H_{15}O_4N_4Cl_2)$.
Constants: Mol wt: 361.1.
Hazard Analysis
Toxicity: Details unknown. See cobalt compounds.
Disaster Hazard: Dangerous; see chlorides.

BIS (1,2-DIMETHYLPROPYL) BORINE
General Information
Synonym: Di-sec-isoceryl borane.
Hazard Analysis and Countermeasures
See boron hydrides.

BIS(DIMETHYLTHIOCARBAMYL) DISULFIDE
General Information
Synonym: Tetramethylthiuram disulfide; disulfiram; TTD; thiram.
Description: Crystals. Insoluble in water. Soluble in alcohol, ether, acetone, chloroform.
Formula: $C_6H_{12}N_2S_4$.
Constants: Mol wt: 296.6, mp: 70°C, d: 1.30.
Hazard Analysis
Toxic Hazard Rating:
Acute Local: Irritant 1; Ingestion 1; Inhalation 2; Allergen 1.
Acute Systemic: Ingestion 2.
Chronic Local: Irritant 1; Allergen 1.
Chronic Systemic: Ingestion 2.
TLV: ACGIH (recommended) 5 milligrams per cubic meter of air.
Toxicology: Acute poisoning in experimental animals produced liver and kidney injury and also brain damage. In the presence of alcohol this compound produces violent nausea, vomiting and collapse. A fungicide.
Disaster Hazard: Dangerous; see oxides of nitrogen and sulfur.
Countermeasures
Ventilation Control: Section 2.
Personal Hygiene: Section 3.
Storage and Handling: Section 7.
MCA warning label.

BIS(DIMETHYLTHIOCARBAMYL) SULFIDE.
See bis(dimethylthiocarbamyl) disulfide.

BIS-2, 2-DIPYRIDYL RHENICHLORIDE
General Information
Description: Green crystals, slightly soluble in water.
Formula: $(C_{10}H_9N_2)_2ReCl_6$.
Constant: Mol wt: 713.4.
Hazard Analysis
Toxicity: Details unknown.
Disaster Hazard: When heated to decomposition it emits highly toxic chloride fumes.

BIS(3, 4-EPOXYBUTYL) ETHER
General Information
Formula: $H_2\overline{COHCH_2CH_2COCH_2CH_2\overline{CHOCH}_2}$.

Constant: Mol wt: 158.

Hazard Analysis
Toxic Hazard Rating:
Acute Local: Irritant 3.
Acute Systemic: Ingestion 2; Skin Absorption: 2.
Chronic Local: U.
Chronic Systemic: U.
Based on limited animal experiments.

BIS(3, 4-EPOXYCYCLOHEXYLMETHYL) ADIPATE
General Information
Formula: $[C_6H_8CH_3O]OOC(CH_2)_4COO[C_6H_8CH_3O]$.
Constant: Mol wt: 366.

Hazard Analysis
Toxic Hazard Rating:
Acute Local: Irritant 1.
Acute Systemic: Ingestion: 1; Inhalation: 1; Skin Absorption: 1.
Chronic Local: U.
Chronic Systemic: U.
Based on limited animal experiments.

BIS(2,3-EPOXYCYCLOPENTYL) ETHER
General Information
Formula: $(C_5H_7O)_2O$.
Constant: Mol wt: 182.

Hazard Analysis
Toxic Hazard Rating:
Acute Local: Irritant: 1.
Acute Systemic: Ingestion 3; Inhalation 1; Skin Absorption 2.
Chronic Local: Allergen 0.
Chronic Systemic: U.
Based on limited animal experiments. See also epoxy resins.

BIS(3, 4-EPOXY-6-METHYLCYCLOHEXYLMETHYL) ADIPATE
Hazard Analysis
Toxic Hazard Rating:
Acute Local: Irritant 1.
Acute Systemic: Ingestion 1; Inhalation 1; Skin Absorption 1.
Chronic Local: Allergen 0.
Chronic Systemic: U.
Based on limited animal experiments.

BIS(2, 3-EPOXY-2-METHYL PROPYL) ETHER
General Information
Formula: $H_2COC(CH_3)CH_2OH_2C(CH_3)COCH_2$.
Constant: Mol wt: 158.

Hazard Analysis
Toxic Hazard Rating:
Acute Local: Irritant 2.
Acute Systemic: Ingestion 2; Inhalation (1; Skin Absorption 2.
Chronic Local: Allergen 2.
Chronic Systemic: U.
Based on limited animal experiments.

2, 3-BIS(2, 3-EPOXYPROPOXY)-1,4-DIOXANE
Hazard Analysis
Toxic Hazard Rating: U. Limited animal experiments suggest high toxicity. See also dioxane.

1, 3-BIS-ETHYLAMINOBUTANE
General Information
Description: Liquid.
Formula: $(C_2H_5)_2C_4H_7NH_2$.

Constants: Mol wt: 129.2, bp: 180°C, flash p: 115°F, d: 0.81, vap. d.: 4.44.

Hazard Analysis
Toxicity: Unknown.
Fire Hazard: Moderate, when exposed to heat or flame; can react with oxidizing materials.

Countermeasures
To Fight Fire: Foam, carbon dioxide, dry chemical or carbon tetrachloride (Section 6).
Storage and Handling: Section 7.

BIS(2-ETHYLHEXYL)-2-ETHYLHEXYL-PHOSPHONATE
General Information
Description: Colorless liquid, very mild odor.
Formula: $C_8H_{17}P(O)(OC_8H_{17})_2$.
Constants: Mol wt: 418.6, bp: 161°C at 0.25 mm, flash p: 419°F (C.O.C.), d: 0.908 at 20°/4°C, vap. d.: 14.4.

Hazard Analysis
Toxicity: Unknown.
Fire Hazard: Slight, when exposed to heat or flame.
Disaster Hazard: Dangerous; see phosphates; can react with oxidizing materials.

Countermeasures
Storage and Handling: Section 7.
To Fight Fire: Foam, carbon dioxide, dry chemical or carbon tetrachloride (Section 6).

2, 2-BIS-(p-FLUOROPHENYL)-1, 1, 1-TRI-CHLOROETHANE. See DFDT.

2, 3-BIS(GLYCIDYLOXY)-1,4-DIOXANE
Hazard Analysis
Toxic Hazard Rating:
Acute Local: Irritant 3.
Acute Systemic: Ingestion 2; Inhalation 1; Skin Absorption 2.
Chronic Local: U.
Chronic Systemic: U.
Based on limited animal experiments.

2, 2,-BIS p-(2, 3-GLYCIDYLOXY) PHENYL PROPANE
Hazard Analysis
Toxic Hazard Rating:
Acute Local: U.
Acute Systemic: Ingestion 1; Skin Absorption 1.
Chronic Local: Allergen 3.
Chronic Systemic: U.
Based on limited animal experiments.

BISH. See aconitum ferox.

BIS-(2-HYDROXY-5-CHLOROPHENYL) SULFIDE
General Information
Formula: $(C_6H_3OHCl)_2S$.
Constant: Mol wt: 287.2.

Hazard Analysis
Toxicity: Details unknown; a fungicide. See chlorinated phenols.
Fire Hazard: Unknown.
Disaster Hazard: Dangerous; see chlorides and oxides of sulfur.

Countermeasures
Storage and Handling: Section 7.

BIS-HYDROXYCOUMARIN
General Information
Synonym: Dicoumarol.
Description: Very small crystals; slight pleasant odor, bitter taste, soluble in alkali.

TOXIC HAZARD RATING CODE (For detailed discussion, see Section 1.)

0 NONE: (a) No harm under any conditions; (b) Harmful only under unusual conditions or overwhelming dosage.

1 SLIGHT: Causes readily reversible changes which disappear after end of exposure.

2 MODERATE: May involve both irreversible and reversible changes; not severe enough to cause death or permanent injury.

3 HIGH: May cause death or permanent injury after very short exposure to small quantities.

U UNKNOWN: No information on humans considered valid by authors.

Formula: $C_{19}H_{12}O_6$.
Constants: Mol wt: 336.3, mp: 287–293°C.
Hazard Analysis
Toxic Hazard Rating:
  Acute Local: 0.
  Acute Systemic: Ingestion 3.
  Chronic Local: 0.
  Chronic Systemic: Ingestion 3.
Toxicology: An anticoagulant. Excessive doses can cause hemorrhages. Antidote: Vitamin K.

BIS-(2-HYDROXY-3, 5-DICHLOROPHENYL) SULFIDE.
  See bithionol.

3, 3-BIS-(p-HYDROXYPHENYL) PHTHALIDE. See phenolphthalein.

2,2-BIS(4-HYDROXYPHENYL) PROPANE
General Information
Synonym: Bisphenol A; p,p'-isopropylidene diphenol.
Description: White flakes; mild phenolic odor; insoluble in water; soluble in alcohol and dilute alkalies; slightly soluble in CCl_4.
Formula: $(CH_3)_2C(C_6H_4OH)_2$
Constant: Mol wt: 228.
Hazard Analysis
Toxic Hazard Rating: U. See phenol.

BIS(ISOPROPYLAMIDO) FLUOROPHOSPHATE
General Information
Synonym: Isopestox, mipafox.
Hazard Analysis
Toxicity: Highly toxic. See parathion.
Disaster Hazard: Very dangerous; see fluorides and phosphates.
Countermeasures
See parathion.
Shipping Regulations: Section 11.
  IATA: Other restricted articles, class A, no limit (passenger), no limit (cargo).

BISMARCK BROWN 331
Hazard Analysis
Toxic Hazard Rating:
  Acute Local: Allergen 1.
  Acute Systemic: U.
  Chronic Local: Allergen 1.
  Chronic Systemic: U.

2, 2-BIS(p-METHOXYPHENYL)-1, 1, 1-TRI-CHLOROETHANE. See methoxychlor.

BIS(a-METHYL BENZYL) AMINE. See a-methyl benzylamine.

BIS-alpha-METHYL BENZYL ETHER. See methyl benzyl ether.

N,N′-BIS(1-METHYL HEPTYL) ETHYLENE DIAMINE
General Information
Description: Insoluble in water.
Formula: $HC(CH_3)(C_6H_{13})NHCH_2CH_2NHCH\cdot(CH_3)(C_6H_{13})$.
Constant: Mol wt: 284, d: 0.8, bp: 424°F at 43 mm, flash p: > 400°F.
Hazard Analysis
Toxic Hazard Rating: U.
Fire Hazard: Slight, when exposed to heat or flame.
Countermeasures
To Fight Fire: Water, foam, (Section 6).
Storage and Handling: Section 7.

BISMUTH
General Information
Description: Hexagonal silver-white or reddish metallic crystals.
Formula: Bi.

Constants: At wt: 209.00, mp: 271.3°C, bp: 1420–1560°C, d: 9.80, vap. press.: 1 mm at 1021°C.
Hazard Analysis
Toxicity: See bismuth compounds.
Radiation Hazard: Section 5. For permissible levels, see Table 5, p. 150.
  Artificial isotope ^{206}Bi half life 6.3 d. Decays to stable ^{206}Pb by electron capture. Emits gamma rays of 0.11–1.0 MeV and x-rays.
  Artificial isotope ^{207}Bi half life 28 y. Decays to stable ^{207}Pb by electron capture. Emits gamma rays of 0.57–1.77 MeV and x-rays.
  Natural isotope ^{210}Bi (Radiom E, Uranium Series) half life 5.0 d. Decays to radioactive ^{210}Po by emitting beta particles of 1.16 MeV.
  Natural isotope ^{212}Bi (Thorium-C, Thorium Series), half life 60.5 m. Decays to radioactive ^{208}Tl by emitting alpha particles (36%) of 6.05, 6.09 MeV. Also decays to radioactive ^{212}Po by emitting beta particles of 0.63 (2%), 1.55 (5%), 2.25 (55%) MeV.
Fire Hazard: Moderate; when exposed to flame and by chemical reaction with oxidizing agents.
Disaster Hazard: Moderately dangerous; can react with acid or acid fumes to emit toxic fumes.
Countermeasures
Ventilation Control: Section 2.
Personal Hygiene: Section 3.
Storage and Handling: Section 7.

BISMUTH ACETATE
General Information
Description: White crystals.
Formula: $Bi(C_2H_3O_2)_3$.
Constants: Mol wt: 386.13, mp: decomposes.
Hazard Analysis
Toxicity: See bismuth compounds.

BISMUTH (COLLOIDAL SOLUTION)
Hazard Analysis
Toxicity: Details unknown. See also bismuth compounds.

BISMUTH o-ARSENATE
General Information
Description: Monoclinic crystals.
Formula: $BiAsO_4$.
Constants: Mol wt: 347.91, d: 7.14.
Hazard Analysis and Countermeasures
See arsenic compounds.

BISMUTH BENZOATE
General Information
Description: White powder.
Formula: $Bi(C_7H_5O_2)_3$.
Constant: Mol wt: 572.33.
Hazard Analysis
Toxicity: See bismuth compounds.
Fire Hazard: Slight; when heated. (Section 6).
Countermeasures
Storage and Handling: Section 7.

BISMUTH CARBONATE, BASIC
General Information
Synonyms: Bismuth oxycarbonate, bismutospherite.
Description: White powder.
Formula: $Bi_2O_2CO_3$.
Constants: Mol wt: 510.01, mp: decomposes, d: 6.86.
Hazard Analysis
Toxicity: See bismuth compounds.

BISMUTH CITRATE
General Information
Description: White crystals.
Formula: $BiC_6H_5O_7$.
Constants: Mol wt: 398.10, mp: decomposes, d: 3.458.
Hazard Analysis
Toxicity: See bismuth compounds.

BISMUTH COMPOUNDS
Hazard Analysis
Toxic Hazard Rating:
Acute Local: Irritant 1; Ingestion 1.
Acute Systemic: Ingestion 2.
Chronic Local: U.
Chronic Systemic: Ingestion 1; Inhalation 1.
Toxicology: Bismuth and its salts can cause kidney damage, although the degree of such damage is usually mild. Large doses can be fatal. Industrially it is considered one of the less toxic of the heavy metals, although intoxication has occurred from its use in medicine. The similarity between the pharmacologic and toxic behavior of lead and bismuth has been pointed out in the literature. Like lead, bismuth may be liberated from tissue deposits during periods of acidosis. Serious and sometimes fatal poisoning may occur from the injection of large doses into closed cavities and from extensive application to burns.

Death of animals from bismuth nephritis following injections of soluble salts occurs within several hours to 24 days, the time being generally inversely proportional to the dose, and it appears to be in the order of 5 to 10 times higher than the dose by slow intravenous injection for rabbits. It is stated that the administration of bismuth should be stopped when gingivitis appears, for otherwise serious ulcerative stomatitis is likely to result. Other toxic results may develop, such as malaise, albuminuria, diarrhea, skin reactions, and sometimes serous exodermatitis.

Industrial bismuth poisoning has not been reported, although bismuth absorbed in industrial cases may complicate a diagnosis of plumbism, since the dark line in the gums which is often present in lead poisoning is also produced by bismuth. All bismuth compounds do not have equal toxicity. See individual entries.

Countermeasures
Treatment and Antidotes: Personnel showing some of the symptoms noted above which might indicate that they were absorbing too much bismuth into the body should be removed from exposure as soon as possible. Get medical advice.

Bismuth and its salts should be labeled as poison. Personnel should be cautioned against careless handling of these materials.
Ventilation Control: Section 2.
Personal Hygiene: Section 3.
Storage and Handling: Section 7.

BISMUTH DICHROMATE, BASIC
General Information
Description: Yellow or red crystals.
Formula: $(BiO)_2Cr_2O_7$.
Constant: Mol wt: 666.
Hazard Analysis
Toxicity: See chromium compounds and bismuth compounds.
Fire Hazard: Moderate, by chemical reaction with easily combustible materials; can react with reducing materials; a powerful oxidizer.
Countermeasures
Storage and Handling: Section 7.

BISMUTH ETHYL CHLORIDE
General Information
Description: Powder.
Formula: $BiHC_2H_5Cl$.
Constant: Mol wt: 274.5.

Hazard Analysis
Toxicity: See bismuth compounds.
Fire Hazard: Dangerous, by spontaneous chemical reactions; also spontaneously flammable in air (Section 6).
Explosion Hazard: Unknown.
Disaster Hazard: Dangerous; see clorides; can react with oxidizing materials.
Countermeasures
Storage and Handling: Section 7.

BISMUTH FLUORIDE. See bismuth trifluoride.

BISMUTH GALLATE, BASIC
General Information
Synonyms: Subgallate; dermatol (com'l).
Description: Yellow amorphous crystals.
Formula: $Bi(OH)_2C_7H_5O_5$ (approx.).
Constants: Mol wt: 412.13, mp: decomposes.
Hazard Analysis
Toxicity: See bismuth compounds.

BISMUTH GLANCE. See bismuth trisulfide.

BISMUTH HEPTADIENE-CARBOXYLATE
Hazard Analysis
Toxicity: See bismuth compounds.
Fire Hazard: Slight; when heated.
Countermeasures
Storage and Handling: Section 7.

BISMUTH HEPTOXIDE
Hazard Analysis
Toxicity: See bismuth compounds.
Fire Hazard: Moderate, by chemical reaction with easily oxidized materials; can react with reducing materials; a powerful oxidizer (Section 6).
Countermeasures
Storage and Handling: Section 7.

BISMUTH HYDROXIDE
General Information
Description: White amorphous powder.
Formula: $Bi(OH)_3$.
Constants: Mol wt: 260.02; $-H_2O$ at 100°C; $-1\frac{1}{2}H_2O$ at 400°C; decomp. at 415°C.
Hazard Analysis
Toxicity: See bismuth compounds.

m-BISMUTHIC ACID. See bismuth pentoxide.

BISMUTHINE. See hydrogen bismuthide.

BISMUTH IODATE
General Information
Description: White crystals.
Formula: $Bi(IO_3)_3$.
Constant: Mol wt: 733.76.
Hazard Analysis
Toxicity: See bismuth compounds and iodates.
Fire Hazard: Moderate, by spontaneous chemical reaction with easily oxidized materials; a powerful oxidizer (Section 6).
Disaster Hazard: Moderately dangerous; see iodates; can react with reducing materials.
Countermeasures
Storage and Handling: Section 7.

BISMUTH dl-LACTATE
General Information
Description: Prisms.
Formula: $Bi(C_6H_9O_6)\cdot 7H_2O$.

TOXIC HAZARD RATING CODE (For detailed discussion, see Section 1.)

0 NONE: (a) No harm under any conditions; (b) Harmful only under unusual conditions or overwhelming dosage.

1 SLIGHT: Causes readily reversible changes which disappear after end of exposure.

2 MODERATE: May involve both irreversible and reversible changes; not severe enough to cause death or permanent injury.

3 HIGH: May cause death or permanent injury after very short exposure to small quantities.

U UNKNOWN: No information on humans considered valid by authors.

Constant: Mol wt: 512.25.
Hazard Analysis
Toxicity: See bismuth compounds.
Fire Hazard: Slight; when heated.
Countermeasures
Storage and Handling: Section 7.

BISMUTH MOLYBDATE
General Information
Description: Tetragonal needles.
Formula: $Bi_2(MoO_4)_3$.
Constants: Mol wt: 897.85, mp: 643°C, d: 6.07.
Hazard Analysis
Toxicity: See bismuth compounds and molybdenum compounds.

BISMUTH MONOSULFIDE
General Information
Description: Gray crystals.
Formula: BiS.
Constants: Mol wt: 241.07, mp: 685°C, d: 7.7.
Hazard Analysis
Toxicity: See bismuth compounds and sulfides.
Fire Hazard: See sulfides.
Explosion Hazard: See sulfides.
Disaster Hazard: Dangerous; see oxides of sulfur; will react with water or steam to produce toxic and flammable vapors.
Countermeasures
Storage and Handling: Section 7.

BISMUTHNITE. See bismuth trisulfide.

BISMUTH NITRATE
General Information
Description: Triclinic, colorless, slightly hygroscopic crystals.
Formula: $Bi(NO_3)_3 \cdot 5H_2O$.
Constants: Mol wt: 485.10; $-5H_2O$ at 80°C, d: 2.83.
Hazard Analysis and Countermeasures
See bismuth compounds and nitrates.

BISMUTH OXALATE
General Information
Description: White powder.
Formula: $Bi_2(C_2O_4)_3$.
Constant: Mol wt: 682.06.
Hazard Analysis and Countermeasures
See bismuth compounds and oxalates.

BISMUTH OXYBROMIDE
General Information
Description: Colorless crystals or white powder.
Formula: BiOBr.
Constants: Mol wt: 304.92, d: 8.08.
Hazard Analysis and Countermeasures
See bismuth compounds and bromides.

BISMUTH OXYCARBONATE. See bismuth carbonate, basic.

BISMUTH OXYCHLORIDE
General Information
Description: White lustrous crystalline powder.
Formula: BiOCl.
Constants: Mol wt: 260.46, mp: red heat, d: 7.72.
Hazard Analysis and Countermeasures
See bismuth compounds and chlorides.

BISMUTH OXYFLUORIDE
General Information
Description: White crystals or powder.
Formula: BiOF.
Constants: Mol wt: 244.00, mp: decomposes, d: 7.5.
Hazard Analysis and Countermeasures
See bismuth compounds and fluorides.

BISMUTH OXYIODIDE
General Information
Description: Red powder.
Formula: BiOI.
Constants: Mol wt: 351.92, mp: decomposes, d: 7.92.
Hazard Analysis and Countermeasures
See bismuth compounds and iodides.

BISMUTH OXYIODOGALLATE
General Information
Description: Grayish-green, odorless, tasteless powder.
Formula: $C_6H_2(OH)_3COO(BiOHI)$.
Constant: Mol wt: 522.0.
Hazard Analysis and Countermeasures
See bismuth compounds and iodides.

BISMUTH OXYNITRATE
General Information
Description: Hexagonal plates or white powder.
Formula: $BiONO_3 \cdot H_2O$.
Constants: Mol wt: 305.02, mp: decomposes at 260°C, d: 4.928 at 15°C.
Hazard Analysis and Countermeasures
See bismuth compounds and nitrates.

BISMUTH PENTAFLUORIDE
General Information
Description: Crystals which react violently with water and petrolatum above 50°C.
Formula: BiF_5.
Constants: Mol wt: 304, sublimes at 550°C.
Hazard Analysis
Toxic Hazard Rating:
   Acute Local: Irritant 3; Ingestion 3; Inhalation 3.
   Acute Systemic: Ingestion 3; Inhalation 3.
   Chronic Local: U.
   Chronic Systemic: U.
Toxicology: Highly toxic and irritating. Decomposes readily on contact with moisture to yield ozone and bismuth trifluoride. See fluorides. See ozone.
Disaster Hazard: Very dangerous. When heated to decomposition it emits highly toxic fluoride fumes. In contact with moisture, acids, etc., it reacts violently liberating much heat and ozone.
Countermeasures
Personnel Protection: Section 3.
Storage and Handling: Section 7.

BISMUTH PENTOXIDE
General Information
Synonym: m-Bismuthic acid.
Description: Brown or dark red crystals.
Formula: Bi_2O_5.
Constants: Mol wt: 498.00, -0 at 150°C, -20 at 357°C, d: 5.10.
Hazard Analysis
Toxicity: See bismuth compounds.
Fire Hazard: Slight, when exposed to heat and oxidizable materials or by chemical reaction with reducing agents; when heated evolves oxygen (Section 6).

BISMUTH o-PHOSPHATE
General Information
Description: Monoclinic white crystals.
Formula: $BiPO_4$.
Constants: Mol wt: 303.98, mp: decomposes, d: 6.323 at 15°C.
Hazard Analysis and Countermeasures
See bismuth compounds and phosphates.

BISMUTH PROPIONATE, BASIC
General Information
Description: White powder; faint odor of propionic acid.
Formula: $BiOC_3H_5O_2$.
Constant: Mol wt: 298.07.

Hazard Analysis
Toxicity: See bismuth compounds.
Fire Hazard: Slight; when heated.
Countermeasures
Storage and Handling: Section 7.

BISMUTH SALICYLATE, BASIC
General Information
Synonym: Subsalicylate (commercial).
Description: White microscopic crystals.
Formula: $Bi(C_7H_5O_3)_3 \cdot Bi_2O_3$ (approx.).
Constant: Mol wt: 1086.33.
Hazard Analysis
Toxicity: A fungicide. See bismuth compounds.
Fire Hazard: Slight; when heated.
Countermeasures
Storage and Handling: Section 7.

BISMUTH SUBCARBONATE
General Information
Description: White, odorless, tasteless powder. Insoluble in water or alcohol.
Formula: $(BiO)_2CO_3 \cdot \frac{1}{2}H_2O$.
Constants: Mol wt: 519, d: 6.86.
Hazard Analysis
Toxicity: See bismuth compounds. Has been used as an opaque medium in x-ray work.

BISMUTH SUBNITRATE
General Information
Synonym: Magistery of bismuth.
Description: White, heavy powder.
Formula: $4BiNO_3(OH)_2 \cdot BiO(OH)$.
Constants: Mol wt: 1462.11, mp: decomposes at 260°C, d: 4.928.
Hazard Analysis and Countermeasures
See bismuth compounds and nitrates.

BISMUTH SUBSALICYLATE. See bismuth salicylate, basic.

BISMUTH SULFATE
General Information
Description: White needles.
Formula: $Bi_2(SO_4)_3$.
Constants: Mol wt: 706.20, mp: decomposes, d: 5.08 at 15°C.
Hazard Analysis and Countermeasures
See bismuth compounds and sulfates.

BISMUTH TARTRATE
General Information
Description: White powder.
Formula: $Bi_2(C_4H_6O_6)_3 \cdot 6H_2O$.
Constants: Mol wt: 970.31, $-3H_2O$ at 105°C, d: 2.595 at 25°C.
Hazard Analysis
Toxicity: See bismuth compounds.

BISMUTH TELLURATE
General Information
Synonym: Montanite.
Description: Powder or crystals.
Formula: $Bi_2TeO_6 \cdot 2H_2O$.
Constants: Mol wt: 677.64, d: 3.79.
Hazard Analysis and Countermeasures
See bismuth compounds and tellurium compounds.

BISMUTH TETRACHLORIDE
General Information
Description: Colorless crystals.
Formula: $BiCl_4$.
Constants: Mol wt: 350.83, mp: 225°C.
Hazard Analysis and Countermeasures
See bismuth compounds and hydrochloric acid.

BISMUTH TETRAOXIDE
General Information
Description: Heavy yellowish-brown powder.
Formula: Bi_2O_4.
Constants: Mol wt: 482, mp: 305°C, d: 5.6.
Hazard Analysis
Toxicity: See bismuth compounds.
Fire Hazard: Slight, by chemical reaction with reducing agents; an oxidizer (Section 6).
Countermeasures
Storage and Handling: Section 7.

BISMUTH TITANATE
General Information
Formula: $BiTiO_3$.
Constant: Mol wt: 305.
Hazard Analysis
Toxic Hazard Rating: U. Animal experiments show low toxicity. See also bismuth and titanium compounds.

BISMUTH TRIBROMIDE
General Information
Description: Yellow crystalline deliquescent powder.
Formula: $BiBr_3$.
Constants: Mol wt: 448.75, mp: 218°C, bp: 461°C, d: 5.7, vap. press.: 10 mm at 282°C.
Hazard Analysis and Countermeasures
See bismuth compounds, bromides and hydrobromic acid..

BISMUTH TRICHLORIDE
General Information
Description: White, deliquescent crystals.
Formula: $BiCl_3$.
Constants: Mol wt: 315.37, mp: 230°C, bp: 441°C, d: 4.75, vap. press.: 10 mm at 264°C.
Hazard Analysis and Countermeasures
See bismuth compounds and hydrochloric acid.

BISMUTH TRIETHYL. See triethyl bismuthine.

BISMUTH TRIFLUORIDE
General Information
Synonym: Bismuth fluoride.
Description: White, cubic crystals, insoluble in water.
Formula: BiF_3.
Constants: Mol wt: 266, d: 8.3, mp: 725–730°C.
Hazard Analysis and Countermeasures
See fluorides and bismuth compounds.

BISMUTH TRIIODIDE
General Information
Description: Hexagonal reddish, brown-gray-blue crystals.
Formula: BiI_3.
Constants: Mol wt: 589.76, mp: 439°C, bp: decomposes at 500°C, d: 5.7.
Hazard Analysis and Countermeasures
See bismuth compounds and iodides.

BISMUTH TRIMETHYL. See trimethyl bismuthine.

BISMUTH TRIOXIDE
General Information
Description: Rhombic yellow crystals.
Formula: Bi_2O_3.
Constants: Mol wt: 466.00, mp: 820°C, bp: 1890°C, d: 8.9.

TOXIC HAZARD RATING CODE (For detailed discussion, see Section 1.)

0 NONE: (a) No harm under any conditions; (b) Harmful only under unusual conditions or overwhelming dosage.

1 SLIGHT: Causes readily reversible changes which disappear after end of exposure.

2 MODERATE: May involve both irreversible and reversible changes; not severe enough to cause death or permanent injury.

3 HIGH: May cause death or permanent injury after very short exposure to small quantities.

U UNKNOWN: No information on humans considered valid by authors.

Hazard Analysis
Toxicity: See bismuth compounds.

BISMUTH TRIPHENYL. See triphenyl bismuthine.

BISMUTH TRISELENIDE
General Information
Synonym: Guanajuatite.
Description: Rhombic black crystals.
Formula: Bi_2Se_3.
Constants: Mol wt: 654.88, mp: 710°C, bp: decomposes, d: 6.82.
Hazard Analysis
Toxicity: See bismuth compounds and selenium compounds.
Fire Hazard: Moderate, by chemical reaction with powerful oxidizers. On contact with moisture a flammable gas is evolved (Section 6).
Explosion Hazard: Slight, by chemical reaction with powerful oxidizers and moisture.
Disaster Hazard: See selenium compounds.
Countermeasures
Storage and Handling: Section 7.

BISMUTH TRISULFIDE
General Information
Synonyms: Bismuthnite; bismuth glance.
Description: Rhombic black crystals.
Formula: Bi_2S_3.
Constants: Mol wt: 514.20, mp: 685°C decomposes, d: 7.39.
Hazard Analysis and Countermeasures
See bismuth compounds and sulfides.

BISMUTH TRITELLURIDE
General Information
Synonym: Tetradymite.
Description: Crystals.
Formula: Bi_2Te_3.
Constants: Mol wt: 800.83, mp: 573°C, d: 7.7.
Hazard Analysis
Toxicity: See bismuth compounds and tellurium compounds.
Fire Hazard: Moderate by spontaneous chemical reaction with powerful oxidizers; reacts with moisture to evolve a gas (Section 6).
Explosion Hazard: Slight, by chemical reaction with powerful oxidizers; it reacts with moisture.
Disaster Hazard: See tellurium compounds.
Countermeasures
Storage and Handling: Section 7.

BIS(5-OXY-2,8-DITHIOOCTANE) GERMANIUM.
General Information
Description: Colorless crystals, soluble in benzene.
Formula: $[(SCH_2CH_2)_2O]_2Ge$.
Constants; Mol wt: 345.1, mp: 159.5°C.
Hazard Analysis
Toxicity: Details unknown. See germanium compounds.
Disaster Hazard: Dangerous; see sulfates.

BISPHENOL A. See 2,2-bis(4-hydroxyphenol) propane.

BIS-PROPIONYL ACETONE GERMANIUM DICHLORIDE
General Information
Description: White crystals.
Formula: $GeCl_2[CHC(C_2H_5)CH_3O_2]_2$.
Constants: Mol wt: 369.8, mp: 129°C.
Hazard Analysis
Toxicity: Details unknown. See germanium compounds.
Disaster Hazard: Dangerous; see chlorides.

BIS-TRIBENZYL GERMANYL SULFIDE
General Information
Description: Colorless crystals, soluble in alcohol.
Formula: $SGe_2(C_6H_5CH_2)_6$.
Constants: Mol wt: 724, mp: 124°C.
Hazard Analysis
Toxicity: Details unknown. See germanium compounds.
Disaster Hazard: Dangerous; see sulfates.

BIS-TRIBIPHENYLYL GERMANYL SULFIDE
General Information
Description: Colorless crystals, soluble in organic solvents.
Formula: $[(C_6H_5 \cdot C_6H_4)_3Ge]_2S$.
Constants: Mol wt: 1096.4, mp: 238°C.
Hazard Analysis
See bis-tribenzyl germanyl sulfide.

BIS-TRICHLOROGERMANYL METHANE
General Information
Description: Colorless liquid, hydrolyzed by water.
Formula: $CH_2(GeCl_3)_2$.
Constants: Mol wt: 372, bp: 110°C at 18 mm Hg.
Hazard Analysis
See bis-propionyl acetone germanium dichloride.

BIS-TRICYCLOHEXYL GERMANIUM DISULFIDE
General Information
Description: Colorless crystals insoluble in water.
Formula: $[(C_6H_{11})_3Ge]_2S_2$.
Constants: Mol wt: 708.2, mp: 88°C.
Hazard Analysis
See bis-tribenzyl germanyl sulfide.

BIS(TRIDECYL) SODIUM SULFOSUCCINATE
General Information
Description: White waxy pellets.
Formula: $(C_{13}H_{27}CO_2)_2CH_2SCHO_3Na$.
Constants: Mol wt: 584, mp: 70–75°C, d: 1.0 at 25°C.
Hazard Analysis
Toxicity: Unknown.
Fire Hazard: Slight. (Section 6).
Countermeasures
Storage and Handling: Section 7.

BIS-TRIETHYL GERMANYL SULFIDE
General Information
Description: Colorless oily liquid soluble in organic solvents.
Formula: $[(C_2H_5)_3Ge]_2S$.
Constants: Mol wt: 351.6, bp: 150°C at 12 mm Hg.
Hazard Analysis
See bis-tribenzyl germanyl sulfide.

BIS(TRIFLUOROMETHYL) BENZENE. See xylene hexafluoride.

2,2,-BIS-(2,2,2,-TRIMETHYL ETHYL) OXIRANE
Hazard Analysis
Toxic Hazard Rating: U. Limited animal experiments suggest low toxicity. See also epoxy resins.

BIS(2,2,4-TRIMETHYL PENTANEDIOL MONOISO-BUTYRATE) DIGLYCOLLATE
General Information
Formula: $C_{28}H_{27}O_9$.
Constants: Mol wt: 507, d: 1.1, bp: 639°F, flash p: 383°F (O.C.).
Hazard Analysis
Toxic Hazard Rating: U.
Fire Hazard: Slight, when exposed to heat or flame.
Countermeasures
To Fight Fire: Water, foam (Section 6).
Storage and Handling: Section 7.

BIS-TRIPHENYL GERMANYL SULFIDE
General Information
Description: Colorless crystals soluble in organic solvents.
Formula: $[(C_6H_5)_3Ge]_2S$.
Constants: Mol wt: 640, mp: 138°C.
Hazard Analysis
See bis-tribenzyl germanyl sulfide.

BIS-TRITOLYL GERMANYL SULFIDE
General Information
Description: Colorless crystals soluble in organic solvents.
Formula: $[(C_6H_4CH_3)_3Ge]_2S$.
Constants: Mol wt: 724, mp: 157°C.

Hazard Analysis
See bis-tribenzyl germanyl sulfide.

BISULFITE DRY POWDER SODA. See sodium bisulfite.

BISULFITE OF SODIUM. See sodium bisulfite.

BISULFITES
Hazard Analysis
Toxic Hazard Rating:
  Acute Local: Irritant 3; Ingestion 3; Inhalation 1.
  Acute Systemic: U.
  Chronic Local: Irritant 2.
  Chronic Systemic: U.
Toxicology: See individual compounds.
Disaster Hazard: Dangerous; when heated to decomposition
  or on contact with acid or acid fumes, it emits highly
  toxic fumes of sulfur dioxide; will react with water or
  steam to produce toxic and corrosive fumes.
Countermeasures
Ventilation Control: Section 2.
Personnel Protection: Section 3.
First Aid: Section 1.
Storage and Handling: Section 7.

BITHIONOL
General Information
Synonyms: 2,2-Thiobis (4,6-dichlorophenol);
  bis-(2-hydroxy-3,5-dichlorophenyl) sulfide.
Description: White crystalline powder; very faint phenolic
  odor.
Formula: $(C_6H_2Cl_2OH)_2S$.
Constants: Mol wt: 356.1, mp: 187–188°C, d: 1.61 at 25°C,
  vap. press.: 1.1×10^{-9} mm at 37°C.
Hazard Analysis
Toxic Hazard Rating:
  Acute Local: Irritant 2; Ingestion 2; Inhalation 2.
  Acute Systemic: Ingestion 2; Inhalation 2
  Chronic Local: Irritant 1.
  Chronic Systemic: U.
A food additive permitted in feed and drinking water of ani-
  mals and/or for the treatment of food producing ani-
  mals. Also a food additive permitted in food for human
  consumption. Section 10.
Disaster Hazard: Dangerous; when heated to decomposition
  or on contact with acid or acid fumes, it emits highly
  toxic fumes of oxides of sulfur.
Countermeasures
Ventilation Control: Section 2.
Personnel Protection: Section 3.
Personal Hygiene: Section 3.
Storage and Handling: Section 7.

3,3 -BITOLYLENE-4,4′-DIISOCYANATE
General Information
Description: Small pale, yellowish flakes.
Formula: $[C_6H_3(OCN)(CH_3)]_2$.
Constants: Mol wt: 264.3, fp: 69.6°C, d: 1.197 at
  80°/4°C.
Hazard Analysis
Toxicity: Details unknown, but probably an irritant. See also
  2,4-tolylene diisocyanate.
Disaster Hazard: Moderately dangerous; when heated to
  decomposition, it emits toxic fumes.
Countermeasures
Storage and Handling: Section 7.

BITTER APPLE. See colocynth.

BITTER CUCUMBER. See colocynth.

BITTER WOOD TREE. See quassia.

BITUMEN. See asphalt.

BLACK BLASTING POWDER. See explosives, low.

BLACK EBONY WOOD
General Information
Description: A wood.
Hazard Analysis
  Acute Local: Irritant 1; Allergen 1.
  Acute Systemic: 0.
  Chronic Local: Allergen 1.
  Chronic Systemic: 0.
Fire Hazard: Slight, in the form of dust when exposed to heat
  or flame; can react with oxidizing materials (Section 6).
Countermeasures
Personal Hygiene: Section 3.
Storage and Handling: Section 7.

BLACK LEAD. See graphite.

BLACK POWDER. See charcoal, sulfur, potassium nitrate
  or sodium nitrate.
Shipping Regulations: Section 11.
  I.C.C.: Class A explosive—not accepted.
  Coast Guard Classification: Class A explosive.
  IATA: Explosive, not acceptable (passenger and cargo).

**BLACK POWDER IGNITERS WITH EMPTY
  CARTRIDGE BAGS.** See explosives, low.
Shipping Regulations: Section 11.
  I.C.C.: Class C explosive, 150 pounds.
  Coast Guard Classification: Class C explosive.
  IATA: Explosive, explosive label, 25 kilograms (pas-
    senger), 70 kilograms (cargo).

BLACK PRECIPITATE. See mercurous nitrate, ammoniated.

BLADEX. See tetraethyl pyrophosphate.

BLANC FIXE. See barium sulfate.

BLAST FURNACE GAS. See carbon monoxide.

BLASTING CAPS
Shipping Regulations: Section 11.
  IATA: Explosive, not acceptable (passenger and cargo).

BLASTING CAPS, 1000 OR LESS. See explosives, high.
Shipping Regulations: Section 11.
  I.C.C.: Class C explosive.
  Coast Guard Classification: Class C explosive.

BLASTING CAPS, MORE THAN 1000. See explosives,
  high.
  I.C.C.: Class A explosive—not accepted.
  Coast Guard Clasification: Class A explosive.

**BLASTING CAPS WITH METAL CLAD MILD
  DETONATING FUSE—1000, OR LESS.**
Shipping Regulations: Section 11.
  I.C.C.: Class C explosive.

**BLASTING CAPS WITH METAL CLAD MILD
  DETONATING FUSE—MORE THAN 1000**
Shipping Regulations: Section 11.
  I.C.C.: Explosive A, not accepted.

**BLASTING CAPS WITH SAFETY FUSE, 1000 OR
  LESS CAPS.** See explosives, high.

TOXIC HAZARD RATING CODE (For detailed discussion, see Section 1.)

0 NONE: (a) No harm under any conditions; (b) Harmful only under unusual conditions or overwhelming dosage.

1 SLIGHT: Causes readily reversible changes which disappear after end
  of exposure.

2 MODERATE: May involve both irreversible and reversible changes;
  not severe enough to cause death or permanent injury.

3 HIGH: May cause death or permanent injury after very short exposure
  to small quantities.

U UNKNOWN: No information on humans considered valid by authors.

Shipping Regulations: Section 11.
  I.C.C.: Class C explosive.
  Coast Guard Classification: Class C explosive.

BLASTING CAPS WITH SAFETY FUSE—MORE THAN 1000
Shipping Regulations: Section 11.
  I.C.C.: Explosive A, not accepted.

BLASTING GELATIN. See explosives, high.
Synonym: SNG.
Shipping Regulations: Section 11.

BLASTING OIL. See nitroglycerine.

BLASTING POWDERS. See explosives, low.
Shipping Regulations: Section 11.
  I.C.C. Classification: Class A explosive—not accepted.
  Coast Guard Classification: Class A explosive.

BLAU GAS
General Information
Description: A gas.
Composition: Typical gas will analyze 51.9% illuminants; 0.1% carbon monoxide; 2.7% hydrogen; 44.1% methane; 1.2% nitrogen, etc.
Hazard Analysis
Toxic Hazard Rating:
  Acute Local: 0.
  Acute Systemic: Inhalation 2.
  Chronic Local: 0.
  Chronic Systemic: Inhalation 1.
Fire Hazard: Dangerous, when exposed to flame (Section 6).
Explosion Hazard: Slight, when exposed to flame (Section 7).
Disaster Hazard: Moderately dangerous; when heated, it burns and it can react with oxidizing materials.
Countermeasures
Ventilation Control: Section 2.
Storage and Handling: Section 7.
Shipping Regulations: Section 11.
  IATA: Flammable gas, red label, not acceptable (passenger), 140 kilograms (cargo).

BLEACHING POWDER.
General Information
Synonym: Calcium hypochlorite, chlorinated lime, calcium chloride hypochlorite.
Description: White powder.
Formula: $CaCl(ClO) \cdot 4H_2O$.
Constant: Mol wt: 199.0.
Hazard Analysis
Toxic Hazard Rating:
  Acute Local: Irritant 2; Ingestion 3; Inhalation 3.
  Acute Systemic: U.
  Chronic Local: Irritant 2, Inhalation 2.
  Chronic Systemic: U.
Toxicology: Can cause severe irritation of skin and emit fumes capable of causing pulmonary edema.
Fire Hazard: Moderate by chemical reaction with combustible materials (Section 6).
Caution: A powerful oxidizer. Deflagration occurs in contact with combustible substances.
Explosion Hazard: Moderate, in its solid form when heated (Section 7).
Explosive Range: When heated suddenly above 212° F.
Disaster Hazard: Dangerous, when heated to decomposition or on contact with acid or acid fumes, it emits highly toxic fumes and explodes; it will react with water or steam to produce toxic and corrosive fumes; it can react vigorously with reducing materials.
Countermeasures
Ventilation Control: Section 2.
Personnel Protection: Section 3.
First Aid: Section 1.
Storage and Handling: Section 7.
Shipping Regulations: Section 11.
  Coast Guard Classification: Hazardous material.

BLEACH LEATHER
Countermeasures
Shipping Regulations: Section 11.
  IATA: Flammable liquid, red label, 1 liter (passenger), 40 liters (cargo).

BLISTERING BEETLES. See cantharides.

BLISTERING FLIES. See cantharides.

BLUE LEAD. See lead sulfate.

BLUE STONE. See copper sulfate.

BLUE VITRIOL. See copper sulfate.

BOILER COMPOUND LIQUID
Description: A caustic solution.
Toxicity: See sodium hydroxide.
Shipping Regulations: Section 11.
  I.C.C.: Corrosive liquid; white label, 10 gallons.
  Coast Guard Classification: Corrosive liquid; white label.

BOLIDEN SALT. See chromated zinc arsenate.
Bombs, practice, non-explosive, with electric primers or electric squibs.
Shipping Regulations: Section 11.
  I.C.C.: Section 73.55.
  IATA: Not restricted (passenger and cargo).

BOMBS, SAND-LOADED OR EMPTY. See explosives, high.
Shipping Regulations: Section 11.
  IATA: Not restricted (passenger and cargo).

BONE ASH. See calcium phosphate, tribasic.

BONE DUST
General Information
Composition: Phosphates and silicates of calcium, magnesium, etc.
Hazard Analysis
Toxic Hazard Rating:
  Acute Local: Irritant 1; Inhalation 1.
  Acute Systemic: 0.
  Chronic Local: Irritant 1.
  Chronic Systemic: Inhalation 1.
TLV: ACGIH (recommended); 50 million particles per cubic foot of air. See also nuisance dusts.
Countermeasures
Ventilation Control: Section 2.
Personal Hygiene: Section 3.

BOOSTERS (EXPLOSIVE). See explosives, high.
Shipping Regulations: Section 11.
  I.C.C.: Class A explosive—not accepted.
  Coast Guard Classification: Class A explosive.
  IATA: Explosive, not acceptable (passenger and cargo).

BORACIC ACID. See boric acid.

BORANES. See boron hydrides.

BORAX
General Information
Synonyms: Tincal; tinkal.
Description: White gray, bluish or greenish white streak, vitreous or dull luster.
Formula: $Na_2B_4O_7 \cdot 10H_2O$.
Constants: Mol wt: 381.44, mp: 75°C; $-8H_2O$ at 60°C; $-10H_2O$ at 200°C, d: 1.69–1.72.
Hazard Analysis
Toxicity: A herbicide. See boron compounds.

BORAZINE. See S-triazaborane.

BORAZOLE
General Information
Description: Colorless liquid.
Formula: $B_3N_3H_6$.

Constants: Mol wt: 80.53, mp: $-58°$C, bp: $53°$C, d: 0.824 at $0°$C.

Hazard Analysis

Toxic Hazard Rating:
  Acute Local: Irritant 3; Ingestion 3; Inhalation 3.
  Acute Systemic: U.
  Chronic Local: Irritant 2.
  Chronic Systemic: U.
Fire Hazard: Dangerous by chemical reaction to produce flammable or even spontaneously flammable gases (Section 6).
Caution: Hydrolyzes in water to evolve boron hydrides.
Explosion Hazard: Details unknown. See also boron hydrides.
Disaster Hazard: Moderate; when heated to decomposition it emits toxic fumes; and upon reaction with water it can evolve toxic and flammable gases.

Countermeasures

Ventilation Control: Section 2.
Personnel Protection: Section 3.
First Aid: Section 1.

BORDEAUX, ARSENITES, LIQUID. See arsenic and copper compounds.

Shipping Regulations: Section 11.
  I.C.C.: Poison B; poison label, 55 gallons.
  Coast Guard Classification: Poison B; poison label.
  IATA: Poison B, poison label, 1 liter (passenger), 220 liters (cargo).

BORDEAUX, ARSENITES, SOLID. See arsenic and copper compounds.

Shipping Regulations: Section 11.
  I.C.C.: Poison B; poison label, 200 pounds.
  Coast Guard Classification: Poison B; poison label.
  IATA: Poison B, poison label, 25 kilograms (passenger), 95 kilogram (cargo).

BORDEAUX MIXTURE. See copper compounds.

BORIC ACETIC ANHYDRIDE. See boron acetate.

BORIC ACID

General Information

Synonyms: Boracic acid; o-boric acid.
Description: White crystals or powder.
Formula: H_3BO_3.
Constants: Mol wt: 61.84, mp: $185°$C (decomp.) $-1\frac{1}{2}H_2O$ at $300°$C, d: 1.435 at $15°$C.

Hazard Analysis

Toxicity: See boron compounds.

BORIC ACID OINTMENT. See boric acid.

BORINE CARBONYL

General Information

Description: Colorless, unstable gas.
Formula: BH_3CO.
Constants: Mol wt: 41.85, mp: $-137.0°$C, bp: $-64°$C.

Hazard Analysis

Toxicity: See carbon monoxide, boron compounds and boron hydrides.
Fire Hazard: Dangerous, when exposed to heat or flame or by chemical reaction (Section 6). See also boron hydrides and carbon monoxide which are readily evolved by this material upon contact with heat or moisture.
Explosion Hazard: See boron hydrides and carbon monoxide.
Disaster Hazard: Dangerous; when heated to decomposi-

tion, emits highly toxic fumes and it will react with water or steam to produce toxic and flammable vapors.

Countermeasures

Storage and Handling: Section 7.

BORINOAMINOBORINE

General Information

Description: Colorless liquid.
Formula: B_2H_7N.
Constants: Mol wt: 42.70, mp: $-66.5°$C, bp: $76.2°$C, vap d: 14.7.

Hazard Analysis

Toxicity: See boron compounds.
Fire Hazard: It may evolve boron hydrides upon reaction with water or acids (Section 6).
Disaster Hazard: Dangerous; when heated to decomposition, it can emit toxic and flammable decomposition products.

BORNEO CAMPHOR. See d-borneol.

d-BORNEOL

General Information

Synonyms: α-Camphanol; borneo camphor.
Description: Hexagonal crystals, peppery odor and burning taste.
Formula: $C_{10}H_{17}OH$.
Constants: Mol wt: 154.24, mp: $208°$C, bp: $212°$C, flash p: $150°$F, d: 1.01 at $20°/4°$C, vap d: 5.31.

Hazard Analysis

Toxic Hazard Rating:
  Acute Local: Irritant 1.
  Acute Systemic: Ingestion 2.
  Chronic Local: Irritant 1.
  Chronic Systemic: Ingestion 1.
Toxicity: Can cause nausea, vomiting, mental disturbances and convulsions.
Fire Hazard: Slight, when exposed to heat or flame; it can react with oxidizing material.

Countermeasures

To Fight Fire: Water, carbon dioxide, dry chemical or carbon tetrachloride (Section 6).
Storage and Handling: Section 7.
Shipping Regulations: Section 11.
  IATA: Flammable solid, yellow label, 12 kilograms (passenger), 45 kilograms (cargo).

BORNEOL ACETATE. See bornyl acetate.

BORNEOL THIOCYANOACETATE. See bornyl thiocyanoacetate.

BORNYL ACETATE

General Information

Synonym: Borneol acetate.
Description: Crystals; very slightly soluble in water; soluble in alcohol and ether.
Formula: $C_{10}H_{17}OCOCH_3$.
Constants: Mol wt: 196.28, mp: $29°$C, bp: $225–226°$C.

Hazard Analysis

Toxic Hazard Rating: U. See borneol.

BORNYL THIOCYANOACETATE

General Information

Synonym: Borneol thiocyanoacetate.
Description: Crystals.
Formula: $C_{10}H_{17}OOCCH_2CNS$.
Constant: Mol wt: 253.4.

TOXIC HAZARD RATING CODE (For detailed discussion, see Section 1.)

0 NONE: (a) No harm under any conditions; (b) Harmful only under unusual conditions or overwhelming dosage.

1 SLIGHT: Causes readily reversible changes which disappear after end of exposure.

2 MODERATE: May involve both irreversible and reversible changes; not severe enough to cause death or permanent injury.

3 HIGH: May cause death or permanent injury after very short exposure to small quantities.

U UNKNOWN: No information on humans considered valid by authors.

Hazard Analysis
Toxicity: Unknown; an insecticide. See also isobornyl-thiocyanoacetate.
Disaster Hazard: Dangerous; when heated to decomposition or on contact with acid or acid fumes, it emits highly toxic fumes of cyanide.
Countermeasures
Storage and Handling: Section 7.

BOROBUTANE. See dihydrotetraborane.

BOROCAINE.
General Information
Synonym: Procaine borate.
Description: Crystals.
Formula: $C_{13}H_{20}N_2O_2 \cdot 5HBO_2$.
Constants: Mol wt: 455.5, mp: 165–166°C.
Hazard Analysis
Toxic Hazard Rating:
Acute Local: U.
Acute Systemic: Ingestion 2.
Chronic Local: Irritant 1.
Chronic Systemic: Ingestion 1; Inhalation 1; Skin Absorption 1.
Fire Hazard: Slight, when heated.
Countermeasures
Personal Hygiene: Section 3.
Storage and Handling: Section 7.

BOROETHANE. See diborane.

BORON
General Information
Description: Monoclinic crystals, yellow or brown amorphous powder.
Formula: B.
Constants: At wt: 10.82, mp: 2300°C, bp: 2550°C, d: 3.33 at 20°C.
Hazard Analysis
Toxicity: See boron compounds.
Fire Hazard: Moderate in the form of dust when exposed to air or by chemical reaction with oxidizing materials. See also powdered metals.
Explosion Hazard: An explosion hazard in the form of dust, which ignites on contact with air. See also powdered metals.
Countermeasures
Storage and Handling: Section 7.

BORON ACETATE
General Information
Synonym: Boric acetic anhydride.
Description: White to cream, hygroscopic crystals.
Formula: $(CH_3CO_2)_3B$.
Constants: Mol wt: 188.0, mp: 135–144°C.
Hazard Analysis
Toxicity: See boron compounds and acetic acid.
Countermeasures
Storage and Handling: Section 7.

BORON BROMIDE. See boron tribromide.

BORON BROMIDE DIIODIDE
General Information
Description: Colorless liquid.
Formula: $BBrI_2$.
Constants: Mol wt: 344.58, bp: 180°C.
Hazard Analysis
Toxicity: See boron compounds, bromides and iodides.
Disaster Hazard: Dangerous; see bromides and iodides.

BORON BROMIDE PENTAHYDRIDE
General Information
Description: Colorless gas.
Formula: B_2H_5Br.
Constants: Mol wt: 106.60, mp: −104°C, bp: 10°C.

Hazard Analysis
Toxicity: See boron compounds and bromides.
Fire Hazard: Details unknown; this material can evolve a boron hydride.
Disaster Hazard: Dangerous; when heated to decomposition, it emits highly toxic fumes of boron hydrides; it will react with water or steam to produce toxic and corrosive fumes; it can react vigorously on contact with oxidizing materials.
Countermeasures
Storage and Handling: Section 7.

BORON CARBIDE
General Information
Description: Black crystals.
Formula: B_4C.
Constants: Mol wt: 55.29, mp: 2450°C, bp: >3500°C, d: 2.50.
Hazard Analysis
Toxicity: See nuisance dusts.
Countermeasures
Ventilation Control: Section 2.

BORON CHLORIDE. See boron trichloride.

BORON CHLORIDE PENTAHYDRIDE
General Information
Description: Colorless gas, highly unstable.
Formula: B_2H_5Cl.
Constants: Mol wt: 62.14, bp: −78°C at 18 mm.
Hazard Analysis
Toxicity: See boron compounds and hydrochloric acid.
Fire Hazard: See boron hydrides.
Explosion Hazard: See boron hydrides.
Disaster Hazard: Dangerous; when heated to decomposition, it emits highly toxic fumes of boron hydrides and hydrochloric acid; it will react with water or steam to produce toxic and corrosive fumes; can react vigorously with oxidizing material.
Countermeasures
Storage and Handling: Section 7.

BORON COMPOUNDS
Hazard Analysis
Toxic Hazard Rating:
Acute Local: Ingestion 2; Inhalation 2.
Acute Systemic: U.
Chronic Local: 0.
Chronic Systemic: Ingestion 2; Inhalation 2; Skin Absorption 2.
Toxicology: Not highly toxic and therefore not considered an industrial poison. Used in medicine as sodium borate, boric acid, or borax, which is also a common cleaner. Fatal poisoning of children has been caused in some instances by the accidental substitution of boric acid for powdered milk. The medical literature reveals many instances of accidental poisoning due to boric acid, oral ingestion of borates or boric acid, and presumably absorption of boric acid from wounds and burns. The fatal dose of orally ingested boric acid for an adult is somewhat more than 15 or 20 grams and for an infant 5 to 6 grams.
Boron is one of a group of elements, such as lead, manganese and arsenic, which affects the central nervous system. Boron poisoning causes depression of the circulation, persistent vomiting and diarrhea, followed by profound shock and coma. The temperature is subnormal and a scarletina-form rash may cover the entire body. Boric acid intoxication can come about from absorbing toxic quantities from ointments applied to burned areas or to wounds involving loss or damage to such areas of skin, but it is not absorbed from intact skin. When a 5% boric acid solution is used to irrigate body cavities most of the boric acid is absorbed by the tissues. Continuous irrigation of the body

cavities with solutions containing boron can be dangerous.

Countermeasures

Treatment and Antidotes: Large intravenous doses of isotonic salt solution and plasma have been shown to act as an antidote.

Care should be observed in applying ointments and dressings which contain boron over large areas of the body where the skin has been destroyed. It can be absorbed by the body in this way with the toxic effects noted above. Containers of boric acid should be plainly labeled and should differ radically from those which contain powdered milk, particularly in institutions such as hospitals. The careless use of borax as a skin cleaner should be discouraged as well as the continuous irrigation of body cavities with solutions containing boron.

Ventilation Control: Section 2.

Personnel Protection: Section 3.

BORON DIBROMIDE IODIDE

General Information

Description: Colorless liquid.

Formula: BBr_2I.

Constants: Mol wt: 297.57, bp: $125°C$, vap d: 10.3.

Hazard Analysis

Toxicity: See boron compounds, bromides and iodides.

Disaster Hazard: Dangerous; see bromides and iodides; it will react with water or steam to produce toxic and corrosive fumes.

Countermeasures

Storage and Handling: Section 7.

BORON FLUORIDE. See boron trifluoride.

BORON FLUORIDE ETHERATE. See boron trifluoride etherate.

BORON FLUORIDE MONOETHYLAMINE. See boron, fluorides, and amines.

BORON HYDRIDES

General Information

Diborane; dihydrotetraborane; pentaborane (stable); pentaborane (unstable); hexaborane; decaborane.

Hazard Analysis

Toxic Hazard Rating:

Acute Local: Irritant 3; Inhalation 3.

Acute Systemic: Inhalation 3.

Chronic Local: Inhalation 3.

Chronic Systemic: Inhalation 3.

Toxicology: Experimental studies in which dogs were exposed to diborane by inhalation resulted in severe irritation of the lungs and pulmonary edema. Injuries to CNS, liver and kidneys have also been produced in experimental animals. Similar observations have been reported in humans resulting at times in a reaction resembling metal fume fever. Human exposure to pentaborane has produced signs of severe central nervous system irritation such as drowsiness, dizziness, visual disturbances, muscle twitching and in severe cases painful muscle spasm.

Fire Hazard: Dangerous, when exposed to heat or flame or by chemical reaction. Pentaborane (stable) is spontaneously flammable in air. On contact with moisture, hydrogen is usually evolved (Section 6).

Explosion Hazard: Moderate, when exposed to heat or flame or by chemical reaction. Diborane reacts explosively with chlorine. Other boron hydrides evolve hydrogen

upon contact with moisture or can propagate a flame rapidly enough to cause an explosion.

Disaster Hazard: Severe. Heat can cause these materials to decompose violently or at least to evolve hydrogen; these materials react with water or steam to evolve hydrogen. Powerful oxidizing agents such as chlorine gas, etc., can react violently with boron hydrides.

Countermeasures

Storage and Handling: Section 7.

BORON METHYL. See boron trimethyl.

BORON MONOXIDE. See boron compounds.

BORON NITRIDE

General Information

Description: Hexagonal, white crystals.

Formula: BN.

Constants: Mol wt: 24.83, mp: $3000°C$ (sublimes), d: 2.20.

Hazard Analysis

Toxicity: Details unknown. This material is nearly insoluble and possibly therefore, a nuisance dust (see also boron compounds).

BORON OXIDE

General Information

Description: Vitreous, colorless crystals.

Formula: B_2O_3.

Constants: Mol wt: 69.64, mp: $450°C$ (approx.), bp: $1500°C$, d: 1.844.

Hazard Analysis

Toxicity: Animal experiments suggest relatively low toxicity. A herbicide. See boron compounds.

TLV: (Accepted) 15 mg per cubic meter of air.

BORON PENTASULFIDE

General Information

Description: White crystals.

Formula: B_2S_5.

Constants: Mol wt: 181.97, mp: $390°C$, d: 1.85.

Hazard Analysis

Toxicity: See boron compounds and sulfides.

Fire Hazard: See sulfides.

Disaster Hazard: See sulfides.

Countermeasures

Storage and Handling: Section 7.

BORON PHOSPHATE

General Information

Description: A white powder.

Formula: BPO_4.

Constants: Mol wt: 105.8, mp: $>1200°C$, d: 2.52.

Hazard Analysis

Toxicity: See boron compounds.

BORON PHOSPHIDE

General Information

Description: Maroon powder.

Formula: BP.

Constants: Mol wt: 41.80, mp: ign. $200°C$.

Hazard Analysis

Toxicity: See boron compounds and phosphides.

Fire Hazard: Ignites at $200°C$. See also phosphides.

Explosion Hazard: See phosphides.

Disaster Hazard: See phosphides.

Countermeasures

Storage and Handling: Section 7.

TOXIC HAZARD RATING CODE (For detailed discussion, see Section 1.)

0 NONE: (a) No harm under any conditions; (b) Harmful only under unusual conditions or overwhelming dosage.

1 SLIGHT: Causes readily reversible changes which disappear after end of exposure.

2 MODERATE: May involve both irreversible and reversible changes; not severe enough to cause death or permanent injury.

3 HIGH: May cause death or permanent injury after very short exposure to small quantities.

U UNKNOWN: No information on humans considered valid by authors.

BORON TRIBROMIDE
General Information
Synonym: Boron bromide.
Description: Colorless, fuming liquid.
Formula: BBr_3.
Constants: Mol wt: 250.57, mp: $-45°C$, bp: $91.7°C$, d: 2.650 at $0°C$, vap. press.: 40 mm at $14.0°C$, 100 mm at $33.5°C$.
Hazard Analysis
Toxicity: See boron compounds and hydrobromic acid.
Disaster Hazard: Dangerous; when heated to decomposition it can explode and it emits toxic fumes of bromides and it will react with water or steam to produce toxic and corrosive fumes.
Countermeasures
Storage and Handling: Section 7.
Shipping Regulations: Section 11.
  IATA: Corrosive liquid, white label, not accepted (passenger), 1 liter (cargo).

BORON TRICHLORIDE
General Information
Synonym: Boron chloride.
Description: Colorless, fuming liquid.
Formula: BCl_3.
Constants: Mol wt: 117.19, mp: $-107°C$, bp: $12.5°C$, d: 1.434 at $0°C$, vap pres. 1 atm at $12.7°C$, vap d: 4.03.
Hazard Analysis
Toxicity: See boron compounds and hydrochloric acid.
Disaster Hazard: Dangerous; when heated to decomposition, it emits toxic fumes of chlorides; it will react with water or steam to produce heat, toxic and corrosive fumes.
Countermeasures
Storage and Handling: Section 7.
Shipping Regulations: Section 11.
  I.C.C.: Corrosive liquid; white label, 1 quart.
  Coast Guard Classification: Corrosive liquid; white label.
  IATA: Corrosive liquid, white label, not acceptable (passenger), 1 liter (cargo).

BORON TRICYCLOHEXYL. See tricyclohexyl borine.

BORON TRIETHYL
General Information
Synonyms: Boron ethyl; triethyl borine; triethyl boron.
Description: Colorless liquid.
Formula: $B(C_2H_5)_3$.
Constants: Mol wt: 98.0, mp: $-93°C$, bp: $95°C$, d: 0.6961 at $23°C$.
Hazard Analysis
Toxic Hazard Rating:
  Acute Local: Inhalation 3.
  Acute Systemic: Inhalation 3.
  Chronic Local: U.
  Chronic Systemic: Ingestion 2.
Toxicity: Animal experiments show high vapor toxicity and moderate toxicity leading to pulmonary irritation and convulsions.
Fire Hazard: Dangerous, by spontaneous chemical reaction with oxidizers (Section 6).
Caution: Spontaneously flammable in air.
Explosion Hazard: Unknown.
Disaster Hazard: Highly dangerous, when heated to decomposition or upon contact with air it emits toxic fumes; it will react with water or steam to produce toxic and flammable vapors; it can react vigorously with oxidizing materials.
Countermeasures
Storage and Handling: Section 7.
To Fight Fire: Do not use halogenated extinguishing agents. (Section 6).

BORON TRIFLUORIDE
General Information
Synonym: Boron fluoride.
Description: Colorless gas.
Formula: BF_3.
Constants: Mol wt: 67.82, mp: $-126.8°C$, bp: $-110.7°C$, d: 2.99 g/liter.
Hazard Analysis
Toxic Hazard Rating:
  Acute Local: Irritant 3; Inhalation 3.
  Acute Systemic: U.
  Chronic Local: Irritant 3; Inhalation 3.
  Chronic Systemic: U.
A strong irritant.
TLV: ACGIH (recommended) 1 ppm in air; 3 mg/mi.
Disaster Hazard: Dangerous; when heated to decomposition or upon contact with water or steam, it will produce toxic and corrosive fumes.
Countermeasures
Storage and Handling: Section 7.
Shipping Regulations: Section 11.
  I.C.C.: Nonflammable gas; green label, 300 pounds.
  Coast Guard Classification: Noninflammable gas; green gas label.
  IATA: Nonflammable gas, green label, not acceptable (passenger), 140 kilograms (cargo).

BORON TRIFLUORIDE—ACETIC ACID COMPLEX
Shipping Regulations: Section 11.
  IATA: Corrosive liquid, white label, 1 liter (passenger), 5 liters (cargo).

BORON TRIFLUORIDE ETHERATE
General Information
Composition: Boron trifluoride ether.
Formula: $CH_3CH_2O(BF_3)CH_2CH_3$.
Constants: Mol wt: 142, d: 1.1, bp: $259°F$, flash p: $147°F$ (O.C.).
Hazard Analysis
Toxicity: See boron compounds, fluorides and ether.
Fire Hazard: See ether.
Explosion Hazard: See ether.
Disaster Hazard: Dangerous; see fluorides; it will react with water or steam to produce toxic, corrosive and flammable vapors; it can react vigorously with oxidizing materials.
Countermeasures
Storage and Handling: Section 7.

BORON TRIIODIDE
General Information
Description: Colorless, hygroscopic plates.
Formula: BI_3.
Constants: Mol wt: 391.58, mp: $43°C$, bp: $210°C$, d: 3.35 at $50°C$.
Hazard Analysis
Toxicity: See boron compounds and iodides.
Disaster Hazard: Dangerous; see iodides.
Countermeasures
Storage and Handling: Section 7.

BORON TRIMETHYL
General Information
Synonyms: Boron methyl; trimethyl borine.
Description: Colorless gas.
Formula: $B(CH_3)_3$.
Constants: Mol wt: 55.9, mp: $-161.5°C$, bp: $-20°C$, d: 1.91 g/liter (gas); 0.625 (solid at $-100°C$).
Hazard Analysis
Toxicity: Probably very toxic. See boron compounds.
Fire Hazard: Dangerous, when exposed to flame or by chemical reaction with oxidizing agents (Section 6).
Caution: Spontaneously flammable gas.
Explosion Hazard: Unknown.
Disaster Hazard: Highly dangerous; can react vigorously with oxidizing materials.

Countermeasures
Storage and Handling: Section 7.

BORON TRIOXIDE. See boron oxide.

BORON TRIPHENYL
General Information
Synonym: Triphenyl boron.
Description: Colorless crystals decomposed by water.
Formula: $(C_6H_5)_3B$.
Constants: Mol wt: 242.1, mp: 136°C, bp: 203°C at 15 mm Hg.
Hazard Analysis
Toxicity: Details unknown. See phenol and boron compounds.

BORON TRIPHENYL AMMINE
General Information
Description: Colorless crystals soluble in alcohol.
Formula: $(C_6H_5)_3BNH_3$.
Constants: Mol wt: 259.2, mp: 216°C (decomp).
Hazard Analysis
Toxicity: Details unknown. See phenols and boron compounds.

BORON TRISELENIDE
General Information
Description: Yellow gray powder.
Formula: B_2Se_3.
Constant: Mol wt: 258.52.
Hazard Analysis
Toxicity: See boron compounds and selenium compounds.
Disaster Hazard: See selenium compounds.
Countermeasures
Storage and Handling: Section 7.

BORON TRISULFIDE
General Information
Description: White crystals.
Formula: B_2S_3.
Constants: Mol wt: 117.84, mp: 310°C, d: 1.55.
Hazard Analysis and Countermeasures
See boron compounds and sulfides.

BOROPHOSPHORIC ACID
General Information
Description: White crystals.
Formula: $BPO_4 \cdot H_2O$.
Constants: Mol wt: 123.8, mp: decomposes, d: 1.873.
Hazard Analysis and Countermeasures
See boron compounds and phosphoric acid.

BOROTUNGSTIC ACID
General Information
Description: Tetragonal, colorless crystals.
Formula: $H_5BW_{12}O_{40} \cdot 30H_2O$.
Constants: Mol wt: 3403.39, mp: 45–51°C, d: 3.0.
Hazard Analysis
Toxicity: See boron compounds and tungsten compounds.

BOTTLE GAS. See liquefied hydrocarbon gas.

BOX TOE GUM
General Information
Description: Liquid.
Hazard Analysis
Toxicity: Unknown.
Fire Hazard: Dangerous, when exposed to heat or flame; can react with oxidizing materials (Section 6).

Countermeasures
Storage and Handling: Section 7.
Shipping Regulations: Section 11.
 I.C.C.: Flammable liquid; red label, 10 gallons.
 Coast Guard Classification: Inflammable liquid; red label.
 IATA: Flammable liquid, red label, 1 liter (passenger), 40 liters (cargo).

BOXWOOD
General Information
Description: A wood.
Hazard Analysis
Toxic Hazard Rating:
 Acute Local: Irritant 1; Allergen 2.
 Acute Systemic: 0.
 Chronic Local: Allergen 2.
 Chronic Systemic: 0.
Fire Hazard: Slight, when exposed to heat or flame; it can react with oxidizing materials (Section 6).
Countermeasures
Personnel Protection: Section 3.

BRAKE FLUID, HYDRAULIC
Shipping Regulations: Section 11.
 IATA: Flammable liquid, red label, 1 liter (passenger), 40 liters (cargo).

BRASS (MASSIVE)
General Information
Description: Pale gold to red metal alloy.
Composition: $Zn + Cu$ (mainly), Pb (small amounts), As, Sb, P, Sn, CN^- (traces).
Constants: Mp: 940–1030°C, d: 8.40–8.75.
Hazard Analysis
Toxicity: See brass scrapings.

BRASS POWDER. See brass scrapings.

BRASS SCRAPINGS
General Information
Synonym: Brass powder.
Description: Finely divided particles of metal.
Composition: Zinc, copper (mainly), lead (minor constituent), arsenic, phosphorus, tin, antimony, cyanides (trace quantities).
Hazard Analysis
Toxicity: See specific element, particularly zinc, copper, lead. Where quantities of brass scrapings can come in contact with acid or acid fumes, significant quantities of toxic fumes can be evolved. See arsenic, antimony, phosphorus, tin and cyanides.
Fire Hazard: Moderate, when exposed to heat or flame. See also powdered metals.
Explosion Hazard: Slight, when exposed to flame. See also powdered metals.
Disaster Hazard: Moderately dangerous; it can react with oxidizing materials, and on contact with acid or acid fumes, it can emit toxic fumes.
Countermeasures
Storage and Handling: Section 7.

BRAZIL NUT
Hazard Analysis
Toxic Hazard Rating:
 Acute Local: Allergen 1.
 Acute Systemic: U.
 Chronic Local: Allergen 1.
 Chronic Systemic: U.

TOXIC HAZARD RATING CODE *(For detailed discussion, see Section 1.)*

0 NONE: (a) No harm under any conditions; (b) Harmful only under unusual conditions or overwhelming dosage.

1 SLIGHT: Causes readily reversible changes which disappear after end of exposure.

2 MODERATE: May involve both irreversible and reversible changes; not severe enough to cause death or permanent injury.

3 HIGH: May cause death or permanent injury after very short exposure to small quantities.

U UNKNOWN: No information on humans considered valid by authors.

Toxicology: These nuts have been found to contain a concentration of about 1000 times as much ^{226} Ra and ^{222}Th and their associated daughter products as usual foods. These nuts are also relatively high in barium content.

Fire Hazard: Slight, when exposed to heat or flame (section 6).

Countermeasures
Personal Hygiene: Section 3.
Storage and Handling: Section 7.

BRAZIL WAX. See carnauba wax.

BRAZIL WOOD
General Information
Synonyms: Redwood; pernambuco; fernambuco.
Description: A wood.
Toxic Hazard Rating:
  Acute Local: Allergen 1.
  Acute Systemic: U.
  Chronic Local: Allergen 1.
  Chronic Systemic: U.
Fire Hazard: Slight, when exposed to heat or flame; it can react with oxidizing materials (Section 6).
Countermeasures
Personal Hygiene: Section 3.
Storage and Handling: Section 7.

BREITHAUPTITE. See nickel antimonide.

BRILLIANT CRESYL BLUE BB (L) 877
General Information
Description: A dye.
Hazard Analysis
Toxic Hazard Rating:
  Acute Local: Allergen 1.
  Acute Systemic: U.
  Chronic Local: Allergen 1.
  Chronic Systemic: U.
Disaster Hazard: Moderately dangerous; when heated to decomposition, it emits toxic fumes.
Countermeasures
Personal Hygiene: Section 3.
Storage and Handling: Section 7.

BRIMSTONE. See sulfur.
Shipping Regulations: Section 11.
  Coast Guard Classification: Hazardous article.

BRITISH ANTI-LEWISITE. See 2,3-dimercapto-1-propanol.

BROMACETIC ACID
General Information
Synonym: Bromoacetic acid.
Description: Hygroscopic crystals; very soluble in water or alcohol.
Formula: $CH_2BrCOOH$.
Constants: Mol wt: 158.96, d: 1.93, mp: 50°C, bp: 208°C.
Hazard Analysis
Toxic Hazard Rating:
  Acute Local: Irritant 3; Ingestion 3; Inhalation 3.
  Acute Systemic: U.
  Chronic Local: U.
  Chronic Systemic: U.
A strong irritant and corrosive.
Disaster Hazard: Dangerous. See bromides.
Countermeasures
Shipping Regulations: Section 11.
  IATA: Other restricted articles, class B; no label required, 12 kilograms (passenger), 45 kilograms (cargo).

BROMACETIC ACID, SOLUTION
Shipping Regulations: Section 11.
  IATA: Corrosive liquid, white label, 1 liter (passenger and cargo).

BROMACETYLENE. See bromoacetylene.

BROMATES
Hazard Analysis
Toxic Hazard Rating:
  Acute Local: U.
  Acute Systemic: Ingestion 2.
  Chronic Local: U.
  Chronic Systemic: Ingestion 2.
Toxicology: Generally considered to be more toxic than chlorates, causing central nervous system paralysis. They may form methemoglobin, but less actively than chlorates. See also specific compounds as listed.
Fire Hazard: Moderate; in the form of gas, vapor or dust by chemical reaction with reducing agents (Section 6).
Caution: These are powerful oxidizing agents. They may react with oxidizable materials more or less violently, very often causing combustion.
Disaster Hazard: Dangerous; when heated to decomposition, they emit toxic fumes of bromine and they can react with reducing material.
Countermeasures
Personal Hygiene: Section 3.
Storage and Handling: Section 7.
Shipping Regulations: Section 11.
  IATA (n.o.s.): Oxidizing material, yellow label, 12 kilograms (passenger), 45 kilograms (cargo).

p-BROMBENZYL CHLORIDE
General Information
Synonyms: p-Bromo-α-chlorotholuene; α-chloro-4-bromotoluene.
Description: Needles; freely soluble in hot alcohol.
Formula: $C_6H_4BrCH_2Cl$.
Constants: Mol wt: 206, mp: 40–41°C, bp: 105–115°C (12 mm).
Hazard Analysis
See benzyl chloride.

BROMELLITE. See beryllium oxide.

BROMIC ACID
General Information
Description: Colorless or slightly yellow liquid; turns yellow on exposure.
Formula: $HBrO_3$.
Constants: Mol wt: 128.92, mp: decomposes 100°C, d: 3.1883.
Hazard Analysis
Toxic Hazard Rating:
  Acute Local: Irritant 3; Ingestion 3; Inhalation 3.
  Acute Systemic: U.
  Chronic Local: Irritant 2.
  Chronic Systemic: U.
Fire Hazard: See bromates.
Disaster Hazard: See bromates.
Countermeasures
Ventilation Control: Section 2.
Personnel Protection: Section 3.
First Aid: Section 1.
Storage and Handling: Section 7.

BROMIC ETHER. See ethyl bromide.

BROMIDES
Hazard Analysis
Toxic Hazard Rating:
  Acute Local: Irritant 1; Ingestion 2; Inhalation 2.
  Acute Systemic: Ingestion 2.
  Chronic Local: Irritant 1.
  Chronnic Systemic: Ingestion 2; Inhalation 2.
Toxicology: The most common inorganic bromides are sodium, potassium, ammonium, calcium and magnesium bromides. Methyl and ethyl bromides are among the most common organic bromides.

The inorganic bromides produce depression, emaciation and in severe cases, psychoses and mental deterioration. Bromide rashes (bromoderma) especially of the face, and resembling acne and furunculosis, often occur when bromide inhalation or administration is prolonged.

Organic bromides such as methyl bromide and ethyl bromide, are volatile liquids of relatively high toxicity. See also specific compounds.

Disaster Hazard: When strongly heated, they emit highly toxic fumes.

Countermeasures

Ventilation Control: Section 2.

Personnel Protection: Section 3.

BROMINE
General Information

Description: Rhombic crystals or dark-red liquid.

Formula: Br_2.

Constants: Mol wt: 159.83, fp: $-7.3°C$, bp: $58.73°C$, d: 2.928 at 59°C; 3.12 at 20°C, vap. press.: 175 mm at 21°C; 1 atm at 58.2°C, vap. d.: 5.5.

Hazard Analysis

Toxic Hazard Rating:

Acute Local: Irritant 3; Ingestion 3; Inhalation 3.

Acute Systemic: U.

Chronic Local: Irritant 2.

Chronic Systemic: Ingestion 2; Inhalation 2.

TLV: ACGIH (recommended): 0.1 parts per million in air; 0.7 mg/cubic meter of air.

Toxicology: The action of bromine is essentially the same as that of chlorine, being an irritant to the mucous membranes of the eyes and upper respiratory tract. Severe exposures may result in pulmonary edema. Usually, however, the irritating qualities of the chemical force the workman to leave the exposure before serious poisoning can result. Chronic exposure similar to therapeutic ingestion of excessive bromides. See also bromides. Regular physical examinations should be made upon people who work with bromine.

Radiation Hazard: Section 5. For permissible levels, see Table 5, p. 150.

Artificial isotope ^{82}Br, half life 36 h. Decays to stable ^{82}Kr by emitting beta particles of 0.44 MeV and others. Also emits gamma rays of 0.55–1.47 MeV.

Fire Hazard: Moderate; in the form of liquid or vapor by spontaneous chemical reaction with reducing materials (Section 6).

Caution: May ignite a combustible material upon contact. A very powerful oxidizer.

Disaster Hazard: Highly dangerous; when heated, it emits highly toxic fumes; it will react with water or steam to produce toxic and corrosive fumes; and it can react vigorously with reducing materials.

Countermeasures

Ventilation Control: Section 2.

Personnel Protection: Section 3.

Storage and Handling: Section 7.

Shipping Regulations: Section 11.

I.C.C.: Corrosive liquid; white label, 1 quart.

Coast Guard Classification: Corrosive liquid; white label.

IATA: Corrosive liquid, white label, not acceptable (passenger), 1 liter (cargo).

MCA warning label.

BROMINE ANALOG OF DDT. See 1-trichloro-2,2-bis (p-bromophenyl)ethane.

BROMINE AZIDE
General Information

Synonym: Bromoazide.

Description: Crystals or red liquid.

Formula: BrN_3.

Constants: Mol wt: 121.94, mp: 45°C, bp: explodes.

Hazard Analysis

Toxicity: See bromine and azides.

Fire Hazard: Moderate, in the form of vapor by chemical reaction (Section 6).

Caution: A powerful oxidant. See also bromine.

Explosion Hazard: Moderate, when exposed to heat (Section 7).

Disaster Hazard: Dangerous; when heated to decomposition, it emits highly toxic fumes of bromine and explodes; it will react with water or steam to produce toxic and corrosive fumes and it can react on contact with reducing materials.

Countermeasures

Storage and Handling: Section 7.

BROMINE CHLORIDE
General Information

Description: Reddish-yellow liquid or gas.

Formula: BrCl.

Constants: Mol wt: 115.37, bp: decomposes 10°C.

Hazard Analysis

Toxicity: See bromine, chlorine.

Fire Hazard: Moderate, in the form of vapor by spontaneous chemical reaction.

Caution: A powerful oxidant. See also bromine.

Disaster Hazard: Dangerous; when heated to decomposition it emits highly toxic fumes of bromine and chlorine; it will react with water or steam to produce toxic and corrosive fumes and on contact with reducing material, it can react vigorously.

Countermeasures

Storage and Handling: Section 7.

BROMINE CYANIDE. See cyanogen bromide.

BROMINE DIOXIDE
General Information

Description: Light-yellow crystals.

Formula: BrO_2.

Constants: Mol wt: 111.92, mp: 0°C (decomposes).

Hazard Analysis

Toxicity: See bromine.

Fire Hazard: Moderate, in the form of vapor by chemical reaction with reducing agents (Section 6).

Caution: A strong oxidant.

Disaster Hazard: Dangerous; when heated to decomposition it emits highly toxic fumes of bromine and will react with water or steam to produce toxic and corrosive fumes; can react vigorously with reducing materials.

Countermeasures

Storage and Handling: Section 7.

BROMINE HYDRATE
General Information

Description: Red crystals.

Formula: $Br_2 \cdot 10H_2O$.

Constants: Mol wt: 340.0, mp: 6.8°C(decomposes).

Hazard Analysis

Toxicity: See bromine.

Fire Hazard: Moderate, in the form of vapor by chemical reaction with reducing materials (Section 6).

TOXIC HAZARD RATING CODE (For detailed discussion, see Section 1.)

0 NONE: (a) No harm under any conditions; (b) Harmful only under unusual conditions or overwhelming dosage.

1 SLIGHT: Causes readily reversible changes which disappear after end of exposure.

2 MODERATE: May involve both irreversible and reversible changes; not severe enough to cause death or permanent injury.

3 HIGH: May cause death or permanent injury after very short exposure to small quantities.

U UNKNOWN: No information on humans considered valid by authors.

Caution: A strong oxidant.
Disaster Hazard: Dangerous; see bromine and on contact with water or steam, it will react to produce toxic and corrosive fumes; can react vigorously with reducing materials.
Countermeasures
Storage and Handling: Section 7.

BROMINE MONOXIDE
General Information
Description: Dark-brown crystals.
Formula: Br_2O.
Constants: Mol wt: 175.83, mp: −17 to −18°C (decomposes).
Hazard Analysis
Toxicity: See bromine.
Fire Hazard: Moderate, by chemical reaction with reducing agents (Section 6).
Caution: A strong oxidant.
Disaster Hazard: Dangerous; see bromine; it will react with water or steam to produce toxic and corrosive fumes; can react vigorously with reducing materials.
Countermeasures
Storage and Handling: Section 7.

BROMINE OCTOXIDE
General Information
Description: White crystals.
Formula: Br_3O_8.
Constants: Mol wt: 367.75, mp: stable at −40°C.
Hazard Analysis
Toxicity: See bromine.
Fire Hazard: Moderate, upon contact with reducing agents (Section 6).
Caution: A strong oxidant.
Disaster Hazard: Dangerous; see bromine; it will react with water or steam to produce toxic and corrosive fumes; can react vigorously with reducing materials.
Countermeasures
Storage and Handling: Section 7.

BROMINE PENTAFLUORIDE
General Information
Description: Colorless fuming liquid.
Formula: BrF_5.
Constants: Mol wt: 174.92, mp: −61.3°C, bp: 40.5°C, d: 2.466 at 25°C, vap. d.: 6.05.
Hazard Analysis
Toxic Hazard Rating:
Acute Local: Irritant 3; Ingestion 3; Inhalation 3.
Acute Systemic: U.
Chronic Local: Irritant 3.
Chronic Systemic: U.
Toxicity: See bromine and fluorides.
Disaster Hazard: Dangerous; see bromine and fluorine; it will react with water or steam to produce toxic and corrosive fumes.
Countermeasures
Storage and Handling: Section 7.
Shipping Regulations: Section 11.
I.C.C.: Corrosive liquid; white label, 100 pounds.
Coast Guard Classification: Corrosive liquid; white label.
IATA: Corrosive liquid, white label, not acceptable (passenger), 45 kilograms (cargo).

BROMINE TRIFLUORIDE
General Information
Description: Colorless, fuming liquid.
Formula: BrF_3.
Constants: Mol wt: 136.92, mp: 8.8°C, bp: 127°C, d: 2.84.
Toxicity: See bromine pentafluoride.
Disaster Hazard: Very dangerous; see bromides and fluorides. Very reactive.
Countermeasures
Storage and Handling: Section 7.

Shipping Regulations: Section 11.
I.C.C.: Corrosive liquid; white label, 100 pounds.
Coast Guard Classification: Corrosive liquid; white label.
IATA: Corrosive liquid, white label, not acceptable (passenger), 45 kilograms (cargo).

N-BROMOACETAMIDE
General Information
Synonym: NBA.
Description: White powder, odor similar to bromine.
Formula: $CH_3CONHBr$.
Constants: Mol wt: 138.0, mp: 105–108°C.
Hazard Analysis
Toxicity: Unknown.
Disaster Hazard: Dangerous; see bromine.
Countermeasures
Storage and Handling: Section 7.

BROMOACETIC ACID. See bromacetic acid

BROMOACETONE
General Information
Description: Colorless liquid when pure; pungent odor. Slightly soluble in water. Soluble in alcohol and acetone.
Formula: $CH_2BrCOCH_3$.
Constants: Mol wt: 136.99, mp: −54°C, bp: 136°C, d: 1.631 at 0°C.
Hazard Analysis
Toxic Hazard Rating:
Acute Local: Irritant 3; Inhalation 3.
Acute Systemic: U.
Chronic Local: U.
Chronic Systemic: U.
Toxicology: Intensely irritating. Powerful lacrimator. A chemical warfare agent.
Disaster Hazard: Dangerous; see bromides.
Countermeasures
Storage and Handling: Section 7.
Shipping Regulations: Section 11.
I.C.C.: Poison A; poison gas label, not accepted.
Coast Guard Classification: Poison A; poison gas label.
IATA: Poison A, not acceptable (passenger and cargo).

BROMOACETOPHENONE
General Information
Synonym: Phenacyl bromide.
Description: White rhombic prisms changing to greenish color under the influence of light.
Formula: $BrCH_2COC_6H_5$.
Constants: Mol wt: 199.05, mp: 50°C, bp: 140°C at 12 mm, d: 1.647 at 20°/4°C.
Hazard Analysis
Toxicity: Unknown. See bromines and phenyl methyl ketone.
Disaster Hazard: Dangerous; see bromides.
Countermeasures
Storage and Handling: Section 7.

BROMOACETYLENE
General Information
Synonym: Bromoethyne, bromacetylene.
Description: Gas.
Formula: CHCBr.
Constants: Mol wt: 104.9, bp: −2°C, vap. d.: 4.684.
Hazard Analysis
Toxicity: Unknown. Probably similar to dibromoacetylene.
Fire Hazard: Dangerous, by spontaneous chemical reaction (Section 6).
Caution: A spontaneously flammable gas.
Explosion Hazard: Unknown.
Disaster Hazard: Dangerous; when heated to decomposition, it burns and emits toxic fumes; it can react with oxidizing materials.
Countermeasures
Storage and Handling: Section 7.

BROMOANILINE
General Information
Synonym: o-Bromoaniline.
Description: Crystals.
Formula: $BrC_6H_4NH_2$.
Constants: Mol wt: 172.0, mp: 32°C, bp: 229°C.
Hazard Analysis
Toxicity: Details unknown. See also aniline.
Disaster Hazard: Dangerous; when heated to decomposition or on contact with acid or acid fumes, it emits toxic fumes.
Countermeasures
Storage and Handling: Section 7.

BROMOAURIC ACID
General Information
Description: Red-brown crystals.
Formula: $HAuBr_4 \cdot 5H_2O$.
Constants: Mol wt: 607.95, mp: 27°C.
Hazard Analysis
Toxicity: See bromides and gold compounds.
Disaster Hazard: Dangerous, see bromine.
Countermeasures
Storage and Handling: Section 7.

BROMOAZIDE. See bromine azide.

p-BROMOAZOBENZENE
General Information
Formual: $BrC_6H_4NNC_6H_5$.
Constant: Mol wt: 261.1.
Hazard Analysis
Toxicity: Unknown; an insecticide. See also bromobenzene.
Disaster Hazard: Dangerous; see bromides.
Countermeasures
Storage and Handling: Section 7.

BROMOBENZENE
General Information
Synonym: Phenyl bromide.
Description: Clear, colorless, mobile liquid.
Formual: C_6H_5Br.
Constants: Mol wt: 157.02, mp: −30.7°C, bp: 156.2°C, flash p: 124°F, d: 1,497, vap. press.: 10 mm at 40°C, vap. d.: 5.41, autoign. temp.: 1051°F.
Hazard Analysis
Toxic Hazard Rating:
Acute Local: Irritant 2.
Acute Systemic: Ingestion 1; Inhalation 1.
Chronic Local: U.
Chronic Systemic: U.
Fire Hazard: Moderate, when exposed to heat or flame.
Disaster Hazard: Dangerous; see bromides. It can react with oxidizing materials.
Countermeasures
Storage and Handling: Section 7.
To Fight Fire: Water, foam, carbon dioxide, dry chemical, or carbon tetrachloride (Section 6).
Shipping Regulations: Section 11.
Coast Guard Classification: Inflammable liquid.

p-BROMOBENZENETETRATHIO-o-STANNATE
General Information
Description: Solid.
Formula: $Sn(SC_6H_4Br)_4$.
Constants: Mol wt: 871.0, mp: 217°C.
Hazard Analysis
Toxicity: Details unknown. See tin compounds.
Disaster Hazard: Dangerous. Sulfur and bromine fumes.

p-BROMOBENZONITRILE
General Information
Description: Crystals.
Formula: BrC_6H_4CN.
Constants: Mol wt: 182, mp: 113°C.
Hazard Analysis
Toxicity: An insecticide. See also nitriles.
Disaster Hazard: See nitriles.
Countermeasures
Storage and Handling: Section 7.

p-BROMOBENZYL BROMIDE
General Information
Synonym: p,α-Dibromotoluene.
Description: Crystals; aromatic odor; soluble in water and cold alcohol; more soluble in hot alcohol, ether, carbon disulfide, benzene, and glacial acetic acid.
Formula: $C_7H_6Br_2$.
Constants: Mol wt: 249.5, mp: 61°C, bp: 115–124°C at 12 mm.
Hazard Analysis
See benzyl bromide.

BROMOBENZYL CYANIDE
General Information
Synonym: o-Bromo-2-phenylacetonitrile; BBC; bromobenzyl nitride.
Description: (Pure) yellowish-white crystals; (tech) brown, oily liquid; pungent odor of soured fruit.
Formula: $C_6H_4CH_2CNBr$.
Constants: Mol wt: 196.1, mp: 29°C, bp: 242°C, fp: 25.5°C, d: 1.5160 at 20°C, flash p: none, vap. d.: 6.8, vap. press.: .011 mm at 20°C.
Hazard Analysis
Toxic Hazard Rating:
Acute Local: Irritant 3; Inhalation 3.
A strong lachrymator. See cyanides.
Disaster Hazard: See cyanides.
Countermeasures
Storage and Handling: Section 7.
Shipping Regulations: Section 11.
I.C.C.: Poison C; tear gas label, 20 pounds.
Coast Guard Classification: Poison C; tear gas label.
IATA: Poison C, poison label, not acceptable (passenger), 10 kilograms (cargo).

o-BROMOBENZYL CYANIDE. See bromobenzyl cyanide.

BROMOBENZYL NITRILE. See bromobenzyl cyanide.

BROMOBUTANE. See butyl bromide.

α-BROMOBUTYRIC ACID
General Information
Description: Colorless, oily liquid; soluble in alcohol and ether; sparingly soluble in water.
Formula: $CH_3CH_2CHBrCOOH$.
Constants: Mol wt: 167, d: 1.54, bp: 181°C at 250 mm. mp: −4°C.
Hazard Analysis
Toxic Hazard Rating: U. Animal experiments suggest toxicity greater than that of α-bromoisobutyric acid.
Disaster Hazard: Dangerous. See bromides.

2-BROMO-4-tert-BUTYL PHENOL
General Information
Description: Straw-colored liquid.
Formula: $(CH_3)_3CC_6H_3BrOH$.

TOXIC HAZARD RATING CODE (For detailed discussion, see Section 1.)

0 NONE: (a) No harm under any conditions; (b) Harmful only under unusual conditions or overwhelming dosage.

1 SLIGHT: Causes readily reversible changes which disappear after end of exposure.

2 MODERATE: May involve both irreversible and reversible changes; not severe enough to cause death or permanent injury.

3 HIGH: May cause death or permanent injury after very short exposure to small quantities.

U UNKNOWN: No information on humans considered valid by authors.

Constants: Mol wt: 229.1, bp: 104–105° C at 5 mm,
   flash p: 240° F, fp: < −20° C, d: 1.319 at 25° /25° C.
Hazard Analysis
Toxicity: Unknown. See also phenol.
Fire Hazard: Moderate, when exposed to heat or flame.
Disaster Hazard: Moderately dangerous; when heated to de-
   composition, it emits toxic fumes; it can react with
   oxidizing materials.
Countermeasures
Storage and Handling: Section 7.
To Fight Fire: Water, foam, carbon dioxide, dry chemical
   or carbon tetrachloride (Section 6).

α-BROMO-n-CAPROIC ACID
General Information
Synonym: 2-Bromohexanoic acid.
Description: Liquid; soluble in alcohol, ether.
Formula: $(CH_2)_3CHBrCOOH$.
Constant: Mol wt: 195.06, bp: 240° C.
Hazard Analysis
Toxic Hazard Rating: U. Animal experiments suggest high
   toxicity.
Disaster Hazard: Dangerous, See bromides.

1-BROMO-4-CHLOROBENZENE
General Information
Description: White crystals.
Formula: C_6H_4BrCl.
Constants: Mol wt: 191.5, mp: 67.4, bp: 196.3° C, fp: 64.5° C.
Hazard Analysis
Toxicity: Unknown.
Disaster Hazard: Dangerous; see bromine.
Countermeasures
Storage and Handling: Section 7.

BROMOCHLORO TRIFLUOROETHANE
General Information
Synonym: Fluothane; halothane; 2-bromo-2-chloro,1,1,1,-
   trifluoro ethane.
Description: Nonflammable liquid; highly volatile; sweet
   odor; slightly soluble in water; miscible with many
   organic solvents.
Formula: $CF_3CHBrCl$.
Constant: Mol wt: 197.39, d: 1.86 (20/4° C), bp: 50.7° C
   (760 mm).
Hazard Analysis
Toxic Hazard Rating:
   Acute Local: Inhalation 1.
   Acute Systemic: Inhalation 3.
   Chronic Local: U.
   Chronic Systemic: U.
Has been used as a surgical anesthetic.
Disaster Hazard: Dangerous. See bromides, chlorides and
   fluorides.

1-BROMO-2-CHLOROETHANE. See ethylene chloro-
bromide.

BROMOCHLOROMETHANE. See methylene chloro-
bromide.

BROMOCUMENE. See bromoisopropyl benzene.

BROMOCYCLOHEXANE
General Information
Description: Yellowish liquid.
Formula: $C_6H_{11}Br$.
Constants: Mol wt: 163.1, bp: 165.8–167.3° C, flash p: 145° F,
   d: 1.337.
Hazard Analysis
Toxicity: Unknown. Probably narcotic in high concentra-
   tions. See also cyclohexane.
Fire Hazard: Moderate, when exposed to heat or flame.
Disaster Hazard: Dangerous; see bromides, it can react with
   oxidizing materials.
Countermeasures
Storage and Handling: Section 7.

To Fight Fire: Water, foam, carbon dioxide, dry chemical or
   carbon tetrachloride (Section 6).

BROMODICHLOROMETHANE
General Information
Description: Colorless liquid.
Formula: $CHBrCl_2$.
Constants: Mol wt: 163.8, bp: 89.2–90.6° C, d: 1.971 at
   25° /25° C.
Hazard Analysis
Toxicity: Unknown. Probably narcotic in high con-
   centrations. See also methylene chloride.
Disaster Hazard: Dangerous; see bromides and chlorides.
Countermeasures
Storage and Handling: Section 7.

3-BROMO-5,5-DIMETHYLHYDANTOIN
General Information
Description: White to buff powder; mild odor but like
   bromine.
Formula: $C_5H_7N_2O_2Br$.
Constant: Mol wt: 207.03.
Hazard Analysis
Toxicity: Unknown. See also dimethyl hydantoin.
Disaster Hazard: Dangerous; see bromine and bromides.
Countermeasures
Storage and Handling: Section 7.

4-BROMODIPHENYL
General Information
Description: White crystals.
Formula: $C_6H_5C_6H_4Br$.
Constants: Mol wt: 233.1, mp: 89.8–90.3° C, bp: 310.8° C,
   flash p: 290° F.
Hazard Analysis
Toxicity: Unknown.
Fire Hazard: Moderate, when exposed to heat or flame.
Disaster Hazard: Dangerous; when heated to decomposition
   it emits toxic fumes; it can react with oxidizing mate-
   rials.
Countermeasures
Storage and Handling: Section 7.
To Fight Fire: Water, foam, carbon dioxide, dry chemical or
   carbon tetrachloride (Section 6).

BROMOETHANE. See ethyl bromide.

BROMOETHENE. See vinyl bromide.

2-BROMOETHYL ACETATE
General Information
Formula: $BrC_2H_4O_2C_2H_3$.
Constant: Mol wt: 165.
Hazard Analysis
Toxicity: See esters.
Disaster Hazard: Dangerous; see bromine and bromides.
Countermeasures
Storage and Handling: Section 7.

BROMOETHYLBENZENE
General Information
Description: Colorless to pale-yellow liquid.
Formula: $BrC_6H_4CH_2CH_3$.
Constants: Mol wt: 185.1, mp: −65° C, bp: 201° C,
   d: 1.35–1.4 at 25° /25° C, flash p: 205° F, vap. d.: 6.4.
Hazard Analysis
Toxicity: Unknown. See also bromobenzene.
Fire Hazard: Moderate, when exposed to heat or flame.
Disaster Hazard: Dangerous; see bromides, it can react with
   oxidizing materials.
Countermeasures
Storage and Handling: Section 7.
To Fight Fire: Water, foam, carbon dioxide, dry chemical
   or carbon tetrachloride (Section 6).

BROMOETHYL CHLOROSULFONATE
General Information
Description: Liquid.

Formula: $BrCH_2CH_2OSO_2Cl$.
Constants: Mol wt: 223.49, bp: 100–105°C at 18 mm.
Hazard Analysis
Toxic Hazard Rating:
  Acute Local: Irritant 3; Inhalation 3.
  Acute Systemic: U.
  Chronic Local: U.
  Chronic Systemic: U.
Disaster Hazard: Dangerous; when heated to decomposition
  or on contact with acid or acid fumes, it emits toxic
  fumes of oxides of sulfur and bromides; it will react with
  water or steam to produce toxic and corrosive fumes.
Countermeasures
Ventilation Control: Section 2.
Personnel Protection: Section 3.
Storage and Handling: Section 7.

BROMOETHYLENE. See vinyl bromide.

2-BROMOETHYL ETHYL ETHER
General Information
Description: Liquid.
Formula: $BrCH_2CH_2OCH_2CH_3$.
Constants: Mol wt: 153, vap. d: 5.25.
Hazard Analysis
Toxicity: Unknown; an insecticide. See also ethers.
Fire Hazard: See ethers.
Explosion Hazard: See ethers.
Disaster Hazard: Dangerous; see bromides, it can react vigor-
  ously with oxidizing materials.
Countermeasures
Storage and Handling:Section 7.

BROMOETHYNE. See bromoacetylene.

BROMOFORM
General Information
Synonym: Tribromomethane.
Description: Colorless liquid or hexagonal crystals.
Formula: $CHBr_3$.
Constants: Mol wt: 252.77, mp: 6–7°C, bp: 149.5°C,
  flash p: none, d: 2.890 at 20°/4°C.
Hazard Analysis
Toxic Hazard Rating:
  Acute Local: Ingestion 2; Inhalation 2.
  Acute Systemic: Ingestion 2; Inhalation 2.
  Chronic Local: Irritant 2.
  Chronic Systemic: Ingestion 2; Inhalation 2.
  TLV: ACGIH (recommended) 0.5 part per million; 5
  milligrams per cubic meter of air. May be absorbed
  through the skin.
Toxicology: This material causes lachrymation. It can damage the liver to a serious degree and cause death. It has
  been said that its medicinal application has resulted in
  numerous poisonings. It has anesthetic properties similar to those of chloroform, but it is not sufficiently volatile for inhalation purposes and is far too toxic to be recommended. In addition to its narcotic effects, it is a
  metabolic poison. Petroleum geologists working in
  closed rooms with a large number of funnels or open
  separatory flasks, as required for routine procedure in
  the separation of minerals, can be subjected to appreciable concentrations of bromoform as an atmospheric
  contaminant. The inhalation of small amounts of this
  material causes irritation, provoking the flow of tears
  and saliva and reddening of the face. In dogs, 29,000
  ppm caused a deep narcosis after 8 minutes, and death
  after 1 hour; in 30 minutes there was deep narcosis and
  recovery on the next day.

Disaster Hazard: Dangerous; when heated to decomposition
  it emits highly toxic fumes.
Countermeasures
Ventilation Control: Section 2.
Treatment and Antidotes: Remove patient to fresh air and if
  breathing has stopped administer artificial respiration.
  Give oxygen if necessary. Call a physician.
Personnel Protection: Section 3.
Storage and Handling: Section 7.
Shipping Regulations: Section 11.
  IATA: Poison B, poison label, 1 liter (passenger), 220 liters
  (cargo).

BROMOGERMANE
General Information
Description: Colorless liquid.
Formula: GeH_3Br.
Constants: Mol wt: 155.54, mp:−32.0°C, bp: 52.0°C,
  d: 2.34 at 29.5°C.
Hazard Analysis
Toxicity: See germanium compounds and bromides.
Fire Hazard: Moderate, when exposed to heat or flame (Sec-
  tion 6).
Disaster Hazard: Dangerous; see bromine; it can react vigor-
  ously with oxidizing materials.
Countermeasures
Storage and Handling: Section 7.

p-BROMOHYDROAZOBENZENE
Hazard Analysis
Toxicity: Unknown; an insecticide.
Disaster Hazard: Dangerous; see bromides.
Countermeasures
Storage and Handling: Section 7.

2-BROMOISOBUTYRIC ACID
General Information
Description: Crystals; sparingly soluble in cold water; de-
  composed by hot water into the hydroxy acid; soluble
  in alcohol and ether.
Formula: $(CH_3)_2CBrCOOH$.
Constants: Mol wt: 167.01, d: 1.52, mp: 48–49°C, bp: 198–
  200°C.
Hazard Analysis
Toxic Hazard Rating: U. Limited animal experiments sug-
  gest low toxicity.
Disaster Hazard: Dangerous. See bromides.

BROMOISOPROPYL BENZENE.
General Information
Synonym: ar-Bromo cumene.
Description: Colorless liquid.
Formula: $BrC_6H_4CH(CH_3)_2$.
Constants: Mol wt: 199.1, mp: −20°C, bp: 212–216°C,
  flash p: 207°F, d: 1.27–1.3 at 25°/25°C, vap. d.: 6.90.
Hazard Analysis
Toxicity: Unknown. See also bromobenzene.
Fire Hazard: Moderate, when exposed to heat or flame; it
  can react with oxidizing materials.
Countermeasures
To Fight Fire: Water, foam, carbon dioxide, dry chemical
  or carbon tetrachloride (Section 6).
Storage and Handling: Section 7.

5-BROMO-3-ISOPROPYL-6-METHYL URACIL
General Information
Synonym: Isocil, hyvar.
Description: Crystals, soluble in absolute alcohol.

Formula: $OCCBrC(CH_3)NHC(O)NCH(CH_3)_2$.
Constants: Mol wt: 247, mp: 158°C.
Hazard Analysis
Toxic Hazard Rating: U. A herbicide. Animal experiments show low oral toxicity.
Disaster Hazard: Dangerous. See bromides.

BROMOISOVALERIC ACID
General Information
Synonym: 2-Bromo-3-methyl butyric acid.
Descriptions: 2 isometric forms (α and β)
(1) α: Lustrous crystals; sparingly soluble in water; soluble in alcohol and ether.
(2) β: Needles; slightly soluble in water; soluble in alcohol, benzene, and ether.
Formula: (1) α: $(CH_3)_2CHCHBrCOOH$.
Formula: (2) β: $(CH_3)_2CBrCH_2COOH$.
Constants: Mol wt: 181.04 (α and β), mp (α): 44°C, (β): 73.74°C, bp: (α): $\approx$ 230°C (slight decomposition).
Hazard Analysis
Toxic Hazard Rating: U. Limited animal experiments suggest low toxicity.
Disaster Hazard: Dangerous. See bromides.

BROMOL. See tri bromo phenol.

BROMOMETHANE. See methyl bromide.

2-BROMO-3-METHYL BUTYRIC ACID. See α-bromoisovaleric acid.

BROMO METHYLETHYL KETONE
General Information
Description: Colorless to pale-yellowish liquid.
Formula: $BrCH_2COC_2H_5$.
Constants: Mol wt: 135.01, bp: 145–146°C decomposes, d: 1.43.
Hazard Analysis
Toxic Hazard Rating:
Acute Local: Irritant 3; Ingestion 3; Inhalation 3.
Acute Systemic: U.
Chronic Local: U.
Chronic Systemic: U.
Disaster Hazard: Dangerous; when heated to decomposition it emits toxic fumes of bromides; it can react with oxidizing materials.
Countermeasures
Ventilation Control: Section 2.
Personnel Protection: Section 3.
Storage and Handling: Section 7.

1-BROMO-3-NITROBENZENE
General Information
Description: Crystals.
Formula: $BrC_6H_4NO_2$.
Constants: Mol wt: 202.0, mp: 56°C, bp: 256.5°C, d: 1.70 at 20°/4°C.
Hazard Analysis
Toxicity: Details unknown. See also nitrobenzene.
Disaster Hazard: Dangerous; when heated to decomposition, it emits highly toxic fumes of oxides of nitrogen and bromides; it can react vigorously with oxidizing materials.
Countermeasures
Storage and Handling: Section 7.

BROMOPHENOLS (m,p,o)
General Information
Description: (m-) Crystals; insoluble in water; soluble in alcohol, ether and alkalies.
(p-) Crystals; slightly soluble in water; soluble in alcohol, ether, chloroform, and glacial acetic acid.
(o-) Yellow to oily red liquid; unpleasant odor; insoluble in water; soluble in alcohol, ether, and chloroform.
Formula: $HO(C_6H_4)Br$.
Constants: Mol wt: 173 (m,p,o). d (p): 1.840 (15°C), 1.5875

(80°C), d (o): 1.5, mp (m): 33°C, d (p): 64°C, d (o): 6°C, bp (m): 236°C, bp (p): 238°C, bp (o): 194°C.
Hazard Analysis
Toxic Hazard Rating: U. Similar to pentachlorophenol (q.v.).
Disaster Hazard: Dangerous. See bromides.

o-BROMO-2-PHENYL ACETONITRILE. See bromobenzyl cyanide.

2-BROMO-4-PHENYLPHENOL
General Information
Description: Crystals.
Formula: $C_{12}H_9BrO$.
Constants: Mol wt: 249.1, mp: 95°C, bp: 195°C decomp., flash p: 405°F, vap. press.: 1 mm at 100.0°C.
Hazard Analysis
Toxicity: Unknown. See also phenol.
Disaster Hazard: Dangerous; see bromides.
Countermeasures
Storage and Handling: Section 7.

BROMOPICRIN
General Information
Synonyms: Nitrobromoform; tribomonitromethane.
Description: Prismatic crystals.
Formula: BCr_3NO_2.
Constants: Mol wt: 297.77, mp: 103°C, bp: 127°C, d: 2.79 at 18°C.
Hazard Analysis
Toxicity: Powerful irritant.
Fire Hazard: Unknown.
Explosion Hazard: Severe, when heated rapidly (Section 7).
Disaster Hazard: Dangerous; see oxides of nitrogen and bromides. It explodes.
Countermeasures
Storage and Handling: Section 7.

BROMOPLATINIC ACID
General Information
Description: Monoclinic, deliquescent red crystals.
Formula: $H_2PtBr \cdot 9H_2O$.
Constants: Mol wt: 838.9, mp: < 100°C decomposes.
Hazard Analysis
Toxicity: See bromides and platinum compounds.
Disaster Hazard: Dangerous; see bromides.
Countermeasures
Storage and Handling: Section 7.

1-BROMOPROPANE. See propyl bromide.

BROMOSUCCINIMIDE
General Information
Description: White to pale buff, fine crystalline powder with faint odor of bromine.
Formula: $C_4H_4O_2NBr$.
Constants: Mol wt: 178.0, mp: 174–178°C, d: 2.098.
Hazard Analysis
Toxic Hazard Rating:
Acute Local: Irritant 3; Ingestion 3; Inhalation 3.
Acute Systemic: U.
Chronic Local: U.
Chronic Systemic: U.
Caution: A powerful irritant.
Disaster Hazard: Dangerous; see bromides.
Countermeasures
Ventilation Control: Section 2.
Personnel Protection: Section 3.
Storage and Handling: Section 7.

α-BROMOTOLUENE. See benzyl bromide.

m-BROMOTOLUENE
General Information
Synonyms: 3-Bromotoluene; 3-bromo-1-methyl benzene; m-tolyl bromide.

Description: Liquid; soluble in alcohol, ether, and benzene.
Formula: $BrC_6H_4CH_3$.
Constants: Mol wt: 171.04, d: 1.4099 (20/4°C), mp: −39.8°C, bp: 183.7°C.
Hazard Analysis
Toxic Hazard Rating:
Acute Local: Irritant 1; Inhalation 2.
Acute Systemic: Inhalation 2.
Chronic Local: U.
Chronic Systemic: U.
An irritant and narcotic.
Disaster Hazard: Dangerous. See bromides.

o-BROMOTOLUENE
General Information
Description: Colorless liquid.
Formula: $CH_3C_6H_4Br$.
Constants: Mol wt: 171.0, bp: 180–182°C, fp: −27°C, flash p: 175°F, d: 1.422 at 25°/25°C, vap. d.: 5.9.
Hazard Analysis
Toxicity: Unknown. Limited animal experiments suggest low toxicity. See m-bromotoluene.
Fire Hazard: Moderate; when exposed to heat or flame.
Disaster Hazard: Dangerous; see bromides, can react with oxidizing materials.
Countermeasures
Storage and Handling: Section 7.
To Fight Fire: Water, foam, carbon dioxide, dry chemical or carbon tetrachloride (Section 6).

p-BROMOTOLUENE
General Information
Synonym: p-Tolyl bromide.
Description: White crystals.
Formula: $CH_3C_6H_4Br$.
Constants: Mol wt: 171.0, bp: 183.3–184.1°C, freezing p: 25.5°C, flash p: 185°F, d: 1.400 at 27°/25°C, vap. d.: 5.9.
Hazard Analysis
Toxicity: Unknown. See m-bromotoluene.
Fire Hazard: Moderate, when exposed to heat or flame.
Disaster Hazard: Dangerous; See Bromides; it can react with oxidizing materials.
Countermeasures
Storage and Handling: Section 7.
To Fight Fire: Water, foam, carbon dioxide, dry chemical or carbon tetrachloride (Section 6).

BROMOTRICHLOROMETHANE
General Information
Description: Colorless liquid.
Formula: $CBrCl_3$.
Constants: Mol wt: 198.3, bp: 103.8–105.1°C, d: 1.997 at 25°/25°C.
Hazard Analysis
Toxicity: Unknown. Probably narcotic in high concentrations. See also chloroform.
Disaster Hazard: Dangerous; when heated to decomposition it emits toxic fumes.
Countermeasures
Storage and Handling: Section 7.

BROMOTRIFLUOROETHYLENE
General Information
Synonym: BFE.
Description: Colorless gas.
Formula: $BrFC:CF_2$.
Constant: Mol wt: 160.94.

Hazard Analysis
Toxic Hazard Rating: U.
Fire Hazard: Flammable gas or liquid.
Disaster Hazard: Dangerous. Reacts with powerful oxidizer. When heated to decomposition, it emits lightly toxic fumes and bromine and fluorine.
Countermeasures
Storage and Handling: Section 7.
Shipping Regulations: Section 11.
IATA: Flammable liquid, red label, not acceptable (passenger), 40 liters (cargo).

BROMOTRIFLUOROMETHANE. See trifluoro bromomethane.

BRONZE. See copper and tin and zinc.

BRONZE LIQUID PAINT
General Information
Composition: Liquid vehicle and bronze.
Constant: Flash p: below 100°F.
Hazard Analysis
Toxicity: See specific vehicle employed.
Fire Hazard: Dangerous, when exposed to heat or flame; can react with oxidizing materials.
Explosion Hazard: Unknown.
Explosive Range: Depends upon components used.
Countermeasures
Storage and Handling: Section 7.
To Fight Fire: Foam, carbon dioxide, dry chemical or carbon tetrachloride (Section 6).
Shipping Regulations: Section 11.
See also paint, enamel, lacquer, stain, shellac, varnish, etc.
Coast Guard Classification: Inflammable liquid; red label.

BRUCINE
General Information
Description: Monoclinic prisms.
Formula: $C_{23}H_{26}N_2O_4$.
Constants: Mol wt: 394.45, mp: 178°C.
Hazard Analysis
Toxic Hazard Rating:
Acute Local: U.
Acute Systemic: Ingestion 3; Inhalation 3.
Chronic Local: U.
Chronic Systemic: U.
Caution: An alkaloid like strychnine but one-sixth as toxic.
Disaster Hazard: Dangerous; when heated, it emits toxic fumes.
Countermeasures
Ventilation Control: Section 2.
Personal Hygiene: Section 3.
First Aid: Section 1.
Storage and Handling: Section 7.
Shipping Regulations: Section 11.
I.C.C.: Poison B; poison label, 200 pounds.
Coast Guard Classification: Poison B; poison label.
IATA: Poison B, poison label, 25 kilograms (passenger), 95 kilograms (cargo).

BRUCINE COMPOUNDS. See brucine.

BRUCINE NITRATE
General Information
Description: White powder.
Formula: $C_{23}H_{26}N_2O_4 \cdot HNO_3 \cdot 2H_2O$.
Constants: Mol wt: 493.51, mp: anh. 230°C decomposes.
Hazard Analysis
Toxicity: See brucine.

TOXIC HAZARD RATING CODE (For detailed discussion, see Section 1.)

0 NONE: (a) No harm under any conditions; (b) Harmful only under unusual conditions or overwhelming dosage.

1 SLIGHT: Causes readily reversible changes which disappear after end of exposure.

2 MODERATE: May involve both irreversible and reversible changes; not severe enough to cause death or permanent injury.

3 HIGH: May cause death or permanent injury after very short exposure to small quantities.

U UNKNOWN: No information on humans considered valid by authors.

Fire Hazard: Moderate by chemical reaction with reducing agents (Section 6).
Caution: A powerful oxidant.
Disaster Hazard: See nitrates.
Countermeasures
Storage and Handling: Section 7.

BRUNSWICK GREEN. See copper oxychloride.

BTC. See alkyl dimethyl benzyl ammonium chloride.

BULAN. See 1,1-bis(p-chlorophenyl)-2-nitrobutane.

BUNSENITE. See nickel monoxide.

BUQUINOLATE
General Information
Synonym: Ethyl-4-hydroxy-6,7-diisobutoxy-3-quinoline-carboxylate.
Hazard Analysis
Toxic Hazard Rating: U. Used as a food additive permitted in the feed and drinking water of animals and/or for the treatment of food producing animals. Also permitted in food for human consumption. (Section 10).

BURNT COTTON (NOT REPICKED)
Hazard Analysis
Toxicity: Unknown.
Fire Hazard: Moderate when exposed to heat or flame; can react with oxidizing materials (Section 6).
Countermeasures
Storage and Handling: Section 7.
Shipping Regulations: Section 11.
  I.C.C.: Flammable solid; yellow label, not accepted.
  Coast Guard Classification: Inflammable solid; yellow label.
  IATA: Flammable solid, not accepted.

BURNT FIBER
Hazard Analysis
Toxicity: Unknown.
Fire Hazard: Moderate, when exposed to heat or flame or powerful oxidizers (Section 6).
Countermeasures
Storage and Handling: Section 7.
Shipping Regulations: Section 11.
  I.C.C.: Flammable solid; yellow label, not accepted.
  Coast Guard Classification: Inflammable solid; yellow label.
  IATA: Flammable solid, not accepted (passenger and cargo).

BURNT LIME. See calcium oxide.

BURSTERS, EXPLOSIVE
Shipping Regulations: Section 11.
  ICC: Class A explosive, not accepted.
  IATA: Explosive, not acceptable.

BURWEED MARSH ELDER
Hazard Analysis
Toxic Hazard Rating:
  Acute Local: Allergen 1.
  Acute Systemic: 0.
  Chronic Local: Allergen 1.
  Chronic Systemic: 0.
Fire Hazard: Slight, when exposed to heat or flame.
Countermeasures
Personal Hygiene: Section 3.
Storage and Handling: Section 7.

BUTADIENE-1,3
General Information
Synonym: Erythrene.
Description: Colorless gas.
Formula: $CH_2CHCHCH_2$.
Constants: Mol wt: 54.09, bp: $-4.5°C$, mp: $-113°C$, fp: $-108.9°C$, flash p: $<20°F$, lel: 2.0%, uel: 11.5%, d: 0.621 at $20°/4°C$, autoign. temp.: $804°F$, vap. d.: 1.87, vap. press.: 1840 mm at $21°C$.

Hazard Analysis
Toxic Hazard Rating:
  Acute Local: Irritant 2; Ingestion 2; Inhalation 2.
  Acute Systemic: Inhalation 2.
  Chronic Local: Irritant 1.
  Chronic Systemic: U.
Toxicology: The vapors are irritating to eyes and mucous membranes. Inhalation of high concentrations can cause unconsciousness and death. If spilled on skin or clothing can cause burns or frost bite (due to rapid vaporization). Chronic systemic poisoning in humans has not been reported.
TLV: ACGIH (recommended); 1000 parts per million in air; 2200 milligrams per cubic meter of air.
Fire Hazard: Dangerous; when exposed to heat, flame or powerful oxidizers.
Spontaneous Heating: No.
Explosion Hazard: Moderate, when exposed to heat or flame; may form explosive peroxides upon exposure to air.
Disaster Hazard: Moderately dangerous; when heated, it emits acrid fumes; can react with oxidizing materials.
Countermeasures
Personal Hygiene: Section 3.
To Fight Fire: Carbon dioxide, dry chemical or water spray (Section 6).
Storage and Handling: Section 7.
Shipping Regulations: Section 11.
  I.C.C. Classification: Red label.

BUTADIENE DIOXIDE
General Information
Formula: $O_2HC:CHCH:CH_2$.
Constant: Mol wt: 85.
Hazard Analysis
Toxic Hazard Rating:
  Acute Local: Irritant 3.
  Acute Systemic: Ingestion 3; Inhalation 3; Skin Absorption 3.
  Chronic Local: Allergen 3.
  Chronic Systemic: U.
Based on limited animal experiments.

BUTADIENE, INHIBITED. See butadiene.
Shipping Regulations: Section 11.
  I.C.C.: Flammable gas; red gas label, 300 pounds.
  Coast Guard Classification: Inflammable gas; red gas label.
  IATA: Flammable gas, red label, not acceptable (passenger), 140 kilograms (cargo).
  MCA warning label.

BUTADIENE MONOXIDE
General Information
Synonym: Vinylethylene oxide,
Description: Liquid.
Formula: $CH_2CHCHCH_2O$
Constants: Mol wt: 70.09, mp: $-135°C$, bp: $67°C$, flash p: $<-58°F$ (C.C.), d: 0.869, autoign. temp.: $806°F$, vap. d.: 2.41.
Hazard Analysis
Toxicity: See butadiene.
Fire Hazard: Dangerous, when exposed to heat or flame.
Spontaneous Heating: No.
Disaster Hazard: Moderately dangerous; when heated to decomposition it emits acrid fumes; it can react with oxidizing materials.
Countermeasures
Storage and Handling: Section 7.
To Fight Fire: Carbon dioxide, dry chemical, carbon tetrachloride or water spray (Section 6).
Shipping Regulations: Section 11.

BUTADIENE, UNINHIBITED
Shipping Regulations: Section 11.
  IATA: Flammable gas, not acceptable (passenger and cargo).

BUTADIENYL ACETATE
General Information
Formula: $CH_3COOHC:CHHC:CH_2$.
Constant: Mol wt: 112.
Hazard Analysis
Toxic Hazard Rating: U. Limited animal experiments suggest toxicity higher than that of most esters.

BUTADIINE. See diacetylene.

BUTADIYNE. See diacetylene.

BUTANAL. See butyraldehyde.

BUTANAL OXIDE. See butyraldoxime.

BUTANE
General Information
Synonyms: N-Butane; methylethylmethane; butyl hydride.
Description: Colorless gas.
Formula: $CH_3CH_2CH_2CH_3$.
Constants: Mol wt: 58.1, bp: $-0.5°C$, fp: $-138.6°C$, lel: 1.9%, uel: 8.5%, flash p: $-76°F$ (C.C.), d: 0.599, autoign. temp.: $761°F$, vap. press.: 2 atm at $18.8°C$, vap. d.: 2.046.
Hazard Analysis
Toxic Hazard Rating:
Acute Local: 0.
Acute Systemic: Inhalation 2.
Chronic Local: U.
Chronic Systemic: Inhalation 1.
A general purpose food additive (Section 10).
Caution: Produces drowsiness. Simple asphyxiant.
Fire Hazard: Dangerous; when exposed to heat or flame.
Spontaneous Heating: No.
Explosion Hazard: Moderate, when exposed to flame (Section 7).
Disaster Hazard: Moderately dangerous; when heated it emits acrid fumes; can react with oxidizing materials.
Countermeasures
Ventilation Control: Section 2.
To Fight Fire: Carbon dioxide, dry chemical or water spray (Section 6).
Storage and Handling: Section 7.
Shipping Regulations: Section 11.
  I.C.C.: Flammable gas; red gas label, 300 pounds.
  IATA: Flammable gas, red label, not acceptable (passenger), 140 kilograms (cargo).

BUTANE
Butane mixtures, and mixtures having similar properties contained in lighters, candles, heating devices, refill containers and similar devices each *not* exceeding 65 grams (2–3ozs)
Shipping Regulations: Section 11.
  IATA: Flammable gas, red label, 0.6 kilogram (passenger), 12 kilograms (cargo).

n-BUTANE. See butane.

1,4-BUTANEDICARBOXYLIC ACID. See adipic acid.

1,2,4-BUTANEDICARBOXYLLIC ACID, TRI(2-ETHYL HEXYL)ESTER
Hazard Analysis
Toxic Hazard Rating: U.
Limited animal experiments suggest low toxicity. See also esters.

BUTANEDINITRILE. See succinonitrile.

BUTANEDIOIC ACID. See succinic acid.

BUTANEDIOIC ANHYDRIDE. See succinic anhydride.

BUTANEDIOIC PEROXIDE. See succinic acid peroxide.

BUTANEDIOL. See 1,4-butanediol.

1,2-BUTANEDIOL
General Information
Synonym: Dihydroxybutane-1,2; ethyl ethylene glycol.
Formula: $CH_3CH_2CHOHCH_2OH$.
Constants: Mol wt: 90.1, d: 1.0, vap. d.: 3.1, bp: $381°F$, flash p: $104°F$.
Hazard Analysis
Fire Hazard: Moderate, when exposed to heat or flame.
Countermeasures
To Fight Fire: Alcohol foam (Section 6).
Storage and Handling: Section 7.

1,3-BUTANEDIOL. See 1,3-butylene glycol.

1,4-BUTANEDIOL
General Information
Synonym: Butanediol; 1,4-butylene glycol; tetramethylene glycol.
Description: Nearly odorless, colorless, viscid liquid.
Formula: $HOCH_2CH_2CH_2CH_2OH$.
Constants: Mol wt: 90.1, bp: $228°C$, fp: $20.9°C$, flash p: $>250°F$ (O.C.), d: 1.0154 at $25°/4°C$, vap. d.: 3.1.
Hazard Analysis
Toxic Hazard Rating:
Acute Local: U.
Acute Systemic: Ingestion 2.
Chronic Local: U.
Chronic Systemic: U.
Fire Hazard: Moderate, when exposed to heat or flame; it can react with oxidizing materials.
Countermeasures
To Fight Fire: Water, foam, carbon dioxide, dry chemical or carbon tetrachloride. Section 6.
Personal Hygiene: Section 3.
Storage and Handling: Section 7.

2,3-BUTANEDIOL. See 2,3-butylene glycol.

BUTANEDIOL DICAPRYLATE
General Information
Description: A clear liquid.
Formula: $(CH_2)_2(CH_2OCOC_7H_{15})_2$.
Constants: Mol wt: 342, bp: $220°C$ at 5 mm, fp: $10.5°C$, flash p: $390°F$, d: 0.929 at $20°/20°C$, vap. d.: 11.
Hazard Analysis
Toxicity: Unknown.
Fire Hazard: Slight, when exposed to heat or flame; can react with oxidizing materials (Section 6).
Countermeasures
Storage and Handling: Section 7.

BUTANEDIONE. See diacetyl.

BUTANENITRILE. See N-butyronitrile.

1-BUTANETHIOL. See n-butyl mercaptan.

2-BUTANETHIOL. See sec-butyl mercaptan.

BUTANETRIOL. See 1,2,4-butanetriol.

1,2,4-BUTANETRIOL
General Information
Synonym: Butanetriol.
Description: Colorless, nearly odorless, thick liquid.

TOXIC HAZARD RATING CODE (For detailed discussion, see Section 1.)

0 NONE: (a) No harm under any conditions; (b) Harmful only under unusual conditions or overwhelming dosage.

1 SLIGHT: Causes readily reversible changes which disappear after end of exposure.

2 MODERATE: May involve both irreversible and reversible changes; not severe enough to cause death or permanent injury.

3 HIGH: May cause death or permanent injury after very short exposure to small quantities.

U UNKNOWN: No information on humans considered valid by authors.

Formula: $HOCH_2CHOHCH_2CH_2OH$.
Constants: Mol wt: 106.1, bp: 312°C, fp: supercools, flash p: 332°F (C.O.C.), d: 1.184 at 4°C.
Hazard Analysis
Toxicity: Practically nontoxic. Resembles glycerine.

BUTANOIC ANHYDRIDE. See butyric anhydride.

BUTANOL. See butyl alcohol.

2-BUTANOL. See sec-butyl alcohol.

2-BUTANOL ACETATE. See sec-butyl acetate.

BUTANOLAL. See aldol.

2-BUTANONE
General Information
Synonym: Methylethyl ketone.
Description: Colorless liquid, acetone-like odor.
Formula: $CH_3COCH_2CH_3$.
Constants: Mol wt: 72.10, bp: 79.57°C, fp: −85.9°C, lel: 1.8%, uel: 10%, flash p: 22°F (T.O.C.), d: 0.80615 at 20°/20°C, vap. press.: 71.2 mm at 20°C, autoign. temp.: 960°F, vap. d.: 2.41.
Hazard Analysis
Toxic Hazard Rating:
Acute Local: Irritant 1; Ingestion 1; Inhalation 1.
Acute Systemic: Inhalation 2.
Chronic Local: Irritant 1.
Chronic Systemic: U.
Toxicology: Produces local irritation and narcosis. See ketones.
TLV: ACGIH (recommended): 200 parts per million in air; 590 milligrams per cubic meter of air.
Fire Hazard: Dangerous, when exposed to heat or flame; it can react with oxidizing materials.
Underwriters Lab Classification: 85–90.
Spontaneous Heating: No.
Disaster Hazard: Highly dangerous upon exposure to heat or flame.
Explosion Hazard: Moderate, when exposed to flame (Section 7).
Countermeasures
Ventilation Control: Section 2.
Personal Hygiene: Section 3.
To Fight Fire: Alcohol foam, carbon dioxide, dry chemical or carbon tetrachloride (Section 6).
Storage and Handling: Section 7.
Shipping Regulations: Section 11.
I.C.C.: Flammable liquid; red label, 10 gallons.
Coast Guard Classification: Inflammable liquid; red label.
IATA: Flammable liquid, red label, 1 liter (passenger), 40 liters (cargo).

BUTANOYL CHLORIDE. See butyryl chloride.

2-BUTENAL. See crotonaldehyde.

1-BUTENE. See α-butylene.

cis-BUTENE-2
General Information
Synonyms: Dimethylethylene; pseudo-butylene.
Description: Colorless gas.
Formula: $CH_3CHCHCH_3$.
Constants: Mol wt: 56.1, bp: 1°C, fp: −139°C, flash p: >20°F, d: 0.627 at 15.5°/15.5°C, vap. press.: 1410 mm at 21°C, autoign. temp.: 615°F, lel: 1.7% uel: 9.0%, vap. d.: 1.9.
Hazard Analysis
Toxicity: Details unknown. May act as a simple asphyxiant.
Fire Hazard: Dangerous, when exposed to heat or flame; it can react with oxidizing materials (Section 6).
Explosion Hazard: Unknown.

Countermeasures
Ventilation Control: Section 2.
Storage and Handling: Section 7.
To Fight Fire: Stop flow of gas.

trans-BUTENE-2
General Information
Description: A gas.
Formula: C_4H_8.
Constants: Mol wt: 56.1, bp: 2.5°C, fp: −105.6°C, flash p: <20°F, d: 0.613 at 15.5°/15.5°C, vap. d.: 1.95, vap. press.: 1592 mm at 21°C, autoign. temp.: 615°F, lel: 1.8%, uel: 9.7%, vap. d.: 1.9.
Hazard Analysis
Toxicity: Unknown. May act as a simple asphyxiant.
Fire Hazard: Dangerous, when exposed to heat or flame; it can react with oxidizing materials (Section 6).
Explosion Hazard: Unknown.
Countermeasures
Storage and Handling: Section 7.
To Fight Fire: Stop flow of gas.

trans-BUTENEDIOIC ACID. See fumaric acid.

cis-BUTENEDIOIC ANHYDRIDE. See maleic anhydride.

2-BUTENE-1,4-DIOL
General Information
Description: Colorless, odorless liquid.
Formula: $HOCH_2CHCHCH_2OH$.
Constants: Mol wt: 88.1, bp: 234°C, fp: 12.5°C, flash p: 263°F (C.O.C.), d: 1.07 at 25°/15°C, vap. d.: 3.04.
Hazard Analysis
Toxicity: Unknown.
Caution: This material is an irritant to the skin.
Fire Hazard: Moderate, when exposed to heat or flame; it can react with oxidizing materials.
Countermeasures
To Fight Fire: Water, foam, carbon dioxide, dry chemical or carbon tetrachloride (Section 6).
Storage and Handling: Section 7.

trans-BUTENEDIOYL CHLORIDE. See fumaryl chloride.

2-BUTENE-NITRILE. See crotonitrile.

3-BUTENE-2-ONE. See methyl vinyl ketone.

2-BUTEN-1-OL. See α-methallyl alcohol.

3-(2-BUTENYL)-4-METHYL-2-OXO-3-CYCLOPENTEN-1-YL CHRYSANTHEMUM MONO CARBO-XYLATE. See cinerin I.

3-(2-BUTENYL)-4-METHYL-2-OXO-3-CYCLOPENTEN-1-YL ESTER OF CHRYSANTHEMUM DI-CARBOXYLIC ACID MONOMETHYL ESTER. See cinerin II.

BUTESIN
General Information
Synonym: Butyl-p-aminobenzoate.
Description: Yellow, amorphous powder.
Formula: $C_6H_4NH_2COOC_4H_9$.
Constants: Mol wt: 193.24, mp: 55°C.
Hazard Analysis
Toxic Hazard Rating:
Acute Local: Allergen 1.
Acute Systemic: U.
Chronic Local: Allergen 1.
Chronic Systemic: U.
Fire Hazard: Slight.
Countermeasures
Personal Hygiene: Section 3.
Storage and Handling: Section 7.

BUTESIN PICRATE OINTMENT
General Information
Description: A yellow ointment.
Hazard Analysis
Toxic Hazard Rating:
  Acute Local: Allergen 1.
  Acute Systemic: U.
  Chronic Local: ·Allergen 1.
  Chronic Systemic: U.
Caution: Has been known to cause dermatitis (Section 9).
Fire Hazard: Slight, when exposed to heat or flame (Section 6).
Disaster Hazard: Dangerous; See nitrates.
Countermeasures
Personal Hygiene: Section 3.
Storage and Handling: Section 7.

BUTOPYRONOXYL
General Information
Synonym: Butyl mesityl oxide.
Description: Yellow to reddish liquid.
Formula: $C_{12}H_{18}O_4$.
Constants: Mol wt: 226.3, bp: 256–270°C, d: 1.052–1.060 at 25°/25°C.
Hazard Analysis
Toxic Hazard Rating:
  Acute Local: Irritant 1.
  Acute Systemic: U.
  Chronic Local: Irritant 1.
  Chronic Systemic: U.
Toxicology: Mildly irritating to skin. Experimentally it has produced liver necrosis in animals.
Fire Hazard: Slightly dangerous (Section 6).
Countermeasures
Storage and Handling: Section 7.

1-BUTOXYBUTANE. See butyl ether.

2-BUTOXY-N-(2-DIETHYLAMINOETHYL) CINCHONINAMIDE HYDROCHLORIDE. See dibucaine hydrochloride.

BUTOXYETHANOL. See butyl "cellosolve."

2,β-BUTOXY ETHOXY ETHYL CHLORIDE
General Information
Formula: $C_4H_9C_2H_4OC_2H_4Cl$.
Constants: Mol wt: 165, d: 1.0, vap. d.: 6.1, bp: 392–437°F, flash p: 190°F.
Hazard Analysis
Fire Hazard: Moderate, when exposed to heat or flame.
Disaster Hazard: Dangerous. See chlorides.
Countermeasures
Storage and Handling: Section 7.

1-(BUTOXY ETHOXY)-2-PROPANOL
General Information
Description: Soluble in water.
Formula: $CH_3CH(OH)CH_2OC_2H_4OC_2H_4C_2H_5$.
Constants: Mol wt: 176, d: 0.9310(20/20°C), bp: 230.3°C, fp: −90°C, flash p: 250°F(O.C.).
Hazard Analysis
Fire Hazard; Slight, when exposed to heat or flame.
Countermeasures
To Fight Fire: Alcohol foam, water, foam (Section 6).
Storage and Handling: Section 7.

BUTOXY ETHYL ACETATE
General Information
Synonym: Ethylene glycol monobutyl ether acetate.

Description: Colorless liquid, fruity odor; soluble in hydrocarbon and organic solvents; insoluble in water.
Formula: $C_4H_9OCH_2CH_2OOCCH_3$.
Constants: Mol wt: 160, bp: 192.3°C, d: 0.9424(20/20°C), fp: −63.5°C, flash p: 190°F.
Hazard Analysis
Toxic Hazard Rating: U. Limited animal experiments suggest low toxicity. See also esters.
Fire Hazard: Moderate, when exposed to heat or flame.
Countermeasures
Storage and Handling: Section 7.

BUTOXYETHYL ACRYLATE
General Information
Description: Liquid.
Formual: $CH_2CHCO_2CH_2CH_2OC_4H_9$.
Constants: Mol wt: 172.2, bp: 62°C, at 2 mm, flash p: 185°F (O.C.), d: 0.948 at 20°C, vap. d.: 5.93.
Hazard Analysis
Toxicity: See esters, and ethyl acrylate.
Fire Hazard: Moderate, when exposed to heat or flame; it can react with oxidizing materials.
Countermeasures
To Fight Fire: Foam, carbon dioxide, dry chemical or carbon tetrachloride (Section 6).
Storage and Handling: Section 7.

BUTOXY ETHYL DIGLYCOL CARBONATE. See diethylene glycol bis (2-butoxy ethyl carbonate).

BUTOXYETHYL STEARATE
General Information
Description: Light-colored liquid with mild odor.
Formula: $CH_3(CH_2)_{16}COOC_2H_4OC_4H_9$.
Constants: Mol wt: 384, mp: 10–16°C, bp: 215–245°C at 4 mm, flash p: 420°F, d: 0.882 at 20°/20°C, vap. press.: 0.03 mm at 150°C, vap. d.: 13.2.
Hazard Analysis
Toxicity: See esters.
Fire Hazard: Slight, when exposed to heat or flame; it can react with oxidizing materials.
Countermeasures
To Fight Fire: Foam, carbon dioxide, dry chemical or carbon tetrachloride (Section 6).
Storage and Handling: Section 7.

BUTOXYL. See methoxybutyl acetate.

p-BUTOXY PHENOL
General Information
Description: White to faint yellow crystalline powder; soluble in acetone, alcohol, ether, benzene, aqueous alkali; insoluble in water.
Formula: $OHC_6H_4C_4H_8OH$.
Constants: Mol wt: 166, mp: 61–65°C.
Hazard Analysis
Toxic Hazard Rating: U. See phenol.

BUTOXY POLYPROPYLENE GLYCOL
General Information
Description: Colorless liquid.
Hazard Analysis
Toxicity: See glycols.

β-BUTOXY-β-THIOCYANODIETHYL ETHER
General Information
Synonym: Lethane 384; β,β-tubatoxythiocyanodiethyl ether.

TOXIC HAZARD RATING CODE   (For detailed discussion, see Section 1.)

0 NONE: (a) No harm under any conditions; (b) Harmful only under unusual conditions or overwhelming dosage.

1 SLIGHT: Causes readily reversible changes which disappear after end of exposure.

2 MODERATE: May involve both irreversible and reversible changes; not severe enough to cause death or permanent injury.

3 HIGH: May cause death or permanent injury after very short exposure to small quantities.

U UNKNOWN: No information on humans considered valid by authors.

Hazard Analysis
Toxic Hazard Rating:
  Acute Local: Irritant 2; Ingestion 2; Inhalation 2.
  Acute Systemic: Ingestion 2; Inhalation 2.
  Chronic Local: U.
  Chronic Systemic: U.
An insecticide.
Toxicology: Irritating to skin and mucous membrane. High
  concentration can cause CNS depression. See thio-
  cyanates. More toxic than lauric acid-2-thiocyanato-
  ethyl ester.
Fire Hazard: Unknown.
Disaster Hazard: Dangerous; See cyanides.
Countermeasures
Storage and Handling: Section 7.

BUTTER OF ANTIMONY. See antimony trichloride.

BUTTER OF ZINC. See zinc chloride.

BUTYL ACETAMIDE
General Information
Description: Liquid.
Formula: $CH_3CONHC_4H_9$.
Constants: Mol wt: 117.2, bp: 234°C, flash p: 240°F,
  d: 0.89.
Hazard Analysis
Toxicity: Unknown.
Probably has low toxicity. See amides.
Fire Hazard: Moderate, when exposed to heat or flame; it
  can react with oxidizing materials (Section 6).
Countermeasures
Storage and Handling: Section 7.
To Fight Fire: Water, foam (Section 6).

BUTYLACETANILIDE
General Information
Description: Slightly yellow liquid.
Formula: $CH_3(CH_2)_3N(C_6H_5)C(O)CH_3$.
Constants: Mol wt: 191.3, mp: 20.8°C, bp: 277–281°C,
  flash p: 286°F, d: 0.992 at 25°/25°C, vap. d.: 6.6.
Hazard Analysis
Toxicity: Unknown. See also acetanilide.
Fire Hazard: Moderate, when exposed to heat or flame; it
  can react with oxidizing materials.
Countermeasures
To Fight Fire: Foam, carbon dioxide, dry chemical or
  carbon tetrachloride (Section 6).
Storage and Handling: Section 7.

BUTYL ACETATE
General Information
Synonym: Butyl ethanoate.
Description: Colorless liquid.
Formula: $CH_3COOC_4H_9$.
Constants: Mol wt: 116.16, bp: 126°C. fp: −73.5°C, ulc:
  50–60, lel: 1.7%, uel: 7.6%, flash p: 72°F, d: 0.88 at
  20°/20°C, autoign. temp.: 790°F, vap. d.: 0.88, vap.
  press.: 15 mm at 25°C.
Hazard Analysis
Toxic Hazard Rating:
  Acute Local: Irritant 1; Allergen 1; Ingestion 1; Inhala-
  tion 1.
  Acute Systemic: Ingestion 2; Inhalation 2; Skin Absorp-
  tion 1.
  Chronic Local: Allergen 1; Irritant 1.
  Chronic Systemic: Inhalation 1; Skin Absorption 1; Ingestion 1.
Toxicology: High concentrations are irritating to eyes and
  respiratory tract and cause narcosis. Evidence of chronic
  systemic toxicity is inconclusive.
TLV: ACGIH (recommended) 150 parts per million; 710
  milligrams per cubic meter of air. See also esters,
  n-butyl alcohol and acetic acid.
Fire Hazard: Moderate, when exposed to heat or flame; can
  react with oxidizing materials.

Spontaneous Heating: No.
Explosion Hazard: Moderate, when exposed to flame (Sec-
  tion 7).
Countermeasures
Ventilation Control: Section 2.
Personal Hygiene: Section 3.
To Fight Fire: Alcohol foam, carbon dioxide, dry chemical
  or carbon tetrachloride (Section 6).
Storage and Handling: Section 7.
Shipping Regulations: Section 11.
  Coast Guard Classification: Inflammable liquid.
  ICC: Flammable liquid, red label, 10 gallons.
  IATA: Combustible liquid, no label required, 220 liters
  (passenger and cargo).
  MCA warning label.

sec-BUTYL ACETATE
General Information
Synonyms: Acetic acid secondary butyl ester; 2-butanol
  acetate.
Description: Colorless liquid; mild odor.
Formula: $CH_3COOC_4H_9$.
Constants: Mol wt: 116.1, bp: 112°C, flash p: 88°F (O.C.),
  lel: 1.7%, d: 0.862–0.866 at 20°/20°C, vap. d.: 4.00.
Hazard Analysis
Toxic Hazard Rating:
  Acute Local: Irritant 1; Allergen 1; Ingestion 2; Inhala-
  tion 0.
  Acute Systemic: Ingestion 2; Inhalation 2; Skin Absorp-
  tion 1.
  Chronic Local: Allergen 1.
  Chronic Systemic: Ingestion 1; Inhalation 1; Skin Absorp-
  tion 1.
TLV: ACGIH (recommended) 200 parts per million; 950
  milligrams per cubic meter of air.
Fire Hazard: Dangerous, when exposed to heat or flame; it
  can react with oxidizing materials.
Explosion Hazard: Moderate, when exposed to flame (Sec-
  tion 7).
Countermeasures
Ventilation Control: Section 2.
Personnel Protection: Section 3.
To Fight Fire: Alcohol foam, carbon dioxide, dry chemical
  or carbon tetrachloride (Section 6).
Storage and Handling: Section 7.
Shipping Regulations: Section 11.
  IATA: Flammable liquid, red label, 1 liter (passenger),
  40 liters (cargo).

BUTYL ACETOACETATE
General Information
Description: Liquid.
Formula: $CH_3COCH_2COOCH_2CH_2CH_2CH_3$.
Constants: Mol wt: 160.2, bp: 214°C, flash p: 185°F, d:
  0.96, vap. d.: 5.55.
Hazard Analysis
Toxicity: Details unknown. See esters, acetoacetic acid and
  butyl alcohol.
Fire Hazard: Moderate, when exposed to heat or flame;
  it can react with oxidizing materials.
Countermeasures
To Fight Fire: Foam, carbon dioxide, dry chemical or car-
  bon tetrachloride (Section 6).
Storage and Handling: Section 7.

BUTYL ACETYL RICINOLEATE
General Information
Description: Yellow, oily liquid.
Formula: $C_{18}H_{33}O_3C_4H_8CH_3CO$.
Constants: Mol wt: 396.6. mp: −32°C, bp: 220°C, flash
  p. 230°F (O.C.), d: 0.940, autoign. temp.: 725°F, vap.
  d.: 13.7.
Hazard Analysis
Toxicity: Details unknown. See esters.
Fire Hazard: Moderate, when exposed to heat or flame.

Countermeasures
To Fight Fire: Foam, carbon dioxide, dry chemical or carbon tetrachloride (Section 6).
Storage and Handling: Section 7.

N-BUTYLACID PHOSPHATE. See acid butyl phosphate.

N-tert-BUTYLACRYLAMIDE
General Information
Description: White, crystalline solid.
Formula: $C_7H_{13}ON$.
Constants: Mol wt: 127.18, mp: 128–130°C with polymerization, d: 1.015 at 30°C.
Hazard Analysis
Toxic Hazard Rating:
  Acute Local: Irritant 2; Ingestion 2.
  Acute Systemic: Ingestion 2; Inhalation 2.
  Chronic Local: U.
  Chronic Systemic: U.
Fire Hazard: Slight.
Countermeasures
Ventilation Control: Section 2.
Personnel Protection: Section 3.
Storage and Handling: Section 7.

BUTYL ACRYLATE
General Information
Description: Water-white extremely reactive monomer.
Formula: $C_7H_{12}O_2$.
Constants: Mol wt: 128.2, bp: 69°C at 50 mm, fp: –64.6°C, flash p: 120°F (O.C.), d: 0.894 at 25°/25°C, vap. press.: 10 mm at 35.5°C, vap. d.: 4.42.
Hazard Analysis
Toxic Hazard Rating:
  Acute Local: Irritant 2; Ingestion 2.
  Acute Systemic: Ingestion 2.
  Chronic Local: U.
  Chronic Systemic: U.
Fire Hazard: Moderate, when exposed to heat or flame; it can react with oxidizing materials.
Countermeasures
Personnel Protection: Section 3.
Storage and Handling: Section 7.
To Fight Fire: Foam, carbon dioxide, dry chemical or carbon tetrachloride (Section 6).

BUTYL ALCOHOL
General Information
Synonym: n-Butanol.
Description: Colorless liquid.
Formula: $CH_3(CH_2)_2CH_2OH$.
Constants: Mol wt: 74.12, bp: 117.5°C, ulc: 40, lel: 1.4%, uel: 11.2%, fp: –88.9°C, flash p: 84°F, d: 0.80978 at 20°/4°C, autoign. temp.: 689°F, vap. press.: 5.5 mm at 20°C, vap. d.: 2.55.
Hazard Analysis
Toxic Hazard Rating:
  Acute Local: Irritant 1; Ingestion 2.
  Acute Systemic: Ingestion 2; Inhalation 2; Skin Absorption 2.
  Chronic Local: U.
  Chronic Systemic: Ingestion 1; Inhalation 1; Skin Absorption 1.
TLV: ACGIH (recommended); 100 parts per million in air; 303 mg per cubic meter of air.
Toxicology: Though animal experiments have shown the butyl alcohols to possess toxic properties, they have produced few cases of poisoning in industry because of

their low volatility. The use of normal butyl alcohol is reported to have resulted in irritation of the eyes, with corneal inflammation, slight headache and dizziness, slight irritation of the nose and throat, and dermatitis about the fingernails and along the sides of the fingers (Section 9). Keratitis has also been reported.
Fire Hazard: Dangerous, when exposed to heat or flame.
Spontaneous Heating: No.
Explosion Hazard: Moderate, when exposed to flame (Section 7).
Disaster Hazard: Moderately dangerous; when heated to decomposition it emits toxic fumes; it can react with oxidizing materials.
Countermeasures
Ventilation Control: Section 2.
To Fight Fire: Water spray, carbon dioxide, dry chemical or carbon tetrachloride (Section 6).
Personnel Protection: Section 3.
Storage and Handling: Section 7.
Shipping Regulations: Section 11.
  MCA warning label.

sec-BUTYL ALCOHOL
General Information
Synonyms: 2-Butanol; ethylmethyl carbinol.
Description: Colorless liquid.
Formula: $CH_3CH_2CHOHCH_3$.
Constants: Mol wt: 74.12, mp: –89°C, bp: 99.5°C, flash p: 75°F (C.C.), d: 0.808 at 20°/4°C, autoign. temp.: 763°F, vap. press.: 10 mm at 20°C, vap. d.: 2.55, lel: 1.7% at 212°F, uel: 9.8% at 212°F.
Hazard Analysis
Toxic Hazard Rating:
  Acute Local: Irritant 1; Ingestion 2.
  Acute Systemic: Ingestion 2; Inhalation 2; Skin Absorption 2.
  Chronic Local: U.
  Chronic Systemic: Ingestion 1; Inhalation 1; Skin Absorption 1.
TLV: ACGIH (tentative) 150 parts per million; 450 milligrams per cubic meter of air.
Fire Hazard: Dangerous, when exposed to heat or flame; can react with oxidizing materials.
Explosion Hazard: Unknown.
Countermeasures
Ventilation Control: Section 2.
To Fight Fire: Water spray, alcohol foam, carbon dioxide, dry chemical or carbon tetrachloride (Section 6).
Personnel Protection: Section 3.
Storage and Handling: Section 7.
Shipping Regulations: Section 11.
  IATA: Flammable liquid, red label, 1 liter (passenger), 40 liters (cargo).
  MCA warning label.

tert-BUTYL ALCOHOL
General Information
Synonym: 2-Methyl-2-propanol.
Description: Colorless liquid or rhombic prisms or planes.
Formula: $(CH_3)_3COH$.
Constants: Mol wt: 74.12, mp: 25.3°C, bp: 82.8°C, flash p: 52°F (C.C.), d: 0.7887 at 20°/4°C, autoign. temp.: 892°F, vap. press.: 40 mm at 24.5°, vap. d.: 2.55, lel: 2.4%, uel: 8.0%.
Hazard Analysis
Toxic Hazard Rating:
  Acute Local: Irritant 1.

TOXIC HAZARD RATING CODE (For detailed discussion, see Section 1.)

0 NONE: (a) No harm under any conditions; (b) Harmful only under unusual conditions or overwhelming dosage.

1 SLIGHT: Causes readily reversible changes which disappear after end of exposure.

2 MODERATE: May involve both irreversible and reversible changes; not severe enough to cause death or permanent injury.

3 HIGH: May cause death or permanent injury after very short exposure to small quantities.

U UNKNOWN: No information on humans considered valid by authors.

Acute Systemic: Ingestion 2; Inhalation 2; Skin Absorption 2.

Chronic Local: U.

Chronic Systemic: U.

TLV: ACGIH (recommended) 100 parts per million in air; 300 mg per cubic meter of air.

Fire Hazard: Dangerous, when exposed to heat or flame; it can react with oxidizing materials.

Spontaneous Heating: No.

Explosion Hazard: Moderate, in the form of vapor when exposed to flame.

Countermeasures

Ventilation Control: Section 2.

To Fight Fire: Alcohol foam, carbon dioxide, dry chemical or carbon tetrachloride (Section 6).

Personnel Protection: Section 3.

Storage and Handling: Section 7.

Shipping Regulations: Section 11.

MCA warning label.

BUTYL ALDEHYDE. See butyraldehyde.

BUTYLAMINE
General Information
Synonym: 1-Aminobutane.

Description: Liquid, ammonia-like odor.

Formula: $C_4H_9NH_2$.

Constants: Mol wt: 73.1, mp: $-50°C$, bp: $77°C$, flash p: $45°$ (O.C.), $10°F$ (C.C.), d: 0.74–0.76 at $20°/20°C$, autoign. temp.: $594°F$, vap. d.: 2.52, lel: 1.7%, uel: 9.8%.

Hazard Analysis
Toxic Hazard Rating:

Acute Local: Irritant 2; Ingestion 2; Inhalation 2.

Acute Systemic: U.

Chronic Local: U.

Chronic Systemic: U.

See also amines.

TLV: ACGIH (recommended); 5 parts per million of air; 15 milligrams per cubic meter of air.

Absorbed via the skin.

Fire Hazard: Dangerous, when exposed to heat or flame; it can react with oxidizing materials.

Countermeasures
To Fight Fire: Alcohol foam, carbon dioxide, dry chemical or carbon tetrachloride (Section 6).

Ventilation Control: Section 2.

Personnel Protection: Section 3.

Storage and Handling: Section 7.

Shipping Regulations: Section 11.

IATA: Flammable liquid, red label, 1 liter (passenger), 40 liters (cargo).

sec-BUTYLAMINE
General Information
Synonym: 2-Aminobutane.

Description: Liquid.

Formula: $CH_3CH_2CH(NH_2)CH_3$.

Constants: Mol wt: 73.1, mp: $-104°C$, bp: $63°C$, flash p: $15°F$, d: 0.724 at $20°C$.

Hazard Analysis
Toxic Hazard Rating:

Acute Local: Irritant 2; Ingestion 2; Inhalation 2.

Acute Systemic: U.

Chronic Local: U.

Chronic Systemic: U.

See also amines.

Fire Hazard: Dangerous, when exposed to heat or flame; it can react with oxidizing materials.

Explosion Hazard: Unknown.

Countermeasures
Ventilation Control: Section 2.

To Fight Fire: Carbon dioxide, dry chemical or carbon tetrachloride (Section 6).

Personnel Protection: Section 3.

Storage and Handling: Section 7.

tert-BUTYLAMINE
General Information
Synonyms: 2-Aminoisobutane; trimethylaminomethane.

Description: Colorless liquid.

Formula: $(CH_3)_3CNH_2$.

Constants: Mol wt: 73.14, mp: $-67.5°C$, bp: $44–46°C$, d: 0.700 at $15°C$, lel: 1.7% at $212°F$, uel: 8.9% at $212°F$, vap. d.: 2.5.

Hazard Analysis
Toxic Hazard Rating:

Acute Local: Irritant 2; Ingestion 2; Inhalation 2.

Acute Systemic: U.

Chronic Local: U.

Chronic Systemic: U.

See also amines.

Fire Hazard: Moderate, when exposed to heat or flame; it can react with oxidizing materials (Section 6).

Explosion Hazard: Unknown.

Countermeasures
Ventilation Control: Section 2.

Personnel Protection: Section 3.

Storage and Handling: Section 7.

To Fight Fire: Alcohol foam (Section 6).

BUTYLAMINE OLEATE
General Information
Synonym: Monobutylamine oleate.

Description: Liquid.

Formula: $C_{17}H_{33}COONH_3C_4H_9$.

Constants: Mol wt: 355.6, flash p: $150°F$ (O.C.), d: 0.891.

Hazard Analysis
Toxicity: Details unknown. See esters and butylamine.

Fire Hazard: Moderate, when exposed to heat or flame; it can react with oxidizing materials.

Countermeasures
To Fight Fire: Alcohol foam, carbon dioxide, dry chemical or carbon tetrachloride (Section 6).

Storage and Handling: Section 7.

BUTYL AMINOBENZOATE. See butesin.

2-BUTYLAMINOETHANOL. See butylethanolamine.

tert-BUTYLAMINO ETHYL METHACRYLATE
General Information
Description: Liquid.

Formula: $C_{10}H_{19}NO_2$.

Constants: Mol wt: 185, bp: $100–105°C$, d: 0.914, fp: $205°F$ (O.C.).

Hazard Analysis
Fire Hazard: Moderate, when exposed to heat or flame.

Countermeasures
Storage and Handling: Section 7.

N-n-BUTYL-p-AMINOPHENOL
General Information
Description: Crystals.

Formula: $HOC_6H_4HNCH_2CH_2CH_2CH_3$.

Constants: Mol wt: 165.2, mp: $-33°C$, flash p: $61°F$.

Hazard Analysis
Toxicity: Unknown.

Caution: May be an irritant or skin sensitizer.

Fire Hazard: Dangerous, when exposed to heat or flame (Section 6).

Explosion Hazard: Unknown.

Disaster Hazard: Moderately dangerous; when heated to decomposition, it emits toxic fumes; it can react with oxidizing materials.

Countermeasures
Storage and Handling: Section 7.

BUTYL ANILINE
General Information
Description: Colorless liquid.

Formula: (C_4H_9) NHC_6H_5.

Constants: Mol wt: 149.23, bp: $241°C$, flash p: $225°F$

(C.O.C.), fp: −15.1°C, d: 0.9288 at 20°/20°C, vap. press.: 0.02 mm at 20°C, vap. d.: 5.15.

Hazard Analysis
Toxicity: Details unknown, probably highly toxic. See also aniline.
Fire Hazard: Moderate, when exposed to heat or flame.
Disaster Hazard: Dangerous; when heated to decomposition it emits highly toxic fumes of aniline; it can react vigorously with oxidizing materials.

Countermeasures
Storage and Handling: Section 7.
To Fight Fire: Alcohol foam, foam, carbon dioxide, dry chemical or carbon tetrachloride (Section 6).

p-tert-BUTYL ANILINE
General Information
Synonym: 1-Amino-4-tert-butyl benzene.
Description: Oil.
Formula: $(CH_3)_3CC_6H_4NH_2$.
Constants: Mol wt: 149.2, mp: 17°C, bp: 241°C, d: 0.9525 at 15°/4°C.

Hazard Analysis
Toxicity: Details unknown. Probably toxic. See also aniline.
Disaster Hazard : Dangerous; when heated to decomposition, it emits highly toxic fumes of aniline.

Countermeasures
Storage and Handling: Section 7.

BUTYLATED HYDROXYANISOLE
General Information
Synonym: BHA.
Formula: $C_{11}H_{16}O_2$.
Constant: Mol wt: 130.2.

Hazard Analysis
Toxicology: Very low order of toxicity. Used as an antioxidant in foods.

BUTYLATED HYDROXYTOLUENE. See ditert-butyl-p-cresol.

BUTYL BENZENE
General Information
Synonym: 1-Phenyl butane.
Description: Colorless liquid.
Formula: $C_6H_5CH_2CH_2CH_2CH_3$.
Constants: Mol wt: 134.2, mp: −81.2°C, bp: 182.1°C, fp: −88.2°C, flash p: 160°F (T.O.C.), d: 0.8601 at 20°/4°C, vap. press.: 1 mm at 22.7°C, Autoign. temp.: 774°F, lel: 0.8%, uel: 5.8%, vap. d.: 4.6.

Hazard Analysis
Toxicity: Details unknown. See ethyl benzene.
Fire Hazard: Moderate, when exposed to heat or flame; it can react with oxidizing materials.

Countermeasures
To Fight Fire: Foam, carbon dioxide, dry chemical or carbon tetrachloride (Section 6).
Storage and Handling: Section 7.

sec-BUTYL BENZENE
General Information
Synonym: 2-Phenyl butane.
Description: Colorless liquid.
Formula: $C_6H_5CH(CH_3)C_2H_5$.
Constants: Mol wt: 134.2, mp: −82.7°C, bp: 173.5°C, fp: −75.8°C, flash p: 126°F (T.O.C.), d: 0.8621 at 20°C, vap. press.: 1 mm at 18.6°C, vap. d.: 4.62, autoign. temp.: 784°F, lel: 0.8%, uel: 6.9%.

Hazard Analysis
Toxicity: Details unknown. See ethyl benzene.
Fire Hazard: Moderate, when exposed to heat or flame; it can react with oxidizing materials (Section 6).
Spontaneous Heating: No.

Countermeasures
To Fight Fire: Foam, carbon dioxide, dry chemical or carbon tetrachloride (Section 6).
Storage and Handling: Section 7.

tert-BUTYL BENZENE
General Information
Synonym: 2-Methyl-2-phenyl propane.
Description: Colorless liquid.
Formula: $C_6H_5C(CH_3)_3$.
Constants: Mol wt: 134.2, bp: 168.2°C, fp: −58°C, flash p: 140°F (T.O.C.), d: 0.8665 at 20°C, vap. press.: 1 mm at 13.0°C, vap. d.: 4.62, autoign. temp.: 842°F, lel: 0.7% at 212°F, uel: 5.7% at 212°F.

Hazard Analysis
Toxicity: Details unknown. See ethyl benzene.
Fire Hazard: Moderate, when exposed to heat or flame; it can react with oxidizing materials.

Countermeasures
To Fight Fire: Foam, carbon dioxide, dry chemical or carbon tetrachloride (Section 6).
Storage and Handling: Section 7.

p-BUTYL BENZENE TETRATHIO-o-STANNATE
General Information
Description: Solid.
Formula: $(SC_6H_4C_4H_9)_4Sn$.
Constants: Mol wt: 779.8, mp: 106°C.

Hazard Analysis
Toxicology: Details unknown. See tin compounds.
Disaster Hazard: Dangerous; see sulfur compounds.

BUTYL BENZOATE
General Information
Description: Liquid.
Formula: $C_6H_5COOC_4H_9$.
Constants: Mol wt: 178.22, mp: −21.5°C, bp: 250°C, flash p: 225°F (O.C.), d: 1.0073 at 20°/20°C, vap. press.: <0.01 mm at 20°C, vap. d.: 6.15.

Hazard Analysis
Toxicity: Details unknown. See also esters, benzoic acid and n-butyl alcohol.
Fire Hazard: Moderate, when exposed to heat or flame; it can react with oxidizing materials.

Countermeasures
To Fight Fire: Foam, carbon dioxide, dry chemical or carbon tetrachloride (Section 6).
Storage and Handling: Section 7.

p-tert-BUTYL BENZOIC ACID
General Information
Synonym: p-t-BBA.
Description: Colorless, fine crystalline powder.
Formula: $HOOCC_6H_4C(CH_3)_3$.
Constants: Mol wt: 178.11, mp: 166.3°C, d: 1.142 at 20°/4°C.

Hazard Analysis
Toxic Hazard Rating:
Acute Local: Irritant 2; Ingestion 2; Inhalation 2.
Acute Systemic: U.
Chronic Local: U.
Chronic Systemic: U.

TOXIC HAZARD RATING CODE (For detailed discussion, see Section 1.)

0 NONE: (a) No harm under any conditions; (b) Harmful only under unusual conditions or overwhelming dosage.

1 SLIGHT: Causes readily reversible changes which disappear after end of exposure.

2 MODERATE: May involve both irreversible and reversible changes; not severe enough to cause death or permanent injury.

3 HIGH: May cause death or permanent injury after very short exposure to small quantities.

U UNKNOWN: No information on humans considered valid by authors.

Toxicology: Data based on animal experiments show low toxicity. Mild local irritation has been reported in humans.

Fire Hazard: Slight; when exposed to heat or flame; it can react with oxidizing materials.

Countermeasures

To Fight Fire: Foam, carbon dioxide, dry chemical or carbon tetrachloride (Section 6).

Storage and Handling: Section 7.

BUTYL BENZYL PHTHALATE
General Information
Description: Clear, oily liquid.
Formula: $C_4H_9COOC_6H_4COOCH_2C_6H_5$.
Constants: Mol wt: 312.4, mp: $< -35°C$, bp: 370°C, flash p: 390°F, d: 1.116 at 25°/25°C, vap. d.: 10.8.
Hazard Analysis
Toxicity: Details unknown. See also esters.
Fire Hazard: Slight, when exposed to heat or flame; it can react with oxidizing materials.
Countermeasures
To Fight Fire: Water, foam, carbon dioxide, dry chemical or carbon tetrachloride (Section 6).
Storage and Handling: Section 7.

BUTYL BENZYL SEBACATE
General Information
Description: Liquid.
Formula: $(CH_2)_8CO_2C_4H_9CO_2C_6H_5CH_2$.
Constants: Mol wt: 348.5, bp: 245–285°C at 10 mm, flash p: 395°F, d: 1.004 at 20°/20°C, vap. d.: 12.0.
Hazard Analysis
Toxicity: See esters.
Fire Hazard: Slight, when exposed to heat or flame; can react with oxidizing materials.
Countermeasures
To Fight Fire: Foam, carbon dioxide, dry chemical or carbon tetrachloride (Section 6).
Storage and Handling: Section 7.

BUTYL BORIC ACID
General Information
Description: Colorless crystals.
Formula: $C_4H_9B(OH)_2$.
Constants: Mol wt: 101.95, mp: 92–94°C, bp: decomposes.
Hazard Analysis
Toxicity: Details unknown. See boron compounds and butyl alcohol.
Fire Hazard: Slight; when heated, it emits toxic fumes. It may have flammable decomposition products (Section 6).
Countermeasures
Storage and Handling: Section 7.

tert-BUTYL BORIC ACID
General Information
Description: White crystals.
Formula: $C_4H_9B(OH)_2$.
Constants: Mol wt: 101.95, mp: 105°C decomposes, bp: decomposes.
Hazard Analysis
Toxicity: Unknown. See boron compounds and butyl alcohol.
Fire Hazard : Slight; when heated, it emits toxic fumes. It may emit flammable products of decomposition (Section 6).
Countermeasures
Storage and Handling: Section 7.

n-BUTYL BORON OXIDE
General Information
Description: Colorless liquid hydrolyzed by water. Soluble in ether.
Formula: $(C_4H_9)_3B_3O_3$.
Constants: Mol wt: 251.8, bp: 154°C at 30 mm Hg.
Toxicity: Details unknown. See boron compounds.

tert-BUTYL BORON OXIDE
General Information
Description: Colorless liquid hydrolyzed by water. Soluble in ether.
Formula: $(C_4H_9)_3B_3O_3$.
Constants: Mol wt: 251.8, mp: 20°C, bp: 67°C at 5 mm Hg.
Hazard Analysis
Toxicity: Details unknown. See boron compounds.

BUTYL BROMIDE
General Information
Synonym: 1-Bromobutane.
Description: Colorless to pale straw-colored liquid.
Formula: $CH_3(CH_2)_2CH_2Br$.
Constants: Mol wt: 137.03, mp: $-112.4°C$, bp: 101.4°C, flash p: 65°F (O.C.), d: 1.274 at 25°/25°C, autoign. temp.: 509°F, vap. d.: 4.72, lel: 2.6% at 212°F, uel: 6.6% at 212°F.
Hazard Analysis
Toxicity: Details unknown. See also chlorinated hydrocarbons, aliphatic.
Fire Hazard: Dangerous, when exposed to heat or flame.
Disaster Hazard: Dangerous; see bromides; it can react with oxidizing materials.
Countermeasures
Storage and Handling: Section 7.
To Fight Fire: Carbon dioxide, dry chemical or carbon tetrachloride (Section 6).
Shipping Regulations: Section 11.
  IATA: Flammable liquid, red label, 1 liter (passenger), 40 liters (cargo).

sec-BUTYL BROMIDE
General Information
Description: Colorless liquid.
Formula: $C_2H_5CHBrCH_3$.
Constants: Mol wt: 137.0, fp: $< -50°C$, bp: 91.4°C, flash p: 70°F, d: 1.257 at 25°/25°C.
Hazard Analysis
Toxicity: Unknown. Narcotic in high concentrations. See also chlorinated hydrocarbons, aliphatic.
Fire Hazard: Dangerous, when exposed to heat or flame.
Explosion Hazard: Unknown.
Disaster Hazard: Dangerous; when heated to decomposition, it emits toxic fumes; it can react with oxidizing materials.
Countermeasures
Storage and Handling: Section 7.
To Fight Fire: Water, foam, carbon dioxide, dry chemical or carbon tetrachloride (Section 6).

tert-BUTYL BROMIDE
General Information
Description: Colorless liquid.
Formula: $(CH_3)_3CBr$.
Constants: Mol wt: 137.0, mp: $-20°C$, bp: 73.3°C, fp: $-18°C$, d: 1.215 at 25°/25°C.
Hazard Analysis
Toxicity: Unknown. See also chlorinated hydrocarbons, aliphatic.
Disaster Hazard: Moderately dangerous; when heated to decomposition, it emits toxic fumes.
Countermeasures
Storage and Handling: Section 7.

BUTYL BUTANOATE. See n-butyl butyrate.

n-BUTYL BUTYRATE
General Information
Synonym: Butyl butanoate.
Description: Liquid.
Formula: $CH_3CH_2CH_2CO_2CH_2CH_2CH_2CH_3$.
Constants: Mol wt: 144.2, bp: 166°C, flash p: 128°F (O.C.), d: 0.874, vap. d.: 5.0.

Hazard Analysis
Toxic Hazard Rating:
Acute Local: Irritant 2.
Acute Systemic: Inhalation 2.
Chronic Local: Irritant 1.
Chronic Systemic: U.
Toxicology: Irritating and narcotic in high concentrations.
Fire Hazard: Moderate, when exposed to heat or flame;
it can react with oxidizing materials.
Countermeasures
Ventilation Control: Section 2.
Personnel Protection: Section 3.
To Fight Fire: Alcohol foam, foam, carbon dioxide, dry
chemical or carbon tetrachloride (Section 6).
Storage and Handling: Section 7.

BUTYL CARBINOL. See amyl alcohol.

M-BUTYL CARBINAL. See l-pentanol.

sec-BUTYL CARBINOL. See sec-amyl alcohol.

tert-BUTYL CARBINOL. See tert-amyl alcohol.

BUTYL "CARBITOL"
General Information
Synonym: Diethylene glycol monobutyl ether.
Description: Colorless liquid.
Formula: $C_4H_9OCH_2CH_2OCH_2CH_2OH$.
Constants: Mol wt: 162.2, fp: $-68.1°C$, bp: 230.6°C,
flash p: 172°F, d: 0.9553 at 20°/4°C, autoign. temp.:
442°F, vap. press.: 0.02 mm at 20°C, vap. d.: 5.58.
Hazard Analysis
Toxicity: See glycols.
Fire Hazard: Moderate, when exposed to heat or flame,
it emits degradation products; can react with oxidizing
materials.
Spontaneous Heating: No.
Countermeasures
Storage and Handling: Section 7.
To Fight Fire: Alcohol foam, carbon dioxide, dry chemical,
or carbon tetrachloride (Section 6).

BUTYL "CARBITOL" ACETATE
General Information
Synonym: Diethylene glycol monobutyl ether acetate.
Description: Colorless liquid.
Formula: $C_4H_9O(CH_2)_2O(CH_2)_2OOCCH_3$.
Constants: Mol wt: 204.26, fp: $-32.2°C$, bp: 247°C,
flash p: 240°F (O.C.), d: 0.981 at 20°/20°C, autoign.
temp. 570°F, vap. press.: <0.01 mm at 20°C.
Hazard Analysis
Toxicity: See glycols.
Fire Hazard: Moderate, when exposed to heat or flame it
emits degradation products; can react with oxidizing
materials.
Countermeasures
To Fight Fire: Foam, carbon dioxide, dry chemical, or
carbon tetrachloride (Section 6).
Storage and Handling: Section 7.

BUTYL "CARBITOL" THIOCYANATE
General Information
Formula: $C_4H_9O(CH_2)_2O(CH_2)_2CNS$.
Constant: Mol wt: 203.3.
Hazard Analysis
Toxicity: Details unknown. See also thiocyanates. Experi-
mental data show moderately high toxicity for mouse,
rat, and cat.

Disaster Hazard: See thiocyanates.
Countermeasures
Storage and Handling: Section 7.

BUTYL CARBITYL 6-PROPYL PIPERONYL ETHER.
See piperonyl butoxide.

p-tert-BUTYL CATECHOL
General Information
Description: White crystalline solid.
Formula: $(CH_3)_3CC_6H_3(OH)_2$.
Constants: Mol wt: 166.2, fp: 52°C, flash p: 265°F,
bp: 285°C, d: 1.049 at 60°/25°C.
Hazard Analysis
Toxic Hazard Rating:
Acute Local: Irritant 2.
Acute Systemic: U.
Chronic Local: U.
Chronic Systemic: U.
Fire Hazard: Moderate, when exposed to heat or flame.
Disaster Hazard: Moderately dangerous; when heated to
decomposition it emits toxic fumes; can react with
oxidizing materials.
Countermeasures
Personnel Protection: Section 3.
To Fight Fire: Foam, carbon dioxide, dry chemical or
carbon tetrachloride (Section 6).
Storage and Handling: Section 7.

BUTYL "CELLOSOLVE"
General Information
Synonyms: Glycol monobutyl ether; 2-butoxy ethanol.
Description: Colorless liquid.
Formula: $C_4H_9OCH_2CH_2OH$.
Constants: Mol wt: 118.17, mp: $<-40°C$, bp: 171.2°C,
flash p: 141°F (C.C.), d: 0.9027 at 20°/4°C, autoign.
temp.: 472°F, vap. press.: 0.6 mm at 20°C, vap. d.:
4.07.
Hazard Analysis
Toxicity: See Glycols.
TLV: ACGIH (recommended): 50 parts per million in air;
240 milligrams per cubic meter of air. May be absorbed
via skin.
Fire Hazard: Moderate, when exposed to heat or flame; can
react with oxidizing materials.
Spontaneous Heating: No.
Countermeasures
To Fight Fire: Carbon dioxide, dry chemical or carbon
tetrachloride (Section 6).
Storage and Handling: Section 7.

BUTYL "CELLOSOLVE" ACETATE
General Information
Synonym: Glycol monobutyl ether acetate.
Description: Liquid.
Formula: $C_4H_9O(CH_2)_2OOCCH_3$.
Constants: Mol wt: 160.20, bp: 188°C, flash p: 180°F
(O.C.), d: 0.943, vap. d.: 5.5.
Hazard Analysis
Toxicity: See Glycols.
Fire Hazard: Moderate, when exposed to heat or flame, it
emits toxic degradation products; can react with
oxidizing materials.
Countermeasures
To Fight Fire: Foam, carbon dioxide, dry chemical or
carbon tetrachloride (Section 6).
Storage and Handling: Section 7.

TOXIC HAZARD RATING CODE (For detailed discussion, see Section 1.)

0 NONE: (a) No harm under any conditions; (b) Harmful only under unusual conditions or overwhelming dosage.

1 SLIGHT: Causes readily reversible changes which disappear after end
of exposure.

2 MODERATE: May involve both irreversible and reversible changes;
not severe enough to cause death or permanent injury.

3 HIGH: May cause death or permanent injury after very short exposure
to small quantities.

U UNKNOWN: No information on humans considered valid by authors.

BUTYL CHLORIDE
General Information
Synonym: 1-Chlorobutane.
Description: Colorless liquid.
Formula: $CH_3(CH_2)_2CH_2Cl$.
Constants: Mol wt: 92.57, mp: $-123.1°C$, bp: $78°C$, lel. 1.9%, uel: 10.1%, flash p: 15°F, (O.C.), d: 0.884, autoign. temp.: 860°F, vap. d.: 3.20.
Hazard Analysis
Toxicity: Details unknown. See chlorinated hydrocarbons, aliphatic. Limited animal experiments suggest low toxicity.
Fire Hazard: Dangerous, when exposed to heat or flame.
Explosion Hazard: Moderate, when exposed to flame (Section 7).
Disaster Hazard: Dangerous; when heated to decomposition, it emits highly toxic fumes of phosgene; it can react vigorously with oxidizing materials.
Countermeasures
Storage and Handling: Section 7.
To Fight Fire: Foam, carbon dioxide, dry chemical or carbon tetrachloride (Section 6).
Shipping Regulations: Section 11.
   IATA: Flammable liquid, red label, 1 liter (passenger), 40 liters (cargo).

4-tert-BUTYL-2-CHLOROPHENOL
General Information
Formula: $ClC_6H_3(C_4H_9)C(CH_3)_3$.
Constants: Mol wt: 184.5, bp: 453–484°F, d: 1.1, flash p: 225°F.
Hazard Analysis
Fire Hazard: Moderate, when exposed to heat or flame.
Disaster Hazard: Dangerous. See chlorides.
Countermeasures
To Fight Fire: Water, foam (Section 6).
Storage and Handling: Section 7.

tert-BUTYL CHROMATE
Hazard Analysis
Toxic Hazard Rating:
   Acute Local: Irritant 3.
   Acute Systemic: U.
   Chronic Local: U.
   Chronic Systemic: U.
Toxicity: Highly toxic. See chromium compounds.
TLV: ACGIH (recommended) 0.1 milligrams per cubic meter of air. Absorbed via skin.

BUTYL CITRATE. See tributyl citrate.

p-tert-BUTYL-o-CRESOL
General Information
Formula: $(OH)C_6H_3CH_3C(CH_3)_3$.
Constants: Mol wt: 164, d: 1.0–, bp: 278–280°F, flash p: 244°F.
Hazard Analysis
Fire Hazard: Slight, when exposed to heat or flame.
Countermeasures
To Fight Fire: Water, foam.
Storage and Handling: Section 7.

tert-BUTYL-m-CRESOL
General Information
Synonym: MBMC; 6-tert-butyl-methyl cresol.
Description: Clear liquid, soluble in organic solvents and aqueous potassium hydroxide.
Formula: $C_6H_3(C_4H_9)(CH_3)OH$.
Constants: Mol wt: 164, fp: 23.1°C, bp: 244°C, d: 0.922 (80°C), flash p: 116°F.
Hazard Analysis
Fire Hazard: Moderate, when exposed to heat or flame,
Countermeasures
Storage and Handling: section 7.

BUTYL CROTONATE
General Information
Description: Water-white liquid, pleasant persistent odor.
Formula: $CH_3CHCHCOOC_4H_9$.
Constants: Mol wt: 142.2 bp: 180.5°C, d: 0.9037 at 20°/20°C, vap. d.: 4.9.
Hazard Analysis
Toxicity: Details unknown. See also esters.
Fire Hazard: Dangerous, when exposed to heat or flame; it can react with oxidizing materials (Section 6).
Countermeasures
Storage and Handling: Section 7.

n-BUTYL CYCLOHEXYL AMINE
General Information
Formula: $C_4H_9C_6H_{10}NH$.
Constants: Mol wt: 154, flash p: 200°F (O.C.), d: 0.8, bp: 409°F.
Hazard Analysis
Toxic Hazard Rating: U. Limited animal experiments suggest high toxicity and moderate irritation. See also amines.
Fire Hazard: Moderate, when exposed to heat or flame.
Countermeasures
To Fight Fire: Alcohol foam (Section 6).
Storage and Handling: Section 7.

BUTYL CYCLOHEXYLCAPROATE
Hazard Analysis
Toxicity: Details unknown. See esters and butyl alcohol.
Fire Hazard: A combustible material; it can react with oxidizing materials (Section 6).
Countermeasures
Storage and Handling: Section 7.

BUTYL DECALIN
General Information
Description: Liquid.
Formula: $C_4H_9C_{10}H_{17}$.
Constants: Mol wt: 194.35, flash p: 500°F (C.C.), vap. d.: 6.7.
Hazard Analysis
Toxicity: Unknown. See also decalin.
Fire Hazard: Slight, when exposed to heat or flame; it can react with oxidizing materials.
Countermeasures
To Fight Fire: Water, carbon dioxide, dry chemical or carbon tetrachloride (Section 6)
Storage and Handling: Section 7.

tert-BUTYL DECALIN
General Information
Description: Liquid.
Formula: $C_4H_9C_{10}H_{17}$.
Constants: Mol wt: 194.4, flash p: 640°F (C.C.), vap. d.: 6.7.
Hazard Analysis
Toxicity: Unknown.
Fire Hazard: Slight, when exposed to heat or flame; it can react with oxidizing materials.
Countermeasures
To Fight Fire: Water, carbon dioxide, dry chemical or carbon tetrachloride (Section 6).
Storage and Handling: Section 7.

BUTYL DIAMYLAMINE
General Information
Description: Liquid.
Formula: $(C_4H_9)(C_5H_{11})_2N$.
Constants: Mol wt: 213.4, bp: 229°C, flash p: 200°F, d: 0.78, vap. d.: 7.3.
Hazard Analysis
Toxicity: Details unknown. See also amines.
Fire Hazard: Moderate, when exposed to heat or flame, it can react with oxidizing materials.

Countermeasures

To Fight Fire: Foam, carbon dioxide, dry chemical or carbon tetrachloride (Section 6).

Storage and Handling: Section 7.

BUTYL DICHLOROARSINE

General Information

Description: Oily liquid, somewhat agreeable odor.

Formula: $C_4H_9AsCl_2$.

Constants: Mol wt: 202.945, bp: 194°C.

Hazard Analysis

Toxicity: See arsenic compounds. A highly toxic military poison gas.

Disaster Hazard: See arsenic compounds.

Countermeasures

Storage and Handling: Section 7.

BUTYL 2,4-DICHLOROPHENOXYACETATE

General Information

Synonyms: 2,4-D butyl ester.

Description: Light-brown liquid.

Formula: $C_6H_3(Cl)_2OCH_2COO(CH_2)_3CH_3$.

Constants: Mol wt: 277.15, d: 1.235–1.245 at 25°/25°C.

Hazard Analysis

Toxicity: Unknown. See 2,4-dichlorophenoxyacetic acid.

Disaster Hazard: Dangerous; when heated to decomposition it emits toxic fumes.

Countermeasures

Storage and Handling: Section 7.

BUTYL DIETHANOLAMINE

General Information

Description: Liquid.

Formula: $C_4H_9N(CH_2CH_2OH)_2$.

Constants: Mol wt: 161.24, bp: 262°C, flash p: 245°F (O.C.), d: 0.97, vap. d.: 5.55.

Hazard Analysis

Toxicity: Details unknown. See amines.

Fire Hazard: Slight; when exposed to heat or flame, it can react with oxidizing materials.

Countermeasures

To Fight Fire: Alcohol foam, foam, carbon dioxide, dry chemical or carbon tetrachloride (Section 6).

Storage and Handling: Section 7.

tert-BUTYL DIETHANOLAMINE

General Information

Synonym: [2,2-(t-Butylimino)-diethanol].

Formula: $C_8H_{10}NO_2$.

Constants: Mol wt: 152, mp: 117°F, bp: 329–338°F at 33 mm, d: 1.0, flash p: 285°F (O.C.).

Hazard Analysis

Fire Hazard: Slight, when exposed to heat or flame.

Countermeasures

To Fight Fire: Water, foam, alcohol foam (Section 6).

Storage and Handling: Section 7.

BUTYL DIETHYLHEPTYLATE

General Information

Description: Liquid.

Formula: $C_4H_8[COOCH_2CH(C_2H_5)C_4H_8]_2$.

Constants: Mol wt: 370.56, mp: −60°C, bp: 214°C at 5 mm, flash p: 385°F (O.C.), d: 0.9268 at 20°/20°C, vap. d.: 12.8.

Hazard Analysis

Toxicity: Details unknown. See also esters.

Fire Hazard: Slight; when exposed to heat or flame, it can react with oxidizing materials.

Countermeasures

To Fight Fire: Foam, carbon dioxide, dry chemical or carbon tetrachloride (Section 6)

Storage and Handling: Section 7.

BUTYL DIGLYCOL CARBONATE. See diethylene glycol bis(butyl carbonate).

2-sec-BUTYL-4,6-DINITROPHENOL

General Information

Synonym: DNBP; dinoseb.

Description: Crystals.

Formula: $(C_4H_9)(NO_2)_2C_6H_2OH$.

Constant: Mol wt: 195.2.

Hazard Analysis

Toxic Hazard Rating:

Acute Local: Irritant 3; Ingestion 3.

Acute Systemic: Ingestion 3; Inhalation 3; Skin Absorption 3.

Chronic Local: Irritant 2.

Chronic Systemic: Ingestion 3; Inhalation 3; Skin Absorption 3.

Toxicity: Details unknown. See also butyl alcohol and dinitrophenol. Data based on animal experiments.

Fire Hazard: Details unknown. See nitrates.

Disaster Hazard: See nitrates.

Countermeasures

Storage and Handling: Section 7.

Shipping Regulations: Section 11.

IATA: Poison B, poison label, 1 liter (passenger), 220 liters (cargo).

n-BUTYL DISULFIDE

General Information

Synonym: 1-Butyldithiobutane.

Formula: $[CH_3(CH_2)_3]SS[CH_3(CH_2)_3]$.

Constants: Mol wt: 178.4, bp: 103°C at 15 mm Hg.

Hazard Analysis

Toxicity: See alkyl disulfides.

Fire Hazard: Probably moderate.

Disaster Hazard: Dangerous; when heated to decomposition, it emits highly toxic fumes of sulfides. Probably reacts strongly with powerful oxidizers.

Countermeasures

To Fight Fires: Water spray, foam, carbon dioxide, dry chemicals (Section 6).

Storage and Handling: Section 7.

α-BUTYLENE

General Information

Synonym: 1-Butene.

Description: Gas.

Formula: $CH_3CH_2CHCH_2$.

Constants: Mol wt: 56.10, bp: −6.3°C, fp: −185.3°C, lel: 1.7%, uel: 9.0%, flash p: −112°F, d: 0.668 at 0°/1°C, vap. d.: 1.93, vap. press.: 3480 mm at 21°C, antoign. temp.: 723°F.

Hazard Analysis

Toxic Hazard Rating:

Acute Local: 0.

Acute Systemic: Inhalation 2.

Chronic Local: U.

Chronic Systemic: U.

Caution: An anesthetic and asphyxiant.

Fire Hazard: Dangerous, when exposed to heat or flame.

Spontaneous Heating: No.

Explosion Hazard: Moderate, when exposed to flame (Section 7).

TOXIC HAZARD RATING CODE (For detailed discussion, see Section 1.)

0 NONE: (a) No harm under any conditions; (b) Harmful only under unusual conditions or overwhelming dosage.

1 SLIGHT: Causes readily reversible changes which disappear after end of exposure.

2 MODERATE: May involve both irreversible and reversible changes; not severe enough to cause death or permanent injury.

3 HIGH: May cause death or permanent injury after very short exposure to small quantities.

U UNKNOWN: No information on humans considered valid by authors.

Disaster Hazard: Moderately dangerous; it can react with oxidizing materials.

Countermeasures

Ventilation Control: Section 2.

To Fight Fire: Carbon dioxide, dry chemical or water spray. Stop flow of gas. (Section 6).

Storage and Handling: Section 7.

Shipping Regulations: Section 11.

IATA: Flammable gas, red label, not acceptable (passenger), 140 kilograms (cargo).

β-BUTYLENE. See 2-butene.

γ-BUTYLENE. See isobutylene.

BUTYLENE CHLORIDE
General Information

Description: Colorless liquid.

Formula: C_4H_7Cl.

Constants: Mol wt: 90.6, lel: 2.3%, uel: 9.3%, bp: 72°C, d: 0.926, vap. d.: 3.13.

Hazard Analysis

Toxicity: Details unknown. See also chlorinated hydrocarbons, aliphatic.

Fire Hazard: Dangerous, when exposed to heat or flame.

Explosion Hazard: Moderate, when exposed to flame (Section 7).

Disaster Hazard: Dangerous; when heated to decomposition, it emits highly toxic fumes of phosgene; it can react vigorously with oxidizing materials.

Countermeasures

Storage and Handling: Section 7.

To Fight Fire: Foam, carbon dioxide, dry chemical or carbon tetrachloride (Section 6).

α-BUTYLENE DIBROMIDE
General Information

Synonym: 1,2-Dibromo butane,

Description: Yellowish liquid; insoluble in water; miscible with water.

Constants: d: 1.820 (20/4°C), mp: −65°C, bp: 166°C.

Hazard Analysis

Toxic Hazard Rating:

Acute Local: Irritant 1.

Acute Systemic: Inhalation 1.

Chronic Local: U.

Chronic Systemic: U.

An irritant and narcotic. See also bromides.

Disaster Hazard: Dangerous. See bromides.

α-BUTYLENE GLYCOL. See 1,2-butylene glycol.

β-BUTYLENE GLYCOL. See 1,3-butylene glycol.

1,2-BUTYLENE GLYCOL
General Information

Synonym: α-Butylene glycol.

Description: Liquid.

Formula: $CH_3CH_2CH(OH)CH_2OH$.

Constants: Mol wt: 90.1, mp: −114°C, bp: 192°C, d: 1.019 at 0°/4°C, vap. d.: 3.1.

Hazard Analysis

Toxicity: Details unknown. See also glycols.

Fire Hazard: Moderate; when exposed to heat or flame, it can react with oxidizing materials (Section 6).

Countermeasures

To Fight Fire: Water, carbon dioxide, dry chemical or carbon tetrachloride (Section 6).

Storage and Handling: Section 7.

1,3-BUTYLENE GLYCOL
General Information

Synonyms: 1,3-Butanediol; β-butylene glycol.

Description: Viscous liquid.

Formula: $CH_3CH(OH)CH_2CH_2OH$.

Constants: Mol wt: 90.12, bp: 207.5°C fp: < −50°C,

flash p: 250°F, d: 1.006 at 20°/20°C, autoign. temp. 741°F, vap. press.: 0.06 mm at 20°C, vap. d.: 3.2.

Hazard Analysis

Toxicity: Details unknown. See also glycols. A food additive permitted in food for human consumption (Section 10).

Fire Hazard: Slight; when exposed to heat or flame, it can react with oxidizing materials.

Spontaneous Heating: No.

Countermeasures

To Fight Fire: Foam, alcohol foam, carbon dioxide, dry chemical or carbon tetrachloride (Section 6).

Storage and Handling: Section 7.

1,4-BUTYLENE GLYCOL. See 1,4 butanediol.

2,3-BUTYLENE GLYCOL
General Information

Synonyms: 2,3-Butanediol; pseudobutylene glycol.

Description: Colorless liquid.

Formula: $CH_3CH(OH)CH(OH)CH_3$.

Constants: Mol wt: 90.12, bp: 180°C, fp: 19°C, flash p: 185°F (T.O.C.), d: 1.0095 at 20°/20°C, autoign. temp.: 756°F, vap. press.: 0.17 mm at 20°C, vap. d.: 3.1.

Hazard Analysis

Toxicity: Details unknown. See also glycols.

Fire Hazard: Moderate; when exposed to heat or flame, it can react with oxidizing materials (Section 6).

Countermeasures

To Fight Fire: Alcohol foam, carbon dioxide, dry chemical or carbon tetrachloride (Section 6).

Storage and Handling: Section 7.

BUTYLENE OXIDES
General Information

Description: A liquid mixture of isomeric butylene oxides.

Formula: C_4H_8O.

Constants: Mol wt: 72.1, bp: 62–65°C, lel: 1.5%, uel: 18.3%, flash p: 5°F, pour p: −150°C, d: 0.826 at 25°/25°C, vap. d.: 2.49.

Hazard Analysis

Toxic Hazard Rating:

Acute Local: Irritant 3; Inhalation 3.

Acute Systemic: Ingestion 3; Inhalation 3.

Chronic Local: U.

Chronic Systemic: U.

Caution: Highly toxic concentrations of vapor are readily obtained at room temperature.

Fire Hazard: Dangerous; when exposed to heat or flame, reacts with oxidizing materials.

Explosion Hazard: Moderate, in the form of vapor when exposed to flame (Section 7).

Countermeasures

Ventilation Control: Section 2.

Personnel Protection: Section 3.

To Fight Fire: Foam, carbon dioxide, dry chemical or carbon tetrachloride (Section 6).

First Aid: Section 1.

Storage and Handling: Section 7.

N-BUTYLENE PYRROLIDINE
General Information

Description: Colorless to light yellow liquid; penetrating amine-like odor.

Formula: $C_4H_8NC_4H_7$.

Constants: Mol wt: 125.2, bp: 154°C, fp: −75°C, flash p: 93°F, d: 0.837.

Hazard Analysis

Toxicology: A strongly alkaline material. See ammonia. Probably highly toxic. An irritant.

Fire Hazard: Dangerous; when exposed to sparks, heat, open flame or powerful oxidizers.

Disaster Hazard: Dangerous; when heated to decomposition, it emits highly toxic fumes.

Countermeasures
To Fight Fire: Spray, foam, carbon dioxide, dry chemicals (Section 6).
Ventilation Control: See Section 2.
Personnel Protection: Section 3.
Storage and Handling: Section 7.

BUTYL ESTER,2,4-D. See butyl 2,4-dichlorophenoxy acetate.

BUTYL ETHANEDIOATE. See butyl oxalate.

BUTYL ETHANOATE. See butyl acetate.

BUTYLETHANOLAMINE
General Information
Synonym: 2-Butylaminoethanol.
Description: Liquid.
Formula: $CH_3(CH_2)_3NHCH_2CH_2OH$.
Constants: Mol wt: 117.19, bp: 192°C, flash p: 170°F (O.C.), d: 0.89, vap. d.: 4.03.
Hazard Analysis
Toxicity: Details unknown. See also amines.
Fire Hazard: Moderate; when exposed to heat or flame, it can react with oxidizing materials (Section 6).
Spontaneous Heating: No.
Countermeasures
To Fight Fire: Alcohol foam, foam, carbon dioxide, dry chemical or carbon tetrachloride (Section 6).
Storage and Handling: Section 7.

BUTYL ETHER
General Information
Synonyms: 1-Butoxybutane; n-dibutyl ether.
Description: Colorless liquid.
Formula: $CH_3(CH_2)_3O(CH_2)_3CH_3$.
Constants: Mol wt: 130.23, mp: −95°C, bp: 142°C, flash p: 77°F, d: 0.769 at 20°/20°C, autoign. temp.: 382°F, vap. d.: 4.48, lel: 1.5%, uel: 7.6%.
Hazard Analysis
Toxic Hazard Rating:
  Acute Local: Irritant 1; Ingestion 1; Inhalation 1.
  Acute Systemic: Inhalation 2.
  Chronic Local: U.
  Chronic Systemic: U.
Fire Hazard: Moderate. See ethers.
Spontaneous Heating: No.
Explosion Hazard: Moderate. See also ethers (Section 7).
Disaster Hazard: Moderately dangerous; when heated, it emits acrid fumes; it can react with oxidizing materials.
Countermeasures
Ventilation Control: Section 2.
To Fight Fire: Foam, carbon dioxide, dry chemical or carbon tetrachloride (Section 6).
Personal Hygiene: Section 3.
Storage and Handling: Section 7.
Shipping Regulations: Section 11.
  MCA warning label.

BUTYL ETHYL ACETALDEHYDE. See 2-ethylhexaldehyde.

BUTYL ETHYL ACETIC ACID. See 2-ethylhexoic acid.

BUTYL ETHYL "CELLOSOLVE"
General Information
Description: Liquid.
Formula: $C_4H_9OCH_2CH_2OC_2H_5$.

Constants: Mol wt: 146.22, mp: −90°C, bp: 164.2°C, d: 0.8389.
Hazard Analysis
Toxicity: See glycols and "cellosolves."
Fire Hazard: Slight; can react with oxidizing materials.
Countermeasures
Storage and Handling: Section 7.

BUTYL ETHYLENE. See hexene-1.

BUTYL ETHYL ETHER. See ethyl n-butylether.

BUTYL FORMAL. See formaldehyde and butyl alochol.

BUTYL FORMATE
General Information
Synonym: Butyl methanoate.
Description: Colorless liquid.
Formula: $HCOOCH_2CH_2CH_2CH_3$.
Constants: Mol wt: 101.12, mp: −90°C, bp: 106.0°C, flash p: 64°F (C.C.), d: 0.911, autoign. temp.: 612°F, vap. press.: 40 mm at 31.6°C, vap. d.: 3.52, lel: 1.7%, uel: 8%.
Hazard Analysis
Toxic Hazard Rating:
  Acute Local: Irritant 2.
  Acute Systemic: Inhalation 2.
  Chronic Local: U.
  Chronic Systemic: U.
Toxicology: Irritant and narcotic in high concentrations. See esters, butyl alcohol and formic acid.
Fire Hazard: Dangerous, when exposed to heat or flame; it can react with oxidizing materials.
Spontaneous Heating: No.
Countermeasures
To Fight Fire: Alcohol foam, foam, carbon dioxide, dry chemical or carbon tetrachloride (Section 6).
Storage and Handling: Section 7.
Shipping Regulations: Section 11.
  IATA: Flammable liquid, red label, 1 liter (passenger), 40 liters (cargo).

n-BUTYL GLYCIDYL ETHER
General Information
Synonym: BGE.
Hazard Analysis
Toxic Hazard Rating:
  Acute Local: Irritant 1.
  Acute Systemic: Ingestion: 2; Inhalation: 2; Skin Absorption: 2.
  Chronic Local: Allergen 3.
  Chronic Systemic: U.
Based on limited animal experiments.
TLV: ACGIH (recommended) 50 parts per million in air; 270 mg/cubic meter of air.

tert-BUTYL HYDROPEROXIDE
General Information
Description: Water white liquid; slightly soluble in water; very soluble in esters and alcohols.
Formula: $(CH_3)_3COOH$.
Constants: Mol wt: 90.12, flash p: 100°F or above, fp: −35°C, d: 0.860, vap. d.: 2.07.
Hazard Analysis
Toxicology: Limited animal experiments indicate moderate toxicity via oral administration. The LD_{50} for white mice was 710 mg/kg calculated according to the method of Litchfield and Wilcoxson. At highest dosage levels

---

TOXIC HAZARD RATING CODE (For detailed discussion, see Section 1.)

0 NONE: (a) No harm under any conditions; (b) Harmful only under unusual conditions or overwhelming dosage.

1 SLIGHT: Causes readily reversible changes which disappear after end of exposure.

2 MODERATE: May involve both irreversible and reversible changes; not severe enough to cause death or permanent injury.

3 HIGH: May cause death or permanent injury after very short exposure to small quantities.

U UNKNOWN: No information on humans considered valid by authors.

symptoms noted were severe depression, incoordination and cyanosis. Death was due to respiratory arrest.

Fire Hazard: Dangerous, when exposed to heat or flame, or by spontaneous chemical reaction; can react with reducing materials.

Spontaneous Heating: No.

Countermeasures

To Fight Fire: Alcohol foam, carbon dioxide, dry chemical or carbon tetrachloride (Section 6).

Storage and Handling: Section 7.

Shipping Regulations: Section 11.

IATA (Solution, exceeding 60% in a nonvolatile solvent): Oxidizing material, not acceptable (passenger and cargo).

(Solution, not exceeding 60% in a nonvolatile solvent): Oxidizing material, yellow label, 1 liter (passenger and cargo).

BUTYL 12-HYDROXY-9-OCTADECENOATE. See butyl ricinoleate.

BUTYL HYDRIDE. See butane.

p,p-sec-BUTYLIDENEDIPHENOL

General Information

Description: Tan, granular solid.

Formula: $(HOC_6H_4)_2C(CH_3)(C_2H_5)$.

Constants: Mol wt: 242.3, mp: 118.9–121.7°C.

Hazard Analysis

Toxicity: Details unknown. See also phenols.

Fire Hazard: Slight; when heated, it emits acrid fumes; can react with oxidizing materials.

Countermeasures

Storage and Handling: Section 7.

n-BUTYL ISOCYANATE

General Information

Description: Colorless liquid.

Formula: $CH_3CH_2CH_2CH_2NCO$.

Constants: Mol wt: 99.1, bp: 115°C, d: 0.880 at 20°/4°C.

Hazard Analysis

Toxicology: Details unknown. A powerful irritant to eyes, skin and mucous membranes

Disaster Hazard: See cyanates.

Countermeasures

Storage and Handling: Section 7.

Shipping Regulations: Section 11.

IATA: Poison C, poison label, not acceptable (passenger), 40 liters (cargo).

tert-BUTYL ISOPROPYL BENZENE HYDRO-PEROXIDE

General Information

Description: Crystals.

Formula: $(C_4H_9)(C_6H_4)(C_3H_6OOH)$.

Constant: Mol wt: 208.3.

Hazard Analysis

Toxicity: Unknown.

Fire Hazard: Moderate when exposed to heat or flame or by chemical reaction; it can react with oxidizing or reducing materials (Section 6).

Countermeasures

Storage and Handling: Section 7.

Shipping Regulations: Section 11.

IATA (Solution, exceeding 60% in a nonvolatile solvent): Oxidizing material not acceptable (passenger and cargo).

(Solution, not exceeding 60% in a nonvolatile solution): Oxidizing material, yellow label, 1 liter (passenger and cargo).

BUTYL ISOTHIOCYANATE

General Information

Synonym: Butyl mustard oil.

Formula: $C_2H_5CH_2CH_2NCS$.

Constants: Mol wt: 115, d: 1.0-; vap. d.: 4.0, flash p.: 150°F, mp: 342–347°F.

Hazard Analysis

Fire Hazard: Moderate, when exposed to heat or flame.

Disaster Hazard: Dangerous. See thiocyanates.

Countermeasures

Storage and Handling: Section 7.

BUTYL LACTATE

General Information

Description: Liquid.

Formula: $CH_3CH(OH)COOC_4H_9$.

Constants: Mol wt: 146.18, mp: −43°C, bp: 188°C, flash p: 160°F (O.C.), d: 0.968, autoign. temp.: 720°F, vap. d.: 5.04, vap. press.: 0.4 mm at 20°C.

Hazard Analysis

Toxicity: See esters, butyl alcohol and lactic acid.

Fire Hazard: Moderate; when exposed to heat or flame, it can react with oxidizing materials.

Spontaneous Heating: No.

Countermeasures

To Fight Fire: Alcohol foam, foam, carbon dioxide, dry chemical or carbon tetrachloride (Section 6).

Storage and Handling: Section 7.

BUTYL LITHIUM

General Information

Formula: $CH_3CH_2CH_2CH_2Li$.

Constant: Mol wt: 63.94.

Hazard Analysis

Toxic Hazard Rating: U. Probably toxic.

Fire Hazard: Very dangerous. Extremely flammable. Ignites on contact with moist air.

Disaster Hazard: Very dangerous. Heat or moisture can cause it to ignite and burn rapidly.

Countermeasures

Storage and Handling: Section 7.

To Fight Fire: Dry chemical, see special instructions of manufacturer.

Shipping Regulations: Section II.

MCA warning label.

n-BUTYL MAGNESIUM CHLORIDE

General Information

Description: Colorless liquid soluble in ether and tetrahydrofuran.

Formula: C_4H_9MgCl.

Constants: Mol wt: 116.9, d: 0.88.

Hazard Analysis

Toxic Hazard Rating: Unknown. Probably toxic.

Fire Hazard: Dangerous. Flammable.

Disaster Hazard: Dangerous. Heat, source of ignition or powerful oxidizer can cause fires. See chlorides.

Countermeasures

Storage and Handling: Section 7.

To Fight Fire: Water spray, alcohol foam and dry chemical.

Shipping Regulations: Section 11.

I.C.C.: Flammable liquid, red label.

Coast Guard: Inflammable liquid, red label.

BUTYL MERCAPTAN

General Information

Synonyms: 1-Butanethiol; n-butyl thioalcohol.

Description: Colorless liquid; skunk-like odor.

Formula: $CH_3(CH_2)_2CH_2SH$.

Constants: Mol wt: 90.18, mp: −116°C, bp: 98°C, d: 0.8365 at 25°/4°C, flash p: 35°F, vap. d.: 3.1.

Hazard Analysis

Toxic Hazard Rating:

Acute Local: Irritant 2; Ingestion 2; Inhalation 2.

Acute Systemic: U.

Chronic Local: U.

Chronic Systemic: Inhalation 2.

TLV: ACGIH (recommended); 10 parts per million in air. 35 mg/cubic meter of air.

Fire Hazard: Moderate, when exposed to heat or flame (Section 6).

Disaster Hazard: Dangerous; when heated to decomposition or on contact with acid or acid fumes; it emits highly toxic fumes; it can react vigorously with oxidizing materials.

Countermeasures
To Fight Fire: Alcohol foam (Section 6).
Ventilation Control: Section 2.
Personnel Protection: Section 3.
Storage and Handling: Section 7.
Shipping Regulations: Section 11.
  Coast Guard Classification: Inflammable liquid; red label.
  I.C.C.: Flammable liquid, red label, 10 gallons.
  IATA: Flammable liquid, red label, 1 liter (passenger), 40 liters (cargo).

sec-BUTYL MERCAPTAN
General Information
Synonym: 2-Butanethiol.
Description: Mobile liquid, skunk-like odor.
Formula: $C_4H_{10}S$.
Constants: Mol wt: 90.2, mp: $-165°$C, bp: $85°$C, d: 0.83 at $17°$C.
Hazard Analysis
Toxic Hazard Rating:
  Acute Local: Irritant 2; Ingestion 2; Inhalation 2.
  Acute Systemic: U.
  Chronic Local: U.
  Chronic Systemic: Inhalation 2.
Fire Hazard: Moderate, when exposed to heat or flame (Section 6).
Disaster Hazard: Dangerous; when heated to decomposition or on contact with acid or acid fumes, it emits highly toxic fumes; it can react vigorously with oxidizing materials.
Countermeasures
Ventilation Control: Section 2.
Personnel Protection: Section 3.
Storage and Handling: Section 7.

tert-BUTYL MERCAPTAN
General Information
Synonym: 2-Methyl-2-propanethiol.
Description: Liquid; skunk-like odor.
Formula: $C_4H_{10}S$.
Constants: Mol wt: 90.2, mp: $-0.5°$C, d: 0.79–0.82 at $15.5°/15.5°$C, flash p: $-15°$F, bp: $62–67°$C, vap. d.: 3.1.
Hazard Analysis
Toxic Hazard Rating:
  Acute Local: Irritant 2; Ingestion 2; Inhalation 2.
  Acute Systemic: U.
  Chronic Local: U.
  Chronic Systemic: Inhalation 2.
Fire Hazard: Moderate, when exposed to heat or flame (Section 6).
Disaster Hazard: Dangerous; when heated to decomposition or on contact with acid or acid fumes, it emits highly toxic fumes; it can react vigorously with oxidizing materials.
Countermeasures
Ventilation Control: Section 2.
Personnel Protection: Section 3.
Storage and Handling: Section 7.

BUTYL MESITYL OXIDE. See butopyronoxyl.

n-BUTYL MESITYL OXIDE OXALATE. See α,α dimethyl-α'-carbobutoxyhydro-γ-pyrone.

BUTYL METHACRYLATE, MONOMER
General Information
Description: Colorless liquid.
Formula: $CH_2C(CH_2)COOC_4H_9$.
Constants: Mol wt: 142.19, bp: $163°$C, flash p: $130°$F (O.C.), d: 0.895 at $20°/4°$C, vap. d.: 4.8.
Hazard Analysis
Toxic Hazard Rating:
  Acute Local: Irritant 2; Ingestion 2; Inhalation 2.
  Acute Systemic: Ingestion 3; Inhalation 2.
  Chronic Local: U.
  Chronic Systemic: U.
Fire Hazard: Moderate; when exposed to heat or flame, it can react with oxidizing materials (Section 6).
Explosion Hazard: Some when exposed to heat, flame or sparks.
Countermeasures
Ventilation Control: Section 2.
Personnel Protection: Section 3.
First Aid: Section 1.
Storage and Handling: Section 7.

BUTYL METHANOATE. See butyl formate.

N-BUTYL-a-METHYLBENZYLAMINE
General Information
Formula: $C_4H_9C_6H_5CH(CH_3)NH_2$.
Constant: Mol wt: 178.
Hazard Analysis
Toxic Hazard Rating: U. Limited animal experiments suggest high toxicity and moderate irritation.

BUTYL METHYL KETONE. See methyl butyl ketone.

BUTYL MONOETHANOLAMINE
General Information
Description: Liquid.
Formula: $C_4H_9NHC_2H_4OH$.
Constants: Mol wt: 117.19, bp: $192°$C, flash p: $170°$F (O.C.), d: 0.89, vap. d.: 4.03.
Hazard Analysis
Toxicity: Details unknown. See also amines.
Fire Hazard: Moderate; when exposed to heat or flame, it can react with oxidizing materials.
Countermeasures
To Fight Fire: Alcohol foam, foam, carbon dioxide, dry chemical or carbon tetrachloride (Section 6).
Storage and Handling: Section 7.

BUTYL NAPHTHALENE
General Information
Description: Liquid.
Formula: $C_4H_9C_{10}H_7$.
Constants: Mol wt: 184.3, flash p: $680°$F, (C.C.), vap. d.: 6.2.
Hazard Analysis
Toxicity: Details unknown. See also naphthalene.
Fire Hazard: Slight; when exposed to heat or flame, it can react with oxidizing materials.
Countermeasures
To Fight Fire: Water, carbon dioxide, dry chemical or carbon tetrachloride (Section 6).
Storage and Handling: Section 7.

BUTYL NICOTINATE
General Information
Description: Colorless liquid.
Formula: $C_{10}H_{13}NO_2$.
Constants: Mol wt: 179.3, bp: $122–123°$C at 8 mm, d: 1.0471 at $25°/4°$C.

TOXIC HAZARD RATING CODE *(For detailed discussion, see Section 1.)*

0 NONE: (a) No harm under any conditions; (b) Harmful only under unusual conditions or overwhelming dosage.

1 SLIGHT: Causes readily reversible changes which disappear after end of exposure.

2 MODERATE: May involve both irreversible and reversible changes; not severe enough to cause death or permanent injury.

3 HIGH: May cause death or permanent injury after very short exposure to small quantities.

U UNKNOWN: No information on humans considered valid by authors.

Hazard Analysis
Toxicity: Details unknown. See also esters.
Disaster Hazard: Dangerous; when heated to decomposition, it emits highly toxic fumes; it can react vigorously on contact with oxidizing materials.
Countermeasures
Storage and Handling: Section 7.

BUTYL NITRATE
General Information
Description: Liquid.
Formula: $CH_3(CH_2)_3ONO_2$.
Constants: Mol wt: 119.1, bp: 136°C, flash p: 97°F, d: 1.048 at 0°/4°C, vap. d.: 4.0.
Hazard Analysis
Toxicity: Details unknown. See also nitrates and butyl alcohol.
Fire Hazard: Moderate, when exposed to heat or flame or by spontaneous chemical reaction (Section 6).
Caution: An oxidizer.
Explosion Hazard: See nitrates.
Disaster Hazard: Dangerous; see nitrates; it can react vigorously with oxidizing or reducing materials.
Countermeasures
Storage and Handling: Section 7.

sec-BUTYL NITRATE
General Information
Synonym: α-Methyl propyl nitrate.
Description: Liquid.
Formula: $C_2H_5CH(CH_3)ONO_2$.
Constants: Mol wt: 119.1, bp: 124°C, d: 1.0382 at 0°/4°C, vap. d.: 4.0.
Hazard Analysis
Toxicity: Details unknown. See nitrates and sec-butyl alcohol.
Fire Hazard: Moderate, when exposed to heat or flame or by spontaneous chemical reaction (Section 6).
Caution: An oxidizer.
Explosion Hazard: See nitrates.
Disaster Hazard: See nitrates.
Countermeasures
Storage and Handling: Section 7.

BUTYL NITRITE
General Information
Description: Oily liquid; characteristic odor; miscible in alcohol and ether.
Formula: $CH_3(CH_2)_3ONO$.
Constants: Mol wt: 103.1, bp: 75°C, d: 0.9114 at 0°/4°C, vap. d.: 3.5.
Hazard Analysis
Toxic Hazard Rating:
  Acute Local: Irritant 1.
  Acute Systemic: Ingestion 2; Inhalation 2.
  Chronic Local: Irritant 1.
  Chronic Systemic: Inhalation 1.
Toxicology: Resembles amyl nitrite in causing fall in blood pressure, headache, throbbing and weakness. See also nitrites and sec-butyl alcohol.
Fire Hazard: Moderate, when exposed to heat or flame or by spontaneous chemical reaction (Section 6).
Caution: An oxidizer.
Explosion Hazard: See nitrites.
Disaster Hazard: See nitrites.
Countermeasures
Storage and Handling: Section 7.

sec-BUTYL NITRITE
General Information
Synonym: α-Methyl propyl nitrite.
Description: Liquid.
Formula: $C_2H_5CH(CH_3)ONO$.
Constants: Mol wt: 103.1, bp: 68°C, d: 0.8981 at 0°/4°C, vap. d.: 3.5.
Hazard Analysis
Toxicity: Details unknown. See butyl nitrite.

Fire Hazard: Moderate, when exposed to heat or flame or by spontaneous chemical reaction (Section 6).
Caution: An oxidizer.
Explosion Hazard: See nitrites.
Disaster Hazard: See nitrites.
Countermeasures
Storage and Handling: Section 7.

tert-BUTYL NITRITE
General Information
Synonyms: α,α-Dimethyl ethyl nitrite.
Description: Yellowish liquid.
Formula: $(CH_3)_3CONO$.
Constants: Mol wt: 103.1, bp: 63°C, d: 0.8941 at 0°/4°C, vap. d.: 3.5.
Hazard Analysis
Toxicity: Details unknown. See butyl nitrite.
Fire Hazard: Moderate, when exposed to heat or flame or by spontaneous chemical reaction (Section 6).
Caution: An oxidizer.
Explosion Hazard: See nitrites.
Disaster Hazard: See nitrites.
Countermeasures
Storage and Handling: Section 7.

2-BUTYL OCTANOL
General Information
Description: Liquid.
Formula: $C_6H_{13}CH(C_4H_9)CH_2OH$.
Constants: Mol wt: 186.33, mp: −80°C, flash p: 230°F (O.C.), bp: 253.3°C, d: 0.8355 at 20°/20°C, vap. d:: 6.42.
Hazard Analysis
Toxicity: Details unknown. See also alcohols.
Fire Hazard: Slight, when exposed to heat or flame; it can react with oxidizing materials.
Countermeasures
To Fight Fire: Foam, carbon dioxide, dry chemical or carbon tetrachloride (Section 6).
Storage and Handling: Section 7.

BUTYL OLEATE
General Information
Description: Liquid.
Formula: $C_{17}H_{33}COOC_4H_9$.
Constants: Mol wt: 338.56, bp: 173°C, flash p: 356°F (O.C.), d: 0.873, vap. d.: 11.3.
Hazard Analysis
Toxicity: Details unknown. See also esters, butyl alcohol and oleic acid.
Fire Hazard: Slight, when exposed to heat or flame; it can react with oxidizing materials.
Countermeasures
To Fight Fire: Foam, carbon dioxide, dry chemical or carbon tetrachloride (Section 6).
Storage and Handling: Section 7.

BUTYL OXALATE
General Information
Synonym: Butyl ethane dioate.
Description: Liquid.
Formula: $(COOC_4H_9)_2$.
Constants: Mol wt: 202.24, flash p: 265°F (O.C.), d: 0.989–0.993, vap. d.: 7.0.
Hazard Analysis
Toxicity: Details unknown. See also esters, butyl alcohol and oxalic acid.
Fire Hazard: Slight, when exposed to heat or flame, it can react with oxidizing materials.
Countermeasures
To Fight Fire: Foam, carbon dioxide, dry chemical or carbon tetrachloride (Section 6).
Storage and Handling: Section 7.

BUTYL OXYETHYL SALICYLATE
General Information
Formula: $OCH_6H_4COOCH_2CH_2OC_4H_9$.
Constants: Mol wt: 183, d: 1.0+, bp: 367–378° F, flash p: 315° F.
Hazard Analysis
Fire Hazard: Slight, when exposed to heat or flame.
Countermeasures
To Fight Fire: Water, foam (Section 6).
Storage and Handling: Section 7.

tert-BUTYL PERACETATE (solution in benzene).
General Information
Description: Solution is clear, colorless, insoluble in water. Soluble in organic solvents.
Formula: $CH_3CO(O_2)C(CH_3)_3$ + benzene.
Constants: Mol wt: 132.2, d: 0.923, vap. press.: 50 mm Hg at 26° C, flash p: < 80° F (C.O.C.).
Hazard Analysis and Countermeasures
See peroxides, organic, and benzene.
Shipping Regulations:
I.C.C. Classification: Peroxides, organic; solution, liquid N.O.S.
Freight and Express: Red label, chemicals N.O.I.B.N.
Rail Express: Maximum wt.-2 lbs. per shipping case.
Parcel Post: Prohibited.

tert-BUTYL PERBENZOATE
General Information
Description: Colorless to slight yellow liquid, mild aromatic odor. Insoluble in water. Soluble in organic solvents.
Formula: $C_6H_5COOOC(CH_3)_3$.
Constants: Mol wt: 194.2, bp: 112° C (decomp.), flash p: 190° F (O.C.), fp: 8° C, vap. press.: 0.33 mm Hg at 50° C, d: 1.0+.
Hazard Analysis
Toxicology: Details unknown. See peroxides, organic.
Fire Hazard: See peroxides, organic.
Explosion Hazard: See peroxides, organic.
Disaster Hazard: See peroxides, organic.
Countermeasures
To Fight Fire: See peroxides, organic.
Storage and Handling: See peroxides, organic.
Shipping Regulations:
I.C.C.: Peroxides, organic, liquid, N.O.S.
Freight and Express: Yellow label as chemicals N.O.I.B.N.
IATA: Oxidizing material, yellow label, 1 liter (passenger and cargo).
Parcel Post Shipment: Prohibited.

tert-BUTYL PEROXYISOBUTYRATE
General Information
Description: Solution is colorless to yellow liquid, insoluble in water or glycerine. Soluble in alcohols, hydrocarbons, esters, ethers, ketones.
Formula: $C_8H_{16}O_3$ + benzene.
Constants: Mol wt (of solute): 160.2, d: 0.90 at 25° C, flash p: <80° F (micro O.C.).
Hazard Analysis
Toxicity: See benzene.
Fire Hazard: Highly flammable. See benzene and peroxides, organic.
Explosion Hazard: See peroxides, organic.
Disaster Hazard: See peroxides, organic.
Countermeasures
To Fight Fire: Foam, spray, dry chemicals.
Storage and Handling: See peroxides, organic.

Shipping Regulations:
I.C.C. Classification: Peroxides, organic, solution, liquid N.O.S.
Freight and Express: Red label; refrigeration required. Special permit.

tert-BUTYL PEROXYPIVALATE
General Information
Description: Colorless liquid, insoluble in water and ethylene glycol; soluble in most organic solvents.
Formula: $(CH_3)_3COOCOC(CH_3)_3$.
Constants: Mol wt: 159, d: 0.854 (25° /25° C), fp: < 19° C, flash p: >155° F (O.C.), rapid decomposition at 70° F.
Hazard Analysis
Explosion Hazard: Explodes on heating.

o-sec-BUTYLPHENOL
General Information
Description: Colorless liquid.
Formula: $(CH_3CHC_2H_5)C_6H_4OH$.
Constants: Mol wt: 150.2, bp: 226–228° C at 25 mm, fp: 12° C, flash p: 225° F, d: 0.981 at 25° /25° C.
Hazard Analysis
Toxic Hazard Rating:
Acute Local: Irritant 2.
Acute Systemic: U.
Chronic Local: Irritant 2.
Chronic Systemic: U.
Toxicology: See phenol.
Fire Hazard: Moderate, when exposed to heat or flame; it can react with oxidizing materials.
Countermeasures
To Fight Fire: Foam, carbon dioxide, dry chemical or carbon tetrachloride (Section 6).
Storage and Handling: Section 7.

p-sec-BUTYLPHENOL
General Information
Description: Nearly white flakes.
Formula: $(CH_3CHC_2H_5)C_6H_4OH$.
Constants: Mol wt: 150.2, bp: 135.4–136.5° C at 25 mm, fp: 51° C, flash p: 240° F, d: 0.963 at 60° /60° C.
Hazard Analysis
Toxicity: See o-sec-butylphenol.
Fire Hazard: Slight, when exposed to heat or flame.
Disaster Hazard: Moderate; when heated to decomposition, it emits toxic fumes; it can react with oxidizing materials.
Countermeasures
To Fight Fire: Foam, carbon dioxide, dry chemical or carbon tetrachloride (Section 6).
Storage and Handling: Section 7.

p-tert-BUTYLPHENOL
General Information
Description: Crystals or practically white flakes.
Formula: $C_4H_9C_6H_4OH$.
Constants: Mol wt: 150.2, bp: 238° C, fp: 97° C, d: 0.9081 at 114° /4° C, vap. press.: 1 mm at 70.0° C, vap. d.: 5.1.
Hazard Analysis
Toxicity: See o-sec-butyl phenol.
Disaster Hazard: Dangerous; when heated to decomposition, it emits toxic fumes.
Countermeasures
Storage and Handling: Section 7.

2-(p-sec-BUTYLPHENOXY) ETHANOL. See ethylene glycol p-sec-butyl phenyl ether.

TOXIC HAZARD RATING CODE (For detailed discussion, see Section 1.)

0 NONE: (a) No harm under any conditions; (b) Harmful only under unusual conditions or overwhelming dosage.

1 SLIGHT: Causes readily reversible changes which disappear after end of exposure.

2 MODERATE: May involve both irreversible and reversible changes; not severe enough to cause death or permanent injury.

3 HIGH: May cause death or permanent injury after very short exposure to small quantities.

U UNKNOWN: No information on humans considered valid by authors.

2-(p-tert-BUTYLPHENOXY) ETHYL ACETATE. See ethylene glycol p-tert-butyl phenyl ether acetate.

1-(0-sec-BUTYLPHENOXY)-2-PROPANOL. See propylene glycol sec-butyl phenyl ether.

1-(p-sec-BUTYLPHENOXY)-2-PROPANOL. See propylene glycol p-sec-butyl phenyl ether.

1-(p-tert-BUTYLPHENOXY)-2-PROPANOL. See propylene glycol tert-butyl phenyl ether.

p-tert-BUTYLPHENYL DIPHENYL PHOSPHATE. See diphenyl mono-(p-tert-butyl phenyl) phosphate.

BUTYL PHENYL ETHER
General Information
Description: Liquid.
Formula: $C_4H_9OC_6H_5$.
Constants: Mol wt: 150.1, flash p: 180° F, d: 0.929 at 20° C, vap. d.: 5.6.
Hazard Analysis
Toxicity: Details unknown. See also Ethers.
Fire Hazard: Moderate, when exposed to heat or flame; it can react with oxidizing materials. See also Ethers.
Explosion Hazard: Slight, by spontaneous chemical reaction. See also Ethers.
Countermeasures
Storage and Handling: Section 7.
To Fight Fire: Foam, carbon dioxide, dry chemical or carbon tetrachloride (Section 6).

4-tert-BUTYL-2 PHENYL-PHENOL
General Information
Formula: $C_6H_5C_6H_3OHC(CH_3)_3$.
Constants: Mol wt: 226, d: 1.0+, bp: 385–388° F, flash P: 320° F.
Hazard Analysis
Fire Hazard: Slight, when exposed to heat or flame.
Countermeasures
To Fight Fire: Water, foam (Section 6).
Storage and Handling: Section 7.

N-BUTYL PHOSPHORIC ACID. See acid butyl phosphate.

BUTYL PHTHALYL BUTYL GLYCOLATE
General Information
Description: Liquid.
Formula: $C_6H_4(CO_2)_2(C_4H_9)CH_2CO_2C_4H_9$.
Constants: Mol wt: 331.3, bp: 345° C, flash p: 390° F (O.C.), d: 1.097, vap. d.: 11.6.
Hazard Analysis
Toxicity: Unknown.
Fire Hazard: Slight, when exposed to heat or flame; it can react with oxidizing materials.
Spontaneous Heating: No.
Countermeasures
To Fight Fire: Water, foam, carbon dioxide, dry chemical or carbon tetrachloride (Section 6).
Storage and Handling: Section 7.

BUTYL PIPERONYLAMIDE
General Information
Description: Crystals.
Formula: $CH_2(O_2)C_6H_2(C_4H_9)CONH_2$.
Constants: Mol wt: 221.3, vap. d.: 7.4.
Hazard Analysis
Toxicity: Unknown; an insecticide.
Disaster Hazard: Moderately dangerous; when heated to decomposition it emits toxic fumes.
Countermeasures
Storage and Handling: Section 7.

BUTYL PROPANOATE. See butyl propionate.

BUTYL PROPIONATE
General Information
Synonym: Butyl propanoate.

Description: Water-white liquid, apple-like odor.
Formula: $C_2H_5CO_2C_4H_9$.
Constants: Mol wt: 130.2, mp: −89.6° C, bp: 145.4° C, flash p: 90° F., d: 0.875 at 20° C, autoign. temp.: 800° F, vap. d.: 4.49.
Hazard Analysis
Toxic Hazard Rating:
  Acute Local: Irritant 2.
  Acute Systemic: U.
  Chronic Local: Irritant 1.
  Chronic Systemic: U.
Toxicity: See also esters, n-butyl alcohol and propionic acid.
Fire Hazard: Dangerous, when exposed to heat or flame; it can reach with oxidizing materials.
Spontaneous Heating: No.
Countermeasures
To Fight Fire: Foam, carbon dioxide, dry chemical or carbon tetrachloride (Section 6).
Storage and Handling: Section 7.

BUTYL RICINOLEATE
General Information
Synonym: Butyl-12-hydroxy-9-octadecenoate.
Description: Liquid.
Formula: $C_{18}H_{33}O_3C_4H_9$.
Constants: Mol wt: 354.56, bp: 275° C at 13 mm, flash p: 230° F, d: 0.906, vap. d.: 12.2.
Hazard Analysis
Toxicity: Details unknown. See also esters and butyl alcohol.
Fire Hazard: Slight, when exposed to heat or flame; it can react with oxidizing materials.
Countermeasures
To Fight Fire: Foam, carbon dioxide, dry chemical or carbon tetrachloride (Section 6).
Storage and Handling: Section 7.

BUTYL SEBECATE. See dibutyl sebacate.

BUTYL STEARAMIDE
General Information
Description: Liquid.
Formula: $C_4H_9C_{17}H_{34}CONH_2$.
Constants: Mol wt: 339.6, bp: 195° C, flash p: 430° F, vap. d.: 11.7.
Hazard Analysis
Toxicity: Unknown. See also butyl stearate and amides.
Fire Hazard: Slight, when exposed to heat or flame; it can react with oxidizing materials.
Countermeasures
To Fight Fire: Foam, carbon dioxide, dry chemical or carbon tetrachloride (Section 6).
Storage and Handling: Section 7.

BUTYL STEARATE
General Information
Description: Liquid.
Formula: $C_{17}H_{35}COOC_4H_9$.
Constants: Mol wt: 340.57, mp: 19.5° C, bp: 220–225° C at 25 mm Hg, flash p: 320° F (C.C.), d: 0.855 at 25°/25° C, vap. d.: 11.4, autoign. temp.: 671° F.
Hazard Analysis
Toxicity: Details unknown. Limited animal experiments suggest low toxicity. See also esters and n-butyl alcohol.
Fire Hazard: Slight, when exposed to heat or flame; it can react with oxidizing materials.
Countermeasures
To Fight Fire: Foam, carbon dioxide, dry chemical or carbon tetrachloride (Section 6).
Storage and Handling: Section 7.

BUTYL SULFIDE
General Information
Synonyms: Dibutyl sulfide; butyl thiobutane.
Description: Liquid.
Formula: $(CH_3CH_2CH_2CH_2)_2S$.

Constants: Mol wt: 146.3, mp: $-80°$ C, bp: 182°C, d: 0.839 at 16°/0°C, vap. d.: 4.9.

Hazard Analysis

Toxicity: Unknown. See also alkyl disulfides.

Fire Hazard: Moderate, when exposed to heat or flame or on contact with acid or acid fumes.

Disaster Hazard: Dangerous; when heated to decomposition, it emits highly toxic fumes of oxides of sulfur; it can react vigorously with oxidizing materials.

Countermeasures

Storage and Handling: Section 7.

BUTYL TARTRATE. See dibutyl tartrate.

tert-BUTYL TETRALIN

General Information

Description: Liquid.

Formula: $C_4H_9C_{10}H_{11}$.

Constants: Mol wt: 188.30, flash p: 680°F (C.C.), vap. d.: 6.3.

Hazard Analysis

Toxicity: Unknown. See also tetrahydronaphthaline.

Fire Hazard: Slight, when exposed to heat or flame; it can react with oxidizing materials.

Countermeasures

To Fight Fire: Water, carbon dioxide, dry chemical or carbon tetrachloride (Section 6).

Storage and Handling: Section 7.

n-BUTYL TETRATHIO-o-STANNATE

General Information

Description: Solid.

Formula: $(SC_4H_9)_4Sn$.

Constants: Mol wt: 475.4, bp: 136°C at 0.001 mm Hg.

Hazard Analysis

Toxicity: Details unknown. See tin compounds.

Disaster Hazard: Dangerous. See sulfur compounds.

sec-BUTYL TETRATHIO-o-STANNATE

General Information

Description: Solid.

Formula: $(SC_4H_9)_4Sn$.

Constants: Mol wt: 475.4, bp: 111°C at 0.001 mm Hg.

Hazard Analysis

Toxicity: Details unknown. See tin compounds.

Disaster Hazard: Dangerous. See sulfur compounds.

BUTYL TITANATE

General Information

Synonym: Titanium butylate.

Description: Colorless to light-yellow liquid, odor of butanol.

Formula: $Ti(OC_4H_9)_4$.

Constants: Mol wt: 340.4, mp: $-55°$ C, bp: 312°C, flash p: 170°F, vap. d.: 11.5.

Hazard Analysis

Toxicity: See butyl alcohol and titanium compounds.

Fire Hazard: Moderate, when exposed to heat or flame; it can react with oxidizing materials (Section 6).

Countermeasures

Storage and Handling: Section 7.

sec-BUTYL TITANATE

General Information

Description: A clear liquid.

Formula: $Ti[OCH(CH_3)(C_2H_5)]_4$.

Constants: Mol wt: 340, mp: -25 to $-30°$ C, bp: 138°C at 10 mm, d: 0.93.

Hazard Analysis

Toxicity; See esters and sec-butanol.

Fire Hazard: Moderate, when exposed to heat or flame; it can react with oxidizing materials.

Countermeasures

Storage and Handling: Section 7.

n-BUTYL THIOALCOHOL. See n-butyl mercaptan.

BUTYL THIOBUTANE. See butyl sulfide.

p-tert-BUTYLTOLUENE

General Information

Description: Colorless liquid.

Formula: $C_4H_9C_6H_4CH_3$.

Constant: Mol wt: 148.24.

Hazard Analysis

Toxic Hazard Rating:

Acute Local: Irritant 3.

Acute Systemic: Ingestion 3; Inhalation 2; Skin Absorption 2.

Chronic Local: Irritant 2.

Chronic Systemic: Ingestion 3; Inhalation 3; Skin Absorption 2.

Toxicology: Inhalation of vapors causes irritation of lungs and depression of CNS. Prolonged exposure may result in damage to liver and kidneys.

TLV: ACGIH (recommended); 10 parts per million in air; 60 milligrams per cubic meter of air.

Fire Hazard: Moderate, when exposed to heat or flame.

Disaster Hazard: Dangerous; when heated, it emits highly toxic fumes; it can react with oxidizing materials.

Countermeasures

Storage and Handling: Section 7.

BUTYL TRICHLOROSILANE

General Information

Description: Liquid.

Formula: $C_4H_9SiCl_3$.

Constants: Mol wt: 191.6, vap. d.: 6.4, flash p: 130°F (O.C.), d: 1.2.

Hazard Analysis

Toxicity: Details unknown. See also chlorosilanes.

Disaster Hazard: Dangerous; when heated to decomposition, it emits highly toxic fumes of chlorides; it will react with water or steam to produce heat, toxic and corrosive fumes.

Countermeasures

Storage and Handling: Section 7.

Shipping Regulations: Section 11.

I.C.C.: Corrosive liquid; white label, 10 gallons.

Coast Guard Classification: Corrosive liquid; white label.

IATA: Corrosive liquid, white label, not acceptable (passenger), 40 liters (cargo).

BUTYL TRIPHENYL GERMANIUM

General Information

Description: Colorless crystals; insoluble in water.

Formula: $Ge(C_4H_9)(C_6H_5)_3$.

Constant: Mol wt: 361.

Hazard Analysis

Toxicity: Details unknown. See germanium compounds.

BUTYL URETHANE

General Information

Synonym: Carbonic acid butyl ethyl ester; ethyl n-butyl-carbamate.

Formula: $CH_3(CH_2)_3NHCOOC_2H_5$.

Constants: Mol wt: 145, d: 0.9, vap. d.: 5.0, bp: 396–397°F, flash p: 197°F.

TOXIC HAZARD RATING CODE (For detailed discussion, see Section 1.)

0 NONE: (a) No harm under any conditions; (b) Harmful only under unusual conditions or overwhelming dosage.

1 SLIGHT: Causes readily reversible changes which disappear after end of exposure.

2 MODERATE: May involve both irreversible and reversible changes; not severe enough to cause death or permanent injury.

3 HIGH: May cause death or permanent injury after very short exposure to small quantities.

U UNKNOWN: No information on humans considered valid by authors.

Hazard Analysis
Fire Hazard: Moderate, when exposed to heat or flame.
Countermeasures
Storage and Handling: Section 7.

BUTYL VINYL ETHER
General Information
Description: Liquid.
Formula: $CHCH_2OC_4H_9$.
Constants: Mol wt: 100.2, mp: $-92°C$, bp: $93.3°C$, flash p: $30°F$, d: 0.77, vap. d.: 3.4.
Hazard Analysis
Toxicity: Details unknown. See also ethers.
Explosion Hazard: Moderate, by spontaneous chemical reaction (Section 7). See also ethers.
Fire Hazard: Dangerous. See ethers.

BUTYN
General Information
Synonym: p-Aminobenzoyl-γ-di-n-butyl amino-propanol sulfate.
Description: Colorless, odorless powder.
Formula: $[NH_2C_6H_4COO(CH_2)_3N(C_4H_9)_2]H_2SO_4$.
Constants: Mol wt: 714.8, mp: $98-100°C$.
Hazard Analysis
Toxicity: A weak allergen.
Fire Hazard: Slight; (Section 6).
Countermeasures
Storage and Handling: Section 7.

2-BUTYNE. See crotonylene.

2-BUTYNE-1,4-DIOL
General Information
Description: Straw to amber crystals.
Formula: $C_4H_6O_2$.
Constants: Mol wt: 86.1, mp: $57.5°C$, bp: $194°C$ at 100 mm.
Hazard Analysis
Toxic Hazard Rating:
  Acute Local: Irritant 3; Ingestion 3.
  Acute Systemic: Ingestion 3; Inhalation 3.
  Chronic Local: U.
  Chronic Systemic: U.
Explosion Hazard: Moderate, when exposed to heat or by spontaneous chemical reaction in contact with certain materials.
Explosive Hazard: Upon contamination with mercury salts, strong acids and alkali earth hydroxides and halides at high temperatures.
Disaster Hazard: Dangerous; when heated to decomposition it emits acrid fumes and may explode.
Countermeasures
Ventilation Control: Section 2.
Personnel Protection: Section 3.
First Aid: Section 1.
Storage and Handling: Section 7.

BUTYRALDEHYDE
General Information
Synonyms: Butanal; butyric aldehyde; n-butyl aldehyde.
Description: Colorless liquid.
Formula: $CH_3(CH_2)_2CHO$.
Constants: Mol wt: 72.1, mp: $-100°C$, bp: $75.7°C$, flash p: $20°F$ (C.C.), d: 0.817 at $20°/4°C$, autoign. temp.: $446°F$, vap. d.: 2.5, lel: 2.5%.
Hazard Analysis
Toxic Hazard Rating:
  Acute Local: Irritant 2; Ingestion 2; Inhalation 1.
  Acute Systemic: Ingestion 2; Inhalation 2.
  Chronic Local: Irritant 1.
  Chronic Systemic: U.
Fire Hazard: Dangerous, when exposed to heat or flame; it can react with oxidizing materials.
Spontaneous Heating: No.

Countermeasures
To Fight Fire: Foam, carbon dioxide, dry chemical or carbon tetrachloride (Section 6).
Ventilation Control: Section 2.
Personnel Protection: Section 3.
Storage and Handling: Section 7.
Shipping Regulations: Section 11.
  I.C.C.: Flammable liquid; red label, 10 gallons.
  Coast Guard Classification: Inflammable liquid; red label.
  IATA: Flammable liquid, red lable, 1 liter (passenger), 40 liters (cargo).
  MCA warning label.

BUTYRALDEHYDE ANILINE
General Information
Description: Liquid.
Formula: $CH_3(CH_2)_2CONHC_6H_5$.
Constants: Mol wt: 163.2, vap. d.: 5.5.
Hazard Analysis
Toxicity: Details unknown; may resemble aniline. See also aniline.
Fire Hazard: Slight, when exposed to heat or flame (Section 6).
Disaster Hazard: Dangerous; when heated to decomposition it emits highly toxic fumes of aniline; it can react vigorously with oxidizing materials.
Countermeasures
Storage and Handling: Section 7.

BUTYRALDOL
General Information
Description: Slightly soluble in water.
Formula: $C_8H_{16}O_2$.
Constants: Mol wt: 144, d: 0.9, bp: $280°F$ at 50 mm, flash p: $165°F$ (O.C.).
Hazard Analysis
Fire Hazard: Moderate, when exposed to heat or flame.
Countermeasures
To Fight Fire: Alcohol foam (Section 6).
Storage and Handling: Section 7.

BUTYRALDOXIME
General Information
Synonym: Butanal oxime.
Description: Liquid.
Formula: C_4H_8NOH.
Constants: Mol wt: 87.1, mp: $-29.5°C$, bp: $152°C$, flash p: $136°F$ (C.C.), d: 0.923, vap. d.: 3.01.
Hazard Analysis
Toxicity: Unknown.
Fire Hazard: Moderate, when exposed to heat or flame; it can react with oxidizing materials.
Countermeasures
To Fight Fire: Foam, carbon dioxide, dry chemical or carbon tetrachloride (Section 6).
Storage and Handling: Section 7.

BUTYRIC ACID
General Information
Synonyms: Butanoic acid; n-butyric acid; ethyl acetic acid; propyl formic acid.
Description: Liquid.
Formula: $CH_3(CH_2)_2COOH$.
Constants: Mol wt: 88.10, mp: $-7.9°C$, bp: $163.5°C$, flash p: $161°F$, fp: $-5.5°C$, d: 0.9590 at $20°/20°C$, autoign. temp.: $846°F$, vap. press.: 0.43 mm at $20°C$, vap. d.: 3.04, lel: 2.0%, uel: 10.0%.
Hazard Analysis
Toxic Hazard Rating:
  Acute Local: Irritant 1; Ingestion 1; Inhalation 1.
  Acute Systemic: Ingestion 1; Inhalation 1.
  Chronic Local: Irritant 1.
  Chronic Systemic: U.
A synthetic flavoring substance and adjuvant (Section 10).

Fire Hazard: Moderate, when exposed to heat or flame; it can react with oxidizing materials.

Spontaneous Heating: No.

Countermeasures

To Fight Fire: Alcohol foam, carbon dioxide, dry chemical or carbon tetrachloride (Section 6).

Ventilation Control: Section 2.

Personal Hygiene: Section 3.

Storage and Handling: Section 7.

BUTYRIC ALDEHYDE. See butyraldehyde.

BUTYRIC ANHYDRIDE

General Information

Synonym: Butanoic anhydride.

Description: Liquid, decomposes in water.

Formula: $[CH_3(CH_2)_2CO]_2O$.

Constants: Mol wt: 158.19, mp: $-73.3°C$, bp: $198°C$, flash p: $190°F$ (C.C.), d: 0.978, vap. d.: 5.4.

Hazard Analysis

Toxic Hazard Rating:

Acute Local: Irritant 1; Ingestion 1; Inhalation 1.

Acute Systemic: Ingestion 1; Inhalation 1.

Chronic Local: Irritant 1.

Chronic Systemic: U.

Fire Hazard: Moderate; when exposed to heat or flame; it can react with oxidizing materials.

Countermeasures

To Fight Fire: Carbon dioxide, dry chemical or carbon tetrachloride (Section 6).

Ventilation Control: Section 2.

Personal Hygiene: Section 3.

Storage and Handling: Section 7.

BUTYRIC ETHER. See ethyl butyrate.

γ-BUTYROLACTONE

General Information

Description: Colorless liquid; mild odor.

Formula: $C_4H_6O_2$.

Constants: Mol wt: 86, mp: $-44°C$, bp: $206°C$, flash p: $209°F$ (O.C.), d: 1.124 at $25°/4°C$, vap. d.: 3.0.

Hazard Analysis

Toxic Hazard Rating:

Acute Local: Irritant 1.

Acute Systemic: Ingestion 2.

Chronic Local: Irritant 1.

Chronic Systemic: U.

Less toxic then β-propiolactone.

Fire Hazard: Moderate, when exposed to heat or flame; it can react with oxidizing materials.

Countermeasures

To Fight Fire: Water, foam, alcohol foam, carbon dioxide, dry chemical or carbon tetrachloride (Section 6).

Personal Hygiene: Section 3.

Storage and Handling: Section 7.

BUTYRONE

General Information

Synonym: 4-Heptanone; dipropyl ketone.

Description: Colorless, refractive liquid.

Formula: $(C_3H_7)_2CO$.

Constants: Mol wt: 114.18, bp: $144°C$, mp: $-326°C$, vap. press.: 5.2 mm Hg at $20°C$, flash p: $120°F$ (C.C.), d: 0.815, vap. d.: 3.93.

Hazard Analysis

Toxicity: Details unknown. See also ketones.

Fire Hazard: Moderate, when exposed to heat or flame; it can react with oxidizing materials.

Spontaneous Heating: No.

Countermeasures

To Fight Fire: Carbon dioxide, dry chemical or carbon tetrachloride (Section 6).

Storage and Handling: Section 7.

n-BUTYRONITRILE

General Information

Synonym: Butanenitrile; n-propyl cyanide.

Description: Colorless liquid; slightly soluble in water; soluble in alcohol and ether.

Formula: $CH_3(CH_2)CN$.

Constants: Mol wt: 69, d: 0.796 $(15°C)$, mp: $112.6°C$, bp: $116-117°C$, flash p: $79°F$ (O.C.).

Hazard Analysis

Toxic Hazard Rating: U. Animal experiments show high toxicity similar to that of cyanides. See nitriles and cyanides.

Fire Hazard: Dangerous, when exposed to heat or flame.

Disaster Hazard: Dangerous. See nitriles.

Countermeasures

To Fight Fire: Alcohol foam (Section 6).

Storage and Handling: Section 7.

BUTYRYL CHLORIDE

General Information

Synonym: Butanoyl chloride.

Description: Clear, colorless liquid with sharp odor.

Formula: C_3H_7COCl.

Constants: Mol wt: 106.6, mp: $-89°C$, bp: $100-110°C$, d: 1.028 at $20°/4°C$, vap. d.: 3.67.

Hazard Analysis

Toxic Hazard Rating:

Acute Local: Irritant 3; Ingestion 3; Inhalation 3.

Acute Systemic: U.

Chronic Local: Irritant 1.

Chronic Systemic: U.

Disaster Hazard: Dangerous; when heated to decomposition, it emits highly toxic fumes of chlorides; it will react with water or steam to produce toxic and corrosive fumes; can react vigorously with oxidizing materials.

Countermeasures

Ventilation Control: Section 2.

Personnel Protection: Section 3.

Storage and Handling: Section 7.

TOXIC HAZARD RATING CODE *(For detailed discussion, see Section 1.)*

0 NONE: (a) No harm under any conditions; (b) Harmful only under unusual conditions or overwhelming dosage.

1 SLIGHT: Causes readily reversible changes which disappear after end of exposure.

2 MODERATE: May involve both irreversible and reversible changes; not severe enough to cause death or permanent injury.

3 HIGH: May cause death or permanent injury after very short exposure to small quantities.

U UNKNOWN: No information on humans considered valid by authors.

C

CA. See cellulose acetate.

CABLE CUTTERS, EXPLOSIVE
Shipping Regulations: Section 11.
  IATA: Explosive, explosive label, 25 kilograms (passenger), 70 kilograms (cargo).

CACODYL
General Information
Synonyms: Dicacodyl; tetramethyl diarsyl.
Description: Oily liquid; colorless to yellow, slightly soluble in water.
Formula: $(CH_3)_2As-As(CH_3)_2$.
Constants: Mol wt: 210.0, bp: 170°C, fp: −6°C, d: 1.15.
Hazard Analysis
Toxicity: Highly toxic. See arsenic compounds.
Fire Hazard: Dangerous; by spontaneous chemical reaction. Ignites spontaneously in dry air (Section 6).
Spontaneous Heating: Yes.
Explosion Hazard: Unknown.
Disaster Hazard: Dangerous; see arsenic; it can react vigorously with oxidizing materials.
Countermeasures
First Aid: Section 1.
Personal Hygiene: Section 3.
Storage and Handling: Section 7.

CACODYL BROMIDE. See dimethyl bromarsine.

CACODYL CHLORIDE. See dimethyl chlorarsine.

CACODYL DIOXIDE
General Information
Description: A liquid.
Hazard Analysis
Toxicity: See arsenic compounds.
Fire Hazard: Dangerous, by spontaneous chemical reaction. Ignites spontaneously in air (Section 6).
Spontaneous Heating: Yes.
Explosion Hazard: Unknown.
Disaster Hazard: Dangerous; see arsenic; can react vigorously with oxidizing materials.
Countermeasures
First Aid: Section 1.
Personal Hygiene: Section 3.
Storage and Handling: Section 7.

CACODYL HYDRIDE. See dimethyl arsine.

CACODYLIC ACID
General Information
Synonyms: Dimethylarsenic acid; alkargen.
Description: Colorless crystals, odorless and soluble in water.
Formula: $(CH_3)_2AsOOH$.
Constants: Mol wt: 138.0, mp: 200°C.
Hazard Analysis
Toxicity: See arsenic compounds.
Disaster Hazard: See arsenic compounds.
Countermeasures
Storage and Handling: Section 7.
Shipping Regulations: Section 11.
  I.C.C.: Poison B; poison label, 200 pounds.
  Coast Guard Classification: Poison B; poison label.
  IATA: Poison B, poison label, 25 kilograms (passenger), 95 kilograms (cargo).

CACODYL OXIDE
General Information
Synonym: Dicacodyl oxide.
Description: Colorless liquid, slightly soluble in water.
Formula: $O[(CH_3)_2As]_2$.
Constants: Mol wt: 226, d: 1.486 at 15°C, mp: −25°C, bp: 150°C.
Hazard Analysis
Toxicity: Highly toxic. See arsenic compounds.
Disaster Hazard: See cacodyl dioxide.

CACODYL SULFIDE
General Information
Synonym: Dicacodyl sulfide.
Description: Oily liquid slightly soluble in water.
Formula: $[(CH_3)_2As]_2S$.
Constants: Mol wt: 242, bp: 211°C.
Hazard Analysis
Toxicity: See arsenic compounds and sulfides.
Fire Hazard: Dangerous, when exposed to heat or by spontaneous chemical reaction. Ignites spontaneously in air (Section 6).
Spontaneous Heating: Yes.
Explosion Hazard: Unknown.
Disaster Hazard: Dangerous; see arsenic and oxides of sulfur; it can react vigorously with oxidizing materials.
Countermeasures
First Aid: Section 1.
Personal Hygiene: Section 3.
Storage and Handling: Section 7.

CADAVERINE. See pentamethylene diamine.

CADE OIL
General Information
Synonym: Juniper tar.
Description: Dark brown, viscous, volatile oil.
Constant: D: 0.950–1.055 at 25°/25°C.
Hazard Analysis
Toxic Hazard Rating:
  Acute Local: Allergen 1.
  Acute Systemic: U.
  Chronic Local: Allergen 1.
  Chronic Systemic: U.
Fire Hazard: A combustible material; can react with oxidizing materials (Section 6).
Countermeasures
Personal Hygiene: Section 3.
Storage and Handling: Section 7.

CADMIUM
General Information
Description: Hexagonal crystals; silver-white malleable metal.
Formula: Cd.
Constants: At wt: 112.41, mp: 320.9°C, bp: 767 ± 2°C, d: 8.642, vap. press.: 1 mm at 394°C.
Hazard Analysis
Radiation Hazard: Section 5. For permissible levels, see Table 5, p. 150.
  Artificial isotope ^{109}Cd, half life 470 d. Decays to stable ^{109}Ag by electron capture. Emits gamma rays of 0.09 MeV and X-rays.

Artifical isotope ^{115m}Cd, half life 43 d. Decays to radio-active ^{115}In by emitting beta particles of 1.63 MeV.

Artificial isotope ^{115}Cd, half life 2.3 d. Decays to radio-active ^{115}In by emitting beta particles of 0.59, 0.63, 1.11 MeV. Also emits gamma rays of 0.49, 0.52 MeV.

TLV: ACGIH (tentative) 0.2 milligrams per cubic meter of air (metal dust and soluble salts).

Toxicity: Cadmium plating of food and beverage containers has resulted in a number of outbreaks of gastro-enteritis (food poisoning). See cadmium compounds.

Fire Hazard: Moderate, in the form of dust when exposed to heat or flame or by chemical reaction with oxidizing agents (Section 6). See also powdered metals.

Explosion Hazard: Slight, in the form of dust when exposed to flame (Section 7).

Disaster Hazard: Dangerous; see cadmium; it can react vigorously with oxidizing materials.

Countermeasures
Ventilation Control: Section 2.
Personal Hygiene: Section 3.
First Aid: Section 1.
Storage and Handling: Section 7.

CADMIUM ACETATE
General Information
Description: Monoclinic colorless crystals, odor of acetic acid.
Formula: $Cd(C_2H_3O_2)_2$.
Constants: Mol wt: 230.50, mp: 256°C, bp: decomposes, d: 2.341.
Hazard Analysis and Countermeasures
See cadmium compounds.

CADMIUM AMIDE
General Information
Description: White solid.
Formula: $Cd(NH_2)_2$.
Constants: Mol wt: 144.46, mp: decomposes at 120°C, d: 3.05 at 25°C.
Hazard Analysis and Countermeasures
See cadmium compounds and ammonia.

CADMIUM AMMONIUM BROMIDE
General Information
Synonyms: Ammonium-cadmium bromide.
Description: Colorless crystals.
Formula: $CdBr_2 \cdot 4NH_4Br$.
Constant: Mol wt: 664.00.
Hazard Analysis and Countermeasures
See cadmium compounds and bromides.

CADMIUM ARSENIDE
General Information
Description: Dark-gray cubes.
Formula: Cd_3As_2.
Constants: Mol wt: 487.05, mp: 721°C, d: 6.21 at 15°/4°C.
Hazard Analysis
Toxicity: See As compounds and Cd compounds.
Fire Hazard: Moderate, when exposed to heat or flame. May evolve arsine upon contact with moisture or acids (Section 6).
Explosion Hazard: Moderate when exposed to flame (Section 7). See also arsine.
Disaster Hazard: Dangerous; when heated to decomposition or on contact with acids, it emits highly toxic fumes which will react violently with water, steam or oxidizing materials.

Countermeasures
First Aid: Section 1.
Personal Hygiene: Section 3.
Storage and Handling: Section 7.

CADMIUM BENZOATE
General Information
Description: White solid.
Formula: $Cd(C_7H_5O_2)_2 \cdot 2H_2O$.
Constant: Mol wt: 390.66.
Hazard Analysis and Countermeasures
See cadmium compounds.

CADMIUM BOROTUNGSTATE
General Information
Description: Yellow, triclinic crystals.
Formula: $Cd_5(BW_{12}O_{40})_2 \cdot 18H_2O$.
Constants: Mol wt: 6602.06, mp: 75°C.
Hazard Analysis and Countermeasures
See cadmium compounds and boron compounds.

CADMIUM BROMATE
General Information
Description: Rhombic, white crystals.
Formula: $Cd(BrO_3)_2 \cdot H_2O$.
Constants: Mol wt: 386.26, mp: decomposes, d: 3.758.
Hazard Analysis
Toxicity: See cadmium compounds and bromates.
Fire Hazard: Moderate, by chemical reaction with reducing agents (Section 6).
Caution: A powerful oxidizing agent.
Disaster Hazard: Dangerous; see cadmium, bromine and bromides; it can react vigorously with reducing materials.
Countermeasures
Storage and Handling: Section 7.

CADMIUM BROMIDE
General Information
Description: Yellow crystals.
Formula: $CdBr_2$.
Constants: Mol wt: 272.24, mp: 567°C, bp: 963°C, d: 5.192 at 25°C.
Hazard Analysis and Countermeasures
See cadmium compounds and bromides.

CADMIUM CARBONATE
General Information
Description: Trigonal, white crystals.
Formula: $CdCO_3$.
Constants: Mol wt: 172.42, mp: decomposes <500°C, d: 4.258 at 4°C.
Hazard Analysis and Countermeasures
See cadmium compounds.

CADMIUM CHLORATE
General Information
Description: Colorless, deliquescent prisms.
Formula: $Cd(ClO_3)_2 \cdot 2H_2O$.
Constants: Mol wt: 315.36, mp: 80°C, d: 2.28 at 18°C.
Hazard Analysis
Toxicity: See cadmium compounds and chlorates.
Fire Hazard: Moderate, by chemical reaction with reducing agents (Section 6).
Caution: A powerful oxidizing agent. See also chlorates.
Explosion Hazard: Moderate, when shocked or exposed to heat. See also chlorates.
Disaster Hazard: Dangerous; heat and shock will explode it;

TOXIC HAZARD RATING CODE (For detailed discussion, see Section 1.)

0　NONE: (a) No harm under any conditions; (b) Harmful only under unusual conditions or overwhelming dosage.

1　SLIGHT: Causes readily reversible changes which disappear after end of exposure.

2　MODERATE: May involve both irreversible and reversible changes; not severe enough to cause death or permanent injury.

3　HIGH: May cause death or permanent injury after very short exposure to small quantities.

U　UNKNOWN: No information on humans considered valid by authors.

see cadmium, chlorine, and chlorides; it can react vigorously with reducing materials.

Countermeasures
First Aid: Section 1.
Personal Hygiene: Section 3.
Storage and Handling: Section 7.

CADMIUM CHLORIDE
General Information
Description: Hexagonal, colorless crystals.
Formula: $CdCl_2$.
Constants: Mol wt: 183.32, mp: 568°C, d: 4.047 at 25°C, vap. press.: 10 mm at 656°C.
Hazard Analysis and Countermeasures
See cadmium compounds and chlorides.

CADMIUM COBALTINITRITE. See cadmium nitrocobaltate (III).

CADMIUM COMPOUNDS
Hazard Analysis
Toxic Hazard Rating:
Acute Local: Irritant 3; Ingestion 3; Inhalation 3.
Acute Systemic: Ingestion 3; Inhalation 3.
Chronic Local: Variable.
Chronic Systemic: Ingestion 3; Inhalation 3.
Toxicology: The inhalation of fumes or dusts of cadmium primarily affects the respiratory tract; the kidneys may also be affected. Even brief exposure to high concentrations may result in pulmonary edema and death. Usually the edema is not massive, with little pleural effusion. In fatal cases, fatty degeneration of the liver, and acute inflammatory changes in the kidneys have been noted. Ingestion of cadmium results in a gastro-intestinal type of poisoning resembling food poisoning in its symptoms.

Inhalation of dust or fumes may cause dryness of the throat, cough, headache, a sense of constriction in the chest, shortness of breath (dyspnea) and vomiting. More severe exposure results in marked lung changes, with persistent cough, pain in the chest, severe dyspnea and prostration which may terminate fatally. X-ray changes are usually similar to those seen in bronchopneumonia. The urine is frequently dark. These symptoms are usually delayed for some hours after exposure, and fatal concentrations may be breathed without sufficient discomfort to warn the workman to leave the exposure. See also cadmium.

Ingestion of cadmium results in sudden nausea, salivation, vomiting and diarrhea and abdominal pain and discomfort. Symptoms begin almost immediately after ingestion.

A yellow discoloration of the teeth has been reported in workers exposed to cadmium. Cadmium oxide fumes can cause metal fume fever resembling that caused by zinc oxide fumes.
Countermeasures
Ventilation Control: Section 2.
Personal Hygiene: Section 3.
First Aid: Section 1.
Storage and Handling: Section 7.

CADMIUM CYANIDE
General Information
Description: Crystals.
Formula: $Cd(CN)_2$.
Constants: Mol wt: 164.45, mp: > 200°C decomposes.
Hazard Analysis and Countermeasures
See cyanides and cadmium compounds.

CADMIUM DIHYDROGEN PHOSPHATE
General Information
Description: Triclinic crystals.
Formula: $Cd(H_2PO_4)_2 \cdot 2H_2O$.
Constants: Mol wt: 342.4, mp: decomposes 100°C, d: 2.74 at 15°/4°C.

Hazard Analysis and Countermeasures
See cadmium compounds and phosphoric acid.

CADMIUM 9,10-EPOXYSTEARATE
Hazard Analysis
Toxic Hazard Rating:
Acute Local: U.
Acute Systemic: Ingestion 1.
Chronic Local: U.
Chronic Systemic: U.
Based on Limited animal experiments.
Disaster Hazard: Dangerous. See cadmium compounds.

CADMIUM ETHYLENE BISDITHIOCARBAMATE
Hazard Analysis
Toxicity: See cadmium compounds.
Disaster Hazard: Dangerous; see cadmium and oxides of sulfur.
Countermeasures
Storage and Handling: Section 7.

CADMIUM FERROCYANIDE
General Information
Description: Solid.
Formula: $Cd_2Fe(CN)_6 \cdot xH_2O$.
Constant: Mol wt: 454.8.
Hazard Analysis and Countermeasures
See cadmium compounds and ferrocyanides.

CADMIUM FLUOGALLATE
General Information
Description: Colorless crystals.
Formula: $[Cd(H_2O)_6](GaF_5H_2O)$.
Constants: Mol wt: 403.24, mp: $-5H_2O$ at 110°C, d: 2.79.
Hazard Analysis and Countermeasures
See fluorides and cadmium compounds.

CADMIUM FLUORIDE
General Information
Description: Cubic white crystals.
Formula: CdF_2.
Constants: Mol wt: 150.41, mp: 1100°C, bp: 1751°C, d: 6.64, vap. press.: 1 mm at 1112°C.
Hazard Analysis and Countermeasures
See fluorides and cadmium compounds.

CADMIUM FLUOSILICATE
General Information
Description: Hexagonal, colorless crystals.
Formula: $CdSiF_6 \cdot 6H_2O$.
Constant: Mol wt: 362.6.
Hazard Analysis and Countermeasures
See cadmium compounds and fluosilicates.

CADMIUM FORMATE
General Information
Description: Monoclinic crystals.
Formula: $Cd(CHO_2)_2 \cdot 2H_2O$.
Constants: Mol wt: 238.48, mp: decomposes, d: 2.44.
Hazard Analysis and Countermeasures
Toxicity: See cadmium compounds and formic acid.

CADMIUM FUMARATE
General Information
Description: Solid.
Formula: $CdC_4H_2O_4$.
Constant: Mol wt: 226.47.
Hazard Analysis and Countermeasures
See cadmium compounds.

CADMIUM HYDROGEN ARSENATE
General Information
Description: Solid.
Formula: $CdHAsO_4 \cdot H_2O$.
Constants: Mol wt: 270.34, d: 4.164 at 15°/4°C.

Hazard Analysis and Countermeasures
See arsenic compounds and cadmium compounds.

CADMIUM HYDROXIDE
General Information
Description: Trigonal or amorphous white crystals.
Formula: $Cd(OH)_2$.
Constants: Mol wt: 146.43, mp: decomposes 300°C, d: 4.79 at 15°C.
Hazard Analysis and Countermeasures
See cadmium compounds.

CADMIUM IODATE
General Information
Description: White crystals.
Formula: $Cd(IO_3)_2$.
Constants: Mol wt: 462.25, mp: decomposes, d: 6.43.
Hazard Analysis and Countermeasures
See cadmium compounds and iodates.

CADMIUM IODIDE
General Information
Description: Hexagonal, brownish crystals.
Formula: $CdI_2(\alpha)$; $CdI_2(\beta)$.
Constants: Mol wt: 366.25; 366.25, mp: 385°C, bp: 713°C, d: 5.670 at 30°C, vap. press.: 1 mm at 416°C.
Hazard Analysis and Countermeasures
See cadmium compounds and iodides.

CADMIUM LACTATE
General Information
Description: Needles.
Formula: $Cd(C_2H_5O_3)_2$.
Constant: Mol wt: 290.55.
Hazard Analysis and Countermeasures
See cadmium compounds.

CADMIUM MALEATE
General Information
Description: Solid.
Formula: $CdC_4H_2O_4 \cdot 2H_2O$.
Constant: Mol wt: 262.50.
Hazard Analysis and Countermeasures
See cadmium compounds.

CADMIUM MOLYBDATE
General Information
Description: A solid.
Formula: $CdMoO_4$.
Constants: Mol wt: 272.36, d: 5.347.
Hazard Analysis and Countermeasures
See cadmium compounds and molybdenum compounds.

CADMIUM NITRATE
General Information
Description: White, prismatic needles; hygroscopic.
Formula: $Cd(NO_3)_2$.
Constants: Mol wt: 236.43, mp: 350°C.
Hazard Analysis and Countermeasures
See cadmium compounds and nitrates.

CADMIUM NITROCOBALTATE (III)
General Information
Synonym: Cadmium cobaltinitrite.
Description: Yellow crystals.
Formula: $Cd_3[Co(NO_2)_6]_2$.
Constants: Mol wt: 1007.21, mp: decomposes 175°C.
Hazard Analysis and Countermeasures
See cadmium, cobalt compounds, and nitrates.

CADMIUM ORANGE. See cadmium selenide.

CADMIUM OXALATE
General Information
Description: Colorless crystals.
Formula: CdC_2O_4.
Constants: Mol wt: 200.43, mp: decomposes at 340°C, d: 3.32 at 18°/4°C.
Hazard Analysis and Countermeasures
See cadmium compounds and oxalates.

CADMIUM OXIDE
General Information
Description: (1) Amorphous, brown crystals; (2) cubic, brown crystals.
Formula: CdO.
Constants: Mol wt: 128.41, mp: (1) >1426°C, (2) 900°C decomposes; bp: 1559°C; d: (1) 6.95, (2) 8.15; vap. press.: 1 mm at 1000°C.
Hazard Analysis and Countermeasures
See cadmium compounds.
TLV: ACGIH (accepted) for cadmium oxide fumes 0.1 milligrams per cubic meter of air.

CADIUM PERCHLORATE
General Information
Description: Crystals.
Formula: $Cd(ClO_4)_2 \cdot 6H_2O$.
Constants: Mol wt: 419.9, mp: 129.4°C.
Hazard Analysis and Countermeasures
See cadmium compounds and perchlorates.

CADMIUM PERMANGANATE
General Information
Description: Violet crystals.
Formula: $Cd(MnO_4)_2 \cdot 6H_2O$.
Constants: Mol wt: 458.4, mp: decomposes at 95°C, d: 2.81.
Hazard Analysis and Countermeasures
See cadmium compounds and manganese compounds.

CADMIUM PHOSPHATE
General Information
Description: Amorphous or colorless crystals.
Formula: $Cd_3(PO_4)_2$.
Constants: Mol wt: 527.19, mp: 1500°C.
Hazard Analysis and Countermeasures
See cadmium compounds and phosphates.

CADMIUM PICRATE
General Information
Description: Yellow solid.
Formula: $Cd[C_6H(NO_2)_3OH]_2$.
Constant: Mol wt: 568.6.
Hazard Analysis and Countermeasures
See cadmium compounds and picric acid.

CADMIUM POTASSIUM CYANIDE
General Information
Description: Cubic crystals; soluble in 3 parts cold, 1 part hot water; insoluble in alcohol.
Formula: $Cd(CN)_2 \cdot 2KCN$.
Constant: Mol wt: 294.67.
Hazard Analysis and Countermeasures
See cyanides.

CADMIUM POTASSIUM IODIDE. See potassium cadmium iodide.

CADMIUM RED. See cadmium selenide.

TOXIC HAZARD RATING CODE (For detailed discussion, see Section 1.)

0 NONE: (a) No harm under any conditions; (b) Harmful only under unusual conditions or overwhelming dosage.

1 SLIGHT: Causes readily reversible changes which disappear after end of exposure.

2 MODERATE: May involve both irreversible and reversible changes; not severe enough to cause death or permanent injury.

3 HIGH: May cause death or permanent injury after very short exposure to small quantities.

U UNKNOWN: No information on humans considered valid by authors.

CADMIUM RICINOLEATE
General Information
Description: Fine white powder.
Formula: $Cd[CO_2(CH_2)_7CHCHCH_2CHOH(CH_2)_5CH_3]_2$.
Constants: Mol wt: 702.5, mp: 103°C, d: 1.11 at 25°/25°C.
Hazard Analysis and Countermeasures
See cadmium compounds.

CADMIUM SALICYLATE
General Information
Description: White needles.
Formula: $Cd(C_7H_5O_3)_2 \cdot H_2O$.
Constant: Mol wt: 404.65.
Hazard Analysis and Countermeasures
See cadmium compounds.

CADMIUM SELENATE, DIHYDRATE
General Information
Description: Rhombic crystals.
Formula: $CdSeO_4 \cdot 2H_2O$.
Constants: Mol wt: 291.40, mp: $-H_2O$ at 100°C, d: 3.632.
Hazard Analysis and Countermeasures
See cadmium compounds and selenium compounds.

CADMIUM SELENIDE
General Information
Description: Red powder, or it may be gray to brown.
Formula: CdSe.
Constants: Mol wt: 191.37, mp: >1350°C, d: 5.81 at 15°/4°C.
Hazard Analysis
Toxicity: See cadmium compounds and selenium compounds.
Fire Hazard: Moderate when exposed to heat or flame (Section 6).
Caution: See hydrogen selenide, which can be evolved on contact with acids or moisture.
Explosion Hazard: Moderate, when exposed to flame (Section 7).
Disaster Hazard: Dangerous; see cadmium; it can emit hydrogen selenide and selenium compounds on contact with moisture, acid or acid fumes.
Countermeasures
Storage and Handling: Section 7.

CADMIUM m-SILICATE
General Information
Description: Colorless, rhombic crystals.
Formula: $CdSiO_3$.
Constants: Mol wt: 188.47, mp: 1242°C, d: 4.93.
Hazard Analysis and Countermeasures
See cadmium compounds and silicates.

CADMIUM SOAP
Hazard Analysis and Countermeasures
See cadmium compounds.

CADMIUM SULFATE
General Information
Description: Rhombic, white crystals.
Formula: $CdSO_4$.
Constants: Mol wt: 208.48, mp: 1000°C, d: 4.691.
Hazard Analysis and Countermeasures
See cadmium compounds and sulfates.

CADMIUM SULFIDE
General Information
Synonym: Greenockite.
Description: Hexagonal, yellow-orange crystals.
Formula: CdS.
Constants: Mol wt: 144.48, mp: 1750° at 100 atm, bp: sublimes in N_2, d: 4.82.
Hazard Analysis and Countermeasures
See cadmium compounds and sulfides.

CADMIUM SULFITE
General Information
Description: Crystals.
Formula: $CdSO_3$.
Constants: Mol wt: 192.48, mp: decomposes.
Hazard Analysis and Countermeasures
See cadmium compounds and sulfites.

CADMIUM TARTRATE
General Information
Description: White, crystalline powder.
Formula: $CdC_4H_4O_6$.
Constant: Mol wt: 260.48.
Hazard Analysis and Countermeasures
See cadmium compounds.

CADMIUM TELLURIDE
General Information
Description: Black, cubic crystals.
Formula: CdTe.
Constants: Mol wt: 240.02, mp: 1041°C, d: 6.20 at 15°C.
Hazard Analysis
Toxicity: See cadmium compounds and tellurium compounds.
Fire Hazard: Moderate when exposed to heat or flame (Section 6).
Caution: See hydrogen telluride which can be evolved on contact with acids or moisture.
Explosion Hazard: Moderate, when exposed to flame (Section 7).
Disaster Hazard: Dangerous; see cadmium; it can emit hydrogen telluride and tellurium compounds on contact with acid or acid fumes, or moisture.
Countermeasures
Storage and Handling: Section 7.

CADMIUM TRICHLOROACETATE
General Information
Description: Rhombic crystals.
Formula: $Cd(C_2Cl_3O_2)_2 \cdot 1\frac{1}{2}H_2O$.
Constants: Mol wt: 464.22, d: 2.093 at 25°C.
Hazard Analysis and Countermeasures
See cadmium compounds and trichloroacetic acid.

CADMIUM TUNGSTATE
General Information
Description: Yellow crystals.
Formula: $CdWO_4$.
Constant: Mol wt: 360.33.
Hazard Analysis and Countermeasures
See cadmium and tungsten compounds.

"CADMOLITH". See barium sulfate and cadmium sulfide.

CAFFEIC ACID
General Information
Synonym: 3,4-Dihydroxycinnamic acid.
Description: Yellow crystals. Soluble in hot water, alcohol.
Formula: $C_9H_8O_4$.
Constants: Mol wt: 180.2, mp: 195°C.
Hazard Analysis
Toxic Hazard Rating:
Acute Local: Allergen 2.
Acute Systemic: Ingestion 1.
Chronic: Allergen 2.
Chronic Systemic: Ingestion 1.

CAFFEINE
General Information
Synonym: Theine.
Description: White, fleecy masses.
Formula: $C_8H_{10}N_4O_2 \cdot H_2O$.
Constants: Mol wt: 212.11, mp: 236.8°C.
Hazard Analysis
Toxic Hazard Rating:
Acute Local: 0.

Acute Systemic: Ingestion 2.
Chronic Local: 0.
Chronic Systemic: Ingestion 2.
A general purpose food additive (Section 10).
Toxicology: Large doses (above 1.0 gram) cause palpitation, excitement, insomnia, dizziness, headache and vomiting. Continued excessive use of caffeine in tea or coffee may lead to digestive disturbances, constipation, palpitations, shortness of breath and depressed mental states.

Countermeasures
Treatment and Antidotes: Evacuate stomach with emetic or stomach tube. Call a physician.
Fire Hazard: Slight; when heated to decomposition, it emits toxic fumes.

CAKE ALUM. See aluminum sulfate, dry.

CALABAR BEAN. See physostigma.

CALABARINE. See physostigmine.

CALCIC LIVER OF SULFUR. See calcium sulfide.

CALCIFEROL. See vitamin D_2.

CALCIMINE
General Information
Description: A white powder.
Composition: Chalk and glue.
Hazard Analysis
Toxic Hazard Rating:
    Acute Local: Inhalation 1.
    Acute Systemic: 0.
    Chronic Local: Inhalation 1.
    Chronic Systemic: 0.
Countermeasures
Ventilation Control: Section 2.

CALCITE. See calcium carbonate.

CALCIUM
General Information
Description: Silver-white, soft metal.
Formula: Ca.
Constants: At wt: 40.08, mp: 842°C, bp: 1240°C, d: 1.54 at 20°C, vap. press.: 10 mm at 983°C.
Hazard Analysis
Radiation Hazard: Section 5. For permissible levels, see Table 5, p. 150.
    Artificial isotope ^{45}Ca, half life 165 d. Decays to stable ^{45}Sc by emitting beta particles of 0.25 MeV.
    Artificial isotope ^{47}Ca, half life 4.7 d. Decays to radioactive ^{47}Sc by emitting beta particles of 0.69 (82%), 2.0 (18%) MeV. Also emits gamma rays of 1.30 MeV and others.
Toxicology: See calcium compounds.
Fire Hazard: Moderate, when heated or in intimate contact with oxidizing agents. Contact with moisture or acids evolves hydrogen. See also hydrogen.
Explosion Hazard: Moderate, in intimate contact with very powerful oxidizing agents (Section 7).
Disaster Hazard: Dangerous; reacts with moisture or acids to liberate large quantities of hydrogen; it can develop explosive pressure in containers. See also hydrogen.
Countermeasures
Ventilation Control: Section 2.
To Fight Fire: Special mixtures of dry chemical (Section 6).
Personal Hygiene: Section 3.

Storage and Handling: Section 7.
Shipping Regulations: Section 11.
    I.C.C. Classification: Flammable solid; yellow label.
    Coast Guard Classification: Inflammable solid; yellow label.

CALCIUM ACETATE
General Information
Synonyms: Vinegar salts; gray acetate; lime acetate; calcium diacetate.
Description: Brown, gray, or white powder; slight odor of acetic acid; soluble in water; slightly soluble in alcohol.
Formula: $Ca(C_2H_3O_2)_2 \cdot H_2O$.
Constant: Mol wt: 176.
Hazard Analysis
Toxic Hazard Rating: U. A sequestrant food additive (Section 10).

CALCIUM ACRYLATE
General Information
Description: White crystals.
Formula: $Ca(CO_2CHCH_2)_2$.
Constant: Mol wt: 182.
Hazard Analysis
Toxicity: See calcium compounds and acrylic acid.
Fire Hazard: Slight; it will react with water or steam to produce heat (Section 6).
Countermeasures
Storage and Handling: Section 7.

CALCIUM ACRYLATE DIHYDRATE
General Information
Description: White powder.
Formula: $(CH_2CHCOO)_2Ca \cdot 2H_2O$.
Constant: Mol wt: 218.22.
Hazard Analysis and Countermeasures
See calcium compounds and acrylic acid.

CALCIUM ALGINATE
General Information
Description: White or cream colored powder, or filaments, grains, or granules; slight characteristic odor. It is a colloidal substance, having a mol wt of 32,000–35,000. Insoluble in water, acids; soluble in alkaline solutions.
Hazard Analysis
Toxic Hazard Rating: U. A Stabilizer food additive (Section 10).

CALCIUM ALUMINUM ANHYDRIDE. See hydrides.

CALCIUM ARSENATE
General Information
Synonyms: Tricalcium o-arsenate; calcium o-arsenate.
Description: White, amorphous powder.
Formula: $Ca_3(AsO_4)_2$.
Constant: Mol wt: 398.06.
Hazard Analysis
Toxicity: See arsenic compounds.
TLV: ACGIH (accepted); 1 milligram per cubic meter of air.
Countermeasures
Storage and Handling: Section 7.
Shipping Regulations: Section 11.
    I.C.C.: Poison B; poison label, 200 pounds.
    IATA: Poison B, poison label, 25 kilograms (passenger), 95 kilograms (cargo).
    Coast Guard Classification: Poison B; poison label.

CALCIUM-o-ARSENATE. See calcium arsenate.

TOXIC HAZARD RATING CODE (*For detailed discussion, see Section 1.*)

0 NONE: (a) No harm under any conditions; (b) Harmful only under unusual conditions or overwhelming dosage.

1 SLIGHT: Causes readily reversible changes which disappear after end of exposure.

2 MODERATE: May involve both irreversible and reversible changes; not severe enough to cause death or permanent injury.

3 HIGH: May cause death or permanent injury after very short exposure to small quantities.

U UNKNOWN: No information on humans considered valid by authors.

CALCIUM ARSENIDE
General Information
Description: Red Crystals.
Formula: Ca_3As_2.
Constants: Mol wt: 270.06, mp: decomposes, d: 3.031 at 25°C.
Hazard Analysis
Toxicity: See arsenic compounds and arsine. Highly toxic.
Fire Hazard: Moderate; on contact with moisture, acid, or acid fumes, it will evolve arsine. See also arsine.
Explosion Hazard: Moderate; on contact with moisture, acid or acid fumes, it will evolve arsine. See also arsine.
Disaster Hazard: Dangerous; see arsenic; it will react with water, steam, acid or acid fumes to produce toxic and flammable vapors of arsine.
Countermeasures
Storage and Handling: Section 7.

CALCIUM ARSENITE
General Information
Description: White, granular powder.
Formula: $CaAsO_3H$.
Constant: Mol wt: 164.00.
Hazard Analysis
Toxicity: See arsenic compounds.
Disaster Hazard: See arsenic compounds.
Countermeasures
Storage and Handling: Section 7.
Shipping Regulations: Section 11.
  I.C.C.: Poison B; poison label, 200 pounds.
  Coast Guard Classification: Poison B; poison label.
  MCA warning label.
  IATA: Poison B, poison label, 25 kilograms (passenger), 95 kilograms (cargo).

CALCIUM ASCORBATE
General Information
Description: A white to yellow crystalline powder; odorless; soluble in water; slightly soluble in alcohol; insoluble in ether.
Formula: $Ca(C_6H_7O_6)_2 \cdot 2H_2O$.
Constant: Mol wt: 426.
Hazard Analysis
Toxic Hazard Rating: U. A chemical preservative food additive. (Section 10).

CALCIUM AZIDE
General Information
Description: Rhombic, colorless crystals.
Formula: $Ca(N_3)_2$.
Constants: Mol wt: 124.13, mp: explodes 144–156°C.
Hazard Analysis and Countermeasures
See azides.

CALCIUM BIPHOSPHATE. See calcium phosphate, monobasic.

CALCIUM BISULFITE
General Information
Synonym: Calcium hydrogen sulfite.
Description: Colorless or slightly yellowish liquid; strong sulfur dioxide odor.
Formula: $Ca(HSO_3)_2$.
Constants: Mol wt: 202.21, d: 1.06.
Hazard Analysis and Countermeasures
See bisulfites.
Shipping Regulations: Section 11.
  IATA: Other restricted articles, class B, no label, no limit (passenger and cargo).

CALCIUM m-BORATE
General Information
Description: Colorless, rhombic or long, flat plates.
Formula: $Ca(BO_2)_2$.
Constants: Mol wt: 125.72, mp: 1154°C.
Hazard Analysis and Countermeasures
See boron compounds.

CALCIUM BORIDE
General Information
Description: Cubic, black crystals.
Formula: CaB_6.
Constants: Mol wt: 105.00, d: 2.3 at 20°C.
Hazard Analysis
Toxicity: See calcium compounds and boron compounds.
Fire Hazard: Moderate; on contact with moisture, acid or acid fumes, it will evolve boron hydrides, some of which are highly toxic and flammable.
Countermeasures
Storage and Handling: Section 7.

CALCIUM BROMATE
General Information
Description: Monoclinic crystals.
Formula: $Ca(BrO_3)_2 \cdot H_2O$.
Constants: Mol wt: 313.9, mp: $-H_2O$ at 180°C, d: 3.329.
Hazard Analysis and Countermeasures
See bromates.

CALCIUM BROMIDE
General Information
Description: Deliquescent needles.
Formula: $CaBr_2$.
Constants: Mol wt: 199.91, mp: 765°C, d: 3.353 at 25°C, bp: 806–812°C.
Hazard Analysis and Countermeasures
See bromides.

CALCIUM CARBAMATE
General Information
Description: Crystals. Soluble in water; insoluble in alcohol.
Formula: $Ca(NH_2CO_2)_2$.
Constant: Mol wt: 160.2.
Hazard Analysis
Toxicology: Details unknown. In general, carbamic acid salts have low toxicity.

CALCIUM CARBIDE
General Information
Description: Rhombic, gray crystals.
Formula: CaC_2.
Constants: Mol wt: 64.10, mp: approx. 2300°C, d: 2.22.
Hazard Analysis
Toxicity: Acetylene is evolved when calcium carbide is in contact with moisture. See also calcium hydroxide and acetylene.
Fire Hazard: Moderate; on contact with moisture, acid or acid fumes it evolves heat or flammable vapors. See also acetylene.
Explosion Hazard: Moderate. See also acetylene.
Countermeasures
Storage and Handling: Section 7.
Shipping Regulations: Section 11.
  Coast Guard Classification: Hazardous article.
  IATA: Flammable solid, yellow label, not acceptable (passenger), 12 kilograms (cargo).

CALCIUM CARBIMIDE. See calcium cyanamide.

CALCIUM CARBONATE
General Information
Synonyms: Calcite; aragonite.
Description: White powder.
Formula: $CaCO_3$.
Constants: Mol wt: 100.09, mp: decomposes at 825°C; 1339°C at 1025 atm., d: 2.7–2.95.
Hazard Analysis and Countermeasures
See calcium compounds. A nutrient and/or dietary supplement food additive. Also a general purpose food additive (Section 10). Calcium carbonate is a common air contaminant (Section 4).

CALCIUM CHLORATE
General Information
Description: Monoclinic, white-yellowish, deliquescent crystals.

Formula: Ca(ClO$_3$)$_2$·2H$_2$O.
Constants: Mol wt: 243.03, mp:−H$_2$O > 100°C, d: 2.711.
Hazard Analysis
Toxicity: See chlorates and calcium compounds.
Fire Hazard: See chlorates.
Explosion Hazard: See chlorates.
Disaster Hazard: See chlorates.
Countermeasures
Storage and Handling: Section 7.
Shipping Regulations: Section 11.
    I.C.C.: Oxidizing material; yellow label, 100 pounds.
    Coast Guard Classification: Oxidizing material; yellow label.
    IATA: Oxidizing material, yellow label, 12 kilograms (passenger), 45 kilograms (cargo).

CALCIUM CHLORIDE
General Information
Description: Cubic colorless, deliquescent crystals.
Formula: CaCl$_2$.
Constants: Mol wt: 110.99, mp: 772°C, bp: > 1600°C, d: 2.512 at 25°C.
Hazard Analysis and Countermeasures
See calcium compounds.
A sequestrant food additive, also a general purpose food additive and a substance migrating to food from packaging materials (Section 10).

CALCIUM CHLORIDE FLUORIDE o-PHOSPHATE
General Information
Description: Colorless crystals.
Formula: CaClF·3Ca$_3$(PO$_4$)$_2$.
Constants: Mol wt: 1025.14, mp: 1270°C, d: 3.14.
Hazard Analysis and Countermeasures
See fluorides and chlorides.

CALCIUM CHLORIDE HYPOCHLORITE. See bleaching powder.

CALCIUM CHLORITE
General Information
Description: White solid.
Formula: CaClO$_2$.
Constant: Mol wt: 107.6.
Hazard Analysis
Toxicity: See chlorites.
Fire Hazard: See chlorites.
Disaster Hazard: See chlorites.
Countermeasures
Storage and Handling: Section 7.
Shipping Regulations: Section 11.
    I.C.C.: Oxidizing material; yellow label, 100 pounds.
    Coast Guard Classification: Oxidizing material; yellow label.
    IATA: Oxidizing material, yellow label, not acceptable (passenger), 45 kilograms (cargo).

CALCIUM CITRATE
General Information
Synonym: Lime citrate; tricalcium citrate.
Description: White powder; odorless; almost insoluble in water; insoluble in alcohol.
Formula: Ca$_3$(C$_6$H$_5$O$_7$)$_2$·4H$_2$O.
Constant: Mol wt: 570.
Hazard Analysis
Toxic Hazard Rating: U. A nutrient and/or dietary supplement food additive. Also a sequestrant food additive and a general purpose food additive (Section 10).

CALCIUM CHROMATE
General Information
Description: Monoclinic prisms, yellow color.
Formula: CaCrO$_4$·2H$_2$O.
Constants: Mol wt: 192.12, mp: −2H$_2$O at 200°C.
Hazard Analysis and Countermeasures
See chromium compounds.

CALCIUM COMPOUNDS
Hazard Analysis
Toxic Hazard Rating:
    Acute Local: Irritant 1; Ingestion 1; Inhalation 1.
    Acute Systemic: U.
    Chronic Local: Irritant 1.
    Chronic Systemic: U.
Toxicology: The fumes evolved by burning calcium in air are composed of calcium oxide (quick lime). This material is irritating to the skin, eyes and mucous membranes. Many calcium compounds are used medicinally. Generally speaking, calcium compounds should be considered toxic only when they contain a toxic component (such as arsenic, etc.) or as calcium oxide or hydroxide. Calcium compounds are common air contaminants (Section 4).
Countermeasures
Treatment and Antidotes: Any calcium residue left on the body or clothing should be brushed off immediately.

CALCIUM CYANAMIDE
General Information
Synonym: Calcium carbimide.
Description: Hexagonal, rhombohedral, colorless crystals.
Formula: CaCN$_2$.
Constants: Mol wt: 80.11, mp: 1300°C; sublimes > 1500°C.
Hazard Analysis
Toxic Hazard Rating:
    Acute Local: Irritant 2; Inhalation 2.
    Acute Systemic: Ingestion 2; Inhalation 2.
    Chronic Local: Irritant 2.
    Chronic Systemic: U.
Toxicology: A herbicide. Calcium cyanamide acts locally on the skin as a primary irritant, and the lesions produced vary from erythema to acute and subacute eczema. Usually the moist skin areas are attacked first, but the material is spread by scratching and parts of the body not ordinarily exposed may be affected. In severe cases ulceration may develop; the ulcers are usually covered by a black, necrotic crust. There is frequently irritation of the conjunctivae and of the mucous membranes of the nose and throat, with production of conjunctivitis, inflamed ulcers in the nose and throat, rhinitis and gingivitis. Systemically, headache, flushing of the skin of the head and neck, shortness of breath, vasodilation with lowered blood pressure, and rapid pulse have been described among exposed persons who have consumed alcohol. No fatalities have been reported. Calcium cyanamide is not believed to have a cumulative action. The fatal dose, by ingestion, is probably around 20 to 30 grams for an adult. It does not have a cyanide effect. See also amides.
Disaster Hazard: See cyanides.
Countermeasures
Ventilation Control: Section 2.
Personnel Protection: Section 3.
Personal Hygiene: Section 3.
Storage and Handling: Section 7.
Shipping Regulations: Section 11.

TOXIC HAZARD RATING CODE (*For detailed discussion, see Section 1.*)

0 NONE: (a) No harm under any conditions; (b) Harmful only under unusual conditions or overwhelming dosage.

1 SLIGHT: Causes readily reversible changes which disappear after end of exposure.

2 MODERATE: May involve both irreversible and reversible changes; not severe enough to cause death or permanent injury.

3 HIGH: May cause death or permanent injury after very short exposure to small quantities.

U UNKNOWN: No information on humans considered valid by authors.

Coast Guard Classification: (Not hydrated); hazardous article label.
Coast Guard Classification: Poison B; poison label.
IATA (containing more than 0.5% calcium carbide): Flammable solid, yellow label, 12 kilograms (passenger), 95 kilograms (cargo).

CALCIUM CYANIDE
General Information
Description: Rhombohedron crystals; white powder.
Formula: $Ca(CN)_2$.
Constants: Mol wt: 92.12, mp: decomposes > 350°C.
Hazard Analysis
See cyanides.
Countermeasures
Storage and Handling: Section 7.
Shipping Regulations: Section 11.
I.C.C.: Poison B; poison label, 200 pounds.
Coast Guard Classification: Poison B; poison label.
MCA warning label.
IATA: Poison B, poison label, 12 kilograms (passenger), 95 kilograms (cargo).

CALCIUM CYANOPLATINITE
General Information
Description: Rhombic, yellow-green fluorescent crystals.
Formula: $CaPt(CN)_4 \cdot 5H_2O$.
Constant: Mol wt: 429.46.
Hazard Analysis
See cyanides.
Countermeasures
Storage and Handling: Section 7.

CALCIUM CYCLAMATE
General Information
Synonym: Calcium cyclohexyl sulfamate; calcium cyclohexane sulfanate.
Description: White crystalline powder; almost odorless; freely soluble in water; practically insoluble in alcohol, benzene, chloroform and ether.
Formula: $(C_6H_{11}NHSO_3)_2Ca \cdot 2H_2O$.
Constant: Mol wt: 432.
Hazard Analysis
Toxic Hazard Rating: U. A non-nutritive sweetener food additive (Section 10).

CALCIUM DIACETATE. See calcium acetate.

CALCIUM DICYANOGUANIDINE
General Information
Description: Crystals.
Formula: $Ca[NH_2CN(NCN)CN]_2 \cdot 4H_2O$.
Constants: Mol wt: 328.3, mp: > 300°C.
Hazard Analysis and Countermeasures
See cyanides.

CALCIUM DISODIUM EDETATE. See calcium disodium EDTA.

CALCIUM DISODIUM EDTA
General Information
Synonym: Calcium disodium edetate; edathamil calcium disodium; calcium disodium ethylenediamene tetraacetate.
Description: White, odorless powders or flakes; soluble in water; insoluble in organic solvents.
Formula: $CaNa_2C_{10}H_{12}N_2O_8 \cdot xH_2O$.
Hazard Analysis
Toxic Hazard Rating: U. A food additive permitted in food for human consumption (Section 10).

CALCIUM DISODIUM ETHYLENEDIAMINETETRA-ACETATE. See calcium disodium EDTA.

CALCIUM 2,3-EPOXY-2-ETHYLHEXANOATE
Hazard Analysis
Toxic Hazard Rating:
Acute Local: U.
Acute Systemic: Ingestion 1.
Chronic Local: U.
Chronic Systemic: U.
Based on limited animal experiments.

CALCIUM 9,10-EPOXY STEARATE
Hazard Analysis
Toxic Hazard Rating:
Acute Local: U.
Acute Systemic: Ingestion 1.
Chronic Local: U.
Chronic Systemic: U.
Based on limited animal experiments.

CALCIUM ETHYLENE BISDITHIOCARBAMATE
General Information
Description: Solid.
Formula: $Ca[(C_2H_3)_2NCS_2]_2$.
Constant: Mol Wt: 330.5.
Hazard Analysis
Toxicity: Details unknown; a fungicide. See bis(dimethyl-thiocarbamyl)disulfide.
Disaster Hazard: Moderately dangerous; when heated to decomposition it emits toxic fumes.
Countermeasures
Storage and Handling: Section 7.

CALCIUM FERRICYANIDE
General Information
Description: Red, deliquescent needles.
Formula: $Ca_3[Fe(CN)_6]_2 \cdot 12H_2O$.
Constant: Mol wt: 760.4.
Hazard Analysis and Countermeasures
See ferricyanides and calcium compounds.

CALCIUM FERROCYANIDE
General Information
Description: Yellow, triclinic crystals.
Formula: $Ca_2Fe(CN)_6 \cdot 11H_2O$.
Constants: Mol wt: 490.3, mp: decomposes, d: 1.68.
Hazard Analysis and Countermeasures
See calcium compounds and ferrocyanides.

CALCIUM FLUORIDE
General Information
Description: Cubic, colorless crystals; luminous with heat.
Formula: CaF_2.
Constants: Mol wt: 78.1, mp: 1360°C, d: 3.180.
Hazard Analysis
See fluorides.
Countermeasures
See fluorides.

CALCIUM FLUOROPHOSPHATE
General Information
Synonym: Calcium monofluorophosphate.
Description: Colorless monoclinic crystals. Slightly water soluble. Insoluble in organic solvents.
Formula: $CaPO_3F$.
Constant: Mol wt: 138.1.
Hazard Analysis
Toxicology: See fluorides.
Disaster Hazard: Dangerous; see fluorides and phosphates.

CALCIUM FLUOSILICATE
General Information
Description: White, crystalline powder.
Formula: $CaSiF_6$.
Constants: Mol wt: 182.14, d: 2.662 at 17.5°C.
Hazard Analysis and Countermeasures
See fluosilicates.

CALCIUM GLUCONATE
General Information
Description: White fluffy powder or granules; odorless; soluble in hot water; less soluble in cold water; insoluble in alcohol, acetic acid and other organic solvents.

Formula: $Ca(C_6H_{11}O_7)_2 \cdot H_2O$.
Constant: Mol wt: 448, loses H_2O at 120°C.
Hazard Analysis
Toxic Hazard Rating: U. A sequestrant food additive also a general purpose food additive (Section 10).

CALCIUM GLYCERINOPHOSPHATE. See calcium glycerophosphate.

CALCIUM GLYCEROPHOSPHATE
General Information
Synonym: Calcium glycerino phosphate.
Description: White, crystalline powder; odorless; slightly soluble in water; insoluble in alcohol.
Formula: $CaC_3H_7O_2PO_4$.
Constant: Mol wt: 210, decomposes $> 170°$C.
Hazard Analysis
Toxic Hazard Rating: U. A nutrient and/or dietary supplement food additive (Section 10).
Disaster Hazard: Dangerous. See Phosphates.

CALCIUM HEXAMETA PHOSPHITE
Hazard Analysis
Toxic Hazard Rating: U. A sequestrant food additive.
Disaster Hazard: Dangerous. See phosphates.

CALCIUM HYDRIDE
General Information
Synonym: Hydrolith.
Description: Gray-white crystal; powder.
Formula: CaH_2.
Constants: Mol wt: 42.10, mp: 816°C in H; decomposes at about 600°C, d: 1.8 approx.
Hazard Analysis and Countermeasures
    See calcium compounds and hydrides.
Shipping Regulations: Section 11.
  IATA: Flammable solid, yellow label, not acceptable (passenger), 12 kilograms (cargo).

CALCIUM HYDROGEN SULFITE. See calcium bisulfite.

CALCIUM HYDROSULFIDE. See calcium sulfhydrate.

CALCIUM HYDROXIDE
General Information
Synonyms: Hydrate lime; slaked lime.
Description: Rhombic, trigonal, colorless crystals.
Formula: $Ca(OH)_2$.
Constants: Mol wt: 74.10, mp: $-H_2O$ at 580°C, bp decomposes, d: 2.343.
Hazard Analysis
Toxicity: A general purpose food additive, also a substance migrating to food from packaging materials (Section 10). See calcium compounds.
Toxicology: Calcium hydroxide has a caustic reaction and therefore is irritating to the skin and respiratory system. In the form of dust it is considered to be an important industrial hazard. It can cause dermatitis, irritation of the eyes and mucous membranes (Section 9). It is a common air contaminant (Section 4).
Countermeasures
Treatment and Antidotes: Irrigate any areas which have come in contact with this material. If the eyes are involved, they should be washed at once with copious amounts of warm water. If the skin is involved, a shower is recommended. See also calcium compounds.
Ventilation Control: Section 2.
Personal Hygiene: Section 3.

CALCIUM HYPOCHLORITE. See bleaching power.

CALCIUM HYPOCHLORITE COMPOUNDS, DRY, containing more than 8.80% available oxygen, 39% available chlorine. See hypochlorites.
Shipping Regulations: Section 11.
  I.C.C.: Oxidizing material; yellow label, 100 pounds.
  Coast Guard Classification: Oxidizing material; yellow label.
  IATA: Oxidizing material, yellow label, 25 kilograms (passenger), 45 kilograms (cargo).

CALCIUM IODATE
General Information
Synonym: Autarite.
Description: Triclinic crystals.
Formula: $Ca(IO_3)_2$.
Constants: Mol wt: 389.9, mp: decomposes, d: 4.519 at 15°C.
Hazard Analysis and Countermeasures
A trace mineral added to animal feed (Section 10). See iodates andd calcium compounds.

CALCIUM IODOBEHENATE
General Information
Description: White or yellowish powder; odorless or slightly fatty odor; contains 24% iodine; soluble in warm chloroform; slightly soluble in alcohol and ether; insoluble in water.
Formula: $Ca(OOCC_{21}H_{42}I)_2$.
Constant: Mol wt: 960.
Hazard Analysis
Toxic Hazard Rating: U. A trace mineral added to animal feeds (Section 10).
Disaster Hazard: Dangerous. See iodides.

CALCIUM LACTATE
General Information
Description: White powder; almost odorless; soluble in water; practically insoluble in alcohol.
Formula: $Ca(C_3H_5O_3)_2 \cdot 5H_2O$.
Constant: Mol wt: 308, loses H_2O at 120°C.
Hazard Analysis
Toxic Hazard Rating: U. A general purpose food additive (Section 10).

CALCIUM LACTOBIONATE
General Information
Description: The calcium salt of lactobionic acid, produced by the oxidation of lactone.
Hazard Analysis
Toxic Hazard Rating: U. A food additive permitted in food for human consumption (Section 10).

CALCIUM LIGNOSULFATE
Hazard Analysis
Toxic Hazard Rating: U. A food additive permitted in food for human consumption (Section 10).

CALCIUM METALLIC
Shipping Regulations: Section 11.
  I.C.C.: Flammable solid, yellow label, 100 pounds.
  IATA: Flammable solid, yellow label, 12 kilograms (passenger), 425 kilograms (cargo).

CALCIUM METALLIC, CRYSTALLINE
Shipping Regulations: Section 11.
  I.C.C.: Flammable solid, no exemption, yellow label, 25 pounds.
  IATA: Flammable solid, yellow label, not acceptable (passengers), 12 kilograms (cargo).

CALCIUM MONOFLUOROPHOSPHATE. See calcium fluorophosphate.

TOXIC HAZARD RATING CODE (For detailed discussion, see Section 1.)

0 NONE: (a) No harm under any conditions; (b) Harmful only under unusual conditions or overwhelming dosage.

1 SLIGHT: Causes readily reversible changes which disappear after end of exposure.

2 MODERATE: May involve both irreversible and reversible changes; not severe enough to cause death or permanent injury.

3 HIGH: May cause death or permanent injury after very short exposure to small quantities.

U UNKNOWN: No information on humans considered valid by authors.

CALCIUM NITRATE
General Information
Description: Cubic, colorless, hygroscopic crystals.
Formula: $Ca(NO_3)_2$.
Constants: Mol wt: 164.10, mp: 561°C, d: 2.36.
Hazard Analysis
Toxicity: See nitrates and calcium compounds.
Fire Hazard: See nitrates.
Disaster Hazard: See nitrates.
Countermeasures
Storage and Handling: Section 7.
Shipping Regulations: Section 11.
  I.C.C.: Oxidizing material; yellow label, 100 pounds.
  Coast Guard Classification: Oxidizing material.
  IATA: Oxidizing material, yellow label, 12 kilograms (passenger), 45 kilograms (cargo).

CALCIUM NITRIDE
General Information
Description: Brown crystals.
Formula: Ca_3N_2.
Constants: Mol wt: 148.3, mp: 900°C, d: 2.63 at 17°C.
Hazard Analysis and Countermeasures
See calcium compounds and ammonia.
Caution: Ammonia can be evolved upon contact with moisture.

CALCIUM NITRITE
General Information
Description: Hexagonal, colorless to yellowish crystals.
Formula: $Ca(NO_2)_2 \cdot H_2O$.
Constants: Mol wt: 150.11, mp: $-H_2O$ at 100°C, d: 2.23 at 34°C (anh.); 2.53 at 30°C.
Hazard Analysis and Countermeasures
See nitrites and calcium compounds.

CALCIUM OXALATE
General Information
Description: Cubic, colorless crystals.
Formula: CaC_2O_4.
Constants: Mol wt: 128.10, mp: decomposes, d: 2.2 at 4°C.
Hazard Analysis and Countermeasures
See oxalates and calcium compounds.

CALCIUM OXIDE
General Information
Synonyms: Unslaked lime; quick-lime; burnt lime; calx.
Description: Cubic, colorless crystals.
Formula: CaO.
Constants: Mol wt: 56.08, mp: 2580°C, bp: 2850°C, d: 3.37.
Hazard Analysis
Toxicity: A nutrient and/or dietary supplement food additive. See calcium hydroxide.
TLV: ACGIH (recommended) 5 milligrams per cubic meter of air.
Toxicology: A common air contaminant (Section 4).
Disaster Hazard: Slightly dangerous; on contact with water, steam, acid or acid fumes, it will react to produce heat.
Countermeasures
Storage and Handling: Section 7.
Shipping Regulations: Section 11.
  IATA: Other restricted articles, class B, no label required, 12 kilograms (passenger), 45 kilograms (cargo).

CALCIUM PANTOTHENATE
General Information
Description: White, slightly hygroscopic, powder; odorless; soluble in water and glycerol; insoluble in alcohol, chloroform, and ether.
Formula: $(C_9H_{16}NO_5)_2Ca$.
Constant: Mol wt: 476, mp: 170–172°C, decomposes at 195–196°C.
Hazard Analysis
Toxic Hazard Rating: A nutrient and/or dietary supplement food additive (Section 10).

CALCIUM PERCHLORATE
General Information
Description: Colorless crystals.
Formula: $Ca(ClO_4)_2$.
Constant: Mol wt: 238.09.
Hazard Analysis and Countermeasures
See perchlorates and calcium compounds.
Shipping Regulations: Section 11.
  IATA: Oxidizing material, yellow label, 12 kilograms (passenger), 45 kilograms (cargo).

CALCIUM PERMANGANATE
General Information
Synonym: Acerdol.
Description: Violet, deliquescent crystals.
Formula: $Ca(MnO_4)_2 \cdot 5H_2O$.
Constants: Mol wt: 368.02, mp: decomposes, d: 2.4.
Hazard Analysis and Countermeasures
See calcium compounds and permanganates.
Shipping Regulations: Section 11.
  I.C.C.: Oxidizing material; yellow label, 100 pounds.
  Coast Guard Classification: Oxidizing material; yellow label.
  IATA: Oxidizing material, yellow label, 12 kilograms (passenger), 45 kilograms (cargo).

CALCIUM PEROXIDE
General Information
Synonym: Calcium superoxide.
Description: Yellow crystals, powder or white crystals.
Formula: CaO_2.
Constants: Mol wt: 72.08, mp: decomposes 275°C.
Hazard Analysis
Toxicology: Although this material is used in dentifrices, the concentrated form would be an irritant. It would react with moisture to form slaked lime. See also calcium compounds, calcium hydroxide and peroxides, inorganic.
Fire Hazard: Moderate, if hot and mixed with finely divided combustible material. Decomposes rapidly above 200°C. Like calcium oxide, it is a strong alkali. An oxidizer. See also peroxides, inorganic.
Explosion Hazard: An explosion hazard when intimately mixed with finely divided reducing agents such as organic matter (Section 7).
Disaster Hazard: See peroxides, inorganic.
Countermeasures
Storage and Handling: Section 7.
Shipping Regulations: Section 11.
  I.C.C.: Oxidizing material; yellow label, 100 pounds.
  Coast Guard Classification: Oxidizing material; yellow label.
  IATA: Oxidizing material, yellow label, 12 kilograms (passenger), 45 kilograms (cargo).

CALCIUM PHENATE. See calcium phenoxide.

CALCIUM-1-PHENOL-4-p-SULFONATE
General Information
Description: White to pinkish powder.
Formula: $Ca[C_6H_4(OH)SO_3]_2 \cdot H_2O$.
Constant: Mol wt: 404.43.
Hazard Analysis and Countermeasures
See phenol sulfonates.

CALCIUM PHENOXIDE
General Information
Synonym: Calcium phenate.
Description: Powder.
Formula: $Ca(OC_6H_5)_2$.
Constant: Mol wt: 226.28.
Hazard Analysis and Countermeasures
See phenol.

sec-CALCIUM PHOSPATE See calcium phosphate, dibasic.

CALCIUM PHOSPHATE, DIBASIC
General Information
Synonym: Dicalcium o-phosphate; bicalcium phosphate; sec-calcium phosphate.
Description: White crystalline powder; odorless; soluble in dilute hydrochloric, nitric, and acetic acids; insoluble in alcohol; slightly soluble in water.
Formula: (1)$CaHPO_4$. (2) $CaHPO_4 \cdot 2H_2O$.
Constant: Mol wt (1): 136, mol wt (2): 154.
Hazard Analysis
Toxic Hazard Rating: U. A nutrient and/or dietary supplement food additive. Also a general purpose food additive (Section 10).
Disaster Hazard: Dangerous. See phosphates.

CALCIUM PHOSPHATE, MONOBASIC
General Information
Synonym: Calcium biphosphate; acid calcium phosphate; calcium phosphate, primary; monocalcium.
Description: Colorless, pearly scales or powder; soluble in water and acids.
Formula: $CaH_4(PO_4)_2 \cdot H_2O$.
Constants: Mol wt: 252, mp: loses H_2O at 100°C, decomposes at 200°C, d: 2.20.
Hazard Analysis
Toxic Hazard Rating: U. A nutrient and/or dietary supplement food additive. Also a general purpose food additive and a sequestrant food additive (Section 10).
Disaster Hazard: Dangerous. See phosphates.

CALCIUM PHOSPHATE, PRIMARY. See caclium phosphate, monobasic.

CALCIUM PHOSPHATE, TRIBASIC
General Information
Synonyms: Bone ash; tricalcium phosphate.
Description: Amorphous, odorless powder.
Formula: $Ca_3(PO_4)_2$.
Constants: Mol wt: 310.2, mp: 1670°C, d: 3.14.
Hazard Analysis and Countermeasures
A nutrient and/or dietary supplement food additive. Also a general purpose food additive (Section 10). See calcium compounds and phosphates.

CALCIUM-m-PHOSPHATE
Hazard Analysis
Toxic Hazard Rating: U. A sequestrant food additive (Section 10).
Disaster Hazard: Dangerous. See phosphates.

CALCIUM PHOSPHIDE
General Information
Description: Red crystals.
Formula: Ca_3P_2.
Constants: Mol wt: 182.2, mp: >1600°C, d: 2.238 at 25°C.
Hazard Analysis and Countermeasures
See phosphides.
Shipping Regulations: Section 11.
I.C.C.: Flammable solid; yellow label, 125 pounds.
IATA: Flammable solid, yellow label, not acceptable (passenger), 12 kilograms (cargo).
Coast Guard Classification: Inflammable solid; yellow label.

CALCIUM PHYTATE
General Information
Synonym: Hexacalcium phytate.
Description: Free flowing white powder; slightly soluble in water.

Formula: $C_6H_6(CaPO_4)_6$.
Constant: Mol wt: 788.
Hazard Analysis
Toxic Hazard Rating: U. A sequestrant food additive (Section 10).
Disaster Hazard: Dangerous. See phosphates.

CALCIUM o-PLUMBATE
General Information
Description: Reddish-brown crystals.
Formula: Ca_2PbO_4.
Constants: Mol wt: 351.37, mp: decomposes, d: 5.71.
Hazard Analysis and Countermeasures
See lead compounds.

CALCIUM POLYSULFIDES. See calcium sulfide.

CALCIUM PROPIONATE
General Information
Description: White powder; soluble in water; slightly soluble in alcohol.
Formula: $Ca(OOCCH_2CH_3)_2$ (occurs also with 1 H_2O).
Constant: Mol wt: 188.
Hazard Analysis
Toxic Hazard Rating: Used as a chemical preservative food additive. (Section 10).

CALCIUM PYROPHOSPHATE
General Information
Description: White powder; soluble in dilute hydrochloric acid and nitric acid; insoluble in water.
Formula: $Ca_2P_2O_7$.
Constants: Mol wt: 254, d: 3.09, mp: 1230°C.
Hazard Analysis
Toxic Hazard Rating: U. A nutrient and/or dietary supplement food additive.(Section 10).
Disaster Hazard: Dangerous. See phosphates.

CALCIUM RESINATE
General Information
Description: Yellowish white amorphous powder or lumps.
Formula: $Ca(C_{44}H_{62}O_4)_2$.
Constant: Mol wt: 1349.50.
Hazard Analysis
Toxicity: Unknown.
Fire Hazard: Slight; when heated, it can react with oxidizing materials (Section 6).
Countermeasures
Storage and Handling: Section 7.
Shipping Regulations: Section 11.
I.C.C.: Flammable solid; yellow label, 125 pounds.
Coast Guard Classification: Inflammable solid; yellow label.
IATA: Flammable solid, yellow label, not acceptable (passenger), 60 kilograms (cargo).

CALCIUM RESINATE, FUSED. See calcium resinate.
Shipping Regulations: Section 11.
I.C.C.: Flammable solid; yellow lable, 125 pounds.
Coast Guard Classification: Inflammable solid; yellow label.
IATA: Flammable solid, yellow label, not acceptable (passenger), 60 kilograms (cargo).

CALCIUM RICINOLEATE
General Information
Description: Fine, white powder.
Formula: $Ca[CO_2(CH_2)_7CHCHCH_2CHOH(CH_2)_5CH_3]_2$.
Constants: Mol wt: 630, mp: 84°C, d: 1.09 at 25°C.

TOXIC HAZARD RATING CODE (For detailed discussion, see Section 1.)

0 NONE: (a) No harm under any conditions; (b) Harmful only under unusual conditions or overwhelming dosage.

1 SLIGHT: Causes readily reversible changes which disappear after end of exposure.

2 MODERATE: May involve both irreversible and reversible changes; not severe enough to cause death or permanent injury.

3 HIGH: May cause death or permanent injury after very short exposure to small quantities.

U UNKNOWN: No information on humans considered valid by authors.

Hazard Analysis and Countermeasures
See calcium compounds.

CALCIUM SACCHARIN
General Information
Description: White crystalline powder, odorless or faint aromatic odor; soluble in water.
Formula: $Ca(C_6H_4COSO_2N)_2$.
Constant: Mol wt: 464.
Hazard Analysis
Toxic Hazard Rating: U. A non-nutritive sweetener food additive (Section 10).
Disaster Hazard: Dangerous. See sulfates.

CALCIUM SELENATE
General Information
Description: Colorless crystals.
Formula: $CaSeO_4$.
Constants: Mol wt: 183.04, d: 2.93.
Hazard Analysis and Countermeasures
See selenium compounds.

CALCIUM SELENIDE
General Information
Description: Simple cubic crystals.
Formula: CaSe.
Constants: Mol wt: 119.04, d: 7.593.
Hazard Analysis
Toxicity: See selenium compounds.
Fire Hazard: Moderate, when heated. See hydrogen selenide, which can be evolved on contact with acids or moisture (Section 6).
Explosion Hazard: Moderate, when exposed to flame (Section 7).
Disaster Hazard: Dangerous; it can emit highly toxic and flammable fumes of hydrogen selenide of selenium compounds on contact with water, steam, acid or acid fumes.
Countermeasures
Storage and Handling: Section 7.

CALCIUM SILICATE
General Information
Description: White to cream colored, free flowing powder.
Formula: Ca_2SiO_4, $Ca_3Si_2O_7$, $Ca_3(Si_3O_9)$, $Ca_4(H_2Si_4O_{13})$.
Constant: D: 2.10 at $25°/4°C$.
Hazard Analysis
Toxic Hazard Rating:
  Acute Local: Inhalation 2; Irritant 1.
  Acute Systemic: Ingestion 1.
  Chronic Local: Inhalation 1; Irritant 1.
  Chronic Systemic: U.
An anti-caking agent food additive. Also a food additive permitted in feed and drinking water of animals and/or for the treatment of food producing animals. (Section 10).

CALCIUM SILICIDE
General Information
Description: Glassy solid.
Formula: $CaSi_2$.
Constants: Mol wt: 96.2, d: 2.5.
Hazard Analysis and Countermeasures
See calcium hydroxide and silanes.
Shipping Regulations: Section 11.
  IATA: Flammable solid, yellow label, 12 kilograms (passenger), 45 kilograms (cargo).

CALCIUM SILICOFLUORIDE. See fluosilicates.

CALCIUM SILICON
Shipping Regulations: Section 11.
  IATA: Flammable solid, yellow label, 12 kilograms (passenger), 45 kilograms (cargo).

CALCIUM SORBATE
General Information
Formula: $Ca(OOC_5H_7)_2$.
Constant: Mol wt: 262.
Hazard Analysis
Toxic Hazard Rating: U. Used as a chemical preservative food additive (Section 10).

CALCIUM STEARYL-2-LACTYLATE
General Information
Synonym: Verv-Ca.
Description: Free flowing white powder, sparingly soluble in water.
Formula: $(C_{24}H_{43}O_6)_2Ca$.
Constant: Mol wt: 895.
Hazard Analysis
Toxicity: Practically non-toxic. A food additive.

CALCIUM SULFATE
General Information
Description (pure anhydrous): White powder or crystals; odorless.
Formula: $CaSO_4$.
Constants: Mol wt: 136, d: 2.964, mp: $1450°C$.
Hazard Analysis
Toxic Hazard Rating: U. A nutrient and/or dietary supplement food additive (Section 10).
Disaster Hazard: Dangerous. See sulfates.

CALCIUM SULFHYDRATE.
General Information
Synonym: Calcium hydrosulfide.
Description: Colorless, transparent crystals; decomposes in air.
Formula: $Ca(HS)_2$.
Constant: Mol wt: 106.21.
Hazard Analysis and Countermeasures
See sulfides.

CALCIUM SULFIDE
General Information
Synonyms: Oldhamite; hepar calcis; calcic liver of sulfur.
Description: Cubic, colorless crystals.
Formula: CaS.
Constants: Mol wt: 72.14, bp: decomposes, d: 2.18 at $15°C$.
Hazard Analysis and Countermeasures
See sulfides.

CALCIUM SULFITE
General Information
Description: Hexagonal, colorless crystals.
Formula: $CaSO_3 \cdot 2H_2O$.
Constants: Mol wt: 156.18, mp: $-2H_2O$ at $100°C$.
Hazard Analysis and Countermeasures
See sulfites.

CALCIUM SUPEROXIDE. See calcium peroxide.

CALCIUM SUPERPHOSPHATE
General Information
Synonym: Calcium tri-o-phosphate (fertilizer grade).
Description: Amorphous, white powder.
Formula: $Ca_3(PO_4)_2$.
Constants: Mol wt: 310.28, mp: $1670°C$, d: 3.14.
Hazard Analysis and Countermeasures
See calcium compounds and phosphates.

CALCIUM TELLURIDE
General Information
Description: Simple cubic crystals.
Formula: CaTe.
Constants: Mol wt: 167.69, d: 7.593.
Hazard Analysis
Toxicity: See tellurium compounds.
Fire Hazard: Moderate, when heated. See hydrogen telluride, which can be evolved on contact with acids or moisture (Section 6).

Explosion Hazard: Slight, when exposed to flame.

Disaster Hazard: Dangerous; it emits highly toxic and flammable fumes of hydrogen telluride and tellurium compounds on contact with water, steam, acid or acid fumes.

Countermeasures

Storage and Handling: Section 7.

CALCIUM TELLURITE

General Information

Description: White flakes.

Formula: $CaTeO_3$.

Constants: Mol wt: 215.69, mp: > 960° C.

Hazard Analysis and Countermeasures

Toxicity: See tellurium compounds.

CALCIUM TETRABORATE

General Information

Description: White solid.

Formula: CaB_4O_7.

Constants: Mol wt: 195.36, mp: 986° C.

Hazard Analysis and Countermeasures

See boron compounds.

CALCIUM THIOCYANATE

General Information

Description: White crystals, deliquescent.

Formula: $Ca(SCN)_2 \cdot 3H_2O$.

Constant: Mol wt: 210.30.

Hazard Analysis and Countermeasures

See thiocyanates.

Shipping Regulations: Section 11.

IATA: Poison B, poison label, 1 liter (passenger), 220 liters (cargo).

CALCIUM TITANATE

General Information

Description: Powder.

Formula: $CaTiO_3$.

Constants: Mol wt: 136, d: 3.98, mp: 1800° C.

Hazard Analysis

Toxic Hazard Rating: U. Animal experiments show low toxicity. See also titanium compounds.

CALCIUM TRI-o-PHOSPHATE. See calcium super-phosphate.

CALCIUM ZINC ARSENATE MONOHYDRATE

General Information

Synonym: Austinite.

Description: A solid.

Formula: $2CaO \cdot 2ZnO \cdot As_2O_5 \cdot H_2O$.

Constant: Mol wt: 523.

Hazard Analysis

Toxicity: Highly toxic. See arsenic compounds.

Disaster Hazard: Dangerous; see arsenic.

CALCO URANITE. See autunite.

CALGON. See sodium hexa-m-phosphate.

CALIFORNIUM

General Information

Description: A synthetic radioactive element. At number 98. The chemical properties of Californium have been studied by tracer techniques and are similar to those of the other transuranium elements.

Hazard Analysis

Radiation Hazard: Section 5. For permissible levels, see Table 5, p. 150.

Artificial isotope ^{249}Cf, half life about 400 y. Decays to radioactive ^{245}Cm by emitting alpha particles of 5.8 MeV. Also emits gamma rays of 0.06–0.4 MeV.

Artificial isotope ^{250}Cf, half life about 10 y. Decays to radioactive ^{246}Cm by emitting alpha particles of 6.0 MeV.

Artificial isotope ^{251}Cf, half life about 800 y. Decays to radioactive ^{247}Cm by emitting alpha particles of uncertain energy.

Artificial isotope ^{252}Cf, half life about 2.5 y. Decays to radioactive ^{248}Cm by emitting alpha particles of 6.1 MeV.

Artificial isotope ^{253}Cf, half life about 18 d. Decays to radioactive ^{253}Es by emitting beta particles of 0.27 MeV.

Artificial isotope ^{254}Cf, half life 56 d. Decays by spontaneous fission.

CALOMEL. See mercurous chloride.

CALX. See calcium oxide.

CAMBOGIA. See gamboge.

α-CAMPHANOL. See d-borneol.

CAMPHENE

General Information

Synonym: 3,3-Dimethyl-2-methylene norcamphone.

Description: Cubic crystals.

Formula: $C_{10}H_{16}$.

Constants: Mol wt: 136.2, mp: 50–51° C, bp: 159° C, d: 0.842 at 54° /4° C.

Hazard Analysis

Toxicity: Unknown.

Fire Hazard: Slight, yields flammable vapors when heated; it can react with oxidizing materials (Section 6).

Countermeasures

Storage and Handling: Section 7.

Shipping Regulations: Section 11.

Coast Guard Classification: Hazardous article.

2-CAMPHONONE. See camphor.

CAMPHOR

General Information

Synonyms: 2-Camphonone; gum camphor; laurel camphor.

Description: White, transparent, crystalline masses, pentrating odor, pungent, aromatic taste.

Formula: $C_{10}H_{16}O$.

Constants: Mol wt: 152.23, mp: 174–177° C, bp: 204° C, lel: 0.6%, uel: 3.5%, flash p: 150° F (C.C.), d: 0.992 at 25° /4° C, autoign. temp.: 871° F, vap. d.: 5.24.

Hazard Analysis

Toxic Hazard Rating:

Acute Local: Irritant 2; Allergen 1; Ingestion 2.

Acute Systemic: Ingestion 2.

Chronic Local: Allergen 1.

Chronic Systemic: U.

Toxicology: Locally, camphor is an irritant. When swallowed it causes nausea, vomiting, dizziness, excitation and convulsions.

TLV: ACGIH (tentative) 5 mg/m³ of air.

Fire Hazard: Moderate, when exposed to heat or flame; can react with oxidizing materials.

Spontaneous Heating: No.

Explosion Hazard: Moderate, in the form of vapor when exposed to heat or flame (Section 7).

TOXIC HAZARD RATING CODE (For detailed discussion, see Section 1.)

0 NONE: (a) No harm under any conditions; (b) Harmful only under unusual conditions or overwhelming dosage.

1 SLIGHT: Causes readily reversible changes which disappear after end of exposure.

2 MODERATE: May involve both irreversible and reversible changes; not severe enough to cause death or permanent injury.

3 HIGH: May cause death or permanent injury after very short exposure to small quantities.

U UNKNOWN: No information on humans considered valid by authors.

Countermeasures

Personal Hygiene: Section 3.

Treatment and Antidotes: Evacuate stomach with emetic or stomach tube. Give sedatives if necessary. Call a physician.

Storage and Handling: Section 7.

To Fight Fire: Foam, carbon dioxide, dry chemical or carbon tetrachloride (Section 6).

Shipping Regulations: Section 11.
    Coast Guard Classification:
        Hazardous article.

CAMPHORIC ANHYDRIDE
General Information

Description: Acicular, odorless, crystals; almost insoluble in water; freely soluble in alcohol, benzene, ether, carbon disulfide, chloroform, ethyl acetate.

Formula: $C_{10}H_{14}O_3$.

Constants: Mol wt: 182.21, d: 1.19, mp: 216–217°C, bp: 270°C.

Hazard Analysis

See camphor.

CAMPHOR OIL (LIGHT)
General Information

Synonym: Liquid camphor.

Description: An oily, fragrant liquid.

Constants: Bp: 175–200°C, flash p: 117°F (C.C.), d: 0.88.

Hazard Analysis

Toxicity: See camphor.

Fire Hazard: Moderate, when exposed to heat or flame; it can react with oxidizing materials.

Spontaneous Heating: No.

Countermeasures

To Fight Fire: Foam, carbon dioxide, dry chemical or carbon tetrachloride (Section 6).

Storage and Handling: Section 7.

Shipping Regulations: Section 11.
    Coast Guard Classification: Combustible liquid.

CANADA BALSAM
General Information

Synonyms: Canada turpentine; balm of Gilead.

Description: Yellowish to greenish viscid, transparent, fluorescent, aromatic liquid.

Constant: D: 0.987–0.994.

Hazard Analysis

Toxic Hazard Rating:

Acute Local: Irritant 1; Allergen 1; Ingestion 1.

Acute Systemic: U.

Chronic Local: Allergen 1.

Chronic Systemic: U.

Countermeasures

Personal Hygiene: Section 3.

CANADA TURPENTINE. See canada balsam.

CANNABIS
General Information

Synonyms: Cannabin resin; Indian hemp; Indian cannabis; hashish; guaza; marihuana.

Description: A resinous, bitter substance from Cannabis Sativa, greenish black mass.

Hazard Analysis

Toxic Hazard Rating:

Acute Local: Allergen 1.

Acute Systemic: Inhalation 2.

Chronic Local: Allergen 1.

Chronic Systemic: Inhalation 2.

Toxicology: Cannabis is a narcotic analgesic and sedative, it causes the pupils to be widely dilated and to barely react to light. When ingested or inhaled as smoke can cause euphoria, delirium, hallucinations, drowsiness, weakness and hyporeflexia. An overdose can cause coma and death. Like the other narcotics, this material can be habit-forming. It can be grown in the temperate zones and acquired illegally for narcotic purposes.

Fire Hazard: Slight, because the dried material can burn; it can react with oxidizing materials (Section 6).

Countermeasures

Treatment and Antidotes: Removal from exposure or cessation of ingestion of this material will generally cause the symptoms to disappear. In case of a large dose, or if overdose is suspected, a physician should be consulted at once.

Ventilation Control: Section 2.

Storage and Handling: Section 7.

CANNABIS RESIN. See cannabis.

CANNON PRIMERS. See explosives, low and high.

Shipping Regulations: Section 11.

I.C.C.: Explosive C, 150 pounds.

Coast Guard Classification: Explosive C.

IATA: Explosive, explosive label, 25 kilograms (passenger), 70 kilograms (cargo).

CANTHARIDES
General Information

Synonyms: Blistering flies; blistering beetles; Spanish fly.

Description: Brown to black powder and scales.

Formula: $C_{10}H_{12}O_4$.

Constants: Mol wt: 196.15, mp: 218°C, bp: sublimes at 90°C.

Hazard Aanalysis

Toxic Hazard Rating:

Acute Local: Irritant 3; Allergen 1; Ingestion 3; Inhalation 3.

Acute Systemic: Ingestion 3.

Chronic Local: Irritant 3; Allergen 1.

Chronic Systemic: Ingestion 3.

Toxicology: Extremely irritating to the eyes. Can cause conjuctivitis, keratitis, blepharitis, slight swelling of cornea and inflammation of iris. It is a powerful irritant. Used externally chiefly as a blistering agent. It is often and mistakenly used as an aphrodisiac, but is much too dangerous and irritant a material for this purpose. Symptoms of intoxication are extreme irritation of intestine and kidneys.

Disaster Hazard: Slight; on decomposition it emits toxic fumes.

Countermeasures

Ventilation Control: Section 2.

Treatment and Antidotes: If it has been ingested, a stomach syphon is recommended, or emetics, demulcents but not oils. Administer morphine, stimulants or poultices to the abdomen. Call a physician.

Personnel Protection: Section 3.

Personal Hygiene: Section 3.

First Aid: Section 1.

Storage and Handling: Section 7.

CAOUTCHOUC. See rubber, crude.

CAPRIC ETHER. See ethyl caprate.

CAPRINIC ETHER. See ethyl caprate.

CAPROIC ACID
General Information

Synonym: Hexanoic acid.

Description: Colorless liquid, odor of limburger cheese.

Formula: $C_6H_{12}O_2$.

Constants: Mol wt: 116.2, bp: 205.0°C, fp: −5.4°C, flash p: 215°F (C.O.C.), d: 0.9295 at 20°/20°C, vap. press.: 0.18 mm at 20°C, vap. d.: 4.0.

Hazard Analysis

Toxic Hazard Rating:

Acute Local: Irritant 3.

Acute Systemic: Ingestion 1; Inhalation 1.

Chronic Local: U.

Chronic Systemic: U.
Data based on animal experiments.
Fire Hazard: Moderate, when exposed to heat or flame; it
can react with oxidizing materials.
Countermeasures
To Fight Fire: Foam, carbon dioxide, dry chemical, or
carbon tetrachloride (Section 6).
Ventilation Control: Section 2.
Personal Hygiene: Section 3.
Storage and Handling: Section 7.

ε-CAPROLACTAM
General Information
Synonym: 2-Oxohexamethylenimine.
Description: White crystals.
Formula: $C_6H_{11}NO$.
Constants: Mol wt: 113.16, mp: 69°C, vap. press.: 6 mm
at 120°C.
Hazard Analysis
Toxicity: May be an irritant.
Disaster Hazard: Moderately dangerous; when heated to
decomposition, it emits toxic fumes.
Countermeasures
Storage and Handling: Section 7.

CAPRYL ALCOHOL
General Information
Synonym: 2-Octanol.
Description: Clear, colorless, oily liquid, pungent, aromatic
odor.
Formula: $CH_3HCOH(CH_2)_5CH_3$.
Constants: Mol wt: 130.2, bp: 178.5°C, fp: −38.6°C,
d: 0.8193 at 25°/4°C, vap. d.: 4.48, flash p: 190°F.
Hazard Analysis
Toxicity: See alcohols.
Fire Hazard: Moderate, when exposed to heat or flame; it
can react with oxidizing materials.
Countermeasures
Storage and Handling: Section 7.

CAPRYLALDEHYDE. See octyl aldehyde.

1-CAPRYLENE. See 1-octene.

CAPRYLIC ACID
General Information
Synonym: Octanoic acid.
Description: Colorless, oily liquid; unpleasant odor; burn-
ing rancid taste.
Formula: $CH_3(CH_2)_6COOH$.
Constants: Mol wt: 144.2, d: 0.91 at 20°C, bp: 237°C,
mp: 16°C.
Hazard Analysis
Toxicology: Details unknown. Yields irritating vapors
which can cause coughing. Experimental data suggest
low toxicity.
Used as a chemical preservative food additive (Section 10).

CAPRYLIC ALDEHYDE. See caprylaldehyde.

CAPRYLYL CHLORIDE. See ethyl hexanoyl chloride.

CAPRYLYL PEROXIDE SOLUTION
Hazard Analysis
Toxicity: See peroxides, organic.
Fire Hazard: See peroxides, organic.
Disaster Hazard: See peroxides, organic.
Countermeasures
Storage and Handling: Section 7.

Shipping Regulations: Section 11.
I.C.C: Oxidizing material; yellow label, 1 quart.
Coast Guard Classification: Oxidizing material; yellow
label.
IATA: Oxidizing material, yellow label, 1 liter (passenger),
1 liter (cargo).

CAPRYLYL PEROXIDE, solution in mineral oil (50%).
General Information
Description: Colorless to light yellow liquid, insoluble in
water; very soluble in organic solvents.
Formula: $[CH_3(CH_2)_6CO]_2O_2$ + mineral oil.
Constants: D: 0.88, fp: −16°C.
Hazard Analysis
See peroxides, organic.
Countermeasures
See peroxides, organic.

CAPSICUM
General Information
Synonym: Tincture of tabasco pepper.
Description: An alcoholic solution of capsicum.
Hazard Analysis
Toxic Hazard Rating:
Acute Local: Irritant 2; Ingestion 3.
Acute Systemic: Ingestion 2.
Chronic Local: Irritant 2.
Chronic Systemic: U.
Fire Hazard: Moderate, when exposed to heat. See also
alcohol.
Disaster Hazard: See alcohol.
Countermeasures
Ventilation Control: Section 2.
Personnel Protection: Section 3.
First Aid: Section 1.
Storage and Handling: Section 7.

CAPTAN. See N-trichloromethylthiotetrahydrophthali-
mide.

CARAMEL
General Information
Synonym: Sugar coloring; burnt sugar.
Description: Deliquescent, dark brown powder or thick
liquid; burnt sugar odor; soluble in water and dilute
alcohol; immiscible with most organic solvents.
Constant: D: ≈1.35.
Hazard Analysis
Toxic Hazard Rating: U. A general purpose food additive
(Section 10).

CARBAMIC ACID BUTYLETHYL ESTER. See butyl-
urethane.

CARBAMIDE
General Information
Synonym: Urea crystal.
Description: White crystals.
Formula: $(NH_2)_2CO$.
Constants: Mol wt: 60.1, mp: 132.7°C, bp: decomposes,
d: (solid) 1.335.
Hazard Analysis
See urea and amides.
Countermeasures
Ventilation Control: Section 2.
Storage and Handling: Section 7.

TOXIC HAZARD RATING CODE (For detailed discussion, see Section 1.)

0 NONE: (a) No harm under any conditions; (b) Harmful only under unusual conditions or overwhelming dosage.

1 SLIGHT: Causes readily reversible changes which disappear after end
of exposure.

2 MODERATE: May involve both irreversible and reversible changes;
not severe enough to cause death or permanent injury.

3 HIGH: May cause death or permanent injury after very short exposure
to small quantities.

U UNKNOWN: No information on humans considered valid by authors.

CARBAMIDE PHOSPHORIC ACID
General Information
Description: White crystals.
Formula: $CO(NH_2)_2 \cdot H_3PO_4$.
Constant: Mol wt: 158.
Hazard Analysis
Toxicity: See phosphoric acid, urea and amides.
Disaster Hazard: Dangerous; when heated to decomposition, it emits highly toxic fumes of cyanides or oxides of phosphorus.
Countermeasures
Storage and Handling: Section 7.

CARBAMYL CHLORIDE
General Information
Synonym: Chloroformamide.
Description: Liquid; offensive odor.
Formula: H_2NCOCl.
Constants: Mol wt: 79.5, mp: 50°C (approx.), bp: 61-62°C (decomposes).
Hazard Analysis
Toxicity: See hydrochloric acid which is evolved upon standing.
Disaster Hazard: Dangerous; see chlorides.
Countermeasures
Storage and Handling: Section 7.

CARBARYL
General Information
Synonym: 1-Naphthyl-N-methyl carbamate; sevin.
Description: Solid.
Formula: $C_{10}H_7OOCNHCH_3$.
Constants: Mol wt: 201, mp: 142°C.
Hazard Analysis
Toxic Hazard Rating:
Acute Local: Irritant 2.
Acute Systemic: Ingestion 2; Inhalation 2; Skin Absorption 2.
Chronic Local: Irritant 1.
Chronic Systemic: U.
A reversible cholinesterase inhibitor. Symptoms similar to those due to parathion, but much less severe from equal doses
TLV: ACGIH (recommended): 5 milligrams per cubic meter of air.

CARBAZOLE
General Information
Synonym: Dibenzopyrrole.
Description: White crystals.
Formula: $(C_6H_4)_2NH$.
Constants: Mol wt: 167.20, mp: 244.8°C, bp: 354.8°C, d: 1.10 at 18°/4°C, vap. press.: 400 mm at 323.0°C.
Hazard Analysis
Toxic Hazard Rating:
Acute Local: Allergen 1.
Acute Systemic: U.
Chronic Local: Allergen 1.
Chronic Systemic: U.
Fire Hazard: Slight (Section 6).
Countermeasures
Personal Hygiene: Section 3.
Storage and Handling: Section 7.

CARBAZOTIC ACID. See picric acid.

CARBIDES. See acetylides.

"CARBITOL"
General Information
Synonym: Diethylene glycol monoethyl ether.
Description: Colorless liquid; mild pleasant odor.
Formula: $C_2H_5OCH_2CH_2OCH_2CH_2OH$.
Constants: Mol wt: 134.17, bp: 201.9°C, flash p: 205°F (O.C.), d: 0.9902 at 20°/4°C, vap. d.: 4.62.

Hazard Analysis
Toxic Hazard Rating:
Acute Local: Irritant 1.
Acute Systemic: Ingestion 3.
Chronic Local: U.
Chronic Systemic: Ingestion 1; Inhalation 1; Skin Absorption 1.
Toxicity: See also glycols.
Fire Hazard: Moderate, when exposed to heat; can react with oxidizing materials.
Spontaneous Heating: No.
Countermeasures
To Fight Fire: Alcohol foam, carbon dioxide, dry chemical, or carbon tetrachloride (Section 6).
Ventilation Control: Section 2.
Personal Hygiene: Section 3.
Storage and Handling: Section 7.

"CARBITOL" ACETATE
General Information
Synonym: Diethylene glycol monoethyl ether acetate.
Description: Liquid.
Formula: $C_2H_5O(CH_2)_2O(CH_2)_2OOCH_3$.
Constants: Mol wt: 176.21, bp: 217.4°C, fp: −25°C, flash p: 225°F (C.C.), d: 1.0114 at 20°/20°C, vap. press.; 0.05 mm at 20°C, vap. d.: 6.07.
Hazard Analysis
Toxicity: See glycols.
Fire Hazard; moderate, when exposed to heat; it can react with oxidizing materials.
Spontaneous Heating: No.
Countermeasures
To Fight Fire: Alcohol foam, water, carbon dioxide, dry chemical or carbon tetrachloride (Section 6).
Storage and Handling: Section 7.

"CARBITOL" PHTAHALATE
General Information
Synonym: Dicarbitol phthalate.
Description: Liquid.
Formula: $C_6H_1(COOC_2H_4OC_2H_4OC_2H_5)_2$.
Constants: Mol wt: 398.44, bp: 200-260°C, flash p: 405°F (C.C.), d: 1.121.
Hazard Analysis
Toxicity: See glycols.
Fire Hazard: Slight, when exposed to heat; can react with oxidizing materials.
Spontaneous Heating: No.
Countermeasures
To Fight Fire: Water, foam, carbon dioxide, dry chemical, or carbon tetrachloride (Section 6).

CARBODIMIDE. See cyanimide.

CARBOLIC ACID, LIQUID (liquid tar acid containing over 50% benzophenol). See phenol.
Shipping Regulations: Section 11.
I.C.C.: Poison B; poison label, 55 gallons.
Coast Guard Classification: Poison B; poison label.
IATA: Poison B, poison label, 1 liter (passenger), 220 liters (cargo).
MCA warning label.

CARBOLIC ACID, SOLID. See phenol.
Shipping Regulations: Section 11.
I.C.C. Poison B; poison label, 250 pounds.
Coast Guard Classification: Poison B; poison label.
IATA: Poison B, poison label, 25 kilograms (passenger), 115 kilograms (cargo).

CARBOLINEUM
General Information
Synonym: Chlorinated anthracene oil.
Hazard Analysis
Toxic Hazard Rating:
Acute Local: Irritant 2.

Actue Systemic: U.
Chronic Local: Irritant 2.
Chronic Systemic: U.
Toxicology: Action resembles that of phenol. May contain carcinogenic impurities.
Disaster Hazard: See chlorinated hydrocarbons, aromatic.

CARBOMETHANE. See ketene.

2-CARBOMETHOXY-1-METHYL VINYL DIMETHYL PHOSPHATE. See phosdrin.

CARBOMETHOXY PHENYL TRICHLORO-STANNANE
General Information
Description: A solid.
Formula: $(CH_3OCOC_6H_4)SnCl_3$.
Constants: Mol wt: 360, mp: 164° C.
Hazard Analysis
Toxicity: Details unknown. See tin compounds.
Disaster Hazard: Dangerous; see chlorides.

CARBON
General Information
Description: Black crystals, powder or diamond form.
Formula: C.
Constants: At wt: 12.01, mp: 3652–3697°C (sublimes), bp: approx. 4200°C, density amorphous 1.8-2.1; graphite 2.25; diamond 3.51, vap. press.: 1 mm at 3586°C.
Hazard Analysis
Toxic Hazard Rating:
Acute Local: Inhalation 1.
Acute Systemic: 0.
Chronic Local: Irritant 1.
Chronic Systemic: 0.
Toxicology: In the form of graphite (which is one of the common forms of carbon) it can cause a dust irritation, particularly to the eyes. Carbon in the form of soot can cause conjunctivitis. It can also cause epithelial hyperplasia of cornea, as well as eczematous inflammation of eyelids. Some forms of carbon dust can cause irritation of eyes and mucous membranes.
Radiation Hazard: Section 5. For permissible levels, see Table 5, p. 150.
    Artificial and natural isotope ^{14}C, half life 5570 y. Decays to stable ^{14}N by emitting beta particles of 0.155 MeV. ^{14}C occurs in nature as a result of cosmic ray bombardment of ^{14}N. The natural activity in living things or atmospheric CO_2 is 12.5 dpm per gram of carbon.
It is a common air contaminant. Section 4.
Fire Hazard: Slight, when exposed to heat (Section 6).
Explosion Hazard: Slight, in the form of dust when exposed to heat or flame (Section 7).
Countermeasures
Shipping Regulations: Section 11.
Coast Guard Classification: Oxidizing materials.

CARBON, ACTIVATED. See carbon.
Shipping Regulations: Section 11.
Coast Guard Classification: Oxidizing materials.

CARBON BISULFIDE. See carbon disulfide.

CARBON DIBROMIDE. See dicarbon tetrabromide.

CARBON DICHLORIDE. See perchloroethylene.

CARBON DIOXIDE
General Information
Synonyms: Carbonic acid; carbonic anhydride.
Description: Colorless, odorless gas.
Formula: CO_2.
Constants: Mol wt: 44.01, mp: −57.7° (sublimes), bp: −78.2° C, vap. d.:'1.53.
Hazard Analysis
Toxic Hazard Rating:
Acute Local: 0.
Acute Systemic: Inhalation 1.
Chronic Local: 0.
Chronic Systemic: Inhalation 1.
A general purpose food additive (Section 10).
TLV: ACGIH (accepted); 5000 parts per million in air; 9000 milligrams per cubic meter of air.
Toxicology: Carbon dioxide is generally regarded as a simple asphyxiant, symptoms resulting only when such high concentrations are reached that there is insufficient oxygen in the atmosphere to support life. The signs and symptoms are those which precede asphyxia, namely, headache, dizziness, shortness of breath, muscular weakness, drowsiness, and ringing in the ears. Removal from exposure results in rapid recovery. Contact of carbon dioxide snow with the skin may cause a "burn." See also discussion of simple asphyxiants under Argon.
Disaster Hazard: Slightly dangerous.
Countermeasures
Ventilation Control: Section 2.
Shipping Regulations; section 11.
I.C.C.: Nonflammable gas; green label, 300 pounds.
Coast Guard Classification: Nonflammable gas; green gas label.
IATA: Nonflammable gas, green label, 70 kilograms (passenger), 140 kilograms (cargo).

CARBON DIOXIDE-NITROUS OXIDE MIXTURE
General Information
Description: Gas.
Composition: CO_2 + N_2O.
Hazard Analysis
Toxic Hazard Rating:
Acute Local: 0.
Acute Systemic: Inhalation 2.
Chronic Local: 0.
Chronic Systemic: Inhalation 1.
Countermeasures
Ventilation Control: Section 2.
Shipping Regulations: Section 11.
I.C.C.: Nonflammable gas; green label, 300 pounds.
Coast Guard Classification: Nonflammable gas; green gas label.
IATA: Nonflammable gas, green label, 70 kilograms (passenger), 140 kilograms (cargo).

CARBON DIOXIDE-OXYGEN MIXTURE
General Information
Description: Gas.
Composition: CO_2 + O_2.
Hazard Analysis
Toxicity: None.
Fire Hazard: Slight (Section 6).
Caution: An oxidizing mixture.
Disaster Hazard: Moderately dangerous; it can react with reducing materials.
Countermeasures
Storage and Handling: Section 7.

TOXIC HAZARD RATING CODE *(For detailed discussion, see Section 1.)*

0 NONE: (a) No harm under any conditions; (b) Harmful only under unusual conditions or overwhelming dosage.

1 SLIGHT: Causes readily reversible changes which disappear after end of exposure.

2 MODERATE: May involve both irreversible and reversible changes; not severe enough to cause death or permanent injury.

3 HIGH: May cause death or permanent injury after very short exposure to small quantities.

U UNKNOWN: No information on humans considered valid by authors.

Shipping Regulations: Section 11.
  I.C.C.: Nonflammable gas; green label, 300 pounds.
  Coast Guard Classification: Nonflammable gas; green gas label.
  IATA: Nonflammable gas, green label, 70 kilograms (passenger), 140 kilograms (cargo).

CARBON DIOXIDE, SOLID
General Information
Synonym: Dry Ice.
Formula: CO_2.
Description: White, snow-like solid.
Constants: Mol wt: 44, mp: $-56.6°C$ at 5.2 atm (sublimes), bp: $-78.5°C$, d: 1.56 at 70°C.
Hazard Analysis
Toxic Hazard Rating:
  Acute Local: Irritant 3.
  Acute Systemic: 0.
  Chronic Local: 0.
  Chronic Systemic: 0.
Countermeasures
Ventilation Control: Section 2.
Shipping Regulations: Section 11.
  Coast Guard Classification: Hazardous article. MCA warning label.
  IATA: Other restricted articles, class A, no label required, no limit (passenger and cargo).

CARBON DISELENIDE
General Information
Description: Greenish-yellow liquid.
Formula: CSe_2.
Constants: Mol wt: 169.9, bp: 90°C (approx.).
Hazard Analysis
Toxicity: Probably high toxic. See selenium compounds and carbon disulfide.

CARBON DISULFIDE
General Information
Synonym: Carbon bisulfide.
Description: Clear, colorless liquid, nearly odorless when pure.
Formula: CS_2.
Constants: Mol wt: 76.1, mp: $-110.8°C$, bp: 46.5°C, lel. 1.3%, uel. 44%, flash p: $-22°F$, (C.C.), d: 1.261 at 20/20°C, autoign. temp.: 212°F, vap. press.: 400 mm at 28°C, vap. d.: 2.64.
Hazard Analysis
Toxic Hazard Rating:
  Acute Local: Irritant 1; Ingestion 1; Inhalation 1.
  Acute Systemic: Ingestion 3; Inhalation 3; Skin Absorption 3.
  Chronic Local: U.
  Chronic Systemic: Ingestion 3; Inhalation 3; Skin Absorption 3.
TLV: ACGIH (accepted); 20 parts per million in air; 60 milligrams per cubic meter of air. Can be absorbed via intact skin.
Toxicology: An insecticide. The chief toxic effect is on the central nervous system, acting as a narcotic and anesthetic in acute poisoning with death following from respiratory failure. The anesthetic action is much more powerful than that of chloroform. In chronic poisoning, the effect on the nervous system is one of central and peripheral damage, which may be permanent if the damage has been severe. Sensory symptoms usually precede motor involvement. A secondary anemia may be damaged.

  In acute poisoning, early excitation of the central nervous system resembling alcoholic intoxication occurs, followed by depression, with stupor, restlessness, unconsciousness, and possibly death. If recovery occurs, the patient usually passes through the after-stage of narcosis, with nausea, vomiting, headache, etc. In chronic poisoning, the picture is that of involvement

of the nervous system, with neuritis and disturbance of vision being the commonest early changes. Sensory changes, such as a crawling sensation in the skin, sensations of heaviness and coldness, and visually, "veiling" of objects so that they appear indistinct, are noticed first. Often there is pain in the affected parts, particularly the limbs. These symptoms are followed by gradually increasing loss of strength. Wasting of the muscles may occur. Mental symptoms vary from simple excitation or depression and irritability in the mild cases to mental deterioration, Parkinsonian paralysis, and even insanity. These changes are accompanied by insomnia, loss of memory, and personality changes. Chronic fatigue is a very common complaint.
Fire Hazard: Highly dangerous when exposed to heat; flame, sparks or friction.
Spontaneous Heating: No.
Explosion Hazard: Severe, when exposed to heat or flame (Section 7).
Disaster Hazard: Highly dangerous; when heated to decomposition, it emits highly toxic fumes of oxides of sulfur; it can react vigorously with oxidizing materials.
Countermeasures
Ventilation Control: Section 2.
To Fight Fire: Water, carbon dioxide, dry chemical or carbon tetrachloride (Section 6).
Personnel Protection: Section 3.
First Aid: Section 1.
Storage and Handling: Section 7.
Shipping Regulations: Section 11.
  I.C.C.: Flammable liquid; red lebel, not accepted.
  Coast Guard Classification: Inflammable liquid; red label.
  MCA warning label.
  IATA: Flammable liquid, not acceptable (passenger and cargo).

CARBON HEXACHLORIDE. See hexachloroethane.

CARBONIC ACID. See carbon dioxide.

CARBONIC ANHYDRIDE. See carbon dioxide.

CARBONIC ETHER. See diethyl carbonate.

CARBONITES. See explosives, high.

CARBON MONOSULFIDE
General Information
Description: Red powder.
Formula: CS, or (CS)x.
Constants: Mol wt: 44.08, mp: decomposes at 200°C, d: 1.66.
Hazard Analysis and Countermeasures
See sulfides.

CARBON MONOXIDE
General Information
Description: Colorless, odorless gas.
Formula: CO.
Constants: Mol wt: 28.01, mp: $-207°C$, bp: $-191.3°C$, lel: 12.5%, uel: 74.2%, d: (gas) 1.250 grams per liter at 0°C; (liquid) 0.793, autoign. temp.: 1128°F.
Hazard Analysis
Toxic Hazard Rating:
  Acute Local: 0.
  Acute Systemic: Inhalation 3.
  Chronic Local: 0.
  Chronic Systemic: Inhalation 1.
Toxicology: Carbon monoxide has an affinity for hemoglobin 210 times that of oxygen, and by combining with the hemoglobin, renders the latter incapable of carrying oxygen to the tissues. The effect on the body is therefore predominantly one of asphyxia. In addition to this action, the presence of CO-hemoglobin in the

blood interferes with the dissociation of the remaining oxyhemoglobin, so that the tissues are further deprived of oxygen.

A concentration of 400 to 500 ppm in the air can be inhaled without appreciable effect for 1 hour. An hour's exposure to 600 to 700 ppm will cause barely appreciable effects, and a similar exposure to 1,000 to 1,200 ppm is dangerous; concentrations of 4,000 ppm and over are fatal in less than an hour.

Carbon monoxide is eliminated through the lungs when air free from CO is inhaled. Over half the CO is eliminated in the first hour, when the exposure has been moderate.

With concentrations up to 10% of CO-hemoglobin in the blood, there rarely are any symptoms. Concentrations of 20 to 30% cause shortness of breath on moderate exertion and slight headache. Concentrations from 30 to 50% cause severe headache, mental confusion and dizziness, impairment of vision and hearing, and collapse and fainting on exertion. With concentrations of 50 to 60%, unconsciousness results, and death may follow if exposure is long. Concentrations of 80% result in almost immediate death.

Acute cases of poisoning, resulting from brief exposures to high concentrations, seldom result in any permanent disability if recovery takes place. Chronic effects as the result of repeated exposure to lower concentrations have been described, particularly in the Scandinavian literature. Auditory disturbances and contraction of the visual fields have been demonstrated. Glycosuria does occur, and heart irregularities have been reported. Other workers have found that where the poisoning has been relatively long and severe, cerebral congestion and edema may occur, resulting in long-lasting mental or nervous damage. Repeated exposure to low concentrations of the gas, up to 100 ppm in air, is generally believed to cause no signs of poisoning or permanent damage. Industrially, sequelae are rare, as exposure though often severe, is usually brief.

It is a common air contaminant (Section 4).

TLV: ACGIH (tentative) 50 parts per million; 55 milligrams per cubic meter of air.

Fire Hazard: Dangerous, when exposed to flame.

Spontaneous Heating: No.

Explosion Hazard: Severe, when exposed to heat or flame (Section 7).

Countermeasures

Ventilation Control: Section 2.

To Fight Fire: Stop flow of gas. Carbon dioxide, dry chemical or water spray (Section 6).

Storage and Handling: Section 7.

Shipping Regulations: Section 11.

I.C.C.: Flammable gas; red gas label, 150 pounds.

Coast Guard Classification: Flammable gas; red gas label.

IATA: Flammable gas, red label, not acceptable (passenger), 70 kilograms (cargo).

CARBON OXYBROMIDE. See carbonyl bromide.

CARBON OXYCHLORIDE. See phosgene.

CARBON OXYCYANIDE. See carbonyl cyanide.

CARBON OXYSULFIDE
General Information
Synonym: Carbonyl sulfide.
Description: Gas or liquid.
Formula: COS.

Constants: Mol wt: 60.07, mp: −138°C, bp: 49.9°C, lel: 12%, uel: 28.5%, d: liquid 1.24 at −87°C, vap. d.: 2.1.

Hazard Analysis
Toxic Hazard Rating:
Acute Local: Irritant 1; Ingestion 1; Inhalation 1.
Acute Systemic: Inhalation 3.
Chronic Local: Irritant 1, Inhalation 1.
Chronic Systemic: Inhalation 3.

Toxicology: Narcotic in high concentrations. May liberate highly toxic hydrogen sulfide upon decomposition.

Fire Hazard: Dangerous, when exposed to heat or flame.

Spontaneous Heating: No.

Explosion Hazard: Moderate, when exposed to heat or flame (Section 7). See also sulfides.

Disaster Hazard: Dangerous; see sulfides. It can react vigorously with oxidizing materials.

Countermeasures
Ventilation Control: Section 2.

To Fight Fire: Stop flow of gas. Carbon dioxide, dry chemical or water spray (Section 6). See also sulfides.

Storage and Handling: Section 7.

Shipping Regulations: Section 11.

IATA: Flammable gas, red label, not acceptable (passenger), 140 kilograms (cargo).

CARBON PAPER
Hazard Analysis
Toxic Hazard Rating:
Acute Local: Allergen 1.
Acute Systemic: 0.
Chronic Local: Allergen 1.
Chronic Systemic: 0.

CARBON REMOVER, LIQUID
General Information
Constant: Flash p: < 80°F.
Hazard Analysis
Toxicity: Unknown.
Fire Hazard: Dangerous, when exposed to heat or flame; it can react with oxidizing materials.

Countermeasures
To Fight Fire: Carbon dioxide, dry chemical or carbon tetrachloride (Section 6).

Storage and Handling: Section 7.

Shipping Regulations: Section 11.

I.C.C.: Flammable liquid; red label, 10 gallons.

Coast Guard Classification: Inflammable liquid; red label.

IATA: Flammable liquid, red label, 1 liter (passenger), 40 liters (cargo).

CARBON SELENOSULFIDE
General Information
Description: Yellow, oily liquid.
Formula: CSeS.
Constants: Mol wt: 123.04, mp: −75.2°C, bp: 85.6°C, d: 1.9874, vap. press.: 100 mm at 28.3°C.

Hazard Analysis and Countermeasures
See selenium compounds and sulfides.

CARBON SUBOXIDE
General Information
Description: Colorless gas or liquid.
Formula: C_3O_2.
Constants: Mol wt: 68.03, mp: −111.3°C, bp: 6.3°C, d: liq. 1.114 at 0°C.

Hazard Analysis
Toxic Hazard Rating:
Acute Local: Inhalation 3.

TOXIC HAZARD RATING CODE (For detailed discussion, see Section 1.)

0 NONE: (a) No harm under any conditions; (b) Harmful only under unusual conditions or overwhelming dosage.

1 SLIGHT: Causes readily reversible changes which disappear after end of exposure.

2 MODERATE: May involve both irreversible and reversible changes; not severe enough to cause death or permanent injury.

3 HIGH: May cause death or permanent injury after very short exposure to small quantities.

U UNKNOWN: No information on humans considered valid by authors.

Acute Systemic: U.
Chronic Local: U.
Chronic Systemic: U.

CARBON SUBSULFIDE
General Information
Description: Red liquid.
Formula: C_3S_2.
Constants: Mol wt: 100.16, mp: $0.4°C$, d: 1.274, vap. press.: 1 mm at $14.0°C$.
Hazard Analysis and Countermeasures
See carbon disulfide.

CARBON SULFIDE TELLURIDE
General Information
Description: Yellow-red crystals.
Formula: CSTe.
Constants: Mol wt: 171.69, mp: $-54°C$, bp: decomposes $> -54.0°C$, d: 2.9 at $50°C$.
Hazard Analysis and Countermeasures
See carbon disulfide. See also tellurium compounds and sulfides.

CARBON TETRABROMIDE
General Information
Synonym: Tetrabromomethane.
Description: Colorless monoclinic tablets.
Formula: CBr_4.
Constants: Mol wt: 331.67, mp: α $48.4°C$; β $90.1°C$, bp: $189.5°C$, d: 3.42, vap. press.: 40 mm at $96.3°C$.
Hazard Analysis
Toxic Hazard Rating:
Acute Local: Ingestion 2; Inhalation 2.
Acute Systemic: Inhalation 2.
Chronic Local: Irritant 1.
Chronic Systemic: U.
Toxicology: Narcotic in high concentrations.
Disaster Hazard: See chlorinated hydrocarbons, aliphatic.
Countermeasures
Storage and Handling: Section 7.

CARBON TETRACHLORIDE
General Information
Synonym: Tetrachloromethane.
Description: Colorless liquid; heavy, ethereal odor.
Formula: CCl_4.
Constants: Mol wt: 153.84, mp: $-22.6°C$, bp: $76.8°C$, fp: $-22.9°C$, flash p: none, d: 1.597 at $20°C$, vap. press.: 100 mm at $23.0°C$.
Hazard Analysis
Toxic Hazard Rating:
Acute Local: 0.
Acute Systemic: Ingestion 3; Inhalation 3; Skin Absorption 1.
Chronic Local: Irritant 1.
Chronic Systemic: Ingestion 3; Inhalation 3; Skin Absorption 3.
TLV: ACGIH (accepted); 10 parts per million in air; 65 milligrams per cubic meter of air. Can be absorbed via intact skin.
Toxicology: Carbon tetrachloride has a narcotic action resembling that of chloroform, though not as strong. Following exposures to high concentrations, the workman may become unconscious, and if exposure is not terminated, death can follow from respiratory failure. In cases of narcosis that recover, the aftereffects are more serious than those of delayed chloroform poisoning, usually taking the form of damage to the kidneys, liver and lungs. Exposure to lower concentrations, insufficient to produce unconsciousness, usually results in severe gastro-intestinal upset, and may progress to serious kidney and hepatic damage. The kidney lesion is an acute nephrosis; the liver involvement consists of an acute degeneration of the central portions of the lobules. Where recovery takes place, there may be no permanent disability. Marked variation in individual susceptibility to carbon tetrachloride exists, some persons appearing to be unaffected by exposure which seriously poison their fellow-workers. Alcoholism and previous liver and kidney damage seem to render the individual more susceptible. Concentrations of the order of 1,000 to 1,500 ppm are sufficient to cause symptoms if exposure continues for several hours. Repeated daily exposure to such concentrations may result in poisoning.

Though the common form of poisoning following industrial exposure is usually one of gastrointestinal upset, which may be followed by renal damage, other cases have been reported in which the central nervous system has been affected, with production of polyneuritis, narrowing of the visual fields, and other neurological changes. Prolonged exposure to small amounts of carbon tetrachloride has also been reported as causing cirrhosis of the liver.

Locally, a dermatitis may be produced following long or repeated contact with the liquid. The skin oils are removed, and the skin becomes red, cracked and dry (Section 9). The effect of carbon tetrachloride on the eyes either as a vapor or as a liquid, is one of irritation with lacrimation and burning.

Industrial poisoning is usually acute, with malaise, headache, nausea, dizziness, and confusion, which may be followed by stupor and sometimes loss of consciousness. Symptoms of liver and kidney damage may follow later, with development of dark urine, sometimes jaundice and liver enlargement, followed by scanty urine, albumenuria and renal casts, uremia may develop and cause death. Where the exposure has been less acute, the picture is usually one of headache, dizziness, nausea, vomiting, epigastric distress, loss of appetite, and fatigue. Visual disturbances (blind spots, spots before the eyes, a visual "haze" and restriction of the visual fields), secondary anemia, and occasionally a slight jaundice may occur. Dermatitis may be noticed on the exposed parts.
Disaster Hazard: Dangerous; when heated to decomposition, it emits highly toxic fumes of phosgene.
Countermeasures
Ventilation Control: Section 2.
Personal Hygiene: Section 3.
Storage and Handling: Section 7.
Shipping Regulations: Section 11.
MCA warning label.
IATA: Poison B, poison label, 1 liter (passenger), 220 liters (cargo).

CARBON TETRAFLUORIDE
General Information
Synonym: Tetrafluoromethane.
Description: Colorless gas.
Formula: CF_4.
Constants: Mol wt: 88.01, mp: $-184°C$, bp: $-127.7°C$, d: 1.96 at $-184°C$.
Hazard Analysis
Toxic Hazard Rating:
Acute Local: Inhalation 2.
Acute Systemic: Inhalation 2.
Chromic Local: U.
Chronic Systemic: U.
Toxicology: Appears to be less toxic than carbon tetrachloride. See halogenated hydrocarbons.
Disaster Hazard: See halogenated hydrocarbons, aliphatic.
Countermeasures
Storage and Handling: Section 7.
Shipping Regulations; section 11.
IATA: Nonflammable gas, green label, 70 kilograms (passenger), 140 kilograms (cargo).

CARBON TETRAIODIDE
General Information
Synonym: Tetraiodomethane.

Description: Octahedral, red crystals.
Formula: CI₄.
Constants: Mol wt: 519.69, mp: 171°C decomposes, d: 4.32.
Hazard Analysis and Countermeasures
Toxicity Unknown. See also iodoform.

CARBON TRICHLORIDE. See hexachloroethane.

CARBONYL BROMIDE
General Information
Synonym: Carbon oxybromide.
Description: Liquid.
Formula: COBr₂.
Constants: Mol wt: 187.84, bp: 64.5°C, d: 2.44.
Hazard Analysis
Toxic Hazard Rating:
  Acute Local: Irritant 3; Inhalation 3.
  Acute Systemic: Inhalation 3.
  Chronic Local: U.
  Chronic Systemic: U.
Fire Hazard: See carbon monoxide and bromine.
Explosion Hazard: See carbon monoxide.
Disaster Hazard: Dangerous; when heated to decomposition, it emits highly toxic fumes; it will react with water or steam to produce toxic and flammable vapors.
Countermeasures
Ventilation Control: Section 2.
Personnel Protection: Section 3.
Personal Hygiene: Section 3.
Storage and Handling: Section 7.

CARBONYL CHLORIDE. See phosgene.

CARBONYL CYANIDE
General Information
Synonym: Carbon oxycyanide.
Description: Colorless liquid.
Formula: CO(CN)₂.
Constants: Mol wt: 80.05, bp: 65.5°C, d: 1.124.
Hazard Analysis and Countermeasures
See cyanides and carbon monoxide.

CARBONYL FLUORIDE
General Information
Synonym: Fluoroformyl fluoride.
Description: Colorless gas, pungent, hygroscopic.
Formula: COF₂.
Constants: Mol wt: 66.01, mp: −114°C, bp: −83°C, d: 1.139 at −114°C.
Hazard Analysis
Toxic Hazard Rating:
  Acute Local: Irritant 3; Inhalation 3.
  Acute Systemic: U.
  Chronic Local: U.
  Chronic Systemic: U.
Toxicology: A powerful irritant. See hydrofluoric acid and fluorine. It hydrolyzes instantly upon contact with moisture.
Fire Hazard: See carbon monoxide.
Explosion Hazard: See carbon monoxide.
Disaster Hazard: See fluorides and carbon monoxide.
Countermeasures
Storage and Handling: Section 7.

CARBONYLS
General Information
Description: The (CO)⁻ group with a metal.

Hazard Analysis
Toxic Hazard Rating:
  Acute Local: Inhalation 3.
  Acute Systemic: Inhalation 3.
  Chronic Local: U.
  Chronic Systemic: Inhalation 3.
Toxicology: Most carbonyls are highly toxic. The toxicity of carbonyls depends in part, but not always entirely, on their ready decomposition which releases carbon monoxide. Symptoms are due in part to carbon monoxide and in part to the direct irritating action of the carbonyl. See also specific carbonyl in question.
Fire Hazard: Moderate, when exposed to heat or flame. More or less readily evolves carbon monoxide. See also carbon monoxide and powdered metals.
Explosion Hazard: Moderate, when exposed to heat or flame (Section 7).
Explosive Range: See carbon monoxide.
Caution: Powerful oxidizers; can react violently. Heat can cause carbonyls to decompose.
Disaster Hazard: Dangerous; when heated to decomposition, it emits highly toxic fumes of carbon monoxide; it will react with water or steam to produce toxic and flammable vapors; it can react vigorously with oxidizing materials.
Countermeasures
Storage and Handling: Section 7.

CARBONYL SULFIDE. See carbon oxysulfide.

CARBOPHENOTHION
General Information
Synonym: S-[{(p-Chlorophenyl) thio} methyl]-o,o-diethyl phosphorodithioate; o,o-diethyl S-(p-chlorophenyl thio methyl) phosphorodithioate; S-(4-chlorophenyl thio-methyl) diethyl phosphorophorothithionate.
Description: Amber liquid; essentially insoluble in water; miscible in common solvents.
Formula: (C₂H₅O)₂ P(S)SCH₂S(C₆H₄)Cl.
Constants: Mol wt: 343, bp: 82°C at 0.1 mm; d: 1.29 (20°C).
Hazard Analysis
Toxic Hazard Rating: U. A food additive permitted in feed and drinking water of animals and/or for the treatment of food producing animals. (Section 10).
Disaster Hazard: Dangerous. See sulfur compounds, phosphates and chlorides.

CARBOPOLS. See Carboxypolymethylene.

CARBOPROPOXIDE, STABLIZED
General Information
Description: Liquid.
Hazard Analysis
Toxic Hazard Rating:
  Acute Local: Irritant 3; Ingestion 3; Inhalation 3.
  Acute Systemic: U.
  Chronic Local: U.
  Chronic Systemic: U.
Fire Hazard: Moderate, when exposed to heat or flame (Section 6).
Disaster Hazard: Moderately dangerous because when heated to decomposition, it emits toxic fumes; it can react with oxidizing materials.
Countermeasures
Ventilation Control: Section 2.
Personnel Protection: Section 3.
Storage and Handling: Section 7.

TOXIC HAZARD RATING CODE *(For detailed discussion, see Section 1.)*

0 NONE: (a) No harm under any conditions; (b) Harmful only under unusual conditions or overwhelming dosage.

1 SLIGHT: Causes readily reversible changes which disappear after end of exposure.

2 MODERATE: May involve both irreversible and reversible changes; not severe enough to cause death or permanent injury.

3 HIGH: May cause death or permanent injury after very short exposure to small quantities.

U UNKNOWN: No information on humans considered valid by authors.

Shipping Regulations: Section 11.
See isopropyl percarbonate, stabilized.

CARBOPROPOXIDE, UNSTABILIZED
General Information
Description: Solid.
Hazard Analysis
Toxic Hazard Rating:
  Acute Local: Irritant 3; Ingestion 3; Inhalation 3.
  Acute Systemic: U.
  Chronic Local: U.
  Chronic Systemic: U.
Fire Hazard: Dangerous, when exposed to heat (Section 6).
Disaster Hazard: Moderately dangerous when heated to decomposition, it emits toxic fumes; it can react with oxidizing materials.
Countermeasures
Ventilation Control: Section 2.
Personnel Protection: Section 3.
Storage and Handling: Section 7.
Shipping Regulations: section 11.
  See isopropyl percarbonate, unstabilized.

"CARBORUNDUM." See silicon carbide.

"CARBOWAX" METHOXY POLYETHYLENE CLYCOL 350
General Information
Description: Slightly viscous liquid.
Formula: $CH_3OCH_2(CH_2OCH_2)_xCH_2OH$.
Constants: Mol wt: 330–370, mp: $-5°$ to $-10°C$, flash p: 440° F (O.C.), d: 1.09 at 20° /20° C, vap. press.: <50 mm at 37° C.
Hazard Analysis
Toxicity: See glycols.
Fire Hazard: Slight, when exposed to heat or flame; it can react with oxidizing materials.
To Fight Fire: Water, foam, carbon dioxide, dry chemical or carbon tetrachloride (Section 6).
Storage and Handling: Section 7.

"CARBOWAX" METHOXY POLYETHYLENE GLYCOL 550
General Information
Description: Soft, wax-like solid.
Formula: $CH_3OCH_2(CH_2OCH_2)_xCH_2OH$.
Constants: Mol wt: 525–575, mp: 15°–25°C, flash p: 460°F (O.C.), d: 1.07 at 55°/20°C, vap. press.: < 50 mm at 37°C.
Hazard Analysis
Toxicity: See glycols.
Fire Hazard: Slight, when exposed to heat or flame; it can react with oxidizing materials.
Countermeasures
To Fight Fire: Water, foam, carbon dioxide, dry chemical or carbon tetrachloride (Section 6).
Storage and Handling: Section 7.

CARBOXIDE
General Information
Description: Liquid.
Formula: 1 part ethylene oxide + 9 parts carbon dioxide.
Hazard Analysis
Toxic Hazard Rating:
  Acute Local: Irritant 3; Inhalation 3.
  Acute Systemic: Inhalation 3.
  Chronic Local: U.
  Chronic Systemic: U.
Toxicology: A powerful irritant. May cause CNS damage including hemorrhages into the brain; hemorrhagic nasal discharge, unsteadiness, dyspnea and death following sufficient exposure. 30,000 ppm may be tolerated for perhaps 1 hour.

CARBOXY METHYL MERCAPTOSUCCINIC ACID
General Information
Description: White powder.

Formula: $C_6H_8O_6S$.
Constants: Mol wt: 208.2, mp: 128°–136°C.
Hazard Analysis
Toxicity: Details unknown. Preliminary tests indicate toxicity is somewhat less than citric acid.
Disaster Hazard: See sulfur compounds.
Countermeasures
Storage and Handling: Section 7.

CARBOXYPOLYMETHYLENE
General Information
Synonym: Carbopol.
Description: White powder; a vinyl polymer with active carboxyl groups.
Hazard Analysis
Toxic Hazard Rating: U. Animal experiments show low toxicity.

CARBOYS, ACID, EMPTY
Hazard Analysis
Toxicity: See specific material shipped.
Disaster Hazard: Slightly dangerous; when heated, they may emit toxic fumes.
Countermeasures
Storage and Handling: Section 7.
Shipping Regulations: Section 11.
  I.C.C. Classification: See Section 11.
  Coast Guard Classification: Hazardous article.

CARBURIZING SALT. See cyanides.

CARITOL. See carotene.

CARNAUBA WAX
General Information
Synonym: Brazil wax.
Constants: Flash p: 540° F (C.C.), mp: 85°C.
Hazard Analysis
Toxic Hazard Rating:
  Acute Local: 0.
  Acute Systemic: 0.
  Chronic Local: Allergen 1.
  Chronic Systemic: 0.
A general purpose food additive (Section 10).
Fire Hazard: Slight, when exposed to heat; it can react with oxidizing materials.
Countermeasures
To Fight Fire: Water, carbon dioxide, dry chemical or carbon tetrachloride (Section 6).
Personal Hygiene: Section 3.
Storage and Handling: Section 7.

CAROB BEAN GUM
General Information
Synonym: Locust bean gum; carob seed gum.
Description: A galactomannan polysaccharide with a mol wt of about 310,000. In powdered form, nearly pure white. Insoluble in most organic solvents.
Hazard Analysis
Toxic Hazard Rating: U. A stabilizer food additive. Also a substance migrating to food from packaging materials. (Section 10).

CAROB SEED GUM. See carob bean gum.

CARO'S ACID. See peroxysulfuric acid.

CAROTENE
General Information
Synonym: Caritol; provitamin A.
Description: Ruby red crystals; insoluble in water; slightly soluble in alcohol and ether; soluble in oils, chloroform, benzene. It is a precursor of vitamin A occurring naturally in plants. It consists of 3 isomers (15% α, 85% β, 0.1% γ).
Formula: $C_{40}H_{56}$.

Constants: Mol wt: 536, mp (α): 188°C, mp (β): 184°C, mp (γ): 178°C.

Hazard Analysis
Toxic Hazard Rating: U. A nutrient and/or dietary supplement food additive (Section 10).

CARRAGEENIN. See chondrus extract.

CARTRIDGE BAGS, EMPTY, WITH BLACK POWDER IGNITERS. See explosives, low.
Shipping Regulations: Section 11.
I.C.C.: Explosive C, 150 pounds.
Coast Guard Classification: Explosive C.
IATA: Explosive, explosive label, 25 kilograms (passenger), 70 kilograms (cargo).

CARTRIDGE CASES, EMPTY, PRIMED. See explosives, low.
Shipping Regulations: Section 11.
I.C.C.: Explosive C, 150 pounds.
Coast Guard Classification: Explosive C.
IATA: Explosive, explosive label, 25 kilograms (passenger), 70 kilograms (cargo).
Cartridges, oil well, containing more than 13 grams of propellant powder each or equipped with ignition device or element. IATA: Explosive, not acceptable (passenger or cargo).

CARTRIDGES, actuating, explosive, for fire extinguisher or valve.
Shipping Regulations: Section 11.
IATA: Explosive, explosive label, 25 kilograms (passenger), 70 kilograms (cargo).

CARVONE
General Information
Description: Optically active ketone occurring in d- and l-forms. Pale yellowish or colorless liquid; strong characteristic odor; soluble in alcohol, ether, chloroform, propyleneglycol and mineral oils; insoluble in glycerin and water.
Formula: $CH_3\overset{|}{C}:CHCH_2CH[C(CH_3):CH_2]CH_2\overset{|}{C}O$.
Constants: Mol wt: 150, d: 0.960 at 20°C, bp: 227–230°C.
Hazard Analysis
Toxic Hazard Rating: U. A synthetic flavoring substance and adjuvant. (Section 10).

CARYOPHYLLIC ACID. See eugenol.

CASE-HARDENING COMPOUNDS. See cyanides.

CASEIN
General Information
Description: The principle protein in milk; a phospho protein. White odorless, amorphous solid; soluble in dilute alkalies and concentrated acids; almost insoluble in water.
Composition: ~85% phosphorous and 0.76% sulfur. Consists of ~ 15 amino acids.
Constants: Mol wt: from 75,000–375,000, d: 1.25–1.31.
Hazard Analysis
Toxic Hazard Rating: U. A substence migrating to food from packaging materials. (Section 10).

CASEIN-SODIUM. See sodium caseinate.

CASHEW NUT SHELL OIL
General Information
Composition: Mostly anacardic acid.

Hazard Analysis
Toxic Hazard Rating:
Acute Local: Irritant 3.
Acute Systemic: U.
Chronic Local: Allergen 3.
Chronic Systemic: U.
Fire Hazard: Slight; when heated.
Countermeasures
Personnel Protection: Section 3.
Storage and Handling: Section 7.

CASSEL'S GREEN. See barium manganate.

CASSEL YELLOW LAURIONITE. See lead oxychlorides.

CASTOR-BEAN MEAL
Hazard Analysis
Toxicity: This material contains the very toxic materials ricin and ricinine. It is also a very potent allergen. See Section 9.

CASTOR OIL
General Information
Synonyms: Ricinus oil.
Description: A colorless to pale-yellow viscous liquid, characteristic odor.
Constants: Mp: -12°C, bp: 313°C, flash p: 445°F (C.C.), d: 0.96, autoign. temp.: 840°F.
Hazard Analysis
Toxic Hazard Rating:
Acute Local: Irritant 1; Allergen 1.
Acute Systemic: Ingestion 2.
Chronic Local: Allergen 1.
Chronic Systemic: U.
A food additive permitted in food for human consumption.
Fire Hazard: Slight, when exposed to heat.
Spontaneous Heating: Yes.
Countermeasures
To Fight Fire: Foam, carbon dioxide, dry chemical or carbon tetrachloride (Section 6).
Personal Hygiene: Section 3.
Storage and Handling: Section 7.

CASTOR OIL, HYDROGENATED
General Information
Description: Insoluble in water.
Formula: $(C_{18}H_{35}O_3)_3C_3H_5$.
Constant: Mol wt: 938, flash p: 401°F.
Hazard Analysis
Fire Hazard: Slight, when exposed to heat or flame.
Countermeasures
To Fight Fire: Water, foam (Section 6).
Storage and Handling: Section 7.

CASTOR OIL PLANT DUST
Hazard Analysis
Toxic Hazard Rating:
Acute Local: Irritant 1.
Acute Systemic: U.
Chronic Local: Allergen 1.
Chronic Systemic: U.
Fire Hazard: Moderate, when exposed to heat or flame; it can react vigorously with oxidizing materials (Section 6).
Explosion Hazard: Slight, when exposed to flame (Section 7).
Countermeasures
Personnel Protection: Section 3.
Personal Hygiene: Section 3.

TOXIC HAZARD RATING CODE (For detailed discussion, see Section 1.)

0 NONE: (a) No harm under any conditions; (b) Harmful only under unusual conditions or overwhelming dosage.

1 SLIGHT: Causes readily reversible changes which disappear after end of exposure.

2 MODERATE: May involve both irreversible and reversible changes; not severe enough to cause death or permanent injury.

3 HIGH: May cause death or permanent injury after very short exposure to small quantities.

U UNKNOWN: No information on humans considered valid by authors.

CASTRIX. See 2-chloro-4-dimethylamino-6-methyl pyrimidine.

CATECHOL. See pyrocatechol.

CATIONIC AMINE 200
General Information
Synonym: 1-Hydroxyethyl-1,2-heptadecenyl glyoxaldene.
Formula: $C_{17}H_{33}CH(CH_2)_2NHC_2CH_2OH$.
Constants: Mol wt: 350.58, flash p: 465° F (O.C.).
Hazard Analysis
Toxicity: Unknown. See also amines.
Fire Hazard: Slight, when exposed to heat.
Countermeasures
To Fight Fire: Water, foam, carbon dioxide, dry chemical or carbon tetrachloride (Section 6).
Storage and Handling: Section 7.

CATIONIC SP. See stereamidopropyldimethyl-β-hydroxy-ethylammonium phosphate.

CATTLE FEED DUST
Hazard Analysis
Toxic Hazard Rating:
  Acute Local: Allergen 1.
  Acute Systemic: 0.
  Chronic Local: Allergen 1.
  Chronic Systemic: Inhalation 1.
Fire Hazard: Moderate, when exposed to heat or flame; it can react with oxidizing materials (Section 6).
Spontaneous Heating: Yes; dangerous when very dry or very moist.
Explosion Hazard: Slight, when exposed to flame (Section 7).
Countermeasures
Ventilation Control: Section 2.
Personal Hygiene: Section 3.

CAUSTIC ALCOHOL. See sodium ethylate.

CAUSTIC ALCOHOL SOLUTION. See sodium hydroxide and ethyl alcohol.

CAUSTIC ALKALI SOLUTION. See alkaline caustic liquid N.O.S.

CAUSTIC BARYTA. See barium hydroxide.

CAUSTIC BARLEY. See sabadilla seed.

CAUSTIC POTASH. See potassium hydroxide.
Shipping Regulations: Section 11.
  Coast Guard Classification: Hazardous article.
  MCA warning label.

CAUSTIC POTASH, LIQUID. See potassium hydroxide.
Shipping Regulations: Section 11.
  I.C.C.: Corrosive liquid; white label, 10 gallons.
  Coast Guard Classification: Corrosive liquid; white label.
  MCA warning label.

CAUSTIC SODA. See sodium hydroxide.
Shipping Regulations: Section 11.
  Coast Guard Classification: Hazardous article.
  MCA warning label.

CAUSTIC SODA, LIQUID. See sodium hydorxide.
Shipping Regulations: Section 11.
  I.C.C.: Corrosive liquid; white label, 10 gallons.
  Coast Guard Classification: Corrosive liquid; white label.
  MCA warning label.

CBP. See chlorobromopropene.

CDAA. See N,N-diallyl-2-chloroacetamide.

CEDAR
General Information
Description: A wood.

Hazard Analysis
Toxic Hazard Rating:
  Acute Local: Inhalation 1.
  Acute Systemic: U.
  Chronic Local: Allergen 1.
  Chronic Systemic: U.
Toxicology: The wood dust contains chemical irritants which can cause conjunctivitis, intense lachrymation, iridocyclitis, keratitis as well as dermatitis (Section 9).
Fire Hazard: Moderate, when exposed to heat.
Exposion Hazard: Slight in the form of dust when exposed to flame. See also dust.
Countermeasures
Ventilation Control: Section 2.
Treatment and Antidotes: Removal from exposure will cause the symptoms to disappear.
Personal Hygiene: Section 3.
Storage and Handling: Section 7.

CEDAR WOOD OIL
General Information
Description: Colorless or slightly yellow, viscid liquid.
Composition: Cedrene and cedrol.
Constant: D: 0.940–0.950 at 20°/20° C.
Hazard Analysis
Toxic Hazard Rating:
  Acute Local: Irritant 1; Allergen 1.
  Acute Systemic: U.
  Chronic Local: Allergen 1.
  Chronic Systemic: U,
Fire Hazard: Slight; when heated.
Countermeasures
Personal Hygiene: Section 3.
Storage and Handling: Section 7.

"CELANESE SOLVENT 203"
General Information
Description: Liquid; odor of isobutanol.
Constants: Bp: 110–120° C, flash p: 100° F (T.O.C.), d: 0.805–0.815 at 20°/4° C.
Hazard Analysis
Toxicity: See isobutyl alcohol, n-butyl alcohol and amyl alcohol.
Fire Hazard: Moderate; when exposed to heat or flame, it can react with oxidizing materials.
Countermeasures
To Fight Fire: Foam, carbon dioxide, dry chemical or carbon tetrachloride (Section 6).
Storage and Handling: Section 7.

"CELANESE SOLVENT 301"
General Information
Description: Liquid.
Constants: Bp: 97–100° C, flash p: 90° F (T.O.C.), d: 0.804-0.808 at 20°/4° C.
Hazard Analysis
Toxicity: See n-propanol, sec-butanol.
Fire Hazard: Dangerous; when exposed to heat or flame, can react with oxidizing materials.
Countermeasures
To Fight Fire: Foam, carbon dioxide, dry chemical or carbon tetrachloride (Section 6).
Storage and Handling: Section 7.

"CELANESE SOLVENT 601"
General Information
Description: Liquid, ether-like odor.
Constants: Bp: 74–84° C, flash p: 10° F (T.O.C.), d: 0.840–0.850 at 20°/4° C.
Hazard Analysis
Toxicity: Unknown.
Fire Hazard: Dangerous; when exposed to heat or flame, it can react with oxidizing materials.

Countermeasures

To Fight Fire: Foam, carbon dioxide, dry chemical or carbon tetrachloride (Section 6).

Storage and Handling: Section 7.

"CELANESE SOLVENT 901"

General Information

Description: Liquid.

Constants: Bp: 125–155°C, flash p: 115°F (T.O.C.;, d: 0.835–0.840, 840 at 20°/4°C.

Hazard Analysis

Toxicity: Unknown.

Fire Hazard: Moderate; when exposed to heat or flame, it can react with oxidizing materials.

Countermeasures

To Fight Fire: Foam, carbon dioxide, dry chemical or carbon tetrachloride (Section 6).

Storage and Handling: Section 7.

CELERY SEED

General Information

Description: Volatile and fixed oil, bitter extractive and resin.

Hazard Analysis

Toxic Hazard Rating:

  Acute Local: Allergen 1.

  Acute Systemic: U.

  Chronic Local: Allergen 1.

  Chronic Systemic: U.

Fire Hazard: Slight, when exposed to heat (Section 6).

Countermeasures

Personal Hygiene: Section 3.

Storage and Handling: Section 7.

CELLOIDIN. See nitrocellulose.

"CELLOSOLVE" ACETATE

General Information

Synonym: Ethylene glycol monoethyl ether acetate; ethoxy acetate.

Description: Colorless liquid with a mild, pleasant ester-like odor.

Formula: $CH_3COOCH_2CH_2OC_2H_5$.

Constants: Mol wt: 132.17, bp: 156.4°C, flash p: 120°F (C.O.C.), lel: 1.7%, fp: −61.7°C, d: 0.9748 at 20°/20°C, autoign. temp.: 715°F, vap. press.: 1.2 mm at 20°C, vap. d.: 4.72.

Hazard Analysis

Toxic Hazard Rating:

  Acute Local: Irritant 1.

  Acute Systemic: Ingestion 2.

  Chronic Local: U.

  Chronic Systemic: Ingestion 2; Inhalation 2.

  See glycols.

TLV: ACGIH, 100 parts per million; 540 milligrams per cubic meter of air. May be absorbed via the skin.

Fire Hazard: Moderate; when exposed to heat or flame, can react with oxidizing materials.

Spontaneous Heating: No.

Explosion Hazard: Moderate in the form of vapor when heated (Section 7).

Countermeasures

Ventilation Control: Section 2.

Personnel Protection: Section 3.

To Fight Fire: Alcohol foam, carbon dioxide, dry chemical, or carbon tetrachloride (Section 6).

Storage and Handling: Section 7.

"CELLOSOLVE" SOLVENT

General Information

Synonym: Ethylene glycol monoethyl ether; 2-ethoxy ethanol.

Description: Colorless liquid; practically odorless.

Formula: $CH_2OHCH_2OC_2H_5$.

Constants: Mol wt: 90.12, bp: 135.1°C, lel: 2.6%, uel: 15.7%, flash p: 106°F (C.C.), d: 0.9360 at 15°/15°C, autoign. temp.: 460°F, vap. press.: 3.8 mm at 20°C, vap. d.: 3.10.

Hazard Analysis

Toxic Hazard Rating:

  Acute Local: Irritant 1; Ingestion 1.

  Acute Systemic: Ingestion 2.

  Chronic Local: U.

  Chronic Systemic: Ingestion 1; Inhalation 1.

  See also Glycols.

TLV: ACGIH, 200 parts per million; 740 milligrams per cubic meter of air. May be absorbed via the skin.

Toxicology: Animal experiments indicate that exposure to air saturated with "Cellosolve" vapor (0.6%) for periods of 18 to 24 hours may produce congestion and edema of the lungs and congestion of the kidneys. Exposure of humans to the same concentration for a few seconds resulted in irritation of the eyes. No cases of poisoning have so far been reported in industry.

Fire Hazard: Moderate, when exposed to heat or flame, can react with oxidizing materials.

Spontaneous Heating: No.

Explosion Hazard: Moderate; in the form of vapor when exposed to heat or flame (Section 7).

Countermeasures

Ventilation Control: Section 2.

Personnel Protection: Section 3.

To Fight Fire: Carbon dioxide, dry chemical or carbon tetrachloride (Section 6).

Storage and Handling: Section 7.

"CELLULOID"

General Information

Description: Clear or colored cellulose nitrate.

Constant: D: 1.35–1.60.

Hazard Analysis

Toxicity: Little or no toxicity unless burned or in solution.

Fire Hazard: Moderate, when exposed to heat (Section 6).

Disaster Hazard: Dangerous; when heated to decomposition, it emits toxic fumes; it can react with oxidizing materials.

COUNTERMEASURES

Storage and Handling: Section 7.

CELLULOSE ACETATE

General Information

Synonym: CA.

Description: White flakes or powder; soluble in acetone, ethyl acetate, cyclohexanol, nitropropane, ethylene dichloride.

Constant: Melts ∼ 260°C, d: 1.27–1.34.

Hazard Analysis

Toxic Hazard Rating: U. A substance migrating to food from packaging materials (Section 10).

CELLULOSE DINITRATE

General Information

Synonym: Cellulose tetranitate.

Description: White, amorphous solid.

Formula: $C_{12}H_{16}(ONO_2)_4O_6$.

Constants: Mol wt: 504.3, d: 1.66.

TOXIC HAZARD RATING CODE (For detailed discussion, see Section 1.)

0 NONE: (a) No harm under any conditions; (b) Harmful only under unusual conditions or overwhelming dosage.

1 SLIGHT: Causes readily reversible changes which disappear after end of exposure.

2 MODERATE: May involve both irreversible and reversible changes; not severe enough to cause death or permanent injury.

3 HIGH: May cause death or permanent injury after very short exposure to small quantities.

U UNKNOWN: No information on humans considered valid by authors.

Hazard Analysis
Toxicity: See nitrocellulose.
Fire Hazard: See nitrocellulose.
Explosion Hazard: See nitrocellulose.
Disaster Hazard: See nitrocellulose.

CELLULOSE GUM. See sodium carboxymethyl cellulose.

CELLULOSE HEXANITRATE. See nitrocellulose.

CELLULOSE METHYL ETHER. See methyl cellulose.

CELLULOSE NITRATE. See nitrocellulose.

CELLULOSE PENTANITRATE. See nitrocellulose.

CELLULOSE TETRANITRATE. See cellulose dinitrate and nitrocellulose.

CELLULOSE TRINITRATE. See nitrocellulose.

CEMENT, LEATHER
Hazard Analysis
Fire Hazard: Dangerous; when exposed to heat or flame, it can react with oxidizing materials (Section 6).
Explosion Hazard: Unknown.
Countermeasures
Storage and Handling: Section 7.
Shipping Regulations: Section 11.
  I.C.C.: Flammable liquid; red label, 12 gallons.
  Coast Guard Classification: Inflammable liquid; red label.
  IATA: Flammable liquid, red label, 1 liter (passenger), 40 liters (cargo).

CEMENT, LINOLEUM, TILE, WALLBOARD, OR CONTAINER, LIQUID.
Shipping Regulations: Section 11.
  ICC: Flammable liquid, red label, 15 gallons.

CEMENT, LIQUID, N.O.S.
Hazard Analysis
Fire Hazard: Dangerous; when exposed to heat or flame, it can react with oxidizing materials (Section 6).
Countermeasures
Storage and Hanlding: Section 7.
Shipping Regulations: Section 11.
  I.C.C.: Flammable liquid; red label, 15 gallons.
  Coast Guard Classification: Inflammable liquid; red label.
  IATA: Flammable liquid, red label, 1 liter (passenger), 40 liters (cargo).

CEMENT, PORTLAND
General Information
Synonym: Cement, hydraulic.
Description: Fire gray powder composed of compounds of lime, aluminum, silica, and iron oxide as $(4CaO \cdot Al_2O_3 \cdot Fe_2O_3)$, $(3CaOAl_2O_3)$, $(3CaO \cdot SiO_2)$, and $(2CaO \cdot SiO_2)$ abbreviated as C_4AF, C_3A, C_3C, C_2S. Small amounts of magnesia, sodium, potassium, and sulfur are also present in combined form.
Hazard Analysis
Toxic Hazard Rating:
  Acute Local: Irritant 2; Allergen 1; Inhalation: 1.
  Acute Systemic: U.
  Chronic Local: Irritant 2; Allergen 1; Inhalation: 1.
  Chronic Systemic: U.
Signficant pulmonory fibrosis due to cement dust occurs rarely if at all. Eczema is by no means rare and is believed to be due to chromium in the cement. Cement dust is a common air contaminant (Section 4).

CEMENT, PYROXYLIN
Hazard Analysis
Fire Hazard: Dangerous; when exposed to heat or flame; it can react with oxidizing materials (Section 6).
Countermeasures
Shipping Regulations:
  I.C.C.: Flammable liquid; red label, 15 gallons.

Coast Guard Classification: Inflammable liquid; red label.
IATA: Flammable liquid, red label, 1 liter (passenger), 60 liters (cargo).

CEMENT, ROOFING, LIQUID
Hazard Analysis
Fire Hazard: Dangerous; when exposed to heat or flame, it can react with oxidizing materials (Section 6).
Countermeasures
Storage and Handling: Section 7.
Shipping Regulations: Section 11.
  I.C.C.: Flammable liquid; red label, 12 gallons.
  Coast Guard Classification: Inflammable liquid; red label.
  IATA: Flammable liquid, red label, 1 liter (passenger), 40 liters (cargo).

CEMENT, RUBBER
General Information
Constant: Flash p: 50° F or less.
Hazard Analysis
Toxicity: Often contains benzene or other toxic solvents. See specific constituent.
Fire Hazard: Dangerous; when exposed to heat or flame, it can react with oxidizing materials (Section 6).
Countermeasures
Storage and Handling: Section 7.
Shipping Regulations: Section 11.
  I.C.C.: Flammable liquid; red label, 15 gallons.
  Coast Guard Classification: Inflammable liquid; red label.
  IATA: Flammable liquid, red label, 1 liter (passenger), 60 liters (cargo).

CERA ALBA. See beeswax, white.

CERESIN WAX
General Information
Synonym: Earth wax; mineral wax.
Description: White or yellow waxy material.
Constants: Mp: 68–72°C, d: 0.92–0.94, flash p: 236°F.
Hazard Analysis
Toxic Hazard Rating:
  Acute Local: 0.
  Acute Systemic: 0.
  Chronic Local: Allergen 1.
  Chronic Systemic: U.
Fire Hazard: Slight, when exposed to heat or flame.
Countermeasures
Personal Hygiene: Section 3.
To Fight Fire: Water, foam (Section 6).
Storage and Handling: Section 7.

CERIC AMMONIUM NITRATE
General Information
Synonym: Cerium-ammonium nitrate.
Description: Small prismatic, orange-red crystals.
Formula: $Ce(NO_3)_4 \cdot 2NH_4NO_3 + 1\frac{1}{2}H_2O$.
Constant: Mol wt: 575.40.
Hazard Analysis and Countermeasures
See cerium compounds and nitrates.

CERIC FLUROIDE
General Information
Synonym: Cerium tetrafluoride.
Description: Tiny crystals. Nearly insoluble in water.
Formula: CeF_4.
Constants: Mol wt: 216.1, d: 4.77, mp: > 650°C.
Hazard Analysis
Toxicology: Details unknown. Low solubility reduces toxicity. See fluorides.
Disaster Hazard: Dangerous; See fluorides.

CERIC IODATE
General Information
Description: Colorless crystals.
Formula: $Ce(IO_3)_4$.

Constant: Mol wt: 839.81.
Hazard Analysis
Toxicity: See cerium compounds and iodates.

CERIC NITRATE, BASIC
General Information /
Description: Long, red needles.
Formula: $Ce(OH)(NO_3)_3 \cdot 3H_2O$.
Constant: Mol wt: 397.21.
Hazard Analysis and Countermeasures
See cerium compounds and nitrates.

CERIUM
General Information
Description: Cubic or hexagonal, steel-gray crystals.
Formula: Ce.
Constants: At wt: 140.13, mp: 640°C, bp: 1400°C, density cubic form 6.90; hexagonal form 6.7.
Hazard Analysis
Toxic Hazard Rating:
Acute Local: 0.
Acute Systemic: Ingestion 1.
Chronic Local: 0.
Chronic Systemic: Ingestion 1; Inhalation 1.
Toxicology: Cerium resembles aluminum in its pharmacological action as well as in its chemical properties. The insoluble salts such as the oxalate are stated to be nontoxic even in large doses. It is used to prevent vomiting in pregnancy. The average dose is from 0.05 to 0.5 g. Cerium tartrate has been found to produce a direct injurious action on the hearts of small animals. The effect on the nervous system of the rare-earth metals following inhalation may preclude welding operations with these materials to any large extent. Cerium is stated to produce polycythemia but is useless in the treatment of anemia owing to its toxic effects. The salts of cerium increase the blood coagulation rate.
Radiation Hazard: Section 5. For permissible levels, see Table 5, p. 150.
Artificial isotope ^{141}Ce, half life 33 d. Decays to stable ^{141}Pr by emitting beta particles of 0.44 (70%), 0.58 (30%) MeV. Also emits gamma rays of 0.14 MeV.
Artificial isotope ^{143}Ce, half life 33 h. Decays to radioactive ^{143}Pr by emitting beta particles of 0.22–1.38 MeV. Also emits gamma rays of 0.29–0.72 MeV.
Artificial isotope ^{144}Ce, half life 284 d. Decays to radioactive ^{144}Pr by emitting beta particles of 0.19 (20%), 0.24 (8%), 0.32 (72%) MeV. Also emits gamma rays of 0.03–0.13 MeV. ^{144}Ce exists in radioactive equilibrium with its daughter, and permissible levels include the daughter activity.
Fire Hazard: Moderate, ignites spontaneously in air at 150–180°C. A strong reducing agent. See also powdered metals.
Explosion Hazard: Moderate, in the form of dust when exposed to flame. See also powdered metals.
Countermeasures
Ventilation Control: (Section 2).
Personal Hygiene: Section 3.

CERIUM AMMONIUM NITRATE. See ceric ammonium nitrate.

CERIUM CARBIDE
General Information
Description: Hexagonal, red crystals.
Formula: CeC_2.
Constants: Mol wt: 164.15, d: 5.23.

Hazard Analysis and Countermeasures
Toxicity: See cerium and acetylides.

CERIUM CHLORIDE
General Information
Synonym: Cerous chloride.
Description: Colorless, deliquescent crystals.
Formula: $CeCl_3$.
Constants: Mol wt: 246.50, mp: 848°C, bp: decomposes, d: 3.92.
Hazard Analysis and Countermeasures
See cerium and chlorides.

CERIUM COMPOUNDS
Hazard Analysis
Toxicity: The toxicity of cerium compounds may be taken to be that of cerium, except when the anion has a toxicity of its own.

CERIUM TARTRATE
General Information
Description: White powder.
Formula: $Ce_2(C_4H_4O_6)_3$.
Constants: Mol wt: 724.5.
Hazard Analysis and Countermeasures
See cerium.

CERIUM TETRABORIDE
General Information
Description: Tetragonal crystals.
Formula: CeB_4.
Constants: Mol wt: 183.41, d: 5.74.
Hazard Analysis and Countermeasures
See cerium and boron compounds.

CERIUM TETRAFLUORIDE. See ceric fluoride.

CERIUM TRIFLUORIDE. See cerous fluoride.

CEROUS AMMONIUM NITRATE
General Information
Description: Monoclinic, white, transparent crystals.
Formula: $2NH_4NO_3 \cdot Ce(NO_3)_3 \cdot 4H_2O$.
Constants: Mol wt: 588.32, mp: 74°C.
Hazard Analysis and Countermeasures
See cerium and nitrates.

CEROUS BROMATE
General Information
Description: Hexagonal, reddish-white crystals.
Formula: $Ce(BrO_3)_3 \cdot 9H_2O$.
Constants: Mol wt: 686.02, mp: 49°C.
Hazard Analysis and Countermeasures
See cerium and bromates.

CEROUS CHLORIDE. See cerium chloride.

CEROUS CYNANOPLATINITE
General Information
Description: Monoclinic, yellow-blue, lustrous crystals.
Formula: $Ce[Pt(CN)_4]_3 \cdot 18H_2O$.
Constants: Mol wt: 1502.45, d: 2.657.
Hazard Analysis and Countermeasures
See cyanides and cerium.

CEROUS FLUORIDE
General Information
Synonym: Cerium trifluoride.
Description: Colorless crystals or powder nearly insoluble in water.

TOXIC HAZARD RATING CODE (For detailed discussion, see Section 1.)

0 NONE: (a) No harm under any conditions; (b) Harmful only under unusual conditions or overwhelming dosage.

1 SLIGHT: Causes readily reversible changes which disappear after end of exposure.

2 MODERATE: May involve both irreversible and reversible changes; not severe enough to cause death or permanent injury.

3 HIGH: May cause death or permanent injury after very short exposure to small quantities.

U UNKNOWN: No information on humans considered valid by authors.

Formula: CeF$_3$.
Constants: Mol wt: 197.13, mp: 1460°C, d: 6.16.
Hazard Analysis
Toxicity: See fluorides.
Disaster Hazard: See fluorides.
Countermeasures
Storage and Handling: Section 7.

CEROUS HYDRIDE
General Information
Description: Amorphous, dark blue powder.
Formula: CeH$_3$.
Constants: Mol wt: 143.15, mp: ignites.
Hazard Analysis and Countermeasures
See cerium and hydrides.

CEROUS IODIDE
General Information
Description: Reddish-white crystals.
Formula: CeI$_3 \cdot$9H$_2$O.
Constant: Mol wt: 683.03.
Hazard Analysis and Countermeasures
See cerium and iodides.

CEROUS NITRATE
General Information
Description: Colorless or reddish deliquescent crystals.
Formula: Ce(NO$_3$)$_3 \cdot$6H$_2$O.
Constants: Mol wt: 434.25, mp: $-$3H$_2$O at 150°C, bp: decomposes 200°C.
Hazard Analysis and Countermeasures
See cerium and nitrates.

CEROUS OXYCHLORIDE
General Information
Description: Purple, leaf-like crystals.
Formula: CeOCl.
Constant: Mol wt: 191.6.
Hazard Analysis and Countermeasures
See cerium and hydrochloric acid.

CEROUS SELENATE
General Information
Description: Rhombic crystals.
Formula: Ce$_2$(SeO$_4$)$_3$.
Constants: Mol wt: 709.14, d: 4.456.
Hazard Analysis and Countermeasures
See selenium and cerium.

CEROUS SULFIDE
General Information
Description: Red crystals; brown, dark purple powder.
Formula: Ce$_2$S$_3$.
Constants: Mol wt: 376.46, mp: decomposes, d: 5.020 at 11°C.
HAZARD Analysis and Countermeasures
See sulfides and cerium.

CERUSSITE. See lead carbonate.

CESIUM
General Information
Description: Hexagonal crystals; silver-white, ductile metal.
Formula: Cs.
Constants: At wt: 132.91, mp: 28.5°C, bp: 670°C, d: 1.873, vap. press.: 1 mm at 279°C.
Hazard Analysis
Toxic Hazard Rating:
Acute Local: U.
Acute Systemic: Ingestion 1.
Chronic Local: U.
Chronic Systemic: Ingestion 1.
Toxicology: Cesium is quite similar to potassium in its elemental state. It has been shown, however, to have pronounced physiological action in experiments with animals. Hyperirritability, including marked spasms, has been shown to follow the administration of cesium in amounts equivalent to the potassium content of the diet. It has been found that, by replacing the potassium in the diet of rats with cesium, death occurred after 10 days to 17 days.
Radiation Hazard: Section 5. For permissible levels, see Table 5, p. 150.
    Artificial isotope ^{131}Cs, half life 10 d. Decays to stable ^{131}Xe by electron capture. Emits X-rays.
    Artificial isotope ^{134m}Cs, half life 2.9 h. Decays to radioactive ^{134}Cs by emitting gamma rays of 0.01, 0.13, 0.14 MeV.
    Artificial isotope ^{134}Cs, half life 2.2 y. Decays to stable ^{134}Ba by emitting beta particles of 0.09 (27%), 0.41 (9%), 0.66 (61%) MeV. Also emits gamma rays of 0.48–1.36 MeV.
    Artificial isotope ^{135}Cs, half life 2×10^6 y. Decays to stable ^{135}Ba by emitting beta particles of 0.21 MeV.
    Artificial isotope ^{136}Cs, half life 13 d. Decays to stable ^{136}Ba by emitting beta particles of 0.34 (93%), 0.66 (7%), MeV. Also emits gamma rays of 0.07–1.26 MeV.
    Artificial isotope ^{137}Cs, half life 30 y. Decays to stable ^{137}Ba by emitting beta particles of 0.51 (95%), 1.18 (5%) MeV. Also emits gamma rays of 0.66 MeV.
Fire Hazard: Dangerous, by chemical reaction; reacts with oxidizing materials. Can ignite spontaneously in moist air. See also sodium.
Explosion Hazard: Moderate, by chemical reaction. Reacts with moisture to liberate hydrogen (Section 7).
Disaster Hazard: Moderately dangerous; it will react with water or steam to produce heat and hydrogen; on contact with oxidizing materials, it can react vigorously.
Countermeasures
Personal Hygiene: Section 3.
Storage and Handling: Section 7.

CESIUM 137
Shipping Regulations: Section 11.
    ICC: Poison D, radioactive materials label, 300 curies.
    IATA: Radioactive material, radioactive (red) label, 300 curies (passenger and cargo).

CESIUM BROMATE
General Information
Description: Solid.
Formula: CsBrO$_3$.
Constant: Mol wt: 260.83.
Hazard Analysis and Countermeasures
See bromates and cesium.

CESIUM BROMOCHLOROIODIDE
General Information
Description: Rhombic, yellow-red crystals.
Formula: CsBrClI.
Constants: Mol wt: 375.20, mp: 235°C, bp: decomposes 290°C.
Hazard Analysis and Countermeasures
See bromides, chlorides, iodides and cesium.

CESIUM BROMODIIODIDE
General Information
Formula: CsI$_2$Br.
Constants: Mol wt: 466.67, mp: 195.5°C.
Hazard Analysis and Countermeasures
See bromides, iodides and cesium.

CESIUM CHLORATE
General Information
Description: White crystals.
Formula: CsClO$_3$.
Constants: Mol wt: 216.37, d: 3.57.
Hazard Analysis and Countermeasures
See chlorates and cesium.

CESIUM CHLOROAURATE
General Information
Description: Monoclinic, yellow crystals.
Formula: $CsAuCl_4$.
Constant: Mol wt: 471. 94.
Hazard Analysis and Countermeasures
See gold compounds and cesium.

CESIUM CHLORODIBROMIDE
General Information
Description: Yellow crystals.
Formula: $CsBr_2Cl$.
Constants: Mol wt: 328.20, mp: 191°C.
Hazard Analysis and Countermeasures
See bromides, chlorides and cesium.

CESIUM CHLOROPLATINATE
General Information
Description: Cubic, yellow crystals.
Formula: Cs_2PtCl_6.
Constants: Mol wt: 673.79, mp: decomposes.
Hazard Analysis and Countermeasures
See platinum compounds, cesium and chlorides.

CESIUM CHLOROSTANNATE
General Information
Description: Cubic, white crystals.
Formula: Cs_2SnCl_6.
Constants: Mol wt: 597.26, d: 3.33.
Hazard Analysis and Countermeasures
See tin compounds and chlorides.

CESIUM CHROMATE
General Information
Description: Alpha: yellow prisms; beta: yellow, rhombic
crystals.
Formula: Cs_2CrO_4.
Constants: Mol wt: 381.83, d: 4.237.
Hazard Analysis and Countermeasures
See chrominum compounds and cesium.

CESIUM COMPOUNDS
Hazard Analysis
Toxicity: Probably have the toxicity of cesium unless they
contain a more toxic radical. Toxicity is of minor im-
portance in industry. See also cesium.
Countermeasures
Personal Hygiene: Section 3.

CESIUM CYANIDE
General Information
Description: Colorless crystals.
Formula: $CsCN$.
Constant: Mol wt:158.93.
Hazard Analysis and Countermeasures
See cyanides.

CESIUM DIBROMOIODIDE
General Information
Description: Rhombic crystals.
Formula: $CsIBr_2$.
Constants: Mol wt: 419.66, mp: 248°C, bp: decomposes
320°C.
Hazard Analysis and Countermeasures
See bromides, iodides and cesium.

CESIUM DICHLOROBROMIDE
General Information
Description: Crystals.

Formula: $CsBrCl_2$.
Constants: Mol wt: 283.74, mp: 205°C.
Hazard Analysis and Countermeasures
See bromides, chlorides and cesium.

CESIUM DICHLOROIODIDE
General Information
Description: Rhombic, pale orange crystals.
Formula: $CsICl_2$.
Constants: Mol wt: 330.74, mp: 230°C, bp: decomposes
290°C, d: 3.86.
Hazard Analysis and Countermeasures
See iodides, cesium and chlorides.

CESIUM DISULFIDE
General Information
Description: Amorphous or dark-red crystals.
Formula: Cs_2S_2.
Constants: Mol wt: 329.95, bp: >800°C, mp: 460°C.
Hazard Analysis and Countermeasures
See sulfides.

CESIUM FLUOGERMANATE
General Information
Description: Isotropic crystals; regular; octahedral.
Formula: Cs_2GeF_6.
Constants: Mol wt: 452.42, d: 4.10.
Hazard Analysis and Countermeasures
See cesium and fluorides.

CESIUM FLUORIDE
General Information
Description: Cubic, colorless crystals.
Formula: CsF.
Constants: Mol wt: 151.91, mp: 683°C, bp: 1251°C,
d: 3.586, vap. press.: 1 mm at 712°C.
Hazard Analysis and Countermeasures
See fluorides.

CESIUM FLUOSILICATE
General Information
Description: Cubic, white crystals.
Formula: Cs_2SiF_6.
Constants: Mol wt: 407.88, d: 3.372 at 17°C.
Hazard Analysis and Countermeasures
See fluosilicates.

CESIUM FLUOTELLURITE
General Information
Description: Colorless needles.
Formula: $CsTeF_5$.
Constant: Mol wt: 355.52.
Hazard Analysis and Countermeasures
See tellurium compounds and fluorides.

CESIUM GALLIUM SELENATE
General Information
Description: Colorless crystals.
Formula: $CsGa(SeO_4)_2 \cdot 12H_2O$.
Constant: Mol wt: 704.74.
Hazard Analysis and Countermeasures
See selenium and gallium compounds.

CESIUM GALLIUM SULFATE
General Information
Description: Cubic, colorless crystals.
Formula: $CsGa(SO_4)_2 \cdot 12H_2O$.
Constants: Mol wt: 610.95, d: 2.113.

TOXIC HAZARD RATING CODE *(For detailed discussion, see Section 1.)*

0 NONE: (a) No harm under any conditions; (b) Harmful only under unusual conditions or overwhelming dosage.

1 SLIGHT: Causes readily reversible changes which disappear after end
of exposure.

2 MODERATE: May involve both irreversible and reversible changes;
not severe enough to cause death or permanent injury.

3 HIGH: May cause death or permanent injury after very short exposure
to small quantities.

U UNKNOWN: No information on humans considered valid by authors.

Hazard Analysis and Countermeasures
See cesium, gallium compounds and sulfates.

CESIUM GRAPHITE
General Information
Description: Violet to black platelets. Hightly reactive with water, air, alcohols.
Formula: CsC_x.
Constant: Mol wt: 228.9.
Hazard Analysis
Toxicology: See cesium. For all other properties and associated hazards see cesium.

CESIUM HEXASULFIDE
General Information
Description: Bright, red crystals.
Formula: Cs_2S_6.
Constants: Mol wt: 458.22, mp: 186°C.
Hazard Analysis and Countermeasures
See sulfides and cesium.

CESIUM HYDRIDE
General Information
Description: White crystals.
Formula: CsH.
Constants: Mol wt: 133.92, mp: decomposes, d: 2.7.
Hazard Analysis and Countermeasures
See cesium and hydrides.

CESIUM HYDROXIDE
General Information
Description: Colorless, yellowish, very deliquescent crystals.
Formula: CsOH.
Constants: Mol wt: 149.92, mp: 272.3°C, d: 3.675.
Hazard Analysis and Countermeasures
See cesium compounds.

CESIUM IODATE
General Information
Description: Monoclinic, white crystals.
Formula: $CsIO_3$.
Constants: Mol wt: 307.83, d: 4.85.
Hazard Analysis and Countermeasures
See iodates and cesium.

CESIUM MERCURIC BROMIDE
General Information
Description: Rhombic crystals.
Formula: $CsBr \cdot 2HgBr_2$.
Constant: Mol wt: 933.71.
Hazard Analysis and Countermeasures
See mercury compounds, inorganic, bromides and cesium.

CESIUM MERCURIC CHLORIDE
General Information
Description: Cubic or rhombic, colorless crystals.
Formula: $CsCl \cdot HgCl_2$.
Constant: Mol wt: 439.89.
Hazard Analysis and Countermeasures
See mercury compounds, inorganic and chlorides.

CESIUM METAL
Shipping Regulations: Section 11.
IATA: Flammable solid, yellow label, not acceptable (passenger), 12 kilograms (cargo).

CESIUM METAL IN CARTRIDGES
Shipping Regulations: Section 11.
IATA: Flammable solid, yellow label, $\frac{1}{2}$ kilogram (passenger), 12 kilograms (cargo).

CESIUM m-PERIODATE
General Information
Description: Rhombic, white plates.
Formula: $CsIO_4$.
Constants: Mol wt: 323.83, d: 4.259.

Hazard Analysis and Countermeasures
See iodates and cesium.

CESIUM MONOBROMIDE
General Information
Description: Cubic, colorless crystals.
Formula: CsBr.
Constants: Mol wt: 212.83, mp: 636°C, bp: 1300°C, d: 4.44; 3.04 at 700°C.
Hazard Analysis and Countermeasures
See bromides and cesium.

CESIUM MONOIODIDE
General Information
Description: Cubic, colorless crystals.
Formula: CsI.
Constants: Mol wt: 259.83, mp: 621°C, bp: 1280°C, d: 4.510.
Hazard Analysis and Countermeasures
See iodides and cesium.

CESIUM NITRATE
General Information
Description: Colorless, hexagonal or cubic, glittering crystalline powder.
Formula: $CsNO_3$.
Constants: Mol wt: 194.92, mp: 414°C, bp: decomposes, d: 3.685; 2.71 at 500°C.
Hazard Analysis and Countermeasures
See cesium and nitrates.
Shipping Regulations: Section 11.
IATA: Oxidizing material, yellow label, 12 kilograms (passenger), 45 kilograms (cargo).

CESIUM NITRITE
General Information
Description: Yellow crystals.
Formula: $CsNO_2$.
Constant: Mol wt: 178.92.
Hazard Analysis and Countermeasures
See nitrites.

CESIUM OXALATE
General Information
Description: White solid.
Formula: $Cs_2C_2O_4$.
Constant: Mol wt: 353.84.
Hazard Analysis and Countermeasures
See oxalates.

CESIUM PENTASULFIDE
General Information
Description: Solid.
Formula: Cs_2S_5.
Constants: Mol wt: 426.15, mp: 210°C, d: 2.806 at 15°C.
Hazard Analysis and Countermeasures
See sulfides.

CESIUM PERCHLORATE
General Information
Description: Rhombic, colorless crystals.
Formula: $CsClO_4$.
Constants: Mol wt: 232.37, mp: decomposes, d: 3.327.
Hazard Analysis and Countermeasures
See perchlorates and cesium.

CESIUM PERMANGANATE
General Information
Description: A violet solid.
Formula: $CsMnO_4$.
Constants: Mol wt: 251.84, mp: decomposes, d: 3.597.
Hazard Analysis and Countermeasures
See permanganates.

CESIUM SULFIDE
General Information
Description: White, deliquescent crystals.

Formula: $Cs_2S \cdot 4H_2O$.
Constant: Mol wt: 369.95.
Hazard Analysis and Countermeasures
See sulfides.

CESIUM TETRASULFIDE
General Information
Description: Yellow crystals.
Formula: Cs_2S_4.
Constants: Mol wt: 394.08, mp: 160° C decomposes.
Hazard Analysis and Countermeasures
See sulfides.

CESIUM TRIBROMIDE
General Information
Description: Rhombic crystals.
Formula: $CsBr_3$.
Constants: Mol wt: 372.66, mp: 180° C.
Hazard Analysis and Countermeasures
See bromides and cesium.

CESIUM TRISULFIDE
General Information
Description: Yellow leaf.
Formula: Cs_2S_3.
Constants: Mol wt: 362.02, mp: 217° C, bp: 780° C.
Hazard Analysis and Countermeasures
See sulfides.

CETAB. See cetyl trimethyl ammonium bromide.

CETANE. See n-hexadecane.

CETYL ALCOHOL
General Information
Synonym: Hexadecanol.
Description: Solid or leaf-like crystals.
Formula: $C_{16}H_{33}OH$.
Constants: Mol wt: 242.4, mp: 49.3° C, bp: 344° C, d: 0.8176 at 50° /4° C.
Hazard Analysis
Toxicity: Practically non-toxic.
Fire Hazard: Moderate, when exposed to heat and/or flame; it can react with oxidizing materials.
Countermeasures
To Fight Fire: Foam, carbon dioxide, dry chemical or carbon tetrachloride (Section 6).
Storage and Handling: Section 7.

CETYL BROMIDE
General Information
Description: Dark yellow liquid.
Formula: $CH_3(CH_2)_{15}Br$.
Constants: Mol wt: 305.3, bp: 186–197° C at 10 mm, fp: 15° C, flash p: 350° F, d: 0.991 at 25° /25° C.
Hazard Analysis
Toxicity: Unknown.
Fire Hazard: Slight, when exposed to heat or flame.
Disaster Hazard: Dangerous; when it can react with oxidizing materials. See bromides.
Countermeasures
Storage and Handling: Section 7.

CETYL ISOQUINOLINIUM BROMIDE
Hazard Analysis
Toxicity: Details unknown; a fungicide.
Fire Hazard: Slight. (Section 6).
Disaster Hazard: Dangerous. See bromides.

Countermeasures
Storage and Handling: Section 7.

CETYL TRIMETHYL AMMONIUM BROMIDE
General Information
Synonym: Cetab.
Description: Creamy white powder. Water soluble.
Formula: $C_{16}H_{33}(CH_3)_3NBr$(at least 80%).
Hazard Analysis
Toxicology: See also benzethonium chloride.
Disaster Hazard: Dangerous; when strongly heated, it emits highly toxic fumes.

CETYLTRIMETHYLAMMONIUM PENTACHLORO-PHENATE
Hazard Analysis
See pentachlorophenol.

CEVADILLA. See sabadilla seed.

CEVADINE. See veratrine.

CG. See phosgene.

CHALCOCITE. See cuprous sulfide.

CHAMBER CRYSTALS. See nitrosyl sulfuric acid.

CHAMOMILE OIL
General Information
Description: Blue liquid, turning brownish-yellow.
Composition: Amyl and butyl esters of angelic and tiglic acids, butyric acid, etc.
Constant: D: 0.905–0.915 at 15° /15° C.
Hazard Analysis
Toxic Hazard Rating:
Acute Local: 0.
Acute Systemic: U.
Chronic Local: Allergen 1.
Chronic Systemic: U.
Fire Hazard: Slight; when heated.
Countermeasures
Personal Hygiene: Section 3.
Storage and Handling: Section 7.

CHARCOAL
General Information
Description: Black amorphous solid.
Composition: C + impurities.
Constants: Mol wt: 12.0, mp: > 3500° C, bp: 4200° C, d: 3.51.
Hazard Analysis
Toxic Hazard Rating:
Acute Local: 0.
Acute Systemic: 0.
Chronic Local: Inhalation 1.
Chronic Systemic: 0.
Caution: Carbon itself has no toxic action, but if it contains impurities, these may be toxic.
Fire Hazard: Moderate; reacts with oxidizing materials.
Spontaneous Heating: Yes, particularly when wet, freshly calcined or tightly packed.
Explosion Hazard: Slight, when exposed to heat or flame.
Countermeasures
Ventilation Control: Section 2.
To Fight Fire: Water (Section 6).
Storage and Handling: Section 7.

CHARCOAL, ACTIVATED. See activated carbon.

TOXIC HAZARD RATING CODE *(For detailed discussion, see Section 1.)*

0 NONE: (a) No harm under any conditions; (b) Harmful only under un-usual conditions or overwhelming dosage.

1 SLIGHT: Causes readily reversible changes which disappear after end of exposure.

2 MODERATE: May involve both irreversible and reversible changes; not severe enough to cause death or permanent injury.

3 HIGH: May cause death or permanent injury after very short exposure to small quantities.

U UNKNOWN: No information on humans considered valid by authors.

CHARCOAL BRIQUETTES. See charcoal.
Shipping Regulations: Section 11.
I.C.C.: Flammable solid, yellow label, 200 pounds.
  Coast Guard Classification: Inflammable solid.
  IATA: Flammable solid, yellow label, 25 kilograms (passenger), 25 kilograms (cargo).

CHARCOAL SHELL. See charcoal.
Shipping Regulations: Section 11.
  I.C.C.: Flammable solid, yellow label, 200 pounds.
  Coast Guard Classification: Inflammable solid.
  IATA: Flammable solid, yellow label, 12 kilograms (passenger), 95 kilograms (cargo).

CHARCOAL, WET. See also charcoal.
Shipping Regulations: Section 11.
  Coast Guard Classification: Not permitted.
  I.C.C.: Not accepted.
  IATA: Flammable solid, not acceptable (passenger and cargo).

CHARCOAL, WOOD (GROUND, CRUSHED, GRANULATED OR PULVERIZED). See also charcoal.
Shipping Regulations: Section 11.
  I.C.C.: Flammable solid, yellow label, 200 pounds.
  Coast Guard Classification: Inflammable solid; yellow label.
  IATA: Flammable solid, yellow label, 12 kilograms (passenger), 95 kilograms (cargo).

CHARCOAL, WOOD (LUMP). See also charcoal.
Shipping Regulations: Section 11.
  I.C.C.: Flammable solid, yellow label, 100 pounds.
  Coast Guard Classification: Inflammable solid.
  IATA: Flammable solid, yellow label, 25 kilograms (passenger), 25 kilograms (cargo).

CHARCOAL, WOOD SCREENINGS, MADE FROM "PINON". see also charcoal.
Shipping Regulations: Section 11.
  I.C.C.:Flammable solid, yellow label, 200 pounds.
  Coast Guard Classification: Inflammable solid.
  IATA: Flammable solid, yellow label, 12 kilograms (passenger), 95 kilograms (cargo).

CHARCOAL, WOOD SCREENINGS, OTHER THAN "PINON". See also charcoal.
Shipping Regulations: Section 11.
  I.C.C.:Flammable solid, not accepted.
  Coast Guard Classification: Inflammable solid; yellow label.
  IATA: Flammable solid, not acceptable (passenger and cargo).

CHARCOAL WOOD SCREENINGS, WET. See also charcoal.
Shipping Regulations: Section 11.
  Coast Guard Classification: Not permitted.

CHARGED OIL WILL JET PERFORATING GUNS
Shipping Regulations: Section 11.
  I.C.C. (total explosive contents in guns exceeding 20 pounds per motor vehicle): Explosive A, not accepted.
  I.C.C. (total explosive contents in guns not exceeding 20 pounds per motor vehicle): Class C explosive, not accepted.
  IATA: Explosive, not acceptable (passenger and cargo).

CHELIDONINE
General Information
Description: White crystalline powder.
Formula: $C_{20}H_{19}O_5N \cdot H_2O$.
Constants: Mol wt: 371.38, mp: 135–136°C.
Hazard Analysis
Toxic Hazard Rating:
  Acute Local: U.

Acute Systemic: Ingestion 3; Inhalation 3.
Chronic Local: U.
Chronic Systemic: U.
Toxicology: A central nervous system depressant, causing sleepiness, depression, slowing of pulse, and in large doses, coma and circulatory failures.
Fire Hazard: Slight. (Section 6).
Countermeasures
Ventilation Control: Section 2.
Treatment and Antidotes: Stimulants, epinephrine, pilocarpine. Call a physician.
Personal Hygiene: Section 3.
First Aid: Section 1.
Storage and Handling: Section 7.

CHELIDONINE HYDROCHLORIDE. See chelidonine.

CHELIDONINE PHOSPHATE. See chelidonine.

CHELIDONINE TANNATE. See chelidonine.

"CHEM-D". See DDT.

CHEM-HEX. See 1,2,3,4,5,6-hexachlorocyclohexane.

CHEMICAL AMMUNITION (CONTAINING CLASS A POISONS, LIQUIDS OR GASES)
Hazard Analysis
Toxicity: All class A poisons are highly toxic. See individual components.
Countermeasures
Shipping Regulations: Section 11.
  I.C.C.: Poison A, poison gas label, not accepted.
  Coast Guard Classification: Poison A; poison gas label.
  IATA: Poison A, not acceptable (passenger and cargo).

CHEMICAL AMMUNITION (CONTAINING CLASS B POISONS, LIQUIDS OR GASES)
Hazard Analysis
Toxicity: All class B poisons are toxic. See individual components.
Countermeasures
Shipping Regulations: Section 11.
  I.C.C.: Poison B, poison label, 55 gallons.
  Coast Guard Classification: Poison B; poison label.
  IATA: Poison B, poison label, not acceptable (passenger), 220 liters (cargo).

CHEMICAL AMMUNITION (CONTAINING CLASS C POISONS, LIQUIDS OR SOLIDS)
Hazard Analysis
Toxicity: All class C poisons are moderately toxic. See individual components.
Countermeasures
Shipping Regulations: Section 11.
  I.C.C.: Poison C, tear gas label, 20 pounds.
  Coast Guard Classification: Poison C; tear gas label.
  IATA: Poison C, poison label, not acceptable (passenger), 10 kilograms (cargo).

CHEMICAL AMMUNITION, EXPLOSIVE. See explosives.

CHEMICAL KITS
Shipping Regulations: Section 11.
  Coast Guard Classification: Corrosive liquid.
  IATA: Corrosive liquid, white label, 1 liter (passenger and cargo).

CHEWING GUM BASE
General Information
Description: Made up of a number of substances both natural and synthetic. It is used in the manufacture of chewing gum.
Hazard Analysis
Toxic Hazard Rating: U. A food additive permitted in food for human consumption (Section 10).

CHINA CLAY. See kaolin.

CHINA GREEN. See malachite green.

CHINA WOOD OIL. See tung oil.

CHINESE BEAN OIL. See soy bean oil.

CHINESE WAX
General Information
Description: White to yellowish-white solid.
Composition: Excretion of an insect. Primarily ceryl cerotate.
Constants: Mp: 92°C, d: 0.93.
Hazard Analysis
Toxic Hazard Rating:
  Acute Local: Allergen 1.
  Acute Systemic: U.
  Chronic Local: Allergen 1.
  Chronic Systemic: U.
Fire Hazard: Slight, when exposed to heat.
Countermeasures
To Fight Fire: Foam, carbon dioxide, dry chemical or carbon tetrachloride (Section 6).
Personal Hygiene: Section 3.
Storage and Handling: Section 7.

CHINESE WHITE. See zinc oxide.

CHINOLINE. See quinoline.

CHINONE. See quinone.

CHINOSOL. See quinosol.

CHIOLITE. See sodium aluminum fluoride.

CHLORACETOPHENONE. See chloroacetophenone.

CHLORAL
General Information
Synonym: Trichloracetaldehyde.
Description: Oily liquid; irritating odor.
Formula: CCl_3CHO.
Constants: Mol wt: 147.4, mp: −57.5°C, bp: 97.8°C, flash p: none, d: 1.510 at 20°/4°C, vap. d.: 5.1.
Hazard Analysis
Toxic Hazard Rating:
  Acute Local: Irritant 1.
  Acute Systemic: Ingestion 3.
  Chronic Local: U.
  Chronic Systemic: Ingestion 2.
Disaster Hazard: Dangerous; when heated to decomposition, it emits toxic fumes.
Countermeasures
Personal Hygiene: Section 3.
First Aid: Section 1.
Storage and Handling: Section 7.
MCA warning label.

CHLORAL ALCOHOLATE
General Information
Synonym: Chloral ethylalcoholate; trichloracetaldehyde monoethylacetal.
Description: Crystals, less soluble in water than chloral hydrate; soluble in organic solvents.
Formula: $C_4H_7Cl_3O_2$.
Constants: Mol wt: 193.47, d: 1.143, mp: 47.5°C, bp: 116°C.

Hazard Analysis
Toxic Hazard Rating:
  Acute Local: Irritant 1.
  Acute Systemic: Ingestion 1; Inhalation 1.
  Chronic Local: U.
  Chronic Systemic: U.

CHLORAL HYDRATE
General Information
Description: Transparent, colorless crystals; aromatic, penetrating, slightly acrid odor and slightly bitter, caustic taste.
Formula: $CCl_3CH(OH)_2$.
Constants: Mol wt: 165.41, mp: 52°C, bp: 97.5°C, d: 1.901.
Hazard Analysis
Toxic Hazard Rating:
  Acute Local: Irritant 2; Ingestion 2; Inhalation 2.
  Acute Systemic: Ingestion 3.
  Chronic Local: U.
  Chronic Systemic: Ingestion 2.
Toxicology: Chloral hydrate is a poisonous drug popularly called "knockout drops." The liquid and vapors are both very dangerous to the eyes. If this material is ingested, the pupils contract during sleep, but dilate upon waking. A number of incidents have been reported where it was given in alcoholic beverages and death was caused when only its hypnotic effect was desired. Recently there has been much addiction to chloral hydrate. These addicts are mostly between the ages of 21 and 40 years. In the majority of cases they are also addicted to alcohol or opium. Occasionally this material may produce excitement and delirium. Nausea and vomiting may be the result of local action of the drug upon the stomach. The respiration is irregular and shallow and the pulse is scarcely perceptible. The pupils are moderately contracted, but rarely dilated. The face is cyanotic or in the early stages, flushed. On account of the dilation of the vessels, the extremities are cold and the blood pressure and temperature are slightly lower, but little more so than in natural sleep. If the dose exceeds 2 to 3 grams, the patient passes into stupor and coma with complete muscular relaxation. The action is very similar to that of chloroform. Death is ordinarily caused by paralysis of the respiratory center. The average fatal dose is placed at about 10 grams. Use caution in giving over the therapeutic dose.
Fire Hazard: Slight; when heated.
Countermeasures
Personal Hygiene: Section 3.
First Aid: Section 1.
Treatment and Antidotes: A physician should be called at once. Wash out the stomach. Administer strychnine hypodermically, also give caffeine, and caffeine with sodium benzoate. Maintain the temperature of the patient with the use of electric pads, hot water bottles and blankets if necessary.
Storage and Handling: Section 7.

CHLORAL HYDROCYANIDE. See chlorocyanohydrin.

CHLORAMINE. See chloramine-T.

CHLORAMINE-T
General Information
Synonyms: Chloramine; sodium p-toluene sulfon chloramide.
Description: White or faintly yellow crystals. Slight chlorine odor; water soluble.

TOXIC HAZARD RATING CODE (For detailed discussion, see Section 1.)

0 NONE: (a) No harm under any conditions; (b) Harmful only under unusual conditions or overwhelming dosage.

1 SLIGHT: Causes readily reversible changes which disappear after end of exposure.

2 MODERATE: May involve both irreversible and reversible changes; not severe enough to cause death or permanent injury.

3 HIGH: May cause death or permanent injury after very short exposure to small quantities.

U UNKNOWN: No information on humans considered valid by authors.

Formula: $C_7H_7ClNNaO_2S \cdot 3H_2O$.
Constant: Mol wt: 281.
Hazard Analysis
Toxic Hazard Rating:
Acute Local: Irritant 1.
Acute Systemic: U.
Chronic Local: Allergen 1.
Chronic Systemic: U.
Toxicology: Inhalation of vapors can cause vasomotor rhinitis and asthma.
Disaster Hazard: Dangerous; see sulfonates and chlorides.
Countermeasures
Personal Hygiene: Section 3.
Storage and Handling: Section 7.

CHLORANIL. See tetrachlorobenzoquinone.

cis-CHLORAQUOTETRAMMINECOBALT (III) CHLORIDE
General Information
Description: Rhombic violet crystals.
Formula: $[Co(NH_3)_4(H_2O)Cl]Cl_2$.
Constants: Mol wt: 251.40, mp: decomposes, d: 1.847.
Hazard Analysis
Toxicity: See cobalt compounds.

"CHLORASOL" FUMIGANT. See ethylene dichloride and carbon tetrachloride.

CHLORATE AND BORATE MIXTURES. See also chlorates, N.O.S. and boron compounds.
Shipping Regulations: Section 11.
I.C.C.: Oxidizing material; yellow label, 100 pounds.
Coast Guard Classification: Oxidizing material; yellow label.
IATA (mixtures containing more than 28% chlorate): Oxidizing material, yellow label, 12 kilograms (passenger), 45 kilograms (cargo).

CHLORATE AND MAGNESIUM CHLORIDE MIXTURE
Shipping Regulations: Section 11.
I.C.C.: Oxidizing material, yellow label, 100 pounds.
IATA (mixtures containing more than 28% chlorate): Oxidizing material, yellow label, 12 kilograms (passenger), 45 kilograms (cargo).

CHLORATE OF POTASH. See potassium chlorate.
Shipping Regulations: Section 11.
I.C.C.: Oxidizing material; yellow label, 100 pounds.
Coast Guard Classification: Oxidizing material; yellow label.
IATA: Oxidizing material, yellow label, 12 kilograms (passenger), 45 kilograms (cargo).

CHLORATE OF SODA. See sodium chlorate.
Shipping Regulations: Section 11.
I.C.C.: Oxidizing material; yellow label, 100 pounds.
Coast Guard Classification: Oxidizing material; yellow label.
IATA: Oxidizing material, yellow label, 12 kilograms (passenger), 45 kilograms (cargo).

CHLORATES, N.O.S.
General Information
Description: Chlorates are a combination of a metal or hydrogen and $^-ClO_3$ monovalent radical. They are crystalline and somewhat deliquescent.
Hazard Analysis
Toxic Hazard Rating:
Acute Local: Irritant 1.
Acute Systemic: Ingestion 2.
Chronic Local: Irritant 1.
Chronic Systemic: Ingestion 2.
Toxicology: The principal toxic effects of chlorates are the production of methemoglobin in the blood and destruc-
tion of red blood corpuscles. The latter may lead to irritation of the kidneys. Damage to heart muscle has been reported.
Fire Hazard: Moderate, in contact with flammable matter. When contaminated with oxidizable materials, they are particularly sensitive to friction, heat and shock; they are powerful oxidizing agents (Section 6).
Explosion Hazard: Moderate, when shocked, exposed to heat or rubbed, particularly when contaminated with sugar, charcoal, shellac, sulfur, starch, sawdust, sulfuric acid, ammonium compounds, cyanides, phosphorus or antimony sulfide.

Chlorates when mixed with combustible materials may form explosive mixtures. For instance, potassium chlorate, when mixed with sulfur or with other combustible substances explodes on friction. Pure chlorates which have been spilled on the floor, or mixed with small amounts of impurities, become very sensitive to shock and friction. Water is considered the best agent for fighting fires involving chlorates. In the explosive industry, chlorates are used as oxidizing agents in the primer caps in combination with mercury fulminates, phosphorus, antimony sulfide and other combustible substances. They are used in pyrotechnic mixtures, as a component of airplane flares and aerial bombs. They are also used as a component of permissible explosives. Chlorates are used extensively in the manufacture of chlorate explosives. The chief constituent of such an explosive is from 60 to 80 percent chlorate. This can be the chlorate of ammonium, sodium or potassium. The other ingredients in such a mixture are combustible materials, such as metallic powders, powdered sulfur, powdered charcoal or possibly mixtures of organic matter. Nitro derivatives of benzene, toluene, and other aromatic compounds are also added. Paraffin may be added as a desensitizer. Recently, similar mixtures were used in Europe but with the addition of small amounts of nitroglycerin or collodion cotton. Chlorate explosives are more sensitive than modern permissible explosives, and therefore not as safe as for instance the perchlorate explosives, or the permissibles. Plastic mixtures of chlorate explosives (containing nitroglycerin) are somewhat less sensitive to shock and friction, in spite of the nitroglycerin present, than the dryer explosives with no nitroglycerin. In this case the nitroglycerin or "explosive oil," as it is known, serves to wet the rest of the mixture. Barium chlorate is shipped and stored in wooden boxes, barrels, or kegs. It should have isolated storage in a cool, ventilated place, away from acute fire hazards and should not be stored in the same building with combustible materials, acids, sulfur, powdered magnesium or powdered aluminum. Examples of chlorates used in the explosive industry, would be potassium chlorate, sodium chlorate and barium chlorate.
Disaster Hazard: Moderate; shock will explode them; when heated to decomposition, they can emit toxic fumes and explode; can react with reducing materials.
Countermeasures
Personal Hygiene: Section 3.
Storage and Handling: Section 7.
Shipping Regulations: Section 11.
I.C.C.: Oxidizing material; yellow label, 200 pounds.
IATA: Oxidizing material, yellow label, 12 kilograms (passenger), 45 kilograms (cargo).

CHLORATES, N.O.S. WET. See also chlorates, N.O.S.
Shipping Regulations: Section 11.
I.C.C.: Oxidizing material; yellow label, 200 pounds.
Coast Guard Classification: Oxidizing material; yellow label.
IATA: Oxidizing material, yellow label, 12 kilograms (passenger), 95 kilograms (cargo).

CHLORAURIC ACID. See gold compounds.

CHLORAZENE. See chloramine.

CHLORDAN. See chlordane.

CHLORDANE
General Information
Synonyms: 1,2,4,5,6,7,8,8 Octachloro-4,7-methano-3a,4,7,-7a-tetrahydroindane; chlordan; octachlorotetrahydro methano indane; "Octa Klor"; "1068"; "Velsicol 1068"; "Dowklor"; "Ortho-Klor"; and other trade names.
Description: Colorless to amber, odorless, viscous liquid.
Formula: $C_{10}H_6Cl_8$.
Constants: Mol wt: 409.75, bp: 175°C, d: 1.57–1.67 at 60°/60°F.
Hazard Analysis
Toxic Hazard Rating:
Acute Local: U.
Acute Systemic: Ingestion 3; Inhalation 3; Skin Absorption 3.
Chronic Local: U.
Chronic Systemic: Ingestion 3; Inhalation 3; Skin Absorption 2.
Toxicity: Probably toxic; an insecticide.
TLV: ACGIH (accepted); 0.5 milligrams per cubic meter of air. May be absorbed via the skin.
Toxicology: Chlordane is readily absorbed through the skin as well as through other portals. It is a central nervous system stimulant whose exact mode of action is unknown, but it may involve microsomal enzyme stimulation. Animals poisoned by this and related compounds show an extremely marked loss of appetite and neurological symptoms. The fatal dose to man is unknown. It has been estimated to be between 6 and 60 grams (1/5 to 2 oz.). One person receiving an accidental skin application of 25 percent solution (amounting to something over 30 g. of technical chlordane) developed symptoms within about 40 minutes and died before medical attention was obtained. In two patients, death followed exposure to low oral doses of chlordane (2–4 g.); on microscopic examination both patients showed severe chronic fatty degeneration of the liver, characteristic of chronic alcoholism. Although these two fatalities cannot be attributed exclusively to chlordane, they are entirely consistent with previous observations that the toxicity of other chlorinated hydrocarbons is much enhanced in the presence of chronic liver damage. The dangerous chronic dose in man is unknown.

One person poisoned by chlordane developed convulsions within 40 minutes of gross skin contamination and died, apparently of respiratory failure, before medical aid could be obtained.

Acutely poisoned experimental animals show similar signs. Experimental animals exposed to repeated small doses exhibit hyperexcitability, tremors, and convulsions, and those which survive long enough show marked anorexia and loss of weight. Symptoms in animals frequently occur within an hour of the administration of a large dose, but death often is delayed for several days depending on the dosage and route of administration. In any event, symptoms are of longer duration whith chlordane than with DDT under similar conditions.

Laboratory findings are essentially normal, except that the insecticide may be demonstrated in tissues of poisoned animals by means of bioassay. A method for specific, quantitative chemical analysis for chlordane is now available using small amounts of subcutaneous fat. Chronically poisoned animals show degenerative changes in the liver and kidney tubules.
Disaster Hazard: Dangerous; see chlorides.
Countermeasures
Storage and Handling: Section 7.
Treatment of Poisoning: Removal of the poison from the skin or the alimentary tract should be attempted. Oil laxatives should be avoided. The nervous symptoms may best be combatted with pentobarbital or phenobarbital.
Shipping Regulations: Section 11.
MCA warning label.
IATA: Other restricted articles, class A, no label required, no limit (passenger and cargo).

CHLORETHYL BENZENE
General Information
Description: Liquid.
Formula: $C_6H_5ClC_2H_5$.
Constant: Mol wt: 141.6.
Hazard Analysis
Toxicity: See chlorinated hydrocarbons, aromatic.
Fire Hazard: Moderate; when exposed to heat or flame (Section 6).
Explosion Hazard: Unknown.
Disaster Hazard: Dangerous; when heated to decomposition, it emits toxic fumes; reacts with oxidizing materials.
Countermeasures
Storage and Handling: Section 7.

CHLOREX. See dichloroethyl ether.

CHLORGUANIDE
General Information
Synonyms: 1-(p-Chlorophenyl)-5-isopropylibiguanide hydrochloride.
Description: White powder.
Formula: $ClC_6H_4C_3H_7C_2H_5N_3HCl$.
Constant: Mol wt: 290.2.
Hazard Analysis
Toxicity: Unknown.
Disaster Hazard: Dangerous; see chlorides.
Countermeasures
Storage and Handling: Section 7.

CHLORHYDROL
General Information
Synonym: Aluminum chlorohydroxide complex.
Formula: $Al(OH)_2Cl$.
Constant: Mol wt: 96.4.
Hazard Analysis
Toxicology: No details. Probably has low toxicity.

CHLORIC ACID
General Information
Description: Colorless solution.
Formula: $HClO_3 \cdot 7H_2O$.
Constants: Mol wt: 210.58, mp: $< -20°C$, bp: decomposes 40°C, d: 1.282 at 14.2°C.
Hazard Analysis
Toxic Hazard Rating: (See also chlorates).
Acute Local: Irritant 3; Ingestion 3; Inhalation 3.
Acute Systemic: U.
Chronic Local: U.
Chronic Systemic: U.
Fire Hazard: Dangerous; ignites organic matter upon contact; a very powerful oxidizing agent.

TOXIC HAZARD RATING CODE (For detailed discussion, see Section 1.)

0 NONE: (a) No harm under any conditions; (b) Harmful only under unusual conditions or overwhelming dosage.

1 SLIGHT: Causes readily reversible changes which disappear after end of exposure.

2 MODERATE: May involve both irreversible and reversible changes; not severe enough to cause death or permanent injury.

3 HIGH: May cause death or permanent injury after very short exposure to small quantities.

U UNKNOWN: No information on humans considered valid by authors.

Explosion Hazard: Slight.

Disaster Hazard: Dangerous; see chlorides; reacts vigorously with reducing materials.

Countermeasures

Ventilation Control: Section 2.

Personnel Protection: Section 3.

Storage and Handling: Section 7.

Shipping Regulations: Section 11.

IATA: Oxidizing material, not acceptable (passenger and cargo).

CHLORIC ETHER
General Information

Description: A liquid solution of 60 cc chloroform and 940 cc alcohol.

Hazard Analysis

Toxic Hazard Rating: (See also ethyl alcohol and chloroform).

Acute Local: U.

Acute Systemic: Ingestion 3; Inhalation 3; Skin Absorption 2.

Chronic Local: Irritant 2; Ingestion 2; Inhalation 2.

Chronic Systemic: Ingestion 1; Inhalation 1; Skin Absorption 1.

Fire Hazard: Moderate, when exposed to heat or flame (Section 6).

Disaster Hazard: Dangerous; when heated to decomposition, it emits highly toxic fumes of phosgene; can react vigorously with oxidizing materials.

Countermeasures

Ventilation Control: Section 2.

Personnel Protection: Section 3.

Storage and Handling: Section 7.

CHLORIDE OF LIME. See bleaching powder.

CHLORIDES
Hazard Analysis

Toxicity: Varies widely. Sodium chloride (table salt) has very low toxicity, while carbonyl chloride (phosgene) is lethal in small doses.

Disaster Hazard: Dangerous; when heated to decomposition or on contact with acids or acid fumes they evolve highly toxic chloride fumes. Some organic chlorides decompose to yield phosgene.

Countermeasures

Storage and Handling: Section 7.

CHLORINATED ANTHRACENE OIL. See carbolineum.

CHLORINATED BIPHENOLS. See chlorinated diphenyls.

CHLORINATED CAMPHENE. See "Toxaphene."

CHLORINATED DIPHENYL OXIDE
Hazard Analysis

Toxic Hazard Rating:

Acute Local: Irritant 2.

Acute Systemic: Ingestion 3; Inhalation 3.

Chronic Local: Irritant 3.

Chronic Systemic: Inhalation 3; Skin Absorption 2.

TLV: ACGIH (accepted); 0.5 milligrams per cubic meter of air.

Disaster Hazard: Dangerous; when heated to decomposition, it emits highly toxic fumes.

Countermeasures

Storage and Handling: Section 7.

Shipping Regulations: Section 11.

MCA warning label.

CHLORINATED DIPHENYLS
General Information

Description: Colorless mobile liquid.

Constants: Bp: 340–375° C, flash p: 383° F (C.O.C), d: 1.44 at 30° C.

Hazard Analysis

Toxic Hazard Rating:

Acute Local: Irritant 2.

Acute Systemic: Ingestion 3; Inhalation 3.

Chronic Local: Irritant 3.

Chronic Systemic: Inhalation 3; Skin Absorption 2.

TLV for chlorodiphenyl (42% chlorine) ACGIH (accepted): 1 milligram per cubic meter of air. Can be absorbed via intact skin.

TLV for chlorodiphenyl (54% chlorine) ACGIH (accepted): 0.5 milligram per cubic meter of air. Can be absorbed via intact skin.

Toxicology: Like the chlorinated naphthalenes, the chlorinated diphenyls have two distinct actions on the body, namely, a skin effect and a toxic action on the liver. The lesion produced in the liver is an acute yellow atrophy. This hepato toxic action of the chlorinated diphenyls appears to be increased if there is exposure to carbon tetrachloride at the same time. The higher the chlorine content of the diphenyl compound, the more toxic is it liable to be. Oxides of chlorinated diphenyls are more toxic than the unoxidized materials.

The skin lesion is known as chloracne, and consists of small pimples and dark pigmentation of the exposed areas, initially. Later, comedones and pustules develop. In persons who have suffered systemic intoxication the usual signs and symptoms are nausea, vomiting, loss of weight, jaundice, edema and abdominal pain. Where the liver damage has been severe, the patient may pass into coma and die.

Fire Hazard: Slight, when exposed to heat or flame (Section 6).

Disaster Hazard: Dangerous; when heated to decomposition, they emit highly toxic fumes.

Countermeasures

Ventilation Control: Section 2.

Personnel Protection: Section 3.

First Aid: Section 1.

Storage and Handling: Section 7.

Shipping Regulations: Section 11.

MCA warning label.

CHLORINATED HYDROCARBONS, ALIPHATIC
Hazard Analysis

Toxicology: The subsitution of a chlorine (or other halogen) atom for a hydrogen greatly increases the anesthetic action of a member of the aliphatic hydrocarbons. In addition, the chlorine derivative is usually less specific in its action and may affect other tissues of the body in addition to those of the central nervous system; in many cases the chlorine derivative is quite toxic. Thus, chloroform, in addition to its narcotic qualties, may cause liver, heart, and kidney damage.

As a general rule, the unsaturated chlorine derivatives are highly toxic but less toxic than the saturated derivatives, thus causing degenerative changes in the liver and kidneys less frequently. In the saturated group, the narcotic effect is enhanced with an increase in the number of chlorine atoms. However, there is less relationship between the number of chlorine atoms present and the toxicity of the compound.

In dealing with these chlorinated hydrocarbons, it must be remembered that a toxic action may result from repeated exposure to concentrations which are too low to produce a narcotic effect, and which, consequently, are too low to give warning of danger. Individual susceptibility is also important when poisoning by this group of solvents is being considered. Certain workmen may be seriously affected by concentrations that seem to have no effect on fellow employees in the same exposure.

Disaster Hazard: Dangerous; when heated to decomposition, they emit highly toxic fumes of phosgene; they can react with oxidizing materials.

Countermeasures

Storage and Handling: Section 7.

CHLORINATED HYDROCARBONS, AROMATIC
Hazard Analysis
Toxicology: In most instances it is difficult to predict the toxicity of these compounds. However, in the case of most aromatic chlorine compounds, their toxicity is usually no greater, and frequently is less, than that of the corresponding aromatic hydrocarbons, with the notable exception of naphthalene.
Fire Hazard: Unknown.
Explosion Hazard: Unknown.
Disaster Hazard: Dangerous; when heated to decomposition, they emit toxic fumes; they can react with oxidizing materials.
Countermeasures
Storage and Handling: Section 7.

CHLORINATED HYDROCHLORIC ETHER. See ethylidene chloride.

CHLORINATED LIME. See bleaching powder.

CHLORINATED NAPHTHALENES
Hazard Analysis
Toxic Hazard Rating: Acute Local: Irritant 3.
  Acute Systemic: Ingestion 3; Inhalation 3.
  Chronic Local: Irritant 3.
  Chronic Systemic: Ingestion 3; Inhalation 3; Skin Absorption 3.
Toxicology: The action of the chlorinated naphthalenes on the body is quite similar to that of the chlorinated diphenyls, the chief effects being the production of chloracne of the skin and, systemically, an acute yellow atrophy of the liver. See also chlorinated diphenyls.
Disaster Hazard: Dangerous; see chlorides.
Countermeasures
Ventilation Control: Section 2.
Personnel Protection: Section 3.
Personal Hygiene: Section 3.
First Aid: Section 1.
Storage and Handling: Section 7.
Shipping Regulations: Section 11.
  MCA warning label (for tri and higher).

CHLORINATED PHENOLS
Hazard Analysis
Toxic Hazard Rating:
  Acute Local: Irritant 3, Ingestion 3; Inhalation 3.
  Acute Systemic: Ingestion 3; Inhalation 3; Skin Absorption 3.
  Chronic Local: U.
  Chronic Systemic: Ingestion 3; Inhalation 3; Skin Absorption 3.
Disaster Hazard: Dangerous; when heated to decomposition, they emit highly toxic fumes.
Countermeasures
Storage and Handling: Section 7.
Ventilation and Indicated Hygiene: Section 2.
Respiratory Protection: Section 3.

CHLORINATED TRIPHENYLS
Hazard Analysis
Toxic Hazard Rating:
  Acute Local: Irritant 2.
  Acute Systemic: Ingestion 2; Inhalation 2; Skin Absorption 2.
  Chronic Local: Irritant 3.
  Chronic Systemic: Ingestion 2; Inhalation 2; Skin Absorption 2.

See also chlorinated dyphenyls.
Disaster Hazard: Dangerous; when heated to decomposition, they emit highly toxic fumes of chlorides.
Countermeasures
Ventilation Control: Section 2.
Personnel Protection: Section 3.
Personal Hygiene: Section 3.
Storage and Handling: Section 7.

CHLORINE
General Information
Description: Greenish-yellow gas, liquid, or rhombic crystals.
Formula: Cl_2.
Constants: Mol wt: 70.914, mp: $-101°C$, bp: $-34.5°C$, d: (liquid) 1.47 at $0°C$ (3.65 atmos.), vap. press.: 4800 mm at $20°C$, vap. d.: 2.49.
Hazard Analysis
Toxic Hazard Rating:
  Acute Local: Irritant 3; Inhalation 3.
  Acute Systemic: 0.
  Chronic Local: U.
  Chronic Systemic: U.
TLV: ACGIH (tentative); 1 part per million in air; 3 milligrams per cubic meter of air.
Toxicology: Chlorine is extremely irritating to the mucous membranes of the eyes and respiratory tract. It combines with moisture to liberate nascent oxygen and form hydrochloric acid. Both these substances, if present in quantity, cause inflammation of the tissues with which they come in contact. If the lung tissues are attacked, pulmonary edema may result. A concentration of 3.5 ppm produces a detectable odor; 15 ppm causes immediate irritation of the throat. Concentrations of 50 ppm are dangerous for even short exposures. 1,000 ppm may be fatal, even where the exposure is brief.
  Because of its intensely irritating properties, severe industrial exposure seldom occurs, as the workman is forced to leave exposure before he can be seriously affected. In cases where this is impossible, the initial irritation of the eyes and mucous membranes of the nose and throat is followed by cough, a feeling of suffocation, and later, pain and a feeling of constriction in the chest. If exposure has been severe, pulmonary edema may follow, with rales being heard over the chest. It is a common air contaminant (Section 4).
Radiation Hazard: Section 5. For permissible levels, see Table 5, p. 150.
  Artificial isotope ^{36}Cl, half life 3×10^5 y. Decays to stable ^{36}A by emitting beta particles of 0.71 MeV.
  Artificial isotope ^{38}Cl, half life 38 m. Decays to stable ^{38}A by emitting beta particles of 1.11 (31%), 2.77 (16%), 4.81 (53%) MeV. Also emits gamma rays of 1.60, 2.16 MeV.
Fire Hazard: Moderate; can react to cause fires or explosions upon contact with turpentine, ether, ammonia gas, illuminating gas, hydrocarbons, hydrogen and powdered metals (Section 6).
Explosion Hazard: Slight, by reaction with reducing agents.
Disaster Hazard: Dangerous; when heated, it emits highly toxic fumes; will react with water or steam to produce toxic and corrosive fumes of hydrogen chloride.
Countermeasures
Ventilation Control: Section 2.
Personnel Protection: Section 3.
Personal Hygiene: Section 3.

TOXIC HAZARD RATING CODE *(For detailed discussion, see Section 1.)*

.0 NONE: (a) No harm under any conditions; (b) Harmful only under unusual conditions or overwhelming dosage.

1 SLIGHT: Causes readily reversible changes which disappear after end of exposure.

2 MODERATE: May involve both irreversible and reversible changes; not severe enough to cause death or permanent injury.

3 HIGH: May cause death or permanent injury after very short exposure to small quantities.

U UNKNOWN: No information on humans considered valid by authors.

First Aid: Section 1.
Storage and Handling: Section 7.
Shipping Regulations: Section 11.
  I.C.C.: Nonflammable gas; green label, 150 pounds.
  Coast Guard Classification: Nonflammable gas; green gas label.
  MCA warning label.
  IATA: Nonflammable gas, green label, not acceptable (passenger), 70 kilograms (cargo).

CHLORINE AZIDE
General Information
Synonym: Chlor(o)azide.
Description: Gas.
Formula: ClN_3.
Constant: Mol wt: 77.48
Hazard Analysis
Toxic Hazard Rating:
  Acute Local: Irritant 3; Inhalation 3.
  Acute Systemic: Inhalation 3.
  Chronic Local: U.
  Chronic Systemic: U.
Explosion Hazard: Severe, when shocked or exposed to heat or flame.
Disaster Hazard: Dangerous; shock can explode; when heated to decomposition, it emits highly toxic fumes of chlorine and oxides of nitrogen; will react with water or steam to produce toxic and corrosive fumes of hydrogen chloride.
Countermeasures
Ventilation Control: Section 2.
Personnel Protection: Section 3.
Personal Hygiene: Section 3.
First Aid: Section 1.
Storage and Handling: Section 7.

CHLORINE CYANIDE. See cyanogen chloride.

CHLORINE DIOXIDE
General Information
Description: Red-yellow gas or orange-red crystals.
Formula: ClO_2.
Constants: Mol wt: 67.5, mp: $-59°C$, bp: $9.9°C$ at 731 mm, d: 309 g/liter at 11°C.
Hazard Analysis
Toxic Hazard Rating:
  Acute Local: Irritant 3; Ingestion 3; Inhalation 3.
  Acute Systemic: U.
  Chronic Local: Irritant 3.
  Chronic Systemic: U.
TLV: ACGIH (accepted) 0.1 parts per million in air; 0.3 mg/cubic meter of air.
Fire Hazard: Dangerous; a powerful oxidizer (Section 6).
Explosion Hazard: Severe, when heated to 100°C or by chemical reaction.
Disaster Hazard: Dangerous; shock will explode it; when heated to decomposition, it emits highly toxic fumes of chlorine; will react with water or steam to produce toxic and corrosive fumes of hydrochloric acid; can react vigorously with reducing materials.
Countermeasures
Ventilation Control: Section 2.
Personnel Protection: Section 3.
Personal Hygiene: Section 3.
First Aid: Section 1.
Storage and Handling: Section 7.

CHLORINE DIOXIDE HYDRATE, FROZEN
Shipping Regulations: Section 11.
  I.C.C.: Oxidizing material, yellow label, not accepted.

CHLORINE HEPTAOXIDE
General Information
Description: Colorless oil.
Formula: Cl_2O_7.

Constants: Mol wt: 182.91, mp: $-91.5°C$, bp: 82°C, vap. press.: 100 mm at 29.1°C.
Hazard Analysis
Toxic Hazard Rating:
  Acute Local: Irritant 3; Ingestion 3; Inhalation 3.
  Acute Systemic: U.
  Chronic Local: U.
  Chronic Systemic: U.
Fire Hazard: Dangerous; a very powerful oxidizing agent (Section 6).
Explosion Hazard: Severe, when shocked or exposed to heat or flame.
Disaster Hazard: Dangerous; shock or heat will explode it; on decomposition, it emits highly toxic fumes of chlorine; will react with water or steam to produce toxic and corrosive fumes.
Countermeasures
Ventilation Control: Section 2.
Personnel Protection: Section 3.
Personal Hygiene: Section 3.
First Aid: Section 1.
Storage and Handling: Section 7.

CHLORINE HYDRATE
General Information
Description: Rhombic light yellow crystals.
Formula: $Cl_2 \cdot 8H_2O$.
Constants: Mol wt: 215.04, mp: decomposes 9.6°C, d: 1.23.
Hazard Analysis
Toxic Hazard Rating:
  Acute Local: Irritant 3; Ingestion 3; Inhalation 3.
  Acute Systemic: U.
  Chronic Local: U.
  Chronic Systemic: U.
Disaster Hazard: Dangerous; see chlorine; will react with water or steam to produce toxic and corrosive fumes.
Countermeasures
Ventilation Control: Section 2.
Personnel Protection: Section 3.
Personal Hygiene: Section 3.
First Aid: Section 1.
Storage and Handling: Section 7.

CHLORINE MONOFLUORIDE
General Information
Description: Nearly colorless gas.
Formula: ClF.
Constants: Mol wt: 54.46, mp: $-154 \pm 0.5°C$, bp: $-100.8°C$, d: 1.62 at $-100°C$.
Hazard Analysis and Countermeasures
See fluorides and chlorine.

CHLORINE MONOXIDE
General Information
Description: Yellow-red gas or red-brown liquid.
Formula: Cl_2O.
Constants: Mol wt: 86.91, mp: $-20°C$, bp: 2.2°C, d: 3.89 g/liter at 0°C, lel: 23.5%, uel: 100%.
Hazard Analysis
Toxicity: See chlorine.
Explosion Hazard: Severe, when shocked or exposed to heat.
Disaster Hazard: Dangerous; see chlorine; will react with water or steam to produce toxic and corrosive fumes. When heated it explodes.
Countermeasures
Storage and Handling: Section 7.

CHLORINE TETROXIDE
General Information
Synonym: Hypochlorous anhydride.
Description: Gas.
Formula: ClO_4 or Cl_2O_8.
Constants: Mol wt: 99.46, bp: decomposes.

Hazard Analysis

Toxicity: See chlorine.

Fire Hazard: Moderate; a powerful oxidizer.

Explosion Hazard: Moderate, when exposed to heat.

Disaster Hazard: Dangerous; see chlorine; will react with water or steam to produce toxic and corrosive fumes; on contact with acid fumes, it can emit flammable vapors.

Countermeasures

Storage and Handling: Section 7.

CHLORINE TETROXYFLUORIDE. See fluorine perchlorate.

CHLORINE TRIFLUORIDE

General Information

Description: Colorless gas to yellow liquid, sweet odor.

Formula: ClF_3.

Constants: Mol wt: 92.46, mp: $-83°C$, bp: 11.8C, d: 1.77 at 13°C.

Hazard Analysis

Toxicity: See fluorides, chlorine and fluorine.

TLV: ACGIH (accepted); 0.1 parts per million of air; 0.4 milligrams per cubic meter of air.

Fire Hazard: Dangerous. Spontaneously flammable. It ignites on contact with many organic compounds, and reacts violently with oxidizable materials, metals, etc. (Section 6).

Disaster Hazard: Dangerous; when heated to decomposition or on contact with acid or acid fumes, it emits highly toxic fumes; will react with water or steam to produce much heat and toxic and corrosive fumes; reacts vigorously with reducing materials.

Countermeasures

Ventilation Control: Section 2.

Storage and Handling: Section 7.

Shipping Regulations: Section 11.

  I.C.C.: Corrosive liquid; white label, 25 pounds.

  Coast Guard Classification: Corrosive liquid; white label.

  IATA: Corrosive liquid, white label, not acceptable (passenger), 45 kilograms (cargo).

CHLOROACETALDEHYDE

General Information

Synonym: Chloroaldehyde.

Description: Clear, colorless liquid; pungent odor.

Formula: C_2H_3OCl.

Constants: Mol wt: 78.5, bp: 90.0–100.1°C (40% sol.), fp: $-16.3°C$ (40% sol.), flash p: 190°F, d: 1.19 at 25°/25°C (40% sol.), vap. press.: 100 mm at 45°C (40% sol.).

Hazard Analysis

Toxicity: See aldehydes.

TLV: ACGIH (accepted) 1 part per million in air; 3 milligrams per cubic meter of air.

Fire Hazard: Moderate, when exposed to heat or flame.

To Fight Fire: Water, foam, carbon dioxide, dry chemical or carbon tetrachloride (Section 6).

Disaster Hazard: Dangerous; see chlorides; reacts with oxidizing materials.

Countermeasures

Storage and Handling: Section 7.

CHLOROACETIC ACID. See monochloroacetic acid.

CHLOROACETONE

General Information

Synonym: Chlorinated acetone.

Description: Colorless liquid; pungent odor.

Formula: CH_3COCH_2Cl.

Constants: Mol wt: 92.53, mp: $-44.5°C$, bp: 119°C, d: 1.162.

Hazard Analysis

Toxic Hazard Rating:

  Acute Local: Irritant 3; Inhalation 3.

  Acute Systemic: Inhalation 3.

  Chronic Local: U.

  Chronic Systemic: U.

A lachrymator poison gas.

Toxicity: See chlorinated hydrocarbons, aliphatic, and acetone.

Fire Hazard: Moderate, when exposed to heat or flame (Section 6).

Disaster Hazard: Dangerous; when heated to decomposition it emits highly toxic fumes of phosgene; can react vigorously with oxidizing materials.

Countermeasures

Storage and Handling: Section 7.

Shipping Regulations: Section 11.

  IATA (stabilized): Poison C, poison label, not acceptable (passenger), 20 liters (cargo).

  IATA (unstabilized): Poison C, not acceptable (passenger and cargo).

CHLOROACETONITRILE

General Information

Description: Crystals.

Formula: CH_2ClCN.

Constant: Mol wt: 75.5.

Hazard Analysis

Toxicity: See nitriles.

Disaster Hazard: Dangerous; when heated to decomposition, it emits highly toxic fumes; will react with water, steam, acid or acid fumes to produce toxic and flammable vapors.

Countermeasures

Storage and Handling: Section 7.

CHLOROACETOPHENONE

General Information

Synonyms: Phenacylchloride; phenylchloromethyl ketone.

Description: Pale straw-colored liquid or white crystals; fragrant, non-persistent odor.

Formula: C_8H_7ClO.

Constants: Mol wt: 154.6, mp: 56°C, bp: 237-247°C, fp: 59°C, d: 1.19 at 25°/25°C, vap. press.: 0.012 mm at 0°C, vap. d.: 5.2.

Hazard Analysis

Toxic Hazard Rating:

  Acute Local: Irritant 2; Inhalation 2.

  Acute Systemic: Inhalation 2.

  Chronic Local: U.

  Chronic Systemic: U.

Data based on animal experiments.

TLV: ACGIH (tentative) 0.05 parts per million; 0.3 milligrams per cubic meter of air.

Caution: A lachrymator type of military poison.

Disaster Hazard: Dangerous; when heated to decomposition, it emits toxic fumes; will react with water or steam to produce toxic and corrosive fumes.

Countermeasures

Ventilation Control: Section 2.

Personnel Protection: Section 3.

Personal Hygiene: Section 3.

First Aid: Section 1.

Storage and Handling: Section 7.

TOXIC HAZARD RATING CODE (*For detailed discussion, see Section 1.*)

0 NONE: (a) No harm under any conditions; (b) Harmful only under unusual conditions or overwhelming dosage.

1 SLIGHT: Causes readily reversible changes which disappear after end of exposure.

2 MODERATE: May involve both irreversible and reversible changes; not severe enough to cause death or permanent injury.

3 HIGH: May cause death or permanent injury after very short exposure to small quantities.

U UNKNOWN: No information on humans considered valid by authors.

Shipping Regulations: Section 11.
    I.C.C.: Poison C; tear gas label (gas, liquid or solid), 75 pounds.
    Coast Guard Classification: Poison C; tear gas label (gas, liquid or solid).
    IATA: Poison C, poison label, not acceptable (passenger), 35 kilograms (cargo).

CHLOROACETYL CHLORIDE
General Information
Description: Water-white or slightly yellow liquid.
Formula: $CH_2ClCOCl$.
Constants: Mol wt: 112.95, bp: 105–106°C, fp: −22.5°C, flash p: none, d: 1.495 at 0°C.
Hazard Analysis
Toxicity: See hydrochloric acid.
Caution: A lachrymator (tear gas).
Disaster Hazard: Dangerous; when heated to decomposition, it emits highly toxic fumes of chlorides, will react with water or steam to produce toxic and corrosive fumes.
Countermeasures
Storage and Handling: Section 7.
Shipping Regulations: Section 11.
    I.C.C.: Corrosive liquid; white label, 1 quart.
    Coast Guard Classification: Corrosive liquid; white label.
    IATA: Corrosive liquid, white label, not acceptable (passenger), 1 liter (cargo).

CHLOROACETYL ISOCYANATE
Hazard Analysis
Toxic Hazard Rating: U. An irritant. See also cyanates.
Disaster Hazard: Dangerous. See chlorides and isocyanates.

CHLOROACROLEIN
General Information
Description: Colorless liquid.
Formula: $CH_2CClCHO$.
Constants: Mol wt: 90.52, bp: 30°C, d: 1.205 at 15°C.
Hazard Analysis
Toxic Hazard Rating:
    Acute Local: Irritant 3, Ingestion 3; Inhalation 3.
    Acute Systemic: U.
    Chronic Local: U.
    Chronic Systemic: U.
Caution: A poison gas.
Disaster Hazard: Dangerous; when heated to decomposition, it emits highly toxic fumes; will react with water or steam to produce toxic and corrosive fumes.
Countermeasures
Ventilation Control: Section 2.
Personnel Protection: Section 3.
First Aid: Section 1.
Storage and Handling: Section 7.

CHLOROALDEHYDE. See chloroacetaldehyde.

2-CHLOROALLYL DIETHYL DITHIOCARBAMATE.
    See vegadex.

2-CHLORO-4-tert-AMYL PHENOL
General Information
Description: Liquid.
Formula: $HOC_6H_3ClC_5H_{11}$.
Constants: Mol wt: 198.7, bp: 253–265°C, flash p: 225°F, d: 1.11.
Hazard Analysis
Toxicity: Details unknown. See chlorinated phenols.
Fire Hazard: Slight, when exposed to heat or flame (Section 6).
Disaster Hazard: Dangerous, see chlorides; reacts with oxidizing materials.
Countermeasures
Storage and Handling: Section 7.
To Fight Fire: Water, foam (Section 6).

CHLORO-4-tert-AMYL PHENYL METHYL ETHER
General Information
Formula: $C_5H_{11}C_6H_3ClOCH_3$.
Constant: Mol wt: 212.5, d: 1.1, vap. d.: 7.3, bp: 518–529°F, flash p: 230°F.
Hazard Analysis
Fire Hazard: Slight, when exposed to heat or flame.
Disaster Hazard: Dangerous. See chlorides and ethers.
Countermeasures
To Fight Fire: Water, foam (Section 6).
Storage and Handling: Section 7.

m-CHLOROANILINE
General Information
Description: Liquid.
Formula: C_6H_6ClN.
Constants: Mol wt: 127.6, mp: −10°C, bp: 228.5°C, d: 1.223 at 15°/15°C, vap. press.: 1 mm at 63.5°C.
Hazard Analysis
Toxic Hazard Rating:
    Acute Local: Allergen 1.
    Acute Systemic: Ingestion 3; Inhalation 3; Skin Absorption 3.
    Chronic Local: Allergen 1.
    Chronic Systemic: Ingestion 3; Inhalation 3; Skin Absorption 3.
Toxicology: See also aniline.
Disaster Hazard: Dangerous; see chlorides and aniline.
Countermeasures
Ventilation Control: Section 2.
Personnel Protection: Section 3.
First Aid: Section 1.
Storage and Handling: Section 7.
Shipping Regulations: Section 11.
    IATA: Poison B, poison label, 1 liter (passenger), 220 liters (cargo).

o-CHLOROANILINE
General Information
Description: Liquid.
Formula: C_6H_6ClN.
Constants: Mol wt: 127.6, mp: −14°C or −3.5°C (2 forms), bp: 209°C, d: 1.213 at 20°/4°C, vap. press.: 1 mm at 46.3°C.
Hazard Analysis
Toxic Hazard Rating:
    Acute Local: Allergen 1.
    Acute Systemic: Ingestion 3; Inhalation 3; Skin Absorption 3.
    Chronic Local: Allergen 1.
    Chronic Systemic: Ingestion 3; Inhalation 3; Skin Absorption 3.
Toxicology: See also aniline.
Disaster Hazard: Dangerous; see aniline and chlorides.
Countermeasures
Ventilation Control: Section 2.
Personnel Protection: Section 3.
First Aid: Section 1.
Storage and Handling: Section 7.
Shipping Regulations: Section 11.
    IATA: Poison B, poison label, 1 liter (passenger), 220 liters (cargo).

p-CHLOROANILINE
General Information
Description: Crystalline solid.
Formula: C_6H_6ClN.
Constants: Mol wt: 127.6, mp: 70–71°C, bp: 230.5°C, d: 1.427, vap. press.: 1 mm at 59.3°C.
Hazard Analysis
Toxic Hazard Rating:
    Acute Local: Allergen 1.
    Acute Systemic: Ingestion 3; Inhalation 3.
    Chronic Local: Allergen 1.
    Chronic Systemic: Ingestion 3; Inhalation 3.

Toxicology: See also aniline.
Disaster Hazard: Dangerous; see aniline and chlorides.
Countermeasures
Ventilation Control: Section 2.
Personal Hygiene: Section 3.
First Aid: Section 1.
Storage and Handling: Section 7.
Shipping Regulations: Section 11.
IATA: Poison B, poison label, 25 kilograms (passenger), 95 kilograms (cargo).

β-(o-CHLOROANILINO) PROPIONITRILE
General Information
Description: Colorless to red (on aging) liquid.
Formula: $ClC_6H_4NHCH_2CH_2CN$.
Constants: Mol wt: 180.6, bp: 139–141°C at 0.3 mm, d: 1.2103 at 25°/25°C, vap. d.: 6.23.
Hazard Analysis
Toxicity: See nitriles.
Fire Hazard: Moderate, when exposed to heat or flame (Section 6).
Disaster Hazard: Dangerous; see hydrogen cyanide; will react with water, steam, acid or acid fumes to produce toxic and flammable vapors of cyanides.
Countermeasures
Storage and Handling: Section 7.
Ventilation Control: Section 2.

CHLOROAURIC ACID
General Information
Description: Bright yellow needles; deliquescent.
Formula: $HAuCl_4 \cdot 4H_2O$.
Constants: Mol wt: 412.10, mp: decomposes.
Hazard Analysis and Countermeasures
See gold compounds and hydrochloric acid.

CHLOROAZIDE. See chlorine azide.

CHLOROAZODIN. See azochloramide.

CHLOROBENZENE
General Information
Synonyms: Phenyl chloride; monochlorobenzene; chlorobenzol.
Description: Clear, colorless liquid.
Formula: C_6H_5Cl.
Constants: Mol wt: 112.56, bp: 131.7°C, lel: 1.3%, uel: 7.1% at 150°C, fp: −55.6°C, flash p: 85°F (C.C.), d: 1.113 at 15.5°/15.5°C, autoign. temp.: 1180°F, vap. press.: 10 mm at 22.2°C, vap. d.: 3.88.
Hazard Analysis
Toxic Hazard Rating:
Acute Local: Irritant 1; Ingestion 1; Inhalation 1.
Acute Systemic: Ingestion 2; Inhalation 2; Skin Absorption 1.
Chronic Local: 0.
Chronic Systemic: Ingestion 2; Inhalation 2; Skin Absorption 2.
TLV: ACGIH (accepted); 350 milligrams per cubic meter of air; 75 parts per million in air.
Toxicology: Monochlorobenzol is a fairly strong narcotic and possesses only slight irritant qualities. For cats, concentrations of 1,200 ppm are quite narcotic, and concentrations of 3,700 ppm are fatal after several hours. The dichlorobenzols are strongly narcotic, 1,000 ppm causing narcosis in guinea pigs followed by death after 20 hours exposure. Knowledge of the effects on man of repeated exposure to subnarcotic concentrations is meager. In general, it appears that the chlorobenzols

are not as toxic as benzol. Some of the symptoms described (methemoglobinemia) suggest that other substances, such as nitrobenzol, may have been partially responsible for the few cases of industrial illness reported. It is possible that prolonged exposure to chlorobenzol may cause kidney and liver damage.
Somnolence, loss of consciousness, twitchings of the extremities, cyanosis, deep, rapid respirations and a small, irregular pulse are the chief symptoms occurring in acute exposures. The urine may be burgundy red, and the red blood cells show degenerative and regenerative changes.
Fire Hazard: Dangerous, when exposed to heat or flame.
Spontaneous Heating: No.
Explosion Hazard: Moderate, when exposed to heat or flame.
Disaster Hazard: Dangerous; see chlorine compounds; can react vigorously with oxidizing materials.
Countermeasures
Ventilation Control: Section 2.
To Fight Fire: Foam, carbon dioxide, dry chemical or carbon tetrachloride (Section 6).
Personnel Protection: Section 3.
Storage and Handling: Section 7.
Shipping Regulations: Section 11.
I.C.C. Classification: Flammable liquid; red label.
Coast Guard Classification: Inflammable liquid.
MCA warning label.

p-CHLOROBENZENE SULFONAMIDE
General Information
Description: Crystals.
Formula: $ClC_6H_4SO_2NH_2$.
Constant: Mol wt: 191.6.
Hazard Analysis
Toxicity: See chlorinated hydrocarbons, aromatic.
Disaster Hazard: Dangerous; see sulfonates and chlorides.
Countermeasures
Storage and Handling: Section 7.

4-CHLOR BENZENE SULFONIC ACID, CHLOROPHENYLESTER
Hazard Analysis
Toxic Hazard Rating: U. An insecticide. Relatively low toxicity to experimental animals.
Disaster Hazard: Dangerous. See chlorides and sulfonates.

p-CHLOROBENZENE SULFONYL FLUORIDE
General Information
Description: White solid.
Formula: $ClC_6H_4SO_2F$.
Constants: Mol wt: 195, mp: 46–51°C, bp: 229–230°C, flash p: 340°F, d: 1.475, vap. press.: 11 mm at 100°C, vap. d.: 6.73.
Hazard Analysis
Toxicity: Does not appear to be highly toxic to test animals by inhalation on an acute basis. May be slightly irritating to the skin.
Fire Hazard: Slight, when exposed to heat or flame.
Disaster Hazard: Dangerous; see fluorides, chlorides and sulfonates; can react vigorously with oxidizing materials.
Countermeasures
Storage and Handling: Section 7.
To Fight Fire: Water, foam, carbon dioxide, dry chemical or carbon tetrachloride (Section 6).

p-CHLOROBENZENE TETRATHIO-o-STANNATE
General Information
Description: Solid.

TOXIC HAZARD RATING CODE (*For detailed discussion, see Section 1.*)

0 NONE: (a) No harm under any conditions; (b) Harmful only under unusual conditions or overwhelming dosage.

1 SLIGHT: Causes readily reversible changes which disappear after end of exposure.

2 MODERATE: May involve both irreversible and reversible changes; not severe enough to cause death or permanent injury.

3 HIGH: May cause death or permanent injury after very short exposure to small quantities.

U UNKNOWN: No information on humans considered valid by authors.

Formula: $(SC_6H_4Cl)_4Sn$.
Constants: Mol wt: 693.1, mp: 189°C.
Hazard Analysis
Toxicology: Details unknown. See tin compounds.
Disaster Hazard: Dangerous; see sulfates and chlorides.

CHLOROBENZILATE
Toxicity: Similar to DDT.

CHLOROBENZOL. See chlorobenzene.

CHLOROBENZONITRILES (o.p)
General Information
Description: Crystals.
Formula: ClC_6H_4CN.
Constant: Mol wt: 137.6.
Hazard Analysis
Toxicity: See cyanides.
Disaster Hazard: Dangerous; see cyanides and chlorides;
they will react with water, steam, acid or acid fumes to
produce toxic fumes.
Countermeasures
Storage and Handling: Section 7.

m-CHLOROBENZOTRIFLUORIDE
General Information
Synonym: m-Chlorotrifluoromethylbenzene.
Description: Water-white aromatic liquid.
Formula: $ClC_6H_4CF_3$.
Constants: mol wt: 180.56, mp: $-56°C$, bp: 138°C,
d: 1.351, vap. d.: 6.24.
Hazard Analysis
Toxicology: Only slightly toxic to experimental animals by
skin exposure and oral ingestion.
Disaster Hazard: Dangerous, see chlorides and fluorides.
Countermeasures
Storage and Handling: Section 7.

O-CHLOROBENZOTRIFLUORIDE
General Information
Synonym: o-Chlorotrifluoro methylbenzene; o-chloro-
α,α-trifluorotoluene.
Description: Colorless liquid; aromatic odor.
Formula: $ClC_6H_4CF_3$.
Constants: Mol wt: 180.56, d: 1.379 (15.5/15.5°C),
bp: 152°C, fp: $-7.4°C$, flash p: 138°F, vap. d.: 6.2.
Hazard Analysis
Fire Hazard: Moderate, when exposed to heat or flame.
Disaster Hazard: Dangerous; see chlorides and fluorides.
Countermeasures
Storage and Handling: Section 7.

p-CHLOROBENZOTRIFLUORIDE
General Information
Synonym: p-Chlorotrifluoromethylbenzene.
Description: Clear water-white liquid.
Formula: $ClC_6H_4CF_3$.
Constants: Mol wt: 180.56, mp: $-36°C$, bp: 139.3°C,
flash p: 116°F, d: 1.353 at 15.5°/15.5°C, vap. d.: 6.24.
Hazard Analysis
Toxicology: See m-Chlorobenzotrifluoride.
Fire Hazard: Moderate, when exposed to heat or flame.
Disaster Hazard: Dangerous; see chlorides and fluorides;
can react vigorously with oxidizing materials.
Countermeasures
Storage and Handling: Section 7.
To Fight Fire: Water, foam, carbon dioxide, dry chemical or
carbon tetrachloride (Section 6).

p-CHLOROBENZOYL PEROXIDE
General Information
Synonym: Luperco BDB.
Description: A white granular material; insoluble in water;
soluble in organic solvents.
Formula: $(ClC_6H_4CO)_2O_2$.
Constant: Mol wt: 311.1.

Hazard Analysis
Toxicology: Limited animal experiments indicate a moderate toxicity via oral administration. Probably an irritant
to skin and mucous membranes. See peroxides, organic.
Fire Hazard: Dangerous; a powerful oxidizer. Store in a cool
place away from fire hazards, sparks, open flames and
out of the direct rays of the sun.
Explosion Hazard: Dangerous; this material may be caused
to explode by heat (over 38°C) or contamination. Any
contaminant which acts as an accelerator to the polymerization or on decomposition of this material can
cause an explosion.
Disaster Hazard: Dangerous; heat or contact with certain
fumes or mists can cause it to explode.
Countermeasures
To Fight Fire: For small quantities, carbon dioxide, or foam
extinguishers may be used. Water spray or mist may
also be used. Dry chemical is effective.
Storage and Handling: Must be kept below 38°C in even
cool storage. This material is dangerous when involved
in a fire. Consult manufacturer's literature on storage
and handling. See Section 7.
Shipping Regulations:
I.C.C.: Oxidizing material N.O.S., yellow label, 25
pounds.
Freight: Yellow label chemicals N.O.I.B.N.
Rail Express: 25 lbs. maximum per shipping case.
Parcel Post: Prohibited.
IATA: Oxidizing material, yellow label, not acceptable
(passenger), 12 kilograms (cargo).

o-CHLOROBENZYL CHLORIDE
General Information
Description: A colorless liquid or crystals.
Formula: $ClC_6H_4CH_2Cl$.
Constants: Mol wt: 161.08, bp: 216–222°C, fp: $< -30°C$,
d: 1.270–1.280 at 25°/15°C, vap. d.: 5.55.
Hazard Analysis
Toxicity: Details unknown. See also benzyl chloride.
Disaster Hazard: Dangerous; see phosgene.
Countermeasures
Storage and Handling: Section 7.

p-CHLOROBENZYL CHLORIDE
General Information
Synonym: α-4-Dichlorotoluene.
Description: Needle-like crystals or colorless liquid.
Formula: $ClC_6H_4CH_2Cl$.
Constants: Mol wt: 161.0, mp: 29°C, bp: 222°C, d: 1.250–
1.260 at 25°/15°C, vap. d.: 5.55.
Hazard Analysis
Toxicity: Details unknown. See also benzyl chloride.
Disaster Hazard: Dangerous; see phosgene.
Countermeasures
Storage and Handling: Section 7.

p-CHLOROBENZYL p-CHLOROPHENYL SULFIDE
General Information
Synonyms: Chlorosulfacide, mitox, chlorbenside.
Description: Crystals; almond like odor; insoluble in water;
soluble in most organic solvents.
Formula: $ClC_6H_4CH_2SC_6H_4Cl$.
Constants: Mol wt: 269, mp: 75–76°C.
Hazard Analysis
Toxic Hazard Rating: U. Has caused liver and kidney in-
jury; also skin irritation in experimental animals.
Disaster Hazard: Dangerous. See chlorides and sulfur com-
pounds.

o-CHLOROBENZYLIDENE MALONITRILE
General Information
Synonym: OCBM
Hazard Analysis
Toxic Hazard Rating:
Acute Local: Irritant 3; Inhalation 3.

Acute Systemic: Inhalation 1.
Chronic Local: Irritant 2; Inhalation 2.
Chronic Systemic: Inhalation 1.
Human exposure data suggest relatively low systemic toxicity but intense irritation of eyes, skin mucous membranes. See nitriles.
TLV: ACGIH (tenative) 0.05 part per million; 0.4 milligrams per cubic meter of air.
Disaster Hazard: Dangerous. See chlorides and nitriles.

2-CHLORO-4,6-BIS-(DIETHYLAMINE)-3-TRIAZINE
General Information
Description: Clear liquid; slight odor.
Formula: $C_8H_{20}N_5Cl$.
Constants: Mol wt: 221.75, mp: 27°C, d: 1.0956.
Hazard Analysis
Toxicity: Details unknown.
Disaster Hazard: Dangerous; see chlorides.
Countermeasures
Storage and Handling: Section 7.

CHLOROBROMOMETHANE. See methylene chlorobromide.

CHLOROBROMOPHOSGENE.
General Information
Description: Liquid.
Formula: COClBr.
Constants: Mol wt: 143.39, bp: 25°C, d: 1.82 at 15°C.
Hazard Analysis
Toxicity: See carbon oxychloride.
Disaster Hazard: Dangerous; see phosgene.
Countermeasures
Storage and Handling: Section 7.

CHLOROBROMOPROPENE.
General Information
Synonym: CBP.
Description: Liquid.
Formula: $ClBrC_3H_4$.
Constant: Mol wt: 155.4.
Hazard Analysis
Toxic Hazard Rating:
Acute Local: Irritant 3; Inhalation 3.
Acute Systemic: U.
Chronic Local: U.
Chronic Systemic: U.
Disaster Hazard: Dangerous; see chlorides and bromides.

2-CHLORO-1,3-BUTADIENE. See chloroprene.

1-CHLOROBUTANE. See butyl chloride.

CHLOROBUTANOL
General Information
Synonym: Acetone chloroform.
Description: Crystals, camphor odor.
Formula: $Cl_3CC(CH_3)_2OH$.
Constants: Mol wt: 177.5, mp: 97°C, bp: 167°C.
Hazard Analysis
Toxicity: Details unknown. A food additive permitted in food for human consumption. See chloral hydrate which acts similarly.
Fire Hazard: Slight; when exposed to heat or flame (Section 6).
Disaster Hazard: Dangerous; see phosgene; can react with oxidizing materials.
Countermeasures
Storage and Handling: Section 7.

2-CHLOROBUTENE-2. See methallyl chloride.

1-CHLORO-2-(β-CHLOROETHOXY)ETHANE. See 2,2'-dichloroethyl ether.

o-CHLOROCINNAMIC ACID
General Information
Description: Light tan powder.
Formula: $ClC_6H_4CHCHCOOH$.
Constants: Mol wt: 182.6, mp: 207–212°C.
Hazard Analysis
Toxicity: Details unknown.
Disaster Hazard: Dangerous; see phosgene.
Countermeasures
Storage and Handling: Section 7.

p-CHLORO-m-CRESOL
Synonym: 4-Chloro-3-hydroxytoluene; PCMC.
Description: Odorless crystals (when pure). Somewhat soluble in water. Very soluble in organic solvents.
Formula: C_7H_7ClO.
Constants: Mol wt: 142.6, mp: 66°C, bp: 235°C.
Hazard Analysis
Toxicity: Details unknown. An allergen. See cresol.
Disaster Hazard: Dangerous; see phosgene.
Countermeasures
Storage and Handling: Section 7.

CHLOROCYANOHYDRIN
General Information
Synonym: Chloral hydrocyanide; trichloroacetonitrile.
Description: Crystals; odor of chloral and hydrogen cyanide.
Formula: $CCl_3CH(OH)CN$.
Constants: Mol wt: 174.4, mp: 61°C, bp: 220°C.
Hazard Analysis
Toxicity: See cyanides.
Disaster Hazard: Dangerous; see cyanides; will react with water, steam, acid or acid fumes to produce toxic fumes.
Countermeasures
Storage and Handling: Section 7.

2-CHLORO-4,6-di-tert-AMYL PHENOL
General Information
Formula: $(C_5H_{11})_2C_6H_2ClOH$.
Constants: Mol wt: 268.5, bp: 320–354°F at 22 mm, flash p: 250°F.
Hazard Analysis
Toxic Hazard Rating: U.
Fire Hazard Rating: Slight when exposed to heat or flame.
Disaster Hazard: Dangerous. See chlorides.
Countermeasures
To Fight Fire: Water, foam (Section 6).
Storage and Handling: Section 7.

2-CHLORO-2-DIETHYLCARBAMYL-1-METHYL VINYL-DIMETHYL PHOSPHATE
General Information
Synonym: Phosphamidon.
Description: Colorless liquid; soluble in water and organic solvents.
Formula: $(CH_3O)_2P(O)OC(CH_3):C(Cl)C(O)N(C_2H_5)_2$.
Constants: Mol wt: 299.5, bp: 162°C at 1.5 mm.
Hazard Analysis
Toxic Hazard Rating:
Acute Local: U.
Acute Systemic: Ingestion 3; Inhalation 3; Skin Absorption 3.
Chronic Local: U.
Chronic Systemic: U.

TOXIC HAZARD RATING CODE (For detailed discussion, see Section 1.)

0 NONE: (a) No harm under any conditions; (b) Harmful only under unusual conditions or overwhelming dosage.

1 SLIGHT: Causes readily reversible changes which disappear after end of exposure.

2 MODERATE: May involve both irreversible and reversible changes; not severe enough to cause death or permanent injury.

3 HIGH: May cause death or permanent injury after very short exposure to small quantities.

U UNKNOWN: No information on humans considered valid by authors.

A highly toxic organic phosphate insecticide. See also parathion.

Disaster Hazard: Dangerous. See chlorides and phosphates.

1-CHLORO-1,1-DIFLUOROETHANE
General Information
Description: Gas.
Formula: $ClF_2C_2H_3$.
Constants: Mol wt: 100.50.
Hazard Anslysis
Toxic Hazard Rating:
Acute Local: Irritant 1; Inhalation 1.
Acute Systemic: Inhalation 2.
Chronic Local: Irritant 1.
Chronic Systemic: U.
Toxicology: In high concentrations it acts as a simple asphyxiant.
Disaster Hazard: Dangerous; see chlorides and fluorides.
Countermeasures
Storage and Handling: Section 7.

CHLORODIFLUOROMETHANE
General Information
Synonym: Freon-22.
Description: Gas.
Formula: $ClHCF_2$.
Constants: Mol wt: 86.5, d: 3.87 that of air at 0°C, mp: −146°C, bp: −40.8°C.
Hazard Analysis
Toxicity: Asphyxiant in high concentrations.
Disaster Hazard: Dangerous; see chlorides and fluorides.
Countermeasures
Shipping Regulations: Section 11.
IATA: Nonflammable gas, green label, 70 kilograms (passenger), 140 kilograms (cargo).

2-CHLORO-4-DIMETHYLAMINO-6-METHYL PYRIMIDINE
General Information
Synonym: Castrix.
Description: Brownish waxlike. Very slightly water soluble.
Formula: $C_7H_{10}ClN_3$.
Constant: Mol wt: 171.6.
Hazard Analysis
Toxic Hazard Rating:
Acute Local: U.
Acute Systemic: Ingestion 3.
Chronic Local: U.
Chronic Systemic: U.
Toxicology: Can cause CNS damage and convulsions. A rodenticide.
Disaster Hazard: Dangerous; see chlorides.

CHLORODIMETHYLARSINE. See dimethyl chloroarsine.

2,1,4-CHLORODINITROBENZENE. See dinitrochlorobenzene.

CHLORODIPHENYL. See chlorinated diphenyls.

CHLORODIPHENYL OXIDE. See chlorinated diphenyl oxide.

1-CHLORO-2,3-EPOXYPROPANE. See epichlorohydrin.

CHLOROETHANE. See ethyl chloride.

CHLOROETHANOIC ACID. See monochloroacetic acid.

2-CHLOROETHANOL. See ethylene chlorohydrin.

CHLOROETHENE. See vinyl chloride.

CHLOROETHYL ACETATE. See ethyl chloroacetate.

β-CHLOROETHYL ALCOHOL. See ethylene chlorohydrin.

2-CHLORO-4-ETHYLAMINO-6-ISOPROPYLAMINO -s-TRIAZINE
General Information
Synonym: Atrazine.
Hazard Analysis
See 3-amino-1 H-1,2,4-triazole.

1-CHLORO-4-ETHYLBENZENE
General Information
Description: Clear, colorless liquid.
Formula: $ClC_6H_1CH_2CH_3$.
Constants: Mol wt: 140.6, mp: −62.6°C, bp: 184.3°C, flash p: 147°F, d: 1.05 at 25°/25°C, vap. press.: 1 mm at 19.2°C, vap. d.: 4.86.
Hazard Analysis
Toxicity: Details unknown. See chlorinated hydrocarbons, aromatic and chlorobenzene.
Fire Hazard: Moderate, when exposed to heat or flame; reacts with oxidizing materials.
Disaster Hazard: Dangerous. See chlorides.
Countermeasures
To Fight Fire: Foam, carbon dioxide, dry chemical or carbon tetrachloride (Section 6).
Storage and Handling: Section 7.

β-CHLOROETHYLCHLOROFORMATE
General Information
Description: Colorless liquid.
Formula: $COClOCH_2CH_2Cl$.
Constants: Mol wt: 142.99, bp: 152.5°C at 752 mm, d: 1.3825 at 20°C.
Hazard Analysis
Toxic Hazard Rating:
Acute Local: Irritant 2; Ingestion 2; Inhalation 2.
Acute Systemic: U.
Chronic Local: U.
Chronic Systemic: U.
Disaster Hazard: Dangerous; see chlorides.
Countermeasures
Ventilation Control: Section 2.
Personnel Protection: Section 3.
Storage and Handling: Section 7.

β-CHLOROETHYL CHLOROSULFONATE
General Information
Description: Chloropicrin-like odor.
Formula: $ClCH_2CH_2OSO_2Cl$.
Constants: Mol wt: 179.04, bp: 101°C at 23 mm.
Hazard Analysis
Toxic Hazard Rating:
Acute Local: Irritant 3; Ingestion 3; Inhalation 3.
Acute Systemic: Ingestion 3; Inhalation 3.
Chronic Local: U.
Chronic Systemic: U.
Disaster Hazard: Dangerous; see chlorides and sulfonates.
Countermeasures
Ventilation Control: Section 2.
Personnel Protection: Section 3.
First Aid: Section 1.
Storage and Handling: Section 7.

CHLOROETHYLENE. See vinyl chloride.

2-CHLOROETHYL ETHER. See dichloroethyl ether.

6-CHLOROETHYLHEXANOATE
General Information
Formula: $CH_3(CH_2)_3COOCH_3CHCl$.
Constant: Mol wt: 178.5.
Hazard Analysis
Toxic Hazard Rating: Limited animal experiments suggest moderate toxicity and low irritation. See also esters.
Disaster Hazard: Dangerous. See chlorides.

2-CHLOROETHYL METHYL SULFIDE. See hemisulfur mustard.

2-CHLOROETHYL VINYL ETHER
General Information
Description: Liquid.
Formula: $CH_2ClCH_2OCHCH_2$.
Constants: Mol wt: 106.55, bp: 104–108° C.
Hazard Analysis
Toxic Hazard Rating:
Acute Local: Irritant 1; Ingestion 2; Inhalation 2.
Acute Systemic: Ingestion 2; Inhalation 2; Skin Absorption 2.
Chronic Local: Irritant 1.
Chronic Systemic: Ingestion 2; Inhalation 2; Skin Absorption 2.
Fire Hazard: Moderate, when exposed to heat or flame (Section 6).
Explosion Hazard: See ethers.
Disaster Hazard: Dangerous; see chlorides; can react with oxidizing materials.
Countermeasures
Ventilation Control: Section 2.
Personnel Protection: Section 3.
Personal Hygiene: Section 3.
First Aid: Section 1.
Storage and Handling: Section 7.

2-CHLOROETHYL-2-XENYL ETHER.
General Information
Formula: $C_6H_5C_6H_4OCH_2CH_2Cl$.
Constants: Mol wt: 232.5, flash p: 320° F, d: 1.1, bp: 613° F.
Hazard Analysis
Toxic Hazard Rating: U.
Fire Hazard: Slight, when exposed to heat or flame.
Disaster Hazard: Dangerous. See chlorides and ethers.
Countermeasures
To Fight Fire: Water, foam, alcohol foam (section 6).
Storage and Handling: Section 7.

CHLOROETHYNE. See acethylene chloride.

2-CHLOROFLUORENE
Hazard Analysis
Toxicity: Details unknown.
Disaster Hazard: Dangerous; see chlorides.
Countermeasures
Storage and Handling: Section 7.

3-CHLORO-2-FLUORO-1-PROPENE
General Information
Formula: $CH_2ClCFCH_2$.
Constant: Mol wt: 94.5.
Hazard Analysis
Toxic Hazard Rating: U. Limited animal experiments suggest high oral, skin, and inhalation toxicity.
Disaster Hazard: Dangerous. See chlorides and fluorides.

CHLOROFORM
General Information
Synonym: Trichloromethane.
Description: Colorless liquid; heavy, ethereal odor.
Formula: $CHCl_3$.
Constants: Mol wt: 119.39, mp: $-63.5°$C, bp: 61.26°C, fp: $-63.5°$C, flash p: none, d: 1.49845 at 15°C, vap. press.: 100 mm at 10.4° C, vap. d.: 4.12.
Hazard Analysis
Toxic Hazard Rating:
Acute Local: Irritant 1.
Acute Systemic: Inhalation 3.
Chronic Local: U.

Chronic Systemic: U.
TLV: ACGIH (accepted): 50 parts per million in air; 240 miligrams per cubic meter of air.
Toxicology: Chloroform causes irritation of the conjunctiva. Upon inhalation, it causes dilation of the pupils with reduced reaction to light, as well as reduced intraocular pressure (experimental). The material is well known as an anesthetic. In the initial stages there is a feeling of warmth of the face and body, then an irritation of the mucous membranes and skin followed by nervous aberration. Prolonged inhalation will bring on paralysis accompanied by cardiac and respiratory failure and finally death.

It has been widely used as an anesthetic. However, due to its toxic effects, this use is being abandoned. 68,000 to 82,000 ppm kill most animals in a few minutes. 14,000 ppm is dangerous to life after an exposure of from 30 to 60 minutes. 5,000 to 6,000 ppm can be tolerated by animals for one hour without serious disturbances. The maximum concentration tolerated for several hours or for prolonged exposure with slight symptoms is 2,000 to 2,500 ppm. The harmful effects are narcosis, and damage to the liver and heart. Prolonged administration as an anesthetic may lead to such serious effects as profound toxemia and damage to the liver, heart and kidneys. Experimentally prolonged but light anesthesia in dogs produces a typical hepatitis. Inhalation of the concentrated chloroform vapor results in irritation of the mucous surfaces exposed to it. The narcosis is ordinarily preceded by a stage of excitation which is followed by loss of reflexes, sensation, and consciousness.
Fire Hazard: Slight, when exposed to high heat; otherwise practically nonflammable (Section 6).
Disaster Hazard: Dangerous; see phosgene.
Countermeasures
Ventilation Control: (Section 2.
Personal Hygiene: Section 3.
Treatment and Antidotes: If it has been ingested, or there has been great overexposure, the following antidotes may be applied: emetics, stomach syphon, friction, cold douche, fresh air, strychnine (hypodermically— from 1/120 to 1/60 grain), rubefactions, artificial respiration, etc. If during exposure to unknown amounts of chloroform vapor, the patient should feel any of the symptoms noted above, he should immediately be moved to fresh air and kept under observation until the symptoms disappear.
Storage and Handling: Section 7.
Shipping Regulations: Section 11.
MCA warning label.
IATA: Other restricted articles, class A no label, 40 liters (passenger), 220 liters (cargo).

CHLOROFORMAMIDE. See carbamyl chloride.

CHLOROFORMOXIME
General Information
Description: Needles; odor resembling hydrocyanic acid.
Formula: $CClHNOH$.
Constant: Mol wt: 79.5.
Hazard Analysis
Toxic Hazard Rating:
Acute Local: Irritant 3; Inhalation 3.
Acute Systemic: U.
Chronic Local: U.
Chronic Systemic: U.

TOXIC HAZARD RATING CODE (For detailed discussion, see Section 1.)

0 NONE: (a) No harm under any conditions; (b) Harmful only under unusual conditions or overwhelming dosage.

1 SLIGHT: Causes readily reversible changes which disappear after end of exposure.

2 MODERATE: May involve both irreversible and reversible changes; not severe enough to cause death or permanent injury.

3 HIGH: May cause death or permanent injury after very short exposure to small quantities.

U UNKNOWN: No information on humans considered valid by authors.

Disaster Hazard: Dangerous; see chlorides.
Countermeasures
Ventilation Control: (Section 2).
Personnel Protection: (Section 3).
First Aid: Section 1.
Storage and Handling: Section 7.

CHLOROGERMANE
General Information
Description: Colorless liquid.
Formula: GeH$_3$Cl.
Constants: Mol wt: 111.08, mp: $-52.0°$C, bp: $28.0°$C,
   d: 1.75 at $-52°$C.
Hazard Analysis and Countermeasures
See hydrochloric acid and germanium compounds.

1-CHLOROHEXANE
General Information
Synonym: n-Hexyl chloride.
Description: Mobile liquid; insoluble in water.
Formula: C$_6$H$_{13}$Cl.
Constants: Mol wt: 120.62, d: 0.8780 (20/4°C), bp: 134°C
   vap. d.: 4.2, flash p: 95°F.
Hazard Analysis
Toxic Hazard Rating: U.
Fire Hazard: Moderate, when exposed to heat or flame.
Disaster Hazard: Dangerous. See chlorides.
Countermeasures
To Fight Fire: Water may be ineffective (Section 6).
Storage and Handling: Section 7.

6-CHLOROHEXANOIC ACID
General Information
Formula: CH$_2$Cl(CH$_2$)$_4$COOH.
Constant: Mol wt: 150.5.
Hazard Analysis
Toxic Hazard Rating: U. Limited animal experiments
   suggest moderate toxicity and high irritation.
Disaster Hazard: Dangerous. See chlorides.

CHLOROHYDRIC ACID. See hydrochloric acid.

α-**CHLOROHYDRIN.** See 1-chloropropane-2,3-diol.

2-CHLORO-5-HYDROXY-1,3-DIMETHYL BENZENE.
   See p-chloro-m-xylenol.

2-CHLORO-4-(HYDROXYMERCURI)PHENOL. See
   hydroxymercurichlorophenol.

4-CHLORO-3-HYDROXYTOLUENE. See p-chloro-m-
   cresol.

CHLOROISOPROPYL ALCOHOL. See propylene
   chlorohydrin.

6-CHLORO-4-ISOPROPYL-1-METHYL-3-PHENOL.
   See chlorothymol.

CHLOROMANGANOKALITE. See potassium manganous
   chloride.

CHLOROMERCURIPHENOL. See o-hydroxyphenol
   mercuric chloride.

CHLOROMETHANE. See methyl chloride.

2-CHLORO-1-METHYLBENZENE. See o-chlorotoluene.

3-CHLORO-1-METHYLBENZENE. See m-chlorotoluene.

4-CHLORO-1-METHYLBENZENE. See p-chlorotoluene.

CHLOROMETHYL CHLOROFORMATE
General Information
Description: Mobile, colorless liquid; penetrating ir-
   ritating odor.
Formula: ClCOOCH$_2$Cl.
Constants: Mol wt: 129, bp: 106.5–107°C, d: 1.465 at 15°C.

Hazard Analysis
Toxic Hazard Rating:
   Acute Local: Irritant 3; Ingestion 3; Inhalation 3.
   Acute Systemic: Ingestion 3; Inhalation 3.
   Chronic Local: U.
   Chronic Systemic: U.
Disaster Hazard: Dangerous; see chlorides.
Countermeasures
Ventilation Control: (Section 2).
Personnel Protection: (Section 3).
First Aid: Section 1.
Storage and Handling: Section 7.

CHLOROMETHYL-4-CHLOROPHENYL SULFONE
General Information
Description: Crystals.
Formula: ClC$_6$H$_4$SO$_2$CH$_2$Cl.
Constant: Mol wt: 225.1.
Hazard Analysis
Toxicity: Details unknown. See also chloromethyl chloro-
   sulfonate.
Disaster Hazard: Dangerous; see chlorides and sulfonates;
   will react with water or steam to produce toxic and cor-
   rosive fumes.
Countermeasures
Storage and Handling: Section 7.

CHLOROMETHYL CHLOROSULFONATE
General Information
Description: Colorless liquid.
Formula: ClCH$_2$OClSO$_2$.
Constants: Mol wt: 165.01, bp: 49–50°C at 14 mm Hg,
   d: 1.63.
Hazard Analysis
Toxic Hazard Rating:
   Acute Local: Irritant 3; Ingestion 3; Inhalation 3.
   Acute Systemic: Ingestion 3; Inhalation 3.
   Chronic Local: U.
   Chronic Systemic: U.
Disaster Hazard: Dangerous; see chlorides and sulfonates;
   will react with water or steam to produce toxic and cor-
   rosive fumes.
Countermeasures
Ventilation Control: Section 2.
Personnel Protection: Section 3.
First Aid: Section 1.
Storage and Handling: Section 7.

4-CHLORO-2-METHYLPHENOXYACETIC ACID. See
   methoxone.

CHLOROMETHYLSILICANE
General Information
Description: Gas.
Formula: CH$_3$SiH$_2$Cl.
Constants: Mol wt: 80.6, d: 0.935 at $-80°$C, mp:
   $-134.1°$C, bp: 7°C.
Hazard Analysis and Countermeasures
See silanes and chlorides.

CHLOROMETHYL SULFONYL TRICHLORANILIDE
Hazard Analysis
Toxic Hazard Rating: U. Increased basal metabolism,
   sweating, weight loss reported in humans. See also
   aniline.
Disaster Hazard: Dangerous. See chlorides, sulfonates and
   aniline.

2-CHLOROMETHYLTHIOPHENE
General Information
Description: Crystals.
Formula: C$_4$H$_3$SCH$_2$Cl.
Constant: Mol wt: 132.6.
Hazard Analysis
Toxicity: Details unknown. See also hydrochloric acid

which is liberated by this material upon storage. See also thiophene.

Fire Hazard: Moderate, when exposed to heat or flame (Section 6).

Explosion Hazard: Severe, when shocked, exposed to heat or by spontaneous chemical reaction.

Disaster Hazard: Dangerous, shock will explode it; see chlorides and oxides of sulfur; can react vigorously with oxidizing materials.

Countermeasures

Storage and Handling: Section 7.

3'-CHLORO-2-METHYL-p-VALEROTOLUIDIDE
General Information

Synonym: Solan; N-(3-chloro-4-methyl phenyl)-2-methyl-pentanamide.

Description: Solid; insoluble in water; soluble in pine oil, diisobutyl ketone; isophorone, xylene.

Formula: $H_3CC_6H_3(Cl)NHCOCH(CH_3)CH_2CH_2CH_3$.

Constants: Mol wt: 239.5, mp: 86°C.

Hazard Analysis

Toxic Hazard Rating: U. A herbicide. Animal experiments show low toxicity.

Disaster Hazard: Dangerous. See chlorides.

CHLORONAPHTHALENE
General Information

Synonym: α-Chloronaphthalene; 1-Chloronaphthalene.

Description: Oily liquid; volatile with steam; soluble in benzene, petroleum ether, alcohol; insoluble in water.

Formula: $C_{10}H_7Cl$.

Constants: Mol wt: 162.61, flash p: 270°F (O.C.), autoign. temp.: > 1036°F, d: 1.19382 (20/4°C), mp: −2.5°C, bp: 505°F.

Hazard Analysis

Toxic Hazard Rating: See chlorinated naphthalenes.

Fire Hazard: Slight, when exposed to heat or flame.

Disaster Hazard: Dangerous. See chlorides.

Countermeasures

To Fight Fire: Water, foam (Section 6).

Storage and Handling: Section 7.

o-CHLORO-p-NITROANILINE
General Information

Synonym: OCPN.

Description: Yellow crystalline powder.

Formula: $O_2NC_6H_3(Cl)NH_2$.

Constants: Mol wt: 172.6, mp: 108.4°C.

Hazard Analysis

Toxic Hazard Rating:

Acute Local: Allergen 1.

Acute Systemic: Ingestion 3; Inhalation 3.

Chronic Local: Allergen 1.

Chronic Systemic: Ingestion 3; Inhalation 3.

Fire Hazard: Slight, when exposed to heat or flame (Section 6).

Disaster Hazard: Dangerous; when heated to decomposition, or on contact with acid or acid fumes, it emits highly toxic fumes; it can react with oxidizing materials.

Countermeasures

Ventilation Control: Section 2.

Personnel Protection: Section 3.

First Aid: Section 1.

Storage and Handling: Section 7.

p-CHLORO-o-NITROANILINE
General Information

Description: Orange crystalline powder.

Formula: $ClC_6H_3(NO_2)NH_2$.

Constants: Mol wt: 172.6, mp: 116.3°C.

Hazard Analysis

Toxic Hazard Rating:

Acute Local: Allergen 1.

Acute Systemic: Ingestion 3; Inhalation 3.

Chronic Local: Allergen 1.

Chronic Systemic: Ingestion 3; Inhalation 3.

Fire Hazard: Slight, when exposed to heat or flame (Section 6).

Disaster Hazard: Dangerous; when heated to decomposition or on contact with acid or acid fumes, it emits highly toxic fumes; can react with oxidizing materials.

Countermeasures

Ventilation Control: Section 2.

Personal Hygiene: Section 3.

First Aid: Section 1.

Storage and Handling: Section 7.

2-CHLORO-4-NITROBENZAMIDE
General Information

Synonym: "Aklomide."

Formula: $C_7H_5O_3N_2Cl$.

Constants: Mol wt: 201, minimum bp: 170°C.

Hazard Analysis

Toxic Hazard Rating: U. A food additive permitted in the feed and drinking water of animals and/or for the treatment of food producing animals (Section 10). Also a food additive permitted in food for human consumption.

Disaster Hazard: Dangerous. See chlorides and nitrates.

1-CHLORO-3-NITROBENZENE. See m-chloronitrobenzene.

4-CHLORO-1-NITROBENZENE. See p-chloronitrobenzene.

m-CHLORONITROBENZENE
General Information

Synonym: 1-Chloro-3-nitrobenzene.

Description: Yellowish crystals.

Formula: $ClC_6H_4NO_2$.

Constants: Mol wt: 157.6, mp: 46°C, bp: 236°C, d: 1.534 at 20°/4°C.

Hazard Analysis

Toxic Hazard Rating:

Acute Local: U.

Acute Systemic: Ingestion 3; Inhalation 3.

Chronic Local: U.

Chronic Systemic: Ingestion 3; Inhalation 3.

Toxicology: Intoxication from this material can be serious. When absorbed, it forms methemoglobin and gives rise to cyanosis and blood changes. Its effects are analogous to those of nitrobenzene. It can cause poisoning by the pulmonary route and its effects are cumulative. Chemically, it is probably reduced in the body to chloroaniline, which is also poisonous. The para compound is thought to be somewhat less toxic than the ortho compound. In industry it is the dust of this material that is most often the source of intoxication.

Fire Hazard: Slight, when exposed to heat or flame (Section 6).

Disaster Hazard: Dangerous; see nitrates and phosgene; can react with oxidizing materials.

TOXIC HAZARD RATING CODE (For detailed discussion, see Section 1.)

0 NONE: (a) No harm under any conditions; (b) Harmful only under unusual conditions or overwhelming dosage.

1 SLIGHT: Causes readily reversible changes which disappear after end of exposure.

2 MODERATE: May involve both irreversible and reversible changes; not severe enough to cause death or permanent injury.

3 HIGH: May cause death or permanent injury after very short exposure to small quantities.

U UNKNOWN: No information on humans considered valid by authors.

Countermeasures

Ventilation Control: Section 2.

Treatment and Antidotes: Removal from exposure is important as soon as the symptoms appear. If cyanosis is evident, administer oxygen and call a physician as soon as possible. If breathing has stopped, give artificial respiration.

Personal Hygiene: Section 3.

First Aid: Section 1.

Storage and Handling: Section 7.

Shipping Regulations: Section 11.

　IATA (solid): Poison B, poison label, 25 kilograms (passenger), 95 kilograms (cargo).

o-CHLORONITROBENZENE

General Information

Synonym: o-Nitrochlorobenzene.

Description: Yellow crystals.

Formula: $C_6H_4ClNO_2$.

Constants: Mol wt: 157.6, mp: 32–33°C, bp: 245–246°C, d: 1.368, flash p: 261°F.

Hazard Analysis

See m-chloronitrobenzene.

Countermeasures

Storage and Handling: Section 7.

To Fight Fire: Water, foam (Section 6).

Shipping Regulations: Section 11.

　IATA (liquid): Poison B, poison label, 1 liter (passenger), 220 liters (cargo).

p-CHLORONITROBENZENE.

General Information

Synonym: 4-Chloro-1-nitrobenzene.

Description: Yellow crystals.

Formula: $C_6H_4ClNO_2$.

Constants: Mol wt: 157.6, mp: 83°C, bp: 242°C, flash p: 257°F, d: 1.520.

Hazard Analysis

See m-chloronitrobenzene.

Countermeasures

Storage and Handling: Section 7.

To Fight Fire: Water, foam, carbon dioxide, dry chemical or carbon tetrachloride (Section 6).

Shipping Regulations: Section 11.

　IATA (solid): Poison B, poison label, 25 kilograms (passenger), 95 kilograms (cargo).

2-CHLORO-5-NITROBENZOTRIFLUORIDE

General Information

Description: Liquid.

Formula: $C_7H_3F_3ClNO_2$.

Constants: Mol wt: 225.6, bp: 231.9°C, flash p: 275°F, d: 1.504 at 30°/4°C, vap. d.: 7.8.

Hazard Analysis

Toxicity: See m-chloronitrobenzene and fluorides.

Fire Hazard: Moderate, when exposed to heat or flame.

Disaster Hazard: Dangerous; see chlorides, nitrates and fluorides; can react with oxidizing materials.

Countermeasures

Storage and Handling: Section 7.

To Fight Fire: Foam, carbon dioxide, dry chemical or carbon tetrachloride (Section 6).

4-CHLORO-3-NITROBENZOTRIFLUORIDE

General Information

Description: Liquid.

Formula: $C_7H_3F_3ClNO_2$.

Constants: Mol wt: 225.6, bp: 222.6°C, mp: −2.5°C.

Hazard Analysis

Toxicity: See m-chloronitrobenzene and fluorides.

Disaster Hazard: Dangerous; see chlorides, nitrates and fluorides.

Countermeasures

Storage and Handling: Section 7.

1-CHLORO-1-NITROETHANE

General Information

Description: Liquid.

Formula: $CH_3CHCl(NO_2)$.

Constants: Mol wt: 109.5, bp: 129°C, flash p: 133°F, d: 1.258 at 20°/20°C, vap. d.: 3.77.

Hazard Analysis

Toxic Hazard Rating:

　Acute Local: Irritant 2; Ingestion 3; Inhalation 3.

　Acute Systemic: Inhalation 3.

　Chronic Local: Irritant 2.

　Chronic Systemic: U.

Toxicology: High concentrations can cause pulmonary edema and narcosis.

Fire Hazard: Moderate, when exposed to heat.

Explosion Hazard: Dangerous, when shocked or exposed to heat or flame.

Disaster Hazard: Dangerous; shock will explode it; See phosgene and oxides of nitrogen; can react vigorously with oxidizing materials.

Countermeasures

Storage and Handling: Section 7.

To Fight Fire: Alcohol foam, water, carbon dioxide, dry chemical or carbon tetrachloride (Section 6).

1-CHLORO-1-NITROPROPANE

General Information

Description: liquid.

Formula: $CH_3CH_2CH(NO_2)Cl$.

Constants: Mol wt: 123.5, bp: 139.5°C, flash p: 144°F, d: 1.209 at 20°/20°C, vap. d.: 4.26.

Hazard Analysis

Toxic Hazard Rating:

　Acute Local: Irritant 2; Ingestion 2; Inhalation 2.

　Acute Systemic: Ingestion 3; Inhalation 3.

　Chronic Local: U.

　Chronic Systemic: U.

Toxicology: In experiments it has produced injury to kidneys, liver and cardiovascular system.

TLV: ACGIH (accepted); 20 parts per million in air; 101 milligrams per cubic meter of air.

Fire Hazard: Moderate, when exposed to heat.

Explosion Hazard: Moderate, when exposed to heat.

Disaster Hazard: Dangerous; See chlorides; can react with oxidizing materials. When heated it explodes.

Countermeasures

Ventilation Control: Section 2.

To Fight Fire: Alcohol foam, water, carbon dioxide, dry chemical or carbon tetrachloride (Section 6).

Personnel Protection: Section 3.

First Aid: Section 1.

Storage and Handling: Section 7.

2-CHLORO-2-NITROPROPANE

General Information

Description: Liquid.

Formula: $CH_3CCl(NO_2)CH_3$.

Constants: Mol wt: 123.5, bp: 134°C, flash p: 135°F, d: 1.197 at 20°/20°C, vap. d.: 4.26.

Hazard Analysis

Toxic Hazard Rating:

　Acute Local: Irritant 3; Ingestion 3; Inhalation 3.

　Acute Systemic: Ingestion 3; Inhalation 3.

　Chronic Local: U.

　Chronic Systemic: U.

Fire Hazard: See nitrates.

Explosion Hazard: See nitrates.

Disaster Hazard: See nitrates.

Countermeasures

Personnel Protection: Section 3.

To Fight Fire: Water, carbon dioxide, dry chemical or carbon tetrachloride (Section 6).

First Aid: Section 1.

Storage and Handling: Section 7.

α-CHLORO-p-NITROSTYRENE
General Information
Formula: $C_6H_4ClCH:CHNO_2$.
Constant: Mol wt: 183.5.
Hazard Analysis
Toxic Hazard Rating: U. Limited animal experiments suggest high toxicity and irritation. See also phenyl ethylene (styrene).
Disaster Hazard: Dangerous. See chlorides and nitrates.

2-CHLORO-6-NITROTOLUENE
General Information
Description: Liquid.
Formula: $C_7H_6NO_2Cl$.
Constant: Mol wt: 171.6.
Hazard Analysis
Toxicity: See trinitrotoluene.
Fire Hazard: See nitrates.
Disaster Hazard: See nitrates and chlorides.
Countermeasures
Storage and Handling: Section 7.

2-CHLORO-5-NITROTRIFLUOROMETHYL-BENZENE.
See m-chloronitrobenzene and fluorides.

CHLOROPENTAFLUOROETHANE
General Information
Synonym: Monochloropenta fluoroethane; fluorocarbon 115; propellant 115; Refrigerant 115.
Description: Colorless gas; insoluble in water; soluble in alcohol and ether.
Formula: $CClF_2CF_3$.
Constants: Mol wt: 155; bp: $-38.7°C$; mp: $-106°C$.
Hazard Analysis
Toxic Hazard Rating: U. A food additive permitted in food for human consumption (Section 10).
Disaster Hazard: Dangerous. See chlorides and fluorides.
Countermeasures
Shipping Regulations: Section 11.
  ICC: Nonflammable gas, green label, 300 pounds.
  IATA: Nonflammable gas, green label, 70 kilograms (passenger), 140 kilograms (cargo).

CHLOROPENTAMMINE CHROMIUM (III) CHLORIDE
General Information
Description: Red crystals.
Formula: $[Cr(NH_3)_5Cl]Cl_2$.
Constants: Mol wt: 243.54, d: 1.696.
Hazard Analysis
Toxicity: See chromium compounds.
Fire Hazard: Slight; when heated.
Disaster Hazard: Dangerous. See chlorides.
Countermeasures
Storage and Handling: Section 7.

CHLOROPENTAMMINECOBALT (III) CHLORIDE
General Information
Synonym: Purpureo.
Description: Rhombic dark-red violet crystals.
Formula: $[Co(NH_3)_5Cl]Cl_2$.
Constants: Mol wt: 250.47, mp: decomposes, d: 1.819 at $25°/25°C$.
Hazard Analysis
Toxicity: See cobalt compounds.

1-CHLOROPENTANE. See amyl chloride.

β-CHLOROPHENETOLE
General Information
Synonym: Phenoxy ethyl chloride.
Formula: $C_6H_5OCH_2CH_2C$.
Constants: Mol wt: 133, flash p: $225°F$, d: 1.1, bp: $306-311°F$.
Hazard Analysis
Toxic Hazard Rating: U.
Fire Hazard: Moderate, when exposed to heat or flame.
Disaster Hazard: Dangerous. See chlorides.
Countermeasures
To Fight Fire: Water, foam, alcohol foam. Section 6.

m-CHLOROPHENOL
General Information
Description: Crystals.
Formula: C_6H_4ClOH.
Constants: Mol wt: 128.6, mp: $32.5°C$, bp: $214°C$, d: 1.245, vap. press.: 1 mm at $44.2°C$.
Hazard Analysis
Toxic Hazard Rating:
  Acute Local: Irritant 3; Ingestion 3; Inhalation 3.
  Acute Systemic: Ingestion 3; Inhalation 3.
  Chronic Local: Irritant 2.
  Chronic Systemic: Ingestion 3; Inhalation 3.
Disaster Hazard: Dangerous; when heated to decomposition, it emits highly toxic fumes.
Countermeasures
Ventilation Control: Section 2.
Personnel Protection: Section 3.
First Aid: Section 1.
Storage and Handling: Section 7.
Shipping Regulations: Section 11.
  MCA warning label.

o-CHLOROPHENOL
General Information
Description: Light amber liquid.
Formula: C_6H_4OHCl.
Constants: Mol wt: 128.6, bp: $174.5°C$, fp: $7°C$, d: 1.256 at $25°/25°C$, flash p: $147°F$, vap. press.: 1 mm at $12.1°C$.
Hazard Analysis
Toxicity: See chlorinated phenols.
Fire Hazard: Moderate, when exposed to heat (Section 6).
Disaster Hazard: Dangerous; see phenol and chlorides; it can react with oxidizing materials.
Countermeasures
Ventilation Control
Storage and Handling: Section 7.
To Fight Fire: Water, foam (Section 6).
Shipping Regulations: Section 11.
MCA warning label.

p-CHLOROPHENOL
General Information
Description: Needle-like, white to straw-colored crystals; unpleasant odor.
Formula: C_6H_4ClOH.
Constants: Mol wt: 128.6, bp: $220°C$, fp: $42.8°C$, flash p: $250°F$, d: 1.246 at $60°/25°C$, vap. press.: 1 mm at $49.8°C$.
Hazard Analysis
Toxicity: See chlorinated phenols.
Fire Hazard: Moderate, when exposed to heat or flame (Section 6).
Disaster Hazard: Dangerous; see chlorides and phenol.

TOXIC HAZARD RATING CODE (For detailed discussion, see Section 1.)

0 NONE: (a) No harm under any conditions; (b) Harmful only under unusual conditions or overwhelming dosage.

1 SLIGHT: Causes readily reversible changes which disappear after end of exposure.

2 MODERATE: May involve both irreversible and reversible changes; not severe enough to cause death or permanent injury.

3 HIGH: May cause death or permanent injury after very short exposure to small quantities.

U UNKNOWN: No information on humans considered valid by authors.

Countermeasures
Ventilation Control: Section 2.
Storage and Handling: Section 7.
Shipping Regulations: Section 11.
MCA warning label.

CHLOROPHENOTHANE. See DDT.

2-(4-CHLOROPHENOXY) ETHANOL
General Information
Formula: $C_6H_4Cl(O)CH_2CH_2OH$.
Constants: Mol wt: 172.5.
Hazard Analysis
Toxic Hazard Rating: Limited animal experiments suggest
high toxicity and irritation. See also 2,4-dichlorophen-
oxyacetic acid.
Disaster Hazard: Dangerous. See chlorides.

p-CHLOROPHENOXYETHOXYETHYL CHLORIDE
Hazard Analysis
Toxicity: See chlorinated hydrocarbons, aromatic.
Disaster Hazard: Dangerous; see chlorides.
Countermeasures
Storage and Handling: Section 7.

1-(o-CHLOROPHENOXY)-2-PROPANOL. See propylene
glycol chlorophenyl ether.

**3-(α, p-CHLOROPHENYL-β-ACETYLETHYL)-4-
HYDROXYCOUMARIN.** See tomorin.

CHLOROPHENYL CARBAMATE
General Information
Description: Crystals.
Formula: $ClC_6H_4NHCOOH$.
Constants: Mol wt: 171.6.
Hazard Analysis
Toxicity: Details unknown.
Disaster Hazard: Dangerous; see chlorides.
Countermeasures
Storage and Handling: Section 7.

**p-CHLOROPHENYL-p-CHLOROBENZENE
SULFONATE**
General Information
Synonyms: K-101; kolker acaricide; ovex; ovotran.
Description: White to cream, free flowing powder.
Formula: $C_{12}H_8Cl_2O_3S$.
Constants: Mol wt: 303.2, mp: 87°C.
Hazard Analysis
Toxic Hazard Rating:
Acute Local: Irritant 2.
Acute Systemic: Ingestion 3.
Chronic Local: U.
Chronic Systemic: U.
Toxicology: An insecticide. Has produced kideny and liver
damage in animal experiments. See chloro phenyls.
Disaster Hazard: Dangerous; see chlorides and phosphates;
Countermeasures
Storage and Handling: Section 7.
Shipping Regulations: Section 11.
MCA warning label.

p-CHLOROPHENYL-p-CHLOROBENZYL SULFIDE
General Information
Formula: $C_{13}H_{10}Cl_2S$.
Constant: Mol wt: 269.2.
Hazard Analysis
Toxic Hazard Rating:
Acute Local: Irritant 2.
Acute Systemic: Ingestion 3.
Chronic Local: U.
Chronic Systemic: U.
Toxicology: Experimentally, has caused liver and kidney
injury.
Disaster Hazard: Dangerous; see chlorides and sulfates.

3-(p-CHLOROPHENYL)-1,1-DIMETHYLUREA. See
monuron.

**3-(p-CHLOROPHENYL)-1,1-DIMETHYLUREA   TRI-
CHLOROACETATE.** See urox.

o-CHLOROPHENYL DIPHENYL PHOSPHATE
General Information
Description: Clear, pale straw colored mobile liquid.
Formula: $C_{18}H_{14}O_4ClP$.
Constants: Mol wt: 360.73, mp: $<0°C$, bp: 240–255°C at
5 mm, flash p: $>419°F$, d: 1.3 at 25°/25°C, vap.
d.: 12.5.
Hazard Analysis
Toxicity: Details unknown.
Fire Hazard: Slight, when exposed to heat or flame.
Disaster Hazard: Dangerous; see chlorides and phosphates;
it can react with oxidizing materials.
Countermeasures
Storage and Handling: Section 7.
To Fight Fire: Water, foam, carbon dioxide, dry chemical
or carbon tetrachloride (Section 6).

o-CHLOROPHENYL ISOCYANATE
General Information
Description: Colorless liquid; soluble in organic solvents.
Formula: C_7H_4ClNO.
Constants: Mol wt: 153.6, bp: 106°C at 30 mm Hg.
Hazard Analysis
Toxicology: A powerful irritant to skin and mucous mem-
brane. A very strong lachrymator.
Disaster Hazard: Dangerous. See chlorides and cyanates.
Countermeasures
Storage and Handling: Section 7.

m-CHLOROPHENYL ISOCYANATE
General Information
Description: Water white liquid. Soluble in organic solvents.
Formula: C_7H_4ClNO.
Constants: Mol wt: 153.6, mp: $-4°C$, bp: 101°C at 30
mm Hg.
Hazard Analysis
Toxicology: See o-chlorophenyl isocyanate.
Disaster Hazard: Dangerous. See chlorides and cyanates.
Countermeasures
Storage and Handling: Section 7.

p-CHLOROPHENYL ISOCYANATE
General Information
Description: White solid; soluble in organic solvents.
Formula: C_7H_4ClNO.
Constants: Mol wt: 153.6, mp: 28°C, bp: 106.5°C at 30
mm Hg.
Hazard Analysis
Toxicology: See o-chlorophenyl isocyanate.
Disaster Hazard: Dangerous. See chlorides and cyanates.
Countermeasures
Storage and Handling: Section 7.

**1-(p-CHLOROPHENYL)-5-ISOPROPYLBIGUANIDE
HYDROCHLORIDE.** See chlorguanide.

CHLORO-2-PHENYLPHENOL. See   2-chloro-4-phenyl-
phenol.

2-CHLORO-4-PHENYLPHENOL
General Information
Synonym: Dowicide 4.
Description: White flakes.
Formula: $C_6H_5C_6H_3OHCl$.
Constants: Mol wt: 204.6, bp: 322°C (decomposes), fp:
74.2°C, flash p: 345°F, d: <1, mp: 172–176°F.
Hazard Analysis
Toxicity: See phenol.
Fire Hazard: Slight, when exposed to heat or flame.
Disaster Hazard: Dangerous, see chlorides.

Countermeasures
Storage and Handling: Section 7.
To Fight Fire: Water, alcohol foam, foam, carbon dioxide,
dry chemical or carbon tetrachloride (Section 6).

4 and 6-CHLORO-2-PHENYLPHENOL
General Information
Description: Clear, colorless to straw-colored, viscous
liquid.
Formula: $C_6H_5C_6H_3OHCl$.
Constants: Mol wt: 204.6, bp: 162–178°C at 10 mm, flash
p: 325°F, d: 1.234 at 25°/25°C, vap. d.: 7.07.
Hazard Analysis
Toxicity: Details unknown. Toxic where taken orally. See
phenols.
Fire Hazard: Moderate, when exposed to heat or flame.
Disaster Hazard: Dangerous; see chlorides.
Countermeasures
Storage and Handling: Section 7.
To Fight Fire: Water, foam, carbon dioxide, dry chemical
or carbon tetrachloride (Section 6).

p-CHLOROPHENYL PHENYL SULFONE. See sul-
phenone.

CHLOROPHENYLTRICHLOROSILANE
General Information
Description: Colorless to pale yellow liquid; readily hydro-
lyzed by moisture, with the liberation of hydrochloric
acid.
Formula: $ClC_6H_4SiCl_3$ (a mixture of 3 isomers).
Constants: Mol wt: 246, bp: 230°C, d: 1.439 (25/25°C),
flash p: 255°F (C.O.C.).
Hazard Analysis
Toxic Hazard Rating: U. See also chlorosilanes.
Fire Hazard: Slight, when exposed to heat or flame.
Disaster Hazard: Dangerous. See chlorides.
Countermeasures
Shipping Regulations: Section 11.
IATA: Corrosive liquid, white label, not acceptable (pas-
senger), 40 liters (cargo).

CHLOROPICRIN
General Information
Synonym: Nitrotrichloromethane; trichloronitromethane,
nitrochloroform.
Description: Slightly oily, colorless liquid.
Formula: CCl_3NO_2.
Constants: Mol wt: 164.39, mp: −64°C, bp: 112°C, d:
1.692, vap. press.: 40 mm at 33.8°C, vap. d.: 5.69.
Hazard Analysis
Toxic Hazard Rating:
Acute Local: Irritant 3; Ingestion 3; Inhalation 3.
Acute Systemic: Ingestion 3; Inhalation 3.
Chronic Local: U.
Chronic Systemic:U.
An insecticide.
Toxicology: Chloropicrin is a powerful irritant and affects
all body surfaces. It causes lachrymation, vomiting,
bronchitis, and pulmonary edema. It is not only a lachry-
mator but also is irritating to skin, gastro-intestinal,
and respiratory tracts. An additional toxic effect is its
reaction with SH-groups in hemoglobin thus interfering with oxygen transport. Photochemical transformation of chloropicrin into phosgene (carboxy chloride,
COCl²) has been reported. A concentration of 1 ppm
causes a smarting pain in the eyes and therefore in
itself constitutes a good warning of exposure. It causes

vomiting probably due to swallowing saliva in which
small amounts of chloropicrin have dissolved. It is
called vomiting gas and has been extensively used by
the military. Its primary lethal effect is to produce lung
injury and it is a difficult gas to protect oneself against
because it is chemically inert and does not react with
the usual chemicals used in gas masks. Four ppm is sufficient to render a man unfit for action and 20 ppm,
when breathed from 1 to 2 minutes, causes definite
bronchial or pulmonary lesions. Industrially it is used as
a warning agent in commercial fumigants. It is more
toxic than chlorine but less so than phosgene.
TLV: ACGIH (accepted); 0.1 part per million in air; 0.7
mg/cu meter of air.
Disaster Hazard: Dangerous; when heated to decomposition,
it emits highly toxic fumes.
Countermeasures
Ventilation Control: Section 2.
Treatment and Antidotes: Removal from exposure is an
immediate necessity. If exposure has been severe, con-
sult a physician.
Personnel Protection: Section 3.
First Aid: Section 1.
Storage and Handling: Section 7.
Shipping Regulations: Section 11.
I.C.C.: Poison B, poison label, 24 pounds.
Coast Guard Classification: Poison B; poison label.
MCA warning label.
IATA: Poison A, not acceptable (passenger and cargo).

CHLOROPICRIN, ABSORBED
Shipping Regulations: Section 11.
I.C.C.: Poison B, poison label, 75 pounds.
IATA: Poison A, not acceptable (passenger and cargo).

CHLOROPICRIN AND METHYL CHLORIDE
    MIXTURES
Shipping Regulations: Section 11.
ICC: Poison A, poison gas label, not acceptable.
IATA: Poison A, not acceptable (passenger and cargo).

CHLOROPICRIN AND METHYL BROMIDE
    MIXTURES
Shipping Regulations: Section 11.
IATA: Poison A, not acceptable (passenger and cargo).

CHLOROPICRIN AND NONFLAMMABLE, NON-
    LIQUEFIED COMPRESSED GAS MIXTURES
Shipping Regulations: Section 11.
ICC: Poison A, poison gas label, not accepted.
IATA: Poison A, not acceptable (passenger and cargo).

CHLOROPICRIN MIXTURES (CONTAINING NO
    COMPRESSED GAS OR POISON LIQUID,
    CLASS A)
Shipping Regulations: Section 11.
ICC: Poison B, poison label 75 pounds.

CHLOROPICRIN MIXTURES, N.O.S.
Shipping Regulations: Section 11.
IATA: Poison A, not acceptable :passenger and cargo).

CHLOROPLATINIC ACID
General Information
Description: Red-brown, deliquescent prisms.
Formual: $H_2PtCl_6 \cdot 6H_2O$.
Constants: Mol wt: 518.08, mp: 60°C, d: 2.431.
Hazard Analysis and Countermeasures
See platinum compounds and chlorides.

TOXIC HAZARD RATING CODE   *(For detailed discussion, see Section 1.)*

0 NONE: (a) No harm under any conditions; (b) Harmful only under unusual conditions or overwhelming dosage.

1 SLIGHT: Causes readily reversible changes which disappear after end
of exposure.

2 MODERATE: May involve both irreversible and reversible changes;
not severe enough to cause death or permanent injury.

3 HIGH: May cause death or permanent injury after very short exposure
to small quantities.

U UNKNOWN: No information on humans considered valid by authors.

Shipping Regulations: Section 11.
   IATA (solid): Other restricted articles, class B, no label required, 12 kilograms (passenger), 45 kilograms (cargo).

CHLOROPRENE
General Information
Synonyms: 2-Chloro-1,3-butadiene; chlorobutadiene.
Description: Colorless liquid.
Formula: $CH_2CHCClCH_2$.
Constants: Mol wt: 88.54, bp: 59.4°C, d: 0.9583, flash p: −4°F, lel: 4.0%, uel: 20.0%, vap. d.: 3.0.
Hazard Analysis
Toxic Hazard Rating:
   Acute Local: Irritant 2; Inhalation 2.
   Acute Systemic: Ingestion 3; Inhalation 3.
   Chronic Local: U.
   Chronic Systemic: Ingestion 3; Inhalation 3.
TLV: ACGIH (accepted); 25 parts per million of air; 90 milligrams per cubic meter of air.
Toxicology: Animal experiments have shown that a concentration of 250 ppm in air is toxic, and a concentration of 75 ppm may be toxic with continued exposure. Exposure to the vapor first causes irritation of the respiratory tract, followed by depression of respiration and, if exposure is continued, asphyxia. The vapor is a central system depressant; in animals it causes severe degenerative changes in the vital organs, particularly the liver and kidneys. Blood pressure is lowered. Lung changes accompany exposure to the higher concentrations. Humans exposed to chloroprene have been reported to develop dermatitis, conjunctivitis, corneal necrosis, anemia, temporary loss of hair, nervousness and irritability.
Disaster Hazard: Dangerous; See chlorides.
Fire Hazard: Dangerous, when exposed to heat or flame.
Countermeasures
Ventilation Control: Section 2.
Personnel Protection: Section 3.
First Aid: Section 1.
Storage and Handling: Section 7.
To Fight Fire: Alcohol foam.
Storage and Handling: Section 7.

1-CHLOROPROPANE. See n-propyl chloride.

2-CHLOROPROPANE. See isopropyl chloride.

1-CHLOROPROPANE-2,3-DIOL
General Information
Synonym: α-Chlorohydrin.
Description: Colorless liquid.
Formula: $CH_2ClCHOHCH_2OH$.
Constants: Mol wt: 110.54, bp: 213°C decomposes, d: 1.326.
Hazard Analysis
Toxicity: Details unknown.
Fire Hazard: Slight, when exposed to heat or flame (Section 6).
Countermeasures
Storage and Handling: Section 7.

1-CHLORO-2-PROPANOL. See propylenechlorohydrin.

2-CHLORO-1-PROPONOL. See propylene chlorohydrin—primary.

2-CHLOROPROPENE
General Information
Synonyms: Isopropenyl chloride.
Formula: $CH_3CCL:CH_2$.
Description: Colorless liquid.
Constants: Mol wt: 76.53, bp: 22.65°C, fp: −137.4°C, d: 0.918 at 9°C.
Hazard Analysis
Toxic Hazard Rating: U.

Fire Hazard: Flammable.
Disaster Hazard: Dangerous. Reacts with powerful oxidizers. See chlorides.
Countermeasures
Storage and Handling: Section 7.
Shipping Regulations: Section 11.
   IATA: Flammable liquid, red label, not acceptable (passenger), 40 liters (cargo).

3-CHLOROPROPENE. See allyl chloride.

α-CHLOROPROPIONIC ACID
General Information
Synonym: 2-Chloropropionic acid.
Description: Soluble in water.
Formula: $CH_3CHClCOOH$.
Constants: Mol wt: 108.5, d: 1.260–1.268 (20°C), bp: 183–187°C, flash p: 225°C.
Hazard Analysis
Fire Hazard: Moderate, when exposed to heat or flame.
Disaster Hazard: Dangerous. See chlorides.
Countermeasures
To Fight Fire: Water, foam, alcohol foam (Section 6).

2-CHLOROPROPIONIC ACID. SEE α-chloropropionic acid.

β-CHLOROPROPIONITRILE
General Information
Description: Colorless liquid.
Formula: $ClCH_2CH_2CN$.
Constants: Mol wt: 89.5, mp: −51°C, bp: 176°C decomposes, flash p: 168°F (C.C.), d: 1.1363 at 25°C, vap. press.: 6 mm at 50°C, vap. d.: 3.09.
Hazard Analysis
Toxicity: See nitriles.
Fire Hazard: Moderate in its liquid form when exposed to heat or flame.
Countermeasures
To Fight Fire: Alcohol foam, water, foam, carbon dioxide or dry chemical or carbon tetrachloride.
Storage and Handling: Section 7.

β-CHLORPROPYL ALCOHOL. See propylene chlorohydrin—primary.

3-CHLOROPROPYLENE-1,2-OXIDE. See epichlorohydrin.

CHLOROPROPYLENE SULFIDE
General Information
Formula: $S \cdot CH_2CHCH_2Cl$.
Constants: Mol. wt: 108.5.
Hazard Analysis
Toxic Hazard Rating:
   Acute Local: Irritation 2; Inhalation 3.
   Acute Systemic: Ingestion 3; Inhalation 2.
   Chronic Local: U.
   Chronic Systemic: U.
Based on animal experiments.
Disaster Hazard: Dangerous. See chlorides and sulfur compounds.

γ-CHLOROPROPYLTRICHLOROSILANE
General Information
Description: Liquid.
Formula: $ClC_3H_6SiCl_3$.
Constants: Mol wt: 212, bp: 180°C, d: 1.336 at 25°C.
Hazard Analysis
Toxicity: See hydrochloric acid and chlorosilanes.
Disaster Hazard: Dangerous; see chlorosilanes.
Countermeasures
Storage and Handling: Section 7.

3-CHLORO-1-PROPYNE. See propargyl chloride.

2-CHLOROPYRIDINE
General Information
Description: Colorless oily liquid.
Formula: C_5H_4NCl.
Constants: Mol wt: 113.6, bp: 170°C, d: 1.205 at 15°C vap. press.: 1 mm at 13.3°C, vap. d.: 3.93.
Hazard Analysis
Toxicity: Details unknown. See also pyridine.
Fire Hazard: Slight, when exposed to heat or flame (Section 6).
Disaster Hazard: Dangerous; see phosgene; it can react with oxidizing materials.
Countermeasures
Storage and Handling: Section 7.

CHLOROSILANES
General Information
Description: Compounds of silicon, chlorine and hydrogen where the total number of atoms of chlorine and hydrogen add up to 4.
Formula: SiH Cl (chlorosilane).
Hazard Analysis
Toxic Hazard Rating:
Acute Local: Irritant 3; Ingestion 3, Inhalation 3.
Acute Systemic: U.
Chronic Local: Irritant 2.
Chronic Systemic: U.
Note: Toxicity based on hydrochloric acid which is formed upon hydrolysis of a chlorosilane.
Disaster Hazard: Dangerous; when heated to decomposition, they emit highly toxic fumes of chlorides; they will react with water or steam to produce heat, toxic and corrosive fumes of hydrochloric acid.
Countermeasures
Ventilation Control: Section 2.
Personnel Protection: Section 3.
Storage and Handling: Section 7.

CHLOROSOL. See ethylene dichloride, carbon tetrachloride.

CHLOROSTANNIC ACID
General Information
Description: Colorless leaf-like crystals or liquid.
Formula: $H_2SnCl_6 \cdot 6H_2O$.
Constants: Mol wt: 441.6, mp: 9°C, d: 1.93.
Hazard Analysis
Toxicity: See tin compounds and hydrochloric acid.
Disaster Hazard: Dangerous; see chlorides; will react with water or steam to produce heat and toxic and corrosive fumes of hydrochloric acid.
Countermeasures
Storage and Handling: Section 7.

N-CHLOROSUCCINIMIDE
General Information
Synonym: Succinchlorimide.
Description: White powder; mild odor of chlorine.
Formula: $C_4H_4O_2NCl$.
Constants: Mol wt: 133.5, mp: 148–149°C.
Hazard Analysis
Toxicity: See hypochlorous acid.
Disaster Hazard: See hypochlorites.
Countermeasures
Storage and Handling: Section 7.

CHLOROSULFACIDE. See p-chlorobenzyl p-chlorophenyl sulfide.

CHLOROSULFONIC ACID
General Information
Synonym: Sulfuric chlorohydrin.
Description: Clear to cloudy, colorless to pale yellow liquid; sharp odor.
Formula: $ClSO_3H$.
Constants: Mol wt: 116.53, mp: −80°C, bp: 151.0°C, d: 1.766 at 18°C, vap. press.: 1 mm at 32.0°C, vap. d.: 4.02.
Hazard Analysis
Toxicity: See sulfuric acid.
Toxicology: Chlorosulfonic acid can cause severe acid burns and is very irritating to the eyes, lungs and mucous membranes. It can cause acute toxic effects either in the liquid or vapor state. Inhalation of concentrated vapor may cause loss of consciousness with serious damage to lung tissue. Contact of liquid with the eyes can cause severe burns if not immediately and completely removed. It also causes severe skin burns due to its highly corrosive action. Upon ingestion, it will irritate the mouth, esophagus and stomach to a serious degree and on contact with skin cause dermatitis. Even in the vapor form it may cause conjunctivitis.
Disaster Hazard: Dangerous. See sulfuric acid and hydrochloric acid and sulfonates.
Countermeasures
Ventilation Control: Section 2.
Storage and Handling: Section 7.
Shipping Regulations: Section 11.
I.C.C.: Corrosive liquid; white label, 1 quart.
Coast Guard Classification: Corrosive liquid; white label.
MCA warning label.
IATA: Corrosive liquid, white label, 1 liter (passenger and cargo).

CHLOROSULFONIC ACID-SULFUR TRIOXIDE MIXTURE. See chlorosulfonic acid and sulfur trioxide.
Shipping Regulations: Section 11.
I.C.C.: Corrosive liquid; white label, 1 quart.
Coast Guard Classification: Corrosive liquid; white label.
IATA: Corrosive liquid, white label, 1 liter (passenger and cargo).

CHLOROSULFURIC ACID. See sulfuryl chloride.

2-CHLORO-2-TETRACHLOROPHENOXYDIETHYL ETHER
General Information
Description: Colorless, odorless liquid.
Formula: $Cl(C_2H_4)_2O_2Cl_4$.
Constants: Mol wt: 338.4, mp: 31°C, bp: 170–176°C at 1 mm, flash p: 392°F, d: 1.506, vap. d.: 11.6.
Hazard Analysis
Toxicity: Details unknown.
Fire Hazard: Slight, when exposed to heat or flame.
Countermeasures
To Fight Fire: Water, foam, carbon dioxide, dry chemical or carbon tetrachloride.
Storage and Handling: Section 7.

CHLOROTETRACYCLINE. See chlortetracycline.

CHLOROTETRAFLUOROETHANE. See monochlorotetrafluoroethane.

CHLOROTHYMOL
General Information
Synonym: 6-Chloro-4-isopropyl-1-methyl-3-phenol.
Description: Crystals.

TOXIC HAZARD RATING CODE (For detailed discussion, see Section 1.)

0 NONE: (a) No harm under any conditions; (b) Harmful only under unusual conditions or overwhelming dosage.

1 SLIGHT: Causes readily reversible changes which disappear after end of exposure.

2 MODERATE: May involve both irreversible and reversible changes; not severe enough to cause death or permanent injury.

3 HIGH: May cause death or permanent injury after very short exposure to small quantities.

U UNKNOWN: No information on humans considered valid by authors.

Formula: $C_{10}H_{13}ClO$.
Constants: Mol wt: 184.7 mp: 62–64°C.
Hazard Analysis
Toxic Hazard Rating:
Acute Local: Irritant 3; Inhalation 3.
Acute Systemic: U.
Chronic Local: U.
Chronic Systemic: U.
Toxicology: Concentrated solutions are strong irritants.
Fire Hazard: Slight, when exposed to heat or flame (Section 6).
Disaster Hazard: Dangerous; see chlorides.
Countermeasures
Storage and Handling: Section 7.

(4-CHLORO-o-TOLOXY) ACETIC ACID. See 2,4-dichlorophenoxy acetic acid.

α-CHLOROTOLUENE. See benzyl chloride.

m-CHLOROTOLUENE
General Information
Synonym: 3-Chloro-1-methylbenzene.
Description: Liquid.
Formula: C_7H_7Cl.
Constants: Mol wt: 126.6, mp: −48°C, bp: 162.3°C, d: 1.0797 at 13.90°/4°C, vap. press.: 10 mm at 43.2°C.
Hazard Analysis
Toxicity: Details unknown. Narcotic in high concentrations. See also chlorinated hydrocarbons, aromatic.
Fire Hazard: Slight, when exposed to heat or flame (Section 6).
Countermeasures
Storage and Handling: Section 7.
Ventilation Control: Section 2.
Shipping Regulations: Section 11.
MCA warning label.

o-CHLOROTOLUENE
General Information
Synonym: 2-Chloro-1-methylbenzene.
Description: Liquid.
Formula: C_7H_7Cl.
Constants: Mol wt: 126.6, mp: −35°C, bp: 159.3°C, d: 1.1018 at 0°/4°C, vap. press. 10 mm at 43.2°C.
Hazard Analysis
Toxicity: Details unknown. Vapor toxic. Avoid prolonged breathing of vapor. Use with adequate ventilation. See m-chlorotoluene.
Fire Hazard: Slight, when exposed to heat or flame (Section 6).
Countermeasures
Ventilation Control: Section 2.
Storage and Handling: Section 7.
Shipping Regulations: Section 11.
MCA warning label.

p-CHLOROTOLUENE
General Information
Synonym: 4-Chloro-1-methylbenzene.
Description: Liquid.
Formula: C_7H_7Cl.
Constants: Mol wt: 126.6, mp: 7.3°C, bp: 162.3°C, d: 1.0651 at 24.4°/4°C, vap. press.: 10 mm at 43.8°C.
Hazard Analysis
Toxicity: Details unknown. Vapor is toxic. Avoid prolonged breathing of vapor. Use with adequate ventilation. See also m-chlorotoluene.
Fire Hazard: Slight, when exposed to heat or flame (Section 6).
Countermeasures
Ventilation Control: Section 2.
Storage and Handling: Section 7.
Shipping Regulations: Section 11.
MCA warning label.

CHLOROTOLUIDINE
General Information
Synonym: p-Toluidine hydrochloride.
Description: Grayish-white crystals.
Formula: $C_7H_9N \cdot HCl$.
Constants: Mol wt: 143.6, mp: 243°C, bp: 257.5 (sublimes).
Hazard Analysis
Toxicity: See toluidine.
Toxicology: May cause cyanosis, tachycardia, hematuria, albuminuria.
Disaster Hazard: Dangerous; see chlorides.
Countermeasures
Storage and Handling: Section 7.

4-CHLORO-o-TOLUIDENE HYDROCHLORIDE
General Information
Description: Solid.
Formula: $CH_3C_6H_3(Cl)NH_2 \cdot HCl$.
Constant: Mol wt: 178.
Hazard Analysis
Toxic Hazard Rating: U.
Disaster Hazard: Dangerous. See chlorides.
Countermeasures
Shipping Regulations: Section 11.
ICC: Poison B, poison label, 1 quart.
IATA: Poison B, poison label, not acceptable (passenger), 1 liter (cargo).

2-CHLORO-1,3,3-TRIETHOXYPROPANE
General Information
Description: Liquid.
Formula: $C_9H_{19}O_3Cl$.
Constants: Mol wt: 210.70.
Hazard Analysis
Toxicity: Details unknown. Limited experiments show low toxicity for rabbits and moderate toxicity for rats.
Fire Hazard: Slight, when exposed to heat or flame (Section 6).
Countermeasures
Storage and Handling: Section 7.

2-CHLORO-1,1,2-TRIFLUORO-3-CYANOCYCLOBUTANE
General Information
Description: Liquid.
Formula: $C_5H_3F_3ClN$.
Constant: Mol wt: 169.5.
Hazard Analysis
Toxicity: Details unknown.
Disaster Hazard: Dangerous; see chlorides and cyanides.

CHLOROTRIFLUOROETHYLENE
General Information
Formula: $ClFCCF_2$.
Constants: Mol wt: 116.46, lel: 8.4%, uel: 38.7%, −18°F.
Hazard Analysis
Toxicity: Details unknown. See fluorides. Limited experiments with animals show moderate degree of toxicity. See also chlorinated hydrocarbons, aliphatic.
Disaster Hazard: Dangerous; see chlorides and fluorides.
Countermeasures
Storage and Handling: Section 7.
To Fight Fire: Stop flow of gas (Section 6).
Shipping Regulations: Section 11.
IATA (Inhibited): Flammable gas, red label, not acceptable (passenger), 140 kilograms (cargo).
IATA (uninhibited): Flammable gas, no label, not acceptable (passenger or cargo).

CHLOROTRIFLUOROGERMANE
General Information
Description: Colorless gas.
Formula: GeF_3Cl.
Constants: Mol wt: 165.06, mp: −66.2°C, bp: −20.3°C.
Hazard Analysis and Countermeasures
See fluorides, germanium compounds, and chlorides.

CHLOROTRIFLUOROMETHANE. See monochlorotri-
fluoromethane.

m-CHLOROTRIFLUOROMETHYLBENZENE. See m-
chlorobenzotrifluoride.

o-CHLOROTRIFLUOROMETHYLBENZENE. See o-
chlorobenzotrifluoride.

p-CHLOROTRIFLUOROMETHYLBENZENE. See p-
chlorobenzotrifluoride.

**2-CHLORO-1,1,2-TRIFLUORO-3-METHYL-3-VINYLCY-
CLOBUTANE**
General Information
Description: Liquid.
Formula: $C_7H_8F_3Cl$.
Constant: Mol wt: 184.5.
Hazard Analysis
Toxicity: Details unknown.
Disaster Hazard: Dangerous; when heated to decomposi-
tion, it emits highly toxic halide fumes.

**3-CHLORO-1,1,2-TRIFLUORO-3-PHENYLCYCLOBU-
TANE**
General Information
Description: Liquid.
Formula: $C_{10}H_8F_3Cl$.
Constant: Mol wt: 220.5.
Hazard Analysis
Toxicity: Details unknown.
Disaster Hazard: Dangerous; when heated to decomposi-
tion, it emits highly toxic halide fumes.

O-CHLORO-2,α-TRIFLUORO TOLUENE. See o-chloro-
benzotrifluoride.

**2-CHLORO-1,1,2-TRIFLUORO-3-VINYLCYCLOBU-
TANE**
General Information
Description: Liquid.
Formula: $C_6H_6F_3Cl$.
Constant: Mol wt: 170.5.
Hazard Analysis
Toxicity: Details unknown.
Disaster Hazard: Dangerous; see chlorides and fluorides.

2-CHLORO-1,3,5-TRINITROBENZENE. See picryl
chloride.

β-CHLOROVINYLDICHLOROARSINE. See dichloro-
(2-chlorovinyl) arsine.

β-CHLOROVINYLMETHYLCHLOROARSINE
General Information
Description: Liquid.
Formula: $CH_3AsClCHCHCl$.
Constants: Mol wt: 187.0, bp: 112–115°C.
Hazard Analysis
Toxic Hazard Rating:
  Acute Local: Irritant 3; Ingestion 3; Inhalation 3.
  Acute Systemic: Ingestion 3; Inhalation 3; Skin Absorp-
  tion 3.
  Chronic Local: Irritant 3.
  Chronic Systemic: Ingestion 3; Inhalation 3; Skin Absorp-
  tion 3.
Disaster Hazard: Dangerous! See arsenic compounds.
Countermeasures
Storage and Handling: Section 7.

6-CHLORO-o-XENOL. See 2-phenyl-6-chlorophenol.

p-CHLORO-m-XYLENOL
General Information
Synonym: 2-Chloro-5-hydroxy-1,3-dimethlybenzene.
Description: Crystals. Phenolic odor. Slightly water soluble.
Formula: C_8H_9OCl.
Constants: Mol wt: 156.6, mp: 115.5°C, bp: 246°C.
Hazard Analysis
Toxic Hazard Rating:
  Acute Local: Irritant 2; Ingestion 3.
  Acute Systemic: Ingestion 3; Skin Absorption 2.
  Chronic Local: Irritant 2.
  Chronic Systemic: U.
Toxicology: See phenol.
Disaster Hazard: Dangerous; see chlorides.
Countermeasures
Storage and Handling: Section 7.

CHLORTETRACYCLINE
General Information
Synonym: CTC, chlorotetracycline.
Description: Golden yellow crystals; slightly soluble in
water. Very soluble in aqueous solution ph 76.5; freely
soluble in the "cellosolves," dioxane, "Carbitol";
slightly soluble in methanol, ethanol, butanol, acetone,
ethyl acetate, and benzene; insoluble in ether and
petroleum ether.
Formula: $C_{22}H_{23}ClN_2O_8$.
Constants: Mol wt: 479, mp: 168–169°C.
Hazard Analysis
Toxic Hazard Rating: U. A food additive permitted in the
feed and drinking water of animals; and/or for the
treatment of food producing animals. Also a food additive permitted in food for human consumption (Section
10).
Disaster Hazard: Dangerous. See chlorides.

CHLORTHION
General Information
Synonym: Compound 22/190.
Formula: Phosphoric acid ester containing chlorine.
Hazard Analysis
Toxicity: See parathion. This is a very toxic organic phos-
phate cholinesterase inhibitor.
Disaster Hazard: Dangerous; when heated to decomposi-
tion, it emits highly toxic fumes.

CHOLECALCIFEROL. See vitamin D_3.

5,7-CHOLESTADUN-3-β-ol. See vitamin D_3.

CHOLIC ACID
General Information
Description: The most abundant bile acid, the monohydrate
crystallizes in plates from dilute acetic acid; soluble in
glacial acetic acid, acetone, and alcohol; slightly sol-
uble in chloroform; practically insoluble in water and
benzene.
Formula: $C_{24}H_{40}O_5$.
Constant: Mol. wt: 408.
Hazard Analysis
Toxic Hazard Rating: U. An emulsifying agent food addi-
tive (Section 10).

CHOLINE BITARTRATE
General Information
Description: White crystalline powder, odorless or faint
trimethylamine like odor; soluble in water and alcohol;
insoluble in ether, chloroform and benzene.

TOXIC HAZARD RATING CODE (For detailed discussion, see Section 1.)

0 NONE: (a) No harm under any conditions; (b) Harmful only under unusual conditions or overwhelming dosage.

1 SLIGHT: Causes readily reversible changes which disappear after end
of exposure.

2 MODERATE: May involve both irreversible and reversible changes;
not severe enough to cause death or permanent injury.

3 HIGH: May cause death or permanent injury after very short exposure
to small quantities.

U UNKNOWN: No information on humans considered valid by authors.

Formula: $(C_5H_{14}NO\cdot C_4H_5O_8)$.
Constant: Mol wt: 253.
Hazard Analysis
Toxic Hazard Rating: U. A nutrient and/or dietary supplement food additive (Section 10).

CHOLINE XANTHATE
Hazard Analysis
A food additive permitted in the feed and drinking water of animals and/or for the treatment of food producing animals (Section 10).
Disaster Hazard: Dangerous. See sulfur compounds.

CHONDRUS EXTRACT
General Information
Synonym: Carrageenan; Irish moss; chondrus; Irish gum; pig wrack; rock salt moss.
Description: Dried plant of sea weed Chondrus crispus; yellow white when powdered; insoluble in organic solvents.
Hazard Analysis
Toxic Hazard Rating: U. A stabilizer food additive (Section 10).

CHROMATED ZINC ARSENATE
General Information
Synonym: Boliden salt.
Description: Powder.
Formula: Arsenic acid + sodium arsenate + sodium dichromate + zinc sulfate.
Hazard Analysis
Toxicity: Highly toxic. See arsenic compounds and chromium compounds.
Disaster Hazard: Dangerous! See arsenic compounds.

CHROMATES. See chromium compounds.

CHROME ALUM
General Information
Synonym: Chromium-potassium sulfate.
Description: Dark violet-red crystals.
Formula: $CrK(SO_4)_2\cdot 12H_2O$.
Constants: Mol wt: 499.3, d: 1.813.
Hazard Analysis and Countermeasures
See chromium compounds.

CHROME YELLOW. See lead chromate.

CHROMIC ACETATE
General Information
Description: Gray-green powder or bluish-green pasty mass.
Formula: $Cr(C_2H_3O_2)_3\cdot H_2O$.
Constant: Mol wt: 247.16.
Hazard Analysis and Countermeasures
See chromium compounds.

CHROMIC ACID
General Information
Synonyms: Chromic anhydride; chromium trioxide.
Description: Dark, purple-red crystals.
Formula: CrO_3.
Constants: Mol wt: 100.01, mp: 196°C, d: 2.70.
Hazard Analysis
Toxicity: See chromium compounds.
Toxicology: It is a common air contaminant (Section 4).
Caution: This material is usually caustic in its action on skin, mucous membranes or organic matter in general.
TLV: ACGIH (accepted); 0.1 milligrams per cubic meter of air.
Fire Hazard: Dangerous; a very powerful oxidizing agent. In contact with organic matter or reducing agents in general it causes violent reactions (Section 6).
Explosion Hazard: Upon intimate contact with powerful reducing agents it can cause violent explosions.
Countermeasures
First Aid: Section 1.

Storage and Handling: Section 7.
Shipping Regulations: Section 11.
    I.C.C.: Oxidizing material; yellow label, 100 pounds.
    Coast Guard Classification: Oxidizing material; yellow label.
    MCA warning label.
    IATA: Oxidizing material, yellow label, 12 kilograms (passenger), 45 kilograms (cargo).

CHROMIC ACID SOLUTION. See also chromic acid.
Shipping Regulations: Section 11.
    I.C.C.: Corrosive liquid; white label, 1 gallon.
    Coast Guard Classification: Corrosive liquid: white label.
    IATA: Corrosive liquid, white label, 1 liter (passenger), 5 liters (cargo).

CHROMIC ANHYDRIDE. See chromic acid.

CHROMIC BROMIDE
General Information
Description: Hexagonal olive-green crystals.
Formula: $CrBr_3$.
Constants: Mol wt: 291.76, bp: sublimes, d: 4.25 at 0°C.
Hazard Analysis and Countermeasures
See chromium compounds and bromides.

CHROMIC CHLORIDE
General Information
Description: Crystals.
Formula: $CrCl_3$.
Constants: Mol wt: 158.38, bp: 1300°C (sublimes), d: 2.76 at 15°C.
Hazard Analysis and Countermeasures
See chromium compounds and chlorides.

CHROMIC FLUORIDE
General Information
Description: Rhombic green crystals.
Formula: CrF_3.
Constants: Mol wt: 109.01, mp: >1000°C, bp: sublimes, d: 3.8.
Hazard Analysis and Countermeasures
See chromium compounds and fluorides.
Shipping Regulations: Section 11.
    IATA (solid): Other restricted articles, class B, no label required, 12 kilograms (passenger), 45 kilograms (cargo).
    IATA (liquid): Corrosive liquid, white label, 1 liter (passenger), 5 liters (cargo).

CHROMIC NITRATE
General Information
Description: Monoclinic brown crystals.
Formula: $Cr(NO_3)_3\cdot 7\frac{1}{2}H_2O$.
Constants: Mol wt: 373.15, mp: 100°C.
Hazard Analysis and Countermeasures
See chromium compounds and nitrates.

CHROMIC o-PHOSPHATE
General Information
Description: Violet crystals.
Formula: $Cr(PO_4)\cdot 2H_2O$.
Constants: Mol wt: 183.02, d: 2.42 at 32.5°C.
Hazard Analysis and Countermeasures
See chromium compounds and phosphates.

CHROMIC SESQUISULFIDE
General Information
Description: Brown-black powder.
Formula: Cr_2S_3.
Constants: Mol wt: 200.02, d: 3.77 at 19°C.
Hazard Analysis and Countermeasures
See chromium compounds and sulfides.

CHROMIC SULFATE
General Information
Description: Violet or red powder.
Formula: $Cr_2(SO_4)_3$.

Constants: Mol wt: 392.22, d: 3.012.
Hazard Analysis and Countermeasures
See chromium compounds and sulfates.

CHROMIC SULFITE
General Information
Description: Greenish-white crystals.
Formula: $Cr_2(SO_3)_3$.
Constants: Mol wt: 344.22, mp: decomposes, d: 2.2.
Hazard Analysis and Countermeasures
See chromium compounds and sulfites.

CHROMIUM
General Information
Description: Very hard metal; cubic steel-gray crystals.
Formula: Cr.
Constants: At wt: 52.01, mp: 1890°C, bp: 2200°C, d: 7.20, vap. press.: 1 mm at 1616°C.
Hazard Analysis
Toxicity: Essentially nontoxic.
Radiation Hazard: Section 5. For permissible levels, see Table 5, p. 150.
   Artificial isotope ^{51}Cr, half life 28 d. Decays to stable ^{51}V by electron capture. Emits gamma rays of 0.32 MeV and X-rays.
Fire Hazard: Moderate in form of dust.

CHROMIUM CARBONYL
General Information
Description: Colorless crystals.
Formula: $Cr(CO)_6$.
Constants: Mol wt: 220.07, mp: sublimes at room temp.; sinters at 90°C. Decomp at 130°C. Explodes at 210°C, bp: 151.0°C, d: 1.77, vap. press.: 1 mm at 36.0°C, vap. d.: 7.6.
Hazard Analysis and Countermeasures
See chromium compounds and carbonyls.

CHROMIUM CHLORIDE. See chromic compounds.

CHROMIUM COMPOUNDS
Hazard Analysis
Toxic Hazard Rating:
   Acute Local: Irritant 3; Ingestion 3; Inhalation 3.
   Acute Systemic: U.
   Chronic Local: Irritant 3; Ingestion 3; Inhalation 3.
   Chronic Systemic: Ingestion 3; Inhalation 3.
TLV: ACGIH (tentative) 0.5 milligram per cubic meter of air. (Soluble chromic, chromous salts as Cr); 1 milligram per cubic meter of air (metallic and insoluble salts).
Toxicology: Chromic acid and its salts have a corrosive action on the skin and mucous membranes. The lesions are confined to the exposed parts, affecting chiefly the skin of the hands and forearms and the mucous membranes of the nasal septum. The characteristic lesion is a deep, penetrating ulcer, which, for the most part, does not tend to suppurate, and which is slow in healing.
   Small ulcers, about the size of a matchhead or end of a lead pencil may be found, chiefly around the base of the nails, on the knuckles, dorsum of the hands and forearms. These ulcers tend to be clean, and progress slowly. They are frequently painless, even though quite deep. They heal slowly, and leave scars. On the mucous membrane of the nasal septum the ulcers are usually accompanied by purulent discharge and crusting. If exposure continues, perforation of the nasal septum may result, but produces no deformity of the nose.

Chromate salts have been associated with cancer of the lungs.
   Hexavalent compounds are said to be more toxic than the trivalent. Eczematous dermatitis due to trivalent chromium compounds has been reported.
Countermeasures
Ventilation Control: Section 2.
Storage and Handling: Section 7.

CHROMIUM DIFLUORIDE. See chromous fluoride.

CHROMIUM FORMATE
General Information
Description: Crystals.
Formula: $Cr(CHO_2)_3$.
Constant: Mol wt: 187.1.
Hazard Analysis
Toxicity: See chromium compounds.
Countermeasures
Storage and Handling: Section 7.

CHROMIUM MONOARSENIDE
General Information
Description: Gray hexagonal crystals.
Formula: CrAs.
Constants: Mol wt: 126.92, d: 6.35 at 16°C.
Hazard Analysis
Toxicity: See arsenic compounds and chromium compounds.
Fire Hazard: See arsine.
Explosion Hazard: See arsine.
Disaster Hazard: Dangerous; when heated to decomposition or on contact with water, stean, acid or acid fumes, it will react to produce toxic and flammable vapors of arsine.
Countermeasures
Storage and Handling: Section 7.

CHROMIUM MONOBORIDE
General Information
Description: Orthorhombic silver crystals.
Formula: CrB.
Constants: Mol wt: 62.83, mp: 2760°C, d: 6.17.
Hazard Analysis
Toxicity: See chromium compounds and boron compounds.
Fire Hazard: See boron hydrides.
Explosion Hazard: See boron hydrides.
Disaster Hazard: Dangerous; on contact with water, steam, acid or acid fumes, it will react to produce toxic and flammable vapors or boron hydrides.
Countermeasures
Storage and Handling: Section 7.

CHROMIUM MONOPHOSPHIDE
General Information
Description: Gray-black crystals.
Formula: CrP.
Constants: Mol wt: 82.99, d: 5.7 at 15°C.
Hazard Analysis
Toxicity: See chromium compounds and phosphides.
Fire Hazard: Dangerous; upon contact with moisture, acid or acid fumes, phosphine is evolved. See phosphine.
Explosion Hazard: See phosphides and phosphine.
Disaster Hazard: Dangerous; See phosphides.
Countermeasures
Storage and Handling: Section 7.

TOXIC HAZARD RATING CODE (For detailed discussion, see Section 1.)

0 NONE: (a) No harm under any conditions; (b) Harmful only under unusual conditions or overwhelming dosage.

1 SLIGHT: Causes readily reversible changes which disappear after end of exposure.

2 MODERATE: May involve both irreversible and reversible changes; not severe enough to cause death or permanent injury.

3 HIGH: May cause death or permanent injury after very short exposure to small quantities.

U UNKNOWN: No information on humans considered valid by authors.

CHROMIUM OXYCHLORIDE
General Ifnormation
Synonym: Chromyl chloride,
Description: Dark red liquid; musty burning odor.
Formula: CrO_2Cl_2.
Constants: Mol wt: 154.92, mp: $-96.5°C$, bp: $115.7°C$, d: 1.9145 at $25°/4°C$, vap. press.: 20 mm at $20°C$.
Hazard Analysis
Toxic Hazard Raging:
  Acute Local: Irritant 3, Inhalation 3.
  Acute Systemic: U.
  Chronic Local: U.
  Chronic Systemic: Inhalation 3.
A strong irritant. Hydrolyzes to form chromic and hydrochloric acids. Suspected as having a role in lung cancer in chromate workers.
Disaster Hazard: Dangerous. See chlorides.
Countermeasures
Shipping Regulations: Section 11.
  IATA: Corrosive liquid, white label, not acceptable (passenger), 5 liters (cargo).
  ICC: Corrosive liquid, white label, 1 gallon.

CHROMIUM 2,4-PENTANEDIONE DERIVATIVE
General Information
Synonym: Acetylacetonate of chromium.
Description: A solid.
Formula: $Cr(C_5H_7O_2)_3$.
Constants: Mol wt: 349.33, mp: $216°C$, bp: $340°C$.
Hazard Analysis and Countermeasures
See chromium compounds.

CHROMIUM PICRATE
General Information
Description: Solid.
Formula: $Cr[C_6H_2OH(NO_2)_3]_3$.
Constant: Mol wt: 739.4.
Hazard Analysis
Toxicity: See chromium compounds and picric acid.
Fire Hazard: See nitrates.
Explosion Hazard: See explosives, high and nitrates.
Disaster Hazard: See nitrates.
Countermeasures
Storage and Handling: Section 7.

CHROMIUM POTASSIUM SULFATE. See chrome alum.

CHROMIUM SULFATE. See chromic sulfate.

CHROMIUM TETRAFLUORIDE
General Information
Description: Brown amorphous, hygroscopic mass; soluble in water with hydrolysis.
Formula: CrF_4.
Constants: Mol wt: 128.01, d: 2.89, mp: $200°C$, bp: $\simeq 400°C$ evolving intensely blue fumes.
Hazard Analysis
Toxic Hazard Rating:
  Acute Local: Irritant 3; Ingestion 3; Inhalation 3.
  Acute Systemic: Ingestion 3.
  Chronic Local: U.
  Chronic Systemic: U.
  See also chromium compounds.
Disaster Hazard: Dangerous. See fluorides.

CHROMIUM TRIOXIDE. See chromic acid.

CHROMOUS ACETATE
General Information
Description: Red crystals.
Formula: $Cr(C_2H_3O_2)_2$.
Constant: Mol wt: 170.10.
Hazard Analysis and Countermeasures
See chromium compounds.

CHROMOUS BROMIDE
General Information
Description: White crystals.
Formula: $CrBr_2$.
Constants: Mol wt: 211.84, mp: $842°C$, d: 4.356.
Hazard Analysis and Countermeasures
See chromium compounds and bromides.

CHROMOUS CHLORIDE
General Information
Description: White deliquescent needles.
Formula: $CrCl_2$.
Constants: Mol wt: 122.92, mp: $824°C$, d: 2.75.
Hazard Analysis and Countermeasures
See chromium compounds and chlorides.

CHROMOUS FLUORIDE
General Information
Synonym: Chromium difluoride.
Description: Green crystals.
Formula: CrF_2.
Constants: Mol wt: 90.01, mp: $1100°C$, bp: $>1300°C$, d: 4.11.
Hazard Analysis
Toxicity: See chromium compounds and fluorides. A powerful irritant.
Disaster Hazard: See fluorides.
Countermeasures
Storage and Handling: Section 7.

CHROMOUS HYDROXIDE
General Information
Description: Yellow-brown crystals.
Formula: $Cr(OH)_2$.
Constant: Mol wt: 86.03.
Hazard Analysis and Countermeasures
See chromium compounds.

CHROMOUS IODIDE
General Information
Description: Grayish powder.
Formula: CrI_2.
Constants: Mol wt: 305.85, d: 5.196.
Hazard Analysis and Countermeasures
See chromium compounds and iodides.

CHROMOUS MONOSULFIDE
General Information
Description: Black powder.
Formula: CrS.
Constants: Mol wt: 84.08, d: 4.1.
Hazard Analysis and Countermeasures
See chromium compounds and sulfides.

CHROMOUS MONOXIDE
General Information
Description: Black crystals.
Formula: CrO.
Constants: Mol wt: 68.01.
Hazard Analysis and Countermeasures
See chromium compounds.

CHROMOUS OXLATE
General Information
Description: Yellow crystalline powder.
Formula: $CrC_2O_4 \cdot H_2O$.
Constant: Mol wt: 158.05.
Hazard Analysis and Countermeasures
See chromium compounds and oxalates.

CHROMOUS SULFATE
General Information
Description: Blue crystals.
Formula: $CrSO_4 \cdot 7H_2O$.
Constant: Mol wt: 274.19.
Hazard Analysis and Countermeasures
See chromium compounds.

CHROMYL   CHLORIDE. See chromium oxychloride.

CHROMYL FLUORIDE
General Information
Description: Exists in two modifications; appears first as a reddish black solid. It polymerizes on exposure to light into a dirty white solid (m. 200°C), forming reddish brown vapors on melting.
Formula: CrO_2F_2.
Constant: Mol wt: 122.01.
Hazard Analysis and Countermeasures
See chromium compounds and fluorides.

CHRYSAROBIN
General Information
Synonym: Goa powder.
Description: Brownish to orange-yellow crystals.
Hazard Analysis
Toxic Hazard Rating:
  Acute Local: Allergen 1.
  Acute Systemic: U.
  Chronic Local: Allergen 1.
  Chronic Systemic: U.
Fire Hazard: Slight; when heated, it emits smoke (Section 6).
Countermeasures
Storage and Handling: Section 7.

CHRYSENE
General Information
Synonym: 1,2-Benzphenanthrene.
Description: Crystals; slightly soluble in ether, alcohol, glacial acetic acid; insoluble in water.
Formula: $C_{18}H_{12}$.
Constants: D: 1.274 (20/4°C), mp: 254°C, bp: 448°C.
Hazard Analysis
A polycyclic hydrocarbon air pollutant. Section 4.

CHRYSOPHANIC ACID
General Information
Synonym: 1,8-Dihydroxy-3-methyl anthraquinone.
Description: Microcrystalline orange-yellow powder.
Formula: $C_6H_3(OH)(CO)_2(C_6H_2)(OH)CH_3$.
Constants: Mol wt: 254.2, mp: 196°C.
Hazard Analysis
Toxic Hazard Rating:
  Acute Local: Irritant 1.
  Acute Systemic: U.
  Chronic Local: U.
  Chronic Systemic: U.
Countermeasures
Personal Hygiene: Section 3.

CHRYSOTILE. See asbestos particles.

CHYMOSIN. See rennet.

CICUTA. See coniine.

CICUTINE. See coniine.

CIGAR AND CIGARETTE LIGHTER FLUID
Countermeasures
Shipping Regulations: Section 11.
  ICC: Flammable liquid, red label, 10 gallons.
  IATA: Flammable liquid, red label, 1 liter (passenger), 40 liters (cargo).

CIGARETTE LOADS
Shipping Regulations: Section 11.
  ICC: Class C explosive, not accepted.

  IATA: Explosive, explosive label, 25 kilograms (passenger), 70 kilograms (cargo).

CIGARETTES, SELF LIGHTING
Shipping Regulations: Section 11.
  IATA: Flammable solid, not acceptable (passenger and cargo).

CINENE. See dipentene.

CINERIN I
General Information
Synonym: 3-(2-Butenyl)-4-methyl-2-oxo-3-cyclopenten-1-yl ester of chrysanthemum monocarboxylic acid.
Description: Viscous liquid.
Formula: $C_{20}H_{28}O_3$.
Constants: Mol wt: 316.4, bp: 200°C at 0.1 mm w. decomposes.
Hazard Analysis
Toxic Hazard Rating:
  Acute Local: U.
  Acute Systemic: Ingestion 3.
  Chronic Local: U.
  Chronic Systemic: U.
Toxicology: Large doses can cause diarrhea and convulsions as well as damage to kidneys and liver; prostration and death from respiratory paralysis. See also Pyretrin I.
Countermeasures
Storage and Handling: Section 7.

CINERIN II
General Information
Synonym: 3-(2-Butenyl)-4-methyl-2-oxo-3-cyclopenten-1-yl ester of chrysanthemum dicarboxylic acid monomethyl ester.
Description: A viscous liquid.
Formula: $C_{21}H_{28}O_5$.
Constants: Mol wt: 360.4, bp: 200°C at 0.1 mm.
Hazard Analysis
Toxicity: Details unknown; an insecticide. See cinerin I.
Fire Hazard: Slight; when heated.
Countermeasures
Storage and Handling: Section 7.

CINNABAR. See mercuric sulfide.

CINNAMALDEHYDE
General Information
Synonym: Cinnamic aldehyde; 3-phenyl propenal, cinnamyl aldehyde.
Description: Yellowish oil; cinnamon odor; soluble in 5 volumes of 60% alcohol; very slightly soluble in water.
Formula: $C_6H_5CH:CHO$.
Constants: Mol wt: 115, d: 1.048–1.052, mp: −8°C, bp: 248°C.
Hazard Analysis
Toxic Hazard Rating: U. Synthetic flavoring substance and adjuvant (Section 10).

CINNAMAMIDE
General Information
Description: Solid.
Formula: $C_6H_5CHCHCONH_2$.
Constants: Mol wt: 147.2, mp: 147°C.
Hazard Analysis
Toxicity: Details unknown; an insecticide.
Fire Hazard: Slight.
Countermeasures
Storage and Handling: Section 7.

TOXIC HAZARD RATING CODE (For detailed discussion, see Section 1.)

0 NONE: (a) No harm under any conditions; (b) Harmful only under unusual conditions or overwhelming dosage.

1 SLIGHT: Causes readily reversible changes which disappear after end of exposure.

2 MODERATE: May involve both irreversible and reversible changes; not severe enough to cause death or permanent injury.

3 HIGH: May cause death or permanent injury after very short exposure to small quantities.

U UNKNOWN: No information on humans considered valid by authors.

CINNAMEIN. See benzyl cinnamate.

CINNAMENE. See phenyl ethylene.

CINNAMIC ACID
General Information
Description: White crystalline scales.
Formula: $C_6H_5CHCHCOOH$.
Constants: Mol wt: 148.15, mp: 133°C, bp: 300°C, vap. press.: 1 mm at 127.5°C.
Hazard Analysis
Toxic Hazard Rating:
 Acute Local: U.
 Acute Systemic: U.
 Chronic Local: Allergen 1.
 Chronic Systemic: U.
Fire Hazard: Slight; Section 6.
Countermeasures
Personal Hygiene: Section 3.
Storage and Handling: Section 7.

CINNAMIC ALDEHYDE. See cinnamaldehyde.

CINNAMOYL CHLORIDE
General Information
Description: Yellow crystals.
Formula: $C_6H_5CH:CHCOCl$.
Constants: Mol wt: 166.6, mp: 35–36°C, bp: 170–171°C at 58 mm Hg., d: 1.617 at 45.3/4°C.
Hazard Analysis
Toxic Hazard Rating:
 Acute Local: Irritant 2.
 Acute Systemic: Ingestion 1.
 Chronic Local: U.
 Chronic Systemic: U.
Toxicology: May cause dermatitis.
Disaster Hazard: Dangerous; see chlorides.

CINNAMYL ALDEHYDE. See cinnamaldehyde.

CIPC. See isopropyl-n-(3-chlorophenyl) carbamate.

CITRAL
General Information
Synonym: Geranial; neral; geranialdehyde; 2,6-dimethyl-2,6-octadienal.
Description: Mobile, pale yellow liquid; strong lemon odor; soluble in 5 volumes of 60% alcohol; soluble in all proportions of benzyl benzoate, diethyl phthalate, glycerin, propylene glycol, mineral oil, fixed oils and 95% alcohol; insoluble in water.
Formula: $(CH_3)_2C:CHC_2H_4C(CH_3):CHCHO$.
Constant: Mol wt: 152, d: 0.891–0.897 (15°C).
Hazard Analysis
Toxic Hazard Rating: U. A synthetic flavoring substance and adjuvant (Section 10).

CITRAZINC ACID
General Information
Synonyms: CZA; 2,6-dihydroxy-4-carboxypyridine.
Description: Buff to grey powder.
Formula: $C_6H_4NO_4$.
Constant. Mol wt: 155.12.
Hazard Analysis
Toxic Hazard Rating:
 Acute Local: Irritant 1; Ingestion 1; Inhalation 1.
 Acute Systemic: U.
 Chronic Local: Irritant 1.
 Chronic Systemic: U.
Countermeasures
Ventilation Control: Section 2.
Personal Hygiene: Section 3.
Storage and Handling: Section 7.

CITRIC ACID
General Information
Synonym: β-Hydroxytricarballylic acid.

Description: Colorless, odorless crystals.
Formula: $C_3H_4(OH)(COOH)_3 \cdot H_2O$.
Constants: Mol wt: 192.12, mp: 153°C, bp: decomposes, d: 1.542.
Hazard Analysis
Toxic Hazard Rating:
 Acute Local: Allergen 1.
 Acute Systemic: 0.
 Chronic Local: Allergen 1.
 Chronic Systemic: 0.
A sequestrant food additive. Also a general purpose food additive (Section 10).
Fire Hazard: Slight; when heated.
Countermeasures
Personal Hygiene: Section 3.
Storage and Handling: Section 7.

CITRONELLA
General Information
Synonym: Oil of citronella (Ceylon).
Description: Colorless to pale-yellow liquid; turns red on standing; pleasant odor.
Composition: 60% Geraniol; 15% citronellol; 15% camphene and dipentene; also linalool and borneol.
Constnat: D: 0.897–0.912.
Hazard Analysis
Toxic Hazard Rating:
 Acute Local: Allergen 1.
 Acute Systemic: U.
 Chronic Local: Allergen 1.
 Chronic Systemic: U.
Fire Hazard: Slight; when heated.
Countermeasures
Storage and Handling: Section 7.

CK. See cyanogen chloride.

CLAUSTHALITE. See lead selenide.

CLEANING FLUID
General Information
Description: Insoluble in water.
Constant: Flash p: < 80°F.
Hazard Analysis
Fire Hazard: Dangerous, when exposed to heat or flame.
Countermeasures
Storage and Handling: Section 7.
Shipping Regulations: Section 11.
 ICC: Flammable liquid, red label, 10 gallons.
 IATA: Flammable liquid, red label, 1 liter (passenger), 40 liters (cargo).

CLEANING SOLVENTS (KEROSENE CLASS)
See Stoddard solvent.

CLEANING SOLVENTS, 140°F CLASS
General Information
Description: Insoluble in H_2O.
Constants: Flash p: 138.2°F or higher, autoign. temp.: 453.2°F or higher; lel: 0.8% at 302°F; bp (initial); 357.8°F or higher.
Hazard Analysis
Fire Hazard: Moderate, when exposed to heat or flame.
Countermeasures
Storage and Handling: Section 7.

"CLOROX." See hypochlorites.

CLYSOLATE. See 4-chloro benzenesulfonic acid, chlorophenyl ester.

CMC. See sodium carboxymethyl cellulose.

CM CELLULOSE. See sodium carboxymethyl cellulose.

CMU. See monuron.

COAL BRIQUETTES, HOT. See also carbon.
Shipping Regulations: Section 11.
   I.C.C. Classification: Not accepted.
   Coast Guard Classification: Not permitted.

COAL DUST. See anthricite particles.
Countermeasures
Shipping Regulations: Section 11.
   IATA: No label required, no limit (passenger and cargo).

COAL GAS
General Information
Description: Contains hydrogen, methane, carbon monoxide, etc.
Constants: Lel: 5.3%, uel: 31%, autoign. temp.: 1200° F.
Hazard Analysis
Toxicity: See carbon monoxide.
Fire Hazard: Dangerous. See hydrogen.
Explosion Hazard: Moderate, when exposed to heat or flame.
Disaster Hazard: Dangerous. See hydrogen.
Countermeasures
Storage and Handling: Section 7.
To Fight Fire: Stop flow of gas. Carbon dioxide, dry chemical or water spray (Section 6).
Shipping Regulations: Section 11.
   I.C.C.: Flammable gas; red gas label, 300 pounds.
   IATA: Flammable gas, red label, not acceptable (passenger), 140 kilograms (cargo).

COAL, GROUND BITUMINOUS, SEA COAL, COAL FACINGS, ETC.
General Information
Description: Black powder or chunks.
Hazard Analysis
Toxic Hazard Rating:
   Acute Local: Inhalation 1.
   Acute Systemic: 0.
   Chronic Local: Inhalation 1.
   Chronic Systemic: Inhalation 1.
TLV: ACHIH (accepted) 50 million particles per cubic foot.
Fire Hazard: Moderate, when exposed to heat; can react with oxidizing materials (Section 6).
Spontaneous Heating: Moderate.
Explosion Hazard: Slight, when exposed to flame.
Countermeasures
Ventilation Control: Section 2.
Storage and Handling: Section 7.
Shipping Regulations: Section 11.
   Coast Guard Classification: Inflammable solid; yellow label.
   ICC: Section 73.165.

COALITES. See explosives, high.

COAL NAPHTHA. See benzene.

COAL SPECIALS. See explosives, high.

COAL TAR
General Information
Description: Black, viscous liquid.
Composition: Benzene, toluene, naphthalene, anthracene, xylene, phenol, cresol, ammonia, pyridine, thiophene, etc.
Hazard Analysis
Toxic Hazard Rating:
   Acute Local: Irritant 1; Allergen 1; Ingestion 1; Inhalation 1.

   Acute Systemic: Inhalation 2.
   Chronic Local: Irritant 1; Allergen 1.
   Chronic Systemic: Inhalation 3.
   May contain carcinogenic compounds.
Fire Hazard: Moderate, when exposed to heat (Section 6).
Explosion Hazard: Moderate, when vapor is exposed to heat or flame.
Disaster Hazard: Dangerous; when heated to decomposition, it emits highly toxic fumes; can react with oxidizing materials.
Countermeasures
Ventilation Control: Section 2.
Personnel Protection: Section 3.
Storage and Handling: Section 7.

COAL TAR DISTILLATE
Hazard Analysis
Toxic Hazard Rating:
   Acute Local: Allergen 1; Ingestion 2, Inhalation 2.
   Acute Systemic: Inhalation 2.
   Chronic Local: Allergen 2.
   Chronic Systemic: Inhalation 3; Skin Absorption 3.
Fire Hazard: Dangerous, when exposed to heat or flame; it can react vigorously with oxidizing materials (Section 6).
Explosion Hazard: Unknown.
Countermeasures
Ventilation Control: Section 2.
Personnel Protection: Section 3.
First Aid: Section 1.
Storage and Handling: Section 7.
Shipping Regulations: Section 11.
   I.C.C.: Flammable liquid; red label, 10 gallons.
   Coast Guard Classification: Inflammable liquid; red label.
   IATA: Flammable liquid, red label, 1 liter (passenger), 40 liters (cargo).

COAL TAR DYES
Hazard Analysis
Toxicity: Many of the coal tar dyes are quite harmless and are permitted for foods, drugs and cosmetics. Some of them may be allergens (Section 9).

COAL TAR LIGHT OIL
General Information
Constants: Lel: 1.3%, uel: 8%, d.: <1, flash point: 60–77° F.
Hazard Analysis
Toxic Hazard Rating:
   Acute Local: Allergen 1; Ingestion 1; Inhalation 1.
   Acute Systemic: Inhalation 1.
   Chronic Local: Allergen 2.
   Chronic Systemic: Inhalation 3; Skin Absorption 3.
Fire Hazard: Dangerous, when exposed to heat or flame.
Explosion Hazard: Moderate, when exposed to heat or flame.
Disaster Hazard: Dangerous; when heated to decomposition, it emits highly toxic fumes; can react vigorously with oxidizing materials.
Countermeasues
Ventilation Control: Section 2.
To Fight Fire: Foam, carbon dioxide, dry chemical or carbon tetrachloride (Section 6).
Personal Hygiene: Section 3.
Storage and Handling: Section 7.
Shipping Regulations: Section 11.
   I.C.C.: Flammable liquid; red label, 10 gallons.
   Coast Guard Classification: Inflammable liquid; red label.
   IATA: Flammable liquid, red label, 1 liter (passenger), 40 liters (cargo).

TOXIC HAZARD RATING CODE (For detailed discussion, see Section 1.)

0 NONE: (a) No harm under any conditions; (b) Harmful only under unusual conditions or overwhelming dosage.

1 SLIGHT: Causes readily reversible changes which disappear after end of exposure.

2 MODERATE: May involve both irreversible and reversible changes; not severe enough to cause death or permanent injury.

3 HIGH: May cause death or permanent injury after very short exposure to small quantities.

U UNKNOWN: No information on humans considered valid by authors.

COAL TAR NAPHTHA. See naphtha, coal tar.
Shipping Regulations: Section 11.
 I.C.C.: Flammable liquid; red label, 10 gallons.
 Coast Guard Classification: Inflammable liquid; red label.
 IATA: Flammable liquid, red label, 1 liter (passenger), 40 liters (cargo).

COAL TAR OIL. See coal tar light oil.

COAL TAR PITCH
General Information
Description: A black to brown tarry mass.
Constant: Flash p: 405° F (C.C.), d: > 1.
Hazard Analysis
Toxic Hazard Rating:
 Acute Local: Irritant 1; Allergen 2; Inhalation 1.
 Acute Systemic: U.
 Chronic Local: Allergen 2.
 Chronic Systemic: Inhalation 3.
Fire Hazard: Slight, when exposed to heat.
Disaster Hazard: Dangerous; when heated to decomposition, it emits toxic fumes.
Countermeasures
Ventilation Control: Section 2.
To Fight Fire: Water, foam, carbon dioxide, dry chemical or carbon tetrachloride (Section 6).
Personal Hygiene: Section 3.
Storage and Handling: Section 7.

COAL TAR RESINS. See coumarone-indene resins.

COATING SOLUTION
Countermeasures
Shipping Regulations: Section 11.
 ICC: Flammable liquid, red label, 15 gallons.
 IATA: Flammable liquid, red label, 1 liter (passenger), 60 liters (cargo).

COBALAMIN. See vitamin B_{12}.

COBALT
General Information
Description: Silver-gray metal.
Formula: Co.
Constants: At wt: 58.9, mp: 1495°C, bp: 2900°C, d: 8.9.
Hazard Analysis
Toxicity: See cobalt compounds.
Radiation Hazard: Section 5. For permissible levels, see Table 5, p. 150.
 Artificial isotope ^{57}Co, half life 270 d. Decays to stable ^{57}Fe by electron capture. Emmits gamma rays of 0.01, 0.12 MeV and X-rays.
 Artificial isotope ^{58m}Co, half life 9.0 h. Decays to radioactive ^{58}Co by emitting gamma rays of 0.025 MeV.
 Artificial isotope ^{58}Co, half life 71 d. Decays to stable ^{58}Fe by emitting positrons (15%) of 0.48 MeV. Also decays by electron capture. Also emits gamma rays of 0.81, 1.65 MeV.
 Artificial isotope ^{60}Co, half life 5.3 y. Decays to stable ^{60}Ni; by emitting beta particles of 0.32 MeV. Also emits gamma rays of 1.17, 1.33 MeV.
TLV: (Tenetative) 0.1 milligrams per cubic meter of air.
Caution: ^{60}Co is more likely to be an external hazard than an internal hazard, since the majority of ^{60}Co is used as a source of gamma radiation in the form of massive peices.
Fire Hazard: Moderate, when exposed to heat or flame or by spontaneous chemical reaction. See also powdered metals (Section 6).

COBALT 60. see cobalt.
Shipping Regulations: Section 11.
 ICC: Poison D, radioactive materials, red label, 300 curies.

IATA: Radioactive materials, radioactive (red) label, 300 curies (passenger and cargo).

COBALT ACETATE
General Information
Synonym: Cobaltous acetate.
Description: Monoclinic, red-violet, deliquescent crystals.
Formula: $Co(C_2H_3O_2)_2 \cdot 4H_2O$.
Constants: Mol wt: 249.09; $-4H_2O$ at 140°C, d: 1.705.
Hazard Analysis and Countermeasures
A trace mineral added to animal feeds (Section 10).
See cobalt compounds.

COBALT ALUMINATE
General Information
Description: Cubic blue crystals.
Formula: $CoAl_2O_4$.
Constant: Mol wt: 176.88.
Hazard Analysis and Countermeasures
See cobalt compounds and aluminum compounds.

COBALT ARSENIC SULFIDE
General Information
Synonym: Cobaltite.
Description: Gray-reddish crystals.
Formula: CoAsS.
Constants: Mol wt: 165.92, mp: decomposes, d: 6.2–6.3.
Hazard Analysis
Toxicity: See arsenic compounds and sulfides.
Fire Hazard: See sulfides and arsine.
Explosion Hazard: See sulfides and arsine.
Disaster Hazard: Dangerous; when heated to decomposition, it emits highly toxic fumes of arsine and oxides of sulfur; it will react with water, steam, acid or acid fumes to produce toxic and flammable vapors of arsine and hydrogen sulfide.
Countermeasures
Storage and Handling: Section 7.

COBALT BROMOPLATINATE
General Information
Description: Crystals.
Formula: $CoPtBr_6 \cdot 12H_2O$.
Constants: Mol wt: 949.86, d: 2.762.
Hazard Analysis and Countermeasures
See cobalt compounds, bromides and platinum compounds.

COBALT CARBONYL
General Information
Synonyms: Cobalt tricarbonyl; tetracobalt dodecacarbonyl.
Description: Black crystals.
Formula: $[Co(CO)_3]_4$ or $Co_4(CO)_{12}$.
Constant: Mol wt: 571.88.
Hazard Analysis
Toxicity: See carbon monoxide, carbonyls and cobalt compounds.
Fire Hazard: See carbonyls and carbon monoxide.
Explosion Hazard: See carbonyls and carbon monoxide.
Disaster Hazard: Dangerous; see carbonyls.
Countermeasures
Storage and Handling: Section 7.

COBALT CHLORIDE
General Information
Description: Blue powder.
Formula: $CoCl_2$.
Constants: Mol wt: 129.86, mp: sublimes, bp: 1049°C, d: 3.348.
Hazard Analysis and Countermeasures
A trace mineral added to animal feeds (Section 10). See cobalt compounds and chlorides.

COBALT CHLOROPLATINATE
General Information
Description: Crystals.
Formula: $CoPtCl_6 \cdot 6H_2O$.

Constants: Mol wt: 575.01, mp: decomposes, d: 2.699.
Hazard Analysis and Countermeasures
See cobalt compounds, platinum compounds and chlorides.

COBALT CHLOROSTANNATE
General Information
Description: Rhombic or trigonal crystals.
Formula: $CoSnCl_6 \cdot 6H_2O$.
Constants: Mol wt: 498.48, mp: decomposes 100°C.
Hazard Analysis and Countermeasures
See cobalt compounds, tin compounds and chlorides.

COBALT COMPOUNDS
Hazard Analysis
Toxic Hazard Rating:
Acute Local: Allergen 1.
Acute Systemic: Ingestion 1; Inhalation 1.
Chronic Local: Irritant 1; Allergen 1.
Chronic Systemic: Ingestion 1; Inhalation 1.
Toxicology: Experimental evidence shows that the toxicity of cobalt by mouth is low. In animals, administration of cobalt salts produces polycythemia. In humans, a single case of poisoning liver and kidney damage has been attributed to cobalt. Locally, cobalt has been shown to produce dermatitis and certain investigators have been able to demonstrate a hypersensitivity of the skin to cobalt. There have also been reports of hematologic, digestive, and pulmonary changes in humans (Section 9).
Countermeasures
Ventilation Countrol: Section 2.
Personal Hygiene: Section 3.

COBALT HEXAMETHYLENE TETRAMINE
General Information
Description: Ultramarine blue crystals, soluble in water.
Formula: $CoC_6H_{12}N_4$.
Constant: Mol wt: 199.
Hazard Analysis and Countermeasures
See chlorides and cobalt compounds and hexamethylene tetramine.

COBALT HYDROXYQUINONE
General Information
Description: Ruby red crystals.
Formula: $Co(C_{10}H_5O_3)_2$.
Constants: Mol wt: 405.2, mp: 210–215°C (decomposes).
Hazard Analysis
See cobalt compounds.

COBALTIC ACETATE
General Information
Description: Green crystals.
Formula: $Co(C_2H_3O_2)_3$.
Constant: Mol wt: 236.07.
Hazard Analysis and Countermeasures
See cobalt compounds.

COBALTIC CHLORIDE
General Information
Description: Red crystals.
Formula: $CoCl_3$.
Constants: Mol wt: 165.31, mp: sublimes, d: 2.94.
Hazard Analysis and Countermeasures
See cobalt compounds and chlorides.

COBALTIC FLUORIDE
General Information
Description: Hexagonal brown crystals.
Formula: CoF_3.

Constants: Mol wt: 115.94, d: 3.88.
Hazard Analysis and Countermeasures
See fluorides.

COBALTIC HYDROXIDE
General Information
Description: Black-brown powder.
Formula: $Co(OH)_3$.
Constants: Mol wt: 109.96, mp: decomposes; $-1\frac{1}{2}H_2O$ at 100°C.
Hazard Analysis and Countermeasures
See cobalt compounds.

COBALTIC OXIDE
General Information
Description: Black-gray powder.
Formula: $Co2O_3$.
Constants: Mol wt: 165.88, mp: decomposes 895°C, d: 5.18.
Hazard Analysis and Countermeasures
See cobalt compounds.

COBALTIC SESQUISULFIDE
General Information
Description: Black crystals.
Formula: Co_2S_3.
Constants: Mol wt: 214.08, d: 4.8.
Hazard Analysis and Countermeasures
See sulfides and cobalt compounds.

COBALTIC SULFATE
General Information
Description: Blue needles.
Formula: $Co_2(SO_4)_3 \cdot 18H_2O$.
Constant: Mol wt: 730.37.
Hazard Analysis and Countermeasures
See cobalt compounds and sulfates.

COBALTICYANIC ACID. See cyanocobaltic III acid.

COBALT IODOPLATINATE
General Information
Description: Crystals.
Formula: $CoPtI_6 \cdot 9H_2O$.
Constants: Mol wt: 1177.83, d: 3.618.
Hazard Analysis and Countermeasures
See cobalt compounds, iodides and platinum compounds.

COBALTITE. See cobalt arsenic sulfide.

COBALT MONOBORIDE
General Information
Description: Prisms.
Formula: CoB.
Constants: Mol wt: 69.76, d: 7.25 at 18°C.
Hazard Analysis
Toxicity: See cobalt compounds and boron compounds.
Fire Hazard: See boron hydrides.
Explosion Hazard: See boron hydrides.
Disaster Hazard: Dangerous; it will react with water, steam, acid or acid fumes to produce toxic and flammable vapors of boron hydrides.
Countermeasures
Storage and Handling: Section 7.

COBALT MONOSELENIDE
General Information
Description: Yellow crystals.
Formula: CoSe.
Constants: Mol wt: 137.90, mp: red heat, d: 7.65.

TOXIC HAZARD RATING CODE (For detailed discussion, see Section 1.)

0 NONE: (a) No harm under any conditions; (b) Harmful only under unusual conditions or overwhelming dosage.

1 SLIGHT: Causes readily reversible changes which disappear after end of exposure.

2 MODERATE: May involve both irreversible and reversible changes; not severe enough to cause death or permanent injury.

3 HIGH: May cause death or permanent injury after very short exposure to small quantities.

U UNKNOWN: No information on humans considered valid by authors.

Hazard Analysis
Toxicity: See selenium and cobalt compounds.
Fire Hazard: See hydrogen selenide.
Explosion Hazard: See hydrogen selenide.
Disaster Hazard: Dangerous; when heated or on contact
with water, steam, acid or acid fumes, it will react to
produce highly toxic and flammable vapors of hydro-
gen selenide.
Countermeasures
Storage and Handling: Section 7.

COBALT MONOSULFIDE
General Information
Description: Reddish, silver-white crystals.
Formula: CoS.
Constants: Mol wt: 91.01, mp: $> 1116°$C, d: 5.45.
Hazard Analysis and Countermeasures
See sulfides and cobalt compounds.

COBALT NAPHTHA
General Information
Synonym: Cobalt naphthenate; cobaltous naphthenate.
Description: Brown amorphous powder or bluish-red solid,
insoluble in water; soluble in alcohol, ether, and oils.
Composition: Indefinite.
Constants: Flash p: 121°F, autoign. temp.: 529°F, d: 0.9.
Hazard Analysis
Fire Hazard: Moderate, when exposed to heat or flame.
Countermeasures
Storage and Handling: Section 7.

COBALTOUS ACETATE. See cobalt acetate.

COLBALTOUS o-ARSENATE
General Information
Synonym: Erythrite.
Description: Monoclinic violet-red crystals.
Formula: $Co_3(AsO_4)_2 \cdot 8H_2O$.
Constants: Mol wt: 598.77, mp: decomposes, d: 2.948.
Hazard Analysis and Countermeasures
See arsenic compounds and cobalt compounds.

COBALTOUS BENZOATE
General Information
Description: Gray-red leaf.
Formula: $Co(C_7H_5O_2)_2 \cdot 4H_2O$.
Constants: Mol wt: 373.22, mp: $-4H_2O$ at 115°C.
Hazard Analysis and Countermeasures
See cobalt compounds.

COBALTOUS BROMATE
General Information
Description: Red crystals.
Formula: $Co(BrO_3)_2 \cdot 6H_2O$.
Constant: Mol wt: 422.87.
Hazard Analysis and Countermeasures
See cobalt compounds and bromates.

COBALTOUS BROMIDE
General Information
Description: Green deliquescent crystals.
Formula: $CoBr_2$.
Constants: Mol wt: 218.77, mp: decomposes, d: 4.909 at
25°C.
Hazard Analysis and Countermeasures
See cobalt compounds and bromides.

COBALTOUS CARBONATE
General Information
Synonym: Spherocobaltite.
Description: Red crystals.
Formula: $CoCO_3$.
Constants: Mol wt: 118.95, mp: decomposes, d: 4.13.
Hazard Analysis and Countermeasures
A trace mineral added to annimal feeds (Section 10). See
cobalt compounds.

COBALTOUS CHLORATE
General Information
Description: Cubic, red, deliquescent crystals.
Formula: $Co(ClO_3)_2 \cdot 6H_2O$.
Constants: Mol wt: 333.95, mp: 61°C, bp: decomposes at
100°C, d: 1.92.
Hazard Analysis
Toxicity: See cobalt compounds and chlorates.
Fire Hazard: See chlorates.
Caution: Can explode easily when contaminated.
Explosion Hazard: See chlorates.
Disaster Hazard: See chlorates.
Countermeasures
Storage and Handling: Section 7.

COBALTOUS CHLORIDE
General Information
Description: Black crystals.
Formula: $CoCl_2$.
Constants: Mol wt: 129.85, mp: 735°C, bp: 1050°C, d:
3.356, vap. pres.: 40 mm at 770°C.
Hazard Analysis and Countermeasures
See cobalt compounds and chlorides.

COBALTOUS CHROMATE
General Information
Description: Gray-black crystals.
Formula: $CoCrO_4$.
Constants: Mol wt: 174.95, mp: decomposes.
Hazard Analysis and Countermeasures
See chromium compounds and cobalt compounds.

COBALTOUS CITRATE
General Information
Description: Rose-red crystals.
Formula: $Co_3(C_6H_5O_7)_2 \cdot 2H_2O$.
Constants: Mol wt: 591.05, mp: $-2H_2O$ at 150°C.
Hazard Analysis and Countermeasures
See cobalt compounds.

COBALTOUS CYANIDE
General Information
Description: Buff color; anhydrous, a blue violet powder.
Formula: $Co(CH)_2 \cdot 2H_2O$.
Constants Mol wt: 147.01, mp: $-2H_2O$ at 280°C, d:
1.872 at 25°C.
Hazard Analysis
Toxicity: See Cyanides.
Fire Hazard: See cyanides.
Explosion Hazard: See cynadies.
Disaster Hazard: Dangerous; see cyanides.
Countermeasures
Storage and Handling: Section 7.

COBALTOUS FERRICYANIDE
General Information
Description: Red needles.
Formula: $Co_3[Fe(CN)_6]_2$.
Constant: Mol wt: 600.74.
Hazard Analysis and Countermeasures
See ferricyanides and cobalt compounds.

COBALTOUS FERROCYANIDE
General Information
Description: Gray-green crystals.
Formula: $Co_2Fe(CN)_6 \cdot xH_2O$.
Hazard Analysis and Countermeasures
See ferrocyanides and cobalt compounds.

COBALTOUS FLUOGALLATE
General Information
Description: Pink crystals.
Formula: $[Co(H_2O)_6](GaF_5H_2O)$.
Constants: Mol wt: 349.8, mp: $-5H_2O$ at 110°C,
d: 2.35.
Hazard Analysis and Countermeasures
See fluogallates and cobalt compounds.

COBALTOUS FLUORIDE
General Information
Description: Monoclinic, rose-red crystals.
Formula: $CoF_2 \cdot 2H_2O$.
Constants: Mol wt: 132.97, d: 4.46.
Hazard Analysis and Countermeasures
See fluorides and cobalt compounds.

COBALTOUS FLUOSILICATE
General Information
Description: Pink crystals.
Formula: $CoSiF_6 \cdot 6H_2O$.
Constants: Mol wt: 309.10, d: 2.113 at 19°C.
Hazard Analysis and Countermeasures
See fluosilicates and cobalt compounds.

COBALTOUS FORMATE
General Information
Description: Red crystals.
Formula: $Co(CHO_2)_2 \cdot 2H_2O$.
Constants: Mol wt: 185.01, mp: $-2H_2O$ at 140°C, bp: anhydrous decomposes 175°C, d: 2.129 at 22°C.
Hazard Analysis and Countermeasures
See cobalt compounds and formic acid.

COBALTOUS HYDROXIDE
General Information
Description: Rhombic rose-red crystals.
Formula: $Co(OH)_2$.
Constants: Mol wt: 92.96, mp: decomposes, d: 3.597 at 15°C.
Hazard Analysis and Countermeasures
See cobalt compounds.

COBALTOUS IODATE
General Information
Description: Black-violet needles.
Formula: $Co(IO_3)_2$.
Constants: Mol wt: 408.78, d: 5.008 at 18°C.
Hazard Analysis and Countermeasures
See cobalt compounds and iodates.

α-COBALTOUS IODIDE
General Information
Description: Hexagonal black cyrstals.
Formula: CoI_2.
Constants: Mol wt: 312.78, d: 5.68.
Hazard Analysis and Countermeasures
See cobalt compounds and iodides.

COBALTOUS LINOLEATE
General Information
Description: Brown amorphous mass.
Formula: $Co(C_{18}H_{31}O_2)_2$.
Constant: Mol wt: 617.80.
Hazard Analysis and Countermeasures
See cobalt compounds.

COLBALTOUS NITRATE
General Information
Description: Red crystals.
Formula: $Co(NO_3)_2 \cdot 6H_2O$.
Constants: Mol wt: 291.05, mp: $<100°C$; $-3H_2O$ at 55°C, d: 1.87.
Hazard Analysis and Countermeasures
See nitrates and cobalt compounds.

COLBALTOUS OLEATE
General Information
Description: Brown amorphous powder.

Formula: $Co(C_{18}H_{33}O_2)_2$.
Constant: Mol wt: 621.83.
Hazard Analysis and Countermeasures
See cobalt compounds.

COBALTOUS OXALATE
General Information
Description: Reddish white crystals.
Formula: CoC_2O_4.
Constants: Mol wt: 146.96, d: 3.021 at 25°C.
Hazard Analysis and Countermeasures
See oxalates and cobalt compounds.

COBALTOUS OXIDE
General Information
Description: Cubic green-brown crystals.
Formula: CoO.
Constants: Mol wt: 74.94, mp: 1800°C decomposes, d: 5.7-6.7.
Hazard Analysis and Countermeasures
See cobalt compounds.. A trace mineral added to animal feeds (Section 10).

COBALTOUS PERCHLORATE
General Information
Description: Red needles.
Formula: $Co(ClO_4)_2$.
Constants: Mol wt: 257.85, d: 3.327.
Hazard Analysis and Countermeasures
See perchlorates and cobalt compounds.

COBALTOUS PERRHENATE
General Information
Description: Dark pink crystals.
Formula: $Co(ReO_4)_2 \cdot 5H_2O$.
Constants: Mol wt: 649.64, mp: decomposes.
Hazard Analysis
Toxicity: See cobalt compounds and rhenium compounds.
Fire Hazard: Dangerous; an oxidizing agent.
Disaster Hazard: Dangerous; keep away from combustible materials.
Countermeasures
Storage and Handling: Section 7.

COBALTOUS o-PHOSPHATE
General Information
Description: Reddish crystals.
Formula: $Co_3(PO_4)_2$.
Constant: Mol wt: 366.78.
Hazard Analysis and Countermeasures
See cobalt compounds and phosphates.

COBALTOUS PROPIONATE
General Information
Description: Dark red crystals.
Formula: $Co(C_3H_5O_2)_2 \cdot 3H_2O$.
Constants: Mol wt: 259.13, mp: ca. 250°C.
Hazard Analysis and Countermeasures
See cobalt compounds.

COBALTOUS RESINATE
General Information
Description: Brown-red powder.
Formula: $Co(C_{44}H_{62}O_4)_2$.
Constant: Mol wt: 1368.81.
Hazard Analysis
Toxicity: See cobalt compounds.
Fire Hazard: Dangerous; spontaneously flammable in air; reacts vigorously with oxidizing materials (Section 6).

TOXIC HAZARD RATING CODE (For detailed discussion, see Section 1.)

0 NONE: (a) No harm under any conditions; (b) Harmful only under unusual conditions or overwhelming dosage.

1 SLIGHT: Causes readily reversible changes which disappear after end of exposure.

2 MODERATE: May involve both irreversible and reversible changes; not severe enough to cause death or permanent injury.

3 HIGH: May cause death or permanent injury after very short exposure to small quantities.

U UNKNOWN: No information on humans considered valid by authors.

Countermeasures
Storage and Handling: Section 7.

COBALTOUS SELENATE
General Information
Description: Ruby-red crystals.
Formula: $CoSeO_4 \cdot 5H_2O$.
Constants: Mol wt: 291.98, mp: decomposes, d: 2.512.
Hazard Analysis and Countermeasures
See selenium compounds and cobalt compounds.

COBALTOUS o-SILICATE
General Information
Description: Violet crystals.
Formula: Co_2SiO_4.
Constants: Mol wt: 209.94, d: 4.63.
Hazard Analysis and Countermeasures
See cobalt compounds and silicates.

COBALTOUS SULFATE
General Information
Description: Red, water-soluble powder.
Formula: $CoSO_4$.
Constants: Mol wt: 155, mp: 989°C, d: 3.71 at 25°/25°C.
Hazard Analysis and Countermeasures
See cobalt compounds and sulfates. A trace mineral added
to animal feeds. (Section 10).

COBALTOUS SULFITE
General Information
Description: Red crystals.
Formula: $CoSO_3 \cdot 5H_2O$.
Constant: Mol wt: 229.08.
Hazard Analysis and Countermeasures
See cobalt compounds and sulfites.

COBALTOUS TARTRATE
General Information
Description: Monoclinic reddish crystals.
Formula: $CoC_4H_4O_6$.
Constant: Mol wt: 207.01.
Hazard Analysis and Countermeasures
See cobalt compounds.

COBALTOUS THIOCYANATE
General Information
Description: Rhombic violet crystals.
Formula: $Co(SCN)_2 \cdot 3H_2O$.
Constants: Mol wt: 229.16, mp: $-3H_2O$ at 105°C.
Hazard Analysis and Countermeasures
See thiocyanates and cobalt compounds.

COBALTOUS TUNGSTATE
General Information
Description: Monoclinic blue-green crystals.
Formula: $CoWO_4$.
Constants: Mol wt: 306.86, d: 8.42.
Hazard Analysis and Countermeasures
See cobalt compounds and tungsten compounds.

COBALT OXIDE
General Information
Description: Brown or black powder.
Formula: CoO.
Constants: Mol wt: 74.94, mp: 1800°C, d: 5.6–5.75.
Hazard Analysis and Countermeasures
See cobalt compounds.

COBALT PALMITATE
General Information
Description: A solid.
Formula: $Co(C_{16}H_{31}O_2)_2$.
Constants: Mol wt: 569.76, mp: 70.5°C.
Hazard Analysis and Countermeasures
See cobalt compounds.

COBALT PHOSPHIDE
General Information
Description: Small gray needles.
Formula: Co_2P.
Constants: Mol wt: 148.9, mp: 138.6°C.
Hazard Analysis
Toxicity: See phosphides and cobalt compounds.
Fire Hazard: Dangerous; it decomposes to evolve phosphine
(Section 6).
Explosion Hazard: See phosphides.
Disaster Hazard: Dangerous; see phosphides.
Countermeasures
Storage and Handling: Section 7.

COBALT RESINATE, PRECIPITATE
Hazard Analysis
Toxicity: See cobalt compounds.
Fire Hazard: Dangerous; spontaneously flammable in air;
reacts with oxidizing materials (Section 6).
Countermeasures
Shipping and Handling: Section 7.
Shipping Regulations: Section 11.
I.C.C.: Flammable solid; Yellow lable, 125 pounds.
Coast Guard Classification: Inflammable solid; yellow
label.
IATA: Flammable solid, yellow label, not acceptable
(passenger), 60 kilograms (cargo).

COBALT   TETRACARBONYL. See dicobalt octacar-
bonyl.

COBALT o-TITANATE
General Information
Description: Cubic greenish-black crystals.
Formula: Co_2TiO_4.
Constants: Mol wt: 229.78, d: 5.07–5.12.
Hazard Analysis and Countermeasures
See cobalt compounds and titanium compounds.

COBALT TRICARBONYL. See cobalt carbonyl.

COCAINE
General Information
Synonym: Methyl benzoylecgonine.
Description: Colorless to white crystals.
Formula: $C_{17}H_{21}NO_4$.
Constants: Mol wt: 303.35, mp: 98°C.
Hazard Analysis
Toxic Hazard Rating:
Acute Local: Allergen 1.
Acute Systemic: Ingestion 3; Inhalation 3.
Chronic Local: Allergen 1.
Chronic Systemic: Ingestion 3; Inhalation 3.
Toxicology: A poisonous and habit-forming alkaloid. It can
cause dilation and immobility of the pupils, blepharitis
and blindness. Fatal cases run a very rapid course with
anxiety, sudden fainting, extreme pallor, dyspnea, some-
times brief convulsions, arrest of respiration and death,
generally in a few minutes. If death does not occur in a
few minutes to a half hour, the patient almost always
recovers.
Disaster Hazard: Dangerous; when heated to decomposition,
it emits highly toxic fumes.
Countermeasures
Ventilation Control: Section 2.
Treatment and Antidotes: Wash stomach out immediately
with water and sodium bicarbonate (60 grains to 1 pint).
If a solution of permanganate is handy (1 crystal to
8 oz. of water), use it to wash out stomach. Give inhalations of ammonia. Allay convulsions with chloral
or chloroform. If breathing is disturbed, give artificial
respiration and cardiac massage after respiration fails.
Inhalations of oxygen plus 5% carbon dioxide should
be given.
Personal Hygiene: Section 3.

First Aid: Section 1.
Storage and Handling: Section 7.

COCCULUS SOLID
General Information
Synonym: Fish berry.
Description: Dried ripe fruit of a woody climbing plant.
Hazard Analysis
Toxic Hazard Rating:
  Acute Local: U.
  Acute Systemic: Ingestion 3.
  Chronic Local: U.
  Chronic Systemic: U.
Caution: Contains picrotoxin, a convulsant poison.
Fire Hazard: Slight, when exposed to heat (Section 6).
Countermeasures
Personal Hygiene: Section 3.
First Aid: Section 1.
Storage and Handling: Section 7.
Shipping Regulations: Section 11.
  I.C.C.: Poison B; Poison label, 200 pounds.
  Coast Guard Classification: Poison B; poison label.
  IATA: Poison B, poison label, 25 kilograms (passenger), 95 kilograms (cargo).

COCOA
General Information
Description: Brown powder.
Hazard Analysis
Toxic Hazard Rating:
  Acute Local: Allergen 2.
  Acute Systemic: Ingestion 1; Inhalation 1.
  Chronic Local: Allergen 1.
  Chronic Systemic: Ingestion 1; Inhalation 1.
Fire Hazard: Slight; when heated.
Countermeasures
Personal Hygiene: Section 3.
Storage and Handling: Section 7.

COCOA BEAN SHELL TANKAGE
General Information
Description: A hygroscopic material.
Hazard Analysis
Toxicity: See cocoa.
Fire Hazard: Moderate, when exposed to heat.
Spontaneous Heating: Moderate; extreme caution must be observed to keep the moisture content low; on contact with oxidizing materials, it can react vigorously.
Countermeasures
Storage and Handling: Section 7.

COCOANUT OIL
General Information
Synonym: Coconut oil; coconut palm oil; coconut butter.
Description: White, semi-solid, lard like fat; characteristic odor. Soluble in alcohol, ether, chloroform, and carbon disulfide.
Constants: D: 0.92, mp: 20–25°C, flash p: 420°F (crude), 548°F (refined).
Hazard Analysis
Fire Hazard: Slight, when exposed to heat or flame.
Storage and Handling: Section 7.
To Fight Fire: Water, foam (Section 6).

COCO BOLO
General Information
Description: A wood.

Hazard Analysis
Toxic Hazard Rating:
  Acute Local: Allergen 1.
  Acute Systemic: 0.
  Chronic Local: Allergen 1.
  Chronic Systemic: 0.
Fire Hazard: Slight; when exposed to heat or flame (Section 6).
Countermeasures
Personal Hygiene: Section 3.
Storage and Handling: Section 7.

COCONUT BUTTER. See cocoanut oil.

COCONUT OIL. See cocoanut oil.

COCONUT PALM OIL. See cocoanut oil.

CODEINE
General Information
Synonym: Methylmorphine.
Formula: $C_{18}H_{21}NO_3 \cdot H_2O$.
Constants: Mol wt: 317.37, mp: 154.9°C, d: 0.9, vap. d.: 3.5, bp: 239°F, flash p: 75°F.
Hazard Analysis
Toxic Hazard Rating:
  Acute Local: 0.
  Acute Systemic: 3.
  Chronic Local: 0.
  Chronic Systemic: 3.
Fire Hazard: Dangerous, when exposed to heat or flame.
Countermeasures
Shipping Regulations: Subject to Federal narcotic restrictions.
To Fight Fire: Alcohol foam (Section 6).

COD LIVER OIL
General Information
Synonym: Morrhua oil.
Description: Pale yellow liquid, fixed, nondrying oil; characteristic odor; soluble in ether, chloroform, ethyl acetate, petroleum ether, carbon disulfide; slightly soluble in alcohol.
Constants: D: 0.9, flash p: 412°F.
Hazard Analysis
Fire Hazard: Slight, when exposed to heat or flame.
Countermeasures
Storage and Handling: Section 7.
To Fight Fire: Water, foam (Section 6).

COFFEE
Hazard Analysis
Toxic Hazard Rating:
  Acute Local: Allergen 1.
  Acute Systemic: Ingestion 1.
  Chronic Local: Allergen 1.
  Chronic Systemic: Ingestion 1.
Remarks: Coffee tasters can have amblyopia and coffee has been known to cause temporary blindness.
Disaster Hazard: Slightly dangerous in dry state; when heated, it burns.
Countermeasures
Storage and Handling: Section 7.

COKE, HOT
General Information
Description: A black amorphous mass.
Composition: Carbon and impurities.

TOXIC HAZARD RATING CODE (For detailed discussion, see Section 1.)

0 NONE: (a) No harm under any conditions; (b) Harmful only under unusual conditions or overwhelming dosage.

1 SLIGHT: Causes readily reversible changes which disappear after end of exposure.

2 MODERATE: May involve both irreversible and reversible changes; not severe enough to cause death or permanent injury.

3 HIGH: May cause death or permanent injury after very short exposure to small quantities.

U UNKNOWN: No information on humans considered valid by authors.

Hazard Analysis
Toxicity: See carbon and charcoal.
Fire Hazard: See carbon.
Caution: While still hot it may ignite easily in air or by contact with spark or flame.
Disaster Hazard: See carbon and charcoal.
Countermeasures
Storage and Handling: Section 7.
Shipping Regulations: Section 11.
 I.C.C. Classification: Not accepted.
 Coast Guard Classification: Not permitted.

COLCHICINE
General Information
Description: Yellow, crystalline alkaloid, amorphous powder.
Formula: $C_{22}H_{25}NO_6$.
Constants: Mol wt: 399.43, mp: 143.7°C, anhyd.
Hazard Analysis
Toxic Hazard Rating:
 Acute Local: Irritant 2; Inhalation 2.
 Acute Systemic: Ingestion 3.
 Chronic Local: U.
 Chronic Systemic: Ingestion 3.
Toxicology: A poisonous alkaloid. No symptoms result from small doses. Large doses cause diarrhea with griping in susceptible persons. Symptoms take several hours and arise from the alimentary tract. Pain in the gastric region is followed by salivation, nausea, vomiting and diarrhea. Depression, apathy and collapse follow. In experimental poisoning, paralysis starts with the posterior limbs and increases upward until the forelimbs and respiratory muscles are involved and death occurs from asphyxia. Externally, it is a local irritant, which causes redness and smarting when applied to the skin. Upon inhalation, the dust causes sneezing and conjunctival hyperemia and a burning sensation of mouth and throat. 0.02 g is very likely to cause death in about 24 hours from a single dose. Repeated doses may be cumulative.
Fire Hazard: Slight (Section 6).
Countermeasures
Ventilation Control: Section 2.
Treatment and Antidotes: Give a cathartic and an emetic at once, also large quantities of water for the kidneys. Tannic acid should be given in large amounts. Opium and stimulants should also be given to counteract depression.
Personal Hygiene: Section 3.
First Aid: Section 1.
Storage and Handling: Section 7.

COLCHICINE TANNATE. See colchicine.

COLD STARTERS
Shipping Regulations: Section 11.
 IATA: Flammable solid, yellow label, 12 kilograms (passenger and cargo).

COLLIERS. See explosives, high.

COLLODION
General Information
Description: Solution of nitrated cellulose in ether-alcohol.
Formula: $C_{12}H_{16}O_6(NO_3)_4C_{13}H_{17}O_7(NO_3)_3$
Constants: Mol wt: 975, flash p: < 0°F.
Hazard Analysis
Fire Hazard: Dangerous, when exposed to heat or flame.
Countermeasures
To Fight Fire: Alcohol foam (Section 6).
Shipping Regulations: Section 11.
 I.C.C.: Flammable liquid, red label, 10 gallons.
 IATA: Flammable liquid, red label, 1 liter (passenger), 40 liters (cargo).

COLLODION COTTON. See nitrocelluose.

COLLODION WOOL. See nitrocellulose.

COLLOIDAL CLAY. See bentonite.

COLOCYNTH
General Information
Synonyms: Bitter apple; bitter cucumber.
Description: Dried pulp.
Hazard Analysis
Toxic Hazard Rating:
 Acute Local: U.
 Acute Systemic: Ingestion 3.
 Chronic Local: U.
 Chronic Systemic: U.
Fire Hazard: Slight (Section 6).
Countermeasures
Personal Hygiene: Section 3.
First Aid: Section 1.
Storage and Handling: Section 7.

COLOGNE SPIRITS (ALCOHOL). See ethyl alcohol.
Shipping Regulations: Section 11.
 I.C.C: Flammable liquid; red label, 10 gallons.
 Coast Guard Classification: Inflammable liquid; red label.
 IATA: Flammable liquid, red label, 1 liter (passenger), 40 liters (cargo).

"COLONIAL SPIRITS". See methyl alcohol.

COLOPHONY, POWDER. See rosin.

COLUMBIAN SPIRITS. See methyl alcohol.
Shipping Regulations: Section 11.
 Coast Guard Classification: Inflammable liquid; red label.
 IATA: Flammable liquid, red label, 1 liter (passenger), 40 liters (cargo).
 I.C.C.: Flammable liquid, red label, 10 gallons.

COLUMBIUM. See niobium.

COLUMBIUM CHLORIDE
General Information
Synonym: Columbium pentachloride.
Description: Yellow-white, deliquescent powder.
Formula: $CbCl_5$.
Constants: Mol wt: 270.20, mp: 194°C, bp: 240.5°C.
Hazard Analysis and Countermeasures
See niobium and chlorides.

COLUMBIUM PENTACHLORIDE. See columbium chloride.

COLZA OIL. See rapeseed oil.

COMBINATION FUZES. See explosives, high.
Shipping Regulations: Section 11.
 I.C.C.: Explosive C, 150 pounds.
 Coast Guard Classification: Explosive C.
 IATA: Explosive, explosive label, 25 kilograms (passenger), 70 kilograms (cargo).

COMBINATION PRIMERS. See explosives, high.
Shipping Regulations: Section 11.
 I.C.C.: Explosive C, 150 pounds.
 Coast Guard Classification: Explosive C.
 IATA: Explosive, explosive label, 25 kilograms (passenger), 70 kilograms (cargo).

COMBUSTIBLE LIQUID (N.O.S.)
Shipping Regulations: Section 11.
 IATA: Combustible liquid, no label, 220 liters (passenger and cargo).

COMBUSTION PRODUCT GAS
General Information
Description: A food additive manufactured by the con-

trolled combustion in air of butane, propane, or natural gas.

Hazard Analysis

Toxic Hazard Rating: U. A food additive permitted in food for human consumption. Section 10.

COMMON FIREWORKS

Shipping Regulations: Section 11.
I.C.C.: Explosive C, 150 pounds.
IATA: Explosive, explosive label, 25 kilograms (passenger), 95 kilograms (cargo).

COMMON MALIC ACID. See malic acid.

COMPOSITIONS, EXPLOSIVE, N.O.S.

Shipping Regulations: Section 11.
IATA: Explosive, not acceptable (passenger and cargo).

COMPOUND 22/190. See chlorthion.

COMPOUND 118. See aldrin.

COMPOUND 497. See dieldrin.

COMPOUND 1836. See diethyl-2-chlorovinyl phosphate.

COMPOUND 3422. See parathion.

COMPOUND 3956. See toxaphene.

COMPOUND G 23, 922. See ethyl-4,4'-dichlorobenzilate.

COMPOUNDS, CLEANING, LIQUID

Shipping Regulations: Section 11.
I.C.C. (containing hydrochloric acid): Corrosive liquid, white label, 10 pints.
IATA (containing hydrochloric acid): Corrosive liquid, white label, 1 liter (passenger), 5 liters (cargo).
I.C.C. (containing hydrofluoric acid): Corrosive liquid, white label, 10 pints.
IATA (containing hydrofluoric acid): Corrosive liquid, white label, 1 liter (passenger), 5 liters (cargo).
I.C.C (corrosive): Corrosive liquid, white label, 1 quart.
IATA (corrosive): Corrosive liquid, white label, 1 liter (passenger and cargo).

COMPOUNDS, ENAMEL

Shipping Regulations: Section 11.
I.C.C.: Flammable liquid, red label, 55 gallons.
IATA: Flammable liquid, red label, 1 liter (passenger), 220 liters (cargo).

COMPOUNDS, FLAME RETARDANT, LIQUID.

Shipping Regulations: Section 11.
IATA: Corrosive liquid, white label, 1 liter (passenger), 40 liters (cargo).

COMPOUNDS, IRON OR STEEL RUST PREVENTING OR REMOVING

Shipping Regulations: Section 11.
I.C.C.: Corrosive liquid, white label, 1 gallon.
IATA: Corrosive liquid, white label, 1 liter (passenger), 5 liters (cargo).

COMPOUNDS, LACQUER, PAINT, OR VARNISH etc., REMOVING, LIQUID, CORROSIVE

Shipping Regulations: Section 11.
I.C.C.: Corrosive liquid, white label, 1 gallon.
IATA: Corrosive liquid, white label, 1 liter (passenger), 5 liters (cargo).

COMPOUNDS, LACQUER, PAINT, OR VARNISH ETC., REMOVING, REDUCING, OR THINNING, LIQUID, FLAMMABLE.

Shipping Regulations: Section 11.
I.C.C.: Flammable liquid, red label, 55 gallons.
IATA: Flammable liquid, red label, 1 liter (passenger), 220 liters (cargo).

COMPOUNDS, MOTOR FUEL, ANTI-KNOCK

Shipping Regulations: Section 11.
IATA: Poison B, poison label, not acceptable (passenger), 220 liters (cargo).

COMPOUNDS, POLISHING, LIQUID.

Shipping Regulations: Section 11.
I.C.C.: Flammable liquid, red label, 55 gallons.
IATA: Flammable liquid, red label, 1 liter (passenger), 220 liters (cargo).

COMPOUNDS, TREE OR WEED KILLING

Shipping Regulations: Section 11.
1) Flammable liquid:
I.C.C.: Flammable liquid, red label, 10 gallons.
IATA: Flammable liquid, red label, 1 liter (passenger), 40 liters (cargo).
2) Oxidizing material, solid:
I.C.C.: Oxidizing material, yellow label, 100 pounds.
IATA: Oxidizing material, yellow label, 12 kilograms (passenger), 45 kilograms (cargo).
3) Poisonous liquid:
I.C.C.: Poison B, poison label, 55 gallons.
IATA: Poison B, poison label, 1 liter (passenger), 220 liters (cargo).
4) Poisonous solid:
IATA: Poison B, poison label, 25 kilograms (passenger), 95 kilograms (cargo).

COMPOUNDS, TYPE CLEANING FLUID

Shipping Regulations: Section 11.
I.C.C.: Flammable liquid, red label, 10 gallons.
IATA: Flammable liquid, red label, 1 liter (passenger), 40 liters (cargo).

COMPOUNDS, VULCANIZING, LIQUID

Shipping Regulations: Section 11.
I.C.C. (corrosive): Corrosive liquid, white label, 1 quart.
IATA (corrosive): Corrosive liquid, white label, 1 liter (passenger and cargo).
I.C.C. (flammable): Flammable liquid, red label, 10 gallons.
IATA (flammable): Flammable liquid, red label, 1 liter (passenger), 40 liters (cargo).

COMPOUNDS, WATER TREATMENT, LIQUID

Shipping Regulations: Section 11.
IATA: Corrosive liquid, white label, 1 liter (passenger), 40 liters (cargo).

COMPOUNDS, WATER PURIFYING

Shipping Regulations: Section 11.
IATA: Oxidizing material, yellow label, 12 kilograms (passenger), 45 kilograms (cargo).

COMPRESSED GASES (FLAMMABLE). See specific gas.

Shipping Regulations: Section 11.
I.C.C.: Flammable gas; red gas label, 300 pounds.
IATA: Flammable gas, red label, not acceptable (passenger), 140 kilograms (cargo).

TOXIC HAZARD RATING CODE (For detailed discussion, see Section 1.)

0 NONE: (a) No harm under any conditions; (b) Harmful only under unusual conditions or overwhelming dosage.

1 SLIGHT: Causes readily reversible changes which disappear after end of exposure.

2 MODERATE: May involve both irreversible and reversible changes; not severe enough to cause death or permanent injury.

3 HIGH: May cause death or permanent injury after very short exposure to small quantities.

U UNKNOWN: No information on humans considered valid by authors.

Coast Guard Classification: Inflammable gas; red gas label.

COMPRESSED GASES, N.O.S. See specific gas.
Shipping Regulations: Section 11.
I.C.C.: Nonflammable gas; green label, 300 pounds.
Coast Guard Classification: Noninflammable gas; green gas label.
IATA: Nonflammable gas, green label, 70 kilograms (passenger), 140 kilograms (cargo).

CONDURANGIN
General Information
Description: Yellowish, amorphous, bitter powder.
Formula: $C_{35}H_{54}O_{14}(OCH_3)_2$.
Constant: Mol wt: 760.9.
Hazard Analysis
Toxic Hazard Rating:
Acute Local: Irritant 1.
Acute Systemic: Ingestion 3.
Chronic Local: U.
Chronic Systemic: U.
Toxicology: May cause convulsions and then paralysis.
Fire Hazard: Slight; when heated.
Countermeasures
Personal Hygiene: Section 3.
Storage and Handling: Section 7.

CONHYDRINE
General Information
Description: Colorless, crystalline alkaloid.
Formula: $C_8H_{17}NO$.
Constants: Mol wt: 143.23, mp: 121°C, bp: 226°C.
Hazard Analysis
Toxic Hazard Rating:
Acute Local: 0.
Acute Systemic: Ingestion 3; Inhalation 3.
Chronic Local: U.
Chronic Systemic: U.
Toxicology: Poisonous alkaloid. Toxic doses cause weakness and drowsiness but not actual sleep. Movements are weak and unsteady and gait is staggering. There are usually nausea and vomiting with profuse salivation. Intelligence usually remains clear. The pupils are ordinarily somewhat dilated and ptosis also occurs, indicating oculomotor paralysis. Speech is thick and hearing is imperfect. Tremors and occasionally convulsions occur. Breathing becomes weaker and slower and finally stops, causing death. It is said to have a depressant effect on bruised surfaces of the skin. One-half to one grain is considered a poisonous dose.
Fire Hazard: Slight; when heated.
Countermeasures
Ventilation Control: Section 2.
Treatment and Antidotes: Administer emetics and wash stomach out. Then give tannic acid freely and wash stomach out again. Strychnine and other stimulants are given hyperdermically. Keep patient warm and administer artificial respiration. A physician should be summoned.
Personal Hygiene: Section 3.
First Aid: Section 1.
Storage and Handling: Section 7.

CONICINE. See coniine.

CONIINE
General Information
Synonyms: Cicutine; cicuta; conicine.
Description: Colorless, oily liquid with mousy odor.
Formula: $C_8H_{17}N$.
Constants: Mol wt: 127.23, bp: 166.5°C, fp: −2.5°C, d: 0.844–0.848 at 20°/4°C.
Hazard Analysis
Toxic Hazard Rating:
Acute Local: 0.

Acute Systemic: Ingestion 3; Inhalation 3.
Chronic Local: U.
Chronic Systemic: U.
Toxicology: Produces paralysis of the nervous system. In small doses it is a sedative. Poisoning is treated by evacuating the stomach and administering tannic acid.
Fire Hazard: Slight; when heated.
Countermeasures
Ventilation Control: Section 2.
Personal Hygiene: Section 3.
First Aid: Section 1.
Storage and Handling: Section 7.

CONIINE HYDROCHLORIDE. See coniine.

CONIUM. See coniine.

CONIUM MACULATIUM. See coniine.

CONTAINERS, EMPTY. see acid carboys, empty; bottles, empty; drums, empty; cylinders, empty.

COPAIBA
General Information
Synonym: Balsam copaiba.
Description: Transparent, viscid yellow liquid. Peculiar odor. Bitter, acrid nauseous taste.
Constituents: Volatile oil, resin.
Hazard Analysis
Toxicology: Large doses cause vomiting and diarrhea. Can also cause dermatitis and kidney damage.

COPAL
General Information
Synonyms: Resin copal; gum copal.
Description: Yellowish-brown pieces.
Hazard Analysis
Toxic Hazard Rating:
Acute Local: Allergen 1.
Acute Systemic: U.
Chronic Local: Allergen 1.
Chronic Systemic: U.
Fire Hazard: Slight; when heated.
Countermeasures
Personal Hygiene: Section 3.
Storage and Handling: Section 7.

COPPER
General Information
Description: Distinct reddish color.
Formula: Cu.
Constants: At wt: 63.54, mp: 1083°C, bp: 2324°C, d: 8.92, vap. press.: 1 mm at 1628°C.
Hazard Analysis
Toxicity: See copper compounds.
Radiation Hazard: Section 5. For permissible levels, see Table 5, p. 150.
Artificial isotope ^{64}Cu, half life 12.8 h. Decays to stable ^{64}Ni by emitting positrons of 0.66 MeV. Also decays to stable ^{64}Zn by emitting beta particles of 0.57 MeV. Also emits gamma rays of 1.4 MeV.
Fire Hazard: See powdered metals.

COPPER ABIETINATE
General Information
Synonym: Cupric abietinate.
Description: Green scales.
Formula: $Cu(C_{19}H_{27}O_2)_2$.
Constant: Mol wt: 637.69.
Hazard Analysis
Toxicity: See copper compounds.
Fire Hazard: Slight; when heated.
Countermeasures
Storage and Handling: Section 7.

COPPER ACETATE
General Information
Synonyms: Cupric acetate; neutral verdigris.
Description: Greenish-blue, fine powder or small crystals.
Formula: $Cu(C_2H_3O_2)_2 \cdot H_2O$.
Constants: Mol wt: 199.64, mp: 115°C, bp: 240°C decomp.
    d: 1.882; (anhydrous) 1.93.
Hazard Analysis and Countermeasures
See copper compounds.

COPPER ACETATE, BASIC
General Information
Synonym: Verdigris.
Description: Greenish-blue powder.
Formula: $Cu(C_2H_3O_2)_2 \cdot CuO \cdot 6H_2O$.
Constant: Mol wt: 369.33.
Hazard Analysis and Countermeasures
See copper compounds.

COPPER ACETOARSENITE
General Information
Synonyms: Emerald green; imperial; king's green; moss
    green; Vienna green.
Description: Emerald green powder.
Formula: $(CuO)_3As_2O_3 \cdot Cu(C_2H_3O_2)_2$.
Constant: Mol wt: 618.15.
Hazard Analysis
See arsenic compounds and copper compounds.
Countermeasures
Storage and Handling: Section 7.
Shipping Regulations: Section 11.
    I.C.C.: Poison B; poison label, 200 pounds.
    Coast Guard Classification: Poison B; poison label.
    IATA: Poison B, poison label, 25 kilograms (passenger),
    95 kilograms (cargo).

COPPER ACETONATE
Hazard Analysis and Countermeasures
See copper compounds.

COPPER ACETYLIDE. See cuprous acetylide.

COPPER AMMONIUM SULFATE
General Information
Description: Crystals.
Formula: $CuSO_4 \cdot 4NH_3 \cdot H_2O$.
Constant: Mol wt: 245.8.
Hazard Analysis and Countermeasures
See copper compounds and sulfates.

COPPER ARSENATE, BASIC
General Information
Synonym: Cuprous arsenate, basic.
Description: A green solid.
Formula: $Cu(CuOH)AsO_4$.
Constant: Mol wt: 283.0.
Hazard Analysis and Countermeasures
See arsenic compounds and copper compounds.

COPPER ARSENIDE
General Information
Description: Black crystals.
Formula: Cu_5As_2.
Constants: Mol wt: 467.52, mp: decomposes, d: 7.56.
Hazard Analysis
See arsenic compounds and copper compounds.
Fire Hazard: See arsine.
Explosion Hazard: See arsine.
Disaster Hazard: Dangerous; see arsenides.

Countermeasures
Storage and Handling: Section 7.

COPPER ARSENITE
General Information
Synonyms: Cupric arsenite; Sheele's mineral.
Description: Yellowish-green powder.
Formula: $CuHAsO_3$.
Constants: Mol wt: 187.5, mp: decomposes.
Hazard Analysis
Toxicity: See arsenic compounds and copper compounds.
Disaster Hazard: See arsenic compounds and copper com-
    pounds.
Countermeasures
Storage and Handling: Section 7.
Shipping Regulations: Section 11.
    I.C.C.: Poison B; poison label, 200 pounds.
    Coast Guard Classification: Poison B; poison label.
    IATA: Poison B, poison label, 25 kilograms (passenger),
    95 kilograms (cargo).

COPPERAS. See ferrous sulfate.

COPPER BORIDE
General Information
Synonym: Cupric boride.
Description: Yellow crystals.
Formula: Cu_3B_2.
Constants: Mol wt: 212.26, d: 8.116.
Hazard Analysis
Toxicity: See copper compounds and boron compounds.
Fire Hazard: See boron hydrides.
Explosion Hazard: See boron hydrides.
Disaster Hazard: Moderately dangerous; on contact with
    acid, acid fumes, water or steam, it will react to produce
    toxic and flammable vapors of boron hydrides.
Countermeasures
Storage and Handling: Section 7.

COPPER CARBONATE
General Information
Synonym: Cupric carbonate.
Description: Green powder.
Formula: $CuCO_3 \cdot Cu(OH)_2$.
Constants: Mol wt: 221.17, mp: decomposes 200°C, d: 4.0.
Hazard Analysis and Countermeasures
A fungicide. Also a trace mineral added to animal feeds
    (Section 10). See copper compounds.

COPPER CHLORATE. See cupric chlorate.

COPPER CHLORIDE
General Information
Synonym: Cupric chloride.
Description: Yellowish-brown, hygroscopic powder.
Formula: $CuCl_2$.
Constants: Mol wt: 134.48, mp: 498°C, d: 3.054.
Hazard Analysis and Countermeasures
Used as a fungicide. Also, a trace mineral added to
    animal feeds (Section 10). See copper compounds and
    chlorides.
Shipping Regulations: Section 11.
    IATA: Other restricted articles, class B, no label required,
    12 kilograms (passenger), 45 kilograms (cargo).

COPPER-γ-CHLOROACETOACETANILIDE
General Information
Description: Solid.

TOXIC HAZARD RATING CODE (For detailed discussion, see Section 1.)

0 NONE: (a) No harm under any conditions; (b) Harmful only under un-
    usual conditions or overwhelming dosage.

1 SLIGHT: Causes readily reversible changes which disappear after end
    of exposure.

2 MODERATE: May involve both irreversible and reversible changes;
    not severe enough to cause death or permanent injury.
3 HIGH: May cause death or permanent injury after very short exposure
    to small quantities.
U UNKNOWN: No information on humans considered valid by authors.

Formula: $Cu(C_8H_7ClNO)_2$.
Constant: Mol wt: 400.7.
Hazard Analysis and Countermeasures
See copper compounds, acetanilde, and chlorides.

COPPER CHROMATE, BASIC. See cupric chromate, basic.

COPPER COMPOUNDS
Hazard Analysis
Toxic Hazard Rating:
Acute Local: Irritant 1; Allergen 1; Ingestion 1; Inhalation 1.
Acute Systemic: Ingestion 2; Inhalation 2.
Chronic Local: Allergen 1.
Chronic Systemic: Ingestion 1; Inhalation 1.
Toxicology: As the sublimed oxide, copper may be responsible for one form of metal fume fever. Inhalation of copper dust has caused, in animals, hemolysis of the red blood cells, deposition of hemofuscin in the liver and pancreas, and injury to the lung cells; injection of the dust has caused cirrhosis of the liver and pancreas, and a condition closely resembling hemochromatosis, or bronzed diabetes. However, considerable trial exposure to copper compounds has not resulted in such disease.

As regards local effect, copper chloride and sulfate have been reported as causing irritation of the skin and conjunctivae which may be on an allergic basis (Section 9). Cuprous oxide is irritating to the eyes and upper respiratory tract. Discoloration of the skin is often seen in persons handling copper, but this does not indicate any actual injury from copper.

"In man the ingestion of a large quantity of copper sulfate has caused vomiting, gastric pain, dizziness, exhaustion, anemia, cramps, convulsions, shock, coma and death. Symptoms attributed to damage to the nervous system and kidney have been recorded, jaundice has been observed and, in some cases, the liver has been enlarged. Deaths have been reported to have occurred following the ingestion of so little as 27 g of the salt, while other victims have recovered after having taken much larger amounts up to 120 g." "Copper: The Metal, Its Alloys and Compounds," ed. Allison Butts, p. 857, Reinhold, 1954.
Many copper containing compounds are used as fungicides.
Countermeasures
Ventilation Control: Section 2.
Personnel Protection: Section 3.
Personal Hygiene: Section 3.
Shipping Regulations: (10% and over as metallic copper): Section 11. MCA warning label.

COPPER-8-CUNILATE
Hazard Analysis and Countermeasures
See copper compounds.

COPPER CYANIDE
General Information
Synonym: Cupric cyanide.
Description: Yellowish-green powder.
Formula: $Cu(CN)_2$.
Constants: Mol wt: 115.61, mp: decomposes before melting.
Hazard Analysis and Countermeasures
See cyanides and copper compounds.
Shipping Regulations: Section 11.
I.C.C.: Section 73.370.
IATA: Poison B, poison label, 12 kilograms (passenger), no limit (cargo).

COPPER DIAZOAMINOBENZENE
General Information
Description: Orange crystals insoluble in water. Soluble in benzene.
Formula: $CuN_3(C_6H_5)_2$.

Constants: Mol wt: 259.8, mp: 270°C (decomposes).
Hazard Analysis
Toxicology: Details unknown. See copper compounds.

COPPER DICHLOROBENZOATE
General Information
Synonym: CDCB.
Hazard Analysis
Toxic Hazard Rating: U. Used as a fungicide.
Disaster Hazard: Dangerous. See chlorides.

COPPER DIMETHYLDITHIOCARBAMATE
Hazard Analysis
Toxicity: See ferbam.
Disaster Hazard: Dangerous; when heated to decomposition, it emits toxic fumes.
Countermeasures
Storage and Handling: Section 7.

COPPER ETHYL XANTHOGENATE. See copper xanthate.

COPPER FLUORIDE
General Information
Synonym: Cupric fluoride.
Description: Monoclinic blue crystals.
Formula: $CuF_2 \cdot 2H_2O$.
Constants: Mol wt: 137.60, d: 2.93.
Hazard Analysis and Countermeasures
See fluorides and copper compounds.

COPPER GLUCONATE
General Information
Synonym: Cupric gluconate.
Description: Light blue, fine crystalline powder; soluble in water; insoluble in acetone, alcohol, and ether.
Formula: $[CH_2OH(CHOH)_4COO]_2Cu$.
Constant: Mol wt: 453.5.
Hazard Analysis
Toxic Hazard Rating: U. A nutrient and/or dietary supplement food additive. Also a trace mineral added to animal feeds. (Section 10).

COPPER HYDRIDE
General Information
Description: Red-brown crystals.
Formula: CuH.
Constants: Mol wt: 64.55, mp: decomposes 60°C.
Hazard Analysis and Countermeasures
See copper compounds and hydrides.

COPPER HYDROSELENITE
General Information
Description: Bluish-green, tiny prisms.
Formula: $Cu(HSeO_3)_2$.
Constant: Mol wt: 319.5.
Hazard Analysis and Countermeasures
See selenium compounds and copper compounds.

COPPER HYDROXIDE
General Information
Synonym: Cupric hydroxide.
Description: Blue gelatinous or amorphous powder.
Formula: $Cu(OH)_2$.
Constants: Mol wt: 97.59, d: 3.368.
Hazard Analysis and Countermeasures
A trace mineral added to animal feeds (Section 10). Also used as a fungicide. See copper compounds.

COPPER 8-HYDROXYQUINOLINE. See copper 8-quinolinolate.

COPPER MERCURY IODIDE. See mercuric cuprous iodide.

COPPER NAPHTHENATE
General Information
Description: A solid.

Formula: $(C_6H_5COO)_2Cu$.
Constants: Mol wt: 221.9, flash p: 100° F, d: 1.055.
Hazard Analysis
Toxicity: See copper compounds.
Fire Hazard: Moderate, when exposed to heat or flame; it can react with oxidizing materials.
Countermeasures
To Fight Fire: Foam, carbon dioxide, dry chemical or carbon tetrachloride (Section 6).
Storage and Handling: Section 7.

COPPER NITRATE
General Information
Synonym: Cupric nitrate.
Description: Blue, deliquescent crystals.
Formula: $Cu(NO_3)_2 \cdot 3H_2O$.
Constants: Mol wt: 241.63, mp: 114.5°C, d: 2.047.
Hazard Analysis and Countermeasures
See copper compounds and nitrates.

COPPER NITRIDE
General Information
Description: Dark-green powder.
Formula: Cu_3N.
Constants: Mol wt: 204.63, mp: decomposes 300°C.
Hazard Analysis and Countermeasures
See copper compounds and nitrides.

COPPER NITRODITHIOACETATE
General Information
Description: Solid.
Hazard Analysis
Toxicity: See copper compounds.
Disaster Hazard: Dangerous; when heated to decomposition, it emits toxic fumes.
Countermeasures
Storage and Handling: Section 7.

COPPER OLEATE
General Information
Synonym: Cupric oleate.
Description: Brown powder or greenish-blue mass.
Formula: $Cu(C_{18}H_{33}O_2)_2$.
Constant: Mol wt: 626.46.
Hazard Analysis and Countermeasures
Toxicity: Used as a fungicide. See copper compounds and oleic acid.

COPPER OXALATE
General Information
Synonym: Cupric oxalate.
Description: Solid; light bluish-green powder.
Formula: $CuC_2O_4 \cdot \frac{1}{2}H_2O$.
Constant: Mol wt: 160.57.
Hazard Analysis and Countermeasures
See oxalates and copper compounds.

COPPER OXIDE
General Information
Synonym: Cupric oxide; paramelaconite.
Description: Fine, black powder.
Formula: CuO.
Constants: Mol wt: 79.5, bp: decomposes at 1026°C, d: 6.4.
Hazard Analysis and Countermeasures
Used as a fungicide. Also a trace mineral added to animal feeds (Section 10). See copper compounds.

COPPER OXIDE, RED. See cuprous oxide.

COPPER OXYCHLORIDE
General Information
Synonyms: Brunswick green; cupric oxychloride.
Description: Emerald green to greenish-black powder.
Formula: $CuCl_2 \cdot 2CuO \cdot 4H_2O$.
Constants: Mol wt: 365.60, mp: $-3H_2O$ at 140°C.
Hazard Analysis and Countermeasures
See copper compounds and chlorides.

COPPER 2,4-PENTANEDIONE DERIVATIVE
General Information
Synonym: Acetylacetonate of copper.
Description: Blue crystals.
Formula: $Cu(C_6H_7O_2)_2$.
Constants: Mol wt: 261.75, mp: $> 230°C$, bp: sublimes.
Hazard Analysis and Countermeasures
See copper compounds.

COPPER PERCHLORATE
General Information
Description: Crystalline.
Formula: $Cu(ClO_4)_2 \cdot 6H_2O$.
Constants: Mol wt: 370.6, mp: 60°C.
Hazard Analysis and Countermeasures
See perchlorates and copper compounds.

COPPER PEROXIDE
General Information
Description: Brown or brownish-black crystals.
Formula: $CuO_2 \cdot H_2O$.
Constant: Mol wt: 113.56.
Hazard Analysis and Countermeasures
See copper compounds and peroxides.

COPPER 3-PHENYL SALICYLATE
General Information
Description: An odorless, nonvolatile crystalline material.
Formula: $C_{26}H_{18}CuO_6$.
Constants: Mol wt: 490.0, mp: 145°C.
Hazard Analysis
Toxicity: See copper compounds.
Disaster Hazard: Moderately dangerous; when heated to decomposition, it emits toxic fumes.
Countermeasures
Storage and Handling: Section 7.

COPPER PHOSPHATE
General Information
Synonym: Cupric phosphate.
Description: Solid; bluish-green powder.
Formula: $Cu_3(PO_4)_2$.
Constant: Mol wt: 380.6.
Hazard Analysis and Countermeasures
A trace mineral added to animal feeds Section 10. See copper compounds and phosphates.

COPPER PHOSPHIDE
General Information
Synonym: Cupric phosphide.
Description: Solid.
Formula: Cu_3P_2.
Constants: Mol wt: 252.6, mp: decomposes, d: 6.67.
Hazard Analysis and Countermeasures
See phosphides and copper compounds.

COPPER PICRATE
General Information
Description: A solid.
Formula: $Cu(C_6H_2N_3O_7)_2$.

TOXIC HAZARD RATING CODE (For detailed discussion, see Section 1.)

0 NONE: (a) No harm under any conditions; (b) Harmful only under unusual conditions or overwhelming dosage.

1 SLIGHT: Causes readily reversible changes which disappear after end of exposure.

2 MODERATE: May involve both irreversible and reversible changes; not severe enough to cause death or permanent injury.

3 HIGH: May cause death or permanent injury after very short exposure to small quantities.

U UNKNOWN: No information on humans considered valid by authors.

Constant: Mol wt: 519.7.
Hazard Analysis and Countermeasures
See picrates and copper compounds.

COPPER POLYSULFIDE. See cuprous sulfide.

COPPER PROPARGYLATE
General Information
Description: A solid.
Hazard Analysis
Toxicity: See copper compounds.
Explosion Hazard: Severe, when shocked or exposed to heat. See also explosives, high.
Disaster Hazard: Dangerous; shock and heat will explode it.
Countermeasures
Storage and Handling: Section 7.

COPPER PROPIONYL ACETATE
General Information
Description: Crystals.
Formula: CuC_5H_9O.
Constant: Mol wt: 148.7.
Hazard Analysis and Countermeasures
See copper compounds.

COPPER PYROPHOSPHATE
Hazard Analysis
Toxic Hazard Rating: U. A trace mineral added to animal feeds (Section 10).
Disaster Hazard: Dangerous. See phosphates.

COPPER-8-QUINOLINOLATE
General Information
Synonyms: Copper-8-hydroxyquinoline.
Description: Yellow-green powder.
Formula: $C_{18}H_{12}N_2O_2Cu$.
Constants: Mol wt: 351.83, decomposes at 210°C.
Hazard Analysis and Countermeasures
Used as a fungicide. See copper compounds.

COPPER RESINATE
General Information
Description: Green powder.
Formula: $Cu(C_{20}H_{29}O_2)_2$.
Constant: Mol wt: 666.43.
Hazard Analysis and Countermeasures
See copper compounds.

COPPER RICINOLEATE
General Information
Description: A green, plastic solid.
Formula: $Cu[CO_2(CH_2)_7CHCHCH_2CHOH(CH_2)_5 \cdot CH_3]_2$.
Constants: Mol wt: 654, softening p: 64°C.
Hazard Analysis and Countermeasures
See copper compounds.

COPPER SEBACATE
General Information
Description: A solid.
Formula: $Cu(CH_2)_8C_2O_4$.
Constant: Mol wt: 263.8.
Hazard Analysis and Countermeasures
See copper compounds.

COPPER SILICATE
General Information
Description: Greenish crystals.
Formula: $CuSiO_3$.
Constant: Mol wt: 139.6.
Hazard Analysis and Countermeasures
See copper compounds and silicates.

COPPER SILICIDE
General Information
Description: White metallic crystals.
Formula: Cu_4Si.
Constants: Mol wt: 282.22, mp: 850°C, d: 7.53.

Hazard Analysis and Countermeasures
See copper compounds and silanes.

COPPER SILICOFLUORIDE. See cupric fluosilicate.

COPPER STEARATE
General Information
Synonym: Cupric stearate.
Description: Light-blue, amorphous powder.
Formula: $Cu(C_{18}H_{35}O_2)_2$.
Constants: Mol wt: 630.50, mp: 125°C.
Hazard Analysis and Countermeasures
See copper compounds.

COPPER SUBOXIDE
General Information
Description: Olive-green crystals.
Formula: Cu_4O.
Constants: Mol wt: 270.16, mp: decomposes.
Hazard Analysis and Countermeasures
See copper compounds.

COPPER SUBSULFATE
General Information
Synonym: Cupric sulfate, basic.
Description: Light-blue powder.
Formula: $4CuO \cdot SO_3$.
Constant: Mol wt: 398.2.
Hazard Analysis and Countermeasures
See copper compounds and sulfates.

COPPER SULFATE
General Information
Synonyms: Blue vitriol; blue stone; Roman vitriol.
Description: Blue crystals or blue, crystalline granules or powder.
Formula: $CuSO_4 \cdot 5H_2O$.
Constants: Mol wt: 249.71, mp: $-4H_2O$ at 110°C, d: 2.284.
Hazard Analysis and Countermeasures
Used as a trace mineral added to animal feeds. It is a substance which migrates to food from packaging materials (Section 10). It is also used as a herbicide. See copper compounds and sulfuric acid.

COPPER SULFATE, AMMONIATED
General Information
Synonyms: Cupric sulfate, ammoniated.
Description: Dark-blue crystals.
Formula: $CuSO_4 \cdot 4NH_3 \cdot H_2O$.
Constant: Mol wt: 245.8.
Hazard Analysis and Countermeasures
See copper compounds, ammonia and sulfates.

COPPER SULFATE, BASIC
Hazard Analysis
Toxic Hazard Rating: U. Used as a fungicide.
Disaster Hazard: Dangerous. See sulfates.

COPPER SULFATE, MONOHYDRATE
Hazard Analysis
Toxic Hazard Rating: U. Used as a fungicide.
Disaster Hazard: Dangerous. See sulfates.

COPPER SULFIDE
General Information
Synonym: Cupric sulfide.
Description: Black powder or crystals.
Formula: CuS.
Constants: Mol wt: 95.6, mp: transition at 103°C, bp: decomposes at 220°C, d: 4.6.
Hazard Analysis and Countermeasures
See copper compounds and sulfides.

COPPER TELLURITE
General Information
Description: Green solid.
Formula: $CuTeO_3$.
Constant: Mol wt: 239.15.

Hazard Analysis and Countermeasures
See tellurium compounds and copper compounds.

COPPER TETRAZOL
Hazard Analysis
Toxicity: See copper compounds.
Explosion Hazard: Severe, when exposed to heat or shock.
Disaster Hazard: Highly dangerous; shock and heat will explode it; when heated to decomposition, it emits toxic fumes.
Countermeasures
Storage and Handling: Section 7.

COPPER THIOCYANATE
General Information
Synonym: Cuprous thiocyanate.
Description: White to yellowish powder.
Formula: CuCNS.
Constants: Mol wt: 121.6, mp: 1084°C, d: 2.85.
Hazard Analysis and Countermeasures
See copper compounds and thiocyanates.

"COPPER THIURAM"
Hazard Analysis
Toxicity: See copper compounds.
Disaster Hazard: Dangerous; when heated to decomposition, it emits toxic fumes.
Countermeasures
Storage and Handling: Section 7.

COPPER TRICHLOROPHENATE
General Information
Description: A crystalline solid.
Formula: $Cu(Cl_3C_6H_2O)_2$.
Constant: Mol wt: 456.5.
Hazard Analysis and Countermeasures
See chlorinated phenols and copper compounds.

COPPER XANTHATE
General Information
Synonym: Copper ethylxanthogenate.
Description: Yellow precipitate.
Formula: $Cu(C_3H_5OS_2)_2$.
Constants: Mol wt: 305.94, mp: decomposes.
Hazard Analysis
Toxicity: See copper compounds.
Disaster Hazard: Dangerous; see sulfates.
Countermeasures
Storage and Handling: Section 7.

COPPER ZINC CHROMATE
General Information
Description: Variable in composition.
Hazard Analysis
Toxic Hazard Rating: U. Used as a fungicide.

COPPER ZINC SULFATE
Hazard Analysis
Toxic Hazard Rating: U. Used as a fungicide.
Disaster Hazard: Dangerous. See sulfates.

COPRA
Hazard Analysis
Toxicity: Unknown.
Fire Hazard: Moderate, when exposed to heat; dangerous if stored wet and hot (Section 6).
Spontaneous Heating: Slight.
Countermeasures
Storage and Handling: Section 7.

Shipping Regulations: Section 11.
    Coast Guard Classification: Hazardous article.

CORDEAU DETONANT FUSE. See explosives, high.
Shipping Regulations: Section 11.
    I.C.C.: Explosive C, 300 pounds.
    Coast Guard Classification: Explosive C.
    IATA: Explosive, explosive label, 25 kilograms (passenger), 140 kilograms (cargo).

CORDITE. See smokeless powder.

CORN MEAL FEEDS
Hazard Analysis
Toxicity: Weak allergens. Local contact can cause contact dermatitis. Inhalation or ingestion may provoke bronchial asthma, eczema, hives, angioedema, conjunctivitis, rhinorrhea, etc. in already sensitized persons (Section 9).
Fire Hazard: Slight, when exposed to heat (Section 6).
Spontaneous Heating: High. A safe moisture content must be retained. Presence of oil may be dangerous during storage.
Countermeasures
Storage and Handling: Section 7.

CORN OIL
General Information
Description: Light-yellow, clear, oily liquid; faint characteristic odor.
Constants: Mp: −10°C, flash p: 490°F (C.C.), d: 0.92, autoign. temp.: 740°F.
Hazard Analysis
Toxicity: May be an allergen.
Fire Hazard: Slight, when exposed to heat or flame. Dangerous when stored if leakage impregnates rags, waste, etc. (Section 6).
Spontaneous Heating: Moderate.
Countermeasures
To Fight Fire: Foam, carbon dioxide, dry chemical or carbon tetrachloride. Section 6.
Storage and Handling: Section 7.

CORN SUGAR. See dextrose.

CORONENE
Hazard Analysis
A polycyclic hydrocarbon air pollutant.

CORROSIVE LIQUID, N.O.S.
Hazard Analysis
Toxicity: See sulfuric acid or sodium hydroxide.
Countermeasures
Personnel Protection: Section 3.
First Aid: Section 1.
Shipping Regulations: Section 11.
    I.C.C.: Corrosive liquid, white label, 5 pints.
    Coast Guard Classification: Corrosive liquid, white label.
    IATA: Corrosive liquid, white label, 1 liter (passenger and cargo).

CORROSIVE SUBLIMATE. See mercuric chloride.

CORYNINE. See yohimbine.

CORUNDUM. See emery.

COTARNINE CHLORIDE
General Information
Synonym: Stypticin.
Description: Light-yellow powder.

TOXIC HAZARD RATING CODE (*For detailed discussion, see Section 1.*)

0 NONE: (a) No harm under any conditions; (b) Harmful only under unusual conditions or overwhelming dosage.

1 SLIGHT: Causes readily reversible changes which disappear after end of exposure.

2 MODERATE: May involve both irreversible and reversible changes; not severe enough to cause death or permanent injury.

3 HIGH: May cause death or permanent injury after very short exposure to small quantities.

U UNKNOWN: No information on humans considered valid by authors.

Formula: $C_{12}H_{13}NO_3HCl$.
Constant: Mol wt: 255.64.
Hazard Analysis
Toxic Hazard Rating:
  Acute Local: Irritant 1; Ingestion 1; Inhalation 1.
  Acute Systemic: Ingestion 3; Inhalation 3.
  Chronic Local: U.
  Chronic Systemic: Ingestion 1; Inhalation 1.
Toxicology: Large doses cause paralysis of the central
  nervous system and death due to depression of the
  respiratory center.
Disaster Hazard: Dangerous; see chlorides and nitrates.
Countermeasures
Ventilation Control: Section 2.
Personnel Protection: Section 3.
First Aid: Section 1.
Storage and Handling: Section 7.

COTTON DUST
Hazard Analysis
Toxic Hazard Rating:
  Acute Local: Inhalation 2.
  Acute Systemic: Inhalation 2.
  Chronic Local: Inhalation 2.
  Chronic Systemic: Inhalation 1.
TLV: ACGIH; 1 milligram per cubic meter of air.
Toxicology: Can cause a mild febrile condition of the lungs
  known as byssinosis or Monday fever. This resembles
  metal fume fever and is prevalent in plants where the
  dusts of such fibers are found. Immunity can be acquired
  after a few days of exposure. It does not ordinarily cause
  any fibrosis.
    Coarser grades of cotton contain more dust than the
  finer varieties, and therefore constitute a greater hazard.
  It is considered an inert dust and is so within the meaning of that term. However, it can cause some illness, due
  to the allergens or fungi in the cotton or on the dust.
  Workers in processing rooms may develop conjunctivitis or blepharitis from the burned products of the
  gassing of the double yarn. It is a mild allergen. Inhalation may produce bronchial asthma, sneezing and
  eczema in sensitized persons (Section 9).
Fire Hazard: Moderate, when exposed to heat or flame; can
  react with oxidizing materials (Section 6).
Explosion Hazard: Moderate, when exposed to heat or flame
  (Section 7).
Countermeasures
Ventilation Control: Section 2.
Treatment and Antidotes: Removal from exposure will usually cause symptoms to disappear. Medical attention
  should be given to ascertain that there is no fibrosis of
  the lungs.

COTTON SEED
Hazard Analysis
Toxicity: A very powerful allergen. Inhalation or ingestion
  may produce bronchial asthma, sneezing, rhinorrhea,
  conjunctivitis, eczema and hives in persons already
  sensitized to this material (Section 9).
Fire Hazard: Slight, when exposed to heat (Section 6).
Spontaneous Heating: Low. If piled or stored wet and hot, it
  can generate dangerous amounts of heat.
Countermeasures
Storage and Handling: Section 7.
Shipping Regulations: Section 11.
  Coast Guard Classification: Hazardous article.
Cotton seed, cut linters, hull fibers, pulp, waste, and shavings,
  with animal or vegetable oil.
Shipping Regulations: Section 11.
  IATA: Flammable solid, not acceptable (passenger and
  cargo).

**COTTON SEED, CUT LINTERS, PULP, WASTE AND
SHAVINGS WITH ANIMAL OR VEGETABLE
OIL.**

Shipping Regulations: Section 11.
  IATA: Flammable solid, not acceptable (passenger and
  cargo).

COTTONSEED OIL (REFINED)
General Information
Description: Pale oily, yellow, nearly odorless liquid from
  seeds of species of Gossypium.
Constants: Flash p.: 486° F (C.C.), fp: 0 to −5°C,
  d: 0.915–0.921 at 25°/25°C, autoign. temp.: 650° F.
Hazard Analysis
Toxic Hazard Rating:
  Acute Local: Allergen 1.
  Acute Systemic: 0.
  Chronic Local: Allergen 1.
  Chronic Systemic: 0.
Fire Hazard: Slight, when exposed to heat or flame. How-
  ever, if allowed to impregnate rags or oily waste, it can
  be a dangerous hazard (Section 6).
Spontaneous Heating: Moderate.
Countermeasures
To Fight Fire: Foam, carbon dioxide, dry chemical or car-
  bon tetrachloride. Section 6.
Personal Hygiene: Section 3.
Storage and Handling: Section 7.

COTTON WASTE, OILY, with more than 5% animal or
  vegetable oil.
Hazard Analysis
Toxicity: No rating.
Fire Hazard: Moderate; when exposed to heat or flame; it
  can react with oxidizing materials (Section 6).
Countermeasures
Storage and Handling: Section 7.
Shipping Regulations: Section 11.
  I.C.C.: Flammable solid; yellow label, not accepted.
  Coast Guard Classification: Inflammable solid; yellow
  label.
  IATA: Flammable solid, not acceptable (passenger and
  cargo).

COTUNNITE. See lead chloride.

COUMARIN
General Information
Synonym: 1,2-benzopyrone.
Description: Crystals; fragrant, pleasant odor; burning taste.
Formula: $C_9H_6O_2$.
Constants: Mol wt: 146.1, mp: 70°C, bp: 291.0°C,
  vap. press.: 1 mm at 106.0°C.
Hazard Analysis
Toxic Hazard Rating:
  Acute Local: Allergen 1.
  Acute Systemic: Ingestion 1; Inhalation 1.
  Chronic Local: Allergen 1.
  Chronic Systemic: U.
Toxicology: The parent substance of dicoumarol, which
  causes disturbances in the clotting mechanism of the
  blood, and hence can lead to spontaneous bleeding.
Fire Hazard: Slight, when exposed to heat or flame (Section
  6).
Countermeasures
Ventilation Control: Section 2.
Personal Hygiene: Section 3.
Storage and Handling: Section 7.

COUMARONE-INDENE RESINS
General Information
Synonyms: Coal tar resins; indene resins; polycoumarone
  resins; polyindene resins.
Description: Vary from fairly viscous liquids to hard resins;
  color—pale yellow to nearly black; soluble in hydrocarbon solvents; pyridine, acetone, carbon disulfide, and
  carbon tetrachloride; insoluble in water and alcohol.
Hazard Analysis
Toxic Hazard Rating: U. A food additive permitted in food
  for human consumption (Section 10).

CRAB-E-RAD. See disodium monomethylarsonate.

CRAG HERBICIDE 974. See 3,5-dimethyltetrahydro-
-1,3,5-2H-thiodiazine-2-thione.

CRAG I. See dichlorophenoxyethyl sulfate.

CRAG FRUIT FUNGICIDE 341. See glyodin.

CRAG FUNGICIDE 974. See mylone.

CREAM OF TARTAR. See potassium acid tartrate.

CREOSOL
General Information
Synonym: 2-methoxy-4-methyl phenol; 4-methylguaiacol;
    2-methoxy-p-cresol.
Description: Colorless to yellow liquid, slightly soluble in
    water; soluble in alcohol, benzene, chloroform, ether,
    acetic acid.
Formula: $CH_3O(CH_3)C_6H_3OH$.
Constants: Mol wt: 138.16, d: 1.092 (25°/4°C),
    mp: 5.5°C, bp: 220°C.
Hazard Analysis and Countermeasures
See phenol.

CREOSOTE, COAL TAR
General Information
Synonym: Creosote oil.
Description: Colorless or yellow clear, oily liquid.
Composition: A mixture of phenols from coal tar.
Constants: Bp: 200–250°C, flash p: 165°F (C.C.), d: 1.07,
    autoign. temp.: 637°F.
Hazard Analysis
Toxic Hazard Rating:
    Acute Local: Irritant 2; Ingestion 2; Inhalation 2.
    Acute Systemic: Ingestion 2; Inhalation 2.
    Chronic Local: Irritant 2; Allergen 2.
    Chronic Systemic: Ingestion 2; Inhalation 2.
Fire Hazard: Moderate, when exposed to heat or flame.
Disaster Hazard: Dangerous; when heated to decomposi-
    tion, it emits toxic fumes.
Countermeasures
Ventilation Control: Section 2.
Personnel Protection: Section 3.
Storage and Handling: Section 7.
Shipping Regulations: Section 11.
    Coast Guard Classification: Combustible liquid.
    MCA warning label.

CREOSOTE (CRESOL AND PHENOL) MIXTURE. See
    cresol and phenol.

CREOSOTE OIL. See creosote, coal tar.

CRESOL
General Information
Synonyms: Cresylic acid, cresylol, tricresol.
Description: (U.S.P. XVI) mixture of isomeric cresols ob-
    tained from coal tar; colorless to yellowish to brown
    yellow or pinkish liquid, phenol-like odor.
Formula: $C_6H_4OHCH_3$.
Constants: Mol wt: 108.10, mp: 10.9–35.5°C, bp: 191–
    203°C, flash p: 110°F, d: 1.030–1.038 at 25°/25°C,
    vap. press.: 1 mm at 38–53°C, vap. d.: 3.72.
Hazard Analysis
Toxic Hazard Rating:
    Acute Local: Irritant 2; Allergen 1; Ingestion 2; Inhala-
    tion 2.
    Acute Systemic: Ingestion 2; Inhalation 2; Skin Absorp-
    tion 2.

Chronic Local: Irritant 3; Allergen 1.
Chronic Systemic: Ingestion 2; Inhalation 2; Skin Ab-
    sorption 2.
TLV: ACGIH (accepted); 22 milligrams per cubic meter of
    air; 5 parts per million in air. May be absorbed via in-
    tact skin.
Toxicology: Cresol is similar to phenol in its action on the
    body, but it is less severe in its effects. It has corrosive
    action on the skin and mucous membranes. Systemic
    poisoning has rarely been reported, but it is possible
    that absorption may result in damage to the kidneys,
    liver and nervous system. The main hazard accompany-
    ing its use in industry lies in its action on the skin and
    mucous membranes, with production of severe chemi-
    cal burns and dermatitis (Section 9).
Fire Hazard: Moderate, when exposed to heat or flame.
Explosion Hazard: Slight, in the form of vapor when exposed
    to heat or flame (Section 7).
Explosive Range: 1.35% at 300°F.
Disaster Hazard: Dangerous; when heated to decomposi-
    tion, it emits highly toxic fumes; it can react vigorously
    with oxidizing materials.
Countermeasures
Ventilation Control: Section 2.
Personnel Protection: Section 3.
First Aid: Section 1.
Storage and Handling: Section 7.
To Fight Fire: Foam, carbon dioxide, dry chemical or carbon
    tetrachloride (Section 6).
Shipping Regulations: Section 11.
    Coast Guard Classification: Inflammable liquid.
    MCA warning label.
    IATA (liquid: Poison B, poison label, 1 liter (passenger),
    220 liters (cargo).

m-CRESOL
General Information
Synonym: m-Methylphenol.
Description: Colorless to yellowish liquid; phenolic odor.
Formula: C_7H_8O.
Constants: Mol wt: 108.1, mp: 10.9°C, bp: 202.8°C,
    lel: 1.1% at 302°F, flash p: 202°F, d: 1.034 at
    20°/4°C, autoign. temp.: 1038°F, vap. press.: 1 mm
    at 52.0°C, vap. d.: 3.72.
Hazard Analysis
Toxicity: See cresol.
TLV: ACGIH (accepted); 22 milligrams per cubic meter of
    air; 5 parts per million in air. May be absorbed via intact
    skin.
Fire Hazard: See cresol.
Explosion Hazard: Moderate, in the form of vapor when ex-
    posed to heat or flame (Section 7).
Disaster Hazard: See cresol.
Countermeasures
Storage and Handling: Section 7.
    MCA warning label.

o-CRESOL
General Information
Synonyms: o-Cresylic acid; o-hydroxytoluene.
Description: Crystals or liquid darkening with exposure to
    air and light.
Formula: C_7H_8O.
Constants: Mol wt: 108.1, mp: 30.8°C, bp: 190.8°C,
    flash p: 178°F, density: 1.047 at 20°/4°C,
    autoign. temp.: 1110°F, vap. press.: 1 mm at 38.2°C,
    vap. d.: 3.72, lel: 1.4% at 300°F.

TOXIC HAZARD RATING CODE *(For detailed discussion, see Section 1.)*

.0 NONE: (a) No harm under any conditions; (b) Harmful only under un-
    usual conditions or overwhelming dosage.

1 SLIGHT: Causes readily reversible changes which disappear after end
    of exposure.

2 MODERATE: May involve both irreversible and reversible changes;
    not severe enough to cause death or permanent injury.

3 HIGH: May cause death or permanent injury after very short exposure
    to small quantities.

U UNKNOWN: No information on humans considered valid by authors.

Hazard Analysis
Toxicity: See cresol.
TLV: ACGIH (accepted); 22 milligrams per cubic meter of air; 5 parts per million in air. Absorbed via intact skin.
Fire Hazard: See cresol.
Explosion Hazard: See cresol.
Disaster Hazard: See cresol.
Countermeasures
Storage and Handling: Section 7.
Shipping Regulations: Section 11.
  Coast Guard Classification: Inflammable liquid.
  MCA warning label.

p-CRESOL
General Information
Synonyms: 4-Cresol.
Description: Crystals, phenolic odor.
Formula: C_7H_8O.
Constants: Mol wt: 108.1, mp: 35.5°C, bp: 201.8°C, lel: 1.1% at 302°F, flash p: 202°F, d: 1.0341 at 20°/4°C, autoign. temp.: 1038°F, vap. press.: 1 mm at 53.0°C, vap. d.: 3.72.
Hazard Analysis
Toxicity: See cresol.
TLV: ACGIH (accepted); 5 parts per million in air; 22 milligrams per cubic meter of air; absorbed via intact skin.
Fire Hazard: Moderate, when exposed to heat or flame.
Spontaneous Heating: No.
Explosion Hazard: Moderate, in the form of vapor when exposed to heat or flame (Section 7).
Disaster Hazard: See cresol.
Countermeasures
Storage and Handling: Section 7.
To Fight Fire: Carbon dioxide, dry chemical or carbon tetrachloride (Section 6).
MCA warning label.

4-CRESOL. See p-cresol.

CRESOLITE. See 2,4,6-trinitro-m-cresol.

CRESOTIC ACID. See o-cresotinic acid.

o-CRESOTINIC ACID
General Information
Synonym: Cresotic acid; hydroxytoluic acid; homosalicylic acid.
Description: White to yellowish needle-like crystals.
Formula: $C_6H_3(CH_3)(OH)COOH$.
Constants: Mol wt: 152.1, mp: 166°C sublimes.
Hazard Analysis
Toxicity: Unknown. For symptoms, see salicylic acid.
Fire Hazard: Slight (Section 6).
Countermeasures
Storage and Handling: Section 7.

CRESYL DIPHENYL PHOSPHATE
General Information
Description: Liquid.
Formula: $(CH_3C_6H_4)(C_6H_5)_2PO_4$.
Constants: Mol wt: 403.3, bp: 734°F, flash p: 450°F, d: 1.208, vap. d.: 11.7.
Toxicity: See tri-o-cresyl phosphate.
Fire Hazard: Slight, when exposed to heat or flame.
Disaster Hazard: Dangerous; see phosphates.
Countermeasures
Storage and Handling: Section 7.
To Fight Fire: Water, foam, carbon dioxide, dry chemical or carbon tetrachloride (Section 6).

CRESYLIC ACID. See cresol.

o-CRESYLIC ACID. See o-cresol.

CRISTOBALITE. See quartz.

CROCOITE. See lead chromate.

CROTONALDEHYDE
General Information
Synonyms: 2-Butenal; crotonic aldehyde; β-methylacrolein.
Description: Water-white mobile liquid; pungent suffocating odor.
Formula: $CH_3CHCHCHO$.
Constants: Mol wt: 70.09, bp: 104°C, fp: −76.0°C, lel: 2.1%, uel: 15.5%, flash p: 55°F, d: 0.853 at 20°/20°C, vap. d.: 2.41.
Hazard Analysis
Toxic Hazard Rating:
  Acute Local: Irritant 3; Allergen 2; Ingestion 3; Inhalation 3.
  Acute Systemic: U.
  Chronic Local: Allergen 2.
  Chronic Systemic: U.
TLV: ACGIH (tentative) 2 parts per million of air; 6 milligrams per cubic meter of air.
Toxicology: A lachrymating material which is very dangerous to the eyes. Can cause corneal burns and is irritating to the skin.
Fire Hazard: Dangerous, when exposed to heat or flame; can react with oxidizing materials.
Spontaneous Heating: No.
Disaster Hazard: Dangerous; keep away from heat and open flame.
Countermeasures
Ventilation Control: Section 2.
Treatment and Antidotes: In case of contact, immediately flush the skin or eyes with water for at least 15 minutes. Get medical attention.
Personnel Protection: Section 3.
First Aid: Section 1.
Storage and Handling: Section 7.
To Fight Fire: Alcohol foam, carbon dioxide, dry chemical or carbon tetrachloride (Section 6).
Shipping Regulations: Section 11.
  I.C.C. Classification: Flammable liquid; red label, 10 gallons.
  Coast Guard Classification: Inflammable liquid; red label.
  IATA: Flammable liquid, red label, 1 liter (passenger), 40 liters (cargo).

CROTONALIC ACID. See tiglic acid.

CROTONIC ACID
General Information
Synonym: β-Methacrylic acid.
Description: Colorless needle-like crystals.
Formula: $CH_3CHCHCOOH$.
Constants: Mol wt: 86.09, bp: 172°C, fp: 72°C, flash p: 190°F (C.O.C.), d: 1.018 at 20°/4°C, vap. press.: 0.19 mm at 20°C, vap. d.: 2.97.
Hazard Analysis
Toxicity: Unknown. Experimental data show moderate toxicity in rats and higher toxicity in guinea pigs.
Fire Hazard: Slight, when exposed to heat or flame; it can react with oxidizing materials.
Countermeasures
To Fight Fire: Alcohol foam, carbon dioxide, dry chemical or carbon tetrachloride (Section 6).
Storage and Handling: Section 7.

CROTONIC ACID-2,4-DINITRO-6(1-METHYL-HEPTYL) PHENYL ESTER
Hazard Analysis
Toxicity: Details unknown. Limited observations show an irritant action. See also dinitrophenol.
Disaster Hazard: When heated to decomposition, it emits highly toxic fumes.

CROTONIC ALDEHYDE. See crotonaldehyde.

CROTONIC ANHYDRIDE
Hazard Analysis
Toxic Hazard Rating: U. Limited animal experiments suggest moderate toxicity and high irritation.

CROTONITRILE
General Information
Synonym: 2-Butene nitrile.
Formula: C_4H_5N.
Constants: Mol wt: 67, flash p: <212° F, d: 0.8, vap. d.:2.3, bp: 230–240.8° F.
Hazard Analysis
Toxic Hazard Rating: U. Probably very toxic.
Fire Hazard: Moderate when exposed to heat or flame.
Disaster Hazard: Dangerous. See nitriles.
Countermeasures
Storage and Handling: Section 7.

CROTON OIL
General Information
Synonym: Tiglium oil.
Description: Brownish-yellow, viscid oil, slight offensive odor.
Composition: Croton resin, glycerides of fatty acids and crotin.
Constant: D: 0.935–0.950 at 25° /25° C.
Hazard Analysis
Toxic Hazard Rating:
  Acute Local: Irritant 2; Allergen 1; Ingestion 3.
  Acute Systemic: Ingestion 3.
  Chronic Local: Allergen 1.
  Chronic Systemic: U.
Fire Hazard: Slight (Section 6).
Countermeasures
Personal Hygiene: Section 3.
First Aid: Section 1.
Storage and Handling: Section 7.

CROTONYLENE
General Information
Synonym: 2-Butyne.
Description: Liquid.
Formula: CH_3CCCH_3.
Constants: Mol wt: 54.09, bp: 27° C, flash p: 64° F, lel: 1.4%, d: 0.688 at 25° C, vap. d.: 1.91.
Hazard Analysis
Toxicity: A simple asphyxiant. See also argon for action of such asphyxiants.
Fire Hazard: Dangerous, when exposed to heat or flame; it can react with oxidizing materials.
Explosion Hazard: Moderate, in the form of vapor when exposed to heat or flame.
Countermeasures
Storage and Handling: Section 7.
To Fight Fire: Foam, carbon dioxide, dry chemicals or carbon tetrachloride (Section 6).
Shipping Regulations: Section 11.
  IATA: Flammable liquid, red label, 1 liter (passenger), 40 liters (cargo).

CROTOXIN
General Information
Synonym: Rattlesnake venom.
Hazard Analysis
Toxicity: Highly toxic.
Toxicology: When injected as in a snake bite, local pain results as well as inflammation, hemorrhage, necrosis and clouding of sensorium. Effects are vertigo, impairment of motor activity and collapse with signs of shock. May be fatal.

CROTYLIDENE DICROTONATE
Hazard Analysis
Toxic Hazard Rating: Limited animal experiments suggest moderate toxicity. See also esters.

CRUDE NITROGEN FERTILIZER SOLUTION
See compressed gases, N.O.S.
Shipping Regulations: Section 11.
  I.C.C.: Nonflammable gas; green label, 300 pounds.
  Coast Guard Classification: Noninflammable gas; green label.
  IATA: Nonflammable gas, green label, 70 kilograms (passenger), 140 kilograms (cargo).

CRUDE OIL (PETROLEUM)
General Information
Synonyms: Earth oil; Seneca oil.
Description: Thick, flammable liquid, dark-yellow to brown or green-black.
Composition: A mixture of hydrocarbons.
Constants: D: 0.780–0.970, flash p: 20–90° F.
Hazard Analysis
Toxic Hazard Rating:
  Acute Local: Irritant 1; Allergen 1; Ingestion 2; Inhalation 1.
  Acute Systemic: U.
  Chronic Local: Irritant 2; Allergen 1.
  Chronic Systemic: U.
Fire Hazard: Moderate, when exposed to heat or flame.
Disaster Hazard: Moderately dangerous; when heated to decomposition, it emits toxic fumes; it can react with oxidizing materials.
Countermeasures
Personnel Protection: Section 3.
To Fight Fire: Foam, carbon dioxide, dry chemical or carbon tetrachloride (Section 6).
Storage and Handling: Section 7.
Shipping Regulations: Section 11.
  I.C.C.: Flammable liquid; red label, 10 gallons.
  Coast Guard Classification: Inflammable liquid; red label.
  IATA: Flammable liquid, red label, 1 liter (passenger), 40 liters (cargo).

CRYOLITE. See sodium aluminum fluoride.

CRYPTOHALITE. See ammonium fluosilicate.

CRYPTOPINE
General Information
Description: White, crystalline powder.
Formula: $C_{21}H_{23}NO_5$.
Constants: Mol wt: 369.4, mp: 217° C decomposes.
Hazard Analysis
Toxic Hazard Rating:
  Acute Local: U.
  Acute Systemic: Ingestion 3; Inhalation 3.
  Chronic Local: U.
  Chronic Systemic: Ingestion 3; Inhalation 3.
Caution: Related to opium in derivation and action.
Disaster Hazard: Dangerous; when heated to decomposition it emits highly toxic fumes.
Countermeasures
Ventilation Control: Section 2.
Personnel Protection: Section 3.
First Aid: Section 1.
Storage and Handling: Section 7.

CRYPTOPINE HYDROCHLORIDE. See cryptopine.

CHRYSOBERYL. See beryllium aluminate.

TOXIC HAZARD RATING CODE *(For detailed discussion, see Section 1.)*

.0 NONE: (a) No harm under any conditions; (b) Harmful only under unusual conditions or overwhelming dosage.

1 SLIGHT: Causes readily reversible changes which disappear after end of exposure.

2 MODERATE: May involve both irreversible and reversible changes; not severe enough to cause death or permanent injury.

3 HIGH: May cause death or permanent injury after very short exposure to small quantities.

U UNKNOWN: No information on humans considered valid by authors.

CRYSTAL AMMONIA. See ammonium carbonate.

CRYSTAL CARBONATE. See sodium carbonate.

C-STUFF
General Information
Description: Liquid.
Formula: 50% hydrazine hydrate + 50% methanol.
Hazard Analysis
Toxicology: See hydrazine and methanol.
Fire Hazard: This material is a high energy rocket propellant. Special instructions on storage and handling and fire fighting must be obtained before having anything to do with this material.
Explosion Hazard: See fire hazard.
Disaster Hazard: See fire hazard.

CUMENE
General Information
Synonyms: Isopropylbenzene; 2-phenylpropane; cumol.
Description: Colorless liquid.
Formula: $C_6H_5CH(CH_3)_2$.
Constants: Mol wt: 120.19, mp: $-96.0°C$, bp: $152°C$, flash p: $111°F$, d: 0.864 at $20°/4°C$, vap. press.: 10 mm at $38.3°C$, autoign. temp.: $795°F$, lel: 0.9%, uel: 6.5%, vap. d.: 4.1.
Hazard Analysis
Toxic Hazard Rating:
Acute Local: Irritant 1.
Acute Systemic: Ingestion 3; Inhalation 3; Skin Absorption 3.
Chronic Local: Irritant 2.
Chronic Systemic: Ingestion 2; Inhalation 2; Skin Absorption 2.
TLV: ACGIH (tentative) 50 parts per million of air; 245 milligrams per cubic meter of air. May be absorbed via the skin.
Toxicology: Cumene has a potent narcotic action characterized by a slow induction period, although the effects are of long duration. It is a depressant to the central nervous system and the minimum lethal concentration for mice is 2,000 ppm. The long duration of its action indicates a possible slow rate of elimination, meaning that possible cumulative effects must be considered. It is thought to have a greater acute toxicity than benzene or toluene; there is no apparent difference between the toxicity of pure cumene or that derived from petroleum. See also benzene and toluene.
Fire Hazard: Moderate, when exposed to flame; it can react with oxidizing materials.
Countermeasures
To Fight Fire: Foam, carbon dioxide, dry chemical or carbon tetrachloride (Section 6).
Ventilation Control: Section 2.
Personnel Protection: Section 3.
First Aid: Section 1.
Storage and Handling: Section 7.

CUMENE HYDROPEROXIDE
General Information
Synonym: α-Dimethyl benzyl hydroperoxide.
Description: Colorless to pale-yellow liquid.
Constants: Bp: $153°C$, flash p: $175°F$, d: 1.05.
Hazard Analysis
Toxic Hazard Rating:
Acute Local: Irritant 3; Ingestion 3; Inhalation 3.
Acute Systemic: Ingestion 3; Inhalation 3; Skin Absorption 3.
Chronic Local: U.
Chronic Systemic: U.
Fire Hazard: Moderate; when exposed to heat or flame, it can react with reducing materials.
Caution: A strong oxidizing agent. See peroxides.
Countermeasures
To Fight Fire: Foam, carbon dioxide, dry chemical or carbon tetrachloride (Section 6).

Ventilation Control: Section 2.
Personnel Protection: Section 3.
First Aid: Section 1.
Storage and Handling: Section 7.
Shipping Regulations: Section 11.
I.C.C.: Oxidizing material; yellow label, 1 quart.
Coast Guard Classification: Oxidizing material; yellow label.
IATA (solution, exceeding 96%): Oxidizing material, not acceptable (passenger and cargo).
IATA (solution, *not* exceeding 96% in a non-violatile solvent): Oxidizing material, yellow label, 1 liter (passenger and cargo).

o-CUMENOL. See isopropyl phenol.

CUMOL. See cumene.

p-α-CUMYL PHENOL
General Information
Description: White to light tan crystals.
Formula: $C_6H_5C(CH_3)_2C_6H_4OH$.
Constants: Mol wt: 212.3, bp: $187°C$ at 10 mm, fp: $70°C$, flash p: $320°F$.
Hazard Analysis
Toxicity: Unknown.
Fire Hazard: Slight; when exposed to heat or flame; it can react with oxidizing materials (Section 6).
To Fight Fire: Foam, carbon dioxide, dry chemical or carbon tetrachloride. Section 6.
Storage and Handling: Section 7.

CUPRAMATE. See copper dimethyldithiocarbamate.

CUPRIC ABIETINATE. See copper abietinate.

CUPRIC ACETATE. See copper acetate.

CUPRIC ACETATE m-ARSENATE
General Information
Synonym: Paris green.
Description: Emerald green powder.
Formula: $Cu(C_2H_3O_2)_2 \cdot 3Cu(AsO_2)_2$ (approx.).
Constant: Mol wt: 1013.71.
Hazard Analysis
Toxicity: See arsenic compounds and copper compounds.
Disaster Hazard: Dangerous, see Arsenic.
Personnel Protection: Section 3.
First Aid: Section 1.
Storage and Handling: Section 7.
Shipping Regulations: Section 11.
I.C.C.: Poison B; poison label.
Coast Guard Classification: Poison B; poison label.
MCA warning label.

CUPRIC o-ARSENATE
General Information
Description: Bluish-green crystals.
Formula: $Cu_3(AsO_4)_2 \cdot 4H_2O$.
Constant: Mol wt: 540.50.
Hazard Analysis
See arsenic compounds and copper compounds.
Disaster Hazard: Dangerous; see Arsenic.
Countermeasures
Ventilation Control: Section 2.
Personnel Protection: Section 3.
First Aid: Section 1.
Storage and Handling: Section 7.

CUPRIC ARSENATE, BASIC
General Information
Description: Green powder.
Formula: $Cu_3(AsO_4)_2Cu(OH)_2$.
Constant: Mol wt: 566.
Hazard Analysis
See arsenic compounds and copper compounds.
Disaster Hazard: Dangerous; see arsenic.

Countermeasures
Ventilation Control: Section 2.
Personnel Protection: Section 3.
First Aid: Section 1.
Storage and Handling: Section 7.

CUPRIC ARSENITE. See copper arsenite.

CUPRIC m-ARSENITE
General Information
Description: Powder.
Formula: $Cu(AsO_2)_2 \cdot H_2O$.
Constant: Mol wt: 295.4.
Hazard Analysis
See arsenic compounds and copper compounds.
Disaster Hazard: Dangerous; see arsenic.
Countermeasures
Storage and Handling: Section 7.

CUPPIC BENZOATE
General Information
Description: Light-blue, crystalline powder.
Formula: $Cu(C_7H_5O_2)_2 \cdot 2H_2O$.
Constants: Mol wt: 341.8, mp: $-2H_2O$ at $110°C$.
Hazard Analysis and Countermeasures
See copper compounds.

CUPRIC m-BORATE
General Information
Description: Bluish-green, crystalline powder.
Formula: $Cu(BO_2)_2$.
Constants: Mol wt: 149.18, d: 3.859.
Hazard Analysis and Countermeasures
See copper compounds and boron compounds.

CUPRIC BORIDE. See copper boride.

CUPRIC BROMATE
General Information
Description: Cubic blue-green crystals.
Formula: $Cu(BrO_3)_2 \cdot 6H_2O$.
Constants: Mol wt: 427.47, mp: decomposes $180°C$;
$-6H_2O$ at $200°C$, d: 2.583.
Hazard Analysis and Countermeasures
See bromates and copper compounds.

CUPRIC BROMIDE
General Information
Description: Monoclinic black, deliquescent crystals.
Formula: $CuBr_2$.
Constants: Mol wt: 223.37, mp: $498°C$.
Hazard Analysis and Countermeasures
See bromides and copper compounds.

CUPRIC BUTYRATE
General Information
Description: Dark-green crystals, odor of butyric acid.
Formula: $Cu(C_4H_7O_2)_2 \cdot 2H_2O$.
Constant: Mol wt: 273.8.
Hazard Analysis and Countermeasures
See copper compounds.

CUPRIC CARBONATE. See copper carbonate.

CUPRIC CHLORATE
General Information
Description: Cubic green, deliquescent crystals.
Formula: $Cu(ClO_3)_2 \cdot 6H_2O$.
Constants: Mol wt: 338.55, mp: $65°C$, bp: decomposes
at $100°C$.

Hazard Analysis and Countermeasures
See chlorates and copper compounds.

CUPRIC CHLORIDE. See copper chloride.

CUPRIC CHROMATE, BASIC
General Information
Description: Yellow-brown crystals.
Formula: $CuCrO_4 \cdot 2CuO \cdot 2H_2O$.
Constants: Mol wt: 374.66, mp: $-2H_2O$ at $260°C$.
Hazard Analysis and Countermeasures
See chromium compounds and copper compounds.

CUPRIC CITRATE
General Information
Description: Bluish-green powder.
Formula: $2Cu_2C_6H_4O_7 \cdot 5H_2O$.
Constant: Mol wt: 720.42.
Hazard Analysis and Countermeasures
See copper compounds.

CUPRIC CYANIDE. See copper cyanide.

CUPRIC DICHROMATE
General Information
Description: Black, deliquescent crystals.
Formula: $CuCr_2O_7 \cdot 2H_2O$.
Constants: Mol wt: 315.59, mp: $-2H_2O$ at $100°C$, d: 2.283.
Hazard Analysis and Countermeasures
See chromium compounds and copper compounds.

CUPRIC DIHYDROGEN o-ARSENATE
General Information
Description: Blue crystals.
Formula: $Cu_5H_2(AsO_4)_4 \cdot 2H_2O$.
Constant: Mol wt: 911.39.
Hazard Analysis and Countermeasures
See arsenic compounds and copper compounds.

CUPRIC ETHYLACETOACETATE
General Information
Description: Green needles.
Formula: $Cu(C_6H_9O_3)_2$.
Constants: Mol wt: 321.80, mp: $192°C$, bp: sublimes.
Hazard Analysis and Countermeasures
See copper compounds.

CUPRIC FERRICYANIDE
General Information
Description: Yellow-green crystals.
Formula: $Cu_3[Fe(CN)_6]_2$.
Constant: Mol wt: 870.76.
Hazard Analysis and Countermeasures
See ferricyanides and copper compounds.

CUPRIC FLUORIDE. See copper fluoride.

CUPRIC FLUOSILICATE
General Information
Description: Monoclinic prisms.
Formula: $CuSiF_6 \cdot 4H_2O$.
Constants: Mol wt: 277.66, d: 2.158.
Hazard Analysis and Countermeasures
See fluosilicates and copper compounds.

CUPRIC FORMATE
General Information
Description: Monoclinic, blue crystals.
Formula: $Cu(CHO_2)_2$.
Constants: Mol wt: 153.58, d: 1.831.

TOXIC HAZARD RATING CODE   *(For detailed discussion, see Section 1.)*

0 NONE: (a) No harm under any conditions; (b) Harmful only under unusual conditions or overwhelming dosage.

1 SLIGHT: Causes readily reversible changes which disappear after end of exposure.

2 MODERATE: May involve both irreversible and reversible changes; not severe enough to cause death or permanent injury.

3 HIGH: May cause death or permanent injury after very short exposure to small quantities.

U UNKNOWN: No information on humans considered valid by authors.

Hazard Analysis and Countermeasures
See copper compounds.

CUPRIC GLUCONATE. See copper gluconate.

CUPRIC GLYCINE DERIVATIVE
General Information
Description: Blue needles.
Formula: $Cu(C_2H_4NO_2)_2 \cdot 2H_2O$.
Constants: Mol wt: 229.68, mp: $-H_2O$ at 130°C.
Hazard Analysis and Countermeasures
See copper compounds.

CUPRIC HYDROXIDE. See copper hydroxide.

CUPRIC IODATE
General Information
Description: Monoclinic, green crystals.
Formula: $Cu(IO_3)_2$.
Constants: Mol wt: 413.38, mp: decomposes, d: 5.241 at 15°C.
Hazard Analysis and Countermeasures
See iodates and copper compounds.

CUPRIC LACTATE
General Information
Description: Monoclinic, dark blue crystals.
Formula: $Cu(C_3H_5O_3)_2 \cdot 2H_2O$.
Constant: Mol wt: 277.71.
Hazard Analysis and Countermeasures
See copper compounds.

CUPRIC LAURATE
General Information
Description: Light-blue powder.
Formula: $Cu(C_{12}H_{23}O_2)_2$.
Constants: Mol wt: 462.15, mp: 111–113°C.
Hazard Analysis and Countermeasures
See copper compounds.

CUPRIC NITRATE. See copper nitrate.

CUPRIC NITROPRUSSIDE
General Information
Description: White-greenish powder.
Formula: $CuFe(CN)_5NO \cdot 2H_2O$.
Constant: Mol wt: 315.52.
Hazard Analysis and Countermeasures
See cyanides and copper compounds.

CUPRIC OLEATE. See copper oleate.

CUPRIC OXALATE. See copper oxalate.

CUPRIC OXIDE. See copper oxide.

CUPRIC OXYCHLORIDE. See copper oxychloride.

CUPRIC PALMITATE
General Information
Description: Green-blue powder.
Formula: $Cu(C_{16}H_{31}O_2)_2$.
Constants: Mol wt: 574.36, mp: 120°C.
Hazard Analysis and Countermeasures
See copper compounds.

CUPRIC p-PERIODATE
General Information
Description: Green powder.
Formula: Cu_2HIO_6.
Constants: Mol wt: 351.01, mp: decomposes 110°C.
Hazard and Analysis and Countermeasures
See copper compounds and iodates.

CUPRIC PHENOLSULFONATE
General Information
Description: Bluish-green crystals.
Formula: $Cu[C_6H_4(OH)SO_3]_2 \cdot 6H_2O$.

Constant: Mol wt: 518.0.
Hazard Analysis and Countermeasures
See copper compounds and phenosulfonates.

CUPRIC o-PHOSPHATE. See copper phosphate.

CUPRIC PHOSPHIDE. See copper phosphide.

CUPRIC SALICYLATE
General Information
Description: Blue-green needles.
Formula: $Cu(C_7H_5O_3)_2 \cdot 4H_2O$.
Constant: Mol wt: 409.8.
Hazard Analysis and Countermeasures
See copper compounds.

CUPRIC SELENATE
General Information
Description: Blue, triclinic crystals.
Formula: $CuSeO_4 \cdot 5H_2O$.
Constants: Mol wt: 296.58, d: 2.559.
Hazard Analysis and Countermeasures
See selenium compounds and copper compounds.

CUPRIC STEARATE. See copper stearate.

CUPRIC SULFATE. See copper sulfate.

CUPRIC SULFATE, AMMONIATED. See copper sulfate ammoniated.

CUPRIC SULFATE, BASIC. See copper subsulfate.

CUPRIC SULFIDE. See copper sulfide.

CUPRIC TARTRATE
General Information
Description: Light-blue powder.
Formula: $CuC_4H_4O_6$.
Constant: Mol wt: 211.6.
Hazard Analysis and Countermeasures
See copper compounds.

CUPRIC THIOCYANATE
General Information
Description: Black crystals.
Formula: $Cu(SCN)_2$.
Constants: Mol wt: 179.71, mp: decomposes at 100°C.
Hazard Analysis and Countermeasures
See thiocyanates and copper compounds.

CUPRIC TUNGSTATE
General Information
Description: Light-green crystals.
Formula: $CuWO_4 \cdot 2H_2O$.
Constants: Mol wt: 347.49, mp: at red heat.
Hazard Analysis and Countermeasures
See copper compounds and tungsten compounds.

CUPRIETHYLENE-DIAMINE SOLUTION
Countermeasures
Shipping Regulations: Section 11.
 I.C.C.: Corrosive liquid, white label, 1 gallon.
 IATA: Corrosive liquid, white label, 1 liter (passenger), 5 liters (cargo).

CUPRITE. See cuprous oxide.

CUPROUS ACETYLIDE
General Information
Description: Amorphous red powder.
Formula: Cu_2C_2.
Constants: Mol wt: 150.10, mp: explodes.
Hazard Analysis
Toxicity: See copper compounds.
Explosion Hazard: Severe when shocked or exposed to heat (Section 7).
Disaster Hazard: Highly dangerous; shock and heat will explode it.

Countermeasures
Storage and Handling: Section 7.

CUPROUS ARSENATE, BASIC. See copper arsenate, basic.

CUPROUS BROMIDE
General Information
Description: White crystals.
Formula: CuBr.
Constants: Mol wt: 143.66, mp: 504°C, bp: 1355°C, d: 4.718, vap. press.: 1 mm at 572°C.
Hazard Analysis and Countermeasures
See bromides and copper compounds.

CUPROUS CARBONATE
General Information
Description: Yellow crystals.
Formula: Cu_2CO_3.
Constants: Mol wt: 187.1, mp: decomposes, d: 4.40.
Hazard Analysis and Countermeasures
See copper compounds.

CUPROUS CHLORIDE
General Information
Synonym: Nantokite.
Description: Cubic, white crystals.
Formula: CuCl.
Constants: Mol wt: 99.00, mp: 422°C, bp: 1366°C, d: 3.53, vap. press.: 1 mm at 546°C.
Hazard Analysis and Countermeasures
See chlorides and copper compounds.

CUPROUS CYANIDE
General Informat on
Description: Monoclinic, white prisms.
Formula: CuCN.
Constants: Mol wt: 89.56, mp: 473°C in N_2, bp: decomposes, d: 2.92.
Hazard Analysis and Countermeasures
See cyanides and copper compounds.

CUPROUS FERRICYANIDE
General Information
Description: Brown-red crystals.
Formula: $Cu_3Fe(CN)_6$.
Constant: Mol wt: 402.59.
Hazard Analysis and Countermeasures
See ferricyanides and copper compounds.

CUPROUS FLUOGALLATE
General Information
Description: Pale-blue crystals.
Formula: $[Cu(H_2O)_6](GaF_5 \cdot H_2O)$.
Constants: Mol wt: 354.37, mp: $-5H_2O$ at 110°C, d: 2.20.
Hazard Analysis and Countermeasures
See fluogallates and copper compounds.

CUPROUS FLUORIDE
General Information
Description: Red crystals.
Formula: CuF.
Constants: Mol wt: 82.54, mp: 908°C, bp: sublimes at 1100°C.
Hazard Analysis and Countermeasures
See fluorides and copper compounds.

CUPROUS FLUOSILICATE
General Information
Synonym: Cuprous silicofluoride.

Description: Red powder.
Formula: Cu_2SiF_6.
Constants: Mol wt: 269.14, bp: decomposes.
Hazard Analysis and Countermeasures
See fluosilicates and copper compounds.

CUPROUS HYDRIDE
General Information
Description: Yellow-brown crystals; insoluble in cold water; reacts in hot water.
Formula: CuH.
Constants: Mol wt: 64.6, mp: decomposes slowly at 25°C, d: 6.29.
Hazard Analysis and Countermeasures
See copper compounds and hydrides.

CUPROUS HYDROXIDE
General Information
Description: Yellow crystals.
Formula: CuOH.
Constants: Mol wt: 80.55, mp: $-\frac{1}{2}H_2O$ at 360°C, d: 3.37.
Hazard Analysis and Countermeasures
See copper compounds.

CUPROUS IODIDE
General Information
Synonym: Marshite.
Description: Cubic, white crystals.
Formula: CuI.
Constants: Mol wt: 190.46, mp: 605°C, bp: 1290°C, d: 5.62, vap. press.: 10 mm at 656°C.
Hazard Analysis and Countermeasures
A nutrient and/or dietary supplement food additive. Also a trace mineral added to animal feeds (Section 10). See iodides and copper compounds.

CUPROUS OXIDE
General Information
Synonym: Cuprite.
Description: Octahedral, cubic red crystals.
Formula: Cu_2O.
Constants: Mol wt: 143.14, mp: 1235°C, bp: $-O_2$ at 1800°C, d: 6.0.
Hazard Analysis and Countermeasures
See copper compounds.

CUPROUS SELENIDE
General Information
Description: Black cubes.
Formula: Cu_2Se.
Constants: Mol wt: 206.04, mp: 1113°C, d: 6.749 at 30°/4°C.
Hazard Analysis and Countermeasures
See selenium compounds and copper compounds.

CUPROUS SILICOFLUORIDE. See cuprous fluosilicate.

CUPROUS SULFATE
General Information
Description: Gray powder.
Formula: Cu_2SO_4.
Constants: Mol wt: 223.15, mp: + O at 200°C.
Hazard Analysis and Countermeasures
See copper compounds and sulfates.

CUPROUS SULFIDE
General Information
Synonym: Chalcocite.

TOXIC HAZARD RATING CODE (For detailed discussion, see Section 1.)

0 NONE: (a) No harm under any conditions; (b) Harmful only under unusual conditions or overwhelming dosage.

1 SLIGHT: Causes readily reversible changes which disappear after end of exposure.

2 MODERATE: May involve both irreversible and reversible changes; not severe enough to cause death or permanent injury.

3 HIGH: May cause death or permanent injury after very short exposure to small quantities.

U UNKNOWN: No information on humans considered valid by authors.

Description: Rhombic, black crystals.
Formula: Cu_2S.
Constants: Mol wt: 159.15, mp: 1100° C, d: 5.6.
Hazard Analysis and Countermeasures
See sulfides and copper compounds.

CUPROUS SULFITE
General Information
Description: Red prisms.
Formula: $Cu_2SO_3 \cdot H_2O$.
Constants: Mol wt: 225.16, d: 4.46 at 15°C.
Hazard Analysis and Countermeasures
See sulfites and copper compounds.

CUPROUS THIOCYANATE. See copper thiocyanate.

CURARE
General Information
Description: Brown, brittle, resinous mass.
Toxic Hazard Rating:
  Acute Local: U.
  Acute Systemic: Ingestion 3; Inhalation 3; Skin Absorption 3.
  Chronic Local: U.
  Chronic Systemic: U.
Hazard Analysis
Toxicology: Extremely poisonous. Visual disturbances, choking sensation, then generalized paralysis.
Disaster Hazard: Dangerous; when heated to decomposition, it emits highly toxic fumes.
Countermeasures
Ventilation Control: Section 2.
Treatment and Antidotes: Artificial respiration, physostigmine. Call a physician immediately.
Personnel Protection: Section 3.
First Aid: Section 1.
Storage and Handling: Section 7.

CURIUM
General Information
Description: An artificially produced radioactive element.
Formula: Cm.
Constant: At wt: 242.
Hazard Analysis
Toxicity: Unknown except in relation to radioactivity.
Radiation Hazard: Section 5. For permissible levels, see Table 5, p. 150.
  Artificial isotope ^{242}Cm, half life 163 d. Decays to radioactive ^{238}Pu by emitting alpha particles of 6.1 MeV.
  Artificial isotope ^{243}Cm, half life 35 y. Decays to radioactive ^{239}Pu by emitting alpha particles of 5.8 MeV.
  Artificial isotope ^{244}Cm, half life about 18 y. Decays to radioactive ^{240}Pu by emitting alpha particles of 5.8 MeV.
  Artificial isotope ^{245}Cm, half life about 10^4y. Decays to radioactive ^{241}Pu by emitting alpha particles of 5.4 MeV.
  Artificial isotope ^{246}Cm, half life about 5000 y. Decays to radioactive ^{242}Pu by emitting alpha particles of 5.4 MeV.
  Artificial isotope ^{247}Cm, half life about 4×10^7 y. Decays to radioactive ^{243}Pu by emitting alpha particles of unknown energy.
  Artificial isotope ^{248}Cm, half life about 5×10^5 y. Decays to radioactive ^{244}Pu by emitting alpha particles (90%) of 5.0 MeV. The other 10% of the disintegrations are by spontaneous fission.
  Artificial isotope ^{249}Cm, half life 64 m. Decays to radioactive ^{249}Bk by emitting beta particles of 0.9 MeV.
Countermeasures
Shipping Regulations: Section 11.

CURLED MINT OIL. See spearmint oil.

CUTTING OILS
Hazard Analysis
Toxic Hazard Rating:
  Acute Local: Irritant 1.
  Acute Systemic: U.
  Chronic Local: Irritant 1.
  Chronic Systemic: U.
Toxicology: This oil is the cause of "cutting oil" dermatitis. Although it is generally caused by an insoluble oil, it can occasionally be caused by a soluble one. Many have looked for a causative factor other than the oil itself. Bacteria have frequently been blamed, although insoluble oil is usually sterile while the soluble oils may contain bacteria. The metal slivers which occur in these oils after use have also been thought to be a reason for their irritant properties, as well as the sulfur, chlorine and inhibitors which they contain. The oil itself can plug the pores forming boils (Section 9).
Fire Hazard: Slight, when exposed to heat or flame (Section 6).
Countermeasures
Personal Hygiene: Section 3.
Treatment and Antidotes: Disinfectant agents have been added to the oils to cut down the infectious effects. Filters have been installed in the cutting oil lines to filter out the tiny metal slivers which may irritate and cause infection of the skin. General housecleaning measures are also thought advisable to remove promptly any excess oil which gets onto the hands and skin. It has been found that occasionally changing from one oil to another may bring relief from the irritation. This would also indicate that the specific irritant is not present in all oils.
Storage and Handling: Section 7.

CYAMELIDE
General Information
Synonym: Insoluble cyanuric acid.
Description: White powder.
Formula: $C_3H_3N_3O_3$.
Constant: Mol wt: 129.1.
Hazard Analysis
Toxic Hazard Rating:
  Acute Local: U.
  Acute Systemic: Ingestion 3; Inhalation 3.
  Chronic Local: U.
  Chronic Systemic: Ingestion 3; Inhalation 3.
See also cyanides.
Disaster Hazard: Dangerous; when heated to decomposition or on contact with acid or acid fumes, it emits highly toxic fumes.
Countermeasures
Ventilation Control: Section 2.
Personal Hygiene: Section 3.
First Aid: Section 1.
Storage and Handling: Section 7.

CYANAMIDE
General Information
Synonyms: Carbodiimide; cyanogenamide.
Description: Deliquescent crystals.
Formula: HNCNH.
Constants: Mol wt: 42.05, mp: 42° C, bp: 260° C, flash p: 285° F, d: 1.073, vap. d.: 1.45.
Hazard Analysis
Toxic Hazard Rating:
  Acute Local: U.
  Acute Systemic: Ingestion 2; Inhalation 2.
  Chronic Local: U.
  Chronic Systemic: Ingestion 1; Inhalation 1.
Toxicology: Does not contain free cyanide. Causes increase in respiration and pulse rate, lowered blood pressure and dizziness. There may be a flushed appearance of the face.
Fire Hazard: Slight, when exposed to heat or flame.

Disaster Hazard: Moderately dangerous; when heated to decomposition or on contact with acid or acid fumes, it emits toxic fumes.

Countermeasures

Ventilation Control: Section 2.

To Fight Fire: Carbon dioxide, dry chemical or carbon tetrachloride (Section 6).

Personal Hygiene: Section 3.

Storage and Handling: Section 7.

CYANATES

Hazard Analysis

Toxicity: Variable.

Disaster Hazard: Dangerous; when heated to decomposition or on contact with acid or acid fumes, they emit toxic fumes.

Countermeasures

Storage and Handling: Section 7.

CYANIC ACID

General Information

Synonym: Isocyanic acid.

Description: A white solid.

Formula: HOCN.

Constants: Mol wt: 43.01, bp: decomposes, d: 1.140 at 0°/0°C.

Hazard Analysis

Toxicity: Highly toxic. See cyanides.

Explosion Hazard: Severe.

Disaster Hazard: Dangerous; can explode; when heated to decomposition or on contact with acid or acid fumes, it emits highly toxic fumes and flammable vapors.

Countermeasures

Storage and Handling: Section 7.

CYANIDE OF CALCIUM OR CYANIDE OF CALCIUM, MIXTURE, SOLID

Shipping Regulations: Section 11.

  I.C.C.: Poison B, poison label, 200 pounds.

  IATA: Poison B, poison label, 12 kilograms (passenger), 95 kilograms (cargo).

CYANIDE OF POTASSIUM. See potassium cyanide.

Shipping Regulations: Section 11.

  I.C.C.: Poison B; poison label, 55 gallon (liquid).

  I.C.C.: Poison B, poison label, 200 pounds (solid).

  Coast Guard Classification: Poison B; poison label.

  IATA: Poison B, poison label, 1 liter (passenger), 220 liters (cargo) (liquid).

  IATA: Poison B, poison label, 12 kilograms (passenger), 95 kilograms (cargo) (solid).

CYANIDE OF SODIUM. See sodium cyanide.

Shipping Regulations: Section 11.

  I.C.C. (liquid): Poison B, poison, 55 gallons (liquid).

  I.C.C. (solid): Poison B, poison label, 200 pounds (solid).

  Coast Guard Classification: Poison B; poison label.

  IATA (liquid): Poison B, poison label, 1 liter (passenger), 220 liters (cargo).

  IATA (solid): Poison B, poison label, 12 kilograms (passenger), 95 kilograms (cargo).

CYANIDE SOLUTIONS, N.O.S.

Shipping Regulations: Section 11.

  IATA: Poison B, poison label, 1 liter (passenger), 220 liters (cargo).

CYANIDES

Hazard Analysis

Toxic Hazard Rating:

  Acute Local: Irritant 1.

  Acute Systemic: Ingestion 3; Inhalation 3.

  Chronic Local: Irritant 2.

  Chronic Systemic: Ingestion 1; Inhalation 1.

TLV: ACGIH (recommended); 5 milligrams per cubic meter of air. Absorbed via intact skin.

Toxicology: The volatile cyanides resemble hydrocyanic acid physiologically, inhibiting tissue oxidation and causing death through asphyxia. Cyanogen is probably as toxic as hydrocyanic acid; the nitriles are generally considered somewhat less toxic, probably because of their lower volatility. The non-volatile cyanide salts appear to be relatively nontoxic systemically, so long as they are not ingested and care is taken to prevent the formation of hydrocyanic acid. Workers, such as electroplaters and picklers, who are daily exposed to cyanide solutions may develop a "cyanide" rash, characterized by itching, and by macular, papular, and vesicular eruptions. Frequently there is secondary infection. Exposure to small amounts of cyanide compounds over long periods of time is reported to cause loss of appetite, headache, weakness, nausea, dizziness, and symptoms of irritation of the upper respiratory tract and eyes. See also specific compounds.

Fire Hazard: Moderate, by chemical reaction with heat, moisture, acid; emit hydrocyanic acid (Section 6).

Caution: Many cyanides evolve hydrocyanic acid rather easily. This is a flammable gas and is highly toxic. Carbon dioxide from the air is sufficiently acidic to liberate hydrocyanic acid from cyanide solutions. See also hydrocyanic acid.

Explosion Hazard: See hydrocyanic acid.

Disaster Hazard: Dangerous; on contact with acid, acid fumes, water or steam, they will produce toxic and flammable vapors.

Countermeasures

Ventilation Control: Section 2.

Personal Hygiene: Section 3.

Storage and Handling: Section 7.

Shipping regulations: Section 11.

  Coast Guard Classification: Poison B; poison label.

  MCA warning label.

CYANIDE OR CYANIDE MIXTURES, DRY

Shipping Regulations: Section 11.

  I.C.C.: Poison B, poison label, 200 pounds.

  IATA: Poison B, poison label, 12 kilograms (passenger), 95 kilograms (cargo).

CYANOACETAMIDE

General Information

Synonyms: Nitrilomalonamide; propionamide nitrile.

Description: White powder.

Formula: $CNCH_2CONH_2$.

Constants: Mol wt: 84.08, mp: 119°C, bp: decomposes.

Hazard Analysis and Countermeasures

See cyanides.

CYANOACETIC ACID

General Information

Synonym: Malonic acid mononitrile.

Description: White, deliquescent needles.

Formula: $CNCH_2COOH$.

Constants: Mol wt: 85.06, mp: 66°C, bp: 108°C at 0.15 mm.

TOXIC HAZARD RATING CODE (For detailed discussion, see Section 1.)

0 NONE: (a) No harm under any conditions; (b) Harmful only under unusual conditions or overwhelming dosage.

1 SLIGHT: Causes readily reversible changes which disappear after end of exposure.

2 MODERATE: May involve both irreversible and reversible changes; not severe enough to cause death or permanent injury.

3 HIGH: May cause death or permanent injury after very short exposure to small quantities.

U UNKNOWN: No information on humans considered valid by authors

Hazard Analysis and Countermeasures
See cyanides.

CYANOACETONITRILE
Hazard Analysis
Toxicology: Details unknown. Animal experiments show low toxicity since the CN group here is stable.
Disaster Hazard: Dangerous; when heated to decomposition, it emits highly toxic fumes.

CYANOAURIC ACID
General Information
Synonym: Auric cyanide.
Description: Tablets.
Formula: $HAu(CN)_4 \cdot 3H_2O$
Constants: Mol wt: 356.33, mp: 50°C, bp: decomposes.
Hazard Analysis and Countermeasures
Highly toxic. See cyanides and gold compounds.

CYANOCOBALTIC III ACID
General Information
Synonym: Cobalticyanic acid.
Description: Colorless needles; deliquescent.
Formula: $[H_3Co(CN)_6]_2H_2O$
Constants: Mol wt: 454.16, mp: decomposes at 100°C.
Hazard Analysis and Countermeasures
See cyanides and cobalt compounds.

N-CYANODIALLYLAMINE. See diallyl cyanamide.

N-CYANODIETHYLAMINE. See diethyl cyanamide.

CYANODIMETHYLAMINOETHOXYPHOSPHINE OXIDE. See tabun.

2-(2-CYANO ETHOXY) ETHYL ACETATE
Hazard Analysis
Toxic Hazard Rating: U. Limited animal experiments suggest high toxicity. See also nitriles.
Disaster Hazard: Dangerous. See nitriles.

4-CYANOETHOXY-2-METHYL-2-PENTANOL
Hazard Analysis
Toxic Hazard Rating: U. Limited animal experiments suggest toxicity less than that of cyanides. See also nitriles.
Disaster Hazard: Dangerous. See nitriles.

CYANOETHYL ACRYLATE
General Information
Description: Soluble in water.
Formula: $CH_2COOCH_2CH_2CN$.
Constants: Mol wt: 80, d: 1.069, bp: polymerizes when heated, fp: −16.9°C, flash p: 255°F (C.O.C.), vap. d.: 4.3.
Hazard Analysis
Toxic Hazard Rating: U. Limited animal experiments suggest high toxicity and irritation. See also nitriles.
Fire Hazard: Slight, when exposed to heat or flame.
Disaster Hazard: Dangerous. See nitriles.
Countermeasures
To Fight Fire: Water, foam (Section 6).

n-(2-CYANO ETHYL) CYCLO HEXYL AMINE
General Information
Description: Insoluble in water.
Formula: $C_6H_{11}NHC_2H_4CN$.
Constants: Mol wt: 152, flash p: 255°F (O.C.), d: 0.9, vap. d.: 5.2.
Hazard Analysis
Fire Hazard: Slight, when exposed to heat or flame.
Disaster Hazard: Dangerous. See nitriles.
Countermeasures
To Fight Fire: Water, foam (Section 6).

3-CYANOETHYL PROPIONATE
General Information
Formula: $CH_3CH_2COOCH_2CH_2CN$.

Constant: Mol wt: 127.
Hazard Analysis
Toxic Hazard Rating: U. Limited animal experiments suggest moderate toxicity. See also nitriles.
Disaster Hazard: Dangerous. See nitriles.

CYANOFORMIC CHLORIDE
General Information
Description: Oily liquid.
Formula: $COCNCl$.
Constants: Mol wt: 89.49, bp: 126–128°C at 750 mm.
Hazard Analysis and Countermeasures
See cyanides and chlorides.

CYANOGEN
General Information
Synonyms: Ethane dinitrile; prussite.
Description: Colorless gas, pungent odor.
Formula: $NCCN$.
Constants: Mol wt: 52.04, mp: −34.4°C, bp: −21.0°C, d: 0.866 at 17°/4°C, lel: 6%, uel: 32%, vap. d.: 1.8.
Hazard Analysis
Toxicity: Highly toxic. See cyanides.
Disaster Hazard: Dangerous; when heated to decomposition or on contact with acid, acid fumes, water or steam, it will react to produce highly toxic fumes.
Countermeasures
To Fight Fire: Stop flow of gas (Section 6).
Storage and Handling: Section 7.
Shipping Regulations: Section 11.
 I.C.C.: Poison A; poison gas label, not accepted.
 Coast Guard Classification: Poison A; poison gas label.
 IATA: Poison A, not acceptable (passenger and cargo).

CYANOGENAMIDE. See cyanamide.

CYANOGEN BROMIDE
General Information
Synonym: Bromine cyanide.
Description: Colorless needles.
Formula: $BrCN$.
Constants: Mol wt: 105.93, mp: 52°C, bp: 61.6°C, d: 2.015 at 20°/4°C, vap. press.: 100 mm at 22.6°C.
Hazard Analysis and Countermeasures
See cyanides and bromides.
Shipping Regulations: Section 11.
 I.C.C.: Poison B, poison label, 25 pounds.
 IATA: Poison B, poison label, 1 kilogram (passenger), 12 kilograms (cargo).

CYANOGEN CHLORIDE
General Information
Synonym: Chlorine cyanide; CK.
Description: Colorless liquid or gas; lacrimatory and irritating odor.
Formula: $CNCl$.
Constants: Mol wt: 61.48, mp: −6.5°C, bp: 13.1°C, d: 1.218 at 4°/4°C, vap. press.: 1010 mm at 20°C, vap. d.: 1.98.
Hazard Analysis
Toxicity: An insecticide. Highly toxic. See cyanides and hydrochloric acid.
Disaster Hazard: Highly dangerous; when heated to decomposition or on contact with water or steam, it will react to produce highly toxic and corrosive fumes.
Countermeasures
Storage and Handling: Section 7.
Shipping Regulations: Section 11.
 I.C.C.: Poison A; poison gas label, not accepted.
 Coast Guard Classification: Poison A; poison gas label.
 IATA: Poison A, not acceptable (passenger and cargo).

CYANOGEN FLUORIDE
General Information
Description: Colorless gas.
Formula: CNF.

Constants: Mol wt: 45.02, mp: sublimes at $-72°$C, bp: $-72.6°$C.

Hazard Analysis and Countermeasures
See cyanides and fluorides.

CYANOGEN IODIDE
General Information
Synonym: Iodine cyanide.
Description: Colorless needles.
Formula: CNI.
Constants: Mol wt: 152.94, mp: 146.5°C, vap. press.: 1 mm at 25.2°C.

Hazard Analysis and Countermeasures
See cyanides and iodides.

CYANOGEN SULFIDE
General Information
Synonym: Cyanogen thiocyanate.
Description: Crystals.
Formula: NCSCN.
Constants: Mol wt: 84.1, mp: 65°C, bp: sublimes at 30–40°C.

Hazard Analysis and Countermeasures
See cyanides and thiocyanates.

CYANOGEN THIOCYANATE. See cyanogen sulfide.

CYANOGUANIDINE. See dicyandiamide.

6-CYANOHEXANOIC ACID
General Information
Formula: $CH_2CN(CH_2)_4COOH$.
Constant: Mol wt: 141.

Hazard Analysis
Toxic Hazard Rating: U. Limited animal experiments suggest low toxicity. See also nitriles.
Disaster Hazard: Dangerous. See nitriles.

CYANOMETHYL ACETATE
General Information
Synonym: Methyl cyanoethanoate.
Description: Colorless liquid.
Formula: $CNCH_2COOCH_3$.
Constants: Mol wt: 99.09, mp: $-22.5°$C, bp: 200°C, d: 1.123 at 15°C.

Hazard Analysis and Countermeasures
See cyanides.

N-CYANOMETHYL MORPHOLINE
General Information
Formula: $OCH_2CH_2NHCHCHCH_2CN$.
Constant: Mol wt: 125.

Hazard Analysis
Toxic Hazard Rating: U. Limited animal experiments suggest moderate toxicity. See also nitriles and morpholine.
Disaster Hazard: Dangerous. See nitriles.

3-CYANOPYRIDINE
General Information
Description: Gray crystals.
Formula: $C_6H_4N_2$.
Constants: Mol wt: 104.1, mp: 47–49°C, bp: 83–84°C at 10 mm.

Hazard Analysis
Toxicity: Details unknown; probably highly toxic.
Fire Hazard: Slight, when exposed to heat or flame (Section 6).

Disaster Hazard: Dangerous; see cyanides; it can react with oxidizing materials.

Countermeasures
Storage and Handling: Section 7.

CYANOTRIAMIDE. See melamine.

CYANOVINYL ACETATE
General Information
Formula: $CH_3COO(CNCH:CH_2)$.
Constant: Mol wt: 112.

Hazard Analysis
Limited animal experiments suggest high toxicity. See also nitriles.
Disaster Hazard: Dangerous. See nitriles.

CYANURIC ACID
General Information
Synonyms: Sym-triazinetriol; tricyanic acid.
Description: Crystals.
Formula: $C_3H_3N_3O_3$.
Constants: Mol wt: 129.1, mp: $>360°$C, d: 2.500 at 20°/4°C.

Hazard Analysis
Toxicity: Highly toxic. See cyanides.
Disaster Hazard: Dangerous; see cyanides.

Countermeasures
Storage and Handling: Section 7.

CYANURIC CHLORIDE
General Information
Synonyms: Trichloro-s-triazine; trichlorocyanidine; tricyanogen chloride.
Description: Monoclinic, colorless crystals; pungent odor.
Formula: $C_3Cl_3N_3$.
Constants: Mol wt: 184.43, mp: 145.8°C, bp: 190°C, d: 1.32 at 20°/4°C, vap. press.: 2 mm at 70°C, vap. d.: 6.36.

Hazard Analysis
Toxic Hazard Rating:
  Acute Local: Irritant 2; Allergen 2; Ingestion 2; Inhalation 2.
  Acute Systemic: U.
  Chronic Local: Allergen 2.
  Chronic Systemic: U.
Toxicity: Has been reported as causing irritation of mucous membranes and disturbed rhythm of the heart in humans.
Disaster Hazard: Dangerous; see chlorides and nitriles.

Countermeasures
Ventilation Control: Section 2.
Personnel Protection: Section 3.
Storage and Handling: Section 7.

CYANURIC TRIAZIDE
Hazard Analysis
Toxicity: Details unknown. See also cyanides and azides.
Explosion Hazard: Severe, when shocked or exposed to heat.
Disaster Hazard: Dangerous, shock and heat will explode it; when heated to decomposition or on contact with acid or acid fumes, it emits highly toxic fumes.

Countermeasures
Storage and Handling: Section 7.

CYCLETHRIN. See 3-(2-cyclopentenyl)-2-methyl-4-oxo-2-cyclopentenyl ester of chrysanthemum-monocarboxylic acid.

TOXIC HAZARD RATING CODE (For detailed discussion, see Section 1.)

0 NONE: (a) No harm under any conditions; (b) Harmful only under unusual conditions or overwhelming dosage.

1 SLIGHT: Causes readily reversible changes which disappear after end of exposure.

2 MODERATE: May involve both irreversible and reversible changes; not severe enough to cause death or permanent injury.

3 HIGH: May cause death or permanent injury after very short exposure to small quantities.

U UNKNOWN: No information on humans considered valid by authors.

"CYCLINE OIL." See mineral oil, vegetable oil.

CYCLOBUTANE
General Information
Synonym: Tetramethylene.
Description: A gas.
Formula: C_4H_8.
Constants: Mol wt: 56.10, mp: $-50°C$, bp: 12.9°C, flash p: below 50°F (C.C.), d: 0.708 at 11°C, vap. d.: 1.93.
Hazard Analysis
Toxicity: Details unknown. May be a simple asphyxiant. See also cycloparaffins.
Fire Hazard: Dangerous when exposed to heat or flame; can react with oxidizing materials.
Spontaneous Heating: No.
Countermeasures
To Fight Fire: Stop flow of gas. Carbon dioxide, dry chemical or water spray (Section 6).
Storage and Handling: Section 7.

CYCLOBUTENE
General Information
Synonym: Cyclobutylene.
Description: Gas.
Formula: C_4H_6.
Constants: Mol wt: 54.1, bp: 2.4°C, d: 0.733 at 0°/4°C.
Hazard Analysis
Toxicity: Details unknown. May be a simple asphyxiant.
Fire Hazard: Dangerous, when exposed to heat or flame; it can react with oxidizing materials.
Countermeasures
Storage and Handling: Section 7.

CYCLOBUTYLENE. See cyclobutene.

CYCLOHEPTANE
General Information
Synonym: Suberane.
Description: An oil.
Formula: C_7H_{14}.
Constants: Mol wt: 98.2, mp: $-12°C$, bp: 244°C, flash p: below 100°F, d: 0.8099 at 20°/4°C, vap. d.: 3.3.
Hazard Analysis
Toxicity: Details unknown. See cycloparaffins.
Fire Hazard: Dangerous, when exposed to heat or flame; it can react with oxidizing materials.
Countermeasures
To Fight Fire: Foam, carbon dioxide, dry chemical or carbon tetrachloride (Section 6).
Storage and Handling: Section 7.

CYCLOHEPTANONE
General Information
Synonym: Suberone.
Description: Liquid; nearly insoluble in water.
Formula: $C_7H_{12}O$.
Constants: Mol wt: 112.2, bp: 181°C, d: 0.9490 at 20°/4°C.
Hazard Analysis
Toxicity: Unknown.
Toxicology: Limited experimental studies show moderate toxicity; mainly CNS depression.

CYCLOHEXANE
General Information
Synonyms: Hexahydrobenzene; hexamethylene.
Description: Colorless mobile liquid, pungent odor.
Formula: C_6H_{12}.
Constants: Mol wt: 84.16, mp: 6.5°C, bp: 80.7°C, fp: 4.6°C, flash p: $-4°F$, ulc: 90–95, lel: 1.3%, uel: 8.4%, d: 0.7791 at 20°/4°C, autoign. temp.: 500°F, vap. press.: 100 mm at 60.8°C, vap. d.: 2.90.
Hazard Analysis
Toxic Hazard Rating:
Acute Local: Irritant 2.
Acute Systemic: Inhalation 2.

Chronic Local: U.
Chronic Systemic: Inhalation 1.
Toxicology: Can cause skin irritation.
TLV: ACGIH (recommended) 300 parts per million of air; 1050 milligrams per cubic meter of air.
Caution: May act as a simple asphyxiant. See also cycloparaffins.
Fire Hazard: Dangerous; when exposed to heat or flame, it can react with oxidizing materials.
Spontaneous Heating: No.
Explosion Hazard: Moderate, in the form of vapor when exposed to flame (Section 7).
Countermeasures
Ventilation Control: Section 2.
Storage and Handling: Section 7.
To Fight Fire: Foam, carbon dioxide, dry chemical or carbon tetrachloride (Section 6).
Shipping Regulations: Section 11.
  I.C.C.: Flammable liquid; red label, 10 gallons.
  Coast Guard Classification: Inflammable liquid; red label.
  MCA warning label. ·
  IATA: Flammable liquid, red label, 1 liter (passenger), 40 liters (cargo).

cis-1,2-CYCLOHEXANEDICARBOXYLIC ANHYDRIDE.
See cis-hexahydrophthalic anhydride.

CYCLOHEXANESULFAMIC ACID
General Information
Synonym: N-Cyclohexylsulfamic acid.
Description: Crystals. Sweet-sour taste. Slightly soluble in water.
Formula: $C_6H_{13}NO_3S$.
Constants: Mol wt: 179.2, mp: 170°C.
Hazard Analysis
Toxic Hazard Rating:
Acute Local: U.
Acute Systemic: Ingestion 1.
Chronic Local: U.
Chronic Systemic: Ingestion 1.
Toxicology: Large doses by mouth have a laxative effect.
Disaster Hazard: Dangerous; see sulfonates.

CYCLOHEXANOL
General Information
Synonyms: Hexahydrophenol.
Description: Colorless needles or viscous liquid; hygroscopic; camphor-like odor.
Formula: $C_6H_{11}OH$.
Constants: Mol wt: 100.16, mp: $-9°C$ (sets to a glass), bp: 161.5°C, flash p: 154°F (C.C.), d: 0.9449 at 25°/4°C, vap. press.: 1 mm at 21.0°C, vap. d.: 3.45, autoign. temp.: 572°F.
Hazard Analysis
Toxic Hazard Rating:
Acute Local: Irritant 1; Inhalation 1.
Acute Systemic: Ingestion 2; Inhalation 2; Skin Absorption 1.
Chronic Local: Irritant 1; Ingestion 1; Inhalation 1.
Chronic Systemic: U.
Toxicity: Narcotic in high concentrations. Has caused damage to kidneys, liver, and blood vessels in experimental animals.
TLV: ACGIH (recommended); 200 milligrams per cubic meter of air; 50 parts per million in air.
Fire Hazard: Moderate; when exposed to heat or flame, it can react with oxidizing materials.
Spontaneous Heating: No.
Countermeasures
Ventilation Control: Section 2.
Personnel Protection: Section 3.
To Fight Fire: Alcohol Foam, foam, carbon dioxide, dry chemical or carbon tetrachloride (Section 6).
Storage and Handling: Section 7.
MCA warning label.

CYCLOHEXANOL ACETATE
General Information
Description: Pale-yellow liquid.
Formula: $C_6H_{11}OOCCH_3$.
Constants: Mol wt: 142.19, bp: 177°C, d: 0.996, vap. d.: 4.9.
Hazard Analysis
Toxicity: Details unknown. See cycloparaffins.
Fire Hazard: Slight; (Section 6.)
Countermeasures
Storage and Handling: Section 7.

CYCLOHEXANONE
General Information
Synonym: Ketohexamethylene; pimelic ketone.
Description: Colorless liquid, acetone-like odor.
Formula: $CO(CH_2)_4CH_2$.
Constants: Mol wt: 98.14, mp: −45.0°C, bp: 155.6°C,
  ulc: 35–40, lel: 1.1% at 100°C, flash p: 111°F, d:
  0.9478 at 20°/4°C, autoign. temp.: 788°F, vap.
  press.: 10 mm at 38.7°C, vap. d.: 3.4.
Hazard Analysis
Toxic Hazard Rating:
  Acute Local: Irritant 2.
  Acute Systemic: Ingestion 1; Inhalation 1.
  Chronic Local: Irritant 1.
  Chronic Systemic: Ingestion 1.
Toxicology: Experimentally, animals exposed to 0.4% of
  cyclohexanone may be killed after prolonged exposure.
  The oral M.L.D. for rabbits is between 1.0 and 1.6
  grams per kilogram. In man, 50 ppm can cause irrita-
  tion of the eyes and throat. Mild narcotic properties
  have also been described. See also cycloparaffins.
TLV: ACGIH (recommended); 50 parts per million in air;
  200 milligrams per cubic meter of air.
Fire Hazard: Moderate, when exposed to heat or flame;
  can react vigorously with oxidizing materials.
Spontaneous Heating: No.
Explosion Hazard: Slight, in its vapor form, when exposed
  to flame.
Countermeasures
Ventilation Control: Section 2.
To Fight Fire: Alcohol foam, dry chemical, carbon tetra-
  chloride or carbon dioxide. Section 6.
Personnel Protection: Section 3.
Storage and Handling: Section 7.
Shipping Regulations: Section 11.
  MCA warning label.

CYCLOHEXANONE-Δ
General Information
Description: Liquid.
Formula: C_6H_8O.
Constants: Mol wt: 96.12, bp: 155.5°C, flash p: 93°F
  (C.C.), vap. d.: 3.31, vap. press.: 4 mm at 20°C.
Hazard Analysis
Toxicity: Details unknown. Can cause a dermatitis upon
  skin contact (Section 9). Irritating to eyes, skin and
  mucous membranes. Can damage the liver and kidneys.
  See also cycloparaffins.
Fire Hazard: Moderate, when exposed to flame and heat;
  it can react with oxidizing materials.
Spontaneous Heating: No.
Countermeasures
To Fight Fire: Carbon dioxide, dry chemical or carbon
  tetrachloride (Section 6).
Ventilation Control: Section 2.

Shipping Regulations: Section 11.
  MCA warning label.

CYCLOHEXANONE PEROXIDE (paste with dibutyl phthalate).
General Information
Description: Off white thick paste, not soluble in water;
  soluble in common organic solvents.
Formula: $C_{12}H_{22}O_5$ + dibutyl phthalate.
Constant: Mol wt: 246.3.
Hazard Analysis and Countermeasures
See peroxides, organic.
Shipping Regulations:
  I.C.C.: Oxidizing material N.O.S., yellow label, 25
    pounds.
  Express and Freight: Yellow label; chemicals N.O.I.B.N.
  Rail Express: Maximum weight 25 pounds per shipping
    case.
  Parcel Post: Prohibited.
  IATA (not over 50% concentration): Oxidizing ma-
    terial, yellow label, 1 kilogram (passenger), 12 kilo-
    grams (cargo).
  IATA (Over 50% but not exceeding 85% concentration):
    Oxidizing material, yellow label, not acceptable (pas-
    senger), 12 kilograms (cargo).
  IATA (Over 85% concentration): Oxidizing material,
    not acceptable (passenger and cargo).

CYCLOHEXENE
General Information
Synonym: 1,2,3,4-Tetrahydrobenzene.
Description: Colorless liquid.
Formula: $CH_2CH_2CH_2CH_2CHCH$.
Constants: Mol wt: 82.14, bp: 83°C, fp: −103.7°C,
  flash p: < 20°F, d: 0.8102 at 20°/4°C, vap. press.:
  160 mm at 38°C, vap. d.: 2.8.
Hazard Analysis
Toxic Hazard Rating:
  Acute Local: U.
  Acute Systemic: Inhalation 2.
  Chronic Local: U.
  Chronic Systemic: U.
TLV: ACGIH (recommended) 300 parts per million of air;
  1015 milligrams per cubic meter of air.
Fire Hazard: Dangerous, when exposed to flame; can react
  with oxidizing materials.
Disaster Hazard: Dangerous! Keep away from heat and
  open flame!
Countermeasures
Ventilation Control: Section 2.
Storage and Handling: Section 7.
To Fight Fire: Foam, carbon dioxide, dry chemical or
  carbon tetrachloride (Section 6).

4-CYCLOHEXENE-1-CARBOXALDEHYDE
General Information
Synonym: 1,2,5,6-Tetrahydrobenzaldehyde.
Description: Liquid, slightly soluble in water.
Formula: $CH_2CH:CHCH_2CH_2CHCHO$.
Constants: D: 0.9721, bp: 164.2°C, fp: −108°C, flash
  p: 135°F.
Hazard Analysis
Toxic Hazard Rating: U. Limited animal experiments sug-
  gest moderate toxicity. See also aldehydes.
Fire Hazard: Moderate, when exposed to heat or flame.

TOXIC HAZARD RATING CODE (For detailed discussion, see Section 1.)

0 NONE: (a) No harm under any conditions; (b) Harmful only under unusual conditions or overwhelming dosage.

1 SLIGHT: Causes readily reversible changes which disappear after end
of exposure.

2 MODERATE: May involve both irreversible and reversible changes;
not severe enough to cause death or permanent injury.

3 HIGH: May cause death or permanent injury after very short exposure
to small quantities.

U UNKNOWN: No information on humans considered valid by authors.

CYCLOHEXENE OXIDE
General Information
Description: Clear liquid.
Formula: $C_6H_{10}O$.
Constants: Mol wt: 98.14, bp: 129.5°C, flash p: 81°F, d: 0.9678 at 25°/4°C, vap. d.: 3.5.
Hazard Analysis
Toxicity: Details unknown. See also cyclohexene.
Fire Hazard: Dangerous, when exposed to heat or flame; it can react with oxidizing materials (Section 6).
Explosion Hazard: Unknown.
Countermeasures
Storage and Handling: Section 7.

CYCLOHEXENYL TRICHLOROSILANE
Shipping Regulation: Section 11.
IATA: Corrosive liquid, white label, not acceptable (passenger), 40 liters (cargo).
I.C.C.: Corrosive liquid, white label, 10 gallons.

CYCLOHEXIMIDE
General Information
Synonym: Actidione.
Description: Crystals; moderately soluble in water. Soluble in chloroform, ether and acetone.
Formula: $C_{15}H_{23}NO_4$.
Constants: Mol wt: 281.3, mp: 120°C.
Hazard Analysis
Toxic Hazard Rating:
Acute Local: Irritant 2.
Acute Systemic: U.
Chronic Local: U.
Chronic Systemic: U.
Toxicology: LD_{50} intravenously in mice 150 mg/kg.

CYCLOHEXYL ACETATE
General Information
Synonym: Hexalin acetate.
Formula: $CH_3COOC_6H_{11}$.
Constants: Mol wt: 142.19, bp: 177°C, flash p: 136°F, vap. d.: 4.9, autoign. temp.: 633°F, d: 1.0−.
Hazard Analysis
Toxic Hazard Rating:
Acute Local: Irritant 1; Inhalation 1.
Acute Systemic: Inhalation 2.
Chronic Local: Irritant 1.
Chronic Systemic: Inhalation 1.
Toxicology: Is said to be considerably more toxic than amyl acetate.
Fire Hazard: Moderate, when exposed to flame; it can react with oxidizing materials.
Spontaneous Heating: No.
Countermeasures
To Fight Fire: Foam, carbon dioxide, dry chemical or carbon tetrachloride (Section 6).
Storage and Handling: Section 7.

CYCLOHEXYL ALCOHOL. See cyclohexanol.

CYCLOHEXYLAMINE
General Information
Synonyms: Hexahydroaniline; aminocyclohexane.
Description: Liquid, strong fishy odor.
Formula: $C_6H_{13}N$.
Constants: Mol wt: 99.2, mp: −17.7°C, bp: 134.5°C, flash p: 90°F (O.C.), d: 0.865 at 25°/25°C, autoign. temp.: 560°F. vap. d.: 3.42.
Hazard Analysis
Toxic Hazard Rating:
Acute Local: Irritant 2.
Acute Systemic: Ingestion 3; Inhalation 3; Skin Absorption 3.
Chronic Local: Irritant 2.
Chronic Systemic: Ingestion 2; Inhalation 2; Skin Absorption 2.
Toxicology: May cause dermatitis and convulsions.

Fire Hazard: Moderate, when exposed to heat or flame.
Disaster Hazard: Dangerous; when heated to decomposition, it emits highly toxic fumes, it can react vigorously with oxidizing materials.
Countermeasures
Ventilation Control: Section 2.
Personnel Protection: Section 3.
To Fight Fire: Alcohol foam, carbon dioxide, dry chemical or carbon tetrachloride (Section 6).
Storage and Handling: Section 7.

CYCLOHEXYLBENZENE. See phenyl cyclohexane.

N-CYCLOHEXYL-2-BENZOTHIAZOLE SULFENAMIDE
General Information
Description: Light tan of buff powder.
Formula: $C_6H_4NSCSNHC_6H_{11}$.
Constants: Mol wt: 264.4, mp: 94°C, d: 1.27 at 25°C.
Hazard Analysis
Toxicity: Details unknown.
Disaster Hazard: Dangerous; see sulfonates.
Countermeasures
Storage and Handling: Section 7.

CYCLOHEXYL CHLORIDE
General Information
Description: Clear liquid.
Formula: $C_6H_{11}Cl$.
Constants: Mol wt: 118.61, mp: −43°C, bp: 142°C, flash p: 89°F, d: 0.9923 at 25°/4°C, vap. d.: 4.0.
Hazard Analysis
Toxicity: See cyclohexane.
Fire Hazard: Moderate, when exposed to heat or flame (Section 6).
Disaster Hazard: Moderately dangerous; see chlorides, it can react with oxidizing materials.
Countermeasures
Ventilation Control (Section 2).
Storage and Handling: Section 7.

2-CYCLOHEXYL CYCLOHEXANOL
General Information
Description: Colorless liquid.
Formula: $C_6H_{11}C_6H_{10}OH$.
Constants: Mol wt: 182.3, bp: 271–277°C, flash p: 270°F, d: 0.977 at 25°/25°C.
Hazard Analysis
Toxicity: See alcohols.
Fire Hazard: Slight, when exposed to heat or flame; it can react with oxidizing materials.
Countermeasures
To Fight Fire: Foam, carbon dioxide, dry chemicals or carbon tetrachloride (Section 6).
Storage and Handling: Section 7.

2-CYCLOHEXYL-4,6-DINITROPHENOL
General Information
Description: Crystals.
Formula: $C_6H_{11}C_6H_2OH(NO_2)_2$.
Constant: Mol wt: 266.23.
Hazard Analysis
Toxicity: See dinitrophenol.
Fire Hazard: See nitrates.
Disaster Hazard: Dangerous; see nitrates, it can react vigorously with oxidizing materials.
Countermeasures
Storage and Handling: Section 7.

CYCLOHEXYL FORMATE
General Information
Formula: $HCOOC_6H_{11}$.
Constant: Mol wt: 128.2.
Hazard Analysis
Toxicology: Is said to resemble amyl acetate in type and intensity of toxicity. See amyl acetate.

CYCLOHEXYL ISOCYANATE
General Information
Formula: $C_6H_{11}NCO$.
Constant: Mol wt: 125.
Hazard Analysis
Toxic Hazard Rating: U.
Disaster Hazard: Dangerous. See isocyanates.
Countermeasures
Shipping Regulations: Section 11.
   IATA: Poison C, poison label, not accepted (passenger), 40 liters (cargo).

CYCLOHEXYLMETHANE. See methyl cyclohexane.

o-CYCLOHEXYLPHENOL
General Information
Description: Nearly white crystals.
Formula: $C_6H_{11}C_6H_4OH$.
Constants: Mol wt: 176.2, mp: 50°C, bp: 169°C at 25 mm, flash p: 270°F, d: 1.018 at 60°/25°C.
Hazard Analysis
Toxicity: Details unknown. See phenol.
Fire Hazard: Slight, when exposed to heat or flame; it can react with oxidizing materials.
Countermeasures
To Fight Fire: Alcohol foam, foam, carbon dioxide, dry chemical or carbon tetrachloride (Section 6).
Storage and Handling: Section 7.

N-CYCLOHEXYLSULFAMIC ACID. See cyclohexane-sulfamic acid.

CYCLOHEXYLTETRATHIO-o-STANNATE
General Information
Description: Solid.
Formula: $(SC_6H_{11})_4Sn$.
Constants: Mol wt: 579.5, mp: 54°C.
Hazard Analysis
Toxicology: Details unknown. See tin compounds.
Disaster Hazard: Dangerous; emits sulfur fumes.

N-CYCLOHEXYL-p-TOLUENE SULFONAMIDE
General Information
Description: Fine white crystals.
Formula: $CH_3C_6H_4SO_2NHC_6H_{11}$.
Constants: Mol wt: 253, mp: 86°C, bp: 350°C, d: 1.125.
Hazard Analysis
Toxicity: Details unknown.
Disaster Hazard: Dangerous; see sulfonates.
Countermeasures
Storage and Handling: Section 7.

CYCLOHEXYLTRICHLOROSILANE
General Information
Formula: $C_6H_{11}SiCl_3$.
Constants: Mol wt: 218.6, bp: 208°C, flash p: 196°F (O.C.), d: 1.2, vap. d.: 7.5.
Hazard Analysis
Toxic Hazard Rating:
   Acute Local: Irritant 2; Ingestion 2; Inhalation 2.
   Acute Systemic: U.
   Chronic Local: U.
   Chronic Systemic: U.
Fire Hazard: Moderate, when exposed to flame.
Disaster Hazard: Dangerous; See chlorosilanes.
Countermeasures
Storage and Handling: Section 7.
To Fight Fire: Foam, carbon dioxide, dry chemical or carbon tetrachloride (Section 6).

Shipping Regulations: Section 11.
   Coast Guard Classification: Corrosive liquid; white label.
   I.C.C.: Corrosive liquid, white label, 10 gallons.
   IATA: Corrosive liquid, white label, not acceptable (passenger), 40 liters (cargo).

CYCLONITE. See cyclotrimethylene trinitramine.

CYCLOOCTATETRAENE
General Information
Description: Liquid.
Formula: C_8H_8.
Constants: Mol wt: 104.14, mp: −4.7°C, bp: 140.6°C, fp: −4.7°C, vap. press.: 7.9 mm at 25°C.
Hazard Analysis
Toxicity: Details unknown. May be a simple asphyxiant.
Fire Hazard: Moderate, when exposed to heat or flame; it can react with oxidizing materials.
Countermeasures
Storage and Handling: Section 7.

CYCLOPARAFFINS
Hazard Analysis
Toxicology: Both the saturated and unsaturated members of the cycloparaffin series are narcotic, and may cause death through respiratory paralysis. For most of the members there appears to be little range between the concentrations causing deep narcosis and those causing death. There is very little information in the literature regarding the chronic effects resulting from exposure of humans to the cycloparaffins. Experimental work with rabbits indicates that barely demonstrable changes in the liver and kidneys may result from exposure to 786 ppm of cyclohexane for 6 hours daily, repeated for 50 days.
Countermeasures
Ventilation Control: Section 2.

CYCLOPENTADIENE-1,3
General Information
Description: Colorless liquid.
Formula: C_5H_6.
Constants: Mol wt: 66.1, mp: −85°C, bp: 42.5°C, d: 0.80475 at 19°/4°C.
Hazard Analysis
Toxicity: Details unknown. Animal experiments suggest moderate toxicity. An insecticide and fungicide.
TLV: ACGIH; 75 parts per million of air; 200 milligrams per cubic meter of air.
Fire Hazard: Moderate, when exposed to heat or flame; it can react with oxidizing materials (Section 6).
Explosion Hazard: Moderate, in the form of gas when exposed to heat or by chemical reaction. It decomposes violently at high temperatures and pressures (Section 7).
Countermeasures
Storage and Handling: Section 7.

CYCLOPENTAMETHYLENE GERMANIUM DICHLORIDE
General Information
Description: Colorless liquid.
Formula: $(CH_2)_5GeCl_2$.
Constants: Mol wt: 213.7, bp: 55–60°C at 12 mm Hg.
Hazard Analysis
Toxicology: Details unknown. See germanium compounds.
Disaster Hazard: Dangerous; see chlorides.

TOXIC HAZARD RATING CODE **(For detailed discussion, see Section 1.)**

0 NONE: (a) No harm under any conditions; (b) Harmful only under unusual conditions or overwhelming dosage.

1 SLIGHT: Causes readily reversible changes which disappear after end of exposure.

2 MODERATE: May involve both irreversible and reversible changes; not severe enough to cause death or permanent injury.

3 HIGH: May cause death or permanent injury after very short exposure to small quantities.

U UNKNOWN: No information on humans considered valid by authors.

CYCLOPENTANE

General Information

Synonym: Pentamethylene.

Description: Colorless liquid.

Formula: $(CH_2)_5$.

Constants: Mol wt: 70.08, bp: 49.3°C, fp: −93.7°C, flash p: < 20°F, d: 0.745 at 20°/4°C, vap. press.: 400 mm at 31.0°C, vap. d.: 2.42.

Hazard Analysis

Toxic Hazard Rating:

Acute Local: U.

Acute Systemic: Ingestion 2; Inhalation 2.

Chronic Local: U.

Chronic Systemic: U.

Toxicology: High concentrations have narcotic action.

Fire Hazard: Dangerous, when exposed to flame; can react with oxidizing materials.

Explosion Hazard: Unknown.

Disaster Hazard: Dangerous! Keep away from heat and open flame!

Countermeasures

Ventilation Control: Section 2.

To Fight Fire: Foam, carbon dioxide, dry chemical or carbon tetrachloride (Section 6).

Personal Hygiene: Section 3.

Storage and Handling: Section 7.

Shipping Regulations: Section 11.

I.C.C.: Flammable liquid; red label, 10 gallons.

Coast Guard Classification: Inflammable liquid; red label.

IATA: Flammable liquid, red label, 1 liter (passenger), 40 liters (cargo).

CYCLOPENTANOL

General Information

Description: Clear liquid.

Formula: C_5H_9OH.

Constants: Mol wt: 86.13, bp: 139.5°C, flash p: 124°F, d: 0.9422 at 25°/4°C.

Hazard Analysis

Toxicity: Details unknown. See also alcohols.

Fire Hazard: Moderate, in the presence of heat or flame; it can react with oxidizing materials.

Countermeasures

Storage and Handling: Section 7.

CYCLOPENTANONE

General Information

Synonyms: Dumasin; ketocyclopentane.

Description: Liquid.

Formula: C_5H_8O.

Constants: Mol wt: 84.1, mp: −58.2°C, bp: 130.6°C, flash p: 79°F, d: 0.9509 at 18°/4°C, vap. d.: 2.3.

Hazard Analysis

Toxicity: See cycloparaffins.

Fire Hazard: Moderate, when exposed to flame; it can react with oxidizing materials.

Countermeasures

To Fight Fire: Alcohol foam, foam, carbon dioxide, dry chemical or carbon tetrachloride (Section 6).

Storage and Handling: Section 7.

CYCLOPENTANONE OXIME

General Information

Description: Solid.

Formula: C_5H_8NOH.

Constants: Mol wt: 99.13, mp: 57.5°C.

Hazard Analysis

Toxicity: Unknown.

Fire Hazard: Slight.

Countermeasures

Storage and Handling: Section 7.

CYCLOPENTENE

General Information

Description: Liquid.

Formula: C_5H_8.

Constants: Mol wt: 68.1, mp: −93.3°C, bp: 44.242°C, fp: −135.2°C, flash p: −20°F, d: 0.77199 at 20°C.

Hazard Analysis

Toxicity: See cycloparaffins.

Fire Hazard: Dangerous, when exposed to flame or heat; it can react with oxidizing materials.

Disaster Hazard: Dangerous! Keep away from heat and open flame!

Countermeasures

Ventilation Control: Section 2.

To Fight Fire: Foam, carbon dioxide, dry chemical or carbon tetrachloride (Section 6).

2-CYCLOPENTENE-1-OL

General Information

Formula: $OHCHCH:CHCH_2CH_2$.

Constant: Mol wt: 84.

Hazard Analysis

Toxic Hazard Rating: U. Limited animal experiments suggest high toxicity.

3-(2-CYCLOPENTENYL)-2-METHYL-4-OXO-CYCLOPENTENYL ESTER OF CHRYSANTHE-MUM-MONOCARBOXYLIC ACID

General Information

Synonym: Cyclethrin.

Hazard Analysis

Toxicity: See pyrethrin I.

CYCLOPENTYL BROMIDE

General Information

Description: Liquid.

Formula: C_5H_9Br.

Constants: Mol wt: 149.04, bp: 137.5°C, flash p: 108°F, d: 1.3866 at 25°/4°C, vap. d.: 5.

Hazard Analysis

Toxicity: See bromides.

Fire Hazard: Moderate, when exposed to heat or flame (Section 6).

Disaster Hazard: Dangerous; see bromides; it can react with oxidizing materials.

Countermeasures

Storage and Handling: Section 7.

CYCLOPENTYL CHLORIDE

General Information

Description: Liquid.

Formula: C_5H_9Cl.

Constants: Mol wt: 104.58, bp: 113.5°C, flash p: 60°F, d: 1.0024 at 25°/4°C, vap. d.: 3.5.

Hazard Analysis

Toxicity: See chlorinated hydrocarbons, aliphatic and aromatic.

Fire Hazard: Dangerous; when exposed to heat or flame (Section 6).

Explosion Hazard: Unknown.

Disaster Hazard: Dangerous; see chlorides; it can react with oxidizing materials.

Countermeasures

Ventilation Control: Section 2.

CYCLOPENTYL ETHER

General Information

Formula: $(C_5H_9)_2O$.

Constant: Mol wt: 154.

Hazard Analysis

Toxic Hazard Rating: U. Limited animal experiments suggest high toxicity. See also ethers.

Disaster Hazard: Unknown. See ethers.

CYCLOPHOSPHAMIDE. See endoxan.

CYCLOPROPANE

General Information

Synonym: Trimethylene.

Description: Colorless gas.

Formula: $CH_2CH_2CH_2$.
Constants: Mol wt: 42.08, mp: $-126.6°C$, bp: $-33.5°C$, lel: 2.4%, uel: 10.4%, d: 1.879 g/l at $0°C$, autoign. temp.: $928°F$.

Hazard Analysis
Toxic Hazard Rating:
  Acute Local: 0.
  Acute Systemic: Inhalation 2.
  Chronic Local: 0.
  Chronic Systemic: U.
Toxicology: High concentrations have narcotic action. Used as a surgical anesthetic.
Fire Hazard: Dangerous; when exposed to heat or flame; can react with oxidizing materials.
Spontaneous Heating: No.
Explosion Hazard: Moderate, in the form of vapor when exposed to heat or flame (Section 7).
Disaster Hazard: Dangerous! Keep away from heat and open flame!

Countermeasures
Ventilation Control: Section 2.
To Fight Fire: Stop flow of gas. Carbon dioxide, dry chemical or water spray (Section 6).
Storage and Handling: Section 7.
Shipping Regulations: Section 11.
  I.C.C.: Flammable gas; red gas label, 300 pounds.
  Coast Guard Classification: Inflammable gas; red gas label.
  IATA: Flammable gas, red label, not acceptable (passenger), 140 kilograms (cargo).

CYCLOPROPYL ETHYL ETHER
General Information
Description: Liquid.
Formula: $C_3H_5OC_2H_5$.
Constant: Mol wt: 86.1.
Hazard Analysis and Countermeasures
See ethers.

CYCLOPROPYL METHYL ETHER
General Information
Synonym: Cypronic ether.
Description: Liquid.
Formula: $C_3H_5OCH_3$.
Constants: Mol wt: 72.1, mp: $-119°C$, bp: $44.7°C$, d: 0.786 at $25°/4°C$.
Hazard Analysis and Countermeasures
See ethers.

CYCLOPROPYL PROPYL ETHER
General Information
Description: Liquid.
Formula: $C_3H_5OC_3H_7$.
Constant: Mol wt: 100.2.
Hazard Analysis and Countermeasures
See ethers.

CYCLOTETRAMETHYLENE OXIDE. See tetrahydrofuran.

CYCLOTRIMETHYLENE TRINITRAMINE
General Information
Synonyms: RDX; cyclonite; hexogen.
Description: White, crystalline powder.
Formula: $C_3H_6N_6O_6$.
Constants: Mol wt: 222.15, mp: $202°C$.
Hazard Analysis
Toxic Hazard Rating:
  Acute Local: Irritant 1.
  Acute Systemic: U.
  Chronic Local: Irritant 1.
  Chronic Systemic: Inhalation 2.
Toxicology: Cases of epileptiform convulsions have been reported from exposure to it.
Fire Hazard: See nitrates.
Explosion Hazard: It is one of the most powerful high explosives in use today. See explosives, high. Has more shattering power than TNT. Is often used mixed with TNT as a bursting charge for aerial bombs, mines and torpedoes. Because it is easily initiated by mercury fulminate it may be used as a booster.
Disaster Hazard: See nitrates.
Countermeasures
Ventilation Control: Section 2.
Personal Hygiene: Section 3.
Storage and Handling: Section 7.
Shipping Regulations: Section 11.
  IATA: Explosive, not acceptable (passenger and cargo).

CYLINDERS, EMPTY
Shipping Regulations: Section 11.
  I.C.C. Classification: See Section 11, § 73.29.
  Coast Guard Classification: Hazardous article.

p-CYMENE
General Information
Synonym: Isopropyl toluene.
Description: Liquid.
Formula: $CH_3C_6H_4CH(CH_3)_2$.
Constants: Mol wt: 134.21, mp: $-68.2°C$, bp: $176°C$, lel: 0.7% at $100°C$, ulc: 30–35, flash p: $117°F$ (C.C.), d: 0.86, autoign. temp.: $817°F$, vap. d.: 4.62, vap. press.: 1 mm at $17.3°C$, flash p (technical): $833°F$, uel (technical): 5.6%.
Hazard Analysis
Toxicity: See toluene.
Fire Hazard: Moderate, when exposed to heat or flame.
Spontaneous Heating: No.
Explosion Hazard: Slight, in the form of vapor (Section 7).
Disaster Hazard: Moderately dangerous; it can react with oxidizing materials.
Countermeasures
Storage and Handling: Section 7.
To Fight Fire: Foam, carbon dioxide, dry chemical or carbon tetrachloride (Section 6).

CYMOGENE. See butane.

CYPREX. See n-dodecyl guanidine acetate.

CYPRONIC ETHER. See cyclopropyl methyl ether.

CYSTEINE
General Information
Synonyms: α-Amino-β-thiolpropionic acid; β-mercapto-alanine.
Description: An amino acid derived from cystine, occurring naturally in the L-form which will be considered here. Colorless crystals; soluble in water, ammonium hydroxide, and acetic acid; insoluble in ether, acetone, benzene, carbon disulfide, and carbon tetrachloride.
Formula: $HSCH_2CH(NH_2)COOH$.
Constant: Mol wt: 121.
Hazard Analysis
Toxic Hazard Rating: U. A nutrient and/or dietary supplement food additive (Section 10).

TOXIC HAZARD RATING CODE (For detailed discussion, see Section 1.)

0 NONE: (a) No harm under any conditions; (b) Harmful only under unusual conditions or overwhelming dosage.

1 SLIGHT: Causes readily reversible changes which disappear after end of exposure.

2 MODERATE: May involve both irreversible and reversible changes; not severe enough to cause death or permanent injury.

3 HIGH: May cause death or permanent injury after very short exposure to small quantities.

U UNKNOWN: No information on humans considered valid by authors.

CYSTINE
General Information
Synonym: β,β'-Dithiobisalanine; di(α-amino-β-thiol-propionic acid).

Description: The chief sulfur containing amino acid of protein white crystalline plates; soluble in water; insoluble in alcohol. Occurrs in Dl, L, and D form. We consider the L and DL forms here.

Formula: $HOOCCH(NH_2)CH_2SSCH_2CH(NH_2)COOH$.

Constants: Mol wt: 240, mp (DL): 260°C, mp (L): 258–261°C.

Hazard Analysis
Toxic Hazard Rating: U. A nutrient and/or dietary supplement food additive (Section 10).

CYTISUS
General Information
Description: A wood dust.

Hazard Analysis
Toxic Hazard Rating:
  Acute Local: Irritant 1; Allergen 1.
  Acute Systemic: U.
  Chronic Local: Allergen 1.
  Chronic Systemic: U.

Fire Hazard: Moderate, when exposed to heat or flame (Section 6).

Explosion Hazard: Slight, when exposed to flame (Section 7).

Countermeasures
Personal Hygiene: Section 3.

CZA. See citrazinic acid.

D

2,4-D. See 2,4-dichlorophenoxyacetic acid.

DACTIN. See 1,3-dichloro-5,5-dimethylhydantoin.

DALMATIAN INSECT POWDER. See pyrethrum flowers.

DAP. See ammonium phosphate, dibasic.

DAPHNIN. See 7,8-dihydroxycoumarin-7-β-D-glucoside.

DATURINE. See atropine.

DBH. See 1,2,3,4,5,6-hexachlorocyclohexane.

DBMC. See di-tert-butyl-m-cresol.

DCMX. See 2,4-dichloro-3,5-xylenol.

DBPC. See di-tert-butyl-p-cresol.

"DC 200 FLUID"
General Information
Synonym: Hexamethyldisiloxane.
Description: Viscous liquid.
Formula: $(CH_3)_3Si-O-Si(CH_3)_3$.
Constants: Mol wt: 162.4, vap. d.: 5.5.
Hazard Analysis
Toxic Hazard Rating:
  Acute Local: Ingestion 2.
  Acute Systemic: U.
  Chronic Local: U.
  Chronic Systemic: U.
Toxicology: See also silanes.
Fire Hazard: Slight; (Section 6).
Countermeasures
Personal Hygiene: Section 3.
Storage and Handling: Section 7.

DDD
General Information
Synonym: 2,2-Bis-(p-chlorophenyl)-1,1-dichloroethane; dichloro diphenyl dichloro ethane.
Description: Crystalline solid.
Formula: $(ClC_6H_4)_2HCCHCl_2$.
Constants: Mol wt: 320.1, mp: 110°C, vap. d.: 11.
Hazard Analysis
Toxic Hazard Rating:
  Acute Local: Ingestion 2.
  Acute Systemic: Ingestion 2; Inhalation 1.
  Chronic Local: U.
  Chronic Systemic: Ingestion 2; Inhalation 1.
Toxicology: Insecticide similar in action to DDT but considered less toxic to mammals. Used as a food additive permitted in the food and drinking water of animals

and/or for the treatment of food producing animals (Section 10).. See also DDT.
Disaster Hazard: Dangerous; when heated to decomposition, it emits highly toxic fumes of chlorides.
Countermeasures
Personnel Protection: Section 3.
Storage and Handling: Section 7.

DDT
General Information
Synonyms: Dichloro-diphenyl-trichloroethane; chlorophenothane, dicophane, 1,1,1-trichloro-2,2-bis (p-chlorophenyl) ethane, "Gesarol"; "Neocoid"; "Persisto Spray"; "Suntobane" and many other trade names.
Description: Colorless crystals or white to slightly off-white powder. Odorless or with slight aromatic odor.
Formula: $(ClC_6H_4)_2CHCCl_3$.
Constants: Mol wt: 354.5, mp: 108.5–109°C.
Hazard Analysis
TLV: ACGIH (recommended) 1 milligram per cubic meter of air. Absorbed via intact skin.
Toxicity: Used as a food additive permitted in the food and drinking water of animals and/or for the treatment of food producing animals. Also a food additive permitted in food for human consumption (Section 10). Note: DDT is a common air contaminant (Section 4).
Toxicology: DDT is readily absorbed from the intestinal tract and, if it occurs in the air in the form of an aerosol or dust, it may be taken into the lung and readily absorbed. DDT is not, however, absorbed from the skin unless it is in solution. Solutions are absorbed from the skin and, by the same token, emulsions are absorbed to some extent. Likewise, fats and oils from whatever source increase the absorption of DDT from the intestine. DDT acts on the central nervous system, but the exact mechanism of this action either in man or in animals has not been elucidated. Large doses of DDT also induce nausea and/or diarrhea in man; however, whether this is a central or local action is not yet clear. Chronically, DDT produces microscopic changes in the liver and kidneys in some experimental animals. This has not been demonstrated in man. DDT is secreted in the milk and, as an acid derivative is excreted in the urine of rabbits, dogs and man. DDT and certain of its degradation products, particularly DDE, are stored in fat. Such storage results either from a single large dose or from repeated small doses. DDT stored in the fat is at least largely inactive since a greater total dose may be stored in an experimental animal than is sufficient as a lethal dose for that same animal if given at one time. A study based on 75 human cases reported an average of 5.3 ppm of DDT stored in the fat. A higher content of

---

TOXIC HAZARD RATING CODE *(For detailed discussion, see Section 1.)*

0 NONE: (a) No harm under any conditions; (b) Harmful only under unusual conditions or overwhelming dosage.

1 SLIGHT: Causes readily reversible changes which disappear after end of exposure.

2 MODERATE: May involve both irreversible and reversible changes; not severe enough to cause death or permanent injury.

3 HIGH: May cause death or permanent injury after very short exposure to small quantities.

U UNKNOWN: No information on humans considered valid by authors.

DDT and its derivatives (up to 434 ppm of DDE and 648 ppm of DDT) was found in workers who had very extensive exposure. Without exception, the samples were taken from persons who were either asymptomatic or suffering from some disease completely unrelated to DDT. Careful hospital examination of workers who had been very extensively exposed and who had volunteered for examination revealed no abnormality which could be attributed to DDT. Much higher levels than have been found in man have been observed in the fat of experimental animals which were apparently asymptomatic. DDT stored in the fat is eliminated only very gradually when further dosage is discontinued. After a single dose, the secretion of DDT in the milk and its excretion in the urine reach their height within a day or two and continue at a lower level thereafter.

Dangerous Acute Dose in Man. A dose of 20 g has proved highly dangerous though not fatal to man. This dose was taken by five persons who vomited an unknown portion of the material and even so recovered only incompletely after 5 weeks. Smaller doses produced less important symptoms with relatively rapid recovery. Experimental ingestion of 1.5 g resulted in great discomfort and moderate neurological changes including paraesthesia, tremor, moderate ataxia, exaggeration of part of the reflexes, headache, and fatigue. Vomiting followed only after 11 hours. Recovery was complete on the following day. The fatal dose of DDT for man is not known. Judging from the literature, no one has ever been killed by DDT in the absence of other insecticides and/or a variety of toxic solvents. However, these common solvent formulations are highly fatal when taken in small doses, partly because of the toxicity of the solvent, and perhaps because of the increased absorbability of the DDT; several fatal cases in man have been reported.

Dangerous Chronic Dose in Man. Even less is known of the hazard of chronic DDT poisoning. It is known that certain experimental animals fed diets containing one part of DDT per million store the compound in their fat. The storage of DDT in man has been mentioned above. The exact significance of these findings is not known and their further investigation is of the greatest importance. The consistent lack of any demonstrable pathology in those occasional cases where high levels of DDT are present in fat would suggest that chronic exposure to low DDT residues on food would not be likely to lead to tissue damage. Human volunteers have ingested up to 35 mg/day for 21 months with no ill effects.

Signs and Symptoms of Poisoning in Man. In patients who ate substantial doses of DDT in flour, the symptoms observed were vomiting, numbness and partial paralysis of the extremities, mild convulsions, loss of proprioception and vibratory sensation of the extremities, and hyperactive knee jerk reflexes. Symptoms appeared in 30 to 60 minutes after eating the DDT. The paralysis and numbness were most evident in the most distal portions of the extremities, and their intensity was directly proportional to the amount of DDT ingested. All the patients were apprehensive and excited; respiration was moderately rapid; pulse remained slow to normal. The immediate protective mechanism in man, following substantial doses, is vomiting. With smaller doses, nausea and vomiting are less prominent, but diarrhea has been observed. Signs and symptoms of chronic poisoning in man are unknown, although, judging from the observed microscopic changes in experimental animals, liver and kidney dysfunctions should be looked for. The primary irritancy of DDT is practically nil, and it has little or no tendency to produce allergy. Dermatitis induced by DDT has occasionally been reported, but these reports are unconfirmed; nevertheless,

the phenomenon should be expected to occur in rare instances (Section 9).

Laboratory Findings: Laboratory findings are essentially negative except for the presence of DDT which may be quantitatively measured in stomach contents, urine, or tissues.

Countermeasures

Treatment of Poisoning: Depending on the condition of the patient, attention should first be given to the sedation or to the removal of poison which may have been taken internally. Stomach lavage and saline laxatives may be used. Oil laxatives should be avoided; they promote absorption of DDT and of many organic solvents. The five drugs of choice, arranged roughly in order of their effectiveness, are phenobarbital, pentobarbital, paraldehyde, urethane, and calcium gluconate. Phenobarbital, which has been used in doses up to 0.7 g per day in epilepsy, and pentobarbital (0.25 to 0.5 g) are the barbiturates known to control convulsions of central origin. Paraldehyde (average dosage 15 cc orally, 1 cc undiluted intravenously, 35 cc rectally in normal saline) controls the convulsions of DDT-poisoned animals. Urethane (human dosage 1 to 4 g) has proved very effective in rats, but it should be remembered that the hypnotic and narcotic effects of urethane are not correspondingly high in man. Urethane has an added advantage, however, of being tolerated in the young and the aged. The object of sedation is not to induce sleep but to restore a relative calm; however, the proper dosage in the presence of poisoning may be so large that it would induce anesthesia if poisoning were not present.

Calcium gluconate has been used less than the other antidotes, but it is reported to control DDT-induced convulsions in several animals. Since its mechanism of action is entirely different, it may be used in addition to sedatives. Epinephrine is contraindicated.

Shipping Regulations: Section 11.
IATA: Other restricted articles, class A; no label required; no limit (passenger and cargo).

DDVP. See dimethyldichlorovinyl phosphate.

DEAC. See diethyl aluminum chloride.

DEADLY NIGHTSHADE
Source of the alkaloids, atropine and belladonine.

DEAPASIL. See p-aminosalicylic acid.

DECABORANE
General Information
Synonyms: Boron hydride; decaboron tetradecahydride.
Description: Colorless needles.
Formula: $B_{10}H_{14}$.
Constants: Mol wt: 122.3, mp: 99.7°C, bp: 213°C, d: 0.94 (solid), 0.78 (liquid at 100°C), vap. press.: 19 mm at 100°C.
Hazard Analysis and Countermeasures
See boron hydrides.
TLV: ACGIH (accepted); 0.05 parts per million of air; 0.3 milligrams per cubic meter of air. Absorbed via intact skin.
Shipping Regulations: Section 11.
I.C.C.: Flammable solid, yellow label, 25 pounds.
IATA: Flammable solid, yellow label, not acceptable (passenger), 12 kilograms (cargo).

DECABORON TETRADECAHYDRIDE. See decaborane.

DECAHYDRONAPHTHALENE. See decalin.

"DECALIN"
General Information
Synonym: Decahydronaphthalene.
Description: Water-white liquid.

Formula: $C_{10}H_{18}$.
Constants: Mol wt: 138.3, mp (cis) $-43.3°C$, (trans) $-30.7°C$, bp (cis) 194.6°C, (trans) 186.7°C, flash p: 136°F (C.C.), autoign. temp.: 482°F, vap. press.: (cis) 1 mm at 22.5°C, (trans) 10 mm at 47.2°C, d: 0.8963, vap. d.: 4.76, lel: 6.7% at 212°F, uel: 4.9% at 212°F.

Hazard Analysis
Toxic Hazard Rating:
  Acute Local: Irritant 2; Ingestion 2.
  Acute Systemic: Ingestion 2.
  Chronic Local: Allergen 1.
  Chronic Systemic: U.
Toxicology: Irritating to skin, eyes and mucous membranes. Has caused kidney damage in experimental animals.
Fire Hazard: Moderate, when exposed to heat or flame; it can react with oxidizing materials.
Spontaneous Heating: No.

Countermeasures
To Fight Fire: Foam, carbon dioxide, dry chemical or carbon tetrachloride (Section 6).
Ventilation Control: Section 2.
Personal Hygiene: Section 3.
Storage and Handling: Section 7.

1-DECANAL
General Information
Synonyms: Caprylaldehyde; capric aldehyde; n-decylaldehyde; aldehyde-C-10.
Description: Colorless to light yellow liquid; floral fatty odor; soluble in 80% alcohol, fixed oils, volatile oils, mineral oils; insoluble in water and glycerol.
Formula: $CH_3(CH_2)_8CHO$.
Constants: Mol wt: 156, d: 0.831–0.838 (15°C).

Hazard Analysis
Toxic Hazard Rating: U. Limited animal experiments suggest low toxicity and irritation. See also aldehydes. Used as a synthetic flavoring substance and adjuvant (Section 10).

DECANE
General Information
Synonym: Decyl hydride.
Description: Liquid.
Formula: $CH_3(CH_2)_8CH_3$.
Constants: Mol wt: 142.3, mp: $-29.7°C$, bp: 174.1°C, lel: 0.8%, uel: 5.4%, flash p.: 115°F (C.C.), d: 0.730 at 20°/4°C, autoign. temp.: 482°F, vap. press.: 1 mm at 16.5°C, vap. d.: 4.90.

Hazard Analysis
Toxic Hazard Rating:
  Acute Local: U.
  Acute Systemic: Inhalation 1.
  Chronic Local: U.
  Chronic Systemic: U.
Toxicology: A simple asphyxiant. Narcotic in high concentrations.
Fire Hazard: Moderate, when exposed to heat or flame; can react with oxidizing materials.
Spontaneous Heating: No.
Explosion Hazard: Moderate, in its vapor form.

Countermeasures
Ventilation Control: Section 2.
To Fight Fire: Foam, carbon dioxide, dry chemical or carbon tetrachloride (Section 6).
Personal Hygiene: Section 3.
Storage and Handling: Section 7.

DECANOIC ACID
General Information
Synonyms: Decoic acid; decylic acid.
Description: White crystals; unpleasant odor; soluble in most organic solvents and in dilute nitric acid; insoluble in water.
Formula: $CH_3(CH_2)_8COOH$.
Constants: D: 0.0886, bp: 270°C, mp: 315°C.

Hazard Analysis
Toxic Hazard Rating: U. Limited animal experiments suggest moderate toxicity and irritation.

1-DECANOL. See n-decyl alcohol.

1-DECENE
General Information
Synonym: n-Decylene.
Description: Colorless liquid.
Formula: $H_2CCH(CH_2)_7CH_3$.
Constants: Mol wt: 140.26, mp: $-66.3°C$, bp: 172°C, d: 0.7396 at 20°/4°C, vap. press.: 1 mm at 95.7°C, vap. d.: 4.83.

Hazard Analysis
Toxicity: Unknown. Compounds in this group generally have irritant and narcotic action. See also hexene-1.
Fire Hazard: Moderate; when exposed to heat or flame; can react with oxidizing materials.
To Fight Fire: Foam, carbon dioxide, dry chemical or carbon tetrachloride (Section 6).
Explosion Hazard: Unknown.

Countermeasures
Storage and Handling: Section 7.

DECYL ACRYLATE
General Information
Description: Very slightly soluble in water.
Formula: $CH_3(CN_2)_9OCOCN:CH_2$.
Constants: Mol wt: 459, flash p: 441°F (O.C.), d: 0.9, bp: 316°F at 50 mm.

Hazard Analysis
Toxic Hazard Rating: U. Limited animal experiments suggest low toxicity. See also esters.
Fire Hazard: Slight, when exposed to heat or flame.

Countermeasures
To Fight Fire: Water, foam (Section 6).
Storage and Handling: Section 7.

n-DECYL ALCOHOL
General Information
Synonym: 1-Decanol; nonyl carbinol.
Description: Viscous, refractive liquid.
Formula: $CH_3(CH_2)_8CH_2OH$.
Constants: Mol wt: 158.3, mp: 7°C, bp: 231.0°C, flash p: 180°F, d: 0.8297 at 20°/4°C, vap. press.: 1 mm at 69.5°C, vap. d.: 5.3.

Hazard Analysis
Toxicity: Details unknown. Animal experiments suggest low toxicity. See also alcohols.
Fire Hazard: Moderate, when exposed to heat or flame; can react with oxidizing materials.

Countermeasures
To Fight Fire: Foam, carbon dioxide, dry chemical or carbon tetrachloride (Section 6).
Storage and Handling: Section 7.

DECYLAMINE
General Information
Description: Liquid.

TOXIC HAZARD RATING CODE *(For detailed discussion, see Section 1.)*

0 NONE: (a) No harm under any conditions; (b) Harmful only under unusual conditions or overwhelming dosage.

1 SLIGHT: Causes readily reversible changes which disappear after end of exposure.

2 MODERATE: May involve both irreversible and reversible changes; not severe enough to cause death or permanent injury.

3 HIGH: May cause death or permanent injury after very short exposure to small quantities.

U UNKNOWN: No information on humans considered valid by authors.

Formula: $CH_3(CH_2)_9NH_2$.
Constants: Mol wt: 157.3, mp: 17°C, bp: 95°C at 10 mm, flash p: 210°F, d: 0.79 at 20°C, vap. d.: 5.5.
Hazard Analysis
Toxicity: Details unknown. Limited animal experiments suggest high toxicity and irritation. See also amines and fatty amines.
Fire Hazard: Moderate, when exposed to heat or flame; can react with oxidizing materials.
Countermeasures
To Fight Fire: Alcohol foam, foam, carbon dioxide, dry chemical or carbon tetrachloride (Section 6).
Storage and Handling: Section 7.

DECYL BENZENE
General Information
Description: Insoluble in water.
Formula: $C_{10}H_{21}C_6H_5$.
Constants: D: 0.9, bp: 491–536°F, flash p: 225°F.
Hazard Analysis
Toxic Hazard Rating: U.
Fire Hazard: Moderate, when exposed to heat or flame.
Countermeasures
To Fight Fire: Water, foam (Section 6).
Storage and Handling: Section 7.

DECYL CHLORIDE
General Information
Formula: $C_{10}H_{21}Cl$.
Constant: Mol wt: 176.5.
Hazard Analysis
Toxic Hazard Rating: U. Limited animal experiments suggest low toxicity.
Disaster Hazard: Dangerous. See chlorides.

n-DECYLENE. See 1-decene.

DECYL HYDRIDE. See decane.

tert-DECYL MERCAPTAN
General Information
Formula: $C_{10}H_{21}SH$.
Constants: Mol wt: 174, d: 0.9, vap. d.: 6.0, bp: 410–424°F, flash p: 190°F.
Hazard Analysis
Toxic Hazard Rating: U.
Fire Hazard: Moderate, when exposed to heat or flame.
Disaster Hazard: Dangerous. See sulfur compounds.
Countermeasures
Storage and Handling: Section 7.
To Fight Fire: Foam, fog. dry chemical (Section 6).

DECYL NAPHTHALENE
General Information
Description: Liquid.
Formula: $C_{10}H_{21}C_{10}H_7$.
Constants: Mol wt: 268.5, bp: 330–440°C, flash p: 350°F, d: 1.0, vap. d.: 9.6.
Hazard Analysis
Toxicity: Details unknown. See also naphthalene.
Fire Hazard: Slight; when exposed to heat or flame; can react with oxidizing materials.
Countermeasures
To Fight Fire: Foam, carbon dioxide, dry chemical or carbon tetrachloride (Section 6).

DECYL NITRATE
General Information
Description: Insoluble in water.
Formula: $CH_3(CH_2)_9ONO_2$.
Constants: Mol wt: 203, d: 1.0−, bp: 261°F at 11 mm, flash p: 235°F (O.C.).
Hazard Analysis
Toxic Hazard Rating: U.
Fire Hazard: Slight, when exposed to heat or flame.
Disaster Hazard: Dangerous. See nitrates, organic.

Countermeasures
Storage and Handling: Section 7.
To Fight Fire: Water, foam, mist (Section 6).

DEGUELIN
General Information
Description: Pale-green crystals.
Formula: $C_{23}H_{22}O_6$.
Constants: Mol wt: 394.41, mp: 171°C.
Hazard Analysis
Toxic Hazard Rating:
Acute Local: Irritant 3; Ingestion 3; Inhalation 3.
Acute Systemic: Ingestion 2.
Chronic Local: U.
Chronic Systemic: U.
Toxicology: A strong irritant. Inhalation can produce pulmonary edema.
Fire Hazard: Slight; when heated, it emits acrid fumes (Section 6).
Countermeasures
Storage and Handling: Section 7.

de HAENS SALT. See antimony salt.

DEHYDRITE. See magnesium perchlorate.

DEHYDROABIETYLAMINE
General Information
Synonym: "Rosinamine-D."
Description: Amber, viscous liquid.
Formula: $C_{20}H_{25}N$.
Constants: Mol wt: 279.4, bp: 344°C decomposes, flash p: 375°F (O.C.), d: 1.001 at 25°/15.6°C, autoign. temp.: 430°F, vap. d.: 10.
Hazard Analysis
Toxic Hazard Rating:
Acute Local: Irritant 2; Ingestion 2; Inhalation 2.
Acute Systemic: U.
Chronic Local: U.
Chronic Systemic: U.
Fire Hazard: Slight, when exposed to heat or flame; can react with oxidizing materials.
Countermeasures
To Fight Fire: Foam, carbon dioxide, dry chemical or carbon tetrachloride (Section 6).
Ventilation Control: Section 2.
Personnel Protection: Section 3.
Storage and Handling: Section 7.

DEHYDROABIETYLAMINE ACETATE
General Information
Description: Amber paste or solution.
Formula: $C_{20}H_{25}NCH_3COO$.
Constants: Mol wt: 338.5, d: 1.029 at 25°/15.6°C, vap. d.: 11.5.
Hazard Analysis
Toxicity: See dehydroabietylamine.
Fire Hazard: Slight (Section 6).
Countermeasures
Storage and Handling: Section 7.

DEHYDROABIETYLAMINE NAPHTHENATE
General Information
Description: Viscous liquid.
Formula: $C_{20}H_{25}NOOCC_6H_5$.
Constant: Mol wt: 400.6.
Hazard Analysis
Toxicity: Unknown. See also dehydroabietylamine.
Fire Hazard: Slight, when exposed to heat or flame; can react with oxidizing materials (Section 6).
Countermeasures
Storage and Handling: Section 7.

DEHYDROABIETYLAMINE PENTACHLORO-PHENATE
General Information
Description: Crystals.

Formula: $C_{20}H_{25}NC_6Cl_5$.
Constants: Mol wt: 528.8, vap. d.: 18.
Hazard Analysis
Toxicity: Details unknown. See also pentachlorophenol.
Disaster Hazard: Dangerous; when heated to decomposition or on contact with acid or acid fumes, it emits highly toxic fumes of chlorides.
Countermeasures
Storage and Handling: Section 7.

DEHYDROABIETYL NITRILE
General Information
Description: Light amber, partially crystallized solid.
Hazard Analysis and Countermeasures
See Nitriles.

DEHYDROACETIC ACID
General Information
Synonyms: 3-acetyl-6-methyl-1,2-pyran-2-2,4(3H)-dione.
Description: Crystals. Moderately soluble in water and organic solvents.
Formula: $C_8H_8O_4$.
Constants: Mol wt: 168.2, mp: 109°C, bp: 269.0°C, vap. press.: 1 mm at 91.7°C, vap. d.: 5.8.
Hazard Analysis
Toxic Hazard Rating:
　Acute Local: U.
　Acute Systemic: Ingestion 3.
　Chronic Local: U.
　Chronic Systemic: Ingestion 1.
Toxicity: Large internal doses have produced vomiting, ataxia and convulsions in experimental animals. Can depress kidney function.
A food additive permitted in food for human consumption (Section 10).
Fire Hazard: Slight (Section 6).
Countermeasures
Storage and Handling: Section 7.

DEHYDROCHOLESTEROL,　ACTIVATED. See vitamin D₃.

DEICING FLUIDS
Countermeasures
Shipping Regulations: Section 11.
　IATA: Flammable liquid, red label, 1 liter (passenger), 40 liters (cargo).

DEKANITROCELLULOSE. See nitrocellulose.

DELAY ELECTRIC IGNITERS. See explosives, high.

m-DELPHENE. See N,N-dimethyl-m-toluamide.

DELPHININE
General Information
Description: White, crystalline, rhombic plates.
Formula: $C_{34}H_{47}NO_9$.
Constants: Mol wt: 613.73, mp: 191°C decomposes.
Hazard Analysis
Toxic Hazard Rating:
　Acute Local: U.
　Acute Systemic: Ingestion 3; Inhalation 3.
　Chronic Local: U.
　Chronic Systemic: U.
Caution: An alkaloid poison.
Disaster Hazard: Dangerous; when heated to decomposition, it emits highly toxic fumes of oxides of nitrogen.

Countermeasures
Ventilation Control: Section 2.
Personal Hygiene: Section 3.
First Aid: Section 1.
Storage and Handling: Section 7.

DELPHINIUM
General Information
Synonym: Larkspur; stagger weed.
Description: Dried ripe seeds.
Hazard Analysis
Toxic Hazard Rating:
　Acute Local: Allergen 1.
　Acute Systemic: Ingestion 3; Inhalation 3.
　Chronic Local: Irritant 2; Allergen 1.
　Chronic Systemic: U.
Fire Hazard: Slight; when exposed to heat or flame (Section 6).
Disaster Hazard: Dangerous; when heated to decomposition, it emits highly toxic fumes and it can react with oxidizing materials.
Countermeasures
Ventilation Control: Section 2.
Personal Hygiene: Section 3.
First Aid: Section 1.
Storage and Handling: Section 7.

DEMETON
General Information
Synonyms: o, o-diethyl-o-2-(ethylthio)ethyl thiophosphate; "E-1059"; "systox."
Hazard Analysis
Hazardous Properties: An insecticide. It is believed that most of the physiological actions of demeton resemble, generally, those of parathion, TEPP, and other related organic phosphorus poisons. The actions of this compound and its metabolites are based principally upon the inhibition of the enzyme cholinesterase, thus allowing the accumulation of large amounts of acetylcholine.
TLV: ACGIH (recommended) 0.1 milligram per cubic meter of air. May be absorbed via skin.
Dangerous Acute and Chronic Doses: Little is known regarding the acute and chronic dosages which would be dangerous to man. However, the acute dose is believed to be approximately equal to that of parathion (estimated at from 12 to 20 mg or one-fifth to one-third grain). The compound is only slightly less toxic by the dermal route. Doses of organic phosphorus insecticides tend to be cumulative in their effect. However, if illness occurs, it is acute in nature, whether caused by a single large dose or by repeated exposure.
Signs and Symptoms of Poisoning: Persons poisoned with demeton may be expected to show the following symptoms: headache, giddiness, blurred vision, weakness, nausea, diarrhea, and discomfort in the chest. In addition to these symptoms, sweating, miosis, muscular fasciculation, incoordination, tearing, salivation, pulmonary edema, cyanosis, papilledema, convulsions, coma, and loss of reflexes and sphincter control may occur. There is one result of demeton poisoning in rats which may or may not be peculiar to the species. Shortly after other signs of poisoning appeared, rats fed 50 or 100 ppm of demeton in their regular diet developed severe exophthalmos, and perhaps associated changes, leading to drying and ulceration of the cornea so severe that some animals lost both eyes through necrosis.

TOXIC HAZARD RATING CODE *(For detailed discussion, see Section 1.)*

0　NONE: (a) No harm under any conditions; (b) Harmful only under unusual conditions or overwhelming dosage.

1　SLIGHT: Causes readily reversible changes which disappear after end of exposure.

2　MODERATE: May involve both irreversible and reversible changes; not severe enough to cause death or permanent injury.

3　HIGH: May cause death or permanent injury after very short exposure to small quantities.

U　UNKNOWN: No information on humans considered valid by authors.

Countermeasures

Treatment of Poisoning: Keep the patient fully atropinized, using 1 to 2 mg of atropin sulfate per dose and up to 10 to 20 mg per day. Remove any remaining poison. Administer oxygen and other supportive measures early and be prepared for the immediate use of mechanical artificial respiration. Watch the patient continuously. See also parathion.

Shipping Regulations: Section 11.

See Organic Phosphates.

DENATURANTS

Hazard Analysis

Toxicity: These materials are almost always toxic. See also specific denaturant.

DENATURED ALCOHOL. Ethyl alcohol to which a toxic ingredient has been added (denaturant).

General Information

Constants: Flash p: 60° F, autoign. temp.: 750° F, d: 0.8, vap. d.: 1.6, bp: 175° F.

Hazard Analysis

Toxic Hazard Rating: Almost always toxic.

Fire Hazard: Dangerous, when exposed to heat or flame.

Countermeasures

To Fight Fire: Alcohol foam (Section 6).

Storage and Handling: Section 7.

Shipping Regulations: Section 11.

See alcohol, N.O.S.

DENATURED SPIRITS. See alcohol, denatured.

"DEPENDIP"

General Information

Description: Water-white liquid.

Constants: Flash p: 52° F, d: 0.758 at 15.5° C.

Hazard Analysis

Toxic Hazard Rating:

Acute Local: Irritant 1.

Acute Systemic: Inhalation 1.

Chronic Local: U.

Chronic Systemic: U.

Fire Hazard: Dangerous, when exposed to heat or flame; it can react vigorously with oxidizing materials.

Disaster Hazard: Dangerous; when heated, it emits acrid fumes.

Countermeasures

Ventilation Control: Section 2.

To Fight Fire: Foam, carbon dioxide, dry chemical or carbon tetrachloride (Section 6).

Personal Hygiene: Section 3.

Storage and Handling: Section 7.

DERMATOL (COMMERCIAL). See bismuth gallate, basic.

DERRIS. See rotenone.

DESMODUR. See 1,5-naphthyl diisocyanate.

DESOXYCHOLIC ACID

General Information

Synonym: Deoxycholic acid.

Description: A bile acid; crystals, practically insoluble in water and benzene. Slightly soluble in chloroform and ether; soluble in acetone and solutions of alkali hydroxides and carbonates; freely soluble in alcohol.

Formula: $C_{24}H_{40}O_4$.

Constants: Mol wt: 392, mp: 172–173° C.

Hazard Analysis

Toxic Hazard Rating: U. An emulsifying agent food additive (Section 10).

DETONATING FUSES. See explosives, high.

DETONATING PRIMERS. See explosives, high.

DEUTERIUM. See hydrogen.

DEXTRANS

General Information

Synonym: Mucroses.

Description: Certain polymers of glucose which have chain like structures and very high molecular weights (up to 200,000 or higher).

Hazard Analysis

Toxic Hazard Rating: U. General purpose food additives (those with mol wt < 100,000). See Section 10.

DEXTRIN

General Information

Synonym: Starch gum; artificial gum; vegetable gum; tapioca; dextran.

Description: An intermediate product formed by the hydrolysis of starches yellow or white powder or granules. Soluble in water; insoluble in alcohol and ether. It is a colloidal in properties and describes a class of substances, hence it has no formula.

Hazard Analysis

Toxic Hazard Rating: U. A substance migrating to food from packaging material (Section 10).

DEXTROSE

General Information

Synonym: Dextroglucose; grape sugar; corn sugar.

Description: Colorless crystals or white crystalline or granular powder; odorless; soluble in water; slightly soluble in alcohol.

Formula: $C_6H_{12}O_6 \cdot H_2O$.

Constants: Mol wt: 198, d: 1.544, mp: 146° C.

Hazard Analysis

Toxic Hazard Rating: U. A substance migrating to food from packaging materials (Section 10).

DEXTROSE NITRATE. See nitrolevulose.

DFDD

General Information

Synonym: Difluorodiphenyldichloroethane.

Description: Crystals.

Formula: $C_{14}H_{10}Cl_2F_2$.

Constants: Mol wt: 287.2, mp: 77.5° C, vap. d.: 10.

Hazard Analysis

Toxicity: Details unknown; a contact insecticide. See DDT.

Disaster Hazard: Dangerous; when heated to decomposition, it emits highly toxic fumes of fluorides and chlorides.

Countermeasures

Storage and Handling: Section 7.

DFDT

General Information

Synonym: 2,2-bis(p-fluorophenyl)-1,1,1 trichloroethane, difluorodiphenyl trichloroethane.

Description: Crystals.

Formula: $(FC_6H_4)_2CHCCl_3$.

Constants: Mol wt: 321.6, mp: 45.5° C, vap. d.: 10.5.

Hazard Analysis

Toxicity: Details unknown; an insecticide. Similar in action to DDT but apparently less toxic. See DDT.

Disaster Hazard: Dangerous; when heated to decomposition, it emits highly toxic fumes of fluorides and chlorides.

Countermeasures

Storage and Handling: Section 7.

DFP. See diisopropyl fluorophosphate.

DGE. See diglycidyl ether.

DHS. See dihydrostreptomycin.

DIACETIC ETHER. See ethyl acetoacetate.

DIACETIN. See glyceryl diacetate.

DIACETONE ALCOHOL
General Information
Synonym: 4-hydroxy-4-methylpentanone-2.
Description: Liquid; faint pleasant odor.
Formula: $(CH_3)_2C(OH)CH_2COCH_3$.
Constants: Mol wt: 116.2, mp: -47 to $-54°C$, bp:
167.9°C, flash p: 151°F, d: 0.9306 at 25°/4°C,
autoign. temp.: 1118°F, vap. d.: 4.00, vap. press.: 1.1
mm at 20°C, lel: 1.8%, uel: 6.9%, flash p (acetone
free): 136°F.
Hazard Analysis
Toxic Hazard Rating:
  Acute Local: Irritant 2; Inhalation 2.
  Acute Systemic: Inhalation 2.
  Chronic Local: Irritant 1.
  Chronic Systemic: U.
Toxicity: Irritating to eyes and mucous membranes. Narcotic
in high concentration. Experimentally has caused
anemia and damage to kidneys and liver.
TLV: ACGIH (recommended) 50 parts per million of air;
240 milligrams per cubic meter of air.
Fire Hazard: Moderate, when exposed to heat or flame; it
can react with oxidizing materials.
Countermeasures
To Fight Fire: Alcohol foam, foam, carbon dioxide, dry
chemical or carbon tetrachloride (Section 6).
Ventilation Control: Section 2.
Personal Hygiene: Section 3.
Storage and Handling: Section 7.
Shipping Regulations: Section 11.
  IATA: Flammable liquid, red label, 1 liter (passenger),
40 liters (cargo).

DIACETONE ALCOHOL (TECHNICAL)
General Information
Description: Liquid, pleasant odor.
Formula: $CH_3COCH_2C(CH_3)_2OH$.
Constants: Mol wt: 116.2, bp: 163°C, flash p.: 48°F (O.C.),
d: 0.931–0.940, autoign. temp.: 1118°F, vap. press.:
1.1 mm at 20°C, vap. d.: 4.00, fp: $-42.8°C$.
Hazard Analysis and Countermeasures
See diacetone alcohol.

1,1-DIACETOXY-2,3-DICHLOROPROPANE
Hazard Analysis
Toxic Hazard Rating: U. Limited Animal experiments
suggest high toxicity and irritation.
Disaster Hazard: Dangerous. See chlorides.

DIACETYL
General Information
Synonym: Butanedione.
Description: Greenish-yellow liquid, strong odor.
Formula: $CH_3COCOCH_3$.
Constants: Mol wt: 86.09, bp: 88°C, flash p: $< 80°F$, d:
0.9904 at 15°/15°C, vap. d.: 3.00.
Hazard Analysis
Toxicity: Details unknown. Probably has irritant and
narcotic action. Note: Used as a synthetic flavoring
substance and adjuvant (Section 10).
Fire Hazard: Dangerous, when exposed to heat or flame.
Disaster Hazard: Dangerous! Keep away from heat and
open flame.
Countermeasures
Ventilation Control: Section 2.
To Fight Fire: Foam, carbon dioxide, dry chemical or car-
bon tetrachloride (Section 6).
Storage and Handling: Section 7.

Shipping Regulations: Section 11.
  IATA: Flammable liquid, red label, 1 liter (passenger),
40 liters (cargo).

DIACETYLENE
General Information
Synonyms: Butadiyne; butadiine.
Description: Gas.
Formula: $HC:CC:CH$.
Constants: Mol wt: 50.1, mp: $-36.4°C$, bp: 10.3°C, d:
2.233.
Hazard Analysis
Toxic Hazard Rating:
  Acute Local: 0.
  Acute Systemic: Inhalation 2.
  Chronic Local: 0.
  Chronic Systemic: U.
Toxicology: In high concentrations it acts as a simple as-
phyxiant.
Fire Hazard: Moderate, when exposed to heat or flame,
or by chemical reaction with oxidizers. Can ignite
spontaneously in contact with moist silver salts.
Explosion Hazard: Unknown.
Countermeasures
Storage and Handling: Section 7.

DIACETYLENE CARBONIC ACID
Hazard Analysis
Toxicity: Unknown.
Fire Hazard: Unknown.
Explosion Hazard: Moderate, when exposed to heat at
367°F.
Countermeasures
Storage and Handling: Section 7.

1,2-DIACETYLETHANE. See acetonyl acetone.

DIACETYL MORPHINE
General Information
Synonyms: Diamorphine; heroin.
Description: White, odorless, bitter, crystals or crystalline
powder.
Formula: $C_{17}H_{17}(OOCCH_3)_2NO$.
Constants: Mol wt: 369.40, mp: 171–173°C.
Hazard Analysis
Toxic Hazard Rating:
  Acute Local: U.
  Acute Systemic: Ingestion 3; Inhalation 3.
  Chronic Local: U.
  Chronic Systemic: Ingestion 2.
Toxicology: Fatal dose between 1/6 and 2 grains. Resembles
morphine in its general results, but acts more strongly
on the respiration and is therefore more poisonous. Its
depressant effects on the cerebrum appear to be greater
than that of codeine. Large doses cause excitement
and convulsions in animals and man. The more common symptoms are headache, disturbance of vision,
slow, small regular pulse, restlessness, cramps in the
extremities, slight cyanosis, respiration slow and deep,
and death from respiratory paralysis. Poisonous habit-
forming drug.
Disaster Hazard: Dangerous; when heated to decomposition, it emits toxic fumes of oxides of nitrogen.
Countermeasures
Ventilation Control: Section 2.
Treatment and Antidotes: Use the same treatment as for
morphine, with special attention to control of the
respiratory paralysis.

TOXIC HAZARD RATING CODE (For detailed discussion, see Section 1.)

0 NONE: (a) No harm under any conditions; (b) Harmful only under unusual conditions or overwhelming dosage.

1 SLIGHT: Causes readily reversible changes which disappear after end
of exposure.

2 MODERATE: May involve both irreversible and reversible changes;
not severe enough to cause death or permanent injury.

3 HIGH: May cause death or permanent injury after very short exposure
to small quantities.

U UNKNOWN: No information on humans considered valid by authors.

Personal Hygiene: Section 3.
First Aid: Section 1.
Storage and Handling: Section 7.

DIACETYLMORPHINE HYDROCHLORIDE. See diacetyl morphine.

DIACETYL PEROXIDE. See acetyl peroxide.

DIACETYL PEROXIDE SOLUTION. See acetyl peroxide 25% solution in dimethylphthalate.

DIACETYL TARTARIC ACID ESTERS OF MONO- and DI-GLYCERIDES FROM THE GLYCEROLS OF EDIBLE FATS OR OILS
Hazard Analysis
Toxic Hazard Rating: Used as Emulsifying agent food additives. (Section 10).

DIALLATE. See 2,3-dichloroallyl diisopropyl thiocarbamate.

DIALLYLAMINE
General Information
Synonym: di-2-propenylamine.
Description: Liquid; soluble in water.
Formula: $(CH_2:CHCH_2)_2NH$.
Constants: Mol wt: 103, d: 0.7889 at 20°C, bp: 112°C, fp: −100°C.
Hazard Analysis
Toxic Hazard Rating: U. Limited animal experiments suggest moderate toxicity and irritation. See also allylamine.

N,N-DIALLYL-2-CHLOROACETAMIDE
General Information
Synonym: CDAA; "Randox"; α-chloro-n,n-diallyl acetamide.
Description: Amber liquid; slightly soluble in water; soluble in alcohol, hexane, and xylene.
Formula: $ClCH_2CON(CH_2CH:CH_2)_2$.
Constants: Mol wt: 183.5, bp: 74°C at 0.3 mm
Hazard Analysis
Toxic Hazard Rating:
  Acute Local: Irritant 3; Inhalation 3.
  Acute Systemic: Ingestion 3; Skin Absorption 3.
  Chronic Local: U.
  Chronic Systemic: U.
Disaster Hazard: Dangerous. See chlorides.

DIALLYLCYANAMIDE
General Information
Synonym: N-cyanodiallylamine.
Description: Colorless, mobile liquid when pure.
Formula: $(CN_2CHCH_2)_2NCN$.
Constants: Mol wt: 122.17, mp: $< -70°C$, bp: 222°C (slight decomposition), d: 0.9021, vap. d.: 4.1.
Hazard Analysis
Toxicity: It is said to be more toxic than cyanamide. See cyanamide and amines.
Disaster Hazard: Dangerous; when heated to decomposition or on contact with acid or acid fumes, it emits highly toxic fumes of cyanides.
Countermeasures
Ventilation Control: Section 2.
Personnel Protection: Section 3.
First Aid: Section 1.
Storage and Handling: Section 7.

DIALLYL DISULFIDE. See allyl sulfide.

DIALLYL ETHER. See allyl ether.

DIALLYL ETHYLENE BIS-GLYCOLATE
General Information
Description: Liquid.
Formula: $(CH_2OCH_2CO_2CH_2CHCH_2)_2$.

Constants: Mol wt: 258.3, bp: 130–133°C at 1 mm, flash p: 320°F (O.C.), d: 1.1227 at 25°/25°C, vap. d.: 8.9.
Hazard Analysis
Toxicity: Details unknown.
Fire Hazard: Slight, when exposed to heat or flame; it can react with oxidizing materials.
Countermeasures
To Fight Fire: Water, foam, carbon dioxide, dry chemical or carbon tetrachloride (Section 6).
Storage and Handling: Section 7.

DI-2-ALLYL-4-HYDROXY-3 METHYL-2-CYCLOPEN-TEN-1-ONE ESTER OF CIS AND TRANS-DI-CHRYSANTHEMUM MONOCARBOXYLIC ACID. See allethrin.

DIALLYL MALEATE
General Information
Synonym: Allyl maleate.
Description: Liquid.
Formula: $(CHCOOCH_2CHCH_2)_2$.
Constants: Mol wt: 196.2, vap. d.: 6.6.
Hazard Analysis
Toxic Hazard Rating:
  Acute Local: Irritant 2.
  Acute Systemic: Ingestion 2.
  Chronic Local: U.
  Chronic Systemic: U.
Fire Hazard: Slight (Section 6).
Countermeasures
Ventilation Control: Section 2.
Personnel Protection: Section 3.
Storage and Handling: Section 7.

DIALLYL MELAMINE
General Information
Description: White, crystalline solid.
Formula: $(CH_2CHCH_2)_2N(CN)_3(NH_2)_2$.
Constants: Mol wt: 226.3, mp: 142°C, d: 1.242 at 30°C, vap. d.: 7.6.
Hazard Analysis
Toxic Hazard Rating:
  Acute Local: Irritant 2; Ingestion 2.
  Acute Systemic: U.
  Chronic Local: U.
  Chronic Systemic: U.
Disaster Hazard: Dangerous; when heated to decomposition or on contact with acid or acid fumes, it emits highly toxic fumes of cyanides.
Countermeasures
Personnel Protection: Section 3.
Storage and Handling: Section 7.

DIALLYL PHTHALATE
General Information
Description: Nearly colorless, oily liquid.
Formula: $C_6H_4(CO_2CH_2CHCH_2)_2$.
Constants: Mol wt: 246.3, bp: 157°C, flash p: 330°F, d: 1.120 at 20°/20°C, vap. d.: 8.3.
Hazard Analysis
Toxicity: See phthalic acid and esters.
Fire Hazard: Slight, when exposed to heat or flame; it can react with oxidizing materials.
Countermeasures
To Fight Fire: Water, foam, carbon dioxide, dry chemical or carbon tetrachloride (Section 6).
Storage and Handling: Section 7.

DIALLYL SULFIDE. See allyl sulfide.

DIALLYL TRISULFIDE. See allyl trisulfide.

DIAMINE. See hydrazine.

p-DIAMINOAZOBENZENE
General Information
Synonym: p-azoaniline; 4,4'-azodianiline; 4,4'-diamino-azobenzene.
Description: Golden yellow needles; slightly soluble in water, benzene, petroleum ether; freely soluble in alcohol.
Formula: $C_{12}H_{12}N_4$.
Constants: Mol wt: 212.25, mp: 250°C.
Hazard Analysis
Toxic Hazard Rating: U. See aniline.

DI-p-AMINOAZOBENZENE FLUOSILICATE
General Information
Description: Brown crystals slightly soluble in water.
Formula: $(NH_2C_6H_4N_2C_6H_5)_2 \cdot H_2SiF_6$.
Constants: Mol wt: 538.5, mp: 220°C (decomposes).
Hazard Analysis and Countermeasures
See fluosilicates.

p-DIAMINOBENZENE. See p-phenylenediamine.

m-DIAMINOBENZENE. See m-phenylene diamine.

o-DIAMONOBENZENE. See o-phenylene diamine.

3,5-DIAMINOBENZOIC ACID
General Information
Description: Needlelike crystals. Slightly water soluble. Soluble in alcohol and ether.
Formula: $C_7H_8N_2O_2 \cdot H_2O$.
Constants: Mol wt: 170.2, mp: 228°C; $-H_2O$ at 110°C.
Hazard Analysis and Countermeasures
See p-aminobenzoic acid.

DI-p-AMINOBENZOIC ACID FLUOSILICATE
General Information
Description: White Crystals.
Formula: $(C_6H_4NH_2COOH)_2 \cdot H_2SiF_6$.
Constants: Mp: 242°C, mol wt: 418.4.
Hazard Analysis and Countermeasures
See fluosilicates.

1,3-DIAMINOBUTANE
General Information
Description: Liquid.
Formula: $C_4H_8(NH_2)_2$.
Constants: Mol wt: 88.2, bp: 142–150°C, flash p: 125°F, d: 0.85, vap. d.: 3.04.
Hazard Analysis
Toxicity: Details unknown. See also fatty amines.
Fire Hazard: Moderate, when exposed to heat or flame; it can react with oxidizing materials.
Countermeasures
To Fight Fire: Alcohol foam, foam, carbon dioxide, dry chemical or carbon tetrachloride (Section 6).
Ventilation Control: Section 2.
Storage and Handling: Section 7.

α-epsilon—DIAMINOCAPROIC ACID. See lysine.

3-DIAMINO-4-DIHYDROXY-1-ARSENOBENZENE HYDROCHLORIDE. See arsphenamine.

p-DIAMINODIPHENYL. See benzidine.

4,4'-DIAMINODIPHENYLAMINE
General Information
Description: Crystals.
Formula: $C_{12}H_{13}N_3$.
Constants: Mol wt: 199.3, mp: 158°C.

Hazard Analysis and Countermeasures
See diphenylamine.

DIAMINODIPHENYL SULFONE
General Information
Synonym: p,p'-sulfonyl dianiline.
Description: Crystals. Nearly insoluble in water. Soluble in acetone, alcohol.
Formula: $C_{12}H_{12}O_2N_2S$.
Constants: Mol wt: 248.3, mp: 176°C, vap. d.: 8.3.
Hazard Analysis
Toxic Hazard Rating:
    Acute Local: U.
    Acute Systemic: Ingestion 2.
    Chronic Local: U.
    Chronic Systemic: Ingestion 2.
Toxicology: Moderate doses can cause neuritis, dermatitis and hepatitis. Used in treating leprosy and in veterinary medicine.
Disaster Hazard: Dangerous; when heated to decomposition, it emits highly toxic fumes of oxides of sulfur.
Countermeasures
Storage and Handling: Section 7.

1,2-DIAMINOETHANE. See ethylenediamine.

2,4-DIAMINO-4'-ETHOXYAZOBENZENE. See ethoxazene.

1,3-DIAMINOISOPROPANOL
General Information
Description: Soluble in water.
Formula: $NH_2CH_2CHOHCH_2NH_2$.
Constants: Mol wt: 90, d: 1.1, bp: 266°F, flash p: 27°F.
Hazard Analysis
Toxic Hazard Rating: U. See also amines.
Fire Hazard: Dangerous, when exposed to heat or flame.
Countermeasures
Storage and Handling
To Fight Fire: Water, foam, alcohol foam (Section 6).
Storage & Handling: Section 7.

1,8-DIAMINO-p-MENTHONE
General Information
Formula: $C_{10}H_{16}O(NH_2)_2$.
Constant: Mol wt: 184.
Hazard Analysis
Toxic Hazard Rating: U. Limited animal experiments suggest high toxicity and irritation. See also amines.

1,8-DIAMINONAPHTHALENE. See 1,8-naphthalenediamine.

DIAMINOPHENOL HYDROCHLORIDE. See amidol.

1,3-DIAMINOPROPANE
General Information
Description: Liquid.
Formula: $C_3H_6(NH_2)_2$.
Constants: Mol wt: 74.1, bp: 133°C, flash p: 75°F, d: 0.88, vap. d.: 2.5.
Hazard Analysis
Toxicity: Details unknown. See also fatty amines.
Fire Hazard: Dangerous, when exposed to heat or flame.
Explosion Hazard: Unknown.
Disaster Hazard: Dangerous! Keep away from heat and open flame!
Countermeasures
Ventilation Control: Section 2.

TOXIC HAZARD RATING CODE (For detailed discussion, see Section 1.)

0 NONE: (a) No harm under any conditions; (b) Harmful only under unusual conditions or overwhelming dosage.

1 SLIGHT: Causes readily reversible changes which disappear after end of exposure.

2 MODERATE: May involve both irreversible and reversible changes; not severe enough to cause death or permanent injury.

3 HIGH: May cause death or permanent injury after very short exposure to small quantities.

U UNKNOWN: No information on humans considered valid by authors.

To Fight Fire: Alcohol foam, foam, carbon dioxide, dry chemical or carbon tetrachloride (Section 6).
Storage and Handling: Section 7.

2,6-DIAMINOPYRIDINE
General Information
Description: Crystals.
Formula: $NC_5H_3(NH_2)_2$.
Constants: Mol wt: 109.08, mp: 120.8°C, bp: 285°C.
Hazard Analysis
Toxicity: Details unknown. See also pyridine.
Disaster Hazard: Dangerous; when heated to decomposition, it emits highly toxic fumes of nitrogen compounds.
Countermeasures
Storage and Handling: Section 7.

Di(α-AMINO-β-THIOLPROPIONIC ACID). See cystine.

2,5-DIAMINOTOLUENE. See 2,5-toluenediamine.

DIAMMINECOBALT (II) CHLORIDE
General Information
Description: Rose crystals.
Formula: $CoCl_2 \cdot 2NH_3$.
Constants: Mol wt: 163.92, mp: 273°C, d: 2.097.
Hazard Analysis and Countermeasures
See cobalt compounds, chlorides and ammonia.

DIAMMINECOPPER (II) ACETATE
General Information
Description: Violet-blue crystals.
Formula: $Cu(C_2H_3O_2)_2 \cdot 2NH_3$.
Constants: Mol wt: 215.69, mp: decomposes at approx. 175°C.
Hazard Analysis and Countermeasures
See copper compounds and ammonia.

DIAMMONIUM HYDROGEN PHOSPHATE. See ammonium phosphate, dibasic.

DIAMMONIUM PHOSPHATE. See ammonium phosphate, dibasic.

DIAMORPHINE. See diacetylmorphine.

DIAMYL ACETAMIDE
General Information
Description: Liquid.
Formula: $(C_5H_{11})_2CHCONH_2$.
Constants: Mol wt: 199.3, flash p: 235°F, vap. d.: 6.7.
Hazard Analysis
Toxicity: Details unknown.
Fire Hazard: Slight, when exposed to heat or flame; it can react with oxidizing materials.
Countermeasures
To Fight Fire: Foam, carbon dioxide, dry chemical or carbon tetrachloride (Section 6).
Storage and Handling: Section 7.

DIAMYLAMINE
General Information
Synonym: di-n-amylamine.
Description: Water-white liquid.
Formula: $(C_5H_{11})_2NH$.
Constants: Mol wt: 157.26, bp: 202°C, flash p: 124°F, d: 0.777 at 20°/20°C, vap. d.: 5.42.
Hazard Analysis
Toxicity: Unknown. Limited animal experiments suggest high toxicity. See also amines.
Fire Hazard: Moderate, when exposed to heat or flame; it can react with oxidizing materials.
Countermeasures
To Fight Fire: Alcohol foam, foam, carbon dioxide, dry chemical or carbon tetrachloride (Section 6).
Ventilation Control: Section 2.
Storage and Handling: Section 7.

DI-n-AMYLAMINE. See diamylamine.

DIAMYL ANILINE
General Information
Description: Liquid.
Formula: $C_6H_5N(C_5H_{11})_2$.
Constants: Mol wt: 233.4, bp: 277°C, flash p: 260°F, d: 0.89, vap. d.: 7.8.
Hazard Analysis
Toxicity: Details unknown. See also aniline.
Fire Hazard: Slight, when exposed to heat or flame.
Disaster Hazard: Moderately dangerous; when heated to decomposition, it emits toxic fumes and it can react with oxidizing materials.
Countermeasures
Storage and Handling: Section 7.
To Fight Fire: Foam, carbon dioxide, dry chemical or carbon tetrachloride (Section 6).

DIAMYL BENZENE
General Information
Description: Liquid.
Formula: $C_6H_4(C_5H_{11})_2$.
Constants: Mol wt: 218.4, bp: 265°C, flash p: 225°F, d: 0.85, vap. d.: 7.3.
Hazard Analysis
Toxicity: Details unknown. See also benzene.
Fire Hazard: Slight, when exposed to heat or flame; it can react with oxidizing materials.
Countermeasures
To Fight Fire: Foam, carbon dioxide, dry chemical or carbon tetrachloride (Section 6).
Storage and Handling: Section 7.

DIAMYL BIPHENYL
General Information
Description: Liquid.
Formula: $C_5H_{11}(C_6H_4)_2C_5H_{11}$.
Constants: Mol wt: 294.5, mp: −30°C, flash p: 340°F, bp: 355–385°C, d: 0.938 at 20°/20°C, vap. d.: 10.2.
Hazard Analysis
Toxicity: Details unknown. See also diphenyl and amyl alcohol.
Fire Hazard: Slight, when exposed to heat or flame; it can react with oxidizing materials.
Countermeasures
To Fight Fire: Foam, carbon dioxide, dry chemical or carbon tetrachloride (Section 6).
Storage and Handling: Section 7.

DIAMYL CHLORONAPHTHALENE
General Information
Description: Liquid.
Formula: $(C_5H_{11})_2C_{10}H_5Cl$.
Constants: Mol wt: 302.9, bp: 350°C, flash p: 330°F, d: 1.06, vap. d.: 10.4.
Hazard Analysis
Toxicity: Details unknown. See also chloronaphthalenes.
Fire Hazard: Slight, when exposed to heat or flame.
Disaster Hazard: Dangerous; see chlorides; it can react vigorously with oxidizing materials.
Countermeasures
Storage and Handling: Section 7.
To Fight Fire: Water, foam, carbon dioxide, dry chemical or carbon tetrachloride (Section 6).

Di-tert-AMYLCYCLOHEXANOL
General Information
Description: Insoluble in water.
Formula: $(C_5H_{11})_2C_6H_9OH$.
Constants: D: 0.9, bp: 554–572°F, flash p: 270°F. Mol wt: 240.4.
Hazard Analysis
Toxic Hazard Rating: U.
Fire Hazard: Slight, when exposed to heat or flame.
Countermeasures
Storage and Handling: Section 7.
To Fight Fire: Water, foam (Section 6).

DIAMYLENE
General Information
Description: Liquid.
Formula: $C_{10}H_{20}$.
Constants: Mol wt: 140.26, mp: below $-50°C$, bp: $150°C$, flash p: $118°F$ (O.C.), d: 0.77–0.78, vap. d.: 4.7.
Hazard Analysis
Toxicity: Details unknown. Compounds of this type have irritant and narcotic action.
Fire Hazard: Moderate, when exposed to heat or flame; it can react with oxidizing materials.
Countermeasures
To Fight Fire: Foam, carbon dioxide, dry chemical or carbon tetrachloride (Section 6).
Ventilation Control: Section 2.
Storage and Handling: Section 7.

DIAMYL ETHER. See amyl ether.

DIAMYL MALEATE
General Information
Description: Liquid.
Formula: $(CHCO_2C_5H_{11})_2$.
Constants: Mol wt: 256.3, flash p: $270°F$, d: 0.981, vap. d.: 8.6, bp: $500–572°F$.
Hazard Analysis
Toxicity: Details unknown. See also esters and amyl alcohol.
Fire Hazard: Slight, when exposed to heat or flame; it can react with oxidizing materials.
Countermeasures
To Fight Fire: Foam, carbon dioxide, dry chemical or carbon tetrachloride (Section 6).
Storage and Handling: Section 7.

DI-n-AMYL MERCURY
General Information
Description: Crystals.
Formula: $Hg(C_5H_{11})_2$.
Constants: Mol wt: 342.9, d: 1.6369, bp: $133°C$ at 10 mm.
Hazard Analysis and Countermeasures
See mercury compounds, organic.

DIAMYL NAPHTHALENE
General Information
Description: Liquid.
Formula: $C_{10}H_6(C_5H_{11})_2$.
Constants: Mol wt: 268.43, bp: $326°C$, flash p: $315°F$ (O.C.), d: 0.93–0.94, vap. d.: 9.3, vap. press.: 0.00124 mm at $20°C$.
Hazard Analysis
Toxicity: Details unknown. See also naphthalene.
Fire Hazard: Slight, when exposed to heat or flame; it can react with oxidizing materials.
Countermeasures
To Fight Fire: Foam, carbon dioxide, dry chemical or carbon tetrachloride (Section 6).

DIAMYL β-NAPHTHOL
General Information
Description: Liquid.
Formula: $(C_5H_{11})_2C_{10}H_5OH$.
Constants: Mol wt: 284.4, bp: $205°C$, flash p: $345°F$, d: 0.97, vap. d.: 9.6.
Hazard Analysis
Toxicity: Details unknown. See also β-naphthol.
Fire Hazard: Slight, when exposed to heat or flame; it can react with oxidizing materials.

Countermeasures
To Fight Fire: Foam, carbon dioxide, dry chemical or carbon tetrachloride (Section 6).
Storage and Handling: Section 7.

DIAMYL OXALATE
General Information
Description: Liquid.
Formula: $(C_5H_{11}CO_2)_2$.
Constants: Mol wt: 230.3, flash p: $257°F$, vap. d.: 7.7, d: 1.0
Hazard Analysis
Toxicity: See oxalates and amyl alcohol.
Fire Hazard: Slight, when exposed to heat or flame; it can react with oxidizing materials.
Countermeasures
To Fight Fire: Foam, carbon dioxide, dry chemical or carbon tetrachloride (Section 6).
Storage and Handling: Section 7.

2,4-DIAMYL PHENOL
General Information
Description: Liquid.
Formula: $(C_5H_{11})_2C_6H_3OH$.
Constants: Mol wt: 234.37, bp: $278°C$, flash p: $260°F$ (O.C.), d: 0.93–0.94, vap. d.: 8.1.
Hazard Analysis
Toxicity: Details unknown. See also phenol. Highly toxic to rats.
Fire Hazard: Slight, when exposed to heat or flame.
Disaster Hazard: Moderately dangerous; when heated to decomposition, it emits toxic fumes; it can react with oxidizing materials.
Countermeasures
Storage and Handling: Section 7.
To Fight Fire: Foam, carbon dioxide, dry chemical or carbon tetrachloride (Section 6).

DI-sec-AMYL PHENOL
General Information
Description: Liquid.
Formula: $(C_5H_{11})_2C_6H_3OH$.
Constants: Mol wt: 234.4, bp: $280°C$, flash p: $260°F$ (O.C.), d: 0.91, vap. d.: 8.1.
Hazard Analysis
Toxicity: Probably toxic. See also phenol.
Fire Hazard: Slight, when exposed to heat or flame; it can react with oxidizing materials.
Countermeasures
To Fight Fire: Foam, carbon dioxide, dry chemical or carbon tetrachloride (Section 6).
Storage and Handling: Section 7.

DI-tert-AMYLPHENOXYETHANOL
General Information
Description: Liquid.
Formula: $C_6H_3(C_5H_{11})_2OCH_2CH_2OH$.
Constants: Mol wt: 278.4, mp: $-35°C$, flash p: $300°F$ (O.C.), bp: $321°C$, d: 0.959, vap. d.: 9.6, vap. press.: < 0.01 mm.
Hazard Analysis
Toxicity: Details unknown. See also alcohols.
Fire Hazard: Slight, when exposed to heat or flame; it can react with oxidizing materials.
Countermeasures
To Fight Fire: Carbon dioxide, dry chemical or carbon tetracholoride (Section 6).
Storage and Handling: Section 7.

TOXIC HAZARD RATING CODE (For detailed discussion, see Section 1.)

0 NONE: (a) No harm under any conditions; (b) Harmful only under unusual conditions or overwhelming dosage.

1 SLIGHT: Causes readily reversible changes which disappear after end of exposure.

2 MODERATE: May involve both irreversible and reversible changes; not severe enough to cause death or permanent injury.

3 HIGH: May cause death or permanent injury after very short exposure to small quantities.

U UNKNOWN: No information on humans considered valid by authors.

DIAMYL PHTHALATE
General Information
Description: Liquid.
Formula: $C_6H_4(COOC_5H_{11})_2$.
Constants: Mol wt: 306.4, bp: 333°C, flash p: 357°F (C.C.), vap. d.: 10.5.
Hazard Analysis
Toxicity: See esters, amyl alcohol and phthalic acid.
Fire Hazard: Slight, when exposed to heat or flame; it can react with oxidizing materials.
Countermeasures
To Fight Fire: Water, foam, carbon dioxide, dry chemical or carbon tetrachloride (Section 6).
Storage and Handling: Section 7.

DI-n-AMYL PHTHALATE
General Information
Synonym: Amyl phthalate.
Description: Liquid.
Formula: $C_6H_4(CO_2C_5H_{11})_2$.
Constants: Mol wt: 306.4, bp: 243–255°C at 50 mm, flash p: 245°F (C.C.), d: 1.023, vap. d.: 10.5.
Hazard Analysis
Toxicity: See esters, amyl alcohol and phthalic acid.
Fire Hazard: Slight, when exposed to heat or flame, it can react with oxidizing materials.
Countermeasures
To Fight Fire: Water, foam, carbon dioxide, dry chemical or carbon tetrachloride (Section 6).
Storage and Handling: Section 7.

DIAMYL SODIUM SULFOSUCCINATE
General Information
Description: White, waxy solid.
Formula: $C_5H_{11}CO_2CH_2CHCO_2C_5H_{11}SO_3Na$.
Constant: Mol wt: 360.
Hazard Analysis
Toxic Hazard Rating:
Acute Local: Irritant 1; Ingestion 1; Inhalation 1.
Acute Systemic: U
Chronic Local: Irritant 1.
Chronic Systemic: U.
Disaster Hazard: Dangerous; see sulfonates.
Countermeasures
Ventilation Control: Section 2.
Personal Hygiene: Section 3.
Storage and Handling: Section 7.

DIAMYL SULFIDE. See amyl sulfide.

DIANILINE CALCIUM
General Information
Description: White crystals decomposed by water.
Formula: $(NHC_6H_5)_2Ca$.
Constant: Mol wt: 224.3.
Hazard Analysis and Countermeasures
See aniline.

DIANILINE FLUOSILICATE
General Information
Description: White crystals, water soluble.
Formula: $(C_6H_5NH_2)_2 \cdot H_2SiF_6$.
Constants: Mol wt: 330, mp: sublimes at 230°C.
Hazard Analysis and Countermeasures
See aniline and fluosilicates.

DIANISIDINE
General Information
Synonym: Dimethoxybenzidine.
Description: Liquid.
Formula: $[NH_2(OCH_3)C_6H_3]_2$.
Constants: Mol wt: 244.29, mp: 137–138°C, flash p: 403°F, vap. d.: 8.5.
Hazard Analysis
Toxicity: Details unknown. See also benzidine and anisidine.

Fire Hazard: Slight, when exposed to heat or flame; it can react with oxidizing materials.
Countermeasures
To Fight Fire: Water, carbon dioxide, dry chemical or carbon tetrachloride (Section 6).
Storage and Handling: Section 7.

DI-o-ANISYL DICHLOROSTANNANE
General Information
Description: Solid.
Formula: $(CH_3OC_6H_4)_2SnCl_2$.
Constants: Mol wt: 404, mp: 113°C.
Hazard Analysis and Countermeasures
See chlorides and tin compounds.

2,2-DI-p-ANISYL-1,1,1-TRICHLOROETHANE.
See methoxychlor.

DIATOMACEOUS EARTH
General Information
Synonyms: Infusorial earth; kieselguhr.
Description: A soft, earthy rock.
Formula: SiO_2, K_2O, Al_2O_3, Fe_2O_3, CaO.
Constant: D: 0.24–0.34.
Hazard Analysis
Toxic Hazard Rating:
Acute Local: Inhalation 2.
Acute Systemic: U.
Chronic Local: Inhalation 2.
Chronic Systemic: Inhalation 2.
Toxicology: Diatomaceous earth dust may cause disabling fibrosis of the lungs, but it is less likely to do so than SiO_2 in the crystalline form.
Note: It is a substance which migrates to food from packaging materials (Section 10).
Note: Before roasting or calcining it is less dangerous than after.
Countermeasures
Ventilation Control: Section 2.

DIAZINON
General Information
Synonyms: G-24,480; O,O-diethyl-O,2-isopropyl-4-methyl-pyrimidyl thiophosphate.
Description: Liquid with faint ester-like odor. Miscible in organic solvents.
Formula: $C_{12}H_{21}N_2O_3PS$.
Constants: Mol wt: 304.4, bp: 84°C at 0.002 mm, d: 1.116 at 20/4°C.
Hazard Analysis and Countermeasures
Action less severe than parathion. See parathion.
Shipping Regulations: Section 11.
IATA: Other restricted articles, class A, no label required, no limit (passenger and cargo).

DIAZOACETIC ESTER
General Information
Synonyms: Ethyl diazoacetate; ethyl diazoethanoate.
Description: Yellow oil, pungent odor, very volatile.
Formula: $C_4H_6N_2O_2$.
Constants: Mol wt: 114.1, mp: −22°C, bp: 141°C at 720 mm, d: 1.0852 at 17.6°/4°C, vap. d.: 3.9.
Hazard Analysis
Toxicity: Details unknown. See also esters.
Explosion Hazard: Moderate, when exposed to heat, upon contact with concentrated sulfuric acid, or upon distillation.
Disaster Hazard: Dangerous; explodes when heated; on contact with acid or acid fumes, it emits toxic fumes.
Countermeasures
Storage and Handling: Section 7.

α-DIAZOAMIDOBENZOL
General Information
Synonyms: Diazoaminobenzene; 1,3-diphenyltriazine.
Description: Golden yellow crystals.

Formula: $C_6H_5NNNHC_6H_5$.
Constants: Mol wt: 197.2, mp: 98–99°C, bp: explodes, vap. d.: 6.8.
Hazard Analysis
Toxicity: Details unknown. An insecticide. See also benzene.
Explosion Hazard: Severe, when shocked or exposed to heat (Section 7).
Disaster Hazard: Dangerous; shock or heat will explode it.
Countermeasures
Storage and Handling: Section 7.

β-DIAZOAMIDOBENZOL
General Information
Synonym: Benzene diazoanilide.
Description: Yellow crystals.
Formula: $C_6H_5NNNHC_6H_5$.
Constants: Mol wt: 197.2, mp: 80–81°C, bp: explodes, vap. d.:. 6.8.
Hazard Analysis
Toxicity: Details unknown. See also aniline.
Explosion Hazard: Severe, when shocked or exposed to heat (Section 7).
Disaster Hazard: Dangerous; explodes when heated and emits toxic fumes.
Countermeasures
Storage and Handling: Section 7.

1,1′-DIAZOAMIDONAPHTHALENE
General Information
Synonym: 1,3-di-1-naphthyltriazine.
Description: Yellow leaflets.
Formula: $C_{10}H_7NNNHC_{10}H_7$.
Constants: Mol wt: 297.4, mp: explodes at 100°C, vap. d.: 10.25.
Hazard Analysis
Toxicity: Details unknown. See also naphthalene.
Explosion Hazard: Severe, when shocked or exposed to heat.
Disaster Hazard: Dangerous; explodes when heated or shocked; emits toxic fumes.
Countermeasures
Storage and Handling: Section 7.

DIAZOAMINOBENZENE. See α-diazoamidobenzol.

DIAZOBENZENE CHLORIDE
General Information
Synonym: Benzene diazonium chloride.
Description: Colorless needles.
Formula: $C_6H_5N(:N)Cl$.
Constants: Mol wt: 140.6, mp: decomposes, bp: explodes, vap. d.: 4.9.
Hazard Analysis
Toxicity: Unknown.
Explosion Hazard: Severe, when shocked or exposed to heat (Section 7).
Disaster Hazard: Dangerous; shock and heat will explode it; emits toxic fumes.
Countermeasures
Storage and Handling: Section 7.

DIAZOBENZENEIMIDE. See phenyl azoimide.

DIAZOBENZENE NITRATE
General Information
Synonym: Benzene diazonium nitrate.
Description: Colorless needles.
Formula: $C_6H_5N(:N)NO_3$.

Constants: Mol wt: 167.1, mp: explodes at 90°C, d: 1.37 at 20°/4°C, vap. d.: 5.8.
Hazard Analysis
Toxicity: Unknown. See also nitrates.
Fire Hazard: See nitrates.
Explosion Hazard: Severe; explodes when slightly shocked or exposed to heat (Section 7). See also nitrates.
Disaster Hazard: See nitrates.
Countermeasures
Storage and Handling: Section 7.

DIAZOBENZENE SULFATE. See diazobenzol sulfate.

p-DIAZOBENZENESULFONIC ACID
General Information
Synonym: Sulfanilic acid diazide.
Description: White paste (white or slightly red crystals).
Formula: $C_6H_4NSO_3N$.
Constants: Mol wt: 184.17, vap. d.: 6.3.
Hazard Analysis
Toxicity: Details unknown.
Explosion Hazard: Severe, when shocked, exposed to heat, or by chemical reaction (Section 7).
Disaster Hazard: Dangerous; shock and heat will explode it; see sulfonates.
Countermeasures
Storage and Handling: Section 7.

DIAZOBENZOIC NITRATE. See diazobenzene nitrate.

DIAZOBENZOL ANILIDE. See diazoamidobenzol.

DIAZOBENZOL CHLORIDE. See diazobenzene chloride.

DIAZOBENZOL CHROMATE
General Information
Synonym: Benzene diazonium chromate.
Description: Crystals.
Formula: $[C_6H_5N(:N)]_2CrO_4$.
Constant: Mol wt: 326.3.
Hazard Analysis
Toxicity: See chromium compounds.
Fire Hazard: See chromates.
Explosion Hazard: Severe, when shocked or exposed to heat.
Disaster Hazard: Dangerous; shock and heat will explode it; emits toxic fumes.
Countermeasures
Storage and Handling: Section 7.

DIAZOBENZOLIMIDE. See phenyl azoimide.

DIAZOBENZOL NITRATE. See diazobenzene nitrate.

DIAZOBENZOL SULFATE
General Information
Synonym: Diazobenzene sulfate.
Description: Crystals.
Formula: $[C_6H_5N(:N)]_2SO_4$.
Constants: Mol wt: 306.3, mp: explodes at 100°C, vap. d.: 10.6.
Hazard Analysis
Toxicity: Details unknown.
Explosion Hazard: Severe, when shocked or exposed to heat (section 7).
Disaster Hazard: Dangerous; shock and heat will explode it; see sulfates.
Countermeasures
Storage and Handling: Section 7.

TOXIC HAZARD RATING CODE (For detailed discussion, see Section 1.)

0 NONE: (a) No harm under any conditions; (b) Harmful only under unusual conditions or overwhelming dosage.

1 SLIGHT: Causes readily reversible changes which disappear after end of exposure.

2 MODERATE: May involve both irreversible and reversible changes; not severe enough to cause death or permanent injury.

3 HIGH: May cause death or permanent injury after very short exposure to small quantities.

U UNKNOWN: No information on humans considered valid by authors.

DIAZOBENZOLSULFONIC ACID. See p-diazobenzene-sulfonic acid.

DIAZODINITROPHENOL
General Information
Synonym: Dinol.
Description: Crystals.
Formula: $HOC_6H_3(NO_2)_2N(:N)$.
Constants: Mol wt: 212.1, mp: explodes at 180°C, d: 1.63, vap. d.: 7.3.
Hazard Analysis
Toxicity: Unknown. See also dinitrophenol.
Fire Hazard: See nitrates.
Explosion Hazard: Severe; when shocked or exposed to heat or friction. A less sensitive high explosive to impact than lead azide or mercury fulminate. It is a powerful initiator (Section 7).
Countermeasures
Storage and Handling: Section 7.

DIAZOMETHANE
General Information
Synonym: Azimethylene.
Description: Yellow gas at ordinary temperature.
Formula: $CH_2 = N^+ = N^-$.
Constants: Mol wt: 42.04, mp: $-145°C$, bp: $-23°C$, d: 1.45.
Hazard Analysis
Toxic Hazard Rating:
  Acute Local: Irritant 3; Inhalation 3.
  Acute Systemic: U.
  Chronic Local: Irritant 2; Inhalation 2; Allergen 3.
  Chronic Systemic: Inhalation 3.
  TLV: ACGIH (recommended) 0.2 part per million of air; 0.4 milligrams per cubic meter of air.
Toxicology: A highly toxic, powerful irritant. It can cause pulmonary edema and frequently causes hypersensitivity leading to asthmatic symptoms.
Explosion Hazard: Severe, when shocked, exposed to heat or by chemical reaction. Undiluted liquid or gas may explode on contact with alkali metals, rough surfaces, heat (100°C) and shock (Section 7).
Disaster Hazard: Highly dangerous; shock and heat will explode it; when heated to decomposition or on contact with acid or acid fumes, it emits highly toxic fumes.
Countermeasures
Ventilation Control: Section 2.
Personnel Protection: Section 3.
Storage and Handling: Section 7.

1-DIAZO-2-NAPHTHOL-4-SULFONIC ACID
General Information
Description: Yellow needles in paste or dry form.
Formula: $C_{10}H_5N_2OSO_3H$.
Constants: Mol wt: 250.23, bp: decomp. by heat above 100°C, vap. d.: 8.6.
Hazard Analysis
Toxicity: Unknown.
Explosion Hazard: Moderate, when exposed to heat (Section 7).
Explosive Range: Above 100°C.
Disaster Hazard: Dangerous; when heated to decomposition, it emits highly toxic fumes of oxides of sulfur and explodes.
Countermeasures
Storage and Handling: Section 7.

DIAZONITROPHENOL
General Information
Description: Crystals.
Formula: $HOC_6H_4(NO_2)N(:N)$.
Constants: Mol wt: 167.1, mp: explodes at 180°C, vap. d.: 5.8.
Hazard Analysis
Toxicity: Unknown.

Fire Hazard: See nitrates.
Explosion Hazard: Severe, when shocked, exposed to heat or by chemical reaction (Section 7).
Disaster Hazard: See nitrates.
Countermeasures
Storage and Handling: Section 7.

5-DIAZOSALICYLIC ACID
General Information
Description: Yellow crystals or powder.
Formula: $C_7H_4N_2O_3$.
Constants: Mol wt: 164.12, mp: explodes at 155°C.
Hazard Analysis
Toxicity: Unknown. See also salisylic acid.
Explosion Hazard: Moderate, when exposed to heat (Section 7).
Disaster Hazard: Dangerous, it emits toxic fumes, and explodes when heated.
Countermeasures
Storage and Handling: Section 7.

DIAZOTIZING SALTS. See sodium nitrite.

DIBAL. See diisobutyl aluminum hydride.

DIBASIC SODIUM o-ARSENATE
General Information
Description: White powder.
Formula: $Na_2HAsO_4 \cdot 7H_2O$.
Constants: Mol wt: 312.0, d:1.87.
Hazard Analysis and Countermeasures
See arsenic compounds.

DIBASIC SODIUM o-ARSENITE
General Information
Description: White powder.
Formula: Na_2HAsO_3.
Constant: Mol wt: 169.9.
Hazard Analysis and Countermeasures
See arsenic compounds.

DIBENAMINE. See dibenzyl chloroethylamine.

DIBENZANTHRENE. See perylene.

DIBENZOFURAN. See diphenylene oxide.

DIBENZOPYRAZINE. See phenazine.

DIBENZOPYRROLE. See carbazole.

2,2'-DIBENZOTHIAZYL DISULFIDE
General Information
Synonym: 2,2'-dithio bis-benzothiazole.
Description: Light-yellow crystals.
Formula: $C_{14}H_8N_2S_4$.
Constants: Mol wt: 332.5, mp: 180°C, bp: decomposes, d: 1.50 at 20°/4°C.
Hazard Analysis
Toxicity: Unknown.
Disaster Hazard: Dangerous; when heated or on contact with acid or acid fumes, it emits highly toxic fumes of oxides of sulfur.
Countermeasures
Storage and Handling: Section 7.

DI-(2-BENZOTHIAZYL-THIOMETHYL)-UREA
General Information
Synonym: 1,3-bis(2-benzothiazolyl mercaptomethyl) urea.
Description: Cream-colored powder.
Formula: $(C_6H_4NSCSCH_2NH)_2CO$.
Constants: Mol wt: 418.6, mp: 220°C, d: 1.29 at 20°C.
Hazard Analysis
Toxicity: Unknown. See also thiourea
Disaster Hazard: Dangerous; when heated to decomposition, it emits highly toxic fumes of oxides of sulfur.
Countermeasures
Storage and Handling: Section 7.

DIBENZOTHIOXIN. See phenothioxin.

DIBENZOXYHYDROQUINONE. See dibenzyl ether of hydroquinone.

DIBENZOYL. See benzil.

DIBENZOYL DIETHYLENE GLYCOL ESTER
Hazard Analysis
Toxic Hazard Rating: U. Limited animal experiments suggest low toxicity. See also glycols.

DIBENZOYL DIPROPYLENE GLYCOL ESTER
Hazard Analysis
Toxic Hazard Rating: U. Limited animal experiments suggest low toxicity. See also glycols.

DIBENZOYL PEROXIDE. See benzoyl peroxide.

DIBENZPHENANTHRENE. See picene.

DIBENZYL AMINE
General Information
Description: An oil with ammonia-like odor, insoluble in water, soluble in alcohol and ether.
Formula: $C_{14}H_{15}N$.
Constants: Mol wt: 197.3, bp: 300°C (partial decomp.).
Hazard Analysis and Countermeasures
See diphenyl amine.

DIBENZYLCHLOROETHYLAMINE
Hazard Analysis
Toxic Hazard Rating:
    Acute Local: Irritant 1.
    Acute Systemic: Ingestion 2.
    Chronic Local: Irritant 1.
    Chronic Systemic: Ingestion 2.
Can cause leukopenia.
Disaster Hazard: Dangerous. See chlorides.

DIBENZYL DIETHYL STANNANE
General Information
Description: A liquid soluble in organic solvents.
Formula: $(C_6H_5CH_2)_2Sn(C_2H_5)_2$.
Constants: Mol wt: 359.1, d: > 1, mp:< 20°C, bp: 224°C at 20 mm.
Hazard Analysis and Countermeasures
See tin compounds.

DIBENZYL ETHER
General Information
Description: Liquid.
Formula: $(C_6H_5CH_2)O(C_6H_5CH_2)$.
Constants: Mol wt: 198.25, mp: 5°C, bp: 298°C, flash p: 275°F (C.C.), d: 1.036, vap. d.: 6.84.
Hazard Analysis
Toxicity: Details unknown. Probably an irritant. Vapors probably narcotic in high concentrations. See also ethers.
Fire Hazard: Slight, when exposed to heat or flame; it can react with oxidizing materials.
Spontaneous Heating: No.
Explosion Hazard: Moderate, by spontaneous chemical reaction. See also ethers.
Countermeasures
Storage and Handling: Section 7.
To Fight Fire: Water, foam, carbon dioxide, dry chemical or carbon tetrachloride (Section 6).

DIBENZYL ETHER OF HYDROQUINONE
General Information
Synonym: Dibenzoxyhydroquinone.
Description: Powder.
Formula: $C_6H_4(OC_6H_5CH_2)_2$.
Constants: Mol wt: 290.3, vap. d.: 10.
Hazard Analysis
Toxicity: Details unknown. See also hydroquinones and ethers.
Fire Hazard: Moderate, when exposed to heat or flame; it can react with oxidizing materials (Section 6).
Countermeasures
Storage and Handling: Section 7.

DIBENZYL ETHYL PROPYL STANNANE
General Information
Description: Liquid miscible in organic liquids.
Formula: $(C_6H_5CH_2)_2Sn(C_2H_5)(C_3H_7)$.
Constants: Mol wt: 373.1, mp: > 0°C, bp: 225°C at 15 mm.
Hazard Analysis and Countermeasures
See tin compounds.

DIBENZYL MERCURY
General Information
Description: Colorless crystals soluble in organic solvents.
Formula: $(C_7H_7)_2Hg$.
Constants: Mol wt: 382.9.
Hazard Analysis and Countermeasures
See mercury compounds, organic.

DIBENZYL SEBACATE
General Information
Description: Plastic material or light-straw colored liquid.
Formula: $[(CH_2)_4CO_2C_6H_5CH_2]_2$.
Constants: Mol wt: 382.5, bp: 265°C at 4 mm, flash p: 457°F, d: 1.055 at 30°/20°C.
Hazard Analysis
Toxicity: Details unknown. See also esters, benzyl alcohol and sebacic acid.
Fire Hazard: Slight, when exposed to heat or flame; it can react with oxidizing materials.
Countermeasures
To Fight Fire: Water, foam, carbon dioxide, dry chemical or carbon tetrachloride (Section 6).
Storage and Handling: Section 7.

DIBENZYL TIN ACETATE
General Information
Description: Colorless crystals soluble in acetone and benzene.
Formula: $(C_6H_5CH_2)_2Sn(CH_3CO_2)_2$.
Constants: Mol wt: 419, mp: 137°C.
Hazard Analysis and Countermeasures
See tin compounds.

DIBENZYL TIN DIBROMIDE
General Information
Description: Colorless crystals soluble in organic solvents.
Formula: $(C_6H_5CH_2)_2SnBr_2$.
Constants: Mol wt: 460.8, mp: 130°C.
Hazard Analysis and Countermeasures
See tin compounds and bromides.

DIBENZYL TIN DICHLORIDE
General Information
Description: Colorless crystals soluble in organic solvents.
Formula: $(C_6H_5CH_2)_2SnCl_2$.
Constants: Mol wt: 371.9, mp: 164°C.

TOXIC HAZARD RATING CODE (*For detailed discussion, see Section 1.*)

0 NONE: (a) No harm under any conditions; (b) Harmful only under unusual conditions or overwhelming dosage.

1 SLIGHT: Causes readily reversible changes which disappear after end of exposure.

2 MODERATE: May involve both irreversible and reversible changes; not severe enough to cause death or permanent injury.

3 HIGH: May cause death or permanent injury after very short exposure to small quantities.

U UNKNOWN: No information on humans considered valid by authors.

Hazard Analysis and Countermeasures
See tin compounds and chlorides.

DIBENZYL TIN DIIODIDE
General Information
Description: Colorless crystals soluble in organic solvents.
Formula: $(C_6H_5CH_2)_2SnI_2$.
Constants: Mol wt: 555, mp: 87°C.
Hazard Analysis and Countermeasures
See tin compounds and iodides.

DIBORANE
General Information
Synonyms: boroethane; boron hydride.
Description: Colorless gas, sickly sweet odor.
Formula: B_2H_6.
Constants: Mol wt: 27.7, mp: −165.5°C, bp: −92.5°C, d: 0.447 (liquid at −112°C); 0.577 (solid at −183°C), vap. press.: 224 mm at −112°, autoign. temp.: 100–125°F, lel: 0.9%, uel: 98%, flash p: −130°F.
Hazard Analysis
Toxic Hazard Rating:
  Acute Local: Irritant 3; Inhalation 3.
  Acute Systemic: Inhalation 3.
  Chronic Local: U.
  Chronic Systemic: U.
TLV: ACGIH (recommended); 0.1 part per million in air, 0.1 milligram per cubic meter of air.
Toxicology: Inhalation causes irritation of lungs and can lead to pulmonary edema. Severe exposures produce symptoms resembling metalfume fever. See also boron hydrides.
Fire Hazard: Dangerous; when exposed to flame. See also boron hydrides.
Explosion Hazard: Slight, by chemical reaction.
Disaster Hazard: Dangerous; when heated to decomposition, it emits toxic fumes of boron oxides; it will react with water or steam to produce hydrogen; and it can react explosively with oxidizing materials.
Countermeasures
To Fight Fire: Stop the flow of gas. Caution! Diborane reacts violently with halogenated extinguishing agents.
Ventilation Control: Section 2.
Personnel Protection: Section 3.
First Aid: Section 1.
Storage and Handling: Section 7.

DIBORANIDE. See potassium diborane.

DIBORON TETRACHLORIDE
General Information
Formula: B_2Cl_4.
Constant: Mol wt: 164.
Hazard Analysis
Toxic Hazard Rating: U. May liberate highly irritating fumes.
Disaster Hazard: Dangerous. See chlorides.

DIBORON TETRAHYDROXIDE
Hazard Analysis
See boron compounds.

DIBROMETHYNE. See dibromoacetylene.

DIBROM (R). See dimethyl 1,2-dibromo-2,2-di-chloro-ethyl phosphate.

DIBROMOACETYLENE
General Information
Synonym: Dibromethyne.
Description: Liquid.
Formula: BrC:CBr.
Constants: Mol wt: 183.9, mp: 76°C (approx.), bp: explodes, d: 2 (approx.), vap. d.: 6.35.
Hazard Analysis
Toxic Hazard Rating:

Acute Local: Irritant 3; Ingestion 3; Inhalation 3.
Acute Systemic: Inhalation 3.
Chronic Local: U.
Chronic Systemic: U.
Explosion Hazard: Severe, when exposed to heat or by spontaneous chemical reaction. Explodes easily with a trace of oxygen (Section 7).
Disaster Hazard: Highly dangerous; when heated, it emits toxic fumes of bromides; can react explosively with oxidizing materials.
Countermeasures
Ventilation Control: Section 2.
Personnel Protection: Section 3.
First Aid: Section 1.
Storage and Handling: Section 7.

p-DIBROMOBENZENE
General Information
Description: Crystals, odor of xylene.
Formula: BrC_6H_4Br.
Constants: Mol wt: 235.9, mp: 87°C, bp: 219°C, d: 2.261, vap. press.: 1 mm at 61.0°C, vap. d.: 8.1.
Hazard Analysis
Toxicity: Details unknown. A fumigant. Limited experiments with animals suggest moderate toxicity.
Disaster Hazard: Moderately dangerous; when heated to decomposition, it emits toxic fumes of bromides.
Countermeasures
Storage and Handling: Section 7.

DIBROMOCHLOROMETHANE
General Information
Description: Colorless to pale yellow, heavy liquid.
Formula: $CHClBr_2$.
Constants: Mol wt: 208.3, bp: 118–122°C, fp: < −20°C, d: 2.440 at 25°/25°C.
Hazard Analysis
Toxicity: Unknown. Compounds of this type are generally irritant and narcotic. See also bromoform and chloroform.
Disaster Hazard: Moderately dangerous; when heated to decomposition, it emits toxic fumes.
Countermeasures
Storage and Handling: Section 7.

1,2-DIBROMO-3-CHLOROPROPANE
General Information
Synonym: Nemagon.
Formula: $C_3H_5Br_2Cl$.
Constants: Mol wt: 236.4, bp: 196°C.
Hazard Analysis
Toxicology: Details unknown. Irritating to skin and mucous membranes. Narcotic in high concentrations.
Disaster Hazard: Dangerous. See bromides and chlorides.

DIBROMODIETHYL SULFIDE
General Information
Description: White crystals.
Formula: $(CH_2CH_2Br)_2S$.
Constants: Mol wt: 232.00; mp: 31–34°C, bp: 240°C, decomp., d: 2.05 at 15°C, vap. d.: 8.0.
Hazard Analysis
Toxic Hazard Rating:
  Acute Local: Irritant 3; Ingestion 3; Inhalation 3.
  Acute Systemic: U.
  Chronic Local: U.
  Chronic Systemic: U.
Toxicology: A military poison.
Fire Hazard: Moderate, when exposed to heat or flame.
Disaster Hazard: Dangerous; when heated to decomposition or on contact with acid or acid fumes, it emits highly toxic fumes of oxides of sulfur and bromides; will react with water or steam to produce toxic and corrosive fumes; can react vigorously with oxidizing materials.
Countermeasures
Ventilation Control: Section 2.

To Fight Fire: Carbon dioxide, dry chemical or carbon tetrachloride (Section 6).
Personnel Protection: Section 3.
Storage and Handling: Section 7.

DIBROMODIFLUOROMETHANE
General Information
Synonym: Difluorodibromomethane.
Description: Colorless, heavy liquid.
Formula: CF_2Br_2.
Constants: Mol wt: 209.8, bp: 23.2°C, fp: −141°C, d: 2.288 at 15°/4°C.
Hazard Analysis
Toxic Hazard Rating:
  Acute Local: Irritant 2.
  Acute Systemic: Irritant 2.
  Chronic Local: Inhalation 1.
  Chronic Systemic: Inhalation 2.
Rating based on animal experiments.
TLV: ACGIH (recommended) 100 parts per million in air; 860 milligrams per cubic meter of air.
Disaster Hazard: Dangerous; when heated to decomposition, it emits toxic fumes.
Countermeasures
Storage and Handling: Section 7.
Shipping Regulations: Section 11.
  IATA: Other restricted articles, class A, no label required, 40 liters (passenger), 220 liters (cargo).

1,1-DIBROMOETHANE
General Information
Synonym: Ethylidene bromide.
Description: Liquid, insoluble in water, soluble in organic solvents.
Formula: CH_3CHBr_2.
Constants: Mol wt: 188, d: 2.089 at 20.5/4°C, bp: 110°C.
Hazard Analysis
Toxic Hazard Rating:
  Acute Local: Irritant 2; Inhalation 2.
  Acute Systemic: Inhalation 2.
  Chronic Local: U.
  Chronic Systemic: U.
Toxicology: Irritant and narcotic material.
Disaster Hazard: See bromides.
Countermeasures
Storage and Handling: See Section 7.

1,2-DIBROMOETHANE. See ethylene dibromide.

1,2-DIBROMOETHYL BENZENE
General Information
Synonym: Styrene dibromide.
Description: Colorless liquid.
Formula: $C_6H_5(Br_2C_2H_3)$.
Constants: Mol wt: 264.2, bp: 254.0°C, fp: −43°C, flash p: none, d: 1.744 at 25°/25°C, vap. press.: 1 mm at 86.0°C, vap. d.: 9.1.
Countermeasures
Storage and Handling: Section 7.

DIBROMOFORMOXIME
General Information
Description: Crystals.
Formula: CBr_2NOH.
Constants: Mol wt: 202.87, mp: 70–71°C, vap. d.: 7.0.
Hazard Analysis
Toxic Hazard Rating:
  Acute Local: Irritant 3; Ingestion 3; Inhalation 3.

Acute Systemic: U.
Chronic Local: U.
Chronic Systemic: U.
Toxicology: A military poison.
Disaster Hazard: Dangerous; when heated, it emits highly toxic fumes.
Countermeasures
Ventilation Control: Section 2.
Personnel Protection: Section 3.
Storage and Handling: Section 7.

DIBROMOGERMANE
General Information
Description: Colorless liquid.
Formula: GeH_2Br_2.
Constants: Mol wt: 234.45, mp: −15.0°C, bp: 89.0°C, d: 2.80 at 0°C, vap. d.: 8.1.
Hazard Analysis and Countermeasures
See germanium compounds and bromides.

DIBROMOKETONE
General Information
Formula: $(Br)_2C = O$.
Constants: Mol wt: 188, vap. d.: 6.5.
Hazard Analysis
Toxic Hazard Rating:
  Acute Local: Irritant 3; Ingestion 3; Inhalation 3.
  Acute Systemic: U.
  Chronic Local: U.
  Chronic Systemic: U.
Disaster Hazard: Dangerous; when heated it emits highly toxic fumes; it will react with water or steam to produce toxic and corrosive fumes.
Countermeasures
Ventilation Control: Section 2.
Personnel Protection: Section 3.
Personal Hygiene: Section 3.
Storage and Handling: Section 7.

DIBROMOMETHANE. See methylene bromide.

DIBROMOMETHYL ETHER
General Information
Description: Colorless liquid.
Formula: $(CH_2Br)_2O$.
Constants: Mol wt: 203.90, mp: −34°C, bp: 154–155°C, d: 2.2, vap. d.: 7.0.
Hazard Analysis
Toxic Hazard Rating:
  Acute Local: Irritant 3; Ingestion 3; Inhalation 3.
  Acute Systemic: U.
  Chronic Local: U.
  Chronic Systemic: U.
Toxicology: A lacrimator type of military poison.
Fire Hazard: Details unknown. See also ethers.
Explosion Hazard: Details unknown. See also ethers.
Disaster Hazard: Dangerous; when heated it emits highly toxic fumes; it can react vigorously with oxidizing materials.
Countermeasures
Ventilation Control: Section 2.
Personnel Protection: Section 3.
Personnel Hygiene: Section 3.
Storage and Handling: Section 7.

DIBROMOMETHYL SULFIDE
General Information
Formula: Br_2CS.
Constant: Mol wt: 204.

TOXIC HAZARD RATING CODE (For detailed discussion, see Section 1.)

0 NONE: (a) No harm under any conditions; (b) Harmful only under unusual conditions or overwhelming dosage.

1 SLIGHT: Causes readily reversible changes which disappear after end of exposure.

2 MODERATE: May involve both irreversible and reversible changes; not severe enough to cause death or permanent injury.

3 HIGH: May cause death or permanent injury after very short exposure to small quantities.

U UNKNOWN: No information on humans considered valid by authors.

Hazard Analysis
Toxicology: A strong irritant. Details unknown.
Disaster Hazard: Dangerous. See sulfides.

DIBROMO NITROETHYLBENZENE
General Information
Formula: $C_6H_3(Br_2)(C_2H_4NO_2)$.
Constants: Mol wt: 309.2, vap. d.: 10.7.
Hazard Analysis
Toxicity: Details unknown. An insecticide.
Fire Hazard: See nitrates.
Disaster Hazard: See nitrates.
Countermeasures
Storage and Handling: Section 7.

DIBUCAINE HYDROCHLORIDE
General Information
Synonym: 2-butoxy-N-(2-diethylaminoethyl) cinchoninamide
hydrochloride; nupercaine hydrochloride; percaine.
Description: Crystals.
Formula: $C_{20}H_{29}N_3O_2 \cdot HCl$.
Constants: Mol wt: 379.9, mp: decomposes at 90–98°C,
vap. d.: 13.1.
Hazard Analysis
Toxicity: Appears to be more toxic than procaine. See
procaine hydrochloride.
Disaster Hazard: Dangerous; when heated to decomposi-
tion, it emits highly toxic fumes of chlorides.
Countermeasures
Storage and Handling: Section 7.

DIBUTOXYETHANE
General Information
Formula: $CH_3CH_2(OC_4H_9)_2$.
Constant: Mol wt: 175.
Hazard Analysis
Toxic Hazard Rating: U. Animal experiments suggest low
toxicity.

DIBUTOXYETHYL PHTHALATE
General Information
Description: Light-colored liquid.
Formula: $C_6H_4(COOC_2H_4OC_4H_9)_2$.
Constants: Mol wt: 366, mp: −50°C (becomes viscous),
bp: 210–233°C at 4 mm, flash p: 407°F, d: 1.063 at
20°/20°C, vap. press.: 0.05 mm at 150°C.
Hazard Analysis
Toxicity: See esters.
Fire Hazard: Slight, when exposed to heat or flame; it
can react with oxidizing materials.
Countermeasures
To Fight Fire: Water, foam, carbon dioxide, dry chemical
or carbon tetrachloride (Section 6).
Storage and Handling: Section 7.

DIBUTOXYMETHANE
General Information
Description: Liquid.
Formula: $CH_2(OC_4H_9)_2$.
Constants: Mol wt: 160.25, mp: 60°C, bp: 165–188°C,
flash p: 140°F (C.C.), d: 0.838, vap. d.: 5.5.
Hazard Analysis
Toxicity: Details unknown. See also dibutoxyethane.
Fire Hazard: Moderate, when exposed to heat or flame; it
can react with oxidizing materials.
Countermeasures
To Fight Fire: Foam, carbon dioxide, dry chemical or car-
bon tetrachloride (Section 6).
Storage and Handling: Section 7.

DIBUTOXYTETRAGLYCOL
General Information
Description: Practically colorless liquid with characteristic
odor.
Formula: $(C_2H_4OC_2H_4OC_4H_9)_2O$.
Constants: Mol wt: 306.44, mp: −20°C, bp: 237°C at

50 mm, flash p: 305°F (O.C.), d: 0.9436 at 20°/20°C,
vap. d.: 10.6.
Hazard Analysis
Toxicity: Details unknown. See also glycols.
Fire Hazard: Slight, when exposed to heat or flame; it can
react with oxidizing materials.
Countermeasures
To Fight Fire: Water, alcohol foam, foam, carbon dioxide,
dry chemical or carbon tetrachloride (Section 6).
Storage and Handling: Section 7.

DIBUTYL ACETAMIDE
General Information
Description: Liquid.
Formula: $(C_4H_9)_2CHCONH_2$.
Constants: Mol wt: 171.3, bp: 242°C, flash p: 225°F,
d: 0.89, vap. d.: 5.9.
Hazard Analysis
Toxicity: Details unknown. See also amines.
Fire Hazard: Moderate, when exposed to heat or flame;
it can react with oxidizing materials.
Countermeasures
To Fight Fire: Foam, carbon dioxide, dry chemical or
carbon tetrachloride (Section 6).
Storage and Handling: Section 7.

DIBUTYLAMINE
General Information
Description: Liquid.
Formula: $(C_4H_9)_2NH$.
Constants: Mol wt: 129.2, mp: −51.1°C, bp: 159°C,
flash p: 125°F (O.C.), d: 0.76, vap. d.: 4.46, vap. press.:
2 mm at 20°C.
Hazard Analysis
Toxic Hazard Rating:
Acute Local: Irritant 3; Ingestion 3; Inhalation 3.
Acute Systemic: Ingestion 2; Inhalation 3.
Chronic Local: U.
Chronic Systemic: U.
Rating based on animal experiments.
Toxicity: See amines.
Fire Hazard: Moderate, when exposed to heat or flame;
it can react with oxidizing materials.
Countermeasures
To Fight Fire: Alcohol foam, foam, carbon dioxide, dry
chemical or carbon tetrachloride (Section 6).
Storage and Handling: Section 7.

DI-sec-BUTYLAMINE
General Information
Description: Liquid.
Formula: $(C_4H_9)_2NH$.
Constants: Mol wt: 129.2, bp: 134°C, flash p: 75°F,
d: 0.75, vap. d.: 4.5.
Hazard Analysis
Toxicity: See Dibutylamine.
Fire Hazard: Dangerous; when exposed to heat or flame; it
can react with oxidizing materials.
Countermeasures
To Fight Fire: Alcohol foam, foam, carbon dioxide, dry
chemical or carbon tetrachloride (Section 6).
Ventilation Control: Section 2.
Storage and Handling: Section 7.

DIBUTYLAMINOETHANOL
General Information
Description: Liquid.
Formula: $(C_4H_9)_2NCH_2CH_2OH$.
Constants: Mol wt: 173.3, bp: 222°C, flash p: 200°F,
d: 0.85, vap. d.: 6.0.
Hazard Analysis
Toxicity: Details unknown. See also alcohols.
Fire Hazard: Moderate, when exposed to heat or flame;
can react with oxidizing materials.

Countermeasures
To Fight Fire: Carbon dioxide, dry chemical or carbon tetrachloride.
Storage and Handling: Section 7.

3-(DIBUTYL AMINO) PROPYLAMINE
General Information
Formula: $(C_4H_9)_2NCH_2CH_2NH_2$.
Constant: Mol wt: 186.
Hazard Analysis
Toxic Hazard Rating: U. Limited animal experiments suggest high toxicity and irritation. See also amines.

DIBUTYLANILINE
General Information
Description: Liquid.
Formula: $C_6H_5N(C_4H_9)_2$.
Constants: Mol wt: 205.3, bp: 266°C, flash p: 230°F, d: 0.94, vap. d.: 7.1.
Hazard Analysis
Toxicity: Details unknown. See also aniline.
Fire Hazard: Slight, when exposed to heat or flame (Section 6).
Disaster Hazard: Dangerous; when heated to decomposition, it emits highly toxic fumes of nitrogen compounds; can react vigorously with oxidizing materials.
Countermeasures
Storage and Handling: Section 7.
To Fight Fire: Water, foam (Section 6).

DIBUTYL 1,2-BENZENEDICARBOXYLATE. See dibutyl o-phthalate.

2,5-DI-tert-BUTYL BENZOQUINONE
General Information
Description: Crystals.
Formula: $(O)_2 \cdot C_6H_2 \cdot [C(CH_3)_3]_2$.
Constants: Mol wt: 220.3, mp: 150°C, vap. d.: 7.6.
Hazard Analysis
Toxicity: Details unknown. See also quinone.
Fire Hazard: Slight (Section 6).
Countermeasures
Storage and Handling: Section 7.

DIBUTYL BERYLLIUM
General Information
Description: Colorless liquid.
Formula: $Be(C_4H_9)_2$.
Constants: Mol wt: 123.25, bp: 170°C at 25 mm, vap. d.: 4.2.
Hazard Analysis
Toxicity: See beryllium compounds.
Fire Hazard: Moderate, when in contact with heat or flame (Section 6).
Disaster Hazard: Dangerous; when heated to decomposition, it emits highly toxic fumes of beryllium and it can react vigorously with oxidizing materials.
Countermeasures
Storage and Handling: Section 7.

DI-tert-BUTYL BERYLLIUM
General Information
Description: A clear, colorless, mobile liquid.
Formula: $Be(C_4H_9)_2$.
Constants: Mol wt: 123.25, mp: −16°C, d: 0.65, vap. press.: 35 mm at 25°C, vap. d.: 4.2.
Hazard Analysis
Toxicity: See beryllium compounds.

Fire Hazard: Moderate, when exposed to heat or flame (Section 6).
Disaster Hazard: Dangerous; when heated to decomposition, it emits highly toxic fumes; it can react vigorously with oxidizing materials.
Countermeasures
Storage and Handling: Section 7.

DIBUTYL BUTYL PHOSPHONATE
General Information
Description: Colorless liquid, mild odor.
Formula: $C_4H_9P(O)(OC_4H_9)_2$.
Constants: Mol wt: 250.3, bp: 128°C at 2.5 mm, flash p: 311°F (C.O.C.), d: 0.948 at 20°/4°C, vap. d.: 8.62.
Hazard Analysis
Toxicity: Unknown.
Fire Hazard: Slight, when exposed to heat or flame.
Disaster Hazard: Dangerous, when heated to decomposition, it emits highly toxic fumes of oxides of phosphorus; it can react vigorously with oxidizing materials.
Countermeasures
Storage and Handling: Section 7.
To Fight Fire: Foam, carbon dioxide, dry chemical or carbon tetrachloride (Section 6).

DIBUTYL CADMIUM
General Information
Description: Oily liquid decomposed by water.
Formula: $(C_4H_9)_2Cd$.
Constants: Mol wt: 227, d: 1.3056, mp: −48°C, bp: 104°C at 13 mm.
Hazard Analysis
Toxicity: See cadmium compounds.
Fire Hazard: Details unknown. Probably dangerous when exposed to heat or flame. See Section 6.
Disaster Hazard: Dangerous. See cadmium compounds.
Countermeasures
Storage and Handling: See Section 7.

DIBUTYL "CARBITOL." See diethylene glycol dibutyl ether.

DIBUTYL "CELLOSOLVE." See glycols.

DI(BUTYL "CELLOSOLVE") ETHYLENE BISGLYCOLATE
General Information
Description: Liquid.
Formula: $(CH_2OCH_2CO_2CH_2CH_2OC_4H_9)_2$.
Constants: Mol wt: 378.5, bp: 199–207°C at 1.3 mm, flash p: 410°F (O.C.), d: 1.0667 at 25°/25°C, vap. d.: 13.1.
Hazard Analysis
Toxicity: Details unknown. See also glycols.
Fire Hazard: Slight, when exposed to heat or flame; it can react with oxidizing materials.
Countermeasures
To Fight Fire: Foam, carbon dioxide, dry chemical or carbon tetrachloride (Section 6).
Storage and Handling: Section 7.

DIBUTYL "CELLOSOLVE" PHTHALATE
Hazard Analysis
Toxicity: Details unknown. See also glycols.
Fire Hazard: Slight; when heated, it can react with oxidizing materials (Section 6).
Countermeasures
Storage and Handling: Section 7.

TOXIC HAZARD RATING CODE (For detailed discussion, see Section 1.)

0 NONE: (a) No harm under any conditions; (b) Harmful only under unusual conditions or overwhelming dosage.

1 SLIGHT: Causes readily reversible changes which disappear after end of exposure.

2 MODERATE: May involve both irreversible and reversible changes; not severe enough to cause death or permanent injury.

3 HIGH: May cause death or permanent injury after very short exposure to small quantities.

U UNKNOWN: No information on humans considered valid by authors.

DIBUTYL "CELLOSOLVE" SUCCINATE
Hazard Analysis
Toxicity: Details unknown. See also glycols.
Fire Hazard: Slight; when heated, it can react with oxidizing materials (Section 6).
Countermeasures
Storage and Handling: Section 7.

DIBUTYL CHLOROPHOSPHATE
General Information
Description: Water-white liquid.
Formula: $(C_4H_9O)_2POCl$.
Constants: Mol wt: 228.5, bp: 103–106°C at 1.5 mm, d: 1.0742 at 20°/20°C.
Hazard Analysis
Toxicity: Unknown.
Disaster Hazard: Dangerous; when heated to decomposition, it emits highly toxic fumes of chlorides and oxides of phosphorus.
Countermeasures
Storage and Handling: Section 7.

DI-tert-BUTYL-m-CRESOL
General Information
Synonym: DBMC.
Description: Yellow, crystalline solid.
Formula: $C_{15}H_{24}O$.
Constants: Mol wt: 220.3, mp: 62.1°C, bp: 282°C, flash p: 262°F (O.C.), d: 0.912 at 80°/4°C.
Hazard Analysis
Toxicity: Details unknown. See di-tert-butyl-p-cresol.
Fire Hazard: Slight, when exposed to heat or flame.
Disaster Hazard: Moderately dangerous; when heated to decomposition, it emits toxic fumes; it can react with oxidizing materials.
Countermeasures
Storage and Handling: Section 7.
To Fight Fire: Foam, carbon dioxide, dry chemical or carbon tetrachloride (Section 6).

DI-tert-BUTYL-p-CRESOL
General Information
Synonyms: DBPC; 4-methyl-2, 6-di-tert-butyl phenol; BHT; butylated hydroxy toluene; o,o'-di-tert-butyl-p-cresol.
Description: White, crystalline solid.
Formula: $C_6H_2(C_4H_9)_2(CH_3)OH$.
Constants: Mol wt: 220.3, bp: 265°C, fp: 68°C, flash p: 260°F (T.O.C.), d: 1.048 at 20°/4°C, vap. d.: 7.6.
Hazard Analysis
Toxic Hazard Rating:
  Acute Local: Irritant 1.
  Acute Systemic: U.
  Chronic Local: Ingestion 1.
  Chronic Systemic: U.
Toxicology: Limited animal experiments suggest low toxicity. Used as a food additive.
Fire Hazard: Slight, when exposed to heat or flame.
Disaster Hazard: Moderately dangerous; when heated to decomposition, it emits toxic fumes; it can react with oxidizing materials.
Countermeasures
Storage and Handling: Section 7.
To Fight Fire: Water, foam, carbon dioxide, dry chemical or carbon tetrachloride (Section 6).

o,o'-DI-tert-BUTYL-p-CRESOL. See di-tert-butyl-p-cresol.

DIBUTYL-1,2-CYCLOHEXANEDICARBOXYLIC ACID. See dibutyl hexahydrophthalate.

DIBUTYL DIACETOXYSTANNANE. See dibutyltin diacetate.

DIBUTYL DICHLOROSTANNANE. See dibutyltin dichloride.

DIBUTYL-2,3-DIHYDROXYBUTANEDIOATE. See dibutyl tartrate.

DI-tert-BUTYL DIPERPHTHALATE (solution in dibutyl phthalate).
General Information
Description: Solution is a clear liquid, immiscible in water but miscible in common organic solvents.
Formula: $C_{16}H_{22}O_6$.
Constants: Mol wt: 310.3, d: 1.056 at 25°C, flash p: 145°F (M.O.C.).
Hazard Analysis
See peroxides, organic; also dibutyl phthalate.
Countermeasures
See peroxides, organic.
Shipping Regulations: Section 11.
  I.C.C. Classification: Peroxides; organic solution liquid N.O.S.
  Express and Freight: Yellow label-chemicals N.O.I.B.N.
  Parcel Post: Prohibited.

N,N-DIBUTYL ETHANOLAMINE. See dibutylaminoethanol.

n-DIBUTYL ETHER. See n-butyl ether.

DIBUTYL ETHYLENE BIS-GLYCOLATE
General Information
Description: Liquid.
Formula: $(CH_2OCH_2CO_2C_4H_9)_2$.
Constants: Mol wt: 290.4, bp: 145–150°C at 0.6 mm, flash p: 248°F (O.C.), d: 1.0391 at 25°/25°C, vap. d.: 10.0.
Hazard Analysis
Toxicity: Unknown.
Fire Hazard: Slight, when exposed to heat or flame; it can react with oxidizing materials.
Countermeasures
To Fight Fire: Foam, carbon dioxide, dry chemical or carbon tetrachloride (Section 6).
Storage and Handling: Section 7.

DIBUTYL FUMARATE
General Information
Description: Colorless, clear, mobile, liquid. Typical odor.
Formula: $C_4H_9OCOCHCHCOOC_4H_9$.
Constants: Mol wt: 228.28, bp: 285.1°C, fp: −19°C, flash p: 300°F (O.C.), d: 0.986 at 20°/20°C, vap. d.: 7.88.
Hazard Analysis
Toxicity: Practically non-toxic by ingestion or skin contact. Can cause slight irritation of eyes and mucous membranes, based upon animal experiments. See also esters and butyl alcohol.
Fire Hazard: Slight, when exposed to heat or flame; it can react with oxidizing materials.
Countermeasures
To Fight Fire: Foam, carbon dioxide, dry chemical or carbon tetrachloride (Section 6).
Storage and Handling: Section 7.

DIBUTYL HEXAHYDROPHTHALATE
General Information
Synonym: dibutyl-1,2-cyclohexanedicarboxylic acid.
Description: Liquid.
Formula: $C_6H_{10}(CO_2C_4H_9)_2$.
Constants: Mol wt: 284.4, bp: 185°C, flash p: 305°F, d: 1.0, vap. d.: 9.8.
Hazard Analysis
Toxicity: Unknown.
Fire Hazard: Slight, when exposed to heat or flame; can react with oxidizing materials.
Countermeasures
To Fight Fire: Foam, carbon dioxide, dry chemical or carbon tetrachloride (Section 6).
Storage and Handling: Section 7.

N,N-DIBUTYL-(2-HYDROXYPROPYL)AMINE
General Information
Formula: $CH_3CHOHCH_2NH(C_4H_9)_2$.
Constant: Mol wt: 187.
Hazard Analysis
Toxic Hazard Rating: U. Limited animal experiments suggest moderate toxicity and irritation. See also amines.

DIBUTYL ISOPHTHALATE. See dibutyl-m-phthalate.

DIBUTYL ISOPROPANOLAMINE
General Information
Synonym: 2-propanol-1-N,N-dibutylamine.
Description: Liquid.
Formula: $CH_3CHOHCH_2N(C_4H_9)_2$.
Constants: Mol wt: 187.3, mp: $-80°C$, bp: $229°C$, flash p.: $205°F$ (O.C.), d: 0.8419 at $20°/20°C$, vap. press.: <0.1 mm, vap. d.: 6.5.
Hazard Analysis
Toxicity: Details unknown. See also amines.
Fire Hazard: Moderate, when exposed to heat or flame; can react with oxidizing materials.
Countermeasures
To Fight Fire: Alcohol foam, foam, carbon dioxide, dry chemical or carbon tetrachloride (Section 6).
Storage and Handling: Section 7.

DIBUTYL LAURAMIDE
General Information
Description: Liquid.
Formula: $(C_4H_9)_2CH(CH_2)_{10}CONH_2$.
Constants: Mol wt: 311.5, bp: $200°C$, flash p: $375°F$, d: 0.86, vap. d.: 10.7.
Hazard Analysis
Toxicity: Unknown.
Fire Hazard: Slight, when exposed to heat or flame; can react with oxidizing materials.
Countermeasures
To Fight Fire: Foam, carbon dioxide, dry chemical or carbon tetrachloride (Section 6).
Storage and Handling: Section 7.

DIBUTYL MALEATE
General Information
Description: Liquid.
Formula: $C_4H_9OOCHCCHCOOC_4H_9$.
Constants: Mol wt: 228.3, mp: $-85°C$ (sets to a glass), bp: $281°C$, flash p: $285°F$ (O.C.), d: 0.9964 at $20°/20°C$, vap. d.: 7.9.
Hazard Analysis
Toxicity: Slightly toxic by ingestion or skin exposure. Upon skin it causes moderate irritation causing slight swelling and redness. See also esters and butyl alcohol.
Fire Hazard: Slight, when exposed to heat or flame; can react with oxidizing materials.
Countermeasures
To Fight Fire: Foam, carbon dioxide, dry chemical or carbon tetrachloride (Section 6).
Storage and Handling: Section 7.

DIBUTYL MALYL DIOXYSTANNANE. See dibutyltin maleate.

DIBUTYL MERCURY
General Information
Description: Liquid.
Formula: $Hg(C_4H_9)_2$.

Constants: Mol wt: 314.8, bp: $105°C$ at 10 mm, d: 1.7779, vap. d.: 10.8.
Hazard Analysis
Toxicity: Highly toxic. See mercury compounds, organic.
Fire Hazard: Moderate, when exposed to heat or flame (Section 6).
Disaster Hazard: Dangerous; when heated to decomposition or on contact with acid or acid fumes, it emits highly toxic fumes of mercury; can react vigorously with oxidizing materials.
Countermeasures
Storage and Handling: Section 7.

DIBUTYL METHIONATE
General Information
Description: Liquid.
Formula: $CH_2(SO_3C_4H_9)_2$.
Constants: Mol wt: 288.3, vap. d.: 10.0.
Hazard Analysis
Toxicity: Unknown.
Disaster Hazard: Dangerous; when heated to decomposition, it emits highly toxic fumes of oxides of sulfur; it will react with water or steam to produce toxic and corrosive fumes.
Countermeasures
Storage and Handling: Section 7.

N,N-DIBUTYLMETHYLAMINE
General Information
Description: Colorless liquid; amine odor; insoluble in water; soluble in alcohol and ether. Miscible with hydrocarbons.
Formula: $(C_4H_9)_2NH$.
Constant: Mol wt: 129, d: 0.7613 (20/20°C), bp: $159.6°C$, fp: $-62°C$, flash p: $125°F$ (O.C.).
Hazard Analysis
Toxic Hazard Rating: U. Limited animal experiments suggest high toxicity and moderate irritation. See also amines.
Fire Hazard: Moderate, when exposed to heat or flame.
Countermeasures
Storage and Handling: Section 7.

2,6-DI-tert-BUTYL-4-METHYL PHENOL
General Information
Synonym: Ionol.
Description: Granular crystals.
Formula: $(CH_3)_3C-C_6H_2(OH)(CH_3)-C(CH_3)_3$.
Constants: Mol wt: 220.3, mp: $70°C$, bp: $265°C$, flash p: $260°F$ (T.O.C.), d: 0.899 at $80°/4°C$, vap. d.: 7.61.
Hazard Analysis
Toxic Hazard Rating:
  Acute Local: 0.
  Acute Systemic: Ingestion 1.
  Chronic Local: 0.
  Chronic Systemic: Ingestion 1.
Toxicology: Animal experiments indicate low toxicity.
Fire Hazard: Slight, when exposed to heat or flame.
Disaster Hazard: Dangerous; when heated to decomposition, it emits highly toxic fumes of phenol; can react with oxidizing materials.
Countermeasures
Storage and Handling: Section 7.
To Fight Fire: Foam, carbon dioxide, dry chemical or carbon tetrachloride (Section 6).

DIBUTYLOLEAMIDE
General Information
Description: Liquid.

TOXIC HAZARD RATING CODE (For detailed discussion, see Section 1.)

0 NONE: (a) No harm under any conditions; (b) Harmful only under unusual conditions or overwhelming dosage.

1 SLIGHT: Causes readily reversible changes which disappear after end of exposure.

2 MODERATE: May involve both irreversible and reversible changes; not severe enough to cause death or permanent injury.

3 HIGH: May cause death or permanent injury after very short exposure to small quantities.

U UNKNOWN: No information on humans considered valid by authors.

Formula: $C_8H_{17}CHCH(CH_2)_7CON(C_4H_9)_2$.
Constants: Mol wt: 393.7, bp: 220°C, d: 0.85, vap. d.: 13.6.
Hazard Analysis
Toxicity: Unknown.
Fire Hazard: Slight, when exposed to heat or flame; can react with oxidizing materials (Section 6).
Countermeasures
Storage and Handling: Section 7.

DIBUTYL OXALATE
General Information
Description: Water-white, high-boiling liquid, mild odor.
Formula: $(COOC_4H_9)_2$.
Constants: Mol wt: 202.3, mp: −30°C, bp: 246°C, flash p: 220°F (C.C.), d: 0.989–0.993 at 20°/20°C, vap. d.: 7.0.
Hazard Analysis
Toxicity: See oxalic acid.
Fire Hazard: Slight, when exposed to heat or flame; can react with oxidizing materials.
Spontaneous Heating: No.
Countermeasures
To Fight Fire: Foam, carbon dioxide, dry chemical or carbon tetrachloride (Section 6).
Storage and Handling: Section 7.

DIBUTYL OXOSTANNANE. See dibutyltin oxide.

DI-tert-BUTYL PEROXIDE
General Information
Description: Clear, water-white liquid.
Formula: $(C_4H_9O)_2$.
Constants: Mol wt: 146.2, mp: −40°C, bp: 80°C at 284°F, flash p: 65°F (C.C.), d: 0.79, vap. press.: 19.51 mm at 20°C, vap. d.: 5.03.
Hazard Analysis
Toxic Hazard Rating:
Acute Local: Irritant 2; Ingestion 1; Inhalation 1.
Acute Systemic: Inhalation 2.
Chronic Local: Irritant 3.
Chronic Systemic: Ingestion 2; Inhalation 2.
Fire Hazard: See peroxides, organic.
Explosion Hazard: See peroxides, organic.
Disaster Hazard: See peroxides, organic.
Countermeasures
To Fight Fire: Foam, carbon dioxide, dry chemical or carbon tetrachloride (Section 6).
Ventilation Control: Section 2.
Storage and Handling: Section 7.
Shipping Regulations: Section 11.
I.C.C. Classification: Peroxides, organic liquid, N.O.S.
Freight and Rail Express: Red label as chemicals N.O.I.B.N.
Parcel Post: Prohibited.

DI-sec-BUTYL PHENOL
General Information
Description: Amber liquid.
Formula: $(CH_3CHC_2H_5)_2C_6H_3OH$.
Constants: Mol wt: 206.3, bp: 152–165°C at 25 mm, fp: −50°C, flash p: 280°F, d: 0.936 at 25°/4°C.
Hazard Analysis
Toxicity: Unknown. See also phenol.
Fire Hazard: Slight, when exposed to heat or flame; can react with oxidizing materials.
Countermeasures
To Fight Fire: Foam, carbon dioxide, dry chemical or carbon tetrachloride (Section 6).
Storage and Handling: Section 7.

2,4-DI-tert-BUTYL PHENOL
General Information
Description: Tan crystals.
Formula: $HOC_6H_3(C_4H_9)_2$.
Constants: Mol wt: 206.3, mp: 51°C, bp: 260.8°C, flash p:

265°F, d: 0.907 at 60°/4°C, vap. press.: 1 mm at 84.5°C.
Hazard Analysis
Toxicity: Details unknown. See also phenol.
Fire Hazard: Slight, when exposed to heat or flame.
Disaster Hazard: Moderately dangerous; when heated to decomposition, it emits toxic fumes; can react with oxidizing materials.
Countermeasures
Storage and Handling: Section 7.
To Fight Fire: Foam, carbon dioxide, dry chemical or carbon tetrachloride (Section 6).

N,N'-DI-sec-BUTYL-p-PHENYLENEDIAMINE
General Information
Description: Liquid.
Formula: $C_6H_4(CH_3CHNHCH_2CH_3)_2$.
Constants: Mol wt: 220.4, mp: 17.8°C, flash p: 285°F (O.C.), d: 0.94–0.95 at 24°/24°C.
Hazard Analysis
Toxic Hazard Rating:
Acute Local: Irritant 3.
Acute Systemic: Ingestion 3; Inhalation 3; Skin Absorption 3.
Chronic Local: U.
Chronic Systemic: U.
Toxicity: A mild allergen. Individual susceptibility to these materials may vary. Should be handled with care. See also p-phenylenediamine.
Toxicology: Can cause severe burns of skin. Systemic symptoms are sweating, flushing, shortness of breath and slow pulse.
Fire Hazard: Slight, when exposed to heat or flame; can react with oxidizing materials.
Countermeasures
To Fight Fire: Foam, carbon dioxide, dry chemical or carbon tetrachloride (Section 6).
Storage and Handling: Section 7.

DI-(p-tert-BUTYL PHENYL) MONO PHENOL PHOSPHATE
General Information
Description: Insoluble in water.
Formula: $(C_4H_9C_6H_4O)_2POOC_6H_5$.
Constants: Mol wt: 438, d: 1.1, bp: 500–527°F at 5 mm, flash p: 482°F.
Hazard Analysis
Toxic Hazard Rating: U.
Fire Hazard: Slight, when exposed to heat or flame.
Disaster Hazard: Dangerous. See phosphorus compounds.
Countermeasures
To Fight Fire: Water, foam (Section 6).
Storage and Handling: Section 7.

DIBUTYL PHOSPHATE
Hazard Analysis
Toxic Hazard Rating: U. Animal experiments suggest high toxicity.
TLV: ACGIH (tentative) 1 part per million of air; 5 milligrams per cubic meter of air.
Disaster Hazard: Dangerous. See phosphorus compounds.

DIBUTYL PHOSPHITE
General Information
Description: Liquid.
Formula: $(C_4H_9O)_2P(O)H$.
Constants: Mol wt: 194.2, bp: 115°C at 10 mm, flash p: 120°F, d: 0.971 at 35°/4°C, vap. press.: < 1 mm at 20°C, vap. d.: 6.7.
Hazard Analysis
Toxicity: Unknown. See also phosphine.
Fire Hazard: Slight, when exposed to heat or flame or by chemical reaction. Many phosphites decompose to evolve phosphine when heated (Section 6).
Explosion Hazard: See phosphine.

Disaster Hazard: Dangerous; when heated to decomposition or on contact with acid or acid fumes, it emits highly toxic fumes of oxides of phosphorus; can react vigorously with oxidizing materials.

Countermeasures

Storage and Handling: Section 7.

To Fight Fire: Foam, carbon dioxide, dry chemical or carbon tetrachloride (Section 6).

DIBUTYL-m-PHTHALATE
General Information

Synonyms: Dibutyl isophthalate, dibutyl phthalate.

Description: Liquid.

Formula: $(C_6H_4)(CO_2C_4H_9)_2$.

Constants: Mol wt: 278.3, flash p: 322°F (C.C.), vap. d.: 9.58.

Hazard Analysis

Toxic Hazard Rating:

Acute Local: Ingestion 2.

Acute Systemic: U.

Chronic Local: Ingestion 1.

Chronic Systemic: U.

TLV: ACGIH (tentative) 5 milligrams per cubic meter of air.

Toxicity: See esters, butyl alcohol and phthalic acid.

Toxicology: Limited animal experiments show low toxicity.

Fire Hazard: Slight, when exposed to heat or flame; can react with oxidizing materials.

Countermeasures

To Fight Fire: Water, foam, carbon dioxide, dry chemical or carbon tetrachloride (Section 6).

Storage and Handling: Section 7.

DIBUTYL o-PHTHALATE
General Information

Synonym: Dibutyl-1,2-benzenedicarboxylate, n-butyl phthalate.

Description: Oily liquid, mild odor.

Formula: $C_6H_4(COOC_4H_9)_2$.

Constants: Mol wt: 278.3, bp: 340°C, fp: −35°C, flash p: 315°F (C.C.), d: 1.047–1.049 at 20°/20°C, autoign. temp.: 757°F, vap. d.: 9.58.

Hazard Analysis

Toxic Hazard Rating:

Acute Local: Ingestion 2.

Acute Systemic: U.

Chronic Local: U.

Chronic Systemic: U.

Toxicity: See esters, phthalic acid and butyl alcohol.

Fire Hazard: Slight, when exposed to heat or flame; can react with oxidizing materials.

Countermeasures

To Fight Fire: Water, foam, carbon dioxide, dry chemical or carbon tetrachloride (Section 6).

Storage and Handling: Section 7.

3,5-DIBUTYLPYRIDINE
General Information

Description: Colorless liquid with mild pyridine-like odor.

Formula: $C_5H_3N(C_4H_9)_2$.

Constants: Mol wt: 191.3, bp: 271°C, fp: −75°C, flash p: 252°F, d: 0.882.

Hazard Analysis

Toxicology: Probably somewhat toxic. A mild irritant. See also pyridine.

Fire Hazard: Low but can be ignited.

Disaster Hazard: Dangerous; when heated to decomposition it emits highly toxic fumes.

Countermeasures

To Fight Fire: Spray, foam, carbon dioxide or dry chemical.

Ventilation Control: Section 2.

Personnel Protection: Section 3.

Storage and Handling: Section 7.

DIBUTYL SEBACATE
General Information

Description: Clear liquid.

Formula: $[(CH_2)_4COOC_4H_9]_2$.

Constants: Mol wt: 314.45, bp: 180°C at 3 mm; fp: −11°C, flash p: 353°F (C.O.C.), d: 0.936 at 20°/20°C, vap. d.: 10.8.

Hazard Analysis

Toxicity: See esters and butyl alcohol.

Fire Hazard: Slight, when exposed to heat or flame; can react with oxidizing materials.

Countermeasures

To Fight Fire: Water, carbon dioxide, foam, dry chemical or carbon tetrachloride (Section 6).

Storage and Handling: Section 7.

DIBUTYL STEARAMIDE
General Information

Description: Liquid.

Formula: $CH_3(CH_2)_{16}CON(C_4H_9)_2$.

Constants: Mol wt: 395.7, bp: 175°C, flash p: 420°F, d: 0.86, vap. d.: 13.7.

Hazard Analysis

Toxicity: Unknown.

Fire Hazard: Slight, when exposed to heat or flame; can react with oxidizing materials.

Countermeasures

To Fight Fire: Foam, carbon dioxide, dry chemical or carbon tetrachloride (Section 6).

Storage and Handling: Section 7.

DIBUTYL SUCCINATE
General Information

Synonym: Tabutrex.

Description: Liquid.

Formula: $C_{12}H_{22}O_4$.

Constants: Mol wt: 230.3, bp: 275°C, mp: −29°C, d: 0.9760 at 20/4°C.

Hazard Analysis

Toxicity: Details unknown. Limited animal experiments show low toxicity.

DIBUTYL SULFIDE. See butyl sulfide.

DIBUTYL TARTRATE
General Information

Synonym: dibutyl-d-2,3-dihydroxybutanedioate.

Description: Light tan liquid, mild odor.

Formula: $(COOC_4H_9)_2(CHOH)_2$.

Constants: Mol wt: 262.30, mp: 21°C, bp: 204°C at 26 mm, flash p: 195°F (C.C.), d: 1.087–1.093 at 20°/20°C, autoign. temp.: 544°F, vap. d.: 9.03.

Hazard Analysis

Toxicity: See esters, butyl alcohol and tartaric acid.

Fire Hazard: Moderate, when exposed to heat or flame; can react with oxidizing materials.

Spontaneous Heating: No.

Countermeasures

To Fight Fire: Water, foam, carbon dioxide, dry chemical or carbon tetrachloride (Section 6).

Storage and Handling: Section 7.

TOXIC HAZARD RATING CODE (For detailed discussion, see Section 1.)

0 NONE: (a) No harm under any conditions; (b) Harmful only under unusual conditions or overwhelming dosage.

1 SLIGHT: Causes readily reversible changes which disappear after end of exposure.

2 MODERATE: May involve both irreversible and reversible changes; not severe enough to cause death or permanent injury.

3 HIGH: May cause death or permanent injury after very short exposure to small quantities.

U UNKNOWN: No information on humans considered valid by authors.

DI-n-BUTYL TELLURIDE
General Information
Description: Yellow oil.
Formula: $(C_4H_9)_2Te$.
Constants: Mol wt: 241.8, d: 1.334 at 40°C, bp: 132–135°C.
Hazard Analysis
Toxicity: Highly toxic. See tellurium compounds.
Disaster Hazard: Dangerous. See tellurium compounds.

1,3-DIBUTYLTHIOUREA
General Information
Description: White to light tan powder.
Formula: $C_4H_9NHCSNHC_4H_9$.
Constants: Mol wt: 188.4, mp: 60°C, vap. d.: 6.5.
Hazard Analysis
Toxicity: Details unknown. See also thiourea.
Disaster Hazard: Dangerous; see sulfur compounds.
Countermeasures
Storage and Handling: Section 7.

DIBUTYLTIN DIACETATE
General Information
Synonym: Dibutyldiacetoxystannane.
Description: Clear, colorless liquid with a slight acetic acid odor.
Formula: $(C_4H_9)_2Sn(CH_3CO_2)_2$.
Constants: Mol wt: 351.0, bp: decomposes, fp: 5–10°C, flash p: 290°F (O.C.), d: 1.31 at 25°C, vap. d.: 12.1.
Hazard Analysis
Toxicity: See tin compounds and butyl alcohol.
Fire Hazard: Slight, when exposed to heat or flame; can react with oxidizing materials.
Countermeasures
To Fight Fire: Water, foam, carbon dioxide, dry chemical or carbon tetrachloride (Section 6).
Storage and Handling: Section 7.

DIBUTYLTIN DIBROMIDE
General Information
Description: Small crystals.
Formula: $(C_4H_9)_2SnBr_2$.
Constants: Mol wt: 392.8, mp: 20°C.
Hazard Analysis
Toxicity: Details unknown. See tin compounds, bromides and butyl alcohol.
Disaster Hazard: Dangerous. See bromides.

DIBUTYLTIN DICHLORIDE
General Information
Synonym: Dibutyldichlorostannane.
Description: White, crystalline solid.
Formula: $(C_4H_9)_2SnCl_2$.
Constants: Mol wt: 303.8, mp: 43°C, bp: 135°C at 10 mm, flash p: 335°F (O.C.), d: 1.36 at 50°C, vap. press.: 2 mm at 100°C, vap. d.: 10.5.
Hazard Analysis
Toxicity: See tin compounds, butyl alcohol, hydrochloric acid.
Fire Hazard: Slight, when exposed to heat or flame.
Disaster Hazard: Dangerous; it emits highly toxic fumes of hydrochloric acid; will react with water or steam to produce heat and toxic fumes; can react vigorously with oxidizing materials.
Countermeasures
Storage and Handling: Section 7.
To Fight Fire: Water, foam, carbon dioxide, dry chemical or carbon tetrachloride (Section 6).

DIBUTYLTIN DILAURATE
General Information
Description: Pale yellow liquid to colorless solid (when pure).
Formula: $(C_4H_9)_2Sn[O_2C(CH_2)_{10}CH_3]_2$.
Constants: Mol wt: 631.5, mp: 27°C, bp: nondistillable at 10 mm, flash p: 455°F (O.C.), d: 1.066 at 20°/20°C, vap. d.: 21.8.

Hazard Analysis
Toxicity: Details unknown. Used against tapeworm in chickens in doses of 100 mg. See tin compounds and butyl alcohol.
Caution: The vapors evolved when this material is heated should be avoided. A skin irritant.
Fire Hazard: Slight, when exposed to heat or flame; can react with oxidizing materials.
Countermeasures
To Fight Fire: Water, foam, carbon dioxide, dry chemical or carbon tetrachloride (Section 6).
Storage and Handling: Section 7.

DIBUTYLTIN MALEATE
General Information
Synonym: Dibutylmalyldioxystannane.
Description: A white, amorphous powder or glass-like resin.
Formula: $(C_4H_9)_2Sn(CHCO_2)_2$.
Constants: Mol wt: 347.0, mp: 110°C, flash p: 400°F (O.C.), bulk density: 0.36, vap. d.: 12.0.
Hazard Analysis
Toxicity: See tin compounds.
Caution: Avoid vapors emitted when this material is hot.
Fire Hazard: Slight, when exposed to heat or flame.
Disaster Hazard: Moderately dangerous; when heated to decomposition, it emits toxic fumes; can react with oxidizing materials.
Countermeasures
Storage and Handling: Section 7.
To Fight Fire: Foam, carbon dioxide, dry chemical or carbon tetrachloride (Section 6).

DIBUTYLTIN OXIDE
General Information
Synonym: Dibutyloxostannane.
Description: White, amorphous powder.
Formula: $(C_4H_9)_2SnO$.
Constants: Mol wt: 248.9, mp: decomposes without melting, bulk density: 0.5, vap. d.: 8.6.
Hazard Analysis
Toxicity: See tin compounds and butyl alcohol.
Fire Hazard: Ignites when exposed to flame; can react with oxidizing materials (Section 6).
Countermeasures
Storage and Handling: Section 7.

n,n-DIBUTYLTOLUENE SULFONAMIDE
General Information
Formula: $CH_3C_6H_4SO_3N(C_4H_9)_2$.
Constants: Mol wt: 299, d: 1.1, bp: 392°C at 10 mm.
Hazard Analysis
Toxic Hazard Rating: U.
Fire Hazard: Moderate when exposed to heat or flame.
Disaster Hazard: Dangerous. See sulfonates.
Countermeasures
Storage and Handling: Section 7.
To Fight Fire: Water, foam (Section 6).

DIBUTYL UREA
General Information
Description: Liquid.
Formula: $C_4H_9NHCONHC_4H_9$.
Constants: Mol wt: 172.3, mp: 25°C, bp: 119°C, flash p: 279°F, vap. d.: 5.95.
Hazard Analysis
Toxicity: Unknown.
Fire Hazard: Slight, when exposed to heat or flame; can react with oxidizing materials.
Countermeasures
To Fight Fire: Foam, carbon dioxide, dry chemical or carbon tetrachloride (Section 6).
Storage and Handling: Section 7.

DI-n-BUTYL ZINC
General Information
Description: Liquid—decomposed by cold water.
Formula: $(C_4H_9)_2Zn$.
Constants: Mol wt: 179.6, bp: 82°C at 9 mm.
Hazard Analysis
Toxicity: Details unknown. See zinc compounds.

DICACODYL. See cacodyl.

DICACODYL OXIDE. See cacodyl oxide.

DICACODYL SULFIDE. See cacodyl sulfide.

DICALCIUM o-PHOSPHATE. See calcium phosphate, dibasic.

DICAPRYL ADIPATE
General Information
Description: Liquid.
Formula: $(CH_2CH_2CO_2C_8H_{17})_2$.
Constants: Mol wt: 370.6, bp: 211–217°C at 4 mm, flash p: 352°F, d: 0.916 at 20°/20°C, vap. d.: 12.8.
Hazard Analysis
Toxicity: See esters.
Fire Hazard: Slight, when exposed to heat or flame; can react with oxidizing materials.
Countermeasures
To Fight Fire: Foam, carbon dioxide, dry chemical or carbon tetrachloride (Section 6).
Storage and Handling: Section 7.

DICAPRYL PHTHALATE
General Information
Description: Clear liquid.
Formula: $C_6H_4(CO_2C_8H_{17})_2$.
Constants: Mol wt: 390.6, bp: 222–230 at 4 mm, fp: −60°C, flash p: 394°F (C.O.C.), d: 0.978 at 20°/20°C, vap. d.: 9.8.
Hazard Analysis
Toxicity: See esters.
Fire Hazard: Slight, when exposed to heat or flame; can react with oxidizing materials.
Countermeasures
To Fight Fire: Foam, carbon dioxide, dry chemical or carbon tetrachloride (Section 6).
Storage and Handling: Section 7.

DICAPRYL SEBACATE
General Information
Description: Liquid.
Formula: $[(CH_2)_4CO_2C_8H_{17}]_2$.
Constants: Mol wt: 426.7, bp: 230–240°C at 4 mm, flash p: 445°F, d: 0.907 at 20°/20°C, vap. d.: 14.7.
Hazard Analysis
Toxicity: See esters.
Fire Hazard: Slight, when exposed to heat or flame; can react with oxidizing materials.
Disaster Hazard: Slight; can react with oxidizing materials.
Countermeasures
Storage and Handling: Section 7.
To Fight Fire: Foam, carbon dioxide, dry chemical or carbon tetrachloride (Section 6).

DICAPTHON
General Information
Synonyms: p-nitro-chlorophenyl dimethyl thiono phosphate; o-(2-chloro-4-nitrophenyl) o,o-dimethyl phosphorothioate; p-n-o-chlorophenyl dimethyl thiono phosphate.

Description: White solid, insoluble in water, soluble in acetone, cyclohexane, ethyl acetate, toluene, and xylene.
Formula: $(CH_3O)_2P(:S)OC_6H_3(Cl)NO_2$.
Constants: Mol wt: 297.5, mp: 51°–52°C.
Hazard Analysis and Countermeasures
A cholinesterase inhibitor. See parathion, chlorides and nitrates.

DICARBITOL PHTHALATE. See "Carbitol" phthalate.

DICARBON HEXABROMIDE
General Information
Synonym: Hexabromoethane.
Description: Rhombic crystals.
Formula: C_2Br_6.
Constants: Mol wt: 503.52, mp: 149°C decomposes, bp: 210°C, d: 3.823, vap. d.: 18.2.
Hazard Analysis
Toxicity: Details unknown. See also bromides.
Disaster Hazard: Moderately dangerous; when heated to decomposition, it emits toxic fumes of bromides and will react with water or steam to produce toxic and corrosive fumes.
Countermeasures
Storage and Handling: Section 7.

DICARBON HEXACHLORIDE. See hexachloroethane.

DICARBON TETRABROMIDE
General Information
Synonyms: Carbon dibromide; tetrabromethylene.
Description: Solid.
Formula: C_2Br_4.
Constants: Mol wt: 343.68, mp: 57.5°C, bp: 227°C, vap. d.: 11.9.
Hazard Analysis
Toxicity: Details unknown. See also bromides.
Disaster Hazard: Dangerous; see bromides.
Countermeasures
Storage and Handling: Section 7.

DICARBON TETRACHLORIDE. See perchloroethylene.

DICHLONE. See 2,3-dichloro-1,4-naphthoquinone.

4,4-DICHLORO-a-(TRICHLOROMETHYL) BENZHYDRAL
General Information
Synonym: DTMC, "Kelthane."
Formula: $C_6H_4Cl_2CHOHC_6H_4CHCl_3$.
Constant: Mol wt: 372.5.
Hazard Analysis
Toxic Hazard Rating: U. Animal experiments suggest low toxicity.
Disaster Hazard: Dangerous. See chlorides.

DICHLORETHANAL. See dichloroacetaldehyde.

DICHLORINATED METHYL OXIDE. See dichloromethyl ether.

DICHLOROACETALDEHYDE
General Information
Synonym: 2,2-dichlorethanal.
Description: Colorless liquid, polymerizes slowly to white solid.
Formula: $CHCl_2CHO$.
Constants: Mol wt: 112.95, bp: 88°C, fp: −50°C, flash p: 140°F (C.C.), d: 1.436 at 25°/4°C, vap. press.: 50 mm at 20°C, vap. d.: 3.9.

TOXIC HAZARD RATING CODE (For detailed discussion, see Section 1.)

0 NONE: (a) No harm under any conditions; (b) Harmful only under unusual conditions or overwhelming dosage.

1 SLIGHT: Causes readily reversible changes which disappear after end of exposure.

2 MODERATE: May involve both irreversible and reversible changes; not severe enough to cause death or permanent injury.

3 HIGH: May cause death or permanent injury after very short exposure to small quantities.

U UNKNOWN: No information on humans considered valid by authors.

Hazard Analysis
Toxic Hazard Rating:
 Acute Local: Irritant 3; Ingestion 3; Inhalation 3.
 Acute Systemic: U.
 Chronic Local: U.
 Chronic Systemic: U.
Fire Hazard: Moderate, when exposed to heat or flame.
Disaster Hazard: Dangerous; see chlorides; can react vigorously with oxidizing materials.
Countermeasures
Ventilation Control: Section 2.
To Fight Fire: Water, foam, carbon dioxide, dry chemical or carbon tetrachloride (Section 6).
Personnel Protection: Section 3.
Storage and Handling: Section 7.

DICHLOROACETIC ACID
General Information
Description: Colorless corrosive liquid, pungent odor.
Formula: $CHCl_2COOH$.
Constants: Mol wt: 129.0, mp a: 10°C, b: −4°C, bp: 194°C, d: 1.5634 at 20°/4°C, vap. press.: 1 mm at 44.0°C, vap. d.: 4.45.
Hazard Analysis
Toxic Hazard Rating:
 Acute Local: Irritant 3; Ingestion 3; Inhalation 3.
 Acute Systemic: U.
 Chronic Local: U.
 Chronic Systemic: U.
Caution: The LD_{50} for dosage by mouth is of the same order of magnitude as acetic acid or trichloroacetic acid. It is corrosive and irritating to the skin and mucous membranes.
Disaster Hazard: Dangerous; when heated to decomposition, it emits highly toxic fumes of chlorides; it will react with water or steam to produce toxic and corrosive fumes.
Countermeasures
Ventilation Control: Section 2.
Personnel Protection: Section 3.
First Aid: Section 1.
Storage and Handling: Section 7.
Shipping Regulations: Section 11.
 IATA: Corrosive liquid, white label, 1 liter (passenger and cargo).

1,1-DICHLOROACETONE
General Information
Synonym: 1,1-dichloro-2-propanone.
Description: Oily liquid.
Formula: $CH_3COCHCl_2$.
Constants: Mol wt: 127.0, bp: 120°C, d: 1.305 at 18°/15°C, vap. d.: 4.38.
Hazard Analysis
Toxicity: Details unknown. See 1,3-dichloroacetone and ketones.
Disaster Hazard: Dangerous; see chlorides; can react vigorously with oxidizing materials.
Countermeasures
Storage and Handling: Section 7.

1,3-DICHLOROACETONE
General Information
Synonym: 1,3-dichloro-2-propanone.
Description: Crystals.
Formula: $CH_2ClCOCICH_2$.
Constants: Mol wt: 127.0, mp: 45°C, bp: 173°C, d: 1.3826 at 46°/4°C, vap. d.: 4.38.
Hazard Analysis
Toxic Hazard Rating:
 Acute Local: Irritant 3; Ingestion 3; Inhalation 3.
 Acute Systemic: U.
 Chronic Local: U.
 Chronic Systemic: U.

Disaster Hazard: Dangerous; when heated to decomposition, it emits highly toxic fumes of chlorides.
Countermeasures
Ventilation Control: Section 2.
Personnel Protection: Section 3.
Storage and Handling: Section 7.

DICHLOROACETONITRILE
General Information
Formula: $CHCl_2CN$.
Constants: Mol wt: 110.0, vap. d.: 3.8.
Hazard Analysis and Countermeasures
See nitriles and chlorides.

α,α-DICHLOROACETOPHENONE
General Information
Description: Crystals.
Formula: $C_6H_5COCHCl_2$.
Constants: Mol wt: 189.05, mp: 21°C, bp: 247°C decomposes, d: 1.34 at 15°C, vap. d.: 6.5.
Hazard Analysis
Toxic Hazard Rating:
 Acute Local: Irritant 3; Ingestion 3; Inhalation 3.
 Acute Systemic: U.
 Chronic Local: U.
 Chronic Systemic: U.
Disaster Hazard: Dangerous; when heated to decomposition, it emits highly toxic fumes of chlorides.
Countermeasures
Ventilation Control: Section 2.
Personnel Protection: Section 3.
Storage and Handling: Section 7.

2,2-DICHLOROACETYL CHLORIDE
General Information
Synonym: Dichloroethanoyl chloride.
Description: Fuming liquid. Acrid odor. Miscible in ether.
Formula: $CHCl_2COCl$.
Constants: Mol wt: 147.4, d: 1.5315 at 16°/4°C, bp: 108°C, flash p: 151°F, vap. d.: 5.1.
Hazard Analysis
Toxic Hazard Rating: (based on limited animal experiments).
 Acute Local: Irritant 2; Ingestion 2; Inhalation 2.
 Acute Systemic: Ingestion 1.
 Chronic Local: U.
 Chronic Systemic: U.
Fire Hazard: Moderate, when exposed to heat or flame.
Disaster Hazard: Dangerous; see chlorides.
Countermeasures
Shipping Regulations: Section 11.
 IATA: Corrosive liquid, white label, 1 liter (passenger), 5 liters (cargo).

DICHLOROACETYLENE
General Information
Synonym: Dichloracetylene.
Formula: ClC:CCl.
Constant: Mol wt: 94.94.
Hazard Analysis
Toxicity: Details unknown. See also chlorinated hydrocarbons, aliphatic.
Explosion Hazard: Severe, when shocked or exposed to heat (Section 7).
Disaster Hazard: Highly dangerous; shock and heat will explode it; when heated to decomposition or on contact with acid or acid fumes, it emits highly toxic fumes of chlorides; can react vigorously with oxidizing materials.
Countermeasures
Storage and Handling: Section 7.

2,3-DICHLOROALLYL DIISOPROPYL THIOCARBAMATE
General Information
Synonym: Diallate.

Description: Brown liquid; slightly soluble in water; soluble in organic solvents.
Formula: $[(CH_3)_2CH]_2NCOSCH_2C(Cl):CHCl$.
Constants: Mol wt: 270, bp: 159°C at 9 mm.

Hazard Analysis
Toxic Hazard Rating: U. Animal experiments suggest high toxicity. Mainly on nervous system. It can be absorbed via the skin. See also bis (dimethyl-thiocarbamyl) disulfide.
Disaster Hazard: Dangerous. When strongly heated it gives off highly toxic fumes.

Countermeasures
Storage and Handling: Section 7.

2,5-DICHLOROANILINE
General Information
Description: Needle-like crystals.
Formula: $Cl_2C_6H_3NH_2$.
Constants: Mol wt: 162.0, bp: 251°C, mp: 50°C, vap. d.: 5.6.

Hazard Analysis
Toxicity: Details unknown. See also aniline.
Disaster Hazard: Dangerous; when heated to decomposition, it emits toxic fumes.

Countermeasures
Storage and Handling: Section 7.

3,4-DICHLOROANILINE
General Information
Description: Crystals; insoluble in water; soluble in most organic solvents.
Formula: $Cl_2C_6H_3NH_2$.
Constants: Mol wt: 162.0, mp: 68–72°C, bp: 272°C, flash p: 331°F.

Hazard Analysis
Fire Hazard: Slight, when exposed to heat or flame.
Disaster Hazard: Dangerous. See chlorides and aniline.

Countermeasures
To Fight Fire: Water, foam (Section 6).
Storage and Handling: Section 7.

DICHLORO ANILINE, LIQUID
Shipping Regulations: Section 11.
IATA: Poison B, poison label, 1 liter (passenger), 220 liters (cargo).

1,3-DICHLOROBENZENE. See m-dichlorobenzene.

1,4-DICHLOROBENZENE. See p-dichlorobenzene.

m-DICHLOROBENZENE
General Information
Synonym: 1,3-dichlorobenzene.
Description: Colorless liquid.
Formula: $C_6H_4Cl_2$.
Constants: Mol wt: 147.0, mp: −24.8°C, bp: 173°C, d: 1.288 at 20°/4°C, vap. press.: 1 mm at 12.1°C, vap. d.: 5.08.

Hazard Analysis
Toxicity: An insecticide and fumigant. See also o-dichlorobenzene and chlorobenzene.
Fire Hazard: Moderate.
Disaster Hazard: Dangerous; see chlorides; can react vigorously with oxidizing materials.

Countermeasures
Ventilation Control: Section 2.
Personnel Protection: Section 3.

To Fight Fire: Water, foam, carbon dioxide, dry chemical or carbon tetrachloride (Section 6).
Storage and Handling: Section 7.

o-DICHLOROBENZENE
General Information
Description: Clear liquid.
Formula: $C_6H_4Cl_2$.
Constants: Mol wt: 147.0, mp: −17.5°C, bp: 180–183°C, fp: −22°C, flash p: 151°F, d: 1.307 at 20°/20°C, vap. d.: 5.05, autoign. temp.: 1198°F, lel: 2.2%, uel: 9.2%.

Hazard Analysis
Toxic Hazard Rating:
Acute Local: Irritant 3; Ingestion 2; Inhalation 2.
Acute Systemic: Inhalation 2.
Chronic Local: Irritant 2; Inhalation 2.
Chronic Systemic: Inhalation 2.
TLV: ACGIH (recommended); 50 parts per million in air; 300 milligrams per cubic meter of air.
Toxicology: See chlorobenzene. The o-isomer is probably more toxic than the m- or p- forms. It is irritating to skin and mucous membrane. Experimentally it has produced liver and kidney injury.
Fire Hazard: Moderate, when exposed to heat or flame.
Disaster Hazard: Dangerous; see chlorides; can react vigorously with oxidizing materials.

Countermeasures
Ventilation Control: Section 2.
To Fight Fire: Water, foam, carbon dioxide, dry chemical or carbon tetrachloride (Section 6).
Personal Hygiene: Section 3.
Storage and Handling: Section 7.
Shipping Regulations: Section 11.
IATA: Other restricted articles, class A, no label required, no limit (passenger and cargo).

p-DICHLOROBENZENE
General Information
Synonym: 1,4-dichlorobenzene.
Description: White crystals, penetrating odor.
Formula: $C_6H_4Cl_2$.
Constants: Mol wt: 147.0, mp: 53°C, bp: 173.4°C, flash p: 150°F (C.C.), d: 1.4581 at 20.5°/4°C, vap. press.: 10 mm at 54.8°C, vap. d.: 5.08.

Hazard Analysis
Toxic Hazard Rating:
Acute Local: Irritant 1; Allergen 1; Ingestion 2; Inhalation 1.
Acute Systemic: Ingestion 2.
Chronic Local: Irritant 2; Allergen 1.
Chronic Systemic: Ingestion 1; Inhalation 1.
Toxicology: An insecticide. Has been reported to cause liver injury in humans.
TLV: ACGIH (recommended); 75 parts per million in air; 450 milligrams per cubic meter of air.
Fire Hazard: Moderate, when exposed to heat or flame.
Spontaneous Heating: No.
Disaster Hazard: Dangerous; see chlorides; can react vigorously with oxidizing materials.

Countermeasures
Ventilation Control: Section 2.
Personnel Protection: Section 3.
To Fight Fire: Water, foam, carbon dioxide, dry chemical or carbon tetrachloride (Section 6).
Storage and Handling: Section 7.
Shipping Regulations: Section 11.

TOXIC HAZARD RATING CODE *(For detailed discussion, see Section 1.)*

0 NONE: (a) No harm under any conditions; (b) Harmful only under unusual conditions or overwhelming dosage.

1 SLIGHT: Causes readily reversible changes which disappear after end of exposure.

2 MODERATE: May involve both irreversible and reversible changes; not severe enough to cause death or permanent injury.

3 HIGH: May cause death or permanent injury after very short exposure to small quantities.

U UNKNOWN: No information on humans considered valid by authors.

IATA: Other restricted articles, class A; no label required, no limit (passenger and cargo).

2,2'-DICHLOROBENZIDINE
General Information
Description: Crystalline solid. Needle-like; insoluble in water. Soluble in alcohol and ether.
Formula: $C_6H_3ClNH_2C_6H_3ClNH_2$.
Constants: Mol wt: 253.14, mp: 133°C, vap. d.: 8.73.
Hazard Analysis
Toxic Hazard Rating:
  Acute Local: Allergen 2.
  Acute Systemic: U.
  Chronic Local: Allergen 2.
  Chronic Systemic: U.
Toxicity: Details unknown. See also benzidine.
Disaster Hazard: Dangerous; see chlorides.
Countermeasures
Storage and Handling: Section 7.

3,3'-DICHLOROBENZIDINE
General Information
Description: Crystals; insoluble in water, soluble in alcohol, benzene and glacial acetic acid.
Formula: $C_{12}H_{10}Cl_2N_2$.
Constants: Mol wt: 253.1, mp: 133°C.
Hazard Analysis
Toxicology: See 2,2'-dichlorobenzidine.
Disaster Hazard: See 2,2'-dichlorobenzidine and benzidine.

4,4'-DICHLOROBENZILIC ACID ETHYL ESTER.
See ethyl 4,4'-dichlorobenzilate.

2,6-DICHLOROBENZONITRILE
General Information
Synonym: Dichlobenil.
Description: White solid; almost insoluble in water; soluble in organic solvents.
Formula: $Cl(Cl)C_6H_3CN$.
Constants: Mol wt: 172, mp: 144°C.
Hazard Analysis
Toxic Hazard Rating: U. Does not hydrolyze to HCN in body. Less toxic than most aliphatic nitriles. See also benzonitrile.
Disaster Hazard: Dangerous. See chlorides and nitriles.

2,4-DICHLOROBENZOYL PEROXIDE PASTE WITH DIBUTYL PHTHALATE
General Information
Synonym: Luperco CCC.
Description: Thick white paste insoluble in water. Soluble in organic nonpolar solvents.
Formula: $(Cl_2C_6H_3CO)_2O_2$ + dibutyl phthalate.
Hazard Analysis
Toxicity: Details unknown. See peroxides, organic.
Fire Hazard: See peroxides, organic.
Explosion Hazard: See peroxides, organic.
Countermeasures
See peroxides, organic.
Shipping Regulations:
  I.C.C.: Oxidizing material N.O.S.
  Freight and Express: Yellow label, chemicals N.O.I.B.N.
  Rail Express: Max. wt.: 25 pounds per shipping case.
  Parcel Post: Prohibited.

3,4-DICHLOROBENZYL ALCOHOL
General Information
Description: Crystals.
Formula: $Cl_2C_6H_3CH_2OH$.
Constants: Mol wt: 177.1, vap. d.: 6.1.
Hazard Analysis
Toxicity: Details unknown; an insecticide. See also alcohols.
Disaster Hazard: Dangerous; see chlorides.
Countermeasures
Storage and Handling: Section 7.

2,4-DICHLOROBENZYL CHLORIDE
General Information
Description: Liquid.
Formula: $Cl_2C_6H_3CH_2Cl$.
Constants: Mol wt: 195.5, vap. d.: 6.76.
Hazard Analysis
Toxicity: Details unknown; an insecticide. See also p-dichlorobenzene.
Disaster Hazard: Dangerous; see chlorides.
Countermeasures
Storage and Handling: Section 7.

3,4-DICHLOROBENZYL CHLORIDE
General Information
Description: Liquid.
Formula: $Cl_2C_6H_3CH_2Cl$.
Constants: Mol wt: 195.5, bp: 255°C, vap. d.: 6.76.
Hazard Analysis
Toxicity: Details unknown, an insecticide. See also p-dichlorobenzene.
Disaster Hazard: Dangerous; see chlorides.
Countermeasures
Storage and Handling: Section 7.

1,1-DICHLORO-2,2-BIS(p-CHLOROPHENYL) ETHANE
General Information
Synonyms: tetrachlorodiphenylethane; TDE.
Description: Crystals.
Formula: $C_{14}H_{10}Cl_4$.
Constants: Mol wt: 320.05, mp: 110°C, vap. d.: 11.0.
Hazard Analysis
Toxic Hazard Rating:
  Acute Local: Irritant 1; Allergen 1.
  Acute Systemic: Ingestion 2.
  Chronic Local: Allergen 1.
  Chronic Systemic: Ingestion 2.
See also DDT.
Disaster Hazard: Dangerous; see chlorides.
Countermeasures
Personal Hygiene: Section 3.
Storage and Handling: Section 7.

1,1-DICHLORO-2,2-BIS-(p-ETHYLPHENYL) ETHANE.
See perthane.

1,1-DICHLOROBUTANE
General Information
Formula: $Cl_2CHCH_2CH_2CH_3$.
Constant: Mol wt: 127.0.
Hazard Analysis
Toxicity: Details unknown. Animal experiments suggest moderate toxicity. See also chlorinated hydrocarbons, aliphatic.

1,4-DICHLOROBUTANE
General Information
Synonym: sym-dichlorobutane.
Description: Colorless, mobile liquid with mild, pleasant odor.
Formula: $CH_2ClCH_2CH_2CH_2Cl$.
Constants: Mol wt: 127.02, bp: 155°C, flash p: 126°F (C.C.), d: 1.141, vap. d.: 4.4, vap. press.: 4 mm at 20°C.
Hazard Analysis
Toxicity: Details unknown. See 1,1-dichlorobutane.
Fire Hazard: Moderate, when exposed to heat or flame.
Disaster Hazard: Dangerous; see chlorides; can react vigorously with oxidizing materials.
Countermeasures
Storage and Handling: Section 7.
To Fight Fire: Foam, carbon dioxide, dry chemical or carbon tetrachloride (Section 6).

2,3-DICHLOROBUTANE
General Information
Description: Liquid.
Formula: $CH_3CHClCHClCH_3$.

Constants: Mol wt: 127, mp: −80.4°C, bp: 116.0°C, flash p: 194°F (O.C.), vap. press.: 40 mm at 35.0°C, vap. d.: 4.4, d: 1.1.

Hazard Analysis

Toxicity: Details unknown. See also chlorinated hydrocarbons, aliphatic and 1,1-dichlorobutane.

Fire Hazard: Dangerous, when exposed to heat or flame.

Explosion Hazard: Unknown.

Disaster Hazard: Dangerous; see chlorides; can react vigorously with oxidizing materials.

Countermeasures

Ventilation Control: Section 2.

To Fight Fire: Foam, carbon dioxide, dry chemical or carbon tetrachloride (Section 6).

Storage and Handling: Section 7.

sym-DICHLOROBUTANE. See 1,4-dichlorobutane.

1,3-DICHLORO-2-BUTENE

General Information

Formula: $CH_2ClCHCClCH_3$.

Constants: Mol wt: 125.01, bp: 123°C, flash p: 80°F (C.C.), vap. d.: 4.31.

Hazard Analysis

Toxicity: Details unknown. See also chlorinated hydrocarbons, aliphatic. Limited animal experiments suggest moderate toxicity.

Fire Hazard: Dangerous, when exposed to heat or flame.

Spontaneous Heating: No.

Explosion Hazard: Unknown.

Disaster Hazard: Dangerous; see chlorides; can react with oxidizing materials.

Countermeasures

Ventilation Control: Section 2.

To Fight Fire: Foam, carbon dioxide, dry chemical or carbon tetrachloride (Section 6).

Storage and Handling: Section 7.

DICHLORO-(2-CHLOROVINYL) ARSINE

General Information

Synonyms: β-chlorovinyl dichlorarsine; lewisite.

Description: Liquid, faint odor of geranium.

Formula: $C_2H_2AsCl_3$.

Constants: Mol wt: 207.32, bp: 190°C decomposes, fp: −13°C, d: 1.888 at 20°/4°C, vap. press.: 0.4 mm at 20°C, vap. d.: 7.15.

Hazard Analysis

Toxic Hazard Rating:

Acute Local: Irritant 3; Ingestion 3; Inhalation 3.

Acute Systemic: Ingestion 3; Inhalation 3; Skin Absorption 3.

Chronic Local: U.

Chronic Systemic: Ingestion 3; Inhalation 3; Skin Absorption 3.

Toxicology: A blistering type military poison. Has a delayed action similar to distilled mustard gas. This gas exhibits a systemic poisoning effect. To decontaminate, use 2,3-dimercapto-1-propanol (BAL). For poisoning give BAL intramuscularly. This material is absorbed through skin. As little as 2 ml can cause death.

Disaster Hazard: Dangerous; see arsenic.

Countermeasures

Ventilation Control: Section 2.

Personnel Protection: Section 3.

First Aid: Section 1.

Storage and Handling: Section 7.

Shipping Regulations: Section 11.

I.C.C.: Poison A, poison gas label, Not accepted.

Coast Guard Classification: Poison A, poison gas label.

IATA: Poison A, not acceptable (passenger and cargo).

DICHLORODIETHYL SILANE. See diethyldichlorosilane.

DICHLORODIETHYL SULFIDE

General Information

Synonyms: Distilled mustard gas; HD.

Description: Colorless (if pure), to light yellow, oily liquid.

Formula: $S(CH_2CH_2Cl)_2$.

Constants: Mol wt: 159.1, bp: 228°C, fp. 14.4°C, flash p: 221°F, d: 1.2741 at 20°/4°C, vap. d.: 5.4, vap. press.: 0.09 mm at 30°C.

Hazard Analysis

Toxic Hazard Rating:

Acute Local: Irritant 3; Ingestion 3; Inhalation 3.

Acute Systemic: Ingestion 3; Inhalation 3; Skin Absorption 3.

Chronic Local: U.

Chronic Systemic: U.

Toxicology: A "blistering" gas. Highly irritant to eyes, skin and lungs. Pulmonary lesions are often fatal. However, effects upon eyes and skin are likewise serious. There can also be severe gastric disturbances. Only slight evidence for systemic damage. This material penetrates deeply and injures blood vessels which can result in a severe inflammatory reaction and tissue necrosis. Exposure, even to minute traces, can lead to intense inflammation after some hours. It produces marked leucopenia with involution of the lymph nodes; secondary infections are common.

Treatment of local lesions is mainly by cleanliness and emollients, similar to that of burns. Oils protect the skin only slightly. Immediately after exposure, the poison may be partly removed by scrubbing the victim with kerosene, but penetration is so rapid that this treatment is not successful if delayed for 15 to 30 minutes.

Fire Hazard: Slight, when exposed to heat or flame; can be ignited by a large explosive charge.

Disaster Hazard: Dangerous; when heated to decomposition or on contact with acid or acid fumes, it emits highly toxic fumes of oxides of sulfur and chlorides; it will react with water or steam to produce toxic and corrosive fumes; it can react vigorously with oxidizing materials.

Countermeasures

Ventilation Control: Section 2.

To Fight Fire: Water, foam, carbon dioxide, dry chemical or carbon tetrachloride (Section 6).

Personnel Protection: Section 3.

First Aid: Section 1.

Storage and Handling: Section 7.

Shipping Regulations: Section 11.

IATA: Poison A, not acceptable (passenger and cargo).

DICHLORODIETHYLSULFONE

General Information

Description: Colorless crystals.

Formula: $(CH_2CH_2Cl)_2O_2S$.

Constants: Mol wt: 191.08, mp: 52°C, bp: 179–181°C at 14–15 mm, vap. d.: 6.6.

Hazard Analysis

Toxic Hazard Rating:

Acute Local: Irritant 3; Ingestion 3; Inhalation 3.

Acute Systemic: U.

TOXIC HAZARD RATING CODE (For detailed discussion, see Section 1.)

0 NONE: (a) No harm under any conditions; (b) Harmful only under unusual conditions or overwhelming dosage.

1 SLIGHT: Causes readily reversible changes which disappear after end of exposure.

2 MODERATE: May involve both irreversible and reversible changes; not severe enough to cause death or permanent injury.

3 HIGH: May cause death or permanent injury after very short exposure to small quantities.

U UNKNOWN: No information on humans considered valid by authors.

Chronic Local: U.
Chronic Systemic: U.
Disaster Hazard: Dangerous; see chlorides and sulfur compounds.
Countermeasures
Ventilation Control: Section 2.
Personnel Protection: Section 3.
First Aid: Section 1.
Storage and Handling: Section 7.

1,1-DICHLORO-2,2-DIFLUOROETHYLENE
General Information
Description: Liquid.
Formula: Cl_2CCF_2.
Constants: Mol wt: 133, vap. d.: 4.6.
Hazard Analysis
Toxic Hazard Rating:
Acute Local: U.
Acute Systemic: Ingestion 2; Inhalation 2; Skin Absorption 2.
Chronic Local: U.
Chronic Systemic: Ingestion 1; Inhalation 1; Skin Absorption 1.
Disaster Hazard: Dangerous; when heated to decomposition, it emits highly toxic fumes of fluorides and chlorides; it will react with water or steam to produce toxic and corrosive fumes.
Countermeasures
Ventilation Control: Section 2.
Personnel Protection: Section 3.
Storage and Handling: Section 7.
Shipping Regulations: Section 11.
IATA: Other restricted articles, class A, no label, 40 liters (passenger), 220 liters (cargo).

DICHLORODIFLUOROGERMANE
General Information
Description: Colorless gas.
Formula: $GeCl_2F_2$.
Constants: Mol wt: 181.51, mp: $-51.8°C$, bp: $-2.8°C$.
Hazard Analysis and Countermeasures
See fluorides and germanium compounds.

DICHLORODIFLUOROMETHANE
General Information
Synonym: "Freon-12."
Description: Colorless, almost odorless gas.
Formula: CCl_2F_2.
Constants: Mol wt: 120.92, mp: $-158°C$, bp: $-29°C$, vap. press.: 5 atm at $16.1°C$.
Hazard Analysis
Toxic Hazard Rating:
Acute Local: 0.
Acute Systemic: Inhalation 1.
Chronic Local: U.
Chronic Systemic: Inhalation 1.
Toxicology: Narcotic in high concentrations.
TLV: ACGIH (recommended); 1000 parts per million in air; 4950 milligrams per cubic meter of air.
Disaster Hazard: Dangerous; when heated to decomposition, it emits highly toxic fumes of phosgene and fluorides.
Countermeasures
Ventilation Control: Section 2.
Storage and Handling: Section 7.
Shipping Regulations: Section 11.
I.C.C.: Nonflammable gas; green label, 300 pounds.
Coast Guard Classification: Noninflammable gas; green label.
IATA: Nonflammable gas, green label, 70 kilograms (passenger), 140 kilograms (cargo).

DICHLORODIFLUOROMETHANE AND DIFLUORO ETHANE, CONSTANT BOILING MIXTURE
Shipping Regulations: Section 11.

I.C.C.: Nonflammable gas, green label, 200 pounds.
IATA: Nonflammable gas, green label, 70 kilograms (passenger), 140 kilograms (cargo).

DICHLORODIFLUORO-METHANE-DICHLORO-TETRAFLUOROETHANE MIXTURE
Countermeasures
Shipping Regulations: Section 11.
I.C.C.: Nonflammable gas, green label, 300 pounds.
IATA: Nonflammable gas, green label, 70 kilograms (passenger), 140 kilograms (cargo).

DICHLORODIFLUOROMETHANE-MONOCHLORO-DIFLUORO METHANE MIXTURE
Countermeasures
Shipping Regulations: Section 11.
I.C.C.: Nonflammable gas, green label, 300 pounds.
IATA: Nonflammable gas, green label, 70 kilograms (passenger), 140 kilograms (cargo).

DICHLORODIFLUOROMETHANE MONOFLUORO-TRICHLORO METHANE MIXTURE
Countermeasures
Shipping Regulations: Section 11.
I.C.C: Nonflammable gas, green label, 300 pounds.
IATA: Nonflammable gas, green label, 70 kilograms (passenger), 140 kilograms (cargo).

DICHLORODIFLUOROMETHANE-TRICHLORO-MONOFLUOROMETHANE-MONOCHLORODI-FLUOROMETHANE MIXTURE
Shipping Regulations: Section 11.
I.C.C.: Nonflammable gas, green label, 300 pounds.
IATA: Nonflammable gas, green label, 70 kilograms (passenger), 140 kilograms (cargo).

DICHLORODIFLUOROMETHANE-TRICHLORO TRIFLUORO ETHANE MIXTURE
Shipping Regulations: Section 11.
I.C.C.: Nonflammable gas, green label, 300 pounds.
IATA: Nonflammable gas, green label, 70 kilograms (passenger), 140 kilograms (cargo).

2,2-DICHLORO-1,1-DIFLUORO-3-VINYL-CYCLO-BUTANE
General Information
Formula: $C_6H_6F_2Cl_2$.
Constant: Mol wt: 187.
Hazard Analysis
Toxicity: Unknown.
Disaster Hazard: Dangerous; when heated to decomposition, it emits highly toxic fumes.

1,3-DICHLORO-5,5-DIMETHYL HYDANTOIN
General Information
Synonyms: Dichlorodimethyl hydantoin, dactin, halane.
Description: Crystals. Liberates chlorine on contact with hot water.
Formula: $C_5H_6N_2Cl_2O_2$.
Constants: Mol wt: 197, mp: $132°C$, sublimes at $100°C$, conflagrates at $212°C$, d: 1.5 at $20°C$, vap. d.: 6.8.
Hazard Analysis
Toxicity: Details unknown. Toxicity to warm-blooded animals appears to be low. Avoid excessive contact because of effects of active chlorine to irritate the skin. Some of the hydantoins are central nervous system depressants. Readily releases chlorine on decomposition. See also chlorine.
TLV: ACGIH (recommended); 0.2 milligrams per cubic meter of air.
Disaster Hazard: Dangerous; see chlorides and it will react with water or steam to produce toxic and corrosive fumes.
Countermeasures
Storage and Handling: Section 7.

DICHLORODIMETHYLSILANE
General Information
Description: Liquid.
Formula: $C_2H_6Cl_2Si$.
Constants: Mol wt: 129.02, mp: below $-70°C$, bp: $70.5°C$, d: 1.07 at $25°/25°C$, vap. d.: 4.45.
Hazard Analysis
Toxic Hazard Rating:
Acute Local: Irritant 3; Ingestion 3; Inhalation 3.
Acute Systemic: U.
Chronic Local: U.
Chronic Systemic: U.
Disaster Hazard: Dangerous; see chlorides and it will react with water or steam to produce heat, toxic and corrosive fumes.
Countermeasures
Ventilation Control: Section 2.
Personnel Protection: Section 3.
Storage and Handling: Section 7.

1,8-DICHLORO-3,6-DINITROCARBAZOLE
General Information
Description: Crystals.
Formula: $(NO_2ClC_6H_2NH)_2$.
Constant: Mol wt: 341.1.
Hazard Analysis
Toxicity: Details unknown. An insecticide. See also nitro compounds of aromatic hydrocarbons and chlorides.
Disaster Hazard: Dangerous; see chlorides and oxides of nitrogen; can react vigorously with oxidizing materials.
Countermeasures
Storage and Handling: Section 7.

DICHLORO DIPHENYL DICHLOROETHANE See DDD.

DICHLORODIPHENYLENE OXIDE. See dichlorodiphenyl oxide.

DICHLORODIPHENYL OXIDE
General Information
Synonym: dichlorodiphenylene oxide.
Description: Liquid.
Formula: $Cl_2C_{12}H_6O$.
Constants: Mol wt: 237.1, vap. d.: 8.2.
Hazard Analysis and Countermeasures
See chlorides.

DICHLORODIPHENYLTRICHLOROETHANE. See DDT.

DICHLORO DI-m-TOLYLSTANNANE
General Information
Description: Solid.
Formula: $(C_6H_4CH_3)_2SnCl_2$.
Constants: Mol wt: 371.9, mp: 40°C.
Hazard Analysis
Toxicity: Details unknown. See tin compounds and chlorides.
Disaster Hazard: Dangerous; see chlorides.

1,4-DICHLORO-2,3-EPOXYBUTANE
General Information
Formula: $CH_2ClHCOCHCH_2Cl$.
Constant: Mol wt: 131.
Hazard Analysis
Toxic Hazard Rating:
Acute Local: Irritant 1.
Acute Systemic: Ingestion 2; Inhalation 2; Skin Absorption 1.

Chronic Local: U.
Chronic Systemic: U.
Based on limited animal experiments.
Disaster Hazard: Dangerous. See chlorides.

1,2-DICHLOROETHANE. See ethylene dichloride.

1,1-DICHLOROETHANE. See ethylidene chloride.

DICHLOROETHANOYL CHLORIDE. See 2,2-dichloroacetyl chloride.

1,2-DICHLOROETHYL ACETATE
General Information
Description: Water-white liquid.
Formula: $CH_3COOCHClCH_2Cl$.
Constants: Mol wt: 157, mp: $< -32°C$, bp: $58-65°C$ at 13 mm, flash p: $307°F$, d: 1.296 at $20°C$, vap. d.: 5.42.
Hazard Analysis
Toxicity: Details unknown. See also esters, alcohols and acetic acid.
Fire Hazard: Slight, when exposed to heat or flame.
Disaster Hazard: Dangerous; see chlorides; can react vigorously with oxidizing materials.
Countermeasures
Storage and Handling: Section 7.
To Fight Fire: Water, foam, carbon dioxide, dry chemical or carbon tetrachloride (Section 6).

α,β-DICHLOROETHYL ALCOHOL. See 1,2-dichloroethyl alcohol.

1,2-DICHLOROETHYL ALCOHOL
General Information
Synonym: α,β-dichloroethyl alcohol.
Description: Liquid.
Formula: $CH_2ClCHClOH$.
Constants: Mol wt: 115.0, vap. d.: 3.97.
Hazard Analysis
Toxicity: Details unknown. See also alcohols.
Disaster Hazard: Dangerous; see chlorides.
Countermeasures
Storage and Handling: Section 7.

2,2′ DICHLOROETHYLAMINE. See nitrogen mustard.

DICHLOROETHYLARSINE. See ethyldichloroarsine.

ar-DICHLOROETHYLBENZENE
General Information
Description: Colorless liquid.
Formula: $Cl_2C_6H_3CH_2CH_3$.
Constants: Mol wt: 175.1, mp: $< -70°C$, bp: $220-224°C$, flash p: $205°F$, d: 1.208 at $25°/25°C$, vap. d.: 6.05.
Hazard Analysis
Toxicity: Details unknown. See also chlorinated hydrocarbons, aromatic.
Fire Hazard: Moderate, when exposed to heat or flame.
Disaster Hazard: Dangerous; see chlorides; can react vigorously with oxidizing materials.
Countermeasures
Storage and Handling: Section 7.
To Fight Fire: Water, foam, carbon dioxide, dry chemical or carbon tetrachloride (Section 6).

1,2-DICHLOROETHYLBENZENE
General Information
Description: Clear, colorless liquid.
Formula: $Cl_2C_6H_3CH_2CH_3$.

TOXIC HAZARD RATING CODE (For detailed discussion, see Section 1.)

0 NONE: (a) No harm under any conditions; (b) Harmful only under unusual conditions or overwhelming dosage.

1 SLIGHT: Causes readily reversible changes which disappear after end of exposure.

2 MODERATE: May involve both irreversible and reversible changes; not severe enough to cause death or permanent injury.

3 HIGH: May cause death or permanent injury after very short exposure to small quantities.

U UNKNOWN: No information on humans considered valid by authors.

Constants: Mol wt: 175.1, mp: $< -70°C$, bp: 224–226°C, flash p: 401°F, d: 1.21 at 25°/25°C, vap. d.: 6.05.

Hazard Analysis

Toxicity: Details unknown. See also chlorinated hydrocarbons, aromatic.

Fire Hazard: Slight, when exposed to heat or flame.

Disaster Hazard: Dangerous; see chlorides; can react vigorously with oxidizing materials.

Countermeasures

Storage and Handling: Section 7.

To Fight Fire: Water, foam, carbon dioxide, dry chemical or carbon tetrachloride (Section 6).

1,1-DICHLOROETHYLENE. See vinylidene chloride.

1,2-DICHLOROETHYLENE. See cis- and trans-dichloroethylene.

cis-DICHLOROETHYLENE

General Information

Synonyms: 1,2-dichloroethylene; acetylene dichloride.

Description: Colorless liquid, pleasant odor.

Formula: ClCHCHCl.

Constants: Mol wt: 97.0, mp: $-80.5°C$, bp: 59°C, lel: 9.7%, uel: 12.8%, flash p: 39°F, d: 1.2743 at 25°/4°C, vap. press.: 400 mm at 41.0°C, vap. d.: 3.34.

Hazard Analysis

Toxic Hazard Rating:

Acute Local: Inhalation 2.

Acute Systemic: Ingestion 2; Inhalation 2; Skin Absorption 2.

Chronic Local: Irritant 1.

Chronic Systemic: Ingestion 1; Inhalation 1.

TLV: ACGIH (recommended); 200 parts per million in air; 794 milligrams per cubic meter of air. See also chlorinated hydrocarbons, aliphatic.

Toxicology: In high concentrations it is irritant and narcotic. Has produced liver and kidney injury in experimental animals.

Fire Hazard: Dangerous, when exposed to heat or flame.

Spontaneous Heating: No.

Explosion Hazard: Moderate, in the form of vapor when exposed to flame (Section 7).

Disaster Hazard: Dangerous; see chlorides; can react vigorously with oxidizing materials.

Countermeasures

Ventilation Control: Section 2.

To Fight Fire: Water spray, foam, carbon dioxide, dry chemical or carbon tetrachloride (Section 6).

Personnel Protection: Section 3.

Storage and Handling: Section 7.

Shipping Regulations: Section 11.

I.C.C.: Flammable liquid; red label, 10 gallons.

Coast Guard Classification: Inflammable liquid; red label.

IATA: Flammable liquid, red label, 1 liter (passenger), 40 liters (cargo).

trans-DICHLOROETHYLENE

General Information

Synonym: Acetylene dichloride.

Description: Colorless liquid, pleasant odor.

Formula: ClCHCHCl.

Constants: Mol wt: 97.0, mp: $-50°C$, bp: 48°C, flash p: 36°F, lel: 9.7%, uel: 12.8%, d: 1.2743 at 25°/4°C, vap. press.: 400 mm at 30.8°C, vap. d.: 3.34.

Hazard Analysis

Toxic Hazard Rating:

Acute Local: Inhalation 1.

Acute Systemic: Ingestion 2; Inhalation 2; Skin Absorption 2.

Chronic Local: Irritant 1.

Chronic Systemic: U.

Toxicology: Exposure to high concentrations of vapor can cause nausea, vomiting, weakness, tremor and cramps. Recovery is usually prompt following removal from

exposure. Dermatitis may result from defatting action on skin.

Fire Hazard: Dangerous, when exposed to heat or flame.

Spontaneous Heating: No.

Explosion Hazard: Moderate in the form of vapor when exposed to flame (Section 7).

Disaster Hazard: Dangerous; see chlorides; can react vigorously with oxidizing materials.

Countermeasures

Ventilation Control: Section 2.

To Fight Fire: Water, foam, carbon dioxide, dry chemical or carbon tetrachloride (Section 6).

Personnel Protection: Section 3.

Storage and Handling: Section 7.

Shipping Regulations: Section 11.

I.C.C.: Flammable liquid; red label, 10 gallons.

Coast Guard Classification: Inflammable liquid; red label.

IATA: Flammable liquid, red label, 1 liter (passenger), 40 liters (cargo).

2,2'-DICHLOROETHYL ETHER

General Information

Synonyms: Chlorex; 1-chloro-2-(β-chloroethoxy) ethane.

Description: Colorless, stable liquid.

Formula: ClCH$_2$CH$_2$OCH$_2$CH$_2$Cl.

Constants: Mol wt: 143.0, bp: 178.5°C, fp: $-51.9°C$, flash p: 131°F (C.C.), d: 1.2220 at 20°/20°C, autoign. temp.: 696°F, vap. press.: 0.7 mm at 20°C, vap. d.: 4.93.

Hazard Analysis

Toxic Hazard Rating:

Acute Local: Irritant 3; Inhalation 3.

Acute Systemic: Ingestion 3; Inhalation 3.

Chronic Local: U.

Chronic Systemic: Ingestion 3; Inhalation 3.

TLV: ACGIH (recommended); 15 parts per million in air; 88 milligrams per cubic meter of air. May be absorbed via the skin.

Toxicology: The vapor is irritant to the mucous membranes of the eyes and nose. It affects the kidneys and liver in varying degrees, and is a mild narcotic. Guinea pigs cannot be killed immediately by exposure to concentrations which can be attained at ordinary room temperature, but exposure to 1,000 ppm for 30 to 60 minutes may produce death after several days. Autopsy shows congestion of the lungs and upper respiratory tract, pulmonary edema, and congestion of the liver, brain and kidneys. The pulmonary edema apparently develops after a latent period of several hours, similar to the action of "nitrous fumes." In humans, exposure to 500 to 1,000 ppm causes severe irritation of the eyes and nose after brief exposure, and deep inhalation is nauseating and intolerable. A concentration of 100 ppm produces slight nausea and irritation; concentrations of 35 ppm are practically free from irritation, though the odor is easily detectable. No cases of industrial poisoning have been reported. A fumigant. See also ethers.

Fire Hazard: Moderate, when exposed to heat or flame.

Explosion Hazard: See also ethers.

Disaster Hazard: Dangerous; when heated to decomposition, it emits highly toxic fumes; it reacts with water or steam to evolve toxic and corrosive fumes; can react vigorously with oxidizing materials.

Countermeasures

Ventilation Control: Section 2.

To Fight Fire: Water, foam, carbon dioxide, dry chemical or carbon tetrachloride (Section 6).

Personnel Protection: Section 3.

First Aid: Section 1.

Storage and Handling: Section 7.

DI-2-CHLOROETHYL FORMAL

General Information

Synonym: di-[2-chloroethyl] formal.

Description: Liquid.

Formula: $CH_2(OCH_2CH_2Cl)_2$.

Constants: Mol wt: 173.05, bp: 217.5°C, flash p: 230°F (O.C.), d: 1.23, vap. d.: 5.9.

Hazard Analysis

Toxic Hazard Rating:

Acute Local: Irritant 3; Ingestion 3; Inhalation 3.

Acute Systemic: U.

Chronic Local: U.

Chronic Systemic: U.

Fire Hazard: Slight, when exposed to heat or flame.

Disaster Hazard: Dangerous; see chlorides; can react vigorously with oxidizing materials.

Countermeasures

Ventilation Control: Section 2.

To Fight Fire: Water, foam, carbon dioxide, dry chemical or carbon tetrachloride (Section 6).

Personnel Protection: Section 3.

First Aid: Section 1.

Storage and Handling: Section 7.

2-(1,2-DICHLOROETHYL)-4-METHYL-1,3-DIOXOLANE

Hazard Analysis

Toxic Hazard Rating: U. Limited animal experiments suggest high toxicity and low irritation.

Disaster Hazard: Dangerous. See chlorides.

DICHLOROETHYL SULFIDE. See dichlorodiethyl sulfide.

DICHLOROFLUOROMETHANE

General Information

Synonym: "Freon 21."

Description: Heavy, colorless gas.

Formula: $HCCl_2F$.

Constants: Mol wt: 103, mp: −135°C, bp: 8.9°C, d: 1.48, vap. press.: 2 atm at 28.4°C, vap. d.: 3.82.

Hazard Analysis

Toxic Hazard Rating:

Acute Local: 0.

Acute Systemic: Inhalation 1.

Chronic Local: 0.

Chronic Systemic: Inhalation 1.

TLV: ACGIH (recommended); 1000 parts per million in air; 4200 milligrams per cubic meter of air.

Disaster Hazard: Dangerous; see chlorides and fluorides.

Countermeasures

Storage and Handling: Section 7.

DICHLOROFORMOXIME

General Information

Description: Colorless, prismatic crystals, disagreeable penetrating odor.

Formula: CCl_2NOH.

Constants: Mol wt: 113.94, mp: 39–40°C, bp: 53–54°C at 28 mm, vap. d.: 4.8.

Hazard Analysis

Toxic Hazard Rating:

Acute Local: Irritant 3; Ingestion 3; Inhalation 3.

Acute Systemic: U.

Chronic Local: U.

Chronic Systemic: U.

Toxicology: A military poison.

Disaster Hazard: Dangerous; see chlorides.

Countermeasures

Ventilation Control: Section 2.

Personnel Protection: Section 3.

Storage and Handling: Section 7.

DICHLOROGERMANE

General Information

Description: Colorless liquid.

Formula: GeH_2Cl_2.

Constants: Mol wt: 145.53, mp: −68.0°C, bp: 69.5°C, d: 1.90 at −68°C, vap. d.: 5.0.

Hazard Analysis and Countermeasures

See hydrochloric acid and germanium compounds.

1,3-DICHLORO-2,4-HEXADIENE

General Information

Description: Liquid.

Formula: $CH_2ClHCCClHCCHCH_3$.

Constants: Mol wt: 151.04, flash p: 168°F (O.C.), vap. d.: 5.2.

Hazard Analysis

Toxicity: See chlorinated hydrocarbons, aliphatic.

Fire Hazard: Moderate, when exposed to heat or flame.

To Fight Fire: Water, foam, carbon dioxide, dry chemical or carbon tetrachloride (Section 6).

Disaster Hazard: Dangerous; see chlorides. Will react with water or steam to produce toxic and corrosive fumes; can react vigorously with oxidizing materials.

Countermeasures

Storage and Handling: Section 7.

α-DICHLOROHYDRIN. See 1,3-dichloropropanol-2.

DICHLOROISOCYANURIC ACID

General Information

Description: White crystals, chlorine odor, moderately soluble in water.

Formula: $Cl_2H(NCO)_3$.

Constants: Mol wt: 198, mp: 225°C.

Hazard Analysis

Toxicology: Details unknown. Limited animal experiments suggest slight to moderate toxicity.

Disaster Hazard: Dangerous; when heated to decomposition it emits chlorides and carbon monoxide.

Countermeasures

Storage and Handling: An oxidizing material (Section 7).

Shipping Regulations: Section 11.

I.C.C.: Oxidizing material, yellow label, 200 pounds. Dry, containing more than 39% available chlorine.

IATA: Oxidizing material, yellow label, 25 kilograms (passenger), 45 kilograms (cargo).

DICHLOROISOPROPYL ALCOHOL. See 1,3-dichloropropanol-2.

DICHLOROISOPROPYL ETHER

General Information

Description: Colorless liquid.

Formula: $(CH_2ClCH_3CH)_2O$.

Constants: Mol wt: 171.09, bp: 187.8°C, fp: < −20°C, flash p: 185°F (O.C.), d: 1.11 at 25/25°C, vap. d.: 6.0, vap. press.: 0.10 mm at 20°C.

Hazard Analysis

Toxicity: Details unknown. See also ethers. Animal experiments show variable toxicity depending on species and route of administration.

Fire Hazard: Moderate, when exposed to heat or flame.

Explosion Hazard: Details unknown. See also ethers.

Disaster Hazard: Dangerous; when heated to decomposition it emits highly toxic fumes of chlorides; can react vigorously with oxidizing materials.

Countermeasures

Storage and Handling: Section 7.

TOXIC HAZARD RATING CODE (For detailed discussion, see Section 1.)

0 NONE: (a) No harm under any conditions; (b) Harmful only under unusual conditions or overwhelming dosage.

1 SLIGHT: Causes readily reversible changes which disappear after end of exposure.

2 MODERATE: May involve both irreversible and reversible changes; not severe enough to cause death or permanent injury.

3 HIGH: May cause death or permanent injury after very short exposure to small quantities.

U UNKNOWN: No information on humans considered valid by authors.

To Fight Fire: Water, foam, carbon dioxide, dry chemical or carbon tetrachloride (Section 6). See also ethers.

DICHLOROMALEALDEHYDRIC ACID. See mucochloric acid.

DICHLOROMETHANE. See methylene chloride.

3,9-DICHLORO-7-METHOXYACRIDINE
General Information
Synonyms: Halocrin.
Description: Needlelike crystals soluble in organic solvents.
Formula: $C_{14}H_9Cl_2NO$.
Constants: Mol wt: 278, mp: 161°C.
Hazard Analysis
Toxic Hazard Rating:
　Acute Local: Irritant 3; Ingestion 3; Inhalation 3.
　Acute Systemic: U.
　Chronic Local: U.
　Chronic Systemic: U.
Disaster Hazard: Dangerous; when heated to decomposition it emits highly toxic chloride and oxide of nitrogen fumes.

3',4'-DICHLORO-2-METHYL ACRYL ANILIDE
General Information
Synonym: Dicryl.
Description: Solid; insoluble in water; soluble in acetone, alcohol, isophorone, dimethyl sulfoxide.
Formula: $Cl_2C_6H_3NHCOC(CH_3):CH_2$.
Constants: Mol wt: 230, mp: 128°C.
Hazard Analysis
Toxic Hazard Rating: U. Animal experiments show moderately low toxicity.
Disaster Hazard: Dangerous. See chlorides.

DICHLOROMETHYL ARSINE
General Information
Synonyms: Methylarsenic dichloride; MD.
Description: Colorless liquid.
Formula: CH_3AsCl_2.
Constants: Mol wt: 160.86, bp: 134.5°C, fp: −59°C, flash p: >221°F, d: 1.838 at 20°/4°C, vap. press.: 10 mm at 24.3°C, vap. d.: 5.40.
Hazard Analysis
Toxic Hazard Rating:
　Acute Local: Irritant 3; Ingestion 3; Inhalation 3.
　Acute Systemic: Ingestion 3; Inhalation 3.
　Chronic Local: U.
　Chronic Systemic: Ingestion 3; Inhalation 3.
Toxicology: This is a blistering type of military poison. It is rapidly detoxified in the body. A moderately persistent gas. For effects and antidotes see dichloro-(2-chloro-vinyl) arsine. See also arsenic compounds.
Fire Hazard: Slight, when exposed to heat or flame.
Disaster Hazard: Dangerous; when heated to decomposition or on contact with acid or acid fumes, it emits highly toxic fumes of chlorides and arsenic; can react vigorously with oxidizing materials.
Countermeasures
Ventilation Control: Section 2.
To Fight Fire: Water, foam, carbon dioxide, dry chemical or carbon tetrachloride (Section 6).
Personnel Protection: Section 3.
First Aid: Section 1.
Storage and Handling: Section 7.
Shipping Regulations: Section 11.
　I.C.C. Classification: Poison A, poison gas label.
　Coast Guard Classification: Poison A, poison gas label.

DICHLOROMETHYLCHLOROFORMATE
General Information
Description: Colorless liquid.
Formula: $ClCOOCHCl_2$.
Constants: Mol wt: 163.40, bp: 110–111°C, d: 1.56 at 15°C, vap. d.: 5.63.

Hazard Analysis
Toxic Hazard Rating:
　Acute Local: Irritant 3; Ingestion 3; Inhalation 3.
　Acute Systemic: U.
　Chronic Local: U.
　Chronic Systemic: U.
Disaster Hazard: Dangerous; see chlorides, will react with water or steam to produce toxic and corrosive fumes.
Countermeasures
Ventilation Control: Section 2.
Personnel Protection: Section 3.
Storage and Handling: Section 7.

2,2-DICHLORO-N-METHYL DIETHYLAMINE HYDROCHLORIDE. See mechlorethamine hydrochloride.

DICHLOROMETHYL ETHER
General Information
Synonym: Dichlorinated methyl oxide.
Description: Volatile liquid.
Formula: $O(CH_2Cl)_2$.
Constants: Mol wt: 115, bp: 105°C, d: 1.315 at 20°C, vap. d.: 4.0.
Hazard Analysis
Toxic Hazard Rating:
　Acute Local: Irritant 3; Ingestion 3; Inhalation 3.
　Acute Systemic: U.
　Chronic Local: U.
　Chronic Systemic: U.
Fire Hazard: Details unknown. See also ethers and chlorides.
Explosion Hazard: Details unknown. See also ethers.
Disaster Hazard: Dangerous; see chlorides.
Countermeasures
Ventilation Control: Section 2.
Personnel Protection: Section 3.
Storage and Handling: Section 7.

α,β-DICHLOROMETHYLETHYL KETONE
General Information
Description: Liquid.
Formula: $ClCH_2COCH_2CH_2Cl$.
Constants: Mol wt: 141, bp: 65°C at 3 mm, vap. d.: 4.9.
Hazard Analysis
Toxic Hazard Rating:
　Acute Local: Irritant 3; Ingestion 3; Inhalation 3.
　Acute Systemic: U.
　Chronic Local: U.
　Chronic Systemic: U.
Toxicology: A lacrimator type of military poison.
Disaster Hazard: Dangerous; when heated to decomposition, it burns and emits highly toxic fumes of chlorides.
Countermeasures
Ventilation Control: Section 2.
Personnel Protection: Section 3.
Storage and Handling: Section 7.

2,3-DICHLORO-2-METHYL-PROPIONALDEHYDE
General Information
Formula: CH_2ClCCH_3ClCHO.
Constant: Mol wt: 129.
Hazard Analysis
Toxic Hazard Rating: U. Limited animal experiments suggest high toxicity. See also aldehydes.
Disaster Hazard: Dangerous. See chlorides.

DICHLOROMETHYL SILICANE
General Information
Description: A liquid.
Formula: $HSiCl_2CH_3$.
Constants: Mol wt: 115, d: 0.93 at 0°C, mp: −93°C.
Hazard Analysis and Countermeasures
See silanes and chlorides.

DICHLOROMONOFLUOROMETHANE. See dichlorofluoromethane.

2,3-DICHLORO-1,4-NAPHTHOQUINONE
General Information
Synonyms: Dichlone, phygon.
Description: Golden yellow crystals, insoluble in water, moderately soluble in organic solvents.
Formula: $C_{10}H_4O_2Cl_2$.
Constants: Mol wt: 227.1, mp: 193°C, vap. d.: 7.8.
Hazard Analysis
Toxic Hazard Rating:
Acute Local: Irritant 2; Inhalation 2.
Acute Systemic: Ingestion 2; Inhalation 2.
Chronic Local: U.
Chronic Systemic: U.
Toxicology: A fungicide. Irritating to skin, eyes and mucous membranes. Large doses can cause CNS depression.
Disaster Hazard: Dangerous; see chlorides.
Countermeasures
Storage and Handling: Section 7.

1,2,4-DICHLORONITROBENZENE
General Information
Description: Liquid.
Formula: $C_6H_3Cl_2NO_2$.
Constants: Mol wt: 192.0, vap. d.: 6.6.
Hazard Analysis
Toxicity: Details unknown. See also chlorinated hydrocarbons, aromatic and nitrobenzene.
Fire Hazard: See nitrates.
Disaster Hazard: Dangerous; when heated to decomposition, it emits highly toxic fumes of oxides of nitrogen and chlorides; can react vigorously with oxidizing materials.
Countermeasures
Storage and Handling: Section 7.

1,4,2-DICHLORONITROBENZENE
General Information
Description: Liquid.
Formula: $C_6H_3Cl_2NO_2$.
Constants: Mol wt: 192.0, vap. d.: 6.6.
Hazard Analysis
Toxicity: Details unknown. See also chlorinated hydrocarbons, aromatic and nitrobenzene.
Fire Hazard: See nitrates.
Disaster Hazard: Dangerous; when heated to decomposition, it emits highly toxic fumes of oxides of nitrogen and chlorides; can react vigorously with oxidizing materials.
Countermeasures
Storage and Handling: Section 7.

1,1-DICHLORO-1-NITROETHANE
General Information
Synonym: Ethide.
Description: Liquid.
Formula: $H_3CC(Cl)_2NO_2$.
Constants: Mol wt: 143.96, bp: 124°C, flash p: 168°F (O.C.), d: 1.4153 at 20°/20°C, vap. d.: 4.97.
Hazard Analysis
Toxic Hazard Rating:
Acute Local: Irritant 3; Ingestion 3; Inhalation 3.
Acute Systemic: U.
Chronic Local: U.
Chronic Systemic: U.
Toxicology: Strong irritant. Inhalation causes pulmonary edema.
TLV: ACGIH (recommended); 10 parts per million in air; 59 milligrams per cubic meter of air.

Fire Hazard: Moderate, when exposed to heat or flame.
Spontaneous Heating: No.
Disaster Hazard: Dangerous; when heated to decomposition, it emits highly toxic fumes of chlorides and oxides of nitrogen; can react vigorously with oxidizing materials.
Countermeasures
Ventilation Control: Section 2.
To Fight Fire: Water, carbon dioxide, dry chemical or carbon tetrachloride (Section 6).
Personnel Protection: Section 3.
First Aid: Section 1.
Storage and Handling: Section 7.

1,1-DICHLORO-1-NITROPROPANE
General Information
Description: Liquid.
Formula: $C_2H_5CCl_2NO_2$.
Constants: Mol wt: 158.00, bp: 141°C, flash p: 151°F (O.C.), d: 1.314, vap. d.: 5.45.
Hazard Analysis
Toxicity: Details unknown. See also chlorinated hydrocarbons, aliphatic.
Fire Hazard: Moderate, when exposed to heat or flame.
Disaster Hazard: Dangerous; when heated to decomposition, it emits highly toxic fumes of chlorides and oxides of nitrogen; can react vigorously with oxidizing materials.
Countermeasures
Storage and Handling: Section 7.
To Fight Fire: Water, carbon dioxide, dry chemical or carbon tetrachloride (Section 6).

1,5-DICHLOROPENTANE
General Information
Synonym: Amylene chloride, pentamethylene dichloride.
Description: Insoluble in water.
Formula: $CH_2Cl(CH_2)_3CH_2Cl$.
Constants: Mol wt: 141, d: 1.1, vap. d.: 4.9, bp: 352–358°F, flash p: 780°F (O.C.).
Hazard Analysis
Toxic Hazard Rating: U.
Fire Hazard: Moderate, when exposed to heat or flame.
Disaster Hazard: Dangerous. See chlorides.
Countermeasures
To Fight Fire: Water ineffective except as a blanket (Section 6).
Storage and Handling: Section 7.

DICHLOROPENTANES (MIXED)
General Information
Description: Clear, light yellow colored liquid.
Formula: $C_5H_{10}Cl_2$.
Constants: Mol wt: 141.04, bp: 130°C, flash p: 106°F (O.C.), d: 1.06–1.08 at 20°C, autoign. temp.: 115°F, vap. d.: 4.86.
Hazard Analysis
Toxicity: Details unknown. See also chlorinated hydrocarbons, aliphatic.
Fire Hazard: Moderate, when exposed to heat or flame.
Disaster Hazard: Dangerous; when heated to decomposition, it emits highly toxic fumes of phosgene; can react vigorously with oxidizing materials.
Countermeasures
Ventilation Control: Section 2.
To Fight Fire: Water, foam, carbon dioxide, dry chemical or carbon tetrachloride (Section 6).
Storage and Handling: Section 7.

TOXIC HAZARD RATING CODE *(For detailed discussion, see Section 1.)*

0 NONE: (a) No harm under any conditions; (b) Harmful only under unusual conditions or overwhelming dosage.

1 SLIGHT: Causes readily reversible changes which disappear after end of exposure.

2 MODERATE: May involve both irreversible and reversible changes; not severe enough to cause death or permanent injury.

3 HIGH: May cause death or permanent injury after very short exposure to small quantities.

U UNKNOWN: No information on humans considered valid by authors.

DICHLOROPHENE. See 2,2'-methylene-bis-(4-chlorophenol).

2,4-DICHLOROPHENOL
General Information
Description: Colorless crystals.
Formula: $C_6H_3OHCl_2$.
Constants: Mol wt: 163.0, mp: 45°C, bp: 210°C, flash p: 237°F, d: 1.383 at 60°/25°C, vap. d.: 5.62, vap. press.: 1 mm at 53.0°C.
Hazard Analysis
Toxicity: Details unknown. Limited animal experiments show moderate toxicity. See also chlorinated phenols.
Fire Hazard: Slight, when exposed to heat or flame.
Disaster Hazard: Dangerous; when heated to decomposition, or on contact with acid or acid fumes, it emits highly toxic fumes of chlorides; can react vigorously with oxidizing materials.
Countermeasures
Storage and Handling: Section 7.
To Fight Fire: Water, foam, carbon dioxide, dry chemical or carbon tetrachloride (Section 6).

DICHLORO (PHENOL DIPHENYL ETHER). See dichloro(phenol xenyl ether).

DICHLORO(PHENOLXENYL ETHER)
General Information
Synonym: Dichloro(phenol diphenyl ether).
Description: Light, viscous, straw-colored liquid.
Formula: $C_{18}H_{12}OCl_2$.
Constants: Mol wt: 315.2, mp: $< 0°C$, bp: 222°C at 10 mm, flash p: 399°F, d: 1.233 at 25°/25°C, vap. d.: 10.9.
Hazard Analysis
Toxicity: Details unknown. See also chlorinated hydrocarbons, aromatic and chlorinated phenols.
Fire Hazard: Slight, when exposed to heat or flame.
Disaster Hazard: Dangerous; see chlorides; can react with oxidizing materials.
Countermeasures
Storage and Handling: Section 7.
To Fight Fire: Water, foam, carbon dioxide, dry chemical or carbon tetrachloride (Section 6).

2,4-DICHLOROPHENOXYACETIC ACID
General Information
Synonym: 2,4-D.
Description: White powder.
Formula: $C_8H_6Cl_2O_3$.
Constants: Mol wt: 221, vap. d.: 7.63.
Hazard Analysis
Toxic Hazard Rating:
Acute Local: Irritant 2; Ingestion 2; Inhalation 2.
Acute Systemic: Ingestion 2.
Chronic Local: Irritant 2.
Chronic Systemic: Ingestion 2.
Toxicity: Can cause nausea, vomiting, and CNS depression. Liver and kidney injury have been reported in experimental animals. It is used as a herbicide.
TLV: ACGIH (recommended); 10 milligrams per cubic meter of air.
Disaster Hazard: Dangerous; see chlorides.
Countermeasures
Storage and Handling: Section 7.
Shipping Regulations: Section 11.
IATA: Other restricted articles, class A, no label required, no limit (passenger and cargo).

2,4-DICHLOROPHENOXYETHYL BENZOATE
General Information
Synonym: Sesin.
Description: Crystals.
Formula: $C_{15}H_{12}O_3Cl_2$.
Constant: Mol wt: 311.2.

Hazard Analysis
Toxicology: Details unknown. Limited animal experiments suggest fairly low toxicity.
Disaster Hazard: Dangerous; see chlorides.

DICHLOROPHENOXYETHYL SULFATE
General Information
Synonym: Crag I.
Formula: $C_6H_3Cl_2OC_2H_5SO_4$.
Constant: Mol wt: 287.1.
Hazard Analysis
Toxic Hazard Rating:
Acute Local: Irritant 2.
Acute Systemic: Ingestion 2.
Chronic Local: Irritant 2.
Chronic Systemic: Ingestion 2.
Toxicology: Has produced liver and kidney damage in experimental animals.
TLV: ACGIH (recommended); 15 milligrams per cubic meter of air.
Disaster Hazard: Dangerous; see sulfates.

DI-(4-CHLOROPHENOXY) METHANE. See bis(p-chlorophenoxymethane).

1-(2,4-DICHLOROPHENOXY)-2-PROPANOL. See propylene glycol-2,4-dichlorophenyl ether.

DICHLOROPHENYL BENZENE SULFONATE
General Information
Formula: $C_{12}H_8Cl_2SO_3$.
Constant: Mol wt: 303.2.
Hazard Analysis
Toxicology: Details unknown. Limited animal experiments suggest relatively low toxicity. No specific human toxicologic data is available. Probably highly irritating to skin and mucous membranes. Animal experiments have shown effects similar to DDT. It is used as an insecticide.
Disaster Hazard: Dangerous. See dichlorophenoxyethyl sulfate.

α,α-DI-(4-CHLOROPHENYL) DIETHYL ETHER
General Information
Description: Crystals.
Formula: $C_{16}H_{16}OCl_2$.
Constants: Mol wt: 295, vap. d.: 11.3.
Hazard Analysis
Toxicity: Details unknown. See also chlorinated hydrocarbons, aromatic.
Disaster Hazard: Dangerous; see chlorides.
Countermeasures
Storage and Handling: Section 7.

2,4-DICHLOROPHENYL-O,O-DIETHYL PHOSPHOROTHIOATE. See V-CB nemacide.

3-(3,4-DICHLOROPHENYL)-1,1-DIMETHYL UREA. See diuron.

2,5-DICHLOROPHENYL ISOCYANATE
General Information
Description: White to light green solid, soluble in organic solvents.
Formula: C_7H_3ClNO.
Constants: Mol wt: 189.1, mp: 27°C, bp: 126°C at 30 mm.
Hazard Analysis
Toxicology: A powerful irritant to the skin and mucous membranes. Very strong lacrimator.
Disaster Hazard: Dangerous. See chlorides and cyanates.
Countermeasures
Storage and Handling: Section 7.

3,4-DICHLOROPHENYL ISOCYANATE
General Information
Description: White to light brown solid. Soluble in organic solvents.

Formula: $C_7H_3Cl_2NO$.
Constants: Mol wt: 189.1, mp: $42°C$, bp: $133°C$ at 30 mm.
Hazard Analysis and Countermeasures
See 2,5 dichlorophenyl isocyanate.

3-(3,4-DICHLOROPHENYL)-1-METHOXY-1-METHYL UREA.
General Information
Synonym: "Horox" *, Linuron.
Description: Solid; slightly soluble in water; partially soluble in acetone and alcohol.
Formula: $C_6H_3ClNHC(O)N(OCH_3)CH_3$.
Constants: Mol wt: 214, mp: $93–94°C$.
Hazard Analysis
See Monuron, which has similar properties.

DI(p-CHLOROPHENYL) METHYL CARBINOL
General Information
Synonym: Dimite, DMC.
Formula: $C_{14}H_{12}Cl_2O$.
Constant: Mol wt: 267.2.
Hazard Analysis
Toxicology: Details unknown. An insecticide. Probably similar to DDT, but less toxic. See DDT.
Disaster Hazard: Dangerous. See chlorides.

3,4-DICHLOROPHENYL-N-METHYL SULFONAMIDE
General Information
Description: Crystals.
Formula: $Cl_2C_7H_5O_3NS$.
Constants: Mol wt: 254.2, vap. d.: 6.4.
Hazard Analysis
Toxicity: See chlorinated hydrocarbons, aromatic.
Disaster Hazard: Dangerous; see sulfonates and chlorides.
Countermeasures
Storage and Handling: Section 7.

DICHLOROPHENYL PHOSPHINE. See phosphenyl chloride.

DICHLOROPHENYL TRICHLOROSILANE
General Information
Description: Straw colored liquid; soluble in benzene, perchloroethylene.
Formula: $Cl_2C_6H_3SiCl_3$ (mixture of isomers).
Constants: Mol wt: 280.5, d.: 1.562, bp: $260°C$, flash p: $286°F$.
Hazard Analysis
Toxic Hazard Rating: U.
To Fight Fire: Slight when exposed to heat or flame.
Disaster Hazard: Dangerous. See chlorides and silanes.
Countermeasrues
Shipping Regulations: Section 11.
  IATA: Corrosive liquid, white label, not acceptable (passenger), 40 liters (cargo).

DICHLOROPHTHALIC ANHYDRIDE
General Information
Description: White flakes.
Formula: $C_6H_2Cl_2(CO)_2O$.
Constants: Mol wt: 217, mp: $80–140°C$, bp: $220°C$ at 40 mm.
Hazard Analysis
Toxicity: Details unknown; probably an irritant.
Disaster Hazard: Dangerous; see chlorides; it will react with water or steam to produce heat and toxic fumes.
Countermeasures
Storage and Handling: Section 7.

1,2-DICHLOROPROPANE
General Information
Synonym: propylene dichloride.
Description: Colorless liquid.
Formula: $CH_2ClCHClCH_3$.
Constants: Mol wt: 113.0, bp: $96.8°C$, flash p: $60°F$, d.: 1.1593 at $20°/20°C$, vap. press.: 40 mm at $19.4°C$, vap. d.: 3.9, autoign. temp.: $1035°F$, lel: 3.4%, uel: 14.5%.
Hazard Analysis
Toxic Hazard Rating:
Acute Local: U.
Acute Systemic: Ingestion 3; Inhalation 3; Skin Absorption 3.
Chronic Local: Irritant 1.
Chronic Systemic: Ingestion 3; Inhalation 3; Skin Absorption 3.
TLV: ACGIH (recommended); 75 parts per million in air; 346 milligrams per cubic meter of air.
Toxicology: 1,2-dichloropropane can cause dermatitis, and is regarded as one of the more toxic chlorinated hydrocarbons. A suggested order of increasing toxicity is: dichloromethane, trichloroethylene, carbon tetrachloride, dichloropropane, dichloroethane. Animals, exposed to high concentrations often showed marked visceral congestion, fatty degeneration of the liver, kidney, and less frequently of the heart. They also showed areas of coagulation and necrosis of the liver. There was found to be a heavy mortality among mice exposed to 400 ppm concentrations.
Fire Hazard: Dangerous, when exposed to heat or flame.
Disaster Hazard: Dangerous; see chlorides; can react vigorously with oxidizing materials.
Countermeasures
Ventilation Control: Section 2.
To Fight Fire: Water, foam, carbon dioxide, dry chemical or carbon tetrachloride (Section 6).
Personnel Protection: Section 3.
Storage and Handling: Section 7.

1,3-DICHLOROPROPANE
General Information
Synonym: trimethylene chloride.
Description: Colorless liquid.
Formula: $CH_2ClCH_2CH_2Cl$.
Constants: Mol wt: 113.0, bp: $125°C$, d.: 1.201 at $15°C$, vap. d.: 3.90.
Hazard Analysis
Toxicity: See chlorinated hydrocarbons, aliphatic and 1,2-dichloropropane.
Disaster Hazard: Dangerous; when heated to decomposition, it emits highly toxic fumes of phosgene.
Countermeasures
Storage and Handling: Section 7.

2,2-DICHLOROPROPANE
General Information
Synonym: Acetone dichloride.
Description: Liquid.
Formula: $CH_3CCl_2CH_3$.
Constants: Mol wt: 113.0, mp: $-34.6°C$, bp: $69.7°C$, d.: 1.093 at $20°/20°C$, vap. d.: 3.9.
Hazard Analysis
Toxicity: See chlorinated hydrocarbons, aliphatic and 1,2-dichloropropane.
Disaster Hazard: Dangerous; when heated to decomposi-

TOXIC HAZARD RATING CODE (For detailed discussion, see Section 1.)

0 NONE: (a) No harm under any conditions; (b) Harmful only under unusual conditions or overwhelming dosage.

1 SLIGHT: Causes readily reversible changes which disappear after end of exposure.

2 MODERATE: May involve both irreversible and reversible changes; not severe enough to cause death or permanent injury.

3 HIGH: May cause death or permanent injury after very short exposure to small quantities.

U UNKNOWN: No information on humans considered valid by authors.

tion, it emits highly toxic fumes of phosgene; can react vigorously with oxidizing materials.

Countermeasures

Storage and Handling: Section 7.

1,3-DICHLOROPROPANOL-2

General Information

Synonym: dichloroisopropyl alcohol, α-dichlorohydrin.

Description: Colorless liquid.

Formula: $CH_2ClCHOHCH_2Cl$.

Constants: Mol wt: 129.0, bp: 174°C, d: 1.367 at 20°/4°C, vap. press.: 1 mm at 28.0°C, vap. d.: 4.45, flash p: 165°F (O.C.).

Hazard Analysis

Toxic Hazard Rating:

  Acute Local: Irritant 1; Ingestion 2; Inhalation 2.

  Acute Systemic: Inhalation 2; Ingestion 2.

  Chronic Local: U.

  Chronic Systemic: U.

Toxicology: Action may be similar to carbon tetrachloride but more irritating to mucous membranes.

Disaster Hazard: Dangerous; when heated to decomposition, it emits highly toxic fumes of phosgene.

Fire Hazard: Moderate, when exposed to heat or flame.

Countermeasures

Ventilation Control: Section 2.

Storage and Handling: Section 7.

To Fight Fire: Alcohol foam.

1,1-DICHLORO-2-PROPANONE. See 1,1-dichloro-acetone.

1,3-DICHLORO-2-PROPANONE. See 1,3-dichloro-acetone.

1,2-DICHLOROPROPENE. See 1,2-dichloropropylene.

1,3-DICHLOROPROPENE

General Information

Synonym: 1,3-dichloropropylene.

Description: Liquid.

Formula: $C_3H_4Cl_2$.

Constants: Mol wt: 111.0, bp: 103–110°C, flash p: 95°F, d: 1.22, vap. d.: 3.8.

Hazard Analysis

Toxic Hazard Rating:

  Acute Local: Irritant 3; Ingestion 3; Inhalation 3.

  Acute Systemic: Ingestion 2.

  Chronic Local: U.

  Chronic Systemic: U.

Toxicology: A strong irritant. Has produced liver and kidney injury in experimental animals.

Fire Hazard: Moderate, when exposed to heat or flame.

Disaster Hazard: Dangerous; see chlorides; can react vigorously with oxidizing materials.

Countermeasures

Ventilation Control: Section 2.

To Fight Fire: Water, foam, carbon dioxide, dry chemical or carbon tetrachloride (Section 6).

Storage and Handling: Section 7.

2,3-DICHLOROPROPIONALDEHYDE

General Information

Description: Liquid.

Formula: $CH_2ClCHClCHO$.

Constants: Mol wt: 127, vap. d.: 4.4.

Hazard Analysis

Toxic Hazard Rating:

  Acute Local: Irritant 3; Ingestion 3; Inhalation 3.

  Acute Systemic: U.

  Chronic Local: U.

  Chronic Systemic: U.

Disaster Hazard: Dangerous; see chlorides.

Countermeasures

Ventilation Control: Section 2.

Personnel Protection: Section 3.

Storage and Handling: Section 7.

3,4-DICHLOROPROPIONANILIDE.

General Information

Synonym: Propanil.

Description: Light brown solid (pure compound); liquid (technical compound).

Formula: $Cl_2C_6H_3NHCOCH_2CH_3$.

Constants: Mol wt: 218, mp (pure compound): 85–89°C, bp (technical compound): 91–95°C.

Hazard Analysis

Toxic Hazard Rating: U. Animal experiments suggest moderate toxicity.

Disaster Hazard: Dangerous. See chlorides.

2,2-DICHLOROPROPIONIC ACID

General Information

Description: White to tan powder.

Formula: CH_3CCl_2COOH.

Constant: Mol wt: 143.0.

Hazard Analysis

Toxic Hazard Rating:

  Acute Local: Irritant 2; Ingestion 2; Inhalation 2.

  Acute Systemic: U.

  Chronic Local: U.

  Chronic Systemic: U.

Disaster Hazard: Dangerous; see chlorides.

Countermeasures

Ventilation Control: Section 2.

Personnel Protection: Section 3.

Storage and Handling: Section 7.

1,2-DICHLOROPROPYLENE

General Information

Synonym: 1,2-dichloropropene.

Description: Liquid.

Formula: $CHClCClCH_3$.

Constants: Mol wt: 111.0, bp: 75°C, vap. d.: 3.83.

Hazard Analysis

Toxicity: See chlorinated hydrocarbons, aliphatic.

Disaster Hazard: Dangerous; see chlorides.

Countermeasures

Storage and Handling: Section 7.

1,3-DICHLOROPROPYLENE. See 1,3-dichloropropene.

2,3-DICHLOROQUINOLINE

General Information

Description: Crystals.

Formula: $C_9H_5Cl_2N$.

Constants: Mol wt: 198.1, mp: 105°C, vap. d.: 6.83.

Hazard Analysis

Toxicity: Unknown.

Disaster Hazard: Dangerous; see chlorides.

Countermeasures

Storage and Handling: Section 7.

DICHLOROSTYRENE

General Information

Description: Liquid.

Formula: $C_6H_5CClCHCl$.

Constants: Mol wt: 173.0, flash p: 225°F (O.C.), vap. d.: 6.0.

Hazard Analysis

Toxicity: Unknown. See also phenyl ethylene (styrene).

Fire Hazard: Slight, when exposed to heat or flame.

Disaster Hazard: Dangerous; see chlorides; can react vigorously with oxidizing materials.

Countermeasures

Storage and Handling: Section 7.

To Fight Fire: Water, foam, carbon dioxide, dry chemical or carbon tetrachloride (Section 6).

DICHLOROTETRAFLUOROETHANE

General Information

Synonym: F-114.

Description: Colorless gas.

Formula: CCl_2FCF_3.

Constants: Mol wt: 171, bp: 3.5°C.

Hazard Analysis

Toxic Hazard Rating:

Acute Local: Irritant 1; Inhalation 2.

Acute Systemic: Inhalation 1.

Chronic Local: U.

Chronic Systemic: U.

Toxicology: Narcotic in high concentrations.

TLV: ACGIH (recommended); 1000 parts per million in air; 7000 milligrams per cubic meter of air.

Caution: An asphyxiant.

Disaster Hazard: Dangerous; see fluorides and chlorides.

Countermeasures

Storage and Handling: Section 7.

α,4-DICHLOROTOLUENE. See p-chlorobenzyl chloride.

2,2'-DICHLOROTRIETHYLAMINE. See nitrogen mustard.

β,β-DICHLOROVINYLCHLOROARSINE

General Information

Description: Yellow or yellowish brown liquid, when pure; darker when impure.

Formula: (ClCHCH)$_2$AsCl.

Constants: Mol wt: 233.35, bp: 230°C decomposes, d: 1.702 at 20°C, vap. d.: 8.05.

Hazard Analysis

Toxic Hazard Rating:

Acute Local: Irritant 3; Ingestion 3; Inhalation 3.

Acute Systemic: Ingestion 3; Inhalation 3.

Chronic Local: U.

Chronic Systemic: Ingestion 3; Inhalation 3.

Toxicology: See also dichloro-(2-chloro vinyl) arsine.

Disaster Hazard: Dangerous; see arsenic and chlorides.

Countermeasures

Ventilation Control: Section 2.

Personnel Protection: Section 3.

First Aid: Section 1.

Storage and Handling: Section 7.

Shipping Regulations: Section 11.

IATA: Poison A, not acceptable (passenger and cargo).

β,β-DICHLOROVINYLMETHYLARSINE

General Information

Description: Liquid.

Formula: (CHCHCl)$_2$CH$_3$As.

Constants: Mol wt: 212.9, bp: 140–145°C at 10 mm, vap. d.: 7.35.

Hazard Analysis

Toxic Hazard Rating:

Acute Local: Irritant 3; Ingestion 3; Inhalation 3.

Acute Systemic: Ingestion 3; Inhalation 3.

Chronic Local: U.

Chronic Systemic: Ingestion 3; Inhalation 3.

Toxicology: See also dichloro-(2-chlorovinyl) arsine.

Disaster Hazard: Dangerous; see arsenic and chlorides.

Countermeasures

Ventilation Control: Section 2.

Personnel Protection: Section 3.

First Aid: Section 1.

Storage and Handling: Section 7.

DICHLOROVOS

General Information

Synonym: DDVP; 2,2-dichlorovinyl dimethyl phosphate.

Description: Liquid; slightly soluble in water and gylcerine; miscible with aromatic and chlorinated hydrocarbon solvents and alcohol.

Formula: (CH$_3$O)$_2$P(O)OCH:CCl$_2$.

Constants: Mol wt: 221, bp: 120°C at 14 mm.

Hazard Analysis

Toxic Hazard Rating: Highly toxic!

Disaster Hazard: Dangerous. See chlorides and phosphorous compounds.

Countermeasures

Shipping Regulations: Section 11.

IATA: Poison B, poison label, 1 liter (passenger), 220 liters (cargo).

DICHLOROXYLENOL. See 2,4-dichloro-3,5-xylenol.

2,4-DICHLORO-3,5-XYLENOL

General Information

Synonym: Dichloroxylenol, DCMX.

Formula: (CH$_3$)$_2$(Cl)$_2$OHC$_6$H.

Constant: Mol wt: 191.

Hazard Analysis and Countermeasures

See chlorinated phenols.

DICOBALT OCTACARBONYL

General Information

Synonym: Cobalt tetracarbonyl.

Description: Orange crystals.

Formula: Co$_2$(CO)$_8$.

Constants: Mol wt: 342.0, mp: 51°C, bp: decomposes at 52°C, d: 1.73, vap. press.: 0.07 mm at 15°C.

Hazard Analysis

Toxic Hazard Rating:

Acute Local: U.

Acute Systemic: Ingestion 2; Inhalation 2.

Chronic Local: U.

Chronic Systemic: Ingestion 1; Inhalation 1.

Toxicology: See also carbonyls.

Fire Hazard: Moderate, when exposed to heat or flame. See also carbonyls.

Explosion Hazard: Moderate, when exposed to heat or flame. See also carbon monoxide and carbonyls.

Disaster Hazard: Dangerous; when heated to decomposition, it emits highly toxic fumes; can react vigorously with oxidizing materials.

Countermeasures

Ventilation Control: Section 2.

Storage and Handling: Section 7.

DICOPHANE. See DDT.

DICOUMAROL. See bishydroxy coumarin.

DI-m-CRESYLTRICHLOROETHANE

General Information

Description: Crystals.

Formula: Cl$_3$C$_2$H(C$_6$H$_4$CH$_3$)$_2$.

Constants: Mol wt: 313.7, vap. d.: 10.8.

Hazard Analysis

Toxicity: Details unknown. A fungicide. See also chlorinated hydrocarbons, aromatic and cresol.

Disaster Hazard: Dangerous; see chlorides.

Countermeasures

Storage and Handling: Section 7.

DICRYL. See 3',4'-dichloro-2-methyl acryl anilide.

DICUMYL PEROXIDE, SOLID

General Information

Formula: [C$_6$H$_5$C(CH$_3$)$_2$O]$_2$.

Constant: Mol wt: 270.

TOXIC HAZARD RATING CODE *(For detailed discussion, see Section 1.)*

0 NONE: (a) No harm under any conditions; (b) Harmful only under unusual conditions or overwhelming dosage.

1 SLIGHT: Causes readily reversible changes which disappear after end of exposure.

2 MODERATE: May involve both irreversible and reversible changes; not severe enough to cause death or permanent injury.

3 HIGH: May cause death or permanent injury after very short exposure to small quantities.

U UNKNOWN: No information on humans considered valid by authors.

Countermeasures
Shipping Regulations: Section 11.
   I.C.C.: Oxidizing material, yellow label, 25 pounds.
   IATA: Oxidizing material, yellow label, 1 kilogram
   (passenger), 12 kilograms (cargo).

DICUMYL PEROXIDE, SOLUTION
   I.C.C. (50% solution): Oxidizing material, yellow label,
   1 quart.
   IATA (solution): Oxidizing material, yellow label,
   1 liter (passenger and cargo).

DICYANDIAMIDE
General Information
Synonym: Cyanoguanidine.
Description: Pure white crystals.
Formula: $H_2NC(NH)NHCN$.
Constants: Mol wt: 84.1, mp: 208°C, d: 1.400 at 25°C.
Hazard Analysis
Toxicity: Details unknown. See cyanamides.
Disaster Hazard: Dangerous; see cyanides.
Countermeasures
Storage and Handling: Section 7.

o-DICYANOBENZENE
General Information
Synonyms: 1,2-Benzene dicarbonitrile; phthalonitrile.
Description: Colorless crystals. Insoluble in water; soluble
   in acetone and benzene.
Formula: $C_6H_4(CN)_2$.
Constants: Mol wt: 128.1, mp: 138°C, bp: sublimes, vap. d.:
   4.42.
Hazard Analysis
Toxicity: An insecticide. Probably highly toxic. See also
   nitriles.
Disaster Hazard: Dangerous; see cyanides.
Countermeasures
Ventilation Control: Section 2.
Storage and Handling: Section 7.

p-DICYANOBENZENE
General Information
Description: Crystals.
Formula: $C_6H_4(CN)_2$.
Constants: Mol wt: 128.1, vap. d.: 4.42.
Hazard Analysis
Toxicity: Details unknown; a fumigant. Probably highly
   toxic. See also nitriles.
Disaster Hazard: See cyanides.
Storage and Handling: Section 7.

1,4-DICYANOBUTANE. See adiponitrile.

DICYANOETHYLAMINE
General Information
Description: Liquid.
Formula: $C_4H_5N_3$.
Constants: Mol wt: 95.1, vap. d.: 3.3.
Hazard Analysis
Toxicity: Details unknown. Limited animal experiments
   suggest low toxicity. See also amines.
Disaster Hazard: Dangerous; see cyanides; can react vig-
   orously with oxidizing materials.
Countermeasures
Storage and Handling: Section 7.

DICYANOETHYLSULFIDE
General Information
Description: Liquid.
Formula: $C_4H_3N_2S$.
Constants: Mol wt: 111.2, vap. d.: 3.82.
Hazard Analysis
Toxicity: Details unknown. Limited animal experiments
   suggest low toxicity.
Disaster Hazard: Dangerous; see cyanides and oxides of
   sulfur.

Countermeasures
Storage and Handling: Section 7.

DICYCLOHEXYL. See bicyclohexyl.

DICYCLOHEXYLAMINE
General Information
Description: Liquid; fishy odor.
Formula: $(C_6H_{11})_2NH$.
Constants: Mol wt: 181.32, mp: 20°C, bp: 256°C, flash p:
   210°F (O.C.), d: 0.910, vap. d.: 6.27.
Hazard Analysis
Toxic Hazard Rating:
   Acute Local: Irritant 3; Inhalation 3.
   Acute Systemic: Inhalation 3.
   Chronic Local: Irritant 3.
   Chronic Systemic: U.
Toxicity: Animal experiments show intense local irritation
   and nervous excitation. See also cyclohexylamine.
Fire Hazard: Moderate, when exposed to heat or flame; can
   react with oxidizing materials.
Countermeasures
To Fight Fire: Alcohol foam, carbon dioxide, dry chemi-
   cal or carbon tetrachloride (Section 6).
Storage and Handling: Section 7.

DICYCLOHEXYLAMINE NITRITE
Hazard Analysis
Toxic Hazard Rating: U. Animal experiments show mod-
   erate toxicity. See amines and nitriles.
Fire Hazard: Dangerous. See nitrates.
Disaster Hazard: Dangerous. See nitrates.
Countermeasures
See nitrates.

DICYCLOHEXYL PHTHALATE
General Information
Description: White solid.
Formula: $C_6H_4(COOC_6H_{11})_2$.
Constants: Mol wt: 330.5, mp: 58°C, bp: 200–235°C at
   4 mm, flash p: 405°F, d: 1.148 at 20°/20°C, vap.
   press.: 0.1 mm at 150°C.
Hazard Analysis
Toxicity: Unknown.
Fire Hazard: Slight, when exposed to heat or flame; can
   react with oxidizing materials.
Countermeasures
To Fight Fire: Water, foam, carbon dioxide, dry chemical
   or carbon tetrachloride (Section 6).
Storage and Handling: Section 7.

DICYCLOPENTADIENE
General Information
Description: Colorless crystals.
Formula: $C_{10}H_{12}$.
Constants: Mol wt: 132.2, mp: 32.9°C, bp: 166.6°C,
   d: 0.976 at 35°C, vap. press.: 10 mm at 47.6°C, vap.
   d: 4.55, flash p: 90°F (O.C.).
Hazard Analysis
Toxicity: Unknown. Limited animal experiments suggest
   high toxicity.
Fire Hazard: Moderate, when exposed to heat or flame;
   can react with oxidizing materials (Section 6).
Countermeasures
Storage and Handling: Section 7.

DICYCLOPENTADIENYL IRON. See ferrocene.

DICYCLOPENTENYL ALCOHOL
General Information
Description: Liquid.
Formula: $C_{10}H_{17}O$.
Constants: Mol wt: 153.2, bp: 238°C, flash p: 426°F,
   d: 1.07, vap. d.: 5.3.
Hazard Analysis
Toxicity: Details unknown. See also alcohols.

Fire Hazard: Slight, when exposed to heat or flame; can react with oxidizing materials.

Countermeasures
To Fight Fire: Water, foam, carbon dioxide, dry chemical or carbon tetrachloride (Section 6).
Storage and Handling: Section 7.

DI-(DECANOYL) TRIETHYLENE GLYCOL ESTER
Hazard Analysis
Toxic Hazard Rating: Limited animal experiments suggest low toxicity. See also glycols.

DIDECYL ADIPATE
General Information
Description: A clear liquid.
Formula: $(C_{10}H_{21}CO_2)_2C_4H_8$.
Constants: Mol wt: 426, bp: 240°C at 4 mm, fp: −72°C, flash p: 425°F (C.O.C.), d: 0.916–0.922 at 20°/20°C.
Hazard Analysis
Toxicity: Details unknown. Limited animal experiments suggest low toxicity. See also esters and adipic acid.
Fire Hazard: Slight, when exposed to heat or flame; can react with oxidizing materials.
Countermeasures
To Fight Fire: Foam, carbon dioxide, dry chemical or carbon tetrachloride (Section 6).
Storage and Handling: Section 7.

DI-n-DECYLAMINE
General Information
Description: Liquid.
Formula: $(C_{10}H_{21})_2NH$.
Constants: Mol wt: 297.5, bp: 195°C at 12 mm, vap. d.: 10.3.
Hazard Analysis
Toxicity: Details unknown. See also amines.
Fire Hazard: Moderate, when exposed to heat or flame; can react with oxidizing materials (Section 6).
Countermeasures
Storage and Handling: Section 7.

DIDECYL PHTHALATE
General Information
Description: A clear liquid.
Formula: $(C_{10}H_{21}CO_2)_2C_6H_4$.
Constants: Mol wt: 446, bp: 252°C at 4 mm, fp: −53°C, flash p: 450°F (C.O.C.), d: 0.964–0.968 at 20°/20°C.
Hazard Analysis
Toxicity: Details unknown. Limited animal experiments suggest low toxicity. See also esters and phthalic acid.
Fire Hazard: Slight, when exposed to heat or flame; can react with oxidizing materials.
Countermeasures
To Fight Fire: Foam, carbon dioxide, dry chemical or carbon tetrachloride (Section 6).
Storage and Handling: Section 7.

DIDIPHENYLAMINE FLUOSILICATE
General Information
Description: White crystals.
Formula: $[(C_6H_5)_2NH]_2 \cdot H_2SiF_6$.
Constants: Mol wt: 482.5, mp: 169°C.
Hazard Analysis
Toxicology: See fluosilicates.
Disaster Hazard: Dangerous. See fluosilicates.

DIDODECYLAMINE
General Information
Description: White solid, slight ammoniacal odor.
Formula: $(C_{12}H_{25})_2NH$.

Constants: Mol wt: 253.7, mp: 50°C, bp: 210°C at 1 mm.
Hazard Analysis
Toxicity: Details unknown. See also amines.
Fire Hazard: Slight; (Section 6).
Countermeasures
Storage and Handling: Section 7.

DIDYMIUM NITRATE
General Information
Description: Violet-red, hygroscopic crystals.
Composition: A mixture of praseodmyium and neodymium nitrates.
Hazard Analysis
Toxicity: See nitrates.
Radiation Hazard: Didymium salts are usually slightly radioactive from the presence of thorium as an impurity (Section 5).
Fire Hazard: See nitrates.
Disaster Hazard: See nitrates.
Countermeasures
Storage and Handling: Section 7.
Shipping Regulations: Section 11.
   IATA: Oxidizing material, yellow label, 12 kilograms (passenger), 45 kilograms (cargo).

DIELDRIN
General Information
Synonym: 1, 2, 3, 4, 10, 10-hexachloro-6, 7-epoxy-1, 4, 4a, 5, 6, 7, 8, 8a-octahydro-1, 4, 5, 8-dimethano-naphthalene; compound 497; Octalox, and other trade names.
Description: Crystalline. Insoluble in water. Soluble in common organic solvents.
Formula: $C_{12}H_{10}OCl_6$.
Constants: Mol wt: 383, mp: approx. 150°C, vap. d.: 13.2.
Hazard Analysis
Toxic Hazard Rating:
   Acute Local: 0.
   Acute Systemic: Ingestion 3; Inhalation 3; Skin Absorption 3.
   Chronic Local: U.
   Chronic Systemic: U.
TLV: ACGIH (recommended); 0.25 milligrams per cubic meter of air. May be absorbed via intact skin.
Toxicology: Used as an insecticide. Dieldrin is absorbed readily from the skin as well as through other portals. It acts as a central nervous system stimulant, but the exact mechanism of this action is entirely unknown. It also greatly reduces or eliminates appetite, apparently by an action on the central nervous system. Either nervous symptoms or anorexia may appear first. However, appetite may occasionally return in animals are extremely sick and which eventually die.
Dangerous Acute Dose in Man: The effects of dieldrin and aldrin are similar both quantitatively and qualitatively in animals as well as in man. Persons exposed to oral dosages which exceed 10 mg/kg frequently become acutely ill. Symptoms may appear within 20 minutes and in no instance has a latent period of more than 12 hours been confirmed. No death or permanent sequelae have been reported following known poisoning by aldrin or dieldrin in man. In an attempted suicide by ingestion of dieldrin the dosage was estimated at 25.6 mg/kg.
   The oral LD_{50} of dieldrin for rats is 40–50 mg/kg indicating a toxicity roughly five times that of DDT. The dermal LD_{50} of dieldrin in xylene for rats is only slightly less than the oral toxicity (60 mg/kg for the

---

TOXIC HAZARD RATING CODE (For detailed discussion, see Section 1.)

0 NONE: (a) No harm under any conditions; (b) Harmful only under unusual conditions or overwhelming dosage.

1 SLIGHT: Causes readily reversible changes which disappear after end of exposure.

2 MODERATE: May involve both irreversible and reversible changes; not severe enough to cause death or permanent injury.

3 HIGH: May cause death or permanent injury after very short exposure to small quantities.

U UNKNOWN: No information on humans considered valid by authors.

female and 90 mg/kg for the male) indicating an acute dermal toxicity roughly 40 times that of DDT. Tests with certain other solvents indicate a factor of only six times.

Dangerous Chronic Dose: Nothing is known with certainty about the chronic toxicity of dieldrin for man but poisoning has occurred from use of 0.5–2.5% suspensions. Experimental animals show a wide species variation in their susceptibility to dieldrin. Repeated dermal applications of 10 mg or even 20 mg/kg are tolerated by rats, whereas rabbits are killed by both of these dosages. Animals have shown convulsions as much as 120 days following the last dose of dieldrin indicating that dieldrin or its derivatives and/or residual toxicant-induced injury may persist in the body for a long time once severe poisoning has occurred.

Signs and Symptoms of Poisoning in Man: Early symptons include headache, nausea, vomiting, general malaise, and dizziness. With more severe poisoning, clonic and tonic convulsions ensue or they may appear without the premonitory symptoms just mentioned. Coma may or may not follow the convulsions. Hyperexcitability and hyperirritability are common findings.

Disaster Hazard: Dangerous; when heated to decomposition, it emits highly toxic fumes of chlorides.

Countermeasures

Storage and Handling: Section 7.

Treatment of Poisoning: Every effort should be made to remove dieldrin from the skin by thorough washing with soap and water or from the alimentary tract by the use of lavage and/or saline laxatives. Oil laxatives should be avoided. Experiments with dogs and monkeys indicate that phenobarbital is effective as an antidote. It has been necessary to give the drug in large doses over a period of 2 weeks or more. The dosage which is required to keep poisoned animals from showing hyperexcitability or convulsions and which enables them to eat and behave normally is often a dosage which would induce sleep or even anesthesia in a normal animal of the same species. In human beings the dosage should be adjusted to the symptoms.

Shipping Regulations: Section 11.

IATA: Other restricted articles, class A, no label required, no limit (passenger and cargo).

DIENESTRAL DIACETATE
General Information
Synonym: 3,4,bis (p-acetoxy phenyl) 2,4,-hexadiene.
Hazard Analysis
Toxic Hazard Rating: U. A food additive permitted in the feed and drinking water of animals, and/or for the treatment of food producing animals. Also a food additive permitted in food for human consumption (Section 10).

1,2-DIEPOXYOCTANE
Hazard Analysis
Toxic Hazard Rating: Limited animal experiments suggest moderate toxicity and initation. See also epoxy resins.

1,2,7,8-DIEPOXYOCTANE
General Information
Formula: $H_2COCH(CH_2)_4HCOCH_2$.
Constant: Mol wt: 142.
Hazard Analysis
Toxic Hazard Rating:
  Acute Local: Irritant 3.
  Acute Systemic: Ingestion 2; Inhalation 3; Skin Absorption 3.
  Chronic Local: U.
  Chronic Systemic: U.
Based on limited animal experiments.

1,2,8,9-DIEPOXY LIMONENE
Hazard Analysis
Toxic Hazard Rating: U. Limited animal experiments sug-

gest moderate toxicity and initation. See also epoxy resins.

DIESEL OIL
General Information
Synonyms: Fuel Oil No. 2.
Description: Brown, slightly viscous liquid.
Constants: Flash p: 100°F, d: <1, autoign. temp.: 494°F.
Hazard Analysis
Toxicity: Moderate. See kerosene.
Fire Hazard: Moderate, when exposed to heat or flame (Section 6).
Explosion Hazard: See kerosene.
Disaster Hazard: See kerosene.
Countermeasures
Storage and Handling: Section 7.
To Fight Fire: Foam, carbon dioxide, dry chemical or carbon tetrachloride (Section 6).
Shipping Regulations: Section 11.
  Coast Guard Classification: Inflammable liquid.

DIETHANOLAMINE
General Information
Synonym: Di(2-hydroxyethyl) amine.
Description: A faintly colored, viscous liquid.
Formula: $(HOCH_2CH_2)_2NH$.
Constants: Mol wt: 105.14, mp: 28°C, bp: 269.1°C (decomp.), flash p: 305°F (O.C.), d: 1.0919 at 30°/20°C, autoign. temp.: 1224°F, vap. press.: 5 mm at 138°C, vap. d.: 3.65.
Hazard Analysis
Toxic Hazard Rating:
  Acute Local: Irritant 1; Ingestion 1; Inhalation 1.
  Acute Systemic: U.
  Chronic Local: U.
  Chronic Systemic: Ingestion 1; Inhalation 1.
Toxicology: See also amines.
Fire Hazard: Slight, when exposed to heat or flame; can react with oxidizing materials.
Spontaneous Heating: No.
Countermeasures
To Fight Fire: Alcohol foam, water, carbon dioxide, dry chemical or carbon tetrachloride (Section 6).
Personal Hygiene: Section 3.
Storage and Handling: Section 7.

DIETHOXYBORON CHLORIDE
General Information
Description: Colorless liquid; decomposes upon contact with moisture.
Formula: $(C_2H_5O)_2BCl$.
Constants: Mol wt: 136.4, bp: 112°C.
Hazard Analysis
Toxicity: Details unknown. See boron compounds.
Disaster Hazard: Dangerous. See chlorides.

DIETHOXYCHLOROSILANE
General Information
Description: Liquid.
Formula: $(C_2H_5O)_2SiHCl$.
Constants: Mol wt: 154.7, vap. d.: 5.33.
Hazard Analysis
Toxic Hazard Rating:
  Acute Local: Irritant 3; Ingestion 3; Inhalation 3.
  Acute Systemic: U.
  Chronic Local: U.
  Chronic Systemic: U.
Disaster Hazard: Dangerous; See chlorides; will react with water or stream to produce heat, toxic and corrosive fumes.
Countermeasures
Ventilation Control: Section 2.
Personnel Protection: Section 3.
Storage and Handling: Section 7.

DIETHOXYDIMETHYLSILANE
General Information
Description: Liquid.
Formula: $(C_2H_5O)_2Si(CH_3)_2$.
Constants: Mol wt: 148.2, bp: 113.5°C, d: 0.834, vap.
press.: 10 mm at 13.3°C, vap. d.: 5.1.
Hazard Analysis
Toxic Hazard Rating:
Acute Local: Irritant 2; Ingestion 2; Inhalation 2.
Acute Systemic: U.
Chronic Local: U.
Chronic Systemic: U.
Fire Hazard: Moderate, when exposed to heat or flame; can
react with oxidizing materials.
Countermeasures
Ventilation Control: Section 2.
Personnel Protection: Section 3.
Storage and Handling: Section 7.

1,1-DIETHOXYETHANE. See acetal.

1,2-DIETHOXYETHYLENE. See diethyl cellosolve.

DIETHOXY ETHYL PHTHALATE
General Information
Description: Light-colored liquid with a mild odor.
Formula: $C_6H_4(COOC_2H_4OC_2H_5)_2$.
Constants: Mol wt: 310, mp: 31°C, bp: 198–211°C at
4 mm, flash p: 356°F, d: 1.12, vap. d.: 10.7.
Hazard Analysis
Toxicity: Unknown. See esters.
Fire Hazard: Slight, when exposed to heat or flame; can
react with oxidizing materials.
Countermeasures
To Fight Fire: Water, foam, carbon dioxide, dry chemical
or carbon tetrachloride (Section 6).
Storage and Handling: Section 7.

DIETHOXY TETRAHYDROFURAN
General Information
Description: Liquid.
Formula: $C_2H_5OCH(CH_2)OCH(CH_2)OC_2H_5$.
Constants: Mol wt: 160.21, mp: −26.9°C, bp: 173.1°C,
flash p: 160°F (O.C.), d: 0.9686 at 20°/20°C.
Hazard Analysis
Toxicity: Unknown.
Fire Hazard: Moderate, when exposed to heat or flame;
can react with oxidizing materials.
Countermeasures
To Fight Fire: Foam, carbon dioxide, dry chemical or
carbon tetrachloride (Section 6).
Storage and Handling: Section 7.

DIETHYL ACETAL. See acetal.

DIETHYLACETALDEHYDE. See 2-ethylbutyraldehyde.

DIETHYLACETAMIDE
General Information
Description: Liquid.
Formula: $(C_2H_5)_2CHCONH_2$.
Constants: Mol wt: 115.2, mp: < −65°C, bp: 180°C,
flash p: 170°F, d: 0.92, vap. d.: 4.0.
Hazard Analysis
Toxicity: Unknown.
Fire Hazard: Moderate, when exposed to heat or flame;
can react with oxidizing materials (Section 6).
Countermeasures
Storage and Handling: Section 7.

DIETHYLACETIC ACID. See 2-ethylbutyric acid.

N,N-DIETHYLACETOACETAMIDE
General Information
Description: Liquid; soluble in water.
Formula: $CH_3COCH_2CON(C_2H_5)$.
Constants: Mol wt: 157, d: 0.995 at 20/20°C, bp: decom-
poses, fp: − 70°C, flash p: 250°F (C.O.C.).
Hazard Analysis
Toxic Hazard Rating: Limited animal experiments suggest
low toxicity.
Fire Hazard: Slight, when exposed to heat or flame.
Countermeasures
To Fight Fire: Water, foam, alcohol foam (Section 6).
Storage and Handling: Section 7.

DIETHYL ACETOACETATE
General Information
Description: Very slightly soluble in water.
Formula: $CH_3COC(C_2H_5)_2COOC_2H_5$.
Constants: Mol wt: 186, d: < 1, vap. d.: 6.4, bp: 412–424°F
(decomposes), flash p: 170°F.
Hazard Analysis
Toxic Hazard Rating: U.
Fire Hazard: Moderate, when exposed to heat or flame.
Countermeasures
Storage and Handling: Section 7.
To Fight Fire: Foam, mist (Section 6).

DIETHYL ALUMINUM CHLORIDE
General Information
Synonym: Aluminum diethyl monochloride; DEAC.
Description: Colorless, pyrophoric liquid.
Formula: $(C_2H_5)_2AlCl$.
Constants: Mol wt: 120.5, bp: 208°C, fp: −85.4°C.
Hazard Analysis
Toxic Hazard Rating: U. Probably toxic.
Fire Hazard: Dangerous, ignites spontaneously in air.
Disaster Hazard: Dangerous. See chlorides.
Countermeasures
To Fight Fire: Do not use water, foam, or halogenated ex-
tinguishing agents (Section 6).
Shipping Regulations: Section 11.
ICC: Flammable liquid, Red label, 2 ounces.
IATA: Flammable liquid, not acceptable (passenger and
cargo).

DIETHYL ALUMINUM HYDRATE
General Information
Description: A pyrophoric mixture with triethyl aluminum.
Formula: $(C_2H_5)_2AlH$.
Constant Mol wt: 86.
Hazard Analysis
Toxic Hazard Rating:
Acute Local: Irritant 3; Ingestion 3; Inhalation 3.
Acute Systemic: U.
Chronic Local: Irritant 3; Inhalation 3.
Chronic Systemic: U.
Fire Hazard: Dangerous, ignites spontaneously in air.
Countermeasures
Storage and Handling: Section 7.
To Fight Fire: Do not use water, foam, or halogenated
extinguishing agents. Section 6.

DIETHYL ALUMINUM MALONATE
General Information
Description: White crystals. Insoluble in water. Soluble in
organic solvents.

---

0 NONE: (a) No harm under any conditions; (b) Harmful only under unusual conditions or overwhelming dosage.

1 SLIGHT: Causes readily reversible changes which disappear after end
of exposure.

2 MODERATE: May involve both irreversible and reversible changes;
not severe enough to cause death or permanent injury.

3 HIGH: May cause death or permanent injury after very short exposure
to small quantities.

U UNKNOWN: No information on humans considered valid by authors.

Formula: $Al(C_7H_{11}O_4)_3$.
Constants: Mol wt: 504.5, mp: 98°C, d: 1.084 at 100°C.

Hazard Analysis
Toxicity: Details unknown.

DIETHYLAMINE
General Information
Description: Colorless liquid, ammoniacal odor.
Formula: $(C_2H_5)_2NH$.
Constants: Mol wt: 73.1, mp: −38.9°C, bp: 55.5°C, flash p: < 0°F, d: 0.7108 at 20°/20°C, autoign. temp.: 594°F, vap. press.: 400 mm at 38.0°C, vap. d.: 2.53, lel: 1.8% uel: 10.1%

Hazard Analysis
Toxic Hazaard Rating:
  Acute Local: Irritant 2; Ingestion 3.
  Acute Systemic: U.
  Chronic Local: U.
  Chronic Systemic: U.
Toxicity: See also amines.
TLV: ACGIH (recommended); 25 parts per million in air; 75 milligrams per cubic meter of air.
Fire Hazard: Dangerous, when exposed to heat or flame; can react with oxidizing materials.
Spontaneous Heating: No.
Explosion Hazard: Unknown.

Countermeasures
Ventilation Control: Section 2.
To Fight Fire: Alcohol foam, carbon dioxide, dry chemical or carbon tetrachloride (Section 6).
Personnel Protection: Section 3.
Storage and Handling: Section 7.
Shipping Regulations: Section 11.
  I.C.C.: Flammable liquid; red label, 10 gallons.
  Coast Guard Classification: Inflammable liquid, red label.
  IATA: Flammable liquid, red label, 1 liter (passenger), 40 liters (cargo).
  MCA warning label.

DIETHYL AMINE HYDROCHLORIDE
General Information
Description: White crystals; soluble in water, alcohol; slightly soluble in chloroform; insoluble in ether.
Formula: $(C_2H_5)_2NHHCl$.
Constants: Mol wt: 109.60, d: 1.048, mp: 223–224°C, bp: 320–330°C.

Hazard Analysis
Toxic Hazard Rating: U. See diethylamine and amines.
Disaster Hazard: Dangerous. See chlorides.

DIETHYLAMINOETHANOL. See diethylethanolamine.

N,N,-DIETHYL AMINO ETHYL ACRYLATE
General Information
Formula: $CH_2CHCOOH_2CH_2CN(H_5C_2)_2O$
Constants: Mol wt: 200, d: 0.9, vap. d.: 5.9, flash p: 195°F (O.C.), bp: decomposes.

Hazard Analysis
Toxic Hazard Rating: U. Limited animal experiments suggest high toxicity. See also ethyl acrylate.
Fire Hazard: Moderate, when exposed to heat or flame.

Countermeasures
To Fight Fire: Decomposes in water. Use alcohol foam or dry chemical.
Storage and Handling: Section 7.

3-DIETHYLAMINOPROPYLAMINE
General Information
Description: Liquid.
Formula: $(C_2H_5)_2NC_3H_6NH_2$.
Constants: Mol wt: 130.3, bp: 165–170°C, flash p: 138°F (O.C.), d: 0.82, vap. d.: 4.48.

Hazard Analysis
Toxicity: Details unknown. See also amines. Animal ex-

periments show irritating and sensitizing properties, also slight systemic toxicity.
Fire Hazard: Moderate, when exposed to heat or flame; can react with oxidizing materials.

Countermeasures
To Fight Fire: Foam, carbon dioxide, dry chemical or carbon tetrachloride (Section 6).
Storage and Handling: Section 7.

DIETHYLANILINE
General Information
Description: Colorless to yellow liquid.
Formula: $(C_2H_5)_2NC_6H_5$.
Constants: Mol wt: 149.23, mp: −38°C, flash p: 185°F, bp: 215.5°C, d: 0.9351, vap. press.: 1 mm at 49.7°C, vap. d.: 5.15, autoign. temp.: 630°F.

Hazard Analysis
Toxicity: Details unknown. See also aniline.
Fire Hazard: Moderate, when exposed to heat or flame.
Disaster Hazard: Dangerous; see aniline.

Countermeasures
Storage and Handling: Section 7.
To Fight Fire: Foam, carbon dioxide, dry chemical or carbon tetrachloride (Section 6).

DIETHYLANILINE FLUOSILICATE
General Information
Description: White crystals.
Formula: $(C_6H_5NHC_2H_5)_2 \cdot H_2SiF_6$.
Constants: Mol wt: 386.4, mp: 165°C.

Hazard Analysis
Toxicity: Details unknown. See aniline and fluosilicates.
Disaster Hazard: Dangerous. See fluorides.

DIETHYLBENZENE (mixture of isomers)
General Information
Description: Colorless, mobile liquid.
Formula: $C_6H_4(C_2H_5)_2$.
Constants: Mol wt: 134.21, bp: 183.8°C, flash p: 132°F, d: 0.868 at 25°/25°C, autoign. temp.: 806°F, vap. press.: 1 mm at 20.7°C, vap. d.: 4.62.

Hazard Analysis
Toxicity: Details unknown. See also ethyl benzene.
Fire Hazard: Moderate, when exposed to heat or flame; can react with oxidizing materials.

Countermeasures
To Fight Fire: Carbon dioxide, dry chemical or carbon tetrachloride (Section 6).
Storage and Handling: Section 7.

DIETHYLBERYLLIUM
General Information
Description: Colorless liquid.
Formula: $Be(C_2H_5)_2$.
Constants: Mol wt: 67.14, mp: 12°C, bp: 110°C at 15 mm, vap. d.: 2.3.

Hazard Analysis
Toxicity: See beryllium compounds.
Fire Hazard: Moderate, when exposed to heat or flame (Section 6).
Disaster Hazard: Dangerous; when heated to decomposition, it emits highly toxic fumes of beryllium; can react vigorously with oxidizing materials.

Countermeasures
Storage aand Handling: Section 7.

DI(2-ETHYLBUTYL) AZELATE
General Information
Description: Liquid.
Formula: $(CH_2)_7(CO_2C_6H_{13})_2$.
Constants: Mol wt: 356.5, flash p: 385°F, d: 0.93, vap. d.: 12.3.

Hazard Analysis
Toxicity: Unknown.
Fire Hazard: Slight, when exposed to heat or flame; can react with oxidizing materials.

Countermeasures
To Fight Fire: Foam, carbon dioxide, dry chemical or carbon tetrachloride (Section 6).
Storage and Handling: Section 7.

DI(2-ETHYLBUTYL) PHTHALATE
General Information
Description: Clear, oily, slightly aromatic liquid.
Formula: $C_6H_4[COOCH_2CH(C_2H_5)_2]_2$.
Constants: Mol wt: 334, mp: $-50°C$, bp: $350°C$, flash p: $381°F$ (O.C.), d: 1.01–1.016 at $20°/20°C$, vap. d.: 11.5.
Hazard Analysis
Toxicity: Details unknown. See also esters.
Fire Hazard: Slight, when exposed to heat or flame; can react with oxidizing materials.
Countermeasures
To Fight Fire: Water, foam, carbon dioxide, dry chemical or carbon tetrachloride (Section 6).
Storage and Handling: Section 7.

DIETHYLCADMIUM
General Information
Description: An oil, decomposed by moisture.
Formula: $(C_2H_5)_2Cd$.
Constants: Mol wt: 170.5, d: 1.6562, mp: $-21°C$, bp: $64°C$.
Hazard Analysis
Toxicity: Probably highly toxic. See cadmium compounds.
Fire Hazard: Details unknown. Probably dangerous when exposed to heat or flame (Section 6).
Disaster Hazard: When heated to decomposition it emits highly toxic fumes of cadmium.
Countermeasures
Storage and Handling: Section 7.

DIETHYL CARBAMYL CHLORIDE
General Information
Description: Liquid.
Formula: $(C_2H_5)_2NCOCl$.
Constants: Mol wt: 119.6, mp: $-44°C$, bp: 190–195°C, vap. d.: 4.1.
Hazard Analysis
Toxicity: Details unknown. Probably toxic.
Disaster Hazard: Dangerous; when heated to decomposition, it emits highly toxic fumes of chlorides; will react with water or steam to produce toxic and corrosive fumes.
Countermeasures
Storage and Handling: Section 7.

DIETHYL CARBAMAZINE
Hazard Analysis
A food additive permitted in the food and drinking water of animals, and/or for the treatment of food producing animals (Section 10).

N,N'-DIETHYLCARBANILIDE
General Information
Synonym: N,N'-Diethyl-N,N'-diphenyl urea.
Description: Colorless crystals.
Formula: $CO[N(C_2H_5)(C_6H_5)]_2$.
Constants: Mol wt: 268.4, mp: 73°C, d: 1.12, bp: 326°C, flash p: 302°F (C.C.), vap. d.: 9.3.
Hazard Analysis
Toxicity: Unknown.
Explosion Hazard: Severe, when shocked or exposed to heat (Section 7).
Disaster Hazard: Highly dangerous; shock and heat will explode it; when heated to decomposition, it emits highly toxic fumes.
Countermeasures
Storage and Handling: Section 7.

DIETHYL CARBINOL. See sec-amyl alcohol.

DIETHYL "Carbitol." See "Carbitol."

DIETHYL CARBONATE
General Information
Synonyms: Ethyl carbonate; carbonic ether.
Description: Colorless liquid, mild odor.
Formula: $(C_2H_5)_2CO_3$.
Constants: Mol wt: 118.13, mp: $-43°C$, bp: 125.8°C, flash p: 77°F (O.C.), d: 0.975 at $20°/4°C$, vap. press.: 10 mm at 23.8°C, vap. d.: 4.07.
Hazard Analysis
Toxic Hazard Rating:
    Acute Local: Irritant 2; Ingestion 2; Inhalation 2.
    Acute Systemic: U.
    Chronic Local: U.
    Chronic Systemic: U.
Fire Hazard: Moderate, when exposed to heat or flame; can react with oxidizing materials.
Spontaneous Heating: No.
Countermeasures
To Fight Fire: Foam, carbon dioxide, dry chemical or carbon tetrachloride (Section 6).
Ventilation Control: Section 2.
Personnel Protection: Section 3.
Storage and Handling: Section 7.

DIETHYL "CELLOSOLVE"
General Information
Synonym: 1,2-Diethoxyethylene.
Description: Colorless liquid, slight ethereal odor.
Formula: $C_2H_5OCH_2CH_2OC_2H_5$.
Constants: Mol wt: 119.17, mp: $-74°C$, bp: 121.4°C flash p: 95°F (O.C.), d: 0.8417 at $20°/20°C$, autoign. temp.: 406°F, vap. d.: 6.56, vap. press.: 9.4 mm.
Hazard Analysis
Toxicity: Details unknown. See also glycols and "Cellosolve."
Fire Hazard: Moderately dangerous; when exposed to heat or flame; can react with oxidizing materials.
Spontaneous Heating: No.
Countermeasures
To Fight Fire: Carbon dioxide, dry chemical or carbon tetrachloride (Section 6).
Ventilation Control: Section 2.
Storage and Handling: Section 7.

DIETHYL-β-CHLOROETHYLAMINE
General Information
Description: Liquid.
Formula: $(C_2H_5)_2NC_2H_4Cl$.
Constants: Mol wt: 135.64, vap. d.: 4.69.
Hazard Analysis
Toxicity: Details unknown. See also amines. Limited animal experiments suggest high toxicity.
Disaster Hazard: Dangerous; see chlorides.
Countermeasures
Storage and Handling: Section 7.

O,O-DIETHYL-3-p-CHLOROPHENYL THIOMETHYL PHOSPHORODITHIOATE. See trithion.

TOXIC HAZARD RATING CODE   *(For detailed discussion, see Section 1.)*

0 NONE: (a) No harm under any conditions; (b) Harmful only under unusual conditions or overwhelming dosage.

1 SLIGHT: Causes readily reversible changes which disappear after end of exposure.

2 MODERATE: May involve both irreversible and reversible changes; not severe enough to cause death or permanent injury.

3 HIGH: May cause death or permanent injury after very short exposure to small quantities.

U UNKNOWN: No information on humans considered valid by authors.

DIETHYL CHLOROPHOSPHATE
General Information
Description: Water-white liquid.
Formula: $(C_2H_5O)_2POCl$.
Constants: Mol wt: 172, bp: 60°C at 2 mm, d: 1.1915 at 25°/25°C, vap. d.: 5.94.
Hazard Analysis and Countermeasures
Toxic Hazard Rating: U. Limited animal experiments suggest high toxicity. Readily penetrates skin. See also parathion, phosphates, and chlorides.

DIETHYL-2-CHLOROVINYL PHOSPHATE
General Information
Synonym: Compound 1836.
Description: Powder.
Formula: $C_6H_{12}PO_4Cl$.
Constant: Mol wt: 214.6.
Hazard Analysis
Toxicity: Highly toxic. An organic phosphate insecticide and cholinesterase inhibitor. See parathion.
Disaster Hazard: Dangerous. See phosphates.

DIETHYL CYANAMIDE
General Information
Synonym: N-Cyanodiethylamine.
Description: Liquid.
Formula: $CNN(C_2H_5)_2$.
Constants: Mol wt: 98.15, mp: −80.5°C, bp: 186°C, flash p: 176°F, d: 0.8591 at 30°C, vap. d.: 3.4.
Hazard Analysis
Toxic Hazard Rating:
  Acute Local: Irritant 3; Ingestion 3; Inhalation 3.
  Acute Systemic: U.
  Chronic Local: U.
  Chronic Systemic: U.
Fire Hazard: Moderate, when exposed to heat or flame.
Disaster Hazard: Dangerous; when heated to decomposition or on contact with acid or acid fumes, it emits highly toxic fumes of cyanides; will react with water or stream to produce toxic and corrosive fumes; can react vigorously with oxidizing materials.
Countermeasures
Ventilation Control: Section 2.
To Fight Fire: Foam, carbon dioxide, dry chemical or carbon tetrachloride (Section 6).
Personnel Protection: Section 3.
Storage and Handling: Section 7.

DIETHYL CYCLOHEXANE
General Information
Description: Liquid; insoluble in water.
Formula: $(C_2H_5)_2C_6H_{10}$.
Constants: Mol wt: 140, d: 0.8037 (20/20°C), bp: 174°C, fp: −100°C, flash p: 125°F (O.C.), autoign. temp.: 465°F, lel: 0.8% at 140°F, uel: 6.0% at 230°F.
Hazard Analysis
Toxic Hazard Rating: U.
Fire Hazard: Moderate, when exposed to heat or flame.
Countermeasures
Storage and Handling: Section 7.
To Fight Fire: Foam, mist, dry chemical (Section 6).

N,N-DIETHYL CYCLOHEXYLAMINE
General Information
Description: Clear, colorless liquid.
Formula: $(C_2H_5)_2N(C_6H_{11})$.
Constants: Mol wt: 155.3, bp: 194.5°C, vap. d.: 5.36.
Hazard Analysis
Toxicity: Details unknown. See also amines.
Fire Hazard: Moderate, when exposed to heat or flame; can react with oxidizing materials.
Countermeasures
To Fight Fire: Foam, carbon dioxide, dry chemical or carbon tetrachloride (Section 6).
Storage and Handling: Section 7.

1,1-DIETHYL CYCLOPENTAMETHYLENE GERMANIUM
General Information
Description: Colorless liquid.
Formula: $(CH_2)_5Ge(C_2H_5)_2$.
Constants: Mol wt: 200.9, bp: 52°C at 13 mm.
Hazard Analysis
Toxicity: Details unknown. See germanium compounds.
Fire Hazard: Probably moderately dangerous. Details unknown.

DIETHYL DIBROMODIPYRIDINE TIN
General Information
Description: Solid.
Formula: $(C_2H_5)_2 \cdot SnBr_2 \cdot (C_5H_5N)_2$.
Constants: Mol wt: 494.9, mp: 140°C.
Hazard Analysis
Toxicity: Details unknown. See tin compounds.
Disaster Hazard: Dangerous. Emits bromides and nitrogen oxide fumes when heated to decomposition.

DIETHYL DICHLOROSILANE
General Information
Synonym: Dichlorodiethylsilane.
Description: Liquid.
Formula: $C_4H_{10}Cl_2Si$.
Constants: Mol wt: 157.05, mp: −96°C, bp: 131.0°, flash p: 70°F, d: 1.05, vap. d.: 5.41.
Hazard Analysis
Toxicity: See chlorosilanes.
Fire Hazard: Dangerous, when exposed to heat or flame.
Explosion Hazard: Unknown.
Disaster Hazard: Dangerous; see chlorides and will react with water or steam to produce heat, toxic and corrosive fumes; can react vigorously with oxidizing materials.
Countermeasures
Ventilation Control: Section 2.
To Fight Fire: Foam, carbon dioxide, dry chemical or carbon tetrachloride (Section 6).
Storage and Handling: Section 7.
Shipping Regulations: Section 11.
  I.C.C.: Corrosive liquid—no exemptions; white label, 10 gallons.
  Coast Guard Classification: Corrosive liquid; white label.
  IATA: Corrosive liquid, white label, not acceptable (passenger), 40 liters (cargo).

O,O-DIETHYL-S-(β-DIETHYLAMINO) ETHYL PHOSPHOROTHIOLATE HYDROGEN OXALATE.
See tetram.

DIETHYL DIISOAMYL TIN
General Information
Description: Solid.
Formula: $(C_2H_5)_2Sn(C_5H_{11})_2$.
Constants: Mol wt: 319.1, d: 1.0725 at 19°C, bp: 131°C at 13.5 mm.
Hazard Analysis
Toxicity: Details unknown. See tin compounds.

DIETHYL DIISOBUTYL TIN
General Information
Description: Solid.
Formula: $(C_2H_5)_2Sn(C_4H_9)_2$.
Constants: Mol wt: 291.1, d: 1.1030, bp: 108°C at 13 mm.
Hazard Analysis
Toxicity: Details unknown. See tin compounds.

DIETHYLDIMETHYL METHANE. See 2,3-dimethyl-pentane.

DIETHYL DIPHENYL GERMANIUM
General Information
Description: Colorless liquid; insoluble in water.
Formula: $(C_2H_5)_2Ge(C_6H_5)_2$.
Constants: Mol wt: 284.9, bp: 316°C.

Hazard Analysis
Toxicity: Details unknown. See germanium compounds.
Fire Hazard: Probably moderately dangerous. Details unknown.

DIETHYL DIPHENYL TIN
General Information
Description: Solid.
Formula: $(C_2H_5)_2Sn(C_6H_5)_2$.
Constants: Mol wt: 331.0, bp: 156°C at 4 mm.
Hazard Analysis
Toxicity: Details unknown. See tin compounds.

DIETHYL DIPHENYL UREA. See N,N'-diethylcarbanilide.

DIETHYL DISULFIDE. See 3,4-dithiahexane.

DIETHYLENE DIAMINE. See piperazine.

DIETHYLENE DIOXIDE. See dioxan.

1,3-DIETHYLENE DISULFIDE, 2,2-DIPHENYL. See 1,3-dithiane, 2,2-diphenyl.

DIETHYLENE GLYCOL
General Information
Synonym: Diglycol.
Description: Clear, colorless, practically odorless, syrupy liquid.
Formula: $CH_2OHCH_2OCH_2CH_2OH$.
Constants: Mol wt: 106.12, bp: 245.8°C, fp: $-8°C$, flash p: 255°F, d: 1.1184 at 20°/20°C, autoign. temp.: 444°F, vap. press.: 1 mm at 91.8°C, vap. d.: 3.66.
Hazard Analysis
Toxic Hazard Rating:
  Acute Local: Irritant 1; Ingestion 0; Inhalation 0.
  Acute Systemic: Ingestion 2.
  Chronic Local: Irritant 1.
  Chronic Systemic: Ingestion 1; Inhalation 1.
Toxicology: See glycols.
Fire Hazard: Slight, when exposed to heat or flame; can react with oxidizing materials.
Spontaneous Heating: No.
Countermeasures
To Fight Fire: Alcohol foam, water, carbon dioxide, dry chemical or carbon tetrachloride (Section 6).
Personal Hygiene: Section 3.
Storage and Handling: Section 7.

DIETHYLENE GLYCOL BIS(2-BUTOXYETHYL CARBONATE)
General Information
Synonym: Butoxyethyl diglycol carbonate.
Description: Slightly slouble in water.
Formula: $[CH_3(CH_2)_3O(CH_2)_2OOCOC_2H_4]_3$.
Constants: Mol wt: 378, flash p: 379°F, d: 1.1, bp: 327°C at 2mm.
Hazard Analysis
Toxic Hazard Rating: U.
Fire Hazard: Slight when exposed to heat or flame.
Countermeasures
To Fight Fire: Water, foam, alcohol foam (Section 6).
Storage and Handling: Section 7.

DIETHYLENE GLYCOL BIS(BUTYL CARBONATE)
General Information
Synonym: Butyl diglycol carbonate.
Description: Colorless liquid; slightly soluble in water; widely soluble in organic solvents.

Formula: $[CH_3(CH_2)_3OOCOCH_2CH_2]_2O$.
Constants: Mol wt: 306, d: 1.07 (20/4°C), boiling range: 164°–166°C at 2 mm, flash p: 372°F.
Hazard Analysis
Toxic Hazard Rating: U.
Fire Hazard: Slight, when exposed to heat or flame.
Countermeasures
To Fight Fire: Water, foam, alcohol foam (Section 6).
Storage and Handling: Section 7.

DIETHYLENE GLYCOL BIS PHENYL CARBONATE
General Information
Synonym: Phenyl diglycol carbonate.
Description: Colorless solid; insoluble in water; widely soluble in organic solvents.
Formula: $(C_6H_5OOCOCH_2CH_2)_2O$.
Constants: Mol wt: 346, d: 1.23 (20/4°C), mp: 40°C, bp: 225–229°C at 2 mm, flash p: 460°C.
Hazard Analysis
Toxic Hazard Rating: U.
Fire Hazard: Slight, when exposed to heat or flame.
Countermeasures
To Fight Fire: Water, foam (Section 6).
Storage and Handling: Section 7.

DIETHYLENE GLYCOL n-BUTYL ETHER
General Information
Description: Clear, mobile liquid, pleasant odor.
Formula: $C_4H_9(OCH_2CH_2)_2OH$.
Constants: Mol wt: 162.2, bp: 226.7–230.4°C, fp: $-76°C$ (supercools), flash p: 225°F (C.O.C.), d: 0.9546 at 20°/20°C, vap. press.: 30 mm at 130°C.
Hazard Analysis
Toxicity: Unknown. See also glycols.
Fire Hazard: Slight, when exposed to heat or flame; can react with oxidizing materials.
Countermeasures
To Fight Fire: Foam, carbon dioxide, dry chemical or carbon tetrachloride (Section 6).
Storage and Handling: Section 7.

DIETHYLENE GLYCOL CHLOROHYDRIN
General Information
Description: Water-white liquid.
Formula: $ClCH_2CH_2OCH_2CH_2OH$.
Constants: Mol wt: 124.57, mp: $-90°C$ (sets to a glass), bp: 198.9°C, flash p: 437°F, d: 1.1753 at 20°/20°C, vap. d.: 4.3.
Hazard Analysis
Toxicity: Unknown. See also glycols and 1-chloropropane.
Fire Hazard: Slight, when exposed to heat or flame.
Disaster Hazard: Dangerous; See chlorides. Can react with oxidizing materials.
Countermeasures
Storage and Handling: Section 7.
To Fight Fire: Water, foam, carbon dioxide, dry chemical or carbon tetrachloride (Section 6).

DIETHYLENE GLYCOL DIABIETATE (20% XYLENE)
General Information
Synonym: Diglycol diabietate.
Description: Liquid mixture.
Formula: $(C_{19}H_{29}COOCH_2CH_2)_2O$.
Constants: Mol wt: 622.6, flash p: 97°F.
Hazard Analysis
Toxicity: Details unknown. See also xylene and glycols.
Fire Hazard: Moderate, when exposed to heat or flame, can react with oxidizing materials.

Countermeasures

To Fight Fire: Foam, carbon dioxide, dry chemical or carbon tetrachloride (Section 6). See also xylene.

Storage and Handling: Section 7.

DIETHYLENE GLYCOL DIACETATE
General Information

Synonym: Diglycol diacetate.

Description: Colorless liquid.

Formula: $(CH_3COOCH_2CH_2)_2O$.

Constants: Mol wt: 190.20, mp: 19.1°C, bp: 250°C, flash p: 275°F, d: 1.1159 at 20°/20°C, vap. d.: 6.56, vap. press.: 0.02 mm at 20°C.

Hazard Analysis

Toxicity: Details unknown. See also glycols.

Fire Hazard: Slight, when exposed to heat or flame; can react with oxidizing materials.

Countermeasures

To Fight Fire: Alcohol foam, water, foam, carbon dioxide, dry chemical or carbon tetrachloride (Section 6).

Storage and Handling: Section 7.

DIETHYLENE GLYCOL DIBENZOATE
General Information

Description: Liquid.

Formula: $C_6H_5COOCH_2CH_2OCH_2CH_2OCOC_6H_5$.

Constants: Mol wt: 314.32, mp: 15.9°C, bp: 236°C at 5 mm, flash p: 450°F, d: 1.1765 at 20°/20°C, vap. press.: 0.89 mm at 200°C.

Hazard Analysis

Toxicity: See esters, glycols and benzoic acid.

Fire Hazard: Slight, when exposed to heat or flame; can react with oxidizing materials.

Countermeasures

To Fight Fire: Alcohol foam, water, foam, carbon dioxide, dry chemical or carbon tetrachloride (Section 6).

Storage and Handling: Section 7.

DIETHYLENE GLYCOL DIBUTYL ETHER
General Information

Description: Practically colorless liquid; characteristic odor; slightly soluble in water.

Formula: $C_4H_9O(C_2H_4O)_2C_4H_9$.

Constants: Mol wt: 218, d: 0.8853 (20/20°C), bp: 256°C fp: -60.2°C, flash P: 245°F.

Hazard Analysis

Toxic Hazard Rating: U.

Fire Hazard: Slight, when exposed to heat or flame.

Countermeasures

To Fight Fire: Water, foam, alcohol foam (Section 6).

Storage and Handling: Section 7.

DIETHYLENE GLYCOL DI-2-ETHYL BENZOATE
General Information

Description: Liquid.

Formula: $C_6H_5COOCH_2CH_2OCH_2CH_2OCOC_6H_5$.

Constants: Mol wt: 317, mp: 15.9°C, bp: 236°C at 5 mm, flash p: 455°F (O.C.), d: 1.1765 at 20°/20°C, vap. d.: 10.0.

Hazard Analysis

Toxicity: Details unknown. See also glycols.

Fire Hazard: Slight, when exposed to heat or flame; can react with oxidizing materials.

Countermeasures

To Fight Fire: Water, foam, carbon dioxide, dry chemical or carbon tetrachloride (Section 6).

Storage and Handling: Section 7.

DIETHYLENE GLYCOL DIETHYL ETHER
General Information

Description: Colorless liquid, soluble in hydrocarbons and water.

Formula: $CH_3(CH_2OCH_2)_3CH$.

Constants: Mol wt: 162, d: 0.9082 (20/20°C), bp: 189°C, fp: - 44°C, flash p: 180°F (O.C.), vap. d.: 5.6.

Hazard Analysis

Toxic Hazard Rating: U.

Fire Hazard: Moderate, when exposed to heat or flame.

Countermeasures

Storage and Handling: Section 7.

To Fight Fire: Alcohol foam (Section 6).

DIETHYLENE GLYCOL DIGLYCOLATE
General Information

Description: Yellow liquid, faint odor.

Formula: $(CH_2CH_2OOCCH_2OH)_2O$.

Constants: Mol wt: 222.2, d: 1.30 at 30°C, vap. d.: 7.65.

Hazard Analysis

Toxicity: See glycols.

Fire Hazard: Slight; Section 6.

Countermeasures

Storage and Handling: Section 7.

DIETHYLENE GLYCOL DILEVULINATE
General Information

Description: Amber liquid, pleasant odor.

Formula: $[CH_2CH_2OOC(CH_2)_2COCH_3]_2O$.

Constants: Mol wt: 270.4, flash p: 340°F (C.C.), d: 1.145 at 25°C, vap. d.: 9.33.

Hazard Analysis

Toxicity: Details unknown. See also glycols.

Fire Hazard: Slight, when exposed to heat or flame, can react with oxidizing materials.

Countermeasures

To Fight Fire: Water, foam, carbon dioxide, dry chemical or carbon tetrachloride (Section 6).

Storage and Handling: Section 7.

DIETHYL GLYCOL DIPROPIONATE
General Information

Description: Liquid.

Formula: $CH_3O(CH_2CH_2O)_2CH_3$.

Constants: Mol wt: 134.17, bp: 162°C, mp: $< -75°C$, flash p: 158°F (O.C.), d: 0.9440 at 20°C, vap. d.: 4.62.

Hazard Analysis

Toxicity: Details unknown. See glycols.

Fire Hazard: Moderate, when exposed to heat or flame; can react with oxidizing materials.

Explosion Hazard: Details unknown. See also ethers.

Countermeasures

To Fight Fire: Foam, carbon dioxide, dry chemical or carbon tetrachloride (Section 6).

Storage and Handling: Section 7.

DIETHYLENE GLYCOL DINITRATE
General Information

Synonym: Dinitroglycol.

Description: Liquid.

Formula: $C_4H_8O_3(NO_2)_2$.

Constants: Mol wt: 196.1, vap. d.: 6.76.

Hazard Analysis

Toxic Hazard Rating:

Acute Local: U.

Acute Systemic: Ingestion 2; Inhalation 2; Skin Absorption 1.

Chronic Local: U.

Chronic Systemic: Inhalation 1; Skin absorption 1.

Caution: This compound can cause a drop in blood pressure and possibly various cardiac disturbances. See also glycols and nitrates.

Fire Hazard: See nitrates and explosives, high.

Explosion Hazard: Severe, when shocked or exposed to heat. Used in low freezing dynamites and some permissible explosives. See also explosives, high.

Disaster Hazard: Dangerous; shock and heat will explode it; when heated, it emits toxic fumes of oxides of nitrogen; can react vigorously with oxidizing or reducing materials.

Countermeasures

Ventilation Control: Section 2.

Storage and Handling: Section 7.
Shipping Regulations: Section 11.
  Coast Guard Classification: Prohibited.
  ICC: Section 73, 51 (d).
  IATA: Explosive, not acceptable (passenger and cargo).

DIETHYLENE GLYCOL DIOLEATE
General Information
Description: Pale yellow liquid.
Formula: $(C_{17}H_{33}CO_2C_2H_4)_2O$.
Constants: Mol wt: 635.0, d: 0.9319 at 20°/4°C.
Hazard Analysis
Toxicity: Details unknown. See also glycols.
Fire Hazard: Slight, when exposed to heat or flame; can react with oxidizing materials (Section 6).
Countermeasures
Storage and Handling: Section 7.

DIETHYLENE GLYCOL DIPELARGONATE
General Information
Synonym: Plastolein X-55.
Description: Liquid.
Formula: $(C_8H_{17}COOC_2H_4)_2O$.
Constants: Mol wt: 362, mp: −15°C, bp: 229°C at 5 mm, flash p: 410°F, d: 0.966 at 20°C, vap. d.: 12.5.
Hazard Analysis
Toxic Hazard Rating:
  Acute Local: Irritant 1.
  Acute Systemic: Ingestion 1.
  Chronic Local: U.
  Chronic Systemic: U.
Toxicology: Animal experiments suggest low toxicity. See also glycols.
Fire Hazard: Slight, when exposed to heat or flame; can react with oxidizing materials.
Countermeasures
To Fight Fire: Foam, carbon dioxide, dry chemical or carbon tetrachloride (Section 6).
Storage and Handling: Section 7.

DIETHYL GLYCOL DIPROPIONATE
General Information
Description: Slightly soluble in H_2O.
Formula: $(C_2H_5COOC_2H_4)_2O$.
Constants: Mol wt: 218, d: 1.1, bp: 491–529°F, flash p: 260°F.
Hazard Analysis
Toxic Hazard Rating: U.
Fire Hazard: Slight, when exposed to heat or flame.
Countermeasures
To Fight Fire: Water, foam, alcohol foam (Section 6).
Storage and Handling: Section 7.

DIETHYLENE GLYCOL DIVINYL ETHER
General Information
Formula: $C_8H_{14}O_3$.
Constant: Mol wt: 158.
Hazard Analysis
Toxic Hazard Rating: U. Limited animal experiments suggest low toxicity. See also glycols.

DIETHYLENE GLYCOL ETHYL ETHER
General Information
Description: A clear colorless liquid; mild odor.
Formula: $C_2H_5(OCH_2CH_2)_2OH$.
Constants: Mol wt: 134.2, bp: 200–202°C, flash p: 205°F (C.O.C.), d: 0.9855 at 25°/25°C, vap. press.: 10 mm at 90°C.

Hazard Analysis
Toxicity: Details unknown. See glycols.
Caution: Avoid excessive breathing of vapor.
Fire Hazard: Moderate, when exposed to heat or flame; can react with oxidizing materials.
Countermeasures
To Fight Fire: Foam, carbon dioxide, dry chemical or carbon tetrachloride (Section 6).
Storage and Handling: Section 7.

DIETHYLENE GLYCOL ETHYL VINYL ETHER
Hazard Analysis
Toxic Hazard Rating: U. Limited animal experiments suggest low toxicity. See also glycols.

DIETHYLENE GLYCOL LAURATE
General Information
Synonym: Diglycol laurate.
Description: Light straw-colored, oily liquid.
Formula: $HOCH_2CH_2OCH_2CH_2OOC(CH_2)_{10}CH_3$.
Constants: Mol wt: 228.4, mp: 17°C, bp: 315–325°C, flash p: 290°F, d: 0.965–0.975, vap. d.: 10.0.
Hazard Analysis
Toxicity: Details unknown. See also glycols.
Fire Hazard: Slight, when exposed to heat or flame; can react with oxidizing materials.
Countermeasures
To Fight Fire: Water, foam, carbon dioxide, dry chemical or carbon tetrachloride (Section 6).
Storage and Handling: Section 7.

DIETHYLENE GLYCOL METHYL ETHER. See methyl "Carbitol."

DIETHYLENE GLYCOL METHYL ETHYL ETHER
General Information
Formula: $CH_3O(C_2H_4O)_2C_2H_5$.
Constant: Mol wt: 148.
Hazard Analysis
Toxic Hazard Rating: U. Limited animal experiments suggest low toxicity. See also glycols.

DIETHYLENE GLYCOL MONOBUTYL ETHER. See butyl "Carbitol."

DIETHYLENE GLYCOL MONOBUTYL ETHER ACETATE. See butyl "Carbitol" acetate.

DIETHYLENE GLYCOL MONO-2-CYANOETHYL ETHER
General Information
Formula: $(CH_2CH_2CN)OCH_2CH_2OCH_2CH_2OH$.
Constant: Mol wt: 159.
Hazard Analysis
Toxic Hazard Rating: Limited animal experiments suggest moderate toxicity. See also nitriles.

DIETHYLENE GLYCOL MONOETHYL ETHER. See "Carbitol."

DIETHYLENE GLYCOL MONOLAURATE
General Information
Description: Light yellow, oily liquid.
Formula: $C_{11}H_{23}COOC_2H_4OC_2H_4OH$.
Constants: Mol wt: 288.42, mp: 16–19°C, bp: >280°C, d: 0.97 at 25°C.
Hazard Analysis
Toxicity: Details unknown. See also glycols.
Fire Hazard: Slight, when exposed to heat or flame; can react with oxidizing materials.

TOXIC HAZARD RATING CODE (For detailed discussion, see Section 1.)

0 NONE: (a) No harm under any conditions; (b) Harmful only under unusual conditions or overwhelming dosage.

1 SLIGHT: Causes readily reversible changes which disappear after end of exposure.

2 MODERATE: May involve both irreversible and reversible changes not severe enough to cause death or permanent injury.

3 HIGH: May cause death or permanent injury after very short exposure to small quantities.

U UNKNOWN: No information on humans considered valid by authors.

Countermeasures
Storage and Handling: Section 7.

DIETHYLENE GLYCOL MONOMETHYL ETHER.
See methyl "Carbitol."

DIETHYLENE GLYCOL MONOMETHYL ETHER ACETATE. See methyl "Carbitol" acetate.

DIETHYLENE GLYCOL MONO-2-METHYL PENTYL ETHER
General Information
Formula: $C_{10}H_{22}O_3$.
Constant: Mol wt: 190.
Hazard Analysis
Toxic Hazard Rating: Limited animal experiments suggest low toxicity. See also glycols.

DIETHYLENE GLYCOL MONOPHENYL ETHER.
See phenyl "Carbitol."

DIETHYLENE GLYCOL MYRISTATE
General Information
Synonym: Diglycol myristate.
Description: Light colored wax; faint odor.
Formula: $HOCH_2CH_2OCH_2CH_2OOC(CH_2)_{12}CH_3$.
Constants: Mol wt: 316.5, mp: 37°C, flash p: 290°F, d: 0.938 at 38°C, vap. d.: 10.9.
Hazard Analysis
Toxicity: Details unknown. See also glycols.
Fire Hazard: Slight, when exposed to heat or flame; can react with oxidizing materials.
Countermeasures
To Fight Fire: Foam, carbon dioxide, dry chemical or carbon tetrachloride (Section 6).
Storage and Handling: Section 7.

DIETHYLENE OXIDE. See dioxan.

DIETHYLENETRIAMINE
General Information
Description: Yellow, viscous liquid; mild ammoniacal odor.
Formula: $(NH_2C_2H_4)_2NH$.
Constants: Mol wt: 103.17, mp: −39°C, bp: 207°C, flash p: 215°F (O.C.), d: 0.9586 at 20°/20°C, autoign temp.: 750°F, vap. press.: 0.22 mm at 20°C, vap. d.: 3.48.
Hazard Analysis
Toxic Hazard Rating:
  Acute Local: Irritant 3; Ingestion 3; Inhalation 3.
  Acute Systemic: Ingestion 2; Inhalation 2.
  Chronic Local: Allergen 3.
  Chronic Systemic: U.
Toxicology: High concentration of vapors causes irritation of respiratory tract, nausea and vomiting. Repeated exposures can cause asthma and sensitization of skin.
Fire Hazard: Moderate, when exposed to heat or flame; can react with oxidizing materials.
Countermeasures
Ventilation Control: Section 2.
Personnel Protection: Section 3.
Storage and Handling: Section 7.
Shipping Regulations: Section 11.
  MCA warning label.
To Fight Fire: Alcohol foam, foam, water (Section 6).

DIETHYLENIMIDE OXIDE. See morpholine.

DIETHYL ETHANEDIOATE. See ethyl oxalate.

DIETHYLETHANOLAMINE
General Information
Synonym: Diethyl amino ethanol.
Description: Colorless, hygroscopic liquid.
Formula: $(C_2H_5)_2NCH_2CH_2OH$.
Constants: Mol wt: 117.2, bp: 162°C, flash p: 140°F

(O.C.), d: 0.8851 at 20°/20°C, vap. press.: 1.4 mm at 20°C, vap. d.: 4.03.
Hazard Analysis
Toxic Hazard Rating:
  Acute Local: Irritant 3.
  Acute Systemic: Ingestion 2; Inhalation 3.
  Chronic Local: U.
  Chronic Systemic: U.
See also amines and monoethanol amine.
TLV: ACGIH (recommended) 10 parts per million of air; 50 milligrams per cubic meter of air. May be absorbed via the skin.
Fire Hazard: Moderate, when exposed to heat or flame; can react with oxidizing materials.
Countermeasures
To Fight Fire: Foam, carbon dioxide, dry chemical or carbon tetrachloride (Section 6).
Storage and Handling: Section 7.

DIETHYL ETHER. See ethyl ether.

DIETHYL ETHOXYMETHYLENE MALONATE
General Information
Description: Liquid.
Formula: $C_2H_5OCH:C(COOC_2H_5)_2$.
Constants: Mol wt: 217, vap. d.: 7.48.
Hazard Analysis
Toxicity: Details unknown. See also esters.
Fire Hazard: Slight; when heated, it emits acrid fumes; can react with oxidizing materials (Section 6).
Countermeasures
Storage and Handling: Section 7.

DIETHYL ETHYLENE BIS-GLYCOLATE
General Information
Description: Liquid.
Formula: $(CH_2OCH_2CO_2C_2H_5)_2$.
Constants: Mol wt: 234.2, bp: 152–155°C at 5 mm, flash p: 255°F (O.C.), d: 1.1199 at 25°/25°C, vap. d.: 8.09.
Hazard Analysis
Toxicity: Details unknown. See also esters and ethyl alcohol.
Fire Hazard: Slight, when exposed to heat or flame; can react with oxidizing materials.
Countermeasures
To Fight Fire: Water, foam, carbon dioxide, dry chemical or carbon tetrachloride (Section 6).
Storage and Handling: Section 7.

N,N-DIETHYL ETHYLENE DIAMINE
General Information
Description: Liquid.
Formula: $(C_2H_5)_2NC_2H_4NH_2$.
Constants: Mol wt: 116.2, bp: 60°C at 40 mm, flash p: 115°F (O.C.), d: 0.82 at 20°/20°C, vap. d.: 4.00.
Hazard Analysis
Toxicity: Details unknown. See also amines.
Fire Hazard: Moderate, when exposed to heat or flame; can react with oxidizing materials.
Countermeasures
To Fight Fire: Foam, carbon dioxide, dry chemical or carbon tetrachloride (Section 6).
Storage and Handling: Section 7.

DIETHYL ETHYL PHOSPHONATE
General Information
Description: Colorless liquid; sweet odor.
Formula: $C_2H_5P(O)(OC_2H_5)_2$.
Constants: Mol wt: 166.2, bp: 83°C at 11 mm, flash p: 221°F (C.O.C.), d: 1.025 at 20°/4°C, vap. d.: 5.73.
Hazard Analysis
Toxicity: Details unknown.
Fire Hazard: Moderate, when exposed to heat or flame.
Disaster Hazard: Dangerous; See phosphorus compounds; can react with oxidizing materials.
Countermeasures
Storage and Handling: Section 7.

To Fight Fire: Foam, carbon dioxide, dry chemical or carbon tetrachloride (Section 6).

O,O-DIETHYL-S, 2-(ETHYLTHIO) ETHYL PHOS-PHODITHIOATE
General Information
Synonym: Disulfoton.
Description: Pure compound; yellow liquid; technical compound: brown liquid; insoluble in most organic solvents.
Formula: $(C_2H_5O)_2P(S)SCH_2CH_2SCH_2CH_3$.
Constants: Mol wt: 174, bp: 62° C at 0.001 mm, d: 1.144 at 20° C.
Hazard Analysis
Toxic Hazard Rating: U. An insecticide. A food additive permitted in the feed and drinking water of animals and/or for the treatment of food producing animals. (Section 10).

O,O-DIETHYL-O, 2-(ETHYLTHIO)ETHYL THIO-PHOSPHATE. See demeton.

DIETHYL FLUOROPHOSPHATE
General Information
Description: A liquid with a sweet or fruity odor.
Formula: $(C_2H_5O)_2POF$.
Constants: Mol wt: 156.10, mp: low, bp: 170° C, d: 1.15 (approx.), vap. d.: 5.38,
Hazard Analysis
Toxic Hazard Rating:
  Acute Local: Irritant 3; Ingestion 3; Inhalation 3.
  Acute Systemic: Ingestion 3; Inhalation 3.
  Chronic Local: U.
  Chronic Systemic: U.
Disaster Hazard: Dangerous; when heated to decomposition or on contact with acid or acid fumes, it emits highly toxic fumes of fluorides and oxides of phosphorus.
Countermeasures
Ventilation Control: Section 2.
Personnel Protection: Section 3.
First Aid: Section 1.
Storage and Handling: Section 7.

DIETHYL FUMARATE
General Information
Description: White crystals or liquid.
Formula: $C_2H_5OCOCHCHCOOC_2H_5$.
Constants: Mol wt: 172.2, mp: 0.6° C, bp: 218.5° C, flash p: 220° F, d: 1.0529 at 20°/20° C, vap. press.: 1 mm at 53.2° C, vap. d.: 5.93.
Hazard Analysis
Toxicity: Details unknown. Limited animal experiments suggest moderate toxicity. See esters, fumaric acid and ethyl alcohol.
Fire Hazard: Moderate, when exposed to heat or flame; can react with oxidizing materials.
Countermeasures
To Fight Fire: Water, alcohol foam, foam, carbon dioxide dry chemical or carbon tetrachloride (Section 6).
Storage and Handling: Section 7.

DIETHYL GERMANIUM BROMIDE
General Information
Description: Colorless liquid, decomposed by water.
Formula: $(C_2H_5)_2GeBr_2$.
Constants: Mol wt: 290.6, mp: $< -33°$ C, bp: 202° C.
Hazard Analysis

Toxicity: Details unknown. See germanium compounds and bromides.
Disaster Hazard: Dangerous. See bromides.

DIETHYL GERMANIUM CHLORIDE
General Information
Description: Colorless liquid decomposed by water.
Formula: $(C_2H_5)_2GeCl_2$.
Constants: Mol wt: 201.6, mp: $-38°$ C, bp: 175° C.
Hazard Analysis
Toxicity: Details unknown. See germanium compounds and chlorides.
Disaster Hazard: Dangerous; see chlorides.

DIETHYL GERMANIUM IMINE
General Information
Description: Colorless liquid.
Formula: $(C_2H_5)_2GeNH$.
Constants: Mol wt: 145.7, bp: 100° C at 0.01 mm.
Hazard Analysis
Toxicity: Details unknown. See germanium compounds.

DIETHYL GERMANIUM IODIDE
General Information
Description: Colorless liquid, decomposed by water.
Formula: $(C_2H_5)_2GeI_2$.
Constants: Mol wt: 384.6, mp: $-2°$ C, bp: 252° C.
Hazard Analysis
Toxicity: Details unknown. See germanium compounds and iodides.
Disaster Hazard: Dangerous; see iodides.

DIETHYL GERMANIUM OXIDE
General Information
Description: Colorless liquid, insoluble in water.
Formula: $[(C_2H_5)_2GeO]_3$.
Constants: Mol wt: 440.2, mp: 18° C.
Hazard Analysis
Toxicity: Details unknown. See germanium compounds.

DIETHYL GLYCOCOLL-p-AMINO-o-OXYBENZOIC METHYL ESTER
General Information
Description: Colorless, prismatic crystals.
Formula: $C_6H_3OHCOOCH_3NHCOCH_2N(C_2H_5)_2$.
Constants: Mol wt: 280.3, mp: 185° C, vap. d.: 9.65.
Hazard Analysis
Toxicity: Details unknown. See also esters.
Fire Hazard: Slight (Section 6).
Countermeasures
Storage and Handling: Section 7.

DIETHYL GLYCOCOLLGUAIACOL HYDRO-CHLORIDE
General Information
Description: White prisms.
Formula: $CH_3OC_6H_4OCOCH_2N(C_2H_5)_2 \cdot HCl$.
Constants: Mol wt: 273.70, vap. d.: 9.45.
Hazard Analysis
Toxicity: Details unknown.
Disaster Hazard: Dangerous; see chlorides.
Countermeasures
Storage and Handling: Section 7.

DIETHYL GLYCOL PHTHALATE
General Information
Description: Liquid.
Formula: $C_6H_4[COO(CH_2)_2OC_2H_5]_2$.
Constants: Mol wt: 310.34, flash p: 343° F, d: 1.11.

TOXIC HAZARD RATING CODE (For detailed discussion, see Section 1.)

0 NONE: (a) No harm under any conditions; (b) Harmful only under unusual dosage or overwhelming dosage.

1 SLIGHT: Causes readily reversible changes which disappear after end of exposure.

2 MODERATE: May involve both irreversible and reversible changes not severe enough to cause death or permanent injury.

3 HIGH: May cause death or permanent injury after very short exposure to small quantities.

U UNKNOWN: No information on humans considered valid by authors

Hazard Analysis
Toxicity: See diethyl-o-phthalate and esters.
Fire Hazard: Slight, when exposed to heat or flame; can react with oxidizing materials.
Countermeasures
To Fight Fire: Alcohol foam, water, carbon dioxide, dry chemical or carbon tetrachloride (Section 6).
Storage and Handling: Section 7.

DI(2-ETHYLHEXYL) ADIPATE
General Information
Description: Liquid.
Formula: $[C_4H_9(C_2H_5)CHCH_2OCOC_2H_4]_2$.
Constants: Mol wt: 370.56, mp: $-60°C$, bp: $214°C$ at 5 mm, flash p: $385°F$, d: 0.9268 at $20°/20°C$, vap. press.: 2.6 mm at $200°C$, vap. d.: 12.8.
Hazard Analysis
Toxicity: See esters, adipic acid and ethyl hexanol.
Fire Hazard: Slight, when exposed to heat or flame; can react with oxidizing materials.
Countermeasures
To Fight Fire: Foam, carbon dioxide, dry chemical or carbon tetrachloride (Section 6).
Storage and Handling: Section 7.

DI(2-ETHYLHEXYL) AMINE
General Information
Synonym: Dioctyl amine.
Description: Water-white liquid with slightly ammoniacal odor.
Formula: $[C_4H_9CH(C_2H_5)CH_2]_2NH$.
Constants: Mol wt: 241.45, bp: $281.1°C$, flash p: $270°F$ (O.C.), d: 0.8062 at $20°/20°C$, vap. d.: 8.35.
Hazard Analysis
Toxicity: Details unknown. See also amines. Limited animal experiments suggest low toxicity.
Fire Hazard: Slight, when exposed to heat or flame; can react with oxidizing materials.
Countermeasures
To Fight Fire: Water, alcohol foam, foam, carbon dioxide, dry chemical or carbon tetrachloride (Section 6).
Storage and Handling: Section 7.

DI(2-ETHYLHEXYL) AZELATE
General Information
Description: Liquid.
Formula: $C_{25}H_{48}O_4$.
Constants: Mol wt: 412.63, mp: $-67.8°C$, bp: $237°C$ at 5 mm, flash p: $415°F$, d: 0.918 at $20°C$, vap. d.: 18.7.
Hazard Analysis
Toxicity: Details unknown. See also esters.
Fire Hazard: Slight, when exposed to heat or flame; can react with oxidizing materials.
Countermeasures
To Fight Fire: Foam, carbon dioxide, dry chemical or carbon tetrachloride (Section 6).
Storage and Handling: Section 7.

DI(2-ETHYLHEXYL)4,5-EPOXYCYCLOHEXANE-1,2-DICARBOXYLATE
General Information
Formula: $C_6H_8O[COOC_4H_9CH(C_2H_5)CH_2]_2$.
Constant: Mol wt: 327.
Hazard Analysis
Toxic Hazard Rating:
Acute Local: Irritant 1.
Acute Systemic: Ingestion 1; Inhalation 1; Skin Absorption 1.
Chronic Local: U.
Chronic Systemic: U.
Based on limited animal experiments.

DI(2-ETHYLHEXYL) ETHANOLAMINE
General Information
Description: Liquid.
Formula: $[C_4H_9CH(C_2H_5)CH_2]_2NC_2H_4OH$.

Constants: Mol wt: 285.5, mp: $-60°C$, bp: $216°C$ at 50 mm, flash p: $280°F$ (O.C.), d: 0.8573 at $20°/20°C$, vap. press.: < 0.01 mm at $20°C$, vap. d.: 9.87.
Hazard Analysis
Toxicity: Details unknown. See also amines.
Fire Hazard: Slight, when exposed to heat or flame; can react with oxidizing materials.
Countermeasures
To Fight Fire: Water, alcohol foam, foam, carbon dioxide, dry chemical or carbon tetrachloride (Section 6).
Storage and Handling: Section 7.

DI(2-ETHYLHEXYL) ETHYLENE BIS-GLYCOLATE
General Information
Description: Liquid.
Formula: $(CH_2OCH_2CO_2C_8H_{17})_2$.
Constants: Mol wt: 402.6, bp: $192-202°C$ at 0.3 mm, flash p: $390°F$ (O.C.), d: 0.9760 at $25°/25°C$, vap. d.: 13.9.
Hazard Analysis
Toxicity: Details unknown. See also glycols and esters.
Fire Hazard: Slight, when exposed to heat or flame; can react with oxidizing materials.
Countermeasures
To Fight Fire: Foam, carbon dioxide, dry chemical or carbon tetrachloride (Section 6).
Storage and Handling: Section 7.

DI(2-ETHYLHEXYL) HEXAHYDROPHTHALATE
General Informaton
Description: Liquid.
Formula: $C_6H_{10}[COOCH_2CH(C_2H_5)C_4H_9]_2$.
Constants: Mol wt: 396.59, mp: $-53°C$, bp: $216°C$ at 5 mm, flash p: $425°F$, d: 0.9586 at $20°/20°C$, vap. press.: 2.2 mm at $200°C$, vap. d.: 13.7.
Hazard Analysis
Toxicity: Details unknown. See also esters.
Fire Hazard: Slight, when exposed to heat or flame; can react with oxidizing materials.
Countermeasures
To Fight Fire: Foam, carbon dioxide, dry chemical or carbon tetrachloride (Section 6).
Storage and Handling: Section 7.

DI(2-ETHYLHEXYL) FUMARATE. See dioctyl fumarate.

DI(2-ETHYLHEXYL) MALEATE
General Information
Description: Liquid.
Formula: $[CHCO_2CH_2CH(C_2H_5)C_4H_9]_2$.
Constants: Mol wt: 340.5, mp: $-60°C$, bp: $164°C$ at 10 mm, flash p: $365°F$, d: 0.9436 at $20°/20°C$, vap. d.: 11.7.
Hazard Analysis
Toxicity: Details unknown. See also esters.
Fire Hazard: Slight, when exposed to heat or flame; can react with oxidizing materials.
Countermeasures
Storage and Handling: Section 7.
To Fight Fire: Water, foam.

DI(2-ETHYLHEXYL) PHOSPHITE
General Information
Description: Liquid.
Formula: $(C_8H_{17}O)_2PHO$.
Constants: Mol wt: 306, bp: $150-155°C$ at 2-3 mm, d: 0.929 at $25°/4°C$, vap. d.: 10.6.
Hazard Analysis
Toxicity: Details unknown.
Disaster Hazard: Dangerous; see phosphorus compounds.
Countermeasures
Storage and Handling: Section 7.

DI(2-ETHYLHEXYL) PHOSPHORIC ACID
General Information
Synonym: Dioctyl phosphoric acid.

Description: Liquid; insoluble in water; soluble in organic solvents.

Formula: $[C_4H_9CH(C_2H_5)CH_2]_2HPO_4$.

Constants: Mol wt: 322, d: 0.973 (25/25°C), fp: −60°C, flash p: 385°F (O.C.).

Hazard Analysis

Toxic Hazard Rating: U. Limited animal experiments suggest moderate toxicity and high irritation.

Fire Hazard: Slight, when exposed to heat or flame.

Disaster Hazard: Dangerous. See phosphorous compounds.

Countermeasures

To Fight Fire: Water, foam (Section 6).

Storage and Handling: Section 7.

DI(2-ETHYLHEXYL) PHTHALATE

General Information

Synonym: Dioctyl phthalate; di-sec-octyl phthalate.

Description: Stable, light colored liquid; mild odor.

Formula: $C_6H_4[CO_2CH_2CH(C_2H_5)C_4H_9]_2$.

Constants: Mol wt: 390.6, bp: 230°C at 5 mm, fp: −55°C, flash p: 425°F (O.C.), d: 0.9861 at 20°/20°C, vap. press.: 1.2 mm at 200°C, vap. d.: 16.0.

Hazard Analysis

Toxic Hazard Rating:

Acute Local: Irritant 0; allergen 0; Ingestion 1; Inhalation 0.

Acute Systemic: Inhalation 1.

Chronic Local: Irritant 0; Allergen 0.

Chronic Systemic: Ingestion 1; Inhalation 1; Skin Absorption 0.

Data based on human and animal observations.

TLV: ACGIH; 5 milligrams per cubic meter of air

Fire Hazard: Slight, when exposed to heat or flame; can react with oxidizing materials.

Countermeasures

To Fight Fire: Foam, carbon dioxide, dry chemical or carbon tetrachloride (Section 6).

Storage and Handling: Section 7.

DI(2-ETHYLHEXYL) SEBACATE

General Information

Description: Light, clear liquid; mild odor.

Formula: $(CH_2)_8(CO_2C_8H_{17})_2$.

Constants: Mol wt: 426, bp: 248°C at 9 mm, fp: −55°C, flash p: 410°F, d: 0.913 at 25°/25°C, vap. d.: 14.7.

Hazard Analysis

Toxicity: See esters and ethyl hexyl alcohol.

Fire Hazard: Slight, when exposed to heat or flame; can react with oxidizing materials.

Countermeasures

To Fight Fire: Foam, carbon dioxide, dry chemical or carbon tetrachloride (Section 6).

Storage and Handling: Section 7.

DI(2-ETHYLHEXYL) SUCCINATE

General Information

Description: Liquid.

Formula: $C_8H_{17}OCOCH_2CH_2COOC_8H_{17}$.

Constants: Mol wt: 342.5, mp:. −60°C, bp: 257°C at 50 mm, flash p: 315°F (O.C.), d: 0.9346 at 20°/20°C, vap. d.: 11.8.

Hazard Analysis

Toxicity: Details unknown. See also esters.

Fire Hazard: Slight, when exposed to heat or flame; can react with oxidizing materials.

Countermeasures

To Fight Fire: Water, alcohol foam, foam, carbon dioxide, dry chemical or carbon tetrachloride (Section 6).

Storage and Handling: Section 7.

DI(2-ETHYLHEXYL) TETRAHYDROPHTHALATE

General Information

Description: Liquid.

Formula: $C_6H_8[COOCH_2CH(C_2H_5)C_4H_9]_2$.

Constants: Mol wt: 394.58, mp: −50°C, bp: 219°C at 5 mm, flash p: 350°F (O.C.), d: 0.9685 at 20°/20°C, vap. d.: 13.6.

Hazard Analysis

Toxicity: Details unknown. See also esters.

Fire Hazard: Slight, when exposed to heat or flame; can react with oxidizing materials.

Countermeasures

To Fight Fire: Foam, carbon dioxide, dry chemical or carbon tetrachloride (Section 6).

Storage and Handling: Section 7.

DI(2-ETHYLHEXYL) TIN DICHLORIDE

General Information

Synonym: Di-n-octyltin dichloride.

Description: Crystals.

Formula: $(C_8H_{17})_2SnCl_2$.

Constant: Mol wt: 656.2.

Hazard Analysis

Toxicity: See tin compounds.

Disaster Hazard: Dangerous; when heated to decomposition it emits highly toxic fumes of chlorides.

DIETHYL ISOAMYLTIN BROMIDE

General Information

Description: Solid.

Formula: $(C_2H_5)_2(C_5H_{11})SnBr$.

Constants: Mol wt: 327.9, bp: 138°C at 17 mm, d: 1.4881 at 17°C.

Hazard Analysis

Toxicity: Details unknown. See tin compounds and bromides.

Disaster Hazard: Dangerous. See bromides.

DIETHYL ISOAMYLTIN CHLORIDE

General Information

Description: Solid.

Formula: $(C_2H_5)_2(C_5H_{11})SnCl$.

Constants: Mol wt: 283.4, d: 1.2994 at 20°C, bp: 126°C at 13 mm.

Hazard Analysis

Toxicity: Details unknown. See tin compounds.

Disaster Hazard: Dangerous. See chlorides.

DIETHYL ISOBUTYLTIN BROMIDE

General Information

Description: Solid.

Formula: $(C_2H_5)_2(C_4H_9)SnBr$.

Constants: Mol wt: 313.9, d: 1.5108, bp: 122°C at 17 mm

Hazard Analysis

Toxicity: Details unknown. See tin compounds.

Disaster Hazard: Dangerous, see bromides.

O,O-DIETHYL-S-ISOPROPYLMERCAPTOMETHYL PHOSPHORODITHIOATE. See thimet.

O,O-DIETHYL-O-2-ISOPROPYL-4-METHYL 6-PYRIMIDYL THIOPHOSPHATE. See diazinon.

DIETHYL KETONE

General Information

Synonyms: 3-Pentanone; metacetone; propione; ethyl propionyl.

Description: Colorless, mobile liquid; acetone-like odor.

Formula: $C_2H_5COC_2H_5$.

TOXIC HAZARD RATING CODE (For detailed discussion, see Section 1.)

0 NONE: (a) No harm under any conditions; (b) Harmful only under unusual conditions or overwhelming dosage.

1 SLIGHT: Causes readily reversible changes which disappear after end of exposure.

2 MODERATE: May involve both irreversible and reversible changes not severe enough to cause death or permanent injury.

3 HIGH: May cause death or permanent injury after very short exposure to small quantities.

U UNKNOWN: No information on humans considered valid by authors

Constants: Mol wt: 86.13, mp: $-42°C$, bp: $101°C$, flash p: $55°F$, d: 0.8159 at $19°/4°C$, vap. d.: 2.96, autoign. temp.: $846°F$.

Hazard Analysis

Toxicity: Details unknown. See also ketones. Animal experiments suggest slight toxicity.

Fire Hazard: Dangerous, when exposed to heat or flame; can react vigorously with oxidizing materials.

Explosion Hazard: Unknown.

Countermeasures

Ventilation Control: Section 2.

To Fight Fire: Alcohol foam, foam, carbon dioxide, dry chemical or carbon tetrachloride (Section 6).

Storage and Handling: Section 7.

Shipping Regulations: Section 11.

IATA: Flammable liquid, red label, 1 liter (passenger), 40 liters (cargo).

DIETHYL LAURAMIDE

General Information

Description: Liquid.

Formula: $(C_2H_5)_2NOC(CH_2)_{10}CH_3$.

Constants: Mol wt: 255, bp: $165°C$, flash p: $> 150°F$, d:0.86, vap. d.: 8.8.

Hazard Analysis

Toxicity: Details unknown.

Fire Hazard: Moderate, when exposed to heat or flame.

Disaster Hazard: Dangerous; when heated to decomposition, it emits highly toxic fumes of oxides of nitrogen; can react with oxidizing materials.

Countermeasures

Storage and Handling: Section 7.

To Fight Fire: Foam, carbon dioxide, dry chemical or carbon tetrachloride (Section 6).

DIETHYL MALEATE

General Information

Description: Water-white liquid.

Formula: $(HCCOOC_2H_5)_2$.

Constants: Mol wt: 172.2, mp: $-11.5°C$, bp: $225.0°C$, flash p: $200°F$ (O.C.), d: 1.0687 at $20°C$, vap. press.: 1 mm at $57.3°C$, vap. d.: 5.93.

Hazard Analysis

Toxicity: See esters, ethyl alcohol and maleic acid.

Fire Hazard: Moderate, when exposed to heat or flame; can react with oxidizing materials.

Countermeasures

To Fight Fire: Foam, carbon dioxide, dry chemical or carbon tetrachloride (Section 6).

Storage and Handling: Section 7.

DIETHYL MALONATE

General Information

Synonym: Ethyl malonate

Description: Clear, colorless liquid.

Formula: $CH_2(COOC_2H_5)_2$.

Constants: Mol wt: 160.17, bp: $198.9°C$, fp: $-49.8°C$, flash p: $200°F$ (O.C.), d: 1.055 at $25°/25°C$, vap. press.: 1 mm at $40.0°C$, vap. d.: 5.52.

Hazard Analysis

Toxicity: See esters and ethyl alcohol.

Fire Hazard: Moderate, when exposed to heat or flame; can react with oxidizing materials.

Spontaneous Heating: No.

Countermeasures

To Fight Fire: Foam, carbon dioxide, dry chemical or carbon tetrachloride (Section 6).

Storage and Handling: Section 7.

DIETHYL MERCURY

General Information

Description: Colorless liquid; hazel-like odor.

Formula: $Hg(C_2H_5)_2$.

Constants: Mol wt: 258.73, bp: $159°C$, d: 2.4660 at $20°C$.

Hazard Analysis

Toxicity: Highly toxic. See mercury compounds.

Fire Hazard: Moderate, when exposed to heat or flame.

Disaster Hazard: Dangerous; when heated to decomposition or on contact with acid or acid fumes, it emits highly toxic fumes of mercury; can react with oxidizing materials.

Countermeasures

Storage and Handling: Section 7.

DIETHYLMETHYLMETHANE. See 3-methyl-pentane.

O,O-DIETHYL-O-(3-METHYL-5-PYRAZOLYL) PHOSPHATE

Hazard Analysis

Toxicity: An experimental insecticide. A cholinesterase inhibitor. See parathion.

Disaster Hazard: See phosphorous compounds.

O,O-DIETHYL O-(3-METHYL-5-PYRAZOLYL) PHOSPHOROTHIOATE

Hazard Analysis

Toxicity: An experimental insecticide. A cholinesterase inhibitor. See parathion.

Disaster Hazard: Dangerous. See phosphorous compounds.

DIETHYL-p-NITROPHENYLPHOSPHATE

General Information

Synonym: Para-oxon.

Description: Oily liquid; slight odor; slightly water soluble; freely soluble in organic solvents.

Formula: $C_{10}H_{14}NO_6P$.

Constants: Mol wt: 275.2, bp: $170°C$ at 1 mm, d: 1.2736 at $20/4°C$.

Hazard Analysis

Toxicity: Highly toxic. Cholinesterase inhibitor. See parathion.

Disaster Hazard: Dangerous; see phosphorous compounds.

O,O-DIETHYL-O,p-NITROPHENYL THIOPHOSPHATE. See parathion.

DIETHYL OXALATE. See ethyl oxalate.

3,3-DIETHYLPENTANE

General Information

Description: Liquid.

Formula: $CH_3CH_2C(C_2H_5)_2CH_2CH_3$.

Constants: Mol wt: 128.3, autoign. temp.: $554°F$, vap. d.: 4.42, lel: 0.7%, uel: 7.7%.

Hazard Analysis

Toxicity: Details unknown.

Probably has irritant and narcotic effects.

Fire Hazard: Dangerous, when exposed to heat or flame; can react vigorously with oxidizing materials.

Countermeasures

To Fight Fire: Foam, carbon dioxide, dry chemical or carbon tetrachloride (Section 6).

Ventilation Control: Section 2.

DIETHYL PEROXIDE

General Information

Formula: $C_2H_5OOC_2H_5$.

Constants: Mol wt: 90, lel: 2.3%, d: 0.8, vap. d.: 7.7, bp: $149°F$.

Hazard Analysis

Toxic Hazard Rating: U.

Disaster Hazard: Dangerous, explodes on heating.

Fire Hazard: Dangerous. Powerful oxidizer.

DIETHYLPHOSPHINE

General Information

Description: Colorless liquid.

Formula: $(C_2H_5)_2PH$.

Constants: Mol wt: 90.1, bp: $85°C$, d: < 1, vap. d.: 3.11.

Hazard Analysis

Toxic Hazard Rating:

Acute Local: U.
Acute Systemic: Ingestion 3; Inhalation 3.
Chronic Local: U.
Chronic Systemic: Ingestion 3; Inhalation 3.
Fire Hazard: Dangerous, when exposed to heat or flame; may be spontaneously flammable in air.
Explosion Hazard: Unknown.
Disaster Hazard: Dangerous; see phosphorus; can react vigorously with oxidizing materials.

Countermeasures
Ventilation Control: Section 2.
To Fight Fire: Foam, carbon dioxide, dry chemical or carbon tetrachloride (Section 6).
Personal Hygiene: Section 3.
First Aid: Section 1.
Storage and Handling: Section 7.

DIETHYL PHOSPHITE
General Information
Description: Liquid.
Formula: $(C_2H_5O)_2PHO$.
Constants: Mol wt: 138, bp: 138°C, d: 1.071 at 20°/20°C, vap. d.: 4.76.

Hazard Analysis
Toxicity: Details unknown.
Fire Hazard: Moderate, when exposed to heat or flame (Section 6).
Disaster Hazard: Dangerous; see phosphorous compounds, can react with oxidizing materials.

Countermeasures
Storage and Handling: Section 7.

DIETHYL o-PHTHALATE
General Information
Synonym: Ethyl phthalate.
Description: Clear, colorless liquid.
Formula: $C_6H_4(CO_2C_2H_5)_2$.
Constants: Mol wt: 222.2, mp: −40.5°C, bp: 302°C, flash p: 325°F (O.C.), d: 1.110, vap. d.: 7.66.

Hazard Analysis
Toxic Hazard Rating:
Acute Local: Irritant 2; Ingestion 2; Inhalation 2.
Acute Systemic: Inhalation 2.
Chronic Local: U.
Chronic Systemic: U.
Toxicology: Irritant to mucous membranes. Narcotic in high concentrations.
Fire Hazard: Slight, when exposed to heat or flame; can react with oxidizing materials.

Countermeasures
To Fight Fire: Water, foam, carbon dioxide, dry chemical or carbon tetrachloride (Section 6).
Storage and Handling: Section 7.

DIETHYL p-PHTHALATE
General Information
Description: Liquid.
Formula: $C_6H_4(COOC_2H_5)_2$.
Constants: Mol wt: 222.23, mp: −5°C, bp: 296°C, flash p: 243°F (C.C.), d: 1.117–1.121 at 20°/20°C, vap. d: 7.66.

Hazard Analysis and Countermeasures
See diethyl-o-phthalate.

N,N-DIETHYL PIPERONYLAMIDE
General Information
Description: Crystals.
Formula: $CH_2O_2:C_6H_3CON(C_2H_5)_2$.

Constants: Mol wt: 237.25, vap. d.: 9.44.

Hazard Analysis
Toxicity: Details unknown, but piperonyl compounds are generally of a low order of toxicity; an insecticide.
Fire Hazard: Slight; Section 6.

Countermeasures
Storage and Handling: Section 7.

2,2-DIETHYL-1,3-PROPANEDIOL
General Information
Synonym: Prenderol.
Description: Crystals.
Formula: $HOCH_2C(C_2H_5)_2CH_2OH$.
Constants: Mol wt: 132.2, bp: 125°C at 10 mm, fp: 61.3°C, flash p: 215°F (O.C.), d: 1.052 at 20°C.

Hazard Analysis
Toxicology: Large doses can cause drowsiness, vertigo, nausea and vomiting.
Fire Hazard: Moderate, when exposed to heat or flame; can react with oxidizing materials.

Countermeasures
To Fight Fire: Water, alcohol foam, foam, carbon dioxide, dry chemical or carbon tetrachloride (Section 6).
Storage and Handling: Section 7.

DIETHYL-n-PROPYLTIN BROMIDE
General Information
Description: A powder.
Formula: $(C_2H_5)_2(C_3H_7)SnBr$.
Constants: Mol wt: 300, d: 1.5910 at 21°C, bp: 112°C at 16 mm.

Hazard Analysis and Countermeasures
See tin compounds and bromides.

DIETHYL-n-PROPYLTIN CHLORIDE
General Information
Description: A powder.
Formula: $(C_2H_5)_2(C_3H_7)SnCl$.
Constants: Mol wt: 255, d: 1.3848 at 16°C, bp: 108°C at 17 mm Hg.

Hazard Analysis and Countermeasures
See tin compounds and chlorides.

DIETHYL-n-PROPYLTIN FLUORIDE
General Information
Description: A powder soluble in alcohol.
Formula: $(C_2H_5)_2(C_3H_7)SnF$.
Constants: Mol wt: 239, mp: 271°C.

Hazard Analysis and Countermeasures
See tin compounds and fluorides.

O,O-DIETHYL-O-(2-PYRAZINYL) PHOSPHORO-THIOATE. See zinophos.

DIETHYL PYROCARBONATE
General Information
Description: Clear, colorless liquid; sweet ester like odor miscible with ethanol and methanol.
Formula: $C_2H_5OC(O)OC(O)OC_2H_5$.
Constant: Mol wt: 162.

Hazard Analysis
Toxic Hazard Rating: U. A food additive permitted in food for human consumption (Section 10).

DIETHYL SELENIDE
General Information
Description: Liquid.
Formula: $(C_2H_5)_2Se$.

TOXIC HAZARD RATING CODE *(For detailed discussion, see Section 1.)*

0 NONE: (a) No harm under any conditions; (b) Harmful only under unusual conditions or overwhelming dosage.

1 SLIGHT: Causes readily reversible changes which disappear after end of exposure.

2 MODERATE: May involve both irreversible and reversible change not severe enough to cause death or permanent injury.

3 HIGH: May cause death or permanent injury after very short exposure to small quantities.

U UNKNOWN: No information on humans considered valid by author

Constants: Mol wt: 137.08, bp: 108.0°C, d: 1.23, lel: 2.5%, vap. press.: 40 mm at 31.2°C, vap. d.: 4.73.

Hazard Analysis

Toxicity: See selenium and its compounds.

Fire Hazard: Dangerous, when exposed to heat or flame.

Spontaneous Heating: No.

Explosion Hazard: Moderate, when exposed to flame.

Disaster Hazard: See selenium compounds.

Countermeasures

Storage and Handling: Section 7.

To Fight Fire: Water, foam, carbon dioxide, dry chemical or carbon tetrachloride (Section 6).

DIETHYLSILANE

General Information

Description: Liquid. Stable when pure.

Formula: $(C_2H_5)_2SiH_2$.

Constants: Mol wt: 88.2, d: 0.6843 at 20/4°C, bp: 56°C at 741 mm.

Hazard Analysis and Countermeasures

Toxic Hazard Rating:

Acute Local: Irritant 3; Inhalation 3.

Acute Systemic: U.

Chronic Local: U.

Chronic Systemic: U.

DIETHYL STEARAMIDE

General Information

Description: Liquid.

Formula: $(C_2H_5)_2NOC(CH_2)_{16}CH_3$.

Constants: Mol wt: 339.6, bp: 140°C, flash p: 375°F, d: 0.86, vap. d.: 11.7.

Hazard Analysis

Toxicity: Unknown. See also amides.

Fire Hazard: Slight; when exposed to heat or flame; can react with oxidizing materials.

Countermeasures

To Fight Fire: Water, foam, carbon dioxide, dry chemical or carbon tetrachloride (Section 6).

Storage and Handling: Section 7.

DIETHYLSTILBESTROL

General Information

Synonym: Stilbestrol.

Description: Small crystals.

Formula: $C_{18}H_{20}O_2$.

Constants: Mol wt: 268.3, mp: 169–172°C.

Hazard Analysis

Toxic Hazard Rating:

Acute Local: Irritant 2; Allergen 1.

Acute Systemic: U.

Chronic Local: Allergen 1.

Chronic Systemic: Ingestion 2; Inhalation 2; Skin Absorption 2.

Note: A food additive permitted in the feed and drinking water of animals, and/or for the treatment of food producing animals. Also a food additive permitted in food for human consumption (Section 10).

Fire Hazard: Slight; Section 6.

Countermeasures

Ventilation Control: Section 2.

Personnel Protection: Section 3.

Storage and Handling: Section 7.

DIETHYLSTILBESTROL DIPROPIONATE

General Information

Description: Crystals.

Formula: $C_{24}H_{28}O_4$.

Constants: Mol wt: 380.4, mp: 104°C.

Hazard Analysis

Toxicity: See diethyl stilbestrol.

Fire Hazard: Slight (Section 6).

Countermeasures

Storage and Handling: Section 7.

DIETHYL SUCCINATE

General Information

Synonym: Ethyl succinate.

Description: Essentially colorless, clear, mobile liquid; slight odor.

Formula: $C_8H_{14}O_4$.

Constants: Mol wt: 174.2, mp: −20.8°C, bp: 216.5°C, flash p: 230°F (O.C.), d: 1.037 at 25°/25°C, vap. press.: 1 mm at 54.6°C, vap. d.: 6.0.

Hazard Analysis

Toxicity: Probably very low. See also succinic acid.

Fire Hazard: Slight, when exposed to heat or flame; can react with oxidizing materials.

Countermeasures

To Fight Fire: Water, alcohol foam, foam, carbon dioxide, dry chemical or carbon tetrachloride (Section 6).

Storage and Handling: Section 7.

DIETHYL SULFATE

General Information

Synonym: Ethyl sulfate.

Description: Colorless, oily liquid; faint ethereal odor.

Formula: $(C_2H_5)_2SO_4$.

Constants: Mol wt: 154.18, mp: −25.0°C, bp: decomposes to ethyl ether, flash p: 220°F (C.C.), d: 1.172 at 25°/4°C, autoign. temp.: 817°F, vap. press.: 1 mm at 47.0°C, vap. d.: 5.31.

Hazard Analysis

Toxic Hazard Rating:

Acute Local: Irritant 3; Ingestion 3; Inhalation 3.

Acute Systemic: U.

Chronic Local: U.

Chronic Systemic: U.

Fire Hazard: Moderate, when exposed to heat or flame.

Spontaneous Heating: No.

Disaster Hazard: Dangerous; see sulfates. It can react with oxidizing materials.

Countermeasures

Ventilation Control: Section 2.

To Fight Fire: Alcohol foam, water, foam, carbon dioxide, dry chemical or carbon tetrachloride (Section 6).

Personnel Protection: Section 3.

Storage and Handling: Section 7.

Shipping Regulations: Section 11.

IATA: Poison B, poison label, 1 liter (passenger), 220 liters (cargo).

DIETHYL SULFIDE. See ethyl sulfide.

DIETHYL TARTRATE

General Information

Description: Colorless, thick, oily liquid soluble in water and alcohol.

Formula: $CHOHCO_2(C_2H_5)_2$.

Constants: Mol wt: 108, bp: 280°C, mp: 17°C, d: 1.204 (20/4°C), flash p: 200°F.

Hazard Analysis

Toxic Hazard Rating: U.

Fire Hazard: Moderate, when exposed to heat or flame.

Countermeasures

To Fight Fire: Alcohol foam (Section 6).

Storage and Handling: Section 7,

DIETHYL TELLURIDE

General Information

Description: Liquid.

Formula: $(C_2H_5)_2Te$.

Constants: Mol wt: 186, bp: 138°C.

Hazard Analysis and Countermeasures

See tellurium compounds.

DIETHYLTHIOCARBAMYL CHLORIDE

General Information

Description: Crystals.

Formula: $(C_2H_5)_2NC(S)Cl$.

Constants: Mol wt: 151.7, mp: 46.2°C, bp: 97–103°C at 5 mm.
Hazard Analysis
Toxicity: Details unknown. See also chlorides.
Disaster Hazard: Dangerous; see chlorides and sulfates.
Countermeasures
Storage and Handling: Section 7.

DIETHYLTIN
General Information
Description: Yellowish oily liquid not miscible with water.
Formula: $(C_2H_5)_2SN$.
Constants: Mol wt: 177, d: 1.654, mp: $< -12°C$, bp: 150°C (decomp).
Hazard Analysis
Toxicity: Probably toxic. Details unknown. See tin compounds.
Fire Hazard: Dangerous. A powerful reducing agent. See Section 6.
Countermeasures
Storage and Handling: Section 7.

DIETHYLTIN DIBROMIDE
General Information
Description: Colorless water soluble crystals.
Formula: $(C_2H_5)_2SNBr_2$.
Constants: Mol wt: 337, d: 2.068 at 74°C, mp: 63°C, bp: 233°C.
Hazard Analysis and Countermeasures
See tin compounds and bromides.

DIETHYLTIN DICHLORIDE
General Information
Description: White water soluble cyrstals.
Formula: $(C_2H_5)_2SNCl_2$.
Constants: Mol wt: 248, mp: 85°C, bp: 220°C.
Hazard Analysis and Countermeasures
See tin compounds and chlorides.

DIETHYLTIN DIFLUORIDE
General Information
Description: Crystals slightly soluble in alcohol.
Formula: $(C_2H_5)_2SnF_2$.
Constants: Mol wt: 215, mp: 229°C.
Hazard Analysis and Countermeasures
See tin compounds and fluorides.

DIETHYLTIN DIIODIDE
General Information
Description: Very slightly soluble white crystals.
Formula: $(C_2H_5)_2SnI_2$.
Constants: Mol wt: 431, mp: 45°C, bp: 240–245°C (decomposes).
Hazard Analysis and Countermeasures
See tin compounds and iodides.

DIETHYLTIN OXIDE
General Information
Description: White, insoluble powder.
Formula: $(C_2H_5)_2SnO$.
Constants: Mol wt: 193; mp: infusible.
Hazard Analysis
Toxicity: See tin compounds.

DIETHYL VALERAMIDE. See valyl.

DIETHYL XANTHOGEN SULFIDE. See ethyl xanthogen disulfide.

DIFLUOROBENZENE
General Information
Description: Liquid; pungent odor.
Formula: $C_6H_4F_2$.
Constants: Mol wt: 114.09, mp: $-23.7°C$, d: 1.17006 (20°C), bp: 88.82°C.
Hazard Analysis
Toxic Hazard Rating: U. Vapors may be irritant and narcotic.
Disaster Hazard: Dangerous. See fluorides.

1,1,1-DIFLUOROCHLOROETHANE. See difluoromonochloroethane.

DIFLUOROCHLOROMETHYLMETHANE. See difluorochloroethane.

DIFLUORODIBROMOMETHANE. See dibromo-difluoromethane.

DIFLUORODIPHENYLDICHLOROETHANE. See DFDD.

DIFLUORODIPHENYL DISULFIDE
General Information
Description: Crystals.
Formula: $C_{12}H_6S_2F_2$.
Constants: Mol wt: 252.3, vap. d.: 8.72.
Hazard Analysis
Toxicity: An insecticide. See also fluorides.
Disaster Hazard: Dangerous; see fluorides and oxides of sulfur; it will react with water or steam to produce toxic fumes.
Countermeasures
Storage and Handling: Section 7.

DIFLUORODIPHENYL TRICHLOROETHANE. See DFDT.

DIFLUOROETHANE. See 1,1-difluoroethane.

1,1-DIFLUOROETHANE
General Information
Synonyms: Difluoroethane: ethylene fluoride; ethylidene difluoride.
Description: Colorless gas.
Formula: $F_2C_2H_4$.
Constants: Mol wt: 66.1, mp: $-117.0°C$, bp: $-26.5°C$, d: 1.004 at 25°C, vap. d.: 2.28.
Hazard Analysis
Toxic Hazard Rating:
  Acute Local: Irritant 1.
  Acute Systemic: Inhalation 2.
  Chronic Local: U.
  Chronic Systemic: U.
Toxicology: Irritant to lungs. Narcotic in high concentrations.
Fire Hazard: Dangerous, when exposed to heat or flame (Section 6).
Explosion Hazard: Unknown.
Disaster Hazard: Dangerous; see fluorides; it can react vigorously with oxidizing materials.
Countermeasures
Ventilation Control: Section 2.
Storage and Handling: Section 7.
Shipping Regulations: Section 11.
  I.C.C.: Flammable gas; red gas label, 300 pounds.
  Coast Guard Classification: Inflammable gas; red gas label.

TOXIC HAZARD RATING CODE (For detailed discussion, see Section 1.)

0 NONE: (a) No harm under any conditions; (b) Harmful only under unusual conditions or overwhelming dosage.

1 SLIGHT: Causes readily reversible changes which disappear after end of exposure.

2 MODERATE: May involve both irreversible and reversible changes; not severe enough to cause death or permanent injury.

3 HIGH: May cause death or permanent injury after very short exposure to small quantities.

U UNKNOWN: No information on humans considered valid by authors

IATA: Flammable gas, red label, not acceptable (passenger), 140 kilograms (cargo).

1,1-DIFLUOROETHYLENE
General Information
Description: Colorless gas.
Formula: CH_2CF_2.
Constants: Mol wt: 64.04, bp: $< -70°C$.
Hazard Analysis
Toxic Hazard Rating:
  Acute Local: U.
  Acute Systemic: Inhalation 1.
  Chronic Local: U.
  Chronic Systemic: U.
Disaster Hazard: Dangerous; when heated to decomposition, it emits toxic fumes of fluorides.
Countermeasures
Ventilation Control: Section 2.
Storage and Handling: Section 7.

DIFLUOROMONOCHLOROETHANE
General Information
Synonym: 1,1,1,-Difluorochloroethane.
Description: Gas.
Formula: $C_2H_3F_2Cl$.
Constants: Mol wt: 100.5, mp: $-131°C$, bp: $-9.5°C$, d: 1.19.
Hazard Analysis
Toxic Hazard Rating:
  Acute Local: U.
  Acute Systemic: Inhalation 1.
  Chronic Local: U.
  Chronic Systemic: U.
Fire Hazard: Dangerous, when exposed to heat or flame (Section 6).
Explosion Hazard: Unknown.
Disaster Hazard: Dangerous; when heated to decomposition, it emits toxic fumes of fluorides and chlorides; can react vigorously with oxidizing materials.
Countermeasures
Ventilation Control: Section 2.
Storage and Handling: Section 7.
Shipping Regulations: Section 11.
  I.C.C.: Flammable gas; red gas label, 300 pounds.
  Coast Guard Classification: Inflammable gas; red gas label.
  IATA: Flammable gas, red label, not acceptable (passenger), 140 kilograms (cargo).

DIFLUOROMONOCHLOROMETHANE
General Information
Description: Gas.
Formula: $CHClF_2$.
Constants: Mol wt: 86.48, bp: $-41°C$, autoign. temp.: 1170°F.
Hazard Analysis
Toxic Hazard Rating:
  Acute Local: 0.
  Acute Systemic: Inhalation 1.
  Chronic Local: U.
  Chronic Systemic: Inhalation 1.
Fire Hazard: Slight, when exposed to heat or flame.
Disaster Hazard: Dangerous; see fluorides and chlorides; it can react with oxidizing materials.
Countermeasures
To Fight Fire: Carbon dioxide, dry chemical or water spray (Section 6).
Storage and Handling: Section 7.

DIFLUOROPHOSPHORIC ACID, ANHYDROUS
General Information
Description: Mobile, strongly fuming, colorless liquid.
Formula: HPO_2F_2.
Constants: Mol wt: 102, mp: $-75°C$, bp: 116°C, d: 1.583 at 25°/4°C, vap. d.: 3.52.

Hazard Analysis
Toxicity: See fluorides and phosphoric acid.
Disaster Hazard: Dangerous; see fluorides and oxides of phosphorous; it will react with water or steam to produce toxic and corrosive fumes.
Countermeasures
Storage and Handling: Section 7.
Shipping Regulations: Section 11.
  I.C.C.: Corrosive liquid; white label, 1 gallon.
  Coast Guard Classification: Corrosive liquid; white label.
  IATA: Corrosive liquid, white label, not acceptable (passenger), 5 liters (cargo).

DIFOLATAN
General Information
Synonym: N-(1,1,2,2-Tetrachloroethyl thio)-4-cyclohexene-1,2-dicarboximide.
Description: White solid; insoluble in water; slightly soluble in most organic solvents.
Formula: $C_{10}H_9Cl_4NO_2S$.
Constant: Mol wt: 349.
Hazard Analysis
Toxic Hazard Rating: U. See captan.
Disaster Hazard: Dangerous. See chlorides and sulfur compounds.

N,N'-DIFURFURYLIDENE-2-FURANMETHANE DIAMINE. See hydrofuramide.

DIGALLANE. See gallium hydride.

DIGERMANE
General Information
Synonym: Germanium hydride.
Description: Liquid or gas.
Formula: Ge_2H_6.
Constants: Mol wt: 151.25, bp: 29°C, mp: $-109°C$, d (gas): 6.74 g/l at 20°C, (liquid): 1.98 at $-109°C$.
Hazard Analysis and Countermeasures
See hydrides and germanium compounds.

DIGITALIN. See digitalis.

DIGITALIS
General Information
Synonyms: Foxglove; purple foxglove; fairy gloves.
Description: Dried leaves of digitalis purpurea.
Composition: Digitoxin (0.2–0.4%), etc.
Hazard Analysis
Toxic Hazard Rating:
  Acute Local: Allergen 1.
  Acute Systemic: Ingestion 3.
  Chronic Local: Allergen 1.
  Chronic Systemic: Ingestion 2.
Toxicology: An overdose can be fatal. 2.5 g or 30 cc of the tincture is considered a toxic dose. It contains digitalin, digitalein, digitonin and digitoxin, the most toxic component. An overdose causes the eyes to become prominent, the pupils dilated and the sclera blue. Furthermore, it seems to affect the digestive system, causing nausea, vomiting, cardiac irregularities, and excessive doses cause heart failure. In addition to the nausea, great malaise and often headache are noted. If this material is taken for medical purposes and the above symptoms are noted, the dosage should be discontinued until medical advice can be obtained.
Countermeasures
Treatment and Antidotes: Wash stomach with tannic acid or strong tea. Keep patient lying down. Administer stimulants such as caffeine or ammonia, atropine if pulse falls below 50 per minute, morphine for irritability. Call a physician.
Personal Hygiene: Section 3.
First Aid: Section 1.
Storage and Handling: Section 7.

DIGITOXIN. See digitalis.

DIGLYCEROL TETRANITRATE
General Information
Description: Liquid.
Hazard Analysis
Toxicity: Details unknown. See also TNT.
Fire Hazard: See nitrates.
Explosion Hazard: Severe, when shocked or exposed to heat. See explosives, high.
Disaster Hazard: Dangerous; shock will explode it; when heated to decomposition, it emits highly toxic fumes of oxides of nitrogen; it can react vigorously with oxidizing and reducing materials.
Countermeasures
Ventilation Control: Section 2.
Storage and Handling: Section 7.

DIGLYCIDYL ETHER
General Information
Synonym: DGE.
Hazard Analysis
Toxic Hazard Rating:
Acute Local: Irritant 3.
Acute Systemic: Ingestion 2; Inhalation 3.
Chronic Local: U.
Chronic Systemic: Ingestion 3; Inhalation 3; Skin Absorption 3.
Toxicology: Animal experiments show severe irritation of skin. Repeated exposures caused depression of bone marrow in rats, rabbits and dogs.
TLV: ACGIH (recommended) 0.5 ppm of air; 2.8 milligrams per cubic meter of air.

DIGLYCOL. See diethylene glycol.

DIGLYCOL CHLOROFORMATE
General Information
Description: Liquid.
Formula: $ClCOOCH_2CH_2OCH_2CH_2OH$.
Constants: Mol wt: 191.0, bp: 125°C at 5 mm, flash p: 295°F (O.C.), vap. d.: 5.83.
Hazard Analysis
Toxicity: Details unknown. See also glycols.
Fire Hazard: Slight, when exposed to heat or flame.
Disaster Hazard: Dangerous; see chlorides; it will react with water or steam to produce toxic and corrosive fumes; it can react with oxidizing materials.
Countermeasures
Storage and Handling: Section 7.
To Fight Fire: Water, foam, carbon dioxide, dry chemical or carbon tetrachloride (Section 6).

DIGLYCOLCHLOROHYDRIN
General Information
Description: Colorless liquid.
Formula: $ClCH_2CH_2OCH_2CH_2OH$.
Constants: Mol wt: 124.57, bp: 196.8°C, flash p: 225°F (O.C.), d: 1.1698, vap. press.: 0.17 mm at 20°C.
Hazard Analysis
Toxic Hazard Ratings:
Acute Local: Irritant 2; Ingestion 2; Inhalation 2.
Acute Systemic: U.
Chronic Local: U.
Chronic Systemic: U.
Fire Hazard: Slight, when exposed to heat or flame.
Disaster Hazard: Dangerous; see chlorides; it will react with water or steam to produce toxic and corrosive fumes; it can react with oxidizing materials.

Countermeasures
Ventilation Control: Section 2.
To Fight Fire: Water; foam; carbon dioxide; dry chemical or carbon tetrachloride (Section 6).
Personnel Protection: Section 3.
Storage and Handling: Section 7.

DIGLYCOL DIABIETATE. See diethylene glycol diabietate (20% xylene).

DIGLYCOL DIACETATE. See diethylene glycol diacetate.

DIGLYCOL DILEVULINATE. See diethylene glycol dilevulinate.

DIGLYCOLIC ACID
General Information
Description: White; crystalline solid.
Formula: $O(CH_2COOH)_2$.
Constants: Mol wt: 134.1, mp: 148°C, bp: decomposes.
Hazard Analysis
Toxicity: Details unknown. See also glycols and acetic acid.
Fire Hazard: Slight; Section 6.
Countermeasures
Storage and Handling: Section 7.

DIGLYCOL LAURATE. See diethylene glycol laurate.

DIGLYCOL MYRISTATE. See diethylene glycol myristate.

DIGOXIN
General Information
Synonym: Lanoxin.
Description: White, crystalline powder.
Formula: $C_{41}H_{64}O_{14}$.
Constants: Mol wt: 780.9, mp: 265°C.
Hazard Analysis
Toxicity: Details unknown. See also digitalis.
Fire Hazard: Slight; Section 6.
Countermeasures
Storage and Handling: Section 7.

DI-n-HEPTYLAMINE
General Information
Description: Liquid.
Formula: $(C_7H_{15})_2NH$.
Constants: Mol wt: 213.4, vap. d.: 7.35.
Hazard Analysis
Toxicity: Details unknown. See also amines.
Fire Hazard: Moderate, when exposed to heat or flame; can react with oxidizing materials.
Countermeasures
Storage and Handling: Section 7.

DIHEXYL. See dodecane.

DIHEXYL ADIPATE
General Information
Synonym: Hexyl adipate.
Description: Liquid.
Formula: $(CH_2CH_2CO_2C_6H_{13})_2$.
Constants: Mol wt: 314.5, bp: 168–170°C at 4 mm, flash p: 325°F, d: 0.926 at 20°/20°C, vap. d.: 10.85.
Hazard Analysis
Toxicity: Details unknown. See also esters.
Fire Hazard: Slight, when exposed to heat or flame; can react with oxidizing materials.
Countermeasures
To Fight Fire: Foam; carbon dioxide; dry chemical or carbon tetrachloride (Section 6).
Storage and Handling: Section 7.

TOXIC HAZARD RATING CODE *(For detailed discussion, see Section 1.)*

0 NONE: (a) No harm under any conditions; (b) Harmful only under unusual conditions or overwhelming dosage.

1 SLIGHT: Causes readily reversible changes which disappear after end of exposure.

2 MODERATE: May involve both irreversible and reversible changes; not severe enough to cause death or permanent injury.

3 HIGH: May cause death or permanent injury after very short exposure to small quantities.

U UNKNOWN: No information on humans considered valid by authors.

DI-n-HEXYLAMINE
General Information
Description: Liquid.
Formula: $(C_6H_{13})_2NH$.
Constants: Mol wt: 185.4, bp: 233–243°C, flash p: 220°F, d: 0.78, vap. d.: 6.38.
Hazard Analysis
Toxic Hazard Ratings:
  Acute Local: Irritant 1; Ingestion 1: Inhalation 1.
  Acute Systemic: U.
  Chronic Local: U.
  Chronic Systemic: U.
Fire Hazard: Moderate, when exposed to heat or flame; can react with oxidizing materials.
Countermeasures
To Fight Fire: Water; foam; carbon dioxide; dry chemical or carbon tetrachloride (Section 6).
Ventilation Control: Section 2.
Personal Hygiene: Section 3.
Storage and Handling: Section 7.

DIHEXYL ETHER. See hexyl ether.

DIHEXYL MALEATE
General Information
Synonym: Di-n-hexyl maleate.
Description: Liquid; solubility in water less than 0.01% by weight at 20°C.
Formula: $C_6H_3OOCCH:CHCOOC_6H_3$.
Constants: Mol. wt: 274, d: 0.9602 (20/20°C), bp: 179°C at 10 mm, vap. press.: <0.01 mm at 20°C.
Hazard Analysis
Toxic Hazard Rating: U. Limited animal experiments suggest low toxicity. See also esters and maleic acid.

DI-n-HEXYLMERCURY
General Information
Description: Liquid.
Formula: $Hg(C_6H_{13})_2$.
Constants: Mol wt: 370.9, bp: 158°C at 10 mm, d: 1.5361, vap. d.: 12.8.
Hazard Analysis
Toxicity: See mercury compounds, organic.
Fire Hazard: Moderate, when exposed to heat or flame (Section 6).
Disaster Hazard: Dangerous; see mercury; it can react with oxidizing materials.
Countermeasures
Storage and Handling: Section 7.

DIHEXYL PHTHALATE
General Information
Description: Liquid.
Formula: $C_6H_4(CO_2C_6H_{13})_2$.
Constants: Mol wt: 334.4, mp: −58°C, bp: 210°C at 5 mm, flash p: 350°F, d: 0.995 at 20°/20°C, vap. d.: 11.5.
Hazard Analysis
Toxicity: Details unknown. See phthalic acid and esters.
Fire Hazard: Slight, when exposed to heat or flame; can react with oxidizing materials.
Countermeasures
To Fight Fire: Foam; carbon dioxide; dry chemical or carbon tetrachloride (Section 6).
Storage and Handling: Section 7.

DIHEXYL SEBACATE
General Information
Description: Liquid.
Formula: $[(CH_2)_4CO_2C_6H_{13}]_2$.
Constants: Mol wt: 370.6, bp: 184°C at 1 mm, flash 415°F, d: 0.911 at 20°/20°C, vap. d.: 12.8.
Hazard Analysis
Toxicity: Details unknown. See also esters.
Fire Hazard: Slight, when exposed to heat or flame; can react with oxidizing materials.

Countermeasures
To Fight Fire: Foam; carbon dioxide; dry chemical or carbon tetrachloride (Section 6).
Storage and Handling: Section 7.

DIHEXYL SODIUM SULFOSUCCINATE
General Information
Description: Clear, viscous liquid.
Formula: $C_6H_{13}CO_2CH_2CH(SO_3Na)CO_2C_6H_{13}$.
Constant: Mol wt: 388.
Hazard Analysis
Toxic Hazard Rating:
  Acute Local: Irritant 1; Ingestion 1.
  Acute Systemic: U.
  Chronic Local: Irritant 1.
  Chronic Systemic: U.
Fire Hazard: Moderate; Section 6.
Countermeasures
Personal Hygiene: Section 3.
Storage and Handling: Section 7.

DIHYDRAZINE SULFATE
General Information
Description: White, crystalline flakes.
Formula: $(N_2H_4)_2 \cdot H_2SO_4$.
Constants: Mol wt: 162.2, mp: 104°C, bp: 180°C decomposes.
Hazard Analysis
Toxic Hazard Rating:
  Acute Local: Irritant 2; Ingestion 2; Inhalation 2.
  Acute Systemic: U.
  Chronic Local: U.
  Chronic Systemic: U.
See also hydrazine.
Disaster Hazard: Dangerous. See sulfates.
Countermeasures
Storage and Handling: Section 7.

9,10-DIHYDROANTHRACENE
General Information
Description: Crystals.
Formula: $C_6H_4CH_2C_6H_4CH_2$.
Constants: Mol wt: 180.2, vap. d.: 6.21.
Hazard Analysis
Toxicity: Details unknown; an insecticide.
Fire Hazard: Slight, when exposed to heat; can react with oxidizing materials (Section 6).
Countermeasures
Storage and Handling: Section 7.

1,2-DIHYDRO-6-ETHOXY-2,2,4-TRIMETHYL-QUINOLINE
Hazard Analysis
Toxic Hazard Rating: U. Animal experiments suggest toxicity.
Disaster Hazard: Moderate. When strongly heated it emits acrid, possibly toxic fumes.
Countermeasures
Storage and Handling: Section 7.

2,3-DIHYDRO-2-FORMYL-1,4-PYRAN
General Information
Description: Liquid.
Formula: $OHCCH(CH_2)_2CHCHO$.
Constants: Mol wt: 112.12, mp: −90°C, bp: 150.6°C, d: 1.0776 at 20°/20°C, vap. press.: 2.4 mm at 20°C, vap. d.: 3.87.
Hazard Analysis
Toxicity: Details unknown. See also dihydropyran.
Fire Hazard: Slight; Section 6.
Countermeasures
Storage and Handling: Section 7.

DIHYDROGEN POTASSIUM ARSENATE
General Information
Description: White solid.

Formula: KH_2AsO_4.
Constant: Mol wt: 180, mp: 288°C.
Hazard Analysis and Countermeasures
See arsenic compounds.

1,2-DIHYDRO-4-METHOXY-1-METHYL-2-OXONICOTINONITRILE. See ricinine.

DIHYDROPENTABORANE. See pentaborane, unstable.

DIHYDROPYRAN
General Information
Description: Colorless, mobile liquid; ethereal odor.
Formula: C_5H_8O.
Constants: Mol wt: 84.1, bp: 85.6°C, flash p: 0°F, d: 0.923 at 20°/4°C, vap. d.: 2.90.
Hazard Analysis
Toxic Hazard Rating:
  Acute Local: Irritant 2; Ingestion 2; Inhalation 2.
  Acute Systemic: U.
  Chronic Local: U.
  Chronic Systemic: U.
Fire Hazard: Dangerous, when exposed to heat or flame; can react vigorously with oxidizing materials.
Explosion Hazard: Unknown.
Disaster Hazard: Dangerous! Keep away from heat and open flame!
Countermeasures
Ventilation Control: Section 2.
To Fight Fire: Foam, carbon dioxide, dry chemical or carbon tetrachloride (Section 6).
Personnel Protection: Section 3.
Storage and Handling: Section 7.
Shipping Regulations: Section 11.
  IATA: Flammable liquid, red label, 1 liter (passenger), 40 liters (cargo).

1,2-DIHYDRO-3,6-PYRIDAZINEDIONE. See maleic hydrazide.

2,5-DIHYDROPYRROLE. See pyrroline.

DIHYDROROTENONE
General Information
Formula: $C_{23}H_{24}O_6$.
Constant: Mol wt: 396.4.
Hazard Analysis
Toxicity: Details unknown. Said to be more toxic than rotenone. See also rotenone.

DIHYDRO STREPTOMYCIN
General Information
Synonym: DHS.
Description: A derivative of streptomycin; has anesthetic properties.
Formula: $C_{21}H_{41}N_7O_{12}$.
Constant: Mol wt: 583.
Hazard Analysis
Toxic Hazard Rating: U. A food additive permitted in food for human consumption (Section 10).

DIHYDRO STREPTOMYCIN SULFATE
General Information
Description: White or practically white powder; odorless or slight odor; freely soluble in water; very slightly soluble in alcohol; practically insoluble in chloroform.
Formula: $(C_{21}H_{41}N_7O_{12})_2 \cdot 3H_2SO_4$.
Constant: Mol wt: 1386.

Hazard Analysis
Toxic Hazard Rating: U. A food additive permitted in the feed and drinking water of animals, and/or for the treatment of food producing animals (Section 10).
Disaster Hazard: Dangerous. See sulfates.

DIHYDROTETRABORANE
General Information
Synonyms: Borobutane; boron hydride; tetraborondecahydride. tetraborane.
Description: Colorless gas; repulsive odor.
Formula: B_4H_{10}.
Constants: Mol wt: 53.4, mp: −120°C, bp: 18°C, d: 0.59 (liquid at −70°C); 0.56 (liquid at −35°C), vap. press.: 580 mm at 6°C, vap. d.: 1.8.
Hazard Analysis
Toxic Hazard Rating:
  Acute Local: Inhalation 3.
  Acute Systemic: Inhalation 3.
  Chronic Local: U.
  Chronic Systemic: U.
Toxicology: See also boron hydrides.
Fire Hazard: See boron hydrides.
Explosion Hazard: See boron hydrides.
Disaster Hazard: See boron hydrides.
Countermeasures
Ventilation Control: Section 2.
Storage and Handling: Section 7.

5,6-DIHYDRO-2,3,6-TRIMETHYL-1,3,5-DITHIAZINE. See thialdine.

1,2-DIHYDRO-1,4,6-TRIMETHYL-2-OXO-NICOTINONITRILE
Hazard Analysis
Toxicity: Probably highly toxic. See also cyanides and nicotine.
Disaster Hazard: See cyanides.
Countermeasures
Storage and Handling: Section 7.

1,2-DIHYDRO-2,2,4-TRIMETHYLQUINOLINE, POLYMERIZED
General Information
Description: Light tan powder.
Formula: $C_6H_4C(CH_3)CHC(CH_3)_2NH$.
Constants: Mol wt: 173.3, mp: 120°C, d: 1.08 at 25°C.
Hazard Analysis
Toxicity: Details unknown. See also quinoline.
Fire Hazard: Slight; Section 6.
Countermeasures
Storage and Handling: Section 7.

DIHYDROXYACETONE
General Information
Synonym: Man-tan.
Description: Crystalline powder. Hygroscopic, characteristic odor, sweet cooling taste. Very water soluble. Soluble in alcohol, ether, and acetone.
Formula: $HO \cdot CH_2COCH_2OH$.
Constants: Mol wt: 90.1, mp: 75–80°C.
Hazard Analysis
Toxicity: Details unknown. Has been reported as causing dermatitis in man.

1,8-DIHYDROXYANTHRANOL. See anthralin.

3,4-DIHYDROXYANTHRANOL. See anthrarobin.

TOXIC HAZARD RATING CODE (For detailed discussion, see Section 1.)

0 NONE: (a) No harm under any conditions; (b) Harmful only under unusual conditions or overwhelming dosage.

1 SLIGHT: Causes readily reversible changes which disappear after end of exposure.

2 MODERATE: May involve both irreversible and reversible changes; not severe enough to cause death or permanent injury.

3 HIGH: May cause death or permanent injury after very short exposure to small quantities.

U UNKNOWN: No information on humans considered valid by authors.

1,2-DIHYDROXYANTHRAQUINONE. See alizarin.

1,4-DIHYDROXYANTHRAQUINONE
General Information
Description: Crystals.
Formula: $C_{14}H_8O_4$.
Constants: Mol wt: 240.2, mp: 194°C, bp: 450.0°C, vap. press.: 1 mm at 196.7°C, vap. d.: 8.3.
Hazard Analysis
Toxicity: Details unknown. A weak allergen.
Fire Hazard: Slight; Section 6.
Countermeasures
Personal Hygiene: Section 3.
Storage and Handling: Section 7.

1,5-DIHYDROXYANTHRAQUINONE
General Information
Description: Green to yellow crystals.
Formula: $C_{14}H_8O_4$.
Constants: Mol wt: 240.2, mp: 280°C, bp: sublimes, vap. d.: 8.3.
Hazard Analysis
Toxicity: Details unknown. A weak allergen.
Fire Hazard: Slight; Section 6.
Countermeasures
Storage and Handling: Section 7.

1,8-DIHYDROXYANTHRAQUINONE
General Information
Description: Crystals.
Formula: $C_{14}H_8O_4$.
Constants: Mol wt: 240.2, mp: 193°C, vap. d.: 8.3.
Hazard Analysis
Toxicity: Details unknown. A weak allergen.
Fire Hazard: Slight; Section 6.
Countermeasures
Storage and Handling: Section 7.

m-DIHYDROXYBENZENE. See resorcinol.

o-DIHYDROXYBENZENE. See pyrocatechol.

DIHYDROXYBENZYL DIMETHYL METHANE
General Information
Description: Flake material.
Formula: $HOC_6H_4CH_2C(CH_3)_2CH_2C_6H_4OH$.
Constants: Mol wt: 256.3, mp: 152.5°C, bp: 220°C at 4 mm.
Hazard Analysis
Toxicity: Details unknown.
Fire Hazard: Slight; Section 6.
Countermeasures
Storage and Handling: Section 7.

2,6-DIHYDROXY-4-CARBOXYPYRIDINE.
See citrazinic acid.

3,4-DIHYDROXYCINNAMIC ACID. See caffeic acid.

7,8-DIHYDROXYCOUMARIN-7-β-D-GLUCOSIDE
General Information
Synonym: Daphnin.
Description: Crystals.
Formula: $C_{15}H_{16}O_9$.
Constant: Mol wt: 340.3.
Hazard Analysis
Toxicity: Details unknown. See warfarin.

DIHYDROXYDIAMINOMERCUROBENZENE
General Information
Formula: $OHNH_2C_6H_3HgC_6H_3OHNH_2$.
Constants: Mol wt: 416.9, vap. d.: 14.4.
Hazard Analysis and Countermeasures
See mercury compounds, organic.

2,2'-DIHYDROXY-5,5'-DICHLORODIPHENYL-METHANE. See 2,2'-methylene bis (4-chlorophenol).

2,4-DIHYDROXY-3,3-DIMETHYL BUTYRONITRILE
General Information
Formula: $NCCHOHC(CH_3)_2CH_2OH$.
Constant: Mol wt: 129.
Hazard Analysis
Toxic Hazard Rating: U. Limited animal experiments suggest moderate toxicity. See also nitriles.

DI(2-HYDROXY ETHYL) AMINE. See diethanolamine.

N,N'-DIHYDROXYETHYLETHYLENE DIAMINE
General Information
Description: Crystal.
Formula: $HOC_2H_4NHCH_2CH_2NHC_2H_4OH$.
Constants: Mol wt: 148.2, mp: 98.1°C, bp: 196°C at 10 mm, flash p: 355°F, vap. d.: 5.1.
Hazard Analysis
Toxicity: Details unknown. See also amines.
Fire Hazard: Slight, when exposed to heat or flame; can react with oxidizing materials.
Countermeasures
To Fight Fire: Foam, carbon dioxide, dry chemical or carbon tetrachloride (Section 6).

DIHYDROXYETHYL NITRAMINE DINITRATE
Hazard Analysis
See also Tetryl.
Toxicity: Details unknown.
Fire Hazard: See nitrates.
Explosion Hazard: Severe, when shocked or exposed to heat.
Disaster Hazard: Dangerous; shock will explode it; when heated to decomposition, it emits highly toxic fumes of oxides of nitrogen; can react vigorously with reducing materials.
Storage and Handling: Section 7.

1,4-DI(2-HYDROXYETHYL) PIPERAZINE
General Information
Formula: $N(C_2H_4OH)CH_2CH_2N(C_2H_4OH)CH_2CH_2$.
Constant: Mol wt: 174.
Hazard Analysis
Toxic Hazard Rating: U. Limited animal experiments suggest low toxicity and moderate irritation. See also piperazine.

DI-(HYDROXYETHYL)-o-TOLYLAMINE
General Information
Synonym: o-Tolyldiethanol amine.
Hazard Analysis
Toxic Hazard Rating: U. Limited animal experiments suggest moderate toxicity and low irritation. See also amines.

2,2'-DIHYDROXY-3,5,6,3',5',6'-HEXACHLORODI-PHENYLMETHANE. See hexachlorophene.

1,6-DIHYDROXYHEXANE. See 1,6-hexanediol.

2,2'-DIHYDROXYISOPROPYL ETHER. See dipropylene glycol.

1,8-DIHYDROXY-3-METHYL-ANTHRAQUINONE. See chrysophanic acid.

1,3-DIHYDROXYNAPHTHALENE. See 1,3-naphthalenediol.

DIHYDROXY OCTACHLORODIPHENYL
General Information
Formula: $C_{12}H_2O_2Cl_8$.
Constants: Mol wt: 461.8, vap. d.: 15.9.
Hazard Analysis
Toxic Hazard Rating:
　Acute Local: Irritant 3.
　Acute Systemic: Ingestion 3; Inhalation 3.
　Chronic Local: U.

Chronic Systemic: Ingestion 3; Inhalation 3; Skin Absorption 3.
Disaster Hazard: Dangerous; when heated to decomposition, it emits toxic fumes of chlorides.
Countermeasures
Ventilation Control: Section 2.
Personnel Protection: Section 3.
First Aid: Section 1.
Storage and Handling: Section 7.

1,2-DIHYDROXYPROPANE. See propylene glycol.

DIIODOACETYLENE
General Information
Synonym: Diiodoethyne.
Description: Colorless, rhombic, crystals; unpleasant odor.
Formula: IC:CI.
Constants: Mol wt: 277.86, mp: 78–82°C, bp: 80–100°C decomposes, vap. d.: 9.6.
Hazard Analysis
Toxic Hazard Rating:
  Acute Local: Irritant 3; Inhalation 3.
  Acute Systemic: Ingestion 3; Inhalation 3.
  Chronic Local: U.
  Chronic Systemic: U.
Toxicology: A military poison.
Explosion Hazard: Details unknown; many acetylene compounds are unstable.
Disaster Hazard: Dangerous; when heated to decomposition, it emits toxic fumes of iodine.
Countermeasures
Storage and Handling: Section 7.

DIIODODIACETYLENE
Hazard Analysis
Toxicity: Details unknown. Probably has irritant and narcotic effects. See also chlorinated hydrocarbons, aliphatic for similar properties.
Explosion Hazard: Severe, when exposed to heat.
Disaster Hazard: Dangerous; see iodides.
Countermeasures
Storage and Handling: Section 7.

DIIODODIETHYL SULFIDE
General Information
Description: Bright yellow prisms.
Formula: (CH₂CH₂I)₂S.
Constants: Mol wt: 215.08, mp: 62°C, vap. d.: 7.41.
Hazard Analysis
Toxic Hazard Rating:
  Acute Local: Irritant 3; Ingestion 3; Inhalation 3.
  Acute Systemic: U.
  Chronic Local: U.
  Chronic Systemic: U.
Disaster Hazard: Dangerous; when heated to decomposition or on contact with acid or acid fumes, it emits highly toxic fumes of iodides, oxides of sulfur.
Countermeasures
Ventilation Control: Section 2.
Personnel Protection: Section 3.
First Aid: Section 1.
Storage and Handling: Section 7.

1,1-DIIODOETHANE. See ethylidine diiodide.

DIIODOETHYNE. See diiodoacetylene.

DIIODOFORMOXIME
General Information
Description: Crystals.

Formula: CI₂NOH.
Constants: Mol wt: 296.9, mp: 69°C, vap. d.: 10.2.
Hazard Analysis
Toxic Hazard Rating:
  Acute Local: Irritant 3; Ingestion 3; Inhalation 3.
  Acute Systemic: U.
  Chronic Local: U.
  Chronic Systemic: U.
Disaster Hazard: Dangerous; see iodides and oxides of nitrogen.
Countermeasures
Ventilation Control: Section 2.
Personnel Protection: Section 3.
Storage and Handling: Section 7.

DIIODOMETHANE. See methylene iodide.

3,5-DIIODO SALICYLIC ACID
General Information
Description: White to pale pink crystalline powder, slightly soluble in water.
Formula: I₂C₆H₂(OH)COOH.
Constant: Mol wt: 390.
Hazard Analysis
Toxic Hazard Rating: U. A trace mineral added to animal feeds (Section 10).

DIIRON PHOSPHIDE
General Information
Description: Blue-gray crystals or powder.
Formula: Fe₂P.
Constants: Mol wt: 142.68, mp: 1290°C, d: 6.56.
Hazard Analysis and Countermeasures
See phosphides.

DIISOAMYLAMINE
General Information
Synonym: Isodiamylamine.
Description: Liquid; soluble in water, alcohol, ether, chloroform.
Formula: HNCH₂CH₂CH(CH₃)₂.
Constants: Mol wt: 193.76, d: 0.767 (20/4°C), bp: 188°C, mp: −44°C.
Hazard Analysis
Toxic Hazard Rating:
  Acute Local: Irritant 2; Ingestion 2; Inhalation 2.
  Acute Systemic: Ingestion 2.
  Chronic Local: U.
  Chronic Systemic: U.
May cause flushing and rise in blood pressure.

DI-sec-ISOAMYLBORANE. See bis(1,2-dimethylpropyl) borine.

DIISOAMYLMERCURY
General Information
Description: Liquid.
Formula: Hg(C₅H₁₁)₂.
Constants: Mol wt: 342.9, bp: 125°C at 10 mm, d: 1.6397, vap. d.: 11.8.
Hazard Analysis and Countermeasures
See mercury compounds, organic.

DIISOAMYLOXYBORON CHLORIDE
General Information
Description: Colorless liquid which decomposes upon contact with moisture.
Formula: (C₅H₁₁O)₂BCl.
Constants: Mol wt: 188.6, bp: 110–115°C at 14 mm.

TOXIC HAZARD RATING CODE (For detailed discussion, see Section 1.)

0 NONE: (a) No harm under any conditions; (b) Harmful only under unusual conditions or overwhelming dosage.

1 SLIGHT: Causes readily reversible changes which disappear after end of exposure.

2 MODERATE: May involve both irreversible and reversible changes; not severe enough to cause death or permanent injury.

3 HIGH: May cause death or permanent injury after very short exposure to small quantities.

U UNKNOWN: No information on humans considered valid by authors.

Hazard Analysis
Toxicity: Details unknown. See boron compounds.
Disaster Hazard: Dangerous; see chlorides.

DIISOAMYLCADMIUM
General Information
Description: An oil.
Formula: $(C_5H_{11})_2Cd$.
Constants: Mol wt: 254.7, d: 1.2209, mp: $-115°C$, bp: 121.5°C at 15 mm.
Hazard Analysis
Toxicity: Probably highly toxic. Details unknown. See cadmium compounds.
Fire Hazard: Details unknown. Probably dangerous. See Section 6.
Disaster Hazard: Dangerous; see cadmium.
Countermeasures
Storage and Handling: Section 7.

DIISOAMYLTIN DIBROMIDE
General Information
Description: Liquid.
Formula: $(C_5H_{11})_2SnBr_2$.
Constants: Mol wt: 420.8, mp: $-24°C$.
Hazard Analysis
Toxicity: Details unknown. See tin compounds.
Disaster Hazard: See bromides.

DIISOAMYLTIN DICHLORIDE
General Information
Description: Solid.
Formula: $(C_5H_{11})_2SnCl_2$.
Constants: Mol wt: 331.9, mp: 28°C.
Hazard Analysis
Toxicity: Details unknown. See tin compounds.
Disaster Hazard: See chlorides.

DIISOAMYLTIN DIIODIDE
General Information
Description: Oily liquid.
Formula: $(C_5H_{11})_2SnI_2$.
Constants: Mol wt: 514.8, bp: 202–205°C at 8 mm.
Hazard Analysis
Toxicity: Details unknown. See tin compounds.
Disaster Hazard: Dangerous. See iodides.

DIIOSBUTYL ALUMINUM CHLORIDE
General Information
Synonym: DIBAC.
Description: Colorless liquid.
Formula: $[(CH_3)_2CHCH_2]_2AlCl$.
Constants: Mol wt: 176.5, d: 0.905, fp: $-39.5°C$.
Hazard Analysis
Toxic Hazard Rating:
    Acute Local: Irritant 3; Ingestion 3; Inhalation 3.
    Acute Systemic: Inhalation 3.
    Chronic Local: U.
    Chronic Systemic: U.
Animal experiments show strong irritant and corrosive action on skin and lungs.

DIISOBUTYL ALUMINUM HYDRIDE
General Information
Synonym: DIBAL-H.
Description: Colorless, pyrophoric liquid; miscible in hydrocarbon solvents.
Formula: $[(CH_3)_2CHCH_2]_2AlH$.
Constants: Mol wt: 142, fp: $-80°C$, d: 0.798, bp: 105°C at .2 mm.
Hazard Analysis
Toxic Hazard Rating: U. See also hydrides.
Fire Hazard: Dangerous, ignites spontaneously in air.
Countermeasures
To Fight Fire: Do not use water, foam, or halogenated extinguishing agents (Section 6).

DIISOBUTYLAMINE
General Information
Description: Water-white liquid; amine odor.
Formula: $[(CH_3)_2CHCH_2]_2NH$.
Constants: Mol wt: 129.24, mp: $-70°C$, bp: 139.5°C, flash p: 85°F, d: 0.745 at 20°/4°C, vap. press.: 10 mm at 30.6°C, vap. d.: 4.46.
Hazard Analysis
Toxicity: Details unknown. See amines.
Fire Hazard: Moderate, when exposed to heat or flame; can react vigorously with oxidizing materials.
Explosion Hazard: Unknown.
Countermeasures
Ventilation Control: Section 2.
To Fight Fire: Foam, carbon dioxide, dry chemical or carbon tetrachloride (Section 6).
Storage and Handling: Section 7.

DIISOBUTYLCADMIUM
General Information
Description: An oil, decomposed by water.
Formula: $(C_4H_9)_2Cd$.
Constants: Mol wt: 226.6, d: 1.2690, mp: $-37°C$, bp: 90.5°C at 20 mm.
Hazard Analysis
Toxicity: Probably highly toxic. Details unknown. See cadmium compounds.
Fire Hazard: Details unknown. Probably dangerous. (See Section 6).
Disaster Hazard: Dangerous; see cadmium.
Countermeasures
Storage and Handling: Section 7.

DIISOBUTYL CARBINOL
General Information
Synonym: Nonyl alcohol; 2,6-dimethyl heptanol-4.
Description: Colorless liquid.
Formula: $(CH_3)_2CHCH_2CHOHCH_2CH(CH_3)_2$.
Constants: Mol wt: 144.3, bp: 173.3°C, fp: $-65°C$, flash p: 165°F, d: 0.8121 at 20°/20°C, vap. press.: 0.3 mm at 20°C, vap. d.: 4.98, lel: 0.8% at 212°F; uel: 6.1% at 212°F.
Hazard Analysis
Toxicity: Details unknown. See also alcohols. Limited animal experiments suggest low toxicity. Has caused central nervous and liver injury in animals.
Fire Hazard: Moderate, when exposed to heat or flame; can react with oxidizing materials.
Countermeasures
To Fight Fire: Alcohol foam, foam, carbon dioxide, dry chemical or carbon tetrachloride (Section 6).
Storage and Handling: Section 7.

DIISOBUTYLENE
General Information
Synonym: 2,6-Dimethyl-4-heptanone.
Description: Liquid.
Formula: C_8H_{16}.
Constants: Mol wt: 112.2, mp: $-101°C$, bp: 102°C, flash p: 20°F (C.C.), d: 0.7227 at 15.6°C, vap. d.: 3.97.
Hazard Analysis
Toxic Hazard Rating:
    Acute Local: Irritant 2; Inhalation 1.
    Acute Systemic: Inhalation 1.
    Chronic Local: U.
    Chronic Systemic: Inhalation 1.
Irritant and narcotic in high concentrations. Has caused liver and kidney damage in experimental animals.
Fire Hazard: Dangerous, when exposed to heat or flame; can react vigorously with oxidizing materials.
Disaster Hazard: Dangerous! Keep away from heat and open flame!
Explosion Hazard: Unknown.
Countermeasures
Ventilation Control: Section 2.

To Fight Fire: Foam, carbon dioxide, dry chemical or carbon tetrachloride (Section 6).
Storage and Handling: Section 7.

DIISOBUTYLENE OXIDE
General Information
Synonym: (2,4,4-Trimethyl pentene-1) oxide.
Hazard Analysis
Toxic Hazard Rating:
Acute Local: Irritant 1.
Acute Systemic: Ingestion 1; Inhalation 3; Skin Absorption 1.
Chronic Local: U.
Chronic Systemic: U.
Based on limited animal experiments.

DIISOBUTYL FUMARATE
Hazard Analysis
Toxic Hazard Rating: U. Limited animal experiments suggest low toxicity. See also esters and fumaric acid.

DIISOBUTYL KETONE
General Information
Synonym: Isovalerone; 2,6-dimethyl-4-heptanone.
Description: Liquid.
Formula: $[(CH_3)_2CHCH_2]_2CO$.
Constants: Mol wt: 142.2, bp: 166°C, flash p: 140°F, d: 0.81, vap. d.: 4.9, lel: 0.8% at 212°F; uel: 6.2% at 212°F.
Hazard Analysis
Toxic Hazard Rating:
Acute Local: Irritant 1.
Acute Systemic: Inhalation 1.
Chronic Local: Irritant 1.
Chronic Systemic: Inhalation 1.
A mild irritant. Narcotic in high concentrations.
TLV: ACGIH (recommended); 50 parts per million in air; 290 milligrams per cubic meter of air.
Fire Hazard: Moderate, when exposed to heat or flame; can react with oxidizing materials.
Countermeasures
To Fight Fire: Carbon dioxide, dry chemical or carbon tetrachloride (Section 6).
Storage and Handling: Section 7.
Shipping Regulations: Section 11.
Coast Guard Classification: Combustible liquid.
MCA warning label.
IATA: Other restricted articles, class A, no label required, 40 liters (passenger), 220 liters (cargo).

DIISOBUTYLMERCURY
General Information
Description: Colorless liquid.
Formula: $Hg(C_4H_9)_2$.
Constants: Mol wt: 314.8, mp: volatile at 100°C, bp: 207°C, d: 1.7678, vap. d.: 10.8.
Hazard Analysis and Countermeasures
See mercury compounds, organic.

DIISOBUTYL PHENOL. See octyl phenol.

DIISOBUTYL PHTHALATE
General Information
Description: Liquid.
Formula: $C_6H_4[COOCH_2CH(CH_3)_2]_2$.
Constants: Mol wt: 278, mp: −64°C, flash p.: 385°F, d: 1.039–1.043, vap. d.: 9.59.
Hazard Analysis
Toxicity: Details unknown. See phthalic acid and esters.

Fire Hazard: Slight.
Countermeasures
To Fight Fire: Foam, carbon dioxide, dry chemical, or carbon tetrachloride (Section 6).
Storage and Handling: Section 7.

DIISOBUTYL SODIUM SULFOSUCCINATE
General Information
Synonym: Aerosol IB; Alphasol IB; dibutyl sodium sulfosuccinate.
Description: Available as 3 esters of sulfosuccinic acid. The mixture is white, powder like; soluble in water, glycerol, pine oil, oleic acid; insoluble in acetone, kerosene, liquid petrolatum, carbon tetrachloride, ethanol, benzene, olive oil.
Formula: $C_{12}H_{21}NaO_7S$.
Constant: Mol wt: 332.35.
Hazard Analysis
Toxic Hazard Rating:
Acute Local: Irritant 2; Ingestion 2; Inhalation 2.
Acute Systemic: U.
Chronic Local: U.
Chronic Systemic: U.
Disaster Hazard: Dangerous. See sulfates.

DIISOBUTYLTIN DIIODIDE
General Information
Description: Solid.
Formula: $(C_4H_9)_2SnI_2$.
Constants: Mol wt: 486.8, bp: 290–295°C.
Hazard Analysis
Toxicity: Details unknown. See tin compounds.
Disaster Hazard: Dangerous. See iodides.

p,p′-DIISOCYANATODIPHENYLMETHANE
General Information
Description: Crystals.
Formula: $OCNC_6H_4CH_2C_6H_4NCO$.
Constants: Mol wt: 250.3, bp: 194–199°C at 5 mm, fp: 37.2°C, d: 1.19 at 50°C, vap. d: 8.63.
Hazard Analysis
Toxic Hazard Rating:
Acute Local: Irritant 2; Ingestion 2; Inhalation 2.
Acute Systemic: U.
Chronic Local: U.
Chronic Systemic: U.
Disaster Hazard: Dangerous. See cyanides.
Countermeasures
Ventilation Control: Section 2.
Personnel Protection: Section 3.
Storage and Handling: Section 7.

2,4-DIISOCYANOTOLUENE. See 2,4-tolylene diisocyanate.

DIISODECYL ADIPATE
General Information
Description: Clear, mobile, oily liquid.
Formula: $(CH_2CH_2CO_2C_{10}H_{21})_2$.
Constants: Mol wt: 426, bp: 240°C at 4 mm, flash p: 225°F (O.C.), vap. d.: 14.7.
Hazard Analysis
Toxicity: Details unknown. See also esters.
Fire Hazard: Slight, when exposed to heat or flame; can react with oxidizing materials.
Countermeasures
To Fight Fire: Foam, carbon dioxide, dry chemical or carbon tetrachloride (Section 6).
Storage and Handling: Section 7.

TOXIC HAZARD RATING CODE (For detailed discussion, see Section 1.)

0 NONE: (a) No harm under any conditions; (b) Harmful only under unusual conditions or overwhelming dosage.

1 SLIGHT: Causes readily reversible changes which disappear after end of exposure.

2 MODERATE: May involve both irreversible and reversible changes not severe enough to cause death or permanent injury.

3 HIGH: May cause death or permanent injury after very short exposure to small quantities.

U UNKNOWN: No information on humans considered valid by authors

DI-(ISODECYL)4,5-EPOXYCYCLOHEXANE-1,2-DICARBOXYLATE

Hazard Analysis
Toxic Hazard Rating:
Acute Local: Irritant 1.
Acute Systemic: Ingestion 1; Inhalation 1; Skin Absorption 1.
Chronic Local: U.
Chronic Systemic: U.
Based on limited animal experiments.

DIISODECYL PHTHALATE

General Information
Description: Clear, oily liquid; milk characteristic odor.
Formula: $(C_{10}H_{21}CO_2)_2C_6H_4$.
Constants: Mol wt: 446.7, flash p: 450° F, d: 0.966 at 25°/25°C, vap. d.: 15.4.
Hazard Analysis
Toxic Hazard Rating:
Acute Local: Irritant 2; Ingestion 2.
Acute Systemic: U.
Chronic Local: U.
Chronic Systemic: U.
Fire Hazard: Slight; can react with oxidizing materials. Section 6.
Countermeasures
Personnel Protection: Section 3.
Storage and Handling: Section 7.

DIISOOCTYL ACID PHOSPHATE

General Information
Description: A corrosive liquid.
Formula: $(C_8H_{17})_2HPO_4$.
Constant: Mol wt: 322.
Hazard Analysis
Toxic Hazard Rating: U. An irritant.
Disaster Hazard: Dangerous. See phosphorous compounds.
Countermeasures
Storage and Handling: Section 7.
Shipping Regulations: Section 11.
I.C.C.: Corrosive liquid, white label, 1 quart.
IATA: Corrosive liquid, white label, 1 liter (passenger and cargo).

DIISOOCTYL ADIPATE

General Information
Description: Liquid.
Formula: $[(CH_2)_2CO_2C_8H_{17}]_2$.
Constants: Mol wt: 370.6, bp: 207–213°C at 4 mm, flash p: 370° F, d: 0.930 at 20°/20°C, vap. d.: 12.8.
Hazard Analysis
Toxicity: See esters.
Fire Hazard: Slight, when exposed to heat or flame; can react with oxidizing materials.
Countermeasures
To Fight Fire: Foam, carbon dioxide, dry chemical or carbon tetrachloride (Section 6).
Storage and Handling: Section 7.

DIISOOCTYL AZELATE

General Information
Description: Liquid.
Formula: $(CH_2)_7[CO_2(CH_2)_5CH(CH_3)_2]_2$.
Constants: Mol wt: 412, bp:235°C at 5 mm, mp: < −59.5°C, flash p: 415° F, d: 0.92 at 20°C, vap. d.: 18.7.
Hazard Analysis
Toxicity: Details unknown. See also esters.
Fire Hazard: Slight, when exposed to heat or flame; can react with oxidizing materials.
Countermeasures
To Fight Fire: Foam, carbon dioxide, dry chemical or carbon tetrachloride (Section 6).
Storage and Handling: Section 7.

DIISOOCTYL PHENYL PHOSPHONATE

General Information
Description: Liquid.
Formula: $C_6H_5PO(OC_8H_{17})_2$.
Constants: Mol wt: 382, mp: < 0°C, bp: 180–190°C at 1 mm, d: 0.973 at 25°/25°C, vap. d.: 13.2.
Hazard Analysis and Countermeasures
See esters, phosphates.

DIISOOCTYL PHTHALATE

General Information
Description: Liquid; mild odor.
Formula: $C_6H_4(CO_2C_8H_{17})_2$.
Constants: Mol wt: 390.5, bp: 230–240°C at 4 mm, fp: −45°C, flash p.: 450° F, d: 0.986 at 20°/20°C, vap. d.: 13.5.
Hazard Analysis
Toxicity: Details unknown. See phthalic acid and esters.
Fire Hazard: Slight, when exposed to heat or flame; can react with oxidizing materials.
Countermeasures
To Fight Fire: Foam, carbon dioxide, dry chemical or carbon tetrachloride (Section 6).
Storage and Handling: Section 7.

DIISOOCTYL SEBACATE

General Information
Description: Liquid
Formula: $[(CH_2)_4CO_2C_8H_{17}]_2$.
Constants: Mol wt: 426.7, bp: 248–255°C at 4 mm, flash p: 470° F, d: 0.917 at 20°/20°C, vap. d.: 14.7.
Hazard Analysis
Toxicity: Details unknown. See esters.
Fire Hazard: Slight, when exposed to heat or flame; can react with oxidizing materials.
Countermeasures
To Fight Fire: Foam, carbon dioxide, dry chemical or carbon tetrachloride (Section 6).
Storage and Handling: Section 7.

DIISOOCTYL STYRYLPHOSPHONATE

General Information
Description: Liquid.
Formula: $C_6H_5C_2H_2PO(OC_8H_{17})_2$.
Constants: Mol wt: 408, mp: < 0°C, bp: 185–190°C at 1 mm, d: 0.977 at 25°/25°C.
Hazard Analysis and Countermeasures
See esters and phosphates.

DIISOPROPANOLAMINE

General Information
Formula: $[CH_3CH(OH)CH_2]_2NH$.
Constants: Mol wt: 133.19, mp: 42°C, bp: 249°C, flash p: 260° F (O.C.), d: 0.9890 at 45°/ 20°C, vap. d.: 4.59.
Hazard Analysis
Toxicity: See amines.
Fire Hazard: Slight, when exposed to heat or flame; can react with oxidizing materials.
Countermeasures
To Fight Fire: Alcohol foam, carbon dioxide dry chemical or carbon tetrachloride (Section 6).
Storage and Handling: Section 7.

DIISOPROPYL. See 2,3-dimethylbutane.

DIISOPROPYLAMINE

General Information
Description: Colorless liquid.
Formula: $[(CH_3)_2CH]_2NH$.
Constants: Mol wt: 101.19, bp: 83–84°C, flash p: 30° F (O.C.), d: 0.722 at 22.0°C, vap. d.: 3.5.
Hazard Analysis
Toxic Hazard Rating:
Acute Local: Irritant 3; Ingestion 3; Inhalation 3.
Acute Systemic: Inhalation 3.
Chronic Local: U.

Chronic Systemic: U.

TLV: ACGIH (tentative) 5 parts per million of air; 20 milligrams per cubic meter of air; may be absorbed via the skin.

Toxicology: Inhalation of fumes can cause pulmonary edema. See also amines.

Fire Hazard: Dangerous, when exposed to heat or flame; can react vigorously with oxidizing materials.

Explosion Hazard: Unknown.

Disaster Hazard: Dangerous! Keep away from heat and open flame!

Countermeasures

To Fight Fire: Alcohol foam, foam, carbon dioxide, dry chemical or carbon tetrachloride (Section 6).

Ventilation Control: Section 2.

Personnel Protection: Section 3.

Storage and Handling: Section 7.

Shipping Regulations: Section 11.

IATA: Flammable liquid, red label, 1 liter (passenger), 40 liters (cargo).

DIISOPROPYLBENZENE
General Information

Description: Clear, colorless liquid.

Formula: $[(CH_3)_2CH]_2C_6H_4$.

Constants: Mol wt: 162.26, mp: $< -55°C$, bp: 205°C, flash p: 170°F (O.C.), d: 0.863–0.867 at 25°/25°C, autoign. temp.: 840°F, vap. d.: 5.6.

Hazard Analysis

Toxicity: Details unknown. See also cumene.

Fire Hazard: Moderate, when exposed to heat or flame; can react with oxidizing materials.

Countermeasures

To Fight Fire: Foam, carbon dioxide, dry chemical or carbon tetrachloride (Section 6).

Storage and Handling: Section 7.

DIISOPROPYL BENZENE HYDROPEROXIDE
General Information

Description: Colorless to pale yellow liquid.

Hazard Analysis

Toxic Hazard Rating: U. Probably toxic.

Fire Hazard: A powerful oxidizer.

Countermeasures

Storage and Handling: Section 7.

Shipping Regulations: Section 11.

I.C.C.: Oxidizing material, yellow label, 1 quart.

IATA (solution exceeding 60%): Oxidizing material, no label, not acceptable (passenger and cargo).

(solution not exceeding 60% in non-volatile solvent): Oxidizing material, yellow label, 1 liter (passenger and cargo).

DIISOPROPYL CARBINOL
General Information

Synonym: 2,4-Dimethyl-3-pentanol.

Description: Colorless liquid.

Formula: $(C_3H_7)_2CHOH$.

Constants: Mol wt: 116.2, mp: $-70°C$, bp: 140°C, flash p: 120°F, d: 0.8288 at 20°/4°C, vap. d.: 4.0.

Hazard Analysis

Toxicity: See alcohols.

Fire Hazard: Moderate, when exposed to heat or flame; can react with oxidizing materials.

Countermeasures

To Fight Fire: Foam, carbon dioxide, dry chemical or carbon tetrachloride (Section 6).

Storage and Handling: Section 7.

DIISOPROPYL CYANAMIDE
General Information

Description: Colorless, mobile liquid; characteristic odor.

Formula: $(C_3H_7)_2NCN$.

Constants: Mol wt: 126.20, mp: $-27.3°C$, bp: 207°C, flash p: 179.5°F, d: 0.8451 at 30°C, vap. press.: 9 mm at 80°C, vapor density: 4.34.

Hazard Analysis

Toxic Hazard Rating:

Acute Local: Irritant 3; Ingestion 3; Inhalation 3.

Acute Systemic: Ingestion 3; Inhalation 3.

Chronic Local: U.

Chronic Systemic: U.

Fire Hazard: Moderate, when exposed to heat or flame.

Disaster Hazard: Dangerous; when heated to decomposition or on contact with acid or acid fumes, it emits highly toxic fumes of cyanides; will react with water or steam to produce toxic and corrosive fumes; can react with oxidizing materials.

Countermeasures

Ventilation Control: Section 2.

To Fight Fire: Foam, carbon dioxide, dry chemical or carbon tetrachloride (Section 6).

Personnel Protection: Section 3.

Storage and Handling: Section 7.

DIISOPROPYL DIXANTHOGEN
General Information

Description: Yellow to greenish pellets.

Formula: $(C_4H_7OS_2)_2$.

Constants: Mol wt: 270.5, mp: 52°C, d: 1.28, vap. d.: 9.35.

Hazard Analysis

Toxic Hazard Rating:

Acute Local: Irritant 3; Ingestion 3; Inhalation 3.

Acute Systemic: U.

Chronic Local: U.

Chronic Systemic: U.

Fire Hazard: Moderate, when exposed to heat or flame (Section 6).

Disaster Hazard: Dangerous; when heated to decomposition or on contact with acid or acid fumes, it emits highly toxic fumes of oxides of sulfur; will react with water or steam to produce toxic fumes; can react with oxidizing materials.

Countermeasures

Ventilation Control: Section 2.

Personnel Protection: Section 3.

Storage and Handling: Section 7.

N,N-DIISOPROPYL ETHANOLAMINE
General Information

Synonym: n,n-Diisopropylamino ethanol.

Description: Colorless liquid, slightly soluble in water.

Formula: $[(CH_3)_2CH]_2NC_2H_4OH$.

Constants: Mol wt: 145, d: 0.8742 (20°C); vap. p.: 0.08 mm at 20°C, fp: $-39.3°C$, bp: 191°C, flash p: 175°F (O.C.).

Hazard Analysis

Toxic Hazard Rating: U. See also amines.

Fire Hazard: Moderate, when exposed to heat or flame.

Countermeasures

Storage and Handling: Section 7.

DIISOPROPYL ETHER. See isopropyl ether.

DIISOPROPYL FLUOROPHOSPHATE
General Information

Description: Oily liquid.

TOXIC HAZARD RATING CODE (For detailed discussion, see Section 1.)

0 NONE: (a) No harm under any conditions; (b) Harmful only under unusual conditions or overwhelming dosage.

1 SLIGHT: Causes readily reversible changes which disappear after end of exposure.

2 MODERATE: May involve both irreversible and reversible change not severe enough to cause death or permanent injury.

3 HIGH: May cause death or permanent injury after very short exposure to small quantities.

U UNKNOWN: No information on humans considered valid by author

ormula: $C_6H_{14}FPO_3$.
Constants: Mol wt: 208.17, mp: $-82°C$, bp: $46°C$ at 5 mm, d: 1.07 (approx.), vap. d.: 5.24.

Hazard Analysis
Toxic Hazard Rating:
  Acute Local: Irritant 3; Inhalation 3.
  Acute Systemic: Ingestion 3; Skin Absorption 2.
  Chronic Local: Irritant 2.
  Chronic Systemic: Ingestion 2.
Toxicology: An insecticide. Ingestion can cause damage to eyes, nausea, vomiting, diarrhea and CNS disturbances.
Disaster Hazard: Dangerous; see fluorides and phosphorus compounds.

Countermeasures
Storage and Handling: Section 7.

DIISOPROPYL MALEATE
General Information
Description: Insoluble in water.
Formula: $(CH_3)_2CHOCOCH:COOCH(CH_3)_2$.
Constants: Mol wt: 187, d: 1.0+, bp: 444°F, flash p: 220°F (O.C.).

Hazard Analysis
Toxic Hazard Rating: U.
Fire Hazard: Moderate, when exposed to heat or flame.

Countermeasures
To Fight Fire: Water, foam, alcohol foam (Section 6).
Storage and Handling: Section 7.

DIISOPROPYLMERCURY
General Information
Description: Liquid.
Formula: $Hg(C_3H_7)_2$.
Constants: Mol wt: 286.8, bp: 63°C at 10 mm d: 2.0024, vap. d.: 9.9.

Hazard Analysis and Countermeasures
See mercury compounds, organic.

DIISOPROPYL PEROXYDICARBONATE
General Information
Synonym: Isopropyl percarbonate; isopropyl peroxydi-carbonate; IPP.
Description: Colorless, crystalline solid; almost insoluble in water; miscible with aliphatic and aromatic hydrocarbons, esters, ethers, chlorinated hydrocarbons.
Formula: $(CH_3)_2CHOCOOCOOCH(CH_3)_2$.
Constants: Mol wt: 120, rapid decomposition at 53°F, mp: 8–10°C, d: 1.080 (15.5/4°C).

Hazard Analysis
Toxic Hazard Rating: U.
Explosion Hazard: Dangerous, explodes on heating.

Countermeasures
Storage and Handling: Section 7.
Shipping Regulations: Section 11.
  I.C.C. (stabilized): Corrosive liquid, white label, not accepted.
       (unstabilized): Flammable solid, yellow label, not accepted.
  IATA (stabilized): Corrosive liquid, not acceptable (passenger and cargo).
       (unstabilized): Flammable solid, not acceptable (passenger and cargo).

2,6-DIISOPROPYL PHENOL
General Information
Description: A colorless liquid or solid.
Formula: $HC(CH_3)_2C_6H_3(OH)(CH_3)_2CH$.
Constants: Mol wt: 178.3, bp: 242.4°C, fp: 17.9°C, flash p: 235°F (C.C.), d: 0.955 at 20°/4°C.

Hazard Analysis
Toxicity: Unknown. See also phenol.
Fire Hazard: Slight; when exposed to heat or flame; can react with oxidizing materials.

Countermeasures
To Fight Fire: Foam, carbon dioxide, dry chemical or carbon tetrachloride (Section 6).
Storage and Handling: Section 7.

DIISOPROPYL PHOSPHO FLUORIDATE
General Information
Description: Odorless, colorless liquid.
Constants: Bp: 183°C, fp: $-82°C$.

Hazard Analysis
Toxicity: Highly toxic; used by Germans in World War II as the basis of nerve gases. See tabun and sarin.

DIISOPROPYLTIN DIBROMIDE
General Information
Description: Pale yellow crystals, decomposed by water.
Formula: $(C_3H_7)_2SnBr_2$.
Constants: Mol wt: 364.7, mp: 54°C.

Hazard Analysis
Toxicity: Details unknown. See tin compounds.
Disaster Hazard: Dangerous. See bromides.

DIISOPROPYLTIN DICHLORIDE
General Information
Description: Colorless crystals, soluble in water.
Formula: $(C_3H_7)_2SnCl_2$.
Constants: Mol wt: 275.8, mp: 84°C.

Hazard Analysis
Toxicity: Details unknown. See tin compounds.
Disaster Hazard: Dangerous. See chlorides.

DIISOPROPYLTIN OXIDE
General Information
Description: Solid, insoluble in water.
Formula: $(C_3H_7)_2SnO$.
Constant: Mol wt: 220.9.

Hazard Analysis
Toxicity: Details unknown. See tin compounds.

DIKERYL BENZENE-12
General Information
Description: Liquid.
Constant: Flash p: 320°F.

Hazard Analysis
Toxicity: Details unknown. Probably irritant and narcotic.
Fire Hazard: Slight, when exposed to heat or flame; can react with oxidizing materials.

Countermeasures
To Fight Fire: Foam, carbon dioxide, dry chemical or carbon tetrachloride (Section 6).
Storage and Handling: Section 7.

DIKETENE
General Information
Synonym: Acetyl ketene.
Description: Colorless, non-hygroscopic liquid; pungent odor; decomposes in water.
Formula: $C_4H_4O_2$.
Constants: Mol wt: 84.07, mp: $-6.5°C$, bp: 127.4°C, d: 1.0897, vap. d.: 2.9, flash p: 115°F (O.C.).

Hazard Analysis
Toxic Hazard Rating:
  Acute Local: Irritant 2; Ingestion 2; Inhalation 2.
  Acute Systemic: Ingestion 2; Inhalation 2.
  Chronic Local: U.
  Chronic Systemic: U.
Fire Hazard: Moderate, when exposed to heat or flame; can react with oxidizing materials (Section 6).

Countermeasures
Ventilation Control: Section 2.
Personnel Protection: Section 3.
Storage and Handling: Section 7.

DILAN
General Information
Description: Dilan is a mixture of 2-nitro-1,1-bis(p-chloro-

phenyl)propane (1 part) and 2-nitro-1,1-bis(p-chloro-phenyl) butane (2 parts). The first of the two components is known as Prolan, the second as Bulan. (The former is not to be confused with the pituitary hormone of the same name.)

Hazard Analysis

Toxic Hazard Rating: U. An insecticide.

Toxicity: Exact physiologic action of this compound is unknown. Toxicity tests in lower animals indicate that it can be absorbed from the digestive tract. Dilan does not seem to be absorbed through the skin to an appreciable extent, although it may be.

Dangerous Acute Dose: The acute dose in man in unknown. The LD_{50} to mice by oral administration about 1100 mg/kg for Dilan as compared to 135 mg/kg for DDT and about 3000 mg/kg for methoxychlor. In rats is has been found that neither Prolan nor Bulan is irritating to the skin. The LD_{50} by the dermal route for each is 74000 mg/kg. The dermal LD_{50} for Dilan in white rats is 5900 mg/kg for females, and 6900 mg/kg for males.

Dangerous Chronic Dose: In man it is unknown.

Signs and Symptoms: No signs and symptoms are as yet known in man. In animals they simulate those caused by other chlorinated hydrocarbons

Disaster Hazard: Dangerous. See chlorides and nitrates.

Countermeasures

Treatment of Poisoning: The treatment of exposed individuals should be directed along similar lines to those recommended for chlorinated hydrocarbons (Section 1).

DILAURYL THIO DIPROPIONATE

General Information

Synonym: Didodecyl 3,3'-thio dipropionate; thiodipropionic acid dilauryl ester.

Description: White flakes, sweet ester odor; soluble in most organic solvents.

Formula: $(C_{12}H_{25}OOCCH_2CH_2)_2S$.

Constants: Mol wt: 504, d (solid, 25°C): 0.975, mp: 40°C.

Hazard Analysis

Toxic Hazard Rating: U. Used as a chemical preservative food additive. (Section 10).

Disaster Hazard: Dangerous. See sulfur compounds.

DIMAGNESIUM PHOSPHATE. See magnesium phosphate, dibasic.

DIMAGNESIUM o-PHOSPHATE. See magnesium phosphate, dibasic.

DIMANGANESE ARSENIDE

General Information

Description: Solid.

Formula: Mn_2As.

Constants: Mol wt: 184.77, mp: 1400°C.

Hazard Analysis

Toxicity: Highly toxic. See arsenic compounds and manganese compounds.

Fire Hazard: Dangerous; when exposed to heat or by chemical reaction.

Explosion Hazard: See arsine.

Disaster Hazard: Dangerous; see arsenic; will react with water or acid to produce arsine; can react vigorously with oxidizing materials.

Countermeasures

Ventilation Control: Section 2.

Storage and Handling: Section 7.

DIMEFOX. See tetramethyl phosphorodiamine fluoride.

1,3-DIMERCAPTOPROPANE. See 1,3 propanedithiol.

2,3-DIMERCAPTO-1-PROPANOL

General Information

Synonyms: BAL; British anti-lewisite.

Description: Viscous, oily liquid; pungent odor.

Formula: $C_3H_8OS_2$.

Constants: Mol wt: 124.2, bp: 140°C at 40 mm, vap. d.: 4.3, d: 1.2385 at 25°/4°C.

Hazard Analysis

Toxic Hazard Rating:

Acute Local: Irritant 2.

Acute Systemic: Ingestion 2.

Chronic Local: U.

Chronic Systemic: U.

Toxicology: Applied locally to the skin it causes redness and swelling, but does not produce blisters or ulcers. It is intensely irritating to the eyes and mucous membranes. Systemic symptoms are caused by injection of the drug and usually occur within a few minutes, reaching their maximum in 15 to 30 minutes. At first there is a feeling of warmth with tingling sensations in the nose, mouth and skin. There may be nausea, vomiting, restlessness, weakness, rapid pulse and rise in blood pressure. In severe cases there may be tremors and convulsions.

Treatment is symptomatic. Epinephrine and antihistamines may be helpful. It may be necessary to give large doses of barbiturates or even anesthetics to control convulsions.

Disaster Hazard: Dangerous; see sulfur compounds.

Countermeasures

Personnel Protection: Section 3.

Storage and Handling: Section 7.

DIMERCURIC AMMONIUM OXIDE

Hazard Analysis

Toxicity: See mercury compounds, inorganic.

Explosion Hazard: Severe, when shocked or exposed to heat.

Disaster Hazard: Dangerous; shock or heat will explode it; when heated to decomposition, it emits highly toxic fumes of mercury.

Countermeasures

Storage and Handling: Section 7.

DIMETAN

General Information

Synonym: 5,5-Dimethyldihydroresorcinoldimethyl carbamate; G-19258.

Description: Crystals. Soluble in water and organic solvents.

Formula: $C_{11}H_{17}NO_3$.

Constants: Mol wt: 211.3, mp: 46°C, bp: 170–180°C at 11 mm.

Hazard Analysis

Toxic Hazard Rating:

Acute Local: U.

Acute Systemic: Inhalation 3; Skin Absorption 3.

Chronic Local: U.

Chronic Systemic: Ingestion 3.

Causes symptoms similar to parathion, but less severe.

Disaster Hazard: Dangerous; when heated to decomposition, it emits highly toxic fumes.

TOXIC HAZARD RATING CODE (For detailed discussion, see Section 1.)

0 NONE: (a) No harm under any conditions; (b) Harmful only under unusual conditions or overwhelming dosage.

1 SLIGHT: Causes readily reversible changes which disappear after end of exposure.

2 MODERATE: May involve both irreversible and reversible changes not severe enough to cause death or permanent injury.

3 HIGH: May cause death or permanent injury after very short exposure to small quantities.

U UNKNOWN: No information on humans considered valid by authors

DIMETHICONE
General Information
Description: Water-white, viscous, oil-like liquid. Not miscible with water.
Formula: A mixture of dimethyl siloxane polymers.
Hazard Analysis
Toxicity: Details unknown. Limited animal experiments suggest low toxicity.

DIMETHOATE. See o,o-dimethyl S-(N-methyl carbamoyl methyl) phosphorodithioate.

DIMETHOXANE. See 2,6-dimethyl-m-dioxan-4-yl acetate.

DIMETHOXYBENZALDEHYDE. See veratraldehyde.

1,4-DIMETHOXYBENZENE. See hydroquinone dimethyl ether.

DIMETHOXYBENZIDINE. See dianisidine.

3,4-DIMETHOXYBENZYL ALCOHOL
General Information
Description: Colorless liquid.
Formula: $(CH_3O)_2C_6H_3CH_2OH$.
Constants: Mol wt: 168.2, mp: 22°C, bp: 174°C at 15 mm, d: 1.17 at 25°/25°C, vap. d.: 5.8.
Hazard Analysis
Toxicity: Details unknown. See also alcohols.
Fire Hazard: Moderate; can react with oxidizing materials.
Countermeasures
To Fight Fire: Water, foam, carbon dioxide, dry chemical or carbon tetrachloride (Section 6).
Storage and Handling: Section 7.

DIMETHOXYBORINE
General Information
Description: Colorless, unstable liquid which decomposes upon contact with moisture.
Formula: $(CH_3O)_2BH$.
Constants: Mol wt: 73.9, mp: −131°C, bp: 26°C.
Hazard Analysis
Toxicity: Details unknown. See boron compounds.

DIMETHOXYBORON CHLORIDE
General Information
Description: Colorless liquid which decomposes upon contact with moisture.
Formula: $(CH_3O)_2BCl$.
Constants: Mol wt: 108.4, mp: −87.5°C, bp: 74.7°C.
Hazard Analysis
Toxicity: Details unknown. See boron compounds.
Disaster Hazard: Dangerous; when heated to decomposition, it emits highly toxic chloride fumes.

2,3-DIMETHOXYBUTANE
General Information
Formula: $CH_3OCH_2CH_2CH(OCH_3)CH_3$.
Constant: Mol wt: 118.
Hazard Analysis
Toxic Hazard Rating: U. Limited animal experiments suggest moderate toxicity.

DIMETHOXYCHLOROBENZENE-2,5
General Information
Description: Slightly soluble in water.
Formula: $C_8H_9ClO_2$.
Constants: Mol wt: 172.5, vap. d.: 5.9, bp: 460–467°F, flash p: 243°F.
Hazard Analysis
Toxic Hazard Rating: U.
Fire Hazard: Slight, when exposed to heat or flame.
Countermeasures
To Fight Fire: Water, foam, alcohol foam (Section 6).
Storage and Handling: Section 7.

DIMETHOXYDIPHENYL TRICHLOROETHANE
General Information
Description: Crystals (methoxy analog of DDT).
Formula: $Cl_3C_{14}H_9(OCH_3)_2$.
Constants: Mol wt: 345.7, vap. d.: 11.9.
Hazard Analysis
Toxic Hazard Rating:
    Acute Local: Irritant 2; Ingestion 2; Inhalation 2.
    Acute Systemic: Ingestion 2; Inhalation 2.
    Chronic Local: U.
    Chronic Systemic: U.
Disaster Hazard: Dangerous; see chlorides.
Countermeasures
Ventilation Control: Section 2.
Personnel Protection: Section 3.
Storage and Handling: Section 7.

1,2-DIMETHOXYETHANE. See ethylene glycol dimethyl ether.

DIMETHOXYETHYL PHTHALATE
General Information
Description: Light colored, clear liquid; mild aromatic odor.
Formula: $C_6H_4(COOC_2H_4OCH_3)_2$.
Constants: Mol wt: 282, mp: −40°C (forms gel), bp: 190–210°C at 4 mm, flash p: 360°F, d: 1.171 at 20°/20°C, vap. press.: 0.3 mm at 150°C, vap. d.: 9.75.
Hazard Analysis
Toxicity: Details unknown. See also esters.
Fire Hazard: Slight, when exposed to heat or flame; can react with oxidizing materials.
Countermeasures
To Fight Fire: Water, foam, carbon dioxide, dry chemical or carbon tetrachloride (Section 6).
Storage and Handling: Section 7.

β-3,4-DIMETHOXYPHENYLETHYLAMINE
General Information
Description: Colorless to pale yellow liquid.
Formula: $(CH_3O)_2C_6H_3CH_2CH_2NH_2$.
Constants: Mol wt: 181.2, mp: 15°C, bp: 156°C at 10 mm, d: 1.08 at 28°/4°C, vap. d.: 6.25.
Hazard Analysis
Toxicity: Details unknown. See also amines.
Fire Hazard: Slight; Section 6.
Countermeasures
Storage and Handling: Section 7.

1,2-DIMETHOXYPROPANE
General Information
Description: Liquid.
Formula: $CH_3CHOCH_2CH_2OCH_3$.
Constants: Mol wt: 104.15, bp: 95°C, d: 0.8461, vap. d.: 3.59.
Hazard Analysis
Toxicity: Details unknown. Probably irritant and narcotic in high concentrations.
Fire Hazard: Moderate, when exposed to heat or flame; can react with oxidizing materials.
Explosion Hazard: Unknown.
Countermeasures
Storage and Handling: Section 7.
To Fight Fire: Foam, carbon dioxide, dry chemical or carbon tetrachloride (Section 6).

DIMETHOXYTETRAETHYLENE GLYCOL. See dimethoxytetraglycol.

DIMETHOXYTETRAGLYCOL
General Information
Synonyms: Tetraethylene glycol dimethyl ether; dimethoxytetraethylene glycol.
Description: Water-white, practically odorless liquid.
Formula: $(CH_3OC_2H_4OC_2H_4)_2O$.
Constants: Mol wt: 222.28, bp: 275.3°C, fp: −29.7°C,

flash p: 285° F (O.C.), d: 1.0132 at 20°/20° C, vap. press.: 0.01 mm at 20° C, vap. d.: 7.7.

Hazard Analysis

Toxicity: See glycols.

Fire Hazard: Slight, when exposed to heat or flame; can react with oxidizing materials.

Countermeasures

To Fight Fire: Foam, carbon dioxide, dry chemical or carbon tetrachloride (Section 6).

Storage and Handling: Section 7.

2,5-DIMETHOXYTETRAHYDROFURAN

General Information

Description: Colorless liquid.

Formula: $(CH_3O)_2C_4H_6O$.

Constants: Mol wt: 132.16, bp: 35° C at 10 mm, vap. d: 4.56.

Hazard Analysis

Toxicity: Details unknown. See also tetrahydrofuran.

Fire Hazard: Slight; Section 6.

Countermeasures

Storage and Handling: Section 7.

DIMETHYL. See ethane.

DIMETHYLACETAL

General Information

Synonym: Ethylidene dimethyl ether.

Description: Colorless liquid; strong aromatic odor.

Formula: $CH_3(OCH_3)_2CH$.

Constants: Mol wt: 90.12, bp: 61.8° C, flash p: <80° F, d: 0.848 at 25° C, vap. d.: 3.1.

Hazard Analysis

Toxic Hazard Rating:

Acute Local: U.

Acute Systemic: Ingestion 2; Inhalation 2.

Chronic Local: U.

Chronic Systemic: U.

Fire Hazard: Dangerous, when exposed to heat or flame; can react vigorously with oxidizing materials.

Explosion Hazard: Unknown.

Countermeasures

Ventilation Control: Section 2.

To Fight Fire: Foam, carbon dioxide, dry chemical or carbon tetrachloride (Section 6).

Personal Hygiene: Section 3.

Storage and Handling: Section 7.

N,N-DIMETHYL ACETAMIDE

General Information

Description: Liquid.

Formula: $CH_3CON(CH_3)_2$.

Constants: Mol wt: 87.12, mp: −20° C, bp: 165° C, d: 0.9448 at 15.5° C, vap. d.: 3.01, vap. press.: 1.3 mm at 25° C, flash p: 171° F (T.O.C.). lel: 2.0%; uel: 11.5% at 740 mm and 160° C.

Hazard Analysis

Toxicity: Details unknown.

Toxic Hazard Rating:

Acute Local: Irritant 2.

Acute Systemic: Ingestion 2; Inhalation 2; Skin Absorption 2.

Chronic Local: Irritant 2; Ingestion 2; Inhalation 2.

Chronic Systemic: Ingestion 3; Inhalation 3; Skin Absorption 3.

TLV: ACGIH (tentative) 10 parts per million in air; 35 milligrams per cubic meter of air may be absorbed via intact skin.

Toxicology: Somewhat less toxic than dimethyl formamide based upon animal tests. It appears to be from slightly to moderately toxic by inhalation, skin exposure or ingestion. Data based on animal experiments.

Fire Hazard: Slight; Section 6.

Countermeasures

Storage and Handling: Section 7.

N,N-DIMETHYLACETOACETAMIDE

General Information

Description: Liquid; miscible in water and organic solvents.

Formula: $CH_3COCH_2CON(CH_3)_2$.

Constants: Mol wt: 129, bp: 220° C, d: 1.049–1.052 (20/20° C), flash p: 252° F (C.O.C.).

Hazard Analysis

Toxic Hazard: U. Limited animal experiments suggest low toxicity. See also N,N-dimethylacetamide.

Fire Hazard: Slight, when exposed to heat or flame.

Countermeasures

Storage and Handling: Section 7.

DIMETHYL ACETONYL CARBONYL

Hazard Analysis

Toxic Hazard Rating:

Acute Local: U.

Acute Systemic: Ingestion 2; Inhalation 2.

Chronic Local: U.

Chronic Systemic: U.

Fire Hazard: Moderate; can react with oxidizing materials.

Countermeasures

Ventilation Control: Section 2.

Personal Hygiene: Section 3.

Storage and Handling: Section 7.

DIMETHYL ACROLEIN. See senecidaldehyde.

DIMETHYLAMINE, ANHYDROUS

General Information

Description: Colorless gas.

Formula: $(CH_3)_2NH$.

Constants: Mol wt: 45.08, bp: 6.88° C, flash p: 21° F (C.C.), fp: −92.19° C, d: 0.6804 at 0°/4° C, autoign temp.: 806° F, vap. d.: 1.55, lel: 2.8%, uel: 18.4%.

Hazard Analysis

Toxic Hazard Rating:

Acute Local: Irritant 2; Ingestion 2; Inhalation 2.

Acute Systemic: U.

Chronic Local: U.

Chronic Systemic: U.

TLV: ACGIH; 10 parts per million of air; 18 milligrams per cubic meter of air.

Fire Hazard: Dangerous, when exposed to heat or flame can react vigorously with oxidizing materials.

Explosion Hazard: Moderate, when exposed to flame.

Disaster Hazard: Dangerous! Keep away from heat and open flame!

Countermeasures

Ventilation Control: Section 2.

To Fight Fire: Stop flow of gas, foam, carbon dioxide dry chemical or carbon tetrachloride (Section 6).

Personnel Protection: Section 3.

Storage and Handling: Section 7.

Shipping Regulations: Section 11.

I.C.C.: Flammable gas; red gas label, 300 pounds.

Coast Guard Classification: Inflammable gas; red gas label.

MCA warning label.

TOXIC HAZARD RATING CODE (For detailed discussion, see Section 1.)

0 NONE: (a) No harm under any conditions; (b) Harmful only under unusual conditions or overwhelming dosage.

1 SLIGHT: Causes readily reversible changes which disappear after end of exposure.

2 MODERATE: May involve both irreversible and reversible change not severe enough to cause death or permanent injury.

3 HIGH: May cause death or permanent injury after very short exposu to small quantities.

U UNKNOWN: No information on humans considered valid by autho

IATA: Flammable gas, red label, not acceptable (passenger), 140 kilograms (cargo).

DIMETHYLAMINE, AQUEOUS SOLUTION
Shipping Regulations: Section 11.
  I.C.C.: Flammable liquid; red label.
  Coast Guard Classification: Inflammable liquid; red label.
  IATA: Flammable liquid, red label, 1 liter (passenger), 40 liters (cargo).
  MCA warning label.

DIMETHYLAMINOACETONITRILE
General Information
Formula: $(CH_3)_2NCH_2CN$.
Constant: Mol wt: 84.
Hazard Analysis
Toxic Hazard Rating: U. Limited animal experiments suggest high toxicity. See also nitriles.
Disaster Hazard: Dangerous. See cyanides.

DIMETHYLAMINOANTIPYRINE. See aminopyrine.

DIMETHYLAMINOBENZALDEHYDE
General Information
Synonym: 4-Dimethylaminobenzenecarbinol; Ehrlich's reagent.
Description: Small, granular lemon colored crystals (may turn pink upon exposure to light), slightly soluble in water; soluble in alcohol, ether, chloroform, acetic acid and many other organic solvents.
Formula: $C_6H_4[N(CH_3)_2]CHO$.
Constants: Mol wt: 149.19, mp: 73°C, bp: 176–177°C at 17 mm.
Hazard Analysis
Toxic Hazard Rating: U. Not highly toxic. See benzaldehyde.

DIMETHYLAMINOCHLOROPROPANE
General Information
Description: Liquid.
Formula: $C_5H_{12}NCl$.
Constants: Mol wt: 121.6, vap. d.: 4.20.
Hazard Analysis
Toxic Hazard Rating:
  Acute Local: Irritant 2; Ingestion 2; Inhalation 2.
  Acute Systemic: U.
  Chronic Local: U.
  Chronic Systemic: U.
Disaster Hazard: Dangerous; see chlorides; can react with oxidizing materials.
Countermeasures
Ventilation Control: Section 2.
Personnel Protection: Section 3.
Storage and Handling: Section 7.

2-DIMETHYLAMINOETHANOL
General Information
Synonym: Dimethylethanolamine.
Description: Liquid.
Formula: $(CH_3)_2NCH_2CH_2OH$.
Constants: Mol wt: 89.14, bp: 131°C, flash p: 105°F, d: 0.8866 at 20°/4°C, vap. d.: 3.03.
Hazard Analysis
Toxicity: Details unknown. See also amines. Used medically as a CNS stimulant.
Fire Hazard: Moderate, when exposed to heat or flame; can react vigorously with oxidizing materials.
Explosion Hazard: Unknown.
Countermeasures
Ventilation Control: Section 2.
To Fight Fire: Alcohol foam, foam, carbon dioxide, dry chemical or carbon tetrachloride (Section 6).
Storage and Handling: Section 7.

DIMETHYLAMINOETHYL METHACRYLATE
General Information
Description: Liquid; soluble in water and organic solvents.
Formula: $CH_2(CH_3)CCO_2(CH_2)_2N(CH_3)_2$.
Constants: Mol wt: 157, d: 0.933 at 25°C, bp: 182–190°C, flash p: 165°F (T.O.C.), vap. d.: 5.4.
Hazard Analysis
Toxicity: Details unknown. It is an irritant to skin, eyes and mucous membrane. A powerful lachrymator.
Fire Hazard: Moderate, when exposed to sparks, heat or open flame.
Countermeasures
To Fight Fire: Water, foam, dry chemical, spray. Section 6.
Personal Hygiene: Section 3.
Storage and Handling: Section 7.

1-[2-(N,N-DIMETHYLAMINO)ETHYL]-4-METHYL PIPERAZINE
Hazard Analysis
Toxic Hazard Rating: U. Limited animal experiments suggest high toxicity and irritation. See also piperazine.

DIMETHYLAMINOPHENYL MERCURIC ACETATE
General Information
Description: Colorless crystals; insoluble in water.
Formula: $C_6H_4N(CH_3)_2HgO_2C_2H_3$.
Constants: Mol wt: 379.8, mp: 165°C.
Hazard Analysis and Countermeasures
Highly toxic. See mercury compounds organic.

1-DIMETHYLAMINO-2-PROPANOL
General Information
Description: Clear, amber-colored, volatile liquid.
Formula: $(CH_3)_2NCH_2CHOHCH_3$.
Constants: Mol wt: 103.2, bp: 122.5°–126.2°C, flash p: 90°F, fp: $< -20°C$, d: 0.850 at 25°/25°C, vap. d.: 3.52.
Hazard Analysis
Toxicity: Unknown. See also amines.
Fire Hazard: Moderate, when exposed to heat or flame; can react vigorously with oxidizing materials.
Explosion Hazard: Unknown.
Countermeasures
Ventilation Control: Section 2.
To Fight Fire: Foam, carbon dioxide, dry chemical or carbon tetrachloride (Section 6).
Storage and Handling: Section 7.

DIMETHYLAMINOPROPIONITRILE
General Information
Description: Liquid.
Formula: $(CH_3)_2NCH_2CH_2CN$.
Constants: Mol wt: 98.2, mp: $-43°C$, bp: 170°C, flash p: 147°F, d: 0.8617, vap. d.: 3.35.
Hazard Analysis
Toxicity: Highly toxic. See nitriles.
Fire Hazard: Moderate, when exposed to heat or flame.
Disaster Hazard: Dangerous; when heated to decomposition, it emits highly toxic fumes; can react with oxidizing materials.
Countermeasures
Storage and Handling: Section 7.
To Fight Fire: Foam, carbon dioxide, dry chemical or carbon tetrachloride (Section 6).

3-DIMETHYLAMINOPROPYLAMINE
General Information
Description: Colorless liquid.
Formula: $(CH_3)_2NCH_2CH_2CH_2NH_2$.
Constants: Mol wt: 102.18, mp: $< -70°C$, bp: 123°C, flash p: 95°F (T.C.C.), d: 0.8100 at 30°C, vap. press.: 10 mm at 30°C, vap. d.: 3.52.
Hazard Analysis
Toxic Hazard Rating:

Acute Local: Irritant 3; Ingestion 3; Inhalation 3.
Acute Systemic: U.
Chronic Local: U.
Chronic Systemic: U.
Fire Hazard: Moderate, when exposed to heat or flame.
Explosion Hazard: U.
Disaster Hazard: Dangerous; reacts vigorously with oxidizing materials, and emits toxic fumes when heated.

Countermeasures
Ventilation Control: Section 2.
To Fight Fire: Foam, carbon dioxide, dry chemical or carbon tetrachloride (Section 6).
Personnel Protection: Section 3.
Storage and Handling: Section 7.

DIMETHYL AMMONIUM DIMETHYLCARBAMATE
General Information
Description: Liquid or crystals.
Formula: $(CH_3)_2NCOONH_2(CH_3)_2$.
Constants: Mol. wt. 134.17, mp: $-38°C$, bp: $60.2°C$, d: 1.026 at $25°/4°C$, vap. d.: 4.62.

Hazard Analysis
Toxicity: Unknown.
Disaster Hazard: Slight; on decomposition it emits toxic fumes.

Countermeasures
Storage and Handling: Section 7.

DI(METHYL AMYL) MALEATE
General Information
Description: Insoluble in water.
Formula: $[(CH_3)_2CHCH_2CHCH_3OCOCH:]_2$.
Constants: Mol. wt. 284, d: 0.9, bp: 394°F at 50 mm, flash p: 290°F (O.C.).

Hazard Analysis
Toxic Hazard Rating: U.
Fire Hazard: Slight, when exposed to heat or flame.

Countermeasures
To Fight Fire: Water, foam (Section 6).
Storage and Handling: (Section 7).

N,N-DIMETHYLANILINE
General Information
Description: Liquid.
Formula: $C_6H_5N(CH_3)_2$.
Constants: Mol. wt. 121.18, mp: 2.5°C, bp: 193.1°C, flash p: 145°F (C.C.), d: 0.9557 at $20°/4°C$, ulc: 20–25, autoign. temp.: 700°F, vap. press.: 1 mm at 29.5°C, vap. d.: 4.17.

Hazard Analysis
Toxic Hazard Rating:
  Acute Local: 0.
  Acute Systemic: Ingestion 3; Inhalation 3; Skin Absorption 3.
  Chronic Local: U.
  Chronic Systemic: Ingestion 3; Inhalation 3; Skin Absorption 3.
TLV: ACGIH (recommended); 5 parts per million in air; 25 milligrams per cubic meter of air. May be absorbed via intact skin.
Toxicology: Its physiological action is similar to that of aniline, although it is believed to be more toxic. It acts as a depressant on the central nervous system. Oral or subcutaneous administration of 2 g/kg of body weight in guinea pigs has caused weakness, tremors, tonic and clonic convulsions, slowing of the respiration, and finally death due to respiratory paralysis. The hazard associated with this material is that untrained personnel

will disregard small splashes of it upon the shoes, clothing, or body and will use improperly ventilated equipment. Industrial accidents involving dimethyl aniline are dangerous in that they can release sudden massive quantities of the oil or its vapor from breaks in the pipes of a closed system. See also aniline.
Fire Hazard: Moderate, when exposed to heat or flame.
Spontaneous Heating: No.
Disaster Hazard: Dangerous; when heated to decomposition, it emits highly toxic fumes of aniline; can react with oxidizing materials.

Countermeasures
Ventilation Control: Section 2.
To Fight Fire: Foam, carbon dioxide, dry chemical or carbon tetrachloride (Section 6).
Personnel Protection: Section 3.
First Aid: Section 1.
Storage and Handling: Section 7.
Shipping Regulations: Section 11.
  MCA warning label.

DIMETHYL ANALOG OF PARATHION. See O,O-dimethyl O-p-nitrophenyl thiophosphate.

DIMETHYLANILINE FLUOSILICATE
General Information
Description: White crystals, soluble in hot alcohol.
Formula: $(C_6H_5NHCH_3)_2 \cdot H_2SiF_6$.
Constants: Mol. wt. 358.4.

Hazard Analysis
Toxicity: See fluosilicates.
Disaster Hazard: Dangerous; see fluorides.

DIMETHYLANILINE MERCURY
General Information
Description: Colorless crystals, soluble in chloroform.
Formula: $Hg[C_6H_4N(CH_3)_2]_2$.

Hazard Analysis and Countermeasures
Highly toxic. See mercury compounds, organic.

DIMETHYLARSINE
General Information
Synonym: Cacodyl hydride.
Description: Colorless liquid.
Formula: $(CH_3)_2AsH$.
Constants: Mol. wt. 106.0, bp: 36°C, d: 1.213 at $29°/4°C$, vap. d.: 3.65.

Hazard Analysis
Toxicity: Highly toxic. See arsenic compounds.
Fire Hazard: Dangerous, ignites spontaneously in air (Section 6).
Explosion Hazard: Unknown.
Disaster Hazard: Dangerous. See arsenic compounds.
Spontaneous Heating: Yes.

Countermeasures
To Fight Fire: Water, foam, dry chemical, carbon dioxide and water spray.
Storage and Handling: Section 7.

DIMETHYLARSINIC ACID. See cacodylic acid.

DIMETHYLBENZENE. See xylene.

p,α-DIMETHYL BENZYL ALCOHOL
General Information
Synonym: p-Tolylmethyl carbinol; methyl-p-tolylcarbinol, 4-(α-hydroxyethyl)toluene; 4-methyl-α-phenethyl alcohol; 1-p-tolyl-1-ethanol.
Description: Viscous liquid; menthol like odor; very spar-

TOXIC HAZARD RATING CODE (For detailed discussion, see Section 1.)

0 NONE: (a) No harm under any conditions; (b) Harmful only under unusual conditions or overwhelming dosage.

1 SLIGHT: Causes readily reversible changes which disappear after end of exposure.

2 MODERATE: May involve both irreversible and reversible changes; not severe enough to cause death or permanent injury.

3 HIGH: May cause death or permanent injury after very short exposure to small quantities.

U UNKNOWN: No information on humans considered valid by authors.

ingly soluble in water; miscible with absolute alcohol, ether. Also soluble in isopropanol; liquid petrolatum.
Formula: $H_3C(C_6H_4)CH(CH_3)(OH)$.
Constants: Mol wt: 136.19, d: 0.9668 (15.5/4°C), bp: 219°C at 756 mm.
Hazard Analysis
Toxic Hazard Rating: U. Probably irritant and narcotic. See also toluene.

DIMETHYLBENZYLCETYL AMMONIUM CHLORIDE
General Information
Formula: $(CH_3)_2(C_6H_5CH_2)C_{16}H_{33}NCl$.
Constants: Mol wt: 436.1, vap. d.: 15.
Hazard Analysis
Toxic Hazard Rating:
  Acute Local: Irritant 2; Ingestion 2.
  Acute Systemic: U.
  Chronic Local: U.
  Chronic Systemic: U.
Fire Hazard: Slight; (Section 6).
Countermeasures
Ventilation Control: Section 2.
Personnel Protection: Section 3.
Storage and Handling: Section 7.

α-DIMETHYL BENZYL HYDROPEROXIDE. See cumene hydroperoxide.

DIMETHYLBERYLLIUM
General Information
Description: White needles.
Formula: $Be(CH_3)_2$.
Constants: Mol wt: 39.09, bp: sublimes at 200°C.
Hazard Analysis
Toxicity: Highly toxic. See beryllium compounds.
Fire Hazard: Moderate, when exposed to heat or flame (Section 6).
Disaster Hazard: Dangerous; when heated to decomposition, it emits highly toxic fumes of beryllium oxide; can react with oxidizing materials.
Countermeasures
Storage and Handling: Section 7.

DIMETHYLBORIC ACID
General Information
Synonym: Dimethyl hydroxyborine.
Description: Colorless liquid miscible with water.
Formula: $(CH_3)_2BOH$.
Constants: Mol wt: 57.9, bp: 0°C at 36 mm.
Hazard Analysis
Toxicity: Details unknown. See boron compounds.

DIMETHYLBORIC ANHYDRIDE
General Information
Description: Colorless crystals which hydrolyze in water.
Formula: $(CH_3)_2BOB(CH_3)_2$.
Constants: Mol wt: 97.8, mp: −37.3°C, bp: 43°C.
Hazard Analysis
Toxicity: Details unknown. See boron compounds.

DIMETHYLBORINE TRIMETHYL AMMINE
General Information
Description: Colorless liquid which decomposes in water.
Formula: $(CH_3)_2HB \cdot N(CH_3)_3$.
Constants: Mol wt: 101.0, mp: −18°C, bp: 172°C (decomp.).
Hazard Analysis
Toxicity: Details unknown. See boron compounds.

DIMETHYLBORON BROMIDE
General Information
Description: Colorless liquid or gas decomposed by moisture.
Formula: $(CH_3)_2BBr$.
Constants: Mol wt: 120.8, mp: −123.4°C, bp: 22°C.

Hazard Analysis
Toxicity: Details unknown. See boron compounds and bromides.
Disaster Hazard: Dangerous; see bromides.

DIMETHYLBORON IODIDE
General Information
Description: Colorless liquid decomposed by water.
Formula: $(CH_3)_2BI$.
Constants: Mol wt: 167.8, mp: −111°C, bp: 65°C.
Hazard Analysis
Toxicity: Details unknown. See boron compounds.
Disaster Hazard: Dangerous; see iodides.

DIMETHYLBROMARSINE
General Information
Synonym: Cacodyl bromide.
Description: Yellow oily liquid.
Formula: $(CH_3)_2AsBr$.
Constants: Mol wt: 184.9, bp: 130°C.
Hazard Analysis
Toxicity: Highly toxic. See arsenic compounds.
Disaster Hazard: Dangerous; see arsenic bromides.

2,2-DIMETHYLBUTANE
General Information
Synonym: Neohexane.
Description: Liquid.
Formula: $(CH_3)_3CCH_2CH_3$.
Constants: Mol wt: 86.2, bp: 49.7°C, mp: −98.2°C, flash p: −54°F, fp: −101.9°C, d: 0.649, autoign. temp.: 797°F, vap. press.: 400 mm at 31.0°C, vap. d.: 3.00, lel: 1.2%, uel: 7.0%.
Hazard Analysis
Toxicity: Details unknown. Probably is irritant and narcotic in high concentrations.
Fire Hazard: Dangerous when exposed to heat or flame; can react vigorously with oxidizing meterials.
Explosion Hazard: Unknown.
Disaster Hazard: Dangerous! Keep away from heat or open flame.
Ventilation Control: Section 2.
Countermeasures
To Fight Fire: Foam, carbon dioxide, dry chemical or carbon tetrachloride (Section 6).
Storage and Handling: Section 7.
Shipping Regulations: Section 11.
  I.E.C. Classification: Flammable liquid. Red Label.
  Coast Guard Classification: Inflammable liquid. Red label.

2,3-DIMETHYLBUTANE
General Information
Synonym: Diisopropyl.
Description: Liquid.
Formula: $(CH_3)_2CHCH(CH_3)_2$.
Constants: Mol wt: 86.17, mp: −135.1°C, bp: 58.0°C, flash p: −20°F, d: 0.662 at 20°/4°C, autoign. temp.: 788°F, vap. press.: 400 mm at 39.0°C, vap. d.: 3.0, lel: 1.2%, uel: 7.0%.
Hazard Analysis
Toxicity: Unknown. Probably irritant and narcotic in high concentrations.
Fire Hazard: Dangerous, when exposed to heat or flame; can react vigorously with oxidizing materials.
Explosion Hazard: Unknown.
Disaster Hazard: Dangerous; Keep away from heat and open flame!
Countermeasures
Ventilation Control: Section 2.
To Fight Fire: Foam, carbon dioxide, dry chemical or carbon tetrachloride (Section 6).
Storage and Handling: Section 7.
Shipping Regulations: Section 11.
  IATA: Flammable liquid, red label, 1 liter (passenger), 40 liters (cargo).
  I.C.C.: Flammable liquid, red label, 10 gallons.

2,2-DIMETHYL-1,3-BUTANEDIOL
General Information
Description: Liquid; very soluble in water.
Formula: $CH_3CH(OH)C(CH_3)_2CH_2OH$.
Constants: Mol wt: 119, d: 0.9700, bp: 202.4°C, fp: −12.8°C.
Hazard Analysis
Toxic Hazard Rating: U. Limited animal experiments suggest moderate toxicity.

2,2-DIMETHYL BUTANOL
General Information
Formula: $CH_3CH_2C(CH_3)_2CH_2OH$.
Constant: Mol wt: 102.
Hazard Analysis
Toxic Hazard Rating: U. Limited animal experiments suggest high toxicity and high irritation.

1,3-DIMETHYLBUTYL ACETATE
General Information
Description: Slightly soluble in water.
Formula: $CH_3COOCH(CH_3)CH_2CH(CH_3)_2$.
Constants: Mol wt: 144, d: 0.9, vap. d.: 5.0, bp: 284–297°F, flash p: 113°F.
Hazard Analysis
Toxic Hazard Rating: U.
Fire Hazard: Moderate, when exposed to heat or flame.
Countermeasures
To Fight Fire: Alcohol foam (Section 6).
Storage and Handling: Section 7.

1,3-DIMETHYL BUTYLAMINE
General Information
Description: A liquid.
Formula: $CH_3CH(NH_2)CH_2CH(CH_3)_2$.
Constants: Mol wt: 101.2, bp: 106°–109°C, flash p: 55°F, d: 0.750 at 20°/20°C.
Hazard Analysis
Toxicity: See amines.
Fire Hazard: Dangerous, when exposed to heat or flame, can react vigorously with oxidizing materials.
Explosion Hazard: Unknown.
Countermeasures
Ventilation Control: Section 2.
To Fight Fire: Foam, carbon dioxide, dry chemical or carbon tetrachloride (Section 6).
Storage and Handling: Section 7.

3,3-DIMETHYL BUTYRIC ACID. See tert-butyl acetic acid.

DIMETHYLCADMIUM
General Information
Description: Oil decomposed by water.
Formula: $(CH_3)_2Cd$.
Constants: Mol wt: 142.5, d: 1.984, mp: −45°C, bp: 106°C.
Hazard Analysis and Countermeasures
See cadmium compounds.

DIMETHYL-α-CAPROLACTONE
Hazard Analysis
Toxic Hazard Rating: U. Limited animal experiments suggest low toxicity and irritation. See also lactones.

DIMETHYL CARBAMYL CHLORIDE
General Information
Synonym: N,N-Dimethyl carbamyl chloride.
Formula: $(CH_3)_2NCOCl$.
Description: Liquid.

Constants: Mol wt: 107.5, mp: −33°C, bp: 165°–167°C, d: 1.678 at 20°/4°C, vap. d.: 3.73.
Hazard Analysis
Toxicity: Lachrymator; strong local irritant action.
Disaster Hazard: Dangerous; see chlorides; it will react with water or steam to produce toxic and corrosive fumes.
Countermeasures
Storage and Handling: Section 7.

DIMETHYL CARBATE
General Information
Synonym: Cis-bicyclo (2,2,1)-5-heptene-2,3-dicarboxylic acid dimethyl ester.
Hazard Analysis
Toxic Hazard Rating:
  Acute Local: Irritant 1; Ingestion 1; Inhalation 1.
  Acute Systemic: U.
  Chronic Local: U.
  Chronic Systemic: U.
Fire Hazard: Slight; Section 6.
Countermeasures
Personal Hygiene: Section 3.
Storage and Handling: Section 7.

DIMETHYLCARBINOL. See isopropyl alcohol.

3-DIMETHYLCARBINOLPROPYLAMINE. See 3-isopropoxypropylamine.

α,α-DIMETHYL-α'-CARBOBUTOXYDIHYDRO-γ-PYRONE
General Information
Synonym: n-Butyl mesityl oxide oxalate.
Description: Amber liquid.
Formula: $OC(CH_3)_2CH_2COCHC(CO_2C_4H_9)$.
Constants: Mol wt: 226.3, bp: 113°C, flash p: 315°F, d: 1.08 at 20°/20°C, vap. d.: 7.8.
Hazard Analysis
Toxicity: Details unknown. Insect repellent.
Fire Hazard: Slight, when exposed to heat or flame; can react with oxidizing materials.
Countermeasures
To Fight Fire: Foam, carbon dioxide, dry chemical or carbon tetrachloride (Section 6).
Storage and Handling: Section 7.

DIMETHYL-1-CARBOMETHOXY-1-PROPENYL-2-PHOSPHATE
Hazard Analysis
Toxicity: Highly toxic. An organophosphate cholinesterase inhibitor. See parathion.
Disaster Hazard: Dangerous; see phosphorus compounds.

DIMETHYL CARBONATE
General Information
Synonym: Methyl carbonate.
Description: Colorless liquid; pleasant odor; miscible with acids and alkalies, soluble in most organic solvents; insoluble in water.
Formula: $CO(OCH_3)_2$.
Constant: Mol wt: 90.
Hazard Analysis
Toxic Hazard Rating:
  Acute Local: Irritation 3; Inhalation 3.
  Acute Systemic: Inhalation 3.
  Chronic Local: Irritation 3; Inhalation 3.
  Chronic Systemic: U.

TOXIC HAZARD RATING CODE (For detailed discussion, see Section 1.)

0 NONE: (a) No harm under any conditions; (b) Harmful only under unusual conditions or overwhelming dosage.

1 SLIGHT: Causes readily reversible changes which disappear after end of exposure.

2 MODERATE: May involve both irreversible and reversible changes; not severe enough to cause death or permanent injury.

3 HIGH: May cause death or permanent injury after very short exposure to small quantities.

U UNKNOWN: No information on humans considered valid by authors.

Countermeasures
Storage and Handling: Section 7.
To Fight Fire: Alcohol foam (Section 6).
Shipping Regulations: Section 11.
   IATA: Flammable liquid, red label, 1 liter (passenger), 40 liters (cargo).

DI(METHYL "CELLOSOLVE") MALEATE
General Information
Description: Liquid.
Formula: $CH_3O(CH_2)_2OCOCH:CHCOO(CH_2)_2OCH_3$.
Constants: Mol wt: 232.23, mp: $-50°C$, bp: $152°C$ at 5 mm, d: 1.1413 at $20°/20°C$, vap. d.: 8.01.
Hazard Analysis
Toxicity: Details unknown. See also glycols.
Fire Hazard: Slight, when heated, can react with oxidizing materials (Section 6).
Countermeasures
Storage and Handling: Section 7.

DIMETHYL "CELLOSOLVE" PHTHALATE. See dimethylglycol phthalate.

DIMETHYLCHLORACETAL
General Information
Description: Colorless liquid.
Formula: $ClCH_2CH(OCH_3)_2$.
Constants: Mol wt: 124.6, bp: $126°-132°C$, flash p: $110°F$, d: 1.082–1.092 at $25°/4°C$, autoign. temp.: $450°F$, vap. d.: 4.3.
Hazard Analysis
Toxicity: Details unknown. As an aldehyde it may have irritant and narcotic action.
Fire Hazard: Moderate, when exposed to heat or flame.
Disaster Hazard: Dangerous; see chlorides; can react with oxidizing materials.
Countermeasures
To Fight Fire: Foam, carbon dioxide, dry chemical or carbon tetrachloride (Section 6).
Storage and Handling: Section 7.

DIMETHYLCHLOROARSINE
General Information
Synonyms: Cacodyl chloride; chlorodimethylarsine.
Description: Colorless liquid; insoluble in water.
Formula: $(CH_3)_2AsCl$.
Constants: Mol wt: 140.44, mp: $<-45°C$, bp: $106.5°C$, d: >1, vap. d.: 4.84.
Hazard Analysis
Toxicity: Highly toxic. See arsenic compounds. See also dichloro-(2-chlorvinyl)arsine.
Fire Hazard: Flammable liquid. Keep away from powerful oxidizers.
Spontaneous Heating: Yes.
Disaster Hazard: See arsenic compounds.
Countermeasures
To Fight Fire: Water, water spray, foam, dry chemical carbon dioxide (Section 6).
Storage and Handling: Section 7.

DIMETHYL-β-CHLOROETHYLAMINE
General Information
Description: Liquid.
Formula: $(CH_3)_2NC_2H_4Cl$.
Constants: Mol wt: 107.6, vap. d.: 3.72.
Hazard Analysis
Toxic Hazard Rating:
  Acute Local: Irritant 3; Ingestion 3; Inhalation 3.
  Acute Systemic: U.
  Chronic Local: U.
  Chronic Systemic: U.
Disaster Hazard: Dangerous; when heated to decomposition, it emits highly toxic fumes of chlorides.
Countermeasures
Ventilation Control: Section 2.

Personnel Protection: Section 3.
Storage and Handling: Section 7.

O,O-DIMETHYL-O-(3-CHLORO-4-NITROPHENYL) THIOPHOSPHATE. See chlorthion.

DIMETHYL CYANAMIDE
General Information
Description: Colorless, mobile liquid.
Formula: $(CH_3)_2NCN$.
Constants: Mol wt: 74.09, mp: $-41.0°C$, bp: $160°C$, flash p: $160°F$ (T.C.C.), d: 0.8768 at $30°C$, vap. press.: 40 mm at $80°C$, vap. d.: 2.55.
Hazard Analysis
Toxicity: Details unknown. See also cyanides.
Fire Hazard: Moderate, when exposed to heat or flame.
Disaster Hazard: Dangerous; see cyanides; it will react with water or steam to produce toxic and flammable vapors; can react with oxidizing materials.
Countermeasures
Storage and Handling: Section 7.
To Fight Fire: Foam, carbon dioxide, dry chemical or carbon tetrachloride (Section 6).

p-DIMETHYLCYCLOHEXANE
General Information
Synonym: Hexahydroxylene.
Description: Liquid.
Formula: $(CH_3)_2C_6H_{10}$.
Constants: Mol wt: 112.21, mp: $-86°C$, bp: $119.5°C$, flash p: $52°F$ (C.C.), d: 0.77, vap. press.: 10 mm at $10.2°C$, vap. d.: 3.86.
Hazard Analysis
Toxicity: See cycloparaffins.
Fire Hazard: Dangerous, when exposed to heat or flame; can react vigorously with oxidizing materials.
Spontaneous Heating: No.
Explosion Hazard: Unknown.
Disaster Hazard: Dangerous. Keep away from heat and open flame.
Countermeasures
Ventilation Control: Section 2.
To Fight Fire: Foam, carbon dioxide, dry chemical or carbon tetrachloride (Section 6).
Storage and Handling: Section 7.
Shipping Regulations: Section 11.
   IATA: Flammable liquid, red label, 1 liter (passenger), 40 liters (cargo).

DIMETHYL 1-CYCLOHEXENE-1,2-DICARBOXYLIC ACID. See dimethyl tetrahydrophthalate.

DIMETHYL DECALIN
General Information
Formula: $C_{12}H_{18}$.
Constants: Mol wt: 162, bp: $455°F$, d: 1.0, flash p: $184°F$, autoign. temp.: $453°F$, lel: 0.7% at $200°F$, uel: 5.3% at $300°F$.
Hazard Analysis
Toxic Hazard Rating: U.
Fire Hazard: Moderate, when exposed to heat or flame.
Countermeasures
Storage and Handling: Section 7.

1,1-DIMETHYLDIBORANE
General Information
Description: Colorless gas decomposed by water.
Formula: $B_2H_4(CH_3)_2$.
Constants: Mol wt: 55.7, mp: $-150°C$, bp: $-2.6°C$.
Hazard Analysis
Toxicity: Details unknown. See boron compounds and boron hydrides.
Fire Hazard: Details unknown. See also boron hydrides.
Countermeasures
Storage and Handling: Section 7.

1,2-DIMETHYLDIBORANE
General Information
Description: Colorless gas decomposed by water.
Formula: $B_2H_4(CH_3)_2$.
Constants: Mol wt: 55.7, mp: $-125°C$, bp: $-49°C$.
Hazard Analysis and Countermeasures
Details unknown. See boron hydrides.

DIMETHYL 1,2-DIBROMO-2,2-DICHLORETHYL PHOSPHATE
General Information
Synonym: "DiBrom".
Hazard Analysis
Toxic Hazard Rating: U. Limited animal experiments suggest high toxicity.
TLV: ACGIH (recommended): 3 milligrams per cubic meter of air.
Disaster Hazard: Dangerous. See phosphorus compounds and chlorides.

DIMETHYL DIBROMO DIPYRIDINE TIN
General Information
Description: Solid.
Formula: $(CH_3)_2SnBr_2(C_5H_5N)_2$.
Constants: Mol wt: 466.8, mp: $172°C$.
Hazard Analysis
Toxicity: Details unknown. See tin compounds.
Disaster Hazard: Dangerous. See bromides.

2,5-DIMETHYL-2,5-DI-(tert-BUTYLPEROXY) HEXANE
General Information
Description: Colorless to light yellow liquid. Insoluble in water, soluble in many organic solvents.
Formula: $C_{16}H_{34}O_4$.
Constants: Mol wt: 290.5, d: 0.85, fp: $8°C$, flash p: $>180°F$ (MOC), bp: $250°C$.
Hazard Analysis and Countermeasures
Details unknown. See peroxides, organic.

2,5-DIMETHYL-2,5-DI(tert-BUTYL PEROXY) HEXYNE-3
General Information
Description: Contains 50% inert filler, free-flowing white powder.
Formula: $C_{16}H_{30}O_4$.
Constants: Mol wt: 286.4, fp: 9.7, bp: $240°C$.
Hazard Analysis and Countermeasures
See peroxides, organic.
Shipping Regulations:
I.C.C. Classification: Oxidizing materials N.O.S.
Rail and Freight: Yellow label chemicals N.O.I.B.N.
Parcel Post: Prohibited.

DIMETHYLDICHLOROSILANE
General Information
Description: Liquid; decomposes in water.
Formula: $(CH_3)_2SiCl_2$.
Constants: Mol wt: 129.1, mp: $-86°C$, bp: $70°C$, lel: 3.4%, uel: $>9.5\%$, vap. d.: 4.44, d: 1.1.
Hazard Analysis
Toxicity: Probably high. See chlorosilanes.
Fire Hazard: Dangerous, when exposed to heat or flame (Section 6).
Explosion Hazard: Moderate, when exposed to flame.
Disaster Hazard: Dangerous; see chlorides; it will react with water or steam to produce heat and toxic and corrosive fumes.
Countermeasures
Ventilation Control: Section 2.

Personnel Protection: Section 3.
Storage and Handling: Section 7.
Shipping Regulations: Section 11.
Coast Guard Classification: Inflammable liquid; red label.
I.C.C.: Flammable liquid, red label, 40 liters.
IATA: Flammable liquid, red label, not acceptable (passenger), 40 liters (cargo).

DIMETHYLDICHLOROVINYL PHOSPHATE
General Information
Synonym: DDVP.
Hazard Analysis
Toxicity: Highly toxic. A cholinesterase inhibitor. See parathion.
TLV: ACGIH (recommended) 1 milligram per cubic meter of air. May be absorbed via the skin.
Disaster Hazard: Dangerous. See phosphates.
Countermeasures
See parathion.

DIMETHYL DICHLORO DIPYRIDINE TIN
General Information
Description: Solid.
Formula: $(CH_3)_2(C_5H_5N)_2SnCl_2$.
Constants: Mol wt: 377.9, mp: $163°C$.
Hazard Analysis
Toxicity: Details unknown. See tin compounds.
Disaster Hazard: Dangerous. See chlorides.

DIMETHYLDIDODECYLAMMONIUM CHLORIDE
General Information
Description: Solid.
Formula: $(CH_3)_2(C_{12}H_{25})_2NCl$.
Constants: Mol wt: 418.2, vap. d.: 14.4.
Hazard Analysis
Toxic Hazard Rating:
Acute Local: Irritant 2; Ingestion 2; Inhalation 2.
Acute Systemic: U.
Chronic Local: U.
Chronic Systemic: U.
Disaster Hazard: Dangerous; see chlorides.
Countermeasures
Ventilation Control: Section 2.
Personnel Protection: Section 3.
Storage and Handling: Section 7.

DIMETHYLDIETHYLTIN
General Information
Description: Colorless liquid insoluble in water.
Formula: $(CH_3)_2Sn(C_2H_5)_2$.
Constants: Mol wt: 206.9, d: 1.2319 at $19°C$, mp: $< -13°C$, bp: $146°C$.
Toxicity: Details unknown. See tin compounds.

5,5-DIMETHYLDIHYDRORESORCINOL DIMETHYLCARBAMATE. See dimetan.

DIMETHYLDIISOBUTYLTIN
General Information
Description: Solid.
Formula: $(CH_3)_2Sn(C_4H_9)_2$.
Constants: Mol wt: 263.0, d: 1.1179 at $20°C$, bp: $85°C$ at 16.5 mm.
Hazard Analysis
Toxicity: Details unknown. See tin compounds.

TOXIC HAZARD RATING CODE (For detailed discussion, see Section 1.)

0 NONE: (a) No harm under any conditions; (b) Harmful only under unusual conditions or overwhelming dosage.

1 SLIGHT: Causes readily reversible changes which disappear after end of exposure.

2 MODERATE: May involve both irreversible and reversible changes; not severe enough to cause death or permanent injury.

3 HIGH: May cause death or permanent injury after very short exposure to small quantities.

U UNKNOWN: No information on humans considered valid by authors.

o,o-DIMETHYL-o,p-(DIMETHYL SULFAMOYL) PHENYL PHOSPHOROTHIOATE
General Infirmation
Synonym: Famphur.
Description: Crystalline powder; very soluble in chloroform and carbon tetrachloride; slightly soluble in water.
Formula: $(CH_3O)_2P(S)OC_6H_4SO_2N(CH_3)_2$.
Constants: Mol wt: 325, mp: 55°C.
Hazard Analysis
Toxic Hazard Rating: U. A cholinesterase inhibitor. See parathion.
Disaster Hazard: Dangerous. See phosphates.

DIMETHYLDINITROOXAMIDE
General Information
Description: Crystals.
Formula: $(CH_3)_2NCOCON(NO_2)_2$.
Constant: Mol wt: 206.1.
Hazard Analysis
Toxicity: Details unknown. See amides and nitrates.
Fire Hazard: See nitrates.
Explosion Hazard: Severe, when shocked or exposed to heat. See also explosives, high.
Disaster Hazard: Dangerous; shock will explode it; when heated, it emits highly toxic fumes; can react vigorously with reducing materials.
Countermeasures
Storage and Handling: Section 7.

DIMETHYLDIOXANE
General Information
Description: Water-white liquid.
Formula: $OCH(CH_3)CH_2OCH_2CH(CH_3)$.
Constants: Mol wt: 116.16, bp: 117.5°C, flash p: 75°F, d: 0.9268, vap. press.: 15.4 mm at 20°C, vap. d.: 4.0.
Hazard Analysis
Toxic Hazard Rating:
  Acute Local: Irritant 2; Ingestion 2; Inhalation 2.
  Acute Systemic: Ingestion 2; Inhalation 2.
  Chronic Local: U.
  Chronic Systemic: U.
Fire Hazard: Dangerous, when exposed to heat or flame; can react vigorously with oxidizing materials.
Countermeasures
To Fight Fire: Foam, carbon dioxide, dry chemical or carbon tetrachloride (Section 6).
Ventilation Control: Section 2.
Personnel Protection: Section 3.
Storage and Handling: Section 7.

3,6-DIMETHYL-2,5-p-DIOXANEDIONE
General Information
Synonym: Lactide.
Description: Pale yellow, crystalline solid.
Formula: $C_6H_8O_4$.
Constants: Mol wt: 144.1, mp: 96–104°C.
Hazard Analysis
Toxicity: Unknown.
Disaster Hazard: Slight; Section 6.
Countermeasures
Storage and Handling: Section 7.

2,6-DIMETHYL-m-DIOXAN-4-yl-ACETATE
General Information
Synonym: Dimethoxane; 6-acetoxy-2,4-dimethyl-m-dioxane.
Description: Clear yellow to light amber liquid; soluble in or miscible with water and organic solvents.
Formula: $CH_3COOC_4H_5O_2(CH_3)_2$.
Constants: Mol wt: 174, d: 1.068–1.075 (25/25°C), bp: 66–68°C at 3 mm, fp: < −25°C.
Hazard Analysis
Toxic Hazard Rating: U. Low toxicity and no irritant effects except in undiluted form.

N,N-DIMETHYL-2,2-DIPHENYL ACETAMIDE
General Information
Synonym: Diphenamid; Dymid.
Description: White solid; very slightly soluble in water; moderately soluble in acetone, dimethyl formamide, and phenyl cellosolve.
Formula: $(C_6H_5)_2CHCON(CH_3)_2$.
Constants: Mol wt: 303, mp: 134.5–135.5°C.
Hazard Analysis
Toxic Hazard Rating: U. Animal experiments show moderately high toxicity. See also amides.

1,3-DIMETHYL 1,3-DIPHENYL CYCLOBUTANE
General Information
Description: Insoluble in water.
Formula: $(C_6H_5CH_3)_2(CH_2)_2$.
Constant: Mol wt: 236, mp: 120°F, d: 1.0 at 122°F, bp: 585–588°F, flash p: 289°F.
Hazard Analysis
Toxic Hazard Rating: U.
Fire Hazard: Slight, when exposed to heat or flame.
Countermeasures
To Fight Fire: Water, foam (Section 6).
Storage and Handling: Section 7.

1,1'-DIMETHYL-4,4'-DIPYRIDINIUM   DICHLORIDE.
See paraquat.

DIMETHYL DISULFIDE. See 2,3-dithiabutane.

DIMETHYL DITHIOPHOSPHATE
Hazard Analysis
Toxic Hazard Rating: An insecticide. Similar to parathion. See parathion.

DIMETHYL DITHIOPHOSPHATE OF DIETHYL MERCAPTOSUCCINATE
General Information
Synonym: Malathion.
Description: Brown to yellow liquid. Characteristic odor. Miscible in organic solvents, slightly water soluble.
Formula: $C_{10}H_{19}O_6PS_2$.
Constants: Mol wt: 330, d: 1.23 at 25°/4°C, mp: 2.9°C, bp: 156°C at 0.7 mm.
Hazard Analysis
Toxicology: Highly toxic. Has caused allergic sensitization of the skin. Details unknown. An organic phosphate cholinesterase inhibitor. Less toxic than parathion. Used as a food additive permitted in the feed and drinking water of animals and or for the treatment of food producing animals (Section 10).
TLV: ACGIH (recommended) 15 milligrams per cubic meter of air. May be absorbed via intact skin.
Disaster Hazard: Dangerous. See phosphates.
Countermeasures
See parathion.

DIMETHYLENEIMINE. See ethylenimine.

DIMETHYLENE METHANE. See allene.

DIMETHYL ETHANOLAMINE. See dimethylaminoethanol.

DIMETHYL ETHANOL OCTADECYL AMMONIUM CHLORIDE
General Information
Description: Solid.
Formula: $(CH_3)_2(C_2H_5O)(C_{18}H_{37})NCl$.
Constants: Mol wt: 378.1, vap. d.: 13.0.
Hazard Analysis
Toxic Hazard Rating:
  Acute Local: Irritant 2; Ingestion 2; Inhalation 2.
  Acute Systemic: U.
  Chronic Local: U.
  Chronic Systemic: U.
Disaster Hazard: Dangerous, see chlorides.

Countermeasures
Ventilation Control: Section 2.
Personnel Protection: Section 3.
Storage and Handling: Section 7.

DIMETHYL ETHER
General Information
Synonyms: Methyl ether; methyl oxide.
Description: Colorless liquid.
Formula: CH_3OCH_3.
Constants: Mol wt: 46.07, mp: $-138.5°C$, bp: $-23.7°C$,
   lel: 3.5%, uel: 18%, flash p: $-42°F$ (C.C.), d: 0.661,
   autoign. temp.: 662°F, vap. d.: 1.59.
Hazard Analysis
Toxic Hazard Rating:
   Acute Local: Ingestion 2; Inhalation 2.
   Acute Systemic: Inhalation 2; Skin Absorption 2.
   Chronic Local: U.
   Chronic Systemic: U.
Toxicology: Effects similar to those of ethyl ether.
Fire Hazard: Highly dangerous when exposed to heat or
   flame.
Spontaneous Heating: No.
Explosion Hazard: Dangerous, when exposed to flame,
   sparks, etc.
Disaster Hazard: Highly dangerous. Keep in closed con-
   tainer away from heat and open flame.
Countermeasures
Ventilation Control: Section 2.
To Fight Fire: Carbon dioxide, dry chemical or carbon
   tetrachloride (Section 6).
Storage and Handling: Section 7.
Shipping Regulations: Section 11.
   I.C.C.: Flammable gas; red gas label, 300 pounds.
   Coast Guard Classification: Inflammable gas; red gas
   label.
   IATA: Flammable gas, red label, not acceptable (pas-
   senger), 140 kilograms (cargo).
MCA warning label.

DIMETHYL ETHYL CARBINOL. See tert-amyl alcohol.

DIMETHYLETHYLENE. See cis-butene-2.

DIMETHYLETHYLENE BIS-GLYCOLATE
General Information
Description: Liquid.
Formula: $(CH_2OCH_2CO_2CH_3)_2$.
Constants: Mol wt: 206.2, bp: 145°–146°C at 4 mm, flash
   p: 290°F (O.C.), d: 1.1994 at 25°/25°C, vap. d.: 7.11.
Hazard Analysis
Toxicity: Details unknown. See esters.
Fire Hazard: Slight, when exposed to heat or flame; can
   react with oxidizing materials.
Countermeasures
To Fight Fire: Water, foam, carbon dioxide, dry chemical
   or carbon tetrachloride (Section 6).
Storage and Handling: Section 7.

α,α-**DIMETHYLETHYL NITRITE.** See tert-butyl nitrite.

2,3-DIMETHYL-3-ETHYLPENTANE
General Information
Description: Liquid.
Formula: C_9H_{20}.
Constants: Mol wt: 128.25, bp: 142°C, flash p: 47°F, d:
   0.754 at 20°C, vap. d.: 4.43.

Hazard Analysis
Toxicity: Details unknown. Probably irritant and narcotic
   in high concentrations.
Fire Hazard: Dangerous, when exposed to heat or flame;
   can react vigorously with oxidizing materials.
Explosion Hazard: Unknown.
Disaster Hazard: Dangerous! Keep away from heat and
   open flame!
Countermeasures
Ventilation Control: Section 2.
To Fight Fire: Foam, carbon dioxide, dry chemical or car-
   bon tetrachloride (Section 6).
Storage and Handling: Section 7.

DIMETHYL ETHYL PROPYL TIN
General Information
Description: A powder.
Formula: $(CH_3)_2(C_2H_5)(C_3H_7)Sn$.
Constants: Mol wt: 221, d: 1.2014 at 20/20°C, bp: 150°C.
Hazard Analysis and Countermeasures
See tin compounds.

DIMETHYL ETHYL TIN IODIDE
General Information
Description: Crystals.
Formula: $(CH_3)_2(C_2H_5)SnI$.
Constants: Mol wt: 305, d: 2.026 at 20/20°C, bp: 78°C at
   11 mm.
Hazard Analysis and Countermeasures
See tin compounds and iodides.

DIMETHYL FLUOROPHOSPHATE
General Information
Description: Liquid.
Formula: $(CH_3O)_2POF$.
Constants: Mol wt: 128.1, mp: low, bp: 149°C, d: 1.28, vap.
   d.: 4.42.
Hazard Analysis
Toxicity: Details unknown. See fluorides.
   Animal experiments show high toxicity.
Disaster Hazard: Dangerous. See fluorides and phosphates.
Countermeasures
Storage and Handling: Section 7.

DIMETHYL FORMAMIDE
General Information
Description: Colorless, mobile liquid.
Formula: $(CH_3)_2NCHO$.
Constants: Mol wt: 73.1, bp: 152.8°C, lel: 2.2% at 100°C,
   uel: 15.2% at 100°C, flash p: 136°F, fp: $-61°C$, d:
   0.9445 at 25°/4°C, autoign. temp.: 833°F, vap.
   press. 3.7 mm at 25°C, vap. d.: 2.51.
Hazard Analysis
Toxic Hazard Rating:
   Acute Local: Irritant 2; Inhalation 2.
   Acute Systemic: Ingestion 1; Inhalation 2; Skin Absorp-
   tion 2.
   Chronic Local: Irritant 2.
   Chronic Systemic: Ingestion 3; Inhalation 3; Skin Ab-
   sorption 3.
Toxicology: Highly irritating. Prolonged inhalation of 100
   ppm has produced liver damage in experimental animals.
TLV: ACGIH (recommended) 10 parts per million of air;
   30 milligrams per cubic meter of air. May be absorbed
   via the skin.
Fire Hazard: Moderate, when exposed to heat or flame; can
   react with oxidizing materials.

TOXIC HAZARD RATING CODE (For detailed discussion, see Section 1.)

0 NONE: (a) No harm under any conditions; (b) Harmful only under unusual conditions or overwhelming dosage.

1 SLIGHT: Causes readily reversible changes which disappear after end
   of exposure.

2 MODERATE: May involve both irreversible and reversible changes;
   not severe enough to cause death or permanent injury.

3 HIGH: May cause death or permanent injury after very short exposure
   to small quantities.

U UNKNOWN: No information on humans considered valid by authors.

Explosion Hazard: Moderate, when exposed to flame.

Caution: Avoid contact with halogenated hydrocarbons and inorganic nitrates.

Countermeasures

Ventilation Control: Section 2.

To Fight Fire: Foam, carbon dioxide, dry chemical or carbon tetrachloride (Section 6).

Personnel Protection: Section 3.

Storage and Handling: Section 7.

DIMETHYL FORMOCARBOTHIALDINE. See mylone.

2,5-DIMETHYLFURAN

General Information

Description: Colorless liquid.

Formula: $C_4H_2O(CH_3)_2$.

Constants: Mol wt: 96.12, bp: 94°C, flash p: 45°F, d: 0.9026 at 17.7°/4°C, vap. d.: 3.31.

Hazard Analysis

Toxic Hazard Rating:

Acute Local: Irritant 1; Ingestion 1; Inhalation 1.

Acute Systemic: U.

Chronic Local: U.

Chronic Systemic: U.

Fire Hazard: Dangerous, when exposed to heat or flame; can react vigorously with oxidizing materials.

Explosion Hazard: Unknown.

Disaster Hazard: Dangerous! Keep away from heat and open flame!

Countermeasures

Ventilation Control: Section 2.

To Fight Fire: Alcohol foam, foam, carbon dioxide, dry chemical or carbon tetrachloride (Section 6).

Personal Hygiene: Section 3.

Storage and Handling: Section 7.

DIMETHYL GALLIUM AMIDE

General Information

Description: White crystals.

Formula: $Ga(CH_3)_2NH_2$.

Constant: Mol wt: 115.8.

Hazard Analysis and Countermeasures

See gallium compounds and amides.

DIMETHYL GALLIUM CHLORIDE DIAMMINE

General Information

Description: White crystals, decomposed by water.

Formula: $Ga(CH_3)_2Cl \cdot 2NH_3$.

Constants: Mol wt: 169.3, mp: 112°C.

Hazard Analysis and Countermeasures

See gallium compounds and chlorides.

DIMETHYL GALLIUM CHLORIDE MONAMMINE

General Information

Description: White crystals, decomposed by water.

Formula: $Ga(CH_3)_2Cl \cdot NH_3$.

Constants: Mol wt: 152.3, mp: 54°C.

Hazard Analysis and Countermeasures

See gallium compounds and chlorides.

DIMETHYLGLYCOL PHTHALATE

Genejal Information

Synonym: Dimethyl "cellosolve" phthalate.

Description: Liquid.

Formula: $C_6H_4[COO(CH_2)_2OCH_3]_2$.

Constants: Mol wt: 282.3, bp: 230°C, flash p: 369°F (C.C.), d: 1.8, vap. d.: 9.72.

Hazard Analysis

Toxicity: Details unknown. See also glycols.

Fire Hazard: Slight, when exposed to heat or flame; can react with oxidizing materials.

Spontaneous Heating: No.

Countermeasures

To Fight Fire: Water, carbon dioxide, dry chemical or carbon tetrachloride (Section 6).

Storage and Handling: Section 7.

2,6-DIMETHYL-4-HEPTANONE. See diisobutyl ketone.

2,6-DIMETHYL-2-5-HEPTADIEN-4-ONE. See phorone.

2,5-DIMETHYLHEPTANE

General Information

Description: Liquid.

Formula: C_9H_{20}.

Constants: Mol wt: 128.25, bp: 136°C, flash p: 73.5°F, d: 0.715 at 20°C, vap. d.: 4.42.

Hazard Analysis

Toxicity: Unknown.

Fire Hazard: Dangerous, when exposed to heat or flame; can react vigorously with oxidizing materials.

Explosion Hazard: Unknown. Probably irritant and narcotic in high concentrations.

Countermeasures

Ventilation Control: Section 2.

To Fight Fire: Foam, carbon dioxide, dry chemical or carbon tetrachloride (Section 6).

Storage and Handling: Section 7.

3,5-DIMETHYLHEPTANE

General Information

Description: Liquid.

Formula: C_9H_{20}.

Constants: Mol wt: 128.25, bp: 136°C, flash p: 73.5°F, d: 0.723 at 20°C, vap. press.: 9.5 mm at 25°C, vap. d.: 4.42.

Hazard Analysis

Toxicity: Unknown. Probably irritant and narcotic in high concentrations.

Fire Hazard: Dangerous, when exposed to heat or flame; can react vigorously with oxidizing materials.

Explosion Hazard: Unknown.

Countermeasures

Ventilation Control: Section 2.

To Fight Fire: Carbon dioxide, dry chemical or carbon tetrachloride (Section 6).

Storage and Handling: Section 7.

4,4-DIMETHYLHEPTANE

General Information

Description: Liquid.

Formula: C_9H_{20}.

Constants: Mol wt: 128.25, bp: 135.2°C, flash p: 73.5°F, d: 0.72 at 25°/4°C, vap. press.: 10.4 mm at 25°C, vap. d.: 4.42.

Hazard Analysis

Toxicity: Unknown. Probably irritant and narcotic in high concentrations.

Fire Hazard: Dangerous, when exposed to heat or flame; can react vigorously with oxidizing materials.

Explosion Hazard: Unknown.

Countermeasures

Ventilation Control: Section 2.

To Fight Fire: Carbon dioxide, dry chemical or carbon tetrachloride (Section 6).

Storage and Handling: Section 7.

2,6-DIMETHYLHEPTANOL-4. See diisobutylcarbinol.

2,6-DIMETHYLHEPTENE-3 (cis and trans ISOMERS)

General Information

Description: A clear liquid.

Formula: C_9H_{18}.

Constants: Mol wt: 126.23, bp: 128.5–129°C, flash p: 70°F (T.O.C.), d: 0.722 at 15.5°/15.5°C, vap. press.: 28.4 mm at 38°C, vap. d.: 4.38.

Hazard Analysis

Toxicity: Details unknown. Probably irritant and narcotic in high concentrations.

Fire Hazard: Dangerous, when exposed to heat or flame; can react vigorously with oxidizing materials.

Countermeasures

To Fight Fire: Foam, carbon dioxide, dry chemical or carbon tetrachloride (Section 6).
Ventilation Control: Section 2.
Storage and Handling: Section 7.

2,5-DIMETHYLHEXADIENE-2,4

General Information

Description: Liquid.
Formula: $(CH_3)_2CCHCHC(CH_3)_2$.
Constants: Mol wt: 110.12, mp: $-91.3°C$, bp: $102.5°C$, d: 0.762 at $20°/20°C$, vap. d.: 3.8.

Hazard Analysis

Toxicity: Details unknown. Probably irritant and narcotic in high concentrations.
Fire Hazard: Moderate; can react with oxidizing materials.

Countermeasures

To Fight Fire: Foam, carbon dioxide, dry chemical or carbon tetrachloride (Section 6).
Storage and Handling: Section 7.

2,3-DIMETHYLHEXANE

General Information

Description: A clear liquid.
Formula: C_8H_{18}.
Constants: Mol wt: 114.23, bp: $116°C$, flash p: $45°F$, d: 0.716 at $15.5°/15.5°C$, vap. d.: 4.1, autoign. temp.: $820°F$.

Hazard Analysis

Toxicity: Details unknown. Probably irritant and narcotic in high concentrations.
Fire Hazard: Moderate, when exposed to heat or flame; can react vigorously with oxidizing materials.
Explosion Hazard: Unknown.

Countermeasures

To Fight Fire: Foam, carbon dioxide, dry chemical or carbon tetrachloride (Section 6).
Storage and Handling: Section 7.

2,4-DIMETHYLHEXANE

General Information

Description: A liquid.
Formula: C_8H_{18}.
Constants: Mol wt: 114.23, bp: $109°C$, flash p: $50°F$, d: 0.705 at $15.5°/15.5°C$, vap. d.: 3.9.

Hazard Analysis

Toxicity: Unknown. Probably irritant and narcotic in high concentrations.
Fire Hazard: Dangerous, when exposed to heat or flame; can react vigorously with oxidizing materials.

Countermeasures

To Fight Fire: Foam, carbon dioxide, dry chemical or carbon tetrachloride (Section 6).
Ventilation Control: Section 2.
Storage and Handling: Section 7.

2,5-DIMETHYL HEXANE-2,5-DIHYDROPEROXIDE

General Information

Description: Fine white crystals. Insoluble in hydrocarbons; slightly soluble in water, esters, glycerine; soluble in other organic solvents.
Formula: $C_8H_{18}O_4$.
Constants: Mol wt: 178.2, mp: $104°C$.

Hazard Analysis and Countermeasures

See peroxides, organic.
Shipping Regulations.
   I.C.C.: Oxidizing material, yellow label, 25 pounds.

Freight and Rail Express: Yellow label, chemicals N.O.I.B.N.
Parcel Post: Prohibited.
   IATA: Oxidizing material, yellow label, not acceptable (passenger), 12 kilograms (cargo).

2,5-DIMETHYLHEXANE-2,5-DIOL

General Information

Description: Crystals.
Formula: $(CH_3)_2(OH)C(CH_2)_2C(OH)(CH_3)_2$.
Constants: Mol wt: 146.14, mp: $88-89°C$, vap. d.: 5.03.

Hazard Analysis

Toxicity: Details unknown.

DIMETHYLHEXYNEDIOL

Geneeal Information

Description: White crystals.
Formula: $C_8H_{14}O_2$.
Constants: Mol wt: 142.2, mp: $94-95°C$, bp: $205-206°C$, vap. d.: 4.9.

Hazard Analysis

Toxicity: Details unknown.
Fire Hazard: Slight; Section 6.

Countermeasures

Storage and Handling: Section 7.

2,5-DIMETHYL-1,2,6-HEXANETRIOL

General Information

Formula: $CH_2(OH)CH(CH_3)CH_2CH_2C(CH_3)(OH)\cdot CH_2(OH)$.
Constant: Mol wt: 158.

Hazard Analysis

Toxic Hazard Rating: U. Limited animal experiments suggest low toxicity.

DIMETHYLHEXYNOL

General Information

Synonym: 3,5-Dimethyl-1-hexyn-3-ol.
Description: Colorless liquid; camphor-like odor.
Formula: $CH_3CH(CH_3)CH_2C(CH_3)(OH)C\colon CH$.
Constants: Mol wt: 126.5, bp: $150-151°C$, fp: $-68°C$, flash p: $134°F$ (T.O.C.), d: 0.8545 at $20°/20°C$.

Hazard Analysis

Toxicity: Unknown.
Fire Hazard: Moderate, when exposed to heat or flame; can react with oxidizing materials.

Countermeasures

To Fight Fire: Foam, carbon dioxide, dry chemical or carbon tetrachloride (Section 6).
Storage and Handling: Section 7.

DIMETHYL HYDANTOIN

General Information

Synonym: α-Ureido-isobutyric acid lactam.
Description: White crystalline solid.
Formula: $C_5H_8O_2N_2$.
Constants: Mol wt: 128.13, mp: $178°C$, vap. d.: 4.4.

Hazard Analysis

Toxicity: Details unknown. Hydantoins are related to barbiturates and have a depressant action on the central nervous system.
Disaster Hazard: Slight; Section 6.

Countermeasures

Storage and Handling: Section 7.

DIMETHYL HYDANTOIN-FORMALDEHYDE RESIN

General Information

Description: Colorless to lightly yellow, brittle lumps; odorless or faint caramel-like odor.

TOXIC HAZARD RATING CODE (For detailed discussion, see Section 1.)

0 NONE: (a) No harm under any conditions; (b) Harmful only under unusual conditions or overwhelming dosage.

1 SLIGHT: Causes readily reversible changes which disappear after end of exposure.

2 MODERATE: May involve both irreversible and reversible changes; not severe enough to cause death or permanent injury.

3 HIGH: May cause death or permanent injury after very short exposure to small quantities.

U UNKNOWN: No information on humans considered valid by authors.

Constants: Mol wt: 240–300 (average), mp: 60°C (minimum), d: 1.30.

Hazard Analysis

Toxic Hazard Rating:

Acute Local: Irritant 1; Allergen 1; Ingestion 1; Inhalation 1.

Acute Systemic: U.

Chronic Local: Allergen 1.

Chronic Systemic: U.

Toxicology: This resin contains a small percentage of formaldehyde (about 0.3% maximum) which may cause dermatitis in individuals sensitive to formaldehyde. When crushed or ground it gives a fine dust which may pass through ordinary respirators irritating mucous membranes and leading to sore throat, coughing and occasionally vomiting.

In case of contract with the body, individuals are advised to flush skin or eyes with water. In grinding or crushing operations, adequate ventilation should be provided.

Disaster Hazard: Slight; Section 6.

Countermeasures

Personal Hygiene: Section 3.

Storage and Handling: Section 7.

asym-DIMETHYLHYDRAZINE
General Information

Synonym: 1,1-Dimethylhydrazine.

Description: Colorless liquid, ammonia-like odor.

Formula: $(CH_3)_2NNH_2$.

Constants: Mol wt: 60.1, bp: 63.3°C, fp: −58°C, flash p: 5°F, d: 0.782 at 25°/4°C, vap. press.: 157 mm at 25°C, vap. d.: 1.94, autoign. temp.: 480°F, lel: 2%, uel: 95%.

Hazard Analysis

Toxic Hazard Rating:

Acute Local: Irritant 3; Ingestion 3; Inhalation 3.

Acute Systemic: Ingestion 3; Inhalation 3.

Chronic Local: U.

Chronic Systemic: Ingestion 2; Inhalation 2.

TLV: ACGIH (recommended); 0.5 parts per million in air; 1 milligram per cubic meter of air. May be absorbed via intact skin.

Fire Hazard: Dangerous, when exposed to heat or flame.

Explosion Hazard: Unknown.

Disaster Hazard: Highly dangerous; when heated to decompoistion, it emits highly toxic fumes; can react vigorously with oxidizing materials.

Countermeasures

Ventilation Control: Section 2.

To Fight Fire: Alcohol foam, carbon dioxide, dry chemical or carbon tetrachloride (Section 6).

Personnel Protection: Section 3.

Storage and Handling: Section 7.

Shipping Regulations: Section 11.

I.C.C.: Flammable liquid, red label, 5 pints.

IATA: Flammable liquid, red label, not acceptable (passenger), 2½ liters (cargo).

N,N-DIMETHYLHYDROXYACETAMIDE
General Information

Description: Crystals.

Formula: $HOCH_2CON(CH_3)_2$.

Constants: Mol wt: 103.12, mp: 45°C, bp: 213°C, d: 1.076 at 50°/4°C.

Hazard Analysis

Toxicity: Details unknown. See also amides.

Disaster Hazard: Slight; when heated, it emits toxic fumes.

Countermeasures

Storage and Handling: Section 7.

2,5-DIMETHYL-2-HYDROXYADIPALDEHYDE
General Information

Formula: $OHCCH(CH_3)(CH_2)_2C(CH_3)OHCHO$.

Constant: Mol wt: 158.

Hazard Analysis

Toxic Hazard Rating: U. Limited animal experiments suggest low toxicity. See also aldehydes.

DIMETHYL HYDROXYBORINE. See dimethylboric acid.

O,O-DIMETHYL-1-HYDROXY-2,2,2-TRI-CHLOROETHYL PHOSPHONATE. See dipterex.

DIMETHYLISOPROPANOLAMINE
General Information

Description: Liquid.

Formula: $C_3H_7ON(CH_3)_2$.

Constants: Mol wt: 103.2, mp: −85°C, bp: 125.8°C, flash p: 95°F (O.C.), d: 0.86, vap. press.: 9.0 mm at 20°C, vap. d.: 3.55.

Hazard Analysis

Toxicity: Details unknown. See also amines.

Fire Hazard: Moderate, when exposed to heat or flame; can react with oxidizing materials.

Countermeasures

To Fight Fire: Alcohol foam, carbon dioxide, dry chemical or carbon tetrachloride (Section 6).

Ventilation Control: Section 2.

Storage and Handling: Section 7.

N,N-DIMETHYL-m-ISOPROPYLPHENYL CARBAMATE
General Information

Formula: $C_6H_3(CH_3)_2NHCOOHC(CH_3)_2$.

Constant: Mol wt: 207.

Hazard Analysis

Toxic Hazard Rating: U. Limited animal experiments suggest high toxicity.

Disaster Hazard: Dangerous. See carbamates.

DIMETHYL KETOL. See acetoin.

DIMETHYL KETONE. See acetone.

DIMETHYL MALEATE
General Information

Description: Liquid.

Formula: $(:CHCOOCH_3)_2$.

Constants: Mol wt: 144.12, mp: −17.5°C, bp: 205.0°C, flash p: 235°F (O.C.), d: 1.153, vap. press.: 1 mm at 45.7°C, vap. d.: 4.97.

Hazard Analysis

Toxicity: Details unknown. Limited animal experiments suggest moderate toxicity. See also esters and maleic acid.

Fire Hazard: Slight, when exposed to heat or flame; can react with oxidizing materials.

Countermeasures

To Fight Fire: Water, foam, carbon dioxide, dry chemical or carbon tetrachloride (Section 6).

Storage and Handling: Section 7.

DIMETHYLMERCURY
General Information

Description: Colorless liquid; sweet odor, soluble in alcohol.

Formula: $Hg(CH_3)_2$.

Constants: Mol wt: 230.7, d: 3.069, bp: 96°C.

Hazard Analysis and Countermeasures

See mercury compounds, organic.

DIMETHYLMETHANE. See propane.

O,O-DIMETHYL S-(N-METHYL CARBAMOYL METHYL) PHOSPHORODITHIOATE.
General Information

Synonym: Dimethoate; S-methylcarbamoyl methyl o,o-dimethyl phosphorodithioate.

Description: White solid; moderately soluble in water; soluble in most organic solvents except hydrocarbons.

Formula: $(CH_3O)_2PSSCH_2CONHCH_3$.

Constant: Mol wt: 197, mp: 51–52°C.

Hazard Analysis
Toxic Hazard Rating: U. An organic phosphate ester insecticide. A cholinesterase inhibitor. See parathion.
Disaster Hazard: Dangerous. See phosphates.

3,3-DIMETHYL-2-METHYLENENORCAMPHONE.
See camphene.

O,O-DIMETHYL O-[4-(METHYL THIO)m-TOLYL] PHOSPHORTHIOATE
General Information
Synonym: Tenthion; Baytex.
Description: Brown liquid; insoluble in water; soluble in most organic solvents.
Formula: $(CH_3O)_2P(S)OC_6H_3(CH_3)SCH_3$.
Constants: Mol wt: 278, bp: 105°C at 0.01 mm.
Hazard Analysis
Toxic Hazard Rating: U. A cholinesterase inhibitor. See parathion.
Disaster Hazard: Dangerous. See phosphates and sulfur compounds.

2,6-DIMETHYL MORPHOLINE
General Information
Description: Liquid; very soluble in water.
Formula: $OCH(CH_3CH_2NHCH_2CH(CH_3)$.
Constants: Mol wt: 115, d: 0.9346, bp: 146.6°C, fp: −85°C, flash p: 112°F, vap. d.: 4.0.
Hazard Analysis
Toxic Hazard Rating: Limited animal experiments suggest moderate toxicity. See also morpholine.
Fire Hazard: Moderate, when exposed to heat or flame.
Countermeasures
To Fight Fire: Alcohol foam (Section 6).
Storage and Handling: Section 7.

O,O-DIMETHYL-O,p-NITROPHENYL THIOPHOSPHATE
General Information
Synonym: Dimethyl analog of parathion.
Description: Crystals.
Formula: $(CH_3O)_2SPOC_6H_4NO_2$.
Constants: Mol wt: 263.2, vap. d.: 9.1.
Hazard Analysis
Toxic Hazard Rating:
Acute Local: Irritant 2.
Acute Systemic: Ingestion 3; Inhalation 3.
Chronic Local: U.
Chronic Systemic: Ingestion 3; Inhalation 3.
Toxicology: See also parathion.
Disaster Hazard: Dangerous; when heated to decomposition or on contact with acid or acid fumes, it emits highly toxic fumes.
Countermeasures
Ventilation Control: Section 2.
Personnel Protection: Section 3.
First Aid: Section 1.
Storage and Handling: Section 7.

DIMETHYL NITROSAMINE
Hazard Analysis
Toxic Hazard Rating: U. Limited animal experiments show strong irritation and liver injury; also tumors of liver, lung and kidney.

DIMETHYL-p-NITROSOANILINE. See p-nitrosodimethylaniline.

2,6-DIMETHYL-2,6-OCTADIENOL. See citral.

3,7-DIMETHYL-1,6-OCTADIEN-3-OL. See linalool.

3,7-DIMETHYL-2,6-OCTADIEN-1-OL. See geraniol.

3,7-DIMETHYL-3,6-OCTADIEN-1-OL. See geraniol.

DIMETHYL OCTYNEDIOL
General Information
Synonym: 3,6-Dimethyl-4-octyne-3,6-diol.
Description: White crystals.
Formula: $C_{10}H_{18}O_2$.
Constants: Mol wt: 170.3, mp: 55–56°C, bp: 135°C at 20 mm, vap. d.: 5.9.
Hazard Analysis
Toxicity: Details unknown; probably toxic.
Fire Hazard: Slight when heated.
Countermeasures
Storage and Handling: Section 7.

2,2-DIMETHYLOLPROPANOL-1. See trimethylolethane.

O,O-DIMETHYL S-(4-OXOBENZOTRIAZINO-3-METHYL) PHOSPHORODITHIOATE. See guthion.

2,3-DIMETHYL PENTALDEHYDE
General Information
Description: Liquid; slightly soluble in water.
Formula: $CH_3CH_2CH(CH_3)CH(CH_3)CHO$.
Constants: Mol wt: 114, d: 0.8293, bp: 140.5°C, fp: −110°C, flash p: 94°F (O.C.).
Hazard Analysis
Toxic Hazard Rating: U.
Fire Hazard: Moderate, when exposed to heat or flame.
Countermeasures
To Fight Fire: Foam, mist, dry chemical.
Storage and Handling: Section 7.

2,3-DIMETHYL-4-PENTANAL
General Information
Formula: $CH_3CH(CH_3)CH(CH_3)OCHOH$.
Constant: Mol wt: 115.
Hazard Analysis
Toxic Hazard Rating: U. Limited animal experiments suggest low toxicity and irritation. See also aldehydes.

2,3-DIMETHYLPENTANE
General Information
Synonym: Diethyldimethylmethane.
Description: Liquid.
Formula: C_7H_{16}.
Constants: Mol wt: 100.20, mp: −135°C, bp: 89.8°C, d: 0.69 at 15.5/15.5°C, autoign. temp.: 639°F, flash p: <20°F, vap. press.: 40 mm at 13.9°C, vap. d.: 3.45, lel: 1.1%, uel: 6.7%.
Hazard Analysis
Toxicity: Unknown. Probably irritant and narcotic in high concentrations.
Fire Hazard: Dangerous, when exposed to heat or flame.
Explosion Hazard: Unknown.
Disaster Hazard: Dangerous; keep away from heat and open flame; can react vigorously with oxidizing materials.
Countermeasures
Ventilation Control: Section 2.
Storage and Handling: Section 7.
To Fight Fire: Foam, carbon dioxide, dry chemical, or carbon tetrachloride (Section 6).

TOXIC HAZARD RATING CODE *(For detailed discussion, see Section 1.)*

0 NONE: (a) No harm under any conditions; (b) Harmful only under unusual conditions or overwhelming dosage.

1 SLIGHT: Causes readily reversible changes which disappear after end of exposure.

2 MODERATE: May involve both irreversible and reversible changes; not severe enough to cause death or permanent injury.

3 HIGH: May cause death or permanent injury after very short exposure to small quantities.

U UNKNOWN: No information on humans considered valid by authors.

2,4-DIMETHYLPENTANE
General Information
Description: Formula: C_7H_{16}. A clear liquid.
Constants: Mol wt: 100.2, mp: $-123.4°C$, bp: $80.3°C$, fp: $-119.4°C$, flash p: $<20°F$, d: 0.6728 at $20°/4°C$, vap. press.: 8.2 mm at $21°C$, vap. d.: 3.48.
Hazard Analysis
Toxicity: Unknown.
Fire Hazard: Dangerous, when exposed to heat or flame.
Explosion Hazard: Unknown. Probably irritant and narcotic in high concentrations.
Disaster Hazard: Dangerous. Keep away from heat and open flame; can react vigorously with oxidizing materials.
Countermeasures
Ventilation Control: Section 2.
To Fight Fire: Foam, carbon dioxide, dry chemical or carbon tetrachloride (Section 6).
Storage and Handling: Section 7.

2,3-DIMETHYL PENTANOL
General Information
Formula: $CH_3CH(CH_3)CH(CH_3)CH_2CH_2OH$.
Constant: Mol wt: 116.
Hazard Analysis
Toxic Hazard Rating: U. Limited animal experiments suggest moderate toxicity and high eye irritation.

2,4-DIMETHYL-3-PENTANOL. See diisopropyl carbinol.

3,5-DIMETHYLPHENOL. See 3,5-xylenol.

DIMETHYL-p-PHENYLENEDIAMINE
General Information
Synonym: p-Aminodimethyl aniline.
Description: Crystalline mass.
Formula: $C_6H_4NH_2N(CH_3)_2$.
Constants: Mol wt: 136.20, mp: $53°C$, bp: $150°C$ at 17 mm, d: 1.036 at $20/4°C$, vap. d.: 4.69.
Hazard Analysis
Toxic Hazard Rating:
　Acute Local: Irritant 2.
　Acute Systemic: Ingestion 3; Inhalation 3.
　Chronic Local: U.
　Chronic Systemic: Ingestion 3; Inhalation 3.
Disaster Hazard: Moderate, when heated to decomposition it emits toxic fumes.
Countermeasures
Ventilation Control: Section 2.
Personnel Protection: Section 3.
First Aid: Section 1.
Storage and Handling: Section 7.

2,4-DIMETHYL PHENYL MALEIMIDE
Hazard Analysis
Toxic Hazard Rating: U. Limited animal experiments suggest high toxicity and irritation.

1,1-DIMETHYL PHENYL UREA
General Information
Synonym: Fenuron; 3-phenyl-1,1-dimethyl urea.
Description: White crystalline solid; almost insoluble in water; sparingly soluble in hydrocarbon solvents.
Formula: $C_6H_5NHCON(CH_3)_2$.
Constants: Mol wt: 164, mp: 127–129°C.
Hazard Analysis
Toxic Hazard Rating: U. If hydrolyzed it can liberate aniline. See aniline.

DIMETHYLPHOSPHINE
General Information
Description: Colorless liquid.
Formula: $(CH_3)_2PH$.
Constants: Mol wt: 62.1, bp: $25°C$, d: <1, vap. d.: 2.14.
Hazard Analysis
Toxic Hazard Rating:

　Acute Local: U.
　Acute Systemic: Ingestion 3; Inhalation 3.
　Chronic Local: U.
　Chronic Systemic: Ingestion 3; Inhalation 3.
Fire Hazard: Dangerous, when exposed to heat or flame; spontaneously flammable in air.
Explosion Hazard: Unknown.
Disaster Hazard: Dangerous; see phosphates; can react vigorously with oxidizing materials.
Countermeasures
Ventilation Control: Section 2.
To Fight Fire: Foam, carbon dioxide, dry chemical or carbon tetrachloride (Section 6).
Personnel Protection: Section 3.
First Aid: Section 1.
Storage and Handling: Section 7.

o-DIMETHYL PHTHALATE
General Information
Synonym: DMP.
Description: Colorless, odorless liquid.
Formula: $C_6H_4(COOCH_3)_2$.
Constants: Mol wt: 194.18, bp: $283.7°C$, flash p: $295°F$ (C.C.), d: 1.189 at $25°/25°C$, autoign. temp.: $1032°F$, vap. d.: 6.69, vap. press.: 1 mm at $100.3°C$.
Hazard Analysis
Toxic Hazard Rating:
　Acute Local: Irritant 2; Ingestion 2; Inhalation 2.
　Acute Systemic: Ingestion 2.
　Chronic Local: U.
　Chronic Systemic: U.
TLV: ACGIH (tentative) 5 milligrams per cubic meter of air.
Toxicology: Can cause CNS depression as well as local irritation.
Fire Hazard: Slight, when exposed to heat or flame; can react with oxidizing materials.
Spontaneous Heating: No.
Countermeasures
To Fight Fire: Water, foam, carbon dioxide, dry chemical or carbon tetrachloride (Section 6).
Personal Hygiene: Section 3.
Storage and Handling: Section 7.

1,4-DIMETHYLPIPERAZINE
General Information
Description: Colorless, mobile liquid.
Formula: $CH_3N(CH_2CH_2)_2NCH_3$.
Constants: Mol wt: 114.2, d: 0.8565 at $20°/4°C$, flash p: $85°F$ (T.O.C.), bp: approx. $130°C$.
Hazard Analysis
Toxicity: Details unknown. See also trans-2,5-Dimethylpiperazine.
Fire Hazard: Dangerous, when exposed to heat, sparks, powerful oxidizers.
Countermeasures
To Fight Fire: Foam, spray, carbon dioxide, dry chemical.
Storage and Handling: Section 7.

cis-2,5-DIMETHYLPIPERAZINE
General Information
Description: A liquid; typical amine odor.
Formula: $C_6H_{14}N_2$.
Constants: Mol wt: 114.2, mp: $17.5°C$, bp: $164.5°C$ at 746 mm, flash p: $154.5°F$ (C.O.C.), d: 0.9195 at $25°/25°C$.
Hazard Analysis
Toxicology: Details unknown. See also trans-2,5-dimethylpiperazine.
Fire Hazard: Moderate, when exposed to heat or flame; can react with oxidizing materials.
Countermeasures
To Fight Fire: Foam, carbon dioxide, dry chemical or carbon tetrachloride (Section 6).
Storage and Handling: Section 7.

trans-2,5-DIMETHYLPIPERAZINE
General Information
Description: Crystals; typical amine odor.
Formula: HNCH$_2$CH(CH$_3$)NHCH$_2$CH(CH$_3$).
Constants: Mol wt: 114.19, mp: 117.5°C, bp: 161.9°C at 746 mm, flash p: 210°F (O.C.).
Hazard Analysis
Toxicology: Details unknown. Animal experiments show low toxicity. See also piperazine.
Fire Hazard: Moderate, when exposed to heat or flame; can react with oxidizing materials.
Countermeasures
To Fight Fire: Foam, carbon dioxide, dry chemical or carbon tetrachloride (Section 6).
Storage and Handling: Section 7.

2,2-DIMETHYLPROPANE
General Information
Synonym: Neopentane.
Description: Liquid.
Formula: (CH$_3$)$_4$C.
Constants: Mol wt: 72.2, bp: 9.5°C, fp: −18.2°C, flash p: < 20°F, d: 0.590 at 20°/4°C, autoign. temp.: 842°F, vap. press.: 1100 mm at 21°C, vap. d.: 2.48, lel: 1.4%, uel: 7.5%.
Hazard Analysis
Toxicity: Unknown. Propably irritant and narcotic in high concentrations.
Fire Hazard: Dangerous, when exposed to heat or flame; can react vigorously with oxidizing materials.
Explosion Hazard: Unknown.
Countermeasures
Ventilation Control: Section 2.
To Fight Fire: Foam, carbon dioxide, dry chemical or carbon tetrachloride (Section 6).
Storage and Handling: Section 7.
Shipping Regulations: Section 11.
IATA: Flammable liquid, red label, 1 liter (passenger), 40 liters (cargo).

2,2-DIMETHYL-1,3-PROPANEDIOL. See neopentyl glycol.

DIMETHYLPROPYLMETHANE. See 2-methylpentane.

p-(α,α-DIMETHYLPROPYL) PHENOL. See p-tert-amyl phenol.

2,5-DIMETHYLPYRAZINE
General Information
Description: A liquid; pyridine odor.
Formula: C$_6$H$_8$N$_2$.
Constants: Mol wt: 108, mp: 15°C, bp: 154°C at 742 mm, flash p: 147.3°F (C.O.C.), d: 0.9873 at 25°/25°C.
Hazard Analysis
Toxicity: Unknown.
Fire Hazard: Moderate, when exposed to heat or flame; can react with oxidizing materials.
Countermeasures
To Fight Fire: Foam, carbon dioxide, dry chemical or carbon tetrachloride (Section 6).
Storage and Handling: Section 7.

2,6-DIMETHYLPYRIDINE. See 2,6-lutidine.

N-(4,6-DIMETHYL-2-PYRIMIDYL)SULFANILAMIDE. See sulfamethazine.

7,8-DIMETHYL-10-(1²-D-RIBITYL) ISSOALLOXAZINE. See riboflavin.

DIMETHYL SEBACATE
General Information
Synonyms: Methyl sebacate.
Description: Liquid.
Formula: [(CH$_2$)$_4$CO$_2$CH$_3$]$_2$.
Constants: Mol wt: 230.3, mp: 38°C, bp: 293.5°C, flash p: 293°F, d: 0.986 at 30°/25°C, vap. press.: 1 mm at 104.0°C, vap. d.: 7.95.
Hazard Analysis
Toxicity: See esters and methyl alcohol.
Fire Hazard: Slight, when exposed to heat or flame; can react with oxidizing materials.
Countermeasures
To Fight Fire: Foam, carbon dioxide, dry chemical or carbon tetrachloride (Section 6).
Storage and Handling: Section 7.

DIMETHYL SELENIDE
General Information
Synonym: Methyl selenide.
Description: Liquid.
Formula: (CH$_3$)$_2$Se.
Constants: Mol wt: 109.0, bp: 58°C, d: 1.4077 at 14.6°/4°C, vap. d.: 3.75.
Hazard Analysis
Toxicity: See selenium and its compounds.
Fire Hazard: Dangerous, when exposed to heat or flame.
Explosion Hazard: Unknown.
Disaster Hazard: Dangerous; see selenium; will react with water, steam, acid or acid fumes to produce toxic fumes; can react vigorously with oxidizing materials.
Countermeasures
First Aid: Section 1.
To Fight Fire: Foam, carbon dioxide, dry chemical or carbon tetrachloride (Section 6).
Storage and Handling: Section 7.

DIMETHYLSILICANE
General Information
Description: Gas.
Formula: H$_2$Si(CH$_3$)$_2$.
Constants: Mol wt: 60.1, d: 0.68 at −80°C, mp: −150°C, bp: −20°C.
Hazard Analysis and Countermeasures
See silanes.

DIMETHYL SULFATE
General Information
Synonym: Methyl sulfate.
Description: Colorless liquid.
Formula: (CH$_3$)$_2$SO$_4$.
Constants: Mol wt: 126.13, mp: −31.8°C, bp: 188°C, flash p: 182°F (C.C.), d: 1.3322 at 20°/4°C, vap. d.: 4.35.
Hazard Analysis
Toxic Hazard Rating:
Acute Local: Irritant 3; Inhalation 3.
Acute Systemic: Ingestion 3; Inhalation 3; Skin Absorption 3.
Chronic Local: Irritant 3; Inhalation 3.
Chronic Systemic: Ingestion 3; Inhalation 3; Skin Absorption 3.
TLV: ACGIH (recommended); 1 part per million in air; 5 milligrams per cubic meter of air. May be absorbed via intact skin.
Toxicology: Contact of the skin and mucous membranes with the liquid or vapor, even for short periods, results in intense irritation of these tissues several hours

---

TOXIC HAZARD RATING CODE (For detailed discussion, see Section 1.)

0 NONE: (a) No harm under any conditions; (b) Harmful only under unusual conditions or overwhelming dosage.

1 SLIGHT: Causes readily reversible changes which disappear after end of exposure.

2 MODERATE: May involve both irreversible and reversible changes; not severe enough to cause death or permanent injury.

3 HIGH: May cause death or permanent injury after very short exposure to small quantities.

U UNKNOWN: No information on humans considered valid by authors.

later. There is no odor or initial irritation to give warning of exposure. On brief, mild exposures, conjunctivitis, catarrhal inflammation of the mucous membranes of the nose, throat, larynx and trachea and possibly some reddening of the skin develop after the latent period. With longer, heavier exposures, the cornea shows clouding, the irritative changes of the nasopharynx are more marked and after 6 to 8 hours pulmonary edema may develop. Death may occur in 3 or 4 days. The liver and kidneys are frequently damaged. Spilling of the liquid on the skin can cause ulceration and local necrosis. The fatal concentration for cats and monkeys is in the range of 25 to 200 ppm of the vapor in air.

After a latent period of several hours, there is severe lacrimation, conjunctivitis, photophobia, coughing and hoarseness, followed, in the case of more severe exposures, by chest pain, dyspnea, cyanosis and possibly death. In patients surviving severe exposures, there may be serious injury of the liver and kidneys, with suppression of urine, jaundice, albuminuria and hematuria appearing. Death, resulting from the kidney or liver damage, may be delayed for several weeks.

Fire Hazard: Moderate, when exposed to heat or flame.
Spontaneous Heating: No.
Disaster Hazard: Dangerous; see sulfates. Can react with oxidizing materials.

Countermeasures
Ventilation Control: Section 2.
To Fight Fire: Water, foam, carbon dioxide, dry chemical or carbon tetrachloride (Section 6).
Personnel Protection: Section 3.
First Aid: Section 1.
Storage and Handling: Section 7.
Shipping Regulations: Section 11.
  I.C.C.: Corrosive liquid; white label, 1 quart.
  Coast Guard Classification: Corrosive liquid; white label.
  MCA warning label.
  IATA: Corrosive liquid, white label, not acceptable (passenger), 1 liter (cargo).

DIMETHYL SULFIDE. See methyl sulfide.

DIMETHYL SULFITE. See methyl sulfite.

DIMETHYLSULFOLANE
General Information
Description: Solid.
Formula: $C_6H_{12}SO_2$.
Constants: Mol wt: 148.22, bp: 280°C, flash p: 290°F (O.C.), d: 1.1362 at 20°/4°C, vap. press.: 0.006 mm at 20°C.

Hazard Analysis
Toxicity: Unknown. Limited animal experiments suggest moderate toxicity.
Fire Hazard: Slight, when exposed to heat or flame.
Disaster Hazard: Dangerous; see sulfates; can react with oxidizing materials.

Countermeasures
Storage and Handling: Section 7.
To Fight Fire: Water, foam, carbon dioxide, dry chemical or carbon tetrachloride (Section 6).

DIMETHYL SULFOXIDE
General Information
Synonym: DMSO.
Description: Clear, water-white, hygroscopic liquid.
Formula: $(CH_3)_2SO$.
Constants: Mol wt: 78.13, mp: 6°C, bp: decomp. at 100°C, flash p: 203°F (O.C.), d: 1.100 at 20°C, vap. press.: 0.37 mm at 20°C, lel: 2.6%, uel: 28.5%.

Hazard Analysis
Toxic Hazard Rating:
  Acute Local: Irritant 2; Allergen 2; Inhalation 1.
  Acute Systemic: Ingestion 2; Skin Absorption 2.
  Chronic Local: Irritant 1; Allergen 2.

Chronic Systemic: Ingestion 1; Skin Absorption 2.
Freely penetrates skin. Acts as a primary irritant on skin causing redness, burning, itching, and scaling, also causes urticaria: systemic symptoms are nausea, vomiting, chills, cramps, and lethargy. A case of anaphylactic reaction has been reported. Has caused corneal opacity in experimental animals.
Fire Hazard: Moderate, when exposed to heat or flame.
Disaster Hazard: Moderately dangerous; when heated to decomposition, it emits toxic fumes; can react with oxidizing materials.

Countermeasures
Storage and Handling: Section 7.
To Fight Fire: Water, foam, alcohol foam, carbon dioxide, dry chemical or carbon tetrachloride (Section 6).

DIMETHYL TELLURIDE
General Information
Synonym: Methyl telluride.
Description: Yellowish oil with odor of garlic.
Formula: $(CH_3)_2Te$.
Constants: Mol wt: 157.7, bp: 82°C, vap. d.: 5.45.
Hazard Analysis and Countermeasures
See tellurium compounds.

α-DIMETHYL TELLURONIUM DIBROMIDE
General Information
Description: Orange crystals; soluble in alcohol.
Formula: $C_2H_6Br_2Te$.
Constants: Mol wt: 317.5, mp: 142°C (decomp.).
Hazard Analysis and Countermeasures
See tellurium compounds and bromides.

α-DIMETHYL TELLURONIUM DICHLORIDE
General Information
Description: Crystals; soluble in water.
Formula: $C_2H_6Cl_2Te$.
Constants: Mol wt: 228.6, mp: 92°C.
Hazard Analysis and Countermeasures
See tellurium compounds and chlorides.

β-DIMETHYL TELLURONIUM DICHLORIDE
General Information
Description: Crystals. Soluble in alcohol and ether.
Formula: $C_2H_6Cl_2Te$.
Constants: Mol wt: 228.6, mp: 134°C.
Hazard Analysis and Countermeasures
See tellurium compounds and chlorides.

α-DIMETHYL TELLURONIUM DIIODIDE
General Information
Description: Red crystals; insoluble in cold water.
Formula: $C_2H_6I_2Te$.
Constants: Mol wt: 411.5, mp: 125°C (decomposes).
Hazard Analysis and Countermeasures
See tellurium compounds and iodides.

DIMETHYL   2,3,5,6-TETRACHLOROTEREPHTHA-LATE
General Information
Description: Crystals; insoluble in water; slightly soluble in acetone and benzene.
Formula: $C_6Cl_4(COOCH_3)_2$.
Constants: Mol wt: 324, mp: 156°C.
Hazard Analysis
Toxic Hazard Rating: U. Animal experiments show low oral toxicity.
Disaster Hazard: Dangerous. See chlorides.

DIMETHYL TETRAHYDROPHTHALATE
General Information
Synonym: Dimethyl-1-cyclohexene-1,2-dicarboxylic acid.
Description: Crystals.
Formula: $C_6H_8(COOCH_3)_2$.
Constants: Mol wt: 198.2, vap. d.: 6.83.

Hazard Analysis
Toxicity: Probably slight. See phthalic acid.
Fire Hazard: Slight, when exposed to heat or flame; can react with oxidizing materials (Section 6).
Countermeasures
Personal Hygiene: Section 3.
Storage and Handling: Section 7.

2,6-DIMETHYL TETRAHYDRO-1,4-PYRONE
General Information
Formula: $C_7H_{14}O_2$.
Constants: Mol wt: 130.2, vap. d.: 4.48.
Hazard Analysis
Toxic Hazard Rating:
  Acute Local: U.
  Acute Systemic: Ingestion 2; Inhalation 2.
  Chronic Local: U.
  Chronic Systemic: U.
Fire Hazard: Slight; Section 6.
Countermeasures
Ventilation Control: Section 2.
Personal Hygiene: Section 3.
Storage and Handling: Section 7.

3,5-DIMETHYLTETRAHYDRO-1,3,5,2-H-THIO-DIAZINE-2-THIONE
General Information
Synonym: Crag herbicide 974.
Description: White crystalline powder, slightly soluble in water or alcohol; soluble in acetone.
Formula: $C_5H_{10}N_2S_2$.
Constants: Mol wt: 162.2, mp: 156°C, vap. d.: 5.59, d: 1.3.
Hazard Analysis
Toxic Hazard Rating:
  Acute Local: Irritant 2; Ingestion 2; Inhalation 2.
  Acute Systemic: U.
  Chronic Local: U.
  Chronic Systemic: U.
Toxicity: Details unknown. A fungicide, a herbicide, and a food additive resulting from contact with containers or equipment (Section 10). Limited animal experiments show slight irritant effects.
TLV: ACGIH (recommended); 15 milligrams per cubic meter of air.
Disaster Hazard: Dangerous; when heated to decomposition, it emits highly toxic fumes of oxides of sulfur and nitrogen.
Countermeasures
Ventilation Control: Section 2.
Personnel Protection: Section 3.
Storage and Handling: Section 7.

2,4-DIMETHYL THIOPHENE. See p-thioxene.

DIMETHYLTIN
General Information
Description: Yellow solid, insoluble in water.
Formula: A polymer of $(CH_3)_2Sn$.
Constants: Mol wt: a multiple of 148.8.
Hazard Analysis and Countermeasures
See tin compounds.

DIMETHYLTIN DIBROMIDE
General Information
Description: Colorless crystals, soluble in water and organic solvents.
Formula: $(CH_3)_2SnBr_2$.
Constants: Mol wt: 308.6, mp: 76°C, bp: 208–213°C.
Hazard Analysis and Countermeasures
See tin compounds and bromides.

DIMETHYLTIN DICHLORIDE
General Information
Description: Solid; water soluble.
Formula: $(CH_3)_2SnCl_2$.
Constants: Mol wt: 219.7, mp: 90°C, bp: 190°C.
Hazard Analysis and Countermeasures
See tin compounds and chlorides.

DIMETHYLTIN DIFLUORIDE
General Information
Description: White crystals, water soluble.
Formula: $(CH_3)_2SnF_2$.
Constants: Mol wt: 186.8, bp: decomposes $< 360°C$.
Hazard Analysis and Countermeasures
See fluorides and tin compounds.

DIMETHYLTIN DIIODIDE
General Information
Description: White crystals. Soluble in hot water.
Formula: $(CH_3)_2SnI_2$.
Constants: Mol wt: 402.6, d: 2.872, mp: 43°C, bp: 228°C.
Hazard Analysis and Countermeasures
See tin compounds and iodides.

DIMETHYLTIN OXIDE
General Information
Description: White powder. Insoluble in water.
Formula: $(CH_3)_2SnO$.
Constant: Mol wt: 164.8.
Hazard Analysis
Toxicity: Details unknown. See tin compounds.

DIMETHYLTIN SULFIDE
General Information
Description: Solid.
Formula: $(CH_3)_2SnS$.
Constants: Mol wt: 180.8, mp: 148°C.
Hazard Analysis and Countermeasures
See sulfides and tin compounds.

N,N-DIMETHYL-m-TOLUAMIDE
General Information
Synonym: m-Delphene.
Formula: $C_6H_2(CH_3)_3CONH_2$.
Constant: Mol wt: 163.
Hazard Analysis
Toxic Hazard Rating:
  Acute Local: Irritant 1.
  Acute Systemic: U.
  Chronic Local: Irritant 1.
  Chronic Systemic: Ingestion 1.
Irritates eyes and mucous membranes. Ingestion can cause central nervous depression.

DIMETHYL TRIBORINE TRIAMINE (B)
General Information
Description: Colorless liquid; hydrolyzed by water.
Formula: $(CH_3)_2B_3N_3H_4$.
Constants: Mol wt: 108.6, mp: $-48°C$, bp: 107°C.
Hazard Analysis
Toxicity: Details unknown. See boron compounds. and amines.

DIMETHYL TRIBORINE TRIAMINE (N)
General Information
Description: Colorless liquid, hydrolyzed by water.
Formula: $(CH_3)_2B_3N_3H_4$.
Constants: Mol wt: 108.6, bp: 108°C.

TOXIC HAZARD RATING CODE (For detailed discussion, see Section 1.)

0 NONE: (a) No harm under any conditions; (b) Harmful only under unusual conditions or overwhelming dosage.

1 SLIGHT: Causes readily reversible changes which disappear after end of exposure.

2 MODERATE: May involve both irreversible and reversible changes; not severe enough to cause death or permanent injury.

3 HIGH: May cause death or permanent injury after very short exposure to small quantities.

U UNKNOWN: No information on humans considered valid by authors.

Hazard Analysis
Toxicity: Details unknown. See also boron compounds and boron hydrides and amines.

DIMETHYLTRITHIOCARBONATE
General Information
Description: Crystals.
Formula: $CS_3(CH_3)_2$.
Constants: Mol wt: 139.26, vap. d.: 4.8.
Hazard Analysis
Toxic Hazard Rating:
  Acute Local: Irritant 3; Inhalation 3.
  Acute Systemic: Inhalation 3.
  Chronic Local: U.
  Chronic Systemic: U.
Fire Hazard: Moderate, when exposed to heat (Section 6).
Disaster Hazard: Dangerous; see sulfates; it can react with oxidizing materials.
Countermeasures
Ventilation Control: Section 2.
Personnel Protection: Section 3.
First Aid: Section 1.
Storage and Handling: Section 7.

2,3-DIMETHYL VALERALDEHYDE
General Information
Synonym: 2,3-Dimethyl pentaldehyde.
Description: Liquid; slightly soluble in water.
Formula: $CH_3CH_2CH(CH_3)CH(CH_3)CHO$.
Constants: Mol wt: 111, d: 0.8293, bp: 140.5°C, fp: 110°C, flash p: 94°F.
Hazard Analysis
Toxic Hazard Rating: U. Limited animal experiments suggest low toxicity and irritation. See also aldehydes.
Fire Hazard: Moderate, when exposed to heat or flame.

β,β-DIMETHYLVINYL CHLORIDE
General Information
Synonym: Isocrotyl chloride.
Description: Liquid.
Formula: C_4H_7Cl.
Constants: Mol wt: 90.6, d: 0.919 at 20°/4°C, bp: 68°C.
Hazard Analysis
Toxic Hazard Rating:
  Acute Local: Irritant 2; Ingestion 2; Inhalation 2.
  Acute Systemic: Inhalation 2.
  Chronic Local: U.
  Chronic Systemic: U.
Toxicology: Causes local irritation and is narcotic in high concentrations.
Disaster Hazard: Dangerous; when heated to decomposition, it emits highly toxic chloride fumes.

1,3-DIMETHYLXANTHINE. See theophylline.

3,6-DIMETHYLXANTHINE. See theobromine.

DIMETRIDIAZOLE
General Information
Synonym: 1,2-Dimethyl-5-nitroimidazole.
Formula: $C_5H_7O_2N_5$.
Constant: Mol wt: 147.
Hazard Analysis
Toxic Hazard Rating: U. A food additive permitted in the feed and drinking water of animals and/or for the treatment of food producing animals. Also a food additive permitted in food for human consumption (Section 10).

"DIMITE." See di(p-chlorophenyl) methyl carbinol.

DIMORPHOLINETHIURAM DISULFIDE
Hazard Analysis
Toxicity: Details unknown; a fungicide. See disulfiram.
Disaster Hazard: Dangerous; see sulfates.
Countermeasures
Storage and Handling: Section 7.

DI-α-NAPHTHYLAMINE FLUOSILICATE
General Information
Description: White crystals; slightly soluble in alcohol.
Formula: $(C_{10}H_7NH_2)_2 \cdot H_2SiF_6$.
Constants: Mol wt: 430.4, mp: 218°C.
Hazard Analysis and Countermeasures
See fluosilicates.

DI-β-NAPHTHYLAMINE FLUOSILICATE
General Information
Description: White crystals.
Formula: $(C_{10}H_7NH_2)_2 \cdot H_2SiF_6$.
Constants: Mol wt: 430.4, mp: 236°C.
Hazard Analysis and Countermeasures
See fluosilicates.

DINAPHTHYLMERCURY
General Information
Description: White crystals. Insoluble in cold water.
Formula: $(C_{10}H_7)Hg(C_{10}H_7)$.
Constants: Mol wt: 454.9, d: 1.929, mp: 188°C, bp: 249°C.
Hazard Analysis and Countermeasures
See mercury compounds, organic.

DI-β-NAPHTHYL-p-PHENYLENE DIAMINE
General Information
Description: Solid.
Formula: $(C_{10}H_7)_2NC_6H_4NH_2$.
Constants: Mol wt: 360.42, vap. d.: 12.4.
Hazard Analysis
Toxicity: Details unknown; probably toxic. See p-phenylene diamine.
Disaster Hazard: Moderately dangerous; when heated to decomposition, it emits toxic fumes.
Countermeasures
Storage and Handling: Section 7.

DI-α-NAPHTHYLTIN
General Information
Description: A powder.
Formula: $(C_{10}H_7)_2Sn$.
Constants: Mol wt: 373, mp: 200°C, bp: decomposes at 255°C.
Hazard Analysis and Countermeasures
See tin compounds.

1,3-DI-1-NAPHTHYLTRIAZINE. See   1,1'-diazoamido-naphthalene.

DINICKEL PHOSPHIDE
General Information
Description: Gray crystals.
Formula: NI_2P.
Constants: Mol wt: 148.36, mp: 1112°C, d: 6.31 at 15°C.
Hazard Analysis and Countermeasures
See phosphine, phosphides, and nickel compounds.

DI-m-NITRANILINE FLUOSILICATE
General Information
Description: White crystals slightly soluble in alcohol.
Formula: $(C_6H_4NH_2NO_2)_2 \cdot H_2SiF_6$.
Constants: Mol wt: 420.3, mp: 200°C.
Hazard Analysis and Countermeasures
See fluosilicates and nitrates, organic and m-nitroaniline.

"DINITRO." See 4,6-dinitro-o-cresol.

DINITROAMINOPHENOL. See picramic acid.

DINITRO-o, sec-AMYLPHENOL
General Information
Formula: $(NO_2)_2C_6H_2OHC_5H_{11}$.
Constant: Mol wt: 253.2.
Hazard Analysis and Countermeasures
A herbicide. See dinitrophenol and nitrates.

2,4-DINITROANILINE
General Information
Synonym: 2,4-Dinitrophenylamine.
Description: Yellow, needlelike crystals, insoluble in water.
Formula: $(NO_2)_2C_6H_3NH_2$.
Constants: Mol wt: 183.13, mp: 180°C, bp: 56.7°C, flash p: 435°F (C.C.), d: 1.615, vap. d.: 6.31.
Hazard Analysis
Toxic Hazard Rating:
  Acute Local: Irritant 2; Ingestion 2; Inhalation 2.
  Acute Systemic: Ingestion 3.
  Chronic Local: Irritant 2.
  Chronic Systemic: Ingestion 3.
Toxicology: A powerful poison. See also nitroanilines.
Fire Hazard: Slight, when exposed to heat or flame.
Disaster Hazard: Dangerous; when heated to decomposition, it emits highly toxic fumes; it can react with oxidizing materials.
Countermeasures
Ventilation Control: Section 2.
To Fight Fire: Water; carbon dioxide, dry chemical or carbon tetrachloride (Section 6).
Personal Hygiene: Section 3.
First Aid: Section 1.
Storage and Handling: Section 7.

DINITROANILINE, LIQUID
Shipping Regulations: Section 11.
  IATA: Poison B, poison label, 1 liter (passenger), 220 liters (cargo).

2,4-DINITROANISOLE
General Information
Synonym: 2,4-Dinitrophenylmethyl ether.
Description: Colorless to yellow crystals.
Formula: $CH_3OC_6H_3(NO_2)_2$.
Constants: Mol wt: 198.1, mp: 89°C, bp: sublimes, d: 1.341, at 20°/4°C, vap. d.: 6.83.
Hazard Analysis and Countermeasures
See nitro compounds of aromatic hydrocarbons and nitrates.

3,5-DINITROBENZAMIDE
General Information
Formula: $C_7H_5N_3O_5$.
Constants: Mol wt: 211.13, mp range: 180–184°C.
Hazard Analysis
Toxic Hazard Rating: U. A food additive permitted in the feed and drinking water of animals and/or for the treatment of food producing animals (Section 10).
Disaster Hazard: Dangerous. See nitrates.

1,2-DINITROBENZENE
General Information
Synonym: o-Dinitrobenzol.
Description: Colorless or yellowish needles or plates.
Formula: $C_6H_4(NO_2)_2$.
Constants: Mol wt: 168.11, mp: 118°C, bp: 302.8°C at 770 mm, flash p: 302°F (C.C.), d: 1.571 at 0°/4°C, vap. d.: 5.79.
Hazard Analysis
Toxic Hazard Rating:
  Acute Local: U.
  Acute Systemic: Ingestion 3; Inhalation 3; Skin Absorption 3.
  Chronic Local: U.
  Chronic Systemic: Ingestion 3; Inhalation 3; Skin Absorption 3.
Toxicology: Produces a wide variety of pathological changes

including anemia, jaundice, enlarged liver or yellow atrophy, degeneration of kidneys and injury to CNS.
TLV: ACGIH (recommended) 1 milligram per cubic meter of air. May be absorbed via intact skin.
Fire Hazard: Slight, when exposed to heat or flame.
Explosion Hazard: Severe, when shocked or exposed to heat or flame. This compound is a high explosive used in bursting charges and to fill artillery shells. It is a useful industrial explosive when mixed with more powerful explosives or with oxygen carriers, such as inorganic nitrates or chlorates or with ammonium nitrate.
Disaster Hazard: Dangerous; when heated to decomposition, it emits highly toxic fumes of oxides of nitrogen and explodes; it can react vigorously with oxidizing materials.
Countermeasures
Ventilation Control: Section 2.
To Fight Fire: Water; carbon dioxide; dry chemical or carbon tetrachloride (Section 6).
Personnel Protection: Section 3.
First Aid: Section 1.
Storage and Handling: Section 7.
Shipping Regulations: Section 11.
  Coast Guard Classification: Poison B (liquid, solid); poison label required.
  IATA (liquid): Poison B, poison label, 1 liter (passenger), 220 liters (cargo).
      (solid): Poison B, poison label, 25 kilograms (passenger), 95 kilograms (cargo).
  I.C.C. (solid): Poison B, poison label, 200 pounds.
      (liquid): Poison B, poison label, 55 gallons.

2,4-DINITROBENZENE SULFENYL CHLORIDE
General Information
Description: Bright yellow, crystalline solid.
Formula: $C_6H_3SCl(NO_2)_2$.
Constants: Mol wt: 234, mp: 95°–96°C, vap. d.: 8.08.
Hazard Analysis
Toxicity: Details unknown. See also nitro compounds of aromatic hydrocarbons.
Fire Hazard: See nitrates.
Explosion Hazard: Slight, when shocked.
Disaster Hazard: Dangerous; shock will explode it; when heated to decomposition, it emits highly toxic fumes of oxides of sulfur and nitrogen; it will react with water or steam to produce toxic and corrosive fumes; it can react with reducing materials.
Countermeasures
Storage and Handling: Section 7.

o-DINITROBENZOL. See 1,2-dinitrobenzene.

3,5-DINITROBENZOYL CHLORIDE
General Information
Description: Yellow crystals.
Formula: $(NO_2)_2C_6H_3COCl$.
Constants: Mol wt: 230.6, mp: 69°C, bp: 196°C at 12 mm, vap. d.: 7.96.
Hazard Analysis
Toxic Hazard Rating:
  Acute Local: Irritant 3; Ingestion 3.
  Acute Systemic: U.
  Chronic Local: U.
  Chronic Systemic: U.
Disaster Hazard: Dangerous; see chlorides.
Countermeasures
Ventilation Control: Section 2.
Personal Hygiene: Section 3.

TOXIC HAZARD RATING CODE (For detailed discussion, see Section 1.)

0 NONE: (a) No harm under any conditions; (b) Harmful only under unusual conditions or overwhelming dosage.

1 SLIGHT: Causes readily reversible changes which disappear after end of exposure.

2 MODERATE: May involve both irreversible and reversible changes; not severe enough to cause death or permanent injury.

3 HIGH: May cause death or permanent injury after very short exposure to small quantities.

U UNKNOWN: No information on humans considered valid by authors.

First Aid: Section 1.
Storage and Handling: Section 7.

DINITRO-o-sec-BUTYL PHENOL
General Information
Description: Crystals.
Formula: $(NO_2)_2C_4H_7C_6H_4OH$.
Constants: Mol wt: 240.3, vap. d.: 7.73.
Hazard Analysis
Toxic Hazard Rating:
Acute Local: Irritant 2.
Acute Systemic: Ingestion 3; Inhalation 3; Skin Absorption 3.
Chronic Local: Irritant 2.
Chronic Systemic: Ingestion 3; Inhalation 3; Skin Absorption 3.
Toxicology: An insecticide and a herbicide. See also dinitrophenol.
Fire Hazard: See nitrates.
Disaster Hazard: See nitrates.
Countermeasures
Storage and Handling: Section 7.

DINITROCAPRYL CROTONATE
General Information
Description: Crystals.
Formula: $(NO_2)_2CH(CH_2)_6CH_2O_2CCH{:}CHCH_3$.
Constants: Mol wt: 288.3, vap. d.: 9.95.
Hazard Analysis
Toxic Hazard Rating:
Acute Local: Irritant 3; Ingestion 3; Inhalation 3.
Acute Systemic: U.
Chronic Local: U.
Chronic Systemic: U.
A fungicide.
Fire Hazard: See nitrates.
Disaster Hazard: See nitrates.
Countermeasures
Ventilation Control: Section 2.
Personnel Protection: Section 3.
Storage and Handling: Section 7.

4,6-DINITRO(2-CAPRYL) PHENYL CROTONATE
General Information
Synonym: Karathane.
Description: Liquid.
Formula: $C_{18}H_{24}N_2O_6$.
Constant: Mol wt: 364.4
Hazard Analysis
Toxic Hazard Rating:
Acute Local: Irritant 2.
Acute Systemic: U.
Chronic Local: U.
Chronic Systemic: U.
Toxicology: A fungicide and an insecticide. Limited animal experiments show moderate toxicity.
Countermeasures
See nitrates.

2,4-DINITROCHLOROBENZENE
General Information
Synonym: 1-Chloro-2,4-dinitrobenzene.
Description: Yellow, rhombic crystals; insoluble in water.
Formula: $(NO_2)_2C_6H_3Cl$.
Constants: Mol wt: 202.56, mp alpha: 53.4°C, beta: 43°C, gamma: 27°C, bp: 315°C, lel: 2.0%, uel: 22%; flash p: 382°F (C.C.); d alpha: 1.687 at 22°C, beta: 1.680 at 20°/4°C; vap. d.: 6.98.
Hazard Analysis
Toxic Hazard Rating:
Acute Local: Irritant 2; Allergen 2.
Acute Systemic: Ingestion 3; Inhalation 3,
Chronic Local: Allergen 2.
Chronic Systemic: Ingestion 3; Inhalation 3; Skin Absorption 3.

Toxicology: Acts as a primary irritant as well as a sensitizer of skin. See also dinitrobenzene and nitro compounds of aromatic hydrocarbons.
Fire Hazard: Slight, when exposed to heat or flame.
Explosion Hazard: Moderate, when exposed to flame, sparks or in a fire.
Disaster Hazard: See nitrates.
Countermeasures
Ventilation Control: Section 2.
Personnel Protection: Section 3.
To Fight Fire: Water; carbon dioxide; dry chemical or carbon tetrachloride (Section 6).
First Aid: Section 1.
Storage and Handling: Section 7.
Shipping Regulations: Section 11.
I.C.C.: Poison B; poison label, 200 pounds.
Coast Guard Classification: Poison B; poison label.
IATA: Poison B, poison label, 25 kilograms (passenger), 95 kilograms (cargo).

DINITROCHLOROHYDRIN
General Information
Formula: $CH_2ClCHNO_2CH_2NO_2$.
Constants: Mol wt: 168.6, bp: 180°C, d: 1.5, vap. d.: 5.83.
Hazard Analysis
Toxic Hazard Rating:
Acute Local: Irritant 3; Ingestion 3; Inhalation 3.
Acute Systemic: U.
Chronic Local: U.
Chronic Systemic: U.
Fire Hazard: See nitrates.
Explosion Hazard: Severe, when shocked.
Disaster Hazard: Dangerous; shock will explode it; when heated to decomposition, it burns and emits highly toxic fumes; it can react vigorously with reducing materials.
Countermeasures
Ventilation Control: Section 2.
Personnel Protection: Section 3.
Storage and Handling: Section 7.

4,6-DINITRO-o-CRESOL
General Information
Synonym: 2-Methyl-4,6-dinitrophenol; "dinitro;" DNC; DNOC.
Description: Yellow, prismatic crystals.
Formula: $(NO_2)_2C_6H_2(CH_3)OH$.
Constants: Mol wt: 198.1, mp: 85.8°C, vap. d.: 6.82.
Hazard Analysis
Toxic Hazard Rating:
Acute Local: Irritant 3; Ingestion 2; Inhalation 2.
Acute Systemic: Ingestion 3; Inhalation 3.
Chronic Local: Irritant 3.
Chronic Systemic: Ingestion 3; Inhalation 3.
TLV: ACGIH (recommended); 0.2 milligram per cubic meter of air. May be absorbed via intact skin.
Toxicology: An insecticide and a herbicide. Appears to be somewhat less toxic than the p-form, but is still highly toxic. See also dinitrophenol.
Fire Hazard: See nitrates.
Disaster Hazard: See nitrates.
Countermeasures
Ventilation Control: Section 2.
Personnel Protection: Section 3.
First Aid: Section 1.
Storage and Handling: Section 7.
Shipping Regulations: Section 11.
IATA: Poison B, poison label, 25 kilograms (passenger), 95 kilograms (cargo).

3,5-DINITRO-p-CRESOL
General Information
Description: Crystals.
Formula: $(NO_2)_2C_6H_2CH_3OH$.
Constant: Mol wt: 198.1.

Hazard Analysis
Toxic Hazard Rating:
Acute Local: Irritant 3; Ingestion 3; Inhalation 3.
Acute Systemic: Ingestion 3; Inhalation 3; Skin Absorption 3.
Chronic Local: Irritant 3.
Chronic Systemic: Ingestion 3; Inhalation 3; Skin Absorption 3.
Toxicology: Can cause brain damage, as well as damage to liver and kidneys. See also dinitrophenol.
Countermeasures
See 4,6-Dinitro-o-cresol.

3,5-DINITRO-o-CRESYL ACETATE
General Information
Formula: $CH_3COOC_6H_2CH_3(NO_2)_2$.
Constants: Mol wt: 240.2, vap. d.: 8.28.
Hazard Analysis and Countermeasures
See cresol and nitro compounds of aromatic hydrocarbons.

2,4-DINITRO-6-CYCLOHEXYL PHENOL
General Information
Synonym: 2,4-Dinitro-o-cyclohexyl phenol.
Description: Crystals.
Formula: $(NO_2)_2(C_6H_{11})C_6H_2OH$.
Constants: Mol wt: 266.3, vap. d.: 8.62.
Hazard Analysis
Toxicity: An insecticide. Details unknown; probably toxic. See also nitro compounds of aromatic hydrocarbons. Animal experiments indicate toxicity similar to dinitrophenol.
Fire Hazard: See nitrates.
Disaster Hazard: See nitrates.
Countermeasures
Storage and Handling: Section 7.
Shipping Regulations: Section 11.
IATA: Other restricted articles, class A; no label required, no limit (passenger and cargo).

DINITRODICHLOROBENZENE
General Information
Synonym: Parazol.
Description: Crystalline material.
Formula: $C_6H_2(NO_2)_2Cl_2$.
Constants: Mol wt: 237.0, vap. d.: 8.17.
Hazard Analysis
Toxic Hazard Ratings:
Acute Local: Irritant 3.
Acute Systemic: Ingestion 3; Inhalation 3.
Chronic Local: Irritant 2.
Chronic Systemic: Ingestion 3; Inhalation 3.
Toxicology: See also dinitrochlorobenzene and nitro compounds of aromatic hydrocarbons.
Fire Hazard: See nitrates.
Disaster Hazard: Dangerous; when heated to decomposition it emits highly toxic fumes of oxides and chlorides; it can react vigorously with oxidizing materials.
Countermeasures
Ventilation Control: Section 2.
Personnel Protection: Section 3.
First Aid: Section 1.
Storage and Handling: Section 7.

2,4-DINITRO-1-FLUOROBENZENE. See 1-fluoro-2,4-dinitrobenzene.

DINITROGEN TRIOXIDE
General Information
Synonym: Nitrous anhydride.

Description: Red-brown gas; blue solid or liquid.
Formula: N_2O_3.
Constants: Mol wt: 76.02, mp: $-102°C$, bp: $3.5°$ decomposes, d: 1.447 at 20°C.
Hazard Analysis
Toxic Hazard Rating:
Acute Local: Irritant 3; Inhalation 3.
Acute Systemic: Inhalation 3.
Chronic Local: U.
Chronic Systemic: Inhalation 3.
Disaster Hazard: Dangerous. See nitric oxide.
Countermeasures
Ventilation Control: Section 2.
Personnel Protection: Section 3.
First Aid: Section 1.
Storage and Handling: Section 7.

DINITROGLYCOL. See diethylene glycol dinitrate.

DINITROMONOCHLORHYDRIN
General Information
Description: Dinitromonochlorhydrin is a high explosive miscible with nitroglycerine and is used as a component in low-freezing dynamites and also in permissible gelatinous explosives.
Formula: $(NO_2)_2C_3H_5O_2Cl$.
Constant: Mol wt: 200.6.
Hazard Analysis
Toxicity: Unknown. See also dinitrochlorohydrin.
Fire Hazard: See nitrates.
Explosion Hazard: See explosives, high and nitrates.
Disaster Hazard: See explosives, high and nitrates.
Countermeasures
Storage and Handling: Section 7.

1,5-DINITRONAPHTHALENE. See dinitronaphthalene.

1,8-DINITRONAPHTHALENE. See dinitronaphthalene.

DINITRONAPHTHALENE
General Information
Synonyms: (a) 1,5-Dinitronaphthalene; (b) 1,8-dinitronaphthalene.
Description: (a) yellowish needles; (b) yellow, rhombic crystals.
Formula: $C_{10}H_6(NO_2)_2$.
Constants: Mol wt: 218.16, mp: (a) 217.5°C (b) 173–173.5°C, bp: (a) sublimes,; (b) decomposes, vap. d.: 7.51.
Hazard Analysis
Toxicity: Details unknown. See also nitro compounds of aromatic hydrocarbons.
Fire Hazard: Moderate, when exposed to heat or flame (Section 6).
Explosion Hazard: Moderate, when shocked or exposed to heat. It is used mixed with chlorates and perchlorates and in combination with picric acid. Dinitronaphthalene is an ingredient of permissible explosives and is also used in combination with ammonium nitrate.
Disaster Hazard: See explosives, high and nitrates.
Countermeasures
Storage and Handling: Section 7.

2,4-DINITRO-NAPHTHOL-1
General Information
Description: Yellow needles or leaflets.
Formula: $(NO_2)_2C_{10}H_5OH$.
Constants: Mol wt: 234.16, mp: 138°C, vap. d.: 8.08.

Hazard Analysis
Toxicity: See nitro compounds of aromatic hydrocarbons.
Fire Hazard: See nitrates.
Explosion Hazard: See nitrates.
Disaster Hazard: See nitrates.
Countermeasures
Storage and Handling: Section 7.

DINITRONAPHTHOL SALTS
Hazard Analysis
Toxicity: See nitro compounds of aromatic hydrocarbons.
Fire Hazard: See nitrates.
Explosion Hazard: See nitrates.
Disaster Hazard: See nitrates.
Countermeasures
Storage and Handling: Section 7.

2,4-DINITROPHENETOLE
General Information
Description: Crystals.
Formula: $(NO_2)_2C_6H_3OC_2H_5$.
Constants: Mol wt: 212.2, vap. d.: 7.32.
Hazard Analysis and Countermeasures
See nitro compounds of aromatic hydrocarbons and nitrates.

2,3-DINITROPHENOL
General Information
Synonym: 1-Hydroxy-2,3-dinitrobenzene.
Description: Yellow needles.
Formula: $(NO_2)_2C_6H_3OH$.
Constants: Mol wt: 184.11, mp: 144°C, d: 1.681 at 20°C, vap. d.: 6.35.
Hazard Analysis
Toxic Hazard Rating:
Acute Local: Irritant 3; Allergen 1; Inhalation 2.
Acute Systemic: Ingestion 3; Inhalation 3; Skin Absorption 3.
Chronic Local: Irritant 2; Allergen 2; Inhalation 2.
Chronic Systemic: Ingestion 3; Inhalation 3; Skin Absorption 3.
Toxicology: The harmful effects, which can be fatal, are damage to the liver and induced fever. Fatal cases have been reported in the literature from the inhalation of the dust in a concentration estimated at approximately 40 mg/cu. meter. It is a powerful stimulant of metabolism. It is the excessive oxidation effects of this material upon the metabolism and nutrition that damage the liver and kidney cells. Like other nitrated phenols, it is an irritant to the skin and has been known to cause dermatitis (Section 9). Ingestion will cause dilation of the pupils or posterior subcapsular opacities or cataracts. It also exhibits some allergic manifestations. Wood preservative. See also nitro compounds of aromatic hydrocarbons.
Fire Hazard: See nitrates.
Explosion Hazard: Severe, when exposed to heat. A high explosive used as a component of some shell and bomb charges. See also explosives, high.
Disaster Hazard: See explosives, high and nitrates.
Countermeasures
Ventilation Control: Section 2.
Personnel Protection: Section 3.
First Aid: Section 1.
Storage and Handling: Section 7.

2,4-DINITROPHENOL
General Information
Description: Yellow crystals.
Formula: $(NO_2)_2C_6H_3OH$.
Constants: Mol wt: 184.11, mp: 112°C, d: 1.683 at 24°C, vap. d.: 6.35.
Hazard Analysis
Toxicity: See 2,3-Dinitrophenol.
Fire Hazard: See nitrates.
Explosion Hazard: Moderate, when exposed to heat.

Disaster Hazard: See nitrates.
Countermeasures
Storage and Handling: Section 7.

2,6-DINITROPHENOL
General Information
Description: Yellow crystals.
Formula: $(NO_2)_2C_6H_3OH$.
Constants: Mol wt: 184.11, mp: 63°C, vap. d.: 6.35.
Hazard Analysis
Toxicity: See 2,3-Dinitrophenol.
Fire Hazard: See nitrates.
Explosion Hazard: Moderate, when exposed to heat.
Disaster Hazard: See nitrates.
Countermeasures
Storage and Handling: Section 7.

DINITROPHENOL, dry or wet with less than 15% water.
Shipping Regulations: Section 11.
IATA: Explosive, not acceptable (passenger and cargo).

DINITROPHENOL, wet with not less than 15% water.
IATA: Flammable solid, yellow label, 12 kilograms (passenger and cargo).

DINITROPHENOL, Solutions
Shipping Regulations: Section 11.
I.C.C.: Poison B, poison label, 65 pounds.
IATA: Poison B, poison label, 1 liter (passenger), 30 kilograms (cargo).

2,4-DINITROPHENYL ACETATE
General Information
Description: Crystals.
Formula: $CH_3CO_2C_6H_3(NO_2)_2$.
Constants: Mol wt: 228.2, vap. d.: 7.87.
Hazard Analysis and Countermeasures
See nitro compounds of aromatic hydrocarbons and nitrates.

DINITROPHENYLAMINE. See 2,4-dinitroaniline.

DINITROPHENYLHYDRAZINE
General Information
Description: Red crystalline powder; slightly soluble in water and alcohol.
Formula: $(NO_2)_2C_6H_3NHNH_2$.
Constants: Mol wt: 198.2, mp: 200°C (approx.).
Hazard Analysis
Toxic Hazard Rating: U.
Explosion Hazard Rating: Dangerous explosive.
Countermeasures
Storage and Handling: Section 7.
Shipping Regulations: Section 11.
I.C.C. and CG: Explosive A.
IATA: Explosive, not acceptable (passenger and cargo).

2,4-DINITROPHENYL METHYL ETHER. See 2,4-dinitroanisole.

2,4-DINITRORESORCINOL
General Information
Synonyms: 2,4-Dinitro-1,3-benzendiol; styphnic acid.
Description: Yellow crystals.
Formula: $(NO_2)_2C_6H_2(OH)_2$.
Constants: Mol wt: 200.11, mp: 148°C, bp: sublimes and explodes, vap. d.: 6.79.
Hazard Analysis
Toxicity: Probably toxic. See nitro compounds of aromatic hydrocarbons.
Fire Hazard: See nitrates.
Explosion Hazard: Severe, when shocked or exposed to heat. Dinitroresorcinol and its lead salt are used in commercial priming compositions and blasting caps. It is also used to facilitate the ignition of lead azide. See also explosives, high.
Disaster Hazard: See explosives, high and nitrates.

Countermeasures
Ventilation Control: Section 2.
Storage and Handling: Section 7.

DINITROSO DIPHENYL AMINE FLUOSILICATE
General Information
Description: Indigo colored crystals, slightly soluble in alcohol.
Formula: $[(C_6H_5)_2N:NO]_2 \cdot H_2SiF_6$.
Constants: Mol wt: 540.5, mp: 124.5°C.
Hazard Analysis and Countermeasures
See fluosilicates and nitro compounds of aromatic hydrocarbons.

1,4-DINITROSOPIPERAZINE
General Information
Description: White crystals.
Formula: $C_4H_8N_2(NO_2)_2$.
Constants: Mol wt: 144.1, mp: 158°C, vap. d.: 4.97.
Hazard Analysis
Toxicity: Details unknown; a stomach insecticide.
Fire Hazard: See nitrates.
Disaster Hazard: See nitrates.
Countermeasures
Storage and Handling: Section 7.

4,4'-DINITRO-2,2'-STILBENE DISULFONIC ACID
General Information
Description: Yellow paste or brownish crystals.
Constant: Mol wt: 430.
Hazard Analysis
Toxicity: Unknown.
Fire Hazard: See nitrates.
Disaster Hazard: Dangerous; when heated to decomposition, it emits highly toxic fumes of oxides of nitrogen and sulfur; can react vigorously with reducing materials.
Countermeasures
Storage and Handling: Section 7.

3,5-DINITRO-o-TOLUAMIDE. See zoalene.

2,4-DINITROTOLUENE
General Information
Synonym: Dinitrotoluol.
Description: Yellow needles.
Formula: $(NO_2)_2C_6H_3CH_3$.
Constants: Mol wt: 182.13, mp: 69.5°C, bp: 300°C, d: 1.521 at 15°C, vap. d.: 6.27.
Hazard Analysis
Toxic Hazard Rating:
Acute Local: Irritant 1; Allergen 2; Inhalation 3.
Acute Systemic: Ingestion 3; Inhalation 3; Skin Absorption 3.
Chronic Local: Allergen 2.
Chronic Systemic: Ingestion 3; Inhalation 3; Skin Absorption 3.
Toxicology: Can cause anemia, methemoglobinemia, cyanosis and liver damage. See also trinitrotoluene.
TLV: ACGIH (recommended); 1.5 milligrams per cubic meter of air. May be absorbed via intact skin.
Fire Hazard: Moderate, when exposed to heat or flame; an oxidizer.
Explosion Hazard: Moderate, when exposed to heat.
Disaster Hazard: See explosives, high and nitrates.
Countermeasures
Ventilation Control: Section 2.
To Fight Fire: Water, carbon dioxide, dry chemical or carbon tetrachloride (Section 6).

Personnel Protection: Section 3.
First Aid: Section 1.
Storage and Handling: Section 7.

DINITROTOLUENES, liquid.
Shipping Regulations: Section 11.
IATA: Poison B, poison label, 1 liter (passenger), 220 liters (cargo).

DINITROTOLUOL. See 2,4-dinitrotoluene.

DINOL. See diazodinitrophenol.

DINONYL NAPHTHALENE
General Information
Description: Dark straw-colored viscous liquid.
Formula: $(C_9H_{19})_2C_{10}H_6$.
Constants: Mol wt: 380.6, bp: 200–270°C at 20 mm, d: 0.92–0.95 at 30°/20°C, vap. d.: 13.1.
Hazard Analysis
Toxicity: Unknown. See also naphthalene.
Disaster Hazard: Slight; when heated, it emits acrid fumes; can react with oxidizing materials.
Countermeasures
Storage and Handling: Section 7.

DINONYL PHENOL
General Information
Description: Clear liquid.
Formula: $C_6H_3OH(C_9H_{19})_2$.
Constants: Mol wt: 346.6, bp: 180–220°C at 10 mm, d: 0.914 at 25°C, vap. d.: 12.0.
Hazard Analysis
Toxicity: Details unknown. See also phenols.
Fire Hazard: Moderate, when exposed to heat or flame; can react with oxidizing materials.
Countermeasures
To Fight Fire: Foam, carbon dioxide, dry chemical or carbon tetrachloride. Section 6.
Storage and Handling: Section 7.

DINOSEB. See 2-sec-butyl-6,4-dinitrophenol.

DIOCTYL ADIPATE
General Information
Description: A clear liquid.
Formula: $(C_8H_{17}CO_2)_2C_4H_8$.
Constants: Mol wt: 370, bp: 214°C at 5 mm, fp: −79°C, flash p: 400°F (C.O.C.), d: 0.924–0.930 at 20°/20°C.
Hazard Analysis
Toxicity: See esters and adipic acid.
Disaster Hazard: Slight; when heated, it emits acrid fumes; can react with oxidizing materials.
Countermeasures
To Fight Fire: Foam, carbon dioxide, dry chemical or carbon tetrachloride (Section 6).
Storage and Handling: Section 7.

DIOCTYLAMINE. See di-2-ethylhexylamine.

DIOCTYLCHLOROPHOSPHATE
General Information
Description: Liquid.
Formula: $(C_8H_{17}O)_2POCl$.
Constants: Mol wt: 341, bp: 125°C at 0.2 mm, d: 0.9839 at 25°/25°C.
Hazard Analysis
Toxicity: Details unknown. An insecticide.
Disaster Hazard: Dangerous; when heated to decomposi-

TOXIC HAZARD RATING CODE (For detailed discussion, see Section 1.)

0 NONE: (a) No harm under any conditions; (b) Harmful only under unusual conditions or overwhelming dosage.

1 SLIGHT: Causes readily reversible changes which disappear after end of exposure.

2 MODERATE: May involve both irreversible and reversible changes; not severe enough to cause death or permanent injury.

3 HIGH: May cause death or permanent injury after very short exposure to small quantities.

U UNKNOWN: No information on humans considered valid by authors.

tion, it emits highly toxic fumes of oxides of phosphorus and chlorides.
Countermeasures
Storage and Handling: Section 7.

DI(n-OCTYL-n-DECYL) PHTHALATE
General Information
Description: Clear, oily liquid.
Formula: $C_{26}H_{42}O_2$.
Constants: Mol wt: 418.3, bp: 232–267°C at 5 mm, fp: −30°C, flash p: 426°F (O.C.), d: 0.968–0.977 at 25°/25°C, vap. d.: 14.4.
Hazard Analysis
Toxicity: See esters.
Fire Hazard: Slight, when exposed to heat or flame; can react with oxidizing materials.
Countermeasures
To Fight Fire: Foam, carbon dioxide, dry chemical or carbon tetrachloride (Section 6).
Storage and Handling: Section 7.

DIOCTYL FUMARATE
General Information
Synonym: Di-2-ethylhexyl fumarate.
Description: Clear, mobile liquid; mild odor.
Formula: $(C_8H_{17}CO_2)_2C_2H_2$.
Constants: Mol wt: 340, bp: 211–220°C, flash p: 365°F (C.O.C.), d: 0.942 at 20°/20°C.
Hazard Analysis
Toxicity: See esters and fumaric acid.
Fire Hazard: Slight; can react with oxidizing materials.
Countermeasures
To Fight Fire: Foam, carbon dioxide, dry chemical or carbon tetrachloride (Section 6).
Storage and Handling: Section 7.

DIOCTYL PHOSPHITE
General Information
Formula: $(C_8H_{17}O)_2PHO$.
Constants: Mol wt: 306.3, bp: 128°C at 0.5 mm, d: 0.9291 at 25°/25°C.
Hazard Analysis
Toxicity: Unknown.
Disaster Hazard: Dangerous; when heated to decomposition, it emits highly toxic fumes of oxides of phosphorus.
Countermeasures
Storage and Handling: Section 7.

DIOCTYL PHOSPHORIC ACID. See di-2-(ethylhexyl) phosphoric acid.

DIOCTYL PHTHALATE. See di-2-ethylhexyl phthalate.

DI-SEC-OCTYL PHTHALATE. See di(2-ethylhexyl) phthalate.

DIOCTYL SEBACATE
General Information
Description: Liquid.
Formula: $[(CH_2)_4CO_2C_8H_{17}]_2$.
Constants: Mol wt: 426.7, mp: −55°C, bp: 248°C at 4 mm, flash p: 465°F, d: 0.916 at 20°/20°C.
Hazard Analysis
Toxicity: See esters.
Fire Hazard: Slight, when exposed to heat or flame; can react with oxidizing materials.
Countermeasures
To Fight Fire: Foam, carbon dioxide, dry chemical or carbon tetrachloride (Section 6).
Storage and Handling: Section 7.

DIOCTYL SODIUM SULFOSUCCINATE
General Information
Description: White solid or clear, viscous liquid.

Formula: $C_8H_{17}CO_2CH_2CH(CO_2C_8H_{17})SO_3Na$.
Constant: Mol wt: 444.
Hazard Analysis
Toxic Hazard Rating:
Acute Local: Irritant 1; Ingestion 1; Inhalation 1.
Acute Systemic: U.
Chronic Local: Irritant 1; Ingestion 1; Inhalation 1.
Chronic Systemic: U.
Note: Used as a food additive permitted in food for human consumption (Section 10).
Disaster Hazard: Moderately dangerous; when heated to decomposition, it emits toxic fumes.
Countermeasures
Personal Hygiene: Section 3.
Storage and Handling: Section 7.

DIOCTYL TETRAHYDROPHTHALATE
General Information
Description: Crystals.
Formula: $C_6H_8(COOC_8H_{17})_2$.
Constants: Mol wt: 394.6, vap. d.:. 13.6.
Hazard Analysis
Toxic Hazard Rating:
Acute Local: Irritant 1; Ingestion 1; Inhalation 1.
Acute Systemic: U.
Chronic Local: U.
Chronic Systemic: U.
Fire Hazard: Slight, when exposed to heat or flame; can react with oxidizing materials (Section 6).
Countermeasures
Personal Hygiene: Section 3.
Storage and Handling: Section 7.

DI-n-OCTYLTIN DICHLORIDE. See di-2-ethylhexyltin dichloride.

DIONIN
General Information
Synonym: Ethylmorphine hydrochloride.
Description: White, microscopic, crystalline powder.
Formula: $C_{19}H_{23}NO_3 \cdot HCl \cdot 2H_2O$.
Constants: Mol wt: 385.88, mp: 125°C decomposes, vap. d.: 13.3.
Hazard Analysis
Toxic Hazard Rating:
Acute Local: Allergen 1.
Acute Systemic: Ingestion 3.
Chronic Local: Allergen 1.
Chronic Systemic: U.
Caution: May cause habituation. See also codeine.
Disaster Hazard: Dangerous. See Chlorides. (Section 6).
Countermeasures
Personal Hygiene: Section 3.
Storage and Handling: Section 7.

1,4-DIOXANE
General Information
Synonyms: p-Dioxane; diethylene oxide; diethylene dioxide.
Description: Colorless liquid.
Formula: $OCH_2CH_2OCH_2CH_2$.
Constants: Mol wt: 88.10, mp: 10°C, bp: 101.1°C, lel: 2.0%, uel: 22.2%, flash p: 54°F (C.C.), d: 1.0353 at 20°/4°C, autoign. temp.: 356°F, vap. press.: 40 mm at 25.2°C, vap. d.: 3.03.
Hazard Analysis
Toxic Hazard Rating:
Acute Local: U.
Acute Systemic: Ingestion 3; Inhalation 3; Skin Absorption 2.
Chronic Local: U.
Chronic Systemic: Ingestion 3; Inhalation 3; Skin Absorption 2.
TLV: ACGIH (recommended); 100 parts per million in air; 360 milligrams per cubic meter of air.
Toxicology: Exposure of animals to concentrations of 0.1 to

3% of dioxane vapor causes irritation of the eyes and nose, followed by narcosis and/or pulmonary edema and death. The irritative effects probably provide sufficient warning, in acute exposures, to enable the workman to leave exposure before he is seriously affected. On the other hand, repeated exposure to low concentrations has resulted in human fatalities, the organs chiefly affected being the liver and kidneys. Death resulted from acute hemorrhagic nephritis. The hepatic lesion consists of an acute central necrosis of the lobules. The brain and lungs may show acute edema.

In acute exposures, the signs and symptoms consist of irritation of the eyes and naso-pharynx, which may later subside, to be followed by headache, drowsiness, dizziness, and occasionally nausea and vomiting. In chronic exposures, there may be loss of appetite, nausea and vomiting, pain and tenderness in the abdomen and lumbar region, malaise, and enlargement of the liver without jaundice. There may be changes in the blood picture. Further exposure may result in suppression of urine, followed by uremia and death.

Fire Hazard: Dangerous, when exposed to heat or flame; can react vigorously with oxidizing materials.

Explosion Hazard: Moderate, when exposed to flame or by chemical reaction with oxidizers.

Countermeasures
Ventilation Control: Section 2.
To Fight Fire: Alcohol foam, carbon dioxide, dry chemical or carbon tetrachloride (Section 6).
Personnel Protection: Section 3.
First Aid: Section 1.
Storage and Handling: Section 7.
Shipping Regulations: Section 11.
  MCA warning label.
  IATA: Flammable liquid, red label, 1 liter (passenger), 40 liters (cargo).

DIOXOLANE
General Information
Description: Water-white liquid.
Formula: $OCH_2CH_2OCH_2$.
Constants: Mol wt: 74.08, mp: $-26.4°C$, bp: $74°C$, flash p: $35°F$ (O.C.), d: 1.065, vap. press.: 70 mm at $20°C$, vap. d.: 2.6.
Hazard Analysis
Toxic Hazard Rating:
  Acute Local: U.
  Acute Systemic: Ingestion 2; Inhalation 2.
  Chronic Local: U.
  Chronic Systemic: U.
Fire Hazard: Dangerous, when exposed to heat or flame; can react vigorously with oxidizing materials.
Explosion Hazard: Unknown.
Disaster Hazard: Dangerous! Keep away from heat and open flame!
Countermeasures
Ventilation Control: Section 2.
To Fight Fire: Alcohol foam, carbon dioxide, dry chemical or carbon tetrachloride (Section 6).
Personal Hygiene: Section 3.
Storage and Handling: Section 7.
Shipping Regulations: Section 11.
  IATA: Flammable liquid, red label, 1 liter (passenger), 40 liters (cargo).

DIOXIN. See (2,6-dimethyl-m-dioxan-4-yl)acetate.

DIOXOLONE-2. See ethylene carbonate.

DIOXYGEN DIFLUORIDE. See fluorine dioxide.

DIPAXIN. See 2-diphenylacetyl-1,3-indandione.

DIPENTAERYTHRITOL
General Information
Description: White powder; odorless.
Formula: $C_{10}H_{22}O_7$.
Constants: Mol wt: 254.3, mp: $212-220°C$, d: 1.33 at $25°/4°C$, vap. d.: 8.77.
Hazard Analysis
Toxicity: Details unknown.
Fire Hazard: Slight; Section 6.
Countermeasures
Storage and Handling: Section 7.

DIPENTENE
General Information
Synonym: Cinene.
Description: Colorless liquid, pleasant, lemon-like odor.
Formula: $C_{10}H_{16}$.
Constants: Mol wt: 136.23, bp: $174.6°C$, flash p: $113°F$, d: 0.865 at $18°C$; 0.845 at $20°C$, vap. press.: 1 mm at $14.0°C$, vap. d.: 4.66, autoign. remp.: $458°F$, lel: 0.7% at $302°F$, uel: 6.1% at $503°F$.
Hazard Analysis
Toxic Hazard Rating:
  Acute Local: Irritant 2; Allergen 1; Ingestion 2; Inhalation 1.
  Acute Systemic: U.
  Chronic Local: Allergen 1.
  Chronic Systemic: U.
Fire Hazard: Moderate, when exposed to heat or flame; can react with oxidizing materials.
Countermeasures
To Fight Fire: Foam, carbon dioxide, dry chemical or carbon tetrachloride (Section 6).
Ventilation Control: Section 2.
Personnel Protection: Section 3.
Storage and Handling: Section 7.

DIPENTYL MALEATE
General Information
Synonym: Diamyl maleate.
Description: Liquid; water white; faintly alcoholic odor.
Formula: $(CHCOOC_5H_{11})_2$.
Constants: Mol wt: 256, d: 0.981 at $20°C$, boiling range: $263-300°C$, flash p: $270°F$.
Hazard analysis
Toxic Hazard Rating: U. Limited animal experiments suggest low toxicity. See also esters and maleic acid.
Fire Hazard: Slight, when exposed to heat or flame.

DIPHENADIONE. See 2-diphenylacetyl-1,3-indandione.

DIPHENAMIDE. See N,N-dimethyl-2,2-diphenylacetamide.

DI-p-PHENETYL DITELLURIDE
General Information
Description: Orange-brown crystals.
Formula $(C_2H_5OC_6H_4)_2Te_2$.
Constants: Mol wt: 497.5, d: 1.666, mp: $108°C$.
Hazard Analysis and Countermeasures
See tellurium compounds.

DIPHENOLIC ACID
General Information
Synonym: DPA.

TOXIC HAZARD RATING CODE *(For detailed discussion, see Section 1.)*

0 NONE: (a) No harm under any conditions; (b) Harmful only under unusual conditions or overwhelming dosage.

1 SLIGHT: Causes readily reversible changes which disappear after end of exposure.

2 MODERATE: May involve both irreversible and reversible changes not severe enough to cause death or permanent injury.

3 HIGH: May cause death or permanent injury after very short exposure to small quantities.

U UNKNOWN: No information on humans considered valid by authors

Description: Crystals. Soluble in water, acetone, alcohol and acetic acid.
Formula: $C_{17}H_{18}O_4$.
Constants: Mol wt: 286.3, mp: 172° C.
Hazard Analysis
Toxicity: Details unknown. See phenol.

DIPHENYL
General Information
Synonym: Biphenyl.
Description: White scales, pleasant odor.
Formula: $C_6H_5C_6H_5$.
Constants: Mol wt: 154.2, mp: 70° C, bp: 255° C, flash p: 235° F (C.C.), d: 0.991 at 75°/4° C, autoign. temp. 1004° F, vap. d.: 5.31, lel: 0.6% at 232°, uel: 5.8% at 331° F.
Hazard Analysis
Toxic Hazard Rating:
  Acute Local: U.
  Acute Systemic: Ingestion 3; Inhalation 3.
  Chronic Local: U.
  Chronic Systemic: Ingestion 2; Inhalation 2.
TLV: ACGIH (tentative) 0.2 part per million; 1 milligram per cubic meter.
Toxicology: Experimentally has caused paralysis and convulsions in animals.
Fire Hazard: Slight, when exposed to heat or flame; can react with oxidizing materials.
Spontaneous Heating: No.
Countermeasures
To Fight Fire: Carbon dioxide, dry chemical or carbon tetrachloride (Section 6).
Ventilation Control: Section 3.
Personal Hygiene: Section 3.
Storage and Handling: Section 7.

DIPHENYLACETIC ACID
General Information
Synonym: Diphenylmethane-α-carboxylic acid.
Description: White crystals.
Formula: $(C_6H_5)_2CHCOOH$.
Constants: Mol wt: 212.2, mp: 147–148.2° C, bp: sublimes, vap. d.:. 7.3.
Hazard Analysis
Toxicity: Details unknown. See also diphenyl.
Fire Hazard: Slight; Section 6.
Countermeasures
Storage and Handling: Section 7.

DIPHENYL ACETYLENE
General Information
Synonym: Tolan; diphenyl ethyne.
Description: Monoclinic pseudorhombic crystals; insoluble in water; freely soluble in ether, hot alcohol.
Formula: $(C_6H_5)C:C(C_6H_5)$.
Constants: Mol wt: 178.22, mp: 60–61° C, bp: 300° C, bp: 170° C at 19 mm, d: 0.966 (100/4° C).
Hazard Analysis
See diphenyl ethylene.

2-DIPHENYLACETYL-1,3-INDANDIONE
General Information
Synonym: Diphenadione; dipaxin.
Description: Pale yellow crystals. Soluble in acetone, and acetic acid.
Formula: $C_{23}H_{16}O_3$.
Constants: Mol wt: 340.4, mp: 147° C.
Hazard Analysis
Toxic Hazard Rating:
  Acute Local: U.
  Acute Systemic: Ingestion 2.
  Chronic Local: U.
  Chronic Systemic: Ingestion 3.
Toxicology: Inhibits blood clotting leading to hemorrhages. Action similar to that of warfarin.

DIPHENYLAMINE
General Information
Synonym: Phenylaniline; anilinobenzene.
Description: Crystals, floral odor. Soluble in benzene, ether and carbon disulfide.
Formula: $(C_6H_5)_2NH$.
Constants: Mol wt: 169.24, mp: 52.9° C, bp: 302.0° C, flash p: 307° F (C.C.), d: 1.16, autoign. temp. 1173° F, vap. press.: 1 mm at 108.3° C, vap. d.: 5.82.
Hazard Analysis
Toxic Hazard Rating:
  Acute Local: U.
  Acute Systemic: Ingestion 3; Inhalation 3; Skin Absorption 3.
  Chronic Local: U.
  Chronic Systemic: Ingestion 3; Inhalation 3; Skin Absorption 3.
Toxicity: Action similar to aniline but less severe. See also aniline and amines.
TLV: ACGIH (tentative) 10 milligrams per cubic meter of air.
Fire Hazard: Slight, when exposed to heat or flame.
Spontaneous Heating: No.
Disaster Hazard: Dangerous; when heated to decomposition, it emits highly toxic fumes; can react with oxidizing materials.
Countermeasures
Ventilation Control: Section 2.
To Fight Fire: Water, carbon dioxide, dry chemical or carbon tetrachloride (Section 6).
Personnel Protection: Section 3.
First Aid: Section 1.
Storage and Handling: Section 7.

DIPHENYLAMINE ARSENIOUS OXIDE
Hazard Analysis and Countermeasures
See arsenic compounds.

DIPHENYLAMINECHLOROARSINE
General Information
Synonym: Adamsite.
Description: Light Yellow to green granules, irritant odor.
Formula: $NH(C_6H_4)_2AsCl$.
Constants: Mol wt: 277.6, mp: 195° C, bp: 410° C (decomposes), d: 1.65, vap. press.: very low at 20° C, vap. d.: 9.6.
Hazard Analysis
Toxic Hazard Rating:
  Acute Local: Irritant 3; Ingestion 3; Inhalation 3.
  Acute Systemic: Ingestion 3; Inhalation 3.
  Chronic Local: U.
  Chronic Systemic: Ingestion 3; Inhalation 3.
Caution: A vomiting type of military poison and a nonpersistent military gas. See also arsenic compounds.
Disaster Hazard: Dangerous; when heated it emits highly toxic fumes of arsenic.
Countermeasures
Ventilation Control: Section 2.
Personnel Protection: Section 3.
First Aid: Section 1.
Storage and Handling: Section 7.
Shipping Regulations: Section 11.
  I.C.C.: Poison C (gas, liquid or solid); tear gas label, 75 pounds.
  Coast Guard Classification: Poison C (gas, liquid or solid); tear gas label.
  IATA (solid): Poison C, poison label, not acceptable (passenger), 10 kilograms (cargo).
  IATA (gas or liquid): Poison C, poison label, not acceptable (passenger), 35 kilograms (cargo).

DIPHENYL AMINE HYDROCHLORIDE
General Information
Description: White crystals, become blue in air; freely soluble in water and alcohol.

Formula: $(C_6H_5)_2NHHCl$.
Constant: Mol wt: 205.68.
Hazard Analysis
See diphenyl amine.
Disaster Hazard: Dangerous. See chlorides.

DIPHENYL AMINE SULFATE
General Information
Description: White to yellowish powder, insoluble in water, soluble in alcohol, sulfuric acid.
Formula: $(C_6H_5)_2NHH_2SO_4$.
Constants: Mol wt: 267.30, mp: 123-125°C.
Hazard Analysis
See diphenyl amine.
Disaster Hazard: Dangerous. See sulfates.

1,2-DIPHENYLBENZENE. See o-terphenyl.

DIPHENYLBORIC ACID
General Information
Synonym: Diphenylhydroxyborine.
Description: Colorless crystals, insoluble in water.
Formula: $(C_6H_5)_2BOH$.
Constants: Mol wt: 182.0, mp: 226°C, bp: 215-235°C at 17 mm.
Hazard Analysis
Toxicity: Details unknown. See also boron compounds.

DIPHENYLBORON BROMIDE
General Information
Description: Colorless thick liquid or crystals; decomposed by water.
Formula: $(C_6H_5)_2BBr$.
Constants: Mol wt: 244.9, mp: 25°C, bp: 150-160°C at 8 mm.
Hazard Analysis
Toxicity: Details unknown. See also boron compounds.
Disaster Hazard: Dangerous. See bromides.

DIPHENYLBORON CHLORIDE
General Information
Description: Colorless liquid; decomposed by water.
Formula: $(C_6H_5)_2BCl$.
Constants: Mol wt: 200.5, bp: 271°C.
Hazard Analysis
Toxicity: Details unknown. See boron compounds.
Disaster Hazard: Dangerous; See chlorides.

DIPHENYLBROMOARSINE
General Information
Description: Crystals.
Formula: $(C_6H_5)_2AsBr$.
Constants: Mol wt: 309.1, mp: 54°C, vap. d.: 10.7.
Hazard Analysis
Toxic Hazard Rating:
Acute Local: Irritant 3; Allergen 1; Ingestion 3; Inhalation 3.
Acute Systemic: Ingestion 3; Inhalation 3.
Chronic Local: Allergen 1.
Chronic Systemic: Ingestion 3; Inhalation 3.
Fire Hazard: Moderate, when exposed to heat or flame (Section 6).
Disaster Hazard: Dangerous; See arsenic.
Countermeasures
Ventilation Control: Section 2.
Personnel Protection: Section 3.
First Aid: Section 1.
Storage and Handling: Section 7.

DIPHENYL CARBINOL. See benzhydrol.

DIPHENYLCHLOROARSINE
General Information
Description: Colorless crystals when pure; technical product is dark brown liquid.
Formula: $(C_6H_5)_2AsCl$.
Constants: Mol wt: 264.57, bp: 333°C (decomposes), fp: 44°C, d: 1.363 at 40°C (solid): 1.358 at 45°C (liquid), vap. press.: 0.00049 mm at 20°C, vap. d.: 9.15.
Hazard Analysis
Toxic Hazard Rating:
Acute Local: Irritant 3; Ingestion 3; Inhalation 3.
Acute Systemic: Ingestion 3; Inhalation 3.
Chronic Local: U.
Chronic Systemic: Ingestion 3; Inhalation 3.
Caution: A powerful irritant military poison. Can be decontaminated by use of chlorine or caustic soda in confined spaces.
Toxicology: A non-persistent gas. Exposure yields cold-like symptoms, plus headache, vomiting and nausea. Relatively nontoxic to eyes and skin, mainly an irritant.
Disaster Hazard: Dangerous; see arsenic.
Countermeasures
Ventilation Control: Section 2.
Personnel Protection: Section 3.
First Aid: Section 1.
Storage and Handling: Section 7.
Shipping Regulations: Section 11.
I.C.C.: Poison C; tear gas label, 20 pounds.
Coast Guard Classification (solid): Poison C; tear gas label.
IATA: Poison C, poison label, not acceptable (passenger), 10 kilograms (cargo).

DIPHENYLCYANOARSINE
General Information
Description: Colorless prisms, characteristic odor resembling that of a mixture of bitter almonds and garlic.
Formula: $(C_6H_5)_2AsCN$.
Constants: Mol wt: 255.0, bp: 213°C at 21 mm, fp: 31°C, d: 1.33 at 20°C, vap. press.: 0.0002 mm at 20°C, vap. d.: 8.75.
Hazard Analysis
Toxic Hazard Rating:
Acute Local: Irritant 3; Ingestion 3; Inhalation 3.
Acute Systemic: Ingestion 3; Inhalation 3.
Chronic Local: U.
Chronic Systemic: Ingestion 3; Inhalation 3.
Toxicology: Acts as an irritant only to the eyes. Exposure yields cold-like symptoms, plus headache, vomiting and nausea. A vomiting gas of the non-persistent variety. Decontamination needed only in confined spaced by alkali solution. See also cyanides and arsenic compounds.
Disaster Hazard: Dangerous; see cyanides and arsenic.
Countermeasures
Ventilation Control: Section 2.
Personnel Protection: Section 3.
First Aid: Section 1.
Storage and Handling: Section 7.

DIPHENYL DECYL PHOSPHITE
General Information
Description: Nearly water white liquid; insoluble in water.
Formula: $(C_6H_5O)_2POC_{10}H_{21}$.
Constants: Mol wt: 374, d: 1.023 (25/15.5°C) mp: 18°C, flash p: 425°F (O.C.).
Hazard Analysis
Toxic Hazard Rating: U.
Fire Hazard: Slight when exposed to heat or flame.

TOXIC HAZARD RATING CODE (For detailed discussion, see Section 1.)

0 NONE: (a) No harm under any conditions; (b) Harmful only under unusual conditions or overwhelming dosage.

1 SLIGHT: Causes readily reversible changes which disappear after end of exposure.

2 MODERATE: May involve both irreversible and reversible changes; not severe enough to cause death or permanent injury.

3 HIGH: May cause death or permanent injury after very short exposure to small quantities.

U UNKNOWN: No information on humans considered valid by authors.

Disaster Hazard: Dangerous. See phosphates.
Countermeasures
To Fight Fire: Water, foam (Section 6).
Storage and Handling: Section 7.

DIPHENYLDICHLOROSILANE
General Information
Description: Colorless liquid.
Formula: $(C_6H_5)_2SiCl_2$.
Constants: Mol wt: 245.22, mp: $-22°C$, bp: $303°C$,
 d: 1.19 at 20° C, vap. d.: 8.45.
Hazard Analysis
Toxic Hazard Rating:
 Acute Local: Irritant 3; Ingestion 3; Inhalation 3.
 Acute Systemic: U.
 Chronic Local: U.
 Chronic Systemic: U.
Toxicology: See also chlorosilanes.
Disaster Hazard: Moderately dangerous; when heated to
 decomposition or on contact with acid or acid fumes, it
 emits toxic fumes; will react with water or steam to
 produce heat, toxic and corrosive fumes; can react
 vigorously with oxidizing materials.
Countermeasures
Ventilation Control: Section 2.
Personnel Protection: Section 3.
Storage and Handling: Section 7.
Shipping Regulations: Section 11.
 I.C.C.: Corrosive liquid; white label, 10 gallons.
 IATA: Corrosive liquid, white label, not acceptable (pas-
 senger), 40 liters (cargo).
 Coast Guard Classification: Corrosive liquid; white label.

DIPHENYL DISULFIDE
General Information
Description: White powder.
Formula: $C_6H_5-S-S-C_6H_5$.
Constants: Mol wt: 218.3, mp: 59.5–60.0° C, bp: 310° C.
Hazard Analysis
Toxicity: Details unknown. See also sulfides and alkyl disul-
 fides.
Fire Hazard: Moderate, when exposed to heat or flame
 (Section 6).
Disaster Hazard: Dangerous; see sulfur compounds. Can
 react with oxidizing materials.
Countermeasures
Storage and Handling: Section 7.

DIPHENYLENE OXIDE
General Information
Synonym: Dibenzofuran.
Description: Colorless crystals.
Formula: $C_6H_4OC_6H_4$.
Constants: Mol wt: 168.2, mp: 87° C, bp: 288° C, vap. d.:
 5.8.
Hazard Analysis
Toxicity: Details unknown; an insecticide.
Fire Hazard: Slight; Section 6.
Countermeasures
Storage and Handling: Section 7.

1,1-DIPHENYLETHANE
General Information
Description: Pale, yellowish liquid, aromatic odor.
Formula: $(C_6H_5)_2CHCH_3$.
Constants: Mol wt: 182.3, mp: $-20°C$, bp: 272°C,
 flash p: 264° F, d: 0.987 at 25° /25° C, vap. d.: 6.28.
Hazard Analysis
Toxicity: Details unknown. Probably irritant and narcotic
 in high concentration.
Fire Hazard: Slight, when exposed to heat or flame.
Disaster Hazard: Moderately dangerous; can react with
 oxidizing materials.
Countermeasures
Storage and Handling: Section 7.

To Fight Fire: Foam, carbon dioxide, dry chemical or
 carbon tetrachloride (Section 6).

DIPHENYL ETHER. See diphenyl oxide.

DI-β-PHENYLETHYLAMINE
General Information
Description: Colorless to slightly yellow liquid.
Formula: $(C_6H_5CH_2CH_2)_2NH$.
Constants: Mol wt: 225.3, bp: 204° C at 22 mm, d: 1.0 at
 25° /25° C, vap. d.: 7.76.
Hazard Analysis
Toxicity: Details unknown. See also amines.
Fire Hazard: Slight; Section 6.
Countermeasures
Storage and Handling: Section 7.

DIPHENYL ETHYLENE
General Information
Synonym: Stibene: tolulyne: trans form of α, β-diphenyl
 ethylene.
Description: Colorless or slightly yellow crystals; in-
 soluble in water; soluble in 90 parts cold alcohol, 13
 parts boiling alcohol; freely soluble in benzene, ether.
Formula: $C_6H_5CH:CHC_6H_5$.
Constants: Mol wt: 180.24, mp: 124–125° C, bp: 306–
 307° C, d: 0.9707.
Hazard Analysis
Toxic Hazard Rating: U. Animal experiments show low
 oral and inhalant toxicity but greater effects from skin
 absorption.

DIPHENYLGERMANIUM
General Information
Description: White crystals; insoluble in water.
Formula: $[(C_6H_5)_2Ge]_4$.
Constants: Mol wt: 907.2, mp: 295° C.
Hazard Analysis and Countermeasures
See germanium compounds.

DIPHENYLGERMANIUM DIBROMIDE
General Information
Description: Colorless liquid, hydrolyzed by water.
Formula: $(C_6H_5)_2GeBr_2$.
Constants: Mol wt: 386.6, bp: 207° C at 512 mm.
Hazard Analysis and Countermeasures
See germanium compounds and bromides.

DIPHENYLGERMANIUM DICHLORIDE
General Information
Description: Colorless liquid; hydrolyzed by water.
Formula: $(C_6H_5)_2GeCl_2$.
Constants: Mol wt: 297.7, d: 0.71, mp: 9° C, bp: 223° C
 at 12 mm.
Hazard Analysis and Countermeasures
See germanium compounds and chlorides.

DIPHENYLGERMANIUM DIFLUORIDE
General Information
Description: Colorless liquid; hydrolyzed by water.
Formula: $(C_6H_5)_2GeF_2$.
Constants: Mol wt: 264.8, bp: 100° C at 0.007 mm.
Hazard Analysis and Countermeasures
See fluorides and germanium compounds.

DIPHENYLGLYCOLIC ACID. See benzilic acid.

DIPHENYLGUANIDINE
General Information
Description: White powder.
Formula: $NH:C(NHC_6H_5)_2$.
Constants: Mol wt: 211.1, mp: 145° C, d: 1.15 at 25° C.
Hazard Analysis
Toxicity: Details unknown. Animal experiments suggest
 high toxicity.
Disaster Hazard: Moderately dangerous; when heated to
 decomposition, it emits toxic fumes.

Countermeasures
Storage and Handling: Section 7.

DIPHENYLGUANIDINE PHTHALATE
General Information
Description: White to light gray power.
Formula: $C_6H_4(COOH)_2 \cdot [HNC(HNC_6H_5)_2]_2 \cdot 1/2H_2O$.
Constants: Mol wt: 604.7, mp: 178°C, d: 1.2 at 25°C.
Hazard Analysis
Toxicity: Details unknown. See also esters and diphenyl guanidine. Possibly a skin sensitizer.
Fire Hazard: Slight; Section 6.
Countermeasures
Storage and Handling: Section 7.

DIPHENYLHYDROXYBORINE. See diphenylboric acid.

DIPHENYLINE. See 2,4′-biphenyl diamine.

DIPHENYL KETENE
General Information
Synonym: Diphenylethenone.
Description: Reddish yellow liquid.
Formula: $(C_6H_5)_2C:CO$.
Constants: Mol wt: 194.22, d: 1.1107 (13.7/4°C), bp: 265–276°C (decomposition).
Hazard Analysis
Toxic Hazard Rating: U. Vapors may have a narcotic effect.

DIPHENYL KETONE. See benzophenone.

DIPHENYLMERCURY
General Information
Description: White crystals, insoluble in water.
Formula: $(C_6H_5)_2Hg$.
Constants: Mol wt: 354.8, d: 2.318, mp: 122°C (sublimes). bp: 204°C at 10.5 mm.
Hazard Analysis and Countermeasures
See mercury compounds, organic.

DIPHENYLMETHANE
General Information
Synonym: Benzylbenzene.
Description: Liquid.
Formula: $(C_6H_5)_2CH_2$.
Constants: Mol wt: 168.23, mp: 26.5°C, bp: 264.5°C, flash p: 266°F (C.C.), d: 1.006, vap. press.: 1 mm at 76.0°C, vap. d.: 5.79, autoign. temp.: 907°F.
Hazard Analysis
Toxicity: Details unknown. Probably irritant; narcotic in high concentrations.
Fire Hazard: Slight, when exposed to heat or flame; can react with oxidizing materials.
Spontaneous Heating: No.
Countermeasures
To Fight Fire: Foam, carbon dioxide, dry chemical or carbon tetrachloride (Section 6).
Storage and Handling: Section 7.

DIPHENYLMETHANE-α-CARBOXYLIC ACID. See diphenylacetic acid.

DIPHENYLMETHANE DIISOCYANATE. See methylene bis(4-phenyl isocyanate).

DIPHENYL METHYL BROMIDE
General Information
Synonym: Benzhydryl bromide.

Description: Solid; decomposes in hot water; soluble in alcohol; very soluble in benzene.
Formula: $BrCH(C_6H_5)_2$.
Constants: Mol wt: 247, mp: 45°C, bp: 193°C at 26 mm.
Hazard Analysis
Toxic Hazard Rating: U.
Disaster Hazard: Dangerous. See bromides.
Countermeasures
Shipping Regulations: Section 11.
IATA: Corrosive liquid, white label, 1 liter (passenger), 5 liters (cargo).

DIPHENYL MONO(p-tert-BUTYLPHENYL) PHOSPHATE
General Information
Synonym: p-tert-Butyl phenyl diphenyl phosphate.
Description: Colorless to pale straw-colored liquid.
Formula: $(C_6H_5)_2[(CH_3)_3CC_6H_4]PO_4$.
Constants: Mol wt: 382.4, bp: 245–260°C at 5 mm, fp: 0°C, flash p: 435°F, d: 1.16 at 25°/25°C.
Hazard Analysis
Toxicity: Details unknown.
Fire Hazard: Slight, when exposed to heat or flame.
Disaster Hazard: Moderately dangerous; when heated to decomposition, it emits toxic fumes; can react with oxidizing materials.
Countermeasures
Storage and Handling: Section 7.
To Fight Fire: Water, foam, carbon dioxide, dry chemical or carbon tetrachloride (Section 6).

DIPHENYL MONO(O-XENYL) PHOSPHATE
General Information
Formula: $(C_6H_5O)_2PO(OC_6H_4C_6H_5)$.
Constants: Mol wt: 402, d: 1.2, bp: 482–545°F at 5 mm, flash p: 437°F.
Hazard Analysis
Toxic Hazard Rating: U.
Fire Hazard: Slight when exposed to heat or flame.
Disaster Hazard: Dangerous. See phosphates.
Countermeasures
To Fight Fire: Water, foam (Section 6).
Storage and Handling: Section 7.

DIPHENYL OXIDE
General Information
Synonym: Diphenyl ether.
Description: Colorless crystals, geranium odor.
Formula: $(C_6H_5)_2O$.
Constants: Mol wt: 170.20, mp: 28°C, bp: 257°C, flash p: 239°F, d: 1.0728 at 20°C, vap. d.: 5.86, autoign. temp.: 1144°F.
Hazard Analysis
Toxic Hazard Rating:
Acute Local: Irritant 1; Ingestion 1; Inhalation 1.
Acute Systemic: U.
Chronic Local: U.
Chronic Systemic: U.
Fire Hazard: Moderate, when exposed to heat or flame; can react with oxidizing materials (Section 6).
Spontaneous Heating: No.
Explosion Hazard: See ethers.
Countermeasures
Ventilation Control: Section 2.
To Fight Fire: Water, foam, carbon dioxide, dry chemical or carbon tetrachloride. See also ethers. (Section 6).
Personal Hygiene: Section 3.
Storage and Handling: Section 7.

TOXIC HAZARD RATING CODE (For detailed discussion, see Section 1.)

0 NONE: (a) No harm under any conditions; (b) Harmful only under unusual conditions or overwhelming dosage.

1 SLIGHT: Causes readily reversible changes which disappear after end of exposure.

2 MODERATE: May involve both irreversible and reversible changes not severe enough to cause death or permanent injury.

3 HIGH: May cause death or permanent injury after very short exposure to small quantities.

U UNKNOWN: No information on humans considered valid by author.

N,N-DIPHENYL-p-PHENYLENEDIAMINE
General Information
Description: Solid.
Formula: $(C_6H_5)_2NC_6H_4NH_2$.
Constants: Mol wt: 260.3, d: 1.20, vap. d.: 9.0.
Hazard Analysis
Toxicity: A weak allergen. See also p-phenylenediamine.
Fire Hazard: Slight, when exposed to heat or flame (Section 6).
Disaster Hazard: Moderately dangerous; when heated to decomposition, it emits toxic fumes; can react with oxidizing materials.
Countermeasures
Storage and Handling: Section 7.

DIPHENYL PHTHALATE
General Information
Description: Solid.
Formula: $C_6H_4(COOC_6H_5)_2$.
Constants: Mol wt: 318.3, mp: 70.7°C, bp: 405°C, flash p: 435°F (C.C.), d: 1.28.
Hazard Analysis
Toxicity: Moderate. See phenol and phthalic acid.
Fire Hazard: Slight, when exposed to heat or flame; can react with oxidizing materials.
Countermeasures
To Fight Fire: Water, carbon dioxide, dry chemical or carbon tetrachloride (Section 6).
Storage and Handling: Section 7.

N,N'-DIPHENYL PIPERAZINE
General Information
Description: Crystals.
Formula: $C_6H_5N(CH_2)_4NC_6H_5$.
Constants: Mol wt: 238.3, mp: 167°C, vap. d.: 8.22.
Hazard Analysis
Toxicity: Details unknown; piperazines generally have a low order of toxicity.
Countermeasures
Storage and Handling: Section 7.

DIPHENYL-sec-PROPYLGERMANIUM BROMIDE
General Information
Description: Colorless liquid.
Formula: $(C_6H_5)_2(C_3H_7)GeBr$.
Constants: Mol wt: 349.8, bp: 215–250°C at 13 mm.
Hazard Analysis and Countermeasures
See bromides and germanium compounds.

3,5-DIPHENYLPYRIDINE
General Information
Description: Pale yellow to green solid. Insoluble in water.
Formula: $(C_6H_5)_2C_5H_3N$.
Constants: Mol wt: 231.3, mp: 137°C.
Hazard Analysis
Toxicology: Based upon animal experiments, it appears to be practically not toxic.
Disaster Hazard: Dangerous; when heated to decomposition, it emits toxic fumes.

1,4-DIPHENYLSEMICARBAZIDE
General Information
Description: Crystals.
Formula: $C_6H_5NHNHCONHC_6H_5$.
Constants: Mol wt: 227.3, vap. d.: 7.8.
Hazard Analysis
Toxicity: Details unknown; an insecticide,
Disaster Hazard: Dangerous; when heated to decomposition, it emits toxic fumes.
Countermeasures
Storage and Handling: Section 7.

DIPHENYL SULFOXIDE
General Information
Description: Crystals.
Formula: $(C_6H_5)_2SO$.

Constants: Molecular weight: 202.3, vap. d.: 7.0.
Hazard Analysis
Toxicity: Details unknown. A fungicide. See also diphenyl.
Disaster Hazard: Dangerous; see sulfates.
Countermeasures
Storage and Handling: Section 7.

DIPHENYL THIOCARBAZONE
General Information
Synonym: Octhiozone.
Description: Bluish black, crystalline powder; insoluble in water; sparingly soluble in alcohol; freely soluble in carbon tetrachloride and chloroform.
Formula: $C_6H_5N:NCSNHNHC_6H_5$.
Constant: Mol wt: 256.32.
Hazard Analysis
Toxic Hazard Rating: U. Has caused eye injury and glycosuria in animals.
Disaster Hazard: Dangerous, when heated to decomposition it emits highly toxic fumes.
Countermeasures
Storage and Handling: Section 7.

DIPHENYLTHIOL DIPHENYL STANNANE
General Information
Description: Solid.
Formula: $(C_6H_5)_2Sn(C_6H_5S)_2$.
Constants: Mol wt: 491.2, mp: 65°C.
Hazard Analysis and Countermeasures
See tin compounds and sulfides.

N,N'-DIPHENYLTHIOUREA. See thiocarbanilide.

1,3-DIPHENYL-2-THIOUREA. See thiocarbanilide.

1,3-DIPHENYLTRIAZINE. See α-diazoamidobenzol.

DIPHENYLTIN
General Information
Description: Yellow powder, insoluble in water.
Formula: $(C_6H_5)_2Sn$.
Constants: Mol wt: 272.9, mp: 226°C.
Hazard Analysis
Toxicity: Details unknown. See tin compounds.

DIPHENYLTIN BROMIDE
General Information
Description: Colorless crystals soluble in alcohol and ether.
Formula: $(C_6H_5)_2SnBr_2$.
Constants: Mol wt: 432.7, mp: 38°C, bp: 230°C at 42 mm.
Hazard Analysis and Countermeasures
See tin compounds and bromides.

DIPHENYLTIN DICHLORIDE
General Information
Description: Colorless crystals decomposed by water.
Formula: $(C_6H_5)_2SnCl_2$.
Constants: Mol wt: 343.8, mp: 42°C, bp: 333–337°C (decomposes).
Hazard Analysis and Countermeasures
See tin compounds and chlorides.

DIPHENYLTIN DIFLUORIDE
General Information
Description: Solid.
Formula: $(C_6H_5)_2SnF_2$.
Constants: Mol wt: 310.9, mp: 360°C.
Hazard Analysis and Countermeasures
See fluorides and tin compounds.

DIPHENYLTIN DIIODIDE
General Information
Description: Colorless crystals, insoluble in water.
Formula: $(C_6H_5)_2SnI_2$.
Constants: Mol wt: 526.7, mp: 72°C, bp: 176–182°C at 2 mm.

Hazard Analysis and Countermeasures
See tin compounds and iodides.

DIPHENYLTIN HYDROXY CHLORIDE
General Information
Description: White powder, insoluble in water.
Formula: $(C_6H_5)_2Sn(OH)(Cl)$.
Constants: Mol wt: 325.4, mp: 187° C.
Hazard Analysis and Countermeasures
See tin compounds and chlorides.

DIPHENYLTIN OXIDE
General Information
Description: Colorless powder insoluble in water.
Formula: $(C_6H_5)_2SnO$.
Constants: Mol wt: 288.9, mp: infusible.
Hazard Analysis
Toxicity: Details unknown. See tin compounds.

DIPHENYLZINC
General Information
Description: White crystals decomposed by cold water.
Formula: $(C_6H_5)_2Zn$.
Constants: Mol wt: 219.6, mp: 107° C.
Hazard Analysis
Toxicity: Details unknown. See zinc compounds and phenol.

DIPHOSGENE
General Information
Synonym: Trichloromethyl chloroformate.
Description: Colorless liquid, odor of phosgene.
Formula: $ClCO_2CCl_3$.
Constants: Mol wt: 197.9, mp: −57° C, bp: 128° C, d: 1.653 at 14°/4° C, vap. press.: 10.3 mm at 20° C, vap. d.: 6.9.
Hazard Analysis
Toxic Hazard Rating:
Acute Local: Irritant 3; Ingestion 3; Inhalation 3.
Acute Systemic: Ingestion 3; Inhalation 3.
Chronic Local: U.
Chronic Systemic: U.
Toxicology: A military poison. A lung irritant. It is but slightly lachrimatory. Its physiological action like phosgene is delayed. To decontaminate in enclosed spaces, use ammonia or steam. It is moderately persistent. See also phosgene.
Disaster Hazard: Dangerous; when heated it emits highly toxic fumes; will react with water or steam to produce toxic and corrosive fumes.
Countermeasures
Ventilation Control: Section 2.
Personnel Protection: Section 3.
Storage and Handling: Section 7.
Shipping Regulations: Section 11.
I.C.C.: Poison A; poison gas label, not accepted.
Coast Guard Classification: Poison A; poison gas label.
IATA: Poison A, not acceptable (passenger and cargo).

DIPHOSPHORUS PENTASELENIDE. See phosphorus pentaselenide.

DIPICRYLAMINE. See 2,4,6,2′,4′,6′-hexanitrodiphenylamine.

DIPICRYL SULFIDE. See hexanitrodiphenyl sulfide.

1,10-DIPIPERIDINEDECANE
General Information
Description: Solid.

Formula: $(CH_2)_5NCH_2(CH_2)_8CH_2N(CH_2)_5$.
Constants: Mol wt: 308.5, vap. d.: 10.7.
Hazard Analysis
Toxicity: Details unknown. Piperidines usually have a low toxicity.
Countermeasures
Storage and Handling: Section 7.

DIPOTASSIUM NITROACETATE
General Information
Description: Solid.
Formula: $K_2(NO_2)C_2O_2H$.
Constant: Mol wt: 181.2.
Hazard Analysis
Toxicity: Unknown.
Explosion Hazard: Severe, when shocked or exposed to heat (Section 7).
Disaster Hazard: Highly dangerous; shock and heat will explode it: when heated to decomposition, it emits highly toxic fumes.
Countermeasures
Storage and Handling: Section 7.

DIPOTASSIUM-o-PHOSPHATE. See potassium phosphate, dibasic.

DIPPING ACID. See sulfuric acid.

DIPROPARGYL
General Information
Synonym: 1,5-Hexadiyne.
Description: Colorless liquid.
Formula: $CH:CCH_2CH_2C:CH$.
Constants: Mol wt: 78.1, bp: 86° C, fp: −6° C, d: 0.8049 at 20°/4° C, vap. d.: 2.69.
Hazard Analysis
Toxicity: Details unknown.
Fire Hazard: Moderate, when exposed to heat or flame (Section 6).
Explosion Hazard: Moderate, when exposed to heat (Section 7).
Disaster Hazard: Moderately dangerous; when heated to decomposition it explodes; can react vigorously with oxidizing materials.
Countermeasures
Storage and Handling: Section 7.

DIPROPYL ALUMINUM HYDRIDE
General Information
Formula: $(C_3H_7)_2AlH$.
Constant: Mol wt: 114.
Hazard Analysis
Toxic Hazard Rating: U. See also hydrides.
Fire Hazard: Dangerous, ignites spontaneously in air.
Countermeasures
To Fight Fire: Do not use water, foam, or halogenated extinguishing agents (Section 6).
Storage and Handling: Section 7.

DI-n-PROPYLAMINE
General Information
Synonym: Dipropylamine.
Description: Water-white, amine odor.
Formula: $(C_3H_7)_2NH$.
Constants: Mol wt: 101.19, mp: −40° C, bp: 105° C flash p: 63° F (O.C.), d: 0.741 at 20° C, vap. d.: 3.5.
Hazard Analysis
Toxicity: Details unknown. Limited animal experiment

TOXIC HAZARD RATING CODE *(For detailed discussion, see Section 1.)*

0 NONE: (a) No harm under any conditions; (b) Harmful only under unusual conditions or overwhelming dosage.

1 SLIGHT: Causes readily reversible changes which disappear after end of exposure.

2 MODERATE: May involve both irreversible and reversible change not severe enough to cause death or permanent injury.

3 HIGH: May cause death or permanent injury after very short exposu to small quantities.

U UNKNOWN: No information on humans considered valid by author

show high oral toxicity and low inhalation toxicity and moderate irritation. See also amines.

Fire Hazard: Dangerous, when exposed to heat or flame.

Explosion Hazard: Unknown.

Disaster Hazard: Dangerous! Keep away from heat and open flame!

Countermeasures

Ventilation Control: Section 2.

To Fight Fire: Foam, carbon dioxide, dry chemical or carbon tetrachloride (Section 6).

Storage and Handling: Section 7.

DIPROPYLBERYLLIUM
General Information

Description: Liquid.

Formula: $Be(C_3H_7)_2$.

Constants: Mol wt: 95.19, mp: $< -17°C$, bp: 245°C, vap. d.: 3.29.

Hazard Analysis

Toxicity: See beryllium compounds.

Fire Hazard: Moderate, when exposed to heat or flame (Section 6).

Disaster Hazard: Dangerous; when heated to decomposition, it emits highly toxic fumes of beryllium oxide; can react vigorously with oxidizing materials.

Countermeasures

Ventilation Control: Section 2.

Personal Hygiene: Section 3.

First Aid: Section 1.

Storage and Handling: Section 7.

DIPROPYLCADMIUM
General Information

Description: An oil decomposed by water.

Formula: $(C_3H_7)_2Cd$.

Constants: Mol wt: 198.6, d: 1.420, mp: $-83°C$, bp: 84°C at 21.5 mm.

Hazard Analysis and Countermeasures

See cadmium compounds.

DI-n-PROPYL DIBROMO DIPYRIDINE TIN
General Information

Description: Solid.

Formula: $(C_3H_7)_2SnBr_2(C_5H_5N)_2$.

Constants: Mol wt: 522.9, mp: 128°C.

Hazard Analysis and Countermeasures

See tin compounds and bromides.

DIPROPYLENE GLYCOL
General Information

Synonym: 2,2'-Dihydroxyisopropyl ether.

Description: Colorless, slightly viscous liquid. Practically no odor.

Formula: $(CH_3CHOHCH_2)_2O$.

Constants: Mol wt: 134.17, bp: 231.8°C, flash p: 280°F (O.C.), d: 1.0252 at 20°/20°C, vap. press.: 1 mm at 73.8°C, vap. d.: 4.63.

Hazard Analysis

Toxic Hazard Rating:

Acute Local: U.

Acute Systemic: Ingestion 2.

Chronic Local: U.

Chronic Systemic: Ingestion 1; Inhalation 1.

Toxicology: See also glycols.

Fire Hazard: Slight, when exposed to heat or flame; can react with oxidizing materials.

Spontaneous Heating: No.

Countermeasures

To Fight Fire: Water, foam, alcohol foam, carbon dioxide, dry chemical or carbon tetrachloride (Section 6).

Ventilation Control: Section 2.

Personal Hygiene: Section 3.

Storage and Handling: Section 7.

DIPROPYLENE GLYCOL DIBENZOATE
General Information

Description: Liquid.

Formula: $CH_3CHC_6H_5CO_2CH_2OCH_2CHC_6H_5CO_2CH_3$.

Constants: Mol wt: 342.4, mp: $< -30°C$ (sets to a glass), bp: 250°C at 10 mm, d: 1.1260 at 20°/20°C, vap. press.: 1.2 mm at 200°C, vap. d.: 11.8.

Hazard Analysis

Toxicity: See glycols and benzoic acid.

Fire Hazard: Slight, when exposed to heat or flame; can react with oxidizing materials (Section 6).

Countermeasures

Storage and Handling: section 7.

DIPROPYLENE GLYCOL DIMETHYL ETHER
General Information

Description: Liquid.

Formula: $CH_3OC_3H_6OC_3H_6OCH_3$.

Constants: Mol wt: 148.2, bp: 190°C, d: 0.951, vap. d.: 5.11.

Hazard Analysis

Toxicity: Details unknown. See also glycols.

Fire Hazard: Moderate, when exposed to heat or flame; can react with oxidizing materials (Section 6).

Countermeasures

Storage and Handling: Section 7.

DI(PROPYLENE GLYCOL) 4,4'-ISOPROPYLIDENE BISPHENYL
General Information

Description: Very viscous liquid.

Formula: $(CH_3CHOHCH_2OC_6H_4)_2C_3H_6$.

Constants: Mol wt: 344.4, mp: 54°C, bp: decomposes, flash p: 450°F, d: 1.058 at 80°/4°C.

Hazard Analysis

Toxicity: Unknown.

Fire Hazard: Slight, when exposed to heat or flame; can react with oxidizing materials.

Countermeasures

To Fight Fire: Water, foam, carbon dioxide, dry chemical or carbon tetrachloride (Section 6).

Storage and Handling: Section 7.

DIPROPYLENE GLYCOL MONOMETHYL ETHER
General Information

Synonym: Dowanol 50B.

Description: Colorless liquid.

Formula: $CH_3OC_3H_6OC_3H_6OH$.

Constants: Mol wt: 148.2, bp: 189°C, flash p: 185°F, d: 0.950 at 25°/4°C, vap. d.: 5.11.

Hazard Analysis

Toxicity: See glycols.

TLV: ACGIH (recommended) 100 parts per million of air; 600 mg per cubic meter of air. May be absorbed via the skin.

Fire Hazard: Moderate, when exposed to heat or flame; can react with oxidizing materials.

Countermeasures

To Fight Fire: Foam, carbon dioxide, dry chemical or carbon tetrachloride (Section 6).

Storage and Handling: Section 7.

DIPROPYLENE GLYCOL PHENYL ETHER
General Information

Description: Slightly yellow liquid.

Formula: $C_6H_5OC_3H_6OC_3H_6OH$.

Constants: Mol wt: 210.2, mp: $< -25°C$, bp: 285.7°C, flash p: 315°F, d: 1.044 at 25°/25°C, vap. d.: 7.25.

Hazard Analysis

Toxicity: See glycols.

Fire Hazard: Slight, when exposed to heat or flame; can react with oxidizing materials.

Explosion Hazard: Details unknown. See also ether.

Countermeasures

Storage and Handling: Section 7.

To Fight Fire: Carbon dioxide, dry chemical or carbon tetrachloride (Section 6). See also ethers.

DIPROPYLENE TRIAMINE
General Information
Description: Water-white liquid.
Formula: $(CH_3)_2(CH_2)_2(CH)_2(NH_2)_2NH.$
Constants: Mol wt: 131.2, flash p: 190°F, d: 0.96, vap. d.: 4.6.
Hazard Analysis
Toxicity: Details unknown. See also amines.
Fire Hazard: Moderate,when exposed to heat or flame; can react with oxidizing materials.
Countermeasures
To Fight Fire: Foam, carbon dioxide, dry chemical or carbon tetrachloride (Section 6).
Storage and Handling: Section 7.

DIPROPYL ETHER. See propyl ether.

DIPROPYL KETONE. See butyrone.

DIPROPYLMERCURY
General Information
Description: Colorless liquid, insoluble in water.
Formula: $(C_3H_7)_2Hg.$
Constants: Mol wt: 286.8, d: 2.0208, bp: 190°C.
Hazard Analysis and Countermeasures
See mercury compounds, organic.

DIPROPYLMETHANE. See heptane.

DI-n-PROPYLPHOSPHOTHIONIC CHLORIDE
General Information
Formula: $(n-C_3H_7O)_2PSCl.$
Constants: Mol wt: 216.67, bp: 70–75°C at 1 mm, d: 1.123 at 25°/4°C.
Hazard Analysis
Toxicity: Details unknown.
Disaster Hazard: Dangerous; when heated to decomposition or on contact with acid or acid fumes, it emits highly toxic fumes of oxides of phosphorus and sulfur.
Countermeasures
Storage and Handling: Section 7.

DIPROPYL PHTHALATE
General Information
Description: Liquid.
Formula: $C_6H_4(COOC_3H_7)_2.$
Constants: Mol wt: 250.28, bp: 129–132°C at 1 mm, d: 1.071 at 25°C.
Hazard Analysis
Toxicity: See esters, phthalic acid and propyl alcohol.
Fire Hazard: Slight, when exposed to heat or flame; can react with oxidizing materials.
Countermeasures
To Fight Fire: Water, foam, carbon dioxide, dry chemical, or carbon tetrachloride (Section 6).
Storage and Handling: Section 7.

DI-n-PROPYL SULFATE. See propyl sulfate.

DIPROPYLTIN DIBROMIDE
General Information
Description: Colorless crystals, soluble in organic solvents.
Formula: $(C_3H_7)_2SnBr_2.$
Constants: Mol wt: 364.7, mp: 49°C.
Hazard Analysis and Countermeasures
See tin compounds and bromides.

DIPROPYLTIN DICHLORIDE
General Information
Description: Colorless crystals, soluble in organic solvents.

Formula: $(C_3H_7)_2SnCl_2.$
Constants: Mol wt: 275.8, mp: 81°C.
Hazard Analysis and Countermeasures
See tin compounds and chlorides.

DIPROPYLTIN DIFLUORIDE
General Information
Description: Crystals, nearly insoluble in water.
Formula: $(C_3H_7)_2SnF_2.$
Constants: Mol wt: 242.9, mp: 205°C.
Hazard Analysis and Countermeasures
See fluorides and tin compounds.

DIPROPYLTIN DIIODIDE
General Information
Description: Colorless oily liquid; insoluble in wate soluble in organic solvents.
Formula: $(C_3H_7)_2SnI_2.$
Constants: Mol wt: 458.7, mp: $< -15°C$, bp: 273°C.
Hazard Analysis and Countermeasures
See tin compounds and iodides.

DI-n-PROPYLZINC
General Information
Description: Liquid, decomposed in cold water.
Formula: $(C_3H_7)_2Zn.$
Constants: Mol wt: 151.6, bp: 146°C.
Hazard Analysis
Toxicity: Details unknown. See zinc compounds and prop alcohol.

DIPTEREX
General Information
Synonym: O,O-dimethyl-l-hydroxy-2,2,2-trichloroethyl phosphonate.
Description: Crystals, soluble in water, chloroform, ether.
Formula: $C_4H_8Cl_3O_4P.$
Constants: Mol wt: 257.5, d: 1.73, mp: 84°C.
Hazard Analysis
Toxicology: Highly toxic. A cholinesterase inhibitor. I believed to be less toxic than parathion. See relate compound parathion.
Disaster Hazard: Dangerous, when heated to decompositio it emits highly toxic fumes of chloride and oxides c phosphorus.

A,A'-DIPYRIDYL
General Information
Synonym: 2,2'-bipyridine; 2,2'-dipyridyl.
Description: White crystals; soluble in 2200 parts wate very soluble in alcohol, ether, benzene, chloroform, an petroleum ether.
Formula: $C_{10}H_8N_2$
Constants: Mol wt: 156.18, mp: 69.7°C, bp: 272–273°C
Hazard Analysis
See pyridine.

2,2'-DIPYRIDYLAMINE
General Information
Description: Crystals.
Formula: $C_{10}H_9N_3.$
Constants: Mol wt: 171.2, mp: 95.3°C, bp: 222°C.
Hazard Analysis
Toxicity: Details unknown. See also amines.
Disaster Hazard: Moderately dangerous; when heated t decomposition, it emits toxic fumes.
Countermeasures
Storage and Handling: Section 7.

TOXIC HAZARD RATING CODE *(For detailed discussion, see Section 1.)*

0 NONE: (a) No harm under any conditions; (b) Harmful only under unusual conditions or overwhelming dosage.

1 SLIGHT: Causes readily reversible changes which disappear after end of exposure.

2 MODERATE: May involve both irreversible and reversible change not severe enough to cause death or permanent injury.

3 HIGH: May cause death or permanent injury after very short exposu to small quantities.

U UNKNOWN: No information on humans considered valid by author

DIPYRIDYLETHYL SULFIDE
General Information
Description: Liquid.
Formula: $H_4NC_5(CH_2)_2S(CH_2)_2C_5NH_4$.
Constants: Mol wt: 241.4, mp: 1.5°C, d: 1.113 at 25°C, vap. d.: 8.31.
Hazard Analysis
Toxicity: Details unknown; an insecticide.
Disaster Hazard: Dangerous; when heated to decomposition or on contact with acid or acid fumes, it emits highly toxic fumes of oxides of sulfur and nitrogen.
Countermeasures
Storage and Handling: section 7.

2,2-DIPYRIDYL PERRHENATE
General Information
Description: Colorless crystals, water soluble.
Formula: $(C_5H_4N)_2HReO_4$.
Constant: Mol wt: 407.5.
Hazard Analysis
Toxicity: Details unknown.

2,2'-DIPYRIDYL RHENICHLORIDE
General Information
Description: Yellow crystals, slightly water soluble.
Formula: $(C_5H_4N)_2ReCl_6$.
Constant: Mol wt: 557.3.
Hazard Analysis and Countermeasures
See chlorides.

DIQUAT
General Information
Synonym: 6,7-Dihydrodipyrido   (1,2-a:2',1'-C)   pyrazidinium salt; 1,1'-ethylene-2,2'-dipyridinium dibromide.
Description: Yellow crystals; soluble in water.
Formula: $(C_5H_4NCH_2-)_2Br_2$.
Constants: Mol wt: 344, mp: 355°C.
Hazard Analysis
See paraquat.

DISALICYLAL PROPYLENE DIIMINE
General Information
Synonym: Schiff's base.
Formula: $HOC_6H_4CHNCH_2CHCH_3NHCC_6H_4OH$.
Constants: Mol wt: 282.3, mp: −23.3°C.
Hazard Analysis
Toxic Hazard Rating:
  Acute Local: Irritant 1; Allergen 1.
  Acute Systemic: U.
  Chronic Local: Allergen 1.
  Chronic Systemic: U.
Fire Hazard: Slight; Section 6.
Countermeasures
Personal Hygiene: Section 3.
Storage and Handling: Section 7.

DISILANE
General Information
Synonym: Silicoethane.
Description: Gas, repulsive odor.
Formula: Si_2H_6.
Constants: Mol wt: 62.2, mp: −132.5°C, bp: −14.5°C, d: 0.686 at −25°/4°C.
Hazard Analysis
Toxic Hazard Rating:
  Acute Local: Inhalation 3.
  Acute Systemic: U.
  Chronic Local: U.
  Chronic Systemic: U.
Toxicity: See also hydrides and silanes.
Fire Hazard: Dangerous, when exposed to heat or flame or by chemical reaction; can react with oxidizing materials. Ignites spontaneously in air (Section 6).
Explosion Hazard: Unknown.
Countermeasures
Storage and Handling: Section 7.

DISILICON HEXABROMIDE
General Information
Synonym: Hexabromodisilane.
Description: Rhombic, white crystals.
Formula: Si_2Br_6.
Constants: Mol wt: 535.62, mp: 95°C, bp: 240°C, vap. d.: 18.5.
Hazard Analysis
Toxicity: See bromides and silanes.
Disaster Hazard: Dangerous; See bromides; will react with water or steam to produce heat, toxic and corrosive fumes.
Countermeasures
Storage and Handling: Section 7.

DISILICON HEXACHLORIDE
General Information
Synonym: Hexachlorodisilane.
Description: Colorless liquid.
Formula: Si_2Cl_6.
Constants: Mol wt: 268.86, mp: −1°C, bp: 139°C, d (liquid): 1.58 at 0°C, vap. d.: 9.29.
Hazard Analysis
Toxicity: Details unknown. See also silanes.
Disaster Hazard: Dangerous; See chlorides; will react with water or steam to produce toxic and corrosive fumes.
Countermeasures
Storage and Handling: Section 7.

DISILICON HEXAFLUORIDE
General Information
Synonym: Hexafluorodisilane.
Description: Gas.
Formula: Si_2F_6.
Constants: Mol wt: 170.12, mp: −18.7°C, bp: −19.1°C sublimes, vap. d.: 5.87.
Hazard Analysis and Countermeasures
See fluorides and silanes.

DISILICON HEXAIODIDE
General Information
Synonym: Hexaiododisilane.
Description: Hexagonal, colorless crystals.
Formula: Si_2I_6.
Constants: Mol wt: 817.64, mp: 250°C, bp: decomposes, vap. d.: 28.2.
Hazard Analysis and Countermeasures
See iodides and silanes.

DISILICON TETRAIODIDE
General Information
Synonym: Tetraiododisilane-ethylene.
Description: Orange-red crystals.
Formula: Si_2I_4.
Constants: Mol wt: 563.80, vap. d.: 19.4.
Hazard Analysis
Toxicity: Details unknown. See iodides and silanes.
Disaster Hazard: Dangerous; See iodides. Will react with water or steam to produce toxic and corrosive fumes.
Countermeasures
Storage and Handling: Section 7.

DISILOXANE. See silicyl oxide.

DISINFECTANT, LIQUID, CORROSIVE
Shipping Regulations: Section 11.
  IATA: Corrosive liquid, white label, 1 liter (passenger), 40 liters (cargo).

DISINFECTANT, LIQUID, POISONOUS
Shipping Regulations: Section 11.
  IATA: Poison B, poison label, 1 liter (passenger), 220 liters (cargo).

DISINFECTANTS, SOLID
Shipping Regulations: Section 11.

IATA: Poison B, poison label, 25 kilograms (passenger), 9 kilograms (cargo).

DISODIUM ACETOARSENATE
General Information
Synonym: Aricyl.
Description: White, crystalline powder.
Formula: COONaCH₂AsOOHONa.
Constant: Mol wt: 227.95.
Hazard Analysis and Countermeasures
Highly toxic. See arsenic compounds.

DISODIUM DIHYDROGEN PYROPHOSPHATE. See sodium acid phosphate.

DISODIUM DIPHOSPHATE. See sodium acid pyrophosphate.

DISODIUM ETHYLENE BIS(DITHIOCARBAMATE). See nabam.

DISODIUM ETHYLENEDIAMINE TETRAACETATE
General Information
Synonyms: Sequestrene; ethylenediamine tetracetic acid.
Description: White crystals.
Formula: $C_{10}H_{14}O_8Na_2 \cdot 2H_2O$.
Constants: Mol wt: 372, vap. d.: 12.8.
Hazard Analysis
Toxic Hazard Rating:
Acute Local: Irritant 1; Ingestion 1; Inhalation 1.
Acute Systemic: U.
Chronic Local: U.
Chronic Systemic: Ingestion 1.
Toxicology: There have been reports of kidney injury in man following use of this type of compound in treating lead poisoning. Used as a food additive permitted in the feed and drinking water of animals and/or for the treatment of food producing animals. Also used as a food additive permitted as food for human consumption (Section 10).
Countermeasures
Personal Hygiene: Section 3.
Storage and Handling: Section 7.

DISODIUM GUANYLATE. See sodium guanylate.

DISODIUM HYDROGEN PHOSPHATE. See sodium phosphate, dibasic.

DISODIUM HYDROXYMERCURY SALICYLOXY ACETATE. See "Mercurosal."

DISODIUM INOSINATE. See sodium inosinate.

DISODIUM MONOMETHYLARSONATE
General Information
Synonym: Crab-e-rad.
Description: Crystals, water soluble.
Formula: $CH_3AsO_3Na_2$.
Constants: Mol wt: 183.9.
Hazard Analysis and Countermeasures
See arsenic compounds.

DISODIUM PHOSPHATE. See sodium phosphate, dibasic.

DISODIUM-o-PHOSPHATE. See sodium phosphate, dibasic.

DISODIUM PYROPHOSPHATE. See sodium acid pyrophosphate.

DISODIUM TARTRATE. See sodium tartrate.

DISPERSANT GAS, N.O.S.
General Information
Description: Gas.
Hazard Analysis
Toxicity: Unknown.
Countermeasures
Shipping Regulations: Section 11.
I.C.C.: Nonflammable gas; green label.
Coast Guard Classification: Noninflammable gas; green gas label.

DISTILLATE
Shipping Regulations: Section 11.
ICC: Flammable liquid, red label, 10 gallons.
IATA: Flammable liquid, red label, 1 liter (passenger) 40 liters (cargo).

DISTILLED MUSTARD GAS. See dichlorodiethyl sulfide

DISULFIRAM. See bis(diethylthiocarbamyl) disulfide.

DISULFOTON. See o,o-diethyl-s-2-(ethylthio) ethyl phosphorodithioate.

DISULFUR DECAFLUORIDE. See sulfur pentafluoride.

DISULFURIC ACID. See pyrosulfuric acid.

DISULFURYL CHLORIDE. See pyrosulfuryl chloride.

DITAINE
General Information
Synonym: Echitamine.
Description: White, thick, glistening, crystalline alkaloid.
Formula: $C_{22}H_{28}N_2O_4 \cdot 4H_2O$.
Constants: Mol wt: 456.53, mp: 206° C decomposes.
Hazard Analysis
Toxic Hazard Rating:
Acute Local: U.
Acute Systemic: Ingestion 3; Inhalation 3.
Chronic Local: U.
Chronic Systemic: U.
Disaster Hazard: Dangerous; when heated to decomposition, it emits toxic fumes.
Countermeasures
Ventilation Control: Section 2.
Personal Hygiene: Section 3.
Storage and Handling: Section 7.

DITCH POWDER. See explosives, high.

DITELLUROMETHANE
General Information
Description: Dark red solid, insoluble in water.
Formula: TeCH₂Te.
Constants: Mol wt: 269.3, mp: 214° C (decomp).
Hazard Analysis and Countermeasures
See tellurium compounds.

2,3-DITHIABUTANE
General Information
Synonym: Dimethyl disulfide.
Description: Liquid.
Formula: CH_3-S-S-CH_3.
Constants: Mol wt: 94.2, bp: 109.7° C, d: 1.0569 at 25° C, vap. press.: 28.6 mm at 25° C, vap. d.: 3.24.
Hazard Analysis
Toxicity: Details unknown. See also sulfides and alkyl disulfides.

TOXIC HAZARD RATING CODE *(For detailed discussion, see Section 1.)*

0 NONE: (a) No harm under any conditions; (b) Harmful only under unusual conditions or overwhelming dosage.

1 SLIGHT: Causes readily reversible changes which disappear after end of exposure.

2 MODERATE: May involve both irreversible and reversible changes; not severe enough to cause death or permanent injury.

3 HIGH: May cause death or permanent injury after very short exposure to small quantities.

U UNKNOWN: No information on humans considered valid by authors.

Fire Hazard: Moderate (Section 6).
Disaster Hazard: Dangerous; Can react vigorously with oxidizing materials. See sulfates.
Countermeasures
Storage and Handling: Section 7.

3,4-DITHIAHEXANE
General Information
Synonym: Diethyl disulfide.
Description: Liquid.
Formula: C_2H_5-S-S-C_2H_5.
Constants: Mol wt: 122.3, bp: 154°C, fp: −101.5°C, d: 0.99267 at 20/4°C, vap. d.: 4.22, vap. press.: 4.28 mm at 25°C.
Hazard Analysis
Toxicity: Details unknown. See also sulfides and alkyl disulfides.
Fire Hazard: Moderate (Section 6).
Disaster Hazard: Dangerous; can react vigorously with oxidizing materials. See sulfates.
Countermeasures
Storage and Handling: Section 7.
To Fight Fire: Foam, carbon tetrachloride, dry chemical (Section 6).

1,3-DITHIANE-3,2-DIPHENYL
General Information
Synonym: 1,3-Diethylene disulfide-2,2-diphenyl.
Description: Crystals.
Formula: $S[CH_2C(C_6H_5)_2CH_2CH_2]S$.
Constants: Mol wt: 272.4, vap. d.: 9.4.
Hazard Analysis
Toxicity: Details unknown. See also sulfides and alkyl disulfides.
Disaster Hazard: Dangerous; see sulfates.
Countermeasures
Storage and Handling: Section 7.

β,β-DITHIO BIS ALANINE. See cystine.

2,2'-DITHIOBIS(BENZOTHIAZOLE). See 2,2'-dibenzothiazyl disulfide.

2,4-DITHIOBIURET
General Information
Description: Crystals.
Formula: $H_2NC(S)NHC(S)NH_2$.
Constants: Mol wt: 135.20, mp: 181°C, bp: decomposes, d: 1.522 at 30°C.
Hazard Analysis
Toxic Hazard Rating:
Acute Local: U.
Acute Systemic: Ingestion 3; Inhalation 3.
Chronic Local: U.
Chronic Systemic: Ingestion 2; Inhalation 2.
Toxicology: Details unknown; but animal experiments suggest high toxicity with respiratory paralysis.
Disaster Hazard: Dangerous; when heated to decomposition, it emits highly toxic fumes.
Countermeasures
Ventilation Control: Section 2.
Personal Hygiene: Section 3.
Storage and Handling: Section 7.

DITHIOCARBAMIC ACID
General Information
Synonym: Aminodithioformic acid.
Description: Colorless needles.
Formula: NH_2CS_2H.
Constants: Mol wt: 93.2, vap. d.: 3.21.
Hazard Analysis
Toxic Hazard Rating:
Acute Local: Irritant 1; Ingestion 1; Inhalation 1.
Acute Systemic: U.
Chronic Local: U.
Chronic Systemic: Ingestion 1; Inhalation 1.

Disaster Hazard: Dangerous; when heated to decomposition, it emits toxic fumes.
Countermeasures
Personal Hygiene: Section 3.
Storage and Handling: Section 7.

β-DITHIOCYANOETHYL ETHER
General Information
Formula: $O[C(CNS)_2CH_3]_2$.
Constants: Mol wt: 302.5, vap. d.: 10.4.
Hazard Analysis
Toxicity: Details unknown; an insecticide.
Fire Hazard: Moderate (Section 6).
Disaster Hazard: Dangerous; when heated to decomposition or on contact with acid or acid fumes, it emits toxic fumes; can react with oxidizing materials.
Countermeasures
Storage and Handling: Section 7.

4,4-DITHIO MORPHOLINE
General Information
Synonym: Sulfasan; 4,4-dithiodimorpholine.
Description: Tan to gray powder.
Formula: $C_4H_8ONSSNOC_4H_8$.
Constants: Mp: 122°C minimum, d: 1.36 at 25°C.
Hazard Analysis
Toxicity: See morpholine.
Disaster Hazard: Dangerous, when heated to decomposition it emits highly toxic fumes.
Countermeasures
Storage and Handling: Section 7.

DITHIONE
General Information
Synonym: 7-Hydroxy-3, 4-tetramethylene coumarin O,O-diethylthiophosphate.
Description: Crystals, nearly insoluble in water.
Formula: $C_{17}H_{21}O_5PS$.
Constants: Mol wt: 368.4, mp: 88°C.
Hazard Analysis
Toxicity: Details unknown. An organic phosphate insecticide. Probably similar in effect to parathion.
Disaster Hazard: Dangerous; when heated to decomposition it emits highly toxic fumes of oxides of sulfur and phosphorous.

DITHIOZONE. See diphenyl thiocarbazone.

DITHRANOL. See anthralin.

DI-o-TOLUIDINE FLUOSILICATE
General Information
Description: White crystals, soluble in hot alcohol.
Formula: $(C_6H_4NH_2CH_3)_2 \cdot H_2SiF_6$.
Constant: Mol wt: 358.4.
Hazard Analysis and Countermeasures
See fluosilicates.

DI-m-TOLUIDINE FLUOSILICATE
General Information
Description: White crystals, soluble in hot alcohol.
Formula: $(C_6H_4NH_2CH_3)_2 \cdot H_2SiF_6$.
Constant: Mol wt: 358.4.
Hazard Analysis and Countermeasures
See fluosilicates.

DI-p-TOLUIDINE FLUOSILICATE
General Information
Description: White crystals.
Formula: $(C_6H_4NH_2CH_3)_2 \cdot H_2SiF_6$.
Constant: Mol wt: 358.4.
Hazard Analysis and Countermeasures
See fluosilicates.

DI-p-TOLYL BORIC ANHYDRIDE
General Information
Description: White powder, insoluble in water, soluble in ether.

Formula: $(C_7H_7)_2BOB(C_7H_7)_2$.
Constants: Mol wt: 402.2, mp: 78°C.
Hazard Analysis
Toxicity: Details unknown. See also boron compounds.

DI-p-TOLYLGERMANIUM DIBROMIDE
General Information
Description: Yellowish liquid hydrolyzed by water.
Formula: $(CH_3C_6H_4)_2GeBr_2$.
Constants: Mol wt: 414.7, bp: 230–233°C at 13 mm.
Hazard Analysis and Countermeasures
See germanium compounds and bromides.

DI-o-TOLYLGUANIDINE
General Information
Description: White crystals.
Formula: $C_{15}H_{17}N_3$.
Constants: Mol wt: 239.3, mp: 179°C, d: 1.10 at 20°/4°C, vap. d.: 8.24.
Hazard Analysis
Toxic Hazard Rating:
  Acute Local: Allergen 1.
  Acute Systemic: U.
  Chronic Local: Allergen 1.
  Chronic Systemic: U.
Disaster Hazard: Dangerous; when heated to decomposition, it emits highly toxic fumes of oxides of nitrogen.
Countermeasures
Personal Hygiene: Section 3.
Storage and Handling: Section 7.

o-DITOLYLMERCURY
General Information
Description: White crystals, insoluble in water.
Formula: $Hg(C_7H_7)_2$.
Constants: Mol wt: 382.9, mp: 107°C, bp: 219°C at 14 mm.
Hazard Analysis and Countermeasures
See mercury compounds, organic.

m-DITOLYLMERCURY
General Information
Description: Colorless to light yellow crystals, insoluble in water.
Formula: $(C_7H_7)_2Hg$.
Constants: Mol wt: 382.9, mp: 102°C.
Hazard Analysis and Countermeasures
See mercury compounds, organic.

p-DITOLYLMERCURY
General Information
Description: Crystals, insoluble in water.
Formula: $(C_7H_7)_2Hg$.
Constants: Mol wt: 382.9, mp: 238°C.
Hazard Analysis and Countermeasures
See mercury compounds, organic.

DI-p-TOLYL PHENYL GERMANIUM BROMIDE
General Information
Description: Colorless crystals.
Formula: $(CH_3C_6H_4)_2(C_6H_5)GeBr$.
Constants: Mol wt: 411.9, mp: 119°C.
Hazard Analysis and Countermeasures
See bromides and germanium compounds.

DI-o-TOLYLTHIOUREA
General Information
Synonym: sym-Di-o-tolylthiourea.
Description: Crystals.

Formula: $CS(NHC_6H_4CH_3)_2$.
Constants: Mol wt: 256.4, mp: 178°C. vap. d.: 8.85.
Hazard Analysis
Toxic Hazard Rating:
  Acute Local: Allergen 1.
  Acute Systemic: U.
  Chronic Local: Allergen 1.
  Chronic Systemic: U.
Disaster Hazard: Dangerous; when heated to decomposition, it emits highly toxic fumes.
Countermeasures
Personal Hygiene: Section 3.
Storage and Handling: Section 7.

DI-p-TOLYLTIN
General Information
Description: Orange-yellow powder, soluble in benzene.
Formula: $(CH_3C_6H_4)_2Sn$.
Constants: Mol wt: 301.0, mp: 111.5°C.
Hazard Analysis
Toxicity: Details unknown. See tin compounds.

DI-o-TOLYLTIN DICHLORIDE
General Information
Description: Solid.
Formula: $(CH_3C_6H_4)_2SnCl_2$.
Constants: Mol wt: 371.9, mp: 50°C.
Hazard Analysis and Countermeasures
See tin compounds and chlorides.

DI-p-TOLYLTIN DICHLORIDE
General Information
Description: Solid.
Formula: $(CH_3C_6H_4)_2SnCl_2$.
Constants: Mol wt: 371.9, mp: 50°C.
Hazard Analysis and Countermeasures
See tin compounds and chlorides.

DI-m-TOLYLTIN OXIDE
General Information
Description: White powder, insoluble in water.
Formula: $(CH_3C_6H_4)_2SnO$.
Constants: Mol wt: 317.0.
Hazard Analysis
Toxicity: Details unknown. See tin compounds.

DI-m-TOLYLTIN SULFIDE
General Information
Description: Solid, soluble in chloroform and benzene.
Formula: $(CH_3C_6H_4)_2SnS$.
Constants: Mol wt: 333.0, mp: 122°C.
Hazard Analysis and Countermeasures
See tin compounds and sulfides.

2,2-DI-p-TOLYL-1,1,1-TRICHLOROETHANE
General Information
Description: Crystals.
Formula: $(C_6H_4CH_3)_2C_2HCl_3$.
Constants: Mol wt: 313.7, vap. d.: 10.8.
Hazard Analysis
Toxicity: Probably toxic. An insecticide. See also chlorinated hydrocarbon insecticides and DDT.
Disaster Hazard: Moderately dangerous; when heated to decomposition, it emits toxic fumes.
Countermeasures
Storage and Handling: Section 7.

TOXIC HAZARD RATING CODE (For detailed discussion, see Section 1.)

0 NONE: (a) No harm under any conditions; (b) Harmful only under unusual conditions or overwhelming dosage.

1 SLIGHT: Causes readily reversible changes which disappear after end of exposure.

2 MODERATE: May involve both irreversible and reversible changes; not severe enough to cause death or permanent injury.

3 HIGH: May cause death or permanent injury after very short exposure to small quantities.

U UNKNOWN: No information on humans considered valid by authors.

DI-o-TOLYLZINC
General Information
Description: White crystals, soluble in xylene, petroleum ether.
Formula: $(C_7H_7)_2Zn$.
Constants: Mol wt: 247.6, mp: 210°C.
Hazard Analysis
Toxicity: Details unknown. See zinc compounds.

DI-(TRIDECYL) AMINE
Hazard Analysis
Toxic Hazard Rating: U. Limited animal experiments show low toxicity and moderate irritation. See also amines.

DI-(TRIDECYL) PHTHALATE
General Information
Synonym: DTDP.
Formula: $C_6H_4(COOC_{13}H_{27})_2$.
Constants: Mol wt: 538, d: 0.951 (20/20°C), bp: >285°C at 5 mm, flash p: 470°F(O.C.).
Hazard Analysis
Toxic Hazard Rating: U. Limited animal experiments suggest low toxicity.
Fire Hazard: Slight when exposed to heat or flame.
Countermeasures
To Fight Fire: Water, foam (Section 6).
Storage and Handling: Section 7.

DITRIPHENYL GERMANYL METHANE
General Information
Description: Colorless crystals, insoluble in water.
Formula: $[(C_6H_5)_3Ge]_2CH_2$.
Constants: Mol wt: 621.8, mp: 133°C.
Hazard Analysis
Toxicity: Details unknown. See germanium compounds.

DITRIPHENYL STANNYL METHANE
General Information
Description: White crystals, soluble in benzene, ether and chloroform.
Formula: $[(C_6H_5)_3Sn]_2CH_2$.
Constants: Mol wt: 714.0 mp: 104.5°C.
Hazard Analysis
Toxicity: Details unknown. See tin compounds.

DIURON
General Information
Synonym: 3-(3,4-Dichlorophenyl)-1,1-dimethyl urea.
Description: Crystals, low solubility in water and hydrocarbon solvents.
Formula: $C_9H_{10}Cl_2N_2O$.
Constants: Mol wt: 233.1, mp: 159°C.
Hazard Analysis
Toxicity: Details unknown. Limited animal experiments suggest low to moderate toxicity. Used as a food additive permitted in the feed and drinking water of animals and or for the treatment of food producing animals. (Section 10).
Disaster Hazard: Dangerous; when heated to decomposition, it emits highly toxic fumes.

1,2-DIVALERATE DIETHYL SULFIDE
General Information
Description: Liquid.
Formula: $[CH_2(COOC_4H_9)CH(COOC_4H_9)]_2S$.
Constants: Mol wt: 490.64, mp: −45°C, bp: 246°C at 5 mm, flash p: 430°F (O.C.), d: 1.0543 at 20°/20°C, vap. d.: 16.95.
Hazard Analysis
Toxicity: Details unknown. See also sulfides and alkyl disulfides.
Fire Hazard: Slight, when exposed to heat or flame.
Disaster Hazard: Dangerous; see sulfates. Can react with oxidizing materials.
Countermeasures
Storage and Handling: Section 7.

To Fight Fire: Water, foam, carbon dioxide, dry chemical or carbon tetrachloride (Section 6).

DIVINYL-b. See 1,3-butadiene.

m-DIVINYLBENZENE
General Information
Synonym: Vinylstyrene.
Description: Water-white liquid.
Formula: $C_6H_4(CHCH_2)_2$.
Constants: Mol wt: 130.1, bp: 199.5°C, fp: −66.9°C, lel: 0.3%, flash p: 165°F, d: 0.9289 at 20°C, vap. press.: 1 mm at 32.7°C, vap. d.: 4.48.
Hazard Analysis
Toxic Hazard Rating:
Acute Local: U.
Acute Systemic: Inhalation 2.
Chronic Local: U.
Chronic Systemic: Inhalation 2.
Fire Hazard: Moderate, when exposed to heat or flame or by spontaneous chemical reaction; can react with oxidizing materials.
Spontaneous Heating: Yes.
Explosion Hazard: Moderate, in the form of vapor when exposed to flame (Section 7).
Countermeasures
Ventilation Control: Section 2.
To Fight Fire: Foam, carbon dioxide, dry chemical or carbon tetrachloride (Section 6). Commercial DVB without inhibitor may start to polymerize and gel. Heat can then build up rapidly to possibly 300°C and the vapor pressure may attain dangerous levels. In storage this material should be kept below 90°F, possibly by water spray.
Storage and Handling: Section 7.

DIVINYL ETHER. See vinyl ether.

DI(VINYL OCTYLATE) CAPRYLAMIDE
General Information
Description: Liquid.
Formula: $(C_7H_{15}COOC_2H_3)_2NCOC_7H_{15}$.
Constants: Mol wt: 483.71, mp: −33°C, bp: 256°C at 5 mm, flash p: 420°F, d: 0.9564 at 20°/20°C, vap. d.: 16.7.
Hazard Analysis
Toxicity: Unknown. See also amides.
Fire Hazard: Slight when exposed to heat or flame; can react with oxidizing materials.
Countermeasures
To Fight Fire: Foam, carbon dioxide, dry chemical or carbon tetrachloride (Section 6).
Storage and Handling: Section 7.

DIVINYL OXIDE. See vinyl ether.

DIVINYL PYRETHRIN I
General Information
Synonyms: DVPy; allyl homolog of Cinerin I.
Hazard Analysis
Toxic Hazard Rating:
Acute Local: Ingestion 1.
Acute Systemic: U.
Chronic Local: U.
Chronic Systemic: U.
Disaster Hazard: Slight; when heated, it emits acrid fumes.
Countermeasures
Personal Hygiene: Section 3.
Storage and Handling: Section 7.

DIVINYL SULFIDE
General Information
Description: Mobile liquid. characteristic odor.
Formula: $(CHCH_2)_2S$.
Constants: Mol wt: 86.2, bp: 86°C, d: 0.9174 at 15°C, vap. d.: 2.97.

Hazard Analysis
Toxicity: Details unknown. See also sulfide and alkyl disufides.
Fire Hazard: Moderate, when exposed to heat or flame (Section 6).
Disaster Hazard: Dangerous; when heated to decomposition, it emits highly toxic fumes of oxides of sulfur; on contact with oxidizing materials, it can react vigorously.

Countermeasures
Storage and Handling: Section 7.

DI(o-XENYL) MONOPHENYL PHOSPHATE
General Information
Description: Insoluble in water.
Formula: $(C_6H_5C_6H_4)_2PO(OC_6H_5)$.
Constants: Mol wt: 446, d: 1.20 at 60°C, boiling range: 285–330°C at 5 mm, flash p: 482°F.

Hazard Analysis
Toxic Hazard Rating: U.
Fire Hazard: Slight when exposed to heat or flame.
Disaster Hazard: Dangerous. See phosphates.

Countermeasures
To Fight Fire: Water, foam (Section 6).
Storage and Handling: Section 7.

DI-p-XYLYLTIN
General Information
Description: Solid.
Formula: $[(CH_3)_2C_6H_3]_2Sn$.
Constants: Mol wt: 329.0, mp: 157°C, bp: 240°C (decomp).

Hazard Analysis
Toxicity: Details unknown. See tin compounds.

DKP. See potassium phosphate, dibasic.

"DM". See diphenylamine chloroarsine.

DMC. See di-(p-chlorophenyl)methyl carbinol.

DMDT. See methoxychlor.

DMF. See dimethylformamide.

DMP. See dimethyl phthalate.

DNBP. See 2-sec-butyl-6,4-dinitrophenol.

DNC. See dinitro-o-cresol.

DNOC. See dinitro-o-cresol.

DNTP. See parathion.

DODECANE
General Information
Synonym: Dihexyl.
Description: Liquid.
Formula: $C_{12}H_{26}$.
Constants: Mol wt: 170.3, mp: −12°C, bp: 216.2°C, lel: 0.6%, flash p: 165°F, d: 0.750, vap. press.: 1 mm at 47.8°C, vap. d.: 5.96, autoign. temp.: 399°F.

Hazard Analysis
Toxicity: Unknown. Probably irritant and narcotic in high concentrations.
Fire Hazard: Moderate, when exposed to heat or flame.
Spontaneous Heating: No.
Explosion Hazard: Moderate, in the form of vapor when exposed to flame (Section 7).

Disaster Hazard: Dangerous; keep away from heat and open flame.

Countermeasures
Storage and Handling: Section 7.
To Fight Fire: Foam, carbon dioxide, dry chemical or carbon tetrachloride (Section 6).

DODECANETHIOL. See lauryl mercaptan.

DODECANOIC ACID. See lauric acid.

l-DODECANOL
General Information
Synonym: Dodecyl alcohol; lauryl alcohol; 1°-n-lauryl alcohol.
Description: Colorless liquid; floral odor; soluble in 2 parts of 70% alcohol, insoluble in water.
Formula: $CH_3(CH_2)_{11}OH$.
Constants: Mol wt: 186.33, d: 0.8201 (24/4°C), mp: 24°C, bp: 259°C, flash p: 260°F.

Hazard Analysis
Toxic Hazard Rating: U. Limited animal experiments suggest low toxicity. See also alcohols.
Fire Hazard: Slight when exposed to heat or flame.

Countermeasures
To Fight Fire: Water, foam (Section 6).
Storage and Handling: Section 7.

DODECANOYL PEROXIDE. See lauryl peroxide.

DODECENE
General Information
Synonym: Dodecylene.
Description: Colorless liquid.
Formula: $C_{12}H_{24}$.
Constants: Mol wt: 168.31, mp: −31.5°C, bp: 213°C, d: 0.76 at 20°/4°C, vap. press.: 1 mm at 47.2°C, vap. d.: 5.81.

Hazard Analysis
Toxicity: Details unknown. Probably irritant and narcotic in high concentrations.
Fire Hazard: Moderate, when exposed to heat or flame; can react with oxidizing materials.

Countermeasures
To Fight Fire: Foam, carbon dioxide, dry chemical or carbon tetrachloride (Section 6).
Storage and Handling: Section 7.

DODECENYL SUCCINIC ANHYDRIDE
General Information
Description: Light, yellow, clear viscous oil.
Formula: $C_{12}H_{23}CHOOOCCH_2$.
Constants: Mol wt: 266, bp: 180–182°C at 5 mm, flash p: 352°F (C.O.C.), d: 1.002 at 25°/4°C.

Hazard Analysis
Toxicity: Details unknown. Animal experiments show irritating and sensitizing properties as well as slight systemic toxicity.
Fire Hazard: Slight, when exposed to heat or flame; can react with oxidizing materials.

Countermeasures
To Fight Fire: Foam, carbon dioxide, dry chemical or carbon tetrachloride (Section 6).
Storage and Handling: Section 7.

DODECYL ALCOHOL. See 1-dodecanol.

TOXIC HAZARD RATING CODE (For detailed discussion, see Section 1.)

.0 NONE: (a) No harm under any conditions; (b) Harmful only under unusual conditions or overwhelming dosage.

1 SLIGHT: Causes readily reversible changes which disappear after end of exposure.

2 MODERATE: May involve both irreversible and reversible changes; not severe enough to cause death or permanent injury.

3 HIGH: May cause death or permanent injury after very short exposure to small quantities.

U UNKNOWN: No information on humans considered valid by authors.

DODECYLAMINE
General Information
Description: Oil, amine odor.
Formula: $C_{12}H_{25}NH_2$.
Constants: Mol wt: 185, fp: 28.3°C, vap. d.: 64 mm at 170°C.
Hazard Analysis
Toxicity: Details unknown. See fatty amines.

DODECYLBENZENE, CRUDE
General Information
Description: Liquid.
Formula: $C_{12}H_{25}C_6H_5$.
Constants: Mol wt: 246.4, bp: 290–410°C, flash p: 285°F, d: 0.9, vap. d.: 8.47.
Hazard Analysis
Toxicity: Details unknown. See also benzene.
Fire Hazard: Slight, when exposed to heat or flame; can react with oxidizing materials.
Countermeasures
To Fight Fire: Foam, carbon dioxide, dry chemical or carbon tetrachloride (Section 6).
Storage and Handling: Section 7.

DODECYL BROMIDE. See lauryl bromide.

DODECYL CHLORIDE. See lauryl chloride.

DODECYLENE. See dodecene.

N-DODECYL GUANIDINE ACETATE
General Information
Synonym: Dodine, cyprex.
Description: Crystals; soluble in hot water and alcohol.
Formula: $C_{12}H_{25}NHC(:NH)NH_2CH_3COOH$.
Constants: Mol wt: 287, mp: 136°C.
Hazard Analysis
Toxic Hazard Rating:
Acute Local: Irritant 2; Ingestion 2; Inhalation 2.
Acute Systemic: Ingestion 2.
Chronic Local: U.
Chronic Systemic: U.
A fungicide.

DODECYL MERCAPTAN. See lauryl mercaptan.

tert-DODECYL MERCAPTAN
General Information
Description: White to light yellow liquid.
Formula: $C_{12}H_{25}SH$.
Constants: Mol wt: 202.4, bp: 200–235°C, flash p: 205°F, d: 0.85 at 25°/25°C, vap. d.: 6.98.
Hazard Analysis
Toxicity: Probably toxic. See also mercaptans.
Fire Hazard: Moderate, when exposed to heat or flame.
Disaster Hazard: Dangerous; see sulfates. Can react vigorously with oxidizing materials.
Countermeasures
Storage and Handling: Section 7.
To Fight Fire: Foam, carbon dioxide, dry chemical or carbon tetrachloride (Section 6).

DODECYL PHENOL
General Information
Description: Straw-colored liquid, phenolic odor.
Formula: $C_{12}H_{25}C_6H_4OH$.
Constants: Mol wt: 262.4, bp: 154°–168°C, flash p. 325°F (O.C.), d: 0.93 at 20°/20°C, vap. d.: 9.04.
Hazard Analysis
Toxicity: Details unknown. Limited animal experiments suggest moderate toxicity and high irritation. See phenol.
Fire Hazard: Slight, when exposed to heat or flame.
Disaster Hazard: Moderately dangerous; when heated to decomposition, it emits toxic fumes; can react with oxidizing materials.

Countermeasures
Storage and Handling: Section 7.
To Fight Fire: Foam, carbon dioxide, dry chemical or carbon tetrachloride (Section 6).

DODECYL PYRIDINIUM CHLORIDE. See lauryl pyridinium chloride.

DODECYL TRICHLOROSILANE
General Information
Description: Colorless to yellow liquid; readily hydrolyzed by moisture with the production of hydrochloric acid.
Formula: $C_{12}H_{25}SiCl_3$.
Constants: Mol wt: 303.5, bp: 288°C, d: 1.026 (25/25°C).
Hazard Analysis
Toxic Hazard Rating: U. See silanes.
Disaster Hazard: Dangerous. See chlorides.
Countermeasures
Storage and Handling: Section 7.
Shipping Regulations: Section 11.
ICC: Corrosive liquid, white label, 10 gallons.
IATA: Corrosive liquid, white label, not acceptable (passenger), 40 liters (cargo).

DODINE. See N-dodecyl guanidine acetate.

DOMEYKITE. See tricopper arsenide.

DORMISON. See methylpentynol.

DOWANOL-50B. See dipropylene glycol monomethyl ether.

DOWICIDE 1. See o-phenylphenol.

DOWICIDE 4. see 2-chloro-4-phenylphenol.

DOWKLOR. See chlordane.

DOWLAP. See 3,4,6-trichloro-2-nitrophenol.

DPA. See diphenolic acid.

DRESSING, LEATHER
Shipping Regulations: Section 11.
ICC: Flammable liquid, red label, 10 gallons.
IATA: Flammable liquid, red label, 1 liter (passenger), 40 liters (cargo).

DRIED GRASS. See hay.

DRILL CARTRIDGES. See explosives, high.
Shipping Regulations: Section 11.
Coast Guard Classification: No restrictions.

DRUGS, MEDICINES, OR COSMETICS, n.o.s., COR-ROSIVE LIQUID
Shipping Regulations: Section 11.
ICC: Corrosive liquid, white label, 1 quart.
IATA: Corrosive liquid, white label, 1 liter (passenger and cargo).

DRUGS, MEDICINES, OR COSMETICS, n.o.s., FLAM-MABLE LIQUID
Shipping Regulations: Section 11.
ICC: Flammable liquid, red label, 10 gallons.
IATA: Flammable liquid, red label, 1 liter (passenger), 40 liters (cargo).

DRUGS, MEDICINES OR COSMETICS, n.o.s., FLAM-MABLE SOLID
Shipping Regulations: Section 11.
ICC: Flammable solid, yellow label, 100 pounds.
IATA: Flammable solid, yellow label, 12 kilograms (passenger), 45 kilograms (cargo).

DRUGS, MEDICINES, OR COSMETICS, n.o.s., OXI-DIZING MATERIAL
Shipping Regulations: Section 11.

ICC: Oxidizing material, yellow label, 100 pounds.
IATA: Oxidizing material, yellow label, 12 kilograms (passenger), 45 kilograms (cargo).

DRUGS, MEDICINES, OR COSMETICS, n.o.s., POISONOUS LIQUID
Shipping Regulations: Section 11.
ICC: Poison B, poison label, 55 gallons.
IATA: Poison B, poison label, 1 liter (passenger), 220 liters (cargo).

DRUGS, MEDICINES OR COSMETICS, n.o.s., POISONOUS SOLID
Shipping Regulations: Section 11.
ICC: Poison B, poison label, 200 pounds.
IATA: Poison B, poison label, 25 kilograms (passenger), 95 kilograms (cargo).

DRUMS, EMPTY (USED)
Shipping Regulations: Section 11.
Coast Guard Classification: Hazardous material.
ICC: Section 73.29.

DRY ICE. See carbon dioxide, solid.

"DRYOLENE." See naphtha, VM & P.

DRYORTH. See sodium o-silicate.

DSP. See sodium phosphate, dibasic.

DTMC. See 4,4'-dichloro-a-(trichloromethyl) benzhydrol.

DUBOISINE. See hyoscyamine.

DULCITOL HEXANITRATE
General Information
Synonym: 1,2,3,4,5,6-Hexane hexanitrate.
Description: Crystals.
Formula: $C_6H_8(NO_2)_6$.
Constants: Mol wt: 356.2, mp: explodes at 205° C.
Hazard Analysis
Toxicity: Details unknown. See also nitrates.
Fire Hazard: Dangerous, when exposed to heat or flame or by chemical reaction. A powerful oxidizing agent (Section 6).
Explosion Hazard: Severe, when shocked or exposed to heat (Section 7).
Disaster Hazard: Highly dangerous; shock and heat will explode it; when heated to decomposition it emits highly toxic fumes of oxides of nitrogen; can react vigorously with oxidizing materials.
Countermeasures
Storage and Handling: Section 7.

DUMASIN. See cyclopentanone.

DUMMY CARTRIDGES
Shipping Regulations: Section 11.
Coast Guard Classification: No restrictions.

DUPHAR. See tedion.

DUPONOL. See sodium lauryl sulfate.

DURASET
General Information
Synonym: N-m-Tolylphthalamic acid.
Hazard Analysis
Toxicity: Details unknown. Hydrolyzes to yield m-toluidine. See m-toluidine.

DUSTS, METALLIC. See specific metal and powdered metals.

DUST, TOTAL (below 5% FREE SiO₂)
Hazard Analysis
Toxic Hazard Rating:
Acute Local: Inhalation 1.
Acute Systemic: Variable.
Chronic Local: Inhalation 1.
Chronic Systemic: Variable.
TLV: ACGIH (recommended); 50 million particles per cubic foot of air.
Toxicology: The effects of dust vary according to the composition. If it is organic dust, it is primarily a nuisance, but may produce asthma by being an allergen. If the dust contains toxic compounds, either organic or inorganic, the hazard is that of the toxic component. A common air contaminant (Section 4).
Fire and Explosion Hazard: Dangerous; both organic and inorganic dusts can explode or burn violently if exposed to sparks or open flame. See Section 6.

DVPy. See divinyl pyrethrin I.

DYE INTERMEDIATES, n.o.s., FLAMMABLE LIQUID
Shipping Regulations: Section 11.
IATA: Flammable liquid, red label, 1 liter (passenger), 40 liters (cargo).

DYE INTERMEDIATE, n.o.s., POISONOUS LIQUID
Shipping Regulations: Section 11.
IATA: Poison B, poison label, 1 liter (passenger), 220 liters (cargo).

DYE INTERMEDIATE, n.o.s., POISONOUS SOLID
Shipping Regulations: Section 11.
IATA: Poison B, poison label, 25 kilograms (passenger), 95 kilograms (cargo).

DYMID. See N,N-dimethyl-2,2-diphenyl acetamide.

DYNALTONE. See aralkonium chloride.

DYNAMAGNITE. See dynamite.

DYNAMITE
General Information
Description: Dynamite is a high explosive used industrially in construction and mining. The name generally refers to a mixture containing as its principal explosive ingredient either glyceryl trinitrate (nitroglycerin) or ammonium nitrate, suitably sensitized. It does not apply to black blasting powders, chlorate powders, and other deflagrating mixtures.

An ordinary blasting cap or an electric blasting cap is used for detonating a charge of dynamite. The various classes and grades of dynamites are made from mixtures composed of an explosive compound or a mixture of explosive compounds, a dope, and an antacid. If any of the explosive ingredients are in a liquid state they are referred to as the "explosive oil," which is usually composed of glyceryl trinitrate (nitroglycerin) and about 25 to 30 percent of ethylene glycol dinitrate. The latter compound depresses the freezing point of the nitroglycerin and renders the dynamite low-freezing. Other compounds may also be used as freezing point depressants. The expsive oil is absorbed by carbonaceous materials that have entirely replaced kieselguhr (diatomaceous earth), formerly used ex-

---

0 NONE: (a) No harm under any conditions; (b) Harmful only under unusual conditions or overwhelming dosage.

1 SLIGHT: Causes readily reversible changes which disappear after end of exposure.

2 MODERATE: May involve both irreversible and reversible changes; not severe enough to cause death or permanent injury.

3 HIGH: May cause death or permanent injury after very short exposure to small quantities.

U UNKNOWN: No information on humans considered valid by authors.

clusively as the absorbent or dope in dynamites. This type of dope does not enter into the explosive reaction. Wood pulp is now most commonly used as the absorber, either alone or mixed in suitable proportions with flour, starch, etc.

The absorbents may be mixed with an oxidizer such as sodium nitrate, in which case an active dope is formed. For neutralizing any acid that may be present, about 1% of an antacid (calcium carbonate or zinc oxide) is added to the mixture. The explosive oil is mixed into the dope. The strength of a kieselguhr dynamite, when detonated, is derived only from the explosive oil, since kieselguhr is inert. A mixture of this kind is known as a straight dynamite. See dynamite, straight. On the other hand, an active dope, (an admixture of carbonaceous absorbents with an oxidizer), furnishes explosive strength in addition to that derived from the explosive ingredients.

By replacing a part of the explosive oil of a straight dynamite with ammonium nitrate, so that the latter becomes the principal explosive ingredient, a mixture known as an ammonia dynamite is obtained. See dynamite, ammonia.

When the explosive oil is gelatinized the explosive is known as a gelatin or an ammonia gelatin dynamite. See also dynamite, ammonia gelatin and dynamite, gelatin.

Blasting gelatin is a gelatinized mass of an elastic nature obtained by incorporating nitrocotton with an explosive oil into which is mixed about 1% of antacid. See also dynamite, blasting gelatin.

Dynamites may be in bulk form (bag powder) or put up in cartridge form, the most common size being $1\frac{1}{4}$ inches in diameter and 8 inches long, although for holes of small diameter, cartridges as small as $\frac{7}{8}$ inch in diameter are also used. In large diameter well-drill holes for quarry blasting, cartridge diameters up to 10 inches, and lengths up to 30 inches may be used. These upper limits or 50 pounds in weight of each cartridge are imposed by the ICC Regulations, and the maximum length of 30 inches applies to all cartridge diameters between 4 and 10 inches.

An integral part of a stick of dynamite is the paraffined paper wrapper that not only holds the ingredients together but enters into the explosive reaction.

The wrapper also affords some measure of protection from moderate exposure to dampness. For blasting in wet operations, a gelatinized dynamite which resists the absorption of water should be used.

The strength of a straight dynamite is graded by its explosive oil content (percentage by weight), while for any other class of dynamite, the strength is determined experimentally in comparison with the various grades of the strength dynamites. For example, a 40% straight dynamite is one which contains 40% of explosive oil; a 40% strength ammonia dynamite, as determined by tests, equals a 40% straight dynamite in strength. In other words a 40% strength ammonia dynamite will release the same energy as an equivalent weight of a 40% straight dynamite.

Hazard Analysis

Toxicity: See nitrates.

Fire Hazard: See explosives, high and nitrates.

Explosion Hazard: While this material is a powerful explosive when detonated by shock or heat, it is only moderately hazardous. See also explosives, high and nitrates.

Disaster Hazard: Dangerous; shock and heat will explode it; when heated to decomposition it emits highly toxic fumes of oxides of nitrogen, and carbon monoxide, etc. It can react vigorously with oxidizing materials.

Countermeasures

Storage and Handling: Section 7.

Shipping Regulations: See explosives, high.

DYNAMITE, AMMONIA

General Information

Description: This class of dynamite contains ammonium nitrate as the principal explosive ingredient, which may be considered as replacing some of the sodium nitrate and at least 60 percent of the explosive oil as found in the straight dynamites.

A suitable sensitizer must be added because the firing of an ordinary No. 6 blasting cap embedded in pure ammonium nitrate is not capable of bringing about its detonation. If the sensitizer in the dynamite is a liquid ingredient, carbonaceous absorbent materials are added in amounts adequate to avoid leakage of the liquid ingredients and to obtain a suitable oxygen balance. Only a few of the various substances which have been proposed as sensitizers for ammonium nitrate have really been successful. The most successful one takes the form of an explosive oil, like that used in the straight dynamites. Other sensitizers are in solid form and may be either of an explosive nature or a non-explosive nature. Certain of these solid sensitizers are explosive and may be dissolved by the nitroglycerin. Among the solid sensitizers are the organic compounds such as diphenyl and diphenylamine, both of which are nonexplosive; or nitrated organic compounds, such as nitrostarch (as used in nitrostarch dynamites), nitrotoluenes, nitronaphthalenes, etc., which are explosive. Inorganic solid substances like calcium silicide, ferrosilicon, aluminum, and sulfur have also been used as sensitizers, although none of them are explosive except when mixed with other substances.

The following is a typical formula for a 40 percent ammonia dynamite: 1% moisture; 15% explosive oil; 31% ammonium nitrate; 38% sodium nitrate; 1% antacid; 4% sulfur; 9% carbonaceous material. For a $1\frac{1}{4}$ by 8 inch cartridge, the weight of the wrapper per 100 grams of explosive ingredient averages 7 grams.

The ingredients of ammonia dynamites are of a granular nature that have very little cohesion and do not readily resist the absorption of water. Owing to the hygroscopic nature of ammonium nitrate, ammonia dynamites are not usually used in wet holes. Under this condition the gelatinized types of dynamites (ammonia gelatin or gelatin dynamites) should be used and for very wet work it is advisable to use only the gelatin dynamites.

The regular grades of ammonia dynamites range from 15 to 60 percent strength. The ammonia gelatin, or gelatin dynamites of 40 to 60 percent grades are used to greatest extent in blasting operations.

High count powders are modifications of ammonia dynamites. The cartridge count per 50 pound case for the regular grades of ammonia dynamite is about 100 in comparison with 115 to 172 for the high count ammonia dynamites. Those having a cartridge count of 115 have a bulk strength of 60 percent and those of the 172 count have a strength of 20 percent on the same basis. In weight strength all these powders are approximately equivalent to a 60 percent straight dynamite.

Hazard Analysis

Toxicity: See nitrates.

Fire Hazard: Dangerous; when exposed to heat or flame (Section 6). See also explosives, high.

Explosion Hazard: Moderate when shocked or exposed to heat or flame (Section 7). See also explosives, high.

Disaster Hazard: Highly dangerous; shock and heat will explode it; when heated to decomposition, it emits highly toxic fumes of oxides of nitrogen.

Countermeasures

Storage and Handling: Section 7.

DYNAMITE, AMMONIA GELATIN

General Information

Description: An ammonia gelatin dynamite is a gelatinized

explosive mixture that contains ammonium nitrate. At least 60 percent of the ammonium nitrate and all of the explosive oil entering into the composition of an ammonia dynamite can be considered as being replaced by a plastic material which then becomes the main explosive ingredient. This material is obtained by gelatinizing about 0.7 percent of a low-nitrated cellulose with an explosive oil which is warmed slightly to produce the jelly-like mass. The other ingredients are then mixed with the gelatinized material. The final product is dense, plastic and more water-resistant than an ammonia dynamite. The regular grades of ammonia gelatin dynamites range from 30 to 90 percent strength.

The following formula is typical of a 40 percent ammonia gelatin: 1% moisture; 26% gelatinized ingredient; 8% ammonium nitrate; 50% sodium nitrate; 1% antacid; 6% sulfur; 8% carbonaceous combustible material.

For a 1¼ by 8 inch cartridge the wrapper per 100 grams of explosive ingredient averages 4.5 grams. The cartridge count per 50 pound case averages 100.

When detonated, this class of dynamite emits a low volume of poisonous gases and is therefore recommended along with the gelatin dynamites for underground blastings. On the other hand, quarry gelatin which is a modified type of ammonia gelatin dynamite should be used only in open work and should never be used underground becasue it emits large volumes of poisonous gases on detonation.

Hazard Analysis
Toxicity: See nitrates.

DYNAMITE, BLASTING GELATIN
General Information
Description: Blasting gelatin, which is the most powerful explosive used industrially, is a translucent material of an elastic texture. It is obtained by incorporating about 7 percent of a low-nitrated cellulose with an explosive oil which is slightly warmed during the process of gelatinization. This explosive is employed only where its high density and water-resisting properties can be advantageously used, such as in submarine work and other operations that are very wet. There is only one grade and it is rated as 100 percent.

Hazard Analysis
Toxicity: See nitrates.

DYNAMITE, GELATIN
General Information
Description: In the gelatin dynamites all the explosive oil contained in a straight dynamite is replaced by a gelatinized material prepared from 0.7 percent of soluble cellulose nitrate. The latter substance is dissolved in the explosive oil to form a jelly-like mass. Wood pulp or other carbonaceous materials, sodium nitrate, and an antacid are mixed together to form the dope.

The grades of gelatin dynamites range from 20 to 90 percent, depending on the relative proportions of the gelatinized ingredient and the sodium nitrate.

Gelatin dynamites have a higher density than the ammonia dynamites and are more suitable for blasting in hard ground. They have the further advantage of being plastic and can therefore be packed solidly in the hole so that the explosive is concentrated to best advantage with the least loss of force. Gelatin dynamites are the most suitable dynamites to use in wet holes due to their superior water-resisting properties. On detona-

tion they emit a very low volume of poisonous gases which makes them suitable for use in underground operations where trouble might otherwise be experienced from fumes.

DYNAMITE, SEMI-GELATIN
General Information
Description: The principal explosive ingredient of a semi-gelatin dynamite is ammonium nitrate, sensitized by a small quantity of a gelatinized explosive oil. For semi-gelatin dynamite only about 0.3 percent of a solution nitrocellulose is used for the gelatinization process. Wood pulp and other carbonaceous materials, sodium nitrate, and an antacid are also added to the dope.

The two most commonly used grades correspond to a bulk strength of 45 and 60 percent and have cartridge count per 50 pound case of 105 and 120, respectively, for the 1¼ by 8 inch size.

These dynamites are cohesive and are superior in water-resisting properties to the ammonia dynamites. They also emit a smaller volume of poisonous gases. Semi-gelatin dynamites are usually used for blasting rocks and ores or moderate hardness, such as limestones and gypsum.

DYNAMITE, STRAIGHT
General Information
Description: A straight dynamite contains no ammonium nitrate, nitroglycerin being the explosive ingredient. Straight dynamites have good water-resisting properties and their mixed ingredients are definitely cohesive. Their present use is confined mainly to propagated ditch blasting and submarine blasting. However, for virtually all other purposes, this class of dynamite has been replaced by the ammonia dynamites and the gelatin types of dynamites. See also dynamite, ammonia and dynamite, gelatin. On account of the large volume of poisonous gases emitted on detonation the straight dynamites are not recommended for blasting operations in underground workings. For blasting ditches, 50 percent straight dynamite (containing 50 percent explosive oil) has been found to be generally suitable and for other purposes, straight dynamites ranging from 15 to 60 percent can be procured. A typical formula for a 40 percent straight dynamite is as follows: 1% moisture; 40% explosive oil; 44% sodium nitrate; 1% antacid; 14% carbonaceous material. The wrapper of straight dynamite constitutes from 5 to 6 percent of the weight of a 1¼ by 8 inch cartridge which weighs nearly one-half pound.

DYPNONE
General Information
Description: Liquid.
Synonym: β-Methylchalcone.
Formula: $C_6H_5COCH{:}C(CH_3)C_6H_5$.
Constants: Mol wt: 222.27, mp: $-30°C$, bp: $246°C$ at 50 mm, flash p: $350°F$ (O.C.), d: 1.093 at $20°/20°C$, vap. press.: < 0.01 at $20°C$, vap. d.: 7.67.

Hazard Analysis
Toxicity: Details unknown. Limited animal experiments suggest low toxicity.
Fire Hazard: Slight, when exposed to heat or flame; can react with oxidizing materials.

Countermeasures
To Fight Fire: Water, foam, alcohol foam, carbon dioxide, dry chemical or carbon tetrachloride (Section 6).
Storage and Handling: Section 7.

TOXIC HAZARD RATING CODE (For detailed discussion, see Section 1.)

0 NONE: (a) No harm under any conditions; (b) Harmful only under unusual conditions or overwhelming dosage.

1 SLIGHT: Causes readily reversible changes which disappear after end of exposure.

2 MODERATE: May involve both irreversible and reversible changes not severe enough to cause death or permanent injury.

3 HIGH: May cause death or permanent injury after very short exposure to small quantities.

U UNKNOWN: No information on humans considered valid by authors

DYSPROSIUM
General Information
Formula: Dy.
Constant: At wt: 162.46.
Hazard Analysis
Toxicity: Unknown. May have anticoagulant effect.
Fire Hazard: Moderate. An active reducing agent.
Radiation Hazard: Section 5. For permissible levels, see Table 5, p. 150.

Artificial isotope ^{165}Dy, half life 2.4 h. Decays to stable ^{165}Ho by emitting beta particles of 1.20 (15%), 1.29 (82%) MeV. Also emits gamma rays of 0.095 MeV.

Artificial isotope ^{166}Dy, half life 81.5 h. Decays to radioactive ^{166}Ho by emitting beta particles of 0.40, 0.48 MeV. Also emits gamma rays of 0.08 MeV.

DYSPROSIUM BROMATE
General Information
Description: Yellow, hexagonal needles.
Formula: $Dy(BrO_3)_3 \cdot 9H_2O$.
Constants: Mol wt: 708.35, mp: 78°C, $-6H_2O$ at 110°C.
Hazard Analysis
Toxicity: Details unknown. See bromates.
Fire Hazard: Moderate, by spontaneous chemical reaction. A powerful oxidizing agent (Section 6).
Disaster Hazard: Dangerous; see bromides; can react vigorously with reducing materials.
Countermeasures
Storage and Handling: Section 7.

DYSPROSIUM CHROMATE
General Information
Description: Yellow crystals.
Formula: $Dy_2(CrO_4)_3 \cdot 10H_2O$.
Constants: Mol wt: 853.11, mp: $-3\frac{1}{2}H_2O$ at 150°C, bp: decomposes.

Hazard Analysis
Toxicity: See chromium compounds.
Fire Hazard: Slight, by chemical reaction; an oxidizing agent; can react with reducing materials (Section 6).
Countermeasures
Storage and Handling: Section 7.

DYSPROSIUM NITRATE
General Information
Description: Yellow crystals.
Formula: $Dy(NO_3)_3 \cdot 5H_2O$.
Constants: Mol wt: 438.56, d: 88.6.
Hazard Analysis
Toxicity: Details unknown. See nitrates.
Fire Hazard: Moderate, by chemical reaction. A powerful oxidizing agent (Section 6).
Disaster Hazard: Dangerous; when heated to decomposition, it emits highly toxic fumes of oxides of nitrogen; can react vigorously with reducing materials.
Countermeasures
Storage and Handling: Section 7.

DYSPROSIUM OXALATE
General Information
Description: Prisms.
Formula: $Dy_2(C_2O_4)_3 \cdot 10H_2O$.
Constant: Mol wt: 769.14.
Hazard Analysis
Toxicity: See oxalates.

DYSPROSIUM SELENATE
General Information
Description: Yellow needles.
Formula: $Dy_2(SeO_4)_3 \cdot 8H_2O$.
Constants: Mol wt: 897.93, mp: $-8H_2O$ at 200°C.
Hazard Analysis and Countermeasures
See selenium and its compounds.

E

E-60. See ethylene glycol dibenzoate.

E-605. See parathion.

E-1059. See demeton.

EADC. See ethyl aluminum dichloride.

EASC. See ethyl aluminum sesquichloride.

EARTHNUT OIL. See peanut oil.

EARTH OIL. See crude oil (petroleum).

EARTH WAX. See ceresin wax.

ECGONINE
General Information
Synonym: β-Oxymethyl-β-pyridylpropionic acid; tropine-carboxylic acid.
Description: White crystalline alkaloid; slightly bitter taste.
Formula: $C_9H_{15}NO_3 \cdot H_2O$.
Constants: Mol wt: 203.24, mp: 205°C, d: 1.370 at 12°/4°C, vap. d.: 7.0.
Hazard Analysis
Toxic Hazard Rating:
Acute Local: Allergen 1.
Acute Systemic: Ingestion 3; Inhalation 3.
Chronic Local: Allergen 1.
Chronic Systemic: U.
Disaster Hazard: Dangerous; when heated to decomposition, it emits highly toxic fumes of oxides of nitrogen.
Countermeasures
Ventilation Control: Section 2.
Personnel Protection: Section 3.
First Aid: Section 1.

ECHITAMINE. See ditaine.

EDATHMIL CALCIUM DISODIUM. See calcium disodium edta.

EDTA. See ethylenediaminetetraacetic acid.

EDTN. See ethylenediaminetetraacetonitrile.

EINSTEINIUM
General Information
Description: Element 99; a synthetic radioactive element; chemical properties similar to holmium.
Symbol: Es.
Constant: At wt: 254
Hazard Analysis
Radiation Hazard: (Section 5). For permissible levels, see Table 5, p. 150.
Artificial isotope ^{253}Es, half life 20 d. Decays to radioactive ^{249}Bk by emitting alpha particles of 6.6 MeV.
Artificial isotope ^{254m}Es, half life 480 d. Decays to radioactive ^{250}Bk by emitting alpha particles of 6.4 MeV.
Artificial isotope ^{254}Es, half life 37 h. Decays to radioactive ^{254}Fm by emitting beta particles of 1.0 MeV. Also emits gamma rays of 0.66 MeV.
Artificial isotope ^{255}Es, half life 24 d. Decays to radioactive ^{255}Fm by emitting beta particles of unknown energy.

EKABORON. See scandium.

ELAIDIC ACID. See oleic acid.

ELAYL. See ethylene.

ELECTRIC BLASTING CAPS (1,000 OR LESS). See explosives, high.
Shipping Regulations: Section 11.
I.C.C.: Explosive C.
Coast Guard Classification: Explosive C.
IATA: Explosive, not acceptable (passenger and cargo).

ELECTRIC BLASTING CAPS (MORE THAN 1,000). See explosives, high.
Shipping Regulations: Section 11.
I.C.C.: Explosive A—not accepted.
IATA: Explosive, not acceptable (passenger and cargo).
Coast Guard Classification: Explosive A.

ELECTRIC SQUIBS. See explosives, high.
Shipping Regulations: Section 11.
I.C.C.: Explosive C, 150 pounds.
Coast Guard Classification: Explosive C.
IATA: Explosive, explosive label, 25 kilograms (passenger), 70 kilograms (cargo).

ELECTROLYTE (ACID) BATTERY FLUID. See sulfuric acid.
Shipping Regulations: Section 11.
I.C.C.: Corrosive liquid; white label, 5 gallons.
Coast Guard Classification: Corrosive liquid; white label.
IATA (not over 47% strength): Corrosive liquid, white label, 1 liter (passenger), 20 liters (cargo).

ELECTROLYTE (ALKALINE) CORROSIVE BATTERY FLUID
Shipping Regulations: Section 11.
IATA: Corrosive liquid, white label, 1 liter (passenger), 20 liters (cargo).

ELECTROLYTE (ACID) OR ALKALINE CORROSIVE BATTERY FLUID PACKED WITH STORAGE BATTERIES
Shipping Regulations: Section 11.
I.C.C.: Corrosive liquid, white label, 6 quarts.

TOXIC HAZARD RATING CODE (For detailed discussion, see Section 1.)

0 NONE: (a) No harm under any conditions; (b) Harmful only under unusual conditions or overwhelming dosage.

1 SLIGHT: Causes readily reversible changes which disappear after end of exposure.

2 MODERATE: May involve both irreversible and reversible changes; not severe enough to cause death or permanent injury.

3 HIGH: May cause death or permanent injury after very short exposure to small quantities.

U UNKNOWN: No information on humans considered valid by authors.

ELECTROLYTE (ACID OR ALKALINE) CORROSIVE BATTERY FLUID PACKED WITH ACTUATING DEVICE, BATTERY CHARGER, ELECTRONIC EQUIPMENT, RADIO CURRENT SUPPLY DEVICE, OR EMPTY STORAGE BATTERY.
Shipping Regulations: Section 11.
  I.C.C.: Corrosive liquid, white label, 6 quarts.
  IATA: Corrosive liquid, white label, not acceptable (passenger), 2½ liters (cargo).

ELIXIR OF VITRIOL. See sulfuric acid, aromatic.

EMERALD GREEN. See copper acetoarsenite.

EMERY
General Information
Synonym: Corundum.
Description: A varicolored mineral.
Formula: Al_2O_3.
Constant: D: 3.95–4.10.
Hazard Analysis
Toxic Hazard Rating:
  Acute Local: Inhalation 1.
  Acute Systemic: 0.
  Chronic Local: Inhalation 1.
  Chronic Systemic: 0.
TLV: ACGIH (accepted); 50 million particles per cubic foot of air.

EMETINE
General Information
Description: White powder or lumps; bitter taste; darkens on exposure.
Formula: $C_{29}H_{40}O_4N_2$.
Constants: Mol wt: 480.63, mp: 74°C.
Hazard Analysis
Toxic Hazard Rating:
  Acute Local: U.
  Acute Systemic: Ingestion 3; Inhalation 3.
  Chronic Local: U.
  Chronic Systemic: U.
Toxicology: This is one of the two potent alkaloids obtained from the Brazilian plant ipecac. The therapeutic use of various ipecac preparations has given rise to many cases of poisoning, in some instances with fatal results. The toxic effects are particularly prominent if the drug is given intravenously. Special care should therefore be exercised when administering it in this manner. The symptoms of intoxication are gastrointestinal irritation and salivation, as well as general edema, which follows renal insufficiency, hemoptysis, flaccid paralysis, peripheral neuritis, aphonia, difficulties in swallowing, delirium, coma and failure of the heart. The fatal dose is considered to be approximately 2 g, whether administered over a short or relatively long period. The drug seems to have a cumulative effect.
Disaster Hazard: Dangerous; when heated to decomposition, it emits highly toxic fumes of oxides of nitrogen.
Countermeasures
Personnel Protection: Section 3.
First Aid: Section 1.
Storage and Handling: Section 7.
Treatment and Antidotes: Since poisoning occurs generally only after repeated dosage, discontinuing use of the drug usually stops the symptoms and recovery follows. When acute intoxication occurs, the remedial measures are purely symptomatic. Heart depression is the most serious symptom and is most to be guarded against.

EMPTY CARTRIDGE BAGS, BLACK POWDER IGNITERS. See explosives, low.
Shipping Regulations: Section 11.
  I.C.C.: Explosive C, 150 pounds.
  Coast Guard Classification: Explosive C.

EMPTY CARTRIDGE CASES, PRIMED. See explosives, high.
Shipping Regulations: Section 11.
  I.C.C.: Explosive C, 150 pounds.
  Coast Guard Classification: Explosive C.

ENDO-cis-BICYCLO (2,2,1)-5-HEPTENE-2,3-DICARBOXYLIC ANHYDRIDE
General Information
Description: White crystals.
Formula: $C_9H_8O_3$.
Constants: Mol wt: 164.2, mp: 164–165°C.
Hazard Analysis
Toxicity: Unknown.
Fire Hazard: Slight; it will react with water or steam to produce heat (Section 6).
Countermeasures
Storage and Handling: Section 7.

ENDO-DICYCLOPENTADIENE DIOXIDE
General Information
Description: White crystal powder; slightly soluble in water; soluble in acetone and benzene.
Formula: $C_{10}H_{12}O$.
Constants: Mol wt: 164, mp: 180–184°C, d: 1.331 at 25°C.
Hazard Analysis
Toxic Hazard Rating:
  Acute Local: U.
  Acute Systemic: Ingestion 3; inhalation 1, Skin Absorption 2.
  Chronic Local: Allergen 1.
  Chronic Systemic: U.
Based on limited animal experiments.

2,5-ENDOMETHYLENE CYCLOHEXANE CARBOXYLIC ACID, ETHYL ESTER
Hazard Analysis
Toxic Hazard Rating: U. Limited animal experiments suggest low toxicity. See also esters.

(2,5-ENDOMETHYLENE CYCLOHEXYL METHYL) AMINE
Hazard Analysis
Toxic Hazard Rating: U. Limited animal experiments suggest moderate toxicity and irritation. See also amines.

ENDOTHAL
General Information
Synonym: 3,6-Endoxohexahydrophthalic acid, disodium salt.
Hazard Analysis
Toxic Hazard Rating:
  Acute Local: Irritant 3; Ingestion 3; Inhalation 3.
  Acute Systemic: U.
  Chronic Local: U.
  Chronic Systemic: U.
Toxicology: Very irritating to eyes, skin and mucous membrane.

ENDOXAN
General Information
Synonym: Cyclophosphamide.
Description: Crystals. Water soluble, slightly soluble in organic solvents.
Formula: $C_7H_{15}Cl_2N_2O_2P$.
Constants: Mol wt: 261.1, mp: 41–45°C.
Hazard Analysis
Toxic Hazard Rating:
  Acute Local: U.
  Acute Systemic: Ingestion 2.
  Chronic Local: U.
  Chronic Systemic: Ingestion 2.
Toxicology: Can cause gastrointestinal disturbances and leukopenia.

Disaster Hazard: When heated to decomposition it emits highly toxic fumes of oxides of phosphorus and nitrogen.

3,6-ENDOXOHEXAHYDROPHTHALIC ACID DISODIUM SALT. See endothal.

ENDRIN
General Information
Synonym: 1,2,3,4,10,10-Hexachloro-6,7-epoxy-1,4,4a,5,6, 7,8,8a-octahydro-1,4,5,8-endo-endo-dimethanonaphthalene.
Hazard Analysis
Toxicity: An insecticide; similar to aldrin and dieldrin. See dieldrin.
TLV: ACGIH (recommended) 0.1 milligrams per cubic meter of air.
Disaster Hazard: Dangerous; when heated to decomposition, it emits highly toxic fumes.
Countermeasures
Shipping Regulations: Section 11.
  IATA: Poison B, poison label, 25 kilograms (passenger), 95 kilograms (cargo).

ENGINES, INTERNAL COMBUSTION
Shipping Regulations: Section 11.
  I.C.C.: See § 73.120.
  IATA: No limit (passenger and cargo).

ENGINE STARTING FLUID
Shipping Regulations: Section 11.
  I.C.C.: Flammable gas, red label, 60 pounds.
  IATA: (charged with flammable gas): Flammable gas, red label, not acceptable (passenger), 30 kilograms (cargo).

EPHEDRINE
General Information
Synonym: 1-Phenyl-2-methylaminopropanol.
Description: White granules.
Formula: $C_{10}H_{15}NO$.
Constants: Mol wt: 165.23, mp: 34-40°C, bp: 255°C decomposes.
Hazard Analysis
Toxic Hazard Rating:
  Acute Local: Allergen 1.
  Acute Systemic: Ingestion 2; Inhalation 2.
  Chronic Local: Allergen 1.
  Chronic Systemic: Ingestion 1; Inhalation 1.
Toxicology: Causes rapid pulse, rise in blood pressure, and other actions similar to epinephrine. Has been known to cause allergic sensitization (Section 9).
Disaster Hazard: Slight; when heated, it emits toxic fumes (Section 6).
Countermeasures
Personnel Protection: Section 3.
Storage and Handling: Section 7.

EPICHLOROHYDRIN
General Information
Synonym: 1-Chlor-2,3-epoxypropane.
Description: Colorless, mobile liquid; irritating chloroform-like odor.
Formula: C_3H_5OCl.
Constants: Mol wt: 92.52, bp: 117.9°C, fp: −57.1°C, flash p: 105°F (O.C.), d: 1.1761 at 20°/20°C, vap. press.: 10 mm at 16.6°C, vap. d.: 3.29.
Hazard Analysis
Toxic Hazard Rating:
  Acute Local: Irritant 3; Ingestion 3; Inhalation 3; Allergen 2.

Acute Systemic: Ingestion 3; Inhalation 3; Skin Absorption 3.
Chronic Local: Irritant 3; Allergen 2; Inhalation 3.
Chronic Systemic: Ingestion 3; Inhalation 3; Skin Absorption 3.
Toxicology: In acute poisoning death may be caused by respiratory paralysis. In chronic poisoning there is kidney damage. Inflammatory changes in the eyes and lungs have been observed. Primary irritation and sensitization of skin has been described.
TLV: ACGIH (tentative) 5 parts per million in air; 19 milligrams per cubic meter of air; may be absorbed through intact skin.
Fire Hazard: Moderate, when exposed to heat or flame.
Disaster Hazard: Dangerous; when heated to decomposition, it emits highly toxic fumes of phosgene; can react with oxidizing materials.
Countermeasures
Ventilation Control: Section 2.
To Fight Fire: Water, foam, alcohol foam, carbon dioxide, dry chemical or carbon tetrachloride (Section 6).
Personnel Protection: Section 3.
Storage and Handling: Section 7.
MCA warning label.

EPINEPHRINE. See "Adrenaline."

EPN
General Information
Synonym: Ethyl-p-nitrophenyl thionobenzene phosphate.
Formula: $C_{14}H_{14}O_5NPS$.
Constant: Mol wt: 339.3.
Hazard Analysis
Toxicity: An insecticide. Highly toxic. A cholinesterase inhibitor. See related compound, parathion.
TLV: ACGIH (recommended) 0.5 milligrams per cubic meter of air. Absorbed through intact skin.
Disaster Hazard: Dangerous; when heated to decomposition, it emits highly toxic fumes of oxides of sulfur and phosphorus.

EPON RESINS, CURED. See epoxy resins, cured.

EPON RESINS, UNCURED. See epoxy resins, uncured.

1,2-EPOXYBUTANE
General Information
Synonym: 1,2-Butylene oxide.
Description: Colorless liquid; soluble in water; miscible with most organic solvents.
Formula: $H_2COCHCH_2CH_3$.
Constants: Mol wt: 72, d: 0.8312 (20/20°C); bp: 63°C; flash p: < 0°F.
Hazard Analysis
Toxic Hazard Rating:
  Acute Local: Irritant 2.
  Acute Systemic: Ingestion 2, Inhalation 2, Skin Absorption 2.
  Chronic Local: U.
  Chronic Systemic: U.
Based on limited animal experiments. See also butylene oxides.
Fire Hazard: Dangerous, when exposed to heat or flame.
Countermeasures
Storage and Handling: See Section 7.

2,3-EPOXYBUTYRIC ACID, BUTYL ESTER
Hazard Analysis
Toxic Hazard Rating: U. Limited animal experiments sug-

TOXIC HAZARD RATING CODE (For detailed discussion, see Section 1.)

0 NONE: (a) No harm under any conditions; (b) Harmful only under unusual conditions or overwhelming dosage.

1 SLIGHT: Causes readily reversible changes which disappear after end of exposure.

2 MODERATE: May involve both irreversible and reversible changes; not severe enough to cause death or permanent injury.

3 HIGH: May cause death or permanent injury after very short exposure to small quantities.

U UNKNOWN: No information on humans considered valid by authors.

gest moderate toxicity and low irritant. See also epoxy resins and esters.

4,5-EPOXYCYCLOHEXANE-1,2-DICARBOXYLIC ACID; DI(DECYL) ESTER
Hazard Analysis
Toxic Hazard Rating: U. Limited animal experiments suggest low toxicity and irritant. See also epoxy resins and esters.

3,4-EPOXYCYCLOHEXANECARBONITRILE
General Information
Description: Liquid, soluble in water.
Formula: $O(C_6H_9)CN$
Constants: Mol wt: 123; d: 1.0929 (19/4°C), bp: 244.5°C; flash p: −33°C.
Hazard Analysis
Toxic Hazard Rating:
  Acute Local: Irritant 2.
  Acute Systemic: Ingestion 2, Inhalation 1, Skin Absorption 3.
  Chronic Local: U.
  Chronic Systemic: U.
Limited Animal experiments suggest high toxicity. See also epoxy resins and nitriles.

3,4-EPOXYCYCLOHEXYLMETHYL-3,4-EPOXY-CYCLOHEXANE CARBOXYLATE
Hazard Analysis
Toxic Hazard Rating:
  Acute Local: Irritant 1.
  Acute Systemic: Ingestion 1, Skin Absorption 1.
  Chronic Local: U.
  Chronic Systemic: U.
Based on limited animal experiments.

1,2-EPOXYETHANE. See ethylene oxide.

3-(a,b-EPOXYETHYL)-5,6-EPOXYBENZENE
Hazard Analysis
Toxic Hazard Rating: U. Limited animal experiments suggest moderate toxicity and irritant. See also epoxy resins.

2,3-EPOXY-2-ETHYL HEXANOL
General Information
Formula: $C_3H_7CHOC(C_2H_5)CH_2OH$
Constant: Mol wt: 144.
Hazard Analysis
Toxic Hazard Rating:
  Acute Local: Irritant 2.
  Acute Systemic: Ingestion 1, Inhalation 1, Skin Absorption 1.
  Chronic Local: U.
  Chronic Systemic: U.
Based on limited animal experiments. See also epoxy resins.

2,3-EPOXY-2-ETHYLHEXYL-9,10-EPOXYSTEARATE
Hazard Analysis
Toxic Hazard Rating:
  Acute Local: U.
  Acute Systemic: Ingestion 1.
  Chronic Local: U.
  Chronic Systemic: U.
Based on limited animal experiments.

4-EPOXY-6-METHYLCYCLOHEXANE CAR-BOXYLIC ACID, ALLYL ESTER
Hazard Analysis
Toxic Hazard Rating: U. Limited animal experiments suggest moderate toxicity and low irritant. See also epoxy resins and esters.

4-EPOXY-6-METHYLCYCLOHEXYL METHYL-3,4-ACRYLATE
Hazard Analysis
Toxic Hazard Rating: U. Limited animal experiments suggest low toxicity and irritant. See also epoxy resins.

3,4-EPOXY-6-METHYL CYCLOHEXYL METHYL-3,4-EPOXY-6-METHYL CYCLOHEXANE CARBOXYL-ATE
Hazard Analysis
Toxic Hazard Rating:
  Acute Local: Irritant 1.
  Acute Systemic: Ingestion 1, Skin Absorption 1.
  Chronic Local: O.
  Chronic Systemic: U.
Based on Limited animal experiments.

1,2-EPOXYPROPANE. See propylene oxide.

1,3-EPOXYPROPANE. See trimethylene oxide.

2,3-EPOXY-1-PROPANOL. See glycidol.

2,3-EPOXYPROPYL ACRYLATE
General Information
Formula: $H_2C-CHCH_2OOCHC:CH_2$.
      $\underset{O}{}$
Constant: Mol wt: 123.
Hazard Analysis
Toxic Hazard Rating: U. Limited animal experiments suggest high toxicity and irritation. See also epoxy resins.

N-(2,3-EPOXYPROPYL) DIETHYL AMINE
Hazard Analysis
Toxic Hazard Rating: U. Limited animal experiments suggest high toxicity and irritation. See also epoxy resins and amines.

2,3-EPOXYPROPYL ETHER
General Information
Formula: $H_2COCHOCH_3$.
Constant: Mol wt: 74.
Hazard Analysis
Toxic Hazard Rating: U. Limited animal experiments suggest moderate toxicity. See also epoxy resins and ethers.

EPOXY RESINS, CURED
Hazard Analysis
Toxic Hazard Rating:
  Acute Local: Irritant 1, Inhalation 1.
  Acute Systemic: Ingestion 0–1.
  Chronic Local: 0.
  Chronic Systemic: Ingestion 0–1.
Toxicology: Most cured resins have little or no toxic effects. If curing is incomplete there may be residues of highly toxic curing agents such as the organic amines: m-phenylene diamine; diethylene triamine, tetraethylene pentamine; hexamethylene tetramine; also phthalic anhydride and related compounds.
Disaster Hazard: Dangerous. When heated to decomposition they emit highly toxic fumes.

EPOXY RESINS, UNCURED
General Information
Synonym: Polymers of epichlorohydrin and 2,2-bis(4-hydroxy-phenyl) piperazine
Hazard Analysis
Toxic Hazard Rating:
  Acute Local: Irritant 3, allergen 2, inhalation 1.
  Acute Systemic: Ingestion 1, inhalation 2.
  Chronic Local: Irritant 3.
  Chronic Systemic: Ingestion 1, inhalation 1.
Toxicology: Animal experiments have shown disturbed blood formation. The degree of toxicity of uncured epoxy resins varies, and is partly dependent on the extent of unreacted curing agents. See also epoxy resins, cured and specific agents.
Disaster Hazard: Dangerous. When heated to decomposition they emit highly toxic fumes.

9,10-EPOXY STEARIC ACID, ALLYL ESTER
Hazard Analysis
Toxic Hazard Rating: U. Limited animal experiments suggest moderate toxicity. See also epoxy resins and esters.

9,10-EPOXY STEARIC ACID, 2 ETHYLHEXYL ESTER
Hazard Analysis
Toxic Hazard Rating: U. Limited animal experiments suggest low toxicity. See also epoxy resins and esters.

1,2-EPOXY-4-VINYLCYCLOHEXANE
Hazard Analysis
Toxic Hazard Rating: U. Limited Animal experiments suggest moderate toxicity. See also epoxy resins.

EPSOM SALTS. See magnesium sulfate.

ERADICATOR (PAINT OR GREASE), LIQUID
Hazard Analysis
Fire Hazard: Dangerous, when exposed to heat or flame; can react vigorously with oxidizing materials (Section 6).
Explosion Hazard: Unknown.
Countermeasures
Ventilation Control: Section 2.
Storage and Handling: Section 7.
Shipping Regulations: Section 11.
I.C.C.: Flammable liquid; red label, 10 gallons.
Coast Guard Classification: Combustible liquid.
IATA: Flammable liquid, red label, 1 liter (passenger), 40 liters (cargo).
Coast Guard Classification: Inflammable liquid; red label.

ERASOL. See mechlorethamine hydrochloride.

ERBIUM
General Information
Description: Dark gray powder.
Formula: Er.
Constants: At wt: 167.20, mp: 1520°C, bp: 2600°C, d: 9.15, vap. press.: 2 mm at 1530°C.
Hazard Analysis
Toxicity: Unknown.
Radiation Hazard: Section 5. For permissible levels, see Table 5, p. 150.
Artificial isotope ^{169}Er, half life 9.3 d. Decays to stable. ^{169}Tm by emitting beta particles of 0.33–0.34 MeV. Also emits gamma rays of 0.01 MeV.
Artificial isotope ^{171}Er, half life 7.5 h. Decays to radioactive ^{171}Tm by emitting beta particles of 1.06 MeV and others. Also emits gamma rays of 0.11–0.31 MeV.
Fire Hazard: Moderate, when exposed to heat or flame by chemical reaction with oxidizers (Section 6).
Explosion Hazard: Moderate, when exposed to flame or spark.

ERBIUM NITRATE
General Information
Description: Reddish crystals.
Formula: $Er(NO_3)_3 \cdot 5H_2O$.
Constant: Mol wt: 443.30.
Hazard Analysis and Countermeasures
See nitrates.

ERBIUM OXALATE
General Information
Description: Reddish powder.
Formula: $Er_2(C_2O_4)_3 \cdot 10H_2O$.
Constants: Mol wt: 778.62, mp: decomposes at 575°C.

Hazard Analysis
Toxicity: See oxalates.

ERBON
General Information
Synonyms: 2-(2,4,5-Trichlorophenoxy) ethyl-2,2-dichloropropionate.
Description: Crystals; insoluble in water, soluble in acetone, alcohol, kerosene and xylene.
Formula: $C_{11}H_9Cl_5O_3$.
Constants: Mol wt: 366.5, mp: 50°C, bp: 161–164°C at 0.5 mm.
Hazard Analysis
Toxicity: Details unknown. Animal experiments show effects similar to those of 2,4-D. See 2,4-dichlorophenoxyacetic acid.
Disaster Hazard: Dangerous; when heated to decomposition it emits highly toxic fumes of chlorides.

ERGAMINE. See histamine.

ERGOCALCIFEROL. See vitamin D_2.

ERGOTININE
General Information
Synonym: Ergotoxine.
Description: White, crystalline alkaloid.
Formula: $C_{35}H_{39}N_5O_5$.
Constants: Mol wt: 609.53, mp: 205°C.
Hazard Analysis
Toxic Hazard Rating:
Acute Local: U.
Acute Systemic: Ingestion 3.
Chronic Local: U.
Chronic Systemic: Ingestion 2.
Toxicology: Acute poisoning by ergot is very rare and is caused accidentally, except in some cases where it has been employed to produce abortion. The symptoms are vomiting, burning pains in the abdomen, tingling of the extremities, great thirst, diarrhea, collapse with weak, rapid pulse, cold skin, hemorrhage from the uterus, and abortion. There is ecchymosis in many organs and in the subcutaneous tissues, suppression of urine, prostration, coma and death from respiratory and cardiac failure. Death may occur in a few hours or be delayed a few days. In cases of recovery, abnormal symptoms persist for a few days. Sometimes a cataract forms in the eyes.

Chronic poisoning is very rarely caused by the medicinal use of ergot. Some epidemics have occurred from eating bread made with grain containing ergot. A severe epidemic causing 11,319 cases of poisoning occurred in Russia. Of these, 1,618 were admitted to the hospital and 93 died. There are two types of symptoms from ergot poisoning, one which causes a nervous disorder and one which causes gangrene. In some cases both types have been found, but as a rule one prevails. The gangrenous type generally develops in the fingers and toes; sometimes the entire leg or arm becomes cold and numb, dark, dry, hard and shrunken and falls off with little or no pain and no hemorrhage. Gangrene also occurs in the internal organs. Cataracts are common. The symptoms of spasmodic ergotism are drowsiness, depression, weakness, giddiness, headache, and cramps in the limbs, with itching. In severe cases, convulsions generally clonic and often epileptiform occur. There is mental weakness and sometimes complete dementia. In animals, restlessness, salivation

---

TOXIC HAZARD RATING CODE *(For detailed discussion, see Section 1.)*

0 NONE: (a) No harm under any conditions; (b) Harmful only under unusual conditions or overwhelming dosage.

1 SLIGHT: Causes readily reversible changes which disappear after end of exposure.

2 MODERATE: May involve both irreversible and reversible change not severe enough to cause death or permanent injury.

3 HIGH: May cause death or permanent injury after very short exposure to small quantities.

U UNKNOWN: No information on humans considered valid by author

vomiting and purging, depression, ataxia, clonic convulsions and death by paralysis of the respiratory center occur. The fatal dose cannot be stated accurately. However, 30 grains has caused severe poisoning, but recovery has followed 150 grains (9.72 g). Gangrene and death have been said to follow 10 grains. Death is more likely to follow prolonged use of small medicinal doses than one large dose.

Disaster Hazard: Dangerous; when heated to decomposition, it emits highly toxic fumes of oxides of nitrogen.

Countermeasures

Ventilation Control: Section 2.

Treatment and Antidotes: Wash out the stomach with water containing 80 to 100 grains of sodium bicarbonate. Call a physician. Give 0.5 to 1.0 g of sodium tetrathionate dissolved in 10 cc of sterile distilled water intravenously daily. Empty the bowels by soap-suds enemas and purgatives, such as calomel and castor oil. Treat symptomatically. Give stimulants, such as strychnine and coffee. If gangrene sets in, the parts should be bathed with warm water and wrapped in soothing ointment.

Personnel Protection: Section 3.
Storage and Handling: Section 7.

ERGOTOXINE. See ergotinine.

ERIOCHALCITE. See copper chloride.

ERYTHORBIC ACID
General Information
Synonym: d-erythro-ascorbic acid.
Description: Shiny granular crystals; soluble in water, alcohol, pyridine; moderately soluble in acetone; slightly soluble in glycerol.
Formula: $C_6H_8O_6$.
Constant: Mol wt: 176.
Hazard Analysis
Toxic Hazard Rating: U. Used as a chemical preservative food additive. See section 10.

ERYTHRENE. See butadiene-1,3.

ERYTHRITE. See cobaltous o-arsenate.

ERYTHROMYCIN
General Information
Description: An antibiotic; white or slightly yellow, crystalline powder, odorless; freely soluble in alcohol, chloroform, and ether, very slightly soluble in water.
Formula: $C_{37}H_{67}NO_{13}$.
Constants: Mol wt: 733, mp: 133–138°C.
Hazard Analysis
Toxic Hazard Rating: U. A food additive permitted in food for human consumption. See section 10.

ERYTHROMYCIN THIOCYANATE
General Information
Description: The thiocyanate salt of the antibiotic substance produced by the growth of streptomyces erythreus.
Hazard Analysis
Toxic Hazard Rating: U. A food additive permitted in the feed and drinking water of animals and/or for the treatment of food producing animals. See section 10.

ESERINE. See physostigmine.

ESSENTIAL OILS, OLEORESINS (SOLVENT FREE), AND NATURAL EXTRACTIVES (INCLUDING DISTILLATES). See Food Additives Section 10.

ESSENTIAL SALT OF LEMON. See potassium binoxalate.

ESTER GUM
General Information
Synonym: Rosin ester.

Description: Hard resin.
Constants: Mp: 141°C, bp: 257°C, flash p: 375°F (C.C.).
Hazard Analysis
Toxicity: Unknown.
Fire Hazard: Slight, when exposed to heat or flame; can react with oxidizing materials.
Spontaneous Heating: No.
Countermeasures
To Fight Fire: Water, carbon dioxide, dry chemical or carbon tetrachloride (Section 6).
Storage and Handling: Section 7.

ESTERS
General Information
Description: A large group of compounds which correspond structurally to salts in inorganic chemistry. They are considered as being derived from acids by the replacement of hydrogen by an organic alkyl radical.
Hazard Analysis
Toxicology: No general statement can be made as to the toxicity of esters. Many of them are highly volatile and hence can act as asphyxiants or narcotics. Skin absorption, as well as inhalation, may be an important route of absorption for those esters which are volatile and have high solvent action. The degree of toxicity of the esters covers a wide range, from slight to high. Esters generally hydrolyze upon contact with moisture; hence a rough guide to the toxicity of a given ester may be the sum of the toxicities of the products of hydrolysis.

ESTRADIOL BENZOATE
General Information
Description: White or slightly yellow to brownish crystalline powder; odorless. Almost insoluble in water; soluble in alcohol, acetone, and dioxane; sparingly soluble in vegetable oils; slightly soluble in ether.
Formula: $C_{18}H_{23}O \cdot C_7H_5O_2$.
Constants: Mol wt: 376; mp: 191–196°C.
Hazard Analysis
Toxic Hazard Rating: U. A food additive permitted in the feed and drinking water of animals and/or for the treatment of food producing animals. See section 10.

ESTRADIOL MONOPALMITATE
General Information
Synonym: 1,3,5(10)-estratriene-3,17β-diol-17-palmitate.
Formula: $C_{34}H_{54}O_3$.
Constants: Mol wt: 510, mp: 81–85°C.
Hazard Analysis
Toxic Hazard Rating: U. A food additive permitted in feed and drinking water of animals and/or for the treatment of food producing animals. Also a food additive permitted in food for human consumption. See section 10.

ETCHING ACID LIQUID
Shipping Regulations: Section 11.
I.C.C. (N.O.S.): Corrosive liquid, white label, 10 pounds.
IATA (containing a mixture of nitric and hydrochloric acids).: Corrosive liquid, white label, not acceptable (passenger), 5 kilograms (cargo).

ETHANE
General Information
Synonyms: Bimethyl; methylmethane; dimethyl; ethyl hydride.
Description: Colorless, odorless gas.
Formula: C_2H_6.
Constants: Mol wt: 30.07, mp: −172°C, bp: −88.6°C, lel: 3.0%, uel: 12.5%, fp: −183.2°C, d: 0.446 at 0°C (liquid), autoign. temp.: 959°F, vap. d.: 1.04.
Hazard Analysis
Toxic Hazard Rating:
Acute Local: 0.
Acute Systemic: Inhalation 2.

Chronic Local: 0.
Chronic Systemic: Inhalation 1.
Toxicology: A simple asphyxiant. See argon for properties of simple asphyxiants.
Fire Hazard: Dangerous, when exposed to heat or flame; can react vigorously with oxidizing materials.
Spontaneous Heating: No.
Explosion Hazard: Moderate, when exposed to flame.
Disaster Hazard: Dangerous; upon exposure to heat or flame.
Countermeasures
Ventilation Control: Section 2.
To Fight Fire: Stop flow of gas. Carbon dioxide or dry chemical (Section 6).
Storage and Handling: Section 7.
Shipping Regulations: Section 11.
    I.C.C.: Flammable gas; red gas label, 300 pounds.
    Coast Guard Classification: Inflammable gas; red gas label.
    IATA: Flammable gas, red label, not acceptable (passenger), 140 kilograms (cargo).

ETHANEARSONIC ACID
General Information
Synonym: Ethylarsonic acid.
Description: Crystals, water soluble.
Formula: $C_2H_5AsO(OH)_2$.
Constants: Mol wt: 154.0, mp: 99.5°C.
Hazard Analysis and Countermeasures
See arsenic compounds.

1,2-ETHANEDIAMINE. See ethylenediamine.

ETHANEDIOIC ACID. See oxalic acid.

ETHANEDIAL. See glyoxal.

1,2-ETHANEDIOL. See ethylene glycol.

ETHANEDIOYL BROMIDE. See oxalyl bromide.

ETHANEDIOYL CHLORIDE. See oxalyl chloride.

ETHANEDINITRILE. See cyanogen.

ETHANE HEXAMERCARBIDE
General Information
Description: Yellowish-white powder. Insoluble in water.
Formula: $C_2Hg_6O_2(OH)_2$.
Constants: Mol wt: 1293.7, mp: explodes at 230°C, vap. d.: 44.6.
Hazard Analysis
Toxicity: See mercury compounds, organic. Highly toxic.
Explosion Hazard: Dangerous. When shocked or heated to 230°C it explodes.
Disaster Hazard: Dangerous; when heated to decomposition or on contact with acid or acid fumes, below 230°C, it emits highly toxic fumes of mercury; can react vigorously with oxidizing materials.
Countermeasures
Storage and Handling: Section 7.

ETHANENITRILE. See methyl cyanide.

ETHANETETRACARBONITRILE. See tetracyanoethylene.

ETHANETHIOL. See ethyl mercaptan.

ETHANETHIOLIC ACID. See thioacetic acid.

ETHANOIC ACID. See acetic acid.

ETHANOIC ANHYDRIDE. See acetic anhydride.

ETHANOL. See ethyl alcohol.

ETHANOLAMINE. See monoethanolamine.

ETHANOL FORMAMIDE
General Information
Description: Liquid.
Formula: $HOCH_2CH_2NHCHO$.
Constants: Mol wt: 89.09, mp: −72°C, bp: 143°C at 2.5 mm, flash p: 165°C, d: 1.17 at 25°/4°C, vap. press.: 0.1 mm at 20°C, vap. d.: 3.07.
Hazard Analysis
Toxicity: Details unknown; probably slight. See also amides.
Fire Hazard: Moderate, when exposed to heat or flame; can react with oxidizing materials.
Countermeasures
To Fight Fire: Water, foam, carbon dioxide, dry chemical or carbon tetrachloride (Section 6).
Storage and Handling: Section 7.

ETHANOL MERCURIC CHLORIDE
Hazard Analysis and Countermeasures
See mercury compounds, organic.

ETHANOYL BROMIDE. See acetyl bromide.

ETHANOYL CHLORIDE. See acetyl chloride.

ETHANOYL IODIDE. See acetyl iodide.

ETHANOYL PEROXIDE. See acetyl peroxide.

ETHENE. See ethylene.

ETHENOL. See vinyl alcohol.

ETHENONE. See ketene.

ETHENOXYETHANE. See ethyl vinyl ether.

ETHER. See ethyl ether.

ETHERS
General Information
Description: Organic compounds in which an oxygen atom is interposed between two carbon atoms in the structure of the molecule.
Hazard Analysis
Toxicology: The simpler ethers such as ethyl ether, isopropyl ether, etc. are powerful narcotics which in large doses can cause death. The danger from ethers is usually acute and seldom chronic.
    There are seldom after-effects to ether intoxication although continued exposure to small concentrations (not enough to cause an overt symptom) has been known to cause loss of appetite, excessive thirst and fatigue.
Fire Hazard: The most common ethers such as ethyl and methyl are particularly dangerous fire hazards. The common ones are easily ignited and have low flash points. It is necessary to control smoking, open flames or even the use of hot plates in areas where low molecular weight ethers are apt to reach 1% concentration or more in air. Only electrical equipment of explosion-proof type (Group C classification) is permitted to be operated in ether areas. Ethers should not be stored near powerful oxidizers or in areas of high fire hazard. They should be kept cool and the containers electrically grounded to avoid sparks (Section 6).

TOXIC HAZARD RATING CODE (For detailed discussion, see Section 1.)

0 NONE: (a) No harm under any conditions; (b) Harmful only under unusual conditions or overwhelming dosage.

1 SLIGHT: Causes readily reversible changes which disappear after end of exposure.

2 MODERATE: May involve both irreversible and reversible changes; not severe enough to cause death or permanent injury.

3 HIGH: May cause death or permanent injury after very short exposure to small quantities.

U UNKNOWN: No information on humans considered valid by authors.

Explosion Hazard: Dangerous; when heated or exposed to flame or sparks. Besides the risk of explosion from air mixtures of ether vapors, ethers tend to form peroxides upon standing. When ethers containing peroxides are heated they can detonate. See also ethyl ether.

Disaster Hazard: Dangerous; shock or heat can cause gaseous ethers to escape from their containers; they can react vigorously with oxidizing materials.

Countermeasures
Ventilation Control: Section 2.
Storage and Handling: Section 7.
Shipping Regulations: Section 11.
  Coast Guard Classification: Inflammable liquid; red label.

ETHERIN. See ethylene.

ETHIDE. See 1,1-dichloro-1-nitroethane.

ETHINE. See acetylene.

ETHINYL TRICHLORIDE. See trichloroethylene.

ETHION
General Information
Synonym: O,O,O',O'-Tetraethyl-S,S-methylene di-phosphorodithioate.
Description: Liquid, slightly soluble in water, soluble in xylene, chloroform, acetone.
Formula: $C_9H_{22}O_4P_2S_4$.
Constants: Mol wt: 384.5, mp: $-13°C$, d: 1.220 at 20/4°C.

Hazard Analysis
Toxicity: Details unknown. Limited animal experiments show moderate toxicity. A cholinesterase inhibiting phosphate ester insecticide. For effects see parathion.
Note: Used also as a food additive permitted in the feed and drinking water of animals and/or for the treatment of food for producing animals; also permitted in food for human consumption. See section 10.
Disaster Hazard: Dangerous; when heated to decomposition, it emits highly toxic fumes of oxides of sulfur and phosphorus.

ETHOPABATE
General Information
Synonym: Methyl-4-acetamide-2-ethoxy benzoate.
Hazard Analysis
Toxic Hazard Rating: U. Used as a food additive permitted in food for human consumption. See section 10.

ETHOXAZENE
General Information
Synonym: 2,4-Diamino-4'-ethoxyazobenzene.
Description: Crystals.
Formula: $C_{14}H_{16}N_4O$.
Constant: Mol wt: 256.3.
Hazard Analysis
Toxicity: Details unknown. Large doses are said to cause irritation of urinary tract.

ETHOXYACETYLENE
General Information
Description: Insoluble in water.
Formula: C_4H_6O.
Constants: Mol wt: 70, flash p: $< 20°F$, d: 0.8, vap. d.: 2.4, bp: 124°F.
Hazard Analysis
Fire Hazard: Dangerous, when exposed to heat or flame.
Countermeasures
Storage and Handling: Section 7.

ETHOXYBORON DICHLORIDE
General Information
Description: Colorless liquid; decomposed by water.
Formula: $C_2H_5OBCl_2$.
Constants: Mol wt: 126.8, bp: 78°C.

Hazard Analysis and Countermeasures
See chlorides and boron compounds.

2-ETHOXY-3,4-DIHYDRO-2,2-PYRAN
General Information
Synonym: 2-ethoxy-3,4-dihydro-2H-pyran.
Description: Liquid, very slightly soluble in water.
Formula: $OCH:CHCH_2CH_2CHOC_2H_5$.
Constants: Mol wt: 128, d: 1.0, bp: 289°F, flash p: 111°F (O.C.).
Hazard Analysis
Toxic Hazard Rating: U. Limited animal experiments suggest moderate toxicity.
Fire Hazard: Moderate, when exposed to heat or flame.
Countermeasures
Storage and Handling: Section 7.

6-ETHOXY-1,2-DIHYDRO-2,2,4-TRIMETHYLQUINO-LINE
General Information
Description: Clear, light yellow liquid.
Formula: $C_{14}H_{19}NO$.
Constants: Mol wt: 217.3, mp: $< 0°C$, bp: 125°C at 2 mm, vap. d.: 7.48, d: 1.030 at 25°C.
Hazard Analysis
Note: Used as a food additive permitted in the feed and drinking water of animals and/or for the treatment of food producing animals. Also permitted in food for human consumption. See Section 10.
Fire Hazard: Slight; when heated to decomposition, it emits toxic fumes; can react with oxidizing materials.
Countermeasures
Personal Hygiene: Section 3.
Storage and Handling: Section 7.

2-ETHOXY ETHANOL. See "Cellosolve" solvent.

2-ETHOXY ETHYL ACETATE. See "Cellosolve" acetate.

2-ETHOXY-4-METHYL-3,4-DIHYDROPYRAN
Hazard Analysis
Toxic Hazard Rating: U. Limited animal experiments suggest moderate toxicity.

2-ETHOXYMETHYL-2,4-DIMETHYL PENTANE-DIOL-1,5
General Information
Description: Liquid.
Formula: $HOCH_2CH(CH_3)CH_2C(CH_3)(CH_2CH_2OCH_3)\cdot CH_2OH$.
Constants: Mol wt: 190.3, mp: $-40°C$, bp: 208°C at 100 mm, flash p: 295°F (O.C.), d: 1.0165, vap. d.: 6.08.
Hazard Analysis
Toxicity: Unknown.
Fire Hazard: Slight, when exposed to heat or flame; can react with oxidizing materials.
Countermeasures
To Fight Fire: Foam, carbon dioxide, dry chemical or carbon tetrachloride (Section 6).
Storage and Handling: Section 7.

3-ETHOXYPROPIONALDEHYDE
General Information
Description: Liquid.
Formula: $C_2H_5OCH_2CH_2CHO$.
Constants: Mol wt: 102.13, mp: $-69.4°C$, bp: 135.2°C, flash p: 100°F (O.C.), d: 0.918 at 20°/20°C, vap. d.: 3.63, vap. press.: 5.5 mm at 20°C.
Hazard Analysis
Toxic Hazard Rating:
  Acute Local: U.
  Acute Systemic: Ingestion 2; Inhalation 2.
  Chronic Local: U.
  Chronic Systemic: U.

Fire Hazard: Moderate, when exposed to heat or flame; can react with oxidizing materials.

Countermeasures

To Fight Fire: Foam, carbon dioxide, dry chemical or carbon tetrachloride (Section 6).

Ventilation Control: Section 2.

Personal Hygiene: Section 3.

Storage and Handling: Section 7.

3-ETHOXYPROPIONIC ACID

General Information

Description: Liquid.

Formula: $C_2H_5OCH_2CH_2COOH$.

Constants: Mol wt: 118.13, mp: $-10.7°C$, bp: $219°C$, flash p: $225°F$ (O.C.), d: 1.0474, vap. d.: 4.08.

Hazard Analysis

Toxicity: Unknown.

Fire Hazard: Slight, when exposed to hear or flame; can react with oxidizing materials.

Countermeasures

To Fight Fire: Water, alcohol foam, foam, carbon dioxide, dry chemical or carbon tetrachloride (Section 6).

Storage and Handling: Section 7.

3-(ETHOXY PROPYL) MERCURY BROMIDE

General Information

Synonym: Aagrano.

Hazard Analysis

Toxic Hazard Rating: A fungicide. See mercury compounds, organic.

N'-6-(ETHOXY-3-PYRIDAZINYL) SULFANILAMIDE.
See sulfaethoxypyridazine.

ETHOXYTRIGLYCOL

General Information

Formula: $C_2H_5O(C_2H_4O)_3H$.

Constants: Mol wt: 178.21, bp: $255.4°C$, flash p: $275°F$ (C.C.), d: 1.0208 at $20°/20°C$, vap. press.: 0.01 mm at $20°C$.

Hazard Analysis

Toxicity: See glycols.

Fire Hazard: Slight, when exposed to heat or flame; can react with oxidizing materials.

Countermeasures

To Fight Fire: Water, foam, alcohol foam, carbon dioxide, dry chemical or carbon tetrachloride (Section 6).

Personal Hygiene: Section 3.

Storage and Handling: Section 7.

ETHOXYTRIMETHYLSILANE

General Information

Formula: $C_5H_{14}OSi$.

Constants: Mol wt: 118.3, bp: $75.7°C$, vap. press.: 100 mm at $22.1°C$, vap. d.: 4.1.

Hazard Analysis

Toxic Hazard Rating:

Acute Local: Irritant 2; Ingestion 2; Inhalation 2.

Acute Systemic: U.

Chronic Local: U.

Chronic Systemic: U.

Toxicology: See also silanes.

Fire Hazard: Moderate; when heated, it can ignite; can react with oxidizing materials.

Countermeasures

Ventilation Control: Section 2.

Personnel Protection: Section 3.

Storage and Handling: Section 7.

ETHYL ABIETATE

General Information

Synonym: Abietic acid, ethyl ester.

Description: Amber-colored, viscous liquid.

Formula: $C_{19}H_{29}CO_2C_2H_5$.

Constants: Mol wt: 330.4, mp: $45°C$, bp: $350°C$, flash p: $352.4°F$, d: 1.02, vap. d.: 11.4.

Hazard Analysis

Toxicity: See abietic acid and ethyl alcohol.

Fire Hazard: Slight, when exposed to heat or flame; can react with oxidizing materials.

Countermeasures

To Fight Fire: Foam, carbon dioxide, dry chemical or carbon tetrachloride (Section 6).

Storage and Handling: Section 7.

ETHYL ACETAMIDE

General Information

Description: Water-white liquid.

Formula: $CH_3CONHC_2H_5$.

Constants: Mol wt: 87.1, mp: $-32°C$, bp: $206-208°C$, flash p: $230°F$, d: 0.920 at $20°/20°C$, vap. d.: 3.0.

Hazard Analysis

Toxicity: Unknown. See also amides.

Fire Hazard: Slight, when exposed to heat or flame; can react with oxidizing materials (Section 6).

Countermeasures

Storage and Handling: Section 7.

To Fight Fire: Water, foam, alcohol foam.

ETHYL ACETANILIDE

General Information

Synonym: Ethyl phenyl acetamide.

Description: White crystals; faint odor.

Formula: $CH_3CON(C_2H_5)(C_6H_5)$.

Constants: Mol wt: 163.21, mp: $54°C$, bp: $258°C$, flash p: $126°F$, d: 0.994, vap. d.: 5.62.

Hazard Analysis

Toxicity: Details unknown. See also acetanilide.

Fire Hazard: Slight, when exposed to heat or flame; can react with oxidizing materials.

Countermeasures

To Fight Fire: Foam, carbon dioxide, dry chemical or carbon tetrachloride (Section 6).

Storage and Handling: Section 7.

ETHYL ACETATE

General Information

Synonym: Acetic ether; ethyl ester; ethyl ethanoate.

Description: Colorless liquid; fragrant odor.

Formula: $CH_3COOC_2H_5$.

Constants: Mol wt: 88.10, mp: $-83.6°C$, bp: $77.15°C$, ulc: 85-90, lel: 2.5%, uel: 9%, flash p: $24°F$, d: 0.8946 at $25°C$, autoign. temp.: $800°F$, vap. press.: 100 mm at $27.0°C$, vap. d.: 3.04.

Hazard Analysis

Toxic Hazard Rating:

Acute Local: Irritant 1.

Acute Systemic: Ingestion 2; Inhalation 2; Skin Absorption 2.

Chronic Local: Irritant 1.

Chronic Systemic: Ingestion 1; Inhalation 1; Skin Absorption 1.

TLV: ACGIH (recommended); 400 parts per million in air; 1400 milligrams per cubic meter of air.

Toxicology: Ethyl acetate is irritating to mucous surfaces, particularly the eyes, gums, and respiratory passages

---

TOXIC HAZARD RATING CODE (For detailed discussion, see Section 1.)

0 NONE: (a) No harm under any conditions; (b) Harmful only under unusual conditions or overwhelming dosage.

1 SLIGHT: Causes readily reversible changes which disappear after end of exposure.

2 MODERATE: May involve both irreversible and reversible changes; not severe enough to cause death or permanent injury.

3 HIGH: May cause death or permanent injury after very short exposure to small quantities.

U UNKNOWN: No information on humans considered valid by authors.

and is also mildly narcotic. On repeated or prolonged exposures, it causes conjunctival irritation and corneal clouding. It can cause dermatitis. High concentrations have a narcotic effect and can cause congestion of the liver and kidneys. Chronic poisoning has been described as producing secondary anemia, leucocytosis and cloudy swelling, and fatty degeneration of the viscera. Note: Used as a synthetic flavoring substance and adjuvant (Section 10).

Fire Hazard: Dangerous, when exposed to heat or flame; can react vigorously with oxidizing materials.

Spontaneous Heating: No.

Explosion Hazard: Moderate, when exposed to flame.

Disaster Hazard: Dangerous, upon exposure to heat or flame.

Countermeasures

Ventilation Control: Section 2.

To Fight Fire: Carbon dioxide, dry chemical or carbon tetrachloride (Section 6).

Personnel Protection: Section 3.

Storage and Handling: Section 7.

Shipping Regulations: Section 11.

　I.C.C.: Flammable liquid; red label, 10 gallons.

　Coast Guard Classification: Inflammable liquid; red label.

　MCA warning label.

　IATA: Flammable liquid, red label, 1 liter (passenger), 40 liters (cargo).

N-ETHYLACETOACETAMIDE

General Information

Formula: $CH_3CONH_2CO_2C_2H_5$.

Constant: Mol wt: 132.14.

Hazard Analysis

Toxic Hazard Rating: U. Limited animal experiments suggest low toxicity.

ETHYL ACETOACETATE

General Information

Synonym: Diacetic ether; acetoacetic ester.

Description: Colorless liquid; fruity odor.

Formula: $CH_3COCH_2COOC_2H_5$.

Constants: Mol wt: 130.14, bp: 180.8°C, fp: −45°C, flash p: 185°F (C.O.C.), d: 1.0261 at 20°/20°C, vap. press.: 1 mm at 28.5°C, vap. d.: 4.48.

Hazard Analysis

Toxic Hazard Rating:

　Acute Local: Irritant 2; Ingestion 2.

　Acute Systemic: Ingestion 2; Inhalation 2.

　Chronic Local: U.

　Chronic Systemic: U.

Fire Hazard: Moderate, when exposed to heat or flame can react with oxidizing materials.

Spontaneous Heating: No.

Countermeasures

To Fight Fire: Alcohol foam, carbon dioxide, dry chemical or carbon tetrachloride (Section 6).

Ventilation Control: Section 2.

Personnel Protection: Section 3.

Storage and Handling: Section 7.

ETHYL ACRYLATE

General Information

Synonym: Ethyl propenoate.

Description: Colorless liquid.

Formula: $CH_2CHCOOC_2H_5$.

Constants: Mol wt: 100.11, bp: 99.8°C, fp: < −72°C, lel: 1.8%, flash p: 60°F (O.C.), d: 0.924 at 20°/4°C, vap. press.: 29.3 mm at 20°C, vap. d.: 3.45.

Hazard Analysis

Toxic Hazard Rating:

　Acute Local: Irritant 3.

　Acute Systemic: Ingestion 3; Inhalation 3; Skin Absorption 3.

　Chronic Local: U.

　Chronic Systemic: U.

TLV: ACGIH (recommended); 25 parts per million in air; 100 milligrams per cubic meter of air. Can penetrate skin.

Toxicology: Oral administration of 0.42 g or more per kilogram of body weight in the case of rabbits resulted in fatal poisoning. This was characterized in its terminal stages by dyspnea, cyanosis, and convulsive movements. It caused severe local irritation of the gastro-enteric tract; and toxic degenerative changes of cardiac, hepatic, renal, and splenic tissues were observed. It gave no evidence of cumulative effects in rabbits. When applied to the intact skin of rabbits, the ethyl ester caused marked local irritation, erythema, edema, thickening, and vascular damage. Animals subjected to a fairly high concentration of these esters suffered irritation of the mucous membranes of the eyes, nose, and mouth as well as lethargy, dyspnea, and convulsive movements. Note: A substance migrating to food from packaging materials. See Section 10.

Fire Hazard: Dangerous, when exposed to heat or flame; can react vigorously with oxidizing materials.

Explosion Hazard: Dangerous, when exposed to a source of heat or open flame or sparks.

Disaster Hazard: Dangerous, upon exposure to heat or flame.

Countermeasures

Ventilation Control: Section 2.

To Fight Fire: Alcohol foam, foam, carbon dioxide, dry chemical or carbon tetrachloride (Section 6).

Personnel Protection: Section 3.

Storage and Handling: Section 7.

Shipping Regulations: Section 11.

　MCA warning label.

　IATA (inhibited): Flammable liquid, red label, 1 liter (passenger), 40 liters (cargo).

　IATA (uninhibited): Flammable liquid, not acceptable (passenger and cargo).

ETHYL ALCOHOL

General Information

Synonyms: Ethanol; methyl carbinol; spirit of wine.

Description: Clear, colorless, fragrant liquid.

Formula: CH_3CH_2OH.

Constants: Mol wt: 46.07, bp: 78.32°C, ulc: 70, lel: 4.3%, uel: 19%, fp: −114.1°C, flash p: 55°F, d: 0.7893 at 20°/4°C, autoign. temp.: 793°F, vap. press.: 40 mm at 19°C, vap. d.: 1.59.

Hazard Analysis

Toxic Hazard Rating:

　Acute Local: Irritant 1.

　Acute Systemic: Ingestion 2; Inhalation 2; Skin Absorption 1.

　Chronic Local: Irritant 1.

　Chronic Systemic: Ingestion 1; Inhalation 1; Skin Absorption 1.

TLV: ACGIH (recommended); 1000 parts per million in air; 1880 milligrams per cubic meter of air.

Toxicology: The systemic effect of ethyl alcohol differs from that of methyl alcohol. Ethyl alcohol is rapidly oxidized in the body to carbon dioxide and water, and in contrast to methyl alcohol, no cumulative effect occurs. Though ethyl alcohol possesses narcotic properties, concentrations sufficient to produce this effect are not reached in industry. Exposure to concentrations of 5,000 to 10,000 ppm results in irritation of the eyes and mucous membranes of the upper respiratory tract. If continued for an hour, stupor and drowsiness may result. Concentrations below 1,000 ppm usually produce no signs of intoxication. There is no concrete evidence that repeated exposure to ethyl alcohol vapor results in cirrhosis of the liver. The main effect of ethyl alcohol is due to its irritant action on the mucous membranes of the eyes and upper respiratory tract.

Exposure to concentrations of over 1,000 ppm may cause headache, irritation of the eyes, nose and throat, and, if long continued, drowsiness and lassitude, loss of appetite and inability to concentrate.

Fire Hazard: Dangerous, when exposed to heat or flame; can react vigorously with oxidizing materials.

Disaster Hazard: Dangerous, when exposed to heat or flame.

Spontaneous Heating: No.

Explosion Hazard: Moderate, when exposed to flame.

Countermeasures

Ventilation Control: Section 2.

To Fight Fire: Alcohol foam, carbon dioxide, dry chemical or carbon tetrachloride (Section 6).

Storage and Handling: Section 7.

Shipping Regulations: Section 11.

   I.C.C.: Flammable liquid; red label, 10 gallons.

   Coast Guard Classification: Inflammable liquid; red label.

   MCA warning label.

ETHYL ALDEHYDE. See acetaldehyde.

ETHYL α-ALLYL ACETOACETATE
General Information

Description: Liquid.

Formula: $CH_3COCH(CH_2CHCH_2)CO_2C_2H_5$.

Constants: Mol wt: 170.23, bp: 208–218°C, d: 0.989 at 25°/20°C, vap. d.: 5.88.

Hazard Analysis

Toxicity: Details unknown. See also esters.

Fire Hazard: Slight; can react with oxidizing materials (Section 6).

Countermeasures

Storage and Handling: Section 7.

ETHYL ALUMINUM DICHLORIDE
General Information

Synonym: EADC.

Description: Clear, yellow, pyrophoric liquid.

Formula: $C_2H_5AlCl_2$.

Constants: Mol wt: 127, bp (extrapolated): 194°C, fp: 32°C, d: 1.222.

Hazard Analysis

Fire Hazard: Dangerous; fumes vigorously in air, may ignite spontaneously.

Countermeasures

To Fight Fire: Do not use foam, water, or halogenated extinguishing agents.

Storage and Handling: Section 7.

Shipping Regulations: Section 11.

   I.C.C.: Flammable liquid, red label, 2 ounces.

   IATA: Flammable liquid, not acceptable (passenger and cargo).

ETHYL ALUMINUM SESQUICHLORIDE
General Information

Synonym: EASC.

Description: Clear, yellow liquid.

Formula: $(C_2H_5)_3Al_2Cl_3$.

Constant: Mol wt: 247.5.

Hazard Analysis

Fire Hazard: Dangerous; ignites spontaneously in air.

Countermeasures

To Fight Fire: Do not use water, foam or halogenated extinguishing agents.

Storage and Handling: Section 7.

Shipping Regulations: Section 11.

   I.C.C.: Flammable liquid, red label, 2 ounces.

   IATA: Flammable liquid, not acceptable (passenger and cargo).

ETHYLAMINE
General Information

Synonyms: Aminoethane; monoethylamine.

Description: Colorless liquid; strong ammoniacal odor.

Formula: $C_2H_5NH_2$.

Constants: Mol wt: 45.08, mp: −80.6°C, bp: 16.6°C, flash p: <0°F, d: 0.7059 at 0°/4°C, autoign. temp.: 723°F, vap. press.: 400 mm at 2.0°C, vap. d.: 1.56, lel: 3.5%, uel: 14.0%.

Hazard Analysis

Toxic Hazard Rating:

   Acute Local: Irritant 3; Ingestion 3; Inhalation 3.

   Acute Systemic: U.

   Chronic Local: U.

   Chronic Systemic: U.

Toxicology: See also amines.

TLV: ACGIH (recommended); 10 ppm of air; 18 milligrams per cubic meter of air.

Fire Hazard: Dangerous, when exposed to heat or flame; can react vigorously with oxidizing materials.

Disaster Hazard: Dangerous; keep away from heat and open flame!

Countermeasures

Ventilation Control: Section 2.

To Fight Fire: Alcohol foam, carbon dioxide, dry chemical or carbon tetrachloride (Section 6).

Personnel Protection: Section 3.

Storage and Handling: Section 7.

Shipping Regulations: Section 11.

   I.C.C.: Flammable liquid, red label, 10 gallons.

   IATA: Flammable liquid, red label, 1 liter (passenger), 40 liters (cargo).

ETHYL p-AMINOBENZOATE. See anesthesin.

ETHYL AMYL KETONE
General Information

Synonym: 5-Methyl-3-heptanone.

Description: Liquid with mild fruity odor. Soluble in many organic solvents.

Formula: $C_8H_{16}O$.

Constants: Mol wt: 128.1, bp: 157–162°C, d: 0.822 at 20°/20°C, flash p: 138°F.

Hazard Analysis

Toxic Hazard Rating:

   Acute Local: Irritant 2; Inhalation 1.

   Acute Systemic: Inhalation 2.

   Chronic Local: Irritant 2.

   Chronic Systemic: U.

TLV: ACGIH (recommended) 25 ppm of air; 130 milligrams per cubic meter of air.

Toxicology: Lower concentrations, about 100 ppm can cause headache and nausea. High concentrations are narcotic. It is moderately irritating to eyes and mucous membranes.

N-ETHYLANILINE
General Information

Synonym: Ethyl phenyl amine.

Description: Clear, yellow-brown oil.

Formula: $C_2H_5NH(C_6H_5)$.

Constants: Mol wt: 121.2, mp: −63.5°C, bp: 204°C, d: 0.958 at 25°/25°C, vap. press.: 1 mm at 38.5°C, vap. d.: 4.18.

TOXIC HAZARD RATING CODE *(For detailed discussion, see Section 1.)*

0 NONE: (a) No harm under any conditions; (b) Harmful only under unusual conditions or overwhelming dosage.

1 SLIGHT: Causes readily reversible changes which disappear after end of exposure.

2 MODERATE: May involve both irreversible and reversible changes; not severe enough to cause death or permanent injury.

3 HIGH: May cause death or permanent injury after very short exposure to small quantities.

U UNKNOWN: No information on humans considered valid by authors.

Hazard Analysis
Toxic Hazard Rating:
Acute Local: Allergen 1.
Acute Systemic: Ingestion 3; Inhalation 3; Skin Absorption 3.
Chronic Local: Allergen 1.
Chronic Systemic: Ingestion 3; Inhalation 3; Skin Absorption 3.
Toxicity: See also aniline and amines.
Fire Hazard: Unknown.
Disaster Hazard: Highly dangerous; on decomposition or on contact with acid or acid fumes, it emits highly toxic fumes of aniline and oxides of nitrogen; can react with oxidizing materials.

Countermeasures
Ventilation Control: Section 2.
Personnel Protection: Section 3.
First Aid: Section 1.
Storage and Handling: Section 7.

o-ETHYLANILINE
General Information
Description: Yellow liquid; darkens upon standing.
Formula: $C_6H_4NH_2C_2H_5$.
Constants: Mol wt: 121.2, mp: $-63.5°C$, bp: $215°C$, flash p: $185°F$ (O.C.), d: 0.98 at $25°/25°C$, vap. d.: 4.17.

Hazard Analysis
Toxic Hazard Rating:
Acute Local: Allergen 1.
Acute Systemic: Ingestion 3; Inhalation 3; Skin Absorption 3.
Chronic Local: Allergen 1.
Chronic Systemic: Ingestion 3; Inhalation 3; Skin Absorption 3.
Toxicity: See also aniline and amines.
Fire Hazard: Moderate, when exposed to heat or flame.
Disaster Hazard: Dangerous; when heated to decomposition, it emits highly toxic fumes of aniline and oxides of nitrogen; can react with oxidizing materials.

Countermeasures
Ventilation Control: Section 2.
To Fight Fire: Foam, carbon dioxide, dry chemical or carbon tetrachloride (Section 6).
Personnel Protection: Section 3.
First Aid: Section 1.
Storage and Handling: Section 7.

ETHYL ARSENIOUS OXIDE
General Information
Description: Colorless oil; garlic-like odor.
Formula: C_2H_5AsO.
Constants: Mol wt: 120.0, bp: $158°C$ at 10 mm, d: 1.802 at $11°C$, vap. d.: 4.14.

Hazard Analysis and Countermeasures
See arsenic compounds.

ETHYLBENZENE
General Information
Synonyms: Ethylbenzol; phenylethane.
Description: Colorless liquid; aromatic odor.
Formula: $C_6H_5C_2H_5$.
Constants: Mol wt: 106.16, bp: $136.2°C$, fp: $-94.9°C$, flash p: $59°F$, d: 0.8669 at $20°/4°C$, autoign. temp.: $810°F$, vap. press.: 10 mm at $25.9°C$, vap. d.: 3.66.

Hazard Analysis
Toxic Hazard Rating:
Acute Local: Irritant 2.
Acute Systemic: Ingestion 2; Inhalation 2; Skin Absorption 2.
Chronic Local: U.
Chronic Systemic: U.
TLV: ACGIH (recommended) 100 ppm of air; 435 milligrams per cubic meter of air.
Toxicology: The liquid is an irritant to the skin and mucous membranes. A concentration of 0.1% of the vapor in air is an irritant to the eyes of humans, and a concentration of 0.2% is extremely irritating at first, then causes dizziness, irritation of the nose and throat and a sense of constriction of the chest. Exposure of guinea pigs to 1% concentrations has been reported as causing ataxia, loss of consciousness, tremor of the extremities and finally death through respiratory failure. The pathological findings were congestion of the brain and lungs, with edema. No data are available regarding the effect of chronic exposure.

Erythema and inflammation of the skin may result from contact of the skin with the liquid (Section 9). Exposure to the vapor causes lachrymation and irritation of the nose and throat, dizziness, and a sense of constriction of the chest. The irritant properties are sufficient to cause workers to leave an atmosphere containing 0.5% of the vapor.
Fire Hazard: Moderate, when exposed to heat or flame; can react vigorously with oxidizing materials.
Spontaneous Heating: No.
Disaster Hazard: Dangerous; keep away from heat and open flame!

Countermeasures
Ventilation Control: Section 2.
To Fight Fire: Foam, carbon dioxide, dry chemical or carbon tetrachloride (Section 6).
Personnel Protection: Section 3.
Storage and Handling: Section 7.
Shipping Regulations: Section 11.
Coast Guard Classification: Combustible liquid.
IATA: Flammable liquid, red label, 1 liter (passenger), 40 liters (cargo).

ETHYL BENZOATE
General Information
Synonym: Benzoic ether.
Description: Colorless, aromatic liquid.
Formula: $C_6H_5CO_2C_2H_5$.
Constants: Mol wt: 150.2, mp: $-34.6°C$, bp: $213.4°C$, flash p: $>204°F$, d: 1.048 at $20°/20°C$, vap. press.: 1 mm at $44.0°C$, vap. d.: 5.17.

Hazard Analysis
Toxicity: See ethyl alcohol, benzoic acid, and esters. Animal experiments show low toxicity.
Fire Hazard: Moderate, when exposed to heat or flame; can react with oxidizing materials.

Countermeasures
To Fight Fire: Foam, carbon dioxide, dry chemical or carbon tetrachloride (Section 6).
Storage and Handling: Section 7.

ETHYLBENZOL. See ethylbenzene.

ETHYL BENZOYLACETATE
General Information
Description: Clear liquid.
Formula: $C_6H_5COCH_2COOC_2H_5$.
Constants: Mol wt: 192.21, bp: $265.0°C$ decomposes, flash p: $285°F$ (O.C.), d: 1.11, vap. press.: 1 mm at $107.6°C$, vap. d.: 6.63.

Hazard Analysis
Toxicity: Details unknown. See also esters.
Fire Hazard: Slight, when exposed to heat or flame; can react with oxidizing materials.

Countermeasures
To Fight Fire: Water, carbon dioxide, dry chemical or carbon tetrachloride (Section 6).
Storage and Handling: Section 7.

ETHYL BENZYL ANILINE
General Information
Description: Clear, colorless oil.
Formula: $C_6H_5N(C_2H_5)CH_2C_6H_5$.
Constants: Mol wt: 211.3, bp: $286°C$, d: 1.034, vap. d.: 7.28.

Hazard Analysis and Countermeasures
See aniline.

ETHYL BORATE
General Information
Synonyms: Triethyl borate; triethoxyborine.
Description: Colorless liquid; mild odor, decomposes in water.
Formula: $B(OC_2H_5)_3$.
Constants: Mol wt: 146.00, bp: 120°C, flash p: 52°F (C.C.), d: 0.864 at 26.5°C, vap. d.: 5.04.
Hazard Analysis
Toxicity: Details unknown. See also boron compounds.
Fire Hazard: Dangerous, when exposed to heat or flame; will react with water or steam to produce flammable vapors; can react vigorously with oxidizing materials.
Spontaneous Heating: No.
Disaster Hazard: Dangerous; keep away from heat and open flame!
Countermeasures
Ventilation Control: Section 2.
To Fight Fire: Carbon dioxide, dry chemical or carbon tetrachloride (Section 6).
Storage and Handling: Section 7.
Shipping Regulations: Section 11.
    I.C.C.: Flammable liquid; red label.
    IATA: Flammable liquid, red label, 1 liter (passenger), 40 liters (cargo).

ETHYL BORIC ACID
General Information
Description: White crystals, water soluble.
Formula: $(C_2H_5)B(OH)_2$.
Constants: Mol wt: 73.9, mp: sublimes at 40°C.
Hazard Analysis
Toxicity: Details unknown. See also boron compounds.

ETHYL BROMIDE
General Information
Synonyms: Bromoethane; hydrobromic ether; bromic ether.
Description: Colorless, volatile liquid.
Formula: CH_3CH_2Br.
Constants: Mol wt: 108.98, mp: −119°C, bp: 38.4°C, lel: 6.7%, uel: 11.3%, fp: −119.0°C, flash p: none, d: 1.430 at 20°/4°C, autoign. temp.: 952°F, vap. press.: 400 mm at 21°C, vap. d.: 3.76.
Hazard Analysis
Toxic Hazard Rating:
    Acute Local: Irritant 2.
    Acute Systemic: Ingestion 3; Inhalation 3; Skin Absorption 3.
    Chronic Local: U.
    Chronic Systemic: Ingestion 2; Inhalation 2; Skin Absorption 2.
TLV: ACGIH (recommended); 200 ppm in air; 892 milligrams per cubic meter of air.
Toxicology: It readily decomposes into volatile toxic products, such as hydrobromic acid and bromine, particularly in the presence of hot surfaces or open flame. Physiologically, it is an anesthetic and narcotic. Its vapors are markedly irritating to the lungs when inhaled for even short periods. It can produce acute congestion and edema. Liver and kidney damage in humans has been reported. It is much less toxic than methyl bromide, but more toxic than ethyl chloride.
Fire Hazard: Dangerous, when exposed to heat or flame.
Spontaneous Heating: No.

Explosion Hazard: Moderate, when exposed to flame.
Disaster Hazard: Dangerous; when heated to decomposition, it emits highly toxic fumes of bromine; will react with water or steam to produce toxic and corrosive fumes; can react vigorously with oxidizing materials.
Countermeasures
Ventilation Control: Section 2.
To Fight Fire: Water, carbon dioxide, dry chemical or carbon tetrachloride (Section 6).
Personnel Protection: Section 3.
First Aid: Section 1.
Storage and Handling: Section 7.
Shipping Regulations: Section 11.
    MCA warning label.
    IATA: Other restricted articles, class A, no label required, 40 liters (passenger), 220 liters (cargo).

ETHYL BROMOACETATE
General Information
Synonym: Ethyl bromoethanoate.
Description: Colorless to straw-colored liquid.
Formula: $CH_2BrCOOC_2H_5$.
Constants: Mol wt: 167.01, bp: 158.8°C, fp: < −20°C, flash p: none, d: 1.514 at 13°/4°C, vap. d.: 5.8.
Hazard Analysis
Toxic Hazard Rating:
    Acute Local: Irritant 3; Ingestion 3; Inhalation 3.
    Acute Systemic: Ingestion 3; Inhalation 3; Skin Absorption 3.
    Chronic Local: U.
    Chronic Systemic: U.
Disaster Hazard: Dangerous; when heated to decomposition or on contact with acid or acid fumes, it emits highly toxic fumes of bromides; will react with water or steam to produce toxic and corrosive fumes.
Countermeasures
Ventilation Control: Section 2.
Personnel Protection: Section 3.
First Aid: Section 1.
Storage and Handling: Section 7.
Shipping Regulations: Section 11.
    IATA: Poison A, not acceptable (passenger and cargo).

ETHYL BROMOETHANOATE. See ethyl bromoacetate.

ETHYL BUTANOATE. See ethyl n-butyrate.

2-ETHYLBUTANOIC ACID. See 2-ethyl-butyric acid.

2-ETHYLBUTANOL
General Information
Synonyms: 2-Ethylbutyl alcohol.
Description: Clear liquid.
Formula: $(C_2H_5)_2CHCH_2OH$.
Constants: Mol wt: 102.17, bp: 148.9°C, flash p: 135°F (C.O.C.), d: 0.8328, vap. press.: 0.9 mm at 20°C, vap. d.: 3.4.
Hazard Analysis
Toxicity: See alcohols.
Fire Hazard: Moderate, when exposed to heat or flame; can react with oxidizing materials (Section 6).
Countermeasures
Ventilation Control: Section 2.
Personnel Protection: Section 3.
Storage and Handling: Section 7.

2,2-ETHYL BUTOXY ETHANOL. See 2-ethylbutyl "Cellosolve."

TOXIC HAZARD RATING CODE (For detailed discussion, see Section 1.)

0 NONE: (a) No harm under any conditions; (b) Harmful only under unusual conditions or overwhelming dosage.

1 SLIGHT: Causes readily reversible changes which disappear after end of exposure.

2 MODERATE: May involve both irreversible and reversible changes; not severe enough to cause death or permanent injury.

3 HIGH: May cause death or permanent injury after very short exposure to small quantities.

U UNKNOWN: No information on humans considered valid by authors.

3-(2-ETHYL BUTOXY) PROPIONIC ACID
General Information
Description: Water white liquid; insoluble in water.
Formula: $CH_3CH_2CH(C_2H_5)CH_2OCH_2CH_2COOH$.
Constants: Mol wt: 174, d: 0.96 (20/20°C) bp: 200°C at 100 mm, vap. press.: <0.1 mm at 20°C, flash p: 280°F (O.C.).
Hazard Analysis
Fire Hazard: Slight, when exposed to heat or flame.
Countermeasures
To Fight Fire: Water, foam.
Storage and Handling: Section 7.

2-ETHYLBUTYL ACETATE
General Information
Description: Colorless liquid; mild odor.
Formula: $(C_2H_5)_2CHCH_2OOCCH_3$.
Constants: Mol wt: 144.21, bp: 163°C, flash p: 130°F (O.C.), d: 0.875–0.881 at 20°/20°C, vap. d.: 5.0.
Hazard Analysis
Toxicity: Unknown. See esters.
Fire Hazard: Moderate, when exposed to heat or flame; can react with oxidizing materials.
Countermeasures
To Fight Fire: Foam, carbon dioxide, dry chemical or carbon tetrachloride (Section 6).
Ventilation Control: Section 2.
Storage and Handling: Section 7.
Shipping Regulations: Section 11.
  Coast Guard Classification: Combustible liquid.

2-ETHYLBUTYL ACRYLATE
General Information
Description: Clear, colorless liquid.
Formula: $H_2C = CHCOOCH_2C(C_2H_5)HC_2H_5$.
Constants: Mol wt: 156.22, bp: 82°C at 10 mm, fp: −70°C, flash p: 125°F (O.C.), d: 0.8964 at 20°/20°C, vap. press.: 1.7 mm at 20°C.
Hazard Analysis
Toxicity: Details unknown; a moderate irritant. See also ethyl acrylate.
Fire Hazard: Moderate, when exposed to heat or flame; can react with oxidizing materials.
Countermeasures
To Fight Fire: Foam, carbon dioxide, dry chemical or carbon tetrachloride (Section 6).
Ventilation Control: Section 2.
Storage and Handling: Section 7.

ETHYL BUTYL ALCOHOL. See amyl carbinol.

2-ETHYLBUTYL ALCOHOL. See 2-ethylbutanol.

ETHYL n-BUTYLAMINE
General Information
Description: Water-white liquid.
Formula: $C_2H_5NHCH_2CH_2CH_2CH_3$.
Constants: Mol wt: 101.2, bp: 110–113°C, flash p: 65°F, d: 0.739 at 20°/20°C, vap. d.: 3.5.
Hazard Analysis
Toxicity: Details unknown. See also amines.
Fire Hazard: Dangerous; when exposed to heat or flame; can react vigorously with oxidizing materials (Section 6).
Explosion Hazard: Unknown.
Disaster Hazard: Dangerous; keep away from heat and open flame!
Countermeasures
Ventilation Control: Section 2.
Storage and Handling: Section 7.

ETHYL-n-BUTYL CARBAMATE. See butyl urethane.

ETHYLBUTYL CARBONATE
General Information
Description: Liquid.
Formula: $(C_2H_5)(C_4H_9)CO_3$.

Constants: Mol wt: 146.48, bp: 135°C, flash p: 122°F (C.C.); d: 0.92, vap. d.: 5.03.
Hazard Analysis
Toxicity: Unknown.
Fire Hazard: Moderate, when exposed to heat or flame; can react with oxidizing materials.
Spontaneous Heating: No.
Countermeasures
To Fight Fire: Foam, carbon dioxide, dry chemical or carbon tetrachloride (Section 6).
Ventilation Control: Section 2.
Storage and Handling: Section 7.

2-ETHYLBUTYL "CELLOSOLVE"
General Information
Synonym: 2,2-Ethylbutoxy ethanol.
Description: Liquid.
Formula: $(C_2H_5)_2CHCH_2OC_2H_4OH$.
Constants: Mol wt: 146.2, bp: 197°C, fp: −90°C, flash p: 180°F (O.C.), d: 0.8954 at 20°/20°C, vap. press.: 0.17 mm at 20°C, vap. d.: 5.04.
Hazard Analysis
Toxicity: Details unknown. See also glycols.
Fire Hazard: Moderate, when exposed to heat or flame; can react with oxidizing materials.
Countermeasures
To Fight Fire: Foam, carbon dioxide, dry chemical or carbon tetrachloride (Section 6).
Storage and Handling: Section 7.

ETHYL n-BUTYL ETHER
General Information
Synonym: Butyl ethyl ether.
Description: Liquid.
Formula: $C_2H_5OC_4H_9$.
Constants: Mol wt: 102.2, bp: 92°C, mp: −124°C, flash p: 40°F, d: 0.7528 at 20°/20°C, vap. d.: 3.52.
Hazard Analysis
Toxicity: Details unknown. See also ethers.
Fire Hazard: Dangerous, when exposed to heat or flame; can react vigorously with oxidizing materials. See also ethers.
Explosion Hazard: Details unknown. See also ethers.
Disaster Hazard: Dangerous; keep away from heat and open flame!
Countermeasures
Ventilation Control: Section 2.
To Fight Fire: Foam, carbon dioxide, dry chemical or carbon tetrachloride (Section 6)
Storage and Handling: Section 11.
Shipping Regulations: Section 11.
  IATA: Flammable liquid, red label, 1 liter (passenger), 40 liters (cargo).

ETHYL BUTYL KETONE
General Information
Description: Clear liquid.
Formula: $(C_2H_5)(C_4H_9)CO$.
Constants: Mol wt: 114.2, mp: −36.7°C, bp: 148°C, flash p: 115°F (O.C.), d: 0.8198 at 20°/20°C, vap. d.: 3.93.
Hazard Analysis
Toxic Hazard Rating:
  Acute Local: Irritant 2; Ingestion 2; Inhalation 2.
  Acute Systemic: Ingestion 2; Inhalation 2.
  Chronic Local: U.
  Chronic Systemic: U.
Toxicity: See also ketones.
TLV: ACGIH (recommended) 50 ppm in air; 230 milligrams per cubic meter of air.
Fire Hazard: Moderate, when exposed to heat or flame; can react with oxidizing materials.
Countermeasures
To Fight Fire: Foam, carbon dioxide, dry chemical or carbon tetrachloride (Section 6).
Ventilation Control: Section 2.

Personnel Protection: Section 3.
Storage and Handling: Section 7.

2-ETHYL-2-BUTYL-1,3-PROPANEDIOL
General Information
Description: Crystals or liquid.
Formula: $HOCH_2C(C_2H_5)(C_4H_9)CH_2OH$.
Constants: Mol wt: 160.25, bp: 178°C at 50 mm, fp: 40.1°C, flash p: 280°F (O.C.), d: 0.931 at 50°/20°C, vap. press.: < 0.01 mm at 20°C, vap. d.: 5.53.
Hazard Analysis
Toxicity: Unknown.
Fire Hazard: Slight, when exposed to heat or flame; can react with oxidizing materials.
Countermeasures
To Fight Fire: Foam, carbon dioxide, dry chemical or carbon tetrachloride (Section 6).
Storage and Handling: Section 7.

2-ETHYLBUTYL TITANATE
General Information
Synonym: Hexyl titanate.
Description: A liquid.
Formula: $Ti[OCH_2CH(C_2H_5)_2]_4$.
Constants: Mol wt: 453, mp: −55°C, bp: 208°C at 10 mm, d: 0.96.
Hazard Analysis
Toxicity: See titanium compounds and organometals.
Fire Hazard: Moderate, when exposed to heat or flame; can react with oxidizing materials (Section 6).
Countermeasures
Storage and Handling: Section 7.

2-ETHYLBUTYRALDEHYDE
General Information
Synonyms: 2-Ethylbutyric aldehyde, diethylacetaldehyde.
Description: Colorless liquid.
Formula: $(C_2H_5)_2CHCHO$.
Constants: Mol wt: 100.16, bp: 116.8°C, flash p: 70°F (O.C.), fp: −89°C, d: 0.8164 at 20°/20°C, vap. press.: 13.7 mm at 20°C, vap. d.: 3.45, lel: 1.2%, uel: 7.7%.
Hazard Analysis
Toxicity: Details unknown. See also aldehydes.
Fire Hazard: Dangerous, when exposed to heat or flame; can react vigorously with oxidizing materials.
Disaster Hazard: Dangerous, keep away from heat and open flame!
Countermeasures
Ventilation Control: Section 2.
To Fight Fire: Alcohol foam, carbon dioxide, dry chemical or carbon tetrachloride (Section 6).
Storage and Handling: Section 7.
Shipping Regulations: Section 11.
   IATA: Flammable liquid, red label, 1 liter (passenger), 40 liters (cargo).

ETHYL n-BUTYRATE
General Information
Synonyms: Butyric ether, ethyl butanoate.
Description: Colorless, volatile liquid; pineapple-like odor.
Formula: $C_3H_7CO_2C_2H_5$.
Constants: Mol wt: 116.16, mp: −93.3°C, bp: 121.0°C, flash p: 78°F (C.C.), d: 0.8788, vap. press.: 10 mm at 15.3°C, vap. d.: 4.0, autoign. temp.: 865°F.
Hazard Analysis
Toxicity: Details unknown. See esters. Irritating to mucous membranes and narcotic in high concentrations. Note:

Used as a synthetic flavoring substance and adjuvant (Section 10).
Fire Hazard: Dangerous, when exposed to heat or flame; can react vigorously with oxidizing materials.
Spontaneous Heating: No.
Explosion Hazard: Unknown.
Disaster Hazard: Dangerous; keep away from heat and open flame!
Countermeasures
Ventilation Control: Section 2.
To Fight Fire: Carbon dioxide, dry chemical or carbon tetrachloride (Section 6).
Storage and Handling: Section 7.
Shipping Regulations: Section 11.
   Coast Guard Classification: Combustible liquid.
   IATA: Flammable liquid, red label, 1 liter (passenger), 40 liters (cargo).

2-ETHYLBUTYRIC ACID
General Information
Synonyms: 2-Ethyl butanoic acid, diethylacetic acid.
Description: Water-white liquid.
Formula: $(C_2H_5)_2CHCOOH$.
Constants: Mol wt: 116.16, mp: −15°C, bp: 193.4°C, flash p: 210°F (C.O.C.), fp: −9.4°C, d: 0.9225 at 20°/20°C, vap. press.: 0.06 mm at 20°C, vap. d.: 4.01.
Hazard Analysis
Toxicity: Unknown. Animal experiments show low toxicity and moderate irritation. It can be absorbed via the skin.
Fire Hazard: Moderate, when exposed to heat or flame; can react vigorously with oxidizing materials.
Countermeasures
To Fight Fire: Foam, carbon dioxide, dry chemical or carbon tetrachloride (Section 6).
Storage and Handling: Section 7.

2-ETHYLBUTYRIC ALDEHYDE. See 2-ethyl-butyraldehyde.

ETHYL CACODYL. See arsenic diethyl.

ETHYL CALCIUM IODIDE
General Information
Description: A powder, decomposed by water.
Formula: C_2H_5CaI.
Constant: Mol wt: 196.1
Hazard Analysis and Countermeasures
See iodides.

ETHYL CAPRATE
General Information
Synonyms: Capric ether; caprinic ether.
Description: Colorless, fragrant liquid.
Formula: $C_9H_{19}CO_2C_2H_5$.
Constants: Mol wt: 200.3, bp: 243°C, d: 0.862, vap. d.: 6.9.
Hazard Analysis
Toxicity: See esters and ethers.
Fire Hazard: Slight; can react with oxidizing materials.
Countermeasures
Storage and Handling: Section 7.

ETHYL CARBAMATE
General Information
Synonym: Urethane; ethyl urethane.
Description: Colorless, odorless crystals.
Formula: $CO(NH_2)OC_2H_5$.
Constants: Mol wt: 89.1, mp: 49°C, bp: 184°C, d: 0.9862, vap. press.: 10 mm at 77.8°C, vap. d.: 3.07.

TOXIC HAZARD RATING CODE (For detailed discussion, see Section 1.)

0 NONE: (a) No harm under any conditions; (b) Harmful only under unusual conditions or overwhelming dosage.

1 SLIGHT: Causes readily reversible changes which disappear after end of exposure.

2 MODERATE: May involve both irreversible and reversible changes; not severe enough to cause death or permanent injury.

3 HIGH: May cause death or permanent injury after very short exposure to small quantities.

U UNKNOWN: No information on humans considered valid by authors.

Hazard Analysis
Toxic Hazard Rating:
    Acute Local: 0.
    Acute Systemic: Ingestion 3.
    Chronic Local: 0.
    Chronic Systemic: Ingestion 2.
Toxicology: Causes depression of bone marrow and occasionally focal degeneration in the brain. Can also produce CNS depression, nausea and vomiting.
Disaster Hazard: Moderate; when heated, it emits toxic fumes.
Countermeasures
Personal Hygiene: Section 3.
Storage and Handling: Section 7.

ETHYL CARBINOL. See n-propyl alcohol.

ETHYL CARBONATE. See diethyl carbonate.

ETHYL CELLULOSE
General Information
Description: White, granular solid.
Formula: Variable.
Constant: Mp: 240+°C.
Hazard Analysis
Toxicity: Unknown. Used as a food additive permitted in the feed and drinking water of animals and/or for the treatment of food producing animals. It migrates to food from packaging materials (Section 10).
Fire Hazard: Moderate, when exposed to heat or flame or by chemical reaction with oxidizing agents. Flammability varies with degree of replacement of OH⁻ radicals of cellulose by ethyl radical (Section 6).
Countermeasures
Storage and Handling: Section 7.

ETHYL CHLORIDE
General Information
Synonyms: Chloroethane; hydrochloric ether; muriatic ether.
Description: Colorless liquid or gas; ether-like odor, burning taste.
Formula: CH_3CH_2Cl.
Constants: Mol wt: 64.52, bp: 12.3°C, lel: 3.8%, uel: 15.4%, fp: −139°C, flash p: −58°F (C.C.), d: 0.9214 at 0°/4°C, autoign. temp.: 966°F, vap. press.: 1000 mm at 20°C, vap. d.: 2.22.
Hazard Analysis
Toxic Hazard Rating:
    Acute Local: Irritant 1; Ingestion 2; Inhalation 2.
    Acute Systemic: Ingestion 2; Inhalation 2.
    Chronic Local: U.
    Chronic Systemic: U.
TLV: ACGIH (recommended); 1000 ppm in air 2600 milligrams per cubic meter of air.
Toxicology: The liquid is harmful to the eyes and can cause some irritation. In the case of guinea pigs, the symptoms attending exposure are similar to those caused by methyl chloride, except that the signs of lung irritation are not as pronounced. It gives some warning of its presence because it is irritating, but it is possible to tolerate exposure to it until one becomes unconscious. It is the least toxic of all of the chlorinated hydrocarbons. It can cause narcosis, although the effects are usually transient. Animal experiments show some evidence of kidney irritation and accumulation of fat due to this material in the kidneys, cardiac muscles and liver.
Fire Hazard: Highly dangerous, when exposed to heat or flame.
Spontaneous Heating: No.
Explosion Hazard: Severe, when exposed to flame.
Disaster Hazard: Highly dangerous! Keep away from heat and open flame; forms phosgene on combustion; reacts with water or steam to produce toxic and corrosive fumes; can react vigorously with oxidizing materials.

Countermeasures
Ventilation Control: Section 2.
To Fight Fire: Carbon dioxide, dry chemical or carbon tetrachloride (Section 6).
Personnel Protection: Section 3.
Storage and Handling: Section 7.
Shipping Regulations: Section 11.
    I.C.C.: Flammable liquid; red label, 300 pounds in cylinders, 15 pounds in other containers.
    Coast Guard Classification: Inflammable liquid; red label.
    MCA warning label.
    IATA: Flammable liquid, red label, not acceptable (passenger), 140 kilograms (cargo).

ETHYL CHLOROACETATE
General Information
Synonym: Ethyl chloroethanoate.
Description: Colorless liquid; fruity odor.
Formula: $CH_2ClCOOC_2H_5$.
Constants: Mol wt: 122.55, bp: 143.6°C, fp: −26.6°C, flash p: 179°F, d: 1.159 at 20°/4°C, vap. press.: 10 mm at 37.5°C, vap. d.: 4.3.
Hazard Analysis
Toxic Hazard Rating:
    Acute Local: Irritant 3; Ingestion 3; Inhalation 3.
    Acute Systemic: U.
    Chronic Local: U.
    Chronic Systemic: U.
Fire Hazard: Moderate, when exposed to heat or flame.
Disaster Hazard: Dangerous; when heated to decomposition, it emits highly toxic fumes of chlorides; will react with water or steam to produce toxic and corrosive fumes; can react with oxidizing materials.
Countermeasures
Ventilation Control: Section 2.
To Fight Fire: Water, foam, carbon dioxide, dry chemical or carbon tetrachloride (Section 6).
Personnel Protection: Section 3.
Storage and Handling: Section 7.
Shipping Regulations: Section 11.
    Coast Guard Classification: Combustible liquid.

ETHYL CHLOROACETAL
General Information
Description: Water-white liquid; pleasant odor.
Formula: $ClCH_2CH(OC_2H_5)_2$.
Constants: Mol wt: 156.6, mp: −32°C, bp: 151°C, flash p: 117°F, d: 1.022 at 20°C, vap. d.: 5.41.
Hazard Analysis
Toxicity: Details unknown. See aldehydes.
Fire Hazard: Moderate, when exposed to heat or flame.
Disaster Hazard: Dangerous; when heated to decomposition, it emits highly toxic fumes of chlorides; will react with water or steam to produce toxic and corrosive fumes; can react with oxidizing materials.
Countermeasres
Ventilation Control: Section 2.
To Fight Fire: Foam, carbon dioxide, dry chemical or carbon tetrachloride (Section 6).
Storage and Handling: Section 7.

ETHYL CHLOROCARBONATE
General Information
Synonyms: Ethyl chloromethanoate, ethyl chloroformate.
Description: Colorless liquid; decomposes in water.
Formula: $ClCOOC_2H_5$.
Constants: Mol wt: 108.53, mp: −80.6°C, bp: 94°C, flash p: 61°F (C.C.), d: 1.138 at 20°/4°C, vap. d.: 3.74.
Hazard Analysis
Toxic Hazard Rating:
    Acute Local: Irritant 3; Ingestion 3; Inhalation 3.
    Acute Systemic: U.
    Chronic Local: U.
    Chronic Systemic: U.

Fire Hazard: Dangerous; when exposed to heat or flame.

Explosion Hazard: Unknown.

Disaster Hazard: Dangerous; when heated to decomposition, it emits highly toxic fumes of chlorides; will react with water or steam to produce toxic and corrosive fumes; can react vigorously with oxidizing materials.

Countermeasures

Ventilation Control: Section 2.

To Fight Fire: Carbon dioxide, dry chemical or carbon tetrachloride (Section 6).

Personnel Protection: Section 3.

Storage and Handling: Section 7.

Shipping Regulations: Section 11.

   I.C.C.: Corrosive liquid; white label, 5 pints.

   Coast Guard Classification: Corrosive liquid; white label.

   IATA: Corrosive liquid, white label, not acceptable (passenger), 2½ liters (cargo).

ETHYL CHLOROETHANOATE. See ethyl chloroacetate.

ETHYL CHLOROFORMATE. See ethyl chlorocarbonate.

ETHYL CHLOROMETHANOATE. See ethyl chlorocarbonate.

ETHYLCHLOROSTANNIC ACID

General Information

Description: Colorless deliquescent crystals, decomposed by water.

Formula: $H_2SnC_2H_5Cl_5$.

Constants: Mol wt: 327.1.

Hazard Analysis and Countermeasures

See tin compounds and chlorides.

ETHYL CHLOROSULFONATE

General Information

Description: Colorless oily liquid; pungent odor.

Formula: $C_2H_5ClSO_3$.

Constants: Mol wt: 144.57, bp: 152–153°C, d: 1.379 at 0°C, vap. d.: 5.0.

Hazard Analysis

Toxic Hazard Rating:

   Acute Local: Irritant 3; Ingestion 3; Inhalation 3.

   Acute Systemic: U.

   Chronic Local: U.

   Chronic Systemic: U.

Disaster Hazard: Dangerous; when heated, it emits highly toxic fumes of phosgene; will react with water or steam to produce toxic and corrosive fumes.

Countermeasures

Ventilation Control: Section 2.

Personnel Protection: Section 3.

Storage and Handling: Section 7.

ETHYL CROTONATE

General Information

Description: Colorless, monoclinic prisms, or water-white liquid; pungent odor.

Formula: $CH_3CHCHCOOC_2H_5$.

Constants: Mol wt: 114.14, mp: 45°C (solid), bp: 209°C (solid), 139.0°C (liq.), flash p: 36.0°F, d: 0.9207 at 20°/20°C, vap. d.: 3.93.

Hazard Analysis

Toxic Hazard Rating:

   Acute Local: Irritant 3; Ingestion 3; Inhalation 3.

   Acute Systemic: U.

   Chronic Local: U.

   Chronic Systemic: U.

Fire Hazard: Dangerous, when exposed to heat or flame; can react vigorously with oxidizing materials.

Spontaneous Heating: No.

Disaster Hazard: Dangerous, upon exposure to heat or flame.

Countermeasures

Ventilation Control: Section 2.

To Fight Fire: Foam, carbon dioxide, dry chemical or carbon tetrachloride (Section 6).

Personnel Protection: Section 3.

First Aid: Section 1.

Storage and Handling: Section 7.

Shipping Regulations: Section 11.

   IATA: Flammable liquid, red label, 1 liter (passenger), 40 liters (cargo).

ETHYL CYANIDE

General Information

Synonyms: Propionitrile; propanenitrile.

Description: Colorless liquid; ethereal odor.

Formula: CH_3CH_2CN.

Constants: Mol wt: 55.08, mp: −103.5°C, bp: 97.1°C, d: 0.783 at 21°/4°C, vap. d.: 1.9.

Hazard Analysis and Countermeasures

See nitriles.

ETHYL CYANOACETATE

General Information

Synonym: Malonic ethyl ester nitrile.

Description: Colorless to pale straw-colored liquid.

Formula: $CNCH_2CO_2C_2H_5$.

Constants: Mol wt: 113.1, mp: −22.5°C, bp: 206°C, flash p: 230°F, d: 1.06 at 25°/25°C, vap. press.: 1 mm at 67.8°C, vap. d.: 3.9.

Hazard Analysis

Toxicity: Highly toxic. See nitriles.

Fire Hazard: Slight, when exposed to heat or flame.

Disaster Hazard: Dangerous; when heated to decomposition or on contact with acid or acid fumes, it emits highly toxic fumes of cyanides; will react with water or steam to produce toxic and flammable vapors; it can react with oxidizing materials.

Countermeasures

Storage and Handling: Section 7.

To Fight Fire: Foam, carbon dioxide, dry chemical or carbon tetrachloride (Section 6).

ETHYL CYCLOBUTANE

General Information

Description: Insoluble in water.

Formula: $C_2H_5C_4H_7$.

Constants: Mol wt: 84, autoign. temp.: 414°F, lel: 1.2%, uel: 7.7%.

ETHYL CYCLOHEXANE

General Information

Description: Colorless liquid; insoluble in water.

Formula: $C_2H_5C_6H_{11}$.

Constants: Mol wt: 112, d: 0.787, bp: 131.8°C, flash p: 95°F, autoign. temp.: 504°F, lel: 0.9%, uel: 6.6%, vap. d.: 3.9.

Hazard Analysis

Toxic Hazard Rating: U.

Fire Hazard: Moderate, when exposed to heat or flame.

Countermeasures

Storage and Handling: Section 7.

TOXIC HAZARD RATING CODE   *(For detailed discussion, see Section 1.)*

0 NONE: (a) No harm under any conditions; (b) Harmful only under unusual conditions or overwhelming dosage.

1 SLIGHT: Causes readily reversible changes which disappear after end of exposure.

2 MODERATE: May involve both irreversible and reversible changes; not severe enough to cause death or permanent injury.

3 HIGH: May cause death or permanent injury after very short exposure to small quantities.

U UNKNOWN: No information on humans considered valid by authors.

N-ETHYL(CYCLOHEXYL) AMINE
General Information
Description: Slightly soluble in water.
Formula: $C_2H_5(C_6H_{10})NH_2$.
Constants: Mol wt: 126, flash p: 86°F (O.C.), d: 0.8, vap. d.: 4.4.
Hazard Analysis
Toxic Hazard Rating: U. Limited animal experiments suggest high oral, inhalation and irritant toxicity. See also amines.
Fire Hazard: Moderate, when exposed to heat or flame.
Countermeasures
To Fight Fire: Alcohol foam.
Storage and Handling: Section 7.

ETHYL CYCLOPENTANE
General Information
Formula: $C_2H_5C_5H_9$.
Constants: Mol wt: 98, autoign. temp.: 504°F, lel: 1.1%, uel: 6.7%, d: 0.8, vap. d.: 3.4, bp: 218°F.

ETHYL DIAZOACETATE. See diazoacetic ester.

ETHYL DIAZOETHANOATE. See diazoacetic ester.

ETHYLDICHLOROARSINE
General Information
Synonym: Dichloroethylarsine.
Description: Colorless liquid; fruity, biting, irritating odor.
Formula: $C_2H_5AsCl_2$.
Constants: Mol wt: 174.89, mp: −65°C, bp: 156°C decomposes, d: 1.742 at 14°C, vap. press.: 2.29 mm at 21.5°C, vap. d.: 6.03.
Hazard Analysis
Toxic Hazard Rating:
  Acute Local: Irritant 3; Ingestion 3; Inhalation 3.
  Acute Systemic: Ingestion 3; Inhalation 3; Skin Absorption 3.
  Chronic Local: U.
  Chronic Systemic: U.
Toxicity: See arsenic compounds. Extremely irritating. Used as a military poison gas.
Disaster Hazard: Dangerous; on contact with acid or acid fumes, it emits highly toxic fumes of arsenic and phosgene; will react with water or steam to produce toxic and corrosive fumes; can react with oxidizing materials.
Countermeasures
Storage and Handling: Section 7.
Shipping Regulations: Section 11.
  I.C.C.: Poison A; poison gas label, not accepted.
  Coast Guard Classification: Poison A; poison gas label.
  IATA: Poison A, not accepted (passenger and cargo).

ETHYL 4,4′-DICHLOROBENZILATE
General Information
Synonym: 4,4′-Dichlorobenzilic acid ethyl ester; compound G23, 922.
Description: Viscous liquid, sometimes yellow; slightly soluble in water.
Formula: $(C_6H_4Cl)_2C(OH)COOC_2H_5$.
Constants: Mol wt: 325.2, bp: 156–158°C, vap. press.: 2.2×10^{-6} mm at 20°C.
Hazard Analysis and Countermeasures
Dangerous. See related compound DDT.

ETHYLDICHLOROSILANE
General Information
Description: Liquid.
Formula: $C_2H_5SiHCl_2$.
Constants: Mol wt: 129.1, vap. d.: 4.45.
Hazard Analysis
Toxic Hazard Rating:
  Acute Local: Irritant 3; Ingestion 3; Inhalation 3.
  Acute Systemic: U.
  Chronic Local: U.
  Chronic Systemic: U.

Toxicology: See also chlorosilanes.
Disaster Hazard: Dangerous; when heated, it emits highly toxic fumes of phosgene; will react with water or steam to produce heat, toxic and corrosive fumes.
Countermeasures
Ventilation Control: Section 2.
Personnel Protection: Section 3.
Storage and Handling: Section 7.
Shipping Regulations: Section 11.
  I.C.C.: Flammable liquid, red label, 10 gallons.
  IATA: Flammable liquid, red label, not acceptable (passenger), 40 liters (cargo).

ETHYL DIETHANOLAMINE
General Information
Description: White, clear liquid.
Formula: $C_2H_5N(C_2H_4OH)_2$.
Constants: Mol wt: 133.19, bp: 240°C, flash p: 280°F (O.C.), d: 1.015 at 20°C, vap. d.: 4.59.
Hazard Analysis
Toxicity: Details unknown. See also amines.
Fire Hazard: Slight, when exposed to heat or flame; can react with oxidizing materials.
Disaster Hazard: Moderately dangerous; when heated to decomposition it emits toxic fumes of nitrogen oxides.
Countermeasures
Storage and Handling: Section 7.
To Fight Fire: Alcohol foam, carbon dioxide, dry chemical or carbon tetrachloride (Section 6).

ETHYL DIISOAMYL TIN BROMIDE
General Information
Description: Solid.
Formula: $(C_2H_5)(C_5H_{11})_2SnBr$.
Constants: Mol wt: 370.0, d: 1.3650, bp: 155°C at 16 mm.
Hazard Analysis and Countermeasures
See tin compounds and bromides.

ETHYL DIISOBUTYL TIN BROMIDE
General Information
Formula: $(C_2H_5)(C_4H_9)_2SnBr$.
Constants: Mol wt: 341.9, d: 1.4089 at 19.5°C, bp: 131°C at 13 mm.
Hazard Analysis and Countermeasures
See tin compounds and bromides.

ETHYL DIMETHYLCHLOROSILANE
General Information
Description: Liquid.
Formula: $C_2H_5(CH_3)_2SiCl$.
Constants: Mol wt: 122.7, bp: 89.2°C, vap. d.: 4.24.
Hazard Analysis
Toxic Hazard Rating:
  Acute Local: Irritant 3; Ingestion 3; Inhalation 3.
  Acute Systemic: U.
  Chronic Local: U.
  Chronic Systemic: U.
Disaster Hazard: Dangerous; when heated, it emits highly toxic fumes of phosgene; will react with water or steam to produce heat and toxic and corrosive fumes.
Countermeasures
Ventilation Control: Section 2.
Personnel Protection: Section 3.
Storage and Handling: Section 7.

ETHYL DIMETHYL-7-OCTADECENYLAMMONIUM BROMIDE
General Information
Synonym: Onyxide.
Hazard Analysis
Toxic Hazard Rating: U. Limited animal experiments suggest low toxicity.

ETHYL DIPHENYL ETHER
General Information
Description: Light green to yellow liquid; aromatic odor.

Formula: $C_6H_5OC_6H_4C_2H_5$.
Constants: Mol wt: 198.3, mp: $< -20°C$, flash p: 295° F, d: 1.032 at 25°/25°C, vap. d.: 6.84.
Hazard Analysis
Toxicity: Details unknown. See also ethers.
Fire Hazard: Slight, when exposed to heat or flame; can react with oxidizing materials.
Explosion Hazard: Slight, by chemical reaction with oxidizing agents.
Countermeasures
Storage and Handling: Section 7.
To Fight Fire: Foam, carbon dioxide, dry chemical or carbon tetrachloride (Section 6).

ETHYLDIPHENYL PHOSPHINE
General Information
Description: Liquid.
Formula: $(C_6H_5)_2PC_2H_5$.
Constants: Mol wt: 214.2, bp: 293°C, vap. d.: 7.39.
Hazard Analysis
Toxic Hazard Rating:
  Acute Local: U.
  Acute Systemic: Ingestion 3; Inhalation 3.
  Chronic Local: U.
  Chronic Systemic: Ingestion 3; Inhalation 3.
Fire Hazard: Unknown.
Explosion Hazard: Unknown.
Disaster Hazard: Dangerous; when heated, it emits highly toxic fumes of phosphine and oxides of phosphorus, can react with oxidizing materials.
Countermeasures
Ventilation Control: Section 2.
Personnel Protection: Section 3.
Storage and Handling: Section 7.

ETHYL DISULFIDE
General Information
Synonym: Diethyl disulfide; ethyl dithioethane.
Description: Oily liquid; very slightly soluble in water. Soluble in alcohol and ether.
Formula: $C_2H_5SSC_2H_5$.
Constants: Mol wt: 122.2, bp: 154°C, d: 0.99267 at 20°/4°C.
Hazard Analysis
Toxicity: See alkyl disulfides.
Fire Hazard: Probably moderate.
Disaster Hazard: Dangerous; when exposed to heat or decomposition, it emits highly toxic fumes of sulfides. Probably reacts strongly with powerful oxidizers.
Countermeasures
To Fight Fire: Water, water spray, carbon dioxide, foam, dry chemicals (Section 6).
Storage and Handling: Section 7.

ETHYL DODECANOATE. See ethyl laurate.

ETHYLENE
General Information
Synonyms: Ethene; elayl; etherin.
Description: Colorless gas; sweet odor and taste.
Formula: CH_2CH_2.
Constants: Mol wt: 28.05, mp: $-169.4°C$, bp: $-103.9°C$, lel: 3.1%, uel: 32%, d: 0.610 at 0°C, autoign. temp.: 842°F, vap. d.: 0.98.
Hazard Analysis
Toxic Hazard Rating:
  Acute Local: 0.
  Acute Systemic: Inhalation 2.

Chronic Local: 0.
Chronic Systemic: 0.
Toxicology: High concentrations cause anesthesia. A simple asphyxiant. A common air contaminant (Section 4).
Fire Hazard: Dangerous, when exposed to heat or flame; can react vigorously with oxidizing materials.
Explosion Hazard: Moderate, when exposed to flame.
Disaster Hazard: Dangerous! Flammable gas!
Countermeasures
Ventilation Control: Section 2.
To Fight Fire: Stop flow of gas, carbon dioxide, dry chemical or fine water spray (Section 6).
Storage and Handling: Section 7.
Shipping Regulations: Section 11.
  I.C.C.: Flammable gas; red gas label, 300 pounds.
  Coast Guard Classification: Inflammable gas; red gas label.
  IATA: Flammable gas, red label, not acceptable (passenger), 140 kilograms (cargo).

ETHYLENE ACETATE. See ethylene glycol diacetate.

ETHYLENE ALCOHOL. See ethylene glycol.

ETHYLENE BIS-DITHIOCARBAMATE DERIVATIVE
Hazard Analysis
Toxicity: Details unknown, but probably toxic; a fungicide. See ferbam, maneb, nabam and zineb.
Disaster Hazard: Moderately dangerous; when heated to decomposition, it emits toxic fumes.
Countermeasures
Storage and Handling: Section 7.

ETHYLENE BIS(GLYCOLIC ACID)
General Information
Description: Light-colored, viscous liquid.
Formula: $(CH_2OCH_2COOH)_2$.
Constants: Mol wt: 178.14, mp: 30–35°C, d: 1.3 at 75°/25°C, vap. d.: 6.14.
Hazard Analysis
Toxicity: Unknown.
Fire Hazard: Slight; (Section 6).
Countermeasures
Storage and Handling: Section 7.

ETHYLENE BROMIDE. See ethylene dibromide.

ETHYLENE BUTYRATE. See ethylene dibutyrate.

ETHYLENE CARBONATE
General Information
Synonyms: Glycol carbonate; dioxolone-2.
Description: Colorless liquid or crystalline solid.
Formula: OCH_2CH_2OCO.
Constants: Mol wt: 88.06, bp: 244°C at 740 mm, fp: 35.7°C, flash p: 290°F(O.C.), d: 1.322 at 40°/20°C, vap. press.: 0.01 mm at 20°C, vap. d.: 3.04.
Hazard Analysis
Toxicity: Details unknown, but probably low as ordinarily used.
Fire Hazard: Slight, when exposed to heat or flame; can react with oxidizing materials.
Countermeasures
To Fight Fire: Water, foam, carbon dioxide, dry chemical or carbon tetrachloride (Section 6).
Storage and handling: Section 7.

ETHYLENE CHLORIDE. See ethylene dichloride.

TOXIC HAZARD RATING CODE (For detailed discussion, see Section 1.)

0 NONE: (a) No harm under any conditions; (b) Harmful only under unusual conditions or overwhelming dosage.

1 SLIGHT: Causes readily reversible changes which disappear after end of exposure.

2 MODERATE: May involve both irreversible and reversible changes; not severe enough to cause death or permanent injury.

3 HIGH: May cause death or permanent injury after very short exposure to small quantities.

U UNKNOWN: No information on humans considered valid by authors.

ETHYLENE CHLOROBROMIDE
General Information
Synonym: 1-Bromo-2-chloroethane.
Description: Colorless, volatile liquid; sweet chloroform-like odor.
Formula: $BrCH_2CH_2Cl$.
Constants: Mol wt: 143.4, bp: 106.1°C, fp: -18.4°C, flash p: none, d: 1.7272 at 25°/4°C, vap. press.: 40 mm at 29.7°C, vap. d.: 4.94.
Hazard Analysis
Toxic Hazard Rating:
Acute Local: Irritant 3; Ingestion 3; Inhalation 3.
Acute Systemic: Ingestion 3; Inhalation 3; Skin Absorption 3.
Chronic Local: Irritant 2.
Chronic Systemic: Ingestion 2; Inhalation 2; Skin Absorption 2.
Toxicology: Data based on animal experiments show action similar to that of ethylene dichloride. May cause injury to liver and kidneys.
Disaster Hazard: Dangerous; when heated to decomposition, it emits highly toxic fumes of chlorides and bromides.
Countermeasures
Storage and Handling: Section 7.

ETHYLENE CHLOROHYDRIN
General Information
Synonyms: β-Chloroethyl alcohol; glycol chlorohydrin; 2-chloroethanol.
Description: Colorless liquid; faint ethereal odor.
Formula: CH_2ClCH_2OH.
Constants: Mol wt: 80.52, mp: -69°C, bp: 128.8°C, flash p: 140°F (O.C.), d: 1.213 at 20°/4°C, autoign. temp.: 797°F, vap. press.: 10 mm at 30.3°C, vap. d.: 2.78, lel: 4.9%, uel: 15.9%.
Hazard Analysis
Toxic Hazard Rating:
Acute Local: Irritant 2.
Acute Systemic: Ingestion 3; Inhalation 3; Skin Absorption 3.
Chronic Local: U.
Chronic Systemic: U.
TLV: ACGIH (recommended); 5 ppm in air; 16 milligrams per cubic meter of air. Can penetrate the skin.
Toxicology: Ethylene chlorohydrin is a narcotic poison affecting the nervous system and the liver, spleen and lungs. Exposure to the vapor may result in irritation of the mucous membranes, followed by sleepiness, drowsiness, giddiness, nausea and vomiting. The initial symptoms may be slight. After a latent period of several hours, dyspnea, severe headache, stupor, cyanosis, and pain over the heart may develop. Autopsy shows pulmonary edema, ulceration of the mucous membranes of the larger bronchi, and acute liver and kidney lesions. Fatal amounts of ethylene chlorohydrin may be absorbed through the skin.
Fire Hazard: Moderate, when exposed to heat or flame.
Spontaneous Heating: No.
Disaster Hazard: Dangerous; when heated to decomposition, it emits highly toxic fumes of phosgene; will react with water or steam to produce toxic and corrosive fumes; can react with oxidizing materials.
Countermeasures
Ventilation Control: Section 2.
To Fight Fire: Water, alcohol foam, carbon dioxide, dry chemical or carbon teterachloride (Section 6).
Personnel Protection: Section 3.
First Aid: Section 1.
Storage and Handling: Section 7.
MCA warning label.
IATA: Poison B, poison label, 1 liter (passenger), 220 liters (cargo).

ETHYLENE CYANIDE. See succinontrile.

ETHYLENE CYANOHYDRIN
General Information
Synonyms: β-Hydroxypropionitrile; hydracrylonitrile; glycol cyanohydrin.
Description: Colorless to straw-colored liquid.
Formula: $HOCH_2CH_2CN$.
Constants: Mol wt: 71.08, bp: 228°C decomposes, fp: -46°C, flash p: 265°F (O.C.), d: 1.0404 at 25°C, vap. press.: 0.08 mm at 25°C, vap. d.: 2.45.
Hazard Analysis
Toxicity: See nitriles. See also acrylonitrile.
Fire Hazard: Dangerous, when exposed to heat or flame.
Disaster Hazard: Dangerous; when heated or on contact with acid or acid fumes, it emits highly toxic fumes of cyanides; will react with water or steam to produce toxic and flammable vapors; can react vigorously with oxidizing materials.
Countermeasures
Ventilation Control: Section 2.
To Fight Fire: Carbon dioxide, dry chemical or carbon tetrachloride (Section 6).
Personnel Protection: Section 3.
First Aid: Section 1.
Storage and Handling: Section 7.

ETHYLENE DIACETATE. See ethylene glycol diacetate.

ETHYLENEDIAMINE
General Information
Synonyms: 1,2-Ethanediamine; 1,2-diaminoethane.
Description: Volatile, colorless, hygroscopic liquid; ammonia-like odor.
Formula: $NH_2CH_2CH_2NH_2$.
Constants: Mol wt: 60.10, mp: 8.5°C, bp: 117.2°C, flash p: 93°F (C.C.), d: 0.8994 at 20°/4°C, vap. press.: 10.7 mm at 20°C, vap. d.: 2.07.
Hazard Analysis
Toxic Hazard Rating:
Acute Local: Irritant 2; Ingestion 2; Inhalation 2.
Acute Systemic: Ingestion 2; Inhalation 2; Skin Absorption 2.
Chronic Local: Irritant 2; Allergen 2.
Chronic Systemic: Ingestion 2; Inhalation 2; Skin Absorption 2.
Toxicology: In addition to being an irritant it can cause sensitization leading to allergic dermatitis asthma. Note: Used as a food additive permitted in food for human consumption.
TLV: ACGIH (recommended); 10 ppm in air; 30 milligrams per cubic meter of air.
Fire Hazard: Moderate, when exposed to heat or flame; can react with oxidizing materials.
Spontaneous Heating: No.
Countermeasures
To Fight Fire: Carbon dioxide, dry chemical or carbon tetrachloride (Section 6).
Ventilation Control: Section 2.
Personal Hygiene: Section 3.
Storage and Handling: Section 7.
Shipping Regulations: Section 11.
MCA warning label.
IATA: Other restricted articles, class A, no label required, no limit (passenger and cargo).

ETHYLENE DIAMINE DIHYDROIODIDE
General Information
Formula: $(-CH_2NH_2)_2 \cdot 2HI$.
Constant: Mol wt: 316.
Hazard Analysis
Toxic Hazard Rating: U. A trace material added to animal feeds for iodine content.

ETHYLENEDIAMINETETRAACETIC ACID
General Information
Synonym: EDTA.

Description: Colorless crystals. Slightly water soluble. Insoluble in common organic solvents.

Formula: (HOOCCH$_2$)$_2$NCH$_2$CH$_2$N(CH$_2$COOH)$_2$.

Constants: Mol wt: 292.3, decomposes at 240° C.

Hazard Analysis

Toxicity: See disodium ethylenediaminetetraacetate. The calcium disodium salt of EDTA is used as a chelating agent in treating lead poisoning.

ETHYLENE DIAMINE TETRAACETONITRILE
General Information

Synonym: EDTN.

Description: White to cream colored powder.

Formula: (NCCH$_2$)$_2$NCH$_2$CH$_2$N(CH$_2$CN)$_2$.

Constants: Mol wt: 216, mp: 132° C, d: 0.5 at 25° C.

Hazard Analysis

Toxicity: Highly toxic. See nitriles.

ETHYLENEDIAMINO DIBUTYL GOLD BROMIDE
General Information

Description: Colorless crystals, water soluble.

Formula: (CH$_2$NH$_2$)$_2$Au(C$_4$H$_9$)$_2$Br.

Constants: Mol wt: 451.4, mp: 190° C (decomp).

Hazard Analysis and Countermeasures

Dangerous. See bromides and gold compounds.

ETHYLENEDIAMINO DIPROPYL GOLD BROMIDE
General Information

Description: Colorless crystals.

Formula: (CH$_2$NH$_2$)$_2$Au(C$_3$H$_7$)$_2$Br.

Constants: Mol wt: 423.4, mp: volatile at 130° C, decomposes at 190° C.

Hazard Analysis and Countermeasures

Dangerous. See bromide and gold compounds.

ETHYLENE DIBENZOATE
General Information

Description: Crystals.

Formula: (C$_6$H$_5$CO$_2$)$_2$C$_2$H$_4$.

Constants: Mol wt: 270.3, mp: 73–74° C, bp: 360° C decomposes, vap. d.: 9.35.

Hazard Analysis

Toxicity: Probably moderate. See ethylene glycol and benzoic acid.

Fire Hazard: Slight; can react with oxidizing materials.

Countermeasures

Storage and Handling: Section 7.

ETHYLENE DIBROMIDE
General Information

Synonyms: 1,2-Dibromoethane; glycol dibromide.

Description: Colorless, heavy liquid; sweet odor.

Formula: CH$_2$BrCH$_2$Br.

Constants: Mol wt: 187.88, bp: 131.4° C, fp: 9.3° C, flash p: none, d: 2.1701 at 25°/4° C, vap. press.: 17.4 mm at 30° C, vap. d.: 6.48.

Hazard Analysis

Toxic Hazard Rating:

Acute Local: Irritant 3; Ingestion 3; Inhalation 3.

Acute Systemic: Ingestion 3; Inhalation 3; Skin Absorption 3.

Chronic Local: Irritant 2.

Chronic Systemic: Ingestion 2; Inhalation 2; Skin Absorption 2.

Toxicology: An insecticide. Action resembles that of ethylene dichloride.

TLV: ACGIH (recommended) 25 ppm in air; 190 milligrams per cubic meter of air. May be absorbed via the skin.

Disaster Hazard: Dangerous; when heated to decomposition, it emits highly toxic fumes of bromides.

Countermeasures

Ventilation Control: Section 2.

Personnel Protection: Section 3.

First Aid: Section 1.

Storage and Handling: Section 7.

Shipping Regulations: Section 11.

MCA warning label.

IATA: Poison B, poison label, 1 liter (passenger), 220 liters (cargo).

ETHYLENE DIBUTYRATE
General Information

Synonym: Ethylene butyrate.

Description: Liquid.

Formula: (CH$_2$O$_2$CCH$_2$CH$_2$CH$_3$)$_2$.

Constants: Mol wt: 202.3, bp: 240° C, d: 1.024 at 0°/4° C, vap. d.: 6.96.

Hazard Analysis

Toxicity: Details unknown. See esters.

Fire Hazard: Slight; can react with oxidizing materials.

Countermeasures

Storage and Handling: Section 7.

ETHYLENEDICARBOXYLIC ACID. See succinic acid.

ETHYLENE DICHLORIDE
General Information

Synonyms: Ethylene chloride; 1,2-dichloroethane.

Description: Colorless liquid.

Formula: CH$_2$ClCH$_2$Cl.

Constants: Mol wt: 99.0, bp: 83.5° C, ulc: 60–70, lel: 6.2%, uel: 15.9%, fp: −35.7° C, flash p: 56° F, d: 1.257 at 20°/4° C, autoign. temp.: 775° F, vap. press.: 100 mm at 29.4° C, vap. d.: 3.35.

Hazard Analysis

Toxic Hazard Rating:

Acute Local: Irritant 3; Ingestion 3; Inhalation 3.

Acute Systemic: Ingestion 3; Inhalation 3.

Chronic Local: Irritant 2.

Chronic Systemic: Ingestion 3; Inhalation 3; Skin Absorption 2.

TLV: ACGIH (recommended) 50 ppm of air; 200 milligrams per cubic meter of air.

Toxicology: Ethylene dichloride has a distinctive odor and strong local irritating effects, which give warning of its presence in relatively safe concentrations. There is irritation of the eyes and upper respiratory passages. Ethylene dichloride has a specific effect on the cornea. Exposure to the vapor, or, in animals, injection under the skin, produces a clouding which may progress to endothelial necrosis and infiltration of the cornea by lymphocytes and connective tissue cells. The narcotic action of the compound is strong, probably of the same order as chloroform. Its toxic effects upon the liver and kidneys are less than that of carbon tetrachloride, but animal experiments indicate that these organs may show congestion and fatty degeneration. Edema of the lungs has also been reported in animals. Dermatitis in man has been observed (Section 9).

In short exposures to high concentrations, the picture is one of irritation of the eyes, nose and throat, followed by dizziness, nausea, vomiting, increasing stupor, cyanosis, rapid pulse, and loss of consciousness.

Chronic poisoning, where exposure has occurred over a period of several months, may cause loss of appetite, nausea and vomiting, epigastric distress, tre-

---

TOXIC HAZARD RATING CODE (For detailed discussion, see Section 1.)

0 NONE: (a) No harm under any conditions; (b) Harmful only under unusual conditions or overwhelming dosage.

1 SLIGHT: Causes readily reversible changes which disappear after end of exposure.

2 MODERATE: May involve both irreversible and reversible changes; not severe enough to cause death or permanent injury.

3 HIGH: May cause death or permanent injury after very short exposure to small quantities.

U UNKNOWN: No information on humans considered valid by authors.

mors, nystagmus, leucocytosis, low blood sugar levels, and possibly dermatitis if there has been skin contact. A soil fumigant. Used as a food additive permitted in food for human consumption.

Fire Hazard: Dangerous, if exposed to heat or flame.

Spontaneous Heating: No.

Explosion Hazard: Moderate, in the form of vapor when exposed to flame (Section 7).

Disaster Hazard: Dangerous; when heated to decomposition, it emits highly toxic fumes of phosgene; can react vigorously with oxidizing materials.

Countermeasures

Ventilation Control: Section 2.

To Fight Fire: Water, foam, carbon dioxide, dry chemical or carbon tetrachloride (Section 6).

Personnel Protection: Section 3.

First Aid: Section 1.

Storage and Handling: Section 7.

Shipping Regulations: Section 11.

    I.C.C.: Flammable liquid; red label, 10 gallons.

    Coast Guard Classification: Inflammable liquid; red label. MCA warning label.

    IATA: Flammable liquid, red label, 1 liter (passenger), 40 liters (cargo).

ETHYLENE DIFLUORIDE. See 1,1-difluoroethane.

ETHYLENE DIFORMATE. See ethylene glycol diformate.

ETHYLENE DIIODIDE

General Information

Synonym: 1,2-Diodoethane.

Description: Yellow crystals.

Formula: CH_2ICH_2I.

Constants: Mol wt: 281.9, mp: 82°C, bp: decomposes, d: 2.132 at 10°/4°C, vap. d.: 9.72.

Hazard Analysis and Countermeasures

See iodides.

ETHYLENE DILAURATE

General Information

Synonym: Ethylene laurate.

Description: Solid.

Formula: $(C_{11}H_{23}CO_2CH_2)_2$.

Constants: Mol wt: 426.7, mp: 52°C, bp: 188°C at 20 mm, vap. d.: 14.8.

Hazard Analysis

Toxicity: Details unknown.

Disaster Hazard: Slight; when heated, it emits acrid fumes; can react with oxidizing materials.

Countermeasures

Storage and Handling: Section 7.

ETHYLENE DINITRATE

General Information

Synonym: Ethylene nitrate.

Description: Yellow liquid.

Formula: $C_2H_4(ONO_2)_2$.

Constants: Mol wt: 152.07, mp: −20°C, bp: explodes at 114°C, d: 1.483 at 8°C, vap. d.: 5.25.

Hazard Analysis

Toxic Hazard Rating:

    Acute Local: U.

    Acute Systemic: Ingestion 3; Inhalation 3.

    Chronic Local: U.

    Chronic Systemic: U.

Fire Hazard: See nitrates.

Disaster Hazard: See nitrates.

Explosion Hazard: See nitrates.

Countermeasures

Ventilation Control: Section 2.

Personal Hygiene: Section 3.

First Aid: Section 1.

Storage and Handling: Section 7

ETHYLENE DINITRITE

General Information

Synonym: Ethylene nitrite.

Description: Liquid.

Formula: $C_2H_4(ONO)_2$.

Constants: Mol wt: 120.1, mp: < −15°C, bp: 98°C, d: 1.2156 at 0°/4°C, vap. d.: 4.13.

Hazard Analysis and Countermeasures

See nitrites.

ETHYLENE DITHIOCYANATE

General Information

Description: Colorless crystals.

Formula: $(CH_2SCN)_2$.

Constants: Mol wt: 144.2, mp: 90°C, bp: decomposes, vap. d.: 4.98.

Hazard Analysis

Toxicity: Probably low. See also thiocyanates.

Disaster Hazard: Dangerous! Will decompose when heated to give off highly toxic fumes.

Countermeasures

Storage and Handling: Section 7.

ETHYLENE FLUORIDE. See 1,1-difluoroethane.

ETHYLENE FORMATE. See ethylene glycol diformate.

ETHYLENE GLYCOL

General Information

Synonyms: 1,2-Ethanediol; glycol; ethylene alcohol; glycol alcohol.

Description: Colorless, sweet-tasting liquid.

Formula: CH_2OHCH_2OH.

Constants: Mol wt: 62.1, bp: 197.5°C, lel: 3.2%, fp: −13°C, flash p: 232°F (C.C.), d: 1.113 at 25°/25°C, autoign. temp.: 775°F, vap. press.: 0.05 mm at 20°C, vap. d.: 2.14.

Hazard Analysis

Toxic Hazard Rating:

    Acute Local: 0.

    Acute Systemic: Ingestion 2.

    Chronic Local: Irritant 1.

    Chronic Systemic: Ingestion 1.

    (Lethal dose for man reported to be 100 ml.).

Toxicology: If ingested it causes initial CNS stimulation followed by depression. Later it causes kidney damage which can terminate fatally.

Fire Hazard: Slight, when exposed to heat or flame; can react with oxidizing materials.

Spontaneous Heating: No.

Explosion Hazard: Moderate, when exposed to flame.

Countermeasures

Personal Hygiene: Section 3.

To Fight Fire: Alcohol foam, water, foam, carbon dioxide, dry chemical or carbon tetrachloride (Section 6).

Storage and Handling: Section 7.

ETHYLENE GLYCOL BIS-(2,3-EPOXY-2-METHYL PROPYL) ETHER

Hazard Analysis

Toxic Hazard Rating:

    Acute Local: Irritant 2.

    Acute Systemic: Ingestion 1; Skin Absorption 2.

    Chronic Local: O.

    Chronic Systemic: U.

Based on limited animal experiments. See also glycols and epoxy resins.

ETHYLENE GLYCOL n-BUTYL ETHER

General Information

Description: Clear, mobile liquid; pleasant odor.

Formula: $C_4H_9OCH_2CH_2OH$.

Constants: Mol wt: 118.2, bp: 168.4–170.2°C, fp: −74.8°C, flash p: 160°F (C.O.C.), d: 0.9012 at 20°/20°C, vap. press.: 300 mm at 140°C.

Hazard Analysis
Toxicity: Unknown. See glycols.
Fire Hazard: Moderate, when exposed to heat or flame; can react with oxidizing materials.
Countermeasures
To Fight Fire: Foam, carbon dioxide, dry chemical or carbon tetrachloride (Section 6).
Storage and Handling: Section 7.

ETHYLENE GLYCOL-p-sec-BUTYLPHENYL ETHER
General Information
Synonym: 2-(p-sec-Butylphenoxy) ethanol.
Description: Straw-colored liquid.
Formula: $C_2H_5CH(CH_3)C_6H_4OCH_2CH_2OH$.
Constants: Mol wt: 194.3, mp: $< -20°C$, bp: 151–161°C at 10 mm, d: 1.008 at 25°/25°C, flash p: 248°F (O.C.), vap. d.: 6.7.
Hazard Analysis
Toxicity: See glycols.
Fire Hazard: Slight, when exposed to heat or flame; can react with oxidizing materials.
Countermeasures
To Fight Fire: Foam, carbon dioxide, dry chemical or carbon tetrachloride (Section 6).
Storage and Handling: Section 7.

ETHYLENE GLYCOL-p-tert-BUTYLPHENYL ETHER
General Information
Description: Light yellow liquid.
Formula: $(CH_3)_3CC_6H_4OCH_2CH_2OH$.
Constants: Mol wt: 194.3, mp: 12.2°C, bp: 289–328°C at 10 mm, flash p: 314.6°F, d: 1.017 at 25°/25°C, vap. d.: 6.7.
Hazard Analysis
Toxicity: Details unknown. See also glycols.
Fire Hazard: Slight, when exposed to heat or flame; can react with oxidizing materials.
Countermeasures
To Fight Fire: Foam, carbon dioxide, dry chemical or carbon tetrachloride (Section 6).
Storage and Handling: Section 7.

ETHYLENE GLYCOL-p-tert-BUTYLPHENYL ETHER ACETATE
General Information
Synonyms: 2-(p-tert Butylphenoxy) ethyl acetate.
Description: Slightly yellow liquid.
Formula: $(CH_3)_3CC_6H_4OCH_2CH_2OOCCH_3$.
Constants: Mol wt: 236.3, mp: $< -20°C$, bp: 208°C, flash p: 327°F, d: 1.029 at 25°/25°C, vap. d.: 8.15.
Hazard Analysis
Toxicity: Details unknown. See also glycols.
Fire Hazard: Slight, when exposed to heat or flame; can react with oxidizing materials.
Countermeasures
To Fight Fire: Foam, carbon dioxide, dry chemical or carbon tetrachloride (Section 6).
Storage and Handling: Section 7.

ETHYLENE GLYCOL DIABIETATE
General Information
Synonym: Glycol diabietate.
Description: Liquid.
Formula: $(CH_2CO_2C_{19}H_{29})_2$.
Constants: Mol wt: 630.9, flash p: 520°F, vap. d.: 21.8.
Hazard Analysis
Toxicity: Details unknown. See also glycols.

Fire Hazard: Slight, when exposed to heat or flame; can react with oxidizing materials.
Countermeasures
To Fight Fire: Foam, carbon dioxide, dry chemical or carbon tetrachloride (Section 6).
Storage and Handling: Section 7.

ETHYLENE GLYCOL DIACETATE
General Information
Synonyms: Ethylene diacetate; glycol diacetate.
Description: Colorless liquid or crystals.
Formula: $(CH_2OOCCH_3)_2$.
Constants: Mol wt: 146.14, mp: $-31°C$, bp: 191°C, flash p: 205°F (O.C.), d: 1.128 at 0°/4°C, vap. press.: 1 mm at 38.3°C, vap. d.: 5.04.
Hazard Analysis
Toxicity: Details unknown. See also glycols. Animal experiments suggest low toxicity.
Fire Hazard: Moderate when exposed to heat or flame; can react with oxidizing materials.
Countermeasures
To Fight Fire: Water, foam, carbon dioxide, dry chemical or carbon tetrachloride (Section 6).
Storage and Handling: Section 7.

ETHYLENE GLYCOL DIBENZOATE
General Information
Synonym: Benzoflex; E-60.
Description: Crystalline material.
Formula: $(CH_2O_2CC_6H_5)_2$.
Constants: Mol wt: 270.2, bp: 208–211°C, fp: 69–71°C, flash p: 365°F, vap. d.: 9.38.
Hazard Analysis
Toxicity: Details unknown. See also glycols.
Fire Hazard: Slight, when exposed to heat or flame; can react with oxidizing materials.
Countermeasures
To Fight Fire: Water, foam, carbon dioxide, dry chemical or carbon tetrachloride (Section 6).
Storage and Handling: Section 7.

ETHYLENE GLYCOL DIFORMATE
General Information
Synonyms: Ethylene diformate; glycol diformate.
Description: Liquid; decomposes in water.
Formula: $HCOOCH_2CH_2OOCH$.
Constants: Mol wt: 118.09, mp: $-10°C$, bp: 177°C, flash p: 200°F, d: 1.2277 at 20°/20°C, vap. d.: 4.07.
Hazard Analysis
Toxicity: Details unknown. More toxic than ethyl formate in animal experiments. See ethyl formate.
Fire Hazard: Moderate, when exposed to heat or flame; can react with oxidizing materials.
Countermeasures
To Fight Fire: Water, foam, carbon dioxide, dry chemical or carbon tetrachloride (Section 6).
Storage and Handling: Section 7.

ETHYLENE GLYCOL DIMETHYL ETHER
General Information
Synonym: 1,2-Dimethoxy ethane.
Description: Water-white liquid; ethereal odor.
Formula: $CH_3OCH_2CH_2OCH_3$.
Constants: Mol wt: 90.1, bp: 85.2°C, flash p: 104°F, fp: $< -60°C$, d: 0.8672 at 20°/20°C, vap. press.: 48 mm at 20°C, vap. d.: 3.11.
Hazard Analysis
Toxicity: Details unknown. See also glycols.

TOXIC HAZARD RATING CODE (For detailed discussion, see Section 1.)

0 NONE: (a) No harm under any conditions; (b) Harmful only under unusual conditions or overwhelming dosage.

1 SLIGHT: Causes readily reversible changes which disappear after end of exposure.

2 MODERATE: May involve both irreversible and reversible changes; not severe enough to cause death or permanent injury.

3 HIGH: May cause death or permanent injury after very short exposure to small quantities.

U UNKNOWN: No information on humans considered valid by authors.

Fire Hazard: Dangerous, when exposed to heat or flame; can react vigorously with oxidizing materials.

Explosion Hazard: See ethers.

Disaster Hazard: Dangerous; upon exposure to heat or flame.

Countermeasures

Ventilation Control: Section 2.

To Fight Fire: Foam, carbon dioxide, dry chemical or carbon tetrachloride (Section 6).

Storage and Handling: Section 7.

ETHYLENE GLYCOL DINITRATE

Toxic Hazard Rating:

    Acute Local: Irritant 1.

    Acute Systemic: Ingestion 2; Skin Absorption 1.

    Chronic Local: Irritant 1.

    Chronic Systemic: Ingestion 2; Skin Absorption 2.

TLV: ACGIH (recommended) 0.02 parts per million of air; 0.1 milligram per cubic meter of air. May be absorbed via the skin.

Can cause lowered blood pressure leading to headache, dizziness, and weakness.

ETHYLENE GLYCOL ETHYL ETHER

General Information

Description: Colorless liquid; mild pleasant odor.

Formula: $C_2H_5OCH_2CH_2OH$.

Constants: Mol wt: 90.1, bp: 134.5°C, flash p: 110°F (C.O.C.), d: 0.9275 at 25°/25°C, vap. press.: 40 mm at 60°C, vap. d.: 3.1.

Hazard Analysis

Toxic Hazard Rating:

    Acute Local: Irritant 1; Ingestion 2; Inhalation 2.

    Acute Systemic: Ingestion 2; Inhalation 2.

    Chronic Local: U.

    Chronic Systemic: U.

Fire Hazard: Moderate, when exposed to heat or flame; can react with oxidizing materials (Section 6).

Countermeasures

To Fight Fire: Foam, carbon dioxide, dry chemical or carbon tetrachloride (Section 6).

Ventilation Control: Section 2.

Personnel Protection: Section 3.

Storage and Handling: Section 7.

ETHYLENE GLYCOL METHYL BUTYL ETHER

General Information

Description: Liquid.

Formula: $C_4H_9OCH_2CH_2OCH_3$.

Constants: Mol wt: 132.2, bp: 147°C, d: 0.8487, vap. d.: 4.57.

Hazard Analysis and Countermeasures

See glycols.

ETHYLENE GLYCOL METHYL ETHER. See ethylene glycol monomethyl ether.

ETHYLENE GLYCOL METHYL n-HEXYL ETHER

General Information

Description: Liquid.

Formula: $C_6H_{13}OCH_2CH_2OCH_3$.

Constants: Mol wt: 160.25, bp: 190°C, d: 0.855, vap. d.: 5.53.

Hazard Analysis

Toxicity: Details unknown. See also glycols.

Fire Hazard: Moderate; can react with oxidizing materials.

Countermeasures

To Fight Fire: Foam, carbon dioxide, dry chemical or carbon tetrachloride (Section 6).

Storage and Handling: Section 7.

ETHYLENE GLYCOL METHYL PHENYL ETHER

General Information

Description: Liquid.

Formula: $C_6H_5OCH_2CH_2OCH_3$.

Constants: Mol wt: 152.19, bp: 215°C, d: 1.03.

Hazard Analysis

Toxicity: Details unknown. See also glycols.

Countermeasures

Storage and Handling: Section 7.

ETHYLENE GLYCOL MONOACETATE

General Information

Synonym: Glycol monoacetate.

Description: Colorless, almost odorless liquid.

Formula: $CH_3COOCH_2CH_2OH$.

Constants: Mol wt: 104.10, bp: 182°C, flash p: 215°F (O.C.), d: 1.108 at 15°C, vap. d.: 3.59.

Hazard Analysis

Toxic Hazard Rating:

    Acute Local: Irritant 1; Inhalation 1.

    Acute Systemic: Ingestion 1; Inhalation 1.

    Chronic Local: U.

    Chronic Systemic: U.

Toxicology: Animal experiments show low toxicity. See also glycols.

Fire Hazard: Moderate, when exposed to heat or flame; can react with oxidizing materials.

Spontaneous Heating: No.

Countermeasures

To Fight Fire: Alcohol foam, foam, water, carbon dioxide, dry chemical or carbon tetrachloride (Section 6).

Storage and Handling: Section 7.

ETHYLENE GLYCOL MONOBENZYL ETHER. See benzyl "Cellosolve."

ETHYLENE GLYCOL MONOBUTYL ETHER. See butyl "Cellosolve."

ETHYLENE GLYCOL MONOETHYL ETHER. See "Cellosolve" solvent.

ETHYLENE GLYCOL MONOETHYL ETHER ACETATE. See "Cellosolve" acetate.

ETHYLENE GLYCOL MONOISOPROPYL ETHER

General Information

Description: Colorless liquid; mild, agreeable odor.

Formula: $(CH_3)_2CHOCH_2CH_2OH$.

Constants: Mol wt: 104.14, bp: 139°C, d: 0.906, vap. d.: 3.6.

Hazard Analysis

Toxicity: Details unknown. See also glycols.

Fire Hazard: Moderate; can react with oxidizing materials.

Countermeasures

Storage and Handling: Section 7.

ETHYLENE GLYCOL MONOMETHYL ETHER

General Information

Synonym: 2-Methoxyethanol.

Description: Colorless liquid; mild agreeable odor.

Formula: $CH_3OCH_2CH_2OH$.

Constants: Mol wt: 76.09, bp: 124.5°C, fp: −86.5°C, flash p: 115°F (O.C.), d: 0.9660 at 20°/4°C, autoign. temp.: 551°F, vap. press.: 6.2 mm at 20°C, vap. d.: 2.62.

Hazard Analysis

Toxic Hazard Rating:

    Acute Local: U.

    Acute Systemic: Ingestion 2.

    Chronic Local: Irritant 1.

    Chronic Systemic: Ingestion 2; Inhalation 2.

TLV: ACGIH (recommended) 25 parts per million of air; 80 milligrams per cubic meter of air. May be absorbed via the skin.

Toxicology: When used under conditions which do not require the application of heat, this material probably presents little hazard to health. However, in the manufacture of fused collars, which require pressing with a hot iron, cases have been reported showing disturbance of the hemopoietic system with or without neurological signs and symptoms. The blood picture may resemble that produced by exposure to benzene. Two cases reported had severe aplastic anemia with tremors and marked mental dullness. One case had multiple neuritis and four others had abnormal reflexes. The commonest change in the blood picture was the finding of immature neutrophils (shift to the left); in other cases there were a reduction in number of the blood platelets and a macrocytic anemia. The persons affected had been exposed to vapors of methyl "Cellosolve" (76 ppm), ethyl and methyl alcohol, ethyl acetate and petroleum naphtha.

The first signs of poisoning are probably abnormalities found in the blood picture, as mentioned above. Reflexes may be exaggerated or abnormal in character, and may be accompanied by complaints of drowsiness and fatigue. Tremors may be present. Severe damage probably takes the form of aplastic anemia.

Fire Hazard: Moderate, when exposed to heat or flame; can react with oxidizing materials.

Spontaneous Heating: No.

Countermeasures

To Fight Fire: Alcohol foam, carbon dioxide, dry chemical or carbon tetrachloride (Section 6).

Ventilation Control: Section 2.

Personnel Protection: Section 3.

Storage and Handling: Section 7.

Shipping Regulations: Section 11.

Coast Guard Classification: Combustible liquid.

MCA warning label.

ETHYLENE GLYCOL MONOMETHYL ETHER ACETAL

General Information

Description: Liquid.

Formula: $CH_3CH(OCH_2CH_2OCH_3)_2$.

Constants: Mol wt: 178.2, mp: $-85°C$ (glass), bp: $207.2°C$, flash p: $205°F$, d: 0.9773 at $20°/20°C$, vap. press.: 0.1 mm at $20°C$, vap. d.: 6.16.

Hazard Analysis

Toxicity: Details unknown. See also glycols.

Fire Hazard: Moderate, when exposed to heat or flame; can react with oxidizing materials.

Countermeasures

To Fight Fire: Foam, carbon dioxide, dry chemical or carbon tetrachloride (Section 6).

Storage and Handling: Section 7.

Shipping Regulations: Section 11.

Coast Guard Classification: Combustible liquid.

ETHYLENE GLYCOL MONOMETHYL ETHER ACETATE. See methyl "Cellosolve" acetate.

ETHYLENE GLYCOL MONO-2-METHYL PENTYL ETHER

General Information

Formula: $HOCH_2CH_2OCH_3C_5H_{10}$.

Constant: Mol wt: 146.

Hazard Analysis

Toxic Hazard Rating: U. Limited animal experiments suggest moderate toxicity. See also glycols.

ETHYLENE GLYCOL MONOOCTYL ETHER

General Information

Description: Colorless, odorless liquid.

Formula: $C_4H_9CHC_2H_5CH_2OCH_2CH_2OH$.

Constants: Mol wt: 174.28, bp: $228.3°C$, flash p: $230°F$, d: 0.8859 at $20°C$, vap. press.: 0.02 mm at $20°C$.

Hazard Analysis

Toxicity: Unknown. See also glycols.

Fire Hazard: Slight, when exposed to heat or flame; can react with oxidizing materials.

Explosion Hazard: See ethers.

Countermeasures

Storage and Handling: Section 7.

To Fight Fire: Foam, carbon dioxide, dry chemical or carbon tetrachloride (Section 6).

ETHYLENE GLYCOL MONO-2,6,8-TRIMETHYL-4-NONYL ETHER

General Information

Formula: $HOCH_2CH_2OCH_2CH_3CHCH_2CH_2CH_2$-$CH_3CHCH_2CH_3CHCH_3$.

Constant: Mol wt: 230.

Hazard Analysis

Toxic Hazard Rating: U. Limited animal experiments suggest low toxicity. See also glycols.

ETHYLENE GLYCOL PHENYL ETHER. See phenyl "Cellosolve."

ETHYLENE GLYCOL PHENYL ETHER

(commercial grade)

General Information

Synonyms: 2-Phenoxyethanol; phenyl "Cellosolve."

Description: Pale straw-colored liquid.

Formula: $C_6H_5OCH_2CH_2OH$.

Constants: Mol wt: 138.2, bp: $245°C$, fp: $12°C$, flash p: $250°F$, d: 1.105 at $25°/25°C$, vap. d.: 4.76.

Hazard Analysis

Toxicity: See glycols.

Fire Hazard: Slight, when exposed to heat or flame; can react with oxidizing materials.

Explosion Hazard: See ethers.

Countermeasures

Storage and Handling: Section 7.

To Fight Fire: Water, foam, carbon dioxide, dry chemical or carbon tetrachloride (Section 6).

ETHYLENE GLYCOL PHENYL ETHER ACETATE

General Information

Synonyms: 2-Phenoxyethyl acetate.

Description: Colorless, mobile liquid.

Formula: $C_6H_5OCH_2CH_2OOCCH_3$.

Constants: Mol wt: 180.2, mp: $< -20°C$, bp: $231°C$, d: 1.103 at $25°/25°C$, flash p: $275°F$, vap. d.: 6.2.

Hazard Analysis

Toxicity: Details unknown. See also glycols.

Fire Hazard: Slight, when exposed to heat or flame; can react with oxidizing materials.

Countermeasures

To Fight Fire: Water, foam, carbon dioxide, dry chemical or carbon tetrachloride (Section 6).

Storage and Handling: Section 7.

ETHYLENE IMINE

General Information

Synonyms: Ethylenimine; dimethylenimine.

Description: Oil; pungent, ammoniacal odor; water-white liquid.

Formula: $NHCH_2CH_2$.

Constants: Mol wt: 43.07, bp: $55-56°C$, fp: $-71.5°C$, flash p: $12°F$, d: 0.832 at $20°/4°C$, autoign. temp.: $612°F$, vap. press.: 160 mm at $20°C$, vap. d.: 1.48, lel: 3.6%, uel: 4.6%.

TOXIC HAZARD RATING CODE (For detailed discussion, see Section 1.)

0 NONE: (a) No harm under any conditions; (b) Harmful only under unusual conditions or overwhelming dosage.

1 SLIGHT: Causes readily reversible changes which disappear after end of exposure.

2 MODERATE: May involve both irreversible and reversible changes; not severe enough to cause death or permanent injury.

3 HIGH: May cause death or permanent injury after very short exposure to small quantities.

U UNKNOWN: No information on humans considered valid by authors.

Hazard Analysis
Toxic Hazard Rating:
Acute Local: Irritant 3; Ingestion 3; Inhalation 3.
Acute Systemic: Ingestion 3; Inhalation 3; Skin Absorption 3.
Chronic Local: Allergen 2.
Chronic Systemic: U.
TLV: ACGIH (recommended) 0.5 part per million; 1 milligram per cubic meter of air. May be absorbed via the skin.
Toxicology: Intensely irritating and can also cause allergic sensitization of skin. Readily absorbed through skin.
Caution: Causes opaque cornea, keratoconus, and necrosis of cornea (experimental). Has been known to cause severe human eye injury. Drinking of carbonated beverages is recommended as antidote to this material in stomach.
Fire Hazard: Dangerous, when exposed to heat or flame.
Disaster Hazard: Dangerous. Heat and/or the presence of catalytically active metals or chloride ions can cause a violent exothermic reaction. This material should be handled as per instructions of the manufacturer.
Countermeasures
Ventilation Control: Section 2.
To Fight Fire: Alcohol foam, carbon dioxide, dry chemical or carbon tetrachloride (Section 6).
Personnel Protection: Section 3.
Storage and Handling: Section 7.
Shipping Regulations: Section 11.
I.C.C. (inhibited): Flammable liquid, red label, 5 pints.
IATA (inhibited): Flammable liquid, red label, not acceptable (passenger), 2½ liters (cargo).
IATA (uninhibited): Flammable liquid, not acceptable (passenger and cargo).

ETHYLENE LAURATE. See ethylene dilaurate.

ETHYLENE MONOCHLOROCHLORIDE
General Information
Synonym: Monochloroethylene chloride.
Formula: $CH_2ClCHCl_2$.
Constant: Mol wt: 134.4, vap. d.: 4.8.
Hazard Analysis
Toxicity: Details unknown, but probably toxic, a fumigant.
Disaster Hazard: Dangerous; when heated to decomposition, it emits highly toxic fumes of chlorides; will react with water or steam to produce toxic and corrosive fumes.
Countermeasures
Storage and Handling: Section 7.

ETHYLENE NITRATE. See ethylene dinitrate.

ETHYLENE NITRITE. See ethylene dinitrite.

ETHYLENE OXIDE
General Information
Synonyms: 1,2-Epoxyethane; oxirane.
Description: Colorless gas at room temperature.
Formula: $(CH_2)_2O$.
Constants: Mol wt: 44.05, mp: −111.3°C, bp: 10.7°C, ulc: 100, lel: 3.0%, ucl: 100%, flash p: <0°F, d: 0.8711 at 20°/20°C, autoign. temp.: 804°F, vap. press.: 1095 mm at 20°C, vap. d.: 1.52.
Hazard Analysis
Toxic Hazard Rating:
Acute Local: Irritant 3; Inhalation 2.
Acute Systemic: Inhalation 2.
Chronic Local: Irritant 2.
Chronic Systemic: U.
Toxicology: Irritating to eyes and mucous membranes of respiratory tract. High concentrations can cause pulmonary edema.
TLV: ACGIH (recommended); 50 parts per million in air; 90 milligrams per cubic meter of air.

Fire Hazard: Dangerous, when exposed to heat or flame; can react with oxidizing materials.
Spontaneous Heating: No.
Explosion Hazard: Severe, when exposed to flame.
Disaster Hazard: Highly dangerous, upon exposure to heat or flame.
Countermeasures
Ventilation Control: Section 2.
To Fight Fire: Alcohol foam, carbon dioxide, dry chemical or carbon tetrachloride (section 6).
Personnel Protection: Section 3.
Storage and Handling: Section 7.
Shipping Regulations: Section 11.
I.C.C.: Flammable liquid, red label, 300 pounds in cylinder, 15 pounds in other containers.
Coast Guard Classification: Inflammable liquid; red label.
MCA warning label.
IATA: Flammable liquid, red label, not acceptable (passenger), 140 kilograms (cargo).

ETHYLENE OXIDE POLYMER
General Information
Description: The polymer of ethylene oxide.
Hazard Analysis
Toxic Hazard Rating: U. A food additive permitted in foods for human consumption (Section 10).

ETHYLENESUCCINIC ACID. See succinic acid.

ETHYLENE SULFIDE
General Information
Synonym: Thiirane.
Description: Colorless liquid.
Formula: $(CH_2)_2S$.
Constants: Mol wt: 60.11, mp: decomposes, bp: 55–56°C, d: 1.0368 at 0°/4°C, vap. d.: 2.07.
Hazard Analysis
Toxic Hazard Rating:
Acute Local: Irritant 2; Inhalation 3.
Acute Systemic: Ingestion 3; Inhalation 2.
Chronic Local: U.
Chronic Systemic: U.
Based on animal experiments.
Fire Hazard: Unknown.
Disaster Hazard: Dangerous; when heated to decomposition or on contact with acid or acid fumes, it emits highly toxic fumes of oxides of sulfur; can react with oxidizing materials.
Countermeasures
Personnel Protection: Section 3.
Storage and Handling: Section 7.

ETHYLENE TETRACHLORIDE. See perchloroethylene.

ETHYLENE TRICHLORIDE. See trichloroethylene.

ETHYLENE TRIFLUORIDE
General Information
Formula: CH_3CF_3.
Constant: Mol wt: 84.
Hazard Analysis
Toxic Hazard Rating: U. Animal experiments show low irritant and narcotic effects.

ETHYLENE TRITHIOCARBONATE
General Information
Description: Bright yellow to brown, crystalline solid.
Formula: $C_3H_4S_3$.
Constants: Mol wt: 136.3, mp: 34.5°C, bp: 117.5°C at 0.6 mm, flash p: 330°F (T.C.C.), d: 1.48, vap. d.: 4.70.
Hazard Analysis
Toxicity: Details unknown; avoid excessive skin contact, inhalation or ingestion of material.
Fire Hazard: Slight, when exposed to heat or flame.
Disaster Hazard: Dangerous; when heated to decomposi-

tion, it emits highly toxic fumes of oxides of sulfur; can react with oxidizing materials.

Countermeasures

Storage and Handling: Section 7.

To Fight Fire: Water, carbon dioxide, dry chemical or carbon tetrachloride (Section 6).

ETHYLENIMINE. See ethylene imine.

ETHYL-2,3-EPOXYBUTYRATE

General Information

Formula: $CH_3HCOCHCH_2OOC_2H_5$.

Constant: Mol wt: 132.

Hazard Analysis

Toxic Hazard Rating:

  Acute Local: Irritant 1.

  Acute Systemic: Ingestion 2; Inhalation 1; Skin Absorption 2.

  Chronic Local: U.

  Chronic Systemic: U.

Based on limited animal experiments.

ETHYL ESTER. See ethyl acetate.

ETHYL ETHANEDIOATE. See ethyl oxalate.

ETHYL ETHANOATE. See ethyl acetate.

ETHYL ETHANOLAMINE-1

General Information

Description: Liquid.

Formula: $C_2H_5NHC_2H_4OH$.

Constants: Mol wt: 89.14, mp: $-7.8°C$, bp: $167°C$, flash p: $160°F$ (O.C.), d: 0.9182 at $20°/20°C$, vap. press.: 0.4 mm at $20°C$, vap. d.: 3.00.

Hazard Analysis

Toxicity: Details unknown. See also amines.

Fire Hazard: Moderate, when exposed to heat or flame; can react with oxidizing materials.

Countermeasures

To Fight Fire: Foam, carbon dioxide, dry chemical or carbon tetrachloride (Section 6).

Storage and Handling: Section 7.

ETHYL ETHER

General Information

Synonyms: Sulfuric ether; anesthesia ether; ether; ethyl oxide.

Description: A clear, volatile liquid.

Formula: $C_2H_5OC_2H_5$.

Constants: Mol wt: 74.12, mp: $-116.2°C$, bp: $34.6°C$, ulc: 100, lel: 1.85%, uel: 48%, flash p: $-49°F$, d: 0.7135 at $20°/4°C$, autoign. temp.: $356°F$, vap. press.: 442 mm at $20°C$, vap. d.: 2.56.

Hazard Analysis

Toxic Hazard Rating:

  Acute Local: Ingestion 2; Inhalation 1.

  Acute Systemic: Inhalation 2; Skin Absorption 2.

  Chronic Local: U.

  Chronic Systemic: U.

TLV: ACGIH (recommended); 400 parts per million in air; 1212 milligrams per cubic meter of air.

Toxicology: Ether is not corrosive or dangerously reactive. However, it must not be considered safe for individuals to inhale or ingest. It is not toxic in the sense of being a poison. It is, however, a depressant of the central nervous system and is capable of producing intoxication, drowsiness, stupor, and unconsciousness. Death

due to respiratory failure may result from severe and continued exposure.

Fire Hazard: Dangerous, when exposed to heat or flame; can react vigorously with oxidizing materials. See ethers.

Explosion Hazard: Severe, when exposed to heat or flame.

Disaster Hazard: Highly dangerous, in the presence of heat or flame. See ethers.

Countermeasures

Ventilation Control: Section 2.

To Fight Fire: Alcohol foam, carbon dioxide, dry chemical or carbon tetrachloride (Section 6).

Treatment and Antidotes: Removal from exposure almost always produces rapid and complete recovery.

Personal Hygiene: Section 3.

Storage and Handling: Section 7.

Shipping Regulations: Section 11.

  I.C.C.: Flammable liquid; red label, 10 gallons.

  Coast Guard Classification: Inflammable liquid; red label.

  MCA warning label.

  IATA: Flammable liquid, red label, 1 liter (passenger), 40 liters (cargo).

ETHYL ETHER OF PROPYLENE GLYCOL

General Information

Description: Liquid.

Formula: $CH_2OC_2H_5CHOHCH_3$.

Constants: Mol wt: 104.2, vap. d.: 3.59.

Hazard Analysis

Toxic Hazard Rating:

  Acute Local: U.

  Acute Systemic: Ingestion 2; Inhalation 1.

  Chronic Local: U.

  Chronic Systemic: Ingestion 1; Inhalation 1.

See also glycols.

Fire Hazard: Slight; can react with oxidizing materials.

Countermeasures

Ventilation Control: Section 2.

Personal Hygiene: Section 3.

Storage and Handling: Section 7.

ETHYL-3-ETHOXYPROPIONATE

General Information

Description: Liquid.

Formula: $C_2H_5OCH_2CH_2COOC_2H_5$.

Constants: Mol wt: 146.18, mp: $-100°C$, bp: $170.1°C$, flash p: $180°F$ (O.C.), d: 0.9496 at $20°/20°C$, vap. d.: 5.03.

Hazard Analysis

Toxicity: Unknown. See esters.

Fire Hazard: Moderate, when exposed to heat or flame; can react with oxidizing materials.

Countermeasures

To Fight Fire: Foam, carbon dioxide, dry chemicals or carbon tetrachloride (Section 6).

Storage and Handling: Section 7.

ETHYLETHYLENE ACETATE

General Information

Formula: $CH_2(CO_2CH_3)CH_2CHCH_2$.

Constants: Mol wt: 114.1, vap. d.: 3.93.

Hazard Analysis

Toxic Hazard Rating:

  Acute Local: Irritant 1.

  Acute Systemic: U.

  Chronic Local: U.

  Chronic Systemic: U.

TOXIC HAZARD RATING CODE (For detailed discussion, see Section 1.)

0 NONE: (a) No harm under any conditions; (b) Harmful only under unusual conditions or overwhelming dosage.

1 SLIGHT: Causes readily reversible changes which disappear after end of exposure.

2 MODERATE: May involve both irreversible and reversible changes; not severe enough to cause death or permanent injury.

3 HIGH: May cause death or permanent injury after very short exposure to small quantities.

U UNKNOWN: No information on humans considered valid by authors.

Disaster Hazard: Slight, (Section 6).
Countermeasures
Personal Hygiene: Section 3.
Storage and Handling: Section 7.

ETHYL FLUORIDE
General Information
Synonym: Fluoroethane.
Description: Colorless gas.
Formula: CH_3CH_2F.
Constants: Mol wt: 48.06, mp: $-143.2°C$, bp: $-37.7°C$, d: 0.8158 at $-37.7°C$, vap. d.: 1.66.
Hazard Analysis
Toxic Hazard Rating:
  Acute Local: U.
  Acute Systemic: Inhalation 1.
  Chronic Local: U.
  Chronic Systemic: U.
Toxicology: Narcotic in high concentrations.
Disaster Hazard: Dangerous; when heated to decomposition, it emits toxic fumes of fluorides.
Countermeasures
Ventilation Control: Section 2.
Storage and Handling: Section 7.

ETHYL FLUOROFORMATE
General Information
Description: Liquid.
Formula: $FCOOC_2H_5$.
Constants: Mol wt: 92.07, bp: $57°C$, d: 1.11 at $33°C$, vap. d.: 3.18.
Hazard Analysis
Toxic Hazard Rating:
  Acute Local: Irritant 3; Ingestion 3; Inhalation 3.
  Acute Systemic: U.
  Chronic Local: U.
  Chronic Systemic: U.
Disaster Hazard: Dangerous, when heated to decomposition, or on contact with acid fumes, it emits toxic fumes of fluorides; will react with water or steam to produce toxic and corrosive fumes.
Countermeasures
Ventilation Control: Section 2.
Personnel Protection: Section 3.
Storage and Handling: Section 7.

ETHYL p-FLUOROPHENYL SULFONE
General Information
Synonym: Fluoresone.
Hazard Analysis
Toxic Hazard Rating: U. Has shown moderately low toxicity in experimental animals.

ETHYL FLUOROSULFONATE
General Information
Description: Liquid, ethereal odor.
Formula: $C_2H_5SO_3F$.
Constants: Mol wt: 128.1, vap. d.: 4.42.
Hazard Analysis
Toxic Hazard Rating:
  Acute Local: Irritant 3; Ingestion 3; Inhalation 3.
  Acute Systemic: U.
  Chronic Local: U.
  Chronic Systemic: U.
Disaster Hazard: Dangerous; when heated to decomposition or on contact with acid or acid fumes, it emits highly toxic fumes of fluorides and oxides of sulfur; will react with water or steam to produce toxic and corrosive fumes.
Countermeasures
Ventilation Control: Section 2.
Personnel Protection: Section 3.
Storage and Handling: Section 7.

ETHYL FORMATE
General Information
Synonyms: Formic ether; ethyl methanoate.
Description: Water-white liquid; pleasant, aromatic odor.
Formula: $HCOOC_2H_5$.
Constants: Mol wt: 74.08, mp: $-79°C$, bp: $54.3°C$, lel: 2.7%, uel: 13.5%, flash p: $-4°F$ (C.C.), d: 0.9236 at $20°/20°C$, autoign. temp.: $851°F$, vap. press.: 100 mm at $5.4°C$, vap. d.: 2.55.
Hazard Analysis
Toxic Hazard Rating:
  Acute Local: Irritant 2; Inhalation 2.
  Acute Systemic: Inhalation 2.
  Chronic Local: U.
  Chronic Systemic: U.
Toxicology: Irritating to skin and mucous membranes. Narcotic in high concentrations. A general purpose food additive (Section 10).
TLV: ACGIH (recommended); 100 parts per million in air; 303 milligrams per cubic meter of air.
Fire Hazard: Dangerous, when exposed to heat or flame; can react vigorously with oxidizing materials.
Spontaneous Heating: No.
Explosion Hazard: Severe, when exposed to flame.
Disaster Hazard: Highly dangerous; upon exposure to heat or flame.
Countermeasures
Ventilation Control: Section 2.
To Fight Fire: Alcohol foam, foam, carbon dioxide, dry chemical or carbon tetrachloride (Section 6).
Personnel Protection: Section 3.
Storage and Handling: Section 7.
Shipping Regulations: Section 11.
  I.C.C.: Flammable liquid, red label, 10 gallons.
  Coast Guard Classification: Inflammable liquid; red label.
  MCA warning label.
  IATA: Flammable liquid, red label, 1 liter (passenger), 40 liters (cargo).

ETHYL FORMATE, ANHYDROUS. See ethyl
formate.

ETHYL FORMYL PROPIONATE
General Information
Description: Liquid; somewhat soluble in water.
Formula: $C_2H_5OOCC_2H_4C(O)H$.
Constants: Mol wt: 130, d: 1.0625 at $20/20°C$, flash p: $200°F$.
Hazard Analysis
Toxic Hazard Rating: U. Limited animal experiments suggest low toxicity. See also esters and ethyl formate.
Fire Hazard: Moderate, when exposed to heat or flame.

ETHYL GERMANIUM OXIDE
General Information
Description: Colorless crystals, water soluble.
Formula: $(C_2H_5GeO)_2O$.
Constants: Mol wt: 251.3, mp: $> 300°C$.
Hazard Analysis
Toxicity: Details unknown. See germanium compounds.

ETHYLGERMANIUM TRIBROMIDE
General Information
Description: Colorless liquid decomposed by water.
Formula: $(C_2H_5)GeBr_3$.
Constants: Mol wt: 341.4, mp: $< -33°C$, bp: $200°C$ at 763 mm.
Hazard Analysis and Countermeasures
Dangerous. See bromides and germanium compounds.

ETHYLGERMANIUM TRICHLORIDE
General Information
Description: Colorless liquid, decomposed by water.
Formula: $(C_2H_5)GeCl_3$.

Constants: Mol wt: 208.0, mp: < −33°C, bp: 144°C at 762 mm.
Hazard Analysis and Countermeasures
Dangerous. See chlorides and germanium compounds.

ETHYLGERMANIUM TRIFLUORIDE
General Information
Description: Colorless liquid decomposed by water.
Formula: $(C_2H_5)GeF_3$.
Constants: Mol wt: 158.7, mp: −16°C, bp: 112°C at 750 mm.
Hazard Analysis and Countermeasures
Highly toxic. See fluorides and germanium compounds.

ETHYLGERMANIUM TRIIODIDE
General Information
Description: Yellow liquid decomposed by water.
Formula: $(C_2H_5)GeI_3$.
Constants: Mol wt: 482.2, mp: −2°C, bp: 281°C at 755 mm.
Hazard Analysis and Countermeasures
Dangerous. See iodides and germanium compounds.

ETHYL GLYCOL. See "Cellosolve."

ETHYL GLYCOL ACETATE. See "Cellosolve" acetate.

ETHYLHEPTANOIC ACID
General Information
Synonym: Pelargonic acid; nonanoic acid.
Description: Oily colorless liquid. Very slightly soluble in water.
Formula: $CH_3(CH_2)_7COOH$.
Constants: Mol wt: 158.2, bp: 254°C, mp: 12°C, d: 0.9055 at 20°/4°C.
Hazard Analysis
Toxicity: Details unknown. Limited data suggests low systemic toxicity but high irritation of skin.

2-ETHYLHEXALDEHYDE
General Information
Synonym: Butyl ethyl acetaldehyde; 2-ethylhexanal.
Description: Liquid.
Formula: $C_4H_9CH(C_2H_5)CHO$.
Constants: Mol wt: 128.21, bp: 163.4°C, flash p: 125°F (O.C.), d: 0.8205, vap. press.: 1.8 mm at 20°C, vap. d.: 4.42.
Hazard Analysis
Toxicity: Details unknown. See also aldehydes.
Fire Hazard: Moderate, when exposed to heat or flame; can react with oxidizing materials.
Countermeasures
To Fight Fire: Foam, carbon dioxide, dry chemical or carbon tetrachloride (Section 6).
Storage and Handling: Section 7.

2-ETHYLHEXANAL. See 2-ethylhexaldehyde.

2,2-(2-ETHYLHEXANAMIDO) DIETHYL DI-(2-ETHYLHEXANOATE)
General Information
Description: Light brown liquid.
Formula: $(C_7H_{15}OCOC_2H_4)_2NCOC_7H_{15}$.
Constants: Mol wt: 483.71, mp: −33°C, bp: 255°C at 5 mm, flash p: 420°F (O.C.), d: 0.9564 at 20°/20°C, vap. press.: 0.31 mm at 200°C, vap. d.: 16.7.
Hazard Analysis
Toxicity: Unknown.
Fire Hazard: Slight; can react with oxidizing materials.

Countermeasures
To Fight Fire: Foam, carbon dioxide, dry chemical or carbon tetrachloride (Section 6).
Storage and Handling: Section 7.

2-ETHYL-1,3-HEXANDIO BIS(9,10-EPOXYSTEARATE)
Hazard Analysis
Toxic Hazard Rating:
Acute Local: U.
Acute Systemic: Ingestion 1.
Chronic Local: U.
Chronic Systemic: U.
Based on limited animal experiments.

2-ETHYLHEXANE-DIOL-1,3
General Information
Description: Practically colorless, somewhat viscous, odorless liquid.
Formula: $C_3H_7CH(OH)CH(C_2H_5)CH_2OH$.
Constants: Mol wt: 146.22, bp: 243.1°C, flash p: 260°F (O.C.), fp: −40°C, d: 0.9422 at 20°/20°C, vap. press.: < 0.01 mm at 20°C, vap. d.: 5.03.
Hazard Analysis
Toxic Hazard Rating:
Acute Local: Irritant 1; Ingestion 1; Inhalation 1.
Acute Systemic: U.
Chronic Local: U.
Chronic Systemic: U
Fire Hazard: Slight, when exposed to heat or flame; can react with oxidizing materials.
Countermeasures
To Fight Fire: Alcohol foam, foam, carbon dioxide, dry chemical or carbon tetrachloride (Section 6).
Ventilation Control: Section 2.
Personal Hygiene: Section 3.
Storage and Handling: Section 7.

2-ETHYLHEXANOIC ACID. See 2-ethylhexoic acid.

2-ETHYL HEXANOIC ACID, 2-ETHYL HEXYL ESTER
General Information
Formula: $C_4H_9CH(C_2H_5)COO(CH_2)_3CH(C_2H_5)CH_3$.
Constant: Mol wt: 242.
Hazard Analysis
Toxic Hazard Rating: U. Limited animal experiments suggest low toxicity. See also esters.

2-ETHYLHEXANOIC ANHYDRIDE
General Information
Synonym: Octoic anhydride.
Formula: $[CH_3(CH_2)_3CH(C_2H_5)CO]_2O$.
Constants: Mol wt: 270.39, vap. d.: 7.82.
Hazard Analysis
Toxic Hazard Rating:
Acute Local: Irritant 1.
Acute Systemic: U.
Chronic Local: U.
Chronic Systemic: U.
Fire Hazard: Slight; can react with oxidizing materials.
Countermeasures
Personal Hygiene: Section 3.
Storage and Handling: Section 7.

2-ETHYLHEXANOL. See 2-ethylhexyl alcohol.

ETHYLHEXANOYL CHLORIDE
General Information
Synonym: Caprylyl chloride, octanoyl chloride.

TOXIC HAZARD RATING CODE (For detailed discussion, see Section 1.)

0 NONE: (a) No harm under any conditions; (b) Harmful only under unusual conditions or overwhelming dosage.

1 SLIGHT: Causes readily reversible changes which disappear after end of exposure.

2 MODERATE: May involve both irreversible and reversible changes; not severe enough to cause death or permanent injury.

3 HIGH: May cause death or permanent injury after very short exposure to small quantities.

U UNKNOWN: No information on humans considered valid by authors.

Description: Liquid, clear water-white to straw-colored, decomposes in water.
Formula: $C_7H_{15}COCl$.
Constants: Mol wt: 162.66, d: 0.9671 at $0°/4°C$, mp: $-6°C$, flash p: $180°F$ (C.C.), vap. d.: 5.63, bp: $195.55°C$.

Hazard Analysis
Toxicity: An irritant material and a source of hydrochloric acid. See hydrochloric acid.
Fire Hazard: Moderate when exposed to heat or flame.
Disaster Hazard: Dangerous; when heated to decomposition, it evolves toxic chlorides.

Countermeasures
Personal Hygiene: Section 3.
Storage and Handling: Section 7.
To Fight Fire: Foam, carbon dioxide, dry chemical or carbon tetrachloride (Section 6).

2-ETHYL-1-HEXENE
General Information
Description: Colorless liquid.
Formula: C_8H_{16}.
Constants: Mol wt: 112.21, bp: $120°C$, d: 0.7270 at $20°/20°C$, vap. d.: 3.87.

Hazard Analysis
Toxic Hazard Rating:
 Acute Local: U.
 Acute Systemic: Ingestion 2; Inhalation 2.
 Chronic Local: U.
 Chronic Systemic: U.
Fire Hazard: Slight; can react with oxidizing materials.

Countermeasures
Ventilation Control: Section 2.
Personal Hygiene: Section 3.
Storage and Handling: Section 7.

2-ETHYLHEXOIC ACID
General Information
Synonyms: 3-Heptanecarboxylic acid; butylethylacetic acid.
Description: Mild-odored, colorless liquid.
Formula: $C_4H_9CH(C_2H_5)COOH$.
Constants: Mol wt: 144.21, bp: $227.6°C$, fp: $-118.4°C$, flash p: $260°F$ (C.O.C.), d: 0.903 at $20°/4°C$, vap. press.: $< .01$ mm at $20°C$, vap. d.: 4.98.

Hazard Analysis
Toxic Hazard Rating:
 Acute Local: Irritant 1.
 Acute Systemic: U.
 Chronic Local: U.
 Chronic Systemic: U.
Fire Hazard: Slight, when exposed to heat or flame; can react with oxidizing materials.

Countermeasures
To Fight Fire: Foam, carbon dioxide, dry chemical or carbon tetrachloride (Section 6).
Personal Hygiene: Section 3.
Storage and Handling: Section 7.

2-ETHYLHEXYL ACETATE
General Information
Synonyms: Octyl acetate; β-ethylhexyl acetate.
Description: Water-white, stable liquid.
Formula: $CH_3COOCH_2CHC_2H_5C_4H_9$.
Constants: Mol wt: 172.26, mp: $-93°C$, bp: $199.3°C$, flash p: $190°F$ (O.C.), d: 0.872 at $20°/4°C$, vap. press.: 0.4 mm at $20°C$, vap. d.: 5.93.

Hazard Analysis
Toxic Hazard Rating:
 Acute Local: Irritant 1.
 Acute Systemic: Ingestion 1; Inhalation 1.
 Chronic Local: U.
 Chronic Systemic: U.
Toxicity: Slight, based on animal experiments. See octyl alcohol and acetic acid.
Fire Hazard: Moderate, when exposed to heat or flame; can react with oxidizing materials.

Countermeasures
To Fight Fire: Foam, carbon dioxide, dry chemical or carbon tetrachloride (Section 6).
Personal Hygiene: Section 3.
Storage and Handling: Section 7.

2-ETHYLHEXYL ACRYLATE
General Information
Synonym: Octyl acrylate.
Description: Liquid.
Formula: $CH_2CHCO_2CH_2CH(C_2H_5)C_4H_9$.
Constants: Mol wt: 184.3, bp: $130°C$ at 50 mm, fp: $-90°C$, flash p: $180°F$ (O.C.), d: 0.8869 at $20°/20°C$, vap. press.: 1 mm at $50.0°C$, vap. d.: 6.35.

Hazard Analysis
Toxicity: Slight. Animal experiments show lower toxicity than for ethyl acrylate. See acrylic acid.
Fire Hazard: Moderate, when exposed to heat or flame; can react with oxidizing materials.

Countermeasures
To Fight Fire: Foam, carbon dioxide, dry chemical or carbon tetrachloride (Section 6).
Personal Hygiene: Section 3.
Storage and Handling: Section 7.

2-ETHYLHEXYL ALCOHOL
General Information
Synonym: 2-Ethyl-1-hexanol; octyl alcohol.
Description: Liquid.
Formula: $C_4H_9CH(C_2H_5)CH_2OH$.
Constants: Mol wt: 130.23, bp: $179-185.5°C$, mp: $< -76°C$, flash p: $178°F$, d: 0.834 at $20°/20°C$, vap. press.: 0.2 mm at $20°C$, vap. d.: 4.49.

Hazard Analysis
Toxicity: Slight. See alcohols.
Fire Hazard: Moderate, when exposed to heat or flame; can react with oxidizing materials.

Countermeasures
To Fight Fire: Foam, carbon dioxide, dry chemical or carbon tetrachloride (Section 6).
Storage and Handling: Section 7.

2-ETHYLHEXYLAMINE
General Information
Description: A clear, miscible liquid.
Formula: $C_4H_9CH(C_2H_5)CH_2NH_2$.
Constants: Mol wt: 129.24, bp: $169.2°C$, flash p: $140°F$ (O.C.), d: 0.7894 at $20°/20°C$, vap. press.: 1.2 mm at $20°C$, vap. d.: 4.45.

Hazard Analysis
Toxic Hazard Rating:
 Acute Local: U.
 Acute Systemic: Ingestion 2; Inhalation 2.
 Chronic Local: U.
 Chronic Systemic: U.
Fire Hazard: Moderate, when exposed to heat or flame; can react with oxidizing materials.

Countermeasures
To Fight Fire: Carbon dioxide, dry chemical or carbon tetrachloride (Section 6).
Ventilation Control: Section 2.
Personal Hygiene: Section 3.
Storage and Handling: Section 7.

2-ETHYLHEXYL-3-AMINOPROPYL ETHER
General Information
Description: Liquid.
Formula: $C_4H_9CH(C_2H_5)CH_2OC_3H_6NH_2$.
Constants: Mol wt: 187.32, bp: $239°C$, mp: $-90°C$, flash p: $210°F$ (O.C.), d: 0.8483, vap. d.: 6.47.

Hazard Analysis
Toxicity: Details unknown. See also ethers.
Fire Hazard: Moderate, when exposed to heat or flame; can react with oxidizing materials.

Countermeasures
To Fight Fire: Foam, carbon dioxide, dry chemical or carbon tetrachloride (Section 6).
Storage and Handling: Section 7.

N-2-ETHYLHEXYL ANILINE
General Information
Description: Liquid, mild odor.
Formula: $C_6H_5NHCH_2CH(C_2H_5)(C_4H_9)$.
Constants: Mol wt: 205.3, bp: 194°C at 50 mm, fp: < −70°C, flash p: 325°F (C.O.C.), d: 0.9119 at 20°/20°C, vap. press.: < 0.01 at 20°C, vap. d.: 6.9.
Hazard Analysis
Toxic Hazard Rating:
  Acute Local: Irritant 2; Ingestion 2; Inhalation 2.
  Acute Systemic: Ingestion 2; Inhalation 2; Skin Absorption 1.
  Chronic Local: U.
  Chronic Systemic: Ingestion 3; Inhalation 3; Skin Absorption 3.
Fire Hazard: Slight, when exposed to heat or flame.
Disaster Hazard: Dangerous; when heated to decomposition, it emits highly toxic fumes; it will react with water or steam to produce toxic fumes; can react with oxidizing materials.
Countermeasures
Ventilation Control: Section 2.
Personnel Protection: Section 3.
Storage and Handling: Section 7.
To Fight Fire: Water, foam.

ETHYL HEXYL AZELATE
Hazard Analysis
Toxic Hazard Rating: U. Limited animal experiments suggest low toxicity. See also esters.

2-ETHYLHEXYL "CELLOSOLVE"
General Information
Description: Water-white liquid.
Formula: $C_4H_9CH(C_2H_5)CH_2OCH_2CH_2OH$.
Constants: Mol wt: 174.28, bp: 228.3°C, flash p: 230°F, d: 0.8859, vap. press.: 0.02 mm at 20°C, vap. d.: 6.02.
Hazard Analysis
Toxicity: Details unknown. See also glycols.
Fire Hazard: Slight, when exposed to heat or flame; can react with oxidizing materials.
Countermeasures
To Fight Fire: Foam, carbon dioxide, dry chemical or carbon tetrachloride (Section 6).
Storage and Handling: Section 7.

2-ETHYLHEXYL-1-CHLORIDE
General Information
Description: Colorless liquid.
Formula: $C_4H_9CH(C_2H_5)CH_2Cl$.
Constants: Mol wt: 148.67, bp: 172.9°C, fp: −135°C, flash p: 140°F (O.C.), d: 0.8833 at 20°C, vap. d.: 5.14.
Hazard Analysis
Toxic Hazard Rating:
  Acute Local: U.
  Acute Systemic: Ingestion 2; Innalation 2.
  Chronic Local: U.
  Chronic Systemic: U.
Fire Hazard: Moderate, when exposed to heat or flame.
Disaster Hazard: Dangerous; when heated to decomposition, it emits highly toxic fumes of phosgene; can react with oxidizing materials.

Countermeasures
Ventilation Control: Section 2.
To Fight Fire: Foam, carbon dioxide, dry chemical or carbon tetrachloride (Section 6).
Personal Hygiene: Section 3.
Storage and Handling: Section 7.

N-(2-ETHYL HEXYL) CYCLOHEXYLAMINE
General Information
Description: Insoluble in water.
Formula: $C_4H_9CH(C_2H_5)CH_2C_6H_{11}NH_2$.
Constants: Mol wt: 211, d: 0.8, bp: 342°F at 50 mm, flash p: 265°F.
Hazard Analysis
Toxic Hazard Rating: U. Limited animal experiments suggest high toxicity and moderate irritation. See also amines.
Fire Hazard: Slight, when exposed to heat or flame.
Countermeasures
To Fight Fire: Water, foam.

2-ETHYL HEXYL 9,10-EPOXYSTEARATE
Hazard Analysis
Toxic Hazard Rating:
  Acute Local: Irritant 1.
  Acute Systemic: Ingestion 1; Inhalation 1; Skin Absorption 1.
  Chronic Local: Allergen 1.
  Chronic Systemic: U.
Based on limited animal studies.

2-ETHYLHEXYL ETHER
General Information
Description: Liquid.
Formula: $[C_4H_9CH(C_2H_5)CH_2]_2CO$.
Constants: Mol wt: 242.43, mp: −95°C, bp: 269.4°C, flash p: 235°F (O.C.), d: 0.8121 at 20°/20°C, vap. d.: 8.36.
Hazard Analysis
Toxicity: Details unknown. See also ethers.
Fire Hazard: Slight, when exposed to heat or flame; can react with oxidizing materials.
Explosion Hazard: Slight, when exposed to heat, spark or flame.
Countermeasures
Storage and Handling: Section 7.
To Fight Fire: Foam, carbon dioxide, dry chemical or carbon tetrachloride (Section 6).

2-ETHYLHEXYL-2-ETHYLHEXANOATE
General Information
Description: Liquid.
Formula: $C_4H_9CH(C_2H_5)CH_2CO_2(C_2H_5)CHC_4H_9$.
Constants: Mol wt: 256.4, vap. d.: 8.85.
Hazard Analysis
Toxic Hazard Rating:
  Acute Local: Irritant 1.
  Acute Systemic: U.
  Chronic Local: U.
  Chronic Systemic: U.
Fire Hazard: Slight; can react with oxidizing materials.
Countermeasures
Personal Hygiene: Section 3.
Storage and Handling: Section 7.

2-ETHYLHEXYL MERCAPTAN
General Information
Description: Colorless liquid.

TOXIC HAZARD RATING CODE (For detailed discussion, see Section 1.)

0 NONE: (a) No harm under any conditions; (b) Harmful only under unusual conditions or overwhelming dosage.

1 SLIGHT: Causes readily reversible changes which disappear after end of exposure.

2 MODERATE: May involve both irreversible and reversible changes not severe enough to cause death or permanent injury.

3 HIGH: May cause death or permanent injury after very short exposure to small quantities.

U UNKNOWN: No information on humans considered valid by authors

Formula: $CH_3CH(C_2H_5)(CH_2)_4SH$.
Constants: Mol wt: 146.3, bp: 90°C at 35 mm, d: 0.8543 at 20°/20°C, vap. d.: 5.04.
Hazard Analysis
Toxicity: Details unknown. See also mercaptans.
Disaster Hazard: Dangerous; when heated to decomposition, it emits highly toxic fumes of oxides of sulfur; can react with oxidizing materials.
Countermeasures
Storage and Handling: Section 7.

2-ETHYLHEXYL TITANATE
General Information
Synonym: Octyl titanate.
Description: Liquid.
Formula: $Ti[OCH_2CH(C_2H_5)(C_4H_9)]_4$.
Constants: Mol wt: 565, mp: −55°C, bp: 254°C, d: 0.93.
Hazard Analysis
Toxicity: See titanium compounds.
Fire Hazard: Moderate, when exposed to heat or flame; can react with oxidizing materials.
Countermeasures
Storage and Handling: Section 7.

ETHYL HYDRIDE. See ethane.

ETHYL HYDROGEN SULFATE. See ethylsulfuric acid.

ETHYL HYDROSULFIDE. See ethyl mercaptan.

ETHYL p-HYDROXYBENZOATE
General Information
Formula: $C_6H_4(OH)COOC_2H_5$.
Constant: Mol wt: 156.
Hazard Analysis
Toxic Hazard Rating:
  Acute Local: Irritant 1; Allergen 1.
  Acute Systemic: U.
  Chronic Local: Allergen 2.
  Chronic Systemic: U.

ETHYL-2-HYDROXY PROPANOATE. See ethyl lactate.

ETHYLIDENE ACETONE. See 1-pentene-2-one.

ETHYLIDENE CHLORIDE
General Information
Synonyms: Ethylidene dichloride; chlorinated hydrochloric ether; 1,1-dichlorethane.
Description: Colorless liquid; aromatic, ethereal odor, hot saccharin taste.
Formula: CH_3CHCl_2.
Constants: Mol wt: 99.0, mp: −96.7°C, lel: 5.6%, uel: 11.4%, bp: 57.3°C, flash p: 22°F (C.C.), d: 1.174 at 20°/4°C, vap. press.: 230 mm at 25°C, vap. d.: 3.44, autoign. temp.: 856°F.
Hazard Analysis
Toxic Hazard Rating:
  Acute Local: Irritant 1; Inhalation 2.
  Acute Systemic: Ingestion 2; Inhalation 2.
  Chronic Local: U.
  Chronic Systemic: Inhalation 2.
TLV: ACGIH (recommended); 100 parts per million in air; 405 milligrams per cubic meter.
Toxicology: Limited data available. Liver injury has been reported in experimental animals.
Fire Hazard: Dangerous, when exposed to heat or flame.
Explosion Hazard: Moderate, when exposed to heat or flame.
Disaster Hazard: Dangerous; when heated to decomposition, it emits highly toxic fumes of phosgene; can react vigorously with oxidizing materials.
Countermeasures
Ventilation Control: Section 2.
To Fight Fire: Alcohol foam, water, foam, carbon dioxide, dry chemical or carbon tetrachloride (Section 6).

Personal Hygiene: Section 3.
First Aid: Section 1.
Storage and Handling: Section 7.

ETHYLIDENE CYCLOHEXANE
General Information
Formula: $C_6H_{11}CH_3CH_2$.
Constants: Mol wt: 112.2, vap. d.: 4.
Hazard Analysis
Toxic Hazard Rating: U. Animal experiments show high narcotic action on acute inhalation of vapors and injury to livers and kidneys on chronic exposure.

ETHYLIDENE DICHLORIDE. See ethylidene chloride.

ETHYLIDENE DIETHYL ETHER. See acetal.

ETHYLIDENE DIFLUORIDE. See 1,1-difluoroethane.

ETHYLIDENE DIIODIDE
General Information
Synonym: 1,1-Diiodoethane.
Description: Liquid.
Formula: CH_3CHI_2.
Constants: Mol wt: 281.9, bp: 179°C, d: 2.84 at 0°/4°C, vap. d.: 9.74.
Hazard Analysis
Toxicity: Details unknown. Probably irritant and narcotic in high concentrations. See also iodides.
Disaster Hazard: Dangerous; when heated to decomposition or on contact with acid or acid fumes, it emits highly toxic fumes of iodides; will react with water or steam to produce toxic and corrosive fumes.
Countermeasures
Storage and Handling: Section 7.

ETHYLIDENE DIMETHYL ETHER. See dimethyl acetal.

ETHYLIDENE LACTIC ACID. See lactic acid.

ETHYL IODIDE
General Information
Synonyms: Hydriodic ether; iodoethane.
Description: Clear, colorless liquid; turns brown on exposure to light.
Formula: C_2H_5I.
Constants: Mol wt: 156.0, mp: −105°C, bp: 72.4°C, d: 1.90–1.93 at 25°/25°C, vap. press.: 100 mm at 18.0°C, vap. d.: 5.38.
Hazard Analysis
Toxic Hazard Rating:
  Acute Local: U.
  Acute Systemic: Inhalation 2.
  Chronic Local: Irritant 1.
  Chronic Systemic: Inhalation 2; Skin Absorption 2.
Toxicology: Narcotic in high concentrations.
Fire Hazard: Moderate, when exposed to heat or flame.
Disaster Hazard: Dangerous; when heated to decomposition, it emits highly toxic fumes of iodides; will react with water or steam to produce toxic and corrosive fumes; can react vigorously with oxidizing materials.
Countermeasures
Ventilation Control: Section 2.
To Fight Fire: Water, carbon dioxide, dry chemical or carbon tetrachloride (Section 6).
Storage and Handling: Section 7.

ETHYL IODOACETATE
General Information
Description: Dense, colorless liquid.
Formula: $CH_2ICOOC_2H_5$.
Constants: Mol wt: 214.01, bp: 179°C, d: 1.80, vap. press.: 0.54 mm at 20°C, vap. d.: 7.4.
Hazard Analysis
Toxic Hazard Rating:
  Acute Local: Irritant 3; Ingestion 3; Inhalation 3.

Acute Systemic: U.

Chronic Local: U.

Chronic Systemic: U.

Disaster Hazard: Dangerous; when heated to decomposition or on contact with acid or acid fumes, it emits highly toxic fumes of iodides; will react with water or steam to produce toxic and corrosive fumes.

Countermeasures

Ventilation Control: Section 2.

Personnel Protection: Section 3.

Storage and Handling: Section 7.

5-ETHYL-5-ISOAMYLBARBITURIC ACID. See amobarbital.

ETHYL ISOBUTYRATE

General Information

Synonyms: Isobutyric ether; ethyl-2-methyl-propanoate.

Description: Colorless, volatile liquid.

Formula: $(CH_3)_2CHCOOC_2H_5$.

Constants: Mol wt: 116.16, mp: $-88°C$, bp: $110-111°C$, d: 0.870, vap. press.: 40 mm at $33.8°C$, vap. d.: 4.01.

Hazard Analysis

Toxicity: See isobutyric acid.

Fire Hazard: Moderate, when exposed to heat or flame; can react vigorously with oxidizing materials.

Countermeasures

To Fight Fire: Foam, carbon dioxide, dry chemical or carbon tetrachloride (Section 6).

Storage and Handling: Section 7.

ETHYL ISOCYANATE

General Information

Formula: C_2H_5NCO.

Constants: Mol wt: 71.1, bp: $60°C$, d: 0.90 at $20°/4°C$, vap. d.: 2.45.

Hazard Analysis and Countermeasures

See isocyanates.

Shipping Regulations: Section 11.

IATA: Poison C, poison label, not acceptable (passenger), 40 liters (cargo).

ETHYL ISOCYANIDE

General Information

Synonym: Ethyl isonitrile.

Description: Colorless liquid, soluble in water and organic solvents.

Formula: C_2H_5NC.

Constants: Mol wt: 55.1, d: 0.7402 at $20/4°C$, mp: $< -66°C$, bp: $79°C$.

Hazard Analysis and Countermeasures

See cyanides.

ETHYL ISONITRILE. See ethyl isocyanide.

ETHYL ISOTHIOCYANATE

General Information

Synonym: Ethyl thiocarbimide.

Description: Colorless liquid; pungent odor.

Formula: C_2H_5NCS.

Constants: Mol wt: 87.2, mp: $-5.9°C$, bp: $131°C$, d: 1.004 at $15°/4°C$, vap. press.: 10 mm at $22.8°C$, vap. d.: 3.01.

Hazard Analysis

Toxic Hazard Rating:

Acute Local: Irritant 3; Ingestion 3; Inhalation 3.

Acute Systemic: U.

Chronic Local: U.

Chronic Systemic: U.

Toxicology: Intensely irritating. A military poison gas.

Disaster Hazard: Dangerous; when heated to decomposition or on contact with acid or acid fumes, it emits highly toxic fumes of oxides of sulfur and cyanides; can react with oxidizing materials.

Countermeasures

Ventilation Control: Section 2.

Personnel Protection: Section 3.

Storage and Handling: Section 7.

ETHYL LACTATE

General Information

Synonym: Ethyl-2-hydroxypropanoate.

Description: Colorless liquid; mild odor.

Formula: $CH_3CHOHCOOC_2H_5$.

Constants: Mol wt: 118.13, bp: $154°C$, ulc: 30–35, lel: 1.55% at $212°F$, flash p: $115°F$ (C.C.), flash p (technical): $131°F$, d: 1.020–1.036 at $20°/20°C$, autoign. temp.: $752°F$, vap. d.: 4.07.

Hazard Analysis

Toxicity: See esters.

Fire Hazard: Moderate, when exposed to heat or flame; can react with oxidizing materials.

Spontaneous Heating: No.

Explosion Hazard: Slight, when exposed to flame.

Countermeasures

Storage and Handling: Section 7.

To Fight Fire: Foam, carbon dioxide, dry chemical or carbon tetrachloride (Section 6).

Shipping Regulations: Section 11.

Coast Guard Classification: Combustible liquid.

ETHYL LAURATE

General Information

Synonym: Ethyl dodecanoate.

Description: Oily liquid, insoluble in water.

Formula: $CH_3(CH_2)_{10}COOC_2H_5$.

Constants: Mol wt: 228.4, d: 0.8615 at $20/4°C$, mp: $-10.7°C$, bp: $269°C$.

Hazard Analysis

Toxicity: Very low, based upon animal experiments.

ETHYL LEVULINATE

General Information

Description: Colorless liquid.

Formula: $CH_3CO(CH_2)_2CO_2C_2H_5$.

Constants: Mol wt: 144.2, bp: $206.2°C$, d: 1.012, vap. press.: 1 mm at $47.3°C$, vap. d.: 4.97.

Hazard Analysis

Toxicity: Unknown. See esters.

Fire Hazard: Slight; can react with oxidizing materials.

Countermeasures

Storage and Handling: Section 7.

ETHYLLITHIUM. See lithium ethyl.

ETHYLMAGNESIUM BROMIDE

General Information

Description: Solid.

Formula: C_2H_5MgBr.

Constant: Mol wt: 133.3.

Hazard Analysis

Toxicity: See bromides and magnesium compounds.

Fire Hazard: Moderate; it can evolve flammable ethane (Section 6).

Disaster Hazard: Moderately dangerous; when heated to decomposition, it emits toxic fumes of bromides; will react with water or steam to produce flammable vapors.

TOXIC HAZARD RATING CODE (*For detailed discussion, see Section 1.*)

0 NONE: (a) No harm under any conditions; (b) Harmful only under unusual conditions or overwhelming dosage.

1 SLIGHT: Causes readily reversible changes which disappear after end of exposure.

2 MODERATE: May involve both irreversible and reversible changes; not severe enough to cause death or permanent injury.

3 HIGH: May cause death or permanent injury after very short exposure to small quantities.

U UNKNOWN: No information on humans considered valid by authors.

Countermeasures
Storage and Handling: Section 7.

ETHYLMAGNESIUM CHLORIDE
General Information
Description: Solid.
Formula: C_2H_5MgCl.
Constant: Mol wt: 88.8.
Hazard Analysis
Toxicity: See magnesium compounds and chlorides.
Fire Hazard: Moderate; it can react with moisture to evolve ethane (Section 6).
Disaster Hazard: Dangerous; when heated to decomposition, it emits highly toxic fumes of chlorides; will react with water or steam to produce flammable vapors.
Countermeasures
Storage and Handling: Section 7.

N-ETHYL MALEIMIDE
Hazard Analysis
Toxic Hazard Rating: U.
Disaster Hazard: Dangerous, on heating liberates highly irritating vapors.

ETHYL MALONATE. See diethyl malonate.

ETHYL MANDELATE
General Information
Description: Crystals.
Formula: $C_6H_5CH(OH)CO_2CH_2CH_3$.
Constant: Mol wt: 180.2.
Hazard Analysis
Toxicity: Details unknown. See esters.
Fire Hazard: Slight; can react with oxidizing materials.
Countermeasures
Personal Hygiene: Section 3.
Storage and Handling: Section 7.

ETHYL MERCAPTAN
General Information
Synonyms: Ethanethiol; ethyl hydrosulfide; ethyl thio-alcohol; ethyl sulfhydrate.
Description: Colorless liquid; penetrating garlic-like odor.
Formula: C_2H_5SH.
Constants: Mol wt: 62.13, mp: $-121°C$, bp: $36.2°C$, lel: 2.8%, uel: 18.2%, flash p: $< 80°F$ (C.C.), d: 0.83907 at $20°/4°C$, autoign. temp.: $570°F$, vap. d.: 2.14.
Hazard Analysis
Toxic Hazard Rating:
  Acute Local: Irritant 2; Ingestion 2; Inhalation 2.
  Acute Systemic: Ingestion 2; Inhalation 2.
  Chronic Local: U.
  Chronic Systemic: U.
TLV: ACGIH (recommended); 10 parts per million of air; 25 milligrams per cubic meter of air.
Fire Hazard: Dangerous, when exposed to heat or flame (Section 6).
Explosion Hazard: Moderate, when exposed to spark or flame.
Disaster Hazard: Dangerous; when heated to decomposition or on contact with acid or acid fumes, it emits highly toxic fumes of oxides of sulfur; will react with water or steam to produce toxic and flammable vapors; can react vigorously with oxidizing materials.
Countermeasures
Ventilation Control: Section 2.
To Fight Fire: Carbon dioxide, dry chemical or carbon tetrachloride (Section 6).
Personnel Protection: Section 3.
Storage and Handling: Section 7.
Shipping Regulations: Section 11.
  Coast Guard Classification: Inflammable liquid; red label.
  I.C.C.: Flammable liquid, red label, 10 gallons.
  IATA: Flammable liquid, red label, 1 liter (passenger), 40 liters (cargo).

ETHYL MERCURIC CHLORIDE. See ethylmercury chloride.

ETHYLMERCURY CHLORIDE
General Information
Synonym: Ethyl mercuric chloride; lignasan.
Description: Silvery, irridescent leaflets.
Formula: C_2H_5HgCl.
Constants: Mol wt: 265.13, mp: $192.5°C$.
Hazard Analysis and Countermeasures
A fungicide. See mercury compounds, organic.

ETHYLMERCURY CRESOL
General Information
Formula: $C_2H_5HgC_6H_4CH_3$.
Constants: Mol wt: 320.8, vap. d.: 11.05.
Hazard Analysis and Countermeasures
See mercury compounds, organic.

ETHYLMERCURY HYDROXIDE
General Information
Description: Silvery crystals, insoluble in water.
Formula: C_2H_5HgOH.
Constants: Mol wt: 246.7, mp: $37°C$.
Hazard Analysis and Countermeasures
Highly toxic. See mercury compounds, organic.

ETHYLMERCURY IODIDE
General Information
Description: Crystals; soluble in alcohol.
Formula: C_2H_5HgI.
Constants: Mol wt: 356.6, mp: $186°C$.
Hazard Analysis and Countermeasures
Highly toxic. See mercury compounds, organic.

ETHYLMERCURY ISOTHIOCARBAMIDE
General Information
Description: Crystals.
Formula: $C_2H_5HgNHCSNH_2$.
Constants: Mol wt: 304.8, vap. d.: 10.5.
Hazard Analysis and Countermeasures
See mercury compounds, organic.

ETHYLMERCURY NITROPHENOL
General Information
Description: Crystals.
Formula: $C_2H_5HgC_6H_4NO_2$.
Constants: Mol wt: 351.8, vap. d.: 12.1.
Hazard Analysis and Countermeasures
See mercury compounds, organic.

ETHYLMERCURY PHOSPHATE
General Information
Description: Solid.
Formula: $C_2H_5HgH_2PO_4$.
Constant: Mol wt: 326.7.
Hazard Analysis and Countermeasures
A fungicide. See mercury compounds, organic and phosphates.

ETHYLMERCURY SULFATE
General Information
Description: Crystals.
Formula: $(C_2H_5Hg)_2SO_4$.
Constant: Mol wt: 555.4.
Hazard Analysis and Countermeasures
See mercury compounds, organic and sulfates.

ETHYLMERCURY-p-TOLUENE SULFONANILIDE
General Information
Synonym: Ceresan M.
Description: Crystals, pungent, garlic-like odor. Water insoluble.
Formula: $C_{15}H_{17}HgNO_2S$.
Constant: Mol wt: 476.0.
Hazard Analysis and Countermeasures
See mercury compounds, organic and sulfonates.

ETHYL METHACRYLATE
General Information
Description: A liquid.
Formula: $H_2CCCH_3COOC_2H_5$.
Constants: Mol wt: 114.07, mp: $< -75°C$, bp: 119°C, lel: 1.8%, uel: saturation, flash p: 68°F (O.C.), d: 0.911 at 25°/25°C, vap. d.: 3.94.
Hazard Analysis
Toxicity: Animal experiments show moderate toxicity and irritation. Less toxic than ethyl acrylate.
Fire Hazard: Moderate, when exposed to heat or flame; can react with oxidizing materials.
Explosion Hazard: Dangerous when exposed to heat, sparks or flame.
Countermeasures
To Fight Fire: Carbon dioxide, dry chemical or carbon tetrachloride (Section 6).
Storage and Handling: Section 7.

ETHYL METHANOATE. See ethyl formate.

N-ETHYL METHYL (α-METHYL BENZYL) AMINE
General Information
Formula: $C_2H_5NH_2C_6H_4CH_3$.
Constant: Mol wt: 136.
Hazard Analysis
Toxic Hazard Rating: U. Limited animal experiments suggest high toxicity and irritation. See also amines.

ETHYL METHYL CARBINOL. See sec-butyl alcohol.

2-ETHYL-2-METHYL-1,3-DIOXOLANE
General Information
Formula: $H_2COCC_2H_5CH_3OCH_2$.
Constant: Mol wt: 128.
Hazard Analysis
Toxic Hazard Rating: U. Limited animal experiments suggest moderate toxicity.

ETHYL METHYL ETHER
General Information
Synonym: Methoxyethane.
Description: Colorless liquid or gas.
Formula: $CH_3OC_2H_5$.
Constants: Mol wt: 60.09, bp: 7.5°C, lel: 2.0%, uel: 10.1%, flash p: $-35°F$ (C.C.), d: 0.7260 at 0°/4°C, autoign. temp.: 374°F, vap. d.: 2.07.
Hazard Analysis
Toxicity: Details unknown; Has anesthetic properties. See also ethers.
Fire Hazard: Highly dangerous; when exposed to heat or flame; can react vigorously with oxidizing materials. See also ethers.
Explosion Hazard: Moderate, when exposed to flame. See also ethers.
Disaster Hazard: Highly dangerous, upon exposure to heat or flame.
Countermeasures
Storage and Handling: Section 7.
To Fight Fire: Alcohol foam, carbon dioxide, dry chemical or carbon tetrachloride (Section 6).
Shipping Regulations: Section 11.
  I.C.C.: Flammable liquid; red label, 10 gallons.
  Coast Guard Classification: Inflammable liquid; red label.
  IATA: Flammable liquid, red label, 1 liter (passenger), 40 liters (cargo).

7-ETHYL-2-METHYL-4-HENDECANAL
General Information
Description: Very slightly soluble in water.
Formula: $C_4H_9CH(C_2H_5)C_2H_4CHOHCH_2CH(CH_3)_2$.
Constants: Mol wt: 214, d: 0.8, bp: 507°F, flash p: 285°F.
Hazard Analysis
Fire Hazard: Slight, when exposed to heat or flame.
Countermeasures
To Fight Fire: Water, foam.

ETHYL METHYL KETONE. See butanone.

ETHYL METHYL KETONE PEROXIDE (in dimethyl phthalate)
General Information
Description: Water-white liquid; insoluble in water, soluble in benzene, alcohols, esters, ethers.
Constants: D: 1.091, flash p: 122°F (C.O.C.), decomposes mildly at 130°C.
Hazard Analysis and Countermeasures
See peroxides, organic.
Shipping Regulations: Section 11.
  I.C.C. Classification: Peroxide organic solution liquid N.O.S.
  Freight and Express: Yellow label, chemicals N.O.I.B.N.
  Rail Express: Max. wt: 1 lb. per shipping case.
  Parcel Post: Prohibited.
  IATA (containing not more than 60% by weight of peroxide): Oxidizing material, yellow label, 1 liter (passenger and cargo).
  IATA (containing > 60% peroxide): Oxidizing material not acceptable (passenger and cargo).

2-ETHYL-4-METHYL PENTANOL
General Information
Formula: $CH_3CHC_2H_5CH_2CHOHCH_3$.
Constant: Mol wt: 104.
Hazard Analysis
Toxic Hazard Rating: U. Limited animal experiments suggest moderate toxicity.

ETHYL METHYL PHENYL GLYCIDATE
General Information
Synonym: 3-Methyl-3-phenyl glycidic acid ethyl ester; "strawberry aldehyde"; aldehyde C-16.
Description: Colorless to yellowish liquid; strawberry like odor; soluble in 3 volumes of 60% alcohol.
Formula: $CH_3(C_6H_5)COCHCOOC_2H_5$.
Constants: Mol wt: 206, d: 1.104–1.123.
Hazard Analysis
Toxic Hazard Rating: U. A synthetic flavoring substance and adjuvant (Section 10).

ETHYL-2-METHYLPROPANOATE. See ethyl isobutyrate.

ETHYL METHYL PROPYL TIN IODIDES
General Information
Description: Crystals.
Formula: $(C_2H_5)(CH_3)(C_3H_7)SnI$.
Constants: Mol wt: 332.8, d: 1.8182 at 20°/20°C, bp: 110°C at 11 mm.
Hazard Analysis and Countermeasures
See tin compounds and iodides.

5-ETHYL-2-METHYL PYRIDINE
General Information
Synonym: Aldehydine.
Description: Liquid.

TOXIC HAZARD RATING CODE (For detailed discussion, see Section 1.)

0 NONE: (a) No harm under any conditions; (b) Harmful only under unusual conditions or overwhelming dosage.

1 SLIGHT: Causes readily reversible changes which disappear after end of exposure.

2 MODERATE: May involve both irreversible and reversible changes; not severe enough to cause death or permanent injury.

3 HIGH: May cause death or permanent injury after very short exposure to small quantities.

U UNKNOWN: No information on humans considered valid by authors

Formula: $(C_2H_5)(CH_3)C_5H_3N$.
Constants: Mol wt: 121.2, bp: 174°C, d: 0.9184 at 23°/4°C.
Hazard Analysis
Toxicity: Details unknown. See also pyridine.
Disaster Hazard: Dangerous; when heated to decomposition, it emits toxic fumes; can react with oxidizing materials.
Countermeasures
Storage and Handling: Section 7.

-ETHYL-2 METHYL PYRIDINE-1-OXIDE
General Information
Formula: $C_5H_3(CH_3)(C_2H_5)NO$.
Constant: Mol wt: 136.
Hazard Analysis
Toxic Hazard Rating: U. Limited animal experiments suggest moderate toxicity. See also pyridine.

ETHYL METHYL TELLUROPHETONE
General Information
Description: Dark yellow oil.
Formula: $C_2H_5CTeCH_3$.
Constants: Mol wt: 183.7, d: 1.8711, bp: 63–66°C.
Hazard Analysis and Countermeasures
Highly toxic. See tellurium compounds.

ETHYLMONOBROMOACETATE
General Information
Description: Insoluble in water.
Formula: $BrCH_2COOC_2H_5$.
Constants: Mol wt: 167, d: 1.5, bp: 318°F, flash p: 118°F.
Hazard Analysis
Fire Hazard: Moderate, when exposed to heat or flame.
Disaster Hazard: Dangerous. When heated to decomposition it emits highly toxic fumes.
Countermeasures
Storage and Handling: Section 7.
To Fight Fire: Water may be used to blanket fire.

ETHYLMONOCHLOROACETATE
General Information
Description: Insoluble in water.
Formula: $ClCH_2COOC_2H_5$.
Constants: Mol wt: 112.5, d: 1.2, bp: 295°F, flash p: 100°F.
Hazard Analysis
Fire Hazard: Moderate, when exposed to heat or flame.
Disaster Hazard: Dangerous. When heated to decomposition emits highly toxic fumes.
Countermeasures
Storage and Handling: Section 7.
To Fight Fire: Water may be used to blanket fire.

ETHYL MORPHINE HYDROCHLORIDE. See dionin.

ETHYL N-MORPHOLINE
General Information
Description: Colorless liquid.
Formula: $CH_2CH_2OCH_2CH_2NCH_2CH_3$.
Constants: Mol wt: 115.2, bp: 138°C, flash p: 89.6°F (O.C.), d: 0.916 at 20°/20°C, vap. d.: 4.00.
Hazard Analysis
Toxicity: Unknown.
MLV: ACGIH (tentative) 20 parts per million of air; 94 milligrams per cubic meter in air. May be absorbed via the skin.
Fire Hazard: Moderate when exposed to heat or flame; can react with oxidizing materials.
Explosion Hazard: Unknown.
Countermeasures
Storage and Handling: Section 7.
To Fight Fire: Alcohol foam, foam, carbon dioxide, dry chemical or carbon tetrachloride (Section 6).

ETHYL MUSTARD OIL. See dichlorodiethyl sulfide.

ETHYL NITRATE
General Information
Synonym: Nitric ether.
Description: Colorless liquid; pleasant odor; sweet taste.
Formula: $C_2H_5ONO_2$.
Constants: Mol wt: 91.07, mp: −112°C, bp: 88.7°C, explodes at 185°F, lel: 3.8%, flash p: 50°F (C.C.), d: 1.105 at 20°/4°C, vap. d.: 3.14.
Hazard Analysis
Toxic Hazard Rating:
    Acute Local: U.
    Acute Systemic: Ingestion 2; Inhalation 2.
    Chronic Local: U.
    Chronic Systemic: U.
Fire Hazard: Dangerous, when exposed to heat or flame.
Spontaneous Heating: No.
Explosion Hazard: Moderate, when exposed to heat.
Disaster Hazard: See nitrates.
Countermeasures
Ventilation Control: Section 2.
To Fight Fire: Foam, carbon dioxide, dry chemical or carbon tetrachloride (Section 6).
Personal Hygiene: Section 3.
Storage and Handling: Section 7.
Shipping Regulations: Section 11.
    I.C.C.: Flammable liquid; red label, 10 gallons.
    Coast Guard Classification: Inflammable liquid; red label.
    IATA: Flammable liquid, not acceptable (passenger and cargo).

ETHYL NITRITE
General Information
Synonyms: Nitrous ether; hyponitrous ether.
Description: Colorless or yellowish liquid; highly aromatic; ethereal odor.
Formula: C_2H_5ONO.
Constants: Mol wt: 75.07, bp: 16.4°C, lel: 4.1%–3.0%, uel: 50%, explodes at 194°F, flash p: −31°F (C.C.), d: 0.900 at 15.5°C, autoign. temp.: 194°F, vap. d.: 2.59.
Hazard Analysis
Toxic Hazard Rating:
    Acute Local: U.
    Acute Systemic: Inhalation 2.
    Chronic Local: U.
    Chronic Systemic: Inhalation 2.
Toxicology: Narcotic in high concentrations and lowers blood pressure. Methemoglobinemia has been reported.
Fire Hazard: Highly dangerous when exposed to heat or flame.
Spontaneous Heating: No.
Explosion Hazard: Severe.
Disaster Hazard: Highly dangerous; when heated to decomposition or on contact with acid or acid fumes, it emits highly toxic fumes of oxides of nitrogen; can react vigorously with oxidizing materials.
Countermeasures
Ventilation Control: Section 2.
To Fight Fire: Foam, carbon dioxide, dry chemical, carbon tetrachloride or water spray (Section 6).
Personnel Protection: Section 3.
Storage and Handling: Section 7.
Shipping Regulations: Section 11.
    I.C.C.: Flammable liquid; red label, 10 gallons.
    Coast Guard Classification: Inflammable liquid; red label.
    IATA: Flammable liquid, red label, 1 liter (passenger), 40 liters (cargo).

ETHYL NITRITE SPIRIT. See spirits of niter (sweet).

1-ETHYL-2-NITROBENZENE. See o-ethylnitrobenzene.

o-ETHYLNITROBENZENE
General Information
Synonym: 1-Ethyl-2-nitrobenzene.

Description: Clear yellow to green liquid.
Formula: $C_6H_4NO_2C_2H_5$.
Constants: Mol wt: 151.16, mp: 23°C, bp: about 228°C, fp: −13°C, d: 1.1 at 25°/25°C, vap. d.: 5.2.

Hazard Analysis
Toxicity: Details unknown; may resemble nitrobenzene. See nitrobenzene.
Fire Hazard: See nitrates.
Explosion Hazard: See nitrates.
Disaster Hazard: See nitrates.

Countermeasures
See nitrates.
To Fight Fire: Water, foam, carbon dioxide, dry chemical or carbon tetrachloride (Section 6).

ETHYL p-NITROBENZOATE. See nitrobenzoic acid ethyl ester.

ETHYL-p-NITROPHENYL THIONO BENZENE PHOSPHATE. See EPN.

ETHYL OLEATE
General Information
Description: Light-colored oleaginous liquid.
Formula: $C_{17}H_{33}COOC_2H_5$.
Constants: Mol wt: 310.5, mp: −32°C (approx.), bp: 205°C, flash p: 347.5°F, d: 0.867.

Hazard Analysis
Toxicity: Details unknown. See esters.
Fire Hazard: Slight; can react with oxidizing materials (Section 6).

Countermeasures
Storage and Handling: Section 7.

ETHYL OXALATE
General Information
Synonyms: Diethyl oxalate; diethyl ethanedioate; oxalic ether.
Description: Colorless, oily aromatic liquid; decomposes in water.
Formula: $(COOC_2H_5)_2$.
Constants: Mol wt: 146.14, mp: −40.6°C, bp: 185.4°C, flash p: 168°F (C.C.), d: 1.08426 at 15°C, 1.0785 at 20°/4°C, vap. d.: 5.04.

Hazard Analysis
Toxic Hazard Rating: Toxic. See oxalates.
Fire Hazard: Moderate, when exposed to heat or flame; can react with oxidizing materials.
Spontaneous Heating: No.
Disaster Hazard: On combustion, it will give off toxic fumes.

Countermeasures
Storage and Handling: Section 7.
To Fight Fire: Foam, carbon dioxide, dry chemical or carbon tetrachloride (Section 6).

ETHYL OXIDE. See ethyl ether.

ETHYL PARABEN. See ethyl p-hydroxy benzoate.

ETHYL PARASEPT. See ethyl p-hydroxy benzoate.

o-ETHYL PHENOL
General Information
Synonym: Phloral.
Description: Colorless liquid; phenol odor; insoluble in water; freely soluble in alcohol, benzene, and glacial acetic acid.
Formula: $C_8H_{10}O$.

Constants: Mol wt: 122.16, d: 1.037 at 12°C, bp: 207–208°C, solidifies < 18°C.
Hazard Analysis
Toxic Hazard Rating: U. See phenol.

p-ETHYL PHENOL
General Information
Description: Colorless needles; soluble in alcohol or ether; slightly soluble in water.
Formula: $HOC_6H_4C_2H_5$.
Constants: Mol wt: 122.6, mp: 46°C, bp: 219°C, d: 1.0– at 140°F.

Hazard Analysis
Toxic Hazard Rating: U. See phenol.
Fire Hazard: Moderate when exposed to powerful oxidizers, heat or source of ignition.
Countermeasures
To Fight Fire: Water, foam, alcohol foam.
Storage and Handling: Section 7.

ETHYL PHENYL ACETAMIDE. See ethyl acetanilide.

ETHYL PHENYLACETATE
General Information
Description: Colorless liquid.
Formula: $C_6H_5CH_2COOC_2H_5$.
Constants: Mol wt: 164.2, bp: 276°C, d: 1.033 at 20°C, vap. d.: 5.67.

Hazard Analysis
Toxicity: Details unknown. A weed killer.
Disaster Hazard: Moderately dangerous; when heated to decomposition, it may emit toxic fumes.

ETHYL PHENYLAMINE. See N-ethylaniline.

ETHYL PHENYL DICHLOROSILANE
General Information
Description: Liquid.
Formula: $C_2H_5SiCl_2C_6H_5$.
Constants: Mol wt: 205.06, vap. d.: 7.07.

Hazard Analysis
Toxic Hazard Rating:
    Acute Local: Irritant 3; Ingestion 3; Inhalation 3.
    Acute Systemic: U.
    Chronic Local: U.
    Chronic Systemic: U.
Disaster Hazard: Dangerous; when heated to decomposition, it emits highly toxic fumes of chlorides and phenol; will react with water or steam to produce toxic and corrosive fumes; can react with oxidizing materials.

Countermeasures
Ventilation Control: Section 2.
Personnel Protection: Section 3.
First Aid: Section 1.
Storage and Handling: Section 7.
Shipping Regulations: Section 11.
    I.C.C.: Corrosive liquid; white label, 10 gallons.
    Coast Guard Classification: Corrosive liquid; white label.
    IATA: Corrosive liquid, white label, not acceptable (passenger), 40 liters (cargo).

ETHYL PHENYL DI-p-TOLYL GERMANIUM
General Information
Description: White crystals.
Formula: $(C_2H_5)(C_6H_5)Ge(C_6H_4CH_3)_2$.
Constants: Mol wt: 361.0, mp: 55°C.
Hazard Analysis and Countermeasures
See germanium compounds.

TOXIC HAZARD RATING CODE (For detailed discussion, see Section 1.)

0 NONE: (a) No harm under any conditions; (b) Harmful only under unusual conditions or overwhelming dosage.

1 SLIGHT: Causes readily reversible changes which disappear after end of exposure.

2 MODERATE: May involve both irreversible and reversible changes not severe enough to cause death or permanent injury.

3 HIGH: May cause death or permanent injury after very short exposure to small quantities.

U UNKNOWN: No information on humans considered valid by authors

ETHYL PHENYL ETHER. See phenyl ethyl ether.

ETHYL PHENYL KETONE. See propiophenone.

ETHYL PHOSPHINE
General Information
Synonym: Phosphinoethane.
Description: Colorless liquid.
Formula: $C_2H_5PH_2$.
Constants: Mol wt: 62.1, bp: 25°C, d: < 1, vap. d.: 2.14.
Hazard Analysis
Toxicity: Details unknown. See also phosphine.
Fire Hazard: Dangerous, when exposed to heat or flame.
Explosion Hazard: Unknown.
Disaster Hazard: Dangerous; when heated to decomposition, it emits highly toxic fumes of oxides of phosphorus; can react vigorously with oxidizing materials.
Countermeasures
Storage and Handling: Section 7.
To Fight Fire: Foam, carbon dioxide, dry chemical or carbon tetrachloride (Section 6).

ETHYL PHTHALATE. See diethyl o-phthalate.

ETHYL PHTHALYL ETHYL GLYCOLLATE
General Information
Description: Liquid.
Formula: $C_2H_5OCOC_6H_4OCOCH_2OCOC_2H_5$.
Constants: Mol wt: 280.3, bp: 320°C, flash p: 365°F (C.C.), d: 1.180, vap. d.: 9.6.
Hazard Analysis
Toxicity: Unknown. See also esters.
Fire Hazard: Slight, when exposed to heat or flame; can react with oxidizing materials.
Countermeasures
To Fight Fire: Water, foam. alcohol foam, carbon dioxide, dry chemical or carbon tetrachloride (Section 6).
Storage and Handling: Section 7.

ETHYL PICOLINES
General Information
Synonyms: 5-Ethyl-2-picoline; 4-ethyl-2-picoline; 3-ethyl-2-picoline.
Description: Liquids with aromatic or pungent odors. Volatile with steam.
Formula: $C_8H_{11}N$.
Constant: Mol wt: 121.2.
Hazard Analysis
Toxicity: Variable. Some highly toxic, some not.
Disaster Hazard: Dangerous; when heated to decomposition, they emit highly toxic fumes.

ETHYL PROPENOATE. See ethyl acrylate.

ETHYL 1-PROPENYL ETHER
General Information
Formula: $CH_3CH:CHOCH_2CH_3$.
Constants: Mol wt: 86, d: 0.8, bp: 158°F, flash p: > 19°F (O.C.).
Hazard Analysis
Toxic Hazard Rating: U. Limited animal experiments suggest low toxicity. See also ethers.
Fire Hazard: Dangerous, when exposed to heat or flame.
Storage and Handling: Section 7.

ETHYL PROPIONATE
General Information
Synonym: Propionic ether.
Description: Water-white liquid; pineapple-like odor.
Formula: $C_2H_5COOC_2H_5$.
Constants: Mol wt: 102.11, mp: −72.6°C, bp: 99°C, flash p: 54°F (C.C.), d: 0.895 at 15.5°C, autoign. temp.: 890°F, vap. press.: 40 mm at 27.2°C, vap. d.: 3.52, lel: 1.9%, uel: 11%.
Hazard Analysis
Toxicity: Details unknown. See esters.

Fire Hazard: Dangerous, when exposed to heat or flame; can react vigorously with oxidizing materials.
Spontaneous Heating: No.
Explosion Hazard: Unknown.
Disaster Hazard: Dangerous, upon exposure to heat or flame.
Countermeasures
Ventilation Control: Section 2.
To Fight Fire: Foam, carbon dioxide, dry chemical or carbon tetrachloride (Section 6).
Storage and Handling: Section 7.
Shipping Regulations: Section 11.
IATA: Flammable liquid, red label, 1 liter (passenger), 40 liters (cargo).

ETHYL PROPIONYL. See diethyl ketone.

2-ETHYL-3-PROPYL ACROLEIN
General Information
Synonym: 2-Ethylhexenal.
Description: Colorless liquid; powerful odor.
Formula: $CH_3(CH_2)_2CHC(C_2H_5)CHO$.
Constants: Mol wt: 126.19, bp: 175°C, flash p: 155°F (O.C.), d: 0.848 at 20°/4°C, vap. d.: 4.35, vap. press.: 1.0 mm at 20°C.
Hazard Analysis
Toxic Hazard Rating:
Acute Local: Irritant 2; Ingestion 2; Inhalation 2.
Acute Systemic: U.
Chronic Local: U.
Chronic Systemic: U.
Fire Hazard: Moderate, when exposed to heat or flame; can react with oxidizing materials.
Countermeasures
To Fight Fire: Foam, carbon dioxide, dry chemical or carbon tetrachloride (Section 6).
Ventilation Control: Section 2.
Personnel Protection: Section 3.
Storage and Handling: Section 7.

2-ETHYL-3-PROPYL ACRYLIC ACID
General Information
Description: Liquid.
Formula: $C_3H_7CHC(C_2H_5)COOH$.
Constants: Mol wt: 142.19, mp: −7.8°C, bp: 232.1°C, d: 0.948 at 20°/20°C, vap. d.: 4.91, flash p: 330°F (O.C.).
Hazard Analysis
Toxicity: Unknown.
Fire Hazard: Slight; Section 6.
Countermeasures
Storage and Handling: Section 7.
To Fight Fire: Water, foam, alcohol foam.

ETHYL n-PROPYL DIISOAMYL TIN
General Information
Formula: $(C_2H_5)(C_3H_7)(C_5H_{11})_2Sn$.
Constants: Mol wt: 333.1, d: 1.0654 at 22°C, bp: 141°C at 17 mm.
Hazard Analysis
Toxicity: Details unknown. See tin compounds.

ETHYL sec-PROPYL DIPHENYL GERMANIUM
General Information
Description: Liquid.
Formula: $(C_2H_5)(C_3H_7)Ge(C_6H_5)_2$.
Constants: Mol wt: 299.0, mp: 175–190°C.
Hazard Analysis
Toxicity: Details unknown. See germanium compounds.

ETHYL PROPYL ETHER
General Information
Synonym: Ethoxy propane-1.
Description: Soluble in water.
Formula: $C_2H_5OC_3H_7$.

Constants: Mol wt: 88, lel: 1.9%, uel: 2.4%, d: 0.8, bp: 147° F.

Hazard Analysis
Toxic Hazard Rating: U. See ethers.
Fire Hazard: Dangerous when exposed to heat or open flame. See ethers.

Countermeasures
Storage and Handling: Section 7.
To Fight Fire: Alcohol foam.

1 ETHYL-2-PROPYLETHYLENE. See 1,3-heptene (mixture of cis and trans isomers).

ETHYL PROPYL KETONE. See 3-hexanone.

ETHYL PROPYL TIN DICHLORIDE
General Information
Description: Crystals; water soluble.
Formula: $(C_2H_5)(C_3H_7)SnCl_2$.
Constants: Mol wt: 261.8, mp: 58° C.
Hazard Analysis and Countermeasures
See tin compounds and chlorides.

2-(5'-ETHYL PYRID-2'-yl) ETHYL ACRYLATE
General Information
Description: Liquid, very slightly soluble in water.
Formula: $CH_2:CHOOC_2H_4C_5H_3NC_2H_5$.
Constants: Mol wt: 193, d: 1.0458 at 20° C, bp: 181° C at 50 mm, fp: −75° C.
Hazard Analysis
Toxic Hazard Rating: U. Limited animal experiments suggest moderate toxicity. See also esters and ethyl acrylate.

4-ETHYL PYRIDINE
General Information
Description: Liquid.
Formula: $CH_3CH_2C_5H_4N$.
Constants: Mol wt: 107.1, bp: 168.3° C, d: 0.9460 at 20°/20° C, vap. d.: 3.70.
Hazard Analysis
Toxicity: Details unknown. See also pyridine.
Fire Hazard: Moderate.
Explosion Hazard: Unknown.
Disaster Hazard: Dangerous; when heated to decomposition, it emits highly toxic fumes of oxides of nitrogen; can react with oxidizing materials.
Countermeasures
Storage and Handling: Section 7.
To Fight Fire: Foam, carbon dioxide, dry chemical or carbon tetrachloride (Section 6).

ETHYL SELENIDE. See diethyl selenide.

ETHYL SILICATE
General Information
Synonym: Tetraethyl-o-silicate.
Description: Colorless liquid; faint odor; decomposes in water.
Formula: $(C_2H_5)_4SiO_4$.
Constants: Mol wt: 208.30, bp: 165.5° C, flash p: 125° F (O.C.), d: 0.933 at 20°/4° C, vap. press.: 1.0 mm at 20° C, vap. d.: 7.22.
Hazard Analysis
Toxic Hazard Rating:
  Acute Local: Irritant 3; Ingestion 3; Inhalation 3.
  Acute Systemic: Ingestion 2; Inhalation 2.
  Chronic Local: U.
  Chronic Systemic: U.

TLV: ACGIH (recommended); 100 parts per million in air; 851 milligrams per cubic meter of air.
Toxicology: Ethyl silicate is irritating to the eyes and upper respiratory tract. Concentrations of 2,500 ppm produce narcosis in guinea pigs after 1½ hours exposure. This is the maximum concentration obtainable at ordinary room temperatures by evaporation. The maximum concentration which these animals can inhale for 60 minutes without the production of serious disturbance is 2,000 ppm. For an exposure of several hours, concentrations of 500 ppm or more are required to cause serious injury. Animals dying after acute exposures show pulmonary edema, sometimes with secondary pneumonia, acute nephritis and evidence of injury to the liver. For humans, concentrations of 85 ppm produce a detectable odor, and 700 ppm cause stinging of the eyes and nose. No cases of human poisoning have been reported nor have experiments dealing with the effects of prolonged exposure been reported.

Irritation of the eyes and nose, tremors, respiratory difficulty or irregularity, anemia, leucocytosis, and narcosis have been reported in animal experiments. Brief exposures of humans to 3,000 ppm cause only extreme irritation of the eyes and nose.
Fire Hazard: Moderate, when exposed to heat or flame.
Disaster Hazard: Dangerous; when heated to decomposition, it emits toxic fumes; can react with oxidizing materials.

Countermeasures
Ventilation Control: Section 2.
To Fight Fire: Carbon dioxide, dry chemical or carbon tetrachloride (Section 6).
Personnel Protection: Section 3.
Storage and Handling: Section 7.
Shipping Regulations: Section 11.
  Coast Guard Classification: Combustible liquid.

ETHYLSTANNIC ACID
General Information
Description: White amorphous powder; insoluble in water.
Formula: C_2H_5SnOOH.
Constant: Mol wt: 180.8.
Hazard Analysis
Toxicity: Details unknown. See tin compounds.

ETHYL STYRENE. See m-ethylvinylbenzene.

ETHYL SUCCINATE. See diethyl succinate.

ETHYL SULFATE. See diethyl sulfate.

ETHYL SULFHYDRATE. See ethyl mercaptan.

ETHYL SULFIDE
General Information
Synonyms: Diethyl sulfide; thioethyl ether.
Description: Liquid; garlic-like odor.
Formula: $(C_2H_5)_2S$.
Constants: Mol wt: 90.2, mp: −102° C, bp: 92–93° C, d: 0.837 at 20°/4° C, vap. d.: 3.11.
Hazard Analysis
Toxicity: Details unknown, but probably highly toxic.
Fire Hazard: Moderate. See also sulfides (Section 6).
Disaster Hazard: Dangerous; when heated to decomposition, or on contact with acid or acid fumes, it emits highly toxic fumes of oxides of sulfur; will react with water or steam to produce toxic and flammable vapors; can react with oxidizing materials.

TOXIC HAZARD RATING CODE (For detailed discussion, see Section 1.)

0 NONE: (a) No harm under any conditions; (b) Harmful only under unusual conditions or overwhelming dosage.

1 SLIGHT: Causes readily reversible changes which disappear after end of exposure.

2 MODERATE: May involve both irreversible and reversible changes; not severe enough to cause death or permanent injury.

3 HIGH: May cause death or permanent injury after very short exposure to small quantities.

U UNKNOWN: No information on humans considered valid by authors.

Countermeasures

Storage and Handling: Section 7.

2-(ETHYL SULFONYL) ETHANOL
General Information

Synonym: Ethylsulfonylethyl alcohol.

Description: Hygroscopic crystals. Water soluble.

Formula: $CH_3CH_2SO_2CH_2CH_2OH$.

Constants: Mol wt: 138.2, d: 1.25 at 20°/4°C, mp: 43°C, bp: 153°C at 2.5 mm, flash p: 371°F.

Hazard Analysis

Toxicity: Details unknown. Limited animal experiments suggest low toxicity.

Disaster Hazard: Dangerous; when heated to decomposition, it emits highly toxic fumes of oxides of sulfur.

Fire Hazard: Slight, when exposed to heat or flame.

Countermeasures

To Fight Fire: Water, foam, carbon dioxide, dry chemicals, carbon tetrachloride.

Storage and Handling: Section 7.

ETHYLSULFONYLETHYL ALCOHOL. See 2-(ethylsulfonyl)ethanol.

ETHYLSULFURIC ACID
General Information

Synonyms: Ethyl hydrogen sulfate; acid ethyl sulfate.

Description: Colorless, oily liquid.

Formula: $C_2H_5OSO_3H$.

Constants: Mol wt: 126.13, bp: 280°C decomposes, d: 1.316 at 17°/4°C.

Hazard Analysis

Toxic Hazard Rating:

Acute Local: Irritant 3; Ingestion 3; Inhalation 3.

Acute Systemic: U.

Chronic Local: U.

Chronic Systemic: U.

Disaster Hazard: Dangerous; when heated to decomposition, it emits highly toxic fumes of oxides of sulfur; will react with water or steam to produce heat.

Countermeasures

Ventilation Control: Section 2.

Personnel Protection: Section 3.

Storage and Handling: Section 7.

ETHYL TETRAPHOSPHATE. See hexaethyl tetraphosphate.

ETHYL TETRATHIO-o-STANNATE
General Information

Description: Solid.

Formula: $(SC_2H_5)_4Sn$.

Constants: Mol wt: 363.2, bp: 105°C at 0.001 mm.

Hazard Analysis and Countermeasures

See tin and sulfur compounds.

ETHYL THIOALCOHOL. See ethyl mercaptan.

ETHYL THIOCARBIMIDE. See ethyl isothiocyanate.

O-[2-(ETHYLTHIO) ETHYL]-O,O-DIMETHYL PHOSPHOROTHIOATE. See demeton.

ETHYLTIN TRIBROMIDE
General Information

Description: Colorless crystals; water soluble.

Formula: $(C_2H_5)SnBr_3$.

Constants: Mol wt: 387.5, mp: 310°C.

Hazard Analysis and Countermeasures

See tin compounds and bromides.

ETHYLTIN TRIIODIDE
General Information

Description: Crystals.

Formula: $(C_2H_5)SnI_3$.

Constants: Mol wt: 528.5, bp: 181–184°C at 19 mm.

Hazard Analysis and Countermeasures

See tin compounds and iodides.

ETHYL-p-TOLUENE SULFONAMIDE
General Information

Description: Liquid.

Formula: $C_7H_7SO_2NHC_2H_5$.

Constants: Mol wt: 199.3, bp: 98°C at 745 mm, flash p: 260°F (C.C.), d: 1.253, vap. d.: 5.6.

Hazard Analysis

Toxicity: Unknown. See also amides.

Fire Hazard: Slight, when exposed to heat or flame.

Disaster Hazard: Dangerous; when heated to decomposition, it emits highly toxic fumes of oxides of sulfur; can react with oxidizing materials.

Countermeasures

Storage and Handling: Section 7.

To Fight Fire: Water, carbon dioxide, dry chemical or carbon tetrachloride (Section 6).

ETHYL-p-TOLUENE SULFONATE
General Information

Description: Liquid.

Formula: $C_7H_7SO_3C_2H_5$.

Constants: Mol wt: 200.25, mp: 33°C, bp: 221.3°C, flash p: 316°F (C.C.), d: 1.17, vap. d.: 6.98.

Hazard Analysis

Toxicity: Unknown.

Fire Hazard: Slight, when exposed to heat or flame.

Disaster Hazard: Dangerous; when heated to decomposition, it emits highly toxic fumes of oxides of sulfur; can react with oxidizing materials.

Countermeasures

Storage and Handling: Section 7.

To Fight Fire: Water, foam, carbon dioxide, dry chemical or carbon tetrachloride (Section 6).

ETHYL TRIBENZYL GERMANIUM
General Information

Description: Colorless crystals; soluble in methyl alcohol.

Formula: $(C_2H_5)Ge(CH_2C_6H_5)_3$.

Constants: Mol wt: 375.0, mp: 57°C.

Hazard Analysis

Toxicity: Details unknown. See germanium compounds.

ETHYL TRI-n-BUTYL TIN
General Information

Description: Crystals.

Formula: $(C_2H_5)(C_4H_9)_3Sn$.

Constants: Mol wt: 319.1, d: 1.0783, bp: 129°C at 10 mm.

ETHYL TRICHLOROSILANE
General Information

Description: Liquid.

Formula: $C_2H_5Cl_3Si$.

Constants: Mol wt: 163.47, mp: −105.6°C, bp: 99.5°C, flash p: 72°F (O.C.) d: 1.24 at 25°/25°C, vap. d.: 5.6.

Hazard Analysis

Toxic Hazard Rating:

Acute Local: Irritant 3; Ingestion 3; Inhalation 3.

Acute Systemic: U.

Chronic Local: U.

Chronic Systemic: U.

Fire Hazard: Dangerous, when exposed to heat or flame.

Explosion Hazard: Unknown.

Disaster Hazard: Dangerous; when heated to decomposition, it emits highly toxic fumes of phosgene; will react with water or steam to produce heat and toxic and corrosive fumes; can react vigorously with oxidizing materials.

Countermeasures

Ventilation Control: Section 2.

To Fight Fire: Carbon dioxide, dry chemical or carbon tetrachloride (Section 6).

Personnel Protection: Section 3.

First Aid: Section 1.
Storage and Handling: Section 7.
Shipping Regulations: Section 11.
I.C.C.: Flammable liquid; red label, 10 gallons.
Coast Guard Classification: Inflammable liquid; red label.
IATA: Flammable liquid, red label, not acceptable (passenger), 40 liters (cargo).

ETHYL TRIPHENYL GERMANIUM
General Information
Description: Colorless powder; insoluble in water.
Formula: $(C_2H_5)Ge(C_6H_5)_3$.
Constants: Mol wt: 333.0, mp: 78.5° C.
Hazard Analysis
Toxicity: Details unknown. See germanium compounds.

ETHYL TRIPHENYL SILANE
General Information
Description: Crystals; insoluble in water.
Formula: $(C_6H_5)_3Si(C_2H_5)$.
Constants: Mol wt: 288.4, mp: 76° C.
Hazard Analysis
Toxicity: Details unknown. See silanes.

ETHYL TRI-n-PROPYL TIN
General Information
Formula: $(C_2H_5)Sn(C_3H_7)_3$.
Constants: Mol wt: 277.0, mp: 101° C at 10 mm.
Hazard Analysis
Toxicity: Details unknown. See tin compounds.

ETHYL TRIS-p-BIPHENYL GERMANIUM
General Information
Description: Colorless crystals.
Formula: $(C_6H_5C_6H_4)_3(C_2H_5)Ge$.
Constants: Mol wt: 561.2, mp: 155° C.
Hazard Analysis
Toxicity: Details unknown. See germanium compounds.

ETHYLUREA
General Information
Description: Crystals.
Formula: $NH_2CONHC_2H_5$.
Constants: Mol wt: 88.1, mp: 92° C, flash p: > 200° F, d: 1.213 at 18° C.
Hazard Analysis
Toxicity: Details unknown, but may be toxic.
Fire Hazard: Moderate, when exposed to heat or flame.
Disaster Hazard: Dangerous; when heated to decomposition, it emits highly toxic fumes of cyanides; can react with oxidizing materials.
Countermeasures
Storage and Handling: Section 7.
To Fight Fire: Foam, carbon dioxide, dry chemical or carbon tetrachloride (Section 6).

ETHYL URETHANE. See ethyl carbamate.

ETHYL VANILLIN
General Information
Description: Fine, crystalline needles.
Formula: $C_6H_3(OH)(CHO)(OC_2H_5)$.
Constants: Mol wt: 166.17, mp: 76.5° C.
Hazard Analysis
Toxicity: Unknown. Animal experiments show low toxicity. Used as a synthetic flavoring substance and adjuvant. It is a substance which migrates to food from packaging materials. (Section 10).

Disaster Hazard: Slight; (Section 6).
Countermeasures
Storage and Handling: Section 7.

m-ETHYLVINYLBENZENE
General Information
Synonym: Ethylstyrene.
Description: Water-white liquid.
Formula: $C_2H_5C_6H_4CHCH_2$.
Constants: Mol wt: 132.1, bp: 191.5° C, fp: −127° C, d: 0.8955 at 20° C, vap. press.: 2.17 mm at 40° C, vap. d.: 4.56.
Hazard Analysis
Toxicity: Details unknown. See also phenyl ethylene (styrene).
Fire Hazard: Moderate; can react with oxidizing materials.
Countermeasures
To Fight Fire: Foam, carbon dioxide, dry chemical or carbon tetrachloride (Section 6).
Storage and Handling: Section 7.

ETHYL VINYL ETHER
General Information
Synonym: Ethenoxyethane.
Description: Liquid.
Formula: $CH_2CHOC_2H_5$.
Constants: Mol wt: 72.1, mp: −115° C, bp: 35.5° C, flash p: 0° F, d: 0.763 at 15°/18° C, vap. d.: 2.49.
Hazard Analysis
Toxicity: Details unknown. Limited animal experiments show anaesthetic effects. See also ethers.
Fire Hazard: Dangerous, when exposed to heat or flame; can react vigorously with oxidizing materials.
Explosion Hazard: Details unknown. See also ethers.
Disaster Hazard: Highly dangerous, upon exposure to heat or flame.
Countermeasures
Ventilation Control: Section 2.
To Fight Fire: Foam, carbon dioxide, dry chemical or carbon tetrachloride (Section 6).
Storage and Handling: Section 7.

ETHYLXANTHOGEN DISULFIDE
General Information
Synonym: Diethylxanthogen sulfide.
Hazard Analysis
Toxic Hazard Rating:
Acute Local: Irritant 2.
Acute Systemic: Ingestion 2; Skin Absorption 2.
Chronic Local: Allergen 2.
Chronic Systemic: U.
Toxicology: Data based upon limited animal experiments.
Disaster Hazard: Dangerous; when heated to decomposition, it emits highly toxic fumes of oxides of sulfur.

ETHYNE. See acetylene.

2-ETHYNYL-2-BUTANOL. See methyl pentynol.

1-ETHYNYLCYCLOHEXAN-1-OL. See ethynylcyclohexanol.

ETHYNYLCYCLOHEXANOL
General Information
Synonym: 1-Ethynylcylohexan-1-ol.
Description: Liquid or crystal.
Formula: $C_8H_{12}O$.
Constants: Mol wt: 124.18, mp: 32° C, bp: 180° C, vap. d.: 3.73.

TOXIC HAZARD RATING CODE (For detailed discussion, see Section 1.)

0 NONE: (a) No harm under any conditions; (b) Harmful only under unusual conditions or overwhelming dosage.

1 SLIGHT: Causes readily reversible changes which disappear after end of exposure.

2 MODERATE: May involve both irreversible and reversible changes; not severe enough to cause death or permanent injury.

3 HIGH: May cause death or permanent injury after very short exposure to small quantities.

U UNKNOWN: No information on humans considered valid by authors.

Hazard Analysis

Toxicity: Details unknown. Limited animal experiments suggest high tocicity and eye irritation. See also alcohols.

Fire Hazard: Moderate; can react with oxidizing materials.

Countermeasures

To Fight Fire: Foam, carbon dioxide, dry chemical or carbon tetrachloride (Section 6).

Storage and Handling: Section 7.

ETIOLOGIC AGENTS

General Information

Description: Cultures and infectious materials causative of serious communicable diseases.

Countermeasures

Shipping Regulations: Section 11.

  IATA: Other restricted articles, class C, no label required, 5 liters (passenger and cargo).

EUCALYPTUS OIL. See Oil of Eucalyptus.

EUCLASE. See beryllium aluminum silicate.

EUGENIC ACID. See eugenol.

EUGENOL

General Information

Synonym: 4-Allyl-2-methoxy phenol; coryophyllic acid; eugenic acid.

Description: Colorless or yellowish liquid; oily, spicy odor; soluble in alcohol, chloroform, ether, and volatile oils; very slightly soluble in water.

Formula: $C_3H_5C_6H_3(OH)OCH_3$.

Constants: Mol wt: 164, d: 1.064–1.070, bp: 253.5°C.

Hazard Analysis

Toxic Hazard Rating: U. A synthetic flavoring substance and adjuvant. (Section 10).

EUROPIUM

General Information

Description: Gray crystals.

Formula: Eu.

Constant: At wt: 152.00.

Hazard Analysis

Toxicity: Unknown.

Radiation Hazard: Section 5. For permissible levels, see Table 5, p. 150.

  Artificial isotope ^{152}Eu, half life 12.4 y. Decays to stable ^{152}Sm by electron capture. Also decays to radioactive ^{152}Gd by emitting beta particles of 0.70, 1.48 MeV and others. Also emits gamma rays of 0.34, 0.41, 0.78 MeV, other gammas and X-rays.

  Artificial isotope ^{154}Eu, half life 16 y. Decays to stable ^{154}Gd by emitting beta particles of 0.26–1.86 MeV. Also emits gamma rays of 0.12–1.28 MeV.

  Artificial isotope ^{155}Eu, half life 1.8 y. Decays to stable ^{155}Gd by emitting beta particles of 0.14–0.25 MeV. Also emits gamma rays of 0.04–0.11 MeV.

EXPELLERS, SEED OR NUT, WET

Countermeasures

Shipping Regulations: Section 11.

  IATA: Flammable solid, not acceptable (passenger and cargo).

EXPLOSIVE AUTO ALARMS

Countermeasures

Shipping Regulations: Section 11.

  I.C.C.: Class C explosive, 150 pounds.

EXPLOSIVE BOMB

Shipping Regulations: Section 11.

  I.C.C.: Class A explosive, not acceptable.

  IATA: Explosive, not acceptable (passenger and cargo).

EXPLOSIVE CABLE CUTTERS

Shipping Regulations: Section 11.

  I.C.C.: Class C explosive, 150 pounds.

  IATA: Explosive, explosive label, 25 kilograms (passenger), 70 kilograms (cargo).

EXPLOSIVES

Shipping Regulations: Section 11.

  I.C.C.: Section 73.53 (Class A).

         Section 73.88 (Class B).

         Section 73.100 (Class C).

  IATA (n.o.s.): Explosive, not acceptable (passenger and cargo).

EXPLOSIVE COMPOSITIONS

Shipping Regulations: Section 11.

  I.C.C.: Class A or B explosive, 10 pounds.

  IATA: Explosive, not acceptable (passenger and cargo).

EXPLOSIVE MINE

Shipping Regulations: Section 11.

  I.C.C.: Explosive A, not accepted.

  IATA: Explosive, not acceptable (passenger and cargo).

EXPLOSIVE POWER DEVICE, CLASS B, CLASS C

Shipping Regulations: Section 11.

  I.C.C. (Class B): Class B explosive, red label, 150 pounds.

  I.C.C. (Class C): Class C explosive, not accepted.

  IATA: Explosive, not acceptable (passenger and cargo).

EXPLOSIVE PROJECTILE

Shipping Regulation: Section 11.

  I.C.C.: Class A explosive, not accepted.

  IATA: Explosive, not acceptable (passenger and cargo).

EXPLOSIVE RELEASE DEVICE

Shipping Regulations: Section 11.

  I.C.C.: Class C explosive, 150 pounds.

EXPLOSIVE RIVETS

Shipping Regulations: Section 11.

  I.C.C.: Class C explosive, 150 pounds.

  IATA: Explosive, explosive label, 25 kilograms (passenger), 70 kilograms (cargo).

EXPLOSIVES, HIGH

General Information

  Description: High explosives (HE) are those which decompose by detonation. This is a very rapid (nearly instantaneous) process; hence the action is fast and violent. An explosion may be initiated by sudden shock, by high temperature or by a combination of the two. For many explosives the conditions under which they will explode are well known, for example:

  An explosion may be initiated by elevated temperature alone:

  (1) In the case of mercury fulminate 15 seconds exposure to 200°C or 1 second exposure to 340°C will set if off.

  (2) Trinitrotoluene will be set off by exposure to 500°C for 1 second.

  (3) Tetryl will detonate in 1000 seconds at 160°C or in 0.1 second at 500°C.

  (4) Picric acid will detonate in 9 seconds at 300°C or 1 second at 355°C.

  An explosion of HE may also be initiated by severe shock. Sensitivity of explosives to shock may be measured in several ways, such as the impact pendulum method and the drop test. The impact pendulum test operates by allowing a heavy pendulum to swing down over a sample of explosive in a dished, inclined container so arranged that there is very little clearance between the pendulum and the sample. Thus the effect of contact between the sample and the pendulum bob is one of a combination of shock and rubbing. The height from which the pendulum is allowed to swing to explode the sample is a measure of the sensitivity of the sample to this test. The drop test consists of placing a sample upon an anvil and allowing a 5 pound weight to drop on it. The height from which the weight must drop to explode the sample is a measure of the sample's sensitivity to shock.

Below is a table of the results of a drop test upon several samples. These results must be considered as relative and not by any means absolute. Solid explosive in a tightly fitting container is much more sensitive to shock.

(1) mercury fulminate
      = 2 inches at 5 pounds
(2) nitroglycerin
      = 4 inches at 5 pounds
(3) tetryl = 8 inches at 5 pounds
(4) picric acid
      = 14 inches at 5 pounds
(5) trinitrotoluene
      = 20 inches at 5 pounds
(6) black powder (a low explosive)
      = 30 inches at 5 pounds*

Another test for explosives is the speed at which a detonation travels. This speed is usually in the range of thousands of meters per second. Speed of detonation is found to be a function of kind of explosive and state of compaction. There is an optimum state of compaction beyond which the explosive tends to become "deadpressed," in which state it is difficult to make the whole sample explode. Below the point of optimum compaction the rate of detonation is found to be directly proportional to the density of the sample. Below are listed some maximum detonation rates, in meters/second, for some common explosives:

(1) nitroglycerin	8500
(2) PETN	8100
(3) tetryl	7700
(4) picric acid	7500
(5) trinitrotoluene	7400
(6) lead azide	4900
(7) mercury fulminate	4800
(8) ammonium nitrate	1100
(9) low explosives	1000

It has been found that upon detonation an explosive can cause a nearby sample of explosive to detonate "sympathetically." The distance over which one charge can detonate another is a function of the amount of energy produced by the first explosion and the medium through which the shock wave is propagated to the second charge of explosive. For instance the relationship for air (very approximately) would be expected to be: Weight of explosive in pounds/(distance in feet)^3 = 4. Thus to calculate the maximum distance for a possible sympathetic detonation of 40,000 pounds of explosive, the calculation is:

$$D^3 = (40,000)/4$$
$$D^3 = 10,000$$
$$D = 22 \text{ feet (approximately)}.$$

According to C. S. Robinson the formula is more nearly:
      weight of explosive = 4 × (distance)$^{2.25}$.

The power of the shock wave is much more rapidly attenuated in water, wood, etc., than in air, which means that if a shield of water or wood is interposed between piles of explosive the distance between them may

*From "Explosions, Their Anatomy and Destructiveness," by C. S. Robinson (McGraw-Hill).

be lessened. (See also American Table of Distances in Section 7, p. 186.)

Liquid Oxygen. Though not by itself explosive, liquid oxygen can be dangerous when blended with highly flammable or carbonaceous materials. In this combination it is used in coal mining, quarrying, strip mining, and open-cut ore mining, and in rocket fuels. Its use underground or in confined places is not recommended by the U. S. Bureau of Mines, but it evolves a large percentage of carbon monoxide (See carbon monoxide). This type of explosive has many safety advantages. For instance, it is not itself an explosive until mixed with a flammable absorbent which can be done at the last moment before firing. However, once the explosive has been made up, it is very flammable and when it catches fire it will usually detonate. Liquid oxygen explosives are not stored, as they deteriorate rapidly and lose a great deal of their explosive power in a short time.

Hazard Analysis

General Fire Hazard: Severe, when exposed to heat by chemical reaction with powerful oxidizing or reducing agents.

General Explosion Hazard: Moderate to dangerous when severely shocked or heated, depending upon kind of explosive, state of compaction, degree of confinement, etc. Practically all high explosives used commercially require a detonator or cap to set them off, as compared to an igniter needed to set off black blasting powder (See also Explosives, Permissible; Dynamite; Nitroglycerin; Ammonium Nitrate; Nitrates, Trinitrotoluene).

Countermeasures

Storage and Handling: See detailed description in Section 7.

Detonating Devices: To develop the desired disruptive effect of an explosive, some means must be adopted to "set off," "fire," or "detonate" it without killing or maiming the persons doing the blasting. Several devices or methods are being utilized, all with a view to having this work done as safely and efficiently as possible. There are two general types of devices or methods of getting explosives into action, namely, igniters and detonators. The former merely convey a flame to the explosive mass and ignite it, while the latter transmit (originally through ignition of a small quantity of highly explosive substance by an arc, a flame, or spark) a sharp blow that causes the explosive to dissociate, or detonate, or burn with very great rapidity. Igniters are squibs (plain and electric), fuse, and delay igniters; detonators are blasting caps (plain and electric), delay electric blasting caps, delay electric igniters with caps, and Cordeau-Bickford detonating fuse.

The squib is a small-diameter tube of straw or paper filled with a quick-burning powder and having a relatively slow burning "match head" attached to one end; the latter is ignited or lighted by an ordinary match or other flame, and its relatively slow burning allows the person handling the ignition to retire before the fire is communicated to the quick-burning material in the tube. Squibs are by no means either safe or efficient, even though still used to a considerable extent, especially in coal mining. Electric squibs are somewhat similar to ordinary squibs, except that the ignition is accomplished by means of an electric arc; electric squibs are much more satisfactory from a safety viewpoint than ordinary squibs.

A fuse (or, as it is sometimes called, "safety fuse") consists of a fine-grained black powder core covered with cotton hemp or jute to form a rope-like material about 3/16 inch in diameter; one end of the fuse is brought in contact with the powder charge or with a detonating "cap," and the other end (usually several feet away from the explosive) is lighted by a flame from a match or open light. The fine-grained black powder burns gradually and somewhat slowly (about 30 to 40 seconds to the linear foot of fuse) until it reaches the explosive (black powder) or the detonating cap (if some form of dynamite is used), giving the blaster time to get in the clear before the main explosion takes place. Fuses are much safer than squibs, but have their own hazards and must be used with care.

Delay electric igniters usually are a combination of electric igniters and fuses, the latter being lighted by the igniters within the blasting hole, the fuse transmitting the ignition to the explosive. Delay igniters usually are much safer than fuse, particularly for coal-mine use; but they, too, have their hazards. Delay blasting is by no means a safe procedure in coal mining, though it is a standard and relatively safe practice in metal mining and tunneling, if sensible precautions are taken.

Blasting caps or detonators are metallic cylinders (usually copper) closed at one end, about 3/16 inch in diameter, and usually less than 2 inches in length, partly filled with a small amount of a relatively easily fired or "detonated" compound, the resultant shock or blow when fired being sufficient, when embedded in dynamite, to fire or detonate the dynamite mass. Ordinary blasting caps usually are fired or detonated by the flame of the fuse, the end of the latter being inserted into the open end of the detonator or cap and placed in contact with the highly explosive material in the interior of the metallic capsule or cap. Caps are extremely hazardous to handle, as they are likely to be detonated by heat, friction, or a relatively moderate blow; however, they are relatively safe if handled carefully. Partial proof of this is in the fact that they are manufactured and shipped by the thousands daily and accidents are decidedly rare, primarily because the caps are at all times handled with utmost care.

Electric blasting caps are somewhat similar to ordinary caps or detonators, but the cap is fired by electricity. The electric wires are so placed in the capsule or cap that when attached to an electric current an arc is formed within the cap, which detonates the sensitive explosive material in the cap. A hazard in the use of electric blasting caps is unexpected explosions due to radio or radar induced electric currents which may activate the cap.

Delay electric blasting caps or detonators are somewhat similar to ordinary electric blasting caps, except that several time intervals in blasting are obtained by having the electric arc ignite a short piece of fuse or some slow-burning substance before it reaches the highly sensitive detonating material in the capsule or cap. Numerous time-interval delays are obtained; in general delay electric blasting caps are relatively safe and effective even in wet holes, though they ought not be used in coal mining if explosions of gas or dust are to be avoided. Delay electric igniters with caps or detonators are a combination of electric igniter and blasting cap, usually with suitable lengths of fuse between to give the desired delay; they have some advantages but are relatively unsafe and should not be used in coal mining.

The Cordeau-Bickford detonating fuse is a combined fuse and detonator in the form of a lead tube about 1/4 inch in diameter filled throughout its length with a high explosive, trinitrotoluene (TNT). It is fired by fuse and an ordinary detonator or cap or by an electric cap; when fired, it detonates throughout its length (which may be up to or over 100 feet) almost instantaneously, the explosion wave traveling at a rate of about 17,500 feet

per second. Although somewhat expensive, it is relatively safe to handle and is particularly effective in deepwell drill holes in quarry and similar work, as it detonates simultaneously throughout its length, adding effectiveness to the main body of explosive which it detonates. It fires black powder as well as high explosives (dynamite, etc.) and is obtainable in lengths of approximately 500 feet, wound on spools.

Transportation of Explosives. In the transportation of explosives, the manufacturers and the railroads have set a most excellent example in safety, as is proved by the fact that accidents in rail transportation of explosives were so rare as to have been almost nonexistent in the past decade. The problem of transporting explosives from the railroad car to the users' magazines and from the magazines to the places of use (the mine, quarry, tunnel) is decidedly serious and should be given much more consideration than it usually receives. The Interstate Commerce Commission has issued numerous explicit regulations on the transportation of explosives, probably the most complete being "Regulations for the Transportation of Explosives and Other Dangerous Articles by Freight and Express and in Baggage Service"; the most recent issue was in 1960.

Users of explosives who transport them by truck would profit by familiarizing themselves with the regulations established by the Interstate Commerce Commission. Railroad companies formulate and issue their own regulations for the transportation of explosives and other dangerous articles, and the restrictions of some of the carriers are drastic in the extreme. The Bureau of Explosives of the American Railway Association has issued much interesting information relating to the transportation of dangerous substances, including explosives of various kinds. The Bureau of Mines and other organizations have also issued information on the transportation of explosives, but much careless explosives-transportation practice occurs, not only in mining work but to an even greater extent in other activities in which explosives are used. State laws on transportation of explosives usually are confusing and inadequate; in fact, state laws on this subject usually are almost nonexistent or are phrased so broadly as to be almost meaningless.

EXPLOSIVES, INITIATING
Shipping Regulations: Section 11.
    IATA: Explosive, not acceptable (passenger and cargo).

EXPLOSIVES, LOW
General Information
Synonyms: Black blasting powder; gunpowder; "A" blasting powder; "B" blasting powder.
Composition: Black powder is composed of saltpeter, charcoal and sulfur in the approximate proportions of 6:1:1. ("A" blasting powder uses KNO_3 and "B" blasting powder uses $NaNO_3$). Low explosives are explosives which deflagrate; this differentiates them both in composition and properties from high explosives, which detonate. A deflagrating explosive is one that burns progressively over a relatively sustained period of time in comparison with a detonating explosive, which decomposes almost instantaneously. See also Section 7.

Hazard Analysis
Toxicity: Variable.
Fire Hazard: Dangerous, when exposed to heat or flame or by chemical reaction.

    Although most safety men now look upon black blasting powder with disfavor, it is one of the oldest and most generally used explosives in commercial work. It burns with extreme rapidity instead of detonating as high explosives do; it is highly sensitive to flame or sparks or friction and gives off much flame, which is hot and of great length of duration. These properties make it extremely hazardous for use in mines (especially coal mines) and quarries. The gases given off in detonation are not only very hot but frequently contain harmful consitutents. Notwithstanding its numerous deficiencies, from a safety standpoint, it has action characteristics that make it valuable in both coal mining and quarrying, though it has relatively little utility in metal mining. It is difficult to use effectively in wet places and this is its main disadvantage from an efficiency standpoint. Pellet powder is black blasting powder in consolidated (pellet or stick) form rather than in grains or granules, and it has few if any real advantages over black blasting powder, notwithstanding the fairly prevalent idea that it is a "safe" explosive.

Smokeless powders have a somewhat different composition from that of black blasting powder and are used chiefly for sportsmen's ammunition and, more widely, for military purposes. They are decidedly sensitive to flame and impact but ordinarily are so packaged that if reasonable judgment is used they are relatively harmless.

Most black powder fires start from sparks. Ignition results in an explosion so quickly that no attempt can be made to fight the fire. Every effort should be made to prevent fires from reaching stores of black powder; but if this fails, fire-fighting forces should be withdrawn at least 800 feet from the fire and should protect themselves against an explosion by seeking any cover available or by lying flat on the ground. If an explosion does occur, every effort should be made to prevent flames from spreading to neighboring magazines. Fire-fighting forces should be cautioned against approaching a fire which may involve black powder to avoid being trapped or injured by an explosion. Most explosions from black powder originate from sparks and the following safety rules should be strictly enforced and obeyed.

(1) Open no containers in a magazine in which explosives or ammunition are stored. This should be done only in a building free from all other explosives or ammunitions; or in suitable weather in the open, at least 100 feet from the nearest magazine. The quantity at or near such an operation should be limited to 100 pounds. Safety tools only should be used in opening or closing containers or in other operations involving black powder. Safety shoes (non-insulating) should be worn in all rooms where black powder is handled and by all persons engaged in handling black powder. The wearing of all non-conductive shoes, such as rubber is prohibited. If the handling of black powder is carried on over a concrete floor, the floor should be covered with a tarpaulin or other suitable material.

Loose black powder is extremely dangerous. Whenever it is necessary to handle loose black powder, not over 50 pounds should be permitted at or near such operations. If black powder is spilled on benches or floors, all work should be stopped until it has been removed and the explosive hazard of any remaining dust or particles has been neutralized with water. Rooms or buildings in which black powder is handled should be inspected frequently for dust, and all such dust should be immediately removed with water. See also Section 7.

Explosive Hazard: Severe, when exposed to heat or flame or by chemical reaction. Black powder is the slowest acting of all explosives. It has a shearing and heaving action, tending to blast materials in large, firm fragments. The action derives from a relatively slow development of gas pressure so that it must be carefully loaded and closely confined. The application is limited because it disintegrates in water and therefore cannot be used in wet work, except with special precautions. It also produces more smoke and fumes than most dynamites.

Precautions in Handling: Black powder is the most treacherous explosive material used today and it is regarded as one of the worst known explosive hazards. When ignited unconfined it burns with explosive violence and will explode if ignited under even slight confinement. It can be ignited easily by very small sparks, heat and friction. It is subject to rapid deterioration in the presence of moisture, but if kept dry it retains its explosive properties for many years. It is used to ignite smokeless powder, propelling charges, airplane flares and bursting charges of hand grenades, as a bursting charge in shrapnel, practice bombs, practice trench-mortar shells, in saluting charges, smoke-puff charges, time and percussion fuses, pellets, primers and primer detonators, and in expelling charges of pyrotechnic signals.

When black powder is shipped or received, each container should be inspected for holes and weak spots. It should be examined particularly for small holes, such as those made by nails, which are visible only upon close examination. Damaged containers should not be repaired. The contents should instead be transferred to new or serviceable containers. Metal containers for export shipment should be crated. Usually two containers are packed in each crate. Repainting of containers and repacking of black powder, contained in damaged or unserviceable containers, constitutes the principal maintenance activity for such stocks. Black powder containers are also subject to sweating, which rusts metal drums or kegs, so that repainting is necessary to keep containers serviceable. Repainting should not be done in a magazine in which explosives or ammunition are stored. It may be done in a nearly empty magazine, or in clear weather in the open, at least 100 feet from the nearest magazine. The quantity of black powder, at or near such operations should be limited to 100 pounds.

The marking on repainted containers should be checked carefully to see that it is a facsimile of the old. Furthermore, the metal caps on certain types of black powder containers deteriorate in storage. Replacement of these caps may be allowed, but the same safety precautions as outlined above for repainting containers should be followed. Operations, such as removal of black powder from containers and its transfer from unserviceable to serviceable drums should be conducted in strict compliance with applicable portions of the above outlined safety regulations. Black powder operations of any kind should be conducted in special buildings which are not used for other purposes at the same time. The floors of such buildings should be covered with suitable materials. Intraplant-quantity-distance requirements for high explosives as given in Section 7 should be followed. Absolute cleanliness should be maintained at all times in and around each operation. Non-insulating safety shoes should be worn by personnel in all assembly operations. All equipment should be electrically grounded and this should be determined by tests. Empty metal containers which have held black powder should be thoroughly washed inside with water before they are disposed of. Serious explosions

---

TOXIC HAZARD RATING CODE (For detailed discussion, see Section 1.)

0 NONE: (a) No harm under any conditions; (b) Harmful only under unusual conditions or overwhelming dosage.

1 SLIGHT: Causes readily reversible changes which disappear after end of exposure.

2 MODERATE: May involve both irreversible and reversible changes; not severe enough to cause death or permanent injury.

3 HIGH: May cause death or permanent injury after very short exposure to small quantities.

U UNKNOWN: No information on humans considered valid by authors.

have occurred from supposedly empty cans. Wooden containers should be destroyed by burning. Safety tools only should be used in opening or closing containers or in handling black powder. Processes should be so laid out as to bring about frequent grounding of all operators handling this material.

Destruction: Black powder is best destroyed by dumping it into water, preferably a stream or large body of water, but some states have anti-stream pollution laws that may be violated in this manner. It is well to be informed on this subject and to be certain that no damage will be done to water supplies.

If it is inconvenient or inadvisable to destroy black powder thus, it may be burned. If that is done, the contents of only one container holding 25 pounds or less should be burned at a time. The powder should be poured on the ground in a trail not wider than 2 inches, and no part of such a trail should parallel another part at a distance of less than 10 feet. A train of combustible material should be laid to the explosive for igniting it, as with dynamite.

"Pellet powder" (black powder in compressed stick form) may be destroyed either by throwing it into water after removing the wrapper, or as described for dynamite. If dry and in good condition, black powder burns rapidly, especially in small grain size, with a yellow or pinkish-blue flame and dense smoke.

The empty powder containers should be washed out, as explosions are said to have occurred from "empty" containers.

Countermeasures
Shipping Regulations: Class A explosive. Not accepted as freight.

EXPLOSIVES OR EXPLOSIVE DEVICES, NEW
Shipping Regulations: Section 11.
   IATA: Explosive, not acceptable (passenger and cargo).

EXPLOSIVES, PERMISSIBLE
"Permissible" explosives are essentially high explosives (dynamite) modified by the introduction of "dopes." The function of the dopes in general is to decrease flame temperature and to a smaller extent, the length and duration of flame when the explosive is converted from a solid into a gas; in other words, when it is fired or detonated. The designation "permissible" is given to an explosive of modified dynamite type after it has passed certain tests made by the Federal Bureau of Mines. The permissible character of such explosives depends not only upon the ingredients in the explosive, but also on certain well-defined specifications as to handling and use. As with the dynamites, there are several different types and grades; "permissibles," hydrated "permissibles," organic nitrate "permissibles," nitroglycerin "permissibles," ammonium perchlorate "permissibles," and gelatin "permissible." Essentially all those now used to any extent are in either the ammonium nitrate or the gelatin classes. (See also dynamite.)

The ammonium nitrate "permissible" explosives contain relatively little nitroglycerin and relatively large proportions of ammonium nitrate. The latter is an explosive but one less sensitive to impact, sparks, and flames than nitroglycerin. This type of "permissible" explosive is now used extensively, as it has a rather wide range of strength, rate of detonation, density, size of cartridge, etc., and can be utilized not only in dry but also to some extent in fairly wet holes if charged carefully and fired promptly. Gelatin "permissible," explosives are more suitable than ammonium nitrate "permissibles," for wet holes, and in general are stronger and more violent than the ammonium-nitrate types.

All "permissible" explosives are strong, must be used in relatively small quantities (less than 1½ pounds) per hole to retain their permissibility, give off considerable quantities of toxic gases on detonation, and, while much safer than black blasting powder or dynamite, must be stored, handled, and used with care.

Classification Upon Basis of Toxic Gases: All "permissible" explosives, when detonated, emit some toxic gases and a much larger volume of nontoxic gases. In order that the toxic products may not become a menace to the life or health of miners, no explosive is now or can become "permissible" if upon detonation it evolves more than 158 liters (5½ cubic feet) of toxic gases per 1½ pound charge as determined by tests in the Bichel pressure gage.

Classification upon the basis of the volume of toxic gases produced by 680 grams (1½ pounds) of explosive is as follows: Class A, not more than 53 liters; Class B, between 53 and 106 liters; and Class C, between 106 and 158 liters. (These classifications are not to be confused with the I.C.C. Classification of explosives).

Field tests were made with a 1½ pound charge of a "permissible explosive that produced, in the Bichel gage, the maximum allowable quantity of poisonous gases (158 liters per 1½ pounds); these tests indicated that in a narrow entry, without artificial ventilation, 1800 ppm of carbon monoxide (the only poisonous gas present) was produced, as shown by analysis of an air sample taken 2 minutes after the shot. Another sample of the air taken 2 minutes later contained 800 ppm of carbon monoxide. Under no conditions should miners or shot firers return to the face until the poisonous gases have been removed by adequate ventilation.

It is provided further that, in accordance with the provisions and conditions, explosives enumerated on the "permissible" lists of the Bureau of Miners are "permissible" in use only when they satisfy the following requirements:

1. That the explosive be in all respects similar to the sample submitted by the manufacturer for test, and that the diameters of the cartridges used must be those that have been approved.

2. That electric detonators (not fuze and detonators) be used of not less efficiency than No. 6, the detonation charge of which shall consist of a 1 gram mixture of 80 parts of mercury fulminate and 20 parts of potassium chlorate (or their equivalents), and that the required electric firing must be done by means of a "permissible" type blasting unit.

3. That the explosive be stored in surface magazines under proper conditions, so that it will not undergo change in character, and that after taking underground it be used in less than 36 hours.

4. That the coal to be blasted be undercut or equivalently relieved; that, to prevent blow-through, all portions of the borehole must be at least 18 inches from relief in any direction; that, to prevent blowouts, the charge be properly confined with not less than 2 feet of clay (if the length of the hole will not permit the charge desired and 2 feet of stemming, at least half the length of the hole shall be filled with stemming) or other incombustible stemming and not be on the solid; that, to prevent the hole being on the solid, it shall be at least 6 inches shorter than the depth of the undercut or equivalent relief, and, when placed adjacent to the roof, ribs, or floor, all but 12 inches at the rear of the hole must be at least 6 inches from the adjacent surface as projected into the coal to be blasted, and all parts of the hole shall be free from the adjacent surface as projected into the coal to be blasted; that the shot be not a dependent shot; and that the shot hole be cleaned before charging.

5. That the quantity used for a shot (1) be not in excess of 680 grams (1½ pounds) when fired in accordance with these requirements and (2) when used under certain

additional requirements or restrictions be not in excess of 1,361 grams (3 pounds). The use of charges over $1\frac{1}{2}$ pounds and not exceeding 3 pounds is approved tentatively pending further investigation. For charges of over $1\frac{1}{2}$ pounds, the following additional requirements must be observed; (a) Shot holes must be 6 feet or more in length. (b) Explosive must be charged in a continuous train, with no cartridges deliberately deformed or crushed, with all cartridges in contact with each other, and with the end cartridges touching the rear of the hole and the stemming, respectively. (c) Examinations for gas must be made in the blasting area before and after a shot is fired. (d) The "permissible" explosive must be one showing toxic gas emission that will place it either in Class A or Class B.

6. That the region in which the blasting is done be kept well-protected by rock dust or otherwise in accordance with Bureau of Mines inspection standards.

7. That the shot not be fired in the presence of a dangerous percentage of firedamp. Examination for firedamp to be made at the blasting area before shooting in a gassy mine.

See also articles on dynamite, nitroglycerin, trinitrotoluene, pentaerythritol tetranitrate, nitrates, ammonium nitrates, picrates, azides, fulminates.

EXPLOSIVES, SAMPLES FOR LABORATORY. EXAMINATION
Shipping Regulations: Section 11.

EXPLOSIVE TORPEDO
Shipping Regulation: Section 11.
    I.C.C.: Class A explosive, not accepted.
    IATA: Explosive, not acceptable (passenger and cargo).

EXTRACTS, LIQUID FLAVORING
General Information
Constant: Flash p: < 80° F.
Hazard Analysis
Toxicity: Variable; an allergen.
Fire Hazard: Dangerous, when exposed to heat or flame; can react vigorously with oxidizing materials (Section 6).
Explosion Hazard: Unknown.
Countermeasures
Ventilation Control: Section 2.
Storage and Handling: Section 7.
Shipping Regulations: Section 11.
    I.C.C.: Flammable liquid; red label, 10 gallons.
    Coast Guard Classification: Combustible liquid; inflammable liquid; red label.
    IATA: Flammable liquid, red label, 1 liter (passenger), 40 liters (cargo).

I.C.C.: Section 73.86.
IATA: Explosive, not acceptable (passenger and cargo).

TOXIC HAZARD RATING CODE (For detailed discussion, see Section 1.)

.0 NONE: (a) No harm under any conditions; (b) Harmful only under unusual conditions or overwhelming dosage.

1 SLIGHT: Causes readily reversible changes which disappear after end of exposure.

2 MODERATE: May involve both irreversible and reversible changes; not severe enough to cause death or permanent injury.

3 HIGH: May cause death or permanent injury after very short exposure to small quantities.

U UNKNOWN: No information on humans considered valid by authors.

F

F-114. See dichlorotetrafluoroethane.

FAIRY GLOVES. See digitalis.

FATTY ACIDS
General Information
Description: Monobasic organic acids derived from natural fats and oils. The term is also applied to all the monobasic acids of the general formula: $C_nH_{2n+1}COOH.$, e.g., stearic acid, oleic acid, linoleic acid etc.
Hazard Analysis
Toxic Hazard Rating: U. Food additives permitted in foods for human consumption.

FATTY AMINES
General Information
Synonym: Aliphatic amines.
Description: Look like ordinary oils or fats.
Formula: $R-NH_2$ where R = 8-22 C atoms. Can be primary, secondary or tertiary.
Hazard Analysis
Toxicity: As organic bases they are mild irritants to skin and mucous membrane. Regard these compounds as mild alkalies. See also amines.

FENAC. See 2,3,6-trichlorophenylacetic acid.

FENTHION. See o,o-dimethyl-o[(4-methylthio)m-tolyl] phosphorothioate.

FENURON. See 1,1-dimethyl-3-phenyl urea.

FERACONITINE. See pseudoaconitine.

FERBAM
General Information
Synonym: Ferric dimethyldithiocarbamate.
Description: Black solid; slightly soluble in water.
Formula: $Fe[(CH_3)_2NCS_2]_3$.
Constants: Mol wt: 416.5, mp: decomposes.
Hazard Analysis
Toxic Hazard Rating:
 Acute Local: Irritant 2; Inhalation 1.
 Acute Systemic: Ingestion 2.
 Chronic Local: Irritant 1.
 Chronic Systemic: U.
Toxicology: A fungicide. See also bis(diethylthiocarbamyl) disulfide.
TLV: ACGIH (recommended) 15 milligrams per cubic meter of air.
Disaster Hazard: Dangerous; when heated to decomposition it emits highly toxic fumes.

FERMATE
General Information
Description: A wettable powder containing 76% ferbam.
Hazard Analysis
See ferbam.

FERMIUM
General Information
Description: A synthetic radioactive element. Properties similar to those of erbium.
Formula: Fm.
Constant: At wt: 252.

Hazard Analysis
Radiation Hazard: Section 5. For permissible levels, see Table 5, p. 150.
 Artificial isotope ^{254}Fm, half life 3.2 h. Decays to radioactive ^{250}Cf by emitting alpha particles of 7.7 MeV.
 Artificial isotope ^{255}Fm, half life 22 h. Decays to radioactive ^{251}Cf by emitting alpha particles of 7.1 MeV.
 Artificial isotope ^{256}Fm, half life 3–4 h. Decays by spontaneous fission.

FERNAMBUCO. See brazil wood.

FERRIC AMMONIUM CITRATE. See iron ammonium citrate.

FERRIC ARSENATE
General Information
Synonym: Scorodite.
Description: Rhombic, green crystals.
Formula: $FeAsO_4 \cdot 2H_2O$.
Constants: Mol wt: 230.78, mp: decomposes, d: 3.18.
Hazard Analysis
Toxicity: Highly toxic. See arsenic compounds.
Disaster Hazard: See arsenic compounds.
Countermeasures
Storage and Handling: Section 7.
Shipping Regulations: Section 11.
 I.C.C.: Poison B; poison label, 200 pounds.
 Coast Guard Classification: Poison B: poison label.
 IATA: Poison B, poison label, 25 kilograms (passenger), 95 kilograms (cargo).

FERRIC ARSENIDE
General Information
Synonym: Iron arsenide.
Description: White powder.
Formula: FeAs.
Constants: Mol wt: 130.8, mp: 1020°C, d: 7.83.
Hazard Analysis and Countermeasures
See arsenic compounds and arsine.

FERRIC ARSENITE
General Information
Description: Brown-yellow powder.
Formula: $2FeAsO_3 \cdot Fe_2O_3 \cdot 5H_2O$.
Constants: Mol wt: 607.26, mp: decomposes.
Hazard Analysis
Toxicity: Highly toxic. See arsenic compounds.
Disaster Hazard: See arsenic compounds.
Countermeasures
Storage and Handling: Section 7.
Shipping Regulations: Section 11.
 I.C.C.: Poison B; poison label, 200 pounds.
 Coast Guard Classification: Poison B; poison label.
 IATA: Poison B, poison label, 25 kilograms (passenger), 95 kilograms (cargo).

FERRIC BROMIDE
General Information
Description: Dark red-brown, deliquescent crystals.
Formula: $FeBr_3$.
Constants: Mol wt: 295.60, mp: sublimes with decomposition.

Hazard Analysis
Toxic Hazard Rating:
  Acute Local: Irritant 1; Ingestion 1.
  Acute Systemic: Ingestion 1.
  Chronic Local: U.
  Chronic Systemic: Ingestion 1.
Toxicity: See also bromides.
Fire Hazard: Slight; (Section 6).
Countermeasures
Personal Hygiene: Section 3.
Storage and Handling: Section 7.

FERRIC CACODYLATE
General Information
Description: Yellowish-brown, odorless powder.
Formula: $Fe[(CH_3)_2AsO_2]_3$.
Constant: Mol wt: 466.8.
Hazard Analysis and Countermeasures
See arsenic compounds.

FERRIC CHLORIDE
General Information
Description: Black-brown solid.
Formula: $FeCl_3$.
Constants: Mol wt: 162.2, mp: 282°C, bp: 319.0°C, d: 2.804 at 11°C, vap. press.: 1 mm at 194.0°C.
Hazard Analysis
Toxic Hazard Rating:
  Acute Local: Irritant 1; Ingestion 1.
  Acute Systemic: U.
  Chronic Local: Irritant 1.
  Chronic Systemic: U.
Note: Used as a trace mineral added to animal feeds. (Section 10).
Disaster Hazard: Dangerous; when heated to decomposition, it emits highly toxic fumes of hydrochloric acid; will react with water to produce toxic and corrosive fumes.
Countermeasures
Personal Hygiene: Section 3.
Storage and Handling: Section 7.
Shipping Regulations: Section 11.
  IATA: Other restricted articles, class B, no label required, 12 kilograms (passenger), 45 kilograms (cargo).

FERRIC DIARSENIDE
General Information
Synonym: Iron diarsenide.
Description: Cubic, silver-gray crystals.
Formula: $FeAs_2$.
Constants: Mol wt: 205.7, mp: 990°C, d: 7.4.
Hazard Analysis and Countermeasures
See arsenic compounds and arsine.

FERRIC DICHROMATE
General Information
Description: Red-brown granules.
Formula: $Fe_2(Cr_2O_7)_3$.
Constant: Mol wt: 759.8.
Hazard Analysis
Toxicity: See chromium compounds.
Fire Hazard: Moderate, by chemical reaction with reducing agents (Section 6).
Countermeasures
Storage and Handling: Section 7.

FERRIC DIMETHYLDITHIOCARBAMATE. See ferbam.

FERRIC ETHYLENE BIS-DITHIOCARBAMATE
Hazard Analysis
Toxicity: Details unknown; a fungicide. See ferbam and bis(diethylthiocarbamyl) disulfide.
Disaster Hazard: Dangerous; when heated to decomposition, it emits highly toxic fumes of oxides of sulfur.
Countermeasures
Storage and Handling: Section 7.

FERRIC FERRICYANIDE
General Information
Synonym: Berlin green.
Description: Cubic crystals.
Formula: $Fe[Fe(CN)_6]$.
Constant: Mol wt: 267.8.
Hazard Analysis and Countermeasures
See ferricyanides.

FERRIC FERROCYANIDE
General Information
Description: Dark blue crystals.
Formula: $Fe_4[Fe(CN)_6]_3$.
Constants: Mol wt: 859.3, mp: decomposes.
Hazard Analysis and Countermeasures
See ferrocyanides.

FERRIC FLUORIDE
General Information
Synonym: Iron fluoride.
Description: Green crystals.
Formula: FeF_3.
Constants: Mol wt: 112.9, d: 3.18.
Hazard Analysis and Countermeasures
See fluorides.

FERRIC FLUOSILICATE
General Information
Description: Gelatinous, flesh-colored solid.
Formula: $Fe_2(SiF_6)_3$.
Constant: Mol wt: 537.9.
Hazard Analysis and Countermeasures
See fluosilicates.

FERRIC FORMATE
General Information
Synonym: Iron formate.
Description: Red crystals.
Formula: $Fe(CHO_2)_3$.
Constant: Mol wt: 190.9.
Hazard Analysis and Countermeasures
See formic acid.

FERRIC METHANEARSENATE
General Information
Description: Reddish-brown, lustrous scales.
Formula: $Fe_2(CH_3AsO_3)_3$.
Constant: Mol wt: 525.5.
Hazard Analysis and Countermeasures
See arsenic compounds.

FERRIC NITRATE
General Information
Description: Crystals.
Formula: $Fe(NO_3)_3 \cdot 6H_2O$.
Constants: Mol wt: 349.96, mp: 35°C, d: 1.68.
Hazard Analysis
Toxic Hazard Rating:
  Acute Local: Irritant 2; Ingestion 1.
  Acute Systemic: Ingestion 1; Inhalation 1.

TOXIC HAZARD RATING CODE (*For detailed discussion, see Section 1.*)

0 NONE: (a) No harm under any conditions; (b) Harmful only under unusual conditions or overwhelming dosage.

1 SLIGHT: Causes readily reversible changes which disappear after end of exposure.

2 MODERATE: May involve both irreversible and reversible changes; not severe enough to cause death or permanent injury.

3 HIGH: May cause death or permanent injury after very short exposure to small quantities.

U UNKNOWN: No information on humans considered valid by authors.

Chronic Local: Irritant 1.
Chronic Systemic: Ingestion 1; Inhalation 1.
Fire Hazard: See nitrates.
Disaster Hazard: See nitrates.
Countermeasures
Ventilation Control: Section 2.
Personal Hygiene: Section 3.
Storage and Handling: Section 7.
Shipping Regulations: Section 11.
IATA: Oxidizing material, yellow label, 12 kilograms (passenger), 45 kilograms (cargo).

FERRIC OXALATE
General Information
Description: Yellow powder.
Formula: $Fe_2(C_2O_4)_3 \cdot 5H_2O$.
Constants: Mol wt: 465.8, mp: decomposes at 100°C.
Hazard Analysis
Toxic Hazard Rating:
Acute Local: Irritant 1.
Acute Systemic: Ingestion 2.
Chronic Local: Irritant 1.
Chronic Systemic: Ingestion 1.
Toxicity: See also oxalates.
Fire Hazard: Slight; (Section 6).
Countermeasures
Ventilation Control: Section 2.
Personnel Protection: Section 3.
Storage and Handling: Section 7.

FERRIC PHOSPHATE
General Information
Synonym: Iron phosphate.
Description: Yellowish to white powder; insoluble in water; soluble in acids.
Formula: $FePo_4 \cdot 2H_2O$.
Constants: Mol wt: 187, d: 2.87.
Hazard Analysis
Toxic Hazard Rating: U. A nutrient and/or dietary supplement food additive (section 10). Also used as a trace mineral added to animal feeds.

FERRIC PYROPHOSPHATE
General Information
Synonym: Iron pyrophosphate.
Description: Yellowish white powder; insoluble in water; soluble in dilute acids.
Formula: $Fe_4(P_2O_7)_3 \cdot xH_2O$.
Hazard Analysis
Toxic Hazard Rating: U. A nutrient and/or dietary supplement food additive. Also used as a trace mineral added to animal feeds (Section 10).

FERRIC SODIUM PYROPHOSPHATE
Hazard Analysis
Toxic Hazard Rating: U. Used as a nutrient and/or dietary supplement food additive (Section 10).

FERRIC SULFATE
General Information
Description: Rhombic yellow crystals. Slightly water soluble.
Formula: $Fe_2(SO_4)_3$.
Constants: Mol wt: 399.9, d: 3.097 at 18°C, mp: 480°C (decomposes).
Hazard Analysis
Toxic Hazard Rating:
Acute Local: Irritant 1.
Acute Systemic: Ingestion 1.
Chronic Local: Irritant 1.
Chronic Systemic: 0.
Toxicology: May cause local irritation. Practically non-toxic systemically. See iron compounds. It is a substance which migrates to food from packaging materials (Section 10).

FERRIC SULFIDE
General Information
Description: Yellow-green crystals.
Formula: Fe_2S_3.
Constants: Mol wt: 207.90, mp: decomposes, d: 4.3.
Hazard Analysis
Toxic Hazard Rating:
Acute Local: Irritant 1; Ingestion 2; Inhalation 1.
Acute Systemic: Ingestion 1; Inhalation 2.
Chronic Local: U.
Chronic Systemic: U.
Fire Hazard: See sulfides.
Explosion Hazard: See sulfides.
Disaster Hazard: See sulfides.
Countermeasures
Personal Hygiene: Section 3.
Storage and Handling: Section 7.

FERRIC THIOCYANATE
General Information
Description: Cubic, black-red, deliquescent crystals.
Formula: $Fe(SCN)_3$.
Constants: Mol wt: 230.10, mp: decomposes.
Hazard Analysis
Toxic Hazard Rating:
Acute Local: Irritant 1.
Acute Systemic: Ingestion 1.
Chronic Local: 0.
Chronic Systemic: 0.
Disaster Hazard: Dangerous; when heated to decomposition or on contact with acid or acid fumes, it emits highly toxic fumes of cyanides.
Countermeasures
Personal Hygiene: Section 3.
Storage and Handling: Section 7.

FERRIC m-VANADATE
General Information
Synonym: Iron m-vanadate.
Description: Grayish-brown powder.
Formula: $Fe(VO_3)_3$.
Constant: Mol wt: 352.70.
Hazard Analysis
Toxicity: See vanadium compounds.

FERRICYANIC ACID
General Information
Description: Green-brown, deliquescent needles.
Formula: $H_3Fe(CN)_6$.
Constants: Mol wt: 215.0, mp: decomposes.
Hazard Analysis
Toxic Hazard Rating:
Acute Local: U.
Acute Systemic: Ingestion 1.
Chronic Local: U.
Chronic Systemic: Ingestion 1.
Disaster Hazard: Dangerous; when heated to decomposition or on contact with acid or acid fumes, it emits highly toxic fumes of cyanides.
Countermeasures
Personal Hygiene: Section 3.
Storage and Handling: Section 7.

FERRICYANIDES
Hazard Analysis
Toxic Hazard Rating:
Acute Local: Irritant 1.
Acute Systemic: Ingestion 1.
Chronic Local: U.
Chronic Systemic: Ingestion 1.
Toxicology: Ferricyanides as such are of low toxicity since the CN is bound. It has been stated but not conclusively proven that HCN can be liberated in the stomach as a result of contact with gastric acidity.

Disaster Hazard: Dangerous; when heated to decomposition or on contact with acid or acid fumes, they emit highly toxic fumes of cyanides.

Countermeasures
Personal Hygiene: Section 3.
Storage and Handling: Section 7.

FERROCENE
General Information
Synonym: Dicyclopentadienyl iron.
Description: Orange crystals; camphor odor, insoluble in water, soluble in alcohol and ether.
Formula: $C_{10}H_{10}Fe$.
Constants: Mol wt: 186.0, mp: 174°C, sublimes >100°C, volatile in steam.
Hazard Analysis
Toxicity: Details unknown.
Fire Hazard: Moderate.
Disaster Hazard: Dangerous; when heated to decomposition, it emits toxic fumes.

FERROCERIUM
General Information
Description: An alloy of iron and misch metal.
Countermeasures
Shipping Regulations: Section 11.
    IATA: Flammable solid, yellow label, 12 kilograms (passenger), 45 kilograms (cargo).

FERROCYANIC ACID
General Information
Description: White needles turning blue in moist air.
Formula: $H_4Fe(CN)_6$.
Constants: Mol wt: 216.0, mp: decomposes.
Hazard Analysis
Toxic Hazard Rating:
    Acute Local: Irritant 1.
    Acute Systemic: Ingestion 1.
    Chronic Local: U.
    Chronic Systemic: Ingestion 1.
Disaster Hazard: Dangerous; when heated to decomposition or on contact with acid or acid fumes, it emits highly toxic fumes of cyanides.
Countermeasures
Personal Hygiene: Section 3.
Storage and Handling: Section 7.

FERROCYANIDES
Hazard Analysis
Toxic Hazard Rating:
    Acute Local: Irritant 1.
    Acute Systemic: Ingestion 1.
    Chronic Local: U.
    Chronic Systemic: Ingestion 1; Inhalation 1.
Toxicology: Ferrocyanides as such are of a low order of toxicity. But highly toxic decomposition products can form upon mixing them with hot concentrated acids. Acid, basic or neutral solutions of ferrocyanides liberate hydrocyanic acid upon strong irradiation.
Disaster Hazard: Dangerous; when heated to decomposition or on contact with acid or acid fumes, they emit highly toxic fumes of cyanides.
Countermeasures
Personal Hygiene: Section 3.
Storage and Handling: Section 7.

FERROSILICON
General Information
Description: Crystalline, metallic solid.

Formula: Fe + Si.
Constant: D: 5.4.
Hazard Analysis
Toxicity: Unknown. This material is decomposed by moisture, under which conditions impurities may liberate such poisonous gases as phosphine and arsine.
Fire Hazard: Moderate, by chemical reactcon with moisture (Section 6).
Explosion Hazard: Moderate, by chemical reaction.
Disaster Hazard: Dangerous; it will react with water or steam to produce hydrogen and other flammable vapors; can react with oxidizing materials; and on contact with acid or acid fumes, it can emit toxic fumes.
Countermeasures
Storage and Handling: Section 7.
Shipping Regulations: Section 11.
    Coast Guard Classification: Hazardous Article.
    IATA: (Containing 30% or more but not more than 70% silicon). Flammable solid, yellow label, not accepted (passenger), 12 kilograms (cargo).
                        (Containing less than 30% or more than 70% silicon). Not restricted (passenger and cargo).

FERROUS o-ARSENATE
General Information
Description: Green, amorphous powder.
Formula: $Fe_3(AsO_4)_2 \cdot 6H_2O$.
Constants: Mol wt: 553.5, mp: decomposes.
Hazard Analysis
Toxicity: Highly toxic. See arsenic compounds.
Disaster Hazard: See arsenic compounds.
Countermeasures
Storage and Handling: Section 7.
Shipping Regulations: Section 11.
    I.C.C.: Poison B; poison label, 200 pounds.
    Coast Guard Classification: Poison B; poison label.
    IATA: Poison B, poison label, 25 kilograms (passenger), 95 kilograms (cargo).

FERROUS BROMIDE
General Information
Description: Hexagonal, green-yellow crystals.
Formula: $FeBr_2$.
Constants: Mol wt: 215.7, mp: decomposes, d: 4.636 at 25°C.
Hazard Analysis
Toxic Hazard Rating:
    Acute Local: Irritant 1.
    Acute Systemic: Ingestion 1.
    Chronic Local: U.
    Chronic Systemic: Ingestion 1.
See also bromides.
Disaster Hazard: Dangerous; when heated to decomposition, it emits highly toxic fumes of hydrobromic acid.
Countermeasures
Personal Hygiene: Section 3.
Storage and Handling: Section 7.

FERROUS CHLORIDE
General Information
Synonym: Lawrencite.
Description: Green to yellow, deliquescent crystals.
Formula: $FeCl_2$.
Constants: Mol wt: 126.8, mp: 670–674°C, bp: 1026°C, d: 2.98, vap. press.: 10 mm at 700°C.
Hazard Analysis
Toxic Hazard Rating:
    Acute Local: Irritant 1; Ingestion 1.

TOXIC HAZARD RATING CODE (For detailed discussion, see Section 1.)

0 NONE: (a) No harm under any conditions; (b) Harmful only under unusual conditions or overwhelming dosage.

1 SLIGHT: Causes readily reversible changes which disappear after end of exposure.

2 MODERATE: May involve both irreversible and reversible changes; not severe enough to cause death or permanent injury.

3 HIGH: May cause death or permanent injury after very short exposure to small quantities.

U UNKNOWN: No information on humans considered valid by authors.

Acute Systemic: Ingestion 1.
Chronic Local: Irritant 1.
Chronic Systemic: U.
Disaster Hazard: Dangerous; when heated to decomposition, it emits highly toxic fumes of hydrochloric acid.
Countermeasures
Personal Hygiene: Section 3.
Storage and Handling: Section 7.

FERROUS CHLOROPLATINATE
General Information
Description: Yellow, hexagonal crystals.
Formula: $FePtCl_6 \cdot 6H_2O$.
Constants: Mol wt: 571.9, mp: decomposes, d: 2.714.
Hazard Analysis
Toxicity: See platinum compounds. Chloroplatinates generally tend to be highly irritating to the skin, but systemic poisoning is unknown.
Disaster Hazard: Dangerous; when heated to decomposition, it emits toxic fumes of chlorides.
Countermeasures
Storage and Handling: Section 7.

FERROUS FERRICYANIDE
General Information
Description: Deep blue crystals.
Formula: $Fe[Fe(CN)_6]_2$.
Constants: Mol wt: 591.5, mp: decomposes.
Hazard Analysis and Countermeasures
See ferricyanides.

FERROUS FERROCYANIDE
General Information
Description: Amorphous, blue-white crystals.
Formula: $Fe_2Fe(CN)_6$.
Constant: Mol wt: 323.7.
Hazard Analysis and Countermeasures
See ferrocyanides.

FERROUS FLUORIDE
General Information
Description: Crystals.
Formula: FeF_2.
Constants: Mol wt: 93.85, mp: $> 1000°C$, d: 3.95–4.33.
Hazard Analysis and Countermeasures
See fluorides

FERROUS FLUOSILICATE
General Information
Description: Trigonal colorless crystals.
Formula: $FeSiF_6 \cdot 6H_2O$.
Constants: Mol wt: 306.0, d: 1.961.
Hazard Analysis and Countermeasures
See fluosilicates.

FERROUS FUMARATE
General Information
Description: Reddish-brown, anhydrous powder; odorless; insoluble in alcohol; very slightly soluble in water.
Formula: $FeC_4H_2O_4$.
Constants: Mol wt: 170, mp: $> 280°C$.
Hazard Analysis
Toxic Hazard Rating: U. A food additive permitted in the feed and drinking water of animals and/or for the treatment of food producing animals (Section 10).

FERROUS GLUCONATE
General Information
Synonym: Iron gluconate.
Description: Yellowish gray or pale greenish yellow, fine powder or granules with slight odor; soluble in water and glycerin; insoluble in alcohol.
Formula: $Fe(C_6H_{11}O_7)_2 \cdot 2H_2O$.
Constant: Mol wt: 426.
Hazard Analysis
Toxic Hazard Rating: U. A nutrient and or dietary supple-

ment food additive. Also used as a trace mineral added to animal feeds (Section 10).

FERROUS IODIDE
General Information
Description: Hexagonal gray crystals.
Formula: FeI_2.
Constants: Mol wt: 309.7, mp: 177°C, d: 5.315.
Hazard Analysis and Countermeasures
See iodides.

FERROUS LACTATE
General Information
Synonym: Iron lactate.
Description: Greenish white crystals; slight peculiar odor; moderately soluble in water; slightly soluble in alcohol.
Formula: $Fe(C_3H_5O_3) \cdot 3H_2O$.
Constant: Mol wt: 199.
Hazard Analysis
Toxic Hazard Rating: A nutrient and/or dietary supplement food additive (Section 10).

FERROUS NITRATE
General Information
Description: Rhombic, gray crystals.
Formula: $Fe(NO_3)_2 \cdot 6H_2O$.
Constants: Mol wt: 288.0, mp: 60.5°C decomposes.
Hazard Analysis and Countermeasures
See nitrates.

FERROUS OXALATE
General Information
Description: Pale yellow powder.
Formula: $FeC_2O_4 \cdot 2H_2O$.
Constants: Mol wt: 179.90, mp: decomposes at 160°C, d: 2.28.
Hazard Analysis and Countermeasures
See oxalates.

FERROUS OXIDE
General Information
Description: Black crystals.
Formula: FeO.
Constants: Mol wt: 71.84, mp: 1420°C, d: 5.7.
Hazard Analysis
Toxic Hazard Rating:
    Acute Local: Irritant 1; Inhalation 1.
    Acute Systemic: 0.
    Chronic Local: 0.
    Chronic Systemic: 0.
TLV: ACGIH (recommended) 10 milligrams per cubic meter of air (fume).
Toxicity: A trace mineral added to animal feeds (Section 10).
A common air contaminant (Section 4).
Countermeasures
Personal Hygiene: Section 3.

FERROUS PERCHLORATE
General Information
Description: Green crystals.
Formula: $Fe(ClO_4)_2 \cdot 6H_2O$.
Constants: Mol wt: 362.86, mp: decomposes $> 100°C$.
Hazard Analysis
Toxic Hazard Rating:
    Acute Local: Irritant 1; Inhalation 1.
    Acute Systemic: Ingestion 1.
    Chronic Local: Ingestion 1.
    Chronic Systemic: U.
Fire Hazard: Moderate; a powerful oxidizer (Section 6).
Explosion Hazard: Moderate, when shocked, exposed to heat, or by chemical reaction.
Disaster Hazard: Dangerous; shock or heat will explode it; when heated to decomposition, it emits highly toxic fumes of oxides of chlorides; can react with reducing materials.

Countermeasures
Personal Hygiene: Section 3.
Storage and Handling: Section 7.

FERROUS PYROARSENITE
General Information
Description: Green-white crystals.
Formula: $Fe_2As_2O_5$.
Constant: Mol wt: 341.52.
Hazard Analysis and Countermeasures
See arsenic compounds.

FERROUS REDUCTUM. See iron, reduced.

FERROUS SULFATE
General Information
Synonyms: Iron sulfate; copperas; green vitriol.
Description: Monoclinic crystals.
Formula: $FeSO_4 \cdot H_2O$.
Constants: Mol wt: 169.93, d: 2.99–3.08.
Hazard Analysis
Toxic Hazard Rating:
  Acute Local: Irritant 1; Ingestion 1; Inhalation 1.
  Acute Systemic: 0.
  Chronic Local: 0.
  Chronic Systemic: 0.
Toxicity: Used as a nutrient and/or dietary supplement food
  additive as well as a trace mineral added to animal
  feed; it is a substance migrating to foods from pack-
  aging materials (Section 10).
Countermeasures
Personal Hygiene: Section 3.

FERROUS SULFIDE
General Information
Synonym: Troilite.
Description: Black-brown crystals.
Formula: FeS.
Constants: Mol wt: 87.9, mp: 1193°C, bp: decomposes,
  d: 4.84.
Hazard Analysis and Countermeasures
See sulfides.

FERROUS THIOCYANATE
General Information
Description: Rhombic green crystals.
Formula: $Fe(SCN)_2 \cdot 3H_2O$.
Constants: Mol wt: 226.07, mp: decomposes.
Hazard Analysis and Countermeasures
See thiocyanates.

FERROVANADIUM DUST
General Information
Description: A gray to black dust.
Formula: FeV.
Hazard Analysis
Toxicity: See vanadium compounds and iron compounds.
TLV: ACGIH (accepted); 1 milligram per cubic meter of air.
Fire Hazard: Moderate, when exposed to heat or flame
  (Section 6).
Explosion Hazard: Slight, when exposed to flame.

FERRUM. See iron.

FERTILIZER AMMONIATING SOLUTION CONTAIN-
ING FREE AMMONIA. See ammonia and com-
pressed gasses, N.O.S.
Shipping Regulations: Section 11.
  I.C.C.: Nonflammable gas; green label, 300 pounds.

Coast Guard Classification: Noninflammable gas; green
  gas label.
IATA: Nonflammable gas, green label, not accetpable
  (passenger), 140 kilograms (cargo).

FIBER, BURNT
Hazard Analysis
Toxic Hazard Rating:
  Acute Local: Irritant 1; Inhalation 1.
  Acute Systemi: 0.
  Chronic Local: Inhalation 1.
  Chronic Systemic: 0.
Fire Hazard: Moderate, when exposed to heat or flame; can
  react with oxidizing materials (Section 6).
Explosion Hazard: Slight, when exposed to heat or flame.
Countermeasures
Personal Hygiene: Section 3.
Storage and Handling: Section 7.
Shipping Regulations: Section 11.
  I.C.C.: Flammable solid; yellow label, not accepted.
  Coast Guard Classification: Inflammable solid; yellow
  label.
  IATA: Flammable solid, not acceptable (passenger and
  cargo).

"FIBERGLAS"
General Information
Description: A form of fibrous glass; also applied to pro-
  prietary glass flakes.
Hazard Analysis
Toxic Hazard Rating:
  Acute Local: Irritant 2; Inhalation 1.
  Acute Systemic: 0.
  Chronic Local: Irritant 2.
  Chronic Systemic: U.
TLV: ACGIH (tentative) 5 milligrams per cubic meter of
  air.
Toxicology: This proprietary product is glass in form of
  fine fibers. When handled by workers for a period of
  time, it can cause considerable skin irritation, particularly when in intimate contact with the skin. The possibility of lung injury due to inhalation of fine particles
  of this material has been raised repeatedly, but according to the best evidence available, pulmonary effects do
  not arise from this source. It has been reported that the
  dermatitis due to this material is caused by the mechanical impingement of tiny pieces of glass upon the
  skin during handling. Occasionally, it is thought to be
  due to the tricresyl phosphate which is added to the
  glass wool as a dust adhesive, or it is sometimes blamed
  upon the tar or phenolformaldehyde-resin binder which
  this material may contain. An examination of the
  minute lesions resulting from the handling of glass wool
  shows that these are never of an allergic eczematous
  character. With continued use, most workers become
  hardened to it and the temporary itching, swelling and
  redness subside. Reassurance and treatment with
  calamine lotion and phenol plus other protective
  measures take care of about 95% of the cases.
Countermeasures
Personnel Protection: Section 3.
Shipping Regulations: Section 11.
Fiberglass repair kits.
  IATA: Oxidizing material, yellow label, $\frac{1}{2}$ liter ($\frac{1}{2}$
  kilogram) (passenger and cargo).

TOXIC HAZARD RATING CODE *(For detailed discussion, see Section 1.)*

FIBERS OR FABRICS, WITH ANIMAL OR VEGETABLE OILS
Hazard Analysis
Toxicity: Variable; possible allergen.
Fire Hazard: Moderate, when exposed to heat or flame; can react vigorously with oxidizing materials (Section 6).
Explosion Hazard: Moderate, when finely divided and exposed to flame.
Countermeasures
Storage and Handling: Section 7.
Shipping Regulations: section 11.
  I.C.C.:Flammable solid, yellow label, not accepted.
  Coast Guard Classification: Inflammable solid; yellow label.
  IATA: Flammable solid, not acceptable (passenger and cargo).

FILM, MOTION PICTURE OR X-RAY, NITROCELLU-LOSE, INCLUDING MIXED SHIPMENTS WITH NON-FLAMMABLE OR SLOW BURNING OIL.
Shipping Regulations: Section 11.
  ICC: Flammable solid, yellow label, 200 pounds.
  IATA: Flammable solid, yellow label, 25 kilograms (gross) (passenger); 95 kilograms (gross) (cargo).

FILM, MOTION PICTURE OR X-RAY, NITROCELLU-LOSE, OLD, WORN OUT, OR SCRAP.
Shipping Regulations: Section 11.
  IATA: Flammable solid, not acceptable (passenger and cargo).

FILM, MOTION PICTURE, OR X-RAY, NONFLAM-MABLE OR SLOW BURNING
Shipping Regulations: Section 11.
  IATA: No limit (passenger and cargo).

FIRE   EXTINGUISHER CHARGES. See also sulfuric acid.
Shipping Regulations: Section 11.
  I.C.C.: Corrosive liquid, white label, 1 gallon.
  Coast Guard Classification: Corrosive liquid; white label.
  IATA: (containing > 3.5 grams of propellant explosive per unit). Explosive, not acceptable (passenger and cargo).
    (containing < 3.5 grams of propellant explosive unit): Explosive, explosive label, 25 kilograms (passenger), 70 kilograms (cargo).
    (containing sulfuric acid): Corrosive liquid, white label, 1 liter (passenger), 5 liters (cargo).

FIRE EXTINGUSHERS
Shipping Regulations: Section 11.
  ICC: Nonflammable gas, green label, 300 pounds.
  IATA: Nonflammable gas, green label, 70 kilograms (passenger) 140 kilograms (cargo).

FIRE EXTINGUISHING GASES
Shipping Regulations: Section 11.
  I.C.C. Classification: Nonflammable gas; green label.
  Coast Guard Classification: Noninflammable gas.

FIRE LIGHTERS
Shipping Regulations: Section 11.
  ICC: Flammable solid, yellow label, 12 kilograms (passenger), 45 kilograms (cargo).

FIREWORKS, COMMON. See explosive low.
Shipping Regulations: Section 11.
  I.C.C. Classification: Class C explosive, No label, 200 pounds.
  Coast Guard Classification: Explosive C.

FIREWORKS, n.o.s.
Shipping Regulations: Section 11.
  IATA: Explosive, explosive label, not acceptable (passenger), 95 kilograms (cargo).

FIREWORKS, SPECIAL. See explosives, low.
Shipping Regulations: Section 11.
  I.C.C.: Explosive B; special fireworks label, 200 pounds.
  Coast Guard Classification: Explosive B; special fireworks label.

FISCHER'S SALT. See potassium cobaltinitrate.

FISH BERRY. See cocculus solid.

FISH OILS
General Information
Constant: Flash p: about 420° F.
Hazard Analysis
Toxicity: May act as irritants to the skin or as allergens; insecticides when combined with sodium or potassium base as a soap. A substance migrating to food from packaging materials (Section 10).
Fire Hazard: Slight, when exposed to heat or flame.
Spontaneous Heating: Yes.
Countermeasures
To Fight Fire: Foam, carbon dioxide, dry chemical or carbon tetrachloride (Section 6).
Personal Hygiene: Section 3.
Storage and Handling: Section 7.
Shipping Regulations: Section 11
  Coast Guard Classification: Hazardous article.

FISH SCRAP, DRY
Hazard Analysis
Toxicity: Details unknown, a mild allergen; a nuisance dust.
Fire Hazard: Moderate, when exposed to heat or flame; can react with oxidizing materials (Section 6).
Countermeasures
Storage and Handling: Section 7.
Shipping Regulations: Section 11.
  Coast Guard Classification: Hazardous article.

FISH SCRAP OR FISH MEAL with less than 6% or more than 12% moisture
Hazard Analysis
Toxicity: Details unknown; a mild allergen; a nuisance dust.
Fire Hazard: Moderate, when exposed to heat or flame; can react with oxidizing materials.
Countermeasures
Storage and Handling: Section 7.
Shipping Regulations; section 11.
  I.C.C.: Flammable solid; yellow label, not accepted.
  Coast Guard Classification: Inflammable solid; yellow label.

FISSILE RADIOACTIVE MATERIALS, n.o.s.
Shipping Regulations: Section 11.
  ICC: Class D poison, radioactive materials (red) label; Section 73.393 (g) and (m).

FLAME RETARDANT COMPOUND, LIQ.
Shipping Regulations: Section 11.
  ICC: Corrosive liquid, white label, 10 gallon.
  IATA: Corrosive liquid, white label, 1 liter (passenger), 40 liters (cargo).

FLAMMABLE LIQUIDS, n.o.s.
Shipping Regulations: Section 11.
  ICC: Flammable liquid, red label, 10 gallons.
  IATA: Flammable liquid, red label, 1 liter (passenger), 40 liters (cargo).

FLAMMABLE SOLIDS, n.o.s.
Shipping Regulations: Section 11.
  ICC: Flammable solid, yellow label, 25 pounds.
  IATA: Flammable solid, yellow label, 12 kilograms (passenger and cargo).

FLARES
Shipping Regulations: Section 11.

IATA (regular and signal): Explosive, explosive label, 25 kilograms (passenger), 95 kilograms (cargo).
(aeroplane): Explosive, explosive label, not acceptable (passenger), 95 kilograms (cargo).

FLAVIN MONONUCLEOTIDE. See riboflavin 5'-phosphate.

FLEXIBLE LINEAR SHAPED CHARGES, METAL CLAD
Shipping Regulations: Section 11.
ICC: Class C explosive, 300 pounds.

"FLEXOL" PLASTICIZER CC-55. See hexyldiethyl heptylate.

FLOUR
Hazard Analysis
Toxic Hazard Rating:
Acute Local: Irritant 1; Allergen 1.
Acute Systemic: Inhalation 2.
Chronic Local: Allergen 1; Inhalation 1.
Chronic Systemic: Inhalation 1.
Fire Hazard: Moderate in the form of dust.
Explosion Hazard: Moderate in the form of dust, when exposed to heat or flame (Section 6). Severe flour dust explosions have occurred in grain elevators and processing plants.
Countermeasures
Ventilation Control: Section 2.
Personal Hygiene: Section 3.
Storage and Handling: Section 7.

FLOWERS OF SULFUR. See sulfur.

FLOWERS OF ZINC. See zinc oxide.

FLUE DUST
Hazard Analysis
Toxicity: Variable; depends on composition. A common air contaminant (Section 4).
Fire Hazard: Moderate (Section 6).
Explosion Hazard: Moderate, when exposed to flame or spark.
Countermeasures
Ventilation Control: Section 2.
Shipping Regulations: Section 11.
I.C.C.: Poison B; poison label, 200 pounds.
Coast Guard Classification: Poison B; poison label.
IATA: Poison B, poison label, 25 kilograms (passenger), 95 kilograms (cargo).

FLUOACETATES
Hazard Analysis
Toxic Hazard Rating:
Acute Local: Irritant 2.
Acute Systemic: Ingestion 3.
Chronic Local: Irritant 2.
Chronic Systemic: Ingestion 2; Inhalation 2.
Toxicity: See also sodium fluoacetate.
Disaster Hazard: Dangerous; when heated to decomposition or on contact with acid or acid fumes, they emit highly toxic fumes of fluorides.
Countermeasures
Ventilation Control: Section 2.
Personnel Protection: Section 3.
First Aid: Section 1.
Storage and Handling: Section 7.

FLUOACETIC ACID. See fluoroacetic acid.

FLUOALUMINATES
Hazard Analysis
Toxicity: Details unknown. See also fluorides.
Disaster Hazard: Dangerous; when heated to decomposition or on contact with acid or acid fumes, it emits highly toxic fumes of fluorides.
Countermeasures
Storage and Handling: Section 7.

FLUOANTIMONATES
Hazard Analysis
Toxicity: See fluorides and antimony compounds.
Disaster Hazard: See fluorides and antimony compounds.
Countermeasures
Storage and Handling: Section 7.

FLUOBERYLLATES
Hazard Analysis
Toxic Hazard Rating: Very toxic.
Acute Local: Irritant 3; Inhalation 3.
Acute Systemic: Inhalation 3.
Chronic Local: Inhalation 3.
Chronic Systemic: Ingestion 3; Inhalation 3.
Toxicity: See beryllium compounds.
Disaster Hazard: See fluorides and beryllium compounds.
Countermeasures
Ventilation Control: Section 2.
Personnel Protection: Section 3.
First Aid: Section 1.
Storage and Handling: Section 7.

FLUOBORATES
Hazard Analysis
Toxicity: Highly toxic. See fluorides.
Disaster Hazard: See fluorides.
Countermeasures
Ventilation Control: Section 2.
Personnel Protection: Section 3.
First Aid: Section 1.
Storage and Handling: Section 7.

FLUOBORIC ACID
General Information
Description: Colorless liquid.
Formula: HBF_4.
Constants: Mol wt: 87.8, bp: decomposes at 130°C.
Hazard Analysis
Toxicity: Highly toxic. See fluorides. A very corrosive irritant. See hydrofluoric acid.
Disaster Hazard: See fluorides.
Countermeasures
Storage and Handling: Section 7.
Shipping Regulations: Section 11.
IATA: Corrosive liquid, white label, 1 liter (passenger), 5 liters (cargo).

FLUOGALLATES
Hazard Analysis and Countermeasures
See fluorides.

FLUOGERMANATES
Hazard Analysis and Countermeasures
See fluorides.

FLUOPHOSPHATES
Hazard Analysis and Countermeasures
See fluorides.

TOXIC HAZARD RATING CODE (For detailed discussion, see Section 1.)

0 NONE: (a) No harm under any conditions; (b) Harmful only under unusual conditions or overwhelming dosage.

1 SLIGHT: Causes readily reversible changes which disappear after end of exposure.

2 MODERATE: May involve both irreversible and reversible changes; not severe enough to cause death or permanent injury.

3 HIGH: May cause death or permanent injury after very short exposure to small quantities.

U UNKNOWN: No information on humans considered valid by authors.

FLUOPHOSPHORIC ACIDS
General Information
Description: A term used to designate several acids containing fluorine and phosphorous (from H_2PO_3F to HPF_6).
Hazard Analysis and Countermeasures
See fluorides and phosphates.

FLUORANTHENE
General Information
Description: Colorless solid.
Formula: $C_{16}H_{10}$
Constants: Mol wt: 202.24, mp: 102°C, bp: 367°C, vap. press.: 0.01 mm at 20 °C.
Hazard Analysis
Toxicity: Unknown. A polycyclic hydrocarbon found in air pollution studies. Limited animal experiments suggest moderate toxicity.
Fire Hazard: Slight, when exposed to heat or flame (Section 6).
Countermeasures
Storage and Handling: Section 7.

FLUORENE
General Information
Synonym: o-Biphenylenemethane.
Description: White, shining flakes.
Formula: $C_{13}H_{10}$.
Constants: Mol wt: 166.2, mp: 113°C, bp: 295°C, d: 1.202, vap. press.: 10 mm at 146.0°C.
Hazard Analysis
Toxicity: Unknown.
Fire Hazard: Slight; (Section 6).
Countermeasures
Storage and Handling: Section 7.

FLUORESCEIN
General Information
Synonym: Resorcinol phthalein.
Description: Orange-red, cyrstalline powder.
Formula: $C_{20}H_{12}O_5$.
Constants: Mol wt: 332.3, mp: 314–316°C with decomp.
Hazard Analysis
Toxic Hazard Rating:
  Acute Local: Allergen 1.
  Acute Systemic: U.
  Chronic Local: Allergen 1.
  Chronic Systemic: U.
Fire Hazard: Slight; (Section 6).
Countermeasures
Personal Hygiene: Section 3.
Storage and Handling: Section 7.

FLUORESONE. See ethyl-p-fluorphenyl sulfone.

FLUORIDES
Hazard Analysis
Toxic Hazard Rating:
  Acute Local: Irritant 3; Ingestion 3; Inhalation 3.
  Acute Systemic: Ingestion 3.
  Chronic Local: Irritant 1.
  Chronic Systemic: Ingestion 3; Inhalation 3.
TLV: ACGIH (recommended); 2.5 milligrams per cubic meter of air.
Toxicology: Inorganic fluorides are generally highly irritant and toxic. Acute effects resulting from exposure to fluorine compounds are due to hydrogen fluoride. Chronic fluorine poisoning, or "fluorosis," occurs among miners of cryolite, and consists of a sclerosis of the bones, caused by fixation of the calcium by the fluorine. There may also be some calcification of the ligaments. The teeth are mottled, and there is osteosclerosis and ostemalacia. The bony and ligamentous changes are demonstrable by x-ray.
  Loss of weight, anorexia, anemia, wasting and cachexia, and dental defects are among the common findings in chronic fluorine poisoning. There may be an eosinophilia, and impairment of growth in young workers.
  Organic fluorides are generally less toxic than other halogenated hydrocarbons.
Common air contaminants (Section 4).
Disaster Hazard: Dangerous; when heated to decomposition or on contact with acid or acid fumes, they emit highly toxic fumes.
Countermeasures
Ventilation Control: Section 2.
Personnel Protection: Section 3.
First Aid: Section 1.
Storage and Handling: Section 7.

FLUORINE
General Information
Description: Pale yellow gas.
Formula: F_2.
Constants: Mol wt: 38.0, mp: -223°C, bp: -187°C, d: 1.14 at -200°C; 1.108 at -188°C, vap. d.: 1.695.
Hazard Analysis
Toxic Hazard Rating:
  Acute Local: Irritant 3; Inhalation 3.
  Acute Systemic: U.
  Chronic Local: Irritant 3.
  Chronic Systemic: U.
See also fluorides.
Toxicology: A very powerful caustic irritant.
TLV: ACGIH (recommended); 0.1 part per million in air; 0.2 milligrams per cubic meter of air.
Fire Hazard: Dangerous, by chemical reaction with reducing agents (Section 6).
Disaster Hazard: Highly dangerous; when heated, it emits highly toxic fumes; will react with water or steam to produce heat, toxic and corrosive fumes.
Radiation Hazard: (Section 5). For permissible levels, see Table 5, p. 150.
  Artificial isotope ^{18}F, half life 110 m. Decays to stable ^{18}O by emitting positrons of 0.65 MeV.
Countermeasures
Ventilation Control: Section 2.
Personal Hygiene: Section 3.
First Aid: Section 1.
Storage and Handling: Section 7.
Shipping Regulations: Section 11.
  I.C.C.: Flammable gas; red gas label, 6 pounds.
  Coast Guard Classification: Inflammable gas; red gas label.
  I.A.T.A.: Flammable gas, not acceptable (passenger and cargo).

FLUORINE ANALOG OF DDT. See 1-trichloro-2,2-bis (p-fluorophenyl)ethane.

FLUORINE DIOXIDE
General Information
Synonym: Dioxygen difluoride.
Description: Brown gas; cherry red liquid; orange solid.
Formula: F_2O_2.
Constants: Mol wt: 70.0, mp: -163.5°C, d(of solid): 1.912 at -165°C; (of liquid): 1.45 at -57°C, bp: -57°C, decomposes at -100°C.
Hazard Analysis and Countermeasures
See fluorine.

FLUORINE MONOXIDE
General Information
Synonym: Oxygen fluoride.
Description: Colorless gas, yellowish brown liquid. Reacts slowly with water.
Formula: F_2O.
Constants: Mol wt: 54.0, d (liquid): 1.90 at -224°C, mp: -223.8°C, bp: -144.8°C.

Hazard Analysis
Toxicity: More toxic than fluorine. Attacks lungs with delayed appearance of symptoms. See fluorine.
TLV: ACGIH, 0.05 part per million of air; 0.1 milligram per cubic meter of air.
Fire Hazard: A very powerful oxidizer. Must be kept from contact with reducing agents.
Disaster Hazard: Dangerous; when heated to decomposition, it emits highly toxic fumes of fluorine.
Countermeasures
Storage and Handling: Section 7.

FLUORINE NITRATE
General Information
Synonym: Nitrogen trioxyfluoride.
Description: Colorless gas; acrid odor. Hydrolyzes upon contact with water.
Formula: NO_3F.
Constants: Mol wt: 81.0, d (liquid): 1.507 at $-45.9°C$; (solid): 1.951 at $-193.2°C$, mp: $-175°C$, bp: $-45.9°C$.
Hazard Analysis
Toxicity: Highly toxic. See fluorine.
Fire Hazard: Dangerous. Very powerful oxodizer. Conflagrates upon contact with reducers such as alcohol, ether, aniline.
Explosion Hazard: Very dangerous. Liquid explodes upon slight shock.
Disaster Hazard: Dangerous. Upon warming it emits highly toxic fumes.
Countermeasures
Storage and Handling: Section 7.

FLUORINE PERCHLORATE
General Information
Synonym: Chlorine tetroxyfluoride.
Description: Colorless gas, pungent, acrid odor. Very unstable.
Formula: $FClO_4$.
Constants: Mol wt: 118.5, mp: $-167.3°C$, bp: $-15.9°C$.
Hazard Analysis
Toxicity: Highly toxic. See fluorine.
Fire Hazard: Very powerful oxidizer. See perchlorates.
Explosion Hazard: Very dangerous. Explodes on slightest provocation such as contact with rough surfaces, dirt and grease, heating, melting.
Disaster Hazard: Very dangerous. When warmed it emits highly toxic fumes.
Countermeasures
Storage and Handling: Section 7.

FLUOROACETIC ACID
General Information
Synonym: Fluoroethanoic acid, fluoacetic acid.
Description: Colorless solid, water soluble.
Formula: CH_2FCOOH.
Constants: Mol wt: 78.0, mp: $33°C$, bp: $165°C$.
Hazard Analysis
Toxic Hazard Rating:
Acute Local: Irritant 2; Ingestion 2, Inhalation 2.
Acute Systemic: Ingestion 3.
Chronic Local: Irritant 2.
Chronic Systemic: Ingestion 3.
Toxicity: See also sodium fluoroacetate.
Toxicology: Is said to cause convulsions and ventricular fibrillation.
Disaster Hazard: Dangerous. When heated to decomposition it emits highly toxic fumes of fluorides.

FLUOROACETOPHENONE
General Information
Synonym: Phenacyl fluoride.
Description: Brown liquid; pungent odor.
Formula: $C_6H_5COCH_2F$.
Constant: Bp: $98°C$ at 8 mm.
Hazard Analysis
Toxic Hazard Rating:
Acute Local: Irritant 2; Ingestion 2; Inhalation 2.
Acute Systemic: U.
Chronic Local: U.
Chronic Systemic: U.
Disaster Hazard: See fluorides.
Countermeasures
Ventilation Control: Section 2.
Personnel Protection: Section 3.
Storage and Handling: Section 7.

FLUOROANILINE
General Information
Description: Liquid.
Formula: $FC_6H_4NH_2$.
Constants: Mol wt: 111, d: 1.1524, bp: $187.4°C$, mp: $0.82°C$.
Hazard Analysis
See aniline and fluorides.

FLUOROCARBON 115. See chloropentafluoroethane.

1-FLUORO-2,4-DINITROBENZENE
General Information
Synonym: 2,4-Dinitro-1-fluorobenzene.
Description: Crystals. Soluble in ether, benzene, propylene glycol.
Formula: $C_6H_3F(NO_2)_2$.
Constants: Mol wt: 186.1, mp: $26°C$, bp: $137°C$ at 20 mm.
Hazard Analysis
Toxicity: Details unknown. It is an irritant material. See related compound chlorodinitrobenzene.
Disaster Hazard: Dangerous. When heated to decomposition it emits highly toxic fumes of oxides of nitrogen and fluorides.

FLUOROETHANE. See ethyl fluoride.

FLUOROETHANOIC ACID. See fluoroacetic acid.

FLUOROFORM
General Information
Synonym: Trifluoromethane.
Description: Colorless gas.
Formula: CHF_3.
Constants: Mol wt: 70.02, mp: $-163°C$, bp: $-82.2°C$.
Hazard Analysis
Toxic Hazard Rating:
Acute Local: Inhalation 1.
Acute Systemic: Inhalation 2.
Chronic Local: U.
Chronic Systemic: U.
Toxicology: Irritating to respiratory tract. Narcotic in high concentrations.
Disaster Hazard: Dangerous; when heated to decomposition, it emits highly toxic fumes of fluorides.
Countermeasures
Ventilation Control: Section 2.
Storage and Handling: Section 7.
Shipping Regulations: Section 11.
IATA: Nonflammable gas, green label, 70 kilograms (passenger), 140 kilograms (cargo).

TOXIC HAZARD RATING CODE (For detailed discussion, see Section 1.)

0 NONE: (a) No harm under any conditions; (b) Harmful only under unusual conditions or overwhelming dosage.

1 SLIGHT: Causes readily reversible changes which disappear after end of exposure.

2 MODERATE: May involve both irreversible and reversible changes; not severe enough to cause death or permanent injury.

3 HIGH: May cause death or permanent injury after very short exposure to small quantities.

U UNKNOWN: No information on humans considered valid by authors.

FLUOROFORMYL FLUORIDE. See carbonyl fluoride.

FLUOROISOPROPOXYMETHYLPHOSPHINE OXIDE. See sarin.

FLUOROISOPROPYL ALCOHOL. See propylene fluorohydrin.

FLUOROMETHANE. See methyl fluoride.

FLUOROMETHYLPINACOLYLOXY PHOSPHINE OXIDE. See soman.

FLUOROPHOSPHORIC ACID
General Information
Description: Colorless, viscous liquid; miscible with water.
Formula: H_2PO_3F.
Constants: Mol wt: 100, d: 1.1818 at 25°C.
Hazard Analysis
Toxic Hazard Rating: Very toxic. See fluorides.
Countermeasures
Shipping Regulations: Section 11.
  IATA (anhydrous): Corrosive liquid, white label, not acceptable (passenger), 5 liters (cargo).

3-FLUOROPROPANE. See allyl fluoride.

2-FLUORO-2-PROPENE-1-OL
General Information
Formula: $CH_2OHCFCH_2$.
Constant: Mol wt: 76.
Hazard Analysis
Toxic Hazard Rating: U. Limited animal experiments suggest high toxicity.

FLUOROTRICHLOROMETHANE
General Information
Synonym: "Freon 11," trichlorofluoromethane.
Description: Colorless liquid.
Formula: $FCCl_3$.
Constants: Mol wt: 137.38, mp: $-111°C$, bp: 24.1°C, d: 1.494 at 17.2°C.
Hazard Analysis
Toxic Hazard Rating:
  Acute Local: Irritant 1; Ingestion 1; Inhalation 1.
  Acute Systemic: Ingestion 1; Inhalation 1; Skin Absorption 1.
  Chronic Local: Irritant 1.
  Chronic Systemic: Ingestion 1; Inhalation 1; Skin Absorption 1.
Toxicology: High concentrations cause narcosis and anesthesia.
TLV: ACGIH (recommended); 1000 parts per million in air. 5,600 milligrams per cubic meter of air.
Disaster Hazard: Dangerous; when heated to decomposition, it emits highly toxic fumes of fluorides and chlorides.
Countermeasures
Storage and Handling: Section 7.

2-FLUORYLAMINE
General Information
Formula: $C_6H_4CH_2C_6H_3NH_2$.
Constant: Mol wt: 181.3.
Hazard Analysis
Toxicity: Details unknown; a stomach insecticide. See also amines.
Disaster Hazard: Slight; when heated, it emits acrid fumes (Section 6).
Countermeasures
Storage and Handling: Section 7.

FLUOSILICATES
General Information
Synonym: Silicofluorides.
Hazard Analysis
Toxic Hazard Rating:
  Acute Local: Irritant 3; Ingestion 3; Inhalation 3.

Acute Systemic: Ingestion 3; Inhalation 3.
Chronic Local: Irritant 2; Ingestion 2; Inhalation 2.
Chronic Systemic: Ingestion 3; Inhalation 3.
Disaster Hazard: See fluorides.
Countermeasures
Ventilation Control: Section 2.
Personnel Protection: Section 3.
First Aid: Section 1.
Storage and Handling: Section 7.

FLUOSILICIC ACID. See hydrofluosilicic acid.

FLUOSULFONATES
Hazard Analysis
Toxic Hazard Rating:
  Acute Local: Irritant 3; Ingestion 3; Inhalation 3.
  Acute Systemic: U.
  Chronic Local: U.
  Chronic Systemic: Ingestion 2; Inhalation 2.
Disaster Hazard: Dangerous; when heated to decomposition, they emit highly toxic fumes of fluorides and oxides of sulfur; will react with water or steam to produce toxic and corrosive fumes.
Countermeasures
Ventilation Control: Section 3.
Personnel Protection: Section 3.
First Aid: Section 1.
Storage and Handling: Section 7.

FLUOSULFONIC ACID
General Information
Synonym: Fluosulfuric acid.
Description: Colorless, fuming, highly corrosive liquid.
Formula: HSO_3F.
Constants: Mol wt: 100.07, mp: $-87.3°C$, bp: 165.5°C, d: 1.743 at 15°C.
Hazard Analysis
Toxic Hazard Rating:
  Acute Local: Irritant 3; Ingestion 3; Inhalation 3.
  Acute Systemic: U.
  Chronic Local: U.
  Chronic Systemic: Ingestion 2; Inhalation 2.
Toxicology: A very powerful irritant. See fluorides.
Disaster Hazard: See fluosulfonates.
Countermeasures
Ventilation Control: Section 2.
Personnel Protection: Section 3.
First Aid: Section 1.
Storage and Handling: Section 7.
Shipping Regulations: Section 11.
  I.C.C.: Corrosive liquid; white label, 10 pints.
  Coast Guard Classification: Corrosive liquid; white label.
  IATA: Corrosive liquid, white label, not acceptable (passenger), 5 liters (cargo).

FLUOSULFURIC ACID. See fluosulfonic acid.

FLUOTELLURITES
Hazard Analysis
Toxicity: Details unknown. See also fluorides and tellurium compounds.
Disaster Hazard: See fluorides.
Countermeasures
Storage and Handling: Section 7.

"FLURAL"
General Information
Description: Dry, free-flowing powder.
Formula: $AlFSO_4 \cdot xH_2O$ to $Al_2F_4SO_4 \cdot xH_2O$.
Hazard Analysis and Countermeasures
Highly toxic. See fluorides.

FLUX, BLACK. Dangerous! A low explosive. For details, see explosives, low.

FLUX, WHITE
General Information
Composition: Sodium nitrate and sodium nitrite. See nitrates.
Hazard Analysis
Fire Hazard: Dangerous. This is a strong oxidizing agent.
Disaster Hazard: Keep away from heat and open flame; do not store near reducing agents or easily oxidized materials.
Countermeasures
Storage and Handling: Section 7.

FMN. See riboflavin 5'-phosphate.

FOLACIN. See pteroylglutamic acid.

FOLIC ACID. See pteroylglutamic acid.

FORMAL. See methylal.

FORMALDEHYDE (COMMERCIAL SOLUTIONS)
General Information
Synonyms: Methanal; methyl aldehyde; formalin.
Description: Clear, water-white very slightly acid, gas or liquid; pungent odor. Pure formaldehyde is not available commercially because of its tendency to polymerize. It is sold as aqueous solutions containing from 37% to 50% formaldehyde by weight and varying amounts of methanol. Some alcoholic solutions are used industrially and the physical properties and hazards may be greatly influenced by the solvent.
Formula: HCHO.
Constant: Mol wt: 30.03, lel: 7.0%, uel: 73.0%, autoign. temp.: 806°F, d: 1.0, bp: −3°F, flash p (37% [methanol free]): 185°F, flash p: [15% (methanol free)]: 122°F.
Hazard Analysis
Toxic Hazard Rating:
Acute Local: Irritant 3; Allergen 1; Ingestion 3; Inhalation 3.
Acute Systemic: Ingestion 3.
Chronic Local: Irritant 3; Allergen 1.
Chronic Systemic: U.
Toxicity: Toxic effects are mainly those of irritation. If swallowed it causes violent vomiting and diarrhea which can lead to collapse. A fungicide. A common air contaminant. (Section 4).
TLV: ACGIH (recommended); 5 parts per million in air; 6 milligrams per cubic meter in air.
Caution: Frequent or prolonged exposure can cause hypersensitivity.
Fire Hazard: Moderate; vapors will burn above flash point if exposed to flame, sparks, etc. Should formaldehyde be involved in a fire, irritating gaseous formaldehyde may be evolved.
Spontaneous Heating: No.
Explosion Hazard: When aqueous formaldehyde solutions are heated above their flash points, a potential explosion hazard exists. Higher formaldehyde concentrations or methanol content lower the flash point.
Disaster Hazard: Moderately dangerous; because of irritating vapor which may be in toxic concentrations locally if storage tank is ruptured.
Countermeasures
Ventilation Control: Section 2.
To Fight Fire: Carbon dioxide, dry chemical, foam, or water spray (Section 6).
Personnel Protection: Section 3.

First Aid: Section 1.
Storage and Handling: Section 7.
Shipping Regulations: Section 11.
Coast Guard Classification: Combustible liquid.
MCA warning label.
IATA: Other restricted articles, class A; no label required, 40 liters (passenger), 220 liters (cargo).

FORMALDEHYDE ACETAMIDE. See formicin.

FORMALDEHYDE CYANOHYDRIN. See glycolonitrile.

FORMALDEHYDE DIMETHYL ACETAL. See methylal.

FORMALDEHYDE GAS
General Information
Synonym: Methanal, methyl aldehyde.
Description: Colorless gas; pungent suffocating odor. Very soluble in water.
Formula: HCHO.
Constants: Mol wt: 30.3, d(air = 1.000): 1.067, d(water = 1.000): 0.815 at −20°/4°C, mp: −92°C, bp: −19.5°C, autoign. temp.: 572°F.
Hazard Analysis and Countermeasures
See formaldehyde (commercial solutions).

FORMALIN. See formaldehyde.
Formalin was originally used as a European trade name for 37% formaldehyde solution. It is commonly employed as a synonym for commercial formaldehyde.

FORMAMIDE
General Information
Synonym: Methanamide.
Description: Colorless, hygroscopic and oily liquid.
Formula: HCONH₂.
Constants: Mol wt: 45.04, fp: 2.6°C, vap. press.: 29.7 mm at 129.4°C, flash p: 310°F (COC), bp: 210.7°C, decomposes, d: 1.134 at 20°/4°C; 1.1292 at 25°/4°C.
Hazard Analysis
Toxic Hazard Rating:
Acute Local: Irritant 2; Ingestion 2; Inhalation 2.
Acute Systemic: Ingestion 1; Inhalation 1.
Chronic Local: U.
Chronic Systemic: U.
Toxicology: Based upon animal experiments.
Fire Hazard: Moderate. Vapors will burn in air at temperatures above 310°F.
Disaster Hazard: Moderate. When heated to decomposition, it emits toxic fumes.
Countermeasures
Personnel Protection: Section 3.
Storage and Handling: Section 7.

FORMAMINE. See hexamethylenetetramine.

FORMIC ACID
General Information
Synonyms: Methanoic acid; hydrogen carboxylic acid.
Description: Colorless, fuming liquid; pungent penetrating odor.
Formula: HCOOH.
Constants: Mol wt: 46.03, bp: 100.8°C fp: 8.2°C, flash p: 156°F (O.C.), d: 1.2267 at 15°/4°C; 1.220 at 20°/4°C, autoign. temp.: 1114°F, vap. press.: 40 mm at 24.0°C, vap. d.: 1.59, flash p (90% solution): 122°F, autoign. temp. (90% solution): 813°F, lel (90% solution): 18%, uel (90% solution): 57%.

TOXIC HAZARD RATING CODE *(For detailed discussion, see Section 1.)*

.0 NONE: (a) No harm under any conditions; (b) Harmful only under unusual conditions or overwhelming dosage.

1 SLIGHT: Causes readily reversible changes which disappear after end of exposure.

2 MODERATE: May involve both irreversible and reversible changes; not severe enough to cause death or permanent injury.

3 HIGH: May cause death or permanent injury after very short exposure to small quantities.

U UNKNOWN: No information on humans considered valid by authors.

Hazard Analysis
Toxic Hazard Rating:
  Acute Local: Irritant 3; Ingestion 3; Inhalation 3.
  Acute Systemic: Ingestion 3.
  Chronic Local: Irritant 2.
  Chronic Systemic: Ingestion 1.
TLV: ACGIH (recommended) 5 parts per million of air; 9 milligrams per cubic meter of air.
Toxicity: Note: A substance migrating to food from packaging materials (Section 10).
Fire Hazard: Moderate, when exposed to heat or flame.
Countermeasures
To Fight Fire: Water, carbon dioxide, dry chemical, or carbon tetrachloride (Section 6).
Ventilation Control: Section 2.
First Aid: Section 1.
Personnel Protection: Section 3.
Storage and Handling: Section 7.
Shipping Regulations (regular and solution): Section 11.
  I.C.C.: Corrosive liquid; white label, 5 gallons.
  Coast Guard Classification: Corrosive liquid; white label.
  MCA warning label.
  IATA: Corrosive liquid, white label, 1 liter (passenger), 20 liters (cargo).

FORMIC ETHER. See ethyl formate.

FORMICIN
General Information
Synonym: Formaldehyde acetamide.
Description: Colorless, very hygroscopic mass. Very soluble in water and organic solvents.
Formula: $CH_3CONHCH_2OH$.
Constants: Mol wt: 89.1, d: 1.14–1.18.
Hazard Analysis and Countermeasures
See formaldehyde gas and formaldehyde.

FORMOTHION. See 5-(n-formyl-N-methylcarbamoyl-methyl)-o,o-dimethyl phosphorodithioate.

FORMYL FLUORIDE
General Information
Description: Colorless, mobile liquid.
Formula: HCOF.
Constants: Mol wt: 48.02, bp: $-26°$ C.
Toxic Hazard Rating:
  Acute Local: Irritant 3; Ingestion 3; Inhalation 3.
  Acute Systemic: Ingestion 3.
  Chronic Local: U.
  Chronic Systemic: Ingestion 2; Inhalation 2.
Fire Hazard: Moderate, by chemical reaction (Section 6).
Disaster Hazard: Dangerous; when heated to decomposition or on contact with acid or acid fumes, it emits highly toxic fumes of fluorides; it will react with water or steam to produce toxic fumes, such as carbon monoxide.
Countermeasures
Ventilation Control: Section 2.
Personnel Protection: Section 3.
Storage and Handling: Section 7.

FORMYL FORMIC ACID. See glyoxylic acid.

FORMYL HYDROPEROXIDE. See performic acid.

5-(n-FORMYL-n-METHYL CARBAMOYL METHYL)-o,o-DEMETHYL PHOSPHORO DITHTOATE
Hazard Analysis
Toxicity: Highly toxic! A cholinesterase inhibitor: See parathion.

FOSVEX. See tetraethyl pyrophosphate.

FOX GLOVE. See digitalis.

FP SALT #4. See potassium hexafluorophosphate.

FREMY'S SALT. See potassium bifluoride.

FRENCH CHALK. See soapstone dust.

FRENCH POLISH
General Information
Description: Shellac dissolved in alcohol.
Hazard Analysis
Toxicity: See alcohol and shellac.
Fire Hazard: Moderate, when exposed to heat or flame; can react with oxidizing materials (Section 6).
Countermeasures
Storage and Handling: Section 7.

"FREON-11." See fluorotrichloromethane.

"FREON-12." See dichlorodifluoromethane.

"FREON-21." See dichlorofluoromethane.

"FREON-22." See chlorodifluoromethane.

"FREON-112." See tetrachlorodifluoroethane.

"FREON-113." See trifluorotrichloroethane.

"FREON-114." See 1,2-dichloro-1,1,2,2-tetrafluoroethane.

FUEL, AVIATION, TURBINE ENGINES
Shipping Regulations: Section 11.
  IATA: Flammable liquid, red label, 1 liter (passenger), 40 liters (cargo).

FUEL OIL NO. 1. See kerosene.

FUEL OIL NO. 2. See diesel oil.

FUEL OIL NO. 3.
General Information
Description: Somewhat viscous, brown, odoriferous liquid.
Constants: Flash p: 110–230° F (C.C.), d: < 1, autoign. temp.: 498° F.
Hazard Analysis
Toxicity: Unknown.
Fire Hazard: Moderate, when exposed to heat or flame (Section 6).
Explosion Hazard: Unknown.
Countermeasures
Ventilation Control: Section 2.
Personal Hygiene: Section 3.
Storage and Handling: Section 7.

FUEL OIL NO. 4
General Information
Description: Moderately viscous, dark, odoriferous liquid.
Constants: Flash p: 130 ° F, d: < 1, autoign. temp.: 505° F.
Hazard Analysis
Toxicity: Unknown.
Fire Hazard: Moderate, when exposed to heat or flame (Section 6).
Explosion Hazard: Unknown.
Countermeasures
Ventilation Control: Section 2.
To Fight Fire: Foam, carbon dioxide, dry chemical or carbon tetrachloride (Section 6).
Personal Hygiene: Section 3.
Storage and Handling: Section 7.
Shipping Regulations: Section 11.
  Coast Guard Classification: Combustible liquid.

FUEL OIL NO. 5
General Information
Synonym: Navy special.
Description: Viscous, oily liquid.
Constants: Flash p: 130 + ° F, d: < 1.
Hazard Analysis
Toxicity: Unknown.
Fire Hazard: Moderate, when exposed to heat or flame (Section 6).

Countermeasures

To Fight Fire: Foam, carbon dioxide, dry chemical or carbon tetrachloride (Section 6).

Ventilation Control: Section 2.

Personal Hygiene: Section 3.

Storage and Handling: Sect on 7.

Shipping Regulations: Section 11.

Coast Guard Classification: Combustible liquid.

FUEL OIL NO. 6
General Information

Description: Very viscous, dark colored, odoriferous liquid.

Constants: Flash p: 150+°F, d: < 1, autoign. temp.: 765°F.

Hazard Analysis

Toxicity: Unknown.

Fire Hazard: Moderate, when exposed to heat or flame (Section 6).

Countermeasures

To Fight Fire: Foam, carbon dioxide, dry chemical or carbon tetrachloride (Section 6).

Ventilation Control: Section 2.

Personal Hygiene: Section 3.

Storage and Handling: Section 7.

FUEL, PYROPHORIC
Shipping Regulations: Section 11.

IATA: Flammable liquid, not acceptable (passenger and cargo).

FULMINATE OF MERCURY, DRY
General Information

Description: White solid.

Formula: $HgC_2N_2O_2$.

Constants: Mol wt: 284.7, mp: explodes, d: 4.42.

Hazard Analysis

Toxicity: See mercury compounds, organic.

Fire Hazard: Dangerous! Material should be kept moist till used.

Explosion Hazard: See fulminates.

Disaster Hazard: Highly dangerous. See fulminates.

Countermeasures

Storage and Handling: Section 7.

Shipping Regulations: Section 11.

I.C.C. Classification: Forbidden explosive.

Coast Guard Classification: Prohibited.

IATA: Explosive, not acceptable (passenger and cargo).

FULMINATES
Hazard Analysis

Toxicity: Variable.

Fire Hazard: Dangerous. Keep away from heat and open flame!

Explosion Hazard: Severe, when shocked or exposed to heat or flame (see explosives, high and Section 7).

Storage and Handling: The fulminates are a group of explosives which are very sensitive to heat, impact, and friction when dry. They should be kept moist till ready for use. If compressed beyond 25,000 psi they become what is known as "dead-pressed," i.e., not capable of being exploded by flame. Fulminates are subject to deterioration when stored in hot climates. They decompose completely when they detonate and do so with great violence. They can be ignited with a flame or "spit," with a fuse, or with an electrically heated wire. They are widely used as initiators or primers for bringing about the detonation of high explosives or the ignition of powder. They are commonly used in combination with substances which provide a more prolonged blow and a bigger flame than fulminates alone. In the reinforced type of detonator, fulminates are made more effective by the addition of a more sensitive and powerful high explosive such as tetryl. This material is generally used in the manufacture of caps and detonators for initiating explosions for military, industrial, and sporting purposes. All precautions required for protection of magazines apply to storage of these materials. They should not be handled when frozen. Wet fulminate of mercury or wet floor coverings containing small quantities of fulminates may be burned on windrows of flammable material. Nonexplosive products are formed by neutralizing fulminates with cold sodium thiosulfate. All floors, tables and walls where the dry fulminates have been used should be washed with this solution.

In the manufacture of mercury fulminate, the fumes given off are toxic and flammable. Care is required to prevent fulminate dust from being carried off in the exhaust system; deposits thus made have caused explosions. Much attention should be given to cleanliness as foreign or gritty materials in the product may cause an unexpected explosion. The floors on which fulminates are used, should be covered with 1/16 inch cloth inserted rubber packing or its equal. All cracks and crevices should be covered. The walls of these rooms should be covered with glazed, water-proof material. Frequent washing with neutralizing solution is necessary. In manufacture, the fulminate is dried on muslin squares on a drying table. Drying tables may be heated with hot water or the dryhouse may be heated with an air blower system to between 50 and 60°C. Primer caps and detonators loaded with fulminate of mercury are less sensitive than the dry bulk material but must be handled with great care. Fires involving these assemblies should be treated the same as for the bulk material. They will explode as soon as fire reaches them. Stocks in an assembly or loading room should be kept as small as possible. Examples of fulminates commonly used in the explosives industry are mercury fulminate, copper fulminate and silver fulminate. See also Section 7.

FUMARIC ACID
General Information

Synonym: trans-Butenedioic acid.

Description: Colorless, odorless crystals.

Formula: HOOCCH:CHCOOH.

Constants: Mol wt: 116.1, mp: 287°C, bp: 290°C sublimes, d: 1.635 at 20°/4°C.

Hazard Analysis

Toxic Hazard Rating:

Acute Local: Irritant 1.

Acute Systemic: Ingestion 1.

Chronic Local: U.

Chronic Systemic: U.

Note: A food additive permitted in food for human consumption (Section 10).

Fire Hazard: Slight; (Section 6).

Countermeasures

Storage and Handling: Section 7.

FUMARIN. See 3-(α-acetonylfurfuryl)-hydroxy-coumarin.

FUMARYL CHLORIDE
General Information

Synonym: trans-Butenedioyl chloride.

TOXIC HAZARD RATING CODE (For detailed discussion, see Section 1.)

0 NONE: (a) No harm under any conditions; (b) Harmful only under unusual conditions or overwhelming dosage.

1 SLIGHT: Causes readily reversible changes which disappear after end of exposure.

2 MODERATE: May involve both irreversible and reversible changes not severe enough to cause death or permanent injury.

3 HIGH: May cause death or permanent injury after very short exposure to small quantities.

U UNKNOWN: No information on humans considered valid by authors

Description: Clear, straw-colored liquid.
Formula: (ClOCCH)₂.
Constants: Mol wt: 153, bp: 160.0°C, d: 1.408 at
  20°/4°C.
Hazard Analysis
Toxicity: Probably high. See hydrochloric acid.
Disaster Hazard: Dangerous; when heated to decomposi-
  tion, it emits highly toxic fumes of phosgene and hy-
  drochloric acid; it will react with water or steam to pro-
  duce toxic and corrosive fumes.
Countermeasures
Storage and Handling: Section 7.
Shipping Regulations: Section 11.
  IATA: Corrosive liquid, white label, 1 liter (passenger
  and cargo).

FUMING SULFURIC ACID. See oleum.

FUNGICIDES
Shipping Regulation: Section 11.
  IATA (corrosive liquid): Corrosive liquid, white label,
  1 liter (passenger), 40 liters (cargo).
  IATA (poisonous liquid): Poisonous liquid, poison label,
  1 liter (passenger), 220 liters (cargo).
  IATA (poison solid): Poisonous solid, poison label, 25
  kilograms (passenger), 95 kilograms (cargo).

FURAL. See furfural.

2-FURALDEHYDE. See furfural.

FURALTADONE
Hazard Analysis
Toxic Hazard Rating: U. A food additive permitted in
  food for human consumption (Section 10).

FURAN
General Information
Synonym: Furfurane, oxole.
Description: Water-white liquid.
Formula: C₄H₄O.
Constants: Mol wt: 68.07, mp: −85.65°C, bp: 31.36°C,
  lel: 2.3%, uel: 14.3%, flash p: < 32°F, d: 0.937 at
  20°/4°C, vap. d.: 2.35.
Hazard Analysis
Toxic Hazard Rating:
  Acute Local: U.
  Acute Systemic: Ingestion 3; Inhalation 3; Skin Absorp-
    tion 3.
  Chronic Local: U.
  Chronic Systemic: Ingestion 3; Inhalation 3; Skin Ab-
    sorption 3.
Caution: The exposure concentration limit of 10 ppm to-
  gether with its low boiling point requires that adequate
  ventilation be provided in areas handling this chemical.
  Contact with liquid must be avoided since this chemical
  can be absorbed through the skin. Thorough washing
  with soap and water followed by prolonged rinsing
  should be done immediately after accidental contact.
Fire Hazard: Highly dangerous when exposed to heat or
  flame; can react with oxidizing materials. Unstabilized, it may form unstable peroxides on exposure to air
  and should always be tested before distillation. Washing
  with an aqueous solution of ferrous sulfate slightly
  acidified with sodium bisulfate will remove these peroxides. Confirm by test. Contact with acids can initiate
  a violet exothermic reaction.
Explosion Hazard: Moderate, when exposed to flame. The
  low boiling point of this material makes it easy to obtain explosive concentrations of the vapor in inadequately ventilated areas.
Disaster Hazard: Highly dangerous, upon exposure to heat
  or flame; can react vigorously with oxidizing materials.
Countermeasures
Ventilation Control: Section 2.

To Fight Fire: Carbon dioxide, dry chemical or carbon
  tetrachloride (Section 6).
Personnel Protection: Section 3.
Storage and Handling: Section 7.

2-FURANCARBONAL. See furfural.

2-FURANMETHYLAMINE. See furfurylamine.

FURANYL BORIC ACID
General Information
Description: White crystals, water soluble.
Formula: (C₄H₃O)B(OH)₂.
Constants: Mol wt: 111.9, mp: 110°C (decomposes).
Hazard Analysis
Toxicity: Details unknown. See also boron compounds.

FURAZALIDONE
General Information
Synonym: N-5-Nitro-2-furfurylidene)-3-amino)-2-oxazoli-
  dinone.
Description: Yellow powder; odorless; slightly soluble in
  polyethylene glycol; insoluble in water, alcohol and
  peanut oil.
Formula: C₈H₇N₃O₅.
Constant: Mol wt: 225, mp: 255°C.
Hazard Analysis
Toxic Hazard Rating: U. A food additive permitted in the
  feed and drinking water of animals and/or for the treat-
  ment of food producing animals. Also permitted in food
  for human consumption (Section 10).

FURCELLARON
General Information
Description: Vegetable gum available as an odorless, white
  powder; soluble in warm water.
Hazard Analysis
Toxic Hazard Rating: U. A food additive permitted in food
  for human consumption (Section 10).

FURFURAL
General Information
Synonyms: 2-Furancarbonal; 2-furaldehyde; fural; furale.
Description: Colorless-yellow liquid; almond-like odor.
Formula: C₄H₃OCHO.
Constants: Mol wt: 96.1, mp: −36.5°C, bp: 161.7°C at
  764 mm, lel: 2.1% at 125°C, flash p: 140°F (C.C.), d:
  1.161 at 20°/20°C, autoign. temp.: 600°F, vap. d.:
  3.31.
Hazard Analysis
Toxic Hazard Rating:
  Acute Local: Irritant 2; Ingestion 2; Inhalation 2.
  Acute Systemic: Ingestion 2; Inhalation 2; Skin Ab-
    sorption 2.
  Chronic Local: U.
  Chronic Systemic: U.
TLV: ACGIH (recommended); 5 parts per million in air;
  20 milligrams per cubic meter of air. May be absorbed
  via skin.
Toxicology: The liquid is dangerous to the eyes. The vapor
  is irritant to mucous membranes and is a central
  nervous system poison. However, its low volatility reduces its toxic effect. Furfural which has been ingested
  has produced cirrhosis of the liver in rats. In industry
  there is a tendency to minimize the danger of acute
  effects resulting from exposure to it. This is true,
  particularly, because of its low volatility. Little is
  known concerning the possibility of chronic effects, such
  as nervous disturbances, following prolonged or severe
  exposure to this material.
Fire Hazard: Moderate, when exposed to heat or flame; can
  react with oxidizing materials.
Spontaneous Heating: No.
Explosion Hazard: Moderate, when exposed to heat or
  flame or by chemical reaction. An exothermic resinifica-
  tion of almost explosive violence can occur upon contact
  with strong mineral acids or alkalies.

Disaster Hazard: Moderately dangerous. Keep away from heat and open flame!

Countermeasures
Ventilation Control: Section 2.
To Fight Fire: Water, foam, carbon dioxide, dry chemical or carbon tetrachloride (Section 6).
Personnel Protection: Section 3.
Storage and Handling: Section 7.

FURFURAMIDE. See hydrofuramide.

FURFURANE. See furan.

FURFURYL ACETATE
General Information
Description: Colorless liquid turning brown on exposure to light and air; pungent odor; insoluble in water; soluble in alcohol and ether.
Formula: $C_4H_3OCH_2OOCCH_3$.
Constants: Mol. wt: 136, d: 1.1175 at 20°/4°C, bp: 175–177°C, vap. d.: 4.8, flash p: 185°F.
Hazard Analysis
Toxic Hazard Rating: U.
Fire Hazard: Moderate, when exposed to heat or flame.
Countermeasures
To Fight Fire: Water may be used to blanket fire.

FURFURYL ALCOHOL
General Information
Synonym: 2-Furyl carbinol.
Description: Clear, colorless, mobile liquid.
Formula: $C_4H_3OCH_2OH$.
Constants: Mol wt: 98.1, mp: −31°C, lel: 1.8%, uel: 16.3%, both between 72–122°C, bp: 171°C at 750 mm, flash p: 167°F (O.C.), d: 1.129 at 20°/4°C, autoign. temp.: 915°F, vap. press.: 1 mm at 31.8°C, vap. d.: 3.37.
Hazard Analysis
Toxicity: Details unknown. See also alcohols. Animal experiments suggest moderate toxicity.
TLV: ACGIH (recommended); 50 parts per million in air; 200 milligrams per cubic meter of air.
Fire Hazard: Moderate, when exposed to heat; can react with oxidizing materials.
Explosion Hazard: Moderate, when exposed to heat or flame.
Countermeasures
Storage and Handling: Section 7.
To Fight Fire: Alcohol foam, carbon dioxide, dry chemical or carbon tetrachloride (Section 6).

FURFURYLAMINE
General Information
Synonym: 2-Furanmethylamine.
Description: Light straw-colored liquid.
Formula: $C_4H_3OCH_2NH_2$.
Constants: Mol wt: 97.1, bp: 146°C, flash p: 99°F (O.C.), fp: −70°C, d: 1.0502 at 25°C, vap. d.: 3.35.
Hazard Analysis
Toxic Hazard Rating:
Acute Local: Irritant 2; Ingestion 2; Inhalation 2.
Acute Systemic: U.
Chronic Local: U.
Chronic Systemic: U.
Fire Hazard: Moderate, when exposed to heat or flame; can react with oxidizing materials.
Explosion Hazard: Unknown.

Countermeasures
Ventilation Control: Section 2.
To Fight Fire: Foam, carbon dioxide, dry chemical or carbon tetrachloride (Section 6).
Personnel Protection: Section 3.
Storage and Handling: Section 7.

FURNITURE POLISH
Hazard Analysis
Toxicity: Variable; allergen.
Countermeasures
Shipping Regulations: Section 11.
I.C.C.: Flammable liquid; red label, 10 gallons.
Coast Guard Classification: Combustible liquid; inflammable liquid; red label.
IATA: Flammable liquid, red label, 1 liter (passenger), 40 liters (cargo).

FUROIC ACID
General Information
Synonym: Pyromucic acid.
Description: White solid.
Formula: $C_4H_3OCO_2H$.
Constants: Mol wt: 112.1, mp: 133°C sublimes at 100°C, bp: 230–232°C.
Hazard Analysis
Toxicity: Details unknown; a bactericide.
Fire Hazard: Slight; (Section 6).
Countermeasures
Storage and Handling: Section 7.

FUROYL CHLORIDE
General Information
Description: Colorless liquid.
Formula: C_4H_3OCOCl.
Constants: Mol wt: 130.5, bp: 176°C, mp: −2°C.
Hazard Analysis
Toxic Hazard Rating:
Acute Local: Irritant 3; Ingestion 3; Inhalation 3.
Acute Systemic: U.
Chronic Local: U.
Chronic Systemic: U.
Disaster Hazard: Dangerous; when heated to decomposition, it emits highly toxic fumes of phosgene and chlorides; it will react with water or steam to produce toxic and corrosive fumes.
Countermeasures
Ventilation Control: Section 2.
Personnel Protection: Section 3.
Storage and Handling: Section 7.

FURS
Hazard Analysis
Toxic Hazard Rating:
Acute Local: Allergen 1.
Acute Systemic: 0.
Chronic Local: Allergen 1.
Chronic Systemic: Inhalation 1.
Fire Hazard: Moderate, when exposed to heat or flame (Section 6).

2-FURYL CARBINOL. See furfuryl alcohol.

FUSEES, HIGHWAY
Shipping Regulations: Section 11.
IATA: Explosive, explosive label, 25 kilograms (passenger), 95 kilograms (cargo).

TOXIC HAZARD RATING CODE (For detailed discussion, see Section 1.)

0 NONE: (a) No harm under any conditions; (b) Harmful only under unusual conditions or overwhelming dosage.

1 SLIGHT: Causes readily reversible changes which disappear after end of exposure.

2 MODERATE: May involve both irreversible and reversible changes not severe enough to cause death or permanent injury.

3 HIGH: May cause death or permanent injury after very short exposure to small quantities.

U UNKNOWN: No information on humans considered valid by author.

FUSEES, RAILWAY
Shipping Regulations: Section 11.
  IATA: Explosive, explosive label, 25 kilograms (passenger), 95 kilograms (cargo).

FUSE IGNITERS. See explosives, high.
Shipping Regulations: Section 11.
  I.C.C.: Explosive C, 150 pounds.
  Coast Guard Classification: Explosive C.
  IATA: Explosive, explosive label, 25 kilograms (passenger), 70 kilograms (cargo). Same for metal clad fuse igniters.

FUSE LIGHTERS. See explosives, high.
Shipping Regulations: Section 11.
  I.C.C. Classification: Explosive C, 150 pounds.
  Coast Guard Classification: Explosive C.
  IATA: Explosive, explosive label, 25 kilogram (passenger), 70 kilograms (cargo).

FUSEL OIL. See isoamyl alcohol, primary.
Shipping Regulations: Section 11.
  Coast Guard Classification: Combustible liquid.

FUSES. See explosives, high (detonating devices).

FUZES. Official U.S. Army spelling of "fuses." See explosives, high (detonating devices).

FUZES, COMBINATION PERCUSSION, AND TIME
Shipping Regulations: Section 11.
  I.C.C.: Class C explosive, 150 pounds.
  IATA: Explosive, explosive label, 25 kilograms (passenger), 70 kilograms (cargo).

FUZES, DETONATING
Shipping Regulations: Section 11.
  I.C.C. (Class A): Class A explosive, not accepted.
  I.C.C. (Class C): Class C explosive, 150 pounds.
  IATA: Explosive, not acceptable (passenger and cargo).

FUZES, DETONATING, RADIOACTIVE
Shipping Regulation: Section 11.
  I.C.C. (Class A): Class A explosive, not accepted.
  IATA: Explosive, not acceptable (passenger and cargo).

G

G-4. See 2,2′-methylene-bis(4-chlorophenol).

G-11. See 2,2′-methylene-bis(3,4,6-trichlorophenol).

G19258. See dimetan.

G22008. See pyrolan.

G24480. See diazinon.

GABIAN OIL
Hazard Analysis
Toxicity: Unknown.
Fire Hazard: Moderate, when exposed to heat or flame; can react with oxidizing materials (section 6).
Explosion Hazard: Unknown.
Countermeasures
Storage and Handling: Section 7.

GADOLINIUM
General Information
Description: A metallic element.
Formula: Gd.
Constant: Mol wt: 156.90.
Hazard Analysis
Toxicity: Unknown. As a lanthanon it may have anticoagulant effect on blood.
Radation Hazard: Section 5. For permissible levels, see Table 5, p. 150.
    Artifical isotope ^{153}Gd, half life 242 d. Decays to stable ^{153}Eu by electron capture. Emits gamma rays of 0.07–0.10 MeV and X-rays.
    Artificial isotope ^{159}Gd, half life 18 h. Decays to stable ^{159}Tb by emitting beta particles of 0.59 (13%), 0.89 (24%), 0.95 (6.3%) MeV. Also emits gamma rays of 0.06–0.36 MeV.

GADOLINIUM BROMIDE
General Information
Description: Rhombic plates.
Formula: $GdBr_3 \cdot 6H_2O$.
Constants: Mol wt: 504.74, d: 2.844 at 15°C.
Hazard Analysis
Toxicity: See bromides and gadolinium.

GADOLINIUM FLUORIDE
General Information
Description: White, gelatinous mass.
Formula: GdF_3.
Constant: Mol wt: 213.90.
Hazard Analysis and Countermeasures
See fluorides and gadolinium.

GADOLINIUM NITRATE
General Information
Description: Triclinic crystals.
Formula: $Gd(NO_3)_3 \cdot 5H_2O$.
Constants: Mol wt: 433.00, mp: 92°C, d: 2.406 at 15°C.

Hazard Analysis and Countermeasures
See nitrates and gadolinium.

GADOLINIUM OXALATE
General Information
Description: Monoclinic crystals.
Formula: $Gd_2(C_2O_4)_3 \cdot 10H_2O$.
Constants: Mol wt: 758.02, mp: $-6H_2O$ at 110°C.
Hazard Analysis
Toxicity: Highly toxic. See oxalates and gadolinium.

GADOLINIUM OXIDE
General Information
Description: White to cream colored hygroscopic powder insoluble in water, soluble in acids.
Formula: Gd_2O_3.
Constants: Mol wt: 362, d: 7.41, mp: 2330°C.
Hazard Analysis
Toxic Hazard Rating:
    Acute Local: Inhalation 2.
    Acute Systemic: U.
    Chronic Local: Inhalation 2.
    Chronic Systemic: Inhalation 1.
Data based on animal experiments. See also gadolinium.

GADOLINIUM SELENATE
General Information
Description: Monoclinic, pearly crystals.
Formula: $Gd_2(SeO_4)_3 \cdot 8H_2O$.
Constants: Mol wt: 886.81, mp: $-8H_2O$ at 130°C d: 3.309
Hazard Analysis and Countermeasures
See selenium and its compounds and gadolinium.

GADOLINIUM SULFIDE
General Information
Description: Yellow, hygroscopic mass.
Formula: Gd_2S_3.
Constants: Mol wt: 410.00, d: 3.8.
Hazard Analysis and Countermeasures
See sulfides and gadolinium.

GALENA. See lead compounds.

GALLIC ACID
General Information
Synonym: 3,4,5-Trihydroxybenzoic acid.
Description: White to pale fawn colored, odorless crystals Somewhat water soluble.
Formula: $C_7H_6O_5 \cdot H_2O$.
Constants: Mol wt: 188.1, d: 1.694, mp: 225–250°C (decomp.), $-H_2O$ at 100–120°C.
Hazard Analysis
Toxicity: Details unknown. Has produced weakness an neurological disturbances in experimental animals.

TOXIC HAZARD RATING CODE (For detailed discussion, see Section 1.)

0 NONE: (a) No harm under any conditions; (b) Harmful only under unusual conditions or overwhelming dosage.

1 SLIGHT: Causes readily reversible changes which disappear after end of exposure.

2 MODERATE: May involve both irreversible and reversible change not severe enough to cause death or permanent injury.

3 HIGH: May cause death or permanent injury after very short exposu to small quantities.

U UNKNOWN: No information on humans considered valid by author

GALLIUM
General Information
Description: Silvery-white metal; forms bright, shiny liquid at 30°C.
Formula: Ga.
Constants: At wt: 69.72, mp: 30.0°C, bp: 1983°C, d solid: 5.903 at 25°C; liquid: 6.093 at 32.38°C, vap. press.: 1 mm at 1349°C.
Hazard Analysis
Toxicity: When given to humans it has caused a metallic taste, dermatitis, and depression of bone marrow function. See gallium compounds.
Radiation Hazard: Section 5. For permissible levels, see Table 5, p. 150.
      Artificial isotope ^{72}Ga, half life 14 h. Decays to stable ^{72}Ge by emitting beta particles of 0.66–3.17 MeV. Also emits gamma rays of 0.63–2.50 MeV.
Countermeasures
Ventilation Control: Section 2.

GALLIUM ACETATE, BASIC
General Information
Description: White crystals.
Formula: $4Ga(C_2H_3O_2)_3 \cdot 2Ga_2O_3 \cdot 5H_2O$.
Constants: Mol wt: 1452.4, mp: decomposes < 100°C.
Hazard Analysis
Toxicity: Slight. See gallium compounds and gallium.

GALLIUM ARSENIDE
General Information
Description: Cubic crystals with dark gray metallic sheen.
Formula: GaAs.
Constants: Mol wt: 144.6, hardness: 4.5.
Hazard Analysis and Countermeasures
See arsenic compounds and gallium.

GALLIUM COMPOUNDS
Hazard Analysis
Toxic Hazard Rating:
  Acute Local: U.
  Acute Systemic: Ingestion 1.
  Chronic Local: U.
  Chronic Systemic: Ingestion 1.
Toxicology: Preliminary investigations show that these materials have a low order of toxicity. Work has been done with the oxide, tartrate, benzoate, and anthranilate, which were used by some investigators in the treatment of experimental syphilis. Amounts up to 15 mg/kg of body weight were injected intravenously and were tolerated without harm by laboratory animals. Larger doses produce hemorrhagic nephritis. In the case of gallium lactate, work done at the Naval Medical Research Institute showed that intravenous injections of about 40 mg of gallium per kg of body weight in rats or rabbits was lethal. Metallic gallium as well as the nitrate produced no skin injury and subcutaneous injection of relatively large amounts could be tolerated both by rabbits and rats without evidence of injury. It has, however, been demonstrated that gallium remains in the tissues for long periods of time following intramuscular injection of soluble gallium salts. Tissue distribution experiments indicate that it behaves like bismuth and mercury in that one respect.
Countermeasures
Ventilation Control: Section 2.

GALLIUM FERROCYANIDE
General Information
Description: Gelatinous, white material.
Formula: $Ga_4[Fe(CH)_6]_3$.
Constants: Mol wt: 914.8, bp: decomposes.
Hazard Analysis
Toxicity: Slight, See ferrocyanides and gallium.
Countermeasures
Ventilation Control: Section 2.

GALLIUM HYDRIDE
General Information
Synonyms: Digallane; galloethane.
Description: Colorless liquid.
Formula: Ga_2H_6.
Constants: Mol wt: 145.49, mp: −21.4°C, bp: 139°C; decomp. > 130°C, vap. press.: 2.5 mm at 0°C.
Hazard Analysis and Countermeasures
See hydrides and gallium.

GALLIUM MONOSELENIDE
General Information
Synonym: Gallium selenide.
Description: Dark red-brown, greasy leaf.
Formula: GaSe.
Constants: Mol wt: 148.68, mp: 960°C, d: 5.03.
Hazard Analysis
Toxicity: Probably toxic. See selinium and its compounds and gallium.
Fire Hazard: Moderate, by chemical reaction which liberates hydrogen selenide.
Explosion Hazard: Slight, by chemical reaction.
Disaster Hazard: Moderately dangerous; when heated to decomposition, it emits highly toxic fumes of selenium.
Countermeasures
Storage and Handling: Section 7.

GALLIUM MONOSULFIDE
General Information
Description: Sublimate, light yellow.
Formula: GaS.
Constants: Mol wt: 101.79, mp: 956 ± 10°C, d: 3.86 at 25°C.
Hazard Analysis and Countermeasures
See sulfides and gallium.

GALLIUM MONOTELLURIDE
General Information
Description: Soft black, greasy leaves.
Formula: GaTe.
Constants: Mol wt: 197.33, mp: 824°C, d: 5.44.
Hazard Analysis and Countermeasures
See tellurium compounds and gallium.

GALLIUM NITRATE
General Information
Description: White, deliquescent crystals.
Formula: $Ga(NO_3)_3 \cdot xH_2O$.
Constants: Mp: decomposes at 110°C, bp: $\rightarrow Ga_2O_3$ at 200°C.
Hazard Analysis and Countermeasures
See nitrates and gallium.

GALLIUM OXALATE
General Information
Description: White powder.
Formula: $Ga_2(C_2O_4)_3$.
Constants: Mol wt: 403.50, bp: decomposes at 195°C.
Hazard Analysis
Toxicity: Highly toxic. See oxalates and gallium.

GALLIUM OXIDE
General Information
Description: The sesquioxide Ga_2O_3 and suboxide Ga_2O are known. Both are stable at room temperature. Ga_2O_3 is found in α & β forms; both are white crystals; insoluble in water, soluble in alcohol; very slightly soluble in hot acid.
Formula: Ga_2O_3 (sesquioxide).
Constants: Mol wt: 188, d: 6.44 (α); 5.88 (β); mp: 1900°C; changes to beta at 600°C.
Hazard Analysis
Toxic Hazard Rating: Animal experiments suggest low toxicity for this and other gallium salts. See also gallium compounds and gallium.

GALLIUM OXYCHLORIDE
General Information
Description: Cyrstals.
Formula: $6GaOCl \cdot 14H_2O$.
Constant: Mol wt: 979.3.
Hazard Analysis
Toxicity: See gallium compounds and gallium.
Fire Hazard: Slight; Section 6.
Disaster Hazard: Dangerous. *See* chlorides.
Countermeasures
Storage and Handling: Section 7.

GALLIUM PERCHLORATE
General Information
Description: White, deliquescent crystals.
Formula: $Ga(ClO_4)_3 \cdot 6H_2O$.
Constants: Mol wt: 476.19, mp: decomposes at 175° C.
Hazard Analysis and Countermeasures
See perchlorates and gallium.

GALLIUM PHOSPHIDE
General Information
Description: Pale orange transparent crystals or whiskers up to 2 cm. long.
Formula: GaP
Constant: Mol wt: 101.2.
Hazard Analysis
Toxicity: See gallium compounds & gallium.

GALLIUM SELENATE
General Information
Description: Colorless crystals.
Formula: $Ga_2(SeO_4)_3 \cdot 16H_2O$.
Constant: Mol wt: 856.58.
Hazard Analysis and Countermeasures
See selenium compounds and gallium.

GALLIUM SELENIDE. See gallium monoselenide.

GALLIUM SESQUISELENIDE
General Information
Description: Reddish-black crystals.
Formula: Ga_2Se_3.
Constants: Mol wt: 376.32, mp: 1020° C, d: 4.92.
Hazard Analysis
Toxicity: See selenium compounds and gallium.
Fire Hazard: Moderate; it can liberate a flammable gas upon contact with moisture or acids.
Explosion Hazard: Slight; it can liberate a flammable gas upon contact with moisture or acids.
Disaster Hazard: Dangerous; when heated to decomposition or on contact with acid or acid fumes, it emits highly toxic fumes of selenium.
Countermeasures
Storage and Handling: Section 7.

GALLIUM SESQUISULFIDE
General Information
Description: Yellow crystals; white, amorphous mass.
Formula: Ga_2S_3.
Constants: Mol wt: 235.64, mp: $1255 \pm 10°$ C, d: 3.65 at 25° C.
Hazard Analysis and Countermeasures
See sulfides and gallium.

GALLIUM SESQUITELLURIDE
General Information
Description: Hard, black, brittle crystals.

Formula: Ga_2Te_3.
Constants: Mol wt: 522.3, mp: 790° C, d: 5.57.
Hazard Analysis and Countermeasures
See tellurium compounds and gallium.

GALLIUM SUBOXIDE
General Information
Description: Brown-black powder.
Formula: Ga_2O.
Constants: Mol wt: 155.44, mp: >660° C; sublimes > 500° C, d: 4.77.
Hazard Analysis
Toxicity: See gallium compounds and gallium.

GALLIUM SUBSELENIDE
General Information
Description: Crystals.
Formula: Ga_2Se.
Constants: Mol wt: 218.40, d: 5.02.
Hazard Analysis and Countermeasures
See selenium compounds and gallium.

GALLIUM SUBSULFIDE
General Information
Description: Green crystals or black pwoder.
Formula: Ga_2S.
Constants: Mol wt: 171.51, mp: decomp. >800° C, d: 4.18 at 25° C.
Hazard Analysis and Countermeasures
See sulfides and gallium.

GALLIUM SULFATE
General Information
Description: White powder.
Formula: $Ga_2(SO_4)_3$.
Constants: Mol wt: 427.6, mp: decomp. < 600° C.
Hazard Analysis
Toxicity: See gallium compounds and gallium.

GALLIUM TRIBROMIDE
General Information
Description: Colorless, deliquescent crystals.
Formula: $GaBr_3$.
Constants: Mol wt: 309.47, mp: $121.5 \pm 0.6°$ C, bp: 278.8° C, d: 3.69 at 25° C.
Hazard Analysis and Countermeasures
See bromides and gallium.

GALLIUM TRIFLUORIDE
General Information
Description: White powder.
Formula: GaF_3.
Constants: Mol wt: 126.7, mp: >1000° C, bp: sublimes approx. 950° C, d: 4.47.
Hazard Analysis and Countermeasures
See fluorides and gallium.

GALLIUM TRIFLUORIDE TRIAMMINE
General Information
Description: White powder.
Formula: $GaF_3 \cdot 3NH_3$.
Constants: Mol wt: 177.82, mp: $-NH_3$ at 100° C.
Hazard Analysis and Countermeasures
See fluorides and gallium.

GALLIUM TRIIODIDE
General Information
Description: Colorless to lemon-yellow, hygroscopic needles.

TOXIC HAZARD RATING CODE (For detailed discussion, see Section 1.)

0 NONE: (a) No harm under any conditions; (b) Harmful only under unusual conditions or overwhelming dosage.

1 SLIGHT: Causes readily reversible changes which disappear after end of exposure.

2 MODERATE: May involve both irreversible and reversible changes not severe enough to cause death or permanent injury.

3 HIGH: May cause death or permanent injury after very short exposure to small quantities.

U UNKNOWN: No information on humans considered valid by authors

Formula: GaI$_3$.

Constants: Mol wt: 450.48, mp: 212 ± 1°C, bp: 345°C, sublimes, d: 4.15 at 25°C.

Hazard Analyis and Countermeasures

See iodides and gallium.

GALLOETHANE. See gallium hydride.

GALLOTANNIC ACID. See tannic acid.

GAMBALA

General Information

Description: A wood.

Hazard Analysis

Toxic Hazard Rating:

Acute Local: Irritant 1; Allergen 1.

Acute Systemic: U.

Chronic Local: Allergen 1.

Chronic Systemic: U.

Fire Hazard: Moderate, when exposed to heat or flame; can react with oxidizing materials (Section 6).

Explosion Hazard: Slight, when exposed to heat or flame.

Countermeasures

Personal Hygiene: Section 3.

Storage and Handling: Section 7.

GAMBOGE

General Information

Synonym: Gambogia

Hazard Analysis

Toxic Hazard Rating:

Acute Local: Ingestion 3.

Acute Systemic: Ingestion 3.

Chronic Local: U.

Chronic Systemic: U.

Fire Hazard: Moderate, when exposed to heat or flame.

Countermeasures

Personal Hygiene: Section 3.

GAMMA HEXANE. See 1,2,3,4,5,6-hexachlorocyclohexane.

GAMMA ISOMER OF BHC. See 1,2,3,4,5,6-hexachlorocyclohexane.

GAMMA RAYS

General Information

Description: Electromagnetic radiation emitted by certain radioactive isotopes. Gamma rays are similar to x-rays, but are more energetic and penetrating than the x-rays ordinarily used in medical diagnosis and therapy.

Hazard Analysis

Radiation Hazard: See Section 5, for complete discussion.

GAMMEXANE. See 1,2,3,4,5,6-hexachlorocyclohexane.

GAMOXO. See 1,2,3,4,5,6-hexachlorocyclohexane.

GAMTOX. See 1,2,3,4,5,6-hexachlorocyclohexane.

GARBAGE TANKAGE CONTAINING LESS THAN 5% MOISTURE

Hazard Analysis

Fire Hazard: Dangerous, when exposed to heat or flame; can react vigorously with oxidizing materials.

Explosion Hazard: Unknown.

Countermeasures

Storage and Handling: Section 7.

Shipping Regulations: Section 11.

I.C.C.: Flammable solid; yellow label, not accepted.

Coast Guard Classification: Inflammable solid; yellow label.

GARDENIA OXIDE. See amyl benzyl ether.

GARDINOL. See sodium lauryl sulfate.

GARLIC OIL. See allyl sulfide.

GAS CANDLES CHARGED WITH FLAMMABLE GAS

Shipping Regulations: Section 11.

IATA: Flammable gas; red label, 0.6 kilograms (passenger), 12 kilograms (cargo).

GAS CYLINDERS, EMPTY. See Shipping Regulations Section 11.

GAS DRIPS, HYDROCARBON

Hazard Analysis

Toxicity: Unknown.

Fire Hazard: Dangerous, when exposed to heat or flame; can react vigorously with oxidizing materials (Section 6).

Explosion Hazard: Unknown.

Countermeasures

Storage.and Handling: Section 7.

Shipping Regulations: Section 11.

I.C.C.: Flammable liquid; red label, 10 gallons.

Coast Guard Classification: Combustible liquid.

Coast Guard Classification: Inflammable liquid; red label.

IATA: Flammable liquid, red label, 1 liter (passenger), 40 liters (cargo).

GASHOUSE TANKAGE. See spent oxide.

GAS IDENTIFICATION SETS

Shipping Regulations: Section 11.

I.C.C.: Poison A and C; poison gas label.

Coast Guard Classification: Poison A; poison gas label.

IATA: Poison A, not acceptable (passenger & cargo).

GAS OIL

General Information

Description: Yellow liquid.

Constants: Flash p: 150°F, d: <1, lel: 6.0%, uel: 13.5%, autoign. temp.: 640°F, bp: 230–250°C.

Hazard Analysis

Toxicity: Details unknown. See also kerosene.

Fire Hazard; moderate, when exposed to heat or flame; can react with oxidizing materials.

Explosion Hazard: Moderate, when exposed to heat or flame.

Countermeasures

Storage and Handling: Section 7.

To Fight Fire: Foam, carbon dioxide, dry chemical or carbon tetrachloride (Section 6).

Shipping Regulations: Section 11.

Coast Guard Classification: Combustible liquid.

GASOLINE

General Information

Synonym: Petrol.

Description: Clear, aromatic, volatile liquid.

Formula: A mixture of aliphatic hydrocarbons.

Constants: Flash p: −45°F, d: < 1.0, vap. d.: 3.0–4.0, ulc: 95–100, lel: 1.4%, uel: 7.6%, autoign. temp.: 495°F.

Hazard Analysis

Toxic Hazard Rating:

Acute Local: Irritant 1; Ingestion 1; Inhalation 1.

Acute Systemic: Inhalation 2.

Chronic Local: Irritant 1.

Chronic Systemic: Inhalation 1.

TLV: The composition of gasoline varies greatly and thus a single TLV for all types of gasoline is no longer applicable. In general the aromatic hydrocarbon content will determine what TLV applies. Consequently, the content of benzene, other aromatics and additives should be determined to arrive at the appropriate TLV. (Elkins, et. al. A.I.H.A.J. 24, 99, 1963).

Toxicology: Occurrence of chronic poisoning is questionable. Gasoline can cause hyperemia of the conjunctiva and other disturbances of the eyes. The vapors are not considered to be very poisonous, unless its concentration in air is sufficiently high to reduce the oxygen content below that needed to maintain life, in which case it acts as a simple asphyxiant. Gasoline is a common air contaminant. See Section 4.

Fire Hazard: Dangerous; when exposed to heat or flame; can react vigorously with oxidizing materials.

Explosion Hazard: Moderate, when exposed to heat or flame.

Disaster Hazard: Dangerous, in the presence of heat or flame.

Countermeasures

Ventilation Control: Section 2.

To Fight Fire: Foam, carbon dioxide, dry chemical or carbon tetrachloride (Section 6).

Storage and Handling: Section 7.

Shipping Regulations: Section 11.

I.C.C.: Flammable liquid; red label, 10 gallons.

Coast Guard Classification: Inflammable liquid; red label.

IATA: Flammable liquid, red label, 1 liter (passenger), 40 liters (cargo).

GASOLINE (CASINGHEAD)
General Information

Description: Aromatic, volatile liquid.

Constant: Flash p: 0° F or less.

Hazard Analysis

Toxicity: See gasoline.

Fire Hazard: Dangerous; when exposed to heat or flame; can react vigorously with oxidizing materials (Section 6).

Explosion Hazard: Unknown.

Disaster Hazard: Dangerous, in the presence of heat or flame.

Countermeasures

Ventilation Control: Section 2.

Storage and Handling: Section 7.

GASOLINE VAPORS. See gasoline.

GBH. See benzene hexachloride.

GEL COALITES. See explosives, high.

GELOSE. See agar-agar.

GELSEMININE
General Information

Description: An alkaloid.

Formula: $C_{20}H_{22}N_2O_2$.

Constants: Mol wt: 332.4, mp: 178° C.

Hazard Analysis

Toxic Hazard Rating:

Acute Local: U.

Acute Systemic: Ingestion 3.

Chronic Local: U.

Chronic Systemic: U.

Countermeasures

Personal Hygiene: Section 3.

First Aid: Section 1.

Storage and Handling: Section 7.

GELSEMIUM
General Information

Description: A mixture of alkaloids.

Composition: Chief constituents: Gelsemine, gelseminine and gelsemic acid.

Hazard Analysis

Toxicity: See gelseminine.

Countermeasures

Personal Hygiene: Section 3.

First Aid: Section 1.

Storage and Handling: Section 7.

GELSEMIUM SEMPERVIRENS
General Information

Synonym: Yellow jasmine.

Hazard Analysis

Toxicity: See gelseminine.

Caution: Can seriously affect the eyes.

Countermeasures

Storage and Handling: Section 7.

GEMERINE
General Information

Description: Crystals.

Formula: $C_{37}H_{59}NO_{11}$.

Constants: Mol wt: 693.9, decomposes at 202–205° C.

Hazard Analysis

Toxicity: Details unknown. Limited animal experiment indicate high toxicity. See also veratrine.

Disaster Hazard: Dangerous; when heated to decompositio it emits highly toxic fumes.

GENERALS. See explosives, high.

GENIPHENE. See toxaphene.

GENITHION. See parathion.

GERANIAL. See citral.

GERANIOL
General Information

Synonym: 3,7-Dimethyl-2,6 and 3,6-octadien-1-ol.

Description: Colorless to pale yellow; liquid oil; pleasar geranium odor.

Formula: $(CH_3)_2C:CH(CH_2)_2C(CH_3)$.

Constants: Mol wt: 154, d: 0.870–0.890 at 15° C, mp: 15° C bp: 230° C.

Hazard Analysis

Toxic Hazard Rating: U. A synthetic flavoring substanc and adjuvant. See Section 10.

GERANIOL ACETATE. See geranyl acetate.

GERANYL ACETATE
General Information

Synonym: Geraniol acetate.

Description: Colorless, sweet clear liquid; odor of lavende soluble in alcohol & ether, insoluble in water & glycerol.

Formula: $CH_3COOC_{10}H_{17}$.

Constants: Mol wt: 196, d: 0.907–0.918 at 15° C, b 128–129° C at 16 mm.

Hazard Analysis

Toxic Hazard Rating: U. A synthetic flavoring substance a adjuvant. See Section 10.

GERMANIUM
General Information

Description: Cubic gray-white metal.

Formula: Ge.

Constants: At wt: 72.60, mp: 958.5° C, bp: 2700° C, 5.35 at 20°/20° C.

Hazard Analysis

Toxicity: See germanium compounds.

Radiation Hazard: Section 5. For permissible levels, s Table 5, p. 150.

Artificial isotope [71]Ge, half life 11 d. Decays t stable [71]Ga by electron capture. Emits X-rays.

Fire Hazard: Moderate in the form of dust, when exposed t heat or flame. See also powdered metals (Section 6).

TOXIC HAZARD RATING CODE (For detailed discussion, see Section 1.)

0 NONE: (a) No harm under any conditions; (b) Harmful only under unusual conditions or overwhelming dosage.

1 SLIGHT: Causes readily reversible changes which disappear after end of exposure.

2 MODERATE: May involve both irreversible and reversible change not severe enough to cause death or permanent injury.

3 HIGH: May cause death or permanent injury after very short exposu to small quantities.

U UNKNOWN: No information on humans considered valid by author

Explosion Hazard: See powdered metals.
Countermeasures
Ventilation Control: Section 2.
Personal Hygiene: Section 3.

GERMANIUM BROMOFORM. See tribromogermane and germanium.

GERMANIUM CHLOROFORM. See trichlorogermane and germanium.

GERMANIUM COMPOUNDS
Hazard Analysis
Toxic Hazard Rating:
    Acute Local: U.
    Acute Systemic: Ingestion 2; Inhalation 2.
    Chronic Local: U.
    Chronic Systemic: Ingestion 2; Inhalation 1.
Toxicity: Little is known about the toxicity of organic germanium compounds; but they may resemble other organometals in having higher toxicity than inorganic forms.
Toxicology: Germanium compounds are considered to be of a low order of toxicity, but rare instances of poisoning have been reported in the literature. Interest is high in this material because of its close chemical relationship to arsenic. It has been found that the dioxide stimulates the generation of red blood cells, but it is believed to be relatively nontoxic. When germanium is given in sublethal amounts, it causes a pronounced tolerance to be exhibited. Germanium compounds are considered much less toxic than the corresponding lead and tin compounds. Buffered germanium dioxides in solution have been found to be nonirritating to the skin. Germanium hydride is a hemolytic gas and has been shown to have toxic properties at concentration levels of 100 ppm. It can cause death in concentration of 150 ppm.
Countermeasures
Ventilation Control: Section 2.
Personal Hygiene: Section 3.

GERMANIUM DIBROMIDE
General Information
Description: Colorless needles or plates.
Formula: $GeBr_2$.
Constants: Mol wt: 232.4, mp: 122.0°C, bp: decomposes.
Hazard Analysis
Toxicity: See bromides and germanium.
Disaster Hazard: See bromides.
Countermeasures
Storage and Handling: Section 7.

GERMANIUM DICHLORIDE
General Information
Description: White powder.
Formula: $GeCl_2$.
Constants: Mol wt: 143.5, mp: decomp. to Ge + $GeCl_4$.
Hazard Analysis
Toxicity: See germanium compounds and germanium.
Disaster Hazard: Slight; when heated, it emits toxic fumes.
Countermeasures
Storage and Handling: Section 7.

GERMANIUM DIFLUORIDE
General Information
Description: White, deliquescent crystals.
Formula: GeF_2.
Constants: Mol wt: 110.6, mp: decomp. >350°C, bp: sublimes.
Hazard Analysis
Toxicity: See fluorides and germanium.
Disaster Hazard: See fluorides.
Countermeasures
Storage and Handling: Section 7.

GERMANIUM DIIODIDE
General Information
Description: Yellow, hexagonal crystals.
Formula: GeI_2.
Constants: Mol wt: 326.44, mp: sublimes and decomposes.
Hazard Analysis
Toxicity: See iodides and germanium.
Disaster Hazard: See iodides.
Countermeasures
Storage and Handling: Section 7.

GERMANIUM DIOXIDE (INSOLUBLE)
General Information
Description: Tetragonal crystals.
Formula: GeO_2.
Constants: Mol wt: 104.60, mp: 1086 ± 5°C, d: 6.239.
Hazard Analysis
Toxicity: See germanium compounds and germanium.

GERMANIUM DIOXIDE (SOLUBLE)
General Information
Description: Hexagonal, colorless crystals.
Formula: GeO_2
Constants: Mol wt: 104.60, mp: 1115.0°C, d: 4.703 at 18°C.
Hazard Analysis
Toxicity: See germanium compounds.

GERMANIUM DISULFIDE
General Information
Description: White powder; orthorhombic, white crystals.
Formula: GeS_2.
Constants: Mol wt: 136.73, mp: ca. 800°C, bp: sublimes >600°C, d: 2.94 at 14°C.
Hazard Analysis and Countermeasures
See sulfides and germanium.

GERMANIUM HYDRIDE. See digermane.

GERMANIUM IMIDE
General Information
Description: White, amorphous powder.
Formula: $Ge(NH)_2$.
Constants: Mol wt: 102.63, mp: decomposes at 150°C.
Hazard Analysis
Toxicity: Details unknown. See germanium compounds and germanium.

GERMANIUM MONOHYDRIDE
General Information
Description: Brown powder.
Formula: GeH.
Constants: Mol wt: 73.61, mp: decomposes at 165°C, bp: may explode.
Hazard Analysis and Countermeasures
See hydrides and germanium compounds and germanium.

GERMANIUM MONOSULFIDE
General Information
Description: Yellow-red, amorphous or black crystals.
Formula: GeS.
Constants: Mol wt: 104.67; mp: 530°C; bp: sublimes >430°C; d (amorphous): 3.31, (rhombic): 4.01 at 14°/14°C.
Hazard Analysis and Countermeasures
See sulfides and germanium.

GERMANIUM MONOXIDE
General Information
Description: Black, crystalline powder.
Formula: GeO.
Constants: Mol wt: 88.60, mp: sublimes at 710°C.
Hazard Analysis
Toxicity: See germanium compounds and germanium.

GERMANIUM OXYCHLORIDE
General Information
Description: Colorless liquid.
Formula: GeOCl$_2$.
Constants: Mol wt: 159.51, mp: $-56.0°$C, bp: decomposes $>20°$C.
Hazard Analysis
Toxicity: See hydrochloric acid and germanium.
Disaster Hazard: Dangerous; when heated, it emits toxic fumes of chlorides.
Countermeasures
Storage and Handling: Section 7.

GERMANIUM TETRABROMIDE
General Information
Description: Gray-white crystals.
Formula: GeBr$_4$.
Constants: Mol wt: 392.26, mp: 26.1°C, bp: 186.5°C, d: 3.132 at 29°/29°C.
Hazard Analysis
Toxicity: See bromides and germanium.
Disaster Hazard: Dangerous; when heated, it emits toxic fumes of bromides.
Countermeasures
Storage and Handling: Section 7.

GERMANIUM TETRACHLORIDE
General Information
Description: Colorless liquid.
Formula: GeCl$_4$.
Constants: Mol wt: 214.43, mp: $-49.5°$C, bp: 83.1°C, d: 1.879.
Hazard Analysis
Toxic Hazard Rating:
Acute Local: Irritant 2; Inhalation 1.
Acute Systemic: Inhalation 1.
Chronic Local: U.
Chronic Systemic: Inhalation 1.
Toxicity: Animal experiments show moderate irritant of eyes and skin and slight systemic toxicity. See also germanium compounds, hydrochloric acid, and germanium.
Disaster Hazard: Dangerous; when heated to decomposition, it emits toxic fumes of chlorides; will react with water or steam to produce toxic and corrosive fumes.
Countermeasures
Storage and Handling: Section 7.

GERMANIUM TETRAETHYL. See tetraethyl germanium.

GERMANIUM TETRAFLUORIDE
General Information
Description: Colorless gas or solid; not liquid at atmospheric pressure.
Formula: GeF$_4$.
Constants: Mol wt: 148.60, mp: sublimes, d: 6.65 g/liter.
Hazard Analysis
Toxic Hazard Rating:
Acute Local: Irritant 3; Inhalation 3.
Acute Systemic: U.
Chronic Local: U.
Chronic Systemic: U.
Toxicity: See fluorides, germanium compounds, and germanium.
Disaster Hazard: Dangerous; emits highly toxic fumes of fluorides.
Countermeasures
Storage and Handling: Section 7.

GERMANIUM TETRAHYDRIDE. See monogermane.

GERMANIUM TETRAIODIDE
General Information
Description: Cubic, yellow crystals.
Formula: GeI$_4$.
Constants: Mol wt: 580.28, mp: 144.0°C, bp: decomposes, d: 4.322 at 26°/26°C.
Hazard Analysis
Toxicity: See iodides and germanium.
Disaster Hazard: Moderately dangerous; when heated to decomposition it emits toxic fumes of iodides.
Countermeasures
Storage and Handling: Section 7.

GERMANIUM TETRAPHENYL
General Information
Formula: Ge(C$_6$H$_5$)$_4$.
Constant: Mol wt: 380.6.
Hazard Analysis
Toxic Hazard Rating: U. Animal experiments show inhibition of blood formation. See also germanium compounds and germanium.

GESAROL. See DDT.

GHATTI GUM
General Information
Description: The gummy exudation from the stem of Anogeissus latifolia. Colorless to pale yellow tears; almost odorless; partially soluble in water.
Hazard Analysis
Toxic Hazard Rating: U. Used as a stablizer food additive. See Section 10.

GIBBERELLIC ACID
General Information
Description: A plant growth-promoting hormone, crystals, slightly soluble in water, soluble in methanol, ethanol, acetone, aqueous solutions of sodium bicarbonate & sodium acetate.
Formula: C$_{19}$H$_{22}$O$_6$.
Constants: Mol wt: 346; mp: 233–235°C.
Hazard Analysis
Toxic Hazard Rating: U. A food additive permitted in food for human consumption (Section 10).

GIBBERELLIC ACID, POTASSIUM SALT
General Information
Synonym: "Gibrel"[123]
Hazard Analysis
Toxic Hazard Rating: U. A food additive permitted in food for human consumption. See Section 10.

GILSONITE
General Information
Synonym: Uintaite.
Description: A black solid hydrocarbon mineral formed from petroleum millions of years ago by geologic processes.
Hazard Analysis
Toxic Hazard Rating:
Acute Local: Irritant 2; Inhalation 2.
Acute Systemic: U.
Chronic Local: Irritant 2; Allergen 2; Inhalation 2.
Chronic Systemic: U.
Toxicology: Has been known to cause photosensitization of skin.

TOXIC HAZARD RATING CODE *(For detailed discussion, see Section 1.)*

0 NONE: (a) No harm under any conditions; (b) Harmful only under unusual conditions or overwhelming dosage.

1 SLIGHT: Causes readily reversible changes which disappear after end of exposure.

2 MODERATE: May involve both irreversible and reversible changes; not severe enough to cause death or permanent injury.

3 HIGH: May cause death or permanent injury after very short exposure to small quantities.

U UNKNOWN: No information on humans considered valid by authors.

Fire Hazard: Moderate when exposed to heat or open flame.

Disaster Hazard: Moderate. When heated to decomposition, it emits toxic fumes.

Countermeasures

To Fight Fire: Water, foam, dry chemical and carbon dioxide (Section 6).

Storage and Handling: Section 7.

GIN

General Information

Composition: A 40 to 50% ethyl alcohol beverage made from juniper and coriander berries, etc.

Constants: Bp: 82°C, flash p: 89°F, ulc: 70, lel: 3.3%, uel: 19%, autoign. temp.: 799°F.

Hazard Analysis

Toxic Hazard Rating:
  Acute Local: U.
  Acute Systemic: Ingestion 2.
  Chronic Local: U.
  Chronic Systemic: Ingestion 1.

Fire Hazard: Moderate, when exposed to heat or flame; can react vigorously with oxidizing materials.

Explosion Hazard: Moderate, when heated or exposed to a flame.

Countermeasures

Ventilation Control: Section 2.

Personal Hygiene: Section 3.

To Fight Fire: Carbon dioxide, dry chemical or carbon tetrachloride (Section 6).

Storage and Handling: Section 7.

GINGER

General Information

Synonym: Zingiber.

Description: Irregularly bunched pieces; aromatic odor.

Hazard Analysis

Toxic Hazard Rating:
  Acute Local: Irritant 1; Ingestion 1.
  Acute Systemic: U.
  Chronic Local: Irritant 1; Allergen 1.
  Chronic Systemic: U.

Fire Hazard: Slight.

Countermeasures

Ventilation Control: Section 2.

Personal Hygiene: Section 3.

Storage and Handling: Section 7.

GLASS FIBER, DUST. See "Fiberglas."

GLASS WOOL. See "Fiberglas."

GLUCINUM. See beryllium.

GLUCINUM POTASSIUM FLUORIDE. See beryllium potassium fluoride.

GLUCINUM SODIUM FLUORIDE. See beryllium sodium fluoride.

GLUCOSE PENTAPROPIONATE

General Information

Synonym: Pentapropionyl glucose; tetrapropionyl glucosyl propionate.

Description: Insoluble in water.

Formula: $C_6H_7O_6(COC_2H_5)_5$.

Constants: Mol wt: 460, d: 1.2, bp: 401°F at 2 mm, flash p: 509°F.

Hazard Analysis

Fire Hazard: Slight when exposed to heat or flame.

Countermeasures

To Fight Fire: Water, foam.

GLUE

General Information

Description: A colloidal suspension of proteins in water; from white to brown slabs, flakes or chips.

Hazard Analysis

Toxic Hazard Rating:
  Acute Local: Allergen 1.
  Acute Systemic: U.
  Chronic Local: Allergen 1.
  Chronic Systemic: U.

Disaster Hazard: Slight; when heated, it emits smoke (Section 6).

Countermeasures

Personal Hygiene: Section 3.

Storage and Handling: Section 7.

GLUTAMIC ACID

General Information

Synonym: α-Aminoglutaric acid.

Description: A non-essential amino acid present in all complete proteins; crystals.

Formula: $COOH(CH_2)_2CH(NH_2)COOH$.

Constants: Mol wt: 147, mp(DL form): 194°C, d.(DL form) 1.4601 at 20/4°C; (L form) mp: 224–225°C, d(L form): 1.538 at 20/4°C.

Hazard Analysis

Toxic Hazard Rating: U. A general purpose food additive (Section 10).

GLUTAMIC ACID HYDROCHLORIDE

General Information

Description: White crystal powder; very soluble in water, liberating hydrochloric acid; almost insoluble in alcohol and ether.

Formula: $COOH(CH_2)_2CH(NH_2)COOH$.

Constants: Mol wt: 185; d: 1.525, mp: 202–213°C (decomp.)

Hazard Analysis

Toxic Hazard Rating: U. A general purpose food additive (Section 10).

GLUTARALDEHYDE

General Information

Description: Liquid; completely soluble in water.

Formula: $OHC(CH_2)_3CHO$.

Constants: Mol wt: 84, d: 1.124 at 20/20°C, fp: −14°C, vap. press.: 17 mm at 20°C; flash p.: none.

Hazard Analysis

Toxic Hazard Rating: U. Limited animal experiments suggest moderate toxicity and irritant. See also aldehydes.

GLUTARIC ANHYDRIDE

General Information

Synonym: Pentanedioic acid anhydride; 1,3-propane dicarboxylic acid anhydride.

Description: Soluble in benzene, toluene; highly soluble in water on complete hydrolysis.

Constants: Bp: 144–146°C at 13 mm, d: 0.989.

Hazard Analysis

Toxic Hazard Rating: U. Limited animal experiments suggest moderate toxicity and irritant.

GLUTARONITRILE

General Information

Synonym: Pentanedinitrile.

Description: Colorless liquid. Soluble in water, insoluble in ether.

Formula: $CN(CH_2)_3CN$.

Constants: Mol wt: 94.1, d: 0.995 at 15°/4°C, mp: −29°C, bp: 287.4°C.

Hazard Analysis and Countermeasures

See nitriles.

GLYCERIN DICHLOROHYDRIN

General Information

Description: Insoluble in water.

Formula: $CH_2ClCHClCH_2OH$.

Constants: Mol wt: 129, d: 1.4, bp: 360°F, flash p: 200°F.

Hazard Analysis

Toxic Hazard Rating: U.

Fire Hazard: Moderate, when exposed to heat or flame.
Countermeasures
To Fight Fire: Alcohol foam.
Storage and Handling: Section 7.

GLYCERINE
General Information
Synonyms: 1,2,3-Propanetriol; glycerol.
Description: Colorless or pale yellow liquid; odorless and syrupy; sweet and warm taste.
Formula: $CH_2OHCHOHCH_2OH$.
Constants: Mol wt: 92.09, mp: 17.9°C (solidifies at a much lower temperature), bp: 290°C, ulc: 10–20, flash p: 320°F, d: 1.260 at 20°/4°C, autoign. temp.: 739°F, vap. press.: 0.0025 mm at 50°C, vap. d.: 3.17.
Hazard Analysis
Toxic Hazard Rating:
  Acute Local: Irritant 1; Ingestion 1.
  Acute Systemic: Ingestion 1.
  Chronic Local: 0.
  Chronic Systemic: 0.
Toxicity: A general purpose food additive, it migrates to food from packaging materials (Section 10).
Fire Hazard: Slight, when exposed to heat, flame or powerful oxidizers.
Countermeasures
To Fight Fire: Alcohol foam, water, carbon dioxide, dry chemical or carbon tetrachloride (Section 6).
Personal Hygiene: Section 3.
Storage and Handling: Section 7.

GLYCEROL. See glycerine.

GLYCEROL ESTER OF WOOD ROSIN
Hazard Analysis
Toxic Hazard Rating: U. Used as a food additive permitted in food for human consumption (Section 10).

GLYCEROL MONOSTEARATE. See glyceryl monostearate.

GLYCEROL TRICHLOROHYDRIN. See 1,2,3-trichloropropane.

GLYCEROL TRINITRATE. See nitroglycerine.

GLYCERYL DIACETATE
General Information
Synonym: Diacetin.
Description: Colorless liquid, very soluble in water.
Formula: $C_7H_{12}O_5$.
Constants: Mol wt: 176.2, d: 1.178 at 15°/15°C, bp: 280°C, mp: 40°C.
Hazard Analysis
Toxic Hazard Rating:
  Acute Local: Irritant 2.
  Acute Systemic: Ingestion 1.
  Chronic Local: U.
  Chronic Systemic: U.
Toxicology: Animal experiments suggest low toxicity.

GLYCERYL MONOACETATE
General Information
Synonym: Monacetin, acetin.
Description: Colorless, very hygroscopic liquid. Characteristic odor.
Formula: $C_3H_7O_2OOCCH_3$.
Constants: Mol wt: 134.1, d: 1.206 at 20°/4°C, bp: 158°C at 17 mm.

Hazard Analysis
Toxic Hazard Rating:
  Acute Local: Irritant 2, Inhalation 2.
  Acute Systemic: Inhalation 2.
  Chronic Local: U.
  Chronic Systemic: U.
Toxicology: Some irritant and narcotic action on animals.

GLYCERYL MONOSTEARATE
General Information
Synonym: g.m.s., glycerol monostearate; monostearin.
Description: Pure white or cream colored wax like solid with faint odor. Soluble in (hot) alcohol, oils, and hydrocarbons.
Formula: $(C_{17}H_{35})COOCH_2CHOHCH_2OH$.
Constants: Mol wt: 358, mp: 58–59°C d: 0.97.
Hazard Analysis
Toxic Hazard Rating: U. A general purpose food additive (Section 10).

GLYCERYL TRIACETATE
General Information
Synonym: Triacetin.
Formula: $(C_3H_5)(OOCCH_3)_3$.
Constants: Mol wt: 218.2, mp: −78°C, bp: 258°C, flash p: 280°F (C.C.), d: 1.161, autoign. temp.: 812°F, vap. d.: 7.52.
Hazard Analysis
Toxic Hazard Rating:
  Acute Local: Irritant 1.
  Acute Systemic: Ingestion 1.
  Chronic Local: U.
  Chronic Systemic: Ingestion 1.
Toxicity: Used as a general purpose food additive (Section 10).
Fire Hazard: Slight, when exposed to heat, flame or powerful oxidizers.
Countermeasures
To Fight Fire: Alcohol foam, water, foam, carbon dioxide dry chemical or carbon tetrachloride (Section 6).
Personal Hygiene: Section 3.
Storage and Handling: Section 7.

GLYCERYL TRIBUTYRATE. See tributyrin.

GLYCERYL TRINITRATE. See nitroglycerin.

GLYCIDALDEHYDE
Hazard Analysis
Toxic Hazard Rating:
  Acute Local: Irritant 2; Inhalation 3.
  Acute Systemic: Inhalation 3.
  Chronic Local: U.
  Chronic Systemic: U.
Toxicity: Animal and human observations show severe local and pulmonory irritation. See also aldehydes.

GLYCIDOL
General Information
Synonym: 2,3-Epoxy-1-propanol.
Description: Colorless liquid entirely soluble in water, alcohol and ether.
Formula: OCH_2CHCH_2OH.
Constants: Mol wt: 74.1, d: 1.165 at 0°/4°C, bp: 162°C (decomp).
Hazard Analysis
Toxic Hazard Rating:
  Acute Local: Irritant 2; Ingestion 2; Inhalation 2.

TOXIC HAZARD RATING CODE *(For detailed discussion, see Section 1.)*

0 NONE: (a) No harm under any conditions; (b) Harmful only under unusual conditions or overwhelming dosage.

1 SLIGHT: Causes readily reversible changes which disappear after end of exposure.

2 MODERATE: May involve both irreversible and reversible changes; not severe enough to cause death or permanent injury.

3 HIGH: May cause death or permanent injury after very short exposure to small quantities.

U UNKNOWN: No information on humans considered valid by authors.

Acute Systemic: Ingestion 2; Inhalation 2; Skin Absorption 2.
Chronic Local: U.
Chronic Systemic: U.
Toxicity: See also diglycidyl ether. Animal experiments suggest somewhat lower toxicity than related epoxy compounds. Readily absorbed through the skin. Causes nervous excitation followed by depression.
TLV: ACGIH (recommended); 50 parts per million of air; 150 milligrams per cubic meter of air.

GLYCIDYL ACRYLATE
General Information
Description: Clear liquid practically insoluble in water.
Formula: $CH_2CHCOOCH_2CHCH_2O$.
Constants: Mol wt: 128.3, d: 1.1074 at $20°/20°C$, bp: $57°C$ at 2 mm fp: $-41.5°C$, flash p: $141°F$, vap. d.: 4.4.
Hazard Analysis
Toxic Hazard Rating:
Acute Local: Irritant 3.
Acute Systemic: Ingestion 3; Inhalation 3; Skin Absorption, 3.
Chronic Local: U.
Chronic Systemic: U.
Based on limited animal experiments.

GLYCIDYL ALDEHYDE
General Information
Formula: OCH_2CHCH_2CHO.
Constant: Mol wt: 86.2.
Hazard Analysis
Toxic Hazard Rating:
Acute Local: Irritant 3; Inhalation 3.
Acute Systemic: Ingestion 2, Inhalation 2; Skin Absorption 2.
Chronic Local: U.
Chronic Systemic: Inhalation 2.
Toxicology: Large doses cause convulsions in experimental animals. Repeated injections cause bone marrow depression.

GLYCIDYL BENZOATE
Hazard Analysis
Toxic Hazard Rating:
Acute Local: U.
Acute Systemic: Ingestion 2.
Chronic Local: U.
Chronic Systemic: U.
Based on limited animal experiments.

n-GLYCIDYL DIETHYLAMINE
Hazard Analysis
Toxic Hazard Rating:
Acute Local: Irritant 3.
Acute Systemic: Ingestion 3; Inhalation 3; Skin Absorption 3.
Chronic Local: U.
Chronic Systemic: U.
Based on limited animal experiments. See also amines.

GLYCIDYL OLEATE
Hazard Analysis
Toxic Hazard Rating:
Acute Local: Irritant 2.
Acute Systemic: Ingestion 2; Skin Absorption 2.
Chronic Local: U.
Chronic Systemic: U.
Based on limited animal experiments.

GLYCIDYL PHENYL ETHER
General Information
Description: Liquid.
Formula: $C_6H_4CH_2CHOCH_2$.
Constants: Mol wt: 150.17, mp: $3.5°C$, bp: $245°C$,

flash p: $175°F$ (O.C.), d: 1.1092 at $20°/4°C$, vap. d.: 5.18.
Hazard Analysis
Toxic Hazard Rating:
Acute Local: U.
Acute Systemic: Ingestion 3; Inhalation 3.
Chronic Local: U.
Chronic Systemic: U.
Fire Hazard: Moderate, when exposed to heat or flame; can react with oxidizing materials.
Disaster Hazard: Dangerous; when heated to decomposition, it emits toxic fumes.
Countermeasures
Ventilation Control: Section 2.
Personal Hygiene: Section 3.
To Fight Fire: Water, foam, carbon dioxide, dry chemical or carbon tetrachloride (Section 6).
Storage and Handling: Section 7.

GLYCIDYL SORBATE (DIMER)
Hazard Analysis
Toxic Hazard Rating:
Acute Local: Irritant 1.
Acute Systemic: Ingestion 1.
Chronic Local: U.
Chronic Systemic: U.
Based on limited animal experiments.

GLYCINE
General Information
Synonym: Amino acetic acid; glycocoll.
Description: The principal amino acid in sugar cane. White crystals; odorless; soluble in water; insoluble in alcohol and ether.
Formula: NH_2CH_2COOH.
Constants: Mol wt: 75, mp: $232-236°C$, d: 1.1607.
Hazard Analysis
Toxic Hazard Rating: U. A nutrient and/or dietary supplement food additive (Section 10).

GLYCOCHOLIC ACID
General Information
Synonym: Cholyglycine.
Description: Occurs as sodium salt in bile; practically insoluble in water. The sodium salt is soluble in water and alcohol.
Formula: $C_{26}H_{43}NO_6$.
Constant: Mol wt: 465.
Hazard Analysis
Toxic Hazard Rating: U. An emulsifying agent food additive (Section 10).

GLYCOCOLL. See glycine.

GLYCOCOLL CALCIUM
General Information
Description: Slightly water soluble crystals.
Formula: $Ca(CH_2NHCO_2)$.
Constant: Mol wt: 113.1.
Hazard Analysis
Toxicity: Details unknown. See calcium compounds.

GLYCOL. See ethylene glycol.

GLYCOL ALCOHOL. See ethylene glycol.

GLYCOL BIS(HYDROXYETHYL) ETHER. See triethylene glycol.

GLYCOL CARBONATE. See ethylene carbonate.

GLYCOL CHLOROHYDRIN. See ethylene chlorohydrin.

GLYCOL DIABIETATE. See ethylene glycol diabietate.

GLYCOL BIBROMIDE. See ethylene dibromide.

GLYCOL DIACETATE. See ethlene glycol diacetate.

GLYCOL DIFORMATE. See ethlene glycol diformate.

GLYCOLIC ACID
General Information
Synonyms: Hydroxyethanoic acid; hydroxyacetic acid.
Description: Rhombic leaflets from ether.
Formula: $HOCH_2COOH$.
Constants: Mol wt: 76.05, bp: decomposes, mp: (a) 63°C; (b) 79°C.
Hazard Analysis
Toxic Hazard Rating:
  Acute Local: Irritant 1; Ingestion 2.
  Acute Systemic: Ingestion 1.
  Chronic Local: U.
  Chronic Systemic: U.
Disaster Hazard: Slight; when heated, it emits acrid fumes.
Countermeasures
Personal Hygiene: Section 3.
Storage and Handling: Section 7.

GLYCOL MONOACETATE. See ethylene glycol monoacetate.

GLYCOL MONOBUTYL ETHER. See butyl "Cellosolve."

GLYCOL MONOBUTYL ETHER ACETATE. See butyl "Cellosolve" acetate.

GLYCOLONITRILE
General Information
Synonym: Formaldehyde cyanohydrin.
Description: Mobile, colorless, odorless oil.
Formula: $HOCH_2CN$.
Constants: Mol wt: 57.1, bp: 183°C with slight decomp., fp: −67°C, d: 1.104, vap. d.: 1.97.
Hazard Analysis
Toxicity: Highly toxic. Skin absorption can be significant. See nitriles.
Explosion Hazard: Moderate, when exposed to heat or by spontaneous chemical reaction in the presence of alkalies if uninhibited (polymerizes).
Disaster Hazard: Highly dangerous; emits highly toxic fumes on decomposition.
Countermeasures
Storage and Handling: Section 7.

GLYCOLS
Hazard Analysis
Toxicity: Dihydric alcohols are physically and chemically related to glycerol. Most of the glycols have low volatility and consequently present little danger from inhalation of vapors. Severe and even fatal poisoning has occurred following ingestion of ethylene glycol and severe occupational poisoning has been caused by ethylene glycol monomethyl ether (methyl "Cellosolve"). Under ordinary conditions of industrial use, most of the commonly used glycols are not considered to be toxic.
  There is some evidence that the glycol ethers are more toxic than other members of the group.

GLYCOL URIL. See acetylene diureine.

GLYODIN
General Information
Synonyms: 2-Heptadecyl-glyoxalidine acetate, "Crag," fruit fungicide 341, 2-heptadecyl-2-imidoazoline acetate.
Description: Light orange, soft, waxy crystals. Water insoluble.

Formula: $C_{22}H_{44}N_2O_2$.
Constants: Mol wt: 368.6, mp: 94°C.
Hazard Analysis
Toxic Hazard Rating:
  Acute Local: Irritant 1.
  Acute Systemic: Ingestion 1.
  Chronic Local: Irritant 1.
  Chronic Systemic: Ingestion 1.
Toxicology: Data based on limited animal experiments indicate low toxicity. High concentrations irritating to skin and damaging to cornea.

GLYOXAL
General Information
Synonyms: Ethanedial; oxalaldehyde; biformyl.
Description: Yellow crystal or light yellow liquid.
Formula: $OCHCHO$.
Constants: Mol wt: 58.0, mp: 15°C, bp: 50.4°C, d: 1.14 at 20°/4°C.
Hazard Analysis
Toxic Hazard Rating:
  Acute Local: Irritant 2; Ingestion 2; Inhalation 2.
  Acute Systemic: U.
  Chronic Local: U.
  Chronic Systemic: U.
Fire Hazard: Slight; Section 6.
Countermeasures
Ventilation Control: Section 2.
Personnel Protection: Section 3.
Storage and Handling: Section 7.

GLYOXALIDINE DERIVATIVES. See 1-hydroxethyl-2-heptadecylglyoxalidine.

GLYOXYLIC ACID
General Information
Description: Hemihydrate, $C_2H_2O_3 \cdot \frac{1}{2}H_2O$ from H_2O. Also obtained in anhydrous form as monoclinic crystals from water. Deliquesces quickly and forms a syrup on short exposure to air. Freely soluble in water; insoluble in ether and hydrocarbons.
Formula: $C_2H_2O_3$.
Constants: Mol wt: 74.04, d: 1.42 at 20/4°C.
Hazard Analysis
Toxic Hazard Rating:
  Acute Local: Irritant 3; Ingestion 3; Inhalation 3.
  Acute Systemic: U.
  Chronic Local: U.
  Chronic Systemic: U.
Toxicity: Irritant and corrosive.

GMP. See sodium guanylate.

GMS. See glyceryl monostearate.

GOA POWDER. See chrysarobin.

GOAT HAIR. See mohair.

GOLD
General Information
Description: Cubic yellow, ductile, metallic crystals.
Formula: Au.
Constants: At wt: 197.20, mp: 1063°C, bp: 2966°C, d: 19.3 (liquid): 17.0 at 1063°C, vap. press.: 1 mm at 1869°C.
Hazard Analysis
Toxicity: Negligible; see gold compounds.

TOXIC HAZARD RATING CODE (For detailed discussion, see Section 1.)

0 NONE: (a) No harm under any conditions; (b) Harmful only under unusual conditions or overwhelming dosage.

1 SLIGHT: Causes readily reversible changes which disappear after end of exposure.

2 MODERATE: May involve both irreversible and reversible changes; not severe enough to cause death or permanent injury.

3 HIGH: May cause death or permanent injury after very short exposure to small quantities.

U UNKNOWN: No information on humans considered valid by authors.

Radiation Hazard: Section 5. For permissible levels, see Table 5, p. 150.

Artificial isotope [196]Au, half life 6.2 d. Decays to stable [196]Pt by electron capture (94%). Also decays to stable [196]Hg by emitting beta particles of 0.26 MeV. Also emits gamma rays of 0.33–0.43 MeV and X-rays.

Artificial isotope [198]Au, half life 2.7 d. Decays to stable [198]Hg by emitting beta particles of 0.96 MeV. Also emits gamma rays of 0.41 MeV.

Artificial isotope [199]Au, half life 3.15 d. Decays to stable [199]Hg by emitting beta particles of 0.25 (23%), 0.30 (70%), 0.46 (7%) MeV. Also emits gamma rays of 0.05, 0.16, 0.21 MeV.

GOLD 198
Shipping Regulations: Section 11.
I.C.C.: Class D poison, radioactive materials, red label, 300 curies.

GOLD CHLORIDE. See auric chloride.

GOLD COMPOUNDS
Hazard Analysis
Toxic Hazard Rating:
Acute Local: Allergen 2.
Acute Systemic: Ingestion 2.
Chronic Local: Allergen 2.
Chronic Systemic: Ingestion 2, Inhalation 2.
Toxicology: Salts of gold, particularly gold sodium thiosulfate, have been used extensively in the treatment of some forms of arthritis and for other diseases. Toxic reactions are not infrequent. Various allergic manifestations have occurred, such as urticaria (hives), itching, purpura and other types of skin rashes, some of which may be quite severe (Section 9). Damage to the blood forming organs resulting in aplastic anemia has been described. The liver, kidneys and nervous system may be affected.

If gold therapy is being used, careful watch must be kept for the first signs of toxic reaction and the drug be promptly discontinued. BAL has been reported to be of value in treating gold poisoning.
Countermeasures
Ventilation Control: Section 2.

GOLD CYANIDE. See aurous cyanide.

GOLD DITELLURIDE
General Information
Description: Crystal forms: rhombic; monoclinic; triclinic.
Formula: $AuTe_2$.
Constants: Mol wt: 452.42, mp: 472°C decomposes, d: 8.2–9.3.
Hazard Analysis
Toxicity: See tellurium compounds and gold compounds.
Fire Hazard: See hydrides.
Explosion Hazard: See hydrides.
Disaster Hazard: Dangerous; when heated to decomposition, it emits toxic fumes of tellurium; will react with water, steam or acid to produce toxic and flammable vapors of tellurium hydride.
Countermeasures
Storage and Handling: Section 7.

GOLD IODIDE
General Information
Synonym: Aurous iodide.
Description: Greenish-yellow powder.
Formula: AuI.
Constants: Mol wt: 324.12, mp: 120°C decomposes, d: 8.25.
Hazard Analysis
Toxicity: See iodides.
Disaster Hazard: See iodides.
Countermeasures
Storage and Handling: Section 7.

GOLD NITRATE. See auric hydrogen nitrate.

GOLD OXIDE. See auric oxide.

GOLD PHOSPHIDE
General Information
Description: Gray crystals.
Formula: Au_2P_3.
Constants: Mol wt: 487.34, mp: decomposes, d: 6.67.
Hazard Analysis and Countermeasures
See phosphides.

GOLD POTASSIUM BROMIDE
General Information
Synonym: Potassium auribromide.
Description: Violet crystals.
Formula: $AuBr_3 \cdot KBr \cdot 2H_2O$.
Constant: Mol wt: 592.01.
Hazard Analysis and Countermeasures
See bromides.

GOLD POTASSIUM CHLORIDE
General Information
Synonym: Potassium aurichloride.
Description: Yellow crystals.
Formula: $AuCl_3 \cdot KCl \cdot 2H_2O$.
Constant: Mol wt: 414.17.
Hazard Analysis
Toxicity: See gold compounds.
Disaster Hazard: See chlorides.
Countermeasures
Storage and Handling: Section 7.

GOLD POTASSIUM CYANIDE
General Information
Synonyms: Potassium cyanaurite; potassium aurocyanide.
Description: White, crystalline powder.
Formula: $KAu(CN)_2 \cdot 2H_2O$.
Constant: Mol wt: 324.36.
Hazard Analysis and Countermeasures
See cyanides.

GOLD SELENIDE
General Information
Description: Solid.
Formula: Au_2Se_3.
Constants: Mol wt: 631.28, d: 4.65 at 22°C.
Hazard Analysis
Toxicity: See selenium compounds.
Fire Hazard: Moderate, by chemical reaction with moisture or acids.
Explosion Hazard: Slight. See hydrogen selenide.
Disaster Hazard: Dangerous; when heated, it emits highly toxic fumes of selenium; will react with water, steam, or acid to produce toxic and flammable vapors of hydrogen selenide.
Countermeasures
Storage and Handling: Section 7.

GOLD SODIUM BROMIDE
General Information
Synonym: Sodium auribromide.
Description: Black-brown crystals.
Formula: $AuBr_3 \cdot NaBr \cdot 2H_2O$.
Constant: Mol wt: 575.91.
Hazard Analysis and Countermeasures
See bromides.

GOLD SODIUM CHLORIDE
General Information
Synonyms: Sodium aurichloride; sodium chloraurate.
Description: Yellow crystals.
Formula: $NaAuCl_4 \cdot 2H_2O$.
Constant: Mol wt: 398.05.
Hazard Analysis and Countermeasures
See gold compounds and chlorides.

GOLD SODIUM THIOSULFATE
General Information
Synonym: Aurous sodium thiosulfate.
Description: White crystals, odorless.
Formula: $Na_3Au(S_2O_3)_2 \cdot 2H_2O$.
Constant: Mol wt: 527.47.
Hazard Analysis
Toxicity: See gold compounds.
Disaster Hazard: Moderately dangerous; when heated to decomposition, it emits toxic fumes.
Countermeasures
Storage and Handling: Section 7.

GOLD SULFIDE. See auric sulfide.

GOLD TELLURIDE
General Information
Description: Triclinic crystals.
Formula: Au_2Te.
Constants: Mol wt: 522.0, d: 9.04.
Hazard Analysis
Toxicity: See tellurium compounds.
Fire Hazard: Moderate, on contact with moisture (Section 6).
Explosion Hazard: Slight, upon contact with moisture.
Disaster Hazard: Moderately dangerous in a disaster because, when heated to decomposition, it emits toxic fumes of tellurium; on contact with water, steam or acid, it can emit toxic and flammable fumes of hydrogen telluride.
Countermeasures
Storage and Handling: Section 7.

GOLD TRIBROMIDE
General Information
Description: Gray powder, crystalline.
Formula: $AuBr_3$.
Constants: Mol wt: 437.0, mp: $-Br_2$ at 160°C.
Hazard Analysis and Countermeasures
See bromides.

GRAIN SMUTS
General Information
Description: These are minute, threadlike, parasitic plants, the mycelium of which enters the growing portion of the seedling and grow up with the host. They remain invisible until the heads of grain appear, at which time they partially or wholly destroy the heads, appearing at that stage as masses of dark substance. These masses contain spores which further propagate by getting on or into the seeds where they reenact their life after the seed is planted.
Hazard Analysis
Toxic Hazard Rating:
Acute Local: Allergen 1.
Acute Systemic: U.
Chronic Local: Allergen 1; Inhalation 1.
Chronic Systemic: Inhalation 1.
Toxicology: Due to these smut spores, contact dermatitis, an atopic form of skin allergy, perennial rhinitis, and asthma may occur in those exposed. These conditions may be fairly abundant wherever favorable climate for infestation exists, which can be in the Middle West, the Pacific Northwest, the South, and the Southwest.
Countermeasures
Personal Hygiene: Section 3.

GRANITE DUST
Hazard Analysis
Toxic Hazard Rating:
Acute Local: Irritant 1.
Acute Systemic: Inhalation 2.
Chronic Local: Inhalation 3.
Chronic Systemic: 0.
Toxicity: See silica.
Countermeasures
Ventilation Control: Section 2.

GRAPE SUGAR. See dextrose.

GRAPHITE
General Information
Synonyms: Black lead; plumbago.
Description: A crystalline form of carbon; soft, greasy feel; steel-gray to black color.
Formula: C.
Constants: At wt: 12.00, d: 2.0–2.25.
Hazard Analysis
Toxicity: A nuisance dust. Possibly allergenic.
Cases of pulmonary fibrosis, emphysema and cor pulmonale have resulted from prolonged inhalation of graphite dust.
Countermeasures
Personal Hygiene: Section 3.

GRASSELLIS. See explosives, high.

GRAY ACETATE. See calcium acetate.

GREEK FIRE
General Information
Description: Essentially, this is a mixture of nitrates and pitch, which has been in use since the 8th century. This material is a deflagrating mixture and therefore properly belongs in the class of low explosives.
Hazard Analysis and Countermeasures
See explosives, low.

GREENOCKITE. See cadmium sulfide.

GREEN OIL. See anthracene.

GREEN VITRIOL. See ferrous sulfate.

GRENADES (ALL TYPES). See explosives, high.

GRENADES, EMPTY, PRIMED
Shipping Regulations: Section 11.
ICC: Class C explosive, 150 pounds.
IATA: Explosive, explosive label, 25 kilograms (passenger) 70 kilograms (cargo).

GRIGNARD REAGENTS
Since all Grignard reagents react rapidly with both water and oxygen, contact with these substances must be avoided. So far as is known the ordinary materials of construction are satisfactory for use with Grignard reagents. Adequate ventilation should, of course, be employed.

Because the heat or decomposition of Grignard reagents with water is great and the ether they are dissolved in is highly volatile and flammable, they must be handled with extreme care. Some of these reagents, especially the solution of MeMgBr in ethyl ether, may ignite spontaneously on contact with water or even damp floors; nearly all of these compounds will ignite on a wet rag or similar material.

TOXIC HAZARD RATING CODE (*For detailed discussion, see Section 1.*)

0 NONE: (a) No harm under any conditions; (b) Harmful only under unusual conditions or overwhelming dosage.

1 SLIGHT: Causes readily reversible changes which disappear after end of exposure.

2 MODERATE: May involve both irreversible and reversible changes; not severe enough to cause death or permanent injury.

3 HIGH: May cause death or permanent injury after very short exposure to small quantities.

U UNKNOWN: No information on humans considered valid by authors.

All possible precautions should be taken to eliminate spills, or to curb them if they should occur.

1. Check all drums before use to see that there have been no leaks or damage to the drums. Outdoor storage in a roofed-over area protected from the weather is recommended for all Grignard reagents except phenyl magnesium bromide.

2. Any area where Grignard reagents are to be handled must be well ventilated and free from sources of ignition. All electrical fixtures and wiring must be explosion-proof, Class I Group D, or Class II Group G.

3. Prepare an area where the drums are to be opened that is completely dry, and far away from any possible water splashes. A curbed area or one surrounded by a dry sand dike is recommended for this purpose.

4. All facilities to which the reagent is being transferred should be checked to see that they are completely dry. A supply of dry sand should be close at hand.

5. Connect a solid outlet line to the drum, seeing that all joints are tight and that the outlet bung is free flowing. The inlet bung (also free) should be connected to a source of oil-pumped dry nitrogen, and about 1 to 2 psi. should be applied to the drum when pumping starts.

6. Pump the reagent out of the drums into the reactors through a closed system, using a positive displacement pump, replacing the volume with dry, oil-pumped nitrogen. Do not apply over 5 lbs. nitrogen pressure to the drum.

7. In case of a spill the material should be curbed and covered with dry sand and immediately removed. Should a large spill occur that cannot be taken up with sand, a film of decomposed Grignard will appear on the top of the material. Do not break this film any more than absolutely necessary! By means of a vacuum line or pump remove the spilled reagent to a dry container, and then remove the residue with dry sand.

8. Any solid remaining in the drums is extremely hazardous in contact with air or moisture because it may give off great quantities of heat, and may ignite any solvent remaining with it.

9. Do not attempt to wash drums out with water. It should not be necessary if nitrogen displacement has been complete. If washing is necessary, use butyl ether followed by alcohol.

GRISEOFULVIN
General Information
Synonym: 7-Chloro-2′,4,6-trimethoxy-6′-β-methylspiro [benzofuran-2(3H), 1′-(2)cyclohexene]-3,4′-dione.
Description: White to creamy white powder; odorless; very slightly soluble in water; soluble in acetone and chloroform; sparingly soluble in alcohol.
Formula: $C_{17}H_{17}ClO_6$.
Constant: Mol wt: 352.8, mp: 218°C.
Hazard Analysis
Toxic Hazard Rating: U. A food additive permitted in the feed and drinking water of animals and/or for the treatment of food producing animals (Section 10).

GUAIACOL
General Information
Synonyms: Methyl catechol; 2-methoxyphenol.
Description: Clear, pale yellow liquid or solid.
Formula: $OHC_6H_4OCH_3$.
Constants: Mol wt: 124.1, mp: 28°C, bp: 202–209°C, flash p: 180°F (O.C.), d: 1.097 at 25°/25°C.
Hazard Aanlysis
Toxic Hazard Rating:
  Acute Local : Irritant 1; Ingestion 1.
  Acute Systemic: Ingestion 2, Skin Absorption 2.
  Chronic Local: Irritant 1.
  Chronic Systemic: Ingestion 1, Skin Absorption 1.
Toxicology: Action similar to but less intense than phenol. See also phenol.

Fire Hazard: Moderate, when exposed to heat or flame; can react with oxidizing materials.
Countermeasures
To Fight Fire: Foam, carbon dioxide, dry chemical or carbon tetrachloride (Section 6).
Ventilation Control: Section 2.
Personal Hygiene: Section 3.
Storage and Handling: Section 7.

GUAIACOL VALERATE
General Information
Description: Oily, pale yellow liquid.
Formula: $C_6H_4OCH_3OCOC_4H_9$.
Constants: Mol wt: 208.3, bp: 265°C, d: 1.05.
Hazard Analysis
Toxicity: Details unknown. See also guaiacol and valerian.
Disaster Hazard: Slight; when heated, it emits acrid fumes.
Countermeasures
Storage and Handling: Section 7.

GUANAJUATITE. See bismuth triselenide.

GUANIDINE AMINO VALERIC ACID. See arginine.

GUANIDINE CARBONATE
General Information
Description: Columnar crystals.
Formula: $[H_2NC(NH)NH_2]_2H_2CO_3$.
Constants: Mol wt: 180.2, mp: 333°C, d: 1.24.
Hazard Analysis
Toxicity: Details unknown. See guanidine hydrochloride.
Disaster Hazard: Moderate; when heated to decomposition it emits toxic fumes.
Countermeasures
Storage and Handling: Section 7.

GUANIDINE HYDROCHLORIDE
General Information
Description: White powder.
Formula: $H_2NC(NH)NH_2 \cdot HCl$.
Constants: Mol wt: 95.6, mp: 183°C.
Hazard Analysis
Toxic Hazard Rating:
  Acute Local: 0.
  Acute Systemic: Ingestion 2.
  Chronic Local: 0.
  Chronic Systemic: Ingestion 2.
Toxicology: Can cause nausea, diarrhea and neurological disturbances.
Disaster Hazard: Dangerous; when heated to decomposition, it emits highly toxic fumes of hydrochloric acid.
Countermeasures
Storage and Handling: Section 7.

GUANIDINE NITRATE
General Information
Description: White granules.
Formula: $H_2NC(NH)NH_2 \cdot HNO_3$.
Constants: Mol wt: 122.1, mp: 214°C.
Hazard Analysis
Toxicity: See nitrates and guanidine hydrochloride.
Fire Hazard: See nitrates.
Explosion Hazard: Moderate, when shocked or exposed to heat or flame; a stable, flashless, nonhygroscopic high explosive used as a blasting explosive in combination with charcoal and inorganic nitrates.
Disaster Hazard: Dangerous; keep away from heat and open flame.
Countermeasures
Storage and Handling: Section 7.
Shipping Regulations: Section 11.
  I.C.C.: Oxidizing material; yellow label, 100 pounds.
  Coast Guard Classification: Oxidizing material.
  IATA: Oxidizing material, yellow label, 12 kilograms (passenger), 45 kilograms (cargo).

GUANYL NITROSOAMINO GUANYL TETRAZENE.
  See tetracine.

GUANYL UREA NITRATE
General Information
Description: Crystals.
Formula: $NH_2C(NH)NHCONH_2 \cdot HNO_3$.
Constants: Mol wt: 165.1, mp: >230°C, d: 1.579 at 30°C.
Hazard Analysis and Countermeasures
See nitrates.

GUANYL UREA PHOSPHATE
General Information
Description: Crystals.
Formula: $NH_2C(NH)NHCONH_2 \cdot H_3PO_4$.
Constants: Mol wt: 200.1, mp: 184°C decomposes, d: 1.614.

GUANYL UREA SULFATE
General Information
Description: White powder.
Formula: $NH_2C(NH)NHCONH_2 \cdot H_2SO_4$.
Constants: Mol wt: 302.28, mp: 192°C decomposes, d: 1.609.
Hazard Analysis
Toxicity: Unknown.
Disaster Hazard: Dangerous; when heated to decomposition, it emits highly toxic fumes of oxides of sulfur.
Countermeasures
Storage and Handling: Section 7.

GUAR GUM
General Information
Synonym: Guar flour.
Description: Yellowish white powder; dispersible in hot or cold water; obtained from the ground endosperms of Cyanopsis tetragonoloan.
Hazard Analysis
Toxic Hazard Rating: U. A stabilizer food additive. It migrates to food from packaging materials (Section 10).

GUAR FLOUR. See guar gum.

GUAZA. See cannabis.

GUM ARABIC. See acacia gum.

GUM CAMPHOR. See camphor.

GUM COPAL. See copal.

GUM GUAIAC
General Information
Synonym: Guaiac gum; guaiac resin.
Description: A resin from certain Mexican and West Indian trees; soluble in alcohol, ether, acetone, chloroform and caustic soda.
Hazard Analysis
Toxic Hazard Rating: U. A chemical preservative food additive (Section 10).

GUM OPIUM. See opium.

GUM ROSIN. See rosin.

GUM THUS. See turpentine gum.

GUM TRAGACANTH. See tragacanth.

GUN COTTON. See nitrocellulose.

GUNPOWDER. See explosives, high.

GUTHION
General Information
Synonym: O,O-Dimethyl-S-(4-oxobenzotriazino-3-methyl)-phosphoro dithioate.
Description: Crystals, slightly water soluble but soluble in organic solvents.
Formula: $C_{10}H_{12}N_3O_3PS_2$.
Constants: Mol wt: 317.3, d: 1.44, mp: 74°C.
Hazard Analysis
Toxicity: Highly toxic. A cholinesterase inhibitor. See parathion. A food additive permitted in the food and drinking water of animals and/or for the treatment of food producing animals (Section 10).
Countermeasures
See parathion.

H

HAFNIA. See hafnium oxide.

HAFNIUM
General Information
Description: Gray crystals.
Formula: Hf.
Constants: At wt: 178.60, mp: 2207°C, bp: < 3200°C,
  d: 13.3.
Hazard Analysis
Toxicity: Unknown.
TLV: ACGIH (recommended); 0.5 milligrams per cubic
  meter of air.
Radiation Hazard: Section 5. For permissible levels, see
  Table 5, p. 150.
    Artificial isotope ^{181}Hf, half life 43d. Decays to
  stable ^{181}Ta by emitting beta particles of 0.40 (4%),
  0.41 (89%), 0.55 (5%) MeV. Also emits gamma rays of
  0.13 to 0.48 MeV.

HAFNIUM CARBIDE
General Information
Description: Solid.
Formula: HfC.
Constants: Mol wt: 190.61, mp: 3887°C.
Hazard Analysis
Toxicity: Unknown.
Fire Hazard: Slight; will react with water or steam to pro-
  duce flammable vapors (Section 6).
Countermeasures
Storage and Handling: Section 7.

HAFNIUM COMPOUNDS
Hazard Analysis
Details unknown. Animal experiments show relative non-
toxicity.

HAFNIUM DIOXIDE. See hafnium oxide.

HAFNIUM HYDRIDE
General Information
Description: Metallic crystals insoluble in water.
Formula: HfH$_2$.
Constants: Mol wt: 180.6, d: 11.48, mp: decomp. > 400°C.
Hazard Analysis and Countermeasures
See hafnium compounds and hydrides.

**HAFNIUM METAL, DRY, CHEMICALLY PRO-
  DUCED FINER THAN 20 MESH PARTICLE
  SIZE**
Shipping Regulations: Section 11.
  I.C.C.: Flammable solid, yellow label, 75 pounds.

**HAFNIUM METAL, DRY, MECHANICALLY
  PRODUCED, FINER THAN 270 MESH,
  PARTICLE SIZE.**
Shipping Regulations: Section 11.
  I.C.C.: Flammable solid, yellow label, 75 pounds.

**HAFNIUM METAL, WET, CHEMICALLY PRODUCED,
  FINER THAN 20 MESH PARTICLE SIZE**
Shipping Regulations: Section 11.
  ICC: Flammable solid, yellow label, 150 pounds.

**HAFNIUM METAL, WET, MECHANICALLY PRO-
  DUCED, FINER THAN 270 MESH, PARTICLE SIZE.**
Shipping Regulations: Section 11.
  I.C.C.: Flammable solid, yellow label, 150 pounds.

**HAFNIUM METAL, DRY OR WET WITH LESS
  THAN 25% WATER CHEMICALLY PRODUCED,
  PARTICLE SIZE NOT EXCEEDING 840
  MICRONS (20 MESH).**
Shipping Regulations: Section 11.
  IATA: Flammable solid, yellow label, not acceptable
  (passenger), 35 kilograms (cargo). Same regulations
  IATA for particle size not exceeding 53 microns (270
  mesh).

**HAFNUIM METAL, WET WITH NOT LESS THAN
  25% WATER, CHEMICALLY PRODUCED,
  PARTICLE SIZE NOT EXCEEDING 840
  MICRONS (20 MESH).**
Shipping Regulations: Section 11.
  IATA: Flammable solid, yellow label, not acceptable
  (passenger), 70 kilograms (cargo). Same regulations
  IATA for particle size not exceeding 53 microns
  (270 mesh).

HAFNIUM OXIDE
General Information
Synonyms: Hafnium dioxide; hafnia.
Description: White crystals.
Formula: HfO$_2$.
Constants: Mol wt: 210.60, mp: 2810°C, d: 9.68.
Hazard Analysis
Toxicity: Unknown. See also Hafnium.

HAFNIUM OXYCHLORIDE
General Information
Description: Colorless crystals.
Formula: HfOCl$_2$·8H$_2$O.
Constant: Mol wt: 409.6.
Hazard Analysis
Toxicity: Details unknown. See also hydrochloric acid.
Disaster Hazard: Dangerous; see chlorides.
Countermeasures
Storage and Handling: Section 7.

HAIR
Hazard Analysis
Toxic Hazard Rating:
  Acute Local: Allergen 1.
  Acute Systemic: U.
  Chronic Local: Allergen 1.
  Chronic Systemic: U.
Fire Hazard: Slight, when exposed to heat or flame (Sec-
  tion 6).
Countermeasures
Personal Hygiene: Section 3.
Storage and Handling: Section 7.
Shipping Regulations: Section 11.
  Coast Guard Classification: (Wet) Inflammable solid;
  yellow label.
    I.C.C.: Flammable solid, yellow label, not accepted.

HALANE. See 1,3-dichloro-5,5-dimethyl hydantoin.

HALITE. See sodium chloride.

HALOCARBONS. See polychlorotrifluoroethylene.

HALOCRIN. See 3,9-dichloro-7-methoxyacridine.

HALOWAX. See hexachloronaphthalene.

HAMAMELIA. See witch hazel.

HAMBERGITE. See beryllium o-borate, basic.

HANANE
General Information
Formula: Bis(dimethylamino)fluorophosphine oxide + bis(dimethylamino) phosphorous anhydride.
Hazard Analysis
Toxicity: Highly toxic. A cholinesterase inhibitor. See parathion.
Disaster Hazard: Dangerous. See parathion.

HAND GRENADES
Shipping Regulations: Section 11.
    I.C.C.: Class A explosive, not accepted.
    IATA (cargo and passenger): Explosive, not accepted (with gas or smoke material and with bursting charge).
    IATA: Explosive, explosive label, 25 kilograms (passenger) 70 kilograms (cargo) (with smoke material and without bursting charge).

HAND SIGNAL DEVICES
Shipping Regulations: Section 11.
    I.C.C.: Explosive C, 200 pounds.

HASHISH. See cannabis.

HAT GLAZING LACQUER
Hazard Analysis
Toxicity: Details unknown. See also lacquers.
Fire Hazard: Dangerous, when exposed to heat or flame; can react vigorously with oxidizing materials (Section 6).
Explosion Hazard: Unknown.
Countermeasures
Storage and Handling: Section 7.

HAY
General Information
Synonym: Dried grass.
Description: Yellow, fibrous material.
Hazard Analysis
Toxic Hazard Rating:
    Acute Local: Allergen 1; Inhalation 1.
    Acute Systemic: 0.
    Chronic Local: Allergen 1; Inhalation 1.
    Chronic Systemic: 0.
Fire Hazard: Moderate, when exposed to heat or flame; can react with oxidizing materials (Section 6).
Spontaneous Heating: Occurs when wet or improperly packed.
Countermeasures
Ventilation Control: Section 2.
Personal Hygiene: Section 3.
Storage and Handling: Section 7.
Shipping Regulations: Section 11.
    Coast Guard Classification: Hazardous article.

HCCH. See 1,2,3,4,5,6-hexachlorocyclohexane.

HD. See dichlorodiethyl sulfide.

HEATERS, LIQUID FUEL TYPE
Shipping Regulations: Section 11.
    I.C.C.: Flammable liquid.
    IATA (cargo and passenger): No limit.

HEAZLEWOODITE. See nickel subsulfide.

HELIANTHINE B. See methyl orange.

HELIOTROPIN. See piperonal.

HELIUM
General Information
Description: Colorless, odorless, tasteless, inert gas.
Formula: He.
Constants: At wt: 4.003, mp: $-272.2°C$ at 26 atm., bp: $-268.9°C$, d: (gas) 0.1785 g/liter at 0°C, (liquid) 0.147 at $-270.8°C$.
Hazard Analysis
Toxicity: A simple asphyxiant. For properties see argon. Used as a general purpose food additive (Section 10).
Countermeasures
Storage and Handling: Section 7.
Shipping Regulations: Section 11.
    I.C.C.: Nonflammable gas, green label, 300 pounds.
    Coast Guard Classification: Noninflammable gas; green gas label.
    IATA: Nonflammable gas, green label, 70 kilograms (passenger), 140 kilograms (cargo) (gaseous).

HELIUM, LIQUID. See helium.
Shipping Regulations: Section 11.
    IATA (non-pressurized): Other restricted articles, class C, not acceptable (passenger and cargo).
    IATA (low pressure): Other restricted articles, class C, not acceptable (passenger), 500 kilograms (cargo).
    IATA (pressurized): Nonflammable gas, green label, not acceptable (passenger), 140 kilograms (cargo).

HELIUM-OXYGEN MIXTURE. See helium and oxygen.
Shipping Regulations: Section 11.
    I.C.C.: Nonflammable gas, green label, 300 pounds.
    Coast Guard Classification: Noninflammable gas; green gas label.
    IATA: Nonflammable gas, green label, 70 kilograms (passenger), 140 kilograms (cargo).

HELLEBORE, BLACK. See helleborein.

HELLEBOREIN
General Information
Description: Glucoside crystallizable in yellow prisms.
Formula: $C_{37}H_{56}O_{18}$.
Constants: Mol wt: 788.63, mp: 270°C.
Hazard Analysis
Toxic Hazard Rating:
    Acute Local: Irritant 2; Ingestion 2.
    Acute Systemic: U.
    Chronic Local: U.
    Chronic Systemic: U.
Fire Hazard: Slight; (Section 6).
Countermeasures
Ventilation Control: Section 2.
Personnel Protection: Section 3.
Storage and Handling: Section 7.

HELLEBORIN
General Information
Description: Colorless, lustrous crystals.
Formula: $C_{28}H_{36}O_6$.
Constants: Mol wt: 468.6, mp: $>250°C$.
Hazard Analysis
Toxicity: Details unknown. See also helleborein.
Countermeasures
Storage and Handling: Section 7.

TOXIC HAZARD RATING CODE (For detailed discussion, see Section 1.)

0 NONE: (a) No harm under any conditions; (b) Harmful only under unusual conditions or overwhelming dosage.

1 SLIGHT: Causes readily reversible changes which disappear after end of exposure.

2 MODERATE: May involve both irreversible and reversible changes; not severe enough to cause death or permanent injury.

3 HIGH: May cause death or permanent injury after very short exposure to small quantities.

U UNKNOWN: No information on humans considered valid by authors.

HELVELLIC ACID
Hazard Analysis
Toxic Hazard Rating:
  Acute Local: Irritant 2.
  Acute Systemic: Ingestion 3; Inhalation 3.
  Chronic Local: U.
  Chronic Systemic: U.
Toxicology: This is a poisonous material and a powerful hemolytic. agent. It can cause superficial punctate keratitis. The symptoms of poisoning from this material usually occur within the first few hours after ingestion and consist of nausea, vomiting, gastric pains, and diarrhea. Further, there is stupor, dilation of the pupils, enlargement of the liver and spleen, and tubular damage to the kidneys with albuminuria. In some cases death may follow from an acute yellow atrophy of the liver. In the absence of this last symptom, the prognosis is favorable.
Countermeasures
Treatment and Antidotes: Injections of glucose to protect the liver can be given.
Ventilation Control: Section 2.
Personal Hygiene: Section 3.
First Aid: Section 1.
Storage and Handling: Section 7.

HEMICELLULOSE EXTRACT
General Information
Description: A natural substance occurring in woody tissue.
Hazard Analysis
Toxic Hazard Rating: U. A food additive permitted in the feed and drinking water of animals and/or for the treatment of food producing animals (Section 10).

HEMISULFUR MUSTARD
General Information
Synonym: 2-Chloroethyl methyl sulfide.
Description: Liquid; odor of mustard gas.
Formula: $CH_3SCH_2CH_2Cl$.
Constants: Mol wt: 110.6, d: 1.1155 at 20°/4°C, bp: 140°C.
Hazard Analysis
Toxicity: Highly irritating. Very toxic. See also dichlorodiethyl sulfide.
Disaster Hazard: Dangerous; see chlorides and sulfur compounds.

HEMP
Hazard Analysis
Toxic Hazard Rating:
  Acute Local: Allergen 1; Inhalation 1.
  Acute Systemic: Inhalation 1.
  Chronic Local: Allergen 1.
  Chronic Systemic: Inhalation 1.
Fire Hazard: Slight, when exposed to heat or flame; can react with oxidizing materials (Section 6).
Countermeasures
Personal Hygiene: Section 3.
Storage and Handling: Section 7.
Shipping Regulations: Section 11.
  Coast Guard Classification: Hazardous article.

HENDECANE
General Information
Synonym: Undecane-n.
Description: Colorless liquid; insoluble in water.
Formula: $CH_3(CH_2)_9CH_3$.
Constants: Mol wt: 156, d: 0.7402 at 20/4°C, fp: −25.75°C, bp: 195.6°C, flash p: 149°F (O.C.), vap. d.: 5.4.
Hazard Analysis
Toxic Hazard Rating: U.
Fire Hazard: Moderate, when exposed to heat or flame.
Countermeasures
Storage and Handling: Section 7.

To Fight Fire: Foam, mist, dry chemical (Section 6).

HENDECENAL. See undecanal.

HENDECANOL-2. See undecanol-2.

HENNA
General Information
Description: Dried powdery leaves of Lawsonia alba, Lawsonia inermis and Lawsonia spinosa.
Hazard Analysis
Toxic Hazard Rating:
  Acute Local: Irritant 1; Allergen 1.
  Acute Systemic: Skin Absorption 1.
  Chronic Local: Irritant 1; Allergen 1.
  Chronic Systemic: Skin Absorption 1.
Toxicology: Not a harmful dye. Makes hair brittle and can irritate the skin. May be absorbed via intact skin.

HEPAR CALCIS. See calcium sulfide.

HEPAR SULFURIS. See potassium sulfide.

HEPTACHLOR
General Information
Synonyms: 1,4,5,6,7,8,8-Heptachloro-3a,4,7,7a-tetrahydro-4,7-methanoindene.
Description: Crystals.
Hazard Analysis
Toxicity: Highly toxic upon ingestion, inhalation or skin absorption. Acute exposure causes liver damage. Chronic doses have caused liver damage in experimental animals. See closely related chlordane.
TLV: ACGIH (recommended) 0.5 milligrams per cubic meter of air.
Disaster Hazard: Dangerous; when heated to decomposition, it emits highly toxic fumes.
Countermeasures
Shipping Regulations: Section 11.
  IATA: Other restricted articles, class A, no limit (passenger and cargo).

HEPTACHLOROPROPANE
General Information
Formula: C_3HCl_7.
Constant: Mol wt: 285.3.
Hazard Analysis
Toxicity: Details unknown. See also chlorinated hydrocarbons, aliphatic.
Disaster Hazard: Dangerous; when heated to decomposition, it emits highly toxic fumes of phosgene.
Countermeasures
Storage and Handling: Section 7.

1,4,5,6,7,8,8-HEPTACHLORO-3a,4,7,7a-TETRAHYDRO-4,7-METHANOINDENE. See heptachlor.

HEPTADECANOL
General Information
Synonym: Heptadecyl alcohol.
Formula: $C_{17}H_{35}OH$.
Constants: Mol wt: 256.46, mp: 54°C, bp: 309°C, flash p: 310°F (C.O.C.), d: 0.8475 at 20°/20°C, vap. press.: < 0.01 mm at 20°C, vap. d.: 8.84.
Hazard Analysis
Toxicity: Details unknown. Limited animal experiments suggest low toxicity. See also alcohols.
Fire Hazard: Slight, when exposed to heat or flame; can react with oxidizing materials.
Countermeasures
To Fight Fire: Foam, carbon dioxide, dry chemical or carbon tetrachloride (Section 6).
Storage and Handling: Section 7.

HEPTADECYL ALCOHOL. See heptadecanol.

2-HEPTADECYLGLYOXALIDINE. See 2-heptadecylimi-
doazoline.

2-HEPTADECYL GLYOXALIDINE ACETATE. See
glyodin.

2-HEPTADECYLIMIDOAZOLINE
General Information
Synonym: 2-Heptadecylglyoxalidine.
Description: Waxy solid.
Formula: $C_{20}H_{40}N_2$.
Constants: Mol wt: 308.54, mp: 85°C, bp: 200°C at 2 mm.
Hazard Analysis
Toxic Hazard Rating:
  Acute Local: Irritant 2.
  Acute Systemic: U.
  Chronic Local: U.
  Chronic Systemic: U.
Fire Hazard: Slight, when exposed to heat or flame; can re-
  act with oxidizing materials (Section 6).
Countermeasures
Personal Hygiene: Section 3.
Storage and Handling: Section 7.

2-HEPTADECYL-2-IMIDOAZOLINE ACETATE. See
glyodin.

HEPTAFLUOROBUTYRIC ACID
General Information
Description: Colorless liquid; sharp odor similar to butyric
  acid.
Formula: $CF_3CF_2CF_2COOH$.
Constants: Mol wt: 214.00, bp: 210.0°C at 735 mm.
Hazard Analysis
Toxic Hazard Rating:
  Acute Local: Irritant 3; Ingestion 3; Inhalation 3.
  Acute Systemic: U.
  Chronic Local: Irritant 3.
  Chronic Systemic: U.
Disaster Hazard: Dangerous; when heated to decomposi-
  tion, it emits highly toxic fumes of fluorides; will react
  with water or steam to produce corrosive fumes.
Countermeasures
Storage and Handling: Section 7.

HEPTANE
General Information
Synonyms: Heptyl hydride; dipropyl methane.
Description: Colorless liquid.
Formula: $CH_3(CH_2)_5CH_3$.
Constants: Mol wt: 100.20, bp: 98.52, lel: 1.2% uel: 6.7%,
  fp: −90.5°C, flash p: 25°F (C.C.), d: 0.684 at 20°/4°C,
  autoign. temp.: 433°F, vap. press.: 40 mm at 22.3°C,
  vap. d.: 3.45.
Hazard Analysis
Toxic Hazard Rating:
  Acute Local: Inhalation 2.
  Acute Systemic: Inhalation 2.
  Chronic Local: Irritant 1.
  Chronic Systemic: U.
TLV: ACGIH (recommended); 500 parts per million in air;
  2000 milligrams per cubic meter of air.
Toxicology: Irritating to respiratory tract. Narcotic in high
  concentrations.
Fire Hazard: Dangerous, when exposed to heat or flame.
Spontaneous Heating: No.
Explosion Hazard: Moderate, when exposed to heat or
  flame.

Disaster Hazard: Dangerous, upon exposure to heat or
  flame; can react vigorously with oxidizing materials.
Countermeasures
Ventilation Control: Section 2.
To Fight Fire: Foam, carbon dioxide, dry chemical or car-
  bon tetrachloride (Section 6).
Storage and Handling: Section 7.
Shipping Regulations: Section 11.
  I.C.C.: Flammable liquid, red label, 10 gallons.
  Coast Guard Classification: Inflammable liquid; red label.
  IATA: Flammable liquid, red label, 1 liter (passenger),
  40 liters (cargo).

3-HEPTANECARBOXYLIC ACID. See 2-ethylhexoic acid.

HEPTANOIC ACID
General Information
Synonym: Heptoic acid; oenanthylic acid.
Description: Oily liquid, disagreeable rancid odor; less odor
  when very pure.
Formula: $CH_3(CH_2)_5COOH$.
Constants: Mol wt: 130.2, d: 0.9345 at 0°/4°C, mp:
  −7.5°C, bp: 223.0°C.
Hazard Analysis
Toxicity: Details unknown. Limited animal experiments
  suggest low toxicity.

1-HEPTANOL
General Information
Synonym: Heptyl alcohol.
Description: Liquid.
Formula: $C_7H_{15}OH$.
Constants: Mol wt: 116.2, mp: −34.6°C, bp: 175.8°C, d:
  0.824 at 20°/4°C.
Hazard Analysis
Toxicity: See alcohols.
Fire Hazard: Moderate, when exposed to heat or flame; can
  react with oxidizing materials (Section 6).
Countermeasures
Ventilation Control: Section 2.
Personal Hygiene: Section 3.
Storage and Handling: Section 7.

2-HEPTANOL
General Information
Synonym: Amyl methyl carbinol.
Description: Liquid.
Formula: $CH_3(OH)CH(CH_2)_4CH_3$.
Constants: Mol wt: 116.2, bp: 160.4°C, flash p: 160°F
  (O.C.), d: 0.8344 at 0°C, vap. press.: 1 mm at 14.6°C,
  vap. d.: 4.01.
Hazard Analysis
Toxicity: See alcohols.
Fire Hazard: Moderate, when exposed to heat or flame; can
  react with oxidizing materials.
Explosion Hazard: Unknown.
Countermeasures
Ventilation Control: Section 2.
To Fight Fire: Foam, carbon dioxide, dry chemical or car-
  bon tetrachloride (Section 6).
Personal Hygiene: Section 3.
Storage and Handling: Section 7.

3-HEPTANOL
General Information
Description: Liquid.
Formula: $CH_3CH_2CH(OH)C_4H_9$.
Constants: Mol wt: 116.2, bp: 156.2°C, flash p: 140°F

TOXIC HAZARD RATING CODE (For detailed discussion, see Section 1.)

0 NONE: (a) No harm under any conditions; (b) Harmful only under unusual conditions or overwhelming dosage.

1 SLIGHT: Causes readily reversible changes which disappear after end
  of exposure.

2 MODERATE: May involve both irreversible and reversible changes;
  not severe enough to cause death or permanent injury.

3 HIGH: May cause death or permanent injury after very short exposure
  to small quantities.

U UNKNOWN: No information on humans considered valid by authors.

(C.O.C.), fp: $-70°C$, d: 0.8224 at $20°/20°C$, vap. press.: 0.5 mm at $20°C$, vap. d.: 4.01.

Hazard Analysis
Toxic Hazard Rating:
  Acute Local: U.
  Acute Systemic: Ingestion 2; Inhalation 2.
  Chronic Local: U.
  Chronic Systemic: U.
Fire Hazard: Moderate, when exposed to heat or flame; can react with oxidizing materials.

Countermeasures
To Fight Fire: Foam, carbon dioxide, dry chemical or carbon tetrachloride (Section 6).
Ventilation Control: Section 2.
Personal Hygiene: Section 3.
Storage and Handling: Section 7.

2-HEPTANONE. See methyl amyl ketone.

4-HEPTANONE. See butyrone.

1-HEPTENE. See α-heptylene.

2-HEPTENE
General Information
Description: Clear liquid.
Formula: C_7H_{14}.
Constants: Mol wt: 98.18, bp: $98.2°C$, flash p: $<30°F$, d: 0.709 at $20°/4°C$, vap. d.: 3.4.

Hazard Analysis
Toxic Hazard Rating:
  Acute Local: U.
  Acute Systemic: Ingestion 1; Inhalation 1.
  Chronic Local: Irritant 1.
  Chronic Systemic: U.
Fire Hazard: Dangerous; when exposed to heat or flame (Section 6).
Disaster Hazard: Dangerous, upon exposure to heat or flame; can react vigorously with oxidizing materials.

Countermeasures
Ventilation Control: Section 2.
Personal Hygiene: Section 3.
Storage and Handling: Section 7.

3-HEPTENE (Mixture of cis and trans isomers)
General Information
Synonym: 1-Ethyl-2-propyl ethylene.
Description: Liquid.
Formula: $CH_3CH_2CHCHCH_2CH_2CH_3$.
Constants: Mol wt: 98.18, bp: $96°C$, flash p: $<20°F$, d: 0.705 at $15.5°/25.5°C$, vap. d.: 3.38.

Hazard Analysis
Toxicity: Unknown. Probably irritant and narcotic in high concentrations. See also 2-heptene.
Fire Hazard: Dangerous; when exposed to heat or flame; can react vigorously with oxidizing materials.
Disaster Hazard: Dangerous, upon exposure to heat or flame.

Countermeasures
Ventilation Control: Section 2.
To Fight Fire: Foam, carbon dioxide, dry chemical or carbon tetrachloride (Section 6).
Storage and Handling: Section 7.

HEPTOIC ACID. See heptanoic acid.

HEPTYL ACETATE
General Information
Formula: $CH_3COOCH_3(CH_2)_6$.
Constant: Mol wt: 159.0.

Hazard Analysis
Toxic Hazard Rating:
  Acute Local: Irritant 2.
  Acute Systemic: Ingestion 1; Inhalation 1.
  Chronic Local: U.

Chronic Systemic: U.
Toxicology: Data based on animal experiments.

HEPTYL ALCOHOL. See 1-heptanol.

HEPTYL AMINE
General Information
Synonym: 1-Amino heptane.
Description: Colorless liquid; slightly soluble in water; soluble in alcohol or ether.
Formula: $CH_3(CH_2)_6NH_2$.
Constants: Mol wt: 115, d: 0.727 at $20/4°C$, mp: $-23°C$, bp: $155°C$, vap. d.: 4.0, flash p: $140°F$ (O.C.).

Hazard Analysis
Toxic Hazard Rating: U. See also amines.
Fire Hazard: Moderate; when exposed to heat or flame.

Countermeasures
Storage and Handling: Section 7.
To Fight Fire: Alcohol foam (Section 6).

α-HEPTYLENE
General Information
Synonym: 1-Heptene.
Description: Colorless liquid; insoluble in water; soluble in ether.
Formula: $CH_2CH(CH_2)_4CH_3$.
Constants: Mol wt: 98.2, d: 0.6969 at $20°C$, mp: $-10°C$, bp: $93.6°C$.

Hazard Analysis
Toxic Hazard Rating:
  Acute Local: 0.
  Acute Systemic: Inhalation 2.
  Chronic Local: 0.
  Chronic Systemic: U.
Toxicology: A simple asphyxiant. For effects of exposure see argon.
Fire Hazard: Probably dangerous. Details unknown.
Explosion Hazard: Details unknown.

Countermeasures
To Fight Fire: Foam, dry chemical, carbon dioxide (Section 6).
Storage and Handling: Section 7.

HEPTYL HYDRIDE. See heptane.

N-HEPTYL-p-HYDROXYBENZOATE
Hazard Analysis
Toxic Hazard Rating: U.
Note: Used as a food additive permitted in food for human consumption (Section 10).

HEROIN. See diacetyl morphine.

HESAMITE. See tetraethyl pyrophosphate.

HETP. See hexaethyl tetraphosphate.

HEXAANTIPYRINE CERIUM III PERCHLORATE
General Information
Description: Colorless, hexagonal crystals.
Formula: $[Ce(C_{11}H_{12}N_2O)_6] \cdot (ClO_4)_3$.
Constants: Mol wt: 1567.83, mp: $295-300°C$ decomposes.
Hazard Analysis and Countermeasures
See perchlorates.

HEXAANTIPYRINE LANTHANUM IODIDE
General Information
Description: Yellow crystals, water soluble.
Formula: $La(C_{66}H_{72}O_6N_{12})I_3$.
Constants: Mol wt: 1649.0, mp: $269°C$ (decomp.).
Hazard Analysis and Countermeasures
See iodides and lanthanum compounds.

HEXABENZYL DIGERMANE
General Information
Description: Colorless crystals.

Formula: $Ge_2(C_6H_5CH_2)_6$.
Constants: Mol wt: 692.0, mp: 184°C.
Hazard Analysis
Toxicity: Details unknown. See germanium compounds.

HEXABORANE
General Information
Synonyms: Boron hydride; hexaboron decahydride.
Description: Colorless liquid, turns yellow on standing.
Formula: B_6H_{10}.
Constants: Bp: 0°C at 7.2 mm, mol wt: 75.0, mp: −65.1°C, d: 0.69 at 0°C, vap. press.: 7.2 mm at 0°C, vap. d.: 2.6.
Hazard Analysis and Countermeasures
See boron hydride.

HEXABORON DECAHYDRIDE. See hexaborane.

HEXABROMODISILANE. See disilicon hexabromide.

HEXABROMOETHANE. See dicarbon hexabromide.

HEXACALCIUM PHYTATE. See calcium phytate.

γ-HEXACHLORIDE. See benzene hexachloride.

HEXACHLOROACETONE
General Information
Description: Colorless liquid.
Formula: CCl_3COCCl_3.
Constants: Mol wt: 264.8, bp: 202–204°C, fp: −2°C, vap. d.: 9.2.
Hazard Analysis
Toxic Hazard Rating:
  Acute Local: Irritant 2; Ingestion 2; Inhalation 2.
  Acute Systemic: U.
  Chronic Local: U.
  Chronic Systemic: U.
Disaster Hazard: Dangerous; when heated to decomposition, it emits highly toxic fumes of phosgene.
Countermeasures
Ventilation Control: Section 2.
Personnel Protection: Section 3.
Storage and Handling: Section 7.

HEXACHLOROBENZENE
General Information
Synonym: Perchlorobenzene.
Description: Monoclinic prisms.
Formula: C_6Cl_6.
Constants: Mol wt: 284.80, mp: 230°C, bp: 326°C, flash p: 468°F, d: 1.5, vap. press.: 1 mm at 114.4°C, vap. d.: 9.8.
Hazard Analysis
Toxic Hazard Rating:
  Acute Local: Irritant 1.
  Acute Systemic: Ingestion 1.
  Chronic Local: Irritant 1.
  Chronic Systemic: U.
Toxicology: Limited animal experiments suggest low toxicity. See also chlorinated hydrocarbons, aromatic.
Fire Hazard: Slight, when exposed to heat or flame.
Disaster Hazard: Dangerous; when heated to decomposition, it emits highly toxic fumes of chlorides.
Countermeasures
Ventilation Control: Section 2.
To Fight Fire: Carbon dioxide, dry chemical or carbon tetrachloride (Section 6).
Personal Hygiene: Section 3.

First Aid: Section 1.
Storage and Handling: Section 7.

1,2,3,4,7,7-HEXACHLORO-5,6-BIS(CHLOROMETHYL)-2-NORBORNENE
Hazard Analysis
Toxic Hazard Rating: U.
Toxicity: Limited data suggest toxicity similar to but less than DDT. See also DDT.
Disaster Hazard: Dangerous. See chlorides.

1,2,3,4,5,6-HEXACHLOROCYCLOHEXANE.
General Information
Synonyms: Benzene hexachloride; gammahexane; lindane; streunex; BHC, DBH, HCCH, HCH; "666"; "Gammexane"; "Benzahex"; "Chemhex"; "Gamoxol"; "Hexadon"; and other trade names.
Description: White crystalline powder.
Formula: $C_6H_6Cl_6$.
Constants: Mol wt: 290.84, mp: 157°C, vap. press.: 0.0317 mm at 20°C.
Hazard Analysis
Toxic Hazard Rating:
  Acute Local: Allergen 1.
  Acute Systemic: Ingestion 3; Inhalation 2.
  Chronic Local: Allergen 2.
  Chronic Systemic: Ingestion 2; Inhalation 2.
TLV: ACGIH (recommended); 0.5 milligrams per cubic meter of air. May be absorbed via the skin.
Disaster Hazard: Dangerous. See chlorides.
Toxicology: Hexachlorocyclohexane, a local irritant, may be absorbed through the skin. The several isomers of hexachlorocyclohexane have different actions. The gamma and alpha isomers are CNS stimulants, the principal symptom being convulsions. The beta and delta isomers are depressants of the central nervous system.
  The dangerous acute dose of the technical mixture has been estimated at about 30 g. and the dangerous dose of lindane at about 7 to 15 g. These estimates may be too high, for a young man suffered a serious illness, including convulsions, following a single carefully measured dose of 45 mg intended as a vermifuge. It is true that, in similar studies of the potential use of benzene hexachloride as a drug, other persons have withstood larger doses, especially of undissolved material, apparently without injurious effect. Thus 40 mg of a purified commercial preparation was tolerated daily for 10 days. This mixture of isomers burned the tongue, and unpurified mixtures were found to be very irritating. Doses of 70 mg of the preparation caused attacks of dizziness, slight nausea, headache, and a sensation of pressure in the temples. No change was indicated in the blood or urine. Preparations with more than 25 percent gamma isomer were slightly less toxic. Forty mg of pure gamma isomer were given for 14 days with no bodily disturbance. A daily dose of 180 mg caused diarrhea and a feeling of dizziness. A dose of 30 mg three times a day for a week caused no injurious signs in the patient, and no change in the urine or blood. However, as already mentioned, a single dose of 45 mg (or approximately 0.65 mg/kg) of lindane caused convulsions.
  It is interesting to note that lindane shows a marked difference in toxicity to different species. Its toxic effect on laboratory animals compares favorably with that of

---

TOXIC HAZARD RATING CODE (For detailed discussion, see Section 1.)

0 NONE: (a) No harm under any conditions; (b) Harmful only under unusual conditions or overwhelming dosage.

1 SLIGHT: Causes readily reversible changes which disappear after end of exposure.

2 MODERATE: May involve both irreversible and reversible changes; not severe enough to cause death or permanent injury.

3 HIGH: May cause death or permanent injury after very short exposure to small quantities.

U UNKNOWN: No information on humans considered valid by authors.

DDT, but for several domestic animals, notably calves, lindane is more toxic than DDT or dieldrin.

Four children drank undetermined amounts of a home-made soft drink containing lindane, which had been mixed with sugar. Within less than 6 hours three vomited and had convulsions. The fourth did likewise within about 12 hours. All recovered without treatment. It was not possible to determine the dosage of lindane.

The fatal poisoning of a 5-year old girl weighing 55 pounds was caused by the accidental ingestion of 4.5 g of hexachlorocyclohexane as a 30 percent solution in an unspecified organic solvent. This represents a dosage of 180 mg/kg. Shortly afterward, she developed dyspnea, cyanosis, and clonic-tonic convulsions. In spite of evacuation of her stomach and therapy to restore failing circulation, she died. Autopsy showed pulmonary edema, dilation of the heart, fatty infiltration of the liver, and extensive necrosis of blood vessels in the lungs, kidney and liver.

The use of thermal vaporizers with lindane has caused some clear-cut instances of acute poisoning. For example, two refreshment-stand operators suffered severe headache, nausea, and irritation of eyes, nose, and throat shortly after exposure to lindane vapors from a dispenser in which the insecticide apparently became overheated. The symptoms abated 2 hours after the device was removed. Overheated lindane is more apt to cause respiratory distress than lindane which is vaporized at lower temperatures. This is true because heat releases more of the compound and also causes some splitting of the molecule into highly irritating decomposition products.

Nothing is known of chronic systemic benzene hexachloride poisoning from the study of human subjects. In laboratory animals the gamma isomer has by far the greatest acute toxicity, but its relatively rapid excretion by the kidneys does not permit extensive accumulation in the body and the gamma isomer shows the lowest toxicity on repeated exposure. The highest chronic toxicity and the lowest acute toxicity are shown by the beta isomer. This has no insecticidal importance but forms a part of those formulations prepared from technical grade BHC. The use of lindane is therefore favored not only because it is the most effective form of hexachlorocyclohexane for killing insects, but also because it probably presents a relatively low chronic toxicity to workers.

Dermatitis, and perhaps other manifestations based on sensitivity represent a sort of chronic, though probably not systemic intoxication, which has been observed in human beings. Dermatitis has been reported in workers who came in contact with benzene hexachloride and its precursors during manufacture and without proper hygienic precautions. Shortly after a lindane vaporizer was installed in her place of employment, a 35-year-old woman developed urticaria. The dermatitis improved during weekends, but recurred when she returned to work. Patch tests were positive. Complete elimination of exposure resulted in permanent recovery (Section 9).

The signs and symptoms of confirmed acute poisoning in man have paralleled those of experimental animals. These signs and symptoms are: excitation, hyperirritability, loss of equilibrium, clonic-tonic convulsions, and later depression. There is some evidence that the pulmonary edema and vascular collapse may be of neurogenic origin also. The symptoms in animals systemically poisoned by the gamma isomer alone are essentially similar to those caused by mixtures, although the onset may be earlier. Men acutely exposed to high air concentrations of lindane and its decomposition products show headache, nausea, and irritation of eyes, nose and throat.

Urticaria has followed exposure to lindane vapor in rare instances. Unlike the signs and symptoms already mentioned, this allergic manifestation occurs only in susceptible individuals, and usually only after a period of sensitization.

Laboratory findings are essentially normal, Insecticide may be demonstrated in the tissue of the poisoned animal by the use of bioassay or by a specific chemical test.

Pathology in animals from both acute and chronic poisoning is quite similar to that seen in DDT poisoning see DDT.

Treatment is essentially the same as that for DDT. Wash contaminated skin thoroughly with soap and water. If the poison has been taken internally, clear the alimentary tract with pentobarbital or with phenobarbital. These drugs should be given in large doses and pushed to the limit of therapeutic effectiveness. It is not desirable to depress the patient to the point of hypnosis. However, persons poisoned by stimulants may tolerate large doses of barbiturates without undue depression and with great benefit. Intravenous calcium gluconate has been reported to have some effectiveness as an antidote and may be used in conjunction with the barbiturates.

Disaster Hazard: Dangerous; when heated to decomposition, it emits highly toxic fumes of phosgene.

Countermeasures
Ventilation Control: Section 2.
Personnel Protection: Section 3.
First Aid: Section 1.
Storage and Handling: Section 7.
MCA warning label.

HEXACHLOROCYCLOPENTADIENE
General Information
Description: Yellow to amber colored liquid; pungent odor.
Formula: C_5Cl_6.
Constants: Mol wt: 272.79, mp: 9.9°C, bp: 239°C, fp: −2°C, flash p: none (O.C.), d: 1.715 at 15.5°/15.5°C, vap. d.: 9.42.
Hazard Analysis
Toxicity: This is a toxic material. Readily absorbed via intact skin. Inhalation of vapors causes effects similar to inhalation of carbon tetrachloride.
Disaster Hazard: Dangerous; see chlorides.
Countermeasures
Ventilation Control: Section 2.
Personnel Protection: Section 3.
Storage and Handling: Section 7.

HEXACHLORODIPHENYL OXIDE
General Information
Description: Light yellow, very viscous liquid.
Formula: $C_{12}H_4Cl_6O$.
Constants: Mol wt: 376.9, bp: 230–260°C at 8 mm, d: 1.60 at 20°/60°C.
Hazard Analysis
Toxicity: Unknown. Animal experiments show that all compounds of this class are highly toxic, causing liver injury and acne like skin rashes. See also chlorinated phenols.
Disaster Hazard: Dangerous; see chlorides.
Countermeasures
Ventilation Control: Section 2.
Storage and Handling: Section 7.

HEXACHLORODISILANE. See disilicon hexachloride.

1,2,3,4,10,10-HEXACHLORO-6,7-EPOXY-1,4,4a,5,6,7,8,-8a-OCTAHYDRO-1,4,5,8-ENDO-ENDO-DIMETH-ANONAPHTHALENE. See endrin.

1,2,3,4,10,10-HEXACHLORO-6,7-EPOXY-1,4,4a,5,6,7,8,-8a-OCTAHYDRO-1,4,5,8-ENDO-EXO-DIMETH-ANONAPHTHALENE. See dieldrin.

HEXACHLOROETHANE
General Information
Synonyms: Carbon trichloride; carbon hexachloride.
Description: Rhombic, triclinic or cubic crystals; colorless; camphor-like odor.
Formula: CCl_3CCl_3.
Constants: Mol wt: 236.76, mp: 186.6°C (sublimes), d: 2.091, vap. press.: 1 mm at 32.7°C.
Hazard Analysis
Toxic Hazard Rating:
  Acute Local: Irritant 2; Ingestion 2; Inhalation 2.
  Acute Systemic: Inhalation 2.
  Chronic Local: Irritant 2.
  Chronic Systemic: Ingestion 2.
TLV: ACGIH (recommended); 1 part per million; 10 milligrams per cubic meter of air. May be absorbed via the skin.
Toxicology: Liver injury has been described from exposure to this material. See also chlorinated hydrocarbons.
Explosion Hazard: Slight, by spontaneous chemical reaction. Dehalogenation of this material by reaction with alkalies, metals, etc., will produce spontaneously explosive chloroacetylenes.
Disaster Hazard: Dangerous; when heated to decomposition, it emits highly toxic fumes of phosgene.
Countermeasures
Ventilation Control: Section 2.
Personal Hygiene: Section 3.
Storage and Handling: Section 7.

1,2,3,4,10,10-HEXACHLORO-1,4,4a,8,8a-HEXAHYDRO-1,4,5,8-ENDO-ENDO-DIMETHANONAPHTHALENE
Hazard Analysis
Toxicity: An insecticide. Highly toxic. See closely related compound aldrin.
Disaster Hazard: Dangerous. See chlorides.

1,2,3,4,10,10-HEXACHLORO-1,4,4a,5,8,8a-HEXAHYDRO-1,4,5,8-ENDO-EXO-DIMETHANONAPHTHALENE. See aldrin.

HEXACHLOROMETHYL CARBONATE
General Information
Synonym: Triphosgene.
Description: White crystals.
Formula: $(OCCl_3)_2CO$.
Constants: Mol wt: 296.77, mp: 78–79°C, bp: 205–206°C (partial decomposition), d: 2 (approx.).
Hazard Analysis
Toxic Hazard Rating:
  Acute Local: Irritant 3; Ingestion 3; Inhalation 3.
  Acute Systemic: Inhalation 3.
  Chronic Local: U.
  Chronic Systemic: U.
Disaster Hazard: Dangerous; when heated to decomposition, it emits highly toxic fumes of chlorides.
Countermeasures
Ventilation Control: Section 2.
Personnel Protection: Section 3.
First Aid: Section 1.
Storage and Handling: Section 7.

HEXACHLOROMETHYL ETHER
General Information
Description: Liquid.
Formula: $O(CCl_3)_2$.

Constants: Mol wt: 252.76, bp: 98°C (partial decomp.), d: 1.538 at 18°C.
Hazard Analysis
Toxic Hazard Rating:
  Acute Local: Irritant 3; Ingestion 3; Inhalation 3.
  Acute Systemic: Ingestion 3; Inhalation 3.
  Chronic Local: U.
  Chronic Systemic: U.
Disaster Hazard: Dangerous; when heated to decomposition, it emits highly toxic fumes of chlorides.
Countermeasures
Ventilation Control: Section 2.
Personnel Protection: Section 3.
First Aid: Section 1.
Storage and Handling: Section 7.

HEXACHLORONAPHTHALENE
General Information
Description: White solid.
Formula: $C_{10}H_2Cl_6$.
Constant: Mol wt: 334.9.
Hazard Analysis
Toxic Hazard Rating:
  Acute Local: Irritant 2; Inhalation 2.
  Acute Systemic: Ingestion 3; Inhalation 3.
  Chronic Local: Irritant 3.
  Chronic Systemic: Ingestion 3; Inhalation 3.
TLV: ACGIH (tentative); 0.2 milligrams per cubic meter of air. May be absorbed through the skin.
Toxicology: Causes severe acneform eruptions and toxic narcosis of liver.
Disaster Hazard: Dangerous; see chlorides.
Countermeasures
Ventilation Control: Section 2.
Personnel Protection: Section 3.
First Aid: Section 1.
Storage and Handling: Section 7.

HEXACHLOROPHENE. See 2,2'-methylene bis(3,4,6-trichlorophenol).

HEXACHLOROPROPENE
General Information
Formula: C_3Cl_6.
Constant: Mol wt: 248.8.
Hazard Analysis
Toxic Hazard Rating:
  Acute Local: Irritant 3; Ingestion 3; Inhalation 3.
  Acute Systemic: Ingestion 3; Inhalation 3.
  Chronic Local: U.
  Chronic Systemic: U.
Toxicology: Animal experiments show high toxicity and irritant properties. Is not narcotic.
Disaster Hazard: Dangerous; see chlorides.
Countermeasures
Storage and Handling: Section 7.

n-HEXADECANE
General Information
Synonym: Cetane.
Description: Colorless liquid, soluble in alcohol, acetone, ether; insoluble in water.
Formula: $C_{16}H_{34}$.
Constants: Mol wt: 226, d: 0.77335 at 20/4°C, bp: 286.5°C, mp: 18.14°C, vap. d.: 7.8, autoign. temp.: 401°F.

TOXIC HAZARD RATING CODE (For detailed discussion, see Section 1.)

0 NONE: (a) No harm under any conditions; (b) Harmful only under unusual conditions or overwhelming dosage.

1 SLIGHT: Causes readily reversible changes which disappear after end of exposure.

2 MODERATE: May involve both irreversible and reversible changes; not severe enough to cause death or permanent injury.

3 HIGH: May cause death or permanent injury after very short exposure to small quantities.

U UNKNOWN: No information on humans considered valid by authors.

Hazard Analysis
Toxic Hazard Rating: U.
Fire Hazard: Moderate when exposed to heat or flame.

tert-HEXADECANETHIOL
General Information
Synonym: tert-Hexadecylmercaptan.
Description: Colorless liquid; insoluble in water.
Formula: $C_{16}H_{33}SH$.
Constants: Mol wt: 258, boiling range: 121–149°C at 5 mm, d: 0.874 at 60/60°F, flash p: 265°F (O.C.).
Hazard Analysis
Toxic Hazard Rating: U.
Fire Hazard: Slight, when exposed to heat or flame.
Disaster Hazard: Dangerous, when heated to decomposition it emits highly toxic fumes.
Countermeasures
To Fight Fire: Water or foam (Section 6).
Storage and Handling: Section 7.

HEXADECANOL. See cetyl alcohol.

1-HEXADECINE
General Information
Synonym: 1-Hexadecyne.
Description: Liquid or crystals.
Formula: $CH_3(CH_2)_{13}C:CH$.
Constants: Mol wt: 222.40, mp: 15°C, bp: 274.0°C, d: 0.797 at 20°C, vap. press.: 1 mm at 101.6°C, vap. d.: 7.68.
Hazard Analysis
Toxicity: Unknown.
Fire Hazard: Moderate, when exposed to heat or flame; can react with oxidizing materials.
Countermeasures
To Fight Fire: Foam, carbon dioxide, dry chemical or carbon tetrachloride (Section 6).
Storage and Handling: Section 7.

HEXADECYL ACETYLENE. See 1-octadecene.

tert-HEXADECYL MERCAPTAN. See tert-hexadecanethiol.

HEXADECYL TETRATHIO-o-STANNATE
General Information
Description: Solid.
Formula: $Sn(SC_{16}H_{33})_4$.
Constants: Mol wt: 1148.7, mp: 54°C.
Hazard Analysis and Countermeasures
See tin and sulfur compounds.

HEXADECYL TRICHLOROSILANE
General Information
Description: Colorless to yellow liquid.
Formula: $C_{16}H_{33}SiCl_3$.
Constants: Mol wt: 259.5, d: 0.996 at 25/25°C, bp: 269°C, flash p: 295°F (C.O.C.).
Hazard Analysis
Toxic Hazard Rating: U.
Fire Hazard: Slight, when exposed to heat or flame.
Disaster Hazard: Dangerous. See chlorides.
Countermeasures
Storage and Handling: Section 7.
Shipping Regulations: Section 11.
  I.C.C.: Corrosive liquid, white, 10 gallons.
  IATA: Corrosive liquid, white, not acceptable (passenger), 40 liters (cargo).

1-HEXADECYNE. See 1-hexadecine.

2,4-HEXADIENAL
General Information
Description: Very slightly soluble in water.
Formula: $CH_3CH:CHCH:CHC(O)H$.
Constants: Mol wt: 96, d: 0.9, bp: 339°F, lel: 1.3%, uel: 8.1%, flash p: 154°F (O.C.).

Hazard Analysis
Toxic Hazard Rating: U.
Fire Hazard: Moderate, when exposed to heat or flame.
Countermeasures
Storage and Handling: Section 7.
To Fight Fire: Foam, mist, dry chemical.

1,4-HEXADIENE
General Information
Synonym: Allyl propenyl.
Description: Colorless liquid, insoluble in water.
Formula: $CH_3CH:CHCH_2CH:CH_2$.
Constants: Mol wt: 82, d: 0.6996 at 20/4°C, bp: 64°C at 745 mm, flash p: −6°F, vap. d.: 2.8, lel: 2.0%, uel: 6.1%.
Hazard Analysis
Toxic Hazard Rating: U.
Fire Hazard: Dangerous, when exposed to heat or flame.
Countermeasures
Storage and Handling: Section 7.
To Fight Fire: Foam, fog (Section 6).

2,4-HEXADIENOIC ACID. See sorbic acid.

1,5-HEXADIYNE. See dipropargyl.

"HEXADOW." See 1,2,3,4,5,6-hexachlorocyclohexane.

HEXAETHYL BENZENE
General Information
Description: Colorless crystals.
Formula: $C_6(C_2H_5)_6$.
Constants: Mol wt: 246.42, mp: 130°C, bp: 298°C, d: 0.831 at 130°C, vap. press.: 10 mm at 150.3°C.
Hazard Analysis
Toxicity: Unknown.
Fire Hazard: Moderate, when exposed to heat or flame; can react with oxidizing materials.
Countermeasures
Storage and Handling: Section 7.

HEXAETHYL DIGERMANE
General Information
Description: Colorless liquid, insoluble in water.
Formula: $Ge_2(C_2H_5)_6$.
Constants: Mol wt: 319.6, mp: < −60°C, bp: 266°C.
Hazard Analysis
Toxicity: Details unknown. See germanium compounds.

HEXAETHYL DILEAD. See triethyl lead.

HEXAETHYL DISTANNANE
General Information
Description: Liquid.
Formula: $[Sn(C_2H_5)_3]_2$.
Constants: Mol wt: 411.8, d: 1.412 at 20°C, bp: 160°C at 23 mm.
Hazard Analysis
Toxicity: Details unknown. See tin compounds.

HEXAETHYLENE GLYCOL
General Information
Description: Liquid.
Formula: $(OH)_2(CH_2)_{12}O_5$.
Constant: Mol wt: 282.3.
Hazard Analysis
Toxicity: Details unknown. See also glycols.
Fire Hazard: Moderate, when exposed to heat or flame; can react with oxidizing materials.
Countermeasures
Storage and Handling: Section 7.

HEXAETHYL TETRAPHOSPHATE
General Information
Synonyms: Ethyl tetraphosphate; HETP.
Description: Liquid.
Formula: $(C_2H_5O)_6P_4O_7$.

Constants: Mol wt: 506.4, mp: $-40°$C, bp: decomp.
above 150°C, d: 1.2917 at $27°/4°$C.
Hazard Analysis
Toxic Hazard Rating:
  Acute Local: U.
  Acute Systemic: Ingestion 3; Inhalation 3; Skin Absorption 3.
  Chronic Local: U.
  Chronic Systemic: U.
Toxicology: A cholinesterase inhibitor. An insecticide. See parathion.
Disaster Hazard: See tetraethyl pyrophosphate.
Countermeasures
Ventilation Control: Section 2.
Personnel Protection: Section 3.
First Aid: Section 1.
Storage and Handling: Section 7.
Shipping Regulations: Section 11.
  Coast Guard Classification: Poison B; poison label.
  I.C.C.: Poison B, poison label, 1 quart.
  IATA: Poison B, poison label, not acceptable (passenger), 1 liter (cargo).

HEXAETHYL TETRAPHOSPHATE AND COMPRESSED GAS MIXTURES
Shipping Regulations: Section 11.
  I.C.C.: Poison A, poison gas, not acceptable.
  Coast Guard Classification: Poison A; poison gas label.
  IATA: Poison A, not acceptable (passenger and cargo).

HEXAETHYL TETRAPHOSPHATE MIXTURE, DRY
Shipping Regulations: Section 11.
  I.C.C.: Poison B, poison label, 200 pounds.
  IATA (containing more than 2% hexaethyl tetraphosphate): Poison B, poison label, not acceptable (passenger), 95 kilograms (cargo).
  IATA (containing not more than 2% hexaethyl tetraphosphate(: Poison B, poison label, 25 kilograms (passenger), 95 kilograms (cargo).

HEXAETHYL TETRAPHOSPHATE MIXTURE, LIQUID
Shipping Regulations: Section 11.
  I.C.C.: Poison B, poison label, 1 quart.
  IATA (containing more than 25% hexaethyl tetraphosphate): Poison B, poison label, not acceptable (passenger), 1 liter (cargo).
  IATA (containing not more than 25% hexaethyl tetraphosphate): Poison B, poison label, 1 liter (passenger and cargo).

HEXAFLUORODICHLOROCYCLOPENTENE
General Information
Formula: $C_5F_6Cl_2$.
Constant: Mol wt: 245.
Hazard Analysis
Toxic Hazard Rating:
  Acute Local: Inhalation 3.
  Acute Systemic: Inhalation 3.
  Chronic Local: Inhalation 2.
  Chronic Systemic: Inhalation 2.
Toxicity: Animal experiments show skin irritation and injury to lungs and kidneys.
Disaster Hazard: Dangerous. See fluorides and chlorides.

HEXAFLUORODISILANE. See disilicon hexafluoride.

HEXAFLUOROETHANE
General Information
Synonym: Refrigerant 116.
Description: A colorless stable gas.
Formula: CF_3CF_3.
Constants: Mol wt: 138, bp: $-78.2°$C.
Hazard Analysis
Toxic Hazard Rating: U. Most likely a simple asphyxiant.
Disaster Hazard: Moderate. This material is very stable however, when decomposed by heat it can yield highly toxic fumes.
Countermeasures
Shipping Regulations: Section 11.
  IATA: Nonflammable gas, green label, 70 kilograms (passenger), 140 kilograms (cargo).

HEXAFLUOROPHOSPHORIC ACID
General Information
Description: Corrosive, colorless, clear liquid.
Formula: HPF_6.
Constants: Mol wt: 145.99, mp: 31°C, d: 1.65.
Hazard Analysis
Toxicity: Highly toxic. See fluorides, hydrofluoric acid and phosphoric acid.
Disaster Hazard: Dangerous; when heated to decomposition, it emits highly toxic fumes of fluorides and oxides of phosphorus.
Countermeasures
Ventilation Control: Section 2.
Personnel Protection: Section 3.
First Aid: Section 1.
Storage and Handling: Section 7.
Shipping Regulations: Section 11.
  I.C.C.: Corrosive liquid, white label, 1 gallon.
  Coast Guard Classification: Corrosive liquid; white label.
  IATA: Corrosive liquid, white label, not acceptable (passenger), 5 liters (cargo).

HEXAFLUOROPROPYLENE
General Information
Synonym: Perfluoropropene.
Description: Gas.
Formula: $CF_3CF:CF_2$.
Constants: Mol wt: 112, mp: 156°C, bp: 29°C, d: 1.583 at $-40/4°$C.
Hazard Analysis
Toxic Hazard Rating: U.
Toxicity: Animal experiments suggest moderate toxicity.
Disaster Hazard: Dangerous. See fluorides.
Countermeasures
Shipping Regulations: Section 11.
  I.C.C.: Nonflammable gas, green label, 300 pounds.
  IATA: Nonflammable gas, green label, 70 kilograms (passenger), 140 kilograms (cargo).

HEXAHYDRIC ALCOHOL. See sorbitol.

HEXAHYDROANILINE. See cyclohexylamine.

HEXAHYDROBENZENE. See cyclohexane.

HEXAHYDROBENZOIC ACID. See naphthenic acid.

HEXAHYDROCRESOL. See methyl cyclohexanol.

HEXAHYDRO-1,4-DIAZENE
Hazard Analysis
Toxic Hazard Rating: U.

TOXIC HAZARD RATING CODE (For detailed discussion, see Section 1.)

0 NONE: (a) No harm under any conditions; (b) Harmful only under unusual conditions or overwhelming dosage.

1 SLIGHT: Causes readily reversible changes which disappear after end of exposure.

2 MODERATE: May involve both irreversible and reversible changes; not severe enough to cause death or permanent injury.

3 HIGH: May cause death or permanent injury after very short exposure to small quantities.

U UNKNOWN: No information on humans considered valid by authors.

Toxicity: Limited animal experiments suggest moderate toxicity and high irritation.

HEXAHYDROMETHYLPHENOL. See methylcyclohexanol.

HEXAHYDROPHENOL. See cyclohexanol.

cis-HEXAHYDROPHTHALIC ANHYDRIDE
General Information
Synonym: cis-1,2-Cyclohexane dicarboxylic anhydride.
Description: Clear, colorless, viscous liquid or glassy solid.
Formula: $C_8H_{10}O_3$.
Constants: Mol wt: 154.1, bp: 158°C at 17 mm, mp: 35–36°C, vap. d.: 5.31.
Hazard Analysis
Toxicity: A primary skin irritant and possibly dangerous skin sensitizer. Vapors are irritating to skin and mucous membranes. Inhalation of high concentrations as vapor may result in serious injury to respiratory apparatus. Particularly dangerous to eyes. It appears to have a low systemic toxicity by ingestion.
Fire Hazard: Slight, when exposed to heat or flame; can react with oxidizing materials (Section 6).
Countermeasures
Storage and Handling: Section 7.
Ventilation Control: Section 2.
Personal Hygiene: Section 3.

HEXAHYDROPYRAZINE. See piperazine.

HEXAHYDROPYRIDINE. See piperidine.

HEXAHYDROTHYMOL. See menthol.

HEXAHYDROTOLUENE. See methylcyclohexane.

HEXAHYDRO-1,3,5-TRIPHENYL-SYM-TRIAZINE. See methylene aniline.

HEXAHYDROXYCYCLOHEXANE. See inositol.

HEXAHYDROXYL. See p-dimethyl cyclohexane.

HEXAIODOSILANE. See disilicon hexaiodide.

n-HEXALDEHYDE
General Information
Description: Liquid.
Formula: $C_5H_{11}CHO$.
Constants: Mol wt: 100.16, mp: −56.3°C, bp: 128.7°C, flash p: 90°F (O.C.), d: 0.8156 at 20°/20°C, vap. press.: 8.6 mm at 20°C, vap. d.: 3.45.
Hazard Analysis
Toxicity: Details unknown. Limited animal experiments suggest low toxicity and moderate irritation. See also aldehydes.
Fire Hazard: Moderate, when exposed to heat or flame; can react with oxidizing materials.
Countermeasures
To Fight Fire: Foam, carbon dioxide, dry chemical or carbon tetrachloride (Section 6).
Ventilation Control: Section 2.
Storage and Handling: Section 7.
Shipping Regulations: Section 11.
 Coast Guard Classification: Combustible liquid.

HEXALIN. See cyclohexanal.

HEXALIN ACETATE. See cyclohexyl acetate.

HEXAMETHYL DISILANE
General Information
Description: Liquid.
Formula: $(CH_3)_3$-Si-Si-$(CH_3)_3$.
Constants: Mol wt: 146.3, mp: 13°C, bp: 112.5°C.
Hazard Analysis
Toxicity: Details unknown. See silanes.

HEXAMETHYLDISOLOXANE. See "DC 200 fluid."

HEXAMETHYLENE. See cyclohexanes.

HEXAMETHYLENEDIAMINE
General Information
Synonym: 1,6-Hexanediamine.
Description: Colorless, silk leaf.
Formula: $NH_2(CH_2)_6NH_2$.
Constants: Mol wt: 116.21, mp: 42°C, bp: 205°C.
Hazard Analysis
Toxic Hazard Rating:
 Acute Local: Irritant 2; Ingestion 2; Inhalation 2.
 Acute Systemic: U.
 Chronic Local: U.
 Chronic Systemic: U.
Fire Hazard: Slight, when exposed to heat or flame; can react with oxidizing materials (Section 6).
Countermeasures
Ventilation Control: Section 2.
Personnel Protection: Section 3.
Storage and Handling: Section 7.
Shipping Regulations: Section 11.
 I.C.C.: Corrosive liquid, white label, 10 gallons.
 Coast Guard Classification: Corrosive liquid; white label.
 IATA: Corrosive liquid, white label 1 liter (passenger), 40 liters (cargo).

HEXAMETHYLENE GLYCOL. See hexanediol-1,6.

HEXAMETHYLENETETRAMINE
General Information
Synonyms: Methenamine; formamine; hexamine; urotropin.
Description: Odorless, rhombic crystals from alcohol.
Formula: $(CH_2)_6N_4$.
Constants: Mol wt: 140.19, mp: 280°C, subl. flash p: 482°F, d: 1.331 at −5°C.
Hazard Analysis
Toxic Hazard Rating:
 Acute Local: Irritant 2; Ingestion 2.
 Acute Systemic: Ingestion 2.
 Chronic Local: Irritant 1.
 Chronic Systemic: Ingestion 1.
Toxicology: Some persons suffer a skin rash if they come in contact with this material or the fumes evolved when it is heated (Section 9). Pure hexamethylenetetramine may be taken internally in small amounts and is used in medicine as a urinary antiseptic. Its major industrial use is in the manufacture of phenolic resins. It is combustible and can be readily ignited when a flame is applied directly to its surface. It liberates formaldehyde on decomposition.
Fire Hazard: Slight, when exposed to heat or flame; can react with oxidizing materials.
Countermeasures
Ventilation Control: Section 2.
Personal Hygiene: Section 3.
Storage and Handling: Section 7.
Shipping Regulations: Section 11.
 IATA: Flammable solid, yellow label, 12 kilograms (passenger), 95 kilograms (cargo).

HEXAMETHYLENIMINE
General Information
Synonym: HMI.
Description: Clear colorless liquid, water soluble, ammonia-like odor.
Formula: $C_6H_{13}N$.
Constants: Mol wt: 99.2, bp: 138°C, mp: −37°C.
Hazard Analysis
Toxicity: Details unknown. Probably an irritant and possibly a skin sensitizer.

HEXAMETHYLPHOSPHORAMIDE
General Information
Description: Clear, colorless, mobile liquid; spicy odor.

Formula: $[(CH_3)_2N]_3PO$.
Constants: Mol wt: 179.1, bp: 98–100°C at 6 mm, fp: 4°C, d: 1.024 at 25°/25°C, vap. d.: 6.18.
Hazard Analysis
Toxicity: Unknown.
Disaster Hazard: Dangerous; see phosphorus compounds.
Countermeasures
Storage and Handling: Section 7.

HEXAMINE. See hexamethylenetetramine.

HEXAMMINECHROMIUM III CHLORIDE
General Information
Description: Yellow crystals.
Formula: $[Cr(NH_3)_6]Cl_3 \cdot H_2O$.
Constants: Mol wt: 278.59, d: 1.585.
Hazard Analysis and Countermeasures
See chromium compounds and chlorides.

HEXAMMINE COBALT TRICHLORIDE
Hazard Analysis and Countermeasures
See cobalt and chlorides.

HEXAMMINENICKEL (II) NITRATE
General Information
Description: Cubic, blue crystals.
Formula: $[Ni(NH_3)_6](NO_3)_2$.
Constant: Mol wt: 284.90.
Hazard Analysis and Countermeasures
See nitrates.

HEXANAL. See n-hexaldehyde.

n-HEXANE
General Information
Synonym: Hexyl hydride.
Description: Colorless liquid.
Formula: $CH_3(CH_2)_4CH_3$.
Constants: Mol wt: 86.17, bp: 68.7°C, ulc: 90–95, lel: 1.2%, uel: 7.5%, fp: – 95.6°C, flash p: – 7°F, d: 0.6603 at 20°/4°C, autoign. temp.: 500°F, vap. press.: 100 mm at 15.8°C, vap. d.: 2.97.
Hazard Analysis
Toxic Hazard Rating:
Acute Local: Irritant 1.
Acute Systemic: Ingestion 1; Inhalation 1.
Chronic Local: Irritant 1.
Chronic Systemic: U.
Note: Use as a food additive permitted in food for human consumption (Section 10).
TLV: ACGIH (recommended); 500 parts per million in air; 1760 milligrams per cubic meter of air.
Fire Hazard: Dangerous, when exposed to heat or flame.
Spontaneous Heating: No.
Explosion Hazard: Moderate, when exposed to heat or flame.
Disaster Hazard: Dangerous; when heated or exposed to flame; can react vigorously with oxidizing materials.
Countermeasures
Ventilation Control: Section 2.
To Fight Fire: Carbon dioxide, dry chemical or carbon tetrachloride (Section 6).
Personal Hygiene: Section 3.
Storage and Handling: Section 7.
Shipping Regulations: Section 11.
I.C.C.: Flammable liquid, red label, 10 gallons.
Coast Guard Classification: Inflammable liquid; red label.
IATA: Flammable liquid, red label, 1 liter (passenger), 40 liters (cargo).

1,6-HEXANEDIAMINE. See hexamethylenediamine.

HEXANEDIOIC ACID. See adipic acid.

HEXANEDIOL-1,6
General Information
Synonyms: 1,6-Dihydroxyhexane; hexamethylene glycol.
Description: Clear liquid.
Formula: $CH_2OH(CH_2)_4CH_2OH$.
Constants: Mol wt: 118.2, mp: 42°C, bp: 250°C, flash p: 205°F, d: 0.967 at 0°/4°C, vap. d.: 4.07.
Hazard Analysis
Toxicity: See glycols.
Fire Hazard: Moderate, when exposed to heat or flame; can react with oxidizing materials.
Spontaneous Heating: No.
Countermeasures
To Fight Fire: Foam, carbon dioxide, dry chemical or carbon tetrachloride (Section 6).
Storage and Handling: Section 7.

HEXANEDIOL-2,5
General Information
Description: Liquid.
Formula: $CH_3CH(OH)CH_2CH_2CH(OH)CH_3$.
Constants: Mol wt: 118.17, bp: 220.8°C, flash p: 230°F, d: 0.9617 at 20°/20°C, vap. d.: 4.07.
Hazard Analysis
Toxicity: Details unknown. See also glycols.
Fire Hazard: Moderate, when exposed to heat or flame; can react with oxidizing materials.
Countermeasures
To Fight Fire: Foam, carbon dioxide, dry chemical or carbon tetrachloride (Section 6).
Storage and Handling: Section 7.

HEXANEDIONE-2,5. See acetonyl acetone.

1,2,3,4,5,6-HEXANE HEXANITRATE. See dulcitol hexanitrate.

HEXANETRIOL-1,2,6
General Information
Description: Colorless liquid.
Formula: $HOCH_2CH(OH)(CH_2)_3CH_2OH$.
Constants: Mol wt: 134.17, bp: 178°C at 5 mm, fp: – 20°C, flash p: 380°F (C.O.C.), d: 1.1063 at 20°/20°C, vap. press.: < 0.01 mm at 20°C, vap. d.: 4.63.
Hazard Analysis
Toxicity: Details unknown. See also glycols.
Fire Hazard: Slight, when exposed to heat or flame; can react with oxidizing materials.
Countermeasures
Storage and Handling: Section 7.
To Fight Fire: Alcohol foam.

2,4,6,2′,4′,6′-HEXANITRODIPHENYLAMINE
General Information
Synonyms: Hexil; hexite.
Hazard Analysis
Toxicity: Details unknown. See also nitro compounds of aromatic hydrocarbons.
Fire Hazard: See nitrates.
Explosion Hazard: See nitrates. A powerful and violent explosive used as a booster explosive, in which use it is superior to TNT. It is not as good for this purpose as tetryl, but is extremely stable and much safer to handle.
Disaster Hazard: See nitrates.

Countermeasures
Storage and Handling: Section 7.
Shipping Regulations: Section 11.
  IATA (dry or wet with less than 10% water): Explosive, not acceptable (passenger and cargo).
  IATA (wet with not less than 10% water): Flammable solid, yellow label, ½ kilogram (passenger and cargo).

HEXANITRODIPHENYLAMINOETHYL NITRATE
Hazard Analysis
Toxicity: Details unknown. See also nitro compounds of aromatic hydrocarbons.
Fire Hazard: See nitrates.
Explosion Hazard: See nitrates. This explosive is slightly less sensitive to impact than tetryl, but equally sensitive to detonation by other means. It is considered about as powerful as tetryl. Its explosive strength can be enhanced by addition of oxygen-rich salts, such as potassium chlorate. It may replace tetryl as a booster as well as in detonating compositions.
Disaster Hazard: See nitrates.
Countermeasures
Storage and Handling: Section 7.

HEXANITRODIPHENYL OXIDE
Hazard Analysis
Toxicity: Details unknown. See also nitro compounds of aromatic hydrocarbons.
Fire Hazard: See nitrates.
Explosion Hazard: See nitrates. This high explosive is stable and not very sensitive. It is used in detonating compositions and is considered more powerful than picric acid. See also explosives, high.
Disaster Hazard: See nitrates.
Countermeasures
Storage and Handling: Section 7.

HEXANITRODIPHENYL SULFIDE
General Information
Synonym: Picryl sulfide.
Hazard Analysis
Toxicity: See nitro compounds of aromatic hydrocarbons.
Fire Hazard: See nitrates.
Explosion Hazard: See nitrates. This material is a powerful explosive and has an added military advantage in that its explosion gases contain irritating and very toxic sulfur dioxide. See also explosives, high.
Disaster Hazard: See nitrates.
Countermeasures
Storage and Handling: Section 7.
Shipping Regulations: Section 11.
  IATA: Explosive material, not acceptable (passenger and cargo).

HEXANITRODIPHENYL SULFONE
Hazard Analysis
Toxicity: Details unknown. See also nitro compounds of aromatic hydrocarbons.
Fire Hazard: See nitrates.
Explosion Hazard: See nitrates. This material is a stable, very powerful high explosive, used in detonating compositions. See also explosives, high.
Disaster Hazard: See nitrates.
Countermeasures
Storage and Handling: Section 7.

HEXANITROMANNITOL. See mannitol hexanitrate.

HEXANITROOXANILITE
Hazard Analysis
Toxicity: Details unknown. See also nitro compounds of aromatic hydrocarbons.
Fire Hazard: See nitrates.
Explosion Hazard: See nitrates. This material is about as powerful an explosive as TNT. See also explosives, high.

Disaster Hazard: See nitrates.
Countermeasures
Storage and Handling: Section 7.

HEXANOIC ACID. See caproic acid.

1-HEXANOL
General Information
Synonyms: n-Hexyl alcohol; amylcarbinol.
Description: Colorless liquid.
Formula: $CH_3(CH_2)_4CH_2OH$.
Constants: Mol wt: 102.2, bp: 157.2°C, fp: −44.6°C, flash p: 145°F, d: 0.8186 at 20°/4°C, vap. press.: 1 mm at 24.4°C, vap. d.: 3.52.
Hazard Analysis
Toxicity: See alcohols.
Fire Hazard: Moderate, when exposed to heat or flame; can react with oxidizing materials.
Spontaneous Heating: No.
Countermeasures
To Fight Fire: Alcohol foam, carbon dioxide, dry chemical or carbon tetrachloride (Section 6).
Storage and Handling: Section 7.

2-HEXANONE. See methyl n-butyl ketone.

3-HEXANONE
General Information
Synonym: Ethyl propyl ketone.
Description: Colorless liquid.
Formula: $C_2H_5CO(CH_2)_2CH_3$.
Constants: Mol wt: 100.16, bp: 124°C, d: 0.813 at 21.8°/4°C.
Hazard Analysis
Toxic Hazard Rating:
  Acute Local: Irritant 2; Ingestion 2; Inhalation 2.
  Acute Systemic: Inhalation 2.
  Chronic Local: U.
  Chronic Systemic: U.
Toxicology: See also ketones.
Fire Hazard: Moderate, when exposed to heat or flame; can react with oxidizing materials.
Countermeasures
To Fight Fire: Foam, carbon dioxide, dry chemical or carbon tetrachloride (Section 6).
Ventilation Control: Section 2.
Personnel Protection: Section 3.
Storage and Handling: Section 7.

HEXAPHENYL DIGERMANE
General Information
Description: White crystals insoluble in water.
Formula: $(C_6H_5)_6Ge_2$.
Constants: Mol wt: 607.8, mp: 340°C.
Hazard Analysis
Toxicity: Details unknown. See germanium compounds.

HEXAPHENYL DIGERMANE (TRIBENZENE)
General Information
Description: Colorless crystals, not soluble in water, soluble in benzene.
Formula: $(C_6H_5)_6Ge_2 \cdot (C_6H_6)_3$.
Constants: Mol wt: 842.1, mp: loses benzene upon warming.
Hazard Analysis and Countermeasures
See benzene and germanium compounds.

HEXAPHENYL DITIN
General Information
Description: Crystals.
Formula: $[(C_6H_5)_3Sn]_2$.
Constants: Mol wt: 700, mp: 232.5°C.
Hazard Analysis and Countermeasures
See tin compounds.

HEXA-p-TOLYL DIGERMANE
General Information
Description: Colorless crystals.

HEXA-p-TOLYL DITIN

Formula: $(CH_3C_6H_4)_6Ge_2$.
Constants: Mol wt: 692, mp: 227°C.
Hazard Analysis and Countermeasures
See germanium compounds.

HEXA-p-TOLYL DITIN
General Information
Description: Crystals slightly soluble in some organic solvents.
Formula: $[(C_6H_4CH_3)_3Sn]_2$.
Constants: Mol wt: 784.2, mp: 143.5°C.
Hazard Analysis and Countermeasures
See tin compounds.

HEXA-p-XYLYL DITIN
General Information
Description: Crystals soluble in benzene.
Formula: $[(CH_3C_6H_3CH_3)_3Sn]_2$.
Constants: Mol wt: 868.3, mp: 192.5°C.
Hazard Analysis and Countermeasures
See tin compounds.

HEXENE-1
General Information
Synonyms: Hexene; butylethylene; hexylene.
Description: Colorless liquid.
Formula: $CH_2CH(CH_2)_3CH_3$.
Constants: Mol wt: 84.16, mp: −98.5°C, bp: 63.5°C, fp: −139.9°C, flash p: < 20°F, d: 0.6732 at 20°/4°C, vap. press.: 310 mm at 38°C, vap. d.: 3.0.
Hazard Analysis
Toxic Hazard Rating:
  Acute Local: Irritant 2; Ingestion 2; Inhalation 2.
  Acute Systemic: Inhalation 2.
  Chronic Local: U.
  Chronic Systemic: U.
Fire Hazard: Dangerous; when exposed to heat or flame (Section 6).
Disaster Hazard: Dangerous, upon exposure to heat or flame; can react vigorously with oxidizing materials.
Countermeasures
Ventilation Control: Section 2.
Personnel Protection: Section 3.
Storage and Handling: Section 7.

HEXENE-2
General Information
Description: A liquid.
Formula: C_6H_{12}.
Constants: Mol wt: 84.16, bp: 68.5°C, flash p: < 20°F, d: 0.686 at 15.5°/15.5°C, vap. d.: 2.92.
Hazard Analysis
Toxicity: Details unknown. Limited animal experiments suggest low toxicity. See also hexene-1.
Fire Hazard: Dangerous, when exposed to heat or flame.
Explosion Hazard: Unknown.
Disaster Hazard: Dangerous, upon exposure to heat or flame; can react vigorously with oxidizing materials.
Countermeasures
Ventilation Control: Section 2.
To Fight Fire: Foam, carbon dioxide, dry chemical or carbon tetrachloride (Section 6).
Storage and Handling: Section 7.

4-HEXENE-1-YNE-3-OL
General Information
Formula: $HC:CH_2(OH)C:CHCH_2$.
Constant: Mol wt: 95.

Hazard Analysis
Toxic Hazard Rating: U.
Toxicity: Limited animal experiments suggest high toxicity.

4-HEXENE-1-YNE-3-ONE
Hazard Analysis
Toxic Hazard Rating: U.
Toxicity: Limited animal experiments suggest high toxicity and eye irritant. See also ketones.

5-HEXEN-2-ONE. See allyl acetone.

HEXIL See 2,4,6,2′,4′,6′-hexanitrodiphenylamine.

HEXITE. See 2,4,6,2′,4′,6′-hexanitrodiphenylamine.

HEXOGEN. See cyclotrimethylene trinitramine.

HEXONE. See isobutyl methyl ketone.

sec-HEXYL ACETATE. See methyl amyl acetate.

HEXYL ADIPATE. See dihexyl adipate.

HEXYL ALCOHOL. See hexanol.

n-HEXYL AMINE
General Information
Description: Liquid.
Formula: $C_6H_{13}NH_2$.
Constants: Mol wt: 101.19, mp: −22.9°C, bp: 131.4°C, flash p: 85°F (O.C.), d: 0.7675 at 20°/20°C, vap. d.: 3.49.
Hazard Analysis
Toxicity: Details unknown. See also amines. Limited animal experiments suggest moderate toxicity.
Fire Hazard: Moderate, when exposed to heat or flame; can react with oxidizing materials.
Explosion Hazard: Unknown.
Countermeasures
Ventilation Control: Section 2.
To Fight Fire: Foam, carbon dioxide, dry chemical or carbon tetrachloride (Section 6).
Storage and Handling: Section 7.

N,n-HEXYL-2-AMINOHEPTANE. See N,n-Hexyl-n-heptylamine.

HEXYLBORIC ACID
General Information
Description: White crystals slightly soluble in water.
Formula: $C_6H_{13}B(OH)_2$.
Constants: Mol wt: 130.0, mp: 90°C (decomp).
Hazard Analysis
Toxicity: Details unknown. See also boron compounds.

HEXYLBORIC OXIDE
General Information
Description: Colorless liquid; hydrolyzed by water but soluble in organic solvents.
Formula: $(C_6H_{13})_3B_3O_3$.
Constants: Mol wt: 336.0, d: 0.8876, bp: 180°C at 24 mm.
Hazard Analysis
Toxicity: Details unknown. See also boron compounds.

n-HEXYL "CARBITOL"
General Information
Description: Liquid.
Formula: $C_6H_{13}O(C_2H_4O)_2H$.
Constants: Mol wt: 190.03, bp: 258.2°C, fp: −33.3°C, flash p: 285°F (O.C.), d: 0.9346 at 20°/20°C, vap. press.: 0.01 mm at 20°C.

TOXIC HAZARD RATING CODE (For detailed discussion, see Section 1.)

0 NONE: (a) No harm under any conditions; (b) Harmful only under unusual conditions or overwhelming dosage.

1 SLIGHT: Causes readily reversible changes which disappear after end of exposure.

2 MODERATE: May involve both irreversible and reversible changes; not severe enough to cause death or permanent injury.

3 HIGH: May cause death or permanent injury after very short exposure to small quantities.

U UNKNOWN: No information on humans considered valid by authors.

Hazard Analysis
Toxicity: Details unknown. See also glycols.
Fire Hazard: Slight, when exposed to heat or flame; can react with oxidizing materials.
Countermeasures
To Fight Fire: Foam, carbon dioxide, dry chemical or carbon tetrachloride (Section 6).
Storage and Handling: Section 7.

n-HEXYL "CELLOSOLVE"
General Information
Description: Liquid.
Formula: $C_6H_{13}OCH_2CH_2OH$.
Constants: Mol wt: 146.22, bp: 208.3°C, fp: −45.1°C, flash p: 195°F (O.C.), d: 0.8894 at 20°/20°C, vap. press.: 0.1 mm at 20°C, vap. d.: 5.04.
Hazard Analysis
Toxicity: Details unknown. See also glycols.
Fire Hazard: Moderate, when exposed to heat or flame; can react with oxidizing materials.
Countermeasures
To Fight Fire: Foam, carbon dioxide, dry chemical or carbon tetrachloride (Section 6).
Storage and Handling: Section 7.

HEXYL DIETHYL HEPTYLATE
General Information
Synonym: Flexol Plasticizer CC-55.
Description: Liquid.
Formula: $C_6H_{10}[COOCH_2CH(C_2H_5)C_4H_9]_2$.
Constants: Mol wt: 376.59, mp: −53°C, bp: 216°C at 5 mm, flash p: 425°F (O.C.), d: 0.9586 at 20°/20°C, vap. d.: 13.7.
Hazard Analysis
Toxicity: Details unknown. See also esters.
Fire Hazard: Slight, when exposed to heat or flame; can react with oxidizing materials.
Countermeasures
To Fight Fire: Foam, carbon dioxide, dry chemical or carbon tetrachloride (Section 6).
Storage and Handling: Section 7.

HEXYLENE. See hexene-1.

HEXYLENE GLYCOL
General Information
Synonym: 2-Methylpentanediol-2,4.
Description: Mild odor, colorless liquid, water soluble.
Formula: $CH_3COH(CH_3)CH_2CHOHCH_3$.
Constants: Mol wt: 118.17, bp: 197.1°C, fp: −50°C, flash p: 215°F (O.C.), d: 0.9234 at 20°/20°C, vap. press.: 0.05 mm at 20°C, vap. d.: 4.
Hazard Analysis
Toxic Hazard Rating:
   Acute Local: Irritant 2; Inhalation 1.
   Acute Systemic: Ingestion 1; Inhalation 1.
   Chronic Local: Irritant 1.
   Chronic Systemic: U.
Toxicology: Irritating to skin, eyes and mucous membrane. Large oral doses produce narcosis. Data based upon animal experiments. See also glycols.
Fire Hazard: Moderate, when exposed to heat or flame; can react with oxidizing materials.
Countermeasures
To Fight Fire: Foam, carbon tetrachloride, carbon dioxide or dry chemicals (Section 6).
Storage and Handling: Section 7.

HEXYL ETHER
General Information
Synonym: Dihexyl ether.
Description: Liquid.
Formula: $C_6H_{13}OC_6H_{13}$.
Constants: Mol wt: 186.33, mp: −43.0°C, bp: 440°F, flash p: 170°F (O.C.), d: 0.794, autoign. temp.: 369°F, vap. d.: 6.4.
Hazard Analysis
Toxicity: Details unknown. See also ethers.
Fire Hazard: Moderate, when exposed to heat or flame; can react with oxidizing materials.
Explosion Hazard: Details unknown. See also ethers.
Countermeasures
Storage and Handling: Section 7.
To Fight Fire: Foam, carbon dioxide, dry chemical or carbon tetrachloride (Section 6).

N,n-HEXYL-n-HEPTYL AMINE
General Information
Synonym: N,n-hexyl-2-aminoheptane.
Formula: $(C_6H_{13})_2(C_7H_{15})N$.
Constant: Mol wt: 283.5.
Hazard Analysis
Toxicity: Details unknown. See also amines.
Fire Hazard: Slight, when exposed to heat or flame (section 6).
Countermeasures
Storage and Handling: Section 7.

HEXYL HYDRIDE. See n-hexane.

HEXYL MANDELATE
General Information
Formula: $C_6H_5CHOHCOOC_6H_{13}$.
Constant: Mol wt: 220.
Hazard Analysis
Toxic Hazard Rating: U.
Toxicity: Limited animal experiments suggest low toxicity. See also esters and mandelic acid.

HEXYL METHACRYLATE
General Information
Description: Liquid.
Formula: $C_6H_{13}OOCC(CH_3):CH_2$.
Constants: Mol wt: 170, d: 0.88, vap. d.: 5.9, bp: 67–85°C at 8 mm, flash p: 180°F (O.C.).
Hazard Analysis
Toxic Hazard Rating: U.
Fire Hazard: Moderate, when exposed to heat or flame.

HEXYL METHYL KETONE. See 2-octanone.

n-HEXYLPYRROLIDINE
General Information
Description: Colorless liquid, slightly soluble in water.
Formula: $C_4H_8NC_6H_{13}$.
Constants: Mol wt: 155.3, bp: 201°C, d: 0.835 at 20°C, fp: −75°C, flash p: 154°F.
Hazard Analysis
Toxicology: Highly toxic by ingestion. An irritant.
Fire Hazard: Moderately dangerous when exposed to heat, flame, sparks or powerful oxidizers.
Countermeasures
To Fight Fire: Spray, foam, carbon dioxide or dry chemicals.
Ventilation Control: Section 2.
Personnel Protection: Section 3.
Storage and Handling: Section 7.

HEXYL TITANATE. See 2-ethylbutyl titanate.

HEXYL TRICHLOROSILANE
General Information
Formula: $C_6H_{13}SiCl_3$.
Constant: Mol wt: 219.7.
Hazard Analysis
Toxic Hazard Rating:
   Acute Local: Irritant 3; Ingestion 3; Inhalation 3.
   Acute Systemic: U.
   Chronic Local: U.
   Chronic Systemic: U.

Disaster Hazard: Dangerous; see chlorides; will react with water or steam to produce toxic and corrosive fumes.

Countermeasures

Ventilation Control: Section 2.

Personnel Protection: Section 3.

First Aid: Section 1.

Storage and Handling: Section 7.

Shipping Regulations: Section 11.

I.C.C.: Corrosive liquid; white label, 10 gallons.

Coast Guard Classification: Corrosive liquid; white label.

IATA: Corrosive liquid, white label, not acceptable (passenger), 40 liters (cargo).

HIERATITE. See potassium fluosilicate.

HI-FLASH NAPHTHA. See naphtha (coal tar).

HIGH EXPLOSIVES

Shipping Regulations: Section 11.

I.C.C.: Explosive A, not acceptable.

IATA: Explosive, not acceptable (passenger and cargo).

HIGH TEST HYPOCHLORITES

General Information

Synonym: HTH; calcium hypochlorite.

Description: A dry stable form of calcium hypochlorite. Water soluble.

Formula: 70% available chlorine.

Hazard Analysis

Toxic Hazard Rating:

Acute Local: Irritant 2; Inhalation 2.

Acute Systemic: U.

Chronic Local: Irritant 1; Inhalation 1.

Chronic Systemic: U.

Toxicology: See hypochlorites.

Countermeasures

See hypochlorites.

HISTAMINE

General Information

Synonyms: 4-Imidazole ethylamine; ergamine.

Description: Deliquescent needles.

Formula: $C_5H_9N_3$.

Constants: Mol wt: 111.2, mp: 84°C, bp: 210°C at 18 mm.

Hazard Analysis

Toxic Hazard Rating:

Acute Local: Allergen 2.

Acute Systemic: Ingestion 3; Inhalation 3.

Chronic Local: Allergen 1.

Chronic Systemic: U.

Toxicology: Ingestion or inhalation of this material produces the following effects: flushing followed by pallor, dizziness, fainting, fall in blood pressure, headache, rapid, weak pulse. Allergic effects on skin (hives) may occur (Section 9).

Disaster Hazard: Dangerous; when heated to decomposition it emits highly toxic fumes.

Countermeasures

Ventilation Control: Section 2.

Treatment and Antidotes: If swallowed, wash out stomach. Administer stimulants (epinephrine or ephedrine). Call a physician.

Personal Hygiene: Section 3.

Storage and Handling: Section 7.

HISTAMINE SALTS. See histamine.

HISTIDINE

General Information

Synonym: α-Amino-β-imidazolepropionic acid.

Description: An amino acid essential for rats. Occurs in DL, L, and D forms. The DL, and L forms are considered here: Colorless crystals; Soluble in water; insoluble in alcohol and ether.

Formula: $HOOCCH(NH_2)CH_2C_3H_3N_2$.

Constants: Mol wt: 155, mp(DL): 285–286°C, mp(L): 277°C.

Hazard Information

Toxic Hazard Rating: U.

Toxicity: Note: Used as a nutrient and/or dietary supplement food additive (Section 10).

HMI. See hexamethylenimine.

HN-1. See nitrogen mustard.

HOERNESITE. See magnesium-o-arsenate.

HOLMIUM

General Information

Description: Yellow-colored salts.

Formula: Ho.

Constants: At wt: 164.94, mp: 1497°C, bp: 2327°C, d: 8.80, vap. press.: 2 mm at 1630°C.

Hazard Analysis

Toxicity: Unknown. As a lanthanon it may have anticoagulant action on blood. See also lanthanum.

Radiation Hazard: Section 5. For permissible levels, see Table 5, p. 150.

Artificial isotope ^{166}Ho half life 27 h. Decays to stable ^{166}Er by emitting beta particles of 1.76 (48%), 1.84 (52%) MeV. Also emits gamma rays of 0.08 MeV.

HOMATROPINE AND COMPOUNDS

General Information

Synonyms: Mandelytropeine, homoatropine.

Description: Deliquescent prisms from ether; glistening prisms from alcohol.

Formula: $C_{16}H_{21}NO_3$.

Constants: Mol wt: 275.34, mp: 95.5°–98.5°C.

Hazard Analysis

Toxic Hazard Rating:

Acute Local: 0.

Acute Systemic: Ingestion 3; Inhalation 3.

Chronic Local: 0.

Chronic Systemic: U.

Toxicology: A poisonous alkaloid. Resembles atropine in its action. See also atropine.

Disaster Hazard: Dangerous; when heated to decomposition it emits highly toxic fumes.

Countermeasures

Ventilation Control: Section 2.

Personal Hygiene: Section 3.

Storage and Handling: Section 7.

HORMODIN

General Information

Synonym: Indolebutyric acid.

Description: White crystals or powder. Insoluble in water and chloroform.

Formula: $C_{12}H_{13}NO_2$.

Constants: Mol wt: 203.2, mp: 124°C.

Hazard Analysis

Toxicity: Details unknown. Probably has little or no toxicity, except possibly as skin irritant.

TOXIC HAZARD RATING CODE (For detailed discussion, see Section 1.)

0 NONE: (a) No harm under any conditions; (b) Harmful only under unusual conditions or overwhelming dosage.

1 SLIGHT: Causes readily reversible changes which disappear after end of exposure.

2 MODERATE: May involve both irreversible and reversible changes not severe enough to cause death or permanent injury.

3 HIGH: May cause death or permanent injury after very short exposure to small quantities.

U UNKNOWN: No information on humans considered valid by authors

isaster Hazard: Dangerous. When heated to decomposition, it emits toxic fumes.

CORN DUST
Hazard Analysis
Toxic Hazard Rating:
  Acute Local: Allergen 1; Inhalation 1.
  Acute Systemic: U.
  Chronic Local: Allergen 1; Inhalation 1.
  Chronic Systemic: U.
  TLV: ACGIH (accepted); 50 million particles per cubic foot of air.
Countermeasures
Ventilation Control: Section 2.
Personal Hygiene: Section 3.

CORNSTONE. See magnesium fluosilicate and zinc fluosilicate.

HORSEHAIR. See hair.

HTH. See high test hypochlorite.

HTH-15 A less concentrated formulation of calcium hypochlorite than HTH. See HTH and hypochlorites.

HYAMINE 1622. See benzethonium chloride.

HYCOL
General Information
Synonym: Saponified cresol.
Hazard Analysis
Toxicity: See phenol.

HYDRACRYLONITRILE. See ethylene cyanohydrin.

HYDRASTINE
General Information
Description: Colorless, rhombic prisms; white pulverulent alkaloid.
Formula: $C_{21}H_{21}NO_6$.
Constants: Mol wt: 383.39, mp: 132°C.
Hazard Analysis
Toxic Hazard Rating:
  Acute Local: U.
  Acute Systemic: Ingestion 3; Inhalation 3.
  Chronic Local: U.
  Chronic Systemic: U.
Disaster Hazard: Dangerous; when heated to decomposition, it emits highly toxic fumes.
Countermeasures
Ventilation Control: Section 2.
Personal Hygiene: Section 3.
Storage and Handling: Section 7.

HYDRASTININE
General Information
Description: White-yellowish needles.
Formula: $C_{11}H_{13}NO_3$.
Constants: Mol wt: 207.2, mp: 116–117°C.
Hazard Analysis
Toxic Hazard Rating:
  Acute Local: U.
  Acute Systemic: Ingestion 3; Inhalation 3.
  Chronic Local: U.
  Chronic Systemic: U.
Disaster Hazard: Dangerous; when heated to decomposition, it emits highly toxic fumes.
Countermeasures
Ventilation Control: Section 2.
Personal Hygiene: Section 3.
Storage and Handling: Section 7.

HYDRATED ALUMINA. See aluminum hydroxide.

HYDRATED ALUMINUM HYDROXIDE. See aluminum hydroxide.

HYDRATED ALUMINUM SILICATE. See kaolin.

HYDRATED LEAD OXIDE. See lead hydroxide.

HYDRATED LIME. See calcium hydroxide.

HYDRAZINE (and Compounds)
General Information
Synonyms: Hydrazine base; diamine; hydrazine anhydrous.
Description: Colorless, fuming liquid; white crystals.
Formula: NH_2NH_2.
Constants: Mol wt: 32.05, mp: 1.4°C, bp: 113.5°C, flash p: 126°F (O.C.), d: 1.011 at 15°C (liq.), autoign. temp.: 518°F, vap. d.: 1.1.
Hazard Analysis
Toxic Hazard Rating:
  Acute Local: Irritant 3; Ingestion 3; Inhalation 3.
  Acute Systemic: Ingestion 3; Inhalation 3; Skin Absorption 3.
  Chronic Local: Allergen-2.
  Chronic Systemic: Ingestion 3; Inhalation 3; Skin Absorption 3.
Toxicity: May cause skin sensitization as well as systemic poisoning.
TLV: ACGIH (recommended); 1 part per million in air. 1.3 milligrams per cubic meter of air. Can be absorbed via intact skin.
Toxicology: Hydrazine and some of its derivatives may cause damage to the liver and destruction of red blood cells. See phenyl hydrazine.
Fire Hazard: Moderate, when exposed to heat, flame, or oxidizing agents; a powerful reducing agent.
Explosion Hazard: Severe, when exposed to heat or flame or by chemical reaction with oxidizing agents. It is a powerful explosive, much used in rocket fuels. It is very sensitive and must not be used without full and complete instructions from the manufacturer as to handling, storage and disposal.
Disaster Hazard: Dangerous; when heated to decomposition, it emits highly toxic fumes of nitrogen compounds; may explode by heat or chemical reaction.
Countermeasures
Ventilation Control: Section 2.
To Fight Fire: Foam, carbon dioxide, dry chemical or carbon tetrachloride (Section 6).
Personnel Protection: Section 3.
First Aid: Section 1.
Storage and Handling: Section 7.
Shipping Regulations: Section 11.
  Coast Guard Classification: Corrosive liquid; white label.

HYDRAZINE, ANHYDROUS. See also hydrazine.
General Information
Constants: d: 1.0+, vap. d.: 1.1, bp: 236°F, flash p: 100°F, lel: 4.7%, uel: 100%, ignition temperature varies widely depending on the surface with which this is in contact.
Hazard Analysis
Toxic Hazard Rating: U. Probably highly toxic.
Fire Hazard: Moderate, when exposed to heat or flame.
Countermeasures
Storage and Handling: Section 7.
To Fight Fire: Water, foam, mist (Section 6).
Shipping Regulations: Section 11.
  I.C.C.: Corrosive liquid; white label, 5 pints.
  Coast Guard Classification: Corrosive liquid; white label.
  IATA: Corrosive liquid, white label, not acceptable (passenger), 2½ liters (cargo), hydrate or water solutions containing 50% or less of water.
  Hydrazine solutions containing more than 50% water.
  IATA: Corrosive liquid, white label, 1 liter (passenger) 20 liters (cargo).

HYDRAZINE BASE. See hydrazine (and compounds).

HYDRAZINE DIBORANE
General Information
Description: White crystalline powder. Hygroscopic.
Formula: $B_2N_2H_{10}$.
Constants: Mol wt: 59.7, d: 0.91
Hazard Analysis and Countermeasures
See boron hydrides, boron compounds, and hydrazine (and compounds).

HYDRAZINE NITRATE. See hydrazine.
Shipping Regulations: Section 11.
   IATA: Explosive, not acceptable (passenger and cargo).

HYDRAZINE PERCHLORATE
General Information
Description: Solid; decomposes in water; soluble in alcohol; insoluble in ether, benzene, chloroform, and carbon disulfide.
Formula: $N_2H_4 \cdot HClO_4 \cdot \frac{1}{2}H_2O$.
Constants: Mol wt: 141.5, d: 1.939, mp: 137°C, bp: 145°C.
Hazard Analysis
Toxic Hazard Rating: U. Probably toxic.
Disaster Hazard: An explosive.
Countermeasures
Shipping Regulations: Section 11.
   IATA: Explosive, not acceptable (passenger and cargo).

HYDRAZINE SULFATE
General Information
Description: Colorless crystals. Water soluble, insoluble in alcohol.
Formula: $2N_2H_4 \cdot H_2SO_4$.
Constants: Mol wt: 162.2, mp: 85°C.
Hazard Analysis and Countermeasures
See hydrazine.

HYDRAZINE TARTRATE
General Information
Synonym: Hydrazine acid tartrate.
Description: Crystals. Water soluble.
Formula: $H_2NNH_2 \cdot C_4H_6O_6$.
Constants: Mol wt: 182.1, mp: 183°C.
Hazard Analysis and Countermeasures
See hydrazine.

2-HYDRAZINOETHANOL. See 2-hydroxyethyl hydrazine.

HYDRAZOIC ACID
General Information
Synonyms: Azoimide; hydrogen azide.
Description: Colorless liquid; very soluble in water. Intolerable pungent odor.
Formula: HN_3.
Constants: Mol wt: 43.03, mp: −80°C, bp: 37°C.
Hazard Analysis
Toxic Hazard Rating:
   Acute Local: Irritant 2; Ingestion 3; Inhalation 3.
   Acute Systemic: Ingestion 3; Inhalation 3.
   Chronic Local: Irritant 2; Ingestion 2; Inhalation 2.
   Chronic Systemic: Ingestion 2; Inhalation 2.
Toxicology: Exposure to vapors causes irritation of eyes and mucous membrane. Continued inhalation causes cough, chills and fever. High concentrations can cause fatal convulsions. Chronic exposure has been reported as causing injury to kidneys and spleen.
Explosion Hazard: Dangerous, when shocked or exposed to heat.

Disaster Hazard: Dangerous; shock or heat will explode it.
Countermeasures
Ventilation Control: Section 2.
Personnel Protection: Section 3.
Storage and Handling: Section 7.

HYDRIDES
Hazard Analysis
Toxicity: Variable. The hydrides of phosphorus, arsenic, sulfur, selenium, tellurium and boron which are highly toxic, produce local irritation and destroy red blood cells. They are particularly dangerous because of their volatility and ease of entry into the body. The hydrides of the alkali metals, alkaline earths, aluminum, zirconium and titanium react with moisture to evolve hydrogen and leave behind the hydroxide of the metallic element. This hydroxide is usually caustic. See also sodium hydroxide.
Hydrides, metallic, primary type:
   This group includes the hydrides of calcium, lithium, magnesium, potassium, sodium and strontium. In the presence of moisture they are readily converted to hydroxides which are highly irritating to the skin by caustic and thermal action. Similar effects can occur on contact with eyes and respiratory mucous membranes.
Fire Hazard: The volatile hydrides are flammable, some spontaneously so in air. All hydrides react violently on contact with powerful oxidizing agents. When heated or on contact with moisture or acids an exothermic reaction evolving hydrogen occurs. Often enough heat is evolved to cause ignition. Hydrides require special handling instructions which should be obtained from the manufacturers (Section 6).
Explosion Hazard: The volatile hydrides (such as hydrides of boron, arsenic, phosphorus, selenium, tellurium) form explosive mixtures with air. The nonvolatile hydrides (such as sodium, lithium, calcium) readily liberate hydrogen when heated or on contact with moisture or acids. Furthermore, hydrides form dust clouds which can explode due to contact with flames, sparks, heat or oxidizers.
Disaster Hazard: Highly dangerous; when heated, they can ignite at once or liberate hydrogen: they react with moisture or acids to evolve heat and hydrogen; on contact with powerful oxidizers violent reactions can occur.
Countermeasures
Ventilation Control: Section 2.
Personnel Protection: Section 3.
First Aid: Section 1.
Storage and Handling: Section 7.

HYDRIODIC ACID
General Information
Description: Colorless gas or pale yellow liquid.
Formula: HI.
Constants: Mol wt: 127.93, mp: −50.8°C, bp: −35.38°C at 4 atm., d: 5.66 g/liter at 0°C.
Hazard Analysis
Toxic Hazard Rating:
   Acute Local: Irritant 3; Ingestion 3; Inhalation 3.
   Acute Systemic: Ingestion 2; Inhalation 2.
   Chronic Local: U.
   Chronic Systemic: Ingestion 2; Inhalation 2.
Disaster Hazard: Dangerous; when heated to decomposition, it emits highly toxic fumes of iodides; will react with water or steam to produce toxic and corrosive fumes.

TOXIC HAZARD RATING CODE (For detailed discussion, see Section 1.)

0 NONE: (a) No harm under any conditions; (b) Harmful only under unusual conditions or overwhelming dosage.

1 SLIGHT: Causes readily reversible changes which disappear after end of exposure.

2 MODERATE: May involve both irreversible and reversible changes; not severe enough to cause death or permanent injury.

3 HIGH: May cause death or permanent injury after very short exposure to small quantities.

U UNKNOWN: No information on humans considered valid by authors.

Countermeasures
Ventilation Control: Section 2.
Personnel Protection: Section 3.
Storage and Handling: Section 7.
Shipping Regulations: Section 11.
  I.C.C.: Corrosive liquid; white label, 1 gallon.
  Coast Guard Classification: Corrosive liquid; white label.
  IATA: Corrosive liquid, white label, 1 liter (passenger),
    5 liters (cargo).

HYDRIODIC ETHER. See ethyl iodide.

HYDROABIETYL ALCOHOL
General Information
Description: Colorless, tacky, viscous liquid.
Formula: Mixture of tetra-, di-, and dehydroabietyl alcohols.
Constants: Flash p: 380°F (O.C.), d: 1.007–1.008 at
  20°/20°C, softening p.: 33°C.
Hazard Analysis
Toxicology: Limited animal experiments indicate low order
  of toxicity.
Fire Hazard: Slight, when exposed to heat or flame; can
  react with oxidizing materials.
Countermeasures
To Fight Fire: Foam, carbon dioxide, dry chemical or
  carbon tetrachloride (Section 6).
Storage and Handling: Section 7.

HYDROBERBERINE
General Information
Description: White, crystalline alkaloid.
Formula: $C_{20}H_{21}NO_4$.
Constants: Mol wt: 339.4, mp: 167°C.
Hazard Analysis
Toxic Hazard Rating:
  Acute Local: U.
  Acute Systemic: Ingestion 3; Inhalation 3.
  Chronic Local: U.
  Chronic Systemic: U.
Disaster Hazard: Dangerous; when heated to decomposi-
  tion, it can emit highly toxic fumes of nitrogen oxides.
Countermeasures
Ventilation Control: Section 2.
Personnel Protection: Section 3.
First Aid: Section 1.
Storage and Handling: Section 7.

HYDROBROMIC ACID
General Information
Synonym: Hydrogen bromide.
Description: Colorless gas or pale yellow liquid.
Formula: HBr.
Constants: Mol wt: 80.92, mp: −87°C, bp: −66.5°C,
  d: 3.50 g/liter at 0°C.
Hazard Analysis
Toxic Hazard Rating:
  Acute Local: Irritant 3; Ingestion 3; Inhalation 3.
  Acute Systemic: Ingestion 2; Inhalation 2.
  Chronic Local: Irritant 2.
  Chronic Systemic: Ingestion 2; Inhalation 2.
TLV: ACGIH (recommended); 3 parts per million in air; 10
  milligrams per cubic meter of air.
Disaster Hazard: Dangerous; see bromides; will react with
  water or steam to produce toxic and corrosive fumes.
Countermeasures
Ventilation Control: Section 2.
Personnel Protection: Section 3.
Storage and Handling: Section 7.
Shipping Regulations: Section 11.
  I.C.C.: Corrosive liquid; white label, 1 gallon.
  Coast Guard Classification: Noninflammable gas; green
    gas label.
  IATA (exceeding 49% strength): Corrosive liquid, not
    acceptable (cargo and passenger).

  IATA (not exceeding 49% strength): Corrosive liquid,
    white label, 1 liter (passenger), 5 liters (cargo).

HYDROBROMIC ETHER. See ethyl bromide.

HYDROCARBON GAS, LIQUEFIED. See specific gas.
Shipping Regulations: Section 11.
  I.C.C.: Flammable gas; red gas label, 300 pounds.
  Coast Guard Classification: Inflammable gas; red gas
    label.
  IATA: Flammable gas, red label, not acceptable (pas-
    senger), 140 kilograms (cargo).

HYDROCARBON GAS, NONLIQUEFIED. See specific
material.
Shipping Regulations: Section 11.
  I.C.C.: Flammable gas; red gas label, 300 pounds.
  Coast Guard Classification: Inflammable gas; red gas
    label.
  IATA: Flammable gas, red label, not acceptable (pas-
    senger), 140 kilograms (cargo).

HYDROCHLORIC ACID
General Information
Synonyms: Muriatic acid; chlorohydric acid; hydrogen
  chloride.
Description: Colorless gas or colorless, fuming liquid;
  strongly corrosive.
Formula: HCl.
Constants: Mol wt: 36.47, mp: −114.3°C, bp: −84.8°C,
  d: 1.639 g/liter (gas) at 0°C; 1.194 at −36°C (liquid),
  vap. press.: 4.0 atm at 17.8°C.
Hazard Analysis
Toxic Hazard Rating:
  Acute Local: Irritant 3; Ingestion 3; Inhalation 3.
  Acute Systemic: U.
  Chronic Local: Irritant 2.
  Chronic Systemic: U.
TLV: ACGIH (recommended); 5 parts per million in air;
  7 milligrams per cubic meter of air.
Toxicology: Hydrochloric acid is an irritant to the mucous
  membranes of the eyes and respiratory tract, and a
  concentration of 35 ppm causes irritation of the throat
  after short exposure. Concentrations of 50 to 100 ppm
  are tolerable for 1 hour. More severe exposures result
  in pulmonary edema, and often laryngeal spasm. Concentrations of 1,000 to 2,000 ppm are dangerous, even
  for brief exposures. Mists of hydrochloric acid are considered less harmful than the anhydrous hydrogen chloride, since the droplets have no dehydrating action. In
  general, hydrochloric acid causes little trouble in
  industry, other than from accidental splashes and burns.
  It is used as a general purpose food additive (Section 10).
  It is a common air contaminant (Section 10).
Disaster Hazard: Dangerous; see chlorides; will react with
  water or steam to produce toxic and corrosive fumes.
Countermeasures
Ventilation Control: Section 2.
Personnel Protection: Section 3.
Storage and Handling: Section 7.
First Aid: Section 1.
Shipping Regulations: Section 11.
  I.C.C.: Corrosive liquid; white label, 10 pints.
  Coast Guard Classification: Corrosive liquid; white label.
  Coast Guard (anhydrous) Classification: Noninflammable
    gas; green gas label.
  MCA warning label.
  IATA: Corrosive liquid, white label, 1 liter (passenger),
    5 liters (cargo).

HYDROCHLORIC ACID, MIXTURES. See hydrochloric
acid.
Shipping Regulations: Section 11.
  I.C.C.: Corrosive liquid; white label, 10 pints.
  Coast Guard Classification: Corrosive liquid; white label.

IATA: Corrosive liquid, white label, 1 liter (passenger), 5 liters (cargo).

HYDROCHLORIC ACID SOLUTION, INHIBITED
Shipping Regulations: Section 11.
I.C.C.: Corrosive liquid, white label, 10 pints.
IATA: Corrosive liquid, white label, 1 liter (passenger), 5 liters (cargo).

HYDROCHLORIC ETHER. See ethyl chloride.

HYDROCERUSSITE. See lead carbonate basic.

HYDROCORTISONE ACETATE
General Information
Description: White, odorless, crystalline powder; very slightly soluble in ether; practically insoluble in water; slightly soluble in alcohol and chloroform.
Formula: $C_{23}H_{32}O_6$.
Constants: Mol wt: 404, mp: 216–233°C.
Hazard Analysis
Toxic Hazard Rating: U.
Toxicity: Note: Used as a food additive permitted in food for human consumption (Section 10).

HYDROCORTISONE SODIUM SUCCINATE
General Information
Description: White, odorless, hygoscopic, amorphous, solid; very soluble in water and alcohol; insoluble in chloroform; very slightly soluble in acetone.
Formula: $C_{25}H_{33}NaO_8$.
Constants: Mol wt: 484, mp: 169–171°C.
Hazard Analysis
Toxicity: Details unknown. Used as a food additive permitted in food for human consumption.

HYDROCOTARNINE
General Information
Description: Monoclinic prisms.
Formula: $C_{12}H_{15}NO_3 \cdot \frac{1}{2}H_2O$.
Constants: Mol wt: 230.3, mp: 56°C.
Hazard Analysis
Toxic Hazard Rating:
Acute Local: Irritant 1.
Acute Systemic: Ingestion 3; Inhalation 3.
Chronic Local: U.
Chronic Systemic: U.
Disaster Hazard: Dangerous; when heated to decomposition, it can emit highly toxic fumes of nitrogen compounds.
Countermeasures
Ventilation Control: Section 2.
Personal Hygiene: Section 3.
Storage and Handling: Section 7.

HYDROCYANIC ACID
General Information
Synonyms: Hydrogen cyanide; prussic acid.
Description: Colorless liquid; faint odor of bitter almonds.
Formula: HCN.
Constants: Mol wt: 27.03, mp: −13.2°C, bp: 25.7°C, lel: 6%, uel: 40%, flash p: 0°F (C.C.), d: 0.6876 at 20°/4°C, autoign. temp.: 100°F, vap. press.: 400 mm at 9.8°C, vap. d.: 0.932.
Hazard Analysis
Toxic Hazard Rating:
Acute Local: Irritant 2; Ingestion 2; Inhalation 2.
Acute Systemic: Ingestion 3; Inhalation 3; Skin Absorption 3.

Chronic Local: U.
Chronic Systemic: U.
TLV: ACGIH (recommended); 10 parts per million in air; 11 milligrams per cubic meter of air. Can be absorbed via intact skin.
Toxicology: Hydrocyanic acid and the cyanides are true protoplasmic poisons, combining in the tissues with the enzymes associated with cellular oxidation. They thereby render the oxygen unavailable to the tissues, and cause death through asphyxia. The suspension of tissue oxidation lasts only while the cyanide is present; upon its removal, normal function is restored provided death has not already occurred. Hydrocyanic acid does not combine easily with hemoglobin, but it does combine readily with methemoglobin to form cyanmethemoglobin. This fact is utilized in the treatment of cyanide poisoning, when an attempt is made to induce methemoglobin formation. The presence of cherry-red venous blood in cases of cyanide poisoning is due to the inability of the tissues to remove the oxygen from the blood. Exposure to concentrations of 100 to 200 ppm for periods of 30 to 60 minutes can cause death.

In cases of acute cyanide poisoning, death is extremely rapid; though sometimes breathing may continue for a few minutes. In less acute cases, there is headache, dizziness, unsteadiness of gait, a feeling of suffocation, and nausea. Where the patient recovers, there is rarely any disability. An insecticide.
Fire Hazard: Dangerous, when exposed to heat or flame.
Explosion Hazard: Severe, when exposed to heat or flame or by chemical reaction with oxidizers. Under certain conditions, particularly contact with alkaline materials, hydrogen cyanides can polymerize or decompose explosively. The liquid is commonly stabilized by addition of acids.
Disaster Hazard: Highly dangerous; the gas forms explosive mixtures with air; will react with water, steam, acid or acid fumes to produce highly toxic fumes of cyanides.
Countermeasures
Ventilation Control: Section 2.
To Fight Fire: Carbon dioxide, nonalkaline dry chemical or foam (Section 6).
Personal Hygiene: Section 3.
First Aid: Section 1.
Storage and Handling: Section 7.
Shipping Regulations: Section 11.
I.C.C.: Poison A; poison gas label, not accepted.
Coast Guard Classification: Poison A; poison gas label.
MCA warning label.
IATA: Poison A, not acceptable (passenger and cargo).

HYDROCYANIC ACID SOLUTION. See hydrocyanic acid.
Shipping Regulations: Section 11.
I.C.C.: Poison B; poison label, 25 pounds.
Coast Guard Classification: Poison B; poison gas label.
IATA (solutions exceeding 5%): Poison A, not acceptable (passenger and cargo).
(solutions not exceeding 5%): Poison B, poison label, not acceptable (passenger), 12 kilograms (cargo).

HYDROCYANIC ACID, UNSTABILIZED. See hydrocyanic acid.
Shipping Regulations: Section 11.
I.C.C. Classification: Not accepted.
Coast Guard Classification: Not permitted.
IATA: Poison A, not acceptable (passenger and cargo).

TOXIC HAZARD RATING CODE (For detailed discussion, see Section 1.)

0 NONE: (a) No harm under any conditions; (b) Harmful only under unusual conditions or overwhelming dosage.

1 SLIGHT: Causes readily reversible changes which disappear after end of exposure.

2 MODERATE: May involve both irreversible and reversible changes not severe enough to cause death or permanent injury.

3 HIGH: May cause death or permanent injury after very short exposure to small quantities.

U UNKNOWN: No information on humans considered valid by author

HYDROFERROCYANIC ACID
General Information
Description: White crystalline material.
Formula: $H_4Fe(CN)_6$.
Constant: Mol wt: 216.
Hazard Analysis and Countermeasures
See ferrocyanides.

HYDROFLUORIC ACID
General Information
Synonyms: Hydrogen fluoride; fluorhydric acid.
Description: Clear, colorless, fuming corrosive liquid or gas.
Formula: HF.
Constants: Mol wt: 20.01, mp: $-92.3°C$, bp: $19.4°C$, d: 0.921 g/liter (gas); 0.987 (liquid), vap. press.: 400 mm at 2.5°C.
Hazard Analysis
Toxic Hazard Rating:
Acute Local: Irritant 3; Ingestion 3; Inhalation 3.
Acute Systemic: Ingestion 3; Inhalation 3.
Chronic Local: Irritant 2.
Chronic Systemic: Ingestion 2; Inhalation 2.
TLV: ACGIH (recommended); 3 parts per million in air; 2 milligrams per cubic meter of air.
Toxicology: It is extremely irritating and corrosive to the skin and mucous membranes. Inhalation of the vapor may cause ulcers of the upper respiratory tract. Concentrations at 50 to 250 ppm are dangerous, even for brief exposures. Hydrofluoric acid produces severe skin burns which are slow in healing. The subcutaneous tissues may be affected, becoming blanched and bloodless. Gangrene of the affected areas may follow. See also fluorides. It is a common air contaminant. (Section 4).
Disaster Hazard: Dangerous; when heated, it emits highly corrosive fumes of fluorides; will react with water or steam to produce toxic and corrosive fumes.
Countermeasures
Ventilation Control: Section 2.
Personnel Protection: Section 3.
First Aid: Section 1.
Storage and Handling: Section 7.
Shipping Regulations: Section 11.
I.C.C.: Corrosive liquid; white label, 10 pints.
Coast Guard Classification: Corrosive liquid; white label.
MCA warning label.
IATA: Corrosive liquid, white label, 1 liter (passenger), 5 liters (cargo).

HYDROFLUORIC ACID, ANHYDROUS. See hydrofluoric acid.
Shipping Regulations: Section 11.
I.C.C.: Corrosive liquid; white label, 110 pounds.
Coast Guard Classification: Corrosive liquid; white label.
MCA warning label.
IATA: Corrosive liquid, white label, not acceptable (passenger), 50 kilograms (cargo).

HYDROFLUORIC AND SULFURIC ACIDS, MIXTURES
Shipping Regulations: Section 11.
ICC: Corrosive liquid, white label, 10 pints.
IATA: Corrosive liquid, white label, not acceptable (passenger), 5 liters (cargo).

HYDROFLUOSILICIC ACID
General Information
Synonyms: Fluosilicic acid; silicofluoric acid.
Description: Transparent, colorless, fuming liquid.
Formula: H_2SiF_6.
Constants: Mol wt: 144.08, bp: decomposes.
Hazard Analysis
Toxic Hazard Rating:
Acute Local: Irritant 3; Ingestion 3; Inhalation 3.
Acute Systemic: Ingestion 3; Inhalation 3.
Chronic Local: U.

Chronic Systemic: Ingestion 3; Inhalation 3.
Disaster Hazard: Dangerous; when heated to decomposition, it emits highly toxic and corrosive fumes of fluorides; will react with water or steam to produce toxic and corrosive fumes.
Countermeasures
Ventilation Control: Section 2.
Personnel Protection: Section 3.
First Aid: Section 1.
Storage and Handling: Section 7.
Shipping Regulations: Section 11.
I.C.C.: Corrosive liquid; white label, 10 pints.
Coast Guard Classification: Corrosive liquid; white label.
IATA: Corrosive liquid, white label, 1 liter (passenger), 5 liters (cargo).

HYDROFURAMIDE
General Information
Synonyms: N,N'-Difurfurylidene-2-furan methanediamine; furfuramide.
Description: Light brown crystals.
Formula: $(C_4H_3OCH)_3N_2$.
Constants: Mol wt: 268.26, mp: 117°C, bp: 250°C decomposes.
Hazard Analysis
Toxic Hazard Rating:
Acute Local: Inhalation 3.
Acute Systemic: Inhalation 2.
Chronic Local: Inhalation 3.
Chronic Systemic: Inhalation 2.
Toxicity: A component of fungicides. Causes intense pulmonory irritation and reported to cause liver and kidney damage. See also amines and amides.
Disaster Hazard: Slight; when heated to decomposition, it emits toxic fumes of nitrogen compounds.
Countermeasures
Storage and Handling: Section 7.

HYDROGEN
General Information
Description: Colorless gas.
Formula: H_2.
Constants: Mol wt: 2.0162, mp: $-259.18°C$, bp: $-252.8°C$, lel: 4.1%, uel: 74.2%, d: 0.0899 g/liter, autoign. temp. 1085°F, vap. d.: 0.069.
Hazard Analysis
Toxic Hazard Rating:
Acute Local: 0.
Acute Systemic: Inhalation 1.
Chronic Local: 0.
Chronic Systemic: 0.
Radiation Hazard: Section 5. For permissible levels, see Table 000.
Artificial and natural isotope ^3H (tritium), half life 12.3 y. Decays to stable ^3He by emitting beta particles of 0.018 MeV. Tritium occurs naturally as a result of cosmic ray bombardment of deuterium.
Fire Hazard: Highly dangerous; when exposed to heat or flame.
Explosion Hazard: Severe, when exposed to heat or flame.
Explosive Range: 4.1–74.2%.
Disaster Hazard: Dangerous; can react vigorously with oxidizing materials.
Countermeasures
Ventilation Control: Section 2.
Storage and Handling: Section 7.
To Fight Fire: Carbon dioxide or dry chemical (Section 6).
Shipping Regulations: Section 11.
I.C.C.: Flammable gas; red gas label, 300 pounds.
I.C.C.: (Liquefied) not accepted.
Coast Guard Classification: Inflammable gas; red gas label.
Coast Guard Classification: (Liquefied) not permitted.
IATA: Flammable gas, red label, not acceptable (passenger), 140 kilograms (cargo).

IATA (liquefied): Flammable gas, not accepted (passenger and cargo).

HYDROGEN-3. See tritium.

HYDROGEN ANTIMONIDE. See antimony hydride.

HYDROGEN ARSENIDE (GAS). See arsine.

HYDROGEN ARSENIDE (SOLID)
General Information
Description: Brown powder.
Formula: H_2As_2.
Constants: Mol wt: 151.84, mp: decomposes at 200°C.
Hazard Analysis
Toxic Hazard Rating:
  Acute Local: Irritant 3; Ingestion 3; Inhalation 3.
  Acute Systemic: Ingestion 3; Inhalation 3.
  Chronic Local: U.
  Chronic Systemic: Ingestion 3; Inhalation 3,
Toxicology: See also arsenic compounds.
Fire Hazard: Moderate, by chemical reaction with oxidizers.
Explosion Hazard: Details unknown. See also hydrides.
Disaster Hazard: Dangerous; See arsenic.
Countermeasures
Ventilation Control: Section 2.
Personnel Protecton: Section 3.
First Aid: Section 1.
Storage and Handling: Section 7.

HYDROGEN AZIDE. See hydrazoic acid.

HYDROGEN BISMUTHIDE
General Information
Synonym: Bismuthine.
Description: Liquid.
Formula: H_3Bi.
Constants: Mol wt: 212.02, bp: 22°C.
Hazard Analysis and Countermeasures
See bismuth compounds and hydrides.

HYDROGEN BROMIDE. See hydrobromic acid.
Shipping Regulations: Section 11.
  I.C.C.: Nonflammable gas; green label, 300 pounds.
  Coast Guard Classification: Noninflammable gas; green gas label.
  IATA: Nonflammable gas, green label, not acceptable (passenger), 140 kilograms (cargo).

HYDROGEN BROMIDE (CONSTANT BOILING MIX-TURE)
General Information
Description: Colorless liquid.
Formula: HBr(47%) + H_2O.
Constants: Mp: -11°C, bp: 126°C, d: 1.49.
Hazard Analysis and Countermeasures
See hydrobromic acid.

HYDROGEN BROMIDE, HYDRATED
General Information
Description: Colorless liquid.
Formula: HBr·H_2O.
Constants: Mol wt: 98.94, bp: -3.3°C to -15°C, d: 1.78.
Hazard Analysis and Countermeasures
See hydrobromic acid.

HYDROGEN CARBOXYLIC ACID. See formic acid.

HYDROGEN CHLORIDE. See hydrochloric acid.
Shipping Regulations: Section 11.

I.C.C.: Nonflammable gas; green label, 300 pounds.
Coast Guard Classification: Noninflammable gas; green gas label.
MCA warning label.
IATA: Nonflammable gas, green label, not acceptable (passenger), 140 kilograms (cargo).

HYDROGEN CYANIDE. See hydrocyanic acid.

HYDROGEN DIOXIDE. See hydrogen peroxide.

HYDROGEN DISULFIDE
General Information
Description: Yellow oil.
Formula: H_2S_2.
Constants: Mol wt: 66.15, mp: -89.7°C, bp: 74.5°C, d: 1.376, vap. press.: 100 mm at 22.0°C.
Hazard Analysis and Countermeasures
See hydrogen sulfide and sulfides.

HYDROGEN FLUORIDE. See hydrofluoric acid.

HYDROGEN IODIDE. See hydriodic acid.
Shipping Regulations: Section 11.
  IATA: Nonflammable gas, green label, not acceptable (passenger), 140 kilograms (cargo).

HYDROGEN NITRATE. See nitric acid.

HYDROGEN PENTASULFIDE
General Information
Description: Clear yellow oil.
Formula: H_2S_5.
Constants: Mol wt: 162.35, mp: decomposes, d: 1.67 at 16°C.
Hazard Analysis and Countermeasures
See hydrogen sulfide.

HYDROGEN PEROXIDE
General Information
Synonym: Hydrogen dioxide, T-Stuff.
Description: Colorless, heavy liquid or at low temperatures a crystalline solid.
Formula: H_2O_2.
Constants: Mol wt: 34.016, bp: 107°C for 35 wt%, d: 1.71 at -20°C; 1.46 at 0°C, vap. press.: 1 mm at 15.3°C.
Hazard Analysis
Toxic Hazard Rating: For concentrated solutions:
  Acute Local: Irritant 3; Ingestion 3; Inhalation 3.
  Acute Systemic: U.
  Chronic Local: U.
  Chronic Systemic: U.
TLV: ACGIH (recommended); 1 part per million of air; 1.4 milligrams per cubic meter of air (90 wt% solution).
Toxicology: Pure H_2O_2, its solutions, vapors and mists are irritating to body tissue. This irritation can vary from mild to severe depending upon the concentration of H_2O_2. For instance solutions of H_2O_2 of 35 wt% and over can easily cause blistering of the skin. Irritation caused by H_2O_2 which does not subside upon flushing of the affected part with water should be treated by a physician. The eyes are particularly sensitive to irritation by this material. It is used as a general purpose food additive; it is a substance which migrates to food from packaging materials (Section 10). It is a common air contaminant (Section 4).
Fire Hazard: Dangerous by chemical reaction with flammable materials. H_2O_2 is a powerful oxidizer, particularly in the concentrated state. It is important to keep containers of this material covered because (1) uncovered

---

TOXIC HAZARD RATING CODE (For detailed discussion, see Section 1.)

0 NONE: (a) No harm under any conditions; (b) Harmful only under unusual conditions or overwhelming dosage.

1 SLIGHT: Causes readily reversible changes which disappear after end of exposure.

2 MODERATE: May involve both irreversible and reversible changes; not severe enough to cause death or permanent injury.

3 HIGH: May cause death or permanent injury after very short exposure to small quantities.

U UNKNOWN: No information on humans considered valid by authors

containers are much more prone to react with flammable vapors, gases, etc.; (2) because if uncovered, the water from an H_2O_2 solution can evaporate, concentrating the material and thus increasing the fire hazard of the remainder.

For instance, solutions of H_2O_2 of concentrations in excess of 65 wt% heat up spontaneously when decomposing to $H_2O + \frac{1}{2}O_2$. Thus 90 wt% solutions, when caused to decompose rapidly due to the introduction of a catalytic decomposition agent, can get quite hot and perhaps start fires.

Explosion Hazard: Severe, when highly concentrated or pure H_2O_2 is exposed to heat, mechanical impact, detonation of a blasting cap, or caused to decompose catalytically by metals and their salts, dusts and alkalies.

Although many mixtures of H_2O_2 and organic materials do not explode upon contact, the resultant combination is detonatable either upon catching fire or by impact.

The detonation velocity of aqueous solutions of H_2O_2 has been found to be about 6500 meters/second for solutions of between 96 wt% and 100 wt% H_2O_2.

Another source of H_2O_2 explosions is from sealing the material in strong containers. Under such conditions even gradual decomposition of H_2O_2 to $H_2O + \frac{1}{2}O_2$ can cause large pressures to build up in the containers which may then burst explosively. See also Schumb, Satterfield, and Wentworth "Hydrogen Peroxide," pp. 174, 453.

Disaster Hazard: Highly dangerous because when heated, or shocked or contaminated, the concentrated material can explode or start fires.

Countermeasures
Personnel Protection: Section 3.
Ventilation Control: Section 2.
Storage and Handling: Section 7. (See also Schumb, Satterfield and Wentworth, "Hydrogen Peroxide," pp. 172–176.
Shipping Regulations: Section 11.
  Coast Guard Classification: (Corrosive liquid) Solution in water containing over 8% hydrogen peroxide by weight: white label.
  MCA warning label (for solutions above 7.4%).
  ICC (Solution in water containing over 8% hydrogen peroxide by weight): Corrosive liquid, white label, 1 gallon.
  IATA (Solution containing not more than 8% hydrogen peroxide by weight): Not restricted.
    (Solution containing more than 8% but not more than 40% hydrogen peroxide by weight): Corrosive liquid, white label, 1 liter (passenger), 5 liters (cargo).
    (Solution containing more than 40% hydrogen peroxide by weight): Corrosive liquid, not acceptable.

HYDROGEN PHOSPHIDE
General Information
Description: Colorless liquid.
Formula: H_4P_2.
Constants: Mol wt: 65.99, mp: $< -10°C$, bp: 57.5°C at 735 mm, d: 1.012.
Hazard Analysis and Countermeasures
See phosphine and phosphides.

HYDROGEN PHOSPHIDE. See phosphine.

HYDROGEN PHOSPHIDE (POLYMER)
General Information
Description: Yellow solid.
Formula: $(H_2P_4)_3$.
Constants: Mol wt: 377.81, mp: ignites at 160°C, bp: decomposes, d: 1.83 at 19°C.
Hazard Analysis and Countermeasures
See phosphine and phosphides.

HYDROGEN SELENIDE
General Information
Description: Colorless gas.
Formula: H_2Se.
Constants: Mol wt: 80.98, mp: $-64°C$, bp: $-41.4°C$, d: 3.614 g/liter (gas): 2.12 at $-42°C$ (liquid), vap. press.: 10 atm. at 23.4°C.
Hazard Analysis
Toxic Hazard Rating:
  Acute Local: Irritant 3; Allergen 1; Inhalation 3.
  Acute Systemic: Inhalation 3.
  Chronic Local: Allergen 1.
  Chronic Systemic: Inhalation 3.
TLV: ACGIH (recommended); 0.05 part per million in air; 0.17 milligrams per cubic meter of air.
Toxicology: This material is a hazard compound of selenium which can cause damage to the lungs and liver as well as conjunctivitis. It has been found that repeated 8-hour exposures to concentrations of 0.3 ppm prove fatal to guinea pigs by causing a pneumonitis, as well as injury to the liver and spleen. Concentrations of 0.3 ppm are readily detected by odor, but there is no noticeable irritant effect at that level. Concentrations of 1.5 ppm or higher are strongly irritating to the eyes and nasal passages.

As in the case of hydrogen sulfide, the odor of hydrogen selenide in concentrations below 1 ppm disappears rapidly because of olfactory fatigue. Although the odor and irritating effects are both useful to an experienced investigator for estimating the concentrations, they do not offer a dependable warning to workmen who may be exposed to gradually increasing amounts and therefore become used to it. Due to its extreme toxicity and irritant effects, it seldom is allowed to reach a concentration in which it is flammable in air.

Very few data are available on possible chronic effects of this material, but it is logical to assume that when the concentration of this gas is low enough to avoid the irritant effects, only the systemic effects will be noticeable.

Fire Hazard: Dangerous; will react vigorously with powerful oxidizing agents.
Explosion Hazard: Dangerous; forms explosive mixtures with air. See also hydrides.
Disaster Hazard: Highly dangerous; keep away from heat and open flame.
Countermeasures
Ventilation Control: Section 2.
Personnel Protection: Section 3.
Storage and Handling: Section 7.
Shipping Regulations: Section 11.
  IATA: Poison A, no label, not acceptable (passenger and cargo).

HYDROGEN SULFIDE
General Information
Synonym: Sulfuretted hydrogen.
Description: Colorless; flammable gas, offensive odor.
Formula: H_2S.
Constants: Mol wt: 34.08, mp: $-85.5°C$, bp: $-60.4°C$, lel: 4.3%, uel: 46%, autoign. temp.: 500°F, d: 1.539 g/liter at 0°C, vap. press.: 20 atm at 25.5°C, vap. d.: 1.189.
Hazard Analysis
Toxic Hazard Rating:
  Acute Local: Irritant 3; Inhalation 3.
  Acute Systemic: Inhalation 3.
  Chronic Local: Irritant 3.
  Chronic Systemic: Inhalation 3.
TLV: ACGIH (recommended) 10 parts per million of air; 15 milligrams per cubic meter of air.
Toxicology: Hydrogen sulfide is both an irritant and an asphyxiant. Low concentrations of from 20 to 150 ppm

cause irritation of the eyes; slightly higher concentrations may cause irritation of the upper respiratory tract, and if exposure is prolonged, pulmonary edema may result. The irritant action has been explained on the basis that H_2S combines with the alkali present in moist surface tissues to form sodium sulfide, a caustic.

With higher concentrations the action of the gas on the nervous system becomes more prominent, and a 30-minute exposure to 500 ppm results in headache, dizziness, excitement, staggering gait, diarrhea and dysuria, followed sometimes by bronchitis or bronchopneumonia. The action on the nervous system is, with small amounts, one of depression; in larger amounts, it stimulates, and with very high amounts the respiratory center is paralyzed. Exposures of 800 to 1000 ppm may be fatal in 30 minutes, and high concentrations are instantly fatal. Fatal hydrogen sulfide poisoning may occur even more rapidly than that following exposure to a similar concentration of hydrogen cyanide. H_2S does not combine with the hemoglobin of the blood; its asphyxiant action is due to paralysis of the respiratory center.

With repeated exposures to low concentrations, conjunctivitis, photophobia, corneal bullae, tearing, pain and blurred vision are the commonest findings. High concentrations may cause rhinitis, bronchitis, and occasionally pulmonary edema. Exposure to very high concentrations results in immediate death. Chronic poisoning results in headache, inflammation of the conjunctivae and eyelids, digestive disturbances, loss of weight and general debility. It is a common air contaminant (Section 4).

Fire Hazard: Dangerous, when exposed to heat or flame.
Explosion Hazard: Moderate, when exposed to heat or flame.
Disaster Hazard: Highly dangerous; when heated to decomposition, it emits highly toxic fumes of oxides of sulfur; can react vigorously with oxidizing materials.
Countermeasures
Ventilation Control: Section 2.
Personnel Protection: Section 3.
Storage and Handling: Section 7.
To Fight Fire: Carbon dioxide, dry chemical or water spray (Section 6).
Shipping Regulations: Section 11.
  I.C.C.: Flammable gas; red gas label, 300 pounds.
  Coast Guard Classification: Inflammable gas; red gas label.
  MCA warning label.
  IATA: Flammable gas, red label, not acceptable (passenger), 140 kilograms (cargo).

HYDROGEN TELLURIDE
General Information
Description: Colorless gas or yellow needles.
Formula: H_2Te.
Constants: Mol wt: 129.63, mp: $-49.0°C$, bp: $-2.0°C$, d: 5.81 g/liter.
Hazard Analysis and Countermeasures
See tellurium compounds and hydrides.

HYDROGEN TRISULFIDE
General Information
Description: Bright yellow liquid.
Formula: H_2S_3.
Constants: Mol wt: 98.21, mp: $-52°C$, bp: decomposes at $90°C$, d: 1.496 at $15°C$.

Hazard Analysis and Countermeasures
See hydrogen sulfide.

HYDROHYDRASTININE
General Information
Description: White crystalline alkaloid.
Formula: $C_{11}H_{13}NO_2$.
Constants: Mol wt: 191.22, mp: $66°C$, bp: $303°C$ at 752 mm.
Hazard Analysis
Toxic Hazard Rating:
  Acute Local: U.
  Acute Systemic: Ingestion 3; Inhalation 3.
  Chronic Local: U.
  Chronic Systemic: U.
Disaster Hazard: Dangerous; when heated to decomposition, it emits highly toxic fumes of oxides of nitrogen.
Countermeasures
Ventilation Control: Section 2.
Personal Hygiene: Section 3.
First Aid: Section 1.
Storage and Handling: Section 7.

HYDROLITH. See calcium hydride.

HYDRONIUM PERCHLORATE. See perchloric acid, monohydrate.

HYDROQUINOL. See p-hydroquinone.

p-HYDROQUINONE
General Information
Synonyms: 1,4-Benzenediol; quinol; hydroquinol.
Description: Colorless, hexagonal prisms.
Formula: $C_6H_4(OH)_2$.
Constants: Mol wt: 110.1, mp: $170.5°C$, bp: $286.2°C$, flash p: $329°F$ (C.C.), d: 1.358 at $20°/4°C$, autoign. temp.: $960°F$, vap. press.: 1 mm at $132.4°C$, vap. d.: 3.81.
Hazard Analysis
Toxic Hazard Rating:
  Acute Local : Irritant 2; Allergen 1.
  Acute Systemic: Ingestion 2; Inhalation 2.
  Chronic Local: Allergen 2.
  Chronic Systemic: Ingestion 2; Inhalation 2.
TLV: ACGIH (recommended); 2 milligrams per cubic meter of air.
Toxicology: Absorption of this material by tissues can cause symptoms of illness which resemble those induced by its ortho or meta isomers. For instance, the ingestion of 1 g in an adult or a smaller quantity by a child may induce tinnitis, nausea, dizziness, a sensation of suffocation, an increased rate of respiration, vomiting, pallor, muscular twitchings, headache, dyspnea, cyanosis, delirium, and collapse. The literature contains reports of fatal cases which have been caused by the ingestion of 5 to 12 g. Cases of dermatitis have resulted from skin contact with this material, and have also followed the application of an antiseptic oil which apparently contained traces of hydroquinone added as an antioxidant. The reports also contain cases of keratitis and discoloration of the conjunctiva among personnel exposed to this material in concentrations ranging from 10 to 30 mg of the vapor or dust per cubic meter of air. It is considered to be more toxic than phenol. The inhalation of vapors of this material, particularly when liberated at high temperatures, must be avoided. From toxicity experiments on mice, it has

---

TOXIC HAZARD RATING CODE (For detailed discussion, see Section 1.)

0 NONE: (a) No harm under any conditions; (b) Harmful only under unusual conditions or overwhelming dosage.

1 SLIGHT: Causes readily reversible changes which disappear after end of exposure.

2 MODERATE: May involve both irreversible and reversible changes; not severe enough to cause death or permanent injury.

3 HIGH: May cause death or permanent injury after very short exposure to small quantities.

U UNKNOWN: No information on humans considered valid by authors.

been found that the LD_{50}* via the intraperitoneal route is from 160 to 170 mg/kg of body weight, for the rat via the same route, 170 to 175 mg/kg, and for the rat via the oral route, 370 to 390 mg/kg. Finally for the rabbit via the intraperitoneal route, the LD_{50}* has been found to be 125 to 150 mg/kg.

Fire Hazard: Slight, when exposed to heat or flame; can react with oxidizing materials.

Explosion Hazard: Slight, when exposed to heat.

Countermeasures

Ventilation Control: Section 2.

Treatment and antidotes: When personnel who are working with this material exhibit some of the symptoms listed above, they should immediately be removed to fresh air. If the symptoms do not subside quickly, consult a physician. In cases of determatitis due to this material, removal from exposure will quickly clear up the symptoms. If this material accidentally comes into contact with the skin, it should be removed at once and the affected area washed with plenty of soap and warm water.

Personnel Protection: Section 3.

To Fight Fire: Water, carbon dioxide, dry chemical or carbon tetrachloride (Section 6).

Storage and Handling: Section 7.

HYDROQUINONE (in oil). See antiseptic oil.

HYDROQUINONE DIMETHYL ETHER
General Information
Synonym: 1,4-Dimethoxybenzene.
Description: Colorless leaflets; odor of sweet clover.
Formula: $C_6H_4(OCH_3)_2$.
Constants: Mol wt: 138.16, mp: 56°C, bp: 212.6°C, d: 1.053 at 55°/55°C.

Hazard Analysis
Toxic Hazard Rating:
　Acute Local: Irritant 1; Allergen 1; Ingestion 1; Inhalation 1.
　Acute Systemic: Ingestion 1; Inhalation 1.
　Chronic Local: Allergen 1.
　Chronic Systemic: U.
Fire Hazard: Moderate, when exposed to heat or flame; can react with oxidizing materials (Section 6).
Explosion Hazard: Details unknown. See also ethers.

Countermeasures
Ventilation Control: Section 2.
Personal Hygiene: Section 3.
Storage and Handling: Section 7.

HYDROQUINONE MONOBENZYL ETHER
General Information
Synonym: "Agerite alba."
Formula: $HOC_6H_4OC_6H_5$.
Constant: Mol wt: 186.2.

Hazard Analysis
Toxicity: Details unknown. Can cause permanent depigmentation of skin. Limited animal experiments indicate low systemic toxicity.
Fire Hazard: Moderate, when exposed to heat or flame; can react with oxidizing materials.
Explosion Hazard: Details unknown. See also ethers.

Countermeasures
Ventilation Control: Section 2.
Storage and Handling: Section 7.

HYDROQUINONE MONOMETHYL ETHER
General Information
Synonym: 4-Methoxy phenol; hydroxy anisole.
Description: White, waxy solid.
Formula: $CH_3OC_6H_4OH$.
Constants: Mol wt: 124.13, mp: 52.5°C, bp: 246°C, d: 1.55 at 20°/20°C.

Hazard Analysis
Toxic Hazard Rating:

Acute Local: U.
Acute Systemic: Ingestion 2; Inhalation 2.
Chronic Local: U.
Chronic Systemic: U.
Fire Hazard: Slight, when exposed to heat or flame; can react with oxidizing materials (Section 6).
Explosion Hazard: Details unknown. See also ethers.

Countermeasures
Ventilation Control: Section 2.
Personal Hygiene: Section 3.
Storage and Handling: Section 7.

HYDROUS MAGNESIUM SILICATE. See talc.

HYDROXYACETIC ACID. See glycolic acid.

o-HYDROXYACETOPHENONE
General Information
Description: Greenish-yellow liquid; highly refractive; minty odor.
Formula: $C_6H_4(OH)COCH_3$.
Constants: Mol wt: 136.14, mp: 95°C, vap. d.: 4.69.

Hazard Analysis
Toxicity: Unknown. See also phenyl methyl ketone,
Fire Hazard: Slight; when heated, it emits acrid fumes (Section 6).

Countermeasures
Storage and Handling: Section 7.

2-HYDROXYADIPALDEHYDE (25% AQUEOUS SOLUTION)
General Information
Description: Liquid.
Formula: $OHC(CH_2)_3CH(OH)CHO$.
Constants: Mol wt: 130.1, fp: −3.5°C, d: 1.066 at 20°/20°C, vap. press.: 17 mm at 20°C.

Hazard Analysis
Toxicity: Details unknown. See also aldehydes.
Disaster Hazard: Slight; when heated; it emits acrid fumes (Section 6).

Countermeasures
Storage and Handling: Section 7.

β-HYDROXYALANINE. See serine.

p-HYDROXYANILINE. See p-aminophenol.

HYDROXYANISOLE. See hydroquinone monomethyl ether.

β-HYDROXYANTHRAQUINONE
Hazard Analysis
Toxic Hazard Rating:
　Acute Local: Allergen 1.
　Acute Systemic: U.
　Chronic Local: Allergen 1.
　Chronic Systemic: U.
Disaster Hazard: Slight; when heated, it emits acrid fumes (Section 6).

Countermeasures
Personal Hygiene: Section 3.
Storage and Handling: Section 7.

o-HYDROXYBENZAMIDE. See salicylamide.

p-HYDROXYBENZOATE, METHYL, ETHYL, PROPYL, AND BUTYL ESTERS
Hazard Analysis
Toxic Hazard Rating:
　Acute Local: Allergen 2.
　Acute Systemic: U.
　Chronic Local: Allergen 2.
　Chronic Systemic: U.
Toxicity: Skin sensitivity in humans has been reported.

o-HYDROXYBENZOIC ACID. See salicylic acid.

p-HYDROXYBENZOIC ACID ESTERS
General Information
Description: Esters of p-hydroxybenzoic acid $(C_6H_4(OH) \cdot COOH \cdot H_2O)$.
Hazard Analysis
Toxic Hazard Rating:
Acute Local: Allergen 1.
Acute Systemic: U.
Chronic Local: Allergen 1.
Chronic Systemic: U.
Toxicity: The butyl, ethyl, methyl, and propyl esters have been found to produce allergic sensitization in humans. Systemic toxicity appears to be low.

3-HYDROXYBUTANAL. See aldol.

3-HYDROXY-2-BUTANONE. See acetoin.

β-HYDROXYBUTYRALDEHYDE. See aldol.

4-HYDROXY-3,5-DI-tert-BUTYLTOLUENE. See di-tert-butyl-p-cresol.

1-HYDROXY-3,5-DIMETHYLBENZENE. See 3,5-xylenol.

5-HYDROXY-1,3-DIMETHYLBENZENE. See 3,5-xylenol.

3-HYDROXY-n,n-DIMETHYL-CIS-CROTONAMIDE DI-METHYL SULFIDE
General Information
Synonym: Bidrin.
Description: Brown liquid, ester odor; miscible with water and xylene; very slightly soluble in kerosene and diesel fuel.
Formula: $(CH_3O)_2P(O)OC(CH_3)CHC(O)N(CH_3)_2$.
Constants: Mol wt: 237, bp: 400° C.
Hazard Analysis
Toxicity: Resembles other organic phosphate pesticides. See parathion.
Disaster Hazard: Dangerous. See sulfur compounds.

1-HYDROXY-2,3-DINITROBENZENE. See 2,3-dinitrophenol.

p-HYDROXYDIPHENYLAMINE
General Information
Synonym: Anilinophenol.
Description: Gray solid leaflets.
Formula: $C_6H_5NHC_6H_4OH$.
Constants: Mol wt: 185.2, mp: 70° C, bp: 330° C.
Hazard Analysis
Toxicity: Details unknown. See also amines.
Disaster Hazard: Dangerous; when heated to decomposition, it emits highly toxic fumes of aniline.
Countermeasures
Storage and Handling: Section 7.

(2-HYDROXYDIPHENYL) METHANE. See o-benzylphenol.

(4-HYDROXYDIPHENYL) METHANE. See p-benzylphenol.

HYDROXYETHANOIC ACID. See glycolic acid.

HYDROXYETHYL ACETAMIDE
General Information
Synonym: n-Acetyl ethanolamine.
Description: Brown, viscous liquid.

Formula: $CH_3CONHC_2H_4OH$.
Constants: Mol wt: 103.12, bp: 151° C, flash p: 350° F (O.C.), fp: 15.8° C, d: 1.122 at 20° /20° C.
Hazard Analysis
Toxic Hazard Rating:
Acute Local: Irritant 1; Ingestion 1; Inhalation 1.
Acute Systemic: U.
Chronic Local: U.
Chronic Systemic: U.
Fire Hazard: Slight, when exposed to heat or flame.
Countermeasures
To Fight Fire: Water, foam, alcohol foam, carbon dioxide, dry chemical or carbon tetrachloride (Section 6).
Personal Hygiene: Section 3.
Storage and Handling: Section 7.

β-HYDROXYETHYLANILINE. See 2-anilinoethanol.

n-(2-HYDROXYETHYL) CYCLOHEXYL AMINE
General Information
Description: Soluble in water.
Formula: $C_6H_{11}NHC_2H_4OH$.
Constants: Mol wt: 143, mp: 97–102° F, flash p: 249° F (O.C.).
Hazard Analysis
Fire Hazard: Slight, when exposed to heat or flame.
Countermeasures
Storage and Handling: Section 7.
To Fight Fire: Water or foam (Section 6).

HYDROXYETHYLENE DIAMINE
General Information
Description: Clear liquid.
Formula: $NH_2CHOHCH_2NH_2$.
Constants: Mol wt: 76.1.
Hazard Analysis
Toxicity: Details unknown. See also amines.
Disaster Hazard: Slight; when heated, it emits acrid fumes (Section 6).
Countermeasures
Storage and Handling: Section 7.

2-HYDROXYETHYL ETHYLENE DIAMINE. See 2-amino-ethyl ethanolamine.

2-HYDROXYETHYL-5-ETHYL PYRIDINE
General Information
Formula: $N(CH_3)CC_2H_5CC_2H_5OH$.
Constant: Mol wt: 140.
Hazard Analysis
Toxic Hazard Rating: U.
Toxicity: Limited animal experiments suggest low toxicity. See also pyridine.

1-HYDROXYETHYL-1,2-HEPTADECENYL GLYOXALIDINE. See cationic amine 220.

1-HYDROXYETHYL-2-HEPTADECENYLIMIDAZO-LINE
General Information
Description: Yellow liquid.
Formula: $C_{17}H_{33}C[N(C_2H_4OH)CH_2CH_2N]$.
Constants: Mol wt: 350, bp: 250–280° C at 2mm, d: 0.95 at 20° C.
Hazard Analysis
Toxicity: Details unknown. See also amines.
Disaster Hazard: Slight; when heated, it emits acrid fumes (Section 6).

TOXIC HAZARD RATING CODE *(For detailed discussion, see Section 1.)*

0 NONE: (a) No harm under any conditions; (b) Harmful only under unusual conditions or overwhelming dosage.

1 SLIGHT: Causes readily reversible changes which disappear after end of exposure.

2 MODERATE: May involve both irreversible and reversible changes; not severe enough to cause death or permanent injury.

3 HIGH: May cause death or permanent injury after very short exposure to small quantities.

U UNKNOWN: No information on humans considered valid by authors.

Countermeasures
Storage and Handling: Section 7.

1-HYDROXYETHYL-2-HEPTADECYLGLYOXALIDINE
Hazard Analysis
Toxicity: Details unknown; plant fungicide.
Countermeasures
Storage and Handling: Section 7.

2-HYDROXY-3-ETHYL HEPTANOIC ACID
General Information
Formula: $CH_3CH(OH)CH(C_2H_5)(CH_2)_3COOH$.
Constant: Mol wt: 186.
Hazard Analysis
Toxic Hazard Rating: U.
Toxicity: Limited animal experiments suggest moderate toxicity and irritant.

β-HYDROXYETHYL HYDRAZINE
General Information
Description: Colorless, slightly viscous liquid.
Formula: $HOCH_2CH_2NHNH_2$.
Constants: Mol wt: 76.12, mp: $-70°C$, bp: $145-153°C$ at 25 mm, flash p: $224°F$, vap.d.: 2.63.
Hazard Analysis
Toxicity: Highly toxic. See hydrazine.
Fire Hazard: Moderate, when exposed to heat or flame; can react with oxidizing materials.
Countermeasures
To Fight Fire: Foam, carbon dioxide, dry chemical or carbon tetrachloride (Section 6).
Ventilation Control: Section 2.
Storage and Handling: Section 7.

2-HYDROXYETHYL   MERCAPTAN. See mercaptoethanol.

n-HYDROXYETHYL MORPHOLINE. See morpholine ethanol.

n-(2-HYDROXYETHYL) PHENYL AMINE
General Information
Formula: $C_6H_4NH_2C_2H_4OH$.
Constant: Mol wt: 137.
Hazard Analysis
Toxic Hazard Rating: U.
Toxicity: Limited animal experiments suggest moderate oral toxicity; high toxicity by skin absorption and moderate irritant. See also Amines.

HYDROXYETHYLPIPERAZINE
General Information
Description: Light colored liquid.
Formula: $HOCH_2CH_2N(CH_2)_4NH$.
Constants: Mol wt: 130.2, d: 1.0610, bp: $240°C$, flash p: $255°F(OC)$; vap. d.: 4.5.
Hazard Analysis
Toxicity: Details unknown. Limited animal experiments suggest low toxicity. See also piperazine.
Fire Hazard: Slight, when exposed to heat or flame.
Countermeasures
To Fight Fire: Water, foam, alcohol foam (Section 6).
Storage and Handling: Section 7.

n-HYDROXYETHYL PROPYLENEDIAMINE
General Information
Description: Liquid.
Formula: $CH_3CH(NHC_2H_4OH)CH_2NH_2$.
Constants: Mol wt: 118.18, mp: $-65°C$, bp: $240°C$, flash p.: $260°F$ (O.C.), d: 0.9938, vap. d.: 4.07.
Hazard Analysis
Toxicity: Details unknown. See also amines.
Fire Hazard: Slight, when exposed to heat or flame; can react with oxidizing materials.

Countermeasures
To Fight Fire: Foam, carbon dioxide, dry chemical or carbon tetrachloride (Section 6).
Storage and Handling: Section 7.

N-HYDROXYETHYLPYRROLIDINE
General Information
Description: Colorless liquid.
Formula: $C_4H_8NCH_2CH_2OH$.
Constants: Mol wt: 115.2, bp: $187°C$, fp: $-75°C$, flash p: $160°F$, d: 0.9800 at $20°C$.
Hazard Analysis
Toxicology: This material is moderately toxic by oral ingestion. It is moderately irritating to skin, eyes, and mucous membrane.
Fire Hazard: Moderate; when exposed to heat sources, sparks or open flame.
Disaster Hazard: Dangerous; when heated to decomposition it emits highly toxic fumes.
Countermeasures
To Fight Fire: Foam, spray, dry chemical, or carbon dioxide. Section 6.
Ventilation Control: Section 2.
Personnel Protection: Section 3.
Storage and Handling: Section 7.

HYDROXYETHYLSULFONIC ACID
General Information
Synonym: Isethonic acid.
Hazard Analysis
Toxic Hazard Rating:
    Acute Local: Irritant 3; Ingestion 3.
    Acute Systemic: Ingestion 3.
    Chronic Local: U.
    Chronic Systemic: U.
Disaster Hazard Rating: Dangerous. See sulfonates.

2-HYDROXETHYLTRIMETHYLAMMONIUM BICARBONATE
General Information
Description: A colorless liquid.
Formula: $(CH_3)NCH_2CH_2OH \cdot HCO_3$.
Constants: Mol wt: 165.19, bp: loses CO_2, fp: $-21.3°C$, d: 1.0965 at $25°/4°C$.
Hazard Analysis
Toxicity: Details unknown. See also sodium carbonate.

HYDROXYHEPTYL PEROXIDE (paste with dibutyl phthalate).
General Information
Synonym: "Luperco HDB."
Description: Thick white paste, slightly water soluble; soluble in common organic solvents.
Hazard Analysis
Toxicity: Probably low so far as hydroxyheptyl peroxide is concerned. See dibutyl phthalate and peroxides, organic.
Countermeasures
Shipping Regulations: Section 11.
    I.C.C.: Oxidizing material N.O.S.
    Freight Express: Yellow label, chemicals N.O.I.B.N.
    Rail Express: Maximum weight = 25 lbs per shipping case.
    Parcel Post: Prohibited.

a-HYDROXYISOBUTYRONITRILE. See acetone cyanhydrin.

HYDROXYLAMINE
General Information
Synonym: Oxammonium.
Description: Colorless liquid or white needles.
Formula: NH_2OH.
Constants: Mol wt: 33.02, mp: $34.0°C$, bp: $110.0°C$, flash p: explodes at $265°F$, d: 1.227, vap. press. 10 mm at $47.2°C$.
Hazard Analysis
Toxic Hazard Rating:
    Acute Local: Irritant 2; Ingestion 2; Inhalation 2.

Acute Systemic: Ingestion 2; Inhalation 2.
Chronic Local: Irritant 1; Ingestion 1; Inhalation 1.
Chronic Systemic: Ingestion 2; Inhalation 2.
Toxicology: Locally it is irritating and systemically it can cause methemoglobinemia.
Explosion Hazard: Dangerous; when exposed to heat or open flame.
Disaster Hazard: Dangerous, upon exposure to heat or flame.
Countermeasures
Ventilation Control: Section 2.
Personal Hygiene: Section 3.
Storage and Handling: Section 7.

HYDROXYLAMINE ACID SULFATE
General Information
Description: Crystalline white to brown solid.
Formula: $NH_2OH \cdot H_2SO_4$.
Constants: Mol wt: 131.1, mp: indefinite; bp: decomposes.
Hazard Analysis and Countermeasures
See hydroxylamine and sulfuric acid.

HYDROXYLAMINE FLUOGERMANATE
General Information
Description: Monoclinic prisms.
Formula: $2NH_2OH \cdot H_2GeF_6 \cdot 2H_2O$.
Constants: Mol wt: 290.71, d: 2.229 at 25°/25°C.
Hazard Analysis and Countermeasures
See fluogermanates and hydroxylamine.

HYDROXYLAMINE FLUOSILICATE
General Information
Description: Scales.
Formula: $2NH_2OH \cdot H_2SiF_6 \cdot 2H_2O$.
Constant: Mol wt: 246.17.
Hazard Analysis and Countermeasures
See fluosilicaates, hydroxylamine, and fluorides.

HYDROXYLAMINE HYDROCHLORIDE
General Information
Synonym: Hydroxylammonium chloride.
Description: Colorless, hygroscopic crystals.
Formula: $NH_2OH \cdot HCl$.
Constants: Mol wt: 69.50, mp: 151°C, bp: decomposes, d: 1.67 at 17°C.
Hazard Analysis and Countermeasures
See hydroxylamine and hydrochloric acid.

HYDROXYLAMINE NITRATE
General Information
Description: White crystals.
Formula: $NH_2OH \cdot HNO_3$.
Constants: Mol wt: 96.05, mp: 48°C, bp: <100°C decomposes.
Hazard Analysis and Countermeasures
See hydroxylamine and nitrates.

HYDROXYLAMINE SULFATE
General Information
Synonym: Hydroxylammonium sulfate.
Description: A crystalline material.
Formula: $(NH_2OH)_2 \cdot H_2SO_4$.
Constants: Mol wt: 164.14, mp: 177°C.
Hazard Analysis
Toxicity: Highly toxic. See hydroxylamine and sulfuric acid.
Explosion Hazard: Moderate, when exposed to heat or by chemical reaction. In the presence of alkalies at elevated temperatures, free hydroxylamine is liberated and may decompose explosively.

Disaster Hazard: Dangerous; see sulfur compounds.
Countermeasures
Storage and Handling: Section 7.

HYDROXYLAMMONIUM CHLORIDE. See hydroxylamine hydrochloride.

HYDROXYLAMMONIUM SULFATE. See hydroxylamine sulfate.

HYDROXYMERCURICHLOROPHENOL
General Information
Synonym: 2-Chloro-4-(hydroxymercuri) phenol; semesan.
Description: Solid. Insoluble in water and common organic solvents.
Formula: $C_6H_5ClHgO_2$.
Constant: Mol wt: 345.2.
Hazard Analysis and Countermeasures
A fungicide. See mercury compounds, organic. MCA warning label.

HYDROXYMERCURICRESOL
General Information
Formula: $HOHgC_6H_4OHCH_3$.
Constant: Mol wt: 325.8.
Hazard Analysis and Countermeasures
See mercury compounds, organic.

HYDROXYMERCURINITROPHENOL
General Information
Formula: $HOHgC_6H_3OHNO_2$.
Constant: Mol wt: 356.7.
Hazard Analysis and Countermeasures
A fungicide. See mercury compounds, organic, and nitrates.

1-HYDROXY-4-METHYL-2,6-DI-tert-BUTYL BENZENE. See di-tert-butyl-p-cresol.

1-(HYDROXYMETHYL)-5,5-DIMETHYL HYDANTOIN
General Information
Synonym: MDMH.
Hazard Analysis
Toxic Hazard Rating: U.
Toxicity: Liberates formaldehyde on decomposition. See formaldehyde.

DL-α-HYDROXY-γ-METHYLMERCAPTOBUTYRIC ACID, CALCIUM SALT. See methionine hydroxy analog and its calcium salts.

2-HYDROXYMETHYLNORCAMPHANE
Hazard Analysis
Toxic Hazard Rating: U.
Toxicity: Limited animal experiments suggest high toxicity.

4-HYDROXY-4-METHYL-2-PENTANONE. See diacetone.

2-HYDROXYMETHYL TETRA HYDROPYRAN
Hazard Analysis
Toxic Hazard Rating: U.
Toxicity: Limited animal experiments suggest moderate toxicity.

2-HYDROXY-4-METHYLTHIOBUTYRIC ACID, CALCIUM SALT. See methionine hydroxy analog and its calcium salts.

1-HYDROXYNAPHTHALENE. See α-naphthol.

TOXIC HAZARD RATING CODE (For detailed discussion, see Section 1.)

0 NONE: (a) No harm under any conditions; (b) Harmful only under unusual conditions or overwhelming dosage.

1 SLIGHT: Causes readily reversible changes which disappear after end of exposure.

2 MODERATE: May involve both irreversible and reversible changes; not severe enough to cause death or permanent injury.

3 HIGH: May cause death or permanent injury after very short exposure to small quantities.

U UNKNOWN: No information on humans considered valid by authors.

β-HYDROXYNAPHTHALENE. See β-naphthol.

3-HYDROXY-2-NAPHTHOIC ACID
General Information
Synonyms: β-Oxynaphthoid acid; 2-naphthol-3-carboxylic acid.
Description: Yellow needle-like crystals.
Formula: $HOC_{10}H_6COOH$.
Constants: Mol wt: 188.2, mp: 216°C.
Hazard Analysis
Toxicity: Unknown. Animal experiments show moderate toxicity and irritation.
Fire Hazard: Slight; when heated, it emits acrid fumes (Section 6).
Countermeasures
Storage and Handling: Section 7.

4-HYDROXY-3-NITROBENZENE ARSONIC ACID. See 3-nitro-4-hydroxy phenylarsonic acid.

HYDROXYPENTAMETHYLFLAVAN
General Information
Synonym: 2-Hydroxy-2,4,4,4,7-pentamethylflavan.
Hazard Analysis
Toxicity: Details unknown; an insecticide.
Disaster Hazard: Slight; when heated, it emits acrid fumes (Section 6).
Countermeasures
Storage and Handling: Section 7.

β-p-HYDROXYPHENYL ALANINE. See tyrosine.

o-HYDROXYPHENYLMERCURIC CHLORIDE
General Information
Synonym: o-Chloromercuriphenol.
Description: White to faint pink crystals.
Formula: $C_6H_4OHHgCl$.
Constants: Mol wt: 329.2, mp: 143.5°C.
Hazard Analysis and Countermeasures
See mercury compounds, organic.

2-HYDROXYPROPANENITRILE. See lactonitrile.

β-HYDROXYPROPIONITRILE. See ethylene cyanohydrin

HYDROXYPROPYL ALGINATE. See propylene glycol alginate.

2-HYDROXYPROPYL AMINE. See 1-amino-2-propanol.

HYDROXYPROPYL CELLULOSE
General Information
Description: White powder, soluble in water and organic solvents.
Hazard Analysis
Toxicity: Details unknown. Used as a food additive permitted in food for human consumption see (Section 10).

N-HYDROXYPROPYLDIETHYLENETRIAMINE
Hazard Analysis
Toxic Hazard Rating: U.
Toxicity: Limited animal experiments suggest moderate toxicity and irritation. See also amines.

N-(2-HYDROXYPROPYL) ETHYLENEDIAMINE
General Information
Description: Colorless to faint yellow liquid.
Formula: $H_2NCH_2CH_2NH(CH_2CHOHCH_3)$.
Constants: Mol wt: 118, bp: 180°C at 114 mm, flash p: 254°F (C.O.C.), d: 0.9858 at 25°/25°C.
Hazard Analysis
Toxicity: Details unknown. See also amines.
Fire Hazard: Slight, when exposed to heat or flame; can react with oxidizing materials (Section 6).
Countermeasures
Storage and Handling: Section 7.

HYDROXYPROPYL METHYL CELLULOSE
General Information
Synonyms: Methyl cellulose; propylene glycol ether.
Description: White fibrous or granular powder; insoluble in anhydrous alcohol, ether and chloroform.
Hazard Analysis
Toxicity: Details unknown. Used as a food additive permitted in food for human consumption. (Section 10).

N-(3-HYDROXYPROPYL) 1,2-PROPANE DIAMINE
General Information
Formula: $CH_3CH(CHNH_2CHNH_2CH_3)CH_2CH$.
Constant: Mol wt: 132.
Hazard Analysis
Toxic Hazard Rating: U.
Toxicity: Limited animal experiments suggest low toxicity and moderate irritation.

3-HYDROXYPYRIDINE
General Information
Description: Tan-colored lumps.
Formula: C_5H_5ON.
Constants: Mol wt: 95.1, mp: 126–129°C, bp: 151–153°C at 3 to 4 mm.
Hazard Analysis
Toxicity: Details unknown. See also pyridine.
Fire Hazard: Slight, when exposed to heat or flame (Section 6).
Disaster Hazard: Dangerous; when heated to decomposition, it emits highly toxic fumes of nitrogen oxides; can react vigorously with oxidizing material.
Countermeasures
Storage and Handling: Section 7.

8-HYDROXYQUINOLINE. See 8-quinolinol.

HYDROXYSUCCINIC ACID. See malic acid.

6-HYDROXYTETRAHYDROPYRAN-2-CARBOBOXY-LIC ACID LACTONE
Hazard Analysis
Toxic Hazard Rating: U.
Toxicity: Limited animal experiments suggest low toxicity and irritation. See also lactones.

7-HYDROXY-3,4-TETRAMETHYLENE COUMARIN. See dithione.

HYDROXYTOLUENE. See benzyl alcohol.

α-HYDROXY-o-TOLUIC ACID LACTONE. See phthalide.

β-HYDROXYTRICARBALLYLIC ACID. See citric acid.

HYGROMYCIN B
Hazard Analysis
Toxic Hazard Rating: U.
Note: Used as a food additive permitted in the feed and drinking water of animals and/or for the treatment of food producing animals. Also permitted in food for human consumption (Section 10).

HYOSCINE
General Information
Synonym: Scopolamine.
Description: Thick, colorless, syrupy liquid alkaloid.
Formula: $C_{17}H_{21}NO_4$.
Constants: Mol wt: 303.35, mp: 55°C.
Hazard Analysis
Toxic Hazard Rating:
  Acute Local: U.
  Acute Systemic: Ingestion 3; Inhalation 3.
  Chronic Local: U.
  Chronic Systemic: U.
Toxicology: This material is a poisonous alkaloid. It can produce profound depression of the central nervous

system and occasionally it causes excitement. It can cause hallucinations and the individual affected loses a certain amount of his normal inhibitory control. It is for that reason that it has been called "truth serum." In many cases of poisoning from this material, and even to a certain extent following its medical application, there is retention of the urine caused by paralysis of the bladder, and catheterization is necessary. The fatal dose is variable. Death has occurred from as little as 0.6 mg, while recovery has occurred from doses of 7 to 15 mg. There is even a reported case of recovery following a 500-mg ingestion of the hydrobromide by mouth.

Disaster Hazard: Dangerous; when heated to decomposition, it emits highly toxic fumes of nitrogen oxides.

Countermeasures

Treatment and Antidotes: A physician should be called at once. Wash the stomach with a large volume of warm water containing tannin, then give 15% magnesium sulfate (approximately 8 ounces). A solution of iodine may be used instead of tannin. Charcoal may also be used. If iodine is used, the amount should be roughly 4 times the weight of the hyoscine to be precipitated. The iodine solution should not be allowed to remain in the stomach. In moderate cases of poisoning, morphine is of advantage if given hypodermically. If the respiration is markedly depressed, caffeine and artificial respiration should be employed. Coramine may also be of value. If there is extreme excitement or violent delirium, chloroform or ether may be employed.

Ventilation Control: Section 2.
Personal Hygiene: Section 3.
Storage and Handling: Section 7.

HYOSCINE HYDROBROMIDE. See hyoscine.

HYOSCINE SULFATE. See hyoscine.

HYOSCYAMINE
General Information
Synonyms: l-Hyoscyamine; daturine; duboisine.
Description: White crystalline alkaloid.
Formula: $C_{17}H_{23}NO_3$.
Constants: Mol wt: 289.36, mp: 106–108°C.
Hazard Analysis
Toxic Hazard Rating:
Acute Local: U.
Acute Systemic: Ingestion 3; Inhalation 3.
Chronic Local: U.
Chronic Systemic: U.
Toxicology: This is one of the atropine alkaloids and is very toxic, acting very much like atropine. It has the same effect on the central nervous system but twice the effect on the peripheral nerves. The symptoms of poisoning from this material are those of dryness of the throat and mouth, marked difficulty in swallowing, and a sensation of burning and thirst. The vision becomes impaired through dilation and loss of accommodation, and the eyes present a rather prominent, brilliant, staring appearance. The voice is husky and the tongue is red.
Disaster Hazard: Dangerous; when heated to decomposition, it emits highly toxic fumes of nitrogen oxides.
Countermeasures
Ventilation Control: Section 2.
Personal Hygiene: Section 3.
First Aid: Section 1.
Storage and Handling: Section 7.

HYOSCYAMINE HYDROBROMIDE. See hyoscyamine.

HYOSCYAMINE HYDROCHLORIDE. See hyoscyamine.

HYOSCYAMINE SULFATE. See hyoscyamine.

HY-PHOS. See sodium hexa-m-phosphate.

HYPNONE. See phenyl methyl ketone.

HYPO. See sodium thiosulfate.

HYPOBROMOUS ACID
General Information
Description: Colorless to yellow solution.
Formula: HBrO.
Constants: Mol wt: 96.92, mp: 40°C (in vacuum).
Hazard Analysis
Toxic Hazard Rating:
Acute Local: Irritant 3; Ingestion 3; Inhalation 3.
Acute Systemic: U.
Chronic Local: U.
Chronic Systemic: Ingestion 2; Inhalation 2.
Disaster Hazard: Dangerous; see bromides; will react with water or steam to produce toxic and corrosive fumes.
Countermeasures
Ventilation Control: Section 2.
Personnel Protection: Section 3.
Storage and Handling: Section 7.

HYPOCHLORITES
General Information
Description: Salts of hypochlorous acid.
Hazard Analysis
Toxic Hazard Rating:
Acute Local: Irritant 2; Ingestion 2; Inhalation 2.
Acute Systemic: U.
Chronic Local: U.
Chronic Systemic: U.
Fire Hazard: Moderate, by chemical reaction with reducing agents. These are powerful oxidizers particularly at higher temperatures when chlorine and then oxygen are evolved, or in the presence of moisture or carbon dioxide.
Disaster Hazard: Dangerous; when heated or on contact with acid or acid fumes, it emits highly toxic fumes of chlorine and chlorides; will react with water or steam to produce toxic and corrosive fumes; it can react vigorously with oxidizing materials.
Countermeasures
Ventilation Control: Section 2.
Personnel Protection: Section 3.
Storage and Handling: Section 7.
Shipping Regulations: Section 11.
MCA warning label.

HYPOCHLORITE SOLUTIONS containing more than 7% available chlorine by weight. See hypochlorites.
Shipping Regulations: Section 11.
I.C.C.: Corrosive liquid; white label, 4 gallons.
Coast Guard Classification: Corrosive liquid; white label.
IATA: Corrosive liquid, white label, 1 liter (passenger;, 5 liters (cargo).

HYPOCHLORITE SOLUTIONS containing not more than 7% available chlorine by weight.
Shipping Regulations: Section 11.
IATA: Other restricted articles, class B, no limit (passenger and cargo).

TOXIC HAZARD RATING CODE (For detailed discussion, see Section 1.)

0 NONE: (a) No harm under any conditions; (b) Harmful only under unusual conditions or overwhelming dosage.

1 SLIGHT: Causes readily reversible changes which disappear after end of exposure.

2 MODERATE: May involve both irreversible and reversible changes; not severe enough to cause death or permanent injury.

3 HIGH: May cause death or permanent injury after very short exposure to small quantities.

U UNKNOWN: No information on humans considered valid by authors.

HYPOCHLOROUS ACID
General Information
Description: Aqueous solutions are greenish yellow in color.
Formula: HClO.
Constant: Mol wt: 52.5.
Hazard Analysis
Toxicity: Irritant and corrosive. See hypochlorites and hydrochloric acid.
Countermeasures
See hypochlorites and hydrochloric acid.

HYPOCHLOROUS ANHYDRIDE. See chlorine tetroxide.

HYPOIODOUS ACID
General Information
Synonym: Iodine hydroxide.
Description: Yellow to greenish solution.
Formula: HOI.
Constant: Mol wt: 143.93.
Hazard Analysis
Toxic Hazard Rating:
   Acute Local: Irritant 3; Ingestion 3; Inhalation 3.
   Acute Systemic: U.
   Chronic Local: U.
   Chronic Systemic: Ingestion 2; Inhalation 2.
Disaster Hazard: Dangerous; see iodides; will react with water or steam to produce toxic and corrosive fumes.
Countermeasures
Ventilation Control: Section 2.
Personnel Protection: Section 3.
Storage and Handling: Section 7.

HYPONITROUS ACID
General Information
Description: White solid.
Formula: $H_2N_2O_2$.
Constants: Mol wt: 62.03, mp: explodes.
Hazard Analysis
Toxicity: Details unknown. See also nitric oxide.
Fire Hazard: Unknown.
Explosion Hazard: Severe, when exposed to heat or flame.
Disaster Hazard: Highly dangerous; when heated to decomposition, it emits highly toxic fumes of oxides of nitrogen and may explode.
Countermeasures
Storage and Handling: Section 7.

HYPONITROUS ETHER. See ethyl nitrite.

HYPOPHOSPHITES
General Information
Description: Salts of hypophosphorous acid.
Formula: $(H_2PO_2)^-$
Hazard Analysis
Toxicity: Details unknown; generally have little or no toxicity.
Fire Hazard: Moderate, when exposed to heat (Section 6).
Explosion Hazard: Moderate, when exposed to heat. See also phosphine.
Disaster Hazard: Dangerous; when heated to decomposition, it emits highly toxic fumes or phosphine and oxides of phosphorus.
Countermeasures
Storage and Handling: Section 7.

HYPOPHOSPHORIC ACID
General Information
Description: Crystals.
Formula: $H_4P_2O_6$.
Constants: Mol wt: 161.99, mp: 55°C, bp: decomposes at 100°C.
Hazard Analysis
Toxicity: Details unknown. See also phosphoric acid.
Explosion Hazard: Moderate, when exposed to heat.
Disaster Hazard: Dangerous; see phosphorus compounds.
Countermeasures
Storage and Handling: Section 7.

HYPOPHOSPHOROUS ACID
General Information
Description: Colorless, oily liquid or deliquescent crystals.
Formula: $H(H_2PO_2)$.
Constants: Mol wt: 66.00, mp: 26.5°C, bp: decomposes, d: 1.493 at 19°C.
Hazard Analysis
Toxicity: Details unknown. See also phosphoric acid.
Fire Hazard: Moderate; will react violently with oxidizing agents.
Disaster Hazard: Dangerous; on decomposition it emits highly toxic fumes of phosphine and may explode. Keep away from oxidizing agents.
Countermeasures
Storage and Handling: Section 7.

HYPOVANADIC HYDROCHLORIDE. See vanadium oxydichloride.

HYVAR. See 5-bromo-3-isopropyl-6-methyl uracil.

I

ICTHYOL
General Information
Description: Thick brown liquid, bituminous odor.
Hazard Analysis
Toxic Hazard Rating:
Acute Local: Irritant 1; Allergen 1; Ingestion 1.
Acute Systemic: U.
Chronic Local: Allergen 1.
Chronic Systemic: U.
Fire Hazard: Slight, when exposed to heat or flame; can react with oxidizing materials (Section 6).
Countermeasures
Personal Hygiene: Section 3.
Storage and Handling: Section 7.

IGNITER CORD
Shipping Regulations: Section 11.
I.C.C.: Explosive C, 150 pounds.
IATA: Explosive, explosive label, 25 kilograms (passenger), 70 kilograms (cargo).

IGNITER FUSE-METAL CLAD
Shipping Regulations: Section 11.
I.C.C.: Explosive C, 150 pounds.
IATA: Explosive, explosive label, 25 kilograms (passenger), 70 kilograms (cargo).

IGNITERS
Shipping Regulations: Section 11.
I.C.C.: Explosive C, 150 pounds.
IATA: Explosive, explosive label, 25 kilograms (passenger), 70 kilograms (cargo) (including delay-electric).

IGNITERS, AIRCRAFT ROCKET ENGINE (COMMERCIAL)
Shipping Regulations: Section 11.
IATA: Flammable solid, yellow label, not acceptable (passenger), 12 kilograms (cargo).

IGNITERS, JET THRUST (JATO)
I.C.C. (Class A explosives): Explosive A, not accepted.
I.C.C. (Class B explosives): Explosive B, red label, 550 pounds.
IATA: Explosive, not accepted (passenger and cargo).

IGNITERS, ROCKET MOTOR
I.C.C. (Class A explosives): Explosive A, not accepted.
I.C.C. (Class B explosives): Explosive B, red label, 550 pounds.
IATA: Explosive, no label, not acceptable (passenger or cargo).

ILLUMINATING GAS. See carbon monoxide and methane.

4-IMIDAZOLE ETHYLAMINE. See histamine.

3,3'-IMINO-BIS-PROPYLAMINE
General Information
Description: Colorless liquid.
Formula: $C_6H_{17}N_3$.
Constants: Mol wt: 131.23, mp: $-14.5°C$, bp: 151°C at 5 mm flash p: $> 175°F$ (C.C.), d: 0.9237 g/cc at 25°C, vap. d.: 4.52.
Hazard Analysis
Toxic Hazard Rating:
Acute Local: Irritant 2; Ingestion 2; Inhalation 2.
Acute Systemic: U.
Chronic Local: U.
Chronic Systemic: U.
Toxicity: See also amines.
Fire Hazard: Moderate, when exposed to heat or flame; can react with oxidizing materials.
Countermeasures
To Fight Fire: Foam, carbon dioxide, dry chemical, or carbon tetrachloride (Section 6).
Ventilation Control: Section 2.
Personnel Protection: Section 3.
First Aid: Section 1.
Storage and Handling: Section 7.

IMINO DIPROPIONITRILE
General Information
Description: Colorless liquid.
Formula: $HN(CH_2CH_2CN)_2$.
Constants: Mol wt: 123.16, mp: $-5.50°C$, bp: 173°C at 10 mm, d: 1.0165 at 30°C.
Hazard Analysis and Countermeasures
See nitriles.

IMPERIAL. See copper acetoarsenite.

INDALONE. See butopyronoxyl.

5-INDANOL
General Information
Formula: $C_6H_3(OH)C_3H_6$.
Constant: Mol wt: 124.
Hazard Analysis
Toxic Hazard Rating: U.
Toxicity: Limited animal experiments suggest moderate toxicity and high irritation.

1,2,3-INDANTRIONE HYDRATE. See ninhydrin.

INDENE RESINS. See coumorone-indene resin.

INDIAN ACONITE. See aconitum ferox.

INDIAN CANNABIS. See cannabis.

INDIAN HEMP. See cannabis.

INDIAN TRAGACANTH. See karaya gum.

INDIA RUBBER. See rubber crude.

TOXIC HAZARD RATING CODE (For detailed discussion, see Section 1.)

0 NONE: (a) No harm under any conditions; (b) Harmful only under unusual conditions or overwhelming dosage.

1 SLIGHT: Causes readily reversible changes which disappear after end of exposure.

2 MODERATE: May involve both irreversible and reversible changes; not severe enough to cause death or permanent injury.

3 HIGH: May cause death or permanent injury after very short exposure to small quantities.

U UNKNOWN: No information on humans considered valid by authors.

INDIUM
General Information
Description: Tetragonal, silvery-white, soft metal.
Formula: In.
Constants: At wt: 114.76, mp: 155°C, bp: 2000°C, d: 7.28 at 20°C.
Hazard Analysis
Toxic Hazard Rating:
Acute Local: Irritant 2.
Acute Systemic: Ingestion 3; Inhalation 3.
Chronic Local: U.
Chronic Systemic: U.
Toxicology: When this material is injected into animals either subcutaneously or intravenously, it is found to be highly toxic. It affects the liver, the heart, the kidneys and the blood. However, available data are scanty. Toxicity is based on animal experiments (Section 1).
Radiation Hazard: Section 5. For permissible levels, see Table 5, p. 150.
Artificial isotope ^{113m}In, half life 1.7 h. Decays to stable ^{113}In by emitting gamma rays of 0.39 MeV.
Artificial isotope ^{114m}In, half life 50 d. Decays to stable ^{114}Cd by electron capture (3.5%). Also decays to radioactive ^{114}In by emitting gamma rays of 0.19 MeV.
Artificial isotope ^{114}In, half life 72 s. Decays to stable ^{114}Sn by emitting beta particles of 1.98 MeV.
Artificial isotope ^{115m}In, half life 4.5 h. Decays to stable ^{115}Sn by emitting beta particles (5%) of 0.84 MeV. Also decays to radioactive ^{115}In by emitting gamma rays of 0.34 MeV.
Natural (96%) isotope ^{115}In, half life 6×10^{14} y. Decays to stable ^{115}Sn by emitting beta particles of 0.50 MeV.
Fire Hazard: Moderate in the form of dust when exposed to heat or flame (Section 6).
Countermeasures
Ventilation Control: Section 2.
Personnel Protection: Section 3.

INDIUM ANTIMONIDE
General Information
Description: Crystals.
Formula: InSb.
Constants: Mol wt: 236.58, mp: 535°C.
Hazard Analysis
Toxic Hazard Rating: U.
Toxicity: Animal experiments suggest low toxicity. See also antimony and indium.

INDIUM ARSENIDE
General Information
Description: Crystals.
Formula: InAs.
Constants: Mol wt: 189, mp: 943°C.
Hazard Analysis and Countermeasures
Toxicity: Details unknown. See indium and arsenic compounds.

INDIUM CYANIDE
General Information
Description: White precipitate.
Formula: In(CN)$_3$.
Constant: Mol wt: 192.81.
Hazard Analysis and Countermeasures
See cyanides.

INDIUM DIBROMIDE
General Information
Description: Pale-yellow solid.
Formula: InBr$_2$.
Constants: Mol wt: 274.6, mp: 235°C, bp: 632°C sublimes, d: 4.22 at 25°C.
Hazard Analysis and Countermeasures
See indium and bromides.

INDIUM DICHLORIDE
General Information
Description: White powder.
Formula: InCl$_2$.
Constants: Mol wt: 185.7, mp: 235°C, bp: 550–570°C, d: 3.655 at 25°C.
Hazard Analysis and Countermeasures
See indium and chlorides.

INDIUM DIHYDROGEN SULFATE
General Informat on
Description: Powder.
Formula: In$_2$(SO$_4$)$_3$·H$_2$SO$_4$·7H$_2$O.
Constants: Mol wt: 741.9, mp: $-$H$_2$SO$_4$·7H$_2$O at 250°C.
Hazard Analysis and Countermeasures
See indium and sulfates.

INDIUM DIIODIDE
General Information
Description: Solid.
Formula: InI$_2$.
Constants: Mol wt: 368.6, mp: 212°C, d: 4.71 at 25°C.
Hazard Analysis and Countermeasures
See indium and iodides.

INDIUM FLUORIDE
General Information
Description: Solid.
Formula: InF$_3$.
Constants: Mol wt: 171.8, mp: 1170 $\pm$ 10°C, bp: > 1200°C, d: 4.39 $\pm$ 0.01 at 25°C.
Hazard Analysis and Countermeasures
See indium and fluorides.

INDIUM HYDROXIDE
General Information
Description: White precipitate.
Formula: In(OH)$_3$.
Constants: Mol wt: 165.78, mp: $-$H$_2$O $<$ 150°C.
Hazard Analysis
Toxicity: See indium.

INDIUM IODATE
General Information
Description: White crystals.
Formula: In(IO$_3$)$_3$.
Constants: Mol wt: 639.52, bp: decomposes.
Hazard Analysis
Toxicity: See indium and iodates.
Fire Hazard: Moderate, by chemical reaction with reducing agents (Section 6). A powerful oxidizing agent.
Explosion Hazard: Unknown.
Disaster Hazard: Moderately dangerous; keep away from reducing agents.
Countermeasures
Storage and Handling: Section 7.

INDIUM MONOBROMIDE
General Information
Description: Red-brown crystals.
Formula: InBr.
Constants: Mol wt: 194.7, mp: 220°C, bp: 662°C sublimes, d: 4.96.
Hazard Analysis and Countermeasures
See indium and bromides.

INDIUM MONOCHLORIDE
General Information
Description: White powder.
Formula: InCl.
Constants: Mol wt: 150.2, mp: 225°C, bp: 550°C, d: 4.18 (red); 4.19 (yellow) at 25°C.
Hazard Analysis and Countermeasures
See indium and chlorides.

INDIUM MONOIODIDE
General Information
Description: Brown-red solid.
Formula: InI.
Constants: Mol wt: 241.7, mp: 351°C, bp: 714°C, d: 5.31.
Hazard Analysis
Toxicity: See indium and iodides.

INDIUM MONOXIDE
General Information
Description: Gray-white crystals.
Formula: InO.
Constant: Mol wt: 130.8.
Hazard Analysis
Toxicity: See indium.

INDIUM MONOSULFIDE
General Information
Description: Dark crystals.
Formula: InS.
Constants: Mol wt: 146.8, mp: 692°C, bp: 850°C, d: 5.18 at 25°C.
Hazard Analysis and Countermeasures
See indium and sulfides.

INDIUM PERCHLORATE
General Information
Description: Colorless, deliquescent crystals.
Formula: $In(ClO_4)_3 \cdot 8H_2O$.
Constants: Mol wt: 557.26, mp: ca. 80°C, bp: 200°C.
Hazard Analysis and Countermeasures
See indium and perchlorates.

INDIUM SELENATE
General Information
Description: Deliquescent crystals.
Formula: $In_2(SeO_4)_3 \cdot 10H_2O$.
Constant: Mol wt: 838.56.
Hazard Analysis and Countermeasures
See selenium and indium.

INDIUM SESQUIOXIDE
General Information
Description: Red-brown crystals.
Formula: In_2O_3.
Constants: Mol wt: 277.5, mp: 850°C decomposes, d: 7.179.
Hazard Analysis
Toxicity: See indium.

INDIUM SESQUISULFIDE
General Information
Description: Yellow crystals.
Formula: In_2S_3.
Constants: Mol wt: 325.7, mp: 1050°C, bp: sublimes at 850°C in high vacuum, d: 4.90 at 25°C.
Hazard Analysis and Countermeasures
See sulfides and indium.

INDIUM SUBOXIDE
General Information
Description: Black crystals.
Formula: In_2O.
Constants: Mol wt: 245.5, d: 6.99 at 25°C, mp: sublimes in vacuum at 656°C.
Hazard Analysis
Toxicity: See indium.

INDIUM SUBSULFIDE
General Information
Description: Yellow to black needles.

Formula: In_2S.
Constants: Mol wt: 261.6, mp: 653°C, d: 5.87 at 25°C.
Hazard Analysis and Countermeasures
See sulfides and indium.

INDIUM SULFATE
General Information
Description: Grayish, deliquescent powder.
Formula: $In_2(SO_4)_3$.
Constants: Mol wt: 517.7, d: 3.438.
Hazard Analysis
Toxicity: See indium and sulfates.
Disaster Hazard: Dangerous; see sulfates.
Countermeasures
Storage and Handling: Section 7.

INDIUM SULFITE, BASIC
General Information
Description: Crystals.
Formula: $2In_2O_3 \cdot 3SO_2 \cdot 8H_2O$.
Constants: Mol wt: 891.59, mp: $-3H_2O$ at 100°C; bp: $-8H_2O$ at 260°C.
Hazard Analysis and Countermeasures
See indium compounds and sulfites.

INDIUM TRIBROMIDE
General Information
Description: White to yellow needle-like crystals.
Formula: $InBr_3$.
Constants: Mol wt: 354.5, mp: 436°C, bp: sublimes, d: 4.74 at 25°C.
Hazard Analysis and Countermeasures
See indium and bromides.

INDIUM TRICHLORIDE
General Information
Description: White powder.
Formula: $InCl_3$.
Constants: Mol wt: 221.1, mp: 586°C, bp: volat. at 600°C, d: 3.46.
Hazard Analysis and Countermeasures
See indium and chlorides.

INDIUM TRIIODIDE
General Information
Description: Yellow, deliquescent crystals.
Formula: InI_3.
Constants: Mol wt: 495.5, mp: 210°C, d: 4.69.
Hazard Analysis and Countermeasures
See indium and iodides.

INDIUM TRIMETHYL
General Information
Description: Colorless crystals decomposed by water.
Formula: $In(CH_3)_3$.
Constants: Mol wt: 160.0, d: 1.568 at 19°/19°C, mp: 90°C.
Hazard Analysis
Toxicity: Details unknown. See also organometals.

INDOCENE
General Information
Constants: Bp: 171°C, flash p: 170°F (C.C.), d: 1.064.
Hazard Analysis
Toxicity: Unknown.
Fire Hazard: Moderate, when exposed to heat or flame (Section 6).
Countermeasures
Storage and Handling: Section 7.

TOXIC HAZARD RATING CODE *(For detailed discussion, see Section 1.)*

0 NONE: (a) No harm under any conditions; (b) Harmful only under unusual conditions or overwhelming dosage.

1 SLIGHT: Causes readily reversible changes which disappear after end of exposure.

2 MODERATE: May involve both irreversible and reversible changes; not severe enough to cause death or permanent injury.

3 HIGH: May cause death or permanent injury after very short exposure to small quantities.

U UNKNOWN: No information on humans considered valid by authors.

INDOLE
General Information
Synonym: 1-benzo-β-Pyrrole.
Description: Colorless to yellowish scales; fecal odor; soluble in hot water, alcohol, ether, petroleum ether; insoluble in mineral oil and glyceral.
Formula: CHCHCHCHCHCCCHCHNH.
Constants: Mol wt: 117.14, mp: 52°C, bp: 254°C.
Hazard Analysis
Toxic Hazard Rating: U.
Toxicity: Limited animal experiments suggest low toxicity.

INDOLE-α-AMINOPROPIONIC ACID. See tryptophane.

INDOLEBUTYRIC ACID. See hormodin.

2(3)-INDOLONE
General Information
Synonym: Oxindole.
Description: Colorless, needle-like crystals.
Formula: $C_6H_4NHCOCH_2$.
Constants: Mol wt: 133.1, mp: 120°C, bp: decomposes, flash p: 205–275°F, d: 1.058 at 20°/20°C.
Hazard Analysis
Toxicity: Details unknown, but probably toxic, an insecticide.
Fire Hazard: Slight, when exposed to heat or flame; can react with oxidizing materials.
Countermeasures
To Fight Fire: Water, foam, carbon dioxide, dry chemical or carbon tetrachloride (Section 6).
Storage and Handling: Section 7.

INFUSORIAL EARTH. See diatomaceous earth.

INITIATING EXPLOSIVE
Shipping Regulations: Section 11.
  I.C.C.: Explosive A, not accepted.

INK, PRINTING
General Information
Description: Liquid or paste.
Composition: Carbon, black dye, glue, gum arabic, solvent.
Hazard Analysis
Toxicity: Slight, due to solvent or dye used.
Fire Hazard: Dangerous, when exposed to heat or flame; can react with oxidizing materials(Section 6).
Explosion Hazard: Variable.
Countermeasures
Ventilation Control: Section 2.
Storage and Handling: Section 7.
Shipping Regulations: Section 11.
  I.C.C.: Flammable liquid; red label, 10 gallons.
  Coast Guard Classification: Combustible liquid.
  Coast Guard Classification: Inflammable liquid; red label.
  IATA: Flammable liquid, red label, 1 liter (passenger), 40 liters (cargo).

INSECTICIDE, LIQUEFIED GAS. See specific components.
Shipping Regulations: Section 11.
  I.C.C.: Nonflammable gas; green label, 300 pounds.
  Coast Guard Classification: Noninflammable gas; green gas label.
  IATA: Nonflammable gas, green label, 70 kilograms (passenger), 140 kilograms (cargo).

INSECTICIDES, DRY. See specific components listed on container.
Shipping Regulations: Section 11.
  I.C.C.: Poison B; poison label, 200 pounds.
  Coast Guard Classification: Poison B; poison label.
  IATA: Poison B, poison label, 25 kilograms (passenger), 95 kilograms (cargo).

INSECTICIDES, LIQUID. See specific components.
Shipping Regulations: Section 11.

  I.C.C.: Flammable liquid; red label, 10 gallons. Poison B; poison label, 55 gallons.
  Coast Guard Classification: Combustible liquid. Inflammable liquid; red label. Poison B; poison label. Poison B; (Vermin Exterminator).
  IATA (flammable liquid): Flammable liquid, red label, 1.liter (passenger), 40 liters (cargo).
    (poisonous liquid): Poison B, poison label, 1 liter (passenger), 220 liters (cargo).

INOSITOL
General Information
Synonym: Hexahydroxy cyclohexane.
Description: A constituent of body tissue. There are nine isomeric forms of which the optically inactive 1-inositol, is the one having vitamin activity. White crystals; odorless; soluble in water; insoluble in absolute alcohol and ether.
Formula: $C_6H_6(OH)_6 \cdot 2H_2O$.
Constants: Mol wt: 216, mp: 224–227°C, d: 1.524 (anhydrate).
Hazard Analysis
Toxicity: U. Used as a nutrient and/or dietary supplement food additive (Section 10).

INSOLUBLE CYANURIC ACID. See cyamelide.

INSTANTANEOUS FUSE. See explosives, high.
Shipping Regulations: Section 11.
  I.C.C.: Explosive C - No exemptions, 150 pounds.
  Coast Guard Classification: Explosive C.
  IATA: Explosive, explosive label, 25 kilograms (passenger), 70 kilograms (cargo).

INVERT SUGAR
General Information
Description: A mixture of 50% glucose and 50% fructose obtained by the hydrolysis of sucrose.
Hazard Analysis
Toxicity: Details unknown. It is a substance which migrates to food from packaging materials (Section 10).

IODATES
Hazard Analysis
Toxicity: Variable; similar to bromates and chlorates. See also specific compound.
Fire Hazard: These materials are a moderate fire hazard because they are powerful oxidizers. In contact with flammable or even combustible materials they can start fires (Section 6).
Disaster Hazard: Dangerous; see iodine compounds; they can react vigorously with reducing materials.
Countermeasures
Storage and Handling: Section 7.

IODIC ACID
General Information
Description: Colorless to pale yellow crystals.
Formula: HIO_3.
Constants: Mol wt: 175.9, mp: 110°C, d: 4.629.
Hazard Analysis
Toxic Hazard Rating:
  Acute Local: Irritant 3; Ingestion 3.
  Acute Systemic: U.
  Chronic Local: U.
  Chronic Systemic: U.
Fire Hazard: See iodates.
Explosion Hazard: Unknown.
Disaster Hazard: See Iodides.
Countermeasures
Personnel Protection: Section 3.
Storage and Handling: Section 7.

IODIC m-PERACID
General Information
Description: Rhombic, colorless to pale yellow crystals.

Formula: HIO_4.
Constants: Mol wt: 191.9, bp: sublimes at 110° C.

Hazard Analysis
Toxic Hazard Rating:
Acute Local: Irritant 3; Ingestion 3.
Acute Systemic: U.
Chronic Local: U.
Chronic Systemic: U.
Fire Hazard: See periodates.
Explosion Hazard: See periodates.
Disaster Hazard: See periodates.

Countermeasures
Personnel Protection: Section 3.
Storage and Handling: Section 7.

IODIC o-p-PERACID
General Information
Description: Powder.
Formula: H_5IO_6.
Constants: Mol wt: 228.0, mp: decomposes at 140° C.

Hazard Analysis
Toxic Hazard Rating:
Acute Local: Irritant 3; Ingestion 3.
Acute Systemic: U.
Chronic Local: U.
Chronic Systemic: U.
Fire Hazard: Moderate, by chemical reaction. A powerful oxidizer (Section 6).
Explosion Hazard: Unknown.
Disaster Hazard: Dangerous; see iodine compounds; can react vigorously with reducing materials.

Countermeasures
Personnel Protection: Section 3.
Storage and Handling: Section 7.

IODIDES
Hazard Analysis
Toxic Hazard Rating:
Acute Local: Variable.
Acute Systemic: Ingestion 2; Inhalation 2.
Chronic Local: U.
Chronic Systemic: Ingestion 2; Inhalation 2.
Toxicology: Similar in toxicity to bromides. Prolonged absorption of iodides may produce "iodism" which is manifested by skin rash, running nose, headache and irritation of mucous membranes. In severe cases the skin may show pimples, boils, redness, black and blue spots, hives and blisters (Section 9). Weakness, anemia, loss of weight and general depression may occur.
Disaster Hazard: When heated to decomposition, they can emit highly toxic fumes of iodine and iodine compounds.

Countermeasures
Ventilation Control: Section 2.
Personnel Protection: Section 3.

IODINE
General Information
Description: Rhombic, violet-black crystals, metallic luster.
Formula: I_2.
Constants: Mol wt: 253.84, mp: 112.9° C, bp: 183° C, d: 4.93, vap. press.: 1 mm at 38.7° C.

Hazard Analysis
Toxic Hazard Rating:
Acute Local: Irritant 3; Ingestion 3; Inhalation 3.
Acute Systemic: U.
Chronic Local: U.
Chronic Systemic: Ingestion 3; Inhalation 3.

TLV: ACGIH (recommended); 0.1 part per million in air; 1 milligram per cubic meter of air.
Toxicology: The effect of iodine vapor upon the body is similar to that of chlorine and bromine, but it is more irritating to the lungs. Serious exposures are seldom encountered in industry, due to the low volatility of the solid at ordinary room temperatures. Signs and symptoms are irritation and burning of the eyes, lachrymation, cough, irritation of the nose and throat.
Radiation Hazard: Section 5. For permissible levels, see Table 5, p. 150.
Artificial isotope ^{125}I, half life 60 d. Decays to stable [125]Te by electron capture. Emits gamma rays of 0.035 MeV and X-rays.
Artificial isotope [126]I, half life 13 d. Decays to stable [126]Te by emitting positrons (51%) of 0.46–1.13 MeV. Also decays to stable [126]Te by emitting beta particles of 0.38–1.25 MeV. Also emits gamma rays of 0.39–0.86 MeV.
Artificial isotope [129]I, half life 1.6×10^7y. Decays to stable [129]Xe by emitting beta particles of 0.15 MeV. Also emits gamma rays of 0.04 MeV.
Artificial isotope ^{131}I, half life 8.1 d. Decays to stable [131]Xe by emitting beta particles of 0.33 (9%), 0.61 (88%) MeV. Also emits gamma rays of 0.08–0.72 MeV.
Artificial isotope [132]I, half life 2.3 h. Decays to stable [132]Xe by emitting beta particles of 0.80–2.14 MeV. Also emits gamma rays of 0.24–2.0 MeV.
Artificial isotope [133]I, half life 21 h. Decays to radioactive ^{133}Xe by emitting beta particles of 1.22 MeV. Also emits gamma rays of 0.53 MeV.
Artificial isotope [134]I, half life 53 m. Decays to stable [134]Xe by emitting beta particles of 0.5–2.4 MeV. Also emits gamma rays of 0.12–1.8 MeV.
Artificial isotope [135]I, half life 6.7 h. Decays to radioactive [135]Xe by emitting beta particles of 0.47 (35%), 1.0 (40%), 1.4 (25%) MeV. Also emits gamma rays of 0.42–1.69 MeV.
Disaster Hazard: Dangerous; when heated, it emits highly toxic fumes of iodine and iodine compounds; can react vigorously with reducing materials.

Countermeasures
Ventilation Control: Section 2.
Personnel Protection: Section 3.
First Aid: Section 1.
Storage and Handling: Section 7.

IODINE ACIDS
Hazard Analysis
Toxic Hazard Rating:
Acute Local: Irritant 3; Ingestion 3; Inhalation 3.
Acute Systemic: U.
Chronic Local: U.
Chronic Systemic: Ingestion 3; Inhalation 3.
Explosion Hazard: Unknown.
Disaster Hazard: Dangerous; by decomposition, they can emit highly toxic iodine compounds; they react with water or steam to produce corrosive fumes.

Countermeasures
Ventilation Control: Section 2.
Personnel Protection: Section 3.
First Aid: Section 1.
Storage and Handling: Section 7.

IODINE AZIDE
General Information
Description: Yellow crystals.

TOXIC HAZARD RATING CODE (For detailed discussion, see Section 1.)

0 NONE: (a) No harm under any conditions; (b) Harmful only under unusual conditions or overwhelming dosage.

1 SLIGHT: Causes readily reversible changes which disappear after end of exposure.

2 MODERATE: May involve both irreversible and reversible changes; not severe enough to cause death or permanent injury.

3 HIGH: May cause death or permanent injury after very short exposure to small quantities.

U UNKNOWN: No information on humans considered valid by authors.

Formula: IN_3.
Constant: Mol wt: 168.94.
Hazard Analysis
Toxic Hazard Rating:
Acute Local: Irritant 3; Ingestion 3; Inhalation 3.
Acute Systemic: U.
Chronic Local: Irritant 3; Ingestion 3; Inhalation 3.
Chronic Systemic: U.
Explosion Hazard: Severe, when shocked or exposed to heat (Section 7).
Disaster Hazard: Highly dangerous; shock and heat will explode it; when heated to decomposition, it emits highly toxic fumes of iodine and iodine acids.
Countermeasures
Ventilation Control: Section 2.
Personnel Protection: Section 3.
First Aid: Section 1.
Storage and Handling: Section 7.

IODINE CYANIDE. See cyanogen iodide.

IODINE DIOXIDE. See iodine oxide.

IODINE HEPTAFLUORIDE
General Information
Description: Colorless crystals or liquid.
Formula: IF_7.
Constants: Mol wt: 259.92, mp: 5.5°C, bp: sublimes, d: 2.8 at 6°C (as liquid).
Hazard Analysis
Toxicity: Highly toxic. See iodine and fluorides.
Explosion Hazard: Moderate, when exposed to heat (Section 7).
Disaster Hazard: Dangerous; when heated to decomposition or on contact with acid or acid fumes, it emits highly toxic fumes of iodine and fluorides, and may explode; will react with water or steam to produce toxic and corrosive fumes.
Countermeasures
Ventilation Control: Section 2.
Personnel Protection: Section 3.

IODINE HYDROXIDE. See iodous hypoacid.

IODINE MONOBROMIDE
General Information
Description: Dark gray crystals.
Formula: IBr.
Constants: Mol wt: 206.84, mp: 42°C; sublimes at 50°C, bp: 116°C, d: 4.4157 at 0°C.
Hazard Analysis
Toxic Hazard Rating:
Acute Local: Irritant 3; Ingestion 3; Inhalation 3.
Acute Systemic: U.
Chronic Local: U.
Chronic Systemic: U.
Explosion Hazard: Moderate, when exposed to heat (Section 9).
Disaster Hazard: Dangerous; when heated to decomposition, it emits highly toxic fumes of bromine, iodine and their compounds; will react with water or steam to produce toxic and corrosive fumes.
Countermeasures
Ventilation Control: Section 2.
Personnel Protection: Section 3.
Storage and Handling: Section 7.

β-IODINE MONOCHLORIDE
General Information
Synonyms: α-Iodine monochloride.
Description: Reddish-brown, oily liquid.
Formula: ICl.
Constants: Mol wt: 162.4, mp: α 27°C, β 14°C.
Hazard Analysis
Toxicity: Highly toxic. See iodine and chlorine.
Explosion Hazard: Moderate when exposed to heat.

Disaster Hazard: Dangerous; when heated to decomposition, it emits highly toxic fumes of chlorine and iodine, and may explode; will react with water or steam, to produce toxic and corrosive fumes.
Countermeasures
Ventilation Control: Section 2.
Personnel Protection: Section 3.
Storage and Handling: Section 7.
Shipping Regulations: Section 11.
I.C.C.: Corrosive liquid, white label, 1 quart.
IATA: Corrosive liquid, white label, not acceptable (passenger), 1 liter (cargo).

IODINE NONOXIDE
General Information
Description: Yellow, hygroscopic powder.
Formula: I_4O_9.
Constants: Mol wt: 651.7, mp: 75°C decomposes.
Hazard Analysis
Toxicity: Highly toxic. See iodine.
Fire Hazard: Moderate, by spontaneous chemical reaction. A powerful oxidizing agent (Section 6).
Explosion Hazard: Moderate, when exposed to heat.
Disaster Hazard: Dangerous; when heated to decomposition, it emits highly toxic fumes of iodine and iodine compounds, and may explode; can react vigorously with reducing materials.
Countermeasures
Personnel Protection: Section 3.
Storage and Handling: Section 7.

IODINE OXIDE
General Information
Synonyms: Iodine dioxide; iodine tetroxide.
Description: Lemon-yellow crystals.
Formula: IO_2.
Constants: Mol wt: 158.9, mp: decomposes slowly at 75°C; rapidly at 130°C, d: 4.2 at 10°/10°C.
Hazard Analysis
Toxicity: Highly toxic. See iodine.
Fire Hazard: Moderate; by spontaneous chemical reaction. An oxidizing agent.
Explosion Hazard: Slight, when exposed to heat.
Disaster Hazard: Dangerous; when heated to decomposition, it emits highly toxic fumes of iodine and iodine compounds; can react vigorously with reducing materials.
Countermeasures
Personnel Protection: Section 3.
Storage and Handling: Section 7.

IODINE PENTABROMIDE
General Information
Description: Crystals.
Formula: IBr_5.
Constant: Mol wt: 526.5.
Hazard Analysis
Toxicity: Highly toxic. See iodine and bromine.
Fire Hazard: Dangerous, by spontaneous chemical reaction (Section 6).
Explosion Hazard: Slight, when exposed to heat or flame.
Disaster Hazard: Dangerous; when heated to decomposition, it emits highly toxic fumes of bromine, iodine and their compounds; will react with water or steam to produce toxic and corrosive fumes; can react vigorously with reducing materials.
Countermeasures
Personnel Protection: Section 3.
Storage and Handling: Section 7.

IODINE PENTAFLUORIDE
General Information
Description: Colorless liquid.
Formula: IF_5.

Constants: Mol wt: 221.9, mp: $-8.0°$C, bp: 97°C, d: 3.5, vap. press.: 10 mm at 8.5°C.

Hazard Analysis
Toxicity: Highly toxic. See iodine and fluorides.
Explosion Hazard: Unknown.
Disaster Hazard: Dangerous; when heated to decomposition or on contact with acid or acid fumes, it emits highly toxic fumes of iodine, fluorine and their compounds; will react with water or steam to produce toxic and corrosive fumes.

Countermeasures
Ventilation Control: Section 2.
Personnel Protection: Section 3.
Storage and Handling: Section 7.
Shipping Regulations: Section 11.
    IATA: Corrosive liquid, white label, not acceptable (passenger), 45 kilograms (cargo).

IODINE PENTOXIDE
General Information
Description: White crystals.
Formula: I_2O_5.
Constants: Mol wt: 333.8, mp: 300–350°C, d: 4.799 at 25°C.

Hazard Analysis
Toxicity: Highly toxic. See iodine.
Fire Hazard: Dangerous, by spontaneous chemical reaction (Section 6).
Caution: A powerful oxidizing agent.
Explosion Hazard: Slight, when exposed to heat or flame.
Disaster Hazard: Dangerous; see iodine compounds; will react with water or steam to produce toxic fumes; can react vigorously with reducing materials.

Countermeasures
Personnel Protection: Section 3.
Storage and Handling: Section 7.

IODINE TETRAOXIDE. See iodine oxide.

IODINE TRIBROMIDE
General Information
Description: Dark brown liquid.
Formula: IBr_3.
Constant: Mol wt: 366.7.

Hazard Analysis
Toxicity: See iodine and bromine.
Fire Hazard: Moderate, by spontaneous chemical reaction (Section 6).
Explosion Hazard: Unknown.
Disaster Hazard: Dangerous; when heated to decomposition or on contact with acid or acid fumes, it emits highly toxic fumes and explodes; will react with water or steam to produce toxic and corrosive fumes; can react vigorously with reducing materials.

Countermeasures
Ventilation Control: Section 2.
Personnel Protection: Section 3.
Storage and Handling: Section 7.

IODINE TRICHLORIDE
General Information
Description: Orange-yellow, deliquescent, crystalline powder; pungent, irritating odor.
Formula: ICl_3.
Constants: Mol wt: 233.29, mp: 101°C at 16 atm., bp: decomposes at 77°C, d: 3.117.

Hazard Analysis
Toxicity: Highly toxic. See iodine and chlorine.

Explosion Hazard: Slight, when exposed to heat or flame.
Disaster Hazard: Dangerous; when heated to decomposition or on contact with acid or acid fumes, it emits highly toxic fumes; will react with water or steam to produce toxic and corrosive fumes; can react with reducing materials.

Countermeasures
Personnel Protection: Section 3.
Storage and Handling: Section 7.

IODOACETONE
General Information
Synonym: 1-Iodo-2-propanone.
Formula: CH_2ICOCH_3.
Constant: Mol wt: 183.9.

Hazard Analysis
Toxicity: Details unknown. A strong irritant.
Disaster Hazard: Dangerous; when heated to decomposition, it emits highly toxic fumes.

p-IODOANISOLE. See isoform.

IODOETHANE. See ethyl iodide.

IODOETHENE. See vinyl iodide.

IODOFORM
General Information
Synonym: Triiodomethane.
Description: Yellow powder or crystals. Characteristic disagreeable odor.
Formula: CHI_3.
Constants: Mol wt: 393.8, d: 4.1, mp: 120°C (approx.).

Hazard Analysis
Toxic Hazard Rating:
    Acute Local: Irritant 2; Ingestion 2.
    Acute Systemic: Ingestion 2.
    Chronic Local: Irritant 2.
    Chronic Systemic: U.
Disaster Hazard: Dangerous; see iodides.

IODOMETHANE. See methyl iodide.

IODONITROTETRAZOLIUM
General Information
Synonym: 2-(p-Iodophenyl)-3-(p-nitrophenyl)-5-phenyl-tetrazolium chloride.
Description: Lemon yellow crystals. Slightly soluble in water.
Formula: $C_{19}H_{13}N_5O_2ICl$.
Constants: Mol wt: 505.7, mp: 239°C (decomp.).

Hazard Analysis
Toxicity: Unknown.
Disaster Hazard: Dangerous; when heated to decomposition, it emits highly toxic fumes of chlorides, iodides and oxides of nitrogen.

IODOPLATINIC ACID
General Information
Description: Monoclinic, black, deliquescent crystals.
Formula: $H_2PtI_6 \cdot 9H_2O$.
Constants: Mol wt: 1120.91, mp: $< 100°$C.

Hazard Analysis
Toxicity: Highly toxic. See iodine and platinum compounds.
Disaster Hazard: Dangerous; see iodine compounds.

Countermeasures
Storage and Handling: Section 7.

1-IODOPROPANE. See propyl iodide.

TOXIC HAZARD RATING CODE   *(For detailed discussion, see Section 1.)*

0 NONE: (a) No harm under any conditions; (b) Harmful only under unusual conditions or overwhelming dosage.

1 SLIGHT: Causes readily reversible changes which disappear after end of exposure.

2 MODERATE: May involve both irreversible and reversible changes; not severe enough to cause death or permanent injury.

3 HIGH: May cause death or permanent injury after very short exposure to small quantities.

U UNKNOWN: No information on humans considered valid by authors.

1-IODO-2-PROPANONE. See iodoacetone.

IODOPROPENE. See allyl iodide.

IODOSUCCINIMIDE
General Information
Synonym: NIS.
Description: Off-white, nearly odorless powder.
Formula: $C_4H_4O_2NI$.
Constants: Mol wt: 225.0, mp: 201° C.
Hazard Analysis
Toxicity: Unknown.
Disaster Hazard: Moderately dangerous; when heated to decomposition, or in the presence of acid or acid fumes, it can emit toxic fumes of nitrogen and iodine compounds.
Countermeasures
Storage and Handling: Section 7.

α-**IODOTOLUENE.** See benzyl iodide.

IODOUS HYPOACID
General Information
Synonym: Iodine hydroxide. Description: Yellow to greenish solution.
Formula: HOI.
Constant: Mol wt: 143.9.
Hazard Analysis and Countermeasures
Details unknown. See also iodine and iodides.

IONOL. See 2,6-di-tert-butyl-4-methylphenol.

IPC. See o-isopropyl-N-phenyl carbamate.

IPE. See isopropyl glycidyl ether.

IPECAC
General Information
Synonym: Ipecacuanha.
Description: Dried rhizome and roots of Rio or Brazilian ipecac.
Composition: Emetine, cephaeline, emetamine, ipecacuanic acid, psychotrine, methyl psychotaine, resin.
Hazard Analysis
Toxic Hazard Rating:
  Acute Local: U.
  Acute Systemic: Ingestion 3.
  Chronic Local: U.
  Chronic Systemic: Ingestion 1.
Caution: Can cause conjunctivitis with opacity of the cornea. See also emetine.
Countermeasures
Personal Hygiene: Section 3.
Storage and Handling: Section 7.

IPECACUANHA. See ipecac.

IPP. See Diisopropyl peroxydicarbonate.

IRIDIUM AND IRIDIUM COMPOUNDS.
General Information
Description: Silver-white metal.
Formula: Ir.
Constants: At wt: 193.10, mp: 2440 ± 15°C, bp: 4400°C, d: 22.421.
Hazard Analysis
Toxic Hazard Rating:
  Acute Local: U.
  Acute Systemic: Ingestion 2; Inhalation 2.
  Chronic Local: U.
  Chronic Systemic: U.
Toxicology: Soluble iridium compounds are said to be toxic. However, there are no industrial data available upon which to base a maximum allowable concentration in air.
Radiation Hazard: Section 5. For permissible levels, see Table 5, p. 150.

Artificial isotope [190]Ir, half life 12 d. Decays to stable [190]Os by electron capture. Emits gamma rays of 0.19–0.60 MeV and X-rays.
  Artificial isotope ^{192}Ir, half life 74 d. Decays to stable [192]Pt by emitting beta particles of 0.24 (8%), 0.54 (42%), 0.67 (46%) MeV. Also emits gamma rays of 0.14–0.61 MeV.
  Artificial isotope [194]Ir, half life 19 h. Decays to stable [194]Pt by emitting beta particles of 0.44–2.24 MeV. Also emits gamma rays of 0.29–1.45 MeV.
Fire Hazard: Moderate, in the form of dust when exposed to heat or flame. See also powdered metals.
Explosion Hazard: Moderate in the form of dust when exposed to flame. See also powdered metals.
Countermeasures
Ventilation Control: Section 2.
To Fight Fire: Special mixtures of dry chemical (Section 6).
Personal Hygiene: Section 3.
Shipping Regulations: Section 11. (Iridium-192).
  I.C.C.: Poison D, radioactive materials, red label, 300 curies.
  IATA: Radioactive materials, radioactive (red) label, 300 curies (passenger and cargo).

IRISH MOSS. See chondrus extract.

IROKO
General Information
Description: A wood.
Hazard Analysis
Toxic Hazard Rating:
  Acute Local: Allergen 1.
  Acute Systemic: 0.
  Chronic Local: Allergen 1.
  Chronic Systemic: 0.
Fire Hazard: Moderate, when exposed to heat or flame (Section 6).
Explosion Hazard: Moderate, in the form of dust when exposed to heat or flame (Section 7).
Countermeasures
Personal Hygiene: Section 3,

IRON
General Information
Synonym: Ferrum.
Description: Silvery-white, tenacious, lustrous, ductile metal.
Formula: Fe.
Constants: At wt: 55.8, mp: 1535°C, bp: 3000°C, d: 7.86, vap. press.: 1 mm at 1787°C.
Hazard Analysis
Toxic Hazard Rating:
  Acute Local: 0.
  Acute Systemic: Inhalation 1.
  Chronic Local: 0.
  Chronic Systemic: Inhalation 1.
TLV: ACGIH (recommended); 15 milligrams per cubic meter of air (as iron fume).
Toxicology: Iron dust can cause conjunctivitis, choroiditis, retinitis and siderosis of tissues if iron remains in these tissues. Iron ore dust can cause palpebral conjunctivitis. The mottling of the lungs which has been found in personnel who have been exposed to iron dust or iron oxide fumes in the mining or welding industries is now believed to be a benign condition, which does not predispose the worker to tuberculosis nor impair his ability to work. Also, it is thought that it takes 6 to 10 years to bring this condition about, and then only if the exposure has been severe for that period. An iron oxide fume is generated in welding operations and continued exposure to concentrations above 30 mg/cu meter of air can cause chronic bronchitis. Fresh iron oxide fume can cause metal fume fever.
Radiation Hazard: Section 5. For permissible levels, see Table 5, p. 150.

Artificial isotope ^{55}Fe, half life 2.7 y. Decays to stable ^{55}Mn by electron capture. Emits X-rays.

Artificial isotope ^{59}Fe, half life 45 d. Decays to stable ^{59}Co by emitting beta particles of 0.27 (46%), 0.46 (54%) MeV. Also emits gamma rays of 0.19, 1.10, 1.29 MeV.

Fire Hazard: Moderate in the form of dust when exposed to heat or flame. See also powdered metals.

Explosion Hazard: Moderate, in the form of dust when exposed to heat or flame. See also powdered metals.

Countermeasures
Personal Hygiene: Section 3.
To Fight Fire: Special mixtures of dry chemical (Section 6).

IRON AMMONIUM CITRATE
General Information
Synonym: Ferric ammonium citrate.
Description: Thin, transparent, garnet red scales or granules, or as a brownish yellow powder; odorless or slight ammonia odor; soluble in water; insoluble in alcohol.
Hazard Analysis
Toxicity: Unknown. Used as a trace mineral added to animal feeds (Section 10).

IRON ARSENIDE. See ferric arsenide.

IRON BORIDE
General Information
Description: Gray crystals.
Formula: FeB.
Constant: Mol wt: 66.67.
Hazard Analysis
Toxicity: Details unknown.
Fire Hazard: Moderate. Borides can react with moisture and acids to evolve toxic boron hydrides.
Explosion Hazard: A possible explosion hazard.
Disaster Hazard: Dangerous. Can react with water, steam or acids to evolve toxic and flammable fumes.
Countermeasures
Storage and Handling: Section 7.

IRON CARBONATE
Hazard Analysis
Toxicity: Unknown. Used as a trace mineral added to animal feeds (Section 10).

IRON CARBONYL. See iron pentacarbonyl.

IRON COMPOUNDS. See corresponding ferric and/or ferrous compound.

IRON DIARSENIDE. See ferric diarsenide.

IRON FLUORIDE. See ferric fluoride.

IRON FORMATE. See ferric formate.

IRON GLUCONATE. See ferrous gluconate.

IRON LACTATE. See ferrous lactate.

IRON MASS, SPENT. See iron sponge, spent.

IRON MONOPHOSPHIDE
General Information
Description: Rhombic crystals.
Formula: FeP.
Constants: Mol wt: 86.8, d: 5.2 at 20°C.
Hazard Analysis and Countermeasures
See phosphides.

IRON NONACARBONYL
General Information
Description: Orange, hexagonal crystals.
Formula: $Fe_2(CO)_9$.
Constants: Mol wt: 363.8, mp: decomp. at 100°C, d: 2.805 at 18°C.
Hazard Analysis and Countermeasures
See carbon monoxide and carbonyls.

IRON OXIDE. See ferrous oxide.
Shipping Regulations: Section 11.
Coast Guard Classification: Hazardous material (Wet).

IRON PENTACARBONYL
General Information
Synonym: Iron carbonyl.
Description: Yellow to dark red liquid.
Formula: $Fe(CO)_5$.
Constants: Mol wt: 195.9, mp: −21°C, bp: 105.0°C, flash p: 5°F, d: 1.453 at 25°/4°C, vap. press.: 40 mm at 30.3°C.
Hazard Analysis
Toxic Hazard Rating:
Acute Local: Irritant 1; Inhalation 2.
Acute Systemic: Ingestion 3; Inhalation 3; Skin Absorption 3.
Chronic Local: U.
Chronic Systemic: U.
Toxicology: Inhalation of this material causes dizziness, nausea and vomiting. If continued, unconsciousness follows. Often there is a delayed reaction of chest pain, cough and difficult breathing. There may be cyanosis and circulatory collapse. In fatal cases death occurs from the fourth to eleventh day with pneumonitis and injury to kidneys, liver and brain. Iron carbonyl is less toxic than nickel carbonyl.
Fire Hazard: Dangerous. See carbonyls.
Explosion Hazard: Moderate. See carbonyls.
Disaster Hazard: See carbonyls.
Countermeasures
Ventilation Control: Section 2.
To Fight Fire: Water, foam, carbon dioxide, dry chemical or carbon tetrachloride (Section 6).
Personnel Protection: Section 3.
Storage and Handling: Section 7.

IRON PHOSPHATE. See ferric phosphate.

IRON PYROPHOSPHATE. See ferric pyrophosphate.

IRON REDUCED
General Information
Synonym: Ferrum reductum.
Description: Grayish-black, amorphous, fine granular powder.
Hazard Analysis
Toxicity: Unknown. Used as a trace mineral added to animal feeds and as a nutrient and/or dietary supplement food additive. It migrates to food from packaging materials (Section 10).

IRON SPONGE (NOT PROPERLY OXIDIZED)
Hazard Analysis
Toxicity: Details unknown. See also iron.
Fire Hazard: Dangerous, when exposed to heat or flame; it can react vigorously with oxidizing materials (Section 6).
Countermeasures
Storage and Handling: Section 7.

TOXIC HAZARD RATING CODE (For detailed discussion, see Section 1.)

0 NONE: (a) No harm under any conditions; (b) Harmful only under unusual conditions or overwhelming dosage.

1 SLIGHT: Causes readily reversible changes which disappear after end of exposure.

2 MODERATE: May involve both irreversible and reversible changes; not severe enough to cause death or permanent injury.

3 HIGH: May cause death or permanent injury after very short exposure to small quantities.

U UNKNOWN: No information on humans considered valid by authors.

Shipping Regulations: Section 11.
I.C.C.: Flammable solid; yellow label—not accepted.
Coast Guard Classification: Inflammable solid; yellow label.
IATA: Flammable solid, not acceptable (passenger and cargo).

IRON SPONGE, SPENT
General Information
Composition: Iron + other materials.
Hazard Analysis
Toxicity: Details unknown. See also iron.
Fire Hazard: Dangerous, when exposed to heat or flame; it can react vigorously with oxidizing materials (Section 6).
Countermeasures
Storage and Handling: Section 7.
Shipping Regulations: Section 11.
I.C.C.: Flammable solid; yellow label—not accepted.
Coast Guard Classification: Inflammable solid; (Wet) hazardous material.
IATA: Flammable solid, not acceptable (passenger and cargo).

IRON SULFATE. See ferrous sulfate.

IRON TETRACARBONYL
General Information
Description: Dark green, lustrous crystals.
Formula: $Fe(CO)_4$.
Constants: Mol wt: 167.88, mp: decomp. at 140–150°C, d: 1.996 at 18°C.
Hazard Analysis and Countermeasures
See carbonyls and iron pentacarbonyl.

IRON m-VANADATE. See ferric m-vanadate.

ISANO OIL
General Information
Description: Fatty oil from African tree of same name.
Constant: d: 1.0−.
Hazard Analysis
Toxic Hazard Rating: U.
Explosion Hazard: Dangerous; exothermic reaction above 502°F may become explosive.

ISETHONIC ACID. See hydroxyethyl sulfonic acid.

ISOAMYL ACETATE
General Information
Synonyms: Amyl acetate, iso; banana oil.
Description: Liquid, banana-like odor.
Formula: $CH_3COO(C_5H_{11})$.
Constants: Mol wt: 130.2, bp: 142.0°C, ulc: 55–60, lel: 1% at 212°F, uel: 7.5%, flash p: 77°F, d: 0.876, autoign. temp.: 715°F, vap. d.: 4.49.
Hazard Analysis
Toxic Hazard Rating:
Acute Local: Irritant 2; Ingestion 2; Inhalation 2.
Acute Systemic: Ingestion 2; Inhalation 2.
Chronic Local: Irritant 2.
Chronic Systemic: U.
Toxicology: Exposure to concentrations of about 1,000 ppm for 1 hour can cause headache, fatigue and pulmonary irritation.
TLV: ACGIH (recommended) 100 parts per million of air; 525 milligrams per cubic meter of air.
Fire Hazard: Dangerous; when exposed to heat or flame; can react vigorously with reducing materials (Section 6).
Explosion Hazard: Moderate, when exposed to heat or flame.
Disaster Hazard: Dangerous; keep away from heat and open flame!
Countermeasures
Ventilation Control: Section 2.

To Fight Fire: Alcohol foam, carbon dioxide, dry chemical or carbon tetrachloride (Section 6).
Personnel Protection: Section 3.
Storage and Handling: Section 7.

ISOAMYL ALCOHOL, PRIMARY
General Information
Synonyms: Isobutyl carbinol; fusel oil.
Description: Clear liquid.
Formula: $(CH_3)_2CHCH_2CH_2OH$.
Constants: Mol wt: 88.15, bp: 132°C, ulc: 35–40, lel: 1.2%, uel: 9.0% at 212°F, flash p: 109°F (C.C.), d: 0.813, autoign. temp.: 675°F, vap. d.: 3.04.
Hazard Analysis
Toxicity: See amyl alcohol.
TLV: ACGIH (recommended); 100 parts per million in air; 360 milligrams per cubic meter of air.
Fire Hazard: Moderate, when exposed to heat or flame; can react vigorously with reducing materials.
Explosion Hazard: Slight, when exposed to flame.
Countermeasures
Ventilation Control: Section 2.
To Fight Fire: Alcohol foam, carbon dioxide, dry chemical or carbon tetrachloride (Section 6).
Personnel Protection: Section 3.
Storage and Handling: Section 7.

ISOAMYL ALCOHOL, SECONDARY
General Information
Synonym: Isobutyl carbinol.
Formula: $(CH_3)_2CHCH(OH)CH_3$.
Constants: Mol wt: 88.15, bp: 113°C, flash p: 103°F (C.C.), d: 0.819, vap. d.: 3.04, autoign. temp.: 657°F, lel: 1.2%, uel: 9.0%.
Hazard Analysis
Toxicity: See amyl alcohol.
TLV: ACGIH (recommended); 100 parts per million in air; 360 milligrams per cubic meter of air.
Fire Hazard: Moderate, when exposed to heat or flame; can react vigorously with reducing materials (Section 6).
Explosion Hazard: Unknown.
Countermeasures
Ventilation Control: Section 2.
To Fight Fire: Alcohol foam, carbon dioxide, dry chemical or carbon tetrachloride (Section 6).
Personnel Protection: Section 3.
Storage and Handling: Section 7.

ISOAMYL AMINE
General Information
Formula: $CH_3C_4H_8NH_2$.
Constants: Mol wt: 87.
Hazard Analysis
Toxic Hazard Rating:
Acute Local: Irritant 2; Ingestion 2; Inhalation 2.
Acute Systemic: Ingestion 2.
Chronic Local: U.
Chronic Systemic: U.
Toxicity: Intravenous injection in man has caused flushing and excitability.

ISOAMYL BROMIDE
General Information
Description: Colorless liquid.
Formula: $(CH_3)_2CHC_2H_4Br$.
Constants: Mol wt: 151.1, bp: 119.7–121.8°C, flash p: 90°F, mp: −112°C, d: 1.208 at 25°/25°C, vap. d.: 5.25.
Hazard Analysis
Toxicity: Details unknown. Probably an irritant and narcotic in high concentrations. See also bromine compounds.
Fire Hazard: Moderate, when exposed to heat or flame.
Disaster Hazard: Dangerous; when heated to decomposition, it emits toxic fumes of bromides; can react with oxidizing materials.

Countermeasures
Ventilation Control: Section 2.
To Fight Fire: Water, foam, carbon dioxide, dry chemical or carbon tetrachloride (Section 6).
Storage and Handling: Section 7.

ISOAMYL BUTYRATE
General Information
Description: Water-white liquid.
Formula: $C_5H_{11}COOC_3H_7$.
Constants: Mol wt: 158.23, bp: 178.6°C, d: 0.866–0.868 at 15.5°C, vap. press.: 1 mm at 21.2°C.
Hazard Analysis
Toxicity: Slight. See esters.
Fire Hazard: Moderate, when exposed to heat or flame (Section 6).
Countermeasures
Storage and Handling: Section 7.

ISOAMYL DICHLOROARSINE
General Information
Description: Oily liquid, somewhat agreeable odor, decomposed by water.
Formula: $C_5H_{11}AsCl_2$.
Constants: Mol wt: 217.0, bp: 88.5–91.5°C.
Hazard Analysis and Countermeasures
See arsenic compounds.

α-ISOAMYLENE. See 3-methyl butene-1.

β-ISOAMYLENE. See 2-methyl butene-2.

ISOAMYL FORMATE
General Information
Description: Clear liquid.
Formula: $C_5H_{11}OOCH$.
Constants: Mol wt: 116.16, bp: 123.3°C, d: 0.87 at 20°C, vap. press.: 10 mm at 17.1°C.
Hazard Analysis
Toxic Hazard Rating:
  Acute Local: Irritant 3; Ingestion 3; Inhalation 3.
  Acute Systemic: Ingestion 3; Inhalation 3.
  Chronic Local: U.
  Chronic Systemic: Ingestion 2; Inhalation 2.
Toxicology: This material is very irritating and can cause narcosis. The symptoms are usually transient in nature, but it is possible upon severe or prolonged exposure to have serious consequences.
Fire Hazard: Moderate, when exposed to heat or flame; it can react with oxidizing materials (Section 6).
Countermeasures
Ventilation Control: Section 2.
Personnel Protection: Section 3.
Storage and Handling: Section 7.

ISOAMYL HYDRIDE. See isopentane.

3-ISOAMYL-5-(METHYLENEDIOXYPHENYL)-2-CYCLOHEXENONE. See piperonyl cyclonene.

ISOAMYL NITRITE. See amyl nitrite.

ISOBORNEOL
General Information
Description: A geometrical isomer of borneol. White solid; camphor odor; more soluble in most solvents than borneol.
Formula: $C_{10}H_{17}OH$.
Constants: Mol wt: 154.24, mp: 216°C (sublimes).

Hazard Analysis
See borneol.

ISOBORNYL THIOCYANOACETATE
General Information
Synonym: Thanite.
Description: Yellow, oily liquid.
Formula: $C_{13}H_{19}NO_2S$.
Constants: Mol wt: 253.4, d: 1.1465 at 25°/4°C.
Hazard Analysis
Toxic Hazard Rating:
  Acute Local: Irritant 2; Ingestion 1; Inhalation 1.
  Acute Systemic: Ingestion 1.
  Chronic Local: Irritant 1.
  Chronic Systemic: Ingestion 1.
Toxicity: Animal experiments suggest relatively low toxicity. Highly irritating to eyes, mucous membrane, and skin. See also thiocyanates. An insecticide and fly spray.
Disaster Hazard: See thiocyanates.
Countermeasures
Storage and Handling: Section 7.

ISOBUTANE
General Information
Synonyms: 2-Methyl propane; trimethyl methane.
Description: Colorless gas.
Formula: C_4H_{10}.
Constants: Mol wt: 58.12, bp: −11.7°C, lel: 1.9%, uel: 8.5%, fp: −160°C, d: 0.5572 at 20°C, autoign. temp.: 864°F, vap. d.: 2.01.
Hazard Analysis
Toxic Hazard Rating:
  Acute Local: 0.
  Acute Systemic: Inhalation 1.
  Chronic Local: 0.
  Chronic Systemic: U.
Caution: An asphyxiant. A common air contaminant.
Fire Hazard: Dangerous, when exposed to heat or flame (Section 6).
Explosion Hazard: Severe, when exposed to heat or flame.
Disaster Hazard: Dangerous; on contact with oxidizing materials, it can react vigorously.
Countermeasures
Ventilation Control: Section 2.
To Fight Fire: Carbon dioxide, dry chemical or water spray (Section 6).

ISOBUTANOL. See isobutyl alcohol.

ISOBUTANOLAMINE. See 2-amino-2-methyl-1-n-propanol.

ISOBUTENE. See isobutylene.

ISOBUTYL ACETATE
General Information
Description: Colorless, neutral liquid, fruit-like odor.
Formula: $C_4H_9OOCCH_3$.
Constants: Mol wt: 116.10, mp: −98.9°C, bp: 118.0°C, flash p: 64°F (C.C.), d: 0.8685 at 15°C, vap. press.: 10 mm at 12.8°C, autoign. temp.: 793°F, vap. d.: 4.0.
Hazard Analysis
Toxic Hazard Rating:
  Acute Local: Irritant 2; Ingestion 2; Inhalation 2.
  Acute Systemic: Ingestion 2; Inhalation 2.
  Chronic Local: U.
  Chronic Systemic: U.
TLV: ACGIH (recommended) 150 parts per million of air; 700 milligrams per cubic meter of air.

TOXIC HAZARD RATING CODE (For detailed discussion, see Section 1.)

0 NONE: (a) No harm under any conditions; (b) Harmful only under unusual conditions or overwhelming dosage.

1 SLIGHT: Causes readily reversible changes which disappear after end of exposure.

2 MODERATE: May involve both irreversible and reversible changes; not severe enough to cause death or permanent injury.

3 HIGH: May cause death or permanent injury after very short exposure to small quantities.

U UNKNOWN: No information on humans considered valid by authors.

Caution: Upon absorption it can hydrolyze to acetic acid and isobutanol.

Fire Hazard: Dangerous, when exposed to heat or flame (Section 6).

Spontaneous Heating: No.

Disaster Hazard: Dangerous! Keep away from heat and open flame!

Countermeasures

Ventilation Control:Section 2.

To Fight Fire: Alcohol foam, carbon dioxide, dry chemical or carbon tetrachloride (Section 6).

Personnel Protection: Section 3.

Storage and Handling: Section 7.

Shipping Regulations: Section 11.

   IATA: Flammable liquid, red label, 1 liter (passenger), 40 liters (cargo).

ISOBUTYL ALCOHOL
General Information

Synonyms: Isopropylcarbinol; 2-methyl-propanol-1;iso-butanol.

Description: Clear liquid, sweet odor.

Formula: $(CH_3)_2CHCH_2OH$.

Constants: Mol wt: 74.12, bp: 107.90°C, flash p: 82°F, ulc: 40-45, lel: 1.7%, uel = 10.9% at 212°F, fp: -108°C, d: 0.805 at 20°/4°C, autoign. temp.: 800°F, vap. press.: 10 mm at 21.7°C, vap. d.: 2.55.

Hazard Analysis

Toxic Hazard Rating:

   Acute Local: Irritant 3; Ingestion 3; Inhalation 3.

   Acute Systemic: Ingestion 2; Inhalation 2.

   Chronic Local: U.

   Chronic Systemic: U.

TLV: ACGIH (tentative) 100 parts per million of air; 300 milligrams per cubic meter of air.

Fire Hazard: Moderate, when exposed to heat or flame.

Explosion Hazard: Moderate, in the form of vapor when exposed to heat or flame.

Disaster Hazard: Moderately dangerous; emits toxic fumes when heated. Keep away from heat and open flame.

Countermeasures

Ventilation Control: Section 2.

To Fight Fire: Foam, carbon dioxide, dry chemical or carbon tetrachloride (Section 6).

Personnel Protection: Section 3.

Storage and Handling: Section 7.

ISOBUTYL ALDEHYDE
General Information

Synonyms: Isobutyraldehyde; 2-methylpropanol.

Description: Transparent, colorless, highly refractive liquid; pungent odor.

Formula: $(CH_3)_2CHCHO$.

Constants: Mol wt: 72.10, mp: -65.0°C, bp: 64°C, flash p: -40°F (C.C.), d: 0.7938 at 20°/4°C, autoign. temp: 490°F, lel: 1.6%, uel: 10.6%, vap. d.: 2.5.

Hazard Analysis

Toxicity: See aldehydes.

Fire Hazard: Dangerous, when exposed to heat or flame (Section 6).

Explosion Hazard: Moderate; in the form of vapor when exposed to heat or flame.

Disaster Hazard: Dangerous; keep away from heat and open flame; can react vigorously with reducing materials.

Countermeasures

Ventilation Control: Section 2.

To Fight Fire: Foam, carbon dioxide, dry chemical or carbon tetrachloride (Section 6).

Storage and Handling: Section 7.

ISOBUTYLAMINE
General Information

Synonym: 1-Amino-2-methylpropane.

Description: Colorless liquid.

Formula: $(CH_3)_2CHCH_2NH_2$.

Constants: Mol wt: 73.14, mp: -85.5°C, bp: 68.6°C, flash p: 15°F, d: 0.731 at 20°/20°C, vap. press.: 100 mm at 18.8°C, autoign. temp.: 712°F, vap. d.: 2.5.

Hazard Analysis

Toxic Hazard Rating:

   Acute Local: Irritant 3; Ingestion 3; Inhalation 2.

   Acute Systemic: Inhalation 2.

   Chronic Local: U.

   Chronic Systemic: U.

Toxicity: Skin contact can cause blistering. Inhalation can cause headache and dryness of nose and throat. See also amines.

Fire Hazard: Dangerous, when exposed to heat or flame (Section 6).

Spontaneous Heating: No.

Disaster Hazard: Dangerous; keep away from heat and open flame; can react vigorously with oxidizing materials.

Countermeasures

Ventilation Control: Section 2.

To Fight Fire: Dry chemical, foam, carbon dioxide or carbon tetrachloride (Section 6).

Storage andd Handling: Section 7.

Shipping Regulations: Section 11.

   IATA: Flammable liquid, red label, 1 liter (passenger), 40 liters (cargo).

ISOBUTYL BENZENE
General Information

Description: Liquid.

Formula: $C_4H_9C_6H_5$.

Constants: Mol wt: 134.23, mp: -51.5°C, bp: 172.8°C, flash p: 126°F (C.C.), d: 0.867, autoign. temp.: 802°F, vap. press.: 1 mm at 14.1°C, vap. d.: 4.62, lel: 0.8%, uel: 6.0%.

Hazard Analysis

Toxicity: Details unknown. Probably irritant and narcotic in high concentrations.

Fire Hazard: Moderate, when exposed to heat or flame; can react with oxidizing materials.

Countermeasures

To Fight Fire: Foam, carbon dioxide, dry chemical or carbon tetrachloride (Section 6).

Ventilation Control: Section 2.

Storage and Handling: Section 7.

ISOBUTYLBORIC ACID
General Information

Description: Colorless crystals; soluble in water.

Formula: $C_4H_9B(OH)_2$.

Constants: Mol wt: 102.0, mp: 106-112°C (decomp.).

Hazard Analysis

Toxicity: Details unknown. See also boron compounds.

ISOBUTYL BROMIDE
General Information

Description: Colorless liquid.

Formula: $(CH_3)_2CHCH_2Br$.

Constants: Mol wt: 137.0, mp: -118.5°C, bp: 91-93°C, d: 1.258 at 25°/25°C, vap. d.: 4.76.

Hazard Analysis

Toxicity: Details unknown. See bromides.

Fire Hazard: Moderate, when exposed to heat or flame.

Disaster Hazard: Moderately dangerous; when heated to decomposition, it emits highly toxic fumes of bromine; can react vigorously with oxidizing materials.

Countermeasures

Ventilation Control: Section 2.

To Fight Fire: Water, foam, carbon dioxide, dry chemical or carbon tetrachloride (Section 6).

Storage and Handling: Section 7.

ISOBUTYL CARBINOL. See isoamyl alcohol.

ISOBUTYL CHLORIDE
General Information
Formula: $(CH_3)_2CHCH_2Cl$.
Constants: Mol wt: 92.5, lel: 2.0%, uel: 8.8%, d: 0.9, vap. d.: 3.2, bp: 156° F.
Hazard Analysis
Toxic Hazard Rating: U.
Fire Hazard: Moderate when exposed to heat or flame.
Disaster Hazard: Dangerous. See chlorides.

n-ISOBUTYL-2,6-8-DECATRIENEAMIDE
Hazard Analysis
Toxic Hazard Rating: U.
Toxicity: An insecticide. Limited data suggest low toxicity.

ISOBUTYLENE
General Information
Synonyms: Isobutene; 2-methylpropene.
Description: Volatile liquid or easily liquefied gas.
Formula: $(CH_3)_2CCH_2$.
Constants: Mol wt: 56.1, bp: −6.9° C, fp: −140.3° C, flash p: <20° F, d: 0.600, autoign. temp.: 869° F, vap. press.: 3290 mm at 40.5° C, vap. d.: 1.94, lel: 1.8%, uel: 8.8%.
Hazard Analysis
Toxicity: Details unknown. May have asphyxiant or narcotizing action.
Fire Hazard: Dangerous; when exposed to heat or flame (Section 6).
Disaster Hazard: Dangerous; keep away from heat and open flame; can react vigorously with oxidizing materials.
Countermeasures
Ventilation Control: Section 2.
Personnel Protection: Section 3.
Storage and Handling: Section 7.

ISOBUTYL ISOBUTYRATE
General Information
Description: Liquid with fruity odor.
Formula: $(CH_3)_2CHCOOCH_2CH(CH_3)_2$.
Constants: Mol wt: 144.2, mp: −80.7° C, bp: 147.5° C, d: 0.850–0.860 at 20°/20° C, vap. press.: 10 mm at 39.9° C, vap. d.: 4.97.
Hazard Analysis
Toxicity: Details unknown; an insect repellent. See also isobutyl alcohol.
Fire Hazard: Slight, when exposed to heat or flame (Section 6), can react with oxidizing materials.
Countermeasures
Storage and Handling: Section 7.

ISOBUTYLMETHYL   CARBINOL. See methylamyl alcohol.

ISOBUTYL METHYL KETONE
General Information
Synonyms: Methyl isobutyl ketone; hexone; 4-methyl-2-pentanone; isopropylacetone.
Description: Clear liquid.
Formula: $(CH_3)_2CHCH_2COCH_3$.
Constants: Mol wt: 100.2, bp: 118° C, lel: 1.3% at 122° F, uel: 8.0% at 212° F, flash p: 73° F, d: 0.803, fp: −80.3° C, autoign. temp.: 858° F, vap. press.: 16 mm at 20° C, vap. d.: 3.45.
Hazard Analysis
Toxic Hazard Rating:
Acute Local: Irritant 2; Ingestion 2; Inhalation 2—

Acute Systemic: Ingestion 3; Inhalation 3.
Chronic Local: U.
Chronic Systemic: U.
Toxicology: Irritating to eyes and mucous membranes. Narcotic in high concentrations. See also ketones.
TLV: ACGIH (recommended), 100 parts per million in air; 409 milligrams per cubic meter of air.
Fire Hazard: Dangerous; when exposed to heat or flame.
Explosion Hazard: Moderate, in the form of vapor when exposed to heat or flame.
Disaster Hazard: Dangerous! Keep away from heat and open flame; can react vigorously with reducing materials.
Countermeasures
Ventilation Control: Section 2.
To Fight Fire: Foam, carbon dioxide, dry chemical or carbon tetrachloride (Section 6).
Personnel Protection: Section 3.
Storage and Handling: Section 7.

ISOBUTYL VINYL ETHER
General Information
Description: Liquid.
Formula: $C_4H_9OCHCH_2$.
Constants: Mol wt: 100.2, mp: −112° C, bp: 82.9–83.2° C, flash p: 20° F, d: 0.76 at 25°/4° C, vap. d.: 3.45.
Hazard Analysis
Toxicity: Details unknown. Limited animal experiments suggest low toxicity. See ethers.
Fire Hazard: Dangerous, when exposed to heat or flame. See ethers.
Explosion Hazard: Severe, when exposed to sparks or open flame. See ethers.
Disaster Hazard: Dangerous! Keep away from heat and open flame; can react vigorously with oxidizing materials.
Countermeasures
Ventilation Control: Section 2.
To Fight Fire: Foam, carbon dioxide, dry chemical or carbon tetrachloride (Section 6).
Storage and Handling: Section 7.

ISOBUTYRALDEHYDE. See isobutyl aldehyde.

ISOBUTYRIC ACID
General Information
Description: Colorless liquid, pungent odor.
Formula: $(CH_3)_2CHCOOH$.
Constants: Mol wt: 88.1, mp: −47° C, bp: 154.5° C, flash p: 132° F, d: 0.949 at 20°/4° C, vap. press.: 1 mm at 14.7° C, vap. d.: 3.04, autoign. temp.: 935° F.
Hazard Analysis
Toxic Hazard Rating:
Acute Local: Irritant 2.
Acute Systemic: Ingestion 2.
Chronic Local: Irritant 2.
Chronic Systemic: U.
Data based on limited animal experiments.
Fire Hazard: Moderate, when exposed to heat or flame; can react with oxidizing materials.
Countermeasures
To Fight Fire: Foam, carbon dioxide, dry chemical or carbon tetrachloride (Section 6).
Storage and Handling: Section 7.

ISOBUTYRIC ANHYDRIDE
General Information
Description: Liquid, decomposes in water.

TOXIC HAZARD RATING CODE (For detailed discussion, see Section 1.)

0 NONE: (a) No harm under any conditions; (b) Harmful only under unusual conditions or overwhelming dosage.

1 SLIGHT: Causes readily reversible changes which disappear after end of exposure.

2 MODERATE: May involve both irreversible and reversible changes; not severe enough to cause death or permanent injury.

3 HIGH: May cause death or permanent injury after very short exposure to small quantities.

U UNKNOWN: No information on humans considered valid by authors.

Formula: [(CH₃)₂CHCO]₂O.

Formula: $[(CH_3)_2CHCO]_2O$.

Constants: Mol wt: 158.19, bp: 360°F, d: 0.951–0.956 at 20/20°C, vap. d.: 5.5, flash p: 139°F, autoign. temp.: 665°F.

Hazard Analysis

Toxic Hazard Rating: U.

Fire Hazard: Moderate, when exposed to heat or flame.

Countermeasures

Storage and Handling: Section 7.

To Fight Fire: Foam, fog, dry chemical (Section 6).

ISOBUTYRIC ETHER. See ethyl isobutyrate.

ISOBUTYRONITRILE

General Information

Synonyms: 2-Methyl propanenitrile; isopropyl cyanide.

Description: Colorless liquid; slightly soluble in water; very soluble in alcohol and ether.

Formula: $(CH_3)_2CHCN$.

Constants: Mol wt: 69, d: 0.773 at 20/20°C, bp: 107°C, mp: −75°C.

Hazard Analysis

Toxic Hazard Rating: U.

Toxicity: Animal experiments show high toxicity. See cyanides and nitriles.

Disaster Hazard: Dangerous. See nitriles.

ISOCIL. See 5-bromo-3-isopropyl-6-methyl uracil.

ISOCINCHOMERONIC ACID

General Information

Description: Off-white powder.

Formula: $C_5H_3N(COOH)_2$.

Constants: Mol wt: 167.12, mp: 240°C decomposes.

Hazard Analysis

Disaster Hazard: Slight; when heated it emits acrid fumes.

Toxicity: Unknown.

Countermeasures

Storage and Handling: Section 7.

ISOCROTONIC ACID

General Information

Synonyms: 2-Butenoic acid, β-methacrylic acid.

Description: The cis-isomeric form of crotonic acid.

Formula: $CH_3CH:CHCOOH$.

Constants: Mol wt: 86, mp: 15°C.

Hazard Analysis

Toxic Hazard Rating:

Acute Local: Irritant 3; Ingestion 3; Inhalation 3.

Acute Systemic: U.

Chronic Local: U.

Chronic Systemic: U.

ISOCROTYL CHLORIDE. See β, β-dimethyl vinyl chloride.

ISOCYANIC ACID. See cyanic acid.

ISODECALDEHYDE

General Information

Description: Insoluble in water.

Formula: $C_9H_{19}CHO$.

Constants: Mol wt: 156, d: 0.829, bp: 197.0°C, vap. d.: 5.4, flash p: 185°F (O.C.).

Hazard Analysis

Toxic Hazard Rating: U. See also aldehydes.

Fire Hazard: Moderate, when exposed to heat or flame.

ISODECANOL (MIXED ISOMERS).

General Information

Description: Insoluble in water.

Formula: $C_{10}H_{21}OH$.

Constants: Mol wt: 158.3, d: 0.8395, bp: 220°C, flash p: 220°F (O.C.), vap. d.: 5.5.

Hazard Analysis

Toxic Hazard Rating: U.

Fire Hazard: Moderate, when exposed to heat or flame.

Countermeasures

Storage and Handling: Section 7.

To Fight Fire: Water, foam (Section 6).

ISODECANOIC ACID

General Information

Description: Liquid; very slightly soluble in water.

Formula: $C_9H_{19}COOH$.

Constants: Mol wt: 172, bp: 254°C, d: 0.9019 at 20/20°C, fp: glass below −60°C, vap. d.: 5.9, flash p: <300°F (O.C.).

Hazard Analysis

Toxic Hazard Rating: U.

Fire Hazard: Slight, when exposed to heat or flame.

Countermeasures

Storage and Handling: Section 7.

To Fight Fire: Water, foam (Section 6).

ISODECYL ADIPATE. See dodecyl adipate.

ISODECYL OCTYL ADIPATE

General Information

Description: Liquid, low viscosity, water white.

Formula: $(C_{10}H_{21})(C_8H_{17})(CO_2)_2(CH_2)_4$.

Constants: Mol wt: 398.6, mp: −60°C, bp: 215–240°C at 4 mm, flash p: 400°F, d: 0.924 at 20°/20°C, vap. d.: 13.8.

Hazard Anslysis

Toxicity: Details unknown. See esters.

Fire Hazard: Slight, when exposed to heat or flame; can react with oxidizing materials.

Countermeasures

To Fight Fire: Foam, carbon dioxide, dry chemical or carbon tetrachloride (Section 6).

Personal Hygiene: Section 3.

Storage and Handling: Section 7.

ISODRIN. See 1,2,3,4,10,10-hexachloro-1,4,4a,5,8,8a-hexahydro-1,4,5,8-endo-endo-dimethanonaphthalene.

Shipping Regulations: Section 11.

IATA: Poison B, poison label, 25 kilograms (passenger), 95 kilograms (cargo).

ISOFORM

General Information

Synonym: p-Iodoanisole.

Description: White powder.

Formula: $C_6H_4(OCH_3)IO_2$.

Constant: Mol wt: 265.98.

Hazard Analysis

Toxicity: Unknown. See iodides.

Disaster Hazard: Moderate, see iodides; explodes at 225°C.

Countermeasures

Storage and Handling: Section 7.

ISOHEPTANE. See 2-methylhexane.

ISOHEXANE

General Information

Synonym: Mixture of hexane isomers.

Description: Liquid.

Formula: C_6H_{14}.

Constants: Mol wt: 86.17, lel: 1.0%, uel: 7.0%, bp: 54–60°C, flash p: < −20°F (C.C.), d: 0.669, vap. d.: 3.00.

Hazard Analysis

Toxicity: See hexane.

Fire Hazard: Dangerous, when exposed to heat or flame.

Explosion Hazard: Severe, when exposed to heat or flame.

Disaster Hazard: Dangerous! Keep away from sparks, heat or open flame; can react vigorously with oxidizing materials.

Countermeasures

Ventilation Control: Section 2.

To Fight Fire: Foam, carbon dioxide, dry chemical or carbon tetrachloride (Section 6).

Storage and Handling: Section 7.

1,3-ISOINDOLEDIONE. See phthalimide.

ISOLAN
General Information
Synonym: 1-Isopropyl-3-methyl-5-pyragolyl   dimethylcarbamate.
Formula: $C_{10}H_{17}N_3O_2$.
Constant: Mol wt: 211.3.
Hazard Analysis
Toxicity: Highly toxic. A cholinesterase inhibitor. Although this is not an organic phosphate it resembles that group in action. See parathion for effects.
Disaster Hazard: Dangerous. When heated to decomposition it emits highly toxic fumes.

ISOLEUCINE
General Information
Synonyms: α-Amino-β-methylvaleric   acid;   2-amino-3-methyl-pentanoic acid.
Description: An essential amino acid; many isomeric forms. We consider the DL and L forms: Crystalline; slightly soluble in water; nearly insoluble in alcohol, insoluble in ether.
Formula: $CH_3CH_2CH(CH_3)CH(NH_2)COOH$.
Constants: Mol wt: 131, mp (DL): 292°C (decomposes), mp (L): 283–84°C (decomposes).
Hazard Analysis
Toxic Hazard Rating: U. A nutrient and/or dietary supplement food additive (Section 10).

ISOMER OF YOHIMBINE. See yohimbene.

ISOMETHEPTENE
General Information
Synonym: Octin; 1,5-trimethyl-4-hexenylamine.
Description: Colorless oily liquid. Characteristic amine odor. Water insoluble.
Formula: $(CH_3)_2CCHCH_2CH_2CH(NHCH_3)CH_3$.
Constants: Mol wt: 141.3, d: 0.795, bp: 177°C.
Hazard Analysis
Toxic Hazard Rating:
   Acute Local: Irritant 1.
   Acute Systemic: Ingestion 2.
   Chronic Local: U.
   Chronic Systemic: U.
Toxicology: Animal experiments suggest moderate toxicity. can cause headache, nausea and dizziness.

ISONAPHTHOL. See β-naphthol.

ISOOCTANE
General Information
Synonyms: 2,2,4-Trimethylpentane; 2-methylheptane.
Description: Clear liquid.
Formula: $(CH_3)_2CH(CH_2)_4CH_3$.
Constants: Mol wt: 114.2, bp: 99.2°C, fp: −116°C, flash p: 10°F, d: 0.703 at 20°/4°C, autoign. temp.: 784°F, vap. press.: 40.6 mm at 21°C, vap. d.: 3.93, lel: 1.1%, uel: 6.0%.
Hazard Analysis
Toxic Hazard Rating:
   Acute Local: U.
   Acute Systemic: Ingestion 2; Inhalation 2.
   Chronic Local: U.
   Chronic Systemic: U.
Toxicology: High concentrations can cause narcosis.
Fire Hazard: Dangerous, when exposed to heat or flame.
Spontaneous Heating: No.

Disaster Hazard: Dangerous; keep away from heat and open flame; can react vigorously with reducing materials.
Countermeasures
Ventilation Control: Section 2.
To Fight Fire: Foam, carbon dioxide, dry chemical or carbon tetrachloride (Section 6).
Personnel Protection: Section 3.
Storage and Handling: Section 7.
Shipping Regulations: Section 11.
   I.C.C.: Flammable liquid; red label, 10 gallons.
   Coast Guard Classification: Inflammable liquid; red label.
   IATA: Flammable liquid, red label, 1 liter (passenger), 40 liters (cargo).

ISOOCTANOIC ACID (MIXED ISOMERS).
General Information
Description: Insoluble in water.
Formula: $C_6H_{15}COOH$.
Constants: Mol wt: 132, d: 0.9, vap. d.: 5.0, bp: 428°F (decomposes), flash p: 270°F (O.C.).
Hazard Analysis
Toxic Hazard Rating: U.
Fire Hazard: Slight, when exposed to heat or flame.
Countermeasures
Storage and Handling: Section 7.
To Fight Fire: Water, foam (Section 6).

ISOOCTENES
General Information
Formula: C_8H_{16}.
Constants: Mol wt: 112.1, bp: 102–107°C, flash p: < 20°F, d: 0.723 at 15.5°/15.5°C, vap. press.: 102 mm at 21°C, vap. d.: 3.9.
Hazard Analysis
Toxicity: Unknown. May have narcotic properties.
Fire Hazard: Dangerous; when exposed to heat or flame (Section 6).
Disaster Hazard: Dangerous; keep away from heat and open flame; can react vigorously with oxidizing materials.
Countermeasures
Ventilation Control: Section 2.
Storage and Handling: Section 7.
Shipping Regulations: Section 11.
   I.C.C.: Flammable liquid; red label, 10 gallons.
   Coast Guard Classification: Inflammable liquid; red label.
   IATA: Flammable liquid, red label, 1 liter (passenger), 40 liters (cargo).

ISOOCTYL DECYL PHTHALATE
General Information
Description: Water-white, low viscosity liquid.
Formula: $C_{10}H_{21}O_2CC_6H_4CO_2C_8H_{17}$.
Constants: Mol wt: 418, bp: 232–243°C at 4 mm, pour p: −43°C, flash p: 445°F (C.O.C), d: 0.973 at 20°C, vap. d.: 14.4.
Hazard Analysis
Toxicity: See esters.
Fire Hazard: Slight, when exposed to heat or flame; can react with oxidizing materials.
Countermeasures
To Fight Fire: Foam, carbon dioxide, dry chemical or carbon tetrachloride (Section 6).
Storage and Handling: Section 7.

ISOOCTYL NITRATE
General Information
Description: Insoluble in water.

TOXIC HAZARD RATING CODE (For detailed discussion, see Section 1.)

0 NONE: (a) No harm under any conditions; (b) Harmful only under unusual conditions or overwhelming dosage.

1 SLIGHT: Causes readily reversible changes which disappear after end of exposure.

2 MODERATE: May involve both irreversible and reversible changes; not severe enough to cause death or permanent injury.

3 HIGH: May cause death or permanent injury after very short exposure to small quantities.

U UNKNOWN: No information on humans considered valid by authors.

Formula: $C_8H_{17}NO_3$.
Constants: Bp: 106–109° F at 1 mm, d: 1.0, flash p: 205° F (O.C.). mol wt: 175.
Hazard Analysis
Toxic Hazard Rating: U. See nitrates.
Fire Hazard: Moderate, when exposed to heat or flame.
Disaster Hazard: Dangerous. See nitrates.

ISOOCTYL THIOGLYCOLATE
General Information
Description: A clear, water-white liquid, fruity odor.
Formula: $HSCH_2COOC_8H_{17}$.
Constants: Mol wt: 204.33, bp: 52° C at 17 mm, d: 0.9736 at 25° C.
Hazard Analysis
Toxicity: Probably toxic. See thioglycolic acid.
Disaster Hazard: Dangerous; see oxides of sulfur; can react vigorously with oxidizing materials.
Countermeasures
Storage and Handling: Section 7.

ISOPENTANE
General Information
Synonyms: 2-Methyl butane; isoamyl hydride.
Description: Colorless liquid with pleasant odor.
Formula: $CH_3CHCH_3CH_2CH_3$.
Constants: Mol wt: 72.15, bp: 27.8° C, fp: −160.5° C, flash p: < −60° F (C.C.), d: 0.621 at 20°/4° C, autoign. temp.: 788° F, vap. press.: 595 mm at 21.1° C, vap. d.: 2.48, lel: 1.4%, uel: 7.6%.
Hazard Analysis
Toxicity: See pentane.
Fire Hazard: Highly dangerous when exposed to heat or flame.
Explosion Hazard: Unknown.
Disaster Hazard: Dangerous. Keep away from sparks, heat or open flame; can react vigorously with oxidizing materials.
Countermeasures
Ventilation Control: Section 2.
To Fight Fire: Foam, carbon dioxide, dry chemical or carbon tetrachloride (Section 6).
Storage and Handling: Section 7.
Shipping Regulations: Section 11.
  I.C.C.: Flammable liquid; red label, 10 gallons.
  Coast Guard Classification: Inflammable liquid; red label, IATA: Flammable liquid, red label, 1 liter (passenger), 40 liters (cargo).

ISOPENTYL ALCOHOL. See isoamyl alcohol.

ISOPESTEX. See bis(isopropylamido) fluorophosphate.

ISOPHORONE
General Information
Description: Practically water-white liquid.
Formula: $COCHC(CH_3)CH_2C(CH_3)_2CH_2$.
Constants: Mol wt: 138.1, bp: 215.2° C, flash p: 205° F (O.C.), d: 0.9229, autoign. temp.: 864° F, vap. press.: 1 mm at 38.0° C, vap. d.: 4.77, lel: 0.8%, uel: 3.8%.
Hazard Analysis
Toxic Hazard Rating:
  Acute Local: Irritant 3; Ingestion 3; Inhalation 3.
  Acute Systemic: Ingestion 3; Inhalation 3.
  Chronic Local: U.
  Chronic Systemic: U.
TLV: ACGIH (recommended); 25 parts per million in air; 141 milligrams per cubic meter of air.
Toxicology: Considered to be more toxic than mesityl oxide. However, due to its low volatility, it is not a dangerous industrial hazard. The response of guinea pigs and rats to repeated inhalations of the vapors indicates that it is one of the most toxic of the ketones. It is chiefly a kidney poison. It can cause irritation, lachrymation, possible opacity of the cornea and necrosis of the cornea (experimental). It is irritant at the level of 25 ppm to

humans. In animal experiments death during exposure was usually due to narcosis, but occasionally to irritation of the lungs.
Fire Hazard: Moderate, when exposed to heat or flame; can react with oxidizing materials.
Explosion Hazard: Unknown.
Countermeasures
Ventilation Control: Section 2.
To Fight Fire: Foam, carbon dioxide, dry chemical or carbon tetrachloride (Section 6).
Personnel Protection: Section 3.
Storage and Handling: Section 7.

ISOPRAL. See 1,1,1-trichloroisopropyl alcohol.

ISOPHTHALIC ACID
General Information
Synonym: m-phthalic acid, IPA.
Description: Colorless crystals; slightly soluble in water; soluble in alcohol; acetic acid; insoluble in benzene and petroleum ether.
Formula: $C_6H_4(COOH)_2$.
Constants: Mol wt: 166, mp: 345–348° C.
Hazard Analysis
Toxic Hazard Rating:
  Acute Local: Irritant 1; Ingestion 1; Inhalation 1.
  Acute Systemic: U.
  Chronic Local: U.
  Chronic Systemic: U.

ISOPHTHALOYL CHLORIDE
General Information
Synonym: m-Phthalyl chloride.
Description: White, crystalline solid. Soluble in benzene and carbon tetrachloride.
Formula: $C_8H_4O_2Cl_2$.
Constants: Mol wt: 203.0, bp: 276° C, fp: 43.3° C, d: 1.387 at 46.9° C, flash p: 356° F (C.O.C.), vap. d.: 6.9.
Hazard Analysis
Toxicology: An irritant material. Further details unknown.
Disaster Hazard: Dangerous; see chlorides.

ISOPRENE
General Information
Synonym: 3-Methyl-1,3-butadiene.
Description: Colorless, volatile liquid.
Formula: $CH_2C(CH_3)CHCH_2$.
Constants: Mol wt: 68.11, mp: −146.7° C, bp: 34° C, flash p: −65° F, d: 0.6806 at 20°/4° C, autoign. temp.: 428° F, vap. press.: 400 mm at 15.4° C, vap. d.: 2.35.
Hazard Analysis
Toxic Hazard Rating:
  Acute Local: Irritant 2; Ingestion 2; Inhalation 2.
  Acute Systemic: Inhalation 2.
  Chronic Local: U.
  Chronic Systemic: U.
Toxicology: Isoprene is irritant to mucous membranes of the eyes, nose and upper respiratory passages. Concentrations of 2 percent in air do not narcotize mice but produce bronchial irritation. Concentrations of 5 percent are fatal to mice. There are no data on human exposures.
Fire Hazard: Highly dangerous when exposed to heat or flame.
Disaster Hazard: Dangerous; keep away from sparks, heat or open flame; can react vigorously with reducing materials.
Countermeasures
Ventilation Control: Section 2.
To Fight Fire: Carbon dioxide, dry chemical or carbon tetrachloride (Section 6).
Personnel Protection: Section 3.
Storage and Handling: Section 7.
Shipping Regulations: Section 11.

I.C.C.: Flammable liquid; red label, 10 gallons.

Coast Guard Classification: Inflammable liquid; red label.

IATA: Flammable liquid, red label, 1 liter (passenger), 40 liters (cargo).

ISOPROPANOL. See isopropyl alcohol.

ISOPROPANOLAMINE. See 1-amino-2-propanol.

ISOPROPANOLAMINES, MIXED
General Information
Description: Clear liquid.
Formula: $CH_3CH(OH)CH_2NH_2$.
Constants: Mol wt: 75.1, mp: 29.5°C, bp: 159°C, flash p: 160°F (O.C.), d: 0.962, vap. d.: 2.58.
Hazard Analysis
Toxic Hazard Rating:
  Acute Local: Irritant 1; Ingestion 1; Inhalation 1.
  Acute Systemic: U.
  Chronic Local: U.
  Chronic Systemic: U.
Toxicity: See also amines.
Fire Hazard: Moderate, when exposed to heat or flame; can react with oxidizing materials.
Countermeasures
To Fight Fire: Foam, carbon dioxide, dry chemical or carbon tetrachloride (Section 6).
Ventilation Control: Section 2.
Personal Hygiene: Section 3.
Storage and Handling: Section 7.

ISOPROPENYL ACETATE
General Information
Description: Water-white liquid.
Formula: $CH_3COOC(CH_3)CH_2$.
Constants: Mol wt: 100.06, mp: −92.9°C, bp: 96.6°C at 746 mm, flash p: 60°F (O.C.), d: 0.9226 at 20°/20°C, vap. d.: 3.45.
Hazard Analysis
Toxic Hazard Rating:
  Acute Local: Irritant 1; Ingestion 1; Inhalation 1.
  Acute Systemic: U.
  Chronic Local: U.
  Chronic Systemic: U.
Toxicity: See also esters.
Fire Hazard: Dangerous, when exposed to heat or flame.
Disaster Hazard: Dangerous! Keep away from heat and open flame; can react vigorously with oxidizing materials.
Countermeasures
Ventilation Control: Section 2.
To Fight Fire: Foam, carbon dioxide, dry chemical or carbon tetrachloride (Section 6).
Personal Hygiene: Section 3.
Storage and Handling: Section 7.

ISOPROPENYL ACETYLENE
General Information
Synonym: 2-Methyl-1-buten-3-yne.
Description: Colorless liquid; very slightly soluble in water; miscible with acetone, alcohol, benzene, carbon tetrachloride, and kerosene.
Formula: $CH_2:C(CH_3)C:CH$.
Constants: Mol wt: 66, bp: 33–34°C, fp: 113°C, d: 0.695 at 20/20°C, flash p: < 19°F (O.C.), vap. d.: 2.3.

Hazard Analysis
Toxic Hazard Rating: U.
Fire Hazard: Dangerous; when exposed to heat or flame.
Countermeasures
Storage and Handling: Section 7.
To Fight Fire: Alcohol foam (Section 6).

ISOPROPENYL CHLORIDE. See 2-chloropropene.

ISOPROPENYL CHLOROFORMATE
General Information
Description: Liquid.
Formula: $ClCOOC(CH_3)CH_2$.
Constants: Mol wt: 120.5°C, bp: 93°C at 746 mm, d: 1.103 at 20°C.
Hazard Analysis
Toxic Hazard Rating:
  Acute Local: Irritant 3; Ingestion 3; Inhalation 3.
  Acute Systemic: U.
  Chronic Local: U.
  Chronic Systemic: U.
Disaster Hazard: Dangerous; when heated to decomposition it emits toxic fumes of chlorides.
Countermeasures
Ventilation Control: Section 2.
Personnel Protection: Section 3.
Storage and Handling: Section 7.

2-ISOPROPOXYPROPANE. See isopropyl ether.

β-ISOPROPOXYPROPIONITRILE
General Information
Description: Liquid.
Formula: $C_3H_7OCH_2CH_2CN$.
Constants: Mol wt: 113.16, mp: −67°C, bp: 82–86°C, flash p: 155°F, d: 0.9058, vap. d.: 3.91.
Hazard Analysis
Toxicity: Highly toxic. See nitriles.
Fire Hazard: Moderate, when exposed to heat or flame.
Disaster Hazard: Dangerous; see cyanides.
Countermeasures
Ventilation Control: Section 2.
To Fight Fire: Foam, carbon dioxide, dry chemical or carbon tetrachloride (Section 6).
Personnel Protection: Section 3.
Storage and Handling: Section 7.

3-ISOPROPOXYPROPYLAMINE
General Information
Synonym: 3-Dimethylcarbinolpropylamine.
Description: Colorless liquid.
Formula: $(CH_3)_2CHOCH_2CH_2CH_2NH_2$.
Constants: Mol wt: 118.2, bp: 147°C, d: 0.845 at 20°/20°C, vap. d.: 4.07.
Hazard Analysis
Toxic Hazard Rating:
  Acute Local: Irritant 3; Ingestion 3; Inhalation 3.
  Acute Systemic: Ingestion 3; Inhalation 3.
  Chronic Local: Inhalation 3.
  Chronic Systemic: U.
Toxicity: See also amines.
Disaster Hazard: Dangerous; when heated, it emits toxic fumes; can react with oxidizing materials.
Countermeasures
Ventilation Control: Section 2.
Personnel Protection: Section 3.
Storage and Handling: Section 7.

TOXIC HAZARD RATING CODE (For detailed discussion, see Section 1.)

0 NONE: (a) No harm under any conditions; (b) Harmful only under unusual conditions or overwhelming dosage.

1 SLIGHT: Causes readily reversible changes which disappear after end of exposure.

2 MODERATE: May involve both irreversible and reversible changes; not severe enough to cause death or permanent injury.

3 HIGH: May cause death or permanent injury after very short exposure to small quantities.

U UNKNOWN: No information on humans considered valid by authors.

ISOPROPYL ACETATE
General Information
Description: Colorless aromatic liquid.
Formula: $CH_3COOCH(CH_3)_2$.
Constants: Mol wt: 102.13, mp: $-73°C$, bp: $88.4°C$, lel: 1.8%, uel: 7.8%, fp: $-69.3°C$, flash p: $40°F$, d: 0.874 at $20°/20°C$, autoign. temp.: $860°F$, vap. press.: 40 mm at $17.0°C$, vap. d.: 3.52.
Hazard Analysis
Toxic Hazard Rating:
Acute Local: Irritant 1.
Acute Systemic: Ingestion 2; Inhalation 2.
Chronic Local: U.
Chronic Systemic: Ingestion 1; Inhalation 1.
Toxicology: Narcotic in high concentrations. Chronic exposure can cause liver damage.
TLV: ACGIH (recommended) 250 parts per million of air; 950 milligrams per cubic meter of air.
Fire Hazard: Dangerous, when exposed to heat or flame.
Spontaneous Heating: No.
Explosion Hazard: Moderate, when exposed to heat or flame.
Disaster Hazard: Dangerous! Keep away from heat and open flame; can react vigorously with oxidizing materials.
Countermeasures
Ventilation Control: Section 2.
To Fight Fire: Foam, carbon dioxide, dry chemical or carbon tetrachloride (Section 6).
Personal Hygiene: Section 3.
Storage and Handling: Section 7.
Shipping Regulations: Section 11.
I.C.C.: Flammable liquid, red label, 10 gallons.
IATA: Flammable liquid, red label, 1 liter (passenger), 40 liters (cargo).

ISOPROPYL ACETONE. See isobutyl methyl ketone.

ISOPROPYL ALCOHOL
General Information
Synonyms: Dimethyl carbinol; sec-propyl alcohol; isopropanol.
Description: Clear, colorless liquid.
Formula: $CH_3CHOHCH_3$.
Constants: Mol wt: 60.09, mp: -88.5 to $-89.5°C$, bp: $82.3°C$, lel: 2.0%, uel: 12%, flash p: $53°F$, d: 0.7854 at $20°/4°C$, autoign. temp.: $750°F$, vap. press.: 33.0 mm at $20°C$, vap. d.: 2.07, ulc: .70.
Hazard Analysis
Toxic Hazard Rating:
Acute Local: Irritant 1; Ingestion 1; Inhalation 1.
Acute Systemic: Ingestion 2; Inhalation 2.
Chronic Local: U.
Chronic Systemic: Ingestion 1; Inhalation 1.
TLV: ACGIH (recommended); 400 parts per million in air; 980 milligrams per cubic meter of air.
Toxicology: Acts as a local irritant and in high concentrations as a narcotic. It can cause corneal burns and often eye damage. It has good warning properties because it causes a mild irritation of the eyes, nose and throat at concentration levels of 400 ppm. It may induce a mild narcosis, the effects of which are usually transient, and it is somewhat less toxic than the normal isomer, but twice as volatile. It is not considered an important toxic hazard. There is some evidence that personnel can acquire a slight tolerance to this material, and single or repeated applications of it on the skin of rats, rabbits, dogs or human beings induced no untoward effects. It acts very much like ethanol in regard to absorption, metabolism and elimination but with a stronger narcotic action. Chronic injuries due to it have been detected in animals. Humans have ingested up to 20 ml diluted with water and noticed only a sensation of heat and slight lowering of the blood pressure. There

are, however, reports of serious illness from as little as 10 ml taken internally. A food additive permitted in food for human consumption (Section 10). A common air contaminant (Section 4).
Fire Hazard: Dangerous, when exposed to heat or flame.
Spontaneous Heating: No.
Explosion Hazard: Moderate, when exposed to heat or flame.
Disaster Hazard: Dangerous! Keep away from heat and open flame; can react vigorously with oxidizing materials.
Countermeasures
Ventilation Control: Section 2.
To Fight Fire: Carbon dioxide, dry chemical or carbon tetrachloride (Section 6).
Personal Hygiene: Section 3.
Storage and Handling: Section 7.
Shipping Regulations: Section 11.
I.C.C.: Flammable liquid; red label, 10 gallons.
MCA warning label.

ISOPROPYL ACID PHOSHATE
Shipping Regulations: Section 11.
IATA: Other restricted articles, class B, no label required, no limit (passenger and cargo).

ISOPROPYLAMINE
General Information
Synonym: 2-Aminopropane.
Description: Colorless liquid, amine odor.
Formula: $(CH_3)_2CHNH_2$.
Constants: Mol wt: 59.11, mp: $-101.2°C$, bp: $31.7°C$, flash p: $-35°F$ (O.C.), d: 0.694 at $15°/4°C$, autoign. temp.: $756°F$, vap. d.: 2.03.
Hazard Analysis
Toxic Hazard Rating:
Acute Local: Irritant 3; Allergen 1; Ingestion 3; Inhalation 3.
Acute Systemic: Inhalation 2.
Chronic Local: Irritant 2; Allergen 1.
Chronic Systemic: U.
Toxicology: A strong irritant. Occasionally causes sensitization. Narcotic in high concentrations.
TLV: ACGIH (recommended); 5 parts per million in air, 12 milligrams per cubic meter of air.
Fire Hazard: Dangerous, when exposed to heat or flame.
Disaster Hazard: Dangerous; keep away from sparks, heat or open flame; can react vigorously with oxidizing materials.
Countermeasures
Ventilation Control: Section 2.
To Fight Fire: Alcohol foam, foam, carbon dioxide, dry chemical or carbon tetrachloride (Section 6).
Storage and Handling: Section 7.
Shipping Regulations: Section 11.
IATA: Flammable liquid, red label, 1 liter (passenger), 40 liters (cargo).

β-ISOPROPYLAMINOPROPIONITRILE
General Information
Description: Liquid.
Formula: $C_3H_7NHCH_2CH_2CN$.
Constants: Mol wt: 112.18, mp: $< -20°C$, bp: $86°C$ at 17 mm, flash p: $100°F$, d: 0.864, vap. d.: 3.87.
Hazard Analysis
Toxicity: Highly toxic. See nitriles.
Fire Hazard: Moderate, when exposed to heat or flame.
Disaster Hazard: Dangerous; see cyanides; will react with water or steam to produce toxic and flammable vapors; can react vigorously with oxidizing materials.
Countermeasures
Ventilation Control: Section 2.
To Fight Fire: Foam, carbon dioxide, dry chemical or carbon tetrachloride (Section 6).
Storage and Handling: Section 7.

3-ISOPROPYLAMINOPROPYLAMINE
General Information
Description: Colorless liquid.
Formula: $(CH_3)_2CHNHCH_2CH_2CH_2NH_2$.
Constants: Mol wt: 116.2, bp: 159°C, vap. d.: 4.00, d: 0.827
   at 25°/4°C.
Hazard Analysis
Toxic Hazard Rating:
   Acute Local: Irritant 3; Ingestion 3; Inhalation 3.
   Acute Systemic: Ingestion 3; Inhalation 3.
   Chronic Local: U.
   Chronic Systemic: Ingestion 2; Inhalation 2.
Toxicity: See also amines.
Fire Hazard: Slight, when exposed to heat or flame (section
   6).
Disaster Hazard: Dangerous; see cyanides; can react with
   oxidizing materials.
Countermeasures
Ventilation Control: Section 2.
Personnel Protection: Section 3.
Storage and Handling: Section 7.

ISOPROPYL ANTIMONITE
General Information
Formula: $(C_3H_7O)_3Sb$.
Constants: Mol wt: 299.1, bp: 82°C at 7 mm.
Hazard Analysis and Countermeasures
See antimony compounds.

ISOPROPYL BENZENE. See cumene.

ISOPROPYL BENZOATE
General Information
Description: Liquid.
Formula: $C_6H_5COOCH(CH_3)_2$.
Constants: Mol wt: 164.2, mp: −26.4°C, bp: 219°C,
   flash p: 210°F (O.C.), d: 1.0112, vap. press.: 0.12 mm
   at 20°C, vap. d.: 5.67.
Hazard Analysis
Toxic Hazard Rating:
   Acute Local: Irritant 2; Ingestion 2; Inhalation 2.
   Acute Systemic: U.
   Chronic Local: U.
   Chronic Systemic: U.
Fire Hazard: Moderate, when exposed to heat or flame, can
   react with oxidizing materials.
Countermeasures
To Fight Fire: Water spray, foam, carbon dioxide, dry
   chemical or carbon tetrachloride (Section 6).
Ventilation Control: Section 2.
Personal Hygiene: Section 3.
Storage and Handling: Section 7.

ISOPROPYL BROMIDE
General Information
Description: Colorless liquid.
Formula: $CH_3CHBrCH_3$.
Constants: Mol wt: 123.0, bp: 58.5–60.5°C, fp: −90°C,
   d: 1.304 at 25°/25°C, vap. d.: 4.27.
Hazard Analysis
Toxicity: Details unknown. Animal experiments suggest low
   systemic toxicity. Narcotic in high concentrations.
Disaster Hazard: Dangerous; See bromides.
Countermeasures
Storage and Handling: Section 7.

ISOPROPYLCARBINOL. See isobutyl alcohol.

ISOPROPYL CHLORIDE
General Information
Synonyms: 2-Chloropropane.
Description: Clear liquid.
Formula: $CH_3CHClCH_3$.
Constants: Mol wt: 78.55, bp: 35.3°C, fp: −117.6°C,
   lel: 2.8%, uel: 10.7%, flash p: −26°F (C.C.), d:
   0.858 at 25°/25°C, autoign. temp.: 1100°F, vap. d.:
   2.71.
Hazard Analysis
Toxicity: Details unknown. Animal experiments suggest
   low systemic toxicity. Has been used as a surgical an-
   esthetic.
Fire Hazard: Highly dangerous, when exposed to heat or
   flame.
Explosion Hazard: Moderate, when exposed to heat or
   flame.
Disaster Hazard: Dangerous; keep away from sparks, heat
   or open flame; emits highly toxic fumes of chlorides such
   as phosgene; can react vigorously with oxidizing mate-
   rials.
Countermeasures
Ventilation Control: Section 2.
To Fight Fire: Foam, carbon dioxide, dry chemical or carbon
   tetrachloride (Section 6).
Personnel Protection: Section 3.
Storage and Handling: Section 7.

ISOPROPYL m-CHLOROCARBANILATE. See iso-
   propyl N-(3-chlorophenyl carbamate).

ISOPROPYL CHLOROFORMATE
General Information
Description: Clear colorless liquid; a phosgene derivative.
Formula: $(CH_3)_2CHOOCCl$.
Constant: Mol wt: 122.5.
Hazard Analysis
Toxic Hazard Rating:
   Acute Local: Irritant 3; Inhalation 3.
   Acute Systemic: Inhalation 3.
   Chronic Local: Irritant 3.
   Chronic Systemic: U.
Toxicity: Can cause pulmonary edema.
Disaster Hazard: Dangerous. See chlorides.

ISOPROPYL N-(3-CHLOROPHENYL) CARBAMATE
General Information
Synonyms: Isopropyl m-chlorocarbanilate.
Description: Light brown crystalline solid, faint char-
   acteristic odor.
Formula: $C_{10}H_{12}ClNO_2$.
Constants: Mol wt: 213.66, mp: 36°C.
Hazard Analysis
Toxicity: Details unknown; an insecticide. Limited animal
   experiments suggest low to moderate toxicity.
Disaster Hazard: Dangerous; when heated to decomposi-
   tion, it emits highly toxic fumes of phosgene.
Countermeasures
Ventilation Control: Section 2.
Storage and Handling: Section 7.

ISOPROPYL CITRATE
Hazard Analysis
Toxic Hazard Rating: U.
Note: Used as a sesquestrant food additive (Section 10).

ISOPROPYL CYANIDE. See isobutyronitrile.

TOXIC HAZARD RATING CODE *(For detailed discussion, see Section 1.)*

0 NONE: (a) No harm under any conditions; (b) Harmful only under unusual conditions or overwhelming dosage.

1 SLIGHT: Causes readily reversible changes which disappear after end
   of exposure.

2 MODERATE: May involve both irreversible and reversible changes;
   not severe enough to cause death or permanent injury.

3 HIGH: May cause death or permanent injury after very short exposure
   to small quantities.

U UNKNOWN: No information on humans considered valid by authors.

ISOPROPYL CYCLOHEXYL AMINE
General Information
Description: Insoluble in water.
Formula: $C_6H_{11}NHCHC_2H_6$.
Constants: Mol wt: 141, d: 0.8, vap. d.: 4.9, flash p:
   93° F (O.C.).
Hazard Analysis
Toxic Hazard Rating: U. See also amines.
Fire Hazard: Moderate, when exposed to heat or flame.

ISOPROPYL 2,4-DICHLOROPHENOXYACETATE
General Information
Description: Nearly colorless crystals.
Formula: $C_6H_3Cl_2OCH_2COOCH(CH_3)_2$.
Constants: Mol wt: 263.13, d: 1.255–1.270 at 25°/25°C.
Hazard Analysis
Toxicity: See 2,4-D.
Disaster Hazard: Moderately dangerous; when heated to
   decomposition, it emits toxic fumes of chlorides.
Countermeasures
Ventilation Control: Section 2.
Storage and Handling: Section 7.

ISOPROPYL ETHER
General Information
Synonyms: 2-Isopropoxypropane; diisopropyl ether.
Description: Colorless liquid, ethereal odor.
Formula: $(CH_3)_2CHOCH(CH_3)_2$.
Constants: Mol wt: 102.17, mp: –60°C, bp: 68.5°C,
   fp: –85.6°C, lel: 1.4%, uel: 21.0%, flash p: –18°F
   (C.C.), d: 0.719 at 25°C, autoign. temp.: 830°F,
   vap. press.: 150 mm at 25°C, vap. d.: 3.52.
Hazard Analysis
Toxic Hazard Rating:
   Acute Local: Irritant 3; Ingestion 2; Inhalation 2.
   Acute Systemic: Ingestion 2; Inhalation 2.
   Chronic Local: Irritant 2.
   Chronic Systemic: U.
Toxicology: Irritating to skin and mucous membrane.
   Narcotic in high concentrations.
TLV: ACGIH (recommended); 500 parts per million in air;
   2085 milligrams per cubic meter of air.
Fire Hazard: Highly dangerous, when exposed to heat or
   flame.
Spontaneous Heating: No.
Explosion Hazard: Severe, when exposed to heat or flame.
Disaster Hazard: Dangerous; keep away from heat, sparks
   or open flame; under some conditions shock will ex-
   plode it; emits toxic fumes; reacts vigorously with oxi-
   dizing materials.
Countermeasures
Ventilation Control: Section 2.
To Fight Fire: Alcohol foam, foam, carbon dioxide, dry
   chemical or carbon tetrachloride (Section 6).
Personnel Protection: Section 3.
Storage and Handling: Section 7.
Shipping Regulations: Section 11.
   MCA warning label.

ISOPROPYL ETHYLENE. See 3-methyl butene-1.

ISOPROPYL FORMATE
General Information
Synonyms: Isopropyl methanoate.
Description: Clear liquid.
Formula: $C_3H_7CHO_2$.
Constants: Mol wt: 88.1, bp: 68.3°C, flash p: 22°F (C.C.),
   d: 0.873, autoign. temp.: 905°F, vap. press.: 100 mm
   at 17.8°C, vap. d.: 3.03.
Hazard Analysis
Toxic Hazard Rating:
   Acute Local: Irritant 2; Ingestion 2; Inhalation 2.
   Acute Systemic: Ingestion 2; Inhalation 2.
   Chronic Local: U.
   Chronic Systemic: U.

Toxicity: See also esters.
Fire Hazard: Dangerous, when exposed to heat or flame.
Spontaneous Heating: No.
Disaster Hazard: Dangerous; keep away from heat and open
   flame; can react vigorously with oxidizing materials.
Countermeasures
Ventilation Control: Section 2.
To Fight Fire: Alcohol foam, foam, carbon dioxide, dry
   chemical or carbon tetrachloride (Section 6).
Personnel Protection: Section 3.
Storage and Handling: Section 7.

ISOPROPYL GLYCIDYL ETHER
General Information
Synonym: IGE.
Hazard Analysis
Toxic Hazard Rating:
   Acute Local: Irritant 2; Ingestion 2.
   Acute Systemic: Ingestion 2; Inhalation 2; Skin Absorp-
   tion 1.
   Chronic Local: Allergen 2.
   Chronic Systemic: Inhalation 1.
Toxicity: Data based on animal experiments show moderate
   toxicity, skin irritation and sensitization. Can cause
   central nervous depression.
TLV: ACGIH (recommended); 50 parts per million in air;
   240 milligrams per cubic meter of air.

ISOPROPYL LACTATE
General Information
Synonyms: Isopropyl-2-hydroxy propanoate.
Description: Soluble in water.
Formula: $CH_3CHOHCOOCH(CH_3)_2$.
Constants: Mol wt: 132, d: 1.0–, vap. d.: 4.2, bp: 331–334°F,
   flash p: 130°F (O.C.).
Hazard Analysis
Toxic Hazard Rating: U.
Fire Hazard: Moderate, when exposed to heat or flame.
Countermeasures
Storage and Handling: Section 7.
To Fight Fire: Alcohol foam (Section 6).

ISOPROPYL MANDELATE
General Information
Description: Crystals.
Formula: $C_6H_5CH(OH)COOC_3H_7$.
Constant: Mol wt: 194.2.
Hazard Analysis
Toxic Hazard Rating:
   Acute Local: Irritant 1.
   Acute Systemic: U.
   Chronic Local: U.
   Chronic Systemic: U.
Fire Hazard: Slight (Section 6).
Countermeasures
Storage and Handling: Section 7.

ISOPROPYL MERCAPTAN
General Information
Description: Liquid; with extremely powerful, unpleasant
   odor.
Formula: $(CH_3)_2CH(HS)$.
Constants: Mol wt: 76, d: 0.814 at 60/60°F, boiling range:
   51–55°C, flash p: –30°F.
Hazard Analysis
Toxic Hazard Rating: U.
Fire Hazard: Dangerous, when exposed to heat or flame.
Disaster Hazard: Dangerous; when heated to decomposi-
   tion it emits highly toxic fumes. See mercaptans.
Countermeasures
Storage and Handling: Section 7.
Shipping Regulations: Section 11.
   I.C.C: Flammable liquid, red label, 10 gallons.
   IATA: Flammable liquid, red label, 1 liter (passenger),
   40 liters (cargo).

ISOPROPYL METHANOATE. See isopropyl formate.

1-ISOPROPYL-2-METHYLETHYLENE. See 4-methyl-2-pentene.

ISOPROPYL NITRATE
General Information
Synonym: 2-propanol nitrate.
Description: Liquid.
Formula: $(CH_3)_2CHNO_3$.
Constants: Mol wt: 105, bp: 102°C.
Hazard Analysis
Toxic Hazard Rating: U.
Fire Hazard: See nitrates.
Disaster Hazard: Dangerous. See nitrates.
Countermeasures
Shipping Regulations: Section 11.
    IATA: Flammable liquid, red label, 1 liter (passenger), 40 liters (cargo).

ISOPROPYL NITRITE
General Information
Synonym: 2-Propanol nitrite.
Description: Pale yellow oil.
Formula: $C_3H_7NO_2$.
Constants: Mol wt: 89.1, d: 0.856 at 0°/4°C, bp: 39°C at 752 mm.
Hazard Analysis and Countermeasures
See nitrites and amyl nitrite.

ISOPROPYL PERCARBONATE. See diisopropyl peroxydicarbonate.

ISOPROPYL PEROXYDICARBONATE. See diisopropyl peroxydicarbonate.

o-ISOPROPYL PHENOL
General Information
Synonym: o-Cumenol.
Description: Colorless to amber liquid.
Formula: $(CH_3)_2CHC_6H_4OH$.
Constants: Mol wt: 136.2, mp: 15.5°C, bp: 209°C, flash p: 220°F, d: 0.995 at 25°/25°C, vap. press.: 1 mm at 56.6°C.
Hazard Analysis
Toxicity: Unknown. See also phenol.
Fire Hazard: Moderate, when exposed to heat or flame (Section 6).
Disaster Hazard: Dangerous; when heated to decomposition, it emits highly toxic fumes of phenol; can react with oxidizing materials.
Countermeasures
Storage and Handling: Section 7.

p-ISOPROPYL PHENOL
General Information
Description: Needle-like crystals, white to light tan.
Formula: $(CH_3)_2CHC_6H_4OH$.
Constants: Mol wt: 136.2, mp: 61°C, bp: 228°C, d: 0.983, vap. press.: 1 mm at 67°C.
Hazard Analysis
Toxicity: Unknown. See also phenol.
Disaster Hazard: Dangerous; when heated to decomposition, it emits highly toxic fumes of phenol; can react with oxidizing materials.
Countermeasures
Storage and Handling: Section 7.

o-ISOPROPYL N-PHENYLCARBAMATE
General Information
Synonym: IPC.
Description: White to gray crystals, odorless when pure. Slightly water soluble,.
Formula: $C_6H_5NHCOOCH(CH_3)_2$.
Constants: Mol wt: 155.2, mp: 84°C.
Hazard Analysis
Toxicity: Details unknown. Animal experiments indicate low toxicity. A herbicide.

4-ISOPROPYL PYRIDINE
General Information
Description: Liquid.
Formula: $(CH_3)_2CHC_5H_4N$.
Constants: Mol wt: 121.3, bp: 182.2°C, d: 0.9282 at 20°/20°C, vap. d.: 4.18.
Hazard Analysis and Countermeasures
See pyridine and cyanides.

ISOPROPYL STANNIC ACID
General Information
Description: White powder, insoluble in water.
Formula: C_3H_7SnOOH.
Constant: Mol wt: 194.8.
Hazard Analysis
Toxicity: Details unknown. See tin compounds.

ISOPROPYLTETRATHIO-o-STANNATE
General Information
Description: Solid.
Formula: $(C_3H_7S)_4Sn$.
Constants: Mol wt: 419.3, bp: 92°C at 0.001 mm.
Hazard Analysis and Countermeasures
See tin and sulfur compounds.

ISOPROPYLTIN TRIBROMIDE
General Information
Description: Pale yellow crystals, water soluble.
Formula: $(C_3H_7)SnBr_3$.
Constants: Mol wt: 401.5, mp: 112°C.
Hazard Analysis and Countermeasures
See tin compounds and bromides.

ISOPROPYLTIN TRICHLORIDE
General Information
Description: Solid, insoluble in water.
Formula: $(C_3H_7)SnCl_3$.
Constants: Mol wt: 286.2, bp: 75°C at 16 mm.
Hazard Analysis and Countermeasures
See tin compounds and chlorides.

ISOPROPYL TITANATE
General Information
Synonym: Titanium isopropylate.
Description: Colorless liquid.
Formula: $Ti(OC_3H_7)_4$.
Constants: Mol wt: 284.2, mp: 14.8°C, bp: 104°C at 10 mm, d: 0.954, vap. d.: 9.8.
Hazard Analysis
Toxicity: Details unknown.
Fire Hazard: Moderate, when exposed to heat or flame.
Disaster Hazard: It will react with water or steam to produce flammable vapors; can react with oxidizing materials.
Countermeasures
Storage and Handling: Section 7.

ISOPROPYLTOLUENE. See p-cymene.

TOXIC HAZARD RATING CODE (For detailed discussion, see Section 1.)

0 NONE: (a) No harm under any conditions; (b) Harmful only under unusual conditions or overwhelming dosage.

1 SLIGHT: Causes readily reversible changes which disappear after end of exposure.

2 MODERATE: May involve both irreversible and reversible changes not severe enough to cause death or permanent injury.

3 HIGH: May cause death or permanent injury after very short exposure to small quantities.

U UNKNOWN: No information on humans considered valid by authors

ISOPROPYLTRICHLOROSILANE
General Information
Description: Liquid.
Formula: $C_3H_7SiCl_3$.
Constants: Mol wt: 177.5, mp: $-87.7°C$, bp: 119.4°C, d: 1.196.
Hazard Analysis
Toxic Hazard Rating:
Acute Local: Irritant 3; Ingestion 3; Inhalation 3.
Acute Systemic: U.
Chronic Local: U.
Chronic Systemic: U.
Disaster Hazard: Dangerous; when heated to decomposition it emits highly toxic fumes of chlorides; will react with water or steam to produce toxic and corrosive fumes.
Countermeasures
Ventilation Control: Section 2.
Personnel Protection: Section 3.
Storage and Handling: Section 7.

ISOQUINOLINE
General Information
Synonyms: 2-Benzazine; benzo [c] pyridine.
Description: Liquid; pungent odor; almost insoluble in water; miscible with many organic solvents; soluble in dilute acids.
Formula: $C_6H_4C_3H_3N$.
Constants: Mol wt: 127.15, d: 1.09101 at 30°/4°C, mp: 23°C, bp: 243°C.
Hazard Analysis
Toxic Hazard Rating: U.
Toxicity: Animal experiments show toxicity similar to, but greater than quinoline. See also quinoline.

ISOSAFFROLE n-OCTYL SULFOXIDE. See sulfoxcide.

ISOTHAN. See lauryl isoquinolinium bromide.

ISOTHIOCYANATES. See thiocyanates.

ISOTOC. See 1,2,3,4,5,6-hexachlorocyclohexane.

ISOVALERALDEHYDE. See isovaleric aldehyde.

ISOVALERIC ALDEHYDE
General Information
Synonyms: Isovaleraldehyde; 3-methylbutanal.
Description: Colorless liquid, apple-like odor.
Formula: $(CH_3)_2CHCH_2CHO$.
Constants: Mol wt: 86.13, mp: $-51°C$, bp: 92.5°C, d: 0.803 at 17°/4°C, vap. d: 2.96.
Hazard Analysis
Toxicity: Details unknown. See also aldehydes.
Disaster Hazard: Slight; when heated, it emits acrid fumes (Section 6).
Countermeasures
Storage and Handling: Section 7.

ISOVALERONE. See diisobutyl ketone.

2-ISOVALERYL-1,3-INDANDIONE. See valone.

ITACONIC ACID
General Information
Synonym: Methylene succinic acid.
Description: White, crystalline powder.
Formula: $CH_2C(COOH)CH_2COOH$.
Constants: Mol wt: 130.10, mp: 167.5°C (decomp.), d: 1.63.
Hazard Analysis
Toxic Hazard Rating:
Acute Local: Irritant 1; Ingestion 1; Inhalation 1.
Acute Systemic: U.
Chronic Local: U.
Chronic Systemic: U.
Fire Hazard: Slight.
Countermeasures
Storage and Handling: Section 7.

J

JAPAN (LACQUER)
Hazard Analysis
Toxic Hazard Rating:
Acute Local: Irritant 3; Allergen 1.
Acute Systemic: U.
Chronic Local: Irritant 3; Allergen 1.
Chronic Systemic: U.
Toxicology: Dermatitis is frequently caused by natural Japan lacquer due to a highly irritating chemical urushiol (Section 9). Synthetic Japan lacquer contains linseed oil, lead oxide, pigments and solvents such as kerosene or turpentine.
Fire Hazard: Moderate, when exposed to heat or flame; can react with oxidizing materials (Section 6).
Countermeasures
Ventilation Control: Section 2.
Personnel Protection: Section 3.
Storage and Handling: Section 7.

JAVEL WATER
General Information
Description: A solution of chlorinated potash.
Hazard Analysis
See hypochlorites.

JET FUELS
General Information
Description: Petroleum products similar to kerosene, a number of different types are used.
JP-1: Constants: flash p: 95–145° F, autoign. temp.: 442° F.
JP-4: Composition: 65% gasoline, 35% light petroleum distillate.
Constants: Flash p: −10° to 30° F, autoign. temp.: 468° F.
JP-5: Description: Specially refined kerosene.
Constants: flash p: 95–145° F, autoign. temp.: 475° F (approximate).
JP-6: Description: A higher kerosene cut than JP-4, with fewer impurities.
Constants: Flash p: 100° F (O.C.), autoign. temp.: 435° F, lel: 0.6%, uel: 3.7%, d: 0.8, vap. d.: < 1, bp: 250° F.

Hazard Analysis
Fire Hazard: JP-1, JP-5, JP-6: Moderate, when exposed to heat or flame.
JP-4: Dangerous, when exposed to heat or flame.

JET THRUST UNIT (jato)
Shipping Regulations:
I.C.C. (class A Explosives): Explosive A, not accepted.
(class B Explosives): Explosive B, red label, 55 pounds.
IATA: Explosive, not acceptable (passenger and cargo).

J-O
General Information
Description: A mixture containing white phosphorus.
Hazard Analysis and Countermeasures
See phosphorus, white.

JUNIPER BERRIES
General Information
Description: Dried ripe fruit of Juniperus communis L. pinaceae.
Hazard Analysis
Toxic Hazard Rating:
Acute Local: Irritant 2; Allergen 1; Ingestion 2.
Acute Systemic: Ingestion 2; Inhalation 2.
Chronic Local: Allergen 1.
Chronic Systemic: U.
Toxicology: The oil of juniper berries can be irritating to the skin. If taken internally, a severe kidney irritation similar to that caused by turpentine may result.
Fire Hazard: Unknown.
Disaster Hazard: Slightly dangerous; when heated, it emits toxic fumes.
Countermeasures
Personnel Protection: Section 3.
Storage and Handling: Section 7.

JUNIPER TAR. See cade oil.

JUTE
A combustible fiber. In the form of dust it may be highly flammable. See cotton dust.

TOXIC HAZARD RATING CODE *(For detailed discussion, see Section 1.)*

0 NONE: (a) No harm under any conditions; (b) Harmful only under unusual conditions or overwhelming dosage.

1 SLIGHT: Causes readily reversible changes which disappear after end of exposure.

2 MODERATE: May involve both irreversible and reversible change not severe enough to cause death or permanent injury.

3 HIGH: May cause death or permanent injury after very short exposu to small quantities.

U UNKNOWN: No information on humans considered valid by autho

K

K-101
See p-chlorophenyl-p-chlorobenzene sulfonate.

KAOLIN
General Information
Synonym: China clay, hydrated aluminum silicate.
Description: White or yellowish white, earthy mass or powder, insoluble in water.
Formula: $H_2Al_2Si_2O_8 \cdot H_2O$.
Constant: Mol wt: 258.2.
Hazard Analysis
Toxic Hazard Rating:
  Acute Local: Inhalation 1.
  Acute Systemic: 0.
  Chronic Local: Inhalation 1.
  Chronic Systemic: U.
Toxicology: Prolonged inhalation of high concentrations may cause deposits to form on lungs.

KAPOK
General Information
Description: A fibrous material.
Hazard Analysis
Toxic Hazard Rating:
  Acute Local: Allergen 1; Inhalation 1.
  Acute Systemic: 0.
  Chronic Local: Allergen 1; Inhalation 1.
  Chronic Systemic: 0.
Caution: Gives off a nuisance dust. Inhalation can cause coughing and sneezing.
Fire Hazard: Slight, when exposed to heat or flame; can react with oxidizing materials (Section 6).
Explosion Hazard: Slight, in the form of dust when exposed to flame.
Countermeasures
Ventilation Control: Section 2.
Storage and Handling: Section 7.

KARATHANE. See 4,6-dinitro-(2-capryl) phenyl crotonate.

KARAYA GUM
General Information
Synonym: Indian tragacanth.
Description: Fine white powder; slight odor of acetic acid.
Formula: Dried exudate of the tree Sterculia ureus.
Hazard Analysis
Toxic Hazard Rating:
  Acute Local: Allergen 1.
  Acute Systemic: U.
  Chronic Local: Allergen 1.
  Chronic Systemic: U.
Note: A stabilizer food additive (Section 10).

KARMEX. See diuron.

KATCHUNG OIL. See peanut oil.

KELP
General Information
Description: A large, coarse seaweed.
Composition: Dried kelp contains from 2–4% ammonia, 1–2% phosphoric acid, 15–20% potash, and traces of iodine.
Hazard Analysis
Toxic Hazard Rating: U. A food additive permitted in food for human consumption (Section 10).

KELP DUST
Hazard Analysis
Toxicity: Details unknown. The dried material, if finely divided, may form a nuisance dust.
Fire Hazard: Slight, when exposed to heat or flame (Section 6).
Countermeasures
Ventilation Control: Section 2.

KELTHANE. See 1,1-bis(p-chlorophenyl)-2,2,2-trichloro-ethanol.

KEPONE. See decachlorooctahydro-1,3,4-metheno-2H-cyclobuta (cd) pentalen-2-one.

KERMES MINERAL. See antimony sulfurated.

KEROSENE
General Information
Synonym: Fuel oil No. 1.
Description: A pale yellow to water white oily liquid.
Formula: A mixture of petroleum hydrocarbons, chiefly of the methane series, having 10–16 C atoms.
Constants: Bp: 175–325°C, ulc: 40, flash p: 100–165°F (C.C.), d: 0.80–< 1.0, lel: 0.7%, uel: 5.0%, autoign. temp.: 444°F, vap. d.: 4.5.
Hazard Analysis
Toxic Hazard Rating:
  Acute Local: Irritant 1; Ingestion 2; Inhalation 1.
  Acute Systemic: Ingestion 2; Inhalation 2.
  Chronic Local: Irritant 2.
  Chronic Systemic: Inhalation 1.
Toxicology: Inhalation of high concentrations of vapor can cause headache and stupor. Ingestion causes irritation of the stomach and intestines with nausea and vomiting. Aspiration of vomitus can cause serious pneumonitis, particularly in young children.
Fire Hazard: Moderate, when exposed to heat or flame; can react with oxidizing materials.
Explosion Hazard: Moderate, when exposed to heat or flame.
Countermeasures
Ventilation Control: Section 2.
To Fight Fire: Foam, carbon dioxide, dry chemical or carbon tetrachloride (Section 6).
Personnel Protection: Section 3.
Storage and Handling: Section 7.
Shipping Regulations: Section 11.
  Coast Guard Classification: Combustible liquid.

KEROSENE, DEODERIZED. See ultrasene.

KERYL BENZENE-12
General Information
Synonym: Monododecyl benzene.
Description: Very light straw color; practically odorless.
Formula: $C_{12}H_{25}C_6H_5$.
Constants: Mol wt: 246.4, flash p: 280°F, d: 0.894 at 20°C, b. range: 291–354°C.
Hazard Analysis
Toxicity: Details unknown. Probably irritant and narcotic in high concentrations.
Fire Hazard: Slight when exposed to heat or flame; can react with oxidizing materials (Section 6).

Countermeasures
To Fight Fire: Foam, carbon dioxide, dry chemical or carbon tetrachloride (Section 6).
Storage and Handling: Section 7.

KERYL NAPHTHALENE 10
General Information
Synonym: Monodecyl naphthalene.
Description: Color of straw; odor of naphthalene.
Formula: $C_{10}H_7C_{10}H_{21}$.
Constants: Mol wt: 268.4, flash p: 325° F, d: 0.968 at 20° C, b. range: 341–420° C.
Hazard Analysis
Toxicity: Details unknown. See also naphthalene.
Fire Hazard: Slight when exposed to heat or flame; can react with oxidizing materials.
Countermeasures
To Fight Fire: Foam, carbon dioxide, dry chemical or carbon tetrachloride (Section 6).
Storage and Handling: Section 7.

KETEN. See ketene.

KETENE
General Information
Synonym: Ethenone; carbomethane; keten.
Description: Colorless gas of disagreeable taste. Decomposes in water.
Formula: CH_2CO.
Constants: Mol wt: 42.04, mp: −151° C, bp: −56° C, vap. d.: 1.45.
Hazard Analysis
Toxic Hazard Rating:
    Acute Local: Irritant 3; Ingestion 3.
    Acute Systemic: Inhalation 3.
    Chronic Local: U.
    Chronic Systemic: Inhalation 3.
TLV: ACGIH (recommended) 0.5 parts per million of air; 0.9 milligrams per cubic meter of air.
Toxicology: Can cause pulmonary edema.
Countermeasures
Ventilation Control: Section 2.
Personnel Protection: Section 3.

β-KETOBUTYRANILIDE. See acetoacetanilide.

KETOCYCLOPENTANE. See cyclopentanone.

β-KETOGLUTARIC ACID. See acetone dicarboxylic acid.

KETOHEXAMETHYLENE. See cyclohexanone.

2′-KETO-3,4-IMIDOAZOLIDO-2-TETRAHYDROTHIO-PHENE-n-VALERIC ACID. See biotin.

KETONE PROPANE. See acetone.

KETONES
Organic compounds containing the chemical group = CO derived from secondary alcohols by oxidation. Acetone, which is dimethyl ketone, is the most familiar of this group of compounds. See acetone.

No general statement can be made as to the toxicity of ketones. Some are highly volatile and hence may have narcotic or anesthetic effects. Skin absorption as well as inhalation may be an important route of entry into the body. None of the ketones has been shown to have a high degree of chronic toxicity.

Some of them are dangerous fire hazards. See diethyl ketone.
    Common air contaminants. See Section 4.

KIESELGUHR. See diatomaceous earth.

KILL-TONE POWDER. See arsenic oxide and copper compounds.

KING'S GREEN. See copper acetoarsenite.

KOLKER ACARICIDE. See p-chlorophenyl-p-chlorobenzene sulfonate.

KRYOFIN
General Information
Synonym: Methoxyacetyl-p-phenetidine.
Description: White needles, tasteless but becoming bitter on chewing.
Formula: $CH_3OCH_2CONHC_6H_4OC_2H_5$.
Constants: Mol wt: 209.24, mp: 98° C.
Hazard Analysis
Toxic Hazard Rating:
    Acute Local: U.
    Acute Systemic: Ingestion 2.
    Chronic Local: U.
    Chronic Systemic: Ingestion 2.
Fire Hazard: Slight (Section 6).
Countermeasures
Personal Hygiene: Section 3.

KRYPTON
General Information
Description: Colorless inert gas.
Formula: Kr.
Constants: At wt: 83.8, mp: −156.7° C, bp: −152.0° C, d: 3.708 g per liter at 0° C, (liq) 2.155 at −152.9° C.
Hazard Analysis
Toxicity: A simple asphyxiant in large concentrations. See also argon.
Radiation Hazard: Section 5. For permissible levels, see Table 5, p. 150.
    Artificial isotope ^{85m}Kr, half life 4.4 h. Decays to stable ^{85}Rb by emitting beta particles (81%) of 0.82 MeV. Also decays to radioactive ^{85}Kr by emitting gamma rays of 0.30 MeV. Emits other gamma rays of 0.15 MeV.
    Artificial isotope ^{85}Kr, half life 10.6 y. Decays to stable ^{85}Rb by emitting beta particles of 0.67 MeV. ^{85}Kr, like ^{41}A is produced by activation of natural krypton in air cooled reactors. Also it is a direct fission product.
    Artificial isotope ^{87}Kr, half life 78 m. Decays to radioactive ^{87}Rb by emitting beta particles of 1.25 (25%), 3.3 (10%), 3.8 (65%) MeV. Also emits gamma rays of 0.40, 0.85, 2.57 MeV.
    Artificial isotope ^{88}Kr, half life 2.8 h. Decays to radioactive ^{88}Rb by emitting beta particles of 0.52 (70%), 0.9 (10%), 2.7 (20%) MeV. Also emits gamma rays of 0.17–2.40 MeV.
    ^{88}Rb in turn decays with a half life of 18 minutes to stable ^{88}Sr by emitting beta particles of 2.5 (14%), 3.4 (4%), 5.2 (76%) MeV and gamma rays of 0.91–4.87 MeV.
Countermeasures
Storage and Handling: Section 7.
Shipping Regulations: Section 11.
    IATA: Nonflammable gas, green label, 70 kilograms (passenger), 140 kilograms (cargo).

K-STROPHANTHIN. See strophanthin.

KURON. See 2-(2,4,5-trichlorophenoxy)-propionic acid propylene glycol butyl ether ester.

TOXIC HAZARD RATING CODE *(For detailed discussion, see Section 1.)*

0 NONE: (a) No harm under any conditions; (b) Harmful only under unusual conditions or overwhelming dosage.

1 SLIGHT: Causes readily reversible changes which disappear after end of exposure.

2 MODERATE: May involve both irreversible and reversible changes not severe enough to cause death or permanent injury.

3 HIGH: May cause death or permanent injury after very short exposure to small quantities.

U UNKNOWN: No information on humans considered valid by authors

L

LACQUER BASE OR CHIPS, DRY. See lacquers, nitrocellulose.
Shipping Regulations: Section 11.
  I.C.C.: Flammable solid; yellow label, 100 pounds.
  Coast Guard Classification: Inflammable solid; yellow label.
  IATA: Flammable solid, yellow label, 12 kilograms (passenger), 45 kilograms (cargo).

LACQUER BASE OR CHIPS, PLASTIC (wet with alcohol or solvent). See specific component.
Shipping Regulations: Section 11.
  I.C.C.: Flammable liquid; red label, 25 pounds.
  Coast Guard Classification: Combustible liquid; inflammable liquid; red label.
  IATA: Flammable liquid, red label, 1 liter (passenger), 12 kilograms (cargo).

LACQUER DILUENT
General Information
Description: Insoluble in water.
Constants: d: 0.7, bp: 190–225°F, flash p: 12°F, autoign. temp.: 450–550°F, lel: 1.2%, uel: 6.0%.
Hazard Analysis
Toxic Hazard Rating: U.
Fire Hazard: Dangerous, when exposed to heat or flame.

LACQUERS
General Information
Description: Solutions of resins, gums or plastics in an organic solvent.
Constant: Flash p: 0–80°F.
Hazard Analysis
Toxicity: Variable; they may have allergic effects. Common air contaminants. (See Section 4.)
Fire Hazard: Dangerous, when exposed to heat or flame. A large part of the great fire hazard of lacquers is due to the solvents commonly used. Even when solvent free, however, nitrocellulose lacquers are highly flammable. See lacquers, nitrocellulose.
Explosion Hazard: Severe, in the form of vapor when exposed to flame (Section 7). See also individual solvents as well as individual solid component.
Disaster Hazard: Dangerous; keep away from heat and open flame! They can react vigorously with oxidizing materials.
Explosive Range: Variable.
Countermeasures
Ventilation Control: Section 2.
To Fight Fire: Carbon dioxide, dry chemical, carbon tetrachloride, or water spray (Section 6).
Storage and Handling: Section 7.
Shipping Regulations: Section 11.
  Coast Guard Classification: Combustible material.

LACQUERS, NITROCELLULOSE
General Information
Constant: Flash p: 40°F.
Hazard Analysis
Toxicity: Variable, they may have allergic effects.
Fire Hazard: Highly dangerous, when exposed to heat or flame.
Explosion Hazard: Moderate, when exposed to heat or flame.

Disaster Hazard: Dangerous. On decomposition they emit highly toxic fumes.
Countermeasures
Storage and Handling: Section 7.
To Fight Fire: Carbon dioxide, dry chemical or carbon tetrachloride (Section 6).
Ventilation Control: Section 2.
Shipping Regulations: Section 11.
  Coast Guard Classification: Combustible liquid; inflammable liquid; red label.

LACTIC ACID
General Information
Synonym: Ethylidenelactic acid.
Description: Yellow or colorless, thick liquid.
Formula: $CH_3CHOHCOOH$.
Constants: Mol wt: 90.08, mp: 18°C, bp: 122°C at 15 mm, d: 1.249 at 15°C.
Hazard Analysis
Toxic Hazard Rating:
  Acute Local: Irritant 2; Ingestion 2.
  Acute Systemic: U.
  Chronic Local: U.
  Chronic Systemic: U.
Note: Used as a general purpose food additive (Section 10).
Disaster Hazard: Slight; emits acrid fumes when heated.
Countermeasures
Personnel Protection: Section 3.
Storage and Handling: Section 7.

LACTIDE. See 3,6-dimethyl-2,5-p-dioxanedione.

LACTOL SPIRITS. See naphtha, petroleum.

LACTONES
Lactones are esters of hydroxy acids. Toxicity is variable but in general seems to be relatively low. An exception is β-propiolactone.

LACTONITRILE
General Information
Synonym: 2-Hydroxypropanenitrile.
Description: Straw-colored liquid.
Formula: $CH_3CH(OH)CN$.
Constants: Mol wt: 71.08, mp: −40°C, bp: 103°C at 50 mm, fp: −34°C, flash p: 170°F (T.C.C.), d: 0.9834 at 25°C, vap. d.: 2.45.
Hazard Analysis
Toxicity: Highly toxic. See nitriles.
Caution: In the presence of alkali, it evolves hydrocyanic acid.
Fire Hazard: Moderate, when exposed to heat or flame.
Explosion Hazard: See cyanides.
Disaster Hazard: Dangerous; see cyanides; can react vigorously with oxidizing materials.
Countermeasures
Ventilation Control: Section 2.
Storage and Handling: Section 7.
To Fight Fire: Foam, carbon dioxide, dry chemical or carbon tetrachloride. Section 6.

LAMP BLACK. See carbon.

LANOLIN
General Information
Synonym: Wool grease.

Constants: Mp: 37.9°C, flash p: 460°F (C.C.), d: < 1, autoign. temp.: 833°F.

Hazard Analysis
Toxicity: A mild allergen.
Fire Hazard: Slight when exposed to heat or flame; can react with oxidizing materials. Spontaneous Heating: Yes.

Countermeasures
To Fight Fire: Foam, carbon dioxide, dry chemical or carbon tetrachloride (Section 6).
Storage and Handling: Section 7.

LANOXIN. See digoxin.

LANTHANA. See lanthanum oxide.

LANTHANUM
General Information
Description: Lead-gray metal.
Constants: At wt: 138.92, mp: 826°C, bp: 1800°C, d: 6.15.
Hazard Analysis
Toxic Hazard Rating:
  Acute Local: U.
  Acute Systemic: Inhalation 1, Ingestion 1.
  Chronic Local: U.
  Chronic Systemic: U.
Toxicity: Lanthanum and other lanthanons can cause delayed blood clotting thus leading to hemmorages. Has caused liver injury in experimental animals.
Radiation Hazard: Section 5. For permissible levels, see Table 5, p. 150.
    Artificial isotope ^{140}La, half life 40 h. Decays to stable ^{140}Ce by emitting beta particles of 0.83–2.20 MeV. Also emits gamma rays of 0.32–2.5 MeV. ^{140}La usually exists in equilibrium with its parent, ^{140}Ba. Permissible levels are given for the equilibrium mixture.
Fire Hazard: Dangerous in the form of dust when exposed to flame; can react vigorously with oxidizing materials. See also powdered metals.
Explosion Hazard: Moderate in the form of dust when exposed to flame or by chemical reaction. See also powdered metals.
Countermeasures
Storage and Handling: Section 7.

LANTHANUM BORIDE
General Information
Synonym: Lanthanum hexaboride.
Description: Red-purple, metallic crystals.
Formula: LaB_6.
Constants: Mol wt: 203.8, mp: 2210°C, bp: decomposes, d: 2.61.
Hazard Analysis and Countermeasures
See boron hydrides, and Lanthanum.

LANTHANUM BROMATE
General Information
Description: Crystals.
Formula: $La(BrO_3)_3 \cdot 9H_2O$.
Constants: Mol wt: 684.81, mp: 37.5°C.
Hazard Analysis
Toxic Hazard Rating:
  Acute Local: U.
  Acute Systemic: Ingestion 2; Inhalation 2.
  Chronic Local: 0.
  Chronic Systemic: U.
Toxicity: See also Lanthanum.
Fire Hazard: See bromates.

Explosion Hazard: See bromates.
Disaster Hazard: See bromates.
Countermeasures
Storage and Handling: Section 7.

LANTHANUM BROMIDE
General Information
Description: Colorless crystals.
Formula: $LaBr_3 \cdot 7H_2O$.
Constants: Mol wt: 504.78, mp: anh. 783 ± 3°C, d: 5.057 at 25°C.
Hazard Analysis and Countermeasures
See bromides and lanthanum.

LANTHANUM CARBIDE
General Information
Description: Yellow crystals.
Formula: LaC_2.
Constants: Mol wt: 162.94, d: 5.02.
Hazard Analysis and Countermeasures
See lanthanum and carbides.

LANTHANUM HEXAANTIPYRINE PERCHLORATE
General Information
Description: Colorless, hexagonal crystals.
Formula: $[La(C_{11}H_{12}N_2O)_6] \cdot (ClO_4)_3$.
Constants: Mol wt: 1566.62, mp: 290–305°C decomposes.
Hazard Analysis and Countermeasures
See perchlorates and lanthanum.

LANTHANUM HEXABORIDE. See lanthanum boride.

LANTHANUM IODATE
General Information
Description: Colorless crystals.
Formula: $La(IO_3)_3$.
Constant: Mol wt: 663.68.
Hazard Analysis and Countermeasures
See iodates and lanthanum.

LANTHANUM IODIDE
General Information
Description: Crystals.
Formula: LaI_3.
Constants: Mol wt: 519.68, mp: 761 ± 2°C, d: 5.057 at 25°C.
Hazard Analysis and Countermeasures
See iodides and lanthanum.

LANTHANUM OXIDE
General Information
Synonyms: Lanthana; lanthanum trioxide.
Description: White, amorphous oxide; hisses in moist air like quicklime.
Formula: La_2O_3.
Constants: Mol wt: 325.84, mp: 2315°C, bp: 4200°C, d: 6.51.
Hazard Analysis
Toxic Hazard Rating:
  Acute Local: U.
  Acute Systemic: Inhalation 1.
  Chronic Local: U.
  Chronic Systemic: U.
Toxicity: See also lanthanum.
Countermeasures
Ventilation Control: Section 2.

TOXIC HAZARD RATING CODE *(For detailed discussion, see Section 1.)*

0 NONE: (a) No harm under any conditions; (b) Harmful only under unusual conditions or overwhelming dosage.

1 SLIGHT: Causes readily reversible changes which disappear after end of exposure.

2 MODERATE: May involve both irreversible and reversible changes; not severe enough to cause death or permanent injury.

3 HIGH: May cause death or permanent injury after very short exposure to small quantities.

U UNKNOWN: No information on humans considered valid by authors.

LANTHANUM SULFATE
General Information
Description: White, hygroscopic powder.
Formula: $La_2(SO_4)_3$.
Constants: Mol wt: 566, mp: 1150°C (decomposes), d: 3.60 at 15°C.
Hazard Analysis
Toxic Hazard Rating:
    Acute Local: Irritant 2; Ingestion 2; Inhalation 2.
    Acute Systemic: U.
    Chronic Local: U.
    Chronic Systemic: U.
Toxicity: See also lanthanum.
Caution: Sulfuric acid is formed upon hydrolysis of this material.
Countermeasures
Ventilation Control: Section 2.
Personnel Protection: Section 3.

LANTHANUM SULFIDE
General Information
Description: Red-yellow crystals.
Formula: La_2S_3.
Constants: Mol wt: 374.04, mp: 2110–2150°C (in vacuo), d: 5.0 at 0°/0°C.
Hazard Analysis and Countermeasures
See sulfides and lanthanum.

LANTHANUM TRIOXIDE. See lanthanum oxide.

LARD OIL (COMMERCIAL OR ANIMAL)
General Information
Description: Colorless or pale yellow liquid.
Composition: Olein, stearin.
Constants: Mp: −2°C, ulc: 10–20, flash p: 395°F (C.C.), d: 0.905–0.915, autoign. temp.: 883°F.
Hazard Analysis
Toxicity: A very weak allergen.
Fire Hazard: Slight when exposed to heat or flame; can react with oxidizing materials.
Spontaneous Heating: Yes.
Countermeasures
To Fight Fire: Foam, carbon dioxide, dry chemical or carbon tetrachloride (Section 6).
Storage and Handling: Section 7.

LARD OIL (PURE)
General Information
Synonym: No. 2 mineral oil.
Description: Colorless or pale yellow liquid.
Constants: Mp: −2°C, flash p: 500°F, d: 0.9, ulc = 10–20; flash p: (#1) = 440°F., (#2) = 419°F.
Hazard Analysis
Toxicity: Unknown. A substance migrating to food from packaging materials (Section 10).
Fire Hazard: Slight when exposed to heat or flame; can react with oxidizing materials.
Spontaneous Heating: Yes.
Countermeasures
To Fight Fire: Foam, carbon dioxide, dry chemical or carbon tetrachloride (Section 6).
Storage and Handling: Section 7.

LARKSPUR. See delphinium.

LATEX, RUBBER
Hazard Analysis
Toxic Hazard Rating:
    Acute Local: Irritant 2; Allergen 1; Ingestion 2.
    Acute Systemic: U.
    Chronic Local: Allergen 2.
    Chronic Systemic: U.
Countermeasures
Ventilation Control: Section 2.
Personnel Protection: Section 3.

Storage and Handling: This material is shipped in sealed drums which can build up pressure from within; unless care is used in opening, this can endanger personnel. Latex contains ammonia as a preservative, and caution should be exercised in opening the drums not to inhale the fumes evolved (Section 7).

LAUDANIDINE. See tritopine.

LAUDANINE
General Information
Synonym: dl-Laudanine.
Description: Small trimetric yellowish-white powder.
Formula: $C_{20}H_{25}NO_4$.
Constants: Mol wt: 343.41, mp: 166°C, d: 1.26 at 26°/4°C.
Hazard Analysis
Toxic Hazard Rating:
    Acute Local: U.
    Acute Systemic: Ingestion 3; Inhalation 3.
    Chronic Local: U.
    Chronic Systemic: U.
Fire Hazard: Unknown.
Disaster Hazard: Moderately dangerous; when heated to decomposition it emits toxic fumes.
Countermeasures
Ventilation Control: Section 2.
Personal Hygiene: Section 3.
First Aid: Section 1.
Storage and Handling: Section 7.

LAUDANOSINE
General Information
Synonym: 1-N-Methyltetrahydropapaverine.
Description: White needles.
Formula: $C_{21}H_{27}NO_4$.
Constants: Mol wt: 357.44, mp: 89–90°C.
Hazard Analysis
Toxic Hazard Rating:
    Acute Local: U.
    Acute Systemic: Ingestion 3; Inhalation 3.
    Chronic Local: U.
    Chronic Systemic: U.
Fire Hazard: Unknown.
Disaster Hazard: Dangerous; when heated to decomposition, it emits toxic fumes.
Countermeasures
Ventilation Control: Section 2.
Personal Hygiene: Section 3.
First Aid: Section 1.
Storage and Handling: Section 7.

LAUDANUM
General Information
Description: Brown liquid.
Hazard Analysis
Toxic Hazard Rating:
    Acute Local: U.
    Acute Systemic: Ingestion 3; Inhalation 3.
    Chronic Local: U.
    Chronic Systemic: Ingestion 2; Inhalation 2.
Fire Hazard: Unknown.
Disaster Hazard: Moderately dangerous; when heated to decomposition, it emits toxic fumes.
Countermeasures
Ventilation Control: Section 2.
Personal Hygiene: Section 3.
First Aid: Section 1.
Storage and Handling: Section 7.

LAUGHING GAS. See nitrous oxide.

LAUREL CAMPHOR. See camphor.

LAURIC ACID
General Information
Synonym: Dodecanoic acid.

Description: Colorless, needle-like crystals.

Formula: $CH_3(CH_2)_{10}COOH$.

Constants: Mol wt: 200.3, mp: 48° C, bp: 299.2° C, d: 0.883, vap. press.: 1 mm at 121.0° C.

Hazard Analysis

Toxicity: Animal data suggest low toxicity for lauric acid esters. Details unknown.

Fire Hazard: Slight, when exposed to heat or flame; can react with oxidizing materials (Section 6).

Countermeasures

Storage and Handling: Section 7.

LAURIC ACID 2-THIOCYANOETHYL ESTER
General Information

Synonym: Lethane 60.

Description: Powder.

Hazard Analysis

Toxicity: Details unknown. Irritating to skin; narcotic in high concentrations.

LAURITE. See ruthenium sulfide.

LAUROYL PEROXIDE
General Information

Synonym: Dodecanoyl peroxide. Alperox C.

Description: White, tasteless, coarse powder, faint odor.

Formula: $(C_{11}H_{23}CO)_2O_2$.

Constants: Mol wt: 398.60, mp: 53–55° C.

Hazard Analysis

Toxic Hazard Rating:
  Acute Local: Irritant 3; Ingestion 3; Inhalation 3.
  Acute Systemic: U.
  Chronic Local: U.
  Chronic Systemic: U.

Toxicology: It can irritate and cause burns upon the skin and mucous membranes; it is a powerful oxidizing agent. See also peroxides, organic.

Fire Hazard: See peroxides, organic.

Disaster Hazard: See peroxides, organic.

Countermeasures

Ventilation Control: Section 2.

Personnel Protection: Section 3.

First Aid: Section 1.

Storage and Handling: Section 7. Store away from sources of heat. Keep as cool as possible, but below 27° C. Do not add hot material. See peroxides, organic.

Shipping Regulations: Section 11.
  I.C.C.: Oxidizing material; yellow label. Flammable solid, 25 pounds.
  Coast Guard Classification: Oxidizing material; yellow label.
  IATA: Oxidizing material, yellow label, 1 kilogram (passenger), 12 kilograms (cargo).

LAURYL ALCOHOL. See 1-dodecanol.

LAURYL BROMIDE
General Information

Synonym: Dodecyl bromide.

Description: An amber liquid.

Formula: $CH_3(CH_2)_{10}CH_2Br$.

Constants: Mol wt: 249.2, mp: $< -40°$ C, bp: 136.6–193.5° C at 10 mm, flash p: 291° F, d: 1.021 at 25°/25° C, vap. d.: 8.6.

Hazard Analysis and Countermeasures

See bromides.

LAURYL CHLORIDE
General Information

Synonym: Dodecyl chloride.

Description: Clear, water-white, oily liquid.

Formula: $CH_3(CH_2)_{10}CH_2Cl$.

Constants: Mol wt: 204.8, mp: $-19°$ C, bp: 112–160° C, flash p: 235° F, d: 0.87, vap. d.: 7.07.

Hazard Analysis

Toxicity: Unknown.

Fire Hazard: Slight when exposed to heat or flame.

Disaster Hazard: Moderately dangerous; when heated to decomposition it emits toxic fumes; it can react with oxidizing materials.

Countermeasures

Storage and Handling: Section 7.

To Fight Fire: Foam, carbon dioxide, dry chemical or carbon tetrachloride (Section 6).

LAURYL ISOQUINOLINIUM BROMIDE
General Information

Synonym: Isothan.

Description: Deep amber, water-soluble liquid. Pleasant characteristic odor.

Hazard Analysis

Toxicity: Details unknown. See also bromides. Highly toxic. Animal experiments suggest high toxicity. A fungicide.

Fire Hazard: Slight when exposed to heat or flame (Section 6).

Disaster Hazard: Dangerous; when heated to decomposition, it emits toxic fumes; can react with oxidizing materials.

Countermeasures

Storage and Handling: Section 7.

LAURYL MERCAPTAN
General Information

Synonym: Dodecyl mercaptan.

Description: Water-white to pale yellow liquid.

Formula: $C_{12}H_{25}SH$.

Constants: Mol wt: 202.5, mp: $-7°$ C, bp: 115–177° C, flash p: 262° F, d: 0.849 at 15.5°/15.5° C.

Hazard Analysis and Countermeasures

See mercaptans.

LAURYL PYRIDINIUM CHLORIDE
General Information

Synonym: Dodecyl pyridinium chloride.

Description: Mottled tan semi-solid.

Formula: $C_5H_5NC_{12.6}H_{26.2}Cl$.

Constants: Mol wt: 292.0, flash p: 347° F.

Hazard Analysis

Toxicity: Unknown.

Fire Hazard: Slight when exposed to heat or flame.

Disaster Hazard: Moderately dangerous; when heated to decomposition it emits toxic fumes; can react with oxidizing materials.

Countermeasures

Storage and Handling: Section 7.

To Fight Fire: Foam, carbon dioxide, dry chemical or carbon tetrachloride (Section 6).

LAURYL QUINALDINIUM BROMIDE
Hazard Analysis

Toxicity: Details unknown. See also bromides.

Fire Hazard: Unknown.

Disaster Hazard: Dangerous; see bromides.

TOXIC HAZARD RATING CODE (For detailed discussion, see Section 1.)

0 NONE: (a) No harm under any conditions; (b) Harmful only under unusual conditions or overwhelming dosage.

1 SLIGHT: Causes readily reversible changes which disappear after end of exposure.

2 MODERATE: May involve both irreversible and reversible changes; not severe enough to cause death or permanent injury.

3 HIGH: May cause death or permanent injury after very short exposure to small quantities.

U UNKNOWN: No information on humans considered valid by authors.

Countermeasures
Storage and Handling: Section 7.

LAURYL QUINOLINIUM CHLORIDE
Hazard Analysis
Toxicity: Details unknown; a fungicide.
Fire Hazard: Unknown.
Disaster Hazard: Dangerous; see chlorides.
Countermeasures
Storage and Handling: Section 7.

LAURYL THIOCYANATE
General Information
Formula: $CH_3(CH_2)_{10}CH_2SCN$.
Constant: Mol wt: 227.3.
Hazard Analysis and Countermeasures
See thiocyanates. An insecticide.

LAWRENCITE. See ferrous chloride.

LAWRENCIUM
General Information
Description: A synthetic transuranium element: At. No. 103.
Formula: Lw.
Constant: At wt: 257.
Hazard Analysis
Radioactive.

LEAD
General Information
Synonym: Plumbum.
Description: Bluish-gray, soft metal.
Formula: Pb.
Constants: At wt: 207.21, mp: 327.43°C, bp: 1620°C, d: 11.288 at 20°/20°C, vap. press.: 1 mm at 973°C.
Hazard Analysis
Toxic Hazard Rating:
Acute Local: 0.
Acute Systemic: Inhalation 3.
Chronic Local: 0.
Chronic Systemic: Ingestion 3; Inhalation 3.
Toxicology: See lead compounds. A common air contaminant (Section 4).
TLV: ACGIH (recommended); 0.2 milligrams per cubic meter of air.
Radiation Hazard: Section 5. For permissible levels, see Table 5, p. 150.
Artificial isotope ^{203}Pb, half life 52 h. Decays to stable ^{203}Tl by electron capture. Emits gamma rays of 0.28 MeV and others and X-rays.
Natural isotope ^{210}Pb (Radium-D, Uranium Series), half life 22 y. Decays to radioactive ^{210}Bi by emitting beta particles of 0.015 (80%), 0.061 (20%) MeV. Also emits gamma rays of 0.046 MeV. ^{210}Pb usually exists in equilibrium with its daughters ^{210}Bi and ^{210}Po.
Natural isotope ^{212}Pb (Thorium-B, Thorium Series), half life 10.6 h. Decays to radioactive ^{212}Bi by emitting beta particles of 0.16 (5%), 0.34 (81%), 0.58 (14%) MeV. Also emits gamma rays of 0.24, 0.30 MeV and X-rays.
Fire Hazard: Moderate, in the form of dust when exposed to heat or flame. See also powdered metals.
Explosion Hazard: Moderate, in the form of dust when exposed to heat or flame.
Disaster Hazard: Dangerous; when heated it emits highly toxic fumes; can react vigorously with oxidizing materials.
Countermeasures
Ventilation Control: Section 2.
Personal Hygiene: Section 3.
First Aid: Section 1.
Storage and Handling: Section 7.

LEAD ACETATE
General Information
Synonym: Sugar of lead.

Description: White crystals, soluble in water. Commercial grades are frequently brown or gray lumps.
Formula: $Pb(C_2H_3O_2)_2 \cdot 3H_2O$.
Constants: Mol wt: 379.35, mp: 75°C; anhydrous 280°C, d: 2.55.
Hazard Analysis and Countermeasures
See lead compounds. An insecticide.
Shipping Regulations: Section 11.
IATA: Poison B, poison label; 25 kilograms (passenger), 95 kilograms (cargo) (solid).

LEAD ACETATE, MONOBASIC
General Information
Description: White powder.
Formula: $Pb_2OH(C_2H_3O_2)_3$.
Constant: Mol wt: 608.6.
Hazard Analysis and Countermeasures
See lead compounds.

LEAD ANTIMONATE
General Information
Synonyms: Naples yellow; antimony yellow.
Description: Orange-yellow powder.
Formula: $Pb_3(SbO_4)_2$.
Constant: Mol wt: 993.2.
Hazard Analysis and Countermeasures
See lead and antimony compounds.

LEAD ARSENATES
General Information
Synonyms: Lead o-arsenate; lead di-o-arsenate; lead mono-o-arsenate; lead pyro-arsenate; lead m-arsenate.
Description: White crystals.
Hazard Analysis
Toxicity: Highly toxic. See lead compounds and arsenic compounds.
TLV: ACGIH (recommended); 0.15 milligrams per cubic meter of air.
Disaster Hazard: Dangerous; on heating it emits highly toxic fumes.
Countermeasures
Storage and Handling: Section 7.
Shipping Regulations: Section 11.
I.C.C.: Poison B; poison label, 200 lbs (solid).
Coast Guard Classification: Poison B; poison label.
IATA: Poison B, poison label, 25 kilograms (passenger), 95 kilograms (cargo) (solid).

LEAD-m-ARSENATE. See lead arsenates.

LEAD-o-ARSENATE. See lead arsenates.

LEAD ARSENITES
General Information
Synonyms: Lead o-arsenite; lead m-arsenite.
Description: White powder.
Hazard Analysis
Toxicity: Highly toxic. See lead compounds and arsenic compounds.
Disaster Hazard: Dangerous; on heating it emits highly toxic fumes.
Countermeasures
Storage and Handling: Section 7.
Shipping Regulations: Section 11.
I.C.C.: Poison B; poison label, 200 pounds (solid).
Coast Guard Classification: Poison B; poison label.
IATA: Poison B, poison label, 25 kilograms (passenger), 95 kilograms (cargo) (solid).

LEAD m-ARSENITE. See lead arsenites.

LEAD-o-ARSENITE. See lead arsenites.

LEAD AZIDE
General Information
Description: Colorless needles.

Formula: $Pb(N_3)_2$.
Constant: Mol wt: 291.26.
Hazard Analysis
Toxicity: See lead compounds and azides.
Fire Hazard: Unknown.
Explosion Hazard: Severe, when shocked or exposed to heat or flame. Explodes at 350°C (Section 7).
Disaster Hazard: Highly dangerous; shock and heat will explode it; when heated it emits highly toxic fumes of lead.
Countermeasures
Storage and Handling: Section 7.
Shipping Regulations: Section 11.
Coast Guard Classification: Explosive A.

LEAD BENZOATE
General Information
Description: White crystals.
Formula: $Pb(C_7H_5O_2)_2 \cdot H_2O$.
Constants: Mol wt: 467.5, mp: $-H_2O$ at 100°C.
Hazard Analysis and Countermeasures
See lead compounds and lead.

LEAD m-BORATE
General Information
Description: White powder.
Formula: $Pb(BO_2)_2 \cdot H_2O$.
Constants: Mol wt: 310.87, d: 5.598 (anhydrous).
Hazard Analysis and Countermeasures
See lead and boron compounds.

LEAD BROMATE
General Information
Description: Monoclinic crystals.
Formula: $Pb(BrO_3)_2 \cdot H_2O$.
Constants: Mol wt: 481.06, mp: decomposes 180°C, d: 5.53.
Hazard Analysis and Countermeasures
See lead compounds and bromates.

LEAD CAPRATE
General Information
Formula: $Pb(C_{10}H_{19}O_2)_2$.
Constants: Mol wt: 549.71, mp: 103–104°C.
Hazard Analysis and Countermeasures
See lead compounds and lead.

LEAD CAPROATE
General Information
Description: Crystals.
Formula: $Pb(C_6H_{11}O_2)_2$.
Constants: Mol wt: 437.51, mp: 73–74°C.
Hazard Analysis and Countermeasures
See lead compounds and lead.

LEAD CAPRYLATE
General Information
Description: White leaf.
Formula: $Pb(C_8H_{15}O_2)_2$,
Constants: Mol wt: 493.61, mp: 83.5–84.5°C.
Hazard Analysis and Countermeasures
See lead compounds and lead.

LEAD CARBONATE
General Information
Synonym: Cerussite.
Description: White, powdery crystals.
Formula: $PbCO_3$.

Constants: Mol wt: 267.22, mp: decomposes at 315°C, d: 6.6.
Hazard Analysis and Countermeasures
See lead compounds and lead.

LEAD CARBONATE, BASIC
General Information
Synonyms: White lead; hydrocerussite.
Description: White, amorphous powder.
Formula: $2PbCO_3 \cdot Pb(OH)_2$.
Constants: Mol wt: 775.67, mp: decomposes at 400°C, d: 6.14.
Hazard Analysis and Countermeasures
See lead compounds and lead.

LEAD CEROTATE
General Information
Description: White crystals.
Formula: $Pb(C_{26}H_{51}O_2)_2$.
Constants: Mol wt: 998.55, mp: 113.5°C.
Hazard Analysis and Countermeasures
See lead compounds and lead.

LEAD CHLORATE
General Information
Description: Monoclinic, white crystals.
Formula: $Pb(ClO_3)_2$.
Constants: Mol wt: 374.12, mp: decomposes, d: 3.89.
Hazard Analysis and Countermeasures
Toxicity: See lead compounds, chlorates and lead.

LEAD CHLORIDE
General Information
Synonym: Cotunnite.
Description: White crystals.
Formula: $PbCl_2$.
Constants: Mol wt: 278.12, mp: 501°C, bp: 954°C, d: 5.85, vap. press.: 1 mm at 547°C.
Hazard Analysis and Countermeasures
See lead compounds.
Shipping Regulations: Section 11.
    IATA: Other restricted articles, class B, 12 kilograms (passenger), 45 kilograms (cargo) (solid).

LEAD CHLORITE
General Information
Description: Monoclinic, yellow crystals.
Formula: $Pb(ClO_2)_2$.
Constants: Mol wt: 342.12, mp: explodes at 126°C.
Hazard Analysis and Countermeasures
See lead compounds and chlorites.

LEAD CHROMATE
General Information
Synonyms: Crocoite; chrome yellow.
Description: Yellow crystals.
Formula: $PbCrO_4$.
Constants: Mol wt: 323.22, mp: 844°C, bp: decomposes, d: 6.3.
Hazard Analysis and Countermeasures
See lead and chromium compounds.

LEAD CHROMATE, BASIC
General Information
Description: Red, amorphous or crystals.
Formula: $Pb_2(OH)_2CrO_4$.
Constants: Mol wt: 564.45, mp: 920°C.
Hazard Analysis
Toxicity: See lead and chromium compounds.

TOXIC HAZARD RATING CODE (For detailed discussion, see Section 1.)

0 NONE: (a) No harm under any conditions; (b) Harmful only under unusual conditions or overwhelming dosage.

1 SLIGHT: Causes readily reversible changes which disappear after end of exposure.

2 MODERATE: May involve both irreversible and reversible changes not severe enough to cause death or permanent injury.

3 HIGH: May cause death or permanent injury after very short exposure to small quantities.

U UNKNOWN: No information on humans considered valid by authors

Fire Hazard: Moderate, by chemical reaction with reducing agents.

Disaster Hazard: Dangerous, when heated to decomposition, it emits highly toxic fumes of lead and can react vigorously with reducing material.

Countermeasures

Storage and Handling: Section 7.

LEAD CITRATE

General Information

Description: White, crystalline powder.

Formula: $Pb_3(C_6H_5O_7)_2 \cdot 3H_2O$.

Constant: Mol wt: 1053.88.

Hazard Analysis and Countermeasures

See lead compounds.

LEAD COMPOUNDS

Hazard Analysis

Toxic Hazard Rating:

Acute Local: 0.

Acute Systemic: Ingestion 3; Inhalation 3.

Chronic Local 0.

Chronic Systemic: Ingestion 3; Inhalation 3; Skin Absorption 3.

Toxicology: Lead poisoning is one of the commonest of occupational diseases. The presence of lead-bearing materials or lead compounds in an industrial plant does not necessarily result in exposure on the part of the workman. The lead must be in such form, and so distributed, as to gain entrance into the body or tissues of the workman in measurable quantity, otherwise no exposure can be said to exist.

Mode of entry into body:

1. By inhalation of the dusts, fumes, mists or vapors. (Common air contaminants, Section 4).

2. By ingestion of lead compounds trapped in the upper respiratory tract or introduced into the mouth on food, tobacco, fingers or other objects.

3. Through the skin; this route is of special importance in the case of the organic compounds of lead, as lead tetraethyl. In the case of the inorganic forms of lead, this route is of no practical importance.

Physiological Action and Toxicity: When lead is ingested, much of it passes through the body unabsorbed, and is eliminated in the feces. The greater portion of the lead that is absorbed is caught by the liver and excreted, in part, in the bile. For this reason, larger amounts of lead are necessary to cause poisoning if absorption is by this route, and a longer period of exposure is usually necessary to produce symptoms. On the other hand, when lead is inhaled, absorption takes place easily from the respiratory tract and symptoms tend to develop more quickly. From the point of view of industrial poisoning, inhalation of lead is much more important than is ingestion.

Lead is a cumulative poison. Increasing amounts build up in the body and eventually a point is reached where symptoms and disability occur.

Lead produces a brittleness of the red blood cells so that they hemolyze with but slight trauma; the hemoglobin is not affected. Due to their increased fragility, the red cells are destroyed more rapidly in the body than normally, producing an anemia which is rarely severe. The loss of circulating red cells stimulates the production of new young cells which on entering the blood stream, are acted upon by the circulating lead, with resultant coagulation of their basophilic material. These cells after suitable staining, are recognized as "stippled cells." As regards the effect of lead on the white blood cells, there is no uniformity of opinion. In addition to its effect on the red cells of the blood, lead produces a damaging effect on the organs or tissues with which it comes in contact. No specific or characteristic lesion is produced. Autopsies of deaths attributed to

lead poisoning and experimental work on animals, have shown pathological lesion of the kidneys, liver, male gonads, nervous system, blood vessels and other tissues. None of these changes, however, have been found consistently.

In cases of lead poisoning, the amount of lead found in the blood is frequently in excess of 0.07 mg per 100 cc of whole blood. The urinary lead excretion generally exceeds 0.1 mg per liter of urine.

The toxicity of the various lead compounds appears to depend upon several factors: (1) the solubility of the compound in the body fluids; (2) the fineness of the particles of the compound; solubility is greater, of course, in proportion to the fineness of the particles; (3) conditions under which the compound is being used; where a lead compound is used as a powder, contamination of the atmosphere will be much less where the powder is kept damp. Of the various lead compounds, the carbonate, the monoxide and sulfate are considered to be more toxic than metallic lead or other lead compounds. Lead arsenate is very toxic, due to the presence of the arsenic radical. The toxicity of lead chromate, or "chrome yellow" is less than would be expected, due to its low solubility.

Signs and Symptoms: Industrial lead poisoning commonly occurs following prolonged exposure to lead or its compounds. The common clinical types of lead poisoning may be classified according to their clinical picture as (a) alimentary; (b) neuromotor; and (c) encephalic. Some cases may show a combination of clinical types. The alimentary type occurs most frequently, and is characterized by abdominal discomfort or pain. Severe cases may present actual colic. Other complaints are constipation and/or diarrhea, loss of appetite, metallic taste, nausea and vomiting, lassitude, insomnia, weakness, joint and muscle pains, irritability, headache and dizziness. Pallor, lead line on the gums, pyorrhea, loss of weight, abdominal tenderness, basophilic stippling, anemia, slight albuminuria, increased urinary excretion, and an increase in the lead content of the whole blood, are signs which may accompany the above symptoms.

In the neuromuscular type, the chief complaint is weakness, frequently of the extensor muscles of the wrist and hand, unilateral or bilateral. Other muscle groups which are subject to constant use may be affected. Gastroenteric symptoms are usually present, but are not as severe as in the alimentary type of poisoning. Joint and muscle pains are likely to be more severe. Headache, dizziness and insomnia are frequently prominent. True paralysis is uncommon, and usually is the result of prolonged exposure.

Lead encephalopathy is the most severe but the rarest manifestation of lead poisoning. In the industrial worker it follows rapid and heavy lead absorption. Organic lead compounds, such as tetraethyl lead, are absorbed rapidly through the skin as well as through the lungs, and are selectively absorbed by the central nervous system. The clinical picture in these cases is usually an encephalopathy. With inorganic lead compounds, comparable concentrations in the central nervous system are reached only when the workplace is heavily contaminated with vapor, fume and dust. Encephalopathy begins abruptly, and is characterized by signs of cerebral and meningeal involvement. There is usually stupor, progressing to coma, with or without convulsion, and often terminating in death. Excitation, confusion and mania are less common. In milder cases of shorter duration, there may be symptoms of headache, dizziness, somnolence and insomnia. The cerebrospinal pressure may be increased. See also specific compound.

Diagnosis: A diagnosis of lead poisoning should not be made on the basis of any single clinical or laboratory finding. There must be a history of significant exposure, signs, and symptoms (as described above) compatible with the diagnosis, and confirmatory laboratory tests. Increase of stippled red blood cells, mild anemia, and elevated lead in blood and urine, i.e., more than 0.07 mg/100 ml blood and similar values per liter of urine. An increase of coproporphyrins and certain amino acids in urine may be present. Diagnostic mobilization of lead with calcium EDTA may be useful in questionable cases.

Treatment of Lead Poisoning: It has been found that the chelating agent, calcium ethylenediaminetetracetate, and related compounds are highly efficacious in removing absorbed lead from the tissues of the body. (The therapeutic agents of this group are also known as versene, versenate, edathamil and Ca EDTA.)

Ca EDTA is effective only when administered intravenously. Various dosage schedules have been proposed. An effective regime is 3 to 6 grams of Na Ca EDTA in 300 cc to 500 cc of 5 percent glucose by intravenous drip over a period of 3 to 8 hours. Treatment may be given daily for 5 to 10 days with an interval of one week between courses. Another plan is to give treatment at intervals of 3 to 5 days until deleading has been accomplished.

Disaster Hazard: See lead.
Countermeasures
Ventilation Control: Section 2.
Personal Hygiene: Section 3.
First Aid: Section 1.

LEAD CYANATE
General Information
Description: White needles.
Formula: Pb(OCN)₂.
Constants: Mol wt: 291.25, mp: decomposes.
Hazard Analysis and Countermeasures
See lead compounds and cyanates.

LEAD CYANIDE
General Information
Description: White powder.
Formula: Pb(CN)₂.
Constant: Mol wt: 259.25.
Hazard Analysis and Countermeasures
See lead compounds and cyanides.
Shipping Regulations: Section 11.
　IATA: Poison B, poison label, 12 kilograms (passenger), no limit (cargo).

LEAD DI-o-ARSENATE. See lead arsenates.

LEAD DICHROMATE
General Information
Description: Red crystals.
Formula: PbCr₂O₇.
Constant: Mol wt: 423.23.
Hazard Analysis
Toxicity: See lead and chromium compounds.
Fire Hazard: Moderate by chemical reaction with reducing agents (Section 6).
Disaster Hazard: Dangerous; see lead; can react vigorously with reducing materials.
Countermeasures
Storage and Handling: Section 7.

LEAD DICYANOGUANIDINE
General Information
Description: Crystals.
Formula: Pb[NH₂CN(NCN)CN]₂.

Constants: Mol wt: 423.4, mp: > 300° C.
Hazard Analysis
Toxicity: This material may be quite toxic by skin absorption. See also lead compounds.
Disaster Hazard: Dangerous; see lead and cyanides.
Countermeasures
Storage and Handling: Section 7.

LEAD DIIODIDE
General Information
Description: Golden-yellow crystals or powder.
Formula: PbI₂.
Constants: Mol wt: 461.05, mp: 402° C, bp: 954° C, d: 6.16.
Hazard Analysis and Countermeasures
　See lead compounds and iodides.

LEAD DIMETHYLDITHIOCARBAMATE
General Information
Synonym: Ledate.
Description: Gray solid.
Formula: PbS₂[(CH₃)₂CN]₂.
Constants: Mol wt: 447.6, mp: 258° C, d: 2.5.
Hazard Analysis
Toxicity: See lead compounds and disulfiram. Highly toxic.
Fire Hazard: Slight, when exposed to heat or flame (Section 6).
Disaster Hazard: Dangerous; see lead and oxides of sulfur.
Countermeasures
Storage and Handling: Section 7.

LEAD DIOXIDE
General Information
Synonyms: Plattnerite; lead peroxide.
Description: Brown, hexagonal crystals.
Formula: PbO₂.
Constants: Mol wt: 239.21, mp: decomposes at 290° C, d: 9.375.
Hazard Analysis
Toxicity: See lead compounds.
Fire Hazard: Slight chemical reaction with reducing agents. (Section 6).
Disaster Hazard: Dangerous; see lead; can react with reducing materaial.
Countermeasures
Storage and Handling: Section 7.
Shipping Regulations: Section 11.
　IATA: Oxidizing material, yellow label, 12 kilograms (passenger), 45 kilograms (cargo).

LEAD DI-o-PHOSPHATE
General Information
Description: Colorless crystals.
Formula: PbHPO₄.
Constants: Mol wt: 303.20, mp: decomposes, d: 5.661 at 15° C.
Hazard Analysis and Countermeasures
See lead compounds.

LEAD DITHIONATE
General Information
Description: Crystals.
Formula: PbS₂O₆·4H₂O.
Constants: Mol wt: 439.39, mp: decomposes, d: 3.22.
Hazard Analysis
Toxicity: See lead compounds.
Disaster Hazard: Dangerous; see lead and oxides of sulfur.
Countermeasures
Storage and Handling: Section 7.

TOXIC HAZARD RATING CODE (For detailed discussion, see Section 1.)

0 NONE: (a) No harm under any conditions; (b) Harmful only under unusual conditions or overwhelming dosage.

1 SLIGHT: Causes readily reversible changes which disappear after end of exposure.

2 MODERATE: May involve both irreversible and reversible changes; not severe enough to cause death or permanent injury.

3 HIGH: May cause death or permanent injury after very short exposure to small quantities.

U UNKNOWN: No information on humans considered valid by authors.

LEAD DUST. See lead.

LEAD ENANTHATE
General Information
Description: White leaf-like crystals.
Formula: $Pb(C_7H_{13}O_2)_2$.
Constants: Mol wt: 465.56, mp: 79–80°C.
Hazard Analysis and Countermeasures
See lead compounds.

LEAD FERRICYANIDE
General Information
Description: Black-brown to red crystals.
Formula: $Pb_3[Fe(CN)_6]_2$.
Constant: Mol wt: 1135.65.
Hazard Analysis and Countermeasures
See lead compounds and ferricyanides.

LEAD FERRITE
General Information
Description: Hexagonal crystals.
Formula: $PbFe_2O_4$.
Constants: Mol wt: 382.89, mp: 1530°C (decomp.).
Hazard Analysis and Countermeasures
See lead compounds.

LEAD FERROCYANIDE
General Information
Description: Yellowish-white powder.
Formula: $Pb_2Fe(CN)_6 \cdot 3H_2O$.
Constants: Mol wt: 680.43, mp: $-H_2O$ at 100°C.
Hazard Analysis and Countermeasures
See lead compounds and ferrocyanides.

LEAD FLUORIDE
General Information
Description: Colorless solid.
Formula: PbF_2.
Constants: Mol wt: 245.2, mp: 855°C, bp: 1293°C, d: 8.24,
    vap. press.: 10 mm at 904°C.
Hazard Analysis and Countermeasures
See lead compounds and fluorides.

LEAD FLUOSILICATE
General Information
Description: Monoclinic, colorless powder.
Formula: $PbSiF_6 \cdot 2H_2O$.
Constants: Mol wt: 385.30, mp: decomposes.
Hazard Analysis and Countermeasures
See lead compounds and fluosilicates.

LEAD FORMATE
General Information
Description: Rhombic, white, lustrous crystals.
Formula: $Pb(CHO_2)_2$.
Constants: Mol wt: 297.25, mp: decomposes at 190°C, d: 4.63.
Hazard Analysis and Countermeasures
See lead compounds.

LEAD HYDRATE. See lead hydroxide.

LEAD HYDROXIDE
General Information
Synonyms: Lead hydrate; hydrated lead oxide.
Description: White, bulky powder.
Formula: $Pb(OH)_2$.
Constants: Mol wt: 241.23, mp: decomposes at 145°C, d: 7.592.
Hazard Analysis and Countermeasures
See lead compounds.

LEAD HYPOSULFITE. See lead thiosulfate.

LEAD IODATE
General Information
Description: White solid.

Formula: $Pb(IO_3)_2$.
Constants: Mol wt: 557.05, mp: decomposes at 300°C.
Hazard Analysis and Countermeasures
See lead compounds and iodates.

LEAD ISOBUTYRATE
General Information
Description: White prisms.
Formula: $Pb(C_4H_7O_2)_2$.
Constants: Mol wt: 381.40, mp: < 100°C.
Hazard Analysis and Countermeasures
See lead compounds.

LEAD LAURATE
General Information
Description: Chalky white powder.
Formula: $Pb(C_{12}H_{23}O_2)_2$.
Constants: Mol wt: 942.4, mp: 117°C.
Hazard Analysis and Countermeasures
See lead compounds.

LEAD LIGNOCERATE
General Information
Description: White powder.
Formula: $Pb(C_{24}H_{47}O_2)_2$.
Constants: Mol wt: 942.22, mp: 117°C.
Hazard Analysis and Countermeasures
See lead compounds.

LEAD LINOLEATE
General Information
Description: Yellowish-white paste.
Formula: $Pb(C_{18}H_{31}O_2)_2$.
Constant: Mol wt: 766.07.
Hazard Analysis and Countermeasures
See lead compounds.

LEAD MELISSATE
General Information
Description: White powder.
Formula: $Pb(C_{31}H_{61}O_2)_2$.
Constants: Mol wt: 1138.81, mp: 115–116°C.
Hazard Analysis and Countermeasures
See lead compounds.

LEAD MERCAPTIDE
Hazard Analysis and Countermeasures
See lead compounds and mercaptans.

LEAD MOLYBDATE
General Information
Synonym: Wulfenite.
Description: Yellow powder.
Formula: $PbMoO_4$.
Constants: Mol wt: 367.16, mp: 1070°C.
Hazard Analysis and Countermeasures
See lead compounds.

LEAD MONO-o-ARSENATE. See lead arsenates.

LEAD MONOIODIDE
General Information
Description: Pale yellow.
Formula: PbI.
Constants: Mol wt: 334.13, mp: decomposes at 300°C.
Hazard Analysis and Countermeasures
See lead compounds and iodides.

LEAD MONOXIDE. See litharge.

LEAD MONO-o-PHOSPHATE
General Information
Description: Needles.
Formula: $Pb(H_2PO_4)_2$.
Constant: Mol wt: 401.28.
Hazard Analysis and Countermeasures
See lead compounds.

LEAD MYRISTATE
General Information
Description: White powder.
Formula: $Pb(C_{14}H_{27}O_2)_2$.
Constants: Mol wt: 661.92, mp: 107° C.
Hazard Analysis and Countermeasures
See lead compounds.

LEAD β-NAPHTHALENESULFONATE
General Information
Description: White, crystalline powder.
Formula: $Pb(C_{10}H_7SO_3)_2$.
Constant: Mol wt: 621.64.
Hazard Analysis
Toxicity: See lead compounds.
Disaster Hazard: Dangerous; see lead and oxides of sulfur.
Countermeasures
Storage and Handling: Section 7.

LEAD NITRATE
General Information
Description: White crystals.
Formula: $Pb(NO_3)_2$.
Constants: Mol wt: 331.23, mp: decomposes at 470° C,
d: 4.53 at 20° C.
Hazard Analysis and Countermeasures
See lead compounds and nitrates.
Shipping Regulations: Section 11.
  ICC: Oxidizing material, yellow label, 100 pounds.
  IATA: Oxidizing material, yellow label, 12 kilograms
  (passenger), 45 kilograms (cargo).

LEAD OLEATE
General Information
Description: White, ointment-like granules or mass.
Formula: $Pb(C_{18}H_{33}O_2)_2$.
Constant: Mol wt: 770.10.
Hazard Analysis and Countermeasures
See lead compounds.

LEAD OXALATE
General Information
Description: Heavy, white powder.
Formula: PbC_2O_4.
Constants: Mol wt: 295.23, mp: decomposes at 300° C, d:
5.28.
Hazard Analysis and Countermeasures
See lead compounds and oxalates.

LEAD OXIDE
General Information
Synonym: Red lead, minium.
Description: Bright red powder.
Formula: Pb_3O_4.
Constants: Mol wt: 685.6, mp: 890° C (decomp.), bp: 1472° C,
d: 8.32–9.16, vap. press.: 1 mm at 943° C.
Hazard Analysis
Toxicity: See lead compounds.
Fire Hazard: Slight, by chemical reaction with reducing
agents (Section 6).
Caution: An oxidizing agent.
Disaster Hazard: See lead.
Countermeasures
Storage and Handling: Section 7.

LEAD OXYCHLORIDES
General Information
Synonyms: Mendipite; Cassel yellow; laurionite; matlockite.

Description: Yellow or white solid.
Composition: $PbCl_2$ + varying amounts of PbO and pos-
sibly some H_2O.
Hazard Analysis and Countermeasures
See lead compounds.

LEAD PALMITATE
General Information
Description: Chalky white powder.
Formula: $Pb(C_{16}H_{31}O_2)_2$.
Constants: Mol wt: 718.03, mp: 112.3° C.
Hazard Analysis and Countermeasures
See lead compounds.

LEAD p-PERIODATE
General Information
Description: Crystals.
Formula: $PbHIO_5$.
Constants: Mol wt: 415.14, mp: decomposes at 130° C.
Hazard Analysis and Countermeasures
See lead compounds and iodates.

LEAD PERCHLORATE
General Information
Description: Crystals.
Formula: $Pb(ClO_4)_2 \cdot 3H_2O$.
Constants: Mol wt: 460.17, mp: decomposes at 100° C, d:
2.6.
Hazard Analysis and Countermeasures
See lead compounds and perchlorates.
Shipping Regulations: Section 11.
  IATA: Oxidizing material, yellow label, 12 kilograms
  (passenger), 45 kilograms (cargo).

LEAD PEROXIDE. See lead dioxide.

LEAD PHENATE
General Information
Description: Yellowish to grayish-white powder.
Formula: $Pb(OH)OC_6H_5$.
Constant: Mol wt: 317.28.
Hazard Analysis and Countermeasures
See lead compounds and phenol.

LEAD m-PHOSPHATE
General Information
Description: Colorless crystals.
Formula: $Pb(PO_3)_2$.
Constants: Mol wt: 365.17, mp: 800° C.
Hazard Analysis and Countermeasures
See lead compounds.

LEAD o-PHOSPHATE
General Information
Description: Hexagonal, colorless or white powder.
Formula: $Pb_3(PO_4)_2$.
Constants: Mol wt: 811.59, mp: 1014° C, d: 6.9–7.3.
Hazard Analysis and Countermeasures
See lead compounds.

LEAD o-PHOSPHITE
General Information
Description: White powder.
Formula: $PbHPO_3$.
Constants: Mol wt: 287.20, mp: decomposes.
Hazard Analysis and Countermeasures
See lead compounds and phosphites.

TOXIC HAZARD RATING CODE (For detailed discussion, see Section 1.)

0 NONE: (a) No harm under any conditions; (b) Harmful only under unusual conditions or overwhelming dosage.

1 SLIGHT: Causes readily reversible changes which disappear after end
of exposure.

2 MODERATE: May involve both irreversible and reversible changes;
not severe enough to cause death or permanent injury.

3 HIGH: May cause death or permanent injury after very short exposure
to small quantities.

U UNKNOWN: No information on humans considered valid by authors.

LEAD PICRATE
General Information
Description: Yellow crystals.
Formula: $Pb(C_6H_2O_7)_2 \cdot H_2O$.
Constants: Mol wt: 681.43, mp: $-H_2O$ at 130°C, bp: explodes, d: 2.831 at 20°C.
Hazard Analysis and Countermeasures
See lead compounds and picric acid.

LEAD PROTOXIDE. See litharge.

LEAD PYROARSENATE. See lead arsenates.

LEAD PYROPHOSPHATE
General Information
Description: White crystals.
Formula: $Pb_2P_2O_7$.
Constants: Mol wt: 588.38, mp: 824°C, d: 5.8.
Hazard Analysis and Countermeasures
See lead compounds.

LEAD, RED. See lead oxide.

LEAD RESINATE
General Information
Description: Yellowish-white paste.
Formula: $Pb(C_{20}H_{29}O_2)_2$.
Constant: Mol wt: 810.07.
Hazard Analysis
Toxicity: See lead compounds.
Fire Hazard: Moderate, when exposed to heat or flame (Section 6).
Disaster Hazard: See lead.
Countermeasures
Storage and Handling: Section 7.

LEAD SELENATE
General Information
Description: White crystals.
Formula: $PbSeO_4$.
Constants: Mol wt: 350.17, mp: decomposes, d: 6.37.
Hazard Analysis and Countermeasures
See lead compounds and selenium compounds.

LEAD SELENIDE
General Information
Synonym: Clausthalite.
Description: Cubic crystals.
Formula: PbSe.
Constants: Mol wt: 286.17, mp: 1065°C, d: 8.10 at 15°C.
Hazard Analysis
Toxicity: See lead and selenium compounds.
Fire Hazard: Moderate, in the form of dust when exposed to flame or by chemical reaction with moisture to evolve the hydride. See also hydrogen selenide.
Explosion Hazard: Slight, by chemical reaction with moisture. See also hydrogen selenide.
Disaster Hazard: See lead and selenium.
Countermeasures
Storage and Handling: Section 7.

LEAD SILICATE. See lead m-silicate.

LEAD m-SILICATE
General Information
Synonym: Alamosite.
Description: White crystalline powder.
Formula: $PbSiO_3$.
Constants: Mol wt: 283.27, mp: 766°C, d: 6.49.
Hazard Analysis and Countermeasures
See lead compounds.

LEAD STEARATE
General Information
Description: White powder.
Formula: $Pb(C_{18}H_{35}O_2)_2$.
Constants: Mol wt: 774.1, mp: 115.7°C.

Hazard Analysis and Countermeasures
See lead compounds.

LEAD STYPHNATE. See lead trinitroresorcinate.

LEAD SULFATE
General Information
Synonym: Anglisite.
Description: White, rhombic crystals.
Formula: $PbSO_4$.
Constants: Mol wt: 303.27, mp: decomposes at 1000°C, d: 6.2.
Hazard Analysis and Countermeasures
See lead compounds.
Shipping Regulations: Section 11.
  IATA (containing more than 3% free acid): Corrosive liquid, white label, 1 liter (passenger), 5 liters (cargo).

LEAD SULFIDE
General Information
Synonyms: Galena, plumbous sulfide.
Description: Silvery, metallic crystals or black powder.
Formula: PbS.
Constants: Mol wt: 239.27, mp: 1114°C, bp: 1281°C (sublimes), d: 7.5, vap. press.: 1 mm at 852°C (sublimes).
Hazard Analysis and Countermeasures
See lead compounds and sulfides.

LEAD SULFITE
General Information
Description: White powder.
Formula: $PbSO_3$.
Constant: Mol wt: 287.28.
Hazard Analysis and Countermeasures
See lead compounds and sulfites.

LEAD SULFOCYANATE
General Information
Synonym: Lead thiocyanate.
Description: Monoclinic, white crystals.
Formula: $Pb(SCN)_2$.
Constants: Mol wt: 323.37, d: 3.82.
Hazard Analysis and Countermeasures
See lead compounds and thiocyanates.

LEAD TARTRATE
General Information
Description: White, crystalline powder.
Formula: $PbC_4H_4O_6$.
Constants: Mol wt: 355.28, d: 2.53 at 19°C.
Hazard Analysis and Countermeasures
See lead compounds.

LEAD TELLURIDE
General Information
Synonym: Altaite.
Description: White, cubic crystals.
Formula: PbTe.
Constants: Mol wt: 334.82, mp: 917°C, d: 8.16.
Hazard Analysis and Countermeasures
See lead compounds and tellurium compounds.

LEAD TETRAACETATE
General Information
Description: Colorless to faintly pink, monoclinic crystals.
Formula: $Pb(CH_3COO)_4$.
Constants: Mol wt: 443.39, mp: 175°C, d: 2.228 at 17°C.
Hazard Analysis and Countermeasures
See lead compounds.

LEAD TETRACHLORIDE
General Information
Description: Yellow, oily liquid.
Formula: $PbCl_4$.
Constants: Mol wt: 349.04, mp: -15°C, bp: explodes at 105°C, d: 3.18 at 0°C.

Hazard Analysis and Countermeasures
See lead and hydrochloric acid.

LEAD TETRAETHYL
General Information
Description: Colorless, oily liquid, pleasant characteristic odor.
Formula: $Pb(C_2H_5)_4$.
Constants: Mol wt: 323.5, mp: decomposes 125–150°C, bp: 198 to 202°C with decomp., d: 1.659 at 18°C, vap. press.: 1 mm at 38.4°C.
Hazard Analysis
Toxicity: See lead compounds.
TLV: ACGIH (recommended); 0.075 milligrams per cubic meter of air. May be absorbed via the skin.
Toxicology: This material is a powerful poison and a solvent for fatty materials. It has some solvent action on rubber as well. The fact that it is a lipoid solvent makes it an industrial hazard, because it can cause intoxication not only by inhalation but also by absorption through the skin. It decomposes when exposed to sunlight or allowed to evaporate; it forms triethyl lead, which is also a poisonous compound, as one of its decomposition products. This liquid lead compound, when handled in undiluted form or concentrated solution as when it is manufactured or in the plants where it is mixed with gasoline, may cause lead exposure intoxication by coming in contact with the skin. Therefore, any open receptacle which contains these liquids in high concentration or any container, article of clothing, or any other object which is not kept clean, particularly in contact with this material, may subject personnel to serious lead exposure.
Fire Hazard: Moderate, when exposed to heat or flame (Section 6).
Disaster Hazard: Dangerous; see lead; can react vigorously with oxidizing materials.
Countermeasuress
Storage and Handling: Section 7.
Shipping Regulations: Section 11.
IATA: Poison B, poison label, not acceptable (passenger), 220 liters (cargo).

LEAD TETRAFLUORIDE
General Information
Synonym: Plumbic fluoride.
Description: White crystals, reacts with moisture.
Formula: PbF_4.
Constants: Mol wt: 283.2, d: 6.7, mp: 600°C (approx.).
Hazard Analysis and Countermeasures
See lead compounds and fluorides.

LEAD TETRAMETHYL
General Information
Synonym: Tetramethyl lead.
Description: Colorless liquid.
Formula: $Pb(CH_3)_4$.
Constants: Mol wt: 267.33, mp: −18°F, lel: 1.8%, bp: 110°C, d: 1.99, vap. d.: 9.2.
Hazard Analysis
Toxicity: See lead compounds and lead tetraethyl. Highly toxic. Available data indicates toxicity similar to lead tetraethyl.
TLV: ACGIH (recommended); 0.075 milligrams per cubic meter of air.
Fire Hazard: Dangerous, when exposed to heat or flame.

Explosion Hazard: Moderate, in the form of vapor when exposed to flame.
Disaster Hazard: Dangerous; see lead; can react vigorously with oxidizing materials.
Countermeasures
Storage and Handling: Section 7.
To Fight Fire: Water, foam, carbon dioxide, dry chemical or carbon tetrachloride (Section 6).

LEAD THIOCYANATE. See lead sulfocyanate.

LEAD THIOSULFATE
General Information
Synonym: Lead hyposulfite.
Description: White crystals.
Formula: PbS_2O_3.
Constants: Mol wt: 319.33, mp: decomposes, d: 5.18.
Hazard Analysis
Toxicity: See lead compounds.
Disaster Hazard: Dangerous; see lead and oxides of sulfur.
Countermeasures
Storage and Handling: Section 7.

LEAD m-TITANATE
General Information
Description: Pale yellow solid.
Formula: $PbTiO_3$.
Constants: Mol wt: 303.11, d: 7.52.
Hazard Analysis and Countermeasures
Toxic Hazard Rating: U.
Toxicity: Animal experiments show low toxicity. See also lead and titanium compounds.

LEAD TRINITRORESORCINATE
General Information
Synonym: Lead styphnate.
Description: Orange-yellow, monoclinic crystals.
Formula: $C_6H(NO_2)_3(O_2Pb)$.
Constants: Mol wt: 450.30, mp: explodes at 311°C, d: 3.1–2.9.
Hazard Analysis
Toxicity: See lead compounds and nitrates.
Fire Hazard: See nitrates and explosives, high.
Explosion Hazard: Severe when heated.
Disaster Hazard: Dangerous; explodes at 311°C.
Countermeasures
Storage and Handling: Section 7.

LEAD TUNGSTATE
General Information
Synonym: Lead wolframate.
Description: Yellowish powder.
Formula: $PbWO_4$.
Constants: Mol wt: 455.13, d: 8.235.
Hazard Analysis and Countermeasures
See lead compounds.

LEAD m-VANADATE
General Information
Description: Yellow powder.
Formula: $Pb(VO_3)_2$.
Constant: Mol wt: 405.11.
Hazard Analysis and Countermeasures
See lead compounds and vanadium compounds.

LEAD WOLFRAMATE. See lead tungstate.

TOXIC HAZARD RATING CODE   *(For detailed discussion, see Section 1.)*

0 NONE: (a) No harm under any conditions; (b) Harmful only under unusual conditions or overwhelming dosage.

1 SLIGHT: Causes readily reversible changes which disappear after end of exposure.

2 MODERATE: May involve both irreversible and reversible changes; not severe enough to cause death or permanent injury.

3 HIGH: May cause death or permanent injury after very short exposure to small quantities.

U UNKNOWN: No information on humans considered valid by authors.

LEATHER
Hazard Analysis
Toxic Hazard Rating:
Acute Local: Irritant 1; Allergen 1.
Acute Systemic: 0.
Chronic Local: Irritant 1: Allergen 1.
Chronic Systemic: 0.
Caution: Handling "green" hides from certain parts of the world can bring about contact with anthrax spores. Furthermore, tanning can introduce materials such as chromium, formaldehyde, etc.
Fire Hazard: Slight, when exposed to heat or flame; can react with oxidizing materials (Section 6).
Countermeasures
Personal Hygiene: Section 3.
Storage and Handling: Section 7.

LEATHER BLEACH
Hazard Analysis
Toxicity: Unknown.
Fire Hazard: Dangerous, when exposed to heat or flame (Section 6).
Explosion Hazard: Unknown.
Disaster Hazard: Moderately dangerous; when heated to decomposition, it emits toxic fumes; can react with oxidizing materials.
Countermeasures
Storage and Handling: Section 7.
To Fight Fire: Foam, carbon dioxide, dry chemical or carbon tetrachloride (Section 6).
Shipping Regulations: Section 11.
I.C.C.: Flammable liquid; red label, 10 gallons.
Coast Guard Classification: Combustible liquid and inflammable liquid; red label.
IATA: Flammable liquid, red label, 1 liter (passenger), 40 liters (cargo).

LEATHER DRESSING
General Information
Constant: Flash p: <80°F.
Hazard Analysis
Toxicity: Details unknown; some dressings may act as irritants or allergens.
Fire Hazard: Dangerous, when exposed to heat or flame; can react with oxidizing materials.
Explosion Hazard: Unknown.
Countermeasures
Storage and Handling: Section 7.
To Fight Fire: Foam, carbon dioxide, dry chemical or carbon tetrachloride (Section 6).
Shipping Regulations: Section 11.
I.C.C.: Flammable liquid; red label, 10 gallons.
Coast Guard Classification: Combustible and inflammable liquid; red label.
IATA: Flammable liquid, red label, 1 liter (passenger), 40 liters (cargo).

LECITHIN
General Information
Description: The lecithins are mixtures of diglycerides of fatty acids linked to the choline ester of phosphoric acid. They are classified as phosphoglycerides or phosphatides.
Formula: $CH_2(R)CH(R')CH_2OPO(OH)O(CH_2)_2N(OH) \cdot (CH_3)_3$, where R and R' are fatty acid groups.
Hazard Analysis
Toxic Hazard Rating: U.
Note: Used as a general purpose food additive. It is a substance which migrates to food from packaging materials (Section 10).

"LEDATE." See lead dimethyldithiocarbamate.

"LETHANE 60." See lauric acid 2-thiocyanatoethyl ester.

"LETHANE 384." See β-butoxy-β-thiocyanodiethyl ether.

"LETHANE B-71"
General Information
Description: Powder.
Hazard Analysis
Toxic Hazard Rating:
Acute Local: Irritant 1; Allergen 1.
Acute Systemic: Ingestion 1; Inhalation 1.
Chronic Local: Allergen 1.
Chronic Systemic: Ingestion 1; Inhalation 1.
Disaster Hazard: See thiocyanates.
Countermeasures
Ventilation Control: Section 2.
Personal Hygiene: Section 3.
Storage and Handling: Section 7.

LEUCINE
General Information
Synonyms: α-Amino-γ-methylvaleric acid; α-amino-iso-caproic acid.
Description: An essential amino acid; occurs in isomeric forms. Below we consider the L and DL forms. White crystals, soluble in water, slightly soluble in alcohol, insoluble in ether.
Formula: $(CH_3)_2CHCH_2CH(NH_2)COOH$.
Constants: Mol wt: 131, mp(DL): 332°C with decomposition, mp(L): 295°C, d: 1.239 at 18/4°C.
Hazard Analysis
Toxic Hazard Rating: U. A nutrient and/or dietary supplement food additive (Section 10).

LEUKOL. See quinoline.

LEWISITE. See dichloro-(2-chlorovinyl) arsine.

"LEYTOSAN"
General Information
Synonym: Phenyl mercury urea.
Hazard Analysis and Countermeasures
See mercury compounds, organic. A fungicide.

LEXONE. 1,2,3,4,5,6-hexachlorocyclohexane.

LIGHTER FLUID, CIGAR AND CIGARETTE, CHARGED WITH FLAMMABLE GAS
Shipping Regulations: Section 11.
IATA: Flammable gas, red label, 0.6 kilograms (passenger), 12 kilograms (cargo).

LIGHTERS, CIGAR AND CIGARETTE, CONTAINING FLAMMABLE LIQUID
Shipping Regulations: Section 11.
IATA: Flammable liquid, not acceptable.

LIGHT LIGROIN. See petroleum spirits.

LIGHT OILS
General Information
Constant: Bp: 110–160°C.
Hazard Analysis
Toxicity: See kerosene.
Fire Hazard: Moderate, when exposed to heat or flame; can react with oxidizing materials (Section 6).
Explosion Hazard: Unknown.
Countermeasures
Storage and Handling: Section 7.

LIGNASAN. See ethyl mercury chloride.

LIGNITE DUST
General Information
Constant: Autoign. temp.: 302°F.
Hazard Analysis
Toxic Hazard Rating:
Acute Local: Inhalation 1.
Acute Systemic: U.

Chronic Local: Inhalation 1.
Chronic Systemic: Inhalation 1.
TLV: ACGIH (recommended); 50 million particles per cubic foot of air.
Fire Hazard: Moderate, when exposed to heat or flame; can react with oxidizing materials (Section 6).
Explosion Hazard: Unknown.
Countermeasures
Ventilation Control: Section 2.
Storage and Handling: Section 7.

LIME. Calcium oxide.

LIME ACETATE. See calcium acetate.

LIME CITRATE. See calcium citrate.

LIME URANITE. See autunite.

LIMONENE
General Information
Description: Colorless liquid; optically active; occurs in D and L forms; the racemic mixture of the two (dl-form) is known as dipentene.
Formula: $C_{10}H_{16}$.
Constants: Mol wt: 136, d: 0.841 at 20°C, bp: 176.4°C.
Hazard Analysis
Toxic Hazard Rating: Unknown. A synthetic flavoring substance and adjuvant (Section 10).

LIMONENE DIOXIDE
General Information
Synonym: Dipentene dioxide.
Description: Liquid, soluble in water.
Formula: $C_{10}H_{16}O_2$.
Constants: Mol wt: 148, d: 1.0287 at 20°C, bp: 242°C, fp: −100°C.
Hazard Analysis
Toxic Hazard Rating:
  Acute Local: Irritant 2.
  Acute Systemic: Ingestion 1; Inhalation 1; Skin Absorption 2.
  Chronic Local: 0.
  Chronic Systemic: U.
Based on limited animal experiments.

LINALOL. See linalool.

LINALOOL
General Information
Synonyms: Linalol; 3,7-dimethyl-1,6-octadien-3-ol.
Description: Colorless liquid; odor similar to that of bergamot oil and French lavender; soluble in alcohol and ether.
Formula: $(CH_3)_2C:CHCH_2CH_2C(CH_3)OHCH:CH_2$.
Constants: Mol wt: 154, d: 0.858–0.868 at 25°C, bp: 195–199°C.
Hazard Analysis
Toxic Hazard Rating: U. Used as a synthetic flavoring substance and adjuvant (Section 10).

LINALYL ACETATE
General Information
Description: Clear, colorless, oily, liquid. Odor of bergamot; soluble in alcohol, ether, diethyl phthalate, benzyl benzoate, mineral oil, fixed oils, alcohol; slightly soluble in propylene glycol; insoluble in water, glycerin.
Formula: $C_{10}H_{17}C_2H_3O_2$.
Constants: Mol wt: 196, bp: 108–110°C, d: 0.908–0.920.

Hazard Analysis
Toxic Hazard Rating: U. Used as a synthetic flavoring substance and adjuvant (Section 10).

LINDANE. See 1,2,3,4,5,6-hexachlorocyclohexane.

LINOLEIC ACID
General Information
Synonym: Linolic acid.
Description: Colorless oil. Easily oxidized by air. Soluble in ether, ethanol.
Formula: $CH_3(CH_2)_4CH=CHCH_2CH=CH(CH_2)_7 \cdot COOH$.
Constants: Mol wt: 280.4, d: 0.9038 at 18°/4°C, mp: −12°C, bp: 230°C at 16 mm.
Hazard Analysis
Toxicity: Details unknown. Ingestion can cause nausea and vomiting. A nutrient and/or dietary supplement food additive. See Section 10.

LINOLIC ACID. See linoleic acid.

LINSEED OIL
General Information
Constants: Bp: 343°C, mp: −19°C, d: 0.93, flash p: raw oil; 432°F (C.C.); boiled: 403°F (C.C.), autoign. temp.: 650°F.
Hazard Analysis
Toxic Hazard Rating:
  Acute Local: Irritant 1; Allergen 1.
  Acute Systemic: U.
  Chronic Local: Allergen 2.
  Chronic Systemic: U.
Fire Hazard: Slight, when exposed to heat or flame; can react with oxidizing materials.
Spontaneous Heating: Yes.
Explosion Hazard: Unknown.
Countermeasures
Personnel Protection: Section 3.
To Fight Fire: Foam, carbon dioxide, dry chemical or carbon tetrachloride (Section 6).
Storage and Handling: Section 7.

LINTOX. See 1,2,3,4,5,6-hexachlorocyclohexane.

LIQUEFIED CARBON DIOXIDE
General Information
Synonym: Liquid carbonic gas.
Description: Heavy gas or liquid under pressure.
Formula: CO_2.
Constants: Mol wt: 44.0, mp: −56.6°C at 3952 mm, bp: −78.5°C sublimes, d: 1.977 g/liter at 0°C; liquid: 1.101 at −37°C.
Hazard Analysis
Toxic Hazard Rating:
  Acute Local: Irritant 2; Inhalation 1.
  Acute Systemic: Inhalation 1.
  Chronic Local: 0.
  Chronic Systemic: Inhalation 1.
Disaster Hazard: Moderately dangerous.
Countermeasures
Ventilation Control: Section 2.
Storage and Handling: Section 7.
Shipping Regulations: Section 11.
  I.C.C.: Nonflammable gas; green label.
  Coast Guard Classification: Noninflammable gas; green gas label.

TOXIC HAZARD RATING CODE (For detailed discussion, see Section 1.)

0 NONE: (a) No harm under any conditions; (b) Harmful only under unusual conditions or overwhelming dosage.

1 SLIGHT: Causes readily reversible changes which disappear after end of exposure.

2 MODERATE: May involve both irreversible and reversible changes; not severe enough to cause death or permanent injury.

3 HIGH: May cause death or permanent injury after very short exposure to small quantities.

U UNKNOWN: No information on humans considered valid by authors.

LIQUEFIED HYDROCARBON GAS
General Information
Synonym: LP gas; bottle gas; **LPG.**
Hazard Analysis
Toxicity: Unknown, may act as a simple asphyxiant.
TLV: ACGIH(recommended); 1000 parts per million of air; 1800 milligrams per cubic meter of air.
Fire Hazard: Dangerous, when exposed to heat or flame (Section 6).
Explosion Hazard: Unknown.
Disaster Hazard: Moderately dangerous; can react with oxidizing materials.
Countermeasures
Ventilation Control: Section 2.
To Fight Fire: Carbon dioxide, dry chemical, or water spray. Section 6.
Storage and Handling: Section 7.
Shipping Regulations: Section 11.
  I.C.C.: Flammable gas; red gas label, 300 pounds.
  Coast Guard Classification: Inflammable gas; red gas label.
  IATA: Flammable gas, red label, not acceptable (passenger), 140 kilograms (cargo).

LIQUEFIED NONFLAMMABLE GASES (CHARGED WITH NITROGEN, CARBON DIOXIDE OR AIR).
See specific component.
Shipping Regulations: Section 11.
  I.C.C.: Nonflammable gas; green label, 300 pounds.
  Coast Guard Classification: Noninflammable gas; green gas label.
  IATA: Nonflammable gas, green label, 15 kilograms (passenger and cargo).

LIQUID CAMPHOR. See camphor oil (light).

LIQUID CARBONIC GAS. See liquefied carbon dioxide.

LIQUID ROSIN. See tall oil.

LITHARGE
General Information
Synonyms: Lead oxide, yellow; plumbous oxide; lead protoxide; lead monoxide.
Description: Tetragonal, yellow crystals.
Formula: PbO.
Constants: Mol wt: 223.21, mp: 888°C, d: 9.53.
Hazard Analysis and Countermeasures
See lead compounds.

LITHIC ACID. See uric acid.

LITHIUM
General Information
Description: Silvery light metal; mixture of isotopes Li^6 and Li^7.
Formula: Li.
Constants: At wt: 6.94, mp: 179°C, bp: 1317°C, d: 0.534 at 25°C, vap. press.: 1 mm at 723°C.
Hazard Analysis
Toxicity: See lithium compounds for a discussion of the toxicity of the lithium ion. See sodium for a discussion which applies to the toxicity of metallic lithium.
Fire Hazard: Dangerous, when exposed to heat or flame or by chemical reaction with moisture, acids or oxidizers.
Explosion Hazard: Dangerous, by chemical reaction with oxidizers, moisture and acids.
Disaster Hazard: Dangerous; when burned, it emits toxic fumes of lithium oxide and hydroxide; will react with water or steam to produce heat and hydrogen; can react vigorously with oxidizing materials. Reacts with nitrogen.
Countermeasures
Ventilation Control: Section 2.
To Fight Fire: Special mixutres of dry chemical, soda ash graphite (Section 6).

Personnel Protection: Section 3.
Storage and Handling: Section 7.
Shipping Regulations: Section 11.
  I.C.C.: Flammable solid; yellow label, 25 pounds.
  Coast Guard Classification: Inflammable solid; yellow label.
  IATA: Flammable solid, yellow label, not acceptable (passenger), 12 kilograms (cargo).

LITHIUM ACETYLSALICYLATE
General Information
Description: Slightly hygroscopic powder, decomposes in moist air.
Formula: LiC$_9$H$_7$O$_4$.
Constant: Mol wt: 186.09.
Hazard Analysis
Toxicity: See acetylsalicylates.
Fire Hazard: Slight (Section 6).
Countermeasures
Storage and Handling: Section 7.

LITHIUM ACID OXALATE
General Information
Description: Colorless crystals.
Formula: LiHC$_2$O$_4 \cdot$ H$_2$O.
Constants: Mol wt: 113.98, mp: decomposes.
Hazard Analysis
Toxicity: See oxalates.

LITHIUM ALUMINATE
General Information
Synonym: Lithium m-aluminate.
Description: White powder.
Formula: LiAlO$_2$.
Constants: Mol wt: 65.91, mp: > 1625°C, d: 2.55 at 25°C.
Hazard Analysis
Toxicity: See lithium and aluminum compounds.

LITHIUM ALUMINUM DEUTERIDE
General Information
Description: White to gray-white micro crystalline lumps or powder.
Formula: LiAlD$_4$.
Constants: Mol wt: 42.0, mp: 150° with decomposition (ignites in air), d: 1.02 at 25°C.
Hazard Analysis and Countermeasures
See lithium and aluminum compounds and hydrides.

LITHIUM ALUMINUM HYDRIDE
General Information
Description: White, microcrystalline lumps; solid stable in dry air at room temperature; above 125°C, decomposes forming Al, H$_2$ and lithium hydride.
Formula: LiAlH$_4$.
Constant: Mol wt: 37.94.
Hazard Analysis and Countermeasures
See aluminum and lithium compounds and hydrides.
Shipping Regulations: Section 11.
  ICC: Flammable solid, yellow label, 25 pounds.
  IATA: Flammable solid, yellow label, not acceptable (passenger), 12 kilograms (cargo).

LITHIUM ALUMINUM HYDRIDE, ETHEREAL
General Information
Description: Liquid solution of LiAlH$_4$ in ether.
Composition: LiAlH$_4$ + an ether.
Hazard Analysis and Countermeasures
See ethers, and lithium aluminum hydride.
Shipping Regulations: Section 11.
  ICC: Flammable liquid, red label, 1 quart.
  IATA: Flammable liquid, red label, not acceptable (passenger), 1 liter (cargo).

LITHIUM AMIDE
General Information
Description: White crystalline solid or powder.

Formula: LiNH$_2$.
Constants: Mol wt: 23.0, mp: 373–375°C, bp: 430°C, d: 1.178 at 17.5°C.
Hazard Analysis
Toxicity: Ammonia is liberated and lithium hydroxide is formed when this compound is exposed to moisture. See also ammonia and lithium hydroxide.
Fire Hazard: See ammonia.
Explosion Hazard: See ammonia.
Disaster Hazard: Moderately dangerous; will react with water or steam to produce toxic and flammable vapors; on contact with oxidizing materials, it can react vigorously, and on contact with acid or acid fumes, it can evolve much heat.
Countermeasures
Storage and Handling: Section 7.
Shipping Regulations: Section 11.
  I.C.C.: Flammable solid; yellow label, 100 pounds.
  Coast Guard Classification: Inflammable solid; yellow label.
  IATA: Flammable solid, yellow label, 12 kilograms (passenger), 45 kilograms (cargo).

LITHIUM ANTIMONIDE
General Information
Description: Solid.
Formula: Li$_3$Sb.
Constants: Mol wt: 142.58, mp: >950°C, d: 3.2 at 17°C.
Hazard Analysis
Toxicity: See antimony compounds.
Fire Hazard: In contact with moisture stibine is liberated. See also antimony hydride.
Explosion Hazard: See also antimony hydride.
Disaster Hazard: Dangerous; when heated to decomposition or on contact with acid or acid fumes, it emits highly toxic fumes of antimony; will react with water or steam to produce toxic and flammable vapors; can react vigorously with oxidizing materials.
Countermeasures
Storage and Handling: Section 7.

LITHIUM ARSENATE
General Information
Synonym: Lithium o-arsenate.
Description: White powder.
Formula: Li$_3$AsO$_4$.
Constants: Mol wt: 159.73, d: 3.07 at 15°C.
Hazard Analysis and Countermeasures
See arsenic compounds.

LITHIUM BENZOATE
General Information
Description: White crystals; water soluble.
Formula: C$_7$H$_5$LiO$_2$.
Constants: Mol wt: 128.1.
Hazard Analysis
Toxicity: See lithium compounds.

LITHIUM   BICHROMATE. See lithium dichromate.

LITHIUM BORATE. See lithium m-borate.

LITHIUM m-BORATE
General Information
Synonym: Lithium borate.
Description: White powder.
Formula: LiBO$_2$.
Constants: Mol wt: 49.8, mp: 845°C.

Hazard Analysis and Countermeasures
See boron and lithium compounds.

LITHIUM m-BORATE DIHYDRATE. See lithium m-borate.

LITHIUM BOROHYDRIDE
General Information
Description: White to grayish microcrystalline powder and lumps.
Formula: LiBH$_4$.
Constants: Mol wt: 21.8, mp: 284°C (decomposes), d: 0.66.
Hazard Analysis and Countermeasures
See boron and lithium compounds and hydrides.
Shipping Regulations: Section 11.
  IATA: Flammable solid, yellow label, not acceptable (passenger), 12 kilograms (cargo).

LITHIUM BROMIDE
General Information
Description: White, hygrosocopic, granular powder.
Formula: LiBr.
Constants: Mol wt: 86.86, mp: 549°C, bp: 1265°C, d: 3.46 at 25°C, vap. press.: 1 mm at 748°C.
Hazard Analysis and Countermeasures
See bromides and lithium compounds.

LITHIUM CACODYLATE
General Information
Description: White powder.
Formula: Li(CH$_3$)$_2$AsO$_2$.
Constant: Mol wt: 143.96.
Hazard Analysis and Countermeasures
See arsenic compounds.

LITHIUM CARBIDE
General Information
Description: Crystals or white powder.
Formula: Li$_2$C$_2$.
Constants: Mol wt: 37.90, d: 1.65 at 18°C.
Hazard Analysis and Countermeasures
See lithium compounds.

LITHIUM CARBOLATE. See lithium phenate.

LITHIUM CARBONATE
General Information
Formula: Li$_2$CO$_3$.
Constants: Mol wt: 73.9, mp: 618°C, d: 2.11 at 17.5°C.
Hazard Analysis
Toxicity: See lithium compounds.

LITHIUM CHLORATE
General Information
Description: White crystals.
Formula: LiClO$_3$.
Constants: Mol wt: 90.4, mp: 124°C, bp: decomposes at 270°C.
Hazard Analysis and Countermeasures
See chlorates and lithium compounds.

LITHIUM CHLORIDE
General Information
Description: Cubic, white deliquescent crystals.
Formula: LiCl.
Constants: Mol wt: 42.4, mp: 613°C, bp: 1383°C, d: 2.068 at 25°C, vap. press.: 1 mm at 547°C.
Hazard Analysis
Toxicity: See lithium compounds.

TOXIC HAZARD RATING CODE (For detailed discussion, see Section 1.)

0 NONE: (a) No harm under any conditions; (b) Harmful only under unusual conditions or overwhelming dosage.

1 SLIGHT: Causes readily reversible changes which disappear after end of exposure.

2 MODERATE: May involve both irreversible and reversible changes; not severe enough to cause death or permanent injury.

3 HIGH: May cause death or permanent injury after very short exposure to small quantities.

U UNKNOWN: No information on humans considered valid by authors.

Toxicology: This material has been recommended and used as a substitute for sodium chloride in "salt free" diets, but cases have been reported in which the ingestion of lithium chloride has produced dizziness, ringing in the ears, visual disturbances, tremors and mental confusion. In most cases, the symptoms disappeared when the use was discontinued.

LITHIUM CHLOROPLATINATE
General Information
Description: Red, deliquescent crystals.
Formula: $Li_2PtCl_6 \cdot 6H_2O$.
Constants: Mol wt: 529.95, bp: $-6H_2O$ at 180°C.
Hazard Analysis
Toxicity: See platinum compounds.
Disaster Hazard: Dangerous; see chlorine compounds.
Countermeasures
Storage and Handling: Section 7.

LITHIUM CHROMATE
General Information
Description: Yellow crystalline, deliquescent powder.
Formula: $Li_2CrO_4 \cdot 2H_2O$.
Constants: Mol wt: 165.92, mp: $-2H_2O$ at 150°C.
Hazard Analysis
Toxicity: See chromium compounds.
Fire Hazard: Slight; an oxidizer. It can react with reducing materials (Section 6).
Countermeasures
Storage and Handling: Section 7.

LITHIUM CITRATE
General Information
Description: White crystals; feebly alkaline taste; deliquescent, water soluble.
Formula: $C_6H_5Li_3O_7 \cdot 4H_2O$.
Constants: Mol wt: 282.0, mp: decomposes: $-4H_2O$ at 105°C.
Hazard Analysis
Toxicity: See lithium compounds.

LITHIUM COBALTITE
General Information
Description: Blue-black powder.
Formula: $LiCoO_2$.
Constant: Mol wt: 97.9.
Hazard Analysis
Toxicity: See cobalt compounds.

LITHIUM COMPOUNDS
Hazard Analysis
Toxic Hazard Rating:
Acute Local: Ingestion 1; Inhalation 1.
Acute Systemic: Ingestion 2; Inhalation 2.
Chronic Local: U.
Chronic Systemic: Ingestion 1; Inhalation 1.
Toxicology: Lithium compounds resemble those of potassium in systemic and local action. Large doses of lithium compounds have caused dizziness and prostration. Can cause kidney damage if sodium intake is limited. Lithium oxide, hydroxide, carbonate, etc., are strong bases and these solutions in water are very caustic. See also potassium compounds.
Countermeasures
Ventilation Control: Section 2.
Personnel Protection: Section 3.

LITHIUM DEUTERIDE
General Information
Description: Bluish-gray, fine powder.
Formula: LiD.
Constants: Mol wt: 9.0, mp: $> 600°C$, d: 0.906 at 25°C.
Hazard Analysis and Countermeasures
See lithium compounds and hydrides.

LITHIUM DICHROMATE
General Information
Synonym: Lithium bichromate.
Description: Yellowish-red, crystalline, deliquescent powder.
Formula: $Li_2Cr_2O_7 \cdot 2H_2O$.
Constants: Mol wt: 265.93, mp: $-2H_2O$ at 130°C, bp: decomposes.
Hazard Analysis and Countermeasures
See chromium compounds.

LITHIUM ETHYL
General Information
Synonym: Ethyllithium.
Description: Transparent crystals; decomposed by water.
Formula: LiC_2H_5.
Constants: Mol wt: 36.00, mp: 95°C.
Hazard Analysis
Toxicity: See lithium compounds.
Fire Hazard: A flammable material; can react with oxidizing materials (Section 6).
Explosion Hazard: Unknown.
Countermeasures
Ventilation Control: Section 2.
Storage and Handling: Section 7.

LITHIUM FERROSILICON
General Information
Description: Dark crystalline brittle metallic lumps or powder, evolves a flammable gas when in contact with moisture.
Countermeasures
Shipping Regulations:
ICC: Flammable solid, yellow label, 25 pounds.
IATA: Flammable solid, yellow label, not acceptable (passenger), 12 kilograms (cargo).

LITHIUM FLUORIDE
General Information
Description: Fine white powder.
Formula: LiF.
Constants: Mol wt: 25.94, mp: 845°C, bp: 1671°C, d: 2.295 at 21.5°C, vap. press.: 1 mm at 1047°C.
Hazard Analysis and Countermeasures
See fluorides.

LITHIUM FLUOSILICATE
General Information
Synonym: Lithium silicofluoride.
Description: Monoclinic, white crystals.
Formula: $Li_2SiF_6 \cdot 2H_2O$.
Constants: Mol wt: 191.97, mp: $-2H_2O$ at 100°C, bp: decomposes, d: 2.33 at 12°C.
Hazard Analysis and Countermeasures
See fluosilicates.

LITHIUM GALLIUM HYDRIDE
General Information
Description: White crystals.
Formula: $LiGaH_4$.
Constant: Mol wt: 80.69.
Hazard Analysis and Countermeasures
See gallium and lithium compounds and hydrides.

LITHIUM GALLIUM NITRIDE
General Information
Description: Light gray powder.
Formula: Li_3GaN_2.
Constants: Mol wt: 118.56, mp: decomposes at 800°C, d: 3.35.
Hazard Analysis and Countermeasures
See gallium and lithium compounds and nitrides.

LITHIUM m-GERMANATE
General Information
Description: Crystals.

Formula: Li_2GeO_3.
Constants: Mol wt: 134.48, mp: 1239°C, d: 3.53 at 21°C.
Hazard Analysis
Toxicity: See germanium and lithium compounds.

LITHIUM HYDRIDE
General Information
Description: White, translucent, crystalline mass.
Formula: LiH.
Constants: Mol wt: 7.95, mp: 680°C, d: 0.82.
Hazard Analysis
TLV: ACGIH (recommended); 0.025 milligrams per cubic
meter of air.
Toxicity: In contact with moisture lithium hydroxide is
formed. Can ignite spontaneously in moist air. See
also lithium compounds.
Fire Hazard: See hydrides.
Disaster Hazard: See hydrides.
Explosion Hazard: See hydrides.
Countermeasures
Storage and Handling: Section 7.
To Fight Fire: Special mixtures of dry chemical.
Shipping Regulations: Section 11.
I.C.C.: Flammable solid; yellow label, 25 pounds.
Coast Guard Classification: Inflammable solid; yellow
label.
IATA: Flammable solid, yellow label, not acceptable
(passenger), 12 kilograms (cargo).
IATA (in solid form): Flammable solid, yellow label, not
acceptable (passenger), 45 kilograms (cargo).

LITHIUM HYDROSULFIDE
General Information
Description: White, hygroscopic powder.
Formula: LiHS.
Constant: Mol wt: 40.01.
Hazard Analysis and Countermeasures
See sulfides.

LITHIUM HYDROXIDE
General Information
Synonym: Lithium hydroxide monohydrate.
Description: Colorless crystals.
Formula: LiOH.
Constants: Mol wt: 23.95, mp: 462°C, bp: decomposes,
d: 2.54 at 20°C.
Hazard Analysis
Toxicity: See lithium compounds and sodium hydroxide.

LITHIUM HYDROXIDE MONOHYDRATE. See lithium
hydroxide.

LITHIUM HYPOCHLORITE COMPOUNDS, DRY
(Containing more than 39% available chlorine)
Shipping Regulations: Section 11.
I.C.C.: Oxidizing material, yellow label, 100 pounds.
IATA: Oxidizing material, yellow label, 25 kilograms
(passenger), 45 kilograms (cargo).

LITHIUM IODATE
General Information
Description: White crystals.
Formula: $LiIO_3$.
Constant: Mol wt: 181.86.
Hazard Analysis and Countermeasures
See iodates and lithium compounds.

LITHIUM MANGANITE
General Information
Description: Reddish-brown powder.
Formula: Li_2MnO_3.
Constant: Mol wt: 116.8.
Hazard Analysis
Toxicity: See manganese and lithium compounds.

LITHIUM METAL, IN CARTRIDGES. See lithium.
Shipping Regulations: Section 11.
I.C.C.: See § 73.206.
Coast Guard Classification: Inflammable solid; yellow
label.
IATA: Flammable solid, yellow label, ½ kilogram
(passenger), 12 kilograms (cargo).

LITHIUM MOLYBDATE
General Information
Description: White, crystalline powder.
Formula: Li_2MoO_4.
Constants: Mol wt: 173.83, mp: 705°C.
Hazard Analysis
Toxicity: See molybdenum and lithium compounds.

LITHIUM NITRATE
General Information
Description: Colorless powder, deliquescent granules.
Formula: $LiNO_3$.
Constants: Mol wt: 69.0, mp: 255°C, d: 2.38.
Hazard Analysis and Countermeasures
See nitrates.

LITHIUM NITRIDE
General Information
Description: Brownish red hexagonal crystals. Slowly de-
composes on contact with moisture.
Formula: Li_3N.
Constants: Mol wt: 34.8, d: 1.3, mp: 845°C.
Hazard Analysis
Toxicity: Upon contact with moisture, it decomposes into
lithium hydroxide, lithium compounds and ammonia.
Fire Hazard: Moderate; at elevated temperatures it ignites
and burns intensely in air.
Countermeasures
To Fight Fire: Dry chemicals, sand, graphite. Avoid use of
water or carbon tetrachloride.
Shipping Regulations: Section 11.
IATA: Flammable solid, yellow label, not acceptable
(passenger), 12 kilograms (cargo).

LITHIUM NITRITE
General Information
Description: Colorless crystals.
Formula: $LiNO_2 \cdot H_2O$.
Constants: Mol wt: 71.0, mp: <100°C, bp: decomposes, d:
1.615 at 0°C.
Hazard Analysis and Countermeasures
See nitrites.

LITHIUM OXALATE
General Information
Description: Colorless crystals.
Formula: $Li_2C_2O_4$.
Constants: Mol wt: 101.90, mp: decomposes, d: 2.121 at
17.5°C.
Hazard Analysis and Countermeasures
See oxalates.

TOXIC HAZARD RATING CODE (*For detailed discussion, see Section 1.*)

0 NONE: (a) No harm under any conditions; (b) Harmful only under unusual conditions or overwhelming dosage.

1 SLIGHT: Causes readily reversible changes which disappear after end
of exposure.

2 MODERATE: May involve both irreversible and reversible changes;
not severe enough to cause death or permanent injury.

3 HIGH: May cause death or permanent injury after very short exposure
to small quantities.

U UNKNOWN: No information on humans considered valid by authors.

LITHIUM OXIDE
General Information
Description: Cubic, white crystals.
Formula: Li_2O.
Constants: Mol wt: 29.9, mp: $>1700°C$, d: 2.013 at 25.2°C.
Hazard Analysis and Countermeasures
See lithium compounds.

LITHIUM PERCHLORATE
General Information
Description: Colorless, deliquescent crystals.
Formula: $LiClO_4$.
Constants: Mol wt: 106.40, mp: 239°C, bp: decomposes at 380°C; d: 2.429.
Hazard Analysis and Countermeasures
See perchlorates and lithium compounds. It is irritating to skin and mucous membrane.

LITHIUM PERMANGANATE
General Information
Description: Purple crystals.
Formula: $LiMnO_4·3H_2O$.
Constants: Mol wt: 179.92, mp: decomposes at 190°C, d: 2.06.
Hazard Analysis
Toxicity: See manganese compounds.
Fire Hazard: Moderate; a powerful oxidizing agent; can react with reducing materials (Section 6).
Countermeasures
Storage and Handling: Section 7.

LITHIUM PEROXIDE
General Information
Description: Fine white powder or a sandy yellow, granular material.
Formula: Li_2O_2.
Constants: Mol wt: 45.88, mp: decomposes, d: 2.14 at 20°C.
Hazard Analysis
Toxicity: See lithium compounds and peroxides.
Fire Hazard: Dangerous, because it is an extremely powerful oxidizing agent. See also peroxides, inorganic.
Disaster Hazard: Dangerous; will react with water or steam to produce heat; on contact with reducing material, it can react vigorously.
Countermeasures
Storage and Handling: Section 7.
Shipping Regulations: Section 11.
   I.C.C.: Oxidizing material; yellow label, 100 pounds.
   Coast Guard Classification : Oxidizing material; yellow label.
   IATA: Oxidizing material, yellow label, 12 kilograms (passenger), 45 kilograms (cargo).

LITHIUM PHENATE
General Information
Synonym: Lithium carbolate, phenol lithium.
Description: White or reddish powder. Water soluble.
Formula: C_6H_5OLi.
Constant: Mol wt: 100.0.
Hazard Analysis
Toxicity: A strong caustic and irritant. See lithium compounds.

LITHIUM SALICYLATE
General Information
Description: White or gray white, odorless, sweet tasting powder. Water soluble.
Formula: $C_7H_5LiO_3$.
Constant: Mol wt: 144.1.
Hazard Analysis
Toxicity: See lithium compounds.

LITHIUM SELENIDE
General Information
Description: Colorless, deliquescent crystals.

Formula: $Li_2Se·9H_2O$.
Constant: Mol wt: 255.0.
Hazard Analysis
See selenium compounds and hydrogen selenide.

LITHIUM SILICATE
General Information
Description: White powder.
Formula: Li_2SiO_3.
Constants: Mol wt: 89.9, mp: 1201°C, d: 2.52 at 25°C.
Hazard Analysis
Toxicity: See silicates.

LITHIUM SILICOFLUORIDE. See lithium fluosilicate.

LITHIUM SILICON
General Information
Description: Solid.
Composition: Li + Si.
Hazard Analysis
Toxicity: See lithium and silicon.
Fire Hazard: Dangerous, in the form of dust when exposed to heat or flame or by chemical reaction with moisture or acids. In contact with water, silane and hydrogen are evolved.
Explosion Hazard: Slight, in the form of dust when exposed to flame. See also powdered metals.
Disaster Hazaad: Moderately dangerous; will react with water or steam to produce flammable vapors; on contact with oxidizing materials, it can react vigorously and on contact with acid or acid fumes, it can emit toxic and flammable fumes.
Countermeasures
Storage and Handling: Section 7.
To Fight Fire: Carbon dioxide, dry chemical or carbon tetrachloride (Section 6).
Shipping Regulations: Section 11.
   I.C.C.: Flammable solid; yellow label, 25 pounds.
   Coast Guard Classification: Inflammable solid; yellow label.
   IATA: Flammable Solid, yellow label, not acceptable (passenger), 12 kilograms (cargo).

LITHIUM SODIUM FLUOALUMINATE
General Information
Description: Cubic crystals.
Formula: $Li_3Na_3(AlF_6)_2$.
Constants: Mol wt: 371.75, mp: 710°C, d: 2.774.
Hazard Analysis
See fluoaluminates.

LITHIUM SULFIDE
General Information
Description: Cubic, white-yellow crystals.
Formula: Li_2S.
Constants: Mol wt: 46.0, d: 1.66.
Hazard Analysis and Countermeasures
See sulfides.

LITHIUM SULFITE
General Information
Description: Needles.
Formula: $Li_2SO_3·H_2O$.
Constants: Mol wt: 112.0, mp: 455°C slight decomp.
Hazard Analysis and Countermeasures
See sulfites.

LITHIUM TETRABORATE
General Information
Description: White, crystalline powder.
Formula: $Li_2B_4O_7·5H_2O$.
Constants: Mol wt: 259.24, mp: $-2H_2O$ at 200°C.
Hazard Analysis
Toxicity: See boron and lithium compounds.

LITHIUM THALLIUM dl-TARTRATE
General Information
Description: Crystals.
Formula: $LiTlC_4H_4O_6 \cdot 2H_2O$.
Constants: Mol wt: 395.43, d: 3.144.
Hazard Analysis and Countermeasures
See thallium compounds.

LITHIUM THIOCYANATE
General Information
Description: Deliquescent, white crystals.
Formula: LiSCN.
Constant: Mol wt: 65.0.
Hazard Analysis and Countermeasures
See thiocyanates.

LITHIUM m-VANADATE
General Information
Description: Yellowish powder.
Formula: $LiVO_3 \cdot 2H_2O$.
Constant: Mol wt: 141.92.
Hazard Analysis
Toxicity: See vanadium compounds.

LITHIUM ZIRCONATE
General Information
Description: Creamy-white powder.
Formula: Li_2ZrO_3.
Constant: Mol wt: 153.10.
Hazard Analysis
Toxicity: See zirconium and lithium compounds.

LITHIUM ZIRCONIUM SILICATE
General Information
Description: White powder.
Formula: $2Li_2O \cdot ZrO_2 \cdot SiO_2$.
Constant: Mol wt: 243.0.
Hazard Analysis
Toxicity: See lithium compounds, zirconium compounds and silicates.

LITHOPONE
General Information
Description: White powder.
Formula: Zinc sulfide, barium sulfate, zinc oxide.
Hazard Analysis
Toxicity: Can liberate hydrogen sulfide upon decomposition by heat, moisture and acids. See hydrogen sulfide.
Disaster Hazard: Dangerous; when heated to decomposition, it emits highly toxic fumes.

LOBELIA
Hazard Analysis
Toxicity: See lobeline.
Disaster Hazard: Dangerous; when heated, it emits highly toxic fumes.
Countermeasures
Storage and Handling: Section 7.

LOBELINE
General Information
Description: Yellow, syrupy liquid.
Formula: $C_{22}H_{27}NO_2$.
Constants: Mol wt: 285.28, mp: 131°C.
Hazard Analysis
Toxici Hazard Rating:
    Acute Local : U.
    Acute Systemic: Ingestion 3; Inhalation 3.

Chronic Local: U.
CHronic Systemic: U.
Toxicology: Causes stimulation which leads to convulsions in severe cases. Nausea and vomiting are frequent.
Disaster Hazard: Dangerous; when heated, it emits highly toxic fumes.
Countermeasures
Ventilation Control: Section 2.
Treatment and Antidotes: Wash out stomach if vomiting has not occurred. Use stimulants.
Personal Hygiene: Section 3.
Storage and Handling: Section 7.

LOBELINE SULFATE. See lobeline.

LOCUST BEAN GUM. See carob bean gum.

LOGWOOD
Hazard Analysis
Toxic Hazard Rating:
    Acute Local: Allergen 1.
    Acute Systemic: U.
    Chronic Local: Allergen 1.
    Chronic Systemic: U.
Fire Hazard: Slight, when exposed to heat or flame (Section 6).
Countermeasures
Personal Hygiene: Section 3.
Storage and Handling: Section 7.

LOMITE #1. See explosives, high.

LONDON PURPLE, SOLID. See arsenic compounds and aniline.
Shipping Regulations: Section 11.
    I.C.C.: Poison B; poison label, 200 pounds.
    Coast Guard Classification: Poison B; poison label.
    IATA: Poison B, poison label, 25 kilograms (passenger), 95 kilograms (Cargo).

LOROX. See   3-(3,4-dichlorophenyl)-1-methoxy-1-methyl urea.

LOW EXPLOSIVES. See explosives, low.
Shipping Regulations: Section 11.
    I.C.C.: Explosive A—not accepted.
    Coast Guard Classification: Explosive A.
    IATA: Explosive, not acceptable (passenger and cargo).

LPG. See liquified hydrocarbon gas.

LP GAS. See liquefied hydrocarbon gas.

LUBRICATING OIL
General Information
Synonym: Straw oil.
Constants: Flash p: 315-366°F, d: <1.00, autoign. temp.: 783°F.
Hazard Analysis
Toxicity: Details unknown. Can cause dermatitis. See also petroleum hydrocarbons.
Fire Hazard: Slight, when exposed to heat or flame; can react with oxidizing materials.
Spontaneous Heating: No.
Countermeasures
To Fight Fire: Spray, foam, carbon dioxide, dry chemical or carbon tetrachloride (Section 6).
Storage and Handling: Section 7.

TOXIC HAZARD RATING CODE (For detailed discussion, see Section 1.)

0 NONE: (a) No harm under any conditions; (b) Harmful only under unusual conditions or overwhelming dosage.

1 SLIGHT: Causes readily reversible changes which disappear after end of exposure.

2 MODERATE: May involve both irreversible and reversible changes; not severe enough to cause death or permanent injury.

3 HIGH: May cause death or permanent injury after very short exposure to small quantities.

U UNKNOWN: No information on humans considered valid by authors.

LUBRICATING OIL, CYLINDER
General Information
Constant: D: <1.
Hazard Analysis
Toxicity: Details unknown. Can cause dermatitis. See also petroleum hydrocarbons.
Fire Hazard: Slight when exposed to heat or flame; can react with oxidizing materials.
Spontaneous Heating: No.
Countermeasures
To Fight Fire: Water spray, foam, carbon dioxide, dry chemical or carbon tetrachloride (Section 6).
Storage and Handling: Section 7.

LUBRICATING OIL (MAINLY MINERAL)
General Information
Constants: D: <1, autoign. temp.: 700°F, flash p: 300–450°F.
Hazard Analysis
Toxicity: Details unknown. Can cause dermatitis. See also petroleum hydrocarbons.
Fire Hazard: Slight, when exposed to heat or flame; can react with oxidizing materials.
Spontaneous Heating: No.
Countermeasures
To Fight Fire: Water spray, foam, carbon dioxide, dry chemical or carbon tetrachloride (Section 6).
Storage and Handling: Section 7.

LUBRICATING OIL, MINERAL
General Information
Constant: D: <1.
Hazard Analysis
Toxicity: Details unknown. Can cause dermatitis. See also petroleum hydrocarbons.
Fire Hazard: Slight, when exposed to heat or flame; can react with oxidizing materials.
Spontaneous Heating: No.
Countermeasures
To Fight Fire: Water spray, foam, carbon dioxide, dry chemical or carbon tetrachloride (Section 6).
Storage and Handling: Section 7.

LUBRICATING OIL, MOTOR
General Information
Constants: D: <1, autoign. temp.: 500–700°C, flash p: 300–450°F.
Hazard Anaysis
Toxicity: Details unknown. Can cause dermatitis. See also petroleum hydrocarbons.
Fire Hazard: Slight, when exposed to heat or flame; can react with oxidizing materials.
Spontaneous Heating: No.
Countermeasures
To Fight Fire: Water spray, foam, carbon dioxide, dry chemical or carbon tetrachloride (Section 6).
Storage and Handling: Section 7.

LUBRICATING OILS, AUTO
Hazard Analysis
Toxic Hazard Rating:
Acute Local: Irritant 1; Allergen 1; Ingestion 1.
Acute Systemic: 0.
Chronic Local: Allergen 1.
Chronic Systemic: 0.
Toxicity: Can cause dermatitis.
Fire Hazard: Slight, when exposed to heat or flame (Section 6).
Countermeasures
Personal Hygiene: Section 3.
Storage and Handling: Section 7.

LUBRICATING OIL, SPINDLE
General Information
Constants: D: <1, autoign. temp.: 478°F, flash p: 169°F.

Hazard Analysis
Toxicity: Details unknown. Can cause dermatitis. See also petroleum hydrocarbons.
Fire Hazard: Moderate, when exposed to heat or flame; can react with oxidizing materials.
Spontaneous Heating: No.
Countermeasures
To Fight Fire: Water spray, foam, carbon dioxide, dry chemical or carbon tetrachloride (Section 6).
Storage and Handling: Section 7.

LUBRICATING OIL, TURBINE
General Information
Constants: D: <1, autoign. temp.: 700°F, flash p: 400°F (O.C.).
Hazard Analysis
Toxicity: Details unknown. Can cause dermatitis. See also petroleum hydrocarbons.
Fire Hazard: Slight, when exposed to heat or flame; can react with oxidizing materials.
Spontaneous Heating: No.
Countermeasures
To Fight Fire: Water spray, foam, carbon dioxide, dry chemical or carbon tetrachloride (Section 6).
Storage and Handling: Section 7.

LUCIDOL. See benzoyl peroxide, dry.

"LUCITE." See methyl methacrylate.

"LUMINAL"
General Information
Synonyms: Phenylethylbarbituric acid; phenylethylmalonyl-urea.
Description: White, shining, crystalline, odorless and stable powder.
Formula: $(C_6H_5)(C_2H_5)C(CONH)_2CO$.
Constants: Mol wt: 232.17, mp: 174–178°C.
Hazard Analysis
Toxic Hazard Rating:
Acute Local: Allergen 1.
Acute Systemic: Ingestion 2.
Chronic Local: Allergen 1.
Chronic Systemic: Ingestion 2.
Fire Hazard: Unknown.
Countermeasures
Ventilation Control: Section 2.
Personal Hygiene: Section 3.
Storage and Handling: Section 7.

LUPANINE
General Information
Description: White needles (dextroform); yellow syrupy liquid (levoform).
Formula: $C_{15}H_{24}N_2O$.
Constants: Mol wt: 248.28, mp: 40°C.
Hazard Analysis
Toxic Hazard Rating:
Acute Local: U.
Acute Systemic: Ingestion 3; Inhalation 3.
Chronic Local: U.
Chronic Systemic: U.
Disaster Hazard: Dangerous; when heated to decomposition it emits highly toxic fumes.
Countermeasures
Ventilation Control: Section 2.
Personal Hygiene: Section 3.
Storage and Handling: Section 7.

"LUPERCO"
General Information
Description: Paste or solution in organic or inorganic solvents.
Hazard Analysis
Toxic Hazard Rating:
Acute Local: Irritant 3; Ingestion 3; Inhalation 3.

Acute Systemic: U.
Chronic Local: U.
Chronic Systemic: U.
Fire Hazard: Moderate; a powerful oxidizing agent; can react with reducing materials (Section 6).
Explosion Hazard: Unknown.
Countermeasures
Ventilation Control: Section 2.
Personnel Protection : Section 3.
Storage and Handling: Section 7.

LUPERCO BDB. See chlorobenzoyl peroxide.

LUPERCO CCC. See 2,4-dichlorobenzoyl peroxide.

LUPERCO HDB. See hydroxyheptyl peroxide (paste with dibutyl phthalate).

"LUPERSOL"
General Information
Description: Solutions.
Hazard Analysis
Toxic Hazard Rating:
Acute Local: Irritant 3; Ingestion 3; Inhalation 3.
Acute Systemic: U.
Chronic Local: U.
Chronic Systemic: U.
Fire Hazard: Moderate; a powerful oxidizing agent; can react with reducing materials (Section 6).
Explosion Hazard: Unknown.
Countermeasures
Ventilation Control: Section 2.
Personnel Protection: Section 3,
Storage and Handling: Section 7.

LUPININE
General Information
Description: White, crystalline alkaloid.
Formula: $C_{10}H_{19}NO$.
Constants: Mol wt: 169.21, mp: 69–71°C, bp: 255–257°C.
Hazard Analysis
Toxic Hazard Rating:
Acute Local: U.
Acute Systemic: Ingestion 3; Inhalation 3.
Chronic Local: U.
Chronic Systemic: U.
Disaster Hazard: Dangerous; when heated, it emits toxic fumes.
Countermeasures
Ventilation Control: Section 2.
Personal Hygiene: Section 3.
Storage and Handling: Section 7.

LUTETIUM
General Information
Formula: Lu.
Constant: At wt: 175.0.
Hazard Analysis
Toxicity: Unknown.
Radiation Hazard: Section 5. For permissible levels, see Table 5, p. 150.

Natural (2.6%) isotope ^{176}Lu, half life 2.2×10^{10} y. Decays to stable ^{176}Hf by emitting beta particles of 0.42 MeV. Also emits gamma rays of 0.20, 0.31 MeV.
Artificial isotope ^{177}Lu, half life 6.7 d. Decays to stable ^{177}Hf by emitting beta particles of 0.18 (12%), 0.38 (7%), 0.50 (80%) MeV. Also emits gamma rays of 0.11, 0.21 MeV.
Fire Hazard: Moderate in the form of dust when exposed to heat or flame or by chemical reaction with oxidizers.
Explosion Hazard: See powdered metals.
Countermeasures
Storage and Handling: Section 7.

2,6-LUTIDINE
General Information
Synonym: 2,6-Dimethyl pyridine.
Description: Liquid.
Formula: $(CH_3)_2C_5H_3N$.
Constants: Mol wt: 107.1, bp: 143.7°C, fp: −6.6°C, d: 0.923 at 20°C, vap. d.: 3.70.
Hazard Analysis
Toxicity: Details unknown. See pyridine.
Disaster Hazard: Moderately dangerous; when heated to decomposition, it emits toxic fumes.
Countermeasures
Storage and Handling: Section 7.

LYCORINE
General Information
Description: White, crystalline alkaloid.
Formula: $C_{16}H_{17}NO_4$.
Constants: Mol wt: 287.31, mp: 280°C (decomp.).
Hazard Analysis
Toxic Hazard Rating:
Acute Local: U.
Acute Systemic: Ingestion 3; Inhalation 3.
Chronic Local: U.
Chronic Systemic: U.
Disaster Hazard: Moderately dangerous; when heated to decomposition, it emits toxic fumes.
Countermeasures
Ventilation Control: Section 2.
Personal Hygiene: Section 3.

LYE. See sodium hydroxide.

LYSINE
General Information
Synonym: α, ϵ-Diamino caproic acid.
Description: An essential amino acid, occurs in isomeric forms; the L and DL forms are considered here: Colorless crystals; soluble in water; slightly soluble in alcohol; insoluble in ether.
Formula: $NH_2(CH_2)_4CH(NH_2)COOH$.
Constant: Mol wt: 146.
Hazard Analysis
Toxic Hazard Rating: U. A nutrient and/or dietary supplement food additive (Section 10).

TOXIC HAZARD RATING CODE (For detailed discussion, see Section 1.)

0 NONE: (a) No harm under any conditions; (b) Harmful only under unusual conditions or overwhelming dosage.

1 SLIGHT: Causes readily reversible changes which disappear after end of exposure.

2 MODERATE: May involve both irreversible and reversible changes not severe enough to cause death or permanent injury.

3 HIGH: May cause death or permanent injury after very short exposure to small quantities.

U UNKNOWN: No information on humans considered valid by authors

M

MACASSAR GUM. See agar.

MACHINES OR APPARATUS
Shipping Regulations: Section 11.
  I.C.C.: Section 73.130, 73.306.
  IATA: No limit (passenger and cargo).

MACQUER'S SALT. See potassium arsenate.

MAGISTERY OF BISMUTH. See bismuth subnitrate.

MAGNESIA ALBA. See magnesium carbonate.

MAGNESIA MAGNA. See magnesium hydroxide.

MAGNESIUM
General Information
Description: Hexagonal, silvery-white crystals of metal.
Formula: Mg.
Constants: At wt: 24.32, mp: 651°C, bp: 1107°C, d: 1.74 at 5°C, vap. press.: 1 mm at 621°C.
Hazard Analysis
Toxicity: See magnesium compounds.
Fire Hazard: Dangerous, in the form of dust or flakes, when exposed to flame, or by violent chemical reaction with oxidizing agents. In solid form, magnesium is difficult to ignite because heat is conducted rapidly away from the source of ignition; it must be heated above its melting point before it will burn. However, in finely divided form it may be ignited by a spark or the flame of a match. Magnesium fires do not flare up violently unless there is moisture present. Therefore, it must be kept away from water, moisture, etc. It may be ignited by a spark, match flame, or even spontaneously when the material is finely divided and damp, particularly with water-oil emulsion. Also, magnesium reacts with moisture, acids, etc. to evolve hydrogen which is a highly dangerous fire and explosion hazard.
Explosion Hazard: Moderate, in the form of dust, when exposed to flame or by violent chemical reaction with powerful oxidizers. See also powdered metals.
Disaster Hazard: Dangerous; when heated, it burns violently in air and emits fumes; will react with water or steam to produce hydrogen; on contact with oxidizing materials, it can react vigorously.
Countermeasures
To Fight Fire: Operators and fire fighters can approach a magnesium fire to within a few feet if no moisture is present. Water and other ordinary extinguishers, such as CO_2, carbon tetrachloride, etc., should not be used on magnesium fires. G-1 powder or powdered talc should be used on open fires.

Magnesium turnings, borings, etc., should be collected frequently during the working hours and always at the end of the shift from machines and surrounding area and placed in clean, dry metal covered containers plainly labeled, "For Magnesium Only." Good housekeeping is essential. If more than 50 cubic feet (6 fifty-gallon drums) of turnings, borings, etc., are allowed to accumulate, it is necessary that they be left in drums and removed to a separate fire resistant storage room or separated from other occupancies by a space of at least 50 feet. Magnesium grinding dust should always be collected in a dust collector approved specifically for this service. The collected material should be removed frequently, and burned. This may be accomplished by spreading it three or four inches thick on a layer of firebrick or hard burnt paving brick and igniting some combustible refuse placed on top. The burning refuse ignites the top surface of the magnesium and the layer dries as it burns. Danger of a sudden flare up exists when the sludge has been allowed to become partially dry.

During machining, magnesium fires result from the use of dull cutting tools, machining with light cuts at high speeds, or rubbing of the tool on the work after the cutting operation is finished. Chip fires on machine tools can best be controlled by the use of a newly developed liquid extinguishant, by flooding with coolant, by using G-1 powder or powdered talc, or, if necessary, by dry sand. These dry materials should be shoveled onto a fire until it is completely blanketed. If the fire is on a combustible floor, best results can be obtained by spreading a layer of extinguishing material on the floor and on the fire, and then shoveling the fire on the layer of extinguishing material. Oven fires can easily be controlled by the use of boron trifluoride or boron trichloride gas.

MAGNESIUM ALUMINUM PHOSPHIDE
Hazard Analysis and Countermeasures
See phosphides.
Shipping Regulations: Section 11.
  IATA: Flammable solid, yellow label, not acceptable (passenger), 12 kilograms (cargo).

MAGNESIUM o-ARSENATE
General Information
Synonym: Hoernesite.
Description: Monoclinic, white crystals.
Formula: $Mg_3(AsO_4)_2 \cdot 8H_2O$.
Constants: Mol wt: 494.91, d: 2.60–2.61.
Hazard Analysis and Countermeasures
See arsenic compounds.
Shipping Regulations: Section 11.
  MCA warning label.

MAGNESIUM ARSENATE, SOLID. See magnesium o-arsenate.
Shipping Regulations: Section 11.
  I.C.C.: Poison B; poison label, 200 pounds.
  Coast Guard Classification: Poison B; poison label.
  MCA warning label.
  IATA: Poison B, poison label, 25 kilograms (passenger), 95 kilograms (cargo).

MAGNESIUM o-ARSENITE
General Information
Description: Solid.
Formula: $Mg_3(AsO_3)_2$.
Constant: Mol wt: 318.78.
Hazard Analysis and Countermeasures
See arsenic compounds.

MAGNESIUM o-BORATE
General Information
Description: Rhombic, colorless crystals.
Formula: $Mg_3(BO_3)_2$.
Constants: Mol wt: 190.60, d: 2.99 at 21°C.

Hazard Analysis
Toxicity: See boron compounds and magnesium.

MAGNESIUM BROMATE
General Information
Description: Colorless crystals.
Formula: $Mg(BrO_3)_2 \cdot 6H_2O$.
Constants: Mol wt: 388.25, mp: $-6H_2O$ at 200°C, bp: decomposes, d: 2.29.
Hazard Analysis and Countermeasures
See bromates.
Shipping Regulations: Section 11.
  IATA: Oxidizing material, yellow label, 12 kilograms (passenger), 45 kilograms (cargo).

MAGNESIUM BROMIDE
General Information
Description: Large lustrous, white, deliquescent crystals.
Formula: $MgBr_2$.
Constants: Mol wt: 184.15, mp: 700°C, d: 3.72.
Hazard Analysis
Toxicity: See bromides.

MAGNESIUM BROMOPLATINATE
General Information
Description: Trigonal crystals.
Formula: $MgPtBr_6 \cdot 12H_2O$.
Constants: Mol wt: 915.24, d: 2.802.
Hazard Analysis and Countermeasures
See platinum compounds and bromides.

MAGNESIUM CARBONATE
General Information
Synonym: Magnesium carbonate, precipitated; magnesia alba.
Description: Very light, odorless, white powder; soluble in acids; insoluble in water and alcohol.
Formula: $MgCO_3$.
Constants: Mol wt: 84, d: 3.04, decomposes at 350°C.
Hazard Analysis
Toxic Hazard Rating: Unknown. A general purpose food additive; it migrates to food from packaging materials (Section 10).

MAGNESIUM CARBONATE, PRECIPITATED. See magnesium carbonate.

MAGNESIUM CHLORATE
General Information
Description: White, deliquescent crystals or powder.
Formula: $Mg(ClO_3)_2 \cdot 6H_2O$.
Constants: Mol wt: 299.33, mp: 35°C, bp: decomposes at 120°C, d: 1.80 at 25°C.
Hazard Analysis and Countermeasures
See magnesium compounds and chlorates.

MAGNESIUM CHLORIDE
General Information
Description: Thin white to gray opaque granules and/or flakes.
Formula: $MgCl_2$.
Constants: Mol wt: 95.2, mp: 708°C, bp: 1412°C, d: 2.325.
Hazard Analysis and Countermeasures
A substance which migrates to food from packaging materials (Section 10). See magnesium compounds and chlorides.

MAGNESIUM CHLORIDE, HYDRATED. See magnesium chloride.

MAGNESIUM CHLOROPALLADATE
General Information
Description: Hexagonal crystals.
Formula: $MgPdCl_6 \cdot 6H_2O$.
Constants: Mol wt: 451.86, mp: decomposes, d: 2.12.
Hazard Analysis and Countermeasures
See palladium compounds, magnesium compounds and chlorides.

MAGNESIUM CHLOROPLATINATE
General Information
Description: Crystals.
Formula: $MgPtCl_6 \cdot 6H_2O$.
Constants: Mol wt: 540.39, mp: decomposes, d: 2.437.
Hazard Analysis and Countermeasures
See platinum compounds, magnesium compounds and chlorides.

MAGNESIUM CHLOROSTANNATE
General Information
Description: Crystals.
Formula: $MgSnCl_6 \cdot 6H_2O$.
Constants: Mol wt: 463.86, mp: decomposes at 100°C, d: 2.08.
Hazard Analysis and Countermeasures
See tin compounds, magnesium compounds and chlorides.

MAGNESIUM CHROMATE
General Information
Description: Rhombic, yellow crystals.
Formula: $MgCrO_4 \cdot 7H_2O$.
Constants: Mol wt: 266.44, d: 1.695.
Hazard Analysis
Toxicity: See chromium compounds and magnesium compounds.
Fire Hazard: Slight, by chemical reaction; can react with reducing materials (Section 6).
Countermeasures
Storage and Handling: Section 7.

MAGNESIUM COMPOUNDS
Hazard Analysis
Toxic Hazard Rating:
  Acute Local: Irritant 1.
  Acute Systemic: Ingestion 1; Inhalation 2.
  Chronic Local: Irritant 2.
  Chronic Systemic: U.
Toxicology: The inhalation of fumes of freshly sublimed magnesium oxide may cause metal fume fever. There is no evidence that magnesium produces true systemic poisoning. Occupational health hazards may exist in magnesium foundries, probably from the presence of atmospheric contaminants such as fluorides, sulfur dioxide, carbon tetrachloride and chromium compounds.
  Particles of metallic magnesium or magnesium alloy which perforate the skin or gain entry through cuts and scratches may produce a severe local lesion characterized by the evolution of gas and acute inflammatory reaction, frequently with necrosis. The condition has been called a "chemical gas gangrene." Gaseous blebs may develop within 24 hours of the injury. The inflammatory response is marked at the site of injury and there may be signs of lymphangitis. The lesion is very slow to heal.
  The most serious hazard presented by magnesium is the danger from burns. Protection necessary for personnel handling and processing magnesium is usually

---

TOXIC HAZARD RATING CODE (For detailed discussion, see Section 1.)

0 NONE: (a) No harm under any conditions; (b) Harmful only under unusual conditions or overwhelming dosage.

1 SLIGHT: Causes readily reversible changes which disappear after end of exposure.

2 MODERATE: May involve both irreversible and reversible changes; not severe enough to cause death or permanent injury.

3 HIGH: May cause death or permanent injury after very short exposure to small quantities.

U UNKNOWN: No information on humans considered valid by authors.

no different from that which is necessary for other metals. It is recommended that smooth clothing and leather or fire resistant, easily removable aprons be worn in grinding operations on magnesium. The toxicity of magnesium compounds is usually that of the anion. Refer to magnesium and anion. See also specific compounds.

MAGNESIUM CYANIDE
General Information
Description: Solid.
Formula: $Mg(CN)_2$.
Constant: Mol wt: 76.4.
Hazard Analysis and Countermeasures
See cyanides.

MAGNESIUM CYCLAMATE
General Information
Synonym: Magnesium cyclohexyl sulfamate.
Hazard Analysis
Toxic Hazard Rating: Details unknown. A non-nutritive sweetener food additive (Section 10). See also sodium cyclamate.
Disaster Hazard: Dangerous. See sulfates.

MAGNESIUM DIAMIDE
General Information
Description: White powder.
Formula: $Mg(NH_2)_2$.
Constants: Mol wt: 56.4, mp: decomposes.
Hazard Analysis
Toxicity: Details unknown. See also ammonia, which is evolved upon contact with water, and amides.
Fire Hazard: Dangerous; by chemical reaction to evolve ammonia; spontaneously flammable in air (Section 6).
Explosion Hazard: Moderate, by spontaneous chemical reaction; reacts violently with water.
Disaster Hazard: Dangerous; will react with water or steam to produce toxic and flammable vapors; can react vigorously with oxidizing materials.
Countermeasures
Ventilation Control: Section 2.
Storage and Handling: Section 7.

MAGNESIUM DIETHYL
General Information
Description: Crystals or liquid.
Formula: $Mg(C_2H_5)_2$.
Constants: Mol wt: 82.4, mp: 0°C.
Hazard Analysis
Toxicity: Details unknown. See also magnesium and organometals.
Fire Hazard: Dangerous; spontaneously flammable in air or in carbon dioxide (Section 6).
Explosion Hazard: Moderate, by spontaneous chemical reaction or upon contact with water.
Disaster Hazard: Dangerous; will react violently with water or steam; can react vigorously with oxidizing materials.
Countermeasures
Storage and Handling: Section 7.

MAGNESIUM DIOXIDE. See magnesium peroxide.

MAGNESIUM DIPHENYL
General Information
Description: Feathery crystals.
Formula: $Mg(C_6H_5)_2$.
Constant: Mol wt: 178.5.
Hazard Analysis
Toxicity: Unknown. See magnesium and organometals.
Fire Hazard: Moderate; spontaneously flammable in moist (but not dry) air (Section 6).
Explosion Hazard: Moderate, when exposed to flame, or upon contact with moisture.
Disaster Hazard: Moderately dangerous; when heated to decomposition or on contact with acid or acid fumes,

it emits toxic fumes; decomposes violently in water; can react with oxidizing materials.
Countermeasures
Storage and Handling: Section 7.

MAGNESIUM DROSS
Shipping Regulations: Section 11.
   I.C.C.: Section 73.173.
   IATA (dry): Not restricted.
   IATA (wet): Flammable solid, not acceptable (passenger and cargo).

MAGNESIUM DUST. See magnesium.

MAGNESIUM 9,10-EPOXYSTEARATE
Hazard Analysis
Toxic Hazard Rating:
   Acute Local: U.
   Acute Systemic: Ingestion 1.
   Chronic Local: U.
   Chronic Systemic: U.
Based on limited animal experiments.

MAGNESIUM ETHYLENE BISDITHIOCARBAMATE
Hazard Analysis
Toxicity: Details unknown; a skin irritant.
Disaster Hazard: Dangerous; when heated to decomposition, it emits highly toxic fumes of oxides of sulfur.
Countermeasures
Storage and Handling: Section 7.

MAGNESIUM FERROCYANIDE
General Information
Description: Pale yellow crystals.
Formula: $Mg_2Fe(CN)_6 \cdot 12H_2O$.
Constants: Mol wt: 476.79, mp: decomposes at approx. 200°C.
Hazard Analysis and Countermeasures
See ferrocyanides.

MAGNESIUM FLUORIDE
General Information
Synonym: Sellaite, Afluon.
Description: Faint violet, luminous crystals.
Formula: MgF_2.
Constants: Mol wt: 62.32, mp: 1396°C, bp: 2239°C, d: 2.9–3.2.
Hazard Analysis and Countermeasures
See fluorides.

MAGNESIUM FLUOSILICATE
General Information
Description: White crystals or powder.
Formula: $MgSiF_6$.
Constant: Mol wt: 166.38.
Hazard Analysis and Countermeasures
See fluosilicates.

MAGNESIUM FORMATE
General Information
Description: Rhombic, colorless crystals.
Formula: $Mg(CHO_2)_2 \cdot 2H_2O$.
Constant: Mol wt: 150.4.
Hazard Analysis
Toxicity: See formic acid.
Fire Hazard: Slight.
Countermeasures
Storage and Handling: Section 7.

MAGNESIUM o-GERMANATE
General Information
Description: White precipitate.
Formula: Mg_2GeO_4.
Constant: Mol wt: 185.24.
Hazard Analysis
Toxicity: See germanium compounds.

MAGNESIUM HYDRATE. See magnesium hydroxide.

MAGNESIUM HYDRIDE
General Information
Description: White crystals; reacts with water; soluble in isopropylamine.
Formula: MgH_2.
Constants: Mol wt: 26.3, d: 1.419, mp: decomposes > 200°C.
Hazard Analysis and Countermeasures
See magnesium compounds and hydrides.

MAGNESIUM HYDROGEN PHOSPHATE. See magnesium phosphate, dibasic.

MAGNESIUM HYDROXIDE
General Information
Synonyms: Magnesium hydrate, magnesia magna.
Description: White powder; odorless; soluble in solutions of ammonium salts and dilute acids; almost insoluble in water and alcohol.
Formula: $Mg(OH)_2$.
Constants: Mol wt: 58, d: 2.36, mp: decomposes at 350°C.
Hazard Analysis
A general purpose food additive; it is a substance which migrates to food from packaging materials (Section 10).

MAGNESIUM HYPOPHOSPHITE
General Information
Description: White crystals.
Formula: $Mg(H_2PO_2)_2 \cdot 6H_2O$.
Constants: Mol wt: 262.36, d: 1.59.
Hazard Analysis
Toxicity: See hypophosphites.
Fire Hazard: Moderate, when exposed to heat.
Explosion Hazard: Moderate, by chemical reaction.
Disaster Hazard: Dangerous; when heated to 100°C it evolves flammable gas and highly toxic fumes of phosphine; can react with oxidizing materials.
Countermeasures
Storage and Handling: Section 7.

MAGNESIUM IODATE
General Information
Description: Colorless crystals.
Formula: $Mg(IO_3)_2 \cdot 4H_2O$.
Constants: Mol wt: 446.22, mp: $-4H_2O$ at 210°C, bp: decomposes, d: 3.3 at 13.5°C.
Hazard Analysis and Countermeasures
See iodates and magnesium compounds.

MAGNESIUM IODIDE
General Infromation
Description: White, deliquescent crystals.
Formula: MgI_2.
Constants: Mol wt: 278.16, mp: > 700°C decomposes, d: 4.244.
Hazard Analysis and Countermeasures
See iodides and magnesium compounds.

MAGNESIUM, METALLIC, POWDERED, PELLETS, TURNINGS OF RIBBON
Shipping Regulations: Section 11.
I.C.C.: Flammable solid, yellow label, 100 pounds.
IATA: Flammable solid, yellow label, 12 kilograms (passenger), 45 kilograms (cargo).

MAGNESIUM NITRATE
General Information
Description: (1) white crystals (prisms); (2) monoclinic, colorless, deliquescent crystals.

Formula: (1) $Mg(NO_3)_2 \cdot 2H_2O$; (2) $Mg(NO_3)_2 \cdot 6H_2O$.
Constants: Mol wt: (1) 184.37, (2) 256.43; d: (1) 2.0256 at 25°C, (2) 1.464; mp: (1) 129.0°C, (2) 95°C; bp: (2) $-5H_2O$ at 330°C.
Hazard Analysis and Countermeasures
See nitrates.
Shipping Regulations: Section 11.
I.C.C.: Oxidizing material; yellow label, 100 pounds.
Coast Guard Classification: Oxidizing material.
IATA: Oxidizing material, yellow label, 12 kilograms (passenger), 45 kilograms (cargo).

MAGNESIUM OXALATE
General Information
Description: White powder.
Formula: $MgC_2O_4 \cdot 2H_2O$.
Constants: Mol wt: 148.37, mp: decomposes, d: 2.45.
Hazard Analysis
Toxicity: See oxalates.

MAGNESIUM OXIDE
General Information
Description: White powder.
Formula: MgO.
Constants: Mol wt: 40.32, mp: 2500–2800°C, d: 3.65–3.75.
Hazard Analysis
Toxicity: A nutrient and/or dietary supplement food additive (Section 10). See magnesium compounds.
TLV: ACGIH (recommended); (fume) 15 milligrams per cubic meter of air.

MAGNESIUM PERBORATE
General Information
Description: White powder.
Formula: $Mg(BO_3)_2 \cdot 7H_2O$.
Constant: Mol wt: 268.02.
Hazard Analysis
Toxicity: See magnesium compounds and boron compounds.
Fire Hazard: Slight, by chemical reaction; an oxidant (Section 6).
Countermeasures
Storage and Handling: Section 7.

MAGNESIUM PERCHLORATE
General Information
Description: White, hygroscopic crystals.
Synonym: Dehydrite.
Formula: $Mg(ClO_4)_2$.
Constants: Mol wt: 223.33, mp: decomposes at 251°C, d: 2.60 at 25°C.
Hazard Analysis
Toxic Hazard Rating:
Acute Local: Irritant 2; Ingestion 2; Inhalation 2.
Acute Systemic: U.
Chronic Local: U.
Chronic Systemic: U.
Toxicology: See also magnesium compounds and perchlorates.
Fire Hazard: See perchlorates.
Explosion Hazard: See perchlorates.
Disaster Hazard: See perchlorates.
Countermeasures
Storage and Handling: Section 7.
Shipping Regulations: Section 11.
I.C.C.: Oxidizing material; yellow label, 100 pounds.
Coast Guard Classification: Oxidizing material; yellow label.

TOXIC HAZARD RATING CODE *(For detailed discussion, see Section 1.)*

0 NONE: (a) No harm under any conditions; (b) Harmful only under unusual conditions or overwhelming dosage.

1 SLIGHT: Causes readily reversible changes which disappear after end of exposure.

2 MODERATE: May involve both irreversible and reversible changes not severe enough to cause death or permanent injury.

3 HIGH: May cause death or permanent injury after very short exposure to small quantities.

U UNKNOWN: No information on humans considered valid by authors

IATA: Oxidizing material, yellow label, 12 kilograms (passenger), 45 kilograms (cargo).

MAGNESIUM PERMANGANATE
General Information
Description: Dark purple, deliquescent needles.
Formula: $Mg(MnO_4)_2 \cdot 6H_2O$.
Constants: Mol wt: 370.28, mp: decomposes, d: 2.18.
Hazard Analysis
Toxicity: See manganese compounds.
Fire Hazard: Moderate, by chemical reaction; a powerful oxidizer (Section 6).
Countermeasures
Storage and Handling: Section 7.

MAGNESIUM PEROXIDE
General Information
Synonym: Magnesium dioxide.
Description: White powder.
Formula: MgO_2(theoretical).
Constant: Mol wt: 56.32.
Hazard Analysis
Toxicity: See peroxides, inorganic, and magnesium compounds.
Fire Hazard: Moderate, by chemical reaction with acidic materials and moisture; an oxidizing agent (Section 6).
Disaster Hazard: Dangerous; can react with reducing agents; will decompose violently in or near a fire.
Countermeasures
Storage and Handling: Section 7.
Shipping Regulations: Section 11.
I.C.C.: Oxidizing material; yellow label, 100 pounds.
Coast Guard Classification: Oxidizing material; yellow label.
IATA: Oxidizing material, yellow label, 12 kilograms (passenger), 45 kilograms (cargo).

MAGNESIUM PHOSPHATE, DIBASIC
General Information
Synonyms: Dimagnesium o-phosphate; dimagnesium phosphate; magnesium phosphate secondary; magnesium hydrogen phosphate.
Description: White, crystalline powder; soluble in dilute acids; slightly soluble in water.
Formula: $MgHPO_4 \cdot 3H_2O$.
Constants: Mol wt: 173, d: 2.13, decomposes at 550–560°C.
Hazard Analysis
Toxic Hazard Rating: Unknown. Used as a nutrient and/or dietary supplement food additive (Section 10).
Disaster Hazard: Dangerous. See phosphates.

MAGNESIUM PHOSPHATE, NEUTRAL. See magnesium phosphate, tribasic.

MAGNESIUM PHOSPHATE, SECONDARY. See magnesium phosphate, dibasic.

MAGNESIUM PHOSPHATE, TRIBASIC
General Information
Synonyms: Magnesium phosphate, neutral; trimagnesium phosphate.
Description: Fine, soft, bulky, white powder; odorless; soluble in acids; insoluble in water.
Formula: $Mg_3(PO_4)_2 \cdot 8H_2O$.
Constants: Mol wt: 406, loses all water at 400°C.
Hazard Analysis
Toxic Hazard Rating: Unknown. Used as a nutrient and/or dietary supplement food additive (Section 10).
Disaster Hazard: Dangerous. See phosphates.

MAGNESIUM RICINOLEATE
General Information
Description: Coarse white granules.
Formula: $Mg[CO_2(CH_2)_7CH{=}CHCH_2CHOH(CH_2)_5{-}CH_3]_2$.
Constants: Mol wt: 614, mp: 95°C, d: 106 at 25°/25°C.

Hazard Analysis
Toxicity: see magnesium compounds.
Countermeasures
Storage and Handling: Section 7.

MAGNESIUM SCRAP (SHAVINGS, BORINGS, OR TURNINGS). See magnesium.
Shipping Regulations: Section 11.
I.C.C.: Flammable solid; yellow label, 100 pounds.
Coast Guard Classification: Inflammable solid; yellow label.
IATA: Flammable solid, not acceptable (passenger and cargo).

MAGNESIUM SELENATE
General Information
Description: Colorless crystals.
Formula: $MgSeO_4 \cdot 6H_2O$.
Constants: Mol wt: 275.38, d: 1.928.
Hazard Analysis and Countermeasures
See selenium compounds.

MAGNESIUM SILICATE HYDRATED. See asbestos.

MAGNESIUM SILICATE, HYDROUS
General Information
Description: Solid.
Formula: $Mg_2Si_3O_8 \cdot 5H_2O$.
Hazard Analysis
Toxicity: An anticaking agent food additive (Section 10). See silicates.

MAGNESIUM STEARATE
General Information
Description: Soft white, light powder; odorless; insoluble in water and alcohol.
Formula: $Mg(C_{18}H_{35}O_2)_2$.
Constants: Mol wt: 590, d: 1.028, mp: 88.5°C (pure), 132°C (technical).
Hazard Analysis
Toxic Hazard Rating: Unknown. A general purpose food additive (Section 10).

MAGNESIUM SULFATE
General Information
Synonym: Epsom salts.
Description: Opaque needles.
Formula: $MgSO_4 \cdot 7H_2O$.
Constants: Mol wt: 246.5, mp: $-7H_2O$ at 200°C, d: 1.68.
Hazard Analysis
Toxicity: A nutrient and/or dietary supplement food additive, it migrates to food from packaging material. See section 10. See magnesium compounds.

MAGNESIUM SULFIDE
General Information
Description: Pale red to reddish-brown.
Formula: MgS.
Constants: Mol wt: 56.39, mp: decomposes, d: 2.79–2.85.
Hazard Analysis and Countermeasures
See magnesium compounds and sulfides.

MAGNESIUM SULFITE
General Information
Description: White crystalline powder.
Formula: $MgSO_3 \cdot 6H_2O$.
Constants: Mol wt: 212.48, mp: $-6H_2O$ at 200°C, bp: decomposes, d: 1.725.
Hazard Analysis and Countermeasures
See magnesium compounds and sulfites.

MAGNESIUM THIOTELLURITE
General Information
Description: Pale yellow, crystalline mass.
Formula: Mg_3TeS_5.
Constant: Mol wt: 3ι0.9.

Hazard Analysis
Toxicity: See sulfides and tellurium compounds.
Fire Hazard: Slight, when exposed to heat or flame (Section 6).
Disaster Hazard: Dangerous; when heated to decomposition or on contact with acid or acid fumes, it emits highly toxic fumes of oxides of sulfur and tellurium; can react with oxidizing materials.

Countermeasures
Storage and Handling: Section 7.

MAGNESIUM—THORIUM ALLOYS IN FORMED SHAPES (not powdered and which shall contain not more than 4% nominal thorium 232).
Shipping Regulations: Section 11.
  I.C.C.: Poison D, radioactive materials, red label, Section 73.393 (L).
  IATA: Radioactive material, radioactive red label, 2000 millicuries (passenger and cargo).

MAGNESIUM TUNGSTATE
General Information
Description: Colorless, monoclinic crystals.
Formula: $MgWO_4$.
Constants: Mol wt: 272.24, d: 5.66.
Hazard Analysis
Toxicity: See magnesium compounds and tungsten compounds.

MAGNUS' SALT. See tetrammine platinum II chloroplatinate.

MALACHITE GREEN
General Information
Synonyms: Aniline green; china green.
Description: Green crystals.
Formula: $C_{23}H_{25}N_2Cl$.
Constant: Mol wt: 364.9.
Hazard Analysis
Toxicity: Details unknown. See also aniline dyes.
Disaster Hazard: Dangerous; when heated to decomposition, it emits highly toxic fumes of aniline and oxides of nitrogen.
Countermeasures
Storage and Handling: Section 7.

MALATHION. See O,O-Dimethyl dithiophosphate of diethyl mercaptosuccinate.

MALEANILIC ACID. See N-phenylmaleamic acid.

MALE FERN. See aspidium.

MALEIC ACID
General Information
Synonyms: Maleinic acid; toxilic acid.
Description: White crystals, faint acidulous odor.
Formula: HOOCCHCHCOOH.
Constants: Mol wt: 116.1, mp: 130.5°C, bp: 135°C decomposes, d: 1.590 at 20°/4°C, vap. d.: 4.0.
Hazard Analysis
Toxic Hazard Rating:
  Acute Local: Irritant 3, Ingestion 2; Inhalation 2.
  Acute Systemic: U.
  Chronic Local: U.
  Chronic Systemic: U.
Toxicity: Believed to be more toxic than its isomer fumaric acid.
Fire Hazard: Slight.

Countermeasures
Ventilation Control: Section 2.
Personnel Protection: Section 3.
Storage and Handling: Section 7.

MALEIC ANHYDRIDE
General Information
Synonym: Toxilic anhydride; cisbutenedioic anhydride.
Description: Fused black or white crystals.
Formula: OCOCHCHCO.
Constants: Mol wt: 98.1, mp: 58°C, bp: 202°C, flash p: 218°F (C.C.), d: 0.934 at 20°/4°C, autoign. temp.: 890°F, vap. press.: 1 mm at 44.0°C, vap. d.: 3.4, lel: 1.4%, uel: 7.1%.
Hazard Analysis
Toxic Hazard Rating:
  Acute Local: Irritant 3; Ingestion 3; Inhalation 3.
  Acute Systemic: U.
  Chronic Local: U.
  Chronic Systemic: U.
TLV: ACGIH (tentative); 0.25 part per million of air; 1 milligram per cubic meter of air.
Toxicology: Inhalation of vapor can cause pulmonary edema.
Fire Hazard: Moderate, when exposed to heat or flame; will react with water or steam to produce heat; emits toxic fumes when heated; can react on contact with oxidizing materials.
Countermeasures
Ventilation Control: Section 2.
Personnel Protection: Section 3.
Storage and Handling: Section 7.
To Fight Fire: Alcohol foam.
Shipping Regulations: Section 11.
  MCA warning label.

MALEIC HYDRAZIDE
General Information
Synonym: 1,2-Dihydro-3,6-pyridazinedione.
Description: Crystals; somewhat soluble in water and alcohol.
Formula: $C_4H_4N_2O_2$.
Constants: Mol wt: 112.1, mp: > 300°C.
Hazard Analysis
Toxicity: Details unknown. Has caused CNS disturbances in experimental animals. Also liver damage from chronic exposure. It is used as a food additive permitted in food for human consumption (Section 10).
Disaster Hazard: Dangerous; when heated to decomposition, it emits highly toxic fumes.

MALEINIC ACID See maleic acid.

MALIC ACID
General Information
Synonyms: Common malic acid; hydroxysuccinic acid; apple acid.
Description: Colorless crystals; very soluble in water and alcohol; slightly soluble in ether. Exhibits isomeric forms (DL, L and D).
Formula: $COOHCH_2CH(OH)COOH$.
Constants: Mol wt: 134, d (DL): 1.601, d (D or L): 1.595 at 20/40°C, mp (DL): 128°C, mp (D or L): 100°C, bp (DL): 150°C, bp (D or L): 140°C (decomposes).
Hazard Analysis
Toxic Hazard Rating: Unknown. A general purpose food additive. Also a synthetic flavoring substance and adjuvant (Section 10).

TOXIC HAZARD RATING CODE (For detailed discussion, see Section 1.)

0 NONE: (a) No harm under any conditions; (b) Harmful only under unusual conditions or overwhelming dosage.

1 SLIGHT: Causes readily reversible changes which disappear after end of exposure.

2 MODERATE: May involve both irreversible and reversible changes; not severe enough to cause death or permanent injury.

3 HIGH: May cause death or permanent injury after very short exposure to small quantities.

U UNKNOWN: No information on humans considered valid by authors.

MALONIC ACID
General Information
Synonyms: Methanedicarbonic acid; methanedicarboxylic acid; propanedioic acid.
Description: Small crystals; soluble in water, alcohol, and ether.
Formula: $CH_2(COOH)_2$.
Constants: Mol wt: 104.06, mp: 132–134°C, bp: decomposes, d: 1.67.
Hazard Analysis
Toxic Hazard Rating:
Acute Local: Irritant 3; Ingestion 3; Inhalation 3.
Acute Systemic: U.
Chronic Local: U.
Chronic Systemic: U.

MALONIC ACID MONONITRILE. See cyanoacetic acid.

MALONIC ETHYL ESTERNITRILE. See ethyl cyanoacetate.

MALONYLUREA. See barbituric acid.

MALT EXTRACT
General Information
Description: A powder.
Composition: Contains diastase, dextrin, dextrose, protein bodies and barley salts.
Hazard Analysis
Toxic Hazard Rating:
Acute Local: Allergen 1; Inhalation 1.
Acute Systemic: Inhalation 1.
Chronic Local: Allergen 1.
Chronic Systemic: U.
Toxicology: Malt has been associated with grain or malt fever, an illness existing among grain workers. This is described in the literature as follows. It occurs several hours after the patient leaves work, possibly during sleep. It causes headache, weakness, fever, chills and cold sweats, as well as nausea, coughing and often vomiting. By morning these symptoms have subsided. The victim feels normal and generally reports back for work. This strange illness occurs most frequently among men never before exposed to grain dust, or even among experienced men at the beginning of a new season or upon returning to work after a short absence. It resembles metal fume fever because only those constantly exposed appear to develop an immunity. No really satisfactory explanation can be offered for this reaction to malt dust, except that it is probably due to the presence of a foreign protein.
Fire Hazard: Slight, when exposed to heat or flame; can react with oxidizing materials (Section 6).
Explosion Hazard: Slight, in the form of dust, when exposed to flame.
Countermeasures
Storage and Handling: Section 7.

MANDELIC ACID
General Information
Synonyms: Amygdalic acid; phenylhydroxacetic acid; phenylglycolic acid, dl-mandelic acid; α-hydroxy-α-toluic acid; racemic acid; uromaline.
Description: Large white crystals or powder; faint odor; soluble in ether; slightly soluble in water and alcohol.
Formula: $C_6H_5CH(OH)COOH$.
Constants: Mol wt: 142, d: 1.30, mp: 117–119°C.
Hazard Analysis
Toxic Hazard Rating:
Acute Local: U.
Acute Systemic: Ingestion 2.
Chronic Local: Ingestion 1.
Chronic Systemic: U.
Toxicity: Continued absorption can cause kidney irritation. Used medicinally. Ingestion of large doses causes nausea, diarrhea and possibly kidney damage.

MANDELONITRILE. See benzaldehyde cyanhydrin.

MANDELYLTROPEINE. See homatropine and compounds.

MANEB
General Information
Synonym: Manganous ethylene bis-dithio carbamate; manzate.
Description: Yellow powder or crystals; water soluble.
Formula: $C_4H_6MnN_2S_4$.
Constants: Mol wt: 265.3.
Hazard Analysis
Toxic Hazard Rating:
Acute Local: Irritant 2; Ingestion 2; Inhalation 2, Allergen 2.
Acute Systemic: Ingestion 2.
Chronic Local: Irritant 1, Allergen 2.
Chronic Systemic: Ingestion 2.
Toxicology: Data based upon animal experiments. Has caused allergic dermatitis in humans.
Disaster Hazard: Dangerous; when heated to decomposition, it emits highly toxic fumes of oxides of nitrogen and sulfur.

MANGANESE
General Information
Description: Reddish-gray or silvery, brittle, metallic element.
Formula: Mn.
Constants: At wt: 54.93, mp: 1260°C, bp: 1900°C, d: 7.20, vap. press.: 1 mm at 1292°C.
Hazard Analysis
Toxic Hazard Rating:
Acute Local: 0.
Acute Systemic: Inhalation 2.
Chronic Local: U.
Chronic Systemic: Inhalation 3.
See also manganese compounds.
TLV: ACGIH (recommended); 5 milligrams per cubic meter of air.
Radiation Hazard: Section 5. For permissible levels, see Table 5, p. 150.
Artificial isotope ^{52}Mn, half life 5.7 d. Decays to stable ^{52}Cr by electron capture. Also decays to ^{52}Cr by emitting positrons (29%) of 0.30, 0.58 MeV. Also emits gamma rays of 0.74, 0.93, 1.43 MeV and other gammas and X-rays.
Artificial isotope ^{54}Mn, half life 314 d. Decays to stable ^{54}Cr by electron capture. Emits gamma rays of 0.84 MeV and X-rays.
Artificial isotope ^{56}Mn, half life 2.6 h. Decays to stable ^{56}Fe by emitting beta particles of 0.75 (15%), 1.05 (24%), 2.86 (60%) MeV. Also emits gamma rays of 0.84, 1.81, 2.12 MeV.
Fire Hazard: Moderate, in the form of dust or powder, when exposed to flame.
Spontaneous Heating: No.
Explosion Hazard: Moderate, in the form of dust, when exposed to flame. See also powdered metals.
Disaster Hazard: Moderately dangerous; will react with water or steam to produce hydrogen; can react with oxidizing materials.
Countermeasures
Ventilation Control: Section 2.
Storage and Handling: Section 7.
To Fight Fire: Special mixtures of dry chemical (Section 6).

MANGANESE ACETATE
General Information
Description: Pale red crystals; very soluble in water and alcohol.
Formula: $Mn(C_2H_3O_2)_2 \cdot 4H_2O$.
Constants: Mol wt: 245, d: 1.54, mp: 80°C.

Hazard Analysis
Toxic Hazard Rating: Unknown. Used as a trace mineral added to animal feeds (Section 10).

MANGANESE ARSENATE
General Information
Description: Reddish-white, crystalline solid.
Formula: $MnHAsO_4$.
Constant: Mol wt: 194.9.
Hazard Analysis and Countermeasures
See arsenic compounds and manganese compounds.

MANGANESE BACITRACIN
Hazard Analysis
Toxic Hazard Rating: Unknown. Used as a food additive permitted in food for human consumption (Section 10). See also manganese compounds.

MANGANESE BENZOATE. See manganous benzoate.

MANGANESE BROMIDE. See manganese dibromide.

MANGANESE CACODYLATE
General Information
Description: Reddish-white crystals.
Formula: $Mn[(CH_3)_2AsO_2]_2$.
Constant: Mol wt: 328.9.
Hazard Analysis and Countermeasures
See arsenic compounds and manganese compounds.

MANGANESE CHLORIDE. See manganese dichloride.

MANGANESE COMPOUNDS
Hazard Analysis
Toxic Hazard Rating:
  Acute Local: U.
  Acute Systemic: Ingestion 2; Inhalation 2.
  Chronic Local: U.
  Chronic Systemic: Ingestion 3; Inhalation 3.
Toxicology: Chronic manganese poisoning is a clearly characterized disease which results from the inhalation of fumes or dusts of manganese. Exposure to heavy concentrations of dusts or fumes for as little as three months may produce the condition, but usually cases develop after one to three years of exposure. The central nervous system is the chief site of damage. If cases are removed from exposure shortly after the appearance of symptoms, some improvement in the patient's condition frequently occurs, though there may be some residual disturbances in gait and speech. When well established, however, the disease results in permanent disability.
  Individuals exposed to dusts and fumes of manganese have been reported by several investigators to suffer from a much higher incidence of upper respiratory infections and pneumonia than does the general population. It has not yet been possible to prove that a definite pneumonitis results in humans from exposure to manganese dusts or fumes under industrial conditions. However, experiments with mice have produced definite and striking lung pathology which varied in intensity with the length of exposure to the dust.
  Chronic manganese poisoning begins usually with complaints of languor and sleepiness. This is followed by weakness in the legs and the development of a stolid, mask-like facies, and the patient speaks with a slow monotonous voice. Then muscular twitchings appear, varying from a fine tremor of the hands to coarse, rhythmical movements of the arms, legs and trunk.

Nocturnal cramps of the legs appear about the same time. There is a slight increase in tendon reflexes, ankle and patellar clonus, and a typical Parkinsonian slapping gait. The handwriting may be quite minute. There are no sensory disturbances, and no eye, gastrointestinal or genitourinary complaints. The urine and spinal fluid are normal, and the blood shows no abnormality or only a slight leucopenia. The symptoms may simulate progressive bulbar paralysis, postencephalitic Parkinsonism, multiple sclerosis, amyotrophic lateral sclerosis and progressive lenticular degeneration (Wilson's Disease). Often, a history of exposure is the only aid in establishing the diagnosis.
  The blood may show increased erythrocyte formation and increased osmotic fragility. Early administration of EDTA can hasten recovery, but it is of little value in cases of long standing.
  Manganese compounds are common air contaminants (Section 4).
Countermeasures
Ventilation Control: Section 2.
Personnel Protection: Section 3.
Storage and Handling: Section 7.

MANGANESE CYCLOPENTADIENYLTRICARBONYL
Hazard Analysis
Toxic Hazard Rating:
  Acute Local: U.
  Acute Systemic: Inhalation 2.
  Chronic Local: U.
  Chronic Systemic: Inhalation 2.
Toxicity: Animal experiments show mild narcotic effect and impaired kidney function. See also manganese compounds.

MANGANESE DIBROMIDE
General Information
Synonym: Manganese bromide.
Description: Rose-red crystals.
Formula: $MnBr_2 \cdot 4H_2O$.
Constants: Mol wt: 286.8, mp: 64°C (decomposes), d: 4.385 at 25°C.
Hazard Analysis and Countermeasures
See manganese compounds and bromides.

MANGANESE DICHLORIDE
General Information
Synonyms: Scacchite; managanese chloride.
Description: Cubic, deliquescent, pink crystals.
Formula: $MnCl_2$.
Constants: Mol wt: 125.84, mp: 650°C, bp: 1190°C d: 2.977 at 25°C.
Hazard Analysis and Countermeasures
A trace mineral added to animal feeds, also a nutrient and/or dietary supplement food additive (Section 10) See manganese compounds and chlorides.

MANGANESE DIFLUORIDE
General Information
Synonym: Manganese fluoride.
Description: Red, tetragonal crystals or reddish powder.
Formula: MnF_2.
Constants: Mol wt: 92.93, mp: 856°C, d: 3.98.
Hazard Analysis and Countermeasures
See fluorides and manganese compounds.

TOXIC HAZARD RATING CODE (For detailed discussion, see Section 1.)

0 NONE: (a) No harm under any conditions; (b) Harmful only under unusual conditions or overwhelming dosage.

1 SLIGHT: Causes readily reversible changes which disappear after end of exposure.

2 MODERATE: May involve both irreversible and reversible changes not severe enough to cause death or permanent injury.

3 HIGH: May cause death or permanent injury after very short exposur to small quantities.

U UNKNOWN: No information on humans considered valid by author

MANGANESE DIIODIDE
General Information
Synonym: Manganese iodide.
Description: Yellowish-brown or pink, deliquescent crystalline mass or white powder.
Formula: $MnI_2 \cdot 4H_2O$.
Constants: Mol wt: 380.8, mp: decomposes at approx. 80°C, d: 5.01.
Hazard Analysis and Countermeasures
See manganese compounds and iodides.

MANGANESE DIOXIDE
General Information
Synonym: Manganese oxide.
Description: Black powder.
Formula: MnO_2.
Constants: Mol wt: 86.93, mp: −O at 535°C, d: 5.0.
Hazard Analysis
Toxicity: See manganese compounds.
Fire Hazard: Moderate, by chemical reaction; a powerful oxidizer. It must not be heated or rubbed in contact with easily oxidizable matter (Section 6).
Disaster Hazard: Moderately dangerous; keep away from heat and flammable materials.
Countermeasures
Storage and Handling: Section 7.
Shipping Regulations: Section 11.
  Coast Guard Classification: Hazardous material.
  IATA: Other restricted articles, B, not restricted (passenger and cargo).

MANGANESE FLUORIDE. See manganese difluoride.

MANGANESE GLUCONATE
General Information
Description: Light pinkish powder or coarse granules; soluble in water; insoluble in alcohol and benzene.
Formula: $Mn(C_6H_{11}O_7)_2 \cdot 2H_2O$.
Constant: Mol wt: 481.
Hazard Analysis
Toxic Hazard Rating: Unknown. Used as a nutrient and/or dietary supplement food additive. A trace mineral added to animal feeds (Section 10). See also manganese compounds.

MANGANESE GLYCEROPHOSPHATE
General Information
Description: Yellow-white or pinkish powder; odorless; soluble in water in presence of citric acid; slightly soluble in water; insoluble in alcohol.
Formula: $CH_2OHCHOHCH_2OP(O)O_2Mn$.
Constant: Mol wt: 195.
Hazard Analysis
Toxic Hazard Rating: Unknown. A nutrient and/or dietary supplement food additive (Section 10). See also manganese compounds.

MANGANESE GREEN. See barium manganate.

MANGANESE HEPTOXIDE
General Information
Description: A dark red oil.
Formula: Mn_2O_7.
Constants: Mol wt: 221.9, mp: < −20°C, bp: explodes at 70°C, d: > 1.84.
Hazard Analysis
Toxicity: See manganese compounds.
Fire Hazard: Dangerous, by chemical reaction a powerful oxidizer (Section 6).
Explosion Hazard: Severe, when shocked or exposed to heat.
Disaster Hazard: Highly dangerous; shock or heat will explode it; can react vigorously with reducing materials.
Countermeasures
Storage and Handling: Section 7.

MANGANESE HYPOPHOSPHITE
General Information
Description: Pink, odorless crystals.
Formula: $Mn(H_2PO_2)_2 \cdot H_2O$.
Constants: Mol wt: 202.9, mp: −H_2O > 150°C.
Hazard Analysis and Countermeasures
A nutrient and or dietary supplement food additive (Section 10). See manganese compounds and hypophosphites.

MANGANESE IODIDE. See manganese diiodide.

MANGANESE IODOPLATINATE
General Information
Description: Crystals.
Formula: $MnPtI_6 \cdot 9H_2O$.
Constants: Mol wt: 1173.82, mp: decomposes, d: 3.60.
Hazard Analysis and Countermeasures
See platinum compounds and manganese compounds and iodides.

MANGANESE LEAD RESINATE
General Information
Description: Dark brown to black mass.
Hazard Analysis
Toxicity: See lead compounds and managanese compounds.
Fire Hazard: Moderate, when exposed to heat or flame.
Disaster Hazard: Dangerous; when heated, it emits highly toxic fumes of lead; can react with oxidizing materials.
Countermeasures
Storage and Handling: Section 7.
To Fight Fire: Water, foam, carbon dioxide, dry chemical or carbon tetrachloride (Section 6).

MANGANESE MONOARSENIDE
General Information
Description: Black, hexagonal crystals.
Formula: MnAs.
Constants: Mol wt: 129.84, mp: decomposes at 400°C, d: 6.2.
Hazard Analysis
Toxicity: Highly toxic. See arsenic compounds and manganese compounds.
Fire Hazard: Moderate, when exposed to flame. It will react with water and acids to form hydrogen arsenide (Section 6).
Disaster Hazard: Dangerous; when heated to decomposition, it emits highly toxic fumes of arsenic; will react with water, steam or acids to produce toxic and flammable vapors; can react with oxidizing materials.
Countermeasures
Storage and Handling: Section 7.

MANGANESE MONOBORIDE
General Information
Description: Crystalline powder.
Formula: MnB.
Constants: Mol wt: 65.75, d: 6.2 at 15°C.
Hazard Analysis
Toxicity: See manganese compounds and boron compounds.
Fire Hazard: Moderate; will react with acids or steam to form boron hydride. It will ignite in presence of powerful oxidizers (Section 6).
Explosion Hazard: Moderate, by chemical reaction. See boron hydride.
Disaster Hazard: Dangerous; will react with water, steam or acids to produce toxic and flammable vapors; it reacts vigorously with oxidizing materials.
Countermeasures
Storage and Handling: Section 7.

MANGANESE MONOPHOSPHIDE
General Information
Description: Dark gray crystals.
Formula: MnP.

Constants: Mol wt: 85.91, mp: 1190°C, d: 5.39 at 21°C.
Hazard Analysis
Toxicity: See phosphides and manganese compounds.
Fire Hazard: Dangerous, when exposed to flame or by chemical reaction (Section 6).
Explosion Hazard: Moderate, by chemical reaction with moisture or acids.
Disaster Hazard: Dangerous; when heated to decomposition, lt emits highly toxic fumes of oxides of phosphorus; will react with water, steam or acids to produce toxic and flammable vapors; can react vigorously with oxidizing materials.
Countermeasures
Storage and Handling: Section 7.

MANGANESE MONOSULFIDE. See manganese sulfide.

MANGANESE MONOXIDE. See manganous oxide.

MANGANESE OXALATE. See manganous oxalate.

MANGANESE OXIDE. See manganese dioxide.

MANGANESE PHOSPHATE, DIBASIC
General Information
Synonyms: Manganous phosphate, acid; manganese hydrogen phosphate; manganous phosphate, secondary; manganese phosphate.
Description: Pink powder; soluble in acids; slightly soluble in water.
Formula: $MnHPO_4 \cdot 3H_2O$.
Constant: Mol wt: 205.
Hazard Analysis
Toxic Hazard Rating: Unknown. A trace mineral added to animal feeds (Section 10).
Disaster Hazard: Dangerous. See phosphates.

MANGANESE PROTOXIDE. See manganous oxide.

MANGANESE PYROPHOSPHATE. See manganous pyrophosphate.

MANGANESE RESINATE
General Information
Description: Dark, brownish-black mass.
Formula: $Mn(C_{20}H_{29}O_2)_2$.
Constant: Mol wt: 657.8.
Hazard Analysis
Toxicity: See manganese compounds.
Fire Hazard: Slight, when exposed to heat or flame or by reaction with powerful oxidizers (Section 6).
Countermeasures
Storage and Handling: Section 7.
Shipping Regulations: Section 11.
  IATA: Flammable solid, yellow label, 12 kilograms (passenger), 60 kilograms (cargo).

MANGANESE SELENATE
General Information
Description: Rhombic crystals.
Formula: $MnSeO_4 \cdot 2H_2O$.
Constants: Mol wt: 233.92, d: 2.95–3.01.
Hazard Analysis and Countermeasures
See manganese compounds and selenium compounds.

MANGANESE SELENIDE
General Information
Description: Gray, cubic crystals.
Formula: MnSe.

Constants: Mol wt: 133.89, d: 5.59 at 15°C.
Hazard Analysis
Toxicity: See manganese compounds and selenium compounds.
Fire Hazard: Moderate, by chemical reaction either with moisture or acids to liberate selenium hydride, or on contact with powerful oxidizers (Section 6).
Explosion Hazard: Moderate, by chemical reaction.
Disaster Hazard: Dangerous; when heated to decomposition, it emits highly toxic fumes of selenium; will react with water or acids to produce toxic and flammable vapors; can react with oxidizing materials.

MANGANESE SELENITE. See manganous selenite.

MANGANESE SULFATE. See manganous sulfate.

MANGANESE TARTRATE. See manganous tartrate.

MANGANIC FLUORIDE
General Information
Description: Red crystals.
Formula: MnF_3.
Constants: Mol wt: 111.93, mp: decomposes, d: 3.54.
Hazard Analysis and Countermeasures
See manganese compounds and fluorides.

MANGANIC m-PHOSPHATE
General Information
Description: Pink crystals.
Formula: $Mn_2(PO_3)_6 \cdot 2H_2O$.
Constant: Mol wt: 619.8.
Hazard Analysis
Toxicity: See manganese compounds.
Countermeasures
Storage and Handling: Section 7.

MANGANIC o-PHOSPHATE
General Information
Description: Greenish-gray crystalline powder.
Formula: $MnPO_4 \cdot H_2O$.
Constant: Mol wt: 167.93.
Hazard Analysis and Countermeasures
See manganese compounds.

MANGANIC SULFATE
General Information
Description: Green, del iquescent crystals.
Formula: $Mn_2(SO_4)_3$.
Constants: Mol wt: 398.06, mp: decomposes at 160°C.
Hazard Analysis and Countermeasures
Toxicity: See manganese compounds and sulfates.

MANGANOCYANIC ACID
General Information
Formula: $H_4Mn(CN)_6$.
Constants: Mol wt: 215.07, mp: decomposes.
Hazard Analysis and Countermeasures
See cyanides.

MANGANOLANGBEINITE. See potassium manganous sulfate.

MANGANOUS ARSENATE. See manganese arsenate.

MANGANOUS BENZOATE
General Information
Synonym: Manganese benzoate.
Description: Flat prisms.
Formula: $Mn(C_7H_5O_2)_2 \cdot 3H_2O$.

TOXIC HAZARD RATING CODE (For detailed discussion, see Section 1.)

0 NONE: (a) No harm under any conditions; (b) Harmful only under unusual conditions or overwhelming dosage.

1 SLIGHT: Causes readily reversible changes which disappear after end of exposure.

2 MODERATE: May involve both irreversible and reversible changes; not severe enough to cause death or permanent injury.

3 HIGH: May cause death or permanent injury after very short exposure to small quantities.

U UNKNOWN: No information on humans considered valid by authors.

Constant: Mol wt: 351.20.
Hazard Analysis
Toxicity: See manganese compounds.
Disaster Hazard: Slight; when heated, it emits acrid fumes.
Countermeasures
Storage and Handling: Section 7.

MANGANOUS CHLOROPLATINATE
General Information
Description: Crystals.
Formula: $MnPtCl_6 \cdot 6H_2O$.
Constants: Mol wt: 571.00, mp: decomposes, d: 2.692.
Hazard Analysis and Countermeasures
See platinum compounds, manganese compounds and chlorides.

MANGANOUS CITRATE
General Information
Description: White-reddish powder.
Formula: $Mn_3(C_6H_5O_7)_2$.
Constant: Mol wt: 542.99.
Hazard Analysis
Toxicity: See manganese compounds. Used as a nutrient and/or dietary supplement food additive. Also as a trace mineral added to animal feeds (Section 10).
Disaster Hazard: Slight; when heated, it emits acrid fumes.
Countermeasures
Storage and Handling: Section 7.

MANGANOUS ETHYLENE BIS-DITHIO CARBAMATE. See maneb.

MANGANOUS FERROCYANIDE
General Information
Description: Greenish-white powder.
Formula: $Mn_2Fe(CN)_6 \cdot 7H_2O$.
Constant: Mol wt: 447.93.
Hazard Analysis and Countermeasures
See manganese compounds and ferrocyanides.

MANGANOUS FLUOGALLATE
General Information
Description: Pink crystals.
Formula: $[Mn(H_2O)_6][GaF_5 \cdot H_2O]$.
Constants: Mol wt: 345.76, mp: decomposes at 230°C, d: 2.22.
Hazard Analysis and Countermeasures
See fluogallates and manganese compounds.

MANGANOUS FLUOSILICATE
General Information
Synonym: Manganous silicofluoride.
Description: Hexagonal, rose-red prisms.
Formula: $MnSiF_6 \cdot 6H_2O$.
Constants: Mol wt: 305.09, mp: decomposes, d: 1.903.
Hazard Analysis and Countermeasures
See fluosilicates and manganese compounds.

MANGANOUS FORMATE
General Information
Description: Rhombic crystals.
Formula: $Mn(CHO_2)_2 \cdot 2H_2O$.
Constants: Mol wt: 181.00, mp: decomposes, d: 1.953.
Hazard Analysis
Toxicity: See manganese compounds.
Disaster Hazard: Slight; when heated, it emits acrid fumes.
Countermeasures
Storage and Handling: Section 7.

MANGANOUS HYDROXIDE
General Information
Synonym: Pyrochroite.
Description: White-pink crystals.
Formula: $Mn(OH)_2$.
Constants: Mol wt: 88.95, mp: decomposes, d: 3.258 at 13°C.

Hazard Analysis
Toxicity: See manganese compounds.

MANGANOUS NITRATE
General Information
Description: Monoclinic rose-white or colorless crystals.
Formula: $Mn(NO_3)_2 \cdot 6H_2O$.
Constants: Mol wt: 287.04, mp: 25.8°C, bp: 129.4°C, d: 1.82.
Hazard Analysis and Countermeasures
See manganese compounds and nitrates.

MANGANOUS OXALATE
General Information
Synonym: Manganese oxalate.
Description: White crystals or powder.
Formula: $MnC_2O_4 \cdot 2H_2O$.
Constants: Mol wt: 179.0, mp: $-2H_2O$ at 100°C, d: 2.43 at 21.7°C.
Hazard Analysis and Countermeasures
See oxalates and manganese compounds.

MANGANOUS OXIDE
General Information
Synonyms: Manganese protoxide; manganese monoxide.
Description: Grass green powder; soluble in acids; insoluble in water.
Formula: MnO.
Constants: Mol wt: 71, d: 5.45, mp: 1650°C, converted to Mn_3O_4 if heated in air.
Hazard Analysis
Toxic Hazard Rating: Unknown. Used as a nutrient and/or dietary supplement food additive as well as a trace mineral added to animal feeds (Section 10). See also manganese compounds.

MANGANOUS o-PHOSPHATE
General Information
Description: Rhombic crystals or pale rose-pink or yellowish-white granules.
Formula: $Mn_3(PO_4)_2 \cdot 3H_2O$.
Constants: Mol wt: 408.80, d: 3.102.
Hazard Analysis and Countermeasures
See manganese compounds.

MANGANOUS PYROPHOSPHATE
General Information
Synonym: Manganese pyrophosphate.
Description: Monoclinic, brown-pink crystals.
Formula: $Mn_2P_2O_7$.
Constants: Mol wt: 283.82, mp: 1196°C, d: 3.707 at 25°C.
Hazard Analysis and Countermeasures
See manganese compounds.

MANGANOUS SELENITE
General Information
Synonym: Manganese selenite.
Description: Crystals.
Formula: $MnSeO_3 \cdot 2H_2O$.
Constant: Mol wt: 217.92.
Hazard Analysis and Countermeasures
See manganese compounds and selenium compounds.

MANGANOUS SILICOFLUORIDE. See manganous fluosilicate.

MANGANOUS SULFATE
General Information
Synonym: Manganese sulfate.
Description: Reddish crystals.
Formula: $MnSO_4$.
Constants: Mol wt: 151.00, mp: 700°C, bp: decomposes at 850°C, d: 3.25.
Hazard Analysis and Countermeasures
Used as a nutrient and/or dietary supplement food additive

and as a trace mineral added to animal feeds (Section 10). See manganese compounds and sulfates.

MANGANOUS SULFIDE
General Information
Synonym: Alabandite, manganese sulfide.
Description: Cubic, green, amorphous crystals.
Formula: MnS.
Constants: Mol wt: 87.00, mp: decomposes, d: 3.99.
Hazard Analysis and Countermeasures
See manganese compounds and sulfides.

MANGANOUS TARTRATE
General Information
Synonym: Manganese tartrate.
Description: White powder.
Formula: $MnC_4H_4O_6$.
Constant: Mol wt: 203.00.
Hazard Analysis
Toxicity: See manganese compounds.
Disaster Hazard: Slight; when heated, it emits acrid fumes.
Countermeasures
Storage and Handling: Section 7.

MANGANOUS THIOCYANATE
General Information
Description: Deliquescent crystals.
Formula: $Mn(SCN)_2 \cdot 3H_2O$.
Constants: Mol wt: 225.15, mp: $-3H_2O$ at 160°C to 170°C.
Hazard Analysis and Countermeasures
See manganese compounds and thiocyanates.

MANNA SUGAR. See mannitol.

MANNITE. See mannitol.

MANNITOL
General Information
Synonym: Manna; mannite.
Description: White crystalline powder; odorless; soluble in water; slightly soluble in lower alcohols and amines; almost insoluble in organic solvents.
Formula: $C_6H_8(OH)_6$ (straight chain hexahydric alcohol).
Constants: Mol wt: 182, d: 1.52, mp: 165–167°C, bp: 290–295°C at 3–5 mm.
Hazard Analysis
Toxic Hazard Rating: Unknown. Used as a nutrient and/or dietary supplement food additive (Section 10).

MANNITOL HEXANITRATE
General Information
Synonyms: Mannitol nitrate, nitromannite, nitromannitol.
Description: Colorless crystals.
Formula: $C_6H_8(NO_3)_6$.
Constants: Mol wt: 452.17, mp: 112°C, bp: explodes at 120°C, d: 1.603 at 0°C.
Hazard Analysis
Toxic Hazard Rating:
Acute Local: U.
Acute Systemic: Ingestion 2; Inhalation 2.
Chronic Local: U.
Chronic Systemic: Ingestion 2; Inhalation 2.
Toxicology: Causes fall in blood pressure, which may result in weakness, headache and dizziness. Chronic exposure may produce methemoglobinemia with cyanosis.
Fire Hazard: See nitrates.

Explosion Hazard: Severe, when shocked or exposed to heat. A very sensitive high explosive used as a substitute for fulminates and in combination with tetracine. See also explosives, high and nitrates.
Disaster Hazard: Highly dangerous! See nitrates and explosives, high.
Countermeasures
Ventilation Control: Section 2.
Personal Hygiene: Section 3.
Storage and Handling: Section 7.

MANNITOL MONOLAURATE
General Information
Description: Colorless solid.
Formula: $C_6H_8(OH)_5CO_2C_{11}H_{23}$.
Constant: Mol wt: 364.5.
Hazard Analysis
Toxicity: Details unknown; a fly spray.
Disaster Hazard: Slight; when heated, it emits acrid fumes; can react with oxidizing materials.
Countermeasures
Storage and Handling: Section 7.

MANNITOL NITRATE. See mannitol hexanitrate.

MAN-TAN. See dihydroxyacetone.

"MANZATE."* See maneb.

MAPP. See methyl acetylene-propadiene mixture.

MAPO. See tris[1-(2-methyl)aziredinyl]-phosphine oxide.

MARIHUANA. See cannabis.

MARSH GAS. See methane.

MARSHITE. See cuprous iodide.

MATCHES (SAFETY). See matches (strike anywhere)
IATA: Other restricted articles, 70K class C, 25 kilograms (passenger and cargo).

MATCHES (strike anywhere)
General Information
Composition: Usually contain phosphorus, antimony and sulfur.
Hazard Analysis
Toxicity: See individual components.
Fire Hazard: Dangerous, when exposed to heat or flame or by spontaneous chemical reaction. Keep away from rodents!
Disaster Hazard: Dangerous; will tend to propagate fires; emit toxic fumes of phosphorus, sulfur and antimony.
Countermeasures
Storage and Handling: Section 7.
Shipping Regulations: Section 11.
I.C.C.: Flammable solid; yellow label, 60 pounds.
Coast Guard Classification: Inflammable solid.

MATCHES, STRIKE ANYWHERE, SHIP'S LIFEBOAT TYPE AND SIMILAR MATCHES
IATA: Flammable solid, yellow label, 25 kilograms (passenger and cargo).

MATCHES (TRICK). See matches (strike anywhere)
Shipping Regulation: Section 11.
IATA: Explosive, not accepted (passenger and cargo).

MATLOCHITE. See lead oxychlorides.

*Registered trademark E. I. Du Pont de Nemours & Co.

TOXIC HAZARD RATING CODE (For detailed discussion, see Section 1.)

0 NONE: (a) No harm under any conditions; (b) Harmful only under unusual conditions or overwhelming dosage.

1 SLIGHT: Causes readily reversible changes which disappear after end of exposure.

2 MODERATE: May involve both irreversible and reversible changes not severe enough to cause death or permanent injury.

3 HIGH: May cause death or permanent injury after very short exposure to small quantities.

U UNKNOWN: No information on humans considered valid by authors.

MATTING ACID.
General Information
Description: A special mixture of acids having properties similar to white acid.
Countermeasures
Shipping Regulations: Section 11.
IATA: Corrosive liquid, white label, 1 liter (passenger), 5 liters (cargo).

MAUVINE HYDROBROMIDE
General Information
Description: Yellowish powder.
Hazard Analysis
Toxic Hazard Rating:
Acute Local: U.
Acute Systemic: Ingestion 3; Inhalationn 3.
Chronic Local: U.
Chronic Systemic: U.
Disaster Hazard: Dangerous; when heated, it emits highly toxic fumes.
Countermeasures
Ventilation Control: Section 2.
Personal Hygiene: Section 3.
Storage and Handling: Section 7.

MBA. See mechlorethamine.

MBDSA. See m-benzenedisulfonic acid.

McABEES. See explosives, high.

MCPA. See methyl chlorophenoxyacetic acid.

MD. See dichloromethyl arsine.

MDMH. See 1-(hydroxymethyl)-5,5-dimethyl hydantoin.

MECHLORETHAMINE
General Information
Synonym: MBA.
Description: Mobile liquid. Faint odor of herring. Very slightly water soluble.
Formula: $CH_3N(CH_2CH_2Cl)_2$.
Constants: Mol wt: 156.1, d: 1.118 at $25/4°C$, mp: $-60°C$, bp: $75°C$ at 10 mm.
Hazard Analysis
Toxicity: This material is a nitrogen mustard, a vesicant and necrotizing irritant. It is highly toxic.
Disaster Hazard: Dangerous; when heated to decomposition, it emits highly toxic fumes of chlorides and oxides of nitrogen.

MECHLORETHAMINE HYDROCHLORIDE
General Information
Synonym: Erasol; 2,2-dichloro-N-methyl diethylamine hydrochloride; nitol.
Description: Hygroscopic leaf-like crystals. Soluble in water and alcohol.
Formula: $CH_3N(CH_2CH_2Cl)_2 \cdot HCl$.
Constant: Mol wt: 192.5.
Hazard Analysis
Toxic Hazard Rating:
Acute Local: Irritant 3; Ingestion 3; Inhalation 3.
Acute Systemic: U.
Chronic Local: U.
Chronic Systemic: U.
Toxicity: See mechlorethamine.
Disaster Hazard: Dangerous; when heated to decomposition, it emits highly toxic fumes of chlorides.
Countermeasures
Storage and Handling: Section 7.

MEDROXYPROGESTERONE ACETATE
General Information
Synonyms: 17-Hydroxy-6,α-methyl-preg-4-ene-3,20-dione 17-acetate.

Description: White to off white, odorless crystalline powder; insoluble in water; freely soluble in chloroform; sparingly soluble in alcohol.
Formula: $C_{24}H_{34}O_4$.
Constants: Mol wt: 386, melting range: 200–205°C.
Hazard Analysis
Toxic Hazard Rating: Unknown. Used as a food additive permitted in the feed and drinking water of animals and/or for the treatment of food producing animals, also permitted in food for human consumption (Section 10).

MELAMINE
General Information
Synonym: 2,4,6-Triamino-s-triazine; cyanurotriamide.
Description: Monoclinic colorless prisms.
Formula: $NC(NH_2)NC(NH_2)NC(NH_2)$.
Constants: Mol wt: 126.13, mp: 250°C, bp: sublimes, d: 1.573 at $20°/4°C$, vap. press.: 50 mm at 315°C, vap. d.: 4.34.
Hazard Analysis
Toxicity: Details unknown. Animal experiments suggest relatively low oral toxicity. Suspected of causing dermatitis in humans.
Disaster Hazard: Dangerous; when heated to decomposition, it emits highly toxic fumes of cyanides.
Countermeasures
Storage and Handling: Section 7.

MEMTETRAHYDROPHTHALIC ANHYDRIDE
General Information
Description: Clear, transparent, slightly viscous liquid. colorless to light yellow.
Hazard Analysis
Toxic Hazard Rating: U. A corrosive material.
Countermeasures
Shipping Regulations: Section 11.
I.C.C.: Corrosive liquid, white label, 1 quart.

MENDIPITE. See lead oxychlorides.

MENHADEN OIL
General Information
Synonyms: Pogy oil; moss bunker oil.
Description: Thick liquid derived from fish.
Constants: Autoign. temp.: 828°F, fp: $-5°C$, flash p: 435°F (C.C.), d: 0.923–0.933.
Hazard Analysis
Toxicity: Unknown.
Fire Hazard: Slight, when exposed to heat or flame; can react with oxidizing materials.
Spontaneous Heating: Yes.
Countermeasures
To Fight Fire: Foam, carbon dioxide, dry chemical or carbon tetrachloride (Section 6).
Storage and Handling: Section 7.

1(7)-2-p-MENTHADIENE
General Information
Synonym: β-Phellandrene.
Description: Oily liquid; slight citrus odor.
Formula: $C_{10}H_{16}$.
Constants: Mol wt: 136.2, bp: 171.2°C, flash p: 120°F (T.C.C.), d: 0.843–0.849 at 75°C, vap. d.: 4.68.
Hazard Analysis
Toxicity: Unknown. May act as a skin irritant.
Fire Hazard: Moderate, when exposed to heat or flame; can react with oxidizing materials.
Countermeasures
To Fight Fire: Foam, carbon dioxide, dry chemical or carbon tetrachloride (Section 6).
Storage and Handling: Section 7.

p-MENTHANE HYDROPEROXIDE
General Information
Description: Clear, pale yellow liquid.

Constant: D: 0.910–0.925 at 15.5/4°C.
Hazard Analysis
Toxic Hazard Rating: U. An irritant.
Fire Hazard: Dangerous. A powerful oxidizer.
Disaster Hazard: Dangerous. A powerful oxider.
Countermeasures
Shipping Regulations: Section 11.
  IATA solution, exceeding 60%: Oxidizing material, not
      acceptable (passenger and cargo).
  (solution, not exceeding 60% in a non-volatile
      solvent): Oxidizing material, yellow label, 1
      liter (passenger and cargo).

MENTHOL
General Information
Synonym: Hexahydrothymol.
Description: Crystals; peppermint taste and odor.
Formula: $C_{10}H_{19}OH$.
Constants: Mol wt: 156.26, mp: 42.5°C, bp: 212.0°C, d:
    0.890 at 15°/15°C, vap. press.: 1 mm at 56.0°C,
    vap. d.: 5.38.
Hazard Analysis
Toxic Hazard Rating:
  Acute Local: Irritant 1; Allergen 1; Ingestion 1; Inhala-
      tion 1.
  Acute Systemic: U.
  Chronic Local: Irritant 1; Allergen 1.
  Chronic Systemic: U.
Fire Hazard: Slight (Section 6).
Countermeasures
Personnel Protection: Section 3.
Storage and Handling: Section 7.

MERBROMIN. See 'Mercurochrome."

MERCAPTAN MIXTURES, ALIPHATIC
Shipping Regulations: Section 11.
  I.C.C.: Flammable liquid, red label, 10 gallons.
  IATA: Flammable liquid, red label, 1 liter (passenger),
      40 liters (cargo).

MERCAPTANS
Hazard Analysis
Toxic Hazard Rating:
  Acute Local: Irritant 3; Inhalation 3.
  Acute Systemic: Inhalation 3.
  Chronic Local: U.
  Chronic Systemic: Inhalation 2.
Toxicology: Mercaptans generally have a very offensive
    odor, which may cause nausea and headache. High con-
    centrations can produce unconsciousness with cyanosis,
    cold extremities and rapid pulse. A common air con-
    taminant. (Section 4).
Disaster Hazard: Dangerous; when heated to decomposi-
    tion, they almost always emit highly toxic fumes of
    oxides of sulfur; they will react with water, steam or
    acids to produce toxic and flammable vapors; they can
    react with oxidizing materials.
Countermeasures
Ventilation Control: Section 2.
Personnel Protection: Section 3.
Storage and Handling: Section 7.
Shipping Regulations: Section 11.
  IATA (not exceeding 3; c.c. in hermetically sealed warn-
      ing or odorizing devices): Other restricted articles,
      class A, no limit (passenger and cargo).

MERCAPTOACETATES. See thioglycolates.

MERCAPTOACETIC ACID. See thioglycolic acid.

β-MERCAPTOALANINE. See cysteine.

2-MERCAPTOBENZOTHIAZOLE
General Information
Description: Light yellow powder.
Formula: C_6H_4SCNSH.
Constants: Mol wt: 167.3, mp: 170°C, d: 1.42 at 25°C.
Hazard Analysis
Toxicity: Details unknown. See mercaptans.
Disaster Hazard: Dangerous; when heated to decomposi-
    tion or on contact with acid or acid fumes, it emits
    highly toxic fumes of sulfur compounds; can react with
    oxidizing materials.
Countermeasures
Storage and Handling: Section 7.

2-MERCAPTOETHANOIC ACID. See thioglycolic acid.

MERCAPTOETHANOL
General Information
Synonym: 2-Hydroxyethylmercaptan.
Description: Water-white, mobile liquid.
Formula: $HSCH_2CH_2OH$.
Constants: Mol wt: 78.13, bp: 157.1°C, flash p: 170°F
    (C.O.C.), d: 1.1168 at 20°/20°C, vap. press.: 1.0 mm
    at 20°C, vap. d.: 2.69.
Hazard Analysis
Toxic Hazard Rating:
  Acute Local: Irritant 2; Ingestion 2; Inhalation 2.
  Acute Systemic: U.
  Chronic Local: U.
  Chronic Systemic: U.
Fire Hazard: Moderate, when exposed to heat or flame.
Disaster Hazard: Dangerous; when heated to decomposi-
    tion, it emits highly toxic fumes of oxides of sulfur; can
    react with oxidizing materials.
Countermeasures
Ventilation Control: Section 2.
To Fight Fire: Alcohol foam, carbon dioxide, dry chemical,
    carbon tetrachloride or water spray (Section 6).
Personnel Protection: Section 3.
Storage and Handling: Section 7.

2-MERCAPTOETHYLAMINE HYDROCHLORIDE
General Information
Description: White, slightly hygroscopic crystals.
Formula: $HSCH_2CH_2NH_2 \cdot HCl$.
Constants: Mol wt: 113.6, mp: 70.2–70.7°C.
Hazard Analysis
Toxic Hazard Rating:
  Acute Local: U.
  Acute Systemic: Ingestion 2.
  Chronic Local: U.
  Chronic Systemic: U.
Disaster Hazard: Dangerous; when heated to decomposi-
    tion, it emits highly toxic fumes of oxides of sulfur; will
    react with water or steam to produce toxic fumes.
Countermeasures
Personnel Protection: Section 3.
Storage and Handling: Section 7.

MERCURIAL LIQUID, N.O.S.
Shipping Regulations: Section 11.
  I.C.C.: Poison B, poison label, 55 gallons.

TOXIC HAZARD RATING CODE (For detailed discussion, see Section 1.)

0 NONE: (a) No harm under any conditions; (b) Harmful only under unusual conditions or overwhelming dosage.

1 SLIGHT: Causes readily reversible changes which disappear after end
of exposure.

2 MODERATE: May involve both irreversible and reversible changes;
not severe enough to cause death or permanent injury.

3 HIGH: May cause death or permanent injury after very short exposure
to small quantities.

U UNKNOWN: No information on humans considered valid by authors.

MERCURIC ACETATE
General Information
Description: White, crystalline powder.
Formula: $Hg(C_2H_3O_2)_2$.
Constants: Mol wt: 318.70, mp: decomposes, d: 3.270.
Hazard Analysis and Countermeasures
See mercury compounds, organic.
Shipping Regulations: Section 11.
  I.C.C.: Poison B; poison label, 200 pounds.
  Coast Guard Classification: Poison B; poison label.
  IATA: Poison B; poison label; 25 kilgorams (passenger); 95 kilograms (cargo).

MERCURIC ACETYLIDE
General Information
Description: White powder.
Formula: $3HgC_2 \cdot H_2O$.
Constants: Mol wt: 691.91, mp: explodes, d: 5.3.
Hazard Analysis
Toxicity: Highly toxic. See mercury compounds, organic.
Fire Hazard: Details unknown. See also acetylides.
Explosion Hazard: Severe, when shocked or exposed to heat.
Disaster Hazard: Dangerous! Heat or shock will cause detonation; when heated, it emits highly toxic fumes of mercury.
Countermeasures
Storage and Handling: Section 7.

MERCURIC AMINOACETATE
General Information
Synonym: Mercuric glycocollate.
Description: Crystals; soluble in water.
Formula: $Hg[CH_2(NH_2)COO]_2$.
Constant: Mol wt: 348.7.
Hazard Analysis and Countermeasures
See mercury compounds, organic.

MERCURIC AMINOPROPIONATE
General Information
Synonym: Mercury alanine.
Description: White, crystalline powder; soluble in 3 parts water.
Formula: $Hg[CH_3CH(NH_2)COO]_2$.
Constant: Mol wt: 374.77.
Hazard Analysis
See mercury compounds, organic.

MERCURIC AMMONIUM CHLORIDE
General Information
Description: White, pulverulent lumps or powder.
Formula: $HgCl_2 \cdot 2NH_4Cl \cdot 2H_2O$.
Constant: Mol wt: 414.5.
Hazard Analysis and Countermeasures
See mercury compounds, inorganic.
Shipping Regulations: Section 11.
  I.C.C.: Poison B; poison label, 200 pounds.
  Coast Guard Classification: Poison B; poison label.
  IATA: Poison B, poison label, 25 kilograms (passenger), 95 kilograms (cargo).

MERCURIC ARSENATE
General Information
Description: Yellow powder.
Formula: $HgHAsO_4$.
Constants: Mol wt: 340.5, mp: decomposes.
Hazard Analysis
Toxicity: Highly toxic. See mercury compounds, inorganic and arsenic compounds.
Disaster Hazard: Dangerous; when heated to decomposition or on contact with acid or acid fumes, it emits highly toxic fumes of mercury and arsenic.
Countermeasures
Storage and Handling: Section 7.
Shipping Regulations: Section 11.

IATA: (solid): Poison B, poison label, 25 kilograms (passenger), 95 kilograms (cargo).

MERCURIC o-ARSENATE
General Information
Description: Yellow crystals.
Formula: $Hg_3(AsO_4)_2$.
Constant: Mol wt: 879.65.
Hazard Analysis and Countermeasures
See arsenic compounds and mercury compounds, inorganic.

MERCURIC BARIUM BROMIDE
General Information
Synonym: Barium-mercury bromide.
Description: Colorless, very hygroscopic crystals.
Formula: $HgBr_2 \cdot BaBr_2$.
Constant: Mol wt: 657.6.
Hazard Analysis and Countermeasures
See mercury compounds, inorganic, barium compounds and bromides.

MERCURIC BARIUM IODIDE
General Information
Description: Reddish or yellow, unstable, deliquescent, crystalline mass.
Formula: $HgI_2 \cdot BaI_2 \cdot 5H_2O$.
Constant: Mol wt: 935.73.
Hazard Analysis and Countermeasures
See mercury compounds, inorganic, iodides and barium compounds.

MERCURIC BENZOATE
General Information
Description: White crystalline powder.
Formula: $Hg(C_7H_5O_2)_2$.
Constants: Mol wt: 442.83, mp: 165°C.
Hazard Analysis and Countermeasures
See mercury compounds, organic.
Shipping Regulations: Section 11.
  I.C.C.: Poison B; poison label, 200 pounds.
  Coast Guard Classification: Poison B; poison label.
  IATA: Poison B, poison label, 25 kilograms (passenger), 95 kilograms (cargo).

MERCURIC BROMATE
General Information
Description: Crystals.
Formula: $Hg(BrO_3)_2 \cdot 2H_2O$.
Constants: Mol wt: 492.47, mp: decomp. at 130–140°C.
Hazard Analysis and Countermeasures
See mercury compounds, inorganic and bromates.
Shipping Regulations: Section 11.
  IATA: Poison B, poison label, 25 kilograms (passenger), 95 kilograms (cargo).

MERCURIC BROMIDE
General Information
Description: Rhombic, colorless crystals.
Formula: $HgBr_2$.
Constants: Mol wt: 360.44, mp: 237°C, bp: 319°C (sublimes), d: 6.109 at 25°C, vap. press.: 1 mm at 136.5°C.
Hazard Analysis and Countermeasures
See mercury compounds, inorganic and bromides.
Shipping Regulations: Section 11.
  I.C.C.: Poison B; poison label, 200 pounds.
  Coast Guard Classification: Poison B; poison label.
  IATA: Poison B, poison label, 25 kilograms (passenger), 95 kilograms (cargo).

MERCURIC BROMIDE, AMMONOBASIC
General Information
Description: White powder.
Formula: $Hg(NH_2)Br$.
Constants: Mol wt: 296.55, mp: decomposes.
Hazard Analysis and Countermeasures
See mercury compounds, inorganic and bromides.

MERCURIC BROMIDE DIAMMINE
General Information
Description: White powder.
Formula: $Hg(NH_3)_2Br_2$.
Constants: Mol wt: 394.51, mp: 180°C.
Hazard Analysis and Countermeasures
See mercury compounds, inorganic and bromides.

MERCURIC BROMIDE IODIDE
General Information
Description: Rhombic, yellow crystals.
Formula: HgBrI.
Constants: Mol wt: 407.45, mp: 229°C, bp: 360°C.
Hazard Analysis and Countermeasures
See mercury compounds, inorganic and bromides.

MERCURIC CACODYLATE
General Information
Description: Hygroscopic, somewhat unstable crystalline
powder; soluble in water and alcohol; insoluble in ether.
Formula: $Hg[(CH_3)_2AsO_2]_2$.
Constant: Mol wt: 474.52.
Hazard Analysis
See mercury compounds, organic and arsenic compounds.

MERCURIC CHLORATE
General Information
Description: Colorless, needle-like crystals.
Formula: $Hg(ClO_3)_2$.
Constants: Mol wt: 367.52, mp: decomposes, d: 4.998.
Hazard Analysis and Countermeasures
See mercury compounds, inorganic, and chlorates.

MERCURIC CHLORIDE
General Information
Synonym: Corrosive sublimate.
Description: White crystals or powder.
Formula: $HgCl_2$.
Constants: Mol wt: 271.52, mp: 277°C, bp: 304°C, d:
5.440 at 25°C, vap. press.: 1 mm at 136.2°C.
Hazard Analysis and Countermeasures
See mercury compounds, inorganic and chlorides.
Shipping Regulations: Section 11.
  MCA warning label.

MERCURIC CHLORIDE DIAMMINE
General Information
Description: Fusible white precipitate.
Formula: $Hg(NH_3)_2Cl_2$.
Constants: Mol wt: 305.59, mp: 300°C.
Hazard Analysis and Countermeasures
See mercury compounds, inorganic and chlorides.

MERCURIC CHLORIDE IODIDE
General Information
Description: Rhombic, red crystals.
Formula: HgClI.
Constants: Mol wt: 362.99, mp: 153°C, bp: 315°C.
Hazard Analysis and Countermeasures
See mercury compounds, inorganic, chlorides and iodides.

MERCURIC CHLOROIODIDE
General Information
Description: White crystals.
Formula: $2HgCl_2 \cdot HgI_2$.
Constant: Mol wt: 997.48.
Hazard Analysis and Countermeasures
See mercury compounds, inorganic.

MERCURIC CHROMATE
General Information
Description: Rhombic, red crystals.
Formula: $HgCrO_4$.
Constants: Mol wt: 316.62, mp: decomposes.
Hazard Analysis
Toxicity: Highly toxic. See mercury compounds, inorganic.
Fire Hazard: Moderate, by chemical reaction; an oxidizer
(Section 6).
Disaster Hazard: Dangerous; when heated to decomposi-
tion, it emits highly toxic fumes of mercury; can react
with reducing materials.
Countermeasures
Storage and Handling: Section 7.

MERCURIC CUPROUS IODIDE
General Information
Synonym: Copper-mercury iodide.
Description: Dark red, crystalline powder.
Formula: $HgI_2 \cdot 2CuI$.
Constant: Mol wt: 835.42.
Hazard Analysis and Countermeasures
See mercury compounds, inorganic.

MERCURIC CYANIDE
General Information
Description: Colorless, transparent prisms, darkened by
light.
Formula: $Hg(CN)_2$.
Constants: Mol wt: 252.65, mp: decomposes, d: 3.996.
Hazard Analysis
Toxicity: Highly toxic. See cyanides and mercury com-
pounds, organic.
Fire Hazard: See cyanides.
Explosion Hazard: See cyanides.
Disaster Hazard: Dangerous; when heated to decomposition
or on contact with acid or acid fumes, it emits highly
toxic fumes of cyanides and mercury; will react with
water or steam to produce toxic fumes.
Countermeasures
Storage and Handling: Section 7.
Shipping Regulations: Section 11.
  I.C.C.: Poison B; poison label, 200 pounds.
  Coast Guard Classification: Poison B; poison label.
  IATA: Poison B, poison label 12 kilograms (passenger),
  95 kilograms (cargo).

MERCURIC ETHYL MERCAPTIDE
General Information
Description: Crystals; insoluble in water.
Formula: $(SC_2H_5)_2Hg$.
Constants: Mol wt: 322.9, mp: 77°C.
Hazard Analysis and Countermeasures
See mercury compounds, organic and sulfides.

MERCURIC FLUORIDE
General Information
Description: Transparent crystals.
Formula: HgF_2.
Constants: Mol wt: 238.61, mp: 645°C decomposes, bp:
650°C, d: 8.95 at 15°C.
Hazard Analysis and Countermeasures
See fluorides and mercury compounds, inorganic.

MERCURIC FLUOSILICATE
General Information
Description: Colorless, deliquescent crystals.
Formula: $HgSiF_6 \cdot 6H_2O$.

TOXIC HAZARD RATING CODE (For detailed discussion, see Section 1.)

0 NONE: (a) No harm under any conditions; (b) Harmful only under unusual conditions or overwhelming dosage.

1 SLIGHT: Causes readily reversible changes which disappear after end
of exposure.

2 MODERATE: May involve both irreversible and reversible changes;
not severe enough to cause death or permanent injury.
3 HIGH: May cause death or permanent injury after very short exposure
to small quantities.
U UNKNOWN: No information on humans considered valid by authors.

Constant: Mol wt: 450.77.
Hazard Analysis and Countermeasures
See mercury compounds, inorganic.

MERCURIC FORMAMIDE
General Information
Description: A mixture of mercuric oxide + formamide.
Hazard Analysis and Countermeasures
See mercury compounds, organic.

MERCURIC GALLATE
General Information
Description: Gray-green powder. Insoluble in water.
Formula: $Hg[C_6H_2(OH)_3COO]_2$.
Constant: Mol wt: 538.8.
Hazard Analysis and Countermeasures
See mercury compounds, organic.

MERCURIC GLYCOCOLLATE. See mercuric amino-
acetate.

MERCURIC GUAIACOLSULFONATE
General Information
Description: Brown crystals, soluble in water.
Formula: $Hg[C_6H_3(OH)(OCH_3)SO_3]_2$.
Constant: Mol wt: 606.99.
Hazard Analysis
See mercury compounds, organic and sulfonates.

MERCURIC IODATE
General Information
Description: White crystals; slightly water soluble.
Formula: $Hg(IO_3)_2$.
Constant: Mol wt: 550.45.
Hazard Analysis and Countermeasures
See mercury compounds, inorganic and iodates.

MERCURIC IODIDE
General Information
Description: (1) Rhombic, yellow crystals; (2) tetragonal
red crystals or powder.
Formula: HgI_2.
Constants: Mol wt: 454.45; d: (1) 6.271, (2) 6.283; mp: (1)
259°C, (2) 126-127°C; bp: 354°C (sublimes); vap.
press.: (1) 1 mm at 157.5°C.
Hazard Analysis and Countermeasures
See mercury compounds, inorganic and iodides.
Shipping Regulations: Section 11.
I.C.C.: Poison B; poison label, 200 pounds. [solid].
Coast Guard Classification: Poison B; poison label.
IATA: Poison B, poison label, 25 kilograms (passenger),
95 kilograms (cargo).

MERCURIC IODIDE AMMONOBASIC
General Information
Description: Dirty white crystals.
Formula: $Hg(NH_2)I$.
Constant: Mol wt: 343.55.
Hazard Analysis and Countermeasures
See mercury compounds, inorganic and iodides.

MERCURIC IODIDE, AQUOBASIC-AMMONO-BASIC
General Information
Description: Yellow to brown crystals.
Formula: OHg_2NH_2I.
Constants: Mol wt: 560.16, mp: > 128°C, bp: explodes.
Hazard Analysis and Countermeasures
See mercury compounds, inorganic and iodides.

MERCURIC IODIDE DIAMMINE
General Information
Description: Colorless or pale yellow powder or needles.
Formula: $Hg(NH_3)_2I_2$.
Constant: Mol wt: 488.51.
Hazard Analysis and Countermeasures
See mercury compounds, inorganic and iodides.

MERCURIC IODIDE SOLUTION. See mercuric iodide.
Shipping Regulations: Section 11.
I.C.C.: Poison B; poison label, 55 gallons.
Coast Guard Classification: Poison B; poison label.
IATA: Poison B, poison label, 1 liter (passenger), 220
liters (cargo).

MERCURIC LACTATE
General Information
Description: White crystalline powder.
Formula: $Hg(C_3H_5O_3)_2$.
Constant: Mol wt: 378.71.
Hazard Analysis and Countermeasures
See mercury compounds, organic.

MERCURIC NAPHTHOLATE
General Information
Synonym: Mercury β-naphthol.
Description: Brown powder. Insoluble in water.
Formula: $Hg(C_{10}H_7O)_2$.
Constant: Mol wt: 486.9.
Hazard Analysis and Countermeasures
See mercury compounds, organic.

MERCURIC NITRATE
General Information
Description: White-yellowish, deliquescent powder.
Formula: $Hg(NO_3)_2 \cdot \frac{1}{2}H_2O$.
Constants: Mol wt: 333.63, mp: 79°C, bp: decomposes, d:
4.39.
Hazard Analysis and Countermeasures
See mercury compounds, inorganic and nitrates.

MERCURIC OLEATE
General Information
Synonym: Mercury oleate.
Description: Yellowish to red liquid; semisolid or solid
mass.
Formula: $(C_{17}H_{33}CO_2)_2Hg$.
Constant: Mol wt: 763.1.
Hazard Analysis and Countermeasures
See mercury compounds, organic.
Shipping Regulations: Section 11.
I.C.C.: Poison B; poison label, 200 pounds.
Coast Guard Classification: Poison B; poison label.
IATA: Poison B, posion label, 25 kilograms (passenger),
95 kilograms (cargo).

MERCURIC OXALATE
General Information
Description: White powder.
Formula: HgC_2O_4.
Constants: Mol wt: 288.63, mp: decomposes.
Hazard Analysis and Countermeasures
See mercury compounds, organic and oxalates.

MERCURIC OXIDE, RED
General Information
Description: Heavy, bright orange-red powder.
Formula: HgO.
Constants: Mol wt: 216.61, mp: decomposes at 500°C, d:
11.14.
Hazard Analysis
Toxicity: Highly toxic. See mercury compounds, inorganic.
Fire Hazard: Slight, by chemical reactions; an oxidizer
(Section 6).
Disaster Hazard: Dangerous; when heated to decomposi-
tion, it emits highly toxic fumes of mercury; can react
with reducing materials.
Countermeasures
Storage and Handling: Section 7.
Shipping Regulations: Section 11.
I.C.C.: Poison B; poison label, 200 pounds.
Coast Guard Classification: Poison B; poison label.
IATA: Poison B, poison label, 25 kilograms (passenger),
95 kilograms (cargo).

MERCURIC OXIDE, YELLOW
General Information
Description: Light, odorless, amorphous, orange-yellow powder.
Formula: HgO.
Constants: Mol wt: 216.61, mp: decomposes at 500°C, d: 11.14.
Hazard Analysis
Toxicity: Highly toxic. See mercury compounds, inorganic.
Fire Hazard: Slight, by chemical reaction; an oxidizer (Section 6).
Disaster Hazard: Dangerous; when heated to decomposition, it emits highly toxic fumes of mercury; can react with reducing materials.
Countermeasures
Storage and Handling: Section 7.
Shipping Regulations: Section 11.
  I.C.C.: Poison B; poison label, 200 pounds.
  Coast Guard Classification: Poison B; poison label.
  IATA: Poison B, poison label, 25 kilograms (passenger), 95 kilograms (cargo).

MERCURIC OXYBROMIDE
General Information
Description: Yellow crystals.
Formula: $HgBr_2 \cdot 3HgO$.
Constant: Mol wt: 1010.27.
Hazard Analysis and Countermeasures
See mercury compounds, inorganic and bromides.

MERCURIC OXYCHLORIDE
General Information
Description: Hexagonal, yellow crystals.
Formula: $HgCl_2 \cdot 3HgO$.
Constants: Mol wt: 921.35, mp: decomposes at 260°C, d: 7.93.
Hazard Analysis and Countermeasures
See mercury compounds, inorganic and chlorides.

MERCURIC OXYCYANIDE
General Information
Description: White, crystalline powder or needles, slightly water soluble.
Formula: $Hg(CN)_2 \cdot HgO$.
Constants: Mol wt: 469.26, mp: explodes, d: 4.437 at 19°C.
Hazard Analysis
Toxicity: Highly toxic. See cyanides and mercury compounds, organic.
Explosion Hazard: Dangerous; the pure material explodes very easily. See fulminate of mercury, dry.
Disaster Hazard: Dangerous; shock and heat will explode it; when heated to decomposition or on contact with acid or acid fumes, it emits highly toxic fumes of cyanides and mercury.
Countermeasures
Storage and Handling: Section 7.
Shipping Regulations: Section 11.
  I.C.C.: Poison B; poison label, 200 pounds.
  Coast Guard Classification: Poison B; poison label.

MERCURIC OXYCYANIDE-MERCURIC CYANIDE MIXTURE, SOLID.
Shipping Regulations: Section 11.
  IATA: Poison B, poison label, 12 kilograms (passenger), 95 kilograms (cargo).

MERCURIC OXYFLUORIDE
General Information
Description: Yellow crystals.
Formula: $HgF_2 \cdot HgO \cdot H_2O$.
Constants: Mol wt: 473.24, mp: decomposes at 100°C.
Hazard Analysis and Countermeasures
See fluorides and mercury compounds, inorganic.

MERCURIC OXYIODIDE
General Information
Description: Yellow-brown crystals.
Formula: $HgI_2 \cdot 3HgO$.
Constant: Mol wt: 1104.28.
Hazard Analysis and Countermeasures
See mercury compounds, inorganic and iodides.

MERCURIC PERCHLORATE
General Information
Description: Colorless, deliquescent crystals.
Formula: $Hg(ClO_4)_2 \cdot 6H_2O$.
Constant: Mol wt: 507.62.
Hazard Analysis and Countermeasures
See mercury compounds, inorganic and perchlorates.

MERCURIC PHENATE
General Information
Synonym: Mercury carbolate.
Description: Gray or reddish gray powder. Nearly insoluble in water.
Formula: $Hg(C_6H_5O)_2$.
Constants: Mol wt: 386.8.
Hazard Analysis and Countermeasures
See mercury compounds, organic.

MERCURIC PHENYL CYANAMIDE
General Information
Description: Powder.
Formula: $C_6H_5HgNCNH$.
Constant: Mol wt: 318.8.
Hazard Analysis and Countermeasures
See mercury compounds, organic and cyanides.

MERCURIC PHENYL MERCAPTIDE
General Information
Description: Light yellow crystals. Insoluble in water.
Formula: $(C_6H_5S)_2Hg$.
Constants: Mol wt: 418.9, mp: 153°C (decomposes).
Hazard Analysis and Countermeasures
See mercury compounds, organic.

MERCURIC PHOSPHATE
General Information
Synonym: Mercuri o-phosphate.
Description: Heavy white or yellowish powder.
Formula: $Hg_3(PO_4)_2$.
Constant: Mol wt: 791.87.
Hazard Analysis and Countermeasures
See mercury compounds, inorganic.

MERCURIC o-PHOSPHATE. See mercuric phosphate.

MERCURIC POTASSIUM CYANIDE
General Information
Description: Colorless crystals.
Formula: $Hg(CN)_2 \cdot 2KCN$.
Constant: Mol wt: 382.85.
Hazard Analysis
Toxicity: Highly toxic. See cyanides and mercury compounds, organic.

TOXIC HAZARD RATING CODE (For detailed discussion, see Section 1.)

0 NONE: (a) No harm under any conditions; (b) Harmful only under unusual conditions or overwhelming dosage.

1 SLIGHT: Causes readily reversible changes which disappear after end of exposure.

2 MODERATE: May involve both irreversible and reversible changes; not severe enough to cause death or permanent injury.

3 HIGH: May cause death or permanent injury after very short exposure to small quantities.

U UNKNOWN: No information on humans considered valid by authors.

Fire Hazard: See cyanides.

Disaster Hazard: Dangerous; when heated to decomposition or on contact with acid or acid fumes, it emits highly toxic fumes of cyanides and mercury; will react with water or steam to produce toxic and flammable vapors.

Explosion Hazard: See cyanides.

Countermeasures

Storage and Handling: Section 7.

Shipping Regulations: Section 11.

    I.C.C.: Poison B; poison label, 200 pounds.

    Coast Guard Classification: Poison B; poison label.

    IATA: Poison B, poison label, 12 kilograms (passenger), 95 kilograms (cargo).

MERCURIC POTASSIUM IODIDE
General Information

Description: Heavy, bright orange-red to yellow powder.

Formula: $HgI_2 \cdot 2KI$.

Constant: Mol wt: 786.5.

Hazard Analysis and Countermeasures

See mercury compounds, inorganic and iodides.

Shipping Regulations: Section 11.

    I.C.C.: Poison B; poison label, 200 pounds.

    Coast Guard Classification: Poison B; poison label.

    IATA: Poison B, poison label, 25 kilograms (passenger), 95 kilograms (cargo).

MERCURIC POTASSIUM THIOSULFATE
General Information

Description: White powder.

Formula: $HgS_2O_3 \cdot K_2S_2O_3$.

Constant: Mol wt: 503.1.

Hazard Analysis and Countermeasures

See mercury compounds, inorganic and sulfates.

MERCURIC RESORCINOL ACETATE
General Information

Synonym: Resorcinol-mercury acetate.

Description: Contains $\cong 69\%$ mercury; yellow crystalline powder; insoluble in water; soluble in hot glacial acetic acid and solutions of fixed alkali hydroxides.

Hazard Analysis

See mercury compounds, organic.

MERCURIC SALICYLATE
General Information

Synonym: Mercury subsalicylate.

Description: White to yellowish powder.

Formula: $C_7H_4O_3Hg$.

Constant: Mol wt: 336.6.

Hazard Analysis and Countermeasures

See mercury compounds, organic.

Shipping Regulations: Section 11.

    I.C.C.: Poison B; poison label, 200 pounds.

    Coast Guard Classification: Poison B; poison label.

    IATA: Poison B, poison label, 25 kilograms (passenger), 95 kilograms (cargo).

MERCURIC SELENIDE
General Information

Synonym: Tiemannite.

Description: Gray plates.

Formula: HgSe.

Constants: Mol wt: 279.57, mp: sublimes, d: 7.1–8.9.

Hazard Analysis

Toxicity: Highly toxic. See mercury compounds, inorganic.

Fire Hazard: This material readily liberates hydrogen selenide upon contact with acids or moisture (Section 6).

Disaster Hazard: Dangerous; when heated to decomposition, it emits highly toxic fumes of mercury; will react with water, steam or acid to produce toxic fumes; can react with oxidizing materials.

Countermeasures

Storage and Handling: Section 7.

MERCURIC SESQUIIODIDE
Hazard Analysis and Countermeasures

See mercury compounds, inorganic and iodides.

MERCURIC SILVER IODIDE
General Information

Description: Deep yellow powder.

Formula: $HgI_2 \cdot 2AgI$.

Constant: Mol wt: 924.1.

Hazard Analysis and Countermeasures

See mercury compounds, inorganic and iodides.

MERCURIC STEARATE
General Information

Description: Yellowish, granular powder.

Formula: $(C_{18}H_{35}O_2)_2Hg$.

Constant: Mol wt: 767.5.

Hazard Analysis and Countermeasures

See mercury compounds, organic.

MERCURIC SUBSULFATE
General Information

Description: Lemon yellow powder.

Formula: $HgSO_4 \cdot 2HgO$.

Constants: Mol wt: 729.90, bp: volatilizes, d: 6.44, vap. d.: 25.2.

Hazard Analysis and Countermeasures

See mercury compounds, inorganic and sulfates.

Shipping Regulations: Section 11.

    I.C.C.: Poison B; poison label, 200 pounds.

    Coast Guard Classification: Poison B; poison label.

    IATA: Poison B, poison label, 25 kilograms (passenger), 95 kilograms (cargo).

MERCURIC SUCCINATE
General Information

Description: White to yellowish crystalline powder; slightly soluble in water; soluble in aqueous sodium chloride.

Formula: $Hg(CH_2COO)_2$.

Constant: Mol wt: 316.68.

Hazard Analysis

See mercury compounds, organic.

MERCURIC SULFATE
General Information

Description: White, crystalline powder.

Formula: $HgSO_4$.

Constants: Mol wt: 296.67, mp: decomposes, d: 6.47.

Hazard Analysis and Countermeasures

See mercury compounds, inorganic and sulfates.

Shipping Regulations: Section 11.

    I.C.C.: Poison B; poison label, 200 pounds.

    Coast Guard Classification: Poison B; poison label.

    IATA: Poison B, poison label, 25 kilograms (passenger), 95 kilograms (cargo).

MERCURIC SULFIDE, BLACK
General Information

Synonym: Metacinnabarite.

Description: Black powder.

Formula: HgS.

Constants: Mol wt: 232.67, mp: 583.5°C, d: 7.73.

Hazard Analysis

Toxicity: Highly toxic. See mercury compounds, inorganic and sulfides.

MERCURIC SULFIDE, RED
General Information

Synonyms: Cinnabar; vermillion.

Description: Hexagonal, red crystals or powder.

Formula: HgS.

Constants: Mol wt: 232.68, vap. d.: 8.0, mp: subl. at 583.5°C, d: 8.10.

Hazard Analysis and Countermeasures

See mercury compounds, inorganic and sulfides.

MERCURIC SULFOCYANATE
General Information
Synonyms: Mercury rhodanide; mercuric thiocyanate.
Description: White powder.
Formula: $Hg(SCN)_2$.
Constants: Mol wt: 316.77, mp: decomposes.
Hazard Analysis and Countermeasures
See mercury compounds, organic and cyanates.
Shipping Regulations: Section 11.
　I.C.C.: Poison B; poison label, 200 pounds.
　Coast Guard Classification: Poison B; poison label.
　IATA: Poison B, poison label, 25 kilograms (passenger), 95 kilograms (cargo).

MERCURIC TELLURATE
General Information
Description: Amber, cubic crystals.
Formula: Hg_3TeO_6.
Constant: Mol wt: 825.44.
Hazard Analysis and Countermeasures
See mercury compounds, inorganic and tellurium.

MERCURIC THALLIUM IODIDE
General Information
Description: Red, crystalline lumps.
Formula: $HgI_2 \cdot TlI$.
Constant: Mol wt: 708.9.
Hazard Analysis and Countermeasures
See mercury compounds, inorganic and thallium.

MERCURIC THIOCYANATE. See mercuric sulfocyanate.

MERCURIC THYMOLATE
General Information
Description: Yellowish, gray powder; variable composition.
Hazard Analysis and Countermeasures
See mercury compounds, organic.

MERCURIC THYMOL NITRATE
General Information
Description: Amorphous, white to reddish-white powder.
Hazard Analysis and Countermeasures
See mercury compounds, organic and nitrates.

MERCURIC THYMOLSALICYLATE
General Information
Description: White to reddish powder.
Hazard Analysis and Countermeasures
See mercury compounds, organic.

MERCURIC THYMOLSULFATE
General Information
Description: White to reddish-white powder.
Hazard Analysis and Countermeasures
See mercury compounds, organic.

MERCURIC TUNGSTATE
General Information
Description: Yellow crystals.
Formula: $HgWO_4$.
Constants: Mol wt: 448.53, mp: decomposes.
Hazard Analysis and Countermeasures
See mercury compounds, inorganic.

MERCURICYANAMID
General Information
Description: A white solid.
Formula: $HgCN_2$.
Constant: Mol Wt: 240.64.

Hazard Analysis and Countermeasures
See mercury compounds, organic and cyanides.
Shipping Regulations: Section 11.
　I.C.C.: Poison B; poison label.
　Coast Guard Classification: Poison B; poison label.

MERCURIC ZINC ACETATE
General Information
Description: White crystals.
Formula: $Hg(C_2H_3O_2)_2 + Zn(C_2H_3O_2)_2$.
Hazard Analysis and Countermeasures
See mercury compounds, organic.

MERCURIC ZINC CYANIDE
General Information
Description: White powder, a mixture of zinc cyanide with varying quantities of mercuric cyanide. Water dissolves the mercuric cyanide.
Hazard Analysis
See mercury compounds and cyanides.

"MERCUROCHROME"
General Information
Synonym: Merbromin.
Formula: $C_{20}H_8Br_2HgNa_2O_6 \cdot 3H_2O$.
Hazard Analysis
Toxicity: Highly toxic if ingested. Relatively non-irritating and non-toxic to damaged skin or tissue. An antiseptic.
See mercury compounds, organic.

MERCUROL, SOLID
General Information
Description: Colorless to brownish powder.
Composition: Contains 20% mercury.
Hazard Analysis and Countermeasures
See mercury compounds, organic.
Shipping Regulations: Section 11.
　I.C.C.: Poison B; poison label, 200 pounds.
　Coast Guard Classification: Poison B; poison label.
　IATA: Poison B, poison label, 25 kilograms (passenger), 95 kilograms (cargo).

"MERCUROSAL"
General Information
Synonym: Disodium hydroxymercury salicyloxyacetate.
Description: White, amorphous powder.
Formula: $(HOHg)NaOOCC_6H_3OCH_2COONa$.
Constant: Mol wt: 456.69.
Hazard Analysis and Countermeasures
See mercury compounds, organic.

MERCUROUS ACETATE
General Information
Description: Colorless scales or plates.
Formula: $Hg_2(C_2H_3O_2)_2$.
Constants: Mol wt: 519.31, mp: decomposes.
Hazard Analysis and Countermeasures
See mercury compounds, organic.

MERCUROUS ARSENITE
General Information
Description: Unstable brown powder.
Hazard Analysis and Countermeasures
See mercury compounds, inorganic and arsenic.

MERCUROUS AZIDE
General Information
Description: White crystals.
Formula: $Hg_2(N_3)_2$.

TOXIC HAZARD RATING CODE (For detailed discussion, see Section 1.)

0 NONE: (a) No harm under any conditions; (b) Harmful only under unusual conditions or overwhelming dosage.

1 SLIGHT: Causes readily reversible changes which disappear after end of exposure.

2 MODERATE: May involve both irreversible and reversible changes not severe enough to cause death or permanent injury.

3 HIGH: May cause death or permanent injury after very short exposure to small quantities.

U UNKNOWN: No information on humans considered valid by authors.

Constants: Mol wt: 485.27, mp: explodes.
Hazard Analysis and Countermeasures
See mercury compounds, inorganic and azides.

MERCUROUS BROMATE
General Information
Description: Crystals.
Formula: $Hg_2(BrO_3)_2$.
Constants: Mol wt: 657.06, mp: decomposes.
Hazard Analysis and Countermeasures
See mercury compounds, inorganic and bromates.

MERCUROUS BROMIDE
General Information
Description: Tetragonal, white-yellow crystals.
Formula: Hg_2Br_2.
Constants: Mol wt: 561.07, mp: sublimes at 345°C, d: 7.307, vap. d.: 19.3.
Hazard Analysis and Countermeasures
See mercury compounds, inorganic and bromides.
Shipping Regulations: Section 11.
  I.C.C.: Poison B; poison label, 200 pounds.
  Coast Guard Classification: Poison B; poison label.
  IATA: Poison B, poison label, 25 kilograms (passenger), 95 kilograms (cargo).

MERCUROUS CARBONATE
General Information
Description: Yellow-brown crystals.
Formula: Hg_2CO_3.
Constants: Mol wt: 461.23, mp: decomp. at 130°C, d: 5.07 at 218°C.
Hazard Analysis and Countermeasures
See mercury compounds, inorganic.

MERCUROUS CHLORATE
General Information
Description: White crystals.
Formula: $HgClO_3$.
Constants: Mol wt: 284.1, mp: decomp. at 250°C, d: 6.409.
Hazard Analysis and Countermeasures
See mercury compounds, inorganic and chlorates.

MERCUROUS CHLORIDE
General Information
Synonym: Calomel.
Description: White, rhombic crystals or crystalline powder.
Formula: Hg_2Cl_2.
Constants: Mol wt: 472.14, mp: sublimes at 400°C, d: 7.150.
Hazard Analysis and Countermeasures
See mercury compounds, inorganic and chlorides.

MERCUROUS CHROMATE
General Information
Description: Red needles or powder.
Formula: Hg_2CrO_4.
Constants: Mol wt: 517.23, mp: decomposes.
Hazard Analysis
Toxicity: Highly toxic. See mercury compounds, inorganic and chromates.
Fire Hazard: Moderate by chemical reaction; an oxidizer (Section 6).
Disaster Hazard: Dangerous; when heated to decomposition, it emits highly toxic fumes of mercury; can react with reducing materials.
Countermeasures
Storage and Handling: Section 7.

MERCUROUS FLUORIDE
General Information
Description: Cubic yellow crystals.
Formula: Hg_2F_2.
Constants: Mol wt: 439.22, mp: 570°C, d: 8.73.
Hazard Analysis and Countermeasures
See fluorides and mercury compounds, inorganic.

MERCUROUS FLUOSILICATE
General Information
Description: Colorless prisms.
Formula: $Hg_2SiF_6 \cdot 2H_2O$.
Constant: Mol wt: 579.31.
Hazard Analysis and Countermeasures
See mercury compounds, inorganic and fluosilicates.

MERCUROUS FORMATE
General Information
Description: Glistening scales.
Formula: $Hg_2(CHO_2)_2$.
Constants: Mol wt: 491.26, mp: decomposes.
Hazard Analysis and Countermeasures
See mercury compounds, organic.

MERCUROUS GLUCONATE
General Information
Description: White solid.
Formula: $HgCO_2C_5H_6(OH)_5$.
Constant: Mol wt: 395.8.
Hazard Analysis and Countermeasures
See mercury compounds, organic.
Shipping Regulations: Section 11.
  I.C.C.: Poison B; poison label, 200 pounds.
  Coast Guard Classification: Poison B; poison label.
  IATA: Poison B, poison label, 25 kilograms (passenger), 95 kilograms (cargo).

MERCUROUS IODATE
General Information
Description: Yellowish crystals.
Formula: $Hg_2(IO_3)_2$.
Constants: Mol wt: 751.06, mp: decomposes.
Hazard Analysis and Countermeasures
See mercury compounds, inorganic and iodates.

MERCUROUS IODIDE
General Information
Description: Yellow, tetragonal crystals or amorphous powder.
Formula: HgI.
Constants: Mol wt: 327.53, mp: sublimes at 140°C; bp: decomp. at 290°C, d: 7.70.
Hazard Analysis and Countermeasures
See mercury compounds, inorganic and iodides.
Shipping Regulations: Section 11.
  I.C.C.: Poison B; poison label, 200 pounds.
  Coast Guard Classification: Poison B; poison label.
  IATA: Poison B, poison label, 25 kilograms (passenger), 95 kilograms (cargo).

MERCUROUS MONOHYDROGEN o-ARSENATE
General Information
Description: Yellow-red crystals.
Formula: Hg_2HAsO_4.
Constant: Mol wt: 541.14.
Hazard Analysis and Countermeasures
See arsenic compounds and mercury compounds, inorganic.

MERCUROUS NITRATE
General Information
Description: Short, colorless, prismatic crystals.
Formula: $Hg_2(NO_3)_2 \cdot 2H_2O$.
Constants: Mol wt: 561.26, mp: 70°C, d: 4.79 at 4°C.
Hazard Analysis and Countermeasures
See mercury compounds, inorganic and nitrates.
Shipping Regulations: Section 11.
  I.C.C.: Poison B; poison label, 200 pounds.
  Coast Guard Classification: Poison B; poison label.
  IATA: Poison B, poison label, 25 kilograms (passenger), 95 kilograms (cargo).

MERCUROUS NITRATE, AMMONIATED
General Information
Synonym: Black precipitate.

Description: Black powder.
Formula: $Hg_2ONH_2Hg_2NO_3$.
Constant: Mol wt: 896.48.
Hazard Analysis and Countermeasures
See mercury compounds, inorganic and nitrates.

MERCUROUS NITRITE
General Information
Description: Yellow crystals.
Formula: $Hg_2(NO_2)_2$.
Constants: Mol wt: 493.24, mp: decomp. at 100°C, d: 7.33.
Hazard Analysis and Countermeasures
Highly toxic. See mercury compounds, inorganic and nitrites.

MERCUROUS OXALATE
General Information
Description: White crystals.
Formula: $Hg_2C_2O_4$.
Constant: Mol wt: 489.24.
Hazard Analysis and Countermeasures
See oxalates and mercury compounds, organic.

MERCUROUS OXIDE, BLACK
General Information
Description: Black to grayish-black powder.
Formula: Hg_2O.
Constants: Mol wt: 417.22, mp: decomp. at 100°C, d: 9.8.
Hazard Analysis
Toxicity: Highly toxic. See mercury compounds, inorganic.
Fire Hazard: Moderate by chemical reaction; an oxidizer (Section 6).
Disaster Hazard: Dangerous; when heated to decomposition, it emits highly toxic fumes of mercury; can react with reducing materials.
Countermeasures
Storage and Handling: Section 7.
Shipping Regulations: Section 11.
    I.C.C.: Poison B; poision label, 200 pounds.
    Coast Guard Classification: Poison B; poison label.
    IATA: Poison B, poison label, 25 kilograms (passenger), 95 kilograms (cargo).

MERCUROUS PHOSPHATE
General Information
Description: Heavy, white powder.
Formula: Hg_3PO_4.
Constant: Mol wt: 696.85.
Hazard Analysis and Countermeasures
See mercury compounds, inorganic.

MERCUROUS SULFATE
General Information
Description: White, crystalline powder.
Formula: Hg_2SO_4.
Constants. Mol wt: 497.28, mp: decomposes, d: 7.56.
Hazard Analysis and Countermeasures
See mercury compounds, inorganic and sulfates.
Shipping Regulations: Section 11.
    I.C.C.: Poison B; poison label, 200 pounds.
    Coast Guard Classification: Poison B; poison label.
    IATA: Poison B, poison label, 25 kilograms (passenger), 95 kilograms (cargo).

MERCUROUS SULFIDE
General Information
Description: Black crystals.
Formula: Hg_2S.
Constants: Mol wt: 433.24, mp: decomposes.

Hazard Analysis and Countermeasures
See mercury compounds, inorganic and sulfides.

MERCUROUS TARTRATE
General Information
Description: Yellowish-white crystalline powder.
Formula: $Hg_2C_4H_4O_6$.
Constant: Mol wt: 549.29.
Hazard Analysis and Countermeasures
See mercury compounds, organic.

MERCURY
General Information
Description: Silvery liquid; metallic element.
Formula: Hg.
Constants: At wt: 200.61, mp: −38.89°C, bp: 356.9°C, d: 13.546, vap. press.: 1 mm at 126.2°C.
Hazard Analysis
Toxicity: See mercury compounds.
TLV: ACGIH (recommended); 0.1 milligram per cubic meter of air. May be absorbed via skin.
Radiation Hazard: Section 5. For permissible levels, see Table 5, p. 150.
    Artificial isotope ^{197m}Hg, half life 24 h. Decays to radioactive ^{197}Hg by emitting gamma rays of 0.13, 0.16 MeV.
    Artificial isotope ^{197}Hg, half life 65 h. Decays to stable ^{197}Au by electron capture. Emits gamma rays of 0.08 MeV and X-rays.
    Artificial isotope ^{203}Hg, half life 47 d. Decays to stable ^{203}Tl by emitting beta particles of 0.21 MeV. Also emits gamma rays of 0.28 MeV.
Disaster Hazard: Dangerous; when heated it emits highly toxic fumes.
Countermeasures
Ventilation Control: Section 2.
Storage and Handling: Section 7.
Shipping Regulations: Section 11.
    IATA: Other restricted articles, class B, no limit (passenger and cargo).

MERCURY ACETAMIDE
General Information
Description: White powder.
Formula: CH_3CONHg.
Constant: Mol wt: 257.7.
Hazard Analysis and Countermeasures
See mercury compounds, organic.

MERCURY ACETATE. See mercurous acetate or mercuric acetate.
Shipping Regulations: Section 11.
    I.C.C.: Poison B; poison label, 200 pounds.
    Coast Guard Classification: Poison B; poison label.
    IATA: Poison B, poison label, 25 kilograms (passenger), 95 kilograms (cargo).

MERCURY ALANINE. See mercury α-aminopropionate.

MERCURY-p-AMINOPHENOL ARSENATE. See mercury atoxylate.

MERCURY α-AMINOPROPIONATE
General Information
Synonym: Mercury alanine.
Description: White crystals. Water soluble.
Formula: $Hg[CH_2CH(NH_2)COO]_2$.
Constant: Mol wt: 374.8.

TOXIC HAZARD RATING CODE (For detailed discussion, see Section 1.)

0 NONE: (a) No harm under any conditions; (b) Harmful only under unusual conditions or overwhelming dosage.

1 SLIGHT: Causes readily reversible changes which disappear after end of exposure.

2 MODERATE: May involve both irreversible and reversible changes not severe enough to cause death or permanent injury.

3 HIGH: May cause death or permanent injury after very short exposure to small quantities.

U UNKNOWN: No information on humans considered valid by authors

Hazard Analysis and Countermeasures
See mercury compounds, organic.

MERCURY, AMMONIATED
General Information
Synonyms: Ammonobasic mercuric chloride; white precipitate.
Description: White lumps or powder.
Formula: $HgNH_2Cl$.
Constant: Mol wt: 252.10.
Hazard Analysis and Countermeasures
See mercury compounds, inorganic and chlorides.

MERCURY ANTIMONY SULFIDE
General Information
Description: Gray-black powder.
Composition: Mixture of equal parts of black mercury sulfide and gray antimony sulfide.
Hazard Analysis and Countermeasures
See mercury compounds, antimony and sulfides.

MERCURY ATOXYLATE
General Information
Synonym: Mercury p-aminophenol arsenate.
Description: White powder.
Formula: $C_{12}H_{14}O_6N_2As_2Hg$.
Constant: Mol wt: 632.71.
Hazard Analysis and Countermeasures
See arsenic compounds and mercury compounds, organic.

MERCURY BENZAMIDE
General Information
Description: White powder.
Formula: C_6H_5CONHg.
Constant: Mol wt: 319.7.
Hazard Analysis and Countermeasures
See mercury compounds, organic.

MERCURY BICHLORIDE, SOLID. See mercuric chloride.
Shipping Regulations: Section 11.
    I.C.C.: Poison B; poison label, 200 pounds.
    Coast Guard Classification: Poison B; poison label.
    IATA: Poison B, poison label, not acceptable (passenger), 12 kilograms (cargo).

MERCURY BISULFATE. See mercuric sulfate.
Shipping Regulations: Section 11.
    I.C.C.: Poison B; poison label, 200 pounds.
    Coast Guard Classification: Poison B; poison label.

MERCURY CARBOLATE. See mercuric phenate.

MERCURY COLLOIDAL. See mercury.

MERCURY COMPOUNDS, INORGANIC
Hazard Analysis
Toxic Hazard Rating:
    Acute Local: Irritant 3; Ingestion 3; Inhalation 2.
    Acute Systemic: Ingestion 2; Inhalation 3.
    Chronic Local: Irritant 2; Allergen 3.
    Chronic Systemic: Ingestion 3; Inhalation 3; Skin Absorption 3.
Toxicology: Mercury is a general protoplasmic poison; after absorption it circulates in the blood and is stored in the liver, kidneys and spleen and bone. It is eliminated in the urine, feces, sweat, saliva and milk. In industrial poisoning, the chief effect is upon the central nervous system and upon the mouth and gums. Colitis has been reported frequently; a nephritis or nephrosis is rarely reported. The organic mercury compounds, like the organic lead compounds, appear to have an affinity for the lipoid-containing organs, resulting in disturbances of the central nervous system resembling those of tetraethyl lead. Fulminate of mercury rarely produces symptoms of systemic poisoning, but frequently causes a dermatitis. Organic mercury compounds may act as vesicants on the skin (Section 9).

The cardinal symptoms of industrial mercury poisoning are stomatitis, tremors, and psychic disturbances. Usually the first complaints are of excessive salivation and pain on chewing; in severe cases there may be gingivitis, with loosening of the teeth, and a dark line on the gum margins, resembling the "lead line." In slow poisoning the salivation may be absent, and the only complaint dryness of the throat and mouth. Tremor and psychic disturbances are commonly seen in the slow, chronic form of the poisoning; the tremor is of the intention type, and may be seen when the patient spreads the outstretched fingers or protrudes the tongue, or attempts to perform specified movements. Muscles of the face, hands and arms are chiefly affected. In more severe cases there may also be convulsive or shaking movements; writing is frequently illegible. Hyperactive kneejerks and scanning speech may be present in advanced cases. The psychic disturbance (so called "erethism") includes such changes as loss of memory, insomnia, lack of confidence, irritability, vague fears and depression.

The dermatitis produced by fulminate of mercury takes the form of small, discrete ulcers on the exposed parts, and is usually accompanied by conjunctivitis and inflammation of the mucuous membranes of the nose and throat.

Elemental mercury is probably not absorbed through the gastrointestinal tract, but many mercury compounds are. A number of mercury compounds, in addition to the fulminate, can cause skin irritations and can be absorbed through the skin; they are strong allergens (Section 9).

These are common air contaminants (Section 4).
Disaster Hazard: Dangerous; when heated to decomposition, they emit highly toxic fumes of mercury.
Countermeasures
Ventilation Control: Section 2.
Personnel Protection: Section 3.
Storage and Handling: Section 7.

MERCURY   COMPOUNDS,   N.O.S.   LIQUID. See mercury.
Shipping Regulations: Section 11.
    IATA: Poison B, posion label, 25 kilograms (passenger), 95 kilograms (cargo).

MERCURY   COMPOUNDS,   N.O.S.   SOLID. See mercury.
Shipping Regulations: Section 11.
    I.C.C.: Poison B; poison label, 200 pounds.
    Coast Guard Classification: Poison B; poison label.
    IATA: Poison B, poison label, 25 kilograms (passenger), 95 kilograms (cargo).

MERCURY COMPOUNDS, ORGANIC
Hazard Analysis
Toxic Hazard Rating:
    Acute Local: Irritant 3; Allergen 2; Ingestion 3; Inhalation 2.
    Acute Systemic: Ingestion 3; Inhalation 3, Skin Absorption 2.
    Chronic Local: Irritant 2; Allergen 3.
    Chronic Systemic: Ingestion 3; Inhalation 3; Skin Absorption 3.
Toxicology: The customary grouping of all organic mercurials in a single category is not justified by the toxicology of the compounds. Alkyl mercurials have very high toxicity, aryl compounds, particularly the phenyls, are much less toxic, and the organomercurials used in therapeutics have very low toxicity. The alkyls and aryls commonly cause skin burns and other forms of irritation, and both can be absorbed through the skin. Fatal poisoning has occurred due to exposure to alkyl mercurials and permanent damage to the brain has been reported. Extensive human observation on exposure

to the phenyl mercurials have shown no greater toxicity than is caused by metallic mercury. In fact up to the present time there has not been an authenticated case of occupational poisoning due to the phenyl mercurials reported in the literature.

These are common air contaminants (Section 4).

TLV: ACGIH (recommended); 0.01 milligrams per cubic meter of air (as mercury). Can be absorbed through the intact skin.

Disaster Hazard: Dangerous; when heated to decomposition, they emit highly toxic fumes of mercury.

Countermeasures

Ventilation Control: Section 2.
Personnel Protection: Section 3.
First Aid: Section 1.
Storage and Handling: Section 7.

MERCURY CYANIDE. See mercuric cyanide.

Shipping Regulations: Section 11.
  I.C.C.: Poison B; poison label, 200 pounds.
  Coast Guard Classification: Poison B; poison label.
  IATA: See mercuric cyanide.

MERCURY ETHYLENE DIAMINE SULFATE. See sublamin.

MERCURY FULMINATE, MERCURIC FULMINATE. See fulminate of mercury, dry.

MERCURY MORPHINE OLEATE
General Information
Description: An oleaginous mass.
Hazard Analysis and Countermeasures
See mercury compounds, organic.

MERCURY NAPHTHENATE
General Information
Description: A dark amber liquid.
Formula: $Hg(C_{10}H_7)_2$.
Constants: Mol wt: 469.0, vap. d.: 15.3.
Hazard Analysis and Countermeasures
See mercury compounds, organic.

MERCURY-β-NAPHTHOL. See mercuric naphtholate.

MERCURY NITRIDE
General Information
Description: Brown powder.
Formula: Hg_3N_2.
Constants: Mol wt: 629.85, mp: explodes.
Hazard Analysis
Toxicity: Highly toxic. See mercury compounds, inorganic.
Explosion Hazard: Severe, when exposed to heat.
Disaster Hazard: Dangerous; when heated to decomposition, it emits highly toxic fumes of mercury and may explode.
Countermeasures
Storage and Handling: Section 7.

MERCURY NUCLEATE. See mercurol, solid.

MERCURY OLEATE. See mercuric oleate.

MERCURY 8-QUINOLATE
Hazard Analysis
Toxicity: U. A fungicide. See mercury compounds, organic.

MERCURY RHODANIDE. See mercuric sulfocyanate.

MERCURY SUBSALICYLATE. See mercuric salicylate.

MESITYLENE
General Information
Synonym: 1,3,5-Trimethylbenzene.
Description: Liquid; peculiar odor; insoluble in water. Soluble in alcohol, benzene.
Formula: C_9H_{12}.
Constants: Mol wt: 120.2, mp: $-44.8°C$, d: 0.8637 at $20/4°C$, bp: $164.7°C$.
Hazard Analysis
Toxic Hazard Rating:
  Acute Local: U.
  Acute Systemic: Inhalation 3.
  Chronic Local: U.
  Chronic Systemic: Inhalation 3.
Toxicology: Data based on animal experiments show narcotic effects and some disturbances of blood. Leucopenia and thrombocytopenia have been reported in experimental animals.

MESITYL OXIDE
General Information
Synonym: 4-Methyl-3-penten-2-one.
Description: Oily, colorless liquid; strong odor.
Formula: $(CH_3)_2CCHCOCH_3$.
Constants: Mol wt: 98.14, mp: $-59°C$, bp: $130.0°C$, flash p: $90°F$ (C.C.), d: 0.8539 at $20°/4°C$, autoign. temp.: $652°F$, vap. press.: 10 mm at $26.0°C$, vap. d.: 3.38.
Hazard Analysis
Toxic Hazard Rating:
  Acute Local: Irritant 3; Ingestion 3; Inhalation 3.
  Acute Systemic: Ingestion 3; Inhalation 3; Skin Absorption 3.
  Chronic Local: Irritant 3; Inhalation 3.
  Chronic Systemic: Inhalation 3; Ingestion 3; Skin Absorption 3.
TLV: ACGIH (recommended); 25 parts per million in air; 100 milligrams per cubic meter of air.
Toxicology: It can cause opaque cornea, keratoconus, and extensive necrosis of cornea. From animal experiments, it has been found that concentrations of 1,000 ppm for 4 hours at a time per day over an observation period of 14 days killed 3 out of 6 white mice, placing it in the category of moderate definite toxic hazard. Guinea pigs and rats were subject to repeated inhalation of the vapors of this material. No effect upon them was found after the concentration had been reduced to 50 ppm, even with 30 eight-hour exposures. Single exposures tend to indicate that this ketone has greater acute and narcotic action than isophorone. It can have harmful effects upon the kidneys and liver, and may damage the eyes and lungs to a serious degree. Animals exposed to 100 ppm for 30 eight-hour exposures showed internal damage as noted above. Exposure to 500 ppm caused their death. This compound is highly irritating to all tissues on contact and its vapors also are irritating. High concentrations are narcotic. Prolonged exposure can injure liver, kidneys and lungs. It is readily absorbed through intact skin.
Fire Hazard: Moderate, when exposed to heat or flame; can react with oxidizing materials.
Countermeasures
To Fight Fire: Alcohol foam, carbon dioxide, dry chemical or carbon tetrachloride (Section 6).
Ventilation Control: Section 2.
Personnel Protection: Section 3.

TOXIC HAZARD RATING CODE (For detailed discussion, see Section 1.)

0 NONE: (a) No harm under any conditions; (b) Harmful only under unusual conditions or overwhelming dosage.

1 SLIGHT: Causes readily reversible changes which disappear after end of exposure.

2 MODERATE: May involve both irreversible and reversible changes not severe enough to cause death or permanent injury.

3 HIGH: May cause death or permanent injury after very short exposure to small quantities.

U UNKNOWN: No information on humans considered valid by authors

First Aid: Section 1.
Storage and Handling: Section 7.
Shipping Regulations: Section 11.
 MCA warning label.

METABORATE PEROXYHYDRATE. See sodium perborate.

METACETONE. See diethyl ketone.

METACIDE. See parathion.

METACINNABARITE. See mercuric sulfide, black.

METALDEHYDE
General Information
Synonyms: m-Acetaldehyde; polymerized acetaldehyde.
Description: Colorless crystals.
Formula: $(C_2H_4O)_4$.
Constants: Mol wt: 176.21, mp: 246°C, sublimes at 112°C, flash p: 97°F (C.C.), vap. d.: 6.06.
Hazard Analysis
Toxic Hazard Rating:
 Acute Local: Irritant 3; Ingestion 3; Inhalation 3.
 Acute Systemic: Ingestion 3; Inhalation 3.
 Chronic Local: Irritant 2.
 Chronic Systemic: Ingestion 2; Inhalation 2.
Toxicology: Irritating to skin and mucous membranes. Can cause kidney and liver damage. See also acetaldehyde.
Note: Used as a food additive permitted in food for human consumption (Section 10).
Fire Hazard: Moderate, when exposed to heat or flame; can react with oxidizing materials.
Spontaneous Heating: No.
Countermeasures
To Fight Fire: Water, carbon dioxide, dry chemical or carbon tetrachloride (Section 6).
Storage and Handling: Section 7.
Shipping Regulations: Section 11.
 IATA: Flammable solid, yellow label, 12 kilograms (passenger), 45 kilograms (cargo).

METAL POLISH LIQUID
General Information
Constant: Flash p: <0°F.
Hazard Analysis
Toxicity: Details unknown; some metal polishes can act as irritants and allergens.
Fire Hazard: Dangerous; when exposed to heat or flame; can react vigorously with oxidizing materials (Section 6).
Explosion Hazard: Unknown.
Countermeasures
Storage and Handling: Section 7.

METANILIC ACID, SODIUM SALT. See m-aniline sodium sulfonate.

META-SYSTOX
General Information
Synonym: O-[2-(ethylthio)ethyl]-O,O-dimethyl phosphorothioate.
Hazard Analysis and Countermeasures
Toxicity: Highly toxic. A cholinesterase inhibitor. See parathion.

METHACROLEIN
General Information
Synonym: α-Methylacrolein.
Description: Colorless liquid.
Formula: $CH_2C(CH_3)CHO$.
Constants: Mol wt: 70.1, bp: 73.5°C, flash p: 35°F (O.C.) d: 0.830 at 20°/4°C, vap. press.: 120 mm at 20°C, vap. d.: 2.42.
Hazard Analysis
Toxic Hazard Rating:
 Acute Local: Irritant 3; Ingestion 3; Inhalation 3.

Acute Systemic: Inhalation 3.
Chronic Local: U.
Chronic Systemic: U.
Fire Hazard: Dangerous, when exposed to heat or flame.
Explosion Hazard: Unknown.
Disaster Hazard: Dangerous; on decomposition, it emits toxic fumes; can react vigorously with oxidizing materials.
Countermeasures
Ventilation Control: Section 2.
To Fight Fire: Foam, carbon dioxide, dry chemical or carbon tetrachloride (Section 6).
Personnel Protection: Section 3.
Storage and Handling: Section 7.

METHACRYLIC ACID
General Information
Synonym: α-Methylacrylic acid.
Description: Corrosive liquid or colorless crystals.
Formula: $(CH_3)(CH_2)CCOOH$.
Constants: Mol wt: 86.1, mp: 15°C, bp: 161°C, flash p: 170°F (C.O.C.), d: 1.014 at 25°C (glacial), vap. press.: 1 mm at 25.5°C.
Hazard Analysis
Toxic Hazard Rating:
 Acute Local: Irritant 3; Ingestion 3; Inhalation 3.
 Acute Systemic: U
 Chronic Local: U.
 Chronic Systemic: U.
Fire Hazard: Moderate, when exposed to heat or flame (Section 6).
Disaster Hazard: Dangerous; when heated to decomposition, it emits toxic fumes.
Countermeasures
Ventilation Control: Section 2.
Personnel Protection: Section 3.
Storage and Handling: Section 7.
To Fight Fire: Alcohol foam.
First Aid: Section 1.

β-METHACRYLIC ACID. See crotonic acid.

METHACRYLONITRILE
General Information
Synonym: 2-Methylpropenitrile.
Description: Colorless liquid.
Formula: $H_2C = C(CH_3)C = N$.
Constants: Mol wt: 67.09, mp: -40°C, bp: 90.3°C, d: 0.805, vap. press.: 40 mm at 12.8°C.
Hazard Analysis and Countermeasures
Toxic Hazard Rating:
 Acute Local: Irritant 1.
 Acute Systemic: Inhalation 3; Skin Absorbtion 3.
 Chronic Local: Irritant 1.
 Chronic Systemic: U.
Toxicity: See also nitriles.

1,4(8)-p-METHADIENE. See terpinolene.

1-METHALLYL ALCOHOL
General Information
Synonym: 2-Buten-1-ol.
Description: Colorless liquid.
Formula: $CH_3CHCHCH_2OH$.
Constants: Mol wt: 72.1, mp: <-30°C, bp: 118°C, flash p: 94°F, d: 0.8726 at 0°/4°C, vap. d.: 2.49.
Hazard Analysis
Toxicity: Details unknown; probably toxic. See also allyl-alcohol.
Fire Hazard: Moderate, when exposed to heat or flame; can react with oxidizing materials.
Countermeasures
To Fight Fire: Alcohol foam, carbon dioxide, dry chemical or carbon tetrachloride (Section 6).
Storage and Handling: Section 7.

METHALLYL CHLORIDE
General Information
Description: Colorless, volatile liquid; disagreeable odor.
Formula: C_4H_7Cl.
Constants: Mol wt: 90.55, bp: 72.17°C, lel: 2.3%, uel: 9.3%, fp: < -80°C, flash p: -3°F, d: 0.9257 at 20°/4°C, vap. press.: 101.7 mm at 20°C, vap. d.: 3.12.
Hazard Analysis
Toxic Hazard Rating:
Acute Local: Irritant 2; Ingestion 2; Inhalation 2.
Acute Systemic: U.
Chronic Local: Irritant 1.
Chronic Systemic: Ingestion 2; Inhalation 2.
Fire Hazard: Dangerous, when exposed to heat or flame.
Explosion Hazard: Moderate, when exposed to heat or flame.
Disaster Hazard: Dangerous; on decomposition, it emits highly toxic fumes of chlorides; can react vigorously with oxidizing materials.
Countermeasures
Ventilation Control: Section 2.
Personnel Protection: Section 3.
To Fight Fire: Alcohol foam, carbon dioxide, dry chemical or carbon tetrachloride (Section 6).
Storage and Handling: Section 7.

METHAM SODIUM. See methyldithiocarbamic acid-sodium salt.

METHANAL. See formaldehyde.

METHANAMIDE. See formamide.

METHANE
General Information
Synonyms: Marsh gas; methyl hydride.
Description: Colorless, odorless, tasteless gas.
Formula: CH_4.
Constants: Mol wt: 16.04, bp: -161.5°C, lel: 5.3%, uel: 14.0%, fp: -183.2°C, d: 0.415 at -164°C; 0.7168 g/liter, autoign. temp.: 1000°F, vap. d.: 0.6.
Hazard Analysis
Toxic Hazard Rating:
Acute Local: 0.
Acute Systemic: Inhalation 1.
Chronic Local: 0.
Chronic Systemic: Inhalation 1.
Caution: A simple asphyxiant. See argon for discussion of simple asphyxiants.
Fire Hazard: Dangerous, when exposed to heat or flame.
Spontaneous Heating: No.
Explosion Hazard: Dangerous, when exposed to heat or flame.
Disaster Hazard: Dangerous.
Countermeasures
Ventilation Control: Section 2.
To Fight Fire: Stop flow of gas, carbon dioxide or dry chemical.
Storage and Handling: Section 7.
Shipping Regulations: Section 11.
I.C.C.: Flammable gas; red gas label, 300 pounds.
Coast Guard Classification: Inflammable gas; red label.
IATA: Flammable gas, red label, not acceptable (passenger), 140 kilograms (cargo).

METHANE ARSONIC ACID
General Information
Synonym: Methyl arsonic acid.

Description: Crystals; water and alcohol soluble.
Formula: $CH_3AsO(OH)_2$.
Constants: Mol wt: 140.0, mp: 161°C.
Hazard Analysis and Countermeasures
See arsenic compounds.

METHANE CARBOTHIOLIC ACID. See thioacetic acid.

METHANE CARBOXYLIC ACID. See acetic acid.

METHANE SULFONIC ACID
General Information
Synonym: Methyl sulfonic acid.
Description: Solid, soluble in water, alcohol, ether.
Formula: CH_3SO_2OH.
Constants: Mol wt: 96.10, d: 1.485 at 20/20°C, mp: 20°C, bp: 167°C at 10 mm.
Hazard Analysis
Toxicity Hazard Rating:
Acute Local: Irritant 3; Ingestion 3; Inhalation 3.
Acute Systemic: U.
Chronic Local: U.
Chronic Systemic: U.
Disaster Hazard: Dangerous. See sulfonates.

METHANETHIOL. See methyl mercaptan.

METHANOIC ACID. See formic acid.

METHANO INDANE. See chlordane.

METHANOL. See methyl alcohol.

METHENAMINE. See hexamethylene tetramine.

N,N'-METHENYL-o-PHENYLENE DIAMINE. See benzimidazole.

METHIONINE
General Information
Synonym: 2-Amino-4(methyl thio) butyric acid.
Description: An essential sulfur-containing amino acid; white crystalline powder or platelets; faint odor; soluble in water, dilute acids, and alkalis; very slightly soluble in alcohol; practically insoluble in ether.
Formula: $CH_3SCH_2CH_2CH(NH_2)COOH$.
Constant: Mol wt: 149.
Hazard Analysis
Toxic Hazard Rating: Unknown. A nutrient and/or dietary supplement food additive (Section 10).
Disaster Hazard: Dangerous. See sulfides.

METHIONINE HYDROXY ANALOG CALCIUM
General Information
Synonym: DL-α-Hydroxy-γ-methyl mercaptobutyricacid, calcium salt; 2-hydroxy-4-methyl thiobutyric acid, calcium salt.
Description: Light tan powder; soluble in water; insoluble in common organic solvents.
Formula: $(CH_3SCH_2CH_2CHOHCOO)_2Ca$.
Constant: Mol wt: 338.
Hazard Analysis
Toxic Hazard Rating: U. A nutrient and/or dietary supplement food additive (Section 10).

METHIOTRIAZAMINE
General Information
Synonym: 4,6-Diamino-1-(4-methylmercaptophenyl)-1,2-dihydro-2,2-dimethyl-1,3,5-triazine hydrochloride.

TOXIC HAZARD RATING CODE *(For detailed discussion, see Section 1.)*

0 NONE: (a) No harm under any conditions; (b) Harmful only under unusual conditions or overwhelming dosage.

1 SLIGHT: Causes readily reversible changes which disappear after end of exposure.

2 MODERATE: May involve both irreversible and reversible changes; not severe enough to cause death or permanent injury.

3 HIGH: May cause death or permanent injury after very short exposure to small quantities.

U UNKNOWN: No information on humans considered valid by authors.

Hazard Analysis
Toxic Hazard Rating: U. Used as a food additive permitted in the feed and drinking water of animals and/or for the treatment of food producing animals. Also permitted in food for human consumption (Section 10).
Disaster Hazard: Dangerous. See chlorides and sulfides.

METHOXONE.
General Information
Synonym: 4-Chloro-2-methylphenoxyacetic acid.
Description: Crystals.
Formula: $C_9H_9O_3Cl$.
Constant: Mol wt: 200.6.
Hazard Analysis
Toxicity: Moderately toxic. Details unknown. See closely related 2,4-dichlorophenoxy acetic acid.
Disaster Hazard: Dangerous; when heated to decomposition, it emits highly toxic fumes of chlorides.

METHOXYACETAL-p-PHENETIDINE. See kryofin.

METHOXYAMINE
General Information
Synonym: Methoxylamine.
Description: Mobile liquid; fishy amine odor; miscible with water.
Formula: CH_3ONH_2.
Constants: Mol wt: 47.1, bp: 50°C.
Hazard Analysis
Toxicity: Highly toxic. Details unknown. See related compound hydroxylamine.

METHOXY ANALOG OF DDT. See 1-trichloro-2,2-bis-(p-methoxyphenyl) ethane.

o-METHOXY ANILINE. See o-anisidine.

p-METHOXY ANILINE. See p-anisidine.

METHOXY BENZALDEHYDE-O
General Information
Synonyms: o-Anisaldehyde; o-anisic aldehyde.
Description: White to light tan solid; slight phenolic odor; slightly soluble in water.
Formula: $CH_3OC_6H_4CHO$.
Constants: Mol wt: 136, bp: 238°C, mp: 38–39°C and 30°C (2 crystalline forms), d (liquid): 1.1274 at 25/25°C, d(solid): 1.258 at 25/25°C, flash p: 244°F (O.C.).
Hazard Analysis
Toxic Hazard Rating: U.
Fire Hazard: Slight, when exposed to heat or flame.
Countermeasures
To Fight Fire: Water, foam.

p-METHOXY BENZOIC ACID. See anisic acid.

METHOXY BORONDICHLORIDE
General Information
Description: Colorless liquid, decomposed by water.
Formula: CH_3OBCl_2.
Constants: Mol wt: 112.8, mp: −15°C, bp: 58°C.
Hazard Analysis and Countermeasures
See boron compounds and chlorides.

METHOXY BORON DIFLUORIDE
General Information
Description: Colorless liquid, decomposed by water.
Formula: CH_3OBF_2.
Constants: Mol wt: 79.9, d: 1.417 at 36°C, bp: 86°C.
Hazard Analysis and Countermeasures
See fluorides and boron compounds.

1-METHOXY-1,3-BUTADIENE
General Information
Formula: $CH_2:CHCH:CH(OCH_3)$.
Constant: Mol wt: 84.

Hazard Analysis
Toxicity: Details unknown. Limited animal experiments suggest moderate toxicity. See also butadiene.

3-METHOXY BUTANOL
General Information
Description: Liquid.
Formula: $CH_3CH(OCH_3)CH_2CH_2OH$.
Constants: Mol wt: 104.15, bp: 161.1°C, fp: −85°C, flash p: 165°F (C.O.C.), d: 0.9229 at 20°/20°C, vap. press.: 0.9 mm at 20°C, vap. d.: 3.59.
Hazard Analysis
Toxicity: Details unknown. See also alcohols.
Fire Hazard: Moderate, when exposed to heat or flame; can react with oxidizing materials.
Countermeasures
To Fight Fire: Foam, carbon dioxide, dry chemical or carbon tetrachloride (Section 6).
Storage and Handling: Section 7.

METHOXY BUTYL ACETATE
General Information
Synonym: Butoxyl.
Description: Liquid; bitter taste, acrid odor.
Formula: $CH_3CO_2C_4H_8OCH_3$.
Constants: Mol wt: 146.2, bp: 135°C, flash p: 170°F, d: 0.952–0.958 at 20°/20°C, vap. d.: 5.05.
Hazard Analysis
Toxicity: Details unknown. See esters.
Fire Hazard: Moderate, when exposed to heat or flame; can react with oxidizing materials.
Countermeasures
To Fight Fire: Foam, carbon dioxide, dry chemical or carbon tetrachloride (Section 6).
Storage and Handling: Section 7.

3-METHOXY BUTYL ALCOHOL. See 3-methoxy butanol.

3-METHOXYBUTYRALDEHYDE
General Information
Formula: $CH_3CH(OCH_3)CH_2CHO$.
Constant: Mol wt: 102.
Hazard Analysis
Toxicity: Details unknown. Limited animal experiments suggest high toxicity and irritation. See also aldehydes.

3-METHOXY BUTYRIC ACID
General Information
Description: Liquid.
Formula: $CH_3OC_3H_6COOH$.
Constants: Mol wt: 118.13, mp: 12°C, bp: 139°C at 50 mm, d: 1.053 at 20°/20°C.
Hazard Analysis
Toxicity: Unknown.
Disaster Hazard: Slight; when heated, it emits acrid fumes; can react with oxidizing materials (Section 6).
Countermeasures
Storage and Handling: Section 7.

METHOXYCHLOR
General Information
Synonym: DMDT; 2,2-bis-(p-methoxyphenyl)-1,1,1-trichloroethane.
Description: Crystals.
Formula: $(CH_3OC_6H_4)_2CHCCl_3$.
Constants: Mol wt: 345.7, mp: 78°C, vap. d.: 12.
Hazard Analysis
Toxic Hazard Rating:
  Acute Local: Allergen 1; Irritant 1.
  Acute Systemic: Ingestion 1; Inhalation 1.
  Chronic Local: Allergen 1; Irritant 1.
  Chronic Systemic: Ingestion 1.
Toxicology: An insecticide. Prolonged exposure may cause kidney injury. See also DDT.
TLV: ACGIH (recommended); 15 milligrams per cubic meter of air.

Disaster Hazard: Dangerous; when heated to decomposition it emits highly toxic fumes of chlorides.
Countermeasures
Personal Hygiene: Section 3.
Storage and Handling: Section 7.

METHOXYETHANE. See ethyl methyl ether.

2-METHOXYETHANOL. See ethylene glycol monomethyl ether.

2-(2-METHOXY ETHOXY) ETHANOL. See methyl carbitol.

METHOXY ETHYL ACETYL RICINOLEATE
General Information
Description: Light straw colored liquid.
Formula: $C_{17}H_{32}(OOCCH_3)COOC_2H_4OCH_3$.
Constants: Mol wt: 398, mp: $-60°C$ (gels), bp: 200–260°C at 4 mm, flash p: 446°F, d: 0.966 at 20°/20°C, vap. press.: <0.01 mm at 150°C, vap. d.: 13.8.
Hazard Analysis
Toxicity: Details unknown. See esters.
Fire Hazard: Slight, when exposed to heat or flame.
Countermeasures
To Fight Fire: Foam, carbon dioxide, dry chemical or carbon tetrachloride (Section 6).
Storage and Handling: Section 7.

METHOXY ETHYL ACRYLATE
General Information
Description: Liquid.
Formula: $CH_2CHCO_2CH_2CH_2OCH_3$.
Constants: Bp: 61°C at 17 mm, flash p: 180°F (O.C.), d: 1.0134 at 20°C, vap. d.: 4.49.
Hazard Analysis
Toxicity: Details unknown. See ethyl acrylate.
Fire Hazard: Moderate, when exposed to heat or flame; can react with oxidizing materials.
Countermeasures
To Fight Fire: Foam, carbon dioxide, dry chemical or carbon tetrachloride (Section 6).
Storage and Handling: Section 7.

METHOXY ETHYL MERCURI ACETATE
General Information
Description: Crystals; water soluble.
Formula: $CH_3OCH_2CH_2HgOOCCH_3$.
Constant: Mol wt: 318.7.
Hazard Analysis and Countermeasures
See mercury compounds, organic.

METHOXY ETHYL MERCURY CHLORIDE
General Information
Description: Crystals.
Formula: $OCH_3CH_2CH_2HgCl$.
Constant: Mol wt: 295.2.
Hazard Analysis and Countermeasures
See mercury compounds, organic and chlorides.

2-METHOXY ETHYL MERCURY SILICATE
General Information
Synonym: Verisan.
Hazard Analysis and Countermeasures
See mercury compounds, organic.

METHOXY ETHYL OLEATE
General Information
Description: Light-colored liquid.

Formula: $C_8H_{17}CHCH(CH_2)_7COOC_2H_4OCH_3$.
Constants: Mol wt: 340, mp: $-20°C$, bp: 188-225°C at 4 mm, flash p: 386°F, d: 0.902 at 20°/20°C, vap. press.: 0.04 mm at 150°C, vap. d.: 11.8.
Hazard Analysis
Toxicity: Details unknown. See esters.
Fire Hazard: Slight, when exposed to heat or flame; can react with oxidizing materials.
Coontermeasures
To Fight Fire: Foam, carbon dioxide, dry chemical or carbon tetrachloride (Section 6).
Storage and Handling: Section 7.

METHOXY ETHYL PHTHALATE
General Information
Description: Liquid.
Formula: $(OCH_3CH_2CH_2CO_2)_2C_6H_4$.
Constants: Mol wt: 282.3, bp: 190–220°C, flash p: 275°F (C.C.), d: 1.17, vap. d.: 9.75.
Hazard Analysis
Toxic Hazard Rating:
Acute Local: Irritant 2.
Acute Systemic: Ingestion 2; Inhalation 2.
Chronic Local: U.
Chronic Systemic: U.
Fire Hazard: Slight, when exposed to heat or flame; can react with oxidizing materials.
Spontaneous Heating: No.
Countermeasures
To Fight Fire: Water, foam, carbon dioxide, dry chemical or carbon tetrachloride (Section 6).
Ventilation Control: Section 2.
Personnel Protection: Section 3.
Storage and Handling: Section 7.

3-METHOXY-4-HYDROXYBENZALDEHYDE. See vanillin.

METHOXYLAMINE. See methoxyamine.

2-METHOXY-4-METHYL PHENOL. See creosol.

2-METHOXY-4-NITROANILINE. See 1-amino-2-methoxy-4-nitrobenzene.

2-METHOXY PHENOL. See guaiacol.

4-METHOXY PHENOL. See hydroquinone monomethyl ether.

m-METHOXY PHENYL BORIC ACID. See m-anisyl boric acid.

o-METHOXY PHENYL BORIC ACID. See o-anisyl boric acid.

p-METHOXY PHENYL BORIC ACID. See p-anisyl boric acid.

1-METHOXY-2-PROPANOL. See monopropylene glycol methyl ether.

p-METHOXY PROPENYL BENZENE. See anethole.

METHOXY PROPIONITRILE
General Information
Description: Liquid.
Formula: $OCH_3CH_2CH_2CN$.
Constants: Mol wt: 85.1, mp: $-63°C$, bp: 160°C, flash p: 149°F, d: 0.92, vap. d.: 2.94.

TOXIC HAZARD RATING CODE (For detailed discussion, see Section 1.)

0 NONE: (a) No harm under any conditions; (b) Harmful only under unusual conditions or overwhelming dosage.

1 SLIGHT: Causes readily reversible changes which disappear after end of exposure.

2 MODERATE: May involve both irreversible and reversible changes; not severe enough to cause death or permanent injury.

3 HIGH: May cause death or permanent injury after very short exposure to small quantities.

U UNKNOWN: No information on humans considered valid by authors.

Hazard Analysis

Toxicity: Highly toxic. See cyanides.

Fire Hazard: Moderate, when exposed to heat or flame. See nitriles.

Disaster Hazard: Dangerous; when heated to decomposition, it emits highly toxic fumes of cyanides; will react with water, steam, or acids to produce toxic and flammable vapors; can react with oxidizing materials.

Countermeasures

Storage and Handling: Section 7.

To Fight Fire: Carbon dioxide, dry chemical or carbon tetrachloride (Section 6).

3-METHOXYPROPYLAMINE
General Information

Description: Colorless liquid.

Formula: $CH_3OCH_2CH_2CH_2NH_2$.

Constants: Mol wt: 89.14, mp: $-75.7°C$, bp: $116°C$, flash p: $90°F$ (T.C.C.), d: 0.8615 at $30°C$, vap. press.: 20 mm at $30°C$, vap. d.: 3.07.

Hazard Analysis

Toxic Hazard Rating:

Acute Local: U.

Acute Systemic: Ingestion 2; Inhalation 2.

Chronic Local: U.

Chronic Systemic: Ingestion 2; Inhalation 2.

Fire Hazard: Moderate, when exposed to heat or flame; can react with oxidizing materials.

Explosion Hazard: Unknown.

Countermeasures

Ventilation Control: Section 2.

To Fight Fire: Carbon dioxide, dry chemical or carbon tetrachloride (Section 6).

Personal Hygiene: Section 3.

Storage and Handling: Section 7.

METHOXY TRIGLYCOL
General Information

Description: Infinitely soluble in water.

Formula: $CH_3O(C_2H_4O)_3H$.

Constants: Mol wt: 122, d: 1.0494, bp: $249°C$, fp: $-44°C$, flash p: $245°F$ (O.C.).

Hazard Analysis

Toxic Hazard Rating: U.

Fire Hazard: Slight, when exposed to heat or flame.

Countermeasures

Storage and Handling: Section 7.

To Fight Fire: Water, foam, alcohol foam.

METHOXYTRIGLYCOL ACETATE
General Information

Synonym: Triethylene glycol methyl ether acetate.

Description: Liquid.

Formula: $CH_3O(CH_2)_2O(CH_2)_2O(CH_2)_2OCOCH_3$.

Constants: Mol wt: 206.23, bp: $130°C$, flash p: $260°F$ (O.C.), d: 1.094, vap. d.: 7.11.

Hazard Analysis

Toxicity: Details unknown. See also glycols.

Fire Hazard: Slight, when exposed to heat or flame; can react with oxidizing materials.

Countermeasures

To Fight Fire: Alcohol foam, carbon dioxide, dry chemical or carbon tetrachloride (Section 6).

Storage and Handling: Section 7.

METHYL ABIETATE
General Information

Description: Liquid.

Formula: $CH_3C_{20}H_{29}O_2$.

Constants: Mol wt: 316.5, bp: $361°C$, flash p: $356°F$ (O.C.), d: 1.02, vap. d.: 10.9.

Hazard Analysis

Toxicity: Probably similar to abietic acid.

Fire Hazard: Slight, when exposed to heat or flame; can react with oxidizing materials.

Countermeasures

To Fight Fire: Foam, carbon dioxide, dry chemical or carbon tetrachloride (Section 6).

Storage and Handling: Section 7.

METHYL-4-ACETAMIDO-2-ETHOXYBENZOATE. See ethopabate.

METHYL ACETATE
General Information

Description: Colorless, volatile liquid.

Formula: $CH_3CO_2CH_3$.

Constants: Mol wt: 74.08, mp: $-98.7°C$, lel: 3.1%, uel: 16%, bp: $57.8°C$, ulc: 85–90, flash p: $14°F$, d: 0.92438, autoign. temp.: $935°F$, vap. press.: 100 mm at $9.4°C$, vap. d.: 2.55.

Hazard Analysis

Toxic Hazard Rating:

Acute Local: Irritant 1.

Acute Systemic: Ingestion 2; Inhalation 2; Skin Absorption 2.

Chronic Local: U.

Chronic Systemic: Ingestion 2; Inhalation 2; Skin Absorption 2.

TLV: ACGIH (recommended); 200 parts per million in air; 610 milligrams per cubic meter of air.

Toxicology: Methyl acetate is narcotic, but is less so than the higher members of the acetate series. It has an irritating effect upon the mucous membranes of the eyes and upper respiratory tract, and in this respect its action is stronger than that of the higher members of the series. The irritant concentration is about 10,000 ppm.

Signs and symptoms are irritation and burning of the eyes, lachrymation, dyspnea, palpitation of the heart, and complaints of depression or dizziness.

Fire Hazard: Dangerous, when exposed to heat or flame.

Spontaneous Heating: No.

Explosion Hazard: Moderate, when exposed to heat or flame.

Disaster Hazard: Dangerous, upon exposure to heat or flame; can react vigorously with oxidizing materials.

Countermeasures

Ventilation Control: Section 2.

To Fight Fire: Alcohol foam, carbon dioxide, dry chemical or carbon tetrachloride (Section 6).

Personnel Protection: Section 3.

Storage and Handling: Section 7.

Shipping Regulations: Section 11.

I.C.C.: Flammable liquid; red label, 10 gallons.

Coast Guard Classification; Inflammable liquid; red label.

IATA: Flammable liquid, red label, 1 liter (passenger), 40 liters (cargo).

METHYL ACETIC ACID. See propionic acid.

METHYL ACETOACETATE
General Information

Description: Colorless liquid.

Formula: $CH_3COCH_2COOCH_3$.

Constants: Mol wt: 116.11, mp: $27.5°C$, bp: $170°C$, flash p: $170°F$, d: 1.077, vap. d.: 4.00.

Hazard Analysis

Toxicity: Details unknown. Animal experiments suggest low toxicity.

Fire Hazard: Moderate, when exposed to heat or flame.

Spontaneous Heating: No.

Countermeasures

To Fight Fire: Foam, carbon dioxide, dry chemical or carbon tetrachloride (Section 6).

Storage and Handling: Section 7.

METHYL ACETONE
General Information

Description: Clear liquid.

Composition: A mixture of acetone, methyl acetate and methyl alcohol.
Constant: Flash p: 0° F (C.C.).
Hazard Analysis
Toxicity: Probably toxic. See components, especially methyl alcohol.
Fire Hazard: Dangerous, when exposed to heat or flame.
Explosion Hazard: Unknown.
Disaster Hazard: Dangerous! Keep away from heat and open flame; can react vigorously with oxidizing materials.
Countermeasures
Ventilation Control: Section 2.
To Fight Fire: Carbon dioxide, dry chemical or carbon tetrachloride (Section 6).
Storage and Handling: Section 7.
Shipping Regulations: Section 11.
  I.C.C.: Flammable liquid; red label, 10 gallons.
  Coast Guard Classification: Inflammable liquid; red label.
  IATA: Flammable liquid, red label, 1 liter (passenger), 40 liters (cargo).

METHYL ACETYLENE. See allylene.

METHYL ACETYLENE-PROPADIENE MIXTURE
General Information
Synonyms: Methylacetylene propadiene, stabilized; MAPP.
Description: Clear, colorless, liquefied gas.
Constants: D (liquid): 0.576 at 15/15° C, boiling range: −39° to −20° C.
Hazard Analysis
Toxic Hazard Rating: U.
TLV: ACGIH (recommended); 1000 parts per million of air; 1800 milligrams per cubic meter of air.
Fire Hazard: Flammable when exposed to heat or flame.
Countermeasures
Storage and Handling: Section 7.
Shipping Regulations: Section 11.
  I.C.C.: Flammable gas, red gas label, 300 pounds.
  IATA: Flammable gas, red label, not acceptable (passenger), 140 kilograms (cargo).

METHYL ACETYLENE-PROPADIENE, STABILIZED.
  See methyl acetylene-propadiene mixture.

METHYLACETYL RICINOLEATE
General Information
Description: Crystals.
Formula: $CH_3CH_2CH_2CO_2(CH_2)_7CHCHCH_2CHOH-(CH_2)_5CH_3$.
Constants: Mol wt: 340.5, vap. d.: 11.9.
Hazard Analysis
Toxic Hazard Rating:
  Acute Local: Irritant 1; Ingestion 1; Inhalation 1.
  Acute Systemic: U.
  Chronic Local: U.
  Chronic Systemic: U.
Fire Hazard: Slight, when exposed to heat or flame; can react with oxidizing materials (Section 6).
Countermeasures
Personal Hygiene: Section 3.
Storage and Handling: Section 7.

α-METHYL ACROLEIN. See methacrolein.

β-METHYL ACROLEIN. See crotonaldehyde.

METHYL ACRYLATE
General Information
Synonym: Acrylic acid methyl ester.
Description: Colorless liquid.
Formula: $CH_2CHCO_2CH_3$.
Constants: Mol wt: 86.1, bp: 80° C, lel: 2.8%, uel: 25%, fp: −75° C, flash p: 27° F (O.C.), d: 0.949 at 25° C, vap. press.: 100 mm at 28° C, vap. d.: 2.97.
Hazard Analysis
Toxic Hazard Rating:
  Acute Local: Irritant 3; Ingestion 3; Inhalation 3.
  Acute Systemic: Ingestion 3.
  Chronic Local: Irritant 2; Inhalation 2.
  Chronic Systemic: Ingestion 3; Inhalation 3.
Toxicology: Chronic exposure has produced injury to lungs, liver and kidneys in experimental animals. It is a substance which migrates to food from packaging material (Section 10).
TLV: ACGIH (recommended) 10 parts per million in air; 35 milligrams per cubic meter of air. Can be absorbed through intact skin.
Fire Hazard: Dangerous, when exposed to heat or flame.
Explosion Hazard: Dangerous, when exposed to heat, sparks or flame.
Disaster Hazard: Dangerous; when heated to decomposition, it emits toxic fumes; can react vigorously with oxidizing materials.
Countermeasures
Ventilation Control: Section 2.
To Fight Fire: Foam, carbon dioxide, dry chemical or carbon tetrachloride (Section 6).
Personnel Protection: Section 3.
Storage and Handling: Section 7.

METHYL   ACRYLATE,   INHIBITED. See methyl acrylate.
Shipping Regulations: Section 11.
  IATA: Flammable liquid, red label, 1 liter (passenger), 40 liters (cargo).

METHYL   ACRYLATE,   UNINHIBITED. See methyl acrylate.
Shipping Regulations: Section 11.
  IATA: Flammable liquid, not acceptable (passenger and cargo).

METHYL ACRYLIC ACID. See methacrylic acid.

METHYLAL
General Information
Synonyms: Formal; methylene dimethyl ether; formaldehyde dimethylacetal.
Description: Colorless liquid; pungent odor.
Formula: $CH_2(OCH_3)_2$.
Constants: Mol wt: 76.1, mp: −104.8° C, bp: 42.3° C, flash p: 0° F (O.C.), d: 0.864 at 20°/4° C, vap. press.: 330 mm at 20° C, vap. d.: 2.63, autoign. temp.: 459° F.
Hazard Analysis
Toxic Hazard Rating:
  Acute Local: 0.
  Acute Systemic: Inhalation 2; Ingestion 2.
  Chronic Local: 0.
  Chronic Systemic: Ingestion 2; Inhalation 2.
Toxicology: Narcotic in high concentrations. Has produced injury to lungs, liver, kidneys and heart in experimental animals.
TLV: ACGIH (recommended); 1000 parts per million in air; 3100 milligrams per cubic meter of air.

TOXIC HAZARD RATING CODE (For detailed discussion, see Section 1.)

0 NONE: (a) No harm under any conditions; (b) Harmful only under unusual conditions or overwhelming dosage.

1 SLIGHT: Causes readily reversible changes which disappear after end of exposure.

2 MODERATE: May involve both irreversible and reversible changes; not severe enough to cause death or permanent injury.

3 HIGH: May cause death or permanent injury after very short exposure to small quantities.

U UNKNOWN: No information on humans considered valid by authors.

Caution: High concentrations of vapors act as an anesthetic.
Fire Hazard: Dangerous, when exposed to heat or flame.
Explosion Hazard: Moderate, when exposed to heat or flame.
Disaster Hazard: Dangerous, upon exposure to heat or flame; can react vigorously with oxidizing materials.
Countermeasures
Ventilation Control: Section 2.
To Fight Fire: Foam, carbon dioxide, dry chemical or carbon tetrachloride (Section 6).
Personnel Protection: Section 3.
Storage and Handling: Section 7.
Shipping Regulations: Section 11.
   IATA: Flammable liquid, red label, 1 liter (passenger), 40 liters (cargo).

METHYL ALCOHOL
General Information
Synonyms: Methanol.
Description: Clear colorless very mobile liquid.
Formula: CH_3OH.
Constants: Mol wt: 32.04, bp: 64.8°C, lel: 7.3%, uel: 36.5%, fp: −97.8°C, flash p: 52°F, d: 0.7913 at 20°/4°C, autoign. temp.: 867°F, vap. press.: 100 mm at 21.2°C, vap. d.: 1.11.
Hazard Analysis
Toxic Hazard Rating:
   Acute Local: Irritant 1; Inhalation 1.
   Acute Systemic: Ingestion 3; Inhalation 2; Skin Absorption 2.
   Chronic Local: Irritant 1; Inhalation 1.
   Chronic Systemic: Ingestion 2; Inhalation 2; Skin Absorption 2.
TLV: ACGIH (recommended); 200 parts per million in air; 262 milligrams per cubic meter of air.
Toxicology: Methyl alcohol possesses distinct narcotic properties. It is also a slight irritant to the mucous membranes. Its main toxic effect is exerted upon the nervous system, particularly the optic nerves and possibly the retinae. The effect upon the eyes has been attributed to optic neuritis, which subsides but is followed by atrophy of the optic nerve. Once absorbed, methyl alcohol is only very slowly eliminated. Coma resulting from massive exposures may last as long as 2 to 4 days. In the body the products formed by its oxidation are formaldehyde and formic acid, both of which are toxic. Because of the slowness with which it is eliminated, methyl alcohol should be regarded as a cumulative poison. Though single exposures to fumes may cause no harmful effect, daily exposure may result in the accumulation of sufficient methyl alcohol in the body to cause illness.
   Severe exposures may cause dizziness, unconsciousness, sighing respiration, cardiac depression, and eventually death. Where the exposure is less severe, the first symptoms may be blurring of vision, photophobia and conjunctivitis, followed by the development of definite eye lesions. There may be headache, gastrointestinal disturbances, dizziness and a feeling of intoxication. The visual symptoms may clear temporarily, only to recur later and progress to actual blindness. Irritation of the mucous membranes of the throat and respiratory tract, peripheral neuritis, and occasionally, symptoms referable to other lesions of the nervous system have been reported. The skin may become dry and cracked due to the solvent action of methyl alcohol.
   Methyl alcohol is a common air contaminant (Section 4). It is used as a food additive permitted in foods for human consumption. Section 10.
Fire Hazard: Dangerous, when exposed to heat or flame.
Spontaneous Heating: No.
Explosion Hazard: Moderate, when exposed to flame.
Disaster Hazard: Dangerous, upon exposure to heat or flame; can react vigorously with oxidizing materials.

Countermeasures
Ventilation Control: Section 2.
To Fight Fire: Carbon dioxide, dry chemical, or carbon tetrachloride (Section 6).
Personnel Protection: Section 3.
Storage and Handling: Section 7.
Shipping Regulations: Section 11.
   I.C.C.: Flammable liquid; red label, 10 gallons.
   Coast Guard Classification: Inflammable liquid; red label.
   MCA warning label.
   IATA: See Alcohol, N.O.S.

METHYL ALDEHYDE. See formaldehyde.

METHYLALLENE. See butadiene-1,3.

METHYL ALLYL CHLORIDE. See methallyl chloride.

METHYL ALUMINUM SESQUIBROMIDE
General Information
Description: Cloudy yellow liquid at 25°C.
Formula: $(CH_3)_3Al_2Br_3$.
Constants: Mol wt: 204, fp: −40°C, bp: 166°C, d: 1.514.
Hazard Analysis
Toxic Hazard Rating: U.
Fire Hazard: Dangerous, ignites spontaneously in air.
Disaster Hazard: Dangerous, spontaneously flammable. See also bromides.
Countermeasures
Storage and Handling: Section 7.
To Fight Fire: Do not use water, foam, or halogenated extinguishing agents.
Shipping Regulations: Section 11.
   I.C.C.: Flammable liquid, red label, 2 ounces.
   IATA: Flammable liquid, not acceptable (passenger and cargo).

METHYL ALUMINUM SESQUICHLORIDE
General Information
Description: Clear colorless liquid at 25°C.
Formula: $(CH_3)_3Al_2Cl_3$.
Constants: Mol wt: 203.5, fp: 22.80°C, bp: 143.7°C, d: 1.1629 at 25°C.
Hazard Analysis
Toxic Hazard Rating: U.
Fire Hazard: Dangerous, ignites spontaneously in air.
Disaster Hazard: Dangerous. Ignites spontaneously. See also chlorides.
Countermeasures
Storage and Handling: Section 7.
To Fight Fire: Do not use water, foam, or halogenated extinguishing agents.
Shipping Regulations: Section 11.
   I.C.C.: Flammable liquid, red label, 2 ounces.
   IATA: Flammable liquid, not acceptable (passenger and cargo).

METHYLAMINE (30% SOLUTION). See monomethylamine.

N-METHYLAMINO ACETIC ACID. See sarcosine.

METHYL-2-AMINOBENZOATE. See methyl anthranilate.

2-METHYL AMINOETHANOL
General Information
Description: Viscous liquid; fishy odor; miscible with water, alcohol, and ether.
Formula: $CH_3NHCH_2CH_2OH$.
Constants: Mol wt: 75, d: 0.9, vap. d.: 2.9, bp: 319°F, flash p: 165°F.
Hazard Analysis
Toxic Hazard Rating:
   Acute Local: Irritant 3; Ingestion 3; Inhalation 3.
   Acute Systemic: U.

Chronic Local: U.
Chronic Systemic: U.
Fire Hazard: Moderate, when exposed to heat or flame.

METHYL p-AMINOPHENOL
General Information
Description: Colorless needles.
Formula: $CH_3NHC_6H_4OH$.
Constants: Mol wt: 123.17, mp: 87°C.
Hazard Analysis
Toxic Hazard Rating:
Acute Local: Irritant 2; Ingestion 2; Inhalation 2.
Acute Systemic: U.
Chronic Local: U.
Chronic Systemic: U.
Disaster Hazard: Moderate; when heated, it emits toxic
fumes (Section 6).
Countermeasures
Ventilation Control: Section 2.
Personnel Protection: Section 3.
Storage and Handling: Section 7.

METHYL p-AMINOPHENOL SULFATE
General Information
Description: Colorless needles.
Formula: $(CH_3NHC_6H_4OH)_2 \cdot H_2SO_4$.
Constants: Mol wt: 344.38, mp: 127°C (decomp.).
Hazard Analysis
Toxic Hazard Rating:
Acute Local: Irritant 2; Ingestion 2; Inhalation 2.
Acute Systemic: U.
Chronic Local: Allergen 2.
Chronic Systemic: U.
Disaster Hazard: Dangerous; when heated to decomposi-
tion, it emits highly toxic fumes of oxides of sulfur.
Countermeasures
Ventilation Control: Section 2.
Personnel Protection: Section 3.
Storage and Handling: Section 7.

METHYL AMYLACETATE
General Information
Synonym: 4-Methyl pentyl 2-acetate; sec-hexyl acetate.
Description: Clear liquid; pleasant odor.
Formula: $(CH_3)_2CH(CH_2)_3OOCCH_3$.
Constants: Mol wt: 144.21, bp: 146.3°C, fp: −63.8°C,
flash p: 110°F (C.O.C.), d: 0.8598 at 20°/20°C,
vap. press.: 3.8 mm at 20°C, vap. d.: 4.97.
Hazard Analysis
Toxic Hazard Rating:
Acute Local: Irritant 1.
Acute Systemic: Ingestion 2; Inhalation 2.
Chronic Local: U.
Chronic Systemic: U.
TLV: ACGIH (recommended); 50 parts per million of air;
295 milligrams per cubic meters of air.
Fire Hazard: Moderate, when exposed to heat or flame; can
react with oxidizing materials.
Countermeasures
To Fight Fire: Foam, carbon dioxide, dry chemical or
carbon tetrachloride (Section 6).
Ventilation Control: Section 2.
Personal Hygiene: Section 3.
Storage and Handling: Section 7.
Shipping Regulations: Section 11.
MCA warning label.

METHYL AMYL ALCOHOL
General Information
Synonyms: Methyl isobutyl carbinol; 4-methyl pentanol-2.
Description: Clear liquid.
Formula: $(CH_3)_2CHCH_2CHOHCH_3$.
Constants: Mol wt: 102.2, bp: 131.8°C, fp: < −90°C
(sets to a glass), flash p: 106°F, d: 0.8079 at 20°/20°C,
vap. press.: 2.8 mm at 20°C, vap. d.: 3.53, lel: 1.0%,
uel: 5.5%.
Hazard Analysis
Toxic Hazard Rating:
Acute Local: Irritant 3; Ingestion 3; Inhalation 3.
Acute Systemic: Inhalation 3.
Chronic Local: Irritant 2.
Chronic Systemic: U.
Toxicology: A strong irritant. High concentrations cause
anesthesia.
TLV: ACGIH (recommended) 25 parts per million in air;
100 milligrams per cubic meter of air. May be absorbed
via the skin.
Fire Hazard: Moderate, when exposed to heat or flame;
can react with oxidizing materials.
Countermeasures
Ventilation Control: Secton 2.
Personnel Protection: Section 3.
Storage and Handling: Section 7.
Shipping Regulations: Section 11.
MCA warning label.

METHYL n-AMYL KETONE
Synonym: Heptanone-2.
Description: Water-white liquid.
Formula: $CH_3(CH_2)_4COCH_3$.
Constants: Mol wt: 114.18, bp: 150.6°C, flash p: 120°F
(O.C.), d: 0.8166 at 20°/20°C, vap. press.: 2.6 mm
at 20°C, autoign. temp.: 991°F.
Hazard Analysis
Toxic Hazard Rating:
Acute Local: Irritant 1; Inhalation 2.
Acute Systemic: Inhalation 2.
Chronic Local: Irritant 1.
Chronic Systemic: U.
Toxicology: Irritating to mucous membranes and narcotic
in high concentrations.
TLV: ACGIH (recommended); 100 parts per million of air;
465 milligrams per cubic meter of air.
Fire Hazard: Moderate, when exposed to heat or flame;
can react with oxidizing materials.
Countermeasures
To Fight Fire: Foam, carbon dioxide, dry chemical or
carbon tetrachloride (Section 6).
Ventilation Control: Section 2.
Personal Hygiene: Section 3.
Storage and Handling: Section 7.
Shipping Regulations: Section 11.
Coast Guard Classification: Combustible liquid.
MCA warning label.

N-METHYLANILINE
General Information
Synonym: Monomethylaniline.
Description: Reddish-brown, oily liquid.
Formula: $C_6H_5NHCH_3$.
Constants: Mol wt: 107.15, mp: −57.0°C, bp: 195.7°C,
d: 0.989 at 20°/4°C, vap. press.: 1 mm at 36.0°C,
vap. d.: 3.70.

TOXIC HAZARD RATING CODE (For detailed discussion, see Section 1.)

0 NONE: (a) No harm under any conditions; (b) Harmful only under unusual conditions or overwhelming dosage.

1 SLIGHT: Causes readily reversible changes which disappear after end
of exposure.

2 MODERATE: May involve both irreversible and reversible changes;
not severe enough to cause death or permanent injury.

3 HIGH: May cause death or permanent injury after very short exposure
to small quantities.

U UNKNOWN: No information on humans considered valid by authors.

Hazard Analysis
Toxic Hazard Rating:
Acute Local: Allergen 2.
Acute Systemic: Ingestion 3; Inhalation 3; Skin Absorption 3.
Chronic Local: Allergen 2.
Chronic Systemic: Ingestion 3; Inhalation 3; Skin Absorption 3.
Toxicity: See also aniline.
TLV: ACGIH (recommended) 2 parts per million of air; 9 milligrams per cubic meter of air. May be absorbed via intact skin.
Caution: Both the liquid and vapor can be absorbed through the skin. See also aniline.
Disaster Hazard: Dangerous; when heated to decomposition, it emits highly toxic fumes of aniline; can react with oxidizing materials.

Countermeasures
Ventilation Control: Section 2.
Personnel Protection: Section 3.
First Aid: Section 1.
Storage and Handling: Section 7.

o-METHYLANILINE. See o-toluidine.

p-METHYLANILINE. See p-toluidine.

METHYL ANTHRANILITE
General Information
Synonyms: Methyl-2-amino benzoate; neroli oil, artificial.
Description: Colorless to pale yellow liquid with light bluish fluorescence; grape type odor; soluble in 5 volumes or more of 60% alcohol; soluble in fixed oils, volatile oils, and propylene glycol; slightly soluble in water and mineral oil; insoluble in glycerin.
Formula: $H_2NC_6H_4CO_2CH_3$.
Constant: Mol wt: 151.
Hazard Analysis
Toxic Hazard Rating: U. Used as a synthetic flavoring substance and adjuvant (Section 10).

METHYLARSENIC DICHLORIDE. See dichloromethyl arsine.

METHYL ARSINE
General Information
Description: Colorless liquid. Insoluble in water, soluble in ether and alcohol.
Formula: CH_3AsH_2.
Constants: Mol wt: 92.0, bp: 2°C.
Hazard Analysis and Countermeasures
See arsenic compounds and arsine.

METHYL ARSONIC ACID. See methane arsonic acid.

METHYL BENZENE. See toluene.

METHYL BENZENE SULFONIC ACID. See o-toluene sulfonic acid.

METHYL BENZOATE
General Information
Synonym: Niobe oil.
Description: Colorless liquid; fragrant odor.
Formula: $C_6H_5COOCH_3$.
Constants: Mol wt: 136.14, mp: −12.5°C, bp: 199.6°C, flash p: 181°F, d: 1.0937, vap. press.: 1 mm at 39.0°C, vap. d.: 4.69.
Hazard Analysis
Toxic Hazard Rating:
Acute Local: Irritant 2; Ingestion 2.
Acute Systemic: Ingestion 3; Inhalation 3.
Chronic Local: U.
Chronic Systemic: U.
Fire Hazard: Moderate, when exposed to heat or flame; can react with oxidizing materials.

Countermeasures
To Fight Fire: Foam, carbon dioxide, dry chemical or carbon tetrachloride (Section 6).
Storage and Handling: Section 7.

METHYL BENZOYL ECGONINE. See cocaine.

α-METHYL BENZYL ALCOHOL. See phenyl methyl carbinol.

α-METHYL BENZYLAMINE
General Information
Description: Liquid.
Formula: $C_6H_5CH(CH_3)NH_2$.
Constants: Mol wt: 121.18, mp: −65°C, bp: 188.5°C, flash p: 175°F (O.C.), d: 0.9535, vap. press.: 0.5 mm at 20°C, vap. d.: 4.18.
Hazard Analysis
Toxicity: Details unknown. See also amines. Limited animal experiments suggest low toxicity and moderate irritation.
Fire Hazard: Moderate, when exposed to heat or flame; can react with oxidizing materials.
Countermeasures
To Fight Fire: Foam, carbon dioxide, dry chemical or carbon tetrachloride (Section 6).
Storage and Handling: Section 7.

α-METHYL BENZYL BENZOIC ACID
General Information
Synonym: Benzoic acid α-methyl benzyl ester.
Description: Crystals.
Formula: $CH_3C_6H_5CHC_6H_4COOH$.
Constant: Mol wt: 226.3.
Hazard Analysis
Toxicity: Details unknown; an insecticide. See also esters and benzoic acid.
Fire Hazard: Slight.
Countermeasures
Storage and Handling: Section 7.

o-METHYLBENZYL BROMIDE. See xylyl bromide.

α-METHYL BENZYL "CELLOSOLVE"
General Information
Description: Liquid.
Formula: $CH_3C_6H_4CH_2OCH_2CH_2OH$.
Constants: Mol wt: 166.21, bp: 253.5°C, fp: −50°C, d: 1.0395 at 20°/20°C, vap. press.: 0.02 mm at 20°C, vap. d.: 5.73.
Hazard Analysis
Toxicity: Details unknown. See also glycols.
Fire Hazard: Slight, when exposed to heat or flame; can react with oxidizing materials.
Countermeasures
To Fight Fire: Foam, carbon dioxide, dry chemical or carbon tetrachloride (Section 6).
Storage and Handling: Section 7.

o-METHYLBENZYL CHLORIDE. See o-xylyl chloride.

α-METHYLBENZYL DIETHANOLAMINE
General Information
Description: Liquid.
Formula: $C_6H_5CH(CH_3)N(C_2H_4OH)_2$.
Constants: Mol wt: 209.28, mp: −7°C, bp: 244°C at 50 mm, flash p: 370°F (O.C.), d: 1.0812 at 20°/20°C, vap. press.: < 0.01 mm at 20°C, vap. d.: 7.2.
Hazard Analysis
Toxicity: Details unknown. See also amines.
Fire Hazard: Slight, when exposed to heat or flame; can react with oxidizing materials.
Countermeasures
To Fight Fire: Foam, carbon dioxide, dry chemical or carbon tetrachloride (Section 6).
Storage and Handling: Section 7.

α-METHYLBENZYL DIMETHYLAMINE
General Information
Description: Liquid.
Formula: $C_6H_5CH(CH_3)N(CH_3)_2$.
Constants: Mol wt: 149.23, mp: $-75°C$, bp: $195°C$, flash p: $175°F$ (O.C.), d: 0.9044 at $20°/20°C$, vap. press.: 0.6 mm at $20°C$, vap. d.: 5.15.
Hazard Analysis
Toxicity: Details unknown. See also amines.
Fire Hazard: Moderate, when exposed to heat or flame; can react with oxidizing materials.
Countermeasures
To Fight Fire: Foam, carbon dioxide, dry chemical or carbon tetrachloride (Section 6).
Storage and Handling: Section 7.

METHYL BENZYL ETHER
General Information
Synonym: Bis-α-methyl benzyl ether.
Description: Water-white liquid.
Formula: $C_6H_5CH(CH_3)O(CH_3)CHC_6H_5$.
Constants: Mol wt: 226.3, mp: $< -30°C$ (sets to a glass), bp: $286.3°C$, flash p: $275°F$, d: 1.0017 at $20°/20°C$, vap. press.: < 0.01 mm at $20°C$, vap. d.: 7.8.
Hazard Analysis
Toxicity: Details unknown. See also ethers.
Fire Hazard: Slight, when exposed to heat or flame; can react with oxidizing materials.
Countermeasures
To Fight Fire: Alcohol foam, foam, carbon dioxide, dry chemical or carbon tetrachloride (Section 6).
Storage and Handling: Section 7.

α-METHYLBENZYL ETHER
General Information
Description: Liquid.
Formula: $C_6H_5CH(CH_3)OCH(CH_3)C_6H_5$.
Constants: Mol wt: 226.3, mp: $-30°C$, bp: $286.3°C$, flash p: $275°F$ (O.C.), d: 1.0017 at $20°/20°C$, vap. press.: < 0.01 mm at $20°C$, vap. d.: 7.82.
Hazard Analysis
Toxicity: Details unknown. See also ethers.
Fire Hazard: Slight, when exposed to heat or flame; can react with oxidizing materials.
Countermeasures
To Fight Fire: Foam, carbon dioxide, dry chemical or carbon tetrachloride (Section 6).
Storage and Handling: Section 7.

α-METHYLBENZYL MONOETHANOLAMINE
General Information
Description: Liquid.
Formula: $C_6H_5CH(CH_3)NHC_2H_4OH$.
Constants: Mol wt: 165.23, mp: $-15°C$, bp: $182°C$ at 50 mm, flash p: $280°F$ (O.C.), d: 1.0327 at $20°/20°C$, vap. press.: < 0.01 mm at $20°C$, vap. d.: 5.7.
Hazard Analysis
Toxicity: Details unknown. See also amines.
Fire Hazard: Slight, when exposed to heat or flame; can react with oxidizing materials.
Countermeasures
To Fight Fire: Foam, carbon dioxide, dry chemical or carbon tetrachloride (Section 6).
Storage and Handling: Section 7.

METHYL BISMUTHINE
General Information
Description: Liquid.

Formula: $CH_3 \cdot BiH_2$.
Constants: Mol wt: 226.05, bp: $110°C$, d: 2.30 at $18°C$, vap. d.: 7.8.
Hazard Analysis
Toxicity: See bismuth compounds.
Fire Hazard: Moderate (Section 6).
Disaster Hazard: Dangerous; when heated to decomposition or on contact with acid or acid fumes, it emits toxic fumes of bismuth; can react with oxidizing materials.
Countermeasures
Storage and Handling: Section 7.

METHYL BORATE
General Information
Synonyms: Trimethyl borate; trimethoxy borine.
Description: Colorless liquid, decomposes in water.
Formula: $B(OCH_3)_3$.
Constants: Mol wt: 103.92, mp: $-29°C$, bp: $68°C$, flash p: $< 80°F$, d: 0.92 at $20°C$, vap. d.: 3.59.
Hazard Analysis
Toxic Hazard Rating: U.
Toxicity: Limited animal experiments suggest low toxicity. See also esters and boron compounds.
Fire Hazard: Dangerous, when exposed to heat or flame (Section 6).
Explosion Hazard: Moderate, when exposed to flame.
Disaster Hazard: Dangerous; will react with water or steam to produce toxic and flammable vapors; can react vigorously with oxidizing materials.
Countermeasures
Ventilation Control: Section 2.
Storage and Handling: Section 7.

METHYLBORIC ACID
General Information
Description: White crystals; slightly water soluble.
Formula: $CH_3B(OH)_2$.
Constants: Mol wt: 59.9, mp: decomposes before melting.
Hazard Analysis
Toxicity: Details unknown. See also boron compounds.

METHYLBORIC ANHYDRIDE
General Information
Description: Colorless liquid; hydrolyzed by water; soluble in ether.
Formula: $(CH_3)_3B_3O_3$.
Constant: Mol wt: 125.6.
Hazard Analysis
Toxicity: Details unknown. See also boron compounds.

METHYLBORINE TRIMETHYL AMMINE
General Information
Description: Colorless liquid; decomposed by water.
Formula: $(CH_3)_3NBH_2CH_3$.
Constants: Mol wt: 87.0, mp: $0.8°C$, bp: $177°C$.
Hazard Analysis
Toxicity: Details unknown. See also boron compounds.

METHYL BROMIDE
General Information
Synonym: Bromomethane.
Description: Colorless, transparent, volatile liquid or gas; burning taste; chloroform-like odor.
Formula: CH_3Br.
Constants: Mol wt: 94.95, bp: $3.56°C$, lel: 10%, uel: 16%, fp: $-93°C$, flash p: none, d: 1.732 at $0°/0°C$, autoign. temp.: $998°F$, vap. d.: 3.27.

TOXIC HAZARD RATING CODE (For detailed discussion, see Section 1.)

0 NONE: (a) No harm under any conditions; (b) Harmful only under unusual conditions or overwhelming dosage.

1 SLIGHT: Causes readily reversible changes which disappear after end of exposure.

2 MODERATE: May involve both irreversible and reversible changes; not severe enough to cause death or permanent injury.

3 HIGH: May cause death or permanent injury after very short exposure to small quantities.

U UNKNOWN: No information on humans considered valid by authors.

Hazard Analysis
Toxic Hazard Rating:
Acute Local: Irritant 3; Ingestion 3; Inhalation 3.
Acute Systemic: Ingestion 3; Inhalation 3.
Chronic Local: U.
Chronic Systemic: Ingestion 2; Inhalation 2; Skin Absorption 2.
TLV: ACGIH (recommended); 20 parts per million in air; 78 milligrams per cubic meter of air. Can be absorbed through the skin.
Toxicology: An insecticide. Methyl bromide is reported to be 8 times more toxic on inhalation than ethyl bromide. Moreover, because of its greater volatility, methyl bromide is a much more frequent cause of poisoning. Death following acute poisoning is usually caused by its irritant effect on the lungs. In chronic poisoning, death is due to injury to the central nervous system. Fatal poisoning has always resulted from exposures to relatively high concentrations of methyl bromide vapors (from 8,600 to 60,000 ppm). Nonfatal poisoning has resulted from exposure to concentrations as low as 100 to 500 ppm. In addition to the lung and central nervous system injury mentioned, the kidneys may be damaged with development of albuminuria and, in fatal cases, cloudy swelling and/or tubular degeneration. The liver may be enlarged. There are no characteristic blood changes.

The onset of symptoms following the inhalation of methyl bromide vapor is usually delayed for 4 to 6 hours, though the latent period may vary from 2 to 48 hours. In fatal poisoning, the early symptoms are headache, visual disturbances, nausea and vomiting, smarting of the eyes, irritation of the skin, listlessness, vertigo and tremor. Progress is nearly always rapid, with the development of convulsions, fever, pulmonary edema, cyanosis, unconsciousness and death. Signs of involvement of the nervous system may be present before death. The clinical picture in nonfatal poisoning is extremely variable. Fatigue, blurred or double vision, nausea and vomiting are frequent; incoordination, tremors, convulsions, exaggeration of the patellar reflexes and a positive Babinski's sign may develop. Nearly every type of nervous disturbance has been reported. The pulmonary symptoms are comparatively slight. Recovery is frequently prolonged and there may be permanent injury, commonly characterized by sensory disturbances, weakness, disturbances of gait, irritability, and blurred vision. Locally, methyl bromide is an extreme irritant to the skin and may produce severe burns. (Section 9).
Fire Hazard:Moderate, when exposed to heat or flame.
Spontaneous Heating: No.
Explosion Hazard: Moderate, when exposed to sparks or flame.
Disaster Hazard: Dangerous; when heated to decomposition, it emits highly toxic fumes of bromides.
Countermeasures
Ventilation Control: Section 2.
To Fight Fire: Water, foam, carbon dioxide, dry chemical or carbon tetrachloride (Section 6).
Personnel Protection: Section 3.
Storage and Handling: Section 7.
Shipping Regulations: Section 11.
I.C.C. (liquid): Poison B; poison label, 55 gallons.
Coast Guard Classification (liquid): Poison B; poison label.
MCA warning label.
IATA (liquid): Poison B, poison label, not acceptable (passenger), 220 liters (cargo).

METHYL BROMIDE AND CHLOROPICRIN MIXTURE, LIQUID
Shipping Regulations: Section 11.
I.C.C.: Poison B, poison label, 55 gallons.

METHYL BROMIDE AND ETHYLENE DIBROMIDE MIXTURE, LIQUID
Shipping Regulations: Section 11.
I.C.C.: Poison B, poison label, 55 gallons.
IATA: Poison B, poison label, not acceptable (passenger), 220 liters (cargo).

METHYL BROMIDE AND NONFLAMMABLE, NONLIQUIFIED COMPRESSED GAS MIXTURES, LIQUID
Shipping Regulations: Section 11.
I.C.C.: Poison B, poison label, 300 pounds.
IATA: Poison B, poison label, not acceptable (passenger), 140 kilograms (cargo).

2-METHYL BUTADIENE-1,3. See isoprene.

3-METHYLBUTANAL. See isovaleric aldehyde.

2-METHYLBUTANE. See isopentane.

2-METHYL-1-BUTANOL
General Information
Synonyms: Amyl alcohol, primary, active; sec-butyl carbinol.
Description: Colorless liquid; slightly soluble in water; miscible with alcohol and ether.
Formula: $CH_3CH_2CH(CH_3)CH_2OH$.
Constants: Mol wt: 88, d: 0.81–0.82 at 20°C, fp: $< -70°C$, bp: 128°C, flash p: 122°F (O.C.), vap. d.: 3.0.
Hazard Analysis
Toxic Hazard Rating: U.
Fire Hazard: Moderate, when exposed to heat or flame.
Countermeasures
Storage and Handling: Section 7.
To Fight Fire: Alcohol foam.

2-METHYL-2-BUTANOL. See tert-amyl alcohol, refined.

METHYL BUTANONE. See butanone.

METHYLBUTENE. See amylene.

2-METHYL BUTENE-1 (TECHNICAL GRADE).
General Information
Description: Colorless extremely volatile liquid or gas, insoluble in water.
Formula: $CH_2:C(CH_3)CH_2CH_3$.
Constants: Mol wt: 70, d: 0.7, vap. d.: 2.4, bp: 88°F, flash p: $< 20°F$.
Hazard Analysis
Toxic Hazard Rating: U.
Fire Hazard: Dangerous, when exposed to heat or flame.
Countermeasures
Storage and Handling: Section 7.

3-METHYL BUTENE-1
General Information
Synonyms: Isopropyl ethylene; isoamylene.
Description: Colorless, very volatile flammable liquid; disagreeable odor; insoluble in water; soluble in alcohol.
Formula: $(CH_3)_2CHCH:CH_2$.
Constants: Mol wt: 70, bp: 31.11°C, d: 0.65 at 20/20°C, fp: $-137.52°C$, flash p: $< 20°F$, vap. d.: 2.4.
Hazard Analysis
Toxic Hazard Rating: U.
Fire Hazard: Dangerous, when exposed to heat or flame.
Countermeasures
Storage and Handling: Section 7.
To Fight Fire: Alcohol foam.

2-METHYL BUTENE-2
General Information
Synonyms: 3-Methyl butene-2; trimethylethylene; β-isoamylene.
Description: Colorless, volatile, flammable liquid, disagreeable odor, soluble in alcohol, insoluble in water.

Formula: $(CH_3)_2C:CHCH_3$.
Constants: Mol wt: 70, bp: 38.51°C, d: 0.6623 at 20/4°C, fp: 133.83°C, flash p: <20°F, vap. d.: 2.4.

Hazard Analysis
Toxic Hazard Rating: U.
Fire Hazard: Dangerous, when exposed to heat or flame.

Countermeasures
Storage and Handling: Section 7.
To Fight Fire: Alcohol foam.

3-METHYL BUTENE-2. see 2-methyl butene-2.

3-METHYL BUTENONITRILE
Hazard Analysis
Toxicity: Details unknown. Limited animal experiments suggest high toxicity. See also nitriles.

n-METHYL BUTYLAMINE
General Information
Description: Liquid, soluble in water.
Formula: $CH_3(CH_2)_3NHCH_3$.
Constants: Mol wt: 87, d: 0.7335, bp: 91.1°C, vap. d.: 3.0, flash p: 55°F (O.C.).

Hazard Analysis
Toxicity: U. Limited animal experiments suggest high toxicity. See also amines.
Fire Hazard: Dangerous, when exposed to heat or flame.

Countermeasures
Storage and Handling: Section 7.
To Fight Fire: Alcohol foam.

METHYL n-BUTYL KETONE
General Information
Synonyms: 2-Hexanone; n-butyl methyl ketone.
Description: Clear liquid.
Formula: $CH_3OC(CH_2)_3CH_3$.
Constants: Mol wt: 100.16, mp: −56.9°C, bp: 127.2°C, lel: 1.22%, uel: 8.0%, flash p: 95°F (O.C.), d: 0.830 at 0°/4°C, vap. press.: 10 mm at 38.8°C, vap. d.: 3.45, autoign. temp.: 991°F.

Hazard Analysis
Toxic Hazard Rating:
  Acute Local: Ingestion 3; Inhalation 3.
  Acute Systemic: Ingestion 3; Inhalation 3.
  Chronic Local: U.
  Chronic Systemic: Inhalation 2.
TLV: ACGIH (recommended); 100 parts per million in air; 409 milligrams per cubic meter of air.
Fire Hazard: Moderate, when exposed to heat or flame; can react with oxidizing materials.
Spontaneous Heating: No.
Explosion Hazard: Moderate, when exposed to heat or flame.
Disaster Hazard: Dangerous fire and explosion hazard.

Countermeasures
Ventilation Control: Section 2.
To Fight Fire: Foam, carbon dioxide, dry chemical or carbon tetrachloride (Section 6).
Personnel Protection: Section 3.
Storage and Handling: Section 7.

METHYL BUTYNOL
General Information
Synonym: 2-Methyl-3-butyn-2-ol.
Description: Colorless liquid.
Formula: $(CH_3)_2OHCCH$.
Constants: Mol wt: 72.1, mp: 2.6°C, bp: 104–105°C, vap. d.: 2.49, d: 0.9, flash p.: 77°F (O.C.).

Hazard Analysis
Toxicity: Details unknown; probably moderately toxic.
Fire Hazard: Slight; can react with oxidizing materials.

Countermeasures
Storage and Handling: Section 7.
To Fight Fire: Alcohol foam.

2-METHYL-3-BUTYN-2-OL. See methyl butynol.

METHYL BUTYRATE
General Information
Description: Liquid.
Formula: $CH_3COOC_3H_7$.
Constants: Mol wt: 102.13, mp: <−97°C, bp: 102.3°C, flash p: 57°F (C.C.), d: 0.898, vap. press.: 40 mm at 29.6°C, vap. d.: 3.53.

Hazard Analysis
Toxic Hazard Rating:
  Acute Local: Irritant 1.
  Acute Systemic: Ingestion 2; Inhalation 2.
  Chronic Local: U.
  Chronic Systemic: U.
Fire Hazard: Dangerous, when exposed to heat or flame.
Spontaneous Heating: No.
Explosion Hazard: Unknown.
Disaster Hazard: Dangerous, upon exposure to heat or flame; can react vigorously with oxidizing materials.

Countermeasures
Ventilation Control: Section 2.
To Fight Fire: Foam, carbon dioxide, dry chemical or carbon tetrachloride (Section 6).
Personnel Protection: Section 3.
Storage and Handling: Section 7.
Shipping Regulations: Section 11.
  IATA: Flammable liquid, red label, 1 liter (passenger), 40 liters (cargo).

METHYL α-CAPROLACTONE
Hazard Analysis
Toxicity: Details unknown. Limited animal experiments suggest low toxicity. See also lactones.

METHYL CARBINOL. See ethyl alcohol.

METHYL "CARBITOL"
General Information
Synonym: Diethylene glycol methyl ether; 2-(2-methoxy-ethoxy)ethanol.
Description: Hygroscopic, water-white liquid.
Formula: $CH_3OCH_2CH_2OCH_2CH_2OH$.
Constants: Mol wt: 120.15, bp: 194.2°C, flash p: 200°F (O.C.), d: 1.0354 at 20°/4°C, vap. press.: 0.2 mm at 20°C, vap. d.: 4.14.

Hazard Analysis
Toxic Hazard Rating:
  Acute Local: Irritant 1.
  Acute Systemic: Ingestion 3; Inhalation 1.
  Chronic Local: 0.
  Chronic Systemic: Ingestion 2; Inhalation 2.
Toxicology: See also glycols.
Fire Hazard: Moderate, when exposed to heat or flame can react with oxidizing materials.

Countermeasures
To Fight Fire: Carbon dioxide, dry chemical or carbon tetrachloride (Section 6).
Ventilation Control: Section 2.
Personal Hygiene: Section 3.
Storage and Handling: Section 7.

TOXIC HAZARD RATING CODE *(For detailed discussion, see Section 1.)*

0 NONE: (a) No harm under any conditions; (b) Harmful only under unusual conditions or overwhelming dosage.

1 SLIGHT: Causes readily reversible changes which disappear after end of exposure.

2 MODERATE: May involve both irreversible and reversible changes; not severe enough to cause death or permanent injury.

3 HIGH: May cause death or permanent injury after very short exposure to small quantities.

U UNKNOWN: No information on humans considered valid by authors.

METHYL "CARBITOL" ACETATE
General Information
Synonym: Diethylene glycol monomethyl ether acetate.
Description: Colorless liquid.
Formula: $CH_3COOC_2H_4OC_2H_4OCH_3$.
Constants: Mol wt: 162.19, bp: 209.1°C, flash p: 180°F
   (O.C.), d: 1.0396 at 20°/20°C, vap. press.: 0.12 mm
   at 20°C.
Hazard Analysis
Toxic Hazard Rating:
   Acute Local: Irritant 1.
   Acute Systemic: Ingestion 3; Inhalation 1.
   Chronic Local: U.
   Chronic Systemic: Ingestion 2; Inhalation 2.
Fire Hazard: Moderate, when exposed to heat or flame; can
   react with oxidizing materials.
Spontaneous Heating: No.
Countermeasures
To Fight Fire: Foam, carbon dioxide, dry chemical or
   carbon tetrachloride (Section 6).
Ventilation Control: Section 2.
Personal Hygiene: Section 3.
Storage and Handling: Section 7.

METHYL "CARBITOL" FORMAL
General Information
Description: Liquid.
Formula: $CH_2(CH_3OCH_2CH_2OCH_2CH_2O)_2$.
Constants: Mol wt: 252.3, mp: −37.4°C, bp: 305°C,
   flash p: 310°F, d: 1.040, vap. d.: 8.7.
Hazard Analysis
Toxicity: Details unknown. See also glycols.
Fire Hazard: Slight, when exposed to heat or flame; can
   react with oxidizing materials.
Countermeasures
To Fight Fire: Foam, carbon dioxide, dry chemical or
   carbon tetrachloride (Section 6).
Storage and Handling: Section 7.

METHYL CARBONATE. See dimethyl carbonate.

METHYL CATECHOL. See guaiacol.

METHYL "CELLOSOLVE." See ethylene glycol monomethyl ether.

METHYL "CELLOSOLVE" ACETAL. See ethylene
   glycol monomethyl ether acetal.

METHYL "CELLOSOLVE" ACETATE
General Information
Synonym: Ethylene glycol monomethyl ether acetate.
Description: Colorless liquid.
Formula: $CH_3COOCH_2CH_2OCH_3$.
Constants: Mol wt: 118.13, bp: 143°C fp: −70°C,
   flash p: 132°F (C.C.), d: 1.005 at 20°/20°C, vap. d.:
   4.07.
Hazard Analysis
Toxic Hazard Rating:
   Acute Local: Irritant 1.
   Acute Systemic: Ingestion 3; Inhalation 1.
   Chronic Local: U.
   Chronic Systemic: Ingestion 2; Inhalation 2.
TLV: ACGIH (recommended); 25 parts per million in air;
   121 milligrams per cubic meter of air. May be absorbed
   through the skin.
Fire Hazard: Moderate, when exposed to heat or flame;
   can react with oxidizing materials.
Spontaneous Heating: No.
Countermeasures
To Fight Fire: Carbon dioxide, dry chemical or carbon
   tetrachloride (Section 6).
Ventilation Control: Section 2.
Personal Hygiene: Section 3.
Storage and Handling: Section 7.

METHYL "CELLOSOLVE" ACETYLRICINOLEATE
General Information
Description: White crystals.
Formula: $CH_3(CH_2)_7CHOHCH_2CHCH(CH_2)_7CO_2$-
   CH_2CH_2OH.
Constants: Mol wt: 370.6, vap. d.: 12.8.
Hazard Analysis
Toxicity: Details unknown. See also glycols and "Cello-
   solve."
Fire Hazard: Slight, when exposed to heat or flame; can
   react with oxidizing materials (Section 6).
Countermeasures
Storage and Handling: Section 7.

METHYL CELLULOSE
General Information
Synonym: Cellulose methyl ether.
Description: Grayish white, fibrous powder; insoluble in
   alcohol, ether and chloroform; soluble in glacial acetic
   acid.
Note: Molecular weights vary from 40,000—180,000. Here
   we consider U.S.P. Methyl cellulose except that the
   methoxy content shall be not less than 27.5% and not
   > 31.5% on a dry weight basis.
Hazard Analysis
Toxic Hazard Rating: U. Used as a general purpose food
   additive (Section 10).

**METHYL   CELLULOSE,   PROPYLENE   GLYCOL
ETHER.** See hydroxypropyl methylcellulose.

β-METHYL CHALCONE. See dypnone.

METHYL CHLORIDE
General Information
Synonyms: Chloromethane.
Description: Colorless gas.
Formula: CH_3Cl.
Constants: Mol wt: 50.49, bp: −23.7°C, lel: 10.7% uel:
   17.2%, fp: −97.7°C, flash p: < 32°F (O.C.), d: 0.918
   at 20°/4°C, autoign. temp.: 1170°F, vap. d.: 1.78.
Hazard Analysis
Toxic Hazard Rating:
   Acute Local: Irritant 1; Inhalation 1.
   Acute Systemic: Ingestion 3; Inhalation 3.
   Chronic Local: U.
   Chronic Systemic: Ingestion 2; Inhalation 2.
TLV: ACGIH (recommended); 100 parts per million in air;
   209 milligrams per cubic meter of air.
Toxicology: Methyl chloride has very slight irritant properties and may be inhaled without noticeable discomfort.
   It has some narcotic action, but this effect is weaker
   than that of chloroform. Acute poisoning, characterized by the narcotic effect, is rare in industry. Repeated exposure to low concentrations causes damage
   to the central nervous system, and, less frequently, to
   the liver, kidneys, bone marrow and cardiovascular
   system. Hemorrhages into the lungs, intestinal tract
   and dura have been reported. Sprayed on the skin,
   methyl chloride produces anesthesia through freezing
   of the tissue as it evaporates.
      In exposures to high concentrations, dizziness,
   drowsiness, incoordination, confusion, nausea and
   vomiting, abdominal pains, hiccoughs, diplopia and
   dimness of vision are followed by delirium, convulsions
   and coma. Death may be immediate, but if the exposure is not fatal, recovery is usually slow, and degenerative changes in the central nervous system are not uncommon. The liver, kidneys, and bone marrow may be
   affected, with resulting acute nephritis and anemia.
   Death may occur several days after exposure, resulting
   from degenerative changes in the heart, liver and
   especially the kidneys. In repeated exposures to lower
   concentrations there is usually fatigue, loss of appetite,
   muscular weakness, drowsiness, and dimness of vision.

After effects are commonly the result of damage to the central nervous system, with visual changes and attacks of depression and other psychic disturbances being reported.

Note: Used as a food additive permitted in food for human consumption.

Fire Hazard: Dangerous, when exposed to heat or flame.

Spontaneous Heating: No.

Explosion Hazard: Moderate, when exposed to heat or flame.

Disaster Hazard: Dangerous; when heated to decomposition, it emits highly toxic fumes of chlorides; can react vigorously with oxidizing materials.

Countermeasures

Ventilation Control: Section 2.

To Fight Fire: Stop flow of gas, carbon dioxide, dry chemical or water spray (Section 6).

Personal Hygiene: Section 3.

Storage and Handling: Section 7.

Shipping Regulations: Section 11.

   I.C.C.: Flammable gas label, 300 pounds.

   Coast Guard Classification: Inflammable gas; red gas label.

   MCA warning label.

   IATA: Flammable gas, red label, not acceptable (passenger), 140 kilograms (cargo).

METHYL CHLORIDE—METHYLENE CHLORIDE MIXTURE

Shipping Regulations: Section 11.

   I.C.C.: Flammable gas, red gas label, 300 pounds.

   IATA: Flammable gas, red label, not acceptable (passenger), 140 kilograms (cargo).

METHYL CHLOROACETATE

General Information

Synonym: Methyl chloroethanoate.

Description: Colorless liquid; sweet pungent odor; slightly soluble in water; miscible with alcohol and ether.

Formula: $CH_2ClCOOCH_3$.

Constants: Mol wt: 108.5, d: 1.236 at 20/4°C, mp: −32.7°C, bp: 131°C, vap. d.: 3.8, flash p: 122°F (O.C.).

Hazard Analysis

Toxic Hazard Rating: U.

Fire Hazard: Moderate, when exposed to heat or flame.

METHYL CHLOROBENZENE. See monochlorotoluene.

METHYL CHLOROCARBONATE. See methyl chloroformate.

METHYL CHLOROETHANOATE. See methyl chloroacetate.

METHYL CHLOROFORM. See α-trichloroethane.

METHYL CHLOROFORMATE

General Information

Synonyms: Methyl chloromethanoate; methyl chlorocarbonate.

Description: Colorless liquid.

Formula: $ClCOOCH_3$.

Constants: Mol wt: 94.50, bp: 71.4°C, d: 1.223 at 20°/4°C, vap. d.: 3.26.

Hazard Analysis

Toxic Hazard Rating:

   Acute Local: Irritant 3; Ingestion 3; Inhalation 3.

   Acute Systemic: U.

   Chronic Local: U.

   Chronic Systemic: U.

Disaster Hazard: Dangerous; when heated to decomposition, it emits highly toxic fumes of methyl chloroformate and phosgene; will react with water or steam to produce toxic and corrosive fumes.

Countermeasures

Ventilation Control: Section 2.

Personnel Protection: Section 3.

Storage and Handling: Section 7.

Shipping Regulations: Section 11.

   I.C.C.: Corrosive liquid; white label, 5 pints.

   Coast Guard Classification: Corrosive liquid; white label.

   IATA: Corrosive liquid, white label, not acceptable (passenger), $2\frac{1}{2}$ liters (cargo).

METHYL CHLOROMETHANOATE. See methyl chloroformate.

METHYL CHLOROMETHOXY STEARATE

General Information

Description: Straw-yellow liquid.

Formula: $C_{20}H_{39}O_3Cl$.

Constants: Mol wt: 354, fp: 8.0°C, flash p: 266°F (O.C.), d: 0.980 at 20°/40°C, vap. d.: 12.2.

Hazard Analysis

Toxicity: Details unknown. See also esters.

Fire Hazard: Slight, when exposed to heat or flame.

Disaster Hazard: Dangerous; when heated to decomposition, it emits highly toxic fumes of chlorides; can react with oxidizing materials.

Countermeasures

To Fight Fire: Foam, carbon dioxide, dry chemical or carbon tetrachloride (Section 6).

Storage and Handling: Section 7.

METHYL CHLOROMETHYL ETHER—ANHYDROUS

General Information

Description: Clear, colorless, liquid; decomposes in water; soluble in alcohol, ether.

Formula: $ClCH_2OCH_3$.

Constants: Mol wt: 76.5, d: 1.0625 at 10/4°C, mp: 103.5°C, bp: 59.5°C.

Hazard Analysis

Toxic Hazard Rating: U.

Fire Hazard: Moderate. See also ether.

Countermeasures

Storage and Handling: Section 7.

Shipping Regulations: Section 11.

   I.C.C.: Flammable liquid, red label, not accepted.

   IATA: Flammable liquid, not acceptable (passenger and cargo).

METHYL CHLOROPHENOXYACETIC ACID

General Information

Synonym: MCPA.

Formula: $ClC_6H_4OCH_2COOH$.

Constant: Mol wt: 186.6.

Hazard Analysis

Toxicity: Moderately toxic. Details unknown. Resembles 2,4-D. See 2,4-dichlorophenoxyacetic acid.

Disaster Hazard: Dangerous; when heated to decomposition it emits highly toxic fumes of chlorides.

TOXIC HAZARD RATING CODE *(For detailed discussion, see Section 1.)*

0 NONE: (a) No harm under any conditions; (b) Harmful only under unusual conditions or overwhelming dosage.

1 SLIGHT: Causes readily reversible changes which disappear after end of exposure.

2 MODERATE: May involve both irreversible and reversible changes; not severe enough to cause death or permanent injury.

3 HIGH: May cause death or permanent injury after very short exposure to small quantities.

U UNKNOWN: No information on humans considered valid by authors.

METHYL CHLOROSULFONATE
General Information
Description: Colorless liquid; pungent odor.
Formula: CH_3OSO_2Cl.
Constants: Mol wt: 130.55, mp: $-70°C$, bp: $135°C$ decomposes, d: 1.492 at $10°C$, vap. d.: 4.51.
Hazard Analysis
Toxic Hazard Rating:
　Acute Local: Irritant 3; Ingestion 3; Inhalation 3.
　Acute Systemic: U.
　Chronic Local: U.
　Chronic Systemic: U.
Toxicology: Used as a military poison gas.
Disaster Hazard: Dangerous; when heated to decomposition, it emits highly toxic fumes of chlorides; will react with water, steam or acids to produce toxic and corrosive fumes.
Countermeasures
Ventilation Control: Section 2.
Personnel Protection: Section 3.
Storage and Handling: Section 7.

3-METHYLCHOLANTHRENE
General Information
Description: Yellow crystals.
Formula: $C_{21}H_{16}$.
Constants: Mol wt: 268.3, bp: $280°C$ at 80 mm, mp: $180°C$, d: 1.28 at $20°C$.
Hazard Analysis
Toxic Hazard Rating:
　Acute Local: Irritant 2.
　Acute Systemic: U.
　Chronic Local: Irritant 3.
　Chronic Systemic: U.
Caution: Thought to be a carcinogenic agent on repeated local application.
Fire Hazard: Slight; can react with oxidizing materials (Section 6).
Countermeasures
Personnel Protection: Section 3.
First Aid: Section 1.
Storage and Handling: Section 7.

METHYL CYANIDE
General Information
Synonyms: Ethanenitrile; acetonitrile.
Description: Colorless liquid; aromatic odor.
Formula: CH_3CN.
Constants: Mol wt: 41.05, mp: $-41°C$, bp: $80.1°C$, flash p: $42°F$ (C.O.C.), d: 0.7868 at $20°/20°C$, vap. d.: 1.42, vap. press.: 100 mm at $27°C$.
Hazard Analysis
Toxicity: Highly toxic. See cyanides.
TLV: ACGIH (recommended); 40 parts per million in air; 70 milligrams per cubic meter of air.
Fire Hazard: Dangerous, when exposed to heat or flame.
Explosion Hazard: See cyanides.
Disaster Hazard: Dangerous; when heated to decomposition, it emits highly toxic fumes of cyanides; will react with water, steam or acids to produce toxic and flammable vapors; can react vigorously with oxidizing materials.
Countermeasures
Ventilation Control: Section 2.
To Fight Fire: Foam, carbon dioxide, dry chemical or carbon tetrachloride (Section 6).
Personnel Protection: Section 3.
First Aid: Section 1.
Storage and Handling: Section 7.
Shipping Regulations: Section 11.
　I.C.C.: Flammable liquid, red label, 10 gallons.
　IATA: Flammable liquid, red label, 1 liter (cargo), 40 liters (passenger).
　MCA warning label.

METHYL CYANOACETATE
General Information
Description: Liquid.
Formula: $CNCH_2COOCH_3$.
Constants: Mol wt: 99.09, mp: $-22.5°C$, bp: $203°C$, vap. d.: 3.41.
Hazard Analysis and Countermeasures
See cyanides.

METHYL CYANOETHANOATE. See cyanomethyl acetate.

METHYL CYANOFORMATE
General Information
Description: Colorless liquid; ethereal odor.
Formula: $COOCH_3CN$.
Constants: Mol wt: 77.06, bp: $100°C$, d: approx 1.00 at $20°C$, vap. d.: 2.66.
Hazard Analysis and Countermeasures
See cyanides.

METHYLCYCLOHEXANE
General Information
Synonyms: Hexahydrotoluene; cyclohexylmethane.
Description: Colorless liquid.
Formula: $CH_3C_6H_{11}$.
Constants: Mol wt: 98.18, mp: $-126.4°C$, lel: 1.2%, bp: $100.3°C$, flash p: $25°F$ (C.C.), d: 0.7864 at $0°/4°C$, 0.769 at $20°/4°C$, vap. press.: 40 mm at $22.0°C$, vap. d.: 3.39, autoign. temp.: $545°F$.
Toxic Hazard Rating:
　Acute Local: U.
　Acute Systemic: Inhalation 2.
　Chronic Local: U.
　Chronic Systemic: Inhalation 1.
TLV: ACGIH (recommended); 500 parts per million in air; 2000 milligrams per cubic meter of air.
Toxicology: The minimum lethal dose for rabbits via the oral route is from 4.0 to 4.5 g/kg of body weight; by inhalation, 15,000 ppm caused the death of rabbits in about 70 minutes. As to chronic effects, these are also present, since all rabbits exposed to 10,000 ppm concentrations for 6 hours a day, 5 days a week, for 2 weeks died. Not enough data are available in the literature to decide what the limits should be for human beings. However, until these limits have been agreed upon, the allowable concentration given above seems safe enough. This material does not cause irritation to the eyes and nose, and even at the level of 500 ppm, exhibits only a very faint odor. Therefore, it cannot be said to have any warning properties. It is believed to be about 3 times as toxic as hexane and can cause death by tetanic spasm as has been noted in animals. In sub-lethal concentrations it causes narcosis and anesthesia.
Fire Hazard: Dangerous, when exposed to heat or flame.
Spontaneous Heating: No.
Explosion Hazard: Moderate! when exposed to heat or flame.
Disaster Control: Dangerous; upon exposure to heat or flame; can react vigorously with oxidizing materials.
Countermeasures
Ventilation Control: Section 2.
To Fight Fire: Foam, carbon dioxide, dry chemical or carbon tetrachloride (Section 6).
Storage and Handling: Section 7.
Shipping Regulations: Section 11.
　IATA: Flammable liquid, red label, 1 liter (passenger), 40 liters (cargo).

METHYL CYCLOHEXANOL
General Information
Synonyms: Hexahydromethyl phenol; hexahydrocresol.
Description: Colorless, viscous liquid; aromatic, menthol-like odor.
Formula: $CH_3C_6H_{10}OH$.

Constants: Mol wt: 114.1, bp: 155–180°C, flash p: 154°F (C.C.), d: 0.924 at 15.5°/15.5°C, vap. d.: 3.93.

Hazard Analysis

Toxic Hazard Rating:

Acute Local: U.

Acute Systemic: Ingestion 3; Inhalation 3; Skin Absorption 3.

Chronic Local: U.

Chronic Systemic: Ingestion 1; Inhalation 1; Skin Absorption 1.

TLV: ACGIH (recommended); 100 parts per million in air; 466 milligrams per cubic meter of air.

Fire Hazard: Moderate, when exposed to heat or flame.

Spontaneous Heating: No.

Disaster Hazard: Dangerous; on heating it emits highly toxic fumes; can react with oxidizing materials.

Countermeasures

Ventilation Control: Section 2.

Personnel Protection: Section 3.

To Fight Fire: Foam, carbon dioxide, dry chemical or carbon tetrachloride (Section 6).

Storage and Handling: Section 7.

Shipping Regulations: Section 11.

MCA warning label.

METHYLCYCLOHEXANOL ACETATE. See methylcyclohexyl acetate.

METHYL CYCLOHEXANONE

General Information

Description: Water-white to pale yellow liquid; acetone-like odor.

Formula: $COCH(CH_3)CH_2CH_2CH_2CH_2$.

Constants: Mol wt: 112.17, bp: 160–170°C, flash p: 118°F (C.C.), d: 0.925 at 15°/5°C, vap. d.: 3.86.

Hazard Analysis

Toxic Hazard Rating:

Acute Local: Irritant 2.

Acute Systemic: Ingestion 3; Inhalation 3; Skin Absorption 3.

Chronic Local: U.

Chronic Systemic: Ingestion 1; Inhalation 1; Skin Absorption 2.

TLV: ACGIH (recommended); 100 parts per million in air; 458 milligrams per cubic meter of air. May be absorbed through the skin.

Toxicology: This is a toxic compound which can damge the kidney and the liver. It is similar to cyclohexanol in its toxic action, although it is somewhat less active. Harmful exposures in industry are rare. From experiments on rabbits, it has been found that they could withstand prolonged exposures to concentrations of from 0.02 to 0.05% by volume in air.

Fire Hazard: Moderate, when exposed to heat or flame.

Spontaneous Heating: No.

Countermeasures

Ventilation Control: Section 3.

To Fight Fire: Foam, carbon dioxide, dry chemical or carbon tetrachloride (Section 6).

Personnel Protection: Section 3.

Storage and Handling: Section 7.

Shipping Regulations: Section 11.

MCA warning label.

4-METHYLCYCLOHEXENE-1

General Information

Description: A clear liquid.

Formula: C_7H_{12}.

Constants: Mol wt: 96.17, bp: 102.5°C, flash p: 30°F (T.O.C.), d: 0.804 at 15.5°/15.5°C, vap. press.: 10.3 mm at 38°C, vap. d.: 3.34.

Hazard Analysis

Toxicity: Unknown. Probably an irritant and narcotic in high concentrations.

Fire Hazard: Dangerous, when exposed to heat or flame; can react vigorously with oxidizing materials.

Countermeasures

Ventilation Control: Section 2.

To Fight Fire: Foam, carbon dioxide, dry chemical or carbon tetrachloride (Section 6).

Storage and Handling: Section 7.

2-METHYL-4-CYCLOHEXENE-1-CARBOXALDEHYDE

General Information

Formula: $CHCHOCH_3CHCH_2CH:CHCH_2$.

Constant: Mol wt: 124.

Hazard Analysis

Toxicity: Details unknown. Limited animal experiments suggest low toxicity and moderate irritation. See also aldehydes.

METHYL CYCLOHEXYL ACETATE

General Information

Synonym: Methyl cyclohexanol acetate.

Description: Liquid.

Formula: $C_9H_{16}O_2$.

Constants: Mol wt: 156.22, flash p: 147°F (C.C.), vap. d.: 5.37, d: 0.9.

Hazard Analysis

Toxicity: See esters.

Fire Hazard: Moderate, when exposed to heat or flame; can react with oxidizing materials.

Spontaneous Heating: No.

Countermeasures

To Fight Fire: Foam, carbon dioxide, dry chemical or carbon tetrachloride (Section 6).

Ventilatinon Control: Section 2.

Storage and Handling: Section 7.

METHYLCYCLOHEXYL FORMATE

General Information

Description: Liquid.

Formula: $H_3CC_6H_{10}COOH$.

Constant: Mol wt: 142.2.

Hazard Analysis

Toxicity: Details unknown. Is said to resemble amyl acetate in type and severity of effects. See esters.

METHYL CYCLOHEXYL LACTATE

General Information

Description: Liquid.

Formula: $C_{10}H_{18}O_3$.

Constants: Mol wt: 186.3, flash p: 208°F, d: 1.02, vap. d.: 6.43.

Hazard Analysis

Toxicity: See esters.

Fire Hazard: Moderate, when exposed to heat or flame; can react with oxidizing materials.

Countermeasures

To Fight Fire: Foam, carbon dioxide, dry chemical or carbon tetrachloride (Section 6).

Ventilation Control: Section 2.

Storage and Handling: Section 7.

TOXIC HAZARD RATING CODE (For detailed discussion, see Section 1.)

0 NONE: (a) No harm under any conditions; (b) Harmful only under unusual conditions or overwhelming dosage.

1 SLIGHT: Causes readily reversible changes which disappear after end of exposure.

2 MODERATE: May involve both irreversible and reversible changes not severe enough to cause death or permanent injury.

3 HIGH: May cause death or permanent injury after very short exposur to small quantities.

U UNKNOWN: No information on humans considered valid by authors

METHYL CYCLOHEXYL STEARATE
General Information
Description: Clear, oily liquid; straw yellow.
Eormula: $C_{17}H_{35}COOC_6H_{10}CH_3$.
Constants: Mol wt: 380, mp: 10°C, bp: 105–116°C at 4 mm, flash p: 338°F, d: 0.890 at 15°/15°C, vap. d.: 13.1.
Hazard Analysis
Toxicity: See esters.
Fire Hazard: Slight, when exposed to heat or flame; can react with oxidizing materials.
Countermeasures
To Fight Fire: Foam, carbon dioxide, dry chemical or carbon tetrachloride (Section 6).
Ventilation Control: Section 2.
Storage and Handling: Section 7.

METHYL CYCLOPENTADIENE
General Information
Formula: C_6H_8.
Constants: Mol wt: 80, d: 0.9, bp: 163°F, flash p: 120°F, autoign. temp.: 834°F, lel: 1.3% at 212°F, uel: 7.6%, at 212°F.
Hazard Analysis
Toxic Hazard Rating: U.
Fire Hazard: Moderate, when exposed to heat or flame.

METHYL CYCLOPENTANE
General Information
Description: Colorless liquid or solid.
Formula: C_6H_{12}.
Constants: Mol wt: 84.16, mp: −142.5°C, bp: 71.8°C, flash p: < 20°F, d: 0.750 at 20°/4°C, vap. press.: 100 mm at 17.9°C, vap. d.: 2.9.
Hazard Analysis
Toxicity: Unknown.
Fire Hazard: Dangerous, when exposed to heat or flame. Probably irritant and narcotic in high concentrations.
Disaster Hazard: Dangerous; upon exposure to heat or flame; can react vigorously with oxidizing materials.
Countermeasures
Ventilation Control: Section 2.
To Fight Fire: Foam, carbon dioxide, dry chemical or carbon tetrachloride (Section 6).
Storage and Handling: Section 7.
Shipping Regulations: Section 11.
  IATA: Flammable liquid, red label, 1 liter (passenger), 40 liters (cargo).

METHYL DEMETON. See m-systox.

METHYL DIACETOACETATE
General Information
Description: Colorless liquid.
Formula: $(CH_3CO)_2CHCO_2CH_3$.
Constants: Mol wt: 158.2, vap. d.: 5.45.
Hazard Analysis
Toxic Hazard Rating:
  Acute Local: Irritant 1.
  Acute Systemic: U.
  Chronic Local: U.
  Chronic Systemic: U.
Fire Hazard: Slight (Section 6).
Countermeasures
Personal Hygiene: Section 3.
Storage and Handling: Section 7.

METHYL DIBORANE
General Information
Description: Colorless gas. Very unstable. Decomposed by water.
Formula: $B_2H_5CH_3$.
Constants: Mol wt: 41.7, bp: −80°C at 50 mm (decomp.).
Hazard Analysis and Countermeasures
See boron hydrides and boron compounds.

4-METHYL-2,6-DI-tert-BUTYLPHENOL. See di-tert-butyl-p-cresol.

METHYL DICHLOROACETATE
General Information
Synonym: Methyl dichloroethanoate.
Description: Colorless liquid; ethereal odor.
Formula: $Cl_2CHCOOCH_3$.
Constants: Mol wt: 143.0, bp: 143.0°C, d: 1.3809 at 19.2°/19.2°C, vap. d.: 4.93.
Hazard Analysis
Toxicity: This material hydrolyzes upon contact with moisture to form a product corrosive to tissue. See dichloracetic acid.
Disaster Hazard: Dangerous; when heated to decomposition, it emits highly toxic fumes of phosgene.
Countermeasures
Storage and Handling: Section 7.

METHYL DICHLOROARSINE
General Information
Synonym: Methylarsenic dichloride.
Description: Colorless, mobile liquid.
Formula: CH_3AsCl_2.
Constants: Mol wt: 160.86, mp: −59°C, bp: 136°C, d: 1.838 at 20°/4°C, vap. press.: 8.5 mm at 20°C.
Toxic Hazard Rating:
  Acute Local: Irritant 3; Ingestion 3; Inhalation 3.
  Acute Systemic: Ingestion 3; Inhalation 3.
  Chronic Local: U.
  Chronic Systemic: Ingestion 3; Inhalation 3.
Toxicology: See also arsenic compounds.
Disaster Hazard: Dangerous; when heated to decomposition, it emits highly toxic fumes of arsenic and chlorine; will react with water, steam or acids to produce toxic and corrosive fumes.
Countermeasures
First Aid: Section 1.
Storage and Handling: Section 7.
Shipping Regulations: Section 11.
  I.C.C.: Poison A; poison gas label, not accepted.
  Coast Guard Classification: Poison A; poison gas label.
  I.A.T.A.: Poison A, not acceptable (passenger and cargo).

METHYL DICHLOROETHANOATE. See methyl dichloroacetate.

METHYL DICHLOROSILANE
General Information
Description: Colorless liquid; soluble in benzene, ether, heptane.
Formula: CH_3SiHCl_2.
Constants: Mol wt: 115, bp: 41°C, d: 1.10 at 27°C, flash p: −26°F.
Hazard Analysis
Toxic Hazard Rating: U. See silanes.
Fire Hazard: Dangerous, when exposed to heat or flame.
Countermeasures
Storage and Handling: Section 7.
Shipping Regulations: Section 11.
  ICC: Flammable liquid, red label, 10 gallons.
  IATA: Flammable liquid, red label, not acceptable (passenger), 40 liters (cargo).

METHYL DICHLORSTEARATE
General Information
Description: Light yellow, oily liquid.
Formula: $C_{17}H_{33}Cl_2CO_2CH_2$.
Constants: Mol wt: 367.40, mp: −5°C to +7°C, bp: 250°C decomposes, flash p: 358°F, d: 0.997 at 15.5°/15.5°C, vap. d.: 12.7.
Hazard Analysis
Toxicity: Probably low. See esters.
Fire Hazard: Slight, when exposed to heat or flame.
Disaster Hazard: Dangerous; when heated to decomposi-

tion, it emits highly toxic fumes of phosgene; can react with oxidizing materials.

Countermeasures
Storage and Handling: Section 7.
To Fight Fire: Foam, carbon dioxide, dry chemical or carbon tetrachloride (Section 6).

METHYL DIETHANOLAMINE
General Information
Description: Clear liquid.
Formula: $CH_3N(CH_2CH_2OH)_2$.
Constants: Mol wt: 119.16, bp: 240°C, flash p: 260°F (O.C.), d: 1.043, vap. press.: 0.01 mm at 20°C.

Hazard Analysis
Toxicity: Details unknown. See amines.
Fire Hazard: Slight, when exposed to heat or flame; can react with oxidizing materials.

Countermeasures
To Fight Fire: Alcohol foam, carbon dioxide, dry chemical or carbon tetrachloride (Section 6).
Storage and Handling: Section 7.

METHYL DIHYDROABIETATE
General Information
Description: Liquid.
Formula: $CH_3CO_2C_{19}H_{32}$.
Constants: Mol wt: 319.5, bp: 365°C, flash p: 361°F (O.C.), d: 1.02, vap. d.: 11.0.

Hazard Analysis
Toxicity: Unknown.
Fire Hazard: Slight, when exposed to heat or flame; can react with oxidizing materials.

Countermeasures
To Fight Fire: Foam, carbon dioxide, dry chemical or carbon tetrachloride (Section 6).
Storage and Handling: Section 7.

2-METHYL-4,6-DINITROPHENOL. See 4,6-dinitro-o-cresol.

2-METHYL-1,3-DIOXOLANE
General Information
Description: Water-white liquid.
Formula: $OCH(CH_3)OCH_2CH_2$.
Constants: Mol wt: 88.10, bp: 82.5°C, d: 1.002 at 0°/4°C, vap. d.: 3.03.

Hazard Analysis
Toxic Hazard Rating:
Acute Local: Irritant 2; Ingestion 2; Inhalation 2.
Acute Systemic: U.
Chronic Local: U.
Chronic Systemic: U.
Fire Hazard: Slight.

Countermeasures
Ventilation Control: Section 2.
Personnel Protection: Section 3.
Storage and Handling: Section 7.

METHYL DISULFIDE
General Information
Synonym: Dimethyl disulfide; methyl dithiomethane.
Description: Liquid, insoluble in water; soluble in alcohol and ether.
Formula: CH_3SSCH_3.
Constants: Mol wt: 94.2, bp: 117°C, d: 1.057 at 16°/4°C.

Hazard Analysis
Toxicity: See alkyl disulfides.
Fire Hazard: Moderate.

Disaster Hazard: Dangerous. See sulfides.
Countermeasures
Storage and Handling: Section 7.
To Fight Fire: Water, spray, carbon dioxide, foam, dry chemical.

METHYL DITHIOCARBAMIC ACID, SODIUM SALT
Hazard Analysis
Toxicity: Details unknown. Irritating to skin and mucous membranes. Accompanied by alcohol intake causes violent vomiting and shock. See also bis(diethyl thiocarbamyl) disulfide.

METHYL DIVINYL ACETYLENE
Hazard Analysis
Toxic Hazard Rating:
Acute Local: 0.
Acute Systemic: Inhalation 1.
Chronic Local: U.
Chronic Systemic: U.
Toxicity: Animal experiments show slight narcotic effect.

METHYLENE ANILINE
General Information
Synonym: Hexahydro-1,3,5-triphenyl-sym-triazine.
Description: White, silky needles.
Formula: $(C_6H_5NCH_2)_3$.
Constants: Mol wt: 315.40, mp: 141°C, bp: 185°C.

Hazard Analysis
Toxic Hazard Rating:
Acute Local: Irritant 3; Ingestion 3; Inhalation 3.
Acute Systemic: Ingestion 3; Inhalation 3; Skin Absorption 3.
Chronic Local: U.
Chronic Systemic: Ingestion 3; Inhalation 3; Skin Absorption 3.
Disaster Hazard: Dangerous; when heated to decomposition, it emits highly toxic fumes of aniline.

Countermeasures
Ventilation Control: Section 2.
Personnel Protection: Section 3.
First Aid: Section 1.
Storage and Handling: Section 7.

N,N'-METHYLENEBISACRYLAMIDE
General Information
Description: Colorless, crystalline; stable, white powder.
Formula: $H_2C(CH_2CHCONH)_2$.
Constants: Mol wt: 154.17, mp: 185°C (with decomp.), d: 1.235 at 30°C, vap. d.: 5.31.

Hazard Analysis
Toxic Hazard Rating:
Acute Local: U.
Acute Systemic: Ingestion 2.
Chronic Local: U.
Chronic Systemic: U.
Disaster Hazard: Slight; when heated, it emits toxic fumes.

Countermeasures
Personal Hygiene: Section 3.
Storage and Handling: Section 7.

2,2'-METHYLENE-BIS(4-CHLOROPHENOL)
General Information
Synonym: Dichlorophene; G-4; 2,2' dihydroxy-5,5' dichlorodiphenylmethane.
Description: Crystals, nearly water insoluble.
Formula: $C_{13}H_{10}Cl_2O_2$.

TOXIC HAZARD RATING CODE (For detailed discussion, see Section 1.)

0 NONE: (a) No harm under any conditions; (b) Harmful only under unusual conditions or overwhelming dosage.

1 SLIGHT: Causes readily reversible changes which disappear after end of exposure.

2 MODERATE: May involve both irreversible and reversible changes not severe enough to cause death or permanent injury.

3 HIGH: May cause death or permanent injury after very short exposure to small quantities.

U UNKNOWN: No information on humans considered valid by authors

Constants: Mol wt: 269.1, mp: 178°C, vap. d.: 9.3, vap. press.: 10^{-4} mm at 100°C.

Hazard Analysis

Toxic Hazard Rating:

Acute Local: U.

Acute Systemic: Ingestion 2.

Chronic Local: U.

Chronic Systemic: U.

Toxicology: Moderately toxic. Strong concentrations are irritating to skin. Large doses can cause cramps and diarrhea, from limited animal experiments.

Disaster Hazard: Dangerous; when heated to decomposition, it emits highly toxic fumes.

2,2'-METHYLENEBIS (4-METHYL-6-tert-BUTYL-PHENOL)

General Information

Description: Pale cream to white crystals.

Formula: $C_{23}H_{33}O_2$.

Constants: Mol wt: 641.5, mp: 120–130°C, d: 1.074 at 30°C.

Hazard Analysis

Toxicity: Unknown.

Disaster Hazard: Slight; when heated, it may emit toxic fumes.

Countermeasures

Storage and Handling: Section 7.

METHYLENEBIS (4-PHENYL ISOCYANATE)

General Information

Synonyms: Diphenyl methane diisocyanate, MDI.

Description: Crystals or yellow fused solid.

Formula: $OCNC_6H_4CH_2C_6H_4NCO$.

Constants: Mol wt: 250.25, mp: 37.2°C, bp: 194–199°C at 5 mm, d: 1.19 at 50°C.

Hazard Analysis

Toxicity: Details unknown; an irritant.

TLV: ACGIH (recommended) 0.02 parts per million of air, 0.2 milligrams per cubic meter of air.

Disaster Hazard: See isocyanates.

Countermeasures

Storage and Handling: Section 7.

2,2'-METHYLENEBIS(3,4,6-TRICHLOROPHENOL)

General Information

Synonyms: G-11; hexachlorophene.

Description: Crystals; water insoluble.

Formula: $C_{13}H_6Cl_6O_2$.

Constants: Mol wt: 406.9, mp: 165°C.

Hazard Analysis

Toxic Hazard Rating:

Acute Local: Irritant 1; Allergen 1.

Acute Systemic: U.

Chronic Local: Irritant 1; Allergen 1.

Chronic Systemic: U.

Toxicology: Strong concentrations may be irritating but ordinary use of 1-2% is not irritating. Note: Used as a food additive permitted in the feed and drinking water of animals and/or for the treatment of food producing animals; also permitted in food for human consumption (Section 10).

Disaster Hazard: Dangerous; when heated to decomposition, it emits highly toxic fumes of chlorides.

METHYLENE BROMIDE

General Information

Synonym: Dibromomethane.

Description: Colorless, heavy liquid.

Formula: CH_2Br_2.

Constants: Mol wt: 173.9, bp: 95.6–97.4°C, fp: $< -50°C$, d: 2.485 at 25°/25°C, vap. d.: 6.05.

Hazard Analysis

Toxicity: See bromides. Details unknown. Limited animal experiments suggest moderate toxicity.

Disaster Hazard: Moderately dangerous; when heated to decomposition, it emits toxic fumes of bromides.

Countermeasures

Storage and Handling: Section 7.

METHYLENE CHLORIDE

General Information

Synonyms: Dichloromethane.

Description: Colorless, volatile liquid.

Formula: CH_2Cl_2.

Constants: Mol wt: 84.94, bp: 40.1°C, lel: 15.5% in O_2, uel: 66.4% in O_2, fp: $-96.7°C$, d: 1.326 at 20°/4°C, autoign. temp.: 1224°F, vap. press.: 380 mm at 22°C, vap. d.: 2.93.

Hazard Analysis

Toxic Hazard Rating:

Acute Local: Irritant 2; Ingestion 2; Inhalation 2.

Acute Systemic: Ingestion 2; Inhalation 3; Skin Absorption 2.

Chronic Local: U.

Chronic Systemic: Ingestion 1; Inhalation 1; Skin Absorption 1.

TLV: ACGIH (recommended); 500 parts per million in air; 1740 mg per cubic meter of air.

Toxicology: This material is very dangerous to the eyes. Except for its property of inducing narcosis, it has very few other toxic effects. Its narcotic powers are quite strong, and in view of its great volatility, care should be taken in its use. It will not form explosive mixtures with air at ordinary temperatures. However, it can be decomposed by contact with hot surfaces and open flame, and it can then yield toxic fumes, which are irritating and will thus give warning of their presence. It has been used as an anesthetic in Europe and is still used there for local anesthesia. Experiments have shown that 25,000 ppm concentrations for 2 hour exposures were not lethal. Concentrations of 7,200 ppm after 8 minutes caused paresthesia of the extremities; after 16 minutes, acceleration of the pulse to 100 and during the first 20 minutes, congestion in the head, a sense of heat and slight irritation of the eyes. At a level of 2,300 ppm, there was no feeling of dizziness during one-hour exposures, but nausea did occur after 30 minutes of exposure. The limit of perception by smell is set at 25-50 ppm concentrations. Can cause a dermatitis upon prolonged skin contact (Section 9). A gas mask for organic vapors and fumes should be worn to avoid excessive inhalation.

Note: Used as a food additive permitted in food for human consumption (Section 10).

Fire Hazard: None.

Explosion Hazard: None under ordinary conditions, but will form explosive mixtures in atmosphere having high oxygen content.

Disaster Hazard: Dangerous; when heated to decomposition, it emits highly toxic fumes of phosgene.

Countermeasures

Ventilation Control: Section 2.

Personnel Protection: Section 3.

Storage and Handling: Section 7.

Shipping Regulations: Section 11.

MCA warning label.

IATA: Other restricted articles, class A, 40 liters (passenger), 220 liters (cargo).

METHYLENE CHLOROBROMIDE

General Information

Synonyms: Bromochloromethane; chlorobromomethane.

Description: Clear, colorless liquid; sweet odor.

Formula: $BrCH_2Cl$.

Constants: Mol wt: 129.4, bp: 67.8°C, fp: $-88°C$, flash p: none, d: 1.930 at 25°/25°C, vap. d.: 4.46.

Hazard Analysis

Toxic Hazard Rating:

Acute Local: Irritant 2; Inhalation 3.

Acute Systemic: Ingestion 1; Skin Absorption 1; Inhalation 3.
Chronic Local: Irritant 1.
Chronic Systemic: Inhalation 2.
Toxicity: Data based upon animal experiments. See methylene chloride. Irritant and narcotic.
TLV: ACGIH (recommended); 200 parts per million of air; 1050 milligrams per cubic meter of air.
Disaster Hazard: Dangerous; when heated to decomposition, it emits highly toxic fumes of halides.
Countermeasures
Storage and Handling: Section 7. It should not be stored in aluminum or magnesium containers.

p,p'-METHYLENE DIANILINE
General Information
Description: Tan flakes or lumps, faint amine like odor.
Formula: $CH_2(C_6H_4NH_2)_2$.
Constants: Mol wt: 198.3, mp: 90°C, flash p: 440°F.
Hazard Analysis
Toxic Hazard Rating:
  Acute Local: Irritant 3; Allergen 2; Ingestion 3; Inhalation 3.
  Acute Systemic: Ingestion 2.
  Chronic Local: Allergen 2.
  Chronic Systemic: Ingestion 2.
Toxicity: It has moderately high acute oral toxicity. Does not seem to be rapidly absorbed through skin in dangerous quantities or to irritate the eyes. No information regarding chronic toxicity or allergy. Tested only on rats.
Disaster Hazard: Dangerous; when heated to decomposition, it emits highly toxic fumes of aniline.
Countermeasures
Storage and Handling: Section 7.

METHYLENE DIISOCYANATE
General Information
Formula: $CH_2(NCO)_2$.
Constants: Mol wt: 98, flash p: 185°F (O.C.).
Hazard Analysis
Toxic Hazard Rating: U.
Fire Hazard: Moderate, when exposed to heat or flame.

METHYLENE DIMETHYL ETHER. See methylal.

3,4-METHYLENE DIOXYBENZALDEHYDE. See piperonal.

2-(3,4-METHYLENE DIOXYPHENOXY)-3,6,9-TRIOXA-UNDECANE. See sesoxane.

4,4-METHYLENE DIPHENOL
General Information
Formula: $C_6H_5CH_2C_6H_5$.
Constant: Mol wt: 174.
Hazard Analysis
Toxic Hazard Rating: U.
Toxicity: Limited animal experiments suggest moderate toxicity. See also phenol.

METHYLENE DISALICYLIC ACID
General Information
Description: Light tan, coarse powder.
Formula: $C_{15}H_{12}O_6$.
Constants: Mol wt: 288.25, mp: 225–238°C, d: 1.430 at 25°/4°C.
Hazard Analysis
Toxic Hazard Rating:
  Acute Local: Irritant 1; Ingestion 1; Inhalation 1.

Acute Systemic: U.
Chronic Local: U.
Chronic Systemic: U.
Fire Hazard: Slight.
Countermeasures
Personal Hygiene: Section 3.
Storage and Handling: Section 7.

METHYLENE IODIDE
General Information
Synonym: Diiodomethane.
Description: Light straw to clear colored liquid.
Formula: CH_2I_2.
Constants: Mol wt: 267.9, mp: 5–6°C, bp: 181°C (decomp.), d: 3.325 at 20°/4°C, vap. d.: 9.25.
Hazard Analysis
Toxicity: Unknown. Probably irritant and narcotic in high concentrations.
Disaster Hazard: Dangerous; when heated to decomposition, it emits toxic fumes of iodides.
Countermeasures
Storage and Handling: Section 7

METHYLENE SUCCINIC ACID. See itaconic acid.

N-METHYL ETHANOLAMINE
General Information
Description: Liquid.
Formula: $CH_3NHC_2H_4OH$.
Constants: Mol wt: 75.11, mp: −4.5°C, bp: 159.5°C, flash p: 165°F (O.C.), d: 0.9414 at 20°/20°C, vap. press.: 0.7 mm at 20°C, vap. d.: 2.59.
Hazard Analysis
Toxicity: Details unknown. See amines.
Fire Hazard: Moderate, when exposed to heat or flame; can react with oxidizing materials.
Countermeasures
To Fight Fire: Foam, carbon dioxide, dry chemical or carbon tetrachloride.
Storage and Handling: Section 7.

METHYL ETHER. See dimethyl ether.

METHYL ETHER OF PROPYLENE GLYCOL. See propylene glycol methyl ether.

METHYL ETHYL CELLULOSE
General Information
Description: A white to pale cream colored fibrous solid or powder. Practically odorless.
Hazard Analysis
Toxic Hazard Rating: Details unknown. Used as a food additive permitted in food for human consumption (Section 10).

2-METHYL-2-ETHYL-1,3-DIOXOLANE
General Information
Description: Insoluble in water.
Formula: $(CH_3)(C_2H_5)COCH_2CH_2O$.
Constants: Mol wt: 116, d: 0.9392, bp: 117.6°C, fp: −81.96°C, vap. d.: 4.0, flash p: 74°F (O.C.).
Hazard Analysis
Toxic Hazard Rating: U.
Fire Hazard: Dangerous, when exposed to heat or flame.

sym-METHYLETHYL ETHYLENE. See 2-pentene.

METHYL ETHYL ETHER. See ethyl methyl ether.

TOXIC HAZARD RATING CODE *(For detailed discussion, see Section 1.)*

0 NONE: (a) No harm under any conditions; (b) Harmful only under unusual conditions or overwhelming dosage.

1 SLIGHT: Causes readily reversible changes which disappear after end of exposure.

2 MODERATE: May involve both irreversible and reversible changes; not severe enough to cause death or permanent injury.

3 HIGH: May cause death or permanent injury after very short exposure to small quantities.

U UNKNOWN: No information on humans considered valid by authors.

3-METHYL-4-ETHYLHEXANE
General Information
Description: Clear, colorless liquid.
Formula: C_9H_{20}.
Constants: Mol wt: 128.25, bp: 140°C, flash p: 75°F, d: 0.738 at 25°/4°C, vap. press.: 8.1 mm at 25°C, vap. d.: 4.43.
Hazard Analysis
Toxicity: Unknown. Probably irritant and narcotic in high concentrations.
Fire Hazard: Dangerous, when exposed to heat or flame.
Explosion Hazard: Unknown.
Disaster Hazard: Dangerous; keep away from heat and open flame; can react with oxidizing materials.
Countermeasures
Ventilation Control: Section 2.
To Fight Fire: Foam, carbon dioxide, dry chemical or carbon tetrachloride (Section 6).
Storage and Handling: Section 7.

METHYLETHYL KETONE. See butanone.
MCA warning label.

METHYL ETHYL KETONE PEROXIDE
General Information
Formula: $C_4H_8O_2$.
Constant: Mol wt: 88.1.
Hazard Analysis
Toxic Hazard Rating:
   Acute Local: Irritant 3.
   Acute Systemic: Ingestion 2; Inhalation 2.
   Chronic Local: Irritant 3.
   Chronic Systemic: Ingestion 2; Inhalation 2.
Toxicology: Data based upon animal experiments.
Fire Hazard: See peroxides, organic.
Countermeasures
Storage and Handling: See peroxides, organic.

METHYLETHYLMETHANE. See butane.

2-METHYL-5-ETHYL PIPERIDINE
General Information
Description: Slightly soluble in water.
Formula: $NHCH(CH_3)CH_2CH_2CH(C_2H_5)CH_2$.
Constants: Mol wt: 127, flash p: 126°F (O.C.), d: 0.8, vap. d.: 4.4, bp: 326°F.
Hazard Analysis
Toxic Hazard Rating: U.
Fire Hazard: Moderate, when exposed to heat or flame.
Countermeasures
Storage and Handling: Section 7.
To Fight Fire: Alcohol foam.

METHYL ETHYL PYRIDINE. See 2-methyl-5-ethyl pyridine.

2-METHYL-5-ETHYL PYRIDINE
General Information
Description: Liquid.
Formula: $NC(CH_3)CHCHC(C_2H_5)CH$.
Constants: Mol wt: 121.17, mp: −70.3°C, bp: 178.3°C, flash p: 165°F (O.C.), d: 0.9215 at 20°/20°C, vap. press.: 0.9 mm at 20°C, vap. d.: 4.18.
Hazard Analysis
Toxicity: Moderately toxic. Details unknown. See ethyl picoline and pyridine.
Fire Hazard: Moderate, when exposed to heat or flame; can react with oxidizing materials.
Countermeasures
To Fight Fire: Foam, carbon dioxide, dry chemical or carbon tetrachloride (Section 6).
Storage and Handling: Section 7.

METHYL FLUORIDE
General Information
Synonym: Fluoromethane.

Description: Colorless gas.
Formula: CH_3F.
Constants: Mol wt: 34.0, mp: −142°C, bp: −78.2°C, d: 0.8774 at −78.6°/4°C.
Hazard Analysis
Toxicity: Details unknown; may act as a simple asphyxiant.
Disaster Hazard: Dangerous, when heated to decomposition it emits highly toxic fumes of fluorides.
Countermeasures
Storage and Handling: Section 7.
Shipping Regulations: Section 11.
   IATA: Flammable gas, red label, not acceptable (passenger), 140 kilograms (cargo).

METHYL FLUOROFORMATE
General Information
Description: Liquid.
Formula: CH_3COOF.
Constants: Mol wt: 78.04, bp: 40°C, d: 1.06 at 33°C, vap. d.: 2.69.
Hazard Analysis
Toxic Hazard Rating:
   Acute Local: Irritant 3; Ingestion 3; Inhalation 3.
   Acute Systemic: U.
   Chronic Local: U.
   Chronic Systemic: U.
Disaster Hazard: Dangerous; when heated to decomposition, it emits highly toxic fumes of fluorides; will react with water, steam or acids to produce toxic and corrosive fumes.
Countermeasures
Ventilation Control: Section 2.
Personnel Protection: Section 3.
First Aid: Section 1.
Storage and Handling: Section 7.

METHYL FLUOROSULFONATE
General Information
Description: Liquid; ethereal odor.
Formula: CH_3OSO_2F.
Constants: Mol wt: 114.09, bp: 92°C, d: 1.427 at 16°C, vap. d.: 3.94.
Hazard Analysis
Toxic Hazard Rating:
   Acute Local: Irritant 3; Ingestion 3; Inhalation 3.
   Acute Systemic: U.
   Chronic Local: U.
   Chronic Systemic: U.
Disaster Hazard: Dangerous; when heated to decomposition, it emits highly toxic fumes of fluorides and oxides of sulfur; will react with water, steam or acids to produce toxic and corrosive fumes.
Countermeasures
Ventilation Control: Section 2.
Personnel Protection: Section 3.
First Aid: Section 1.
Storage and Handling: Section 7.

METHYL FORMATE
General Information
Synonym: Methyl methanoate.
Description: Colorless liquid; agreeable odor.
Formula: $HCOOCH_3$.
Constants: Mol wt: 60.05, mp: −99.8°C, bp: 32.0°C, lel: 5.9%, uel: 20%, flash p: −2°F, d: 0.98149 at 15°/4°C; 0.975 at 20°/4°C, autoign. temp.: −853°F, vap. press.: 400 mm at 16.0°C, vap. d.: 2.07.
Hazard Analysis
Toxic Hazard Rating:
   Acute Local: Irritant 2; Ingestion 2; Inhalation 2.
   Acute Systemic: Inhalation 2.
   Chronic Local: U.
   Chronic Systemic: U.

TLV: ACGIH (recommended) 100 parts per million in air; 250 milligrams per cubic meter of air.

Toxicology: It can cause irritation to the conjunctiva and optic neuritis. Industrial exposures resulting in fatalities from this material are extremely rare, having occurred only in instances where high concentrations are encountered, as in painting the inside of a tank or working in a tank containing a residue of this material.

Exposure of guinea pigs to 5% concentrations of methyl formate vapor in air proved lethal in from 20 to 30 minutes, whereas 1.5 to 2.5% was dangerous in from 30 to 60 minutes. The maximum concentration tolerated without serious disturbance was 0.5% and the maximum concentration tolerated for several hours without serious disturbances was 0.15–0.20% by volume in air.

Note: Used as a food additive permitted in food for human consumption (Section 10).

Fire Hazard: Dangerous, when exposed to heat or flame.

Spontaneous Heating: No.

Explosion Hazard: Moderate, when exposed to heat or flame.

Disaster Hazard: Dangerous, upon exposure to heat or flame; it emits toxic fumes and can react vigorously with oxidizing materials.

Countermeasures

Ventilation Control: Section 2.

To Fight Fire: Alcohol foam, carbon dioxide, dry chemical or carbon tetrachloride (Section 6).

Personnel Protection: Section 3.

Storage and Handling: Section 7.

Shipping Regulations: Section 11.

    I.C.C.: Flammable liquid; red label, 10 gallons.

    Coast Guard Classification: Inflammable liquid; red label.

    IATA: Flammable liquid, red label, 1 liter (passenger), 40 liters (cargo).

2-METHYL FURAN
General Information

Description: Colorless, mobile liquid; ether-like odor.

Formula: C_5H_6O.

Constants: Mol wt: 82.1, bp: 63.7°C, fp: −88.7°C, flash p: −22°F, d: 0.914 at 20°/4°C, vap. press.: 139 mm at 20°C, vap. d.: 2.8.

Hazard Analysis

Toxic Hazard Rating:

    Acute Local: Irritant 2; Ingestion 2.

    Acute Systemic: Inhalation 2.

    Chronic Local: U.

    Chronic Systemic: U.

Fire Hazard: Dangerous, when exposed to heat or flame.

Explosion Hazard: See ethers.

Disaster Hazard: Dangerous, upon exposure to heat or flame; will emit toxic fumes on heating and can react vigorously with oxidizing materials.

Countermeasures

Ventilation Control: Section 2.

To Fight Fire: Carbon dioxide, dry chemical or carbon tetrachloride (Section 6).

Personnel Protection: Section 3.

Storage and Handling: Section 7.

Shipping Regulations: Section 11.

    IATA: Flammable liquid, red label, 1 liter (passenger), 40 liters (cargo).

METHYL GALLIUM DICHLORIDE
General Information

Description: White cyrstals decomposed by water.

Formula: $Ga(CH_3)Cl_2$.

Constants: Mol wt: 155.7, mp: 75°C.

Hazard Analysis

Toxicity: Details unknown. See gallium compounds.

Disaster Hazard: See chlorides.

METHYL GALLIUM DICHLORIDE MONAMMINE
General Information

Description: White crystals decomposed by water.

Formula: $Ga(CH_3)Cl_2 \cdot NH_3$.

Constant: Mol wt: 172.7.

Hazard Analysis and Countermeasures

See gallium compounds and chlorides.

METHYL GALLIUM DICHLORIDE PENTAMMINE
General Information

Description: White crystals decomposed by water.

Formula: $5NH_3 \cdot Cl_2Ga(CH_3)$.

Constants: Mol wt: 240.8, mp: >80°C (decomp.).

Hazard Analysis and Countermeasures

See gallium compounds, chlorides and ammonia.

N-METHYLGLUCAMINE
General Information

Description: White crystals.

Formula: $HOCH_2(CHOH)_4CH_2NHCH_3$.

Constants: Mol wt: 195.21, mp: 128–129°C.

Hazard Analysis

Toxic Hazard Rating:

    Acute Local: Irritant 1; Ingestion 1; Inhalation 1.

    Acute Systemic: U.

    Chronic Local: U.

    Chronic Systemic: U.

Disaster Hazard: Dangerous; when heated to decomposition it emits toxic fumes.

Countermeasures

Personal Hygiene: Section 3.

Storage and Handling: Section 7.

β-METHYLGLYCIDOL
General Information

Description: Liquid.

Formula: $OCHCH_3CHCH_2OH$.

Constants: Mol wt: 88.1, vap. d.: 3.04.

Hazard Analysis

Toxicity: Unknown.

Disaster Hazard: Slight; when heated, it emits acrid fumes; can react with oxidizing materials.

Countermeasures

Storage and Handling: Section 7.

N-METHYLGLYCINE
General Information

Synonym: Sarcosine.

Description: Crystals.

Formula: CH_3NHCH_2COOH.

Constants: Mol wt: 89.1, mp: 210°C (decomp.), bp: decomposes.

Hazard Analysis

Toxicity: Details unknown.

Disaster Hazard: Slight; when heated, it emits acrid fumes; can react with oxidizing materials.

Countermeasures

Storage and Handling: Section 7.

METHYL GLYCOL ACETATE
General Information

Synonym: Propylene glycol acetate.

Description: Liquid.

TOXIC HAZARD RATING CODE *(For detailed discussion, see Section 1.)*

0 NONE: (a) No harm under any conditions; (b) Harmful only under unusual conditions or overwhelming dosage.

1 SLIGHT: Causes readily reversible changes which disappear after end of exposure.

2 MODERATE: May involve both irreversible and reversible changes; not severe enough to cause death or permanent injury.

3 HIGH: May cause death or permanent injury after very short exposure to small quantities.

U UNKNOWN: No information on humans considered valid by authors.

Formula: $CH_2OHCHOHCH_2CO_2CH_3$.
Constants: Mol wt: 134.1, flash p: 111° F (C.C.), vap. d.: 4.62.

Hazard Analysis
Toxic Hazard Rating:
 Acute Local: Irritant 1; Ingestion 1; Inhalation 1.
 Acute Systemic: Ingestion 2; Inhalation 2.
 Chronic Local: U.
 Chronic Systemic: Ingestion 1; Inhalation 1.
Toxicology: See also glycols.
Fire Hazard: Moderate, when exposed to heat or flame; can react with oxidizing materials.

Countermeasures
To Fight Fire: Carbon dioxide, dry chemical or carbon tetrachloride (Section 6).
Personal Hygiene: Section 3.
Storage and Handling: Section 7.

METHYLGLYCOCOLL. See N-methylglycine.

METHYL HEPTADECYL KETONE
General Information
Description: White solid.
Formula: $(CH_3)(C_{17}H_{35})CO$.
Constants: Mol wt: 282, bp: 165° C at 3 mm, mp: 54° C, flash p: 255° F (C.C.), vap. d.: 9.72.

Hazard Analysis
Toxicity: Unknown. See ketones.
Fire Hazard: Slight, when exposed to heat or flame; can react with oxidizing materials.

Countermeasures
To Fight Fire: Foam, carbon dioxide, dry chemical or carbon tetrachloride (Section 6).
Storage and Handling: Section 7.

2-METHYLHEPTANE. See isooctane.

METHYL HEPTANOL
General Information
Formula: $CH_3CH_2CH(CH_3)CH_2CH_2CH_2CH_2OH$.
Constant: Mol wt: 130.

Hazard Analysis
Toxicity: Details unknown. Limited animal experiments suggest moderate toxicity.

5-METHYL-3-HEPTANONE. See ethyl amyl ketone.

METHYL HEPTYL KETONE
General Information
Description: Colorless liquid.
Formula: $(CH_3)(C_7H_{15})CO$.
Constants: Mol wt: 142, mp: −9° C, bp: 194° C, flash p: 160° F (C.C.), vap. d.: 4.9, d: 0.832 at 30° C.

Hazard Analysis
Toxicity: Details unknown. See ketones.
Fire Hazard: Moderate, when exposed to heat or flame; can react with oxidizing materials.

Countermeasures
To Fight Fire: Foam, carbon dioxide, dry chemical or carbon tetrachloride (Section 6).
Storage and Handling: Section 7.

2-METHYLHEXANE
General Information
Synonyms: Ethylisobutylmethane; isoheptane.
Description: Colorless liquid.
Formula: $(CH_3)_2CH(CH_2)_2CH_2CH_3$.
Constants: Mol wt: 100.20, fp: −118.2° C, bp: 90.0° C, flash p: <0° F, d: 0.6789 at 20°/4° C, vap. press.: 40 mm at 14.9° C, vap. d.: 3.45, lel: 1.0%, uel: 6.0%.

Hazard Analysis
Toxic Hazard Rating:
 Acute Local: U.
 Acute Systemic: Ingestion 1; Inhalation 2.
 Chronic Local: U.

Chronic Systemic: U.
Fire Hazard: Unknown.

Countermeasures
To Fight Fire: Foam, carbon dioxide, dry chemical or carbon tetrachloride. (Section 6).
Ventilation Control: Section 2.
Personal Hygiene: Section 3.

5-METHYL-2-HEXANONE
General Information
Synonym: MIAK; methyl isoamyl ketone.
Description: Colorless, stable liquid; pleasant odor; slightly soluble in water; miscible with most organic solvents.
Formula: $CH_3COC_2H_4CH(CH_3)_2$.
Constants: Mol wt: 114, bp: 144° C, d: 0.8132 at 20/20° C, fp: −73.9° C, flash p: 110° F (O.C.).

Hazard Analysis
Toxicity: Details unknown. Limited animal experiments suggest low toxicity and irritation. See also ketones.
Fire Hazard Rating: Moderate, when exposed to heat or flame.

METHYL HEXYL KETONE. See 2-OCTANONE.

METHYL HYDRAZINE
General Information
Synonyms: Monomethyl hydrozine; MMH.
Description: Colorless, hygroscopic liquid; ammonia like odor; slightly soluble in water; soluble in alcohol and ether.
Formula: CH_3NHNH_2.
Constants: Mol wt: 46, d: 0.874 at 20/4° C, mp: −20.9° C, bp: 190° F, vap. d.: 1.6, flash p: <80° F.

Hazard Analysis
Toxic Hazard Rating: Very toxic.
TLV: ACGIH (recommended); 0.2 part per million of air; 0.35 milligrams per cubic meter of air. May be absorbed through the skin.
Fire Hazard: Dangerous, when exposed to heat or flame.

Countermeasures
Storage and Handling: Section 7.
To Fight Fire: Alcohol foam.
Shipping Regulations: Section 11.
 ICC: Flammable liquid, red label, 5 pints.
 IATA: Flammable liquid, red label, not acceptable (passenger), 2½ liters (cargo).

METHYL HYDRIDE. See methane.

METHYL HYDROGEN SULFATE. See methylsulfuric acid.

METHYL-3-HYDROXYBUTYRATE
General Information
Description: Colorless liquid.
Formula: $CH_3CHOHCH_2COOCH_3$.
Constants: Mol wt: 118.13, bp: 174.9° C, flash p: 180° F, d: 1.0559, vap. press.: 0.85 mm at 20° C, vap. d.: 4.1.

Hazard Analysis
Toxic Hazard Rating:
 Acute Local: Irritant 2.
 Acute Systemic: U.
 Chronic Local: U.
 Chronic Systemic: U.
Fire Hazard: Moderate, when exposed to heat or flame; can react with oxidizing materials.

Countermeasures
To fight Fire: Foam, carbon dioxide, dry chemical or carbon tetrachloride (Section 6).
Personnel Protection: Section 3.
Storage and Handling: Section 7.

4-METHYL-7-HYDROXYCOUMARIN DIETHOXY-THIOPHOSPHATE. See potasan.

METHYL-(B-HYDROXYETHYL)AMINE. See methyl-
aminoethanol.

METHYL IODIDE
General Information
Synonym: Iodomethane.
Description: Colorless liquid; turns brown on exposure to
light.
Formula: CH_3I.
Constants: Mol wt: 141.95, mp: $-64.4°C$, bp: 42.5°C,
d: 2.279 at 20°/4°C, vap. press.: 400 mm at 25.3°C,
vap. d.: 4.89.
Hazard Analysis
Toxic Hazard Rating:
Acute Local: Irritant 2; Ingestion 2; Inhalation 2.
Acute Systemic: Ingestion 3; Inhalation 3; Skin Ab-
sorption 3.
Chronic Local: U.
Chronic Systemic: Ingestion 2; Inhalation 2; Skin Ab-
sorption 2.
Toxicology: A strong narcotic and anesthetic.
TLV: ACGIH (recommended); 5 parts per million of air;
28 milligrams per cubic meter of air.
Fire Hazard: Unknown.
Disaster Hazard: Dangerous; when heated to decompo-
sition, it emits highly toxic fumes of iodides.
Countermeasures
Ventilation Control: Section 2.
Personnel Protection: Section 3.
Storage and Handling: Section 7.

METHYL ISOAMYL KETONE. See 5-methyl-2-hex-
anone.

METHYL ISOBUTYL CARBINOL. See methyl amyl
alcohol.

METHYL ISOBUTYL KETONE. See isobutyl methyl
ketone.

METHYL ISOCYANATE
General Information
Description: Liquid, reacts with water.
Formula: CH_3NCO.
Constants: Mol wt: 57, d: 0.9599 at 20/20°C, bp: 39.1°C,
flash p: $<20°F$.
Hazard Analysis
TLV: ACGIH (recommended); 0.02 parts per million of air;
0.05 milligrams per cubic meter of air. May be absorbed
via the skin.
Fire Hazard: Dangerous, when exposed to heat or flames.
Countermeasures
Shipping Regulations: Section 11.
IATA: Poison C, not acceptable (passenger), 40 liters
(cargo).

METHYL ISOCYANIDE
General Information
Synonym: Methyl isonitrile.
Description: Colorless liquid.
Formula: CH_3NC.
Constants: Mol wt: 41.1, mp: $-45°C$, bp: 59.6°C, d:
0.7464 at 20°/4°C.
Hazard Analysis
Toxicity: Highly toxic. See cyanides.
Explosion Hazard: Severe, when shocked or exposed to heat.
Disaster Hazard: Highly dangerous; shock will explode it;
when heated to decomposition, it emits highly toxic
fumes of cyanides.

Countermeasures
Ventilation Control: Section 2.
Storage and Handling: Section 7.

METHYL ISONITRILE. See methyl isocyanide.

METHYL ISOPRENYL KETONE
General Information
Formula: $CH_3COC:CH_2(CH_3)$.
Constants: Mol wt: 84, lel: 1.8%, uel: 9.0%, vap. d.: 2.9,
bp: 208°F.
Hazard Analysis
Toxic Hazard Rating: U. See also ketones.
Fire Hazard: Dangerous, when exposed to heat or flame.
Countermeasures
Storage and Handling: Section 7.
Shipping Regulations: Section 11.
ICC (inhibited): Flammable liquid, red label, 10 gallons.
IATA (inhibited): Flammable liquid, red label, 1 liter
(passenger), 40 liters (cargo).
(uninhibited): Flammable liquid, not acceptable
(passenger and cargo).

METHYL ISOPROPYL CARBINOL
General Information
Description: Clear liquid.
Formula: $CH_3C_3H_7CHOH$.
Constants: Mol wt: 88.2, flash p: 103°F (C.C.), vap. d.:
3.04, d: 0.8.
Hazard Analysis
Toxicity: Details unknown. See also alcohols.
Fire Hazard: Moderate, when exposed to heat or flame; can
react with oxidizing materials.
Countermeasures
To Fight Fire: Alcohol foam, carbon dioxide, dry chemical
or carbon tetrachloride (Section 6).
Ventilation Control: Section 2.
Storage and Handling: Section 7.

METHYL ISOTHIOCYANATE
General Information
Synonym: Methyl mustard oil.
Description: Crystalline, slightly water soluble.
Formula: CH_3NCS.
Constants: Mol wt: 73.1, mp: 36°C, bp: 119°C.
Hazard Analysis
Toxicity: Highly toxic! A powerful irritant. A military
poison.
Disaster Hazard: Very dangerous. See cyanides and
sulfur compounds for effects when heated to decompo-
sition.
Countermeasures
Storage and Handling: Section 7.
Shipping Regulations: Section 11.
IATA: Poison C, not acceptable (passsnger), 40 liters
(cargo).

METHYL LACTATE
General Information
Description: Colorless liquid, soluble in water, decomposes.
Formula: $C_4H_8O_3$.
Constants: Mol wt: 116.16, bp: 144°C, lel: 2.2% at 212°F,
flash p: 121°F (C.C.), d: 1.09, autoign. temp.: 725°F,
vap. d.: 4.02, ulc: 30–35.
Hazard Analysis
Toxic Hazard Rating:
Acute Local: Irritant 1; Ingestion 1; Inhalation 1.
Acute Systemic: Ingestion 1; Inhalation 1.

TOXIC HAZARD RATING CODE (For detailed discussion, see Section 1.)

0 NONE: (a) No harm under any conditions; (b) Harmful only under unusual conditions or overwhelming dosage.

1 SLIGHT: Causes readily reversible changes which disappear after end
of exposure.

2 MODERATE: May involve both irreversible and reversible changes;
not severe enough to cause death or permanent injury.
3 HIGH: May cause death or permanent injury after very short exposure
to small quantities.
U UNKNOWN: No information on humans considered valid by authors.

Chronic Local: U.

Chronic Systemic: U.

Fire Hazard: Moderate, when exposed to heat or flame.

Explosion Hazard: Moderate, when exposed to heat or flame.

Disaster Hazard: Moderately dangerous; vapor-air mixtures may explode; can react with oxidizing materials.

Countermeasures

Ventilation Control: Section 2.

To Fight Fire: Carbon dioxide, dry chemical or carbon tetrachloride (Section 6).

Personal Hygiene: Section 3.

Storage and Handling: Section 7.

METHYL MAGNESIUM BROMIDE IN ETHYL ETHER

General Information

Formula: CH_3MgBr.

Hazard Analysis

Toxic Hazard Rating: U. See bromides and ether.

Fire Hazard: See ether.

Disaster Hazard: See ether and bromides.

Countermeasures

Storage and Handling: Section 7.

Shipping Regulations: Section 11.

ICC (solutions with concentration not over 40%): Flammable liquid, red label, 6 quarts.

IATA: Flammable liquid, not acceptable (passenger and cargo).

METHYL MERCAPTAN

General Information

Synonym: Methanethiol.

Description: Liquid or gas.

Formula: CH_3SH.

Constants: Mol wt: 48.10, bp: 7.6°C, mp: −123.1°C, flash p: 0°F, d: 0.868 at 20°/4°C, vap. d.: 1.66, lel: 3.9%, uel: 21.8%.

Hazard Analysis

Toxic Hazard Rating:

Acute Local: Irritant 2; Ingestion 2; Inhalation 2.

Acute Systemic: Ingestion 2; Inhalation 2.

Chronic Local: U.

Chronic Systemic: U.

TLV: ACGIH (recommended); 10 parts per million in air; 20 milligrams per cubic meter of air.

Note: Methyl mercaptan is a common air contaminant. (Section 4).

Fire Hazard: Dangerous, when exposed to heat or flame.

Explosion Hazard: Unknown.

Disaster Hazard: Dangerous; on decomposition, it emits highly toxic fumes of oxides of sulfur; will react with water, steam or acids to produce toxic and flammable vapors, can react vigorously with oxidizing materials.

Countermeasures

Ventilation Control: Section 2.

To Fight Fire: Foam, carbon dioxide, dry chemical or carbon tetrachloride (Section 6).

Personnel Protection: Section 3.

First Aid: Section 1.

Storage and Handling: Section 7.

Shipping Regulations: Section 11.

I.C.C.: Flammable gas; red gas label, 300 pounds.

Coast Guard Classification: Inflammable gas; red gas label.

IATA: Flammable gas, red label, not acceptable (passenger), 140 kilograms (cargo).

METHYLMERCURIC CHLORIDE

General Information

Description: White crystals, characteristic odor.

Formula: $CH_3 \cdot Hg \cdot Cl$.

Constants: Mol wt: 251, d: 4.063, mp: 170°C.

Hazard Analysis and Countermeasures

See mercury compounds, organic and chlorides.

METHYL MERCURY DICYANDIAMIDE

General Information

Description: Crystals.

Formula: $CH_3HgNHC(NH)NHCN$.

Constants: Mol wt: 298.7, vap. d.: 10.3.

Hazard Analysis and Countermeasures

See mercury compounds, organic.

METHYL MERCURY IODIDE

General Information

Description: Crystals.

Formula: CH_3HgI.

Constants: Mol wt: 342.6, vap. d.: 11.8.

Hazard Analysis and Countermeasures

See mercury compounds, organic and iodides.

METHYL METHACRYLATE (MONOMER)

General Information

Description: Colorless liquid, very slightly soluble in water.

Formula: $CH_2C(CH_3)COOCH_3$.

Constants: Mol wt: 100.11, mp: −50°C, bp: 101.0°C, lel: 2.1%, uel: 12.5%, flash p: 50°F (O.C.), d: 0.936 at 20°/4°C, vap. press.: 40 mm at 25.5°C, vap. d.: 3.45.

Hazard Analysis

Toxic Hazard Rating:

Acute Local: Irritant 1; Ingestion 1; Inhalation 1.

Acute Systemic: Ingestion 1; Inhalation 1.

Chronic Local: U.

Chronic Systemic: Ingestion 1; Inhalation 1.

TLV: ACGIH (recommended); 100 parts per million of air; 410 milligrams per cubic meter of air.

Note: Methyl methacrylate is a common air contaminant (Section 4).

Fire Hazard: Moderate, when exposed to heat or flame; can react with oxidizing materials.

Explosion Hazard: Moderate when exposed to heat, sparks or flame.

Countermeasures

Ventilation Control: Section 2.

To Fight Fire: Foam, carbon dioxide, dry chemical or carbon tetrachloride (Section 6).

Personal Hygiene: Section 3.

Storage and Handlinb: Section 7

Shipping Regulations: Section 11.

I.C.C.: Flammable liquid; red label, 10 gallons.

MCA warning label.

IATA (inhibited): Flammable liquid, red label, 1 liter (passenger), 40 liters (cargo).

(noninhibited): Flammable liquid, not acceptable (passenger and cargo).

METHYLMETHANE. See ethane.

METHYL METHANOATE. See methyl formate.

METHYL MONOCHLOROACETATE

General Information

Description: Clear liquid.

Formula: $CH_2ClCOOCH_3$.

Constants: Mol wt: 108.53, bp: 132°C, flash p: 116°F (C.C.), d: 1.227, vap. d.: 3.74.

Hazard Analysis

Toxic Hazard Rating:

Acute Local: Irritant 2; Ingestion 2; Inhalation 2.

Acute Systemic: U.

Chronic Local: U.

Chronic Systemic: U.

Fire Hazard: Moderate, when exposed to heat or flame.

Disaster Hazard: Dangerous; when heated to decomposition, it emits highly toxic fumes of phosgene; can react with oxidizing materials.

Countermeasures

Ventilation Control: Section 2.

To Fight Fire: Carbon dioxide, dry chemical or carbon tetrachloride (Section 6).

Personnel Protection: Section 3.
Storage and Handling: Section 7.

N-METHYLMORPHINE. See Codeine

METHYL MUSTARD OIL. See methyl isothiocyanate.

1-METHYL NAPHTHALENE
General Information
Description: Colorless liquid; insoluble in water; soluble in alcohol and ether.
Formula: $C_{10}H_7CH_3$.
Constants: Mol wt: 142, d: 1.025, mp: $-32°C$, bp: 240–243°C, autoign. temp.: 984°F.

2-METHYL NAPTHALENE
General Information
Description: Solid; insoluble in water; soluble in alcohol and ether.
Formula: $C_{10}H_7CH_3$.
Constants: Mol wt: 142, d: 0.994 at 40/4°C, bp: 241–242°C, mp: 34°C.
Hazard Analysis
Toxicity: Details unknown. Limited animal experiments suggest high toxicity. See also naphthylamine.

METHYL NITRATE
General Information
Description: Colorless liquid.
Formula: CH_3NO_3.
Constants: Mol wt: 77.04, bp: 65°C (explodes), d: 1.206 at 20°/4°C, vap. d.: 2.66.
Hazard Analysis
Toxicity: Details unknown. See nitrates. Irritating to humans. Narcotic in high concentrations.
Fire Hazard: See nitrates.
Explosion Hazard: Severe, when shocked or exposed to heat. A rocket fuel.
Disaster Hazard: Dangerous! See nitrates.
Ventilation Control: Section 2.
Storage and Handling: Section 7.

METHYL NITRITE
General Information
Description: Gas.
Formula: CH_3ONO.
Constants: Mol wt: 61.04, mp: $-17°C$, bp: $-12°C$, d: 0.991 at 15°C.
Hazard Analysis
Toxicity: Details unknown. See nitrites. Narcotic in high concentrations.
Fire Hazard: Dangerous, when exposed to heat or flame (Section 6).
Explosion Hazard: Severe, when exposed to heat or flame.
Disaster Hazard: Highly dangerous; when heated, it emits highly toxic fumes of oxides of nitrogen; can react vigorously with oxidizing materials.
Countermeasures
Ventilation Control: Section 2.
Storage and Handling: Section 7.
Shipping Regulations: Section 11.
 IATA: Flammable gas, red label, not acceptable (passenger), 140 kilograms (cargo).

METHYL NITROBENZENE. See p-nitrotoluene.

1-METHYL-3-NITRO-1-NITROSO GUANIDINE
General Information
Description: Crystals.

Formula: $C_2H_5N_5O_3$.
Constant: Mol wt: 147.1.
Hazard Analysis and Countermeasures
See nitrates.

METHYL NONYL KETONE
General Information
Description: Colorless liquid, insoluble in water.
Formula: $(CH_3)(C_9H_{19})CO$.
Constants: Mol wt: 170, mp: 12°C, bp: 223°C, flash p 192°F (C.C.), d: 0.829 at 30°C, vap. d.: 5.9.
Hazard Analysis
Toxicity: Details unknown. See ketones.
Fire Hazard: Moderate, when exposed to heat or flame; car react with oxidizing materials.
Countermeasures
To Fight Fire: Carbon dioxide, dry chemical or carbor tetrachloride (Section 6).
Storage and Handling: Section 7.

9-METHYLOL CARBAZOLE
General Information
Description: Crystals.
Formula: $(C_6H_4)_2NCH_2OH$.
Constant: Mol wt: 197.2.
Hazard Analysis
Toxicity: Details unknown; a fumigant. Probably toxic.
Disaster Hazard: Dangerous, when heated, it emits highly toxic fumes of nitrogen oxides; can react on contact with oxidizing materials.
Countermeasures
Storage and Handling: Section 7.

METHYLOL PINENE
General Information
Description: Water-white to light straw-colored liquid.
Formula: $C_{11}H_{17}OH$.
Constants: Mol wt: 146.3, bp: 111°C at 10 mm, d: 0.963 at 25°C, vap. d.: 5.05.
Hazard Analysis
Toxicity: Unknown.
Fire Hazard: Slight, when exposed to heat or flame; can react with oxidizing materials (Section 6).
Countermeasures
Storage ahd Handling: Section 7.

METHYL ORANGE
General Information
Synonyms: Helianthine B; orange III.
Description: Orange-yellow powder.
Formula: $C_{14}H_{14}N_3NaO_3S$.
Constant: Mol wt: 327.3.
Hazard Analysis
Toxicity: Unknown.
Disaster Hazard: Dangerous; when heated to decomposition, it emits highly toxic fumes of oxides of sulfur.
Countermeasures
Storage and Handling: Section 7.

METHYL OXIDE. See dimethyl ether.

METHYLOXIRANE. See propylene oxide.

METHYL PARABEN
General Information
Synonyms: Methyl p-hydroxybenzoate; methyl parasept.
Description: Colorless crystals or white crystalline powder odorless or slight characteristic odor; almost insoluble in water; soluble in alcohol and ether.

Formula: $CH_3OOCC_6H_4OH$.
Constants: Mol wt: 152, mp: 125–128°C.
Hazard Analysis
See p-hydroxybenzoic acid esters.
Note: Used as a chemical preservative food additive. (Section 10).

METHYL PARASEPT. See methyl paraben.

METHYL PARATHION
General Information
Description: A solid or powder.
Hazard Analysis
Toxicity: See parathion.
Explosion Hazard: Moderate, when exposed to 120°C.
Disaster Hazard: Moderately dangerous; when heated, it emits highly toxic fumes and may explode; can react with oxidizing materials.
Countermeasures
Storage and Handling: Section 7.
Shipping Regulations: Section 11.
MCA warning label.
ICC (liquid): Poison B, poison label, 1 quart.
IATA (liquid): Poison B, poison label, not acceptable (passenger), 1 liter (cargo).

METHYL PARATHION MIXTURE. See methyl parathion.
Shipping Regulations: Section 11.
ICC (liquid): Poison B, poison label, 1 quart.
ICC (dry): Poison B, poison label, 200 pounds.
IATA (dry, containing more than 2% methyl parathion): Poison B, poison label, not acceptable (passenger), 95 kilograms (cargo).
IATA (dry, containing not more than 2% methyl parathion): Poison B, poison label, 25 kilograms (passenger), 95 kilograms (cargo).
IATA (liquid, containing more than 25% methyl parathion): Poison B, poison label, not acceptable (passengers), 1 liter (cargo).
IATA (liquid, containing not more than 25% methyl parathion): Poison B, poison label, 1 liter (passenger and cargo).

METHYL PENTADECYL KETONE
General Information
Description: Insoluble in water.
Formula: $C_{15}H_{31}COCH_3$.
Constants: Mol wt: 254, bp: 313°F at 3 mm, flash p: 248°F.
Hazard Analysis
Toxic Hazard Rating: U. See also ketones.
Fire Hazard: Slight, when exposed to heat or flame.
Countermeasures
Storage and Handling: Section 7.
To Fight Fire: Water, foam (Section 6).

METHYL PENTADIENE
General Information
Description: Liquid.
Formula: C_6H_{10}.
Constants: Mol wt: 82.14, bp: 75–77°C, flash p: −30°F, d: 0.7184 at 20°/4°C, vap. d.: 2.83.
Hazard Analysis
Toxicity: Unknown. Probably irritant and narcotic in high concentration.
Fire Hazard: Highly dangerous, when exposed to heat or flame.
Disaster Hazard: Dangerous! Keep away from heat and open flame.
Countermeasures
Ventilation Control: Section 2.
To Fight Fire: Foam, carbon dioxide, dry chemical or carbon tetrachloride (Section 6).
Storage and Handling: Section 7.
Shipping Regulations: Section 11.

IATA: Flammable liquid, red label, 1 liter (passenger), 40 liters (cargo).

METHYL PENTALDEHYDE
General Information
Synonym: Methyl pentanal.
Description: Very slightly soluble in water.
Formula: $CH_3CH_2CH_2C(CH_3)HCHO$.
Constants: Mol wt: 100, d: 0.8092, bp: 118.3°C, fp: −100°C, flash p: 68°F (O.C.).
Hazard Analysis
Toxic Hazard Rating: U. See also aldehydes.
Fire Hazard: Dangerous, when exposed to heat or flame.

METHYL PENTANAL. See methyl pentaldehyde.

2-METHYLPENTANE
General Information
Synonym: Dimethylpropylmethane.
Description: Colorless liquid.
Formula: $(CH_3)_2CH(CH_2)_2CH_3$.
Constants: Mol wt: 86.17, bp: 60°C, fp: −154°C, flash p: <20°F, d: 0.654 at 20°/4°C, vap. press.: 400 mm at 41.6°C, vap. d.: 2.97, autoign. temp.: 583°F, lel: 1.2%, uel: 7.0%.
Hazard Analysis
Toxicity: Details unknown; may have narcotic or anesthetic properties.
Fire Hazard: Dangerous, when exposed to heat or flame. Can react vigorously with oxidizing materials.
Explosion Hazard: Unknown.
Disaster Hazard: Dangerous; keep away from heat and open flame; reacts vigorously with oxidizing agents.
Countermeasures
Ventilation Control: Section 2.
To Fight Fire: Foam, carbon dioxide, dry chemical or carbon tetrachloride (Section 6).
Storage and Handling: Section 7.
Shipping Regulations: Section 11.
IATA: Flammable liquid, red label, 1 liter (passenger), 40 liters (cargo).

3-METHYLPENTANE
General Information
Synonym: Diethylmethylmethane.
Description: Colorless liquid.
Formula: C_6H_{11}.
Constants: Mol wt: 86.17, bp: 63.3°C, fp: −118°C (sets to a glass), flash p: <20°F, d: 0.664 at 20°/4°C, vap. press.: 100 mm at 10.5°C, vap. d.: 2.97.
Hazard Analysis
Toxicity: Details unknown; may have narcotic or anesthetic properties.
Fire Hazard: Dangerous, when exposed to heat or flame. Can react vigorously with oxidizing agents.
Explosion Hazard: Unknown.
Disaster Hazard: Dangerous; keep away from heat and open flame; can react vigorously with oxidizing materials.
Countermeasures
Ventilation Control: Section 2.
To Fight Fire: Foam, carbon dioxide, dry chemical or carbon tetrachloride (Section 6).
Storage and Handling: Section 7.
Shipping Regulations: See 2-Methylpentane.

2-METHYL-1,3-PENTANEDIOL
General Information
Description: Soluble in water.
Formula: $CH_3CH_2CH(OH)CH(CH_3)CH_2OH$.
Constants: Mol wt: 118.17, d: 0.9745, bp: 220.30°C, fp: −30°C, flash p: 230°F.
Hazard Analysis
Toxicity: Unknown. See also alcohols.
Fire Hazard: Slight, when exposed to heat or flame.

Countermeasures
Storage and Handling: Section 7.
To Fight Fire: Water or foam (Section 6).

2-METHYL-2,4-PENTANEDIOL. See hexylene glycol.

3-METHYLPENTANEDIOL-1,5
General Information
Description: Clear liquid.
Formula: $HOCH_2CH_2CH(CH_3)CH_2CH_2OH$.
Constants: Mol wt: 118.17, bp: 248.4°C, fp: −60°C, d: 0.9755 at 20°/20°C, vap. press.: <0.01 mm at 20°C, vap. d.: 4.
Hazard Analysis
Toxicity: Details unknown. See also alcohols.
Fire Hazard: Moderate, when exposed to heat or flame; can react with oxidizing materials.
Countermeasures
To Fight Fire: Foam, dry chemical, carbon dioxide or carbon tetrachloride (Section 6).
Storage and Handling: Section 7.

2-METHYL PENTANOIC ACID
General Information
Description: Water-white liquid; insoluble in water.
Formula: $C_3H_7CH(CH_3)COOH$.
Constants: Mol wt: 116, d: 0.9242 at 20/20°C, bp: 196.4°C, vap. press.: 0.02, mm at 20°C, fp: sets to glass <85°C, flash p: 225°F (O.C.).
Hazard Analysis
Toxic Hazard Rating: U.
Fire Hazard: Moderate, when exposed to heat or flame.
Countermeasures
Storage and Handling: Section 7.
To Fight Fire: Water or foam (Section 6).

2-METHYL PENTANOL-1
General Information
Description: Liquid.
Formula: $CH_3CH_2CH_2CH(CH_3)CH_2OH$.
Constants: Mol wt: 102.17, bp: 147.5°C, flash p: 135°F (O.C.), d: 0.8230 at 20°/20°C, vap. d.: 3.52.
Hazard Analysis
Toxicity: Details unknown. See also alcohols.
Fire Hazard: Moderate, when exposed to heat or flame; can react with oxidizing materials.
Countermeasures
To Fight Fire: Foam, carbon dioxide, dry chemical or carbon tetrachloride (Section 6).
Storage and Handling: Section 7.

4-METHYL PENTANOL-2. See methyl amyl alcohol.

4-METHYL-2-PENTANONE. See methyl isobutyl ketone.

2-METHYL-1-PENTENE
General Information
Synonym: 1-Methyl-1-propyl ethylene.
Description: Liquid.
Formula: $CH_2C(CH_3)CH_2CH_2CH_3$.
Constants: Mol wt: 84.2, mp: −135.8°C, bp: 62°C, flash p: <20°F, d: 0.684 at 15.5°/15.5°C, vap. press.: 326 mm at 37.3°C, vap. d.: 2.9.
Hazard Analysis
Toxicity: Unknown. Probably irritant and narcotic in high concentrations.
Fire Hazard: Dangerous, when exposed to heat or flame.
Explosion Hazard: Unknown.

Disaster Hazard: Dangerous; keep away from heat and open flame; can react vigorously with oxidizing materials.
Countermeasures
Ventilation Control: Section 2.
To Fight Fire: Carbon dioxide, dry chemical or carbon tetrachloride (Section 6).
Storage and Handling: Section 7.

2-METHYLPENTENE-2
General Information
Description: A liquid.
Formula: C_6H_{12}.
Constants: Mol wt: 84.2, bp: 66.9°C, flash p: <20°F, d: 0.690 at 15.5°/15.5°C, vap. press.: 326 mm at 38°C, vap. d.: 2.9.
Hazard Analysis
Toxicity: Unknown.
Fire Hazard: Dangerous, when exposed to heat or flame.
Explosion Hazard: Unknown.
Disaster Hazard: Dangerous, upon exposure to heat or flame; can react vigorously with oxidizing materials.
Countermeasures
Ventilation Control: Section 2.
To Fight Fire: Foam, carbon dioxide, dry chemical or carbon tetrachloride (Section 6).
Storage and Handling: Section 7.

4-METHYLPENTENE
General Information
Description: A liquid.
Formula: C_6H_{12}.
Constants: Mol wt: 84.2, mp: −153.6°C, bp: 54°C, flash p: <20°F, d: 0.668 at 15.5°/15.5°C, vap. press.: 424 mm at 38°C, vap. d.: 2.9.
Hazard Analysis
Toxicity: Unknown. Probably irritant and narcotic in high concentrations.
Fire Hazard: Dangerous, when exposed to heat or flame.
Explosion Hazard: Unknown.
Disaster Hazard: Dangerous, upon exposure to heat or flame; can react vigorously with oxidizing materials.
Countermeasures
Ventilation Control: Section 2.
To Fight Fire: Foam, carbon dioxide, dry chemical or carbon tetrachloride (Section 6).
Storage and Handling: Section 7.

4-METHYL-2-PENTENE
General Information
Synonym: 1-Isopropyl-2-methyl ethylene.
Description: Liquid.
Formula: $CH_3CHCHCH(CH_3)_2$.
Constants: Mol wt: 84.16, mp: −134.4°C, bp: 58°C, d: 0.670 at 20°/4°C, vap. d.: 2.90, flash p: <20°F.
Hazard Analysis
Toxicity: Unknown. Probably irritant and narcotic in high concentrations.
Disaster Hazard: Moderate; when heated, it emits acrid fumes; can react with oxidizing materials (Section 6).
Countermeasures
Storage and Handling: Section 7.

4-METHYL-3-PENTEN-2-ONE. See mesityl oxide.

4-METHYLPENTYL 2-ACETATE. See methylamyl acetate.

3-METHYL-1-PENTYN-3-OL. See methyl pentynol.

TOXIC HAZARD RATING CODE (For detailed discussion, see Section 1.)

0 NONE: (a) No harm under any conditions; (b) Harmful only under unusual conditions or overwhelming dosage.

1 SLIGHT: Causes readily reversible changes which disappear after end of exposure.

2 MODERATE: May involve both irreversible and reversible changes not severe enough to cause death or permanent injury.

3 HIGH: May cause death or permanent injury after very short exposure to small quantities.

U UNKNOWN: No information on humans considered valid by authors.

METHYL PENTYNOL
General Information
Synonyms: 3-Methyl-1-pentyn-3-ol; dormison; 2-ethynyl-2-butanol.
Description: Colorless, mobile liquid. Acrid odor, burning taste. Soluble in water, ether etc.
Formula: HC:CC(OH)CH₃C₂H₅.
Constants: Mol wt: 98.14, mp: −30.6°C, bp: 122°C, flash p: 101°F, d: 0.8688 at 20°C, vap. d.: 3.38.
Hazard Analysis
Toxicity: Details unknown. Used as a soporific. Average doses may produce dermatitis, eructations, psychoses and CNS abnormalities. Overdoses can produce coma and death.
Fire Hazard: Slight; can react with oxidizing materials.
Countermeasures
Storage and Handling: Section 7.
To Fight Fire: Alcohol foam.

3-METHYL-1-PENTYN-3-OL CARBAMATE
General Information
Synonym: Oblivon.
Description: Crystals, slightly water soluble.
Formula: C₇H₁₁NO₂.
Constants: Mol wt: 141.2, mp: 57°C, bp: 121°C at 16 mm.
Hazard Analysis
Toxicity: Used as a tranquilizer. Overdoses may cause CNS depression and death. Potentiates action of alcohol and barbiturates.
Disaster Hazard: Dangerous; when heated to decomposition, it forms highly toxic fumes of oxides of nitrogen.

n-METHYLPHENOL. See m-cresol.

METHYL PHENYLACETATE
General Information
Description: Colorless liquid.
Formula: C₆H₅CH₂COOCH₃.
Constants: Mol wt: 150.17, d: 1.062, vap. d.: 5.18.
Hazard Analysis
Toxicity: Details unknown. See esters.
Disaster Hazard: Moderately dangerous; when heated to decomposition, it emits toxic fumes.
Countermeasures
Storage and Handling: Section 7.

METHYL PHENYL CARBINOL. See phenyl methyl carbinol.

α-METHYL-1-PHENYL ETHYLENE. See α-methyl styrene.

α-METHYL-3-PHENYL GLYCIDIC ACID, ETHYL ESTER. See ethyl methyl phenylglycidate.

METHYL PHENYL KETONE. See phenyl methyl ketone.

2-METHYL-2-PHENYLPROPANE. See tert-butylbenzene.

3-METHYL-1-PHENYL-5-PYRAZOLONE
General Information
Description: White powder.
Formula: C₆H₅N₂COCH₂CCH₃.
Constants: Mol wt: 174.2, bp: 191°C at 7 mm, mp: 128.9°C.
Hazard Analysis
Toxicity: Unknown.
Disaster Hazard: Dangerous; when heated to decomposition, it emits highly toxic fumes of oxides of nitrogen; can react with oxidizing materials.
Countermeasures
Storage and Handling: Section 7.

METHYL PHOSPHINE
General Information
Description: Colorless gas.
Formula: CH₃PH₂.

Constants: Mol wt: 48.0, bp: −14°C.
Hazard Analysis
Toxic Hazard Rating:
   Acute Local: U.
   Acute Systemic: Inhalation 3.
   Chronic Local: U.
   Chronic Systemic: Inhalation 3.
Fire Hazard: Dangerous, when exposed to heat or flame (Section 6).
Explosion Hazard: Unknown.
Disaster Hazard: Dangerous; when heated, it emits highly toxic fumes of oxides of phosphorus; can react vigorously with oxidizing materials.
Countermeasures
Ventilation Control: Section 2.
Storage and Handling: Section 7.

METHYL PHOSPHONIC DICHLORIDE
Hazard Analysis
Toxicity: Details unknown. Animal experiments show strong irritation. See also hydrochloric acid.
Disaster Hazard: Dangerous. See phosphorus compounds and hydrochloric acid.

3-METHYLPHTHALANIC ACID
Hazard Analysis
Toxicity: U. It is hydrolyzed in body to m-toluidine which can cause kidney damage and methemoglobinemia. See also m-toluidine.

METHYL PHTHALATE. See dimethyl phthalate.

METHYL PHTHALYL ETHYL GLYCOLATE
General Information
Description: Liquid.
Formula: CH₃CO₂C₆H₄CO₂CH₂CO₂C₂H₅.
Constants: Mol wt: 266.2, bp: 310°C, flash p: 375°F (C.C.), d: 1.220, vap. d.: 9.16.
Hazard Analysis
Toxicity: Details unknown. See esters.
Fire Hazard: Slight, when exposed to heat or flame; can react with oxidizing materials.
Countermeasures
To Fight Fire: Water, foam, carbon dioxide, dry chemical or carbon tetrachloride (Section 6).
Storage and Handling: Section 7.

METHYL PICRATE. See trinitroanisole.

n-METHYL PIPERAZINE
General Information
Description: A hygroscopic solid; typical amine-like odor.
Formula: C₅H₁₂N₂.
Constants: Mol wt: 100.1, d: 0.9031 at 20°/20°C, mp: 65.5°C, bp: 139°C, flash p: 108°F, vap. d.: 3.5.
Hazard Analysis
Toxicity: Unknown. Limited animal experiments suggest moderate toxicity and high irritation. See also piperazine.
Fire Hazard: Moderate, when exposed to heat or flame; can react with oxidizing materials.
Countermeasures
To Fight Fire: Foam, carbon dioxide, dry chemical or carbon tetrachloride (Section 6).
Storage and Handling: Section 7.

2-METHYLPROPANE. See isobutane.

2-METHYL-2-PROPANETHIOL. See t-butyl mercaptan.

2-METHYLPROPANAL. See isobutyl aldehyde.

2-METHYL-1-PROPANAL. See isobutyl alcohol.

2-METHYL-2-PROPANAL. See tert-butyl alcohol.

2-METHYL PROPANOIC ACID. See isobutyric acid.

2-METHYL PROPENE. See isobutylene.

2-METHYLPROPENITRILE. See methacrylonitrile.

METHYL PROPIONATE
General Information
Description: Colorless liquid.
Formula: $CH_3CH_2COOCH_3$.
Constants: Mol wt: 88.15, mp: $-87.5°C$, bp: $79.8°C$, flash p: $28°F$ (C.C.), d: 0.937 at $4°C$, autoign. temp.: $876°F$, vap. press.: 40 mm at 11.0 C, vap. d.: 3.03, lel: 2.5%, uel: 13%.
Hazard Analysis
Toxic Hazard Rating:
  Acute Local: Irritant 1; Ingestion 1; Inhalation 1.
  Acute Systemic: U.
  Chronic Local: U.
  Chronic Systemic: U.
Fire Hazard: Dangerous, when exposed to heat or flame.
Spontaneous Heating: No.
Explosion Hazard: Unknown.
Disaster Hazard: Dangerous; keep away from heat and open flame; can react vigorously with oxidizing materials.
Countermeasures
Ventilation Control: Section 2.
To Fight Fire: Foam, carbon dioxide, dry chemical or carbon tetrachloride (Section 6).
Storage and Handling: Section 7.
Shipping Regulations: Section 11.
  IATA: Flammable liquid, red label, 1 liter (passenger), 40 liters (cargo).

α-METHYL PROPIONIC ACID. See isobutyric acid.

METHYL PROPYL CARBINOL
General Information
Synonym: dl-Pentanol-2.
Description: Water-white liquid.
Formula: $CH_3CH_2CH_2CH(OH)CH_3$.
Constants: Mol wt: 88.2, bp: $119°C$, flash p: $105°F$ (C.C.), d: 0.8303 at $20°/4°C$, vap. d.: 3.03.
Hazard Analysis
Toxicity: Details unknown. See also alcohols.
Fire Hazard: Moderate, when exposed to heat or flame; can react with oxidizing materials.
Countermeasures
To Fight Fire: Carbon dioxide, dry chemical or carbon tetrachloride (Section 6).
Ventilation Control: Section 2.
Storage and Handling: Section 7.

METHYL PROPYL CARBINYLAMINE. See sec-amylamine.

METHYL PROPYL ETHER
General Information
Synonym: Neothyl.
Description: Mobile liquid; slightly water soluble; miscible in alcohol and ether.
Formula: $C_4H_{10}O$.
Constants: Mol wt: 74.1, d: 0.7494 at $0°/4°C$, bp: $38.5°C$.
Hazard Analysis and Countermeasures
See ethers.

1-METHYL-1-PROPYLETHYLENE. See 2-methyl-1-pentene.

METHYL n-PROPYL KETONE. See pentanone-2.

α-METHYLPROPYL NITRITE. See sec-butyl nitrite.

2-METHYLPYRAZINE
General Information
Description: Liquid, pyridine-like odor.
Formula: $C_5H_6N_2$.
Constants: Mol wt: 94.12, mp: $-29°C$, bp: $133°C$ at 737 mm, flash p: $122°F$ (C.O.C.), d: 1.0224 at $25°/25°C$, vap. d.: 3.2.
Hazard Analysis
Toxicity: Details unknown.
Fire Hazard: Moderate, when exposed to heat or flame (Section 6).
Disaster Hazard: Dangerous; when heated to decomposition, it emits highly toxic fumes of nitrogen oxides; can react with oxidizing materials.
Countermeasures
Storage and Handling: Section 7.

α-METHYLPYRIDINE. See α-picoline.

3-METHYLPYRIDINE. See β-picoline.

4-METHYLPYRIDINE. See γ-picoline.

1-METHYL-2-β-PYRIDYLPYRROLIDINE
General Information
Description: Pale straw-colored liquid.
Formula: $C_{10}H_{10}N_2$.
Constants: Mol wt: 158.2, bp: $247°C$, d: 1.01 at $20°/20°C$, vap. d.: 5.46.
Hazard Analysis
Toxicity: Highly toxic. See closely related compound, nicotine.
Disaster Hazard: Dangerous; when heated to decomposition, it emits highly toxic fumes of oxides of nitrogen; can react with oxidizing materials.
Countermeasures
Storage and Handling: Section 7.

METHYL PYRROLE-n
General Information
Description: Liquid; insoluble in water.
Formula: $N(CH_3)CH:CHCH:CH$.
Constants: Mol wt: 91, d: 0.9, vap. d.: 2.8, bp: $112°C$, fp: $-57°C$, flash p: $61°F$.
Hazard Analysis
Toxic Hazard Rating: U. See pyrrole.
Fire Hazard: Dangerous, when exposed to heat or flame.

N-METHYL PYRROLIDINE
General Information
Description: Colorless to yellow liquid with penetrating amine-like odor.
Formula: $C_4H_8NCH_3$.
Constants: Mol wt: 85.2, bp: $80.5°C$, fp: $-90°C$, d: 0.8054 at $20°/20°C$, flash p: $7°F$, vap. d.: 2.9.
Hazard Analysis
Toxicology: This material is strongly alkaline. Contact of liquid or vapors of this material with skin, eyes or mucous membranes should be avoided. See also ammonia.
Fire Hazard: Dangerous; keep away from sparks, heat sources and powerful oxidizers. Keep in closed containers.
Disaster Hazard: Dangerous; when heated to decomposition, it emits highly toxic fumes.
Countermeasures
Ventilation Control: Section 2.

TOXIC HAZARD RATING CODE (For detailed discussion, see Section 1.)

0 NONE: (a) No harm under any conditions; (b) Harmful only under unusual conditions or overwhelming dosage.

1 SLIGHT: Causes readily reversible changes which disappear after end of exposure.

2 MODERATE: May involve both irreversible and reversible change not severe enough to cause death or permanent injury.

3 HIGH: May cause death or permanent injury after very short exposure to small quantities.

U UNKNOWN: No information on humans considered valid by author

Personnel Protection: Section 3,
Storage and Handling: Section 7.
To Fight Fire: Alcohol foam.

N-METHYL-2-PYRROLIDONE
General Information
Synonym: 1-Methyl-2-pyrrolidone.
Description: Colorless liquid; mild odor.
Formula: $H_3CNCH_2CH_2CH_2CO$.
Constants: Mol wt: 99.13, bp: 202°C, fp: −24°C, flash p: 204°F, d: 1.027 at 25°/4°C, vap. d.: 3.4.
Hazard Analysis
Toxicity: Unknown. See also N-methyl pyrrolidone.
Fire Hazard: Moderate, when exposed to heat or flame; can react with oxidizing materials.
Countermeasures
To Fight Fire: Foam, carbon dioxide, dry chemical or carbon tetrachloride (Section 6).
Storage and Handling: Section 7.

2-METHYL QUINOLINE
General Information
Synonym: Quinaldine.
Description: Colorless, oily liquid; quinoline odor; insoluble in water; soluble in chloroform, ether.
Formula: $C_6H_4(CH_3)C_3H_2N$.
Constants: Mol wt: 143.18, d: 1.06, bp: 246–247°C.
Hazard Analysis
Toxic Hazard Rating:
Acute Local: Irritant 3; Ingestion 3; Inhalation 3.
Acute Systemic: Ingestion 3.
Chronic Local: U.
Chronic Systemic: U.
Toxicity: Data based on animal experiments.

METHYL SALICYLATE
General Information
Synonym: Oil of wintergreen.
Description: Colorless, yellowish or reddish, oily liquid.
Formula: HOC_6H_4-1-$COOCH_3$.
Constants: Mol wt: 152.14, bp: 223.3°C, ulc: 20–25, flash p: 214°F (C.C.), fp: −1.2°C, d: 1.1840 at 20.2°/4°C, autoign. temp.: 850°F, vap. press.: 1 mm at 54.0°C, vap. d.: 5.24.
Hazard Analysis
Toxic Hazard Rating:
Acute Local: Irritant 2; Allergen 1; Ingestion 2.
Acute Systemic: Ingestion 3.
Chronic Local: Allergen 1.
Chronic Systemic: U.
Toxicology: Accidental acute poisoning is not uncommon. Kidney irritation, vomiting and convulsions occur.
Fire Hazard: Slight, when exposed to heat or flame; can react with oxidizing materials.
Spontaneous Heating: No.
Countermeasures
To Fight Fire: Water, foam, carbon dioxide, dry chemical or carbon tetrachloride (Section 6).
Personnel Protection: Section 3.
Storage and Handling: Section 7.

METHYL SEBACATE. See dimethyl sebacate.

METHYL SELENIDE. See dimethyl selenide.

METHYL SILICANE
General Information
Description: Liquid.
Formula: $H_3Si(CH_3)$.
Constants: Mol wt: 46.1, d: 0.62 at −57°C, mp: −156.4°C, bp: 31°C.
Hazard Analysis
Toxicity: Details unknown. See silanes.

METHYL SILICATE
General Information
Description: Clear liquid.

Formula: $Si(OCH_3)_4$.
Constants: Mol wt: 152.2, vap. d.: 5.25.
Hazard Analysis
Acute Local: Irritant 2; Ingestion 2; Inhalation 2.
Acute Systemic: Ingestion 2; Inhalation 2.
Chronic Local: Irritant 2.
Chronic Systemic: Ingestion 2; Inhalation 2.
Toxicology: This material can cause extensive necrosis (experimental), keratoconus and opaque cornea; it also causes severe human eye injuries, as well as necrosis of corneal cells, which progresses long after the exposure has ceased. It is destructive and its effects resist treatment. Permanent blindness is possible from exposure to it. Both the liquid and vapor of this material are dangerous to the eyes. From animal experimentation, the minimum lethal dose by oral administration has been found to be 0.07 ml/100 g of body weight for the rat and 0.01 ml/100 g of body weight intravenously for the rabbit. Administration of this material to animals resulted in death from within a few hours to a few days.

The kidney seems to be the organ which is usually injured, regardless of the mode of administration of the toxic material. In less severe cases, degeneration of the convoluted tubules was found, with complete degeneration of the organ in the more severe cases. Pulmonary edema also occurred in those animals who had received intravenous injections. It has been reported that exposure to methyl silicate vapor under certain conditions of humidity or exposure to the liquid may cause the eye damages noted above. This material was found to be more toxic to animals than either ethyl silicate or silicic acid, although it has been thought that the injury caused is largely due to the action of the silicic acid.
Countermeasures
Ventilation Control: Section 2.
Personnel Protection: Section 3.
Storage and Handling: Section 7.

METHYLSTANNIC ACID
General Information
Description: White powder; insoluble in water.
Formula: $(CH_3)SnOOH$.
Constants: Mol wt: 166.7.
Hazard Analysis
Toxicity: Details unknown. See tin compounds.

METHYL STEARATE
General Information
Description: Liquid to semi-solid.
Formula: $C_{17}H_{35}CO_2CH_3$.
Constants: Mol wt: 298.5, mp: 38°C, bp: 215°C at 15 mm, flash p: 307°F (C.C.), d: 0.860.
Hazard Analysis
Toxicity: Unknown.
Fire Hazard: Slight, when exposed to heat or flame; can react with oxidizing materials.
Countermeasures
To Fire Fire: Foam, carbon dioxide, dry chemical or carbon tetrachloride (Section 6).
Storage and Handling: Section 7.

METHYL STYRENE
General Information
Description: Liquid, insoluble in water.
Formula: C_9H_{10}.
Constants: Mol wt: 118.17, mp: −23.21°C, bp: 165.4°C, lel: 0.7%, autoign. temp.: 921°F, flash p: 134°F (O.C.), d: 0.9062 at 25°/25°C, vap. press.: 1 mm at 7.4°C, vap. d.: 4.08.
Hazard Analysis
Toxic Hazard Rating:
Acute Local: Irritant 2; Inhalation 2.
Acute Systemic: Ingestion 2; Inhalation 2.
Chronic Local: U.
Chronic Systemic: U.

TLV: ACGIH (recommended) 100 parts per million of air; 480 milligrams per cubic meter of air.

Toxicology: From animal studies it appears that immediate deaths occur from primary action on CNS whereas delayed deaths come about from pneumonia.

Fire Hazard: Moderate, when exposed to heat or flame; can react with oxidizing materials.

Explosion Hazard: Moderate, when exposed to heat or flame.

Countermeasures

Ventilation Control: Section 2.

To Fight Fire: Foam, carbon dioxide, dry chemical or carbon tetrachloride (Section 6).

Personnel Protection: Section 3.

Storage and Handling: Section 7.

α-METHYL STYRENE
General Information

Synonym: 1-Methyl-1-phenyl ethylene.

Description: Colorless liquid; insoluble in water.

Formula: $C_6H_5C(CH_3)CH_2$.

Constants: Mol wt: 118.17, d: 0.9062 at 25/25°C, bp: 165.4°C, flash p: 129°F, autoign. temp.: 1066°F, lel: 1.9%, uel: 6.1%.

Hazard Analysis

Toxicity: See methyl styrene.

Fire Hazard: Moderate, when exposed to heat or flame.

Countermeasures

Storage and Handling: Section 7.

To Fight Fire: Water, foam, mist or dry chemical (Section 6).

METHYLSTYRYL PHENYL KETONE
General Information

Description: Liquid.

Formula: $C_{16}H_{14}O$.

Constants: Mol wt: 222.3, vap. d.: 7.67.

Hazard Analysis

Toxic Hazard Rating:

Acute Local: Irritant 1; Ingestion 1; Inhalation 1.

Acute Systemic: U.

Chronic Local: U.

Chronic Systemic: U.

Fire Hazard: Slight, when exposed to heat or flame; can react with oxidizing materials (Section 6).

Countermeasures

Personal Hygiene: Section 3.

Storage and Handling: Section 7.

METHYL SULFATE. See dimethyl sulfate.

METHYL SULFIDE
General Information

Synonyms: Methylthiomethane; dimethyl sulfide.

Description: Colorless liquid; disagreeable odor, soluble in alcohol and ether.

Formula: $(CH_3)_2S$.

Constants: Mol wt: 62.13, mp: −83.2°C, lel: 2.2%, uel: 19.7%, bp: 37.5°–38°C, d: 0.8458 at 21°/4°C, vap. d.: 2.14, autoign. temp.: 403°F.

Hazard Analysis

Toxicity: Details unknown; probably highly toxic. See sulfides.

Fire Hazard: Dangerous, when exposed to heat or flame.

Explosion Hazard: Moderate.

Disaster Hazard: Dangerous; when heated to decomposition, it emits highly toxic fumes of oxides of sulfur,

and may explode; can react vigorously with oxidizing materials.

Countermeasures

Ventilation Control: Section 2.

To Fight Fire: Carbon dioxide, dry chemical or carbon tetrachloride (Section 6).

Storage and Handling: Section 7.

Shipping Regulations: Section 11.

I.C.C.: Flammable liquid; red label, 10 gallons.

Coast Guard Classification: Inflammable liquid; red label.

IATA: Flammable liquid, red label, 1 liter (passenger) 40 liters (cargo).

METHYL SULFITE
General Information

Synonym: Dimethyl sulfite.

Description: Colorless liquid.

Formula: $(CH_3)_2SO_3$.

Constants: Mol wt: 110.1, bp: 126°C, d: 1.242 at 0°/0°C vap. d.: 3.8.

Hazard Analysis

Toxicity: Details unknown, an irritant and probably toxic material.

Disaster Hazard: Dangerous; when heated to decomposition, it emits highly toxic fumes of oxides of sulfur; it will react with water, steam or acids to produce toxic and corrosive fumes; can react with oxidizing materials.

Countermeasures

Storage and Handling: Section 7.

METHYL SULFOXIDE
General Information

Synonyms: USAN; DMSO; dimethyl sulfoxide.

Description: Colorless liquid; nearly odorless; soluble in water, ethonal, acetone, ether, benzene, and chloroform.

Formula: $(CH_3)_2SO$.

Constants: Mol wt: 78, bp: 189°C, mp: 18.45°C, d: 1.1 at 20°/4°C.

Hazard Analysis

Toxicity: Details unknown. Limited animal experiment suggest low toxicity.

Disaster Hazard: Dangerous. See sulfur compounds.

METHYL SULFURIC ACID
General Information

Synonyms: Methyl hydrogen sulfate; hydrogen methyl sulfate; monomethyl sulfate; sulfuric acid monomethyl ether; acid methyl sulfate.

Description: Oily liquid; freely soluble in water; less soluble in alcohol.

Formula: CH_3OSO_2OH.

Constant: Mol wt: 112.10.

Hazard Analysis and Countermeasures

See dimethyl sulfate.

N-METHYLTAURINE. See 2-(N-methylamino) ethan sulfonic acid.

METHYL TELLURIDE. See dimethyl telluride.

2-METHYL TETRAHYDROFURAN
General Information

Synonyms: Tetrahydromethylfuran; tetrahydrosylvan.

Description: Colorless, mobile liquid; ether-like odor.

Formula: $OCH_2CH_2CH_2CHCH_3$.

Constants: Mol wt: 86.13, bp: 80°C, flash p: 12°F, d: 0.853 at 20°/4°C, vap. d.: 2.97.

TOXIC HAZARD RATING CODE *(For detailed discussion, see Section 1.)*

0 NONE: (a) No harm under any conditions; (b) Harmful only under unusual conditions or overwhelming dosage.

1 SLIGHT: Causes readily reversible changes which disappear after end of exposure.

2 MODERATE: May involve both irreversible and reversible change not severe enough to cause death or permanent injury.

3 HIGH: May cause death or permanent injury after very short exposure to small quantities.

U UNKNOWN: No information on humans considered valid by author.

Hazard Analysis
Toxicity: Unknown. See also tetrahydrofuran.
Fire Hazard: Dangerous, when exposed to heat or flame.
Explosion Hazard: Unknown.
Disaster Hazard: Dangerous; keep away from heat and open flame; can react vigorously with oxidizing materials.
Countermeasures
Ventilation Control: Section 2.
To Fight Fire: Alcohol foam, carbon dioxide, dry chemical or carbon tetrachloride (Section 6).
Storage and Handling: Section 7.

METHYL 1,2,5,6-TETRAHYDRO-1-METHYL-NICOTINATE. See arecoline base.

1-N-METHYL TETRAHYDROPAPAVERINE. See laudanosine.

METHYL TETRATHIO-o-STANNATE
General Information
Description: Solid.
Formula: $(CH_3S)_4Sn$.
Constants: Mol wt: 307.1, mp: 31°C, bp: 81°C at 0.001 mm.
Hazard Analysis and Countermeasures
See tin compounds and sulfur compounds.

2-METHYL THIOETHYL ACRYLATE
Hazard Analysis
Toxicity: Details unknown. Limited animal experiments suggest moderate toxicity.
Disaster Hazard: Dangerous. See sulfur compounds.

METHYL THIOMETHANE. See methyl sulfide.

METHYLTIN TRIBROMIDE
General Information
Description: White crystals, soluble in water, ether, alcohol and benzene.
Formula: $(CH_3)SnBr_3$.
Constants: Mol wt: 373.5, mp: 55°C, bp: 211°C at 746 mm.
Hazard Analysis and Countermeasures
See tin compounds, hydrobromic acid and bromides.

METHYLTIN TRICHLORIDE
General Information
Description: Colorless crystals, soluble in water and organic solvents.
Formula: $(CH_3)SnCl_3$.
Constants: Mol wt: 240.1, mp: 43°C.
Hazard Analysis and Countermeasures
See tin compounds, hydrochloric acid and chlorides.

METHYLTIN TRIIODIDE
General Information
Description: Light yellow crystals, soluble in water, ether, alcohol and benzene.
Formula: $(CH_3)SnI_3$.
Constants: Mol wt: 514.5, mp: 86.5°C.
Hazard Analysis and Countermeasures
See tin compounds, hydriodic acid and iodides.

METHYL TOLUENE SULFONATE
General Information
Description: Light brown crystals.
Formula: $CH_3C_6H_4SO_3CH_3$.
Constants: Mol wt: 186.22, d: 1.230–1.238 at 25°/25°C vap. d.: 6.45, mp: 28°C.
Hazard Analysis
Toxic Hazard Rating:
Acute Local: Irritant 3; Inhalation 3.
Acute Systemic: Inhalation 3.
Chronic Local: Irritant 3; Allergen 3.
Chronic Systemic: U.
Toxicology: A vesicant and skin sensitizer.

Disaster Hazard: Dangerous. See sulfonates.
Countermeasures
Storage and Handling: Section 7.

METHYL TRIBORINE TRIAMINE (B)
General Information
Description: Colorless liquid; hydrolyzed by water.
Formula: $CH_3B_3N_3H_5$.
Constants: Mol wt: 94.6, mp: −59°C, bp: 87°C.
Hazard Analysis
Toxicity: Details unknown. See also boron compounds and amines.

METHYL TRIBORINE TRIAMINE (N)
General Information
Description: Colorless liquid; hydrolyzed by water.
Formula: $CH_3B_3N_3H_5$.
Constants: Mol wt: 94.6, bp: 84°C.
Hazard Analysis
Toxicity: Details unknown. See also boron compounds and amines.

METHYL TRIBROMODIPYRIDINE TIN
General Information
Description: Solid.
Formula: $CH_3SnBr_3(C_5H_5N)_2$.
Constants: Mol wt: 531.7, mp: 203°C.
Hazard Analysis and Countermeasures
See tin compounds and bromides.

METHYL TRI-n-BUTYL TIN
General Information
Description: Liquid.
Formula: $(CH_3)Sn(C_4H_9)_3$.
Constants: Mol wt: 305.1, d: 1.0898 at 20/4°C, bp: 121°C at 10 mm.
Hazard Analysis
Toxicity: Details unknown. See tin compounds.

METHYL TRICHLOROSILANE
General Information
Description: A liquid.
Formula: CH_3Cl_3Si.
Constants: Mol wt: 149.46, mp: −90°C, bp: 66.5°C, d: 1.28 at 25°/25°C, vap. d.: 5.17.
Hazard Analysis and Countermeasures
See chlorosilanes.
Shipping Regulations: Section 11.
I.C.C.: Flammable liquid, red label, 10 gallons.
IATA: Flammable liquid, red label, not acceptable (passenger), 40 liters (cargo).

METHYL TRIETHOXYSILANE
General Information
Description: Liquid.
Formula: $C_7H_{18}O_3Si$.
Constants: Mol wt: 178.20, bp: 141°C, d: 0.890, vap. d.: 6.14.
Hazard Analysis
Toxic Hazard Rating:
Acute Local: Irritant 2; Ingestion 2; Inhalation 2.
Acute Systemic: U.
Chronic Local: U.
Chronic Systemic: U.
Fire Hazard: Slight; can react with oxidizing materials.
Countermeasures
Ventilation Control: Section 2.
Personnel Protection: Section 3.
Storage and Handling: Section 7.

METHYL TRIPHENYL GERMANIUM
General Information
Description: Colorless crystals; insoluble in water. Soluble in organic solvents.
Formula: $(CH_3)Ge(C_6H_5)_3$.
Constants: Mol wt: 318.9, mp: 71.0°C.

Hazard Analysis
Toxicity: Details unknown. See germanium compounds and organometals.

METHYLTRIPHENYLSILANE. See methyl triphenyl silicane.

METHYL TRIPHENYLSILICANE
General Information
Synonym: Methyltriphenylsilane.
Description: Crystals.
Formula: $CH_3(C_6H_5)_3Si$.
Constants: Mol wt: 274.4, mp: 67.3°C.
Hazard Analysis
Toxic Hazard Rating:
  Acute Local: Irritant 1; Ingestion 1; Inhalation 1.
  Acute Systemic: U.
  Chronic Local: U.
  Chronic Systemic: U.
Disaster Hazard: Moderately dangerous; when heated to decomposition, it emits toxic fumes; can react vigorously with oxidizing materials.
Countermeasures
Ventilation Control: Section 2.
Personal Hygiene: Section 3.
Storage and Handling: Section 7.

METHYL TRI-n-PROPYL TIN
General Information
Description: Liquid.
Formula: $(CH_3)Sn(C_3H_7)_3$.
Constants: Mol wt: 263.0, bp: 93°C at 10 mm.
Hazard Analysis
Toxicity: Unknown. See tin compounds.

N-METHYL TROPOLINE. See tropine.

METHYL UNDECYL KETONE
General Information
Description: Colorless liquid or white solid.
Formula: $(CH_3)(C_{11}H_{23})CO$.
Constants: Mol wt: 198, mp: 28°C, bp: 120°C at 5 mm, flash p: 225°F (C.C.), vap. d.: 6.84, d: 0.825 at 30°C.
Hazard Analysis
Toxicity: Details unknown. See ketones.
Fire Hazard: Slight, when exposed to heat or flame; can react with oxidizing materials.
Countermeasures
To Fight Fire: Foam, carbon dioxide, dry chemical or carbon tetrachloride (Section 6).
Storage and Handling: Section 7.

METHYLUREA
General Information
Description: Crystals.
Formula: $NH_2CONHCH_3$.
Constants: Mol wt: 74.1, mp: 101°C, bp: decomp., d: 1.205 at 20°/20°C.
Hazard Analysis
Toxicity: Details unknown; probably low.
Fire Hazard: Slight; can react with oxidizing materials.
Countermeasures
Storage and Handling: Section 7.

4-METHYL VALERALDEHYDE
General Information
Formula: $CH_3CH(CH_3)CH_2CH_2CHO$.
Constant: Mol wt: 100.

Hazard Analysis
Toxicity: Details unknown. Limited animal experiments suggest low toxicity and irritation. See also aldehydes.

n-METHYL-n-VINYLACETAMIDE
General Information
Formula: $CH_3CH:CHCH_2CONH_2$.
Constant: Mol wt: 99.
Hazard Analysis
Toxic Hazard Rating: U.
Toxicity: Limted animal experiments suggest low toxicity.

1-METHYL VINYL ACETATE. See isopropenyl acetate.

4-METHYL-2-VINYL-1.3-DIOXOLANE
General Information
Formula: $CH_2OCH(CH:CH_2)CH(CH_3)O$.
Constant: Mol wt: 114.
Hazard Analysis
Toxicity: Details unknown. Limited animal experiments suggest high toxicity.

METHYL VINYL ETHER
General Information
Description: Liquid or gas.
Formula: $CHCH_2OCH_3$.
Constants: Mol wt: 58.1, mp: −122°C, bp: 5°C, flash p: −60°F, vap. d.: 2.0.
Hazard Analysis
Toxicity: Details unknown; probably has narcotic and anesthetic properties. See also ethers.
Fire Hazard: Highly dangerous, when exposed to heat or flame.
Explosion Hazard: Details unknown. See also ethers.
Disaster Hazard: Dangerous; keep away from heat and open flame; can react vigorously with oxidizing materials.
Countermeasures
Ventilation Control: Section 2.
To Fight Fire: Foam, carbon dioxide, dry chemical, carbon tetrachloride or water spray (Section 6).
Storage and Handling: Section 7.

METHYL VINYL KETONE
General Information
Synonym: 3-Butene-2-one.
Description: Colorless liquid, powerfully irritating odor.
Formula: $CH_3COCHCH_2$.
Constants: Mol wt: 70.09, bp: 81.4°C, flash p: 20°F (C.C.), d: 0.8393 at 25°/4°C, vap. d.: 2.41.
Hazard Analysis
Toxicity: Skin irritant and lachrymator.
Fire Hazard: Dangerous; when exposed to heat or flame.
Explosion Hazard: Unknown.
Disaster Hazard: Dangerous, upon exposure to heat or flame; emits toxic and irritant fumes; can react with oxidizing materials.
Countermeasures
Ventilation Control: Section 2.
To Fight Fire: Foam, carbon dioxide, dry chemical or carbon tetrachloride (Section 6).
Storage and Handling: Section 7.
Shipping Regulations: Section 11.
  I.C.C.: (inhibited): Flammable liquid, red label, 10 gallons.
  IATA (inhibited): Flammable liquid, red label, 120 c.c. (passenger), 40 liters (cargo).

TOXIC HAZARD RATING CODE (For detailed discussion, see Section 1.)

0 NONE: (a) No harm under any conditions; (b) Harmful only under unusual conditions or overwhelming dosage.

1 SLIGHT: Causes readily reversible changes which disappear after end of exposure.

2 MODERATE: May involve both irreversible and reversible changes; not severe enough to cause death or permanent injury.

3 HIGH: May cause death or permanent injury after very short exposure to small quantities.

U UNKNOWN: No information on humans considered valid by authors.

(uninhibited): Flammable liquid, not acceptable (passenger and cargo).

METHYL VINYL SULFONE
Hazard Analysis
Toxicity: Limited animal experiments suggest high toxicity.
Disaster Hazard: Dangerous. See sulfur compounds.

MEVINPHOS. See phosdrin.

MGK. See n-octyl bicycloheptene dicarboximide.

MICA DUST (<5% FREE SILICA)
General Information
Description: Light gray to dark flakes or particles.
Hazard Analysis
Toxic Hazard Rating:
Acute Local: Irritant 2; Inhalation 2.
Acute Systemic: 0.
Chronic Local: Inhalation 2.
Chronic Systemic: 0.
TLV: ACGIH (recommended); 20 million particles per cubic foot of air.
Caution: Can cause a chronic fibrosis of the lungs.
Countermeasures
Ventilation Control: Section 2.

MILD DETONATING FUSE, METAL CLAD
Shipping Regulations: Section 11.
I.C.C.: Explosive C, 300 pounds.
IATA: Explosive, explosive label, 25 kilograms (passenger), 95 kilograms (cargo).

MILD SILVER PROTEIN. See argyrol.

MILITARY GUNCOTTON. See cellulose nitrate.

MILL DUST
Hazard Analysis
Toxic Hazard Rating:
Acute Local: Irritant 2; Allergen 1; Inhalation 2.
Acute Systemic: 0.
Chronic Local: Inhalation 2.
Chronic Systemic: Inhalation 1.
TLV: ACGIH (recommended); 50 million particles per cubic foot of air.
Countermeasures
Ventilation Control: Section 2.

MILLERITE. See nickel monosulfide.

MILLON'S BASE
General Information
Description: Crystals.
Formula: $(HO)_2Hg_2NH_2OH$.
Constants: Mol wt: 468.27, d: 4.083 at 18°C.
Hazard Analysis and Countermeasures
See mercury compounds, inorganic.

MINERAL DUSTS
Hazard Analysis
Toxicity: Variable. From the economic and toxicologic standpoint, the most important are those containing free silica, which can cause silicosis if inhaled in sufficient quantity. These include sand, sandstone, quartz and flint. They consist mainly of silica in the form of quartz, diatomaceous earth, which is essentially amorphous silica, and granite, which contains from 20 to 40% quartz. Minerals that contain combined silica in the form of silicates but no free silica are generally less capable of causing silicosis. Asbestos, however, can cause a fibrotic lung condition of its own, known as asbestosis also implicated in lung cancer. (See asbestos). Mica and talc dust are also considered somewhat hazardous (See mica dust). Non-siliceous minerals, like limestone, marble, dolomite, etc., which do not contain toxic elements, do not ordinarily present any significant dust hazard, although minerals containing toxic elements, such as cryolite, which contains fluorine, and pyrolusite, which contains manganese, may cause systemic poisoning if the dust is inhaled or ingested in sufficient quantity. In any event, the minerals are usually less reactive than synthetic compounds of the same elements, and in fact may be relatively inert by comparison.
Note: These are common air contaminants (Section 4).

#2 MINERAL OIL. See lard oil (pure).

MINERAL SEAL OIL
General Information
Synonym: Signal oil.
Description: A viscous liquid.
Constant: Flash p: 170°F (C.C.), d: 0.8.
Hazard Analysis
Toxicity: Details unknown. See also kerosene.
Fire Hazard: Moderate, when exposed to heat or flame; can react with oxidizing materials.
Countermeasures
To Fight Fire: Foam, carbon dioxide, dry chemical or carbon tetrachloride (Section 6).
Storage and Handling: Section 7.

MINERAL SPIRITS NO. 10
General Information
Synonyms: Turpentine substitute; white spirit.
Description: A clear, water-white liquid.
Constants: Bp: 150-190°C, lel: 0.77% at 212°F, flash p: 104°F (C.C.), d: 0.80, autoign. temp.: 473°F, vap. d.: 3.9.
Hazard Analysis
Toxic Hazard Rating:
Acute Local: Irritant 1; Ingestion 1; Inhalation 1.
Acute Systemic: Ingestion 2; Inhalation 2.
Chronic Local: U.
Chronic Systemic: U.
Note: these are common air contaminants (Section 4).
Fire Hazard: Moderate, when exposed to heat or flame.
Explosion Hazard: Moderate, when exposed to heat or flame.
Disaster Hazard: Dangerous; may explode on heating.
Countermeasures
Ventilation Control: Section 2.
To Fight Fire: Foam, carbon dioxide, dry chemical or carbon tetrachloride (Section 6).
Personal Hygiene: Section 3.
Storage and Handling: Section 7.

MINERAL WAX. See ceresin wax.

MINING DEVICE, CONTAINING CARBON DIOXIDE GAS, LIQUEFIED, WITH EXPLOSIVE HEATING ELEMENT
Shipping Regulations: Section 11.
IATA; Nonflammable gas, green label, not acceptable (passenger), 3 kilograms (cargo).

MINIUM. See lead oxide.

MIPA. See 1-amino-2-propanol.

MIPAFOX. See bis(isopropylamido) fluoro-phosphate.

MIRBANE OIL. See nitrobenzene.

MISCHMETAL
General Information
Description: A commercial form of mixed rare earth metals, such as cerium, etc., used for making lighter flints and as an addition to alloys for improving their characteristics.
Countermeasure
Shipping Regulations: Section 11.
IATA: Flammable solid, yellow label, 12 kilograms (passenger), 45 kilograms (cargo).

MITOX. See p-chlorobenzyl and p-chlorophenyl sulfide.

MIXED ACID. See sulfuric acid and nitric acid.
Shipping Regulations: Section 11.
   Coast Guard Classification: Corrosive liquid; white label.
   MCA warning label.

MIXTURES OF HYDROFLUORIC AND SULFURIC ACIDS. See hydrofluoric and sulfuric acids.
Shipping Regulations: Section 11.
   Coast Guard Classification: Corrosive liquid; white label.

MODIFIED HOP EXTRACT
General Information
Description: Manufactured from a hexane extract of hops.
Hazard Analysis
Toxic Hazard Rating: Unknown. A flavoring agent in the brewing of beer (Section 10).

MOHAIR
General Information
Synonym: Goat hair.
Description: A fabric.
Hazard Analysis
Toxic Hazard Rating:
   Acute Local: Allergen 1.
   Acute Systemic: Inhalation 1.
   Chronic Local: Allergen 1.
   Chronic Systemic: Inhalation 1.
Fire Hazard: Slight, when exposed to heat or flame; can react with oxidizing materials (Section 6).
Explosion Hazard: Slight, in the form of dust, when exposed to flame. See also dust explosions.
Countermeasures
Personal Hygiene: Section 3.
Storage and Handling: Section 7.

MOLYBDENITE. See molybdenum sulfide.

MOLYBDENUM
General Information
Description: Cubic, silver-white metallic crystals or gray-black powder.
Formula: Mo.
Constants: At wt: 95.95, mp: 2622°C, bp: 5560°C, d: 10.2, vap. press.: 1 mm at 3102°C.
Hazard Analysis
Radiation Hazard: Section 5. For permissible levels, see Table 5, p. 150.
     Artificial isotope ^{99}Mo, half life 67 h. Decays to radioactive ^{99}Tc by emitting beta particles of 0.45 (14%), 1.23 (85%) MeV. Also emits gamma rays of 0.04–0.78 MeV.
Fire Hazard: Moderate, in the form of dust, when exposed to heat or flame; can react with oxidizing materials. See also powdered metals (Section 6).
Explosion Hazard: Slight, in the form of dust, when exposed to flame. See also powdered metals.
Countermeasures
Storage and Handling: Section 7.

MOLYBDENUM CARBONYL
General Information
Description: Diamagnetic crystals.
Formula: Mo(CO)₆.
Constants: Mol wt: 264.01, mp: decomposes at 150°C, d: 1.96, vap. press.: 2.3 mm at 55°C, vap. d.: 9.1.
Hazard Analysis
Toxicity: See carbonyls.

Caution: In stable form this compound is not highly toxic, but it may liberate carbon monoxide because it decomposes easily.
Fire Hazard: See carbonyls.
Explosion Hazard: See carbonyls.
Disaster Hazard: See carbonyls.
Countermeasures
Storage and Handling: Section 7.

MOLYBDENUM CHLORIDE
General Information
Synonym: Molybdenum dichloride.
Description: Amorphous or yellow crystals.
Formula: MoCl₂.
Constants: Mol wt: 166.86, mp: decomposes, d: 3.714 at 25°C.
Hazard Analysis
Toxicity: See molybdenum compounds.
Disaster Hazard: Dangerous; when heated to decomposition, it emits toxic fumes of chlorides.
Countermeasures
Personal Hygiene: Section 3.
Storage and Handling: Section 7.

MOLYBDENUM COMPOUNDS
Hazard Analysis
Toxic Hazard Rating:
   Acute Local: Irritant 1.
   Acute Systemic: U.
   Chronic Local: U.
   Chronic Systemic: Inhalation 1.
TLV: ACGIH (recommended); soluble compounds, 5 milligrams per cubic meter of air; insoluble compounds, 15 milligrams per cubic meter of air.
Toxicology: Molybdenum and its compounds are said to be somewhat toxic, but in spite of their considerable use in industry, industrial poisoning by molybdenum has yet to be reported. Some studies have been made of its effects, and it is suggested that suitable precautions should be taken against the inhalation of considerable amounts of the more soluble molybdenum compounds. From animal experimentation, it was found that no fatalities occurred to those subjected to molybdic oxide fumes for 25 one-hour exposures at an average concentration of 1.5 mg/cu. foot of air, and only 1 fatality in 24 one-hour exposures to molybdenite dust at an average concentration of 8.1 mg/cu. foot of air. Molybdenum is not stored in the body to any extent because it is rapidly excreted. Experiments with the sodium salts of hexavalent chromium, tungsten and molybdenum have shown that sodium molybdate is the least toxic of the three, and is less toxic following intraperitoneal injection than either sodium chromate or sodium tungstate in equivalent concentrations. All molybdenum compounds can be referred to molybdenum or to more toxic anions if present.
   Recent studies have shown that molybdenum has importance as a trace element in the normal growth and development of certain forms of plant life. It is found also in animal tissue, although its precise function is unknown. It is a common air contaminant. See Section 4.

MOLYBDENUM   DICHLORIDE. See molybdenum chloride.

MOLYBDENUM DISULFIDE. See molybdenum sulfide.

TOXIC HAZARD RATING CODE (For detailed discussion, see Section 1.)

0 NONE: (a) No harm under any conditions; (b) Harmful only under unusual conditions or overwhelming dosage.

1 SLIGHT: Causes readily reversible changes which disappear after end of exposure.

2 MODERATE: May involve both irreversible and reversible changes; not severe enough to cause death or permanent injury.

3 HIGH: May cause death or permanent injury after very short exposure to small quantities.

U UNKNOWN: No information on humans considered valid by authors.

MOLYBDENUM HEXAFLUORIDE
General Information
Description: Colorless crystals or liquid.
Formula: MoF_6.
Constants: Mol wt: 209.95, mp: 17°C, bp: 35°C, d: 2.55.
Hazard Analysis and Countermeasures
See fluorides.

MOLYBDENUM MONOBORIDE
General Information
Description: Crystals.
Formula: MoB.
Constants: Mol wt: 106.77, d: 8.65.
Hazard Analysis
Toxicity: See boron compounds and molybdenum compounds.
Fire Hazard: See borides.
Explosion Hazard: Details unknown. See also borides.
Disaster Hazard: See borides.
Countermeasures
Personal Hygiene: Section 3.
Storage and Handling: Section 7.

MOLYBDENUM OXIDE
General Information
Synonym: Molybdenum sesquioxide.
Description: Gray-black powder.
Formula: Mo_2O_3.
Constant: Mol wt: 239.90.
Hazard Analysis
Toxicity: See molybdenum compounds.
Countermeasures
Ventilation Control: Section 2.

MOLYBDENUM OXYTETRAFLUORIDE
General Information
Description: Colorless-white, deliquescent crystals.
Formula: $MoOF_4$.
Constants: Mol wt: 187.95, mp: 98°C, bp: 180°C, d: 3.0.
Hazard Analysis and Countermeasures
See fluorides.

MOLYBDENUM PENTACHLORIDE
General Information
Description: Green-black solid, dark red as liquid or vapor; hygroscopic, reacting with water and air; soluble in dry ether, dry alcohol, and other anhydrous organic solvents.
Formula: $MoCl_5$.
Constants: Mol wt: 273.5, mp: 194°C, bp: 268°C, d: 2.9.
Hazard Analysis
Toxic Hazard Rating: Toxic. See also hydrochloric acid.
Disaster Hazard: Dangerous. See hydrochloric acid.
Countermeasures
Storage and Handling: Section 7.
Shipping Regulations: Section 11.
  IATA: Other restricted articles, class B, no label required, 12 kilograms (passenger), 45 kilograms (cargo).

MOLYBDENUM PENTASULFIDE
General Information
Description: Dark brown powder.
Formula: Mo_2S_5.
Constant: Mol wt: 352.23.
Hazard Analysis and Countermeasures
See sulfides.

MOLYBDENUM PHOSPHIDE
General Information
Description: Gray-green crystalline powder.
Formula: MoP.
Constants: Mol wt: 126.93, d: 6.167.
Hazard Analysis and Countermeasures
See phosphides.

MOLYBDENUM SESQUIOXIDE. See molybdenum oxide.

MOLYBDENUM SESQUISULFIDE
General Information
Description: Steel-gray needles.
Formula: Mo_2S_3.
Constants: Mol wt: 288.10, mp: decomposes at 1100°C, bp: volat. at 1200°C, d: 5.91 at 15°C.
Hazard Analysis and Countermeasures
See sulfides.

MOLYBDENUM SULFIDE
General Information
Synonyms: Molybdic sulfide; molybdenite.
Description: Black, lustrous powder.
Formula: MoS_2.
Constants: Mol wt: 160.07, mp: 1185°C, bp: decomposes in air, d: 4.80 at 14°C.
Hazard Analysis and Countermeasures
See sulfides.

MOLYBDENUM TETRABROMIDE
General Information
Description: Black, deliquescent needles.
Formula: $MoBr_4$.
Constants: Mol wt: 415.61, mp: decomposes, bp: volatilizes.
Hazard Analysis and Countermeasures
See bromides and molybdenum compounds.

MOLYBDENUM TETRASULFIDE
General Information
Description: Brown powder.
Formula: MoS_4.
Constants: Mol wt: 224.19, mp: decomposes.
Hazard Analysis and Countermeasures
See sulfides.

MOLYBDENUM TRIOXIDE
General Information
Synonym: Molybdenum anhydride.
Description: White or yellow to slightly bluish powder or granules; soluble in 1000 parts water; soluble in concentrated mineral acids and solutions of alkali hydroxides, ammonia, or potassium bitartrate.
Formula: MoO_3.
Constants: Mol wt: 143.95, d: 4.5, mp: 795°C.
Hazard Analysis
See molybdenum compounds.

MOLYBDENUM TRISULFIDE
General Information
Description: Red-dark brown crystals.
Formula: MoS_3.
Constants: Mol wt: 192.13, mp: decomposes.
Hazard Analysis and Countermeasures
See sulfides.

MOLYBDIC ACID
General Information
Description: Hexagonal, white or slight yellowish crystals.
Formula: H_2MoO_4.
Constants: Mol wt: 162.0, mp: decomp. 115°C, d: 3.112.
Hazard Analysis
Toxicity: See molybdenum compounds.

MOLYBDIC ANHYDRIDE. See molybdenum trioxide.

MOLYBDIC SULFIDE. See molybdenum sulfide.

MONACETIN. See glyceryl monoacetate.

MOND GAS. See carbon monoxide and hydrogen.

MONKSHOOD. See aconite.

MONOAMMONIUM GLUTAMATE
General Information
Synonym: Ammonium glutamate.

Hazard Analysis
Toxic Hazard Rating: U. Used as a general purpose food additive (Section 10).

MONO-sec-AMYLAMINE
General Information
Description: Colorless liquid; amine-like odor.
Formula: $CH_3(CH_2)_2CH(CH_3)NH_2$.
Constants: Mol wt: 87.16, bp: 92°C, flash p: 20°F, d: 0.73839 at 20°/0°C, vap. d.: 3.01.
Hazard Analysis
Toxicity: Details unknown. See amines.
Fire Hazard: Dangerous, when exposed to heat or flame.
Explosion Hazard: Unknown.
Disaster Hazard: Dangerous, upon exposure to heat or flame; can react vigorously with oxidizing materials.
Countermeasures
Ventilation Control: Section 2.
To Fight Fire: Foam, carbon dioxide, dry chemical or carbon tetrachloride (Section 6).
Storage and Handling: Section 7.

N-MONOAMYL ANILINE
General Information
Description: Liquid.
Formula: $C_5H_{11}NHC_6H_5$.
Constants: Mol wt: 163.3, bp: 245°C, flash p: 225°F, d: 0.9, vap. d.: 5.63.
Hazard Analysis
Toxicity: Probably toxic. See aniline.
Fire Hazard: Slight, when exposed to heat or flame.
Disaster Hazard: Dangerous; when heated to decomposition, it emits highly toxic fumes of aniline; can react with oxidizing materials.
Countermeasures
Ventilation Control: Section 2.
To Fight Fire: Foam, carbon dioxide, dry chemical or carbon tetrachloride (Section 6).
Storage and Handling: Section 7.

MONOBROMOTRIFLUOROMETHANE
General Information
Synonym: Bromotrifluoro methane.
Description: Colorless, non-corrosive gas.
Formula: $CBrF_3$.
Constants: Mol wt: 149, fp: −168°C, bp: −58°C, gas density (bp): 8.71 grams/liter.
Hazard Analysis
Toxic Hazard Rating: U.
Disaster Hazard: Dangerous. See bromides and fluorides.
Countermeasures
Storage and Handling: Section 7.
Shipping Regulations: Section 11.
I.C.C.: Nonflammable gas, green label, 300 pounds.
IATA: Nonflammable gas, green label, 70 kilograms (passenger), 140 kilograms (cargo).

MONO-n-BUTYLAMINE
General Information
Description: Colorless liquid.
Formula: $CH_3(CH_2)_2CH_2NH_2$.
Constants: Mol wt: 73.14, mp: −50.5°C, flash p: 10°F (O.C.), d: 0.7401 at 20.4°C.
Hazard Analysis
Toxic Hazard Rating:
Acute Local: Irritant 1; Ingestion 1; Inhalation 1.
Acute Systemic: U.
Chronic Local: U.

Chronic Systemic: U.
Fire Hazard: Dangerous, when exposed to heat or flame.
Explosion Hazard: Unknown.
Disaster Hazard: Dangerous; keep away from heat and open flame; can react vigorously with oxidizing materials.
Countermeasures
Ventilation Control: Section 2.
To Fight Fire: Foam, carbon dioxide, dry chemical or carbon tetrachloride (Section 6).
Personal Hygiene: Section 3.
Storage and Handling: Section 7.
Shipping Regulations: Section 11.
MCA warning label.

MONO-sec-BUTYLAMINE
General Information
Description: Colorless liquid.
Formula: $CH_3CH(NH_2)CH_2CH_3$.
Constants: Mol wt: 73.14, mp: −104.5°C, bp: 63°C, flash p: 15°F, d: 0.724 at 20°/4°C.
Hazard Analysis
Toxic Hazard Rating:
Acute Local: Irritant 1; Ingestion 1; Inhalation 1.
Acute Systemic: U.
Chronic Local: U.
Chronic Systemic: U.
Fire Hazard: Dangerous, when exposed to heat or flame.
Explosion Hazard: Unknown.
Disaster Hazard: Dangerous; keep away from heat and open flame; can react vigorously with oxidizing materials.
Countermeasures
Ventilation Control: Section 2.
To Fight Fire: Foam, carbon dioxide, dry chemical or carbon tetrachloride (Section 6).
Personal Hygiene: Section 3.
Storage and Handling: Section 7.

MONOBUTYLAMINE OLEATE. See butylamine oleate.

MONO-tert-BUTYL-m-CRESOL
General Information
Description: Clear liquid.
Formula: $(CH_3)_3CC_6H_3(CH_3)OH$.
Constants: Mol wt: 164.24, mp: 27.0°C, bp: 137°C at 25 mm, d: 0.969 at 30°/25°C, vap. d.: 5.77.
Hazard Analysis
Toxicity: Details unknown. See cresol.
Disaster Hazard: Moderately dangerous; when heated, it emits toxic fumes; can react with oxidizing materials.
Countermeasures
Storage and Handling: Section 7.

MONOBUTYL DIPHENYL SODIUM MONO-SULFONATE
General Information
Synonym: Areseket.
Description: Powder.
Formula: $C_{16}H_{17}O_4SNa$.
Constant: Mol wt: 328.3.
Hazard Analysis
Toxicity: Details unknown. Properties are those of the general group of arylalkyl sulfamates. Can cause dermatitis by defatting of skin.
Disaster Hazard: Dangerous. See sulfonates.

TOXIC HAZARD RATING CODE (For detailed discussion, see Section 1.)

0 NONE: (a) No harm under any conditions; (b) Harmful only under unusual conditions or overwhelming dosage.

1 SLIGHT: Causes readily reversible changes which disappear after end of exposure.

2 MODERATE: May involve both irreversible and reversible changes; not severe enough to cause death or permanent injury.

3 HIGH: May cause death or permanent injury after very short exposure to small quantities.

U UNKNOWN: No information on humans considered valid by authors.

MONOBUTYL PHENYL PHENOL SODIUM SULFATE
General Information
Synonym: Areskap.
Description: Liquid.
Formula: $C_{16}H_{17}SO_4Na$.
Constant: Mol wt: 328.3.
Hazard Analysis
See monobutyl diphenyl sodium monosulfonate.

MONOCALCIUM. See calcium phosphate, monobasic.

MONOCHLORATED ACETONE. See monochloroacetone, stabilized.

MONOCHLORACETIC ACID
General Information
Synonym: Chloroethanoic acid; chloroacetic acid.
Description: Colorless crystals.
Formula: $CH_2ClCOOH$.
Constants: Mol wt: 94.5, mp: $\alpha = 63°C$; $\beta = 56°C$, $\gamma = 50°C$, bp: 189°C, flash p: none, d: 1.58 at 20°/20°C, vap. d.: 3.26.
Hazard Analysis
Toxic Hazard Rating:
Acute Local: Irritant 2; Ingestion 2; Inhalation 2.
Acute Systemic: U.
Chronic Local: U.
Chronic Systemic: U.
Disaster Hazard: Dangerous; when heated to decomposition, it emits highly toxic fumes of phosgene and chlorides.
Countermeasures
Ventilttion Control: Section 2.
Personnel Protection: Section 3.
Storage and Handling: Section 7.

MONOCHLOROACETONE, STABILIZED
General Information
Synonym: Monochlorated acetone.
Description: Colorless liquid; pungent odor.
Formula: CH_3COCH_2Cl.
Constants: Mol wt: 92.52, bp: 119°C, d: 1.162 at 16°C, vap. d.: 3.2.
Hazard Analysis
Toxic Hazard Rating:
Acute Local: Irritant 3; Ingestion 3; Inhalation 3.
Acute Systemic: U.
Chronic Local: U.
Chronic Systemic: U.
Fire Hazard: Unknown.
Disaster Hazard: Dangerous; when heated to decomposition, it emits highly toxic fumes of phosgene; can react with oxidizing materials.
Countermeasures
Ventilation Control: Section 2.
Personnel Protection: Section 3.
Storage and Handling: Section 7.
Shipping Regulations: Section 11.
I.C.C.: Poison C; tear gas label, 5 gallons.
Coast Guard Classification: Poison C; tear gas label.

MONOCHLOROACETONE, UNSTABILIZED. See monochloroacetone, stabilized.
Shipping Regulations: Section 11
I.C.C.: Not accepted.
Coast Guard Classification: Not permitted.

MONOCHLOROBENZENE. See chlorobenzene.

MONOCHLOROBROMOMETHANE. See monochloromonobromomethane.

MONOCHLORODIFLUOROMETHANE
General Information
Description: Colorless gas.

Formula: $CHClF_2$.
Constants: Mol wt: 86.48, mp: $-146°C$, bp: $-40.8°C$, d: 3.87 at 0°C.
Hazard Analysis
Toxic Hazard Rating:
Acute Local: U.
Acute Systemic: Inhalation 2.
Chronic Local: U.
Chronic Systemic: U.
Disaster Hazard: Dangerous; when heated to decomposition, it emits highly toxic fumes of phosgene and fluorine.
Countermeasures
Ventilation Control: Section 2.
Storage and Handling: Section 7.
Shipping Regulations: Section 11.
I.C.C.: Nonflammable gas; green label, 300 pounds.
Coast Guard Classification: Non-inflammable gas; green gas label.
IATA: Nonflammable gas, green label, 70 kilograms (passenger), 140 kilograms (cargo).

MONOCHLOROETHYLENE CHLORIDE. See ethylene monochlorochloride.

MONOCHLOROMALEIC ANHYDRIDE
General Information
Description: Nearly white liquid or solid.
Formula: $ClCCH(CO)_2O$.
Constants: Mol wt: 132.4, mp: 33°C, bp: 196–197.5°C, d: 1.54 at 25°/25°C.
Hazard Analsysi
Toxicity: Unknown.
Disaster Hazard: Dangerous; when heated to decomposition, it emits highly toxic fumes of phosgene.
Countermeasures
Storage and Handling: Section 7.

MONOCHLOROMETHANE. See methyl chloride.

MONOCHLOROMONOBROMO METHANE
General Information
Synonym: Monochlorobromomethane.
Description: Liquid.
Formula: CH_2BrCl.
Constants: Mol wt: 129, bp: 60–90°C, vap. d.: 4.45.
Hazard Analysis
Toxic Hazard Rating:
Acute Local: Irritant 2; Ingestion 2; Inhalation 2.
Acute Systemic: Ingestion 3; Inhalation 3; Skin Absorption 3.
Chronic Local: U.
Chronic Systemic: Inhalation 2; Skin Absorption 2.
Toxicology: This material has a narcotic action of moderate intensity, although of prolonged duration. Animals exposed for several weeks to 1,000 ppm of this substance had blood bromide levels as high as 350 mg/100 grams of blood. Therefore, until further data are available concerning it, it should be considered at least as toxic as carbon tetrachloride, and more than minimal exposure to its vapors should be avoided.
Disaster Hazard: Dangerous; when heated to decomposition, it emits highly toxic fumes of chlorides and bromides.
Countermeasures
Ventilation Control: Section 2.
Personnel Protection: Section 3.
Storage and Handling: Section 7.

MONOCHLORONAPHTHALENE. See chlorinated naphthalenes.

MONOCHLOROPENTAFLUOROETHANE. See chloropentafluoroethane.

MONOCHLOROTETRAFLUOROETHANE
General Information
Description: Gas.
Formula: C_2HF_4Cl.
Constant: Mol wt: 136.49.
Hazard Analysis
Toxicity: Details unknown; probably acts as a simple asphyxiant. See also chlorinated hydrocarbons.
Disaster Hazard: Dangerous; when heated to decomposition it emits highly toxic fumes of chlorides and fluorides.
Countermeasures
Storage and Handling: Section 7.
Shipping Regulations: Section 11.
  I.C.C.: Nonflammable gas; green label, 300 pounds.
  Coast Guard Classification: Non-inflammable gas; green gas label.
  IATA: Nonflammable gas, green label, 70 kilograms (passenger), 140 kilograms (cargo).

MONOCHLOROTOLUENE
General Information
Synonym: Methylchlorobenzene.
Description: Clear, colorless to straw-colored liquid.
Formula: $CH_3C_6H_4Cl$.
Constants: Mol wt: 126.5, bp: 158–165°C, fp: $< -45°$C, flash p: 125°F, d: 1.080 at 15.5°/15.5°C, vap. d.: 4.38.
Hazard Analysis
Toxicity: Details unknown. See chlorinated hydrocarbons, aromatic.
Fire Hazard: Moderate, when exposed to heat or flame.
Explosion Hazard: Unknown.
Disaster Hazard: Dangerous; when heated to decomposition, it emits highly toxic fumes of chlorides; can react with oxidizing materials.
Countermeasures
Storage and Handling: Section 7.
To Fight Fire: Foam, carbon dioxide, dry chemical or carbon tetrachloride (Section 6).
Shipping Regulations: Section 11.
  MCA warning label.

MONOCHLOROTRIFLUOROETHYLENE. See trifluorochlorethylene.

MONOCHLOROTRIFLUOROMETHANE
General Information
Synonym: Chlorotrifluoromethane.
Description: Colorless gas; ethereal odor.
Formula: $CClF_3$.
Constants: Mol wt: 104.47, mp: $-181°$C, bp: $-80°$C.
Hazard Analysis
Toxic Hazard Rating:
  Acute Local: Irritant 1.
  Acute Systemic: Inhalation 2.
  Chronic Local: U.
  Chronic Systemic: U.
Toxicology: Narcotic in high concentrations.
Disaster Hazard: Dangerous; when heated to decomposition, it emits highly toxic fumes of chlorides and fluorides.
Countermeasures
Ventilation Control: Section 2.
Storage and Handling: Section 7.
Shipping Regulations: Section 11.
  I.C.C.: Nonflammable gas; green label, 300 pounds.

Coast Guard Classification: Noninflammable gas; green gas label.
IATA: Nonflammable gas, green label, 70 kilograms (passenger), 140 kilograms (cargo).

MONOCRESYL DIPHENYL PHOSPHATE
General Information
Description: Liquid.
Formula: $CH_3C_6H_4O(C_6H_5O)_2PO_4$.
Constants: Mol wt: 388.3, flash p: 450°F (C.C.), d: > 1, vap. d.: 13.4.
Hazard Analysis
Toxicity: Details unknown; probably toxic and an irritant.
Fire Hazard: Slight, when exposed to heat or flame.
Spontaneous Heating: No.
Disaster Hazard: Dangerous; when heated to decomposition, it emits highly toxic fumes of oxides of phosphorus; can react with oxidizing materials.
Countermeasures
Storage and Handling: Section 7.
To Fight Fire: Foam, carbon dioxide or dry chemical or carbon tetrachloride (Section 6).

MONODECYL NAPHTHALENE. See keryl naphthalene.

MONO AND DIGLYCERIDES FROM THE GLYCEROLYSIS OF EDIBLE FATS AND OILS
Hazard Analysis
Toxic Hazard Rating: U. Used as emulsifying agent food additive (Section 10).

MONODODECYL BENZENE. See keryl benzene-12.

MONOETHANOL AMINE
General Information
Synonym: 2-Aminoethanol, ethanol amine.
Description: Colorless liquid; ammoniacal odor.
Formula: $NH_2CH_2CH_2OH$.
Constants: Mol wt: 61.08, bp: 170.5°C, fp: 10.5°C, flash p: 200°F (O.C.), d: 1.0180 at 20°/4°C, vap. press.: 6 mm at 60°C, vap. d.: 2.11.
Hazard Analysis
Toxic Hazard Rating:
  Acute Local: Irritant 2; Ingestion 2; Inhalation 2.
  Acute Systemic: U.
  Chronic Local: U.
  Chronic Systemic: U.
TLV: ACGIH (recommended); 3 parts per million of air; 6 milligrams per cubic meter of air.
Fire Hazard: Moderate, when exposed to heat or flame; can react with oxidizing materials.
Countermeasures
To Fight Fire: Foam, carbon dioxide, dry chemical or carbon tetrachloride (Section 6).
Ventilation Control: Section 2.
Personnel Protection: Section 3.
Storage and Handling: Section 7.

MONOETHYL AMINE. See ethylamine.

MONOFLUOPHOSPHORIC ACID
General Information
Description: Colorless, viscous liquid.
Formula: H_2PO_3F.
Constants: Mol wt: 100, d: 1.818 at 25°/4°C, vap. d.: 3.45.
Hazard Analysis and Countermeasures
See fluorides and phosphoric acid.

MONOFLUOROPHOSPHORIC ACID ANHYDROUS.
See monofluophosphoric acid.
Shipping Regulations: Section 11.
   I.C.C.: Corrosive liquid; white label, 1 gallon.
   Coast Guard Classification: Corrosive liquid; white label.
   IATA: Corrosive liquid, white label, not acceptable (passenger), 5 liters (cargo).

MONOFLUOROTRICHLOROMETHANE. See fluoro-trichloromethane.

MONOGERMANE
General Information
Synonym: Germanium tetrahydride.
Description: Gas.
Formula: GeH_4.
Constants: Mol wt: 76.63, mp: $-165°C$, bp: $-88.5°C$, d liq: 1.523 at $-142°C$, d gas: 3.43 g/liter.
Hazard Analysis
Toxicity: Details unknown. See also hydrides and germanium compounds.
Fire Hazard: Dangerous. See hydrides.
Explosion Hazard: Dangerous. See hydrides.
Disaster Hazard: See hydrides.
Countermeasures
Storage and Handling: Section 7.

MONO-sec-HEXYLAMINE
General Information
Synonym: 2-Amino-4-methyl pentane.
Description: Water-white liquid; amine-like odor.
Formula: $CH_3CHNH_2C_4H_9$.
Constants: Mol wt: 101.19, bp: $223°C$, flash p: $55°F$, d: 0.746 at $20°C$, vap. d.: 3.49.
Hazard Analysis
Toxicity: Details unknown. See also amines. The heptane analog shows high toxicity in mouse, rat, rabbit and guinea pig.
Fire Hazard: Dangerous, when exposed to heat or flame.
Explosion Hazard: Unknown.
Disaster Hazard: Dangerous, upon exposure to heat or flame; can react vigorously with oxidizing materials.
Countermeasures
Ventilation Control: Section 2.
To Fight Fire: Foam, carbon dioxide, dry chemical or carbon tetrachloride (Section 6).
Storage and Handling: Section 7.

MONOISOPROPANOLAMINE
General Information
Description: Colorless liquid.
Formula: $CH_3CH(OH)CH_2NH_2$.
Constants: Mol wt: 75.11, bp: $159.9°C$, fp: $-2°C$, flash p $160°F$ (O.C.), d: 0.9619 at $20°/20°C$, vap. press.: 0.6 mm at $20°C$, vap. d.: 2.59.
Hazard Analysis
Toxicity: Details unknown. See also amines.
Fire Hazard: Moderate, when exposed to heat or flame; can react with oxidizing materials.
Countermeasures
To Fight Fire: Foam, carbon dioxide, dry chemical or carbon tetrachloride (Section 6).
Storage and Handling: Section 7.

MONOISOPROPYL BICYCLOHEXYL
General Information
Formula: $C_{15}H_{28}$.
Constants: Mol wt: 208, d: 0.9, bp: 530–541°F, flash p: 255°F, autoign. temp.: 442°F, lel: 0.5% at 347°F, uel: 4.1% at 400°F.
Hazard Analysis
Toxic Hazard Rating: U.
Fire Hazard: Slight, when exposed to heat or flame.
Countermeasures
Storage and Handling: Section 7.
To Fight Fire: Water or foam.

2-MONOISOPROPYL BIPHENYL
General Information
Formula: $C_{15}H_{16}$.
Constants: Mol wt: 196, bp: 518°F, flash p: 285°F, autoign. temp.: 513°F, d: 1.0–, lel: 0.5% at 347°F, uel: 3.2% at 392°F.
Hazard Analysis
Toxic Hazard Rating: U.
Fire Hazard: Slight, when exposed to heat or flame.
Countermeasures
Storage and Handling: Section 7.
To Fight Fire: Water or foam.

MONOISOPROPYL CITRATE
Hazard Analysis
Toxicity: Details unknown. A sesquestrant food additive (Section 10).

MONOMETHYLAMINE
General Information
Synonym: Aminomethane.
Description: Colorless gas or liquid; strong ammoniacal odor.
Formula: CH_3NH_2.
Constants: Mol wt: 31.3, bp: $-6.79°C$, lel: 4.95%, uel: 20.75%, fp: $-93.5°C$, flash p: $0°F$ (C.C.), d: 0.662 at $20°/4°C$, autoign. temp.: 806°F, vap. d.: 1.07.
Hazard Analysis
Toxic Hazard Rating:
   Acute Local: Irritant 3; Ingestion 3; Inhalation 3.
   Acute Systemic: U.
   Chronic Local: Irritant 2.
   Chronic Systemic: U.
Toxicology: A strong irritant. See also amines.
TLV: ACGIH (recommended); 10 parts per million of air; 12 milligrams per cubic meter of air.
Fire Hazard: Dangerous, when exposed to heat or flame.
Explosion Hazard: Moderate, when exposed to sparks or flame.
Disaster Hazard: Dangerous; keep away from heat and open flame; can react vigorously with oxidizing materials.
Countermeasures
Ventilation Control: Section 2.
To Fight Fire: Foam, carbon dioxide, dry chemical or carbon tetrachloride (Section 6).
Storage and Handling: Section 7.

MONOMETHYLAMINE, ANHYDROUS. See monomethylamine.
Shipping Regulations: Section 11.
   I.C.C.: Flammable gas; red gas label, 300 pounds.
   Coast Guard Classification: Inflammable gas; red gas label.
   IATA: Flammable gas, red label, not acceptable (passenger), 140 kilograms (cargo).

MONOMETHYLAMINE, AQUEOUS SOLUTION. See monomethyl amine.
Shipping Regulations: Section 11.
   I.C.C.: Flammable liquid; red label, 10 gallons.
   Coast Guard Classification: Inflammable liquid; red label.
   IATA: Flammable liquid, red label, 1 liter (passenger), 40 liters (cargo).

MONOMETHYLANILINE. See n-methylaniline.

MONOMETHYLOL DIMETHYL HYDANTOIN
General Information
Description: Odorless, white crystalline solid.
Formula: $C_6H_{10}N_2O_3$.
Constants: Mol wt: 158.16, mp: 99–103°C, d: 0.78.
Hazard Analysis
Toxic Hazard Rating:
   Acute Local: Irritant 3; Ingestion 3; Inhalation 3.
   Acute Systemic: U.

Chronic Local: U.
Chronic Systemic: U.
Disaster Hazard: Slight; when heated, it emits toxic fumes (Section 6).
Countermeasures
Ventilation Control: Section 2.
Personnel Protection: Section 3.
Storage and Handling: Section 7.

MONONITROTOLUENE. See p-nitrotoluene.

MONONONYL NAPHTHALENE
General Information
Descript on: Light straw-colored viscous liquid.
Formula: $C_9H_{19}C_{10}H_7$.
Constants: Mol wt: 254.4, bp: 300–350°C, d: 0.93–0.94 at 20°C, vap. d.: 8.8.
Hazard Analysis
Toxicity: Details unknown. See also naphthalene.
Fire Hazard: Slight, when exposed to heat or flame; can react with oxidizing materials (Section 6).
Storage and Handling: Section 7.

MONOPOTASSIUM GLUTAMATE
General Information
Synonym: Potassium glutamate.
Description: White, free flowing, crystalline powder; practically odorless; freely soluble in water; slightly soluble in alcohol.
Formula: $KOOC(CH_2)_2CH(NH_2)COOH \cdot H_2O$.
Constant: Mol wt: 203.
Hazard Analysis
Toxic Hazard Rating: Unknown. Used as a general purpose food additive (Section 10).

MONO-n-PROPYLAMINE
General Information
Description: Colorless liquid; amine-like odor.
Formula: $CH_3(CH_2)_2NH_2$.
Constants: Mol wt: 59.11, mp: −83°C, bp: 48.7°C, flash p: <20°F, d: 0.719, vap. d.: 2.04.
Hazard Analysis
Toxicity: Details unknown. See also amines.
Fire Hazard: Dangerous, when exposed to heat or flame.
Explosion Hazard: Unknown.
Disaster Hazard: Dangerous, upon exposure to heat or flame; on decomposition it emits toxic fumes of nitrogen oxides; can react vigorously with oxidizing materials.
Countermeasures
Ventilation Control: Section 2.
To Fight Fire: Foam, carbon dioxide, dry chemical or carbon tetrachloride (Section 6).
Storage and Handling: Section 7.

MONOPROPYLENE GLYCOL METHYL ETHER
General Information
Synonym: 1-Methoxy-2-propanol.
Description: Clear, colorless liquid.
Formula: $CH_3OCH_2CH(OH)CH_3$.
Constants: Mol wt: 90.1, bp: 120°C, flash p: 102°F, d: 0.92 at 25°/20°C, vap. d.: 3.1.
Hazard Analysis
Toxicity: Details unknown. See also glycols.
Fire Hazard: Moderate, when exposed to heat or flame; can react with oxidizing materials.
Explosion Hazard: Details unknown. See also ethers.
Countermeasures
Ventilation Control: Section 2.

To Fight Fire: Foam, carbon dioxide, dry chemical or carbon tetrachloride (Section 6).
Storage and Handling: Section 7.

MONOSTEARIN. See glyceryl monostearate.

MONOTHIOGLYCEROL
General Information
Description: Liquid.
Formula: $C_3H_8O_2S$.
Constants: Mol wt: 108.2, bp: 118°C at 5 mm, d: 1.248 at 20°/20°C.
Hazard Analysis
Toxicity: Details unknown. See also glycerol.
Fire Hazard: Moderate.
Disaster Hazard: Dangerous; when heated to decomposition, it emits highly toxic fumes of oxides of sulfur; can react with oxidizing materials.
Countermeasures
Storage and Handling: Section 7.

MONOURANYL-o-PHOSPHATE. See uranium phosphate.

MONTANITE. See bismuth tellurate.

MONURON
General Information
Synonym: CMU; 3-(p-chlorophenyl)-1, 1-dimethylurea.
Description: Crystals. Nearly water insoluble. Slight odor.
Formula: $C_9H_{11}ClN_2O$.
Constants: Mol wt: 198.7, mp: 171°C, vap. press.: 178 × 10^{-5}mm at 100°C.
Hazard Analysis
Toxicity: Details unknown. Has produced anemia and methemoglobinemia in experimental animals.
Disaster Hazard: See chlorides.

MORPHINE
General Information
Description: White, crystalline alkaloid.
Formula: $C_{17}H_{19}NO_3 \cdot H_2O$.
Constants: Mol wt: 303.35, mp: anh. 254°C decomposes, bp: 191–192°C (vacuum), d: 1.317.
Hazard Analysis
Toxic Hazard Rating:
 Acute Local: U.
 Acute Systemic: Ingestion 3.
 Chronic Local: U.
 Chronic Systemic: Ingestion 3.
Toxicology: Morphine is the constituent of opium most responsible for its poisonous effects. When taken orally, the effects of morphine poisoning begin to appear in 20 to 40 minutes; if taken hypodermically, the symptoms appear much earlier and narcotism is more likely to follow the early symptoms. Individual susceptibility varies greatly and children are more susceptible than adults. The usual symptoms due to an overdose of morphine are a sense of mental exhilaration and physical ease with a quickening and strengthening of the pulse and then a depression of the brain with special reference to its higher functions. Smaller doses which are not sufficient to cause this depression do cause a diminished sensibility to lasting impressions, such as pain, cold, hunger and discomfort. Following this there may be dizziness and heaviness of the head, nausea, languor and drowsiness, the pulse being reduced in force. Individuals with great susceptibility may com-

---

TOXIC HAZARD RATING CODE (For detailed discussion, see Section 1.)

0 NONE: (a) No harm under any conditions; (b) Harmful only under unusual conditions or overwhelming dosage.

1 SLIGHT: Causes readily reversible changes which disappear after end of exposure.

2 MODERATE: May involve both irreversible and reversible changes not severe enough to cause death or permanent injury.

3 HIGH: May cause death or permanent injury after very short exposure to small quantities.

U UNKNOWN: No information on humans considered valid by authors.

plain of itching of the skin and even erythema and as the action continues, it produces sleep. There is a gradual loss of muscular power and a diminished sense of feeling. The pupils become contracted and fail to respond to light. The respirations become less and less frequent, finally being reduced in some cases to four or five a minute with stertorous breathing. If the patient who has taken an overdose of morphine survives for 48 hours, the prognosis is favorable.

Countermeasures

Treatment and Antidotes: A physician should be called at once. The stomach should be washed repeatedly and at short intervals, whether the poison was taken by mouth or injection, since it is excreted into the stomach anyway. Emetics should be given. Recommended is apomorphine hydrochloride, hypodermically. If narcosis has already set in, the effect of the emetic may be interfered with. A dilute solution of potassium permanganate in the form of a lavage which should be 0.5 grams per liter has been used successfully. Tincture of belladonna by mouth, 2 cc by volume, may be repeated every 15 minutes but this antidote must be used cautiously. Solubility of the morphine in the stomach may be decreased by giving tannic acid, coffee, tea and finally, powdered charcoal in water; also an iodine solution may be given. Every effort should be made to arouse the patient and keep him awake, especially by walking him around, pinching the skin, slapping him, and by ammonia inhalations and artificial respiration if necessary. Keep the patient warm under any conditions. The fatal dose may be from 1 to 2 to 4 or 6 grains, depending very much upon the individual.

Ventilation Control: Section 2.
Personal Hygiene: Section 3.
Storage and Handling: Section 7.

MORPHINE ACETATE. See morphine.

MORPHINE BENZYL ETHER HYDROCHLORIDE. See peronine.

MORPHINE HYDROCHLORIDE. See morphine.

MORPHINE MECONATE. See morphine.

MORPHINE NITRATE
General Information
Description: White powder, darkens when exposed to light.
Formula: $C_{17}H_{19}NO_3 \cdot HNO_3$.
Constant: Mol wt: 348.26.
Hazard Analysis and Countermeasures
See morphine and nitrates.

MORPHINE SULFATE. See morphine.

MORPHOLENE THIURAM DISULFIDE
Hazard Analysis
Toxicity: U. Used as a fungicide.
Disaster Hazard: Dangerous. See sulfur compounds.

MORPHOLINE
General Information
Synonyms: Tetrahydro-1, 4-oxazine; diethylenimide oxide.
Description: Colorless, hygroscopic oil.
Formula: $OCH_2CH_2NHCH_2CH_2$.
Constants: Mol wt: 87.12, bp: 128.9°C, fp: −7.5°C, flash p: 100°F (O.C.), d: 0.998 at 25°/25°C, autoign. temp.: 590°F, vap. press.: 10 mm at 23°C, vap. d.: 3.00.
Hazard Analysis
Toxic Hazard Rating:
Acute Local: Irritant 2; Ingestion 2; Inhalation 2.
Acute Systemic: Ingestion 2; Inhalation 2; Skin Absorption 2.
Chronic Local: U.
Chronic Systemic: U.

TLV: ACGIH (recommended); 20 parts per million of air; 70 milligrams per cubic meter of air. May be absorbed through the skin.
Toxicology: Irritating to skin, eyes and mucous membranes. Has produced kidney damage in experimental animals. Note: Used as a food additive permitted in food for human consumption (Section 10).
Fire Hazard: Moderate, when exposed to flame or heat.
Disaster Hazard: Dangerous; when heated to decomposition, it emits highly toxic fumes of oxides of nitrogen; can react with oxidizing materials.
Countermeasures
Ventilation Control: Section 2.
To Fight Fire: Alcohol foam, carbon dioxide, dry chemical or carbon tetrachloride (Section 6).
Personal Hygiene: Section 3.
Storage and Handling: Section 7.
Shipping Regulations: Section 11.
MCA warning label.

MORPHOLINE ETHANOL
General Information
Synonym: N-hydroxy ethyl morpholine.
Description: Colorless liquid.
Formula: $O(CH_2CH_2)_2NCH_2CH_2OH$.
Constants: Mol wt: 131.17, bp: 225.5°C, mp: 1.6°C, vap. press.: 0.1 mm at 20°C, flash p: 210°F, vap. d.: 4.54, d: 1.071.
Hazard Analysis
Toxic Hazard Rating:
Acute Local: Irritant 1; Ingestion 1; Inhalation 1.
Acute Systemic: U.
Chronic Local: U.
Chronic Systemic: U.
Fire Hazard: Moderate, when exposed to heat or flame; can react with oxidizing materials.
Countermeasures
To Fight Fire: Foam, carbon dioxide, dry chemical or carbon tetrachloride (Section 6).
Personal Hygiene: Section 3.
Storage and Handling: Section 7.

MORRHUA OIL. See cod liver oil.

MORTAR STAIN, LIQUID
Hazard Analysis
Toxicity: Variable.
Fire Hazard: Moderate, when exposed to heat or flame; can react with oxidizing materials (Section 6).
Explosion Hazard: Unknown.
Countermeasures
Storage and Handling: Section 7.
Shipping Regulations: Section 11.
I.C.C.: Flammable liquid; red label, 55 gallons.
Coast Guard Classification: Combustible liquid; inflammable liquid, red label.
IATA: Flammable liquid, red label, 1 liter (passenger), 220 liters (cargo).

MOSS BUNKER OIL. See menhaden oil.

MOSS GREEN. See copper acetoarsenite.

MOTH BALLS. See naphthalene.

MOTH FLAKES. See naphthalene.

MOTION PICTURE FILM, NITROCELLULOSE BASE
Hazard Analysis
Fire Hazard: Dangerous, when exposed to heat or flame. See also nitrocellulose (Section 6).
Explosion Hazard: Moderate, when exposed to heat or flame. See also nitrocellulose.
Disaster Hazard: Dangerous; when heated to decomposition, it emits highly toxic fumes of oxides of nitrogen; can react vigorously with oxidizing materials.

Countermeasures
Storage and Handling: Section 7.

MOTION PICTURE FILM, SLOW BURNING
Hazard Analysis
Fire Hazard: Moderate, when exposed to heat or flame (Section 6).
Disaster Hazard: Dangerous; when heated to decomposition, it emits highly toxic fumes of oxides of nitrogen; can react with oxidizing materials.
Countermeasures
Storage and Handling: Section 7.

MOTOR FUEL ANTIKNOCK COMPOUND. See lead tetraethyl.
Shipping Regulations: Section 11.
  I.C.C.: Poison B; poison label, 55 gallons.
  Coast Guard Classification: Poison B; poison label.
  IATA: Poison B, poison label, not acceptable (passenger), 220 liters (cargo).

MOTOR FUEL, N.O.S. See gasoline.
Shipping Regulations: Section 11.
  I.C.C.: Flammable liquid; red label, 10 gallons.
  Coast Guard Classification: Combustible liquid; inflammable liquid; red label.
  IATA: Flammable liquid, red label, 1 liter (passenger), 40 liters (cargo).

MOUNTAIN TOBACCO. See arnica.

MSP. See sodium phosphate, monobasic.

MUCOCHLORIC ACID
General Information
Synonym: α,β-Dichloro-β-formylacrylic acid; 2,3-dichloromaleic aldehyde acid; dichloromalealdehydic acid.
Description: Monoclinic prisms; slightly soluble in cold water; soluble in hot water, hot benzene, and alcohol.
Formula: OHCCCl:CClCOOH.
Constants: Mol wt: 168.97, mp: 127°C.
Hazard Analysis
Toxic Hazard Rating:
  Acute Local: Irritant 3; Ingestion 3; Inhalation 3.
  Acute Systemic: U.
  Chronic Local: U.
  Chronic Systemic: U.
Disaster Hazard: Dangerous. See chlorides.

MUIRA PUAMA
General Information
Description: A wood.
Composition: Aromatic resin, muira puamine, fat.
Hazard Analysis
Toxic Hazard Rating:
  Acute Local: Irritant 2; Ingestion 2.
  Acute Systemic: Ingestion 2.
  Chronic Local: U.
  Chronic Systemic: U.
Fire Hazard: Moderate, when exposed to heat or flame; can react with oxidizing materials (Section 6).
Explosion Hazard: Slight, in the form of dust, when exposed to heat or flame. See also dust explosions.
Countermeasures
Ventilation Control: Section 2.
Personnel Protection: Section 3.
Storage and Handling: Section 7.

MURIATIC ACID. See hydrochloric acid.
Shipping Regulations: Section 11.
  Coast Guard Classification: Corrosive liquid; red label.
  MCA warning label.

MURIATIC ETHER. See ethyl chloride.

MUSTARD GAS. See β,β'-dichloroethyl sulfide.
Shipping Regulations: Section 11.
  I.C.C.: Poison A; poison gas label—not accepted.
  Coast Guard Classification: Poison A; poison gas label.
  IATA: Poison A, not acceptable (passenger or cargo).

MYLONE
General Information
Synonym: Crag fungicide 974, dimethyl formocarbo-thialdine.
Description: Crystals, soluble in alcohol.
Formula: $C_5H_{10}N_2S_2$.
Constants: Mol wt: 162.3, mp: 107°C.
Hazard Analysis
Toxic Hazard Rating:
  Acute Local: Irritant 2; Allergen 2.
  Acute Systemic: U.
  Chronic Local: Irritant 2; Allergen 2.
  Chronic Systemic: U.
Toxicology: Mild, primary skin irritant and sensitizer.
Disaster Hazard: Dangerous. See sulfates.

TOXIC HAZARD RATING CODE (For detailed discussion, see Section 1.)

0 NONE: (a) No harm under any conditions; (b) Harmful only under unusual conditions or overwhelming dosage.

1 SLIGHT: Causes readily reversible changes which disappear after end of exposure.

2 MODERATE: May involve both irreversible and reversible changes; not severe enough to cause death or permanent injury.

3 HIGH: May cause death or permanent injury after very short exposure to small quantities.

U UNKNOWN: No information on humans considered valid by authors.

N

NABAM
General Information
Synonym: Disodium ethylene bis(dithiocarbamate).
Description: Crystals; soluble in water.
Formula: $C_4H_6N_2Na_2S_4$.
Constant: Mol wt: 256.4.
Hazard Analysis
Toxicity: Irritant to skin and mucous membranes. Narcotic in high concentrations. In presence of alcohol can cause violent vomiting.
Disaster Hazard: Dangerous; when heated to decomposition, it emits highly toxic fumes.

NaK
General Information
Description: Liquid or solid, metallic alloy.
Hazard Analysis
Toxicity: See sodium and potassium.
Toxicology: NaK as its name implies is a low-melting alloy of Na and K. Its toxicity is that of either Na or K alone. In contact with moisture this material reacts very violently to evolve hydrogen and much heat, leaving behind a highly caustic residue of NaOH or KOH. Of course the heat would damage tissue as would the caustic residue. Even fine particles of NaK would damage the eyes, irritate the lungs, etc. Finally when NaK burns it evolves quantities of finely divided fumes of Na_2O, K_2O which is a powerful caustic.
Fire Hazard: Dangerous. In the presence of oxygen, moisture, halogens, oxidizers, acids or acid fumes, etc., it will react violently giving off much heat, often either spattering red hot particles or actually flaming.
Explosion Hazard: Servere. It will react explosively under many conditions such as contact with moisture, halogens, acid solutions, mists and fumes, powerful oxidizing agents and many organic compounds containing oxygen or halogen.
Disaster Hazard: Dangerous; when heated, emits highly toxic fumes of sodium and potassium oxide; it will react explosively with water, steam, acid, acid fumes or mists to produce heat, hydrogen, toxic and corrosive fumes; can react very vigorously with oxidizing materials. See also sodium and potassium.
Countermeasures
Ventilation Control: Section 2.
To Fight Fire: G-1 powder, dry sodium chloride or dry soda ash. Never use water, foam or carbon tetrachloride (Section 6).
Storage and Handling: Store in a cool place, preferably under kerosene (Section 7).

NANTOKITE. See cuprous chloride.

NAPHTHACENE. See tetracene.

NAPHTHA (COAL-TAR)
General Information
Synonyms: Hi-flash naphtha; 160° benzol; naphtha, solvent.
Description: Dark, straw-colored to colorless liquid.
Composition: Benzene, toluene, xylene, etc.
Constants: Bp: 149–216°C, flash p: 100°F (C.C.), d: 0.862–0.892, autoign. temp.: 900–950°F.

Hazard Analysis
Toxic Hazard Rating:
Acute Local: Irritant 2; Ingestion 2; Inhalation 1.
Acute Systemic: Ingestion 3; Inhalation 3; Skin Absorption 3.
Chronic Local: Irritant 1.
Chronic Systemic: Ingestion 3; Inhalation 3; Skin Absorption 3.
Toxicity: Note: A common air contaminant (Section 4).
TLV: ACGIH (recommended); 200 parts per million in air; 800 milligrams per cubic meter of air.
Fire Hazard: Moderate, when exposed to heat or flame; can react with oxidizing materials. Keep containers tightly closed.
Explosion Hazard: Slight.
Countermeasures
Ventilation Control: Section 2.
To Fight Fire: Foam, carbon dioxide, dry chemical or carbon tetrachloride (Section 6).
Personnel Protection: Section 3.
Storage and Handling: Section 7.
Shipping Regulations: Section 11.
I.C.C.: Flammable liquid; red label, 10 gallons.
Coast Guard Classification: Combustible liquid; inflammable liquid, red label.
IATA: Flammable liquid, red label, 1 liter (passenger), 40 liters (cargo).

NAPHTHA DISTILLATE
Shipping Regulations: Section 11.
I.C.C.: Flammable liquid, red label, 10 gallons.
IATA: Flammable liquid, red label, 1 liter (passenger) 40 liters (cargo).

NAPHTHALENE
General Information
Synonyms: Moth flakes; white tar; tar camphor.
Description: Aromatic odor; white, crystalline, volatile flakes.
Formula: $C_{10}H_8$.
Constants: Mol wt: 128.16, mp: 80.1°C, bp: 217.9°C (sublimes), flash p: 176°F, d: 1.145, lel: 0.9%, uel: 5.9%, autoign. temp.: 979°F, vap. press.: 1 mm at 52.6°C, vap. d.: 4.42.
Hazard Analysis
Toxic Hazard Rating:
Acute Local: Irritant 2; Ingestion 2; Inhalation 1.
Acute Systemic: Ingestion 2; Inhalation 2.
Chronic Local: Irritant 1.
Chronic Systemic: Ingestion 1; Inhalation 1.
Toxicity: Note: May be used as an insecticide.
TLV: ACGIH (recommended); 10 parts per million of air; 50 milligrams per cubic meter of air.
Fire Hazard: Moderate, when exposed to heat or flame; reacts with oxidizing materials.
Spontaneous Heating: No.
Explosion Hazard: Moderate, in the form of dust, when exposed to heat or flame.
Countermeasures
Ventilation Control: Section 2.
To Fight Fire: Water, carbon dioxide, dry chemical or carbon tetrachloride (Section 6).
Personnel Protection: Section 3.
Storage and Handling: Section 7.

Shipping Regulations: Section 11.
 Coast Guard Classification: Hazardous article.
 IATA (crude or refined): Flammable solid, yellow label, 12 kilograms (passenger), 140 kilograms (cargo).

p-NAPHTHALENE. See anthracene.

α-NAPHTHALENE ACETIC ACID
General Information
Synonym: Naphthyl acetic acid; planofix.
Description: White odorless crystals; only slightly water soluble.
Formula: $C_{10}H_7CH_2COOH$.
Constants: Mol wt: 186.2, mp: 134°C.
Hazard Analysis
Toxic Hazard Rating:
 Acute Local: Irritant 2; Ingestion 2; Inhalation 2.
 Acute Systemic: Ingestion 2.
 Chronic Local: U.
 Chronic Systemic: U.
Toxicology: Skin irritant. Large doses cause CNS depression.

1,8- NAPHTHALENE DIAMINE
General Information
Synonym: 1,8-Diaminonaphthalene.
Description: Crystals. Turn brown upon standing. Slightly water soluble.
Formula: $C_{10}H_{10}N_2$.
Constants: Mol wt: 158.2, mp: 66.5°C, bp: 205°C at 12 mm.
Hazard Analysis and Countermeasures
See amines.

1,3-NAPHTHALENEDIOL. See naphthoresorcinol.

NAPHTHALENE-1,5-DISULFONIC ACID
General Information
Description: Crystals.
Formula: $C_{10}H_6(SO_3H)_2$.
Constant: Mol wt: 288.3.
Hazard Analysis
Toxicity: Details unknown. See also naphthalene.
Disaster Hazard: Dangerous; when heated to decomposition, it emits highly toxic fumes of oxides of sulfur.
Countermeasures
Storage and Handling: Section 7.

NAPHTHALENE ETHYLENE. See acenaphthene.

NAPHTHA (PETROLEUM). See petroleum spirits.

NAPHTHA, SAFETY SOLVENT. See stoddard solvent.

NAPHTHA, SOLVENT. See naphtha (coal-tar).
Shipping Regulations: Section 11.
 I.C.C.: Flammable liquid; red label, 10 gallons.
 Coast Guard Classification: Combustible liquid; inflammable liquid, red label.
 IATA: Flammable liquid, red label, 1 liter (passenger), 40 liters (cargo).

NAPHTHA, V.M.&P.
General Information
Synonyms: Benzine; 76° naphtha.
Description: Volatile liquid.
Constants: Bp: 100–140°C, flash p: 20°F (C.C.), d: 0.67–0.80, lel: 0.9% at 212°F, uel: 6.0% at 212°F, autoign. temp.: 450°F.

Hazard Analysis
Toxicity: See naphtha (petroleum).
Fire Hazard: Dangerous, when exposed to heat or flame.
Explosion Hazard: Moderate, when exposed to flame.
Disaster Hazard: Dangerous, upon exposure to heat or flame; can react vigorously with oxidizing materials.
Countermeasures
Ventilation Control: Section 2.
To Fight Fire: Foam, carbon dioxide, dry chemical or carbon tetrachloride (Section 6).
Storage and Handling: Section 7.

NAPTHA V.M.&P. 50° FLASH
General Information
Description: Insoluble in water.
Constants: Flash p: 50°F,* autoign. temp.: 450°F,* lel: 0.9%, uel: 6.7%, d: < 1, vap. d.: 4.1, bp: 240–290°F.
*Note: Flash p and autoign. temp. will vary depending on the manufacturer.
Hazard Analysis and Countermeasures
See naphtha V.M.&P.

NAPHTHA V.M.&P., HIGH FLASH
General Information
Description: Insoluble in water.
Constants: flash p: 85°F,* autoign. temp.: 450°F,* lel: 1.0%, uel: 6.0%, d: < 1, vap. d.: 4.3, bp: 280–350°F.
*Note: Flash p and autoign. temp. will vary depending on the manufacturer
Hazard Analysis and Countermeasurer
See naphtha V.M.&P.

NAPHTHENIC ACID
General Information
Synonym: Hexahydrobenzoic acid.
Description: Odorless crystals; slightly water soluble.
Formula: $C_7H_{12}O_2$.
Constants: Mol wt: 128.2, d: 1.034, mp: 31°C, bp: 233°C.
Hazard Analysis
Toxicity: Details unknown. See related compound cyclopentane.

NAPHTHOIC ACID
General Information
Synonym: Naphthylene-carboxylic acid.
Description: Two forms: α and β. (1) α: Needles, slightly soluble in hot water; soluble in hot alcohol and ether. (2) β: Plates or needles; slightly soluble in hot water; soluble in alcohol and ether.
Formula: $C_{11}H_8O_2$.
Constants: Mol wt: 172.17, mp (α): 160–161°C, mp(β): 184–185°C, bp (α): 300°C, bp(β): > 300°C.
Hazard Analysis
Toxic Hazard Rating:
 Acute Local: Irritant 2; Ingestion 2; Inhalation 2.
 Acute Systemic: U.
 Chronic Local: U.
 Chronic Systemic: U.

NAPHTHITE. See trinitro naph thalene.

α-NAPHTHOL
General Information
Synonyms: 1-Naphthol; 1-hydroxynaphthalene.
Description: Colorless crystals; disagreeable taste.
Formula: $C_{10}H_7OH$.
Constants: Mol wt: 144.2, mp: 96°C, bp: 282.5°C, d 1.0954 at 98.7°/4°C, vap. press.: 1 mm at 94.0°C.

TOXIC HAZARD RATING CODE (For detailed discussion, see Section 1.)

0 NONE: (a) No harm under any conditions; (b) Harmful only under unusual conditions or overwhelming dosage.

1 SLIGHT: Causes readily reversible changes which disappear after end of exposure.

2 MODERATE: May involve both irreversible and reversible changes not severe enough to cause death or permanent injury.

3 HIGH: May cause death or permanent injury after very short exposure to small quantities.

U UNKNOWN: No information on humans considered valid by authors

Hazard Analysis
Toxic Hazard Rating:
Acute Local: Irritant 2; Ingestion 3; Inhalation 1.
Acute Systemic: Ingestion 3; Inhalation 2.
Chronic Local: Irritant 2.
Chronic Systemic: Ingestion 3; Inhalation 1; Skin Absorption 3.
Toxicology: The naphthols are thought to be more toxic than naphthalene, the β-naphthol being slightly less toxic than the α derivative. They are an irritant to the skin and mucous membranes and may cause dermatitis (Section 9). It is reported that sufficient naphthol may be absorbed through the skin to cause irritation of the kidneys and injury to the cornea and lens of the eye.
Fire Hazard: Slight; (Section 6).

Countermeasures
Ventilation Control: Section 2.
Personnel Protection: Section 3.
Storage and Handling: Section 7.

β-NAPHTHOL
General Information
Synonyms: β-Hydroxynaphthalene; isonaphthol.
Description: White to yellowish-white crystals; slight phenolic odor.
Formula: $C_{10}H_7OH$.
Constants: Mol wt: 144.2, mp: 122.5°C, bp: 288.0°C, flash p: 322°F, d: 1.22, vap. press.: 10 mm at 145.5°C, vap. d.: 4.97.

Hazard Analysis
Toxic Hazard Rating:
Acute Local: Irritant 2; Ingestion 3; Inhalation 1.
Acute Systemic: Ingestion 3; Inhalation 2.
Chronic Local: Irritant 2.
Chronic Systemic: Ingestion 3; Inhalation 1; Skin Absorption 3.
Fire Hazard: Slight, when exposed to heat or flame.

Countermeasures
To Fight Fire: Carbon dioxide, dry chemical or carbon tetrachloride (Section 6).
Ventilation Control: Section 2.
Personnel Protection: Section 3.
Storage and Handling: Section 7.

2-NAPHTHOL-3-CARBOXYLIC ACID. See 3-hydroxy-β-naphthoic acid.

2-NAPHTHOL-3,6-DISULFONIC ACID. See sodium β-naphthol disulfonate.

α-NAPHTHOL-1,2,3,4-TETRAHYDRIDE.
See tetrahydronaphthol.

1,4-NAPHTHOQUINONE
General Information
Description: Greenish-yellow powder.
Formula: $C_{10}H_6O_2$.
Constants: Mol wt: 158.1, mp: 123–126°C, bp: sublimes at 100°C.

Hazard Analysis
Toxic Hazard Rating:
Acute Local: Irritant 2; Allergen 1; Ingestion 2; Inhalation 2.
Acute Systemic: Ingestion 2; Inhalation 2.
Chronic Local: Allergen 1.
Chronic Systemic: U.
Disaster Hazard: Slight; when heated, it emits toxic fumes (Section 6).

Countermeasures
Storage and Handling: Section 7.

NAPHTHORESORCINOL
General Information
Synonym: 1,3-Dihydroxynaphthalene; 1,3-naphthalenediol.
Description: White, crystalline leaves or plates.
Formula: $C_{10}H_8O_2$.

Constants: Mol wt: 160.2, mp: 124–125°C.
Hazard Analysis
Toxicity: Details unknown. See naphthalene.
Disaster Hazard: Slight; when heated, it emits toxic fumes (Section 6).
Countermeasures
Storage and Handling: Section 7.

NAPHTHYL ACETIC ACID. See α-naphthalene acetic acid.

α-NAPHTHYLAMINE
General Information
Description: White crystals.
Formula: $C_{10}H_7NH_2$.
Constants: Mol wt: 143.18, mp: 50°C, bp: 300.8°C, flash p: 315°F, d: 1.131, vap. press.: 1 mm at 104.3°C, vap. d.: 4.93.

Hazard Analysis
Toxic Hazard Rating:
Acute Local: U.
Acute Systemic: Ingestion 3; Inhalation 3; Skin Absorption 3.
Chronic Local: U.
Chronic Systemic: Ingestion 3; Inhalation 3; Skin Absorption 3.
Toxicology: Along with β-naphthylamine and benzidene, it has been incriminated as a cause of bladder cancer in the coal tar dye industry. See also β-Naphthylamine.
Fire Hazard: Slight, when exposed to heat or flame; can react with oxidizing materials; when heated it emits toxic fumes.
Spontaneous Heating: No.

Countermeasures
To Fight Fire: Water, carbon dioxide, dry chemical or carbon tetrachloride (Section 6).
Ventilation Control: Section 2.
Personnel Protection: Section 3.
First Aid: Section 1.
Storage and Handling: Section 7.
Shipping Regulations: Section 11.
IATA: Poison B, poison label, 25 kilograms (passenger), 95 kilograms (cargo).

β-NAPHTHYLAMINE
General Information
Description: White to faint pink, lustrous leaflets; faint aromatic odor.
Formula: $C_{10}H_7NH_2$.
Constants: Mol wt: 143.18, mp: 111.5°C, bp: 306.0°C, d: 1.061 at 98°/4°C, vap. press.: 1 mm at 108.0°C.

Hazard Analysis
Toxic Hazard Rating:
Acute Local: Irritant 1.
Acute Systemic: U.
Chronic Local: U.
Chronic Systemic: Ingestion 3; Inhalation 3; Skin Absorption 3.
TLV: ACGIH (recommended); Because of the extremely high incidence of bladder tumors in workers handling this compound, and the inability to control exposures, β-naphthylamine has been prohibited from manufacture, use and other activities that involve human contact by the state of Pennsylvania.
Toxicology: It is not corrosive or dangerously reactive, but it is a very toxic chemical in any of its physical forms, such as flake, lump, dust, liquid or vapor. It can be absorbed into the body through the lungs, the gastrointestinal tract or skin. Long and continued exposure to even small amounts may produce tumors and cancers of the bladder. It is combustible and at elevated temperatures evolves a vapor which is flammable and explosive. The explosive limit of these vapors has not yet been determined.

Fire Hazard: Moderate, when exposed to heat or flame; when heated it emits toxic fumes (Section 6).

Countermeasures
Ventilation Control: Section 2.
Personnel Protection: Section 3.
First Aid: Section 1.
Storage and Handling: Section 7.

1-NAPHTHYLAMINE-8-SULFONIC ACID.
General Information
Synonyms: Peri acid; S-acid; Schoelkopf's acid; 8-amino-1-naphthalenesulfonic acid.
Description: White needles; soluble in 4800 parts cold water, 240 parts boiling water; freely soluble in glacial acetic acid.
Formula: $NH_2C_{10}H_6SO_3H$ (bicyclic).
Constant: Mol wt: 223.25.
Hazard Analysis
Toxicity: Details unknown. Animal experiments show irritant properties. No definite evidence of carcinogenicity.
Disaster Hazard: Dangerous. See sulfonic acid.

1,5-NAPHTHALENE DIISOCYANATE
General Information
Synonym: Desmodur.
Formula: $C_{10}H_6(NCO)_2$.
Constant: Mol wt: 210.
Hazard Analysis
Toxic Hazard Rating.
 Acute Local: Allergen 3.
 Acute Systemic: U.
 Chronic Local: Allergen 3.
 Chronic Systemic: U.
Toxicity: See also 2,4-toluylene diisocyanate.

α-NAPHTHYL ISOCYANATE
General Information
Formula: $C_{11}H_7NO$.
Constants: Mol wt: 169.2, mp: 4°C, bp: 269–270°C, d: 1.179 at 25°C.
Hazard Analysis and Countermeasures
See cyanates.

α-NAPHTHYL ISOTHIOCYANATE
General Information
Description: White, odorless, tasteless needles.
Formula: $C_{10}H_7NCS$.
Constants: Mol wt: 185.2, mp: 58°C, d: 1.81.
Hazard Analysis
Toxicity: Details unknown; is said to cause dermatitis and kidney injury. Can be absorbed via the intact skin when in solution.
Disaster Hazard: See thiocyanates.
Countermeasures
Storage and Handling: Section 7.

2-NAPHTHYL MERCAPTAN
Hazard Analysis
Toxicity: Details unknown; a mosquito larvicide. See also mercaptans.
Disaster Hazard: Dangerous; when heated to decomposition, it emits highly toxic fumes of oxides of sulfur.
Countermeasures
Storage and Handling: Section 7.

NAPHTHYL MERCURIC ACETATE
General Information
Description: Crystals; insoluble in water.

Formula: $C_{10}H_7HgO_2C_2H_3$.
Constants: Mol wt: 386.8, mp: 154°C.
Hazard Analysis and Countermeasures
See mercury compounds, organic.

NAPHTHYL MERCURIC CHLORIDE
General Information
Description: Silky crystals; insoluble in water.
Formula: $(C_{10}H_7)HgCl$.
Constant: Mol wt: 363.2, mp: 189°C.
Hazard Analysis and Countermeasures
See mercury compounds, organic.

1-NAPHTHYL-N-METHYL   CARBAMATE. See carbaryl.

N-1-NAPHTHYLPHTHALAMIC ACID
General Information
Synonym: α-Naphthylphthalamic acid.
Description: Colorless crystals from alcohol.
Formula: $C_{18}H_{13}NO_3$.
Constants: Mol wt: 291.3, mp: 203°C, d: 1.40 at 20°/4°C.
Hazard Analysis
Toxicity: Unknown. Limited animal experiments indicate low toxicity.
Fire Hazard: Moderate, when exposed to heat or flame.
Countermeasures
To Fight Fire: Water, foam, carbon dioxide, dry chemical, water spray, carbon tetrachloride (Section 6).
Storage and Handling: Section 7.

α-NAPHTHYLTHIOUREA
General Information
Synonyms: Antu; 1-(1-naphthyl)-2-thiourea.
Description: Crystals; bitter taste.
Formula: $C_{11}H_{10}N_2S$.
Constants: Mol wt: 202.3, mp: 198°C.
Hazard Analysis
Toxic Hazard Rating:
 Acute Local: 0.
 Acute Systemic: Ingestion 2.
 Chronic Local: 0.
 Chronic Systemic: Ingestion 2.
Toxicology: A rodenticide used extensively against rats. Death is caused by pulmonary edema. Chronic toxicity probably similar to that of thiouracil, which has been known to cause a decrease in the white blood cells and drug rashes.
TLV: ACGIH (recommended); 0.3 milligrams per cubic meter of air.
Disaster Hazard: Dangerous; on decomposition it emits toxic fumes.
Countermeasures
Ventilation Control: Section 2.
Personnel Protection: Section 3.
Storage and Handling: Section 7.

1-(1-NAPHTHYL)-2-THIOUREA. See α-naphthyl thiourea.

NAPLES YELLOW. See lead antimonate.

NATRIUM. See sodium.

NATURAL EXTRACTIVES (SOLVENT FREE) USED IN CONJUNCTION WITH SPICES, SEASONINGS, AND FLAVORINGS. See Section 10—Food additives.

TOXIC HAZARD RATING CODE (For detailed discussion, see Section 1.)

0 NONE: (a) No harm under any conditions; (b) Harmful only under unusual conditions or overwhelming dosage.

1 SLIGHT: Causes readily reversible changes which disappear after end of exposure.

2 MODERATE: May involve both irreversible and reversible changes not severe enough to cause death or permanent injury.

3 HIGH: May cause death or permanent injury after very short exposure to small quantities.

U UNKNOWN: No information on humans considered valid by authors.

NATURAL GAS
General Information
Formula: 85% methane, 10% ethane + propane, butane, nitrogen.
Constants: autoign. temp.: 900–1170° F, lel: 3.8–6.5%, uel: 13–17%.
Hazard Analysis
Toxic Hazard Rating:
Acute Local: 0.
Acute Systemic: Inhalation 2.
Chronic Local: 0.
Chronic Systemic: Inhalation 2.
Toxicology: A simple asphyxiant. Upon incomplete combustion yields carbon monoxide. See also specific components.
Fire Hazard: Dangerous. See methane.
Explosion Hazard: See methane.
Explosive Range: See methane.
Countermeasures
To Fight Fire: See methane.
Shipping Regulations: See methane.

NAVY SPECIAL. See fuel oil #5.

NBA. See n-bromoacetamide.

NEATSFOOT OIL
General Information
Description: Pale yellow liquid.
Constants: Mp: 29–41°C, flash p: 470° F (C.C.), d: 0.92, autoign. temp.: 828° F.
Hazard Analysis
Toxicity: Details unknown. May cause dermatites in sensitive individuals.
Fire Hazard: Slight, when exposed to heat or flame.
Spontaneous Heating: Yes.
Countermeasures
To Fight Fire: Foam, carbon dioxide, dry chemical or carbon tetrachloride (Section 6).
Storage and Handling: Section 7.

NEMAGON. See 1,2-dibromo-3-chloropropane.

NEOARSPHENAMINE
General Information
Synonym: Neosalvarsan.
Description: Yellow, almost odorless powder.
Formula: (partially) $NH_2OHC_6H_3AsAsC_6H_3OH-NH(CH_2O)OSNa$.
Hazard Analysis and Countermeasures
See arsenic compounds.

NEOCOID. See DDT.

NEODYMIUM
General Information
Description: Yellowish, metallic crystals.
Formula: Nd.
Constants: At wt: 144.27, mp: 840°C, d: 6.9.
Hazard Analysis
Toxicity: Unknown. As a lanthanon it may interfere with the clotting of the blood. See also lanthanum.
Radiation Hazard: Section 5. For permissible levels, see Table 5, p. 150.
Natural (24%) isotope ^{144}Nd, half life 5×10^{15} y. Decays to stable ^{140}Ce by emitting alpha particles of 1.8 MeV.
Artificial isotope ^{147}Nd, half life 11 d. Decays to radioactive ^{147}Pm by emitting beta particles of 0.37 (20%), 0.81 (77%) MeV. Also emits gamma rays of 0.09–0.69 MeV.
Artificial isotope ^{149}Nd, half life 1.9 h. Decays to radioactive ^{149}Pm by emitting beta particles of 0.95 (16%), 1.1 (43%), 1.5 (31%) MeV. Also emits gamma rays of 0.03–0.27 MeV.

Fire Hazard: Moderate, in the form of dust, when exposed to heat or flame. See also powdered metals (Section 6).
Explosion Hazard: Slight, in the form of dust, when exposed to flame. See also powdered metals.

NEODYMIUM BROMATE
General Information
Description: Hexagonal, red crystals.
Formula: $Nd(BrO_3)_3 \cdot 9H_2O$.
Constants: Mol wt: 690.16, mp: 66.7°C, bp: $-9H_2O$ at 150°C.
Hazard Analysis and Countermeasures
See bromates and neodymium.

NEODYMIUM BROMIDE
General Information
Description: Green crystals.
Formula: $NdBr_3$.
Constant: Mol wt: 384.02.
Hazard Analysis and Countermeasures
See bromides and neodymium.

NEODYMIUM CARBIDE
General Information
Description: Hexagonal, yellow leaflets.
Formula: NdC_2.
Constants: Mol wt: 168.29, mp: decomposes, d: 5.15.
Hazard Analysis
Toxicity: Unknown. See neodymium.
Fire Hazard: A moderate fire hazard by chemical reaction. See carbides.
Explosion Hazard: Slight by chemical reaction with moisture and acids.
Disaster Hazard: Slight; it will react with water or steam to produce flammable vapors.
Countermeasures
Storage and Handling: Section 7.

NEODYMIUM HEXAANTIPYRINE IODIDE
General Information
Description: Rose colored crystals; soluble in water.
Formula: $[Nd(COC_{10}H_{12}N_2)_6I_3]$.
Constants: Mol wt: 1654.4, mp: 271°C.
Hazard Analysis and Countermeasures
See iodides and neodymium.

NEODYMIUM IODIDE
General Information
Description: Black, crystalline powder.
Formula: NdI_3.
Constants: Mol wt: 525.03, mp: 775 ± 3°C.
Hazard Analysis and Countermeasures
See iodides and neodymium.

NEODYMIUM NITRATE
General Information
Description: Triclinic crystals.
Formula: $Nd(NO_3)_3 \cdot 6H_2O$.
Constant: Mol wt: 438.39.
Hazard Analysis and Countermeasures
See nitrates and neodymium.

NEODYMIUM NITRIDE
General Information
Description: Black powder.
Formula: NdN.
Constant: Mol wt: 158.28.
Hazard Analysis and Countermeasures
See ammonia, anhydrous and neodymium.

NEODYMIUM OXALATE
General Information
Description: Rose crystals.
Formula: $Nd_2(C_2O_4)_3 \cdot 10H_2O$.
Constant: Mol wt: 732.76.
Hazard Analysis and Countermeasures
See oxalates and neodymium.

NEODYMIUM SULFIDE
General Information
Description: Olive-green powder.
Formula: Nd_2S_3.
Constants: Mol wt: 384.72, mp: decomposes, d: 5.179 at 11°C.
Hazard Analysis and Countermeasures
See sulfides and neodymium.

NEOHEXANE. See 2,2-dimethylbutane.

NEOMYCIN
General Information
Description: An antibiotic.
Hazard Analysis
Toxic Hazard Rating: U. Used as a food additive permitted in foods for human consumption (Section 10).

NEON
General Information
Description: Colorless, wholly inert gas.
Formula: Ne
Constants: At wt: 20.18, mp: $-248.67°C$, bp: $-245.9°C$, d: 0.9002 g/liter at 0°C; d liq: 1.204 at $-245.9°C$.
Hazard Analysis
Toxic Hazard Rating:
Acute Local: 0.
Acute Systemic: Inhalation 1.
Chronic Local: 0.
Chronic Systemic: U.
Toxicology: A simple asphyxiant. See argon.
Countermeasures
Storage and Handling: Section 7.
Shipping Regulations: Section 11.
I.C.C.: Nonflammable gas; green label, 300 pounds.
Coast Guard Classification: Noninflammable gas; green gas label.
IATA: Nonflammable gas, green label, 70 kilograms (passenger), 140 kilograms (cargo).

NEON, LIQUID
Shipping Regulations: Section 11.
IATA (non-pressurized): Other restricted articles, class C, not acceptable (passenger and cargo).
IATA (low pressure): Other restricted articles, class C, not acceptable (passenger), 140 kilograms (cargo).
(pressurized): Nonflammable gas, green label, not acceptable (passenger), 140 kilograms (cargo).

NEONICOTINE. See anabasine.

NEONICOTINE SULFATE. See anabasine sulfate.

NEOPENTANE. See 2,2-dimethylpropane.

NEOPENTYL ALCOHOL. See amyl alcohol.

NEOPENTYL GLYCOL
General Information
Synonym: 2,2-Dimethyl 1,3-propanediol.
Description: White, crystalline solid.
Formula: $C_5H_{12}O_2$.
Constants: Mol wt: 104.2, mp: 121–126°C, bp: 207–212°C.
Hazard Analysis
Toxicity: Details unknown; an insect repellent. See also glycols.
Fire Hazard: Slight; can react with oxidizing materials (Section 6).
Countermeasures
Storage and Handling: Section 7.

NEOPRENE
General Information
Synonym: Duprene.
Description: An oil resistant synthetic rubber made by the polymerization of chloroprene.
Formula: $[CH_3ClC:CHCH_3]_x$.
Hazard Analysis and Countermeasures
See chloroprene.

NEOSALVARSAN. See neoarsphenamine.

NEOTHYL. See methyl propyl ether.

NEOTRAN. See bis(p-chlorophenoxy)methane.

NEPALINE. See pseudoaconitine.

NEPTUNIUM
General Information
Description: A radioactive, artifically produced element.
Formula: Np.
Constants: At wt: 239.
Hazard Analysis
Toxicity: Highly radiotoxic.
Radiation Hazard: Section 5. For permissible levels, see Table 5, p. 150.
Artificial isotope ^{237}Np (Neptunium Series), half life 2.2×10^6 y. Decays to radioactive ^{233}Pa by emitting alpha particles of 4.8 MeV.
Artificial isotope ^{239}Np, half life 2.3 d. Decays to radioactive ^{239}Pu by emitting beta particles of 0.33–0.72 MeV. Also emits gamma rays of 0.04–0.3 MeV.

NERAL. See citral.

NEROLI OIL, ARTIFICIAL. See methyl anthranilate.

NERVE GAS. See tubun and sarin.

NEUTRAL VERDIGRIS. See copper acetate.

NIACIN
General Information
Synonyms: Nicotinic acid; pyridene-3-carboxylic acid.
Description: The anti-pellagra vitamin; colorless needles; odorless; soluble in water and alcohol; insoluble in most lipid solvents.
Formula: C_5H_4NCOOH.
Constants: Mol wt: 123, mp: 236°C, sublimes above mp; d: 1.473.
Hazard Analysis
Toxicity: Details unknown. A nutrient and/or dietary supplement food additive (Section 10).

NIACINIMIDE
General Information
Synonyms: Nicotinamide; nicotinic acid amide.
Description: Colorless needles; very soluble in water, ethyl alcohol and glycerol.
Formula: $C_5H_4NCONH_2$.
Constants: Mol wt: 122, mp: 129°C, d: 1.40.
Hazard Analysis
Toxic Hazard Rating: U. Used as a nutrient and/or dietary supplement food additive (Section 10).

NICCOLITE. See nickel arsenide.

NICKEL
General Information
Description: Cubic, silvery, metallic crystals.
Formula: Ni.

TOXIC HAZARD RATING CODE *(For detailed discussion, see Section 1.)*

0 NONE: (a) No harm under any conditions; (b) Harmful only under unusual conditions or overwhelming dosage.

1 SLIGHT: Causes readily reversible changes which disappear after end of exposure.

2 MODERATE: May involve both irreversible and reversible changes; not severe enough to cause death or permanent injury.

3 HIGH: May cause death or permanent injury after very short exposure to small quantities.

U UNKNOWN: No information on humans considered valid by authors.

Constants: At wt: 58.69, mp: 1452°C, bp: 2900°C, d: 8.90, vap. press.: 1 mm at 1810°C.

Hazard Analysis

Toxicity: See nickel compounds.

Radiation Hazard: Section 5. For permissible levels, see Table 5, p. 150.

Artificial isotope ^{59}Ni, half life 8×10^4 y. Decays to stable ^{59}Co by electron capture. Emits X-rays.

Artificial isotope ^{63}Ni, half life 120 y. Decays to stable ^{63}Cu by emitting beta particles of 0.067 MeV.

Artificial isotope ^{65}Ni, half life 2.6 h. Decays to stable ^{65}Cu by emitting beta particles of 0.60 (23%), 1.01 (8%), 2.10 (69%) MeV. Also emits gamma rays of 0.37, 1.11, 1.49 MeV.

Fire Hazard: Moderate, in the form of dust, when exposed to heat or flame. See also powdered metals (Section 6).

Explosion Hazard: Slight, in the form of dust when exposed to flame. See also powdered metals.

NICKEL ACETATE

General Information

Description: Green prisms.

Formula: $Ni(C_2H_3O_2)_2$.

Constants: Mol wt: 176.8, mp: decomposes, d: 1.798.

Hazard Analysis

Toxicity: See nickel compounds.

NICKEL AMMONIUM NITRATE

General Information

Synonym: Nickel nitrate tetrammine.

Description: Green crystals.

Formula: $Ni(NO_3)_2 \cdot 4NH_3 \cdot 2H_2O$.

Constant: Mol wt: 286.87.

Hazard Analysis and Countermeasures

See nickel compounds and nitrates.

NICKEL AMMONIUM SULFATE

General Information

Synonyms: Nickel salts, double; ammonium nickel sulfate.

Description: Green crystals; soluble in water; less soluble in ammonium sulfate solution; insoluble in alcohol.

Formula: $NiSO_4 \cdot (NH_4)_2SO_4 \cdot 6H_2O$.

Constants: Mol wt: 395, d: 1.923.

Hazard Analysis and Countermeasures

See nickel compounds and sulfates.

NICKEL ANTIMONIDE

General Information

Synonym: Breithauptite.

Description: Hexagonal, light copper-red crystals.

Formula: NiSb.

Constants: Mol wt: 180.45, mp: 1158°C, bp: decomposes at 1400°C, d: 7.54.

Hazard Analysis

Toxicity: Highly toxic. See antimony compounds and nickel compounds.

Fire Hazard: Moderate, by chemical reaction with water to form stibine. See also hydrides (Section 6).

Explosion Hazard: Slight, when exposed to flame; can evolve stibine. See also hydrides.

Disaster Hazard: Dangerous; when heated to decomposition, it emits highly toxic fumes of antimony; it will react with water, steam or acids to produce toxic and flammable vapors; it can react with oxidizing materials.

Countermeasures

Storage and Handling: Section 7.

NICKEL ARSENATE

General Information

Synonym: Nickel o-arsenate.

Description: Yellowish-green powder.

Formula: $Ni_3(AsO_4)_2$.

Constants: Mol wt: 435.89, d: 4.98.

Hazard Analysis and Countermeasures

See arsenic compounds and nickel compounds.

Shipping Regulations: Section 11.

IATA (solid): Poison B, poison label, 25 kilograms (passenger), 95 kilograms (cargo).

NICKEL ARSENIDE

General Information

Synonym: Niccolite.

Description: Hexagonal crystals.

Formula: NiAs.

Constants: Mol wt: 133.60, mp: 968°C, d: 7.57 at 0°C.

Hazard Analysis

Toxicity: Highly toxic. See arsenic compounds and nickel compounds.

Fire Hazard: Moderate, by chemical reaction to form arsine. See also hydrides (Section 6).

Explosion Hazard: Moderate, by chemical reaction to evolve arsine. See also hydrides.

Disaster Hazard: Dangerous; when heated to decomposition, it emits highly toxic fumes of arsenic; it will react with water, steam or acids to produce toxic and flammable vapors; it can react with oxidizing materials.

Countermeasures

Storage and Handling: Section 7.

NICKEL o-ARSENITE, ACID

General Information

Description: Green-white crystals.

Formula: $Ni_3H_6(AsO_3)_4 \cdot H_2O$.

Constants: Mol wt: 691.77, mp: decomposes.

Hazard Analysis and Countermeasures

See arsenic compounds and nickel compounds.

NICKEL BENZENE SULFONATE

General Information

Description: Green, monoclinic crystals.

Formula: $Ni(C_6H_5SO_3)_2 \cdot 6H_2O$.

Constants: Mol wt: 481.11, bp: decomposes, d: 1.628 at 25°C.

Hazard Analysis and Countermeasures

See nickel compounds and sulfonates.

NICKEL BORIDE

General Information

Description: Prisms.

Formula: NiB.

Constants: Mol wt: 69.51, d: 7.39 at 18°C.

Hazard Analysis and Countermeasures

See nickel compounds, boron compounds and boron hydride.

NICKEL BROMATE

General Information

Description: Monoclinic crystals.

Formula: $Ni(BrO_3)_2 \cdot 6H_2O$.

Constants: Mol wt: 422.62, mp: decomposes, d: 2.575.

Hazard Analysis and Countermeasures

See nickel compounds and bromates.

NICKEL BROMIDE

General Information

Synonym: Nickelous bromide.

Description: Yellowish, brown, deliquescent crystals.

Formula: $NiBr_2$.

Constants: Mol wt: 218.52, mp: 963°C, d: 4.64 at 28°C.

Hazard Analysis and Countermeasures

See bromides and nickel compounds.

NICKEL BROMIDE AMMONIA

General Information

Description: Violet powder.

Formula: $NiBr_2 \cdot 6NH_3$.

Constants: Mol wt: 320.72, d: 1.837.

Hazard Analysis and Countermeasures

See bromides and nickel compounds.

NICKEL BROMOPLATINATE

General Information

Description: Trigonal crystals.

Formula: $NiPtBr_6 \cdot 6H_2O$.
Constants: Mol wt: 841.51, d: 3.715.
Hazard Analysis and Countermeasures
See nickel compounds, platinum compounds and bromides.

NICKEL CARBIDE
General Information
Description: Powder.
Formula: Ni_3C.
Constants: Mol wt: 188.08, d: 7.957 at 25°C.
Hazard Analysis and Countermeasures
See nickel compounds and carbides.

NICKEL CARBONATE
General Information
Description: Rhombic, light green crystals.
Formula: $NiCO_3$.
Constants: Mol wt: 118.70, mp: decomposes.
Hazard Analysis
Toxicity: See nickel compounds.

NICKEL CARBONYL
General Information
Description: Colorless, volatile liquid or needles.
Formula: $Ni(CO)_4$.
Constants: Mol wt: 170.69, mp: $-25°C$, bp: 43°C, lel: 2% at 20°C, d: 1.3185 at 17°C, vap. press.: 400 mm at 25.8°C.
Hazard Analysis
Toxic Hazard Rating:
  Acute Local: Inhalation 3.
  Acute Systemic: Inhalation 3.
  Chronic Local: Allergen 1.
  Chronic Systemic: Inhalation 3.
Toxicology: Toxic symptoms from inhalation are believed to be caused both by the nickel and the carbon monoxide when liberated in the lungs. In severe acute cases there is headache, dizziness, nausea, vomiting, fever and difficult breathing. Later there is coughing and cyanosis. Chronic exposure is associated with a high incidence of cancer of the respiratory tract and nasal sinuses. Sensitization dermatitis is fairly common. Note: This is a common air contaminant. (Section 4).
TLV: ACGIH (recommended); 0.001 part per million in air; 0.007 milligrams per cubic meter of air.
Fire Hazard: Dangerous, when exposed to heat or flame.
Explosion Hazard: Moderate, when exposed to heat or flame.
Disaster Hazard: Dangerous; when heated or on contact with acid or acid fumes, it emits highly toxic fumes; it can react with oxidizing materials.
Countermeasures
Ventilation Control: Section 2.
To Fight Fire: Water, foam, carbon dioxide, dry chemical or carbon tetrachloride (Section 6).
Personnel Protection: Section 3.
First Aid: Section 1.
Storage and Handling: Section 7.
Shipping Regulations: Section 11.
  I.C.C.: Flammable liquid; red label, not accepted.
  Coast Guard Classification: Inflammable liquid; red label.
  IATA: Flammable liquid, not acceptable (passenger and cargo).

NICKEL CATALYST, FINELY DIVIDED, ACTIVATED OR SPENT
Shipping Regulations: Section 11.

I.C.C.: Flammable solid, yellow label, 100 pounds.
IATA (dry or wet with less than 40% water or other suitable liquid): Flammable solid, not acceptable (passenger and cargo).
IATA (wet, with not less than 40% water or other equally suitable liquid): Flammable solid, yellow label, not acceptable (passenger), 45 kilograms (cargo).

NICKEL CHLORATE
General Information
Description: Dark red crystals.
Formula: $Ni(ClO_3)_2 \cdot 6H_2O$.
Constants: Mol wt: 333.70, mp: decomp. at 80°C, d: 2.07.
Hazard Analysis and Countermeasures
See nickel compounds and chlorates.

NICKEL CHLORIDE
General Information
Description: A: Yellow, deliquescent scales; B: Monoclinic, green crystals.
Formula: $A:NiCl_2$; $B: NiCl_2 \cdot 6H_2O$.
Constants: Mol wt: A: 129.60, B: 237.70, mp: sublimes (A), bp: 987°C (A), d: 3.55 (A), vap. press.: 1 mm at 671°C.
Hazard Analysis and Countermeasures
See nickel compounds and chlorides.

NICKEL CHLOROPALLADATE
General Information
Description: Hexagonal crystals.
Formula: $NiPdCl_6 \cdot 6H_2O$.
Constants: Mol wt: 486.23, d: 2.353.
Hazard Analysis and Countermeasures
See nickel compounds, palladium compounds and chlorides.

NICKEL CHLOROPLATINATE
General Information
Description: Trigonal crystals.
Formula: $NiPtCl_6 \cdot 6H_2O$.
Constants: Mol wt: 574.76, d: 2.798.
Hazard Analysis and Countermeasures
See platinum compounds, nickel compounds and chlorides.

NICKEL COBALT SULFATE
General Information
Description: Reddish-brown, crystalline mass.
Formula: $NiSO_4 \cdot CoSO_4 \cdot 14H_2O$.
Constant: Mol wt: 561.99.
Hazard Analysis and Countermeasures
See nickel compounds, cobalt compounds and sulfates.

NICKEL COMPOUNDS
Hazard Analysis
Toxic Hazard Rating:
  Acute Local: Irritant 1; Allergen 1; Ingestion 1; Inhalation 1.
  Acute Systemic: Ingestion 1; Inhalation 1.
  Chronic Local: Irritant 2; Allergen 1.
  Chronic Systemic: Inhalation 2.
TLV: ACGIH (recommended); 1 milligram per cubic meter of air.
Toxicology: Nickel and most salts of nickel are not generally considered to cause systemic poisoning. Ingestion of large doses of nickel, as a nickel compound, (1 to 3 mg per kg of weight) has been shown to cause intestinal disorders, convulsions and asphyxia in dogs. Nickel has been found in the hair of persons exposed to nickel oxide dust, but no systemic effects which could be attributed to nickel alone have been reported. In 1938, the

---

TOXIC HAZARD RATING CODE (For detailed discussion, see Section 1.)

0 NONE: (a) No harm under any conditions; (b) Harmful only under unusual conditions or overwhelming dosage.

1 SLIGHT: Causes readily reversible changes which disappear after end of exposure.

2 MODERATE: May involve both irreversible and reversible changes; not severe enough to cause death or permanent injury.

3 HIGH: May cause death or permanent injury after very short exposure to small quantities.

U UNKNOWN: No information on humans considered valid by authors.

British described 10 cases of lung cancer and numerous cases of cancer of the nose and nasopharynx occuring in workers in a nickel refinery. The exact cause of the malignancies was never completely explained but arsenic was involved. The most common effect resulting from exposure to nickel compounds is the development of "nickel itch." This form of dermatitis occurs chiefly in persons doing nickel-plating. There is marked variation in individual susceptibility to the dermatitis. It occurs more frequently under conditions of high temperature and humidity when the skin is moist, and chiefly affects the hands and arms (Section 9). Nickel carbonyl is highly irritating to the lungs and also can produce asphyxia by decomposing with the formation of carbon monoxide. These compounds are common air contaminants (Section 4).

Countermeasures
Ventilation Control: Section 2.
Personnel Protection: Section 3.

NICKEL CYANIDE
General Information
Description: Apple-green plates or powder.
Formula: $Ni(CN)_2$.
Constant: Mol wt: 110.73.
Hazard Analysis and Countermeasures
See cyanides and nickel compounds.
Shipping Regulations: Section 11.
    I.C.C.: Poison B; poison label, 200 pounds.
    Coast Guard Classification: Poison B; poison label.
    IATA: Poison B, poison label, 12 kilograms (passenger), 95 kilograms (cargo).

NICKEL DI-n-BUTYLDITHIOCARBAMATE
General Information
Description: Dark olive-green powder.
Formula: $Ni[(C_4H_9)_2NCSS]_2$.
Constants: Mol wt: 467.44, mp: 89–90°C, d: 1.29.
Hazard Analysis
Toxicity: See nickel compounds.
Disaster Hazard: Dangerous; when heated to decomposition, it emits highly toxic fumes of oxides of sulfur.
Countermeasures
Storage and Handling: Section 7.

NICKEL DIFLUORIDE. See nickel fluoride.

NICKEL DITHIONATE
General Information
Description: Green, triclinic crystals.
Formula: $NiS_2O_6 \cdot 6H_2O$.
Constants: Mol wt: 326.92, mp: decomposes, d: 1.908.
Hazard Analysis
Toxicity: See nickel compounds.
Disaster Hazard: Dangerous; See sulfates.
Countermeasures
Storage and Handling: Section 7.

NICKEL FERROCYANIDE
General Information
Description: Green-white crystals.
Formula: $Ni_2Fe(CN)_6 \cdot xH_2O$.
Constant: D: 1.892.
Hazard Analysis and Countermeasures
See ferrocyanides and nickel compounds.

NICKEL FLUOGALLATE
General Information
Description: Pale green crystals.
Formula: $[Ni(H_2O)_6][GaF_5H_2O]$.
Constants: Mol wt: 349.52, mp: $-5H_2O$ at 110°C, d: 2.45.
Hazard Analysis and Countermeasures
See fluogallates and nickel compounds.

NICKEL FLUORIDE
General Information
Synonym: Nickel difluoride.
Description: Green crystals. Slightly water soluble. Decomposed by boiling water.
Formula: NiF_2.
Constants: Mol wt: 96.69, d: 4.63.
Hazard Analysis and Countermeasures
See fluorides and nickel compounds.

NICKEL FORMATE
General Information
Description: Green crystals.
Formula: $Ni(CHO_2)_2 \cdot 2H_2O$.
Constants: Mol wt: 184.76, mp: decomposes, d: 2.154.
Hazard Analysis
Toxicity: See nickel compounds.
Disaster Hazard: Slight; when heated, it emits acrid fumes (Section 6).
Countermeasures
Storage and Handling: Section 7.

NICKEL HYDROXIDE. See nickelous hydroxide.

NICKEL HYPOPHOSPHITE
General Information
Description: Green crystals.
Formula: $Ni(H_2PO_2)_2 \cdot 6H_2O$.
Constants: Mol wt: 296.78, mp: explodes at 100°C, d: 1.82 at 19.8°C.
Hazard Analysis
Toxicity: See nickel compounds.
Explosion Hazard: Moderate, when exposed to heat.
Disaster Hazard: Dangerous; may explode at 100°C. Keep away from heat and open flame.
Countermeasures
Storage and Handling: Section 7.

NICKELIC HYDROXIDE. See nickelous hydroxide.

NICKELIC OXIDE. See nickel peroxide.

NICKEL IODATE
General Information
Description: Yellow needles.
Formula: $Ni(IO_3)_2$.
Constants: Mol wt: 408.53, d: 5.07.
Hazard Analysis and Countermeasures
See nickel compounds and iodates.

NICKEL IODIDE
General Information
Description: Black, deliquescent crystals.
Formula: NiI_2.
Constants: Mol wt: 312.53, mp: 797°C, d: 5.834.
Hazard Analysis and Countermeasures
See nickel compounds and iodides.

NICKEL MONOSULFIDE
General Information
Synonym: Millerite.
Description: Trigonal or amorphous, black crystals.
Formula: NiS.
Constants: Mol wt: 90.76, mp: 797°C, d: 5.3–5.65.
Hazard Analysis and Countermeasures
See sulfides and nickel compounds.

NICKEL MONOXIDE
General Information
Synonym: Bunsenite.
Description: Cubic, green-black crystals.
Formula: NiO.
Constants: Mol wt: 74.69, mp: 1990°C, D: 7.45.
Hazard Analysis
Toxicity: See nickel compounds.

NICKEL NITRATE
General Information
Description: Green deliquescent crystals.
Formula: $Ni(NO_3)_2 \cdot 6H_2O$.
Constants: Mol wt: 290.8, mp: 56.7°C, bp: 136.7°C, d: 2.05.
Hazard Analysis and Countermeasures
See nickel compounds and nitrates.

NICKEL NITRATE TETRAMMINE. See nickel ammonium nitrate.

NICKEL OLEATE
General Information
Description: Green oil.
Formula: $Ni(C_{18}H_{33}O_2)_2$.
Constants: Mol wt: 621.58, mp: 18–20°C.
Hazard Analysis
Toxicity: See nickel compounds.

NICKELOUS BROMIDE. See nickel bromide.

NICKELOUS HYDROXIDE
General Information
Description: Light green crystals or amorphous.
Formula: $Ni(OH)_2$.
Constants: Mol wt: 92.7, d: 4.1.
Hazard Analysis
Toxicity: See nickel compounds.

NICKELOUS, NICKELIC SULFIDE
General Information
Synonym: Polydymite.
Description: Cubic, gray-black crystals.
Formula: Ni_3S_4.
Constants: Mol wt: 304.33, d: 4.7.
Hazard Analysis and Countermeasures
See nickel compounds and sulfides.

NICKELOUS PHOSPHATE. See nickel phosphate.

NICKEL OXALATE
General Information
Description: Light green powder.
Formula: $NiC_2O_4 \cdot 2H_2O$.
Constants: Mol wt: 182.74.
Hazard Analysis
Toxicity: Highly toxic. See oxalates.
Disaster Hazard: Dangerous; when heated it emits toxic fumes of carbon monoxide (Section 6).
Countermeasures
Storage and Handling: Section 7.

NICKEL OXIDE. See nickel monoxide.

NICKEL PERCHLORATE
General Information
Description: Hexagonal, green needles.
Formula: $Ni(ClO_4)_2 \cdot 6H_2O$.
Constants: Mol wt: 365.70, mp: 140°C.
Hazard Analysis and Countermeasures
See nickel compounds and perchlorates.

NICKEL PEROXIDE
General Information
Synonym: Nickelic oxide.
Description: Gray-black powder.
Formula: Ni_2O_3.
Constants: Mol wt: 165.38, mp: $-O_2$ at 600°C, d: 4.83.

Hazard Analysis and Countermeasures
See nickel compounds and peroxides.

NICKEL PHOSPHATE
General Information
Synonym: Nickelous phosphate.
Description: Apple-green plates or emerald cyrstals or granules.
Formula: $Ni_3(PO_4)_2 \cdot 8H_2O$.
Constants: Mol wt: 510.24, mp: decomposes.
Hazard Analysis and Countermeasures
See nickel compounds and phosphates.

NICKEL POTASSIUM SULFATE
General Information
Description: Blue-green crystals.
Formula: $NiSO_4 \cdot K_2SO_4 \cdot 6H_2O$.
Constants: Mol wt: 437.10, mp: decomp. at < 100°C, d: 2.124.
Hazard Analysis and Countermeasures
See nickel compounds and sulfates.

NICKEL SELENATE
General Information
Description: Green, tetragonal, monoclinic cyrstals.
Formula: $NiSeO_4 \cdot 6H_2O$.
Constants: Mol wt: 341.75, d: 2.314.
Hazard Analysis and Countermeasures
See selenium compounds and nickel compounds.

NICKEL SELENIDE
General Information
Description: Cubic, white or gray crystals.
Formula: NiSe.
Constants: Mol wt: 137.65, d: 8.46.
Hazard Analysis
Toxicity: See selenium compounds and nickel compounds.
Fire Hazard: Dangerous; it can easily evolve hydrogen selenide, which is highly flammable.
Explosion Hazard: Slight; it can evolve hydrogen selenide. See also hydrides.
Disaster Hazard: Dangerous; when heated to decomposition, it emits highly toxic fumes of selenium; will react with water, steam or acids to produce toxic and flammable vapors.
Countermeasures
Storage and Handling: Section 7.

NICKEL SUBSULFIDE
General Information
Synonym: Heazlewoodite.
Description: Pale yellowish, bronze metallic, lustrous crystals.
Formula: Ni_3S_2.
Constants: Mol wt: 240.19, d: 5.82.
Hazard Analysis and Countermeasures
See sulfides and nickel compounds.

NICKEL SULFATE
General Information
Description: Cubic, yellow crystals.
Formula: $NiSO_4$.
Constants: Mol wt: 154.76, mp: $-SO_3$ at 840°C, d: 3.68.
Hazard Analysis and Countermeasures
Used as a food additive permitted in the feed and drinking water of animals and/or for the treatment of food producing animals; also permitted in food for human consumption (Section 10). See nickel compounds and sulfates.

TOXIC HAZARD RATING CODE *(For detailed discussion, see Section 1.)*

0 NONE: (a) No harm under any conditions; (b) Harmful only under unusual conditions or overwhelming dosage.

1 SLIGHT: Causes readily reversible changes which disappear after end of exposure.

2 MODERATE: May involve both irreversible and reversible changes; not severe enough to cause death or permanent injury.

3 HIGH: May cause death or permanent injury after very short exposure to small quantities.

U UNKNOWN: No information on humans considered valid by authors

NICKEL SULFATE, ANHYDROUS. See nickel sulfate.

NICKEL SULFITE
General Information
Description: Tetragonal, green crystals.
Formula: $NiSO_3 \cdot 6H_2O$.
Constant: Mol wt: 246.85.
Hazard Analysis and Countermeasures
See nickel compounds and sulfites.

NICOTINAMIDE. See niacinamide.

NICOTINE
General Information
Synonym: β-Pyridyl-α-methyl pyrrolidine.
Description: In its pure state, a colorless and almost odorless oil; sharp burning taste; alkaloid tobacco.
Formula: $C_{10}H_{14}N_2$.
Constants: Mol wt: 162.23, mp: $< -80°C$, bp: 247.3°C, lel: 0.75% uel: 4.0%, d: 1.0092 at 20°C, autoign. temp.: 471°F, vap. press.: 1 mm at 61.8°C, vap. d.: 5.61.
Hazard Analysis
Toxic Hazard Rating:
Acute Local: Irritant 2.
Acute Systemic: Ingestion 3; Inhalation 3; Skin Absorption 3.
Chronic Local: U.
Chronic Systemic: Ingestion 2; Inhalation 2; Skin Absorption 2.
Toxicology: Causes nausea, vomiting, diarrhea, mental disturbances and convulsions. May be absorbed via the intact skin. Highly toxic.
TLV: ACGIH (recommended); 0.5 milligrams per cubic meter of air. Can be absorbed through the intact skin.
Fire Hazard: Moderate, when exposed to heat or flame.
Explosion Hazard: Moderate, when exposed to heat or flame.
Disaster Hazard: Dangerous; when heated, emits highly toxic fumes; can react with oxidizing materials.
Countermeasures
Ventilation Control: Section 2.
To Fight Fire: Carbon dioxide, dry chemical or carbon tetrachloride (Section 6).
Personnel Protection: Section 3.
First Aid: Section 1.
Storage and Handling: Section 7.
Shipping Regulations: Section 11.
I.C.C.: Poison B; poison label, 55 gallons.
Coast Guard Classification: Poison B; poison label.
IATA: Poison B, poison label, 1 liter (passenger), 220 liters (cargo).

NICOTINE COMPOUNDS, SOLID, AND PREPARATIONS THEREOF
Shipping Regulations: Section 11.
IATA: Poison B, poison label, 25 kilograms (passenger), 95 kilograms (cargo).

NICOTINE CUPROCYANIDE. See nicotine and cyanides and copper compounds.

NICOTINE HYDROCHLORIDE. See nicotine.
Shipping Regulations: Section 11.
I.C.C.: Poison B; poison label, 55 gallons.
Coast Guard Classification: Poison B; poison label.
IATA: Poison B, poison label, 1 liter (passenger), 220 liters (cargo).

NICOTINE ISOMER. See anabasine.

NICOTINE SALICYLATE. See nicotine.
Shipping Regulations: Section 11.
I.C.C.: Poison B; poison label, 200 pounds.
Coast Guard Classification: Poison B; poison label.
IATA: Poison B, poison label, 25 kilograms (passenger), 95 kilograms (cargo).

NICOTINE SHEEP DIPS. See nicotine.
Shipping Regulations: Section 11.
IATA (liquid): Poison B, poison label, 1 liter (passenger), 220 liters (cargo).
IATA (solid): Poison B, poison label, 25 kilograms (passenger), 95 kilograms (cargo).

NICOTINE SULFATE. See nicotine.
Shipping Regulations: Section 11.
I.C.C. (liquid): Poison B; poison label, 55 gallons.
(solid): Poison B, poison label, 200 pounds.
Coast Guard Classification: Poison B; poison label.
IATA (liquid): Poison B, poison label, 1 liter (passenger), 220 liters (cargo).
(solid): Poison B, poison label, 25 kilograms (passenger), 95 kilograms (cargo).

NICOTINE TARTRATE. See nicotine.
Shipping Regulations: Section 11.
I.C.C.: Poison B; poison label, 200 pounds.
Coast Guard Classification: Poison B; poison label.
IATA: Poison B, poison label, 25 kilograms (passenger), 95 kilograms (cargo).

NICOTINIC ACID. See niacin.

NICOTINIC ACID AMIDE. See niacinamide.

NIFOS. See tetraethyl pyrophosphate.

NIHYDRAZONE
Hazard Analysis
Toxic Hazard Rating: U. Used as a food additive permitted in the food and drinking water of animals and/or for the treatment of food producing animals; also permitted in food for human consumption (Section 10).

NINHYDRIN
General Information
Synonym: 1,2,3-Indantrione hydrate.
Description: Crystals. Turns reddish at 125°C; swells at 139°C; decomposes at 240°C.
Formula: $C_9H_4O_3 \cdot H_2O$.
Constant: Mol wt: 178.1.
Hazard Analysis
Toxicity: Details unknown. An irritant poison.

NIOBE OIL. See methyl benzoate.

NIOBIUM
General Information
Synonym: Columbium.
Description: Rhombic, steel-gray, lustrous, metallic crystals.
Formula: Nb.
Constants: MAt wt: 92.91, mp: 1950°C, bp: 2900°C, d: 8.55.
Hazard Analysis
Toxicity: Unknown. Limited animal experiments show high toxicity for some salts of niobium. Animal experiments have also shown toxic effects related to disturbed enzyme action.
Radiation Hazard: Section 5. For permissible levels, see Table 5, p. 150.
Artificial isotope ^{93m}Nb, half life 3.7 y. Decays to stable ^{93}Nb by emitting gamma rays of 0.03 MeV.
Artificial isotope ^{95}Nb, half life 35 d. Decays to stable ^{95}Mo by emitting beta particles of 0.16 MeV. Also emits gamma rays of 0.77 MeV.
Artificial isotope ^{97}Nb, half life 74 m. Decays to stable ^{97}Mo by emitting beta particles of 1.27 MeV. Also emits gamma rays of 0.66 MeV.
Fire Hazard: Moderate, in the form of dust, when exposed to flame or by chemical reaction; can react with oxidizing materials. See also powdered metals (Section 6).
Explosion Hazard: Slight, in the form of dust, when exposed to flame.
Countermeasures
Storage and Handling: Section 7.

NIOBIUM CARBIDE
General Information
Description: Cubic, black or lavender-gray powder.
Formula: NbC.
Constants: Mol wt: 104.91, mp: approx. 3900°C, d: 7.82.
Hazard Analysis and Countermeasures
See carbides and niobium.

NIOBIUM CHLORIDE
General Information
Synonym: Niobium pentachloride.
Description: Yellow-white deliquescent powder.
Formula: NbCl₅.
Constants: Mol wt: 270.20, mp: 194°C, bp: 241°C.
Hazard Analysis and Countermeasures
Toxicity: Animal experiments show kidney injury from small doses. See hydrochloric acid and niobium.

NIOBIUM HYDRIDE
General Information
Description: Gray powder.
Formula: NbH.
Constants: Mol wt: 93.92, mp: infusible, d: 6.6
Hazard Analysis and Countermeasures
See hydrides and niobium.

NIOBIUM HYDROGEN OXALATE
General Information
Description: Monocline, colorless crystals.
Formula: Nb(HC₂O₄)₅.
Constant: Mol wt: 538.05.
Hazard Analysis
Toxicity: Highly toxic. See oxalates.
Disaster Hazard: Dangerous. When heated, it emits toxic fumes (Section 6).
Countermeasures
Storage and Handling: Section 7.

NIOBIUM NITRIDE
General Information
Description: Cubic, black crystals.
Formula: NbN.
Constants: Mol wt: 106.92, mp: 2573°C, d: 8.4.
Hazard Analysis and Countermeasures
See nitrides and niobium.

NIOBIUM OXYBROMIDE
General Information
Description: Yellow crystals.
Formula: NbOBr₃.
Constants: Mol wt: 348.66, mp: sublimes.
Hazard Analysis and Countermeasures
See bromides and niobium.

NIOBIUM   PENTACHLORIDE. See niobium chloride.

NIOBIUM PENTAFLUORIDE
General Information
Description: Colorless, monoclinic prisms.
Formula: NbF₅.
Constants: Mol wt: 187.91, mp: 72°C, bp: 220°C (approx.), d: 3.92.
Hazard Analysis and Countermeasures
See fluorides and niobium.

NIOBIUM PENTOXIDE
General Information
Synonym: Niobium oxide.

Description: White, o-rhombic crystals; insoluble in water; soluble in hydrofluoric acid or hot sulfuric acid.
Formula: Nb₂O₅.
Constants: Mol wt: 187.91, d: 4.6, mp: 1520°C.
Hazard Analysis
See niobium.

NIRAN. See parathion.

NIRVANOL
General Information
Synonym: Phenylethylhydantion.
Description: Colorless, odorless, crystalline powder.
Formula: (NHCO)(CONH)C(C₂H₅)(C₆H₅).
Constants: Mol wt: 204.17, mp: 199–200°C.
Hazard Analysis
Toxic Hazard Rating:
    Acute Local: U.
    Acute Systemic: Ingestion 2; Inhalation 2.
    Chronic Local: U.
    Chronic Systemic: U.
Fire Hazard: Slight; (Section 6).
Countermeasures
Ventilation Control: Section 2.
Personal Hygiene: Section 3.
Storage and Handling: Section 7.

NITER. See potassium nitrate.

NITOL. See mechlorethamine hydrochloride.

NITON. See radon.

NITRAMINE. See tetryl.

NITRAMON. See explosives, high.

NITRANILIC ACID
General Information
Synonyms: 2,5-Dehydroxy-3,6-dinitroquinone.
Description: Flat yellow crystals; soluble in water and alcohol; insoluble in ether.
Formula: C₆O₂(NO₂)₂(OH)₂.
Constant: Mol wt: 230.09.
Hazard Analysis and Countermeasures
Forms hydrogen cyanide on decomposition. See cyanides.

m-NITRANILINE. See m-nitroaniline.

o-NITRANILINE. See o-nitroaniline.

NITRATE OF ALUMINUM. See aluminum nitrate.
Shipping Regulations: Section 11.
    Coast Guard Classification: Oxidizing material.

NITRATE OF SODA AND POTASH
Shipping Regulations: Section 11.
    ICC: Oxidixing material, yellow label, 100 pounds.
    IATA: Oxidizing material, yellow label, 12 kilograms (passenger), 45 kilograms (cargo).

NITRATES, N.O.S.
General Information
Description: Organic nitrates are usually termed nitro compounds. These compounds are a combination of the nitro (⁻NO₂) group and an organic radical. However, this term is often used to denote nitric acid esters of an organic material. Inorganic nitrates are compounds of metals which are combined with the mono-valent ⁻NO₃ radical.

TOXIC HAZARD RATING CODE (For detailed discussion, see Section 1.)

0 NONE: (a) No harm under any conditions; (b) Harmful only under unusual conditions or overwhelming dosage.

1 SLIGHT: Causes readily reversible changes which disappear after end of exposure.

2 MODERATE: May involve both irreversible and reversible changes; not severe enough to cause death or permanent injury.

3 HIGH: May cause death or permanent injury after very short exposure to small quantities.

U UNKNOWN: No information on humans considered valid by authors.

Hazard Analysis
Toxic Hazard Rating:
　Acute Local: U.
　Acute Systemic: Ingestion 2; Inhalation 2.
　Chronic Local: U.
　Chronic Systemic: Ingestion 2; Inhalation 2.
Toxicology: Large amounts taken by mouth may have serious or even fatal effects. The symptoms are dizziness, abdominal cramps, vomiting, bloody diarrhea, weakness, convulsions and collapse. Small, repeated doses may lead to weakness, general depression, headache and mental impairment.
Fire Hazard: Moderate, by spontaneous chemical reaction; practically all nitrates are powerful oxidizing agents (Section 6).
Explosion Hazard: Nitrates may explode when shocked, exposed to heat or flame or by spontaneous chemical reaction (See also explosives, high). All the inorganic nitrates act as oxygen carriers; under proper conditions these can give up their oxygen to other materials, which may in turn detonate. For example, potassium or barium nitrate are added to double-base powders for the purpose of reducing flash and rendering the powder more ignitable. A further use for these materials is to mix them with a smokeless powder which is not completely colloided, for the purpose of granulation. An example of such a powder is "E. C. Powder," used for loading blank cartridges and hand grenades. Sodium and potassium nitrate are also used in black powder as the oxygen carrier to support the combustion of the sulfur and the charcoal.
　Ammonium nitrate has all the properties of the other nitrates, but is also able to detonate by itself under certain conditions. It is therefore a high explosive, although very insensitive to impact and difficult to detonate. In the pure state, it requires a combination of an initiator and a high explosive. This combination is known as a reinforced detonator. Ammonium nitrate in combination with nitro compounds (such, perhaps, as trinitrotoluene) forms one of the major high explosives for military use. Ammonium nitrate is widely used also as the chief component of "ammonia permissibles," and of "ammonia dynamites"; as a component of many pyrotechnic mixtures; and in combination with smokeless powder, as a granular blasting explosive. It is a relatively safe high explosive which, however, must be stored in a cool, ventilated place, away from acute fire hazards and easily oxidized materials. Ammonium nitrate must not be confined, because if a fire should start, confinement can cause detonation with extremely violent results.
Disaster Hazard: Dangerous, due to fire and explosion hazard. On decomposition they emit toxic fumes. They are powerful oxidizing agents which may cause violent reaction with reducing materials. Nitrates should be protected carefully as discussed in detail in Section 7.
Countermeasures
Ventilation Control: Section 2.
Personal Hygiene: Section 3.
Storage and Handling: Section 7.
Shipping Regulations : Section 11.
　I.C.C.: Oxidizing material; yellow label, 100 pounds.
　Coast Guard Classification: Oxidizing material.
　IATA: Oxidizing material, yellow label, 12 kilograms (passenger), 45 kilograms (cargo).

NITRATINE. See sodium nitrate.

NITRATING (MIXED) ACID. See nitric acid and sulfuric acid.
Shipping Regulations: Section 11.
　I.C.C.: Corrosive liquid; white label, 2½ pints.
　Coast Guard Classification: Corrosive liquid; white label.

IATA: Corrosive liquid, white label, not acceptable (passenger), 1 liter (cargo).
MCA warning label.

NITRE. See potassium nitrate,

NITRIC ACID
General Information
Synonyms: Aqua fortis; hydrogen nitrate; azotic acid.
Description: Transparent colorless or yellowish, fuming, suffocating, caustic and corrosive liquid.
Formula: HNO_3.
Constants: Mol wt: 63.02, mp: $-42°C$, bp: $86°C$, d: 1.502.
Hazard Analysis
Toxic Hazard Rating:
　Acute Local: Irritant 3; Ingestion 3; Inhalation 3.
　Acute Systemic: Inhalation 3.
　Chronic Local: Irritant 2.
　Chronic Systemic: U.
TLV: ACGIH (recommended); 2 parts per million of air; 5 milligrams per cubic meter of air.
Toxicology: The exact composition of the "fumes" or vapor produced by nitric acid depends upon such factors as temperatues, humidity, and whether or not the acid comes in contact with other materials such as heavy metals or organic compounds. Depending upon these factors, the vapor will consist of a mixture of the various oxides of nitrogen and of nitric acid vapor. Nitric acid vapor is highly irritant to the mucous membranes of the eyes and respiratory tract and to the skin. It is corrosive to the teeth. Because of its irritant properties, chronic exposure to dangerous concentrations of the acid vapor seldom occur.
Fire Hazard: Moderate, by chemical reaction with reducing agents. It is a powerful oxidizing agent.
Explosion Hazard: Slight, by chemical reaction to evolve oxides of nitrogen; can explode on contact with powerful reducing agents.
Disaster Hazard: Dangerous; when heated to decomposition, it emits highly toxic fumes of oxides of nitrogen and hydrogen nitrate; will react with water or steam to produce heat and toxic and corrosive fumes.
Countermeasures
Ventilation Control: Section 2.
To Fight Fire: Water (Section 6).
Personnel Protection: (Section 3).
First Aid: Section 1.
Storage and Handling: Section 7.
Shipping Regulations: Section 11.
　ICC: Corrosive liquid; white label, 5 pints.
　Coast Guard Classification: Corrosive liquid; white label.
　IATA: Corrosive liquid, white label, not acceptable (passenger), 2½ liters (cargo).
　MCA warning label.

NITRIC ACID, ANHYDROUS. See nitric acid, fuming.

NITRIC ACID, FUMING
General Information
Synonym: Nitric acid, anhydrous.
Description: Colorless to yellow to red corrosive liquid.
Formula: $NHO_3 + N_2O_5$
Constant: D: > 1.480.
Hazard Analysis
Toxic Hazard Rating:
　Acute Local: Irritant 3; Ingestion 3; Inhalation 3.
　Acute Systemic: Inhalation 3.
　Chronic Local: Irritant 2.
　Chronic Systemic: Inhalation 3.
Fire Hazard: Dangerous; very powerful oxidizing agent (Section 6).
Explosion Hazard: Moderate; can react explosively with many reducing agents.
Disaster Hazard: Dangerous; when heated to decomposi-

tion, it emits highly toxic fumes of oxides of nitrogen; will react with water or steam to produce heat and toxic, corrosive and flammable vapors.

Countermeasures
Ventilation Control: Section 2.
Personnel Protection: Section 3.
First Aid: Section 1.
Storage and Handling: Section 7.
Shipping Regulations: Section 11.
MCA warning label.

NITRIC ANHYDRIDE. See nitrogen pentoxide.

NITRIC ETHER. See ethyl nitrate.

NITRIC OXIDE
General Information
Description: Colorless gas; blue liquid and solid.
Formula: NO.
Constants: Mol wt: 30.01, mp: $-161°C$, bp: $-151.8°C$, d: 1.3402 g/liter; liquid: 1.269 at $-150°C$.
Hazard Analysis
Toxic Hazard Rating:
   Acute Local: Irritant 3; Inhalation 3.
   Acute Systemic: Inhalation 3.
   Chronic Local: U.
   Chronic Systemic: Inhalation 2.
TLV: ACGIH (tentative); 25 parts per million; 30 milligrams per cubic meter of air.
Toxicology: Exposure to such fumes may occur whenever nitric acid acts upon organic material, such as wood, sawdust and refuse; it occurs when nitric acid is heated, and when organic nitro compounds are burned, as for example, celluloid, cellulose nitrate (guncotton) and dynamite. The action of nitric acid upon metals, as in metal etching and pickling, also liberates the fumes. In high-temperature welding, as with the oxyacetylene or electric torch, the nitrogen and oxygen of the air unite to form oxides of nitrogen. Exposure may also occur in many manufacturing processes when nitric acid is made or used.

The oxides of nitrogen are somewhat soluble in water, reacting with it in the presence of oxygen to form nitric and nitrous acids. This is the action that takes place deep in the respiratory system. The acids formed are irritants, causing congestion of the throat and bronchi, and edema of the lungs. The acids are neutralized by the alkalis present in the tissues, with the formation of nitrates and nitrites. The latter may cause some arterial dilatation, fall in blood pressure, headache and dizziness, and there may be some formation of methemoglobin. However, the nitrite effect is of secondary importance.

Because of their relatively low solubility in water, the nitrogen oxides are only slightly irritating to the mucous membranes of the upper respiratory tract. Their warning power is therefore low, and dangerous amounts of the fumes may be breathed before the workman notices any real discomfort. Higher concentrations (60 to 150 ppm) cause immediate irritation of the nose and throat, with coughing and burning in the throat and chest. These symptoms often clear up on breathing fresh air, and the workman may feel well for several hours. Some 6 to 24 hours after exposure, he develops a sensation of tightness and burning in the chest, shortness of breath, sleeplessness and restlessness. Dyspnea and air hunger may increase rapidly, with de-

velopment of cyanosis and loss of consciousness, followed by death. In cases which recover from the pulmonary edema, there is usually no permanent disability, but pneumonia may develop later. Concentrations of 100 to 150 ppm are dangerous for short exposures of 30 to 60 minutes. Concentrations of 200 to 700 ppm may be fatal after even very short exposures.

Continued exposure to low coentrations of the fumes, insufficient to cause pulmonary edema, is said to result in chronic irritation of the respiratory tract, with cough, headache, loss of appetite, dyspepsia, corrosion of the teeth and gradual loss of strength.

Exposure to nitrous fumes is always potentially serious, and persons so exposed should be kept under close observation for at least 48 hours.
Disaster Hazard: Dangerous; when heated to decomposition, it emits highly toxic fumes of oxides of nitrogen; will react with water or steam to produce heat and corrosive fumes; can react vigorously with reducing materials.
Countermeasures
Ventilation Control: Section 2.
Personnel Protection: Section 3.
Storage and Handling: Section 7.
Shipping Regulations: Section 11.
   ICC: Poison A, poison gas label, not acceptable.
   IATA: Poison A, not acceptable (passenger and cargo).

NITRIDES
General Information
Description: Compounds of $N \equiv$ as the anion, such as Li_3N, Ca_3N_2, etc.
Hazard Analysis
Toxicity: The details of toxicity of nitrides as a group are unknown. However, many nitrides react with moisture to evolve ammonia. This gas is an irritant to the nose, throat and eyes. See ammonia.
Fire Hazard: To the extent that many nitrides evolve flammable ammonia gas upon contact with moisture, nitrides can be fire hazards. See ammonia.
Explosion Hazard: Moderate. See ammonia.
Disaster Hazard: Moderately dangerous due to fire and explosion hazard. On decomposition they emit toxic fumes of ammonia.

NITRILES
General Information
Description: Organic compounds containing the $(-CN)$ grouping; e.g.: acrylonitrile $(CH_2:CHCN)$.
Hazard Analysis
Nitriles are organic cyanides. Acrylonitrile, propionitrile and some others resemble cyanides in toxicity. Other nitriles such as cyanamides and cyanates do not have a cyanide effect. When heated to decomposition they emit highly toxic cyanide fumes. See specific compounds, also cyanides. The nitriles may be used as insecticides.

NITRILOMALONAMIDE. See cyanoacetamide.

NITRILOTRISILANE. See trisilicylamine.

NITRITES
Hazard Analysis
Toxic Hazard Rating:
   Acute Local: U.
   Acute Systemic: Ingestion 3; Inhalation 3.
   Chronic Local: U.

Chronic Systemic: Ingestion 1; Inhalation 1.

Toxicology: Large amounts taken by mouth may produce nausea, vomiting, cyanosis (due to methemoglobin formation) collapse and coma. Repeated small doses cause a fall in blood pressure, rapid pulse, headache and visual disturbances.

Fire Hazard: Details unknown. They are generally powerful oxidizers. In contact with readily oxidized materials, a violent reaction such as a fire or explosion may ensue.

Explosion Hazard: Details unknown. Organic nitrites may decompose violently.

Disaster Hazard: Dangerous; shock may explode them; when heated to decomposition, they emit highly toxic fumes of oxides of nitrogen; can react vigorously with reducing materials.

Countermeasures

Ventilation Control: Section 2.
Personnel Protection: Section 3.
First Aid: Section 1.
Storage and Handling: Section 7.

p-NITROACETANILIDE
General Information
Description: White crystals.
Formula: $CH_3CONHC_6H_4NO_2$.
Constants: Mol wt: 180.16, mp: 215°C.
Hazard Analysis
Toxicity: Details unknown. See also acetanilide and nitro compounds of aromatic hydrocarbons.
Disaster Hazard: Dangerous; when heated to decomposition, it emits highly toxic fumes of oxides of nitrogen; can react with oxidizing materials.
Countermeasures
Storage and Handling: Section 7.

5-NITRO-2-AMINOANISOLE. See 1-amino-2-methoxy-4-nitrobenzene.

m-NITROANILINE
General Information
Synonym: m-Nitraniline.
Description: Yellow crystals.
Formula: $C_6H_6N_2O_2$.
Constants: Mol wt: 138.1, mp: 114°C, bp: 305.7°C, d: 1.43, vap. press.: 1 mm at 119.3°C.
Hazard Analysis
Toxicity: Highly toxic. See p-nitroaniline.
Fire Hazard: Moderate, when exposed to heat or flame or by chemical reaction with oxidizers. Causes nitration of materials which can then ignite spontaneously.
Explosion Hazard: See nitrates.
Disaster Hazard: See nitrates.
Countermeasures
To Fight Fire: Water, carbon dioxide, dry chemical or carbon tetrachloride (Section 6).
Storage and Handling: Section 7.

o-NITROANILINE
General Information
Synonym: o-Nitraniline.
Description: Orange-yellow crystals.
Formula: $C_6H_6N_2O_2$.
Constants: Mol wt: 138.1, mp: 71°-72°C, bp: 284.5°C, d: 1.442, vap. press.: 1 mm at 104.0°C.
Hazard Analysis
Toxicity: Highly toxic. See p-nitroaniline.
Fire Hazard: See nitrates.
Explosion Hazard: See nitrates.
Disaster Hazard: See nitrates.
Countermeasures
Storage and Handling: Section 7.
To Fight Fire: Water, carbon dioxide, dry chemical or carbon tetrachloride (Section 6).

p-NITROANILINE
General Information
Synonym: 1-Amino-4-nitrobenzene.
Description: Yellow crystals.
Formula: $C_6H_6N_2O_2$.
Constants: Mol wt: 138.1, mp: 146.5°C, bp: 336.0°C, flash p: 390°F (C.C.), d: 1.437, vap. press.: 1 mm at 142.4.
Hazard Analysis
Toxic Hazard Rating:
Acute Local: U.
Acute Systemic: Ingestion 3; Inhalation 3; Skin Absorption 3.
Chronic Local: U.
Chronic Systemic: Ingestion 3; Inhalation 3; Skin Absorption 3.
TLV: ACGIH (recommended) 1 part per million in air, 6 milligrams per cubic meter of air. May be absorbed via intact skin.
Toxicology: Acute symptoms are headache, nausea, vomiting, weakness and stupor. See also aniline.
Fire Hazard: Slight, when exposed to heat or flame (Section 6).
Explosion Hazard: See nitrates.
Disaster Control: See nitrates.
Countermeasures
Ventilation Control: Section 2.
Personnel Protection: Section 3.
First Aid: Section 1.
Storage and Handling: Section 7.
Shipping Regulations: Section 11.
I.C.C.: Poison B; poison label.
Coast Guard Classification: Poison B; poison label.
IATA: Poison B, poison label, 25 kilograms (passenger), 95 kilograms (cargo).

4-NITRO-o-ANISIDINE. See 1-amino-2-methoxy-4-nitrobenzene.

NITROBARITE. See barium nitrate.

p-NITROBENZALDEHYDE
General Information
Description: White to yellow crystals; slightly soluble in water or ether; soluble in alcohol, benzene; glacial acetic acid.
Formula: $NO_2C_6H_4CHO$.
Constants: Mol wt: 151.12, mp: 106–107°C.
Hazard Analysis
Toxicity: Details unknown. Animal experiments show injury to eyes, skin, and liver.
Disaster Hazard: Dangerous. See nitrates, organic.

p-NITROBENZALDOXIME
Hazard Analysis
Toxicity: Details unknown. Animal experiments show injury to eyes, skin, and liver.
Disaster Hazard: Dangerous. See nitrates, organic.

NITROBENZENE
General Information
Synonyms: Oil of mirbane; nitrobenzol.
Description: Bright yellow crystals or yellow, oily liquid.
Formula: $C_6H_5NO_2$.
Constants: Mol wt: 123.11, mp: 5.7°C, bp: 210.9°C, ulc: 20–30, lel: 1.8% at 200°F, flash p: 190°F (C.C.), d: 1.9867 at 25°/4°C, autoign. temp.: 900°F, vap. press.: 1 mm at 44.4°C, vap. d.: 4.25.
Hazard Analysis
Toxic Hazard Rating:
Acute Local: U.
Acute Systemic: Ingestion 3; Inhalation 3; Skin Absorption 3.
Chronic Local: U.
Chronic Systemic: Ingestion 3; Inhalation 3; Skin Absorption 3.

TLV: ACGIH (recommended); 1 part per million in air; 5 milligrams per cubic meter of air. May be absorbed via intact skin.

Toxicology: Causes cyanosis due to formation of methemoglobin. A common air contaminant (Section 4).

Fire Hazard: Moderate, when exposed to heat or flame.

Explosion Hazard: Moderate, when exposed to heat or flame.

Disaster Hazard: See nitrates.

Countermeasures

Ventilation Control: Section 2.

Personnel Protection: Section 3.

First Aid: Section 1.

To Fight Fire: Water, foam, carbon dioxide, dry chemical or carbon tetrachloride (Section 6).

Storage and Handling: Section 7.

Shipping Regulations: Section 11.

I.C.C.: Poison B; poison label, 55 gallons.

Coast Guard Classification: Poison B; poison label.

IATA: Poison B, poison label, 1 liter (passenger), 220 liters (cargo).

MCA warning label.

m-NITROBENZENE SULFONYL FLUORIDE
General Information

Description: Yellow solid.

Formula: $O_2NC_6H_4SO_2F$.

Constants: Mol wt: 205, mp: 45–47°C, bp: 143°C at 7 mm, flash p: 335°F, d: 1.582, vap. press.: 5 mm at 130°C, vap. d.: 7.07.

Hazard Analysis

Toxicity: Apparently low acute inhalation toxicity to test animals. Possibly slightly irritating to skin.

Fire Hazard: Slight, when exposed to heat or flame.

Disaster Hazard: Dangerous; when heated to decomposition, it emits highly toxic fumes of fluorides, oxides of sulfur; can react with oxidizing materials.

Countermeasures

Storage and Handling: Section 7.

To Fight Fire: Water, foam, carbon dioxide, dry chemical or carbon tetrachloride (Section 6).

p-NITROBENZOIC ACID
General Information

Description: Crystals.

Formula: $NO_2C_6H_4COOH$.

Constants: Mol wt: 167.1, mp: 242.4°C, bp: sublimes, d: 1.550 at 32°/4°C.

Hazard Analysis

Toxicity: Unknown. Animal experiments suggest moderate toxicity. See also nitro compounds of aromatic hydrocarbons.

Disaster Hazard: See nitrates.

Countermeasures

Storage and Handling: Section 7.

p-NITROBENZOIC ACID ETHYL ESTER
General Information

Synonym: Ethyl p-nitrobenzoate.

Hazard Analysis

Toxicity: Details unknown; larvicide for European corn borer. See also nitro compounds of aromatic hydrocarbons.

Fire Hazard: See nitrates.

Disaster Hazard: See nitrates.

Countermeasures

Storage and Handling: Section 7.

NITROBENZOL. See nitrobenzene.

m-NITROBENZOTRIFLUORIDE
General Information

Synonym: m-Nitrotrifluoromethyl benzene.

Description: Thin, pale straw, oily liquid with aromatic odor.

Formula: $C_7H_4F_3NO_2$.

Constants: Mol wt: 191.1, mp: −5°C, bp: 202.8°C, flash p: 217°F (O.C.), d: 1.437 at 15.5°/15.5°C, vap. press.: 0.3 mm at 25°C.

Hazard Analysis

Toxicity: Unknown. See also nitro compound of aromatic hydrocarbons.

Disaster Hazard: Dangerous; when heated to decomposition, it emits highly toxic fumes of oxides of nitrogen and fluorides.

Countermeasures

Storage and Handling: Section 7.

m-NITROBENZOYL CHLORIDE
General Information

Description: Yellow to brown liquid, partially crystallized at room temperature.

Formula: $NO_2C_6H_4COCl$.

Constants: Mol wt: 185.5, mp: 28–31°C, bp: 278°C, vap. d.: 6.43.

Hazard Analysis

Toxicity: Details unknown. See also nitro compounds of aromatic hydrocarbons.

Disaster Hazard: Dangerous; when heated to decomposition, it emits highly toxic fumes of chlorides and oxides of nitrogen; will react with water or steam to produce toxic and corrosive fumes.

Countermeasures

Storage and Handling: Section 7.

p-NITROBENZOYL CHLORIDE
General Information

Description: Yellow, crystalline solid.

Formula: $NO_2C_6H_4COCl$.

Constants: Mol wt: 185.5, mp: 70°C, bp: 154°C at 15 mm.

Hazard Analysis

Toxicity: Details unknown. See also nitro compounds of aromatic hydrocarbons.

Disaster Hazard: Dangerous; when heated to decomposition, it emits highly toxic fumes of chlorides and oxides of nitrogen; will react with water or steam to produce toxic and corrosive fumes.

Countermeasures

Storage and Handling: Section 7.

o-NITROBENZYL CHLORIDE
General Information

Description: Crystals.

Formula: $NO_2C_6H_4CH_2Cl$.

Constants: Mol wt: 171.59, mp: 48–49°C.

Hazard Analysis

Toxic Hazard Rating:

Acute Local: Irritant 3; Ingestion 3; Inhalation 3.

Acute Systemic: U.

Chronic Local: U.

Chronic Systemic: U.

Toxicity: See also nitro compounds of aromatic hydrocarbons.

Disaster Hazard: Dangerous. See chlorides and nitrates.

Countermeasures

Ventilation Control: Section 2.

Personnel Protection: Section 3.

Storage and Handling: Section 7.

TOXIC HAZARD RATING CODE (For detailed discussion, see Section 1.)

0 NONE: (a) No harm under any conditions; (b) Harmful only under unusual conditions or overwhelming dosage.

1 SLIGHT: Causes readily reversible changes which disappear after end of exposure.

2 MODERATE: May involve both irreversible and reversible changes; not severe enough to cause death or permanent injury.

3 HIGH: May cause death or permanent injury after very short exposure to small quantities.

U UNKNOWN: No information on humans considered valid by authors.

o-NITROBIPHENYL

General Information

Synonyms: ONB; o-nitrodiphenyl.

Description: Light yellow to reddish-colored liquid or crystalline solid.

Formula: $C_6H_5C_6H_4NO_2$.

Constants: Mol wt: 199.2, mp: 35°C, bp: 330°C, flash p: 290°F, d: 1.189 at 40°/15.5°C, autoign. temp.: 356°F, vap. press.: 2 mm at 140°C, vap. d.: 6.9.

Hazard Analysis

Toxic Hazard Rating:

    Acute Local: U.

    Acute Systemic: Ingestion 2; Inhalation 2.

    Chronic Local: U.

    Chronic Systemic: U.

Fire Hazard: Slight, when exposed to heat or flame.

Disaster Hazard: See nitrates.

Countermeasures

Ventilation Control: Section 2.

Personnel Protection: Section 3.

To Fight Fire: Water, carbon dioxide, dry chemical or carbon tetrachloride (Section 6).

Storage and Handling: Section 7.

NITROBROMOFORM. See bromopicrin.

1-NITROBUTANE

General Information

Description: Liquid.

Formula: $C_4H_9NO_2$.

Constants: Mol wt: 103.12, bp: 151°C.

Hazard Analysis

Toxic Hazard Rating:

    Acute Local: U.

    Acute Systemic: Ingestion 2; Inhalation 2.

    Chronic Local: U.

    Chronic Systemic: U.

Fire Hazard: Moderate, when exposed to heat or flame.

Explosion Hazard: See nitrates.

Disaster Hazard: See nitrates.

Countermeasures

Ventilation Control: Section 2.

Personal Hygiene: Section 3.

To Fight Fire: Foam, carbon dioxide, dry chemical or carbon tetrachloride.

Storage ahd Handling: Section 7.

2-NITROBUTANE

General Information

Description: Liquid.

Formula: $C_4H_9NO_2$.

Constants: Mol wt: 103.1, bp: 139°C.

Hazard Analysis

Toxicity: See 1-nitrobutane.

Fire Hazard: Moderate, when exposed to heat or flame.

Explosion Hazard: See nitrates.

Disaster Hazard: See nitrates.

Countermeasures

Ventilation Control: Section 2.

To Fight Fire: Foam, carbon dioxide, dry chemical or carbon tetrachloride.

Storage and Handling: Section 7.

2-NITRO-1-BUTANOL

General Information

Description: Colorless liquid.

Formula: $CH_3CH_2CH(NO_2)CH_2OH$.

Constants: Mol wt: 119.12, mp: −47°C, bp: 105°C, d: 1.133 at 20°/20°C.

Hazard Analysis

Toxicity: Details unknown. See also alcohols.

Fire Hazard: Moderate, when exposed to heat or flame (Section 6).

Disaster Hazard: See nitrates.

Countermeasures

Storage and Handling: Section 7.

2-NITRO-2-BUTENE

General Information

Formula: $CH_3(NO_2)C:CHCH_3$.

Constant: Mol wt: 101.

Hazard Analysis

Toxic Hazard Rating:

    Acute Local: Inhalation 3.

    Acute Systemic: Inhalation 3.

    Chronic Local: Inhalation 3.

    Chronic Systemic: Inhalation 2.

Toxicity: Animal experiments show marked pulmonary irritation and some central nervous irritation.

Disaster Hazard: Dangerous. See nitrates, organic.

NITRO CARBONITRATE

General Information

Description: A solid explosive.

Composition: A mixture of ammonium nitrate, dinitro cotton, etc.

Hazard Analysis

Toxicity: Unknown. See nitrates.

Fire Hazard: Moderate; an oxidizing agent. See nitrates.

Explosion Hazard: See explosives, high; nitrates.

Disaster Hazard: See explosives, high; nitrates.

Countermeasures

Storage and Handling: Section 7.

Shipping Regulations: Section 11.

    I.C.C.: Oxidizing material; yellow label, 100 pounds.

    Coast Guard Classification: Oxidizing material.

    IATA: Oxidizing material, yellow label, 12 kilgorams (passenger), 45 kilograms (cargo).

NITROCELLULOSE

General Information

Synonym: Cellulose nitrate; cellulose hexanitrate; cellulose pentanitrate; cellulose tetranitrate; cellulose trinitrate; guncotton; nitrocotton.

Description: Dry: White amorphous powder or cotton-like solid; Wet: Water-white liquid.

Formula: $C_{12}H_{14}(ONO_2)_6O_4$ to $C_{12}H_{17}(ONO_2)_3O_7$.

Constants: Mol wt: 594.28 to 459.28, mp: ignites at 160–170°C, d: 1.66 flash p: 55°F.

Hazard Analysis

Fire Hazard: Highly dangerous in the dry state, when exposed to heat, flame or powerful oxidizers. When wet with 35% of denatured ethyl alcohol it is about as hazardous as ethyl alcohol alone or gasoline.

Dry nitrocellulose burns rapidly, with intense heat and ignites very easily (Section 6).

Explosion Hazard: Moderately dangerous. See explosives.

Disaster Hazard: The dry material is highly dangerous; the wet, somewhat less so. Keep away from heat and open flame!

Countermeasures

Ventilation Control: Section 2.

To Fight Fire: Use copious volumes of water; carbon dioxide is effective in extinguishing fires of nitrocellulose solvents (Section 6).

Personnel Protection: Section 3.

Storage and Handling: Section 7.

Shipping Regulations: Section 11.

    I.C.C.: Flammable liquid; red label (for wet product).

NITROCELLULOSE, COLLOIDED, GRANULAR OR FLAKE, DRY OR WET WITH LESS THAN 20% SOLVENT OR WATER. See nitrocellulose.

Shipping Regulations: Section 11.

    IATA: Explosive, not acceptable (passenger and cargo).

NITROCELLULOSE, DRY OR WET WITH LESS THAN 30% SOLVENT OR 20% WATER. See nitrocellulose.

Shipping Regulations: Section 11.

    IATA: Explosive, not acceptable (passenger and cargo).

NITROCELLULOSE FLAKES, DRY OR WET WITH LESS THAN 20% ALCOHOL OR SOLVENT. See nitrocellulose.
Shipping Regulations: Section 11.
 IATA: Explosive, not acceptable (passenger and cargo).

o-NITROCHLOROBENZENE, LIQUID. See o-chloronitrobenzene.
Shipping Regulations: Section 11.
 I.C.C.: Poison B; poison label, 55 gallons.
 Coast Guard Classification: Poison B; poison label.
 IATA: Poison B, poison label, 1 liter (passenger), 220 liters (cargo).

p-NITROCHLOROBENZENE
General Information
Synonym: p-Nitrochlorbenzol.
Description: Liquid.
Formula: $ClC_6H_4NO_2$.
Constants: Mol wt: 157.6, mp: 83°C, bp: 242°C, flash p: 261°F (C.C.), d: 1.368 at 22°/4°C, vap. d.: 5.43.
Hazard Analysis
Toxic Hazard Rating:
 Acute Local: Irritant 1.
 Acute Systemic: Ingestion 3; Inhalation 3; Skin Absorption 3.
 Chronic Local: U.
 Chronic Systemic: Ingestion 3; Inhalation 3; Skin Absorption 3.
TLV: ACGIH (recommended); 1 milligram per cubic meter of air. May be absorbed via the skin.
Toxicology: See also nitrobenzene and nitrocompounds of aromatic hydrocarbons.
Fire Hazard: Slight, when exposed to heat or flame.
Spontaneous Heating: No.
Disaster Hazard: Dangerous; when heated to decomposition, it emits highly toxic fumes of oxides of nitrogen and chlorides.
Countermeasures
Ventilation Control: Section 2.
To Fight Fire: Water, carbon dioxide, dry chemical or carbon tetrachloride (Section 6).
Personnel Protection: Section 3.
First Aid: Section 1.
Storage and Handling: Section 7.
Shipping Regulations: Section 11.
 Meta or Para—Solid (I.C.C. and Coast Guard).
 I.C.C.: Poison B; poison label, 200 pounds.
 Coast Guard Classification: Poison B; poison label.
 IATA: Poison B, poison label, 25 kilograms (passenger), 95 kilograms (cargo).

p-NITROCHLOROBENZOL. See p-nitrochlorobenzene.

3-NITRO-4-CHLOROBENZOTRIFLUORIDE
General Information
Description: Thin, yellow, oily liquid.
Formula: $C_6H_3CF_3(NO_2)(Cl)$.
Constants: Mol wt: 225.6, mp: −7.5°C, bp: 222°C, d: 1.542 at 15.5°/15.5°C.
Hazard Analysis
Toxicity: Details unknown. See also nitro compounds of aromatic hydrocarbons.
Disaster Hazard: Dangerous. See chlorides, fluorides and nitrates.
Countermeasures
Storage and Handling: Section 7.

NITROCHLOROFORM. See chloropicrin.

NITROCHLOROMETHANE. See chloropicrin.

p-NITROCHLOROPHENYL DIMETHYL THIONO-PHOSPHATE. See dicapthon.

NITRO COMPOUNDS OF AROMATIC HYDRO-CARBONS
Hazard Analysis
Toxic Hazard Rating:
 Acute Local: Irritant 1.
 Acute Systemic: Ingestion 3; Inhalation 3; Skin Absorption 3.
 Chronic Local: U.
 Chronic Systemic: Ingestion 3; Inhalation 3; Skin Absorption 3,
Toxicology: The di- and trinitrobenzenes, like the mononitrobezene, are absorbed chiefly through the skin and through inhalation of the dust or vapor when these materials are heated. The dinitrobenzenes are believed to be somewhat more toxic than the mononitrobenzene, and more toxic than aniline. The effect of di- and trinitrobenzene on the body is similar to that of aniline and mononitrobenzene, with reduction of the oxygen-carrying power of the blood and depression of the nervous system being responsible for most of the symptoms following acute exposure. Poisoning with the solid nitro compounds is usually slower and less severe, than is the case with the liquid nitro and amino benzenes since absorption is less rapid. Thus, chronic posioning occurs more frequently than acute, the picture in the chronic form being one of anemia, moderate cyanosis, fatigue, slight dizziness, headache, insomnia and loss of weight. The urine is frequently dark in color; the skin on the exposed parts is often yellowish-brown, and the hair yellowish-red. There may be irritation of the nose and throat, nausea and vomiting, sclerotic icterus, and complaints related to the nervous system. Prolonged chronic exposure may result in damage to the liver and kidneys, with production of acute yellow atrophy, toxic hepatitis, and fatty degeneration of the kidneys.
 The introduction of one or more chlorine atoms into the nitrobenzene ring results in the formation of chloro-nitrobenzene compounds of nitrochlors. The chloro-mono-nitrobenzenes have essentially the same toxic effect as nitrobenzene. The chlorine derivatives of dinitrobenzene, on the other hand, while resembling dinitrobenzene in their system effect, are much more irritating to the skin. They act as direct irritants, and in addition may cause sensitivity. For further information see specific nitro compounds.
Disaster Hazard: Dangerous; many of these compounds are highly flammable and some are explsoive. When heated to decomposition they evolve highly toxic oxides of nitrogen.

NITROCOTTON. See cellulose nitrate.

NITROCRESOL METHYL ETHER
General Information
Description: Pale yellow crystals.
Formula: $C_6H_3CH_3NO_2OCH_3$.
Constants: Mol wt: 167.2, mp: 8.5°C, bp: 274°C.
Hazard Analysis
Toxicity: Details unknown; probably toxic. See also ethers.
Fire Hazard: Details unknown. See ethers and nitrates.
Disaster Hazard: See nitrates.

TOXIC HAZARD RATING CODE (*For detailed discussion, see Section 1.*)

0 NONE: (a) No harm under any conditions; (b) Harmful only under unusual conditions or overwhelming dosage.

1 SLIGHT: Causes readily reversible changes which disappear after end of exposure.

2 MODERATE: May involve both irreversible and reversible changes not severe enough to cause death or permanent injury.

3 HIGH: May cause death or permanent injury after very short exposure to small quantities.

U UNKNOWN: No information on humans considered valid by authors

Countermeasures
Storage and Handling: Section 7.

NITROCYCLOHEXANE
General Information
Description: Straw-colored to water-white mobile liquid; mild odor.
Formula: $O_2NCH(CH_2)_5$.
Constants: Mol wt: 129.2, bp: 205.5°C, decomposes, fp: −35.7°C, flash p: 190.4°F (T.O.C.), d: 1.06656 at 25°C, vap. press.: 0.5 mm at 25°C, vap. d.: 4.46.
Hazard Analysis
Toxicity: Details unknown. See also cyclohexane.
Fire Hazard: Moderate, when exposed to heat or flame.
Explosion Hazard: See nitrates.
Disaster Hazard: See nitrates.
Countermeasures
Storage and Handling: Section 7.
To Fight Fire: Foam, carbon dioxide, dry chemical or carbon tetrachloride (Section 6).

NITRODAN
General Information
Synonym: 3-Methyl-5[(p-nitrophenyl)azo]-rhodanine.
Hazard Analysis
Toxic Hazard Rating: U. Used as a food additive permitted in the feed and drinking water of animals and/or for the treatment of food producing animals (Section 10).
Disaster Hazard: Dangerous. See nitrates, organic.

8-NITRO-1-DIAZO-2-NAPHTHOL-4-SULFONIC ACID
General Information
Description: Yellow solid.
Formula: $C_{10}H_5N_3SO_6$.
Constant: Mol wt: 295.22.
Hazard Analysis
Toxicity: Unknown.
Fire Hazard: See nitrates.
Disaster Hazard: See nitrates.
Countermeasures
Storage and Handling: Section 7.

o-NITRODIPHENYL. See o-nitrobiphenyl.

2-NITRODIPHENYLAMINE
General Information
Description: Crystals.
Formula: $O_2NC_6H_4NHC_6H_5$.
Constants: Mol wt: 214.22, mp: 74.7°C, bp: 223°C at 20 mm, d: 1.366.
Hazard Analysis and Countermeasures
See nitro compounds of aromatic hydrocarbons and nitrates.

4-NITRODIPHENYLAMINE
General Information
Description: White solid.
Formula: $C_6H_5NHC_6H_4NO_2$.
Constant: Mol wt: 214.2.
Hazard Analysis and Countermeasures
See nitro compounds of aromatic hydrocarbons and nitrates.

NITROETHANE
General Information
Description: Colorless liquid.
Formula: $C_2H_5NO_2$.
Constants: Mol wt: 75.07, bp: 114.0°C, fp: −90°C, flash p: 82°F, d: 1.052 at 20°/20°C, autoign. temp.: 778°F, vap. press.: 15.6 mm at 20°C, vap. d.: 2.58, lel: 3.4%.
Hazard Analysis
Toxic Hazard Rating:
Acute Local: Irritant 2; Ingestion 2; Inhalation 2.
Acute Systemic: Ingestion 2; Inhalation 2.
Chronic Local: U.
Chronic Systemic: U.

Toxicology: An irritant. Has caused injury to liver and kidneys in experimental animals.
TLV: ACGIH (recommended), 100 parts per million in air; 307 milligrams per cubic meter of air.
Fire Hazard: Moderate, when exposed to heat or flame.
Explosion Hazard: See nitrates.
Disaster Hazard: See nitrates.
Countermeasures
Ventilation Control: Section 2.
Storage and Handling: Section 7.
To Fight Fire: Alcohol foam, carbon dioxide, dry chemical or carbon tetrachloride (Section 6).

2-NITRO-2-ETHYL-1,3-PROPANEDIOL
General Information
Description: Crystalline material.
Formula: $CH_2OHC(C_2H_5)NO_2CH_2OH$.
Constants: Mol wt: 149.15, mp: 56–57°C, bp: decomposes.
Hazard Analysis and Countermeasures
See nitrates.

NITROFORM. See trinitromethane.

5-NITRO-2-FURALDEHYDE SEMICARBAZONE. See nitrofurazone.

NITROFURAZONE
General Information
Synonym: 5-Nitro-2-furaldehyde semicarbazone.
Description: Odorless, lemon yellow, crystalline powder; fairly soluble in alcohol and propylene glycol; slightly soluble in water and polyethylene glycol mixtures; practically insoluble in chloroform or ether.
Formula: $C_6H_6N_4O_4$.
Constant: Mol wt: 198.
Hazard Analysis
Toxic Hazard Rating: U. Used as a food additive permitted in the food and drinking water of animals and/or for the treatment of food producing animals (Section 10).
Disaster Hazard: Dangerous. See nitrates, organic.

NITROGEN
General Information
Description: Colorless gas, colorless liquid or cubic crystals at low temperatures.
Formula: N_2.
Constants: Mol wt: 28.02, mp: −210.0°C, bp: −195.8°C, d: 1.2506 g/liter at 0°C; d liq: 0.808 at −195.8°C.
Hazard Analysis
Toxicity: None. In high concentrations it is a simple asphyxiant. The release of nitrogen from solution in the blood, with formation of small bubbles, is the cause of most of the symptoms and changes found in compressed air illness (caisson disease). See also argon. It is used as a general purpose food additive (Section 10).
Countermeasures
Storage and Handling: Section 7.
Shipping Regulations: Section 11.
I.C.C.: Nonflammable gas; green label, 300 pounds.
Coast Guard Classification: Noninflammable gas; green gas label.
IATA: Nonflammable gas, green label, 70 kilograms (passenger), 140 kilograms (cargo).

NITROGEN BROMIDE
General Information
Synonym: Nitrogen tribromide.
Description: Crystalline solid.
Formula: NBr_3.
Constant: Mol wt: 253.8.
Hazard Analysis
Toxicity: Details unknown; probably highly toxic. See bromides.
Explosion Hazard: Severe, when shocked, exposed to heat or flame or by spontaneous chemical reaction.

Disaster Hazard: Dangerous; shock will explode it; on decomposition, it emits highly toxic fumes of bromine; can react vigorously with reducing materials.

Countermeasures

Storage and Handling: Section 7.

NITROGEN CHLORIDE

General Information

Synonym: Nitrogen trichloride.

Description: Volatile, yellowish oil or rhombic crystals; pungent odor.

Formula: NCl_3.

Constants: Mol wt: 120.38, mp: $< -40°C$, explodes at $95°C$, bp: $< 71°C$, d: 1.653.

Hazard Analysis

Toxic Hazard Rating:

Acute Local: Irritant 2; Ingestion 2; Inhalation 2.

Acute Systemic: Ingestion 2; Inhalation 2.

Chronic Local: U.

Chronic Systemic: Ingestion 2; Inhalation 2.

Explosion Hazard: Severe, when shocked, or exposed to heat or flame or by spontaneous chemical reaction. Certain common materials catalyze its decomposition. It is particularly sensitive when it contains impurities.

Disaster Hazard: Dangerous; even slight shock will explode it; on decomposition, it emits highly toxic fumes of chlorine; can react vigorously with reducing materials.

Countermeasures

Ventilation Control: Section 2.

Personnel Protection: Section 3.

Storage and Handling: Section 7.

NITROGEN DIOXIDE

General Information

Synonym: Nitrogen tetroxide.

Description: Colorless solid to yellow liquid.

Formula: NO_2.

Constants: Mol wt: 46, mp: $-9.3°C$ (yellow liquid), bp: $21°C$ (red brown gas with decomp.), d: 1.491 at $0°C$, vap. press.: 400 mm at $80°C$.

Hazard Analysis

Toxicity: Highly toxic. See nitric oxide.

TLV: ACGIH (recommended), 5 parts per million in air; 9 milligrams per cubic meter of air.

Disaster Hazard: See nitrogen oxides.

Countermeasures

Ventilation Control: Section 2.

Storage and Handling: Section 7.

Shipping Regulations: Section 11.

I.C.C.: Poison A (liquid); poison gas label, not accepted.

Coast Guard Classification: Poison A (liquid); poison gas label.

IATA: Poison A, not acceptable (passenger and cargo).

NITROGEN FERTILIZER SOLUTION

Shipping Regulations: Section 11.

ICC: Nonflammable gas, green label, 300 pounds.

IATA: Nonflammable gas, green label, 70 kilograms (passenger), 140 killogms (cargo).

NITROGEN FLUORIDE. See nitrogen trifluoride.

NITROGEN IODIDE. See nitrogen triiodide.

NITROGEN, LIQUID

Shipping Regulations: Section 11.

ICC (pressurized): Poison A, poison gas, not accepted.

IATA (nonpressurized): Other restricted articles, class C, no label required, 50 liters (passenger and cargo).

(low pressure): Other restricted articles, class C, no label required, not acceptable (passenger), 140 kilograms (cargo).

(pressurized)(Nonflammable gas, green label, not acceptable (passenger), 140 kilograms (cargo).

MCA warning label.

NITROGEN MUSTARD

General Information

Synonym: HN-1.

Description: Dark liquid; no odor if pure.

Formula: $(ClCH_2CH_2)_2NC_2H_5$.

Constants: Mol wt: 170.08, mp: $1°C$, bp: $85°C$ at 10 mm, d: 1.09 at $25°C$, vap. press.: 0.17 mm at $25°C$, vap. d.: 5.9.

Hazard Analysis

Toxic Hazard Rating:

Acute Local: Irritant 3; Ingestion 3; Inhalation 3.

Acute Systemic: Ingestion 3; Inhalation 3.

Chronic Local: Irritant 3; Ingestion 3; Inhalation 3.

Chronic Systemic: U.

Disaster Hazard: Highly dangerous! A military poison!

Countermeasures

Ventilation Control: Section 2.

NITROGEN OXIDES (except nitrous)

Hazard Analysis

Toxic Hazard Rating:

Acute Local: Irritant 3; Ingestion 3; Inhalation 3.

Acute Systemic: Ingestion 3; Inhalation 3.

Chronic Local: U.

Chronic Systemic: Ingestion 2; Inhalation 2.

Toxicology: See nitric oxide.

Disaster Hazard: Dangerous; when heated they evolve highly toxic fumes; they will react with water or steam to produce heat and corrosive liquids; they can react vigorously with reducing materials.

Countermeasures

Ventilation Control: Section 2.

Personnel Protection: Section 3.

Storage and Handling: Section 7.

NITROGEN OXYCHLORIDE. See nitrosyl chloride.

NITROGEN OXYFLUORIDE. See nitrosyl fluoride.

NITROGEN PENTOXIDE

General Information

Synonym: Nitric anhydride.

Description: Hexagonal (rhombic) white crystals.

Formula: N_2O_5.

Constants: Mol wt: 108.02, mp: $30°C$, bp: $47°C$ (decomp.), d: 1.642 at $18°C$, vap. press.: 400 mm at $24.4°C$.

Hazard Analysis

Toxicity: See nitric oxide.

Fire Hazard: Moderate; when heated it liberates oxygen (Section 6).

Explosion Hazard: Dangerous; at $122°F$ it decomposes violently liberating oxygen.

Disaster Hazard: Dangerous; when heated to decomposition, it emits highly toxic fumes of oxides of nitrogen;

TOXIC HAZARD RATING CODE (For detailed discussion, see Section 1.)

0 NONE: (a) No harm under any conditions; (b) Harmful only under unusual conditions or overwhelming dosage.

1 SLIGHT: Causes readily reversible changes which disappear after end of exposure.

2 MODERATE: May involve both irreversible and reversible changes; not severe enough to cause death or permanent injury.

3 HIGH: May cause death or permanent injury after very short exposure to small quantities.

U UNKNOWN: No information on humans considered valid by authors.

will react with water or steam to produce toxic and corrosive fumes; can react with reducing materials.
Countermeasures
Ventilation Control: Section 2.
Storage and Handling: Section 7.

NITROGEN PEROXIDE. See nitric oxide.
Shipping Regulations: Section 11.
I.C.C.: Poison A; poison gas label, not accepted.
Coast Guard Classification: Poison A; poison gas label.
IATA: See nitrogen tetroxide.

NITROGEN SELENIDE
General Information
Description: Solid material.
Constant: Bp; explodes at 230°C.
Hazard Analysis
Toxicity: Highly toxic. See selenium compounds.
Explosion Hazard: Dangerous, when shocked or exposed to a temperature of 230°C.
Disaster Hazard: Dangerous; shock will explode it; on decomposition or on contact with acid or acid fumes it emits highly toxic fumes of selenium and oxides of nitrogen.
Countermeasures
Storage and Handling: Section 7.

NITROGEN SULFIDE
General Information
Description: An unstable, solid material.
Constant: Mp: explodes at 205°C.
Hazard Analysis
Toxicity: See sulfides.
Fire Hazard: See sulfides.
Explosion Hazard: Severe, when shocked or exposed to heat or flame.
Disaster Hazard: Dangerous; shock or heat (205°C) will explode it; on decomposition or on contact with acid or acid fumes, it emits highly toxic fumes of oxides of nitrogen and oxides of sulfur.
Countermeasures
Storage and Handling: Section 7.

NITROGEN TETROXIDE. See nitrogen dioxide.
Shipping Regulations: Section 11.
Coast Guard Classification: Poison A; poison gas label.
ICC (liquid): Poison A, poison gas label, not accepted.
IATA: Poison A, not acceptable (passenger and cargo).
MCA warning label.

NITROGEN TETROXIDE AND NITRIC OXIDE, MIXTURE
Shipping Regulations: Section 11.
ICC (containing up to 33.2% weight nitric oxide): Poison A, poison gas label, not accepted.
IATA: Poison A, not acceptable (passenger and cargo).

NITROGEN TRIBROMIDE. See nitrogen bromide.

NITROGEN TRICHLORIDE. See nitrogen chloride.

NITROGEN TRIFLUORIDE
General Information
Description: Colorless gas.
Formula: NF_3.
Constants: Mol wt: 71.01, mp: −208.5°C, bp: −129°C, d (liquid): 1.537 at −129°C.
Hazard Analysis
Toxicity: Highly toxic. See fluroides and hydrofluoric acid.
TLV: ACGIH (recommended), 10 parts per million; 29 milligrams per cubic meter of air.
Explosion Hazard: Severe, when shocked, exposed to heat or flame or by spontaneous chemical reaction. Can

react explosively with reducing agents, or spontaneously due to shock or blast when under pressure.
Fire Hazard: Dangerous; a very powerful oxidizer; otherwise inert at normal temperatures and pressures. Reacts violently when ignited with hydrogen. When pure (dry) it does not attack gas or mercury at normal temperatures. Can react violently with grease and oil, etc.
Disaster Hazard: Dangerous; shock will explode it; on decomposition, it emits highly toxic fumes of fluorides; can react vigorously with reducing materials. Particularly hazardous under pressure.
Countermeasures
Ventilation Control: Section 2.
Storage and Handling: Section 7.
Personnel Protection: Section 3.
Shipping Regulations: Section 11.
I.C.C.: Compressed gases, N.O.S.
I.C.C. Label: Flammable gas.

NITROGEN TRIIODIDE
General Information
Description: Black crystals.
Formula: NI_3.
Constants: Mol wt: 394.77, mp: explodes, bp: sublimes in vacuo.
Hazard Analysis
Toxicity: Details unknown. See also iodides.
Explosion Hazard: Severe; when shocked, exposed to heat or flame or by spontaneous chemical reaction. It has no known uses as an explosive, because it is far too sensitive in the dry state to store or handle safely. If this material must be worked with, it should be kept wet. A convenient way of keeping it wet is with ether; when it is needed in the dry state, it simply has to be taken out into the open and the ether will evaporate, leaving it perfectly dry. When dry, it will explode when given the slightest touch, vibration or raise in temperature. Even a puff of air directed into it can cause it to detonate. It is a high explosive and is very violent. It can be destroyed by throwing a quantity of it into large bodies of water, flowing streams or rivers.

NITROGEN TRIIODIDE MONOAMMINE
General Information
Description: Dark red rhombic crystals.
Formula: NI_3NH_3.
Constants: Mol wt: 411.80, mp: decomp. <20°C, bp: explodes, d: 3.5.
Hazard Analysis
Toxicity: Details unknown.
Explosion Hazard: Severe. This material is extremely unstable when dry. The slightest shock or heat will cause it to decompose explosively. It should be kept moist.
Disaster Hazard: Dangerous. On decomposition it emits highly toxic fumes of iodine and ammonia.

NITROGEN TRIOXIDE
General Information
Description: Bluish gas.
Formula: NO_3.
Constants: Mol wt: 62.01, mp: decomp. slightly at ordinary temperatures.
Hazard Analysis and Countermeasures
See nitric oxide.

NITROGEN TRIOXYFLUORIDE. See fluorine nitrate.

NITROGLYCERIN
General Information
Synonyms: Glycerol trinitrate; blasting oil; soup.
Description: Colorless to yellow liquid.
Formula: $C_3H_5(ONO_2)_3$.
Constants: Mol wt: 227.09, mp: 11°C, bp: expl. at 260°C,

flash p: explodes, d: 1.601, vap. press.: 1 mm at 127°C, vap. d.: 7.84, autoign. temp.: 518°F.

Hazard Analysis
Toxic Hazard Rating:
 Acute Local: Irritant 2.
 Acute Systemic: Ingestion 3; Inhalation 3; Skin Absorption 3.
 Chronic Local: Irritant 2.
 Chronic Systemic: Ingestion 2; Inhalation 2; Skin Absorption 2.
Toxicology: The symptoms of nitroglycerin poisoning are headaches and reduced blood pressure, excitement, vertigo, fainting, respiratory rales and cyanosis. If this material is taken internally, it causes respiratory difficulties and death, due to respiratory paralysis. Severe poisoning often manifests itself at first by confusion, pugnaciousness, hallucinations, and maniacal manifestations. The most common complaint is headache which is noted upon commencing work but soon passes off. A break in the work interrupts this acclimatization and workers sometimes resort to the device of moistening their hat bands with nitroglycerin when they are off the job so as to maintain this effect during absence from their occupation. Furthermore it can be absorbed through uninjured skin and may produce eruptions on the palms and intradigital spaces of the hands. In normal manufacture and use of dynamite, the physiological effects of nitroglycerin cause only temporary discomfort and are not injurious to health.
TLV: ACGIH (recommended); 0.2 parts per million in air; 2.0 milligrams per cubic meter in air. May be absorbed via intact skin.
Fire Hazard: Dangerous, when exposed to heat or flame or by spontaneous chemical reaction.
Spontaneous Heating: No.
Explosion Hazard: Severe, when shocked or exposed to heat or flame. Nitroglycerin is a powerful explosive, very sensitive to mechanical shock. Small quantities of it can readily be detonated by a hammer blow on a hard surface, particularly when it has been absorbed in filter paper. Frozen nitroglycerin is somewhat less sensitive than the liquid. However, a half or partially thawed out mixture is more sensitive than either one. See also explosives, high and dynamites.
Disaster Hazard: Highly dangerous; shock, heat and flame will explode it, and toxic fumes evolved on decomposition.
 Countermeasures
Ventilation Control: Section 2.
Personnel Protection: Section 3.
First Aid: Section 1.
Storage and Handling: Section 7.
Shipping Regulations: Section 11.
 I.C.C.: Cannot be shipped by common carrier.
 Coast Guard Classification: Prohibited.
 IATA: Explosive, not acceptable (passenger and cargo).

NITROGLYCERIN, LIQUID, DESENSITIZED
See nitroglycerin.
Shipping Regulations: Section 11.
 Coast Guard Classification: Not permitted.
 IATA: Explosive, not acceptable (passenger and cargo).

NITROGUANIDINE
General Information
Description: Yellow solid.
Formula: $H_2NC(NH)NHNO_2$.

Constants: Mol wt: 104.1, mp: 246°C.
Hazard Analysis
Toxicity: Unknown. See also guanidine hydrochloride.
Fire Hazard: Dangerous, when exposed to heat or flame or by chemical reaction with oxidizers (Section 6).
Explosion Hazard: Severe, when shocked or exposed to heat or flame. Nitroguanidine is known as a flashless or cool explosive. It is about as powerful as TNT and is normally used mixed with colloided nitrocellulose, in which form it yields a propellent powder which gives no flash from the muzzle of the gun thus serving as a great advantage to the military. It has also been used mixed with ammonium nitrate and paraffin wax as a trench mortar ammunition.
Disaster Hazard: Dangerous; shock will explode it; when heated to decomposition, it emits highly toxic fumes; can react vigorously with oxidizing materials.
Countermeasures
Storage and Handling: Section 7.
Shipping Regulations: Section 11.
 IATA (dry or wet with less than 20% water): Explosive, not acceptable (passenger and cargo).
  (wet with not less than 20% water): Explosive, not acceptable (passenger and cargo).

3-NITRO-3-HEXENE
General Information
Formula: $CH_3CH_2NO_2C:CHCH_2CH_3$.
Constant: Mol wt: 129.
Hazard Analysis
Toxic Hazard Rating:
 Acute Local: Irritant 3; Inhalation 3.
 Acute Systemic: Inhalation 3.
 Chronic Local: Inhalation 3.
 Chronic Systemic: Inhalation 3.
Toxicity: Rating based on long term animal inhalation.
Disaster Hazard: Dangerous. See nitrates, organic.

NITROHYDRENE
General Information
Description: An oil.
Composition: Nitroglycerin + nitrosucrose.
Hazard Analysis
Toxicity: Unknown. See nitroglycerin.
Fire Hazard: Dangerous, when exposed to heat or flame or by chemical reaction (Section 6).
Explosion Hazard: Severe, when shocked or exposed to heat. It is a powerful explosive, approximately as powerful as nitroglycerin, and is used to stretch glycerin supplies. It is made up by dissolving up to 25 percent of sucrose in glycerin and nitrating the resulting mixture to give an explosive oil. This procedure saves considerable quantities of glycerin. The material is almost exactly like nitroglycerin. See also explosives, high.
Disaster Hazard: Dangerous; shock and heat will explode it; on decomposition it emits highly toxic fumes of oxides of nitrogen; can react vigorously with oxidizing materials.
Countermeasures
Storage and Handling: Section 7.

NITROHYDROCHLORIC ACID. See aqua regia.
Shipping Regulations: Section 11.
 I.C.C.: Corrosive liquid; white label, 5 pints.
 Coast Guard Classification: Corrosive liquid; white label.

TOXIC HAZARD RATING CODE (For detailed discussion, see Section 1.)

0 NONE: (a) No harm under any conditions; (b) Harmful only under unusual conditions or overwhelming dosage.

1 SLIGHT: Causes readily reversible changes which disappear after end of exposure.

2 MODERATE: May involve both irreversible and reversible changes; not severe enough to cause death or permanent injury.

3 HIGH: May cause death or permanent injury after very short exposure to small quantities.

U UNKNOWN: No information on humans considered valid by authors.

IATA: Corrosive liquid, white label, not acceptable (passenger), $2\frac{1}{2}$ liters (cargo).

NITROHYDROCHLORIC ACID, DILUTED.
See aqua regia.
Shipping Regulations: Section 11.
I.C.C.: Corrosive liquid; white label, 5 pints.
Coast Guard Classification: Corrosive liquid; white label.
IATA: Corrosive liquid, white label, not acceptable (passenger), $2\frac{1}{2}$ liters (cargo).

3-NITRO-4-HYDROXYPHENYLARSONIC ACID
General Information
Synonym: 4-Hydroxy-3-nitrobenzenearsonic acid.
Description: Pale yellow crystals.
Formula: $HOC_6H_3(NO_2)AsO(OH)_2$.
Constant: Mol wt: 263.
Hazard Analysis
Toxic Hazard Rating: U. Used as a food additive permitted in the feed and drinking water of animals and/or for the treatment of food producing animals (Section 10).
Disaster Hazard: Dangerous. See nitrates, organic.

NITRO JUTE
Hazard Analysis
Toxicity: Unknown.
Fire Hazard: Dangerous, when exposed to heat or flame (Section 6).
Explosion Hazard: Severe, when shocked or exposed to heat. See also explosives, high.
Disaster Hazard: Dangerous; shock or heat will explode it; on decomposition it emits highly toxic fumes of oxides of nitrogen; can react vigorously with oxidizing materials.
Countermeasures
Storage and Handling: Section 7.

NITROLEVULOSE
General Information
Synonym: Dextrose nitrate.
Hazard Analysis
Toxicity: Unknown. See nitrates.
Fire Hazard: Dangerous, when exposed to heat or flame (Section 6).
Explosion Hazard: Severe, when shocked or exposed to heat. See also explosives, high and nitrates.
Disaster Hazard: Dangerous; shock or heat will explode it; on decomposition, it emits highly toxic fumes of oxides of nitrogen; can react vigorously with oxidizing materials.
Countermeasures
Storage and Handling: Section 7.

NITROMANNITE. See mannitolhexanitrate.

NITROMANNITOL. See mannitol hexanitrate.

NITROMETHANE
General Information
Description: A liquid.
Formula: CH_3NO_2.
Constants: Mol wt: 61.04, bp: 101° C, lel: 7.3%, fp: $-29°$ C, flash p: 95° F (C.C.), d: 1.130 at 20°/4° C, autoign. temp.: 785° F, vap. press.: 27.8 mm at 20° C, vap. d.: 2.11.
Hazard Analysis
Toxic Hazard Rating:
Acute Local: Irritant 1.
Acute Systemic: Ingestion 3; Inhalation 3.
Chronic Local: U.
Chronic Systemic: U.
TLV: ACGIH (recommended); 100 parts per million in air; 250 milligrams per cubic meter of air.

Fire Hazard: Moderate, when exposed to heat or flame. See also nitroparaffins.
Explosion Hazard: Moderate, when shocked or exposed to heat or flame.
Disaster Hazard: Dangerous; shock or heat will explode it; on decomposition it emits highly toxic fumes of oxides of nitrogen; can react with oxidizing materials.
Countermeasures
Ventilation Control: Section 2.
Storage and Handling: Section 7.
To Fight Fire: Alcohol foam, carbon dioxide, dry chemical or carbon tetrachloride (Section 6).
Shipping Regulations: Section 11.
IATA: Other restricted articles, class A, no label required, 40 liters (passenger), 220 liters (cargo).

2-NITRO-2-METHYL-1,3-PROPANEDIOL
General Information
Description: Crystalline material.
Formula: $CH_2OHC(CH_3)NO_2CH_2OH$.
Constants: Mol wt: 135.12, mp: 147–149° C, bp: decomposes.
Hazard Analysis
Toxicity: Details unknown. Limited animal experiments suggest high toxicity.
Fire Hazard: Moderate, when exposed to heat or flame (Section 6).
Disaster Hazard: Dangerous; on decomposition, it emits highly toxic fumes of oxides of nitrogen; can react with oxidizing materials.
Countermeasures
Storage and Handling: Section 7.
Ventilation Control: Section 2.

2-NITRO-2-METHYL-1-PROPANOL
General Information
Description: Crystalline material.
Formula: $CH_3C(CH_3)NO_2CH_2OH$.
Constants: Mol wt: 119.12, mp: 90–91° C, bp: 95° C at 10 mm.
Hazard Analysis
Toxicity: Unknown.
Fire Hazard: See nitrates.
Disaster Hazard: See nitrates.
Countermeasures
Storage and Handling: Section 7.

NITROMOLASSES
Hazard Analysis
Toxicity: Unknown. See nitrates.
Fire Hazard: Dangerous, when exposed to heat or flame (Section 6).
Explosion Hazard: Moderate, when shocked or exposed to heat or flame. See also explosives, high and nitrates.
Disaster Hazard: Dangerous; shock or heat will explode it; on decomposition, it emits highly toxic fumes of oxides of nitrogen; can react vigorously with oxidizing materials.
Countermeasures
Ventilation Control: Section 2.
Storage and Hanlding: Section 7.

NITRONAPHTHALENE
General Information
Description: Yellow crystals.
Formula: $C_{10}H_7NO_2$.
Constants: Mol wt: 173.16, mp: 58.8° C, bp: 304° C, flash p: 327° F (C.C.), d: 1.331 at 4°/4° C, vap. d.: 5.96.
Hazard Analysis
Toxic Hazard Rating:
Acute Local: Irritant 2.
Acute Systemic: U.
Chronic Local: U.

Chronic Systemic: U.
Fire Hazard: Slight, when exposed to heat.
Explosion Hazard: See nitrates.
Disaster Hazard: See nitrates.
Countermeasures
Personal Hygiene: Section 3.
Storage and Handling: Section 7.
To Fight Fire: Water, carbon dioxide, dry chemical, carbon
tetrachloride or water spray (Section 6).

β-NITRONAPHTHALENE
General Information
Formula: $C_{10}H_7NO_2$.
   Constant: Mol wt: 173.16.
Hazard Analysis
Toxic Hazard Rating:
   Acute Local: Irritant 1; Inhalation 1.
   Acute Systemic: Ingestion 3.
   Chronic Local: Inhalation 1.
   Chronic Systemic: Ingestion 3.
Toxicity: Animal experiments show high oral toxicity and
   low irritation of skin and lungs.
Fire Hazard: See nitronaphthalene.
Explosion Hazard: See nitrates.
Disaster Hazard: See nitrates.
Countermeasures
See nitronaphthalene.

2-NITRO-2-NONENE
General Information
Formula: $CH_3NO_2C:CH(CH_2)_5CH_3$.
Constant: Mol wt: 171.
Hazard Analysis
Toxic Hazard Rating:
   Acute Local: Irritant 3; Inhalation 3.
   Acute Systemic: Inhalation 3.
   Chronic Local: Inhalation 3.
   Chronic Systemic: Inhalation 3.
Rating based on long term animal inhalation studies.
Disaster Hazard: Dangerous. See nitrates, organic.

NITRONIUM PERCHLORATE
General Information
Description: White crystals. Odor of oxides of nitrogen and
   chlorine. Soluble (by reaction) in water.
Formula: (NO_2ClO_4).
Constants: Mol wt: 145.5, mp: decomp. at 120–140°C, d:
   2.25 g/cc, vap. press.: < 0.05 mm at 20°C.
Hazard Analysis
Toxicology: A powerful irritant to skin and mucous mem-
   branes. See nitric and perchloric acids, which are this
   material's products of hydrolysis; i.e. the moisture of
   skin or mucous membranes.
Disaster Hazard: Dangerous. Decomposed by heat or steam
   to highly toxic fumes.
Countermeasures
Ventilation Control: Section 2.
Personnel Protection: Section 3.
Storage and Handling: Section 7.
Shipping Regulations: Section 11.
   I.C.C.: 73.154.

NITROPARAFFINS
General Information
Synonyms: 2-Nitropropane, nitroethane; nitromethane;
   1-nitropropane.
Constant: Flash p: 103–120° F.

Hazard Analysis
Toxic Hazard Rating:
   Acute Local: Irritant 1.
   Acute Systemic: Ingestion 2; Inhalation 2.
   Chronic Local: Irritant 1.
   Chronic Systemic: Ingestion 1; Inhalation 1.
Fire Hazard: Moderate. See nitrates.
Explosion Hazard: Dry salts of nitromethane and inorganic
   bases are explosive. Otherwise the mono nitro-paraffins
   are relatively stable. Impact under confined conditions
   can cause explosions of mono nitromethane. Avoid
   heating it to above 450° F.
Disaster Hazard: See nitrates.
Countermeasures
Ventilation Control: Section 2.
To Fight Fire: Foam, carbon dioxide, dry chemical or car-
   bon tetrachloride (Section 6).
Personal Hygiene: Section 3.
Storage and Handling: Section 7.

m-NITROPHENOL
General Information
Description: Monoclinic prisms.
Formula: $C_6H_5NO_3$.
Constants: Mol wt: 139.1, mp: 97°C, bp: 194°C at 70 mm,
   d: 1.485 at 20°/4°C.
Hazard Analysis
Toxicity: See p-nitrophenol.
Disaster Hazard: See nitrates.
Countermeasures
Ventilation Control: Section 2.
Personal Hygiene: Section 3.
First Aid: Section 1.
Storage and Handling: Section 7.

o-NITROPHENOL
General Information
Description: Light yellow crystals; aromatic odor.
Formula: $C_6H_5NO_3$.
Constants: Mol wt: 139.1, mp: 45°C, bp: 214.5°C, d: 1.657
   at 20°C, vap. press.: 1 mm at 49.3°C.
Hazard Analysis
Toxicity: Has produced kidney and liver injury in experi-
   mental animals. See p-nitrophenol.
Disaster Hazard: See nitrates.
Countermeasures
Storage and Handling: Section 7.

p-NITROPHENOL
General Information
Description: Colorless to slightly yellow crystals.
Formula: $C_6H_5NO_3$.
Constants: Mol wt: 139.1, mp: 113–114°C, bp: 279°C (de-
   comp.), d: 1.270 at 120°/4°C.
Hazard Analysis
Toxic Hazard Rating:
   Acute Local: U.
   Acute Systemic: Ingestion 3; Inhalation 3.
   Chronic Local: U.
   Chronic Systemic: Ingestion 3; Inhalation 3.
Toxicology: Has experimentally produced hyperthermia,
   methemoglobinemia and depression. See also dinitro-
   phenol.
Disaster Hazard: See nitrates.
Countermeasures
Ventilation Control: Section 2.
Personal Hygiene: Section 3.

TOXIC HAZARD RATING CODE (For detailed discussion, see Section 1.)

0 NONE: (a) No harm under any conditions; (b) Harmful only under unusual conditions or overwhelming dosage.

1 SLIGHT: Causes readily reversible changes which disappear after end
of exposure.

2 MODERATE: May involve both irreversible and reversible changes;
not severe enough to cause death or permanent injury.

3 HIGH: May cause death or permanent injury after very short exposure
to small quantities.

U UNKNOWN: No information on humans considered valid by authors.

First Aid: Section 1.
Storage and Handling: Section 7.

m-NITROPHENOL DIAZONIUM PERCHLORATE
General Information
Description: A solid.
Hazard Analysis
Toxicity: Unknown. See perchlorates.
Fire Hazard: Dangerous, when exposed to heat or flame (Section 6).
Explosion Hazard: Severe, when shocked or exposed to heat. See also explosives, high.
Disaster Hazard: Dangerous; shock will explode it; when heated to decomposition, it emits highly toxic fumes of chlorides and oxides of nitrogen; can react vigorously with oxidizing materials.
Countermeasures
Storage and Handling: Section 7.

NITROPHENYL BORIC ACID
General Information
Description: Yellow crystals; slightly water soluble.
Formula: $NO_2C_6H_4B(OH)_2$.
Constants: Mol wt: 166.9, mp: decomposes before melting.
Hazard Analysis and Countermeasures
See boron compounds, nitro compounds of aromatic hydrocarbons, and nitrates.

1-(p-NITROPHENYL)ETHYL NITRATE
General Information
Formula: $(NO_2)C_6H_3(OH)(C_2H_4NO_3)$.
Constant: Mol wt: 214.
Hazard Analysis
Toxicity: Details unknown. Limited animal experiments suggest moderate toxicity. See also nitro compounds of aromatic hydrocarbons.
Disaster Hazard: Dangerous. See nitrates, organic.

p-NITROPHENYL SERINE
Hazard Analysis
Toxicity: Details unknown. Animal experiments show injury to eyes, skin and liver.
Disaster Hazard: Dangerous. See nitrates, organic.

1-NITROPROPANE
General Information
Description: Colorless liquid.
Formula: $CH_3CH_2CH_2NO_2$.
Constants: Mol wt: 89.09, bp: 132°C, fp: −108°C, flash p: 120°F (T.O.C.), d: 1.003 at 20°/20°C, autoign. temp.: 789°F, vap. press.: 7.5 mm at 20°C, vap. d.: 3.06, lel: 2.6%.
Hazard Analysis
TLV: ACGIH (recommended) 25 parts per million of air: 90 milligrams per cubic meter of air.
Toxicity: Moderate. See nitroparaffines, 2-nitropropane.
Fire Hazard: Moderate, when exposed to heat or open flame (Section 6).
Explosion Hazard: See nitrates.
Disaster Hazard: See nitrates.
Countermeasures
Storage and Handling: Section 7.
To Fight Fire: Alcohol foam, carbon dioxide, dry chemical or carbon tetrachloride (Section 6).

2-NITROPROPANE
General Information
Description: Colorless liquid.
Formula: $(CH_3)_2CHNO_2$.
Constants: Mol wt: 89.09, bp: 120°C, fp: −93°C, flash p: 103°F (T.O.C.), d: 0.992 at 20°/20°C, autoign. temp.: 802°F, vap. press.: 10 mm at 15.8°C, vap. d.: 3.06, lel: 2.6%.
Hazard Analysis
Toxic Hazard Rating:
Acute Local: Irritant 1; Ingestion 1; Inhalation 1.

Acute Systemic: Ingestion 2; Inhalation 2.
Chronic Local: U.
Chronic Systemic: Ingestion 2; Inhalation 2.
Toxicology: Can cause gastro intestinal disturbances and injury to liver and kidneys. Large doses produce methemoglobinemia and cyanosis. See also nitrates.
TLV: ACGIH (recommended); 25 parts per million in air; 90 milligrams per cubic meter of air.
Fire Hazard: Moderate, when exposed to heat or open flame (Section 6).
Explosion Hazard: See nitrates.
Disaster Hazard: See nitrates.
Countermeasures
Storage and Handling: Section 7.
To Fight Fire: Alcohol foam, carbon dioxide, dry chemical or carbon tetrachloride (Section 6).

5-NITRO-2-n-PROPOXYANILINE
General Information
Synonym: P-4000; 1-propoxy-2-amino-4-nitrobenzene.
Description: Orange crystals; slightly water soluble. Very sweet.
Formula: $C_9H_{12}N_2O_2$.
Constants: Mol wt: 196.2, mp: 48°C.
Hazard Analysis
Toxicity: Details unknown. A synthetic sweetener; suspected of having carcinogenic activity.
Disaster Hazard: Dangerous. When heated to decomposition it emits toxic fumes.

NITROPRUSSIDES
Hazard Analysis
Toxic Hazard Rating:
Acute Local: U.
Acute Systemic: Ingestion 3; Inhalation 3.
Chronic Local: U.
Chronic Systemic: Ingestion 3; Inhalation 3.
Toxicology: According to some observers, sodium nitroprusside decomposes in the body liberating cyanide; hence the toxicity is that of cyanides. There may be a nitrite effect in nonfatal cases. See also cyanides and nitrites.
Disaster Hazard: Dangerous; when heated to decomposition or on contact with acid or acid fumes; they emit highly toxic fumes of cyanides and oxides of nitrogen.
Countermeasures
Ventilation Control: Section 2.
Personal Hygiene: Section 3.
First Aid: Section 1.
Storage and Handling: Section 7.

NITROPYRIDINE
General Information
Description: Liquid.
Formula: $C_5H_4N_2O_2$.
Constants: Mol wt: 124.1, autoign. temp.: 725°F.
Hazard Analysis
Toxicity: Details unknown; probably highly toxic. See also pyridine.
Fire Hazard: Moderate, when exposed to heat or flame.
Explosion Hazard: Unknown.
Disaster Hazard: Dangerous; when heated to decomposition, it emits highly toxic fumes of oxides of nitrogen; can react with oxidizing materials.
Countermeasures
Storage and Handling: Section 7.
To Fight Fire: Water, carbon dioxide, dry chemical, carbon tetrachloride or water spray (Section 6).

NITROSACCHAROSE
General Information
Synonym: Nitrosugar.
Description: White solid.
Hazard Analysis
Toxicity: Unknown. See nitrates.

Fire Hazard: Dangerous, when exposed to heat or flame (Section 6).
Explosion Hazard: Moderate, when shocked or exposed to heat. See also explosives, high and nitrates.
Disaster Hazard: See nitrates.
Countermeasures
Storage and Handling: Section 7.

N-NITROSODIMETHYLAMINE
General Information
Synonym: Dimethyl nitrosamine.
Description: Yellow liquid. Soluble in water, alcohol, and ether.
Formula: $C_2H_6N_2O$.
Constants: Mol wt: 74.1, bp: 152°C, d: 1.006 at 20°/4°C.
Hazard Analysis
Toxic Hazard Rating.
  Acute Local: Irritant 3, Ingestion 3, Inhalation 3.
  Acute Systemic: Ingestion 3.
  Chronic Local: U.
  Chronic Systemic: Ingestion 3; Inhalation 3.
Toxicity: Very highly toxic. Possibly carcinogenic. Has caused fatal liver disease in humans.
TLV: ACGIH (recommended) because of extremely high toxicity and presumed carcinogenic potential of this compound, contact by any routes should not be permitted.
Disaster Hazard: Dangerous. When heated to decomposition it yields highly toxic fumes. May decompose violently under some disaster conditions.
Countermeasures
Use only with the greatest care and with competent industrial hygiene advice. See Sections 2,3,7.

p-NITROSODIMETHYLANILINE
General Information
Synonyms: Accellerene; dimethyl-p-nitrosoaniline.
Description: Green leaflets.
Formula: $NOC_6H_4N(CH_3)_2$.
Constants: Mol wt: 150.14, mp: 87.8°C.
Hazard Analysis
Toxic Hazard Rating:
  Acute Local: Irritant 2; Allergen 2.
  Acute Systemic: Ingestion 3; Inhalation 3.
  Chronic Local: Irritant 2; Allergen 2.
  Chronic Systemic: Ingestion 3; Inhalation 3.
Fire Hazard: Moderate, when exposed to heat or flame (Section 6).
Explosion Hazard: See nitrates.
Disaster Hazard: See nitrates.
Countermeasures
Ventilation Control: Section 2.
Personnel Protection: Section 3.
First Aid: Section 1.
Storage and Handling: Section 7.
Shipping Regulations: Section 11.
  IATA: Flammable solid, yellow label, 12 kilograms (passenger), 45 kilograms (cargo).

N-NITROSODIPHENYLAMINE
Hazard Analysis
Toxicity: Details unknown. See also amines.
Fire Hazard: Dangerous, when exposed to heat or flame or by chemical reaction (Section 6).
Disaster Hazard: Dangerous; when heated to decomposition, it emits highly toxic fumes of oxides of nitrogen; can react vigorously with oxidizing materials.

Countermeasures
Storage and Handling: Section 7.

NITROSOGUANIDINE
General Information
Description: A solid.
Hazard Analysis
Toxicity: Unknown.
Fire Hazard: Dangerous, when exposed to heat or flame (Section 6).
Explosion Hazard: Severe, when shocked or exposed to heat or flame. See also explosives, high and nitrates.
Disaster Hazard: Dangerous; shock will explode it; on decomposition, it emits highly toxic fumes of oxides of nitrogen; can react vigorously with oxidizing materials.
Countermeasures
Storage and Handling: Section 7.

NITROSOMETHYLUREA
General Information
Description: Powder.
Formula: $CH_3NHCONHNO$.
Constant: Mol wt: 89.1.
Hazard Analysis
Toxic Hazard Rating:
  Acute Local: Allergen 1.
  Acute Systemic: U.
  Chronic Local: Allergen 1.
  Chronic Systemic: U.
Toxicology: Has been known to cause a contact dermatitis of the poison ivy type (Section 9).
Fire Hazard: Moderate, when exposed to heat or flame (Section 6).
Disaster Hazard: See nitrates.
Countermeasures
Personal Hygiene: Section 2.
Storage and Handling: Section 7.

NITROSOMETHYLURETHANE
General Information
Description: Solid.
Formula: $CH_3NNOCOOC_2H_5$.
Constant: Mol wt: 132.2.
Hazard Analysis
Toxic Hazard Rating:
  Acute Local: Irritant 2; Ingestion 2; Inhalation 2.
  Acute Systemic: U.
  Chronic Local: U.
  Chronic Systemic: U.
Fire Hazard: Slight, when exposed to heat or flame (Section 6).
Disaster Hazard: Dangerous; when heated to decomposition, it emits highly toxic fumes of oxides of nitrogen.
Countermeasures
Ventilation Control: Section 2.
Personnel Protection: Section 3.
Storage and Handling: Section 7.

p-NITROSOPHENOL
General Information
Synonyms: Quinone monoxime.
Description: Yellow, rhombic needles.
Formula: NOC_6H_4OH.
Constants: Mol wt: 123.11, mp: 126°C decomposes.
Hazard Analysis
Toxicity: Details unknown. Is said to resemble p-nitrophenol in action. See also dinitrophenol.

TOXIC HAZARD RATING CODE (*For detailed discussion, see Section 1.*)

0 NONE: (a) No harm under any conditions; (b) Harmful only under unusual conditions or overwhelming dosage.

1 SLIGHT: Causes readily reversible changes which disappear after end of exposure.

2 MODERATE: May involve both irreversible and reversible changes; not severe enough to cause death or permanent injury.

3 HIGH: May cause death or permanent injury after very short exposure to small quantities.

U UNKNOWN: No information on humans considered valid by authors.

Fire Hazard: Dangerous, when exposed to heat or flame; burns explosively (Section 6).

Explosion Hazard: Moderate, when exposed to heat. See also nitrates.

Disaster Hazard: See nitrates.

Countermeasures

Storage and Handling: Section 7.

NITROSOPHENOL DIMETHYLPYRAZOLE
General Information
Description: A solid.
Hazard Analysis
Toxicity: Unknown.
Fire Hazard: See nitrates.
Explosion Hazard: Severe, when exposed to heat. See also explosives, high and nitrates.
Disaster Hazard: Dangerous; when heated to decomposition, it emits highly toxic fumes of oxides of nitrogen; can react vigorously with oxidizing materials. A high explosive. See explosives, high.
Countermeasures
Storage and Handling: Section 7.

NITROSO PYRAZOLE
Hazard Analysis
Toxic Hazard Rating:
  Acute Local: Irritant 2; Ingestion 2; Inhalation 2.
  Acute Systemic: U.
  Chronic Local: U.
  Chronic Systemic: U.
Fire Hazard: Slight, when exposed to heat or flame (Section 6).
Disaster Hazard: Dangerous; when heated to decomposition, it emits highly toxic fumes of oxides of nitrogen; can react with oxidizing materials.
Countermeasures
Ventilation Control: Section 2.
Personnnel Protection: Section 3.
Storage and Handling: Section 7.

NITRO STARCH
General Information
Synonym: Starch nitrate.
Description: Solid.
Formula: $C_{12}H_{12}(NO_2)_8O_{10}$.
Constant: Mol wt: 684.30.
Hazard Analysis
Toxicity: Unknown. See nitrates.
Fire Hazard: Dangerous, when exposed to heat or flame (Section 6).
Explosion Hazard: Severe, when shocked or exposed to heat or flame. It is a powerful high explosive. Nitrostarch is not a definite compound, but a mixture of various nitric acid esters of starch of different degrees of nitration. It is used as an ingredient of blasting explosives, for quarrying, and as a demolition explosive. See also nitrates.
Countermeasures
Storage and Handling: Section 7.
Shipping Regulations: Section 11.
  IATA (dry or wet with less than 20% water or 30% solvent): Explosive, not acceptable (passenger and cargo).

NITROSTYRENE
Hazard Analysis and Countermeasures
See phenylethylene and nitrates.

NITROSUGAR. See nitrosaccharose.

NITROSYL BROMIDE
General Information
Description: Brown gas or dark brown liquid.
Formula: NOBr.
Constants: Mol wt: 109.92, mp: −55.5°C, bp: −2°C, d:> 1.0.

Hazard Analysis
Toxic Hazard Rating:
  Acute Local: Irritant 3; Ingestion 3; Inhalation 3.
  Acute Systemic: U.
  Chronic Local: U.
  Chronic Systemic: U.
Disaster Hazard: See bromides.
Countermeasures
Ventilation Control: Section 2.
Personnel Protection: Section 3.
Storage and Handling: Section 7.

NITROSYL CHLORIDE
General Information
Synonym: Nitrogen oxychloride.
Description: Yellow gas or yellow liquid or crystals; irritating odor.
Formula: NOCl.
Constants: Mol wt: 65.47, mp: −64.5°C, bp: −5.8°C, d: (liq.) 1.250 at 30°C, vap. press.: 76 mm at −50°C, vap. d.: 2.3.
Hazard Analysis
Toxic Hazard Rating:
  Acute Local: Irritant 3; Ingestion 3; Inhalation 3.
  Acute Systemic: U.
  Chronic Local: U.
  Chronic Systemic: U.
Toxicology: Intensely irritating. Can cause fatal pulmonary edema.
Disaster Hazard: See chlorides.
Countermeasures
Ventilation Control: Section 2.
Personnel Protection: Section 3.
Storage and Handling: Section 7.
Shipping Regulations: Section 11.
  I.C.C.: Nonflammable gas; green label, 300 pounds.
  Coast Guard Classification: Noninflammable gas; green gas label.
  IATA: Nonflammable gas, green label, not acceptable (passenger), 140 kilograms (cargo).

NITROSYL FLUORIDE
General Information
Synonym: Nitrogen oxyfluoride.
Description: Colorless gas.
Formula: NOF.
Constants: Mol wt: 49.01, mp: −134°C, bp: −56°C, d: 2.176 g/l.
Hazard Analysis and Countermeasures
See fluorides and hydrofluoric acid.

NITROSYL SULFURIC ACID
General Information
Synonym: Chamber crystals.
Description: Crystalline, colorless solid.
Formula: $ONOSO_3H$.
Constants: Mol wt: 127.08, mp: 73°C (for crystalline material).
Hazard Analysis
Toxic Hazard Rating:
  Acute Local: Irritant 3; Ingestion 3; Inhalation 3.
  Acute Systemic: U.
  Chronic Local: U.
  Chronic Systemic: U. .
Fire Hazard: See nitrates.
Disaster Hazard: See nitrates and sulfates.
Countermeasures
Ventilation Control: Section 2.
Personnel Protection: Section 3.
First Aid: Section 1.
Storage and Handling: Section 7.

NITROSYLSULFURIC ANHYDRIDE
General Information
Description: Tetragonal crystals.

Formula: $(NOSO_3)_2O$.
Constants: Mol wt: 236.15, mp: 217°C, bp: 360°C.
Hazard Analysis
Toxic Hazard Rating:
Acute Local: Irritant 3; Ingestion 3; Inhalation 3.
Acute Systemic: U.
Chronic Local: U.
Chronic Systemic: U.
Disaster Hazard: Dangerous; when heated to decomposition, it emits highly toxic fumes of oxides of nitrogen and oxides of sulfur; will react with water or steam to produce toxic and corrosive fumes.
Countermeasures
Ventilation Control: Section 2.
Personnel Protection: Section 3.
First Aid: Section 1.
Storage and Handling: Section 7.

m-NITROTOLUENE
General Information
Description: Liquid.
Formula: $CH_3C_6H_4NO_2$.
Constants: Mol wt: 137.1, mp: 15.1°C, bp: 231.9°C, flash p: 233°F (C.C.), d: 1.1630 at 15°/4°C, vap. press.: 1 mm at 50.2°C, vap. d.: 4.72.
Hazard Analysis
Toxicity: Moderate. See also p-nitrotoluene.
Fire Hazard: Moderate, when exposed to heat or flame.
Disaster Hazard: See nitrates.
Countermeasures
Storage and Handling: Section 7.
To Fight Fire: Water, carbon dioxide, dry chemical or carbon tetrachloride (Section 6).

o-NITROTOLUENE
General Information
Description: Yellowish liquid.
Formula: $C_7H_7NO_2$.
Constants: Mol wt: 137.1, mp: −4.1°C, bp: 222.3°C, flash p: 223°F (C.C.), d: 1.1622 at 19°/15°C, vap. press.: 1 mm at 50°C, vap. d.: 4.72.
Hazard Analysis
Toxicity: Moderate. See p-nitrotoluene.
Fire Hazard: Moderate, when exposed to heat or open flame.
Disaster Hazard: See nitrates.
Countermeasures
Storage and Handling: Section 7.

p-NITROTOLUENE
General Information
Synonym: Methyl nitrobenzene.
Description: Yellowish crystals.
Formula: $C_7H_7NO_2$.
Constants: Mol wt: 137.1, mp: 51.9°C, bp: 238.3°C, flash p: 223°F (C.C.), d: 1.286, vap. press.: 1 mm at 53.7°C, vap. d.: 4.72.
Hazard Analysis
Toxic Hazard Rating:
Acute Local: U.
Acute Systemic: Ingestion 2; Inhalation 2.
Chronic Local: U.
Chronic Systemic: Ingestion 2; Inhalation 2.
TLV: ACGIH (recommended) 5 parts per million in air; 28 milligrams per cubic meter of air. Can be absorbed through the skin.

Fire Hazard: Moderate, when exposed to heat or open flame.
Disaster Hazard: Dangerous. See nitrates.
Countermeasures
Ventilation Control: Section 2.
To Fight Fire: Watr, carbon dioxide, dry chemical or carbon tetrachloride (Section 6).
Personal Hygiene: Section 3.
Storage and Handling: Section 7.
Shipping Regulations: Section 11.
IATA: Poison B, poison label, 25 kilograms (passenger), 95 kilograms (cargo).

NITROTOLUENE, LIQUID. See m- and o-nitrotoluene.
Shipping Regulations: Section 11.
IATA: Poison B, poison label, 1 liter (passenger), 220 liters (cargo).

m-NITRO-p-TOLUIDINE
General Information
Description: Clear liquid.
Formula: $C_7H_6NH_2NO_2$.
Constants: Mol wt: 168.15, mp: 116°C, flash p: 315°F (C.C.), d: 1.312, vap. d.: 5.80.
Hazard Analysis
Toxicity: Details unknown. See also toluidine and nitro compounds of aromatic hydrocarbons.
Fire Hazard: Slight, when exposed to heat or flame.
Disaster Hazard: See nitrates.
Countermeasures
Storage and Handling: Section 7.
To Fight Fire: Water, carbon dioxide, dry chemical or carbon tetrachloride (Section 6).

NITROTRICHLOROMETHANE. See chloropicrin.

6-NITRO-2,4-bis(TRICHLOROMETHYL)-1,3-BENZODIOXANE
Hazard Analysis
Toxicity: Details unknown. See also nitro compounds of aromatic hydrocarbons.
Disaster Hazard: Dangerous; when heated to decomposition, it emits highly toxic fumes of chlorides and oxides of nitrogen.
Countermeasures
Storage and Handling: Section 7.

NITROUREA
Hazard Analysis
Toxicity: Unknown.
Fire Hazard: Dangerous, when exposed to heat or flame.
Explosion Hazard: Severe, when shocked or exposed to heat. See also explosives, high and nitrates.
Disaster Hazard: Dangerous; shock will explode it; when heated to decomposition it emits highly toxic fumes of oxides of nitrogen; can react vigorously with oxidizing materials.
Countermeasures
Storage and Handling: Section 7.

NITROUS ACID
General Information
Description: Pale blue solutions.
Formula: HNO_2.
Constant: Mol wt: 47.02.
Hazard Analysis
Toxicity: See nitrites.

TOXIC HAZARD RATING CODE (For detailed discussion, see Section 1.)

0 NONE: (a) No harm under any conditions; (b) Harmful only under unusual conditions or overwhelming dosage.

1 SLIGHT: Causes readily reversible changes which disappear after end of exposure.

2 MODERATE: May involve both irreversible and reversible changes; not severe enough to cause death or permanent injury.

3 HIGH: May cause death or permanent injury after very short exposure to small quantities.

U UNKNOWN: No information on humans considered valid by authors.

Fire Hazard: Moderate, by chemical reaction; a powerful oxidizer (Section 6).
Disaster Hazard: Dangerous; when heated to decomposition, it emits highly toxic fumes of oxides of nitrogen.
Countermeasures
Storage and Handling: Section 7.

NITROUS ANHYDRIDE. See dinitrogen trioxide.

NITROUS ETHER. See ethyl nitrite.

NITROUS FUMES. See nitric oxide.

NITROUS OXIDE
General Information
Synonym: Laughing gas.
Description: Colorless gas or liquid or cubic crystals.
Formula: N_2O.
Constants: Mol wt: 44.02, mp: $-102.4°C$, bp: $-88.49°C$, d: 1.977 g/l; d liq: 1.226 at $-89°C$.
Hazard Analysis
Toxic Hazard Rating:
  Acute Local: 0.
  Acute Systemic: Inhalation 2.
  Chronic Local: 0.
  Chronic Systemic: U.
Note: May be used as a general purpose food additive (Section 10).
Fire Hazard: Moderate, by chemical reaction; it supports combustion (Section 6).
Explosion Hazard: Moderate; it can form an explosive mixture with air.
Disaster Hazard: Slightly dangerous, can form explosive mixtures with air.
Countermeasures
Ventilation Control: Section 2.
Storage and Handling: Section 7.
Shipping Regulations: Section 11.
  I.C.C.: Nonflammable gas; green label, 300 pounds.
  Coast Guard Classification: Noninflammable gas; green gas label.
  IATA: Nonflammable gas, green label, 70 kilograms (passenger), 140 kilograms (cargo).

m-NITROXYLENE. See m-nitroxylol.

o-NITROXYLENE. See o-nitroxylol.

p-NITROXYLENE. See p-nitroxylol.

m-NITROXYLOL
General Information
Synonym: n-Nitroxylene.
Description: Light yellow liquid.
Formula: $C_6H_3(CH_3)_2NO_2$.
Constants: Mol wt: 151.2, mp: 2°C, bp: 244°C, d: 1.135 at 15°/4°C.
Hazard Analysis
Toxicity: Details unknown; probably toxic. See also nitro compounds or aromatic hydrocarbons. Said to be less toxic than nitrobenzene.
Fire Hazard: Moderate, when exposed to heat or flame (Section 6).
Explosion Hazard: See nitrates.
Disaster Hazard: See nitrates.
Countermeasures
Storage and Handling: Section 7.
Shipping Regulations: Section 11.
  Coast Guard Classification: Poison B; poison label.
  I.C.C.: Poison B; poison label, 55 gallons.
  IATA: Poison B, poison label, 1 liter (passenger), 220 liters (cargo).

o-NITROXYLOL
General Information
Synonym: o-Nitroxylene.

Description: Yellowish liquid or solid.
Formula: $C_6H_3(CH_3)_2NO_2$.
Constants: Mol wt: 151.2, mp: 29°C, bp: 258°C, d: 1.139.
Hazard Analysis and Countermeasures
See nitro compounds of aromatic hydrocarbons and nitrates.
Shipping Regulations: See m-nitroxylol.

p-NITROXYLOL
General Information
Synonym: p-Nitroxylene.
Description: Liquid.
Formula: $C_6H_3(CH_3)_2NO_2$.
Constants: Mol wt: 151.2, bp: 240°C, d: 1.132.
Hazard Analysis and Countermeasures
See nitro compounds of aromatic hydrocarbons and nitrates.
Shipping Regulations: See m-nitroxylol.

NITRYL CHLORIDE
General Information
Description: Pale yellow or brown gas.
Formula: NO_2Cl.
Constants: Mol wt: 81.47, mp: $< -31°C$, bp: 5°C, d: 2.57 g/l; liquid 1.32 at 14°C.
Hazard Analysis
Toxic Hazard Rating:
  Acute Local: Irritant 3; Inhalation 3.
  Acute Systemic: U.
  Chronic Local: U.
  Chronic Systemic: U.
Disaster Hazard: Dangerous; when heated to decomposition, it emits highly toxic fumes of chlorides and nitrogen dioxide; will react with water or steam to produce corrosive fumes.
Countermeasures
Ventilation Control: Section 2.
Personal Hygiene: Section 3.
Storage and Handling: Section 7.

NITRYL FLUORIDE
General Information
Description: Colorless gas, pungent odor.
Formula: NO_2F.
Constants: Mol wt: 65.01, d(solid): 1.924, d (liquid): 1.796 at 72°C, mp: $-166°C$, bp: $-72.4°C$.
Hazard Analysis
Toxic Hazard Rating:
  Acute Local: Irritant 3; Ingestion 3; Inhalation 3.
  Acute Systemic: U.
  Chronic Local: U.
  Chronic Systemic: U.
Toxicity: See fluorides.
Disaster Hazard: Dangerous. See oxides of nitrogen and fluorides.

NOISEMAKERS, TRICK, EXPLOSIVE
Shipping Regulations: Section 11.
  IATA: Explosive, explosive label, 25 kilograms (passenger), 70 kilograms (cargo).

NONAETHYLENE GLYCOL
Hazard Analysis
Toxicity: Details unknown. See also glycols.
Fire Hazard: Slight; when heated, it emits acrid fumes (Section 6).
Countermeasures
Storage and Handling: Section 7.

n-NONANE
General Information
Description: Colorless liquid.
Formula: C_9H_{20}.
Constants: Mol wt: 128.25, mp: $-53.7°C$, bp: 150.7°C, lel: 0.8%, uel: 2.9%, flash p: 88°F (C.C.), d: 0.718 at 20°/4°C, autoign. temp.: 403°F, vap. press.: 10 mm at 38.0°C, vap. d.: 4.41.

Hazard Analysis
Toxic Hazard Rating:
   Acute Local: Inhalation 1.
   Acute Systemic: Inhalation 2.
   Chronic Local: Inhalation 1.
   Chronic Systemic: U.
Toxicology: Irritating to respiratory tract. Narcotic in high concentrations.
Fire Hazard: Moderate, when exposed to heat or flame; can react with oxidizing materials.
Spontaneous Heating: No.
Explosion Hazard: Moderate, in the form of gas, when exposed to flame.
Countermeasures
Ventilation Control: Section 2.
To Fight Fire: Carbon dioxide, dry chemical or carbon tetrachloride (Section 6).
Storage and Handling: Section 7.

NONANEDIOIC ACID. See azelaic acid.

NONANOIC ACID. See ethyl heptanoic acid.

NONETHYLENE GLYCOL HEXARICINOLEATE
General Information
Constant: Flash p: 530° F(O.C.).
Hazard Analysis
Toxicity: Details unknown. See also glycols.
Fire Hazard: Slight, when exposed th heat or flame.
Countermeasures
To Fight Fire: Foam, carbon dioxide, dry chemical or carbon tetrachloride (Section 6).
Storage and Handling: Section 7.

NONETHYLENE GLYCOL MONOSTEARATE
General Information
Constant: Flash p: 505° F (O.C.).
Hazard Analysis
Toxicity: Details unknown. See also glycols.
Fire Hazard: Slight, when exposed to heat or flame.
Countermeasures
To Fight Fire: Foam, carbon dioxide, dry chemical or carbon tetrachloride (Section 6).
Storage and Handling: Section 7.

NONLIQUEFIED HYDROCARBON GAS
Shipping Regulations: Section 11.
   I.C.C.: Flammable gas, red gas label, 300 pounds.

NONYL ACETATE
General Information
Description: Liquid.
Formula: $CH_3COOCH[CH_2CH(CH_3)_2]_2$.
Constants: Mol wt: 186.29, mp: $-48.1°$ C, bp: 192.4° C, flash p: 155° F (O.C.), d: 0.8530 at 20°/20° C, vap. d.: 6.42.
Hazard Analysis
Toxicity: Details unknown. See also esters and acetic acid.
Fire Hazard: Moderate, when exposed to heat or flame, can react with oxidizing materials.
Countermeasures
To Fight Fire: Foam, carbon dioxide, dry chemical or carbon tetrachloride (Section 6).
Storage and Handling: Section 7.

NONYL ALCOHOL. See diisobutyl carbinol.

NONYL ALDEHYDE CYANOHYDRIN
General Information
Description: Liquid.
Formula: $(CH_3)_3CCH_2CH(CH_3)CH_2CH(OH)CN$.
Constants: Mol wt: 169, bp: decomposes to HCN, d: 0.8976, vap. d.: 5.83.
Hazard Analysis and Countermeasures
See cyanides.

n-NONYLAMINE
General Information
Synonym: 1-Amino nonane.
Description: Colorless liquid.
Formula: $CH_3(CH_2)_7CH_2NH_2$.
Constants: Mol wt: 143.3, bp: 202.2° C, d: 0.785.
Hazard Analysis
Toxicity: Details unknown. See also amines.
Fire Hazard: Slight, when exposed to heat or flame; can react with oxidizing materials (Section 6).
Countermeasures
Storage and Handling: Section 7.

NONYL BENZENE. See nonyl benzol.

NONYL BENZOL
General Information
Synonym: Nonyl benzene.
Description: Liquid.
Formula: $C_6H_5C_9H_{19}$.
Constants: Mol wt: 204.3, bp: 245° C, flash p: 210° F (C.C.), d: 0.86.
Hazard Analysis
Toxicity: Details unknown. Probably irritant and narcotic in high concentrations.
Fire Hazard: Moderate, when exposed to heat or flame; can react with oxidizing materials.
Countermeasures
To Fight Fire: Foam, carbon dioxide, dry chemical or carbon tetrachloride (Section 6).
Storage and Handling: Section 7.

NONYL CARBINOL. See n-decyl alcohol.

NONYL NAPHTHALENE
General Information
Description: Liquid.
Formula: $C_{10}H_7C_9H_{19}$.
Constants: Mol wt: 254.4, bp: 330° C, flash p: < 200° F, d: 0.94, vap. d: 8.8.
Hazard Analysis
Toxicity: Details unknown. See also naphthalene.
Fire Hazard: Slight, when exposed to heat or flame; can react with oxidizing materials.
Countermeasures
To Fight Fire: Foam, carbon dioxide, dry chemical or carbon tetrachloride (Section 6).
Storage and Handling: Section 7.

NONYL PHENOL
General Information
Description: Clear, viscous, straw-colored liquid.
Formula: $C_6H_4OHC_9H_{19}$.
Constants: Mol wt: 220.34, bp: 290–302° C, pour p: 2° C, flash p: 285° F, d: 0.949 at 20°/4° C, vap. d.: 7.59.
Hazard Analysis
Toxicity: Unknown. See also phenols.
Fire Hazard: Slight, when exposed to heat or flame; can react with oxidizing materials.

TOXIC HAZARD RATING CODE (For detailed discussion, see Section 1.)

0 NONE: (a) No harm under any conditions; (b) Harmful only under unusual conditions or overwhelming dosage.

1 SLIGHT: Causes readily reversible changes which disappear after end of exposure.

2 MODERATE: May involve both irreversible and reversible changes not severe enough to cause death or permanent injury.

3 HIGH: May cause death or permanent injury after very short exposure to small quantities.

U UNKNOWN: No information on humans considered valid by author.

Countermeasures
To Fight Fire: Foam, carbon dioxide, dry chemical or carbon tetrachloride (Section 6).
Storage and Handling: Section 7.

NONYLTRICHLOROSILANE
General Information
Formula: $C_9H_{19}SiCl_3$.
Constants: Mol wt: 261.7, d: 1.072 at 25°C.
Hazard Analysis and Countermeasures
Highly toxic. See chlorosilanes.
Shipping Regulations: Section 11.
  I.C.C.: Corrosive liquid, white label, 10 gallons.
  IATA: Corrosive liquid, white label, not acceptable (passenger), 40 liters (cargo).

NORDIHYDROGUAIARETIC ACID
General Information
Synonyms: NDGA; 4,4'-(2,3-dimethyl tetramethylene) di-pyrocatechol.
Description: Crystals from acetic acid; soluble in methanol, ethanol, and ether; slightly soluble in hot water, chloroform; nearly insoluble in benzene, petroleum ether.
Formula: $[C_6H_3(OH)_2CH_2CH(CH_3)]_2$.
Constants: Mol wt: 304, mp: 184–185°C.
Hazard Analysis
Toxic Hazard Rating: U. Used as a chemical preservative food additive (Section 10).

NORNICOTINE
General Information
Synonym: β-Pyridyl-α-pyrrolidine.
Description: Hygroscopic, viscous liquid.
Formula: $C_9H_{12}N_2$.
Constants: Mol wt: 148.2, bp: 270°C, d: 1.0737 at 20°/4°C.
Hazard Analysis
Toxic Hazard Rating:
  Acute Local: U.
  Acute Systemic: Ingestion 3; Inhalation 3; Skin Absorption 3.
  Chronic Local: U.
  Chronic Systemic: U.
Toxicology: Effects similar to nicotine but less marked.
Disaster Hazard: Dangerous; when heated to decomposition, it emits highly toxic fumes of oxides of nitrogen.
Countermeasures
Ventilation Control: Section 2.
Personnel Protection: Section 3.

First Aid: Section 1.
Storage and Handling: Section 7.

NOVEX. See 2,2'-thiobis(4-chlorophenol).

NOVOBIOCIN
General Information
Description: Light yellow to white colored antibiotic.
Hazard Analysis
Toxic Hazard Rating: U. Used as a food additive permitted in the feed and drinking water of animals and/or for the treatment of food producing animals; also permitted in food for human consumption (Section 10).

"NOVOCAINE." See procaine hydrochloride.

NUISANCE DUSTS
Hazard Analysis
Toxicity: Variable, depending upon composition. Cause local irritation of eyes, nose, throat and lungs. Some may lead to chronic bronchitis, emphysema and bronchial asthma. Dermatitis may result from short contact. Asthma, angioneurotic edema, hives, etc. may result from short periods of inhalation. Atopic eczema, angioneurotic edema, hives, etc. may also result from prolonged contact. See also Section 9. A common air contaminant (Section 4).
TLV: ACGIH (recommended); 50 million particles per cubic foot.
Countermeasures
Ventilation Control: Section 2.

NUPERCAINE   HYDROCHLORIDE. See dibucaine hydrochloride.

NYSTATIN
General Information
Synonym: Fungiciden.
Description: Yellow to light tan powder; odor suggestive of cereals; sparingly soluble in methanol, ethanol, very slightly soluble in water; insoluble in chloroform, ether, and benzene.
Formula: $C_{46}H_{77}NO_{19}$.
Constant: Mol wt: 947.
Hazard Analysis
Toxic Hazard Rating: U. Used as a food additive permitted in the feed and drinking water of animals and/or for the treatment of food producing animals; also permitted in food for human consumption (Section 10).

O

OAT DUST
Hazard Analysis
Toxic Hazard Rating:
  Acute Local: Irritant 1; Allergen 1; Inhalation 1.
  Acute Systemic: U.
  Chronic Local: Allergen 1; Inhalation 1.
  Chronic Systemic: Inhalation 1.
Fire Hazard: Moderate, when exposed to heat or flame (Section 6).
Explosion Hazard: Slight, when exposed to flame.
Countermeasures
Personal Hygiene: Section 3.
Ventilation Control: Section 2.

OBLIVON. See 3-methyl-1-pentyn-3-ol-carbamate.

OCBM. See o-chlorobenzylidene malonitrile.

OCPN. See o-chloro-p-nitroaniline.

OCTACHLORO CAMPHENE
General Information
Description: Yellow, waxy solid.
Formula: $C_{10}H_{10}Cl_8$(approx.).
Constant: Mol wt: 414.2.
Hazard Analysis
Toxic Hazard Rating:
  Acute Local: U.
  Acute Systemic: Ingestion 3; Inhalation 3.
  Chronic Local: U.
  Chronic Systemic: Ingestion 2; Inhalation 2.
Disaster Hazard: Dangerous; when heated to decomposition, it emits highly toxic fumes of chlorides.
Countermeasures
Ventilation Control: Section 2.
Personal Hygiene: Section 3.
Storage and Handling: Section 7.

1,2,4,5,6,7,8,8-OCTACHLORO-4,7-METHANO-3a,4,7,-7a-TETRAHYDROINDANE. See chlordane.

OCTACHLORONAPHTHALENE
General Information
Formula: $C_{10}Cl_8$.
Constant: Mol wt: 403.8.
Hazard Analysis
Toxic Hazard Rating: Toxic.
Disaster Hazard: Dangerous. When heated to decomposition it emits highly toxic fumes.
TLV: ACGIH (tentative); 0.1 milligram per cubic meter of air. May be absorbed via the skin.
Countermeasures
Storage and Handling: Section 7.

OCTACHLOROTETRAHYDRO INDANE. See chlordane.

OCTACHLOROTRISILANE
General Information
Formula: Si_3Cl_8.

Constants: Mol wt: 368.3, mp: $-67°$C, bp: 211.4°C, vap. press.: 1 mm at 46.3°C.
Hazard Analysis
Toxic Hazard Rating:
  Acute Local: Irritant 3; Ingestion 3; Inhalation 3.
  Acute Systemic: U.
  Chronic Local: U.
  Chronic Systemic: U.
Disaster Hazard: Dangerous; when heated to decomposition it emits highly toxic fumes of chlorides; will react with water or steam to produce toxic and corrosive fumes.
Countermeasures
Ventilation Control: Section 2.
Personnel Protection: Section 3.
Storage and Handling: Section 7.

OCTADECANOIC ACID. See stearic acid.

1-OCTADECANOL. See stearyl alcohol.

1-OCTADECENE
General Information
Synonyms: 1-Octadecyne; hexadecyl acetylene.
Description: Crystals.
Formula: $HC:C(CH_2)_{15}CH_3$.
Constants: Mol wt: 250.5, mp: 26°C, bp: 180°C at 15 mm, d: 0.7884 at 20°/4°C.
Hazard Analysis
Toxicity: Unknown.
Fire Hazard: Slight, when exposed to heat or flame; can react with oxidizing materials (Section 6).
Countermeasures
Storage and Handling: Section 7.

OCTADECENYL ALDEHYDE
General Information
Description: Liquid.
Formula: $C_{18}H_{36}O$.
Constants: Mol wt: 266.5, bp: 167°C, d: 0.847, vap. d.: 9.18.
Hazard Analysis
Toxicity: Details unknown. See also aldehydes.
Fire Hazard: Moderate, when exposed to heat or flame; can react with oxidizing materials (Section 6).
Countermeasures
Storage and Handling: Section 7.

OCTADECYL ALCOHOL
General Information
Description: Crystals.
Formula: $CH_3(CH_2)_{16}CH_2OH$.
Constants: Mol wt: 270.5, mp: 58°C, bp: 349.5°C, d: 0.8124 at 59°/4°C, vap. press.: 1 mm at 150.3°C.
Hazard Analysis
Toxicity: Unknown. See alcohols.
Fire Hazard: Moderate, when exposed to heat or flame; can react with oxidizing materials (Section 6).

TOXIC HAZARD RATING CODE (For detailed discussion, see Section 1.)

0 NONE: (a) No harm under any conditions; (b) Harmful only under unusual conditions or overwhelming dosage.

1 SLIGHT: Causes readily reversible changes which disappear after end of exposure.

2 MODERATE: May involve both irreversible and reversible changes; not severe enough to cause death or permanent injury.

3 HIGH: May cause death or permanent injury after very short exposure to small quantities.

U UNKNOWN: No information on humans considered valid by authors.

Countermeasures
Storage and Handling: Section 7.

N-OCTADECYL DISODIUM SULFOSUCCINAMATE
General Information
Description: Cream-colored soft paste.
Formula: $C_{22}H_{41}O_6NSNa_2$.
Constant: Mol wt: 493.
Hazard Analysis
Toxicity: Unknown.
Disaster Hazard: Dangerous; when heated to decomposition, it emits highly toxic fumes of oxides of sulfur and nitrogen.
Countermeasures
Storage and Handling: Section 7.

OCTADECYLISOCYANATE
General Information
Formula: $C_{18}H_{35}NCO$.
Constants: Mol wt: 295.5, mp: 15–18°C, bp: 150–180°C at 0.75 mm, d: 0.86 at 25°C.
Hazard Analysis and Countermeasures
See cyanates.

OCTADECYL TRICHLOROSILANE
General Information
Formula: $C_{18}H_{37}SiCl_3$.
Constants: Mol wt: 268.01, bp: 223°C at 10 mm, d: 0.984 at 25°C.
Hazard Analysis
Toxic Hazard Rating:
  Acute Local: Irritant 3; Ingestion 3; Inhalation 3.
  Acute Systemic: U.
  Chronic Local: U.
  Chronic Systemic: U.
Disaster Hazard: Dangerous; when heated to decomposition, it emits highly toxic fumes of chlorides; will react with water or steam to produce toxic and corrosive fumes.
Countermeasures
Ventilation Control: Section 2.
Personnel Protection: Section 3.
Storage and Handling: Section 7.
Shipping Regulations: Section 11.
  IC.C.: Corrosive liquid, white label, 10 gallons.
  IATA: Corrosive liquid, white label, not acceptable (passenger), 140 kilograms (cargo).

1-OCTADECYNE. See 1-octadecene.

OCTAFLUOROCYCLOBUTANE
General Information
Description: Colorless, nonflammable gas; odorless.
Formula: C_4F_8(cyclic).
Constants: Mol wt: 200, bp: 6.04°C, fp: −41.4°C, d (liquid): 1.513 at 70°F.
Hazard Analysis
Toxic Hazard Rating: U. A food additive permitted in food for human consumption. See section 10.
Shipping Regulations: Section 11.
  IATA: Nonflammable gas, green label, 70 kilograms (passenger), 140 kilograms (cargo).

OCTAFLUOROPROPANE. See perfluoropropane.

"OCTA-KLOR". See chlordane.

"OCTALENE." See aldrin.

"OCTALOX". See dieldrin.

OCTAMETHYL PYROPHOSPHORAMIDE
General Information
Synonyms: OMPA; schradan; pestox and others.
Description: Water-white liquid.
Formula: $C_8H_{24}N_4P_2O_3$.

Constants: Mol wt: 286.34, mp: 20–21°C, bp: 137–142°C at 2 mm, d: 1.137 at 25°/4°C.
Hazard Analysis
Toxic Hazard Rating:
  Acute Local: U.
  Acute Systemic: Ingestion 3; Inhalation 3; Skin Absorption 3.
  Chronic Local: U.
  Chronic Systemic: U.
Toxicology: A cholinesterase inhibitor. See parathion. An insecticide.
Disaster Hazard: Dangerous. See phosphates.
Countermeasures
Ventilation Control: Section 2.
Personnel Protection: Section 3.
First Aid: Section 1.
Storage and Handling: Section 7.

OCTAMETHYL TRISILOXANE
General Information
Description: Liquid.
Formula: $C_8H_{24}O_2Si_3$.
Constants: Mol wt: 236.5, bp: 150.2°C, fp: −80°C, d: 0.82, vap. press.: 1 mm at 7.4°C; 10 mm at 43.1°C.
Hazard Analysis
Toxicity: Details unknown. See also silicones.
Disaster Hazard: Slightly dangerous; when heated to decomposition, it emits acrid fumes.
Countermeasures
Storage and Handling: Section 7.

OCTANAL. See caprylaldehyde.

OCTANE
General Information
Description: Clear liquid.
Formula: $CH_3(CH_2)_6CH_3$.
Constants: Mol wt: 114.23, bp: 125.8°C, lel: 1.0%, uel: 3.2%, fp: −56.5°C, flash p: 56°F, d: 0.7036 at 20°/4°C, autoign. temp.: 428°F, vap. press.: 10 mm at 19.2°C, vap. d.: 3.86.
Hazard Analysis
Toxic Hazard Rating:
  Acute Local: 0.
  Acute Systemic: Inhalation 2.
  Chronic Local: Irritant 1.
  Chronic Systemic: U.
TLV: ACGIH (recommended); 500 parts per million in air; 2350 milligrams per cubic meter of air.
Caution: May act as a simple asphyxiant. See also argon.
Fire Hazard: Dangerous, when exposed to heat or flame.
Explosion Hazard: Severe, when exposed to heat or flame.
Disaster Hazard: Dangerous, upon exposure to heat or flame; can react vigorously with oxidizing materials.
Countermeasures
Ventilation Control: Section 2.
To Fight Fire: Foam, carbon dioxide, dry chemical or carbon tetrachloride (Section 6).
Storage and Handling: Section 7.
Shipping Regulations: Section 11.
  IATA: Flammable liquid, red label, 1 liter (passenger), 40 liters (cargo).

4,5-OCTANEDIOL. See octylene glycol.

OCTANOIC ACID. See caprylic acid.

2-OCTANOL. See capryl alcohol.

2-OCTANONE
General Information
Synonym: Methyl hexyl ketone; hexyl methyl ketone.
Description: Colorless liquid; pleasant odor; slightly soluble in water; soluble in alcohol, hydrocarbons, ether, esters, etc.
Formula: $CH_3CO(CH_2)_5CH_3$.

Constants: Mol wt: 128, d: 0.82 at 20/4°C, mp: −20.9°C, bp: 173.5°C, vap. d.: 4.4, flash p: 160°F.
Hazard Analysis
Toxic Hazard Rating: U. See ketones.
Fire Hazard: Moderate, when exposed to heat or flame.
Countermeasures
Storage and Handling: Section 7.
To Fight Fire: Foam, alcohol foam.

OCTANOYL CHLORIDE. See ethyl hexanoyl chloride.

OCTAPHENYL CYCLOTETRASILOXANE
General Information
Description: Crystals.
Formula: $[(C_6H_5)_2SiO]_4$.
Constants: Mol wt: 793.12, mp: 202°C, bp: 335°C.
Hazard Analysis
Toxicity: Details unknown. See also silicones.
Disaster Hazard: Moderately dangerous; when heated to decomposition, it emits toxic fumes.
Countermeasures
Storage and Handling: Section 7.

OCTAPHENYL TRIGERMANE
General Information
Description: White crystals; insoluble in water.
Formula: $(C_6H_5)_8Ge_3$.
Constants: Mol wt: 834.6, mp: 248°C.
Hazard Analysis
Toxicity: Details unknown. See germanium compounds.

1-OCTENE
Teneral Information
Synonyms: 1-Octylene; 1-caprylene.
Description: Colorless liquid.
Formula: C_8H_{16}.
Constants: Mol wt: 112.21, mp: −101.9°C, bp: 121.27°C, flash p: 70°F (T.O.C.), d: 0.716 at 20°/4°C, vap. d.: 3.87, vap. press.: 36.2 mm at 38°C.
Hazard Analysis
Toxicity: Unknown.
Fire Hazard: Dangerous, when exposed to heat or flame.
Disaster Hazard: Dangerous, upon exposure to heat or flame; can react vigorously with oxidizing materials.
Countermeasures
Ventilation Control: Section 2.
To Fight Fire: Foam, carbon dioxide, dry chemical or carbon tetrachloride (Section 6).
Storage and Handling: Section 7.

2-OCTENE
General Information
Description: Colorless liquid.
Formula: C_8H_{16}.
Constants: Mol wt: 112.21, mp: −94.04°C, bp: 125.2°C, flash p: 70°F (T.O.C.), d: 0.7192 at 20°/4°C, vap. d.: 3.87.
Hazard Analysis
Toxicity: Unknown.
Fire Hazard: Dangerous, when exposed to heat or flame.
Disaster Hazard: Dangerous; upon exposure to heat or flame; can react vigorously with oxidizing materials.
Countermeasures
Ventilation Control: Section 2.
To Fight Fire: Foam, carbon dioxide, dry chemical or carbon tetrachloride (Section 6).
Storage and Handling: Section 7.

OCTIN. See isometheptene.

OCTOIC ANHYDRIDE. See 2-ethyl hexanoic anhydride.

n-OCTYL ACETATE. See 2-ethyl hexyl acetate.

OCTYL ACRYLATE. See 2-ethyl hexyl acrylate.

OCTYL ALCOHOL. See 2-ethyl hexyl alcohol.

OCTYL ALDEHYDE
General Information
Synonyms: 2-Ethylhexanal; caprylaldehyde; octanal.
Description: Liquid.
Formula: $C_7H_{15}CHO$.
Constants: Mol wt: 128.2, bp: 163.4, flash p: 125°F (C.C.), d: 0.821 at 20°/4°C, vap. d.: 4.41.
Hazard Analysis
Toxic Hazard Rating:
  Acute Local: Irritant 2; Ingestion 2; Inhalation 2.
  Acute Systemic: U.
  Chronic Local: U.
  Chronic Systemic: U.
Toxicity: Limited animal experiments suggest low toxicity and moderate irritation. See also aldehydes.
Fire Hazard: Moderate, when exposed to heat or flame; can react with oxidizing materials.
Countermeasures
To Fight Fire: Foam, carbon dioxide, dry chemical or carbon tetrachloride (Section 6).
Ventilation Control: Section 2.
Personnel Protection: Section 3.
Storage and Handling: Section 7.

OCTYLAMINE
General Information
Description: Water-white liquid; amine-like odor.
Formula: $CH_3(CH_2)_7NH_2$.
Constants: Mol wt: 129.3, bp: 170–179°C, flash p: 140°F, d: 0.779 at 20°/20°C, vap. d.: 4.46.
Hazard Analysis
Toxicity: Details unknown. Administration to humans has caused headache, fall in blood pressure, rapid pulse, urticaria, and itching of skin due to release of histamine. See also amines and fatty amines.
Fire Hazard: Moderate, when exposed to heat or flame; can react with oxidizing materials.
Countermeasures
To Fight Fire: Foam, carbon dioxide, dry chemical or carbon tetrachloride (Section 6).
Storage and Handling: Section 7.

tert-OCTYLAMINE
General Information
Description: Clear liquid; amine-like odor.
Formula: $(CH_3)_3CCH_2C(CH_3)_2NH_2$.
Constants: Mol wt: 129.3, bp: 140°C, flash p: 92°F, d: 1.4213, vap. d.: 4.46.
Hazard Analysis
Toxicity: Details unknown. See octylamine.
Fire Hazard: Moderate, when exposed to heat or flame; can react with oxidizing materials.
Countermeasures
To Fight Fire: Water, foam, carbon dioxide, dry chemical or carbon tetrachloride (Section 6).
Storage and Handling: Section 7.

N-OCTYL BICYCLOHEPTENE DICARBOXIMIDE
General Information
Synonym: MGK.

TOXIC HAZARD RATING CODE (*For detailed discussion, see Section 1.*)

0 NONE: (a) No harm under any conditions; (b) Harmful only under unusual conditions or overwhelming dosage.

1 SLIGHT: Causes readily reversible changes which disappear after end of exposure.

2 MODERATE: May involve both irreversible and reversible changes not severe enough to cause death or permanent injury.

3 HIGH: May cause death or permanent injury after very short exposure to small quantities.

U UNKNOWN: No information on humans considered valid by authors.

Hazard Analysis
Toxicity: Details unknown. Large doses can cause CNS stimulation followed by depression.

n-OCTYL CHLORIDE
General Information
Description: Colorless liquid; soluble in most organic solvents; insoluble in water.
Formula: $CH_3(CH_2)_7Cl$.
Constants: Mol wt: 148.5, d: 0.8697 at 25/25°C, fp: −62°C, vap. d.: 5.1, bp: 181.6°C, flash p: 158°F.
Hazard Analysis
Toxic Hazard Rating: U. See esters.
Fire Hazard Rating: Moderate, when exposed to heat or flame.
Disaster Hazard: Dangerous. See chlorides.
Countermeasures
Storage and Handling: Section 7.
To Fight Fire: Foam, alcohol foam.

OCTYL DECYL PHTHALATE
General Information
Description: A clear liquid.
Formula: $C_6H_4(COOC_8H_{17})(COOC_{10}H_{21})$.
Constants: Mol wt: 418.6, bp: 239°C at 4 mm fp: −50°C, flash p: 455°F (C.O.C.), d: 0.980 at 20°/20°C.
Hazard Analysis
Toxicity: Unknown. See phthalic acid.
Fire Hazard: Slight, when exposed to heat or powerful oxidizing agents (Section 6).
Countermeasures
Storage and Handling: Section 7.

1-OCTYLENE. See 1-octene.

OCTYLENE GLYCOL
General Information
Synonyms: 4,5-Octanediol.
Formula: $[CH_3(CH_2)_2CHOH]_2$.
Constants: Mol wt: 146.2, mp: 63°C, bp: 246°C, flash p: 230°F (C.C.), d: 0.94, autoign. temp.: 635°F, vap. d.: 5.05.
Hazard Analysis
Toxicity: Details unknown. See also glycols.
Fire Hazard: Slight, when exposed to heat or flame; can react with oxidizing materials.
Countermeasures
To Fight Fire: Foam, carbon dioxide, dry chemical or carbon tetrachloride (Section 6).
Storage and Handling: Section 7.

OCTYLENE GLYCOL BIBORATE
General Information
Description: A white solid.
Formula: $C_{24}H_{46}O_6B_2$.
Constants: Mp: 69–104°C, mol wt: 454.3, bp: 68–78°C at 0.4 mm.
Hazard Analysis and Countermeasures
See esters and boron compounds.

tert-OCTYL MERCAPTAN
General Information
Description: Liquid; insoluble in water.
Formula: $C_8H_{17}SH$.
Constants: Mol wt: 146, boiling range: 159–166°C, d: 0.848 at 60/60°F, vap. d.: 5.0, flash p: 115°F (O.C.).
Hazard Analysis
Toxic Hazard Rating: U.
Fire Hazard: Moderate, when exposed to heat or flame.
Disaster Hazard: Dangerous. See mercaptans.
Countermeasures
Storage and Handling: Section 7.
To Fight Fire: Foam, alcohol foam.

OCTYL PHENOL
General Information
Synonym: Diisobutyl phenol.
Description: White or light pink flakes.
Formula: $C_6H_4(C_8H_{17})OH$.
Constants: Mol wt: 206.3, bp: 280–283°C, fp: 72–74°C, d: 0.941 at 24°/4°C.
Hazard Analysis
Toxicity: Unknown. See also phenol.
Fire Hazard: Slight, when exposed to heat or flame (Section 6).
Disaster Hazard: Dangerous; when heated to decomposition, it emits highly toxic fumes of phenol; can react with oxidizing materials.
Countermeasures
Storage and Handling: Section 7.

OCTYL SILICATE
General Information
Formula: $Si(OC_8H_{17})_4$.
Constants: Mol wt: 544.3, bp: 204°C at 3 mm, d: 0.8208 at 20°C.
Hazard Analysis
Toxicity: Details unknown; an insecticide.
Fire Hazard: Slight (Section 6).
Countermeasures
Storage and Handling: Section 7.

OCTYL TITANATE. See 2-ethyl butyl titanate.

OCTYL TRICHLOROSILANE
General Information
Description: Fuming liquid.
Formula: $C_8H_{17}SiCl_3$.
Constant: Mol wt: 247.7.
Hazard Analysis
Toxic Hazard Rating:
Acute Local: Irritant 3; Ingestion 3; Inhalation 3.
Acute Systemic: U.
Chronic Local: U.
Chronic Systemic: U.
Disaster Hazard: Dangerous; when heated to decomposition, it emits highly toxic fumes of chlorides; will react with water or steam to produce toxic and corrosive fumes.
Countermeasures
Ventilation Control: Section 2.
Personnel Protection: Section 3.
Storage and Handling: Section 7.
Shipping Regulations: Section 11.
I.C.C.: Corrosive liquid; white label, 10 gallons.
Coast Guard Classification: Corrosive liquid; white label.
IATA: Corrosive liquid, white label, not acceptable (passenger), 40 liters (cargo).

OENANTHYLIC ACID. See heptanoic acid.

OILED CLOTHING, FABRICS, RAGS, ETC.
Hazard Analysis
Fire Hazard: Moderate, when exposed to heat or flame, by spontaneous chemical reaction or on contact with oxidizing materials (Section 6).
Spontaneous Heating: Yes, particularly when wet.
Countermeasures
Storage and Handling: Section 7.
Shipping Regulations: Section 11.
Coast Guard Classification: Hazardous article.

OILED MATERIALS, NOT PROPERLY DRIED (see article No. 1315 in IATA list of restricted articles)
Shipping Regulations: Section 11.
IATA: Flammable solid, not acceptable (passenger and cargo).

OILED MATERIALS, PROPERLY DRIED (see article No. 1316 in IATA list of restricted articles)
Shipping Regulations: Section 11.

IATA: Other restricted articles, Class C, no label required, no limit (passenger and cargo).

OIL GAS
General Information
Description: A gas derived from petroleum.
Composition: Illuminants 4.2%; carbon monoxide 10.4%; hydrogen 47.6%; methane 27.0%; carbon dioxide 4.6%; nitrogen 5.8%; oxygen 0.4%.
Constants: Lel: 4.8%, uel: 32.5%, autoign. temp.: 637° F.
Hazard Analysis
Toxicity: Highly toxic. See carbon monoxide.
Fire Hazard: Dangerous, when exposed to heat or flame.
Spontaneous Heating: No.
Explosion Hazard: Moderate, when exposed to heat or flame.
Disaster Hazard: Dangerous; can react vigorously with oxidizing materials.
Countermeasures
Storage and Handling: Section 7.
To Fight Fire: Carbon dioxide, dry chemical or water spray (Section 6).
Shipping Regulations: Section 11.
   IATA: Flammable gas, red label, not acceptable (passenger), 140 kilograms (cargo).

OIL (MINERAL)
General Information
Synonym: Petrolatum liquid, mineral oil, paraffin oil.
Description: Colorless, oily liquid; practically tasteless and odorless. Insoluble in water and alcohol. Soluble in benzene, chloroform and ether.
Formula: A mixture of liquid hydrocarbons from petroleum.
Constant: D: 0.83–0.86 (light), 0.875–0.905 (heavy), flash p: 444° F.
Hazard Analysis
Toxicology: A laxative. May cause aspiration pneumonia.
TLV: ACGIH (recommended) 5 milligrams per cubic meter of air.

OIL, N.O.S. See crude oil.

OIL OF AMBER. See amber oil.

OIL OF AMERICAN WORMSEED. See oil of chenopodium.

OIL OF BITTER ALMOND
General Information
Description: Colorless to yellow liquid.
Composition: 95% benzaldehyde, 2–4% HCN.
·Constant: D: 1.038–1.060 at 25°/25° C.
Hazard Analysis and Countermeasures
See cyanides.

OIL OF CHENOPODIUM
General Information
Synonym: Oil of American wormseed.
Description: Colorless or pale yellow liquid.
Composition: 60–70% Ascaridol.
Constant: D: 0.950–0.980 at 25°/25° C.
Hazard Analysis
Toxic Hazard Rating:
   Acute Local: Irritant 3; Ingestion 3.
   Acute Systemic: Ingestion 3.
   Chronic Local: U.
   Chronic Systemic: U.
Fire Hazard: Slight (Section 6).
Countermeasures
Personnel Protection: Section 3.

First Aid: Section 1.
Ventilation Control: Section 2.
Storage and Handling: Section 7.

OIL OF CHERRY LAUREL
General Information
Description: Pale yellow liquid.
Composition: Hydrogen cyanide, benzaldehyde, benzaldehyde cyanhydrin, benzyl alcohol.
Constants: D: 1.054–1.066 at 20°/20° C.
Hazard Analysis
Toxicity: Highly toxic. See cyanides.
Disaster Hazard: Dangerous; when heated it emits highly toxic fumes.
Countermeasures
Ventilation Control: Section 2.
Storage and Handling: Section 7.

OIL OF CITRONELLA (CEYLON). See citronella.

OIL OF CRISP MINT. See spearmint oil.

OIL OF EUCALYPTUS
General Information
Description: Colorless to pale yellow liquid.
Composition: Eucalyptol, aldehydes, d-pinene.
Constants: MP: −15.4° C approx., d: 0.905–0.925 at 25°/25° C.
Hazard Analysis
Toxic Hazard Rating:
   Acute Local: Irritant 2; Allergen 1; Ingestion 2; Inhalation 1.
   Acute Systemic: U.
   Chronic Local: Allergen 1; Inhalation 2.
   Chronic Systemic: U.
Disaster Hazard: Slight; when heated to decomposition it emits toxic fumes.
Countermeasures
Ventilation Control: Section 2.
Personal Hygiene: Section 3.
Storage and Handling: Section 7.

OIL OF GARLIC
General Information
Composition: Allyl propyl disulfide; diallyl disulfide.
Constant: D:1.046–1.057 at 15°/15° C.
Hazard Analysis
Toxicity: Details unknown. See individual components.
Disaster Hazard: Dangerous; when heated to decomposition it emits highly toxic fumes of oxides of sulfur.
Countermeasures
Ventilation Control: Section 2.
Storage and Handling: Section 7.

OIL OF MIRBANE. See nitrobenzene.

OIL OF PEPPERMINT
General Information
Description: Colorless to pale yellow liquid.
Constant: D: 0.896–0.908 at 25°/25° C.
Hazard Analysis
Toxic Hazard Rating:
   Acute Local: Irritant 1; Allergen 1.
   Acute Systemic: U.
   Chronic Local: Allergen 1.
   Chronic Systemic: U.
Fire Hazard: Slight (Section 6).
Countermeasures
Personal Hygiene: Section 3.
Storage and Handling: Section 7.

TOXIC HAZARD RATING CODE (For detailed discussion, see Section 1.)

0 NONE: (a) No harm under any conditions; (b) Harmful only under unusual conditions or overwhelming dosage.

1 SLIGHT: Causes readily reversible changes which disappear after end of exposure.

2 MODERATE: May involve both irreversible and reversible changes not severe enough to cause death or permanent injury.

3 HIGH: May cause death or permanent injury after very short exposure to small quantities.

U UNKNOWN: No information on humans considered valid by authors

OIL OF TANSY. See tansy oil.

OIL OF VITRIOL. See sulfuric acid.

OIL WELL SAMPLING DEVICE, CHARGED
Shipping Regulations: Section 11.
    IATA: Flammable gas, red label, not acceptable (passenger), 5 kilograms (cargo).

OIL OF WINTERGREEN. See methyl salicylate.

OIL WELL CARTRIDGES
Shipping Regulations: Section 11.
    I.C.C.: Explosive C, 150 pounds.

OLDHAMITE. See calcium sulfide.

OLEANDOMYCIN
General Information
Synonym: Amimycin; matromycin; romicil.
Description: White amorphous powder; moderately soluble in water; soluble in dilute acids; freely soluble in methanol, ethanol, butanol, acetone; practically insoluble in hexane, carbon tetrachloride, and dibutyl ether.
Formula: $C_{35}H_{63}NO_{12}$.
Constants: Mol wt: 689.8, mp: 134–135°C.
Hazard Analysis
Toxic Hazard Rating: Details unknown. A food additive permitted in food for human consumption (Section 10).

OLEFINS
Hazard Analysis
Toxic Hazard Rating:
    Acute Local: U.
    Acute Systemic: Inhalation 2.
    Chronic Local: U.
    Chronic Systemic: Inhalation 2.
Toxicology: Unsaturated aliphatic hydrocarbons do not differ greatly from paraffins, particularly insofar as their toxic effect on working personnel is concerned. Ethylene and some of its homologs occur in manufactured and natural gases. Ethylene can be used as an anesthetic, and when inhaled in sufficient quantity it can be an asphyxiant. However, the greatest hazard from its use is the danger of fire and explosion. Prolonged or repeated exposures to high concentrations of various olefins have caused certain toxic effects in animals, such as liver damage and hyperplasia of the bone marrow (due to butene-2), but no corresponding effects have been discoverable in human beings as due to industrial exposures. The diolefins, butadiene and isoprene, are more irritating than paraffins or monoolefins of the same volatility. In general, it may be stated that the olefins are comparatively innocuous materials.
Countermeasures
Ventilation Control: Section 2.

OLEIC ACID
General Information
Synonym: Red oil (commercial grade).
Description: Colorless, odorless liquid when pure.
Formula: $C_{17}H_{33}COOH$.
Constants: Mol wt: 282.45, mp: 14°C, bp: 360.0°C, flash p: 372°F (C.C.), d: 0.895 at 25°/25°C, autoign. temp.: 685°F, vap. press.: 1 mm at 176.5°C.
Hazard Analysis
Toxic Hazard Rating:
    Acute Local: Irritant 1.
    Acute Systemic: U.
    Chronic Local: U.
    Chronic Systemic: Ingestion 1.
Note: A substance which migrates to food from packaging materials. Section 10.
Fire Hazard: Slight, when exposed to heat or flame.

Countermeasures
To Fight Fire: Foam, carbon dioxide, dry chemical or carbon tetrachloride (Section 6).
Personal Hygiene: Section 3.
Storage and Handling: Section 7.

OLEO OIL. See tallow oil.

OLEUM
General Information
Synonym: Fuming sulfuric acid.
Description: Heavy, fuming yellow liquid.
Formula: H_2SO_4 and up to 80% SO_3.
Hazard Analysis
Toxic Hazard Rating:
    Acute Local: Irritant 3; Ingestion 3; Inhalation 3.
    Acute Systemic: U.
    Chronic Local: U.
    Chronic Systemic: U.
Fire Hazard: Dangerous, by chemical reaction with reducing agents and carbohydrates (Section 6).
Explosion Hazard: Severe, by chemical reaction with moisture and some organics.
Disaster Hazard: Dangerous; when heated to decomposition, it emits highly toxic fumes of oxides of sulfur; will react with water or steam to produce heat and toxic and corrosive fumes; can react vigorously with reducing materials.
Countermeasures
Ventilation Control: Section 2.
Personnel Protection: Section 3.
First Aid: Section 1.
Storage and Handling: Section 7.
Shipping Regulations: Section 11.
    MCA warning label.
    IATA: Corrosive liquid, white label, not acceptable (passenger), 2½ liters (cargo).
    I.C.C.: Corrosive liquid, white label, 10 pints.

OLIVE OIL
General Information
Synonym: Sweet oil.
Description: Yellow oil.
Constants: Mp: −6°C, flash p: 437°F (C.C.), d: 0.910, autoign. temp.: 650°F.
Hazard Analysis
Fire Hazard: Slight, when exposed to heat or flame; can react with oxidizing materials.
Spontaneous Heating: Some.
Countermeasures
To Fight Fire: Foam, carbon dioxide, dry chemical or carbon tetrachloride (Section 6).
Storage and Handling: Section 7.

"OMILITE"
General Information
Synonym: Polyethylene polysulfide.
Hazard Analysis
Toxicity: Details unknown; a fungicide. See also sulfides.
Fire Hazard: Details unknown. See also sulfides.
Disaster Hazard: Dangerous; when heated to decomposition, it emits highly toxic fumes of oxides of sulfur.
Countermeasures
Storage and Handling: Section 7.

OMPA. See octamethylpyrophosphoramide.

ONB. See o-nitrobiphenyl.

ONION OIL. See allyl propyl disulfide.

ONYXIDE
General Information
Formula: Alkenyl dimethyl ethyl ammonium bromide.
Hazard Analysis
Toxicity: Details unknown. Limited animal experiments suggest low toxicity.

OPIUM
General Information
Synonym: Gum opium.
Description: A habit forming drug.
Hazard Analysis
Toxic Hazard Rating:
  Acute Local: 0.
  Acute Systemic: Ingestion 3.
  Chronic Local: 0.
  Chronic Systemic: Ingestion 3.
Fire Hazard: Slight, when exposed to heat or flame (Section 6).
Countermeasures
Ventilation Control: Section 2.
Personal Hygiene: Section 3.
First Aid: Section 1.
Storage and Handling: Section 7.

ORANGE III. See methyl orange.

ORDEAL BEAN. See physostigma.

ORGANOMETALS
General Information
Description: Compounds containing carbon and a metal. Ordinary metallic carbonates (calcium carbonate, etc.) are excluded and also metallic salts of common organic acids. Examples of organic metal compounds are Grignard compounds such as methyl magnesium iodide (CH_3MgI) and metallic alkyls such as butyllithium (C_4H_9Li).
Hazard Analysis
Toxicity: This group of compounds is constantly growing in importance but there is relatively little toxicological information on most of them. Alkyl compounds of lead, tin, mercury and aluminum are known to be highly toxic. Less is known about other organometals, but for the most part they are highly reactive chemically and therefore dangerous if only on direct contact. Until specific toxicologic data become available, it is prudent to exercise great caution in handling organometals, particularly the alkyl forms.

ORGANIC PHOSPHATE COMPOUND, LIQUID, N.O.S.
Shipping Regulations: Section 11.
  I.C.C.: Poison B, poison label, 1 quart.
  IATA: Poison B, poison label, not acceptable (passenger), 1 liter (cargo).

ORGANIC PHOSPHATE MIXTURES, LIQUID, N.O.S.
Shipping Regulations: Section 11.
  I.C.C.: Poison B, poison label, 1 quart.
  IATA (containing more than 2% organic phosphate): Poison B, poison label, not acceptable (passenger), 95 kilograms (cargo).
    (containing not more than 25% organic phosphate): Poison B, poison label, 1 liter (passenger and cargo).

ORGANIC PHOSPHATE MIXTURES, DRY, N.O.S.
Shipping Regulations: Section 11.
  I.C.C.: Poison B, poison label, 200 pounds.
  IATA (containing more than 2% organic phosphate): Poison B, poison label, not acceptable (passenger), 95 kilograms (cargo).
    (containing not more than 2% organic phosphate): Poison B, poison label, 25 kilograms (passenger), 95 kilograms (cargo).

ORGANIC PHOSPHATE N.O.S., MIXED WITH COMPRESSED GAS
Shipping Regulations: Section 11.
  I.C.C.: Poison A, poison gas label, not accepted.
  IATA: Poison A, not acceptable (passenger and cargo).

ORPIMENT. See arsenic sulfide.

ORRIS
General Information
Synonym: White flag.
Constituents: Iridin, irone, ionone, resin, starch, volatile oil.
Hazard Analysis
Toxic Hazard Rating:
  Acute Local: Allergen 2.
  Acute Systemic: U.
  Chronic Local: Allergen 2.
  Chronic Systemic: U.
Toxicology: Frequently the cause of allergic reactions.

ORTHOFORMIC ESTER. See triethyl-o-formate.

"ORTHOKLOR". See chlordane.

ORTHOXENOL. See o-phenyl phenol.

ORTOL
General Information
Composition: 2 molecules of methyl-o-aminophenol, 1 molecule of hydroquinone. Effects resemble those of phenol toxicity.
Hazard Analysis
Toxicity: See methyl-o-aminophenol and hydroquinone. See also phenol.

OSMIC ACID. See osmium tetraoxide.

OSMIC ACID ANHYDRIDE. See osmium tetraoxide.

OSMIUM AND OSMIUM COMPOUNDS
General Information
Description: White, bluish metal.
Formula: Os.
Constants: At wt: 190.10, mp: 2700°C, bp: above 5300°C, d: 22.48.
Hazard Analysis
Toxic Hazard Rating:
  Acute Local: Irritant 2.
  Acute Systemic: Ingestion 1; Inhalation 1.
  Chronic Local: Irritant 2.
  Chronic Systemic: U.
Toxicology: The principal effects of exposure are ocular disturbances and an asthmatic condition caused by inhalation. Furthermore, it causes dermatitis and ulceration of the skin upon contact (Section 9). When osmium is heated, it gives off a pungent, poisonous fume of osmium tetraoxide. One case of osmium poisoning reported in the literature resulted from the inhalation of osmium tetraoxide which gave rise to a capillary bronchitis and dermatitis. The vapor has a pronounced and nauseating odor which should be taken as a warning of the toxic concentration in the atmosphere, and personnel should immediately remove to an area of fresh air. Osmium compounds, other than the tetraoxide, are probably safe, particularly as ordinarily handled in industry. The metal itself is not highly toxic.
Radiation Hazard: Section 5. For permissible levels, see Table 5, p. 150.

TOXIC HAZARD RATING CODE (For detailed discussion, see Section 1.)

0 NONE: (a) No harm under any conditions; (b) Harmful only under unusual conditions or overwhelming dosage.

1 SLIGHT: Causes readily reversible changes which disappear after end of exposure.

2 MODERATE: May involve both irreversible and reversible changes not severe enough to cause death or permanent injury.

3 HIGH: May cause death or permanent injury after very short exposure to small quantities.

U UNKNOWN: No information on humans considered valid by authors.

Artificial isotope ^{185}Os, half life 94 d. Decays to stable ^{185}Re by electron capture. Emits gamma rays of 0.65, 0.88 MeV and X-rays.

Artificial isotope ^{191m}Os, half life 14 h. Decays to radioactive ^{191}Os by emitting gamma rays of 0.07 MeV.

Artificial isotope ^{191}Os, half life 15 d. Decays to stable ^{191}Ir by emitting beta particles of 0.14 MeV. Also emits gamma rays of 0.04–0.13 MeV.

Artificial isotope ^{193}Os, half life 32 h. Decays to stable ^{193}Ir by emitting beta particles of 0.52–1.13 MeV. Also emits gamma rays of 0.07–0.56 MeV.

Fire Hazard: Moderate, in the form of dust, when exposed to heat or flame (Section 6).

Explosion Hazard: Slight, in the form of dust when exposed to heat or flame.

Countermeasures
Personal Hygiene: Section 3.

OSMIUM DICHLORIDE
General Information
Description: Dark brown, deliquescent crystals.
Formula: $OsCl_2$.
Constants: Mol wt: 261.11, mp: decomposes.
Hazard Analysis and Countermeasures
See osmium, osmium compounds, and chlorides.

OSMIUM DISULFIDE
General Information
Description: Cubic, black crystals.
Formula: OsS_2.
Constants: Mol wt: 254.3, mp: decomposes.
Hazard Analysis and Countermeasures
See osmium and sulfides.

OSMIUM HEXAFLUORIDE
General Information
Description: Green crystals.
Formula: OsF_6.
Constants: Mol wt: 304.20, mp: > 50°C, bp: 205°C.
Hazard Analysis and Countermeasures
See fluorides.

OSMIUM OCTAFLUORIDE
General Information
Description: Yellow crystals.
Formula: OsF_8.
Constants: Mol wt: 342.20, mp: 34.4°C, bp: 47.3°C.
Hazard Analysis and Countermeasures
See fluorides and osmium.

OSMIUM SULFITE
General Information
Description: Blue-black crystals.
Formula: $OsSO_3$.
Constants: Mol wt: 270.3, mp: decomposes.
Hazard Analysis and Countermeasures
See osmium and sulfites.

OSMIUM TETRACHLORIDE
General Information
Description: Red-brown needles.
Formula: $OsCl_4$.
Constants: Mol wt: 332.03, mp: sublimes.
Hazard Analysis and Countermeasures
See hydrochloric acid and osmium.

OSMIUM TETRAFLUORIDE
General Information
Description: Brown powder.
Formula: OsF_4.
Constant: Mol wt: 266.2.
Hazard Analysis and Countermeasures
See fluorides and osmium.

OSMIUM TETRAOXIDE
General Information
Synonym: Osmic acid.
Description: (A) Monoclinic, colorless crystals; (B) Yellow mass, pungent, chlorine-like odor.
Formula: OsO_4.
Constants: Mol wt: 254.20; mp A: 39.5°C, B: 41°C; bp: 130°C sublimes; d: 4.906 at 22°C; vap. press. A: 10 mm at 26.0°C, B: 10 mm at 31.3°C.
Hazard Analysis
Toxic Hazard Rating:
Acute Local: Irritant 3; Inhalation 3.
Acute Systemic: U.
Chronic Local: Irritant 2.
Chronic Systemic: U.
Toxicity: See osmium compounds.
Disaster Hazard: Dangerous; when heated to decomposition, it emits toxic fumes.
TLV: ACGIH (tentative); 0.002 milligrams per cubic meter of air.
Countermeasures
Storage and Handling: Section 7.
Shipping Regulations: Section 11.
IATA: Poison B, poison label, 1 kilogram (passenger), 12 kilograms (cargo).

OSMIUM TETRASULFIDE
General Information
Description: Brown or black crystals.
Formula: OsS_4.
Constants: Mol wt: 318.4, mp: decomposes.
Hazard Analysis and Countermeasures
See sulfides and osmium.

OSMIUM TRICHLORIDE
General Information
Description: Cubic, brown crystals.
Formula: $OsCl_3$.
Constants: Mol wt: 296.57, mp: decomposes at 560–600°C.
Hazard Analysis and Countermeasures
See osmium and chlorides.

OVOTRAN. See p-chlorophenyl-p-chlorobenzene sulfonate.

OVEX. See p-chlorophenyl-p-chlorobenzenesulfonate.

OXALALDEHYDE. See glyoxal.

OXALATES
General Information
Formula: Salts of oxalic acid.
Hazard Analysis
Toxic Hazard Rating:
Acute Local: Irritant 3; Ingestion 3.
Acute Systemic: Ingestion 3.
Chronic Local: Irritant 1.
Chronic Systemic: Ingestion 1.
Toxicology: Oxalates are corrosive and produce local irritation. When taken by mouth they have a caustic effect on the mouth, esophagus and stomach. The soluble oxalates are readily absorbed from the gastro-intestinal tract and can cause severe damage to the kidneys.
Disaster Hazard: Dangerous; when heated to decomposition, they emit toxic fumes.
Countermeasures
Personnel Protection: Section 3.
Ventilation Control: Section 2.
Storage and Handling: Section 7.

OXALIC ACID
General Information
Synonym: Ethanedioic acid.
Description: Transparent, colorless crystals.
Formula: $COOHCOOH \cdot 2H_2O$.
Constants: Mol wt: 126.1, mp: 101°C, 189°C (anh.), bp: sublimes at 150°C, d: 1.653.

Hazard Analysis
Toxic Hazard Rating:
Acute Local: Irritant 3; Ingestion 3; Inhalation 3.
Acute Systemic: Ingestion 3; Inhalation 3.
Chronic Local: Irritant 3; Ingestion 3; Inhalation 3.
Chronic Systemic: Ingestion 2; Inhalation 2.
TLV: ACGIH (recommended); 1 milligram per cubic meter of air.
Toxicology: Acute oxalic poisoning results from ingestion of a solution of the acid. There is marked corrosion of the mouth, esophagus and stomach, with symptoms of vomiting, burning abdominal pain, collapse and sometimes convulsions. Death may follow quickly. The systemic effects are attributed to the removal by the oxalic acid of the calcium in the blood. The renal tubules become obstructed by the insoluble calcium oxalate, and there is profound kidney disturbance. The inhalation of the dust or vapor may cause symptoms of irritation of the upper respiratory tract, gastro-intestinal disturbances, albuminuria, gradual loss of weight, increasing weakness and nervous system complaints. Oxalic acid has a caustic action on the skin and may cause dermatitis (Section 9); a case of early gangrene of the fingers resembling that caused by phenol has been described.

The chief effects of inhalation of the dusts or vapor are irritation of the eyes and upper respiratory tract, ulceration of the mucous membrane of the nose and throat, epistaxis, headache, irritability and nervousness. More severe cases may show albuminura, chronic cough, vomiting, pain in the back and gradual emaciation and weakness. The skin lesions are characterized by cracking and fissuring of the skin and the development of slow-healing ulcers. The skin may be bluish in color, and the nails brittle and yellow.

Countermeasures
Ventilation Control: Section 2.
Personnel Protection: Section 3.
First Aid: Section 1.
Storage and Handling: Section 7.
Shipping Regulations: Section 11.
MCA warning label.
IATA: Poison B, poison label, 25 kilograms (passenger), 95 kilograms (cargo).

OXALIC ETHER. See ethyl oxalate.

OXALIC SALTS, SOLID
Shipping Regulations: Section 11.
IATA: Poison B, poison label, 25 kilograms (passenger), 95 kilograms (cargo).

OXALYL BROMIDE
General Information
Synonym: Ethanedioyl bromide.
Description: A liquid.
Formula: $(COBr)_2$.
Constant: Mol wt: 215.85.
Hazard Analysis
Toxicity: Highly toxic. See oxalic acid.
Disaster Hazard: Dangerous; when heated to decomposition, it emits highly toxic fumes of bromides; will react with water or steam to produce toxic and corrosive fumes.
Countermeasures
Storage and Handling: Section 7.

OXALYL CHLORIDE
General Information
Synonym: Ethanedioyl chloride.
Description: Colorless, fuming liquid.
Formula: $COClCOCl$.
Constants: Mol wt: 126.93, mp: $-12°C$, bp: $64°C$, d: 1.488 at 13°C.
Hazard Analysis
Toxicity: Highly toxic. See oxalic acid.
Disaster Hazard: Dangerous; when heated to decomposition, it emits highly toxic fumes of chlorides; will react with water or steam to produce toxic and corrosive fumes.
Countermeasures
Storage and Handling: Section 7.

OXAMMONIUM. See hydroxylamine.

OX BILE EXTRACT
Hazard Analysis
Toxic Hazard Rating: U. Used as an emulsifying agent food additive (Section 10).

OXIDIZING MATERIALS, N.O.S.
Shipping Regulations: Section 11.
I.C.C.: Oxidizing material; yellow label, 25 pounds.
Coast Guard Classification: Oxidizing material; yellow label.
IATA: Oxidizing material, yellow label, 12 kilograms (passenger and cargo).

OXINDOLE. See 2(3)-indolone.

OXIRANE. See ethylene oxide.

OXOATE. See amiodoxyl benzoate.

2-OXOHEXAMETHYLENIMINE. See ε-caprolactane.

OXOLE. See furan.

OXYBUTYRIC ALDEHYDE. See aldol.

N-OXYDIETHYLENE BENZOTHIAZOLE-2-SULFENAMIDE
General Information
Description: Tan-colored powder.
Formula: $C_6H_4SNCSN(CH_2)_4O$.
Constants: Mol wt: 252.4, mp: 90°C, d: 1.37.
Hazard Analysis
Toxicity: Unknown.
Disaster Hazard: Dangerous; when heated to decomposition, it emits highly toxic fumes of oxides of sulfur and oxides of nitrogen.
Countermeasures
Storage and Handling: Section 7.

β,β'-OXYDIPROPIONITRILE
General Information
Description: Colorless liquid.
Formula: $O(CH_2CH_2CN)_2$.
Constants: Mol wt: 124.1, mp: $-26.3°C$, bp: 172°C at 10 mm, d: 1.041 at 30°C.
Hazard Analysis and Countermeasures
See nitriles.

"OXYFUME" FUMIGANT. See ethylene oxide and carbon dioxide or ethylene oxide and dichloro-difluoromethane.

OXYGEN
General Information
Description: Colorless, odorless, tasteless gas or liquid or hexagonal crystals.
Formula: O_2.
Constants: Mol wt: 32.00, mp: $-218.4°C$, bp: $-183.0°C$, d liquid: 1.14 at $-183.0°C$, solid: 1.426 at $-252.5°C$, vap. d.: 1.429 at $0°C$.
Hazard Analysis
Toxicity: None, as gas. In liquid form it can cause severe "burns" and tissue damage on contact with the skin.
Fire Hazard: Though itself nonflammable, it is essential to combustion. Exclusion of oxygen from the neighborhood of a fire is one of the principle methods of extinguishment. See Section 6, for details.
Explosion Hazard: Liquid oxygen can explode on contact with readily oxidizable material, especially at high temperature.
Disaster Hazard: Compressed oxygen is shipped in steel cylinders under high pressure. If these containers are broken due to shock or exposed to high temperature, an explosion and fire may result.
Countermeasures
Storage and Handling: Section 7.
Shipping Regulations: Section 11.
  I.C.C. (gas and pressurized liquid): Nonflammable gas; green gas label, 300 pounds.
  Coast Guard Classification: Nonflammable gas; green gas label.
  IATA (gas): Nonflammable gas, green label, 70 kilograms (passenger), 140 kilograms (cargo).
  IATA (liquid, nonpressurized): Oxidizing material, not acceptable (passenger and cargo).
    (liquid, pressurized): Nonflammable gas, not acceptable (passenger and cargo).

OXYGEN DIFLUORIDE. See fluorine monoxide.

OXYGEN FLUORIDE. See fluorine monoxide.

OXYGEN, LIQUID. See explosives, high.

OXYISOBUTYRIC NITRILE. See acetone cyanhydrin.

β-OXYMETHYL-β-PYRIDYL PROPIONIC ACID. See ecgonine.

β-OXYNAPHTHOIC ACID. See 3-hydroxy-2-naphthoic acid.

p-OXYPHENYL-β-NAPHTHYLAMINE
Hazard Analysis
Toxic Hazard Rating: Details unknown. Animal experiments suggest low acute toxicity.

8-OXYQUINOLINE BENZOATE
Hazard Analysis
Toxicity: Details unknown; a fungicide.
Fire Hazard: Slight; (Section 6).
Countermeasures
Storage and Handling: Section 7.

OXYQUINOLINE POTASSIUM SULFATE. See quinosol.

OXYSTEARIN
General Information
Description: A mixture of the glycerides of partially oxidized stearic and other fatty acids.
Hazard Analysis
Toxic Hazard Rating: U. A food additive permitted in food for human consumption (Section 10).

OXYTETRACYCLINE
General Information
Description: An antibiotic. Dull, yellow, odorless crystalline powder; soluble in acids and alkalies; very slightly soluble in acetone, alcohol, chloroform, and water; practically insoluble in ether.
Formula: $C_{22}H_{24}N_2O_9 \cdot 2H_2O$.
Constants: Mol wt: 496, mp: $179–182°C$ (decomposes).
Hazard Analysis
Toxicity: Unknown. Used as a food additive permitted in the feed and drinking water of animals and/or for the treatment of food producing animals; also permitted in food for human consumption (Section 10).

OZOKERITE. See ceresin wax.

OZONE
General Information
Description: Colorless gas or dark blue liquid.
Formula: O_3.
Constants: Mol wt: 48.00, mp: $-251°C$, bp: $-111.1°C$, d: gas: 2.144 g/l; 1.71 at $-183°C$.
Toxic Hazard Rating:
  Acute Local: Irritant 3; Inhalation 3.
  Acute Systemic: Inhalation 3.
  Chronic Local: U.
  Chronic Systemic: U.
TLV: ACGIH (recommended); 0.1 part per million in air; 0.2 milligrams per cubic meter of air.
Toxicology: Ozone has a strong irritant action on the upper respiratory system. Concentrations of 0.015 parts of ozone per million parts of air produce a barely detectable odor. Concentrations of 1 ppm produce a disagreeable sulfur-like odor and may cause headache and irritation of the upper respiratory tract which disappear after leaving the exposure. Exposure of guinea pigs to higher concentrations may cause death from lung congestion and edema. No systemic effects have been reported following industrial exposures.
Note: Ozone is a common air contaminant (Section 4).
Fire Hazard: Dangerous, by chemical reaction with reducing agents or combustibles (Section 6).
Explosion Hazard: Severe, when shocked, exposed to heat or flame or by chemical reaction with powerful reducing agents.
Countermeasures
Ventilation Control: Section 2.
Storage and Handling: Section 7.

OZONIZED ETHER
Hazard Analysis
Toxicity: Unknown.
Fire Hazard: Dangerous, when exposed to heat or flame or by chemical reaction with reducing agents (Section 6).
Explosion Hazard: Unknown.
Countermeasures
Storage and Handling: Section 7.

P

22-PB. See 2,2 bis(tert-butylperoxy) butane.

P-4000. See 5-nitro-2-N-propoxyaniline.

P.A.B.A. See p-aminobenzoic acid.

PAINT
General Information
Description: A fluid mixture of a pigment and a vehicle.
Constant: Flash p: 0–80° F.
Hazard Analysis
Toxicity: For lead-base paints, see lead; usually the toxicity is that of the solvent used. Common air contaminants. See Section 4.
Fire Hazard: Dangerous, when exposed to heat or flame.
Spontaneous Heating: Yes.
Explosion Hazard: Unknown.
Disaster Hazard: Dangerous; when heated, the solvent emits acrid fumes; can react vigorously with oxidizing materials.
Countermeasures
Ventilation Control: Section 2.
To Fight Fire: Foam, carbon dioxide, dry chemical or carbon tetrachloride (Section 6).
Storage and Handling: Section 7.
Shipping Regulations: Section 11.
  Coast Guard Classification: Combustible liquid.

PAINT AND GREASE ERADICATORS
General Information
Constant: Flash p: 0–80° F.
Hazard Analysis
Toxicity: Variable.
Fire Hazard: Dangerous, when exposed to heat or flame.
Explosion Hazard: Unknown.
Disaster Hazard: Dangerous, upon exposure to heat or flame; can react vigorously with oxidizing materials.
Countermeasures
To Fight Fire: Foam, carbon dioxide, dry chemical or carbon tetrachloride (Section 6).
Storage and Handling: Section 7.

PAINT SCRAPINGS
Hazard Analysis
Toxicity: Variable.
Fire Hazard: Moderate, when exposed to heat or flame; avoid large unventilated piles. Danger of spontaneous heating depends upon state of dryness (Section 6).
Spontaneous Heating: Moderate.

PALLADIUM AND PALLADIUM COMPOUNDS
Hazard Analysis
Toxicity: Low. This metal, in the form of palladium chloride, has been administered orally in dosage of about 1 grain daily in the treatment of tuberculosis. These amounts resulted in no toxic effects. Applied locally to the skin, palladium chloride shows little or

no irritation. In experimental animals, palladium chloride has been given by intravenous injection, producing damage to bone marrow, liver and kidneys when the dosage was of the order of 0.5 to 1.0 mg per kg of body weight.
Radiation Hazard: Section 5. For permissible levels, see Table 5, p. 150.
  Artificial isotope ^{103}Pd, half life 17 d. Decays to stable ^{103}Rh by electron capture. Emits gamma rays of 0.04, 0.05 MeV and X-rays.
  Artificial iostope ^{109}Pd, half life 14 h. Decays to stable ^{109}Ag by emitting beta particles of 1.02 MeV. Also emits gamma rays of 0.09 MeV.
Fire Hazard: Slight, in the form of dust, when exposed to heat or flame (Section 6).

PALLADIUM BROMIDE
General Information
Description: Red-brown crystals.
Formula: $PdBr_2$.
Constants: Mol wt: 266.53, mp: decomposes.
Hazard Analysis and Countermeasures
See palladium compounds and bromides.

PALLADIUM CHLORIDE
General Information
Synonyms: Palladous chloride; palladium dichloride.
Description: Dark brown crystals; soluble in water, alcohol, acetone, hydrochloric acid.
Formula: $PdCl_2 \cdot 2H_2O$ or $PdCl_2$(anhydrous salt).
Constants: Mol wt: 213.35, 177.35 (anhydrous salt); d: 4.0 at 18°C, mp: 501°C (decomposes).
Hazard Analysis
Toxic Hazard Rating: See palladium.
Disaster Hazard: Dangerous. When heated to decomposition emits highly toxic fumes.
Countermeasures
Storage and Handling: Section 7.

PALLADIUM CYANIDE
General Information
Description: Yellowish-white crystals.
Formula: $Pd(CN)_2$.
Constants: Mol wt: 158.44, mp: decomposes.
Hazard Analysis and Countermeasures
See cyanides.

PALLADIUM DIFLUORIDE
General Information
Description: Brown crystals.
Formula: PdF_2.
Constants: Mol wt: 144.40, mp: volatilizes, bp: decomp. at red heat.
Hazard Analysis and Countermeasures
See fluorides and palladium.

TOXIC HAZARD RATING CODE (*For detailed discussion, see Section 1.*)

0 NONE: (a) No harm under any conditions; (b) Harmful only under unusual conditions or overwhelming dosage.

1 SLIGHT: Causes readily reversible changes which disappear after end of exposure.

2 MODERATE: May involve both irreversible and reversible changes; not severe enough to cause death or permanent injury.

3 HIGH: May cause death or permanent injury after very short exposure to small quantities.

U UNKNOWN: No information on humans considered valid by authors.

PALLADIUM HYDRIDE
General Information
Description: Silver, metallic crystals.
Formula: Pd_2H.
Constants: Mol wt: 213.81, mp: decomposes, d: 10.76.
Hazard Analysis and Countermeasures
See hydrides and palladium.

PALLADIUM IODIDE
General Information
Description: Black powder.
Formula: PdI_2.
Constants: Mol wt: 360.21, mp: decomp. at 350°C.
Hazard Analysis and Countermeasures
See palladium compounds and iodides.

PALLADIUM MONO- AND DISULFIDES
General Information
Description: Brown-black crystals.
Formula: PdS, PdS_2.
Constants: Mol wt: 138.5; 170.5, mp: decomp. at 950°C.
Hazard Analysis and Countermeasures
See sulfides and palladium.

PALLADIUM MONOXIDE
General Information
Synonym: Palladous oxide; palladium oxide.
Description: Black powder; insoluble in water, acids; slightly soluble in aqua regia; soluble in 48% hydrobromic acid.
Formula: PdO.
Constants: Mol wt: 122.40, d: 83, mp: 750°C (decomposes).
Hazard Analysis
See palladium.

PALLADIUM NITRATE
General Information
Description: Rhombic, brown-yellow deliquescent crystals.
Formula: $Pd(NO_3)_2$.
Constants: Mol wt: 230.41, mp: decomposes.
Hazard Analysis and Countermeasures
See nitrates and palladium.

PALLADIUM SELENIDE
General Information
Description: Dary gray crystals.
Formula: $PdSe$.
Constants: Mol wt: 185.36, mp: <960°C.
Hazard Analysis and Countermeasures
See selenium compounds.

PALLADIUM SUBSULFIDE
General Information
Description: Green-gray crystals.
Formula: Pd_2S.
Constants: Mol wt: 244.86, mp: decomp. at 800°C, d: 7.303 at 15°C.
Hazard Analysis and Countermeasures
See sulfides and palladium.

PALM BUTTER. See palm oil.

PALM KERNEL OIL
General Information
Synonym: Palm nut oil.
Constants: Mp: 25.5–30°C, flash p: 398°F (C.C.), d: 0.95.
Hazard Analysis
Toxic Hazard Rating:
  Acute Local: Irritant 1; Allergen 1.
  Acute Systemic: 0.
  Chronic Local: Allergen 1.
  Chronic Systemic: U.
Fire Hazard: Slight; when exposed to heat or flame; can react with oxidizing materials.
Spontaneous Heating: Yes.
Countermeasures
To Fight Fire: Foam, carbon dioxide, dry chemical or carbon tetrachloride (Section 6).

Personal Hygiene: Section 3.
Storage and Handling: Section 7.

PALM NUT OIL. See palm kernel oil.

PALM OIL
General Information
Synonym: Palm butter.
Description: Reddish-yellow to dirty-red, fatty mass; faint odor of violet.
Composition: Palmitin, stearin, linolein.
Constants: Mp: 26.6–43.3°C, flash p: 323°F d: 0.92, autoign. temp.: 600°F.
Hazard Analysis
Toxic Hazard Rating:
  Acute Local: Irritant 1.
  Acute Systemic: 0.
  Chronic Local: Allergen 1.
  Chronic Systemic: U.
Fire Hazard: Slight, when exposed to heat or flame; can react with oxidizing materials.
Spontaneous Heating: Low.
Countermeasures
To Fight Fire: Foam, carbon dioxide, dry chemical or carbon tetrachloride (Section 6).
Personal Hygiene: Section 3.
Storage and Handling: Section 7.

2-PAM
General Information
Synonym: 2-Pyridine aldoxime methiodide.
Description: Crystals; water soluble.
Formula: $C_7H_9IN_2O$.
Constants: Mol wt: 264.1, mp: 214°C.
Hazard Analysis
Toxicity: Not a toxic compound. Has been proven highly effective as an antidote vs. cholinesterase inhibitors of the parathion group.
Disaster Hazard: Dangerous. When heated to decomposition it emits highly toxic fumes.

D-PANTOTHENAMIDE
Hazard Analysis
Toxicity: Unknown. Used as a food additive permitted in food for human consumption (Section 10).

PANTHENOL. See D-pantothenyl alcohol.

PANTOTHENOL. See D-pantothenyl alcohol.

D-PANTOTHENYL ALCOHOL
General Information
Synonyms: Panthenol; pantothenol; 2,4-dihydroxy-N-(3-hydroxypropyl)3,3-dimethyl butyramide.
Description: A viscous liquid; soluble in water, methanol, and ethanol.
Formula: $HOCH_2C(CH_3)_2CHOHCONH(CH_2)_2CH_2OH$.
Constant: Mol wt: 189.
Hazard Analysis
Toxicity: Unknown. Used as a nutrient and or dietary supplement food additive (Section 10).

PAPAIN
General Information
Synonym: Papayotin.
Description: White to gray slightly hygroscopic powder; soluble in water and glycerin; insoluble in other common organic solvents. The most thermostatic enzyme known. Digests protein.
Hazard Analysis
Toxicity: Unknown. Used as a general purpose food additive (Section 10).

PAPAVERINE
General Information
Description: Colorless, rhombic needles.
Formula: $C_{20}H_{21}NO_4$.

Constants: Mol wt: 339.38, mp: 147°C, bp: decomposes, d: 1.337.

Hazard Analysis
Toxic Hazard Rating:
  Acute Local: U.
  Acute Systemic: Ingestion 3.
  Chronic Local: U.
  Chronic Systemic: Ingestion 1; Inhalation 1.
Toxicology: This material is considered to be a comparatively week poison. Its CNS action is about midway between morphine and codeine and even large doses do not produce the amount of excitement caused by codeine or the soporific action of morphine. Small doses are followed by sleep and slow respiration, but these are not increased with larger doses. The heart beat is slowed, but the blood pressure is scarcely affected. It is thought that doses up to 15 grains are nonfatal. See also morphine.
Fire Hazard: Slight (Section 6).
Countermeasures
Personal Hygiene: Section 3.
First Aid: Section 1.
Storage and Handling: Section 7.

PAPAYOTIN. See papain.

PAPER STOCK, WET. See waste paper, wet.

PAPP. See p-aminopropiophenone.

PARABEN. See p-hydroxy benzoic acid esters.

PARADUST. See parathion.

PARAFLOW. See parathion.

PARAFFIN OIL. See oil (mineral).

PARAFFINS
Hazard Analysis
Toxicity: The effects of the paraffin hydrocarbons vary with the volatility. The gaseous hydrocarbons, such as methane, ethane, etc., have but slight anesthetic effects and are hazardous only when present in sufficient concentration to dilute the oxygen to a point below that which is necessary to sustain life. With the volatile liquid hydrocarbons, or with the next higher fraction, the anesthetic action predominates, and with the higher molecular weights or with the less volatile compounds, the anesthetic increases, but at the same time an irritant action becomes more pronounced. For information concerning the toxic and hazardous properties of these materials, see the individual compounds.
  Paraffins are common air contaminants (Section 4).

PARAFFIN WAX
General Information
Description: Colorless or white, somewhat translucent, odorless, mass; greasy feel.
Constants: Mp: 43.3–60°C, bp: >370°C, flash p: 390°F (C.C.), d: 0.9, autoign. temp.: 473°F.
Hazard Analysis
Toxic Hazard Rating:
  Acute Local: 0.
  Acute Systemic: 0.
  Chronic Local: Irritant 1.
  Chronic Systemic: 0.
Fire Hazard: Slight, when exposed to heat or flame; can react with oxidizing materials.
Spontaneous Heating: No.

Countermeasures
To Fight Fire: Carbon dioxide, dry chemical or carbon tetrachloride (Section 6).
Personal Hygiene: Section 3.
Storage and Handling: Section 7.

PARAFORM. See paraformaldehyde.

PARAFORMALDEHYDE
General Information
Synonyms: Polyoxymethylene; paraform; trioxane; trioxymethylene.
Description: White powder.
Formula: $(CH_2O)_x HOH$.
Constants: Mp: 120–160°C (sealed tube), bp: vaporizes, depolymerizing to formaldehyde gas, flash p: 158°F d: 1.39, autoign. temp.: 572°F, vap. press.: 1.45 mm at 25°C.
Hazard Analysis
Toxic Hazard Rating:
  Acute Local: Irritant 2; Ingestion 3; Inhalation 2.
  Acute Systemic: Ingestion 3; Inhalation 2.
  Chronic Local: Irritant 2.
  Chronic Systemic: Inhalation 2.
Toxicology: Action is principally as a primary irritant. Used as a food additive permitted in food for human consumption (Section 10).
Fire Hazard: Moderate, when exposed to heat or flame.
Spontaneous Heating: No.
Disaster Hazard: Moderately dangerous; when heated, it forms formaldehyde gas and oxides of carbon; can react with oxidizing materials.
Countermeasures
Ventilation Control: Section 2.
To Fight Fire: Water, alcohol foam, carbon dioxide, dry chemical or carbon tetrachloride (Section 6).
Personnel Protection: Section 3.
Storage and Handling: Section 7.
  MCA warning label.

PARALDEHYDE
General Information
Synonyms: p-Acetaldehyde; 2,4,6 trimethyl 1,3,5 trioxane
Description: Colorless liquid.
Formula: $\overline{OCH(CH_3)OCH(CH_3OCHCH_3}$.
Constants: Mol wt: 132.16, mp: 12.6°C, lel: 1.3% bp: 124.4°C at 752 mm, flash p: 96°F, d: 0.9943 at 20°/4°C, (O.C.), autoign. temp.: 460°F, vap. d.: 4.55
Hazard Analysis
Toxic Hazard Rating:
  Acute Local: Irritant 1.
  Acute Systemic: Ingestion 2; Inhalation 2.
  Chronic Local: U.
  Chronic Systemic: Ingestion 1; Inhalation 1.
Toxicology: Paraldehyde has hypnotic and analgesic properties. However, poisoning due to it is very rare. There is a wide range between the hypnotic dose and the toxic dose; 2 to 5 grams produces a soporific effect; recovery has been observed following the ingestion of 50 grams. Continued use of the drug has been known to result in habituation. There have been no cases reported of industrial poisoning.
  Signs and symptoms are incoordination and drowsiness, followed by sleep. With larger doses, the pupils dilate, reflexes are lost, and patient lapses into coma. Weak pulse and shallow respiration are followed by cyanosis. Death results from respiratory paralysis.

TOXIC HAZARD RATING CODE (For detailed discussion, see Section 1.)

0 NONE: (a) No harm under any conditions; (b) Harmful only under unusual conditions or overwhelming dosage.

1 SLIGHT: Causes readily reversible changes which disappear after end of exposure.

2 MODERATE: May involve both irreversible and reversible changes not severe enough to cause death or permanent injury.

3 HIGH: May cause death or permanent injury after very short exposure to small quantities.

U UNKNOWN: No information on humans considered valid by authors.

The symptoms of chronic intoxication from this material are disturbances of the digestion, continual thirst, general emaciation, muscular weakness and mental failure followed by tremors of the hands and tongue. It can cause skin eruptions similar to those caused by chloral hydrate (Section 9).

Fire Hazard: Dangerous, when exposed to heat or flame.

Explosion Hazard: Slight, when exposed to heat or flame.

Disaster Hazard: Dangerous; keep away from heat and open flame; emits toxic fumes on heating; can react vigorously with oxidizing materials.

Countermeasures

Ventilation Control: Section 2.

Treatment and Antidotes: When a toxic dose of this material has been ingested, wash out the stomach and give barbituates or morphine to quiet the nerves. Call a physician.

Personnel Protection: Section 3.

To Fight Fire: Alcohol foam, carbon dioxide, dry chemical or carbon tetrachloride (Section 6).

Storage and Handling: Section 7.

Shipping Regulations: Section 11.

  Coast Guard Classification: Combustible liquid.
  MCA warning label.

PARAMINOL. See p-aminobenzoic acid.

PARAMORPHINE. See thebaine.

PARA-OXON. See diethyl-p-nitrophenyl phosphate.

PARAPERIODIC ACID. See periodic acid.

PARAQUAT
General Information
Synonyms: 1,1'-Dimethyl-4,4'-dipyridium dichloride; 1,1'-dimethyl-4,4'-bipyridium salt.
Description: Yellow solid; soluble in water.
Formula: $[CH_3(C_5H_4N)_2CH_3] \cdot 2CH_3SO_4$.
Constant: Mol wt: 408.
Hazard Analysis
Toxic Hazard Rating:
  Acute Local: Inhalation 3.
  Acute Systemic: Ingestion 3; Inhalation 3.
  Chronic Local: U.
  Chronic Systemic: U.
Toxicity: Has caused fatal poisoning in human with severe injury to lungs.
TLV: ACGIH (tentative), 0.5 milligrams per cubic meter of air. May be absorbed via the skin.

PARASEPT. See p-hydroxybenzoic acid esters.

PARASPRAY. See parathion.

PARATHION
General Information
Synonyms: O,O-Diethyl O-p-nitrophenyl thiophosphate; "Alkron"; compound 3422; DNTP; DPP; E-605; genithion; niran; paradust; paraflow; paraspray; parawet; penphos; phos-kil; thiophos; vapophor and many others.
Description: Yellowish liquid.
Formula: $C_{10}H_{14}NO_5PS$.
Constants: Mol wt: 291.3, bp: 375°C, d: 1.26 at 25°/4°C.
Hazard Analysis
Toxic Hazard Rating:
  Acute Local: U.
  Acute Systemic: Ingestion 3; Inhalation 3; Skin Absorption 3.
  Chronic Local: U.
  Chronic Systemic: Ingestion 3; Inhalation 3; Skin Absorption 3.
Toxicology: Parathion, like the other organic phosphorus poisons, acts as an irreversible inhibitor of the molecules of the enzyme cholinesterase and thus allows the accumulation of large amounts of acetylcholine. When a critical level of cholinesterase depletion is reached, grave symptoms appear. Whether death is acutally caused entirely by cholinesterase depletion or by the disturbance of a number of enzymes is not yet known. Recovery is apparently complete if a poisoned animal or man has time to reform his critical quota of cholinesterase. However, if a second small dose is administered before recovery from the first is complete, the effect is partially additive.

Dangerous Acute Dose in Man: Death has followed splashing of the body and clothing of one worker with technical parathion (approximately 95 percent pure). The amount was sufficiently small that the worker was not soaked or at any rate did not follow the simple instructions for changing clothes and bathing. Two operators have died after rather extensive skin contact with agricultural sprays and the inhalation of dust from 25-percent water-wettable powder. Others have died after what was apparently almost entirely respiratory exposure to 25-percent dust. In no single case of fatal human poisoning by parathion has it been possible to determine the exact dosage which the victim received either on the day on which symptoms appeared or subacutely over a period of days or weeks. However, on the basis of animal experiments the acute lethal oral dose for man has been estimated at 10 to 20 mg (1/5 to 1/3 grains). Based again on animal experiments the acute lethal dose by the dermal route would be only about three times as great.

Dangerous Chronic Dose in Man: Exposure to parathion reduces the cholinesterase level and the organism exposed remains susceptible to relatively low dosages of parathion until the cholinesterase level has regenerated. Small doses at frequent intervals are, therefore, more or less additive. There is, however, at the present time no indication that, when recovery from a given exposure is entirely complete, the exposed organism is prejudiced in any way.

Symptoms of Poisoning: Headache, giddiness, blurred vision, weakness, nausea, cramps, diarrhea, and discomfort in the chest, sweating, miosis, tearing, salivation, pulmonary edema, cyanosis, papilledema, convulsions, coma, and loss of reflexes and sphincter control. The last several signs are seen only in advance cases but do not preclude a favorable outcome if energetic treatment is continued. Poisoned animals show various degrees of heart block and cardiac arrest may occur. In most cases of accidental, chiefly dermal exposure, relatively incapacitating symptoms of nausea, cramps, muscular twitching, etc., follow the initial giddiness, blurred vision, and pinpoint pupils only after a period of 2 to 8 hours. In one such case, the onset of serious symptoms was apparently more rapid. Following oral ingestion associated with murder or suicide, death has been essentially instantaneous.

Laboratory Findings: Laboratory findings are essentially normal except that by special techniques the cholinesterase level of the blood or serum may be shown to be greatly reduced. At autopsy, the same may be demonstrated for the cholinesterase level of the brain or other tissues provided fresh, unfixed tissue is employed.

The mean cholinesterase values of normal persons living without exposure to organic phosphorus insecticides have been found by various workers, using the Michel method, to range as follows:

Red blood cell	0.67–0.86	pH units/hr.
Plasma	0.70–0.97	pH units/hr.

It is believed that cholinesterase values of 0.5 or less for either cells or plasma represent abnormal depressions for most individuals. Nevertheless, people may experience far greater depressions (to 0.2 or less) without the onset of clinical signs or symptoms; this is especially true of workers who are exposed daily over a period of weeks but whose exposure at any one time is kept at minimum.

Treatment for poisoning: Keep the patient fully atropinized. Give 2 to 4 milligrams and repeat until signs of atropinization appear. The intravenous route is most rapid. The dosage of atropine is greater than that conventionally employed for other purposes but is within safe limits. Atropine relieves many of the distressing symptoms, reduces heart block, and dries secretions of the respiratory tract. Never give morphine, theophylline, or theophylline-ethylene-diamine (*Aminophylline*). 2-PAM, alone or with atropine, is also an effective antidote.

If the patient has not yet shown symptoms or they have been allayed by the first dose of atropine, he must be completely and quickly decontaminated. Wearing rubber gloves, remove the patient's clothing and, with due regard for his condition at the moment, bathe him thoroughly with soap and water. If washing soda is available, use it for parathion is hydrolyzed more rapidly in the presence of alkali. Any relatively mild alkali may be used.

If there is any suspicion that parathion has been ingested, induce vomiting, give some neutral material such as milk or water and induce vomiting again. Nausea may, of course, be anticipated on the basis of the systemic action of parathion but if vomiting is not profuse, gastric lavage may be used. If the pulmonary secretions have accumulated before atropine has become effective, the patient must be turned upside down or in some other positions of postural drainage in order to drain out mucus. Use suction and a catheter if necessary. If the stomach is distended, empty it with a Levine tube. Atropine does not protect against muscular weakness. The mechanism of death appears to be respiratory failure. The use of an oxygen tent or even the use of oxygen under slight positive pressure is advisable and should be started early. Watch the patient *constantly, for the need of artificial respiration may appear suddenly.* Equipment for artificial respiration should be placed by the patient's bed in readiness as soon as he is hospitalized. Cyanosis (anoxia) should be prevented by the most suitable means, since it aggravates the other signs of poisoning. Complete recovery may occur even after many hours of artificial respiration have been necessary. The acute emergency lasts 24 to 48 hours, and the patient must be *watched continuously* during that time. Favorable response to one or more doses of atropine does not guarantee against sudden and fatal relapse. Medication must be continued during the entire emergency. Following exposure heavy enough to produce symptoms, further organic phosphorous insecticide exposure of any sort should be avoided. The patient remains susceptible to relatively small exposures of parathion until regeneration of cholinesterase is complete or nearly so. Persons exposed to other organic phosphorous insecticides before complete recovery from a previous exposure are made more susceptible and vice-versa.

TLV: ACGIH (recommended) 0.1 milligrams per cubic meter of air. Can be absorbed through the intact skin.

Disaster Hazard: Highly dangerous; shock can shatter the container, releasing the contents when heated to decomposition, it emits highly toxic fumes of nitrogen oxides, phosphorus and sulfur.

Countermeasures
Ventilation Control: Section 2.
Personnel Protection: Section 3.
First Aid: Section 1.
Storage and Handling: Section 7.
Shipping Regulations: Section 11.
   Coast Guard Classification: Poison B; poison label.
   MCA warning label.
   ICC: Poison B, poison label, 1 quart.
   IATA: Poison B, poison label, not acceptable (passenger), 1 liter (cargo).

PARATHION AND COMPRESSED GAS MIXTURES. See parathion.
Shipping Regulations: Section 11.
   I.C.C.: Poison A; poison gas label, not accepted.
   Coast Guard Classification: Poison A; poison gas label.
   IATA: Poison A, not acceptable (passenger and cargo).

PARATHION MIXTURE, DRY. See parathion.
Shipping Regulations: Section 11.
   ICC: Poison B, poison label, 200 pounds.
   IATA (containing more than 2% parathion): Poison B, poison label, not acceptable (passenger), 95 kilograms (cargo).
   IATA (containing not more than 2% parathion): Poison B, poison, 25 kilograms (passenger), 95 kilograms (cargo).

PARATHION MIXTURE, LIQUID. See parathion.
Shipping Regulations: Section 11.
   ICC: Poison B, poison label, 1 quart.
   IATA (containing more than 25% parathion): Poison B, poison label, not acceptable (passenger), 1 liter (cargo).
   IATA (containing not more than 25% parathion): Poison B, poison label, 1 liter (passenger and cargo).

PARAWET. See parathion.

PARAZOL. See dinitrodichlorobenzene.

PARIS GREEN. See cupric acetate m-arsenate.

PARMELACONITE. See copper oxide.

PAS. See p-aminosalicylic acid.

PCMC. See p-chloro-m-cresol.

PDU. See 3-phenyl-1,1-dimethyl urea.

PE. See polyethylene.

PEANUT OIL
General Information
Synonyms: Katchung oil; earthnut oil.
Description: Straw-yellow to greenish-yellow or nearly colorless oil; nutty odor and bland taste.
Constants: Mp: 2.7° C, flash p: 540° F, d: 0.92, autoign. temp.: 883° F.
Hazard Analysis
Toxic Hazard Rating:
   Acute Local: Irritant 1; Allergen 1.
   Acute Systemic: 0.
   Chronic Local: Irritant 1; Allergen 1.
   Chronic Systemic: U.
Note: A substance which migrates to food from packaging materials (Section 10).
Fire Hazard: Slight, when exposed to heat or flame; can react with oxidizing materials.
Spontaneous Heating: Slight.

TOXIC HAZARD RATING CODE (*For detailed discussion, see Section 1.*)

0 NONE: (a) No harm under any conditions; (b) Harmful only under unusual conditions or overwhelming dosage.

1 SLIGHT: Causes readily reversible changes which disappear after end of exposure.

2 MODERATE: May involve both irreversible and reversible changes; not severe enough to cause death or permanent injury.

3 HIGH: May cause death or permanent injury after very short exposure to small quantities.

U UNKNOWN: No information on humans considered valid by authors.

Countermeasures
To Fight Fire: Foam, carbon dioxide, dry chemical or carbon tetrachloride (Section 6).
Personal Hygiene: Section 3.
Storage and Handling: Section 7.

PEANUT RED SKIN
General Information
Description: A red skin-like layer between nut and shell.
Hazard Analysis
Toxic Hazard Rating:
Acute Local: Irritant 1; Allergen 1; Inhalation 1.
Acute Systemic: Ingestion 1; Inhalation 1.
Chronic Local: Irritant 1; Allergen 1; Inhalation 1.
Chronic Systemic: Ingestion 1; Inhalation 1.
Fire Hazard: Moderate, by chemical reaction, can react with oxidizing materials (Section 6).
Spontaneous Heating: High.
Countermeasures
Ventilation Control: Section 2.
Personal Hygiene: Section 3.
Storage and Handling: Section 7.

PEARL ASH. See potassium carbonate.

PEAR OIL. See amyl acetate.

PELARGONIC ACID. See ethyl heptanoic acid.

PELARGONIC MORPHOLIDE
Hazard Analysis
Toxic Hazard: Rating:
Acute Local: Irritant 3; Inhalation 3.
Acute Systemic: U.
Chronic Local: U.
Chronic Systemic: U.
Toxicology: Human and animal observations show intense irritation to mucous membranes.

PELIGOT'S SALT. See potassium chlorochromate.

PENETEK. See pentaerythritol, technical.

PENICILLIN
General Information
Description: A group of isomeric and closely related antibiotic compounds with outstanding bacterial activity.
Formula: $(CH_3)_2C_5H_3NSO(COOH)NHCOOR$ (bicyclic).
Hazard Analysis
Toxicity: U. Used as a food additive permitted in food for human consumption (Section 10).

PENPHOS. See parathion.

PENTABORANE, STABLE
General Information
Synonyms: Pentaboron enneahydride.
Description: Colorless gas or liquid; bad odor.
Formula: B_5H_9.
Constants: Mol wt: 63.2, mp: $-46.6°C$, bp: $0°C$ at 66 mm, d: 0.61 at $0°C$, vap. d.: 2.2, vap. press.: 66 mm at $0°C$.
Hazard Analysis
Toxicity: See boron hydrides.
TLV: ACGIH (recommended), 0.005 parts per million in air; 0.01 milligrams per cubic meter of air.
Fire Hazard: Dangerous, by chemical reaction; spontaneously flammable in air (Section 6).
Explosion Hazard: Dangerous. Details unknown.
Disaster Hazard: Dangerous; on decomposition, it emits toxic fumes and can react vigorously with oxidizing materials.
Countermeasures
To Fight Fire: Water is ineffective; reacts violently with halogenated extinguishing agents.
Ventilation Control: Section 2.

Storage and Handling: Section 7.
Shipping Regulations: Section 11.
I.C.C.: Flammable liquid; red label.
Coast Guard Classification: Inflammable liquid; red label.
MCA warning label.

PENTABORANE, UNSTABLE
General Information
Synonyms: Dihydropentaborane.
Description: Colorless liquid; turns yellow on standing.
Formula: B_5H_{11}.
Constants: Mol wt: 65.2, bp: $63°C$.
Hazard Analysis
Toxicity: See boron hydrides.
TLV: ACGIH (recommended); 0.005 parts per million in air; 0.01 milligrams per cubic meter of air.
Fire Hazard: Dangerous, by chemical reaction; ignites spontaneously in air (Section 6).
Explosion Hazard: Dangerous. Details unknown.
Disaster Hazard: Dangerous; on decomposition, it emits highly toxic fumes; can react vigorously with oxidizing materials.
Countermeasures
Storage and Handling: Section 7.
Shipping Regulations: Section 11.
I.C.C.: Flammable liquid; red label, not accepted.
IATA: Flammable liquid, not acceptable (passenger and cargo).

PENTABORANIDE. See potassium pentaborane.

PENTABORON ENNEAHYDRIDE. See pentaborane, stable.

"PENT-ACETATE"
General Information
Composition: A mixture of amyl acetates and amyl alcohols.
Description: Colorless liquid; agreeable odors.
Constants: Bp: $125°C$, flash p: $98°F$, d: 0.864, autoign. temp.: $110°F$.
Hazard Analysis
Toxicity: Moderate, see amyl alcohol and amyl acetate.
Fire Hazard: Moderate, when exposed to heat or flame; can react with oxidizing materials.
Countermeasures
To Fight Fire: Foam, carbon dioxide, dry chemical or carbon tetrachloride (Section 6).
Ventilation Control: Section 2.
Storage and Handling: Section 7.
Shipping Regulations: Section 11.
Coast Guard Classification: Combustible liquid.

PENTACHLOROETHANE
General Information
Synonym: Pentalin.
Description: Colorless liquid.
Formula: $CHCl_2CCl_3$.
Constants: Mol wt: 202.3, mp: $-29°C$, bp: $162°C$, d: 1.6728 at $25°/4°C$.
Hazard Analysis
Toxic Hazard Rating:
Acute Local: Irritant 2.
Acute Systemi: Inhalation 3.
Chronic Local: U.
Chronic Systemic: Ingestion 3; Inhalation 3.
Fire Hazard: Moderate, when exposed to heat or flame.
Explosion Hazard: Moderate, by spontaneous chemical reaction. Dehalogenation by reaction with alkalies, metals, etc., will produce spontaneously explosive chloroacetylenes.
Disaster Hazard: Dangerous; when heated to decomposition, it emits highly toxic fumes of chlorides.
Countermeasures
Ventilation Control: Section 2.
To Fight Fire: Water, carbon dioxide, dry chemical or carbon tetrachloride (Section 6).

Personnel Protection: Section 3.
Firet Aid: Section 1.
Storage and Handling: Section 7.
Shipping Regulations: Section 11.
    IATA: Poison B, poison label, 1 liter (passenger), 220 liters (cargo).

PENTACHLOROETHYL BENZENE
General Information
Description: White, crystalline solid.
Formula: $Cl_5C_6CH_2CH_3$.
Constants: Mol wt: 278.4, mp: 53.4°C, bp: 305°C, d: 1.552 at 60°/25°C, vap. press.: 1 mm at 96.2°C, vap. d.: 9.6.
Hazard Analysis
Toxicity: Details unknown, See chlorinated hydrocarbons, aromatic.
Disaster Hazard: Dangerous; when heated to decomposition, it emits highly toxic fumes of chlorides.
Countermeasures
Storage and Handling: Section 7.

PENTACHLOROMETHYL ETHER
General Information
Description: Liquid.
Formal: CCl_3OCHCl_2.
Constants: Mol wt: 218.32, bp: 158.5 to 159.5°C, d: 1.64 at 20°C.
Hazard Analysis
Toxic Hazard Rating:
    Acute Local: Irritant 3; Ingestion 3; Inhalation 3.
    Acute Systemic: U.
    Chronic Local: U.
    Chronic Systemic: U.
Fire Hazard: Moderate, when exposed to heat or flame.
Explosion Hazard: Unknown.
Disaster Hazard: Dangerous; when heated to decomposition, it emits highly toxic fumes of chlorides; can react with oxidizing materials.
Countermeasures
Ventilation Control: Section 2.
To Fight Fire: Water, foam, carbon dioxide, dry chemical or carbon tetrachloride (Section 6).
Personnel Protection: Section 3.
Storage and Handling: Section 7.

PENTACHLORONAPHTHALENE
General Information
Description: White solid.
Formula: $C_{10}H_3Cl_5$.
Constant: Mol wt: 300.41.
Hazard Analysis
Toxic Hazard Rating:
    Acute Local: Irritant 2.
    Acute Systemic: Ingestion 3; Inhalation 3.
    Chronic Local: Irritant 2.
    Chronic Systemic: Ingestion 3; Inhalation 3.
TLV: ACGIH (recommended); 0.5 milligrams per cubic meter of air. May be absorbed via intact skin.
Toxicology: Action similar to chlorinated naphthalanes and chlorinated diphenyls. See also chlorinated diphenyl.
Disaster Hazard: Dangerous; when heated to decomposition, it emits highly toxic fumes of chlorides.
Countermeasures
Ventilation Control: Section 2.
Personnel Protection: Section 3.
Storage and Handling: Section 7.

PENTACHLORONITROBENZENE
General Information
Description: A solid.
Formula: $C_6Cl_5NO_2$.
Constant: Mol wt: 295.4.
Hazard Analysis
Toxicity: Details unknown; probably quite toxic; an insecticide. See nitro compounds of aromatic hydrocarbons.
Disaster Hazard: Dangerous; when heated to decomposition, it emits highly toxic fumes of chlorides and oxides of nitrogen.
Countermeasures
Storage and Handling: Section 7.

PENTACHLOROPHENOL
General Information
Description: Dark-colored flakes and sublimed needle crystals with a characteristic odor.
Formula: Cl_5C_6OH.
Constants: Mol wt: 266.4, mp: 191°C, bp: 310°C (decomposes), d: 1.978, vap. press.: 40 mm at 211.2°C.
Hazard Analysis
Toxic Hazard Rating:
    Acute Local: Irritant 3; Ingestion 3; Inhalation 3.
    Acute Systemic: Ingestion 3; Inhalation 3; Skin Absorption 3.
    Chronic Local: Irritant 2.
    Chronic Systemic: Ingestion 2; Inhalation 2; Skin Absorption 2.
TLV: ACGIH (recommended); 0.5 milligrams per cubic meter of air. May be absorbed via the intact skin.
Toxicology: Acute poisoning is marked by weakness, convulsions and collapse. Chronic exposure can cause liver and kidney injury. See also phenols. A fungicide.
Disaster Hazard: Dangerous; when heated to decomposition, it emits highly toxic fumes of chlorides.
Countermeasures
Ventilation Control: Section 2.
Personnel Protection: Section 3.
First Aid: Section 1.
Storage and Handling: Section 7.
MCA warning label required.
Shipping Regulations: Section 11.
    IATA: Other restricted article, class A, no label required, no limit (passenger and cargo).

3-PENTADECYL CATECHOL
General Information
Synonym: Tetrahydrourushiol.
Description: Crystals; soluble in alcohol, ether, benzene.
Formula: $C_{21}H_{36}O_2$.
Constants: Mol wt: 320.5, mp: 60°C.
Hazard Analysis
Toxicity: Details unknown. Causes a dermatitis similar to that of poison ivy.

3-PENTADECYL PHENOL
General Information
Description: Flaked pink to white waxy solid.
Formula: $C_{15}H_{31}C_6H_4OH$.
Constants: Mol wt: 304.5, mp: 49°–51°C, d: < 1, br: 190–195°C, at 1 mm.
Hazard Analysis
Toxicity: Details unknown. See also phenol.
Fire Hazard: Slight, when exposed to heat or flame (Section 6).

TOXIC HAZARD RATING CODE (For detailed discussion, see Section 1.)

0 NONE: (a) No harm under any conditions; (b) Harmful only under unusual conditions or overwhelming dosage.

1 SLIGHT: Causes readily reversible changes which disappear after end of exposure.

2 MODERATE: May involve both irreversible and reversible changes; not severe enough to cause death or permanent injury.

3 HIGH: May cause death or permanent injury after very short exposure to small quantities.

U UNKNOWN: No information on humans considered valid by authors.

Disaster Hazard: Dangerous; when heated to decomposition, it emits acrid fumes; can react with oxidizing materials.
Countermeasures
Storage and Handling: Section 7.

5-PENTADECYL RESORCINOL
General Information
Description: Pale amber to white crystalline solid.
Formula: $C_{15}H_{31}C_6H_3(OH)_2$.
Constants: Mol wt: 320.5, mp: 91–93°C, bp: 220–225°C at 1 mm, d: >1.
Hazard Analysis
Toxicity: Details unknown; a weak allergen.
Fire Hazard: Slight, when exposed to heat or flame; can react with oxidizing materials (Section 6).
Countermeasures
Storage and Handling: Section 7.

PENTAERYTHRITOL DIACETAL
General Information
Description: White flakes.
Formula: $C(CH_2O)_4(CHCH_3)_2$.
Constants: Mol wt: 188.22, mp: 42°C, bp: 216°C.
Hazard Analysis
Toxicity: Unknown. See aldehydes.
Fire Hazard: Moderate, when exposed to heat or flame; can react with oxidizing materials(Section 6).
Countermeasures
Storage and Handling: Section 7.

PENTAERYTHRITOL DIBUTYRAL
General Information
Description: White flakes.
Formula: $C(CH_2O)_4(CHC_3H_7)_2$.
Constants: Mol wt: 244.3, mp: 59°C, bp: 290°C.
Hazard Analysis
Toxicity: Unknown. See aldehydes.
Fire Hazard: Slight, when exposed to heat or flame; can react with oxidizing materials (Section 6).
Countermeasures
Storage and Handling: Section 7.

PENTAERYTHRITOL DIFORMAL
General Information
Description: White flakes.
Formula: $C(CH_2O)_4(CH_2)_2$.
Constants: Mol wt: 160.17, mp: 50°C, bp: 80–83°C at 1 mm.
Hazard Analysis
Toxicity: Unknown. See aldehydes.
Fire Hazard: Slight, when exposed to heat or flame; can react with oxidizing materials (Section 6).
Countermeasures
Storage and Handling: Section 7.

PENTAERYTHRITOL DIPROPIONAL
General Information
Description: Slightly viscous, colorless liquid.
Formula: $C(CH_2O)_4(CHC_2H_5)_2$.
Constants: Mol wt: 216.3, mp: 19°C, bp: 259°C, vap. d.: 7.47.
Hazard Analysis
Toxicity: Unknown. See aldehydes.
Fire Hazard: Slight, when exposed to heat or flame; can react with oxidizing materials (Section 6).
Countermeasures
Storage and Handling: Section 7.

PENTAERYTHRITOL, TECHNICAL
General Information
Synonyms: Pentek; penetek.
Description: Crystalline material.
Formula: $C(CH_2OH)_4$.
Constants: Mol wt: 136.1, mp: 262°C, d: 1.38 at 25°/4°C.
Hazard Analysis
Toxicity: Unknown.

Fire Hazard: Moderate, when exposed to heat or flame; can react with oxidizing materials (Section 6).
Countermeasures
Storage and Handling: Section 7.

PENTAERYTHRITOL TETRANITRATE
General Information
Synonym: PETN.
Description: Crystals.
Formula: $C(CH_2NO_3)_4$.
Constants: Mol wt: 316.55, mp: 138–140°C, bp: explodes at 205–215°C, d: 1.773 at 20°/4°C.
Hazard Analysis
Toxic Hazard Rating:
  Acute Local: U.
  Acute Systemic: Ingestion 2; Inhalation 2; Skin Absorption 2.
  Chronic Local: U.
  Chronic Systemic: U.
Toxicology: Effects are similar to nitroglycerine, i.e., headache, weakness, and fall in blood pressure.
Fire Hazard: See nitrates.
Explosion Hazard: Severe, when shocked or exposed to heat. One of the most powerful high explosives, it is particularly sensitive to shock. It is used in detonating and priming compositions, as a base-charge in anti-aircraft shells and mixed with TNT (70–30) in mines, explosive bombs and torpedoes. It is a very effective demolition explosive. It is also used in blasting caps combined with lead azide and diazodinitrophenol. It explodes at 215°C (see explosives, high).
Disaster Hazard: Highly dangerous; shock or heat will explode it; on decomposition, it emits highly toxic fumes of oxides of nitrogen; can react vigorously with oxidizing materials.
Countermeasures
Ventilation Control: Section 2.
Personnel Protection: Section 3.
Storage and Handling: Section 7.
Shipping Regulations: Section 11.
  I.C.C.: An explosive; not accepted by express. Initiating explosive label.

PENTAFLUOROPROPIONIC ACID
General Information
Formula: $F_5C_3HO_2$.
Constant: Mol wt: 164.
Hazard Analysis
Toxic Hazard Rating:
  Acute Local: Irritant 3; Ingestion 3; Inhalation 3.
  Acute Systemic: Ingestion 3; Inhalation 3.
  Chronic Local: U.
  Chronic Systemic: Inhalation 2.
Toxicity: Data based on animal experiments.
Disaster Hazard: Dangerous; when heated to decomposition it emits highly toxic fumes of fluorides.
Countermeasures
Storage and Handling: Section 7.

"PENTALARM"
General Information
Composition: Mixture of isomeric amyl mercaptans.
Description: A liquid.
Constants: Bp: 100°C, flash p: 63°F (O.C.), d: 0.83–0.85.
Hazard Analysis
Toxic Hazard Rating:
  Acute Local: Irritant 2; Inhalation 2.
  Acute Systemic: U.
  Chronic Local: U.
  Chronic Systemic: U.
Fire Hazard: Dangerous, when exposed to heat or flame.
Explosion Hazard: Unknown.
Disaster Hazard: Dangerous, keep away from heat and open flame; can react vigorously with oxidizing materials.

Countermeasures
Ventilation Control: Section 2.
To Fight Fire: Carbon dioxide, dry chemical or carbon tetrachloride (Section 6).
Personnel Protection: Section 3.
Storage and Handling : Section 7,

PENTALIN. See pentachloroethane.

PENTAMETHYLENE. See cyclopentane.

PENTAMETHYLENE DIAMINE
General Information
Synonym: Cadaverine.
Description: Colorless, thick liquid; characteristic odor.
Formula: $C_5H_{14}N_2$.
Constants: Mol wt: 102.2, mp: 9°C, bp: 178–180°C, d: 0.873 at 25°/4°C.
Hazard Analysis
Toxic Hazard Rating:
  Acute Local: Irritant 2; Allergen 2.
  Acute Systemic: Ingestion 3, Skin Absorbtion 2.
  Chronic Local: Allergen 2.
  Chronic Systemic: Ingestion 2; Skin Absorption 2.
Toxicology: A toxic ptomaine. Can be absorbed through skin and may cause irritation and sensitization.
Disaster Hazard: Dangerous; when heated to decomposition, it emits highly toxic fumes of nitrogen compounds and cyanides.
Countermeasures
Personal Hygiene: Section 3.
Storage and Handling: Section 7.

PENTAMETHYLENE   GLYCOL. See 1,5-pentanediol.

PENTAMETHYLENE OXIDE. See tetrahydropyran.

PENTANAL. See m-valeraldehyde.

4-PENTANAL
Hazard Analysis
Toxicity: Details unknown. Limited animal experiments suggest high toxicity and moderate irritation. See also aldehydes.

n-PENTANE
General Information
Synonym: Amyl hydride.
Description: Colorless liquid.
Formula: $CH_3(CH_2)_3CH_3$.
Constants: Mol wt: 72.15, bp: 36.1°C, flash p: < −40°F, fp: −129.8°C, d: 0.626 at 20°/4°C, autoign. temp.: 588°F, vap. press.: 400 mm at 18.5°C, vap. d.: 2.48, lel: 1.5%, uel: 7.8%.
Hazard Analysis
Toxic Hazard Rating:
  Acute Local: U.
  Acute Systemic: Inhalation 1.
  Chronic Local: U.
  Chronic Systemic: U.
Toxicology: Narcotic in high concentrations.
TLV: ACGIH (recommended); 1000 parts per million; 2950 milligrams per cubic meter of air.
Fire Hazard: Highly dangerous, when exposed to heat or flame.
Spontaneous Heating: No.
Explosion Hazard: Severe, when exposed to heat or flame.
Explosive Range: 1.4–8.0%.
Disaster Hazard: Highly dangerous; keep away from heat,

sparks or open flame; shock can shatter metal containers and release contents.
Countermeasures
Ventilation Control: Section 2.
To Fight Fire: Floam, carbon dioxide, dry chemical or carbon tetrachloride (Section 6).
Storage and Handling: Section 7.
Shipping Regulations: Section 11.
  I.C.C.: Flammable liquid; red label, 10 gallons.
  Coast Guard Classification: Inflammable liquid; red label.
  IATA: Flammable liquid, red label, 1 liter (passenger), 40 liters (cargo).

1,5-PENTANEDIAMINE. See pentamethylene diamine.

PENTANEDINITRILE. See glutaronitrile.

1,5-PENTANEDIOL
General Information
Synonyms: Pentamethylene glycol.
Description: Colorless, viscous, odorless lqiuid.
Formula: $HOCH_2CH_2CH_2CH_2CH_2OH$.
Constants: Mol wt: 104.15, bp: 242.5°C, flash p: 275°F, fp: −15.6°C, d: 0.994 at 20°/4°C, vap. press.: <0.01 mm at 20°C, vap. d.: 3.59, autoign. temp.: 633°F.
Hazard Analysis
Toxicity: Details unknown. See also glycols. Limited animal experiments suggest very low toxicity.
Fire Hazard: Slight, when exposed to heat or flame; can react with oxidizing materials.
Countermeasures
To Fight Fire: Foam, carbon dioxide, dry chemical or carbon tetrachloride (Section 6).
Storage and Handling: Section 7.

PENTANEDIONE-2,4
General Information
Synonym: Acetylacetone.
Description: Colorless liquid.
Formula: $CH_3COCH_2COCH_3$.
Constants: Mol wt: 100.11, mp: −23.2°C, bp: 139°C at 746 mm, flash p: 105°F (O.C.), d: 0.976, vap. d.: 3.45.
Hazard Analysis
Toxic Hazard Rating:
  Acute Local: Irritant 1.
  Acute Systemic: Inhalation 2.
  Chronic Local: U.
  Chronic Systemic: U.
Fire Hazard Moderate, when exposed to heat or flame: can react with oxidizing materials.
Explosion Hazard: Unknown.
Countermeasures
Ventilation Control: Secton 2.
To Fight Fire: Alochol foam, carbon dioxide, dry chemical or carbon tetrachloride (Section 6).
Personal Hygiene: Section 3.
Storage and Handling: Section 7.

1-PENTANETHIOL. See amyl mercaptan.

PENTANICKEL DIPHOSPHIDE
General Information
Description: Needles, tablets or crystals.
Formula: Ni_5P_2.
Constants: Mol wt: 355.41, mp: 1185°C.
Hazard Analysis and Countermeasures
See nickel compounds and phosphine,

PENTANOIC ACID. See valeric acid.

TOXIC HAZARD RATING CODE (For detailed discussion, see Section 1.)

0 NONE: (a) No harm under any conditions; (b) Harmful only under unusual conditions or overwhelming dosage.

1 SLIGHT: Causes readily reversible changes which disappear after end of exposure.

2 MODERATE: May involve both irreversible and reversible changes; not severe enough to cause death or permanent injury.

3 HIGH: May cause death or permanent injury after very short exposure to small quantities.

U UNKNOWN: No information on humans considered valid by authors.

dl-PENTANOL-2. See methyl propyl carbinol.

PENTANONE-2
General Information
Synonyms: Methyl n-propyl ketone; ethyl acetone.
Description: Slightly soluble in water; water white liquid.
Formula: $CH_3COC_3H_7$.
Constants: Mol wt: 86, d: 0.8, vap. d.: 3.0, bp: 216°F, flash p: 45°F, autoign. temp.: 941°F, lel: 1.5% uel: 8%.
Hazard Analysis
TLV: ACGIH (recommended): 200 parts per million in air; 700 milligrams per cubic meter of air.
Toxicity: See diethyl ketone.
Fire Hazard : Dangerous, when exposed to heat or flame.
Countermeasures
Storage and Handling: Section 7.
To Fight Fire: Alcohol foam.

PENTAPROPIONYL GLUCOSE. See glucose pentapropionate.

"PENTASOL"
General Information
Synonym: Amyl alcohol, synthetic.
Description: Liquid.
Composition: A mixture of isomeric amyl alcohols.
Constants: Bp: 108–140°C, flash p: 113°F (O.C.), d: 0.81–0.82.
Hazard Analysis
Toxic Hazard Rating:
Acute Local: Irritant 2; Ingestion 2; Inhalation 2.
Acute Systemic: Ingestion 2; Inhalation 2; Skin Absorption 2.
Chronic Local: U.
Chronic Systemic: Ingestion 2; Inhalation 2; Skin Absorption 2.
Fire Hazard: Moderate, when exposed to heat or flame; can react with oxidizing materials.
Countermeasures
To Fight Fire: Foam, carbon dioxide, dry chemical or carbon tetrachloride (Section 6).
Ventilation Control: Section 2.
Personnel Protection: Section 3.
Storage and Handling: Section 7.

PENTATHIONIC ACID
Hazard Analysis
Toxicity: Details unknown; a fungicide.
Disaster Hazard: Dangerous; when heated to decomposition, it emits highly toxic fumes of oxides of sulfur.
Countermeasures
Storage and Handling: Section 7.

PENTATRIACONTENE
General Information
Description: A white solid.
Formula: $C_{17}H_{35}CHCHC_{16}H_{33}$.
Constants: Mol wt: 491.0, bp: 260–265°C at 0.5 mm, flash p: 239°F (O.C.), mp: 75°C, d: 0.8.
Hazard Analysis
Toxicity: Unknown.
Fire Hazard: Slight, when exposed to heat or flame; can react with oxidizing materials.
Countermeasures
To Fight Fire: Foam, carbon dioxide, dry chemical or carbon tetrachloride (Section 6).
Storage and Handling: Section 7.

PENTEK. See pentaerythritol, technical.

1-PENTENE. See amylene.

2-PENTENE
General Information
Synonym: sym-Methylethylethylene.
Description: Liquid.

Formula: C_5H_{10}.
Constants: Mol wt: 70.1; mp: cis: −179°C, trans: −135°C; bp: cis; 37°C, trans: 35.85°C; flash p: 0°F, d cis form: 0.6503 at 20°/4°C, trans form: 0.6482 at 20°/4°C; vap. d.: 2.41.
Hazard Analysis
Toxicity: Unknown. Probably narcotic in high concentrations.
Fire Hazard: Dangerous, when exposed to heat or flame.
Explosion Hazard: Unknown.
Disaster Hazard: Dangerous; keep away from heat and open flame; can react vigorously with oxidizing materials.
Countermeasures
Ventilation Control: Section 2.
Storage and Handling: Section 7.
To Fight Fire: Foam, carbon dioxide, dry chemical or carbon tetrachloride (Section 6).

2-PENTENE-3-CARBOXYLIC ACID. See ethyl crotonate.

3-PENTENE-2-ONE
General Information
Synonyms: Ethylidene acetone.
Description: Colorless liquid.
Formula: $CH_3CHCHCOCH_3$.
Constants: Mol wt: 84.11, bp: 122–124°C, d: 0.856, vap. d.: 2.89.
Hazard Analysis
Toxic Hazard Rating:
Acute Local: U.
Acute Systemic: Ingestion 2; Inhalation 2; Skin Absorption 2.
Chronic Local: U.
Chronic Systemic: U.
Toxicology: Based upon animal experimentation.
Fire Hazard: Moderate, when exposed to heat or flame; can react with oxidizing materials.
Countermeasures
To Fight Fire: Foam, carbon dioxide, dry chemical or carbon tetrachloride (Section 6).
Ventilation Control: Section 2.
Personnel Protection: Section 3.
Storage and Handling: Section 7.

PENTRYL
General Information
Synonym: Trinitrophenylnitroaminoethylnitrate.
Hazard Analysis and Countermeasures
See picric acid.

PENTYL ALCOHOL. See amyl alcohol.

n-PENTYL BENZENE. See amyl benzene.

PENTYLOXYPHENITOL
Hazard Analysis
Toxicity: Action similar to but less severe than phenol. See phenol.

PENTYL PHENOLS
Hazard Analysis
Toxicity: Action similar to but less severe than phenol. See phenol.

PENTYL PROPANOATE. See amyl propionate.

PEPPERMINT OIL. See oil of peppermint.

PERACETIC ACID (40% SOLUTION)
General Information
Synonyms: Peroxyacetic acid; acetyl hydroperoxide.
Description: Colorless liquid; strong odor.
Formula: CH_3COOOH.
Constants: Mol wt: 76.05, bp: 105°C; explodes at 110°C, flash p: 105°F (O.C.), d: 1.15 at 20°C.

Hazard Analysis
Toxic Hazard Rating:
Acute Local: Irritant 3; Ingestion 3; Inhalation 3.
Acute Systemic: U.
Chronic Local: U.
Chronic Systemic: U.
Fire Hazard: See peroxides, organic.
Explosion Hazard: Moderate, when exposed to heat or by spontaneous chemical reaction. A powerful oxidizing agent.
Disaster Hazard: Dangerous; keep away from combustible materials.
Countermeasures
Ventilation Control: Section 2.
To Fight Fire: Water, foam, carbon dioxide or carbon tetrachloride (Section 6).
Personnel Protection: Section 3.
Storage and Handling: Section 7.
Shipping Regulations: Section 11.
I.C.C.: Oxidizing material; yellow label, 5 pints.
Coast Guard Classification: Oxidizing material; yellow label.
IATA (exceeding 40% by weight): Oxidizing material, not acceptable (passenger and cargo).
(not exceeding 40% by weight): Oxidizing material, yellow label, 1 liter (passenger), $2\frac{1}{2}$ liters (cargo).

PERBENZOIC ACID
General Information
Synonym: Benzoyl hydroperoxide.
Description: Leaflets.
Formula: C_6H_5COOOH.
Constants: Mol wt: 138.1, mp: 42°C, bp: explodes at 80–100°C.
Hazard Analysis
Toxic Hazard Rating:
Acute Local: Irritant 2; Ingestion 2; Inhalation 2.
Acute Systemic: U.
Chronic Local: U.
Chronic Systemic: U.
Fire Hazard: A powerful oxidizing agent. See peroxides, organic.
Explosion Hazard: Severe, when exposed to heat or flame.
Disaster Hazard: Dangerous; heat will cause it to explode; can react vigorously with reducing materials.
Countermeasures
Ventilation Control: Section 2.
Personnel Protection: Section 3.
Storage and Handling: Section 7.

PERBORATES, N.O.S.
Shipping Regulations: Section 11.
IATA: Oxidizing material, yellow label, 12 kilograms (passenger), 45 kilograms (cargo).

PERCAINE. See dibucane hydrochloride.

PERCHLORATES
General Information
Composition: Combinations with the monovalent $^-ClO_4$ radical.
Hazard Analysis
Toxic Hazard Rating:
Acute Local: Irritant 2; Ingestion 2; Inhalation 2.
Acute Systemic: Ingestion 2.
Chronic Local: U.
Chronic Systemic: Ingestion 2.
Toxicology: Perchlorates are unstable materials, and are ir-

ritating to the skin and mucous membranes of the body wherever they come in contact with it. Avoid skin contact with these materials (Section 9).
Fire Hazard: Moderate, by chemical reaction; powerful oxidizers. See also explosives,high.
Explosion Hazard: Moderate, when shocked, exposed to heat or by chemical reaction. Perchlorates, when mixed with carbonaceous material, form explosive mixtures. They are considered a fire and explosive hazard when associated with carbonaceous materials or finely divided metals. This is also true of the presence of sulfur, powdered magnesium and aluminum. See explosives, high, and Section 7.
Disaster Hazard: Dangerous; shock will explode them; when heated, they emit highly toxic fumes of chlorides; they can react with reducing materials.
Countermeasures
Ventilation Control: Section 2.
To Fight Fire: Water or foam (Section 6).
Personnel Protection: Section 3.
Storage and Handling: Section 7.
Shipping Regulations: Section 11.
I.C.C.: Oxidizing material; yellow label, 100 pounds.
Coast Guard Classification: Oxidizing material; yellow label.
IATA: Oxidizing material, yellow label, 12 kilograms (passenger), 45 kilograms (cargo).

PERCHLORIC ACID
General Information
Description: Colorless, fuming, unstable liquid.
Formula: $HClO_4$.
Constants: Mol wt: 100.47, mp: −112°C, bp: 19°C at 11 mm, d: 1.768 at 22°C.
Hazard Analysis
Toxic Hazard Rating:
Acute Local: Irritant 3; Ingestion 3; Inhalation 3.
Acute Systemic: U.
Chronic Local: Irritant 2; Inhalation 2.
Chronic Systemic: U.
Fire Hazard: See perchlorates.
Explosion Hazard: See perchlorates.
Disaster Hazard: See perchlorates.
Countermeasures
Ventilation Control: Section 2.
Personnel Protection: Section 3.
First Aid: Section 1.
Storage and Handling: Section 7.
MCA warning label.

PERCHLORIC ACID DIHYDRATE. See perchloric acid (over 72%).

PERCHLORIC ACID, MONOHYDRATE
General Information
System: Hydronium perchlorate.
Description: Fairly stable needles.
Formula: $HClO_4 \cdot H_2O$.
Constants: Mol wt: 118.48, mp: 50°C, bp: explodes at 110°C, d: 1.88; d liq: 1.776 at 50°C.
Hazard Analysis and Countermeasures
See perchloric acid, sulfuric acid, and perchlorates.

PERCHLORIC ACID (NOT OVER 72%)
General Information
Description: Clear liquid.
Formula: $HClO_4 \cdot 3H_2O$.

TOXIC HAZARD RATING CODE (For detailed discussion, see Section 1.)

0 NONE: (a) No harm under any conditions; (b) Harmful only under unusual conditions or overwhelming dosage.

1 SLIGHT: Causes readily reversible changes which disappear after end of exposure.

2 MODERATE: May involve both irreversible and reversible changes; not severe enough to cause death or permanent injury.

3 HIGH: May cause death or permanent injury after very short exposure to small quantities.

U UNKNOWN: No information on humans considered valid by authors.

Constants: Mol wt: 154.5, mp: $-18°C$, bp: $200°C$, d: 1.5967 at $25°/4°C$.
Hazard Analysis
Toxicity: See perchloric acid.
Countermeasures
Shipping Regulations: Section 11.
  I.C.C.: Corrosive liquid; white label, 5 pints.
  Coast Guard Classification: Corrosive liquid; white label.
  MCA warning label.
  IATA: Corrosive liquid, white label not acceptable (passenger), $2\frac{1}{2}$ kilograms (cargo).

PERCHLORIC ACID (OVER 72%)
General Information
Synonym: Perchloric acid dihydrate.
Description: Stable liquid.
Formula: $HClO_4 \cdot 2H_2O$.
Constants: Mol wt: 136.5, mp: $-17.8°C$, bp: $200°C$, d: 1.729 at $25°/4°C$.
Hazard Analysis
Toxicity: See perchloric acid.
Countermeasures
Shipping Regulations: Section 11.
  I.C.C.: Section 11.
  Coast Guard Classification: Not permitted.
  MCA warning label.
  IATA: Corrosive liquid, not acceptable (passenger and cargo).

PERCHLOROBENZENE. See hexachlorobenzene.

PERCHLOROETHYLENE
General Information
Synonyms: Tetrachloroethylene; ethylene tetrachloride; carbon dichloride.
Description: Colorless liquid; chloroform-like odor.
Formula: CCl_2CCl_2.
Constants: Mol wt: 165.85, mp: $-23.35°C$, bp: $121.20°C$, flash p: none, d: 1.6311 at $15°/4°C$, vap. press.: 15.8 mm at $22°C$, vap. d.: 5.83.
Hazard Analysis
Toxic Hazard Rating:
Acute Local: Irritant 2.
Acute Systemic: Ingestion 3; Inhalation 2; Skin Absorption 2.
Chronic Local: Irritant 2.
Chronic Systemic: Ingestion 2; Inhalation 2; Skin Absorption 2.
TLV: ACGIH (recommended); 100 parts per million in air; 670 milligrams per cubic meter of air.
Toxicology: Not corrosive or dangerously reactive, but only by inhalation, by prolonged or repeated contact with the skin or mucous membrane, or when ingested by mouth. The liquid can cause injuries to the eyes; however, with proper precautions it can be handled safely. The symptoms of acute intoxication from this material are the result of its effects upon the nervous system.

Exposures to higher concentrations than 200 ppm cause irritation, lachrymation and burning of the eyes and irritation of the nose and throat. There may be vomiting nausea, drowsiness, an attitude of irresponsibility, and even an appearance resembling alcoholic intoxication. This material also acts as an anesthetic, through the inhalation of excessive amounts within a short time. The symptoms of fatal intoxication are irritation of the eyes, nose and throat, then fullness in the head, mental confusion; there may be headache stupefaction, nausea and vomiting, personnel suffering from subacute poisoning may suffer from such symptoms as headache, fatigue, nausea, vomiting, mental confusion and temporary blurring of the vision. This can occur when inadequate ventilation results in concentrations higher then 200 ppm, or where the vapor concentrations are intermittently high due to faulty

handling of the material, or when an individual fails to take adequate precautionary measures.

This material can cause dermatitis, particularly after repeated or prolonged contact with the skin. The dermatitis is preceded by a reddening and burning, and more rarely, a blistering of the skin. In any event, the skin becomes rough and dry, due largely to the removal of skin oils by material. The skin then cracks easily and is readily susceptible to infection (Section 9). Upon ingestion it causes irritation of the gastrointestinal tract which in turn causes nausea, vomiting, diarrhea and bloody stools. However, such effects are usually less severe then the effects of swallowing similar amounts of other chlorinated hydrocarbons.

It may be handled in the presence or absence of air, water, and light with any of the common construction materials at temperatures up to 140°C. This material is extremely stable and resists hydrolysis.

A common air containinant   (Section 4).
Disaster Hazard: Dangerous; when heated to decomposition, it emits highly toxic fumes of chlorides.
Countermeasures
Ventilation Control: Section 2.
Personnel Protection: Section 3.
Storage and Handling: Section 7.
Shipping Regulations: Section 11.
  MCA warning label.
  IATA: Other restricted articles, class A, no label required, 40 liters (passenger), 220 liters (cargo).

PERCHLOROMETHYL MERCAPTAN
General Information
Synonyms: Thiocarbonyl tetrachloride; trichloromethane sulfonyl chloride.
Description: Yellow oily liquid.
Formula: $ClSCCl_3$.
Constants: Mol wt: 185.90, bp: slight decomp. at $149°C$, d: 1.700 at $20°C$, vap. d.: 6.414.
Hazard Analysis
Toxic Hazard Rating:
Acute Local: Irritant 3; Ingestion 3; Inhalation 3.
Acute Systemic: Ingestion 3; Inhalation 3.
Chronic Local: U.
Chronic Systemic: Ingestion 3; Inhalation 3.
TLV: ACGIH (recommended); 0.1 part per million in air; 0.8 milligrams per cubic meter of air.
Disaster Hazard: Dangerous; when heated to decomposition, it emits highly toxic fumes of chlorides.
Countermeasures
Ventilation Control: Section 2.
Personnel Protection: Section 3.
Storage and Handling: Section 7.
Shipping Regulations: Section 11.
  I.C.C.: Poison B; poison label, 10 pounds.
  Coast Guard Classification: Poison B; poison label.
  IATA: Poison B, poison label, not acceptable (passenger), 5 kilograms (cargo).

PERCHLORYL FLUORIDE
General Information
Synonym: PF.
Description: Colorless, non-corrosive gas with a characteristic sweet odor.
Formula: ClO_3F.
Constants: Mol wt: 102.5, mp: $-146 \pm 2°C$, bp: $-46.8°C$.
Hazard Analysis
Toxicity: Moderately toxic. Forms methemoglobin in the body and destroys red cells causing anemia, anorexia and cyanosis. Recovery is said to be rapid, leaving no permanent physiological damage. Can be absorbed through the skin. Its odor can be detected as low as 10 ppm although this cannot be relied upon as an indication of toxic concentrations in air. The chronic effects are not known.

TLV: ACGIH (recommended); 3 parts per million in air; 13.5 milligrams per cubic meter of air.
Fire Hazard: Moderate; while nonflammable, it supports combustion. It is a powerful oxidizer.
Explosion Hazard: Moderate, contact with readily oxidized substances, such as hydrogen sulfide, charcoal, sawdust, lampblack, etc. yield explosive products down to $-78°$C in some cases.
Disaster Hazard: Dangerous; when heated to decomposition it emits highly toxic fumes of chlorides and fluorides.
Countermeasures
Ventilation Control: Section 2.
Personal Hygiene: Section 3.
Storage and Handling: Section 7.
Shipping Regulations: Section 11.
  I.C.C.: Compressed gas N.O.S., green label.

PERCUSSION CAPS AND FUSES. See explosives, high.
Shipping Regulations: Section 11.
  I.C.C. (caps and fuses): Explosive C, 150 pounds.
  Coast Guard Classification (caps and fuses): Explosive C.
  IATA (caps and fuses): Explosive, explosive label, 25 kilograms (passenger), 70 kilograms (cargo).

PERFLURO-2-BUTENE
General Information
Synonym: Octafluoro-2-butene.
Description: Colorless gas.
Formula: $CF_3CF:CFCF_3$.
Constants: Mol wt: 200, bp: $1.2°$C, fp: $-135°$C.
Hazard Analysis
Toxic Hazard Rating: U.
Disaster Hazard: Dangerous. When heated to decomposition it emits toxic fumes.
Countermeasures
Shipping Regulations: Section 11.
  IATA: Other restricted articles, class A, 40 liters (passenger), 220 liters (cargo).

PERFLUOROETHYLENE. See tetrafluoroethylene, inhibited.

PERFLUOROISOBUTYLENE
Hazard Analysis
Toxic Hazard Rating:
  Acute Local: Irritant 3; Inhalation 3.
  Acute Systemic: Inhalation 3.
  Chronic Local: U.
  Chronic Systemic: U.
Toxicity: Acute exposure in man has produced marked irritation of conjunctive, throat, and lungs.

PERFLUOROPROPANE
General Information
Synonym: Octafluoropropane.
Formula: C_3F_8.
Description: Colorless gas.
Constants: Mol wt: 188, bp: $-36.7°$C, fp: $-160°$C.
Hazard Analysis
Toxic Hazard Rating: U. Probably a simple asphyxiant.
Countermeasures
Shipping Regulations: Section 11.
  IATA: Nonflammable gas, green label, 70 kilograms (passenger), 140 kilograms (cargo).

PERFORMIC ACID
General Information
Synonym: Formyl hydroperoxide.

Description: Liquid.
Formula: HCOOOH.
Constant: Mol wt: 62.0.
Hazard Analysis
Toxic Hazard Rating:
  Acute Local: Irritant 3; Ingestion 3; Inhalation 3.
  Acute Systemic: U.
  Chronic Local: U.
  Chronic Systemic: U.
Fire Hazard: See peroxides, organic.
Explosive Hazard: See peroxides, organic.
Disaster Hazard: See peroxides, organic.
Countermeasures
Ventilation Control: Section 2.
Personnel Protection: Section 3.
Storage and Handling: Section 7.

PERFUMES
General Information
Composition: Essential oils and alcohol.
Hazard Analysis
Toxic Hazard Rating:
  Acute Local: Allergen 1.
  Acute Systemic: Ingestion 2; Inhalation 1.
  Chronic Local: Allergen 1.
  Chronic Systemic: Inhalation 1.
Fire Hazard: Moderate. See alcohol.
Disaster Hazard: Moderately dangerous due to flammability of alcohol.
Countermeasures
Personal Hygiene: Section 3.
Storage and Handling: Section 7.

PERI ACID. See 1-naphthylamine 8-sulfonic acid.

PERILLA OIL
General Information
Constants: Mp: $-5°$C, flash p: $522°$C (C.C.), d: 0.93–0.94.
Hazard Analysis
Toxicity: Unknown.
Fire Hazard: Slight, when exposed to heat or flame; can react with oxidizing materials.
Spontaneous Heating: Yes.
Countermeasures
To Fight Fire: Foam, carbon dioxide, dry chemicals or carbon tetrachloride (Section 6).
Storage and Handling: Section 7.

PERIODIC ACID
General Information
Synonym: Paraperiodic acid.
Description: White, deliquescent crystals.
Formula: $HIO_4 \cdot 2H_2O$.
Constants: Mol wt: 228.0, mp: $110°$C, bp: decomposes at $140°$C.
Hazard Analysis
Toxic Hazard Rating:
  Acute Local: Irritant 3; Ingestion 3; Inhalation 3.
  Acute Systemic: U.
  Chronic Local: U.
  Chronic Systemic: U.
Fire Hazard: Moderate, by chemical reaction; an oxidizing agent (Section 6).
Disaster Hazard: Dangerous; when heated to decomposition, it emits highly toxic fumes of iodides; reacts vigorously with reducing materials.

TOXIC HAZARD RATING CODE (For detailed discussion, see Section 1.)

0 NONE: (a) No harm under any conditions; (b) Harmful only under unusual conditions or overwhelming dosage.

1 SLIGHT: Causes readily reversible changes which disappear after end of exposure.

2 MODERATE: May involve both irreversible and reversible changes not severe enough to cause death or permanent injury.

3 HIGH: May cause death or permanent injury after very short exposure to small quantities.

U UNKNOWN: No information on humans considered valid by authors.

Countermeasures
Ventilation Control: Section 2.
Personnel Protection: Section 3.
Storage and Handling: Section 7.

PERMANGANATE OF POTASH. See potassium permanganate.
Shipping Regulations: Section 11.
  I.C.C.: Oxidizing material; yellow label, 100 pounds.
  Coast Guard Classification: Inflammable solid; yellow label.
  IATA: Oxidizing material, yellow label, 12 kilograms (passenger), 45 kilograms (cargo).

PERMANGANATE OF SODA. See sodium permanganate.
Shipping Regulations: Section 11.
  I.C.C.: Oxidizing material; yellow label, 100 pounds.
  Coast Guard Classification: Oxidizing material; yellow label.
  IATA: Oxidizing material, yellow label, 12 kilograms (passenger), 45 kilograms (cargo).

PERMANGANATES
General Information
Composition: Compounds containing a MnO_4^- radical.
Hazard Analysis
Toxicity: Many are strong oxidizing agents, hence can be irritating. See manganese compounds.
Fire Hazard: Moderate by chemical reaction with reducing agents.
Explosion Hazard: Moderate, when shocked or exposed to heat. Silver permanganate and other metallic permanganates may detonate when exposed to high temperatures or when they are involved in fires or severely shocked. They should be stored in a cool, ventilated area, away from acute fire hazards and easily oxidized materials. They may be disposed of by dissolving in water. Practically all permanganates are soluble in water.
Disaster Hazard: Dangerous; shock or heat may explode them; they can react vigorously on contact with reducing materials.
Countermeasures
Ventilation Control: Section 2.
Storage and Handling: Section 7.
Shipping Regulations: Section 11.
  I.C.C.: Oxidizing material; yellow label, 100 pounds.
  Coast Guard Classification: Oxidizing material; yellow label.
  IATA: Oxidizing material, yellow label, 12 kilograms (passenger), 45 kilograms (cargo).

PERMANGANIC ACID
General Information
Description: In solution only.
Formula: $HMnO_4$.
Constant: Mol wt: 119.9.
Hazard Analysis and Countermeasures
See manganese compounds and permanganate.

PERMANGANYL CHLORIDE
Hazard Analysis
Toxicity: Highly toxic. See manganese compounds.
Explosion Hazard: Moderate, when exposed to heat.
Disaster Hazard: Dangerous; when heated to decomposition, it emits highly toxic fumes of chlorides; will react with water or steam to produce toxic and corrosive fumes.
Countermeasures
Ventilation Control: Section 2.
Storage and Handling: Section 7.

PERNAM BUCO. See brazil wood.

PERONINE
General Information
Synonym: Morphine benzyl ether hydrochloride.

Description: White prismatic, crystalline, odorless powder; bitter taste.
Formula: $C_{17}H_{17}NO(OH)OC_7H_7 \cdot HCl$.
Constant: Mol wt: 411.9.
Hazard Analysis
Toxic Hazard Rating:
  Acute Local: U.
  Acute Systemic: Ingestion 3; Inhalation 3.
  Chronic Local: U.
  Chronic Systemic: U.
Disaster Hazard: Dangerous; when heated to decomposition, it emits highly toxic fumes.
Countermeasures
Ventilation Control: Section 2.
Personal Hygiene: Section 3.
Storage and Handling: Section 7.

PEROXIDES, INORGANIC
Hazard Analysis
Toxicity: Variable. They may cause injury on contact with skin or mucous membranes. See also hydrogen peroxide.
Fire Hazard: Moderate to dangerous by chemical reaction with reducing agents and contaminants; strong oxidizing agents; contact with moisture may produce much heat. See also hydrogen peroxide and sodium peroxide (Section 6).
Explosion Hazard: Moderate; heat shock or catalysts can cause violent decomposition. Contact with reducing agents may give rise to explosively violent reactions.
Disaster Hazard: Dangerous; shock heat or moisture may cause explosion; reacts with reducing agents.
Countermeasures
Storage and Handling: Section 7.

PEROXIDES, ORGANIC
Hazard Analysis
Caution: These materials are irritating to the skin, eyes and mucous membranes.
Fire Hazard: Dangerous, by chemical reaction with reducing agents or exposure to heat. They are powerful oxidizers.
Explosion Hazard: Severe, when shocked, exposed to heat or by spontaneous chemical reaction. Many peroxides are very unstable. Upon contact with reducing materials an explosive reaction can occur.
Disaster Hazard: Dangerous; shock or heat may explode them; reacts with reducing materials.
Countermeasures
Storage and Handling: They should be stored in a cool, ventilated, isolated area away from organic or other easily oxidizable materials and away from acute fire hazards. Containers should be kept closed and plainly labelled. Personnel should be cautioned regarding their use. Personnel who must be exposed to these materials should use protective equipment. Recommended are protective clothing to avoid skin contact, chemical safety goggles for protection of the eyes, and a respirator to avoid the inhalation of extensive amounts of these materials in vapor form (Section 7).
Shipping Regulations: Section 11.
  I.C.C.: Flammable liquid, red label, 1 quart; oxidizing material, yellow label, 1 quart.
  Coast Guard Classification: Inflammable liquid, red label; oxidizing material, yellow label.
  IATA (liquid, N.O.S., flammable): Flammable liquid, not acceptable (passenger and cargo).
    (liquid, N.O.S., oxidizing material): oxidizing material not acceptable (passenger and cargo).
    (liquid or solution, N.O.S., with water, non-volatile solvent, or plasticizer, flammable): Flammable liquid, red label, 1 liter (passenger and cargo).
    (liquid or solution, N.O.S., with water, nonvolatile solvent, or plasticizer, oxidizing material): oxidizing material, yellow label, 1 liter (passenger and cargo).

(solid, N.O.S.): Oxidizing material, yellow label, 1 kilogram (passenger), 12 kilograms (cargo).
MCA warning label.

PEROXYACETIC ACID. See peracetic acid (40% solution).

PEROXY DISULFURIC ACID
General Information
Synonym: Persulfuric acid.
Description: Hygroscopic crystals.
Formula: $H_2S_2O_8$.
Constants: Mol wt: 194.15, mp: decomposes at 65°C, bp: decomposes.
Hazard Analysis
Toxic Hazard Rating:
  Acute Local: Irritant 3; Ingestion 3; Inhalation 3.
  Acute Systemic: U.
  Chronic Local: Irritant 2.
  Chronic Systemic: U.
Fire Hazard: Moderate, by chemical reaction; a powerful oxidizer (Section 6).
Disaster Hazard: Dangerous; on decomposition, it emits highly toxic fumes of oxides of sulfur.
Countermeasures
Ventilation Control: Section 2.
Personnel Protection: Section 3.
Storage and Handling: Section 7.

PEROXY SULFURIC ACID
General Information
Synonym: Caro's acid.
Description: White crystals.
Formula: H_2SO_5.
Constants: Mol wt: 114.08, mp: decomposes at 45°C.
Hazard Analysis
Toxic Hazard Rating:
  Acute Local: Irritant 3; Ingestion 3; Inhalation 3.
  Acute Systemic: U.
  Chronic Local: U.
  Chronic Systemic: U.
Fire Hazard: Moderate by chemical reaction; a powerful oxidizer (Section 6).
Explosion Hazard: Powerful oxidizer. Can explode in contact with organic materials.
Disaster Hazard: Dangerous; when heated to decomposition, it emits highly toxic fumes of oxides of sulfur.
Countermeasures
Ventilation Control: Section 2.
Personnel Protection: Section 3.
Storage and Handling: Section 7.

"PERSISTO SPRAY." See DDT.

PERSULFURIC ACID. See peroxy disulfuric acid.

PERSULFUR HEPTOXIDE. See sulfur heptoxide.

PERTHANE
General Information
Synonym: 1,1-Dichloro-2,2-bis (p-ethyl phenyl) ethane.
Description: Crystals, soluble in acetone, kerosene.
Formula: $C_{18}H_{20}Cl$.
Constants: Mol wt: 307.3, mp: 57°C.
Hazard Analysis
Toxicity: For action see related compound DDT.
Disaster Hazard: Dangerous; when heated to decomposition, it emits highly toxic fumes of chlorides.

PERUVIAN BALSAM. See balsam of Peru.

PERYLENE
General Information
Synonym: Dibenzanthrecene; peri-dinaphthalene.
Description: Yellow to colorless crystals; freely soluble in carbon bisulfide and chloroform; moderately soluble in benzene; slightly soluble in ether, alcohol, acetone; very sparingly soluble in petroleum ether; insoluble in water.
Formula: $C_{20}H_{12}$.
Constants: Mol wt: 252.30, d: 1.35, mp: 273–274°C.
Hazard Analysis
Toxic Hazard Rating: Highly toxic. Details unknown. A carcinogen.

PESTOX. See octamethyl pyrophosphoramide.

PETN. See pentaerythritol tetranitrate.

"PETROHOL" 91%
General Information
Synonym: Isopropyl alcohol.
Description: Clear liquid.
Constants: Flash p: 66° F, autoign. temp.: 895° F.
Hazard Analysis
Toxicity: Moderate. See isopropyl alcohol.
Fire Hazard: Dangerous, when exposed to heat or flame.
Disaster Hazard: Dangerous; keep away from heat and open flame; can react on contact with oxidizing materials.
Countermeasures
Ventilation Control: Section 2.
Storage and Handling: Section 7.

"PETROHOL" 98%
General Information
Synonym: Isopropyl alcohol.
Description: Clear liquid.
Constants: Flash p: 59° F, autoign. temp.: 845° F.
Hazard Analysis
Toxicity: Moderate. See isopropyl alcohol.
Fire Hazard: Dangerous, when exposed to heat or flame.
Disaster Hazard: Dangerous, upon exposure to heat or flame; can react vigorously with oxidizing materials.
Countermeasures
Ventilation Control: Section 2.
To Fight Fire: Carbon dioxide, dry chemical or carbon tetrachloride (Section 6).
Storage and Handling: Section 7.

PETROL. See gasoline.

PETROLATUM
General Information
Synonyms: Mineral fat; petroleum jelly; mineral jelly.
Description: Almost colorless to amber colored gelatinous, oily, translucent, semisolid, amorphous mass; soluble in chloroform, ether, benzene, carbon disulfide; benzene and oils; very slightly soluble in alcohol; insoluble in water.
Constants: d: 0.815–0.880 at 60°C, mp: 38–60°C.
Hazard Analysis
Toxicity: Details unknown. Used as a food additive permitted in the feed and drinking water of animals and/or for the treatment of food producing animals; also permitted in food for human consumption (Section 10).

PETROLALUM LIQUID. See oil (mineral).

PETROLEUM. See crude oil.

TOXIC HAZARD RATING CODE (For detailed discussion, see Section 1.)

0 NONE: (a) No harm under any conditions; (b) Harmful only under unusual conditions or overwhelming dosage.

1 SLIGHT: Causes readily reversible changes which disappear after end of exposure.

2 MODERATE: May involve both irreversible and reversible changes; not severe enough to cause death or permanent injury.

3 HIGH: May cause death or permanent injury after very short exposure to small quantities.

U UNKNOWN: No information on humans considered valid by authors

PETROLEUM BENZINE. See petroleum spirits.

PETROLEUM DISTILLATE
Shipping Regulations: Sectionn 11.
  I.C.C.: Flammable liquid, red label, 10 gallons.
  IATA: Flammable liquid, red label, 1 liter (passenger),
    40 liters (cargo).

PETROLEUM ETHER. See benzine.
Shipping Regulations: Section 11.
  I.C.C.: Flammable liquid, red label; combustible liquid,
    10 gallons.
  Coast Guard Classification: Combustible liquid; inflam-
    mable liquid, red label.
  IATA: Flammable liquid, red label, 1 liter (passenger), 40
    liters (cargo).

PETROLEUM NAPHTHA. See petroleum spirits.

PETROLEUM NAPHTHAS AND SOLVENTS
Shipping Regulations: Section 11.
  Coast Guard Classification: Combustible liquid; inflam-
    mable liquid, red label.
  MCA warning label.

PETROLEUM PITCH. See asphalt.

PETROLEUM SPIRITS
General Information
Synonyms: Petroleum benzine; petroleum naphtha; light
  ligroin; petroleum ether.
Description: Volatile, clear, colorless and non-fluorescent
  liquid.
Constants: Mp: $< -73°C$, bp: 40–80°C, ulc: 95–100, lel:
  1.1%, uel: 5.9%, flash p: $< 0°F$, d: 0.635 to 0.660,
  autoign. temp.: 550°F, vap. d.: 2.50.
Hazard Analysis
Toxic Hazard Rating:
  Acute Local: Irritant 1; Ingestion 2; Inhalation 1.
  Acute Systemic: Ingestion 1; Inhalation 2; Skin Absorp-
    tion 1.
  Chronic Local: Irritant 1.
  Chronic Systemic: Ingestion 1; Inhalation 1.
TLV: ACGIH (recommended); 500 parts per million of air;
  2000 milligrams per cubic meter of air.
Toxicology: Ingestion can cause a burning sensation, vomit-
  ing, diarrhea, drowsiness, and in severe cases, pulmon-
  ary edema. Inhalation of concentrated vapors causes
  intoxication resembling that from alcohol, headache,
  nausea, and coma. Hemorrhages into various vital
  organs have been reported.
Fire Hazard: Highly dangerous; when exposed to heat,
  flame, sparks, etc.
Spontaneous Heating: No.
Explosion Hazard: Moderate, when exposed to heat or
  flame.
Disaster Hazard: Highly dangerous; keep away from heat
  or flame!
Countermeasures
Ventilation Control: Section 2.
To Fight Fire: Foam, carbon dioxide, dry chemical or car-
  bon tetrachloride (Section 6).
Personal Hygiene: Section 3.
Storage and Handling: Section 7.

PETROLEUM SULFONATE
General Information
Constant: Flash p: 400°F (O.C.).
Hazard Analysis
Toxicity: Unknown.
Fire Hazard: Slight, when exposed to heat or flame; can re-
  act with oxidizing materials.
Countermeasures
To Fight Fire: Water, foam, carbon dioxide, dry chemical
  or carbon tetrachloride (Section 6).
Storage and Handling: Section 7.

PETRYL. See trinitrophenyl nitramine ethyl nitrate.

PF. See perchloryl fluoride.

PGA. See pteroylglutamic acid.

PGE. See glycidyl phenyl ether.

PHALTAN
General Information
Synonym: n(Trichloromethylthio) phthalamide.
Formula: $C_9H_4Cl_3NO_2S$.
Constant: Mol wt: 296.6.
Hazard Analysis
Toxicity: Probably similar to N-trichloromethylthio tetra-
  hydrophthalimide.
Disaster Hazard: Dangerous; when heated to decomposi-
  tion, it emits highly toxic fumes of oxides of sulfur and
  of nitrogen.

α-PHELLANDRENE
General Information
Synonyms: 4-Isopropyl-1-methyl-1,5-cyclohexadiene; 1,5-p-
  menthadiene.
Description: Colorless oil; insoluble in water; soluble in
  ether; two isomeric forms (d and l).
Formula: $CH_3C:CHCH_2CH_2CH[CH(CH_3)_2]CH:CH$.
Constants: Mol wt: 136.23, d (d): 0.8463 at 25°C, bp (d):
  66–68°C at 16 mm, d (l): 0.8324 at 20°C, bp (l): 59°C
  at 16 mm.
Hazard Analysis
Toxic Hazard Rating:
  Acute Local: Irritant 2; Ingestion 2.
  Acute Systemic: Ingestion 2; Skin Absorption 2.
  Chronic Local: U.
  Chronic Systemic: U.

β-PHELLANDRENE. See 1(7)2,-p-menthadiene.

PHENACAINE HYDROCHLORIDE
General Information
Description: Small, white crystals.
Formula: $(C_2H_5OC_6H_4)_2N_2HCCH_3 \cdot HCl \cdot H_2O$.
Constants: Mol wt: 352.8, mp: 190°C.
Hazard Analysis
Toxic Hazard Rating:
  Acute Local: Anesthetic.
  Acute Systemic: Ingestion 3; Inhalation 3.
  Chronic Local: U.
  Chronic Systemic: U.
Disaster Hazard: Dangerous; when heated to decomposi-
  tion, it emits highly toxic fumes of chlorides.
Countermeasures
Ventilation Control: Section 2.
Personal Hygiene: Section 3.
Storage and Handling: Section 7.

PHENACETIN. See acetophenetidine.

PHENACYL AMINE. See o-aminoacetophenone.

PHENACYL BROMIDE. See bromoacetophenone.

PHENACYL CHLORIDE. See chloroacetophenone.

PHENACYL FLUORIDE. See fluoroacetophenone.

PHENANTHRENE
General Information
Description: Solid or monoclinic crystals.
Formula: $(C_6H_4CH)_2$.
Constants: Mol wt: 178.22, mp: 97°C, bp: 339°C, d: 1.179
  at 25°C, vap. press.: 1 mm at 118.3, vap. d.: 6.14.
Hazard Analysis
Toxic Hazard Rating:
  Acute Local: Irritant 1; Allergen 1.
  Acute Systemic: U.

Chronic Local: Allergen 1.
Chronic Systemic: U.
Toxicity: Causes skin photosensitization and is a carcinogen.
Fire Hazard: Slight, when exposed to heat or flame; can react with oxidizing materials.
Countermeasures
To Fight Fire: Water, foam, carbon dioxide or dry chemical or carbon tetrachloride (Section 6).
Storage and Handling: Section 7.

PHENEANTHRENEQUINONE
General Information
Description: Yellow-orange, needle like crystals; soluble in sulfuric acid, benzene, glacial acetic acid, and hot alcohol; slightly soluble in ether; insoluble in water.
Formula: $C_{14}H_8O_2$.
Constants: Mol wt: 208, d: 1.4045, mp: 206–207°C, bp: sublimes > 360°C.
Hazard Analysis
See phenanthrene.

PHENARSAZINE CHLORIDE. See diphenylamine chlorarsine.

PHENAZINE
General Information
Synonym: Dibenzopyrazine.
Description: Pale yellow crystals.
Formula: $C_{12}H_8N_2$.
Constants: Mol wt: 180.2, mp: 171°C, bp: > 360°C (Sublimes).
Hazard Analysis
Toxicity: Details unknown; a larvicide.
Disaster Hazard: Moderately dangerous; when heated to decomposition, it emits toxic fumes.
Countermeasures
Storage and Handling: Section 7.

PHENAZINE OXIDE
Hazard Analysis
Toxicity: Details unknown; an insecticide
Fire Hazard: Slight; when heated, it emits toxic fumes (Section 6).
Countermeasures
Storage and Handling: Section 7.

PHENCAPTON
Shipping Regulations: Section 11.
   IATA: Other restricted articles, class A, no label required, no limit (passenger and cargo).

PHENETHYL ALCOHOL
General Information
Synonyms: 2-Phenyl ethanol; benzyl carbinol.
Description: Colorless liquid; floral odor of roses.
Formula: $C_6H_5CH_2CH_2OH$.
Constants: Mol wt: 122.14, mp: −27°C, bp: 220°C, flash p: 216°F, d: 1.0245 at 15°C, vap. d.: 4.21.
Hazard Analysis
Toxicology: Reported as causing severe CNS injury in experimental animals.
Fire Hazard: Moderate, when exposed to heat or flame; can react with oxidizing materials.
Countermeasures
To Fight Fire: Foam, carbon dioxide, dry chemical or carbon tetrachloride (Section 6).
Storage and Handling: Section 7.

o-PHENETIDINE
General Information
Synonym: 2-Aminophenetole.
Description: Oily liquid.
Formula: $C_8H_{11}NO$.
Constants: Mol wt: 137.2, mp: < −20°C, bp: 229°C, vap. press.: 1 mm at 67.0°C.
Hazard Analysis
Toxic Hazard Rating:
   Acute Local: U.
   Acute Systemic: Ingestion 3; Inhalation 3.
   Chronic Local: U.
   Chronic Systemic: Ingestion 2; Inhalation 2.
   Disaster Hazard: Dangerous; when heated to decomposition, it emits highly toxic fumes of nitrogen oxides.
Countermeasures
Ventilation Control: Section 2.
Personnel Protection: Section 3.
Storage and Handling: Section 7.
MCA warning label.

p-PHENETIDINE
General Information
Synonym: 4-Aminophenetole.
Description: Colorless liquid.
Formula: $C_8H_{11}NO$.
Constants: Mol wt: 137.2, mp: 3°C, bp: 254°C, flash p 240°F, d: 1.0652 at 16°/4°C, vap. d.: 4.73.
Hazard Analysis
Toxic Hazard Rating:
   Acute Local: U.
   Acute Systemic: Ingestion 3; Inhalation 3.
   Chronic Local: U.
   Chronic Systemic: Ingestion 2; Inhalation 2.
   Fire Hazard: Slight, when exposed to heat or flame.
   Disaster Hazard: Dangerous; when heated to decomposition, it emits highly toxic fumes of nitrogen oxides; re acts vigorously with powerful oxidizers.
Countermeasures
Ventilation Control: Section 2.
Personnel Protection: Section 3.
Storage and Handling: Section 7.
MCA warning label.

PHENETOLE. See phenyl ethyl ether.

PHENIC ACID. See phenol.

PHENIDONE
General Information
Synonym: 1-Phenyl-3-pyrazolidone.
Description: Crystals. Water soluble.
Formula: $C_9H_{10}N_2O$.
Constants: Mol wt: 162.2, mp: 121°C.
Hazard Analysis
Toxicity: Details unknown. Animal experiments show lov toxicity and no skin irritation.
Disaster Hazard: Dangerous; when heated to decomposi tion, it emits highly toxic fumes.

PHENOBARBITAL
General Information
Description: White, shining, crystalline, odorless powder bitter taste.
Formula: $CO(NHCO)_2C(C_2H_5)(C_6H_5)$.
Constants: Mol wt: 232.2, mp: 174–178°C.
Hazard Analysis
Toxic Hazard Rating:

TOXIC HAZARD RATING CODE (For detailed discussion, see Section 1.)

0 NONE: (a) No harm under any conditions; (b) Harmful only under unusual conditions or overwhelming dosage.

1 SLIGHT: Causes readily reversible changes which disappear after end of exposure.

2 MODERATE: May involve both irreversible and reversible change not severe enough to cause death or permanent injury.

3 HIGH: May cause death or permanent injury after very short exposu to small quantities.

U UNKNOWN: No information on humans considered valid by author

Acute Local: U.
Acute Systemic: Ingestion 2.
Chronic Local: U.
Chronic Systemic: Ingestion 1.
Caution: Repeated ingestion may lead to habituation.
Fire Hazard: Slight; when heated to decomposition, it emits toxic fumes (Section 6).
Countermeasures
Ventilation Control: Section 2.
Personal Hygiene: Section 3.
Storage and Handling: Section 7.

PHENOL
General Information
Synonyms: Carbolic acid; phenic acid; phenylic acid.
Description: White, crystalline mass which turns pink or red if not perfectly pure; burning taste, distinctive odor.
Formula: C_6H_5OH.
Constants: Mol wt: 94.11, mp: 40.6°C, bp: 181.9°C, flash p: 175°F (C.C.), d: 1.072, autoign. temp.: 1319°F, vap. press.: 1 mm at 40.1°C, vap. d.: 3.24.
Hazard Analysis
Toxic Hazard Rating:
 Acute Local: Irritant 3; Ingestion 3; Inhalation 3.
 Acute Systemic: Ingestion 3; Inhalation 3; Skin Absorption 3.
 Chronic Local: Irritant 2.
 Chronic Systemic: Ingestion 2; Inhalation 2; Skin Absorption 2.
TLV: ACGIH (recommended); 5 parts per million in air; 19 milligrams per cubic meter of air. Can be absorbed through the intact skin.
Toxicology: In acute phenol poisoning, the main effect is on the central nervous system. Absorption from spilling phenolic solutions on the skin may be very rapid, and death results from collapse within 30 minutes to several hours. Death has resulted from absorption of phenol through a skin area of 64 square inches. Where death is delayed, damage to the kidneys, liver, pancreas, spleen, and edema of the lungs may result. Absorbed phenol is partly excreted by the kidneys, partly oxidized. Part of the excreted portion is combined with sulfuric and glycuronic acids; the remainder is excreted unchanged. The symptoms develop rapidly, frequently within 15 to 20 minutes following spilling of phenol on the skin. Headache, dizziness, muscular weakness, dimness of vision, ringing in the ears, irregular and rapid breathing, weak pulse, and dyspnea may all develop, and may be followed by loss of consciousness, collapse and death. When taken internally, there is also nausea, with or without vomiting, severe abdominal pain, and corrosion of the lips, mouth, throat, esophagus and stomach. There may be perforation. On the skin, the affected area is white, wrinkled and softened, and there is usually no immediate complaint of pain; later, intense burning is felt, followed by local anesthesia and still later, by gangrene.
 Chronic poisoning, following prolonged exposures to low concentrations of the vapor or mist, results in digestive disturbances (vomiting, difficulty in swallowing, excessive salivation, diarrhea, loss of appetite), nervous disorders (headache, fainting, dizziness, mental disturbances) and skin eruptions. Chronic poisoning may terminate fatally in cases where there has been extensive damage to the kidneys or liver. Dermatitis resulting from contact with phenol or phenol-containing products is fairly common in industry. A common air contaminant (Section 4).
Fire Hazard: Moderate, when exposed to heat or flame.
Spontaneous Heating: No.
Disaster Hazard: Dangerous; when heated it emits toxic fumes; can react with oxidizing materials.
Countermeasures
Ventilation Control: Section 2.

To Fight Fire: Alcohol foam, carbon dioxide, dry chemical or carbon tetrachloride (Section 6).
Personnel Protection: Section 3.
First Aid: Section 1.
Storage and Handling: Section 7.
Shipping Regulations: Section 11.
 I.C.C.: Poison B; poison label, 55 gallons.
 Coast Guard Classification: Poison B; poison label (liquid and solid).
MCA warning label.
 IATA: Poison B, poison label, 1 liter (passenger), 220 liters (cargo).

PHENOL LITHIUM. See lithium phenate.

PHENOLPHTHALEIN
General Information
Synonym: 3,3-Bis(p-hydroxyphenyl)phthalide.
Description: Small crystals.
Formula: $C_{20}H_{14}O_4$.
Constants: Mol wt: 318.3, mp: 260°C, d: 1.277.
Hazard Analysis
Toxic Hazard Rating:
 Acute Local: U.
 Acute Systemic: Ingestion 3.
 Chronic Local: U.
 Chronic Systemic: U.
Countermeasures
Personal Hygiene: Section 3.
Storage and Handling: Section 7.

PHENOLS, HIGH-BOILING
General Information
Description: Liquid.
Composition: Alkyl-substituted phenols.
Constants: Mol wt: 150 (average), bp: 238–288°C, fp: < −40°C, flash p: 250°F (O.C.), d: 1.033 at 20°/20°C, vap. press.: 0.01 mm at 20°C, vap. d.: approx. 5.2.
Hazard Analysis
Toxicity: See phenol.
Fire Hazard: Slight, when exposed to heat or flame; can react with oxidizing materials.
Disaster Hazard: Dangerous; when heated it emits toxic fumes.
Countermeasures
To Fight Fire: Foam, carbon dioxide, dry chemical or carbon tetrachloride (Section 6).
Storage and Handling: Section 7.

PHENOL SULFONATE
Hazard Analysis
Toxic Hazard Rating:
 Acute Local: Irritant 2; Ingestion 2.
 Acute Systemic: Ingestion 2; Inhalation 2.
 Chronic Local: Irritant 1.
 Chronic Systemic: Ingestion 2; Inhalation 2.
Disaster Hazard: Dangerous; see sulfonates.
Countermeasures
Ventilation Control: Section 2.
Personnel Protection: Section 3.
Storage and Handling: Section 7.

PHENOLSULFONIC ACID
General Information
Synonym: Sulfocarbolic acid.
Description: Yellowish liquid; a mixture of o- and p-phenolsulfonic acids.
Formula: $C_6H_5SO_3H$.
Constants: Mol wt: 174.17, mp: 50°C, d: 1.155.
Hazard Analysis
Toxicity: Details unknown; less irritating and less toxic than phenol. See phenol and sulfuric acid.
Disaster Hazard: Dangerous; see sulfonates; will react with water or steam to produce heat.

Countermeasures
Storage and Handling: Section 7.
Shipping Regulations: Section 11.
    IATA: Corrosive liquid, white label, 1 liter (passenger), 5 liters (cargo).

PHENOTHIAZINE
General Information
Synonym: Thiodiphenylamine.
Description: Grayish-green powder.
Formula: $C_{12}H_9NS$.
Constants: Mol wt: 199.2, mp: 175–179°C, bp: 371°C decomposes.
Hazard Analysis
Toxic Hazard Rating:
    Acute Local: Irritant 2; Ingestion 2.
    Acute Systemic: Ingestion 2.
    Chronic Local: U.
    Chronic Systemic: U.
Toxicology: An insecticide. Large doses or heavy exposure may cause hemolytic anemia and toxic degeneration of the liver. Can cause skin irritation and photosensitization. Used as a food additive permitted in the feed and drinking water of animals and/or for the treatment of food producing animals. Also permitted in food for human consumption (Section 10).
Disaster Hazard: Dangerous; when heated to decomposition or on contact with acid or acid fumes, it emits highly toxic fumes of oxides of sulfur and nitrogen.
Countermeasures
Personnel Protection: Section 3.
Storage and Handling: Section 7.
Shipping Regulations: Section 11.
MCA warning label.

PHENOTHIOXIN
General Information
Synonym: Dibenzothioxin.
Description: A powder.
Formula: $C_{12}H_8OS$.
Constant: Mol wt: 200.2.
Hazard Analysis
Toxic Hazard Rating:
    Acute Local: U.
    Acute Systemic: Ingestion 2.
    Chronic Local: Irritant 2.
    Chronic Systemic: Ingestion 2.
Toxicology: Animal experiments have shown evidence of liver damage and skin irritation.
Disaster Hazard: Dangerous; when heated to decomposition, it emits highly toxic fumes of sulfur compounds.
Countermeasures
Storage and Handling: Section 7.

2-PHENOXYETHANOL. See ethylene glycol phenyl ether.

2-PHENOXYETHYL ACETATE. See ethylene glycol phenyl ether acetate.

n-(β-PHENOXYETHYL)ANILINE
General Information
Description: Insoluble in water.
Formula: $C_6H_5OC_2H_4OH$.
Constants: Mol wt: 138, d: 1.1, bp: 396°F, flash p: 338°F.
Hazard Analysis
Fire Hazard: Slight, when exposed to heat or flame.
Countermeasures
Storage and Handling: Section 7.
To Fight Fire: Water, foam. Section 6.

PHENOXYETHYL CHLORIDE. See β-chlorophenetole.

PHENOZONE. See antipyrine.

N-PHENYLACETAMIDE. See acetanilide.

PHENYL ACETATE
General Information
Synonym: Acetyl phenol.
Description: Water white liquid; infinitely soluble in alcohol and ether; slightly soluble in water.
Formula: $CH_3COOC_6H_5$.
Constants: Mol wt: 136, d: 1.073 at 25/25°C, bp: 195–196°C, vap. d.: 4.7, flash p: 176°F.
Hazard Analysis
Toxic Hazard Rating: U. See esters and phenol.
Fire Hazard: Moderate, when exposed to heat or flame.
Countermeasures
Storage and Handling: Section 7.
To Fight Fire: Alcohol foam. Section 6.

PHENYLACETIC ACID
General Information
Synonym: α-Toluic acid.
Description: White crystals; sweet, floral honey-like odor.
Formula: $C_6H_5CH_2COOH$.
Constants: Mol wt: 136.14, mp: 76.7°C, bp: 265.5°C, d: 1.228 at 20°/4°C, vap. press.: 1 mm at 97.0°C.
Hazard Analysis
Toxicity: Details unknown; a fungicide; probably highly toxic. See also toluene.
Disaster Hazard: Dangerous; when heated to decomposition, it emits toxic fumes.
Countermeasures
Storage and Handling: Section 7.

PHENYLACETONITRILE
General Information
Synonym: Benzyl cyanide.
Description: Oily liquid; aromatic odor.
Formula: $C_6H_5CH_2CN$.
Constants: Mol wt: 118.2, mp: −23.8°C, bp: 233.5°C, d: 1.0214 at 15°/15°C, vap. press.: 1 mm at 60.0°C.
Hazard Analysis and Countermeasures
See nitriles.

2-PHENYL ACRYLIC ACID
General Information
Synonym: Atropic acid.
Description: Tabular or acicular crystals; soluble in 790 parts water, alcohol, benzene, chloroform, ether, and carbon disulfide.
Formula: $C_6H_5C:CH_2COOH$.
Constants: Mol wt: 148.15, mp: 106–107°C, bp: 267°C decomposes.
Hazard Analysis
See acrylic acid and phenols.

PHENYLALANINE
General Information
Synonyms: α-Amino-β-phenylpropionic acid.
Description: An essential amino acid; occurs in isomeric forms; Leaflets or prisms, soluble in water.
Formula: $C_6H_5CH_2CH(NH_2)COOH$.
Constants: Mol wt: 165, L-form decomposes at 283°C, DL form decomposes at 318–320°C.
Hazard Analysis
Toxicity: Details unknown. A nutrient and/or dietary supplement food additive (Section 10).

TOXIC HAZARD RATING CODE (For detailed discussion, see Section 1.)

0 NONE: (a) No harm under any conditions; (b) Harmful only under unusual conditions or overwhelming dosage.

1 SLIGHT: Causes readily reversible changes which disappear after end of exposure.

2 MODERATE: May involve both irreversible and reversible changes not severe enough to cause death or permanent injury.

3 HIGH: May cause death or permanent injury after very short exposure to small quantities.

U UNKNOWN: No information on humans considered valid by authors

PHENYLAMINE. See aniline.

PHENYLANILINE. See diphenylamine.

PHENYLARSONIC ACID
General Information
Synonym: Benzene arsonic acid.
Description: Colorless crystals. Water soluble.
Formula: $C_6H_5AsO(OH)_2$.
Constants: Mol wt: 202, d: 1.760, mp: 160°C decomposes.
Hazard Analysis and Countermeasures
See arsenic compounds.

p-PHENYLAZOANILINE. See p-aminoazobenzene.

PHENYL AZOIMIDE
General Information
Synonyms: Diazobenzeneimide; triazobenzene.
Description: Yellow oil.
Formula: $C_6H_5N_3$.
Constants: Mol wt: 119.1, bp: 59°C (explodes), d: 1.078 at 22.5°C.
Hazard Analysis
Toxicity: Details unknown.
Fire Hazard: Moderate, when exposed to heat or flame (Section 6).
Explosion Hazard: Severe, when shocked, exposed to heat or flame or by chemical reaction.
Disaster Hazard: Dangerous; shock or heat will explode it; can react with oxidizing materials; on contact with acid or acid fumes it can emit toxic fumes.
Countermeasures
Ventilation Control: Section 2.
Storage and Handling: Section 7.

PHENYL BENZOATE
General Information
Description: Colorless crystals, geranium odor.
Formula: $C_6H_5COOC_6H_5$.
Constants: Mol wt: 198.2, mp: 70°C, bp: 314°C, d: 1.235, vap. press.: 1 mm at 106.8°C.
Hazard Analysis
Toxicity: See phenol and benzoic acid.
Disaster Hazard: Slight; when heated, it emits acrid fumes (Section 6).
Countermeasures
Storage and Handling: Section 7.

1-PHENYLBENZOTHIAZOLE
Hazard Analysis
Toxicity: Details unknown; a larvicide.

PHENYL BENZOYL CARBINOL. See benzoin.

PHENYL BENZYL TIN DICHLORIDE
General Information
Description: Colorless crystals.
Formula: $(C_6H_5)(C_6H_5CH_2)SnCl_2$.
Constants: Mol wt: 357.8, mp: 84°C.
Hazard Analysis and Countermeasures
See tin compounds and chlorides.

PHENYLBIGUANIDE
General Information
Description: Colorless crystals.
Hazard Analysis
Toxicity: Details unknown. See also guanidine hydrochloride.
Disaster Hazard: Dangerous; when heated to decomposition, it emits highly toxic fumes of nitrogen compounds.
Countermeasures
Storage and Handling: Section 7.

PHENYLBIGUANIDE HYDROCHLORIDE. See phenylbiguanide.

PHENYLBIGUANIDE MERCAPTO BENZO-THIAZOLE SALT. See phenylbiguanide.

m-PHENYLBIPHENYL. See m-terphenyl.

PHENYLBORON DIBROMIDE
General Information
Description: Colorless crystals decomposed by water.
Formula: $C_6H_5BBr_2$.
Constants: Mol wt: 247.8, mp: 34°C, bp: 100°C at 20 mm.
Hazard Analysis and Countermeasures
See boron compounds and bromides.

PHENYLBORON DICHLORIDE
General Information
Description: Colorless liquid decomposed by water.
Formula: $C_6H_5BCl_2$.
Constants: Mol wt: 158.8, mp: 0°C, bp: 175°C.
Hazard Analysis and Countermeasures
See boron compounds and chlorides.

PHENYL BROMIDE. See bromobenzene.

1-PHENYLBUTANE. See butylbenzene.

2-PHENYLBUTANE. See sec-butylbenzene.

1-PHENYLBUTENE-2
General Information
Description: A liquid.
Formula: $C_{10}H_{12}$.
Constants: Mol wt: 132.2, bp: 175°C, flash p: 160°F (T.O.C.), d: 0.888 at 15.5°/15.5°C.
Hazard Analysis
Toxicity: Unknown. In high concentrations, probably narcotic.
Fire Hazard: Moderate, in the presence of heat or flame; can react with oxidizing materials.
Countermeasures
To Fight Fire: Foam, carbon dioxide, dry chemical or carbon tetrachloride (Section 6).
Ventilation Control: Section 2.
Storage and Handling: Section 7.

PHENYL CARBINOL. See benzyl alcohol.

PHENYL "CARBITOL"
General Information
Synonym: Diethylene glycol monophenyl ether.
Description: Liquid.
Formula: $C_6H_5OC_2H_4OC_2H_4OH$.
Constants: Mol wt: 182.21, bp: 207°C at 55 mm, fp: −50°C, d: 1.1158 at 20°/20°C, vap. press.: < 0.01 mm at 20°C, vap. d.: 6.28.
Hazard Analysis
Toxicity: Details unknown. See also glycols.
Disaster Hazard: Moderate; when heated, it emits acrid fumes.
Countermeasures
Storage and Handling: Section 7.

PHENYLCARBYLAMINE
General Information
Synonym: Phenyl isocyanide.
Description: Colorless to greenish liquid. Soluble in ether.
Formula: $C_6H_5N≡C$.
Constants: Mol wt: 103.1, d: 0.9775 at 15°C, bp: 166°C decomposes.
Hazard Analysis and Countermeasures
See cyanides.

PHENYL CARBYLAMINE CHLORIDE
General Information
Synonym: Phenyliminophosgene.
Description: Pale yellow, oily liquid.
Formula: $C_6H_5NCCl_2$.

Constants: Mol wt: 174.03, bp: 208 to 210°C, d: 1.30 at 15°C, vap. d.: 6.03.

Hazard Analysis
Toxic Hazard Rating:
 Acute Local: Irritant 3; Ingestion 3, Inhalation 3.
 Acute Systemic: Ingestion 3; Inhalation 3; Skin Absorption 3.
 Chronic Local: U.
 Chronic Systemic: U.
Disaster Hazard: Dangerous; when heated to decomposition, it emits highly toxic fumes of chlorides. A tear gas.

Countermeasures
Ventilation Control: Section 2.
Personnel Protection: Section 3.
First Aid: Section 1.
Storage and Handling: Section 7.
Shipping Regulations: Section 11.
 I.C.C.: Poison A; poison gas label, not accepted.
 Coast Guard Classification: Poison A; poison gas label.
 IATA: Poison A, not acceptable (passenger and cargo).

PHENYL "CELLOSOLVE"
General Information
Synonym: Ethylene glycol phenyl ether.
Description: Clear liquid.
Formula: $C_6H_5O(CH_2)_2OH$.
Constants: Mol wt: 138.2, mp: 14°C, bp: 242°C, flash p: 250°F (O.C.), d: 1.109.

Hazard Analysis
Toxicity: Details unknown. See also glycols.
Fire Hazard: Slight, when exposed to heat or flame.
Spontaneous Heating: No.
Disaster Hazard: Moderate; when heated, it emits acrid fumes.

Countermeasures
Storage and Handling: Section 7.
To Fight Fire: Carbon dioxide, dry chemical or carbon tetrachloride (Section 6).

PHENYL CHLORIDE. See chlorobenzene.

PHENYLCHLOROFORM. See benzotrichloride.

2-PHENYL-6-CHLOROPHENOL
General Information
Synonym: 6-Chlor-o-xenol.
Description: Pale yellow, viscous liquid.
Formula: $C_{12}H_9ClO$.
Constants: Mol wt: 204.7, mp: 6°C, bp: 318°C decomposes, d: 1.24 at 25°/4°C.

Hazard Analysis
Toxicity: Details unknown; a germicide and fungicide. See also chlorophenol.
Disaster Hazard: Dangerous; see chlorides.

Countermeasures
Storage and Handling: Section 7.

PHENYLCHLOROPHENYL TRICHLOROETHANE
General Information
Description: Crystals.
Formula: $C_6H_5ClC_6H_4Cl_3C_2H$.
Constant: Mol wt: 320.1.

Hazard Analysis
Toxicity: Details unknown; an insecticide. See also DDT which is a related compound.
Disaster Hazard: Dangerous; see chlorides.

Countermeasures
Storage and Handling: Section 7.

PHENYL CYANIDE. See benzonitrile.

PHENYLCYCLOHEXANE
General Information
Synonym: Cyclohexylbenzene.
Description: Colorless, oily liquid.
Formula: $C_{12}H_{16}$.
Constants: Mol wt: 160.25, mp: 7.5°C, bp: 240.0°C, flash p: 210°F (O.C.), d: 0.938 ± 0.02 at 25°/15.6°C, autoign. temp.: 220°F, vap. press.: 1 mm at 67.5°C.

Hazard Analysis
Toxicity: Details unknown. See also cyclohexane.
Fire Hazard: Moderate, when exposed to heat or flame; can react with oxidizing materials (Section 6).

Countermeasures
Storage and Handling: Section 7.

2-PHENYLCYCLOHEXANOL
General Information
Description: Colorless to pale, straw-colored liquid.
Formula: $C_6H_{10}C_6H_5OH$.
Constants: Mol wt: 176.2, bp: 276–281°C, flash p: 280°F, d: 1.033 at 25°/25°C, vap. d.: 6.13.

Hazard Analysis
Toxicity: Details unknown. Animal experiments indicate moderate toxicity. Can cause dermatitis.
Fire Hazard: Slight; when exposed to heat or flame; can react with oxidizing materials.

Countermeasures
To Fight Fire: Foam, carbon dioxide, dry chemical or carbon tetrachloride (Section 6).
Storage and Handling: Section 7.

PHENYL CYCLOHEXYL HYDROPEROXIDE
General Information
Description: Available as solution.
Formula: $C_{12}H_{16}O_2$.
Constants: Mol wt: 192, flash p: > 212°F.

Hazard Analysis
Toxicity: Details unknown. See also peroxides, organic.
Fire Hazard: Slight, when exposed to heat or flame or by spontaneous chemical reaction.
Disaster Hazard: Dangerous; when heated, it burns and emits acrid fumes; can react with oxidizing or reducing materials.

Countermeasures
Storage and Handling: Section 7.
To Fight Fire: Foam, carbon dioxide, dry chemical or carbon tetrachloride (Section 6).

PHENYL DIAZO SULFIDE
General Information
Description: Red crystals.
Formula: $(C_6H_5N_2)_2S$.
Constant: Mol wt: 242.3.

Hazard Analysis
Toxicity: Unknown. See also sulfides.
Fire Hazard: Slight, when exposed to heat (Section 6).
Explosion Hazard: Severe, when shocked or exposed to heat; when dry it can explode at room temperature.
Disaster Hazard: Dangerous; shock will explode it; on decomposition it emits highly toxic fumes of oxides of sulfur and nitrogen; can react with oxidizing materials.

Countermeasures
Storage and Handling: Section 7.

TOXIC HAZARD RATING CODE (For detailed discussion, see Section 1.)

0 NONE: (a) No harm under any conditions; (b) Harmful only under unusual conditions or overwhelming dosage.

1 SLIGHT: Causes readily reversible changes which disappear after end of exposure.

2 MODERATE: May involve both irreversible and reversible changes not severe enough to cause death or permanent injury.

3 HIGH: May cause death or permanent injury after very short exposure to small quantities.

U UNKNOWN: No information on humans considered valid by authors

PHENYL DICHLOROARSINE, Liquid
General Information
Description: Colorless gas or liquid; changes to yellow.
Formula: $C_6H_5AsCl_2$.
Constants: Mol wt: 222.92, bp: 255–257°C, fp: −15.6°C, d: 1.654 at 20°C, vap. press.: 0.021 mm at 20°C, vap. d.: 7.7.
Hazard Analysis
Toxic Hazard Rating:
  Acute Local: Irritant 3; Ingestion 3; Inhalation 3.
  Acute Systemic: Ingestion 3; Inhalation 3.
  Chronic Local: U.
  Chronic Systemic: U.
Toxicology: A lacrymator type of military poison gas. See also arsenic compounds.
Disaster Hazard: Dangerous; when heated to decomposition, it emits highly toxic fumes of arsenic; will react with water or steam to produce corrosive fumes.
Countermeasures
Ventilation Control: Section 2.
Personnel Protection: Section 3.
Storage and Handling: Section 7.
Shipping Regulations: Section 11.
  I.C.C.: Poison B; poison label, 300 gallons.
  Coast Guard Classification: Poison B; poison label.
  IATA: Poison B, poison label, not acceptable (passenger), 120 liters (cargo).

PHENYL DICHLOROPHOSPHINE. See benzene phosphorus dichloride.

PHENYL DICHLOROPHOSPHINE OXIDE. See benzene phosphorus oxydichloride.

PHENYL DICHLOROPHOSPHINE SULFIDE. See benzene phosphorus thiodichloride.

PHENYL DIDECYL PHOSPHITE
General Information
Description: Nearly water white liquid; alcohol odor; insoluble in water,
Formula: $C_6H_5OP(OC_{10}H_{21})_2$.
Constants: Mol wt: 433, d: 0.940 at 25/15.5°C, mp: <0°C, flash p: 425°F (O.C.).
Hazard Analysis
Toxic Hazard Rating: U.
Fire Hazard: Slight, when exposed to heat or flame.
Disaster Hazard: Dangerous. See phosphites.
Countermeasures
Storage and Handling: Section 7.
To Fight Fire: Water, foam.

PHENYL DIETHANOLAMINE
General Information
Description: Liquid.
Formula: $C_6H_5N(CH_2CH_2OH)_2$.
Constants: Mol wt: 181.23, mp: 57.8°C, bp: 192°C, flash p: 375°F (O.C.), d: 1.1203, vap. press.: < 0.01 mm at 20°C.
Hazard Analysis
Toxicity: See amines.
Fire Hazard: Slight, when exposed to heat or flame; can react with oxidizing materials.
Countermeasures
To Fight Fire: Water, carbon dioxide, dry chemical or carbon tetrachloride (Section 6).
Storage and Handling: Section 7.

PHENYL DIGLYCOL CARBONATE. See diethylene glycol bis phenyl carbonate.

3-PHENYL-1,1-DIMETHYLUREA
General Information
Synonym: PDU.
Description: Crystals.
Formula: $C_6H_5NHCON(CH_3)_2$.

Constant: Mol wt: 164.
Hazard Analysis
Toxicity: Details unknown. See related compound Monuron.
Disaster Hazard: Dangerous. When heated to decomposition, it emits highly toxic fumes.

PHENYL DI-o-XENYL PHOSPHATE
General Information
Description: Insoluble in water.
Formula: $(C_{12}H_9O)_2POOC_6H_5$.
Constants: Mol wt: 484, d: 1.2, bp: 545–626°F, flash p: 428°F.
Hazard Analysis
Toxic Hazard Rating: U.
Fire Hazard: Slight, when exposed to heat or flame.
Disaster Hazard: Dangerous. See phosphates.
Countermeasures
Storage and Handling: Section 7.
To Fight Fire: Foam or water.

m-PHENYLENEDIAMINE
General Information
Synonym: m-Diaminobenzene.
Description: White crystals.
Formula: $C_6H_4(NH_2)_2$.
Constants: Mol wt: 108.1, mp: 63°C, bp: 286°C, d: 1.139, vap. press.: 1 mm at 99.8°C.
Hazard Analysis
Toxicity: Somewhat less than p-phenylene diamine.
Fire Hazard: Slight; when heated to decomposition, it emits toxic fumes (Section 6).
Countermeasures
Storage and Handling: Section 7.
Shipping Regulations: Section 11.
MCA warning label.
  IATA: Other restricted articles, class A, no label required, no limit (passenger and cargo).

o-PHENYLENEDIAMINE
General Information
Synonym: o-Diaminobenzene.
Description: Brownish crystals.
Formula: $C_6H_4(NH_2)_2$.
Constants: Mol wt: 108.1, mp: 104°C, bp: 257°C.
Hazard Analysis
Toxicity: Somewhat less than p-phenylene diamine.
Fire Hazard: Slight; (Section 6).
Countermeasures
Storage and Handling: Section 7.

p-PHENYLENEDIAMINE
General Information
Synonym: Diaminobenzene; ursol.
Description: Colorless crystals.
Formula: $C_6H_4(NH_2)_2$.
Constants: Mol wt: 108.1, mp: 139.7°C, flash p: 312°F, vap. d.: 3.72, bp: 267°C.
Hazard Analysis
Toxic Hazard Rating:
  Acute Local: Irritant 2; Allergen 1.
  Acute Systemic: Ingestion 2; Inhalation 2.
  Chronic Local: Allergen 2.
  Chronic Systemic: Ingestion 2; Inhalation 2.
TLV: ACGIH (recommended) 0.1 milligram per cubic meter in air.
Toxicology: Of the three phenylene diamines, the p-form has proved to be an especially powerful skin irritant. This material is also responsible for asthmatic symptoms and other respiratory symptoms of workers in the fur dye industry where it is commonly used. The p- form will cause kerato-conjunctivitis, swollen conjunctiva, and eczema of the eyelids. Systemic poisoning from this material is uncommon. At least one fatal case of liver involvement due to the p-form has been reported in the literature. m- and o-phenylene diamine

are somewhat less toxic than the p-isomer. For this reason and possibly because they are much less used in industry, they are not considered to be industrial hazards. All forms are flammable liquids and moderate fire hazards.

Fire Hazard: Slight, when exposed to heat or flame.

Spontaneous Heating: No.

Disaster Hazard: Dangerous; when heated, it burns and emits highly toxic fumes of nitrogen compounds; can react with oxidizing materials.

Countermeasures

Ventilation Control: Section 2.

Treatment and Antidotes: If removal from exposure is prompt, the symptoms will be relieved. If the exposure has been severe, a physician should be called at once. If the material has come in contact with the eyes, mucous membranes, or the skin, the contaminated area or the involved skin areas should be washed promptly with plenty of lukewarm water and soap.

Personnel Protection: Section 3.

To Fight Fire: Water, carbon dioxide, dry chemical or carbon tetrachloride (Section 6).

Storage and Handling: Section 7.

Shipping Regulations: Section 11.

MCA warning label.

IATA: Other restricted articles, class A, no label required, no limit (passenger and cargo).

m-PHENYLENEDIAMINE FLUOSILICATE
General Information

Description: Brown crystals; slightly soluble in alcohol.

Formula: $C_6H_4(NH_2)_2 \cdot H_2SiF_6$.

Constants: Mol wt: 252.2, mp: 274°C.

Hazard Analysis and Countermeasures

See fluosilicates and fluorides.

p-PHENYLENEDIAMINE FLUOSILICATE
General Information

Description: Pink irregular crystals. Slightly soluble in alcohol.

Formula: $C_6H_4(NH_2)_2 \cdot H_2SiF_6$.

Constant: Mol wt: 252.2.

Hazard Analysis and Countermeasures

See fluosilicates.

PHENYLETHANE. See ethyl benzene.

2-PHENYLETHANOL. See phenethyl alcohol.

PHENYLETHANOLAMINE
General Information

Description: Liquid.

Formula: $C_6H_5NHCH_2CH_2OH$.

Constants: Mol wt: 137.18, bp: 285.2°C, flash p: 305°F (O.C.), d: 1.097 at 20°/20°C, vap. press.: < 0.01 mm at 20°C.

Hazard Analysis

Toxic Hazard Rating:

Acute Local: Irritant 2; Ingestion 2; Inhalation 2.

Acute Systemic: U.

Chronic Local: U.

Chronic Systemic: U.

Toxicity: See also amines.

Fire Hazard: Slight, when exposed to heat or flame; can react with oxidizing materials.

Countermeasures

To Fight Fire: Alcohol foam, foam, carbon dioxide, dry chemical or carbon tetrachloride (Section 6).

Ventilation Control: Section 2.

Personnel Protection: Section 3.

Storage and Handling: Section 7.

PHENYL ETHYL ACETATE
General Information

Description: Colorless liquid.

Formula: $C_6H_5C_2H_4OOCCH_3$.

Constants: Mol wt: 164.2, bp: 223.6°C, fp: < −20°C, flash p: 230°F, d: 1.032 at 25°/25°C.

Hazard Analysis

Toxicity: Unknown. See esters.

Fire Hazard: Slight, when exposed to heat or flame.

Disaster Hazard: Moderate; when heated, it emits acrid fumes; can react with oxidizing materials.

Countermeasures

Storage and Handling: Section 7.

To Fight Fire: Foam, carbon dioxide, dry chemical or carbon tetrachloride (Section 6).

β-PHENYLETHYLAMINE
General Information

Description: Colorless to slightly yellow liquid.

Formula: $C_6H_5(CH_2)_2NH_2$.

Constants: Mol wt: 121.2, bp: 200°C, d: 0.96 at 15.5°/15.5°C, vap. d.: 4.18.

Hazard Analysis

Toxicity: Details unknown. A skin irritant and possible sensitizer. Section 9. See also amines.

Fire Hazard: Slight, when exposed to heat or flame; can react with oxidizing materials (Section 6).

Countermeasures

Storage and Handling: Section 7.

PHENYLETHYLBARBITURIC ACID. See "Luminal."

PHENYLETHYL BENZOATE
General Information

Description: Crystals.

Formula: $C_2H_5C_6H_4C_6H_4COOH$.

Constant: Mol wt: 226.3.

Hazard Analysis

Toxicity: An insecticide. See benzoic acid.

Fire Hazard: Slight, when exposed to flame (Section 6).

Disaster Hazard: Slightly dangerous; when heated to decomposition it emits acrid fumes; can react with oxidizing materials.

Countermeasures

Storage and Handling: Section 7.

PHENYL ETHYL CARBAMATE. See phenylurethane.

PHENYL ETHYLENE
General Information

Synonym: Vinylbenzene, styrene (monomer), cinnamene.

Description: Colorless liquid.

Formula: $C_6H_5CHCH_2$.

Constants: Mol wt: 104.14, mp: −31°C, bp: 146°C, lel: 1.1%, uel: 6.1%, flash p: 88°F, d: 0.9074 at 20°/4°C, autoign. temp.: 914°F, vap. d.: 3.6.

Hazard Analysis

Toxic Hazard Rating:

Acute Local: Irritant 2; Ingestion 2; Inhalation 2.

Acute Systemic: Ingestion 2; Inhalation 2.

Chronic Local: U.

Chronic Systemic: Ingestion 2; Inhalation 2.

TLV: ACGIH (recommended); 100 parts per million; 420 milligrams per cubic meter of air.

TOXIC HAZARD RATING CODE (For detailed discussion, see Section 1.)

0 NONE: (a) No harm under any conditions; (b) Harmful only under unusual conditions or overwhelming dosage.

1 SLIGHT: Causes readily reversible changes which disappear after end of exposure.

2 MODERATE: May involve both irreversible and reversible changes; not severe enough to cause death or permanent injury.

3 HIGH: May cause death or permanent injury after very short exposure to small quantities.

U UNKNOWN: No information on humans considered valid by authors

Toxicology: It can cause irritation, violent itching of the eyes, lachrymation, and severe human eye injuries. Its toxic effects are usually transient and result in irritation and possible narcosis. It is not considered a very toxic material, because under ordinary conditions it does not vaporize sufficiently to reach a concentration that can kill animals, such as rats and guinea pigs, in a few minutes. Experimentally it has been found that 10,000 ppm was dangerous to animal life in from 30 to 60 minutes, 2,500 ppm was dangerous to life in 8 hours, while 1,300 ppm was the highest amount which was found to cause no serious systemic disturbances in 8 hours. However, all animals exposed to these amounts did evidence eye and nasal irritation, while those exposed to 2,500 ppm or more showed varying degrees of weakness and stupor, followed by incoordination, tremors and unconsciousness. To produce this unconsciousness required 10 hours at a concentration of 2,500 ppm. From a study to determine the chronic effects of this material, it was discovered that rats exposed to 1,300 ppm for from 7 to 8 hours per day 5 days a week for 26 weeks showed evidence and definite signs of eye and nasal irritation and appeared unkempt, though they made a normal gain in weight and presented no significant microscopic tissue changes or changes in the blood picture. 12 rabbits exposed to 1,300 ppm for the same period of time with one unexplained exception showed similar results.

Fire Hazard: Moderate, when exposed to flame.

Explosion Hazard: Moderate, when exposed to flame.

Disaster Hazard: Dangerous, upon exposure to heat or flame; on decomposition it emits acrid fumes; can react vigorously with oxidizing materials.

Countermeasures

Ventilation Control: Section 2.

Treatment and Antidotes: Personnel who show symptoms of irritation or beginning narcosis due to exposure to this material should be removed from exposure and the symptoms will disappear. If the symptoms persist, consult a physician.

Personnel Protection: Section 3.

To Fight Fire: Foam, carbon dioxide, dry chemical or carbon tetrachloride (Section 6).

Storage and Handling: Section 7.

PHENYL ETHYLENE OXIDE. See styrene oxide.

PHENYL ETHYL ETHANOLAMINE
General Information
Description: Crystals.
Formula: $C_6H_5N(C_2H_5)C_2H_4OH$.
Constants: Mol wt: 165.23, mp: 37.2°C, bp: 268°C at 740 mm, flash p: 270°F (O.C.), d: 1.04 at 20°/20°C, vap. press.: < 0.01 mm at 20°C.
Hazard Analysis
Toxicity: Details unknown. See also amines.
Fire Hazard: Slight, when exposed to heat or flame; it can react with oxidizing materials.
Countermeasures
To Fight Fire: Alcohol foam, foam, carbon dioxide, dry chemical or carbon tetrachloride (Section 6).
Storage and Handling: Section 7.

PHENYL ETHYL ETHER
General Information
Synonyms: Ethyl phenyl ether; phenetole.
Description: Colorless liquid; insoluble in water; soluble in alcohol and ether.
Formula: $C_6H_5OC_2H_5$.
Constants: Mol wt: 122.16, d: 0.967 at 20/4°C, mp: −30°C, bp: 172°C.
Hazard Analysis
Toxicity: Animal experiments show low toxicity and low irritation. See ethers.

PHENYL ETHYL HYDANTOIN. See nervanol.

PHENYL ETHYL MALONYLUREA. See "Luminal."

PHENYL ETHYL sec-PROPYL GERMANIUM BROMIDE
General Information
Description: Colorless oil.
Formula: $(C_6H_5)(C_2H_5)[CH(CH_3)_2]GeBr$.
Constants: Mol wt: 301.8, bp: 130–135°C at 13 mm.
Hazard Analysis and Countermeasures
See germanium compounds and bromides.

PHENYLFORMIC ACID. See benzoic acid.

PHENYLGERMANIUM TRIBROMIDE
General Information
Description: Colorless liquid; hydrolyzed by water.
Formula: $(C_6H_5)GeBr_3$.
Constants: Mol wt: 389.5, bp: 121°C at 13 mm.
Hazard Analysis and Countermeasures
See phenol, germanium compounds and bromides.

PHENYLGERMANIUM TRICHLORIDE
General Information
Description: Colorless liquid; hydrolyzed by water.
Formula: $(C_6H_5)GeCl_3$.
Constants: Mol wt: 256.1, bp: 106°C at 12 mm.
Hazard Analysis and Countermeasures
See phenol, germanium compounds, and chlorides.

PHENYLGERMANIUM TRIIODIDE
General Information
Description: White solid decomposed by actinic rays. Hydrolyzed by water.
Formula: $(C_6H_5)GeI_3$.
Constants: Mol wt: 530.5, mp: 56°C.
Hazard Analysis and Countermeasures
See phenol, germanium compounds and iodides.

PHENYL GLYCIDYL ETHER
General Information
Synonym: PGE.
TLV: ACGIH (recommended); 50 parts per million in air; 310 milligrams per cubic meter of air.
Hazard Analysis
Toxic Hazard Rating:
Acute Local: Irritant 2.
Acute Systemic: Ingestion 1; Skin Absorption 2.
Chronic Local: Allergen 1.
Chronic Systemic: U.
Data based on limited animal experiments.

PHENYL GLYCOLIC ACID. See mandelic acid.

PHENYLHYDRAZINE
General Information
Description: Yellow, monoclinic crystals or oil.
Formula: $C_6H_5NHNH_2$.
Constants: Mol wt: 108.14, mp: 19.6°C, bp: 243.5°C decomposes, flash p: 192°F (C.C.), d: 1.0978 at 20°/4°C, vap. press.: 1 mm at 71.8°C, vap. d.: 3.7.
Hazard Analysis
Toxic Hazard Rating:
Acute Local: Irritant 2; Allergen 1.
Acute Systemic: Ingestion 3; Inhalation 3.
Chronic Local: Irritant 2; Allergen 2.
Chronic Systemic: Ingestion 2; Inhalation 2.
TLV: ACGIH (recommended); 5 parts per million in air; 22 milligrams per cubic meter of air. Can be absorbed through the intact skin.
Toxicology: The ingestion or subcutaneous injection of phenyl hydrazine has been shown to cause hemolysis of the red blood cells, an effect which has been utilized in the treatment of polycythemia. The erythrocytes frequently contain Heinz bodies. Part of the hemoglobin is converted to methemoglobin. Pathological changes

seen in animals include congestion of the spleen with hyperplasia of the reticuloendothelial system, degeneration and necrosis of the liver cells with extensive pigmentation, early damage to the tubules of the kidneys with fatty changes in the cortical portion, and hyperplasia of the bone marrow. The most common effect of occupational exposure is the development of dermatitis which, in sensitized persons, may be quite severe. Systemic effects include anemia and general weakness, gastro-intestinal disturbances and injury to the kidneys.
Fire Hazard: Moderate, when exposed to heat or flame (Section 6).
Disaster Hazard: Dangerous; when heated to decomposition, it emits highly toxic fumes of nitrogen compounds; can react with oxidizing materials.

Countermeasures
Ventilation Control: Section 2.
Personnel Protection: Section 3.
Storage and Handling: Section 7.
To Fight Fire: Alcohol foam.

PHENYLHYDRAZINE HYDROCHLORIDE
General Information
Description: Leaflets.
Formula: $C_6H_5NHNH_2 \cdot HCl$.
Constants: Mol wt: 144.6, mp: 245°C.
Hazard Analysis
Toxicity: See phenylhydrazine.
Disaster Hazard: Dangerous; when heated to decomposition, it emits toxic fumes of nitrogen compounds and chlorides.
Countermeasures
Ventilation Control: Section 2.
Storage and Handling: Section 7.

PHENYL HYDRIDE. See benzene.

PHENYLHYDROXYACETIC ACID. See mandelic acid.

PHENYL-α-HYDROXYBENZYL KETONE. See benzoin.

PHENYLIC ACID. See phenol.

PHENYLIMINOPHOSGENE. See phenyl carbylamine chloride.

PHENYL ISOCYANATE
General Information
Description: Liquid.
Formula: C_6H_5NCO.
Constants: Mol wt: 119.1, mp: −30°C approx., bp: 166°C, d: 1.1 at 20°C, vap. press.: 1 mm at 10.6°C.
Hazard Analysis and Countermeasures
See cyanates.
Shipping Regulations: Section 11.
  IATA: Poison C, poison label, not acceptable (passenger), 40 liters (cargo).

PHENYL ISOCYANIDE. See phenyl carbylamine.

PHENYL ISOTHIOCYANATE. See phenyl mustard oil.

PHENYL KETONE. See benzophenone.

PHENYLMAGNESIUM BROMIDE
General Information
Description: A solid.
Formula: C_6H_5MgBr.
Constant: Mol wt: 181.3.

Hazard Analysis
Toxicity: Probably high. See also bromides and phenol.
Fire Hazard: Dangerous, by chemical reaction.
Explosion Hazard: Moderate, by chemical reaction.
Disaster Hazard: Dangerous; will react with water, steam or acids to produce heat and toxic and flammable vapors; can react vigorously with oxidizing materials; on decomposition it emits toxic fumes of bromides.
Countermeasures
To Fight Fire: Carbon dioxide, dry chemical or carbon tetrachloride (Section 6).
Storage and Handling: Section 7.

PHENYLMAGNESIUM CHLORIDE
General Information
Description: Crystals; soluble in ether.
Formula: C_6H_5MgCl.
Constant: Mol wt: 136.9.
Hazard Analysis and Countermeasures
See grignard reagents.

N-PHENYLMALEAMIC ACID
General Information
Synonym: Maleanilic acid.
Description: Yellow, crystalline solid.
Formula: $C_{10}H_9O_3N$.
Constants: Mol wt: 191.18, mp: 190°C, d: 1.418 at 30°C.
Hazard Analysis
Toxic Hazard Rating:
  Acute Local: Irritant 2; Ingestion 2; Inhalation 2.
  Acute Systemic: U.
  Chronic Local: U.
  Chronic Systemic: U.
Toxicology: Based on animal experiments.
Fire Hazard: Slight (Section 6).
Countermeasures
Ventilation Control: Section 2.
Personnel Protection: Section 3.
Storage and Handling: Section 7.

PHENYL MERCAPTAN
General Information
Synonym: Thiophenol; benzenethiol.
Description: Liquid; repulsive odor.
Formula: C_6H_5SH.
Constants: Mol wt: 110.2, bp: 168.3°C, d: 1.0728 at 25°/4°C.
Hazard Analysis
Toxicity: Details unknown; can cause severe dermatitis and exposure is said to be capable of causing headache and dizziness; mosquito larvicide. See also mercaptans.
Fire Hazard: Unknown.
Disaster Hazard: Dangerous; when heated to decomposition or on contact with acids, it emits toxic fumes of sulfur compounds.
Countermeasures
Storage and Handling: Section 7.

PHENYL MERCAPTOACETIC ACID
General Information
Description: White powder.
Formula: $C_6H_5SCH_2COOH$.
Constants: Mol wt: 168.2, mp: 63°C.
Hazard Analysis
Toxicity: Details unknown; a fungicide and bactericide probably highly toxic. See also mercaptans.
Disaster Hazard: Dangerous; when heated to decomposition

TOXIC HAZARD RATING CODE (For detailed discussion, see Section 1.)

0 NONE: (a) No harm under any conditions; (b) Harmful only under unusual conditions or overwhelming dosage.

1 SLIGHT: Causes readily reversible changes which disappear after end of exposure.

2 MODERATE: May involve both irreversible and reversible changes not severe enough to cause death or permanent injury.

3 HIGH: May cause death or permanent injury after very short exposure to small quantities.

U UNKNOWN: No information on humans considered valid by authors.

or on contact with acids, it emits highly toxic fumes of oxides of sulfur.

Countermeasures
Storage and Handling: Section 7.

PHENYLMERCURIC ACETATE
General Information
Description: Lustrous crystals; slightly soluble in water.
Formula: $(C_6H_5)HgC_2H_3CO_2$.
Constants: Mol wt: 336.8, mp: 149°C.
Hazard Analysis and Countermeasures
A fungicide and herbicide. See mercury compounds, organic.

PHENYLMERCURIC ACETOXYDECANOIC ACID
Hazard Analysis and Countermeasures
A fungicide. See mercury compounds, organic.

PHENYLMERCURIC AMMONIUM ACETATE
Hazard Analysis and Countermeasures
A fungicide. See mercury compounds, organic.

PHENYLMERCURIC BROMIDE
General Information
Description: Lustrous crystals; insoluble in water.
Formula: $(C_6H_5)HgBr$.
Constants: Mol wt: 357.6, mp: 276°C.
Hazard Analysis and Countermeasures
See mercury compounds, organic.

PHENYLMERCURIC CHLORIDE
General Information
Description: Satiny crystals; soluble in some organic solvents.
Formula: $(C_6H_5)HgCl$.
Constants: Mol wt: 313.2, mp: 251°C.
Hazard Analysis and Countermeasures
A fungicide. See mercury compounds, organic.

PHENYLMERCURIC CYANIDE
General Information
Description: Crystals; slightly soluble in hot water.
Formula: $(C_6H_5)HgCN$.
Constants: Mol wt: 303.7, mp: 204°C.
Hazard Analysis and Countermeasures
See mercury compounds, organic and cyanides.

PHENYLMERCURIC FORMAMIDE
Hazard Analysis and Countermeasures
A fungicide. See mercury compounds, organic.

PHENYLMERCURIC HYDROXIDE
General Information
Description: Fine white to cream crystals; slightly soluble in water; soluble in acetic acid, alcohol.
Formula: C_6H_5HgOH.
Constants: Mol wt: 294.6, mp: 197–205°C.
Hazard Analysis
See mercury compounds, organic.
Countermeasures
Shipping Regulations: Section 11.
IATA: Poison B, poison label, 25 kilograms (passenger), 95 kilograms (cargo).

PHENYLMERCURIC IODIDE
General Information
Description: Satiny crystals; insoluble in water.
Formula: $(C_6H_5)HgI$.
Constants: Mol wt: 404.6, mp: 266°C.
Hazard Analysis and Countermeasures
See mercury compounds, organic.

PHENYLMERCURICMONOETHANOL AMMONIUM ACETATE
General Information
Description: White crystalline solid; soluble in water.
Formula: $[(HOC_2H_4)NH_2(C_6H_5Hg)]OOCCH_3$.

Constant: Mol wt: 398.
Hazard Analysis and Countermeasures
A fungicide. See mercury compounds, organic.

PHENYLMERCURIC NAPHTHENATE
General Information
Description: Prepared by interaction of phenylmercuric acetate and naphthenic acid, producing colored solutions.
Hazard Aanalysis and Countermeasures
See mercury compounds, organic. A fungicide.

PHENYLMERCURIC NITRATE
General Information
Description: Crystals; insoluble in cold water.
Formula: $(C_6H_5)HgNO_3$.
Constants: Mol wt: 339.7, mp: 176–186°C.
Hazard Analysis and Countermeasures
See mercury compounds, organic and nitrates.
Shipping Regulations: Section 11.
IATA: Poison B, poison label, 25 kilograms (passenger), 95 kilograms (cargo).

PHENYLMERCURIC OLEATE
General Information
Description: White crystalline powder; insoluble in water; soluble in organic solvents and some oils.
Formula: $C_6H_5HgOOC(CH_2)_7CH:CHC_8H_{17}$.
Constant: Mol wt: 558.
Hazard Analysis and Countermeasures
A fungicide. See mercury compounds, organic, and phenols.

PHENYLMERCURIC PHTHALATE
Hazard Analysis and Countermeasures
A fungicide. See mercury compounds, organic.

PHENYLMERCURIC SALICYLATE
General Information
Formula: $C_6H_4(OH)(COOHgC_6H_5)$.
Constant: Mol wt: 415.
Hazard Analysis and Countermeasures
A fungicide. See mercury compounds, organic.

PHENYLMERCURIC STEARATE
Hazard Analysis and Countermeasures
A fungicide. See mercury compounds, organic.

PHENYLMERCURIC STEARATE
General Information
Formula: $CH_3(CH_2)_{16}COOHgC_6H_5$.
Constant: Mol wt: 561.
Hazard Analysis and Countermeasures
A fungicide. See mercury compounds, organic.

PHENYLMERCURIC TRIETHANOLAMINE
General Information
Formula: $(HOCH_2CH_2)_3NHgC_6H_5$.
Constant: Mol wt: 427.
Hazard Analysis and Countermeasures
A fungicide. See mercury compounds, organic.

PHENYLMERCURIC TRIETHANOL AMMONIUM LACTATE
General Information
Synonym: [Tris(2-hydroxyethyl)(phenylmercuri)ammonium lactate].
Description: White crystalline solid; soluble in water.
Formula: $[(HOC_2H_4)_3NHgC_6H_5]OOCCHOHCH_3$.
Constant: Mol wt: 516.
Hazard Analysis and Countermeasures
A fungicide. See mercury compounds, organic.

PHENYLMERCURY UREA. See "Leytosan."

PHENYLMETHANE. See toluene.

PHENYLMETHANE THIOL. See benzyl mercaptan.

1-PHENYL-2-METHYLAMINOPROPANOL. See ephedrine.

PHENYLMETHYL CARBINOL
General Information
Synonyms: Methylphenylcarbinol; 1-phenylethanol, styralyl alcohol.
Description: Colorless liquid.
Formula: $C_6H_5CH(CH_3)OH$.
Constants: Mol wt: 122.16, bp: 204°C, fp: 21.4°C, d: 1.015 at 20°/20°C, vap. press.: 0.1 mm at 20°C, vap. d.: 4.21, flash p: 205°F (O.C.).
Hazard Analysis
Toxicity: Slight. Details unknown. Limited animal experiments indicate low toxicity.
Fire Hazard: Moderate, when exposed to heat or flame; can react with oxidizing materials.
Countermeasures
To Fight Fire: Alcohol foam, foam, carbon dioxide, dry chemical or carbon tetrachloride (Section 6).
Personnel Protection: Section 3.
Storage and Handling: Section 7.

PHENYL METHYL ETHANOLAMINE
General Information
Description: Liquid.
Formula: $C_6H_5N(CH_3)C_2H_4OH$.
Constants: Mol wt: 151.2, bp: 192°C at 100 mm, mp: −30°C, flash p: 280°F (O.C.), d: 1.0661 at 20°/20°C, vap. press.: < 0.01 mm at 20°C, vap. d.: 5.21.
Hazard Analysis
Toxicity: See amines.
Fire Hazard: Slight, when exposed to heat or flame; can react with oxidizing materials.
Disaster Hazard: Dangerous; on heating to decomposition, it emits highly toxic fumes of nitrogen compounds.
Countermeasures
Storage and Handling: Section 7.
To Fight Fire: Foam, carbon dioxide, dry chemical or carbon tetrachloride (Section 6).

PHENYL METHYL ETHER. See anisole.

PHENYL METHYL KETONE
General Information
Synonyms: Acetophenone; methylphenyl ketone; hypnone; acetylbenzene.
Description: Colorless liquid or plates.
Formula: $CH_3COC_6H_5$.
Constants: Mol wt: 120.14, mp: 19.7°C, bp: 202.3°C, flash p: 180°F (O.C.), d: 1.026 at 20°/4°C, vap. d.: 4.14, vap. press.: 1 mm at 15°C, autoign. temp.: 1060°F.
Hazard Analysis
Toxicity: Details unknown. Narcotic in high concentrations. A hypnotic. Animal experiments suggest low toxicity. See ketones.
Fire Hazard: Slight; when exposed to heat or flame; can react with oxidizing materials.
Spontaneous Heating: No.
Countermeasures
To Fight Fire: Foam, carbon dioxide, dry chemical or carbon tetrachloride (Section 6).
Personal Hygiene: Section 3.
Storage and Handling: Section 7.

1-PHENYL-3-METHYL-5-PYRAZOLYL DIMETHYL CARBAMATE. See pyrolan.

4-PHENYLMORPHOLINE
General Information
Synonym: N-Phenylmorpholine.
Description: Crystals.
Formula: $C_6H_5NC_2H_4OCH_2CH_2$.
Constants: Mol wt: 163.2, mp: 57°C, bp: 270°C, flash p: 220°F (O.C.), d: 1.0599 at 57°/20°C, vap. press.: <0.1 mm at 20°C, vap. d.: 5.63.
Hazard Analysis
Toxicity: Details unknown; an insecticide.
Fire Hazard: Slight, when exposed to heat or flame; can react with oxidizing materials.
Disaster Hazard: Dangerous; on decomposition it emits toxic fumes of nitrogen compounds.
Countermeasures
Storage and Handling: Section 7.
To Fight Fire: Alcohol foam, foam, carbon dioxide, dry chemical or carbon tetrachloride (Section 6).

PHENYL MUSTARD OIL
General Information
Synonyms: Thiocarbanil; phenylisothiocyanate; phenylthiocarbonimide.
Description: Pale yellow liquid.
Formula: C_6H_5NCS.
Constants: Mol wt: 135.18, mp: −21°C, bp: 221°C, d: 1.1382.
Hazard Analysis
Toxic Hazard Rating:
Acute Local: Irritant 3; Ingestion 2; Inhalation 2.
Acute Systemic: U.
Chronic Local: U.
Chronic Systemic: U.
Disaster Hazard: Dangerous; when heated to decomposition or on contact with acid or acid fumes, it emits highly toxic fumes of cyanides and oxides of sulfur.
Countermeasures
Ventilation Control: Section 2.
Personnel Protection: Section 3.
Storage and Handling: Section 7.

PHENYL-α-NAPHTHYLAMINE
General Information
Description: Crystals.
Formula: $C_{10}H_7NHC_6H_5$.
Constants: Mol wt: 219.3.
Hazard Analysis
Toxic Hazard Rating:
Acute Local: Irritant 2; Allergen 1.
Acute Systemic: U.
Chronic Local: Irritant 1; Allergen 1.
Chronic Systemic: Inhalation 2.
Toxicology: Reported as capable of producing dermatitis, nephritis, anemia and cyanosis. Cases of bladder tumor have not been reported.

PHENYL-β-NAPHTHYLAMINE
General Information
Synonyms: N-Phenyl-2-naphthylamine.
Description: Needle-like crystals.
Formula: $C_{10}H_7NHC_6H_5$.
Constants: Mol wt: 219.3, mp: 108°C, bp: 399.5°C, d: 1.20.
Hazard Analysis
Toxicity: Details unknown. Animal experiments suggest low acute toxicity. See phenyl-α-naphthylamine.
Fire Hazard: Moderate, when exposed to heat or flame (Section 6).

TOXIC HAZARD RATING CODE (For detailed discussion, see Section 1.)

0 NONE: (a) No harm under any conditions; (b) Harmful only under unusual conditions or overwhelming dosage.

1 SLIGHT: Causes readily reversible changes which disappear after end of exposure.

2 MODERATE: May involve both irreversible and reversible changes not severe enough to cause death or permanent injury.

3 HIGH: May cause death or permanent injury after very short exposure to small quantities.

U UNKNOWN: No information on humans considered valid by authors.

Disaster Hazard: Dangerous; when heated to decomposition, it emits highly toxic fumes of nitrogen compounds; can react with oxidizing materials.
Countermeasures
Storage and Handling: Section 7.

N-PHENYL-2-NAPHTHYLAMINE. See phenyl-β-naphthylamine.

o-PHENYLPHENOL
General Information
Synonyms: Orthoxenol.
Description: White, flaky crystals, mild odor.
Formula: $C_6H_5C_6H_4OH$.
Constants: Mol wt: 170.2, bp: 286°C, fp: 57.2°C, flash p: 255°F, d: 1.217 at 25°/25°C, vap. press.: 1 mm at 100.0°C.
Hazard Analysis
Toxic Hazard Rating:
  Acute Local: Irritant 1; Ingestion 2.
  Acute Systemic: Ingestion 2.
  Chronic Local: Irritant 1.
  Chronic Systemic: Ingestion 2.
See also phenol.
Fire Hazard: Slight, when exposed to heat or flame; can react with oxidizing materials.
Countermeasures
Personal Hygiene: Section 3.
Storage and Handling: Section 7.
To Fight Fire: Alcohol foam. Section 6.

o-PHENYLPHENOL
General Information
Description: Flaky material.
Formula: $C_{12}H_9OH$.
Constants: Mol wt: 170.2, mp: 164.5°C, bp: 308°C, flash p: 330°F, vap. press.: 10 mm at 176.2°C.
Hazard Analysis
Toxicity: See o-phenylphenol.
Fire Hazard: Slight; when exposed to heat or flame; can react with oxidizing materials (Section 6).
Countermeasures
Storage and Handling: Section 7.

PHENYL PHOSPHINE
General Information
Synonym: Phosphaniline.
Description: Liquid.
Formula: $C_6H_5PH_2$.
Constants: Mol wt: 110.1, bp: 160°C, d: 1.001 at 15°C, vap. d.: 3.79.
Hazard Analysis
Toxicity: Details unknown; probably highly toxic. See also phosphine.
Fire Hazard: Moderate, when exposed to heat or flame.
Disaster Hazard: Dangerous; when heated to decomposition, it emits highly toxic fumes of phosphorus compounds; can react with oxidizing materials.
Countermeasures
Ventilation Control: Section 2.
Storage and Handling: Section 7.

1-PHENYLPIPERAZINE
General Information
Description: Pale yellow oil; insoluble in water; soluble in alcohol and ether.
Formula: $C_6H_5NCH_2CH_2NHCH_2CH_2$.
Constants: Mol wt: 162, d: 1.0621 at 20/4°C, bp: 286.5°C, mp: 18.8°C, flash p: 285°F.
Hazard Analysis
Toxicity: Details unknown. Limited animal experiments suggest high toxicity and irritation. See also piperazine.
Fire Hazard: Slight, when exposed to heat or flame.
Disaster Hazard: Moderate. It supports combustion and decomposes to yield toxic fumes.

Countermeasures
Storage and Handling: Section 7.
To Fight Fire: Water, foam, dry chemical.

1-PHENYLPROPANE. See n-propylbenzene.

2-PHENYLPROPANE. See cumene.

3-PHENYLPROPENAL. See cinnamaldehyde.

1-PHENYL-3-PYRAZOLIDONE. See phenidone.

PHENYL SALICYLATE
General Information
Synonym: Salol.
Description: Small, white crystals; aromatic odor.
Formula: $C_{13}H_{10}O_3$.
Constants: Mol wt: 214.2, mp: 42°C, bp: 173°C at 12 mm, d: 1.25.
 Hazard Analysis
Toxic Hazard Rating:
  Acute Local: Irritant 1.
  Acute Systemic: Ingestion 1.
  Chronic Local: U.
  Chronic Systemic: U.
Toxicity: See also acetylsalicylic acid and phenol.
Fire Hazard: Slight (Section 6).
Countermeasures
Personal Hygiene: Section 3.
Storage and Handling: Section 7.

PHENYL SILICATE
General Information
Description: Solid.
Formula: $Si(OC_6H_5)_4$.
Constants: Mol wt: 401.5, mp: 48°C.
Hazard Analysis
Toxicity: Details unknown. See also phenol.
Fire Hazard: Slight; (Section 6).
Countermeasures
Storage and Handling: Section 7.

PHENYL SULFIDE
General Information
Synonym: Diphenyl sulfide.
Description: Colorless liquid; almost no odor; insoluble in water; soluble in hot alcohol; miscible with benzene, ether, carbon disulfide.
Formula: $(C_6H_5)_2S$.
Constants: Mol wt: 186.27, d: 1.118 at 15/15°C, mp: 40°C, bp: 295–297°C.
Hazard Analysis
Toxicity: Details unknown. Limited animal experiments suggest high toxicity.
Disaster Hazard: Dangerous. When heated to decomposition it emits highly toxic fumes.
Countermeasures
Storage and Handling: Section 7.

PHENYL TETRATHIO-o-STANNATE
General Information
Description: Solid.
Formula: $(SC_6H_5)_4Sn$.
Constants: Mol wt: 555.3, mp: 67°C.
Hazard Analysis and Countermeasures
See tin compounds and sulfur compounds.

PHENYLTHIOCARBAMIDE. See phenylthiourea.

PHENYLTHIO CARBONIMIDE. See phenyl mustard oil.

PHENYLTHIOUREA
General Information
Synonym: Phenylthiocarbamide.
Description: Needle-like crystals; bitter taste.
Formula: $C_6H_5NHCSNH_2$.
Constants: Mol wt: 152.2, mp: 154°C, d: 1.3.

PHENYLTIN TRIBROMIDE (continued)

Hazard Analysis
Toxicity: Details unknown; experiments show high toxicity for rats; a rodenticide. The thioureas in general are moderately toxic.
Disaster Hazard: Dangerous; when heated to decomposition or on contact with acid or acid fumes, it emits highly toxic fumes of oxides of sulfur and nitrogen.
Countermeasures
Storage and Handling: Section 7.

PHENYLTIN TRIBROMIDE
General Information
Description: Solid.
Formula: $(C_6H_5)SnBr_3$.
Constants: Mol wt: 435.6, bp: 183°C at 29 mm.
Hazard Analysis and Countermeasures
See tin compounds and bromides.

PHENYLTIN TRICHLORIDE
General Information
Description: Solid, soluble in water.
Formula: $(C_6H_5)SnCl_3$.
Constants: Mol wt: 302.3, bp: 143°C at 25 mm.
Hazard Analysis and Countermeasures
See tin compounds, hydrochloric acid and chlorides.

PHENYL TRIBENZYL TIN
General Information
Description: A liquid; soluble in organic solvents, except alcohol.
Formula: $(C_6H_5)Sn(C_6H_5CH_2)_3$.
Constants: Mol wt: 469.2, bp: 290°C at 5 mm.
Hazard Analysis
Toxicity: Details unknown. See tin compounds.

PHENYLTRICHLOROSILANE
General Information
Description: Liquid.
Formula: $C_6H_5SiCl_3$.
Constants: Mol wt: 211.6, bp: 201°C, flash p: 195°F, d: 1.321 at 25°C.
Hazard Analysis
Toxic Hazard Rating:
Acute Local: Irritant 3; Ingestion 3; Inhalation 3.
Acute Systemic: U.
Chronic Local: U.
Chronic Systemic: U.
Fire Hazard: Moderate, when exposed to heat or flame (Section 6).
Disaster Hazard: Dangerous; when heated to decomposition, it emits highly toxic fumes of chlorides; will react with water or steam to produce toxic and corrosive fumes; can react with oxidizing materials.
Countermeasures
Ventilation Control: Section 2.
Personnel Protection: Section 3.
Storage and Handling: Section 7.
Shipping Regulations: Section 11.
I.C.C.: Corrosive liquid; white label, 10 gallons.
Coast Guard Classification: Corrosive liquid; white label.
IATA: Corrosive liquid, white label, not acceptable (passenger), 40 liters (cargo).

PHENYL TRI-p-TOLYL GERMANIUM
General Information
Description: White crystals; insoluble in water.
Formula: $(C_6H_5)Ge(C_6H_4CH_3)_3$.
Constants: Mol wt: 423.1, mp: 191°C.
Hazard Analysis
Toxicity: Details unknown. See germanium compounds.

PHENYLURETHANE
General Information
Synonym: Phenyl ethyl carbamate.
Description: Crystals.
Formula: $C_6H_5NHCOOC_2H_5$.
Constants: Mol wt: 165.2, mp: 53°C, bp: 238°C (slight decomp.), d: 1.106.
Hazard Analysis
Toxicity: Details unknown; experiments show moderate toxicity for mice. A citrus fruit fungicide.
Disaster Hazard: Dangerous; when heated to decomposition it emits highly toxic fumes of nitrogen compounds.
Countermeasures
Storage and Handling: Section 7.

PHLOROGLUCINOL. See pyrogallol.

PHLORIDZIN
General Information
Synonym: Phlorizin.
Description: Silky needles.
Formula: $C_{21}H_{24}O_{10} \cdot 2H_2O$.
Constants: Mol wt: 472.44, mp: 170°C decomposes, d 1.4298 at 19°C.
Hazard Analysis
Toxic Hazard Rating:
Acute Local: U.
Acute Systemic: Ingestion 3.
Chronic Local: U.
Chronic Systemic: Ingestion 3.
Fire Hazard: Slight; (Section 6).
Countermeasures
Personal Hygiene: Section 3.
Storage and Handling: Section 7.

PHLORIZIN. See phloridzin.

PHORATE
General Information
Synonym: o,o-Dithyl s-(ethyl thiomethyl)phosphorodithioate.
Description: Liquid; insoluble in water; miscible with carbon tetrachloride, dioxane, and xylene.
Formula: $(C_2H_5O)_2P(S)SCH_2SC_2H_5$.
Constants: Mol wt: 260, bp: 118°-120°C at 0.8 mm.
Hazard Analysis
Toxicity: Unknown. An insecticide. A food additive permitted in the feed and drinking water of animals and/or for the treatment of food producing animals (Section 10).

PHORONE
General Information
Synonym: 2,6,-Dimethyl-2,5-heptadien-4-one.
Description: Solid or liquid.
Formula: $(CH_3)_2CCHCOCHC(CH_3)_2$.
Constants: Mol wt: 138.20, mp: 28°C, bp: 197.2°C, flash p: 185°F (O.C.), d: 0.879, vap. press.: 1 mm at 42.0°C, vap. d.: 4.8.
Hazard Analysis
Toxicity: Unknown.
Fire Hazard: Moderate, when exposed to heat or flame; can react with oxidizing materials.
Countermeasures
To Fight Fire: Foam, carbon dioxide, dry chemical or carbon tetrachloride (Section 6).
Storage and Handling: Section 7.

TOXIC HAZARD RATING CODE (For detailed discussion, see Section 1.)

0 NONE: (a) No harm under any conditions; (b) Harmful only under unusual conditions or overwhelming dosage.

1 SLIGHT: Causes readily reversible changes which disappear after end of exposure.

2 MODERATE: May involve both irreversible and reversible changes not severe enough to cause death or permanent injury.

3 HIGH: May cause death or permanent injury after very short exposure to small quantities.

U UNKNOWN: No information on humans considered valid by authors.

PHOSDRIN
General Information
Synonym: 2-Carbomethoxy-1-methyl vinyl dimethylphosphate; mevinphos.
Description: Crystals.
Hazard Analysis
Toxicity: Highly toxic. An organic phosphate. See parathion.
TLV: ACGIH (recommended); 0.1 milligrams per cubic meter of air. May be absorbed via the skin.
Disaster Hazard: Dangerous; when heated to decomposition, it emits highly toxic fumes.

PHOSGENE
General Information
Synonyms: Carbon oxychloride; carbonyl chloride; CG.
Description: Colorless, gas or colorless, volatile liquid; odor of new mown hay or green corn.
Formula: $COCl_2$.
Constants: Mol wt: 98.92, mp: $-104°C$, bp: $8.3°C$, d: 1.37 at $20°C$, vap. press.: 1180 mm at $20°C$, vap. d.: 3.4.
Hazard Analysis
Toxic Hazard Rating:
  Acute Local: Irritant 3; Inhalation 3.
  Acute Systemic: Inhalation 3.
  Chronic Local: U.
  Chronic Systemic: U.
TLV: ACGIH (recommended); 0.1 part per million in air; 0.4 milligrams per cubic meter of air.
Toxicology: In the presence of moisture, phosgene decomposes to form hydrochloric acid and carbon monoxide. This action takes place within the body, when the gas reaches the bronchioles and the alveoli of the lungs. There is little irritant effect upon the respiratory tract, and the warning properties of the gas are therefore very slight. The liberation of hydrochloric acid in the lung tissues results in the development of pulmonary edema, which may be followed by bronchopneumonia, and occasionally lung abscess. Degenerative changes in the nerves have been reported as later sequelae. Concentrations of 3 to 5 ppm of phosgene in air cause irritation of the eyes and throat, with coughing; 25 ppm is dangerous for exposure lasting 30 to 60 minutes, and 50 ppm is rapidly fatal after even short exposure.

There may be no immediate warning that dangerous concentrations of the gas are being breathed. After a latent period of 2 to 24 hours the patient complains of burning in the throat and chest, shortness of breath and increasing dyspnea. There may be moist rales in the chest. Where the exposure has been severe, the development of pulmonary edema may be so rapid that the patient dies within 36 hours after exposure. In cases where the exposure had been less, pneumonia may develop several days after the occurrence of the accident. In patients who recover, no permanent residual disability is thought to occur. A common air contaminant (Section 4).
Disaster Hazard: Highly dangerous; when heated to decomposition or on contact with water or steam, it will react to produce toxic and corrosive fumes.
Countermeasures
Ventilation Control: Section 2.
Storage and Handling: Section 7.
Shipping Regulations: Section 11.
  I.C.C.: Poison A; poison gas label, not accepted.
  Coast Guard Classification: Poison A; poison gas label.
  IATA: Poison A, not acceptable (passenger and cargo).

PHOS-KIL. See parathion.

PHOSPHAM
General Information
Description: White, amorphous crystals.
Formula: PN_2H.

Constant: Mol wt: 60.00.
Hazard Analysis
Toxicity: Unknown.
Disaster Hazard: Dangerous; on decomposition, it emits highly toxic fumes of phosphorus and nitrogen compounds.

PHOSPHATES
Hazard Analysis
Toxicity: *See* individual compounds
Disaster Hazard: Dangerous. When heated to decomposition can emit highly toxic fumes of oxides of phosphorus.

PHOSPHAMIDON. See 2-chloro-2-diethylcarbamoyl-1-methyl vinyl dimethyl phosphate.

PHOSPHENILINE. See phenylphosphine.

PHOSPHENYL CHLORIDE
General Information
Synonym: Dichlorophenyl phosphine.
Description: Fuming liquid.
Formula: $C_6H_5PCl_2$.
Constants: Mol wt: 179.0, bp: $225°C$, d: 1.319, vap. d.: 6.17.
Hazard Analysis
Toxic Hazard Rating:
  Acute Local: Irritant 3; Ingestion 3; Inhalation 3.
  Acute Systemic: Ingestion 3; Inhalation 3.
  Chronic Local: U.
  Chronic Systemic: U.
Disaster Hazard: Dangerous; when heated to decomposition, it emits highly toxic fumes of oxides of phosphorus and chlorides; will react with water or steam to produce heat, toxic and corrosive fumes.
Countermeasures
Ventilation Control: Section 2.
Personnel Protection: Section 3.
First Aid: Section 1.
Storage and Handling: Section 7.

PHOSPHIDES
General Information
Composition: Combination of a cation + elemental phosphorus.
Hazard Analysis
Toxicity: Phosphides are particularly dangerous because they tend to decompose to phosphine upon contact with moisture or acids. See also phosphine.
Fire Hazard: Dangerous by chemical reaction; they react with water and acids to liberate phosphine.
Explosion Hazard: Moderate. See phosphine.
Disaster Hazard: Dangerous; when heated these may emit highly toxic fumes of oxides of phosphorus; they react with water or steam to produce toxic and flammable vapors; on contact with oxidizing materials, they can react vigorously and on contact with acid or acid fumes, they can emit toxic fumes.
Countermeasures
Ventilation Control: Section 2.
Storage and Handling: Section 7.

PHOSPHINE
General Information
Synonyms: Hydrogen phosphide; phosphoretted hydrogen.
Description: Colorless gas.
Formula: PH_3.
Constants: Mol wt: 34.04, mp: $-132.5°C$, bp: $-87.5°C$, d: 1.529 g/l at $0°C$.
Hazard Analysis
Toxic Hazard Rating:
  Acute Local: Irritant 2; Inhalation 2.
  Acute Systemic: Inhalation 3.
  Chronic Local: U.
  Chronic Systemic: Inhalation 3.
TLV: ACGIH (recommended); 0.3 parts per million in air; 0.4 milligrams per cubic meter of air.
Toxicology: Phosphine is a very toxic gas, but its action on

the body has not been fully worked out. It appears to cause, chiefly, a depression of the central nervous system and irritation of the lungs; autopsy findings in human cases may be entirely negative, or there may be pulmonary edema, dilation of the heart and hyperemia of the visceral organs.

Inhalation of phosphine causes restlessness, followed by tremors, fatigue, slight drowsiness, nausea, vomiting, and frequently, severe gastric pain and diarrhea. There is often headache, thirst, dizziness, oppression in the chest and burning substernal pain; later the patient may become dyspneic and develop cough and sputum. Coma or convulsions may precede death. Mild cases recover without after-effects.

Chronic poisoning, characterized by anemia, bronchitis, gastro-intestinal disturbances and visual, speech and motor disturbances, are said to result from continued exposure to very low concentrations.

Fire Hazard: Dangerous, by spontaneous chemical reaction.
Explosion Hazard: Moderate, when exposed to flame.
Explosive Range: Not known.
Disaster Hazard: Dangerous; when heated to decomposition, it emits highly toxic fumes of oxides of phosphorus; can react vigorously with oxidizing materials.

Countermeasures
Ventilation Control: Section 2.
To Fight Fire: Carbon dioxide, dry chemical or water spray (Section 6).
Personnel Protection: Section 3.
Storage and Handling: Section 7.
Shipping Regulations: Section 11.
IATA: Poison A, no label, not acceptable (passenger and cargo).

PHOSPHINOETHANE. See ethyl phosphene.

PHOSPHOLEUM. See polyphosphoric acid.

PHOSPHONIUM BROMIDE
General Information
Description: Cubic, colorless crystals.
Formula: PH_4Br.
Constants: Mol wt: 114.93, bp: 38.8°C at 794 mm; sublimes approx. 30°C, d: 2.464 g/l, vap. press.: 400 mm at 28.0°C.
Hazard Analysis and Countermeasures
See bromides.

PHOSPHONIUM CHLORIDE
General Information
Description: Cubic, colorless crystals.
Formula: PH_4Cl.
Constants: Mol wt: 70.47, mp: 28°C at 46 atm, bp: sublimes.
Hazard Analysis and Countermeasures
See chlorides.

PHOSPHONIUM IODIDE
General Information
Description: Tetragonal, colorless, deliquescent crystals.
Formula: PH_4I.
Constants: Mol wt: 161.93, mp: sublimes at 61.8°C, bp: 80°C, d: 2.86, vap. press.: 40 mm at 16.1°C.
Hazard Analysis and Countermeasures
See phosphine and hydriodic acid.

PHOSPHONIUM SULFATE
General Information
Description: Colorless, deliquescent crystals.

Formula: $(PH_4)_2SO_4$.
Constant: Mol wt: 166.09.
Hazard Analysis
Toxicity: See phosphorus compounds, inorganic.
Disaster Hazard: Dangerous; when heated to decomposition, it emits highly toxic fumes of sulfur compounds and phosphorus compounds.
Countermeasures
Ventilation Control: Section 2.
Storage and Handling: Section 7.

PHOSPHORAMIDE
General Information
Synonym: Phosphoryl amide.
Description: Amorphous, white crystals.
Formula: $PO(NH_2)_3$.
Constants: Mol wt: 95.05, mp: decomposes.
Hazard Analysis
Toxicity: Unknown. See phosphorus compounds, inorganic, and amides.
Disaster Hazard: Dangerous; when heated to decomposition, it emits highly toxic fumes or oxides of nitrogen and phosphorus.
Countermeasures,
Storage and Handling: Section 7.

PHOSPHORETTED HYDROGEN. See phosphine.

PHOSPHORIC ACID
General Information
Description: Colorless liquid or rhombic crystals.
Formula: H_3PO_4.
Constants: Mol wt: 98.04, mp: 42.35°C, $-\frac{1}{2}H_2O$ at 213°C, fp: 42.4°C, d: 1.864 at 25°C, vap. press.: 0.0285 mm at 20°C.
Hazard Analysis
Toxic Hazard Rating:
Acute Local: Irritant 2; Ingestion 2; Inhalation 2.
Acute Systemic: U.
Chronic Local: Irritant 2; Inhalation 2.
Chronic Systemic: U.
Note: Used as a general purpose food additive (Section 10). It is a common air contaminant (Section 4).
TLV: ACGIH (recommended). 1 milligram per cubic meter of air.
Disaster Hazard: Dangerous; when heated to decomposition, it emits toxic fumes of oxides of phosphorus.
Countermeasures
Ventilation Control: Section 2.
Personnel Protection: Section 3.
Storage and Handling: Section 7.
Shipping Regulations: Section 11.
MCA warning label.
IATA: Other restricted articles, class B, no label required, no limit (passenger and cargo).

PHOSPHORIC ACID ANHYDRIDE. See phosphorus pentoxide.

PHOSPHORIC ANHYDRIDE. See phosphorus pentoxide.
Shipping Regulations: Section 11.
I.C.C.: Flammable solid; yellow label, 100 pounds.
Coast Guard Classification: Inflammable solid; yellow label.
MCA warning label.
IATA: Flammable solid, yellow label, not acceptable (passenger), 45 kilograms (cargo).

TOXIC HAZARD RATING CODE (For detailed discussion, see Section 1.)

0 NONE: (a) No harm under any conditions; (b) Harmful only under unusual conditions or overwhelming dosage.

1 SLIGHT: Causes readily reversible changes which disappear after end of exposure.

2 MODERATE: May involve both irreversible and reversible changes; not severe enough to cause death or permanent injury.

3 HIGH: May cause death or permanent injury after very short exposure to small quantities.

U UNKNOWN: No information on humans considered valid by authors

m-PHOSPHOROUS ACID
General Information
Description: Feather-like crystals.
Formula: HPO_2.
Constant: Mol wt: 63.99.
Hazard Analysis and Countermeasures
See phosphorus compounds, inorganic.

o-PHOSPHOROUS ACID
General Information
Description: Colorless to yellow deliquescent crystals; soluble in water and alcohol.
Formula: $H_2(HPO_3)$.
Constants: Mol wt: 82.0, d: 1.651 at 21°C, mp: 736°C, bp: decomposes at 200°C.
Hazard Analysis
See phosphorous compounds, inorganic.
Countermeasures
See phosphorous compounds, inorganic.
Shipping Regulations: Section 11.
IATA: Other restricted articles, class B, no limit (passenger and cargo).

PHOSPHORUS (AMORPHOUS, RED)
General Information
Synonym: Red phosphorus.
Description: Reddish-brown powder.
Formula: P_4.
Constants: Mol wt: 124.08, bp: 280°C (with ignition), mp: 590°C at 43 atm., d: 2.20, autoign. temp.: 500°F in air, vap. d.: 4.77.
Hazard Analysis
Toxic Hazard Rating:
Acute Local: Ingestion 1.
Acute Systemic: Ingestion 1.
Chronic Local: 0.
Chronic Systemic: 0.
Toxicology: Relatively harmless unless it contains white phosphorus as an impurity.
Fire Hazard: Dangerous, when exposed to heat or by chemical reaction with oxidizers.
Spontaneous Heating: No.
Explosion Hazard: Moderate, by chemical reaction or on contact with organic materials.
Disaster Hazard: Dangerous; when heated, it emits highly toxic fumes of oxides of phosphorus; can react with reducing materials.
Countermeasures
Ventilation Control: Section 2.
To Fight Fire: Water (Section 6).
Personnel Protection: Section 3).
Storage and Handling: Section 7.
Shipping Regulations: Section 11.
I.C.C.: Flammable solid; yellow label, 11 pounds.
Coast Guard Classification: Inflammable solid; yellow label.
IATA: Flammable solid, yellow label, not acceptable (passenger), 5 kilograms (cargo).

PHOSPHORUS (WHITE)
General Information
Description: Cubic crystals; colorless to yellow, wax-like solid.
Formula: P_4.
Constants: Mol wt: 124.08, mp: 44.1°C, bp: 280°C, flash p: spontaneous in air, d: 1.82, autoign temp: 86°F, vap. press.: 1 mm at 76.6°C, vap. d.: 4.42.
Hazard Analysis
Toxic Hazard Rating:
Acute Local: Irritant 3; Ingestion 3.
Acute Systemic: Ingestion 3; Inhalation 3.
Chronic Local: U.
Chronic Systemic: Ingestion 3; Inhalation 3.
TLV: ACGIH (recommended) 0.1 milligrams per cubic meter of air.

Toxicology: This material is dangerously reactive in air and turns red in sunlight. If combustion occurs in a confined space, it will remove the oxygen and render the air unfit to support life. High concentrations of the vapors evolved by burning it are irritating to the nose, throat and lungs as well as the skin, eyes and mucous membranes. If phosphorus is ingested, it can be absorbed from the gastrointestinal tract or through the lungs. The absorption of toxic quantities of phosphorus has an acute effect on the liver and is accompanied by vomiting and marked weakness. The long-continued absorption of small amounts of phosphorus can result in necrosis of the mandible or jaw bone, and is known as "phossy-jaw." Long-continued absorption, particularly through the lungs, and through the gastrointestinal tract can cause a chronic poisoning. This gives rise to a generalized form of weakness attended by anemia, loss of appetite, gastrointestinal weakness and pallor. The most common symptom, however, of chronic phosphorus poisoning is necrosis of the jaw. It can also cause changes in the long bones, and seriously affected bones may become brittle, leading to spontaneous fractures. It is especially hazardous to the eyes and can damage them severely. It also has adverse effects on the teeth, and workers should have periodic dental examinations. The yellow form of phosphorus, when it comes into external contact with the eyes, can cause conjunctivitis with a yellow tint. If the material is inhaled, it can cause photophobia with myosis, dilation of pupils, retinal hemorrhage, congestion of the blood vessels and rarely an optic neuritis.
Radiation Hazard: Section 5. For permissible levels, see Table 5, p. 150.
Artificial isotope ^{32}P, half life 14.5 d. Decays to stable ^{32}S by emitting beta particles of 1.71 MeV.
Fire Hazard: Dangerous, when exposed to heat or by chemical reaction with oxidizers. Ignites spontaneously in air.
Spontaneous Heating: No.
Explosion Hazard: Moderate, by chemical reaction with oxidizing agents.
Disaster Hazard: Dangerous; it emits highly toxic fumes of oxides of phosphorus; can react vigorously with oxidizing materials.
Countermeasures
Ventilation Control: Section 2.
To Fight Fire: Water (Section 6).
Personnel Protection: Section 3.
First Aid: Section 1.
Storage and Handling: Section 7.
Shipping Regulations: Section 11.
White or yellow–dry, in water for both I.C.C. and Coast Guard.
I.C.C.: Flammable solid; yellow label, not accepted (dry), 25 pounds (in water).
Coast Guard Classification: Inflammable solid; yellow label.
MCA warning label.
IATA (white or yellow, dry): Flammable solid, not acceptable (passenger and cargo).
(white or yellow, in water): Flammable solid, yellow label, not acceptable (passenger), 12 kilograms (cargo).

PHOSPHORUS CHLORIDE. See phosphorus trichloride.

PHOSPHORUS CHLORIDE NITRIDE
General Information
Description: Prisms.
Formula: $P_6N_7Cl_9$.
Constants: Mol wt: 603.05, mp: 237.5°C, bp: 251–261°C at 13 mm.
Hazard Analysis
Toxicity: Highly toxic. See hydrochloric acid.

Disaster Hazard: Dangerous; when heated to decomposition, it emits highly toxic fumes of oxides of phosphorus and chlorides; will react with water or steam to produce toxic and corrosive fumes.

Countermeasures
Storage and Handling: Section 7.

PHOSPHORUS COMPOUNDS, INORGANIC
Hazard Analysis
Toxicity: Variable. Most inorganic phosphates, except phosphine, have low toxicity but in large doses they may cause serious disturbances, particularly in calcium metabolism. Metaphosphates may be highly toxic, causing irritation and hemorrhages in the stomach as well as liver and kidney damage. Common air contaminants (Section 4).

PHOSPHORUS CYANIDE
General Information
Description: White needles.
Formula: $P(CN)_3$.
Constants: Mol wt: 110, mp: sublimes at 130°C.
Hazard Analysis and Countermeasures
See cyanides.

PHOSPHOROUS DIBROMIDE TRICHLORIDE
General Information
Description: Orange crystals.
Formula: PBr_2Cl_3.
Constants: Mol wt: 297.18, mp: 35°C (decomposes).
Hazard Analysis
Toxicity: See hydrobromic acid and hydrochloric acid.
Disaster Hazard: Dangerous; when heated to decomposition or on contact with acid or acid fumes, it emits highly toxic fumes of oxides of phosphorus, bromides and chlorides.
Countermeasures
Ventilation Control: Section 2.
Storage and Handling: Section 7.

PHOSPHOROUS DIBROMIDE TRIFLUORIDE
General Information
Description: Pale yellow crystals.
Formula: PBr_2F_3.
Constants: Mol wt: 247.81, mp: $-20°C$, bp: decomposes at 15°C.
Hazard Analysis and Countermeasures
See fluorides and hydrobromic acid.

PHOSPHORUS DICHLORIDE
General Information
Description: Colorless crystals.
Formula: PCl_2.
Constants: Mol wt: 101.89, bp: 180°C, mp: $-28°C$.
Hazard Analysis
Toxicity: Probably toxic. See hydrochloric acid.
Disaster Hazard: Dangerous; when heated to decomposition, it emits highly toxic fumes or oxides of phosphorus and chlorides; will react with water, steam or acids to produce toxic and corrosive fumes.
Countermeasures
Ventilation Control: Section 2.
Storage and Handling: Section 7.

PHOSPHORUS DICHLORIDE TRIFLUORIDE
General Information
Formula: PCl_2F_3.
Constants: Mol wt: 158.89, bp: $-8°C$; decomp. at 200°C.

Hazard Analysis
See fluorides and hydrochloric acid.

PHOSPHORUS DIIODIDE
General Information
Description: Crystals.
Formula: P_2I_4.
Constants: Mol wt: 569.64, mp: 124.5°C, bp: decomposes.
Hazard Analysis and Countermeasures
See phosphorus and iodides.

PHOSPHORUS HEMITRISELENIDE
General Information
Description: Dark red mass.
Formula: P_2Se_3.
Constant: Mol wt: 298.8.
Hazard Analysis
Toxicity: See selenium compounds.
Fire Hazard: Moderate, by chemical reaction.
Explosion Hazard: Moderate, by chemical reaction.
Disaster Hazard: Dangerous; when heated to decomposition or on contact with acid or acid fumes, it emits highly toxic fumes of selenium and oxides of phosphorus.
Countermeasures
Ventilation Control: Section 2.

PHOSPHORUS HEPTABROMIDE DICHLORIDE
General Information
Description: Prisms.
Formula: PBr_7Cl_2.
Constant: Mol wt: 661.31.
Hazard Analysis
Toxicity: Highly toxic. See hydrochloric acid.
Disaster Hazard: Dangerous; when heated to decomposition, it emits highly toxic fumes of bromides and chlorides; will react with water, steam or acids to produce toxic and corrosive fumes.
Countermeasures
Ventilation Control: Section 2.
Storage and Handling: Section 7.

PHOSPHORUS HEPTASULFIDE
General Information
Description: Solid, light yellow crystals; light gray powder or fused solid.
Formula: P_4S_7.
Constants: Mol wt: 348.34, mp: 310°C, bp: 523°C, d: 2.19 at 17°C.
Hazard Analysis and Countermeasures
See sulfides and phosphorus compounds.
Shipping Regulations: Section 11.
    IATA: Flammable solid, yellow label, not acceptable (passenger), 5 kilograms (cargo).

PHOSPHORUS MONOBROMIDE TETRACHLORIDE
General Information
Description: Yellow crystals.
Formula: $PBrCl_4$.
Constant: Mol wt: 252.72.
Hazard Analysis
Toxicity: Highly toxic. See hydrochloric acid and bromides.
Disaster Hazard: Dangerous; when heated to decomposition, it emits highly toxic fumes of chlorides and bromides; will react with water, steam or acids to produce toxic and corrosive fumes.
Countermeasures
Ventilation Control: Section 2.
Storage and Handling: Section 7.

TOXIC HAZARD RATING CODE (For detailed discussion, see Section 1.)

0 NONE: (a) No harm under any conditions; (b) Harmful only under unusual conditions or overwhelming dosage.

1 SLIGHT: Causes readily reversible changes which disappear after end of exposure.

2 MODERATE: May involve both irreversible and reversible changes not severe enough to cause death or permanent injury.

3 HIGH: May cause death or permanent injury after very short exposure to small quantities.

U UNKNOWN: No information on humans considered valid by authors.

PHOSPHORUS NITRIDE
General Information
Description: Amorphous, white crystals.
Formula: P_3N_5.
Constants: Mol wt: 162.98, bp: decomposes at 800°C, d: 2.51 at 18°C.
Hazard Analysis
Toxicity: Probably toxic. See phosphorus compounds, inorganic and nitrides.
Fire Hazard: Moderate, by chemical reaction. See ammonia.
Explosion Hazard: Moderate, by chemical reaction.
Disaster Hazard: Dangerous; when heated to decomposition it emits highly toxic fumes of oxides of nitrogen and phosphorus; will react with water or steam to produce toxic and corrosive fumes.
Countermeasures
Ventilation Control: Section 2.
Storage and Handling: Section 7.

PHOSPHORUS OXYBROMIDE
General Information
Description: Colorless plates.
Formula: $POBr_3$.
Constants: Mol wt: 286.73, mp: 56°C, bp: 193°C, d: 2.882.
Hazard Analysis
Toxicity: Highly toxic. See hydrobromic acid.
Disaster Hazard: Dangerous; when heated to decomposition, it emits highly toxic fumes of bromides and oxides of phosphorus; will react with water or steam to produce toxic and corrosive fumes.
Countermeasures
Ventilation Control: Section 2.
Storage and Handling: Section 7.
Shipping Regulations: Section 11.
  ICC: Corrosive liquid, white label, 1 quart.
  IATA: Corrosive liquid, white label, not acceptable (passenger), 1 liter (cargo).

PHOSPHORUS OXYCHLORIDE
General Information
Synonym: Phosphoryl chloride.
Description: Colorless to slightly yellow fuming liquid.
Formula: $POCl_3$.
Constants: Mol wt: 153.39, mp: 2°C, bp: 105.1°C, d: 1.685 at 15.5°C, vap. press.: 40 mm at 27.3°C, vap. d.: 5.3.
Hazard Analysis
Toxicity: Highly toxic. See hydrochloric acid.
Disaster Hazard: Dangerous; when heated to decomposition, it emits highly toxic fumes of chlorides and oxides of phosphorus; will react with water or steam to produce heat and toxic and corrosive fumes.
Countermeasures
Ventilation Control: Section 2.
Personnel Protection: Section 3.
Storage and Handling: Section 7.
Shipping Regulations: Section 11.
  I.C.C.: Corrosive liquid; white label, 1 quart.
  Coast Guard Classification: Corrosive liquid; white label.
  MCA warning label.
  IATA: Corrosive liquid, white label, not acceptable (passenger), 1 liter (cargo).

PHOSPHORUS OXYFLUORIDE
General Information
Description: Colorless gas.
Formula: POF_3.
Constants: Mol wt: 103.98, mp: −68°C, bp: −39.8°C, d: 4.69 g/l.
Hazard Analysis and Countermeasures
See fluorides, hydrofluoric acid and phosphates.

PHOSPHORUS OXYNITRIDE
General Information
Description: Amorphous, white crystals.

Formula: PON.
Constants: Mol wt: 60.99, mp: red heat.
Hazard Analysis
Toxicity: See nitrides and phosphates.
Fire Hazard: Moderate, by chemical reaction (See ammonia).
Explosion Hazard: Moderate, by chemical reaction.
Disaster Hazard: Dangerous; on decomposition it emits highly toxic fumes of oxides of phosphorus and nitrogen; on contact with acid or acid fumes, it can emit toxic fumes.
Countermeasures
Ventilation Control: Section 2.
Storage and Handling: Section 7.

PHOSPHORUS OXYSULFIDE
General Information
Description: Tetragonal deliquescent crystals.
Formula: $P_4O_6S_4$.
Constants: Mol wt: 348.18, mp: 102°C, bp: 295°C.
Hazard Analysis and Countermeasures
See sulfides and phosphorus.

PHOSPHORUS PENTABROMIDE
General Information
Description: Rhombic, yellow crystals.
Formula: PBr_5.
Constants: Mol wt: 430.56, mp: decomposes above 100°C, bp: decomposes at 106°C.
Hazard Analysis
Toxicity: Highly caustic. See bromides.
Fire Hazard: Moderate by chemical reaction. Contact with moisture can cause a violent reaction and evolution of heat.
Disaster Hazard: Dangerous; when heated to decomposition, it emits highly toxic fumes of bromides; will react with water or steam to produce heat and toxic and corrosive fumes.
Countermeasures
Ventilation Control: Section 2.
Storage and Handling: Section 7.

PHOSPHORUS PENTACHLORIDE
General Information
Description: Yellowish-white, fuming, crystalline mass.
Formula: PCl_5.
Constants: Mol wt: 208.31, mp: 166.8°C, bp: sublimes at 162°C, d: 4.65 g/liter at 296°C, vap. press.: 1 mm at 55.5°C.
Hazard Analysis
Toxicity: Highly caustic. See hydrochloric acid and phosphorus.
TLV: ACGIH (recommended); 1 milligram per cubic meter of air.
Fire Hazard: Moderate, by chemical reaction. Reacts violently with moisture.
Disaster Hazard: Dangerous; when heated to decomposition, it emits highly toxic fumes of chlorides; will react with water or steam to produce heat and toxic and corrosive fumes.
Countermeasures
Ventilation Control: Section 2.
To Fight Fire: Carbon dioxide, dry chemical or carbon tetrachloride (Section 6).
Storage and Handling: Section 7.
Shipping Regulations: Section 11.
  I.C.C.: Flammable solid; yellow label, 5 pounds.
  Coast Guard Classification: Inflammable solid; yellow label.
  MCA warning label.
  IATA: Flammable solid, yellow label, not acceptable (passenger), 2½ kilogram (cargo).

PHOSPHORUS PENTAFLUORIDE
General Information
Description: Colorless gas.

Formula: PF_5.
Constants: Mol wt: 125.98, mp: $-93.7°C$, bp: $-84.5°C$, vap. d.: 5.81 g/l.
Hazard Analysis
Toxicity: Highly toxic. See fluorides.
Disaster Hazard: Dangerous; when heated to decomposition, it emits highly toxic fumes of fluorides and phosphorus compounds; will react with water or steam to produce toxic and corrosive fumes.
Countermeasures
Ventilation Control: Section 2.
Storage and Handling: Section 7.
Shipping Regulations: Section 11.
   IATA: Nonflammable gas, green label, not acceptable (passenger), 140 kilograms (cargo).

PHOSPHORUS PENTASELENIDE
General Information
Synonym: Diphosphorus pentaselenide.
Description: Dark red-black needles.
Formula: P_2Se_5.
Constants: Mol wt: 456.76, mp: decomposes.
Hazard Analysis
See selenium compounds and phosphorus.

PHOSPHORUS PENTASULFIDE
General Information
Description: Gray to yellow-green, crystalline, deliquescent mass.
Formula: P_2S_5.
Constants: Mol wt: 222.34, mp: $276°C$, bp: $514°C$, d: 2.03, autoign. temp.: $548.6°F$.
Hazard Analysis
Toxic Hazard Rating:
   Acute Local: Irritant 2.
   Acute Systemic: Inhalation 3.
   Chronic Local: Irritant 2.
   Chronic Systemic: U.
Toxicology: Readily liberates hydrogen sulfide on contact with moisture. See hydrogen sulfide.
TLV: ACGIH (recommended); 1 milligram per cubic meter of air.
Fire Hazard: Dangerous in the form of dust when exposed to heat or flame. Evolves heat in contact with mosisture,
Spontaneous Heating: Yes, in the presence of moisture.
Explosion Hazard: Moderate, in solid form by spontaneous chemical reaction.
Disaster Hazard: Dangerous; when heated to decomposition, it emits highly toxic fumes of oxides of sulfur and phosphorus; will react with water, steam or acids to produce toxic and flammable vapors; can react vigorously with oxidizing materials.
Countermeasures
Ventilation Control: Section 2.
To Fight Fire: Carbon dioxide, snow, dry chemical or sand.
Storage and Handling: Section 7.
Shipping Regulations: Section 11.
   IATA: Flammable solid, yellow label not acceptable (passenger), 5 kilograms (cargo).
   MCA warning label.

PHOSPHORUS PENTOXIDE
General Information
Synonym: Phosphoric acid anhydride.
Description: A fluffy, crystalline, deliquescent powder.
Formula: P_2O_5.
Constants: Mol wt: 143.0, mp: $563°C$, d: 0.77–1.39, vap. press.: 1 mm at $384°C$.

Hazard Analysis
Toxicity: Highly caustic. See phosphorus compounds, inorganic.
Fire Hazard: Moderate, by chemical reaction. Reacts violently with water to evolve heat (Section 6).
Disaster Hazard: Dangerous; it reacts with water or steam to produce heat; on contact with reducing materials, it can react vigorously.
Countermeasures
Storage and Handling: Section 7.

PHOSPHORUS SESQUISULFIDE
General Information
Synonym: Tetraphosphorus trisulfide.
Description: Yellow, crystalline mass.
Formula: P_4S_3.
Constants: Mol wt: 220.26, mp: $172.5°C$, bp: $407°C$, d: 2.03, autoign. temp.: $212°F$.
Hazard Analysis
Toxicity: Probably toxic. See sulfides.
Fire Hazard: Dangerous, when exposed to heat or flame; ignites by friction.
Spontaneous Heating: No.
Explosion Hazard: Moderate by chemical reaction.
Disaster Hazard: Dangerous; when heated to decomposition, it emits highly toxic fumes of oxides of phosphorus and sulfur; will react with water or steam to produce toxic fumes; can react vigorously with oxidizing materials.
Countermeasures
To Fight Fire: Water (Section 6).
Ventilation Control: Section 2.
Personnel Protection: Section 3.
Storage and Handling: Section 7.
Shipping Regulations: Section 11.
   I.C.C.: Flammable solid; yellow label, 11 pounds.
   Coast Guard Classification: Inflammable solid; yellow label.
   IATA: Flammable solid, yellow label, not acceptable (passenger), 5 kilograms (cargo).

PHOSPHORUS SULFOCHLORIDE. See thiophosphoryl chloride.

PHOSPHORUS SULFOFLUORIDE
General Information
Description: Solid.
Formula: PSF_3.
Constant: Mol wt: 120.
Hazard Analysis
Toxicity: Highly toxic. Details unknown. See phosphorus pentafluoride and also fluorides.
Disaster Hazard: Dangerous; when heated to decomposition it emits highly toxic fumes of oxides of sulfur and phosphorus as well as fluroides.

PHOSPHORUS TETROXIDE
General Information
Description: Rhombic, colorless, deliquescent crystals.
Formula: P_2O_4.
Constants: Mol wt: 125.96, mp: $>100°C$, bp: sublimes at $180°C$, d: 2.54 at $23°C$.
Hazard Analysis
Toxicity: See phosphorus compounds, inorganic.

PHOSPHORUS THIOCYANATE
General Information
Description: Liquid.
Formula: $P(SCN)_3$.

TOXIC HAZARD RATING CODE (For detailed discussion, see Section 1.)

0 NONE: (a) No harm under any conditions; (b) Harmful only under unusual conditions or overwhelming dosage.

1 SLIGHT: Causes readily reversible changes which disappear after end of exposure.

2 MODERATE: May involve both irreversible and reversible changes; not severe enough to cause death or permanent injury.

3 HIGH: May cause death or permanent injury after very short exposure to small quantities.

U UNKNOWN: No information on humans considered valid by authors.

Constants: Mol wt: 205.23, mp: approx. $-4°C$, bp: 265°C, d: 1.625 at 18°C.

Hazard Analysis

Toxicity: Probably toxic. See thiocyanates.

Fire Hazard: Slight, when exposed to heat or flame (Section 6).

Disaster Hazard: Dangerous; when heated to decomposition or on contact with acids, it emits highly toxic fumes of cyanides; can react with oxidizing materials.

Countermeasures

Ventilation Control: Section 2.

Storage and Handling: Section 7.

PHOSPHORUS TRIBROMIDE

General Information

Description: Colorless, fuming liquid.

Formula: PBr_3.

Constants: Mol wt: 270.8, mp: $-40°C$, bp: 175.3°C, d: 2.852 at 15°C, vap. press.: 10 mm at 47.8°C.

Hazard Analysis

Toxicity: See hydrobromic acid.

Disaster Hazard: Dangerous; when heated to decomposition, it emits highly toxic fumes of bromides and oxides of phosphorus; will react with water, steam or acids to produce heat and toxic and corrosive fumes.

Countermeasures

Ventilation Control: Section 2.

Personnel Protection: Section 3.

Storage and Handling: Section 7.

Shipping Regulations: Section 11.

   I.C.C.: Corrosive liquid; white label, 1 quart.

   Coast Guard Classification: Corrosive liquid; white label.

   IATA: Corrosive liquid, white label, not acceptable (passenger), 1 liter (cargo).

PHOSPHORUS TRICHLORIDE

General Information

Synonym: Phosphorus chloride.

Description: Clear, colorless, fuming liquid.

Formula: PCl_3.

Constants: Mol wt: 137.39, mp: $-111.8°C$, bp: 74.2°C, d: 1.574 at 21°C, vap. press.: 100 mm at 21°C, vap. d.: 4.75.

Hazard Analysis

Toxicity: Highly toxic. See hydrochloric acid.

TLV: ACGIH (recommended); 0.5 parts per million in air; 3 milligrams per cubic meter of air.

Fire Hazard: Moderate, by chemical reaction.

Disaster Hazard: Dangerous; when heated to decomposition, it emits highly toxic fumes of chlorides and oxides of phosphorus; will react with water, steam, or acids to produce heat and toxic and corrosive fumes; can react with oxidizing materials.

Countermeasures

Ventilation Control: Section 2.

To Fight Fire: Carbon dioxide, dry chemical or carbon tetrachloride (Section 6).

Personnel Protection: Section 3.

Storage and Handling: Section 7.

Shipping Regulations: Section 11.

   I.C.C.: Corrosive liquid; white label, 1 quart.

   Coast Guard Classification: Corrosive liquid; white label.

   MCA warning label.

   IATA: Corrosive liquid, white label, not acceptable (passenger), 1 liter (cargo).

PHOSPHORUS TRIFLUORIDE

General Information

Description: Colorless gas.

Formula: PF_3.

Constants: Mol wt: 87.98, mp: $-160°C$, bp: $-95°C$, d: 3.907 g/l.

Hazard Analysis

Toxicity: Highly toxic. See hydrofluoric acid, fluorides and phosphorus.

Disaster Hazard: Dangerous; when heated to decomposition it emits highly toxic fumes of fluorides and oxides of phosphorus; will react with water or steam to produce toxic and corrosive fumes.

Countermeasures

Ventilation Control: Section 2.

Storage and Handling: Section 7.

PHOSPHORUS TRIIODIDE

General Information

Description: Hexagonal, red, deliquescent crystals.

Formula: PI_3.

Constants: Mol wt: 411.74, mp: 61°C, bp: decomposes.

Hazard Analysis and Countermeasures

See iodides and phosphorus.

PHOSPHORUS TRIOXIDE

General Information

Description: Monoclinic, colorless crystals or white, deliquescent powder.

Formula: P_2O_3.

Constants: Mol wt: 110.0, mp: 22.5°C, bp: 173°C, d: 2.135 at 21°C, vap. press.: 10 mm at 53.0°C.

Hazard Analysis

Toxicity: Caustic. See phosphorus compounds, inorganic.

Disaster Hazard: Slightly dangerous; will react with water or steam to produce heat.

Countermeasures

Storage and Handling: Section 7.

PHOSPHORUS TRISULFIDE

General Information

Synonym: Tetraphosphorus hexasulfide.

Description: Gray-yellow crystals.

Formula: P_4S_6.

Constants: Mol wt: 316.44, mp: 290°C, bp: 490°C.

Hazard Analysis

Toxicity: Highly toxic. See sulfides.

Fire Hazard: Moderate, when exposed to heat or by chemical reaction (Section 6).

Explosion Hazard: Moderate, by chemical reaction.

Disaster Hazard: Dangerous, when heated to decomposition, it emits highly toxic fumes of oxides of phosphorus and sulfides; will react with water or steam to produce toxic fumes; can react with oxidizing materials.

Countermeasures

Ventilation Control: Section 2.

Storage and Handling: Section 7.

Shipping Regulations: Section 11.

   IATA: Flammable solid, yellow label, not acceptable (passenger), 5 kilograms (cargo).

PHOSPHORYL AMIDE. See phosphoramide.

PHOSPHORYL CHLORIDE. See phosphorus oxychloride.

PHOSPHOTUNGSTIC ACID

General Information

Description: Triclinic, yellow-green crystals.

Formula: $H_3PW_{12}O_{40} \cdot 14H_2O$.

Constant: Mol wt: 3133.27.

Hazard Analysis

Toxic Hazard Rating:

   Acute Local: Irritant 3; Ingestion 3.

   Acute Systemic: U.

   Chronic Local: Irritant 2.

   Chronic Systemic: U.

Toxicology: A strong acid with caustic properties.

Disaster Hazard: Dangerous; when heated to decomposition, it emits highly toxic fumes of phosphorus.

Countermeasures

Storage and Handling: Section 7.

PHOTOGRAPHIC AND X-RAY FILMS, SCRAP
Hazard Analysis
Fire Hazard: Dangerous, when exposed to heat or flame (Section 6).
Disaster Hazard: Dangerous; when heated to decomposition, they emit highly toxic fumes; they can react vigorously with oxidizing materials.
Countermeasures
Storage and Handling: Section 7.
Shipping Regulations: Section 11.
I.C.C.: Flammable solid; yellow label, (Section 73.196).
Coast Guard Classification: Inflammable solid; yellow label.
IATA: Flammable solid, not acceptable (passenger & cargo).

PHTHALANDIONE. See phthalic anhydride.

PHTHALIC ACID
General Information
Synonym: Benzene dicarboxylic acid.
Description: Crystals.
Formula: $C_8H_6O_4$.
Constants: Mol wt: 166.1, mp: 206–208° C, d: 1.59.
Hazard Analysis
Toxic Hazard Rating:
  Acute Local: Allergen 1.
  Acute Systemic: U.
  Chronic Local: Allergen 1.
  Chronic Systemic: U.
Fire Hazard: Slight; when heated, it emits acrid fumes (Section 6).
Countermeasures
Personal Hygiene: Section 3.
Storage and Handling: Section 7.

PHTHALIC ANHYDRIDE
General Information
Synonym: Phthalandione.
Description: White crystalline needles.
Formula: $C_6H_4(CO)_2O$.
Constants: Mol wt: 148.11, mp: 131.2° C, lel: 1.7%, uel: 10.4%, bp: 284.5° C sublimes, flash p: 305° F (C.C.), d: 1.527 at 4° C, autoign. temp.: 1083° F, vap. press.: 1 mm at 96.5° C, vap. d.: 5.10.
Hazard Analysis
Toxic Hazard Rating:
  Acute Local: Irritant 1; Ingestion 1; Inhalation 1.
  Acute Systemic: U.
  Chronic Local: Irritant 1; Inhalation 1.
  Chronic Systemic: U.
TLV: ACGIH (recommended), 2 parts per million, 12 milligrams per cubic meter of air.
Toxicity: Note: A common air contaminant (Section 4).
Fire Hazard: Slight, when exposed to heat or flame; can react with oxidizing materials.
Explosion Hazard: Moderate, in the form of dust, when exposed to flame.
Countermeasures
Personal Hygiene: Section 3.
To Fight Fire: Water, carbon dioxide, dry chemical or carbon tetrachloride (Section 6).
Storage and Handling: Section 7.
Shipping Regulations: Section 11.
MCA warning label.

PHTHALIDE
General Information
Synonym: α-Hydroxy-o-toluic acid lactone.
Description: White crystals.
Formula: $C_6H_4CO_2CH_2$.
Constants: Mol wt: 134.1, mp: 72.7° C, bp: 290–292° C, vap. press.: 1 mm at 95.5° C.
Hazard Analysis
Toxicity: Unknown.
Fire Hazard: Slight; when heated, it emits acrid fumes (Section 6).
Countermeasures
Storage and Handling: Section 7.

PHTHALIMIDE
General Information
Synonym: 1,3-Isoindoledione.
Description: Light tan to white powder.
Formula: $C_6H_4(CO)_2NH$.
Constants: Mol wt: 147.1, mp: 233° C, bp: sublimes.
Hazard Analysis
Toxicity: Unknown.
Disaster Hazard: Slight; when heated, to decomposition it emits toxic fumes (Section 6).
Countermeasures
Storage and Handling: Section 7.

PHTHALODINITRILE. See o-dicyanobenzene.

PHTHALONITRILE. See o-dicyanobenzene.

m-PHTHALYL CHLORIDE. See isophthaloyl chloride.

p-PHTHALYL DICHLORIDE. See terephaloyl chloride.

PHYGON. See 2,3-dichloro-1,4-naphthoquinone.

PHYSOSTIGMA
General Information
Synonyms: Calabar bean; ordeal bean.
Description: Dried, ripe seeds.
Hazard Analysis
Toxic Hazard Rating:
  Acute Local: Allergen 1.
  Acute Systemic: Ingestion 3; Inhalation 3.
  Chronic Local: Allergen 1.
  Chronic Systemic: U.
Fire Hazard: Slight, when exposed to heat or flame (Section 6).
Countermeasures
Ventilation Control: Section 2.
Personal Hygiene: Section 3.
Storage and Handling: Section 7.

PHYSOSTIGMINE
General Information
Synonyms: Eserine; calabrine.
Description: Colorless, hygroscopic crystals.
Formula: $C_{15}H_{21}N_3O_2$.
Constants: Mol wt: 275.3, mp: 105–106° C.
Hazard Analysis
Toxic Hazard Rating:
  Acute Local: Allergen 1.
  Acute Systemic: Ingestion 3.
  Chronic Local: U.
  Chronic Systemic: U.
Toxicology: Poisoning can occur as the result of a mistake in dosage or due to hypersensitivity of the patient

---

TOXIC HAZARD RATING CODE *(For detailed discussion, see Section 1.)*

0 NONE: (a) No harm under any conditions; (b) Harmful only under unusual conditions or overwhelming dosage.

1 SLIGHT: Causes readily reversible changes which disappear after end of exposure.

2 MODERATE: May involve both irreversible and reversible changes not severe enough to cause death or permanent injury.

3 HIGH: May cause death or permanent injury after very short exposure to small quantities.

U UNKNOWN: No information on humans considered valid by authors

Symptoms of intoxication from this material, which usually manifest themselves in from 5 to 25 minutes after injection, are a marked cutaneous hyperaesthesia, vomiting, convulsions and diarrhea, paralysis of the diaphragm, spasm of the glottis with transient dyspnea, increased salivation, sweating, slowing of the respiration and lowering of the temperature. Death usually occurs from respiratory paralysis. It has a central and peripheral nervous system action.

Fire Hazard: Slight; when heated to decomposition, it emits toxic fumes of nitrogen oxides (Section 6).

Countermeasures
Treatment and Antidotes: Wash out stomach, keep body warm. Atropine by injection. Artificial respiration and oxygen if needed.

PICENE
General Information
Synonym: 3,4-Benzchrysene; 1,2-7,8-dibenzphenathrene; dibenzo [a,i]-phenanthrene; β,β-binaphthylene ethene.
Description: Leaflets.
Formula: $C_{22}H_{14}$(polycyclic).
Constants: Mol wt: 278.33, bp: 518–520°C, mp: 364°C.
Hazard Analysis
Toxic Hazard Rating: Toxic. A carcinogen.

α-PICOLINE
General Information
Synonym: Methyl pyridine.
Description: Colorless liquid, strong, unpleasant odor.
Formula: $C_5H_4NCH_3$.
Constants: Mol wt: 93.13, mp: −70°C, bp: 129°C, flash p: 102°F (O.C.), d: 0.95 at 15°/4°C autoign. temp.: 1000°F, vap. press.: 10 mm at 24.4°C vap. d.: 3.2.
Hazard Analysis
Toxicity: Details unknown. See pyridine and ethylpicoline.
Fire Hazard: Moderate, when exposed to heat or flame; can react with oxidizing materials.
Explosion Hazard: Unknown.
Disaster Hazard: Dangerous; when heated to decomposition it emits toxic fumes of nitrogen oxides.
Countermeasures
Ventilation Control: Section 2.
To Fight Fire: Carbon dioxide, dry chemical or carbon tetrachloride (Section 6).
Storage and Handling: Section 7.

β-PICOLINE
General Information
Synonyms: 3-Picoline; 3-methyl pyridine.
Description: Colorless liquid.
Formula: $CH_3C_5H_4N$.
Constants: Mol wt: 93.12, bp: 143.5°C, d: 0.9613 at 15°/4°C, vap. d.: 3.21.
Hazard Analysis
Toxicity: Details unknown. See pyridine and ethylpicoline.
Fire Hazard: Details unknown. Probably moderate to dangerous.
Disaster Hazard: Dangerous, when heated to decomposition it emits toxic fumes of nitrogen oxides.

γ-PICOLINE
General Information
Synonyms: 4-Picoline; 4-methylpyridine.
Description: Colorless liquid.
Formula: $CH_3C_5H_4N$.
Constants: Mol wt: 93.06, bp: 145°C, fp: 3.7°C, d: 0.9571 at 15°/4°C, vap. d.: 3.21, flash p: 134°F (O.C.).
Hazard Analysis
Toxicity: Details unknown. See pyridine and ethylpicoline.
Fire Hazard: Probably moderate to dangerous.
Disaster Hazard: See α-picoline.
Countermeasures
To Fight Fire: Alcohol foam.

3-PICOLINE. See β-picoline.

4-PICOLINE. See γ-picoline.

4-PICOLINE-N-OXIDE
General Information
Description: Crystals.
Formula: C_6H_7NO.
Constants: Mol wt: 109.1, fp: 186.3°C.
Hazard Analysis
Toxicity: Details unknown. See pyridine.
Fire Hazard: Slight, when exposed to heat or flame (Section 6).
Disaster Hazard: Dangerous; when heated to decomposition, it emits highly toxic fumes of oxides of nitrogen; can react with oxidizing materials.
Countermeasures
Storage and Handling: Section 7.

PICRAMIC ACID
General Information
Synonyms: Picraminic acid; dinitroaminophenol.
Description: Red, monoclinic crystals.
Formula: $NH_2(NO_2)_2C_6H_2OH$.
Constants: Mol wt: 199.12, mp: 168–169°C.
Hazard Analysis and Countermeasures
See nitrates and 2,3-dinitrophenol.

PICRAMIDE. See trinitroaniline.

PICRAMINIC ACID. See picramic acid.

PICRATES. See nitrates.

PICRATOL.
Hazard Analysis
Details unknown. See also picric acid.

PICRIC ACID
General Information
Synonym: Picronitric acid; trinitrophenol; carbazotic acid.
Description: Yellow crystals or yellow liquid.
Formula: $(NO_2)_3C_6H_2OH$.
Constants: Mol wt: 229.11, mp: 121.8°C, bp: explodes > 300°C, flash p: 302°F, d: 1.763, autoign. temp.: 572°F, vap. d.: 7.90.
Hazard Analysis
Toxic Hazard Rating:
Acute Local: Irritant 2; Allergen 1.
Acute Systemic: Ingestion 3; Inhalation 3; Skin Absorption 2.
Chronic Local: Irritant 2; Allergen 2.
Chronic Systemic: Ingestion 2; Inhalation 2; Skin Absorption 2.
Toxicology: Can cause allergic as well as irritative dermatitis. Symptoms of systemic poisoning are nausea, vomiting, diarrhea, suppressed urine, yellow discoloration of skin and conjunctiva.
TLV: ACGIH (recommended); 0.1 milligrams per cubic meter in air. Can be absorbed through the intact skin.
Spontaneous Heating: No.
Explosion Hazard: Dangerous, when shocked or exposed to heat (See explosives, high). Picric acid forms salts readily. Many of its salts, known as picrates, are more sensitive explosives than picric acid. This is particularly true of copper, lead and zinc. Therefore, this material must be kept out of contact with metals. Picric acid is a more powerful explosive than TNT. Its ability to form the sensitive picrates in contact with metal has somewhat limited its usefulness as an explosive. In America, its primary use is in the converted form of ammonium picrate in base-fused shells for seacoast cannon and in all armor-piercing shells. Picric acid has also found use as a booster explosive and as a substitute for a part of the mercury fulminate charge in detonators. It has been used extensively in the form of mixtures with

other nitro compounds. Such mixtures, having a lower melting point than picric acid, can be melted and cast at temperatures below 100° C. These mixtures are more generally practical for use because of the hazard involved in melting straight picric acid, which has a relatively high melting temperature. Guanidine picrate is an example of a picrate which is used in shells instead of pure picric acid. See also explosives,high.

Disaster Hazard: Highly dangerous; shock will explode it; on decomposition, it emits highly toxic fumes and explodes; can react vigorously with reducing materials.

Countermeasures
Ventilation Control: Section 2.
To Fight Fire: Water (Section 6).
Personnel Protection: Section 3.
Storage and Handling: Section 7.

PICRIC ACID, DRY OR WET WITH LESS THAN 10% WATER. See picric acid.
Shipping Reguations: Section 11.
IATA: Explosive, not acceptable (passenger and cargo).

PICRIC ACID, WET, NOT EXCEEDING 16 OUNCES. See picric acid.
Shipping Regulations: Section 11.
I.C.C.: See Section 11 §73.192.
Coast Guard Classification: Inflammable solid.

PICRIC ACID, WET WITH NOT LESS THAN 10% WATER. See picric acid.
Shipping Regulations: Section 11.
I.C.C.: Flammable solid; yellow label, 25 pounds.
Coast Guard Classification: Inflammable solid; yellow label.
IATA: Flammable solid, yellow label, ½ kilogram (passenger), 12 kilograms (cargo).

PICRONITRIC ACID. See picric acid.

PICRYL CHLORIDE
General Information
Synonym: 2-Chloro-1,3,5-trinitrobenzene.
Description: White needles.
Formula: $C_6H_2ClN_3O_6$.
Constants: Mol wt: 247.6, mp: 83° C, d: 1.797.
Hazard Analysis and Countermeasures
See picric acid, nitrates and chlorides.

PICRYL CHLORIDE, DRY OR WET WITH LESS THAN 10% water
Shipping Regulations: Section 11.
IATA: Explosive, no label, not acceptable (passenger and cargo).

PICRYL CHLORIDE WET WITH NOT LESS THAN 10% WATER
Shipping Regulations: Section 11.
IATA: Flammable solid, yellow label, ½ kilogram (passenger), ½ kilogram (cargo).

PICRYL SULFIDE. See hexanitrodiphenyl sulfide.

PIECE RESIN. See rosin.

PILOCARPINE
General Information
Description: Colorless or yellow, hygroscopic, needle-like crystals.
Formula: $C_{11}H_{16}N_2O_2$.

Constants: Mol wt: 208.26, mp: 34° C, bp: 260° C at 5 mm.
Hazard Analysis
Toxic Hazard Rating:
Acute Local: U.
Acute Systemic: Ingestion 3; Inhalation 3.
Chronic Local: U.
Chronic Systemic: U.
Toxicology: A very poisonous alkaloid. It can cause contraction of the pupils and lachrymation. The ingestion of toxic quantities of this material causes a marked secretion of saliva, excessive perspiration and tears, nausea, retching and vomiting, pain in the abdomen, violent movement of the intestines with profuse watery evacuation. The pulse is sometimes quickened, sometimes slow and irregular. The respiration is quick and labored with rales over the bronchi. There are giddiness and confusion with tremors and feeble convulsions. Eventually slower respiration and weakness occur, but consciousness remains until breathing ceases. The action of pilocarpine on the sweat glands renders it the most powerful sudorific in the pharmacopeia. It is used to remove excess fluid accumulations in the body. It very rarely causes death but when it does, it is by paralysis of the heart or edema of the lungs.
Disaster Hazard: Dangerous; on heating to decomposition it emits toxic fumes of nitrogen oxides.
Countermeasures
Ventilation Control: Section 2.
Personal Hygiene: Section 3.
Treatment and Antidotes: Give general treatment for alkaloidal poisoning and in addition atropine may be administered.
Storage and Handling: Section 7.

PIMELIC KETONE. See cyclohexanone.

PIMENTA. See allspice.

PINANE
General Information
Formula: $C_{10}H_{18}$.
Constants: Mol wt: 138, lel: 3.7% at 320° F, uel: 7.2% at 320° F, d: 0.8, bp: 336° F.
Hazard Analysis
Toxic Hazard Rating: Unknown.
Fire Hazard: Moderate, when exposed to heat or ignition source.
Countermeasures
Storage and Handling: Section 7.

"PINAP"
General Information
Constant: Flash p: 80° F (C.C.).
Hazard Analysis
Toxicity: Unknown.
Fire Hazard: Dangerous, when exposed to heat or flame (Section 6).
Explosion Hazard: Unknown.
Disaster Hazard: Dangerous, upon exposure to heat or flame; can react vigorously with oxidizing materials.
Countermeasures
Ventilation Control: Section 2.
Storage and Handling: Section 7.

PINDONE. See 2-pivalyl-1,3-indandione.
Shipping Regulations: Section 11.
IATA: Poison B, poison label, 25 kilograms (passenger), 95 kilograms (cargo).

TOXIC HAZARD RATING CODE (For detailed discussion, see Section 1.)

0 NONE: (a) No harm under any conditions; (b) Harmful only under unusual conditions or overwhelming dosage.

1 SLIGHT: Causes readily reversible changes which disappear after end of exposure.

2 MODERATE: May involve both irreversible and reversible change not severe enough to cause death or permanent injury.

3 HIGH: May cause death or permanent injury after very short exposu to small quantities.

U UNKNOWN: No information on humans considered valid by author

α-PINENE
General Information
Synonym: 2,6,6-Trimethylbicyclo-(3.1.1)-2-heptene.
Description: Liquid; odor of turpentine.
Formula: $C_{10}H_{16}$.
Constants: Mol wt: 136.2, mp: −55°C, bp: 155°C, flash p: 91°F, d: 0.8585 at 20°/4°C, vap. press.: 10 mm at 37.3°C, vap. d.: 4.7.
Hazard Analysis
Toxic Hazard Rating:
  Acute Local: Irritant 2; Ingestion 2; Inhalation 2.
  Acute Systemic: Ingestion 2; Inhalation 2; Skin Absorption 2.
  Chronic Local: Irritant 2.
  Chronic Systemic: Ingestion 2; Inhalation 2; Skin Absorption 2.
Toxicology: Irritating to skin and mucous membrane. Can cause dizziness, palpitation, bronchitis and nephritis.
Fire Hazard: Moderate, when exposed to heat or flame; can react with oxidizing materials.
Countermeasures
To Fight Fire: Foam, carbon dioxide, dry chemical or carbon tetrachloride (Section 6).
Ventilation Control: Section 2.
Storage and Handling: Section 7.

PINE OIL
General Information
Description: Pale yellow liquid; penetrating odor.
Constants: Bp: 200–220°C, flash p: 172°F (C.C.), d: 0.86, flash p (steam distilled): 138°F.
Hazard Analysis
Toxicity: A weak allergen. See also turpentine.
Fire Hazard: Moderate, when exposed to heat or flame; can react with oxidizing materials,
Spontaneous Heating: Moderate.
Countermeasures
To Fight Fire: Foam, carbon dioxide, dry chemical or carbon tetrachloride (Section 6).
Ventilation Control: Section 2.
Storage and Handling: Section 7.
Shipping Regulations: Section 11.
  Coast Guard Classification: Combustible liquid.

PINE PITCH
General Information
Constants: Mp: 64.4°C, bp: 255°C, flash p: 285°F (C.C.), d: 1.1.
Hazard Analysis
Toxicity: A weak allergen.
Fire Hazard: Slight, when exposed to heat or flame; can react with oxidizing materials.
Spontaneous Heating: Yes.
Countermeasures
To Fight Fire: Water, carbon dioxide, dry chemical or carbon tetrachloride (Section 6).
Storage and Handling: Section 7.

PINE RESIN. See rosin.

PINE TAR
General Information
Description: Black-borwn, viscous liquid; piny odor.
Constants: Bp: 240–400°C, flash p: 130°F (C.C.), d: > 1, autoign. temp.: 671°F.
Hazard Analysis
Toxic Hazard Rating:
  Acute Local: Irritant 1; Allergen 1.
  Acute Systemic: U.
  Chronic Local: Allergen 1.
  Chronic Systemic: U.
Fire Hazard: Moderate, when exposed to heat or flame; can react with oxidizing materials.
Spontaneous Heating: Yes.

Countermeasures
To Fight Fire: Foam, carbon dioxide, dry chemical or carbon tetrachloride (Section 6).
Personal Hygiene: Section 3.
Ventilation Control: Section 2.
Storage and Handling: Section 7.

PINE TAR OIL
General Information
Synonym: Tar oil rectified.
Description: Dark, reddish-brown liquid.
Constants: Flash p: 144°F (C.C.), d: 0.96–0.99 at 25°/25°C.
Hazard Analysis
Toxic Hazard Rating:
  Acute Local: Irritant 1.
  Acute Systemic: U.
  Chronic Local: Allergen 1.
  Chronic Systemic: U.
Fire Hazard: Moderate, when exposed to heat or flame; can react with oxidizing materials.
Countermeasures
To Fight Fire: Carbon dioxide, dry chemical or carbon tetrachloride (Section 6).
Personal Hygiene: Section 3.
Storage and Handling: Section 7.

PINTSCH GAS
General Information
Typical Composition: 30.0% illuminants, 0.1% carbon monoxide, 13.2% hydrogen, 45.0% methane, 9.0% ethane, 0.2% carbon dioxide, 1.6% nitrogen.
Hazard Analysis
Toxic Hazard Rating:
  Acute Local: U.
  Acute Systemic: Inhalation 2.
  Chronic Local: U.
  Chronic Systemic: Inhalation 1.
Fire Hazard: Moderate, when exposed to flame (Section 6).
Explosion Hazard: Unknown.
Disaster Hazard: Moderately dangerous; when heated, it emits toxic fumes; can react with oxidizing materials.
Countermeasures
Ventilation Control: Section 2.
Storage and Handling: Section 7.

PIPERAZINE
General Information
Synonyms: Hexahydropyrazine; diethylenediamine.
Description: Colorless, rhombic crystals.
Formula: $\overline{NHCH_2CH_2NHCH_2CH_2}$.
Constants: Mol wt: 86.14, mp: 104°C, bp: 145°C, flash p: 190°F (O.C.), d: 1.1, vap. d.: 3.0.
Hazard Analysis
Toxic Hazard Rating:
  Acute Local: Irritant 1.
  Acute Systemic: Ingestion 1.
  Chronic Local: U.
  Chronic Systemic: Ingestion 1.
Toxicology: Excessive absorption can cause urticaria, vomiting, diarrhea, blurred vision and weakness.
Disaster Hazard: Dangerous; when heated to decomposition, it emits highly toxic fumes of oxides of nitrogen.
Countermeasures
Personal Hygiene: Section 3.
Storage and Handling: Section 7.
To Fight Fire: Alcohol foam.

PIPERIDINE
General Information
Synonym: Hexahydropyridine.
Description: Clear, colorless liquid; aminelike odor.
Formula: $C_5H_{11}N$.
Constants: Mol wt: 85.2, mp: −7°C, bp: 106°C, flash p:

61°F, d: 0.8622 at 20°/4°C, vap. press.: 40 mm at 29.2°C, vap. d.: 3.0.

Hazard Analysis
Toxic Hazard Rating:
Acute Local: Irritant 3.
Acute Systemic: Ingestion 2.
Chronic Local: U.
Chronic Systemic: U.
Fire Hazard: Dangerous, when exposed to heat or flame.
Explosion Hazard: Unknown.
Disaster Hazard: Dangerous; when heated to decomposition, it emits highly toxic fumes of oxides of nitrogen; can react vigorously with oxidizing materials.

Countermeasures
Ventilation Control: Section 2.
Storage and Handling: Section 7.
To Fight Fire: Alcohol foam, carbon dioxide, dry chemical or carbon tetrachloride (Section 6).

PIPERIDINIUM PENTAMETHYLENE DITHIOCARBAMATE
General Information
Description: White to cream crystals.
Formula: $C_{11}H_{22}N_2S_2$.
Constants: Mol wt: 246, mp: 170°C, d: 1.13 at 25°C.

Hazard Analysis
Toxic Hazard Rating:
Acute Local: Irritant 3.
Acute Systemic: Ingestion 1.
Chronic Local: U.
Chronic Systemic: U.
Disaster Hazard: Dangerous; when heated to decomposition, it emits highly toxic fumes of oxides of sulfur and nitrogen.

Countermeasures
Personal Hygiene: Section 3.
Storage and Handling: Section 7.

PIPERINE
General Information
Synonym: 1-Piperoyl piperidine.
Description: Crystals.
Formula: $C_{17}H_{19}NO_3$.
Constants: Mol wt: 285.3, mp: 130°C, d: 1.193.

Hazard Analysis
Toxicity: Details unknown; an insecticide. Probably low.
Disaster Hazard: Dangerous; when heated to decomposition, it emits highly toxic fumes of oxides of nitrogen.

Countermeasures
Storage and Handling: Section 7.

PIPERONAL
General Information
Synonyms: 3,4-Methylene-dioxybenzaldehyde; heliotropin.
Description: Colorless, lustrous crystals.
Formula: $C_8H_6O_3$.
Constants: Mol wt: 150.1, mp: 37°C, bp: 263°C, vap. press.: 1 mm at 87.0°C.

Hazard Analysis
Toxicity: Details unknown; a mosquito repellent. Limited animal experiments show low toxicity. Used as a synthetic flavoring substance and adjuvant (Section 10). See also aldehydes.
Fire Hazard: Slight, when exposed to heat or flame; can react with oxidizing materials (Section 6).

Countermeasures
Storage and Handling: Section 7.

PIPERONYL BUTOXIDE
General Information
Synonym: Butylcarbityl (6-propyl piperonyl) ether.
Description: Light brown liquid; mild odor.
Formula: $C_{19}H_{30}O_5$.
Constants: Mol wt: 338.4, bp: 180°C at 1 mm, flash p: 340°F, d: 1.04–1.07 at 20°/20°C.

Hazard Analysis
Toxic Hazard Rating:
Acute Local: Ingestion 1; Inhalation 1.
Acute Systemic: Ingestion 1.
Chronic Local: U.
Chronic Systemic: U.
Toxicology: Limited animal experiments show low toxicity and little irritation. Excessive ingestion can cause gastrointestinal disturbances. Used as a food additive permitted in the feed and drinking water of animals and or for the treatment of food producing animals, also permitted in food for human consumption (Section 10).
Fire Hazard: Slight, when exposed to heat or flame; can react with oxidizing materials.

Countermeasures
To Fight Fire: Foam, carbon dioxide, dry chemical or carbon tetrachloride (Section 6).
Storage and Handling: Section 7.

PIPERONYL CYCLONENE
General Information
Synonym: 3-Isoamyl-5-(methylene dioxyphenyl)-2-cyclohexanone.
Description: Powder.
Formula: $C_5H_{11}CH_2(OC_6H_5)_2C_6H_8O$.
Constants: Mol wt: 367.5, flash p: 290–300°F, d: 1.09–1.20 at 20°/20°C.

Hazard Analysis
Toxicity: Details unknown. See piperonyl butoxide, which is similar in action.
Fire Hazard: Slight, when exposed to heat or flame; can react with oxidizing materials.

Countermeasures
To Fight Fire: Foam, carbon dioxide, dry chemical or carbon tetrachloride (Section 6).
Storage and Handling: Section 7.

1-PIPEROYL PIPERIDINE. See piperine.

PITCH BLENDE
General Information
Description: An ore of uranium, radium, etc.

Hazard Analysis
Toxicity: Continuous exposure by inhalation may cause carcinoma of the lungs. See uranium.

PITCH DUST
Hazard Analysis
Toxic Hazard Rating:
Acute Local: Irritant 2.
Acute System: U.
Chronic Local: Irritant 2.
Chronic Systemic: U.
Fire Hazard: Slight, when exposed to heat or flame; can react with oxidizing materials (Section 6).

Countermeasures
Personnel Protection: Section 3.

PIVAL. See 2-pivalyl-1,3-indandione.

PIVALIC ACID. See trimethylacetic acid.

TOXIC HAZARD RATING CODE (For detailed discussion, see Section 1.)

0 NONE: (a) No harm under any conditions; (b) Harmful only under unusual conditions or overwhelming dosage.

1 SLIGHT: Causes readily reversible changes which disappear after end of exposure.

2 MODERATE: May involve both irreversible and reversible changes not severe enough to cause death or permanent injury.

3 HIGH: May cause death or permanent injury after very short exposure to small quantities.

U UNKNOWN: No information on humans considered valid by authors

2-PIVALYL-1,3-INDANDIONE
General Information
Synonym: Pival.
Description: Crystals. Water soluble in form of sodium salt.
Hazard Analysis
Toxicity: Toxic. Reduces blood-clotting leading to hemorrhages. For symptoms of exposure, see warfarin.
TLV: ACGIH (recommended); 0.1 milligrams per cubic meter of air.

PLANOFIX. See α-naphthaleneacetic acid.

PLASTICIZER SC. See triethylene glycol caprylate.

PLASTIC POWDERS
Hazard Analysis
Toxicity: Variable; a weak allergen.
Fire Hazard: Moderate, when exposed to heat or flame (Section 6).
Disaster Hazard: Moderately dangerous; when heated, they burn and may emit acrid fumes; they may react with oxidizing materials.
Countermeasures
Storage and Handling: Section 7.

PLASTIC SOLVENT, N.O.S.
Shipping Regulations: Section 11.
    IATA: Flammable liquid, red label, 1 liter (passenger), 40 liter (cargo).
    ICC: Flammable liquid, red label, 10 gallons.

PLASTOLEIN X-55. See diethylene glycol dipelargonate.

PLATINIC AMMONIUM CHLORIDE. See ammonium chloroplatinate.

PLATINIC CHLORIDE
General Information
Synonyms: Acid platinic chloride; chloroplatinic acid.
Description: Brownish yellow, very deliquescent, crystalline mass; easily soluble in water and alcohol.
Formula: $H_2PtCl_6 \cdot 6H_2O$.
Constants: Mol wt: 517.94, d: 2.431, mp: 60°C.
Hazard Analysis and Countermeasures
See platinum compounds and chlorides.

PLATINIC DISULFIDE
General Information
Description: Black-brown powder.
Formula: PtS_2.
Constants: Mol wt: 259.36, mp: decomp. 225–250°C, d: 7.22.
Hazard Analysis
See platinum compounds and sulfides.
Countermeasures
See platinum compounds and sulfides.

PLATINIC TETRABROMIDE
General Information
Description: Dark brown crystals.
Formula: $PtBr_4$.
Constants: Mol wt: 514.89, mp: decomp. at 180°C, d: 5.69.
Hazard Analysis and Countermeasures
See platinum compounds and bromides.

PLATINIC TETRACHLORIDE
General Information
Description: Brown-red crystals.
Formula: $PtCl_4$.
Constants: Mol wt: 337.06, mp: decomp. at 370°C.
Hazard Analysis and Countermeasures
See platinum compounds and chlorides.

PLATINIC TETRAFLUORIDE
General Information
Description: Deep, red, fused mass or yellow-light brown, deliquescent crystals.

Formula: PtF_4.
Constants: Mol wt: 271.23, mp: decomposes.
Hazard Analysis and Countermeasures
See fluorides and platinum compounds.

PLATINIC TETRAIODIDE
General Information
Description: Amorphous, brown to black crystals.
Formula: PtI_4.
Constants: Mol wt: 702.91, mp: decomp. at 370°C, d: 6.064 at 25°C.
Hazard Analysis and Countermeasures
See platinum compounds and iodides.

PLATINOUS-AMMONIUM CHLORIDE. See ammonium chloroplatinate.

PLATINOUS CYANIDE
General Information
Description: Yellow-brown crystals.
Formula: $Pt(CN)_2$.
Constant: Mol wt: 247.27.
Hazard Analysis and Countermeasures
See cyanides.

PLATINOUS DIBROMIDE
General Information
Description: Brown crystals.
Formula: $PtBr_2$.
Constants: Mol wt: 355.06, mp: decomp. at 250°C, d: 6.65.
Hazard Analysis and Countermeasures
See platinum compounds and bromides.

PLATINOUS DICHLORIDE
General Information
Description: Olive green crystals.
Formula: $PtCl_2$.
Constants: Mol wt: 266.14, mp: decomp. at 581°C, d: 5.87 at 11°C.
Hazard Analysis and Countermeasures
See platinum compounds and chlorides.

PLATINOUS DIFLUORIDE
General Information
Description: Yellowish-green crystals.
Formula: PtF_2.
Constant: Mol wt: 233.23.
Hazard Analysis and Countermeasures
See fluorides and platinum compounds.

PLATINOUS DIIODIDE
General Information
Description: Black crystals.
Formula: PtI_2.
Constants: Mol wt: 449.07, mp: decomp. 300–350°C, d: 6.4.
Hazard Analysis and Countermeasures
See platinum compounds and iodides.

PLATINOUS MONOSULFIDE
General Information
Description: Black crystals.
Formula: PtS.
Constants: Mol wt: 227.29, mp: decomposes, d: 8.847.
Hazard Analysis and Countermeasures
See platinum compounds and sulfides.

PLATINUM
General Information
Description: Silver gray, lustrous, malleable and ductile metallic element. Can be prepared as a black powder or sponge.
Formula: Pt.
Constants: Mol wt: 195.1, mp: 1773.5°C, bp: 4530°C, d: 21.447.
Hazard Analysis
Toxicity: See platinum compounds.

TLV: ACGIH (recommended), 0.002 milligrams per cubic
meter of air.
Radiation Hazard: (Section 5). For permissible levels, see
Table 5, p. 150.
Artificial isotope ^{191}Pt, half life 3.0 d. Decays to
stable ^{191}Ir by electron capture. Emits gamma rays of
0.04 to 0.62 MeV and X-rays.
Artificial isotope ^{193m}Pt, half life 4.4 d. Decays to
radioactive ^{193}Pt by emitting gamma rays of 0.14 MeV.
The characteristics of ^{193}Pt are not well known.
Artificial isotope ^{197m}Pt, half life 82 m. Decays to
radioactive ^{197}Pt by emitting gamma rays of 0.35 MeV.
Artificial isotope ^{197}Pt, half life 20 h. Decays to
stable ^{197}Au by emitting beta particles of 0.48 (9%),
0.67 (90%) MeV. Also emits gamma rays of 0.08 MeV.

PLATINUM ARSENIDE
General Information
Synonym: Sperrylite.
Description: Cubic, white crystals.
Formula: PtAs$_2$.
Constants: Mol wt: 345.05, mp: >800°C, d: 10.602.
Hazard Analysis
Toxicity: Highly toxic. See arsenic compounds.
Fire Hazard: Moderate, by chemical reaction (Section 6).
Explosion Hazard: Moderate, by chemical reaction.
Disaster Hazard: Dangerous. See arsenic compounds.
Countermeasures
Ventilation Control: Section 2.
Personnel Protection: Section 3.
Storage and Handling: Section 7.

PLATINUM COMPOUNDS
Hazard Analysis
Toxicity: Exposure to complex platinum salts has been
shown to cause symptoms of intoxication such as wheezing, coughing, running of the nose, tightness in the
chest, shortness of breath and cyanosis, whereas exposure to dust of pure metallic platinum causes no intoxication. Furthermore, many people working with
platinum salts are troubled with dermatitis (Section 9).
This seems only to be true of complex platinum salts. It
does not include the complex salts of other precious
metals.
Countermeasures
Ventilation Control: Section 2.
Personnel Protection: Section 3.
Treatment and Antidotes: Removal from exposure effectively causes the symptoms to disappear.

PLATINUM FULMINATE
General Information
Description: Solid.
Formula: Pt(C$_2$N$_2$O$_2$)$_2$.
Constants: Mol wt: 363.3.
Hazard Analysis
Toxicity: See platinum compounds.
Fire Hazard: Unknown.
Explosion Hazard: Severe, when shocked or exposed to
heat. See also explosives, high and fulminates.
Disaster Hazard: Dangerous; shock will explode it; when
heated to decomposition, it emits highly toxic fumes of
oxides of nitrogen.
Countermeasures
Storage and Handling: Section 7.

PLATINUM PYROPHOSPHATE
General Information
Description: Green-yellow crystals.
Formula: PtP$_2$O$_7$.
Constants: Mol wt: 369.27, mp: decomp. at 600°C, d: 4.85.
Hazard Analysis and Countermeasures
See platinum.

PLATINUM SESQUISULFIDE
General Information
Description: Gray crystals (exist quest.).
Formula: Pt$_2$S$_3$.
Constants: Mol wt: 486.66, mp: decomposes, d: 5.52.
Hazard Analysis and Countermeasures
See platinum compounds and sulfides.

PLATTNERITE. See lead dioxide.

PLUMBAGO. See graphite.

PLUMBIC FLUORIDE. See lead tetrafluoride.

PLUMBOUS SULFIDE. See lead sulfide.

PLUMBUM. See lead.

PLUTONIUM
General Information
Description: Radioactive metallic element.
Formula: Pu.
Constant: At wt: 239.
Hazard Analysis
Toxicity: Highly toxic. The permissible levels for plutonium
are the lowest for any of the radioactive elements. This
is occasioned by the concentration of plutonium directly
in the blood-forming sections of the bone, rather than
the more uniform bone distribution shown by other
heavy elements. This increases the possibility of damage
from equivalent activities of plutonium and has led to
the adoption of the extremely low permissible levels
given.
Radiation Hazard: Section 5. For permissible levels, see
Table 5, p. 150.
Artificial isotope ^{238}Pu, half life 86 y. Decays to
radioactive ^{234}U by emitting alpha particles of 5.5
MeV.
Artificial isotope ^{239}Pu, half life 24000 y. Decays to
radioactive ^{235}U by emitting alpha particles of 5.1
MeV.
Artificial isotope ^{240}Pu, half life 6600 y. Decays to
radioactive ^{236}U by emitting alpha particles of 5.1
MeV.
Artificial isotope ^{241}Pu (Neptunium Series), half
life 13 y. Decays to radioactive ^{241}Am by emitting beta
particles of 0.02 MeV.
Artificial isotope ^{242}Pu, half life 3.8 × 10^5 y. Decays to radioactive ^{238}U by emitting alpha particles of
4.9 MeV.
Artificial isotope ^{243}Pu, half life 5 h. Decays to
radioactive ^{243}Am by emitting beta particles of 0.49
(38%), 0.58 (62%) MeV. Also emits gamma rays of
0.08 MeV.
Artificial isotope ^{244}Pu, half life about 7.5 × 10^7 y.
Decays to radioactive ^{240}U by emitting alpha particles
of unknown energy.

PMA. See pyromellitic acid.

TOXIC HAZARD RATING CODE (For detailed discussion, see Section 1.)

0 NONE: (a) No harm under any conditions; (b) Harmful only under unusual conditions or overwhelming dosage.

1 SLIGHT: Causes readily reversible changes which disappear after end
of exposure.

2 MODERATE: May involve both irreversible and reversible changes
not severe enough to cause death or permanent injury.

3 HIGH: May cause death or permanent injury after very short exposure
to small quantities.

U UNKNOWN: No information on humans considered valid by authors

PODOPHYLLIN
General Information
Synonym: Podophyllum resin.
Description: Light yellow powder or small, yellow, bulky, fragile lumps; bitter, acrid taste.
Hazard Analysis
Toxic Hazard Rating:
  Acute Local: Irritant 2; Ingestion 2; Inhalation 2.
  Acute Systemic: U.
  Chronic Local: U.
  Chronic Systemic: U.
Fire Hazard: Slight, when exposed to heat or flame; can react with oxidizing materials (Section 6).
Countermeasures
Storage and Handling: Section 7.

PODOPHYLLUM RESIN. See podophyllin.

POGY OIL. See menhaden oil.

POISON IVY
General Information
Synonyms: Poison oak; poison sumac.
Description: Leafy plant, vine or bush.
Hazard Analysis
Toxic Hazard Rating:
  Acute Local: Irritant 1; Allergen 2.
  Acute Systemic: U.
  Chronic Local: Allergen 2.
  Chronic Systemic: U.
Toxicology: Can cause a dermatitis in susceptible individuals, from bodily contact with the plant, exposure to the smoke from burning plants and contact with material which has been in contact with the plant (Section 9). A common air contaminant. (Section 4).

POISON OAK. See poison ivy.

POISONOUS ARTICLES (IATA)
POISONOUS ARTICLES, CLASS A cover extremely poisonous gases and liquids, and their carriage is not acceptable either by passenger or cargo aircraft.
POISONOUS ARTICLES, CLASS B are substances, liquids or solids (including pastes and semi-solids) other than Classes A or C, which are known to be so toxic to man as to afford a hazard to health during transportation or which, in the absence of adequate data on human toxicity, are presumed to be toxic to man.
The criteria used to determine whether an article warrants treatment as a Poison B are generally based on the standards used by the Public Health Authorities and Transport regulatory organisations in North America, such as the Interstate Commerce Commission of USA and the Board of Transport Commissioners in Canada and are generally referred to as the LD50 (Lethal Dose for 50% of animals tested) values. These criteria are:
  (i) Oral Toxicity—Those which produce death within 48 hours in half or more than half of a group of 10 or more white laboratory rats weighing 200 to 300 grams at a single dose of 50 milligrams or less per kilogram of body weight, when administered orally.
  (ii) Toxicity on inhalation—Those which produce death within 48 hours in half or more than half of a group of 10 or more white laboratory rats weighing 200 to 300 grams, when inhaled continuously for a period of one hour or less at a concentration of 2 milligrams or less per litre of vapour, mist or dust provided such concentration is likely to be encountered by man when the chemical product is used in any reasonable foreseeable manner.
  (iii) Toxicity by skin absorption—Those which produce death within 48 hours in half or more than half of a group of 10 or more rabbits tested at a dosage of 200 milligrams or less per kilogram body weight, when administered by continuous contact with the bare skin for 24 hours or less.
The foregoing categories shall not apply if the physical characteristics or the probable hazards to humans, as shown by experience, indicate that the substance will not cause serious sickness or death.
In the case of vaporizing liquids or gases having toxic properties the Threshold Limit Values such as Maximum Allowable Concentration (MAC) are also taken into account. However, vapour pressure is recognized as an important aspect of this assessment and due regard is paid to this.
POISONOUS ARTICLES, CLASS C generally cover dangerous lachrymatory gases or solids, very few of which are permitted carriage on passenger aircraft, but some of which are permitted on cargo aircraft.

POISONOUS LIQUID OR GAS, N.O.S.
Shipping Regulations: Section 11.
  I.C.C.: Poison A; poison gas label, not acceptable.
  Coast Guard Classification: Poison A; poison gas label.
  IATA: Poison A, not acceptable (passenger and cargo).

POISONOUS LIQUIDS, N.O.S.
Shipping Regulations: Section 11.
  I.C.C. (class B): Poison B, poison label, 55 gallons.
  IATA (class B): Poison B, poison label, 1 liter (passenger), 220 liters (cargo).
  I.C.C. (class C): Poison C, poison label, 75 pounds.
  IATA (class C): Poison C, poison label, not acceptable (passenger), 35 kilograms (cargo).

POISONOUS SOLIDS, N.O.S.
Shipping Regulations: Section 11.
  I.C.C.: Poison B; poison label, 200 pounds; Poison C, tear gas label, 75 pounds.
  Coast Guard Classification: Poison B, poison label; Poison C, tear gas label.
  IATA (class B): Poison B, poison label, 25 kilograms (passenger), 95 kilograms (cargo).
       class C): Poison C; poison label, not acceptable (passenger), 35 kilograms (cargo).

POISON SUMAC. See poison ivy.

POLISHES (metal, stove, furniture), LIQUID
Hazard Analysis
Toxicity: Variable.
Fire Hazard: Dangerous, when exposed to heat or flame; they can react with oxidizing materials (Section 6).
Explosion Hazard: Unknown.
Countermeasures
Storage and Handling: Section 7.
Shipping Regulations: Section 11.
  I.C.C.: Flammable liquid; red label, 55 gallons.
  Coast Guard Classification: Inflammable liquid; red label.
  IATA: Flammable liquid, red label, 1 liter (passenger), 220 liters (cargo).

POLISHING COMPOUNDS, LIQUID. See polishes, liquid.
Shipping Regulations: Section 11.
  Coast Guard Classification: Combustible liquid.

POLONIUM
General Information
Description: Radioactive element.
Formula: Po.
Constant: At wt: 210.
Hazard Analysis
Radiation Hazard: Section 5. For permissible levels, see Table 5, p. 150.
Natural isotope ^{210}Po (Radium-F, Uranium Series), half life 138 d. Decays to stable ^{206}Pb by emitting alpha particles of 5.3 MeV.

POLONIUM CARBONYL
General Information
Description: Radioactive material.
Formula: PoCO.
Constant: Mol wt: 238.
Hazard Analysis
Toxicity: Highly toxic. See carbonyls.
Radiation Hazard: See polonium.
Fire Hazard: Dangerous, when exposed to heat or by chemical reaction with oxidizing agents (Section 6). See carbonyls.
Explosion Hazard: See carbonyls.
Disaster Hazard: Dangerous; when heated to decomposition, it emits highly toxic fumes of carbon monoxide; can react vigorously with oxidizing materials.
Countermeasures
Ventilation Control: Section 2.
Storage and Handling: Section 7.

POLONIUM HYDRIDE
General Information
Description: Volatile, unstable material.
Formula: PoH_2.
Constant: Mol wt: 212.02.
Hazard Analysis and Countermeasures
See hydrides and polonium.

POLONIUM HYDROXIDE
General Information
Description: Radioactive solid.
Formula: $Po(OH)_4$.
Constant: Mol wt: 278.
Hazard Analysis and Countermeasures
See polonium.

POLYACRYLAMIDE
General Information
Description: White solid; water soluble high polymer.
Formula: $(CH_2CHCONH_2)_x$.
Hazard Analysis
Toxicity: Unknown. A food additive permitted in food for human consumption (Section 10).

POLYAMYL NAPHTHALENE
General Information
Description: A mixture of polymers.
Constants: Flash p: 360° F (O.C.), d: 0.92–0.93.
Hazard Analysis
Toxicity: Details unknown. See also naphthalene.
Fire Hazard: Slight, when exposed to heat or flame; can react with oxidizing materials.
Countermeasures
To Fight Fire: Foam, carbon dioxide, dry chemical or carbon tetrachloride (Section 6).
Storage and Handling: Section 7.

POLY-sec-AMYL PHENOL
General Information
Constants: Bp: 305–355° C, flash p: 520° F, d: 0.90–0.92 at 30° /20° C.
Hazard Analysis
Toxicity: Details unknown. See also phenol.
Fire Hazard: Slight, when exposed to heat or flame.
Disaster Hazard: Dangerous; when heated to decomposition, it emits toxic fumes; can react with oxidizing materials.
Countermeasures
Storage and Handling: Section 7.

To Fight Fire: Foam, carbon dioxide, dry chemical or carbon tetrachloride (Section 6).

POLYCHLOROPENTANES
General Information
Constants: Bp: 174° C, flash p: 175: 175° F (O.C.), d: 1.33.
Hazard Analysis
Toxicity: See chlorinated hydrocarbons (aliphatic).
Fire Hazard: Moderate, when exposed to heat or flame; can react with oxidizing materials.
Countermeasures
Storage and Handling: Section 7.

POLYCHLOROTRIFLUOROETHYLENE
General Information
Synonym: Halocarbon.
Description: Oils, greases and waxes which are generally chemically inert except to molten sodium, liquid fluorine, and liquid chlorine trifluoride.
Formula: $(CF_2-CFCl)_n$.
Hazard Analysis
Toxicity: Little or none.
Disaster Hazard: Dangerous. Decomposition to highly toxic volatile fumes occurs rapidly at temperatures above 300° C.

POLYCOUMARONE   **RESIN.** See coumorone-indene resin.

POLYDYMITE. See nickelous, nickelic sulfide.

POLYETHYLENE
General Information
Synonyms: PE.
Description: Odorless. The high molecular weight materials are tough, white, leathery, resinous materials.
Formula: $(C_2H_4)_n$
Hazard Analysis
Toxicity: Unknown. A food additive resulting from contact with packaging materials (Section 10).

POLYETHYLENE GLYCOL 400
General Information
Description: Liquid.
Formula: $HOCH_2(CH_2OCH_2)_nCH_2OH$(n varies from 8–10).
Constants: Mol wt: 285–315, d: 1.110–1.140 at 20° C; mp: 4–10° C; flash p: 224° C.
Hazard Analysis
Toxic Hazard Rating: U. A food additive permitted in food for human consumption (Section 10).
Fire Hazard: Moderate, when exposed to heat or flame.
Countermeasures
Storage and Handling: Section 7.
To Fight Fire: Water, foam, dry chemical.

POLYETHYLENE GLYCOL 6000
General Information
Description: White; waxy solid; soluble in water.
Constants: Mp: 58–62° C, flash p: > 246° C.
Hazard Analysis
Toxicity: Unknown. A food additive permitted in food for human consumption (Section 10).
Fire Hazard: Slight, when exposed to heat or flame.

POLYETHYLENE GLYCOL CHLORIDE 110
General Information
Description: Liquid.
Formula: $ClC_2H_4OC_2H_4OH$.

TOXIC HAZARD RATING CODE   *(For detailed discussion, see Section 1.)*

0 NONE: (a) No harm under any conditions; (b) Harmful only under unusual conditions or overwhelming dosage.

1 SLIGHT: Causes readily reversible changes which disappear after end of exposure.

2 MODERATE: May involve both irreversible and reversible changes; not severe enough to cause death or permanent injury.

3 HIGH: May cause death or permanent injury after very short exposure to small quantities.

U UNKNOWN: No information on humans considered valid by authors.

Constants: Mol wt: 124.57, mp: $-90°C$, bp: 198.9°C, flash p: 225°F (O.C.), d: 1.1753 at 20°/20°C, vap. d.: 4.31.
Hazard Analysis
Toxicity: See glycols.
Fire Hazard: Slight, when exposed to heat or flame.
Disaster Hazard: Dangerous; when heated to decomposition, it emits highly toxic fumes of chlorides; can react with oxidizing materials.
Countermeasures
Storage and Handling: Section 7.
To Fight Fire: Water, foam, carbon dioxide, dry chemical or carbon tetrachloride (Section 6).

POLYETHYLENE GLYCOLS
General Information
Description: Colorless viscous liquids.
Formula: $H(OCH_2CH_2)_nOH$.
Constants: Mol wt: 200–9000, mp: -14 to 25°C, flash p: 360–480°F, d: 1.122–1.130.
Hazard Analysis
Toxicity:Food additives permitted in the feed and drinking water of animals and/or for the treatment of food producing animals (Section 10). See glycols.
Fire Hazard: Slight, when exposed to heat or flame; they can react with oxidizing materials.
Countermeasures
To Fight Fire: Water, foam, carbon dioxide, dry chemical or carbon tetrachloride (Section 6).
Storage and Handling: Section 7.

POLYETHYLENE POLYSULFIDE. See "Omilite."

POLYINDENE RESINS. See coumorone-indene resin.

POLYMERIZABLE MATERIALS
General Information
Description: Any liquid, solid, or gaseous material, which under conditions incident to transportation may polymerize (combine or react with itself) so as to cause dangerous evolution of gas or heat.
Countermeasures
Shipping Regulations: Section 11.
  IATA: Other restricted articles, class C, no label required, 5 liters (passenger), 40 liters (cargo).

POLYMERIZED ACETALDEHYDE. See metaldehyde.

POLYMIXIN
General Information
Description: A series of antibiotic substances; polypeptide (basic); soluble in water.
Hazard Analysis
Toxicity: Unknown. A food additive permitted in food for human consumption (Section 10).

POLYOXY ETHYLENE GLYCOL (400) MONO AND DI-OLEATES
Hazard Analysis
Toxicity: Unknown. Used as food additives permitted in the feed and drinking water of animals and/or for the treatment of food producing animals (Section 10).

POLYOXYETHYLENE(20)SORBITAN TRISTEARATE
General Information
Synonym: Polysorbate 65
Description: Tan colored; waxy solid at 25°C; faint odor; soluble in mineral oil, petroleum ether, acetone, ether, and ethanol.
Formula: $C_{100}H_{194}O_{28}$(approximately).
Constants: Mol wt: 1842 (approximately), d: 1.05 at 25°C.
Hazard Analysis
Toxicity: Unknown. A food additive permitted in food for human consumption (Section 10).

POLYOXYMETHYLENE GLYCOL. See paraformaldehyde.

POLYPHOSPHORIC ACID
General Information
Synonym: Phospholeum.
Description: Viscous liquid. Water soluble with evolution of heat.
Formula: $H_3PO_4 + P_2O_5$.
Hazard Analysis
Toxic Hazard Rating:
  Acute Local: Irritant 2; Ingestion 2.
  Acute Systemic: U.
  Chronic Local: Irritant 2.
  Chronic Systemic: U.
Toxicology: See also phosphorus compounds, inorganic.
Disaster Hazard: Dangerous. When heated to decomposition it emits highly toxic fumes of oxides of phosphorus.

POLYPROPYLENE
General Information
Description: The lightest plastic produced by stereoselective catalysts. Color: White to yellow.
Formula: $(C_3H_6)_n$.
Constant: Mol wt (Commercial material): 40,000 or more, d: 0.90, mp: 165–171°C (isostatic).
Hazard Analysis
Toxicity: Unknown. A food additive permitted in food for human consumption (Section 10).

POLYPROPYLENE GLYCOL
General Information
Description: Liquid.
Formula: $HO(C_3H_6O)_nH$.
Constants: Mol wt: 400–2000, mp: does not crystallize, flash p: 390° $->$ 440°F, d: 1.002–1.007.
Hazard Analysis
Toxicity: See glycols.
Fire Hazard: Slight, when exposed to heat or flame; can react with oxidizing materials.
Countermeasures
To Fight Fire: Foam, carbon dioxide, dry chemical or carbon tetrachloride (Section 6).
Storage and Handling: Section 7.

POLYSORBATE 60
General Information
Synonym: Polyoxyethylene (20) sorbetan monostearate.
Description: Lemon to orange colored oily liquid or semigel (at 25°C); faint odor; soluble in water, ethyl acetate, toluene, insoluble in mineral oil, and vegetable oils.
Formula: $C_{64}H_{126}O_{26}$ (approximately).
Constants: Mol wt: 1308 (approximately); d: 1.1.
Hazard Analysis
Toxicity: Unknown. Used as a food additive permitted in the feed and drinking water of animals and/or for the treatment of food producing animals; also permitted in food for human consumption (Section 10).

POLYSORBATE 65. See polyoxyethylene (20) sorbitan tristearate.

POLYSORBATE 80
General Information
Synonym: Polyoxyethylene (20) sorbitan monooleate.
Description: Lemon to amber colored oily liquid at 25°C; faint characteristic odor; soluble in water, alcohol, ethyl acetate; insoluble in mineral oil and petroleum ether.
Formula: $C_{64}H_{124}O_{26}$ (approximately).
Constants: Mol wt: 1308 (approximately), d: 1.07–1.09.
Hazard Analysis
Toxicity: Unknown. Used as a food additive permitted in the feed and drinking water of animals and/or for the treatment of food producing animals (Section 10).

POLYTETRAFLUOROETHYLENE
General Information
Synonym: Teflon.

Description: Grayish-white tough plastic. Chemically very inert.

Formula: Composed of long chains of linked CF_2 units.

Hazard Analysis

Toxic Hazard Rating:

Acute Local: Inhalation 1.

Acute Systemic: Inhalation 1.

Chronic Local: Irritant 1.

Chronic Systemic: U.

Toxicity: The finished polymerized compound is inert under ordinary conditions. There have been reports of "polymer fume fever" in humans exposed to the unfinished product dust or to pyrolysis products which also are irritants. Smoking should be prohibited in areas where this material is being fabricated or in general where there may be dust from it.

Disaster Hazard: Dangerous; when heated to above 750° F it decomposes to yield highly toxic fumes of fluorides.

POLYVINYL ALCOHOL
General Information

Description: Colorless, amorphous powder.

Constants: M: decomp. over 200°C, flash p: 175°F (O.C.), d: 1.329.

Hazard Analysis

Toxicity: Details unknown. Polyvinyl alcohol is not considered a toxic hazard. However, technical grades should not be used in foods or drugs without careful testing to assure conformity with the Federal Food, Drug and Cosmetic Act and similar state acts.

Fire Hazard: Slight, when exposed to heat or flame; can react with oxidizing materials.

Explosion Hazard: Slight, in the form of dust, when exposed to flame.

Countermeasures

Personnel Protection: Section 3.

To Fight Fire: Alcohol foam, water, carbon dioxide, dry chemical or carbon tetrachloride (Section 6).

Storage and Handling: Section 7.

POLYVINYL CHLORIDE
General Information

Synonym: PVC.

Description: A polyvinyl thermoplastic resin.

Formula: $(-H_2CCHC-)_n$.

Hazard Analysis

Toxic Hazard Rating:

Acute Local: U.

Acute Systemic: U.

Chronic Local: Allergen 2.

Chronic Systemic: U.

Toxicology: Allergic dermatitis has been reported.

Disaster Hazard: When heated to decomposition it emits highly toxic fumes.

POLYVINYL METHYL ETHER
General Information

Description: Light yellow to amber-colored, balsam-like liquid.

Formula: $(C_3H_6O)_x$.

Constants: Mol wt: $(58.1)_x$, d: 1.05 at 25°/4°C.

Hazard Analysis

Toxicity: Details unknown. See also ethers.

Fire Hazard: Moderate, when exposed to heat or flame, can react with oxidizing materials (Section 6).

Explosion Hazard: Details unknown. See also ethers.

Countermeasures

Storage and Handling: Section 7.

POLYVINYLPOLYPYRROLIDONE
General Information

Synonym: PVP.

Description: A free flowing white, amorphous powder; soluble in water, chlorinated hydrocarbons, alcohols, amines, nitro paraffins, and lower molecular weight fatty acids.

Formula: $(C_6H_9NO)_n$.

Constant: d: 1.23–1.29.

Hazard Analysis

Toxicity: Unknown. A food additive permitted in food for human consumption (Section 10).

POLYVINYL ALCOHOL. See poppy seed oil.

POPPY OIL. See poppy seed oil.

POPPY SEED OIL
General Information

Synonym: Poppy oil.

Description: Very pale golden yellow liquid; pleasant odor; soluble in ether, chloroform, petroleum ether and carbon disulfide; insoluble in water.

Constants: D: 0.924–0.928; flash p: 491° F.

Hazard Analysis

Toxic Hazard Rating: U.

Fire Hazard: Slight, when exposed to heat or flame.

Countermeasures

Storage and Handling: Section 7.

To Fight Fire: Water, foam.

PORTLAND CEMENT. See cement, portland.

POTABA. See potassium p-amino benzoate.

POTASAN
General Information

Synonym: 4-Methyl-7-hydroxycoumarin diethoxythiophosphate.

Description: Crystals. Weak aromatic odor.

Formula: $C_{14}H_{17}O_5PS$.

Constants: Mol wt: 328.3, mp: 38°C, bp: 210°C at 1 mm d: 1.260 at 38/4°C.

Hazard Analysis

Toxicity: Highly toxic. A cholinesterase inhibitor. See parathion.

Disaster Hazard: Dangerous. When heated to decomposition, it emits highly toxic fumes.

POTASH. See potassium carbonate.

POTASH MURIATE. See potassium chloride.

POTASSAMIDE. See potassium amide.

POTASSIUM
General Information

Description: Cubic, silver-metallic crystals.

Formula: K.

Constants: At wt: 39.10, mp: 62.3°C, bp: 760°C, d: 0.86 at 20°C; 0.83 at 62°C, vap. press.: 1 mm at 341°C.

Hazard Analysis

Toxic Hazard Rating:

Acute Local: Irritant 3; Ingestion 3; Inhalation 3.

Acute Systemic: U.

Chronic Local: U.

Chronic Systemic: U.

Toxicology: The toxicity of potassium compounds is almost always that of the anion.

Radiation Hazard: Section 5. For permissible levels, see Table 5, p. 150.

Natural (0.01%) isotope ^{40}K, half life 1.27×10^9 y.

TOXIC HAZARD RATING CODE (For detailed discussion, see Section 1.)

0 NONE: (a) No harm under any conditions; (b) Harmful only under unusual conditions or overwhelming dosage.

1 SLIGHT: Causes readily reversible changes which disappear after end of exposure.

2 MODERATE: May involve both irreversible and reversible changes; not severe enough to cause death or permanent injury.

3 HIGH: May cause death or permanent injury after very short exposure to small quantities.

U UNKNOWN: No information on humans considered valid by authors.

Decays to stable ^{40}A by electron capture (11%). Also decays to stable ^{40}Ca by emitting beta particles of 1.32 MeV. Also emits gamma rays of 1.4 MeV and X-rays.

Artificial isotope ^{42}K, half life 12.5 h. Decays to stable ^{42}Ca by emitting beta particles of 2.03 (18%), 3.55 (82%) MeV. Also emits gamma rays of 1.52 MeV.

Fire Hazard: Dangerous! Metallic potassium reacts with moisture to form potassium hydroxide and hydrogen. The reaction evolves much heat causing the potassium to melt and spatter. It also ignites the hydrogen which now burns, or if there is any confinemeent an explosion can occur. Burning potassium is difficult to extinguish; dry powdered soda ash or graphite or special mixtures of dry chemical is recommended. It can ignite spontaneously in moist air (Section 6).

Explosion Hazard: Moderate, by chemical reaction. Potassium metal will form the peroxide (K_2O_2) and the superoxide (KO_2 or K_2O_4) at room temperature even when stored under mineral oil. Metal which has oxidized on storage under oil may explode violently when handled or cut. Oxide coated potassium should be destroyed by burning.

Disaster Hazard: Dangerous; a highly reactive alkali metal. See sodium and lithium. In the presence of moist air it can spontaneously catch fire and burn with great intensity. It may even explode. Reacts violently with moisture, acid fumes and oxidizers.

Countermeasures

Ventilation Control: Section 2.

Storage and Handling(Section 7): Store in inert atmospheres, such as argon or nitrogen, or under liquids which are oxygen-free, such as toluene or kerosene, or in glass capsules which have been filled under vacuum or inert atmosphere and sealed before oxygen or moisture can enter. When quantities of this metal are in use, provision should be made for fireproof garments so that personnel can approach close enough to the fire to fight it.

Large quantities of potassium can be disposed of by simply cutting it into small pieces and placing it in an open waste space to react with the moisture in the air and gradually turn to potassium hydroxide. Potassium may be reacted with ethanol to give a stable product. Do not dispose of it, even in small quantities, by throwing it into sinks or waste containers which may contain combustible materials. It should be stored in a detached building which is fireproof and not where it can come in contact with moisture, powerful oxidizing materials or high temperatures.

Personnel Protection: Section 3.

First Aid: Section 1.

Shipping Regulations: Section 11.

I.C.C.: Flammable solid; yellow label, 25 pounds.

Coast Guard Classification: Inflammable solid; yellow label.

IATA: Flammable solid, yellow label, not acceptable (passenger), 12 kilograms (cargo).

POTASSIUM ACETATE
General Information

Description: White powder.

Formula: CH_3COOK.

Constants: Mol wt: 98.14, mp: 292°C, d: 1.8 at 20°/20°C.

Hazard Analysis

Toxicity: Very low.

Disaster Hazard: Slight; when heated, it emits acrid fumes (Section 6).

Countermeasures

Storage and Handling: Section 7.

POTASSIUM ACID CARBONATE. See potassium bicarbonate.

POTASSIUM ACID FLUORIDE. See potassium bifluoride.

POTASSIUM ACID OXALATE. See potassium binoxalate.

POTASSIUM ACID TARTRATE
General Information

Synonym: Cream of tartar; potassium bitartrate.

Description: White crystals, or powder; soluble in water; insoluble in alcohol.

Formula: $KHC_4H_4O_6$.

Constants: Mol wt: 188, d: 1.984 at 18°C.

Hazard Analysis

Toxicity: Unknown. Used as a general purpose food additive (Section 10).

POTASSIUM ALGINATE
General Information

Synonym: Potassium polymannuronate.

Description: A hydrophilic colloidal substance; occurs in filamentous, grainy, granular, or powder forms; colorless or slightly yellow; slowly soluble in water, insoluble in alcohol.

Formula: $(C_6H_7O_6K)_n$.

Constant: Mol wt range: 32,000–35,000.

Hazard Analysis

Toxic Hazard Rating: U. A stabilizer food additive (Section 10).

POTASSIUM ALUMINUM BORATE, BASIC
General Information

Description: Cubic, white crystals.

Formula: $K(AlO)_2(BO_2)_3$.

Constants: Mol wt: 253.50, mp: < 1800°C, d: 3.415.

Hazard Analysis and Countermeasures

See boron compounds and aluminum compounds.

POTASSIUM ALUMINUM FLUORIDE
General Information

Description: White powder.

Formula: K_3AlF_6.

Constant: Mol wt: 258.3.

Hazard Analysis and Countermeasures

See fluorides.

POTASSIUM ALUMINUM SULFATE. See alum.

POTASSIUM AMALGAM
General Information

Description: Silvery liquid or solid.

Formula: $K + Hg$.

Hazard Analysis

Toxicity: Highly toxic. See potassium and mercury.

Fire Hazard: Moderate, by spontaneous chemical reaction; on contact with moisture hydrogen is liberated. See also potassium.

Explosion Hazard: Moderate; it liberates hydrogen upon contact with moisture, acids, etc. See also potassium.

Disaster Hazard: Dangerous; when heated to decomposition, it emits highly toxic fumes of mercury and oxides of potassium; will react with water, steam or acids to produce hydrogen; can react with oxidizing materials.

Countermeasures

Storage and Handling: Section 7.

POTASSIUM AMIDE
General Information

Synonym: Potassamide.

Description: Colorless-white or yellow-green crystals.

Formula: KNH_2.

Constants: Mol wt: 55.12, mp: 335°C, bp: sublimes at 400°C.

Hazard Analysis

Toxicity: See ammonia.

Fire Hazard: Moderate, by chemical reaction; can react with moisture to liberate ammonia (Section 6).

Explosion Hazard: Moderate, due to the ammonia liberated.

Disaster Hazard: Dangerous; will react with water or steam to produce toxic, corrosive and flammable vapors; can react with oxidizing materials.

Countermeasures

Storage and Handling: Section 7.

POTASSIUM p-AMINOBENZOATE
General Information
Synonym: Potaba.
Description: Crystals. Saline taste. Water soluble.
Formula: $C_7H_6KNO_2$.
Constant: Mol wt: 175.2.
Hazard Analysis
Toxicity: Details unknown. See p-amino-benzoic acid.

POTASSIUM AMMONIUM NITRATE
General Information
Synonym: BasF.
Description: Gray to light or dark brown crystals.
Formula: $KNO_3 + NH_4NO_3$.
Hazard Analysis and Countermeasures
See nitrates.

POTASSIUM AMMONIUM SELENOSULFIDE
General Information
Description: A solid.
Formula: $(KNH_4S)_5Se$.
Constant: Mol wt: 525.01.
Hazard Analysis and Countermeasures
See selenium compounds and sulfides.

POTASSIUM ANTIMONIDE
General Information
Description: Yellow-green crystals.
Formula: K_3Sb.
Constants: Mol wt: 239.05, mp: 812°C.
Hazard Analysis
Toxicity: Highly toxic. See antimony compounds.
Fire Hazard: Moderate, when exposed to heat or flame (Section 6).
Explosion Hazard: Moderate, when exposed to flame.
Disaster Hazard: Dangerous; when heated to decomposition or on contact with moisture or acids, it emits highly toxic fumes of antimony; can react with oxidizing materials.
Countermeasures
Ventilation Control: Section 2.
Personnel Protection: Section 3.
Storage and Handling: Section 7.

POTASSIUM ANTIMONY OXALATE. See antimony potassium oxalate.

POTASSIUM ANTIMONY TARTRATE. See antimony potassium tartrate.

POTASSIUM ARSENATE
General Information
Synonyms: Potassium dihydrogen arsenate; Macquer's salt.
Description: Colorless crystals.
Formula: KH_2AsO_4.
Constants: Mol wt: 180.02, mp: 288°C, d: 2.867.
Hazard Analysis and Countermeasures
See arsenic compounds.
Shipping Regulations: Section 11.
I.C.C.: Poison B; poison label, 200 pounds.

Coast Guard Classification: Poison B; poison label.
IATA: Poison B, poison label, 25 kilograms (passenger), 95 kilograms (cargo).

POTASSIUM ARSENITE
General Information
Description: White powder.
Formula: $KAsO_2 \cdot HAsO_2$.
Constant: Mol wt: 253.8.
Hazard Analysis and Countermeasures
See arsenic compounds.
Shipping Regulations: Section 11.
I.C.C.: Poison B, poison label, 200 pounds.
IATA: Poison B, poison label, 25 kilograms (passenger), 95 kilograms (cargo).

POTASSIUM AURATE
General Information
Description: Light yellow needles.
Formula: $KAuO_2 \cdot 3H_2O$.
Constants: Mol wt: 322.34, mp: decomposes.
Hazard Analysis
Toxicity: See gold compounds.

POTASSIUM AURIBROMIDE. See gold potassium bromide.

POTASSIUM AURICHLORIDE. See gold potassium chloride.

POTASSIUM AUROCYANIDE. See gold potassium cyanide.

POTASSIUM AZIDE
General Information
Description: Colorless crystals.
Formula: KN_3.
Constants: Mol wt: 81.12, mp: 350°C, d: 2.04.
Hazard Analysis and Countermeasures
See azides.

POTASSIUM BERYLLIUM FLUORIDE
General Information
Description: White, crystalline masses.
Formula: $BeF_2 \cdot 2KF$.
Constant: Mol wt: 163.21.
Hazard Analysis and Countermeasures
Toxicity: Highly toxic. See beryllium and fluorides.

POTASSIUM BERYLLIUM SULFATE
General Information
Description: White powder.
Formula: $K_2Be(SO_4)_2$.
Constant: Mol wt: 279.4.
Hazard Analysis and Countermeasures
See beryllium and sulfates.

POTASSIUM BICARBONATE
General Information
Synonym: Potassium acid carbonate.
Description: Colorless, transparent crystals or white powder; odorless; soluble in water; insoluble in alcohol.
Formula: $KHCO_3$.
Constants: Mol wt: 100, d: 2.17, mp: decomposes between 100 and 120°C.
Hazard Analysis
Toxic Hazard Rating: U. Used as a general purpose food additive (Section 10).

TOXIC HAZARD RATING CODE (For detailed discussion, see Section 1.)

0 NONE: (a) No harm under any conditions; (b) Harmful only under unusual conditions or overwhelming dosage.

1 SLIGHT: Causes readily reversible changes which disappear after end of exposure.

2 MODERATE: May involve both irreversible and reversible changes; not severe enough to cause death or permanent injury.

3 HIGH: May cause death or permanent injury after very short exposure to small quantities.

U UNKNOWN: No information on humans considered valid by authors.

POTASSIUM BICHROMATE
General Information
Synonyms: Potassium dichromate; red potassium chromate.
Description: Bright, yellowish-red, transparent crystals, bitter, metallic taste.
Formula: $K_2Cr_2O_7$.
Constants: Mol wt: 294.21, mp: 398°C, bp: decomp. at 500°C, d: 2.69.
Hazard Analysis
Toxicity: See chromium compounds.
Fire Hazard: Moderate, by chemical reaction. A powerful oxidizer (Section 6).
Countermeasures
Storage and Handling: Section 7.
Shipping Regulations: Secton 11.
　　IATA: Other restricted articles, class A, no label required, no limit (passenger and cargo).

POTASSIUM BIFLUORIDE
General Information
Synonyms: Potassium acid fluoride; Fremy's salt.
Description: Colorless crystals.
Formula: KHF_2.
Constants: Mol wt: 78.10, mp: decomposes.
Hazard Analysis and Countermeasures
See fluorides.
Shipping Regulations: Section 11.
　　IATA (solid): Poison B, poison label, 25 kilograms (passenger), 95 kilograms (cargo).
　　　　　　(solution): Corrosive liquid, white label, 1 liter (passenger), 20 liters (cargo).

POTASSIUM BINOXALATE
General Information
Synonyms: Potassium acid oxalate; sal acetosella; salt of sorrel; essential salt of lemon.
Description: White, somewhat hygroscopic crystals; bitter, sharp taste.
Formula: KHC_2O_4.
Constants: Mol wt: 128.12, mp: decomposes, d: 2.0.
Hazard Analysis and Countermeasures
See oxalates.

POTASSIUM BISULFIDE. See potassium hydrosulfide.

POTASSIUM　　BISULFITE. See potassium hydrogen sulfite.

POTASSIUM BITARTRATE. See potassium acid tartrate.

POTASSIUM m-BORATE
General Information
Description: Colorless crystals.
Formula: KBO_2.
Constants: Mol wt: 81.92, mp: 947–950°C.
Hazard Analysis
Toxicity: See boron compounds.

POTASSIUM BOROHYDRIDE
General Information
Description: White crystals, soluble in water by reaction.
Formula: KBH_4.
Constants: Mol wt: 54.0, d: 1.177, mp: decomp. > 400°C.
Hazard Analysis and Countermeasures
See boron compounds and hydrides.
Shipping Regulations: Section 11.
　　IATA: Flammable solid, yellow label, 12 kilograms (passenger), 45 kilograms (cargo).

POTASSIUM BROMATE
General Information
Description: White crystals or crystalline powder.
Formula: $KBrO_3$.
Constants: Mol wt: 167.01, mp: 434°C; decomp. at 370°C, d: 3.27 at 17.5°C.

Hazard Analysis
Toxicity: See bromates. A food additive permitted in food for human consumption (Section 10).
Countermeasures
Shipping Regulations: Section 11.
　　I.C.C.: Oxidizing material; yellow label, 100 pounds.
　　Coast Guard Classification: Oxidizing material; yellow label.
　　IATA: Oxidizing material, yellow label, 12 kilograms (passenger), 45 kilograms (cargo).

POTASSIUM BROMIDE
General Information
Description: Cubic, colorless, slightly hygroscopic crystals.
Formula: KBr.
Constants: Mol wt: 119.01, mp: 730°C, bp: 1380°C, d: 2.75 at 25°C, vap. press.: 1 mm at 795°C.
Hazard Analysis and Countermeasures
See bromides.

POTASSIUM BROMOAURATE
General Information
Description: Rhombic, red-brown crystals.
Formula: $KAuBr_4$.
Constants: Mol wt: 555.96, mp: decomp.
Hazard Analysis and Countermeasures
See gold compounds and bromides.

POTASSIUM BROMOPLATINATE
General Information
Description: Cubic, dark red-brown crystals.
Formula: K_2PtBr_6.
Constants: Mol wt: 752.92, mp: decomp. > 400°C, d: 4.66 at 24°C.
Hazard Analysis and Countermeasures
See platinum compounds and bromides.

POTASSIUM BROMOPLATINITE
General Information
Description: Rhombic, brown crystals.
Formula: K_2PtBr_4.
Constant: Mol wt: 593.09.
Hazard Analysis and Countermeasures
See platinum compounds and bromides.

POTASSIUM BROMOSTANNATE
General Information
Description: Crystals.
Formula: K_2SnBr_6.
Constants: Mol wt: 676.39, d: 3.783.
Hazard Analysis and Countermeasures
See tin compounds and bromides.

POTASSIUM CACODYLATE
General Information
Description: White crystals.
Formula: $K[(CH_3)_2AsO_2] \cdot H_2O$.
Constant: Mol wt: 194.09.
Hazard Analysis and Countermeasures
See arsenic compounds.

POTASSIUM CADMIUM IODIDE. See cadmium potassium iodide.

POTASSIUM CARBONATE
General Information
Synonyms: Potash; pearl ash.
Description: White, deliquescent, granular, translucent powder; soluble in water; insoluble in alcohol.
Formula: (a) K_2CO_3, mol wt: 138, (b) $K_2CO_3 \cdot \frac{1}{2}H_2O$, mol wt: 147.
Constants: d: 2.428 at 19°C, mp: 891°C, bp: decomposes, mol wt: 138.2.
Hazard Analysis
Toxic Hazard Rating:
　　Acute Local: Irritant 3; Ingestion 3.
　　Acute Systemic: Ingestion 3.

Chronic Local: Irritant 2.
Chronic Systemic: Ingestion 2.
Toxicity: A strong caustic. Used as a general purpose food additive (Section 10).

POTASSIUM CARBONYL
General Information
Description: Gray-red solid.
Formula: $(KCO)_6$.
Constants: Mol wt: 402.64, mp: explodes.
Hazard Analysis
Toxicity: Highly toxic. See carbonyls.
Fire Hazard: Dangerous, when exposed to heat or flame (Section 6).
Explosion Hazard: Moderate, when shocked, exposed to heat, on contact with water, or by chemical reaction.
Disaster Hazard: Dangerous; shock will explode it; will react with water or steam to produce heat; can react vigorously with oxidizing materials.
Countermeasures
Ventilation Control: Section 2.
Personnel Protection: Section 3.
Storage and Handling: Section 7.

POTASSIUM CHLORATE
General Information
Synonym: Potassium oxymuriate.
Description: Transparent, colorless crystals or white powder; cooling, saline taste.
Formula: $KClO_3$.
Constants: Mol wt: 122.55, mp: 368.4°C, bp: decomp. at 400°C, d: 2.32.
Hazard Analysis and Countermeasures
See chlorates.
Shipping Regulations: Section 11.
    Coast Guard Classification: Oxidizing material; yellow label.

POTASSIUM CHLORIDE
General Information
Synonyms: Potassium muriate; potash muriate.
Description: Colorless or white crystals or powder; soluble in water; slightly soluble in alcohol; insoluble in absolute alcohol.
Formula: KCl.
Constants: Mol wt: 74.5, d: 1.987, mp: 790°C, sublimes at 1500°C.
Hazard Analysis and Countermeasures
Toxic Hazard Rating: U. A nutrient and or dietary supplement food additive (Section 10). See chlorides.

POTASSIUM CHLOROAURATE
General Information
Description: Monoclinic, yellow crystals.
Formula: $KAuCl_4$.
Constants: Mol wt: 378.12, mp: decomp. at 357°C.
Hazard Analysis and Countermeasures
See gold compounds and chlorides.

POTASSIUM CHLOROCHROMATE
General Information
Synonym: Peligot's salt.
Description: Monoclinic, red crystals.
Formula: $KCrO_3Cl$.
Constants: Mol wt: 174.56, mp: decomposes, d: 2.497.
Hazard Analysis
Toxicity: See chromium compounds.

Fire Hazard: Moderate, by chemical reaction; a powerful oxidizer (Section 6).
Disaster Hazard: Dangerous. See chlorides.
Countermeasures
Storage and Handling: Section 7.

POTASSIUM CHLOROIODATE (III)
General Information
Description: Rhombic, yellow crystals.
Formula: $KICl_4$.
Constants: Mol wt: 307.84, mp: decomposes, d: 1.76 at 45°C.
Hazard Analysis
Toxicity: Details unknown.
Disaster Hazard: Dangerous; when heated to decomposition or on contact with acid or acid fumes, it emits highly toxic fumes of chlorides and iodides.
Countermeasures
Storage and Handling: Section 7.

POTASSIUM CHLOROIRIDATE
General Information
Description: Cubic, black crystals.
Formula: K_2IrCl_6.
Constants: Mol wt: 484.03, mp: decomposes, d: 3.546.
Hazard Analysis and Countermeasures
See iridium compounds and chlorides.

POTASSIUM CHLOROOSMATE (IV)
General Information
Description: Cubic, red crystals.
Formula: K_2OsCl_6.
Constants: Mol wt: 481.1, mp: decomposes.
Hazard Analysis and Countermeasures
See osmium compounds and chlorides.

POTASSIUM CHLOROPALLADATE
General Information
Description: Cubic, red crystals.
Formula: K_2PdCl_6.
Constants: Mol wt: 397.63, mp: decomposes, d: 2.738.
Hazard Analysis and Countermeasures
See palladium compounds and chlorides.

POTASSIUM CHLOROPLATINATE
General Information
Description: Cubic, yellow crystals.
Formula: K_2PtCl_6.
Constants: Mol wt: 486.16, mp: decomp. at 250°C, d: 3.499 at 24°C.
Hazard Analysis and Countermeasures
See platinum compounds and chlorides.

POTASSIUM CHLOROPLUMBATE
General Information
Description: Cubes.
Formula: K_2PbCl_6.
Constants: Mol wt: 498.14, mp: decomp. at 190°C.
Hazard Analysis and Countermeasures
See lead compounds and chlorides.

POTASSIUM CHLORORHENATE (IV)
General Information
Description: Yellow-green crystals.
Formula: K_2ReCl_6.
Constants: Mol wt: 477.24, d: 3.34.
Hazard Analysis and Countermeasures
See chlorides.

TOXIC HAZARD RATING CODE (For detailed discussion, see Section 1.)

0 NONE: (a) No harm under any conditions; (b) Harmful only under unusual conditions or overwhelming dosage.

1 SLIGHT: Causes readily reversible changes which disappear after end of exposure.

2 MODERATE: May involve both irreversible and reversible changes; not severe enough to cause death or permanent injury.

3 HIGH: May cause death or permanent injury after very short exposure to small quantities.

U UNKNOWN: No information on humans considered valid by authors.

POTASSIUM CHLORORUTHENATE (IV)
General Information
Description: Cubic, black crystals.
Formula: K_2RuCl_6.
Constants: Mol wt: 392.63, mp: decomposes.
Hazard Analysis and Countermeasures
See ruthenium compounds and chlorides.

POTASSIUM CHLOROSTANNATE
General Information
Description: Cubic, colorless crystals.
Formula: K_2SnCl_6.
Hazard Analysis and Countermeasures
See tin compounds and chlorides.

POTASSIUM CHLOROTELLURATE
General Information
Description: Pale yellow crystals.
Formula: K_2TeCl_6.
Constant: Mol wt: 418.54.
Hazard Analysis and Countermeasures
See tellurium compounds and chlorides.

POTASSIUM CHROMATE
General Information
Synonym: Tarapacaite.
Description: Rhombic, yellow crystals.
Formula: K_2CrO_4.
Constants: Mol wt: 194.20, mp: 971°C, d: 2.732 at 18°C.
Hazard Analysis and Countermeasures
See chromium compounds and chromates.

POTASSIUM CHROMIC SULFATE
General Information
Description: Cubic, red or green crystals.
Formula: $KCr(SO_4)_2 \cdot 12H_2O$.
Constants: Mol wt: 499.4, mp: 89°C, bp: $-12H_2O$ at 400°C, d: 1.83.
Hazard Analysis and Countermeasures
See chromium compounds and sulfates.

POTASSIUM CHROMIUM CHROMATE, BASIC
General Information
Description: Violet-brown, amorphous powder.
Formula: $K_2CrO_4 \cdot 2Cr(OH)CrO_4$.
Constants: Mol wt: 564.26, mp: 300°C, d: 2.28 at 14°C.
Hazard Analysis
Toxicity: See chromium compounds.
Fire Hazard: Moderate, by chemical reaction; a powerful oxidizer (Section 6).
Countermeasures
Storage and Handling: Section 7.

POTASSIUM CITRATE
General Information
Description: Colorless or white crystals or powder; odorless; deliquescent; soluble in water and glycerol; almost insoluble in alcohol.
Formula: $K_3C_6H_5O_7 \cdot H_2O$.
Constants: Mol wt: 324, d: 1.98, decomposes when heated to 230°C.
Hazard Analysis
Toxic Hazard Rating: U. A sequestrant food additive, also a general purpose food additive (Section 10).

POTASSIUM COBALTINITRITE
General Information
Synonym: Fischer's salt.
Description: Yellow prisms.
Formula: $K_3Co(NO_2)_6$.
Constant: Mol wt: 452.3.
Hazard Analysis and Countermeasures
See cobalt compounds and nitrites.

POTASSIUM COBALTOUS SULFATE
General Information
Description: Red, monoclinic prisms.
Formula: $K_2SO_4 \cdot CoSO_4 \cdot 6H_2O$.
Constants: Mol wt: 437.36, d: 2.218.
Hazard Analysis and Countermeasures
See cobalt compounds and sulfates.

POTASSIUM COPPER CHLORIDE
General Information
Description: Red needles.
Formula: $KCl \cdot CuCl_2$.
Constants: Mol wt: 209, d: 2.86.
Hazard Analysis and Countermeasures
See copper compounds and chlorides.

POTASSIUM CUPROCYANIDE
General Information
Synonyms: Potassium copper cyanide; copper potassium cyanide.
Description: White crystalline, double salt of copper cyanide and potassium cyanide.
Composition: Copper content (Cu) min. 25.8%; free KCN 1.25–3.0%.
Hazard Analysis and Countermeasures
See cyanides.
Shipping Regulations: Section 11.
 IATA: Poison B, poison label, 12 kilograms (passenger), 95 kilograms (cargo).

POTASSIUM CYANATE
General Information
Description: Colorless crystals.
Formula: KOCN.
Constants: Mol wt: 81.11, mp: 700–900°C (decomp.), d: 2.056 at 20°C.
Hazard Analysis and Countermeasures
Toxic Hazard Rating:
 Acute Local: Irritant 1; Ingestion 1.
 Acute Systemic: Ingestion 2.
 Chronic Local: Irritant 1.
 Chronic Systemic: Ingestion 2; Inhalation 1.
Toxicity: A herbicide. Ingestion can cause irritation of the gastro-intestinal tract. It is said to be slowly metabolized in body to cyanide but does not have high toxicity of cyanides.
Shipping Regulations: Section 11.
 MCA warning label.

POTASSIUM CYANIDE
General Information
Description: White, deliquescent crystalline solid; faint odor of bitter almonds.
Formula: KCN.
Constants: Mol wt: 65.11, mp: 634.5°C, d: 1.52 at 16°C.
Hazard Analysis and Countermeasures
See cyanides.
Shipping Regulations: Section 11.
 I.C.C.: Poison B; poison label.
 Coast Guard Classification: Poison B; poison label.

POTASSIUM CYANO-COMPOUNDS
With the exception of the ferri- and ferro-cyanides, any compound of potassium containing the cyanogen radical (CN) is highly toxic and should be handled with adequate ventilation and protective equipment. See cyanides.

POTASSIUM CYCLAMATE
General Information
Synonym: Potassium cyclohexyl sulfamate.
Hazard Analysis
Toxicity: Unknown. Used as a non-nutritive sweetener (Section 10). See also sodium cyclamate.

POTASSIUM DIBORANE
General Information
Synonym: Diboranide.
Description: White crystals.
Formula: $K_2B_2H_6$.
Constants: Mol wt: 105.9, bp: decomp. at 300°C at 1 mm,
  d: 1.18.
Hazard Analysis
Toxicity: Details unknown. Probably highly toxic. See also
  boron compounds.

POTASSIUM DIBROMOIODIDE
General Information
Description: Rhombic crystals.
Formula: $KIBr_2$.
Constants: Mol wt: 325.85, mp: 60°C, bp: decomp. at
  180°C.
Hazard Analysis and Countermeasures
See bromides and iodides.

POTASSIUM DICHLOROIODIDE
General Information
Description: Monoclinic crystals.
Formula: $KICl_2$.
Constants: Mol wt: 236.93, mp: 60°C, bp: decomp. at
  215°C.
Hazard Analysis and Countermeasures
See chlorides and iodides.

POTASSIUM DICHLOROISOCYANURATE
General Information
Description: White, slightly hygroscopic crystalline
  powder or granules.
Formula: $Cl_2K(NCO)_3$(cyclic).
Constant: Mol wt: 236.
Hazard Analysis and Countermeasures
See chlorides and oxidizers.
Shipping Regulations: Section 11.
  I.C.C.: Oxidizing material, yellow label, 100 pounds.
    (dry, containing > 39% chlorine).
  IATA: Oxidizing material, yellow label, 25 kilograms
    (passenger), 45 kilograms (cargo).

POTASSIUM DICHROMATE. See potassium bichromate.

POTASSIUM DICYANOGUANIDINE
General Information
Description: Crystals.
Formula: NCHNC(NH)NCNK.
Constants: Mol wt: 147.2, mp: decomp. at 265°C.
Hazard Analysis
Toxicity: Details unknown. Possibly most toxic by skin
  absorption.
Disaster Hazard: Dangerous; when heated to decomposition
  or on contact with acid or acid fumes, it emits highly
  toxic fumes of cyanides.
Countermeasures
Storage and Handling: Section 7.

POTASSIUM DIFLUOTELLURATE
General Information
Description: Crystals.
Formula: $K_2TeO_3F_2 \cdot 3H_2O$.
Constants: Mol wt: 345.85, mp: decomposes.
Hazard Analysis and Countermeasures
See fluorides and tellurium compounds.

POTASSIUM DIGERMANATE
General Information
Description: White crystals.
Formula: $K_2Ge_2O_5$.
Constants: Mol wt: 303.4, mp: > 83°C, d: 4.31 at 21.5°C.
Hazard Analysis
Toxicity: See germanium compounds.

POTASSIUM DIHYDROGEN o-ARSENATE
General Information
Description: Tetragonal, colorless crystals.
Formula: KH_2AsO_4.
Constants: Mol wt: 180, mp: 288°C, d: 2.867.
Hazard Analysis and Countermeasures
See arsenic compounds.

POTASSIUM DIHYDROXYDIBORANE
General Information
Description: Colorless, cubical crystals.
Formula: $K_2B_2H_6O_2$.
Constants: Mol wt: 137.88, mp: decomposes to K, d: 1.39.
Hazard Analysis and Countermeasures
Toxicity: See boron compounds.

POTASSIUM DIISOPROPYL DITHIOPHOSPHATE
General Information
Description: Colorless, crystalline, nearly odorless solid.
Formula: $[(CH_3)_2CHO]_2PSKS$.
Constants: Mol wt: 252.4, mp: decomp. near 200°C.
Hazard Analysis
Toxicity: Details unknown. Many derivatives of this ma-
  terial are highly toxic.
Disaster Hazard: Dangerous; when heated to decomposi-
  tion, it emits highly toxic fumes of oxides of sulfur and
  phosphorus.
Countermeasures
Storage and Handling: Section 7.

POTASSIUM DISULFIDE
General Information
Description: Red-yellow crystals.
Formula: K_2S_2.
Constants: Mol wt: 142.3, mp: 470°C.
Hazard Analysis and Countermeasures
See sulfides.

POTASSIUM ETHYLXANTHATE. See potassium
  xanthate.

POTASSIUM FERRIC SULFIDE
General Information
Description: Purple, hexagonal crystals.
Formula: $KFeS_2$.
Constants: Mol wt: 159.1, d: 2.563.
Hazard Analysis and Countermeasures
See sulfides.

POTASSIUM FERRICYANIDE
General Information
Synonyms: Red prussiate of potash; red potassium prussiate.
Description: Bright red, lustrous crystals or powder.
Formula: $K_3Fe(CN)_6$.
Constants: Mol wt: 329.24, mp: decomposes, d: 1.894 at 17°C
Hazard Analysis
Toxicity: See ferricyanides. This is not a powerful poison
  as are the simple cyanides.
Disaster Hazard: Dangerous; when heated to decomposi-
  tion or on contact with acid or acid fumes, it emits
  highly toxic fumes of cyanides.

TOXIC HAZARD RATING CODE (For detailed discussion, see Section 1.)

0 NONE: (a) No harm under any conditions; (b) Harmful only under unusual conditions or overwhelming dosage.

1 SLIGHT: Causes readily reversible changes which disappear after end
  of exposure.

2 MODERATE: May involve both irreversible and reversible changes;
  not severe enough to cause death or permanent injury.

3 HIGH: May cause death or permanent injury after very short exposure
  to small quantities.

U UNKNOWN: No information on humans considered valid by authors.

Countermeasures
Ventilation Control: Section 2.
Personal Hygiene: Section 3.
Storage and Handling: Section 7.

POTASSIUM FERROCYANIDE
General Information
Synonym: Yellow prussiate of potash.
Description: Lemon yellow crystals.
Formula: $K_4Fe(CN)_6 \cdot 3H_2O$.
Constants: Mol wt: 422.39, mp: $-3H_2O$ at $70°C$, bp: decomposes, d: 1.85 at $17°C$.
Hazard Analysis
Toxicity: Not as toxic as the simple cyanides. See ferrocyanides.
Disaster Hazard: Dangerous; when heated to decomposition or on contact with acid or acid fumes, it emits highly toxic fumes of cyanides.
Countermeasures
Storage and Handling: Section 7.

POTASSIUM FLUOBERYLLATE
General Information
Description: Rhombic, colorless crystals.
Formula: K_2BeF_4.
Constants: Mol wt: 163.2, mp: red heat.
Hazard Analysis and Countermeasures
See beryllium and fluorides.

POTASSIUM FLUOBORATE
General Information
Synonym: Avogadrite.
Description: Rhombic or cubic, colorless crystals.
Formula: KBF_4.
Constants: Mol wt: 125.9, mp: $530°C$, d: 2.498.
Hazard Analysis and Countermeasures
See fluorides and boron compounds.

POTASSIUM FLUOGERMANATE
General Information
Description: Hexagonal, white crystals.
Formula: K_2GeF_6.
Constants: Mol wt: 264.8, mp: $730°C$, bp: approx. $835°C$.
Hazard Analysis and Countermeasures
See fluorides.

POTASSIUM FLUORIDE
General Information
Description: White, crystalline, deliquescent powder; sharp, saline taste.
Formula: KF.
Constants: Mol wt: 58.1, mp: $880°C$, bp: $1500°C$, d: 2.48, vap. press.: 1 mm at $885°C$.
Hazard Analysis and Countermeasures
See fluorides.
Shipping Regulations: Section 11.
  IATA: Other restricted articles, class A, no label, no limit (passenger and cargo).
       Corrosive liquid, white label, 1 liter (passenger), 20 liters (cargo (solution).

POTASSIUM FLUOSILICATE
General Information
Synonym: Hieratite.
Description: Hexagonal or cubic, colorless crystals.
Formula: K_2SiF_6.
Constants: Mol wt: 220.25; mp: decomposes; d: (hex.): 3.08, (cubic): 2.665 at $17°C$.
Hazard Analysis and Countermeasures
See fluosilicates.

POTASSIUM FLUOSTANNATE
General Information
Description: Monoclinic prisms.
Formula: $K_2SnF_6 \cdot H_2O$.
Constants: Mol wt: 328.9, d: 3.058.

Hazard Analysis and Countermeasures
See fluorides.

POTASSIUM FLUOSULFONATE
General Information
Description: Short, thick prisms.
Formula: $KFSO_3$.
Constants: Mol wt: 138.16, mp: $311°C$.
Hazard Analysis and Countermeasures
See fluorides and sulfonates.

POTASSIUM FLUOTANTALATE
General Information
Description: Rhombic, colorless crystals.
Formula: K_2TaF_7.
Constants: Mol wt: 392.1, d: 4.56; 5.24.
Hazard Analysis and Countermeasures
See fluorides.

POTASSIUM FLUOTHORATE
General Information
Description: Colorless crystals.
Formula: $K_2ThF_6 \cdot 4H_2O$.
Constant: Mol wt: 496.38.
Hazard Analysis and Countermeasures
See fluorides and thorium.

POTASSIUM FLUOTITANATE
General Information
Description: Colorless, small, lustrous leaflets.
Formula: $K_2TiF_6 \cdot H_2O$.
Constants: Mol wt: 258.1, mp: $780°C$, bp: decomposes.
Hazard Analysis and Countermeasures
See fluorides.

POTASSIUM FLUOZIRCONATE
General Information
Description: Monoclinic, colorless crystals.
Formula: K_2ZrF_6.
Constants: Mol wt: 283.4, d: 3.48.
Hazard Analysis and Countermeasures
See fluorides.

POTASSIUM FORMATE
General Information
Description: Colorless, deliquescent crystals.
Formula: HCOOK.
Constants: Mol wt: 84.1, mp: $168°C$, bp: decomposes, d: 1.91.
Hazard Analysis
Toxicity: See formic acid and potassium compounds.
Fire Hazard: Slight; when heated, it emits acrid fumes (Section 6).
Countermeasures
Storage and Handling: Section 7.

POTASSIUM GALLIUM SULFATE
General Information
Description: Colorless crystals.
Formula: $KGa(SO_4)_2 \cdot 12H_2O$.
Constants: Mol wt: 517.14, d: 1.895.
Hazard Analysis and Countermeasures
See gallium compounds and sulfates.

POTASSIUM m-GERMANATE
General Information
Description: White crystals.
Formula: K_2GeO_3.
Constants: Mol wt: 198.79, mp: $823°C$, d: 3.40 at $21.5°C$.
Hazard Analysis
Toxicity: See germanium compounds.

POTASSIUM GLUTAMATE. See monopotassium glutamate.

POTASSIUM GLYCERINOPHOSPHATE. See potassium glycerophosphate.

POTASSIUM GLYCEROPHOSPHATE
General Information
Synonym: Potassium glycerinophosphate.
Description: Pale yellow syrupy liquid; soluble in alcohol; miscible with water.
Formula: $K_2C_3H_5O_2 \cdot H_2PO_4 \cdot 3H_2O$.
Constant: Mol wt: 302.
Hazard Analysis and Countermeasures
Toxicity: Details unknown. A nutrient and or dietary supplement food additive (Section 10). See also phosphorus compounds.

POTASSIUM GRAPHITE
General Information
Description: Brass colored platelets. Highly reactive with water, air, alcohol.
Formula: KC_8.
Constant: Mol wt: 135.1.
Hazard Analysis and Countermeasures
See potassium.

POTASSIUM HEXAFLUOROPHOSPHATE
General Information
Synonym: FP Salt No. 4.
Formula: KPF_6.
Constants: Mol wt: 184.1, mp: about 575°C, bp: decomposes, d: 2.59.
Hazard Analysis
Toxicity: Probably very low.
Disaster Hazard: Dangerous; when heated to decomposition, or on contact with acid or acid fumes, it emits highly toxic fumes of fluorides and oxides of phosphorus.
Countermeasures
Storage and Handling: Section 7.

POTASSIUM HYDRATE. See potassium hydroxide.

POTASSIUM HYDRIDE
General Information
Description: White needles.
Formula: KH.
Constants: Mol wt: 40.1, mp: decomposes, d: 1.43–1.47.
Hazard Analysis
Toxicity: See potassium and hydrides.
Fire Hazard: Dangerous, by chemical reaction (See potassium).
Explosion Hazard: Moderate, when exposed to heat or by spontaneous chemical reactinn.
Disaster Hazard: Dangerous; when heated to decomposition, it emits highly toxic fumes of potassium oxide; will react with water, steam or acids to produce hydrogen; can react vigorously with oxidizing materials.
Countermeasures
Ventilation Control: Section 2.
To Fight Fire: Carbon dioxide, dry chemical or carbon tetrachloride (Section 6).
Personnel Protection: Section 3.
Storage and Handling: Section 7.

POTASSIUM HYDROGEN OXALATE
General Information
Description: Monoclinic, colorless crystals.
Formula: KHC_2O_4.
Constants: Mol wt: 128.1, mp: decomposes, d: 2.0.
Hazard Analysis
Toxicity: Highly toxic. See oxalates.

POTASSIUM HYDROGEN PHOSPHATE. See potassium phosphate, dibasic.

POTASSIUM HYDROGEN SULFATE
General Information
Synonyms: Potassium bisulfate; acid potassium sulfate.
Description: Colorless crystals; soluble in water; decomposes in alcohol.
Formula: $KHSO_4$.
Constants: Mol wt: 136, d: 2.245, mp: 214°C, bp: decomposes.
Hazard Analysis and Countermeasures
See sulfates.
Shipping Regulations: Section 11.
　IATA: Other restricted articles, class B, no label required, 12 kilograms (passenger), 45 kilograms (cargo).

POTASSIUM HYDROGEN SULFITE
General Information
Synonym: Potassium bisulfite.
Description: Colorless crystals.
Formula: $KHSO_3$.
Constants: Mol wt: 120.17, mp: decomp. at 190°C.
Hazard Analysis and Countermeasures
A chemical preservative food additive (Section 10). See sulfites.

POTASSIUM HYDROSULFIDE
General Information
Synonym: Potassium bisulfide, potassium sulfhydrate.
Description: Rhombic, yellow, deliquescent crystals (commercial).
Formula: KHS.
Constants: Mol wt: 72.17, mp: 455°C, d: 2.0.
Hazard Analysis and Countermeasures
See sulfides.

POTASSIUM HYDROXIDE
General Information
Synonym: Potassium hydrate.
Description: White, deliquescent pieces, lumps or sticks having crystalline fracture.
Formula: KOH.
Constants: Mol wt: 56.11, mp: 360°C ± 7°C, bp: 1320°C, d: 2.044, vap. press.: 1 mm at 719°C.
Hazard Analysis
Toxic Hazard Rating:
　Acute Local: Irritant 3; Ingestion 3; Inhalation 3.
　Acute Systemic: Ingestion 3.
　Chronic Local: Irritant 3.
　Chronic Systemic: U.
Toxicity: Highly caustic. Used as a general purpose food additive (Section 10). See potassium.
Fire Hazard: Moderate.
Disaster Hazard: Dangerous; will react with water or steam to produce caustic solution and heat.
Countermeasures
Ventilation Control: Section 2.
Personnel Protection: Section 3.
Storage and Handling: Section 7.
Shipping Regulations: Section 11.
　Coast Guard Classification: Hazardous article.
　I.C.C.: Solution: Corrosive liquid; white label, 10 gallons.
　MCA warning label.
　IATA (solid): Other restricted articles, class B, no label required, 12 kilograms (passenger), 45 kilograms (cargo).

TOXIC HAZARD RATING CODE (For detailed discussion, see Section 1.)

0 NONE: (a) No harm under any conditions; (b) Harmful only under unusual conditions or overwhelming dosage.

1 SLIGHT: Causes readily reversible changes which disappear after end of exposure.

2 MODERATE: May involve both irreversible and reversible changes not severe enough to cause death or permanent injury.

3 HIGH: May cause death or permanent injury after very short exposure to small quantities.

U UNKNOWN: No information on humans considered valid by authors.

(solution): Corrosive liquid, white label, 1 liter (passenger), 20 liters (cargo).

POTASSIUM HYDROXOANTIMONATE
General Information
Synonym: Pyroantimonate.
Description: Granular, white, crystalline powder.
Formula: $KSb(OH)_6 \cdot \frac{1}{2}H_2O$.
Constant: Mol wt: 271.9.
Hazard Analysis and Countermeasures
See antimony compounds.

POTASSIUM HYDROXOPLATINATE
General Information
Description: Rhombic, yellow crystals.
Formula: $K_2Pt(OH)_6$.
Constants: Mol wt: 375.47, mp: decomposes.
Hazard Analysis
Toxicity: See platinum compounds.

POTASSIUM HYDROXOPLUMBATE
General Information
Description: Colorless crystals.
Formula: $K_2Pb(OH)_6$.
Constant: Mol wt: 387.45.
Hazard Analysis and Countermeasures
See lead compounds.

POTASSIUM HYDROXOSTANNATE
General Information
Description: Colorless crystals.
Formula: $K_2Sn(OH)_6$.
Constants: Mol wt: 298.94, d: 3.197.
Hazard Analysis
Toxicity: See tin compounds.

POTASSIUM HYPERCHLORATE. See potassium perchlorate.

POTASSIUM HYPOCHLORITE
General Information
Description: Solution only.
Formula: KClO.
Constants: Mol wt: 90.55, mp: decomposes.
Hazard Analysis and Countermeasures
See hypochlorites.

POTASSIUM HYPONITRITE. See nitrites.

POTASSIUM HYPOPHOSPHITE
General Information
Description: White, opaque, very deliquescent crystals or powder; pungent saline taste.
Formula: KH_2PO_2.
Constants: Mol wt: 104.09, mp: decomposes.
Hazard Analysis and Countermeasures
See hypophosphites.

POTASSIUM IODATE
General Information
Description: Colorless crystals.
Formula: KIO_3.
Constants: Mol wt: 214, mp: 560°C, d: 3.89.
Hazard Analysis and Countermeasures
See iodates. A trace mineral added to animal feeds (Section 10).

POTASSIUM IODIDE
General Information
Description: Colorless or white granules.
Formula: KI.
Constants: Mol wt: 166.02, mp: 723°C, bp: 1420°C, d: 3.13, vap. press.: 1 mm at 745°C.
Hazard Analysis and Countermeasures
See iodides. A trace mineral added to animal feeds; a nutrient and or dietary supplement food additive (Section 10).

POTASSIUM IODOAURATE
General Information
Description: Lustrous, black crystals.
Formula: $KAuI_4$.
Constant: Mol wt: 743.98.
Hazard Analysis and Countermeasures
See gold compounds and iodides.

POTASSIUM IODOCADMATE
Hazard Analysis and Countermeasures
See cadmium compounds.

POTASSIUM IODOIRIDITE
General Information
Description: Green crystals.
Formula: K_3IrI_6.
Constants: Mol wt: 1071.9, mp: decomposes.
Hazard Analysis and Countermeasures
See iridium compounds and iodides.

POTASSIUM IODOMERCURATE (II)
General Information
Synonym: Potassium mercury iodide.
Description: Yellow, deliquescent prisms.
Formula: $KHgI_3$.
Constant: Mol wt: 620.47.
Hazard Analysis and Countermeasures
See mercury compounds and iodides.

POTASSIUM IODOPLATINATE
General Information
Description: Black crystals.
Formula: K_2PtI_6.
Constants: Mol wt: 1034.94, d: 5.176.
Hazard Analysis and Countermeasures
See platinum compounds and iodides.

POTASSIUM LEAD CHLORIDE
General Information
Synonym: Pseudocotunnite.
Description: Yellow crystals.
Formula: $2KCl \cdot PbCl_2$.
Constants: Mol wt: 427.23, mp: 490°C.
Hazard Analysis and Countermeasures
See lead compounds and chlorides.

POTASSIUM MAGNESIUM CHROMATE
General Information
Description: Triclinic crystals.
Formula: $K_2CrO_4 \cdot MgCrO_4 \cdot 2H_2O$.
Constants: Mol wt: 370.56, d: 2.59.
Hazard Analysis
Toxicity: See chromium compounds.
Fire Hazard: Moderate, by chemical reaction; a powerful oxidizer (Section 6).
Disaster Hazard: Dangerous. Keep away from combustible materials.
Countermeasures
Storage and Handling: Section 7.

POTASSIUM MANGANATE
General Information
Description: Rhombic, green crystals.
Formula: K_2MnO_4.
Constants: Mol wt: 197.12, mp: decomp. at 190°C.
Hazard Analysis
Toxicity: See manganese compounds and potassium permanganate.
Fire Hazard: Moderate, by chemical reaction; a powerful oxidizer (Section 6).
Disaster Hazard: Dangerous. Keep away from combustible materials.
Countermeasures
Storage and Handling: Section 7.

POTASSIUM MANGANIC SULFATE
General Information
Description: Violet crystals.
Formula: $KMn(SO_4)_2 \cdot 12H_2O$.
Constant: Mol wt: 502.35.
Hazard Analysis and Countermeasures
See manganese compounds and sulfates.

POTASSIUM MANGANICYANIDE
General Information
Description: Rhombic, red crystals.
Formula: $K_2Mn(CN)_6$.
Constant: Mol wt: 328.33.
Hazard Analysis and Countermeasures
See cyanides and manganese compounds.

POTASSIUM MANGANOCYANIDE
General Information
Description: Deep blue crystals.
Formula: $K_4Mn(CN)_6 \cdot 3H_2O$.
Constant: Mol wt: 421.47.
Hazard Analysis and Countermeasures
See cyanides and manganese compounds.

POTASSIUM MANGANOUS CHLORIDE
General Information
Synonym: Chloromanganokalite.
Description: Crystals.
Formula: $4KCl \cdot MnCl_2$.
Constants: Mol wt: 424.06, d: 2.31.
Hazard Analysis and Countermeasures
See manganese compounds and chlorides.

POTASSIUM MANGANOUS SULFATE
General Information
Synonym: Manganolangbeinite.
Description: Rose-red crystals.
Formula: $K_2SO_4 \cdot 2MnSO_4$.
Constants: Mol wt: 476.23, mp: 850°C, d: 3.02.
Hazard Analysis and Countermeasures
Toxicity: See manganese compounds and sulfates.

POTASSIUM MERCURICYANIDE
General Information
Description: Colorless crystals.
Formula: $K_2Hg(CN)_4$.
Constant: Mol wt: 382.87.
Hazard Analysis and Countermeasures
See cyanides and mercury compounds, organic.

POTASSIUM MERCUROUS TARTRATE
General Information
Description: White, crystalline powder.
Formula: $KHgC_4H_4O_6$.
Constant: Mol wt: 387.78.
Hazard Analysis and Countermeasures
See mercury compounds, organic.

POTASSIUM MERCURY IODIDE. See potassium iodomercurate (II).

POTASSIUM METABISULFITE. See potassium pyrosulfite.

POTASSIUM, METALLIC, LIQUID ALLOY.
See potassium.
Shipping Regulations: Section 11.
   I.C.C.: Flammable solid; yellow label, 1 pound.

Coast Guard Classification: Inflammable solid; yellow label.
IATA: Flammable solid, yellow label, not acceptable (passenger), ½ kilogram (cargo).

POTASSIUM METHAZONATE
Hazard Analysis
Toxicity: Unknown.
Fire Hazard: Unknown.
Explosion Hazard: Severe, when shocked or exposed to heat.
Disaster Hazard: Dangerous; heat or shock will explode it.
Countermeasures
Storage and Handling: Section 7.

POTASSIUM METHYL SULFATE
General Information
Description: White crystals.
Formula: $2KCH_3SO_4 \cdot H_2O$.
Constant: Mol wt: 318.41.
Hazard Analysis
Toxicity: Details unknown. Probably highly toxic. See also dimethyl sulfate.
Fire Hazard: Moderate, when exposed to heat or flame; it decomposes readily to flammable products (Section 6).
Explosion Hazard: Unknown.
Disaster Hazard: Dangerous; when heated to decomposition, it emits highly toxic fumes; can react with oxidizing materials.
Countermeasures
Storage and Handling: Section 7.

POTASSIUM MONOHYDROGEN o-ARSENATE
General Information
Description: Colorless crystals.
Formula: K_2HAsO_4.
Constant: Mol wt: 218.11.
Hazard Analysis and Countermeasures
See arsenic compounds.

POTASSIUM MONOPHOSPHATE. See potassium phosphate, dibasic.

POTASSIUM MONOSULFIDE
General Information
Description: Yellow-brown, deliquescent crystals.
Formula: K_2S.
Constants: Mol wt: 110.26, mp: 840°C, d: 1.805 at 14°C.
Hazard Analysis and Countermeasures
See sulfides.

POTASSIUM MONOXIDE
General Information
Description: Cubic colorless-gray crystals.
Formula: K_2O.
Constants: Mol wt: 94.19, d: 2.32 at 0°C.
Hazard Analysis
Toxicity: See potassium hydroxide.
Fire Hazard: Moderate, by chemical reaction; will react with water or steam to produce heat (Section 6).
Countermeasures
Storage and Handling: Section 7.

POTASSIUM MURIATE. See potassium chloride.

POTASSIUM NICKEL SULFATE
General Information
Description: Blue crystals.
Formula: $K_2SO_4 \cdot NiSO_4 \cdot 6H_2O$.

TOXIC HAZARD RATING CODE (For detailed discussion, see Section 1.)

0 NONE: (a) No harm under any conditions; (b) Harmful only under unusual conditions or overwhelming dosage.

1 SLIGHT: Causes readily reversible changes which disappear after end of exposure.

2 MODERATE: May involve both irreversible and reversible changes; not severe enough to cause death or permanent injury.

3 HIGH: May cause death or permanent injury after very short exposure to small quantities.

U UNKNOWN: No information on humans considered valid by authors.

Constants: Mol wt: 437.11, mp: decomp. $< 100°$C, d: 2.124.

Hazard Analysis and Countermeasures
See nickel compounds and sulfates.

POTASSIUM NIOBATE
General Information
Synonym: Potassium columbate.
Description: Crystalline form of niobic acid solutions when treated with concentrated potassium hydroxide; soluble in water.
Formula: $4K_2O \cdot 3Nb_2O_5 \cdot 16H_2O$.
Constant: Mol wt: 1462.
Hazard Analysis
Toxicity: Details unknown. Animal experiments show high toxicity, primarily involving the kidney.

POTASSIUM NITRATE
General Information
Synonyms: Niter; nitre; saltpeter.
Description: Transparent, colorless or white, crystalline powder or crystals; cooling, pungent, saline taste.
Formula: KNO_3.
Constants: Mol wt: 101.10, mp: $334°$C, bp: decomp. at $400°$C, d: 2.109 at $16°$C.
Hazard Analysis and Countermeasures
See nitrates. A food additive permitted in food for human consumption (Section 10).
Shipping Regulations: Section 11.
I.C.C.: Oxidizing material; yellow label, 100 pounds.
Coast Guard Classification: Oxidizing material.
IATA: Oxidizing material, yellow label, 12 kilograms (passenger), 45 kilograms (cargo).

POTASSIUM NITRATE MIXED (FUSED) WITH SODIUM NITRITE. See potassium nitrate and sodium nitrite.
Shipping Regulations: Section 11.
I.C.C.: Oxidizing material, yellow label, 100 pounds.
Coast Guard Classification: Oxidizing material.

POTASSIUM NITRIDE
General Information
Description: Greenish-black crystals.
Formula: K_3N.
Constants: Mol wt: 131.30, mp: decomposes.
Hazard Analysis and Countermeasures
See ammonia which is readily evolved on contact with moisture.

POTASSIUM NITRITE
General Information
Description: White or slightly yellowish, deliquescent prisms or sticks.
Formula: KNO_2.
Constants: Mol wt: 85.10, mp: $387°$C, bp: decomposes, d: 1.915.
Hazard Analysis
Toxicity: See nitrites.
Fire Hazard: Moderate; an oxidizing material. See nitrites.
Explosion Hazard: Slight, when exposed to heat. It will explode at $1000°$F or when mixed with cyanide salts and heated.
Disaster Hazard: Dangerous. See nitrites.
Countermeasures
Storage and Handling: Section 7.
Shipping Regulations: Section 11.
I.C.C.: Oxidizing material; yellow label, 100 pounds.
Coast Guard Classification: Oxidizing material; yellow label.
IATA: Oxidizing material, yellow label, 12 kilograms (passenger), 45 kilograms (cargo).

POTASSIUM NITROCYANIDE
General Information
Description: A solid.

Constant: Mp: explodes at $400°$C.
Hazard Analysis
Toxicity: Highly toxic. See cyanides.
Explosion Hazard: Moderate; it will explode at $752°$F.
Disaster Hazard: Dangerous; on decomposition or on contact with acid or acid fumes, it emits highly toxic fumes of cyanides. An explosive.
Countermeasures
Ventilation Control: Section 2.
Personal Hygiene: Section 3.
First Aid: Section 1.
Storage and Handling: Section 7.

POTASSIUM NITROMETHANE
Hazard Analysis
Toxicity: See nitroparaffins.
Fire Hazard: Dangerous, when exposed to heat or flame (Section 6).
Explosion Hazard: Severe, when shocked or exposed to heat.
Disaster Hazard: Dangerous; shock will explode it; when heated to decomposition, it emits highly toxic fumes of oxides of nitrogen; can react vigorously with oxidizing materials.
Countermeasures
Storage and Handling: Section 7.

POTASSIUM m-NITROPHENOXIDE
General Information
Description: Flat, orange needles.
Formula: $KOC_6H_4NO_2 \cdot 2H_2O$.
Constants: Mol wt: 213.23, mp: $-2H_2O$ at $130°$C, bp: decomposes, d: 1.691 at $20°$C.
Hazard Analysis
Toxicity: Highly toxic. See phenol.
Disaster Hazard: Dangerous; when heated to decomposition or on contact with acids or acid fumes, it emits highly toxic fumes of oxides of nitrogen.
Countermeasures
Ventilation Control: Section 2.
Personnel Protection: Section 3.
Storage and Handling: Section 7.

POTASSIUM NITROPLATINATE
General Information
Description: Monoclinic, colorless crystals.
Formula: $K_2Pt(NO_2)_4$.
Constants: Mol wt: 457.45, mp: decomposes.
Hazard Analysis and Countermeasures
See platinum compounds and nitrates.

POTASSIUM NITROPRUSSIDE
General Information
Description: Red, hygroscopic crystals.
Formula: $K_2[Fe(NO)(CN)_5] \cdot 2H_2O$.
Constant: Mol wt: 330.17.
Hazard Analysis
Toxicity: Highly toxic. See cyanides.
Disaster Hazard: Dangerous; when heated to decomposition or on contact with acids or acid fumes, it emits highly toxic fumes of cyanides and oxides of nitrogen.
Countermeasures
Ventilation Control: Section 2.
Personnel Protection: Section 3.
First Aid: Section 1.
Storage and Handling: Section 7.

POTASSIUM OSMATE
General Information
Synonym: Potassium perosmate.
Description: Violet, hygroscopic crystals.
Formula: $K_2OsO_4 \cdot 2H_2O$.
Constants: Mol wt: 368.42, mp: $-H_2O > 100°$C.
Hazard Analysis
Toxicity: See osmium compounds.

Fire Hazard: Moderate, by chemical reaction; a powerful oxidizer (Section 6).
Countermeasures
Storage and Handling: Section 7.

POTASSIUM OXALATE
General Information
Description: Colorless, transparent crystals.
Formula: $K_2C_2O_4 \cdot H_2O$.
Constants: Mol wt: 184.23, mp: decomposes, d: 2.127 at 39° C.
Hazard Analysis and Countermeasures
See oxalates.

POTASSIUM OXALATOURANATE (IV)
General Information
Description: Yellow crystals.
Formula: $K_4[U(C_2O_4)_4] \cdot 5H_2O$.
Constants: Mol wt: 836.61, d: 2.563.
Hazard Analysis and Countermeasures
See oxalates and uranium.

POTASSIUM OXYBORATE
General Information
Description: White crystals.
Formula: $KBO_3 \cdot \frac{1}{2}H_2O$.
Constant: Mol wt: 106.92.
Hazard Analysis
Toxicity: See boron compounds.

POTASSIUM OXYMURIATE. See potassium chlorate.

POTASSIUM OXYNIOBATE. See potassium pentafluoniobate.

POTASSIUM PENTABORANE
General Information
Synonym: Pentaboranide.
Description: White powder.
Formula: $K_2B_5H_9$.
Constants: Mol wt: 141.36, mp: decomp. < 180° C.
Hazard Analysis and Countermeasures
See boron hydrides.

POTASSIUM PENTABORATE
General Information
Description: Colorless crystals; fine granular structure.
Formula: $KB_5O_8 \cdot 4H_2O$.
Constants: Mol wt: 449.66, mp: 780° C.
Hazard Analysis
Toxicity: See boron compounds.

POTASSIUM PENTACHLOROAQUORUTHENATE (III)
General Information
Description: Rose prisms.
Formula: $K_2Ru(H_2O)Cl_5$.
Constants: Mol wt: 375.19, mp: $-H_2O$ at 200° C.
Hazard Analysis and Countermeasures
See ruthenium compounds and chlorides.

POTASSIUM PENTACHLOROHYDROXO RUTHENATE (IV)
General Information
Description: Brown-red crystals.
Formula: $K_2Ru(OH)Cl_5$.
Constants: Mol wt: 374.19, mp: decomposes.
Hazard Analysis and Countermeasures
See ruthenium compounds and chlorides.

POTASSIUM PENTACHLORONITROSYL RUTHENATE (III)
General Information
Description: Rhombic, dark red crystals.
Formula: $K_2Ru(NO)Cl_5$.
Constants: Mol wt: 387.19, mp: decomposes.
Hazard Analysis and Countermeasures
See ruthenium compounds and chlorides.

POTASSIUM PENTACHLORORHODITE
General Information
Description: Rhombic, red crystals.
Formula: K_2RhCl_5.
Constants: Mol wt: 358.39, mp: decomposes.
Hazard Analysis and Countermeasures
See rhodium and chlorides.

POTASSIUM PENTAFLUONIOBATE
General Information
Synonym: Potassium oxyniobate.
Description: Monoclinic, colorless leaflets.
Formula: $K_2NbOF_5 \cdot H_2O$.
Constant: Mol wt: 300.12.
Hazard Analysis and Countermeasures
See fluorides.

POTASSIUM PENTASULFIDE
General Information
Description: Orange crystals.
Formula: K_2S_5.
Constants: Mol wt: 238.52, mp: 206° C.
Hazard Analysis and Countermeasures
See sulfides.

POTASSIUM PERCARBONATE
General Information
Description: White, granular mass; soluble in water with the evolution of oxygen.
Formula: $K_2C_2O_6 \cdot H_2O$.
Constants: Mol wt: 216.23, mp: 200–300° C.
Hazard Analysis
Toxic Hazard Rating:
Acute Local: Irritant 3; Ingestion 3.
Acute Systemic: Ingestion 3.
Chronic Local: U.
Chronic Systemic: U.
Toxicity: A strong oxidizer, hence caustic to skin and mucous membranes.

POTASSIUM PERCHLORATE
General Information
Synonym: Potassium hyperchlorate.
Description: Colorless crystals or white, crystalline powder.
Formula: $KClO_4$.
Constants: Mol wt: 138.55, mp: 610° ± 10° C, d: 2.52 at 10° C.
Hazard Analysis
Toxic Hazard Rating:
Acute Local: Irritant 2; Ingestion 2.
Acute Systemic: Ingestion 2.
Chronic Local: Irritant 1; Ingestion 1.
Chronic Systemic: Ingestion 2.
Toxicology: Irritating to skin and mucous membrane. Absorption can cause methemoglobinemia and kidney injury.
Countermeasures
Shipping Regulations: Section 11.
I.C.C.: Oxidizing material; yellow label, 100 pounds.

TOXIC HAZARD RATING CODE (For detailed discussion, see Section 1.)

0 NONE: (a) No harm under any conditions; (b) Harmful only under unusual conditions or overwhelming dosage.

1 SLIGHT: Causes readily reversible changes which disappear after end of exposure.

2 MODERATE: May involve both irreversible and reversible changes not severe enough to cause death or permanent injury.

3 HIGH: May cause death or permanent injury after very short exposure to small quantities.

U UNKNOWN: No information on humans considered valid by authors.

Coast Guard Classification: Oxidizing material; yellow label.

IATA: Oxidizing material, yellow label, 12 kilograms (passenger), 45 kilograms (cargo).

POTASSIUM m-PERIODATE
General Information
Description: Tetragonal, colorless crystals.
Formula: KIO_4.
Constants: Mol wt: 230.0, mp: 582°C, bp: −O at 300°C, d: 3.618 at 15°C.
Hazard Analysis
Toxic Hazard Rating:
Acute Local: Irritant 3; Ingestion 3; Inhalation 3.
Acute Systemic: U.
Chronic Local: U.
Chronic Systemic: U.
Toxicology: A strong irritant. See also iodates.
Fire Hazard: An oxidizing agent and moderate fire hazard (Section 6). See iodates.
Explosion Hazard: Slight, when exposed to heat. Explodes at 1076°F.
Disaster Hazard: Dangerous, when exposed to heat or flame; on decomposition it emits toxic fumes of iodine compounds.
Countermeasures
Storage and Handling: Section 7.

POTASSIUM PERMANGANATE
General Information
Description: Dark purple crystals with a blue metallic sheen; sweetish astringent taste.
Formula: $KMnO_4$.
Constants: Mol wt: 158.03, mp: decomp. < 240°C, d: 2.703.
Hazard Analysis
Toxic Hazard Rating:
Acute Local: Irritant 3; Ingestion 3; Inhalation 3.
Acute Systemic: Ingestion 3.
Chronic Local: Irritant 2.
Chronic Systemic: Ingestion 3; Inhalation 3.
Toxicity: A strong irritant because of oxidizing properties. See also manganese compounds.
Fire Hazard: Moderate, by chemical reaction. A powerful oxidizing agent, Spontaneously flammable on contact with glycerine and ethylene glycol. See also permanganates.
Disaster Hazard: Dangerous; keep away from combustible materials.
Countermeasures
Storage and Handling: Section 7.
Shipping Regulations: Section 11.
Coast Guard Classification: Oxidizing material; yellow label.

POTASSIUM PEROSMATE. See potassium osmate.

POTASSIUM PEROXIDE
General Information
Description: Yellow, amorphous mass (white crystals).
Formula: K_2O_2.
Constants: Mol wt: 110.19, mp: 490°C.
Hazard Analysis
Toxicity: See peroxides, inorganic.
Fire Hazard: Dangerous, by spontaneous chemical reaction. It is a very powerful oxidizer. Fires of this material should be handled like sodium peroxide.
Explosion Hazard: Moderate, by spontaneous chemical reaction.
Disaster Hazard: Dangerous; will react with water or steam to produce heat; on contact with reducing material, it can react vigorously, and on contact with acid or acid fumes, it can emit toxic fumes.
Countermeasures
Storage and Handling: Section 7.
Shipping Regulations: Section 11.

I.C.C.: Oxidizing material; yellow label, 100 pounds.
Coast Guard Classification: Oxidizing material; yellow label.
IATA: Oxidizing material, yellow label, 12 kilograms (passenger), 45 kilograms (cargo).

POTASSIUM PEROXYCHROMATE
General Information
Description: Brown-red crystals.
Formula: K_3CrO_8.
Constants: Mol wt: 297.3, mp: decomp. at 170°C.
Hazard Analysis
Toxicity: Highly toxic. See chromium compounds.
Fire Hazard: Moderate, by chemical reaction; a powerful oxidizer (Section 6).
Disaster Hazard: Dangerous; keep away from combustible materials.
Countermeasures
Storage and Handling: Section 7.

POTASSIUM   PEROXYDISULFATE. See potassium persulfate.

POTASSIUM PERRHENATE
General Information
Description: White crystals.
Formula: $KReO_4$.
Constants: Mol wt: 289.41, mp: 350°C, d: 4.887.
Hazard Analysis
Toxicity: See rhenium compounds.
Fire Hazard: Moderate, by chemical reaction; a powerful oxidizer (Section 6).
Disaster Hazard: Dangerous. Keep away from combustible materials.
Countermeasures
Storage and Handling: Section 7.

POTASSIUM PERRUTHENATE
General Information
Description: Black crystals.
Formula: $KRuO_4$.
Constants: Mol wt: 204.8, mp: decomp. at 440°C.
Hazard Analysis
Toxicity: See ruthenium compounds.
Fire Hazard: Moderate, by chemical reaction; a powerful oxidizer (Section 6).
Disaster Hazard: Dangerous; keep away from combustible materials.
Countermeasures
Storage and Handling: Section 7.

POTASSIUM PERSELENATE
General Information
Description: Crystals.
Formula: $KSeO_4$.
Constant: Mol wt: 182.1.
Hazard Analysis
Toxicity: Highly toxic. See selenium compounds.
Fire Hazard: Moderate, by chemical reaction; a powerful oxidizer (Section 6).
Disaster Hazard: Dangerous; when heated to decomposition on contact with acid or acid fumes, it emits highly toxic fumes of selenium; keep away from combustible materials.
Countermeasures
Ventilation Control: Section 2.
Personnel Protection: Section 3.
Storage and Handling: Section 7.

POTASSIUM PERSULFATE
General Information
Synonym: Anthion; potassium peroxydisulfate.
Description: White, odorless crystals.
Formula: $K_2S_2O_8$.
Constants: Mol wt: 270.3, mp: decomposes < 100°C, d: 2.477.

Hazard Analysis
Toxic Hazard Rating:
 Acute Local: Irritant 2; Allergen 2; Inhalation 2; Ingestion 2.
 Acute Systemic: U.
 Chronic Local: Irritant 1; Allergen 2.
 Chronic Systemic: Ingestion 1; Inhalation 1.
Fire Hazard: Moderate, when exposed to heat or by chemical reaction. It liberates oxygen above 100°C when dry, or at about 50°C when in solution (Section 6).
Disaster Hazard: Dangerous; when heated to decomposition, it emits highly toxic fumes of oxides of sulfur; can react with reducing materials.
Countermeasures
Ventilation Control: Section 2.
Personal Hygiene: Section 3.
Storage and Handling: Section 7.
Shipping Regulations: Section.
 IATA: Oxidizing material, yellow label, 12 kilograms (passenger), 45 kilograms (cargo).

POTASSIUM PHENOL SULFONATE
General Information
Description: Rhombic crystals.
Formula: $KC_6H_4(OH)SO_3$.
Constants: Mol wt: 212.26, mp: > 260°C, d: 1.87.
Hazard Analysis
Toxicity: Highly toxic. See phenol.
Disaster Hazard: Dangerous; See sulfonates.
Countermeasures
Ventilation Control: Section 2.
Personnel Protection: Section 3.
First Aid: Section 1.
Storage and Handling: Section 7.

POTASSIUM PHENYLACETATE
General Information
Description: Dry powder.
Formula: $C_6H_5CH_2COOK$.
Constant: Mol wt: 174.2.
Hazard Analysis
Toxicity: Unknown.
Disaster Hazard: Moderately dangerous; when heated to decomposition, it emits toxic fumes.
Countermeasures
Storage and Handling: Section 7.

POTASSIUM PHENYL SULFATE
General Information
Description: Rhombic leaflets.
Formula: $KC_6H_5SO_4$.
Constants: Mol wt: 212.26, mp: 150–160°C decomposes, bp: decomposes.
Hazard Analysis
Toxicity: Highly toxic. See phenol.
Disaster Hazard: Dangerous; See sulfates.
Countermeasures
Ventilation Control: Section 2.
Personnel Protection: Section 3.
First Aid: Section 1.
Storage and Handling: Section 7.

POTASSIUM PHOSPHATE, DIBASIC
General Information
Synonyms: DKP; potassium hydrogen phosphate; potassium monophosphate; dipotassium-o-phosphate.
Description: Deliquescent, white crystals or powder; very soluble in water and alcohol.
Formula: K_2PO_4.
Constant: Mol wt: 173.
Hazard Analysis
Toxic Hazard Rating: U. A sequestrant food additive (Section 10).

POTASSIUM PHTHALIMIDE
General Information
Description: White powder.
Formula: $C_6H_4(CO)_2NK$.
Constant: Mol wt: 185.23.
Hazard Analysis
Toxicity: Details unknown; an insecticide.
Disaster Hazard: Moderate; when heated to decomposition it emits toxic fumes (Section 6).
Countermeasures
Storage and Handling: Section 7.

POTASSIUM PICRATE
General Information
Description: Yellow-reddish or greenish crystals.
Formula: $KC_6H_2N_3O_7$.
Constants: Mol wt: 267.20, bp: explodes at 310°C, d: 1.852.
Hazard Analysis
Toxicity: Highly toxic. See picric acid.
Fire Hazard: See nitrates.
Explosion Hazard: Moderate, when shocked or exposed to heat.
Disaster Hazard: Dangerous; shock will explode it; on decomposition, it emits highly toxic fumes of oxides or nitrogen; can react vigorously with reducing materials.
Countermeasures
Ventilation Control: Section 2.
Personnel Protection: Section 3.
First Aid: Section 1.
Storage and Handling: Section 7.

POTASSIUM POLYMANNURATE. See potassium alginate.

POTASSIUM POLYSULFIDES. See sulfides.

POTASSIUM PYROSULFITE
General Information
Synonym: Potassium metabisulfite.
Description: Monoclinic plates.
Formula: $K_2S_2O_5$.
Constants: Mol wt: 222.32, mp: decomposes, d: 2.3.
Hazard Analysis and Countermeasures
A chemical preservative food additive (Section 10). See sulfites.
Shipping Regulations: Section 11.
 IATA: Other restricted articles, class B, no label, no limit (passenger and cargo).

POTASSIUM RHODANIDE. See potassium sulfocyanate.

POTASSIUM RHODIUM SULFATE
General Information
Description: Yellow cubes.
Formula: $KRh(SO_4)_3 \cdot 12H_2O$.
Constants: Mol wt: 550.33, d: 2.23.
Hazard Analysis and Countermeasures
See rhodium and sulfates.

POTASSIUM RUTHENATE
General Information
Description: Black crystals.
Formula: $K_2RuO_4 \cdot H_2O$.

Constants: Mol wt: 261.91, mp: $-H_2O$ at 200°C, bp: decomposes at 400°C in vacuo.
Hazard Analysis
Toxicity: See ruthenium compounds.

POTASSIUM SELENATE
General Information
Description: Rhombic, colorless crystals.
Formula: K_2SeO_4.
Constants: Mol wt: 221.15, d: 3.066.
Hazard Analysis and Countermeasures
See selenium compounds.

POTASSIUM SELENIDE
General Information
Description: White crystals; reddens on exposure to air.
Formula: K_2Se.
Constants: Mol wt: 157.15, d: 2.851 at 15°C.
Hazard Analysis and Countermeasures
See selenium compounds.

POTASSIUM SELENITE
General Information
Description: White, deliquescent crystals.
Formula: K_2SeO_3.
Constant: Mol wt: 205.15.
Hazard Analysis and Countermeasures
See selenium compounds.

POTASSIUM SELENOCYANATE
General Information
Description: Deliquescent needles.
Formula: KSeCN.
Constants: Mol wt: 144.1, mp: decomp. at 100°C, d: 2.347.
Hazard Analysis and Countermeasures
See selenium compounds and cyanates.

POTASSIUM SELENOCYANOPLATINATE
General Information
Description: Rhombic crystals.
Formula: $K_2Pt(SeCN)_6$.
Constants: Mol wt: 903.29, mp: decomp. at 80°C, d: 3.378 at 12.5°C.
Hazard Analysis
Toxicity: Highly toxic. For other properties see potassium selenocyanate and platinum compounds.

POTASSIUM SESQUIOXIDE. See potassium trioxide.

POTASSIUM SILICOFLUORIDE
General Information
Description: White, odorless, fine crystaline powder.
Formula: K_2SiF_6.
Constants: Mol wt: 220.25, d: 2.27.
Hazard Analysis and Countermeasures
See fluosilicates.

POTASSIUM SILVER CARBONATE
General Information
Description: Rectangular plates.
Formula: $KAgCO_3$.
Constants: Mol wt: 206.99, mp: decomp., d: 3.769.
Hazard Analysis
Toxicity: See silver compounds.

POTASSIUM SILVER NITRATE
General Information
Description: Monoclinic crystals.
Formula: $KNO_3 \cdot AgNO_3$.
Constants: Mol wt: 270.99, mp: 125°C, d: 3.219.
Hazard Analysis and Countermeasures
See nitrates and silver compounds.

POTASSIUM SODIUM ANTIMONY TARTRATE
General Information
Description: White scales or powder.
Formula: $KNaSbC_4H_3O_7$.

Constant: Mol wt: 346.9.
Hazard Analysis and Countermeasures
See antimony compounds.

POTASSIUM SODIUM TARTRATE. See sodium potassium tartrate.

POTASSIUM SORBATE
General Information
Synonym: Potassium 2,4-hexadienoate.
Description: White powder, soluble in water at 25°C.
Formula: $CH_3CH:CHCH:CHCOOK$.
Constants: Mol wt: 167, mp: 270°C (decomposes); d: 1.36 (25/20°C).
Hazard Analysis
Toxicity: Unknown. A chemical preservative food additive; it is a substance which migrates to food from packaging materials (Section 10).

POTASSIUM STRONTIUM CHLORATE. See strontium potassium chlorate.

POTASSIUM STRONTIUM CHROMIC OXALATE
General Information
Description: Greenish-black crystals.
Formula: $KSrCr(C_2O_4)_3 \cdot 6H_2O$.
Constants: Mol wt: 550.89, d: 2.155 at 13°C.
Hazard Analysis and Countermeasures
See oxalates and chromium compounds.

POTASSIUM STYPHNATE
General Information
Synonym: Potassium trinitroresorcinate.
Description: Yellow, monoclinic prisms.
Formula: $KC_6H_2N_3O_8 \cdot H_2O$.
Constants: Mol wt: 301.21, mp: $-H_2O$ at 120°C, bp: explodes.
Hazard Analysis and Countermeasures
See nitrates and explosives, high.

POTASSIUM SULFHYDRATE. See potassium hydrosolfide.

POTASSIUM SULFATE
General Information
Description: Colorless or white hard crystals or powder; soluble in water; insoluble in alcohol.
Formula: K_2SO_4.
Constants: Mol wt: 176, d: 2.66, mp: 1072°C.
Hazard Analysis
Toxic Hazard Rating:
    Acute Local: Irritant 2; Ingestion 3.
    Acute Systemic: Ingestion 3.
    Chronic Local: Irritant 1.
    Chronic Systemic: Ingestion 1.
Toxicity: A general purpose food additive (Section 10).
Disaster Hazard: Dangerous. See sulfates.

POTASSIUM SULFIDE
General Information
Synonyms: Potassium sulfuret, hepar sulfuris.
Description: Red, crystalline mass; deliquescent in air.
Formula: K_2S.
Constants: Mol wt: 110.25, mp: 471°C, d: 1.805 at 14°C.
Hazard Analysis and Countermeasures
See sulfides.
 Shipping Regulations: Section 11.
    I.C.C.: Flammable solid; yellow label, 300 pounds.
    Coast Guard Classification: Inflammable solid; yellow label.
    IATA: Flammable solid, yellow label, 12 kilograms (passenger), 140 kilograms (cargo).

POTASSIUM SULFITE
General Information
Description: White-yellowish crystals.

Formula: $K_2SO_3 \cdot 2H_2O$.
Constants: Mol wt: 194.29, mp: decomp.
Hazard Analysis and Countermeasures
See sulfites.

POTASSIUM SULFOCARBONATE
General Information
Synonym: Potassium thiocarbonate.
Description: Yellowish-red crystals.
Formula: K_2CS_3.
Constants: Mol wt: 186.4, mp: decomp.
Hazard Analysis
Toxic Hazard Rating:
   Acute Local: Irritant 3; Ingestion 3.
   Acute Systemic: Ingestion 3.
   Chronic Local: U.
   Chronic Systemic: U.
Toxicity: Details unknown; a soil fumigant.
Disaster Hazard: Dangerous; when heated to decomposition or on contact with acid or acid fumes, it emits highly toxic fumes of oxides of sulfur.
Countermeasures
Storage and Handling: Section 7.

POTASSIUM SULFOCYANATE
General Information
Synonyms: Potassium rhodanide, potassium thiocyanate.
Description: Colorless crystals.
Formula: KSCN.
Constants: Mol wt: 97.17, mp: 173.2°C, d: 1.886, bp: decomp. at 500°C.
Hazard Analysis
Toxic Hazard Rating:
   Acute Local: U.
   Acute Systemic: Ingestion 2.
   Chronic Local: U.
   Chronic Systemic: Ingestion 2.
Toxicology: Has been used medically to reduce blood pressure. Large doses can cause skin eruption, psychoses and collapse.
Disaster Hazard: Dangerous; on decomposition it emits highly toxic fumes of cyanides. See sulfates.
Countermeasures
Ventilation Control: Section 2.
Personal Hygiene: Section 3.
Storage and Handling: Section 7.

POTASSIUM SUPEROXIDE
General Information
Description: Yellow leaflets.
Formula: KO_2.
Constants: Mol wt: 71.10, mp: approx. 400°C, bp: decomp.
Hazard Analysis
Toxicity: See peroxides, inorganic.
Fire Hazard: Dangerous; a powerful oxidizing agent (Section 6).
Explosion Hazard: Unknown.
Disaster Hazard: Dangerous; reacts with water or steam to evolve heat; keep away from combustible materials.
Countermeasures
Storage and Handling: Section 7.

POTASSIUM TELLURATE
General Information
Description: Soft, glutinous mass.
Formula: K_2TeO_4.
Constants: Mol wt: 269.80, mp: decomp. at 200°C.

Hazard Analysis and Countermeasures
See tellurium compounds.

POTASSIUM TELLURIDE
General Information
Description: Colorless crystals.
Formula: K_2Te.
Constants: Mol wt: 205.80, d: 2.51.
Hazard Analysis and Countermeasures
See tellurium compounds.

POTASSIUM TELLURITE
General Information
Description: White, deliquescent crystals.
Formula: K_2TeO_3.
Constants: Mol wt: 253.80, mp: 460–470°C decomp.
Hazard Analysis and Countermeasures
See tellurium compounds.

POTASSIUM TETRABORATE
General Information
Description: Colorless crystals.
Formula: $K_2B_4O_7 \cdot 8H_2O$.
Constants: Mol wt: 377.60, mp: decomposes; d anhydrous: 1.74.
Hazard Analysis and Countermeasures
See boron compounds.

POTASSIUM TETRAGERMANATE
General Information
Description: White crystals.
Formula: $K_2Ge_4O_9$.
Constants: Mol wt: 512.59, mp: 1033°C, d: 4.12 at 21.5°C.
Hazard Analysis and Countermeasures
See germanium compounds.

POTASSIUM TETRAIODOMERCURATE (II)
General Information
Description: Yellow, deliquescent crystals.
Formula: K_2HgI_4.
Constant: Mol wt: 786.48.
Hazard Analysis and Countermeasures
See mercury compounds and iodides.

POTASSIUM TETRASULFIDE
General Information
Description: Red-brown crystals.
Formula: K_2S_4.
Constants: Mol wt: 206.46, mp: 145°C, bp: decomp. at 850°C.
Hazard Analysis and Countermeasures
See sulfides.

POTASSIUM THIOANTIMONATE
General Information
Description: Yellow crystals.
Formula: $2K_3SbS_4 \cdot 9H_2O$.
Constant: Mol wt: 896.8.
Hazard Analysis and Countermeasures
See antimony compounds and sulfides.

POTASSIUM THIOARSENATE
General Information
Description: Deliquescent crystals.
Formula: K_3AsS_4.
Constants: Mol wt: 320.46, mp: decomp.
Hazard Analysis and Countermeasures
See arsenic compounds and sulfides.

TOXIC HAZARD RATING CODE *(For detailed discussion, see Section 1.)*

0 NONE: (a) No harm under any conditions; (b) Harmful only under unusual conditions or overwhelming dosage.

1 SLIGHT: Causes readily reversible changes which disappear after end of exposure.

2 MODERATE: May involve both irreversible and reversible changes; not severe enough to cause death or permanent injury.

3 HIGH: May cause death or permanent injury after very short exposure to small quantities.

U UNKNOWN: No information on humans considered valid by authors.

POTASSIUM THIOARSENITE
General Information
Description: A solid.
Formula: K_2AsS_3.
Constants: Mol wt: 288.4, mp: decomposes.
Hazard Analysis and Countermeasures
See arsenic compounds and sulfides.

POTASSIUM THIOCARBONATE. See potassium sulfo-carbonate.

POTASSIUM THIOCYANATE. See potassium sulfo-cyanate.

POTASSIUM THIOPLATINATE
General Information
Description: Blue-gray crystals.
Formula: $K_2Pt_4S_6$.
Constants: Mol wt: 1051.51, mp: decomp. upon ignition, d: 6.44 at 15°C.
Hazard Analysis and Countermeasures
See platinum compounds and sulfides.

POTASSIUM TRIIODIDE
General Information
Description: Monoclinic, dark blue, deliquescent crystals.
Formula: KI_3.
Constants: Mol wt: 419.86, mp: 31°C, bp: decomp. at 225°C, d: 3.498.
Hazard Analysis and Countermeasures
See iodides.

POTASSIUM TRINITRORESORCINATE. See potassium styphnate.

POTASSIUM TRIOXIDE
General Information
Synonym: Potassium sesquioxide.
Description: Red crystals (probably a mixture).
Formula: K_2O_3.
Constants: Mol wt: 126.19, mp: 430°C.
Hazard Analysis
Toxicity: Highly toxic. See potassium hydroxide.

POTASSIUM TRISULFIDE
General Information
Description: Yellow crystals.
Formula: K_2S_3.
Constants: Mol wt: 174.39, mp: 252°C.
Hazard Analysis and Countermeasures
See sulfides.

POTASSIUM m-URANATE
General Information
Description: Orange-yellow crystals.
Formula: K_2UO_4.
Constants: Mol wt: 380.26.
Hazard Analysis and Countermeasures
See uranium.

POTASSIUM URANIUM NITRATE. See uranium po-
• tassium nitrate.

POTASSIUM URANYL ACETATE
General Information
Description: Radioactive crystals.
Formula: $KUO_2(C_2H_3O_2)_2 \cdot H_2O$.
Constants: Mol wt: 504.31, mp: $-H_2O$ at 275°C, d: 2.396 at 15°C.
Hazard Analysis and Countermeasures
See uranium.

POTASSIUM URANYL CARBONATE
General Information
Description: Yellow, radioactive crystals.
Formula: $2K_2CO_3 \cdot UO_2CO_3$.
Constants: Mol wt: 606.48, mp: $-CO_2$ at 300°C.

Hazard Analysis and Countermeasures
See uranium.

POTASSIUM URANYL SULFATE. See uranium potas-sium sulfate.

POTASSIUM m-VANADATE
General Information
Description: Colorless crystals.
Formula: KVO_3.
Constant: Mol wt: 138.05.
Hazard Analysis
Toxicity: See vanadium compounds.

POTASSIUM XANTHATE
General Information
Description: Liquid.
Synonym: Potassium ethylxanthate.
Formula: $C_2H_5OCS_2K$.
Constants: Mol wt: 160.3, mp: 200°C (decomposed), uel: 9.5%, flash p: 401°F (C.C.), d: 1.558.
Hazard Analysis
Toxicity: See carbon disulfide.
Fire Hazard: Moderate, when exposed to heat or flame.
Spontaneous Heating: No.
Explosion Hazard: Moderate, when exposed to flame.
Disaster Hazard: Dangerous; when heated to decomposi-tion, it emits highly toxic fumes of oxides of sulfur; can react with oxidizing materials.
Countermeasures
Ventilation Control: Section 2.
To Fight Fire: Water, carbon dioxide, dry chemical or car-bon tetrachloride (Section 6).
Personnel Protection: Section 3.
Storage and Handling: Section 7.

POWDERED METALS
Hazard Analysis
Toxicity: See specific metal.
Fire Hazard: Dangerous in dispersed form when exposed to flame or sparks or by chemical reaction with oxidizers. Many powdered metals can ignite spontaneously and explode when suspended in air.
Countermeasures
To Fight Fire: Use no water; use powdered graphite, dolo-mite, sodium chloride, etc. Get instructions from the supplier of the powdered metal.

POWDER OF ALGAROTH. See antimony oxychloride.

POWER DEVICE, EXPLOSIVE
Shipping Regulations: Section 11.
IATA: Explosive, explosive label, not acceptable (pas-senger), 70 kilograms (cargo).

PRASEODYMIUM
General Information
Description: Pale yellow metal.
Formula: Pr.
Constants: At wt: 140.92, mp: 940°C, d: 6.5.
Hazard Analysis
Toxicity: Details unknown. As a lanthanon it may depress coagulation of the blood. See also lanthanum. Limited animal experiments suggest low toxicity.
Radiation Hazard: Section 5. For permissible levels, see Table 5, p. 150.
    Artificial isotope [142]Pr, half life 19.3 h. Decays to stable [142]Nd by emitting beta particles of 0.58 (4%), 2.15 (96%) MeV. Also emits gamma rays of 1.57 MeV.
    Artificial isotope [143]Pr, half life 13.8 d. Decays to stable [143]Nd by emitting beta particles of 0.93 MeV.
    Artificial isotope [144]Pr, half life 17 m. Decays to radioactive [144]Nd by emitting beta particles of 2.98 MeV. [144]Pr exists in radioactive equilibrium with its parent [144]Ce. Permissible levels are given for the equilibrium mixture.

Fire Hazard: Moderate, in the form of dust, when exposed to heat or flame or by chemical reaction. Fine dust ignites readily. See also powdered metals.

PRASEODYMIUM BROMATE
General Information
Description: Hexagonal, green crystals.
Formula: $Pr(BrO_3)_3 \cdot 9H_2O$.
Constants: Mol wt: 686.81, mp: 56.5°C, bp: $-7H_2O$ at 100°C.
Hazard Analysis and Countermeasures
See bromates. As a lanthanon it may depress coagulation of the blood. See also lanthanum.

PRASEODYMIUM OXALATE
General Information
Description: Light green crystals.
Formula: $Pr_2(C_2O_4)_3 \cdot 10H_2O$.
Constant: Mol wt: 726.06.
Hazard Analysis and Countermeasures
See oxalates. As a lanthanon it may depress coagulation of the blood. See also lanthanum.

PRASEODYMIUM SELENATE
General Information
Formula: $Pr_2(SeO_4)_3$.
Constants: Mol wt: 710.72, d: 4.30 at 15°C.
Hazard Analysis and Countermeasures
See selenium compounds. As a lanthanon it may depress coagulation of the blood. See also lanthanum.

PRASEODYMIUM SULFIDE
General Information
Description: Brown powder.
Formula: Pr_2S_3.
Constants: Mol wt: 378.04, mp: decomposes, d: 5.042 at 11°C.
Hazard Analysis and Countermeasures
See sulfides. As a lanthanon it may depress coagulation of the blood. See also lanthanum.

PREDNISOLONE
General Information
Synonym: $\Delta^{1,4}$-Pregnadione-11 β, 17 α, 21-triol-3,20-dione.
Description: White crystalline powder; odorless; very slightly soluble in water; soluble in alcohol, chloroform, acetone, methanol, and deoxane.
Formula: $C_{21}H_{28}O_5$.
Constants: Mol wt: 360, mp: 235°C with some decomposition.
Hazard Analysis
Toxicity: Unknown. A food additive permitted in food for human consumption (Section 10).

PREDNISONE
General Information
Synonym: $\Delta^{1,4}$-Pregnadiene-17α, 21-diol-3,11,20-trione.
Description: White, odorless, crystalline powder; very slightly soluble in water; slightly soluble in alcohol, chloroform, methanol, and dioxane.
Formula: $C_{21}H_{26}O_5$.
Constants: Mol wt: 358, mp: 225°C with some decomposition.
Hazard Analysis
Toxicity: Unknown. A food additive permitted in food for human consumption (Section 10).

PRENDEROL. See 2,2-diethyl-1,3-propanediol.

PRIMERS. See explosives, high.
Shipping Regulations: Section 11.
   Coast Guard Classification: Explosive C.

PROCAINE BORATE. See borocaine.

PROCAINE HYDROCHLORIDE
General Information
Description: Crystals.
Formula: $C_{13}H_{20}N_2O_2 \cdot HCl$.
Constants: Mol wt: 272.8, mp: 153–156°C, d: 0.707 at 17°C.
Hazard Analysis
Toxic Hazard Rating:
   Acute Local: Allergen 1.
   Acute Systemic: Ingestion 2; Inhalation 2.
   Chronic Local: Allergen 1.
   Chronic Systemic: U.
Disaster Hazard: Dangerous; when heated to decomposition, it emits highly toxic fumes of hydrochloric acid.
Countermeasures
Ventilation Control: Section 2.
Personal Hygiene: Section 3.
Storage and Handling: Section 7.

PROCAINE NITRATE
General Information
Description: Crystals.
Formula: $C_{13}H_{20}N_2O_2 \cdot HNO_3$.
Constants: Mol wt: 299.3, mp: 101°C.
Hazard Analysis and Countermeasures
See procaine hydrochloride and nitrates.

PROCAINE PENICILLIN
General Information
Description: White, fine crystals or powder; odorless; sparingly soluble in water; slightly soluble in alcohol; fairly soluble in chloroform; an antibiotic.
Formula: $C_{16}H_{18}N_2O_4S \cdot C_{13}H_2N_2O_2 \cdot H_2O$.
Constant: Mol wt: 788.
Hazard Analysis
Toxicity: Unknown. A food additive permitted in the feed and drinking water of animals and/or for the treatment of food producing animals (Section 10).
Disaster Hazard: Dangerous: When heated to decomposition it emits highly toxic fumes.

PRODUCER GAS
General Information
Constants: lel: 20–30%, uel: 70–80%.
Hazard Analysis
Toxicity: Highly toxic. See carbon monoxide, methane and hydrogen. An asphyxiant.
Fire Hazard: Dangerous, when exposed to flame.
Explosive Range: 20.7–73.7%.
Disaster Hazard: Dangerous; can react vigorously with oxidizing materials.
Countermeasures
Ventilation Control: Section 2.
To Fight Fire: Carbon dioxide, dry chemical or water spray (Section 6).
Storage and Handling: Section 7.

PROGESTERONE
General Information
Synonym: $\triangle$-4-Pregnene-3,20-dione.
Description: A female sex hormone. White, crystalline powder; odorless; practically insoluble in water; soluble

---

TOXIC HAZARD RATING CODE (For detailed discussion, see Section 1.)

0 NONE: (a) No harm under any conditions; (b) Harmful only under unusual conditions or overwhelming dosage.

1 SLIGHT: Causes readily reversible changes which disappear after end of exposure.

2 MODERATE: May involve both irreversible and reversible changes; not severe enough to cause death or permanent injury.

3 HIGH: May cause death or permanent injury after very short exposure to small quantities.

U UNKNOWN: No information on humans considered valid by authors.

in alcohol, acetone, and dioxane; sparingly soluble in vegetable oils.

Formula: $C_{21}H_{30}O_2$.

Constant: Mol wt: 314.

Hazard Analysis

Toxicity: Unknown. A food additive permitted in the feed and drinking water of animals and/or for the treatment of food producing animals. Also permitted in food for human consumption (Section 10).

PROLAN. See 1,1-bis(p-chlorophenyl)-2-nitropropane.

PROLINE

General Information

Synonym: 2-Pyrrolidinecarboxylic acid.

Description: A non-essential amino acid; colorless crystals, soluble in water and alcohol; insoluble in ether; occurs in isomeric forms, we consider the L and DL forms.

Formula: C_4H_8NCOOH.

Constants: Mol wt: 115, mp (L): 215–220°C (with decomposition); mp (DL): 220–222°C (with decomposition).

Hazard Analysis

Toxicity: Unknown. A nutrient and/or dietary supplement food additive (Section 10).

PROMAZINE HYDROCHLORIDE

General Information

Synonym: 10-(3-Dimethylamino propyl)-phenotheozine hydrochloride.

Description: White to slightly yellow; practically odorless crystalline powder.

Formula: $C_{17}H_{20}N_2S \cdot HCl$.

Constant: Mol wt: 321.

Hazard Analysis

Toxicity: Unknown. A food additive permitted in food for human consumption; also permitted in the feed and drinking water of animals and/or for the treatment of food producing animals. (Section 10).

Disaster Hazard: See chlorides.

PROMIZOLE. See 2-amino-5-sulfanilylthiazole.

PROMETHIUM

Hazard Analysis

Radiation Hazard: Section 5. For permissible levels, see Table 5, p. 150.

Artificial isotope ^{147}Pm, half life 2.6 y. Decays to radioactive ^{147}Sm by emitting beta particles of 0.22 MeV.

Artificial isotope ^{149}Pm, half life 53 h. Decays to radioactive ^{149}Sm by emitting beta particles of 1.07 (97%), 0.78 (3%) MeV. Also emits gamma rays of 0.29 MeV.

PROPANE

General Information

Synonym: Dimethylmethane.

Description: Colorless gas.

Formula: $CH_3CH_2CH_3$.

Constants: Mol wt: 44.09, bp: −42.1°C, lel: 2.3% uel: 9.5%, fp: −187.1°C, flash p: −156°F, d: 0.5852 at −44.5°/4°C, autoign. temp.: 874°F, vap. d.: 1.56.

Hazard Analysis

Toxic Hazard Rating:

Acute Local: 0.

Acute Systemic: Inhalation 1.

Chronic Local: 0.

Chronic Systemic: U.

TLV: ACGIH (recommended); 1000 parts per million of air, 1800 milligrams per cubic meter of air.

Toxicity: A general purpose food additive (Section 10).

Fire Hazard: Highly dangerous when exposed to heat or flame.

Spontaneous Heating: No.

Explosion Hazard: Severe, when exposed to flame.

Disaster Hazard: Dangerous; can react vigorously with oxidizing materials.

Countermeasures

Ventilation Control: Section 2.

To Fight Fire: Carbon dioxide, dry chemical or water spray (Section 6).

Storage and Handling: Section 7.

1,3-PROPANEDIAMINE

General Information

Description: Water white liquid; amine odor; completely soluble in water, methanol, and ether.

Formula: $NH_2(CH_2)_3NH_2$.

Constants: Mol wt: 74, d: 0.8881 at 20/20°C, bp: 139.7°C, fp: −12°C, flash p: 120°F (T.O.C.).

Hazard Analysis

Toxicity: Details unknown. Limited animal experiments suggest high toxicity and irritation. See also amines.

Fire Hazard: Moderate, when exposed to heat or flame.

1,2-PROPANEDIOL. See propylene glycol.

PROPADIENE. See allene.

1,3-PROPANEDIOL. See trimethylene glycol.

1,3-PROPANEDITHIOL

General Information

Synonym: 1,3-Dimercaptopropane.

Description: Oil with disagreeable odor. Slightly water soluble.

Formula: $HSCH_2CH_2CH_2SH$.

Constants: Mol wt: 108.2, d: 1.0772 at 20/4°C, bp: 170°C.

Hazard Analysis

Toxicity: Details unknown. See related compound 2,3-dimercapto-1-propanol.

Disaster Hazard: Dangerous; when heated to decomposition it emits highly toxic fumes.

PROPANENITRILE. See ethyl cyanide.

1,2,3-PROPANETRIOL. See glycerine.

PROPANOIC ANHYDRIDE. See propionic anhydride.

PROPANOL. See propyl alcohol.

2-PROPANOL-1,N,N-DIBUTYLAMINE. See dibutyl isopropanolamine.

2-PROPANOL NITRITE. See isopropyl nitrite.

4-PROPANOL PYRIDINE

General Information

Description: Crystals.

Formula: $(CH_2CH_2CH_2OH)C_5H_4N$.

Constants: Mol wt: 137.1, bp: 289°C, fp: 36.7°C, d: 1.053 at 40°C.

Hazard Analysis and Countermeasures

See pyridine.

PROPANONE. See acetone.

PROPANOYL CHLORIDE. See propionyl chloride.

PROPARGYL ALCOHOL

General Information

Synonym: 2-Propyn-1-ol.

Description: A moderately volatile liquid; geranium-like odor.

Formula: $HC \equiv CCH_2OH$.

Constants: Mol wt: 56.1, mp: −17°C, bp: 115°C, flash p: 97°F (O.C.), d: 0.9715 at 20°/4°C, vap. press.: 11.6 mm at 20°C, vap. d.: 1.93.

Hazard Analysis

Toxic Hazard Rating:

Acute Local: Irritant 3; Ingestion 3; Inhalation 3.

Acute Systemic: U.

Chronic Local: U.
Chronic Systemic: U.
Toxicology: Based upon animal data.
Fire Hazard: Moderate, when exposed to heat or flame; can react with oxidizing materials.
Countermeasures
To Fight Fire: Foam, carbon dioxide, dry chemical or carbon tetrachloride (Section 6).
Ventilation Control: Section 2.
Personnel Protection: Section 3.
Storage and Handling: Section 7.

PROPARGYL BROMIDE
General Information
Description: An almost colorless liquid; sharp odor.
Formula: $HC \equiv CCH_2Br$.
Constants: Mol wt: 118.97, bp: 88–90°C, fp: −61.07°C, flash p: 65°F (C.O.C.), d: 1.564–1.570, vap. d.: 6.87.
Hazard Analysis
Toxic Hazard Rating:
Acute Local: Irritant 2; Ingestion 3; Inhalation 3.
Acute Systemic: Ingestion 2; Inhalation 3.
Chronic Local: U.
Chronic Systemic: U.
Fire Hazard: Dangerous, when exposed to heat or flame.
Explosion Hazard: Unknown.
Disaster Hazard: Dangerous; when heated to decomposition, it emits highly toxic fumes of bromides; can react vigorously with oxidizing materials,
Countermeasures
Ventilation Control: Section 2.
To Fight Fire: Water, foam, carbon dioxide, dry chemical or carbon tetrachloride (Section 6).
Personnel Protection: Section 3.
First Aid: Section 1.
Storage and Handling: Section 7.

PROPARGYL CHLORIDE
General Information
Synonym: 3-Chloro-1-propyne.
Description: Liquid. Insoluble in water, soluble in organic solvents.
Formula: $HC \equiv CCH_2Cl$.
Constants: Mol wt: 74.5, mp: −78°C, bp: 57°C, flash p: < 60°F (C.C.), d: 1.03 at 25/4°C.
Hazard Analysis
Toxicity: Details unknown. See propargyl bromide.
Fire Hazard: Dangerous when exposed to heat, sparks or open flame (Section 6).
Disaster Hazard: See chlorides.

PROPELLANT DEVICES, TOY
Shipping Regulations: Section 11.
IATA: Explosive, explosive label, 25 kilograms (passenger), 70 kilograms (cargo).

PROPELLANT EXPLOSIVES
Shipping Regulations: Section 11.
IATA: Explosive, not acceptable (passenger and cargo).
I.C.C.: (class A): Explosive A (Section 73.86).
(liquid-class B): Explosive B, Red # label, 10 pounds.
(solid-class B): Explosive B, Red # label, 10 pounds.

PROPELLANT EXPLOSIVES IN WATER (SMOKELESS POWDER FOR CANNON OR SMALL ARMS).

Shipping Regulations: Section 11.
I.C.C.: Explosive B, not accepted.
IATA: Explosive, not acceptable (passenger and cargo).

PROPELLANT 115. See chloropentafluoroethane.

PROPENAL. See acrolein.

PROPENE. See propylene.

PROPENE ACID. See acrylic acid.

1-PROPENE-2-CHLORO-1,3-DIOL DIACETATE
Hazard Analysis
Toxic Hazard Rating:
Acute Local: U.
Acute Systemic: Ingestion 3; Inhalation 3.
Chronic Local: U.
Chronic Systemic: U.
Toxicology: Based on animal data.
Disaster Hazard: Dangerous; when heated to decomposition, it emits highly toxic fumes of chlorides.
Countermeasures
Ventilation Control: Section 2.
Personal Hygiene: Section 3.
Storage and Handling: Section 7.

1-PROPENE-1,3-DIOL DIACETATE
Hazard Analysis
Toxic Hazard Rating (animal data):
Acute Local: U.
Acute Systemic: Ingestion 3; Inhalation 3.
Chronic Local: U.
Chronic Systemic: U.
Fire Hazard: Slight, when heated, it emits acrid fumes (Section 6).
Countermeasures
Ventilation Control: Section 2.
Personal Hygiene: Section 3.
Storage and Handling: Section 7.

PROPENE NITRILE. See acrylonitrile.

PROPENE OXIDE. See propylene oxide.

1,2,3-PROPENETRICARBOXYLIC ACID. See aconitic acid.

2-PROPEN-1-OL. See allyl alcohol.

PROPENYL ACETATE
General Information
Synonym: 1-Methyl vinyl acetate.
Description: Slightly soluble in water.
Formula: $CH_3COOC(CH_3):CH_2$.
Constants: Mol wt: 100, d: 0.9, vap. d.: 3.5, bp: 207°F, flash p: 60°F.
Hazard Analysis
Toxic Hazard Rating: U.
Fire Hazard: Dangerous, when exposed to heat or flame.
Countermeasures
Storage and Handling: Section 7.
To Fight Fire: Alcohol foam.

2-PROPENYLAMINE. See allylamine.

p-PROPENYL ANISOLE. See anethole.

1-PROPENYL-2-BUTENE-1-YL ETHER
General Information
Formula: $CH_3CH:CHCH_2OCH:CHCH_3$.

TOXIC HAZARD RATING CODE (For detailed discussion, see Section 1.)

0 NONE: (a) No harm under any conditions; (b) Harmful only under unusual conditions or overwhelming dosage.

1 SLIGHT: Causes readily reversible changes which disappear after end of exposure.

2 MODERATE: May involve both irreversible and reversible changes; not severe enough to cause death or permanent injury.

3 HIGH: May cause death or permanent injury after very short exposure to small quantities.

U UNKNOWN: No information on humans considered valid by authors.

Constant: Mol wt: 112.
Hazard Analysis
Toxicity: Details unknown. Limited animal experiments suggest low toxicity. See also ethers.

2-PROPENYL ETHANOATE. See allyl acetate.

PROPENYL ETHYL ETHER
General Information
Formula: $CH_3CH:CHOCH_2CH_3$.
Constants: Mol wt: 86, d: 0.8, vap. d.: 1.3, bp: 158° F, flash p: < 20° F (O.C.).
Hazard Analysis
Toxic Hazard Rating: U.
Fire Hazard: Dangerous, when exposed to heat or flame.
Countermeasures
Storage and Handling: Section 7.
To Fight Fire: Foam, alcohol foam, mist.

2-PROPENYL METHANOATE. See allyl formate.

β-PROPIOLACTONE
General Information
Formula: $C_3H_4O_2$.
Constants: Mol wt: 72.1, mp: −33.4° C, bp: 155° C (with rapid decomp.), flash p: 165° F, d: 1.1460 at 20° /4° C, lel: 2.9%.
Hazard Analysis
Toxic Hazard Rating:
  Acute Local: Irritant 3.
  Acute Systemic: Ingestion 3; Inhalation 3.
  Chronic Local: Irritant 3.
  Chronic Systemic: U.
Toxicity: A strong irritant, it has produced skin cancer in experimental animals. It is considered to be the most toxic of the lactones.
TLV: Because of high acute toxicity and demonstrated skin tumor production in animals, contact by any route should be avoided.
Fire Hazard: Moderate, when exposed to heat or flame; can react with oxidizing materials.
Countermeasures
To Fight Fire: Water, foam, carbon dioxide, dry chemical or carbon tetrachloride (Section 6).
Ventilation Control: Section 2.
Personal Hygiene: Section 3.
Storage and Handling: Section 7.

PROPIONALDEHYDE. See propyl aldehyde.

PROPIONAMIDE NITRILE. See cyanoacetamide.

PROPIONE. See diethyl ketone.

PROPIONIC ACID
General Information
Synonym: Methylacetic acid.
Description: Colorless liquid.
Formula: CH_3CH_2COOH.
Constants: Mol wt: 74.1, mp: −22° C, bp: 141° C, d: 0.992, vap. press.: 10 mm at 39.7° C, vap. d.: 2.56., flash p: 130° F.
Hazard Analysis
Toxic Hazard Rating:
  Acute Local: Irritant 2.
  Acute Systemic: Ingestion 1.
  Chronic Local: Irritant 1.
  Chronic Systemic: Ingestion 1.
Toxicology: Data based on animal experiments show low toxicity. A chemical preservative food additive (Section 10). A substance which migrates to food from packaging materials.
Disaster Hazard: Slight; when heated, it emits acrid fumes (Section 6).
Fire Hazard: Moderate, when exposed to heat or flame.

Countermeasures
Storage and Handling: Section 7.
To Fight Fire: Alcohol foam.
Shipping Regulations: Section 11.
  IATA: Other restricted articles, class B; no label required, no limit (passenger and cargo).

PROPIONIC ANHYDRIDE
General Information
Synonym: Propanoic anhydride.
Formula: $(CH_3CH_2CO)_2O$.
Constants: Mol wt: 130.14, mp: −45° C, bp: 167.0° C, flash p: 165° F (O.C.), d: 1.012, vap. press.: 1 mm at 20.6° C, vap. d.: 4.49.
Hazard Analysis
Toxicity: Details unknown. Limited animal experiments suggest low systemic toxicity.
Fire Hazard: Moderate, when exposed to heat or flame; can react with oxidizing materials.
Countermeasures
To Fight Fire: Carbon dioxide, dry chemical or carbon tetrachloride (Section 6).
Storage and Handling: Section 7.

PROPIONIC ETHER. See ethyl propionate.

PROPIONITRILE. See ethyl cyanide.

PROPIONYL CHLORIDE
General Information
Synonym: Propanoyl chloride.
Description: Colorless liquid.
Formula: CH_3CH_2COCl.
Constants: Mol wt: 92.5, mp: −94° C, bp: 80° C, flash p: 53° F, d: 1.065, vap. d.: 3.2.
Hazard Analysis
Toxicity: Highly toxic. See hydrochloric acid.
Fire Hazard: Dangerous, when exposed to heat or flame.
Explosion Hazard: Unknown.
Disaster Hazard: Dangerous; when heated to decomposition, it emits highly toxic fumes of chlorides; will react with water or steam to produce toxic and corrosive fumes; can react vigorously with oxidizing materials.
Countermeasures
Ventilation Control: Section 2.
To Fight Fire: Carbon dioxide, dry chemical or carbon tetrachloride (Section 6).
Storage and Handling: Section 7.

PROPIOPHENONE
General Information
Synonym: Ethyl phenyl ketone.
Description: Water-white to light amber liquid.
Formula: $C_9H_{10}O$.
Constants: Mol wt: 134.2, mp: 21° C, bp: 218.0° C, flash p: 210° F, d: 1.012 at 20° /20° C, vap. press.: 1 mm at 50.0° C.
Hazard Analysis
Toxicity: Details unknown. See also ketones.
Fire Hazard: Slight when exposed to heat or flame; can react with oxidizing materials.
Countermeasures
To Fight Fire: Foam, carbon dioxide, dry chemical or carbon tetrachloride (Section 6).
Storage and Handling: Section 7.

1-PROPOXY-2-AMINO-4-NITROBENZENE. See 5-nitro-2-N-propoxyaniline.

n-PROPYL ACETATE
General Information
Description: Clear, colorless liquid; pleasant odor.
Formula: $CH_3COOC_3H_7$.
Constants: Mol wt: 102.13, mp: −92.5° C, bp: 101.6° C, flash p: 58° F, lel: 2.0%; uel: 8.0%, d: 0.887, autoign.

temp.: 842°F, vap. press.: 40 mm at 28.8°C, vap. d.: 3.52.

Hazard Analysis

Toxic Hazard Rating:

Acute Local: Irritant 1.

Acute Systemic: Ingestion 2; Inhalation 2; Skin Absorption 1.

Chronic Local: U.

Chronic Systemic: Inhalation 1.

TLV: ACGIH (recommended), 200 parts per million in air; 834 milligrams per cubic meter of air.

Toxicology: This material causes narcosis and is somewhat irritating. However, it is not likely to cause chronic poisoning, since there is definite evidence of habituation to this material. The after-effects are slight and recovery quick from even deep narcosis. The symptoms noted are sleepiness, fatigue, slight stupefaction and retarded respiration. Repeated or prolonged inhalations of high concentrations have been shown to produce irritation and narcosis and in certain cases death, although no industrial injury has been reported as occurring to workmen exposed to it. Isopropyl acetate has been shown to have slightly less narcotic potency than normal propyl acetate.

Fire Hazard: Dangerous, when exposed to heat or flame.

Spontaneous Heating: No.

Explosion Hazard: Moderate, when exposed to heat or flame.

Disaster Hazard: Dangerous, upon exposure to heat or flame; can react vigorously with oxidizing materials.

Countermeasures

Ventilation Control: Section 2.

Treatment and Antidotes: Personnel who show the symptoms of irritation or narcosis from exposure to this material should immediately be removed to fresh air. Recovery is quick and complete. If the exposure has been very severe, consult a physician.

Personal Hygiene: Section 3.

To Fight Fire: Foam, carbon dioxide, dry chemical or carbon tetrachloride (Section 6).

Storage and Handling: Section 7.

Shipping Regulations: Section 11.

IATA: Flammable liquid, red label, 1 liter (passenger), 40 liters (cargo).

n-PROPYL ALCOHOL

General Information

Synonym: 1-Propanol; ethyl carbinol.

Description: Clear, odorless liquid; alcohol-like odor.

Constants: Mol wt: 60.1, mp: −127°C, bp: 97.19°C, flash p: 59°F (C.C.), ulc: 55–60, d: 0.8044 at 20°/4°C, lel: 2.1%, uel: 13.5%, autoign. temp.: 700°F, vap. press.: 10 mm at 14.7°C, vap. d.: 2.07.

Hazard Analysis

Toxic Hazard Rating:

Acute Local: Irritant 1; Ingestion 1; Inhalation 1.

Acute Systemic: Inhalation 1.

Chronic Local: U.

Chronic Systemic: Inhalation 1.

TLV: ACGIH (tentative); 200 parts per million; 510 milligrams per cubic meter of air.

Fire Hazard: Dangerous, when exposed to heat or flame.

Spontaneous Heating: No.

Explosion Hazard: Moderate, when exposed to flame.

Disaster Hazard: Dangerous, upon exposure to heat or flame; can react vigorously with oxidizing materials.

Countermeasures

Personnel Protection: Section 3.

Ventilation Control: Section 2.

To Fight Fire: Alcohol foam, carbon dioxide, dry chemical or carbon tetrachloride (Section 6).

Storage and Handling: Section 7.

sec-PROPYL ALCOHOL. See isopropyl alcohol.

PROPYL ALDEHYDE

General Information

Synonyms: Propylic aldehyde; propionaldehyde.

Description: Colorless liquid; suffocating odor.

Formula: CH_3CH_2CHO.

Constants: Mol wt: 58.1, mp: −81°C, bp: 48°C, flash p: 15–19°F (O.C.), d: 0.807 at 20°/4°C, lel: 3.7%, uel: 16.1%, vap. d.: 2.0.

Hazard Analysis

Toxic Hazard Rating:

Acute Local: Irritant 2; Ingestion 2; Inhalation 2.

Acute Systemic: Ingestion 2, Inhalation 2.

Chronic Local: Irritant 1.

Chronic Systemic: Ingestion 2; Inhalation 2.

Toxicity: See also aldehydes.

Fire Hazard: Dangerous, when exposed to heat or flame.

Disaster Hazard: Dangerous fire hazard: reacts vigorously with oxidizers.

Countermeasures

To Fight Fire: Foam, carbon dioxide, dry chemical or carbon tetrachloride (Section 6).

Ventilation Control: Section 2.

Personnel Protection: Section 3.

Storage and Handling: Section 7.

Shipping Regulations: Section 11.

IATA: Flammable liquid, red label, 1 liter (passenger), 40 liters (cargo).

PROPYLAMINE

General Information

Synonym: 1-Amino-propane.

Description: Colorless, alkaline liquid; strong ammonia odor; miscible with water, alcohol, ether.

Formula: $CH_3CH_2CH_2NH_2$.

Constants: Mol wt: 59.11, d: 0.7191 at 20/20°C, mp: −83°C, bp: 48–49°C, vap. press.: 248 mm at 20°C, flash p: −35°F, autoign. temp.: 604°F, lel: 2.0%, uel: 10.4%.

Hazard Analysis

Toxic Hazard Rating:

Acute Local: Irritant 3; Ingestion 3; Inhalation 3.

Acute Systemic: Ingestion 3.

Chronic Local: Allergen 1.

Chronic Systemic: U.

Toxicity: A strong irritant and possibly a skin sensitizer. See also amines.

Fire Hazard: Dangerous, when exposed to heat or flame.

Countermeasures

Storage and Handling: Section 7.

To Fight Fire: Alcohol foam.

Shipping Regulations: Section 11.

IATA: Flammable liquid, red label, 1 liter (passenger), 40 liters (cargo).

n-PROPYLBENZENE

General Information

Synonym: 1-Phenylpropane.

Description: Clear liquid.

Formula: $C_3H_7C_6H_5$.

TOXIC HAZARD RATING CODE (For detailed discussion, see Section 1.)

0 NONE: (a) No harm under any conditions; (b) Harmful only under unusual conditions or overwhelming dosage.

1 SLIGHT: Causes readily reversible changes which disappear after end of exposure.

2 MODERATE: May involve both irreversible and reversible changes; not severe enough to cause death or permanent injury.

3 HIGH: May cause death or permanent injury after very short exposure to small quantities.

U UNKNOWN: No information on humans considered valid by authors.

Constants: Mol wt: 120.2, mp: $-99.5°C$, bp: $159.2°C$, flash p: $86°F$ (C.C.), d: 0.862, vap. press.: 10 mm at $43.4°C$, vap. d.: 4.14.

Hazard Analysis

Toxicity: Details unknown. Limited animal experiments show moderately acute vapor toxicity.

Fire Hazard: Moderate, when exposed to heat or flame; can react with oxidizing materials.

Spontaneous Heating: No.

Explosion Hazard: Unknown.

Countermeasures

Ventilation Control: Section 2.

Storage and Handling: Section 7.

To Fight Fire: Foam, carbon dioxide, dry chemical or carbon tetrachloride (Section 6).

PROPYL BORATE. See tri-n-propyl borate.

PROPYL BROMIDE

General Information

Synonym: 1-Bromopropane.

Description: Liquid.

Formula: $CH_3CH_2CH_2Br$.

Constants: Mol wt: 123.00, mp: $-110°C$, bp: $70.9°C$, d: 1.353 at $20°/4°C$.

Hazard Analysis

Toxic Hazard Rating:

Acute Local: Irritant 3; Ingestion 3; Inhalation 3.

Acute Systemic: Ingestion 3; Inhalation 3; Skin Absorption 2.

Chronic Local: U.

Chronic Systemic: U.

Fire Hazard: Moderate, when heated or exposed to flame.

Disaster Hazard: Dangerous; on decomposition, it emits highly toxic fumes of bromides; can react with oxidizing materials.

Countermeasures

Ventilation Control: Section 2.

To Fight Fire: Water, foam, carbon dioxide, dry chemical or carbon tetrachloride (Section 6).

Personnel Protection: Section 3.

First Aid: Section 1.

Storage and Handling: Section 7.

S-PROPYL BUTYLETHYL THIOCARBAMATE

Hazard Analysis

Toxicology: Causes violent vomiting when accompanied by alcohol ingestion.

Disaster Hazard: Dangerous. When heated to decomposition it emits highly toxic fumes.

Storage and Handling: Section 7.

PROPYL BUTYRATE

General Information

Description: Colorless liquid. Slightly soluble in water.

Formula: $CH_3CH_2CH_2CO_2C_3H_7$.

Constants: Mol wt: 130.2, d: 0.879 at $15/4°C$, mp: $-95°C$, bp: $143°C$.

Hazard Analysis

Toxicity: Details unknown. Irritant to mucous membrane. Narcotic in high concentrations.

PROPYL "CELLOSOLVE"

Hazard Analysis

Toxicity: Details unknown. See also glycols.

Fire Hazard: Moderate; can react with oxidizing materials (Section 6).

Countermeasures

Storage and Handling: Section 7.

n-PROPYL CHLORIDE

General Information

Synonym: 1-Chloropropane.

Description: Colorless liquid; chloroform-like odor.

Formula: $CH_3CH_2CH_2Cl$.

Constants: Mol wt: 78.54, mp: $-122.8°C$, bp: $47.2°C$, lel: 2.6%, uel: 11.0%, flash p: $< 0°F$, d: 0.890, vap. d.: 2.71.

Hazard Analysis

Toxic Hazard Rating:

Acute Local: Irritant 1.

Acute Systemic: Inhalation 2.

Chronic Local: U.

Chronic Systemic: U.

Toxicology: Irritant to mucous membrane. Narcotic in high concentration. See also chlorinated hydrocarbons, aliphatic.

Fire Hazard: Dangerous, when exposed to heat or flame.

Explosion Hazard: Moderate, when exposed to flame.

Disaster Hazard: Dangerous. Keep away from heat and open flame; can react vigorously with oxidizing materials.

Countermeasures

Ventilation Control: Section 2.

To Fight Fire: Carbon dioxide, dry chemical or carbon tetrachloride (Section 6).

Personal Hygiene: Section 3.

Storage and Handling: Section 7.

Shipping Regulations: Section 11.

IATA: Flammable liquid, red label, 1 liter (passenger), 40 liters (cargo).

PROPYL CHLOROSULFONATE

General Information

Description: Liquid.

Formula: $CH_3CH_2CH_2OSO_2Cl$.

Constants: Mol wt: 158.61, bp: $70-72°C$ at 20 mm.

Hazard Analysis

Toxic Hazard Rating:

Acute Local: Irritant 3; Ingestion 3; Inhalation 3.

Acute Systemic: U.

Chronic Local: U.

Chronic Systemic: U.

Disaster Hazard: Dangerous; when heated to decomposition, it emits highly toxic fumes of chlorides and oxides of sulfur. A military poison.

Countermeasures

Ventilation Control: Section 2.

Personnel Protection: Section 3.

Storage and Handling: Section 7.

n-PROPYL CHLOROTHIOLFORMATE

General Information

Description: Insoluble in water.

Formula: C_3H_7SCOCl.

Constants: Mol wt: 122.5, d: 1.1, vap. d.: 4.8, bp: $311°F$, flash p: $145°F$.

Hazard Analysis

Toxic Hazard Rating: U.

Fire Hazard: Moderate when exposed to heat or flame.

Disaster Hazard: Dangerous. When heated to decomposition it emits toxic fumes.

Countermeasures

Storage and Handling: Section 7.

To Fight Fire: Water, foam carbon dioxide and dry chemical.

n-PROPYL CYANIDE. See n-butyronitrile.

PROPYLENE

General Information

Synonym: Propene.

Description: A gas.

Formula: C_3H_6.

Constants: Mol wt: 42.1, mp: $-185°C$, bp: $-47.7°C$, d (liquid): 0.581 at $0°C$, autoign. temp.: $927°F$, vap. press.: 10 atm. at $19.8°C$, lel: 2.0%, uel: 11.1%, vap. d.: 1.5.

Hazard Analysis

Toxic Hazard Rating:

Acute Local: 0.

Acute Systemic: Inhalation 2.
Chronic Local: 0.
Chronic Systemic: 0.
Toxicology: A simple asphyxiant. No irritating effects from high concentrations in gaseous form. When compressed to liquid form it can cause skin burns from refrigerating effects on tissue of rapid evaporation. For effects of simple asphyxiants see argon.
Fire Hazard: Dangerous, when exposed to heat or flame.
Spontaneous Heating: No.
Explosion Hazard: Moderate, when exposed to heat or flame. Under unusual conditions, i.e., 955 atmospheres pressure and 327°C, it has been known to explode.
Disaster Hazard: Dangerous; can react vigorously with oxidizing materials.
Countermeasures
Ventilation Control: Section 2.
To Fight Fire: Carbon dioxide, dry chemical or water spray (Section 6).
Storage and Handling: Section 7.
Shipping Regulations: Section 11.
  I.C.C.: See liquefied petroleum gas.
  Coast Guard Classification: Inflammable gas; red gas label.
  MCA warning label.
  IATA: Flammable gas, red label, not acceptable (passenger), 140 kilograms (cargo).

PROPYLENE CARBONATE
General Information
Description: A clear liquid.
Formula: $CH_3CHCH_2CO_3$.
Constants: Mol wt: 102.09, bp: 242.1°C, fp: −48.8°C, flash p: 275°F (O.C.), d: 1.2069 at 20°/20°C, vap. press.: 0.03 mm at 20°C.
Hazard Analysis
Toxicity: Unknown. See also esters.
Fire Hazard: Slight; when exposed to heat or flame.
Disaster Hazard: Can react with oxidizing materials.
Countermeasures
Storage and Handling: Section 7.
To Fight Fire: Water, foam, carbon dioxide, dry chemical or carbon tetrachloride (Section 6).

PROPYLENE CHLOROBROMIDE
General Information
Description: Colorless liquid.
Formula: $CH_3CHBrCH_2Cl$.
Constants: Mol wt: 157.5, bp: 117–118.5°C, fp: < −20°C, d: 1.540 at 25°/25°C, vap. d.: 5.47.
Hazard Analysis
Toxicity: Probably toxic.
Disaster Hazard: Dangerous; when heated to decomposition, it emits highly toxic fumes of chlorides.
Countermeasures
Storage and Handling: Section 7.

PROPYLENE CHLOROHYDRIN
General Information
Synonyms: Chloroisopropyl alcohol; 1-chloro-2-propanol.
Description: Colorless liquid; mild nonresidual odor.
Formula: $CH_2ClCHOHCH_3$.
Constants: Mol wt: 94.54, bp: 127.0°C, flash p: 125°F (C.C.), d: 1.103 at 20°C, vap. d.: 3.26.
Hazard Analysis
Toxic Hazard Rating:
  Acute Local: Irritant 1.

Acute Systemic: Skin Absorption 1.
Chronic Local: U.
Chronic Systemic: Skin Absorption 1.
Fire Hazard: Moderate, when exposed to heat or flame.
Spontaneous Heating: No.
Disaster Hazard: Dangerous; when heated to decomposition, it emits highly toxic fumes of chlorides; can react with oxidizing materials.
Countermeasures
Personal Hygiene: Section 3.
To Fight Fire: Alcohol foam, carbon dioxide, dry chemical or carbon tetrachloride (Section 6).
Storage and Handling: Section 7.
Shipping Regulations: Section 11.
  MCA warning label.

PROPYLENE CHLOROHYDRIN (PRIMARY)
General Information
Synonyms: β-Chloropropyl alcohol; 2-chloro-1-propanol.
Formula: $CH_3CHClCH_2OH$.
Constants: Mol wt: 94.5, flash p: 125°F, d: 1.1, vap. d.: 3.3, bp: 271–273°F.
Hazard Analysis
Toxicity: See propylene chlorohydrin.
Fire Hazard: Moderate, when exposed to heat or flame.
Disaster Hazard: Dangerous. When heated to decomposition it emits highly toxic fumes.
Countermeasures
Storage and Handling: Section 7.
To Fight Fire: Alcohol foam.

PROPYLENEDIAMINE
General Information
Description: Colorless liquid.
Formula: $CH_3CH(NH_2)CH_2NH_2$.
Constants: Mol wt: 74.13, bp: 119–120°C, flash p: 92°F (O.C.), d: 0.87 at 15°C, vap. d.: 2.56.
Hazard Analysis
Toxicity: Corrosive liquid. See amines and fatty acids.
Fire Hazard: Moderate, when exposed to heat or flame.
Disaster Hazard: Dangerous; on heating to decomposition, it emits toxic fumes of nitrogen oxides; can react with oxidizing materials.
Countermeasures
Storage and Handling: Section 7.
To Fight Fire: Alcohol foam, carbon dioxide, dry chemical or carbon tetrachloride (Section 6).

PROPYLENE DIBROMIDE
General Information
Description: Colorless liquid.
Formula: $CH_3CHBrCH_2Br$.
Constants: Mol wt: 201.9, mp: −55°C, bp: 139.6–142.6°C, fp: < −75°C, d: 1.940 at 25°/25°C, vap. d.: 7.0.
Hazard Analysis
Toxicity: Unknown. Probably irritant and narcotic in high concentrations. See also bromides.
Disaster Hazard: Moderately dangerous; when heated to decomposition, it emits toxic fumes of bromides.
Countermeasures
Storage and Handling: Section 7.

PROPYLENE DICHLORIDE. See 1,2-dichloropropane.
Shipping Regulations: Section 11.
  IATA: Flammable liquid, red label, 1 liter (passenger), 40 liters (cargo).

TOXIC HAZARD RATING CODE (For detailed discussion, see Section 1.)

0 NONE: (a) No harm under any conditions; (b) Harmful only under unusual conditions or overwhelming dosage.

1 SLIGHT: Causes readily reversible changes which disappear after end of exposure.

2 MODERATE: May involve both irreversible and reversible changes; not severe enough to cause death or permanent injury.

3 HIGH: May cause death or permanent injury after very short exposure to small quantities.

U UNKNOWN: No information on humans considered valid by authors.

PROPYLENE FLUOROHYDRIN
General Information
Synonym: Fluorisopropyl alcohol.
Description: Colorless liquid.
Formula: $CH_2FCHOHCH_3$.
Constants: Mol wt: 78.1.
Hazard Analysis
Toxic Hazard Rating:
  Acute Local: U.
  Acute Systemic: Ingestion 2; Inhalation 2.
  Chronic Local: U.
  Chronic Systemic: U.
Disaster Hazard: Dangerous; when heated to decomposition; it emits highly toxic fumes of fluorides.
Countermeasures
Ventilation Control: Section 2.
Personal Hygiene: Section 3.
Storage and Handling: Section 7.

PROPYLENE GLYCOL
General Information
Synonyms: 1,2-Propanediol; 1,2-dihydroxypropane.
Description: Colorless liquid; practically odorless.
Formula: $CH_2OHCHOHCH_3$.
Constants: Mol wt: 76.1, bp: 188.2°C, flash p: 210°F (O.C.), lel: 2.6%, uel: 12.6%, d: 1.0362 at 25°/25°C, autoign. temp.: 790°F, vap. press.: 0.08 mm at 20°C, vap. d.: 2.62.
Hazard Analysis
Toxic Hazard Rating:
  Acute Local: Irritant 1; Allergen 1.
  Acute Systemic: Ingestion 1.
  Chronic Local: Irritant 1; Allergen 1.
  Chronic Systemic: Ingestion 1.
Note: Used as an emulsifying agent, and a general purpose food additive. It is a substance which migrates to food from packaging materials (Section 10).
Fire Hazard: Moderate, when exposed to heat or flame; can react with oxidizing materials.
Spontaneous Heating: No.
Explosion Hazard: Moderate, when exposed to flame.
Countermeasures
Personnel Protection: Section 3.
To Fight Fire: Carbon dioxide, dry chemical or carbon tetrachloride (Section 6).
Storage and Handling: Section 7.

1,3-PROPYLENE GLYCOL. See trimethylene glycol.

PROPYLENE GLYCOL ACETATE. See methyl glycol acetate.

PROPYLENE GLYCOL ALGINATE
General Information
Synonym: Hydroxypropyl alginate.
Description: White powder; odorless; soluble in water and dilute organic acids.
Formula: $(C_9H_{14}O_7)_n$.
Hazard Analysis
Toxicity: Unknown. A food additive permitted in food for human consumption (Section 10).

PROPYLENE GLYCOL 4-BIPHENYLYL ETHER
General Information
Synonyms: 1-(4-Biphenylyloxy)-2-propanol.
Description: White, crystalline solid.
Formula: $C_6H_5C_6H_4OCH_2CH(OH)CH_3$.
Constants: Mol wt: 228.3, mp: 121–123°C.
Hazard Analysis
Toxicity: Details unknown. See also glycols.
Fire Hazard: Slight, when exposed to heat or flame; can react with oxidizing materials (Section 6).
Countermeasures
Storage and Handling: Section 7.

PROPYLENE GLYCOL sec-BUTYLPHENYL ETHER
General Information
Synonyms: 1-(o-sec-Butyl phenoxy)-2-propanol.
Description: Slightly yellow liquid.
Formula: $C_2H_5CH(CH_3)C_6H_4OCH_2CH(OH)CH_3$.
Constants: Mol wt: 208.3, mp: < −20°C, bp: 276.8°C, flash p: 270°F, d: 0.992 at 25°/25°C, vap. d.: 7.2.
Hazard Analysis
Toxicity: Details unknown. See also glycols.
Fire Hazard: Slight, when exposed to heat or flame; can react with oxidizing materials.
Explosion Hazard: Details unknown. See also ethers.
Countermeasures
Storage and Handling: Section 7.
To Fight Fire: Carbon dioxide, dry chemical or carbon tetrachloride (Section 6).

PROPYLENE GLYCOL p-tert-BUTYLPHENYL ETHER
General Information
Synonyms: 1-(p-tert-Butylphenoxy)-2-propanol.
Description: White solid.
Formula: $(CH_3)_3CC_6H_4OCH_2CH(OH)CH_3$.
Constants: Mol wt: 208.3, mp: 33°C, bp: 288.8°C, flash p: 290°F, d: 0.979 at 60°/60°C, vap. d.: 7.2.
Hazard Analysis
Toxicity: Details unknown. See also glycols.
Fire Hazard: Slight, when exposed to heat or flame; can react with oxidizing materials.
Countermeasures
To Fight Fire: Foam, carbon dioxide, dry chemical or carbon tetrachloride (Section 6).
Storage and Handling: Section 7.

PROPYLENE GLYCOL CHLOROPHENYL ETHER
General Information
Synonyms: 1-(o-Chlorophenoxy)-2-propanol.
Description: Slightly yellow liquid.
Formula: $ClC_6H_4OCH_2CH(OH)CH_3$.
Constants: Mol wt: 186.6, mp: < −20°C, bp: 272.3°C, flash p: 270°F, d: 1.201 at 25°/25°C, vap. d.: 6.45.
Hazard Analysis
Toxicity: Details unknown. See also glycols.
Fire Hazard: Slight, when exposed to heat or flame.
Disaster Hazard: Dangerous; on heating to decomposition, it emits toxic fumes of chlorides; can react with oxidizing materials.
Countermeasures
Storeage and Handling: Section 7.
To Fight Fire: Water, foam, carbon dioxide, dry chemical or carbon tetrachloride (Section 6).

PROPYLENE GLYCOL DIACETATE
Hazard Analysis
Toxicity: Details unknown. See glycols.
Disaster Hazard: Slight; when heated, it emits acrid fumes (Section 6).
Countermeasures
Personal Hygiene: Section 3.
Storage and Handling: Section 7.

PROPYLENE GLYCOL 2,4-DICHLOROPHENYL ETHER
General Information
Synonyms: 1-(2,4-Dichlorophenoxy)-2-propanol.
Description: Slightly yellow liquid.
Formula: $Cl_2C_6H_3OCH_2CH(OH)CH_3$.
Constants: Mol wt: 221.1, mp: 9.5°C, bp: 297.7°C, flash p: 335°F, d: 1.309 at 25°/25°C, vap. d.: 7.6.
Hazard Analysis
Toxicity: Details unknown. See glycols.
Fire Hazard: Slight, when exposed to heat or flame; can react with oxidizing materials.

Countermeasures
To Fight Fire: Water, foam, carbon dioxide, dry chemical or carbon tetrachloride (Section 6).
Storage and Handling: Section 7.

PROPYLENE GLYCOL METHYL ETHER
General Information
Synonym: Propylene glycol monomethyl ether; methyl ether of propylene glycol; Dowanal; PM; UCAR; solvent LM.
Description: Colorless liquid.
Formula: $CH_3OCH_2CHOHCH_3$.
Constants: Mol wt: 90.1, mp: $-96.7°C$, bp: $120°C$, flash p: $100°F$, d: 0.919 at $25°/25°C$.
Hazard Analysis
Toxic Hazard Rating:
  Acute Local: Irritant 1.
  Acute Systemic: Ingestion 1.
  Chronic Local: U.
  Chronic Systemic: Inhalation 1.
Toxicology: Rating based on extensive animal tests. No cases of human toxicity known. See also ethylene glycol monomethyl ether and glycols.
Fire Hazard: Moderate, when exposed to heat or flame; can react with oxidizing materials.
Countermeasures
To Fight Fire: Foam, carbon dioxide, dry chemical or carbon tetrachloride (Section 6).
Ventilation Control: Section 2.
Storage and Handling: Section 7.

PROPYLENE GLYCOL MONOACRYLATE
General Information
Synonym: Hydroxypropyl acrylate.
Description: Soluble in water.
Formula: $CH_2:CHCOO(C_3H_6)OH$.
Constants: Mol wt: 130; d: 1.0+, vap. d.: 4.5, bp: $171°F$ at 5 mm, flash p: $210°F$.
Hazard Analysis
Toxic Hazard Rating: U.
Fire Hazard: Moderate, when exposed to heat or flame.
Countermeasures
Storage and Handling: Section 7.
To Fight Fire: Alcohol foam.

PROPYLENE GLYCOL MONOETHYL ETHER
General Information
Description: Clear liquid.
Formula: $CH_3CHOHCH_2OC_2H_5$.
Constants: Mol wt: 104.15, bp: $130°C$.
Hazard Analysis
Toxicity: Details unknown. See glycols.
Fire Hazard: Moderate, when exposed to heat or flame; can react with oxidizing materials (Section 6).
Countermeasures
Ventilation Control: Section 2.
Storage and Handling: Section 7.

PROPYLENE GLYCOL MONOMETHYL ETHER. See propylene glycol methyl ether.

PROPYLENE GLYCOL PHENYL ETHER
General Information
Description: Liquid.
Formula: $C_6H_5OCH_2CH(OH)CH_3$.
Constants: Mol wt: 152.2, bp: $240°C$, fp: $13-18°C$, flash p: $275°F$, d: 1.063 at $25°/4°C$, vap. d.: 5.25.

Hazard Analysis
Toxicity: Details unknown. See glycols.
Fire Hazard: Slight, when exposed to heat or flame; can react with oxidizing materials.
Countermeasures
To Fight Fire: Foam, carbon dioxide, dry chemical or carbon tetrachloride (Section 6).
Storage and Handling: Section 7.

PROPYLENE IMINE
General Information
Description: Liquid.
Formula: $NHCH_2CHCH_3$.
Constants: Mol wt: 58.10, vap. d.: 2.0.
Hazard Analysis
Toxic Hazard Rating:
  Acute Local: Irritant 2.
  Acute Systemic: Ingestion 2; Inhalation 2.
  Chronic Local: U.
  Chronic Systemic: U.
TLV: ACGIH (recommended); 2 parts per million of air; 5 milligrams per cubic meter of air. Can be absorbed through the skin.
Fire Hazard: Moderate, when exposed to heat or flame (Section 6).
Disaster Hazard: Dangerous; when heated to decomposition it emits toxic fumes of oxides of nitrogen; can react with oxidizing materials.
Countermeasures
Ventilation Control: Section 2.
Personal Hygiene: Section 3.
Storage and Handling: Section 7.
Shipping Regulations: Section 11.
  ICC (inhibited): Flammable liquid, red label, 5 pints.
  IATA (inhibited): Flammable liquid, red label, not acceptable (passenger), 2½ liter (cargo).
    (uninhibited): Flammable liquid, not acceptable (passenger and cargo).

PROPYLENE OXIDE
General Information
Synonyms: 1,2-Epoxypropane; propene oxide; methyl oxirane.
Description: Colorless liquid; ethereal odor. Soluble in water, alcohol, and ether.
Formula: OCH_2CHCH_3.
Constants: Mol wt: 58.08, bp: $33.9°C$, lel: 2.1%, uel: 21.5%, fp: $-104.4°C$, flash p: $-35°F$ (T.O.C.), d: 0.8304 at $20°/20°C$, vap. press.: 400 mm at $17.8°C$, vap. d.: 2.0.
Hazard Analysis
Toxic Hazard Rating:
  Acute Local: Irritant 2; Ingestion 2; Inhalation 2.
  Acute Systemic: Ingestion 2; Inhalation 2.
  Chronic Local: Irritant 1.
  Chronic Systemic: U.
Note: A food additive permitted in food for human consumption (Section 10).
TLV: ACGIH (recommended); 100 parts per million in air; 240 milligrams per cubic meter of air.
Fire Hazard: Highly dangerous, when exposed to heat or flame.
Spontaneous Heating: No.
Explosion Hazard: Severe, when exposed to flame.
Disaster Hazard: Dangerous; can react vigorously with oxidizing materials. Keep away from heat and open flame!

TOXIC HAZARD RATING CODE (For detailed discussion, see Section 1.)

0 NONE: (a) No harm under any conditions; (b) Harmful only under unusual conditions or overwhelming dosage.

1 SLIGHT: Causes readily reversible changes which disappear after end of exposure.

2 MODERATE: May involve both irreversible and reversible changes; not severe enough to cause death or permanent injury.

3 HIGH: May cause death or permanent injury after very short exposure to small quantities.

U UNKNOWN: No information on humans considered valid by authors.

Countermeasures
Ventilation Control: Section 2.
To Fight Fire: Alcohol foam, carbon dioxide, dry chemical
or carbon tetrachloride (Section 6).
Personal Hygiene: Section 3.
Storage and Handling: Section 7.
Shipping Regulations: Section 11.
  Coast Guard Classification: Inflammable liquid; red
  label.
  ICC: Flammable liquid, red label, 10 gallons.
  IATA: Flammable liquid, red label, 1 liter (passenger,
  40 liter (cargo).

PROPYLENE SULFIDE
General Information
Formula: $(CH_2:CHCH_2)_2S$.
Constant: Mol wt: 114.
Hazard Analysis
Toxic Hazard Rating:
  Acute Local: Irritant 2; Inhalation 3.
  Acute Systemic: Ingestion 3; Inhalation 2.
  Chronic Local: U.
  Chronic Systemic: U.
Toxicology: Data based on animal experiments.
Disaster Hazard: Dangerous. When heated to decomposi-
  tion it emits highly toxic fumes.
Storage and Handling: Section 7.

PROPYL ETHER
General Information
Synonym: Dipropyl ether.
Description: Colorless liquid.
Formula: $(C_3H_7)_2O$.
Constants: Mol wt: 102.2, mp: $-122°C$, bp: $90°C$, d:
  0.736 at $20°/4°C$.
Hazard Analysis
Toxicity: See ethers.
Fire Hazard: Dangerous, when exposed to heat or flame
  (Section 6).
Explosion Hazard: Details unknown. See ethers.
Disaster Hazard: Dangerous, upon exposure to heat or flame;
  can react vigorously with oxidizing materials.
Countermeasures
Ventilation Control: Section 2.
Storage and Handling: Section 7.

PROPYLETHYLENE. See amylene.

n-PROPYL FORMATE
General Information
Synonym: Propyl methanoate.
Description: Colorless liquid; pleasant odor.
Formula: $C_3H_7CHO_2$.
Constants: Mol wt: 88.1, mp: $-93°C$, bp: $82°C$, flash p:
  $27°F$ (C.C.), d: 0.901 at $20°C$, vap. press.: 100 mm
  at $29.5°C$, vap. d.: 3.03, autoign. temp.: $851°F$.
Hazard Analysis
Toxicity: Details unknown. Irritating to mucous mem-
  branes. Narcotic in high concentrations. See esters.
Fire Hazard: Dangerous; when exposed to heat or flame.
Spontaneous Heating: No.
Explosion Hazard: Unknown.
Disaster Hazard: Dangerous, upon exposure to heat or
  flame; can react vigorously with oxidizing materials.
Countermeasures
Ventilation Control: Section 2.
Storage and Handling: Section 7.
To Fight Fire: Alcohol foam.
Shipping Regulations: Section 11.
  IATA: Flammable lqiuid, red label, 1 liter (passenger),
  40 liters (cargo).

PROPYL GALLATE
General Information
Synonym: Propyl 3,4,5-trihydroxy benzoate.

Description: Odorless, fine, ivory powder.
Formula: $(HO)_3C_6H_2COOC_3H_7$.
Constants: Mol wt: 212.2, mp: $147-149°C$.
Hazard Analysis
Toxicity: Details unknown. Animal experiments suggest
  low level of oral toxicity. Used in food as antioxidant.
  See n-propyl alcohol.
Fire Hazard: Slight, when exposed to heat or flame; can
  react with oxidizing materials (Section 6).
Countermeasures
Storage and Handling: Section 7.

2-PROPYL HEPTANOL
General Information
Formula: $CH_3(CH_2)_4CH(C_3H_7)CH_2OH$.
Constant: Mol wt: 158.
Hazard Analysis
Toxicity: Details unknown. Limited animal experiments
  suggest moderate toxicity.

PROPYLIC ALDEHYDE. See propyl aldehyde.

PROPYL IODIDE
General Information
Synonym: 1-Iodopropane.
Description: Colorless to yellow liquid.
Formula: $CH_3CH_2CH_2I$.
Constants: Mol wt: 170.0, mp: $-98°C$, bp: $103°C$, d:
  1.747 at $20°/4°C$.
Hazard Analysis
Toxicity: Details unknown. Probably toxic. See chlorinated
  hydrocarbons, aliphatic.
Fire Hazard: Moderate, when exposed to heat or flame
  (Section 6).
Disaster Hazard: Dangerous; when heated to decomposi-
  tion, it emits toxic fumes of iodides; can react with
  oxidizing materials.
Countermeasures
Ventilation Control: Section 2.
Storage and Handling: Section 7.

n-PROPYL MERCAPTAN
General Information
Synonym: 1-Propanethiol.
Description: Offensive smelling liquid.
Formula: C_3H_7HS.
Constants: Mol wt: 76, boiling range: $67-73°C$, d: 0.8408
  at $20/4°C$, flash p: $-5°F$.
Hazard Analysis
Toxic Hazard Rating: U. Probably toxic.
Fire Hazard: Dangerous, when exposed to heat or flame.
Disaster Hazard: Dangerous. When heated to decomposi-
  tion it emits highly toxic fumes.
Countermeasures
Storage and Handling: Section 7.
Shipping Regulations: Section 11.
  ICC: Flammable liquid, red label, 10 gallons.
  IATA: Flammable liquid, red label, 1 liter (passenger),
  40 liters (cargo).

PROPYL ISOCYANATE
Shipping Regulations: Section 11.
IATA: Poison C, poison label, not acceptable (passenger),
  40 liters (cargo).

PROPYL METHANOATE. See n-propyl formate.

PROPYL NITRATE
General Information
Description: Liquid.
Formula: $CH_3CH_2CH_2NO_3$.
Constants: Mol wt: 105.1, bp: $110.5°C$, d: 1.058 at
  $20°/4°C$, flash p: $68°F$, autoign. temp.: $350°F$ (in
  air), lel: 2%, uel: 100%.
Hazard Analysis
Toxicity: Highly toxic. See nitrates.

TLV: ACGIH (recommended); 25 parts per million in air; 110 milligrams per cubic meter of air.
Fire Hazard: Dangerous. When exposed to heat or flame.
Explosion Hazard: Dangerous. May explode on heating.
Countermeasures
Storage and Handling: Section 7.
To Fight Fire: Alcohol foam.
Shipping Regulations: Section 11.
  IATA: Combustible liquid, no label required, 220 liters (passenger and cargo) (normal).

PROPYL NITRITE
General Information
Description: Liquid.
Formula: $CH_3CH_2CH_2ONO$.
Constants: Mol wt: 89.1, bp: 57°C, d: 0.935.
Hazard Analysis
Toxic Hazard Rating:
  Acute Local: U.
  Acute Systemic: Ingestion 2; Inhalation 2.
  Chronic Local : U.
  Chronic Systemic: Ingestion 2; Inhalation 2.
Toxicology: Can cause fall in blood pressure similar to amyl nitrate. See also nitrates.

PROPYL PARABEN
General Information
Synonym: Propyl p-hydroxybenzoate.
Description: Small, colorless crystals or white powder; very slightly soluble in water; soluble in alcohol, ether, and acetone.
Formula: $C_{10}H_{12}O_3$.
Constants: Mol wt: 180, mp: 95–98°C.
Hazard Analysis
Toxicity: Unknown. A chemical preservative food additive (Section 10).

n-PROPYL PROPIONATE
General Information
Description: Clear liquid.
Formula: $CH_3CH_2COOCH_2CH_2CH_3$.
Constants: Mol wt: 116.16, mp: −76°C, bp: 122.4°C, flash p: 175°F (O.C.), d: 0.885, vap. press.: 10 mm at 19.4°C, vap. d.: 4.0.
Hazard Analysis
Toxicity: Details unknown. See esters.
Fire Hazard: Moderate, when exposed to heat or flame; can react with oxidizing materials.
Spontaneous Heating: No.
Countermeasures
To Fight Fire: Foam, carbon dioxide, dry chemical or carbon tetrachloride (Section 6).
Storage and Handling: Section 7.

n-PROPYL SILICATE
General Information
Formula: $(C_3H_7O)_4Si$.
Constants: Mol wt: 264.4, bp: 226°C, d: 0.915.
Hazard Analysis
Toxicity: Details unknown. See n-propyl alcohol.
Fire Hazard: Moderate, when exposed to heat or flame; can react with oxidizing materials (Section 6).
Countermeasures
Storage and Handling: Section 7.

PROPYL SULFATE
General Information
Synonym: Di-n-propyl sulfate.

Description: Colorless, oily, liquid.
Formula: $(CH_3CH_2CH_2)_2SO_4$.
Constants: Mol wt: 182.2, mp: 140–170°C decomposes, bp: 120°C at 20 mm, d: 1.11 at 22.5°C, vap. d.: 6.28.
Hazard Analysis
Toxicity: Details unknown. See esters.
Fire Hazard: Moderate, when exposed to heat or flame.
Disaster Hazard: Dangerous; when heated to decomposition, it emits highly toxic fumes of oxides of sulfur; can react with oxidizing materials.
Countermeasures
Ventilation Control: Section 2.
Storage and Handling: Section 7.
To Fight Fire: Water, foam, carbon dioxide, dry chemical or carbon tetrachloride (Section 6).

PROPYL SULFIDE
General Information
Synonym: 1-Propylthiopropane.
Description: Liquid.
Formula: $(C_3H_7)_2S$.
Constants: Mol wt: 118.2, bp: 142°C, fp: −101.9°C, d: 0.814 at 17°C, vap. d.: 4.08.
Hazard Analysis
Toxicity: Details unknown. See sulfides.
Fire Hazard: Moderate, when exposed to heat or flame.
Disaster Hazard: Dangerous; when heated to decomposition, it emits highly toxic fumes of oxides of sulfur; can react with oxidizing materials.
Countermeasures
Ventilation Control: Section 2.
Storage and Handling: Section 7.
To Fight Fire: Foam, carbon dioxide, dry chemical or carbon tetrachloride (Section 6).

PROPYL TETRATHIO-o-STANNATE
General Information
Description: Solid.
Formula: $(SC_3H_7)_4Sn$.
Constants: Mol wt: 419.3, bp: 123°C at 0.001 mm.
Hazard Analysis and Countermeasures
See tin compounds and sulfur compounds.

PROPYL THIOCYANATE
General Information
Description: Clear liquid.
Formula: $CH_3CH_2CH_2SCN$.
Constants: Mol wt: 101.13, vap. d.: 3.49.
Hazard Analysis
Toxicity: Probably slight. See thiocyanates.
Fire Hazard: Moderate, when exposed to heat or flame.
Disaster Hazard: Dangerous; when heated to decomposition, it emits highly toxic fumes of cyanides; can react with oxidizing materials.
Countermeasures
Storage and Handling: Section 7.
To Fight Fire: Foam, carbon dioxide, dry chemical or carbon tetrachloride (Section 6).

1-PROPYLTHIOPROPANE. See propyl sulfide.

PROPYLTIN TRIIODIDE
General Information
Description: Solid.
Formula: $(C_3H_7)SnI_3$.
Constants: Mol wt: 542.6, bp: 200°C at 16 mm (decomp.).
Hazard Analysis and Countermeasures
See tin compounds and iodides.

TOXIC HAZARD RATING CODE (For detailed discussion, see Section 1.)

0 NONE: (a) No harm under any conditions; (b) Harmful only under unusual conditions or overwhelming dosage.

1 SLIGHT: Causes readily reversible changes which disappear after end of exposure.

2 MODERATE: May involve both irreversible and reversible changes; not severe enough to cause death or permanent injury.

3 HIGH: May cause death or permanent injury after very short exposure to small quantities.

U UNKNOWN: No information on humans considered valid by authors.

n-PROPYL TRI-n-AMYL TIN
General Information
Description: Liquid.
Formula: $(C_3H_7)Sn(C_5H_{11})_3$.
Constants: Mol wt: 375.2, d: 1.0368, bp: 163°C at 10 mm.
Hazard Analysis
Toxicity: Details unknown. See tin compounds.

PROPYL TRICHLOROSILANE
General Information
Formula: $C_3H_7SiCl_3$.
Constants: Mol wt: 177.6, vap. d.: 6.15.
Hazard Analysis
Toxic Hazard Rating:
Acute Local: Irritant 3; Ingestion 3; Inhalation 3.
Acute Systemic: U.
Chronic Local: U.
Chronic Systemic: U.
Fire Hazard: Slight, when exposed to heat or flame.
Disaster Hazard: Dangerous; when heated to decomposition, it emits highly toxic fumes of chlorides; will react with water or steam to produce toxic and corrosive fumes; can react with oxidizing materials.
Countermeasures
Ventilation Control: Section 2.
To Fight Fire: Foam, carbon dioxide, dry chemical or carbon tetrachloride (Section 6).
Personnel Protection: Section 3.
Storage and Handling: Section 7.
Shipping Regulations: Section 11.
I.C.C.: Corrosive liquid; white label, 10 gallons.
Coast Guard Classification: Corrosive liquid; white label.
IATA: Corrosive liquid, white label, not acceptable (passenger), 40 liters (cargo).

PROPYL 3,4,5-TRIHYDROXY BENZOATE. See propyl gallate.

PROPYL TRIPHENYL GERMANIUM
General Information
Description: Colorless crystals; insoluble in water. Soluble in organic solvents.
Formula: $(C_3H_7)Ge(C_6H_5)_3$.
Constants: Mol wt: 347.0, mp: 865°C.
Hazard Analysis
Toxicity: Details unknown. See germanium compounds.

PROPYNE. See allylene.

2-PROPYN-1-OL. See propyargyl alcohol.

PROPYNYL ADIPATE
Hazard Analysis
Toxicity: Details unknown. Limited animal experiments suggest higher toxicity than that of most esters. See also esters.

PROTACTINIUM
General Information
Synonyms: Uranium X_2; UX_2.
Description: Shiny metallic mass.
Formula: Pa.
Constant: At wt: 234.
Hazard Analysis
Radiation Hazard: Section 5. For permissible levels, see Table 5, p. 150.
Artificial isotope ^{230}Pa, half life about 18 d. Decays to radioactive ^{230}Th by electron capture (85%). Also decays to radioactive ^{230}U by emitting beta particles of 0.41 MeV. Also emits gamma rays of 0.05–1.0 MeV and X-rays.
Natural isotope ^{231}Pa (Actinium Series), half life 3 × 10^4 y. Decays to radioactive ^{227}Ac by emitting alpha particles of 5.0 MeV.
Artificial isotope ^{233}Pa (Neptunium Series), half life 27 d. Decays to radioactive ^{233}U by emitting beta particles of 0.15 (37%), 0.26 (58%), 0.57 (5%) MeV. Also emits gamma rays of 0.02–0.42 MeV.
Natural isotope ^{234}Pa (Uranium Series), half life 6.7 h. Decays to radioactive ^{234}U by emitting beta particles of 0.16–1.13 MeV. Also emits gamma rays of 0.04–0.8 MeV.
Countermeasures
Ventilation Control: Section 2.
Personal Hygiene: Section 3.
First Aid: Section 1.

PROUSTITE. See silver thioarsenite.

PROVITAMIN A. See carotene.

PRUSSIATE OF SODA. See sodium ferricyanide and sodium ferrocyanide.

PRUSSIC ACID. See hydrocyanic acid.

PRUSSITE. See cyanogen.

PSEUDOACONITINE
General Information
Synomym: Feraconitine; nepaline.
Description: White crystals or syrupy mass.
Formula: $C_{36}H_{51}NO_{12}$.
Constants: Mol wt: 689.78, mp: 212°C (decomp.).
Hazard Analysis
Toxic Hazard Rating:
Acute Local: U.
Acute Systemic: Ingestion 3; Inhalation 3; Skin Absorption 3.
Chronic Local: U.
Chronic Systemic: U.
Disaster Hazard: Dangerous; when heated, it emits highly toxic fumes.
Countermeasures
Ventilation Control: Section 2.
Personal Hygiene: Section 3.
First Aid: Section 1.
Storage and Handling: Section 7.

PSEUDOBUTYLENE. See cis-butene-2.

PSEUDOBUTYLENE GLYCOL. See 2,3,-butylene glycol.

PSEUDOCOTUNNITE. See potassium lead chloride.

PSEUDOCUMENE
General Information
Synonyms: 1,2,4-Trimethylbenzene; pseudocumol.
Description: Liquid; insoluble in water; soluble in alcohol, benzene, and ether.
Formula: $C_6H_3(CH_3)_3$.
Constants: Mol wt: 120.19, d: 0.889(4/4°C), fp: −43.91°C, bp: 168.89°C, flash p: 130°F.
Hazard Analysis
Toxic Hazard Rating:
Acute Local: Irritant 1; Ingestion 1.
Acute Systemic: Inhalation 2.
Chronic Local: U.
Chronic Systemic: Inhalation 2; Skin Absorption 2.
Toxicology: Can cause nervous depression and damage to blood.
Fire Hazard: Moderate where exposed to heat or flame.
Countermeasures
Storage and Handling: Section 7.
To Fight Fire: Foam, alcohol foam, mist.

PSEUDOIONONE
General Information
Synonyms: Citrylideneacetone; 2,6-dimethylenedeca-2,6,8-trien-10-one; 6,10-dimethyl-3,5,9-undecatrien-2-one.
Description: Pale yellow liquid; soluble in alcohol and ether.
Formula: $(CH_3)_2 C:CH(CH_2)_2C(CH_3):CHCH: CHCOCH_3$.

Constants: Mol wt: 192.29, d: 0.8984 at 20°C, bp: 143–145°C at 12mm.
Hazard Analysis
See orris.

PTEROYLGLUTAMIC ACID
General Information
Synonyms: Folic acid; folacin; PGA.
Description: A member of the vitamin B complex. Orange yellow needles or platelets; odorless; slightly soluble in water; insoluble in lipid solvents; soluble in dilute alkali hydroxide and carbonate solutions.
Formula: $C_{19}H_{19}N_7O_6$.
Constants: Mol wt: 441.
Hazard Analysis
Toxicity: U. A food additive permitted in food for human consumption (Section 10).

PTOMAINES
Hazard Analysis
Toxic Hazard Rating:
Acute Local: U.
Acute Systemic: Ingestion 3.
Chronic Local: U.
Chronic Systemic: U.
Toxicology: These are exceedingly toxic compounds commonly formed in putrefying proteins, dead bodies, decayed meat and fish. They have been prepared synthetically and are derivatives of ethers of the polyhydric alcohols. "Ptomaine poisoning" is usually a misnomer for other forms of food poisoning.

PULSATILLA POWDER
Hazard Analysis
Toxic Hazard Rating:
Acute Local: Irritant 1.
Acute Systemic: U.
Chronic Local: U.
Chronic Systemic: U.
Fire Hazard: Moderate, when exposed to heat or flame; can react with oxidizing materials (Section 6).
Explosion Hazard: Slight, when exposed to flame.
Countermeasures
Ventilation Control: Section 2.
Personal Hygiene: Section 3.
Storage and Handling: Section 7.

PURPLE FOXGLOVE. See digitalis.

PURPUREO. See chloropentammine cobalt (III) chloride.

PVP. See polyvinylpolypyrrolidone.

PYRAZOTHION. See o,o-diethyl-o-(3-methyl-5-pyrazolyl) phosphorothioate.

PYRAZOXON. See o,o-diethyl o-3-methyl-5-pyrazolyl) phosphate.

PYRENE
General Information
Synonym: Benzo [d e f] phenanthrene.
Description: Colorless solid; solutions have a slight blue color; insoluble in water; fairly soluble in organic solvents.
Formula: $C_{16}H_{10}$ (a condensed ring hydrocarbon).
Constants: Mol wt: 202.24, mp: 156°C, d: 1.271 at 23°C, bp: 404°C.

Hazard Analysis
Toxicology: A carcinogen.

PYRETHRIN I
General Information
Synonym: Pyrethrolone ester of chrysanthemum monocarboxylic acid.
Description: Viscous liquid.
Formula: $C_{21}H_{28}O_3$.
Constants: Mol wt: 328.4, bp: 170°C at 0.1 mm decomp.
Hazard Analysis
Toxic Hazard Rating:
Acute Local: Irritant 2; Ingestion 2; Inhalation 2.
Acute Systemic: Ingestion 2; Inhalation 3.
Chronic Local: U.
Chronic Systemic: Ingestion 2; Inhalation 3.
Toxicology: Has produced diarrhea, convulsions, collapse and respiratory failure in experimental animals.
Fire Hazard: Slight (Section 6).
Countermeasures
Ventilation Control: Section 2.
Personal Hygiene: Section 3.
Storage and Handling: Section 7.

PYRETHRIN II
General Information
Synonym: Pyrethrolone ester of chrysanthemum dicarboxylic acid, monomethyl ester.
Description: Viscous liquid.
Formula: $C_{22}H_{28}O_5$.
Constants: Mol wt: 372.4, bp: 200°C at 0.1 mm decomp.
Hazard Analysis
Toxicity: Less than pyrethrin I.

PYRETHRINS. See pyrethrin I and II.
Shipping Regulations: Section 11.
IATA: Not restricted (passenger and cargo).

PYRETHROLONE ESTER OF CHRYSANTHEMUM DICARBOXYLIC ACID MONOMETHYL ESTER. See pyrethrin II.

PYRETHROLONE ESTER OF CHRYSANTHEMUM MONOCARBOXYLIC ACID. See pyrethrin I.

PYRETHROSIN
General Information
Description: Crystals; insoluble in water; soluble in hot alcohol, chloroform; slightly soluble in ether or petroleum ether.
Formula: $C_{34}H_{44}O_{10}$.
Constant: Mol wt: 612.73.
Hazard Analysis
See pyrethrin I.

PYRETHRUM FLOWERS
General Information
Synonym: Dalmatian insect powder.
Description: Fine powder.
Hazard Analysis
Toxic Hazard Rating:
Acute Local: Irritant 2; Allergen 2.
Acute Systemic: Ingestion 2; Inhalation 2.
Chronic Local: Irritant 2; Allergen 2.
Chronic Systemic: Ingestion 2; Inhalation 2.
Toxicology: Can cause dermatitis of both allergic and contact types. See also pyrethrin I. Large doses can cause hyper-excitability, incoordination, tremors and muscular paralysis.

TOXIC HAZARD RATING CODE (For detailed discussion, see Section 1.)

0 NONE: (a) No harm under any conditions; (b) Harmful only under unusual conditions or overwhelming dosage.

1 SLIGHT: Causes readily reversible changes which disappear after end of exposure.

2 MODERATE: May involve both irreversible and reversible changes not severe enough to cause death or permanent injury.

3 HIGH: May cause death or permanent injury after very short exposure to small quantities.

U UNKNOWN: No information on humans considered valid by authors

TLV: ACGIH (recommended); 5 milligrams per cubic meter of air.

PYRIDINE
General Information
Description: Colorless liquid; sharp, penetrating, empyreumatic odor; burning taste.
Formula: NCHCHCHCHCH.
Constants: Mol wt: 79.10, bp: 115.3°C lel: 1.8% uel: 12.4%, fp: −42°C, flash p: 68°F (C.C.), d: 0.982, autoign. temp.: 900°F, vap. press.: 10 mm at 13.2°C, vap. d.: 2.73.
Hazard Analysis
Toxic Hazard Rating:
Acute Local: Irritant 1; Ingestion 2; Inhalation 1.
Acute Systemic: Ingestion 2; Inhalation 1.
Chronic Local: U.
Chronic Systemic: Ingestion 2; Inhalation 2.
Toxicoloby: Is mildly irritating to skin and can cause CNS depression. Kidney and liver damage has been reported in experimental animals.
TLV: ACGIH (recommended); 5 parts per million in air; 15 milligrams per cubic meter of air.
Fire Hazard: Dangerous, when exposed to heat or flame.
Spontaneous Heating: No.
Explosion Hazard: Severe, in the form of vapor, when exposed to flame or spark.
Disaster Hazard: Dangerous; when heated to decomposition, it emits highly toxic fumes of cyanides; can react vigorously with oxidizing materials.
Countermeasures
Ventilation Control: Section 2.
To Fight Fire: Carbon dioxide, dry chemical or carbon tetrachloride (Section 6).
Personnel Protection: Section 3.
Storage and Handling: Section 7.
Shipping Regulations: Section 11.
I.C.C.: Flammable liquid; red label, 10 gallons.
Coast Guard Classification: Combustible liquid; inflammable liquid, red label.
MCA warning label.
IATA: Flammable liquid, red label, 1 liter (passenger), 40 liters (cargo).

2-PYRIDINE ALDOXIME METHIODIDE. See 2-PAM.

PYRIDINE-3-CARBOXYLIC ACID. See niacin.

3-PYRIDINE METHANOL
General Information
Synonym: β-Pyridyl carbinol.
Description: Very hygroscopic liquid. Water soluble.
Formula: C_6H_7NO.
Constants: Mol wt: 109.1, bp: 154°C at 28 mm.
Hazard Analysis
Toxicity: Can cause gastrointestinal distress, flushing of skin, dizziness and paresthesia.
Disaster Hazard: Dangerous, when heated to decomposition it emits highly toxic fumes of oxides of nitrogen.

PYRIDINE-N-OXIDE
General Information
Description: Water-soluble crystals.
Formula: C_5H_5NO.
Constants: Mol wt: 95.1, fp: 67.0°C.
Hazard Analysis
Toxicity: Details unknown. See pyridine.
Fire Hazard: Moderate, when exposed to heat or flame (Section 6).
Disaster Hazard: Dangerous; when heated to decomposition, it emits toxic fumes of oxides of nitrogen; can react with oxidizing materials.
Countermeasures
Storage and Handling: Section 7.

PYRIDINIUM PERCHLORATE. See perchlorates.

PYRIDINOTRIBROMOGOLD
General Information
Description: Red crystals, water soluble.
Formula: $(C_5H_5N)AuBr_3$.
Constants: Mol wt: 516.1, mp: 150°C (decomp.).
Hazard Analysis and Countermeasures
See gold compounds and bromides.

PYRIDOXINE HYDROCHLORIDE
General Information
Description: Commercial form of pyridoxine (Vitamin B_6); colorless, white platelets; soluble in water, alcohol, acetone; slightly soluble in other organic solvents.
Formula: $C_8H_{11}O_3N \cdot HCl$.
Constants: Mol wt: 205.5, mp: 204–206°C.
Hazard Analysis
Toxic Hazard Rating: U.
Toxicity: A food additive permitted in food for human consumption (Section 10).

β-PYRIDYL CARBINOL. See 3-pyridinemethanol.

PYRIDYL MERCURIC ACETATE
Hazard Aanlysis
A fungicide, highly toxic. See mercury compounds, organic.

PYRIDYL MERCURIC CHLORIDE. See mercury compounds, organic.

β-PYRIDYL-α-METHYLPYRROLIDINE. See nicotine.

PYRIDYL METHYL STERATE
Hazard Analysis
Toxicity: A fungicide. Details unknown.
Disaster Hazard: Dangerous, when heated to decomposition it emits highly toxic fumes.
Countermeasures
Storage and Handling: See Section 7.

β-PYRIDYL-α-PYRROLIDINE. See nornicotine.

PYROANTIMONATE. See potassium hydroxoantimonate.

PYROARSENIC ACID
General Information
Description: Colorless cyrstals.
Formula: $H_4As_2O_7$.
Constants: Mol wt: 265.9, mp: decomp. at 206°C.
Hazard Analysis and Countermeasures
See arsenic compounds.

PYROCATECHIN. See pyrocatechol.

PYROCATECHOL
General Information
Synonyms: 1,2-Benzendiol; catechol; pyrocatechin; o-dihydroxybenzene.
Description: Colorless crystals.
Formula: $C_6H_4(OH)_2$.
Constants: Mol wt: 110.11, mp: 105°C, bp: 240°C, flash p: 261°F (C.C.), d: 1.371 at 15°C vap. press.: 10 mm at 118.3°C, vap. d.: 3.79.
Hazard Analysis
Toxic Hazard Rating:
Acute Local: Irritant 3; Allergen 1.
Acute Systemic: Ingestion 3; Inhalation 3; Skin Absorption 3.
Chronic Local: Allergen 2.
Chronic Systemic: Ingestion 3; Inhalation 3; Skin Absorption 3.
Toxicology: Can cause convulsions and injury to blood. See also phenol.
Fire Hazard: Slight, when exposed to heat or flame.
Spontaneous Heating: No.
Disaster Hazard: Dangerous; when heated, it emits highly toxic fumes; can react with oxidizing materials.

Countermeasures
Ventilation Control: Section 2.
To Fight Fire: Water, carbon dioxide, dry chemical or carbon tetrachloride (Section 6).
Personnel Protection: Section 3.
First Aid: Section 1.
Storage and Handling: Section 7.

PYROCELLULOSE. See cellulose nitrate.

PYROCHROITE. See manganous hydroxide.

PYROGALLIC ACID. See pyrogallol.

PYROGALLOL
General Information
Synonyms: Pyrogallic acid: 1,2,3-benzentriol; trihydroxybenzene.
Description: White, lustrous crystals.
Formula: $C_6H_3(OH)_3$.
Constants: Mol wt: 126.11, mp: 133–134°C, bp: 309°C, d: 1.453 at 4°/4°C vap. press.: 10 mm at 167.7°C.
Hazard Analysis
Toxic Hazard Rating:
  Acute Local: Irritant 3.
  Acute Systemic: Ingestion 3; Inhalation 3; Skin Absorption 3.
  Chronic Local: U.
  Chronic Systemic: Ingestion 3; Inhalation 3; Skin Absorption 3.
Toxicology: If swallowed can cause vomiting and diarrhea. Is reported also as causing hemolysis, methemoglobinemia, kidney injury and liver damage. Readily absorbed via skin.
Disaster Hazard: Dangerous; when heated it may emit highly toxic fumes.
Countermeasures
Ventilation Control: Section 2.
Personnel Protection: Section 3.
Storage and Handling: Section 7.

PYROLAN
General Information
Synonym: G-22008; 1-phenyl-3-methyl-5-pyrazolyl dimethyl carbamate.
Description: Crystals. Water soluble.
Formula: $C_{13}H_{15}N_3O_2$.
Constants: Mol wt: 245.3, mp: 50°C, bp: 161°C at 0.2 mm.
Hazard Analysis
Toxicity: Highly toxic. Details unknown. A cholinesterase inhibitor. For effects see parathion.
Disaster Hazard: Dangerous; when heated to decomposition, it emits highly toxic fumes.

PYROLIGNEOUS ACID
General Information
Synonym: Wood vinegar.
Description: Yellowish, acidic liquid.
Formula: $HC_2H_3O_2$.
Hazard Analysis
Toxic Hazard Rating:
  Acute Local: Irritant 1; Allergen 1.
  Acute Systemic: U.
  Chronic Local: Irritant 1; Allergen 1.
  Chronic Systemic: U.

PYROMELLITIC ACID
General Information
Synonyms: (PMA); 1,2,4,5-benzene tetracarboxylic acid.

Description: Off-white powder.
Formula: $C_6H_2(COOH)_4$.
Constants: Mol wt: 254.15, mp: 257–265°C decomp., d 0.43.
Hazard Analysis
Toxicity: Unknown.
Fire Hazard: Slight; (Section 6).
Countermeasures
Storage and Handling: Section 7.

PYROMUCIC ACID. See furoic acid.

PYROMUCIC ALDEHYDE. See furfural.

PYROPHOSPHORIC ACID
General Information
Description: Colorless, hygroscopic needles or liquid.
Formula: $H_4P_2O_7$.
Constants: Mol wt: 178.0, mp: 61°C.
Hazard Analysis and Countermeasures
See phosphoric acid.

PYROPHORIC SOLUTIONS, NOS
Shipping Regulations: Section 11.
  ICC: Flammable liquids, red label, 10 gallons.
  IATA: Flammable liquid; not acceptable (passenger and cargo).

PYROPHOSPHOROUS ACID
General Information
Description: Needles.
Formula: $H_4P_2O_5$.
Constants: Mol wt: 145.99, mp: 38°C, bp: decomp. a 139°C.
Hazard Analysis and Countermeasures
See phosphoric acid.

PYROPHYLLITE
General Information
Synonym: Agalmatolete.
Description: White, green gray, brown; found in metamorphic rocks.
Formula: $Al_2Si_4O_{10}(OH)$.
Constants: Mol wt: 293, d: 2.8–2.9.
Hazard Analysis
Toxicity: Unknown. A food additive permitted in the feed and drinking water of animals and for the treatment of food producing animals (Section 10).

PYROSULFURIC ACID
General Information
Synonym: Disulfuric acid.
Description: Colorless to yellowish crystals.
Formula: $H_2S_2O_7$.
Constants: Mol wt: 178.2, mp: 35°C, bp: decomposes d: 1.89.
Hazard Analysis
Toxicity: Very corrosive. See sulfuric acid.
Fire Hazard: Moderate, by chemical reaction. It react violently with organic materials containing hydrogen and oxygen (Section 6).
Disaster Hazard: Dangerous. See sulfuric acid.
Countermeasures
Personnel Protection: Section 3.
Storage and Handling: Section 7.

PYROSULFURYL CHLORIDE
General Information
Synonym: Disulfuryl chloride.
Description: Colorless, mobile, fuming liquid.

TOXIC HAZARD RATING CODE (For detailed discussion, see Section 1.)

0 NONE: (a) No harm under any conditions; (b) Harmful only under unusual conditions or overwhelming dosage.

1 SLIGHT: Causes readily reversible changes which disappear after end of exposure.

2 MODERATE: May involve both irreversible and reversible changes not severe enough to cause death or permanent injury.

3 HIGH: May cause death or permanent injury after very short exposure to small quantities.

U UNKNOWN: No information on humans considered valid by authors

rmula: $S_2O_5Cl_2$.
nstants: Mol wt: 215.03; mp: $-39°C$ to $-37°C$; bp: $140°C$; d: gas = 9.6 g/l, liquid = 1.818 at $11°/4°C$.

azard Analysis and Countermeasures
e hydrochloric acid.
ipping Regulations: Section 11.
.C.C.: Corrosive liquid; white label, 1 quart.
Coast Guard Classification: Corrosive liquid; white label.
IATA: Corrosive liquid, white label, 1 liter (passenger and cargo).

ROXYLIN PLASTIC. See cellulose nitrate.
ipping Regulations: Section 11.
Coast Guard Classification (rolls, rods, sheets, tubes): Inflammable solid; yellow label.
ICC: Flammable solid, yellow   label, 350 pounds (rods, sheets, rolls, tubes).
IATA: Flammable solid, yellow labe, 25 kilograms (passenger), 160 kilograms (cargo) (manufactured articles, rolls, sheets, rods and tubes).

ROXYLIN PLASTIC, SCRAP. See cellulose nitrate.
ipping Regulations: Section 11.
I.C.C.: Flammable solid; yellow label.
Coast Guard Classification: Inflammable solid; yellow label.
IATA: Flammable solid, not acceptable (passenger and cargo).

ROXYLIN SOLUTION. See cellulose nitrate.
ipping Regulations: Section 11.
IC.C.: Flammable liquid; red label, 10 gallons.
Coast Guard Classification: Combustible liquid; inflammable liquid, red label (see solvents N.O.S.).
IATA: Flammable liquid, red label, 1 liter (passenger), 40 liters (cargo).

ROXYLIN SOLVENT, N.O.S.
ipping Regulations: Section 11.
I.C.C.: Flammable liquid; red label, 10 gallons.
Coast Guard Classification: Combustible liquid; inflammable liquid, red label.
IATA: Flammable liquid, red label, 1 liter (passenger), 40 liters (cargo).

RROLE
eneral Information
escription: Colorless liquid; darkens on standing; mild odor.
rmula: C_4H_5N.
nstants: Mol wt: 67.09, bp: $129°C$, fp: $-24°C$, flash p: $102°F$ (T.C.C.), d: 0.968 at $20°/4°C$, vap. d.: 2.31.

azard Analysis
xic Hazard Rating:
Acute Local: U.
Acute Systemic: Ingestion 2; Inhalation 2.
Chronic Local: U.
Chronic Systemic: Ingestion 2; Inhalation 2.
re Hazard: Moderate, when exposed to heat or flame.
isaster Hazard: Dangerous; when heated to decomposition, it emits highly toxic fumes of oxides of nitrogen; can react with oxidizing materials.

ountermeasures
entilation Control: Section 2.
Fight Fire: Foam, carbon dioxide, dry chemical or carbon tetrachloride (Section 6).
rsonal Hygiene: Section 3.
orage and Handling: Section 7.

YRROLIDINE
eneral Information
escription: Colorless, mobile liquid; penetrating; amine-like odor.
rmula: C_4H_9N.
nstants: Mol wt: 71.12, bp: $86-87°C$, fp: $-63°C$, flash p: $37°F$ (T.C.C.), d: 0.8618 at $20°/4°C$, vap. press.:128mm at $39°C$, vap. d.: 2.45.

Hazard Analysis
Toxic Hazard Rating:
Acute Local: U.
Acute Systemic: Ingestion 2; Inhalation 2.
Chronic Local: U.
Chronic Systemic: Ingestion 2; Inhalation 2.
Fire Hazard: Dangerous, when exposed to heat or flame.
Explosion Hazard: Unknown.
Disaster Hazard: Dangerous; when heated to decomposition, it emits highly toxic fumes of oxides of nitrogen; can react vigorously with oxidizing materials.

Countermeasures
Ventilation Control: Section 2.
To Fight Fire: Foam, carbon dioxide, dry chemical or carbon tetrachloride (Section 6).
Personal Hygiene: Section 3.
Storage and Handling: Section 7.
Shipping Regulations: Section 11.
    IATA: Flammable liquid, red label, 1 liter (passenger), 40 liter (cargo).

2-PYRROLIDONE
General Information
Description: A colorless liquid.
Formula: $HNCH_2CH_2CH_2CO$.
Constants: Mol wt: 85.11, mp: $25°C$, bp: $245°C$, flash p: $265°F$ (O.C.).

Hazard Analysis
Toxicity: Unknown.
Fire Hazard: Slight, when exposed to heat or flame.
Disaster Hazard: Dangerous; when heated to decomposition, it emits highly toxic fumes of oxides or nitrogen; can react with oxidizing materials.

Countermeasures
Storage and Handling: Section 7.
To Fight Fire: Foam, carbon dioxide, dry chemical or carbon tetrachloride (Section 6).

2-PYRROLIDINECARBOXYLIC ACID. See proline.

PYRROLINE
General Information
Synonym: 2,5-Dihydropyrrole.
Description: Nearly colorless liquid. Fumes in air with an unpleasant ammonia-like odor. Miscible with water.
Formula: C_4H_7N.
Constants: Mol wt: 69.1, bp: $91°C$ at 748 mm.

Hazard Analysis
Toxic Hazard Rating:
Acute Local: Irritant 2; Ingestion 2; Inhalation 2.
Acute Systemic: Ingestion 2.
Chronic Local: U.
Chronic Systemic: Ingestion 2.
Toxicology: Data based on animal experiments.
Disaster Hazard: Dangerous; when heated to decomposition, it emits highly toxic fumes of oxides of nitrogen.

PYRUVALDEHYDE. See pyruvic aldehyde.

PYRUVIC ALDEHYDE
General Information
Synonym: Pyruvaldehyde.
Description: Yellow, mobile liquid; pungent odor.
Formula: CH_3COCHO.
Constants: Mol wt: 72.06, bp: $72°C$, d: 1.20 at $20/20°C$.

Hazard Analysis
Toxic Hazard Rating:
Acute Local: Irritant 1; Ingestion 1; Inhalation 1.
Acute Systemic: U.
Chronic Local: U.
Chronic Systemic: U.
Toxicity: See also aldehydes.
Fire Hazard: Slight (Section 6).

Countermeasures
Personal Hygiene: Section 3.
Storage and Handling: Section 7.

Q

QUARTZ
General Information
Synonyms: Cristobalite; silicon dioxide.
Description: Cubic, colorless crystals.
Formula: SiO_2.
Constants: Mol wt: 60.06, mp: 1710°C, bp: 2230°C, d: 2.32.
Hazard Analysis
Toxic Hazard Rating:
 Acute Local: Inhalation 2.
 Acute Systemic: 0.
 Chronic Local: Inhalation 3.
 Chronic Systemic: Inhalation 2.
Toxicology: See silica. A food additive permitted in the feed and drinking water of animals and/or for the treatment of food producing animals. Also permitted in food for human consumption (Section 10). A common air contaminant (Section 4).
TLV: ACGIH (recommended); 5 million particles per cubic foot (over 70% quartz); 10 million particles per cubic foot (10–70% quartz).
Countermeasures
Ventilation Control: Section 2.

QUASSIA
General Information
Synonym: Bitter wood tree.
Description: Wood or bark.
Toxic Hazard Rating:
 Acute Local: Ingestion 1.
 Acute Systemic: Ingestion 1.
 Chronic Local: U.
 Chronic Systemic: U.
Hazard Analysis
Toxicology: Large doses by mouth may produce nausea and vomiting; an insecticide.
Fire Hazard: Moderate, when exposed to heat or flame (Section 6).
Explosion Hazard: Slight, in the form of dust, when exposed to heat or flame.
Countermeasures
Personal Hygiene: Section 3.
Storage and Handling: Section 7.

QUEBRACHINE. See yohimbine.

QUENCHING OIL
General Information
Description: An oil.
Constants: Flash p: 365°F (C.C.), d: 0.9.
Hazard Analysis
Toxicity: Unknown.
Fire Hazard: Slight, when exposed to heat or flame.
Countermeasures
To Fight Fire: Foam, carbon dioxide, dry chemical or carbon tetrachloride (Section 6).
Storage and Handling: Section 7.

QUICKLIME. See calcium oxide.
Shipping Regulations: Section 11.
 Coast Guard Classification: Hazardous article.

QUICKSILVER. See mercury.

QUINACRINE HYDROCHLORIDE. See "Atabrine" hydrochloride.

QUINALDINE. See methyl quinoline.

QUINHYDRONE
General Information
Description: Dark green crystals; slightly soluble in water; soluble in alcohol, ether, hot water, ammonia.
Formula: $C_6H_4O_2C_6H_4(OH)_2$.
Constants: Mol wt: 210, d: 1.40, mp: 171°C.
Hazard Analysis
Toxicity: Details unknown. Animal experiments show high toxicity.

QUININE
General Information
Description: Bulky, white amorphous powder or crystals; bitter taste.
Formula: $C_{20}H_{24}N_2O_2$.
Constants: Mol wt: 324.4, mp: 174.9°C.
Hazard Analysis
Toxic Hazard Rating:
 Acute Local: Irritant 2; Allergen 1.
 Acute Systemic: Ingestion 2.
 Chronic Local: Allergen 1.
 Chronic Systemic: Ingestion 2.
Toxicology: Upon contact with this material, the eyes become swollen, watery and exude a sticky, viscous liquid which forms yellowish crusts. Upon ingestion, it causes dilation of the pupils. The optic nerve becomes pale and atrophic and retina shows thready arteries; ptosis and clonic spasms of the lids result. It may cause atrophy of the optic nerve. Vision returns in from 24 to 28 hours and gradually improves. Quinine dermatitis is an occupational hazard to barbers particularly and generally to people who work with quinine tonics, medicaments, or cosmetics. Quinine has no influence upon sound skin, but it is distinctly irritant to mucous membranes and raw surfaces. Internally it can cause a sense of fullness in the head, tinnitis aureum, slight deafness, disorders of vision and sometimes blindness. Its physiological effects vary with the individual. Occasionally, it can cause cutaneous eruptions, such as erythema, urticaria, herpes, purpuria, and even gangrenous affections. Quinine is used as a food additive permitted in food for human consumption (Section 10).
Fire Hazard: Slight; when heated to decomposition it emits toxic fumes of oxides of nitrogen.
Countermeasures
Personnel Protection: Section 3.

TOXIC HAZARD RATING CODE *(For detailed discussion, see Section 1.)*

0 NONE: (a) No harm under any conditions; (b) Harmful only under unusual conditions or overwhelming dosage.

1 SLIGHT: Causes readily reversible changes which disappear after end of exposure.

2 MODERATE: May involve both irreversible and reversible change not severe enough to cause death or permanent injury.

3 HIGH: May cause death or permanent injury after very short exposure to small quantities.

U UNKNOWN: No information on humans considered valid by autho

Treatment and Antidotes: Wash out stomach. There are no
   specific antidotes. Treatment is symptomatic.
Storage and Handling: Section 7.

QUININE ACETATE. See quinine.

QUININE ARSENATE
General Information
Description: Fine needles or crystals.
Formula: $3C_{20}H_{21}N_2O_2 \cdot 2H_3AsO_4 \cdot 5H_2O$.
Constant: Mol wt: 1346.
Hazard Analysis and Countermeasures
See arsenic compounds.

QUININE BISULFATE. See quinine.

QUININE CACODYLATE
General Information
Description: White powder.
Formula: $C_{20}H_{24}N_2O_2 \cdot (CH_3)_2AsO_2H$.
Constant: Mol wt: 462.4.
Hazard Analysis and Countermeasures
See arsenic compounds.

QUININE CHLORATE
General Information
Description: Crystals.
Formula: $C_{20}H_{24}N_2O_2 \cdot HClO_3 \cdot 2H_2O$.
Constants: Mol wt: 444.9, mp: explodes.
Hazard Analysis
Toxicity: Highly toxic. See quinine.
Fire Hazard: Dangerous. See chlorates.
Explosion Hazard : Moderate, when shocked or exposed to
   heat. See chlorates.
Disaster Hazard: Dangerous; shock will explode it; when
   heated to decomposition it emits highly toxic fumes of
   oxides of nitrogen and chlorides; can react vigorously
   with reducing materials.
Countermeasures
Ventilation Control: Section 2.
Storage and Handling: Section 7.

QUININE HYDROBROMIDE. See quinine.

QUININE HYDROCHLORIDE. See quinine.

QUININE SALICYLATE. See quinine.

QUINOL. See p-hydroquinone.

QUINOLINE
General Information
Synonyms: Chinoline; leukol.
Description: Refractive, colorless liquid; peculiar odor.
Formula: $C_6H_4NCHCHCH$.
Constants: Mol wt: 129.2, mp: $-19.5°C$, bp: $237.7°C$,
   d: 1.0900 at $25°/4°C$, autoign. temp.: $896°F$, vap.
   press.: 1 mm at $59.7°C$, vap. d.: 4.45.
Hazard Analysis
Toxic Hazard Rating:
   Acute Local: U.
   Acute Systemic: Ingestion 3; Inhalation 3.
   Chronic Local: U.
   Chronic Systemic: Ingestion 2; Inhalation 2.
Caution: May produce retinitis similar to that caused by
   naphthalene but without causing opacity of the lens.
Fire Hazard: Slight, when exposed to heat (Section 6).
Disaster Hazard: Dangerous; when heated to decomposi-
   tion it emits toxic fumes of nitrogen oxides.
Countermeasures
Ventilation Control: Section 2.
Personal Hygiene: Section 3.
Storage and Handling: Section 7.

8-QUINOLINOL
General Information
Synonym: 8-Hydroxyquinoline.

Description: White crystals or powder. Nearly insoluble in
   water.
Formula: C_9H_7NO.
Constants: Mol wt: 145.2, mp: $76°C$, bp: $267°C$.
Hazard Analysis
Toxicity: Details unknown; a fungicide. Limited animal
   experiments indicate moderate toxicity with stimula-
   tion of CNS.
Fire Hazard: Slight (Section 6).
Disaster Hazard: Dangerous; when heated to decomposi-
   tion, it emits highly toxic fumes of oxides of nitrogen.
Countermeasures
Storage and Handling: Section 7.

8-QUINOLINOL BENZOATE
Hazard Analysis
Toxicity: Details unknown; a fungicide.
Fire Hazard: Slight (Section 6).
Countermeasures
Storage and Handling: Section 7.

QUINOLINOTRIBROMOGOLD
General Information
Description: Deep red crystals; soluble in chloroform.
Formula: $(C_9H_7N)AuBr_3$.
Constants: Mol wt: 566.1, mp: decomp. $>200°C$.
Hazard Analysis
Toxicity: Details unknown. See gold compounds.
Disaster Hazard: Dangerous. See bromides.

QUINONE
General Information
Synonyms: Benzoquinone; chinone.
Description: Yellow crystals; chracteristic, irritating odor.
Formula: OC_6H_4O.
Constants: Mol wt: 108.09, mp: $115.7°C$, bp: sublimes,
   d: 1.318 at $20°/4°C$.
Hazard Analysis
Toxic Hazard Rating:
   Acute Local: Irritant 3; Ingestion 3; Inhalation 3.
   Acute Systemic: Ingestion 3; Inhalation 3.
   Chronic Local: Irritant 2.
   Chronic Systemic: Ingestion 3; Inhalation 3.
TLV: ACGIH (recommended): 0.1 parts per million in air.
   0.4 milligrams per cubic meter of air.
Toxicology: Quinone has a characteristic, irritating odor. It
   can cause severe local damage to the skin and mucous
   membranes by contact with it in the solid state, in solution, or in the form of condensed vapors. Locally it can
   cause discoloration, severe irritation, erythema, swelling and the formation of papules and vesicles, whereas
   prolonged contact may lead to necrosis. When the eyes
   become involved, it can cause dangerous disturbances
   of vision. A case is reported where ulceration of the
   cornea resulted from brief exposure to a high concentration of the vapor. An accepted criterion for regulating
   workroom concentration of this material in the air has
   been the comfort of personnel involved, as judged by
   eye irritation, conjunctivitis, photophobia, moderate
   lachrymation, and burning sensations. It has been found
   that personnel can develop corneal injury of two types
   due to this material. One type is a typical superficial
   greenish-brown stain or grayish-white opacity varying
   in size and involving all the layers of the cornea. In a
   few cases there has been an appreciable loss of vision.
   The eye stain is probably an end product of the oxidation of quinone to hydroquinone and the subsequent
   polymerization of this material. Its odor becomes
   perceptible at or just above 0.1 ppm and is quite definite
   in the region of 0.15 ppm and irritating at 0.5 ppm.
Disaster Hazard: Dangerous; when heated to decomposition,
   it emits toxic fumes.
Countermeasures
Ventilation Control: Section 2.

Treatment and Antidotes: Personnel who show the toxic symptoms noted above when exposed to this material should be removed from exposure and the symptoms will rapidly and completely disappear.

Personnel Protection: Section 3.

Storage and Handling: Section 7.

QUINONE MONOXIME. See p-nitrosphenol.

QUINOSOL

General Information

Synonyms: Chinosol; oxyquinoline potassium sulfate.

Description: Yellow crystalline powder.

Formula: $C_9H_6NOSO_3K \cdot H_2O$.

Hazard Analysis

Toxicity: Details unknown; probably low; a fungicide.

Fire Hazard: Slight (Section 6).

Countermeasures

Storage and Handling: Section 7.

TOXIC HAZARD RATING CODE (For detailed discussion, see Section 1.)

0 NONE: (a) No harm under any conditions; (b) Harmful only under unusual conditions or overwhelming dosage.

1 SLIGHT: Causes readily reversible changes which disappear after end of exposure.

2 MODERATE: May involve both irreversible and reversible changes not severe enough to cause death or permanent injury.

3 HIGH: May cause death or permanent injury after very short exposure to small quantities.

U UNKNOWN: No information on humans considered valid by author

R

RACEMIC ACID. See tartaric acid.

RADIOACTIVE DEVICES.
Shipping Regulations: Section 11.
   I.A.T.A.: Radioactive Material, no label, 200 curies
   (passenger and cargo).

RADIOACTIVE MATERIALS. See individual isotopes.
Shipping Regulations: Section 11.
   I.C.C. (N.O.S.): Poison D, radioactive materials, blue
   or red label.
   Coast Guard Classification:
      Groups I and II: Poison D, Poison radioactive mate-
      rials label (red).
      Group III: Poison D: Poison radioactive materials
      label (blue).
      Low Activity: Poison D.

RADIOACTIVE MATERIALS, GROUP IV
Shipping Regulations: Section 11.
   IATA: Radioactive Material, radioactive (red marked
   "group IV") label, 300 curies (passenger), 5000 curies
   (cargo).
   IATA (Group I, N.O.S., not in solid, nonfriable form,
   exceeding 2 curies, but not including Cesium 137):
   Radioactive material, radioactive (red) label, 50 curies
   (passenger and cargo).
      (Group II, N.O.S., exceeding 2 curies): Radioac-
   tive materials, radioactive (red) label, 50 curies (passenger and cargo).
      (Group III, N.O.S., exceeding 2 curies, but not in-
   cluding Tritium): Radioactive material, radioactive
   (blue) label, 50 curies (passenger and cargo).

**RADIO CURRENT SUPPLY DEVICES PACKED
WITH ELECTROLYTE**
Shipping Regulations: Section 11.
   IATA: Corrosive liquid, white label, not acceptable (pas-
   senger), $2\frac{1}{2}$ liters (cargo).

RADIUM
General Information
Description: Silver-white metal.
Formula: Ra.
Constants: Mol wt: 226.05, mp: 960°C, bp: 1140°C, d:5.
Hazard Analysis
Toxicity: Common air contaminants (Section 4).
Radiation Hazard: Section 5. For permissible levels, see
   Table 5, p. 150.
   Natural isotope ^{223}Ra (Actinium-X, Actinium
Series), half life 11.4 d. Decays to radioactive
^{219}Rn by emitting alpha particles of 5.5–5.7 MeV.
Also emits gamma rays of 0.12–0.44 MeV.
   Natural isotope ^{224}Ra (Thorium-X, Thorium
Series), half life 3.6 d. Decays to radioactive ^{220}Rn by
emitting alpha particles of 5.7 MeV. Also emits gamma
rays of 0.24 MeV.
   Natural isotope ^{226}Ra (Uranium Series), half life
1600 y. Decays to radioactive ^{222}Rn by emitting alpha
particles of 4.8 MeV. Also emits gamma rays of 0.19
MeV.
   Natural isotope ^{228}Ra (Mesothorium-1, Thorium
Series), half life 5.8 y. Decays to radioactive ^{228}Ac by
emitting beta particles of 0.05 MeV.

Radium replaces calcium in the bone structure and is
a source of irradiation to the blood forming organs.
The ingestion of luminous dial paint prepared from
radium was the cause of death of many of the early
dial painters before the hazard was fully understood.
The data on these workers has been the source of
many of the radiation precautions and the maximum
permissible levels for internal emitters which are now
accepted. ^{226}Ra is the parent of radon and the precautions described under ^{222}Rn should be followed.
   ^{228}Ra is a member of the thorium series. It was a
common constituent of luminous paints, and while its
low β energy was not a hazard, its daughters in the
series may have been a causative agent in the deaths of
radium dial painters following World War I. Its metabolism is the same as any other radium isotope and it
is a source of thoron. The precautions recommended
under ^{220}Rn should be followed.
Disaster Hazard: Highly dangerous; must be kept heavily
   shielded with lead and stored away from possible dis-
   semination by explosion, flood, etc.
Countermeasures
Ventilation Control: Sections 5 and 2.
Personnel Protection: Section 3.
First Aid: Section 1.
Shipping Regulations: Section 11.

RADIUM BROMIDE
General Information
Description: Colorless-yellowish crystals.
Formula: RaBr$_2$.
Constants: Mol wt: 385.88, mp: 728°C, d: 5.79.
Hazard Analysis and Countermeasures
See radium and bromides.

RADIUM CARBONATE
General Information
Description: White or slight brownish crystals.
Formula: RaCO$_3$.
Constant: Mol wt: 286.06.
Hazard Analysis and Countermeasures
See radium.

RADIUM CHLORIDE
General Information
Description: Monoclinic, colorless-yellowish crystals.
Formula: RaCl$_2$.
Constants: Mol wt: 296.96, mp: 1000°C, d: 4.91.
Hazard Analysis and Countermeasures
See radium and chlorides.

RADIUM EMANATION. See radon.

RADIUM IODATE
General Information
Description: White powder.
Formula: Ra(IO$_3$)$_2$.
Constant: Mol wt: 575.89.
Hazard Analysis and Countermeasures
See radium and iodates.

RADIUM SULFATE
General Information
Description: Colorless crystals.
Formula: RaSO$_4$.

Constant: Mol wt: 322.12.
Hazard Analysis and Countermeasures
See radium and sulfates.

RADON-222
General Information
Synonyms: Niton; radium emanation.
Description: Colorless gas; opaque crystals.
Formula: ^{222}Rn.
Constants: Mol wt: 222.00, mp: -110°C, bp: -61.8°C,
    d: 9.73 g/l, d liquid = 4.4 at -62°C, d solid = 4.
Hazard Analysis
Toxicity: A common air contaminant. See section 4. See
    radiation hazard below.
Radiation Hazard: Section 5. For permissible levels, see
    Table 5, p. 150.
        Natural isotope ^{220}Rn (Thoron, Thorium Series),
    half life 55 s. Decays to radioactive ^{216}Po by emitting
    alpha particles of 6.3 MeV.
        Natural isotope ^{222}Rn (Uranium Series), half life
    3.8 d. Decays to radioactive ^{218}Po by emitting alpha
    particles of 5.5 MeV.
        Both ^{220}Rn and ^{222}Rn are gaseous. Their daughter
    products are solids, and usually become attached to
    aerosols. Permissible levels for radon include the ex-
    pected daughter activity.
        The permissible levels are given for Rn^{222} in equi-
    librium with its daughters. The chief hazard from this
    isotope is inhalation of the gaseous element and its
    solid daughters which are collected on the normal dust
    of the air. This material is deposited in the lung and has
    been considered to be a major causative agent in the
    high incidence of lung cancer found in German and
    Czechoslovakian uranium miners. Radon and its
    daughters build up to an equilibrium value in about a
    month from radium compounds, while the build-up
    from uranium compounds is negligible. Good ventilation
    of areas where radium is handled or stored is recom-
    mended to prevent accumulation of hazardous con-
    centrations of radon and its daughters.
Countermeasures
Ventilation Control: Section 2.

RAGS, OILY
Hazard Analysis
Toxicity: A variable allergen.
Fire Hazard: Moderate, when exposed to heat or flame;
    can react with oxidizing materials (Section 6).
Countermeasures
Storage and Handling: Section 7.
Shipping Regulations: Section 11.
    I.C.C.: Flammable solid; yellow label, not accepted.
    Coast Guard Classification: Inflammable solid; yellow
    label.

RAGS, WET
Hazard Analysis
Toxicity: A weak allergen.
Fire Hazard: Slight, when exposed to heat or flame or by
    chemical reaction (Section 6).
Countermeasures
Storage and Handling: Section 7.
Shipping Regulations: Section 11.
    I.C.C.: Flammable solid; yellow label, not accepted.
    Coast Guard Classification: Inflammable solid; yellow
    label.

RANDOX®. See N,N-diallyl-2-chloroacetamide.

RAPESEED OIL
General Information
Synonym: Colza oil.
Constants: Flash p: 325°F (C.C.), solidifying p: -2° to
    -10°C, d: 0.915, autoign. temp.: 836°F.
Hazard Analysis
Toxicity: Details unknown; a fungicide for plants.
Fire Hazard: Slight, when exposed to heat or flame; can
    react with oxidizing materials.
Spontaneous Heating: Yes.
Countermeasures
To Fight Fire: Foam, carbon dioxide, dry chemical or car-
    bon tetrachloride (Section 6).
Storage and Handling: Section 7.

RATIONITE. See dimethyl sulfate and chlorosulfonic acid.

RAT POISON
Shipping Regulations: Section 11.
    IATA (Solid): Poison B, poison label, 25 kilogram (pas-
    senger), 95 kilograms (cargo).
        (Liquid): Poison B, poison label, 1 liter (pas-
    senger), 220 liters (cargo).

RATTLE SNAKE VENOM. See crotoxin.

RDX. See cyclotrimethylene trinitramine.

REALGAR. See arsenic bisulfide.

RED LEAD. See lead oxide.

RED OIL (COMMERCIAL GRADE). See oleic acid.

RED PHOSPHORUS. See phosphorus (amorphous red).

RED POTASSIUM CHROMATE. See potassium bichromate.

RED POTASSIUM PRUSSIATE. See potassium ferricyanide.

RED PRUSSIATE OF POTASH. See potassium ferricyanide.

RED PRUSSIATE OF SODA. See sodium ferricyanide.

RED PRUSSIATE OF SODIUM. See sodium ferricyanide.

REDWOOD. See brazil wood.

REFRIGERANT 115. See chloropentafluoroethane.

REFRIGERANT 116. See hexafluoroethane.

REISSERT COMPOUNDS
Hazard Analysis
See nitriles.

RELEASE DEVICES, EXPLOSIVE
Shipping Regulations: Section 11.
    IATA: Explosive, explosive label, 25 kilograms (pas-
    senger), 70 kilogram (cargo).

RENNASE. See rennet.

RENNET
General Information
Synonym: Rennim; rennase; chymosin.
Description: The enzyme secreted by the glands of the
    stomach which causes curdling of milk. A yellowish

TOXIC HAZARD RATING CODE (For detailed discussion, see Section 1.)

0 NONE: (a) No harm under any conditions; (b) Harmful only under unusual conditions or overwhelming dosage.

1 SLIGHT: Causes readily reversible changes which disappear after end
of exposure.

2 MODERATE: May involve both irreversible and reversible change
not severe enough to cause death or permanent injury.

3 HIGH: May cause death or permanent injury after very short exposu
to small quantities.

U UNKNOWN: No information on humans considered valid by author

white powder, peculiar but not unpleasant odor; partially soluble in water and dilute alcohol.

Hazard Analysis

Toxicity: Unknown. A general purpose food additive (Section 10).

RENNIN. See rennet.

RESERPINE

General Information

Description: White or pale buff to slightly yellow powder; odorless; insoluble in water; very slightly soluble in alcohol; soluble in chloroform and acetic acid.

Formula: $C_{33}H_{40}N_2O_9$.

Constants: Mol wt: 572, mp: 264-265°C.

Hazard Analysis

Toxicity: Unknown. Used as a food additive permitted in the feed and drinking water of animals and/or for the treatment of food producing animals. Also permitted in food for human consumption (Section 10).

RESIN COPAL. See copal.

RESIN SOLUTION

Shipping Regulations: Section 11.

ICC: Flammable liquid, red label, 55 gallons.

IATA (flammable liquid): Flammable liquid, red label, 1 liter (passenger), 220 liters (cargo).

(poisonous): Poison B, poison label, 1 liter (passenger), 220 liters (cargo).

RESORCIN. See resorcinol.

RESORCINOL

General Information

Synonyms: 1,3-Benzenediol; resorcin; m-dihydroxybenzene.

Description: Very white crystals; become pink on exposure to light when not perfectly pure; unpleasant sweet taste.

Formula: $C_6H_4(OH)_2$.

Constants: Mol wt: 110.11, mp: 110°C, bp: 276.5°C, flash p: 261°F (C.C.), d: 1.285 at 15°C, autoign. temp.: 648°F, vap. press.: 1 mm at 108.4°C, vap. d.: 3.79.

Hazard Analysis

Toxic Hazard Rating:

Acute Local: Irritant 2; Ingestion 2; Inhalation 2.

Acute Systemic: Ingestion 2; Inhalation 2; Skin Absorption.

Chronic Local: Irritant 2; Allergen 1.

Chronic Systemic: Ingestion 2; Inhalation 2; Skin Absorption 2.

Toxicology: It is primarily a skin irritant. However, it can cause systemic poisoning by acting both as a blood and nerve poison. It may also cause injury to the eyes and dermatitis, particularly to those who are sensitive to it. Such individuals are affected by even very slight traces of it. In a suitable solvent, this material can readily be absorbed through human skin, and can cause local hyperemia, itching, dermatitis, edema, and corrosion associated with enlargement or regional lymph glands as well as serious systemic disorders such as restlessness, methemoglobinemia, cyanosis, convulsions, tachycardia, dyspnea, and death. These same symptoms can be induced by ingestion of the material. For poisoning treat symptomatically.

Fire Hazard: Slight, when exposed to heat or flame; can react with oxidizing materials.

Spontaneous Heating: No.

Countermeasures

To Fight Fire: Water, carbon dioxide, dry chemical or carbon tetrachloride (Section 6).

Ventilation Control: Section 2.

Personnel Protection: Section 3.

Storage and Handling: Section 7.

RESORCINOL PHTHALEIN. See fluorescein.

RETINOL. See rosin oil.

RHENIUM

General Information

Description: Hexagonal, metallic, lustrous crystals or gray to black powder.

Formula: Re.

Constants: At wt: 186.31, mp: 3180 ± 60°C, d: 21.03, bp: 5900°C.

Hazard Analysis

Toxicity: Unknown.

Radiation Hazard: Section 5. For permissible levels, see Table 5, p. 150.

Artificial isotope ^{183}Re, half life 68 d. Decays to stable ^{183}W by electron capture. Emits gamma rays of 0.05–0.25 MeV and X-rays.

Artificial isotope ^{186}Re, half life 90 h. Decays to stable ^{186}Os by emitting beta particles of 0.93 (23%), 1.07 (73%) MeV. Also emits gamma rays of 0.14 MeV.

Natural (63%) isotope ^{187}Re, half life 6 x 10^{10} y. Decays to stable ^{187}Os by emitting beta particles of 0.01 MeV.

Artificial isotope ^{188}Re, half life 17 h. Decays to stable ^{188}Os by emitting beta particles of 1.96 (20%) 2.12 (78%) MeV. Also emits gamma rays of 0.16 MeV.

Fire Hazard: Moderate in the form of dust, when exposed to heat or flame (Section 6).

Countermeasures

Storage and Handling: Section 7.

RHENIUM DISULFIDE

General Information

Description: Black, hexagonal leaf.

Formula: ReS_2.

Constants: Mol wt: 250.44, bp: decomp., d: 7.5.

Hazard Analysis and Countermeasures

See sulfides.

RHENIUM HEPTASULFIDE

General Information

Description: Black crystals.

Formula: Re_2S_7.

Constants: Mol wt: 597.1, bp: decomposes, d: 4.87 at 24.5°C.

Hazard Analysis and Countermeasures

See sulfides.

RHENIUM HEXAFLUORIDE

General Information

Description: Pale yellow crystals.

Formula: ReF_6.

Constants: Mol wt: 300.31; mp: 25.6°C; bp: 47.6°C; d liquid: 6.1573, solid: 4.251.

Hazard Analysis and Countermeasures

See fluorides.

RHENIUM OXYTETRAFLUORIDE

General Information

Description: Colorless crystals.

Formula: $ReOF_4$.

Constants: Mol wt: 278.31, mp: 39.7°C, bp: 62.7°C, d liquid: 5.314; solid: 4.032.

Hazard Analysis and Countermeasures

See fluorides.

RHENIUM TETRAFLUORIDE

General Information

Description: White solid.

Formula: ReF_4.

Constants: Mol wt: 262.31, mp: 124.5°C.

Hazard Analysis and Countermeasures

See fluorides.

RHENIUM TRIOXYBROMIDE

General Information

Description: White crystals.

Formula: ReO_3Br.

Constants: Mol wt: 314.23, mp: 39.5°C, bp: 163°C.

Hazard Analysis and Countermeasures
See bromides.

RHIGOLENE. See pentane and isopentane.

RHODANINE
General Information
Synonym: 2-Thio-4-ketothiazolidine.
Description: Yellow crystals.
Formula: $C_3H_3ONS_2$.
Constants: Mol wt: 133.1, mp: 166°C (decomp.), d: 0.868.
Hazard Analysis
Toxicity: Details unknown. See also thiocyanates.
Disaster Hazard: Dangerous; when heated to decomposition, it emits highly toxic fumes of oxides of sulfur.
Countermeasures
Storage and Handling: Section 7.

RHODIUM
General Information
Description: Cubic gray-white crystals.
Formula: Rh.
Constants: Mol wt: 102.9, mp: 1985°C, bp: >2500°C, d: 12.4.
Hazard Analysis
Toxicity: Unknown.
TLV: ACGIH (recommended); Metal fumes and dusts: 0.1 milligram per cubic meter of air; soluble salts: 0.001 milligram per cubic meter of air.
Radiation Hazard: Section 5. For permissible levels, see Table 5, p. 150.
    Artificial isotope ^{103m}Rh, half life 57 m. Decays to stable ^{103}Rh by emitting gamma rays of 0.04 MeV and X-rays. ^{103m}Rh exists in radioactive equilibrium with its parent ^{103}Ru. Permissible levels are given for the equilibrium mixture.
    Artificial isotope ^{105}Rh, half life 36 h. Decays to stable ^{105}Pd by emitting beta particles of 0.25 (10%), 0.56 (90%) MeV. Also emits gamma rays of 0.32 MeV and others.
    Artificial isotope ^{106}Rh, half life 30 s. Decays to stable ^{106}Pd by emitting beta particles of 2.4 (11%), 3.0 (8%), 3.54 (78%), MeV. Also emits gamma rays of 0.51, 0.62, 1.05 MeV. ^{106}Rh exists in radioactive equilibrium with its parent ^{106}Ru. Permissible levels are given for the equilibrium mixture.
Fire Hazard: Moderate, when exposed to heat or flame (Section 6).
Countermeasures
Storage and Handling: Section 7.

RHODIUM HYDROSULFIDE
General Information
Description: Black crystals.
Formula: $Rh(HS)_3$.
Constants: Mol wt: 202.13, mp: decomposes.
Hazard Analysis and Countermeasures
See sulfides.

RHODIUM MONOSULFIDE
General Information
Description: Gray-black crystals.
Formula: RhS.
Constants: Mol wt: 134.98, mp: decomposes.
Hazard Analysis
See sulfides.

RHODIUM NITRATE
General Information
Description: Brown-yellow crystals.
Formula: $Rh(NO_3)_3$.
Constants: Mol wt: 288.9, mp: decomposes.
Hazard Analysis and Countermeasures
See nitrates.

RHODIUM SESQUISULFIDE
General Information
Description: Black crystals.
Formula: Rh_2S_3.
Constants: Mol wt: 302.02, mp: decomposes.
Hazard Analysis and Countermeasures
See sulfides.

RHODIUM SULFITE
General Information
Description: Yellow crystals.
Formula: $Rh_2(SO_3)_3 \cdot 6H_2O$.
Constants: Mol wt: 544.11, mp: decomposes.
Hazard Analysis and Countermeasures
See sulfites.

RHODIUM TRIFLUORIDE
General Information
Description: Rhombic, red crystals.
Formula: RhF_3.
Constants: Mol wt: 159.91, bp: >600°C sublimes, d: 5.38.
Hazard Analysis and Countermeasures
See fluorides.

RIBOFLAVIN
General Information
Synonyms: Vitamin B_2; 7,8-dimethyl-10-(1_1-D-ribityl isoalloxozine.
Description: Orange to yellow crystals; slightly soluble in water and alcohols; insoluble in lipid solvents.
Formula: $C_{17}H_{20}N_4O_6$.
Constants: Mol wt: 376, mp: 282°C (decomposes).
Hazard Analysis
Toxicity: Unknown. A nutrient and/or dietary supplement food additive (Section 10).

RIBOFLAVIN-5'-PHOSPHATE
General Information
Synonyms: FMN; flavin mononucleotide.
Description (sodium salt): Yellow crystals, soluble in water.
Hazard Analysis
Toxicity: Unknown. A nutrient and/or dietary supplement food additive (Section 10).

RICHE GAS. See carbon monoxide.

RICIN
General Information
Synonym: Agglutinin.
Description: White powder.
Hazard Analysis
Toxic Hazard Rating:
    Acute Local: Ingestion 3.
    Acute Systemic: Ingestion 3; Inhalation 3.
    Chronic Local: O.
    Chronic Systemic: Ingestion 2; Inhalation 2.
Toxicology: Causes violent purging which may lead to collapse and death. Small particle in eye, nose or any skin abrasion may prove fatal. May cause destruction of red blood cells.

TOXIC HAZARD RATING CODE (For detailed discussion, see Section 1.)

0 NONE: (a) No harm under any conditions; (b) Harmful only under unusual conditions or overwhelming dosage.

1 SLIGHT: Causes readily reversible changes which disappear after end of exposure.

2 MODERATE: May involve both irreversible and reversible change not severe enough to cause death or permanent injury.

3 HIGH: May cause death or permanent injury after very short exposure to small quantities.

U UNKNOWN: No information on humans considered valid by author

Disaster Hazard: Dangerous; when heated to decomposition it emits highly toxic fumes.
Countermeasures
Personal Hygiene: Section 3.
Storage and Handling: Section 7.

RICININE
General Information
Synonym: 1,2-Dihydro-4-methoxy-1-methyl-2-oxonicotino-nitrile.
Description: Alkaloid from castor bean plant.
Formula: $C_8H_8N_2O_2$.
Constants: Mol wt: 164.2, mp: 201.5°C, sublimes at 170-180°C at 20 mm.
Hazard Analysis
Toxic Hazard Rating:
Acute Local: U.
Acute Systemic: Ingestion 3; Inhalation 3.
Chronic Local: U.
Chronic Systemic: U.
Disaster Hazard: Dangerous; when heated to decomposition, it emits highly toxic fumes of cyanide.
Countermeasures
Ventilation Control: Section 2.
Personnel Protection: Section 3.
Storage and Handling: Section 7.

RICINUS OIL. See castor oil.

RIFLE GRENADES
Shipping Regulations: Section 11.
ICC: Explosive A, not acceptable.
IATA (with smoke material and without bursting charge): Explosive, explosive label, 25 kilograms (passenger), 70 kilograms (cargo).
(with gas or smoke material and with bursting charge; or, with incendiary material and with or without bursting charge): Explosive, not acceptable (passenger and cargo).

ROCKET AMMUNITION. See explosives, high.
Shipping Regulations: Section 11.
Coast Guard—Several Classifications.

ROCKET ENGINES
Shipping Regulations: Section 11.
ICC (liquid, class B explosives): Explosive B, not accepted.
IATA (commercial aircraft): Flammable solid, yellow label, not acceptable (passenger), 250 kilograms (cargo).
(Liquid fuel, for missiles or spacecraft): Explosive material, no label, not acceptable (passenger and cargo).

ROCKET MOTORS
Shipping Regulations: Section 11.
ICC (Class A explosives): Explosive A, not accepted.
(Class B explosives): Explosive B, red label, 550 lbs.
IATA (Solid fuel for missiles, rockets or spacecraft): Explosive material, no label, not acceptable (passenger and cargo).

ROCKETS, LINE THROWING
Shipping Regulations: Section 11.
IATA: Explosive, explosive label, not acceptable (passenger), 95 kilograms (cargo).

RODINOL. See p-aminophenol.

ROHRBACK'S SOLUTION. See mercuric barium iodide.

ROMAN VITRIOL. See copper sulfate.

RONNEL
General Information
Synonym: o,o-Dimethyl o-(2,4,5-trichlorophenyl) phosphorothioate.

Hazard Analysis
Toxicity: A food additive permitted in the feed and drinking water of animals and/or for the treatment of food producing animals. Also permitted in food for human consumption (Section 10). For symptoms see parathion.
TLV: ACGIH (tentative); 15 milligrams per cubic meter.

ROSANILINE DYES. See triphenylmethane dyes.

ROSIN
General Information
Synonyms: Gum rosin; colophony; piece resin.
Description: Pale yellow to amber, translucent fragments, turpentine odor and taste.
Constants: Mp: 100–150°C, flash p: 370°F (C.C.), d: 1.08.
Hazard Analysis
Toxic Hazard Rating:
Acute Local: Allergen 1.
Acute Systemic: U.
Chronic Local: Allergen 1.
Chronic Systemic: U.
Fire Hazard: Slight, can react with oxidizing materials. May ignite spontaneously in air. (Section 6).
Countermeasures
Storage and Handling: Section 7.
Shipping Regulations: Section 11.
Coast Guard Classification: Hazardous article.
To Fight Fire: Carbon dioxide, dry chemicals or carbon tetrachloride (Section 6).
Personal Hygiene: Section 3.

"ROSINAMINE-D." See dehydroabietyl amine.

ROSIN ESTER. See ester gum.

ROSIN OIL
General Information
Synonym: Retinol.
Description: An oil.
Constants: Bp: >80°C, flash p: 266°F (C.C.), d: 0.98, autoign. temp.: 648°F.
Hazard Analysis
Toxicity: Unknown. May cause dermatitis in sensitive individuals.
Fire Hazard: Slight; when exposed to heat or flame.
Spontaneous Heating: Yes.
Countermeasures
To Fight Fire: Foam, carbon dioxide, dry chemical or carbon tetrachloride (Section 6).
Ventilation Control: Section 2.
Storage and Handling: Section 7.

ROTENONE
General Information
Synonyms: Tubatoxin, derris.
Description: White, odorless crystals derived from derris root.
Formula: $C_{23}H_{22}O_6$.
Constants: Mol wt: 394.4, mp: 163°C, d: 1.27 at 20°C.
Hazard Analysis
Toxic Hazard Rating:
Acute Local: Irritant 1; Allergen 1; Ingestion 2; Inhalation 1.
Acute Systemic: Ingestion 2; Inhalation 2.
Chronic Local: Allergen 1; Irritant 1.
Chronic Systemic: Ingestion 1.
Toxicology: Adverse effects require large doses. Acute poisoning gives nausea, vomiting and tremors. Chronic exposure is said to be capable of causing injury to liver and kidneys. An insecticide.
TLV: (Commercial) ACGIH (recommended); 5 milligrams per cubic meter of air.
Fire Hazard: Slight (Section 6).
Countermeasures
Personal Hygiene: Section 3.
Storage and Handling: Section 7.

ROUGH AMMONIATE TANKAGES
Hazard Analysis
Toxicity: Unknown.
Fire Hazard: Dangerous, when exposed to heat or flame;
can react vigorously with oxidizing materials (Section 6).
Countermeasures
Storage and Handling: Section 7.
Shipping Regulations: Section 11.
I.C.C.: Flammable solid; yellow label, not accepted.
Coast Guard Classification: Inflammable solid; yellow
label.

R SALT. See sodium β-naphtholdisulfonate.

RUBBER CEMENT. See cement, rubber.

RUBBER, CRUDE
General Information
Synonyms: India rubber, caoutchouc.
Description: Light cream to dark amber, amorphous,
elastic, dry loaves, sheets or slabs.
Constant: D: About 0.9.
Hazard Analysis
Toxicity: None.
Fire Hazard: Very slight, but will support combustion (Sec-
tion 6).

RUBBER, CURED
Hazard Analysis
Toxicity: Cured rubber products may have slight toxicity
or cause superficial dermatitis due to the active vul-
canizing agents used, for example, sulfur, diphenylguani-
dine, phenyl-β-naphthylamine, mercaptobenzothiazole
and its derivatives. The rubber itself is not toxic.
Fire Hazard: Slight, when exposed to flame. In the form of
dust or fine particles, hard rubber is a more serious
hazard, and should be kept away from sparks and open
flame (Section 6).
Disaster Hazard: Dangerous in the form of dust. In burning
it will emit toxic fumes of sulfides.
Countermeasures
Shipping Regulations: None.

RUBBER LATEX. See latex, rubber.

RUBBER, RECLAIMED
Hazard Analysis
Toxicity: Slight. A weak allergen.
Fire Hazard: Dangerous, when exposed to flame; it emits
acrid and toxic fumes of sulfides when burning.
Countermeasures
Storage and Handling: Section 7.
Shipping Regulations: Section 11.
I.C.C.: Flammable solid; yellow label.
Coast Guard Classification: Inflammable solid; yellow
label.

RUBBER SCRAP OR BUFFINGS
Hazard Analysis
Toxicity: None.
Fire Hazard: Dangerous, when exposed to heat or flame
(Section 6).
Explosion Hazard: Moderate, in the form of dust when ex-
posed to flame.
Disaster Hazard: Dangerous; when heated to decomposition,
it emits toxic fumes of sulfides.
Countermeasures
Ventilation Control: Section 2.
Storage and Handling: Section 7.

Shipping Regulations: Section 11.
I.C.C.: Flammable solid; yellow label, 10 pounds.
Coast Guard Classification: Inflammable solid; yellow
label.

RUBBER SOLVENT
General Information
Description: A petroleum distillate used in making rubber
cements and in tire manufacture. Insoluble in water.
Constants: Flash p.: $-40°$ F (varies with manufacturer
autoign. temp.: $450°$ F (varies with manufacturer
lel: 1.0%, uel(7.0%, d: <1, bp: $100-280°$ F.
Hazard Analysis
Toxic Hazard Rating: U.
Fire Hazard: Dangerous, when exposed to heat or flame.
Countermeasures
Storage and Handling: Section 7.
To Fight Fire: Foam, alcohol foam.

RUBIDIUM
General Information
Description: Soft, silvery-white metal.
Formula: Rb.
Constants: At wt: 85.48, mp: $38.5°$ C, bp: $700°$ C, d: 1.53
d liquid 1.475 at $38.5°$ C, vap. press.: 1 mm at $297°$ C.
Hazard Analysis
Toxic Hazard Rating:
Acute Local: U.
Acute Systemic: Ingestion 2.
Chronic Local: Ingestion 1.
Chronic Systemic: Ingestion 1.
Radiation Hazard: Section 5. For permissible levels, se
Table 5, p. 150.
Artificial isotope ^{86}Rb, half life 19 d. Decays to stab
^{86}Sr by emitting beta particles of 0.70 (9%), 1.78 (91%
MeV. Also emits gamma rays of 1.08 MeV.
Natural (28%) isotope ^{87}Rb half life 4.7×10^{10}
Decays to stable ^{87}Sr by emitting beta particles
0.27 MeV.
Fire Hazard: Dangerous, when exposed to heat or flame
by chemical reaction with oxidizers. See also sodiu
(Section 6).
Explosion Hazard: Moderate; reacts explosively with moi
ture, acids and oxidizers. See also sodium and NaK.
Disaster Hazard: Dangerous; when heated, it emits tox
fumes of rubidium oxide; it will react with water
steam to produce hydrogen and flammable vapor
reacts vigorously with oxidizing materials.
Countermeasures
Storage and Handling: Section 7.
Shipping Regulations: Section 11.
IATA: Flammable solid, yellow label, not acceptabl
(passenger), 12 kilograms (cargo).
IATA (in metal cartridges): Flammable solid, yello
label, $\frac{1}{2}$ kilograms (passenger), 12 kilograms (cargo).

RUBIDIUM BROMATE
General Information
Formula: RbBrO$_3$.
Constants: Mol wt: 213.40, mp: $430°$ C, d: 3.68.
Hazard Analysis and Countermeasures
See bromates.

RUBIDIUM BROMOCHLOROIODIDE
General Information
Description: Rhombic crystals.
Formula: RbIBrCl.

TOXIC HAZARD RATING CODE *(For detailed discussion, see Section 1.)*

0 NONE: (a) No harm under any conditions; (b) Harmful only under unusual conditions or overwhelming dosage.

1 SLIGHT: Causes readily reversible changes which disappear after end
of exposure.

2 MODERATE: May involve both irreversible and reversible chang
not severe enough to cause death or permanent injury.

3 HIGH: May cause death or permanent injury after very short exposu
to small quantities.

U UNKNOWN: No information on humans considered valid by autho

Constants: Mol wt: 327.8, mp: 205°C, bp: decomp. at 200°C.
Hazard Analysis and Countermeasures
See bromides and iodides.

RUBIDIUM CARBONATE
General Information
Description: White, deliquescent powder.
Formula: Rb_2CO_3.
Constants: Mol wt: 230.97, mp: 837°C, bp: decomp.
Hazard Analysis
Toxicity: See also potassium carbonate.

RUBIDIUM CHLORATE
General Information
Description: Crystals.
Formula: $RbClO_3$.
Constants: Mol wt: 168.94, d: 3.19.
Hazard Analysis and Countermeasures
See chlorates.

RUBIDIUM CHLOROPLATINATE
General Information
Description: Cubic, yellow crystals.
Formula: Rb_2PtCl_6.
Constants: Mol wt: 578.93, mp: decomposes, d: 3.94 at 17.5°C.
Hazard Analysis and Countermeasures
See platinum compounds and clorides.

RUBIDIUM CHROMATE
General Information
Description: Yellow crystals.
Formula: Rb_2CrO_4.
Constants: Mol wt: 286.97, d: 3.518.
Hazard Analysis
Toxicity: See chromium compounds.
Fire Hazard: Moderate, by chemical reaction with reducing agents (Section 6).
Countermeasures
Storage and Handling: Section 7.

RUBIDIUM CHROMIUM SULFATE
General Information
Description: Cubic, violet crystals.
Formula: $RbCr(SO_4)_2 \cdot 12H_2O$.
Constants: Mol wt: 545.81, mp: 107°C, d: 1.946.
Hazard Analysis and Countermeasures
See chromium compounds and sulfates.

RUBIDIUM COPPER SULFATE
General Information
Description: Monoclinic crystals.
Formula: $Rb_2SO_4 \cdot CuSO_4 \cdot 6H_2O$.
Constants: Mol wt: 534.73, d: 2.57.
Hazard Analysis and Countermeasures
See copper compounds and sulfates.

RUBIDIUM DICHROMATE
General Information
Description: Triclinic or monoclinic crystals.
Formula: $Rb_2Cr_2O_7$.
Constants: Mol wt: 386.98, d: 3.02-3.13.
Hazard Analysis
Toxicity: See chromium compounds.
Fire Hazard: Moderate, by chemical reaction with reducing agents (Section 6).
Countermeasures
Storage and Handling: Section 7.

RUBIDIUM DISULFIDE
General Information
Description: Dark red crystals.
Formula: Rb_2S_2.
Constants: Mol wt: 235.09, bp: volat. >850°C, mp: 420°C.
Hazard Analysis and Countermeasures
See sulfides.

RUBIDIUM FLUORIDE
General Information
Description: Colorless crystals.
Formula: RbF.
Constants: Mol wt: 104.48, mp: 760°C, bp: 1410°C, d liquid: 2.88 at 820°C, vap. press.: 1 mm at 921°C.
Hazard Analysis and Countermeasures
See fluorides.

RUBIDIUM FLUOSILICATE
General Information
Description: Cubic crystals.
Formula: Rb_2SiF_6.
Constants: Mol wt: 313.02, d: 3.332.
Hazard Analysis and Countermeasures
See fluosilicates.

RUBIDIUM FLUOSULFONATE
General Information
Description: Needles.
Formula: $RbFSO_3$.
Constants: Mol wt: 184.55, mp: 304°C.
Hazard Analysis and Countermeasures
See fluosulfonates.

RUBIDIUM GALLIUM SULFATE
General Information
Description: Colorless crystals.
Formula: $RbGa(SO_4)_2 \cdot 12H_2O$.
Constants: Mol wt: 563.52, d: 1.962.
Hazard Analysis and Countermeasures
See gallium compounds and sulfates.

RUBIDIUM GRAPHITE
General Information
Description: Violet to black platelets. Highly reactive with water, air, and alcohols.
Formula: RbC_8.
Constant: Mol wt: 181.5.
Hazard Analysis and Countermeasures
See rubidium.

RUBIDIUM HEXASULFIDE
General Information
Description: Brown-red crystals.
Formula: Rb_2S_6.
Constants: Mol wt: 363.36, mp: 201°C.
Hazard Analysis and Countermeasures
See sulfides.

RUBIDIUM HYDRATE. See rubidium hydroxide.

RUBIDIUM HYDRIDE
General Information
Description: Colorless needles. Reacts with water.
Formula: RbH.
Constants: Mol wt: 86.49, mp: decomposes at 300°C, d: 2.60.
Hazard Analysis and Countermeasures
See hydrides.

RUBIDIUM HYDROGEN NITRATE
General Information
Description: Tetragonal crystals.
Formula: $RbNO_3 \cdot HNO_3$.
Constants: Mol wt: 210.50, mp: 62°C.
Hazard Analysis and Countermeasures
See nitrates.

RUBIDIUM HYDROXIDE
General Information
Synonym: Rubidium hydrate.
Description: Grayish-white, deliquescent mass; strong base.
Formula: RbOH.
Constants: Mol wt: 102.49, mp: 300°C, d: 3.203 at 11°C.
Hazard Analysis
Toxicity: See also potassium hydroxide.

RUBIDIUM IODATE
General Information
Description: Cubic crystals.
Formula: $RbIO_3$.
Constants: Mol wt: 260.40, mp: decomposes, d: 4.33 at 19.5°C.
Hazard Analysis and Countermeasures
See iodates.

RUBIDIUM IODIDE
General Information
Description: Colorless crystals.
Formula: RbI.
Constants: Mol wt: 212.40; mp: 642°C; bp: 1300°C; d: 3.55, liquid: 2.87 at 825°C; vap. press.: 1 mm at 748°C.
Hazard Analysis and Countermeasures
See iodides.

RUBIDIUM MONOSULFIDE
General Information
Description: Colorless crystals.
Formula: Rb_2S.
Constants: Mol wt: 203.03, mp: 530°C decomposes, d: 2.912.
Hazard Analysis and Countermeasures
See sulfides.

RUBIDIUM MONOXIDE
General Information
Description: Cubic, colorless-yellow crystals.
Formula: Rb_2O.
Constants: Mol wt: 186.96, mp: decomposes at 400°C, d: 3.72.
Hazard Analysis
Toxicity: See also potassium oxide.

RUBIDIUM NITRATE
General Information
Description: Hexagonal, cubic, rhombic or triclinic crystals.
Formula: $RbNO_3$.
Constants: Mol wt: 147.49, mp: 310°C, d: 3.11; d liquid: 2.395 at 400°C.
Hazard Analysis and Countermeasures
See nitrates.

RUBIDIUM PENTASULFIDE
General Information
Description: Rhombic, red, deliquescent crystals.
Formula: Rb_2S_5.
Constants: Mol wt: 331.29, mp: 225°C, d: 2.618 at 15°C.
Hazard Analysis and Countermeasures
See sulfides.

RUBIDIUM PERCHLORATE
General Information
Description: Rhombic crystals.
Formula: $RbClO_4$.
Constants: Mol wt: 184.94, mp: fuses, bp: decomposes, d: 2.9.
Hazard Analysis and Countermeasures
See perchlorates.

RUBIDIUM m-PERIODATE
General Information
Description: Tetragonal crystals.
Formula: $RbIO_4$.
Constants: Mol wt: 276.40, d: 3.918 at 16°c.

Hazard Analysis and Countermeasures
See iodates.

RUBIDIUM PERMANGANATE
General Information
Description: Crystals.
Formula: $RbMnO_4$.
Constants: Mol wt: 204.41, d: 3.235 at 10.4°C.
Hazard Analysis
Toxicity: See manganese compounds.
Fire Hazard: Moderate, by chemical reaction with reducing agents (Section 6). An oxidizing agent; keep away from flammable materials.
Countermeasures
Storage and Handling: Section 7.

RUBIDIUM PEROXIDE
General Information
Description: Yellow crystals.
Formula: Rb_2O_2.
Constants: Mol wt: 202.96, mp: 600°C, d: 3.65 at 0°C.
Hazard Analysis and Countermeasures
See peroxides, inorganic.

RUBIDIUM PRASEODYMIUM NITRATE
General Information
Description: Greenish, hygroscopic needles.
Formula: $2RbNO_3 \cdot Pr(NO_3)_3 \cdot 4H_2O$.
Constants: Mol wt: 693.98, mp: 63.5; $-4H_2O$ at 60°C, d: 2.50.
Hazard Analysis and Countermeasures
See nitrates.

RUBIDIUM SELENATE
General Information
Description: Colorless, rhombic crystals.
Formula: Rb_2SeO_4.
Constants: Mol wt: 313.92, d: 3.90.
Hazard Analysis and Countermeasures
See selenium compounds.

RUBIDIUM SUPEROXIDE
General Information
Description: Yellow crystals.
Formula: RbO_2.
Constants: Mol wt: 117.48, mp: 280°C, d: 3.05 at 0°C.
Hazard Analysis and Countermeasures
See peroxides, inorganic.

RUBIDIUM TRIBROMIDE
General Information
Description: Rhombic crystals.
Formula: $RbBr_3$.
Constants: Mol wt: 325.23, mp: decomp. at 140°C.
Hazard Analysis and Countermeasures
See bromides.

RUBIDIUM TRIIODIDE
General Information
Description: Rhombic, black crystals.
Formula: RbI_3.
Constants: Mol wt: 466.24, mp: 190°C, d: 4.03.
Hazard Analysis and Countermeasures
See iodides.

RUBIDIUM TRISULFIDE
General Information
Description: Reddish-yellow crystals.
Formula: Rb_2S_3.

TOXIC HAZARD RATING CODE (For detailed discussion, see Section 1.)

0 NONE: (a) No harm under any conditions; (b) Harmful only under unusual conditions or overwhelming dosage.

1 SLIGHT: Causes readily reversible changes which disappear after end of exposure.

2 MODERATE: May involve both irreversible and reversible change not severe enough to cause death or permanent injury.

3 HIGH: May cause death or permanent injury after very short exposu to small quantities.

U UNKNOWN: No information on humans considered valid by author

Constants: Mol wt: 267.16, mp: 213°C.
Hazard Analysis and Countermeasures
See sulfides.

RUM, DENATURED
General Information
Constant: Flash p: 77°F (C.C.).
Hazard Analysis
Toxicity: Details unknown. See ethyl alcohol.
Fire Hazard: Dangerous, when exposed to heat or flame.
Disaster Hazard: Dangerous, upon exposure to heat or flame; can react vigorously with oxidizing materials.
Countermeasures
Storage and Handling: Section 7.
To Fight Fire: Carbon dioxide, dry chemical or carbon tetrachloride (Section 6).
Shipping Regulations: Section 11.
  I.C.C.: Flammable liquid; red label, 10 gallons.
  Coast Guard Classification: Inflammable liquid; red label.
  IATA: Flammable liquid, red label, 1 liter (passenger), 40 liters (cargo).

RUTHENIUM
General Information
Description: I: Black, porous metal. II: Silvery-white, non-ductile metal of the platinum group.
Formula: Ru.
Constants: At wt: 101.70; d I: 8.6, II: 12.06; mp I: > 1950°C, II: 2450°C; bp II: 4150°C.
Hazard Analysis
Toxicity: See ruthenium compounds.
Radiation Hazard: Section 5. For permissible levels, see Table 5, p. 150.
  Artificial isotope ^{97}Ru, half life 2.9 d. Decays to radioactive ^{97}Tc by electron capture. Emits gamma rays of 0.22 MeV and X-rays.
  Artificial isotope ^{103}Ru, half life 40 d. Decays to radioactive $^{103\,m}$Rh by emitting beta particles of 0.10 (7%), 0.21 (89%) MeV. Also emits gamma rays of 0.50, 0.61 MeV. ^{103}Ru exists in radioactive equilibrium with its daughter. Permissible levels are given for the equilibrium mixture.
  Artificial isotope ^{105}Ru, half life 4.4 h. Decays to radioactive ^{105}Rh by emitting beta particles of 0.92–1.87 MeV. Also emits gamma rays of 0.13 to 0.97 MeV.
  Artificial isotope ^{106}Ru, half life 1.0 y. Decays to radioactive ^{106}Rh by emitting beta particles of 0.039 MeV. ^{106}Ru exists in radioactive equilibrium with its daughter. Permissible levels are given for the equilibrium mixture.
Fire Hazard: Moderate, in the form of dust, when exposed to heat or flame (Section 6).

RUTHENIUM COMPOUNDS
Hazard Analysis
Toxicity: Details unknown; probably toxic, but such small amounts are used industrially that it does not constitute a hazard. It resembles osmium in that when it is heated in air, it evolves fumes which are injurious to the eyes and lungs.
Disaster Hazard: Dangerous; when heated to decomposition, it emits toxic fumes of ruthenium oxide.
Countermeasures
Storage and Handling: Section 7.

RUTHENIUM HYDROXIDE
General Information
Description: Black powder.

Formula: Ru(OH)₃.
Constant: Mol wt: 152.72.
Hazard Analysis and Countermeasures
See ruthenium compounds.

RUTHENIUM PENTAFLUORIDE
General Information
Description: Dark green crystals.
Formula: RuF₅.
Constants: Mol wt: 196.70, mp: 101°C, bp: 270°C, d: 2.963 at 16.5°C.
Hazard Analysis and Countermeasures
See fluorides and ruthenium compounds.

RUTHENIUM SULFIDE
General Information
Synonym: Laurite.
Description: Cubic, gray-black crystals.
Formula: RuS₂.
Constants: Mol wt: 165.83, d: 6.99.
Hazard Analysis and Countermeasures
See sulfides and ruthenium compounds.

RUTHENIUM TETROXIDE
General Information
Description: Yellow, volatile crystals; odor of ozone.
Formula: RuO₄.
Constants: Mol wt: 165.70, mp: 25.5°C, bp: approx. 100°C decomp., d: 3.29 at 21°C.
Hazard Analysis
Toxic Hazard Rating:
  Acute Local: Irritant 2; Inhalation 2.
  Acute Systemic: U.
  Chronic Local: U.
  Chronic Systemic: U.
Fire Hazard: Moderate, by chemical reaction with reducing agents (Section 6). A powerful oxidizing agent.
Disaster Hazard: See ruthenium compounds.
Countermeasures
Ventilation Control: Section 2.
Personnel Protection: Section 3.
Storage and Handling: Section 7.

RUTILE. See titanium dioxide.

RYANIA
General Information
Composition: Ground wood of Ryania spociosa.
Hazard Analysis
Toxicity: Details unknown; animal experiments show moderate toxicity with gastrointestinal disturbances, tremors and convulsions. An insecticide. Ryania is not considered to be a hazard to man when used as an insecticide according to accepted agricultural practices. No tolerance limits have been established or proposed by the government for residues on foods.
Fire Hazard: Moderate, when exposed to heat or flame (Section 6).

RYANODINE
General Information
Description: Crystals, soluble in water, alcohol, acetone, chloroform; practically insoluble in benzene and petroleum ether.
Formula: C₂₅H₃₅O₉.
Constant: Mol wt: 493.54.
Hazard Analysis
See ryania.

S

SABADILLA SEED
General Information
Synonyms: Cevadilla; caustic barley.
Description: A powder.
Hazard Analysis
Toxic Hazard Rating:
Acute Local: Irritant 3; Ingestion 3; Inhalation 3.
Acute Systemic: Ingestion 2.
Chronic Local: Irritant 2.
Chronic Systemic: Ingestion 2.
Toxicology: Ingestion causes severe gastrointestinal disturbances, burning in the mouth, vomiting, diarrhea and cramps. Also produces headache, dizziness, slow pulse and weakness. Large doses cause death by circulatory and respiratory failure. It is a powerful irritant to skin and mucous membranes. It is used as an insecticide. For hazard analysis as an insecticide. See ryania.
Disaster Hazard: Dangerous; when heated, it emits toxic fumes.
Countermeasures
Ventilation Control: Section 2.
Personal Hygiene: Section 3.
Storage and Handling: Section 7.

SACCHARIN
General Information
Synonyms: Saxin; benzosulfinide.
Description: Crystals or powder.
Formula: $C_7H_5NO_3S$.
Constants: Mol wt: 183.2, mp: 228°C (decomp.), bp: sublimes.
Hazard Analysis
Toxic Hazard Rating:
Acute Local: U.
Acute Systemic: Ingestion 1.
Chronic Local: U.
Chronic Systemic: U.
Note: A non-nutritive sweetener food additive (Section 10).
Disaster Hazard: Dangerous; when heated to decomposition, it emits highly toxic fumes of oxides of sulfur, oxides of nitrogen.
Countermeasures
Personal Hygiene: Section 3.
Storage and Handling: Section 7.

SAFETY FUSES. See explosives, high.
Shipping Regulations: Section 11.
I.C.C.: See Section 11, § 73.1000 (o).
Coast Guard Classification: Explosive C.

SAFETY LAMP GASOLINE. See gasoline.

SAFETY SOLVENT. See Stoddard solvent.

SAFETY SQUIBS. See explosives, high.
Shipping Regulations: Section 11.
I.C.C.: Explosive C, 150 pounds.

Coast Guard Classification: Explosive C.
IATA: Explosive, explosive label, 25 kilograms (passenger), 70 kilograms (cargo).

SAFFLOWER OIL
General Information
Description: Oil from the seed of Carthamus tinctorious.
Constants: D: 0.9211 at 25/25°C.
Hazard Analysis
Toxicity: Details unknown. If swallowed in large amounts, it causes vomiting.

SAFROLE
General Information
Synonyms: 1-Allyl-3,4-methylenedioxybenzene; shikimole.
Description: Colorless liquid or crystals.
Formula: $C_3H_5C_6H_3O_2CH_2$.
Constants: Mol wt: 162.18, mp: 11°C, bp: 234.5°C, d: 1.0960 at 20°C, vap. press.: 1 mm at 63.8°C.
Hazard Analysis
Toxic Hazard Rating:
Acute Local: Irritant 3.
Acute Systemic: Ingestion 3.
Chronic Local: Irritant 2.
Chronic Systemic: U.
Fire Hazard: Slight (Section 6).
Countermeasures
Storage and Handling: Section 7.

SAL ACETOSELLA. See potassium binoxalate.

SAL AMMONIA. See ammonium chloride.

SALICYL ALDEHYDE
General Information
Synonyms: Salicylic aldehyde.
Description: Clear, colorless, oily liquid; burning taste.
Formula: C_6H_4OHCOH.
Constants: Mol wt: 122.1, bp: 197°C, fp: 1°C, d: 1.167 at 20°/4°C, vap. press.: 1mm at 33.0°C, flash p.: 172°F.
Hazard Analysis
Toxicity: Details unknown; an auxiliary fumigant. Limited animal experiments suggest low toxicity. See also aldehydes.
Fire Hazard: Moderate, when exposed to heat or flame; can react with oxidizing materials (Section 6).
Countermeasures
Storage and Handling: Section 7.
To Fight Fire: Alcohol foam.

SALICYLAMIDE
General Information
Synonym: o-Hydroxybenzamide.
Description: White to slightly pink crystals or powder. Somewhat bitter tasting.
Formula: $C_7H_7NO_2$.

TOXIC HAZARD RATING CODE *(For detailed discussion, see Section 1.)*

0 NONE: (a) No harm under any conditions; (b) Harmful only under unusual conditions or overwhelming dosage.

1 SLIGHT: Causes readily reversible changes which disappear after end of exposure.

2 MODERATE: May involve both irreversible and reversible changes; not severe enough to cause death or permanent injury.

3 HIGH: May cause death or permanent injury after very short exposure to small quantities.

U UNKNOWN: No information on humans considered valid by authors.

Constants: Mol wt: 137.1, mp: 140°C.
Hazard Analysis
Toxicity: Details unknown. Can cause dizziness, drowsiness, nausea, vomiting, epigastric distress, allergic reactions and blood dyscrasias in average to large doses.
Disaster Hazard: Dangerous; when heated to decomposition, it emits highly toxic fumes.

SALICYLICANILIDE
General Information
Synonym: Anasadol, salinidol.
Description: White, odorless crystals.
Formula: $C_6H_5NHCOC_6H_4OH$.
Constants: Mol wt: 213.2, mp: 135°C, bp: decomposes.
Hazard Analysis
Toxic Hazard Rating:
   Acute Local: Irritant 2; Ingestion 2.
   Acute Systemic: Ingestion 2.
   Chronic Local: Irritant 1.
   Chronic Systemic: Ingestion 2.
Toxicology: See also salicylic acid, aniline, and amides.
Disaster Hazard: Dangerous; when heated to decomposition, it emits highly toxic fumes of nitrogen oxides.
Countermeasures
Storage and Handling: Section 7.

SALICYLIC ACID
General Information
Synonym: o-Hydroxybenzoic acid.
Description: White needle crystals or powder; sweetish, afterward acrid taste.
Formula: HOC_6H_4COOH.
Constants: Mol wt: 138.12, mp: 159°C, bp: sublimes at 76°C, flash p: 315°F, d: 1.443 at 20°/4°C, autoign. temp.: 1013°F. vap. press.: 1 mm at 113.7°C, vap. d.: 4.8.
 Hazard Analysis
Toxic Hazard Rating:
   Acute Local: Ingestion 2.
   Acute Systemic: Ingestion 1.
   Chronic Local: U.
   Chronic Systemic: Ingestion 1; Inhalation 1.
Toxicology: Symptoms of poisoning are nausea and vomiting, ringing in the ears, dizziness, headache, dullness, confusion, sweating, rapid pulse and breathing and sometimes skin eruptions. Symptoms disappear when exposure or administration of the drug is terminated. Used as a food additive permitted in food for human consumption (Section 10).
Fire Hazard: Slight, when exposed to heat or flame; can react with oxidizing materials.
Countermeasures
To Fight Fire: Water, foam, carbon dioxide, dry chemical or carbon tetrachloride (Section 6).
Personal Hygiene: Section 3.
Storage and Handling: Section 7.

SALICYLIC ALDEHYDE. See salicyl aldehyde.

SALINIDOL. See salicylamide.

SALOL. See phenyl salicylate.

SALT. See sodium chloride.

SAL TARTAR. See sodium tartrate.

SALT BATHS (NITRATE OR NITRITE)
Hazard Analysis
Toxicity: Details unknown. See also nitrates and nitrites.
Fire Hazard: Dangerous, by spontaneous chemical reaction. These baths are oxidizing in nature. (Section 6).
Explosion Hazard: Moderate, by chemical reaction, due to contamination by cyanides or easily oxidizable materials or when heated to over 1000°F.
Disaster Hazard: Highly dangerous; in molten form will react with water, steam or acids to produce heat, hydrogen, toxic and corrosive fumes; can react vigorously with reducing materials.

SALT OF SORREL. See potassium binoxalate.

SALTPETER. See potassium nitrate.

SAMARIUM
General Information
Description: Hexagonal, gray-white, metallic crystalline element.
Formula: Sm.
Constants: At wt: 150.43, mp: >1300°C, d: 7.7.
Hazard Analysis
Toxicity: Unknown. As a lanthanon it may cause impairment of blood clotting. See also lanthanum.
Radiation Hazard: Section 5. For permissible levels, see Table 5, p. 150.
   Natural (15%) isotope ^{147}Sm, half life 1.2×10^{11} y. Decays to stable ^{143}Nd by emitting alpha particles of 2.12 MeV.
   Natural (14%) isotope ^{149}Sm, half life 4×10^{14} y. Decays to stable ^{145}Nd by emitting alpha particles of 1.84 MeV.
   Artificial isotope ^{151}Sm, half life 93 y. Decays to stable ^{151}Eu by emitting beta particles of 0.076 MeV.
   Artificial isotope ^{153}Sm, half life 47 h. Decays to stable ^{153}Eu by emitting beta particles of 0.64 (33%), 0.70 (46%), 0.80 (20%) MeV. Also emits gamma rays of 0.07, 0.10 MeV and others.
Fire Hazard: Moderate, in the form of dust, when exposed to flame or by spontaneous chemical reaction with oxidizers. See also powdered metals (Section 6).

SAMARIUM BROMATE
General Information
Description: Hexagonal, yellow crystals.
Formula: $Sm(BrO_3)_3 \cdot 9H_2O$.
Constants: Mol wt: 696.32, mp: 75°C; $-9H_2O$ at 150°C.
Hazard Analysis and Countermeasures
See bromates and samarium.

SAMARIUM BROMIDE
General Information
Description: Yellow deliquescent crystals.
Formula: $SmBr_3 \cdot 6H_2O$.
Constants: Mol wt: 498.27, d: 2.971 at 22°C.
Hazard Analysis and Countermeasures
See bromides and samarium.

SAMARIUM CARBIDE
General Information
Description: Yellow, crystalline mass.
Formula: SmC_2.
Constants: Mol wt: 174.45, d: 5.86.
Hazard Analysis and Countermeasures
See carbides and samarium.

SAMARIUM DICHLORIDE
General Information
Description: Dark reddish-brown, crystalline mass.
Formula: $SmCl_2$.
Constants: Mol wt: 221.34, mp: 740°C, d: 3.687 at 22°C.
Hazard Analysis and Countermeasures
See hydrochloric acid and samarium.

SAMARIUM IODIDE
General Information
Description: Orange-yellow crystals.
Formula: SmI_3.
Constants: Mol wt: 531.19, mp: 816–824°C, bp: decomposes.
Hazard Analysis and Countermeasures
See iodides and samarium.

SAMARIUM NITRATE
General Information
Description: Triclinic, pale yellow crystals.
Formula: $Sm(NO_3)_3 \cdot 6H_2O$.
Constants: Mol wt: 444.55, mp: 78–79°C, d: 2.375.
Hazard Analysis
Toxicity: Details unknown. Limited animal experiments suggest low toxicity.
Countermeasures
See nitrates and samarium.

SAMARIUM OXALATE
General Information
Description: Crystals.
Formula: $Sm_2(C_2O_4)_3 \cdot 10H_2O$.
Constant: Mol wt: 745.08.
Hazard Analysis and Countermeasures
See oxalates and samarium.

SAMARIUM SULFIDE
General Information
Description: Yellowish-pink crystals.
Formula: Sm_2S_3.
Constants: Mol wt: 397.06, mp: 1900°C, d: 5.729.
Hazard Analysis and Countermeasures
See sulfides and samarium.

SAMPLES, EXPLOSIVES
Shipping Regulations: Section 11.
    ICC: Section 73.86.
    ICC (New): Section 73.86.
    IATA: Explosive, not acceptable (passenger and cargo).
    IATA (New): Explosive, not acceptable (passenger and cargo).

SANTOBANE. See ddt.

SANTONIC LACTONE. See santonin.

SANTONIN
General Information
Synonym: Santonic lactone.
Description: Glossy, colorless crystals or white powder, turning yellow on exposure to light; odorless, tasteless at first, then bitter.
Formula: $C_{15}H_{18}O_3$.
Constants: Mol wt: 246.30, mp: 170°C, bp: sublimes, d: 1.187.
Hazard Analysis
Toxic Hazard Rating:
    Acute Local: Ingestion 1.
    Acute Systemic: Ingestion 2.
    Chronic Local: U.
    Chronic Systemic: U.
Toxicology: It can cause disturbance of color vision. Objects first show bluish tinge, then yellow which is most prominent. Complete blindness may occur, lasting perhaps for nearly a week. Dizziness, drowsiness and nausea may also occur. Recovery is spontaneous.
Fire Hazard: Slight; when heated it emits acrid fumes (Section 6).
Countermeasures
Personal Hygiene: Section 3.
Storage and Handling: Section 7.

SANTOPHEN. See o-benzyl-p-chlorophenol.

SAPOGLYCOSIDES. See saponins.

SAPONIFIED CRESOL. See cresol.

SAPONINS
General Information
Synonyms: Sapotoxins; sapoglycosides.
Hazard Analysis
Toxic Hazard Rating:
    Acute Local: U.
    Acute Systemic: Ingestion 3.
    Chronic Local: U.
    Chronic Systemic: Ingestion 3.
Toxicity: Highly toxic.
Toxicology: When administered by injection, saponins cause rapid and severe destruction of red blood cells.
Fire Hazard: Slight (Section 6).
Countermeasures
Personal Hygiene: Section 3.
Treatment and Antidotes: Wash out stomach if vomiting has not occurred. Other treatment is supportive and symptomatic.
First Aid: Section 1.
Storage and Handling: Section 7.

SAPOTOXINS. See saponins.

SAPP. See sodium acid pyrophosphate.

SARCOSINE. See N-methylglycine.

SARIN
General Information
Synonym: Fluoroisopropoxymethylphosphine oxide.
Description: Colorless liquid.
Formula: $[(CH_3)_2CHO](CH_3)(F)(O)P$.
Constants: Mol wt: 140.1, bp: 147°C, fp: −38°C, d: 1.100 at 20°C, vap. press.: 1.57 mm at 20°C, vap. d.: 4.86.
Hazard Analysis
Toxic Hazard Rating:
    Acute Local: Irritant 3; Ingestion 3; Inhalation 3.
    Acute Systemic: Ingestion 3; Inhalation 3; Skin Absorption 3.
    Chronic Local: U.
    Chronic Systemic: Ingestion 3; Inhalation 3; Skin Absorption 3.
Toxicology: This is a nerve gas, acting much like tabun. Highly toxic to eyes. A small drop on skin will kill a man within 15 minutes. Liquid does jot injure skin but penetrates it rapidly. See parathion.
Fire Hazard: Slight, when exposed to heat or flame.
Disaster Hazard: Highly dangerous; when heated, it emits highly toxic fumes; will react with water or steam to produce toxic and corrosive fumes; can react with oxidizing materials.
Countermeasures
Ventilation Control: Section 2.
To Fight Fire: Foam, carbon dioxide, dry chemical or carbon tetrachloride (Section 6).
Personnel Protection: Section 3.
First Aid: Section 1.
Storage and Handling: Section 7.

SAWDUST
General Information
Synonym: Wood dust.
Description: Yellowish particles of wood.
Hazard Analysis
Toxic Hazard Rating:
    Acute Local: Allergen 1; Inhalation 1.

TOXIC HAZARD RATING CODE (For detailed discussion, see Section 1.)

0 NONE: (a) No harm under any conditions; (b) Harmful only under unusual conditions or overwhelming dosage.

1 SLIGHT: Causes readily reversible changes which disappear after end of exposure.

2 MODERATE: May involve both irreversible and reversible changes; not severe enough to cause death or permanent injury.

3 HIGH: May cause death or permanent injury after very short exposure to small quantities.

U UNKNOWN: No information on humans considered valid by authors.

Acute Systemic: U.
Chronic Local: Allergen 1; Inhalation 1.
Chronic Systemic: U.
Note: A common air contaminant (Section 4).
Fire Hazard: Slight, when exposed to heat or flame; can react with oxidizing materials (Section 6).
Spontaneous Heating: Possible. Avoid hot, humid storage or contact with drying oils. Particularly dangerous if charred or partially burned.
Explosion Hazard: Slight, when exposed to flame.
Countermeasures
Storage and Handling: Section 7.
Shipping Regulations: Section 11.
  Coast Guard Classification: Hazardous, when dry, clean and oil-free.

SAXIN. See saccharin.

SCACCHITE. See manganese dichloride.

SCANDIUM
General Information
Synonym: Ekaboron.
Description: Silver crystals.
Formula: Sc.
Constants: At wt: 45.0, mp: 1200° C, bp: 2400° C, d: 2.5.
Hazard Analysis
Toxicity; unknown.
Radiation Hazard: Section 5. For permissible levels, see Table 5, p. 150.
    Artificial isotope ^{46}Sc, half life 84 d. Decays to stable ^{46}Ti by emitting beta particles of 0.36 MeV. Also emits gamma rays of 0.89, 1.12 MeV.
    Artificial isotope ^{47}Sc, half life 3.4 d. Decays to stable ^{47}Ti by emitting beta particles of 0.44 (70%), 0.60 (30%) MeV. Also emits gamma rays of 0.16 MeV.
    Artificial isotope ^{48}Sc, half life 1.8 d. Decays to stabel ^{48}Ti by emitting beta particles of 0.65 MeV. Also emits gamma rays of 0.99, 1.04, 1.31 MeV.
Fire Hazard: Moderate, in the form of dust, when exposed to heat or flame or by chemical reaction with oxidizers. See also powdered metals (Section 6).

SCANDIUM BROMIDE
General Information
Description: Crystals.
Formula: ScBr$_3$.
Constants: Mol wt: 284.85, mp: sublimes > 1000° C, d: 3.914.
Hazard Analysis and Countermeasures
See bromides.

SCANDIUM NITRATE
General Information
Description: Colorless crystals.
Formula: Sc(NO$_3$)$_3$.
Constants: Mol wt: 231.12, mp: 150° C.
Hazard Analysis and Countermeasures
See nitrates.

SCANDIUM OXALATE
General Information
Description: Crystals.
Formula: Sc$_2$(C$_2$O$_4$)$_3$·5H$_2$O.
Constants: Mol wt: 444.34, mp: −4H$_2$O at 140° C.
Hazard Analysis and Countermeasures
See oxalates.

SCHEELE'S MINERAL. See copper arsenite.

SCHIFF'S BASE. See disalicylal propylene diimine.

SCHRADAN. See octamethyl pyrophosphoramide.
Shipping Regulations: Section 11.
  IATA: Poison B, poison label, 1 liter (passenger), 220 liters (cargo).

SCOPOLAMINE. See hyoscine.

SCORODITE. See ferric arsenate.

SEA ONION. See squill, red.

SEA SALT. See sodium chloride.

SEBACIC ACID
General Information
Synonym: Decanedioic acid.
Description: Thin, colorless crystals.
Formula: COOH(CH$_2$)$_8$COOH.
Constants: Mol wt: 202.3, mp: 133° C, bp: 295° C at 100 mm, vap. press.: 1 mm at 183.0° C.
Hazard Analysis
Toxicity: Unknown.
Countermeasures
Storage and Handling: Section 7.

SECONDARY BARIUM PHOSPHATE. See barium phosphate, dibasic.

SELENIC ACID
General Information
Description: Colorless, hexagonal prisms.
Formula: H$_2$SeO$_4$.
Constants: Mol wt: 144.98, mp: 58° C, bp: 260° C (decomp.), d solid: 2.951 at 15° C; d liquid: 2.609 at 15° C.
Hazard Analysis and Countermeasures
See selenium compounds.
Shipping Regulations: Section 11.
  IATA (liquid): Corrosive liquid, white label, not acceptable (passenger), 2½ liters (cargo).
        (solid): Poison B, poison label, 25 kilograms (passenger), 95 kilograms (cargo).

SELENOUS ACID
General Information
Description: Transparent, colorless crystals.
Formula: H$_2$SeO$_3$.
Constants: Mol wt: 128.98, mp: decomposes, d: 3.004 at 15°/4° C, vap. press.: 2 mm at 15° C.
Hazard Analysis and Countermeasures
See selenium compounds.

SELENIUM
General Information
Description: Steel-gray, non-metallic element.
Formula: Se$_8$.
Constants: Mol wt: 631.68, np: 170–217° C, bp: 690° C, d: 4.26–4.79, vap. press.: 1 mm at 356° C.
Hazard Analysis
TLV: ACGIH (recommended); 0.1 milligrams per cubic meter of air.
Toxicology: See selenium compounds.
Radiation Hazard: Section 5. For permissible levels, see Table 5, p. 150.
    Artificial isotope ^{75}Se, half life 120 d. Decays to stable ^{75}As by electron capture. Emits gamma rays of 0.1–0.4 MeV and X-rays.
Countermeasures
Ventilation Control: Section 2.
Personnel Protection: Section 3.
First Aid: Section 1.
Storage and Handling: Section 7.

SELENIUM COMPOUNDS
Hazard Analysis
Toxic Hazard Rating:
  Acute Local: Irritant 2; Ingestion 2; Inhalation 2–3.
  Acute Systemic: Ingestion 2; Inhalation 2–3.
  Chronic Local: Irritant 2; Inhalation 2–3.
  Chronic Systemic: Ingestion 2; Inhalation 2; Skin absorption 2.

Toxicology: Selenium in small amounts is essential for normal growth of some animals. Deficiency or excess is associated with serious disease in livestock. Elemental selenium has low systemic toxicity, but dust or fumes can cause serious irritation of the respiratory tract. Hydrogen selenide resembles other hydrides in being highly toxic, and selenium oxychloride is a vessicant. Some organoselenium compounds have the high toxicity of other organometals. Inorganic selenium compounds can cause dermatitis. Garlic odor of breath is a common symptom. Pallor, nervousness, depression, digestive disturbances have been reported in cases of chronic exposure. Selenium compounds are common air contaminants. (Section 4).

TLV: ACGIH (recommended); 0.2 milligrams per cubic meter of air.

SELENIUM DIBUTYLDITHIOCARBAMATE
General Information
Description: Liquid.
Formula: $Se[SC(S)N(C_4H_9)_2]_2$.
Constants: Mol wt: 896.4, mp: $-25°C$, flash p.: $225°F$, d: 1.14 at $20°/20°C$, vap. d.: 30.9.
Hazard Analysis
Toxicity: Highly toxic. See selenium compounds.
Fire Hazard: Moderate, when exposed to heat or flame.
Disaster Hazard: Dangerous. See selenium compounds.
Countermeasures
Storage and Handling: Section 7.
To Fight Fire: Water, foam, carbon dioxide, dry chemical or carbon tetrachloride (Section 6).

SELENIUM DIETHYLDITHIOCARBAMATE
General Information
Description: Orange-yellow color.
Formula: $Se[SC(S)N(C_2H_5)_2]_2$.
Constants: Mol wt: 672.1, d: 1.32 at $20°/20°C$.
Hazard Analysis and Countermeasures
See selenium compounds.

SELENIUM DIMETHYLDITHIOCARBAMATE
General Information
Description: Crystals.
Formula: $Se[SC(S)N(CH_3)_2]_2$.
Constants: Mol wt: 559.9, m range: $63-71°C$.
Hazard Analysis and Countermeasures
See selenium compounds.

SELENIUM DIOXIDE
General Information
Description: White to slightly reddish, lustrous, crystalline powder or needles.
Formula: SeO_2.
Constants: Mol wt: 110.96, mp: $340-350°C$, bp: sublimes, d: 3.95 at $15°/15°C$, vap. press.: 1 mm at $157.0°C$.
Hazard Analysis and Countermeasures
See selenium compounds.

SELENIUM DISULFIDE
General Information
Description: Red-yellow crystals.
Formula: $Se S_2$.
Constants: Mol wt: 143.09, mp: $<100°C$, bp: decomposes.
Hazard Analysis and Countermeasures
See selenium compounds and sulfides.

SELENIUM HEXAFLUORIDE
General Information
Description: Colorless gas.

Formula: SeF_6.
Constants: Mol wt: 192.96, mp: $-39°C$; sublimes at $-46.6°C$, bp: $-34.5°C$, d: 3.25 g/1 at $-25°C$.
Hazard Analysis and Countermeasures
See selenium compounds and fluorides.
TLV: ACGIH (recommended); 0.05 parts per million; 0.4 milligrams per cubic meter of air.

SELENIUM HYDRIDE
General Information
Formula: SeH.
Constant: Mol wt: 80.2.
Hazard Analysis
Toxic Hazard Rating:
Acute Local: Irritant 3; Inhalation 3.
Acute Systemic: Inhalation 3; Skin absorption 3.
Chronic Local: U.
Chronic Systemic: U.
Toxicology: Marked irritation of respiratory tract. See also selenium compounds. See also hydrides.

SELENIUM MONOBROMIDE
General Information
Description: Dark red liquid.
Formula: Se_2Br_2.
Constants: Mol wt: 317.75, bp: $227°C$ (decomp.), d: 3.604 at $15°C$.
Hazard Analysis and Countermeasures
See selenium compounds and bromides.

SELENIUM MONOBROMIDE TRICHLORIDE
General Information
Description: Yellow-brown crystals.
Formula: $SeBrCl_3$.
Constants: Mol wt: 265.25, mp: $190°C$.
Hazard Analysis and Countermeasures
See selenium compounds, bromides and chlorides.

SELENIUM MONOCHLORIDE
General Information
Description: Brown-red liquid.
Formula: Se_2Cl_2.
Constants: Mol wt: 228.83, mp: $-85°C$, bp: $130°C$ (decomp.), d: 2.91 at $17°C$; 2.77 at $23°C$.
Hazard Analysis and Countermeasures
See selenium compounds and chlorides.

SELENIUM MONOSULFIDE
General Information
Description: Orange-yellow tablets or powder.
Formula: SeS.
Constants: Mol wt: 111.03, mp: decomposes at $118-119°C$, d: 3.056 at $0°C$.
Hazard Analysis and Countermeasures
See selenium compounds and sulfides.

SELENIUM NITRIDE
General Information
Description: Amorphous orange-yellow to brick-red hygroscopic crystals.
Formula: Se_4N_4.
Constants: Mol wt: 371.87, mp: explodes at $160-200°C$, bp: decomposes.
Hazard Analysis
Toxicity: See selenium compounds.
Explosion Hazard: Moderate, when exposed to heat.
Disaster Hazard: Dangerous; when heated to decomposition or on contact with acid or acid fumes, it emits highly toxic fumes of selenium.

TOXIC HAZARD RATING CODE (For detailed discussion, see Section 1.)

0 NONE: (a) No harm under any conditions; (b) Harmful only under unusual conditions or overwhelming dosage.

1 SLIGHT: Causes readily reversible changes which disappear after end of exposure.

2 MODERATE: May involve both irreversible and reversible changes; not severe enough to cause death or permanent injury.

3 HIGH: May cause death or permanent injury after very short exposure to small quantities.

U UNKNOWN: No information on humans considered valid by authors.

Countermeasures
Storage and Handling: Section 7.

SELENIUM OXIDE
General Information
Synonyms: Selenium dioxide, selinious anhydride.
Description: Lustrous, tetragonal needles; vapor has a pungent sour smell; soluble in alcohol and water.
Formula: SeO_2.
Constants: Mol wt: 110.96, mp: 340°C, d: 3.954 (15/15°C), vap. press.: 12.5 mm at 70°C.
Hazard Analysis
Toxicity: Details unknown. A case of poisoning by percutaneous absorption has been reported.

SELENIUM OXYBROMIDE
General Information
Description: Red-yellow crystals.
Formula: $SeOBr_2$.
Constants: Mol wt: 254.79, mp: 41.6°C, bp: 217°C at 740 mm (decomp.), d: liquid: 3.38 at 50°C.
Hazard Analysis and Countermeasures
See selenium compounds and bromides.

SELENIUM OXYCHLORIDE
General Information
Description: Colorless, yellowish liquid.
Formula: $SeOCl_2$.
Constants: Mol wt: 165.87, mp: 8.5°C, bp: 176.4°C, d: 2.42 at 22°C, vap. press.: 1 mm at 34.8°C.
Hazard Analysis and Countermeasures
See selenium compounds and chlorides.

SELENIUM OXYFLUORIDE
General Information
Description: Colorless liquid.
Formula: $SeOF_2$.
Constants: Mol wt: 132.96, mp: 4.6°C, bp: 124°C, d: 2.67.
Hazard Analysis and Countermeasures
See selenium compounds and fluorides.

SELENIUM SULFIDE. See selenium monosulfide.

SELENIUM TETRABROMIDE
General Information
Description: Orange-red-brown crystals.
Formula: $SeBr_4$.
Constants: Mol wt: 398.62, mp: decomposes at 75°C.
Hazard Analysis and Countermeasures
See selenium compounds and bromides.

SELENIUM TETRACHLORIDE
General Information
Description: Cubic, white-yellow, deliquescent crystals.
Formula: $SeCl_4$.
Constants: Mol wt: 220.79, mp: 305°C; sublimes at 170–196°C, bp: decomposes at 288°C, d: 3.78, vap. press.: 1 mm at 74.0°C.
Hazard Analysis and Countermeasures
See selenium compounds and chlorides.

SELENIUM TETRAFLUORIDE
General Information
Description: Colorless liquid or white crystals.
Formula: SeF_4.
Constants: Mol wt: 154.96, mp: −13.5°C, bp: 93°C.
Hazard Analysis and Countermeasures
See selenium compounds and fluorides.

SELENIUM TRIOXIDE
General Information
Description: Amorphous, pale yellow, hygroscopic solid.
Formula: SeO_3.
Constants: Mol wt: 126.96, mp: decomposes at 120°C, d: 3.6.
Hazard Analysis and Countermeasures
See selenium compounds.

SELLAITE. See magnesium fluoride.

SEMESAN. See hydroxymercurichlorophenol.

SENECA OIL. See crude oil (petroleum).

SENECIDALDEHYDE
General Information
Synonyms: Senecioaldehyde; 2,2-dimet hylaxolein; 2-methylcrotonaldehyde; 2-methyl-2-buten-4-al.
Description: Liquid, pungent odor.
Formula: $(CH_3)_2C{:}CHCHO$.
Constants: Mol wt: 84.11, d: 0.8722 at 20/4°C, bp: 133°C.
Hazard Analysis
See acrolein.

SEQUESTRENE. See disodium ethylenediamine tetraacetate.

SERINE
General Information
Synonyms: β-Hydroxyalanine; α-amino-β-hydroxypropionic acid.
Description: A non-essential amino acid; colorless crystals; soluble in water; insoluble in alcohol and ether; optically active, here we consider L-and DL- forms.
Formula: $HOCH_2CH(NH_2)COOH$.
Constants: Mol wt: 105, mp(DL): 246°C with decomposition, mp(L): 228°C with decomposition.
Hazard Analysis
Toxicity: Details unknown. A nutrient and/or dieting supplement food additive (Section 10).

SES. See sodium-2,4-dichlorophenoxyethyl sulfate.

SESAMIN
General Information
Description: Crystals.
Formula: $C_{20}H_{18}O_6$.
Constant: Mol wt: 354.3.
Hazard Analysis
Toxicity: Details unknown; a fungicide.
Fire Hazard: Slight, when exposed to heat or flame (Section 6).
Countermeasures
Storage and Handling: Section 7.

SESIN. See 2,4-dichlorophenoxyethyl benzoate.

SESOXANE
General Information
Synonym: 2-(3,4-Methylenedioxyphenoxy)-3,6,9-trioxaundecane.
Description: Crystals. Soluble in kerosene.
Formula: $C_{15}H_{21}O_6$.
Constants: Mol wt: 297.3.
Hazard Analysis
Toxicity: Details unknown. Limited animal experiments suggest fairly low toxicity.

SESQUI. See sodium sesquicarbonate.

SEVIN. See carbaryl.

SHALE OIL
Hazard Analysis
Toxic Hazard Rating:
    Acute Local: Irritant 1; Ingestion 1; Inhalation 1.
    Acute Systemic: U.
    Chronic Local: Irritant 3.
    Chronic Systemic: Inhalation 3.
Disaster Hazard: Dangerous; when exposed to heat or flame, it emits acrid fumes.
Countermeasures
Personnel Protection: Section 3.
Storage and Handling: Section 7.

SHAPED CHARGES
Shipping Regulations: Section 11.
  IATA (Commercial): Explosive, not acceptable (passenger
    and cargo).
        (flexible, linear, metal clad): Explosive, explosive
    label, 25 kilograms (passenger), 140 kilograms (cargo).

SHELLAC, LIQUID
General Information
Constant: Flash p: 40–70° F.
Hazard Analysis
Toxicity: Details unknown. May act as an allergen.
Fire Hazard: Dangerous, when exposed to heat or flame.
  See also ethyl alcohol, the usual solvent.
Explosion Hazard: See ethyl alcohol.
Disaster Hazard: See ethyl alcohol.
Countermeasures
Storage and Handling: Section 7.
To Fight Fire: Foam, carbon dioxide, dry chemical or
  carbon tetrachloride (Section 6).
Shipping Regulations: Section 11.
  I.C.C.: Flammable liquid; red label, 55 gallons.
  Coast Guard Classification: Combustible liquid; inflam-
    mable liquid; red label.
  IATA: Flammable liquid, red label, 1 liter (passenger),
    220 liters (cargo).

SHIKIMOLE. See safrole.

SIGNAL DEVICES, HAND
Shipping Regulations: Section 11.
  IATA: Explosive, explosive label, 25 kilograms (pas-
  senger), 95 kilograms (cargo).

SIGNAL FLARES
Shipping Regulations: Section 11.
  ICC: Explosive C, 200 pounds.

SIGNALS, HIGHWAY
Shipping Regulations: Section 11.
  IATA: Explosive, explosive label, 25 kilograms (pas-
  senger), 95 kilograms (cargo).

SIGNAL OIL. See mineral seal oil.

SILANE
General Information
Synonym: Silicon tetrahydride, silicane.
Description: Gas with repulsive odor. Slowly decomposed
  by water.
Formula: SiH_4.
Constants: Mol wt: 32.1, d: 0.68 at −185° C,
  mp: −185° C, bp: −112° C.
Hazard Analysis and Countermeasures
See silanes.

SILANES
General Information
Synonyms: Silicon hydrides, disilane.
Description: Gas or liquid.
Hazard Analysis
Toxic Hazard Rating:
  Acute Local: Irritant 3; Ingestion 3; Inhalation 3.
  Acute Systemic: U.
  Chronic Local: U.
  Chronic Systemic: U.
Fire Hazard: Dangerous, by chemical reaction with oxi-
  dizers; often ignite spontaneously in air (Section 6).
Explosion Hazard: Variable.

Disaster Hazard: Dangerous; when heated, they can burn or
  explode and emit highly toxic fumes.
Countermeasures
Ventilation Control: Section 2.
Personnel Protection: Section 3.
First Aid: Section 1.
Storage and Handling: Section 7.

SILANOLS
General Information
Description: Silanols are the alcohol devivatives of the
  silanes.
Hazard Analysis
Toxicity: Details unknown. Probably have little or no
  toxicity.
Shipping Regulations: Section 11.
  IATA: Flammable gas, no label, not acceptable (pas-
  senger and cargo).

SILICA
General Information
Synonyms: Silicon dioxide; silicic anhydride; cristobalite.
Description: Colorless crystals.
Formula: SiO_2.
Constants: Mol wt: 60.09; mp: 1710° C; bp: 2230° C;
  d: amorphous 2.2, crystalline 2.6; vap. press.: 10 mm at
  1732° C.
Hazard Analysis
Toxic Hazard Rating:
  Acute Local: Inhalation 2.
  Acute Systemic: 0.
  Chronic Local: Inhalation 3.
  Chronic Systemic: Inhalation 1.
Toxicology: From the point of view of numbers of men ex-
  posed and cases of disability produced, silica is the
  chief cause of pulmonary dust disease. The prolonged
  inhalation of dusts containing free silica may result in
  the development of a disabling pulmonary fibrosis
  known as silicosis. The Committee on Pneumoconiosis
  of the American Public Health Association defines
  silicosis as "a disease due to the breathing of air containing silica (SiO_2), characterized by generalized
  fibrotic changes and the development of miliary nodules
  in both lungs, and clinically by shortness of breath,
  decreased chest expansion, lessened capacity for work,
  absence of fever, increased susceptibility to tuberculosis
  (some or all of which symptoms may be present), and
  characteristic x-ray findings."
    Silica occurs in the pure state in nature as quartz. It
  is the main constituent of sand, sandstone, tripoli and
  diatomaceous earth, and is present in high amounts
  (up to 35%) in granite. Exposure to silica occurs in hard
  rock mining, in foundries, in manufacture of procelain
  and pottery, in the spraying of vitreous enamels, in
  sandblasting, in granite-cutting and tombstone-making,
  in the manufacture of silica firebrick and other refrac-
  tories, in grinding and polishing operations where natural abrasive wheels are used and other occupations.
    The duration of exposure which is associated with
  the development of silicosis varies widely for different
  occupations. Thus, the average duration of exposure
  required for the development of silicosis in sand-blasters
  is 2 to 10 years, in moulders and granite cutters, about
  30 years, and in hard rock miners 10 to 15 years. There
  is, also, much variation in individual susceptibility,
  certain workers showing radiological evidence of the
  disease years before their fellow workmen who are

TOXIC HAZARD RATING CODE (For detailed discussion, see Section 1.)

0 NONE: (a) No harm under any conditions; (b) Harmful only under unusual conditions or overwhelming dosage.

1 SLIGHT: Causes readily reversible changes which disappear after end
of exposure.

2 MODERATE: May involve both irreversible and reversible changes;
not severe enough to cause death or permanent injury.

3 HIGH: May cause death or permanent injury after very short exposure
to small quantities.

U UNKNOWN: No information on humans considered valid by authors.

similarly exposed. Such susceptible individuals are fortunately rather rare.

The action of silica on the lungs results in the production of a diffuse, nodular fibrosis in which the parenchyma and the lymphatic system are involved. This fibrosis is, to a certain extent, progressive, and may continue to increase for several years after exposure is terminated. Where the pulmonary reserve is sufficiently reduced, the worker complains of shortness of breath on exertion. This is the first and most common symptom in cases of uncomplicated silicosis. If severe, it may incapacitate the worker for heavy, or even light, physical exertion, and in extreme cases there may be shortness of breath even while at rest. The most common physical sign of silicosis is a limitation of expansion of the chest. There may be a dry cough, sometimes very troublesome. The characteristic radiographic appearance is one of diffuse, discrete nodulation, scattered throughout both lung fields. Where the disease advances, the shortness of breath becomes worse, and the cough more productive and troublesome. There is no fever or other evidence of systemic reaction. Further progress of the disease results in marked fatigue, extreme dyspnea and cyanosis, loss of appetite, pleuritic pain and total incapacity to work. If tuberculosis does not supervene, the condition may eventually cause death either from cardiac failure or from destruction of lung tissue, with resultant anoxemia. In the later stages, the x-ray may show large, conglomerate shadows, due to the coalescence of the silicotic nodules, with areas of emphysema between them.

Silica is used as a food additive permitted in the feed and drinking water of animals and/or for the treatment of food producing animals. It is also permitted in food for human consumption (Section 10). It is a common air contaminant (Section 4).

TLV: ACGIH (recommended) Cristobalite and crystalline quartz threshold limit calculated from the formula: $250/(\%SIO_2 + 5)$; for amorphous silica, including diatomaceous earth, 20 millions of particles per cubic foot of air.

Countermeasures
Ventilation Control: Section 2.

SILICA AEROGEL
General Information
Description: A finely powdered microcellular silica foam having a minimum silica content of 89.5%.
Hazard Analysis
Toxicity: Unknown. A general purpose food additive (Section 10). See also silica.

SILICA GEL
General Information
Synonym: Silicic acid (precipitated).
Description: White powder or lustrous granules.
Formula: H_2SiO_3.
Constant: Mol wt: 78.1.
Toxicity: Slight. See silica.

SILICANE. See silane.

SILICANES. See silanes.

SILICATES
Hazard Analysis
Toxicity: Soluble alkaline silicates act locally like mild alkalies. The dust of certain silicates such as asbestos (hydrated magnesium silicate) and talc can produce fibrotic changes in the lungs.

SILICIC ACID. See silica gel.

SILICIC ANHYDRIDE. See silica.

SILICIDES OF LIGHT METALS
General Information
Description: Metallic, crystalline materials.

Hazard Analysis
Toxicity: Variable.
Fire Hazard: Moderate, by chemical reaction. See also hydrogen (Section 6).
Explosion Hazard: Moderate, by chemical reaction. See also hydrogen.
Disaster Hazard: Moderately dangerous; they will react with water or steam to produce hydrogen; on contact with acid or acid fumes, they can emit toxic fumes.
Countermeasures
Storage and Handling: Section 7.

SILICOBROMOFORM. See tribromosilane.

SILICOCHLOROFORM. See trichlorosilane.

SILICOETHANE. See disilane.

SILICOFLUORIC ACID. See hydrofluosilicic acid.

SILICOFLUORIDES. See fluosilicates.

SILICON
General Information
Description: Cubic, steel-gray crystals or dark brown powder.
Formula: Si.
Constants: At wt: 28.09, mp: 1420°C, bp: 2600°C d: 2.42 or 2.3 at 20°C, vap. press.: 1 mm at 1724°C.
Hazard Analysis
Toxicity: Details unknown. Does not occur free in nature, but is found as silicon dioxide (silica), and as various silicates. See also silica and silicates.
Fire Hazard: Moderate, when exposed to flame or by chemical reaction with oxidizers. See also powdered metals (Section 6).
Disaster Hazard: Dangerous; when heated, it will react with water or steam to produce hydrogen; can react with oxidizing materials.
Radiation Hazard: Section 5. For permissible levels, see Table 5, p. 150.
Artificial isotope ^{31}Si, half life 2.6 h. Decays to stable ^{31}P by emitting beta particles of 1.48 MeV.
Countermeasures
Storage and Handling: Section 7.
Shipping Regulations: Section 11.
IATA: Flammable solid, yellow label, 12 kilograms (passenger), 45 kilograms (cargo).

SILICON BROMIDE
General Information
Synonym: Tetrabromosilicane; tetrabromosilane.
Description: Colorless, fuming liquid; disagreeable odor.
Formula: $SiBr_4$.
Constants: Mol wt: 347.72, mp: 5°C, bp: 153°C, d: 2.814, vap. d.: 2.82.
Hazard Analysis
Toxic Hazard Rating:
Acute Local: Irritant 3; Ingestion 3; Inhalation 3.
Acute Systemic: U.
Chronic Local: U.
Chronic Systemic: U.
Disaster Hazard: Dangerous; when heated to decomposition, it emits highly toxic fumes of hydrobromic acid; will react with water or steam to produce heat, toxic and corrosive fumes.
Countermeasures
Ventilation Control: Section 2.
Personnel Protection: Section 3.
Storage and Handling: Section 7.

SILICON CARBIDE
General Information
Description: Bluish-black, iridescent crystals.
Formula: SiC.

Constants: Mol wt: 40.10, mp: 2600°C, bp: sublimes above 2000°C; decomposes at 2210°C, d: 3.17.
Hazard Analysis
Toxic Hazard Rating:
  Acute Local: Inhalation 1.
  Acute Systemic: U.
  Chronic Local: Inhalation 1.
  Chronic Systemic: U.

SILICON CHLORIDE
General Information
Synonym: Silicon tetrachloride.
Description: Colorless, fuming liquid; suffocating odor.
Formula: $SiCl_4$.
Constants: Mol wt: 169.89, mp: $-70°C$, bp: 57.57°C, d: 1.482.
Hazard Analysis
Toxic Hazard Rating:
  Acute Local: Irritant 3; Ingestion 3; Inhalation 3.
  Acute Systemic: U.
  Chronic Local: U.
  Chronic Systemic: U.
Disaster Hazard: Dangerous; when heated to decomposition, it emits highly toxic fumes of hydrochloric acid; will react with water or steam to produce heat, toxic and corrosive fumes.
Countermeasures
Ventilation Control: Section 2.
Personnel Protection: Section 3.
Storage and Handling: Section 7.
Shipping Regulations: Section 11.
  I.C.C.: Corrosive liquid; white label, 1 gallon.
  Coast Guard Classification: Corrosive liquid; white label.
  IATA: Corrosive liquid, white label, 1 liter (passenger), 5 liters (cargo).

SILICON DIBROMIDE SULFIDE
General Information
Description: Colorless plates.
Formula: $SiSBr_2$.
Constants: Mol wt: 219.95, mp: 93°C, bp: 150°C at 18.3 mm.
Hazard Analysis
Toxic Hazard Rating:
  Acute Local: Ingestion 3; Inhalation 3.
  Acute Systemic: U.
  Chronic Local: U.
  Chronic Systemic: Ingestion 3; Inhalation 3.
Countermeasures
See sulfides and bromides.

SILICON DICHLORIDE SULFIDE
General Information
Description: Colorless prisms.
Formula: $SiSCl_2$.
Constants: Mol wt: 131.04, mp: 75°C, bp: 92°C at 22.5 mm.
Hazard Analysis
Toxic Hazard Rating:
  Acute Local: Ingestion 3; Inhalation 3.
  Acute Systemic: U.
  Chronic Local: U.
  Chronic Systemic: Ingestion 3; Inhalation 3.
Countermeasures
See sulfides and chlorides.

SILICON DIOXIDE. See silica and quartz.

SILICON DISULFIDE
General Information
Description: White needles.
Formula: SiS_2.
Constants: Mol wt: 92.19, mp: sublimes, bp: white heat.
Hazard Analysis
Toxic Hazard Rating:
  Acute Local: Ingestion 3; Inhalation 3.
  Acute Systemic: U.
  Chronic Local: U.
  Chronic Systemic: Ingestion 3; Inhalation 3.
For other properties, see sulfides.

SILICONES
General Information
Synonym: Siloxanes.
Description: Organosilicon oxide polymers such as $-R_2Si-O-$ where R is a monovalent organic radical.
Hazard Analysis
Toxicity: Generally low. Most of the silicones that have been studied have low toxicity or none at all and little or no irritant effect.

SILICON FLUORIDE. See silicon tetrafluoride.

SILICON HYDRIDES. See silanes.

SILICON MONOSULFIDE
General Information
Description: Yellow needles or black solid.
Formula: SiS.
Constants: Mol wt: 60.13, bp: sublimes, 940°C at 20 mm, d: 1.853 at 15°C.
Hazard Analysis and Countermeasures
See sulfides.

SILICON MONOXIDE
General Information
Description: Solid. Insoluble in water, soluble in alkalis and hydrofluoric acid.
Formula: SiO.
Constants: Mol wt: 44.1, d: 2.2, mp: $< 1700°C$, bp: 1880°C.
Hazard Analysis
Toxicity: Unknown. See silica.

SILICON OXYHYDRIDE
General Information
Description: White powder. Insoluble in water.
Formula: $(Si_2H_2O_3)x$.
Constants: D: 1.6, mp: 350°C (with decomp.).
Hazard Analysis
See silica.

SILICON SESQUIOXIDE
General Information
Description: Tan colored powder. Insoluble in water.
Formula: $(Si_2O_3)_x$.
Constants: Mol wt: $(104.2)_x$, mp: 1635°C.
Hazard Analysis
Toxicity: Unknown.

SILICON TETRAACETATE
General Information
Description: Hygroscopic crystals.
Formula: $Si(C_2H_3O_2)_4$.
Constants: Mol wt: 264.24, mp: 110°C sublimes, bp: 148°C.

TOXIC HAZARD RATING CODE (For detailed discussion, see Section 1.)

0 NONE: (a) No harm under any conditions; (b) Harmful only under unusual conditions or overwhelming dosage.

1 SLIGHT: Causes readily reversible changes which disappear after end of exposure.

2 MODERATE: May involve both irreversible and reversible changes; not severe enough to cause death or permanent injury.

3 HIGH: May cause death or permanent injury after very short exposure to small quantities.

U UNKNOWN: No information on humans considered valid by authors.

Hazard Analysis
Toxic Hazard Rating:
Acute Local: Irritant 1; ingestion 1; Inhalation 1.
Acute Systemic: U.
Chronic Local: U.
Chronic Systemic: U.
Countermeasures
Personal Hygiene: Section 3.

SILICON TETRABROMIDE. See silicon bromide.

SILICON TETRACHLORIDE. See silicon chloride

SILICON TETRAFLUORIDE
General Information
Synonym: Tetrafluorosilane.
Description: Colorless gas.
Formula: SiF_4.
Constants: Mol wt: 104.06, mp: $-77°C$, bp: $-65°C$ at 181 mm, d: 4.67 g/l.
Hazard Analysis and Countermeasures
See fluorides and hydrofluoric acid.
Shipping Regulations: Section 11.
ICC: Nonflammable gas, green label, 300 pounds.
IATA: Nonflammable gas, green label, not acceptable (passenger), 140 kilograms (cargo).

SILICON TETRAIODIDE
General Information
Synonym: Tetraiodosilane.
Description: Cubic, colorless crystals.
Formula: SiI_4.
Constants: Mol wt: 535.74, mp: 120.5°C, bp: 290°C.
Hazard Analysis
Toxic Hazard Rating:
Acute Local: Irritant 2; Ingestion 3; Inhalation 3.
Acute Systemic: U.
Chronic Local: U.
Chronic Systemic: U.
Disaster Hazard: Dangerous; when heated to decomposition, it emits toxic fumes of iodides; will react with water or steam to produce toxic and corrosive fumes.
Countermeasures
Ventilation Control: Section 2.
Personnel Protection: Section 3.
Storage and Handling: Section 7.

SILICON TETRATHIOCYANATE
General Information
Description: Small prisms.
Formula: $SI(SCN)_4$.
Constants: Mol wt: 260.40, mp: 143.8°C, bp: 314.2°C.
Hazard Analysis and Countermeasures
See thiocyanates.

SILICON TRICHLORIDE HYDROSULFIDE
General Information
Description: Colorless liquid.
Formula: $SiCl_3HS$.
Constants: Mol wt: 167.51, bp: 96–100°C, d: 1.45.
Hazard Analysis and Countermeasures
See sulfides and chlorides.

SILICYL OXIDE
General Information
Synonym: Disiloxane.
Description: Colorless gas.
Formula: $(SiH_3)_2O$.
Constants: Mol wt: 78.17, mp: $-144°C$, bp: $-15.2°C$, d: 0.881 at $-80°C$.
Hazard Analysis
Toxicity: Unknown.
Fire Hazard: Details unknown; probably quite flammable.

SILK DUST
Hazard Analysis
Toxic Hazard Rating:

Acute Local: Allergen 1.
Acute Systemic: Inhalation 1.
Chronic Local: Allergen 1; Inhalation 1.
Chronic Systemic: Inhalation 1.
TLV: ACGIH (recommended); 50 million particles per cubic foot of air.
Fire Hazard: Moderate, when exposed to flame; can react with oxidizing materials (Section 6).
Explosion Hazard: Moderate, in the form of a dust cloud, when exposed to flame.
Countermeasures
Personal Hygiene: Section 3.
Storage and Handling: Section 7.

SILOXANES. See silicones.

SILVER
General Information
Synonym: Argentum.
Description: Soft, ductile and malleable, lustrous, white metal.
Formula: Ag.
Constants: At wt: 107.88, mp: 960.5°C, bp: 1950°C, d: 10.5, vap. press.: 1 mm at 1357°C.
Hazard Analysis
Toxicity: See silver compounds.
Radiation Hazard(Section 5. For permissible levels, see Table 5, p. 150.
Artificial isotope ^{105}Ag, half life 40 d. Decays to stable ^{105}Pd by electron capture. Emits gamma rays of 0.06 to 1.09 MeV and X-rays.
Artificial isotope ^{110}Ag, half life 253 d. Decays to stable ^{110}Cd by emitting beta particles of 0.085 (65%), 0.53 (33% MeV. Also decays to ^{110}Cd through short-lived ^{110}Ag (5%). Also emits gamma rays of 0.44 to 1.50 MeV.
Artificial isotope ^{111}Ag, half life 7.5 d. Decays to stable ^{111}Cd by emitting beta particles of 0.69 (6%), 1.05 (93%) MeV. Also emits gamma rays of 0.25, 0.34 MeV.
Fire Hazard: Moderate, in the form of dust, when exposed to flame or by chemical reaction with oxidizers. See also powdered metals (Section 6).

SILVER ACETATE
General Information
Description: White plates.
Formula: $AgC_2H_3O_2$.
Constants: Mol wt: 166.92, mp: decomposes, d: 3.259 at 15°C.
Hazard Analysis
Toxicity: See silver compounds.

SILVER ACETYLIDE
General Information
Description: White precipitate.
Formula: Ag_2C_2.
Constants: Mol wt: 239.78, mp: explodes.
Hazard Analysis
Toxicity: See silver compounds.
Fire Hazard: Unknown.
Explosion Hazard: Severe, when shocked or exposed to heat.
Disaster Hazard: Dangerous, shock or heat will explode it.
Countermeasures
Storage and Handling: Section 7.

SILVER AMALGAMS
General Information
Description: Silvery liquid or solid.
Formula: Ag + Hg.
Hazard Analysis and Countermeasures
See mercury compounds, inorganic and silver compounds.

SILVER AMMONIUM COMPOUNDS
Hazard Analysis
Toxicity: See silver compounds.

Explosion Hazard: Severe, when shocked, exposed to heat, or by chemical reaction.

Disaster Hazard: Dangerous; shock or heat will explode them.

Countermeasures

Storage and Handling: Section 7.

SILVER o-ARSENATE
General Information
Description: Cubic, dark red crystals.
Formula: Ag_3AsO_4.
Constants: Mol wt: 462.55, d: 6.657 at 25°C.
Hazard Analysis and Countermeasures
See arsenic compounds and silver compounds.

SILVER ARSENITE
General Information
Synonym: Silver orthoarsenite.
Description: Fine, yellow powder; sensitive to light.
Formula: Ag_3AsO_3.
Constants: Mol wt: 446.55, mp: 150°C (decomp.).
Hazard Analysis and Countermeasures
See arsenic compounds.
Shipping Regulations: Section 11.
    IATA: Poison B, poison label, 25 kilograms (passenger), 95 kilograms (cargo).

SILVER ARSPHENAMINE
General Information
Synonyms: Silver diaminodihydroxyarsenobenzene; silver salvarsan, silver diarsenal.
Description: Brownish-black powder; contains approximately 20% arsenic, 15% silver.
Hazard Analysis and Countermeasures
See arsenic compounds and silver compounds.

SILVER AZIDE
General Information
Description: White prisms.
Formula: AgN_3.
Constants: Mol wt: 149.90, mp: 2.50°C.
Hazard Analysis
Toxicity: See silver compounds and azides.
Explosion Hazard: Severe, when shocked or exposed to heat.
Explosive Range: At 521°F.
Disaster Hazard: Dangerous; shock or heat will explode it.
Countermeasures
Storage and Handling: Section 7.

SILVER BENZOATE
General Information
Description: White powder.
Formula: $AgC_7H_5O_2$.
Constant: Mol wt: 228.99.
Hazard Analysis
Toxicity: See silver compounds.

SILVER BROMATE
General Information
Description: White powder.
Formula: $AgBrO_3$.
Constants: Mol wt: 235.8, mp: decomposes, d: 5.206.
Hazard Analysis and Countermeasures
See silver compounds and bromates.

SILVER CARBONATE
General Information
Description: Yellow crystalline powder.
Formula: Ag_2CO_3.

Constants: Mol wt: 275.8, mp: 218°C (decomp.), d: 6.077.
Hazard Analysis
Toxicity: See silver compounds.

SILVER CHLORATE
General Information
Description: Tetragonal, white crystals.
Formula: $AgClO_3$.
Constants: Mol wt: 191.34, mp: 230°C, bp: decomposes at 270°C, d: 4.430.
Hazard Analysis and Countermeasures
See silver compounds and chlorates.

SILVER CHLORIDE
General Information
Description: White granular powder.
Formula: $AgCl$.
Constants: Mol wt: 143.34, mp: 455°C, bp: 1550°C, d: 5.561, vap. press.: 1 mm at 912°C.
Hazard Analysis and Countermeasures
See silver compounds.

SILVER CHROMATE
General Information
Description: Red crystals.
Formula: Ag_2CrO_4.
Constants: Mol wt: 331.77, d: 5.625.
Hazard Analysis
Toxicity: See chromium compounds and silver compounds.

SILVER COMPOUNDS
Hazard Analysis
Toxicity: The absorption of silver compounds into the circulation and the subsequent deposition of the reduced silver in various tissues of the body may result in the production of a generalized greyish pigmentation of the skin and mucous membranes—a condition known as argyria. The introduction of fine particles of silver through breaks in the skin produces a local pigmentation at the site of the injury.

    Generalized argyria rarely seen at the present time, was not infrequent in the past. The condtion developed slowly, usually after some 2 to 25 years of exposure. Pigmentation was noticeable first in conjunctivae, and later in the mucous membranes of the mouth and gums and in the skin. There were no constitutional symptoms, and no physical disability. Persons exhibiting the condition, and who subsequently died from unrelated disease, showed, on autopsy, a deposition of silver in the blood vessel walls, kidneys, testes, pituitary, choroid plexus, and mucous membrane of the nose, maxillary antra, trachea and bronchi. Once deposited, there is no known method by which the silver can be eliminated; the pigmentation is permanent.

TLV: ACGIH (recommended): 0.01 milligrams per cubic meter of air.

SILVER CYANATE
General Information
Description: Colorless crystals.
Formula: $AgOCN$.
Constants: Mol wt: 149.90, mp: decomposes, d: 4.00.
Hazard Analysis and Countermeasures
See cyanates and silver compounds.

SILVER CYANIDE
General Information
Description: White, odorless, tasteless powder which darkens on exposure to light.

TOXIC HAZARD RATING CODE *(For detailed discussion, see Section 1.)*

0 NONE: (a) No harm under any conditions; (b) Harmful only under unusual conditions or overwhelming dosage.

1 SLIGHT: Causes readily reversible changes which disappear after end of exposure.

2 MODERATE: May involve both irreversible and reversible changes; not severe enough to cause death or permanent injury.

3 HIGH: May cause death or permanent injury after very short exposure to small quantities.

U UNKNOWN: No information on humans considered valid by authors.

Formula: AgCN.
Constants: Mol wt: 133.90, mp: 320° C (decomp.), d: 3.95.
Hazard Analysis and Countermeasures
See cyanides and silver compounds.
Shipping Regulations: Section 11.
   I.C.C.: See Section 11.
   Coast Guard Classification: Poison B; poison label.
   IATA: Poison B, poison label, 12 kilograms (passenger), no limit (cargo).

SILVER DIAMINODIHYDROXYARSENOBENZENE.
  See silver arsphenamine.

SILVER DIARSENAL. See silver arsphenamine.

SILVER DICHROMATE
General Information
Description: Red crystals.
Formula: $Ag_2Cr_2O_7$.
Constants: Mol wt: 431.78, mp: decomposes, d: 4.770.
Hazard Analysis
Toxicity: See chromium compounds and silver compounds.

SILVER DIFLUORIDE
General Information
Description: Brown powder.
Formula: AgF_2.
Constants: Mol wt: 145.88, mp: 690° C, d: 4.57–4.78.
Hazard Analysis and Countermeasures
See fluorides and silver compounds.

SILVER FLUOGALLATE
General Information
Description: Colorless crystals.
Formula: $Ag_3(GaF_6) \cdot 10H_2O$.
Constants: Mol wt: 687.52, d: 2.90.
Hazard Analysis and Countermeasures
See fluogallates and silver compounds.

SILVER FLUORIDE
General Information
Description: Yellow, crystalline masses.
Formula: AgF.
Constants: Mol wt: 126.88, mp: 435° C, d: 5.852 at 15.5° C.
Hazard Analysis and Countermeasures
See fluorides and silver compounds.

SILVER FLUOSILICATE
General Information
Description: Colorless crystals or white, deliquescent powder.
Formula: $Ag_2SiF_6 \cdot 4H_2O$.
Constants: Mol wt: 429.88, mp: < 100° C, bp: decomposes.
Hazard Analysis and Countermeasures
See fluosilicates and silver compounds.

SILVER FULMINATE
General Information
Description: Small needles.
Formula: $Ag_2C_2N_2O_2$.
Constants: Mol wt: 299.80, mp: explodes.
Hazard Analysis
Toxicity: See fulminates and silver compounds.
Fire Hazard: Unknown.
Explosion Hazard: Severe, when shocked or exposed to heat. See also explosives, high.
Disaster Hazard: Dangerous; shock or heat will explode it; when heated to decomposition, it emits highly toxic fumes.
Countermeasures
Storage and Handling: Section 7.

SILVER HYPONITRITE
General Information
Description: Yellow crystals.
Formula: $Ag_2N_2O_2$.

Constants: Mol wt: 275.77, mp: decomposes at 110° C, d: 5.75 at 30° C.
Hazard Analysis
Toxicity: See nitrites and silver compounds.
Fire Hazard: Unknown.
Explosion Hazard: Moderate, when exposed to heat.
Explosive Range: At 302° F.
Disaster Hazard: Dangerous; when heated to decomposition, it emits highly toxic fumes of oxides of nitrogen; can react with reducing materials. Heat can explode it.
Countermeasures
Storage and Handling: Section 7.

SILVER HYPOPHOSPHITE
General Information
Description: White crystals.
Formula: AgH_2PO_2.
Constant: Mol wt: 172.9.
Hazard Analysis
Toxicity: See silver compounds and phosphorus compounds.
Fire Hazard: Unknown.
Explosion Hazard: Moderate, when exposed to heat.
Disaster Hazard: Dangerous; when heated to decomposition, it emits highly toxic fumes of oxides of phosphorus. Heat can explode it.
Countermeasures
Storage and Handling: Section 7.

SILVER IODATE
General Information
Description: Rhombic, colorless crystals.
Formula: $AgIO_3$.
Constants: Mp: > 200° C, bp: decomposes, d: 5.525, mol wt: 282.80.
Hazard Analysis and Countermeasures
See silver compounds and iodates.

SILVER IODIDE
General Information
Description: Pale yellow powder.
Formula: AgI.
Constants: Mol wt: 234.8, mp: 556° C bp: 1506° C, d: 5.675, vap. press.: 1 mm at 820° C.
Hazard Analysis and Countermeasures
See iodides and silver compounds.

SILVER LACTATE
General Information
Description: White or slightly gray crystalline powder.
Formula: $AgC_3H_5O_3 \cdot H_2O$.
Constant: Mol wt: 214.97.
Hazard Analysis and Countermeasures
See silver compounds.

SILVER NITRATE
General Information
Description: Colorless, transparent, tabular, rhombic, odorless crystals, becoming gray or grayish-black on exposure to light in presence of organic matter; bitter, caustic, metallic taste.
Formula: $AgNo_3$.
Constants: Mol wt: 169.89, mp: 212° C, bp: 444° C (decomp.), d: 4.352 at 19° C.
Hazard Analysis and Countermeasures
A powerful caustic. See also silver compounds and nitrates.
Shipping Regulations: Section 11.
   MCA warning label.
   ICC: Oxidizing material, yellow label, 100 pounds.
   IATA: Oxidizing material, yellow label, 12 kilograms (passenger), 45 kilograms (cargo).

SILVER NITRIDE
General Information
Description: Colorless solid.
Formula: Ag_3N.
Constant: Mol wt: 337.7.

Hazard Analysis
Toxicity: See silver compounds and nitrides.
Explosion Hazard: Severe, when shocked or exposed to heat. See also explosives, high.
Disaster Hazard: Dangerous; shock or heat will expode it.
Countermeasures
Storage and Handling: Section 7.

SILVER NITRITE
General Information
Description: Rhombic, white crystals.
Formula: $AgNO_2$.
Constants: Mol wt: 153.89, mp: decomposes at 140°C, d: 4.453 at 26°C.
Hazard Analysis and Countermeasures
See nitrites and silver compounds.

SILVER NITROPRUSSIDE
General Information
Description: Light pink crystals.
Formula: $Ag_2[FeNo(CN)_5]$.
Constant: Mol wt: 431.71.
Hazard Analysis
Toxicity: Highly toxic. See hydrocyanic acid.
Disaster Hazard: Dangerous; emits highly toxic fumes on heating.
Countermeasures
Storage and Handling: Section 7.

SILVER OXALATE
General Information
Description: Colorless crystals.
Formula: $Ag_2C_2O_4$.
Constants: Mol wt: 303.78, d: 5.029 at 4°C.
Hazard Analysis
Toxicity: Highly toxic. See oxalates.
Explosion Hazard: Moderate, when exposed to heat.
Explosive Range: At 140°C.
Disaster Hazard: Dangerous; emits highly toxic fumes on heating and may explode.
Countermeasures
Storage and Handling: Section 7.

SILVER OXIDE
General Information
Description: Dark brown, odorless powder; metallic taste.
Formula: Ag_2O.
Constants: Mol wt: 231.76, mp: decomposes at 300°C, d: 7.143 at 16.6°C.
Hazard Analysis
Toxicity: See silver compounds.
Fire Hazard: Moderate, by chemical reaction; an oxidizing agent (Section 6).
Countermeasures
Storage and Handling: Section 7.

SILVER PERCHLORATE
General Information
Description: White, deliquescent crystals.
Formula: $AgClO_4$.
Constants: Mol wt: 207.34, mp: decomposes at 486°C, d: 2.806 at 25°C.
Hazard Analysis and Countermeasures
See perchlorates and silver compounds.

SILVER PERMANGANATE
General Information
Description: Violet, crystalline powder.

Formula: $AgMnO_4$.
Constants: Mol wt: 226.81, mp: decomposes.
Hazard Analysis and Countermeasures
See silver compounds and permanganates.

SILVER PHENOLSULFONATE
General Information
Synonym: Silver sulfocarbolate.
Description: White to faintly reddish crystals.
Formula: $AgC_6H_4SO_3OH$.
Constant: Mol wt: 281.1.
Hazard Analysis
Toxicity: See silver compounds.
Disaster Hazard: Dangerous; see sulfonates.
Countermeasures
Storage and Handling: Section 7.

SILVER PHOSPHATE
General Information
Description: Yellow powder.
Formula: Ag_3PO_4.
Constants: Mol wt: 418.6, mp: 849°C, d: 6.37.
Hazard Analysis and Countermeasures
See silver compounds and phosphates.

SILVER PICRATE. See picratol.

SILVER-POTASSIUM CYANIDE
General Information
Description: White crystals; slight odor of HCN.
Formula: $KAg(CN)_2$.
Constant: Mol wt: 199.00.
Hazard Analysis and Countermeasures
See cyanides and silver compounds.

SILVER PROPARGYLATE
General Information
Description: Crystals.
Formula: $CHCCH_2Ag$.
Constant: Mol wt: 147.
Hazard Analysis
Toxicity: See silver compounds.
Explosion Hazard: Moderate, when exposed to heat.
Countermeasures
Storage and Handling: Section 7.

SILVER SALVARSAN. See silver arsphenamine.

SILVER SELENATE
General Information
Description: Crystals.
Formula: Ag_2SeO_4.
Constants: Mol wt: 358.7, d: 5.72.
Hazard Analysis and Countermeasures
See selenium compounds and silver compounds.

SILVER SELENIDE
General Information
Description: Cubic, thin, gray plates.
Formula: Ag_2Se.
Constants: Mol wt: 294.72, mp: 880°C, bp: decomposes, d: 8.0.
Hazard Analysis and Countermeasures
See selenium compounds and silver compounds.

SILVER SELENITE
General Information
Description: Needle-like crystals.
Formula: Ag_2SeO_3.
Constants: Mol wt: 342.7, d: 5.9297.

TOXIC HAZARD RATING CODE (For detailed discussion, see Section 1.)

0 NONE: (a) No harm under any conditions; (b) Harmful only under unusual conditions or overwhelming dosage.

1 SLIGHT: Causes readily reversible changes which disappear after end of exposure.

2 MODERATE: May involve both irreversible and reversible changes; not severe enough to cause death or permanent injury.

3 HIGH: May cause death or permanent injury after very short exposure to small quantities.

U UNKNOWN: No information on humans considered valid by authors.

Hazard Analysis and Countermeasures
See selenium compounds and silver compounds.

SILVER SULFATE
General Information
Description: Rhombic, white crystals.
Formula: Ag_2SO_4.
Constants: Mol wt: 311.82, mp: 652°C, bp: decomposes at 1085°C, d: 5.45 at 29.2°C.
Hazard Analysis and Countermeasures
See silver compounds and sulfates.

SILVER SULFIDE
General Information
Synonym: Acanthite.
Description: Rhombic, gray-black crystals.
Formula: Ag_2S.
Constants: Mol wt: 247.83, mp: tr 175°C, bp: decomposes, d: 7.326.
Hazard Analysis and Countermeasures
See sulfides and silver compounds.

SILVER SULFITE
General Information
Description: White crystals.
Formula: Ag_2SO_3.
Constants: Mol wt: 295.82, mp: decomposes at 100°C.
Hazard Analysis and Countermeasures
See silver compounds and sulfites.

SILVER SULFOCARBOLATE. See silver
phenolsulfonate.

SILVER TETRAZOL
General Information
Description: Solid.
Formula: $AgCHN_4$.
Constant: Mol wt: 176.9.
Hazard Analysis
Toxicity: See silver compounds.
Explosion Hazard: Severe, when exposed to heat.
Disaster Hazard: Dangerous; when heated to decomposition, it emits highly toxic fumes of nitrogen oxides and may explode.
Countermeasures
Storage and Handling: Section 7.

SILVER-THALLIUM NITRATE
General Information
Description: White, crystalline powder.
Formula: $AgNO_3 \cdot TlNO_3$.
Constants: Mol wt: 435.90, mp: 75°C.
Hazard Analysis and Countermeasures
See thallium compounds, silver compounds and nitrates.

SILVER THIOARSENITE
General Information
Synonym: Proustite.
Description: Crystals.
Formula: Ag_3AsS_3.
Constants: Mol wt: 494.73, mp: > 175°C, d: 5.49.
Hazard Analysis and Countermeasures
See arsenic compounds and sulfides.

SIMAZINE
General Information
Synonyms: Simazen; 2-chlor-4,6-bis (ethylamino-s-triazine.)
Description: White solid; insoluble in water; slightly soluble in organic solvents.
Formula: $ClC_3N_3(NHC_2H_5)_2$.
Constants: Mol wt: 201.67, mp: 225°C.
Hazard Analysis
See cyanuric chloride from which simazine is derived.
Disaster Hazard: Dangerous. See chlorides.

SISAL
Hazard Analysis
Toxic Hazard Rating:
  Acute Local: Allergen 1; Inhalation 1.
  Acute Systemic: Inhalation 2.
  Chronic Local: Allergen 1; Inhalation 1.
  Chronic Systemic: Inhalation 1.
TLV: ACGIH (recommended); 50 million particles per cubic foot of air.
Fire Hazard: Moderate, in the form of dust, when exposed to heat or flame; keep cool and dry; partially burned or charred material is dangerous (Section 6).
Countermeasures
Ventilation Control: Section 2.
Storage and Handling: Section 7.

SLAKED LIME. See calcium hydroxide.

SLATE (below 5% free silica)
General Information
Description: A fine-grained green, black or red sedimentary rock.
Hazard Analysis
Toxic Hazard Rating:
  Acute Local: Inhalation 1.
  Acute Systemic: 0.
  Chronic Local: Inhalation 1.
  Chronic Systemic: 0.
TLV: ACGIH (recommended); 50 million particles per cubic foot.
Countermeasures
Ventilation Control: Section 2.
Personal Hygiene: Section 3.

SLUDGE ACID. See sulfuric acid.
Shipping Regulations: Section 11.
  I.C.C.: Corrosive liquid; white label, 1 quart.
  Coast Guard Classification: Corrosive liquid; white label.
  IATA: Corrosive liquid, white label, not acceptable (passenger), 1 liter (cargo).

SMALL ARMS AMMUNITION
Shipping Regulations: Section 11.
  I.C.C.: Explosive C, 150 pounds.
  IATA: Explosive, explosive label, 45 kilograms (passenger), 70 kilograms (cargo).

SMALL ARMS AMMUNITION, TEARGAS CARTRIDGES
Shipping Regulations: Section 11.
  I.C.C.: Explosive, tear gas label, 150 pounds.
  IATA: Explosive, poison label, 25 kilograms (passenger), 70 kilograms (cargo).

SMALL ARMS PRIMERS
Shipping Regulations: Section 11.
  I.C.C.: Explosive C, 150 pounds.
  IATA: Explosive, explosive label, 25 kilograms (passenger), 70 kilograms (cargo).

SMOG
General Information
Description: An atmospheric combination of smoke, fog, and industrial gases (Section 4).
Composition: Contents vary, but sulfur dioxide is a common component. Other sulfides, fluorides, chlorides, carbon particles and various hydrocarbons may be found in smog.
Hazard Analysis
Toxic Hazard Rating:
  Acute Local: Inhalation 1.
  Acute Systemic: Inhalation 1.
  Chronic Local: Inhalation 1.
  Chronic Systemic: Inhalation 2.
Toxicity: A common air contaminant (Section 4).

SMOKE CANDLES
Shipping Regulations: Section 11.
I.C.C.: Explosive C, 200 pounds.
IATA: Explosive, explosive label, 25 kilograms (passenger), 95 kilograms (cargo).

SMOKE DEVICES, TOY
Shipping Regulations: Section 11.
IATA: Explosive, explosive label, 25 kilograms (passenger), 70 kilograms (cargo).

SMOKE GRENADES
Shipping Regulations: Section 11.
I.C.C.: Explosive C, 200 pounds.

SMOKELESS POWDER
General Information
Nitrocellulose containing about 13.1% nitrogen, produced by blending material of somewhat lower (12.6%) and slightly higher (13.2%) nitrogen content, converting to a dough with alcohol-ether mixture, extruding, cutting and drying to a hard horny product. Small amounts of stabilizers (amines) and plasticizers are usually present, as well as various modifying agents (nitrotoluene, nitroglycerine salts). See also nitrocellulose and explosives, high.

SMOKELESS POWDER FOR SMALL ARMS
Shipping Regulations: Section 11.
IATA: Explosive, not acceptable (passenger and cargo).
I.C.C: See propellant explosives, class A or class B.

SMOKE POTS
Shipping Regulations: Section 11.
I.C.C.: Explosive C, 200 pounds.
IATA: Explosive, explosive label, 25 kilograms (passenger), 95 kilograms (cargo).

SMOKE SIGNALS
Shipping Regulations: Section 11.
IATA: Explosive, explosive label, 25 kilograms (passenger), 95 kilograms (cargo).
I.C.C.: Explosive C, 200 pounds.

SOAP POWDERS
Hazard Analysis
Toxic Hazard Rating:
Acute Local: Irritant 1; Allergen 1; Ingestion 1.
Acute Systemic: Irritant 1; Allergen 1; Ingestion 1.
Chronic Local: Irritant 1; Allergen 1.
Chronic Systemic: U.
Fire Hazard: Slight, by chemical reaction (Section 6).
Spontaneous Heating: Moderate.
Countermeasures
Ventilation Control: Section 2.
Storage and Handling: Section 7.

SOAPSTONE DUST. See talc.

SODA CHLORATE. See sodium chlorate.

SODA LIME
General Information
Description: Soduim hydroxide with lime. A mixture of calcium oxide with 5-20% sodium hydroxide and containing 6-18% water. White or gray granules.
Hazard Analysis
Toxic Hazard Rating:
Acute Local: Irritant 3; Ingestion 3; Inhalation 3.
Acute Systemic: Ingestion 3.

Chronic Local: Irritant 2.
Chronic Systemic: U.
Countermeasures
Storage and Handling: Section 7.
Shipping Regulations: Section 11.
IATA: Other restricted articles, class B, no label required, 12 kilograms (passenger), 45 kilograms (cargo).

SODAMIDE. See sodium amide.

SODA MONOHYDRATE. See sodium carbonate.

SODA NITER. See sodium nitrate.

SODIUM
General Information
Synonym: Natrium.
Description: Light, soft, ductile, malleable, silver-white metal.
Formula: Na.
Constants: At wt: 23.0, mp: 97.81°C, bp: 892°C, d: 0.9710 at 20°C, autoign. temp.: above 115°C in dry air, vap. press.: 1.2 mm at 400°C.
Hazard Analysis
Toxic Hazard Rating:
Acute Local: (Metallic Na): Irritant 1; Ingestion 3; Inhalation 3. (Na Smoke): Irritant 2; Ingestion 3.
Acute Systemic: U.
Chronic Local: (Metallic Na): Irritant 1. (Na Smoke): Irritant 2.
Chronic Systemic: U.
Caution: Metallic sodium reacts exothermally with the moisture of body or tissue surfaces, causing thermal and chemical burns due to the reaction with sodium and the sodium hydroxide formed.
Radiation Hazard: Section 5. For permissible levels, see Table 5, p. 150.
Artificial isotope ^{22}Na, half life 2.6 y. Decays to stable ^{22}Ne by electron capture and positron emission of 0.54 MeV. Also emits gamma rays of 1.27 MeV and X-rays.
Artificial isotope ^{24}Na, half life 15 h. Decays to stable ^{24}Mg by emitting beta particles of 1.39 MeV. Also emits gamma rays of 1.37, 2.75 MeV.
Fire Hazard: Dangerous, when exposed to heat or flame, or by chemical reaction with moisture, air, or any oxidizing material; decomposes moisture to evolve hydrogen and heat; reacts exothermally with the halogens, acids and halogenated hydrocarbons; is spontaneously flammable in air. Heated sodium is spontaneously flammable in air.
Spontaneous Heating: No.
Explosion Hazard: Dangerous, when exposed to moisture in any form! Keep dry at all times!
Disaster Hazard: Dangerous, when heated in air, it emits toxic fumes of sodium oxide; will react with water or steam to produce heat, hydrogen, and flammable vapors; can react vigorously to explosively with oxidizing materials. See hydrogen.
Countermeasures
Ventilation Control: Section 2.
To Fight Fire: Soda ash, dry sodium chloride, or graphite in order of preference (Section 6).
Storage and Handling(Section 7): In the absence of moisture, oxygen or halides, sodium is safe to handle. As to indoor storage of drums, the important thing in storing sodium is that the storage area must be kept dry, since explosions may result from the contract of sodium with water. No automatic sprinkler system, or water or steam

---

TOXIC HAZARD RATING CODE (For detailed discussion, see Section 1.)

0 NONE: (a) No harm under any conditions; (b) Harmful only under unusual conditions or overwhelming dosage.

1 SLIGHT: Causes readily reversible changes which disappear after end of exposure.

2 MODERATE: May involve both irreversible and reversible changes; not severe enough to cause death or permanent injury.

3 HIGH: May cause death or permanent injury after very short exposure to small quantities.

U UNKNOWN: No information on humans considered valid by authors.

pipes containing water should be allowed in the room. Sufficient heat should be provided (without the use of open flames) to prevent condensation of moisture in the room due to changes in atmospheric conditions. Empty sodium drums should be stored in this same area.

"Fire extinguishers (preferably color-coded) must be provided in the storage area, but only those containing sodium chloride, sodium carbonate, or graphite may be used. Pails are adequate for storing extinguishent if special care is taken to insure that the materials are dry. Water, carbon dioxide, carbon tetrachloride, soda-acid, or conventional dry chemical (bicarbonate) extinguishers must be avoided, and signs should be posted in the storage area warning against their use.

"Only that amount of sodium immediately needed should be removed from the storage area. Sodium should not be withdrawn for intermediate storage in reaction areas. A special metal container with a tight fitting cover should be used for transporting sodium bricks to other plant areas, once they have been removed from the original container.

Large-scale outdoor storage tanks such as tank cars are unloaded after melting the sodium by circulating hot oil and with-drawing the molten sodium by vacuum to storage tanks similar in construction to sodium tank cars. Although steam may be used to heat the circulating oil for use on both tank cars and storage tanks, steam must not be used directly as the heating agent for sodium tanks."*

Personnel Protection: Section 3.
Shipping Regulations: Section 11.
    I.C.C.: Flammable solid; yellow label, 25 pounds.
    Coast Guard Classification: Inflammable solid; yellow label.
    MCA warning label.
    IATA: Flammable solid, yellow label, not acceptable (passenger), 12 kilograms (cargo).
*Sittig, M., "Sodium, Its Manufacture, Properties and Uses," pp. 145–146, New York, Reinhold Publishing Corp., 1956.

SODIUM ACETATE
General Information
Description: White crystals.
Formula: $NaC_2H_3O_2$.
Constants: Mol wt: 82.0, mp: 324°C, d: 1.528.
Hazard Analysis
Toxic Hazard Rating:
    Acute Local: Irritant 1; Ingestion 1.
    Acute Systemic: 0.
    Chronic Local: 0.
    Chronic Systemic: U.
Note: Used as a general purpose food additive (Section 10). It is a substance which migrates to food from packaging materials.
Countermeasures
Personal Hygiene: Section 3.
Storage and Handling: Section 7.

SODIUM p-ACETYLAMINOPHENYL-ANTIMONATE
General Information
Synonym: Stibenyl.
Description: Light yellow powder; antimony content 35%.
Formula: $CH_3CONHC_6H_4SbO_3HNa \cdot H_2O$.
Constants: Mol wt: 345.93.
Hazard Analysis and Countermeasures
See antimony compounds.

SODIUM ACETYLARSANILATE. See arsacetin (sodium salt).

SODIUM ACID CARBONATE. See sodium bicarbonate.

SODIUM ACID CHROMATE. See sodium dichromate.

SODIUM ACID PHOSPHATE. See sodium phosphate, monobasic.

SODIUM ACID PYROPHOSPHATE
General Information
Synonyms: Sodium pyrophosphate, acid; disodium pyrophosphate; disodium diphosphate; disodium dihydrogen pyrophosphate; SAPP.
Description: White, crystalline powder; soluble in water.
Formula: $Na_2H_2P_2O_7 \cdot 6H_2O$.
Constants: Mol wt: 331, d: 1.862, mp: 220°C (decomposes).
Hazard Analysis
Toxicity: Unknown. A general purpose food additive (Section 10).
Disaster Hazard: Dangerous. See phosphates.

SODIUM ACID SULFATE. See sodium bisulfate.

SODIUM ALGINATE
General Information
Synonym: Sodium polymannurate.
Description: Colorless or slightly yellow solid occurring in filamentous, granular, or powdered forms; forms a viscous colloidal solution with water; insoluble in alcohol, ether, and chloroform.
Formula: $(C_6H_7O_6Na)x$.
Constants: Mol wt: 32,000–250,000.
Hazard Analysis
Toxicity: A stabilizer food additive (Section 10).

SODIUM m-ALUMINATE
General Information
Description: White, hygroscopic powder.
Formula: $NaAlO_2$.
Constants: Mol wt: 82.0, mp: 1650°C.
Hazard Analysis
Toxic Hazard Rating:
    Acute Local: Irritant 1; Ingestion 1.
    Acute Systemic: U.
    Chronic Local: U.
    Chronic Systemic: U.
Note: A substance which migrates to food from packaging materials (Section 10).
Countermeasures
Shipping Regulations: Section 11.
    I.C.C. (liquid): Corrosive liquid, wiite label, 10 gallons.
    IATA (liquid): Corrosive liquid, white label, 1 liter (passenger), 20 liters (cargo).
       (solid): Other restricted articles, class B, no label required, 12 kilograms (passenger), 45 kilograms (cargo).

SODIUM ALUMINOSILICATE
General Information
Synonym: Sodium silico aluminate.
Description: Fine white amorphous powder or beads; odorless and tasteless; insoluble in water and in alcohol and other organic solvents.
Hazard Analysis
Toxicity: Unknown. An anti-caking agent food additive (Section 10).

SODIUM ALUMINUM FLUORIDE
General Information
Synonym: Chiolite, cryolite.
Description: Very white, vitreous masses. Soluble in concentrated sulfuric acid.
Formula: $3NaF \cdot AlF_3$.
Constants: Mol wt: 210.0, d: 2.95, mp: 1000°C.
Hazard Analysis and Countermeasures
See fluorides. An insecticide.

SODIUM ALUMINUM HYDRIDE
General Information
Description: White crystalline material stable in dry air but sensitive to moisture; soluble in tetrahydrofuran.

Formula: $NaAlH_4$.
Constants: Mol wt: 54.0, d: 1.24, mp: 183°C.
Hazard Analysis
See hydrides.
Countermeasures
See hydrides.
Shipping Regulations: Section 11.
   I.C.C.: Flammable solid, yellow label.
   Coast Guard Classification: Inflammable solid, yellow label.
   IATA: Flammable solid, yellow label, not acceptable (passenger), 12 kilograms (cargo).

SODIUM ALUMINUM PHOSPHATE
General Information
Description: White powder.
Formula: $Na_3PO_4 \cdot AlPO_4$.
Constant: Mol wt: 285.92.
Hazard Analysis
Toxic Hazard Rating:
   Acute Local: Irritant 1; Ingestion 1.
   Acute Systemic : U.
   Chronic Local: U.
   Chronic Systemic: U.
Note: A general purpose food additive; it is a substance which migrates to food from packaging materials (Section 10).

SODIUM   ALUMINUM   SULFATE. See aluminum sodium sulfate.

SODIUM AMALGAM
General Information
Description: Silver-white liquid or porous, crystalline mass; contains 2 to 10% metallic sodium; decomposes in water.
Formula: $Na_x Hg_y$.
Hazard Analysis and Countermeasures
See sodium and mercury compounds, inorganic.
Shipping Regulations: Section 11.
   IATA: Flammable solid, yellow label, 12 kilograms (passenger), 45 kilograms (cargo).

SODIUM AMIDE
General Information
Synonym: Sodamide.
Description: White crystalline powder.
Formula: $NaNH_2$.
Constants: Mol wt: 39.02, mp: 210°C, bp: 400°C.
Hazard Analysis
Toxicity: See sodium hydroxide and ammonia, both of which are liberated by this material in the presence of moisture.
Fire Hazard: Moderate, by chemical reaction. See also ammonia (Section 6).
Explosion Hazard: Moderate, when exposed to heat or flame or by chemical reaction with powerful oxidizers. See also ammonia.
Disaster Hazard: Dangerous; when heated to decomposition, it emits highly toxic fumes of ammonia and sodium oxide; will react with water or steam to produce heat and toxic and corrosive fumes; can react with oxidizing materials.
Countermeasures
Storage and Handling: Section 7.
Shipping Regulations: Section 11.
   I.C.C.: Flammable solid; yellow label, 25 pounds.

Coast Guard Classification: Inflammable solid; yellow label.
   IATA: Flammable solid, yellow label, not acceptable (passenger), 12 kilograms (cargo).

SODIUM   AMINOPHENYL   ARSONATE. See sodium arsanilate.

SODIUM ANILINE ARSONATE. See sodium arsanilate.

SODIUM ARSANILATE
General Information
Synonyms: Atoxyl; sodium aniline arsonate; sodium aminophenyl arsonate.
Description: White, crystalline, odorless powder; faint salty taste.
Formula: $C_6H_4NH_2(HOAsOONa) \cdot 4H_2O$.
Constant: Mol wt: 311.12.
Hazard Analysis and Countermeasures
See arsenic compounds. A food additive permitted in the feed and drinking water of animals and/or for the treatment of food producing animals (Section 10).
Shipping Regulations: Section 11.
   IATA: Poison B, poison label, 25 kilograms (passenger), 95 kilograms (cargo).

SODIUM ARSENATE
General Information
Synonym: Sodium o-arsenate.
Description: Clear, colorless crystals; mild alkaline taste.
Formula: $Na_3AsO_4 \cdot 12H_2O$.
Constants: Mol wt: 424.10, mp: 86.3°C, d: 1.762–1.804.
Hazard Analysis and Countermeasures
See arsenic compounds.
Shipping Regulations: Section 11.
   Coast Guard Classification (solid): Poison B; poison label.

SODIUM m-ARSENATE
General Information
Description: Rhombic, efflorescent crystals.
Formula: $NaAsO_3$.
Constants: Mol wt: 145.91, d: 2.301.
Hazard Analysis and Countermeasures
See arsenic compounds.
Shipping Regulations: Section 11.
   I.C.C.: Poison B, poison label, 200 pounds.
   IATA: Poison B, poison label, 25 kilograms (passenger), 95 kilograms (cargo).

SODIUM ARSENITE
General Information
Description: Colorless or grayish-white powder.
Formula: $NaAsO_2$.
Constants: Mol wt: 129.91, d: 1.87.
Hazard Analysis and Countermeasures
See arsenic compounds. A herbicide.
Shipping Regulations: Section 11.
   I.C.C.: Poison B (solution); poison label, 55 gallons.
   Coast Guard Classification: Poison B (solution); poison label.
   IATA (solid): Poison B, poison label, 25 kilograms (passenger), 95 kilograms (cargo).
      (solution): Poison B, poison label, 1 liter (passenger), 220 liters (cargo).

SODIUM ARSPHENAMINE
General Information
Synonyms: Sodium diarsenal; sodium arsphenolamine.

TOXIC HAZARD RATING CODE (For detailed discussion, see Section 1.)

0 NONE: (a) No harm under any conditions; (b) Harmful only under unusual conditions or overwhelming dosage.

1 SLIGHT: Causes readily reversible changes which disappear after end of exposure.

2 MODERATE: May involve both irreversible and reversible changes; not severe enough to cause death or permanent injury.

3 HIGH: May cause death or permanent injury after very short exposure to small quantities.

U UNKNOWN: No information on humans considered valid by authors.

Description: Bright yellow powder; contains not less than 19% arsenic.
Formula: $NaONH_2C_6AsAsC_6H_3NH_2ONa$.
Constant: Mol wt: 410.08.
Hazard Analysis and Countermeasures
See arsenic compounds.

SODIUM ARSPHENOLAMINE. See sodium arsphenamine.

SODIUM ASCORBATE
General Information
Description: White crystals or powder; odorless; soluble in water; insoluble in alcohol.
Formula: $C_6H_7N_6O_6$.
Constants: Mol wt: 198, mp: 218°C (decomposes).
Hazard Analysis
Toxicity: Unknown. A chemical preservative food additive (Section 10).

SODIUM AURIBROMIDE. See gold sodium bromide.

SODIUM AURIDE
General Information
Description: Cubic yellow crystals.
Formula: $NaAu_2$.
Constants: Mol wt: 417.40, mp: decomposes at 700°C.
Hazard Analysis
Toxicity: See gold compounds.

SODIUM AUROCYANIDE. See sodium cyanoaurite.

SODIUM AZIDE
General Information
Description: Colorless, hexagonal crystals.
Formula: NaN_3.
Constants: Mol wt: 65.02, bp: decomposes in vacuum, d: 2.846.
Hazard Analysis and Countermeasures
See azides and sodium hydroxide.
Shipping Regulations: Section 11.
  I.C.C.: Poison B; poison label, 100 pounds.
  Coast Guard Classification: Poison B; poison label.
  IATA: Poison B, poison label, 25 kilograms (passenger), 45 kilograms (cargo).

SODIUM BARBITAL
General Information
Description: White powder.
Formula: $NaC_8H_{11}N_2O_3$.
Constants: Mol wt: 206.18.
Hazard Analysis
Toxicity: Moderate. See barbiturates.
Disaster Hazard: Slight; when heated to decomposition, it emits toxic fumes of nitrogen oxides.
Countermeasures
Storage and Handling: Section 7.

SODIUM BENZOATE
General Information
Synonym: Benzoate of soda.
Description: White, odorless, crystalline solid.
Formula: $NaC_7H_5O_2$.
Constant: Mol wt: 144.1.
Hazard Analysis
Toxic Hazard Rating:
  Acute Local: 0.
  Acute Systemic: Ingestion 1.
  Chronic Local: 0.
  Chronic Systemic: 0.
Toxicology: Large doses of 8 to 10 grams by mouth may cause nausea and vomiting. It is possible to tolerate as much as 50 grams per day. Small doses have little or no effect. A fungicide. See also benzoic acid. A chemical preservative food additive (Section 10).
Disaster Hazard: Slight; when heated to decomposition it emits acrid fumes.

Countermeasures
Storage and Handling: Section 7.

SODIUM BICARBONATE
General Information
Synonyms: Baking soda; sodium acid carbonate.
Description: White powder or crystalline lumps; soluble in water; insoluble in alcohol.
Formula: $NaHCO_3$.
Constants: Mol wt: 84, d: 2.159, mp: loses CO_2 at 270°C.
Hazard Analysis
Toxicity: Unknown. Used as a general purpose food additive; it is a substance which migrates to food from packaging materials (Section 10).

SODIUM BICHROMATE. See sodium dichromate.

SODIUM BIFLUORIDE. See sodium difluoride.

SODIUM BINOXIDE. See sodium peroxide.

SODIUM BIPHOSPHATE. See sodium phosphate, monobasic.

SODIUM BISULFATE
General Information
Synonyms: Sodium hydrogen sulfate; sodium acid sulfate.
Description: Colorless crystals.
Formula: $NaHSO_4$.
Constants: Mol wt: 120.1, mp: > 315°C (decomp.), d: 2.435 at 13°C.
Hazard Analysis
Toxicity: Toxic. See sulfuric acid, which is liberated on contact with moisture.
Disaster Hazard: Dangerous; see sulfates; will react with water or steam to produce heat and toxic fumes.
Countermeasures
Storage and Handling: Section 7.
Shipping Regulations: Section 11.
  IATA (solid): other restricted articles, class B, no label required, 12 kilograms (passenger), 45 kilograms (cargo).
    (solution): Corrosive liquid, white label, 1 liter (passenger), 5 liters (cargo).

SODIUM BISULFITE
General Information
Synonym: Sodium hydrogen sulfite.
Description: White crystals.
Formula: $NaHSO_3$.
Constants: Mol wt: 104.1, 1.48.
Hazard Analysis and Countermeasures
See sulfites. A chemical preservative food additive (Section 10).
Shipping Regulations: Section 11.
  IATA: Other restricted articles, class B, no label required, no limit (passenger and cargo).

SODIUM m-BISULFITE
General Information
Synonym: Sodium pyrosulfite.
Description: Chief constituent of commercial dry sodium bisulfite with which most of its properties are identical.
Formula: $Na_2S_2O_5$.
Constant: Mol wt: 190.
Hazard Analysis
Toxicity: Unknown. A chemical preservative food additive (Section 10). See sulfites.
Countermeasures
Shipping Regulations: Section 11.
  IATA: Other restricted articles, class B, no label required, no limit (passenger and cargo).

SODIUM BORATE. See sodium tetraborate.

SODIUM BORATE PERHYDRATE
General Information
Description: Crystals.
Formula: $NaBO_2 \cdot H_2O_2$.
Constants: Mol wt: 99.8, mp: decomposes at 40°C.
Hazard Analysis and Countermeasures
See peroxides.

SODIUM BOROHYDRIDE
General Information
Description: White to gray-white microcrystalline powder or lumps. Reacts with hot water; soluble in liquid ammonia, "Cellosolve" ether.
Formula: $NaBH_4$.
Constants: Mol wt: 37.85, mp: > 400°C, d: 1.07.
Hazard Analysis
Toxicity: See boron compounds and hydrides.
Fire Hazard: Moderate, when exposed to heat or flame or by chemical reaction with oxidizers.
Disaster Hazard: Dangerous; when heated to decomposition, it emits toxic fumes; will react with water or steam to produce hydrogen; on contact with acid fumes, it can emit flammable vapors.
Countermeasures
Storage and Handling: Section 7.
Shipping Regulations: Section 11.
    IATA: Flammable solid, yellow label, not acceptable (passenger), 12 kilograms (cargo).

SODIUM BROMATE
General Information
Description: White crystals or crystalline powder.
Formula: $NaBrO_3$.
Constants: Mol wt: 150.91, mp: 381°C, d: 3.339 at 17.5°C.
Hazard Analysis and Countermeasures
See bromates.
Shipping Regulations: Section 11.
    Coast Guard Classification: Oxidizing material; yellow label.
    I.C.C.: Oxidizing material, yellow label, 100 pounds.
    IATA: Oxidizing material, yellow label, 12 kilograms (passenger), 45 kilograms (cargo).

SODIUM BROMIDE
General Information
Description: Cubic, colorless crystals.
Formula: NaBr.
Constants: Mol wt: 102.91, mp: 755°C, bp: 1390°C, d: 3.203 at 25°C, vap. press.: 1 miligram at 806°C.
Hazard Analysis and Countermeasures
See bromides.

SODIUM BROMOAURATE
General Information
Description: Brown-black crystals.
Formula: $NaAuBr_4 \cdot 2H_2O$.
Constant: Mol wt: 575.89.
Hazard Analysis and Countermeasures
See gold compounds and bromides.

SODIUM 2-BROMO-4-PHENYLPHENOL
General Information
Formula: $NaOC_6H_3Br(C_6H_5.)$
Constant: Mol wt: 271.
Hazard Analysis and Countermeasures
See phenols and bromides.

SODIUM BROMOPLATINATE
General Information
Description: Dark red crystals.
Formula: $Na_2PtBr_6 \cdot 6H_2O$.
Constants: Mol wt: 838.8, mp: decomposes at 150°C, d: 3.323.
Hazard Analysis and Countermeasures
See platinum compounds and bromides.

SODIUM CACODYLATE
General Information
Synonyms: Sodium dimethyl arsenate; arsysodila.
Description: White, amorphous powder.
Formula: $NaAsC_2H_6O_2 \cdot 3H_2O$.
Constants: Mol wt: 214.02, mp: approx 60°C.
Hazard Analysis and Countermeasures
See arsenic compounds.
Shipping Regulations: Section 11.
    I.C.C.: Poison B; poison label, 200 pounds.
    Coast Guard Classification: Poison B; poison label.
    IATA: Poison B, poison label, 25 kilograms (passenger), 95 kilograms (cargo).

SODIUM CALCIUM ALUMINOSILICATE, HYDRATED
General Information
Synonym: Sodium calcium silicoaluminate.
Hazard Analysis
Toxicity: Unknown. An anticaking agent food additive (Section 10).

SODIUM CARBIDE
General Information
Description: White powder.
Formula: Na_2C_2.
Constants: Mol wt: 70.01, bp: 700°C, d: 1.575 at 15°C.
Hazard Analysis
Toxicity: See sodium hydroxide and acetylene (liberated on contact with water).
Fire Hazard: Moderate, by chemical reaction with oxidizers.
Explosion Hazard: Moderate, by chemical reaction; also on contact with bromine. See acetylene.
Disaster Hazard: See carbides.
Countermeasures
Storage and Handling: Section 7.
To Fight Fire: Carbon dioxide, dry chemical or carbon tetrachloride.

SODIUM CARBONATE
General Information
Synonyms: Soda monohydrate; crystal carbonate.
Description: White, odorless, small crystals or crystalline powder; alkaline taste.
Formula: Na_2CO_3.
Constants: Mol wt: 106.00, mp: 851°C, bp: decomposes, d: 2.509 at 0°C.
Hazard Analysis
Toxic Hazard Rating:
    Acute Local: Irritant 2; Ingestion 2; Inhalation 2.
    Acute Systemic: U.
    Chronic Local: Irritant 1.
    Chronic Systemic: U.
Note: A general purpose food additive, it migrates to food from packaging materials (Section 10).
Countermeasures
Ventilation Control: Section 2.

TOXIC HAZARD RATING CODE (For detailed discussion, see Section 1.)

0 NONE: (a) No harm under any conditions; (b) Harmful only under unusual conditions or overwhelming dosage.

1 SLIGHT: Causes readily reversible changes which disappear after end of exposure.

2 MODERATE: May involve both irreversible and reversible changes; not severe enough to cause death or permanent injury.

3 HIGH: May cause death or permanent injury after very short exposure to small quantities.

U UNKNOWN: No information on humans considered valid by authors.

Personnel Protection: Section 3.
Storage and Handling: Section 7.

SODIUM CARBONATE PEROXIDE
General Information
Description: Fine white powder.
Formula: $2Na_2CO_3 \cdot 3H_2O_2$.
Constant: Mol wt: 314.
Hazard Analysis
Toxicity: See sodium carbonate and hydrogen peroxide.

SODIUM CARBONYL
General Information
Formula: NaCO.
Constant: Mol wt: 51.
Hazard Analysis
Toxicity: Highly toxic. See carbonyls.
Fire Hazard: Moderate, when exposed to heat or by chemical reaction with oxidizers. Heat causes evolution of carbon monoxide.
Explosion Hazard: Moderate, when exposed to heat or by chemical reaction.
Disaster Hazard: Dangerous; when heated to decomposition, it emits highly toxic fumes of sodium oxide and carbon monoxide; it may explode on heating.
Countermeasures
Ventilation Control: Section 2.
Storage and Handling: Section 7.

SODIUM CARBOXYMETHYLCELLULOSE
General Information
Synonyms: CMC; sodium cellulose glycolate; cellulose gum; CM cellulose.
Description: A synthetic cellulose gum (the sodium salt of carboxymethyl-cellulose not < 99.5% on a dry weight basis, with maximum substitution of 0.95 carboxymethyl groups per anhydroglucose unit, and with a minimum viscosity of 25 centipoises for 2% weight aqueous solutions at 25°C). Colorless, odorless, hygroscopic powder or granules. Insoluble in most organic solvents.
Hazard Analysis
Toxic Hazard Rating: U. A general purpose food additive, it is a substance which migrates to food from packaging materials (Section 10).

SODIUM CASEINATE
General Information
Synonym: Casein-sodium.
Description: White, coarse powder; odorless; soluble in water.
Hazard Analysis
Toxicity: Unknown. A general purpose food additive (Section 10).

SODIUM CELLULOSE GLYCOLATE. See sodium carboxymethylcellulose.

SODIUM CHLORATE
General Information
Synonym: Soda chlorate.
Description: Colorless, odorless crystals, cooling, saline taste.
Formula: $NaClO_3$.
Constants: Mol wt: 106.45, mp: 248–261°C, bp: decomposes, d: 2.490 at 15°C.
Hazard Analysis
Toxic Hazard Rating:
Acute Local: Irritant 1; Ingestion 2; Inhalation 2.
Acute Systemic: Ingestion 2.
Chronic Local: Irritant 2.
Chronic Systemic: U.
Toxicology: Can cause local irritation to skin, eyes and mucous membranes. Ingestion of large quantities can be fatal. Symptoms are abdominal pain, nausea, vomiting, cyanosis and collapse. A herbicide.
CAUTION: see also chlorates.

Countermeasures
Shipping Regulations: Section 11.
I.C.C.: Oxidizing material; yellow label, 100 pounds.
Coast Guard Classification: Oxidizing material; yellow label.
MCA warning label.
IATA: Oxidizing material, yellow label, 12 kilograms (passenger), 45 kilograms (cargo).

SODIUM CHLORAURATE. See gold sodium chloride.

SODIUM CHLORIDE
General Information
Synonyms: Salt; halite; sea salt.
Description: Colorless, transparent crystals or white crystalline powder.
Formula: NaCl.
Constants: Mol wt: 58.45, mp: 801°C, bp: 1413°C, d: 2.165, vap. press.: 1 mm at 865°C.
Hazard Analysis
Toxic Hazard Rating:
Acute Local: Irritant 1; Ingestion 1.
Acute Systemic: 0.
Chronic Local: 0.
Chronic Systemic: 0.
Toxicology: When bulk sodium chloride is heated to high temperature, a vapor is emitted which is irritating to the eyes, particularly. Ingestion of large amounts of sodium chloride can cause irritation of the stomach. Improper use of salt tablets may produce this effect. A substance which migrates to food from packaging materials (Section 10).
Countermeasures
Storage and Handling: Section 7.

SODIUM CHLORITE
General Information
Description: White crystals or crystalline powder.
Formula: $NaClO_2$.
Constants: Mol wt: 90.45, bp: decomposes at 175°$\rightarrow O_2$.
Hazard Analysis
Toxicity: Details unknown. May act as an irritant due to oxidizing power.
Fire Hazard: A powerful oxidizing agent; ignited on friction, heat or shock (Section 6).
Explosion Hazard: Moderate, when shocked or exposed to heat or flame.
Disaster Hazard: Dangerous; shock will explode it; when heated, it emits highly toxic fumes of chlorides and may explode; can react vigorously on contact with reducing materials.
Countermeasures
Storage and Handling: Section 7.
Shipping Regulations: Section 11.
I.C.C.: Oxidizing material; yellow label, 100 pounds.
Coast Guard Classification: Oxidizing material; yellow label.
IATA: Oxidizing material, yellow label, not acceptable (passenger), 45 kilograms (cargo).

SODIUM CHLORITE SOLUTION (EXCEEDING 40% SODIUM CHLORITE)
Shipping Regulations: Section 11.
IATA: Corrosive liquid, not acceptable (passenger and cargo).

SODIUM CHLORITE SOLUTION (NOT EXCEEDING 40% SODIUM CHLORITE).
Shipping Regulations: Section 11.
I.C.C.: Corrosive liquid; white label, 4 gallons.
Coast Guard Classification: Corrosive liquid; white label.
IATA: Corrosive liquid, white label, 1 liter (passenger), 15 liters (cargo).

SODIUM CHLOROACETATE
General Information
Description: White, odorless, free-flowing powder.

Formula: $ClCH_2COONa$.
Constants: Mol wt: 116.49, mp: decomposes at 200°C, flash p: none.
Hazard Analysis and Countermeasures
See chlorides.

SODIUM CHLOROAURATE
General Information
Description: Rhombic, yellow crystals.
Formula: $NaAuCl_4 \cdot 2H_2O$.
Constants: Mol wt: 398.06, mp: decomposes.
Hazard Analysis and Countermeasures
See gold compounds and chlorides.

SODIUM CHLOROIRIDATE
General Information
Description: Dull red-black crystals.
Formula: $Na_2IrCl_6 \cdot 6H_2O$.
Constants: Mol wt: 559.93, mp: decomposes at 600°C.
Hazard Analysis and Countermeasures
See iridium compounds and chlorides.

SODIUM 4-CHLORO-2-METHYLPHENOXY-ACETATE
Hazard Analysis
See also phenol.
Toxicity: Details unknown; a fungicide.
Disaster Hazard: Dangerous; see chlorides.
Countermeasures
Storage and Handling: Section 7.

SODIUM CHLORO-2-PHENYLPHENATE
General Information
Synonyms: Sodium 2-chloro-o-phenylphenate.
Hazard Analysis
Toxicity: Details unknown; a fungicide. See also phenol.
Disaster Hazard: Dangerous; see chlorides.
Countermeasures
Storage and Handling: Section 7.

SODIUM CHLOROPLATINATE
General Information
Description: Orange-yellow powder.
Formula: Na_2PtCl_6.
Constants: Mol wt: 453.97, mp: 150–160°C.
Hazard Analysis and Countermeasures
See platinum compounds and chlorides.

SODIUM CHROMATE
General Information
Description: Yellow, rhombic crystals.
Formula: Na_2CrO_4.
Constants: Mol wt: 162.00, d: 2.723 at 25°C, mp: 792°C.
Hazard Analysis and Countermeasures
See chromium compounds and chromates.

SODIUM CITRATE
General Information
Synonym: Trisodium citrate.
Description: White crystals or granular powder; odorless; soluble in water; insoluble in alcohol.
Formula: $C_6H_5O_7Na_3 \cdot 2H_2O$.
Constants: Mol wt: 258, mp: loses water at 150°C, bp: decomposes at red heat.
Hazard Analysis
Toxic Hazard Rating: U.
Toxicity: Used as a sequestrant and general purpose food additive (Section 10).

SODIUM COMPOUNDS
Hazard Analysis
Toxicity: Variable. Sodium ion is practically nontoxic. The toxicity of sodium compounds is frequently, though not always, due to the anion involved. The hydroxide is very corrosive, being strongly basic. Even it is the concentration of hydroxyl ion which is responsible for the caustic action of this material.

SODIUM COPPER POLYPHOSPHATE. See copper compounds.

SODIUM CYANATE
General Information
Description: Colorless needles.
Formula: $NaOCN$.
Constants: Mol wt: 65.92, d: 1.937 at 20°C.
Hazard Analysis and Countermeasures
See cyanates.

SODIUM CYANIDE
General Information
Description: White, deliquescent, crystalline powder.
Formula: $NaCN$.
Constants: Mol wt: 49.02, mp: 563.7°C, bp: 1496°C, vap. press: 1 mm at 817°C.
Hazard Analysis
Toxicity: See cyanides.
Countermeasures
Shipping Regulations: Section 11.
MCA warning label.

SODIUM CYANOAURITE
General Information
Synonym: Sodium aurocyanide.
Description: White, crystalline powder.
Formula: $NaAu(CN)_2$.
Constant: Mol wt: 272.23.
Hazard Analysis and Countermeasures
See cyanides.

SODIUM CYCLAMATE
General Information
Synonym: Sodium cyclohexylsulfamate.
Description: White crystalline powder; practically odorless; practically insoluble in alcohol, benzene, chloroform, and ether; soluble in water.
Formula: $C_6H_{11}NHSO_3Na$.
Constant: Mol wt: 169.
Hazard Analysis
Toxicity: Unknown. A non-nutritive sweetener food additive (Section 10).

SODIUM DIACETATE
General Information
Description: White crystals; acetic odor; soluble in water; slightly soluble in alcohol; insoluble in ether.
Formula: $CH_3COONa \cdot x(CH_3COOH)$, anhydrous; or $CH_3COONa \cdot x(CH_3COOH) \cdot yH_2O$, technical.
Constants: Decomposes above 150°C.
Hazard Analysis
Toxic Hazard Rating: Unknown. A sequestrant food additive (Section 10).

SODIUM DIARSENAL. See sodium arsphenamine.

SODIUM DICHLOROISOCYANURATE
General Information
Description: White crystals; soluble in water; chlorine odor.

TOXIC HAZARD RATING CODE (For detailed discussion, see Section 1.)

0 NONE: (a) No harm under any conditions; (b) Harmful only under unusual conditions or overwhelming dosage.

1 SLIGHT: Causes readily reversible changes which disappear after end of exposure.

2 MODERATE: May involve both irreversible and reversible changes; not severe enough to cause death or permanent injury.

3 HIGH: May cause death or permanent injury after very short exposure to small quantities.

U UNKNOWN: No information on humans considered valid by authors.

Formula: $Cl_2Na(NCO)_3$.
Constants: Mol wt: 220.0, mp: 230–250°C.
Hazard Analysis
Toxicity: Limited animal experiments indicate slight to moderate toxicity.
Disaster Hazard: Dangerous. When heated to decomposition it emits carbon monoxide and chloride fumes.
Countermeasures
Storage and Handling: An oxidizing material (Section 7).
Shipping Regulations: Section 11.
  I.C.C.: Oxidizing material, yellow label, 100 pounds.
  IATA (dry, containing more than 39% available chlorine: Oxidizing material, yellow label, 25 kilograms (passenger), 45 kilograms (cargo).

SODIUM 2,4-DICHLOROPHENOXYETHYL SULFATE
General Information
Synonym: SES.
Description: Crystals.
Formula: $Cl_2C_6H_3OC_2H_4OSO_3Na$.
Constant: Mol wt: 309.11.
Hazard Analysis
Toxic Hazard Rating:
  Acute Local: Irritant 2.
  Acute Systemic: Ingestion 2.
  Chronic Local: Irritant 2.
  Chronic Systemic: Ingestion 2.
Toxicology: Limited animal data indicated a moderate degree of toxicity with kidney and liver injury. Strong solutions irritate the skin. See also 2,4-dichlorophenoxyacetic acid.
Disaster Hazard: Dangerous; see chlorides.
Countermeasures
Storage and Handling: Section 7.

SODIUM DICHLOROPROPIONATE
General Information
Formula: CH_3CCl_2COONa.
Constants: Mol wt: 165.
Hazard Analysis
Toxic Hazard Rating: U. A food additive permitted in the feed and drinking water of animals and/or for the treatment of food producing animals (Section 10).

SODIUM DICHROMATE
General Information
Synonym: Sodium acid chromate.
Description: Red crystals.
Formula: $Na_2Cr_2O_7 \cdot 2H_2O$.
Constants: Mol wt: 298.1, mp: $-2H_2O$ at 100°C; anhyd. at 320°C, bp: decomposes at 400°C, d: 2.52 at 13°C.
Hazard Analysis and Countermeasures
See chromium compounds.
Shipping Regulations: Section 11.
  MCA warning label.
  IATA: Other restricted articles, class A, no label required, no limit (passenger and cargo).

SODIUM DICYANAMIDE
General Information
Description: Colorless crystals.
Formula: $NaN(CN)_2$.
Constants: Mol wt: 89.04, mp: > 315°C (decomp.), d: 1.701 at 30°C.
Hazard Analysis and Countermeasures
See calcium cyanamide and cyanides.

SODIUM DIETHYLDITHIOCARBAMATE
General Information
Description: Liquid.
Formula: $NaSC(S)N(C_2H_5)_2$.
Constants: Mol wt: 171.3, mp: 16–19°C, d: 1.1 at 20°/20°C, vap. d.: 5.9.

Hazard Analysis
Toxicity: Details unknown. See also bis(diethylthiocarbamyl) disulfide.
Disaster Hazard: Dangerous; when heated to decomposition, it emits highly toxic fumes of sulfur and nitrogen oxides.
Countermeasures
Ventilation Control: Section 2.
Storage and Handling: Section 7.

SODIUM DIFLUORIDE
General Information
Description: White powder.
Formula: $NaF \cdot HF$.
Hazard Analysis and Countermeasures
See fluorides and hydrofluoric acid.

SODIUM DI-HYDROGEN o-ARSENATE
General Information
Description: Rhombic or monoclinic, colorless crystals.
Formula: $NaH_2AsO_4 \cdot H_2O$.
Constants: Mol wt: 181.94, mp: $-H_2O$ at 100–130°C, bp: decomposes at 200–280°C, d: 2.53.
Hazard Analysis and Countermeasures
See arsenic compounds.

SODIUM DIHYDROGEN PHOSPHATE. See sodium phosphate, monobasic.

SODIUM 6,7-DIHYDROXY-2-NAPHTHALENE-SULFONATE
General Information
Description: Dry paste.
Formula: $C_{10}H_6(OH)_2SO_3Na$.
Constant: Mol wt: 263.2.
Hazard Analysis
Toxicity: Unknown.
Disaster Hazard: Dangerous; see sulfonates.
Countermeasures
Storage and Handling: Section 7.

SODIUM p-DIMETHYLAMINOBENZENE DIAZO-SULFONATE
Hazard Analysis
Toxicity: Details unknown. A rodenticide of moderate toxicity developed in Germany.
Disaster Hazard: Moderately dangerous; see sulfonates and nitrogen oxides.
Countermeasures
Storage and Handling: Section 7.

SODIUM DIMETHYLARSENATE. See sodium cacodylate.

SODIUM DIMETHYLDITHIOCARBAMATE
General Information
Description: Crystals.
Formula: $(CH_3)_2NCS_2Na$.
Constant: Mol wt: 143.2.
Hazard Analysis
Toxicity: Details unknown. See also bis(diethylthiocarbamyl) disulfide.
Disaster Hazard: Dangerous; see sulfonates and nitrogen oxides.
Countermeasures
Storage and Handling: Section 7.

SODIUM DINITRO-o-BUTYL PHENATE
Hazard Analysis
Toxic Hazard Rating: U. A herbicide (Section 1). See also phenols.
Disaster Hazard: Dangerous. See nitrates.
Countermeasures
Storage and Handling: Section 7.

SODIUM DINITRO-o-CRESYLATE
General Information
Description: Brilliant orange-yellow dye.
Formula: $C_6H_2(ONa)(NO_2)_2(CH_3)$.
Constant: Mol wt: 220.1.
Hazard Analysis
Toxicity: Highly toxic. Details unknown. An insecticide and selective herbicide. See also dinitrocresol.
Fire Hazard: Moderate. See nitrates.
Disaster Hazard: Dangerous. See nitrates.
Countermeasures
Storage and Handling: Section 7.

SODIUM DINITROPHENOL
General Information
Formula: $C_6H_3(ONa)(NO_2)_2$.
Constant: Mol wt: 206.1.
Hazard Analysis
Toxicity: Probably toxic. See dinitrophenol.
Fire Hazard: Moderate. See nitrates.
Explosion Hazard: Severe, when shocked or exposed to heat. See also nitrates.
Explosive Range: At 698°F.
Disaster Hazard: Dangerous; when heated to decomposition it emits toxic fumes and may explode.
Countermeasures
Storage and Handling: Section 7.

SODIUM DIOXIDE. See sodium peroxide.

SODIUM DISPERSIONS
General Information
Description: Finely divided metallic sodium suspended in toluene, xylene, naphtha, kerosene, etc.
Hazard Analysis
Toxicity: Highly toxic. See sodium and individual dispersant.
Fire Hazard: Dangerous, when exposed to heat or flame or by chemical reaction. These are very reactive forms of sodium, which if carelessly handled may catch fire. To extinguish, see sodium. After sodium has been extinguished, the burning organic vapor can be dealt with by very cautious use of a carbon dioxide extinguisher. Do not use carbon tetrachloride (Section 6).
Explosion Hazard: Moderate, by chemical reaction. See also sodium.
Disaster Hazard: Dangerous; when heated, it loses the solvent and emits highly toxic fumes of sodium, sodium oxide, etc.; will react with water or steam to produce heat and hydrogen; on contact with oxidizing materials, it can react vigorously and on contact with acid or acid fumes, it can emit toxic fumes.
Countermeasures
Storage and Handling: Section 7.

SODIUM DITHIONATE. See sodium hydrosulfite.

SODIUM DODECYLBENZENESULFONATE
General Information
Description: White to light yellow flakes, granules, or powder.
Formula: $C_{12}H_{25}C_6H_4SO_3Na$.
Constant: Mol wt: 348.
Hazard Analysis
Toxic Hazard Rating:
Acute Local: Irritant 1; Ingestion 2.
Acute Systemic: U.
Chronic Local: U.

Chronic Systemic: U.
Disaster Hazard: Dangerous. See sulfonates.

SODIUM ETHYLATE
General Information
Synonym: Caustic alcohol.
Description: White powder, sometimes having brownish tinge.
Formula: C_2H_5ONa.
Constant: Mol wt: 68.05.
Hazard Analysis
Toxicity: See sodium hydroxide and ethyl alcohol, into which it readily hydrolyzes.
Fire Hazard: Dangerous; when exposed to heat or flame.
Disaster Hazard: Dangerous; when heated to decomposition, it emits highly toxic fumes; can react vigorously with oxidizing materials.
Countermeasures
Ventilaton Control Section 2.
Personal Hygiene: Section 3.
Storage and Handling: Section 7.

SODIUM 2-ETHYL HEXENYL SULFONATE
Hazard Analysis
Toxic Hazard Rating: Details unknown. Limited animal experiments suggest low toxicity and moderate irritation.
Disaster Hazard: Dangerous. See sulfonates.

SODIUM ETHYL MERCURITHIOSALICYLATE. See mercury compounds, organic.

SODIUM ETHYL SULFATE
General Information
Synonym: Sodium sulfovinate.
Description: White hygroscopic crystalline material.
Formula: $NaC_2H_5SO_4 \cdot H_2O$.
Constant: Mol wt: 166.14.
Hazard Analysis and Countermeasures
See sulfuric acid and ethyl alcohol.

SODIUM ETHYL XANTHATE
General Information
Synonym: Sodium xanthogenate.
Description: Yellowish powder, soluble in water and alcohol.
Formula: $C_2H_5OCSSNa$.
Constant: Mol wt: 144.2.
Hazard Analysis
Toxic Hazard Rating:
Acute Local: Irritant 1; Inhalation 2.
Acute Systemic: Ingestion 2; Inhalation 2; Skin Absorption 2.
Chronic Local: Irritant 1.
Chronic Systemic: U.
Toxicology: Animal experiments have shown moderate toxicity.
Fire Hazard: Moderate, when exposed to heat or flame. See sulfides.
Disaster Hazard: Dangerous; when heated to decomposition or on contact with acid or acid fumes, it emits highly toxic fumes of oxides of sulfur.
Countermeasures
Storage and Handling: Section 7.

SODIUM FERRICYANIDE
General Information
Synonyms: Red prussiate of sodium; red prussiate of soda.
Description: Ruby-red, deliquescent crystals.

TOXIC HAZARD RATING CODE (For detailed discussion, see Section 1.)

0 NONE: (a) No harm under any conditions; (b) Harmful only under unusual conditions or overwhelming dosage.

1 SLIGHT: Causes readily reversible changes which disappear after end of exposure.

2 MODERATE: May involve both irreversible and reversible changes; not severe enough to cause death or permanent injury.

3 HIGH: May cause death or permanent injury after very short exposure to small quantities.

U UNKNOWN: No information on humans considered valid by authors.

Formula: $Na_3Fe(CN)_6 \cdot H_2O$.
Constant: Mol wt: 298.96.
Hazard Analysis and Countermeasures
See ferricyanides.

SODIUM FERROCYANIDE
General Information
Synonym: Yellow prussiate of soda.
Description: Yellow crystals.
Formula: $Na_4Fe(CN)_6 \cdot 10H_2O$.
Constants: Mol wt: 484.1, d: 1.458.
Hazard Analysis and Countermeasures
See ferrocyanides. Use as a food additive permitted in the
feed and drinking water of animals and/or for the treat-
ment of food producing animals, also permitted in food
for human consumption (Section 10).

SODIUM FLUOACETATE. See sodium fluoracetate.

SODIUM FLUOALUMINATE
General Information
Description: Colorless crystals.
Formula: Na_3AlF_6.
Constants: Mol wt: 209.96, mp: 1000°C, d: 2.90.
Hazard Analysis and Countermeasures
See fluorides.

SODIUM FLUOANTIMONATE
General Information
Formula: $NaSbF_6$.
Constants: Mol wt: 258.76, d: 3.375.
Hazard Analysis and Countermeasures
See antimony compounds and fluorides.

SODIUM FLUOBERYLLATE
General Information
Description: Rhombic or monoclinic, white crystals.
Formula: Na_2BeF_4.
Constants: Mol wt: 131.01, mp: decomposes.
Hazard Analysis and Countermeasures
See beryllium compounds and fluorides.

SODIUM FLUOBORATE
General Information
Description: White, rhombic crystals.
Formula: $NaBF_4$.
Constants: Mol wt: 109.82, mp: 384°C (slight decomp.),
bp: decomposes, d: 2.47 at 20°C.
Hazard Analysis and Countermeasures
See fluorides.

SODIUM FLUORIDE
General Information
Synonym: Villiaumite.
Description: Clear, lustrous crystals or white powder or balls.
Formula: NaF.
Constants: Mol wt: 42.00, mp: 980–997°C, bp: 1700°C,
d: 2.558 at 41°C, vap. press.: 1 mm at 1077°C.
Hazard Analysis
Toxicity: Highly toxic. See fluorides. Doses of 25 to 50 mg
can cause severe vomiting, diarrhea and CNS manifes-
tations. An insecticide.
Countermeasures
Storage and Handling: Section 7.
Shipping Regulations: Section 11.
MCA warning label.
IATA (solid): Other restricted articles, class A, no label
no limit (passenger and cargo).
(solution): Corrosive liquid, white label, 1 liter
(passenger), 20 liters (cargo).

SODIUM FLUOROACETATE (also known as 1080)
General Information
Description: Fine, white odorless powder.
Formula: FCH_2COONa.

Hazard Analysis
Toxicity: A rodenticide. This material has a strong effect on
either or both the cardiovascular and nervous systems
in all species and in some species on the skeletal muscles. Man gives a mixed type response with the cardiac
feature predominating. By a direct action on the heart,
notably in the rabbit, contractile power is lost which
leads to declining blood pressure. Ventricular premature
contractions are seen in all species and arrhythmias
are seen in all species and arrhythmias are especially
marked in some species including man. The central
nervous system, notably that of the dog, is directly attacked by sodium fluoroacetate. In man, the action on
the central nervous system produces epileptiform convulsive seizures followed by severe depression.

The dangerous dose for man is 0.5 to 2 mg/kg.
Other species vary considerably in their response to
sodium fluoracetate with primates and birds being the
most resistant and carnivora and rodents being the
most susceptible. Most domestic animals show a susceptibility falling between the two extremes indicated
above.

The first indication of poisoning is nausea and mental
apprehension followed by epileptiform convulsions.
After a period of several hours, pulsus alternans may
exist followed by ventricular fibrillation and death.
Children appear to be more subject to cardiac arrest
than to ventricular fibrillation.
TLV: ACGIH (recommended): 0.05 milligrams per cubic
meter of air. Can be absorbed through the intact skin.
Countermeasures
Treatment: The treatment for sodium fluoroacetate poisoning is mainly symptomatic. Immediate emesis and
stomach lavage followed by oral doses of magnesium
sulfate are useful. Administration of certain compounds
capable of supplying acetate ions has shown antidotal
effects in animals, including monkeys; the choice drugs
being monacetin (glycerol monoacetate) (2 to 4 g/kg)
and a combination of sodium acetate and theonol (2
g/kg of each). A single dose of magnesium sulfate (800
mg/kg) given intramuscularly as a 50 percent solution
has saved the life of rats dosed with lethal amounts of
sodium fluoroacetate. Complete quiet and rest are indicated, but barbiturate anesthesia has proved disappoint-
ing when used as an antidote.

SODIUM FLUOSILICATE
General Information
Description: White powder.
Formula: Na_2SiF_6.
Constants: Mol wt: 188.1, mp: decomposes, d: 2.679.
Hazard Analysis and Countermeasures
See fluosilicates.
Shipping Regulations: Section 11.
MCA warning label.

SODIUM FLUOSULFONATE
General Information
Description: Shiny, hygroscopic leaflets.
Formula: $NaSO_3F$.
Constants: Mol wt: 122.07, mp: decomposes at red heat.
Hazard Analysis and Countermeasures
See fluosulfonates.

SODIUM FORMATE
General Information
Description: White, deliquescent crystals.
Formula: HCOONa.
Constants: Mol wt: 68.0, mp: 253°C, d: 1.92 at 20°C.
Hazard Analysis
Toxicity: See formic acid.
Fire Hazard: Slight (Section 6).
Countermeasures
Storage and Handling: Section 7.

SODIUM m-GERMANATE
General Information
Description: Monoclinic, white, deliquescent crystals.
Formula: Na_2GeO_3.
Constants: Mol wt: 166.59, mp: 1083°C, d: 3.31 at 22°C.
Hazard Analysis
Toxicity: See germanium compounds.

SODIUM GLUCONATE
General Information
Description: White to yellowish crystalline powder; readily soluble in water; sparingly soluble in alcohol.
Formula: $NaC_6H_{11}O_7$.
Constant: Mol wt: 218.
Hazard Analysis
Toxicity: Unknown. A sequestrant food additive (Section 10).

SODIUM GUANYLATE
General Information
Synonym: Disodium guanylate; GMP.
Description: A 5'-nucleotide; crystals; soluble in cold water; very soluble in hot water.
Formula: $Na_2C_{10}H_{12}N_5O_8P \cdot 2H_2O$.
Constant: Mol wt: 443.
Hazard Analysis
Toxic Hazard Rating: U. A food additive permitted in food for human consumption (Section 10).
Disaster Hazard: Dangerous. See sulfates and phosphates.

SODIUM HEXAFLUOROPHOSPHATE
General Information
Description: White solid.
Formula: $NaPF_6 \cdot H_2O$.
Constants: Mol wt: 185.99, d: 2.369 at 19°C.
Hazard Analysis and Countermeasures
See fluorides and phosphates.

SODIUM HEXA-m-PHOSPHATE
General Information
Synonyms: Hy-phos; calgon.
Description: White powder or flakes. Water soluble.
Formula: $(NaPO_3)_6$.
Constant: Mol wt: 611.9.
Hazard Analysis
Toxic Hazard Rating:
Acute Local: Irritant 2: Ingestion 2.
Acute Systemic: Ingestion 1.
Chronic Local: Irritant 2; Ingestion 1.
Chronic Systemic: U.
Toxicology: Low toxicity. Upon ingestion probably causes nausea, vomiting and diarrhea. Systemic effects are slight. A sequestrant food additive; it migrates to food from packaging materials (Section 10).

SODIUM HYDRATE. See sodium hydroxide.

SODIUM HYDRIDE
General Information
Description: Microcrystalline, white to brownish-gray powder. Reacts with water.
Formula: NaH.
Constants: Mol wt: 24.00, mp: 800°C (decomp.), d: 0.92.
Hazard Analysis
Toxicity: Highly toxic. See sodium hydroxide.
Fire Hazard: Moderate, when exposed to heat or flame or by chemical reaction with oxidizers.
Explosion Hazard: Moderate, when exposed to heat or flame or by chemical reaction. See also hydrides.

Disaster Hazard: Dangerous; when heated to decomposition, it emits highly toxic fumes of oxides of sodium; it will react with water or steam to produce heat, sodium hydroxide and hydrogen; can react with oxidizing materials.
Countermeasures
Shipping Regulations: Section 11.
I.C.C.: Flammable solid; yellow label, 25 pounds.
Coast Guard Classification: Inflammable solid; yellow label.
IATA: Flammable solid, yellow label, not acceptable (passenger), 12 kilograms (cargo).
To Fight Fire: Special mixtures of dry chemical (Section 6).
Storage and Handling: Section 7.

SODIUM HYDROGEN FLUORIDE. See sodium difluoride.

SODIUM HYDROGEN OXALATE
General Information
Description: Monoclinic, white crystals.
Formula: $NaHC_2O_4 \cdot H_2O$.
Constant: Mol wt: 130.04.
Hazard Analysis
Toxicity: Highly toxic. See oxalates.

SODIUM HYDROGEN SULFATE. See sodium bisulfate.

SODIUM HYDROGEN SULFITE. See sodium bisulfite.

SODIUM HYDROSULFIDE
General Information
Synonym: Sodium sulfhydrate.
Description: Colorless needles.
Formula: NaSH.
Constants: Mol wt: 56.07, mp: 350°C.
Hazard Analysis and Countermeasures
See sulfides.
Shipping Regulations: Section 11.
MCA warning label.

SODIUM HYDROSULFITE
General Information
Synonyms: Sodium dithionite; sodium hyposulfite.
Description: Light lemon-colored solid in powder or flake form or white to grayish-white crystalline powder.
Formula: $Na_2S_2O_4 \cdot 2H_2O$.
Constants: Mol wt: 210.16, mp: 55°C (decomp.).
Hazard Analysis and Countermeasures
See sulfites. A substance which migrates to food from packaging materials (Section 10).
Shipping Regulations: Section 11.
I.C.C.: Flammable solid; yellow label, 100 pounds.
Coast Guard Classification: Inflammable solid; yellow label.
MCA warning label.
IATA: Flammable solid, yellow label, 12 kilograms (passenger), 45 kilograms (cargo).

SODIUM HYDROXIDE
General Information
Synonyms: Caustic soda; sodium hydrate; lye; white caustic.
Description: White, deliquescent pieces, lumps or sticks.
Formula: NaOH.
Constants: Mol wt: 40.01, mp: 318.4°C, bp: 1390°C, d: 2.120 at 20°/4°C, vap. press.: 1 mm at 739°C.
Hazard Analysis
Toxic Hazard Rating:
Acute Local: Irritant 3; Ingestion 3; Inhalation 2.

TOXIC HAZARD RATING CODE (For detailed discussion, see Section 1.)

0 NONE: (a) No harm under any conditions; (b) Harmful only under unusual conditions or overwhelming dosage.

1 SLIGHT: Causes readily reversible changes which disappear after end of exposure.

2 MODERATE: May involve both irreversible and reversible changes not severe enough to cause death or permanent injury.

3 HIGH: May cause death or permanent injury after very short exposure to small quantities.

U UNKNOWN: No information on humans considered valid by authors.

Acute Systemic: U.
Chronic Local: Irritant 2.
Chronic Systemic: U.

TLV: ACGIH (recommended); 2 milligrams per cubic meter of air.

Toxicology: This material, both solid and in solution, has a markedly corrosive action upon all body tissue. The symptoms of irritation from this material are frequently evident immediately. Its corrosive action on tissue causes burns and frequently deep ulceration, with ultimate scarring. Prolonged contact with dilute solutions has a destructive effect upon tissue. Mists, vapors, and dusts of this compound cause small burns, and contact with the eyes, either in the solid or solution form, rapidly causes severe damage to the delicate tissue. Ingestion either in the solid or solution form causes very serious damage to the mucous membranes or other tissues with which contact is made. It can cause perforation and scarring. Inhalation of the dust or concentrated mist can cause damage to the upper respiratory tract and to lung tissue, depending upon the severity of the exposure. Thus, effects of inhalation may vary from mild irritation of the mucous membranes to a severe pneumonitis. It can cause an irritant dermatitis (Section 9). It is a general purpose food additive; it migrates to food from packaging materials (Section 10).

Disaster Hazard: Dangerous; will react with water or steam to produce heat and will attack living tissue.

Countermeasures

Ventilation Control: Section 2.

Treatment and Antidotes: Speed in removing this caustic from contact with the skin of one who has come in contact with it is important to avoid injury. Remove all contaminated clothing at once and if possible give patient a shower under deluge type of shower using plenty of water. If the eyes are involved, they should be irrigated at once with plenty of warm water for 15 minutes. Persons so injured should be referred to a physician.

Personnel Protection: Section 3.

Storage and Handling: Section 7.

Shipping Regulations: Section 11.
Coast Guard Classification: Hazardous material.
MCA warning label.
IATA: Other restricted articles, class B, 12 kilograms (passenger), 45 kilograms (cargo).

SODIUM HYDROXIDE, SOLUTION. See sodium hydroxide.

Shipping Regulations: Section 11.
Coast Guard Classification: Corrosive liquid; white label.
ICC: Corrosive liquid, white label, 10 gallons.
IATA: Corrosive liquid, white label, 1 liter (passenger), 20 liters (cargo).

SODIUM HYDROXOPLUMBATE
General Information
Description: Light yellow-white, fused, hygroscopic lumps.
Formula: $Na_2Pb(OH)_6$.
Constant: Mol wt: 355.25.
Hazard Analysis and Countermeasures
See lead compounds.

SODIUM HYDROXOSTANNATE
General Information
Description: Hexagonal, colorless, white powder or lumps.
Formula: $Na_2Sn(OH)_6$.
Constants: Mol wt: 266.74, mp: $-3H_2O$ at 140°C.
Hazard Analysis
Toxicity: See tin compounds.

SODIUM 2-HYDROXY-3,6-NAPHTHALENESULFO-NATE. See sodium β-naphtholdisulfonate.

SODIUM HYPOCHLORITE
General Information
Formula: NaClO.

Constants: Mol wt: 74.45, mp: decomposes, bp: decomposes.
Hazard Analysis and Countermeasures
See hypochlorites.

SODIUM HYPONITRITE
General Information
Description: Crystals.
Formula: $Na_2N_2O_2$.
Constants: Mol wt: 106.01, mp: decomposes at 300°C, d: 2.466 at 4°C.
Hazard Analysis and Countermeasures
See nitrites.

SODIUM HYPOPHOSPHITE
General Information
Description: Colorless, pearly, crystalline plates or white, granular powder; bitter sweet, saline taste.
Formula: $NaH_2PO_2 \cdot H_2O$.
Constant: Mol wt: 106.01.
Hazard Analysis
Toxicity: Moderate. See hypophosphites.
Fire Hazare: Moderate. See hypophosphites.
Explosion Hazard: Moderate, when exposed to heat. Sodium hypophosphite may detonate if heated. It must be kept cool, and stored in a cool, ventilated place, away from acute fire hazards. It may be disposed of by dissolving in water (Section 7).
Disaster Hazard: Dangerous; see phosphates.
Countermeasures
Ventilation Control: Section 2.
Storage and Handling: Section 7.

SODIUM HYPOSULFITE. See sodium hydrosulfite.

SODIUM INOSINATE
General Information
Synonym: Disodium inosinate.
Description: A 5'-nucleotide derived from sea tangle (a seaweed) or dried fish.
Formula: $C_{10}H_{11}Na_2N_4O_8P$.
Constant: Mol wt: 392.
Hazard Analysis
Toxic Hazard Rating: U. A food additive permitted in food for human consumption (Section 10).
Disaster Hazard: Dangerous. See phosphates.

SODIUM IODATE
General Information
Description: Rhombic, white crystals.
Formula: $NaIO_3$.
Constants: Mol wt: 197.92, mp: decomposes, d: 4.277 at 17.5°C.
Hazard Analysis and Countermeasures
See iodates. A trace mineral added to animal feeds (Section 10).

SODIUM IODIDE
General Information
Description: Cubic, colorless crystals.
Formula: NaI.
Constants: Mol wt: 149.92, mp: 651°C, bp: 1300°C, d: 3.667, vap. press.: 1 mm at 767°C.
Hazard Analysis and Countermeasures
See iodides. A trace mineral added to animal feeds (Section 10).

SODIUM ISOPROPYLXANTHATE
General Information
Formula: $SC(OC_3H_7)SNa$.
Constant: Mol wt: 158.3.
Hazard Analysis
Toxicity: Unknown. Can cause local irritation to skin and mucous membranes. Animal experiments suggest low toxicity.
Fire Hazard: Moderate, by chemical reaction with oxidizers. See also sulfides.

Disaster Hazard: Dangerous. See sulfides.
Countermeasures
Storage and Handling: Section 7.

SODIUM LAURYL SULFATE
General Information
Synonym: Gardinol; duponol.
Description: White to cream colored crystals, flakes or powder. Water soluble.
Formula: $C_{12}H_{25}NaO_4S$.
Constant: Mol wt: 288.4.
Hazard Analysis
Toxic Hazard Rating:
Acute Local: Irritant 1; Allergen 1.
Acute Systemic: U.
Chronic Local: Irritant 1; Allergen 1.
Chronic Systemic: U.
Note: A food additive permitted in food for human consumption (Section 10).
Disaster Hazard: Dangerous; when heated to decomposition it emits highly toxic fumes. See sulfates.

SODIUM-LEAD ALLOY
General Information
Description: Metallic material.
Composition: $(Na)_x + (Pb)_y$.
Hazard Analysis
Toxicity: Highly toxic. See lead compounds and sodium.
Fire Hazard: Moderate; reacts with moisture and acids to evolve hydrogen and heat; can react with oxidizing materials.
Explosion Hazard: Moderate, by chemical reaction to produce hydrogen.
Disaster Hazard: Dangerous. See lead compounds and sodium. See also hydrogen which is liberated on contact with moisture.
Countermeasures
Storage and Handling: Section 7.
To Fight Fire: Carbon dioxide, dry chemical or carbon tetrachloride (Section 6).

SODIUM LEAD POLYPHOSPHATE
General Information
Description: Dense white powder.
Hazard Analysis and Countermeasures
See lead compounds and phosphates.

SODIUM MANGANATE
General Information
Description: Monoclinic, green crystals.
Formula: $Na_2MnO_4 \cdot 10H_2O$.
Constants: Mol wt: 345.08, mp: 17°C.
Hazard Analysis
Toxicity: See manganese compounds.
Fire Hazard: Moderate, by chemical reaction; an oxidizer (Section 6).
Countermeasures
Storage and Handling: Section 7.

SODIUM, METALLIC, DISPERSION IN ORGANIC SOLVENT. See sodium dispersions.
Shipping Regulations: Section 11.
I.C.C.: Flammable solid; yellow label, 10 pounds.
Coast Guard Classification: Inflammable solid; yellow label.
IATA: Flammable solid, yellow label, not acceptable (passenger), 5 kilograms (cargo).

SODIUM, METALLIC, LIQUID ALLOY
Shipping Regulations: Section 11.
I.C.C.: Flammable solid; yellow label, 1 pound.
Coast Guard Classification: Inflammable solid; yellow label.
IATA: Flammable solid, yellow label, not acceptable (passenger), $\frac{1}{2}$ kilogram (cargo).

SODIUM m-PHOSPHATE
General Information
Description: Sodium m-phosphate is known as a number of different molecular species, some of which exhibit various crystalline forms. The vitreous sodium phosphates having a Na_2O/P_2O_3 mole ratio near unity are classified as sodium m-phosphates. The term also extends to short chain vitreous compositions, the compounds of which exhibit the polyphosphate formula $Na_{n+2}P_nO_{3n+1}$ with n as low as 4–5.
Formula: $(NaPO_3)$; n may be a small integer > 3 (cyclic molecules) or a large number (polymers).
Hazard Analysis
Toxic Hazard Rating: U. A sequestrant food additive (Section 10).
Disaster Hazard: Dangerous. See phosphates.

SODIUM METHANEARSONATE
General Information
Synonym: Sodium methyl arsonate.
Description: White, crystalline powder.
Formula: $Na_2CH_3AsO_3 \cdot 6H_2O$.
Constants: Mol wt: 292.03, mp: 130–140°C.
Hazard Analysis and Countermeasures
See arsenic compounds.

SODIUM N-METHYLAMINEACETATE. See sodium sarcosinate.

SODIUM 2-(N-METHYLAMINO) ETHANE SULFONATE
General Information
Synonym: Sodium N-methyltaurine.
Description: Clear, colorless liquid.
Formula: $HN(CH_3)CH_2CH_2SO_3Na$.
Constants: Mol wt: 161, mp: -26.5 to $-30°C$, d: 1.202–1.218, vap. d.: 5.55.
Hazard Analysis
Toxicity: Unknown.
Disaster Hazard: Dangerous; see sulfonates.
Countermeasures
Storage and Handling: Section 7.

SODIUM METHYLARSONATE. See sodium methanearsonate.

SODIUM METHYLATE
General Information
Synonym: Sodium methoxide.
Description: White amorphous free flowing powder, decomposed by water; soluble in methyl and ethyl alcohol; decomposes in air above 260°F.
Formula: CH_3ONa.
Constant: Mol wt: 54.
Hazard Analysis
Toxic Hazard Rating: U.
Fire Hazard: Moderate when exposed to heat or source of ignition.
Countermeasures
Shipping Regulations: Section 11.

TOXIC HAZARD RATING CODE (For detailed discussion, see Section 1.)

0 NONE: (a) No harm under any conditions; (b) Harmful only under unusual conditions or overwhelming dosage.

1 SLIGHT: Causes readily reversible changes which disappear after end of exposure.

2 MODERATE: May involve both irreversible and reversible changes not severe enough to cause death or permanent injury.

3 HIGH: May cause death or permanent injury after very short exposure to small quantities.

U UNKNOWN: No information on humans considered valid by author

ICC (alcohol mixture): Flammable liquid, red label, 10 gallons.

(dry): Flammable solid, yellow label, 100 pounds.

IATA (alcohol mixture): Flammable liquid, red label, 1 liter (passenger), 40 liters (cargo).

(dry): Flammable solid, yellow label, 12 kilograms (passenger), 45 kilograms (cargo).

SODIUM N-METHYLDITHIOCARBAMATE DIHYDRATE. See "Vapam."

SODIUM METHYL SULFATE
General Information
Description: White, hygroscopic crystals; soluble in water, alcohol, or methanol.
Formula: $NaCH_3SO_4 \cdot H_2O$.
Constant: Mol wt: 152.10.
Hazard Analysis
Toxicity: Unknown. A food additive permitted in food for human consumption (Section 10).

SODIUM N-METHYLTAURINE. See sodium 2-(N-methylamino)ethanesulfonate.

SODIUM MOLYBDATE
Hazard Analysis
Toxic Hazard Rating:
Acute Local: Irritant 2; Inhalation 2.
Acute Systemic: U.
Chronic Local: U.
Chronic Systemic: U.
Toxicity: See also molybdenum compounds.

SODIUM MONOFLUOROACETATE. See sodium fluoroacetate.

SODIUM MONOFLUOROPHOSPHATE
General Information
Description: Colorless crystals.
Formula: Na_2PO_3F.
Constants: Mol wt: 143.97, mp: approx. 625°C.
Hazard Analysis and Countermeasures
See fluorides and phosphates.

SODIUM MONOHYDROGEN o-ARSENATE
General Information
Description: Monoclinic, colorless crystals.
Formula: $Na_2HAsO_4 \cdot 7H_2O$.
Constants: Mol wt: 312.02, mp: 120–130°C; $-H_2O$ at 180°C, d: 1.88.
Hazard Analysis and Countermeasures
See arsenic compounds.

SODIUM MONOSULFIDE. See sodium sulfide.

SODIUM MONOXIDE
General Information
Description: White-grey, deliquescent crystals.
Formula: Na_2O.
Constants: Mol wt: 61.99, bp: 1275°C (sublimes), d: 2.27.
Hazard Analysis and Countermeasures
See sodium hydroxide.
Shipping Regulations: Section 11.
IATA: Other restricted articles, class B, no label required, 12 kilograms (passenger), 45 kilograms (cargo).

SODIUM β-NAPHTHOLDISULFONATE
General Information
Synonym: R salt.
Description: Gray paste.
Formula: $(NaSO_3)_2C_{10}H_5OH$.
Constant: Mol wt: 348.26.
Hazard Analysis
Toxicity: Details unknown. See also β-naphthol.
Disaster Hazard: Dangerous; see sulfonates.
Countermeasures
Storage and Handling: Section 7.

SODIUM NITRATE
General Information
Synonyms: Soda niter; nitratine.
Description: Colorless, transparent, odorless crystals; saline, slightly bitter taste.
Formula: $NaNO_3$.
Constants: Mol wt: 85.01, mp: 306.8°C, bp: decomposes at 380°C, d: 2.261.
Hazard Analysis
Toxicity: See nitrates. A food additive permitted in food for human consumption. (Section 10).
Fire Hazard: Moderate, when mixed with organic matter, it will ignite on friction. See nitrates.
Explosion Hazard: Explodes when heated to over 1000°C or when mixed with cyanides.
Disaster Hazard: Dangerous. See nitrates.
Storage and Handling: Section 7.
Shipping Regulations: Section 11.
I.C.C.: Oxidizing material; yellow label, 100 pounds.
Coast Guard Classification: Oxidizing material; yellow label.
IATA: Oxidizing material, yellow label, 12 kilograms (passenger), 45 kilograms (cargo).

SODIUM NITRIDE
General Information
Description: Dark gray crystals.
Formula: Na_3N.
Constants: Mol wt: 83.00, mp: decomposes at 300°C.
Hazard Analysis
Toxicity: Highly toxic. See sodium hydroxide and ammonia.
Fire Hazard: Moderate, by chemical reaction with water (Section 6).
Explosion Hazard: Moderate, when exposed to heat or by chemical reaction. See ammonia.
Disaster Hazard: Dangerous; when heated to decomposition, it emits highly toxic fumes of sodium; will react with water or steam to produce toxic, corrosive and flammable vapors.
Countermeasures
Storage and Handling: Section 7.

SODIUM NITRITE
General Information
Synonym: Diazotizing salts.
Description: Slightly yellowish or white crystals, sticks or powder.
Formula: $NaNO_2$.
Constants: Mol wt: 69.01, mp: 271°C, bp: decomposes at 320°C, d: 2.168.
Hazard Analysis
Toxicity: See nitrites. A food additive permitted in the feed and drinking water of animals and/or for the treatment of food producing animals, also permitted in food for human consumption. (Section 10).
Fire Hazard: Moderate; a strong oxidizing agent. In contact with organic matter will ignite by friction. See nitrites.
Explosion Hazard: Explodes when heated to over 1000°F or on contact with cyanides.
Disaster Hazard: Dangerous. See nitrites.
Countermeasures
Storage and Handling: Section 7.
Shipping Regulations: Section 11.
I.C.C.: Oxidizing material; yellow label, 100 pounds.
Coast Guard Classification: Oxidizing material; yellow label.
IATA: Oxidizing material, yellow label, 12 kilograms (passenger), 45 kilograms (cargo).

SODIUM NITRITE MIXED (FUSED) WITH POTASSIUM NITRITE. See nitrites.
Shipping Regulations: Section 11.
I.C.C.: Oxidizing material, yellow label, 100 pounds.

SODIUM NITRITE MIXTURES (Sodium nitrate, sodium nitrite, and potassium nitrate)
Shipping Regulations: Section 11.
    IATA: Oxidizing material, yellow label, 12 kilograms (passenger), 45 kilograms (cargo).

SODIUM m-NITROBENZENESULFONATE
General Information
Description: Crystals.
Formula: $NO_2C_6H_4OSO_2Na$.
Constants: Mol wt: 225.2.
Hazard Analysis
Toxicity: Details unknown. See also nitrobenzene.
Disaster Hazard: Dangerous; see sulfonates and nitrites.
Countermeasures
Storage and Handling: Section 7.

SODIUM NITROMETHANE
General Information
Formula: $NaCH_2NO_2$.
Constant: Mol wt: 83.3.
Hazard Analysis
Toxicity: Details unknown. See also nitroparaffins.
Fire Hazard: Moderate, when exposed to heat or flame (Section 6).
Explosion Hazard: Severe, when shocked or exposed to heat.
Disaster Hazard: Dangerous; shock will explode it; when heated to decomposition it emits highly toxic fumes of oxides of nitrogen; can react with reducing materials.
Countermeasures
Ventilation Control: Section 2.
Storage and Handling: Section 7..

SODIUM NITROPHENATE. See sodium p-nitrophenoxide.

SODIUM p-NITROPHENOXIDE
General Information
Synonym: Sodium nitrophenate.
Description: Yellow prisms.
Formula: $NaOC_6H_4NO_2 \cdot 4H_2O$.
Constants: Mol wt: 233.16, mp: $-2H_2O$ at $36°C$; $-4H_2O$ at $120°C$, bp: decomposes.
Hazard Analysis and Countermeasures
See p-nitrophenol sodium hydroxide and nitrates.

SODIUM NITROPRUSSIDE
General Information
Description: Rhombic, red crystals.
Formula: $Na_2Fe(NO)(CN)_5 \cdot 2H_2O$
Constants: Mol wt: 297.97, d: 1.72.
Hazard Analysis
Toxic Hazard Rating:
  Acute Local: U.
  Acute Systemic: Ingestion 3; Inhalation 2.
  Chronic Local: U.
  Chronic Systemic: Ingestion 2; Inhalation 2.
Toxicology: The effects of this material are similar to that of nitrites, causing fall in blood pressure but no formation of methemoglobin. Large amounts, when taken internally, may form cyanide upon being metabolized.
Disaster Hazard: Dangerous; see cyanides.
Countermeasures
Ventilation Control: Section 2.
Personal Hygiene: Section 3.
Storage and Handling: Section 7.

SODIUM OLEATE
General Information
Description: White powder; slight tallow odor.
Formula: $C_{17}H_{33}COONa$.
Constants: Mol wt: 304.5, mp: $232-235°C$.
Hazard Analysis
Toxicity: See oleic acid. A substance that migrates to food from packaging materials. (Section 10).
Fire Hazard: Slight, when exposed to heat or flame (section 6).
Countermeasures
Storage and Handling: Section 7.

SODIUM OXALATE
General Information
Description: White, crystalline powder.
Formula: $Na_2C_2O_4$.
Constants: Mol wt: 134.01, d: 2.34.
Hazard Analysis and Countermeasures
See oxalates.

SODIUM OXIDE. See sodium monoxide.

SODIUM PALMITATE
Hazard Analysis
Toxic Hazard Rating: Unknown. A substance which migrates to food from packaging materials. (Section 10).

SODIUM PECTINATE
Hazard Analysis
Toxic Hazard Rating: A general purpose food additive (Section 10).

SODIUM PENTABORATE. See sodium borate.

SODIUM PENTACHLOROPHENATE
General Information
Description: Tan powder.
Formula: C_6Cl_5ONa.
Constant: Mol wt: 288.4.
Hazard Analysis and Countermeasures
See pentachlorophenol and chlorides. A fungicide.
Shipping Regulations: Section 11.
    IATA: Other restricted article, class A, no label required, no limit (passenger and cargo).

SODIUM PENTACHLOROPHENOLATE. See sodium pentachlorophenate.

SODIUM PENTASULFIDE
General Information
Description: Yellow crystals.
Formula: Na_2S_5.
Constants: Mol wt: 206.32, mp: $251.8°C$.
Hazard Analysis and Countermeasures
See sulfides.

SODIUM PENTOBARBITAL
General Information
Description: White powder.
Formula: $NaC_{11}H_{17}N_2O_3$.
Constant: Mol wt: 248.26.
Hazard Analysis
Toxicity: See barbiturates.

SODIUM PERBORATE
General Information
Synonym: Metaborate peroxyhydrate; sodium perborate tetrahydrate.

TOXIC HAZARD RATING CODE *(For detailed discussion, see Section 1.)*

0 NONE: (a) No harm under any conditions; (b) Harmful only under unusual conditions or overwhelming dosage.

1 SLIGHT: Causes readily reversible changes which disappear after end of exposure.

2 MODERATE: May involve both irreversible and reversible changes not severe enough to cause death or permanent injury.

3 HIGH: May cause death or permanent injury after very short exposure to small quantities.

U UNKNOWN: No information on humans considered valid by authors

Description: White crystals with saline taste, slightly water-soluble.

Formula: $NaBO_2 \cdot H_2O_2 \cdot 3H_2O$.

Constants: Mol wt: 153.9, mp: 62°C.

Hazard Analysis

Toxicity: See boron compounds.

Fire Hazard: Slight, by chemical reaction; an oxidizer. Practically non-hazardous unless mixed with highly combustible or reactive organic compounds (Section 6).

Countermeasures

Storage and Handling: Section 7.

Shipping Regulations: Section 11.

IATA: Oxidizing material, yellow label, 12 kilograms (passenger), 45 kilograms (cargo).

SODIUM PERBORATE TETRAHYDRATE. See sodium perborate.

SODIUM PERBORSILICATE

General Information

Description: White powder.

Composition: Sodium borate, sodium silicate and hydrogen peroxide.

Hazard Analysis

Toxicity: See silicates and boron compounds.

Fire Hazard: Slight, by chemical reaction; can react with reducing materials (Section 6).

Countermeasures

Storage and Handling: Section 7.

SODIUM PERCARBONATE

General Information

Description: Decomposes in aqueous solution to hydrogen peroxide and sodium carbonate.

Formula: $Na_2C_2O_6$ or Na_2CO_4.

Hazard Analysis

Toxicity: Probably toxic. An irritant.

Fire Hazard: Dangerous. A powerful oxidizer.

Countermeasures

Storage and Handling: Section 7.

Shipping Regulations: Section 11.

IATA: Oxidizing material, yellow label, 12 kilograms (passenger), 45 kilograms (cargo).

SODIUM PERCHLORATE

General Information

Description: Colorless, deliquescent crystals.

Formula: $NaClO_4$.

Constants: Mol wt: 122.45, mp: 482°C decomposes.

Hazard Analysis and Countermeasures

See perchlorates.

Shipping Regulations: Section 11.

IATA: Oxidizing material, yellow label, 12 kilograms (passenger), 45 kilograms (cargo).

SODIUM PERMANGANATE

General Information

Description: Purple to reddish-black crystals or powder.

Formula: $NaMnO_4$.

Constants: Mol wt: 141.93, mp: decomposes.

Hazard Analysis

Toxicity: See manganese compounds.

Fire Hazard: Moderate, by chemical reaction; a strong oxidizer (Section 6).

Disaster Hazard: Dangerous; will react vigorously with combustible materials.

Countermeasures

Storage and Handling: Section 7.

Shipping Regulations: Section 11.

I.C.C.: Oxidizing material; yellow label, 100 pounds.

Coast Guard Classification: Oxidizing material; yellow label.

IATA: Oxidizing material, yellow label, 12 kilograms (passenger), 45 kilograms (cargo).

SODIUM PEROXIDE

General Information

Synonyms: Sodium dioxide; sodium superoxide; sodium binoxide.

Description: White powder, turning yellow when heated.

Formula: Na_2O_2.

Constants: Mol wt: 77.99, mp: decomposes at 460°C, bp: decomposes, d: 2.805.

Hazard Analysis

Toxicity: Highly toxic. See sodium hydroxide and peroxides, inorganic.

Fire Hazard: Dangerous, by chemical reaction; a powerful oxidizing agent. See peroxides, inorganic.

Explosion Hazard: Moderate, by chemical reaction with water, acids, powdered metals, etc.

Disaster Hazard: Dangerous; will react with water or steam to produce heat and toxic fumes; can react vigorously with reducing materials.

Countermeasures

Ventilation Control: Section 2.

To Fight Fire: Carbon dioxide or dry chemical. Combustible materials ignited by contact with sodium peroxide should be smothered with soda ash, salt or dolomite mixtures. Chemical fire extinguishers should not be used. If the fire cannot be smothered, it should be flooded with large quantities of water from a hose (Section 6).

Personnel Protection: Section 3.

Storage and Handling: Section 7.

Shipping Regulations: Section 11.

I.C.C.: Oxidizing material; yellow label, 100 pounds.

Coast Guard Classification: Oxidizing material; yellow label.

IATA: Oxidizing material, yellow label, 12 kilograms (passenger), 45 kilograms (cargo).

SODIUM PEROXYCHROMATE

General Information

Description: Orange plates.

Formula: Na_3CrO_8.

Constants: Mol wt: 249.00, mp: decomposes at 115°C.

Hazard Analysis

Toxicity: Highly toxic. See chromium compounds.

Fire Hazard: Moderate, by chemical reaction; a strong oxidizer (Section 6).

Disaster Hazard: Dangerous; will react vigorously with combustible materials.

Countermeasures

Storage and Handling: Section 7.

SODIUM PERSULFATE

General Information

Description: White crystalline powder; soluble in water; decomposed by alcohol.

Formula: $Na_2S_2O_8$.

Constant: Mol wt: 238.13.

Hazard Analysis

Toxic Hazard Rating:

Acute Local: Irritant 3; Ingestion 3; Inhalation 3.

Acute Systemic: U.

Chronic Local: U.

Chronic Systemic: U.

Fire Hazard: Moderate. An oxidizer.

Disaster Hazard: Dangerous. A powerful oxidizer. See sulfates.

Countermeasures

Storage and Handling: Section 7.

SODIUM PHENOBARBITAL

General Information

Description: White crystals.

Formula: $NaC_{12}H_{11}N_2O_3$.

Constant: Mol wt: 254.22.

Hazard Analysis

Toxicity: See barbiturates.

SODIUM 1-PHENOL-4-SULFONATE
General Information
Description: Colorless, slightly efflorescent granules.
Formula: $NaC_6H_4(OH)SO_3 \cdot 2H_2O$.
Constants: Mol wt: 232.20, bp: decomposes.
Hazard Analysis
Toxicity: Unknown. See also phenol.
Disaster Hazard: Dangerous. See sulfonates.
Countermeasures
Storage and Handling: Section 7.

SODIUM PHENOXIDE
General Information
Description: White, deliquescent, crystalline needles.
Formula: $NaOC_6H_5$.
Constant: Mol wt: 116.10.
Hazard Analysis
Toxicity: Highly toxic. See phenol and sodium hydroxide.
Disaster Hazard: Dangerous; when heated to decomposition or on contact with acid or acid fumes, it emits highly toxic fumes.
Countermeasures
Storage and Handling: Section 7.

SODIUM PHENYLACETATE
General Information
Description: Dry powder.
Formula: $C_6H_5CH_2COONa$.
Constant: Mol wt: 158.1.
Toxicity: Details unknown. See also phenol.
Disaster Hazard: Dangerous; when heated to decomposition, it emits toxic fumes.
Countermeasures
Storage and Handling: Section 7.

SODIUM PHENYLGLYCINAMINE-p-ARSONATE.
See tryparsamide.

SODIUM o-PHENYLPHENATE
General Information
Description: Crystals or practically white flakes.
Formula: $NaOC_6H_4C_6H_5$.
Constant: Mol wt: 192.2.
Hazard Analysis
Toxicity: Details unknown; a fungicide. See also phenol. Animal experiments suggest fairly high toxicity.
Disaster Hazard: Dangerous; when heated to decomposition, it emits highly toxic fumes.
Countermeasures
Storage and Handling: Section 7.

SODIUM o-PHENYLPHENOLATE. See sodium o-phenylphenate.

SODIUM PHOSPHATE, DIBASIC
General Information
Synonyms: DSP; disodium phosphate; sodium orthophosphate, secondary; disodium o-phosphate; disodium hydrogen phosphate.
Description: Colorless, translucent crystals or white powder; soluble in water, very slightly soluble in alcohol.
Formula: $Na_2HPO_4 \cdot xH_2O$.
Hazard Analysis
Toxic Hazard Rating: A nutrient and/or dietary food additive, a general purpose food additive, and a sequestrant food additive (Section 10).
Disaster Hazard: Dangerous. See Phosphates.

SODIUM PHOSPHATE, MONOBASIC
General Information
Synonym: Sodium acid phosphate; sodium biphosphate; sodium o-phosphate, primary; MSP; sodium dihydrogen phosphate.
Description and Formula: (a) NAH_2PO_4; white crystalline powder; very soluble in water; Mol wt: 120.
(b) $NaH_2PO_4 \cdot H_2O$ large translucent crystals, insoluble in alcohol mol wt: 138.
Constants: (b)mp: loses water at $100°C$, d: 2.040.
Hazard Analysis
Toxicity: Unknown. A sequestrant and a nutrient and/or dietary supplement food additive (Section 10).
Disaster Hazard: See phosphates.

SODIUM-o-PHOSPHATE. See trisodium phosphate.

SODIUM-o-PHOSPHATE, PRIMARY
See sodium phosphate, monobasic.

SODIUM-o-PHOSPHATE, SECONDARY.
See sodium phosphate, dibasic.

SODIUM PHOSPHIDE
General Information
Description: Red crystals.
Formula: Na_3P.
Constants: Mol wt: 100.0 mp: decomposes.
Hazard Analysis and Countermeasures
See phosphides.

SODIUM PHOSPHOALUMINATE. See sodium aluminum phosphates.

SODIUM PICRAMATE
General Information
Description: Yellow, water-soluble salt.
Formula: $NaOC_6H_2(NO_2)_2NH_2$.
Constant: Mol wt: 221.00.
Hazard Analysis and Countermeasures
See picric acid and explosives, high.

SODIUM PICRAMATE (wet with 20% water)
Shipping Regulations: Section 11.
I.C.C.: Flammable solid; yellow label, 25 pounds.
Coast Guard Classification: Inflammable solid; yellow label.
IATA: Flammable solid, yellow label, not acceptable (passenger), 12 kilograms (cargo).

SODIUM PICRAMATE, DRY OR WET WITH LESS THAN 20% WATER.
Shipping Regulations: Section 11.
IATA: Flammable solid, not acceptable (passenger and cargo).

SODIUM PICRATE, DRY OR WET WITH LESS THAN 10% WATER
General Information
Description: Needle-like crystals; some solubility in water and alcohol.
Formula: $NaC_6H_2O_7 \cdot H_2O$.
Constants: Mol wt: 269.12, mp: $-H_2O$ at $150°C$, explode at $310°C$.
Hazard Analysis
See picric acid.

Countermeasures
See picric acid.
Shipping Regulations: Section 11.
  IATA: Explosive, not acceptable (passenger and cargo).

SODIUM PICRATE, WET WITH NOT LESS THAN 10% WATER
General Information
See sodium picrate.
Hazard Analysis
See picric acid.
Countermeasures
See picric acid.
Shipping Regulations: Section 11.
  IATA: Flammable solid, yellow label, $\frac{1}{2}$ kilograms (passenger), $\frac{1}{2}$ kilograms (cargo).

SODIUM POLONIDE
General Information
Description: Radioactive material.
Formula: Na_2Po.
Constant: Mol wt: 256.
Hazard Analysis and Countermeasures
See polonium.

SODIUM POLYMANNURATE
See sodium alginate.

SODIUM POTASSIUM ALLOYS
shipping Regulations: Section 11.
  ICC: Flammable solid, yellow label, 25 pounds.
  IATA: Flammable solid, yellow label, not acceptable (passenger), 12 kilograms (cargo).

SODIUM POTASSIUM TARTRATE
General Information
Synonyms: Rochelle salt; potassium sodium tartrate.
Description: Colorless, transparent, efflorescent crystals or white powder; soluble in water; insoluble in alcohol.
Formula: $KNaC_4H_4O_6 \cdot 4H_2O$.
Constants: Mol wt: 282, mp: 70–80°C, d: 1.77.
Hazard Analysis
Toxicity: Unknown. A sequestrant and general purpose food additive. (Section 10).

SODIUM PROPIONATE
General Information
Description: Transparent crystals or granules; almost odorless; very soluble in water; slightly soluble in alcohol.
Formula: CH_3CH_2COONa.
Constant: Mol wt: 96 (anhydrous).
Hazard Analysis
Toxicity: Unknown. A chemical preservative food additive. (Section 10).

SODIUM 3-PYRIDINESULFONATE
General Information
Description: Yellowish-white powder.
Formula: $C_5H_4NO_3SNa$.
Constant: Mol wt: 181.2.
Hazard Analysis
Toxicity: Unknown.
Disaster Hazard: Dangerous; when heated to decomposition or on contact with acid or acid fumes, it emits highly toxic fumes of oxides of sulfur and nitrogen.
Countermeasures
Storage and Handling: Section 7.

SODIUM PYROPHOSPHATE. See tetrasodiumpyrophosphate.

SODIUM PYROPHOSPHATE, ACID.
See sodium acid pyrophosphate.

SODIUM PYROPHOSPHATE PEROXIDE
General Information
Description: White powder.
Formula: $Na_4P_2O_7 \cdot 2H_2O_2$.
Constant: Mol wt: 334.
Hazard Analysis
Toxicity: See sodium phosphate and hydrogen peroxide.
Fire Hazard: Moderate, by chemical reaction; an oxidizing material; may ignite upon intimate contact with combustible matter (Section 6).
Countermeasures
Storage and Handling: Section 7.

SODIUM PYROVANADATE
General Information
Description: Colorless, hexagonal plates.
Formula: $Na_4V_2O_7$.
Constants: Mol wt: 305.89, mp: 632–654°C.
Hazard Analysis
Toxicity: See vanadium compounds.

SODIUM RHODANATE. See sodium sulfocyanide.

SODIUM RHODANIDE. See sodium sulfocyanide.

SODIUM SACCHARIN
General Information
Synonyms: Sodium benzosulfimide; gluside, soluble; soluble saccharin.
Description: White crystals or crystalline powder; odorless or faint aromatic odor.
Formula: $C_7H_4NNaO_3S \cdot 2H_2O$.
Constant: Mol wt: 221.
Hazard Analysis
Toxicity: Unknown. A non-nutritive sweetener food additive (Section 10).

SODIUM SARCOSINATE
General Information
Synonym: Sodium N-methylamineacetate.
Description: Clear liquid.
Formula: $HN(CH_3)CH_2COONa$.
Constants: Mol wt: 111.1, vap. d.: 3.83.
Hazard Analysis
Toxic Hazard Rating:
  Acute Local: Irritant 2.
  Acute Systemic: U.
  Chronic Local: Irritant 2.
  Chronic Systemic: U.
Countermeasures
Ventilation Control: Section 2.
Personal Hygiene: Section 3.
Storage and Handling: Section 7.

SODIUM SELENATE
General Information
Description: Colorless rhombic crystals.
Formula: Na_2SeO_4.
Constants: Mol wt: 188.95, d: 3.098.
Hazard Analysis
Toxicity: Details unknown. Resembles arsenic in its effects, causing injury to liver and kidneys. See selenium compounds.

SODIUM SELENIDE
General Information
Description: White to red deliquescent crystals.
Formula: Na_2Se.
Constants: Mol wt: 124.95, mp: >875°C, d: 2.625 at 10°C.
Hazard Analysis and Countermeasures
See selenides.

SODIUM SELENITE
General Information
Description: White crystals.
Formula: $Na_2SeO_3 \cdot 5H_2O$.

Constant: Mol wt: 263.04.
Hazard Analysis and Countermeasures
See selenium compounds.

SODIUM SESQUICARBONATE
General Information
Synonym: Sesqui.
Description: White needle-shaped crystals; soluble in water.
Formula: $Na_2CO_3 \cdot NaHCO_3 \cdot 2H_2O$.
Constants: Mol wt: 210, d: 2.112, mp: decomposes.
Hazard Analysis
Toxic Hazard Rating: U.
    A general purpose food additive. (Section 10).

SODIUM SESQUISILICATE, ANHYDROUS
General Information
Description: Crystals.
Formula: $Na_6Si_2O_7$.
Constant: Mol wt: 306.1
Hazard Analysis
Toxic Hazard Rating:
    Acute Local: Irritant 3; Ingestion 3.
    Acute Systemic: U.
    Chronic Local: Irritant 2.
    Chronic Systemic: U.
Countermeasures
Personnel Protection: Section 3.
Shipping Regulations: Section 11.
    MCA warning label.

SODIUM SESQUISILICATE, HYDRATED
General Information
Description: Crystals.
Formula: $Na_3HSiO_4 \cdot 5H_2O$.
Constant: Mol wt: 252.2.
Hazard Analysis
Toxic Hazard Rating:
    Acute Local: Irritant 2; Ingestion 2.
    Acute Systemic: U.
    Chronic Local: Irritant 2.
    Chronic Systemic: U.
Countermeasures
Personnel Protection: Section 3.

SODIUM SILICATE
General Information
Synonym: Water glass.
Description: Amorphous or colorless, deliquescent crystals.
Formula: $Na_2O \cdot xSiO_2$ = 2-5).
Hazard Analysis
Toxic Hazard Rating:
    Acute Local: Irritant 1; Ingestion 1.
    Acute Systemic: U.
    Chronic Local: Irritant 1.
    Chronic Systemic: U.
A substance which migrates to food from packaging materials (Section 10).
Countermeasures
Personal Hygiene: Section 3.
Shipping Regulations: Section 11.
    IATA: Other restricted articles, class B, no label required, no limit (passenger and cargo).

SODIUM SILICOFLUORIDE. See sodium fluosilicate.

SODIUM SORBATE
General Information
Formula: $CH_3CH:CHCH:CHCOONa$.

Constant: Mol wt: 134.
Hazard Analysis
Toxic Hazard Rating: Unknown. A chemical preservative food additive. (Section 10). It migrates to food from packaging materials.

SODIUM STEARYL FUMARATE
Hazard Analysis
Toxic Hazard Rating: U. A food additive permitted in food for human consumption. (Section 10).

SODIUM SULFACHLOROPYRAZINE MONOHYDRATE
Hazard Analysis
Toxic Hazard Rating: U. A food additive permitted in the food and drinking water of animals, and/or for the treatment of food producing animals (Section 10).

SODIUM SULFATE, ANHYDROUS
General Information
Synonym: sodium sulfate, exsiccated.
Description: White crystals or powder; odorless; soluble in water and glycerol; insoluble in alcohol.
Formula: Na_2SO_4
Constants: Mol wt: 142, d: 2.671, mp: 888°C.
Hazard Analysis
Toxic Hazard Rating: Details unknown. A substance which migrates to food from packaging materials (Section 10).
Disaster Hazard: See sulfates.

SODIUM SULFATE, EXSICCATED. See sodium sulfate anhydrous.

SODIUM SULFHYDRATE. See sodium hydrosulfide.

SODIUM SULFIDE
General Information
Synonym: Sodium monosulfide.
Descriptions: Amorphous, yellow-pink or white deliquescent crystals.
Formula: Na_2S.
Constants: Mol wt: 78.06, mp: 1180°C, d: 1.856 at 14°C.
Hazard Analysis and Countermeasures
See sulfides.
Shipping Regulations: Section 11.
    I.C.C.: Flammable solid; yellow label, 300 pounds.
    Coast Guard Classification: Inflammable solid; yellow label (fused or concentrated, ground or not ground;— may be chipped, flaked or broken).
    MCA warning label.
    IATA: Flammable solid, yellow label, 12 kilograms (passenger), 140 kilograms (cargo).

SODIUM SULFITE
General Information
Description: Hexagonal prisms or white powder.
Formula: Na_2SO_3.
Constants: Mol wt: 126.06, bp: decomposes, d: 2.633 at 15.4°C.
Hazard Analysis and Countermeasures
See sulfites. A chemical preservative food additive (Section 10).

SODIUM SULFOCYANATE. See sodium sulfocyanide.

SODIUM SULFOCYANIDE
General Information
Synonyms: Sodium sulfocyanate; sodium rhodanate; sodium rhodanide; sodium thiocyanate.

TOXIC HAZARD RATING CODE (For detailed discussion, see Section 1.)

0 NONE: (a) No harm under any conditions; (b) Harmful only under unusual conditions or overwhelming dosage.

1 SLIGHT: Causes readily reversible changes which disappear after end of exposure.

2 MODERATE: May involve both irreversible and reversible changes; not severe enough to cause death or permanent injury.

3 HIGH: May cause death or permanent injury after very short exposure to small quantities.

U UNKNOWN: No information on humans considered valid by authors.

Description: Colorless, deliquescent crystals or white powder.
Formula: NaCNS.
Constants: Mol wt: 81.08, mp: 287°C.
Hazard Analysis
Toxic Hazard Rating:
    Acute Local: U.
    Acute Systemic: Ingestion 3.
    Chronic Local: U.
    Chronic Systemic: Ingestion 2.
Toxicology: Large doses internally cause vomiting and convulsions. Chronic poisoning is manifested by weakness, confusion, diarrhea and skin rashes.
Disaster Hazard: See thiocyanates.
Countermeasures
Storage and Handling: Section 7.

SODIUM SULFOVINATE. See sodium ethylsulfate.

SODIUM SUPEROXIDE. See sodium peroxide.

SODIUM TARTRATE
General Information
Synonym: Sal tartar; disodium tartrate.
Description: White crystals or granules; soluble in water; insoluble in alcohol.
Formula: $Na_2C_4H_4O_6 \cdot 2H_2O$.
Constants: Mol wt: 230, d: 1.794, loses $2H_2O$ at 150°C.
Hazard Analysis
Toxic Hazard Rating: U. A sequestrant food additive. (Section 10).

SODIUM TELLURATE
General Information
Description: Hexagonal plates or white powder.
Formula: $Na_2TeO_4 \cdot 2H_2O$.
Constants: Mol wt: 273.64, mp: decomposes.
Hazard Analysis and Countermeasures
See tellurium compounds.

SODIUM TELLURITE
General Information
Description: White powder.
Formula: Na_2TeO_3.
Constant: Mol wt: 221.6.
Hazard Analysis and Countermeasures
See tellurium compounds.

SODIUM TETRABORATE
General Information
Synonym: Sodium borate.
Description: White crystals.
Formula: $Na_2B_4O_7$.
Constants: Mol wt: 201,27, mp: 741°C, bp: 1575°C (decomp),, d: 2.367.
Hazard Analysis
Toxicity: See boron compounds.

SODIUM 2,3,4,6-TETRACHLOROPHENATE.
See sodium 2,3,4,6-tetrachlorophenol.

SODIUM 2,3,4,6-TETRACHLOROPHENOL
General Information
Synonym: Sodium-2,3,4,6-tetrachlorophenate; sodium 2,3, 4,6-tetrachlorophenolate.
Description: Buff to light brown flakes.
Formula: C_6HCl_4ONa.
Constants: Mol wt: 253.9, vap. d.: 9.4.
Hazard Analysis
Toxicity: All chlorophenols are toxic. See also phenol and chlorinated phenols.
Disaster Hazard: Dangerous. See chlorophenols.

SODIUM   2,3,4,6-TETRACHLOROPHENOLATE. See sodium 2,3,4,6-tetrachlorophenol.

SODIUM TETRAHYDROGEN o-TELLURATE
General Information
Description: Hexagonal plates.
Formula: $Na_2H_4TeO_6$.
Constants: Mol wt: 273.64, mp: decomposes.
Hazard Analysis and Countermeasures
See tellurium compounds.

SODIUM TETRAPHENYLBORATE
Hazard Analysis
Toxicity: Unknown. See boron compounds.

SODIUM TETRASULFIDE
General Information
Description: Yellow, cubic, hygroscopic crystals.
Formula: Na_2S_4.
Constants: Mol wt: 174.26, mp: 275°C, bp: decomposes.
Hazard Analysis and Countermeasures
See sulfides.

SODIUM THIOARSENATE
General Information
Description: Monoclinic, yellow crystals.
Formula: $Na_3AsS_4 \cdot 8H_2O$.
Constants: Mol wt: 416.29, mp: decomposes.
Hazard Analysis and Countermeasures
See arsenic compounds.

SODIUM THIOCYANATE. See sodium sulfocyanide.

SODIUM THIOGLYCOLATE
General Information
Description: Hygroscopic crystals.
Formula: $HSCH_2COONa$.
Constant: Mol wt: 114.1.
Hazard Analysis
Toxicity: See sulfides. This material yields hydrogen sulfide on decomposition. The literature contains the report of a death attributed to the absorption of toxic decomposition products from the use of this material in a permanent waving solution.
Disaster Hazard: Dangerous. See sulfides.
Countermeasures
See sulfides.

SODIUM THIOSULFATE
General Information
Synonym: Hypo.
Description: Monoclinic colorless, odorless crystals.
Formula: $Na_2S_2O_3 \cdot 5H_2O$.
Constants: Mol wt: 248.2, mp: 48°C (rapid heating), d: 1.69.
Hazard Analysis
Toxicity: See thiosulfates. Large doses internally have a cathartic action. A sequestrant food additive. It migrates to food from packaging materials (Section 10).
Disaster Hazard: Dangerous. See thiosulfates.
Countermeasures
See thiosulfates.
Shipping Regulations: Section 11.
    IATA: Not restricted (passenger and cargo).

SODIUM   p-TOLUENESULFONCHLORAMINE. See chloramine-T.

SODIUM TRICHLOROACETATE
General Information
Description: Crystals; water soluble.
Formula: $Cl_3CCOONa$.
Constant: Mol wt: 185.4.
Hazard Analysis
Toxicity: Details unknown. Animal experiments suggest moderate toxicity. Large doses cause CNS depression. A herbicide.
Disaster Hazard: Dangerous. See chlorides.

SODIUM-2,4,5-TRICHLOROPHENOLATE
General Information
Description: Buff to brown flakes.
Formula: $Cl_3C_6H_2ONa \cdot 1\frac{1}{2}H_2O$.
Constants: Mol wt: 246.4, vap. d.: 8.5.
Hazard Analysis
Toxicity: Details unknown; used as fungicide. See chlorophenols.
Disaster Hazard: Dangerous; when heated to decomposition it emits highly toxic fumes of chlorides and oxides of sodium.
Countermeasures
Ventilation Control: Section 2.
Personnel Protection: Section 3.
Storage and Handling: Section 7.

SODIUM TRIMETHOXYBOROHYDRIDE
General Information
Description: Microcrystalline, fine white powder and white to grayish lumps.
Formula: $NaBH(OCH_3)_3$.
Constants: Mol wt: 127.9, mp: 230°C (decomp. and evolves methyl borate), d: 1.24.
Hazard Analysis and Countermeasures
See boron compounds and hydrides.

SODIUM TRI-α-NAPHTHYL BORIDE
General Information
Description: Blue crystals; decomposed by water.
Formula: $Na_2B(C_{10}H_7)_3$.
Constants: Mol wt: 438.3, mp: decomposes before melting.
Hazard Analysis
Toxicity: Details unknown. See also boron compounds.

SODIUM TRIPOLYPHOSPHATE
General Information
Description: Crystals.
Formula: $Na_5P_3O_{10}$.
Constant: Mol wt: 367.93.
Hazard Analysis
Toxic Hazard Rating:
  Acute Local: Irritant 1, Ingestion 1.
  Acute Systemic: Ingestion 2.
  Chronic Local: U.
  Chronic Systemic: U.
Toxicology: Ingestion of large doses of sodium phosphates causes catharsis. Sodium m- and pyrophosphates can cause hemorrhages from the intestine if taken internally in large doses. A sequestrant and general purpose food additive. It migrates to food from packaging materials (Section 10).
Countermeasures
Personal Hygiene: Section 3.

SODIUM TUNGSTATE
General Information
Description: White rhombic crystals.
Formula: Na_2WO_4.
Constants: Mol wt: 293.91, mp: 698°C, d: 4.179.
Hazard Analysis
Toxicity: See tungsten compounds.

SODIUM URANATE
General Information
Synonym: Sodium m-uranate.
Description: Gray, yellow or red plates or rhombic prisms or powder.
Formula: Na_2UO_4.

Constant: Mol wt: 348.06.
Hazard Analysis and Countermeasures
See uranium.

SODIUM URANYL ACETATE. See uranium sodium oxyacetate.

SODIUM URANYL CARBONATE
General Information
Description: Yellow crystals.
Formula: $2Na_2CO_3 \cdot UO_2CO_3$.
Constants: Mol wt: 542.09, mp: decomposes at 400°C.
Hazard Analysis and Countermeasures
See uranium.

SODIUM m-VANADATE
General Information
Description: Colorless, monoclinic, prismatic crystals or pale green, crystalline powder.
Formula: $NaVO_3 \cdot 4H_2O$.
Constants: Mol wt: 194.0, mp: 630°C.
Hazard Analysis
Toxicity: See vanadium compounds.

SODIUM o-VANADATE
General Information
Description: Colorless, hexagonal prisms.
Formula: Na_3VO_4.
Constants: Mol wt: 183.94, mp: 850-866°C.
Hazard Analysis
Toxicity: See vanadium compounds.

SODIUM XANTHOGENATE. See sodium ethyl xanthate.

SODIUM ZINC URANYL ACETATE
General Information
Description: Tablets or monoclinic crystals.
Formula: $NaZn(UO_2)_3(C_2H_3O_2)_9 \cdot 9H_2O$.
Constant: Mol wt: 1592.13.
Hazard Analysis and Countermeasures
See uranium.

SODIUM ZIRCONIUM LACTATE. See zirconium sodium lactate.

SOLAN. See 3'-chloro-2-methyl-p-valero toluidide.

SOLDERS
General Information
Description: Ductile, relatively low melting alloys.
Formula: Pb + Sn (from 5% Pb to 81.5% Pb).
Constant: D: 7.5–10.2 (depending on composition), mp: 185–232°C.
Hazard Analysis
Toxicity: Fumes may be toxic on repeated inhalation. See lead.

SOLVENT EDM. See ethylene dichloride; carbon tetrachloride.

SOLVENTS, N.O.S. Specific solvent.
Shipping Regulations: Section 11.
  I.C.C.: Flammable liquid; red label, 15 gallons.
  Coast Guard Classification: Combustible liquid; inflammable liquid, red label.
  IATA: Flammable liquid, red label, 1 liter (passenger), 40 liters (cargo).

SOMAN
General Information
Synonym: Fluoromethylpinacolyloxyphosphine oxide.

TOXIC HAZARD RATING CODE (For detailed discussion, see Section 1.)

0 NONE: (a) No harm under any conditions; (b) Harmful only under unusual conditions or overwhelming dosage.

1 SLIGHT: Causes readily reversible changes which disappear after end of exposure.

2 MODERATE: May involve both irreversible and reversible changes; not severe enough to cause death or permanent injury.

3 HIGH: May cause death or permanent injury after very short exposure to small quantities.

U UNKNOWN: No information on humans considered valid by authors.

Description: Colorless liquid; evolves an odorless vapor.
Formula: $(CH_3)_3CCH(CH_3)OP(CH_3)(O)(F)$.
Constants: Mol wt: 182.2, bp: 167°C, fp: −70°C, d: 1.026 at 20°C, vap. d.: 6.33, vap press: 0.207 mm at 20°C.

Hazard Analysis
Toxic Hazard Rating:
  Acute Local: Irritant 3; Ingestion 3; Inhalation 3.
  Acute Systemic: Ingestion 3; Inhalation 3; Skin Absorption 3.
  Chronic Local: U.
  Chronic Systemic: Ingestion 3; Inhalation 3; Skin Absorption 3.
Toxicology: Hydrofluroic acid is product of hydrolysis. Very dangerous to eyes. Extremely toxic by skin absorption; it penetrates the skin rapidly but does not injure it. Death may occur after 15 minutes of exposure. It acts the same as tabun, but faster and in lower concentrations. See fluorides. See also parathion.
Fire Hazard: Slight, when exposed to heat or flame.
Disaster Hazard: Highly dangerous; when heated to decomposition, it emits highly toxic fumes of oxides of phosphorus and fluorides; will react with water, steam or acids to produce toxic and corrosive fumes; can react with oxidizing materials.

Countermeasures
Ventilation Control: Section 2.
To Fight Fire: Foam, carbon dioxide, dry chemical or carbon tetrachloride (Section 6).
Personnel Protection: Section 3.
First Aid: Section 1.
Storage and Handling: Section 7.

SOOT
General Information
Description: A dark brown to black powdery material.
Composition: Carbon. Can contain toxic and irritant impurities.

Hazard Analysis
Toxic Hazard Rating:
  Acute Local: Inhalation 1.
  Acute Systemic: 0.
  Chronic Local: Irritant 3; Inhalation 1.
  Chronic Systemic: U.
TLV: ACGIH (recommended); 50 million particles per cubic foot of air.
Toxicology: Soot is an obstructive and irritating dust, which may be carcinogenic. It has caused skin cancer. This latter property may be due to some coal tar product which adheres to the soot rather than the soot itself. The commercial product in the form of dust has caused cancer of nasal sinuses and lungs. A common air contaminant. (Section 4).
Fire Hazard: Slight, when exposed to heat or flame (Section 6).

Countermeasures
Ventilation Control: Section 2.
Personnel Protection: Section 3.
First Aid: Section 1.

SORBIC ACID
General Information
Synonym: 2,4-Hexadienoic acid.
Description: Colorless needles.
Formula: $CH_3CHCHCHCHCOOH$.
Constants: Mol wt: 112.12, bp: 228°C (decomp.), mp: 134.5°C, flash p: 260°F (C.O.C.), vap. press.: <0.01 mm at 20°C, vap. d.: 3.87.

Hazard Analysis
Toxic Hazard Rating:
  Acute Local: Irritant 2; Ingestion 1; Inhalation 1.
  Acute Systemic: U.
  Chronic Local: U.
  Chronic Systemic: U.
A chemical preservative food additive. (Section 10).

Fire Hazard: Slight, when exposed to heat or flame, can react with oxidizing materials (Section 6).

Countermeasures
To Fight Fire: Water.
Personal Hygiene: Section 3.
Storage and Handling: Section 7.

SORBITAN MONOSTEARATE
General Information
Description: Cream colored waxy solid; slight odor.
Constants: d.: 1.0 at 25°C, mp: 54°C.

Hazard Analysis
Toxic Hazard Rating: Unknown. A food additive permitted in the feed and drinking water of animals and/or for the treatment of food producing animals; also permitted in food for human consumption (Section 10).

d-SORBITE. See sorbitol.

SORBITOL
General Information
Synonyms: d-Sorbital; d-Sorbite; hexahydric alcohol.
Description: White crystalline powder; odorless; soluble in water; slightly soluble in methanol, ethanol, acetic acid, phenol, and acetamide; almost insoluble in other organic solvents.
Formula: $C_6H_8(OH)_6$.
Constants: Mol wt: 182; d: 1.47 at −5°C; mp: 93°C (metastable form), 97.5°C (stable form).

Hazard Analysis
Toxic Hazard Rating: Unknown. A nutrient and/or dietary supplement food additive; it migrates to food from packaging materials (Section 10).

SORBOSE
General Information
Description: White crystalline powder; soluble in water; slightly soluble in ethyl or isopropyl alcohol; insoluble in ether, acetone, benzene, chloroform.
Formula: $HOCH_2CO(CHOH)_3CH_2OH$.
Constants: Mol wt: 180, mp: 159–161°C.

Hazard Analysis
Toxic Hazard Rating: U. A substance which migrates to food from packaging materials (Section 10).

SOUP. See nitroglycerin.

SOYBEAN OIL
General Information
Synonyms: Soya bean oil; chinese bean oil.
Description: Pale yellow to brownish-yellow liquid.
Composition: Glycerides of fatty acids.
Constants: Mp: 22.2°C, flash p: 540°F (C.C.), autoign temp. 833°F, d: 0.925.

Hazard Analysis
A substance which migrates to food from packaging materials (Section 10).
Fire Hazard: Slight, when exposed to heat or flame.
Spontaneous Heating: Moderate.
To Fight Fire: Foam, carbon dioxide, dry chemical or carbon tetrachloride (Section 6).
Ventilation Control: Section 2.
Storage and Handling: Section 7.

SPANISH FLY. See cantharides.

SPARKLERS
Shipping Regulations: Section 11.
  I.C.C.: See "common fireworks."
  IATA: Explosive, explosive label, 25 kilograms (passenger), 95 kilograms (cargo).

SPARTEINE AND COMPOUNDS
General Information
Description: Colorless oil.
Formula: $C_{15}H_{26}N_2$.

Constants: Mol wt: 234.38, bp: 325°C at 754 mm in H_2; 180-181°C at 20 mm in air, d: 1.023 at 20°/4°C.
Hazard Analysis
Toxic Hazard Rating:
  Acute Local: U.
  Acute Systemic: Ingestion 3; Inhalation 3.
  Chronic Local: U.
  Chronic Systemic: U
Disaster Hazard: Dangerous; when heated, it emits highly toxic fumes of nitrogen oxide.
Countermeasures
Ventilation Control: Section 2.
Personal Hygiene: Section 3.
Storage and Handling: Section 7.

SPEARMINT OIL
General Information
Synonyms: Oil of crisp mint; curled mint.
Description: Colorless, yellow or greenish-yellow liquid.
Composition: 50% Carvone, 1-limonene, pinene.
Constant: D: 0.917–0.934 at 25°/25°C.
Hazard Analysis
Toxic Hazard Rating:
  Acute Local: Allergen 1; Ingestion 1.
  Acute Systemic: U.
  Chronic Local: Allergen 1.
  Chronic Systemic: U.
Fire Hazard: Slight; (Section 6.).
Countermeasures
Personal Hygiene: Section 3.
Storage and Handling: Section 7.

SPECIAL FIREWORKS. See explosives, low.
Shipping Regulations: Section 11.
  I.C.C.: Explosive B; special fireworks label, 200 pounds.
  Coast Guard Classification: Explosive B; special fireworks label.

SPENT ACID. See mixed acid.
Shipping Regulations: Section 11.
  I.C.C.: Corrosive liquid; white label, 1 quart.
  Coast Guard Classification: Corrosive liquid; white label.
  IATA: Corrosive liquid, white label, not acceptable (passenger), 1 liter (cargo).

SPENT OXIDE
General Information
Synonym: Gashouse tankage.
Composition: Iron sponge + 5–10% nitrogen.
Hazard Analysis
Toxicity: Details unknown; can contain toxic materials.
Fire Hazard: Moderate, when exposed to heat or flame; can react with oxidizing materials (Section 6, p. 161).
Countermeasures
Storage and Handling: Section 7.
Shipping Regulations: Section 11.
  IC.C.: Flammable solid; yellow label, not accepted.
  Coast Guard Classification: Inflammable solid.
  IATA: Flammable solid, not acceptable (passenger and cargo).

SPENT SULFURIC ACID. See sulfuric acid.
Shipping Regulations: Section 11.
  I.C.C.: Corrosive liquid; white label, 1 quart.
  Coast Guard Classification: Corrosive liquid; white label.
  IATA: Corrosive liquid, white label, not acceptable (passenger), 1 liter (cargo).

SPERMACETI
General Information
Description: A wax.
Hazard Analysis
Toxic Hazard Rating:
  Acute Local: Allergen 1.
  Acute Systemic: 0.
  Chronic Local: Allergen 1.
  Chronic Systemic: U.
Disaster Hazard: Slight; when heated, it emits acrid fumes (Section 6).
Countermeasures
Personal Hygiene: Section 3.
Storage and Handling: Section 7.

SPERM OIL
General Information
Constants: Flash p: 428°F, d: 0.875-0.884 at 25/25°C, autoign temp: 586°F, flash p.(no. 2): 460°F.
Hazard Analysis
Toxicity: details unknown; probably low.
Fire Hazard: Slight, when exposed to heat or flame.
Countermeasures
To Fight Fire: Foam, carbon dioxide, dry chemical or carbon tetrachloride (Section 6).
Storage and Handling: Section 7.

SPERRYLITE. See platinum arsenide.

SPHEROCOBALTITE. See cobaltous carbonate.

SPICES AND OTHER NATURAL SEASONINGS AND FLAVORINGS (LEAVES, ROOTS, BARKS, BERRIES ETC.). See food additives (Section 10).

SPINDLE OIL. See lubricating oil.

SPIRIT OF GLYCERYL TRINITRATE. See spirits of nitroglycerin.

SPIRIT OF HARTSHORN. See aromatic spirits of ammonia.

SPIRIT OF NITER (SWEET)
General Information
Synonym: Ethyl nitrite spirit.
Description: Pale, straw-colored liquid; fragrant, pungent odor; burning taste.
Composition: Alcohol solution 3.5–4.5%-$C_2H_5NO_2$.
Constant: D: Not over 0.823 at 25°C.
Hazard Analysis
Toxic Hazard Rating:
  Acute Local: Irritant 1; Ingestion 1.
  Acute Systemic: Ingestion 1; Inhalation 1.
  Chronic Local: U.
  Chronic Systemic: Ingestion 1; Inhalation 1.
Toxicology: See also nitrites.
Fire Hazard: Moderate, when exposed to heat or flame (Section 6).
Explosion Hazard: Moderate, when vapors are exposed to heat or flame.
Disaster Hazard: Dangerous; when heated to decomposition, it emits highly toxic fumes of oxides of nitrogen; can react with oxidizing materials.
Countermeasures
Ventilation Control: Section 2.
Personal Hygiene: Section 3.
Storage and Handling: Section 7.

TOXIC HAZARD RATING CODE (For detailed discussion, see Section 1.)

0 NONE: (a) No harm under any conditions; (b) Harmful only under unusual conditions or overwhelming dosage.

1 SLIGHT: Causes readily reversible changes which disappear after end of exposure.

2 MODERATE: May involve both irreversible and reversible changes; not severe enough to cause death or permanent injury.

3 HIGH: May cause death or permanent injury after very short exposure to small quantities.

U UNKNOWN: No information on humans considered valid by authors.

SPIRIT OF NITROGLYCERIN
General Information
Synonym: Spirit of glyceryl trinitrate.
Description: Clear, colorless liquid.
Composition: 1.0-1.1% glycerol trinitrate in alcoholic solution.
Constant: D: 0.814-0.820 at 25°C.
Hazard Analysis
Toxic Hazard Rating:
 Acute Local: U.
 Acute Systemic: Ingestion 2; Inhalation 2.
 Chronic Local: U.
 Chronic Systemic: Ingestion 1; Inhalation 1.
Toxicology: See also nitroglycerin.
Fire Hazard: Dangerous, when exposed to heat or flame. See also ethyl alcohol (Section 6).
Explosion Hazard: See nitroglycerine and ethyl alcohol. If the alcohol evaporates, the residue is nitroglycerin.
Disaster Hazard: Dangerous; when dried out, shock will explode it; when heated to decomposition it emits highly toxic fumes; on contact with oxidizing materials the mixture can react vigorously.
Countermeasures
Ventilation Control: Section 2.
Personal Hygiene: Section 3.
Storage and Handling: Section 7.
Shipping Regulations: Section 11.
 I.C.C.: Flammable liquid; red label, 6 quarts. (not exceeding 1% of nitroglycerin by weight): Flammable liquid, red label, 6 quarts.
 Coast Guard Classification: Inflammable liquid; red label.
 IATA (not exceeding 1% of nitroglycerin by weight): Flammable liquid, red label, 1 liter (passenger), 5 liters (cargo).
    (>5% by weight): Explosive, not acceptable (passenger and cargo).
    (>1%, <5% nitroglycerin by weight): Explosive, not acceptable (passenger and cargo).

SPIRIT OF TURPENTINE. See turpentine oil.

SPIRIT OF WINE. See ethyl alcohol.

SPRENGEL EXPLOSIVES
 This type of explosive is a mixture of nitrobenzene and fuming nitric acid. It is a powerful and cheap explosive and would have many uses except that it is limited by practical disadvantages. The components have to be mixed in glass shortly before the explosive is used. This requires preparation and equipment not always available at the site of the explosion. This material can be destroyed by throwing it into large quantities of water, or possibly by burning in small quantities at a time. See explosives, high.

SQUILL, RED
General Information
Synonym: Sea onion.
Hazard Analysis
Toxic Hazard Rating:
 Acute Local: U.
 Acute Systemic: Ingestion 2.
 Chronic Local: U.
 Chronic Systemic: U.
Toxicology: Large doses cause vomiting and cardiac depression.
Countermeasures
Personal Hygiene: Section 3.
Storage and Handling: Section 7.

STAGGER WEED. See delphinium.

STANNANE. See tin tetrahydride.

STANNIC BIS-ACETYLACETONE DIBROMIDE
General Information
Description: Crystals; soluble in benzene, chloroform and acetone.

Formula: $(C_5H_7O_2)_2SnBr_2$.
Constants: Mol wt: 476.8, bp: 187°C.
Hazard Analysis and Countermeasures
See tin compounds and bromides.

STANNIC BIS-ACETYLACETONE DICHLORIDE
General Information
Description: Crystals; water soluble.
Formula: $(C_5H_7O_2)_2SnCl_2$.
Constants: Mol wt: 387.8, bp: 203°C.
Hazard Analysis and Countermeasures
See tin compounds and chlorides.

STANNIC BIS-BENZOYLACETONE DIBROMIDE
General Information
Description: Yellow powder; slightly soluble in organic solvents.
Formula: $(C_{10}H_9O_2)_2SnBr_2$.
Constants: Mol wt: 608, bp: 214°C.
Hazard Analysis
See tin compounds and bromides.
Countermeasures
See tin compounds and bromides.

STANNIC BIS-BENZOYLMETHANE DIBROMIDE
General Information
Description: Yellow crystals; insoluble in water.
Formula: $(C_{15}H_{10}O_2)_2SnBr_2$.
Constants: Mol wt: 723.0, bp: 278°C.
Hazard Analysis and Countermeasures
See tin compounds and bromides.

STANNIC BIS-3-ETHYL ACETYL ACETONE DIBROMIDE
General Information
Description: Colorless crystals; soluble in chloroform and benzene.
Formula: $(C_7H_{11}O_2)_2SnBr_2$.
Constants: Mol wt: 532.9, mp: 166°C.
Hazard Analysis and Countermeasures
See tin compounds and bromides.

STANNIC BROMIDE
General Information
Synonyms: Tin bromide; tin tetrabromide.
Description: White, crystalline mass.
Formula: $SnBr_4$.
Constants: Mol wt: 438.36, mp: 31°C, bp: 202°C at 34 mm, d liq: 3.340 at 35°C, vap. press: 10 mm at 72.7°C.
Hazard Analysis and Countermeasures
See bromides and tin compounds.

STANNIC CHLORIDE
General Information
Synonyms: Tin chloride; tin tetrachloride.
Description: Colorless, fuming caustic liquid or crystals.
Formula: $SnCl_4$.
Constants: Mol wt: 260 53, mp: −33°C, bp: 114.1°C, d: 2.232, vap. press.: 10 mm at 10°C.
Hazard Analysis
Toxicity: Corrosive. See hydrochloric acid.
Fire Hazard: Slight, by chemical reaction. Upon contact with moisture considerable heat is generated.
Disaster Hazard: Dangerous; hydrochloric acid is liberated on contact with moisture or heat.
Countermeasures
Ventilation Control: Section 2.
Personnel Protection: Section 3.
Storage and Handling: Section 7.

STANNIC CHROMATE
General Information
Description: Brownish-yellow, crystalline powder.
Formula: $Sn(CrO_4)_2$.
Constants: Mol wt: 350.72, mp: decomposes.

Hazard Analysis and Countermeasures
See chromium compounds.

STANNIC OXIDE (CASSITERITE). See tin oxide.

STANNIC PHOSPHIDE. See tin monophosphide.

STANNIC SULFIDE
General Information
Synonym: Artificial gold; tin bisulfide; mosaic gold; tin
   bronze; tin disulfide.
Description: Yellow to brown powder; soluble in concen-
   trated hydrochloric acid, aqua regia, solutions of alkali
   hydroxides or sulfides; insoluble in water or dilute
   acids.
Formula: SnS_2.
Constants: Mol wt: 183.82, d: 4.5, mp: decomposes at
   600°C.
Hazard Analysis
Toxic Hazard Rating:
   Acute Local: Inhalation 1.
   Acute Systemic: U.
   Chronic Local: Inhalation 1.
   Chronic Systemic: U.
Toxicity: See also tin compounds.
Disaster Hazard: Dangerous. See sulfides.

STANNOUS ACETATE
General Information
Description: White to yellowish powder, decomposed by
   water.
Formula: $Sn(C_2H_3O_2)_2$.
Constants: Mol wt: 236.79, mp: 182°C.
Hazard Analysis
Toxic Hazard Rating:
   Acute Local: Irritant 1; Ingestion 1; Inhalation 1.
   Acute Systemic: U.
   Chronic Local U.
   Chronic Systemic: U.
Toxicity: See also tin compounds.

STANNOUS CHLORIDE
General Information
Synonyms: Tin crystals; tin salt; tin dichloride; tin pro-
   tochloride.
Description: Colorless crystals; soluble in less than its own
   weight of water; very souble in hydrochloric acid
   (dilute or concentrated); soluble in alcohol, ethyl ace-
   tate, glacial acetic acid, and sodium hydroxide solution.
Formula: $SnCl_2 \cdot 2H_2O$.
Constants: Mol wt: 225.65, d: 2.71 mp: 37–38°C.
Hazard Analysis
Toxic Hazard Rating:
   Acute Local: Irritant 2; Ingestion 2; Inhalation 2.
   Acute Systemic: U.
   Chronic Local: U.
   Chronic Systemic: U.
Toxicity: A chemical preservative food additive. (Section
   10).
Disaster Hazard: See chlorides.

STANNOUS FLUORIDE. See tin fluoride.

STANNOUS OXIDE. See tin oxide.

STARCH DUST
General Information
Description: White powder or granules.
Formula: $(C_6H_{10}O_5)n$.

Hazard Analysis
Toxic Hazard Rating:
   Acute Local: Allergen 1; Inhalation 1.
   Acute Systemic: Ingestion 1; Inhalation 1.
   Chronic Local: Allergen 1.
   Chronic Systemic: Inhalation 1.
Fire Hazard: Moderate, when exposed to flame; can react
   with oxidizing materials (Section 6).
Explosion Hazard: Moderate, when exposed to flame.
Countermeasures
Ventilation Control: Section 2.
Personal Hygiene: Section 3.
Storage and Handling: Section 7.

STARCH GUM. See dextrin.

STARTING CARTRIDGES, JET ENGINE
Shipping Regulations. Section 11.
   I.C.C. (Class B explosives) Explosive B, Red label,
      200 pounds.
         (Class C explosives): Explosive C, 150 pounds.
   IATA: Explosive, explosive label, 25 kilograms (pas-
      senger), 70 kilograms (cargo).
         (other than annunition, small arms or cartridges,
      safety): Explosive, explosive label, not acceptable
      (passenger), 70 kilograms (cargo).

STARTERS COLD
Shipping Regulations Section 11.
   IATA: Flammable solid, yellow label, 12 kilograms
      (passenger and cargo).

**STARTER FLUID, PISTON ENGINE, CHARGED
   WITH FLAMMABLE GAS**
Shipping Regulations: Section 11.
   IATA: Flammable gas, red label, not acceptable (pas-
      senger), 30 kilograms (cargo).

**STEARAMIDOPROPYLIDIMETHYL-β-HYDROXY-
   ETHYLAMMONIUM PHOSPHATE**
General Information
Synonym: "Catanac" SP.
Hazard Analysis
Toxic Hazard Rating:
   Acute Local: Ingestion 2.
   Acute Systemic: Ingestion 2.
   Chronic Local: U.
   Chronic Systemic: U.
Toxicology: A quaternary ammonium compound. Large
   doses by mouth cause vomiting and diarrhea. See also
   benzalkonium chloride.

STEARIC ACID
General Information
Synonym: Octadecanoic acid.
Description: White, amorphous solid.
Formula: $CH_3(CH_2)_{16}COOH$.
Constants: Mol wt: 284.47, mp: 69.3°C. bp: 383°C,
   flash p: 385°F (C.C.), d: 0.847, autoign. temp.: 743°F,
   vap. press.: 1 mm at 173.7°C, vap. d.: 9.80.
Hazard Analysis
Toxicity: Slight. A substance which migrates to food from
   packaging materials (Section 10).
Fire Hazard: Slight, when exposed to heat or flame.
Spontaneous Heating: Yes.
Countermeasures
To Fight Fire: Foam, carbon dioxide, dry chemical or
   carbon tetrachloride (Section 6).
Storage and Handling: Section 7.

TOXIC HAZARD RATING CODE (For detailed discussion, see Section 1.)

0 NONE: (a) No harm under any conditions; (b) Harmful only under unusual conditions or overwhelming dosage.

1 SLIGHT: Causes readily reversible changes which disappear after end
of exposure.

2 MODERATE: May involve both irreversible and reversible changes;
not severe enough to cause death or permanent injury.

3 HIGH: May cause death or permanent injury after very short exposure
to small quantities.

U UNKNOWN: No information on humans considered valid by authors.

STEAROYL PROPYLENE GLYCOL HYDROGEN SUCCINATE. See succistearin.

STEARYL ALCOHOL
General Information
Synonym: 1-Octadecanol.
Description: Colorless solid or flakes.
Formula: $C_{18}H_{37}OH$.
Constants: Mol wt: 270.5,, mp: 58°C, bp: 202°C at 10 mm, d: 0.8124 at 59°/4°C.
Hazard Analysis
Toxicity: Practically non-toxic. See alcohols.
Fire Hazard: Moderate, when exposed to heat or flame; can react with oxidizing materials (Section 6).
Countermeasures
To Fight Fire: Foam, carbon dioxide, dry chemical or carbon tetrachloride.
Storage and Handling: Section 7.

STEARYL CITRATE
Hazard Analysis
Toxic Hazard Rating: A sequestrant food additive (Section 10).

STEATITE. See talc.

STEEL WOOL
General Information
Description: It is composed of long curls of steel wire. Sometimes contains soap to aid in cleaning.
Hazard Analysis
Toxic Hazard Rating:
    Acute Local: Irritant 1.
    Acute Systemic: 0.
    Chronic Local: U.
    Chronic Systemic: 0.
Countermeasures
Personal Hygiene: Section 3.

STIBENYL. See sodium p-acetylaminophenyl antimonate.

STIBINE. See antimony hydride.

STIBIUM. See antimony.

STILBENE. See diphenyl ethylene.

STILBESTROL. See diethylstilbestrol.

STODDARD SOLVENT
General Information
Synonyms: Safety solvent; varnoline; cleaning solvents (kerosene class); naphtha safety solvent; white spirits.
Description: Clear, colorless liquid.
Constants: Bp: 220-300°C, flash p: 100-110°F, lel: 1.1%, uel: 6%, autoign. temp.: 450°F, d: 1.0.
Hazard Analysis
Toxic Hazard Rating:
    Acute Local: Irritant 1, Ingestion 2; Inhalation 1.
    Acute Systemic: Ingestion 1; Inhalation 1.
    Chronic Local: Irritant 1.
    Chronic Systemic: U.
Toxicology: See also gasoline.
TLV: ACGIH (recommended); 500 parts per million in air. 2900 milligrams per cubic meter of air.
Fire Hazard: Moderate, when exposed to heat or flame.
Explosion Hazard: Moderate, when exposed to flame.
Disaster Hazard: Moderate; when heated to decomposition, it emits acrid fumes and may explode; can react with oxidizing materials.
Countermeasures
Ventilation Control: Section 2.
To Fight Fire: Foam, carbon dioxide, dry chemical or carbon tetrachloride (Section 6).
Personal Hygiene: Section 3.
Storage and Handling: Section 7.
Shipping Regulations: Section 11.
    Coast Guard Classification: Combustible liquid.

STOVE POLISH, LIQUID
General Information
Constant: Flash p: < 80°F.
Hazard Analysis
Toxicity: Variable.
Fire Hazard: Dangerous, when exposed to heat or flame or powerful oxidizers (Section 6).
Explosion Hazard: Unknown.
Disaster Hazard: Dangerous, upon exposure to heat or flame; can react vigorously with oxidizing materials.
Countermeasures
Ventilation Control: Section 2.
Storage and Handling: Section 7.
Shipping Regulations: Section 11.
    Coast Guard Classification: Combustible liquid; inflammable liquid, red label.

STRAW
Hazard Analysis
Toxic Hazard Rating:
    Acute Local: Allergen 1; Inhalation 1.
    Acute Systemic: Inhalation 1.
    Chronic Local: Irritant 1; Allergen 1; Inhalation 1.
    Chronic Systemic: Inhalation 1.
Fire Hazard: Moderate, when exposed to heat or flame.
Spontaneous Heating: Yes, especially when damp.
Explosion Hazard: Moderate, in the form of dust, when exposed to flame.
Countermeasures
Personal Hygiene: Section 3.
To Fight Fire: Water (Section 6).
Storage and Handling: Section 7.
Shipping Regulations: Section 11.
    Coast Guard Classification: Hazardous article.

STRAWBERRY ALDEHYDE. See ethyl methyl phenyl glycidate.

STRAW OIL. See lubricating oil.

STREPTOMYCIN
General Information
Description: An antibiotic; it is a base and readily forms salts with anions.
Formula: $C_{21}H_{39}N_7O_{12}$.
Constant: Mol wt: 581.58.
Hazard Analysis
Toxic Hazard Rating: U. A food additive permitted in food for human consumption (Section 10).

STREUNEX. See benzene hexachloride.

STROBANE. See "Toxaphene."

STRONTIUM
General Information
Description: Pale yellow, soft metal.
Formula: Sr.
Constants: At wt: 87.63, bp: 1150°C, fp: 752°C, d: 2.6, vap. press.: 10 mm at 898°C.
Hazard Analysis
Toxic Hazard Rating:
    Acute Local: Irritant 2; Ingestion 2; Inhalation 2.
    Acute Systemic: 0.
    Chronic Local: Irritant 1.
    Chronic Systemic: 0.
Toxicology: It resembles calcium in its metabolism and behavior. The stable form has low toxicity.
Radiation Hazard: Section 5. For permissible levels, see Table 5, p. 150.
    Artificial isotope ^{85m}Sr, half life 70 m. Decays to radioactive ^{85}Rb by electron capture (14%). Also decays to radioactive ^{85}Sr by emitting gamma rays of 0.01, 0.22 MeV. Also emits gamma rays of 0.15 MeV and X-rays.
    Artificial isotope ^{85}Sr, half life 64 d. Decays to stable

[85]Rb by electron capture. Emits gamma rays of 0.51 MeV and X-rays.

Artificial isotope ^{89}Sr, half life 50 d. Decays to stable [89]Y by emitting beta particles of 1.46 MeV.

Artificial isotope ^{90}Sr, half life 28 y. Decays to radioactive ^{90}Y by emitting beta particles of 0.54 MeV.

Artificial isotope [91]Sr, half life 9.7 h. Decays to radioactive [91]Y by emitting beta particles of 0.61–2.67 MeV. Also emits gamma rays of 0.64–1.41 MeV.

Artificial isotope [92]Sr, half life 2.6 h. Decays to radioactive [92]Y by emitting beta particles of 0.54 (90%), 1.5 (10%) MeV. Also emits gamma rays of 1.37 MeV and others.

Fire Hazard: Moderate, in the form of dust, when exposed to flame. See also powdered metals (Section 6).

Explosion Hazard: Moderate, in the form of dust, by chemical reaction. Reacts with water to evolve hydrogen.

Disaster Hazard: Highly dangerous in form of radioactive isotopes; it will react with water or steam to produce heat and hydrogen; on contact with oxidizing materials, it can react vigorously.

Countermeasures
Personal Hygiene: Section 3.
Storage and Handling: Section 7.

STRONTIUM ACID o-ARSENATE
General Information
Description: Rhombic needles.
Formula: $SrHAsO_4 \cdot H_2O$.
Constants: Mol wt: 245.56, mp: $-H_2O$ at 125°C, d: 3.606 at 15°C.
Hazard Analysis and Countermeasures
See arsenic compounds.

STRONTIUM ALLOYS
Shipping Regulations: Section 11.
    IATA: Flammable solid, yellow label, 12 kilograms (passenger), 45 kilograms (cargo).

STRONTIUM ARSENITE
General Information
Synonym: Strontium o-arsenite.
Description: White powder.
Formula: $Sr_3(AsO_3)_2 \cdot 4H_2O$.
Constant: Mol wt: 580.77.
Hazard Analysis and Countermeasures
See arsenic compounds.
Shipping Regulations: Section 11.
    I.C.C.: Poison B; Poison label, 200 pounds.
    Coast Guard Classification; poison B; Poison label.
    IATA: Poison B, poison label, 25 kilograms (passenger), 95 kilograms (cargo).

STRONTIUM BROMATE
General Information
Description: Monoclinic, colorless-yellowish, hygroscopic crystals.
Formula: $Sr(BrO_3)_2 \cdot H_2O$.
Constants: Mol wt: 361.48, mp: $-H_2O$ at 120°C, bp: decomposes at 240°C.
Hazard Analysis and Countermeasures
See bromates.

STRONTIUM BROMIDE
General Information
Description: White, hygroscopic needles.
Formula: $SrBr_2$.
Constants: Mol wt: 247.46, mp: 643°C, bp: decomposes, d: 4.216 at 24°C,.

Hazard Analysis and Countermeasures
See bromides.

STRONTIUM CHLORATE
General Information
Description: White crystalline powder.
Formula: $Sr(ClO_3)_2$.
Constants: Mol wt: 254.54, mp: 120°C decomposes, d: 3.152.
Hazard Analysis and Countermeasures
See chlorates.
Shipping Regulations: Section 11.
    I.C.C.: Oxidizing material; yellow label, 100 pounds.
    Coast Guard Classification: Oxidizing material; yellow label.
    IATA: Oxidizing material, yellow label, 12 kilograms (passenger), 45 kilograms (cargo).

STRONTIUM CHLORATE, WET. See chlorates.
Shipping Regulations: Section 11.
    I.C.C.: Oxidizing material; yellow label, 200 pounds.
    Coast Guard Classification: Oxidizing material; yellow label.
    IATA: Oxidizing material, yellow label, 12 kilograms (passenger), 95 kilograms (cargo).

STRONTIUM CHLORIDE FLUORIDE
General Information
Description: Tetragonal crystals.
Formula: $SrCl_2 \cdot SrF_2$.
Constants: Mol wt: 284.17, mp: 962°C, d: 4.18.
Hazard Analysis and Countermeasures
See fluorides and chlorides.

STRONTIUM CHROMATE
General Information
Description: Monoclinic, yellow crystals.
Formula: $SrCrO_4$.
Constants: Mol wt: 203.64, d: 3.895 at 15°C.
Hazard Analysis and Countermeasures
See chromium compounds.

STRONTIUM COMPOUNDS
Hazard Analysis
Toxicity: The strontium ion has a low order of toxicity. It is chemically and biologically similar to calcium. The oxides and hydroxides are moderately caustic materials. As with other compounds, the toxicity may be a function of the anion.

STRONTIUM CYANIDE
General Information
Description: White, rhombic, deliquescent crystals.
Formula: $Sr(CN)_2 \cdot 4H_2O$.
Constants: Mol wt: 211.73, mp: decomposes.
Hazard Analysis and Countermeasures
See cyanides.

STRONTIUM DIURANATE. See uranium strontium oxide.

STRONTIUM FLUORIDE
General Information
Description: Cubic, colorless crystals or white powder.
Formula: SrF_2.
Constants: Mol wt: 125.63, mp: 1190°C, d: 4.24.
Hazard Analysis and Countermeasures
See fluorides.

TOXIC HAZARD RATING CODE *(For detailed discussion, see Section 1.)*

0 NONE: (a) No harm under any conditions; (b) Harmful only under unusual conditions or overwhelming dosage.

1 SLIGHT: Causes readily reversible changes which disappear after end of exposure.

2 MODERATE: May involve both irreversible and reversible changes; not severe enough to cause death or permanent injury.

3 HIGH: May cause death or permanent injury after very short exposure to small quantities.

U UNKNOWN: No information on humans considered valid by authors.

STRONTIUM FLUOSILICATE
General Information
Description: Monoclinic crystals.
Formula: $SrSiF_6 \cdot 2H_2O$.
Constants: Mol wt: 265.72, mp: decomposes, d: 2.99 at 17.5°C.
Hazard Analysis and Countermeasures
See fluosilicates.

STRONTIUM HYDRIDE
General Information
Description: White crystals; reacts with water.
Formula: SrH_2.
Constants: Mol wt: 89.7, d: 3.27, mp: decomposes at red heat.
Hazard Analysis and Countermeasures
See strontium compounds and hydrides.

STRONTIUM HYDROSULFIDE
General Information
Description: Crystals.
Formula: $Sr(HS)_2$.
Constants: Mol wt: 153.78, mp: decomposes.
Hazard Analysis and Countermeasures
See sulfides.

STRONTIUM HYDROXIDE
General Information
Description: White, deliquescent crystals.
Formula: $Sr(OH)_2$.
Constants: Mol wt: 121.65, mp: 375°C, d: 3.625.
Hazard Analysis and Countermeasures
See strontium compounds.

STRONTIUM IODATE
General Information
Description: Triclinic crystals.
Formula: $Sr(IO_3)_2$.
Constants: Mol wt: 437.47, d: 5.045 at 15°C.
Hazard Analysis and Countermeasures
See iodates.

STRONTIUM IODIDE
General Information
Description: Colorless plates.
Formula: SrI_2.
Constants: Mol wt: 341.47, mp: 402°C, bp: decomposes, d: 4.549 at 25°C.
Hazard Analysis and Countermeasures
See iodides.

STRONTIUM MONOSULFIDE
General Information
Description: Cubic, light gray crystals.
Formula: SrS.
Constants: Mol wt: 119.70, d: 3.70 at 15°C.
Hazard Analysis and Countermeasures
See sulfides.

STRONTIUM NITRATE
General Information
Description: White powder.
Formula: $Sr(NO_3)_2$.
Constants: Mol wt: 211.65, mp: 570°C, d: 2.986.
Hazard Analysis and Countermeasures
See nitrates.
Shipping Regulations: Section 11.
  I.C.C.: Oxidizing material; yellow label, 100 pounds.
  Coast Guard Classification: Oxidizing material.
  IATA: Oxidizing material, yellow label, 12 kilograms (passenger), 45 kilograms (cargo).

STRONTIUM NITRITE
General Information
Description: Hexagonal crystals.
Formula: $Sr(NO_2)_2 \cdot H_2O$.
Constants: Mol wt: 197.66, mp: $-H_2O$ at $>100°C$, bp: decomposes at 240°C, d: 2.408 at 0°/0°C.
Hazard Analysis and Countermeasures
See nitrites.

STRONTIUM OXALATE
General Information
Description: White, odorless, colorless, crystalline powder.
Formula: $SrC_2O_4 \cdot H_2O$.
Constants: Mol wt: 193.67, mp: $-H_2O$ at 150°C.
Hazard Analysis
See oxalates.
Countermeasures
See oxalates.

STRONTIUM PERCHLORATE
General Information
Description: Colorless crystals.
Formula: $Sr(ClO_4)_2$.
Constant: Mol wt: 286.54.
Hazard Analysis and Countermeasures
See perchlorates.
Shipping Regulations: Section 11.
  IATA: Oxidizing material, yellow label, 12 kilograms (passenger), 45 kilograms (cargo).

STRONTIUM PERMANGANATE
General Information
Description: Cubic, purple crystals.
Formula: $Sr(MnO_4)_2 \cdot 3H_2O$.
Constants: Mol wt: 379.54, mp: decomposes at 175°C, d: 2.75.
Hazard Analysis
Toxicity: See manganese compounds.
Fire Hazard: Moderate; a strong oxidizer.
Disaster Hazard: Moderate; keep away from flammable materials.
Countermeasures
Storage and Handling: Section 7.

STRONTIUM PEROXIDE
General Information
Description: White powder.
Formula: SrO_2.
Constants: Mol wt: 119.63, mp: decomposes, d: 4.56.
Hazard Analysis and Countermeasures
See peroxides, inorganic.
Shipping Regulations: Section 11.
  I.C.C.: Oxidizing material; yellow label, 100 pounds.
  Coast Guard Classification: Oxidizing material; yellow label.
  IATA: Oxidizing material, yellow label, 12 kilograms (passenger), 45 kilograms (cargo).

STRONTIUM POTASSIUM CHLORATE
General Information
Synonym: Potassium-strontium chlorate.
Description: White, crystalline powder.
Formula: $Sr(ClO_3)_2 \cdot 2KClO_3$.
Constant: Mol wt: 499.67.
Hazard Analysis and Countermeasures
See chlorates.

STRONTIUM SELENATE
General Information
Description: Rhombic crystals.
Formula: $SrSeO_4$.
Constants: Mol wt: 230.59, d: 4.23.
Hazard Analysis and Countermeasures
See selenium compounds.

STRONTIUM SULFIDE. See strontium monosulfide.

STRONTIUM SULFITE
General Information
Description: Colorless crystals.

Formula: $SrSO_3$.
Constants: Mol wt: 167.70, mp: decomposes.
Hazard Analysis and Countermeasures
See sulfites.

STRONTIUM TETRASULFIDE
General Information
Description: Reddish crystals.
Formula: $SrS_1 \cdot 6H_2O$.
Constants: Mol wt: 324.0, mp: 25°C.
Hazard Analysis and Countermeasures
See sulfides.

STROPHANTHIN
General Information
Synonym: K-strophanthin.
Description: White or yellowish powder; very bitter taste.
Formula: $C_{29}H_{44}O_{12} \cdot 8H_2O$.
Constant:Mol wt: 728.62.
Hazard Analysis
Toxic Hazard Rating:
Acute Local: U.
Acute Systemic: Ingestion 3.
Chronic Local: U.
Chronic Systemic: Ingestion 2.
Fire Hazard: Slight (Section 6).
Countermeasures
Ventilation Control: Section 2.
Personal Hygiene: Section 3.
Storage and Handling: Section 7.

STRYCHNINE AND COMPOUNDS
General Information
Description: Hard, white, crystalline alkaloid; very bitter taste.
Formula: $C_{21}H_{22}N_2O_2$.
Constants: Mol wt: 334.40, mp: 268°C, bp: 270°C, d: 1.359 at 18°C.
Hazard Analysis
Toxic Hazard Rating:
Acute Local: Allergen 1.
Acute Systemic: Ingestion 3; Inhalation 3.
Chronic Local: Allergen 1.
Chronic Systemic: U.
Toxicology: A very poisonous alkaloid. If it is taken by mouth, the time of action depends upon the condition of the stomach, that is, whether empty or full, and the nature of the food present. If taken by subcutaneous injection, the place of administration of the injection will affect the time of action. The first symptoms are a feeling of uneasiness with a heightened reflex of irritability, followed by muscular twitching in some parts of the body. With larger doses, this is followed by a sense of impending suffocation. Convulsive movements begin which have the effect of mechanically causing the patient to cry out or to shriek; then follow the characteristic spasms which set in with violence. These are at first clonic and then tonic. There are successive attacks of spasms. With each successive attack, the symptoms become more violent, eventually resulting in death. A rodenticide.
TLV: ACGIH (recommended); 0.15 milligram per cubic meter of air.
Disaster Hazard: Dangerous; when heated, it emits highly toxic fumes.
Countermeasures
Ventilation Control: Section 2.
Treatment and Antidotes: Call a physician at once. The stomach should be washed out with potassium permanganate solution which is diluted to the color of port wine; give chloral hydrate per rectum to control convulsions. The use of apomorphine has been recommended and the use of phenobarbital on animals has proved successful. Barbiturates do not act readily. One gram of Merck's carbo medicinolis will bind 580 milligrams of strychnine. When this is available the stomach should be washed out with it immediately. Sodium amytal has been used effectively as has sodium pentobarbital, particularly in animal experiments.
Personal Hygiene: Section 3.
First Aid: Section 1.
Storage and Handling: Section 7.
Shipping Regulations: Section 11.
Salts thereof, solid -
I.C.C.: Poison B; poison label, 200 pounds.
Coast Guard Classification: Poison B; poison label.
MCA warning label.
IATA: Poison B, poison label, 25 kilograms (passenger), 95 kilograms (cargo).

STYPHNIC ACID. See 2, 4-dinitroresorcinol.

STYPTICIN. See cotarnine chloride.

STYRALYL ALCOHOL. See phenylmethyl carbinol.

STYRENE (monomer). See phenylethylene.

STYRENE DIBROMIDE. See dibromoethylbenzene.

STYRENE OXIDE
General Information
Synonym: Phenylethylene oxide.
Description: Colorless liquid.
Formula: $C_6H_5CHOCH_2$.
Constants: Mol wt: 120.1, bp: 194.2°C, flash p: 165°F (O.C.), fp: −36.7°C, d: 1.0469 at 25°/4°C, autoign. temp.: 175°F (O.C.), vap. d.: 4.14.
Hazard Analysis
Toxic Hazard Rating:
Acute Local: Irritant 2; Ingestion 2; Inhalation 2.
Acute Systemic: Inhalation 2.
Chronic Local: U.
Chronic Systemic: U.
Fire Hazard: Moderate, when exposed to heat or flame.
Disaster Hazard: Moderately dangerous; when heated, it emits acrid fumes; it can react with oxidizing materials.
Countermeasures
Ventilation Control: Section 2.
To Fight Fire: Foam, carbon dioxide, dry chemical or carbon tetrachloride (Section 6).
Personnel Protection: Section 3.
Storage and Handling: Section 7.

SUBERANE. See cycloheptane.

SUBERONE. See cycloheptanone.

SUBGALLATE. See bismuth gallate, basic.

SUBLAMIN
General Information
Synonym: Mercury ethylenediaminesulfate.
Description: White, crystalline powder; contains approximately 43% mercury.

TOXIC HAZARD RATING CODE (For detailed discussion, see Section 1.)

0 NONE: (a) No harm under any conditions; (b) Harmful only under unusual conditions or overwhelming dosage.

1 SLIGHT: Causes readily reversible changes which disappear after end of exposure.

2 MODERATE: May involve both irreversible and reversible changes; not severe enough to cause death or permanent injury.

3 HIGH: May cause death or permanent injury after very short exposure to small quantities.

U UNKNOWN: No information on humans considered valid by authors.

Formula: $HgSO_4 \cdot 2(CH_2 \cdot NH_2)_2 \cdot 2H_2O$.
Constant: Mol wt: 452.87.
Hazard Analysis and Countermeasures
See mercury compounds, organic.

SUBSALICYLATE (COM'L). See bismuth salicylate, basic.

SUCCINCHLORIMIDE. See N-chlorosuccinimide.

SUCCINIC ACID
General Information
Synonyms: Ethylenesuccinic acid; ethylene dicarboxylic acid; butanedioic acid.
Description: Colorless crystals.
Formula: $COOH(CH_2)_2COOH$.
Constants: Mol wt: 118.09, mp: 185°C, bp: 235°C (decomposes), d: 1.564 at 15°/4°C.
Hazard Analysis
Toxic Hazard Rating:
  Acute Local: Irritant 2.
  Acute Systemic: U.
  Chronic Local: U.
  Chronic Systemic: U.
Note: A general purpose food additive (Section 10).
Fire Hazard: Slight, when exposed to heat or flame; it can react with oxidizing materials (Section 6).
Countermeasures
Personnel Protection: Section 3.
Storage and Handling: Section 7.

SUCCINIC ACID ANHYDRIDE. See succinic anhydride.

SUCCINIC ACID PEROXIDE
General Information
Synonyms: Butanedioic peroxide; succinyl peroxide.
Description: Fine white powder, odorless with tart taste; moderately soluble in water.
Formula: $(HOOCCH_2CH_2CO)_2O_2$.
Constants: Mol wt: 234.2, mp: 125°C (decomposes).
Hazard Analysis
Toxic Hazard Rating:
  Acute Local: Irritant 2; Ingestion 2; Inhalation 2.
  Acute Systemic: U.
  Chronic Local: U.
  Chronic Systemic: U.
Countermeasures
See peroxides, organic.
Shipping Regulations: Section 11.
  I.C.C.: Succinic acid peroxide, oxidizing material, yellow label, 25 pounds.
  Freight and Express: Yellow label—chemicals N.O.I.B.N.
  Rail Express: Max. wt. = 25 pounds per shipping case.
  Parcel Post: Prohibited.
  IATA: Oxidizing material, yellow label, 1 kilogram (passenger), 12 kilograms (cargo).

SUCCINIC ANHYDRIDE
General Information
Synonyms: Butanedioic anhydride; succinic acid anhydride.
Description: Colorless needles.
Formula: $(CH_2CO)_2O$.
Constants: Mol wt: 100.07, mp: 119.6°C, bp: 261°C, d: 1.104, vap. press.: 1 mm at 92.0°C.
Hazard Analysis
Toxic Hazard Rating:
  Acute Local: Irritant 2; Ingestion 2; Inhalation 2.
  Acute Systemic: U.
  Chronic Local: U.
  Chronic Systemic: U.
Fire Hazard: Slight, when exposed to heat or flame (Section 6).
Countermeasures
Ventilation Control: Section 2.
Personnel Protection: Section 3.
Storage and Handling: Section 7.

SUCCINONITRILE
General Information
Synonyms: Butanedinitrile; ethylene cyanide.
Description: Colorless, odorless, waxy material.
Formula: $CNCH_2CH_2CN$.
Constants: Mol wt: 80.09; mp: 58.1°C, bp: 267°C, flash p: 270°F (ASTM D92-46), d: 1.022 at 25°C, vap. press.: 2 mm at 100°C, vap. d.: 2.1.
Hazard Analysis
Toxicity: Highly toxic. See nitriles.
Fire Hazard: Slight, when exposed to heat or flame.
Disaster Hazard: Dangerous; when heated to decomposition or on contact with acid or acid fumes, it emits highly toxic fumes of cyanides; it can react with oxidizing materials.
Countermeasures
Ventilation Control: Section 2.
To Fight Fire: Foam, carbon dioxide, dry chemical or carbon tetrachloride (Section 6).
Personal Hygiene: Section 3.
Storage and Handling: Section 7.

SUCCINYL PEROXIDE. See succinic acid peroxide.

SUCCISTEARIN
General Information
Synonym: Stearoyl propylene glycol hydrogen succinate.
Description: The reaction product of succinic anhydride fully hydrogenated vegetable oil, and propylene glycol.
Hazard Analysis
Toxic Hazard Rating: Unknown. A food additive permitted in food for human consumption. (Section 10).

SUCROSE OCTAACETATE
General Information
Description: Crystals.
Formula: $C_{12}H_{14}O_3(OOCCH_3)_8$.
Constants: Mol wt: 678.58, mp: 72.3°C, bp: 260°C at 0.1 mm, d: 1.28 at 20°/20°C.
Hazard Analysis
Toxicity: Probably low.
Fire Hazard: Slight.
Countermeasures
Storage and Handling: Section 7.

SUGAR OF LEAD. See lead acetate.

SULFAETHOXYPYRIDAZINE
General Information
Synonym: n-(6-Ethoxy-3-pyridazinyl) sulfanilamide.
Formula: $C_{12}H_{14}N_4O_3S$.
Constants: Mol wt: 294, mp. range: 180–186°C.
Hazard Analysis
Toxic Hazard Rating: Unknown. A food additive permitted in the feed and drinking water of animals and/or for the treatment of food producing animals. Also permitted in food for human consumption. (Section 10).
Disaster Hazard: Dangerous. See sulfates.

SULFAGUANIDINE
General Information
Description: Needle like crystals.
Formula: $C_7H_{10}N_4O_2S \cdot H_2O$.
Constants: Mol wt: 232.3, mp: 190–193°C.
Hazard Analysis
Toxic Hazard Rating:
  Acute Local: Allergen 1.
  Acute Systemic: U.
  Chronic Local: Allergen 2.
  Chronic Systemic: U.
Disaster Hazard: Dangerous; when heated to decomposition, it emits highly toxic fumes of oxides of sulfur and nitrogen.
Countermeasures
Personnel Protection: Section 3.
Storage and Handling: Section 7.

SULFAMETHAZINE
General Information
Synonym: N-(4,6-Dimethyl-2-pyrimidyl) sulfanilamide.
Description: White to yellow white powder; almost odorless; soluble in acetone; slightly soluble in alcohol; very slightly soluble in water and ether.
Formula: $NH_2C_6H_4SO_2NHC_4N_2H(CH_3)_2$.
Constants: Mol wt: 278, mp: 197-200° C.
Hazard Analysis
Toxic Hazard Rating: Unknown. A food additive permitted in the feed and drinking water of animals and for the treatment of food producing animals. Also permitted in food for human consumption. (Section 10).
Disaster Hazard: Dangerous. See sulfates.

SULFAMIC ACID
General Information
Synonym: Amidosulfonic acid.
Description: White, crystalline solid.
Formula: H_2NSO_3H.
Constants: Mol wt: 97.09, mp: 200° C (decomposes), bp: decomposes, d: 2.03 at 12° C.
Hazard Analysis
Toxic Hazard Rating:
  Acute Local: Irritant 2.
  Acute Systemic: U.
  Chronic Local: U.
  Chronic Systemic: U.
Note: A substance which migrates to food from packaging materials (Section 10).
Disaster Hazard: Dangerous. See sulfonates.
Countermeasures
Personnel Protection: Section 3.
Storage and Handling: Section 7.

SULFAMIDE
General Information
Synonym: Sulfuryl amide.
Description: Rhombic plates.
Formula: $SO_2(NH_2)_2$.
Constants: Mol wt: 96.11, mp: 91.5° C, bp: decomposes at 250° C.
Hazard Analysis
Toxicity: Unknown. See amides.
Disaster Hazard: Dangerous; when heated to decomposition, it emits highly toxic fumes of oxides of sulfur; will react with water or steam to produce toxic and corrosive fumes.
Countermeasures
Storage and Handling: Section 7.

SULFANILIC ACID
General Information
Synonym: o-Aminobenzenesulfonic acid.
Description: Colorless crystals.
Formula: $NH_2C_6H_4SO_3H \cdot H_2O$.
Constants: Mol wt: 191.2, mp: 288° C (decomposes).
Hazard Analysis
Toxicity: Unknown. Animal experiments suggest low toxicity and slight irritation.
Disaster Hazard: Dangerous; when heated to decomposition or on contact with acids it emits highly toxic fumes of oxides of nitrogen and sulfur.
Countermeasures
Storage and Handling: Section 7.

SULFANILIC ACID DIAZIDE. See p-diazobenzene-sulfonic acid.

SULFANITRAN. See acetyl-(p-nitrophenyl) sulfanilamide.

SULFASAN. See 4,4-dithiomorpholine.

SULFATES
Hazard Analysis
Toxicity: Variable. In general the toxic qualities of substances containing the sulfate radical is that of the material (cation) with which the sulfate (anion) is combined. See specific compound.
Disaster Hazard: Dangerous. When heated to decomposition they emit highly toxic fumes of oxides of sulfur.

SULFIDES
Hazard Analysis
Toxicity: Variable. The alkaline sulfides (potassium, calcium, ammonium and sodium) are similar in action to alkalies. They cause softening and irritation of the skin. If taken by mouth they are corrosive and irritant through the liberation of hydrogen sulfide and free alkali. Hydrogen sulfide is especially toxic and should be specially referred to (see hydrogen sulfide).
  Sulfides of the heavy metals are generally insoluble and hence have little toxic action except through the liberation of hydrogen sulfide.
  Sulfides are used as fungicides.
Fire Hazard: Moderate, when exposed to flame or by spontaneous chemical reaction. Many sulfides ignite easily in air at room temperature. Others require a higher temperature or the presence of an oxidizer. Upon contact with moisture or acids, hydrogen sulfide is evolved. Many powerful oxidizers on contact with sulfides ignite violently. See also hydrogen sulfide (Section 6).
Explosion Hazard: Many sulfides react violently and explosively on contact with powerful oxidizers. Hydrogen sulfide evolved can form explosive mixtures with air. See also hydrogen sulfide.
Disaster Hazard: Dangerous; when heated to decomposition, they emit highly toxic fumes or oxides of sulfur, they react with water, steam or acids to produce toxic and flammable vapors of hydrogen sulfide.
Countermeasures
Storage and Handling: Section 7.
Ventilation Control: Section 2.

SULFITES
Hazard Analysis
Toxic Hazard Rating:
  Acute Local: Ingestion 2; Inhalation 2.
  Acute Systemic: Ingestion 2; Inhalation 2.
  Chronic Local: Unknown.
  Chronic Systemic: Ingestion 1; Inhalation 1.
Toxicology: Fairly large doses of sulfites can be tolerated since they are rapidly oxidized to sulfates, although if swallowed they may cause irritation of the stomach by liberating sulfurous acid. In experimental animals, large doses of sodium sulfite have been shown to cause retarded growth, nerve irritation, atrophy of bone marrow, depression and paralysis.
Disaster Hazard: Dangerous; when heated to decomposition, they emit highly toxic fumes of sulfur dioxide; they will react with water, steam or acids to produce a toxic and corrosive material.
Countermeasures
Personal Hygiene: Section 3.
Storage and Handling: Section 7.

SULFOCARBOLIC ACID. See phenolsulfonic acid.

TOXIC HAZARD RATING CODE (For detailed discussion, see Section 1.)

0 NONE: (a) No harm under any conditions; (b) Harmful only under unusual conditions or overwhelming dosage.

1 SLIGHT: Causes readily reversible changes which disappear after end of exposure.

2 MODERATE: May involve both irreversible and reversible changes; not severe enough to cause death or permanent injury.

3 HIGH: May cause death or permanent injury after very short exposure to small quantities.

U UNKNOWN: No information on humans considered valid by authors.

1-SULFOCYANO-2, 4-DINITROBENZENE
General Information
Description: Powder.
Formula: $CNSC_6H_3(NO_2)_2$.
Constant: Mol wt: 225.2.
Hazard Analysis and Countermeasures
See nitrates and cyanides.

SULFONATED CASTOR OIL. See turkey red oil.

SULFONATES
Hazard Analysis
Toxicity: Variable. See specific compounds.
Disaster Hazard: Dangerous; when heated to decomposition or on contact with acid or acid fumes, they emit highly toxic fumes of oxides of sulfur.
Countermeasures
Storage and Handling: Section 7.

SULFONETHYLMETHANE
General Information
Description: Lustrous scales.
Formula: $C_8H_{18}O_4S_2$.
Constants: Mol wt: 242.4, mp: 74–76° C.
Hazard Analysis
Toxic Hazard Rating:
  Acute Local: U.
  Acute Systemic: Ingestion 2; Inhalation 1.
  Chronic Local: U.
  Chronic Systemic: Ingestion 1; Inhalation 1.
Fire Hazard: Moderate, when exposed to heat or flame.
Disaster Hazard: Dangerous. See sulfones. Can react with oxidizing materials.
Countermeasures
Ventilation Control: Section 2.
To Fight Fire: Carbon dioxide, dry chemical or carbon tetrachloride (Section 6).
Personal Hygiene: Section 3.
Storage and Handling: Section 7.

SULFONYL CHLORIDE. See sulfuryl chloride.

p,p′-SULFONYL DIANILINE. See diaminodiphenyl sulfone.

SULFOSALICYLIC ACID
General Information
Synonyms: 3-Carboxy-4-hydroxybenzenesulfonic acid; 5-sulfosalicylic acid; 2-hydroxybenzoic-sulfonic acid; salicylsulfonic acid.
Description: White crystalline powder; very soluble in water and alcohol; soluble in ether.
Formula: $C_6H_3(HO_3S)(OH)(COOH)$.
Constants: Mol wt: 254.22, mp: 120° C, decomposes at higher temperatures.
Hazard Analysis
Toxic Hazard Rating:
  Acute Local: Irritant 2; Ingestion 2.
  Acute Systemic: U.
  Chronic Local: U.
  Chronic Systemic: U.
Disaster Hazard: Dangerous. See sulfates.

SULFOX-CIDE
General Information
Synonym: Isosafrole n-octyl sulfoxide.
Hazard Analysis
Toxicity: Details unknown. Animal experiments have shown CNS disturbances.
Disaster Hazard: Dangerous. When heated to decomposition, it emits highly toxic fumes.

SULFUR
General Information
Synonyms: Brimstone; flowers of sulfur; sulfur flour.
Description: Rhombic, yellow crystals or yellow powder.

Formula: S_8.
Constants: Mol wt: 256.48, mp: 112.8° C, bp: 444.6° C, flash p: 405° F (C.C.), d: 2.07; d liquid: 1.803, autoign. temp.: 450° F, vap. press.: 1 mm at 183.8° C.
Hazard Analysis
Toxicity: Very low. See nuisance dusts. A fungicide.
Radiation Hazard: Section 5. For permissible levels, see Table 5, p. 150.
  Artificial isotope ^{35}S, half life 87 d. Decays to stable ^{35}Cl by emitting beta particles of 0.17 MeV.
Fire Hazard: Slight, when exposed to heat or flame or by chemical reaction with oxidizers.
Spontaneous Heating: No.
Explosion Hazard: Moderate, in the form of dust, when exposed to flame.
Disaster Hazard: Dangerous; when heated it burns and emits highly toxic fumes of oxides of sulfur; can react with oxidizing materials.
Countermeasures
Personal Hygiene: Section 3.
To Fight Fire: Water or special mixtures of dry chemical (Section 6).
Storage and Handling: Section 7.
Shipping Regulations: Section 11.
  Coast Guard Classification: Hazardous article.

SULFUR BROMIDE
General Information
Synonym: Sulfur monobromide.
Description: Red liquid.
Formula: S_2Br_2.
Constants: Mol wt: 223.96, mp: −40° C, bp: 54° C at 0.2 mm, d: 2.635.
Hazard Analysis
Toxic Hazard Rating:
  Acute Local: Irritant 3; Ingestion 3; Inhalation 3.
  Acute Systemic: U.
  Chronic Local: U.
  Chronic Systemic: U.
Fire Hazard: Slight when exposed to heat or flame.
Disaster Hazard: Dangerous; when heated to decomposition, it emits highly toxic fumes of oxides or sulfur and bromides; will react with water or steam to produce toxic and corrosive fumes.
Countermeasures
Ventilation Control: Section 2.
Personnel Protection: Section 3.
Storage and Handling: Section 7.

SULFUR CHLORIDE
General Information
Synonym: Sulfur monochloride.
Description: Amber to yellowish-red, oily, fuming liquid, penetrating odor, decomposes in water.
Formula: S_2Cl_2.
Constants: Mol wt: 135.03, mp: −80° C, bp: 138.0° C, flash p: 245° F (C.C.), d: 1.6885 at 15.5°/15.5° C, autoign. temp.: 453° F, vap. press.: 10 mm at 27.5° C, vap. d.: 4.66.
Hazard Analysis
Toxic Hazard Rating:
  Acute Local: Irritant 3; Ingestion 3; Inhalation 3.
  Acute Systemic: U.
  Chronic Local: Irritant 2; Inhalation 2.
  Chronic Systemic: U.
TLV: ACGIH (recommended). 1 part per million in air; 6 milligrams per cubic meter of air.
Toxicology: It is a fuming, corrosive liquid with a penetrating odor which is very irritating to the eyes, lungs and mucous membranes. It decomposes on contact with water to form hydrogen chloride, thiosulfuric acid and sulfur. These decomposition products are highly irritant. Its toxic effects are irritation of the upper respiratory tract, although the results of intoxication are usually transitory in nature. However, if hydrolysis is

not complete in the upper respiratory tract, injury to the bronchioles and alveoli can result. The literature notes that concentrations of 2 to 9 ppm have been found in rubber factories and that these concentrations were observed to be mildly irritating. A concentration of 150 ppm has been stated to be fatal to mice after an exposure of only 1 minute.

Fire Hazard: Slight, when exposed to heat or flame.

Spontaneous Heating: No.

Disaster Hazard: Dangerous; when heated to decomposition, it emits highly toxic fumes of chlorides and oxides of sulfur; will react with water or steam to produce heat and toxic and corrosive fumes; can react with oxidizing materials.

Countermeasures

Ventilation Control: Section 2.

To Fight Fire: Water, carbon dioxide, dry chemical or carbon tetrachloride (Section 6).

Personnel Protection: Section 3.

Storage and Handling: Section 7.

Shipping Regulations: Section 11.

I.C.C.: Corrosive liquid; white label, 1 gallon.

Coast Guard Classification: Corrosive liquid; white label.

MCA warning label.

IATA: Corrosive liquid, white label, not acceptable (passenger), 5 liters (cargo).

SULFUR COMPOUNDS

General Information

Description: Variable.

Formula: Variable.

Hazard Analysis

Toxicity: Variable. See specific material as listed. Common air contaminants (Section 4).

Disaster Hazard: Dangerous. When heated to decomposition such materials can evolve highly toxic fumes containing oxides of sulfur. See sulfides.

SULFUR DICHLORIDE

General Information

Description: Reddish-brown liquid; pungent odor.

Formula: SCl_2.

Constants: Mol wt: 103.0, mp: $-78°C$, bp: $59°C$, d: 1.621 at $15°/15°C$, vap. d.: 3.55.

Hazard Analysis

Toxicity: Corrosive. See sulfur chloride.

Fire Hazard: Moderate, when exposed to heat or flame (Section 6).

Disaster Hazard: Dangerous; when heated to decomposition, it emits highly toxic fumes of chlorides and oxides of sulfur; will react with water or steam to produce heat and toxic and corrosive fumes.

Countermeasures

Ventilation Control: Section 2.

Personnel Protection: Section 3.

Storage and Handling: Section 7.

Shipping Regulations: Section 11.

I.C.C.: Corrosive liquid; white label, 1 gallon.

MCA warning label.

IATA: Corrosive liquid, white label, not acceptable (passenger), 5 liters (cargo).

SULFUR DIOXIDE

General Information

Synonym: Sulfurous acid anhydride.

Description: Colorless gas or liquid; pungent odor.

Formula: SO_2.

Constants: Mol wt: 64.06, mp: $-75.5°C$, bp: $-10.0°C$, d liq: 1.434 at $0°C$, vap d: 2.264 at $0°C$, vap. press 2538 mm at $21.1°C$.

Hazard Analysis

Toxic Hazard Rating:

Acute Local: Irritant 3; Ingestion 3; Inhalation 3.

Acute Systemic: U.

Chronic Local: Irritant 2; Inhalation 2.

Chronic Systemic: U.

TLV: ACGIH (recommended); 5 parts per million in air; 13 milligrams per cubic meter of air.

Toxicology: This gas is dangerous to the eyes, as it causes irritation and inflammation of the conjunctiva. It has a suffocating odor and is a corrosive and poisonous material. In moist air or fogs, it combines with water to form sulfurous acid, but is only very slowly oxidized to sulfuric acid (Section 4). Concentrations of 6 to 12 ppm cause immediate irritation of the nose and throat, while 0.3 to 1 ppm can be detected by the average individual possibly by taste rather than by the sense of smell. 3 ppm has an easily noticeable odor and 20 ppm is the least amount which is irritating to the eyes. 10,000 ppm is an irritant to moist areas of the skin within a few minutes of exposure.

It chiefly affects the upper respiratory tract and the bronchi. It may cause edema of the lungs or glottis, and can produce respiratory paralysis. Concentrations of < 1 ppm are believed to be injurious to plant foliage.

This material is so irritating that it provides its own warning of toxic concentrations. 400 to 500 ppm is immediately dangerous to life and 50 to 100 ppm is considered to be the maximum permissible concentration for exposures of 30 to 60 minutes. Excessive exposures to high enough concentrations of this material can be fatal. Its toxicity is comparable to that of hydrogen chloride. However, less than fatal concentrations can be borne for fair periods of time with no apparent permanent damage. It is used as a fumigant, insecticide and fungicide, and a chemical preservative food additive (Section 10). It is a common air contaminant (Section 4).

Disaster Hazard: Dangerous; will react with water or steam to produce toxic and corrosive fumes.

Countermeasures

Ventilation Control: Section 2.

Treatment and Antidotes: Personnel who have shown toxic symptoms when exposed to this material should immediately be removed to fresh air. If the eyes are involved, they should be irrigated with copious quantities of warm water. If the symptoms persist, call a physician.

Storage and Handling: Section 7.

Shipping Regulations: Section 11.

I.C.C.: Nonflammable gas; green label, 300 pounds.

Coast Guard Classification: Noninflammable gas; green gas label.

IATA: Nonflammable gas, green label, not acceptable (passenger), 140 kilograms (cargo).

SULFURETTED HYDROGEN. See hydrogen sulfide.

SULFUR FLOUR. See sulfur.

SULFUR FLUORIDE

General Information

Synonym: Sulfur monofluoride.

Description: Colorless gas.

Formula: S_2F_2.

TOXIC HAZARD RATING CODE (For detailed discussion, see Section 1.)

0 NONE: (a) No harm under any conditions; (b) Harmful only under unusual conditions or overwhelming dosage.

1 SLIGHT: Causes readily reversible changes which disappear after end of exposure.

2 MODERATE: May involve both irreversible and reversible changes; not severe enough to cause death or permanent injury.

3 HIGH: May cause death or permanent injury after very short exposure to small quantities.

U UNKNOWN: No information on humans considered valid by authors.

Constants: Mol wt: 102.12, mp: $-105.5°C$, bp: $-99°C$, d liq.: 1.5 at $-100°C$.

Hazard Analysis and Countermeasures

See fluorides and hydrofluoric acid.

SULFUR HEPTOXIDE

General Information

Synonym: Per sulfur heptoxide.

Description: Viscous liquid or possibly needle-like crystals.

Formula: S_2O_7.

Constants: Mol wt: 176.1, mp: $0°C$, bp: sublimes at $10°C$.

Hazard Analysis

Toxic Hazard Rating:

Acute Local: Irritant 3; Ingestion 3; Inhalation 3.

Acute Systemic: U.

Chronic Local: U.

Chronic Systemic: U.

Fire Hazard: Moderate, when exposed to heat or flame or by chemical reaction. When heated, or in contact with water or alcohol, it liberates oxygen.

Disaster Hazard: Dangerous; when heated to decomposition, it emits highly toxic fumes of oxides of sulfur; can react with reducing materials.

Countermeasures

Ventilation Control: Section 2.

To Fight Fire: Carbon dioxide, dry chemical or carbon tetrachloride (Section 6).

Storage and Handling: Section 7.

SULFUR HEXAFLUORIDE

General Information

Description: Colorless gas.

Formula: SF_6.

Constants: Mol wt: 146.06, mp: $-50.5°C$, bp: $63.8°C$ (sublimes), vap. d: 6.602, d. (liquid): 1.5 at $-100°C$.

Hazard Analysis

Toxic Hazard Rating:

Acute Local: U.

Acute Systemic: Inhalation 1.

Chronic Local: U.

Chronic Systemic: U.

TLV: ACGIH (recommended); 1000 parts per million in air; 6000 milligrams per cubic meter of air.

Toxicology: This material is chemically inert in the pure state and is considered to be physiologically inert as well. However, as it is ordinarily obtainable, it can contain variable quantities of the lower sulfur fluorides. Some of these are toxic, very reactive chemically, and corrosive in nature. These materials can hydrolyze on contact with water to yield hydrogen fluoride, which is highly toxic and very corrosive. In high concentrations and when pure it may act as a simple asphyxiant.

Disaster Hazard: Dangerous; when heated to decomposition, it emits highly toxic fumes of fluorides and oxides of sulfur.

Countermeasures

Storage and Handling: Section 7.

Shipping Regulations: Section 11.

I.C.C.: Nonflammable gas; green label, 300 pounds.

Coast Guard Classification: Noninflammable gas; green gas label.

IATA: Nonflammable gas, green label, 70 kilograms (passenger), 140 kilograms (cargo).

SULFURIC ACID

General Information

Synonyms: Oil of vitriol; dipping acid.

Description: Colorless, oily liquid.

Formula: H_2SO_4.

Constants: Mol wt: 98.08, mp: $10.49°C$, bp: $330°C$, d: 1.834, vap. press.: 1 mm at $145.8°C$.

Hazard Analysis

Toxic Hazard Rating:

Acute Local: Irritant 3; Ingestion 3; Inhalation 3.

Acute Systemic: U.

Chronic Local: Irritant 2; Inhalation 2.

Chronic Systemic: U.

TLV: ACGIH (recommended); 1 milligram per cubic meter of air.

Toxicology: Contact with the body results in rapid destruction of tissue, causing severe burns. No systemic effects due to continual ingestion of small amounts of this material have been noted. There are systemic effects secondary to tissue damage caused by contact with it. However, repeated contact with dilute solutions can cause a dermatitis, and repeated or prolonged inhalation of a mist of sulfuric acid can cause an inflammation of the upper respiratory tract leading to chronic bronchitis. Sensitivity to sulfuric acid or mists or vapors varies with individuals. Normally 0.125 to 0.50 ppm may be mildly annoying and 1.5 to 2.5 ppm can be definitely unpleasant. 10 to 20 ppm is unbearable.

Workers exposed to low concentrations of the vapor gradually lose their sensitivity to its irritant action. Inhalation of concentrated vapor or mists from hot acid or oleum can cause rapid loss of consciousness with serious damage to lung tissue. In concentrated form it acts as a powerful caustic to the skin destroying the epidermis and penetrating some distance into the skin and subcutaneous tissues, in which it causes necrosis. This causes great pain and if much of the skin is involved, it is accompanied by shock, collapse and symptoms similar to those seen in severe burns. The fumes or mists of this material cause coughing and irritation of the mucous membranes of the eyes and upper respiratory tract. Severe exposure may cause a chemical pneumonitis; erosion of the teeth due to exposure to strong acid fumes has been recognized in industry.

Used as a general purpose food additive; it migrates to food from packaging materials (Section 10). It is a common air contaminant (Section 4).

Fire Hazard: Moderate, by chemical reaction; a powerful oxidizer; can ignite upon contact with combustibles.

Disaster Hazard: Dangerous; when heated, it emits highly toxic fumes; will react with water or steam to produce heat; can react with oxidizing or reducing materials.

Countermeasures

Ventilation Control: Section 2.

Treatment and Antidotes: Speed in removing this material from contact with the body is of primary importance. Start first aid at once. In all cases of contact in any form, delay can result in serious injuries and all persons injured should be referred to a physician. However, immediately give prolonged applications of running water to wash the material off the body. Remove contaminated clothing. Subject patient to a deluge type of shower if this is available. Do not attempt to neutralize the acid in contact with the skin until all areas of contact have been thoroughly irrigated with running water. Then applications of mild alkaline solutions may be in order. Shock symptoms will often be noted in cases of severe or extensive burns. In such a case, put patient on his back, keep him warm but not hot until physician arrives. Do not apply oils or ointments to burned area without instructions from a physician. If eyes are involved, they should immediately be irrigated with copious quantities of warm water for at least 15 minutes.

If the material has been taken internally, it causes burns of the mucous membrane of the throat, esophagus, and stomach. Do not attempt to induce vomiting in patients who have swallowed strong solutions of sulfuric acid. Do not give anything by mouth to an unconscious patient. If he is conscious, encourage him to wash out his mouth with copious amounts of water, then have him drink milk mixed with whites of eggs. If this is not available, have him drink as much water as possible. Get medical help.

Personnel Protection: Section 3.

Storage and Handling: Section 7.

Shipping Regulations: Section 11.
  I.C.C.: Corrosive liquid, white label, 10 pints.
  MCA warning label.
  IATA: Corrosive liquid, white label; 1 liter (passenger),
    5 liters (cargo).

SULFURIC ACID, AROMATIC
General Information
Synonym: Elixir of vitriol.
Description: Clear, reddish-brown liquid; peculiar aromatic
  odor; pleasant acid taste when diluted.
Hazard Analysis
Toxicity: Corrosive. See sulfuric acid.
Fire Hazard: Moderate, when exposed to heat or flame. See
  also ethyl alcohol and sulfuric acid.
Explosion Hazard: Moderate, in the form of vapor (ethyl
  alcohol) when exposed to flame.
Disaster Hazard: Dangerous. See sulfuric acid and ethyl
  alcohol.
Countermeasures
Storage and Handling: Section 7.

SULFURIC ACID, FUMING. See oleum.
Shipping Regulations: Section 11.
  IATA: Corrosive liquid, white label, not acceptable
    (passenger), 2½ liters (cargo).
  ICC: Corrosive liquid, white label, 10 pints.

SULFURIC CHLORIDE. See sulfuryl chloride.

SULFURIC CHLOROHYDRIN. See chlorosulfonic acid.

SULFURIC ETHER. See ethyl ether.

SULFURIC OXYCHLORIDE. See sulfuryl chloride.

SULFURIC OXYFLUORIDE. See sulfuryl fluoride.

SULFUR MONOBROMIDE. See sulfur bromide.

SULFUR MONOCHLORIDE. See sulfur chloride.
Shipping Regulations: Section 11.
  Coast Guard Classification: Corrosive liquid; white label.

SULFUR MONOFLUORIDE. See sulfur fluoride.

SULFUR MONOOXYTETRACHLORIDE
General Information
Description: Dark red liquid.
Formula: S_2OCl_4.
Constants: Mol wt: 221.96, bp: 60–61°C, d: 1.656 at 0°C.
Hazard Analysis
Toxicity: Corrosive. See hydrochloric acid.
Disaster Hazard: Dangerous; when heated to decomposi-
  tion, it emits highly toxic fumes of oxides of sulfur and
  hydrochloric acid; will react with water or steam to pro-
  duce toxic and corrosive fumes.
Countermeasures
Storage and Handling: Section 7.

SULFUROUS ACID
General Information
Description: Colorless liquid; suffocating sulfur odor (in
  solution only).
Formula: H_2SO_3.
Constants: Mol wt: 82.08, d: about 1.03.
Hazard Analysis
Toxic Hazard Rating:
  Acute Local: Irritant 3; Ingestion 3; Inhalation 3.
  Acute Systemic: U.

Chronic Local: Irritant 2; Inhalation 3.
Chronic Systemic: U.
Disaster Hazard: Dangerous; when heated to decomposi-
  tion, it emits highly toxic fumes of sulfur dioxide.
Countermeasures
Ventilation Control: Section 2.
Personnel Protection: Section 3.
Storage and Handling: Section 7.
Shipping Regulations: Section 11.
  IATA: Other restricted articles, class A, no label, 10
    liters (passenger and cargo).

SULFUROUS ACID ANHYDRIDE. See sulfur dioxide.

SULFUROUS ACID 2-(p-tert BUTYLPHENOXY)-1-METHYLETHYL-2-CHLOROETHYL ESTER
General Information
Synonym: Aramite.
Description: Liquid, miscible with many organic solvents.
  Insoluble in water.
Formula: $C_{15}H_{23}ClO_4S$.
Constants: Mol wt: 334.9, d: 1.1450–1.1620, mp: −31.7°C,
  bp: 175°C, at 0.1 mm, vap press.: < 10 mm at 25°C.
Hazard Analysis
Toxic Hazard Rating:
  Acute Local: Irritant 2.
  Acute Systemic: Ingestion 2.
  Chronic Local: Irritant 1.
  Chronic Systemic: Ingestion 2.
Remarks: A pesticide.
Disaster Hazard: Dangerous. When heated to decomposi-
  tion it emits highly toxic fumes of chlorides etc.

SULFUROUS OXYCHLORIDE. See thionyl chloride.

SULFUR PENTAFLUORIDE
General Information
Synonym: Disulfur decafluoride.
Description: Colorless liquid.
Formula: S_2F_{10}.
Constants: Mol wt: 254.12, bp: 29°C, fp: −92°C, d: 2.08
  at 0°/4°C.
Hazard Analysis
Toxicity: Highly toxic; see fluorides.
TLV: ACGIH (recommended); 0.025 parts per million in air,
  0.25 milligrams per cubic meter of air.
Disaster Hazard: Dangerous; when heated to decomposi-
  tion, it emits highly toxic fumes; on contact with water,
  steam or acids it hydrolyzes to evolve highly toxic and
  corrosive fumes.
Countermeasures
Storage and Handling: Section 7.

SULFUR SESQUIOXIDE
General Information
Description: Blue-green crystals.
Formula: S_2O_3.
Constants: Mol wt: 112.13, mp: decomposes at 70–95°C.
Hazard Analysis
Toxicity: Unknown.
Disaster Hazard: Dangerous; when heated to decomposi-
  tion, it emits highly toxic fumes of oxides of sulfur; can
  react with oxidizing materials.
Countermeasures
Storage and Handling: Section 7.

TOXIC HAZARD RATING CODE (For detailed discussion, see Section 1.)

0 NONE: (a) No harm under any conditions; (b) Harmful only under unusual conditions or overwhelming dosage.

1 SLIGHT: Causes readily reversible changes which disappear after end
   of exposure.

2 MODERATE: May involve both irreversible and reversible changes;
   not severe enough to cause death or permanent injury.

3 HIGH: May cause death or permanent injury after very short exposure
   to small quantities.

U UNKNOWN: No information on humans considered valid by authors.

SULFUR TETRACHLORIDE
General Information
Description: Yellow-brown liquid or gas at ordinary temperatures.
Formula: SCl_4.
Constants: Mol wt: 173.89, mp: $-30°C$, bp: $-15°C$, (decomp.).
Hazard Analysis
Toxic Hazard Rating:
  Acute Local: Irritant 3; Ingestion 3; Inhalation 3.
  Acute Systemic: U.
  Chronic Local: U.
  Chronic Systemic: U.
Disaster Hazard: Dangerous; when heated to decomposition, it emits highly toxic fumes of hydrochloric acid and oxides of sulfur; will react with water or steam to produce toxic and corrosive fumes.
Countermeasures
Ventilation Control: Section 2.
Storage and Handling: Section 7.

SULFUR TETRAFLUORIDE
General Information
Description: Gas.
Formula: SF_4.
Constants: Mol wt: 108.06, bp: $-40°C$, mp: $-124°C$.
Hazard Analysis
Toxic Hazard Rating:
  Acute Local: Irritant 3; Inhalation 3.
  Acute Systemic: U.
  Chronic Local: Irritant 3; Inhalation 3.
  Chronic Systemic: U.
Disaster Hazard: Dangerous; when heated to decomposition it emits highly toxic fumes of fluorides; will react with water, steam or acids to produce toxic and corrosive fumes.
Countermeasures
Ventilation Control: Section 2.
Storage and Handling: Section 7.
Shipping Regulations: Section 11.
  IATA: Nonflammable gas, green label, not acceptable (passenger), 140 kilograms (cargo).

SULFUR TRIOXIDE (α)
General Information
Description: Colorless crystals or liquid.
Formula: SO_3.
Constants: Mol wt: 80.07, mp: 16.83°C, bp: 44.8°C, d: 2.75; 1.925 at 13°C, (liquid), vap. press.: 100 mm at 10.5°C, vap. d.: 2.76.
Hazard Analysis
Toxic Hazard Rating:
  Acute Local: Irritant 3; Ingestion 3; Inhalation 3.
  Acute Systemic: U.
  Chronic Local: Irritant 2; Inhalation 2.
  Chronic Systemic: U.
Disaster Hazard: Dangerous; when heated to decomposition it emits highly toxic fumes of oxides of sulfur; will react with water or steam to produce heat and toxic and corrosive fumes of sulfuric acid.
Countermeasures
Ventilation Control: Section 2.
Personnel Protection: Section 3.
Storage and Handling: Section 7.
Shipping Regulations: Section 11.
  IATA (with or without stabilizer): Corrosive liquid, not acceptable (passenger and cargo).

SULFUR TRIOXIDE (β)
General Information
Description: Silky, fibrous, needle-like crystals.
Formula: $(SO_3)_2$.
Constants: Mol wt: 160.1, mp: 62.2°C, bp: 44.6°C (sublimes at 50°C), d: 1.97 (as liquid), vap. press.: 100 mm at 14.3°C, vap. d.: 5.52.

Hazard Analysis
Toxic Hazard Rating:
  Acute Local: Irritant 3; Ingestion 3; Inhalation 3.
  Acute Systemic: U.
  Chronic Local: Irritant 2; Inhalation 2.
  Chronic Systemic: U.
Disaster Hazard: Dangerous; when heated, it emits highly toxic fumes of oxides of sulfur; will react with water or steam to produce heat and toxic and corrosive fumes of sulfuric acid.
Countermeasures
Ventilation Control: Section 2.
Personnel Protection: Section 3.
Storage and Handling: Section 7.
Shipping Regulations: Section 11.
  IATA (not stablized): Corrosive liquid, not acceptable (passenger and cargo).

SULFUR TRIOXIDE, STABILIZED. See sulfur trioxide.
  Shipping Regulations: Section 11.
  I.C.C.: Corrosive liquid; white label, 10 pints.
  Coast Guard Classification: Corrosive liquid; white label.
  IATA (stabilized): Corrosive liquid, white label, 1 liter (passenger), 5 liters (cargo).
  MCA warning label.

SULFUR TRIOXYTETRACHLORIDE
General Information
Description: White crystals.
Formula: $S_2O_3Cl_4$.
Constants: Mol wt: 254.0, mp: 57°C (decomp.).
Hazard Analysis
Toxicity: Details unknown. This material readily decomposes into toxic compounds and should be considered highly toxic.
Disaster Hazard: Dangerous; when heated to decomposition, it emits highly toxic fumes of oxides of sulfur and chlorides; will react with water or steam to produce toxic and corrosive fumes.
Countermeasures
Storage and Handling: Section 7.

SULFURYL AMIDE. See sulfamide.

SULFURYL CHLORIDE
General Information
Synonyms: Chlorosulfuric acid; sulfonyl chloride; sulfuric chloride; sulfuric oxychloride.
Description: Colorless liquid; pungent odor.
Formula: SO_2Cl_2.
Constants: Mol wt: 135.0, mp: $-54.1°C$, bp: 69.1°C, d: 1.6674, vap. press.: 100 mm at 17.8°C, vap. d.: 4.65.
Hazard Analysis
Toxicity: Corrosive. See sulfuric acid and hydrochloric acids which are formed upon hydrolysis.
Disaster Hazard: Dangerous; when heated to decomposition, it emits highly toxic fumes of chlorides and oxides of sulfur; will react with water or steam to produce heat, toxic and corrosive fumes.
Countermeasures
Ventilation Control: Section 2.
Personnel Protection: Section 3.
Storage and Handling: Section 7.
Shipping Regulations: Section 11.
Coast Guard Classification: Corrosive liquid; white label.
MCA warning label.
I. C.C.: Corrosive liquid, white label, 1 quart.
IATA: Corrosive liquid, white label, 1 liter (passenger and cargo).

SULFURYL CHLORIDE FLUORIDE
General Information
Description: Colorless gas.
Formula: SO_2ClF.

Constants: Mol wt: 118.52, mp: $-124.7°C$, bp: $7.1°C$.

Hazard Analysis and Countermeasures

See sulfuric acid, chlorides and fluorides.

SULFURYL FLUORIDE

General Information

Description: Colorless gas.

Synonym: Sulfuric oxyfluoride.

Formula: SO_2F_2.

Constants: Mol wt: 102.07, mp: $-120°C$, bp: $-52°C$, d: 3.72 grams per liter.

Hazard Analysis

Toxic Hazard Rating:

Acute Local: Irritant 2; Inhalation 2.

Acute Systemic: Inhalation 3.

Chronic Local: Irritant 1; Inhalation 1.

Chronic Systemic: Inhalation 2.

Toxicity: Accidental exposure of a human resulted in nausea, vomiting, cramps, and itching. May have narcotic action in high concentrations.

TLV: ACGIH (recommended); 5 parts per million in air; 20 milligrams per cubic meter of air.

Disaster Hazard: Dangerous; when heated to decomposition, it emits highly toxic fumes of fluorides and oxides of sulfur; will react with water or steam to produce toxic and corrosive fumes.

Countermeasures

Ventilation Control: Section 2.

Storage and Handling: Section 7.

Shipping Regulations: Section 11.

I.C.C.: Nonflammable gas, green label, 300 pounds.

IATA: Nonflammable gas, green label, 70 kilograms (passenger), 140 kilograms (cargo).

SULPHENONE

General Information

Synonym: p-Chlorophenyl phenyl sulfone.

Description: Crystals; slight aromatic odor; no taste; insoluble in water.

Formula: $C_{12}H_9ClO_2S$.

Constants: Mol wt: 252.7 mp: $90°-94°C$.

Hazard Analysis

Toxicity: Details unknown. Limited animal experiments suggest low oral toxicity. An insecticide. See chlorophenyls.

Disaster Hazard: Dangerous; when heated to decomposition it emits highly toxic fumes.

SUPERPALITE. See trichloroacetyl chloride.

SUPPLEMENTARY CHARGES, EXPLOSIVE

Shipping Regulations: Section 11.

I.C.C.: Explosive A, not accepted.

IATA: Explosive, not acceptable (passenger and cargo).

SWEET OIL. See olive oil.

SYLVIC ACID. See abietic acid.

SYSTOX. See demeton.

TOXIC HAZARD RATING CODE (For detailed discussion, see Section 1.)

0 NONE: (a) No harm under any conditions; (b) Harmful only under unusual conditions or overwhelming dosage.

1 SLIGHT: Causes readily reversible changes which disappear after end of exposure.

2 MODERATE: May involve both irreversible and reversible changes; not severe enough to cause death or permanent injury.

3 HIGH: May cause death or permanent injury after very short exposure to small quantities.

U UNKNOWN: No information on humans considered valid by authors.

T

2,4,5-T. See 2,4,5-trichlorophenoxyacetic acid.

TABUN
General Information
Synonym: Cyanodimethylaminoethoxyphosphine oxide.
Description: A colorless to brownish liquid.
Formula: $CH_3CH_2OPO(CN)[N(CH_3)_2]$.
Constants: Mol wt: 162.1, bp: decomposes at 238°C,
fp: −49.4°C, flash p: 172°F, d: 1.073 at 25°C, vap.
press.: 0.07 mm at 25°C, vap. d.: 5.63.
Hazard Analysis
Toxic Hazard Rating:
Acute Local: U.
Acute Systemic: Ingestion 3; Inhalation 3; Skin absorption 3.
Chronic Local: U.
Chronic Systemic: U.
Toxicology: A nerve gas. Vapor does not penetrate skin;
liquid does so rapidly. The primary physiological action
is on the sympathetic nerve system, causing a vaso-
paresis. Vapors when inhaled can cause nausea, vomit-
ing and diarrhea, which can be followed by muscular
twitchings and convulsions.
Fire Hazard: Moderate, when exposed to heat or flame.
Disaster Hazard: Highly dangerous; it emits highly toxic
fumes; can react with oxidizing materials.
Countermeasures
Ventilation Control: Section 2.
To Fight Fire : Water, foam, carbon dioxide, dry chemical or
carbon tetrachloride (Section 6).
Personnel Protection: Section 3.
First Aid: Section 1.
Storage and Handling: Section 7.

TABUTREX. See dibutyl succinate.

TAGAYASAN
General Information
Description: A wood.
Hazard Analysis
Toxic Hazard Rating:
Acute Local: Allergen 1.
Acute Systemic: U.
Chronic Local: Allergen 1.
Chronic Systemic: U.
Fire Hazard: Slight, in the form of dust when exposed to heat
or flame (Section 6).
Explosion Hazard: Slight, in the form of dust, when exposed
to flame.
Countermeasures
Personal Hygiene: Section 3.
Storage and Handling: Section 7.

TALC
General Information
Synonyms: Talcum, French chalk; steatite.
Description: White to grayish-white fine, odorless powder.
Formula: Powdered native hydrous magnesium silicate.
Hazard Analysis
Toxic Hazard Rating:
Acute Local: Inhalation 1.
Acute Systemic: U.
Chronic Local: Inhalation 2.
Chronic Systemic: Inhalation 3.

It is a substance which migrates to food from packaging ma-
terials (Section 10). It is a common air contaminant
(Section 4).
TLV: ACGIH (recommended); 20 million particles per
cubic foot of air.
Caution: Can produce a form of pulmonary fibrosis (talc
pneumoconiosis).
Countermeasures
Ventilation Control: Section 2.
Personal Hygiene: Section 3.

TALC DUST. See talc.

TALCUM. See talc.

TALL OIL
General Information
Synonyms: Liquid rosin; tallol.
Description: Flammable liquid, dark brown, acrid odor.
Composition: Rosin acids, oleic and linoleic acids.
Constant: D: 0.95–1.0.
Hazard Analysis
Toxic Hazard Rating:
Acute Local: Irritant 1.
Acute Systemic: U.
Chronic Local: Allergen 1.
Chronic Systemic: U.
Toxicity: A substance which migrates to food from packag-
ing materials (Section 10).
Fire Hazard : Slight, when exposed to heat or flame; can
react with oxidizing materials (Section 6).
Countermeasures
Storage and Handling: Section 7.

TALLOL. See tall oil.

TALLOW
General Information
Description: A solid fat.
Constants: Mp: 88–100°F, flash p: 509°F (C.C.),
d: 0.895.
Hazard Analysis
Toxicity: Low. A substance which migrates to food from
packaging materials. (Section 10).
Fire Hazard: Slight, when exposed to heat or flame; can
react with oxidizing material.
Spontaneous Heating: Yes.
Countermeasures
To Fight Fire: Carbon dioxide, dry chemical or carbon
tetrachloride (Section 6).
Storage and Handling: Section 7.

TALLOW OIL
General Information
Synonym: Oleo oil.
Constants: Mp: 42.8°C, flash p: 492°F, d: 0.914, autoign.
temp.: 980°F.
Hazard Analysis
Toxicity: Unknown.
Fire Hazard: Slight, when exposed to heat or flame; can
react with oxidizing materials (Section 6).
Spontaneous Heating: Yes.
Countermeasures
To Fight Fire: Carbon dioxide, dry chemical or carbon
tetrachloride (Section 6).
Storage and Handling: Section 7.

TANKAGE FERTILIZER
Hazard Analysis
Toxicity: Details unknown; may contain ammonia, sulfides and other irritants.
Fire Hazard: Moderate, when exposed to heat or flame. Presence of or absence of moisture can contribute to spontaneous heating. Avoid storage before cooling or extremes of moisture content; can react with oxidizing materials.
Spontaneous Heating: Variable.
Countermeasures
Storage and Handling: Section 7.
Shipping Regulations: Section 11.
   I.C.C.: Flammable solid; yellow label, not accepted.
   Coast Guard Classification: Hazardous (8% or more moisture); inflammable solid (8% or less moisture), yellow label.
   IATA: Flammable solid, not acceptable (passenger and cargo).

TANKAGE, ROUGH AMMONIATE. See tankage fertilizer.
Shipping Regulations: Section 11.
   I.C.C.: Flammable solid; yellow label, not accepted.
   Coast Guard Classification: Inflammable solid; yellow label.
   IATA: Flammable solid, not acceptable (passenger and cargo).

TANKS, EMPTY
Shipping Regulations: Section 11.
   I.C.C.: See Section 11.
   Coast Guard Classification: Hazardous article.

TANNIC ACID
General Information
Synonyms: Tannin; gallotannic acid.
Description: Yellowish-white, brown, bulky powder or flakes.
Formula: $C_{76}H_{52}O_{46}$.
Constants: Mol wt: 1701.2, mp: 200°C, flash p: 390°F (O.C.), autoign. temp.: 980°F.
Hazard Analysis
Toxic Hazard Rating:
   Acute Local: Irritant 1; Ingestion 1; Inhalation 1.
   Acute Systemic: Ingestion 2; Inhalation 2.
   Chronic Local: Irritant 2.
   Chronic Systemic: U.
Fire Hazard: Slight, when exposed to heat or flame.
Spontaneous Heating: No.
Countermeasures
To Fight Fire: Water.
Personnel Protection: Section 3.
Storage and Handling: Section 7.

TANNIN. See tannic acid.

TANSY OIL
General Information
Synonym: Oil of tansy.
Description: Yellowish liquid; strong odor.
Constant: D: 0.925–0.955 at 15°/15°C.
Hazard Analysis
Toxic Hazard Rating:
   Acute Local: U.
   Acute Systemic: Ingestion 3; Inhalation 1.
   Chronic Local: U.
   Chronic Systemic: U.

Disaster Hazard: Moderately dangerous; when heated, it emits acrid fumes.
Countermeasures
Personnel Protection: Section 3.
Storage and Handling: Section 7.

TANTALUM
General Information
Description: Cubic, gray-black, metallic crystals or powder. A chemical element.
Formula: Ta.
Constants: At wt: 180.88, mp: 3027°C, bp: approx. 4100°C, d metal: 16.6; d powder: 14.5.
Hazard Analysis
Toxicity: The metal itself is inert. See also tantalum compounds.
TLV: ACGIH (recommended); 5 milligrams per cubic meter of air.
Toxicology: Some industrial skin injuries from tantalum have been reported. However, systemic industrial poisoning is apparently not known. So far experimental work has indicated that tantalum does not produce any unfavorable effects upon the body. Tantalum metal embedded in the abdominal wall and in bones of dogs caused no physiological disturbances, and so far the use of tantalum in human surgery has received favorable comment.
Radiation Hazard: Section 5. For permissible levels, see Table 5, p. 150.
   Artificial isotope ^{182}Ta, half life 115 d. Decays to stable ^{182}W by emitting beta particles of 0.18–0.51 MeV. Also emits gamma rays of 0.07–1.2 MeV.
Fire Hazard: Moderate, in the form of dust when exposed to heat or flame or by chemical reaction with oxidizing agents. See also powdered metals.

TANTALUM BROMIDE
General Information
Description: Yellow crystals.
Formula: TaBr$_5$.
Constants: Mol wt: 580.46, mp: 240°C, bp: 320°C, d: 4.67.
Hazard Analysis and Countermeasures
See bromides and tantalum compounds.

TANTALUM CHLORIDE
General Information
Description: Light yellow, crystalline powder.
Formula: Ta Cl$_5$.
Constants: Mol wt: 358.17, mp: 221°C, bp: 242°C, d: 3.68 at 27°C.
Hazard Analysis and Countermeasures
See chlorides and tantalum compounds.

TANTALUM COMPOUNDS
Hazard Analysis
Some tantalum compounds have been suspected of causing skin irritation and mild fibrosis of the lungs.

TANTALUM FLUORIDE
General Information
Description: Colorless crystal.
Formula: TaF$_5$.
Constants: Mol wt: 275.88, mp: 96.8°C, bp: 229.5°C, d: 4.74, vap. press.: 100 mm at 130°C.
Hazard Analysis and Countermeasures
See fluorides and tantalum compounds.

TAPIOCA DEXTRIN. See dextrin.

TOXIC HAZARD RATING CODE (*For detailed discussion, see Section 1.*)

0 NONE: (a) No harm under any conditions; (b) Harmful only under unusual conditions or overwhelming dosage.

1 SLIGHT: Causes readily reversible changes which disappear after end of exposure.

2 MODERATE: May involve both irreversible and reversible changes; not severe enough to cause death or permanent injury.

3 HIGH: May cause death or permanent injury after very short exposure to small quantities.

U UNKNOWN: No information on humans considered valid by authors.

TAR ACIDS (COAL). See phenol and cresol.

TARAPACAITE. See potassium chromate.

TAR CAMPHOR. See naphthalene.

TAR, DEHYDRATED
General Information
Description: Dark brown, thick, viscid liquid.
Hazard Analysis
Toxic Hazard Rating:
  Acute Local: Irritant 1.
  Acute Systemic: U.
  Chronic Local: Irritant 3.
  Chronic Systemic: Ingestion 3; Inhalation 3; Skin Absorption 3.
Disaster Hazard: Moderately dangerous; when heated, it emits toxic fumes.
Countermeasures
Personnel Protection: Section 3.
Storage and Handling: Section 7.

TAR DUST
Hazard Analysis
Toxic Hazard Rating:
  Acute Local: Irritant 2; Inhalation 1.
  Acute Systemic: U.
  Chronic Local: U.
  Chronic Systemic: U.
Caution: Irritating to the eyes.
Fire Hazard: Moderate, when exposed to heat or flame; can react with oxidizing materials (Section 6).
Countermeasures
Personnel Protection: Section 3.

TAR, LIQUID
Hazard Analysis
Toxic Hazard Rating:
  Acute Local: Irritant 1.
  Acute Systemic: U.
  Chronic Local: Irritant 3.
  Chronic Systemic: Ingestion 3; Inhalation 3; Skin Absorption 3.
Fire Hazard: Moderate, when exposed to heat or flame (Section 6).
Disaster Hazard: Moderately dangerous; when heated it emits toxic fumes.
Countermeasures
Personnel Protection: Section 3.
Storage and Handling: Section 7.
Shipping Regulations: Section 11.
  I.C.C.: Flammable liquid; red label, 10 gallons.
  Coast Guard Classification: Combustible liquid; inflammable liquid, red label.
  IATA: Flammable liquid, red label, 1 liter (passenger), 40 liters (cargo).

TAR OIL RECTIFIED. See pine tar oil.

TARTAR EMETIC. See antimony potassium tartrate.

TARTARIC ACID
General Information
Synonym: Racemic acid.
Description: White crystals.
Formula: HOOC(CHOH)$_2$COOH.
Constants: Mol wt: 150.09, mp: 168–170°C flash p: 410°F (O.C.), d: 1.76, autoign. temp.: 802°F.
Hazard Analysis
Toxic Hazard Rating:
  Acute Local: Irritant 1; Ingestion 1; Inhalation 1.
  Acute Systemic: U.
  Chronic Local: U.
  Chronic Systemic: U.
Toxicity: Has little or no systemic toxicity. It is used as a

sequestrant and general purpose food additive; it is a substance which migrates to food from packaging materials (Section 10).
Fire Hazard: Slight, when exposed to heat or flame.
Spontaneous Heating: No.
Countermeasures
To Fight Fire: Water (Section 6).
Personal Hygiene: Section 3.
Storage and Handling: Section 7.

TAR, WATER GAS
Hazard Analysis
Toxic Hazard Rating:
  Acute Local: Irritant 2.
  Acute Systemic: U.
  Chronic Local: U.
  Chronic Systemic: U.
Fire Hazard: Slight, when exposed to heat (Section 6).
Disaster Hazard: Moderately dangerous; when strongly heated, it emits acrid fumes.
Countermeasures
Personnel Protection: Section 3.
Storage and Handling: Section 7.

TAUROCHOLIC ACID
General Information
Synonyms: Cholaic acid, cholyltaurine.
Description: Occurs as sodium salt in the bile; crystals, freely soluble in water; soluble in alcohol; insoluble in ether and ethyl acetate.
Formul: C$_{26}$H$_{45}$NO$_7$S.
Constants: Mol wt: 515, mp: 125°C.
Hazard Analysis
Toxic Hazard Rating: U. An emulsifying agent food additive (Section 10).
Disaster Hazard: Dangerous. See sulfides.

TCM. See trichloromelamine.

TCNE. See tetracyanoethylene.

2,4,5-TCPPA. See 2-(2,4,5-trichlorophenoxy) propionic acid.

TDE. See 1,1-dichloro-2,2-bis(p-chlorophenyl) ethane.

TDI. See 2,4-tolylene diisocyanate.

TEAK
General Information
Description: A wood.
Hazard Analysis
Toxic Hazard Rating:
  Acute Local: Irritant 1; Allergen 1; Inhalation 1.
  Acute Systemic: U.
  Chronic Local: Allergen 1.
  Chronic Systemic: U.
Fire Hazard: Moderate, in the form of dust, when exposed to heat or flame (Section 6).
Explosion Hazard: Slight, in the form of dust when exposed to flame. See also dust.
Countermeasures
Personal Hygiene: Section 3.
Storage and Handling: Section 7.

TEAR GAS MATERIAL, LIQUID OR SOLID, N.O.S.
Hazard Analysis
Toxic Hazard Rating:
  Acute Local: Irritant 3; Inhalation 3.
  Acute Systemic: Inhalation 3.
  Chronic Local: U.
  Chronic Systemic: U.
Disaster Hazard: Dangerous; it emits highly toxic fumes.
Countermeasures
Ventilation Control: Section 2.

Personnel Protection: Section 3.
First Aid: Section 1.
Storage and Handling: Section 7.
Shipping Regulations: Section 11.
   I.C.C.: Poison C; tear gas label, 75 pounds.
   Coast Guard Classification: Poison C; tear gas label.
   IATA: Poison C, poison label, not acceptable (passenger), 35 kilograms (cargo).

TECHNETIUM
General Information
Description: Hexagonal crystal structure. A chemical element.
Formula: Tc.
Constant: At wt: 99.
Hazard Analysis
Toxicity: Unknown.
Radiation Hazard : Section 5. For permissible levels, see Table 5, p. 150.
   Artificial isotope ^{96m}Tc, half life 52 m. Decays to radioactive ^{96}Tc by emitting gamma rays of 0.03 MeV and X-rays.
   Artificial isotope ^{96}Tc, half life 4.3 d. Decays to stable ^{96}Mo by electron capture. Emits gamma rays of 0.77 to 1.12 MeV and X-rays.
   Artificial isotope ^{97m}Tc, half life 91 d. Decays to radioactive ^{97}Tc by emitting gamma rays of 0.09 Me V and X-rays.
   Artificial isotope ^{97}Tc, half life 2.6×10^6 y. Decays to stable ^{97}Mo by electron capture. Emits X-rays.
   Artificial isotope ^{99m}Tc, half life 6 h. Decays to radioactive ^{99}Tc by emitting gamma rays of 0.14 MeV and X-rays.
   Artificial isotope ^{99}Tc, half life 2.1×10^5 y. Decays to stable ^{99}Ru by emitting beta particles of 0.29 MeV.

TEDION
General Information
Synonym: Duphar; 2,4,5,4'-tetrachlorodiphenyl sulfone.
Description: Crystals, nearly water insoluble.
Formula: $C_{12}H_6Cl_4O_2S$.
Constants: Mol wt: 356.1, mp: 147°C.
Hazard Analysis
Toxicity: Details unknown. Limited animal experiments suggest low systemic toxicity. A food additive permitted in food for human consumption. (Section 10).
Disaster Hazard: Dangerous; when heated to decomposition it emits highly toxic fumes of chlorides and oxides of sulfur.

TEFLON. See polytetrafluoroethylene.

o-TELLURIC ACID
General Information
Description: Needles.
Formula: $H_6TeO_6 \cdot 4H_2O$.
Constants: Mol wt: 301.72, mp: $-4H_2O$ at 100°C.
Hazard Analysis and Countermeasures
See tellurium compounds.

TELLURIUM
General Information
Description: Dark gray crystalline or amorphous powder or small cakes. A chemical element.
Formula: Te.
Constants: At wt: 127.61, d: 6.25, mp: 452°C, bp: 1390°C, vap. press.: 1 mm at 520°C.

Hazard Analysis
Toxicity: See tellurium compounds.
TLV: ACGIH (recommended); 0.1 milligrams per cubic meter of air.
Radiation Hazard: Section 5. For permissible levels, see Table 5, p. 150.
   Artificial isotope ^{125m}Te, half life 58 d. Decays to stable ^{125}Te by emitting gamma rays of 0.04, 0.11 MeV.
   Artificial isotope ^{127m}Te, half life 105 d. Decays to radioactive ^{127}Te by emitting gamma rays of 0.09 MeV.
   Artificial isotope ^{127}Te, half life 9.3 h. Decays to stable ^{127}I by emitting beta particles of 0.70 MeV.
   Artificial isotope ^{129m}Te, half life 33 d. Decays to radioactive ^{129}Te by emitting gamma rays of 0.11 MeV.
   Artificial isotope ^{129}Te, half life 67 m. Decays to radioactive ^{129}I by emitting beta particles of 0.99 (16%), 1.45 (80%) MeV. Also emits gamma rays of 0.03, 0.48 MeV.
   Artificial isotope ^{131m}Te, half life 1.2 d. Decays to radioactive ^{131}I by emitting beta particles (81%) of 0.22–2.46 MeV. Also decays to radioactive ^{131}Te by emitting gamma rays of 0.18 MeV. Also emits gamma rays of 0.08–2.24 MeV.
   Artificial isotope ^{131}Te, half life 25 m. Decays to radioactive ^{131}I by emitting beta particles of 1.15–2.14 MeV. Also emits gamma rays of 0.14–1.13 MeV.
Fire Hazard: Moderate, in the form of dust when exposed to heat or flame or by chemical reaction with oxidizing agents. See also powdered metals.
Countermeasures
Personal Hygiene: Section 3.
First Aid: Section 1.
Storage and Handling: Section 7.

TELLURIUM COMPOUNDS
Hazard Analysis
Toxic Hazard Rating:
   Acute Local: U.
   Acute Systemic: Ingestion 2; Inhalation 2.
   Chronic Local: U.
   Chronic Systemic: Ingestion 3; Inhalation 3.
Toxicology: Elemental tellurium has relatively low toxicity. It is converted in the body to dimethyl telluride which imparts a garlic like odor to the breath and sweat. Heavy exposures may, in addition, result in headache, drowsiness, metallic taste, loss of appetite and nausea. Various tellurium salts may also produce similar symptoms. Large doses can be fatal, as was the case following accidental administration of sodium tellurite.
Disaster Hazard: Dangerous; when heated or on contact with acid or acid fumes, they emit highly toxic fumes.
Countermeasures
Ventilation Control: Section 2.
Personal Hygiene: Section 3.
First Aid: Section 1.
Storage and Handling: Section 7.

TELLURIUM DIBROMIDE
General Information
Description: Needles.
Formula: $TeBr_2$.
Constants: Mol wt: 287.44, mp: 210°C, bp: 339°C.
Hazard Analysis and Countermeasures
See tellurium compounds and bromides.

TOXIC HAZARD RATING CODE (For detailed discussion, see Section 1.)

0 NONE: (a) No harm under any conditions; (b) Harmful only under unusual conditions or overwhelming dosage.

1 SLIGHT: Causes readily reversible changes which disappear after end of exposure.

2 MODERATE: May involve both irreversible and reversible changes; not severe enough to cause death or permanent injury.

3 HIGH: May cause death or permanent injury after very short exposure to small quantities.

U UNKNOWN: No information on humans considered valid by authors.

TELLURIUM DICHLORIDE
General Information
Description: Crystals or amorphous, unstable.
Formula: $TeCl_2$.
Constants: Mol wt: 198.52, mp: $209 \pm 5°C$, bp: $327°C$, d: 7.05.
Hazard Analysis and Countermeasures
See tellurium compounds and chlorides.

TELLURIUM DIIODIDE
General Information
Description: Crystals.
Formula: TeI_2.
Constants: Mol wt: 381.45, mp: sublimes.
Hazard Analysis and Countermeasures
See tellurium compounds and iodides.

TELLURIUM HEXAFLUORIDE
General Information
Description: Colorless gas.
Formula: TeF_6.
Constants: Mol wt: 241.61, mp: $-36°C$, bp: $-35.5°C$, d: 3.025 at $-35.5°C$.
Hazard Analysis and Countermeasures
See fluorides and tellurium compounds.
TLV: ACGIH (recommended); 0.02 part per million; 0.20 milligrams per cubic meter of air.

TELLURIUM HYDRIDE
Hazard Analysis and Countermeasures
See hydrides.

TELLURIUM NITRIDE
General Information
Description: Solid.
Hazard Analysis
Toxicity: Highly toxic. See tellurium compounds and nitrides.
Explosion Hazard: Severe, when shocked or exposed to heat.
Disaster Hazard: Dangerous; shock will explode it; when heated or on contact with acid or acid fumes, it emits highly toxic fumes of tellurium and may explode; will react with water or steam to produce toxic fumes.
Countermeasures
Storage and Handling: Section 7.

TELLURIUM SULFIDE
General Information
Description: Amorphous powder.
Formula: TeS_2.
Constant: Mol wt: 191.73.
Hazard Analysis and Countermeasures
See sulfides and tellurium compounds.

TELLURIUM SULFITE
General Information
Description: Amorphous, deep red solid.
Formula: $TeSO_3$.
Constants: Mol wt: 207.67, mp: soft. $30°C$, bp: decomposes.
Hazard Analysis and Countermeasures
See tellurium compounds and sulfites.

TELLURIUM TETRAFLUORIDE
General Information
Description: White crystals.
Formula: TeF_4.
Constants: Mol wt: 203.61, mp: sublimes.
Hazard Analysis and Countermeasures
See fluorides and tellurium compounds.

TELONE. See 1,3-dichloropropene.

TEM. See triethylene melamine.

TEMUR. See tetramethylurea.

"TENAMINE-1"
General Information
Description: Lemon yellow, mobile liquid.
Composition: 48% N-n-butyl-p-aminophenol in isopropyl alcohol.
Constants: D: 0.89–0.91 at $25°/25°C$, flash p: $61°F$, fp: $-33°C$.
Hazard Analysis
Toxicity: Details unknown. See individual components.
Fire Hazard : Dangerous, when exposed to heat or flame.
Explosion Hazard: Unknown.
Disaster Hazard: Dangerous; when heated to decomposition it emits toxic fumes; can react with oxidizing materials.
Countermeasures
Ventilation Control: Section 2.
To Fight Fire: Carbon dioxide, dry chemical or carbon tetrachloride (Section 6).
Storage and Handling: Section 7.

"TENAMINE 60"
General Information
Description: Liquid.
Composition: 20% toluene and 80% disalicylal propylenediimine.
Constants: Fp: $-18.3°C$, flash p: $19°F$ (C.C.), d: 1.07 at $25°/25°C$.
Hazard Analysis
Toxicity: Details unknown. See individual components.
Fire Hazard: Dangerous, when exposed to heat or flame.
Explosion Hazard: Unknown.
Disaster Hazard: Dangerous; keep away from open flame or heat; it emits toxic fumes when heated; can react with oxidizing materials.
Countermeasures
Ventilation Control: Section 2.
To Fight Fire: Water, foam, carbon dioxide, dry chemical or carbon tetrachloride (Section 6).
Storage and Handling: Section 7.

TEP. See tetraethyl pyrophosphate.

TEPP. See tetraethyl pyrophosphate.

TERBIUM
General Information
Description: A rare earth chemical element.
Formula: Tb.
Constant: At wt: 159.20.
Hazard Analysis
Toxicity: Unknown. As a lanthanon it may impair blood coagulation. See also lanthanum.
Fire Hazard: Moderate, in the form of dust when exposed to heat or flame or by chemical reaction with oxidizers. See also powdered metals.
Radiation Hazard: Section 5. For permissible levels, see Table 5, p. 150.
Artificial isotope ^{160}Tb, half life 72 d. Decays to stable ^{160}Dy by emitting beta particles of 0.26–0.86 MeV. Also emits gamma rays of 0.09–1.5 MeV.

TERBIUM NITRATE
General Information
Description: Colorless, monoclinic needles.
Formula: $Tb(NO_3)_3 \cdot 6H_2O$.
Constants: Mol wt: 453.32, mp: $89.3°C$.
Hazard Analysis and Countermeasures
See nitrates and terbium.

TEREPHTHALIC ACID
General Information
Synonym: p-Phthalic acid; TPA; benzene-p-dicarboxylic acid.
Description: White crystals or powder; insoluble in water, chloroform, ether, acetic acid; slightly soluble in alcohol; soluble in alkalies.

Formula: $C_6H_4(COOH)_2$.
Constants: Mol wt: 166.13, d: 1.51, sublimes $> 300°$ C.
Hazard Analysis
Toxic Hazard Rating:
    Acute Local: Irritant 1.
    Acute Systemic: U.
    Chronic Local: U.
    Chronic Systemic: U.

TEREPHTHALOYL CHLORIDE
General Information
Synonym: p-Phthalyl dichloride.
Description: White crystalline material; musty odor.
Formula: $C_8H_4O_2Cl_2$.
Constants: Mol wt: 203.0, bp: 266° C, flash p: 356° F (C.O.C.), fp: 81.4° C.
Hazard Analysis
Toxicity: An irritant material. Further details unknown.
Disaster Hazard: Dangerous. When heated to decomposition it emits highly toxic fumes of chlorides.

TERPENES. See turpentine oil.

m-TERPHENYL
General Information
Synonyms: m-Phenylbiphenyl; benzene-1,3-diphenyl.
Description: Colorless needles.
Formula: $(C_6H_5)_2C_6H_4$.
Constants: Mol wt: 230.3, mp: 86–87° C, bp: 363° C, flash p: 375° F (O.C.), d: 1.164, vap. press: 7.95.
Hazard Analysis
Toxic Hazard Rating:
    Acute Local: U.
    Acute Systemic: U.
    Chronic Local: U.
    Chronic Systemic: Ingestion 3.
Toxicity: Animal feeding experiments show injury to liver and kidneys. Toxicity of terphenyls is proportional to solubility. See also diphenyl.
TLV: ACGIH (tentative); 1 part per million; 9.4 milligrams per cubic meter of air (same for o- and p-isomers).
Fire Hazard: Slight, when exposed to heat or flame.
Countermeasures
To Fight Fire: Water, carbon dioxide, dry chemical or carbon tetrachloride (Section 6).
Storage and Handling: Section 7.

o-TERPHENYL
General Information
Synonym: 1,2-Diphenylbenzene.
Description: A liquid.
Formula: $C_6H_4(C_6H_5)_2$.
Constants: Mol wt: 230.3, bp: 332° C, flash p: 325° F (O.C.), d: 1.14, vap. d.: 7.95.
Hazard Analysis
Toxicity: Details unknown. See m-terphenyl.
Fire Hazard: Slight, when exposed to heat or flame.
Countermeasures
To Fight Fire: Water, carbon dioxide, dry chemical or carbon tetrachloride (Section 6).
Storage and Handling: Section 7.

p-TERPHENYL
General Information
Synonym: Benzene-1,4 diphenyl.
Description: Liquid.
Formula: $(C_6H_5)_2C_6H_4$.
Constants: Mol wt: 230.3, mp: 213° C, bp: 405° C, flash p: 405° F (O.C.), d: 1.236, vap. d.: 7.95.

Hazard Analysis
Toxicity: Details unknown. See m-terphenyl.
Fire Hazard: Slight, when exposed to heat or flame.
Countermeasures
To Fight Fire: Water, carbon dioxide, dry chemical or carbon tetrachloride (Section 6).
Storage and Handling: Section 7.

TERPINOLENE
General Information
Synonym: 1,4(8)-p-Menthadiene.
Description: Colorless liquid.
Formula: $C_{10}H_{16}$.
Constants: Mol wt: 136.23, bp: 185° C, d: 0.855, flash p: 100° F (C.C.).
Hazard Analysis
Toxicity: Unknown.
Fire Hazard: Moderate, when exposed to heat or flame.
Disaster Hazard: Moderately dangerous. Keep away from open flame; it can react with oxidizing materials.
Countermeasures
Ventilation Control: Section 2.
To Fight Fire: Foam, carbon dioxide, dry chemical or carbon tetrachloride (Section 6).
Storage and Handling: Section 7.

TESTOSTERONE
General Information
Synonym: Δ^4-Androsten-17(α)-ol-3-one.
Description: Crystals.
Formula: $C_{19}H_{28}O_2$.
Constants: Mol wt: 288.4, mp: 155° C.
Hazard Analysis
Toxic Hazard Rating:
    Acute Local: U.
    Acute Systemic: Ingestion 3; Inhalation 3; Skin Absorption 3.
    Chronic Local: U.
    Chronic Systemic: Ingestion 3; Inhalation 3; Skin Absorption 3.
Caution: Workers engaged in the manufacture and packaging have shown effects from this hormone. Enlargement of the breasts in male workers has been observed.
Disaster Hazard: Dangerous; when heated to decomposition it emits toxic fumes.
Countermeasures
Ventilation Control: Section 2.
Personal Protection: Section 3.
Storage and Handling: Section 7.

TESTOSTERONE PROPIONATE
General Information
Description: White or creamy white crystals or crystalline powder; odorless; freely soluble in alcohol, dioxane, ether, and other organic solvents; soluble in vegetable oils; insoluble in water.
Formula: $C_{19}H_{27}O \cdot OOCC_2H_5$.
Constant: Mol wt: 344, mp: 118°–123° C.
Hazard Analysis
Toxic Hazard Rating: U. A food additive permitted in the feed and drinking water of animals and/or for the treatment of food producing animals. Also permitted in food for human consumption (Section 10).

TETRAAMINECOPPER SULFATE
Hazard Analysis and Countermeasures
See copper sulfate and amines.

TOXIC HAZARD RATING CODE (For detailed discussion, see Section 1.)

0 NONE: (a) No harm under any conditions; (b) Harmful only under unusual conditions or overwhelming dosage.

1 SLIGHT: Causes readily reversible changes which disappear after end of exposure.

2 MODERATE: May involve both irreversible and reversible changes; not severe enough to cause death or permanent injury.

3 HIGH: May cause death or permanent injury after very short exposure to small quantities.

U UNKNOWN: No information on humans considered valid by authors.

TETRAAMYLBENZENE
General Information
Description: Liquid.
Formula: $C_6H_2(C_5H_{11})_4$.
Constants: Mol wt: 358.6, bp: 320°C, flash p: 295°F, d: 0.89.
Hazard Analysis
Toxicity: Details unknown. Probably irritant and narcotic in high concentrations.
Fire Hazard: Slight, when exposed to heat or flame; it can react with oxidizing materials.
Countermeasures
To Fight Fire: Foam, carbon dioxide, dry chemical or carbon tetrachloride (Section 6).
Storage and Handling: Section 7.

TETRA-n-AMYLTHIOGERMANIUM
General Information
Description: Colorless liquid; soluble in benzene.
Formula: $Ge[S(CH_2)_4CH_3]_4$.
Constants: Mol wt: 485.4, d: 1.0697 at 25°C, bp: 241°C at 4 mm.
Hazard Analysis and Countermeasures
See germanium compounds and organometals.

TETRA-n-AMYLTIN
General Information
Description: Colorless stable liquid.
Formula: $(C_5H_{11})_4Sn$.
Constants: Mol wt: 403.3, d: 1.0206, bp: 181°C at 10 mm.
Hazard Analysis
See tin compounds.

TETRAAQUOSTANNIC BIS-ACETYL ACETONE STANNIC BROMIDE
General Information
Description: Colorless crystals; soluble in benzene.
Formula: $(C_5H_7O_2)_2Sn(OH_2)_4SnBr_6$.
Constants: Mol wt: 987.2, mp: 107°C.
Hazard Analysis and Countermeasures
See tin compounds and bromides.

TETRABENZYLGERMANIUM
General Information
Description: Colorless powder.
Formula: $Ge(CH_2C_6H_5)_4$.
Constants: Mol wt: 437.1, mp: 108°C.
Hazard Analysis
Toxicity: Details unknown. See germanium compounds.

TETRABENZYLSILICANE
General Information
Description: Liquid; insoluble in water.
Formula: $(C_6H_5CH_2)_4Si$.
Constants: Mol wt: 392.6, bp: 127.5°C.
Hazard Analysis
Toxicity: See silanes.

TETRABENZYLTIN
General Information
Description: Colorless crystals; insoluble in water; soluble in common organic solvents.
Formula: $(C_6H_5CH_2)_4Sn$.
Constants: Mol wt: 483.2, mp: 43°C.
Hazard Analysis
Toxicity: Details unknown. See tin compounds.

TETRA-p-BIPHENYLYLGERMANIUM
General Information
Description: White crystals; insoluble in water.
Formula: $Ge(C_6H_4C_6H_5)_4$.
Constants: Mol wt: 685.4, mp: 271°C.
Hazard Analysis
Toxicity: Details unknown. See germanium compounds.

TETRABORANE. See dihydrotetraborane.

TETRABORIC ACID
General Information
Description: White powder.
Formula: $H_2B_4O_7$.
Constant: Mol wt: 157.30.
Hazard Analysis
Toxicity: See boron compounds.

TETRABORONDECAHYDRIDE. See dihydrotetraborane.

TETRABROMETHYLENE. See dicarbon tetrabromide.

3,4,6,7-TETRABROMO-o-CRESOL
General Information
Description: White to buff crystals. Insoluble in water.
Formula: $C_7H_4Br_4O$.
Constants: Mol wt: 423.8, mp: 205–208°C (decomp.).
Hazard Analysis
Toxicity: Details unknown. Animal experiments show corrosive action on skin but low systemic toxicity. See also cresol.
Disaster Hazard: Dangerous. When heated to decomposition it emits highly toxic fumes.

TETRABROMOETHANE. See acetylene tetrabromide.

TETRABROMOMETHANE. See carbon tetrabromide.

TETRA-p-BROMOPHENYL THIOGERMANIUM
General Information
Description: Colorless crystals. Soluble in benzene.
Formula: $Ge(SC_6H_4Br)_4$.
Constants: Mol wt: 824.9, mp: 196°C.
Hazard Analysis
See sulfates, bromides and germanium compounds.
Countermeasures
See sulfates, bromides and germanium compounds, and organometals.

TETRABROMOSILANE. See silicon bromide.

TETRABROMOSILICANE. See silicon bromide.

TETRA-n-BUTYLGERMANIUM
General Information
Description: Colorless oily liquid.
Formula: $Ge(C_4H_9)_4$
Constants: Mol wt: 301.1, bp: 180°C (approx.).
Hazard Analysis
Toxicity: Details unknown. See germanium compounds.

TETRA-p-tert-BUTYLPHENYL THIOGERMANIUM
General Information
Description: Colorless crystals. Soluble in petroleum ether, ether and acetone.
Formula: $Ge[SC_6H_4C(CH_3)_3]_4$.
Constants: Mol wt: 733.7, mp: 156°C.
Hazard Analysis and Countermeasures
See germanium compounds and organometals.

TETRABUTYL THIODISUCCINATE
General Information
Description: Liquid.
Formula: $[CH_2(COOC_4H_9)CH(COOC_4H_9)]_2S$.
Constants: Mol wt: 490.64, mp: −45°C, bp: 246°C at 5 mm, flash p: 430°F (O.C.), d: 1.0543 at 20°/20°C, vap. press.: 0.24 mm at 200°C, vap. d.: 16.9.
Hazard Analysis
Toxicity: Unknown.
Fire Hazard: Slight, when exposed to heat or flame.
Disaster Hazard: Dangerous, when heated to decomposition it burns and emits highly toxic fumes of oxides of sulfur; it can react on contact with oxidizing materials.

Countermeasures
Storage and Handling: Section 7.
To Fight Fire: Water, foam, carbon dioxide, dry chemical or carbon tetrachloride (Section 6).

TETRA-n-BUTYL THIOGERMANIUM
General Information
Description: A liquid.
Formula: $Ge[S(CH_2)_3CH_3]_4$.
Constants: Mol wt: 429.3, bp: 222.5°C at 4.5 mm, d: 1.1072 at 25°C.
Hazard Analysis and Countermeasures
See germanium compounds, organometals and sulfur compounds.

TETRA-see-BUTYL THIOGERMANIUM
General Information
Description: A liquid; soluble in benzene.
Formula: $Ge[SCH(CH_3)(C_2H_5)]_4$.
Constants: Mol wt: 429.3, d: 1.1119 at 25°C, bp: 200.5°C at 4 mm.
Hazard Analysis and Countermeasures
See germanium compounds, sulfur compounds and organometals.

TETRA-tert-BUTYL THIOGERMANIUM
General Information
Description: Crystals; soluble in absolute alcohol.
Formula: $Ge[SC(CH_3)_3]_4$.
Constants: Mol wt: 436.77, mp: 173°C, bp: sublimes at 170°C at 4 mm.
Hazard Analysis and Countermeasures
See germanium compounds, sulfur compounds and organometals.

TETRA-n-BUTYLTIN
General Information
Description: Colorless stable liquid.
Formula: $(C_4H_9)_4Sn$.
Constants: Mol wt: 347.2, d: 1.0572, bp: 145°C at 10 mm.
Hazard Analysis
Toxicity: Details unknown. See tin compounds.

TETRA-n-BUTYL TITANATE
General Information
Description: Colorless liquid.
Formula: $(C_4H_9O)_4Ti$.
Constants: Mol wt: 340, mp: $< -40°C$, bp: 206°C at 10 mm, d: 0.9951 at 25°C, vap. d.: 11.7.
Hazard Analysis
Toxicity: Details unknown. See esters and titanium compounds.
Fire Hazard: Moderate, when exposed to heat or flame; can react with oxidizing materials (Section 6).
Countermeasures
Storage and Handling: Section 7.

TETRABUTYL UREA
General Information
Description: Liquid.
Formula: $(C_4H_9)_2NCON(C_4H_9)_2$.
Constants: Mol wt: 284.5, mp: $< -60°C$, bp: 300–325°C, flash p: 200°F, d: 0.876, vap. d.: 9.83.
Hazard Analysis
Toxicity: Unknown.
Fire Hazard: Moderate, when exposed to heat or flame.
Disaster Hazard: Dangerous; when heated to decomposition it emits toxic fumes; can react with oxidizing materials.

Countermeasures
Storage and Handling: Section 7.
To Fight Fire: Foam, carbon dioxide, dry chemical or carbon tetrachloride (Section 6).

TETRACENE
General Information
Synonym: Naphthacene.
Description: Orange crystals.
Formula: $C_{18}H_{12}$.
Constants: Mol wt: 228.3, mp: 341°C, d: 1.35.
Hazard Analysis
Toxicity: Unknown.
Explosion Hazard: Moderate, when shocked.
Disaster Hazard: Dangerous; shock will explode it; when heated it burns and emits acrid fumes; can react on contact with oxidizing materials.
Countermeasures
Storage and Handling: Section 7.

TETRACETYLENE DICARBONIC ACID
General Information
Description: Solid.
Hazard Analysis
Toxicity: Unknown.
Explosion Hazard: Severe, when shocked, exposed to heat or by chemical reaction.
Disaster Hazard: Highly dangerous; shock or heat will explode it.
Countermeasures
Storage and Handling: Section 7.

TETRACETYLTHIOGERMANIUM
General Information
Description: White crystals. Soluble in organic solvents.
Formula: $Ge[SCH_2(CH_2)_{14}CH_3]_4$.
Constants: Mol wt: 1102.6, mp: 51°C.
Hazard Analysis and Countermeasures
See germanium compounds, organometals and sulfur compounds.

1,2,4,5-TETRACHLOROBENZENE
General Information
Synonym: Benzene tetrachloride.
Formula: $C_6H_2Cl_4$.
Constants: Mol wt: 215.9, mp: 47.5°C, bp: 245°C, flash p: 311°F (C.C.), d: 1.734, vap. press.: < 0.1 mm at 25°C, vap. d.: 7.4.
Hazard Analysis
Toxicity: Details unknown. Limited animal experiments show low toxicity and no sensitizing action. See also chlorinated hydrocarbons, aromatic.
Fire Hazard: Slight, when exposed to heat or flame.
Disaster Hazard: Dangerous, when heated to decomposition it emits highly toxic fumes of chlorides; can react vigorously with oxidizing materials.
Countermeasures
Storage and Handling: Section 7.
To Fight Fire: Water, carbon dioxide, dry chemical or carbon tetrachloride (Section 6).

TETRACHLORO-p-BENZOQUINONE. See tetrachloroquinone.

TETRACHLORODIAMMINE PLATINUM (IV)
General Information
Description: Rhombic or hexagonal orange-yellow plates or needles.
Formula: $[Pt(NH_3)_2Cl_4]$.
Constant: Mol wt: 371.12.

TOXIC HAZARD RATING CODE (For detailed discussion, see Section 1.)

0 NONE: (a) No harm under any conditions; (b) Harmful only under unusual conditions or overwhelming dosage.

1 SLIGHT: Causes readily reversible changes which disappear after end of exposure.

2 MODERATE: May involve both irreversible and reversible changes; not severe enough to cause death or permanent injury.

3 HIGH: May cause death or permanent injury after very short exposure to small quantities.

U UNKNOWN: No information on humans considered valid by authors.

Hazard Analysis and Countermeasures
See platinum compounds and chlorides.

TETRACHLORODIFLUOROETHANE
General Information
Description: Liquid.
Formula: $C_2F_2Cl_4$.
Constants: Mol wt: 203.8, bp: 92.8°C, d: 1.6447 at 25°C, vap. d.: 7.03.
Hazard Analysis
Toxic Hazard Rating:
   Acute Local: 0.
   Acute Systemic: Inhalation 2.
   Chronic Local: U.
   Chronic Systemic: Inhalation 1.
TLV: ACGIH (recommended); 500 parts per million; 4170 milligrams per cubic meter of air.
Disaster Hazard: Dangerous; when heated, it emits highly toxic fumes of fluorides and chlorides.
Countermeasures
Ventilation Control: Section 2.
Storage and Handling: Section 7.

TETRACHLORODINITROETHANE
General Information
Description: Crystals.
Formula: $(CCl_2NO_2)_2$.
Constants: Mol wt: 257.87, bp: decomp. at 130°C → peroxide of nitrogen.
Hazard Analysis
Toxic Hazard Rating:
   Acute Local: Irritant 3; Ingestion 3; Inhalation 3.
   Acute Systemic: U.
   Chronic Local: U.
   Chronic Systemic: U.
Disaster Hazard: Dangerous; when heated, it emits highly toxic fumes; can react vigorously with oxidizing materials.
Countermeasures
Ventilation Control: Section 2.
Personnel Protection: Section 3.
First Aid: Section 1.
Storage and Handling: Section 7.
Shipping Regulations: Section 11.
   IATA: Poison A, not acceptable (passenger and cargo).

TETRACHLORODIPHENYLETHANE. See 1,1-dichloro-2,2-bis(p-chlorophenyl) ethane.

2,4,5,4-TETRACHLORODIPHENYL   SULFONE. See tedion.

1,1,2,2-TETRACHLOROETHANE. See acetylene tetrachloride.

TETRACHLOROETHYLENE. See perchlorethylene.

N-(1,1,2,2-TETRACHLOROETHYLTHIO)-4-CYCLOHEXENE-1,2-DICARBOXIMIDE. See difolatan.

TETRACHLOROHEPTANE
General Information
Formula: $C_7H_{12}Cl_4$.
Constant: Mol wt: 238.
Hazard Analysis
Toxicity: Unknown. Limited animal experiments suggest low oral toxicity.
Disaster Hazard: Dangerous. See chlorides.

TETRACHLOROMETHANE. See carbon tetrachloride.

TETRACHLOROMETHYL ETHER
General Information
Description: Fuming liquid, pungent odor.
Formula: $O(CHCl_2)_2$.
Constants: Mol wt: 183.87, bp: 145°C, d: 1.6537 at 18°C.

Hazard Analysis
Toxic Hazard Rating:
   Acute Local: Irritant 3; Ingestion 3; Inhalation 3.
   Acute Systemic: U.
   Chronic Local: U.
   Chronic Systemic: U.
Disaster Hazard: Dangerous; when heated to decomposition, it emits highly toxic fumes.
Countermeasures
Ventilation Control: Section 2.
Personnel Protection: Section 3.
First Aid: Section 1.
Storage and Handling: Section 7.

TETRACHLORONAPHTHALENE
General Information
Synonym: 1,2,3,4-Tetrachloro-1,2,3,4-tetrahydronaphthalene.
Description: Crystals.
Formula: $C_{10}H_8Cl_4$.
Constants: Mol wt: 269.99, mp: 182°C.
Hazard Analysis
Toxic Hazard Rating:
   Acute Local: Irritant 2.
   Acute Systemic: U.
   Chronic Local: Irritant 2.
   Chronic Systemic: Ingestion 3; Inhalation 3; Skin Absorption 3.
TLV: ACGIH (tentative); 2 milligrams per cubic meter of air. May be absorbed via the skin.
Toxicology: See also chlorinated naphthalenes and chlorinated diphenyls.
Disaster Hazard: Dangerous; when heated to decomposition, it emits highly toxic fumes.
Countermeasures
Ventilation Control: Section 2.
Personnel Protection: Section 3.
First Aid: Section 1.
Storage and Handling: Section 7.

TETRACHLORONAPHTHOQUINONE
Hazard Analysis
Toxicity: Unknown. A fungicide. See also quinones.
Disaster Hazard: Dangerous; when heated to decomposition, it emits highly toxic fumes.
Countermeasures
Storage and Handling: Section 7.

TETRACHLORONONANE
General Information
Formula: $C_9H_{16}Cl_4$.
Constant: Mol wt: 266.
Hazard Analysis
Toxicity: Unknown. Limited animal experiments show low oral toxicity.
Disaster Hazard: Dangerous. See chlorides.

TETRACHLOROPENTANE
General Information
Formula: $C_5H_8Cl_4$
Constant: Mol wt: 210.
Hazard Analysis
Toxicity: Unknown. Limited animal experiments suggest low oral toxicity.
Disaster Hazard: Dangerous. See chlorides.

TETRACHLOROPHTHALODINITRILE
Hazard Analysis
Toxicity: Unknown. Animal experiments show very high toxicity of the m-compound.
Disaster Hazard: Dangerous. See nitriles.

2,3,4,6-TETRACHLOROPHENOL
General Information
Description: Light brown mass; strong odor.
Formula: Cl_4C_6HOH.

Constants: Mol wt: 231.9, mp: 69–70°C, bp: 288°C (decomp.), vap. press.: 1 mm at 100.0°C.

Hazard Analysis
Toxic Hazard Rating:
Acute Local: Irritant 2.
Acute Systemic: U.
Chronic Local: Irritant 2.
Chronic Systemic: Ingestion 3; Inhalation 3.
Disaster Hazard: Dangerous, when heated to decomposition it emits highly toxic fumes.

Countermeasures
Ventilation Control: Section 2.
Personnel Protection: Section 3.
First Aid: Section 1.
Storage and Handling: Section 7.
Shipping Regulations: Section 11.
MCA warning label.

2,4,5,6-TETRACHLOROPHENOL
General Information
Description: Brown solid; phenol odor.
Formula: C_6HCl_4OH.
Constants: Mol wt: 231.92, mp: 50°C, d: 1.65 at 60°/4°C.

Hazard Analysis
Toxic Hazard Rating:
Acute Local: Irritant 2.
Acute Systemic: U.
Chronic Local: U.
Chronic Systemic: Ingestion 3; Inhalation 3.
Disaster Hazard: Dangerous; when heated to decomposition it emits highly toxic fumes.

Countermeasures
Ventilation Control: Section 2.
Personal Hygiene: Section 3.
First Aid: Section 1.
Storage and Handling: Section 7.
Shipping Regulations: Section 11.
MCA warning label.

TETRACHLOROQUINONE
General Information
Synonym: Chloranil; tetrachloro-p-benzo quinone.
Description: Yellow crystals. Insoluble in water.
Formula: $C_6O_2Cl_4$
Constants: Mol wt: 245.9, mp: 290°C.

Hazard Analysis
Toxicity: Details unknown. May be irritating to skin and mucous membrane. See quinones. A fungicide.
Disaster Hazard: Dangerous; when heated to decomposition, it emits highly toxic fumes of chlorides.

Countermeasures
Storage and Handling: Section 7.

TETRACHLORORESORCINOL
General Information
Description: Crystals.
Formula: $C_6(OH)_2Cl_4$.
Constant: Mol wt: 247.9.

Hazard Analysis
Toxicity: Probably high. A seed disinfectant. See resorcinol.
Disaster Hazard: Dangerous; when heated to decomposition, it emits highly toxic fumes of chlorides.

Countermeasures
Storage and Handling: Section 7.

TETRACHLOROSILANE
General Information
Description: Colorless fuming liquid.

Formula: $SiCl_4$.
Constants: Mol wt: 169.89, mp: −70°C, bp: 57.6°C, d: 1.50 at 25°/25°C.

Hazard Analysis
Toxic Hazard Rating:
Acute Local: Irritant 3; Ingestion 3; Inhalation 3.
Acute Systemic: U.
Chronic Local: U.
Chronic Systemic: U.
Disaster Hazard: Dangerous; when heated it emits highly toxic fumes; will react with water or steam to produce toxic and corrosive fumes.

Countermeasures
Ventilation Control: Section 2.
Personnel Protection: Section 3.
Storage and Handling: Section 7.

TETRACHLOROTETRAFLUOROPROPANE
General Information
Description: Liquid.
Formula: $C_3Cl_4F_4$.
Constant: Mol wt: 253.9.

Hazard Analysis
Toxicity: Unknown. Limited animal experiments indicate high toxicity.
Disaster Hazard: Dangerous; see fluorides and chlorides.

Countermeasures
Storage and Handling: Section 7.

1,2,3,4-TETRACHLORO-1,2,3,4-TETRAHYDRO-NAPHTHALENE. See tetrachloronaphthalene.

TETRACHLOROUNDECANE
General Information
Formula: $C_{11}H_{20}Cl_4$.
Constant: Mol wt: 294.

Hazard Analysis
Toxicity: Unknown. Limited animal experiments suggest low oral toxicity.
Disaster Hazard: Dangerous. See chlorides.

TETRACINE
General Information
Synonym: Guanylnitrosoaminoguanyltetrazene.
Description: Crystals.

Hazard Analysis
Toxicity: Unknown.
Fire Hazard: Dangerous. See nitrates and explosives, high.
Explosion Hazard: Severe, when shocked or exposed to heat. It is a high explosive which evolves much flame. It is used in priming compositions and sometimes in combinations with lead azide to lower the flash point of the azide.
Disaster Hazard: Highly dangerous! Shock will explode it; when heated to decomposition it emits highly toxic fumes of oxides of nitrogen and explodes.

Countermeasures
Storage and Handling: Section 7.

TETRACOBALT DODECACARBONYL. See cobalt carbonyl.

TETRACYANOETHYLENE
General Information
Synonym: Ethane tetracarbonitrile; TCNE.
Description: Colorless crystals; sublimes > 120°C.
Formuls: $(CN)_2C:C(CN)_2$.
Constants: Mol wt: 128, mp: 198–200°C, bp: 223°C.

TOXIC HAZARD RATING CODE (*For detailed discussion, see Section 1.*)

0 NONE: (a) No harm under any conditions; (b) Harmful only under unusual conditions or overwhelming dosage.

1 SLIGHT: Causes readily reversible changes which disappear after end of exposure.

2 MODERATE: May involve both irreversible and reversible changes; not severe enough to cause death or permanent injury.

3 HIGH: May cause death or permanent injury after very short exposure to small quantities.

U UNKNOWN: No information on humans considered valid by authors.

Hazard Analysis and Countermeasures
See nitriles.

TETRACYCLOHEXYL THIOGERMANIUM
General Information
Description: Crystals insoluble in water.
Formula: $Ge(SC_6H_{11})_4$.
Constants: Mol wt: 533.4, d: 1.259–1.270, mp: 84–88°C.
Hazard Analysis and Countermeasures
See germanium compounds, sulfur compounds and organo-
metals.

TETRACYCLOHEXYLTIN
General Information
Description: White crystals; insoluble in water.
Formula: $(C_6H_{11})_4Sn$.
Constants: Mol wt: 451.3, mp: 264°C.
Hazard Analysis
Toxicity: Details unknown. See tin compounds.

TETRADECANE
General Information
Description: Liquid.
Formula: $CH_3(CH_2)_{12}CH_3$.
Constants: Mol wt: 198.38, mp: 5.5°C, lel: 0.5%, bp:
    252.5°C, flash p: 212°F, d: 0.765, vap. press.: 1 mm
    at 76.4°C, vap. d.: 6.83, autoign. temp.: 396°F.
Hazard Analysis
Toxicity: Unknown. Probably irritant and narcotic in high
    concentrations. See also decane.
Fire Hazard: Moderate, when exposed to heat or flame.
Spontaneous Heating: No.
Explosion Hazard: Moderate in the form of vapor when
    exposed to flame.
Disaster Hazard: Moderately dangerous; when heated, it
    emits acrid fumes; can react with oxidizing materials.
Countermeasures
Storage and Handling: Section 7.
To Fight Fire: Foam, carbon dioxide, dry chemical or
    carbon tetrachloride (Section 6).

TETRADECANOL
General Information
Synonym: Tetradecyl alcohol.
Description: Opaque leaflets.
Formula: $C_{14}H_{29}OH$.
Constants: Mol wt: 214.38, mp: 37.62°C, bp: 264.1°C,
    flash p: 285°F (O.C.), d: 0.8355 at 20°/20°C, liquid:
    0.8236 at 38°/4°C, vap. press.: 0.01 mm at 20°C,
    vap. d.: 7.39.
Hazard Analysis
Toxicity: Details unknown. See alcohols.
Fire Hazard: Slight, when exposed to heat or flame; can react
    with oxidizing materials.
Spontaneous Heating: No.
Countermeasures
To Fight Fire: Foam, carbon dioxide, dry chemical or car-
    bon tetrachloride (Section 6).
Storage and Handling: Section 7.

1-TETRADECENE
General Information
Synonym: α-Tetradecylene.
Description: Colorless liquid.
Formula: $CH_2CH(CH_2)_{11}CH_3$.
Constants: Mol wt: 196.36, mp: −12°C, bp: 127°C at 15
    mm, d: 0.7737 at 20°/4°C, vap. d.: 6.78, flash p.:
    230°F.
Hazard Analysis
Toxicity: Unknown. Probably irritant and narcotic in high
    concentrations.
Fire Hazard: Moderate. Can react with oxidizing materials.
Countermeasures
To Fight Fire: Foam, carbon dioxide, dry chemical or car-
    bon tetrachloride (Section 6).
Storage and Handling: Section 7.

TETRADECYL ALCOHOL. See tetradecanol.

TETRADECYLAMINE
General Information
Description: Oil; amine odor.
Formula: $C_{14}H_{29}NH_2$.
Constants: Mol wt: 213, fp: 38.2°C, vap. press.: 32 mm at
    180°C.
Hazard Analysis
Toxicity: Details unknown. See fatty amines.

α-**TETRADECYLENE.** See 1-tetradecene.

TETRADYMITE. See bismuth tritelluride.

TETRAETHANOL AMMONIUM HYDROXIDE
General Information
Description: Crystals.
Formula: $(HOCH_2CH_2)_4NOH$.
Constants: Mol wt: 211.18, mp: 123°C, vap. press.: < 0.01
    mm at 20°C, vap. d.: 7.28.
Hazard Analysis
Toxicity: Unknown.
Disaster Hazard: Moderately dangerous; when heated to de-
    composition, it emits toxic fumes.
Countermeasures
Storage and Handling: Section 7.

TETRAETHOXYLGERMANIUM. See tetraethyl ger-
    manate.

TETRAETHOXYPROPANE
General Information
Description: Liquid.
Formula: $(C_2H_5O)_2CHCH_2CH(C_2H_5O)_2$.
Constants: Mol wt: 220.3, mp: −90°C, bp: 219.9°C,
    flash p: 190°F (O.C.), d: 0.9197 at 20°/20°C, vap. d.:
    7.58.
Hazard Analysis
Toxicity: Unknown. Limited animal experiments suggest
    low toxicity.
Fire Hazard: Moderate, when exposed to heat or flame; can
    react with oxidizing materials.
Countermeasures
To Fight Fire: Foam, carbon dioxide, dry chemical or car-
    bon tetrachloride (Section 6).
Storage and Handling: Section 7.

TETRAETHOXYSILANE
General Information
Description: Liquid.
Formula: $C_8H_{20}O_4Si$.
Constants: Mol wt: 208.22, bp: 165.5°C, d: 0.933 at
    25°/25°C, vap. press.: 1 mm at 16.0°C.
Hazard Analysis
Toxicity: Details unknown. Animal experiments suggest
    low toxicity by inhalation and other routes. See silanes.
Countermeasures
Storage and Handling: Section 7.

TETRA(2-ETHYLBUTYL)SILICATE
General Information
Description: Insoluble in water; slightly soluble in methanol;
    miscible with most organic solvents.
Formula: $[C_2H_5CH(C_2H_5)CH_2O]_4Si$.
Constants: Mol wt: 422, d: 0.8920–0.9018 at 20/20°C, mp:
    < −100°C, bp: 238°C at 50 mm.
Hazard Analysis
Toxic Hazard Rating: U.
Fire Hazard: Flammable when exposed to heat and open
    flame.
Countermeasures
Storage and Handling: Section 7.
To Fight Fire: Water or foam.

TETRAETHYL DIARSINE. See arsenic diethyl.

TETRAETHYLDIARSYL. See arsenic diethyl.

TETRAETHYL DITHIONOPYROPHOSPHATE
General Information
Synonym: TEDP.
Hazard Analysis
Toxicity: High. See tetraethyl pyrosphosphate.
TLV: ACGIH (recommended); Can be absorbed through the intact skin. 0.2 milligrams per cubic meter of air.
Disaster Hazard: Dangerous; when heated, it emits toxic fumes.
Countermeasures
Ventilation Control: Section 2.
Storage and Handling: Section 7.

TETRAETHYLDITHIOPYROPHOSPHATE, Liquid
Shipping Regulations: Section 11.
I.C.C.: Poison B, poison label, 1 quart.
IATA: Poison B, poison label, not acceptable (passenger), 1 liter (cargo).

TETRAETHYLDITHIOPYROPHOSPHATE AND COMPRESSED GAS MIXTURE
Shipping Regulations: Section 11.
I.C.C.: Poison A, poison gas label, not accepted.
IATA: Poison A, not acceptable (passenger and cargo).

TETRAETHYLDITHIOPYROPHOSPHATE MIXTURE, DRY
Shipping Regulations: Section 11.
I.C.C.: Poison B, poison label, 200 pounds.
IATA (containing more than 2% tetraethyl dithiopyrophosphate): Poison B, poison label, not acceptable (passenger), 95 kilograms (cargo).
(containing not more than 2% tetraethyl dithiopyrophosphate): Poison B, poison label, 25 kilograms (passenger), 95 kilograms (cargo).

TETRAETHYLDITHIOPYROPHOSPHATE MIXTURE, LIQUID
Shipping Regulations: Section 11.
I.C.C.: Poison B, poison label, 1 quart.
IATA (containing more than 25% tetraethyl dithiopyrophosphate): Poison B, poison label, not acceptable (passenger), 1 liter (cargo).
(containing not more than 25% tetraethyl dithiopyrophosphate): Poison B, poison label, 1 liter (passenger and cargo).

TETRAETHYLENE GLYCOL
General Information
Description: Colorless to pale straw-colored liquid.
Formula: $HO(C_2H_4O)_3C_2H_4OH$.
Constants: Mol wt: 194.22, bp: 327.3°C, fp: -6°C, flash p: 360°F (O.C.), d: 1.1248 at 20°/20°C, vap. press.: 1 mm at 153.9°C.
Hazard Analysis
Toxicity: See glycols.
Fire Hazard: Slight, when exposed to heat or flame; can react with oxidizing materials.
Spontaneous Heating: No.
Countermeasures
To Fight Fire: Alcohol foam, water, carbon dioxide, dry chemical or carbon tetrachloride (Section 6).
Storage and Handling: Section 7.

TETRAETHYLENE GLYCOL DIMETHYL ETHER.
See dimethoxytetraglycol.

TETRAETHYLENEPENTAMINE
General Information
Description: Viscous, hygroscopic liquid.
Formula: $NH_2(CH_2CH_2NH)_3CH_2CH_2NH_2$.
Constants: Mol wt: 189.30, bp: 333°C, flash p: 325°F (O.C.), d: 0.9980 at 20°/20°C, vap. press.: <0.01 mm at 20°C.
Hazard Analysis
Toxicity: Details unknown. See amines.
Fire Hazard: Slight, when exposed to heat or flame.
Disaster Hazard: Dangerous; when heated to decomposition, it emits toxic fumes of oxides of nitrogen; can react with oxidizing materials.
Countermeasures
Storage and Handling: Section 7.
To Fight Fire: Foam, carbon dioxide, dry chemical or carbon tetrachloride (Section 6).
Shipping Regulations: Section 11.
MCA warning label.

TETRAETHYL GERMANATE
General Information
Synonym: Tetraethoxyl germanium.
Description: Colorless liquid.
Formula: $Ge(OC_2H_5)_4$.
Constants: Mol wt: 252.8, mp: -81°C, bp: 186°C.
Hazard Analysis
Toxicity: Details unknown. See germanium compounds.

TETRAETHYLGERMANIUM
General Information
Synonym: Germanium tetraethyl.
Description: Colorless oil; decomposed by water.
Formula: $(C_2H_5)_4Ge$.
Constants: Mol wt: 188.8, d: 1.198 at 0°C, mp: -90°C, bp: 163°C.
Hazard Analysis
Toxicity: Details unknown. Animal experiments show stimulation of blood formation. See also germanium compounds.

TETRA-(2-ETHYLHEXYL)SILICATE
General Information
Description: Insoluble in water.
Formula: $[C_4H_9CH(C_2H_5)CH_2O]_4Si$.
Constants: Mol wt: 544, d: 0.8838; bp: 350–370°C, fp: -90°C, flash p: 390°F (O.C.).
Hazard Analysis
Toxic Hazard Rating: U.
Fire Hazard: Slight, when exposed to heat or flame.
Countermeasures
Storage and Handling: Section 7.
To Fight Fire: Water or foam.

TETRA-2-ETHYLHEXYL TITANATE
General Information
Description: Light yellow liquid.
Formula: $Ti(OC_8H_{17})_4$.
Constants: Mol wt: 564, mp: <-25°C, bp: 194°C at 0.25 mm, d: 1.0711 at 25°C, vap. d.: 19.5.
Hazard Analysis
Toxicity: Details unknown. See titanium compounds.
Fire Hazard: Moderate, when exposed to heat or flame; can react with oxidizing materials.

TOXIC HAZARD RATING CODE (For detailed discussion, see Section 1.)

0 NONE: (a) No harm under any conditions; (b) Harmful only under unusual conditions or overwhelming dosage.

1 SLIGHT: Causes readily reversible changes which disappear after end of exposure.

2 MODERATE: May involve both irreversible and reversible changes; not severe enough to cause death or permanent injury.

3 HIGH: May cause death or permanent injury after very short exposure to small quantities.

U UNKNOWN: No information on humans considered valid by authors.

Countermeasures

To Fight Fire: Water, foam, carbon dioxide, dry chemical or carbon tetrachloride (Section 6).

Storage and Handling: Section 7.

TETRAETHYL LEAD. See lead tetraethyl.

Shipping Regulations: Section 11.
  Coast Guard Classification: Poison B; poison label.
  I.C.C.: Poison B, poison label, 55 gallons.
  IATA: Poison B, poison label, not acceptable (passenger), 220 liters (cargo).

O,O,O′,O′-TETRAETHYL-S,S-METHYLENE DI-PHOSPHORODITHIOATE. See ethion.

TETRAETHYL PYROPHOSPHATE

General Information

Synonym: TEP; Bladex; fosvex; hesamite; nifos; TEPP; tetron; vaptone and others.

Description: Water-white to amber, hygroscopic liquid.

Formula: $(C_2H_5)_4P_2O_7$.

Constants: Mol wt: 290.20, d: 1.20.

Hazard Analysis

Toxic Hazard Rating:
  Acute Local: U.
  Acute Systemic: Ingestion 3; Inhalation 3; Skin Absorption 3.
  Chronic Local: U.
  Chronic Systemic: Ingestion 3; Inhalation 3; Skin Absorption 3.
Toxicology: The action is similar to that of parathion. Briefly, the action results in an irreversible inhibition of the cholinesterase molecules and the consequent accumulation of large amounts of acetylcholine. See also parathion.
TLV: ACGIH (recommended); Can be absorbed through the intact skin. 0.05 milligrams per cubic meter of air.
Dangerous Chronic Dose: Exposure to any organic phosphorous insecticide lowers the cholinesterase level and, until that enzyme has been completely regenerated, the exposed organism remains susceptible to relatively small doses of tetraethyl pyrophosphate. In other words, small doses at frequent intervals are largely additive (see parathion for further detail).
Signs and Symptoms of Poisoning: Findings are similar to those for parathion.
Treatment of Poisoning: Same as for parathion.
Disaster Hazard: Dangerous; when heated it emits highly toxic fumes.

Countermeasures

Ventilation Control: Section 2.
Personnel Protection: Section 3.
First Aid: Section 1.
Storage and Handling: Section 7.
Shipping Regulations: Section 11.
  Coast Guard Classification: Poison B; poison label.
  MCA warning label.
  I.C.C.: Poison B, poison label, 1 quart.
  IATA: Poison B, poison label, not acceptable (passenger), 220 liters (cargo).

TETRAETHYL PYROPHOSPHATE, COMPRESSED GAS MIXTURE

Shipping Regulations: Section 11.
  I.C.C.: Poison A; poison gas label, not accepted.
  Coast Guard Classification: Poison A; poison gas label.
  IATA: Poison A, not acceptable (passenger and cargo).

TETRAETHYLPYROPHOSPHATE, DRY MIXTURE

Shipping Regulations: Section 11.
  I.C.C.: Poison B, poison label, 200 pounds.
  IATA (containing more than 2% tetraethyl pyrophosphate): Poison B, poison label, not acceptable (passenger), 95 kilograms.

(containing not more than 2% tetraethyl pyrophosphate): Poison B, poison label, 25 kilograms (passenger), 95 kilograms (cargo).

TETRAETHYL PYROPHOSPHATE MIXTURE, LIQUID

Shipping Regulations: Section 11.
  I.C.C.: Poison B, poison label, 1 quart.
  IATA (containing more than 25% tetraethyl pyrophosphate): Poison B, poison label, not acceptable (passenger), 1 liter (cargo).
    (containing not more than 25% tetraethyl pyrophosphate): Poison B, poison label, 1 liter (passenger and cargo).

TETRAETHYL SILICANE

General Information

Description: Liquid.
Formula: $(C_2H_5)_4Si$.
Constants: Mol wt: 144.3, d: 0.762 at 25°C, bp: 153°C.

Hazard Analysis

Toxicity: Details unknown. See silanes.

TETRAETHYL-o-SILICATE. See ethyl silicate.

TETRAETHYL THIOGERMANIUM

General Information

Description: A liquid; soluble in benzene.
Formula: $Ge(SC_2H_5)_4$.
Constants: Mol wt: 317.1, d: 1.2574 at 25°C, bp: 165°C at 5 mm.

Hazard Analysis and Countermeasures

See germanium compounds and organometals.

TETRAETHYLTHIURAM DISULFIDE. See bis(diethylthiocarbamyl) disulfide.

TETRAETHYLTHIURAM SULFIDE

General Information

Description: Crystals.
Formula: $[(C_2H_5)_2NCS]_2S$.
Constants: Mol wt: 264, bp: 225–240°C at 3 mm, d: 1.12 at 20°/20°C.

Hazard Analysis

Toxic Hazard Rating:
  Acute Local: U.
  Acute Systemic: Ingestion 2; Inhalation 2.
  Chronic Local: U.
  Chronic Systemic: Ingestion 2; Inhalation 2.
Toxicity: See also bis(diethylthiocarbamyl) disulfide.
Disaster Hazard: Dangerous; when heated to decomposition, it emits highly toxic fumes of oxides of sulfur; can react vigorously with oxidizing materials.

Countermeasures

Ventilation Control: Section 2.
Personal Hygiene: Section 3.
Storage and Handling: Section 7.

TETRAETHYLTIN

General Information

Description: Colorless liquid; insoluble in water; soluble in organic solvents.
Formula: $(C_2H_5)_4Sn$.
Constants: Mol wt: 234.9, d: 1.187 at 23°C, mp: −112°C, bp: 181°C.

Hazard Analysis

Toxicity: Details unknown. See tin compounds.

TETRAFLUOROETHYLENE, INHIBITED

General Information

Synonym: Perfluoroethylene.
Description: Colorless gas.
Formula: CF_2CF_2.
Constants: Mol wt: 100.02, mp: −142.5°C, bp: −78.4°C.

Hazard Analysis
Disaster Hazard: Dangerous; when heated to decomposition, it emits highly toxic fumes of fluorides.
Toxicity: Can act as an asphyxiant and may have other toxic properties. See also compressed gases.
Countermeasures
Storage and Handling: Section 7.
Shipping Regulations: Section 11.
    I.C.C.: Nonflammable gas; green label, 100 pounds.
    Coast Guard Classification: Noninflammable gas; green gas label.
    IATA: Flammable gas, red label, not acceptable (passenger), 140 kilograms (cargo).

TETRAFLUOROETHYLENE,   **UNINHIBITED.** See tetrafluoroethylene, inhibited.
Shipping Regulations: Section 11.
    IATA: Flammable gas, not acceptable (passenger and cargo).

TETRAFLUOROHYDRAZINE
General Information
Description: Colorless gas; colorless liquid or white solid when pure.
Formula: N_2F_4
Constants: Mol wt: 104.0, mp: $-163°C$, bp: $-73°C$, d: liquid at $-100°C = 1.5$.
Hazard Analysis
Toxicity: Probably highly toxic. See hydrofluoric acid.
Fire Hazard: Highly reactive with reducing agents. (section 6).
Explosion Hazard: Can react explosively with reducing agents at normal temperatures. When ignited with hydrogen can explode. At high pressures it can explode due to shock or blast.
Disaster Hazard: Dangerous. When heated to decomposition, it emits highly toxic fumes. Heat, shock or blast can detonate it when under pressure. Can react explosively with reducing agents.
Countermeasures
Ventilation Control: Section 2.
Personnel Protection: Section 3.
Storage and Handling: Section 7.
Shipping Regulations: Section 11.
    I.C.C.: Compressed gases, N.O.S.; I.C.C. label–flammable gas.

TETRAFLUOROMETHANE. See carbon tetrafluoride.

TETRAFLUOROSILANE. See silicon tetrafluoride.

TETRA-n-HEPTYLTIN
General Information
Description: Liquid.
Formula: $(C_7H_{15})_4Sn$.
Constants: Mol wt: 515.5, d: 0.9748, bp: $239°C$ at 10 mm.
Hazard Analysis
Toxicity: Details unknown. See tin compounds.

TETRA-n-HEXYLTIN
General Information
Description: Liquid.
Formula: $(C_6H_{13})_4Sn$.
Constants: Mol wt: 459.4, d: 0.9959, bp: $209°C$ at 10 mm.
Hazard Analysis
Toxicity: Details unknown. See tin compounds.

1,2,3,6-TETRAHYDROBENZALDEHYDE
General Information
Description: Liquid.
Formula: $CH_2CHCH(CH_2)_2CHCHO$.
Constants: Mol wt: 110.15, mp: $-110°C$, bp: $164.5°C$, d: 0.9733 at $20°/20°C$, vap. press.: 1.6 mm at $20°C$, vap. d.: 3.80, flash p: $130°F$ (O.C.).
Hazard Analysis
Toxicity: See aldehydes.
Disaster Hazard: Slightly dangerous; when heated, it emits acrid fumes.
Fire Hazard: Moderate, when exposed to heat or flame.
Countermeasures
Storage and Handling: Section 7.
To Fight Fire: Alcohol foam.

1,2,3,4-TETRAHYDROBENZENE. See cyclohexane.

TETRAHYDRODIMETHYLFURAN
General Information
Description: Liquid.
Formula: $C_6H_{12}O$.
Constant: Mol wt: 100.2.
Hazard Analysis
Toxicity: Unknown. See tetrahydrofuran.
Fire Hazard: Moderate, when exposed to heat or flame; can react with oxidizing materials (Section 6).
Countermeasures
Storage and Handling: Section 7.

TETRAHYDROFURAN
General Information
Synonym: Cyclotetramethylene oxide.
Description: Colorless, mobile liquid; ether-like odor.
Formula: $OCH_2CH_2CH_2CH_2$.
Constants: Mol .wt: 72.10, bp: $65.4°C$, flash p: $1°F$ (T.C.C.), lel: 2.3%, uel: 11.8%, fp: $-108.5°C$, d: 0.888 at $20°/4°C$, vap. press.: 114 mm at $15°C$, vap. d.: 2.5, autoign. temp.: $610°F$.
Hazard Analysis
Toxic Hazard Rating:
    Acute Local: Irritant 3; Ingestion 3; Inhalation 3.
    Acute Systemic: U.
    Chronic Local: U.
    Chronic Systemic: U.
Toxicology: Irritating to eyes and mucous membrane. Narcotic in high concentrations. Reported as causing injury to liver and kidneys. It is a food additive resulting from contact with containers or equipment.
TLV: ACGIH (recommended); 200 parts per million in air; 590 milligrams per cubic meter of air.
Fire Hazard: Dangerous (Section 6).
Explosion Hazard: Moderate, by chemical reaction. In common with other ethers, unstabilized tetrahydrofuran forms thermally explosive peroxides on exposure to air. It must always be tested for peroxides prior to distillation. Peroxides can be removed by treatment with strong ferrous sulfate solution made slightly acidic with sodium bisulfate.
Disaster Hazard: Dangerous; when heated to decomposition it emits toxic fumes; can react with oxidizing materials.
Countermeasures
Ventilation Control: Section 2.
To Fight Fire: Foam, dry chemical, carbon dioxide, or carbon tetrachloride (Section 6).
Personnel Protection: Section 3.

TOXIC HAZARD RATING CODE (For detailed discussion, see Section 1.)

0 NONE: (a) No harm under any conditions; (b) Harmful only under unusual conditions or overwhelming dosage.

1 SLIGHT: Causes readily reversible changes which disappear after end of exposure.

2 MODERATE: May involve both irreversible and reversible changes; not severe enough to cause death or permanent injury.

3 HIGH: May cause death or permanent injury after very short exposure to small quantities.

U UNKNOWN: No information on humans considered valid by authors.

First Aid: Section 1.
Storage and Handling: Section 7.

2,5-TETRAHYDROFURAN DIMETHANOL
General Information
Synonym: THF-glycol.
Description: Hygroscopic liquid. Faint odor. Water soluble.
Formula: $C_6H_{12}O_3$.
Constants: Mol wt: 132.2, d: 1.1719 at $0°/4°C$, mp: $< -50°C$, bp: $265°C$.
Hazard Analysis
Toxicity: Details unknown. A strong irritant. Avoid contact with eyes. See also tetrahydrofuran.

TETRAHYDROFURFURYL ALCOHOL
General Information
Synonym: THFA.
Description: A hygroscopic liquid. Water soluble.
Formula: $C_4H_7OCH_2OH$.
Constants: Mol wt: 102.13, mp: $< -80°C$, lel: 1.5%, uel: 9.7% at 72° to 122°F, bp: 178°C at 743 mm, flash p: 167°F (O.C.), d: 1.0495 at $20°/4°C$, autoign. temp.: 540°F, vap. d.: 3.5.
Hazard Analysis
Toxicity: Details unknown. An irritant material. See tetrahydrofuran and alcohols.
Fire Hazard: Moderate, when exposed to heat or flame; can react with oxidizing materials.
Explosion Hazard: Moderate, in the form of vapor when exposed to flame.
Countermeasures
Storage and Handling: Section 7.
To Fight Fire: Alcohol foam, water, carbon dioxide, dry chemical or carbon tetrachloride (Section 6).

TETRAHYDROFURFURYL OLEATE
General Information
Description: A liquid.
Formula: $C_{23}H_{12}O_3$.
Constants: Mol wt: 366.6, bp: 240°C at 5 mm, flash p: 392°F, d: 0.93, vap. d.: 12.65.
Hazard Analysis
Toxicity: Details unknown. See tetrahydrofuran and esters.
Fire Hazard: Slight, when exposed to heat or flame; can react with oxidizing materials.
Countermeasures
To Fight Fire: Foam, carbon dioxide, dry chemical or carbon tetrachloride (Section 6).
Storage and Handling: Section 7.

1,2,5,6-TETRAHYDRO-o-METHYLBENZOIC ACID
General Information
Description: Crystals.
Formula: $C_8H_{13}O_2$.
Constant: Mol wt: 141.2.
Hazard Analysis
Toxicity: Details unknown; a mite repellent. See also methyl benzoate.
Disaster Hazard: Slightly dangerous; when heated, it emits acrid fumes.
Countermeasures
Storage and Handling: Section 7.

TETRAHYDRONAPHTHALENE
General Information
Description: Colorless liquid.
Synonym: Tetralin.
Formula: $C_{10}H_{12}$.
Constants: Mol wt: 132.20, mp: $-30°C$, bp: 207.2°C, flash p: 171°F (C.C.), d: 0.981, vap. press.: 1 mm at 38.0°C, vap. d.: 4.55, autoign temp.: 723°F, lel: 0.8% at 212°F, uel: 5.0% at 302°F.
Hazard Analysis
Toxic Hazard Rating:
Acute Local: Irritant 2; Ingestion 2; Inhalation 2.

Acute Systemic: Ingestion 2; Inhalation 2.
Chronic Local: U.
Chronic Systemic: Ingestion 2; Inhalation 2.
Toxicology: An irritant. Narcotic in high concentrations. Reported as causing cataracts and kidney injury in experimental animals.
Fire Hazard: Moderate, when exposed to heat or flame; can react with oxidizing materials.
Spontaneous Heating: No.
Countermeasures
To Fight Fire: Water, foam, carbon dioxide, dry chemical or carbon tetrachloride (Section 6).
Storage and Handling: Section 7.

TETRAHYDRONAPHTHOL
General Information
Synonym: α-Naphthol-1,2,3,4-tetrahydride.
Description: Colorless liquid.
Formula: $C_6H_1C_4H_7OH$.
Constants: Mol wt: 148.2, bp: 140°C at 17 mm, d: 1.090.
Hazard Analysis
Toxicity: Unknown. Limited animal experiments show moderate toxicity.
Fire Hazard: Moderate, when exposed to heat or flame; can react with oxidizing materials (Section 6).
Countermeasures
Storage and Handling: Section 7.

TETRAHYDRO-1,4-OXAZINE. See morpholine.

TETRAHYDROPHTHALIC ANHYDRIDE
General Information
Description: White powder.
Formula: $C_8H_8O_3$.
Constants: Mol wt: 152.14, mp: 101.9°C, bp: 195°C at 50 mm, flash p: 315°F (O.C.), d: 1.375 at $25°/20°C$, vap. press.: < 0.01 mm at 20°C, vap. d.: 5.25.
Hazard Analysis
Toxicity: Unknown. See also phthalic anhydride.
Fire Hazard: Slight, when exposed to heat or flame.
Disaster Hazard: Slightly dangerous; when heated, it emits acrid fumes; will react with water or steam to produce heat; can react with oxidizing materials.
Countermeasures
Storage and Handling: Section 7.
To Fight Fire: Water, foam, carbon dioxide, dry chemical or carbon tetrachloride (Section 6).

TETRAHYDROPYRAN
General Information
Synonyms: Pentamethylene oxide.
Description: Colorless, mobile liquid, ether-like odor.
Formula: $(CH_2)_5O$.
Constants: Mol wt: 86.13, bp: 88°C, flash p: $-4°F$, d: 0.8814 at $20°/4°C$, vap. d: 4.0.
Hazard Analysis
Toxic Hazard Rating:
Acute Local: Irritant 2; Ingestion 2; Inhalation 2.
Acute Systemic: U.
Chronic Local: U.
Chronic Systemic: U.
Fire-Hazard: Dangerous, when exposed to heat or flame (Section 6).
Explosion Hazard: Moderate, by chemical reaction. It can form explosive peroxides if stored in uninhibited condition.
Disaster Hazard: Dangerous, upon exposure to heat or flame; can react vigorously with oxidizing materials.
Countermeasures
Ventilation Control: Section 2.
Personnel Protection: Section 3.
Storage and Handling: Section 7.

TETRAHYDROPYRAN-2-METHANOL
General Information
Description: Liquid.

Formula: $OCH_2CH_2CH_2CH_2CHCH_2OH$.
Constants: Mol wt: 116.16, fp: $-70°C$, bp: $187°C$, d: 1.0272 at $20°/20°C$, vap. d.: 4.02, vap. press.: 0.4 mm at $20°C$, flash p.: $200°F$ (O.C.).

Hazard Analysis
Toxicity: Unknown.
Fire Hazard: Moderate, when exposed to heat or flame; can react with oxidizing materials (Section 6).

Countermeasures
Storage and Handling: Section 7.
To Fight Fire: Alcohol foam.

1,2,5,6-TETRAHYDROPYRIDINE

General Information
Description: Clear liquid.
Formula: C_5H_9N.
Constants: Mol wt: 83.1, bp: $115°-120°C$, d: 0.913 at $20°/4°C$, flash p: $61°F$.

Hazard Analysis
Toxicity: Details unknown. See also pyridine.
Fire Hazard: Moderate when exposed to heat, sparks or flame.

Countermeasures
To Fight Fire: Foam, spray, dry chemical (Section 6).
Storage and Handling: Section 7.

TETRAHYDROSYLVAN. See 2-methyltetrahydrofuran.

TETRAHYDROURUSHIOL. See 3-pentadecyl catechol.

TETRAIODODISILANE ETHYLENE. See disilicon tetraiodide.

TETRAIODOMETHANE. See carbon tetraiodide.

TETRAIODOSILANE. See silicon tetraiodide.

TETRAISOAMYLGERMANIUM

General Information
Description: Colorless oily liquid.
Formula: $(C_5H_{11})_4Ge$.
Constants: Mol wt: 357.2, d: 0.9147 at $20°/20°C$, bp: $164°C$.

Hazard Analysis
Toxicity: Details unknown. See germanium compounds.

TETRAISOAMYLTIN

General Information
Description: Liquid.
Formula: $(C_5H_{11})_4Sn$.
Constants: Mol wt: 403.3, d: 1.035, bp: $188°C$ at 24 mm.

Hazard Analysis
Toxicity: Details unknown. See tin compounds.

TETRAISOBUTYLLEAD

General Information
Description: Crystals.
Formula: $Pb[CH_2CH(CH_3)_2]_4$.
Constants: Mol wt: 435.7, d: 1.324, mp: $23°C$.

Hazard Analysis and Countermeasures
See lead compounds.

TETRAISOBUTYL THIOGERMANIUM

General Information
Description: A liquid, soluble in alcohol.
Formula: $Ge[SCH_2CH(CH_3)_2]_4$.
Constants: Mol wt: 429.3, d: 1.0984 at $25°C$, bp: $200°C$ at 5 mm.

Hazard Analysis and Countermeasures
See organometals and germanium compounds.

TETRAISOBUTYLTIN

General Information
Description: Colorless liquid; insoluble in water; soluble in organic solvents.
Formula: $(C_4H_9)_4Sn$.
Constants: Mol wt: 347.2, d: 1.054 at $23°C$, mp: $-13°C$, bp: $267°C$.

Hazard Analysis
Toxicity: Details unknown. See tin compounds.

TETRAISOPROPYLLEAD

General Information
Description: Colorless liquid, insoluble in water; decomposes in air.
Formula: $Pb[CH(CH_3)_2]_4$.
Constants: Mol wt: 379.6, d: 1.4504, mp: $-53.5°C$, bp: $120°C$ at 14 mm.

Hazard Analysis and Countermeasures
See lead compounds.

TETRAISOPROPYL THIOGERMANIUM

General Information
Description: Liquid. Soluble in absolute alcohol.
Formula: $Ge[SCH(CH_3)_2]_4$.
Constants: Mol wt: 373.2, d: 1.478 at $25°C$, mp: $15°C$, bp: $163°C$ at 4 mm.

Hazard Analysis and Countermeasures
See germanium compounds and organometals.

TETRAISOPROPYL TITANATE

General Information
Description: Colorlesss liquid.
Formula: $(C_3H_7O)_4Ti$.
Constants: Mol wt: 284, mp: $20°C$, bp: $232°C$, d: 0.955 at $25°C$, vap. d.: 9.8.

Hazard Analysis
Toxicity: Details unknown. See esters and titanium compounds.
Fire Hazard: Moderate, when exposed to heat or flame; can react with oxidizing materials (Section 6).

Countermeasures
Storage and Handling: Section 6.

TETRALIN. See tetrahydronaphthalene.

TETRALITE. See tetryl.

TETRAM

General Information
Synonym: O,O-Diethyl-3-(β-diethylamino)-ethyl phosphorothiolate hydrogen oxalate.
Description: Liquid.
Formula: $(C_2H_5O)_2POSCH_3CH_2N(C_2H_5)_2$.
Constants: Mol wt: 269.4, bp: $110°C$ at 0.2 mm.

Hazard Analysis
Toxicity: Details unknown. An organic phosphate insecticide, hence a cholinesterase inhibitor. See parathion.
Disaster Hazard: Dangerous; when heated to decomposition, it emits highly toxic fumes.

TETRAMETHOXYDIBORANE

General Information
Description: Colorless liquid; decomposed by water.
Formula: $(CH_3O)_4B_2$.
Constants: Mol wt: 145.8, mp: $-24°C$, bp: $21°C$ at 44 mm.

Hazard Analysis
Toxicity: Details unknown. See also boron compounds and boron hydrides.

TOXIC HAZARD RATING CODE (For detailed discussion, see Section 1.)

0 NONE: (a) No harm under any conditions; (b) Harmful only under unusual conditions or overwhelming dosage.

1 SLIGHT: Causes readily reversible changes which disappear after end of exposure.

2 MODERATE: May involve both irreversible and reversible changes; not severe enough to cause death or permanent injury.

3 HIGH: May cause death or permanent injury after very short exposure to small quantities.

U UNKNOWN: No information on humans considered valid by authors.

TETRAMETHYL AMMONIUM HYDROXIDE
General Information
Description: A liquid.
Formula: $(CH_3)_4NOH$.
Constant: Mol wt: 91.
Hazard Analysis
Toxic Hazard Rating: Toxic. A powerful caustic.
Countermeasures
Shipping Regulations: Section 11.
　　IATA: Corrosive liquid, white label, 1 liter (passenger), 40 liters (cargo).

2,2,4,4-TETRAMETHYL-1,3-CYCLOBUTANEDIOL
General Information
Description: White crystalline solid. Slightly soluble in water; soluble in methanol.
Formula: $(CH_3)_2C(CHOH)_2C(CH_3)_2$.
Constants: Mol wt: 144.2, mp: 125°–135°C, bp: 222°C, vap. press.: 144 mm at 171°C, flash p: 125°–135°F (C.O.C.).
Hazard Analysis
Toxicology: Considerably more toxic than cyclohexanol. Has a convulsant action on experimental animals. Causes moderate skin irritation. A moderately toxic material.

2,2,4,4-TETRAMETHYL-1,3-CYCLOBUTANEDIONE
General Information
Description: White crystalline solid, insoluble in water, soluble in alcohol and acetic acid.
Formula: $(CH_3)_2C(CO)_2C(CH_3)_2$.
Constants: Mol wt: 140.2, mp: 116°C (sublimes), bp: 159°C, vap. press.: 6 mm at 52°C, d: 1.11.
Hazard Analysis
Toxicity: A slight skin irritant. Slightly toxic.

TETRAMETHYL DIARSINE. See cacodyl.

TETRAMETHYL DIARSYL. See cacodyl.

TETRAMETHYLDIBORANE
General Information
Description: Colorless liquid decomposed by water.
Formula: $B_2H_2(CH_3)_4$.
Constants: Mol wt: 83.8, mp: −72.5°C, bp: 68.6°C.
Hazard Analysis and Countermeasures
See boron compounds and boron hydrides.

TETRAMETHYL DIPROPYLENETRIAMINE
General Information
Formula: $H_2NCHCH_3CHCH_3CH_2NHCHCH_3 \cdot CHCH_3CH_2NH_2$.
Constant: Mol wt: 189.
Hazard Analysis
Toxicity: Details unknown. Limited animal experiments suggest high toxicity and irritation. See also amines.

TETRAMETHYLENE. See cyclobutane.

TETRAMETHYLENE CYANIDE. See adiponitrile.

TETRAMETHYL ETHYLENEDIAMINE
General Information
Formula: $(CH_3)_2NH_2CCNH_2(CH_3)_2$.
Constant: Mol wt: 116.2.
Hazard Analysis
Toxicity: Details unknown. See also amines.
Fire Hazard: Moderate, when exposed to heat or flame; can react with oxidizing materials (Section 6).
Countermeasures
Storage and Handling: Section 7.

TETRAMETHYLGERMANIUM
General Information
Description: Colorless liquid, soluble in organic solvents.
Formula: $Ge(CH_3)_4$.

Constants: Mol wt: 132.7, d: 1.006 at 0°C, mp: −88°C, bp: 43.4°C.
Hazard Analysis
Toxicity: Details unknown. See germanium compounds.
Fire Hazard: Probably dangerous; details unknown. Section 6.
Countermeasures
Storage and Handling: Section 7.

TETRAMETHYLLEAD. See lead tetramethyl.

TETRAMETHYLMETHYLENE DIAMINE
General Information
Description: Liquid.
Formula: $(CH_3)_2NCH_2N(CH_3)_2$.
Countermeasures
Shipping Regulations: Section 11.
　　IATA: Other restricted articles, class A, no label required, no limit (passenger and cargo).

2,2,3,3-TETRAMETHYL PENTANE
General Information
Formula: C_9H_{20}.
Constants: Mol wt: 128, autoign temp.: 806°F, lel: 0.8%, uel: 4.9%, d: 0.7, vap. d.: 4.4, bp: 273°F.
Hazard Analysis
Toxic Hazard Rating: U.
Explosion Hazard: A moderately dangerous fire and explosion hazard.

TETRAMETHYL-p-PHENYLENEDIAMINE
General Information
Synonym: Wurster's reagent.
Description: Leaflets'; slightly soluble in cold water; more soluble in hot water; freely soluble in alcohol, chloroform, ether, petroleum ether.
Formula: $(CH_3)_2NC_6H_4N(CH_3)_2$.
Constants: Mol wt: 164.24, mp: 51°C, bp: 260°C.
Hazard Analysis
See p-phenylenediamine.

TETRAMETHYLPHOSPHORODIAMINE FLUORIDE
General Information
Synonym: Dimefox.
Description: Liquid with fishy odor. Water soluble.
Formula: $[(CH_3)_2N]_2POF$.
Constants: Mol wt: 154.1, d: 1.1151 at 20°/4°C, bp: 67°C at 4 mm.
Hazard Analysis
Toxicity: Highly toxic. A cholinesterase inhibitor. See parathion.
Disaster Hazard: Dangerous; when heated to decomposition, it emits highly toxic fumes.
Shipping Regulations: Section 11.
　　IATA: Poison B, poison label, 1 liter (passenger), 220 liters (cargo).

TETRAMETHYLSILICANE
General Information
Description: Colorless liquid, soluble in ether.
Formula: $(CH_3)_4Si$.
Constants: Mol wt: 88.2, d: 0.651 at 15°C, bp: 26.5°C.
Hazard Analysis
Toxicity: Details unknown. See silanes.

TETRAMETHYLSTANNANE. See tin tetramethyl.

TETRAMETHYLSUCCINONITRILE
General Information
Synonym: TSN.
Description: Crystallizes in plates; almost no odor.
Formula: $C_8H_{12}N_2$.
Constants: Mol wt: 136.0, mp: 169°C (sublimes).
Hazard Analysis
TLV: ACGIH (recommended); 0.5 parts per million, 3

milligrams per cubic meter of air. May be absorbed via the intact skin.

Toxicity: In the preparation of sponge rubber, an azo compound is used, which decomposes to form tetramethylsuccinonitrile or TSN. Animal experiments indicate that it is toxic. Rats exposed to a concentration of 90 ppm exhibit their first convulsion after 1.5 to 2 hours or less. Rats exposed to concentrations of 5.5 ppm exhibited their first convulsion in 27 to 31 hours and were dead in from 31 to 46 hours.

Disaster Hazard: Dangerous. See nitriles.

Countermeasures

Ventilation Control: Section 2.

Treatment and Antidotes: It has been pointed out that this nitrile is different from other nitriles in that thiosulfate proved to be a poor antidote for intoxication. Barbiturates proved adequate for the control of convulsions. However, the barbiturates which have a short or medium period of action will relieve the condition only for a time. Afterwards, the symptoms may reappear, and eventually the death of the animal may ensue. This indicates that TSN is slowly detoxified by the body. The fatal dose is thought to be about 25 mg/kg of body weight. See also cyanides.

Personnel Protection: Section 3.

Storage and Handling: Section 7.

TETRAMETHYL THIOGERMANIUM
General Information
Description: A liquid; soluble in alcohol and benzene.
Formula: $Ge(SCH_3)_4$.
Constants: Mol wt: 261.0, d: 1.4364 at 25°C, mp: −3°C, bp: 140°C at 4 mm.
Hazard Analysis and Countermeasures
See germanium compounds and organometals.

TETRAMETHYLTHIURAM DISULFIDE. See bis(dimethyl thiocarbamyl) disulfide.

TETRAMETHYLTHIURAM MONOSULFIDE
General Information
Synonym: Bis(dimethylthiocarbamyl) sulfide.
Description: Yellow powder.
Formula: $[(CH_3)_2NCS]_2S$.
Constant: Mol wt: 208.39.
Hazard Analysis
Toxic Hazard Rating:
  Acute Local: U.
  Acute Systemic: Ingestion 2; Inhalation 2.
  Chronic Local: U.
  Chronic Systemic: Ingestion 2; Inhalation 2.
Toxicity: See also bis(dimethyl thiocarbamyl)disulfide.
Disaster Hazard: Dangerous; when heated to decomposition, it emits highly toxic fumes of oxides of sulfur.
Countermeasures
Ventilation Control: Section 2.
Personal Hygiene: Section 3.
Storage and Handling: Section 7.

TETRAMETHYLTRIBORINE TRIAMINE
General Information
Description: Colorless liquid, hydrolyzed by water.
Formula: $(CH_3)_3B_3N_3H_3$.
Constants: Mol wt: 136.6, bp: 158°C.
Hazard Analysis
Toxicity: Details unknown. See also boron compounds and amines.

TETRAMETHYLUREA
General Information
Formula: $(CH_3)_4CO(NH_2)_2$.
Constants: Mol wt: 106.
Hazard Analysis
Toxicity: Details unknown. Animal experiments show moderately low toxicity.

TETRAMMINE CADMIUM PERRHENATE
General Information
Description: Crystals.
Formula: $[Cd(NH_3)_4](ReO_4)_2$.
Constants: Mol wt: 681.16, d: 3.714 at 25°/4°C.
Hazard Analysis
Toxicity: Details unknown. See also cadmium compounds.
Fire Hazard: Slight, by chemical reaction with reducing agents (Section 6).
Countermeasures
Storage and Handling: Section 7.

TETRAMMINECOPPER II SULFATE
General Information
Description: Rhombic, blue crystals.
Formula: $[Cu(NH_3)_4]SO_4 \cdot H_2O$.
Constants: Mol wt: 245.75, mp: decomposes at 150°C, d: 1.81.
Hazard Analysis and Countermeasures
See copper compounds and sulfates.

TETRAMMINE PLATINUM(II) CHLORIDE
General Information
Description: Tetragonal, colorless crystals.
Formula: $[Pt(NH_3)_4]Cl_2 \cdot H_2O$.
Constants: Mol wt: 352.29, mp: 250°C; −H_2O at 100°C, d: 2.737.
Hazard Analysis and Countermeasures
See platinum compounds and chlorides.

TETRAMMINE PLATINUM (II) CHLORO PLATINATE
General Information
Synonym: Magnus' salt.
Description: Green or red crystals.
Formula: $[Pt(NH_3)_4]Pt Cl_4$.
Constants: Mol wt: 600.42, mp: decomposes, d: <4.1.
Hazard Analysis and Countermeasures
See platinum compounds and chlorides.

TETRANAPHTHENE
General Information
Description: Crystals.
Formula: $C_6H_3C_4H_7C_2H_4$.
Constant: Mol wt: 158.2.
Hazard Analysis
Toxic Hazard Rating:
  Acute Local: Irritant 1; Ingestion 1; Inhalation 1.
  Acute Systemic: U.
  Chronic Local: U.
  Chronic Systemic: U.
Fire Hazard: Slight, when exposed to heat or flame; can react with oxidizing materials (Section 6).
Countermeasures
Personal Hygiene: Section 3.
Storage and Handling: Section 7.

TETRANAPHTHENOYL TRIETHYLENE TETRAMINE
General Information
Description: Liquid.
Constants: Flash p: 325°F, d: 1.01.
Hazard Analysis
Toxicity: Unknown. See also amines.

TOXIC HAZARD RATING CODE *(For detailed discussion, see Section 1.)*

0 NONE: (a) No harm under any conditions; (b) Harmful only under unusual conditions or overwhelming dosage.

1 SLIGHT: Causes readily reversible changes which disappear after end of exposure.

2 MODERATE: May involve both irreversible and reversible changes; not severe enough to cause death or permanent injury.

3 HIGH: May cause death or permanent injury after very short exposure to small quantities.

U UNKNOWN: No information on humans considered valid by authors.

Fire Hazard: Slight, when exposed to heat or flame; can react with oxidizing materials.

Countermeasures

To Fight Fire: Foam, carbon dioxide, dry chemical or carbon tetrachloride (Section 6).

Storage and Handling: Section 7.

TETRANITROANILINE
General Information
Synonym: TNA.
Description: Solid.
Formula: $C_6H_3N_5O_8$.
Constants: Mol wt: 273.12, mp: 170°C, bp: explodes at 237°C.

Hazard Analysis

Toxicity: Highly toxic. See aniline.

Fire Hazard: See nitrates.

Explosion Hazard: Severe, when shocked or exposed to heat. Tetranitroaniline is a powerful and sensitive high explosive, similar to tetryl. It deteriorates in the presence of moisture. It is used as a booster for high explosive shells and in primer and detonating compositions. See also explosives, high and nitrates.

Disaster Hazard: Dangerous; shock or heat will explode it; when heated to decomposition, it emits highly toxic fumes of oxides of nitrogen; can react vigorously with reducing materials.

Countermeasures

Storage and Handling: Section 7.

TETRANITROCARBAZOLE
General Information
Description: Crystals.
Formula: $(NO_2)_4(C_6H_2)_2NH$.
Constant: Mol wt: 347.6.

Hazard Analysis

Toxicity: Details unknown; probably toxic; an insecticide.

Fire Hazard: See nitrates.

Explosion Hazard: See nitrates and explosives, high.

Disaster Hazard: Dangerous; shock will explode it; when heated to decomposition, it emits highly toxic fumes; can react vigorously with oxidizing materials.

Countermeasures

Storage and Handling: Section 7.

TETRANITRODIGLYCERIN
Hazard Analysis

Toxicity: See nitroglycerin.

Fire Hazard: See nitrates.

Explosion Hazard: Severe; when shocked or exposed to heat. Tetranitrodiglycerin resembles nitroglycerin, but is less sensitive. It is a component of low-freezing dynamites due to its own low-freezing point.

Disaster Hazard: Dangerous; shock will explode it; when heated to decomposition, it emits highly toxic fumes of oxides of nitrogen; can react vigorously with oxidizing materials.

Countermeasures

Storage and Handling: Section 7.

TETRANITROMETHANE
General Information
Description: Colorless liquid.
Formula: $C(NO_2)_4$.
Constants: Mol wt: 196.04, mp: 13°C, bp: 125.7°C, d: 1.650 at 13°C, vap. press.: 10 mm at 22.7°C.

Hazard Analysis

Toxic Hazard Rating:

Acute Local: Irritant 3; Ingestion 3; Inhalation 3.

Acute Systemic: Ingestion 3; Inhalation 3; Skin Absorption 3.

Chronic Local: U.

Chronic Systemic: Ingestion 3; Inhalation 3; Skin Absorption 3.

TLV: ACGIH (recommended); 1 part per million in air, 8 milligrams per cubic meter of air.

Toxicology: This material irritates the eyes and respiratory passages and does serious damage to the liver. It occurs as an impurity in crude TNT, and is thought to be mainly responsible for the irritating properties of that material. It can cause pulmonary edema, mild methemoglobinemea, and fatty degeneration of the liver and kidneys. From animal experiments it has been found that concentrations as low as 0.1 ppm have proved rapidly fatal, and that concentrations of 3.3 to 25.2 ppm produced very rapid and marked irritation of mucous membranes of the eyes, mouth, and upper respiratory tract.

Fire Hazard: Dangerous. See nitrates and explosives, high.

Explosion Hazard: Severe, when shocked or exposed to heat. It can form very powerful explosives when mixed with other nitro high explosives which are somewhat oxygen-deficient. Its primary use is in blasting explosives and in detonating compositions.

Disaster Hazard: Highly dangerous; shock will explode it; when heated to decomposition, it emits highly toxic fumes of oxides of nitrogen; can react vigorously with oxidizing materials.

Countermeasures

Ventilation Control: Section 2.

Personnel Protection : Section 3.

First Aid: Section 1

Storage and Handling: Section 7.

Shipping Regulations: Section 11.

Coast Guard Classification: Oxidizing material; yellow label.

ICC: Oxidizing material, yellow label, 25 pounds.

IATA: Oxidizing material, not acceptable (passenger and cargo).

TETRANITRONAPHTHALENE
General Information
Description: Crystals.
Formula: $C_{10}H_4(NO_2)_4$.
Constants: Mol wt: 308.2, mp: 200°C (approx.), bp: explodes.

Hazard Analysis

Toxicity: Unknown.

Fire Hazard: Dangerous. See nitrates.

Explosion Hazard: Severe, when shocked or exposed to heat. Tetranitronaphthalene is a much used high explosive equal to but somewhat less sensitive to impact than TNT. It is used for bursting charges. See also nitrates and explosives, high.

Disaster Hazard: Dangerous; shock or heat will explode it; when heated to decomposition, it emits highly toxic fumes of oxides of nitrogen; can react vigorously with reducing materials.

Countermeasures

Storage and Handling: Section 7.

TETRA-n-OCTYLTIN
General Information
Description: Liquid.
Formula: $(C_8H_{17})_4Sn$.
Constants: Mol wt: 571.6, d: 0.9605, bp: 268°C at 10 mm.

Hazard Analysis

Toxicity: Details unknown. See tin compounds.

TETRAPHENOXYGERMANIUM
General Information
Description: Colorless oil; soluble in benzene.
Formula: $Ge(OC_6H_5)_4$.
Constants: Mol wt: 445.0, bp: 210°–220°C at 0.3 mm.

Hazard Analysis

Toxicity: Details unknown. See germanium compounds.

TETRAPHENYL ARSONIUM BROMIDE
General Information
Description: Crystals.

Formula: $(C_6H_5)_4AsBr \cdot 2H_2O$.
Constants: Mol wt: 499.3, mp: 282°C.
Hazard Analysis and Countermeasures
See arsenic compounds and bromides.

TETRAPHENYL ARSONIUM CHLORIDE
General Information
Description: Crystals.
Formula: $(C_6H_5)_4AsCl \cdot 2H_2O$.
Constants: Mol wt: 454.8, mp: 259°C.
Hazard Analysis and Countermeasures
See arsenic compounds and chlorides.

TETRA(2-PHENYLETHYL) GERMANIUM
General Information
Description: Colorless crystals. Soluble in ether.
Formula: $Ge(C_6H_5C_2H_4)_4$.
Constants: Mol wt: 493.2, mp: 57°C.
Hazard Analysis
Toxicity: Details unknown. See germanium compounds.

TETRAPHENYL GERMANIUM
General Information
Description: Crystals; insoluble in water, soluble in organic solvents.
Formuls: $Ge(C_6H_5)_4$.
Constants: Mol wt: 381.0, mp: 235.7°C, bp: >400°C.
Hazard Analysis and Countermeasures
See germanium compounds and organometals.

TETRA PHENYLLEAD
General Information
Description: White crystals. Soluble in benzene.
Formula: $Pb(C_6H_5)_4$.
Constants: Mol wt: 515.6, mp: 228°C.
Hazard Analysis and Countermeasures
See lead and lead compounds.

TETRAPHENYLSILICANE
General Information
Description: Colorless solid; soluble in acetic anhydride.
Formula: $(C_6H_5)_4Si$.
Constant: Mol wt: 336.5.
Hazard Analysis
Toxicity: Details unknown. See silanes.

TETRAPHENYL THIOGERMANIUM
General Information
Description: Colorless crystals; soluble in organic solvents.
Formula: $Ge(SC_6H_5)_4$.
Constants: Mol wt: 509.2, mp: 101.5°C.
Hazard Analysis and Countermeasures
See organometals and germanium compounds.

TETRAPHOSPHORUS HEPTASULFIDE
General Information
Description: Light yellow crystals.
Formula: P_4S_7.
Constants: Mol wt: 348.39, mp: 310°C, bp: 523°C, d: 2.19 at 17°C.
Hazard Analysis
Toxicity: See sulfides.
Fire Hazard: Moderate, when exposed to heat or flame or by chemical reaction. See also sulfides.
Explosion Hazard: Moderate. See sulfides.
Disaster Hazard: Dangerous; when heated to decomposition, it emits highly toxic fumes of oxides of sulfur and phosphorus; will react with water, steam or acids to produce toxic and flammable vapors; can react with oxidizing materials.
Countermeasures
Ventilation Control: Section 2.
Personnel Protection: Section 3.
Storage and Handling: Section 7.

TETRAPHOSPHORUS HEXASULFIDE. See phosphorus trisulfide.

TETRAPHOSPHORUS TRISELENIDE
General Information
Description: Orange-red crystals.
Formula: P_4Se_3.
Constants: Mol wt: 360.80, mp: 242°C, bp: 360–400°C, d: 1.31.
Hazard Analysis and Countermeasures
See selenium compounds and phosphorus.

TETRAPHOSORUS TRISULFIDE. See phosphorus sesquisulfide.

TETRAPROPIONYL GLYCOSYL PROPIONATE. See glucose pentapropionate.

TETRAPROPYLENE OXIDE
Hazard Analysis
Toxic Hazard Rating:
Acute Local: Irritant 2.
Acute Systemic: Ingestion 2.
Chronic Local: U.
Chronic Systemic: U.
Based on limited animal experiments.

TETRA-n-PROPYLGERMANIUM
General Information
Description: Colorless mobile liquid.
Formula: $Ge(C_3H_7)_4$.
Constants: Mol wt: 245.0, d: 0.9539 at 20/20°C, mp: −73°C, bp: 225°C at 746 mm.
Hazard Analysis
Toxicity: Details unknown. See germanium compounds.

TETRA-n-PROPYLLEAD
General Information
Description: Colorless liquid; soluble in benzene.
Formula: $Pb(C_3H_7)_4$.
Constants: Mol wt: 379.6, d: 1.44, bp: 126°C at 13 mm.
Hazard Analysis and Countermeasures
See lead and lead compounds.

TETRAPROPYL THIOGERMANIUM
General Information
Description: A liquid; soluble in absolute alcohol.
Formula: $Ge(SC_3H_7)_4$.
Constants: Mol wt: 373.2, d: 1.1662 at 25°C, bp: 192°C at 5 mm.
Hazard Analysis and Countermeasures
See germanium compounds and organometals.

TETRAPROPYL TIN
General Information
Description: Colorless liquid; insoluble in water; soluble in organic solvents.
Formula: $(C_3H_7)_4Sn$.
Constants: Mol wt: 291.1, d: 1.1065, bp: 225°C.
Hazard Analysis
Toxicity: Details unknown. See tin compounds.

TOXIC HAZARD RATING CODE (For detailed discussion, see Section 1.)

0 NONE: (a) No harm under any conditions; (b) Harmful only under unusual conditions or overwhelming dosage.

1 SLIGHT: Causes readily reversible changes which disappear after end of exposure.

2 MODERATE: May involve both irreversible and reversible changes; not severe enough to cause death or permanent injury.

3 HIGH: May cause death or permanent injury after very short exposure to small quantities.

U UNKNOWN: No information on humans considered valid by authors.

TETRAPYRIDINECADMIUM FLUOSILICATE
General Information
Description: White crystals.
Formula: $[Cd(C_5H_5N)_4]SiF_6$.
Constants: Mol wt: 570.86, d: 2.282.
Hazard Analysis and Countermeasures
See cadmium compounds and fluosilicates.

TETRAPYRIDINECOPPER II FLUOSILICATE
General Information
Description: Rhombic, purplish-blue crystals.
Formula: $[Cu(C_5H_5N)_4]SiF_6$.
Constants: Mol wt: 521.99, d: 2.108.
Hazard Analysis and Countermeasures
See fluosilicates and copper compounds.

TETRAPYRIDINENICKEL 11 FLUOSILICATE
General Information
Description: Rhombic, blue-green crystals.
Formula: $[Ni(C_5H_5N)_4]SiF_6$.
Constants: Mol wt: 517.14, d: 2.307.
Hazard Analysis and Countermeasures
See silicofluorides and nickel compounds.

TETRA-N-PYRRYLGERMANIUM
General Information
Description: Light yellow crystals. Soluble in chloroform.
Formula: $Ge(C_4H_4N)_4$.
Constants: Mol wt: 336.9, mp:202°C.
Hazard Analysis
Toxicity: Details unknown. See germanium compounds.

TETRASILANE
General Information
Description: Colorless liquid.
Formula: Si_4H_{10}.
Constants: Mol wt: 122.32, mp: −93.5°C, bp: 109°C,
d: 0.825 at 0°C.
Hazard Analysis
Toxicity: Details unknown. See also silanes.
Fire Hazard: Severe. Details unknown.
Explosion Hazard: Severe, by chemical reaction with oxygen; can detonate in air.
Disaster Hazard: Dangerous; on decomposition, it emits toxic fumes; can react vigorously with oxidizing materials.
Countermeasures
Storage and Handling: Section 7.

TETRASODIUM PYROPHOSPHATE
General Information
Synonym: TSPP.
Description: White powder.
Formula: $Na_4P_2O_7$.
Constants: Mol wt: 266, mp: 880°C, d: 2.534.
Hazard Analysis
Toxic Hazard Rating:
Acute Local: Irritant 2; Ingestion 2; Inhalation 2.
Acute Systemic: U.
Chronic Local: U.
Chronic Systemic: Ingestion 1; Inhalation 1.
Toxicity: It is not a cholinesterase inhibitor. It is a substance which migrates to food from packaging materials and a sequestrant food additive (Section 10).
Countermeasures
Ventilation Control: Section 2.
Personnel Protection: Section 3.
Storage and Handling: Section 7.

TETRA-α-THIENYLGERMANIUM
General Information
Description: White crystals; insoluble in water.
Formula: $Ge(C_4H_3S)_4$.
Constants: Mol wt: 405.1, mp: 150°C.
Hazard Analysis and Countermeasures
See germanium compounds and sulfates.

TETRA-m-TOLYLGERMANIUM
General Information
Description: White crystals; insoluble in water.
Formula: $Ge(C_6H_4CH_3)_4$.
Constants: Mol wt: 437.1, mp: 146°C.
Hazard Analysis
Toxicity: Details unknown. See germanium compounds.

TETRA-o-TOLYLGERMANIUM
General Information
Description: White crystals; insoluble in water.
Formula: $Ge(C_6H_4CH_3)_4$.
Constants: Mol wt: 437.1, mp: 176°C.
Hazard Analysis
Toxicity: Details unknown. See germanium compounds.

TETRA-p-TOLYLGERMANIUM
General Information
Description: White crystals; insoluble in water.
Formula: $Ge(C_6H_4CH_3)_4$.
Constants: Mol wt: 437.1, mp: 227°C.
Hazard Analysis
Toxicity: Details unknown. See germanium compounds.

TETRA-p-TOLYL THIOGERMANIUM
General Information
Description: Colorless crystals; soluble in benzene.
Formula: $Ge(SC_6H_4CH_3)_4$.
Constants: Mol wt: 565.4, mp: 111°C.
Hazard Analysis and Countermeasures
See sulfates and germanium compounds.

TETRA-m-TOLYLTIN
General Information
Description: Colorless crystals; insoluble in water; soluble in benzene.
Formula: $(C_6H_4CH_3)_4Sn$.
Constants: Mol wt: 483.2, mp: 128.5°C.
Hazard Analysis
Toxicity: Details unknown. See tin compounds.

TETRA-o-TOLYLTIN
General Information
Description: White crystals; insoluble in water; soluble in ether and benzene.
Formula: $(C_6H_4CH_3)_4Sn$.
Constants: Mol wt: 483.2, mp: 159°C.
Hazard Analysis
Toxicity: Details unknown. See tin compounds.

TETRA-p-TOLYLTIN
General Information
Description: White crystals; insoluble in water; soluble in organic solvents.
Formula: $(C_6H_4CH_3)_4Sn$.
Constants: Mol wt: 483.2, mp: 233°C.
Hazard Analysis
Toxicity: Details unknown. See tin compounds.

TETRA-m-XYLYLTIN
General Information
Description: Crystals; slightly soluble in organic solvents.
Formula: $[C_6H_3(CH_3)_2]_4Sn$.
Constants: Mol wt: 539.3, mp: 219.5°C, bp: 360°C (decomp.).
Hazard Analysis
Toxicity: Details unknown. See tin compounds.

TETRA-p-XYLYLTIN
General Information
Description: Crystals; insoluble in water.
Formula: $[C_6H_3(CH_3)_2]_4Sn$.
Constants: Mol wt: 539.3, mp: 273°C, bp: 360°C (decomp.).
Hazard Aanlysis
Toxicity: Details unknown. See tin compounds.

TETRAZENE
General Information
Synonym: Guanylnitrosaminoguanyl tetrazene.
Description: Solid.
Hazard Analysis and Countermeasures
See nitrates.

TETRON. See tetraethyl pyrophosphate.

TETRYL
General Information
Synonyms: Tetralite; trinitrophenylmethyl nitramine;
    nitramine.
Description: Yellow, monoclinic crystals.
Formula: $(NO_2)_3C_6H_2N(NO_2)CH_3$.
Constants: Mol wt: 287.15, mp: 130°C, bp: explodes at
    187°C, d: 1.57 at 19°C.
Hazard Analysis
Toxic Hazard Rating:
    Acute Local: Irritant 2.
    Acute Systemic: Ingestion 2; Inhalation 2.
    Chronic Local: Irritant 2.
    Chronic Systemic: Ingestion 2; Inhalation 2.
TLV: ACGIH (recommended); 1.5 milligrams per cubic
    meter of air; can be absorbed through the intact skin.
Toxicology: The chief effect produced by exposure to tetryl
    is the development of dermatitis. Conjunctivitis may
    be caused by rubbing the eyes with contaminated hands
    or through exposure to air-borne dust. Iridocyclitis
    and keratitis have developed as a sequel to the conjunc-
    tivitis. Some authorities consider that tetryl may be a
    cause of tracheitis and asthma. Sensitization which
    frequently occurs as a result of exposure to tetryl may
    play a part in all these conditions. Tetryl workers may
    develop gastrointestinal symptoms, though these com-
    plaints are more common among TNT workers.
    Anemia has been reported to occur frequently.
Fire Hazard: Dangerous. See nitrates.
Explosion Hazard: Severe, when shocked or exposed to
    heat or flame. It is a powerful explosive quite sensitive
    to percussion and more sensitive to shock and friction
    than TNT. It can be compressed into pellets for use as a
    booster explosive. It is used in reinforced detonators
    and is considered to be the standard booster charge for
    high explosive shells.
Disaster Hazard: Dangerous. See nitrates.
Countermeasures
Ventilation Control: Section 2.
Personnel Protection: Section 3.
First Aid: Section 1.
Storage and Handling: Section 7.
Shipping Regulations: See explosives, high.

TEXTILE WASTE, WET
Hazard Analysis
Fire Hazard: Moderate, when exposed to heat or by chemi-
    cal reaction with oxidizing agents (Section 6).
Spontaneous Heating: Moderate.
Countermeasures
Ventilation Control: Section 2.
Shipping Regulations: Section 11.
    I.C.C.: Flammable solid; yellow label, not accepted.
    Coast Guard Classification: Inflammable solid; yellow
        label.

"TG-9." See triethylene glycol dipelargonate.

TGP. See triglycidyl phosphate.

THALLIC NITRATE
General Information
Description: Crystals.
Formula: $Tl(NO_3)_3$.
Constant: Mol wt: 390.41.
Hazard Analysis and Countermeasures
See thallium compounds and nitrates.

THALLIC OXIDE
General Information
Description: Hexagonal, black crystals; amorphous prisms.
Formula: Tl_2O_3.
Constants: Mol wt: 456.78, mp: 717° ± 5°C, bp: $-O_2$
    at 875°C, d amorphous: 9.65 at 21°C; d hexagonal:
    10.19 at 22°C.
Hazard Analysis
Toxicity: See thallium compounds.
Fire Hazard: Slight, by chemical reaction. Evolves oxygen
    at 875°C (Section 6).
Disaster Hazard: See thallium compounds.
Countermeasures
Storage and Handling: Section 7.

THALLIC SULFIDE
General Information
Description: Black, amorphous powder.
Formula: Tl_2S_3.
Constant: Mol wt: 504.98.
Hazard Analysis and Countermeasures
See sulfides and thallium compounds.

THALLIUM
General Information
Description: Bluish-white, soft and malleable metal.
Formula: Tl.
Constants: At wt: 204.39, mp: 302°C, bp: 1457 ± 10°C,
    d: 11.85, vap. press.: 1 mm at 825°C.
Hazard Analysis
Toxicity: See thallium compounds. A rodenticide.
Radiation Hazard: Section 5. For permissible levels, see
    Table 5, p. 150.
        Artificial isotope ^{200}Tl, half life 26 h. Decays to
    stable ^{200}Hg by electron capture. Emits gamma rays of
    0.11–1.52 MeV and X-rays.
        Artificial isotope ^{201}Tl, half life 73 h. Decays to
    stable ^{201}Hg by electron capture. Emits gamma rays
    of 0.03, 0.14, 0.17 MeV and X-rays.
        Aritficial isotope ^{202}Tl, half life 12 d. Decays to
    stable ^{202}Hg by electron capture. Emits gamma rays
    of 0.44 MeV and X-rays.
        Artificial isotope ^{204}Tl, half life 3.8 y. Decays to
    stable ^{204}Pb by emitting beta particles of 0.76 MeV.
Fire Hazard: Moderate, in the form of dust, when exposed
    to heat or flame. See also powdered metals (Section 6).
Countermeasures
Storage and Handling: Section 7.
Shipping Regulations: Section 11.
    I.C.C.: Thallium salts, solid: Poison B; poison label, 200
        pounds.
    Coast Guard Classification: Thallium salts, solid: Poison
        B; poison label.
    IATA: Poison B, poison label, 25 kilograms (passenger),
        95 kilograms (cargo).

THALLIUM ACETATE
General Information
Synonym: Thallous acetate.
Description: Silk-white crystals.

TOXIC HAZARD RATING CODE (For detailed discussion, see Section 1.)

0 NONE: (a) No harm under any conditions; (b) Harmful only under un-
    usual conditions or overwhelming dosage.

1 SLIGHT: Causes readily reversible changes which disappear after end
    of exposure.

2 MODERATE: May involve both irreversible and reversible changes;
    not severe enough to cause death or permanent injury.

3 HIGH: May cause death or permanent injury after very short exposure
    to small quantities.

U UNKNOWN: No information on humans considered valid by authors.

Formula: $TlC_2H_3O_2$.
Constants: Mol wt: 263.43, mp: 110°C d: 3.68.
Hazard Analysis and Countermeasures
See thallium compounds.
Shipping Regulations: Section 11.
   IATA: Poison B, poison label, 25 kilograms (passenger), 95 kilograms (passenger).

THALLIUM AZIDE
General Information
Description: Yellow crystals.
Formula: TlN_3.
Constants: Mol wt: 246.41, mp: 334°C.
Hazard Analysis and Countermeasures
See azides and thallium compounds.

THALLIUM BROMATE
General Information
Description: Colorless crystals.
Formula: $TlBrO_3$.
Constant: Mol wt: 332.31.
Hazard Analysis and Countermeasures
See thallium compounds and bromates.

THALLIUM BROMIDE
General Information
Synonym: Thallous bromide.
Description: Yellowish-white powder.
Formula: $TlBr$.
Constants: Mol wt: 284.31, mp: 460°C (approx.), bp: 815°C, d: 7.557, vap. press.: 10 mm at 522°C.
Hazard Analysis and Countermeasures
See thallium compounds and bromides.
Shipping Regulations: Section 11.
   IATA: Poison B, poison label, 25 kilograms (passenger), 95 kilograms (cargo).

THALLIUM CARBONATE
General Information
Description: Monoclinic, colorless crystals.
Formula: Tl_2CO_3.
Constants: Mol wt: 468.79, mp: 273°C, d: 7.11.
Hazard Analysis and Countermeasures
See thallium compounds.
Shipping Regulations: Section 11.
   IATA: Poison B, poison label, 25 kilograms (passenger), 95 kilograms (cargo).

THALLIUM CHLORATE
General Information
Synonym: Thallous chlorate.
Description: Solid.
Formula: $TlClO_3$.
Constants: Mol wt: 287.85, d: 5.047 at 90°C.
Hazard Analysis and Countermeasures
See chlorates and thallium compounds.

THALLIUM CHLORIDE
General Information
Synonym: Thallous chloride.
Description: Colorless or white powder.
Formula: $TlCl$.
Constants: Mol wt: 239.85, mp: 430°C, bp: 720°C, d: 7.00, vap. press.: 10 mm at 517°C.
Hazard Analysis and Countermeasures
See thallium compounds and chlorides.
Shipping Regulations: Section 11.
   IATA: Poison B, poison label, 25 kilograms (passenger), 95 kilograms (cargo).

THALLIUM CHLOROPLATINATE
General Information
Description: Pale orange crystals.
Formula: Tl_2PtCl_6.
Constants: Mol wt: 816.75, d: 5.76 at 17°C.

Hazard Analysis and Countermeasures
See thalium compounds and chlorides.

THALLIUM CHROMATE
General Information
Description: Yellow crystals.
Formula: Tl_2CrO_4.
Constant: Mol wt: 524.8.
Hazard Analysis and Countermeasures
See chromium compounds and thallium compounds.

THALLIUM COMPOUNDS
Hazard Analysis
Toxic Hazard Rating:
   Acute Local: Ingestion 2.
   Acute Systemic: Ingestion 3; Inhalation 2.
   Chronic Local: Ingestion 3; Inhalation 3.
   Chronic Systemic: Ingestion 3; Inhalation 2.
Toxicology: Acute poisoning usually follows the ingestion of toxic quantities of a thallium-bearing depilatory, or accidental or suicidal ingestion of rat poison. Children have been known to tolerate 8 mg of thallium acetate per kilogram of weight, but adults and adolescents have not. Acute poisoning results in swelling of the feet and legs, arthralgia, vomiting, insomnia, hyperesthesia and paresthesia of the hands and feet, mental confusion, polyneuritis with severe pains in the legs and loins, partial paralysis of the legs with reaction of degeneration, angina-like pains, nephritis, wasting and weakness, and lymphocytosis and eosinophilia. About the 18th day, complete loss of the hair of the body and head occurs. Fatal poisoning has been known to occur.
   Industrial poisoning is reported to have caused discoloration of the hair, which later falls out, joint pains, loss of appetite, fatigue, severe pain in the calves of the legs, albuminuria, eosinophilia and lymphocytosis, and optic neuritis followed by atrophy. Cases of industrial poisoning are rare, however.
TLV (soluble compounds): ACGIH (recommended), 0.1 milligrams per cubic meter of air. May be absorbed via the skin.
Disaster Hazard: Dangerous; when heated, they emit highly toxic fumes.
Countermeasures
Ventilation Control: Section 2.
Personnel Protection: Section 3.
Storage and Handling: Section 7.

THALLIUM CYANIDE
General Information
Synonym: Thallous cyanide.
Description: Tablets.
Formula: $TlCN$.
Constants: Mol wt: 230.41, mp: decomposes.
Hazard Analysis and Countermeasures
See cyanides and thallium compounds.

THALLIUM DICHROMATE
General Information
Synonym: Thallous dichromate.
Description: Red crystals.
Formula: $Tl_2Cr_2O_7$.
Constant: Mol wt: 624.80.
Hazard Analysis and Countermeasures
See chromium compounds and thallium compounds.

THALLIUM DITHIONATE
General Information
Synonym: Thallous dithionate.
Description: Monoclinic crystals.
Formula: $Tl_2S_2O_6$.
Constants: Mol wt: 568.91, mp: decomposes, d: 5.57.
Hazard Analysis and Countermeasures
See thallium compounds and sulfates.

THALLIUM ETHOXIDE
General Information
Description: Colorless liquid.
Formula: $[TlOC_2H_5]_4$.
Constants: Mol wt: 997.80, mp: $-3°C$, bp: decomposes at 80°C, d: 3.522.
Hazard Analysis
Toxicity: Highly toxic. See thallium compounds.
Fire Hazard: Moderate, when exposed to heat or flame.
Disaster Hazard: Dangerous; when heated to decomposition it emits highly toxic fumes; can react with oxidizing materials.
Countermeasures
Ventilation Control: Section 2.
Personnel Protection: Section 3.
Storage and Handling: Section 7.

THALLIUM FERROCYANIDE
General Information
Description: Triclinic, yellow crystals.
Formula: $Tl_4Fe(CN)_6 \cdot 2H_2O$.
Constants: Mol wt: 1065.55, d: 4.641.
Hazard Analysis and Countermeasures
See thallium compounds and ferrocyanides.

THALLIUM FLUOGALLATE
General Information
Description: Colorless crystals.
Formula: $Tl_2(GaF_5H_2O)$.
Constants: Mol wt: 591.52, d: 6.44.
Hazard Analysis and Countermeasures
See fluorides and thallium compounds.

THALLIUM FLUOSILICATE
General Information
Description: Hexagonal plates.
Formula: $Tl_2SiF_6 \cdot 2H_2O$.
Constant: Mol wt: 586.87.
Hazard Analysis and Countermeasures
See fluosilicates and thallium compounds.

THALLIUM HYDROGEN SULFATE. See thallous hydrogen sulfate.

THALLIUM HYDROXIDE
General Information
Synonym: Thallous hydroxide.
Description: Pale yellow needles.
Formula: TlOH.
Constants: Mol wt: 221.40, bp: decomposes at 139°C.
Hazard Analysis and Countermeasures
See thallium compounds.
Shipping Regulations: Section 11.
   IATA: Poison B, poison label, 25 kilograms (passenger), 95 kilograms (cargo).

THALLIUM IODIDE
General Information
Synonym: Thallium monoiodide.
Description: Cubic red crystals; yellow powder.
Formula: TlI.
Constants: Mol wt: 331.31, mp: 440°C, bp: 824°C, d: 7.09, vap. press.: 1 mm at 440°C.
Hazard Analysis and Countermeasures
See thallium compounds and iodides.
Shipping Regulations: Section 11.
   IATA: Poison B, poison label, 25 kilograms (passenger), 95 kilograms (cargo).

THALLIUM MONOFLUORIDE
General Information
Synonym: Thallous fluoride.
Description: Cubic, colorless crystals.
Formula: TlF.
Constants: Mol wt: 223.39, bp: 300°C.
Hazard Analysis and Countermeasures
See thallium compounds and fluorides.

THALLIUM MONOIODIDE. See thallium iodide.

THALLIUM MONOSULFIDE. See thallium sulfide.

THALLIUM MONOXIDE
General Information
Synonyms: Thallium oxide; thallous oxide.
Description: Black, deliquescent crystals.
Formula: Tl_2O.
Constants: Mol wt: 424.78, mp: 300°C, bp: 1080°C at 600 mm; $-O_2$ at 1865°C.
Hazard Analysis and Countermeasures
See thallium compounds.
Shipping Regulations: Section 11.
   IATA: Poison B, poison label, 25 kilograms (passenger), 95 kilograms (cargo).

THALLIUM NITRATE
General Information
Synonym: Thallous nitrate.
Description: Cubic crystals.
Formula: $TlNO_3$.
Constants: Mol wt: 266.4, mp: 206°C, bp: 430°C.
Hazard Analysis and Countermeasures
See thallium compounds and nitrates.
Shipping Regulations: Section 11.
   IATA: Poison B, poison label, 25 kilograms (passenger), 95 kilograms (cargo).

THALLIUM OLEATE
General Information
Synonym: Thallous oleate.
Description: White, crystalline clusters.
Formula: $TlC_{18}H_{33}O_2$.
Constants: Mol wt: 485.83, mp: 131–132°C.
Hazard Analysis and Countermeasures
See thallium compounds.

THALLIUM OXALATE
General Information
Synonym: Thallous oxalate.
Description: Monoclinic prisms.
Formula: $Tl_2C_2O_4$.
Constants: Mol wt: 496.80, d: 6.31.
Hazard Analysis and Countermeasures
See oxalates and thallium compounds.

THALLIUM OXIDE. See thallium monoxide.

THALLIUM PERCHLORATE
General Information
Synonym: Thallous perchlorate.
Description: Colorless crystals.
Formula: $TlClO_4$.
Constants: Mol wt: 303.85, mp: 501°C, bp: decomposes, d: 4.89.
Hazard Analysis and Countermeasures
See thallium compounds and perchlorates.

THALLIUM PEROXIDE. See thallic oxide.
Shipping Regulations: Section 11.

TOXIC HAZARD RATING CODE (For detailed discussion, see Section 1.)

0 NONE: (a) No harm under any conditions; (b) Harmful only under unusual conditions or overwhelming dosage.

1 SLIGHT: Causes readily reversible changes which disappear after end of exposure.

2 MODERATE: May involve both irreversible and reversible changes; not severe enough to cause death or permanent injury.

3 HIGH: May cause death or permanent injury after very short exposure to small quantities.

U UNKNOWN: No information on humans considered valid by authors.

IATA: Poison B, poison label, 25 kilograms (passenger;, 95 kilograms (cargo).

THALLIUM PHENOXIDE
General Information
Synonym: Thallous phenoxide.
Description: White crystals.
Formula: $TlOC_6H_5$.
Constants: Mol wt: 297.49, mp: 233°C.
Hazard Analysis and Countermeasures
See thallium compounds and phenol.

THALLIUM o-PHOSPHATE
General Information
Synonym: Thallous o-phosphate.
Description: Colorless needles.
Formula: Tl_3PO_4.
Constants: Mol wt: 708.15, d: 6.89.
Hazard Analysis and Countermeasures
See thallium compounds and phosphates.

THALLIUM PICRATE
General Information
Description: Red or yellow crystals.
Formula: $TlC_6H_2N_3O_7$.
Constants: Mol wt: 432.49; mp: explodes at 273–275°C; d: red 3.164 at 17°C, yellow 2.993 at 17°C.
Hazard Analysis and Countermeasures
See thallium compounds and picric acid.

THALLIUM PYROVANADATE
General Information
Description: Solid.
Formula: $Tl_4V_2O_7$.
Constants: Mol wt: 1031.46, mp: 454°C, d: 8.21 at 19°C.
Hazard Analysis and Countermeasures
See thallium compounds.

THALLIUM SELENATE
General Information
Synonym: Thallous selenate.
Description: Rhombic needles.
Formula: Tl_2SeO_4.
Constants: Mol wt: 551.74, mp: >400°C, d: 6.875.
Hazard Analysis and Countermeasures
See selenium compounds and thallium compounds.

THALLIUM SELENIDE
General Information
Synonym: Thallous selenide.
Description: Gray leaf.
Formula: Tl_2Se.
Constants: Mol wt: 487.74, mp: 340°C.
Hazard Analysis and Countermeasures
See selenium compounds and thallium compounds.

THALLIUM SESQUICHLORIDE
General Information
Description: Hexagonal yellow crystals or yellow powder.
Formula: Tl_2Cl_3.
Constants: Mol wt: 515.15, mp: 400°–500°C, bp: decomposes, d: 5.9.
Hazard Analysis and Countermeasures
See thallium compounds and chlorides.
Shipping Regulations: Section 11.
IATA: Poison B, poison label, 25 kilograms (passenger), 95 kilograms (cargo).

THALLIUM SILVER NITRATE
General Information
Synonym: Thallous silver nitrate.
Description: White crystalline powder.
Formula: $TlNO_3 \cdot AgNO_3$.
Constants: Mol wt: 436.29, mp: 75°C.
Hazard Analysis and Countermeasures
See thallium compounds and nitrates.

THALLIUM STEARATE
General Information
Synonym: Thallous stearate.
Description: Needles.
Formula: $TlC_{18}H_{35}O_2$.
Constants: Mol wt: 487.85, mp: 119°C.
Hazard Analysis and Countermeasures
See thallium compounds.

THALLIUM SULFATE
General Information
Synonym: Thallous sulfate.
Description: Colorless crystals.
Formula: Tl_2SO_4.
Constants: Mol wt: 504.84, mp: 632°C, bp: decomposes, d: 6.77.
Hazard Analysis and Countermeasures
See thallium compounds and sulfates.
Shipping Regulations: Section 11.
I.C.C.: Poison B; poison label, 200 pounds.
Coast Guard Classification: Poison B; poison label.
MCA warning label.
IATA: Poison B, poison label, 25 kilograms (passenger), 95 kilograms (cargo).

THALLIUM SULFIDE
General Information
Synonyms: Thallium monosulfide; thallous sulfide.
Description: Blue-black powder.
Formula: Tl_2S.
Constants: Mol wt: 440.84, mp: 443°C, bp: decomposes, d: 8.0.
Hazard Analysis and Countermeasures
See thallium compounds and sulfides.
Shipping Regulations: Section 11.
IATA: Poison B, poison label, 25 kilograms (passenger), 95 kilograms (cargo).

THALLIUM SULFITE
General Information
Synonym: Thallous sulfite.
Description: Crystals.
Formula: Tl_2SO_3.
Constants: Mol wt: 488.85, d: 6.427.
Hazard Analysis and Countermeasures
See thallium compounds and sulfites.

THALLIUM m-TELLURATE
General Information
Synonym: Thallous m-tellurate.
Description: Heavy white precipitate.
Formula: Tl_2TeO_4.
Constants: Mol wt: 600.39, mp: red heat, d: 6.760 at 17.6°C.
Hazard Analysis and Countermeasures
See thallium compounds and tellurium compounds.

THALLIUM THIOCYANATE
General Information
Synonym: Thallous thiocyanate.
Description: Tetragonal, colorless crystals.
Formula: $TlSCN$.
Constant: Mol wt: 262.47.
Hazard Analysis and Countermeasures
See thallium compounds and thiocyanates.

THALLIUM TRIIODIDE
General Information
Description: Brown needles.
Formula: TlI_3.
Constant: Mol wt: 585.15.
Hazard Analysis and Countermeasures
See thallium compounds and iodides.

THALLOUS ACETATE. See thallium acetate.

THALLOUS BROMIDE. See thallium bromide.

THALLOUS CHLORATE. See thallium chlorate.

THALLOUS CHLORIDE. See thallium chloride.

THALLOUS CYANIDE. See thallium cyanide.

THALLOUS DICHROMATE. See thallium dichromate.

THALLOUS DITHIONATE. See thallium dithionate.

THALLOUS FLUORIDE. See thallium monofluoride.

THALLOUS HYDROGEN SULFATE
General Information
Synonym: Thallium hydrogen sulfate.
Description: Crystals.
Formula: $TlHSO_4$.
Constants: Mol wt: 301.46, mp: 120° C decomposes.
Hazard Analysis and Countermeasures
See thallium compounds and sulfuric acid.

THALLOUS HYDROXIDE. See thallium hydroxide.

THALLOUS NITRATE. See thallium nitrate.

THALLOUS OLEATE. See thallium oleate.

THALLOUS OXALATE. See thallium oxalate.

THALLOUS OXIDE. See thallium oxide.

THALLOUS PERCHLORATE. See thallium perchlorate.

THALLOUS PHENOXIDE. See thallium phenoxide.

THALLOUS o-PHOSPHATE. See thallium o-phosphate.

THALLOUS SELENATE. See thallium selenate.

THALLOUS SELENIDE. See thallium selenide.

THALLOUS SILVER NITRATE. See thallium silver nitrate.

THALLOUS STEARATE. See thallium stearate.

THALLOUS SULFATE. See thallium sulfate.

THALLOUS SULFITE. See thallium sulfite.

THALLOUS m-TELLURATE. See thallium m-tellurate.

THALLOUS THIOCYANATE. See thallium thiocyanate.

THANITE. See isobornyl thiocyanoacetate.

THBP. See 2,4,5-trihydroxybutyrophenone.

THEBAINE
General Information
Synonym: p-morphine.
Description: White to slightly yellowish, lustrous leaflets or prisms.
Formula: $C_{19}H_{21}NO_3$.
Constants: Mol wt: 311.37, mp: 193° C, d: 1.305.
Hazard Analysis
Toxic Hazard Rating:
  Acute Local: U.
  Acute Systemic: Ingestion 3; Inhalation 3.
  Chronic Local: U.
  Chronic Systemic: U.
Disaster Hazard: Dangerous; when heated to decomposition it emits highly toxic fumes.

Countermeasures
Ventilation Control: Section 2.
Personal Hygiene: Section 3.
First Aid: Section 1.
Storage and Handling: Section 7.

THEINE. See caffeine.

THEOBROMINE
General Information
Synonym: 3,7-Dimethylxanthine.
Description: White powder; bitter taste.
Formula: $C_7H_8N_4O_2$.
Constants: Mol wt: 180.17, mp: 337° C, bp: sublimes.
Hazard Analysis
Toxic Hazard Rating:
  Acute Local: U.
  Acute Systemic: Ingestion 1; Inhalation 1.
  Chronic Local: U.
  Chronic Systemic: Ingestion 1; Inhalation 1.
Disaster Hazard: Dangerous; when heated to decomposition, it emits toxic fumes.
Countermeasures
Ventilation Control: Section 2.
Personal Hygiene: Section 3.
Storage and Handling: Section 7.

THEOBROMINE LITHIUM. See theobromose.

THEOBROMINE SODIUM ACETATE. See theobromine.

THEOBROMINE SODIUM FORMATE. See theophorin.

THEOBROMINE SODIUM SALICYLATE. See theobromine.

THEOBROMOSE
General Information
Synonym: Theobromine lithium.
Description: Needle-like crystals.
Formula: $C_7H_7O_2N_4Li$.
Constant: Mol wt: 186.10.
Hazard Analysis and Countermeasures
See also theobromine and lithium compounds.

THEOPHORIN
General Information
Synonym: Theobromine-sodium formate.
Description: White powder.
Formula: $C_7H_7N_4O_2Na \cdot HCOONa \cdot H_2O$.
Constant: Mol wt: 288.13.
Hazard Analysis
Toxic Hazard Rating:
  Acute Local: Irritant 3; Ingestion 3; Inhalation 3.
  Acute Systemic: U.
  Chronic Local: U.
  Chronic Systemic: U.
Disaster Hazard: Dangerous; when heated, it emits highly toxic fumes.
Countermeasures
Ventilation Control: Section 2.
Personnel Protection: Section 3.
First Aid: Section 1.
Storage and Handling: Section 7.

THEOPHYLLINE
General Information
Synonym: 1,3-Dimethylxanthine.

TOXIC HAZARD RATING CODE (For detailed discussion, see Section 1.)

0 NONE: (a) No harm under any conditions; (b) Harmful only under unusual conditions or overwhelming dosage.

1 SLIGHT: Causes readily reversible changes which disappear after end of exposure.

2 MODERATE: May involve both irreversible and reversible changes; not severe enough to cause death or permanent injury.

3 HIGH: May cause death or permanent injury after very short exposure to small quantities.

U UNKNOWN: No information on humans considered valid by authors.

Description: Monoclinic, odorless needles; bitter taste.
Formula: $C_7H_8N_4O_2$.
Constants: Mol wt: 180.17, mp: 269.72°C.
Hazard Analysis
Toxic Hazard Rating:
Acute Local: U.
Acute Systemic: Ingestion 1; Inhalation 1.
Chronic Local: U.
Chronic Systemic: Ingestion 1; Inhalation 1.
Disaster Hazard: Dangerous; when heated to decomposition, it emits toxic fumes.
Countermeasures
Personal Hygiene: Section 3.
Storage and Handling: Section 7.

"THERMIT"
General Information
Formula: $Fe_2O_3 + Al$.
Hazard Analysis
Toxicity: See aluminum compounds and iron compounds.
Fire Hazard: Dangerous, when exposed to heat or flame. The reaction of $Fe_2O_3 + Al$ is typical of a series of oxide-metal reactions. They are very dangerous in that once started they are very difficult to stop, as they supply their own oxygen. They may attain a temperature of about 2500°C (Section 6).
Disaster Hazard: Dangerous; keep away from combustible materials.
Countermeasures
Storage and Handling: Section 7.

THIABENDAZOLE
General Information
Synonym: 2-(4-Thiazolyl) benzimidazole.
Description: White to tan, odorless; insoluble in water; slightly soluble in alcohol and acetone; very slightly soluble in ether and chloroform.
Formula: $C_{10}H_7N_3S$.
Constants: Mol wt: 201, melting range: 296–303°C.
Hazard Analysis
Toxic Hazard Rating: U. A food additive permitted in the feed and drinking water of animals; and/or for the treatment of food producing animals; also permitted in food for human consumption (Section 10).
Disaster Hazard: Dangerous. See sulfides.

THIACETIC ACID. See thioacetic acid.

THFA. See tetrahydrofurfuryl alcohol.

THF-GLYCOL. See 2,5-tetrahydrofuran dimethanol.

THIALDINE
General Information
Synonym: 5,6-Dihydro-2,3,6-trimethyl-1,3,5-dithiazine.
Description: Powder.
Formula: $SCH(CH_3)SCH(CH_3)NHCHCH_3$.
Constants: Mol wt: 163.29, mp: 44.4°C, bp: decomposes, flash p: 200°F (O.C.), d: 1.191, vap. d.: 5.63.
Hazard Analysis
Toxicity: Unknown.
Fire Hazard: Moderate, when exposed to heat or flame.
Disaster Hazard: Dangerous; when heated to decomposition, it emits toxic fumes.
Countermeasures
Storage and Handling: Section 7.
To Fight Fire: Foam, carbon dioxide, dry chemical or carbon tetrachloride (Section 6).

THIAMINE HYDROCHLORIDE
General Information
Description: Small white crystals or crystalline powder; hygroscopic; nut like odor; soluble in water and glycerol; slightly soluble in alcohol; insoluble in ether and benzene.
Formula: $C_{12}H_{17}ClN_4OS \cdot HCl$.

Constants: Mol wt: 337, mp: 248°C (decomposes).
Hazard Analysis
Toxic Hazard Rating: U. A nutrient and/or dietary supplement food additive (Section 10).
Disaster Hazard: Dangerous. See chlorides.

THIAMINE MONONITRATE
General Information
Description: White crystals or crystalline powder; non-hygroscopic; slightly soluble in water, alcohol, and chloroform.
Formula: $C_{12}H_{17}N_5O_4S$.
Constants: Mol wt: 327, mp: 196–200°C (decomposes).
Hazard Analysis
Toxic Hazard Rating: U. A nutrient and/or dietary supplement food additive (Section 10).
Fire Hazard: A powerful oxidizer, see nitrates.
Disaster Hazard: Dangerous. See nitrates.
Countermeasures
Storage and Handling: Section 7.

THIAZOSULFONE. See 2-amino-5-sulfanilylthiazole.

THIENYL-α-PYRROLIDINE
Hazard Analysis
Toxicity: Details unknown; an insecticide.
Disaster Hazard: Dangerous; when heated to decomposition, it emits highly toxic fumes.
Countermeasures
Storage and Handling: Section 7.

THIIRANE. See ethylene sulfide.

THIMET
General Information
Synonym: o,o-diethyl-s-isopropylmercaptomethyl phosphorodithioate.
Hazard Analysis
Toxicity: Details unknown. Being an organic phosphorus compound it probably is a cholinesterase inhibitor. See parathion.
Disaster Hazard: Dangerous. When heated to decomposition it emits highly toxic fumes of oxides of sulfur and phosphorus.

THIOACETAMIDE
General Information
Description: Colorless leaflets.
Formula: CH_3CSNH_2.
Constants: Mol wt: 75.20, mp: 109°C.
Hazard Analysis
Toxic Hazard Rating:
Acute Local: Ingestion 2; Inhalation 2.
Acute Systemic: U.
Chronic Local: U.
Chronic Systemic: U.
Toxicity: Details unknown. Limited animal experiments show moderate toxicity. Liver damage has been reported.
Disaster Hazard: Dangerous. See sulfides.
Countermeasures
Ventilation Control: Section 2.
Personal Hygiene: Section 3.
Storage and Handling: Section 7.

THIOACETIC ACID
General Information
Synonyms: Ethanethiolic acid; methanecarbothiolic acid; thiacetic acid.
Description: Colorless liquid; pungent, disagreeable odor.
Formula: CH_3COSH.
Constants: Mol wt: 76.11, mp: < -17°C, bp: 93°C, d: 1.074 at 10°/4°C.
Hazard Analysis
Toxic Hazard Rating:
Acute Local: Irritant 2; Ingestion 2; Inhalation 2.

Acute Systemic: U.
Chronic Local: U.
Chronic Systemic: U.
Countermeasures
See sulfides.

2,2'-THIOBIS(4-CHLOROPHENOL)
General Information
Synonym: Novex.
Description: Crystals; soluble in solutions of sodium hydroxide.
Formula: $C_{12}H_xCl_2O_2S$.
Constants: Mol wt: 287.2, mp: 175°C.
Hazard Analysis and Countermeasures
See chlorinated phenols.

2,2'-THIOBIS(4,6-DICHLOROPHENOL). See bithionol.

THIOCARBAMIDE. See thiourea.

THIOCARBAMO SULFONAMIDES
Hazard Analysis
Toxicity: Details unknown, but probably toxic. A fungicide. See also amides.
Disaster Hazard: Dangerous; when heated to decomposition, it emits highly toxic fumes of oxides of sulfur.
Countermeasures
Storage and Handling: Section 7.

THIOCARBANIL. See phenyl mustard oil.

THIOCARBANILIDE
General Information
Synonyms: N,N'-diphenyl thiourea; 1,3-diphenyl-2-thiourea.
Description: White to faint gray powder.
Formula: $C_6H_5NHC(S)NHC_6H_5$.
Constants: Mol wt: 228.3, mp: 154°C, bp: decomposes, d: 1.32 at 25°C.
Hazard Analysis
Toxicity: Details unknown. Limited animal experiments suggest low toxicity. See also thiourea.
Disaster Hazard: Dangerous; when heated to decomposition, it emits highly toxic fumes of oxides of sulfur.
Countermeasures
Storage and Handling: Section 7.

THIOCARBONYL CHLORIDE. See thiophosgene.

THIOCARBONYL TETRACHLORIDE. See perchloromethyl mercaptan.

THIOCYANATES
Hazard Analysis
Toxicity: Variable. Thiocyanates are not normally dissociated into cyanide; they have a low acute toxicity. Prolonged absorption may produce various skin eruptions, running nose, and occasionally dizziness, cramps, nausea, vomiting and mild or severe disturbances of the nervous system.
Disaster Hazard: Dangerous; when heated to decomposition or on contact with acid or acid fumes, they emit highly toxic fumes of cyanides.
Countermeasures
Storage and Handling: Section 7.

THIOCYANIC ACID
General Information
Description: Colorless gas or white solid.
Formula: HSCN.
Constants: Mol wt: 59.09, mp: 5°C, bp: decomposes.

Hazard Analysis and Countermeasures
See thiocyanates.

p-THIOCYANOCHLOROBENZENE
General Information
Description: Crystals.
Formula: ClC_6H_4SCN.
Constant: Mol wt: 169.6.
Hazard Analysis
Toxicity: Details unknown; a fumigant. See chlorobenzene and thiocyanates.
Disaster Hazard: Dangerous; when heated to decomposition or on contact with acid or acid fumes, it emits highly toxic fumes of cyanides and chlorides.
Countermeasures
Storage and Handling: Section 7.

THIOCYANOGEN
General Information
Description: Liquid or yellow solid.
Formula: $(SCN)_2$.
Constants: Mol wt: 116.17, mp: −2 to −3°C (decomp.).
Hazard Analysis
Toxicity: Details unknown. Probably highly toxic.
Disaster Hazard: Dangerous; when heated to decomposition or on contact with acid or acid fumes, it emits highly toxic fumes of oxides of sulfur and cyanides.
Countermeasures
Storage and Handling: Section 7.

THIOCYANOPROPYL PHENYL ETHER
Hazard Analysis
Toxicity: Details unknown; an insecticide.
Fire Hazard: Details unknown. See also ethers.
Disaster Hazard: Dangerous; when heated to decomposition, it emits highly toxic fumes.
Countermeasures
Storage and Handling: Section 7.

THIODIETHYLENE GLYCOL
General Information
Synonym: Thiodiglycol.
Description: Syrupy, colorless liquid; characteristic odor.
Formula: $(CH_2CH_2OH)_2S$.
Constants: Mol wt: 122.2, mp: −11.2°C, bp: 282°C, flash p: 320°F (O.C.), d: 1.1847 at 20°/20°C, vap. d.: 4.21.
Hazard Analysis
Toxicity: Details unknown. See also glycols.
Fire Hazard: Slight, when exposed to heat or flame.
Disaster Hazard: Dangerous; when heated to decomposition, it emits highly toxic fumes of oxides of sulfur; can react with oxidizing materials.
Countermeasures
Storage and Handling: Section 7.
To Fight Fire: Water, foam, carbon dioxide, dry chemical or carbon tetrachloride (Section 6).

THIODIGLYCOL. See thiodiethylene glycol.

THIODIGLYCOLIC ACID
General Information
Description: A white powder.
Formula: $HOOCCH_2SCH_2COOH$.
Constants: Mol wt: 150, mp: 128°C.
Hazard Analysis
Toxicity: Details unknown; probably moderately toxic. See also thioglycolic acid.

TOXIC HAZARD RATING CODE (For detailed discussion, see Section 1.)

0 NONE: (a) No harm under any conditions; (b) Harmful only under unusual conditions or overwhelming dosage.

1 SLIGHT: Causes readily reversible changes which disappear after end of exposure.

2 MODERATE: May involve both irreversible and reversible changes; not severe enough to cause death or permanent injury.

3 HIGH: May cause death or permanent injury after very short exposure to small quantities.

U UNKNOWN: No information on humans considered valid by authors.

Disaster Hazard: Moderately dangerous; when heated to decomposition or on contact with acid or acid fumes, it emits toxic fumes.

Countermeasures

Storage and Handling: Section 7

THIODIPHENYLAMINE. See phenothiazine.

THIODIPROPIONIC ACID

General Information

Formula: $S(CH_2CH_2CO_2H)_2$.

Constant: Mol wt: 178.

Hazard Analysis

Toxic Hazard Rating: U. A chemical preservative food additive (Section 10).

β,β'-THIODIPROPIONITRILE

General Information

Description: White crystals.

Formula: $S(CH_2CH_2CN)_2$.

Constants: Mol wt: 140.20; mp: alpha = 28.65°C, beta = 22.10°C; d: 1.1095 at 30°C.

Hazard Analysis and Countermeasures

See cyanides and nitriles.

THIOETHYL ETHER. See ethyl sulfide.

THIOFURAN. See thiophene.

THIOGLYCOLIC ACID

General Information

Synonyms: Mercaptoacetic acid; 2-mercaptoethanoic acid; thiovanic acid.

Description: Liquid.

Formula: $HSCH_2COOH$.

Constants: Mol wt: 92.11, mp: −16.5°C, bp: 104–106°C at 11 mm, d: 1.3253 at 20°/4°C.

Hazard Analysis

Toxic Hazard Rating:

Acute Local: Irritant 2; Ingestion 3; Inhalation 3.

Acute Systemic: Ingestion 3; Inhalation 3; Skin Absorption 2.

Chronic Local: Irritant 2.

Chronic Systemic: Ingestion 3; Inhalation 3; Skin Absorption 2.

For other properties, see hydrogen sulfide, which is readily evolved by this compound.

Countermeasures

Ventilation Control: Section 2.

Personnel Protection: Section 3.

First Aid: Section 1.

Storage and Handling: Section 7.

Shipping Regulations: Section 11.

IATA: Corrosive liquid, white label, 1 liter (passenger), 5 liters (cargo).

THIOGLYCOLLATES

General Information

Synonym: Mercaptoacetates.

Hazard Analysis

Toxic Hazard Rating:

Acute Local: Irritant 2.

Acute Systemic: Ingestion 2.

Chronic Local: Irritant 2.

Chronic Systemic: U.

Toxicology: See also thioglycolic acid.

Disaster Hazard: Dangerous; when heated to decomposition they emit highly toxic fumes.

2-THIO-4-KETO-THIAZOLIDINE. See rhodanine.

THIOMALIC ACID

General Information

Description: Off-white powder.

Formula: $HCO_2CH_2C(SH)HCO_2H$.

Constants: Mol wt: 150.2, mp: 150°C.

Hazard Analysis

Toxicity: Animal experiments show low toxicity. Has been proposed as an antidote for heavy metal poisoning. Allergic dermatitis in humans has been reported.

Disaster Hazard: Dangerous; when heated to decomposition or on contact with acid or acid fumes, it emits toxic fumes.

Countermeasures

Storage and Handling: Section 7.

THIONYL BROMIDE

General Information

Description: Yellow liquid.

Formula: $SOBr_2$.

Constants: Mol wt: 207.90, mp: −52°C to −50°C, bp: 138°C at 773 mm, d: 2.68 at 18°C, vap. press.: 10 mm at 31.0°C.

Hazard Analysis

Toxic Hazard Rating:

Acute Local: Irritant 3; Ingestion 3; Inhalation 3.

Acute Systemic: Ingestion 3; Inhalation 3.

Chronic Local: U.

Chronic Systemic: Inhalation 2.

Disaster Hazard: Dangerous; when heated it emits highly toxic fumes; will react with water, steam or acids to produce toxic and corrosive fumes.

Countermeasures

Ventilation Control: Section 2.

Personnel Protection: Section 3.

First Aid: Section 1.

Storage and Handling: Section 7.

THIONYL CHLORIDE

General Information

Synonym: Sulfurous oxychloride.

Description: Colorless to yellow to red liquid.

Formula: $SOCl_2$.

Constants: Mol wt: 119.0, mp: −105°C, bp: 78.8°C at 746 mm, d: 1.640 at 15.5°/15.5°C, vap. press.: 100 mm at 21.4°C.

Hazard Analysis

Toxicity: This material has a pungent odor similar to that of sulfur dioxide, and fumes upon exposure to air. In the presence of moisture it decomposes into hydrogen chloride and sulfur dioxide. Both these decomposition products are very toxic and constitute serious toxic hazards. The material itself is more toxic than sulfur dioxide. In experiments with animals, it was found that an exposure of 20 minutes to a concentration of 17.5 ppm was fatal to cats. It is classified as a corrosive liquid and can cause burns of the skin, eyes, and mucous membranes wherever it comes in contact with the body. See also hydrogen chloride and sulfur dioxide.

Disaster Hazard: Corrosive. See hydrochloric acid and sulfur dioxide.

Countermeasures

Ventilation Control: Section 2.

Treatment and Antidotes: Personnel exposed should follow the treatment outlined under hydrogen chloride and sulfur dioxide.

Personnel Protection: Section 3.

First Aid: Section 1.

Storage and Handling: Section 7.

Shipping Regulations: Section 11.

I.C.C.: Corrosive liquid; white label, 1 gallon.

IATA: Corrosive liquid, white label, not acceptable (passenger), 5 liters (cargo).

Coast Guard Classification: Corrosive liquid; white label. MCA warning label.

THIONYL CHLORIDE FLUORIDE

General Information

Description: Gas.

Formula: SOClF.

Constants: Mol wt: 102.52, mp: −139.5°C, bp: 12.2°C.

Hazard Analysis
Toxic Hazard Rating:
  Acute Local: Irritant 3; Inhalation 3.
  Acute Systemic: Inhalation 3.
  Chronic Local: Irritant 2; Inhalation 3.
  Chronic Systemic: Inhalation 2.
Disaster Hazard: Dangerous; when heated, it emits highly toxic fumes; will react with water or steam to produce toxic and corrosive fumes.
Countermeasures
Ventilation Control: Section 2.
Personnel Protection: Section 3.
First Aid: Section 1.
Storage and Handling: Section 7.

THIONYL FLUORIDE
General Information
Description: Colorless gas.
Formula: SOF_2.
Constants: Mol wt: 86.07, bp: $-30°C$, d: 2.93, mp: $-110°C$.
Hazard Analysis
Toxic Hazard Rating:
  Acute Local: Irritant 3; Inhalation 3.
  Acute Systemic: Inhalation 3.
  Chronic Local: Inhalation 3.
  Chronic Systemic: Inhalation 3.
Disaster Hazard: Dangerous; when heated, it emits highly toxic fumes; will react with water or steam to produce toxic and corrosive fumes.
Countermeasures
Ventilation Control: Section 2.
Personnel Protection: Section 3.
First Aid: Section 1.
Storage and Handling: Section 7.

THIOPHENE
General Information
Synonym: Thiofuran.
Description: Clear and colorless liquid.
Formula: $SCHCHCHCH$.
Constants: Mol wt: 84.13, bp: $84.1°C$, fp: $-38.3°C$, flash p: $30°F$, d: 1.0583 at $25°/4°C$, vap. press.: 40 mm at $12.5°C$, vap. d.: 2.9.
Hazard Analysis
Toxicity: Details unknown. Animal experiments suggest moderate toxicity on acute exposure.
Fire Hazard: Dangerous, when exposed to heat or flame.
Disaster Hazard: Dangerous; when heated to decomposition, it emits highly toxic fumes of oxides of sulfur; can react vigorously with oxidizing materials.
Countermeasures
Ventilation Control: Section 2.
To Fight Fire: Foam, carbon dioxide, dry chemical or carbon tetrachloride (Section 6).
Storage and Handling: Section 7.

THIOPHENOL. See phenyl mercaptan.

α-THIOPHENYLBORIC ACID
General Information
Description: Colorless crystals; soluble in water.
Formula: $(C_4H_3S)B(OH)_2$.
Constants: Mol wt: 128.0, mp: $134°C$.
Hazard Analysis and Countermeasures
See boron compounds and sulfates, and phenyl mercaptan.

THIOPHOS. See parathion.

THIOPHOSGENE
General Information
Synonyms: Thiocarbonyl chloride; thiocarbonaldehyde.
Hazard Analysis
Toxicity: Details unknown. Probably a strong irritant. See also phosgene.
Disaster Hazard: Dangerous. When heated to decomposition or upon hydrolysis it emits highly toxic fumes.
Countermeasures
Storage and Handling: Section 7.
Shipping Regulations: Section 11.
  Coast Guard Classification: Poison B; poison label.
  ICC: Poison B, poison label, 1 gallon.
  IATA: Poison B, poison label, not acceptable (passenger), 5 liters (cargo).

THIOPHOSPHORAMIDE
General Information
Description: Amorphous, yellow-white powder.
Formula: $PS(NH_2)_3$.
Constants: Mol wt: 111.12, mp: decomposes at $200°C$, d: 1.7 at $13°C$.
Hazard Analysis
Toxicity: Unknown. See also amides.
Disaster Hazard: Dangerous; when heated to decomposition or on contact with acid or acid fumes, it emits highly toxic fumes of oxides of sulfur and phosphorus.
Countermeasures
Storage and Handling: Section 7.

THIOPHOSPHORYL BROMIDE
General Information
Description: Cubic yellow crystals.
Formula: $PSBr_3$.
Constants: Mol wt: 302.8, mp: $38°C$, bp: decomposes at $175°C$, d: 2.85 at $17°C$.
Hazard Analysis
Toxicity: Details unknown; probably toxic. See also bromides and thiophosphoryl chloride.
Disaster Hazard: Dangerous; when heated to decomposition, it emits highly toxic fumes of bromides, oxides of sulfur and phosphorus; will react with water or steam to produce toxic and corrosive fumes.
Countermeasures
Storage and Handling: Section 7.

THIOPHOSPHORYL CHLORIDE
General Information
Synonym: Phosphorus sulfochloride.
Description: Colorless, mobile liquid, pungent odor.
Formula: $PSCl_3$.
Constants: Mol wt: 169.45, bp: $125°C$, fp: $-35°C$, flash p: none, d: 1.63 at $25°/4°C$, vap. press.: 22 mm at $25°C$, vap. d.: 5.86.
Hazard Analysis
Toxic Hazard Rating:
  Acute Local: Irritant 3; Ingestion 3; Inhalation 3.
  Acute Systemic: U.
  Chronic Local: U.
  Chronic Systemic: U.
Disaster Hazard: Dangerous; when heated it emits highly toxic fumes; will react with water or steam to produce toxic and corrosive fumes.
Countermeasures
Ventilation Control: Section 2.
Personnel Protection: Section 3.
First Aid: Section 1.

TOXIC HAZARD RATING CODE (For detailed discussion, see Section 1.)

0 NONE: (a) No harm under any conditions; (b) Harmful only under unusual conditions or overwhelming dosage.

1 SLIGHT: Causes readily reversible changes which disappear after end of exposure.

2 MODERATE: May involve both irreversible and reversible changes; not severe enough to cause death or permanent injury.

3 HIGH: May cause death or permanent injury after very short exposure to small quantities.

U UNKNOWN: No information on humans considered valid by authors.

Storage and Handling: Section 7.
Shipping Regulations: Section 11.
    I.C.C.: Corrosive liquid; white label, 1 gallon.
    Coast Guard Classification: Corrosive liquid; white label.
    IATA: Corrosive liquid, white label, not acceptable (passenger), 1 liter (cargo).

THIOPHOSPHORYL DIBROMIDE CHLORIDE
General Information
Description: Pale green, fuming liquid.
Formula: $PSBr_2Cl$.
Constants: Mol wt: 258.34, mp: $-60°C$, bp: $98°C$ at 60 mm , d liquid: 2.48 at $0°C$.
Hazard Analysis
Toxicity: Highly corrosive and toxic. See thiophosphoryl chloride.
Disaster Hazard: Dangerous; when heated to decomposition, it emits highly toxic fumes; will react with water or steam to produce toxic and corrosive fumes.
Countermeasures
Storage and Handling: Section 7.

THIOPHOSPHORYL FLUORIDE
General Information
Description: Gas.
Formula: PSF_3.
Constants: Mol wt: 120.05, mp: $3.8°C$ at 7.6 atm., bp: decomposes.
Hazard Analysis
Toxic Hazard Rating:
    Acute Local: Irritant 3; Inhalation 3.
    Acute Systemic: Ingestion 3; Inhalation 3.
    Chronic Local: Inhalation 3.
    Chronic Systemic: Inhalation 2.
Disaster Hazard: Dangerous; will react with water or steam to produce toxic and corrosive fumes.
Countermeasures
Ventilation Control: Section 2.
Personnel Protection: Section 3.
First Aid: Section 1.
Storage and Handling: Section 7.

"THIOSEMICARBAZONE"
General Information
Synonym: p-Acetylaminobenzaldehyde thiosemicarbazone.
Description: Pale yellow crystals.
Formula: $C_{10}H_{12}N_4OS$.
Constants: Mol wt: 236.3, mp: $207°C$.
Hazard Analysis
Toxic Hazard Rating:
    Acute Local: Irritant 2; Ingestion 2.
    Acute Systemic: Ingestion 2.
    Chronic Local: Allergen 2.
    Chronic Systmeic: Ingestion 2.
Toxicology: Can cause nausea, vomiting, skin rashes, liver injury and bone marrow depression.
Disaster Hazard: Dangerous; when heated to decomposition or on contact with acid fumes or acid, it emits highly toxic fumes of oxides of sulfur.
Countermeasures
Storage and Handling: Section 7.

THIOSULFATES
Hazard Analysis
Toxicity: Up to 12 grams of sodium thiosulfate can be taken daily by mouth with no ill effect except catharsis. Most of the thiosulfates have low toxicity.
Disaster Hazard: Dangerous; when heated to decomposition, they emit highly toxic fumes of oxides of sulfur.
Countermeasures
Storage and Handling: Section 7.

THIOSULFURIC ACID
General Information
Formula: $H_2S_2O_3$.

Constant: Mol wt: 114.15.
Hazard Analysis
Toxicity: Unknown. Probably irritant and corrosive. See also sulfuric acid.
Disaster Hazard: Dangerous; when heated to decomposition, it emits highly toxic fumes of oxides of sulfur; will react with water or steam to produce heat.
Countermeasures
Storage and Handling: Section 7.

THIOUREA
General Information
Synonym: Thiocarbamide.
Description: White powder or crystals.
Formula: NH_2CSNH_2.
Constants: Mol wt: 76.1, mp: $180–182°C$, bp: decomposes, d: 1.405.
Hazard Analysis
Toxic Hazard Rating:
    Acute Local: U.
    Acute Systemic: Ingestion 2.
    Chronic Local: Irritant 1.
    Chronic Systemic: Ingestion 2.
Toxicology: Is said to cause depression of bone marrow with anemia, leukopenia, and thrombocytopenia. May also cause allergic skin eruptions.
Disaster Hazard: Dangerous; when heated to decomposition, it emits highly toxic fumes of oxides of sulfur.
Countermeasures
Personal Hygiene: Section 3.
Storage and Handling: Section 7.

THIOVANIC ACID. See thioglycolic acid.

THIOVANOL. See monothioglycerol.

1,4-THIOXANE
General Information
Description: Water-white, refractive mobile liquid; characteristic odor.
Formula: $O(CH_2CH_2)_2S$.
Constants: Mol wt: 104.1, bp: $148.7°C$, fp: $-17°C$, flash p: $108°$ F (C.C.), d: 1.117 at $20°C$.
Hazard Analysis
Toxicity: Details unknown; probably moderately toxic. Limited animal experiments suggest low toxicity on acute exposure.
Fire Hazard: Moderate, when exposed to heat or flame.
Disaster Hazard: Dangerous; when heated to decomposition, it emits highly toxic fumes of oxides of sulfur; can react with oxidizing materials.
Countermeasures
Storage and Handling: Section 7.
To Fight Fire: Water, foam, carbon dioxide, dry chemical or carbon tetrachloride (Section 6).

p-THIOXENE
General Information
Synonym: 2,4-Thioxene; 2,4-dimethylthiophene.
Description: Liquid.
Formula: $(CH_3)_2C_4H_2S$.
Constants: Mol wt: 112.2, bp: $138°C$, d: 0.9956 at $20°C$.
Hazard Analysis
Toxic Hazard Rating:
    Acute Local: U.
    Acute Systemic: Inhalation 2; Skin Absorption 2.
    Chronic Local: U.
    Chronic Systemic: U.
Fire Hazard: Moderate, when exposed to heat or flame.
Disaster Hazard: Dangerous; when heated to decomposition, it emits highly toxic fumes of sulfur; can react with oxidizing materials.
Countermeasures
Ventilation Control: Section 2.
To Fight Fire: Foam, carbon dioxide, dry chemical or carbon tetrachloride (Section 6).

Personnel Protection: Section 3.
Storage and Handling: Section 7.

THIRAM. See bis(dimethylthiocarbamyl) disulfide.
Shipping Regulations: Section 11.
    IATA: Other restricted articles, class A, no label required, no limit (passenger and cargo).

THORIUM
General Information
Description: Cubic, gray crystals. A chemical element.
Formula: Th.
Constants: At wt: 232.12, mp: 1845°C, bp: >3000°C, d: 11.2.
Hazard Analysis
Toxicology: Cases of dermatitis due to thorium have been reported.
Radiation Hazard: Section 5. For permissible levels, see Table 5, p. 150.
    Natural isotope ^{228}Th (Radiothorium, Thorium Series), half life 1.9 y. Decays to radioactive ^{224}Ra by emitting alpha particles of 5.3–5.4 MeV. Also emits gamma rays of 0.08–0.22 MeV and X-rays.
    Natural isotope ^{230}Th (Ionium, Uranium Series), half life 8×10^4 y. Decays to radioactive ^{226}Ra by emitting alpha particles of 4.6–4.7 MeV.
    Natural isotope ^{232}Th (Thorium Series), half life 1.4×10^{10} y. Decays to radioactive ^{228}Ra by emitting alpha particles of 4.0 MeV.
    Natural isotope ^{234}Th (UX-1, Uranium Series), half life 24 d. Decays to radioactive ^{234}Pa by emitting beta particles of 0.10 (35%), 0.19 (65%) MeV.
Fire Hazard: Moderate, in the form of dust, when exposed to heat or flame or by chemical reaction with oxidizers. See powdered metals (Section 6).
Countermeasures
Ventilaton Control: Section 2.
Personnel Proection: Section 3.
Shipping Regulationns: Section 11.
    I.C.C.: Flammable solid; yellow label, 25 pounds.
    Coast Guard Classification: Inflammable solid; yellow label.
    IATA: Flammable solid, yellow label, not acceptable (passenger), 12 kilograms (cargo).

THORIUM CHLORIDE
General Information
Synonym: Thorium tetrachloride.
Description: White, odorless crystals; soluble in water, alcohol.
Formula: ThCl$_4$.
Constants: Mol wt: 373.88, d: 4.59, mp: 820°C, bp: 921°C.
Hazard Analysis
Toxic Hazard Rating:
    Acute Local: Irritant 2; Ingestion 2; Inhalation 1.
    Acute Systemic: U.
    Chronic Local: U.
    Chronic Systemic: U.
Disaster Hazard: Dangerous. See chlorides.

THORIUM COMPOUNDS. For radiation hazard see thorium. Toxicity other than due to radiation has not been described.

THORIUM DIHYDRIDE
General Information
Description: Black, metallic crystals; reacts with water.

Formula: ThH$_2$.
Constants: Mol wt: 234.1, d: 8.24, mp: decomposes.
Hazard Analysis and Countermeasures
See thorium compounds and hydrides.

THORIUM NITRATE
General Information
Description: White, crystalline mass; soluble in water and alcohol.
Formula: Th(NO$_3$)$_4$·4H$_2$O.
Constant: Mol wt: 552.1.
Hazard Analysis
Toxic Hazard Rating: See thorium and nitrates.
Disaster Hazard: Dangerous, oxidizing material; when in contact with readily combustible substances it will cause violent combustion or ignition.
Countermeasures
Storage and Handling: Section 7.
Shipping Regulations: Section 11.
    IATA: Oxidizing material, yellow label, 12 kilograms (passenger), 45 kilograms (cargo).

THORIUM PICRATE
General Information
Description: Crystalline powder.
Formula: Th(C$_6$H$_2$N$_3$O$_7$)$_4$·10H$_2$O.
Constants: Mol wt: 1324.68.
Hazard Analysis and Countermeasures
See picric acid and thorium.

THORIUM TETRACHLORIDE. See thorium chloride.

THORIUM TETRAHYDRIDE
General Information
Description: Black, metallic crystals; reacts with water.
Formula: Th$_4$H$_{15}$.
Constants: Mol wt: 943.3, d: 8.24, mp: decomposes at red heat.
Hazard Analysis and Countermeasures
See thorium compounds and hydrides.

THORON
General Information
Description: An inert gaseous element.
Formula: Tn or Rn^{220}.
Constant: at wt: 220.
Hazard Analysis
Toxicity: See radiation hazard under thorium.
Radiation Hazard: See Rn-220.
Countermeasures
Ventilation Control: Section 2.

THREONINE
General Information
Synonym: 2-Amino-β-hydroxy butyric acid.
Description: An essential amino acid; colorless crystals; soluble in water; optically active.
Formula: CH$_3$CH(OH)CH(NH$_2$)COOH.
Constants: Mol wt: 119, mp (DL form): 228–229°C with decomposition, mp (L form): 255–257°C with decomposition.
Hazard Analysis
Toxic Hazard Rating: U. A nutrient and/or dietary supplement food additive (Section 10).

THULIUM
General Information
Description: A rare earth metallic element.
Formula: Tm.

TOXIC HAZARD RATING CODE (For detailed discussion, see Section 1.)

0 NONE: (a) No harm under any conditions; (b) Harmful only under unusual conditions or overwhelming dosage.

1 SLIGHT: Causes readily reversible changes which disappear after end of exposure.

2 MODERATE: May involve both irreversible and reversible changes; not severe enough to cause death or permanent injury.

3 HIGH: May cause death or permanent injury after very short exposure to small quantities.

U UNKNOWN: No information on humans considered valid by authors.

Constant: At wt: 169.40.

Hazard Analysis

Toxicity: Unknown. As a lanthanon it may inhibit blood clotting. See also lanthanum.

Radiation Hazard: Section 5. For permissible levels, see Table 5, p. 150.

     Artificial isotope [171]Tm, half life 1.9 y. Decays to stable [171]Yb by emitting beta particles of 0.10 MeV.

Fire Hazard: Moderate, in the form of dust, when exposed to flame. See also powdered metals (Section 6).

Explosion Hazard: Slight, in the form of dust, when exposed to flame. See also powdered metals.

THYME CAMPHOR. See thymol.

THYME OIL

General Information

Description: Colorless to reddish-brown liquid; pleasant odor; sharp taste.

Constant: D: 0.984–0.930 at 25°/25°C.

Hazard Analysis

Toxic Hazard Rating:

     Acute Local: Irritant 1; Ingestion 1.

     Acute Systemic: U.

     Chronic Local: Allergen 1.

     Chronic Systemic: Ingestion 1; Inhalation 1.

Fire Hazard: Slight; (Section 6).

Countermeasures

Personal Hygiene: Section 3.

Storage and Handling: Section 7.

THYMOL

General Information

Synonym: Thyme camphor.

Description: Colorless, translucent crystals.

Formula: $C_{10}H_{14}O$.

Constants: Mol wt: 150.2, mp: 51°C, bp: 233°C, d: 0.972, vap. press.: 1 mm at 64.3°C.

Hazard Analysis

Toxic Hazard Rating:

     Acute Local: Irritant 1; Ingestion 2; Allergen 1.

     Acute Systemic: Ingestion 2; Inhalation 2.

     Chronic Local: Allergen 1.

     Chronic Systemic: Ingestion 1.

Disaster Hazard: Moderately dangerous; when heated, it emits toxic fumes.

Countermeasures

Personnel Protection: Section 3.

Storage and Handling: Section 7.

THYMOL IODIDE

General Information

Description: Red brown powder or crystals; slight aromatic odor; soluble in ether, chloroform and fixed or volatile oils; slightly soluble in alcohol; insoluble in water.

Formula: Principally dithymol diiodide [$C_6H_2(CH_3)\cdot (OI)(C_3H_7)_2$].

Constant: Mol wt: 550.

Hazard Analysis

Toxicity: Unknown. A trace mineral added to animal feeds (Section 10).

TIBAL. See triisobutyl aluminum.

TIEMANNITE. See mercuric selenide.

TIGLIC ACID

General Information

Synonym: Crotonalic acid.

Description: Thick, syrupy liquid or colorless crystals.

Formula: $CH_3CHC(CH_3)CO_2H$.

Constants: Mol wt: 100.11, mp: 65°C, bp: 198.5°C, d: 0.9641, vap. press.: 1 mm at 52.0°C.

Hazard Analysis

Toxic Hazard Rating:

     Acute Local: Irritant 1; Ingestion 2.

     Acute Systemic: Ingestion 3.

     Chronic Local: U.

     Chronic Systemic: U.

Disaster Hazard: Moderately dangerous; when heated to decomposition, it emits toxic fumes.

Countermeasures

Personnel Protection: Section 3.

Storage and Handling: Section 7.

TIGLIUM OIL. See croton oil.

TIME FUZES. See explosives, high.

Shipping Regulations: Section 11.

     I.C.C.: Explosive C, 150 pounds.

     Coast Guard Classification: Explosive C.

     IATA: Explosive, explosive label, 25 kilograms (passenger), 95 kilograms (cargo).

TIN(ALPHA)

General Information

Description: Cubic gray crystalline metallic element.

Formula: Sn.

Constants: At wt: 118.70, mp: 231.9°C; stabilizes <18°C, bp: 2260°C, d: 5.75, vap. press.: 1 mm at 1492°C.

Hazard Analysis

Toxicity: See tin compounds.

Radiation Hazard: Section 5. For permissible levels, see Table 5, p. 150.

     Artificial isotope [113]Sn, half life 118 d. Decays to radioactive [113m]In by electron capture. Emits gamma rays of 0.26 MeV and X-rays.

     Artificial isotope [125]Sn, half life 9.4 d. Decays to radioactive [125]Sb by emitting beta particles of 2.33 MeV.

Fire Hazard: Slight, in the form of dust, when exposed to heat or by spontaneous chemical reaction. See also powdered metals (Section 6).

TIN BIFLUORIDE. See tin fluoride.

TIN BROMIDE. See stannic bromide.

TINCAL. See borax.

TIN CHLORIDE. See stannic chloride.

TIN COMPOUNDS

Hazard Analysis

Toxicity: Elemental tin is not generally considered toxic. Some inorganic tin salts are irritants or can liberate toxic fumes on decomposition. The latter is particularly true of tin halogens. Alkyl tin compounds may be highly toxic and produce skin rashes. Dust of tin oxides have caused pneumoconiosis which is relatively benign.

TLV: ACGIH (recommended); organic: 0.1 milligrams per cubic meter of air; inorganic (except oxide): 2 milligrams per cubic meter of air.

TINCTURE OF TABASCO PEPPER. See capsicum.

TIN FLUORIDE

General Information

Synonym: Stannous fluoride; tin bifluoride.

Description: White lustrous, crystalline powder.

Formula: SnF_2.

Constant: Mol wt: 156.70.

Hazard Analysis and Countermeasures

See fluorides and tin compounds.

TINKAL. See borax.

TIN MONOPHOSPHIDE

General Information

Synonyms: Tin phosphide; stannic phosphide.

Description: Silver-white crystals.

Formula: SnP.

Constants: Mol wt: 149.68, d: 6.56.
Hazard Analysis and Countermeasures
See phosphides and tin compounds.

TIN OXIDES
General Information
Synonyms: (a) Stannous oxide; (b) stannic oxide (cassiterite).
Description: (a) Tetragonal black powder; (b) white powder.
Formula: (a) SnO; (b)SnO_2.
Constants: Mol wt: (a) 134.70; (b) 150.7, mp: (a) decomposes at 700–950°C; (b) decomposes at 1127°C, d: (a) 6.446 at 0°C; (b) 6.95.
Hazard Analysis
Toxicity: See tin compounds.

TIN PHOSPHIDE. See tin monophosphide.

TIN TETRABROMIDE. See stannic bromide.

TIN TETRACHLORIDE. See stannic chloride.

TIN TETRACHLORIDE, ANHYDROUS
Shipping Regulations: Section 11.
   I.C.C.: Corrosive liquid; white label, 1 quart.
   Coast Guard Classification: Corrosive liquid; white label.
   IATA: Corrosive liquid, white label, 1 liter (passenger and cargo).

TIN TETRAHYDRIDE
General Information
Synonym: Stannane.
Description: Gas.
Formula: SnH_4.
Constants: Mol wt: 122.73, mp: −150°C, bp: −52°C (decomp.).
Hazard Analysis
Toxicity: See hydrides.
Fire Hazard: Moderate, when exposed to flame. See also hydrides (Section 6).
Explosion Hazard: Slight, when exposed to flame. See also hydrides.
Disaster Hazard: Moderately dangerous; when heated, it emits toxic fumes and may explode; can react with oxidizing materials.
Countermeasures
Ventilation Control: Section 2.
Storage and Handling: Section 7.

TIN TETRAMETHYL
General Information
Synonym: Tetramethylstannane.
Description: Colorless liquid.
Formula: $Sn(CH_3)_4$.
Constants: Mol wt: 178.84, bp: 78°C, lel: 1.9%, d: 1.314 at 0°/4°C, vap. d.: 6.2.
Hazard Analysis
Toxicity: See tin compounds.
Fire Hazard: Moderate, when exposed to heat or flame.
Spontaneous Heating: No.
Explosion Hazard: Moderate, when exposed to flame.
Disaster Hazard: Dangerous; when heated, it emits acrid fumes and may explode; can react with oxidizing materials.
Countermeasures
Ventilation Control: Section 2.
To Fight Fire: Water, foam, carbon dioxide, dry chemical or carbon tetrachloride (Section 6).
Storage and Handling: Section 7.

TIN TETRAPHENYL
General Information
Description: Colorless crystals.
Formula: $(C_6H_5)_4Sn$.
Constants: Mol wt: 427.1, mp: 226°C, bp: 424°C, flash p: 450°F (C.C.), d: 1.490.
Hazard Analysis
Toxicity: Highly toxic. See phenol and tin compounds.
Fire Hazard: Slight, when exposed to heat or flame.
Disaster Hazard: Dangerous; when heated, it emits toxic fumes; can react with oxidizing materials.
Countermeasures
Storage and Handling: Section 7.
To Fight Fire: Foam, carbon dioxide, dry chemical or carbon tetrachloride (Section 6).

TIN TRIPHOSPHIDE
General Information
Description: Crystals.
Formula: SnP_3.
Constants: Mol wt: 211.64, mp: decomposes $>415°C$ to Sn_4P_3, d: 4.10 at 0°C.
Hazard Analysis and Countermeasures
See phosphides and tin compounds.

TITANIUM
General Information
Description: Dark gray, amorphous powder or white, lustrous metal.
Formula: Ti.
Constants: At wt: 47.90, mp: 1800°C, bp: $>3000°C$, d: 4.5 at 20°C, autoign. temp.: 700–800°C for massive metal in air; 250°C for powder.
Hazard Analysis
Toxicity: See titanium compounds.
Fire Hazard: Moderate, in the form of dust, when exposed to heat or flame or by chemical reaction. See also powdered metals. Titanium can burn in an atmosphere of carbon dioxide or nitrogen or air. Ordinary extinguishers are often ineffective against titanium fires. Such fires require the special extinguishers designed for metal fires. See magnesium. In airtight enclosures, titanium fires can be controlled by the use of argon or helium. When titanium burns in the absence of moisture, it burns slowly but evolves much heat. The application of water to burning titanium can cause an explosion (Section 6).
Explosion Hazard: Finely divided titanium dust and powders, like most metal powders are potential explosion hazards when exposed to sparks, open flame or high heat sources. See magnesium.
Countermeasures
Shipping Regulations: Section 11.
   Coast Guard Classification: Inflammable solid; yellow label.

TITANIUM BUTYLATE. See butyl titanate.

TITANIUM COMPOUNDS
Hazard Analysis
Toxic Hazard Rating:
   Acute Local: Ingestion 1; Inhalation 1.
   Acute Systemic: U.
   Chronic Local: U.
   Chronic Systemic: U.
Toxicology: This material is considered to be physiologically inert. There are no reported cases in the literature where titanium as such has caused intoxication. The

TOXIC HAZARD RATING CODE (For detailed discussion, see Section 1.)

0 NONE: (a) No harm under any conditions; (b) Harmful only under unusual conditions or overwhelming dosage.

1 SLIGHT: Causes readily reversible changes which disappear after end of exposure.

2 MODERATE: May involve both irreversible and reversible changes; not severe enough to cause death or permanent injury.

3 HIGH: May cause death or permanent injury after very short exposure to small quantities.

U UNKNOWN: No information on humans considered valid by authors.

dusts of titanium or titanium compounds such as titanium oxide, may be placed in the nuisance category. Titanium tetrachloride, however, is an irritating and corrosive material, because, when exposed to moisture, it hydrolyzes to hydrogen chloride. See hydrochloric acid.

TITANIUM DIBROMIDE
General Information
Description: Black powder.
Formula: $TiBr_2$.
Constants: Mol wt: 207.73, mp: decomposes $> 500°C$.
Hazard Analysis and Countermeasures
See bromides.

TITANIUM DICHLORIDE
General Information
Description: Light brown to black, deliquescent solid.
Formula: $TiCl_2$.
Constants: Mol wt: 118.81, mp: sublimes in H_2.
Hazard Analysis and Countermeasures
See titanium compounds and hydrochloric acid.

TITANIUM DIOXIDE
General Information
Synonym: Rutile.
Description: Blue crystals.
Formula: TiO_2.
Constants: Mol wt: 79.90, mp: 1640°C (decomp.), d: 4.26.
Hazard Analysis
Toxicity: See titanium compounds. A common air contaminant (Section 4).
TLV: ACGIH (recommended); 15 milligrams per cubic meter of air.

TITANIUM DISULFIDE
General Information
Description: Yellow scales.
Formula: TiS_2.
Constant: Mol wt: 112.03.
Hazard Analysis and Countermeasures
See sulfides.

TITANIUM HYDRIDE
General Information
Description: Metallic, dark gray powder or crystals.
Formula: TiH_2.
Constants: Mol wt: 49.9, d: 3.76.
Hazard Analysis
Toxicity: See hydrides and titanium compounds.
Fire Hazard: Moderate, when exposed to heat or flame. Burns brilliantly in air. See hydrides.
Explosion Hazard: Moderate, in the form of dust, by chemical reaction.
Disaster Hazard: See hydrides.
Countermeasures
Storage and Handling: Section 7.
Shipping Regulations: Section 11.
  IATA: Flammable solid, yellow label, not acceptable (passenger), 12 kilograms (cargo).

TITANIUM ISOPROPYLATE. See isopropyl titanate.

TITANIUM METAL POWDER, DRY
Shipping Regulations: Section 11.
  ICC: Flammable solid, yellow label, 75 pounds.
  IATA: Flammable solid, yellow label, not acceptable (passenger), 35 kilograms (cargo).

TITANIUM METAL POWDER, WET WITH NOT LESS THAN 20% WATER
Shipping Regulations: Section 11.
  Coast Guard Classification: Inflammable solid; yellow label.
  ICC: Flammable solid, yellow label, 150 pounds.

IATA: Flammable solid, yellow label, not acceptable (passenger), 70 kilograms (cargo).

TITANIUM MONOSULFIDE
General Information
Description: Reddish solid.
Formula: TiS.
Constant: Mol wt: 79.97.
Hazard Analysis and Countermeasures
See sulfides.

TITANIUM NITRIDE
General Information
Description: Brassy crystals.
Formula: TiN.
Constants: Mol wt: 62.0, mp: 2950°C, d: 5.43.
Hazard Analysis and Countermeasures
See titanium compounds and nitrides.

TITANIUM OXALATE
General Information
Description: Yellow prisms.
Formula: $Ti_2(C_2O_4)_3 \cdot 10H_2O$.
Constant: Mol wt: 540.02.
Hazard Analysis and Countermeasures
See oxalates and titanium compounds.

TITANIUM OXIDE. See titanic dioxide.

TITANIUM PHOSPHIDE
General Information
Description: Gray, metallic solid.
Formula: TiP.
Constants: Mol wt: 78.88, d: 3.95 at 25°C.
Hazard Analysis and Countermeasures
See phosphides and titanium compounds.

TITANIUM SESQUISULFATE. See titanous sulfate.

TITANIUM SESQUISULFIDE
General Information
Description: Grayish-black crystals.
Formula: Ti_2S_3.
Constant: Mol wt: 192.00.
Hazard Analysis and Countermeasures
See sulfides and titanium compounds.

TITANIUM TETRACHLORIDE
General Information
Description: Colorless, light yellow liquid; fumes in moist air.
Formula: $TiCl_4$.
Constants: Mol wt: 189.73, mp: $-30°C$, bp: 136.4°C, d: 1.722 at 25°/25°C, vap. press.: 10 mm at 21.3°C.
Hazard Analysis
Toxic Hazard Rating:
  Acute Local: Irritant 3; Inhalation 3.
  Acute Systemic: U.
  Chronic Local: Inhalation 3.
  Chronic Systemic: U.
Toxicology: Highly corrosive because it liberates heat and hydrochloric acid upon contact with moisture. If spilled on skin, wipe off with dry cloth before applying water.
Disaster Hazard: See hydrochloric acid.
Countermeasures
Storage and Handling: Section 7.
Shipping Regulations: Section 11.
  I.C.C.: Corrosive liquid; white label, 10 gallons.
  Coast Guard Classification: Corrosive liquid; white label.
  IATA: Corrosive liquid, white label, 1 liter (passenger), 40 liters (cargo).

TITANIUM TETRAFLUORIDE
General Information
Description: White powder.

Formula: TiF$_4$.
Constants: Mol wt: 123.90, bp: 284°C, d: 2.798 at 20.5°C.
Hazard Analysis and Countermeasures
See fluorides.

TITANIUM TRICHLORIDE
General Information
Description: Dark violet-colored in solution.
Formula: TiCl$_3$.
Constants: Mol wt: 154.28, mp: 440°C (decomp.).
Hazard Analysis and Countermeasures
See hydrochloric acid.

TITANIUM TRIFLUORIDE
General Information
Description: Purple-red or violet crystals.
Formula: TiF$_3$.
Constants: Mol wt: 104.90.
Hazard Analysis and Countermeasures
See fluorides.

TITANOUS SULFATE
General Information
Synonym: Titanium sesquisulfate.
Description: Crystalline, green, deliquescent powder.
Formula: Ti$_2$(SO$_4$)$_3$.
Constant: Mol wt: 383.98.
Hazard Analysis and Countermeasures
See titanium compounds and sulfates.

TNA. See tetranitroaniline.

TNT. See trinitrotoluene.

TNX. See 2,4,6-trinitroxylene.

TOBACCO
General Information
Description: Dried leaves.
Hazard Analysis
Toxic Hazard Rating:
  Acute Local: Irritant 1; Allergen 1; Inhalation 1.
  Acute Systemic (infusions): Ingestion 2; Inhalation 2; Skin Absorption 2.
  Chronic Local: Irritant 1; Allergen 1.
  Chronic Systemic: Inhalation 1.
  Remarks: See nicotine.
Fire Hazard: Slight, when exposed to heat or flame (Section 6).
Countermeasures
Ventilation Control: Section 2.
Personel Hygiene: Section 3.
Storage and Handling: Section 7.

TOBACCO WOOD. See witch hazel.

TOCHLORINE. See sodium p-toluenesulfonchloramine.

α-TOCOPHEROL ACETATE
General Information
Description: A form of vitamin E; yellow, nearly odorless clear viscous oil; insoluble in water; freely soluble in alcohol; miscible with acetone and vegetable oils.
Formula: C$_{29}$H$_{49}$O·OOCCH$_3$.
Constants: Mol wt: 472, d: 0.950–0.964.
Hazard Analysis
Toxic Hazard Rating: U. A nutrient and/or dietary supplement food additive (Section 10).

TOCOPHEROLS
General Information
Synonym: Vitamin E.
Description: A group of related substances: α,β,γ,δ-tocopherol, which constitute vitamin E. These vitamin constituents are viscous oils; soluble in lipid solvents; insoluble in water.
Hazard Analysis
Toxic Hazard Rating: U. Used as chemical preservative, nutrient, and/or dietary supplement food additives (Section 10).

TOE PUFFS
General Information
Description: Toe puffs are box toe boards used in the manufacture of boots and shoes and may consist of several layers of fabric impregnated with celluloid solvent, rosin, and dye.
Hazard Analysis
Toxic Hazard Rating: U.
Fire Hazard: Dangerous, They are liable to spontaneous combustion.
Countermeasures
Shipping Regulations: Section 11.
  IATA: Flammable solid, yellow label, 12 kilograms (passenger), 45 kilograms (cargo).

TOLAMINE. See sodium p-toluenesulfonchloramine.

TOLAN. See diphyl acetylene.

o-TOLIDINE FLUOSILICATE
General Information
Description: Small white crystals; slightly soluble in alcohol.
Formula: (C$_6$H$_3$NH$_2$CH$_3$)$_2$·H$_2$SiF$_6$.
Constants: Mol wt: 356.4, mp: 269°C.
Hazard Analysis and Countermeasures
See fluosilicates and fluorides.

TOLUENE
General Information
Synonyms: Methylbenzene; phenylmethane; toluol.
Description: Colorless liquid; benzol-like odor.
Formula: C$_6$H$_5$CH$_3$.
Constants: Mol wt: 92.13, mp: −95°C to −94.5°C, bp: 110.4°C, flash p: 40°F (C.C.), ulc: 75–80, lel: 1.27%, uel: 7.0%, d: 0.866 at 20°/4°C, autoign. temp.: 947°F, vap. press.: 36.7 mm at 30°C, vap. d.: 3.14.
Hazard Analysis
Toxic Hazard Rating:
  Acute Local: Ingestion 2.
  Acute Systemic: Ingestion 2; Inhalation 2; Skin Absorption 1.
  Chronic Local: Irritant 1.
  Chronic Systemic: Ingestion 2; Inhalation 2; Skin Absorption 2.
TLV: ACGIH (recommended): 200 parts per million in air; 750 milligrams per cubic meter of air.
Toxicology: Toluene is derived from coal tar, and commercial grades usually contain small amounts of benzene as an impurity. Acute poisoning, resulting from exposures to high concentrations of the vapors, are rare with toluene. Inhalation of 200 ppm of toluene for 8 hours may cause impairment of coordination and reaction time; with higher concentrations (up to 800 ppm) these effects are increased and are observed in a shorter time. In the few cases of acute toluene poisoning re-

ported, the effect has been that of a narcotic, the workman passing through a stage of intoxication into one of coma. Recovery following removal from exposure has been the rule. An occasional report of chronic poisoning describes an anemia and leucopenia, with biopsy showing a bone marrow hypoplasia. These effects, however, are less common in people working with toluene, and they are not as severe.

Exposure to concentrations up to 200 ppm produces few symptoms. At 200 to 500 ppm, headache, nausea, loss of appetite, a bad taste, lassitude, impairment of coordination and reaction time are reported, but are not usually accompanied by any laboratory or physical findings of significance. With higher concentrations, the above complaints are increased and in addition, anemia, leucopenia and enlarged liver may be found in rare cases.

A common air contaminant (Section 4).

Fire Hazard: Dangerous, when exposed to heat or flame.

Spontaneous Heating: No.

Explosion Hazard: Moderate, when exposed to flame.

Disaster Hazard: Moderately dangerous; when heated, it emits toxic fumes; can react vigorously with oxidizing materials.

Countermeasures

Ventilation Control: Section 2.

To Fight Fire : Foam, carbon dioxide, dry chemical or carbon tetrachloride (Section 6).

Personnel Protection: Section 3.

Storage and Handling: Section 7.

Shipping Regulations: Section 11.

  I.C.C.: Flammable liquid; red label, 10 gallons.

  Coast Guard Classification: Inflammable liquid; red label.

  MCA warning label.

  IATA: Flammable liquid, red label, 1 liter (passenger), 40 liters (cargo).

2,4-TOLUENEDIAMINE
General Information
Synonym: Tolylenediamine.

Description: Prisms.

Formula: $CH_3C_6H_3(NH_2)_2$.

Constants: Mol wt: 122.17, mp: 99°C, bp: 280°C, vap. press.: 1 mm at 106.5°C.

Hazard Analysis
Toxic Hazard Rating:

  Acute Local: U.

  Acute Systemic: Ingestion 2; Inhalation 2.

  Chronic Local: U.

  Chronic Systemic: Ingestion 2; Inhalation 2.

Toxicology: This material has a marked toxic action upon the liver and can cause fatty degeneration of that organ. It is also thought to be an irritant. When solutions of it come in contact with the skin, it can cause irritation and blisters, particularly to individuals who are sensitive to it.

Disaster Hazard: Moderately dangerous; when heated, it emits toxic fumes.

Countermeasures
Ventilation Control: Section 2.

Personal Hygiene: Section 3.

Storage and Handling: Section 7.

Shipping Regulations: Section 11.

  IATA: Poison B, poison label, 25 kilograms (passenger), 95 kilograms (cargo).

2,5-TOLUENEDIAMINE
General Information
Synonyms: 2,5-Tolylenediamine; 2,5-diaminotoluene.

Description: Colorless, crystalline tablets.

Formula: $CH_3C_6H_3(NH_2)_2$.

Constants: Mol wt: 122.17, mp: 64°C, bp: 274°C.

Hazard Analysis
Toxic Hazard Rating:

  Acute Local: U.

  Acute Systemic: Ingestion 2; Inhalation 2.

Chronic Local: U.

Chronic Systemic: Ingestion 2; Inhalation 2.

Toxicology: This material has a toxic action upon the liver and can cause fatty degeneration of that organ. Its total effect upon the body seems to take place 3 different ways. It is toxic to the central nervous system. It produces jaundice by action on the liver and spleen, and it produces anemia by destruction of the red blood cells. In this action it is quite similar to aniline, although by no means identical with it. Its high boiling point and the fact that the material is solid at room temperature makes it somewhat less hazardous than aniline, particularly at ordinary working temperatures. The literature contains a reference to a permanent injury to an eye due to the use of this material as an eyelash dye. It is considered to be an irritant dye material. It can cause irritation and blisters on the fingers of individuals whose skins are sensitive to it.

Disaster Hazard: Moderately dangerous; when heated, it emits toxic fumes.

Countermeasures
Ventilation Control: Section 2.

Personal Hygiene: Section 3.

Storage and Handling: Section 7.

TOLUENE DIISOCYANATE. See 2,4-tolylene diisocyanate.

TOLUENE SUBSTITUTE
General Information
Description: Composed largely of octanes.

Constants: Bp: 100°C, flash p: 30°F, d: 0.743.

Hazard Analysis
Toxicity: See octane.

Fire Hazard: Dangerous, when exposed to heat or flame.

Explosion Hazard: Unknown.

Disaster Hazard: Dangerous, upon exposure to heat or flame; can react vigorously with oxidizing materials.

Countermeasures
Storage and Handling: Section 7.

To Fight Fire: Foam, carbon dioxide, or dry chemical or carbon tetrachloride (Section 6).

o-TOLUENESULFONIC ACID
General Information
Synonym: Methylbenzenesulfonic acid.

Description: Crystals.

Formula: $CH_3C_6H_4SO_3H$.

Constant: Mol wt: 172.2.

Hazard Analysis
Toxic Hazard Rating:

  Acute Local: Irritant 3; Ingestion 3; Inhalation 3.

  Acute Systemic: U.

  Chronic Local: U.

  Chronic Systemic: U.

Toxicity: See also p-toluene sulfonic acid.

Disaster Hazard: Dangerous; See sulfonates.

Countermeasures
Storage and Handling: Section 7.

p-TOLUENESULFONIC ACID
General Information
Synonym: p-Tolunene sulfonate.

Description: Colorless leaflets; soluble in alcohol, ether, and water.

Formula: $C_6H_4(SO_3H)(CH_3)$.

Constants: Mol wt: 172, mp: 107°C, bp: 140°C at 20 mm.

Hazard Analysis
Toxicity: Details unknown. Animal experiments show moderate systemic toxicity and high irritation.

Disaster Hazard: Dangerous. See sulfonates.

p-TOLUENESULFONYLAMIDE
Hazard Analysis
Toxicity: Details unknown; fungicide. See also amides.

Disaster Hazard: Dangerous; See sulfonates.
Countermeasures
Storage and Handling: Section 7.

α-**TOLUENE THIOL.** See benzyl mercaptan.

TOLUENE TRICHLORIDE. See benzotrichloride.

α-**TOLUIC ACID.** See phenylacetic acid.

m-TOLUIDINE
General Information
Synonym: m-Methylaniline.
Description: Colorless liquid.
Formula: $CH_3C_6H_4NH_2$.
Constants: Mol wt: 107.2, mp: −31.5°C, bp: 203.3°C, d: 0.989 at 20°/4°C, vap. press.: 1 mm at 41°C, vap. d.: 3.90.
Hazard Analysis
Toxicity: See o-toluidine.
Fire Hazard: Moderate, when exposed to heat or flame.
Disaster Hazard: Dangerous, when heated, it emits highly toxic fumes; can react vigorously on contact with oxidizing materials.
Countermeasures
Ventilation Control: Section 2.
To Fight Fire: Foam, carbon dioxide, dry chemical or carbon tetrachloride (Section 6).
Personnel Protection: Section 3.
Storage and Handling: Section 7.
Shipping Regulations: Section 11.
  MCA warning label.

o-TOLUIDINE
General Information
Synonym: o-Methylaniline.
Description: Colorless liquid.
Formula: $CH_3C_6H_4NH_2$.
Constants: Mol wt: 107.2, mp: −16.3°C, bp: 199.7°C, ulc: 20–25, flash p: 185°F (C.C.), d: 1.004 at 20°/4°C, autoign. temp.: 900°F, vap. press.: 1 mm at 44°C, vap. d.: 3.69.
Hazard Analysis
Toxic Hazard Rating:
Acute Local: Irritant 2; Allergen 1; Ingestion 2.
Acute Systemic: Ingestion 3; Inhalation 3; Skin Absorption 1.
Chronic Local: Allergen 1.
Chronic Systemic: Ingestion 2; Inhalation 2; Skin Absorption 2.
TLV: ACGIH (recommended); 5 parts per million in air; 22 milligrams per cubic meter of air. Can be absorbed through the intact skin.
Toxicology: This material can produce severe systemic disturbances. The main portal of entry into the body is the respiratory tract, particularly in cases of industrial exposure. The symptoms produced by intoxication due to this compound are headache, weakness, difficulty in breathing, air-hunger, psychic disturbances, and marked irritation of the kidneys and bladder. The literature does not yield any good data for comparing the toxicities of the o-, m- and p-isomers. Their behavior is generally comparable to that of aniline, and while the most frequent type of exposure is inhalation, a certain amount of exposure occurs by skin contact. It has been determined experimentally that a concentration of approximately 100 ppm is the maximum endurable for an hour without serious consequences, and that from 6 to 23 ppm is endurable for several hours without serious disturbance. See aniline.
Fire Hazard: Moderate, when exposed to heat or flame.
Spontaneous Heating: No.
Disaster Hazard: Dangerous; when heated, it emits highly toxic fumes; can react with oxidizing materials.
Countermeasures
Ventilation Control: Section 2.
To Fight Fire: Foam, carbon dioxide, dry chemical or carbon tetrachloride (Section 6).
Personnel Protection: Section 3.
First Aid: Section 1.
Storage and Handling: Section 7.
Shipping Regulations: Section 11.
  MCA warning label.

p-TOLUIDINE
General Information
Synonym: p-Methylaniline.
Description: Colorless leaflets.
Formula: $CH_3C_6H_4NH_2$.
Constants: Mol wt: 107.2, mp: 44.5°C, bp: 200.4°C, flash p: 188°F (C.C.), d: 1.046 at 20°/4°C, autoign. temp.: 900°F, vap. press.: 1 mm at 42°C, vap. d.: 3.90.
Hazard Analysis
Toxicity: See o-toluidine.
Fire Hazard: Moderate, when exposed to heat or flame.
Spontaneous Heating: No.
Disaster Hazard: Dangerous; when heated, it emits highly toxic fumes; can react vigorously on contact with oxidizing materials.
Countermeasures
Ventilation Control: Section 2.
To Fight Fire: Foam, carbon dioxide, dry chemical or carbon tetrachloride. (Section 6).
Personnel Protection: Section 3.
First Aid: Section 1.
Storage and Handling: Section 7.
Shipping Regulations: Section 11.
  MCA warning label.

p-TOLUIDINE HYDROCHLORIDE. See chlorotoluidine.

TOLUIDINE, LIQUID. See m- and o-toluidine.
Shipping Regulations: Section 11.
  IATA: Poison B, poison label, 1 liter (passenger), 220 liters (cargo).

TOLUOL. See toluene.

o-TOLUOL SULFOACID
General Information
Description: Liquid.
Formula: $C_{14}H_{14}O_3S$.
Constants: Mol wt: 262.32, flash p: 363°F.
Hazard Analysis
Toxicity: Unknown.
Fire Hazard: Slight, when exposed to heat or flame.
Disaster Hazard: Dangerous; when heated to decomposition, it emits highly toxic fumes of oxides of sulfur; can react with oxidizing materials.
Countermeasures
Storage and Handling: Section 7.
To Fight Fire: Foam, carbon dioxide, dry chemical or carbon tetrachloride (Section 6).

TOLYL ACETAMIDE. See p-acetotoluidide.

TOXIC HAZARD RATING CODE (For detailed discussion, see Section 1.)

0 NONE: (a) No harm under any conditions; (b) Harmful only under unusual conditions or overwhelming dosage.

1 SLIGHT: Causes readily reversible changes which disappear after end of exposure.

2 MODERATE: May involve both irreversible and reversible changes; not severe enough to cause death or permanent injury.

3 HIGH: May cause death or permanent injury after very short exposure to small quantities.

U UNKNOWN: No information on humans considered valid by authors

o-TOLYL BIGUANIDE HYDROCHLORIDE
General Information
Description: Colorless crystals.
Formula: CH₃C₆H₄NHC(NH)NHC(NH)NH₂·HCl.
Constants: Mol wt: 227.5, mp: 227°C, d: 1.264 at 30°C.
Hazard Analysis
Toxicity: Unknown.
Disaster Hazard: Dangerous; when heated to decomposition, it emits highly toxic fumes.
Countermeasures
Storage and Handling: Section 7.

p-TOLYL BROMIDE. See p-bromotoluene.

p-TOLYLDIETHANOLAMINE
General Information
Description: Crystals.
Formula: (HOC₂H₄)₂NC₆H₄CH₃.
Constants: Mol wt: 195.25, mp: 63.2°C, flash p: 385°F (O.C.), vap. d.: 6.73.
Hazard Analysis
Toxicity: Details unknown. See also amines.
Fire Hazard: Slight, when exposed to heat or flame; can react with oxidizing materials.
Countermeasures
To Fight Fire: Foam, carbon dioxide, dry chemical or carbon tetrachloride (Section 6).
Storage and Handling: Section 7.

TOLYLENE   DIISOCYANATE. See 2,4-tolylene diisocyanate.

2,4-TOLYLENE DIISOCYANATE
General Information
Synonyms: Tolylene diisocyanate; 2,4-diisocyanotoluene; TDI.
Description: Clear faintly yellow liquid.
Formula: H₃CC₆H₃(NCO)₂.
Constants: Mol wt: 174.16, bp: 118–120°C at 10 mm, fp: 20°C, d: 1.22 at 20°/4°C, flash p: 275°F (O.C.), vap. d.: 6.0, lel: 0.9%, uel: 9.5%.
Hazard Analysis
Toxic Hazard Rating:
  Acute Local: Irritant 3; Ingestion 3; Inhalation 3.
  Acute Systemic: Ingestion 2.
  Chronic Local: Irritant 3; Inhalation 3.
  Chronic Systemic: Ingestion 1; Allergen 3.
Toxicology: Capable of producing severe dermatitis and bronchial spasm. Following inhalation (especially if severe) victim should be observed by a physician. Particularly irritating to the eyes. A common air contaminant (Section 4).
TLV: ACGIH (recommended); 0.02 parts per million in air. 0.14 milligrams per cubic meter of air.
Fire Hazard: Combustible, when exposed to heat or flame.
Disaster Hazard: Dangerous; when heated to decomposition it emits highly toxic fumes.
Countermeasures
Ventilation Control: Section 2.
Personnel Protection: Section 3.
Storage and Handling: Section 7.
To Fight Fire: Water, foam, dry chemical.

o-TOLYLETHANOLAMINE
General Information
Description: Liquid.
Formula: CH₃C₆H₄NHCH₂CH₂OH.
Constants: Mol wt: 151.2, mp: −25°C, bp: 297.1°C, flash p: 290°F (O.C.), d: 1.0723 at 20°/20°C.
Hazard Analysis
Toxicity: Details unknown. See also amines.
Fire Hazard: Slight, when exposed to heat or flame; can react with oxidizing materials.
Countermeasures
To Fight Fire: Foam, carbon dioxide, dry chemical or carbon tetrachloride (Section 6).
Storage and Handling: Section 7.

TOLYLGERMANIUM TRIBROMIDE
General Information
Description: Colorless liquid, hydrolyzed by water.
Formula: (CH₃C₆H₄)GeBr₃.
Constants: Mol wt: 403.5, mp: 156°C at 13 mm.
Hazard Analysis and Countermeasures
See germanium compounds and bromides.

TOLYLGERMANIUM TRICHLORIDE
General Information
Description: Colorless liquid; hydrolyzed by water.
Formula: (CH₃C₆H₄)GeCl₃.
Constants: Mol wt: 270.1, bp: 116°C at 12 mm.
Hazard Analysis and Countermeasures
See germanium compounds and chlorides.

TOLYLGERMANIUM TRIIODIDE
General Information
Description: Colorless crystals; hydrolyzed by water.
Formula: (CH₃C₆H₄)GeI₃.
Constants: Mol wt: 545.0, mp: 72°C.
Hazard Analysis and Countermeasures
See germanium compounds and iodides.

TOLYL HYDRAZINE
Hazard Analysis
Toxicity: Unknown. See hydrazine.

TOLYL MALEIMIDE
Hazard Analysis
Toxicity: Unknown. Limited animal experiments suggest high toxicity and irritation.

p-TOLYL MERCURIC BROMIDE
General Information
Description: Lustrous crystals; soluble in organic solvents.
Formula: (C₇H₇)HgBr.
Constants: Mol wt: 371.7, mp: 228°C,
Hazard Analysis and Countermeasures
See mercury compounds, organic and bromides.

p-TOLYL MERCURIC CHLORIDE
General Information
Description: Silky tablets; insoluble in water.
Formula: (C₇H₇)HgCl.
Constants: Mol wt: 327.2, mp: 233°C.
Hazard Analysis and Countermeasures
See mercury compounds, organic and chlorides.

TOLYL MERCURY SALICYLATE.
Hazard Analysis and Countermeasures
See mercury compounds, organic.

o-TOLYL PHOSPHATE. See tri-o-cresyl phosphate.

N,m-TOLYLPHTHALAMIC ACID. See duraset.

o-TOLYLPROPANOLAMINE
General Information
Description: Liquid.
Formula: CH₃C₆H₄NHCH₂CHOHCH₃.
Constants: Mol wt: 165.2, bp: 170–180°C at 20 mm, vap. d.: 5.7.
Hazard Analysis
Toxicity: Details unknown. See also amines.
Fire Hazard: Moderate, when exposed to heat or flame; can react with oxidizing materials (Section 6).
Countermeasures
Storage and Handling: Section 7.

TOLYL TETRATHIO-o-STANNATE
General Information
Description: Solid.
Formula: (SC₆H₄CH₃)₄Sn.
Constants: Mol wt: 611.5, mp: 100°C.
Hazard Analysis and Countermeasures
See tin compounds and sulfates.

o-TOLYLTIN TRICHLORIDE
General Information
Description: Liquid.
Formula: $(C_6H_4CH_3)SnCl_3$.
Constants: Mol wt: 316.2, d: 1.7619, bp: 158°C at 20 mm.
Hazard Analysis and Countermeasures
See tin compounds, hydrochloric acid and chlorides.

m-TOLYLTIN TRICHLORIDE
General Information
Description: Colorless liquid.
Formula: $(C_6H_4CH_3)SnCl_3$.
Constants: Mol wt: 316.2, d: 1.7516, mp: $< -20°$C, bp: 151°C at 23 mm.
Hazard Analysis and Countermeasures
See tin compounds, hydrochloric acid and chlorides.

p-TOLYLTIN TRICHLORIDE
General Information
Description: Liquid; decomposed by water.
Formula: $(C_6H_4CH_3)SnCl_3$.
Constants: Mol wt: 316.2, d: 1.7522, bp: 157°C at 23 mm.
Hazard Analysis and Countermeasures
See tin compounds, hydrochloric acid and chlorides.

o-TOLYL p-TOLUENESULFONATE
General Information
Description: Liquid.
Formula: $C_{14}H_{14}O_3S$.
Constants: Mol wt: 262.31, flash p: 363°F (C.C.).
Hazard Analysis
Toxicity: Unknown. See also esters.
Fire Hazard: Slight, when exposed to heat or flame; can react with oxidizing materials.
Disaster Hazard: Dangerous. See sulfonates.
Countermeasures
To Fight Fire: Water, carbon dioxide, dry chemical or carbon tetrachloride (Section 6).
Storage and Handling: Section 7.

TOMORIN
General Information
Synonym: 3-(α-p-chlorophenyl-β-acetyl-ethyl)-4-hydroxy-coumarin.
Hazard Analysis
Toxicity: Highly toxic. See warfarin which closely resembles this material.
Disaster Hazard: Dangerous. When heated to decomposition it emits highly toxic fumes of chlorides.

TOXADUST. See toxaphene.

TOXALBUMIN. See abrin.

"TOXAPHENE"
General Information
Synonyms: Chlorinated camphene; compound 3956; alltox; geniphene; toxakil; toxadust and others.
Description: Yellow, waxy solid.
Formula: $C_{10}H_{10}Cl_8$.
Constants: Mol wt: 413.84, mp: 65–90°C.
Hazard Analysis
Toxic Hazard Rating:
Acute Local: Irritant 2.
Acute Systemic: Ingestion 3; Inhalation 3; Skin Absorption 3.
Chronic Local: U.
Chronic Systemic: Ingestion 3; Inhalation 3; Skin Absorption 3.

Caution: The Food and Drug Administration (1955) proposes a tolerance of 7 ppm for residues on fruits and vegetables. See section 10.
TLV: ACGIH (recommended); 0.5 milligrams per cubic meter of air. May be absorbed via the skin.
Toxicology: An insecticide. Toxic and in some instances lethal amounts of toxaphene can enter the body through the mouth, lungs, and skin. Systemic absorption of the insecticide is increased by the presence of digestible oils, and liquid preparations of the insecticide penetrate the skin more readily than do dusts and wettable powders.

Toxaphene resembles chlordane and to some extent camphor in its physiological action. It causes diffuse stimulation of the brain and spinal cord resulting in generalized convulsions of a tonic or clonic character. Death usually results from respiratory failure. Detoxification appears to occur in the liver.

Dangerous Acute and Chronic Dose: The lethal oral dose for man is estimated to be 2 to 7 grams; a toxicity of about four times that of DDT. Toxaphene causes moderate irritation on the skin but little or no sensitization. At least 7 human deaths have been reported due to toxaphene, all in children. Two families have been made ill by eating vegetables containing a large residue of toxaphene.

Signs and Symptoms of Poisoning: The symptoms are excitement followed by epileptiform convulsions associated with depression; these symptoms are aggravated by external stimuli. Death usually results from respiratory failure and may occur as early as 4 hours and as late as 24 hours after poisoning. In calves, sheep, and goats, internal hemorrhage and high temperatures have been reported, the latter also being one of the symptoms of camphor poisoning in man.

Pathology: Degenerative changes in the liver parenchyma and renal tubules were reported in laboratory animals chronically receiving toxaphene.

Treatment of poisoning: The stomach and intestinal tract should be evacuated. Oily laxatives should be avoided. Contaminated clothing should be removed and contaminated areas should be scrubbed with soap and water. Anticonvulsant drugs (bromides, barbiturates, etc.) should be administered. Sodium phenobarbital is the drug of choice to prevent convulsions, and sodium pentobarbitol is best to control those already in evidence. Dosages may have to be quite high in the presence of poisoning in order to control symptoms but should not be high enough to interfere with continuous sleep.
Disaster Hazard: Dangerous; when heated to decomposition it emits highly toxic fumes of chlorides.
Countermeasures
Ventilation Control: Section 2.
Personnal Protection: Section 3.
First Aid: Section 1.
Storage and Handling: Section 7.
Shipping Regulations: Section 11.
IATA: Other restricted articles, class A, no label required, no limit (passenger and cargo).

TOXILIC ACID. See maleic acid.

TOXILIC ANHYDRIDE. See maleic anhydride.

TOY CAPS. See explosives, low.
Shipping Regulations: Section 11.

TOXIC HAZARD RATING CODE (For detailed discussion, see Section 1.)

0 NONE: (a) No harm under any conditions; (b) Harmful only under unusual conditions or overwhelming dosage.

1 SLIGHT: Causes readily reversible changes which disappear after end of exposure.

2 MODERATE: May involve both irreversible and reversible changes not severe enough to cause death or permanent injury.

3 HIGH: May cause death or permanent injury after very short exposure to small quantities.

U UNKNOWN: No information on humans considered valid by authors.

I.C.C.: Explosive C, 150 pounds.
Coast Guard Classification: Explosive C.

TOY PROPELLANT DEVICES AND SMOKE DE-VICES
Shipping Regulations: Section 11.
   I.C.C.: Explosive C, 150 pounds.

TRACER FUZES. See explosives, high.
Shipping Regulations: Section 11.
   I.C.C.: Explosive C, 150 pounds.
   Coast Guard Classification: Explosive C.
   IATA: Explosive, explosive label, 25 kilograms (passenger), 70 kilograms (cargo).

TRACERS. See explosives, low.
Shipping Regulations: Section 11.
   I.C.C.: Explosive C, 150 pounds.
   Coast Guard Classification: Explosive C.
   IATA: Explosive, explosive label, 25 kilograms (passenger), 70 kilograms (cargo).

TRAGACANTH
General Information
Synonym: Gum tragacanth.
Description: Powder is white; pieces are white to pale yellow, translucent and horny.
Hazard Analysis
Toxic Hazard Rating:
   Acute Local: Allergen 1.
   Acute Systemic: Ingestion 1; Inhalation 1.
   Chronic Local: Allergen 1.
   Chronic Systemic: Ingestion 1; Inhalation 1.
Note: A stabilizer food additive (Section 10).
Fire Hazard: Slight, when exposed to heat or flame (Section 6).
Countermeasures
Personal Hygiene: Section 3.
Storage and Handling: Section 7.

TRANSFORMER OIL
General Information
Synonym: Transil oil.
Description: Liquid.
Constants: Flash p: 295° F (O.C.), d: 0.9.
Hazard Analysis
Toxicity: Unknown.
Fire Hazard: Slight, when exposed to heat or flame.
Spontaneous Heating: No.
Disaster Hazard: Slightly dangerous; when heated it emits acrid fumes; can react with oxidizing materials.
Countermeasures
Storage and Handling: Section 7.
To Fight Fire: Foam, carbon dioxide, dry chemical or carbon tetrachloride (Section 6).

TRANSIL OIL. See transformer oil.

"TRANSOTE." See creosote.

TREMETOL
Hazard Analysis
Toxic Hazard Rating:
   Acute Local: U.
   Acute Systemic: Ingestion 3; Inhalation 3.
   Chronic Local: U.
   Chronic Systemic: U.
Caution: Found in Rich-wheat or Snake-root. Can occur in milk of cows that have eaten either of these.

TREMOLITE
General Information
Description: A variety of asbestos; white to light green; vitreous to silky.
Formula: $Ca_2Mg_5Si_8O_{22}(OH)_2$.
Constants: Mol wt: 710, d: 3.0–3.3.

Hazard Analysis
TLV: ACGIH (recommended), 5 million particles per cubic foot of air.

TRIACETIN. See glyceryl triacetate.

TRIALLYLAMINE
General Information
Description: Liquid.
Formula: $(H_2C{:}CHCH_2)_3N$.
Constants: Mol wt: 137, d: 0.800 (20/4° C), mp: < −70° C, bp: 150–151° C, flash p: 103° F (T.O.C.).
Hazard Analysis
Toxicity: Details unknown. Limited animal experiments suggest high toxicity and moderate irritation. See also amines.
Fire Hazard: Dangerous, when exposed to heat or flame.
Countermeasures
Storage and Handling: Section 7.
To Fight Fire: Foam, alcohol foam, fog.

TRIALLYL CYANURATE
General Information
Description: Colorless liquid or solid.
Formula: $C_{12}H_{15}O_3N_3$.
Constants: Mol wt: 249.26, bp: 120° C at 5 mm, fp: 27.3° C, flash p: > 176° F (T.O.C.), d: 1.1133 at 30° C, vap. press.: 1 mm at 100° C.
Hazard Analysis
Toxic Hazard Rating:
   Acute Local: Ingestion 2; Inhalation 2.
   Acute Systemic: Ingestion 2; Inhalation 2.
   Chronnic Local: U.
   Chronic Systemic: U.
Fire Hazard: Moderate, when exposed to heat or flame (Section 6).
Disaster Hazard: Dangerous; when heated to decomposition or on contact with acid or acid fumes, it emits highly toxic fumes of cyanides.
Countermeasures
Ventilation Control: Section 2.
Personal Hygiene: Section 3.
Storage and Handling: Section 7.

2,4,6-TRIAMINO-8-TRIAZINE. See melamine.

TRIAMYLAMINE
General Information
Description: A clear, water-white or pale yellow liquid; amine odor.
Formula: $(C_5H_{11})_3N$.
Constants: Mol wt: 227.42, bp: 232° C, vap. press.: 7 mm at 26° C, flash p: 190° F (O.C.), d: 0.79–0.80 at 20°/20° C, vap. d.: 7.83.
Hazard Analysis
Toxicity: Details unknown. See amines.
Fire Hazard: Moderate, when exposed to heat or flame; can react with oxidizing materials.
Disaster Hazard: Moderate, when heated to decomposition it emits toxic fumes.
Countermeasures
Storage and Handling: Section 7.

TRIAMYLBENZENE
General Information
Description: A clear liquid.
Formula: $(C_5H_{11})_3C_6H_3$.
Constants: Mol wt: 288.50, bp: 300° C, flash p: 270° F (O.C.), d: 0.87.
Hazard Analysis
Toxicity: Details unknown. Probably irritant and narcotic in high concentrations.
Fire Hazard: Slight, when exposed to heat or flame; can react with oxidizing materials.
Spontaneous Heating: No.

Countermeasures
To Fight Fire: Foam, carbon dioxide, dry chemical or carbon tetrachloride (Section 6).
Storage and Handling: Section 7.

TRIAMYL BORATE
General Information
Synonym: Tri-n-amyl borate.
Description: A clear liquid; odor of n-amyl alcohol.
Formula: $B(C_5H_{11}O)_3$.
Constants: Mol wt: 272.23, bp: 110–114°C at 2 mm, flash p: 180°F (O.C.), d: 0.852 at 27°C, vap. d.: 9.4.
Hazard Analysis
Toxicity: See boron compounds and esters.
Fire Hazard: Slight, when exposed to heat or flame; can react with oxidizing materials.
Spontaneous Heating: No.
Countermeasures
To Fight Fire: Foam, carbon dioxide, dry chemical or carbon tetrachloride (Section 6).
Storage and Handling: Section 7.

TRI-n-AMYLTIN BROMIDE
General Information
Description: Liquid.
Formula: $(C_5H_{11})_3SnBr$.
Constants: Mol wt: 412.0, d: 1.2678.
Hazard Analysis and Countermeasures
See tin compounds and bromides.

TRI-p-ANISYL BORON
General Information
Description: White crystals.
Formula: $(C_6H_4OCH_3)_3B$.
Constants: Mol wt: 332.2, mp: 128°C.
Hazard Analysis
Toxicity: Details unknown. See also boron compounds.

s-TRIAZABORANE
General Information
Synonym: Borazine; hexahydro-s-triazaborine; borazane; borazole; triborine triamine; triboron nitride.
Description: Mobile liquid; dissolves in water.
Hazard Analysis and Countermeasures
See boron hydrides.

sym-TRIAZINETRIOL. See n-cyanuric acid.

TRIAZOBENZENE. See phenyl azoimide.

TRIBASIC COPPER SULFATE
General Information
Synonyms: Microgel; copper sulfate, tribasic.
Description: Aqua colored powder of extremely fine particle size; water insoluble.
Formula: $CuSO_4 \cdot 3Cu(OH)_2 \cdot H_2O$.
Constant: Mol wt: 284.
Hazard Analysis
Toxicity: Unknown. A fungicide.
Disaster Hazard: Dangerous. See sulfates.

TRIBENZYLBORINE
General Information
Description: Colorless crystals; insoluble in water.
Formula: $B(C_6H_5CH_2)_3$.
Constants: Mol wt: 284.2, mp: 47°C, bp: 230°C at 13 mm.
Hazard Analysis
Toxicity: Details unknown. See boron compounds.

TRIBENZYL ETHYL TIN
General Information
Description: Colorless crystals.
Formula: $(C_6H_5CH_2)_3(C_2H_5)Sn$.
Constants: Mol wt: 421.1, mp: 32°C.
Hazard Analysis
Toxicity: Details unknown. See tin compounds.

TRIBENZYL GERMANIUM BROMIDE
General Information
Description: Colorless crystals.
Formula: $(C_6H_5CH_2)_3GeBr$.
Constants: Mol wt: 425.9, mp: 145°C.
Hazard Analysis and Countermeasures
See germanium compounds and bromides.

TRIBENZYL GERMANIUM CHLORIDE
General Information
Description: Colorless crystals.
Formula: $(CH_2C_6H_5)_3GeCl$.
Constants: Mol wt: 381.4, mp: 155°C.
Hazard Analysis and Countermeasures
See germanium compounds and chlorides.

TRIBENZYL GERMANIUM FLUORIDE
General Information
Description: Colorless crystals.
Formula: $(CH_2C_6H_5)_3GeF$.
Constants: Mol wt: 365.0, mp: 96°C.
Hazard Analysis and Countermeasures
See fluorides and germanium compounds.

TRIBENZYL GERMANIUM IODIDE
General Information
Description: Colorless crystals.
Formula: $(CH_2C_6H_5)_3GeI$.
Constants: Mol wt: 472.9, mp: 141°C.
Hazard Analysis and Countermeasures
See germanium compounds and iodides.

TRIBENZYLGERMANIUM OXIDE
General Information
Description: Solid; soluble in petroleum ether.
Formula: $[(C_6H_5CH_2)_3Ge]_2O$.
Constants: Mol wt: 708.0, mp: 135°C.
Hazard Analysis
Toxicity: Details unknown. See germanium compounds.

TRIBENZYLTIN CHLORIDE
General Information
Description: White crystals; insoluble in water; soluble in organic solvents.
Formula: $(C_6H_5CH_2)_3SnCl$.
Constants: Mol wt: 427.5, mp: 144°C.
Hazard Analysis and Countermeasures
See tin compounds and chlorides.

TRIBENZYLTIN HYDROXIDE
General Information
Description: Colorless crystals; soluble in hot organic solvents.
Formula: $(C_6H_5CH_2)_3SnOH$.
Constants: Mol wt: 409.1, mp: 121°C.
Hazard Analysis
Toxicity: Details unknown. See tin compounds.

TRIBENZYLTIN IODIDE
General Information
Description: Crystals.

TOXIC HAZARD RATING CODE *(For detailed discussion, see Section 1.)*

0 NONE: (a) No harm under any conditions; (b) Harmful only under unusual conditions or overwhelming dosage.

1 SLIGHT: Causes readily reversible changes which disappear after end of exposure.

2 MODERATE: May involve both irreversible and reversible changes; not severe enough to cause death or permanent injury.

3 HIGH: May cause death or permanent injury after very short exposure to small quantities.

U UNKNOWN: No information on humans considered valid by authors.

Formula: $(C_6H_5CH_2)_3SnI$.
Constants: Mol wt: 519.0, mp: 103°C.
Hazard Analysis and Countermeasures
See tin compounds and iodides.

TRI-2-BIPHENYLYL PHOSPHATE
General Information
Description: White, granular solid.
Formula: $(C_6H_5C_6H_4)_3PO_4$.
Constants: Mol wt: 554.6, mp: 113–115°C.
Hazard Analysis and Countermeasures
See phosphorus compounds and phosphates.

TRIBORON SILICIDE
General Information
Description: Rhombic, black crystals.
Formula: B_3Si.
Constants: Mol wt: 60.52, d: 2.52.
Hazard Analysis
Toxicity: Details unknown. See boron compounds.
Fire Hazard: Moderate, by chemical reaction. Reacts with moisture to liberate hydrogen.
Explosion Hazard: Moderate, in the form of dust by chemical reaction with oxidizers.
Disaster Hazard: Moderately dangerous; will react with water or steam to produce heat and hydrogen.
Countermeasures
Storage and Handling: Section 7.

TRIBROMOACETIC ACID.
General Information
Description: Lustrous leaflets; soluble in water, alcohol, ether; slightly soluble in petroleum ether.
Formula: $CBr_3 \cdot COOH$.
Constants: Mol wt: 296.78, mp: 130–131°C, bp: 245°C with decomposition.
Hazard Analysis
Toxicity: Unknown. See trichloroacetic acid.
Disaster Hazard: Dangerous. See bromides.

2,4,6-TRIBROMOANILINE
General Information
Synonym: Aniline tribromide.
Description: Needles; insoluble in water; soluble in hot alcohol, chloroform, ether; slightly soluble in cold alcohol.
Formula: $C_6H_2Br_3NH_2$.
Constants: Mol wt: 329.85, d: 2.35, mp: 120–122°C, bp: 300°C.
Hazard Analysis
Toxicity: Unknown. See aniline.
Disaster Hazard: Dangerous. See bromides and aniline.

TRIBROMOETHANOL
General Information
Synonym: Avertin.
Description: Crystals having an ethereal odor and aromatic taste. Slightly water soluble, but soluble in alcohol and organic solvents.
Formula: CBr_3CH_2OH.
Constants: Mol wt: 282.8, mp: 79–82°C, bp: 92–93°C at 10 mm.
Hazard Analysis
Toxicity: Details unknown. See bromides and alcohols. Is used as a general anesthetic and narcotic.
Disaster Hazard: Dangerous. See bromides.
Countermeasures
Storage and Handling: Section 7.

TRIBROMOGERMANE
General Information
Synonym: Germanium bromoform.
Description: Colorless liquid.
Formula: $GeHBr_3$.
Constants: Mol wt: 313.36, mp: −24.0°C, bp: decomposes.

Hazard Analysis and Countermeasures
See bromides and germanium compounds.

TRI-n-BROMOMELAMINE
General Information
Description: White powder.
Formula: $Br_3C_3H_3N_6$.
Constant: Mol wt: 362.9.
Hazard Analysis
Toxicity: Details unknown; probably toxic. See also melamine.
Explosion Hazard: Slight, by chemical reaction or on contact with allyl alcohol at room temperature.
Disaster Hazard: Moderately dangerous; when heated to decomposition or on contact with acid or acid fumes, it emits toxic fumes of bromides.
Countermeasures
Storage and Handling: Section 7.

TRIBROMOMETHANE. See bromoform.

1,1,1-TRIBROMO-2-METHYL-2-PROPANOL
General Information
Description: Fine white crystals.
Formula: $CBr_3C(CH_3)OH$.
Constants: Mol wt: 310.8, mp: 176–177°C.
Hazard Analysis
Toxicity: Details unknown; probably toxic.
Disaster Hazard: Dangerous; See bromides.
Countermeasures
Storage and Handling: Section 7.

TRIBROMONITROMETHANE. See bromopicrin.

TRIBROMOPHENOL-2,4,6
General Information
Synonym: Bromol.
Description: Long crystals; soluble in 14,000 parts water at 15°C; soluble in alcohol, chloroform, ether, and glycerols.
Formula: $C_6H_2Br_3OH$.
Constants: Mol wt: 330.83, d: 2.55, mp: 94–96°C, bp: 244°C.
Hazard Analysis
Toxic Hazard Rating:
  Acute Local: Irritant 3; Ingestion 3; Inhalation 3.
  Acute Systemic: Ingestion 3; Inhalation 3; Skin Absorption 3.
  Chronic Local: Irritant 2.
  Chronic Systemic: Ingestion 2; Inhalation 2; Skin Absorption 2.
Disaster Hazard: Dangerous. See bromides and phenol.

TRIBROMOSILANE
General Information
Synonym: Silicobromoform.
Description: Mobil liquid. Spontaneously flammable in air.
Formula: $SiHBr_3$.
Constants: Mol wt: 268.9, d: 2.7 at 17°/4°C, mp: −73.5°C, bp: 112°C, vap. press.: 8.8 mm at 0°C.
Hazard Analysis
Toxicity: Readily hydrolyzes to liberate hydrogen bromide, which is a powerful irritant. See also hydrobromic acid.
Disaster Hazard: Dangerous. When heated to decomposition or brought into contact with acid fumes or moisture it emits highly toxic fumes.

TRIBUTOXYETHYL PHOSPHATE
General Information
Description: Light-colored liquid with butyl-like odor.
Formula: $(C_4H_9OC_2H_4O)_3PO$.
Constants: Mol wt: 398, mp: −70°C, bp: 200–230°C at 4 mm, flash p: 435°F, d: 1.02 at 20°/20°C, vap. press.: 0.03 mm at 150°C, vap. d.: 13.8.
Hazard Analysis
Toxicity: Details unknown. See esters.

Fire Hazard: Slight, when exposed to heat or flame.
Disaster Hazard: Dangerous; See phosphates, can react with oxidizing materials.
Countermeasures
Storage and Handling: Section 7.
To Fight Fire: Water, foam, carbon dioxide, dry chemical or carbon tetrachloride (Section 6).

TRIBUTYLAMINE
General Information
Description: A colorless liquid.
Formula: $(C_4H_9)_3N$.
Constants: Mol wt: 185.35, mp: $-70°C$, bp: $213°C$, flash p: $185°F$ (O.C.), d: 0.78–0.79, vap. d.: 6.38.
Hazard Analysis
Toxicity: Details unknown. See amines.
Fire Hazard: Moderate, when exposed to heat or flame.
Disaster Hazard: Moderate; when heated to decomposition it emits toxic fumes; can react with oxidizing materials.
Countermeasures
Storage and Handling: Section 7.
To Fight Fire: Foam, carbon dioxide, dry chemical or carbon tetrachloride (Section 6).

TRI-n-BUTYL BORATE
General Information
Description: Colorless, mobile liquid. Odor like n-butyl alcohol.
Formula: $B(OC_4H_9)_3$.
Constants: Mol wt: 230.16, bp: $230°C$, fp: $< -70°C$, flash p: $200°F$ (C.O.C.), d: 0.847 at $28°C$, vap. d.: 7.95.
Hazard Analysis
Toxicity: See boron compounds and n-butanol.
Fire Hazard: Moderate, when exposed to heat.
Disaster Hazard: Moderately dangerous; when heated to decomposition or on contact with acid or acid fumes, it can emit toxic fumes; on contact with oxidizing materials, it can react vigorously.
Countermeasures
Storage and Handling: Section 7.
To Fight Fire: Foam, carbon dioxide, dry chemical or carbon tetrachloride (Section 6).

TRI-sec-BUTYL BORATE
General Information
Description: Colorless liquid, odor of sec-butanol.
Formula: $[CH_3CH_2C(CH_3)OH]_3B$.
Constants: Mol wt: 230.16, bp: $184–192°C$, flash p: $165°F$ (C.O.C.), d: 0.829 at $24°C$.
Hazard Analysis
Toxicity: See boron compounds and sec-butanol.
Fire Hazard: Moderate, when exposed to heat or flame; can react with oxidizing materials.
Countermeasures
To Fight Fire: Foam, carbon dioxide, dry chemical or carbon tetrachloride (Section 6).
Storage and Handling: Section 7.

TRI-n-BUTYLBORINE
General Information
Description: Mobile liquid, colorless; insoluble in water.
Formula: $B(C_4H_9)_3$.
Constants: Details unknown. See also boron compounds.

TRI-tert-BUTYLBORINE
General Information
Description: Colorless mobile liquid; not soluble in water.

Formula: $B(C_4H_9)_3$.
Constants: Mol wt: 182.2, bp: $71°C$ at 12 mm.
Hazard Analysis
Toxicity: Details unknown. See also boron compounds.

TRIBUTYL CITRATE
General Information
Synonym: Butyl citrate.
Description: Liquid.
Formula: $(CH_2COOC_4H_9)_2COHCOOC_4H_9$.
Constants: Mol wt: 360.44, mp: $-20°C$, bp: $232°C$, flash p: $315°F$ (C.O.C.), d: 1.042 at $25°/25°C$, autoign. temp.: $695°F$, vap. d.: 12.41.
Hazard Analysis
Toxicity: See butyl alcohol and citric acid.
Fire Hazard: Slight, when exposed to heat or flame; can react with oxidizing materials.
Countermeasures
To Fight Fire: Water, carbon dioxide, dry chemical or carbon tetrachloride (Section 6).
Storage and Handling: Section 7.

TRI-n-BUTYLGERMANIUM
General Information
Formula: $(C_4H_9)_3Ge$.
Constant: Mol wt: 243.6.
Hazard Analysis
Toxicity: Details unknown. Animal experiments suggest low toxicity. See also germanium.

TRIBUTYL PHOSPHATE
General Information
Description: Colorless, odorless liquid.
Formula: $(C_4H_9)_3PO_4$.
Constants: Mol wt: 266.32, mp: $< -80°C$, bp: $292°C$, flash p: $295°F$ (C.O.C.), d: 0.982 at $20°C$, vap. d.: 9.20.
Hazard Analysis
Toxic Hazard Rating:
  Acute Local: Irritant 2; Ingestion 2; Inhalation 2.
  Acute Systemic: Ingestion 2; Inhalation 2.
  Chronic Local: U.
  Chronic Systemic: U.
Toxicity: Causes stimulation of central nervous system.
TLV: ACGIH (tentative); 5 milligrams per cubic meter of air.
Fire Hazard: Slight, when exposed to heat or flame.
Spontaneous Heating: No.
Disaster Hazard: Dangerous; See phosphates.
Countermeasures
Storage and Handling: Section 7.
To Fight Fire: Foam, carbon dioxide, dry chemical or carbon tetrachloride. (Section 6).

TRIBUTYL PHOSPHINE
General Information
Description: Colorless liquid; garlic odor; almost insoluble in water; miscible with ether, methanol, ethanol, and benzene.
Formula: $(C_4H_9)_3P$.
Constants: Mol wt: 202, d: 0.8100 at $25/4°C$, fp: -60 to $-65°C$, bp: $240°C$, flash p: $104°F$, autoign. temp.: $392°F$.
Hazard Analysis
Toxic Hazard Rating: U. See esters.
Fire Hazard: Moderate, when exposed to heat or flame.
Disaster Hazard: Dangerous; when heated to decomposition it emits highly toxic fumes.

TOXIC HAZARD RATING CODE *(For detailed discussion, see Section 1.)*

0 NONE: (a) No harm under any conditions; (b) Harmful only under unusual conditions or overwhelming dosage.

1 SLIGHT: Causes readily reversible changes which disappear after end of exposure.

2 MODERATE: May involve both irreversible and reversible changes not severe enough to cause death or permanent injury.

3 HIGH: May cause death or permanent injury after very short exposure to small quantities.

U UNKNOWN: No information on humans considered valid by author

Countermeasures
Storage and Handling: Section 7.
To Fight Fire: Foam, alcohol foam, fog.

TRIBUTYL PHOSPHINE OXIDE
General Information
Description: Crystals.
Formula: $(C_4H_9)_3PO$.
Constant: Mol wt: 218.3.
Hazard Analysis
Toxicity: Details unknown; possibly quite high; has been known to damage the eyes of experimental animals.
Disaster Hazard: Dangerous; when heated to decomposition it emits highly toxic fumes of oxides of phosphorus.
Countermeasures
Storage and Handling: Section 7.

TRIBUTYL PHOSPHINE SULFIDE
General Information
Formula: $(C_4H_9)_3PS$.
Constant: Mol wt: 234.
Hazard Analysis
Toxicity: Unknown. Limited animal experiments suggest high toxicity.
Disaster Hazard: Dangerous; when heated to decomposition it emits highly toxic fumes of phosphorous and sufur oxides.

TRIBUTYL PHOSPHITE
General Information
Description: Liquid, decomposes in water.
Formula: $(C_4H_9)_3PO_3$.
Constant: Mol wt: 202.3, flash p: 248°F (O.C.), d: 0.9, bp: 244–250°F at 7 mm.
Hazard Analysis
Toxicity: Details unknown. Limited animal experiments show low toxicity in acute exposures. Has been known to damage the eyes of experimental animals. See phosphorus acid and butanol.
Disaster Hazard: See tributyl phosphine oxide.
Fire Hazard: Slight, when exposed to heat or flame.
Countermeasures
Storage and Handling: Section 7.

TRIBUTYL PHOSPHOROTHIOATE
General Information
Description: Colorless liquid, mild characteristic odor.
Formula: $(C_4H_9O)_3PS$.
Constants: Mol wt: 282.3, bp: 142–145°C at 4.5 mm, flash p: 295°F (C.O.C.), d: 0.987 at 20°/4°C.
Hazard Analysis
Toxicity: Details unknown; probably moderate.
Fire Hazard: Slight, when exposed to heat or flame (Section 6).
Disaster Hazard: Dangerous; when heated to decomposition it emits highly toxic fumes of phosphorus and oxides of sulfur; can react vigorously with oxidizing materials.
Countermeasures
Storage and Handling: Section 7.

TRIBUTYL PHOSPHOROTRITHIOATE
General Information
Synonym: "DEF".
Description: Liquid; insoluble in water; soluble in aliphatic, aromatic, and chlorinated hydrocarbons.
Formula: $(C_4H_9S)_3PO$.
Constants: Mol wt: 314, bp: 150°C at 0.3 mm.
Hazard Analysis
Toxicity: Details unknown. Animal experiments show anti-cholinesterase effect. See also parathion.
Disaster Hazard: Dangerous. See phosphates and sulfates.

TRI-tert-BUTYL THIOGERMANIUM CHLORIDE
General Information
Description: Colorless crystals; soluble in organic solvents.

Formula: $Ge[SC(CH_3)_3]_3Cl$.
Constants: Mol wt: 375.6, mp: 67°C, bp: 157°C at 4 mm.
Hazard Analysis and Countermeasures
See germanium compounds and organometals.

TRI-n-BUTYLTIN BROMIDE
General Information
Description: Liquid.
Formula: $(C_4H_9)_3SnBr$.
Constants: Mol wt: 370.0, d: 1.3365.
Hazard Analysis and Countermeasures
See tin compounds and bromides.

TRIBUTYRIN
General Information
Synonym: Glyceryl tributyrate.
Formula: $C_{15}H_{26}O_6$.
Constants: Mol wt: 302.4, mp: −75°C, bp: 312–315°C, d: 1.0356 at 20°/20°C.
Hazard Analysis
Toxicity: Unknown. Animal experiments show practically no toxicity. A synthetic flavoring substance and adjuvant (Section 10).
Fire Hazard: Slight, when exposed to heat or flame; can react with oxidizing materials (Section 6).
Countermeasures
Storage and Handling: Section 7.

TRICALCIUM o-ARSENATE. See calcium arsenate.

TRICALCIUM CITRATE. See calcium citrate.

TRICALCIUM PHOSPHATE. See calcium phosphate, tribasic.

TRICALCIUM SILICATE. See cement, portland and other cement articles.
Note: Used as an anti-caking agent in foods (Section 10).

2,2,3-TRICHLOPROPIONALDEHYDE
Hazard Analysis
Toxicity: Details unknown. Limited animal experiments suggest high toxicity and irritation. See also aldehydes.
Disaster Hazard: Dangerous. See chlorides.

TRICHLOROACETALDEHYDE. See chloral.

TRICHLORACETIC ACID
General Information
Description: Colorless, rhombic deliquescent crystals.
Formula: CCl_3COOH.
Constants: Mol wt: 163.40, bp: 197.5°C, fp: 57.5°C, flash p: none, d: 1.6298 at 61°/4°C, vap. press.: 1 mm at 51.0°C.
Hazard Analysis
Toxic Hazard Rating:
Acute Local: Irritant 3; Ingestion 3; Inhalation 3.
Acute Systemic: U.
Chronic Local: U.
Chronic Systemic: U.
Disaster Hazard: Dangerous; See chlorides.
Countermeasures
Ventilation Control: Section 2.
Personnel Protection: Section 3.
Storage and Handling: Section 7.
Shipping Regulations: Section 11.
IATA (solid): Other restricted articles, class B, no label required, 12 kilograms (passenger), 45 kilograms (cargo).
(solution): Corrosive liquid, white label, 1 liter (passenger and cargo).

TRICHLOROACETONITRILE. See chlorocyanohydrin.

TRICHLOROACETYL CHLORIDE
General Information
Synonym: Superpalite.

Description: Liquid.
Formula: CCl₃COCl.

Constants: Mol wt: 181.86, bp: 118°C, d: 1.629 at 16°C.

Hazard Analysis
Toxic Hazard Rating:
Acute Local: Irritant 3; Ingestion 3; Inhalation 3.
Acute Systemic: U.
Chronic Local: U.
Chronic Systemic: U.
Caution: A military poison.
Disaster Hazard: Dangerous; See chlorides. Will react with water or steam to produce toxic and corrosive fumes.
Countermeasures
Ventilation Control: Section 2.
Personnel Protection: Section 3.
Storage and Handling: Section 7.
Shipping Regulations: Section 11.
IATA: Poison A, not acceptable (passenger and cargo).

TRICHLOROACETYL CHLOROETHYLAMIDE
Hazard Analysis
Toxicity: Details unknown; a mosquito repellent. See also amides.
Disaster Hazard: Dangerous; See chlorides.
Countermeasures
Storage and Handling: Section 7.

TRICHLOROACETYL NITRILE
General Information
Synonym: Tritox.
Formula: CCl₃CN.
Constant: Mol wt: 144.4.
Hazard Analysis and Countermeasures
See nitriles and chlorides.

2,4,6-TRICHLOROANISOLE
General Information
Synonym: Tyrene.
Description: Crystals; faint odor, nearly water insoluble.
Formula: C₇H₅Cl₃O.
Constants: Mol wt: 211.5, mp: 59°C, bp: 132°C at 27 mm.
Hazard Analysis and Countermeasures
See anisole and chlorides.

1,2,3-TRICHLOROBENZENE
General Information
Description: White crystals.
Formula: C₆H₃Cl₃.
Constants: Mol wt: 181.5, mp: 52.6°C, bp: 221°C, flash p: 235°F (C.C.), d: 1.69 at 25°/25°C, vap. press.: 1 mm at 40.0°C, vap. d.: 6.26.
Hazard Analysis
Toxic Hazard Rating:
Acute Local: U.
Acute Systemic: Ingestion 2; Inhalation 2.
Chronic Local: Causes loss of hair of experimental animals.
Chronic Systemic: Ingestion 2; Inhalation 2.
Toxicology: Liver injury has been reported. See also chlorinated hydrocarbons, aromatic.
Fire Hazard: Slight; when exposed to heat or flame.
Disaster Hazard: Dangerous; See chlorides; can react with oxidizing materials.
Countermeasures
Ventilation Control: Section 2.
To Fight Fire: Water, foam, carbon dioxide, dry chemical or carbon tetrachloride (Section 6).

Personal Hygiene: Section 3.
Storage and Handling: Section 7.

1,2,4-TRICHLOROBENZENE
General Information
Synonym: uns-Trichlorobenzene.
Description: Colorless liquid.
Formula: C₆H₃Cl₃.
Constants: Mol wt: 181.5, mp: 17°C, bp: 213°C, flash p: 230°F (C.C.), d: 1.454 at 25°/25°C, vap. press.: 1 mm at 38.4°C, vap. d.: 6.26.
Hazard Analysis
Toxic Hazard Rating:
Acute Local: U.
Acute Systemic: Ingestion 2; Inhalation 2.
Chronic Local: U.
Chronic Systemic: Ingestion 2; Inhalation 2.
Toxicology: See 1,2,3-trichlorobenzene.
Fire Hazard: Slight, when exposed to heat or flame.
Disaster Hazard: Dangerous; See chlorides; can react vigorously with oxidizing materials.
Countermeasures
Ventilation Control: Section 2.
To Fight Fire: Water, foam, carbon dioxide, dry chemical or carbon tetrachloride (Section 6).
Personal Hygiene: Section 3.
Storage and Handling: Section 7.

1,3,5-TRICHLOROBENZENE
General Information
Synonym: sym-Trichlorobenzene.
Description: White crystals.
Formula: C₆H₃Cl₃.
Constants: Mol wt: 181.5, mp: 63.4°C, bp: 208.5°C, flash p: 225°F (C.C.), vap. press.: 10 mm at 78.0°C, vap. d.: 6.26.
Hazard Analysis
Toxic Hazard Rating:
Acute Local: U.
Acute Systemic: Ingestion 2; Inhalation 2.
Chronic Local: Causes loss of hair of experimental animals.
Chronic Systemic: Ingestion 2; Inhalation 2.
Toxicology: See, 1,2,3-trichlorobenzene.
Fire Hazard: Moderate, when exposed to heat or flame.
Disaster Hazard: Dangerous; See chlorides; can react vigorously with oxidizing materials.
Countermeasures
Ventilation Control: Section 2.
To Fight Fire: Water, foam, carbon dioxide, or dry chemical or carbon tetrachloride (Section 6).
Personal Hygiene: Section 3.
Storage and Handling: Section 7.

sym-TRICHLOROBENZENE. See 1,3,5-trichlorobenzene.

uns-TRICHLOROBENZENE. See 1,2,4-trichlorobenzene.

TRICHLOROBENZYL CHLORIDE
General Information
Formula: C₆H₂Cl₃CH₂Cl.
Constant: Mol wt: 230.
Hazard Analysis
Toxicity: Details unknown. Animal experiments suggest moderate toxicity. May cause skin irritation.
Disaster Hazard: Dangerous. See chlorides.

TOXIC HAZARD RATING CODE (For detailed discussion, see Section 1.)

0 NONE: (a) No harm under any conditions; (b) Harmful only under unusual conditions or overwhelming dosage.

1 SLIGHT: Causes readily reversible changes which disappear after end of exposure.

2 MODERATE: May involve both irreversible and reversible changes; not severe enough to cause death or permanent injury.

3 HIGH: May cause death or permanent injury after very short exposure to small quantities.

U UNKNOWN: No information on humans considered valid by authors.

1,1,1-TRICHLORO-2,2-BIS(p-BROMOPHENYL) ETHANE
General Information
Synonym: Bromine analog of DDT.
Description: Crystals.
Formula: $(BrC_6H_4)_2CHCCl_3$.
Constants: Mol wt: 443.4.
Hazard Analysis
Toxicity: Details unknown. A pesticide. See DDT.
Disaster Hazard: Dangerous; See chlorides and bromides.
Countermeasures
Storage and Handling: Section 7.

1,1,1-TRICHLORO-2,2-BIS(p-CHLOROPHENYL) ETHANE. See DDT.

1,1,1-TRICHLORO-2,2-BIS(p-FLUOROPHENYL) ETHANE
General Information
Synonym: Fluorine analog of DDT.
Description: Crystals.
Formula: $(FC_6H_4)_2CHCCl_3$.
Constant: Mol wt: 321.6.
Hazard Analysis
Toxicity: Details unknown; probably highly toxic. A pesticide. See also DDT.
Disaster Hazard: Dangerous; See chlorides and fluorides.
Countermeasures
Storage and Handling: Section 7.

1,1,1-TRICHLORO-2,2-BIS(p-METHOXYPHENYL) ETHANE. See methoxychlor.

TRICHLOROBORINE DIMETHYL ETHERATE
General Information
Description: Colorless crystals; decomposed by water.
Formula: $(CH_3)_2OBCl_3$.
Constants: Mol wt: 163.3, mp: 76°C (decomposes).
Hazard Analysis and Countermeasures
See boron compounds, chlorides and ethers.

TRICHLOROBORINE TRIMETHYLAMMINE
General Information
Description: Colorless crystals; soluble in hot water and alcohol.
Formula: $(CH_3)_3NBCl_3$.
Constants: Mol wt: 176.3, mp: 243°C.
Hazard Analysis and Countermeasures
See boron compounds and chlorides.

TRICHLOROBUTANE
General Information
Description: Liquid.
Formula: $Cl_3C_4H_7$.
Constants: Mol wt: 154.5, bp: 168°C, flash p: 195°F.
Hazard Analysis
Toxicity: Details unknown. See chlorinated hydrocarbons, aliphatic.
Fire Hazard: Moderate, when exposed to heat or flame.
Disaster Hazard: Dangerous; See chlorides; can react vigorously with oxidizing materials.
Countermeasures
Storage and Handling: Section 7.
To Fight Fire: Foam, carbon dioxide, dry chemical or carbon tetrachloride (Section 6).

α,α,β-TRICHLOROBUTYRAMIDE
General Information
Description: Crystals.
Formula: $CH_3CHClCCl_2CONH_2$.
Constant: Mol wt: 190.5.
Hazard Analysis
Toxicity: Details unknown; a pesticide; stomach insecticide. See also amides.
Disaster Hazard: Dangerous; See chlorides.

Countermeasures
Storage and Handling: Section 7.

TRICHLOROCHLOROPHENYL PHENYLETHANE
General Information
Description: Crystals.
Constant: Mol wt: 320.1.
Hazard Analysis
Toxicity: Details unknown; an insecticide. See DDT.
Disaster Hazard: Dangerous; See chlorides.
Countermeasures
Storage and Handling: Section 7.

1-TRICHLORO-2-(p-CHLOROPHENYL)-2-PHENYL-ETHANE
General Information
Formula: $(ClC_6H_4)(C_6H_5)CHCCl_3$.
Constant: Mol wt: 320.1.
Hazard Analysis
Toxicity: Details unknown; a pesticide. See DDT.
Disaster Hazard: Dangerous; See chlorides.
Countermeasures
Storage and Handling: Section 7.

TRICHLOROCRESOLS
General Information
Description: Compounds with formula $[OHC_6HCH_3Cl_3]$ (there are a number of different isomers).
Hazard Analysis and Countermeasures
See cresol and chlorides.

TRICHLOROCYANIDINE. See cyanuric chloride.

1,2,4-TRICHLORO-3,5-DINITROBENZENE
General Information
Formula: $Cl_3C_6H(NO_2)_2$.
Constant: Mol wt: 271.5.
Hazard Analysis
Toxicity: Details unknown; a pesticide. See also chlorobenzene and nitrobenzene.
Fire Hazard: Moderate, when exposed to heat or flame. See also nitrates.
Explosion Hazard: See nitrates.
Disaster Hazard: Dangerous; when heated to decomposition it emits highly toxic fumes of oxides of nitrogen and chlorides; can react vigorously with reducing materials.
Countermeasures
Storage and Handling: Section 7.

α-TRICHLOROETHANE
General Information
Synonyms: 1,1,1-Trichloroethane; methyl chloroform.
Description: Colorless liquid.
Formula: CH_3CCl_3.
Constants: Mol wt: 133.42, bp: 74.1°C, fp: −32.5°C, flash p: none, d: 1.3492 at 20°/4°C, vap. press.: 100 mm at 20.0°C.
Hazard Analysis
Toxic Hazard Rating:
  Acute Local: Irritant 1; Ingestion 1; Inhalation 2.
  Acute Systemic: Inhalation 2; Skin Absorption 1.
  Chronic Local: Irritant 1.
  Chronic Systemic: Ingestion 1, Inhalation 1.
Toxicity: Narcotic in high concentrations.
TLV: ACGIH (recommended); 350 parts per million in air; 1900 milligrams per cubic meter of air.
Disaster Hazard: Dangerous; See chlorides.
Countermeasures
Ventilation Control: Section 2.
Personal Hygiene: Section 3.
Storage and Handling: Section 7.

β-TRICHLOROETHANE
General Information
Synonyms: 1,1,2-Trichloroethane; vinyl trichloride.

Description: Liquid, pleasant odor.
Formula: $CH_2ClCHCl_2$.
Constants: Mol wt: 133.4, bp: 114°C, fp: −35°C, d: 1.4416 at 20°/4°C, vap. press.: 40 mm at 35.2°C.
Hazard Analysis
Toxic Hazard Rating:
  Acute Local: Irritant 1; Ingestion 1; Inhalation 1.
  Acute Systemic: Inhalation 2.
  Chronic Local: Irritant 1.
  Chronic Systemic: Ingestion 2; Inhalation 2.
TLV: ACGIH (recommended); 10 parts per million of air; 45 milligrams per cubic meter of air. May be absorbed via the skin.
Toxicology: Trichloroethane has narcotic properties and acts as a local irritant to the eyes, nose and lungs. It may also be injurious to the liver and kidneys. A fumigant.
Disaster Hazard: Dangerous; See chlorides.
Countermeasures
Ventilation Control: Section 2.
Personal Hygiene: Section 3.
Storage and Handling: Section 7.

TRICHLOROETHANOL
General Information
Description: Liquid.
Formula: CCl_3CH_2OH.
Constants: Mol wt: 149.5, mp: 17.8°C, bp: 150°C at 765 mm, d: 1.54 at 25°/4°C, vap. press.: 1 mm at 20°C, vap. d.: 5.16.
Hazard Analysis
Toxicity: Details unknown; moderately toxic, anesthetic.
Disaster Hazard: Dangerous; See chlorides.
Countermeasures
Storage and Handling: Section 7.

1,1,3-TRI(2-CHLOROETHOXY)PROPANE
General Information
Formula: $OHCH_2CHClC_2H_4CH(OHCH_2CHCl)_2$.
Constant: Mol wt: 279.5.
Hazard Analysis
Toxicity: Details unknown. Limited animal experiments suggest high toxicity.
Disaster Hazard: Dangerous. See chlorides.

TRICHLOROETHYLENE
General Information
Synonyms: Ethinyl trichloride; ethylene trichloride.
Description: Stable, colorless, heavy, mobile liquid, chloroform-like odor.
Formula: $CHClCCl_2$.
Constants: Mol wt: 131.40, mp: −73°C, bp: 87.1°C, fp: −86.8°C, d: 1.45560 at 25°/4°C, autoign. temp.: 770°F, vap press.: 100 mm at 32°C, vap. d.: 4.53, flash p: 90°F, lel: 2.5%, uel: 90%.
Hazard Analysis
Toxic Hazard Rating:
  Acute Local: Irritant 1; Ingestion 1; Inhalation 1.
  Acute Systemic: Ingestion 2; Inhalation 3; Skin Absorption 2.
  Chronic Local: Irritant 1.
  Chronic Systemic: Inhalation 1; Skin Absorption 1.
TLV: ACGIH (recommended); 100 parts per million of air; 535 milligrams per cubic meter of air.
Toxicology: Inhalation of high concentrations causes narcosis and anesthesia. A form of addiction has been observed in exposed workers. Prolonged inhalation of moderate concentrations causes headache and drowsi-

ness. Fatalities following severe, acute exposure have been attributed to ventricular fibrillation resulting in cardiac failure. There is some question as to damage to liver or other organs from chronic exposure. Cases have been reported but are of questionable validity. Determination of the metabolites trichloracetic acid and trichloroethanol in urine reflects the absorption of trichloroethylene.
NOTE: A food additive permitted in food for human consumption (Section 10). A common air contaminant (Section 4).
Fire Hazard: Slight, when exposed to heat or flame. High concentrations of trichloroethylene vapor in high-temperature air can be made to burn mildly if plied with a strong flame. Though such a condition is difficult to produce, flames or arcs should not be used in closed equipment which contains any solvent residue or vapor.
Spontaneous Heating: No.
Disaster Hazard: Dangerous; See chlorides.
Countermeasures
Ventilation Control: Section 2.
Personnel Protection: Section 3.
Storage and Handling: Section 7.
Shipping Regulations: Section 11.
  MCA warning label.
  IATA: Other restricted articles, class A, no label required, 40 liters (passenger and cargo).

TRICHLOROFLUOROGERMANE
General Information
Description: Colorless liquid.
Formula: $GeCl_3F$.
Constants: Mol wt: 197.97, mp: −49.8°C, bp: 37.5°C.
Hazard Analysis and Countermeasures
See fluorides and germanium compounds and chlorides.

TRICHLOROFLUOROMETHANE. See fluorotrichloromethane.

TRICHLOROGERMANE
General Information
Synonym: Germanium chloroform.
Description: Colorless liquid.
Formula: $GeHCl_3$.
Constants: Mol wt: 179.98, mp: −71.0°C, bp: 75.2°C, d: 1.93 at 0°C.
Hazard Analysis and Countermeasures
See hydrochloric acid and germanium compounds.

TRICHLOROISOCYANURIC ACID
General Information
Description: White crystals; chlorine odor; moderately soluble in water.
Formula: $(ClNCO)_3$.
Constants: Mol wt: 232.5, mp: 225°–230°C.
Hazard Analysis
Toxic Hazard Rating:
  Acute Local: Irritant 2; Ingestion 2.
  Acute Systemic: Ingestion 1.
  Chronic Local: U.
  Chronic Systemic: Ingestion 1.
Toxicity: Animal experiments suggest that the "tri" form is slightly more toxic than the "di" form.
Disaster Hazard: Dangerous; when heated to decomposition it emits chloride and carbon monoxide fumes.
Countermeasures
Storage and Handling: An oxidizing material. See Section 7.

TOXIC HAZARD RATING CODE (For detailed discussion, see Section 1.)

0 NONE: (a) No harm under any conditions; (b) Harmful only under unusual conditions or overwhelming dosage.

1 SLIGHT: Causes readily reversible changes which disappear after end of exposure.

2 MODERATE: May involve both irreversible and reversible changes not severe enough to cause death or permanent injury.

3 HIGH: May cause death or permanent injury after very short exposure to small quantities.

U UNKNOWN: No information on humans considered valid by authors

Shipping Regulations: Section 11.
  I.C.C.: Oxidizing material, yellow label; 100 pounds.
  IATA (dry, containing more than 39% available chlorine):
    Oxidizing material, yellow label, 25 kilograms (passenger), 45 kilograms (cargo).

1,1,1-TRICHLOROISOPROPYL ALCOHOL
General Information
Synonym: Isopral; 1,1,1-trichloro-2-propanol.
Description: Crystals; camphor-like odor. Pungent taste. Water soluble.
Formula: $C_3H_5Cl_3O$.
Constants: Mol wt: 163.4, mp: 50°C, bp: 162°C.
Hazard Analysis
Toxicity: Details unknown. Limited animal experiments suggest moderate toxicity. See also chlorinated hydrocarbons, aliphatic.
Disaster Hazard: Dangerous; See chlorides.

TRICHLOROMELAMINE
General Information
Synonym: TCM.
Description: White powder; slightly water soluble.
Formula: $C_3H_3Cl_3N_6$.
Constants: Mol wt: 229.4, autoign. temp.: 320°F.
Hazard Analysis
Toxicity: Details unknown. Animal experiments indicate low order of oral toxicity. See also melamine.
Fire Hazard: Moderate. In the pure state when heated to 320°F or ignited by spark or flame it reacts vigorously to evolve smoke and heat. Keep away from reducing agents until properly diluted. Vendor can supply directions for handling.
Disaster Hazard: Dangerous; when heated to decomposition it emits highly toxic chloride and oxide of nitrogen fumes.
Countermeasures
Storage and Handling: Store in a cool place, away from reducing agent (Section 7).
Shipping Regulations: Section 11.
  I.C.C.:Chemicals N.O.I.B.N.

TRICHLOROMETHANE. See chloroform.

TRICHLOROMETHANE SULFENYL CHLORIDE. See perchloromethyl mercaptan.

TRICHLOROMETHYLCHLOROFORMATE. See diphosgene.

TRICHLOROMETHYL ETHER
General Information
Description: A liquid of pungent odor.
Formula: $CHCl_2OCH_2Cl$.
Constants: Mol wt: 149.42, bp: 130–132°C, d: 1.5066 at 10°C.
Hazard Analysis
Toxic Hazard Rating:
  Acute Local: Irritant 3; Ingestion 3; Inhalation 3.
  Acute Systemic: U.
  Chronic Local: U.
  Chronic Systemic: U.
Disaster Hazard: Dangerous; when heated to decomposition, it emits highly toxic fumes; will react with water or steam to produce toxic and corrosive fumes.
Countermeasures
Ventilation Control: Section 2.
Storage and Handling: Section 7.
Personal Protection : Section 3.

N-(TRICHLOROMETHYLTHIO) PHTHALAMIDE. See n-trichloromethylthiotetrahydrophthalamide.

N-TRICHLOROMETHYLTHIOTETRAHYDRO-PHTHALAMIDE
General Information
Synonym: Captan.

Description: Odorless crystals; insoluble in water, soluble in benzene, chloroform.
Formula: $C_9H_8Cl_3NO_2S$.
Constants: Mol wt: 300.6, d: 1.74, mp: 173°C.
Hazard Analysis
Toxicology: Limited animal experimentation indicates low toxicity. Large ingested doses may cause vomiting and diarrhea. A fungicide. A food additive permitted in food for human consumption (Section 10).
Disaster Hazard: Dangerous; See chlorides and sulfates.
Countermeasures
Storage and Handling: Section 7.

TRICHLORONAPHTHALENE
General Information
Description: A white solid.
Formula: $C_{10}H_5Cl_3$.
Constant: Mol wt: 231.51.
Hazard Analysis
Toxic Hazard Rating:
  Acute Local: Irritant 2.
  Acute Systemic: Ingestion 2; Inhalation 2.
  Chronic Local: Irritant 2.
  Chronic Systemic: Ingestion 2; Inhalation 2; Skin Absorption 2.
Toxicology: See also chlorinated naphthalenes and chlorinated diphenyls.
TLV: ACGIH (recommended); 5 milligrams per cubic meter of air. Can be absorbed through the intact skin.
Disaster Hazard: Dangerous; See chlorides.
Countermeasures
Ventilation Control: Section 2.
Personnel Protection: Section 3.
Storage and Handling: Section 7.

TRICHLORONITROMETHANE. See chloropicrin.

3,4,6-TRICHLORO-2-NITROPHENOL
General Information
Synonyms: Dowlap.
Description: Pale yellow crystals.
Formula: $C_6H_2Cl_3NO_3$.
Constants: Mol wt: 230.5, mp: 93°C.
Hazard Analysis and Countermeasures
See chlorinated phenols.

TRICHLORONITROPROPANOL
General Information
Description: Crystalline solid.
Formula: $CCl_3CHOHCH_2NO_2$.
Constants: Mol wt: 208.4, mp: 40°C, bp: 120°C at 5 mm, flash p: 352°F (O.C.), d: 1.605 at 45°/4°C, vap. press.: 0.1 mm at 20°C.
Hazard Analysis and Countermeasures
See alcohols and chlorinated hydrocarbons, aliphatic.

TRICHLORONITROSOMETHANE
General Information
Description: Dark-blue liquid; unpleasant odor.
Formula: CCl_3NO.
Constants: Mol wt: 148.39, bp: 5°C at 70 mm, d: 1.5 at 20°C.
Hazard Analysis
Toxic Hazard Rating:
  Acute Local: Irritant 3; Ingestion 3; Inhalation 3.
  Acute Systemic: U.
  Chronic Local: U.
  Chronic Systemic: U.
Caution: A lachrymator type military poison.
Disaster Hazard: Dangerous; See chlorides and nitrates; will react with water or steam to produce toxic and corrosive fumes.
Countermeasures
Ventilation Control: Section 2.
Personnel Protection: Section 3.
Storage and Handling: Section 7.

1,1,2-TRICHLORO-1,3,3,3-PENTAFLUOROPROPANE
Hazard Analysis
Toxicity: Details unknown. Limited animal experiments suggest low toxicity.
Disaster Hazard: Dangerous; See chlorides and fluorides.

2,3,5-TRICHLOROPHENOL
General Information
Description: Colorless needles.
Formula: $Cl_3C_6H_2OH$.
Constants: Mol wt: 197.5, mp: 62°C, bp: 253°C.
Hazard Analysis and Countermeasures
See chlorinated phenols.
Shipping Regulations: Section 11.
MCA warning label.

2,4,5-TRICHLOROPHENOL
General Information
Description: Colorless needles or gray flakes.
Formula: $C_6H_2Cl_3OH$.
Constants: Mol wt: 197.5, bp: 252°C, fp: 57.0°C, d: 1.678 at 25°/4°C, vap. press.: 1 mm at 72.0°C.
Hazard Analysis and Countermeasures
See chlorinated phenols.
Shipping Regulations: Section 11.
MCA warning label.

2,4,6-TRICHLOROPHENOL
General Information
Description: Colorless needles or yellow solid; strong phenolic odor.
Formula: $Cl_3C_6H_2OH$.
Constants: Mol wt: 197.5, mp: 68°C, bp: 244.5°C, fp: 62°C, d: 1.490 at 75°/4°C, vap. press.: 1 mm at 76.5°C.
Hazard Analysis and Countermeasures
See chlorinated phenols.
Shipping Regulations: Section 11.
MCA warning label.

2,4,5-TRICHLOROPHENOXYACETIC ACID
General Information
Synonym: 2,4,5-T.
Description: Crystals; light tan solid.
Formula: $Cl_3C_6H_2OCH_2COOH$.
Constants: Mol wt: 255.5, mp: 151–153°C.
Hazard Analysis
Toxicity: Details unknown; an herbicide. See 2,4,D.
TLV: ACGIH (recommended) 10 milligrams per cubic meter of air.
Disaster Hazard: Dangerous; See chlorides.
Countermeasures
Storage and Handling: Section 7.
Shipping Regulations: Section 11.
IATA: Other restricted articles, class A, no label, no limit (passenger and cargo).

2-(2,4,5-TRICHLOROPHENOXY) ETHYL-2,2-DICHLOROPROPIONATE. See erbon.

2-(2,4,5-TRICHLOROPHENOXY) PROPIONIC ACID
General Information
Synonym: 2,4,5-TCPPA.
Description: Crystals. Slightly water soluble.
Formula: $C_9H_7Cl_3O_3$.
Constants: Mol wt: 269.5, mp: 182°C.
Hazard Analysis
Toxicity: Details unknown. Animal experiments have

shown fairly high toxicity with liver and kidney injury. See also 2,4-D.
Disaster Hazard: Dangerous. See chlorides.

2-(2,4,5-TRICHLOROPHENOXY) PROPIONIC ACID PROPYLENE GLYCOL BUTYL ETHER ESTER
General Information
Synonym: Kuron.
Description: Liquid; insoluble in water.
Formula: $C_9H_7Cl_3O_3 + CH_3CHOHCH_2O-CH_2CH_2-CH_3$.
Hazard Analysis
Toxicity: Details unknown. In experimental animals it has caused damage to liver and kidneys. See also 2,4-dichlorophenoxyacetic acid.
Disaster Hazard: Dangerous; when heated to decomposition it emits highly toxic fumes.

2,4,5-TRICHLOROPHENYL ACETATE
General Information
Description: Crystals.
Formula: $Cl_3C_6H_2OOCCH_3$.
Hazard Analysis
Toxicity: Details unknown; a fungicide. See also chlorinated hydrocarbons, aromatic; and 2,4-D.
Disaster Hazard: Dangerous. See chlorides.
Countermeasures
Storage and Handling: Section 7.

2,3,6-TRICHLOROPHENYL ACETIC ACID
General Information
Formula: $C_6H_2Cl_3CH_2COOH$.
Constant: Mol wt: 239.5.
Hazard Analysis and Countermeasures
See 2,4-dichlorophenoxy acetic acid.

TRI-o-CHLOROPHENYL BORATE
General Information
Description: White solid; odor of o-chlorophenol.
Formula: $(ClC_6H_4O)_3B$.
Constants: Mol wt: 399.15, mp: 47–49°C, bp: 264–270°C at 14 mm.
Hazard Analysis and Countermeasures
See o-chlorophenol and boron compounds.

TRICHLOROPHENYL MONOCHLOROACETATE
General Information
Description: Crystals.
Formula: $Cl_3C_6H_2OOCCH_2Cl$.
Constant: Mol wt: 274.0.
Hazard Analysis and Countermeasures
See monochloroacetic acid and trichlorophenol.

1,2,3-TRICHLOROPROPANE
General Information
Synonyms: Glycerol trichlorohydrin; allyl trichloride trichlorohydrin.
Description: Colorless liquid.
Formula: $CH_2ClCHClCH_2Cl$.
Constants: Mol wt: 147.44, mp: -14.7°C, bp: 156.17°C, ulc: 20-25, lel: 3.2%, uel: 12.6% at 150°C, flash p: 174°F (T.O.C.), d: 1.3888 at 20°C, autoign. temp.: 579°F (commercial), vap. press.: 10 mm at 46°C, vap. d.: 5.0.
Hazard Analysis
Toxic Hazard Rating:
Acute Local: Irritant 3; Inhalation 3.

TOXIC HAZARD RATING CODE (For detailed discussion, see Section 1.)

0 NONE: (a) No harm under any conditions; (b) Harmful only under unusual conditions or overwhelming dosage.

1 SLIGHT: Causes readily reversible changes which disappear after end of exposure.

2 MODERATE: May involve both irreversible and reversible changes; not severe enough to cause death or permanent injury.

3 HIGH: May cause death or permanent injury after very short exposure to small quantities.

U UNKNOWN: No information on humans considered valid by authors.

Acute Systemic: Ingestion 2; Inhalation 2; Skin Absorption 2.
Chronic Local: Irritant 3.
Chronic Systemic: Ingestion 2; Inhalation 2; Skin Absorption 2.
TLV: ACGIH (recommended): 50 parts per million in air; 300 milligrams per cubic meter of air.
Caution: A lipoid solvent. Cumulative toxicity. See also chlorinated hydrocarbons.
Fire Hazard: Moderate, when exposed to heat or flame.
Disaster Hazard: Dangerous; when heated to decomposition it burns and emits highly toxic fumes of chlorides; can react vigorously with oxidizing materials.
Countermeasures
Ventilation Control: Section 2.
To Fight Fire: Water, foam, carbon dioxide, dry chemical or carbon tetrachloride (Section 6).
Personnel Protection: Section 3.
Storage and Handling: Section 7.

1,1,1-TRICHLORO-2-PROPANOL. See 1,1,1-trichloro-isopropyl alcohol.

1,2,3-TRICHLOROPROPENE
General Information
Synonym: Allyltrichloride.
Formula: ClCH$_2$CCl:CHCl.
Constants: Mol wt: 145.4, bp: 142°C, d: 1.414 at 20/20°C.
Hazard Analysis
Toxicity: Details unknown: Limited animal experiments suggest high toxicity and moderate irritation.
Disaster Hazard: High. When heated to decomposition it yields highly toxic fumes.

2,2,3-TRICHLOROPROPIONIC ACID
General Information
Formula: ClCH$_2$CCl$_2$COOH.
Constant: Mol wt: 177.
Hazard Analysis
Toxicity: Details unknown. Limited animal experiments suggest moderate toxicity and high irritation.
Disaster Hazard: Dangerous. See chlorides.

TRICHLOROSILANE
General Information
Synonym: Silicochloroform.
Description: Colorless, very volatile liquid; decomposes in water.
Formula: SiHCl$_3$.
Constants: Mol wt: 135.44, mp: −134°C, bp: 31.8°C, flash p: < 20°F (O.C.), d: 1.35 at 0°C, vap. press.: 400 mm at 14.5°C, vap. d.: 4.7.
Hazard Analysis
Toxic Hazard Rating:
Acute Local: Irritant 2; Inhalation 2.
Acute Systemic: Ingestion 2; Inhalation 2.
Chronic Local: U.
Chronic Systemic: U.
Fire Hazard: Dangerous, when exposed to flame or by chemical reaction. Spontaneously flammable in air.
Explosion Hazard: Unknown.
Disaster Hazard: Dangerous; when heated to decomposition it emits highly toxic fumes of chlorides; will react with water or steam to produce heat, toxic and corrosive fumes; on contact with oxidizing materials, it can react vigorously.
Countermeasures
Ventilation Control: Section 2.
To Fight Fire: Carbon dioxide, dry chemical or carbon tetrachloride (Section 6).
Personal Hygiene: Section 3.
Storage and Handling: Section 7.
Shipping Regulations: Section 11.
I.C.C.: Flammable liquid; red label, 10 gallons.
Coast Guard Classification: Inflammable liquid; red label.

IATA: Flammable liquid, red label, not acceptable (passenger), 40 liters (cargo).

α-**TRICHLOROTOLUENE.** See benzotrichloride.

TRICHLORO-s-TRIAZINE. See cyanuric chloride.

TRICHLOROTRIFLUOROETHANE. See trifluorotrichloroethane.

TRICK MATCHES AND TRICK NOISE MAKERS, EXPLOSIVE
Shipping Regulations: Section 11.
I.C.C.: Explosive C, 150 pounds.
IATA: See matches, trick.

TRICOPPER ANTIMONIDE
General Information
Description: Gray crystals.
Formula: Cu$_3$Sb.
Constants: Mol wt: 312.38, mp: 687°C, d: 8.51.
Hazard Analysis
Toxicity: See antimony and copper compounds.
Fire Hazard: Moderate, by chemical reaction with moisture to evolve antimony hydride. See also stibine.
Explosion Hazard: Slight, by chemical reaction with powerful oxidizers. See also stibine.
Disaster Hazard: Dangerous; when heated to decomposition or on contact with acid or acid fumes, it emits highly toxic fumes of antimony; it will react with water or steam to produce toxic and flammable vapors.
Countermeasures
Storage and Handling: Section 7.

TRICOPPER ARSENIDE
General Information
Synonym: Domeykite.
Description: Hexagonal crystals.
Formuls: Cu$_3$As.
Constants: Mol wt: 265.53, mp: 830°C, d: 8.0.
Hazard Analysis and Countermeasures
See arsenic compounds and arsenides.

11-TRICOSENE
General Information
Description: A liquid.
Formula: C$_{10}$H$_{21}$CHCHC$_{11}$H$_{23}$.
Constants: Mol wt: 322, bp: 168–170°C at 2.4 mm, flash p: 284°F, d: 0.80.
Hazard Analysis
Toxicity: Unknown.
Fire Hazard: Slight, when exposed to heat or flame; can react with oxidizing materials.
Countermeasures
To Fight Fire: Foam, carbon dioxide, dry chemical or carbon tetrachloride (Section 6).
Storage and Handling: Section 7.

TRI-m,p-CRESYL BORATE
General Information
Description: Straw-yellow, viscous liquid; characteristic odor.
Formula: (CH$_3$C$_6$H$_4$OH)$_3$B.
Constants: Mol wt: 335.2, bp: 179–210°C at 0.1 mm, flash p: 240°F (C.O.C.), d: 1.053 at 27.6°C, vap. d.: 11.6.
Hazard Analysis
Toxicity: See cresol and boric acid.
Fire Hazard: Slight, when exposed to heat or flame.
Disaster Hazard: Dangerous; when heated to decomposition it emits highly toxic fumes; can react vigorously with oxidizing materials.
Countermeasures
Storage and Handling: Section 7.
To Fight Fire: Water, foam, carbon dioxide, dry chemical or carbon tetrachloride (Section 6).

TRI-o-CRESYL BORATE
General Information
Description: Straw-yellow liquid; odor of o-cresol.
Formula: $(CH_3C_6H_4O)_3B$.
Constants: Mol wt: 332.2, bp: 189–195°C at 2 mm, flash p: 345°F (C.O.C.), d: 1.079 at 22°C, vap. d.: 11.4.
Hazard Analysis and Countermeasures
See o-cresol and boron compounds.

TRI-o-CRESYL PHOSPHATE
General Information
Synonym: o-Tolyl phosphate.
Description: Colorless liquid.
Formula: $(CH_3C_6H_4)PO_4$.
Constants: Mol wt: 368.36, mp: −25 to −30°C, bp: 410°C (slight decomp.), flash p: 437°F, d: 1.17, autoign. temp.: 725°F, vap. d.: 12.7.
Hazard Analysis
Toxic Hazard Rating:
Acute Local: U.
Acute Systemic: Ingestion 3; Inhalation 2; Skin Absorption 2.
Chronic Local: U.
Chronic Systemic: Ingestion 2; Inhalation 2; Skin Absorption 2.
TLV: ACGIH (recommended); 0.1 milligrams per cubic meter of air.
Toxicology: Most of the cases of tri-o-cresyl phosphate poisoning have followed its ingestion. In 1930, some 15,000 persons were affected in the United States, and of these, 10 died. The responsible material was found to be an alcoholic drink known as Jamaica ginger, or "jake." This beverage had been adulterated with about 2% of tri-o-cresyl phosphate. The affected persons developed a polyneuritis, which progressed, in many cases, with degeneration of the peripheral motor nerves, the anterior horn cells and the pyramidal tracts. Sensory changes were absent. Since 1930 there have been several other outbreaks of poisoning following ingestion of the material. Recently 3 cases of polyneuritis occurring in England in connection with the manufacture of the tri-o-cresyl phosphate have been reported. Absorption was probably through the respiratory tract, though there may have been some absorption through the skin. All three men made a good recovery.

From ingestion experiments with cockerels, it appears that tri-o-cresyl phosphate is more toxic than the m– form, and much more so than tri-p-cresyl phosphate or triphenyl phosphate.

Irrespective of whether absorption has been by ingestion or by inhalation or skin absorption, the history is usually one of early, transient gastro-intestinal upset, with nausea, vomiting, diarrhea and abdominal pain. These clear up, and are followed in 1 to 3 weeks by soreness of the lower leg muscles, "numbness" of the toes and fingers, and a few days later by weakness of the toes and bilateral foot-drop. After another week or so, weakness of the fingers and bilateral wristdrop follow. There are no sensory changes. Recovery is slow, and the degree of residual paralysis depends upon the extent of damage to the nervous system. Many cases recover completely. In 1958 several thousand persons in Morocco were poisoned with this material which was present in lubricating oil which had been mixed with edible oils by dishonest merchants. Many of the victims suffered a permanent paralysis.

Fire Hazard: Slight, when exposed to heat or flame.
Spontaneous Heating: No.
Disaster Hazard: Dangerous; when heated to decomposition it emits highly toxic fumes of oxides of phosphorus; can react with oxidizing materials.
Countermeasures
Ventilation Control: Section 2.
To Fight Fire: Water, foam, carbon dioxide, dry chemical or carbon tetrachloride (Section 6).
Personnel Protection: Section 3.
Storage and Handling: Section 7.

TRICYANIC ACID. See n-cyanuric acid.

TRICYANOGEN CHLORIDE. See cyanuric chloride.

TRICYCLOHEXYL BORATE
General Information
Description: Large needle-like white crystals. Nearly odorless.
Formula: $[CH_2(CH_2)_4CHO]_3B$.
Constants: Mol wt: 308.3, mp: 59–61°C, bp: 330°C.
Hazard Analysis
Toxicity: See boron compounds and esters.
Fire Hazard: Slight when exposed to heat or flame.
Countermeasures
Storage and Handling: Section 7.

TRICYCLOHEXYLBORINE
General Information
Synonym: Boron tricyclohexyl.
Description: Colorless crystals; insoluble in water; soluble in ether.
Formula: $(C_6H_{11})_3B$.
Constants: Mol wt: 260.3, mp: 100°C, bp: 194°C at 15 mm.
Hazard Analysis
Toxicity: Details unknown. See also boron compounds.

TRI-(2-CYCLOHEXYLCYCLOHEXYL) BORATE
General Information
Description: White solid; odor of 2-cyclohexylcyclohexanol.
Formula: $(C_{12}H_{21}O)_3B$.
Constants: Mol wt: 554.7, mp: 172–175°C, bp: 230–250°C at 0.3 mm.
Hazard Analysis
Toxicity: See boron compounds and esters.
Fire Hazard: Slight, when exposed to heat or flame; can react with oxidizing materials.
Countermeasures
Storage and Handling: Section 7.

TRICYCLOHEXYLGERMANIUM BROMIDE
General Information
Description: Colorless crystals; hydrolyzed in water.
Formula: $(C_6H_{11})_3GeBr$.
Constants: Mol wt: 402.0, mp: 110°C.
Hazard Analysis and Countermeasures
See germanium compounds and bromides.

TRICYCLOHEXYLGERMANIUM CHLORIDE
General Information
Description: Colorless crystals; hydrolyzed in water.
Formula: $(C_6H_{11})_3GeCl$.
Constants: Mol wt: 357.5, mp: 102°C.
Hazard Analysis and Countermeasures
See germanium compounds and chlorides.

TOXIC HAZARD RATING CODE (For detailed discussion, see Section 1.)

0 NONE: (a) No harm under any conditions; (b) Harmful only under unusual conditions or overwhelming dosage.

1 SLIGHT: Causes readily reversible changes which disappear after end of exposure.

2 MODERATE: May involve both irreversible and reversible changes; not severe enough to cause death or permanent injury.

3 HIGH: May cause death or permanent injury after very short exposure to small quantities.

U UNKNOWN: No information on humans considered valid by authors.

TRICYCLOHEXYLGERMANIUM FLUORIDE
General Information
Description: Colorless crystals; hydrolyzed in water.
Formula: $(C_6H_{11})_3GeF$.
Constants: Mol wt: 341.1, mp: 92°C.
Hazard Analysis and Countermeasures
See fluorides.

TRICYCLOHEXYLGERMANIUM HYDROXIDE
Genral Information
Description: Solid material; soluble in organic solvents.
Formula: $(C_6H_{11})_3GeOH$.
Constants: Mol wt: 339.1, mp: 177°C.
Hazard Analysis
Toxicity: Details unknown. See germanium compounds.

TRICYCLOHEXYLGERMANIUM IODIDE
General Information
Description: Colorless crystals; hydrolyzed by water.
Formula: $(C_6H_{11})_3GeI$.
Constants: Mol wt: 449.0, mp: 100°C.
Hazard Analysis and Countermeasures
See germanium compounds and iodides.

TRIDANE. See tridecyl benzene.

TRIDECANOL
General Information
Synonym: Tridecyl alcohol.
Description: General term for a commercial mixture of
isomers of the formula $C_{12}H_{25}CH_2OH$; water white
liquid; pleasant odor.
Constants: Mol wt: 200, boiling range: 252–272°C, d: 0.845
at 20/20°C, flash p: 180°F (T.O.C), 250°F (O.C.),
vap. d.: 6.9.
Hazard Analysis
Toxicity: Details unknown. Limited animal experiments
suggest low toxicity.
Fire Hazard: Moderate to slight, when exposed to heat
or flame.
Countermeasures
Storage and Handling: Section 7.
To Fight Fire: Water, foam.

TRIDECYL ACRYLATE
General Information
Description: Isoluble in water.
Formula: $H_2CCHCOOC_{13}H_{27}$.
Constants: Mol wt: 254, d: 0.9, bp: 302°F at 10 mm, flash
p: 270°F (O.C.).
Hazard Analysis
Toxicity: Details unknown. Limited animal experiments
suggest low toxicity. See also esters.
Fire Hazard: Slight, when exposed to heat or flame.
Countermeasures
Storage and Handling: Section 7.
To Fight Fire: Water, foam.

TRIDECYL BENZENE
General Information
Synonym: Tridane; 1-phenyl tridecane.
Description: Colorless liquid.
Formula: $C_6H_5(CH_2)_{12}CH_3$.
Constants: Mol wt: 260, d: 0.85–0.86 at 60/60°F.
Hazard Analysis
Toxicity: Unknown. Vapors may have a narcotic effect.

1-TRIDECYL-2-BENZYL-2-HYDROXYETHYLIMIDA-
ZOLIUM CHLORIDE
General Information
Synonym: Alrosept MBC.
Hazard Analysis
Toxic Hazard Rating:
Acute Local: Irritant 2; Ingestion 2; Inhalation 2.
Acute Systemic: U.
Chronic Local: Irritant 1.

Chronic Systemic: U.
Toxicology: Effects similar to other quaternary amines.
Experimentally has produced cholinesterase inhibition
and muscular paralysis.
Disaster Hazard: When heated to decomposition it emits
highly toxic fumes.

1-TRIDECYL-2-METHYL-2 HYDROXYETHYL-
IMIDAZOLINIUM CHLORIDE
General Information
Synonym: Alrosept MM.
Hazard Analysis and Countermeasures
See 1-tridecyl-2-benzyl-2-hydroxyethylimidazolium chloride.

TRIDECYL PHOSPHITE
General Information
Description: Water white liquid; decyl alcohol odor; in-
soluble in water.
Formula: $(C_{10}H_{21}O)_3P$.
Constants: Mol wt: 502, d: 0.892 at 25/15.5°C, mp: < 0°C,
flash p: 455°F (O.C.).
Toxic Hazard Rating: U. See esters.
Fire Hazard: Slight, when exposed to heat or flame.
Disaster Hazard: Dangerous; when heated to decomposi-
tion it emits highly toxic fumes.
Countermeasures
Storage and Handling: Section 7.
To Fight Fire: Water, foam.

TRIDEUTERO AMMONIA. See ammonia-d_3.

TRI-(DIISOBUTYL CARBINYL) BORATE
General Information
Description: White crystals; odor of diisobutyl carbinol.
Formula: $C_{27}H_{57}O_3B$.
Constants: Mol wt: 440.55, mp: 99–100°C, bp: 198–209°C
at 22 mm.
Hazard Analysis and Countermeasures
Toxicity: See diisobutylcarbinol and boron compounds.

TRI-n-DODECYL BORATE
General Information
Description: Light, straw-yellow, oily liquid.
Formula: $[CH_3(CH_2)_{10}CH_2O]_3B$.
Constants: Mol wt: 566.8, bp: 479°C, flash p: 465°F
(C.O.C), d: 0.845 at 26.8°C, vap. d.: 19.6.
Hazard Analysis
Toxicity: See boron compounds and esters.
Fire Hazard: Slight, when exposed to heat or flame; can re-
act with oxidizing materials.
Countermeasures
To Fight Fire: Foam, carbon dioxide, dry chemical or car-
bon tetrachloride (Section 6).
Storage and Handling: Section 7.

TRIETHANOLAMINE
General Information
Description: Pale yellow liquid.
Formula: $(CH_2OHCH_2)_3N$.
Constants: Mol wt: 149.19, mp: 21.2°C, bp: 360°C, flash
p: 355°F (C.C.), d: 1.1258 at 20°/20°C, vap. press.:
10 mm at 205°C, vap. d.: 5.14.
Hazard Analysis
Toxic Hazard Rating:
Acute Local: 0.
Acute Systemic: Ingestion 1; Skin Absorption 1.
Chronic Local: 0.
Chronic Systemic: Ingestion 1; Skin Absorption 1.
Toxicology: Liver and kidney damage has been demon-
strated in animals under chronic exposure.
Fire Hazard: Slight, when exposed to heat or flame.
Spontaneous Heating: No.
Disaster Hazard: Dangerous; when heated to decomposi-
tion, it emits toxic fumes of oxides of nitrogen; can re-
act vigorously with oxidizing materials.

Countermeasures
Personal Hygiene: Section 3.
To Fight Fire: Alcohol foam, water, carbon dioxide, dry chemical or carbon tetrachloride (Section 6).
Storage and Handling: Section 7.

TRIETHANOLAMINE BORATE
General Information
Description: White odorless solid.
Constants: Mol wt: 157, mp: 235.5–238.5°C.
Hazard Analysis and Countermeasures
See triethanolamine.

TRIETHANOLAMINE-o-sec-BUTYL PHENATE
Hazard Analysis
Toxicity: Unknown. A fungicide. See triethanolamine and esters.

TRIETHANOLAMINE TITANATE
General Information
Description: Yellow liquid.
Formula: $Ti[(OCH_2CH_2)NCH_2CH_2OH]_2$.
Constants: Mol wt: 254.14, d: 1.05, vap. d.: 8.78.
Hazard Analysis
Toxicity: See esters and triethanol amine.
Fire Hazard: Slight, when exposed to heat or flame; can react with oxidizing materials (Section 6).
Countermeasures
Storage and Handling: Section 7.

TRIETHOXYBORON. See ethyl borate.

1,1,3-TRIETHOXYBUTANE
General Information
Formula: $CH(OC_2H_5)_2CH_2CH(OC_2H_5)CH_3$.
Constant: Mol wt: 190.
Hazard Analysis
Toxicity: Details unknown. Limited animal experiments suggest moderate toxicity.

1,1,3-TRIETHOXYHEXANE
General Information
Description: Liquid, insoluble in water.
Formula: $CH(OC_2H_5)CH_2CH(OC_2H_5)C_3H_7$.
Constants: Mol wt: 218, d: 0.8746 at 20/20°C, bp: 133°C at 50 mm; fp: −100°C, flash p: 210°F (O.C.); vap. d.: 7.5.
Hazard Analysis
Toxicity: Details unknown. Limited animal experiments suggest low toxicity.
Fire Hazard: Moderate, when exposed to heat or flame.
Countermeasures
Storage and Handling: Section 7.
To Fight Fire: Foam, alcohol foam, fog.

TRIETHOXYMETHANE. See triethyl-o-formate.

1,3,3-TRIETHOXYPROPENE-1
General Information
Formula: $(C_2H_5O)HCCHCH(C_2H_5O)_2$.
Constant: Mol wt: 174.2.
Hazard Analysis
Toxic Hazard Rating:
  Acute Local: U.
  Acute Systemic: Ingestion 2; Inhalation 2.
  Chronic Local: U.
  Chronic Systemic: U.
Countermeasures
Ventilation Control: Section 2.
Personal Hygiene: Section 3.

TRIETHYLALUMINUM
General Information
Synonym: Aluminum triethyl.
Constants: Fp: −52.5°C, d: 0.837 at 20°C, vap. press.: 4 mm at 83°C, flash p: < −52.5°C.
Hazard Analysis
Toxicity: Highly toxic. Extremely destructive to living tissue.
Fire Hazard: Dangerous. Ignites spontaneously in air.
Countermeasures
To Fight Fire: Carbon dioxide, dry sand, dry chemical. See metal fires.
Shipping Regulations: Section 11.
  I.C.C.: Flammable liquid, red label, 2 ounces.
  IATA: Flammable liquid, not acceptable (passenger and cargo).

TRIETHYL ALUMINUM ETHERATE
General Information
Description: Colorless liquid; ether odor.
Formula: $4Al(C_2H_5)_3 \cdot 3(C_2H_5)_2O$.
Constants: Mol wt: 679.0, bp: 112°C at 16 mm.
Hazard Analysis
Toxicity: Details unknown. See triethyl aluminum and organometals.
Fire Hazard: Dangerous. See ethers.
Explosion Hazard: Dangerous. See ethers. Also upon contact with moisture this material explodes evolving ethane.
Disaster Hazard: Dangerous; upon contact with moisture it explodes. Warming can evolve copious fumes of ether. See ethers.
Countermeasures
Storage and Handling: Section 7.
To Fight Fire: See ethers (Section 6).

TRIETHYLAMINE
General Information
Description: Colorless liquid.
Formula: $(C_2H_5)_3N$.
Constants: Mol wt: 101.19, mp: −114.8°C, bp: 89.5°C, flash p: 20°F (O.C.), d: 0.7229 at 25°/4°C, vap. d.: 3.48, lel: 1.2%, uel: 8.0%.
Hazard Analysis
Toxic Hazard Rating:
  Acute Local: Irritant 3; Ingestion 3; Inhalation 3.
  Acute Systemic: Ingestion 3; Inhalation 3.
  Chronic Local: Irritant 2; Inhalation 2.
  Chronic Systemic: Ingestion 3; Inhalation 3.
Toxicity: Experimental animals have shown kidney and liver damage.
TLV: ACGIH (recommended); 25 parts per million in air; 100 milligrams per cubic meter of air.
Fire Hazard: Dangerous, when exposed to heat or flame.
Explosion Hazard: Unknown.
Disaster Hazard: Highly dangerous; keep away from heat or open flame; can react with oxidizing materials.
Countermeasures
Ventilation Control: Section 2.
To Fight Fire: Foam, carbon dioxide, dry chemical or carbon tetrachloride (Section 6).
Storage and Handling: Section 7.
Shipping Regulations: Section 11.
  IATA: Flammable liquid, red label, 1 liter (passenger), 40 liters (cargo).

TOXIC HAZARD RATING CODE (For detailed discussion, see Section 1.)

0 NONE: (a) No harm under any conditions; (b) Harmful only under unusual conditions or overwhelming dosage.

1 SLIGHT: Causes readily reversible changes which disappear after end of exposure.

2 MODERATE: May involve both irreversible and reversible changes; not severe enough to cause death or permanent injury.

3 HIGH: May cause death or permanent injury after very short exposure to small quantities.

U UNKNOWN: No information on humans considered valid by authors.

TRIETHYL-n-AMYLTIN
General Information
Description: Liquid.
Formula: $(C_2H_5)_3Sn(C_5H_{11})$.
Constants: Mol wt: 277.0, bp: 102°C at 10 mm.
Hazard Analysis
Toxicity: Details unknown. See tin compounds.

TRIETHYL ANTIMONITE. See antimony ethoxide.

TRIETHYLARSENIC
General Information
Synonym: Arsenic triethyl.
Description: Colorless liquid.
Formula: $As(C_2H_5)_3$.
Constants: Mol wt: 162.1, bp: 140°C at 736 mm, d: 1.152, vap. d.: 5.59.
Hazard Analysis
Toxicity: Highly toxic. See arsenic compounds.
Fire Hazard: Moderate.
Disaster Hazard: Dangerous; when heated to decomposition or on contact with acid or acid fumes, it emits highly toxic fumes of arsenic; can react vigorously with oxidizing materials.
Countermeasures
Ventilation Control: Section 2.
Storage and Handling: Section 7.

TRIETHYLBENZENE
General Information
Description: Clear, colorless liquid.
Formula: $C_6H_3(CH_2CH_3)_3$.
Constants: Mol wt: 162.3, mp: $< -70°C$, bp: 218–219°C, flash p: 181°F, d: 0.870 at 25°/25°C, vap. d.: 5.6.
Hazard Analysis
Toxicity: Details unknown. Animal experiments suggest high toxicity.
Fire Hazard: Moderate, when exposed to heat or flame; can react with oxidizing materials.
Countermeasures
To Fight Fire: Foam, carbon dioxide, dry chemical or carbon tetrachloride (Section 6).
Storage and Handling: Section 7.

TRIETHYLBISMUTHINE
General Information
Synonym: Bismuth triethyl.
Description: Liquid.
Formula: $Bi(C_2H_5)_3$.
Constants: Mol wt: 296.18, bp: 107°C at 79 mm (can explode), d: 1.82.
Hazard Analysis
Toxicity: See bismuth compounds.
Fire Hazard: Dangerous, when exposed to heat or flame (Section 6).
Explosion Hazard: Moderate, when exposed to heat. Explodes at 302°F.
Disaster Hazard: Moderately dangerous; when heated to decomposition it emits toxic fumes and explodes; can react with oxidizing materials.
Countermeasures
Storage and Handling: Section 7.

TRIETHYL BORANE. See boron triethyl.

TRIETHYL BORATE. See ethyl borate.

TRIETHYLBORON. See boron triethyl.

TRIETHYL CITRATE
General Information
Formula: $C_3H_5O(COOC_2H_5)_3$.
Constants: Mol wt: 276.3, bp: 294.0°C, flash p: 303°F (C.O.C.), d: 1.136 at 25°C, vap. press.: 1 mm at 107.0°C.

Hazard Analysis
Toxicity: Probably low. See citric acid. A general purpose food additive (Section 10).
Fire Hazard: Slight, when exposed to heat or flame (Section 6).
Countermeasures
Storage and Handling: Section 7.

TRI-(1-ETHYL CYCLOHEXYL) BORATE
General Information
Formula: $B(C_8H_{15}O)_3$.
Constant: Mol wt: 392.5.
Hazard Analysis
Toxic Hazard Rating: Probably toxic. See boron compounds and esters.
Fire Hazard: A flammable material when exposed to heat or flame.
Countermeasures
Storage and Handling: Section 7.

TRIETHYL (p-DIMETHYLAMINOPHENYL)TIN
General Information
Description: Liquid.
Formula: $(C_2H_5)_3(CH_3)_2NC_6H_4Sn$.
Constants: Mol wt: 326.1, d: 1.2425, bp: 173°C at 3 mm.
Hazard Analysis
Toxicity: Details unknown. See tin compounds.

TRIETHYLENE DIAMINE
General Information
Formula: $N(CH_2CH_2)_3N$.
Constant: Mol wt: 112.
Hazard Analysis
See amines.

TRIETHYLENE GLYCOL
General Information
Synonyms: 2,2'-Ethylene dioxydiethanol; glycol bis(hydroxyethyl)ether.
Description: Colorless liquid.
Formula: $(CH_2OCH_2CH_2OH)_2$.
Constants: Mol wt: 150.17, bp: 291.2°C, fp: −4.3°C, flash p: 350°F, d: 1.122 at 25°/25°C, lel: 0.9%, uel: 9.2%, autoign. temp.: 700°F, vap. press.: 1 mm at 114°C, vap. d.: 5.17.
Hazard Analysis
Toxicity: A fungicide. See also glycols.
Fire Hazard: Slight, when exposed to heat or flame; can react with oxidizing materials.
Spontaneous Heating: No.
Explosion Hazard: Moderate, in the form of vapor when exposed to flame, spark or heat source.
Countermeasures
Ventilation Control: Section 2.
To Fight Fire: Water, carbon dioxide, dry chemical or carbon tetrachloride (Section 6).
Personnel Protection: Section 3.
Storage and Handling: Section 7.

TRIETHYLENE GLYCOL CAPRYLATE
General Information
Synonym: Plasticizer SC.
Description: Clear liquid.
Formula: $C_7H_{15}CO_2CH_2CH_2OCH_2CH_2OCH_2CH_2CO_2C_7H_{15}$.
Constants: Mol wt: 402.6, d: 0.973 at 20°C, mp: −3°C, bp: 243°C at 5 mm.
Hazard Analysis
Toxicity: Details unknown. Animal experiments suggest very low toxicity and moderate irritant effects on skin. See also glycols.

TRIETHYLENE GLYCOL DIBENZOATE
General Information
Description: Crystals.

Formula: $C_6H_5CO_2CH_2CH_2OCH_2CH_2OCH_2CH_2 \cdot$
   $CO_2C_6H_5$.
Constants: Mol wt: 358.3, bp: 210–223°C, fp: 46°C,
   flash p: 457°F (T.O.C.), d: 1.168 at 25°/4°C.
Hazard Analysis
Toxicity: Details unknown. See glycols.
Fire Hazard: Slight, when exposed to heat or flame; can re-
   act with oxidizing materials.
Countermeasures
To Fight Fire: Water, foam, carbon dioxide, dry chemical
   or carbon tetrachloride (Section 6).
Storage and Handling: Section 7.

TRIETHYLENE GLYCOL DICHLORIDE. See triglycol
   dichloride.

TRIETHYLENE GLYCOL DI-2-ETHYLBUTYRATE
General Information
Description: Colorless liquid.
Formula: $C_5H_{11}COOCH_2(CH_2OCH_2)_2CH_2OOC \cdot$
   C_5H_{11}.
Constants: Mol wt: 346.45, mp: −65°C, bp: 197°C at 5
   mm, flash p: 385°F (O.C.), d: 0.9946 at 20°/20°C,
   vap. press.: 5.8 mm at 200°C, vap. d.: 11.95.
Hazard Analysis
Toxicity: Details unknown. See glycols.
Fire Hazard: Slight, when exposed to heat or flame; can re-
   act with oxidizing materials.
Countermeasures
To Fight Fire: Foam, carbon dioxide, dry chemical or car-
   bon tetrachloride (Section 6).
Storage and Handling: Section 7.

TRIETHYLENE GLYCOL DI-2-ETHYLHEXOATE
General Information
Description: Colorless liquid with mild odor.
Formula: $C_7H_{15}COOCH_2(CH_2OCH_2)_2CH_2OCCC_7H_{15}$.
Constants: Mol wt: 402.56, mp: −58°C, bp: 218°C at 5
   mm, flash p: 405°F, d: 0.9679 at 20°/20°C, vap.
   press.: 1.9 mm at 200°C, vap. d.: 13.9.
Hazard Analysis
Toxicity: Details unknown. Very low toxicity based upon
   animal tests. See glycols.
Fire Hazard: Slight, when exposed to heat or flame; can re-
   act with oxidizing materials.
Countermeasures
To Fight Fire: Foam, carbon dioxide, dry chemical or car-
   bon tetrachloride (Section 6).
Storage and Handling: Section 7.

TRIETHYLENE GLYCOL DIMETHYL ETHER
General Information
Description: Liquid, water white and mild ethereal odor.
Formula: $CH_3(OCH_2CH_2)_3OCH_3$.
Constants: Mol wt: 178.22, bp: 216°C, fp: −46°C, flash
   p: 232°F (O.C.), d: 0.982 at 20°/20°C, autoign.
   temp.: 1166°F, vap. d.: 4.7, vap. press.: 0.9 mm at
   20°C.
Hazard Analysis
Toxicity: See glycols.
Fire Hazard: Slight, when exposed to heat or flame; can re-
   act with oxidizing materials.
Countermeasures
To Fight Fire: Foam, carbon dioxide, dry chemical or car-
   bon tetrachloride (Section 6).
Storage and Handling: Section 7.

TRIETHYLENE GLYCOL DIPELARGONATE
General Information
Synonym: TG-9.
Description: Clear liquid; mild characteristic odor.
Formula: $(CH_2OCH_2)_2(CH_2OCOC_8H_{17})_2$.
Constants: Mol wt: 438, fp: −4 to 1°C, flash p: 420°F,
   bp: 251°C at 5 mm, d: 0.964 at 20°/20°C, vap. d.:
   15.1.
Hazard Analysis
Toxicity: Details unknown. See glycols.
Fire Hazard: Slight, when exposed to heat or flame; can
   react with oxidizing materials.
Countermeasures
To Fight Fire: Foam, carbon dioxide, dry chemical or car-
   bon tetrachloride (Section 6).
Storage and Handling: Section 7.

**TRIETHYLENE GLYCOL METHYL ETHER
   ACETATE.** See methoxytriglycol acetate.

TRIETHYLENE GLYCOL MONOBUTYL ETHER
General Information
Synonym: Butoxy triglycol.
Discription: Liquid; completely soluble in water.
Formula: $C_4H_9O(C_2H_4O)_3H$.
Constant: Mol wt: 206, d: 1.0021 at 20/20°C, bp: decom-
   poses, fp: −47.4°C, flash p: 290°F.
Hazard Analysis
Toxicity: Details unknown. Limited animal experiments
   suggest low toxicity. See also glycols.
Fire Hazard: Slight, when exposed to heat or flame.
Countermeasures
Storage and Handling: Section 7.
To Fight Fire: Water, foam, fog.

TRIETHYLENE GLYCOL MONOMETHYL ETHER
General Information
Synonym: Methoxy triglycol.
Description: Infinitely soluble in water.
Formula: $CH_3OCH_2CH_2OCH_2CH_2OCH_2CH_2OH$.
Constants: Mol wt: 164, d: 1.0494, bp: 249°C, fp: −44°C,
   flash p: 245°F.
Hazard Analysis
Toxicity: Details unknown. Limited animal experiments
   suggest low toxicity. See also ethylene glycol mono-
   methyl ether and glycols.
Fire Hazard: Slight when exposed to heat or flame. See also
   ethers.

TRIETHYLENE MELAMINE
General Information
Synonym: TEM.
Description: Small crystals. Water soluble.
Formula: $C_9H_{12}N_6$.
Constants: Mol wt: 204.2, decomposes at 139°C.
Hazard Analysis
Toxicity: Details unknown. Has been reported as capable of
   causing GI disturbances and bone marrow depression.
Disaster Hazard: Dangerous; when heated to decomposition
   it emits highly toxic fumes of oxides of nitrogen.

TRIETHYLENE PHOSPHORAMIDE. See tris-(1-
   aziridinyl) phosphine oxide.

TRIETHYLENETETRAMINE
General Information
Description: Moderately viscous, yellowish liquid.
Formula: $H_2NCH_2(CH_2NHCH_2)_2CH_2CH_2NH_2$.

TOXIC HAZARD RATING CODE (*For detailed discussion, see Section 1.*)

0 NONE: (a) No harm under any conditions; (b) Harmful only under unusual conditions or overwhelming dosage.

1 SLIGHT: Causes readily reversible changes which disappear after end
   of exposure.

2 MODERATE: May involve both irreversible and reversible changes;
   not severe enough to cause death or permanent injury.

3 HIGH: May cause death or permanent injury after very short exposure
   to small quantities.

U UNKNOWN: No information on humans considered valid by authors.

Constants: Mol wt: 146.24, bp: 278°C, flash p: 275°F, d: 0.982, vap. press.: <0.01 mm at 20°C, autoign. temp.: 640°F.

Hazard Analysis

Toxic Hazard Rating:

Acute Local: Irritant 3; Allergen 2; Ingestion 3; Inhalation 3.

Acute Systemic: U.

Chronic Local: Irritant 2; Allergen 2; Inhalation 2.

Chronic Systemic: U.

Toxicity: Causes skin irritation and sensitization.

Fire Hazard: Slight, when exposed to heat or flame; can react with oxidizing materials.

Spontaneous Heating: No.

Countermeasures

To Fight Fire: Foam, carbon dioxide, dry chemical or carbon tetrachloride (Section 6).

Storage and Handling: Section 7.

Shipping Regulations: Section 11.

MCA warning label.

TRIETHYL o-FORMATE

General Information

Synonyms: o-formic ester; triethoxymethane.

Description: Clear liquid, pungent odor.

Formula: $(C_2H_5O)_3CH$.

Constants: Mol wt: 148.20, bp: 145.9°C, flash p: 86°F, d: 0.895 at 20°/20°C, vap. press.: 10 mm at 40.5°C, vap. d.: 5.11.

Hazard Analysis

Toxicity: Details unknown. See also esters.

Fire Hazard: Moderate, when exposed to heat or flame.

Explosion Hazard: Unknown.

Disaster Hazard: Moderately dangerous when exposed to heat or open flame; can react with oxidizing materials.

Countermeasures

Ventilation Control: Section 2.

To Fight Fire: Foam, carbon dioxide, dry chemical or carbon tetrachloride (Section 6).

Storage and Handling: Section 7.

TRIETHYLGALLIUM

General Information

Description: Colorless liquid decomposed by water.

Formula: $(C_2H_5)_3Ga$.

Constants: Mol wt: 156.9, d: 1.0576 at 30°C, mp: 82.3°C, bp: 142.6 °C.

Hazard Analysis and Countermeasures

See gallium compounds.

TRIETHYLGALLIUM MONAMMINE

General Information

Description: Colorless liquid decomposed by water.

Formula: $(C_2H_5)_3GaNH_3$.

Constant: Mol wt: 173.9.

Hazard Analysis

Toxicity: Details unknown. See gallium compounds.

TRIETHYLGALLIUM MONOETHERATE

General Information

Description: Colorless liquid decomposed by water.

Formula: $(C_2H_5)_3Ga \cdot O \cdot C_2H_5$.

Constant: Mol wt: 231.0.

Hazard Analysis and Countermeasures

See gallium compounds and ethyl ether.

TRIETHYLGERMANIUM ACETATE

General Information

Formula: $(C_2H_5)_3GeOOCH_3$.

Constant: Mol wt: 207.

Hazard Analysis

Toxicity: Details unknown. Animal experiments show toxicity much lower than corresponding lead compound. See also germanium.

TRIETHYLGERMANIUM BROMIDE

General Information

Description: Colorless liquid hydrolyzed by water.

Formula: $(C_2H_5)_3GeBr$.

Constants: Mol wt: 239.7, mp: −33°C, bp: 191°C.

Hazard Analysis and Countermeasures

See germanium compounds and bromides.

TRIETHYLGERMANIUM CHLORIDE

General Information

Description: Colorless liquid; hydrolyzed by water.

Formula: $(C_2H_5)_3GeCl$.

Constants: Mol wt: 195.2, mp: < −50°C, bp: 176°C.

Hazard Analysis and Countermeasures

See germanium compounds and chlorides.

TRIETHYLGERMANIUM FLUORIDE

General Information

Description: Colorless liquid; hydrolyzed by water.

Formula: $(C_2H_5)_3GeF$.

Constants: Mol wt: 178.8, bp: 149°C, at 751 mm.

Hazard Analysis and Countermeasures

See fluorides.

TRIETHYLGERMANIUM HYDRIDE

General Information

Description: Colorless liquid; insoluble in water.

Formula: $(C_2H_5)_3GeH$.

Constants: Mol wt: 160.8, bp: 124°C at 751 mm.

Hazard Analysis and Countermeasures

See germanium compounds and hydrides.

TRIETHYLGERMANIUM IMINE

General Information

Description: Colorless liquid, hydrolyzes in water.

Formula: $[(C_2H_5)_3Ge]_2NH$.

Constants: Mol wt: 334.6, bp: 100°C at 0.1 mm.

Hazard Analysis

Toxicity: Details unknown. See germanium compounds.

TRIETHYLGERMANIUM IODIDE

General Information

Description: Colorless liquid; hydrolyzed by water.

Formula $(C_2H_5)_3GeI$.

Constants: Mol wt: 286.7, mp: < −50°C, bp: 212°C.

Hazard Analysis and Countermeasures

See germanium compounds and iodides.

TRIETHYLGERMANIUM OXIDE

General Information

Description: Colorless liquid; insoluble in water.

Formula: $[(C_2H_5)_3Ge]_2O$.

Constants: Mol wt: 335.6, mp: < −50°C, bp: 254°C.

Hazard Analysis

Toxicity: Details unknown. See germanium compounds.

TRI (2-ETHYLHEXYL) BORATE

General Informaion

Description: Colorless, mobile liquid. Odor of 2-ethylhexanol.

Formula: $[C_4H_9(C_2H_5)CHCH_2O]_3B$.

Constants: Mol wt: 398.5, bp: 350–354°C, flash p: 350°F (C.O.C.), d: 0.857 at 23.6°C, vap. d.: 13.8.

Hazard Analysis

Toxicity: Details unknown. See esters.

Fire Hazard: Slight, when exposed to heat or flame; can react with oxidizing materials.

Countermeasures

To Fight Fire: Foam, carbon dioxide, dry chemical or carbon tetrachloride (Section 6).

Storage and Handling: Section 7.

TRI(2-ETHYLHEXYL) PHOSPHATE

General Information

Description: Light-colored liquid.

Formula: $(C_2H_5C_6H_{12})_3PO_4$.

Constants: Mol wt: 434.6, mp: −74 °C, bp: 216 °C at 5 mm,

flash p: 405° F, d: 0.9262 at 20° /20° C, vap. press.:
0.23 mm at 150° C.

Hazard Analysis
Toxic Hazard Rating:
Acute Local: U.
Acute Systemic: Ingestion 1.
Chronic Local: Irritant 1.
Chronic Systemic: Ingestion 1.
Toxicity: Details unknown. Limited animal experimentation
showed low toxicity in acute exposure. See phosphoric
acid.
Disaster Hazard: Dangerous. See phosphates.
Countermeasures
Storage and Handling: Section 7.

TRIETHYL-o-HYDROXYPHENYLTIN
General Information
Description: Liquid.
Formula: $(C_2H_5)_3HOC_6H_4Sn$.
Constants: Mol wt: 299.0, d: 1.3229 at 25° C, bp: 200° C
at 3 mm.
Hazard Analysis
Toxicity: Details unknown. See tin compounds.

TRIETHYLISOAMYLTIN
General Information
Description: Liquid.
Formula: $(C_2H_5)_3Sn(C_5H_{11})$.
Constants: Mol wt: 277.0, d: 1.1203, bp: 111° C at 18.5
mm.
Hazard Analysis
Toxicity: Details unknown. See tin compounds.

TRIETHYLISOBUTYLTIN
General Information
Description: Liquid.
Formula: $(C_2H_5)_3Sn(C_4H_9)$.
Constants: Mol wt: 263.0, d: 1.139, bp: 96.5 °C at 17 mm.
Hazard Analysis
Toxicity: Details unknown. See tin compounds.

TRIETHYLLEAD
General Information
Synonym: Hexaethyldilead.
Description: A liquid; insoluble in water.
Formula: $Pb_2(C_2H_5)_6$.
Constants: Mol wt: 588.8, d: 1.471.
Hazard Analysis and Countermeasures
See lead compounds.

TRI (7-ETHYL-2-METHYL-4-UNDECYL) BORATE.
See tritetradecyl borate.

TRIETHYLPHENYLGERMANIUM
General Information
Description: Colorless liquid; insoluble in water.
Formula: $Ge(C_2H_5)_3(C_6H_5)$.
Constants: Mol wt: 236.9, bp: 117° C at 13 mm.
Hazard Analysis
Toxicity: Details unknown. See germanium compounds.

TRIETHYLPHENYLSILICANE
General Information
Description: Solid.
Formula: $(C_2H_5)_3(C_6H_5)Si$.
Constants: Mol wt: 192.3, mp: 148° C, bp: 230° C.
Hazard Analysis
Toxicity: Details unknown. See silanes.

TRIETHYLPHENYLTIN
General Information
Description: Colorless liquid; insoluble in water, soluble in
organic solvents.
Formula: $(C_2H_5)_3Sn(C_6H_5)$.
Constants: Mol wt: 283.0, d: 1.2639, bp: 254° C.
Hazard Analysis
Toxicity: Details unknown. See tin compounds.

TRIETHYL PHOSPHATE
General Information
Synonym: TEP.
Description: Liquid; soluble in most organic solvents; in-
soluble in water.
Formula: $(C_2H_5)_3PO_4$.
Constants: Mol wt: 182.2, mp: −56.5° C, bp: 209-218° C,
flash p: 240° F, d: 1.067-1.072 at 20° /20° C, vap.
press.: 1 mm at 39.6° C, vap. d.: 6.28.
Hazard Analysis
Toxicity: Details unknown. Causes cholinesterase inhibi-
tion, but to a lesser extent than parathion. May be ex-
pected to cause nerve injury similar to that of other
phosphate esters. See also tri-o-cresyl phosphate.
Fire Hazard: Slight when exposed to heat or flame.
Disaster Hazard: Dangerous. See phosphates; can react
vigorously with ozidizing materials.
Countermeasures
Storage and Handling: Section 7.
To Fight Fire: Water, foam, carbon dioxide, dry chemical
or carbon tetrachloride (Section 6).

TRIETHYL PHOSPHINE
General Information
Description: Colorless liquid.
Formula: $(C_2H_5)_3P$.
Constants: Mol wt: 118.2, bp: 128° C, d: 0.801 at 20° /4° C.
Hazard Analysis
Toxicity:Details unknown; probably high. See phosphides.
Disaster Hazard: Dangerous. See phosphates and phosphine.
Countermeasures
Storage and Handling: Section 7.

TRIETHYL PHOSPHINE OXIDE
General Information
Description: Colorless, deliquescent crystals.
Formula: $(C_2H_5)_3PO$.
Constants: Mol wt: 134.2, mp: 52.9° C, bp: 242.9° C.
Hazard Analysis
Toxicity: Details unknown; probably high.
Disaster Hazard: Dangerous. See phosphates.
Countermeasures
Storage and Handling: Section 7.

TRIETHYL PHOSPHINE SULFIDE
General Information
Description: Crystals.
Formula: $(C_2H_5)PS$.
Constants: Mol wt: 150.2, mp: 94° C, bp: ignites.
Hazard Analysis
Toxicity: Details unknown; probably high.
Fire Hazard: Moderate, when exposed to heat or flame
(Section 6).
Disaster Hazard: Dangerous. See phosphates and sulfides,
can react vigorously with oxidizing materials.
Countermeasures
Storage and Handling: Section 7.

TOXIC HAZARD RATING CODE (For detailed discussion, see Section 1.)

0 NONE: (a) No harm under any conditions; (b) Harmful only under unusual conditions or overwhelming dosage.

1 SLIGHT: Causes readily reversible changes which disappear after end
of exposure.

2 MODERATE: May involve both irreversible and reversible changes;
not severe enough to cause death or permanent injury.

3 HIGH: May cause death or permanent injury after very short exposure
to small quantities.

U UNKNOWN: No information on humans considered valid by authors.

TRIETHYL PHOSPHOROTHIOATE
General Information
Description: Colorless liquid; strong characteristic odor.
Formula: $(C_2H_5O)_3PS$.
Constants: Mol wt: 198.2, bp: 93.5-94°C at 10 mm, flash p: 225°F (C.O.C.), d: 1.074 at 20°/4°C.
Hazard Analysis
Toxicity: Details unknown; probably moderate.
Fire Hazard: Moderate, when exposed to heat or flame (Section 6).
Disaster Hazard: Dangerous. See phosphates and sulfates, can react with oxidizing materials.
Countermeasures
Storage and Handling: Section 7.

TRIETHYL-n-PROPYLTIN
General Information
Description: Liquid.
Formula: $(C_2H_5)_3Sn(C_3H_7)$.
Constants: Mol wt: 249.0, d: 1.1680, bp: 82°C at 13 mm.
Hazard Analysis
Toxicity: Details unknown. See tin compounds.

TRIETHYL STIBINE. See antimony triethyl.

TRIETHYLTHALLIUM
General Information
Description: Yellow liquid.
Formula: $(C_2H_5)_3Tl$.
Constants: Mol wt: 291.6, d: 1.957 at 23°/23°C, mp: −63°C, bp: 192°C.
Hazard Analysis and Countermeasures
See thallium compounds.

TRIETHYLTIN
General Information
Description: Colorless liquid; insoluble in water.
Formula: $(C_2H_5)_3Sn$.
Constants: Mol wt: 205.9, d: 1.3774, mp: < −75°C, bp: 161°C at 23 mm.
Hazard Analysis
Toxicity: See tin compounds.

TRIETHYLTIN BROMIDE
General Information
Description: Colorless liquid; soluble in organic solvents.
Formula: $(C_2H_5)_3SnBr$.
Constants: Mol wt: 285.8, d: 1.630, mp: −13.5°C, bp: 224°C.
Hazard Analysis and Countermeasures
See tin compounds and bromides.

TRIETHYLTIN CHLORIDE
General Information
Description: Colorless liquid; insoluble in water; soluble in organic solvents.
Formula: $(C_2H_5)_3SnCl$.
Constants: Mol wt: 241.3, d: 1.428 at 8°C, mp: 10°C, bp: 210°C.
Hazard Analysis and Countermeasures
See tin compounds and chlorides.

TRIETHYLTIN ETHOXIDE
General Information
Description: Colorless liquid, decomposed by water; soluble in organic solvents.
Formula: $(C_2H_5)_3Sn(OC_2H_5)$.
Constants: Mol wt: 250.94, d: 1.2634, bp: 190°C.
Hazard Analysis
Toxicity: Details unknown. See tin compounds and ethyl alcohol.

TRIETHYLTIN HYDROXIDE
General Information
Description: Colorless crystals; soluble in water and organic solvents.

Formula: $(C_2H_5)_3SnOH$.
Constants: Mol wt: 222.9, mp: 43°C, bp: 271°C.
Hazard Analysis
Toxicity: Details unknown. See tin compounds.

TRIETHYLTIN IODIDE
General Information
Description: Colorless liquid; soluble in organic solvents.
Formula: $(C_2H_5)_3SnI$.
Constants: Mol wt: 332.8,, d: 1.833, mp: 34.5°C, bp: 225°C.
Hazard Analysis and Countermeasures
See tin compounds and iodides.

TRIETHYL-p-TOLYLGERMANIUM
General Information
Description: Colorless liquid; insoluble in water.
Formula: $(C_2H_5)_3Ge(C_6H_4CH_3)$.
Constants: Mol wt: 250.9, bp: 126°C at 12 mm.
Hazard Analysis
Toxicity: Details unknown. See germanium compounds.

2,2,2-TRIETHYL-1,1,1-TRIPHENYL DIGERMANE
General Informaion
Description: Colorless crystals; insoluble in water.
Formula $(C_2H_5)_3Ge_2(C_6H_5)_3$.
Constants: Mol wt: 463.7, mp: 90°C.
Hazard Analysis
Toxicity: Details unknown. See germanium compounds.

TRI-(1-ETHYNYLCYCLOHEXYL) BORATE
General Information
Description: Pale-yellow liquid; odor of 1-ethynylcyclohexa-nol.
Constants: Mol wt: 380.34, bp: 150-170°C at 0.5 mm, flash p: 190°F (C.O.C.), d: 1.006 at 27°C.
Hazard Analysis
Toxicity: See boron compounds and alcohols.
Fire Hazard: Moderate, when exposed to heat or flame; can react with oxidizing materials.
Countermeasures
To Fight Fire: Water, foam, carbon dioxide, dry chemical or carbon tetrachloride (Section 6).
Storage and Handling: Section 7.

TRIFLUOROACETIC ACID
General Information
Description: Colorless liquid; strong pungent odor.
Formula: CF_3COOH.
Constants: Mol wt: 114, mp: −15.25°C, bp: 71.1°C at 734 mm, d: 1.535 at 0°C.
Hazard Analysis and Countermeasures
See fluorides.

1,1,2-TRIFLUORO-4-BROMOBUTENE
General Information
Formula: $CF_2:CFCH_2CH_2Br$.
Constant: Mol wt: 189.
Hazard Analysis
Toxicity: Unknown. Probably irritant and narcotic in high concentrations.
Disaster Hazard: Dangerous. See fluorides and bromides.

TRIFLUOROBROMOMETHANE
General Information
Description: Gas.
Formula: $BrCF_3$.
Constant: Mol wt: 149.
Hazard Analysis
Toxic Hazard Rating:
Acute Local: Irritant 1.
Acute Systemic: Inhalation 1.
Chronic Local: Irritant 1.
Chronic Systemic: Inhalation 1; Skin Absorption 1.
Toxicity: Details unknown. High concentrations are ir-

ritating to the lungs and produce narcosis. Acts also as an anesthetic. Animal experiments confirm low toxicity.
Disaster Hazard: Dangerous. See fluorides and bromides.
Countermeasures
Shipping Regulations: Section 11.
   IATA: Nonflammable gas, green label, 70 kilograms (passenger), 140 kilograms (cargo).

TRIFLUOROCHLOROETHENE. See trifluorochlorethylene.

TRIFLUOROCHLOROETHYLENE
General Information
Synonyms: Monochlorotrifluoroethylene; trichlorofluoroethene.
Description: Colorless gas.
Formula: C_2F_3Cl.
Constants: Mol wt: 116.48, mp: -157.5 °C, bp: -27.9 °C.
Hazard Analysis
Toxic Hazard Rating:
   Acute Local: U.
   Acute Systemic: Inhalation 1.
   Chronic Local: U.
   Chronic Systemic: U.
Toxicology: Reported as causing liver and kidney injury.
Caution: An asphyxiant.
Fire Hazard: Slight, when exposed to heat or flame (Section 6).
Disaster Hazard: Dangerous. See fluorides and chlorides.
Countermeasures
Storage and Handling: Section 7.
Shipping Regulations: Section 11.
   I.C.C.: Flammable gas; red gas label, 300 pounds.
   Coast Guard Classification: Noninflammable gas; green gas label.
   IATA: Flammable gas, red label, not acceptable (passenger), 140 kilograms (cargo).

α,α,α,-TRIFLUORO-2,6-DINITRO-n,n-DIPROPYL-p-TOLUIDINE
General Information
Synonym: Trifluralin.
Description: Yellowish-orange solid; insoluble in water; soluble in xylene, acetone, and ethanol.
Formula: $(CF_3)(NO_2)_2C_6H_2N(C_3H_7)_2$.
Constants: Mol wt: 335, mp: 48.5-49°C, bp: 139-140°C at 4.2 mm.
Hazard Analysis
Toxicity: Details unknown. A herbicide. Animal experiments suggest relatively low toxicity.
Disaster Hazard: Dangerous. See fluorides and nitrates.
Countermeasures
Storage and Handling: Section 7.

TRIFLUOROMETHANE. See fluoroform.

TRIFLUOROMETHYLBENZENE. See benzotrifluoride.

TRIFLUOROMETHYLHEXACHLOROCYCLOHEXANE
General Information
Formula: $H_2C_6Cl_6F_3CH_3$.
Constant: Mol wt: 342.
Hazard Analysis and Countermeasures
Toxicity: An insecticide. Specific toxicologic data not available. Toxicity may be similar to that of DDT and fluorides. See fluorides and chlorides.

TRIFLUOROMONOBROMOMETHANE
TLV: ACGIH (recommended); 1000 parts per million in air; 6100 milligrams per cubic meter of air.
Toxicity: Details unknown. Probably irritant and narcotic in high concentrations.

TRIFLUORONITROSOMETHANE
General Information
Description: Bright blue gas.
Formula: CF_3NO.
Constants: Mol wt: 99.02, mp: -150°C, bp: -80°C.
Hazard Analysis
Toxic Hazard Rating:
   Acute Local: Irritant 3; Inhalation 3.
   Acute Systemic: U.
   Chronic Local: U.
   Chronic Systemic: U.
Caution: A military poison.
Disaster Hazard: Dangerous. See fluorides and oxides of nitrogen.
Countermeasures
Ventilation Control: Section 2.
Personnel Protection: Section 3.
Storage and Handling: Section 7.

TRIFLUOROPENTACHLOROPROPANE
General Information
Description: Liquid.
Formula: $C_3F_3Cl_5$.
Constants: Mol wt: 270.3, bp: 155°C, flash p: 228°F.
Hazard Analysis
Toxicity: See chlorinated hydrocarbons, aliphatic.
Fire Hazard: Moderate, when exposed to heat or flame.
Disaster Hazard: Dangerous. See fluorides and chlorides; can react vigorously with ozidizing materials.
Countermeasures
Storage and Handling: Section 7.
To Fight Fire: Foam, carbon dioxide, dry chemical or carbon tetrachloride (Section 6).

TRIFLUOROTRICHLOROETHANE
General Information
Description: Colorless gas.
Formula: CCl_3CF_3.
Constants: Mol wt: 187.39, mp: 13.2°C, bp: 45.8°C, d: 1.5702, autoign. temp.: 1256°F.
Hazard Analysis
Toxic Hazard Rating:
   Acute Local: U.
   Acute Systemic: Inhalation 1.
   Chronic Local: U.
   Chronic Systemic: U.
TLV: ACGIH (recommended) 1000 parts per million in air; 7600 milligrams per cubic meter of air.
Fire Hazard: Very slight; when exposed to heat or flame (Section 6).
Disaster Hazard: Dangerous. See chlorides and fluorides.
Countermeasures
Storage and Handling: Section 7.

TRIFLUOROVINYL BROMIDE
General Information
Description: Dense gas.
Formula: $CF_2:CFBr$.
Constants: Mol wt: 161, bp: -2°C.
Hazard Analysis
Toxicity: Unknown.
Disaster Hazard: Dangerous. See fluorides and bromides.

TOXIC HAZARD RATING CODE (For detailed discussion, see Section 1.)

0 NONE: (a) No harm under any conditions; (b) Harmful only under unusual conditions or overwhelming dosage.

1 SLIGHT: Causes readily reversible changes which disappear after end of exposure.

2 MODERATE: May involve both irreversible and reversible changes; not severe enough to cause death or permanent injury.

3 HIGH: May cause death or permanent injury after very short exposure to small quantities.

U UNKNOWN: No information on humans considered valid by authors.

TRIGERMANE
General Information
Synonym: Germanium hydride.
Description: Colorless liquid.
Formula: Ge_3H_8.
Constants: Mol wt: 225.86, mp: $-105.6°C$, bp: $110.5°C$, d: 2.2.
Hazard Analysis and Countermeasures
See hydrides and germanium compounds.

TRIGLYCIDYL PHOSPHATE
General Information
Synonym: TGP.
Hazard Analysis
Toxicity: Probably highly toxic. Details unknown. Animal experiments show bone marrow depression. See also diglycidyl ether.

TRIGLYCOL DICHLORIDE
General Information
Synonym: Triethylene glycol dichloride.
Description: Colorless liquid.
Formula: $Cl(C_2H_4O)_2C_2H_4Cl$.
Constants: Mol wt: 187.09, bp: $240°C$, fp: $-31.5°C$, flash p: $250°F$ (O.C.), d: 1.197, vap. press.: 0.03 mm at $20°C$.
Hazard Analysis
Toxicity: Details unknown. See also glycols.
Fire Hazard: Slight, when exposed to heat or flame.
Disaster Hazard: Dangerous. See chlorides; can react with oxidizing materials.
Countermeasures
Storage and Handling: Section 7.
To Fight Fire: Water, foam, carbon dioxide, dry chemical or carbon tetrachloride (Section 6).

TRI-n-HEXYL BORATE
General Information
Description: Colorless liquid; odor of n-hexanol.
Formula: $[CH_3(CH_2)_5O]_3B$.
Constants: Mol wt: 314.3, bp: $140-146°C$ at 2 mm, flash p: $300°F$, d: 0.847 at $28°C$, vap. d.: 10.8.
Hazard Analysis
Toxicity: See esters and boron compounds.
Fire Hazard: Slight, when exposed to heat or flame; can react with oxidizing materials.
Countermeasures
To Fight Fire: Foam, carbon dioxide, dry chemical or carbon tetrachloride (Section 6).
Storage and Handling: Section 7.

TRIHEXYLENE GLYCOL BIBORATE
General Information
Description: Colorless liquid; odor of hexylene glycol.
Formula: $C_{18}H_{36}O_6B_2$.
Constants: Mol wt: 370.1, bp: $143-149°C$ at 2 mm, flash p: $345°F$, d: 0.982 at $21°C$, vap. d.: 12.8.
Hazard Analysis
Toxicity: See boron compounds and glycols.
Fire Hazard: Slight, when exposed to heat or flame; can react with oxidizing materials.
Countermeasures
To Fight Fire: Foam, carbon dioxide, dry chemical or carbon tetrachloride (Section 6).
Storage and Handling: Section 7.

TRIHEXYL PHOSPHITE
General Information
Description: Mobile, colorless liquid; characteristic odor; decomposes in water.
Formula: $(C_6H_{13})_3PO_3$.
Constants: Mol wt: 334, d: 0.897, bp: $135-141°C$ at 0.2 mm, flash p: $320°F$ (O.C.).
Hazard Analysis
Toxicity: See esters.

Fire Hazard: Slight, when exposed to heat or flame.
Disaster Hazard: Dangerous. See phosphorous compounds.
Countermeasures
Storage and Handling: Section 7.
To Fight Fire: Foam, alcohol foam, mist. (Section 6).

TRIHYDROXYBENZENE. See pyrogallol.

3, 4, 5-TRIHYDROXYBENZOIC ACID. See gallic acid.

2, 4, 5-TRIHYDROXYBUTYROPHENONE
General Information
Synonym: THBP.
Description: Yellow-tan crystals; very slightly soluble in water; soluble in alcohol and propylene glycol.
Formula: $C_6H_2(OH)_3COC_3H_7$.
Constants: Mol wt: 198, mp: $149-153 °C$, d: 6.0 lb/gal at $20°C$.
Hazard Analysis
Toxicity: Details unknown. A food additive permitted in food for human consumption.

TRIIODOMETHANE
General Information
Synonym: Iodoform.
Description: Yellow crystals; iodine odor; very slightly water soluble.
Formula: CHI_3.
Constants: Mol wt: 393.8, d: 4.008 at $20/4°C$, mp: $119°C$, bp: sublimes, decomposes violently at $210°C$.
Hazard Analysis
Toxic Hazard Rating:
Acute Local: Irritant 2.
Acute Systemic: Ingestion 2; Inhalation 2.
Chronic Local: Irritant 2.
Chronic Systemic: U.
Disaster Hazard: Dangerous. See iodides.

TRI IRON PHOSPHIDE
General Information
Description: Gray crystals.
Formula: Fe_3P.
Constants: Mol wt: 198.53, mp: $1100°C$, d: 6.74.
Hazard Analysis and Countermeasures
See phosphides.

TRIISOAMYL BORATE
General Information
Description: Liquid.
Formula: $(C_5H_{11}O)_3B$.
Constants: Mol wt: 272.2, d: 0.872 at $0°C$, bp: $255°C$.
Hazard Analysis
Toxicity: Details unknown. See also boron compounds.

TRIISOAMYLBORON
General Information
Description: Colorless mobile liquid, not soluble in water; soluble in ether.
Formula: $(C_5H_{11})_3B$.
Constants: Mol wt: 224.2, d: 0.72, bp: $119°C$ at 14 mm.
Hazard Analysis
Toxicity: Details unknown. See also boron compounds and esters.

TRIISOAMYLTIN BROMIDE
General Information
Description: Solid.
Formula: $(C_5H_{11})_3SnBr$.
Constants: Mol wt: 412.0, d: 1.2613, mp: $21°C$, bp: $177°C$ at 15 mm.
Hazard Analysis and Countermeasures
See tin compounds and bromides.

TRIISOAMYLTIN CHLORIDE
General Information
Description: Liquid.
Formula: $(C_5H_{11})_3SnCl$.

Constants: Mol wt: 367.6, d: 1.1290 at 34°C, mp: −30.2°C, bp: 174°C at 13 mm.
Hazard Analysis and Countermeasures
See tin compounds and chlorides.

TRIISOAMYLTIN FLUORIDE
General Information
Description: Crystals.
Formula: $(C_5H_{11})_3SnF$.
Constants: Mol wt: 351.1, mp: 288°C.
Hazard Analysis and Countermeasures
See fluorides.

TRIISOAMYLTIN IODIDE
General Information
Description: Liquid.
Formula: $(C_5H_{11})_3SnI$.
Constants: Mol wt: 459.0, mp: −22°C, d: 1.3777 at 26.5°C, bp: 182°C at 13 mm.
Hazard Analysis and Countermeasures
See tin compounds and iodides.

TRIISOBUTYLALUMINUM
General Information
Description: Clear colorless liquid. Ignites on exposure to air.
Formula: $(C_4H_9)_3Al$.
Constants: Mol wt: 198.3, d: 0.7859 at 20°C, vap. press.: 1 mm at 47°C, flash p: <4°C, autoign. temp.: <4°C, fp: 4.3°C, bp: decomposes.
Hazard Analysis
Toxicity: Highly toxic. Extremely destructive of living tissue.
Fire Hazard: Dangerous. Pyrophoric in air. Reacts violently with moisture and oxidizers.
Disaster Hazard: Dangerous. In presence of air or moisture it reacts violently.
Countermeasures
To Fight Fire: Carbon dioxide, dry sand, dry chemical. (Section 6).
Storage and Handling: Section 7.
Shipping Regulations: Section 11—Pyrophoric chemicals N.O.I.B.N.
I C C : Red label.
I A T A : Flammable liquid, not acceptable.

TRIISOBUTYL ALUMINUM CHLORIDE
Hazard Analysis
Toxic Hazard Rating:
   Acute Local: Irritant 3; Ingestion 3; Inhalation 3.
   Acute Systemic: Inhalation 3.
   Chronic Local: U.
   Chronic Systemic: U.
Toxicity: Action similar to but less intense than that of di-isobutyl aluminum chloride.
Disaster Hazard: Dangerous. See chlorides.

TRIISOBUTYL BORATE
General Information
Synonym: Triisobutoxyborine.
Description: Colorless liquid; odor of isobutyl alcohol.
Formula: $[CH_3CH(CH_3)CH_2O]_3B$.
Constants: Mol wt: 230.16, bp: 212°C, flash p: 185°F (C.O.C.), d: 0.843 at 23°C.
Hazard Analysis
Toxicity: See boron compounds and isobutanol.
Fire Hazard: Moderate, when exposed to heat or flame; can react with oxidizing materials.

Countermeasures
To Fight Fire: Foam, carbon dioxide, dry chemical or carbon tetrachloride (Section 6).
Storage and Handling: Section 7.

TRIISOBUTYLBORON
General Information
Description: Colorless mobile liquid; insoluble in water.
Formula: $(C_4H_9)_3B$.
Constants: Mol wt: 182.2, d: 0.74, bp: 188°C.
Hazard Analysis
Toxicity: Details unknown. See also boron compounds.

TRIISOBUTYLENE OXIDE
General Information
Formula: $(C_4H_8)_3O$.
Constant: Mol wt: 184.
Hazard Analysis
Toxic Hazard Rating:
   Acute Local: Irritant 2.
   Acute Systemic: Ingestion 1; Inhalation 1; Skin Absorption 1.
   Chronic Local: 0.
   Chronic Systemic: Unknown.
Toxicity: Based on limited animal experiments.

TRIISOBUTYLETHYLTIN
General Information
Description: Liquid.
Formula: $(C_4H_9)_3Sn(C_2H_5)$.
Constants: Mol wt: 319.1, d: 1.0779 at 21°C, bp: 125°C at 16 mm.
Hazard Analysis
Toxicity: Details unknown. See tin compounds.

TRIISOBUTYLISOAMYLTIN
General Information
Description: Liquid.
Formula: $(C_4H_9)_3Sn(C_5H_{11})$.
Constants: Mol wt: 361.2, d: 1,0356 at 27°C, bp: 152.9°C at 16.5 mm.
Hazard Analysis
Toxicity: Details unknown. See tin compounds.

TRIISOBUTYLTIN BROMIDE
General Information
Description: Liquid.
Formula: $(C_4H_9)_3SnBr$.
Constants: Mol wt: 370.0, d: 1.3523, mp: −26.5°C, bp: 148°C at 13 mm.
Hazard Analysis and Countermeasures
See tin compounds and bromides.

TRIISOBUTYLTIN CHLORIDE
General Information
Description: Solid.
Formula: $(C_4H_9)_3SnCl$.
Constants: Mol wt: 325.5, d: 1.1290 at 34°C, mp: 30.2°C, bp: 174°C at 13 mm.
Hazard Analysis and Countermeasures
See tin compounds and chlorides.

TRIISOBUTYLTIN FLUORIDE
General Information
Description: Crystals; slightly soluble in organic solvents.
Formula: $(C_4H_9)_3SnF$.
Constants: Mol wt: 309.0, mp: 244°C.
Hazard Analysis and Countermeasures
See fluorides and tin compounds.

TOXIC HAZARD RATING CODE (For detailed discussion, see Section 1.)

0 NONE: (a) No harm under any conditions; (b) Harmful only under unusual conditions or overwhelming dosage.

1 SLIGHT: Causes readily reversible changes which disappear after end of exposure.

2 MODERATE: May involve both irreversible and reversible changes; not severe enough to cause death or permanent injury.

3 HIGH: May cause death or permanent injury after very short exposure to small quantities.

U UNKNOWN: No information on humans considered valid by authors.

TRIISOBUTYLTIN IODIDE
General Information
Descritpion: Colorless liquid; soluble in organic solvents.
Formula: $(C_4H_9)_3SnI$.
Constants: Mol wt: 417.0, d: 1.378 at 26.5°C, mp: −22°C, bp: 286°C.
Hazard Analysis and Countermeasures
See tin compounds and iodides.

TRIISOOCTYL AMINE
Hazard Analysis
Toxicity: Details unknown. Limited animal experiments suggest moderate toxicity. See also amines.

TRIISOOCTYL PHOSPHINE
General Information
Formula: $(C_8H_{17}O)_3PH_3$.
Constant: Mol wt: 421.
Hazard Analysis
Toxicity: Details unknown. Limited animal experiments suggest low toxicity.

O, O, O-TRIISOOCTYL PHOSPHOROTHIOATE
General Information
Description: Colorless liquid; mild characteristic odor.
Formula: $(C_8H_{17}O)_3PS$.
Constants: Mol wt: 450.7, bp: 160–170°C at 0.2 mm, flash p: 410°F (C.O.C.), d: 0.933 at 20°/4°C.
Hazard Analysis
Toxicity: Details unknown. May be a cholinesterase inhibitor. See parathion.
Fire Hazard: Slight, when exposed to heat or flame (Section 6).
Disaster Hazard: Dangerous. See phosphates and sulfur compounds; can react vigorously with oxidizing materials.
Countermeasures
Storage and Handling: Section 7.

TRIISOPROPANOLAMINE
General Information
Description: Crystalline, pure white solid.
Formula: $N(C_3H_6OH)_3$.
Constants: Mol wt: 191.27, mp: 45°C, bp: 305°C, flash p: 320°F (O.C.), d: 1.0200 at 20°/20°C, vap. press.: <0.01 mm at 20°C.
Hazard Analysis
Toxic Hazard Rating:
  Acute Local: Irritant 2.
  Acute Systemic: U.
  Chronic Local: U.
  Chronic Systemic: U.
Toxicity: See also amines.
Fire Hazard: Slight, when exposed to heat or flame.
Spontaneous Heating: No.
Disaster Hazard: Moderately dangerous; when heated to decomposition it emits toxic fumes.
Countermeasures
Personnel Protection: Section 3.
Storage and Handling: Section 7.
To Fight Fire: Alcohol foam, water, carbon dioxide, dry chemical or carbon tetrachloride (Section 6).

TRIISOPROPYLBENZENE
General Information
Description: Clear, colorless liquid.
Formula: $C_6H_3[CH(CH_3)_2]_3$.
Constants: Mol wt: 204.3, mp: −15°C, bp: 236–237°C, flash p: 205°F, d: 0.854 at 25°/25°C, vap. d.: 7.0.
Hazard Analysis
Toxicity: Details unknown. See cumene.
Fire Hazard: Moderate, when exposed to heat or flame.
Disaster Hazard: Moderately dangerous; when heated it emits toxic fumes; can react with oxidizing materials.
Countermeasures
Storage and Handling: Section 7.

To Fight Fire: Foam, carbon dioxide, dry chemical or carbon tetrachloride (Section 6).

TRIISOPROPYL BORATE
General Information
Description: Colorless liquid.
Formula: $(C_9H_{21}O_3)B$.
Constants: Mol wt: 188.08, mp: −59°C, bp: 141.0–142.4°C, flash p: 82°F (T.C.C.), d: 0.8138 at 25°C.
Hazard Analysis
Toxicity: Moderate. See esters and boron compounds.
Fire Hazard: Dangerous; when exposed to heat or flame.
Explosion Hazard: Unknown.
Disaster Hazard: Dangerous. Keep away from flame and heat; can react vigorously with oxidizing materials.
Countermeasures
Storage and Handling: Section 7.
To Fight Fire: Foam, carbon dioxide, dry chemical or carbon tetrachloride (Section 6).

TRIISOPROPYLTIN BROMIDE
General Information
Description: Liquid; soluble in organic solvents.
Formula: $(C_3H_7)_3SnBr$.
Constants: Mol wt: 327.9, d: 1.4263 at 25°C, mp: −49°C, bp: 133°C at 12 mm.
Hazard Analysis and Countermeasures
See tin compounds and bromides.

TRIISOPROPYLTIN IODIDE
General Information
Description: Liquid.
Formula: $(C_3H_7)_3SnI$.
Constants: Mol wt: 374.9, d: 1.4378 at 22°C, bp: 151°C at 13 mm.
Hazard Analysis and Countermeasures
See tin compounds and iodides.

TRILAURYL TRITHIOPHOSPHITE
General Information
Description: Pale yellow liquid.
Formula: $[CH_3(CH_2)_{11}S]_3P$.
Constants: Mol wt: 634, d: 0.915 at 25/15°C, mp: 20°C, flash p: 398°F (O.C.).
Hazard Analysis
Toxicity: See esters.
Fire Hazard: Slight, when exposed to heat or flame.
Disaster Hazard: Dangerous. See phosphates and sulfur compounds.
Countermeasures
Storage and Handling: Section 7.
To Fight Fire: Water, foam.

TRIMAGNESIUM PHOSPHATE. See magnesium phosphate, tribasic.

TRIMANGANESE DIPHOSPHIDE
General Information
Description: Dark gray crystals.
Formula: Mn_3P_2.
Constants: Mol wt: 226.75, mp: 1095°C, d: 5.12 at 18°C.
Hazard Analysis and Countermeasures
See phosphides and manganese compounds.

TRIMETHOXYBORINE. See methyl borate.

TRIMETHYLACETIC ACID
General Information
Synonym: Pivalic acid.
Description: Liquid.
Formula: $(CH_3)_3CCOOH$.
Constants: Mol wt: 102.1.
Hazard Analysis
Toxicity: Details unknown. Animal experiments suggest low toxicity.

TRIMETHYLADIPIC ACID
General Information
Description: Powder.
Formula: $C_9H_{16}O_4$.
Constant: Mol wt: 188.2.
Hazard Analysis
Toxic Hazard Rating:
Acute Local: Irritant 1.
Acute Systemic: U.
Chronic Local: U.
Chronic Systemic: U.
Disaster Hazard: Slightly dangerous; when heated it emits acrid fumes.
Countermeasures
Personal Hygiene: Section 3.
Storage and Handling: Section 7.

TRIMETHYLALUMINUM. See aluminum methyl.
Shipping Regulations: Section 11.
IATA: Flammable liquid, not acceptable (passenger and cargo).

TRIMETHYLAMINE
General Information
Description: Colorless gas.
Formula: $(CH_3)_3N$.
Constants: Mol wt: 59.11, bp: 2.87 °C, lel: 2%, uel: 11.6%, fp: −117.1°C, d: 0.662 at −5°C, autoign. temp.: 374°F, vap. d: 2.0.
Hazard Analysis
Toxic Hazard Rating:
Acute Local: Inhalation 2.
Acute Systemic: Inhalation 2.
Chronic Local: U.
Chronic System: U.
Fire Hazard: Moderate, when exposed to flame (Section 6).
Explosion Hazard: Moderate, when exposed to spark or flame.
Disaster Hazard: Moderately dangerous; when heated, it emits toxic fumes; can react with oxidizing materials.
Countermeasures
Ventilation Control: Section 2.
Storage and Handling: Section 7.
Shipping Regulations: Section 11.
ICC Anhydrous: Flammable gas; red gas label, 200 pounds.
Coast Guard Classification: Anhydrous: Inflammable gas; red gas label.
MCA warning label.
IATA (Anhydrous): Flammable gas, red label, not acceptable (passenger), 140 kilograms (cargo).

TRIMETHYLAMINE, AQUEOUS SOLUTION.
See trimethylamine.
Shipping Regulations: Section 11.
ICC: Flammable liquid; red label, 10 gallons.
Coast Guard Classification: Inflammable liquid; red label.
IATA: Flammable liquid, red label, 1 liter (passenger), 40 liters (cargo).

TRIMETHYLAMINOBORINE
General Information
Description: Colorless crystals decomposed by water. Soluble in ether.
Formula: $(CH_3)_3NBH_3$.
Constants: Mol wt: 73.0, mp: 94°C, bp: 172°C.

Hazard Analysis
Toxicity: Details unknown. See also boron compounds.

TRIMETHYLAMINOMETHANE. See tert-butylamine.

TRI(METHYLAMYL) BORATE
General Information
Synonym: Tri(methylisobutylcarbinyl) borate.
Description: Colorless, mobile liquid; characteristic odor.
Formula: $[(CH_3)_2CHCH_2CHOCH_3]_3B$.
Constants: Mol wt: 314.3, bp: 257°C, flash p: 220°F (C.O.C.), d: 0.819 at 29°C, vap. d.: 10.8.
Hazard Analysis
Toxicity: See boric acid and esters.
Fire Hazard: Slight when exposed to heat or flame.
Disaster Hazard: Moderately dangerous; when heated to decomposition it emits toxic fumes; on contact with oxidizing materials, it can react vigorously.
Countermeasures
Storage and Handling: Section 7.
To Fight Fire: Foam, carbon dioxide, dry chemical or carbon tetrachloride (Section 6).

TRIMETHYLARSENIC
General Information
Synonym: Arsenic trimethyl.
Description: Colorless liquid.
Formula: $As(CH_3)_3$.
Constants: Mol wt: 120.0, bp: 70°C, d: 1.124, vap. d.: 4.14.
Hazard Analysis
Toxicity: Highly toxic. See arsenic compounds.
Fire Hazard: Moderate, when exposed to heat or flame (Section 6).
Disaster Hazard: Dangerous; when heated to decomposition or on contact with acid or acid fumes, it emits highly toxic fumes of arsenic; can react vigorously with oxidizing materials.
Countermeasures
Storage and Handling: Section 7.

1,2,4-TRIMETHYLBENZENE. See pseudocumene.

1,3,5-TRIMETHYLBENZENE. See mesitylene.

2,6,6-TRIMETHYLBICYCLO-(3,1,1)-2-HEPTENE. See pinene.

TRIMETHYLBISMUTHINE
General Information
Synonym: Bismuth trimethyl.
Discription: Liquid.
Formula: $Bi(CH_3)_3$.
Constants: Mol wt: 254.10, bp: 110°C, d: 2.300 at 18°C.
Hazard Analysis
Toxicity: Highly toxic. Can cause narcosis and CNS depression. Prolonged exposure can cause encephalopathy similar to that of organic lead compounds. See bismuth compounds.
Fire Hazard: Moderate, when exposed to heat or flame (Section 6).
Explosion Hazard: Unknown.
Disaster Hazard: Dangerous; when heated to decomposition it emits toxic fumes of bismuth; reacts with oxidizing materials.
Countermeasures
Ventilation Control: Section 2.
Personal Hygiene: Section 3.
Storage and Handling: Section 7.

TOXIC HAZARD RATING CODE (For detailed discussion, see Section 1.)

0 NONE: (a) No harm under any conditions; (b) Harmful only under unusual conditions or overwhelming dosage.

1 SLIGHT: Causes readily reversible changes which disappear after end of exposure.

2 MODERATE: May involve both irreversible and reversible changes; not severe enough to cause death or permanent injury.
3 HIGH: May cause death or permanent injury after very short exposure to small quantities.
U UNKNOWN: No information on humans considered valid by authors.

TRIMETHYL BORATE. See methyl borate.

TRIMETHYLBORINE. See boron trimethyl.

TRIMETHYL-α-CAPROLACTONE
Hazard Analysis
Toxicity: Details unknown. See also lactones.

TRIMETHYLCHLOROSILANE
General Information
Description: Colorless liquid; soluble in benzene, ether, and perchloroethylene.
Formula: $(CH_3)_3SiCl$.
Constants: Mol wt: 108.8, bp: 57°C, d: 0.854 at 25/25°C, flash p: −18°F.
Hazard Analysis
Toxic Hazard Rating: U. See silanes.
Fire Hazard: Dangerous, when exposed to heat or flame.
Disaster Hazard: Dangerous. See chlorides.
Countermeasures
Storage and Handling: Section 7.
Shipping Regulations: Section 11.
   ICC: Flammable liquid, red label, 10 gallons.
   Coast Guard Classification: Inflammable liquid, red label.
   IATA: Flammable liquid, red label, not acceptable (passenger), 40 liters (cargo).
To Fight Fire: Foam, alcohol foam, and fog.

3,3,5-TRIMETHYLCYCLOHEXANOL
General Information
Synonym: Trimethylcyclohexanol.
Description: Liquid.
Formula: $C_9H_{18}O$.
Constants: Mol wt: 142.23, fp: 37.0°C, bp: 198°C, flash p: 165°F (O.C.), d: 0.878 at 40°/20°C, vap. press.: 0.1 mm at 20°C. vap. d.: 4.91.
Hazard Analysis
Toxic Hazard Rating:
   Acute Local: Irritant 2.
   Acute Systemic: U.
   Chronic Local: U.
   Chronic Systemic: U.
Fire Hazard: Moderate, when exposed to heat or flame.
Disaster Hazard: Moderately dangerous; when heated, it emits toxic fumes; can react with oxidizing materials.
Countermeasures
Personnel Protection: Section 3.
To Fight Fire: Water, foam, carbon dioxide, dry chemical or carbon tetrachloride (Section 6).
Storage and Handling: Section 7.
Shipping Regulations: Section 11.
   MCA warning label.

TRIMETHYLCYCLOHEXANONE
General Information
Description: Liquid.
Formula: $C_9H_{16}O$.
Constant: Mol wt: 140.2.
Hazard Analysis
Toxic Hazard Rating:
   Acute Local: Irritant 1.
   Acute Systemic: U.
   Chronic Local: U.
   Chronic Systemic: U.
Fire Hazard: Moderate, when exposed to heat or flame; can react with oxidizing materials (Section 6).
Countermeasures
Personal Hygiene: Section 3.
Storage and Handling: Section 7.

TRIMETHYL DEHYDROQUINOLINE POLYMER
General Information
Description: Amber pellets.
Formula: $(C_{12}H_{15}N)_3$

Constants: Mol wt: 519.93 (approx.), mp: softens at 75°C, d: 1.08.
Hazard Analysis
Toxicity: Details unknown; Animal experiments suggest moderate to low acute and chronic toxicity.
Disaster Hazard: Dangerous; when heated to decomposition it emits highly toxic fumes of cyanides.
Countermeasures
Storage and Handling: Section 7.

1,1,2-TRIMETHYLDIBORANE
General Information
Description: Colorless liquid, decomposed by water.
Formula: $B_2H_3(CH_3)_3$.
Constants: Mol wt: 69.8, mp: −123°C, bp: 45.5°C.
Hazard Analysis and Countermeasures
See boron hydride and boron compounds.

TRIMETHYLENE. See cyclopropane.

TRIMETHYLENE BROMIDE
General Information
Description: Colorless liquid.
Formula: $C_3H_6Br_2$.
Constants: Mol wt: 201.9, bp: 166.5°C, fp: −33°C, d: 1.977 at 25°/25°C, vap. d.: 7.0.
Hazard Analysis
Toxicity: Unknown. Probably irritant and narcotic in high concentrations.
Disaster Hazard: Dangerous. See bromides.
Countermeasures
Storage and Handling: Section 7.

TRIMETHYLENE CHLORIDE. See 1,3-dichloropropane.

TRIMETHYLENE CHLOROBROMIDE
General Information
Description: Colorless liquid.
Formula: C_3H_6ClBr.
Constants: Mol wt: 157.5, bp: 143-145°C, flash p: none, d: 1.594 at 25°/25°C, vap. d.: 5.5
Hazard Analysis
Toxicity: Unknown. Probably irritant and narcotic in high concentrations.
Disaster Hazard: Dangerous. See chlorides and bromides.
Countermeasures
Storage and Handling: Section 7.

TRIMETHYLENE GLYCOL
General Information
Synonyms: 1,3-Propanediol; 1,3-propylene glycol.
Description: Colorless, odorless liquid; soluble in water, alcohol, and ether.
Formula: $CH_2OHCH_2CH_2OH$.
Constants: Mol wt: 76, d: 1.0537 at 25°C, bp: 210–211°C, vap. d.: 2.6, autoign. temp.: 752°F.
Hazard Analysis
See glycols.

TRIMETHYLENE GLYCOL DINITRATE
Hazard Analysis
Toxicity: Details unknown. See ethylene glycol dinitrate.
Fire Hazard: See nitrates.
Explosion Hazard: See nitrates.
Explosive Range: At 225°F.
Disaster Hazard: Dangerous. See nitrates.
Countermeasures
Storage and Handling: Section 7.

TRIMETHYLENE OXIDE
General Information
Synonym: 1,3-Epoxypropane.
Description: Oil with agreeable odor.
Formula: C_3H_6O.
Constants: Mol wt: 58.1, d: 0.8930 at 25/4°C, bp: 48°C at 750 mm.

Hazard Analysis
Toxicity: Details unknown. May be narcotic in high concentrations.

TRIMETHYLENE TRINITRAMINE. See cyclotrimethylene trinitramine.

TRIMETHYLETHYLENE. See 2-methyl butene-2.

TRIMETHYLETHYLTIN
General Information
Description: Colorless liquid; insoluble in water; soluble in organic solvents.
Formula: $(CH_3)_3Sn(C_2H_5)$.
Constants: Mol wt: 192.9, bp: 108.2°C.
Hazard Analysis
Toxicity: Details unknown. See tin compounds.

TRIMETHYL o-FORMATE
General Information
Description: Colorless liquid; pungent odor.
Formula: $HC(OCH_3)_3$.
Constants: Mol wt: 106.12, vap. d.: 3.67.
Hazard Analysis
Toxicity: Details unknown; probably slight. See esters.
Fire Hazard: Slight, when exposed to heat or flame; can react with oxidizing materials (Section 6).
Countermeasures
Storage and Handling: Section 7.

TRIMETHYLGALLIUM
General Information
Description: Colorless liquid, decomposed by water.
Formula: $(CH_3)_3Ga$.
Constants: Mol wt: 114.8, mp: −19°C, bp: 55.7°C.
Hazard Analysis
Toxicity: Details unknown. See gallium compounds.
Fire Hazard: Probably dangerous. Details unknown (Section 6).
Countermeasures
Storage and Handling: Section 7.

TRIMETHYLGALLIUM MONAMMINE
General Information
Description: White crystals, decomposed by water.
Formula: $Ga(CH_3)_3 \cdot NH_3$.
Constants: Mol wt: 131.9, mp: 31°C.
Hazard Analysis
Toxicity: Details unknown. See gallium compounds.

TRIMETHYLGALLIUM MONOETHERATE
General Information
Description: Colorless liquid decomposed by water.
Formula: $(CH_3)_3Ga \cdot O \cdot C_2H_5$.
Constants: Mol wt: 188.9, mp: < -76°C, bp: 99°C.
Hazard Analysis
Toxicity: Details unknown. See gallium compounds and ethyl ether.
Note: This material easily evolves ether. See ethyl ether.

TRIMETHYLGERMANIUM BROMIDE
General Information
Description: Oily liquid decomposed by water.
Formula: $(CH_3)_3GeBr$.
Constants: Mol wt: 197.6, d: 1.544 at 18/40°C, mp: −25°C, bp: 113.7°C.
Hazard Analysis and Countermeasures
See germanium compounds and bromides.

2,2,5-TRIMETHYLHEXANE
General Information
Description: A clear liquid.
Formula: C_9H_{20}.
Constants: Mol wt: 128.25, bp: 125°C, fp: −106°C, flash p: 55°F, d: 0.707 at 20°/4°C, vap. press.: 12.9 mm at 21°C, vap. d.: 4.7.
Hazard Analysis
Toxicity: Unknown.
Fire Hazard: Dangerous, when exposed to heat or flame.
Explosion Hazard: Unknown.
Disaster Hazard: Highly dangerous; keep away from heat or open flame; can react vigorously with oxidizing materials.
Countermeasures
Ventilation Control: Section 2.
To Fight Fire: Foam, carbon dioxide, dry chemical or carbon tetrachloride (Section 6).
Storage and Handling: Section 7.

2,3,3-TRIMETHYLHEXANE
General Information
Description: Colorless liquid.
Formula: C_9H_{20}.
Constants: Mol wt: 128.25, bp: 137.7°C, fp: −116.8°C, flash p: 79°F, d: 0.734 at 25°/4°C, vap. press.: 10.1 mm at 25°C, vap. d.: 4.43.
Hazard Analysis
Toxicity: Unknown.
Fire Hazard: Dangerous, when exposed to heat or flame.
Explosion Hazard: Unknown.
Disaster Hazard: Dangerous; keep away from heat or open flame; can react with oxidizing materials.
Countermeasures
Ventilation Control: Section 2.
To Fight Fire: Foam, carbon dioxide, dry chemical or carbon tetrachloride (Section 6).
Storage and Handling: Section 7.

2,3,4-TRIMETHYLHEXANE
General Information
Description: Colorless liquid.
Formula: C_9H_{20}.
Constants: Mol wt: 128.25, bp: 139°C, flash p: 80.8°F, d: 0.737 at 25/4°C, vap. press.: 9.1 mm at 25°C, vap. d.: 4.43.
Hazard Analysis
Toxicity: Unknown. Probably irritant and narcotic in high concentrations.
Fire Hazard: Dangerous, when exposed to heat or flame (Section 6).
Explosion Hazard: Unknown.
Disaster Hazard: Dangerous. Keep away from heat or open flame. It can react vigorously with oxidizing materials.
Countermeasures
Ventilation Control: Section 2.
Storage and Handling: Section 7.

3,3,4-TRIMETHYLHEXANE
General Information
Description: Colorless liquid.
Formula: C_9H_{20}.
Constants: Mol wt: 128.25, bp: 140.5°C, fp: −101.2°C, flash p.: 79°F, d: 0.741 at 25°/4°C, vap. press.: 8.6 mm at 25°C, vap. d.: 4.43.
Hazard Analysis
Toxicity: Unknown. Probably irritant and narcotic in high concentrations.

TOXIC HAZARD RATING CODE (For detailed discussion, see Section 1.)

0 NONE: (a) No harm under any conditions; (b) Harmful only under unusual conditions or overwhelming dosage.

1 SLIGHT: Causes readily reversible changes which disappear after end of exposure.

2 MODERATE: May involve both irreversible and reversible changes; not severe enough to cause death or permanent injury.

3 HIGH: May cause death or permanent injury after very short exposure to small quantities.

U UNKNOWN: No information on humans considered valid by authors.

Fire Hazard: Dangerous, when exposed to heat or flame.
Explosion Hazard: Unknown.
Disaster Hazard: Dangerous. Keep away from heat or open flame; can react with oxidizing materials.
Countermeasures
Ventilation Control: Section 2.
To Fight Fire: Foam, carbon dioxide, dry chemical or carbon tetrachloride (Section 6).
Storage and Handling: Section 7.

3,5,5-TRIMETHYLHEXANOL
General Information
Description: Colorless liquid.
Formula: $CH_3(CH_3)_2CH_2CHCH_3CH_2CH_2OH$.
Constants: Mol wt: 144.25, mp: $-70°C$, bp: 195°C, flash p: 200°F (O.C.), d: 0.824 at 25°/4°C, vap. d.: 5.0.
Hazard Analysis
Toxicity: See alcohols.
Fire Hazard: Moderate, when exposed to heat or flame; can react with oxidizing materials.
Countermeasures
To Fight Fire: Foam, carbon dioxide, dry chemical or carbon tetrachloride (Section 6).
Storage and Handling: Section 7.

1,5-TRIMETHYL-4-HEXENYLAMINE. See isomethept-ene.

TRI(METHYLISOBUTYL CARBINYL) BORATE. See tri(methylamyl) borate.

TRIMETHYLMETHANE. See isobutane.

2,6,8-TRIMETHYLNONANONE-4
General Information
Description: Liquid.
Formula: $C_{12}H_{24}O$.
Constants: Mol wt: 184.2, mp: $-75°C$, bp: 211-219°C, flash p: 196°F (O.C.), d: 0.8165 at 20°/20°C, vap. d.: 6.37
Hazard Analysis
Toxicity: Details unknown. See ketones.
Fire Hazard: Moderate, when exposed to heat or flame; can react with oxidizing materials.
Countermeasures
To Fight Fire: Foam, carbon dioxide, dry chemical or carbon tetrachloride (Section 6).
Storage and Handling: Section 7.

TRIMETHYLNONYL ALCOHOL
General Information
Description: Liquid.
Formula: $CH_3CH(CH_3)CH_2CH(OH)CH_2CH(CH_3)-CH_2CH(CH_3)CH_3$.
Constants: Mol wt: 186.33, bp: 225.2°C, fp: $-60°C$, flash p: 200°F (O.C.), vap. press.: <0.01 mm at 20°C, d: 0.8193 at 20°/20°C, vap. d.: 6.43.
Hazard Analysis
Toxicity: See alcohols.
Fire Hazard: Moderate, when exposed to heat or flame; can react with oxidizing materials.
Countermeasures
To Fight Fire: Foam, carbon dioxide, dry chemical or carbon tetrachloride (Section 6).
Storage and Handling: Section 7.

TRIMETHYLOLETHANE
General Information
Synonyms: 2,2-Dimethylolpropanol-1.
Description: White, odorless, crystalline powder.
Formula: $CH_3C(CH_2OH)_3$.
Constants: Mol wt: 120.15, mp: about 200°C.
Hazard Analysis
Toxicity: Unknown.
Fire Hazard: Slight, when exposed to heat (Section 6).

Countermeasures
Storage and Handling: Section 7.

2,2,4-TRIMETHYLPENTANE. See isooctane.

2,2,4-TRIMETHYL-1,3-PENTANEDIOL
General Information
Description: White, crystalline solid.
Formula: $C_8H_{18}O_2$.
Constants: Mol wt: 146.2, mp: 49 to 51°C, bp: 109 to 111°C at 4 mm.
Hazard Analysis
Toxicity: Details unknown; an insect repellent.
Fire Hazard: Slight; when exposed to heat or flame; can react with oxidizing materials (Section 6).
Countermeasures
Storage and Handling: Section 7.

2,2,4-TRIMETHYLPENTANEDIOL DIISOBUTYRATE
General Information
Formula: $C_{16}H_{30}O_4$.
Constants: Mol wt: 286, d: 0.9, vap. d.: 9.9, bp: 536°F, flash p: 250°F (O.C.).
Hazard Analysis
Toxic Hazard Rating: U.
Fire Hazard: Slight, when exposed to heat or flame.
Countermeasures
Storage and Handling: Section 7.
To Fight Fire: Water, foam. Section 6.

2,2,4-TRIMETHYLPENTANEDIOL MONOISOBUTY-RATE BENZOATE
General Information
Formula: $C_{19}H_{28}O_4$.
Constants: Mol wt: 320, d: 1.0, bp: 167°F at 10 mm.
Hazard Analysis
Toxic Hazard Rating: U.
Fire Hazard: Slight, when exposed to heat or open flame.
Countermeasures
Storage and Handling: Section 7.
To Fight Fire: Water, foam.

2,2,4-TRIMETHYLPENTANOL
General Information
Formula: $CH_3CH(CH_3)CH_2C(CH_3)_2CH_2OH$.
Constants: Mol wt: 130.
Hazard Analysis
Toxicity: Details unknown. Limited animal experiments suggest moderate toxicity.

2,4,4-TRIMETHYLPENTENE-1
General Information
Description: A clear liquid.
Formula: C_8H_{16}.
Constants: Mol wt: 112.2, bp: 101.3°C, fp: $-94°C$, flash p: $<20°F$, d: 0.719 at 15.5°/15.5°C, vap. press.: 801 mm at 38°C, vap. d.: 3.9.
Hazard Analysis
Toxicity: Unknown. Probably irritant and narcotic in high concentrations.
Fire Hazard: Dangerous, when exposed to heat or flame.
Explosion Hazard: Unknown.
Disaster Hazard: Highly dangerous. Keep away from heat or open flame; can react vigorously with oxidizing materials.
Countermeasures
Ventilation Control: Section 2.
To Fight Fire: Foam, carbon dioxide, dry chemical or carbon tetrachloride (Section 6).
Storage and Handling: Section 7.

2,2,4-TRIMETHYLPENTENE-2
General Information
Description: A clear liquid.
Formula: C_8H_{16}.
Constants: Mol wt: 112.2, bp: 104.5°C, flash p: 35°F

(T.O.C.), fp: −106.4°C, d: 0.724 at 15.5°/15.5°C, vap. press.: 77.5 mm at 38°C, vap. d.: 3.9.

Hazard Analysis

Toxicity: Unknown. Probably irritant and narcotic in high concentrations.

Fire Hazard: Dangerous, when exposed to heat or flame.

Explosion Hazard: Unknown.

Disaster Hazard: Highly dangerous. Keep away from heat or open flame; can react vigorously with oxidizing materials.

Countermeasures

Ventilation Control: Section 2.

To Fight Fire: Foam, carbon dioxide, dry chemical or carbon tetrachloride (Section 6).

Storage and Handling: Section 7.

TRIMETHYLPHENYLGERMANIUM

General Information

Description: Colorless liquid; insoluble in water.

Formula: $(CH_3)_3Ge(C_6H_5)$.

Constants: Mol wt: 194.8, bp: 183°C.

Hazard Analysis

Toxicity: Details unknown. See germanium compounds.

TRIMETHYL PHOSPHINE

General Information

Description: Colorless liquid.

Formula: $(CH_3)_3P$.

Constants: Mol wt: 76.1, bp: 42°C, d: < 1.

Hazard Analysis

Toxicity: Details unknown; probably high. See also phosphine.

Fire Hazard: Moderate, when exposed to heat or flame or by chemical reaction. See also phosphine.

Explosion Hazard: Slight, when exposed to flame.

Disaster Hazard: Dangerous; when heated to decomposition or on contact with acid or acid fumes, it emits highly toxic fumes of phosphorus oxides; can react vigorously with oxidizing materials.

Countermeasures

Storage and Handling: Section 7.

TRIMETHYL PHOSPHITE

General Information

Description: Colorless liquid; insoluble in water; soluble in hexane, benzene, acetone, alcohol, ether, carbon tetrachloride, and kerosene.

Formula: $(CH_3O)_3P$.

Constants: Mol wt: 124, d: 1.046 at 20/4°C, vap. d.: 4.3, bp: 232–234°F, flash p: 130°F (O.C.).

Hazard Analysis

Toxic Hazard Rating: U.

Fire Hazard: Moderate, when exposed to heat or flame.

Disaster Hazard: Dangerous. See phosphates.

Countermeasures

Shipping and Handling: Section 7.

To Fight Fire: Water, foam, fog, carbon dioxide.

TRIMETHYLRHENIUM

General Information

Description: Colorless oil.

Formula: $Re(CH_3)_3$.

Constants: Mol wt: 231.4, bp: 60°C.

Hazard Analysis

Toxicity: Details unknown. See rhenium compounds.

TRIMETHYL STANNYL TRIPHENYL GERMANIUM

General Information

Description: White crystals; insoluble in water.

Formula: $(CH_3)_3SnGe(C_6H_5)_3$.

Constants: Mol wt: 467.7, mp: 88°C.

Hazard Analysis

Toxicity: Details unknown. See tin and germanium compounds.

TRIMETHYL STIBINE. See antimony trimethyl.

2,4,6-TRIMETHYL-1,2,3,6-TETRAHYDROBENZALDE-HYDE

General Information

Description: Liquid.

Formula: $CH(CH_3)CHC(CH_3)CH_2CH(CH_3)CHCHO$.

Constants: Mol wt: 152.23, mp: −41.9°C, bp: 204.5°C, flash p: 185°F (O.C.), d: 0.9195 at 20°/20°C, vap. press.: 0.3 mm at 20°C, vap. d.: 5.25.

Hazard Analysis

Toxicity: Details unknown. See aldehydes.

Fire Hazard: Moderate, when exposed to heat or flame; can react with oxidizing materials.

Countermeasures

To Fight Fire: Foam, carbon dioxide, dry chemical or carbon tetrachloride (Section 6).

Storage and Handling: Section 7.

TRIMETHYLTIN

General Information

Description: Colorless liquid. Insoluble in water.

Formula: $(CH_3)_3Sn$.

Constants: Mol wt: 163.8, d: 1.570 at 25°C, mp: 23°C, bp: 182°C.

Hazard Analysis

Toxicity: See tin compounds.

TRIMETHYLTIN BROMIDE

General Information

Description: Colorless crystals, soluble in water and organic solvents.

Formula: $(CH_3)_3SnBr$.

Constants: Mol wt: 243.7, mp: 27°C, bp: 165°C.

Hazard Analysis and Countermeasures

See tin compounds and bromides.

TRIMETHYLTIN CHLORIDE

General Information

Description: Colorless crystals, soluble in water and organic solvents.

Formula: $(CH_3)_3SnCl$.

Constants: Mol wt: 199.3, mp: 37°C.

Hazard Analysis and Countermeasures

See tin compounds and chlorides.

TRIMETHYLTIN FLUORIDE

General Information

Description: Colorless crystals; slightly soluble in organic solvents.

Formula: $(CH_3)_3SnF$.

Constants: Mol wt: 182.8, mp: 360°C.

Hazard Analysis and Countermeasures

See fluorides and tin compounds.

TRIMETHYLTIN HYDRIDE

General Information

Description: Colorless oily liquid; soluble in organic solvents.

Formula: $(CH_3)_3SnH$.

Constants: Mol wt: 164.8, bp: 60°C.

Hazard Analysis and Countermeasures

See tin compounds and hydrides.

TOXIC HAZARD RATING CODE (For detailed discussion, see Section 1.)

0 NONE: (a) No harm under any conditions; (b) Harmful only under unusual conditions or overwhelming dosage.

1 SLIGHT: Causes readily reversible changes which disappear after end of exposure.

2 MODERATE: May involve both irreversible and reversible changes; not severe enough to cause death or permanent injury.

3 HIGH: May cause death or permanent injury after very short exposure to small quantities.

U UNKNOWN: No information on humans considered valid by authors.

TRIMETHYLTIN HYDROXIDE
General Information
Description: Colorless crystals; soluble in water and many organic solvents.
Formula: $(CH_3)_3SnOH$.
Constants: Mol wt: 180.8, mp: 118°C (decomp.).
Hazard Analysis
Toxicity: Details unknown. See tin compounds.

TRIMETHYLTIN IODIDE
General Information
Description: Colorless liquid; soluble in many organic solvents.
Formula: $(CH_3)_3SnI$.
Constants: Mol wt: 290.7, d: 2.1432, mp: 3.4°C, bp: 170°C.
Hazard Analysis and Countermeasures
See tin compounds and iodides.

TRIMETHYLTIN OXIDE
General Information
Description: White powder, insoluble in water and organic solvents.
Formula: $(CH_3)_3Sn-O-Sn(CH_3)_3$.
Constants: Mol wt: 343.6.
Hazard Analysis
Toxicity: Details unknown. See tin compounds.

TRIMETHYLTIN SULFIDE
General Information
Description: Light yellow oil; insoluble in water; soluble in organic solvents.
Formula: $(CH_3)_3Sn-S-Sn(CH_3)_3$.
Constants: Mol wt: 359.7, d: 1.649 at 25°C, mp: 6°C, bp: 233.5°C.
Hazard Analysis and Countermeasures
See tin compounds and sulfides.

TRIMETHYL TRIBORINE TRIAMINE (B)
General Information
Description: Colorless crystals (or liquid) hydrolyzed by water.
Formula: $B_3(CH_3)_3N_3H_3$.
Constants: Mol wt: 122.6, mp: 31.5°C, bp: 129°C.
Hazard Analysis
Toxicity: Details unknown. See also boron compounds and amines.

TRIMETHYL TRIBORINE TRIAMINE (N)
General Information
Description: Colorless liquid; hydrolyzed by water.
Formula: $(CH_3)_3B_3N_3H_3$.
Constants: Mol wt: 122.6, bp: 134°C.
Hazard Analysis
Toxicity: Details unknown. See also boron compounds and amines.

TRIMETHYL TRIBORINE TRIAMINE
General Information
Description: Colorless liquid, hydrolyzed in water.
Formula: $(CH_3)_3B_3N_3H_3$.
Constants: Mol wt: 122.6, bp: 139°C.
Hazard Analysis
Toxicity: Details unknown. See also boron compounds and amines.

$\alpha,\alpha,\alpha,$-**TRIMETHYL TRIMETHYLENE GLYCOL.** See 2-methyl-2,4-pentanediol.

2,4,6-TRIMETHYL-1,3,5-TRIOXANE. See paraldehyde.

TRI-β-NAPHTHYL BORATE
General Information
Description: Colorless crystals, decomposed by water. Soluble in benzene.
Formula: $B(C_{10}H_7O)_3$.
Constants: Mol wt: 440.3, mp: 115°C.

Hazard Analysis
Toxicity: Details unknown. See also β-naphthol and boron compounds.

TRI-α-NAPHTHYLBORINE
General Information
Description: Colorless crystals, insoluble in water; soluble in organic solvents.
Formula: $B(C_{10}H_7)_3$.
Constants: Mol wt: 392.3, mp: 203°C.
Hazard Analysis
Toxicity: Details unknown. See also α-naphthol and boron compounds.

TRINICKEL DIPHOSPHIDE
General Information
Description: Dark green-black crystals.
Formula: Ni_3P_2.
Constants: Mol wt: 238.03,, d: 5.99.
Hazard Analysis and Countermeasures
See phosphides and nickel compounds.

TRINTROACETONITRILE
General Information
Description: Solid.
Formula: $C_2N_4O_6$.
Constant: Mol wt: 176.05.
Hazard Analysis
Toxicity: See nitriles.
Explosion Hazard: Moderate, when exposed to heat.
Explosive Range: At 392°F.
Disaster Hazard: Dangerous; when heated to decomposition it emits highly toxic fumes of cyanides and explodes; can react vigorously with oxidizing materials.
Countermeasures
Storage and Handling: Section 7.

TRINITROANILINE
General Information
Synonym: Picramide.
Description: Crystals.
Formula: $C_6H_4N_4O_6$.
Constants: Mol wt: 228.1, mp: 188°C, bp: explodes, d: 1.762.
Hazard Analysis
Toxicity: See nitrates and aniline.
Fire Hazard: See nitrates.
Explosion Hazard: Severe, when shocked or exposed to heat. A very sensitive high explosive.
Disaster Hazard: Highly dangerous; shock will explode it; when heated to decomposition it emits highly toxic fumes of oxides of nitrogen and explodes; can react vigorously with reducing materials.
Countermeasures
Storage and Handling: Section 7.
Shipping Regulations: Section 11.
IATA: Explosive, no label, not acceptable (passenger and cargo).

TRINITROANISOLE
General Information
Synonym: Methyl picrate.
Description: Crystals.
Formula: $CH_3OC_6H_2(NO_2)_3$.
Constants: Mol wt: 243.13, mp: 68.4°C, d: 1.408 at 20°/4°C.
Hazard Analysis
Toxic Hazard Rating:
Acute Local: Irritant 2; Allergen 1.
Acute Systemic: Ingestion 3; Inhalation 3.
Chronic Local: Allergen 1.
Chronic Systemic: Ingestion 2; Inhalation 2; Skin Absorption 2.
Fire Hazard: See nitrates.
Explosion Hazard: Severe, when shocked or exposed to heat. Trinitroanisole resembles picric acid in its high

explosive properties but does not attack metals provided it is protected from moisture. It has a lower melting point than picric acid which is an advantage in shell loading. It is used as a booster charge.

Disaster Hazard: Highly dangerous; shock will explode it; when heated to decomposition it emits highly toxic fumes of oxides of nitrogen and explodes; can react vigorously with reducing materials.

Countermeasures
Ventilation Control: Section 2.
Personnel Protection: Section 3.
First Aid: Section 1.
Storage and Handling: Section 7.
Shipping Regulations: Section 11.
   IATA: Explosive, no label, not acceptable (passenger and cargo).

TRINITROBENZALDEHYDE
General Information
Description: Liquid.
Formula: $C_6H_2CHO(NO_2)_3$.
Constants: Mol wt: 241.1, mp: 119°C.
Hazard Analysis and Countermeasures
See aldehydes and nitrates.

1,3,5-TRINITROBENZENE
General Information
Description: Yellow crystals.
Formula: $C_6H_3(NO_2)_3$.
Constants: Mol wt: 213.11, mp: 122°C, bp: decomposes, d: 1.688 at 20°/4°C.
Hazard Analysis
Toxic Hazard Rating:
  Acute Local: U.
  Acute Systemic: Ingestion 3; Inhalation 3.
  Chronic Local: U.
  Chronic Systemic: Ingestion 3; Inhalation 3.
Toxicology: See also nitro compounds of aromatic hydrocarbons.
Fire Hazard: See nitrates.
Explosion Hazard: Severe, when shocked or exposed to heat. Trinitrobenzene is considered a powerful high explosive and has more shattering power than TNT. It is less sensitive to impact that TNT. However, it is not used much because it is difficult to produce.
Disaster Hazard: Highly dangerous; shock will explode it; when heated to decomposition it emits highly toxic fumes of oxides of nitrogen and explodes; can react vigorously with reducing materials.

Countermeasures
Ventilation Control: Section 2.
Personal Hygiene: Section 3.
First Aid: Section 1.
Storage and Handling: Section 7.
Shipping Regulations: Section 11.
   IATA (dry or wet with less than 10% water): Explosive, not acceptable (passenger and cargo).
   IATA (wet, with not less than 10% water): Flammable solid, yellow label, ½ kilogram (passenger and cargo).

TRINITROBENZENE, WET (NOT TO EXCEED 16 OZ.). See 1,3,5-trinitrobenzene.
Shipping Regulations: Section 11.
   I.C.C.: Flammable solid.
   Coast Guard: Inflammable solid; yellow label.

2,4,6-TRINITROBENZOIC ACID
General Information
Synonym: Trinitrobenzoic acid.

Description: Orthorhombic crystals; solubility at 25°C: 2.05% in water, 26.6% in alcohol, 14.7% in ether; also soluble in methanol; slightly soluble in benzene.
Formula: $C_6H_2(NO_2)_3COOH$.
Constants: Mol wt: 257.12, mp: 228.7°C.
Hazard Analysis
See trinitrotoluene.
Countermeasures
Shipping Regulations: Section 11.
   ICC (wet with not less than 10% water, >16 ounces but <25 pounds). Flammable solid, yellow label, 25 pounds.
   IATA (dry or wet with less than 10% water): Explosive, not acceptable (passenger and cargo).
      (wet with not less than 10% water): Flammable solid, yellow label, ½ kilogram (passenger), 25 kilograms (cargo).

TRINITROCHLORBENZENE
General Information
Description: Solid.
Formula: $C_6H_2Cl(NO_3)_3$.
Constant: Mol wt: 295.6
Hazard Analysis
Toxicity: See nitro compounds of aromatic hydrocarbons.
Fire Hazard: See nitrates.
Explosion Hazard: Severe when shocked or exposed to heat.
Disaster Hazard: Highly dangerous; shock will explode it; when heated to decomposition it emits highly toxic fumes of phosgene and oxides of nitrogen and explodes; can react vigorously with oxidizing materials.
Countermeasures
Storage and Handling: Section 7.

2,4,6-TRINITRO-m-CRESOL
General Information
Synonym: Cresolite.
Description: Yellow crystals.
Formula: $(NO_2)_3C_6H(CH_3)OH$.
Constants: Mol wt: 243.13, mp: 106°C, bp: explodes at 150°C.
Hazard Analysis
Toxicity: See nitro compounds of aromatic hydrocarbons.
Fire Hazard: See nitrates.
Explosion Hazard: Severe, when shocked or exposed to heat. Trinitrocresol is not as powerful a high explosive as TNT or picric acid. It has been used as a bursting charge and in combination with other high explosives.
Disaster Hazard: Highly dangerous; shock will explode it; when heated to decomposition it emits highly toxic fumes of oxides of nitrogen and explodes; can react vigorously with oxidizing materials.
Countermeasures
Storage and Handling: Section 7.
Shipping Regulation: Section 11.
   IATA: Explosive material, not acceptable (passenger and cargo).

2,4,4'-TRINITRODIPHENYLAMINE
General Information
Description: Crystals.
Formula: $C_6H_3(NO_2)_2NHC_6H_4NO_2$.
Constant: Mol wt: 304.2.
Hazard Analysis
Toxic Hazard Rating:
  Acute Local: U.
  Acute Systemic: Ingestion 2; Inhalation 2.
  Chronic Local: U.

TOXIC HAZARD RATING CODE (For detailed discussion, see Section 1.)

0 NONE: (a) No harm under any conditions; (b) Harmful only under unusual conditions or overwhelming dosage.

1 SLIGHT: Causes readily reversible changes which disappear after end of exposure.

2 MODERATE: May involve both irreversible and reversible changes; not severe enough to cause death or permanent injury.

3 HIGH: May cause death or permanent injury after very short exposure to small quantities.

U UNKNOWN: No information on humans considered valid by authors.

Chronic Systemic: U.
Toxicity: See also amines.
Fire Hazard: See nitrates.
Explosion Hazard: Dangerous. See nitrates.
Disaster Hazard: Dangerous. See nitrates.
Countermeasures
Ventilation Control: Section 2.
Personal Hygiene: Section 3.
Storage and Handling: Section 7.

TRINITROMETHANE
General Information
Formula: $CH_3(NO_2)_3$.
Constant: Mol wt: 163.
Hazard Analysis
Toxic Hazard Rating:
Acute Local: Irritant 1; Inhalation 1.
Acute Systemic: Inhalation 1.
Chronic Local: U.
Chronic Systemic: U.
Toxicity: Contact with eyes and mucous membranes results in irritation. Inhalation can cause headache and nausea.

TRINITRONAPHTHALENE (mixture of isomers)
General Information
Synonym: Naphthite.
Description: White to yellow crystals; insoluble in water; soluble in alcohol.
Formula: $C_{10}H_5(NO_2)_3$.
Constants: Mol wt: 263.16, mp: 113-247°C depending upon isomeric composition.
Hazard Analysis
See nitrates, μos.
Countermeasures
See nitrates, μos.
Shipping Regulations: Section 11.
ICC: Explosive, Class A.
Coast Guard Classification: Explosive, Class A.
IATA: Explosive material, not acceptable (passenger and cargo).

TRINITROPHENOL. See picric acid.

TRINITROPHENYLMETHYLNITRAMINE. See tetryl.

TRINITROPHENYLNITRAMINE ETHYL NITRATE
General Information
Synonym: Petryl.
Hazard Analysis
Toxicity: Unknown.
Fire Hazard: See nitrates.
Explosion Hazard: Severe when shocked or exposed to heat. Explodes when heated; is soluble in nitroglycerin. Its high explosive sensitivity to impact and friction are about the same as that of tetryl, but its shattering power is much greater. It is used as a base charge in detonators.
Disaster Hazard: Extremely dangerous; shock will explode it; when heated to decomposition it emits highly toxic fumes of oxides of nitrogen and explodes; can react vigorously with oxidizing materials.
Countermeasures
Storage and Handling: Section 7.

TRINITROPHLOROGLUCIN
General Information
Synonym: Trinitrophloroglucinol.
Description: Powder.
Formula: $C_6(NO_2)_3(OH)_3$.
Constant: Mol wt: 261.1.
Hazard Analysis
Toxicity: Details unknown. See nitrates.
Fire Hazard: See nitrates.
Explosion Hazard: See nitrates.

Disaster Hazard: Dangerous; shock will explode it; when heated to decomposition it emits highly toxic fumes of oxides of nitrogen and explodes; can react vigorously with oxidizing or reducing materials.
Countermeasures
Storage and Handling: Section 7.

TRINITRORESORCINOL.
Hazard Analysis
Toxicity: Unknown. See nitro compounds of aromatic hydrocarbons.

TRINITROTOLUENE
General Information
Synonyms: TNT; sym-trinitrotoluol; triton.
Description: Colorless, monoclinic crystals.
Formula: $(NO_2)_3C_6H_2CH_3$.
Constants: Mol wt: 227.13, mp: 80.7°C, bp: 240°C explodes, flash p: explodes, d: 1.654.
Hazard Analysis
Toxic Hazard Rating:
Acute Local: Irritant 2; Allergen 2.
Acute Systemic: Ingestion 3; Inhalation 3; Skin Absorption 2.
Chronic Local: Irritant 2; Allergen 2.
Chronic Systemic: Ingestion 3; Inhalation 3; Skin Absorption 3.
TLV: ACGIH (recommended); 1.5 milligrams per cubic meter of air. Can be absorbed through the intact skin.
Fire Hazard: See nitrates.
Explosion Hazard: Moderate; will detonate under strong shock. See explosives, high. It detonates at around 240°C but can be distilled safely under reduced pressure. It is a comparatively insensitive explosive. In small quantities it will burn quietly if not confined. However, sudden heating of any quantity will cause it to detonate; the accumulation of heat when large quantities are burning will cause detonation. In other respects it is one of the most stable of all high explosives and there are but few restrictions to its handling. It is for this reason, from the military standpoint, that TNT is quantitatively the most used. It requires a fall of 130 centimeters for a 2 kilogram weight to detonate it. It is one of the most powerful high explosives. It can be detonated by the usual detonators and blasting caps (at least a No. 6). For full efficiency, the use of a high velocity initiator, such as tetryl, is required. TNT is one of those explosives containing an oxygen deficiency. In other words, the addition of products which are oxygen rich can enhance its explosive power. Also mono- and dinitrotoluene may be added for reduction of the temperature of the explosion and to make the explosion flashless. Various materials are added to TNT to make what is known as permissible explosives. TNT may be regarded as the equivalent of 40 percent dynamite and can be used under water. It is also used in the manufacture of detonator fuse known as Cordeau Detonant. For the military, TNT finds use in all types of bursting charges, including armor-piercing types, although it is somewhat too sensitive to be ideal for this purpose, and has since been replaced to a great extent by ammonium picrate. It is a relatively expensive explosive and does not compete seriously with dynamite for general commercial use.
Disaster Hazard: Highly dangerous; shock will explode it; when heated to decomposition it emits highly toxic fumes of oxides of nitrogen; can react vigorously with reducing materials.
Countermeasures
Ventilation Control: Section 2.
Personnel Protection: Section 3.
First Aid: Section 1.
Storage and Handling: Section 7.
Shipping Regulations: Section 11.

IATA (dry or wet with less than 10% water): Explosive, not acceptable (passenger and cargo).

      (wet, with not less than 10% water): Flammable solid, yellow label, ½ kilogram (passenger and cargo).

TRINITROTOLUENE, WET (NOT TO EXCEED 16 OZ.). See trinitrotoluene.
Shipping Regulations: Section 11.
  I.C.C. Classifications: Flammable solid, 1 gallon.
  Coast Guard Classification: Inflammable solid; yellow label.

2, 4, 6-TRINITROXYLENE
General Information
Synonym: TNX.
Description: Rhombic crystals.
Formula: $(NO_2)_3C_6H(CH_3)_2$.
Constants: Mol wt: 241.16, mp: 181.5°C, d: 1.604 at 19°C.
Hazard Analysis
Toxicity: Details Unknown. It is said to be less toxic than trinitrotoluene. See also nitro compounds of aromatic hydrocarbons.
Fire Hazard: See nitrates.
Explosion Hazard: Severe, when shocked or exposed to heat. This high explosive is not very powerful when used alone. However, the addition of picric acid or other nitro-type of high explosive serves to lower its melting point and to reinforce its explosive power. Is also used in mixtures with ammonium nitrate; in detonating compositions, and mixed with other high explosives, as a bursting charge.
Disaster Hazard: Highly dangerous; shock will explode it; when heated to decomposition it emits highly toxic fumes of oxides of nitrogen and explodes; can react vigorously with oxidizing materials.
Countermeasures
Storage and Handling: Section 7.

TRI-2-OCTYL BORATE
General Information
Description: Colorless liquid; odor of 2-octanol.
Formula: $[CH_3(CH_2)_5C(CH_3)HO]_3B$.
Constants: Mol wt: 398.47, bp: 340-349°C, flash p: 330°F (C.O.C.), d: 0.837 at 24.5°C, vap. d.: 13.8.
Hazard Analysis
Toxicity: See boron compounds and esters.
Fire Hazard: Moderate, when exposed to heat or flame; can react with oxidizing materials.
Countermeasures
To Fight Fire: Foam, carbon dioxide, dry chemical or carbon tetrachloride (Section 6).
Storage and Handling: Section 7.

TRI-n-OCTYL BORATE
General Information
Description: Colorless liquid; odor of octyl alcohol.
Formula: $[CH_3(CH_2)_7O]_3B$.
Constants: Mol wt: 398.5, bp: 192-194°C at 2 mm, flash p: 370°F (C.O.C.), d: 0.846 at 23°C, vap. d.: 13.7.
Hazard Analysis
Toxicity: See esters and boron compounds.
Fire Hazard: Slight, when exposed to heat or flame; can react with oxidizing materials.
Countermeasures
To Fight Fire: Foam, carbon dioxide, dry chemical or carbon tetrachloride (Section 6).
Storage and Handling: Section 7.

TRIOCTYL PHOSPHATE
General Information
Description: Liquid.
Formula: $[C_4H_9CH(C_2H_5)CH_2O]_3P:O$.
Constants: Mol wt: 434.63, mp: −74°C, bp: 216°C at 5 mm, flash p: 405°F (O.C.), d: 0.9262 at 20°/20°C, vap. d.: 14.95.
Hazard Analysis
Toxicity: Details unknown. See esters.
Fire Hazard: Slight, when exposed to heat or flame.
Disaster Hazard: Dangerous. See phosphates; can react with oxidizing materials.
Countermeasures
Storage and Handling: Section 7.
To Fight Fire: Foam, carbon dioxide, dry chemical or carbon tetrachloride (Section 6).

TRIOCTYL PHOSPHITE
General Information
Synonym: Tris (2-ethyl hexyl) phosphite.
Description: Insoluble in water.
Formula: $(C_8H_{17}O)_3P$.
Constants: Mol wt: 418, d: 0.9, bp: 212°F at 0.01 mm; flash p: 340°F.
Hazard Analysis
Toxic Hazard Rating: U. See esters.
Fire Hazard: Slight, when exposed to heat or flame.
Disaster Hazard: Dangerous. See phosphates.
Countermeasures
Storage and Handling: Section 7.
To Fight Fire: Water, foam, mist, carbon dioxide.

TRIOL-230
General Information
Description: Clear, colorless liquid.
Formula: $HOCH_2CH_2OCH_2(CH_3)(CH_2OH)CCH_2(CH_3)$-$CHCH_2OH$.
Constants: Mol wt: 206.28, bp: 196°C at 5 mm, flash p: 395°F (C.O.C.), fp: $< -5°C$, d: 1.081 at 20°/20°C, vap. press.: <0.01 mm at 20°C.
Hazard Analysis
Toxicity: Unknown.
Fire Hazard: Slight, when exposed to heat or flame; can react with oxidizing materials.
Countermeasures
To Fight Fire: Water, foam, carbon dioxide, dry chemical or carbon tetrachloride (Section 6).
Storage and Handling: Section 7.

TRIOLEYL BORATE
General Information
Description: Pale yellow liquid; odor of oleyl alcohol.
Formula: $[CH_3(CH_2)_7CHCH(CH_2)_7CH_2O]_3B$.
Constants: Mol wt: 813.20, bp: 300-330 °C at 0.5 mm, flash p: 495°F (C.O.C.), d: 0.860 at 23.6°C.
Hazard Analysis
Toxicity: See boron compounds and esters.
Fire Hazard: Slight, when exposed to heat or flame; can react with oxidizing materials.
Countermeasures
To Fight Fire: Foam, carbon dioxide, dry chemical, or carbon tetrachloride (Section 6).
Storage and Handling: Section 7.

TRIOXANE
General Information
Synonym: sym-Trioxane; α-Trioxymethylene.
Description: Colorless crystals; odor of ethyl alcohol.

TOXIC HAZARD RATING CODE (For detailed discussion, see Section 1.)

0 NONE: (a) No harm under any conditions; (b) Harmful only under unusual conditions or overwhelming dosage.

1 SLIGHT: Causes readily reversible changes which disappear after end of exposure.

2 MODERATE: May involve both irreversible and reversible changes; not severe enough to cause death or permanent injury.

3 HIGH: May cause death or permanent injury after very short exposure to small quantities.

U UNKNOWN: No information on humans considered valid by authors.

Formula: $(CH_2O)_3$.

Constants: Mol wt: 90.08, mp: 62°C, lel: 3.6%, uel: 28.7%, bp: 114.5°C sublimes, flash p: 113°F (C.C.), d: 1.17 at 65°/20°C, autoign. temp.: 777°F, vap. press.: 13 mm at 25°C, vap. d.: 3.1.

Hazard Analysis

Toxicity: Details unknown. Can evolve formaldehyde when heated strongly or in contact with strong acids. See also formaldehyde.

Fire Hazard: Moderate when exposed to heat or flame.

Explosion Hazard: Moderate, in the form of vapor when exposed to flame.

Disaster Hazard: Moderately dangerous; can explode when heated; reacts with oxidizing materials; and on contact with acid or acid fumes, it can emit toxic fumes.

Countermeasures

Storage and Handling: Section 7.

To Fight Fire: Foam, carbon dioxide, dry chemical or carbon tetrachloride (Section 6).

sym-TRIOXANE. See trioxane.

TRIOXIME

Hazard Analysis

Toxicity: Unknown.

Fire Hazard: Unknown.

Explosion Hazard: Moderate, when exposed to heat.

Explosive Range: At 311°F.

Countermeasures

Storage and Handling: Section 7.

α-TRIOXYMETHYLENE. See trioxane.

TRIPENTAERYTHRITOL

General Information

Description: White to ivory, odorless powder.

Formula: $C_{15}H_{32}O_{10}$.

Constants: Mol wt: 372.4, mp: approx. 240°C, d: 1.30.

Hazard Analysis

Toxicity: See alcohols.

Fire Hazard: Slight; (Section 6).

Countermeasures

Storage and Handling: Section 7.

TRIPHENYLALUMINUM

General Information

Description: White crystals. Decomposes upon contact with moisture.

Formula: $(C_6H_5)_3Al$.

Constants: Mol wt: 258.3, mp: 200°C.

Hazard Analysis and Countermeasures

See organometals and aluminum compounds.

TRIPHENYL ANISYL GERMANIUM

General Information

Description: White powder; soluble in alcohol.

Formula: $(C_6H_5)_3Ge(CH_3OC_6H_4)$.

Constants: Mol wt: 411.0, mp: 159°C.

Hazard Analysis

Toxicity: Details unknown. See germanium compounds.

TRIPHENYLANTIMONY. See triphenylstibine.

TRIPHENYLARSENIC

General Information

Synonym: Arsenic triphenyl.

Description: White crystals.

Formula: $As(C_6H_5)_3$.

Constants: Mol wt: 306.2, mp: 60°C, d: 1.2225 at 48°C, bp: >360 °C (in CO_2).

Hazard Analysis and Countermeasures

See arsenic compounds.

TRIPHENYL BENZYL TIN

General Information

Description: Colorless crystals, sobuble in organic solvents except alcohol.

Formula: $(C_6H_5)_3Sn(C_6H_5CH_2)$.

Constants: Mol wt: 441.1, mp: 90°C, bp: 250°C, at 3 mm.

Hazard Analysis

Toxicity: Details unknown. See tin compounds.

TRIPHENYLBISMUTHINE

General Information

Synonym: Bismuth triphenyl.

Description: Monoclinic crystals.

Formula: $Bi(C_6H_5)_3$.

Constants: Mol wt: 440.30, mp: 78°C, bp: 242°C at 14 mm, d: 1.585.

Hazard Analysis and Countermeasures

See bismuth compounds.

TRIPHENYL BORATE

General Information

Description: White to pink solid; odor of phenol; decomposed by water.

Formula: $(C_6H_5O)_3B$.

Constants: Mol wt: 290.12, mp: 35 °C, bp: >360 °C.

Hazard Analysis

Toxicity: Highly toxic. See phenol and boron compounds.

Fire Hazard: Moderate, when exposed to heat or flame. Hydrolyzes to phenol.

Disaster Hazard: Dangerous; when heated it emits highly toxic fumes; reacts with water or steam to form toxic fumes of phenol; in contact with oxidizing material, it can react vigorously.

Countermeasures

Storage and Handling: Section 7.

To Fight Fire: Foam, carbon dioxide, dry chemical or carbon tetrachloride (Section 6).

TRIPHENYL BORON. See boron triphenyl.

TRIPHENYL-3, 4-DICHLOROPHENYL PHOSPHONIUM CHLORIDE

General Information

Description: A solid.

Constant: Mol wt. 443.7.

Formula: $(C_6H_5)_3(C_6H_3Cl_2)PCl$.

Hazard Analysis

Toxicity: Details unknown; a moth repellent.

Disaster Hazard: Dangerous. See phosphates and chlorides.

Countermeasures

Storage and Handling: Section 7.

TRIPHENYL DIMETHYLAMINOPHENYL GERMANIUM

General Information

Description: White crystals.

Formula: $(C_6H_5)_3GeC_6H_4N(CH_3)_2$.

Constants: Mol wt: 424.1, mp: 141°C.

Hazard Analysis

Toxicity: Details unknown. See germanium compounds.

TRIPHENYL ETHYL TIN

General Information

Description: White powder.

Formula: $(C_6H_5)_3Sn(C_2H_5)$.

Constants: Mol wt: 379.1, d: 1.2953 at 62°C, mp: 56°C.

Hazard Analysis

Toxicity: Details unknown. See tin compounds.

TRIPHENYLGERMANIUM AMIDE

General Information

Description: White solid.

Formula: $(C_6H_5)_3GeNH_2$.

Constants: Mol wt: 319.9, mp: decomposes to evolve NH_3.

Hazard Analysis

Toxicity: Details unknown. See amides and germanium compounds.

TRIPHENYLGERMANIUM BROMIDE

General Information

Description: Crystals; hydrolyzed in hot water.

Formula: $(C_6H_5)_3GeBr$.
Constants: Mol wt: 383.8, mp: 139° C.
Hazard Analysis and Countermeasures
See germanium compounds and bromides.

TRIPHENYLGERMANIUM CHLORIDE
General Information
Description: White crystals, hydrolyzes in hot water.
Formula: $(C_6H_5)_3GeCl$.
Constants: Mol wt: 339.4, mp: 118°C, bp: 285°C at 12 mm.
Hazard Analysis and Countermeasures
See germanium compounds and chlorides.

TRIPHENYLGERMANIUM FLUORIDE
General Information
Description: White crystals; hydrolyzes in hot water.
Formula: $(C_6H_5)_3GeF$.
Constants: Mol wt: 322.9, mp: 77°C.
Hazard Analysis and Countermeasures
See fluorides and phenol.

TRIPHENYLGERMANIUM HYDRIDE
General Information
Description: White crystals, insoluble in water.
Formula: $(C_6H_5)_3GeH$.
Constants: Mol wt: 304.9; mp: 47°C (α), 27°C (β).
Hazard Analysis and Countermeasures
See germanium compounds and phenol.

TRIPHENYLGERMANIUM HYDROXIDE
General Information
Description: White crystals, insoluble in water.
Formula: $(C_8H_5)_3GeOH$.
Constants: Mol wt: 320.9, mp: 134°C.
Hazard Analysis
Toxicity: Details unknown. See germanium compounds and phenol.

TRIPHENYLGERMANIUM IODIDE
General Information
Description: White crystals; hydrolyzed by water.
Formula: $(C_6H_5)_3GeI$.
Constants: Mol wt: 430.8, mp: 157°C.
Hazard Analysis and Countermeasures
See germanium compounds and iodides.

TRIPHENYLGERMANIUM OXIDE
General Information
Description: Colorless crystals; insoluble in water.
Formula: $[(C_6H_5)_3Ge]_2O$.
Constants: Mol wt: 623.8, mp: 184°C.
Hazard Analysis and Countermeasures
See germanium compounds and phenol.

TRIPHENYLGERMANIUM SODIUM
General Information
Description: Yellowish crystals, decomposed by water.
Formula: $(C_8H_5)_3GeNa$.
Constant: Mol wt: 326.9.
Hazard Analysis
Toxicity: Highly toxic. See sodium, germanium compounds and phenol.
Fire and Explosion Hazard: This material reacts with water to evolve hydrogen. See hydrogen. (Section 6).
Countermeasures
Storage and Handling: Section 7.

TRIPHENYLGERMANIUM SODIUM OXIDE
General Information
Description: White solid; decomposed by water.
Formula: $(C_6H_5)_3GeONa$.
Constant: Mol wt: 342.9.
Hazard Analysis
Toxicity: Details unknown. See germanium compounds.

TRIPHENYLGERMANIUM SODIUM TRIAMMINE
General Information
Description: Yellow solid; decomposed by water.
Formula: $(C_6H_5)_3GeNa \cdot 3NH_3$.
Constants: Mol wt: 378.0, mp: decomposes.
Hazard Analysis
Toxicity: Details unknown. See ammonia and germanium compounds.

TRIPHENYLMETHANE DYES
General Information
Synonym: Rosaniline dyes.
Hazard Analysis
Toxicity: Details unknown; fungicide.
Disaster Hazard: Moderately dangerous; when heated they emit toxic fumes.
Countermeasures
Storage and Handling: Section 7.

TRIPHENYL METHYL TIN
General Information
Description: Colorless crystals; soluble in chloroform, benzene and ether.
Formula: $(C_6H_5)_3Sn(CH_3)$.
Constants: Mol wt: 365.0, d: 1.3113 at 64°C, mp: 64°C.
Hazard Aanlysis and Countermeasures
Details unknown. See tin compounds and phenol.

TRIPHENYL-α-NAPHTHYL TIN
General Information
Description: Colorless crystals; soluble in benzene, ether and chloroform.
Formula: $(C_6H_5)_3Sn(C_{10}H_7)$.
Constants: Mol wt: 477.2, mp: 125°C.
Hazard Analysis and Countermeasures
See tin compounus and phenol.

TRIPHENYL PHOSPHATE
General Information
Description: Colorless, odorless, crystalline solid.
Formula: $PO(OC_6H_5)_3$.
Constants: Mol wt: 326.28, mp: 48.5°C, bp: 245°C at 11 mm, flash p: 428°F (C.C.), d: 1.268 at 60°C, vap. press.: 1 mm at 193.5°C.
Hazard Analysis
Toxic Hazard Rating:
Acute Local: U.
Acute Systemic: Ingestion 2; Inhalation 2.
Chronic Local: U.
Chronic Systemic: U.
TLV: ACGIH (recommended); 3 milligrams per cubic meter of air.
Fire Hazard: Slight, when exposed to heat or flame.
Spontaneous Heating: No.
Disaster Hazard: Dangerous. See phosphates.
Countermeasures
Personal Hygiene: Section 3.
To Fight Fire: Water, carbon dioxide, dry chemical or carbon tetrachloride (Section 6).
Storage and Handling: Section 7.

TOXIC HAZARD RATING CODE (For detailed discussion, see Section 1.)

0 NONE: (a) No harm under any conditions; (b) Harmful only under unusual conditions or overwhelming dosage.

1 SLIGHT: Causes readily reversible changes which disappear after end of exposure.

2 MODERATE: May involve both irreversible and reversible changes; not severe enough to cause death or permanent injury.

3 HIGH: May cause death or permanent injury after very short exposure to small quantities.

U UNKNOWN: No information on humans considered valid by authors.

TRIPHENYLPHOSPHINE
General Information
Synonym: Triphenyl phosphorous.
Description: Crystals.
Formula: $(C_6H_5)_3P$.
Constants: Mol wt: 262.3, mp: 79°C, bp: > 360°C, d: 1.194, flash p: 356 °F (O.C.), vap. d.: 9.0.
Hazard Analysis
Toxicity: Probably highly toxic. See phosphine.
Fire Hazard: Moderate, when exposed to heat or flame (Section 6).
Explosion Hazard: Slight, in the form of vapor when exposed to flame.
Disaster Hazard: Dangerous; when heated to decomposition it emits highly toxic fumes of phosphine and oxides of phosphorus; can react vigorously with oxidizing materials.
Countermeasures
Storage and Handling: Section 7.
To Fight Fire: Water, foam, fog, carbon dioxide.

TRIPHENYLPHOSPHINE OXIDE
General Information
Description: White crystals.
Formula: $(C_6H_5)_3PO$.
Constants: Mol wt: 278.3, mp: 156°C, bp: > 360°C, d: 1.2124 at 22.6°C.
Hazard Analysis
Toxicity: Probably highly toxic, See phosphine.
Explosion Hazard: Unknown.
Disaster Hazard: Dangerous. See phosphates.
Countermeasures
Storage and Handling: Section 7.

TRIPHENYL PHOSPHITE
General Information
Description: Water white to pale yellow solid or oily liquid; clean pleasant odor; insoluble in water.
Formula: $(C_6H_5O)_3P$.
Constants: Mol wt: 310, d: 1.184 at 25/25°C, mp: 22–25°C, bp: 155–160°C at 0.1 mm, flash p: 425°F (O.C.).
Hazard Analysis
Toxic Hazard Rating: U.
Fire Hazard: Slight, when exposed to heat or flame.
Disaster Hazard: Dangerous. See phosphates.
Countermeasures
To Fight Fire: Water, foam, mist, dry chemical.

TRIPHENYLPHOSPHOROUS. See triphenyl phosphine.

TRIPHENYLSTANNYL METHANE
General Information
Description: Crystals, insoluble in water; soluble in organic solvents.
Formula: $[(C_6H_5)_3Sn]_3CH$.
Constants: Mol wt: 1063.0, mp: 128°C.
Hazard Analysis
Toxicity: Details unknown. See tin compounds.

TRIPHENYL STIBINE
General Information
Synonym: Triphenylantimony.
Description: Crystals. Water insoluble; soluble in organic solvents.
Formula: $(C_6H_5)_3Sb$.
Constants: Mol wt: 353.1, mp: 50°C, d: 1.4343 at 25°C, bp: > 360°C.
Hazard Analysis
Toxicity: See antimony compounds.
Fire Hazard: Moderate, when exposed to heat or flame (Section 6).
Disaster Hazard: Dangerous. See antimony compounds; can react vigorously with oxidizing materials.

Countermeasures
Storage and Handling: Section 7.
To Fight Fire: Water, foam, mist.

TRIPHENYL TIN
General Information
Description: White powder; insoluble in water.
Formula: $(C_6H_5)_3Sn$.
Constants: Mol wt: 350.0, mp: 232.5°C.
Hazard Analysis
Toxicity: Details unknown. See tin compounds.
Disaster Hazard: Dangerous. See phenol.

TRIPHENYLTIN BROMIDE
General Information
Description: Colorless crystals; insoluble in water; soluble in organic solvents.
Formula: $(C_6H_5)_3SnBr$.
Constants: Mol wt: 429.9, mp: 120.5°C, bp: 249°C at 13.5 mm.
Hazard Analysis and Countermeasures
See tin compounds and bromides.

TRIPHENYLTIN CHLORIDE
General Information
Description: Colorless crystals, insoluble in water. Soluble in organic solvents.
Formula: $(C_6H_5)_3SnCl$.
Constants: Mol wt: 385.5, mp: 106°C, bp: 240°C at 13.5 mm.
Hazard Analysis and Countermeasures
See tin compounds and chlorides.

TRIPHENYLTIN FLUORIDE
General Information
Description: Crystals, slightly water soluble.
Formula: $(C_6H_5)_3SnF$.
Constants: Mol wt: 369.0
Hazard Analysis and Countermeasures
See fluorides and phenol.

TRIPHENYLTIN HYDROXIDE
General Information
Description: Solid.
Formula: $(C_6H_5)_3SnOH$.
Constants: Mol wt: 367.0, mp: 118°C.
Hazard Analysis
Toxicity: Details unknown. See tin compounds.

TRIPHENYLTIN IODIDE
General Information
Description: White crystals; insoluble in water; soluble in organic solvents.
Formula: $(C_6H_5)_3SnI$.
Constants: Mol wt: 476.9, mp: 121°C, bp: 253 at 13.5 mm
Hazard Analysis and Countermeasures
See tin compounds and iodides.

TRIPHENYL-m-TOLYLGERMANIUM
General Information
Description: White crystals; insoluble in water.
Formula: $(C_6H_5)_3Ge(C_6H_4CH_3)$.
Constants: Mol wt: 395.0, mp: 137°C.
Hazard Analysis
Toxicity: Details unknown. See germanium compounds.

TRIPHENYL-p-TOLYLGERMANIUM
General Information
Description: White crystals; soluble in organic solvents.
Formula: $(C_6H_5)_3Ge(C_6H_4CH_3)$.
Constants: Mol wt: 395.0, mp: 124°C.
Hazard Analysis
Toxicity: Details unknown. See germanium compounds.

TRIPHENYL-p-TOLYLTIN
General Information
Description: Crystals; soluble in chloroform, benzene and ether.
Formula: $(C_6H_5)_3Sn(C_7H_7)$.
Constants: Mol wt: 441.1, mp: 124°C.
Hazard Analysis
Toxicity: Details unknown. See tin compounds.

TRI-p-PHENYLYLGERMANIUM BROMIDE
General Information
Description: White crystals; soluble in benzene.
Formula: $(C_6H_5C_6H_4)_3GeBr$.
Constants: Mol wt: 612.1, mp: 242°C.
Hazard Analysis and Countermeasures
See germanium compounds and bromides.

TRIPHENYL-p-XYLYLTIN
General Information
Description: Crystals. Soluble in benzene, ether and chloroform.
Formula: $(C_6H_5)_3Sn[C_6H_3(CH_3)_2]$.
Constants: Mol wt: 456.2, mp: 100.5°C.
Hazard Analysis
Toxicity: Details unknown. See tin compounds.

TRIPHOSGENE. See hexachloromethyl carbonate.

TRIPOLI
General Information
Description: Finely granulated white or gray siliceous rock.
Formula: An amorphous form of SiO_2.
Hazard Analysis
Toxic Hazard Rating:
  Acute Local: 0.
  Acute Systemic: 0.
  Chronic Local: Inhalation 2.
  Chronic Systemic: 0.
Toxicology: Effects due to silica content which may be enough to produce pulmonary fibrosis.

TRIPROPOXYBORON. See tri-n-propyl borate.

TRIPROPYLALUMINUM. See aluminum tripropyl.

TRIPROPYLAMINE
General Information
Description: Liquid; very slightly soluble in water.
Formula: $N(C_3H_7)_3$.
Constants: Mol wt: 143.3, mp: −93.5°C, bp: 156°C, flash p: 105°F, d: 0.75, vap. d.: 4.9.
Hazard Analysis
Toxicity: See amines.
Fire Hazard: Moderate, when exposed to heat or flame.
Disaster Hazard: Moderately dangerous; when heated to decomposition it emits toxic fumes; can react with oxidizing materials.
Countermeasures
Storage and Handling: Section 7.
To Fight Fire: Foam, carbon dioxide, dry chemical or carbon tetrachloride (Section 6).

TRI-n-PROPYL BORATE
General Information
Synonym: Propyl borate.
Description: Colorless liquid; odor of n-propanol.
Formula: $(CH_3CH_2CH_2O)_3B$.
Constants: Mol wt: 188.08, bp: 176–179°C, flash p: 155°F (C.O.C.), d: 0.856 at 24°C.

Hazard Analysis
Toxicity: See esters and boron compounds.
Fire Hazard: Moderate, when exposed to heat or flame; can react with oxidizing materials.
Countermeasures
Storage and Handling: Section 7.

TRI-n-PROPYLBORON
General Information
Description: Colorless liquid; insoluble in water, soluble in ether.
Formula: $(C_3H_7)_3B$.
Constants: Mol wt: 140.1, d: 0.725, bp: 157°C at 20 mm.
Hazard Analysis
Toxicity: Details unknown. See boron compounds.

TRI-sec-PROPYLBORON
General Information
Description: Colorless mobile liquid; insoluble in water; soluble in ether.
Formula: $B(C_3H_7)_3$.
Constants: Mol wt: 140.1, bp: 150°C.
Hazard Analysis
Toxicity: Details unknown. See boron compounds.

TRI-n-PROPYL-n-BUTYLTIN
General Information
Description: Liquid.
Formula: $(C_3H_7)_3Sn(C_4H_9)$.
Constants: Mol wt: 305.1, bp: 121°C at 10 mm.
Hazard Analysis
Toxicity: Details unknown. See tin compounds.

TRIPROPYLENE GLYCOL
General Information
Description: Colorless liquid.
Formula: $HOC_3H_6OC_3H_6OC_3H_6OH$.
Constants: Mol wt: 192.3, mp: does not crystallize, bp: 267°C, flash p: 285°F, d: 1.023 at 25°/25°C, vap. press.: 1 mm at 96.0°C, vap. d.: 6.63.
Hazard Analysis
Toxicity: See glycols.
Fire Hazard: Slight, when exposed to heat or flame; can react with oxidizing materials.
Countermeasures
To Fight Fire: Water, foam, carbon dioxide, dry chemical or carbon tetrachloride (Section 6).
Storage and Handling: Section 7.

TRIPROPYLENE GLYCOL METHYL ETHER
General Information
Synonym: Tripropylene glycol monomethyl ether.
Description: Colorless liquid.
Formula: $CH_3OC_3H_6OC_3H_6OC_3H_6OH$.
Constants: Mol wt: 206.3, bp: 243°C, flash p: 250°F, d: 0.967 at 25°/25°C, vap. d.: 7.1.
Hazard Analysis
Toxicity: See glycols.
Fire Hazard: Slight, when exposed to heat or flame; can react with oxidizing materials.
Countermeasures
To Fight Fire: Foam, carbon dioxide, dry chemical or carbon tetrachloride (Section 6).
Storage and Handling: Section 7.

TRIPROPYLENE GLYCOL MONOMETHYL ETHER
See tripropylene glycol methyl ether.

TOXIC HAZARD RATING CODE (For detailed discussion, see Section 1.)

0 NONE: (a) No harm under any conditions; (b) Harmful only under unusual conditions or overwhelming dosage.

1 SLIGHT: Causes readily reversible changes which disappear after end of exposure.

2 MODERATE: May involve both irreversible and reversible changes not severe enough to cause death or permanent injury.

3 HIGH: May cause death or permanent injury after very short exposure to small quantities.

U UNKNOWN: No information on humans considered valid by authors

TRI-n-PROPYL ETHYL TIN
General Information
Description: Liquid.
Formula: $(C_3H_7)_3Sn(C_2H_5)$.
Constants: Mol wt: 277.0, d: 1.1225 at 22°C, bp: 117.5°C at 23.3 mm.
Hazard Analysis
Toxicity: Details unknown. See tin compounds.

TRI-n-PROPYL ISOBUTYL TIN
General Information
Description: Liquid.
Formula: $(C_3H_7)_3Sn(C_4H_9)$.
Constants: Mol wt: 305.1, d: 1.0841 at 24°C, bp: 128°C at 18 mm.
Hazard Analysis
Toxicity: Details unknown. See tin compounds.

TRI-n-PROPYL PHOSPHATE
General Information
Description: Liquid.
Formula: $(C_3H_7O)_3PO$.
Constants: Mol wt: 224, bp: 97°C at 4 mm, d: 1.002 at 25°/4°C, vap. d.: 7.72.
Hazard Analysis
Toxicity: Unknown.
Fire Hazard: Slight, when exposed to heat or flame (Section 6).
Disaster Hazard: Dangerous. See phosphates; can react with oxidizing materials.
can react with oxidizing materials.
Countermeasures
Storage and Handling: Section 7.

TRIPROPYLTIN CHLORIDE
General Information
Description: Colorless liquid; soluble in organic solvents.
Formula: $(C_3H_7)_3SnCl$.
Constants: Mol wt: 283.4, d: 1.2678 at 28°C, mp: −23.5°C.
Hazard Analysis and Countermeasures
See tin compounds and chlorides.

TRI-n-PROPYLTIN FLUORIDE
General Information
Description: Crystals, soluble in organic solvents.
Formula: $(C_3H_7)_3SnF$.
Constants: Mol wt: 267.0, mp: 275°C.
Hazard Analysis and Countermeasures
See fluorides.

TRI-n-PROPYLTIN IODIDE
General Information
Description: Colorless liquid; soluble in organic solvents.
Formula: $(C_3H_7)_3SnI$.
Constants: Mol wt: 374.9, d: 1.692 at 16°C, mp: −53°C, bp: 262°C.
Hazard Analysis and Countermeasures
See tin compounds and iodides.

2,2',2''-TRIPYRIDYL RHENICHLORIDE
General Information
Description: Pale green crystals, insoluble in water.
Formula: $(C_5H_4N)_3HReCl_6$.
Constant: Mol wt: 634.3.
Hazard Analysis and Countermeasures
See rhenium compounds and chlorides.

TRIS-ACETYLACETONEGERMANIUM CUPRI-BROMIDE
General Information
Description: Green black crystals.
Formula: $[(C_5H_7O_2)_3Ge]CuBr_3$.
Constants: Mol wt: 673.2, mp: 139°C.
Hazard Analysis and Countermeasures
See copper compounds, germanium compounds and bromides.

TRIS-ACETYLACETONEGERMANIUM CUPRO-BROMIDE
General Information
Description: Colorless crystals; soluble in chloroform.
Formula: $[(C_5H_7O_2)_3Ge]CuBr_2$.
Constants: Mol wt: 593.3, mp: 166°C.
Hazard Analysis and Countermeasures
See copper compounds, bromides and germanium compounds.

TRIS-ACETYLACETONEGERMANIUM CUPRO-CHLORIDE
General Information
Description: Colorless crystals; soluble in chloroform.
Formula: $[(C_5H_7O_2)_3Ge]CuCl_2$.
Constants: Mol wt: 504.4, mp: 148°C.
Hazard Analysis and Countermeasures
See copper compounds, germanium compounds and chlorides.

TRIS-ACETYLACETONEGERMANIUM DICUPRO-BROMIDE
General Information
Description: Colorless crystals; soluble in acetone.
Formula: $[(C_5H_7O_2)_3Ge]Cu_2Br_3$.
Constants: Mol wt: 736.8, mp: 195°C (decomposes).
Hazard Analysis and Countermeasures
See copper compounds, germanium compounds and bromides.

TRIS (1-AZIRIDINYL)PHOSPHINE OXIDE
General Information
Synonyms: Triethylenephosphoramide; tepa; APO; 1-aziridinyl phosphine oxide (tris).
Description: Colorless crystals; soluble in water, alcohol and ether.
Formula: $(NCH_2CH_2)_3PO$.
Constant: Mol wt: 173.
Hazard Analysis
Toxicity: Highly toxic.
Disaster Hazard: Dangerous. See phosphorus compounds.
Countermeasures
Shipping Regulations: Section 11.
    ICC: Corrosive liquid, white label, 1 gallon.
    Coast Guard Classification: Corrosive liquid, white label.
    IATA (solution): Corrosive liquid, white label, 1 liter (passenger), 5 liters (cargo).

TRIS (β-CHLOROETHYL) PHOSPHATE
General Information
Description: Clear liquid.
Formula: $(ClCH_2CH_2)_3PO_4$.
Constants: Mol wt: 285.5, flash p: 421°F (C.O.C.), boiling range: 210–220°C at 20 mm, d: 1.425 at 20°/20°C, autoign. temp.: 1115°F, vap. press.: 0.5 mm at 145°C.
Hazard Analysis
Toxicity: Unknown.
Fire Hazard: Slight, when exposed to heat or flame (Section 6).
Disaster Hazard: Dangerous. See phosphates and chlorides.
Countermeasures
Storage and Handling: Section 7.

TRIS-CHLOROETHYL PHOSPHITE
General Information
Description: Liquid; water white.
Formula: $(ClCH_2CH_2O)_3P$.
Constants: Mol wt: 269.51, bp: 125–135°C at 7 mm, flash p: > 280°F (O.C.), d: 1.3348 at 35°/4°C, vap. press.: < 1 mm at 20°C, vap. d.: 9.32.
Hazard Analysis
Toxicity: Unknown.
Fire Hazard: Slight, when exposed to heat or flame.
Disaster Hazard: Dangerous. See phosphates and chlorides; can react vigorously with oxidizing materials; it can isomerize vigorously.

Countermeasures
Storage and Handling: Section 7.
To Fight Fire: Water, foam, carbon dioxide, dry chemical or carbon tetrachloride (Section 6).

TRIS(2-ETHYL HEXYL) PHOSPHITE. See trioctyl phosphite.

1,1,3-TRIS-[p-GLYCIDYLOXY PHENYL]PROPANE
Hazard Analysis
Toxic Hazard Rating:
   Acute Local: U.
   Acute Systemic: Skin absorption 2.
   Chronic Local: Allergen 3.
   Chronic Systemic: U.
Toxicity: Data based on limited animal experiments.

TRIS(HYDROXYMETHYL) AMINOMETHANE
General Information
Description: Crystals.
Formula: $(CH_2OH)_3CNH_2$.
Constants: Mol wt: 121.14, mp: 171–172°C, bp: 219°C at 10 mm.
Hazard Analysis
Toxicity: Details unknown. Irritating to skin and mucous membranes.
Fire Hazard: Slight, when exposed to heat or flame (Section 6).
Disaster Hazard: Moderately dangerous; when heated to decomposition it emits toxic fumes; can react with oxidizing materials.
Countermeasures
Storage and Handling: Section 7.

TRIS(HYDROXYMETHYL) NITROMETHANE
General Information
Description: Crystalline.
Formula: $(CH_2OH)_3CNO_2$.
Constants: Mol wt: 151.12, mp: 165–170°C, bp: decomposes.
Hazard Analysis and Countermeasures
Probably an irritant.
See nitrates.

TRISILANE
General Information
Synonym: Trisilicon octahydride, trisilicopropane.
Description: Liquid.
Formula: Si_3H_8.
Constants: Mol wt: 92.2, mp: −117.4°C, bp: 52.9°C, d: 0.743 at 0°C, vap. press.: 95.5 mm at 0°C.
Hazard Analysis
Toxicity: Details unknown. See silanes.
Fire Hazard: Dangerous, by chemical reaction. Decomposes in water (Section 6).
Explosion Hazard: Severe, by spontaneous chemical reaction. Detonates spontaneously in air.
Disaster Hazard: Dangerous; when heated to decomposition it emits toxic fumes and can explode; will react with water or steam to produce hydrogen and toxic fumes; can react vigorously with oxidizing materials.
Countermeasures
Storage and Handling: Section 7.

TRISILICON OCTAHYDRIDE. See trisilane.

TRISILICOPROPANE. See trisilane.

TRISILICYLAMINE
General Information
Synonym: Nitrilotrisilane.
Description: Liquid.
Formula: $(SiH_3)_3N$.
Constants: Mol wt: 107.26, mp: −105.6°C, bp: 52°C, d: 0.895 at −106°C.
Hazard Analysis
Toxicity: Details unknown. See amines and silanes.
Disaster Hazard: Dangerous; when heated to decomposition it emits highly toxic fumes; will react with water or steam to produce toxic and flammable vapors.
Countermeasures
Storage and Handling: Section 7.

TRIS[1-(2-METHYL)AZIRIDINYL] PHOSPHINE OXIDE
General Information
Synonym: MAPO.
Description: Amber colored liquid; amine odor; miscible with water and all organic solvents.
Formula: $[N(CH_2)_2CH_3]P:O$.
Constants: Mol wt: 215, bp: 118°–125°C at 1 mm, d: 1.079 at 25/25°C.
Hazard Analysis
Toxicology: Highly toxic by skin absorption as well as by ingestion. Animal experiments suggest cholinesterase inhibition, possibly due to metabolic products of this material in body.
Disaster Hazard: Dangerous. See phosphates.
Countermeasures
Ventilation Control: Section 2.
Personnel Protection: Section 3.
Storage and Handling: Section 7.

TRISODIUM-1,3,6-NAPHTHALENE   TRISULFONATE
General Information
Description: Light tan to buff powder or crystals.
Formula: $C_{10}H_5(SO_3Na)_3$.
Constant: Mol wt: 434.4.
Hazard Analysis
Toxicity: Unknown.
Disaster Hazard: Dangerous. See sulfonates.
Countermeasures
Storage and Handling: Section 7.

TRISODIUM NITROPHOSPHATE
General Information
Description: White powder.
Hazard Analysis
Toxicity: Details unknown. See trisodium phosphate.
Disaster Hazard: Dangerous. See nitrates and phosphates.
Countermeasures
Storage and Handling: Section 7.

TRISODIUM PHOSPHATE
General Information
Synonym: Sodium o-phosphate.
Description: Colorless crystals.
Formula: $Na_3PO_4 \cdot 12H_2O$.
Constants: Mol wt: 380.21, mp: 73.3-76.7°C (decomp.), −12H_2O at 100°C, d: 1.62 at 20°C.
Hazard Analysis
Toxic Hazard Rating:
   Acute Local: Irritant 2; Ingestion 2; Inhalation 2.
   Acute Systemic: U.
   Chronic Local: Irritant 2.
   Chronic Systemic: U.

TOXIC HAZARD RATING CODE (For detailed discussion, see Section 1.)

0 NONE: (a) No harm under any conditions; (b) Harmful only under unusual conditions or overwhelming dosage.

1 SLIGHT: Causes readily reversible changes which disappear after end of exposure.

2 MODERATE: May involve both irreversible and reversible changes; not severe enough to cause death or permanent injury.

3 HIGH: May cause death or permanent injury after very short exposure to small quantities.

U UNKNOWN: No information on humans considered valid by authors.

Note: A sequestrant, general purpose, and nutrient and/or dietary supplement food additive (Section 10).
Caution: A strong, caustic material.
Disaster Hazard: Dangerous. See phosphates.
Countermeasures
Ventilation Control: Section 2.
Personnel Protection: Section 3.
Storage and Handling: Section 7.

TRISTATE SPECIAL NO. 1. See explosives, high.

TRISTEARYL BORATE
General Information
Description: White solid; odor of stearyl alcohol.
Formula: $[CH_3(CH_2)_{16}CH_2O]_3B$.
Constants: Mol wt: 819.25, mp: 49.8–54°C, bp: 300–331°C at 0.3 mm.
Hazard Analysis
Toxicity: See boron compounds and stearyl alcohol.
Fire Hazard: Slight, when exposed to heat or flame; can react with oxidizing materials.
Countermeasures
Storage and Handling: Section 7.

TRITETRADECYL BORATE
General Information
Synonym: Tri(7-ethyl-2-methyl-4 undecyl) borate.
Description: Light straw-yellow, viscous liquid. Slight odor.
Formula: $[CH_3(CH_2)_3CH(C_2H_5)\cdot(CH_2)_2CHO\cdot CH_2CH-(CH_3)_3]B$.
Constants: Mol wt: 650.94, bp: 225°C at 0.6 mm, flash p: 395°F (C.O.C.), d: 0.846 at 26°C, vap. d.: 22.5.
Hazard Analysis
Toxicity: See boron compounds and esters.
Fire Hazard: Slight, when exposed to heat or flame; can react with oxidizing materials.
Countermeasures
To Fight Fire: Foam, carbon dioxide, dry chemical or carbon tetrachloride (Section 6).
Storage and Handling: Section 7.

TRI(TETRAHYDROFURFURYL) BORATE
General Information
Description: Yellow, mobile liquid; characteristic odor.
Formula: $[C_4H_7O\cdot CH_2O]_3B$.
Constants: Mol wt: 314.2, bp: 321°C, flash p: 295°F (C.O.C.), d: 1.103 at 26.2°C, vap. d.: 10.8.
Hazard Analysis
Toxicity: See boron compounds and esters.
Fire Hazard: Slight, when exposed to heat or flame; can react with oxidizing materials.
Countermeasures
To Fight Fire: Water, foam, carbon dioxide, dry chemical or carbon tetrachloride (Section 6).
Storage and Handling: Section 7.

TRITHION
General Information
Synonym: o,o-Diethyl-S-p-chlorophenyl thiomethyl phosphorodithioate.
Hazard Analysis
Toxicity: Highly toxic. See closely related parathion.
Disaster Hazard: Dangerous. See phosphates and chlorides.

TRITIUM
General Information
Synonym: Hydrogen-3.
Description: A colorless, radioactive gaseous isotope of hydrogen.
Formula: T_2.
Constant: Mol wt: 6.05.
Hazard Analysis
Toxicity: See hydrogen. For radiological information on tritium see hydrogen.

TRI-m-TOLYLBORON
General Information
Description: Colorless crystals; insoluble in water; soluble in benzene, ether.
Formula: $B(CH_3C_6H_4)_3$.
Constants: Mol wt: 284.2, mp: 175°C, bp: 233°C at 12 mm.
Hazard Analysis
Toxicity: Details unknown. See boron compounds.

TOLYL-m-TOLYLGERMANIUM BROMIDE
General Information
Description: White crystals; soluble in organic solvents.
Formula: $(C_6H_4CH_3)_3GeBr$.
Constants: Mol wt: 425.9, mp: 79°C, bp: 223°C at 1 mm
Hazard Analysis and Countermeasures
See germanium compounds and bromides.

TRI-o-TOLYGERMANIUM BROMIDE
General Information
Description: Colorless oil.
Formula: $BrGe(C_6H_4CH_3)_3$.
Constants: Mol wt: 425.9, bp: 205–210°C at 1 mm.
Hazard Analysis and Countermeasures
See germanium compounds and bromides.

TRI-p-TOLYGERMANIUM BROMIDE
General Information
Description: Colorless crystals, soluble in petroleum ether.
Formula: $(C_6H_4CH_3)_3GeBr$.
Constants: Mol wt: 425.9, mp: 129°C.
Hazard Analysis and Countermeasures
See germanium compounds and bromides.

TRI-m-TOLYLGERMANIUM CHLORIDE
General Information
Description: Small crystals soluble in petroleum ether.
Formula: $(C_6H_4CH_3)_3GeCl$.
Constants: Mol wt: 381.4, mp: 85°C, bp: 221°–224°C at 4 mm.
Hazard Analysis and Countermeasures
See germanium compounds and chlorides.

TRI-o-TOLYLGERMANIUM CHLORIDE
General Information
Description: Colorless oil.
Formula: $(C_6H_4CH_3)_3GeCl$.
Constant: Mol wt: 381.4.
Hazard Analysis and Countermeasures
See germanium compounds and chlorides.

TRI-p-TOLYLGERMANIUM CHLORIDE
General Information
Description: White crystals; soluble in petroleum ether.
Formula: $(C_6H_4CH_3)_3GeCl$.
Constants: Mol wt: 381.4, mp: 121°C.
Hazard Analysis and Countermeasures
See germanium compounds and chlorides.

TRI-o-TOLYLGERMANIUM HYDROXIDE
General Information
Description: White powder.
Formula: $(C_6H_4CH_3)_3GeOH$.
Constants: Mol wt: 363.0, bp: 212°–214°C at 1 mm.
Hazard Analysis
Toxicity: Details unknown. See germanium compounds.

TRI-m-TOLYLGERMANIUM OXIDE
General Information
Description: White crystals; insoluble in water.
Formula: $[(C_6H_4CH_3)_3Ge]_2O$.
Constants: Mol wt: 708.0, mp: 125°C.
Hazard Analysis
Toxicity: Details unknown. See germanium compounds.

TRI-p-TOLYGERMANIUM OXIDE
General Information
Description: White crystals; insoluble in water.

Formula: $[(C_6H_4CH_3)_3Ge]_2O$.
Constants: Mol wt: 708.0, mp: 150°C.
Hazard Analysis
Toxicity: Details unknown. See germanium compounds.

TRITOLYL PHOSPHATE. See tri-o-cresyl phosphate.

TRI-o-TOLYLTIN BROMIDE
General Information
Description: Crystals. Soluble in benzene and ether.
Formula: $(C_7H_7)_3SnBr$.
Constants: Mol wt: 472.0, mp: 99.5°C.
Hazard Analysis and Countermeasures
See tin compounds and bromides.

TRI-p-TOLYLTIN BROMIDE
General Information
Description: Crystals; soluble in benzene and ether.
Formula: $(C_7H_7)_3SnBr$.
Constants: Mol wt: 472.0, mp: 98.5°C.
Hazard Analysis and Countermeasures
See tin compounds and bromides.

TRI-m-TOLYLTIN CHLORIDE
General Information
Description: Crystals.
Formula: $(C_7H_7)_3SnCl$.
Constants: Mol wt: 427.5, mp: 108°C.
Hazard Analysis and Countermeasures
See tin compounds and chlorides.

TRI-o-TOLYLTIN CHLORIDE
General Information
Description: Crystals; soluble in benzene and ether.
Formula: $(C_7H_7)_3SnCl$.
Constants: Mol wt: 427.5, mp: 99.5°C.
Hazard Analysis and Countermeasures
See tin compounds and chlorides.

TRI-p-TOLYLTIN CHLORIDE
General Information
Description: Crystals; slightly soluble in organic solvents.
Formula: $(C_7H_7)_3SnCl$.
Constants: Mol wt: 427.5, mp: 97.5°C.
Hazard Analysis and Countermeasures
See tin compounds and chlorides.

TRI-p-TOLYLTIN FLUORIDE
General Information
Description: Hairlike crystals; soluble in alcohol.
Formula: $(C_7H_7)_3SnF$.
Constants: Mol wt: 411.1, mp: 305°C.
Hazard Analysis and Countermeasures
See fluorides and tin compounds.

TRI-p-TOLYLTIN HYDROXIDE
General Information
Description: Solid.
Formula: $(C_7H_7)_3SnOH$.
Constants: Mol wt: 409.1, mp: 109°C.
Hazard Analysis
Toxicity: Details unknown. See tin compounds.

TRI-o-TOLYLTIN IODIDE
General Information
Description: Crystals; soluble in benzene and ether.
Formula: $(C_7H_7)_3SnI$.
Constants: Mol wt: 519.0, mp: 119.5°C.
Hazard Analysis and Countermeasures
See tin compounds and iodides.

TRI-m-TOLYL-p-TOLYL GERMANIUM
General Information
Description: White powder; soluble in methyl alcohol.
Formula: $(C_6H_4CH_3)_3Ge(C_6H_4CH_3)$.
Constants: Mol wt: 437.1, mp: 100°C.
Hazard Analysis
Toxicity: Details unknown. See germanium compounds.

TRI-p-TOLYL-o-TOLYL GERMANIUM
General Information
Description: White crystals.
Formula: $(C_6H_4CH_3)_4Ge$.
Constants: Mol wt: 437.1, mp: 166°C.
Hazard Analysis
Toxicity: Details unknown. See germanium compounds.

TRITOPINE
General Information
Synonym: Laudanidine.
Description: White, crystalline alkaloid.
Formula: $C_{20}H_{25}NO_4$.
Constants: Mol wt: 343.41, mp: 166°C.
Hazard Analysis
Toxic Hazard Rating:
    Acute Local: U.
    Acute Systemic: Ingestion 3; Inhalation 3.
    Chronic Local: U.
    Chronic Systemic: U.
Disaster Hazard: Dangerous; when heated to decomposition it emits toxic fumes.
Countermeasures
Ventilation Control: Section 2.
Storage and Handling: Section 7.

TRI(TRIPHENYLGERMANIUM) NITRIDE
General Information
Description: Colorless crystals; hydrolyzed by water.
Formula: $[(C_6H_5)_3Ge]_3N$.
Constants: Mol wt: 925.7, mp: 164°C.
Hazard Analysis
Toxicity: Details unknown. See ammonia and germanium compounds.

TRI-p-XYLYLBORON
General Information
Description: Colorless crystals; insoluble in water; soluble in organic solvents.
Formula: $B(CH_3C_6H_3CH_3)_3$.
Constants: Mol wt: 326.3, mp: 147°C, bp: 221°C at 12 mm.
Hazard Analysis
Toxicity: Details unknown. See also boron compounds.

TRI-p-XYLYLTIN BROMIDE
General Information
Description: Crystals; soluble in benzene and ether.
Formula: $(C_8H_9)_3SnBr$.
Constants: Mol wt: 514.1, mp: 151°C.
Hazard Analysis and Countermeasures
See tin compounds and bromides.

TRI-p-XYLYLTIN CHLORIDE
General Information
Description: Crystals; soluble in organic solvents.
Formula: $(C_8H_9)_3SnCl$.
Constants: Mol wt: 469.6, mp: 141.5°C.
Hazard Analysis and Countermeasures
See tin compounds and chlorides.

TOXIC HAZARD RATING CODE (For detailed discussion, see Section 1.)

0 NONE: (a) No harm under any conditions; (b) Harmful only under unusual conditions or overwhelming dosage.

1 SLIGHT: Causes readily reversible changes which disappear after end of exposure.

2 MODERATE: May involve both irreversible and reversible changes; not severe enough to cause death or permanent injury.

3 HIGH: May cause death or permanent injury after very short exposure to small quantities.

U UNKNOWN: No information on humans considered valid by authors.

TRI-m-XYLYLTIN FLUORIDE
General Information
Description: Crystals; soluble in benzene, ether and alcohol.
Formula: $(C_8H_9)_3SnF$.
Constants: Mol wt: 453.2, mp: 205°C.
Hazard Analysis and Countermeasures
See fluorides and tin compounds.

TRI-p-XYLYLTIN FLUORIDE
General Information
Description: Crystals; slightly soluble in benzene.
Formula: $(C_8H_9)_3SnF$.
Constants: Mol wt: 453.2, mp: 247°C.
Hazard Analysis and Countermeasures
See fluorides and tin compounds.

TRI-p-XYLYLTIN IODIDE
General Information
Description: Crystals; soluble in benzene, ether and chloroform.
Formula: $(C_8H_9)_3SnI$.
Constants: Mol wt: 561.1, mp: 159.5°C.
Hazard Analysis and Countermeasures
See tin compounds and iodides.

TROILITE. See ferrous sulfide.

TROJAN COAL POWDERS. See explosives, high.

"TROLUOIL"
General Information
Description: Water-white liquid.
Constants: Bp: 90–96°C, flash p: 25°F, d: 0.741 at 15.5°C.
Hazard Analysis
Toxicity: Details unknown. See gasoline.
Fire Hazard: Dangerous, when exposed to heat or flame.
Explosion Hazard: Unknown.
Disaster Hazard: Dangerous; flammable liquid; can react vigorously with oxidizing materials.
Countermeasures
Storage and Handling: Section 7.
To Fight Fire: Foam, carbon dioxide, dry chemical or carbon tetrachloride (Section 6).

TROPACOCAINE HYDROCHLORIDE
General Information
Description: Crystalline salt.
Formula: $C_{15}H_{19}NO_2 \cdot HCl$.
Constants: Mol wt: 281.78, mp: 271°C.
Hazard Analysis
Toxic Hazard Rating:
　　Acute Local: U.
　　Acute Systemic: Ingestion 3; Inhalation 3.
　　Chronic Local: U.
　　Chronic Systemic: U.
Caution: An alkaloid poison.
Disaster Hazard: Dangerous. See nitrates and chlorides.
Countermeasures
Ventilation Control: Section 2.
Personal Hygiene: Section 3.
Storage and Handling: Section 7.

TROPINE
General Information
Synonym: N-Methyltropoline.
Description: White, crystalline solid.
Formula: $C_8H_{15}NO$.
Constants: Mol wt: 141.21, mp: 63°C, bp: 233°C, d: 1.039 at 76°/4°C.
Hazard Analysis
Toxic Hazard Rating:
　　Acute Local: U.
　　Acute Systemic: Ingestion 3; Inhalation 3.
　　Chronic Local: U.
　　Chronic Systemic: Ingestion 2; Inhalation 2.

Disaster Hazard: Dangerous. See nitrates.
Countermeasures
Ventilation Control: Section 2.
Personnel Protection: Section 3.
Storage and Handling: Section 7.

TROPINE CARBOXYLIC ACID. See ecgonine.

TROPINE PLATINUM HYDROCHLORIDE
General Information
Description: Orange-red, monoclinic tablets.
Formula: $(C_8H_{15}NO \cdot HCl)_2PtCl_4$.
Constants: Mol wt: 692.41, mp: 198–200°C.
Hazard Analysis and Countermeasures
See tropine, platinum compounds and chlorides.

TRUE ARSENIC ACID. See o-arsenic acid.

TRYPARSAMIDE
General Information
Synonym: Sodium phenylglycinamine p-arsonate.
Description: White, crystalline powder.
Formula: $(NH_2COCH_2NHC_6H_4AsOOHONa)_2 \cdot H_2O$.
Constant: Mol wt: 314.12.
Hazard Analysis and Countermeasures
See arsenic compounds.

TRYPTOPHANE
General Information
Synonym: Indole-α-amino propionic acid; 1-α-amino-3-indole propionic acid.
Description: An essential amino acid, occurs in isomeric forms. We consider the L and DL forms, white crystals; DL—slightly soluble in water; L—soluble in water, hot alcohol, alkali hydroxides; insoluble in chloroform.
Formula: $C_6H_4NHCHCHCH_2CHNH_2COOH$.
Constant: Mol wt: 204.
Hazard Analysis
Toxic Hazard Rating: U. A nutrient and/or dietary supplement food additive (Section 10).

TSN. See tetramethylsuccinonitrile.

TSPP. See tetrasodium pyrophosphate.

T-STUFF. See hydrogen peroxide.

TTD. See bis(diethylthiocarbamyl) disulfide.

TUBATOXIN. See rotenone.

β,β-TUBATOXYTHIOCYANODIETHYL ETHER. See β-butoxy-β-thiocyanodiethyl ether.

TURMERIC
Hazard Analysis
Toxic Hazard Rating:
　　Acute Local: Irritant 1; Allergen 1; Ingestion 1; Inhalation 1.
　　Acute Systemic: U.
　　Chronic Local: Allergen 1.
　　Chronic Systemic: U.
Fire Hazard: Slight, when exposed to heat or flame (Section 6).
Countermeasures
Personal Hygiene: Section 3.
Storage and Handling: Section 7.

TUNG NUT MEALS
Hazard Analysis
Toxicity: A weak allergen.
Fire Hazard: Moderate, in the form of dust when exposed to heat or flame; process material and cool thoroughly before storage so as not to overdry; can react with oxidizing materials (Section 6).
Countermeasures
Storage and Handling: Section 7.

TUNG OIL
General Information
Synonym: China wood oil.
Description: Thick, yellowish liquid.
Constants: Mp: 31°C, flash p: 552°F (C.C.), d: 0.94, autoign. temp.: 855 °F.
Hazard Analysis
Toxicity: Details unknown. May act as a weak allergen.
Fire Hazard: Slight, when exposed to heat or flame; can react with oxidizing materials. Avoid contact or leakage with combustibles. Keep cool and well ventilated.
Spontaneous Heating: Moderate.
Countermeasures
To Fight Fire: Carbon dioxide, dry chemical or carbon tetrachloride (Section 6).
Personal Hygiene: Section 3.
Storage and Handling: Section 7.

TUNGSTEN
General Information
Synonym: Wolfram.
Description: Cubic, gray-black crystalline metallic element.
Formula: W.
Constants: At wt: 183.92, mp: 3370°C, bp: 5900°C, d: 19.3 vap. press.: 1 mm at 3990°C.
Hazard Analysis
Toxicity: See tungsten compounds.
Radiation Hazard: Section 5. For permissible levels, see Table 5, p. 150.
     Artificial isotope ^{181}W, half life 130 d. Decays to stable ^{181}Ta by electron capture. Emits gamma rays of 0.01 MeV and X-rays.
     Artificial isotope ^{185}W, half life 74 d. Decays to stable ^{185}Re by emitting beta particles of 0.43 MeV.
     Artificial isotope ^{187}W, half life 24 h. Decays to radioactive ^{187}Re by emitting beta particles of 0.34 (10%), 0.63 (70%), 1.32 (20%) MeV. Also emits gamma rays of 0.22–0.87 MeV.
Fire Hazard: Moderate, in the form of dust when exposed to flame. See also powdered metals.

TUNGSTEN CARBONYL
General Information
Description: Colorless, rhombic crystals.
Formula: $W(CO)_6$.
Constants: Mol wt: 351.98, mp: sublimes 50°C, bp: 175°C, d: 2.65, vap. press.: 1.2 mm at 67°C, vap. d.: 12.1.
Hazard Analysis and Countermeasures
See carbonyls and tungsten compounds.

TUNGSTEN COMPOUNDS
Hazard Analysis
Toxic Hazard Rating:
     Acute Local: U.
     Acute Systemic: Inhalation 1.
     Chronic Local: U.
     Chronic Systemic: Inhalation 1.
Toxicology: Tungsten compounds are considered somewhat more toxic than those of molybdenum. However, industrially, this element does not consititute an important health hazard. Exposure is related chiefly to the dust arising from the crushing and milling of the two chief ores of tungsten, namely, scheelite and wolframite. There is very little published with reference to its toxicity. The feeding of 2, 5, and 10% of diet as tungsten metal over a period of 70 days has been shown to be without marked effect upon the growth of rats, as measured in terms of gain in weight. Ammonium p-tungstate has been found to be much less toxic to rats upon ingestion than either tungstic oxide or sodium tungstate. Recent studies have failed to indicate any serious toxic effect following the inhalation or ingestion of various tungsten compounds, although heavy exposure to the dust or the ingestion of large amounts of the soluble compounds produces a certain rate of mortality in experimental animals.
Countermeasures
Ventilation Control: Section 2.

TUNGSTEN DISULFIDE
General Information
Description: Dark-gray crystals.
Formula: WS_2.
Constants: Mol wt: 248.05, d: 7.5 at 10°C.
Hazard Analysis and Countermeasures
See sulfides and tungsten compounds.

TUNGSTEN HEXAFLUORIDE
General Information
Description: Light-yellow liquid or colorless gas.
Formula: WF_6.
Constants: Mol wt: 297.92, mp: 2.5°C, bp: 19.5°C, d gas: 12.9 g/l; d liquid: 3.44.
Hazard Analysis and Countermeasures
See fluorides and tungsten compounds.

TUNGSTEN OXYCHLORIDE
General Information
Synonym: Tungsten oxytetrachloride.
Description: Red needles.
Formula: $WOCl_4$.
Constants: Mol wt: 341.75, mp: 211°C, bp: 227.5°C.
Hazard Analysis
Toxic Hazard Rating:
     Acute Local: Irritant 2; Ingestion 2; Inhalation 2.
     Acute Systemic: U.
     Chronic Local: U.
     Chronic Systemic: U.
Disaster Hazard: Dangerous. See chlorides.
Countermeasures
Ventilation Control: Section 2.
Personnel Protection: Section 3.
Storage and Handling: Section 7.

TUNGSTEN OXYTETRACHLORIDE. See tungsten oxychloride.

TUUNGSTEN PHOSPHIDE
General Information
Description: Dark-gray prisms.
Formula: W_2P.
Constants: Mol wt: 398.82, d: 5.21.
Hazard Analysis and Countermeasures
See phosphides and tungsten compounds.

TUNGSTEN TRISULFIDE
General Information
Description: Chocolate-brown powder.
Formula: WS_3.
Constant: Mol wt: 280.12.
Hazard Analysis and Countermeasures
See sulfides and tungsten compounds.

TURKEY RED OIL
General Information
Synonym: Sulfonated castor oil.

TOXIC HAZARD RATING CODE (For detailed discussion, see Section 1.)

0 NONE: (a) No harm under any conditions; (b) Harmful only under unusual conditions or overwhelming dosage.

1 SLIGHT: Causes readily reversible changes which disappear after end of exposure.

2 MODERATE: May involve both irreversible and reversible changes; not severe enough to cause death or permanent injury.

3 HIGH: May cause death or permanent injury after very short exposure to small quantities.

U UNKNOWN: No information on humans considered valid by authors.

Description: A reddish, viscid liquid; characteristic odor.
Constants: Flash p: 476 °F (C.C.), d: 0.95, autoign. temp.: 833° F.
Hazard Analysis
Toxicity: Unknown.
Fire Hazard: Slight, when exposed to heat or flame.
Spontaneous Heating: No.
Countermeasures
To Fight Fire: Alcohol foam, foam, carbon dioxide, dry chemical or carbon tetrachloride (Section 6).
Storage and Handling: Section 7.

TURPENTINE (GUM)
General Information
Synonym: Gum thus.
Description: Yellowish, opaque, sticky masses.
Hazard Analysis
Toxicity: A mild allergen. See also turpentine oil.
Fire Hazard: Moderate.
Countermeasures
Storage and Handling: Section 7.

TURPENTINE OIL
General Information
Synonym: Spirit of turpentine.
Description: Colorless liquid; characteristic odor.
Formula: Principally $C_{10}H_{16}$.
Constants: Mol wt: 136, bp: 154–170°C, lel: 0.8%, flash p: 95°F (C.C.), d: 0.854–0.868 at 25°/25°C, autoign. temp.: 488°F, vap. d.: 4.84.
Hazard Analysis
Toxic Hazard Rating:
  Acute Local: Irritant 2; Allergen 1.
  Acute Systemic: Ingestion 3; Inhalation 2; Skin Absorption 2.
  Chronic Local: Irritant 2; Allergen 2.
  Chronic Systemic: Ingestion 1; Inhalation 1; Skin Absorption 1.
Toxicology: Can cause serious irritation of kidneys. A common air contaminant. (Section 4).
TLV: ACGIH (recommended); 100 parts per million in air; 556 milligrams per cubic meter of air.
Fire Hazard: Moderate, when exposed to heat or flame. Avoid impregnation of leakage with combustibles. Keep cool and ventilated.
Spontaneous Heating: Yes.
Explosion Hazard: Moderate, in the form of vapor when exposed to flame.
Disaster Hazard: Moderately dangerous; when heated it emits acrid fumes; can react with oxidizing materials.
Countermeasures
Ventilation Control: Section 3.

To Fight Fire: Foam, carbon dioxide, dry chemical or carbon tetrachloride (Section 6).
Personnel Protection: Section 3.
Storage and Handling: Section 7.
Shipping Regulations: Section 11.
  Coast Guard Classification: Combustible liquid.

TURPENTINE SUBSTITUTES. See mineral spirits #10.
Shipping Regulations: Section 11.
  ICC: Flammable liquid, red label, 10 gallons.
  IATA: Flammable liquid, red label, 1 liter (passenger), 40 liters (cargo).

TYLOSIN
General Information
Description: An antibiotic; crystals.
Constant: Mp: 128–130°C.
Hazard Analysis
Toxic Hazard Rating: U. A food additive permitted in the feed and drinking water of animals and/or for the treatment of food producing animals; also permitted in food for human consumption (Section 10).

TYPE-CLEANING COMPOUNDS, LIQUID
Hazard Analysis
Toxicity: Variable; may contain benzene, carbon tetrachloride or other toxic materials.

TYPEWRITING RIBBONS
Hazard Analysis
Toxic Hazard Rating:
  Acute Local: Allergen 1.
  Acute Systemic: 0.
  Chronic Local: Allergen 1.
  Chronic Systemic: U.
Personal Hygiene: Section 3.

TYRAMINE. See ergot.

TYRENE. See 2,4,6-trichloroanisole.

TYROSINE
General Information
Synonyms: β-p-Hydroxyphenyl alanine; α-amino-β-p-hydroxy phenylpropionic acid.
Description: A non-essential amino acid. White crystals; soluble in water; slightly soluble in alcohol; insoluble in ether; optically active.
Formula: $C_6H_4OHCH_2CHNH_2COOH$.
Constant: Mol wt: 181.
Hazard Analysis
Toxic Hazard Rating: U. A nutrient and/or dietary supplement food additive. (Section 10).

U

UINTAITE. See gilsonite.

ULTRASENE
General Information
Synonym: Kerosene, deodorized.
Description: Insoluble in water.
Constant: Flash p: 175° F.
Hazard Analysis
Toxic Hazard Rating: See kerosene.
Fire Hazard: Moderate, when exposed to heat or flame.
Countermeasures
Storage and Handling: Section 7.
To Fight Fire: Water, foam, mist. Section 6.

ULTRAVIOLET RADIATION
Hazard Analysis
Toxic Hazard Rating:
Acute Local: Irritant 2.
Acute Systemic: Over exposure may result in systemic effects.
Chronic Local: Irritant 2.
Chronic Systemic: 0.
Radiation Hazard: Excessive ultraviolet radiation can cause acute inflammation of the eyes as well as burns of the skin. There is some evidence that exposure to ultraviolet radiation over a period of many years may cause cancer of the skin (Section 5).

UNDECANAL
General Information
Synonyms: Hendecanal.
Description: Colorless liquid.
Formula: $CH_3(CH_2)_9CHO$.
Constants: Mol wt: 170.29, mp: $-4°C$, bp: 117°C at 18 mm, flash p: 235° F (C.O.C.), d: 0.830 at 20°/4°C, vap. press.: 0.04 mm at 20°C, vap. d.: 5.94.
Hazard Analysis
Toxic Hazard Rating:
Acute Local: Irritant 2; Ingestion 2; Inhalation 2.
Acute Systemic: U.
Chronic Local: U.
Chronic Systemic: U.
Fire Hazard: Slight, when exposed to heat or flame; can react with oxidizing materials.
Countermeasures
To Fight Fire: Foam, carbon dioxide, dry chemical or carbon tetrachloride (Section 6).
Ventilation Control: Section 2.
Personnel Protection: Section 3.
Storage and Handling: Section 7.

UNDECANE. See hendecane.

UNDECANOL-2
General Information
Synonym: Hendecanol-2.
Description: Colorless liquid; insoluble in water; soluble in alcohol and ether.

Formula: $C_4H_9CH(C_2H_5)C_2H_4CH(OH)CH_3$.
Constants: Mol wt: 172, d: 0.8363 at 20°C, mp: 12°C, bp: 228–229°C, flash p: 235° F (O.C.).
Hazard Analysis
Toxic Hazard Rating: U.
Fire Hazard: Slight, when exposed to heat or flame.
Countermeasures
Storage and Handling: Section 7.
To Fight Fire: Water, foam. Section 6.

UNDECYLENIC ACID
General Information
Description: Bright, clear, mobile liquid.
Formula: $H_2CCH(CH_2)_8COOH$.
Constants: Mol wt: 184.27, mp: 24.5°C, bp: 160°C at 10 mm, flash p: 295° F (C.O.C.), d: 0.910 at 25°/25°C.
Hazard Analysis
Toxic Hazard Rating:
Acute Local: Ingestion 1.
Acute Systemic: Ingestion 1.
Chronic Local: U.
Chronic Systemic: U.
Toxicology: Ingestion may cause nausea, vomiting and urticaria.
Fire Hazard: Moderate, when exposed to heat or flame; can react with oxidizing materials.
Countermeasures
To Fight Fire: Foam, carbon dioxide, dry chemical or carbon tetrachloride (Section 6).
Storage and Handling: Section 7.

UNSLAKED LIME. See calcium oxide.

URANIUM
General Information
Description: Cubic, silver-white or black crystals.
Formula: U.
Constants: At wt: 238.07, mp: approx. 1133°C, bp: ignites, d: 18.7.
Hazard Analysis
Toxicity: Highly toxic. See discussion under radiation hazard below. A common air contaminant. (Section 4).
TLV: ACGIH (recommended); 0.05, milligrams per cubic meter of air (for soluble compounds); 0.25 milligrams per cubic meter of air (for insoluble compounds).
Radiation Hazard: Section 5. For permissible levels, see Table 5, p. 150.
Natural uranium consists of ^{238}U and ^{234}U in radioactive equilibrium plus 0.7% of ^{235}U. Enriched uranium has increased percentages of the lighter isotopes, ^{234}U and ^{235}U. Its specific activity, and consequently the radiation hazard, are correspondingly increased.
Artificial isotope ^{230}U, half life 21 d. Decays to radioactive ^{226}Th by emitting alpha particles of 5.8–5.9 MeV.

Natural isotope ^{232}U, half life 74 y. Decays to radioactive ^{228}Th by emitting alpha particles of 5.3 MeV. Also emits gamma rays of 0.06 MeV.

Artificial isotope ^{233}U (Neptunium Series), half life 1.6×10^5 y. Decays to radioactive ^{229}Th by emitting alpha particles of 4.8 MeV.

Natural isotope ^{234}U (U-II, Uranium Series), half life 2.5×10^5 y. Decays to radioactive ^{230}Th by emitting alpha particles of 4.7–4.8 MeV.

Natural (0.72%) ^{235}isotope U (Actino-uranium, Actinium Series), half life 7×10^8 y. Decays to radioactive ^{231}Th by emitting alpha particles of 4.3–4.6 MeV.

Artificial isotope ^{236}U, half life 2.4×10^7 y. Decays to radioactive ^{232}Th by emitting alpha particles of 4.5 MeV.

Natural isotope ^{238}U (U-I, Uranium Series), half life 4.5×10^9 y. Decays to radioactive ^{234}Th by emitting alpha particles of 4.2 MeV.

Artificial isotope ^{240}U, half life 14 h. Decays to radioactive ^{240}Np by emitting beta particles of 0.36 MeV. ^{240}Np in turn decays to ^{240}Pu with a half life of 7 min.

The permissible levels for soluble compounds and the level for insoluble compounds in air are based on chemical toxicity, while the permissible body level for insoluble compounds is based on radiotoxicity.

The high chemical toxicity of uranium and its salts is largely shown in kidney damage, and acute necrotic arterial lesions. The rapid passage of soluble uranium compounds through the body tends to allow relatively large amounts to be taken in. The highly toxic effect of insoluble compounds is largely due to lung irradiation by inhaled particles. This material is transferred from the lungs of animals quite slowly.

Fire Hazard: Dangerous, in the form of a solid or dust when exposed to heat or flame (Section 6).

Explosion Hazard: Moderate, in the form of dust when exposed to flame.

Disaster Hazard: For further information refer to Atomic Energy Commission.

Countermeasures
Ventilation Control: Section 2.
Personal Hygiene: Section 3.
First Aid: Section 1.
Storage and Handling: Section 7.
Shipping Regulations: Uranium, normal or depleted in solid metal form (not borings, chips, or pieces).
ICC: Poison D, radioactive materials, red label; see §73.393(L).
IATA: Radioactive material; radioactive (red) label, 2000 millicuaries (passenger and cargo).

URANIUM AMMONIUM FLUORIDE
General Information
Synonym: Ammonium uranium fluoride.
Description: Greenish-yellow crystalline powder.
Formula: $UO_2F_2 \cdot 3NH_4F$.
Constant: Mol wt: 419.2.
Hazard Analysis and Countermeasures
See uranium and fluorides.

URANIUM AMMONIUM PENTAFLUORIDE
General Information
Description: Tetragonal crystals.
Formula: $(NH_4)_3UO_2F_5$.
Constants: Mol wt: 419.2, mp: decomposes.
Hazard Analysis and Countermeasures
See uranium and fluorides.

URANIUM BARIUM OXIDE
General Information
Synonym: Barium diuranate.
Description: Yellow or orange powder.
Formula: BaU_2O_7.

Constant: Mol wt: 725.50.
Hazard Analysis and Countermeasures
See uranium and barium compounds.

URANIUM BORIDE
General Information
Synonym: Uranium diboride.
Description: Hexagonal crystals.
Formula: UB_2.
Constants: Mol wt: 248.9, mp: 2365° C, d: 12.70.
Hazard Analysis
Toxicity: See uranium and boron compounds.
Radiation Hazard: See uranium.
Fire Hazard: Moderate; can react with moisture or acids to evolve boron hydride. See also boron hydrides.
Explosion Hazard: Moderate; can react with moisture or acids to evolve boron hydrides. See also boron hydrides.
Disaster Hazard: Moderately dangerous; it will react with water or steam to produce toxic and flammable vapors; it can react with oxidizing materials; and on contact with acid or acid fumes, it can emit toxic fumes.
Countermeasures
Storage and Handling: Section 7.

URANIUM CARBIDE
General Information
Description: Gray crystals.
Formula: UC_2.
Constants: Mol wt: 262.09, mp: 2260° C, bp: 4100° C, d: 11.28 at 18° C.
Hazard Analysis and Countermeasures
See uranium and carbides.

URANIUM DIBORIDE. See uranium boride.

URANIUM DIOXIDE
General Information
Description: Rhombic or cubic, brown-black crystals.
Formula: UO_2.
Constants: Mol wt: 270.07, mp: 2176° C, d: 10.9.
Hazard Analysis and Countermeasures
See uranium.

URANIUM DISULFIDE
General Information
Description: Tetragonal, gray-black crystals.
Formula: US_2.
Constants: Mol wt: 302.20, mp: > 1100° C, bp: oxidizes.
Hazard Analysis and Countermeasures
See uranium and sulfides.

URANIUM HEXAFLUORIDE
General Information
Description: Monoclinic, colorless to pale yellow, deliquescent crystals.
Formula: UF_6.
Constants: Mol wt: 352.07, mp: 69.2° C at 2 atm, bp: 56.2° C at 764.6 mm, d: 4.68 at 20.7° C, vap. press.: 100 mm at 18.2° C.
Hazard Analysis and Countermeasures
See uranium and fluorides.

URANIUM NITRIDE
General Information
Description: Brown-black crystals.
Formula: U_3N_4.
Constants: Mol wt: 770.24, d: 10.09.
Hazard Analysis and Countermeasures
See uranium and nitrides.

URANIUM OXYACETATE
General Information
Synonym: Uranyl acetate.
Description: Rhombic, yellow crystals.
Formula: $UO_2(C_2H_3O_2)_2 \cdot 2H_2O$.
Constants: Mol wt: 424.19, mp: $-2H_2O$ at 110° C, bp: decomposes 275° C, d: 2.893 at 15° C.

Hazard Analysis and Countermeasures
See uranium.

URANIUM OXYAMMONIUM CARBONATE
General Information
Synonym: Uranyl ammonium carbonate.
Description: Monoclinic, yellow crystals.
Formula: $2(NH_4)_2CO_3 \cdot UO_2CO_3 \cdot 2H_2O$.
Constants: Mol wt: 558.29, mp: decomposes at $100°C$, d: 2.773.
Hazard Analysis and Countermeasures
See uranium.

URANIUM OXYAMMONIUM CHLORIDE
General Information
Synonym: Uranyl ammonium chloride.
Description: Greenish-yellow, deliquescent crystals.
Formula: $UO_2Cl_2 \cdot 2NH_4Cl \cdot 2H_2O$.
Constant: Mol wt: 484.0.
Hazard Analysis and Countermeasures
See uranium and chlorides.

URANIUM OXYBENZOATE
General Information
Synonym: Uranyl benzoate.
Description: Yellow powder.
Formula: $UO_2(C_7H_5O_2)_2$.
Constant: Mol wt: 512.29.
Hazard Analysis and Countermeasures
See uranium.

URANIUM OXYBROMIDE
General Information
Synonym: Uranyl bromide.
Description: Green-yellow, hygroscopic needles.
Formula: UO_2Br_2.
Constant: Mol wt: 429.90.
Hazard Analysis and Countermeasures
See uranium and bromides.

URANIUM OXYCHLORIDE
General Information
Synonym: Uranyl chloride.
Description: Yellow, deliquescent crystals.
Formula: UO_2Cl_2.
Constants: Mol wt: 340.98, mp: < red heat.
Hazard Analysis and Countermeasures
See uranium and chlorides.

URANIUM OXYFORMATE
General Information
Synonym: Uranyl formate.
Description: Yellow crystals.
Formula: $UO_2(CHO_2)_2 \cdot H_2O$.
Constants: Mol wt: 378.12, mp: $-H_2O$ at $110°C$, d: 3.695 at $19°C$.
Hazard Analysis and Countermeasures
See uranium.

URANIUM OXYIODATE
General Information
Synonym: Uranyl iodate.
Description: Rhombic, yellow crystals.
Formula: $UO_2(IO_3)_2$.
Constants: Mol wt: 619.91, mp: decomposes $250°C$, d: 5.2.
Hazard Analysis and Countermeasures
See uranium and iodates.

URANIUM OXYIODIDE
General Information
Synonym: Uranyl iodide.
Description: Red, deliquescent crystals.
Formula: UO_2I_2.
Constants: Mol wt: 523.91, mp: decomposes in air.
Hazard Analysis and Countermeasures
See uranium and iodides.

URANIUM OXYMONOHYDROGEN PHOSPHATE
General Information
Synonym: Uranyl monohydrogen phosphate.
Description: Tetragonal, yellow plates.
Formula: $UO_2HPO_4 \cdot 4H_2O$.
Constant: Mol wt: 438.12.
Hazard Analysis and Countermeasures
See uranium and phosphates.

URANIUM OXYNITRATE
General Information
Synonym: Uranyl nitrate.
Description: Rhombic, yellow, deliquescent crystals.
Formula: $UO_2(NO_3)_2 \cdot 6H_2O$.
Constants: Mol wt: 502.18, mp: $60.2°C$; decomp. $100°C$, d: 2.807 at $13°C$.
Hazard Analysis and Countermeasures
See uranium and nitrates.
Shipping Regulations: Section 11.
   IATA: Oxidizing material, yellow label, 12 kilograms (passenger), 45 kilograms (cargo).

URANIUM OXYOXALATE
General Information
Synonym: Uranyl oxalate.
Description: Yellow crystals.
Formula: $UO_2C_2O_4 \cdot 3H_2O$.
Constants: Mol wt: 412.14, mp: $-H_2O$ at $110°C$.
Hazard Analysis and Countermeasures
See uranium and oxalates.

URANIUM OXYPERCHLORATE
General Information
Synonym: Uranyl perchlorate.
Description: Yellow, deliquescent crystals.
Formula: $UO_2(ClO_4)_2 \cdot 6H_2O$.
Constants: Mol wt: 545.07, mp: $90°C$; decomp. at $110°C$.
Hazard Analysis and Countermeasures
See uranium and perchlorates.

URANIUM OXYSULFATE
General Information
Synonym: Uranyl sulfate.
Description: Yellow-green crystals.
Formula: $UO_2SO_4 \cdot 3H_2O$.
Constants: Mol wt: 420.18, mp: decomposes at $100°C$, d: 3.28 at $16.5°C$.
Hazard Analysis and Countermeasures
See uranium and sulfates.

URANIUM OXYSULFIDE
General Information
Synonym: Uranyl sulfide.
Description: Brown-black, tetragonal crystals.
Formula: UO_2S.
Constants: Mol wt: 302.14, mp: decomposes $40–50°C$.
Hazard Analysis and Countermeasures
See uranium and sulfides.

TOXIC HAZARD RATING CODE (For detailed discussion, see Section 1.)

0 NONE: (a) No harm under any conditions; (b) Harmful only under unusual conditions or overwhelming dosage.

1 SLIGHT: Causes readily reversible changes which disappear after end of exposure.

2 MODERATE: May involve both irreversible and reversible changes; not severe enough to cause death or permanent injury.

3 HIGH: May cause death or permanent injury after very short exposure to small quantities.

U UNKNOWN: No information on humans considered valid by authors.

URANIUM OXYSULFITE
General Information
Synonym: Uranyl sulfite.
Description: Pale-green crystals.
Formula: $UO_2SO_3 \cdot 4H_2O$.
Constant: Mol wt: 422.20.
Hazard Analysis and Countermeasures
See uranium and sulfites.

URANIUM PENTACHLORIDE
General Information
Description: Dark green-gray needles; red by transparent light, deliquescent.
Formula: UCl_5.
Constants: Mol wt: 415.36, mp: decomposes at 120°C.
Hazard Analysis and Countermeasures
See uranium and chlorides.

URANIUM PEROXIDE
General Information
Description: Pale yellow, hygroscopic crystals.
Formula: $UO_4 \cdot 2H_2O$.
Constants: Mol wt: 338.10, mp: decomposes at 115°C.
Hazard Analysis and Countermeasures
See uranium and peroxides, inorganic.

URANIUM PHOSPHATE
General Information
Synonyms: Mono-uranyl o-phosphate; uranyl phosphate.
Description: Tetragonal, yellow plates.
Formula: $UO_2HPO_4 \cdot 4H_2O$.
Constant: Mol wt: 438.16.
Hazard Analysis and Countermeasures
See uranium and phosphates.

URANIUM POTASSIUM NITRATE
General Information
Synonyms: Potassium uranium nitrate; uranyl potassium nitrate.
Description: Yellow, crystalline powder.
Formula: $(KNO_3)_2 \cdot UO_2(NO_3)_2$.
Constant: Mol wt: 394.086.
Hazard Analysis and Countermeasures
See uranium and nitrates.

URANIUM POTASSIUM SULFATE
General Information
Synonym: Potassium uranyl sulfate.
Description: Monoclinic, yellow crystals.
Formula: $K_2SO_4 \cdot UO_2SO_4 \cdot 2H_2O$.
Constants: Mol wt: 576.41, mp: $-2H_2O$ at 120°C, d: 3.363 at 19.1°C.
Hazard Analysis and Countermeasures
See uranium and sulfates.

URANIUM SESQUISULFIDE
General Information
Description: Gray-black needles.
Formula: U_2S_3.
Constants: Mol wt: 572.34, mp: ignites.
Hazard Analysis and Countermeasures
See uranium and sulfides.

URANIUM SODIUM OXYACETATE
General Information
Synonym: Sodium uranyl acetate.
Description: Yellow crystals.
Formula: $NaUO_2(C_2H_3O_2)_3$.
Constants: Mol wt: 470.20, d: 2.56.
Hazard Analysis and Countermeasures
See uranium.

URANIUM STRONTIUM OXIDE
General Information
Synonym: Strontium diuranate.
Description: Yellow powder.
Formula: SrU_2O_7.
Constant: Mol wt: 675.77.
Hazard Analysis and Countermeasures
See uranium.

URANIUM TETRABROMIDE
General Information
Description: Brown, deliquescent leaflets.
Formula: UBr_4.
Constants: Mol wt: 557.73, bp: volatilizes, d: 4.84 at 21°/4°C.
Hazard Analysis and Countermeasures
See uranium and bromides.

URANIUM TETRACHLORIDE
General Information
Description: Cubic, dark green-gray deliquescent crystals.
Formula: UCl_4.
Constants: Mol wt: 379.90, mp: sublimes, bp: 618°C, d: 4.725 at 25°/4°C.
Hazard Analysis and Countermeasures
See uranium and chlorides.

URANIUM TETRAFLUORIDE
General Information
Description: Green, amorphous powder.
Formula: UF_4.
Constants: Mol wt: 314.07, mp: approx. 1000°C.
Hazard Analysis and Countermeasures
See uranium and fluorides.

URANIUM TETRAIODIDE
General Information
Description: Black needles.
Formula: UI_4.
Constants: Mol wt: 745.75, mp: 500°C, d: 5.6 at 15°C.
Hazard Analysis and Countermeasures
See uranium and iodides.

URANIUM TRIBROMIDE
General Information
Description: Dark brown, hygroscopic needles.
Formula: UBr_3.
Constants: Mol wt: 477.82, bp: volatilizes.
Hazard Analysis and Countermeasures
See uranium and bromides.

URANIUM TRICHLORIDE
General Information
Description: Dark red, hygroscopic needles.
Formula: UCl_3.
Constants: Mol wt: 344.44, d: 5.44 at 25°/4°C.
Hazard Analysis and Countermeasures
See uranium and chlorides.

URANIUM TRIHYDRIDE
General Information
Description: Black metallic crystals; reacts with water.
Formula: UH_3.
Constants: Mol wt: 241.1, d: 10.5, mp: decomp. > 300°C.
Hazard Analysis and Countermeasures
See uranium and hydrides.

URANIUM TRIOXIDE
General Information
Synonym: Uranyl oxide.
Description: Yellow-red powder.
Formula: UO_3.
Constants: Mol wt: 286.07, mp: decomposes, d: 7.29.
Hazard Analysis and Countermeasures
See uranium.

URANIUM X_2. See protactinium.

URANOUS SULFATE
General Information
Description: Rhombic green crystals.

Formula: $U(SO_4)_2 \cdot 4H_2O$.
Constants: Mol wt: 502.27, mp: $-4H_2O$ at $300°C$.
Hazard Analysis and Countermeasures
See uranium and sulfates.

URANYL ACETATE. See uranium oxyacetate.

URANYL AMMONIUM CARBONATE. See uranium oxyammonium carbonate.

URANYL AMMONIUM CHLORIDE. See uranium oxyammonium chloride.

URANYL BENZOATE. See uranium oxybenzoate.

URANYL BROMIDE. See uranium oxybromide.

URANYL CHLORIDE. See uranium oxychloride.

URANYL FORMATE. See uranium oxyformate.

URANYL IODATE. See uranium oxyiodate.

URANYL IODIDE. See uranium oxyiodide.

URANYL MONOHYDROGEN PHOSPHATE. See uranium oxymonohydrogen phosphate.

URANYL NITRATE. See uranium oxynitrate.

URANYL OXALATE. See uranium oxyoxalate.

URANYL OXIDE. See uranium trioxide.

URANYL PERCHLORATE. See uranium oxyperchlorate.

URANYL PHOSPHATE. See uranium phosphate.

URANYL POTASSIUM NITRATE. See uranium potassium nitrate.

URANYL SULFATE. See uranium oxysulfate.

URANYL SULFIDE. See uranium oxysulfide.

URANYL SULFITE. See uranium oxysulfite.

UREA
General Information
Synonym: Carbamide.
Description: White crystals or powder.
Formula: $(NH_2)_2CO$.
Constants: Mol wt: 60.1, mp: $132.7°C$, bp: decomposes, d: 1.335.
Hazard Analysis
Toxicity: No importance as an industrial hazard. It is a substance which migrates to food from packaging materials (Section 10).
Disaster Hazard: Slightly dangerous when heated to decomposition.
Countermeasures
Storage and Handling: Section 7.

UREA CRYSTAL. See carbamide.

UREA HYDROGEN PEROXIDE. See urea peroxide.

UREA NITRATE, DRY OR WET WITH LESS THAN 10% WATER
Shipping Regulations: Section 11.
IATA: Explosive, not acceptable (passenger and cargo).

UREA NITRATE, WET, NOT EXCEEDING 16 OZ. WITH NOT LESS THAN 10% WATER

General Information
Formula: $CO(NH_2)_2HNO_3$ (varying with moisture content).
Constant: Mol wt: 123.08.
Hazard Analysis
Toxic Hazard Rating:
  Acute Local: Irritant 1.
  Acute Systemic: U.
  Chronic Local: Irritant 1.
  Chronic Systemic: U.
Fire Hazard: See nitrates.
Explosion Hazard: See nitrates.
Disaster Hazard: See nitrates.
Countermeasures
Ventilation Control: Section 2.
Storage and Handling: Section 7.
Shipping Regulations: Section 11.
  ICC Classification: See Section 11, §73.192.
  Coast Guard Classification: Inflammable solid.

UREA NITRATE, WET, WITH NOT LESS THAN 10% WATER IN EXCESS OF 16 OZ., BUT NOT EXCEEDING 25 LBS. See urea nitrate, wet.
Shipping Regulations: Section 11.
  ICC: Flammable solid; yellow label, 25 pounds.
  Coast Guard Classification: Inflammable solid; yellow label.
  IATA: Flammable solid, yellow label, ½ kilogram (passenger), 12 kilograms (cargo).

UREA PEROXIDE
General Information
Synonym: Urea hydrogen peroxide.
Description: White crystals.
Formula: $CO(NH_2)_2 \cdot H_2O_2$.
Constants: Mol wt: 94.08, mp: $75–85°C$ (decomp.).
Hazard Analysis
Toxic Hazard Rating:
  Acute Local: Irritant 2.
  Acute Systemic: U.
  Chronic Local: U.
  Chronic Systemic: U.
Fire Hazard: See peroxides, organic.
Disaster Hazard: See peroxides, organic.
Countermeasures
Shipping Regulations: Section 11.
  ICC: Oxidizing material; yellow label, 25 pounds.
  Coast Guard Classification: Oxidizing material; yellow label.
  IATA: Oxidizing material, yellow label, 1 kilogram (passenger), 12 kilograms (cargo).

α-UREIDO-ISOBUTYRIC ACID LACTAM. See dimethylhydantoin.

URETHANE. See ethyl carbamate.

URIC ACID
General Information
Synonyms: Lithic acid; uric oxide.
Description: White crystals.
Formula: $CO(NH)_2COC_2CO(NH)_2$.
Constants: Mol wt: 168.11, mp: decomposes to hydrogen cyanide, d: 1.855–1.893.
Hazard Analysis
Toxicity: It can evolve hydrogen cyanide when heated. See also cyanides.

TOXIC HAZARD RATING CODE   (For detailed discussion, see Section 1.)

0 NONE: (a) No harm under any conditions; (b) Harmful only under unusual conditions or overwhelming dosage.

1 SLIGHT: Causes readily reversible changes which disappear after end of exposure.

2 MODERATE: May involve both irreversible and reversible changes; not severe enough to cause death or permanent injury.

3 HIGH: May cause death or permanent injury after very short exposure to small quantities.

U UNKNOWN: No information on humans considered valid by authors.

Disaster Hazard: Dangerous; when heated to decomposition it emits highly toxic fumes of cyanides.
Countermeasures
Storage and Handling: Section 7.

URIC OXIDE. See uric acid.

UROTROPINE. See hexamethylenetetramine.

UROX
General Information
Synonym: 3-(p-chlorophenyl)-1, 1-dimethylurea trichloroacetate.
Formula: $CCl_3COOC_6H_4ClNCON(CH_3)_2$.
Constant: Mol wt: 360.1.
Hazard Analysis
Toxicity: Details unknown. Limited animal experiments on closely related compounds suggest low toxicity.
Disaster Hazard: Dangerous. When heated to decomposition it emits highly toxic fumes.

UROXIN. See alloxantin.

URSOL. See p-phenylenediamine.

URSOL P. See p-aminophenol.

URUSHIOL
General Information
Description: Pale yellow liquid; soluble in alcohol and ether. Main constituent of the irritant oil of poison ivy.
Constituents: A mixture of several derivatives of catechol.
Hazard Analysis
Toxic Hazard Rating:
    Acute Local: Allergen 3.
    Acute Systemic: U.
    Chronic Local: Allergen 3.
    Chronic Systemic: U.
Toxicology: The active irritant of poison ivy. See also 3-pentadecylcatechol.

UX₂. See protactinium.

V

VALERAL. See m-valeraldehyde.

m-VALERALDEHYDE
General Information
Synonyms: Valeral; pentanal; amylaldehyde; valeric aldehyde.
Description: Colorless liquid; slightly soluble in water; soluble in alcohol and ether.
Formula: $CH_3(CH_2)_3CHO$.
Constants: Mol wt: 86.13, d: 0.8095 at 20/4°C, bp: 102–103°C, fp: −91°C, flash p: 54°F (O.C.), vap. d.: 3.0.
Hazard Analysis
Toxic Hazard Rating:
  Acute Local: Irritant 1; Ingestion 1; Inhalation 1.
  Acute Systemic: Ingestion 1; Inhalation 1.
  Chronic Local: U.
  Chronic Systemic: U.
Toxicity: Mildly irritant and narcotic.
Fire Hazard: Dangerous, when exposed to heat or flame.
Countermeasures
Storage and Handling: Section 7.
To Fight Fire: Foam, fog, alcohol foam.

VALERIC ACID
General Information
Synonym: Pentanoic acid.
Description: Colorless liquid with unpleasant odor. Somewhat water soluble.
Formula: $CH_3CH_2CH_2CH_2COOH$.
Constants: Mol wt: 102.1, d: 0.939 at 20/4°C, mp: −34.5°C, bp: 187°C, flash p: 205°F (O.C.).
Hazard Analysis
Toxicity: Details unknown. Animal experiments suggest moderate toxicity and high irritation. See butyric acid.

γ-VALEROLACTONE
General Information
Description: Colorless, mobile liquid.
Formula: $(C_5H_8O_2)$.
Constants: Mol wt: 100.06, mp: −31°C, bp: 205–206.5°C, flash p: 205°F, d: 1.0518 at 25°/25°C, vap. d.: 3.45.
Hazard Analysis
Toxicity: Details unknown; animal experiments suggest low toxicity and no irritation. See also lactones.
Fire Hazard: Moderate, when exposed to heat or flame; can react with oxidizing materials.
Countermeasures
To Fight Fire: Water, foam, carbon dioxide, dry chemical or carbon tetrachloride (Section 6).
Storage and Handling: Section 7.

VALERYL DIETHYLAMIDE. See valyl.

VALINE (L AND DL FORMS).
General Information
Synonym: α-amino isovaleric acid.
Description: White crystalline solid; soluble in water; very slightly soluble in alcohol; insoluble in ether; an essential amino acid.
Formula: $(CH_3)_2CHCH(NH_2)COOH$.
Constant: Mol wt: 117, mp (DL): 298°C with decomposition, mp (L): 293°C with decomposition.
Hazard Analysis
Toxic Hazard Rating: U. A nutrient and/or dietary supplement food additive (Section 10).

VALONE
General Information
Synonym: 2-Isovaleryl-1,3-indandione.
Description: Crystals.
Formula: $C_{14}H_{14}O$.
Constants: Mol wt: 230.3, mp: 68°C.
Hazard Analysis
Toxicity: Details unknown. In animal experiments it has displayed fairly high toxicity. Acts as an anticoagulant leading to hemorrhages. See also coumarin.

VALYL
General Information
Synonyms: Diethyl valeramide; valeryl diethylamide.
Description: Colorless liquid; burning taste.
Formula: $C_4H_9CON(C_2H_5)_2$.
Constants: Mol wt: 157.25, bp: 210°C.
Hazard Analysis
Toxic Hazard Rating:
  Acute Local: U.
  Acute Systemic: Ingestion 1.
  Chronic Local: U.
  Chronic Systemic: Ingestion 1.
Caution: May have a depressant action.
Fire Hazard: Slightly dangerous; when heated, it will burn (Section 6).
Countermeasures
Personal Hygiene: Section 3.
Storage and Handling: Section 7.

VANADIC SULFATE. See vanodyl sulfate.

VANADIUM
General Information
Description: Light-gray crystalline metallic element.
Formula: V.
Constants: At wt: 50.95, mp: 1720 ± 20°C, bp: 3000°C, d: 5.866 at 15°C.
Hazard Analysis
Toxicity: See vanadium compounds.
Radiation Hazard: Section 5. For permissible levels, see Table 5, p. 150.
  Artificial isotope ^{48}V, half life 16 d. Decays to stable ^{48}Ti by electron capture (44%) and by emitting positrons of 0.70 MeV. Also emits gamma rays of 0.99, 1.31 MeV and X-rays.

TOXIC HAZARD RATING CODE (For detailed discussion, see Section 1.)

0 NONE: (a) No harm under any conditions; (b) Harmful only under unusual conditions or overwhelming dosage.

1 SLIGHT: Causes readily reversible changes which disappear after end of exposure.

2 MODERATE: May involve both irreversible and reversible changes; not severe enough to cause death or permanent injury.

3 HIGH: May cause death or permanent injury after very short exposure to small quantities.

U UNKNOWN: No information on humans considered valid by authors.

Fire Hazard: Moderate in the form of dust when exposed to heat or flame (Section 6).

Countermeasures

Storage and Handling: Section 7.

VANADIUM CARBONYL

Hazard Analysis

Toxicity: Unknown. On decomposition it may release carbon monoxide. See also vanadium compounds and carbonyls.

Disaster Hazard: See carbonyls.

VANADIUM COMPOUNDS

Hazard Analysis

Toxicity: Variable. Vanadium compounds act chiefly as irritants to the conjunctivae and respiratory tract. Prolonged exposures may lead to pulmonary involvement. There is still some controversy as to the effects of industrial exposure on other systems of the body. Responses are acute, never chronic.

The first report of vanadium poisoning in humans described rather widespread systemic effects, consisting of polycythemia, followed by red blood cell destruction and anemia, loss of appetite, pallor and emaciation, albuminuria and hematuria, gastrointestinal disorders, nervous complaints and cough, sometimes severe enough to cause hemoptysis. More recent reports describe symptoms which, for the most part, are restricted to the conjunctivae and respiratory system, no evidence being found of disturbances of the gastrointestinal tract, kidneys, blood or central nervous system. Though certain workers believe that it is only the pentoxide which is harmful, other investigators have found that patronite dust (chiefly vanadium sulfide) is quite toxic to animals, causing acute pulmonary edema.

Symptoms and signs of poisoning are pallor, greenish-black discoloration of the tongue, paroxysmal cough, conjunctivitis, dyspnea and pain in the chest, bronchitis, rales and rhonchi, bronchospasm, tremor of the fingers and arms, radiographic reticulation. See also specific compounds.

These are common air contaminants (Section 4).

TLV: ACGIH (recommended); V_2O_5 dust 0.5, V_2O_5 fume 0.1 milligram per cubic meter in air.

Countermeasures

Ventilation Control: Section 2.

Personal Hygiene: Section 3.

VANADIUM DIBORIDE

General Information

Description: Hexagonal crystals.

Formula: VB_2.

Constants: Mol wt: 72.59, d: 5.10.

Hazard Analysis

Toxicity: See vanadium compounds and boron compounds.

Fire Hazard: Moderate; on contact with moisture or acids, hydrides can be formed. See also boron hydrides.

Explosion Hazard: Slight; on contact with moisture or acids, hydrides can be formed. See also boron hydrides.

Disaster Hazard: Moderately dangerous; it will react with water or steam to produce toxic and flammable vapors; can react with oxidizing materials; and on contact with acid or acid fumes, it can emit toxic fumes.

Countermeasures

Storage and Handling: Section 7.

VANADIUM DICHLORIDE

General Information

Synonym: Vanadous chloride.

Description: Hexagonal, green plates; deliquescent.

Formula: VCl_2.

Constants: Mol wt: 121.86, d: 3.23 at 18°C.

Hazard Analysis

Toxicity: See hydrochloric acid and vanadium compounds.

Disaster Hazard: Dangerous, see chlorides; will react with water or steam to produce toxic and corrosive fumes.

Countermeasures

Storage and Handling: Section 7.

VANADIUM DIOXIDE. See vanadium tetraoxide.

VANADIUM IODIDE

General Information

Description: Green, deliquescent crystals.

Formula: $VI_3 \cdot 6H_2O$.

Constant: Mol wt: 539.81.

Hazard Analysis and Countermeasures

See vanadium compounds and iodides.

VANDIUM MONOSULFIDE

General Information

Synonym: Vanadium sulfide.

Description: Black plates.

Formula: VS.

Constants: Mol wt: 83.02, mp: decomposes, d: 4.20.

Hazard Analysis and Countermeasures

See sulfides and vanadium compounds.

VANADIUM MONOXIDE

General Information

Synonym: Vanadium oxide.

Description: Light-gray crystals.

Formula: VO.

Constants: Mol wt: 66.95, mp: ignites, d: 5.758 at 14°C.

Hazard Analysis

Toxicity: See vanadium compounds.

Fire Hazard: Moderate, when heated (Section 6).

Countermeasures

Storage and Handling: Section 7.

VANADIUM OXIDE. See vanadium monoxide.

VANADIUM OXYBROMIDE

General Information

Description: Violet crystals.

Formula: VOBr.

Constants: Mol wt: 146.87, mp: decomposes at 480°C, d: 4.00 at 18°C.

Hazard Analysis and Countermeasures

See vanadium compounds and bromides.

VANADIUM OXYCHLORIDE

General Information

Description: Yellow-brown powder.

Formula: VOCl.

Constants: Mol wt: 102.41, bp: 127°C, d: 3.64 at 20°C.

Hazard Analysis and Countermeasures

See vanadium compounds and chlorides.

VANADIUM OXYDIBROMIDE

General Information

Description: Brown, deliquescent powder.

Formula: $VOBr_2$.

Constants: Mol wt: 226.78, mp: decomposes at 180°C.

Hazard Analysis and Countermeasures

See vanadium compounds and bromides.

VANADIUM OXYDICHLORIDE

General Information

Synonyms: Hypovanadic hydrochloride; vanadyl chloride.

Description: Dark green, syrupy mass.

Formula: $VOCl_2$.

Constants: Mol wt: 137.86, d: 2.88 at 13°C.

Hazard Analysis and Countermeasures

See vanadium compounds and chlorides.

VANADIUM OXYDIFLUORIDE

General Information

Description: Yellow solid.

Formula: VOF_2.

Constants: Mol wt: 104.95, mp: decomposes, d: 3.396 at 19°C.
Hazard Analysis and Countermeasures
See fluorides and vanadium compounds.

VANADIUM OXYTRIBROMIDE
General Information
Description: Red liquid.
Formula: $VOBr_3$.
Constants: Mol wt: 306.70, mp: decomposes at 180°C, bp: 130°C at 100 mm, d: 2.933 at 14.5°C.
Hazard Analysis and Countermeasures
See vanadium compounds and bromides.

VANADIUM OXYTRICHLORIDE
General Information
Description: Yellow, deliquescent liquid.
Formula: $VOCl_3$.
Constants: Mol wt: 173.32, mp: $-77 \pm 2°C$, bp: 126.7°C, d: 1.811 at 32°C.
Hazard Analysis and Countermeasures
See vanadium compounds and hydrochloric acid.
Shipping Regulations: Section 11.
   IATA: Corrosive liquid, white label, not acceptable (passenger), 1 liter (cargo).

VANADIUM OXYTRIFLUORIDE
General Information
Description: Yellow-white, hygroscopic crystals.
Formula: VOF_3.
Constants: Mol wt: 123.95, mp: 300°C, bp: 480°C, d: 2.459 at 19°C.
Hazard Analysis and Countermeasures
See fluorides and vanadium compounds.

VANADIUM PENTAFLUORIDE
General Information
Description: Crystals.
Formula: VF_5.
Constants: Mol wt: 145.95, bp: 111.2°C, d: 2.177 at 19°C.
Hazard Analysis and Countermeasures
See fluorides and vanadium compounds.

VANADIUM PENTASULFIDE
General Information
Description: Black-green powder.
Formula: V_2S_5.
Constants: Mol wt: 262.2, mp: decomposes, d: 3.00.
Hazard Analysis and Countermeasures
See sulfides and vanadium compounds.

VANADIUM PENTOXIDE
General Information
Description: Yellow to red crystalline powder.
Formula: V_2O_5.
Constants: Mol wt: 181.90, mp: 690°C, bp: decomposes at 1750°C, d: 3.357 at 18°C.
Hazard Analysis
Toxicity: See vanadium compounds.

VANADIUM SESQUIOXIDE
General Information
Synonym: Vanadium trioxide.
Description: Black crystals.
Formula: V_2O_3.
Constants: Mol wt: 149.9, mp: 1970°C, d: 4.87 at 18°C.
Hazard Analysis
Toxicity: See vanadium compounds.

VANADIUM SESQUISULFIDE
General Information
Synonym: Vanadium trisulfide.
Description: Green-black plates or powder.
Formula: V_2S_3.
Constants: Mol wt: 198.1, mp: decomposes, d: 4.7 at 21°C.
Hazard Analysis and Countermeasures
See sulfides and vanadium compounds.

VANADIUM SULFATE. See vanadyl sulfate.

VANADIUM SULFIDE. See vanadium monosulfide.

VANADIUM TETRACHLORIDE
General Information
Description: Reddish-brown liquid.
Formula: VCl_4.
Constants: Mol wt: 192.78, mp: $-28 \pm 2°C$, bp: 148.5°C, d: 1.816 at 30°C.
Hazard Analysis and Countermeasures
See hydrochloric acid and vanadium compounds.
Shipping Regulations: Section 11.
   IATA: Corrosive liquid, white label, not acceptable (passenger), 1 liter (cargo).

VANADIUM TETRAFLUORIDE
General Information
Description: Brown-yellow crystals.
Formula: VF_4.
Constants: Mol wt: 127.0, mp: decomposes at 325°C, d: 2.975 at 23°C.
Hazard Analysis and Countermeasures
See fluorides and vanadium compounds.

VANADIUM TETRAOXIDE
General Information
Synonym: Vanadium dioxide.
Description: Black crystals.
Formula: V_2O_4.
Constants: Mol wt: 165.90, mp: 1967°C, d: 4.339.
Hazard Analysis
Toxicity: See vanadium compounds.

VANADIUM TRIBROMIDE
General Information
Description: Green-black, deliquescent crystals.
Formula: VBr_3.
Constants: Mol wt: 290.8, mp: decomposes.
Hazard Analysis and Countermeasures
See vanadium compounds and bromides.

VANADIUM TRICHLORIDE
General Information
Description: Pink crystals.
Formula: VCl_3.
Constants: Mol wt: 157.32, mp: decomposes, d: 3.00 at 18°C.
Hazard Analysis and Countermeasures
See vanadium compounds and hydrochloric acid.

VANADIUM TRIFLUORIDE
General Information
Description: Rhombic green crystals.
Formula: VF_3.
Constants: Mol wt: 108.0, mp: $> 800°C$, bp: sublimes, d: 3.363 at 19°C.
Hazard Analysis and Countermeasures
See fluorides and vanadium compounds.

TOXIC HAZARD RATING CODE *(For detailed discussion, see Section 1.)*

0 NONE: (a) No harm under any conditions; (b) Harmful only under unusual conditions or overwhelming dosage.

1 SLIGHT: Causes readily reversible changes which disappear after end of exposure.

2 MODERATE: May involve both irreversible and reversible changes; not severe enough to cause death or permanent injury.

3 HIGH: May cause death or permanent injury after very short exposure to small quantities.

U UNKNOWN: No information on humans considered valid by authors.

VANADIUM TRIOXIDE. See vanadium sesquioxide.

VANADIUM TRISULFIDE. See vanadium sesquisulfide.

VANADOUS CHLORIDE. See vanadium dichloride.

VANADYL CHLORIDE. See vanadium oxydichloride.

VANADYL SULFATE
General Information
Synonyms: Vanadic sulfate; vanadium sulfate.
Description: Blue crystals.
Formula: $VOSO_4$.
Constant: Mol wt: 163.01.
Hazard Analysis and Countermeasures
See vanadium compounds and sulfates.

VANILLA
General Information
Description: Cured, full-grown unripe fruit of Vanilla planifolia.
Composition: 2–3% vanillin; 4% resin; 10% sugar, etc.
Hazard Analysis
Toxic Hazard Rating:
  Acute Local: Irritant 1; Allergen 1.
  Acute Systemic: U.
  Chronic Local: Allergen 1.
  Chronic Systemic: U.
Fire Hazard: Slight, when exposed to heat or flame (Section 6).
Countermeasures
Personal Hygiene: Section 3.
Storage and Handling: Section 7.

VANILLIC ALDEHYDE. See vanillin.

VANILLIN
General Information
Synonyms: 3-Methoxy-4-hydroxy benzaldehyde; vanillic aldehyde.
Description: White crystalline needles; pleasant odor; soluble in 125 parts water, 20 parts glycerol, 2 parts 95% alcohol; chloroform, and ether.
Formula: $(CH_3O)(OH)C_6H_3CHO$.
Constants: Mol wt: 152, d: 1.056, mp: 81–83°C, bp: 285°C.
Hazard Analysis
Toxic Hazard Rating: U. A synthetic flavoring substance and adjuvant (Section 10).

VAPAM
General Information
Synonym: Sodium n-methyldithiocarbamate dihydrate.
Description: White crystals; smells like carbon disulfide. Water soluble.
Formula: $C_2H_4NNaS_2 \cdot 2H_2O$.
Constant: Mol wt: 165.2.
Hazard Analysis
Toxicity: Details unknown. See bis(diethylthiocarbamyl) disulfide.
Disaster Hazard: Dangerous. When heated to decomposition it emits highly toxic fumes of oxides of nitrogen and sulfur.

VAPOPHOS. See parathion.

VAPOTONE. See tetraethyl pyrophosphate.

VARNISH
General Information
Description: Usually contains oil, resins and solvents.
Constant: Flash p: < 80°F.
Hazard Analysis
Toxicity: Variable as an allergen. See specific components. A common air contaminant (Section 4).
Fire Hazard: Dangerous, when exposed to heat or flame.
Explosion Hazard: Unknown.

Disaster Hazard: Dangerous; keep away from heat and open flame; can react vigorously with oxidizing materials.
Countermeasures
Ventilation Control: Section 2.
To Fight Fire: Foam, carbon dioxide, dry chemical or carbon tetrachloride (Section 6).
Storage and Handling: Section 7.
Shipping Regulations: Section 11.
  Coast Guard Classification: Combustible liquid; inflammable liquid, red label.

VARNISH MAKERS' NAPHTHA. See naphtha.

VARNISH SHELLAC
General Information
Constant: Flash p: 40–70°F.
Hazard Analysis
Toxicity: Details unknown; a variable allergen. See also specific components. A common air contaminant (Section 4).
Fire Hazard: Dangerous, when exposed to heat or flame.
Explosion Hazard: Unknown.
Disaster Hazard: Dangerous; keep away from heat or open flame; can react vigorously with oxidizing materials.
Countermeasures
Ventilation Control: Section 2.
To Fight Fire: Carbon dioxide, dry chemical, or carbon tetrachloride (Section 6).
Storage and Handling: Section 7.

VARNOLINE. See Stoddard solvent.

V-C13 NEMACIDE
General Information
Synonym: o-2,4-Dichlorophenyl,-o,o-diethylphosphorothioate.
Hazard Analysis
Disaster Hazard: Dangerous. See chlorides and phosphates.

VEGADEX
General Information
Synonym: 2-Chloroallyl diethyl dithiocarbamate.
Hazard Analysis
Toxicity: Details unknown. Probably acts similarly to other dithiocarbamates. See disulfiram.
Disaster Hazard: Dangerous. When heated to decomposition it emits highly toxic fumes.

VEGETABLE GUM. See dextrin.

VEGETABLE OIL, HYDROGENATED
General Information
Constants: Flash p: 610°F (O.C.), d: < 1.
Hazard Analysis
Fire Hazard: Slight, when exposed to heat or flame.
Countermeasures
To Fight Fire: Carbon dioxide, dry chemical or carbon tetrachloride (Section 6).
Storage and Handling: Section 7.

VERATRALDEHYDE
General Information
Synonym: 3,4-Dimethoxy benzaldehyde; veratric aldehyde; 3,4 dimethoxy benzene carbonal.
Description: Needles; odor of vanilla beans; slightly soluble in hot water; freely soluble in alcohol and ether.
Formula: $C_6H_3CHO(OCH_3)_2$.
Constants: Mol wt: 166.17, mp: 42–43°C, bp: 281°C.
Hazard Analysis
Toxicity: Unknown. See aldehydes.

VERATRIDINE
General Information
Description: Yellow-white powder.
Formula: $C_{36}H_{51}NO_{11}$.
Constants: Mol wt: 673.8, mp: 180°C.

Hazard Analysis
Toxicity: Details unknown; an insecticide. See also veratrine.
Fire Hazard: Slight; (Section 6).
Countermeasures
Storage and Handling: Section 7.

VERATRINE
General Information
Synonym: Cevadine.
Description: Rhombic prismatic crystals. Slightly soluble in water.
Formula: $C_{32}H_{49}NO_9$.
Constants: Mol wt: 591.7, mp: 205°C.
Hazard Analysis
Toxic Hazard Rating:
    Acute Local: Irritant 3.
    Acute Systemic: Ingestion 3; Skin Absorption 3.
    Chronic Local: Irritant 3.
    Chronic Systemic: U.
Toxicology: Particularly dangerous because absorption via intact skin can lead to fatal intoxication.
Caution: Action is similar to that of digitalis.
Disaster Hazard: Moderately dangerous; when heated to decomposition it emits toxic fumes.
Countermeasures
Ventilation Control: Section 2.
Personal Hygiene: Section 3.
Storage and Handling: Section 7.

VERATRINE SULFATE. See veratrine.

VERDIGRIS. See copper acetate, basic.

VERISAN. See 2-methoxyethylmercury silicate.

VERMICULITE
General Information
Description: A hydrated magnesium-aluminum-iron silicate; monoclinic crystals; pseudo hexagonal character; insoluble in water and organic solvents; dissolves in hot concentrated sulfuric acid.
Composition: $\approx 39\%$ SiO_2, $\approx 21\%$ MgO, 15% Al_2O_3, 9% Fe_2O_3, 5–7% K_2O, 1% CaO, 5–9% H_2O, and small quantities of Cr, Mn, P, S, Cl.
Hazard Analysis
Toxic Hazard Rating:
    Acute Local: Inhalation 1.
    Acute Systemic: 0.
    Chronic Local: Inhalation 1.
    Chronic Systemic: 0.
Toxicity: Can act as a nuisance dust.

VERMILION. See mercuric sulfide, red.

VERV-CA. See calcium stearyl-2-lactylate.

VIENNA GREEN. See copper acetoarsenite.

VIGORITE NO. 5 L. F. See explosives, high.

VILLIAUMITE. See sodium fluoride.

VINEGAR
General Information
Description: Clear to yellow liquid.
Composition: Approximately 6% solution of acetic acid.
Hazard Analysis and Countermeasures
See acetic acid.

VINEGAR SALTS. See calcium acetate.

VINEGAR ACID. See acetic acid.

VINOL. See vinyl alcohol.

VINYL ACETATE
General Information
Description: Colorless, mobile liquid, polymerizes to solid on exposure to light.
Formula: $CH_3COOCHCH_2$.
Constants: Mol wt: 86.05, mp: $-100.2°C$, bp: 73°C, flash p: 18°F, d: 0.9335 at 20°C, autoign. temp.: 800°F, vap. press.: 100 mm at 21.5°C, lel: 2.6%, uel: 13.4%, vap. d.: 3.0.
Hazard Analysis
Toxic Hazard Rating:
    Acute Local: Irritant 1.
    Acute Systemic: Inhalation 1.
    Chronic Local: Irritant 1.
    Chronic Systemic: U.
Toxicology: May act as a skin irritant by its defatting action. High concentrations of vapor are narcotic but are formed only if an inhibitor is present.
Fire Hazard: Highly dangerous when exposed to heat or flame.
Spontaneous Heating: No.
Explosion Hazard: Unknown.
Disaster Hazard: Dangerous; when heated to decomposition it burns and emits acrid fumes; can react with oxidizing materials.
Countermeasures
Ventilation Control: Section 2.
To Fight Fire: Foam, carbon dioxide, dry chemical or carbon tetrachloride (Section 6).
Personal Hygiene: Section 3.
Storage and Handling: Section 7.
Shipping Regulations: Section 11.
    I.C.C.: Flammable liquid; red label, 10 gallons.
    Coast Guard Classification: Inflammable liquid; red label.
    IATA: Flammable liquid, red label, 1 liter (passenger), 40 liters (cargo).

VINYL ACETONITRILE. See allyl cyanide.

VINYL ALCOHOL
General Information
Synonyms: Ethenol; vinol.
Description: An unstable liquid. Isolated only in form of its esters or the polymer, polyvinyl alcohol.
Formula: CH_2CHOH.
Constant: Mol wt: 44.1.
Hazard Analysis
Toxicity: Details unknown. See alcohols.
Fire Hazard: Dangerous, when exposed to heat or flame (Section 6).
Explosion Hazard: Unknown.
Disaster Hazard: Dangerous, when exposed to heat or open flame; can react vigorously with oxidizing materials.
Countermeasures
Ventilation Control: Section 2.
Storage and Handling: Section 7.

VINYL ALLYL ETHER. See allyl vinyl ether.

VINYL BENZENE. See phenylethylene.

VINYL BROMIDE
General Information
Synonyms: Bromoethylene; bromoethene.

TOXIC HAZARD RATING CODE (For detailed discussion, see Section 1.)

0 NONE: (a) No harm under any conditions; (b) Harmful only under unusual conditions or overwhelming dosage.

1 SLIGHT: Causes readily reversible changes which disappear after end of exposure.

2 MODERATE: May involve both irreversible and reversible changes; not severe enough to cause death or permanent injury.

3 HIGH: May cause death or permanent injury after very short exposure to small quantities.

U UNKNOWN: No information on humans considered valid by authors.

Description: A gas.
Formula: CH_2CHBr.
Constants: Mol wt: 107.0, mp: $-138°C$, bp: $15.6°C$, d: 1.51.
Hazard Analysis
Toxic Hazard Rating:
  Acute Local: Inhalation 2.
  Acute Systemic: Inhalation 2.
  Chronic Local: U.
  Chronic Systemic: Inhalation 2.
Fire Hazard: Dangerous, when exposed to flame or heat.
Explosion Hazard: Unknown.
Disaster Hazard: Dangerous; can react vigorously with oxidizing materials. See bromides.
Countermeasures
Ventilation Control: Section 2.
To Fight Fire: Carbon dioxide, dry chemical or water spray (Section 6).
Storage and Handling: Section 7.

VINYL BUTYL "CELLOSOLVE"
General Information
Description: Liquid.
Formula: $CH_2CHOCH_2CH_2OC_4H_9$.
Constants: Mol wt: 144.21, mp: $-71°C$, bp: $88°C$ at 50 mm, d: 0.8654 at $20°/20°C$, vap. d.: 4.98.
Hazard Analysis
Toxicity: See glycols.
Fire Hazard: Moderate, when exposed to heat or flame (Section 6).
Disaster Hazard: Moderately dangerous when exposed to heat or flame; can react with oxidizing materials.
Countermeasures
Storage and Handling: Section 7.

VINYL BUTYL ETHER
General Information
Description: Liquid.
Formula: $CH_2CHOC_4H_9$.
Constants: Mol wt: 100.16, mp: $-112.7°C$, bp: $94.1°C$, flash p: $15°F$ (O.C.), d: 0.7803 at $20°/20°C$, vap. d.: 3.45.
Hazard Analysis
Toxicity: Details unknown. See ethers.
Fire Hazard: Dangerous, when exposed to heat or flame.
Explosion Hazard: Details unknown. See ethers.
Disaster Hazard: Highly dangerous, when exposed to heat or flame; can react with oxidizing materials.
Countermeasures
Ventilation Control: Section 2.
To Fight Fire: Foam, carbon dioxide, dry chemical or carbon tetrachloride (Section 6).
Storage and Handling: Section 7.

VINYL S-(BUTYLMERCAPTOETHYL) ETHER
Hazard Analysis
Toxic Hazard Rating: U. Limited animal experiments suggest moderate toxicity.
Fire Hazard: See ethers.
Disaster Hazard: See ethers and mercaptans.

VINYL BUTYRATE
General Information
Formula: $CH_2:CHOOCCH_2CH_2CH_3$.
Constants: Mol wt: 114.1, d: 0.9, vap. d.: 4.0, bp: $242°F$, flash p: $68°F$ (O.C.), lel: 1.4%, uel: 8.8%.
Hazard Analysis
Toxicity: Details unknown. Limited animal experiments suggest low toxicity. See also esters.
Fire Hazard: Dangerous when exposed to heat or flame.
Countermeasures
To Fight Fire: Alcohol foam.

VINYL CAPROATE. See vinyl 2-hexanoate.

VINYL CARBAZOLE
General Information
Description: Liquid.
Formula: $(C_6H_4)_2NCHCH_2$.
Constant: Mol wt: 193.2.
Hazard Analysis
Toxic Hazard Rating:
  Acute Local: Allergen 1.
  Acute Systemic: U.
  Chronic Local: Irritant 1; Allergen 1.
  Chronic Systemic: U.
Fire Hazard: Moderate, when exposed to heat or flame (Section 6).
Explosion Hazard: Unknown.
Disaster Hazard: Moderately dangerous; when heated to decomposition it emits toxic fumes; can react with oxidizing materials.
Countermeasures
Ventilation Control: Section 2.
Personal Hygiene: Section 3.
Storage and Handling: Section 7.

VINYL 2-CHLORETHYL ETHER
General Information
Description: Liquid.
Formula: $CH_2CHOCH_2CH_2Cl$.
Constants: Mol wt: 106.55, mp: $-70.3°C$, bp: $108.9°C$, flash p: $80°F$ (O.C.), d: 1.0498, vap. d.: 3.67.
Hazard Analysis
Toxicity: Details unknown. See ethers.
Fire Hazard: Dangerous, when exposed to heat or flame. See ethers.
Explosion Hazard: Details unknown. See ethers.
Disaster Hazard: Dangerous; when heated to decomposition it emits highly toxic fumes of phosgene; can react vigorously with oxidizing materials.
Countermeasures
Ventilation Control: Section 2.
To Fight Fire: Water, foam, alcohol foam, carbon dioxide, dry chemical or carbon tetrachloride (Section 6).
Storage and Handling: Section 7.

VINYL CHLORIDE
General Information
Synonyms: Chloroethylene; chloroethene.
Description: Colorless liquid or gas (when inhibited); faintly sweet odor.
Formula: CH_2CHCl.
Constants: Mol wt: 62.50, bp: $-13.4°C$, lel: 4%, uel: 22%, flash p: $-108°F$ (C.O.C.), bp: $-13.9°C$, fp: $-159.7°C$, d liquid: 0.9195 at $15°/4°C$, vap. press.: 2600 mm at $25°C$, vap. d.: 2.15, autoign. temp.: $882°F$.
Hazard Analysis
Toxic Hazard Rating:
  Acute Local: Irritant 2.
  Acute Systemic: Inhalation 2.
  Chronic Local: Irritant 2.
  Chronic Systemic: U.
Toxicology: In high concentrations it acts as an anesthetic. Causes skin burns by rapid evaporation and consequent freezing. Chronic exposure has shown liver injury in rats and rabbits. Circulatory and bone changes in the finger tips reported in workers handling unpolymerized material.
TLV: ACGIH (recommended); 500 parts per million in air; 1290 milligrams per cubic meter of air.
Caution: May cause local irritation or frostbite due to rapid evaporation from skin or tissues.
Fire Hazard: Dangerous, when exposed to heat or flame. Large fires of this material are practically inextinguishable.
Spontaneous Heating: No.

Explosion Hazard: Severe in the form of vapor when exposed to heat or flame.

Disaster Hazard: Very dangerous; when heated to decomposition it emits highly toxic fumes of phosgene; can react vigorously with oxidizing materials. Before storing or handling this material instructions for its use should be obtained from the supplier.

Countermeasures

Ventilation Control: Section 2.

To Fight Fire: Carbon dioxide, dry chemical or carbon tetrachloride (Section 6).

Personnel Protection: Section 3.

Storage and Handling: Section 7.

Shipping Regulations: Section 11.

  I.C.C.: Flammable gas; red gas label, 300 pounds.

  Coast Guard Classification: Inflammable gas; red gas label.

  MCA warning label.

  IATA (inhibited): Flammable gas, red label, not acceptable (passenger), 140 kilograms (cargo).

    (uninhibited): Flammable gas, not acceptable (passenger and cargo).

VINYL CROTONATE
General Information

Description: Slightly soluble in water.

Formula: $CH_2:CHOCOCH:CHCH_3$.

Constants: Mol wt: 112, d: 0.9, vap. d.: 4.0, bp: 273°F, flash p: 78°F (O.C.).

Hazard Analysis

Toxic Hazard Rating: U. Probably toxic.

Fire Hazard: Dangerous, when exposed to heat or flame.

Countermeasures

Storage and Handling: Section 7.

To Fight Fire: Alcohol foam.

VINYL CYANIDE. See acrylonitrile.

VINYL CYCLOHEXANE
General Information

Formula: $CH_3CH(HC:CH_2)C_4H_8$.

Constant: Mol wt: 110.

Hazard Analysis

Toxicity: Details unknown. High concentrations produce narcosis, and lower levels on repeated exposure cause liver and kidney injury in animals.

4-VINYLCYCLOHEXENE-1
General Information

Description: Liquid.

Formula: C_8H_{12}.

Constants: Mol wt: 108.18, bp: 128°C, fp: −109°C, flash p: 70°F (T.O.C.), d: 0.832 at 20°/4°C, autoign. temp.: 517°F, vap. press.: 25.8 mm at 38°C, vap. d.: 3.76.

Hazard Analysis

Toxicity: Unknown. Probably irritant and narcotic in high concentrations.

Fire Hazard: Dangerous, when exposed to heat or flame.

Explosion Hazard: Unknown.

Disaster Hazard: Dangerous; when exposed to heat or open flame; can react with oxidizing materials.

Countermeasures

Ventilation Control: Section 2.

To Fight Fire: Foam, carbon dioxide, dry chemical or carbon tetrachloride (Section 6).

Storage and Handling: Section 7.

VINYL CYCLOHEXENE DIOXIDE
General Information

Description: Colorless liquid.

Formula: $CH_2CHOC_6H_9O$.

Constants: Mol wt: 140, d: 1.098 at 20/20°C, bp: 227°C, flash p: 230°F.

Hazard Analysis

Toxic Hazard Rating:

  Acute Local: Irritant 3.

  Acute Systemic: Ingestion 2; Inhalation 1; Skin Absorption 2.

  Chronic Local: 0.

  Chronic Systemic: U.

Data based on limited animal experiments.

Fire Hazard: Slight when exposed to heat or flame.

Countermeasures

Storage and Handling: Section 7.

To Fight Fire: Water, foam, dry chemical.

VINYL CYCLOHEXENE MONOXIDE
General Information

Description: Liquid; very slightly soluble in water.

Formula: $CH_2:CHC_6H_9O$.

Constants: Mol wt: 136, d: 0.9598 at 20/20°C, bp: 169°C, flash p: 136°F, fp: −100°F.

Hazard Analysis

Toxic Hazard Rating:

  Acute Local: Irritant 2.

  Acute Systemic: Ingestion 2; Inhalation 2; Skin Absorption 2.

  Chronic Local: 0.

  Chronic Systemic: U.

Data based on limited animal experiments.

Fire Hazard: Moderate when exposed to heat or flame.

Countermeasures

Storage and Handling: Section 7.

To Fight Fire: Foam, alcohol foam, mist.

VINYL DECANOATE
General Information

Formula: $CH_3(CH_2)_8COOHC:CH_2$.

Constant: Mol wt: 198.

Hazard Analysis

Toxicity: Details unknown. Limited animal experiments suggest low toxicity. See also esters.

VINYL ETHER
General Information

Synonyms: Divinyl ether; divinyl oxide.

Description: Colorless liquid.

Formula: $(CH_2CH)_2O$.

Constants: Mol wt: 70.1, bp: 39°C, ulc: 100, lel: 1.9%, uel: 36.5%, flash p: −22°F (C.C.), d: 0.774 at 20°/20°C, autoign. temp.: 680°F, vap. d.: 2.41.

Hazard Analysis

Toxic Hazard Rating:

  Acute Local: Irritant 1; Inhalation 1.

  Acute Systemic: Inhalation 3.

  Chronic Local: Irritant 1.

  Chronic Systemic: Inhalation 2.

Toxicology: Has been used as an inhalation anesthetic. Prolonged exposure is said to cause liver injury.

Fire Hazard: Highly dangerous, when exposed to heat or flame.

Explosion Hazard: Severe in the form of vapor when exposed to heat or flame.

Disaster Hazard: Highly dangerous, when exposed to heat

TOXIC HAZARD RATING CODE *(For detailed discussion, see Section 1.)*

0 NONE: (a) No harm under any conditions; (b) Harmful only under unusual conditions or overwhelming dosage.

1 SLIGHT: Causes readily reversible changes which disappear after end of exposure.

2 MODERATE: May involve both irreversible and reversible changes; not severe enough to cause death or permanent injury.

3 HIGH: May cause death or permanent injury after very short exposure to small quantities.

U UNKNOWN: No information on humans considered valid by authors.

or open flame; can react vigorously with oxidizing materials.
Countermeasures
Ventilation Control: Section 2.
To Fight Fire: Carbon dioxide, dry chemical or carbon tetrachloride (Section 6).
Storage and Handling: Section 7.

VINYLETHYLENE OXIDE. See butadiene monoxide.

VINYL ETHYL ETHER
General Information
Description: Colorless liquid.
Formula: $CH_2CHOC_2H_5$.
Constants: Mol wt: 72.104, bp: 35.6°C, flash p: $< -50°F$, fp: $-115°C$, d: 0.754, autoign. temp.: 395°F, vap. press.: 428 mm at 20°C, lel: 1.7%, uel: 28%, vap. d.: 2.5.
Hazard Analysis
Toxic Hazard Rating:
  Acute Local: U.
  Acute Systemic: Ingestion 2; Inhalation 2.
  Chronic Local: U.
  Chronic Systemic: U.
Fire Hazard: Highly dangerous, when exposed to heat or flame.
Explosion Hazard: Details unknown. See ethers.
Disaster Hazard: Severe; when heated or exposed to flame; can react vigorously with oxidizing materials. See ethers.
Countermeasures
Ventilation Control: Section 2.
To Fight Fire: Alcohol foam, foam, carbon dioxide, dry chemical or carbon tetrachloride. See also ethers.
Personal Hygiene: Section 3.
Storage and Handling: Section 7.
Shipping Regulations: Section 11.
  IATA (inhibited): Flammable liquid, red label, 1 liter (passenger), 40 liters (cargo).
    (uninhibited): Flammable liquid, not acceptable (passenger and cargo).

VINYL 2-ETHYL HEXOATE
General Information
Description: Liquid; insoluble in water.
Formula: $CH_2:CHOCOCH(C_2H_5)C_4H_9$.
Constants: Mol wt: 170, flash p: 165°F (O.C.), d: 0.8751, bp: 185.2°C, fp: $-90°C$, vap. d.: 6.0.
Hazard Analysis
Toxic Hazard Rating: U.
Fire Hazard: Moderate, when exposed to heat or flame.
Storage and Handling: Section 7.
To Fight Fire: Foam, alcohol foam, mist.

VINYL 2-ETHYLHEXYL ETHER
General Information
Description: Liquid.
Formula: $CH_2CHO(C_2H_5)C_6H_{12}$.
Constants: Mol wt: 156.3, mp: $-100°C$, bp: 177.5°C, flash p: 135°F (O.C.), d: 0.810, autoign. temp.: 395°F, vap. d.: 5.4.
Hazard Analysis
Toxicity: Details unknown. See ethers.
Fire Hazard: Moderate, when exposed to heat or flame; can react with oxidizing materials.
Explosion Hazard: Details unknown. See ethers.
Countermeasures
Storage and Handling: Section 7.
To Fight Fire: Alcohol foam, foam, carbon dioxide, dry chemical or carbon tetrachloride. See also ethers.

VINYL S(ETHYLMERCAPTOETHYL)ETHER
Hazard Analysis
Toxicity: Details unknown. Limited animal experiments suggest moderate toxicity.
Disaster Hazard: Dangerous. See ethers and mercaptans.

2-VINYL-5-ETHYL PYRIDINE
General Information
Description: Insoluble in water.
Formula: $N:C(CH:CH_2)CH:CHC(C_2H_5):CH$.
Constants: Mol wt: 133, d: 0.9449 at 20/20°C, bp: 138°C at 100 mm, vap. press.: 0.2 mm at 20°C, fp: $-50.9°C$, flash p: 200°F (C.O.C.).
Hazard Analysis
Toxic Hazard Rating: U. Probably toxic.
Fire Hazard: Moderate, when exposed to heat or flame.
Disaster Hazard: Dangerous; see cyanides.
Countermeasures
Storage and Handling: Section 7.
To Fight Fire: Foam, alcohol foam, mist.

VINYL FLUORIDE
General Information
Synonym: Fluoroethylene.
Description: Colorless gas; insoluble in water; soluble in alcohol and ether.
Formula: $CH_2:CHF$.
Constants: Mol wt: 46, bp: $-72°C$.
Hazard Analysis
Toxic Hazard Rating: U.
Fire Hazard: Ignites in presence of heat or source of ignition.
Disaster Hazard: See fluorides.
Countermeasures
Storage and Handling: Section 7.
Shipping Regulations: Section 11.
  ICC (inhibited): Flammable gas, red gas label, 300 pounds.
  Coast Guard Classification: Inflammable gas, red label.
  IATA (inhibited): Flammable gas, red label, not acceptable (passenger), 140 kilograms (cargo).
    (uninhibited): Flammable gas, not acceptable (passenger and cargo).

VINYL FORMATE. See acrylic acid.

VINYL 2-HEXANOATE
General Information
Synonym: Vinyl caproate.
Formula: $CH_2:CHCO_2C_5H_{11}$.
Constant: Mol wt: 142.2.
Hazard Analysis
Toxicity: Details unknown. Animal experiments suggest low toxicity.

VINYLIDENE CHLORIDE
General Information
Synonym: 1,1-Dichloroethylene.
Description: Colorless, volatile liquid.
Formula: CH_2CCl_2.
Constants: Mol wt: 97.0, bp: 31.6°C, lel: 5.6%, uel: 11.4%, fp: $-122°C$, flash p: 14°F (O.C.), d: 1.218 at 20°/4°C, autoign. temp.: 856°F.
Hazard Analysis
Toxicity: Details unknown. See vinyl chloride.
Fire Hazard: Dangerous, when exposed to heat or flame.
Explosion Hazard: Moderate in the form of gas when exposed to heat or flame.
Disaster Hazard: Highly dangerous; see chlorides; can react vigorously with oxidizing materials.
Countermeasures
Ventilation Control: Section 2.
To Fight Fire: Water, foam, carbon dioxide, dry chemical or carbon tetrachloride (Section 6).
Storage and Handling: Section 7.
Shipping Regulations: Section 11.
  ICC: Inhibited: Flammable liquid; red label, 10 gallons.
  Coast Guard Classification: Inhibited: Inflammable liquid; red label.
  IATA (inhibited): Flammable liquid, red label, 1 liter (passenger), 40 liters (cargo).

(uninhibited): Flammable liquid, not acceptable (passenger and cargo).

VINYL IODIDE
General Information
Synonym: Iodoethene.
Description: Liquid; insoluble in water.
Formula: $CH_2:CHI$.
Constants: Mol wt: 154.0, d: 2.08 at 0°C, bp: 56°C.
Hazard Analysis
Toxicity: Details unknown. Irritating to mucous membrane. Narcotic in high concentrations.
Disaster Hazard: Dangerous. See iodides.

VINYL ISOBUTYL ETHER.
General Information
Description: Liquid.
Formula: $CH_2CHOC_4H_9$.
Constants: Mol wt: 100.3, mp: −132.3°C, bp: 83.3°C, flash p: 15°F (O.C.), d: 0.77, vap. press.: 68 mm at 20°C, vap. d.: 3.45.
Hazard Analysis
Toxicity: Details unknown. See also ethers.
Fire Hazard: Dangerous, when exposed to heat or flame.
Explosion Hazard: Details unknown. See ethers.
Disaster Hazard: Dangerous. Keep away from heat or open flame. It can react vigorously with oxidizing materials.
Countermeasures
Ventilation Control: Section 2.
To Fight Fire: Foam, carbon dioxide, dry chemical or carbon tetrachloride. See also ethers. Section 6.
Storage and Handling: Section 7.
Shipping Regulations: Section 11.
　IATA (inhibited): Flammable liquid, red label, 1 liter (passenger), 40 liters (cargo).
　　(uninhibited): Flammable liquid, not acceptable (passenger and cargo).

VINYL ISOOCTYL ETHER
General Information
Description: Insoluble in water.
Formula: $CH_2CHO(CH_2)_5CH(CH_3)_2$.
Constants: Mol wt: 156, flash p: 140°F, d: 0.8, vap. d.: 5.4, bp: 347°F.
Hazard Analysis
Toxic Hazard Rating: U. See ethers.
Fire Hazard: Moderate, when exposed to heat or flame. See ethers.

VINYL ISOPROPYL ETHER
General Information
Description: Liquid.
Formula: $CH_2CHOC_3H_7$.
Constants: Mol wt: 86.2, flash p: −26°F (C.C.), autoign. temp.: 522°F.
Hazard Analysis
Toxicity: Details unknown. See ethers.
Fire Hazard: Highly dangerous, when exposed to heat or flame. See ethers.
Explosion Hazard: Details unknown. See ethers.
Disaster Hazard: Highly dangerous. Keep away from heat or open flame; can react vigorously with oxidizing materials. Section 6.
Countermeasures
Ventilation Control: Section 2.
To Fight Fire: Foam, carbon dioxide, dry chemical or carbon tetrachloride. See also ethers.
Storage and Handling: Section 7.

VINYL 2-METHOXYETHYL ETHER
General Information
Description: Liquid.
Formula: $CH_2CHOCH_2CH_2OCH_3$.
Constants: Mol wt: 102.13, mp: −82.8°C, bp: 108.8°C, flash p: 65°F (O.C.), d: 0.8967, vap. d.: 3.53.
Hazard Analysis
Toxicity: Details unknown. See ethers.
Fire Hazard: Dangerous, when exposed to heat or flame; can react with oxidizing materials.
Explosion Hazard: Details unknown. See ethers.
Disaster Hazard: Dangerous. Keep away from heat and flame.
Countermeasures
Ventilation Control: Section 2.
To Fight Fire: Foam, carbon dioxide, dry chemical or carbon tetrachloride. See also ethers. Section 6.
Storage and Handling: Section 7.

VINYL METHYL ETHER
General Information
Description: Colorless, easily liquefied gas or colorless liquid.
Formula: CH_2CHOCH_3.
Constants: Mol wt: 58.1, bp: 6.0°C, flash p: −60°F, d: 0.7500, vap. d.: 2.0, fp: −121.6°C, vap press: 1052 mm at 20°C.
Hazard Analysis
Toxicity: Details unknown. See ethers.
Fire Hazard: Dangerous, when exposed to heat or flame.
Explosion Hazard: Details unknown. See ethers.
Disaster Hazard: Dangerous; can react vigorously with oxidizing materials.
Countermeasures
Storage and Handling: Section 7.
To Fight Fire: Foam, carbon dioxide, dry chemical, carbon tetrachloride or water spray. See also ethers.
Shipping Regulations: Section 11.
　ICC: Inhibited: Flammable gas; red gas label, 20 pounds.
　Coast Guard Classification: Inhibited: Inflammable gas; red gas label.
　IATA (inhibited): Flammable gas, red label, not acceptable (passenger), 10 kilograms (cargo).
　　(uninhibited): Flammable gas, not acceptable (passenger and cargo).

VINYL NONYL ETHER
General Information
Description: Liquid.
Formula: $CH_2CHOC_9H_{19}$.
Constants: Mol wt: 170.3, bp: 161°C, d: 0.81.
Hazard Analysis
Toxicity: Details unknown. See ethers.
Fire Hazard: Moderate, when exposed to heat or flame; can react with oxidizing materials. See also ethers.
Explosion Hazard: Details unknown. See ethers.
Countermeasures
Storage and Handling: Section 7.

VINYL n-OCTADECYL ETHER
General Information
Description: Insoluble in water.
Formula: $CH_2:CHO(CH_2)_{17}CH_3$.
Constants: Mol wt: 308, mp: 82.4°F, bp: 267–369°F at 5 mm, d: 0.8, flash p: 350°F.
Hazard Analysis
Toxic Hazard Rating: U. See ethers.

TOXIC HAZARD RATING CODE *(For detailed discussion, see Section 1.)*

0 NONE: (a) No harm under any conditions; (b) Harmful only under unusual conditions or overwhelming dosage.

1 SLIGHT: Causes readily reversible changes which disappear after end of exposure.

2 MODERATE: May involve both irreversible and reversible changes; not severe enough to cause death or permanent injury.

3 HIGH: May cause death or permanent injury after very short exposure to small quantities.

U UNKNOWN: No information on humans considered valid by authors.

Fire Hazard: Slight, when exposed to heat or flame.
Countermeasures
Storage and Handling: Section 7.
To Fight Fire: Water, foam.

VINYL OCTANOATE
General Information
Formula: $CH_2:CHOOC(CH_2)_6CH_3$.
Constant: Mol wt: 180.
Hazard Analysis
Toxicity: Unknown. Limited animal experiments suggest low toxicity. See also esters.

VINYL PROPIONATE
General Information
Description: Liquid, almost insoluble in water.
Formula: $CH_2:CHOOCC_2H_5$.
Constants: Mol wt: 100, d: 0.9173 at 20/20°C, bp: 95°C, fp: −81.1°C, flash p: 34°F (O.C.), vap. d.: 3.3.
Hazard Analysis
Toxicity: Details unknown. Limited animal experiments suggest low toxicity. See also esters.
Fire Hazard: Dangerous, when exposed to heat or flame.
Countermeasures
Storage and Handling: Section 7.
To Fight Fire: Alcohol foam.

VINYL PYRIDINE
General Information
Description: Liquid.
Formula: C_7H_7N.
Constants: Mol wt: 105.1, bp: 159°C, d: 0.9746 at 20°C.
Hazard Analysis
Toxicity: Details unknown. Animal experiments suggest high toxicity. See pyridine.
Caution: An irritant to the skin, eyes and respiratory tract.
Fire Hazard: Moderate, when exposed to heat or flame (Section 6).
Explosion Hazard: Unknown.
Disaster Hazard: Dangerous; when heated to decomposition it emits highly toxic fumes of cyanide; can react with oxidizing materials.
Countermeasures
Ventilation Control: Section 2.
Storage and Handling: Section 7.

N-VINYL-2-PYRROLIDONE
General Information
Description: Colorless liquid.
Formula: C_6H_9NO.
Constants: Mol wt: 111.1, bp: 148°C at 100 mm, fp: 13.5°C, flash p: 209°F (C.O.C.), d: 1.04 at 25°C, autoign. temp.: 213°F, vap. d.: 3.8.
Hazard Analysis
Toxicity: Unknown. Probably irritant and narcotic in high concentrations.
Fire Hazard: Moderate, when exposed to heat or flame.
Disaster Hazard: Dangerous, when heated to decomposition it emits highly toxic fumes of oxides of nitrogen; can react vigorously with oxidizing materials.
Countermeasures
Storage and Handling: Section 7.
To Fight Fire: Water, foam, alcohol foam, carbon dioxide, dry chemical or carbon tetrachloride (Section 6).

VINYL RESINS
Hazard Analysis
Toxic Hazard Rating:
  Acute Local: Allergen 1.
  Acute Systemic: U.
  Chronic Local: Allergen 1.
  Chronic Systemic: U.
Fire Hazard: Slight, when exposed to heat or flame; can react with oxidizing materials (Section 6).

Countermeasures
Personal Hygiene: Section 3.
Storage and Handling: Section 7.

VINYL STEARATE
General Information
Description: White, waxy solid.
Formula: $H_2CCHCO_2(CH_2)_{16}CH_3$.
Constants: Mol wt: 310.5, mp: 28–30°C, bp: 180°C at 2 mm, d: 0.881 at 20°/20°C.
Hazard Analysis
Toxicity: Details unknown. See esters.
Fire Hazard: Slight, when exposed to heat or flame; can react with oxidizing materials. Section 6.
Countermeasures
To Fight Fire: Foam, carbon dioxide, dry chemical or carbon tetrachloride (Section 6).
Storage and Handling: Section 7.

VINYLSTYRENE. See m-divinylbenzene.

VINYL SULFONE
Hazard Analysis
Toxicity: Details unknown. Limited animal experiments suggest high toxicity and irritation.
Disaster Hazard: Dangerous. See sulfonates.

VINYLTOLUENE
General Information
Description: Colorless liquid.
Formula: $CH_2CHC_6H_4CH_3$.
Constants: Mol wt: 118.2; bp: 170–171°C; fp: −82.5°C; flash p: 140°F; d: monomer 0.89 at 25°/25°C, polymer 1.027 at 25°/25°C; vap. d.: 4.08.
Hazard Analysis
Toxicity: Details unknown. See toluene.
TLV: ACGIH (recommended) 100 parts per million in air; 480 milligrams per cubic meter of air.
Fire Hazard: Moderate, when exposed to heat or flame; can react with oxidizing materials.
Countermeasures
To Fight Fire: Foam, carbon dioxide, dry chemical or carbon tetrachloride (Section 6).
Storage and Handling: Section 7.

VINYL TRICHLORIDE. See β-trichloroethane.

VINYLTRICHLOROSILANE
General Information
Description: Fuming liquid.
Formula: $CH_2CHSiCl_3$.
Constant: Mol wt: 162.5, bp: 90.6°C, d: 1.265 at 25°/25°C.
Hazard Analysis
Toxic Hazard Rating:
  Acute Local: Irritant 3; Ingestion 3; Inhalation 3.
  Acute Systemic: U.
  Chronic Local: U.
  Chronic Systemic: U.
Fire Hazard: Unknown.
Disaster Hazard: Dangerous; see chlorides; will react with water or steam to produce toxic and corrosive fumes.
Countermeasures
Ventilation Control: Section 2.
Personnel Protection: Section 3.
Storage and Handling: Section 7.
Shipping Regulations: Section 11.
  Coast Guard Classification: Inflammable liquid; red label.
  I.C.C.: Flammable liquid, red label, 10 gallons.
  IATA: Flammable liquid, red label, not acceptable (passenger), 40 liters (cargo).

VINYL TRIMETHYLNONYL ETHER
General Information
Description: Liquid.

Formula: $CH_2CHOCHCH_2CH(CH_3)_2$
$$CH_2CH(CH_3)CH_2CH(CH_3)_2.$$
Constants: Mol wt: 212.36, mp: $-90°C$, bp: $223.5°C$, flash p: $200°F$ (O.C.), d: 0.8075 at $20°/20°C$, vap. d.: 7.33.
Hazard Analysis
Toxicity: Details unknown. See ethers.
Fire Hazard: Moderate, when exposed to heat or flame; it can react with oxidizing materials.
Countermeasures
To Fight Fire: Foam, carbon dioxide, dry chemical or carbon tetrachloride (Section 6).
Storage and Handling: Section 7.

VISHA. See aconitum ferox.

VITAMIN A
General Information
Description: A suitable form of derivative of retinol. In liquid form it is a light yellow to red oil; very soluble in chloroform and ether; soluble in absolute alcohol and vegetable oils; insoluble in glycerin and water.
Hazard Analysis
Toxic Hazard Rating: U. A nutrient and/or dietary supplement food additive (Section 10).

VITAMIN A ACETATE
General Information
Description: Synthetic vitamin A acetic acid ester; finely divided dry, light yellow crystalline powder; odorless.
Formula: $C_{20}H_{29}OOCCH_3$.
Constant: Mol wt: 328.
Hazard Analysis
Toxic Hazard Rating: U. A nutrient and/or dietary supplement food additive (Section 10).

VITAMIN A PALMITATE
General Information
Description: Synthetic vitamin A palmitic acid ester; yellow liquid, odorless.
Formula: $C_{20}H_{29}OOC_{15}H_{31}$.
Constant: Mol wt: 512.

Hazard Analysis
Toxic Hazard Rating: U. A nutrient and/or dietary supplement food additive (Section 10).

VITAMIN B$_2$. See riboflavin.

VITAMIN B$_{12}$
General Information
Synonym: Cobalamin.
Description: The anti-pernicious anemia vitamin; all vitamin B$_{12}$ compounds contain the cobalt atom in its trivalent state. There are at least 3 active forms: cyanocobalamin, hydroxycobalamin, and nitrocobalamin.
Hazard Analysis
Toxic Hazard Rating: U. A nutrient and/or dietary supplement food additive (Section 10).

VITAMIN D$_2$
General Information
Synonym: Ergocalciferol; calciferol.
Description: White crystals; odorless; insoluble in water; soluble in alcohol, chloroform, ether and fatty acids.
Formula: $C_{28}H_{44}O$.
Constants: Mol wt: 396, mp: $115-118°C$.
Hazard Analysis
Toxic Hazard Rating: U. A nutrient and/or dietary supplement food additive (Section 10).

VITAMIN D$_3$
General Information
Synonyms: Cholecalciferol; 5,7-Cholestadien-3-β-ol; 7-dehydrocholesteral, activated.
Description: White colorless crystals; insoluble in water; soluble in alcohol, chloroform, and fatty oils.
Formula: $C_{27}H_{44}O$.
Constants: Mol wt: 384, melting range: $84-88°C$.
Hazard Analysis
Toxic Hazard Rating: U. A nutrient and/or dietary supplement food additive (Section 10).

VITAMIN H. See biotin.

VM & P. See "dryolene."

VULCAN COAL POWDERS. See explosives, high.

TOXIC HAZARD RATING CODE (For detailed discussion, see Section 1.)

0 NONE: (a) No harm under any conditions; (b) Harmful only under unusual conditions or overwhelming dosage.

1 SLIGHT: Causes readily reversible changes which disappear after end of exposure.

2 MODERATE: May involve both irreversible and reversible changes; not severe enough to cause death or permanent injury.

3 HIGH: May cause death or permanent injury after very short exposure to small quantities.

U UNKNOWN: No information on humans considered valid by authors.

W

WALL PLASTER
Hazard Analysis
Toxic Hazard Rating:
Acute Local: Irritant 1; Allergen 1; Inhalation 1.
Acute Systemic: U.
Chronic Local: Allergen 1.
Chronic Systemic: U.
Countermeasures
Antidote: In spite of its slight adverse effects, plaster may be used in an emergency to neutralize a strong acid or poison taken internally (Section 1).
Ventilation Control: Section 2.
Personal Hygiene: Section 3.

WARFARIN
General Information
Synonym: 3-(α-Acetonylbenzene)-4-hydroxycoumarin; compound 42; WARF-12.
Description: Colorless, odorless, tasteless crystals.
Constant: Mp: 161°C.
Hazard Analysis
Toxicology: A rodenticide. Warfarin has two actions—inhibition of prothrombin formation and capillary damage. There is evidence that these two actions are produced by the two moieties of the molecule. Thus, 4-hydroxycoumarin inhibits the formation of prothrombin and reduces the clotting power of the blood, while there is some evidence benzalacetone produces capillary damage and leads to bleeding upon the very slightest trauma. Significantly enough, vitamin K has an antidotal action against both actions of warfarin up to a certain point.

The action of warfarin is similar to that of the common drug dicoumarol, except that capillary damage (the so-called "toxic factor") is greatly increased.
Dangerous Acute and Chronic Dose in Man: Information on the toxicity of warfarin to man is available. Serious illness was induced by ingesting 1.7 mg of warfarin per kg per day for 6 consecutive days with suicidal intent. This would correspond to eating almost 1 pound of warfarin bait (0.025 percent warfarin) each day for 6 days. All signs and symptoms were caused by hemorrhage and, following multiple small transfusions and massive doses of vitamin K, recovery was complete.

Data from animal experiments suggest that a single dose would be harmless. The rat is specifically susceptible to warfarin, yet mortalities are highly irregular and may be low following single doses of 50, 100, or even 150 mg/kg. To obtain a dose of 50 mg/kg the average 150-pound man would have to eat 0.7 kg or 1.5 pounds of the warfarin concentrate available on the market, although the amount of active ingredient would be only 3.5 g. To obtain the same dose with the strongest bait recommended, the same man would have to eat 14 kg or 30 pounds of the rat bait.

However, with repeated daily doses, the effective toxicity of the compound is greatly increased. For example, five daily doses of 1.0 mg/kg each (a total of 5.0 mg/kg) is sufficient to kill all rats which eat it. There is considerable species difference in susceptibility. For example, chickens may be raised to maturity on an adequate growing mash containing an effective rodenticidal concentration of warfarin.

The possibility of human poisoning by warfarin must be kept in mind, although the safety factors make it appear unlikely that poisoning will occur except with suicidal intent or as a result of gross carelessness and ignorance.
Signs and Symptoms of Poisoning in Man: The initial symptoms in an attempted suicide using warfarin were back pain and abdominal pain. The onset occurred the first day after the sixth daily dose. A day after onset vomiting and attacks of nose bleeding occurred. On the second day of illness, when admitted to the hospital, the patient was observed to have a generalized petechial rash. The prothrombin time was greatly prolonged. The coagulation time was definitely increased by the Lee-White method and slightly increased by the capillary tube method. Bleeding time was normal. Urine was normal in appearance but contained many red cells on microscopic examination.
TLV: ACGIH (recommended); 0.1 milligrams per cubic meter of air.
Countermeasures
Treatment: After blood has been taken for prothrombin and other differential diagnostic tests, a blood transfusion should be given at once if there is reasonable assurance that warfarin poisoning has occurred, irrespective of whether or not signs and symptoms are present. Vitamin K in a dose of 65 mg repeated three times on the first day of treatment is suggested irrespective of symptoms. Smaller doses should be continued until the prothrombin time has reached normal. In a more seriously ill patient, small transfusions of carefully matched whole blood should be given daily until the patient has returned to normal. Should it ever be necessary to treat a patient in shock from blood loss resulting from warfarin poisoning, frequent small transfusions and a complete consideration of the blood chemistry would be in order. Any large hematomata should be the subject of surgical consultation, but any surgical action should be taken only after the clotting power of the blood is restored to normal.
Personnel Protection: Section 3.
Shipping Regulations: Section 11.
MCA warning label.
IATA: Poison B, poison label, 25 kilograms (passenger), 95 kilograms (cargo).

WASTE PAPER, WET
Hazard Analysis
Fire Hazard: Moderate, when exposed to heat or by spontaneous chemical reaction; can react with oxidizing materials (Section 6).
Countermeasures
Storage and Handling: Section 7.
Shipping Regulations: Section 11.
I.C.C.: Flammable solid; yellow label, not accepted.
Coast Guard Classification: Inflammable solid; yellow label.
IATA: Flammable solid, not acceptable (passenger and cargo).

WASTE TEXTILE, WET
Hazard Analysis
Fire Hazard: Moderate, when exposed to heat or by spon-

taneous chemical reaction; can react with oxidizing materials (Section 6).

Countermeasures
Storage and Handling: Section 7.
Shipping Regulations: Section 11.
    I.C.C.: Flammable solid; yellow label, not accepted.
    Coast Guard Classification: Inflammable solid; yellow label.

WASTE WOOL, WET
Hazard Analysis
Fire Hazard: Moderate, when exposed to heat or flame or by spontaneous chemical reaction; can react with oxidizing materials (Section 6).

Countermeasures
Storage and Handling: Section 7.
Shipping Regulations: Section 11.
    I.C.C.: Flammable solid; yellow label, not accepted.
    Coast Guard Classification: Inflammable solid; yellow label.

WATER GAS. See carbon monoxide and hydrogen (hazardous components).
Constants: lel: 7.0%, uel: 72%.

WATER GLASS. See sodium silicate.

WATER OF AMMONIA. See ammonium hydroxide.

WATER TREATING COMPOUND, LIQUID. See acids, N.O.S. or alkalies, N.O.S.
Shipping Regulations: Section 11.
    I.C.C.: Corrosive liquid; white label, 10 gallons.
    Coast Guard Classification: Corrosive liquid; white label.
    IATA: Corrosive liquid, white label, 1 liter (passenger), 40 liters (cargo).

WAX, MICROCRYSTALLINE
General Information
Description: Derived from petroleum; white, amber or black solid; odorless.
Constants: Flash p: >400° F, d: 0.9.
Hazard Analysis
Toxic Hazard Rating: U. Probably very low.
Fire Hazard: Slight, when exposed to heat or flame.

WELDING FUMES
Hazard Analysis
Toxic Hazard Rating:
    Acute Local: Irritant 2; Inhalation 2.
    Acute Systemic: Irritant 1; Inhalation 1.
    Chronic Local: Irritant 2; Inhalation 2.
    Chronic Systemic: U.
Toxicology: When welding is done on a surface coated with cadmium, toxic fumes of cadmium can be evolved. When zinc-coated surfaces are welded, toxic quantities of zinc oxide may be liberated. When painted surfaces are welded, lead or other pigment fumes may be liberated. And when fluoride fluxes are used in welding, very toxic fluoride fumes are evolved. When oily surfaces are welded, offensive and toxic fumes can be liberated, and when the welding torch is improperly ignited, carbon monoxide which is very toxic may be evolved. Also, oxides of nitrogen may be formed. It is therefore considered hazardous to inhale excessive amounts of welding fumes. It is also possible to inhale sufficient quantities of iron oxide from welding to cause siderosis which is not, however, considered a serious condition. Metal fume fever is a common reaction. It

is characterized by chills, fever, sweating, and leucocytosis coming on several hours after exposure. Recovery is usually complete in 24–48 hours and there are no significant after-effects.

Safety goggles are required to protect against spatter. Light-filtering goggles are required to shield the eyes against the intense light from the arc.

Countermeasures
Ventilation Control: Section 2.
Personnel Protection: Section 3.

WET NITROCELLULOSE, COLLOIDED BLOCK, WET
    IATA (with less than 25% alcohol): Explosive, not acceptable (passenger and cargo).
        (with not less than 25% alcohol): Flammable liquid, red label, 1 liter (passenger), 12 kilograms (cargo).

WET NITROCELLULOSE (COLLODION COTTON) 30% ALCOHOL (OR SOLVENT). See nitrocellulose (collodion cotton).
Shipping Regulations: Section 11.
    I.C.C.: Flammable liquid; red label, 25 pounds.
    Coast Guard Classification: Inflammable liquid; red label.
    IATA: Flammable liquid, red label, 1 liter (passenger), 40 liters (cargo).

WET NITROCELLULOSE (COLLODION COTTON) 20% WATER. See nitrocellulose (collodion cotton).
Shipping Regulations: Section 11.
    I.C.C.: Flammable solid; yellow label, 100 pounds.
    Coast Guard Classification: Inflammable solid; yellow label.
    IATA: Flammable solid, yellow label, 12 kilograms (passenger), 45 kilograms (cargo).

WET NITROCELLULOSE, COLLOIDED, GRANULAR OR FLAKE, 20% ALCOHOL (OR SOLVENT). See nitrocellulose.
Shipping Regulations: Section 11.
    I.C.C.: Flammable liquid; red label, 25 pounds.
    Coast Guard Classification: Inflammable liquid; red label.
    IATA: Flammable liquid, red label, 1 liter (passenger), 12 kilograms (cargo).

WET NITROCELLULOSE, COLLOIDED, GRANULAR OR FLAKE, 20% WATER. See nitrocellulose.
Shipping Regulations: Section 11.
    I.C.C.: Flammable solid; yellow label, 100 pounds.
    Coast Guard Classification: Inflammable solid; yellow label.
    IATA: Flammable solid, yellow label, 12 kilograms (passenger), 45 kilograms (cargo).

WET NITROCELLULOSE FLAKES—20% ALCOHOL OR SOLVENT. See nitrocellulose.
Shipping Regulations: Section 11.
    ICC: Flammable liquid, red label, 25 pounds.
    IATA: Flammable liquid, red label, 1 liter (passenger), 12 kilograms (cargo).

WET NITROGUANIDINE, 20% WATER. See nitroguanidine.
Shipping Regulations: Section 11.
    I.C.C.: Flammable solid; yellow label, 100 pounds.
    Coast Guard Classification: Inflammable solid; yellow label.
    IATA: Flammable solid, yellow label, 12 kilograms (passenger), 45 kilograms (cargo).

TOXIC HAZARD RATING CODE (*For detailed discussion, see Section 1.*)

0 NONE: (a) No harm under any conditions; (b) Harmful only under unusual conditions or overwhelming dosage.

1 SLIGHT: Causes readily reversible changes which disappear after end of exposure.

2 MODERATE: May involve both irreversible and reversible changes; not severe enough to cause death or permanent injury.

3 HIGH: May cause death or permanent injury after very short exposure to small quantities.

U UNKNOWN: No information on humans considered valid by authors.

WET NITROSTARCH, 30% ALCOHOL (OR SOLVENT). See nitrostarch.
Shipping Regulations: Section 11.
  I.C.C.: Flammable liquid; red label, 25 pounds.
  Coast Guard Classification: Inflammable liquid; red label.
  IATA: Flammable liquid, red label, 1 liter (passenger), 12 kilograms (cargo).

WET NITROSTARCH, 20% WATER. See nitrostarch.
Shipping Regulations: Section 11.
  I.C.C.: Flammable solid; yellow label, 100 pounds.
  Coast Guard Classification: Inflammable solid; yellow label.
  IATA: Flammable solid, yellow label, 12 kilograms (passenger), 45 kilograms (cargo).

WHALE OIL
General Information
Constants: D: 0.925, flash p: 446° F (C.C.), autoign. temp.: 800° F.
Hazard Analysis
Toxicity: Details unknown. Probably very low.
Fire Hazard: Slight, when exposed to heat or flame.
Spontaneous Heating: Yes.
Countermeasures
To Fight Fire: Foam, carbon dioxide, dry chemical or carbon tetrachloride (Section 6).
Storage and Handling: Section 7.

WHEAT OIL
Hazard Analysis
Toxic Hazard Rating:
  Acute Local: Allergen 1.
  Acute Systemic: 0
  Chronic Local: Allergen 1.
  Chronic Systemic: 0
Disaster Hazard: Slightly dangerous; when heated, it emits acrid fumes.
Countermeasures
Personal Hygiene: Section 3.
Storage and Handling: Section 7.

WHISKY
General Information
Description: Light to deep-amber liquid; characteristic odor and taste.
Composition: Approx. 50% ethyl alcohol (by vol.) + acetic acid and ethyl acetate.
Constants: Flash p: 82° F (C.C.), d: 0.935–0.923 at 25° C.
Hazard Analysis
Toxic Hazard Rating:
  Acute Local: 0.
  Acute Systemic: Ingestion 2.
  Chronic Local: 0.
  Chronic Systemic: Ingestion 1.
Fire Hazard: Moderate, when exposed to heat or flame; can react vigorously with oxidizing materials.
Explosion Hazard: Unknown.
Countermeasures
Ventilation Control: Section 2.
To Fight Fire: Carbon dioxide, dry chemical or carbon tetrachloride (Section 6).
Storage and Handling: Section 7.

WHITE ACID
General Information
Description: A mixture of ammonium bifluoride and hydrofluoric acid used for etching glass.
Countermeasures
Shipping Regulations: Section 11.
  IATA: Corrosive liquid, white label, 1 liter (passenger), 5 liters (cargo).

WHITE ARSENIC. See arsenic trioxide.

WHITE CAUSTIC. See sodium hydroxide.

WHITE FLAG. See orris.

WHITE LEAD. See lead compounds.

WHITE PRECIPITATE. See mercury, ammoniated.

WHITE SPIRIT. See Stoddard solvent.

WHITE TAR. See naphthalene.

WHITE WAX. See beeswax, white.

WINES, HIGH
General Information
Description: Light golden to amber color, aromatic, alcoholic liquids.
Constant: Flash p: 60–80° F.
Hazard Analysis
Toxic Hazard Rating:
  Acute Local: 0.
  Acute Systemic: Ingestion 1.
  Chronic Local: 0.
  Chronic Systemic: Ingestion 1.
Fire Hazard: Dangerous; when exposed to heat or flame; can react with oxidizing materials.
Explosion Hazard: Unknown.
Countermeasures
Ventilation Control: Section 2.
To Fight Fire: Carbon dioxide, dry chemical or carbon tetrachloride (Section 6).
Storage and Handling: Section 7.

WINES (SHERRY AND PORT)
General Information
Description: Light golden to amber color, aromatic, alcoholic liquids.
Constant: Flash p: 129° F.
Hazard Analysis
Toxic Hazard Rating:
  Acute Local: 0.
  Acute Systemic: Ingestion 1.
  Chronic Local: 0.
  Chronic Systemic: Ingestion 1.
Fire Hazard: Moderate when exposed to heat or flame; can react with oxidizing materials.
Explosion Hazard: Unknown.
Countermeasures
Storage and Handling: Section 7.
To Fight Fire: Carbon dioxide, dry chemical or carbon tetrachloride (Section 6).

WINTERGREEN OIL. See methyl salicylate.

WITCH HAZEL
General Information
Synonyms: Hamamelis; tobacco wood.
Hazard Analysis
Toxic Hazard Rating:
  Acute Local: Irritant 1; Ingestion 1.
  Acute Systemic: Ingestion 1.
  Chronic Local: U.
  Chronic Systemic: U.
Fire Hazard: Slight, when exposed to heat or flame; can react with oxidizing materials (Section 6).
Explosion Hazard: Unknown.
Countermeasures
Personal Hygiene: Section 3.
Storage and Handling: Section 7.

WOLFRAM. See tungsten.

WOLFSBANE. See aconite.

WOOD DUST. See saw dust.

WOOD GAS. See carbon monoxide.

WOOD TAR ACID
Hazard Analysis
Toxicity: Details unknown; a fungicide, herbicide and insecticide.
Fire Hazard: Unknown.
Disaster Hazard: Slightly dangerous; when heated it emits smoke.
Countermeasures
Storage and Handling: Section 7.

WOOD TAR OIL
Hazard Analysis
Toxicity: Details unknown; a fungicide and insecticide.
Fire Hazard: Moderate, when exposed to heat or flame; can react with oxidizing materials (Section 6).
Countermeasures
Storage and Handling: Section 7.

WOOD VINEGAR. See pyroligneous acid.

WOOL AND WOOL WASTES
Hazard Analysis
Toxic Hazard Rating:
    Acute Local: Allergen 1; Inhalation 1.
    Acute Systemic: 0.
    Chronic Local: Allergen 1; Inhalation 1.
    Chronic Systemic: 0.
Fire Hazard: Moderate, when exposed to heat or flame; can react with oxidizing materials.

Spontaneous Heating: Yes, particularly when wet.
Countermeasures
Personal Hygiene: Section 3.
Storage and Handling: Section 7.
Shipping Regulations: Section 11.
    Coast Guard Classification (See cotton waste): Hazardous; (wet, see rags, wet) inflammable solid; yellow label.

WOOL GREASE. See lanolin.

WORMWOOD. See absinthium.

WORMWOOD OIL
Hazard Analysis
Toxic Hazard Rating:
    Acute Local: Irritant 1.
    Acute Systemic: Ingestion 2.
    Chronic Local: U.
    Chronic Systemic: Ingestion 2.
Caution: This is the essence of absinthe, prolonged use of which can lead to mental deterioration.
Countermeasures
Personal Hygiene: Section 3.
Storage and Handling: Section 7.

WULFENITE. See lead molybdate.

WURTZITE. See zinc sulfide (α).

TOXIC HAZARD RATING CODE (For detailed discussion, see Section 1.)

0 NONE: (a) No harm under any conditions; (b) Harmful only under unusual conditions or overwhelming dosage.

1 SLIGHT: Causes readily reversible changes which disappear after end of exposure.

2 MODERATE: May involve both irreversible and reversible changes; not severe enough to cause death or permanent injury.

3 HIGH: May cause death or permanent injury after very short exposure to small quantities.

U UNKNOWN: No information on humans considered valid by authors.

X

XANTHOGEN DISULFIDE
Hazard Analysis and Countermeasures
Details unknown. See sulfides.

o-XENOL. See o-phenylphenol.

XENON
General Information
Description: Colorless, gaseous element.
Formula: Xe.
Constants: At wt: 131.30, mp: −112°C, bp: −107.1°C, d: 5.851 grams per liter; d liquid: 3.06 at −109°C.
Hazard Analysis
Toxic Hazard Rating:
Acute Local: 0.
Acute Systemic: Inhalation 1.
Chronic Local: 0.
Chronic Systemic: 0.
Caution: A simple asphyxiant. For a discussion of toxic effects see argon. A common air contaminant (Section 4).
Radiation Hazard: Section 5. For permissible levels, see Table 5, p. 150.
Artificial isotope ^{131m}Xe, half life 12 d. Decays to stable ^{131}Xe by emitting gamma rays of 0.16 MeV.
Artificial isotope ^{133m}Xe, half life 2.3 d. Decays to radioactive ^{133}Xe by emitting gamma rays of 0.23 MeV.
Artificial isotope ^{133}Xe, half life 5.3 d. Decays to stable ^{133}Cs by emitting beta particles of 0.35 MeV. Also emits gamma rays of 0.08 MeV. ^{133}Xe is produced by neutron irradiation of stable ^{132}Xe in air-cooled reactors.
Artificial isotope ^{135}Xe, half life 9.1 h. Decays to radioactive ^{135}Cs by emitting beta particles of 0.91 MeV. Also emits gamma rays of 0.25 MeV. ^{135}Xe is produced by neutron irradiation of stable ^{134}Xe in air-cooled reactors.
Countermeasures
Storage and Handling: Section 7.
Shipping Regulations: Section 11.
IATA: Nonflammable gas, green label, 70 kilograms (passenger), 140 kilograms (cargo).

XENYLAMINE. See p-aminodiphenyl.

X-RAY FILM (NITROCELLULOSE BASE)
Hazard Analysis and Countermeasures
See nitrates.
Shipping Regulations: Section 11.
I.C.C.: Flammable solid; yellow label, 200 pounds.
Coast Guard Classification: Inflammable solid; yellow label.
IATA: See film, x-ray.

X-RAY FILM SCRAP (NITROCELLULOSE BASE)
Hazard Analysis and Countermeasures
See nitrates.
Shipping Regulations: Section 11.
Coast Guard Classification: Inflammable solid; yellow label.

X-RAY FILM (NITROCELLULOSE BASE), UNEXPOSED
Shipping Regulations: Section 11.
ICC: Flammable solid; yellow label, 250 pounds.

X-RAY FILM, SCRAP (NITROCELLULOSE BASE) OTHER THAN SAMPLES. See x-ray film (nitrocellulose base).
Shipping Regulations: Section 11.
ICC: Flammable solid; yellow label (not accepted).

X-RAY FILM, SCRAP (NITROCELLULOSE BASE), SAMPLES OF
Hazard Analysis and Countermeasures
See nitrates.
Shipping Regulations: Section 11.
I.C.C.: Flammable solid; yellow label, 25 pounds.
Coast Guard Classification: Inflammable solid; yellow label.

X-RAY FILM, SLOW BURNING
Hazard Analysis
Fire Hazard: Moderate when exposed to heat or flame (Section 6).
Disaster Hazard: Dangerous; see nitrates; can react vigorously with oxidizing materials.
Countermeasures
Storage and Handling: Section 7.
Shipping Regulations: Section 11.
Coast Guard Classification: No restriction.

X-RAYS
General Information
Description: Highly penetrating electromagnetic radiation produced by electrical means.
Hazard Analysis
Radiation Hazard: See gamma rays and Section 5.

m-XYLENE
General Information
Synonym: m-Xylol.
Description: Colorless liquid.
Formula: $C_6H_4(CH_3)_2$.
Constants: Mol wt: 106.2, mp: −47.9°C, bp: 139°C, lel: 1.1%, uel: 7.0%, flash p: 84°F, d: 0.864 at 20°/4°C, vap. press.: 10 mm at 28.3°C, vap. d.: 3.66, autoign. temp.: 982°F.
Hazard Analysis
Toxic Hazard Rating:
Acute Local: Irritant 1.
Acute Systemic: Inhalation 1.
Chronic Local: Irritant 1.
Chronic Systemic: Inhalation 1; Skin Absorption 1.
TLV: ACGIH (recommended); 100 parts per million in air; 435 milligrams per cubic meter of air.
Note: A common air contaminant (Section 4).
Fire Hazard: Dangerous, when exposed to heat or flame; can react with oxidizing materials.
Explosion Hazard: Moderate in the form of vapor when exposed to heat or flame.
Disaster Hazard: Dangerous. Keep away from open flame.
Countermeasures
Ventilation Control: Section 2.
To Fight Fire: Foam, carbon dioxide, dry chemical or carbon tetrachloride (Section 6).
Personnel Protection: Section 3.
Storage and Handling: Section 7.
Shipping Regulations: Section 11.
I.C.C.: Flammable liquid; red label, 10 gallons.

Coast Guard Classification: Combustible liquid; inflammable liquid; red label.

MCA warning label.

IATA: Flammable liquid, red label, 1 liter (passenger), 40 liters (cargo).

o-XYLENE
General Information
Synonym: o-Xylol.
Description: Colorless liquid.
Formula: $C_6H_4(CH_3)_2$.
Constants: Mol wt: 106.2, bp: 144.4°C, fp: −25.5°C, ulc: 40–45, lel: 1.0%, uel: 6.0%, flash p: 90°F, d: 0.880 at 20°/4°C, vap. press.: 10 mm at 32.1°C, vap. d.: 3.66, autoign. temp.: 867°F.

Hazard Analysis
Toxicity: See m-xylene.
TLV: ACGIH (recommended); 100 parts per million in air; 435 milligrams per cubic meter of air.
Note: A common air contaminant (Section 4).
Fire Hazard: Moderate, when exposed to heat or flame; can react with oxidizing materials.
Explosion Hazard: Slight, in the form of vapor when exposed to heat or flame.

Countermeasures
Ventilation Control: Section 2.
To Fight Fire: Foam, carbon dioxide, dry chemical or carbon tetrachloride (Section 6).
Storage and Handling: Section 7.
Shipping Regulations: Section 11.
   I.C.C.: Flammable liquid; red label, 10 gallons.
   Coast Guard Classification: Combustible liquid; inflammable liquid, red label.
   MCA warning label.
   IATA: Flammable liquid, red label, 1 liter (passenger), 40 liters (cargo).

p-XYLENE
General Information
Synonym: p-Xylol.
Description: Clear liquid.
Formula: $C_6H_4(CH_3)_2$.
Constants: Mol wt: 106.2, bp: 138.3°C, lel: 1.1%, uel: 7.0%, fp: 13.2°C, flash p: 103°F (T.O.C.), d: 0.8611 at 20°/4°C, vap. press.: 10 mm at 27.3°C, vap. d.: 3.66.

Hazard Analysis
Toxicity: See m-xylene.
TLV: ACGIH (recommended); 100 parts per million in air; 435 milligrams per cubic meter of air.
Note: A common air contaminant (Section 4).
Fire Hazard: Moderate, when exposed to heat or flame; can react with oxidizing materials.
Explosion Hazard: Moderate in the form of vapor when exposed to heat or flame.

Countermeasures
Ventilation Control: Section 2.
To Fight Fire: Foam, carbon dioxide, dry chemical or carbon tetrachloride (Section 6).
Storage and Handling: Section 7.
Shipping Regulations: Section 11.
   I.C.C.: Flammable liquid; red label, 10 gallons.
   Coast Guard Classification: Combustible liquid; inflammable liquid, red label.
   MCA warning label.
   IATA: Flammable liquid, red label, 1 liter (passenger), 40 liters (cargo).

XYLENE HEXAFLUORIDE
General Information
Synonym: Bis(trifluoromethyl) benzene.
Description: Clear, water-white liquid.
Formula: $C_6H_4(CF_3)_2$.
Constants: Mol wt: 214.11, mp: −40 to −50°C, bp: 115°C, d: 1.395 at 20°/15.5°C.

Hazard Analysis and Countermeasures
See xylene and fluorides.

XYLENES (MIXED m- and p- ISOMERS)
General Information
Description: A clear liquid.
Constants: Bp: 138.5°C, flash p: 100°F (T.O.C.), d: 0.864 at 20°/4°C, vap. press.: 6.72 mm at 21°C.

Hazard Analysis
Toxicity: See m-xylene.
Fire Hazard: Moderate, in the presence of heat or flame; can react with oxidizing materials.

Countermeasures
To Fight Fire: Foam, carbon dioxide, dry chemical or carbon tetrachloride (Section 6).
Storage and Handling: Section 7.

XYLENE SUBSTITUTES
General Information
Description: Liquid.
Constants: Flash p: 45°F, d: 0.760.

Hazard Analysis
Toxicity: Unknown.
Fire Hazard: Dangerous, when exposed to heat or flame.
Explosion Hazard: Unknown.
Disaster Hazard: Dangerous. Keep away from heat and flame; can react with oxidizing materials.

Countermeasures
Storage and Handling: Section 7.
To Fight Fire: Foam, carbon dioxide, dry chemical or carbon tetrachloride (Section 6).

3,5-XYLENOL
General Information
Synonyms: 3,5-Dimethylphenol; 1-hydroxy-3,5-dimethylbenzene; 5-hydroxy-1,3-dimethylbenzene.
Description: White crystals.
Formula: $(CH_3)_2C_6H_3OH$.
Constants: Mol wt: 122.16, mp: 68°C, bp: 219.5°C, d: 1.0362, vap. press.: 1 mm at 62.0°C.

Hazard Analysis
Toxicity: See phenol.
Fire Hazard: Slight, when exposed to heat or flame.

Countermeasures
Ventilation Control: Section 2.
Personnel Protection: Section 3.
Storage and Handling: Section 7.

XYLIDINE
General Information
Synonym: Aminodimethylbenzene.
Description: Usually liquid (except for o-4-xylidine).
Formula: $(CH_3)_2C_6H_3NH_2$.
Constants: Mol wt: 121.2, bp: 213–226°C, flash p: 206°F (C.C.), d: 0.97–0.99, vap. d.: 4.17.

Hazard Analysis
Toxic Hazard Rating:
   Acute Local: U.
   Acute Systemic: Ingestion 3; Inhalation 3; Skin Absorption 3.

TOXIC HAZARD RATING CODE (For detailed discussion, see Section 1.)

0 NONE: (a) No harm under any conditions; (b) Harmful only under unusual conditions or overwhelming dosage.

1 SLIGHT: Causes readily reversible changes which disappear after end of exposure.

2 MODERATE: May involve both irreversible and reversible changes; not severe enough to cause death or permanent injury.

3 HIGH: May cause death or permanent injury after very short exposure to small quantities.

U UNKNOWN: No information on humans considered valid by authors.

Chronic Local: U.

Chronic Systemic: Ingestion 3; Inhalation 3; Skin Absorption 3.

TLV: ACGIH (recommended); 5 parts per million; 25 milligrams per cubic meter of air . May be absorbed via the skin.

Toxicology: This material, which so closely resembles aniline in its toxic effects, is actually twice as toxic as aniline, based on the determination of the LD_{50} for mice by inhalation. It can cause injury to the blood and the liver. It does not necessarily give any alarm or warning, such as cyanosis, headache, and dizziness which characterizes aniline poisoning. Thus it may be considered a more insidious poison than aniline, and severe and possibly fatal intoxication may come about through skin absorption. From animal experimentation it has been further found that the minimum lethal dose for rabbits is 0.28 g/kg of body weight, and that the lethal dose intravenously is 240 mg/kg of body weight for rabbits. The signs of intoxication in animals are loss of weight, dyspnea, prostration, albuminuria, and occasional terminal convulsions. This compound penetrates the intact skin of rabbits in sufficient quantity to cause cyanosis and death. There are no local effects upon the skin. The application of 3.3 grams or more of this material per kilogram of body weight on the intact skin of a rabbit for an hour or more always caused fatal results. A 2% solution of this material caused no harm to 3 rabbits in the course of 50 periods of cutaneous contact.

Fire Hazard: Moderate when exposed to heat or flame.

Disaster Hazard: Dangerous; when heated to decomposition it emits highly toxic fumes; can react vigorously with oxidizing materials.

Countermeasures

Ventilation Control: Section 2.

Treatment and Antidotes: In case of exposure, contact the medical department immediately. When it splashes or spills upon the person, the area of contact should be washed with copious quantities of warm water aided by soap. Remove all contaminated clothing and if possible wash the area affected under a deluge-type of shower. Protective clothing should be washed before reuse. This material must not be used without adequate ventilation as well as personal protective equipment. See also aniline. Personnel engaged working with it should do so under the direct supervision of a medical department.

Personnel Protection: Section 3.

To Fight Fire: Foam, carbon dioxide, dry chemical or carbon tetrachloride (Section 6).

First Aid: Section 1.

Storage and Handling: Section 7.

Shipping Regulations: Section 11.

Coast Guard Classification: Poison B; poison label.

MCA warning label.

IATA: Poison B, poison label, 1 liter (passenger), 220 liters (cargo).

XYLITOL

Hazard Analysis

Toxic Hazard Rating: U. An additive permitted in food for human consumption (Section 10).

XYLOLS. See xylenes—o, m, p.

m-XYLYL BROMIDE

General Information

Description: Colorless liquid.

Formula: C_8H_9Br.

Constants: Mol wt: 185.1, bp: 213°C, d: 1.371 at 23°C.

Hazard Analysis

Toxic Hazard Rating:

Acute Local: Irritant 3; Ingestion 3; Inhalation 3.

Acute Systemic: U.

Chronic Local: U.

Chronic Systemic: U.

Fire Hazard: Moderate, when exposed to heat or flame.

Disaster Hazard: Dangerous; see bromides; can react vigorously with oxidizing materials.

Countermeasures

Ventilation Control: Section 2.

Personnel Protection: Section 3.

Storage and Handling: Section 7.

Shipping Regulations: Section 11.

I.C.C.: Poison C; tear gas label, 75 pounds.

Coast Guard Classification: Poison C; tear gas label.

IATA: Poison C, poison label, not acceptable (passenger), 35 kilograms (cargo).

o-XYLYL BROMIDE

General Information

Synonym: o-Methylbenzyl bromide.

Description: Crystals.

Formula: C_8H_9Br.

Constants: Mol wt: 185.1, mp: 21°C, bp: 217°C, d: 1.381.

Hazard Analysis

Toxicity: See m-xylyl bromide.

Fire Hazard: Slight, when exposed to heat or flame (Section 6).

Disaster Hazard: Dangerous; see bromides.

Countermeasures

Storage and Handling: Section 7.

p-XYLYL BROMIDE

General Information

Description: White crystals.

Formula: C_8H_9Br.

Constants: Mol wt: 185.1, mp: 36°C, bp: 220°C at 740 mm, d: 1.324.

Hazard Analysis and Countermeasures

See m-xylyl bromide.

m-XYLYL CHLORIDE

General Information

Description: Colorless liquid.

Formula: $CH_3C_6H_4CH_2Cl$.

Constants: Mol wt: 140.61, bp: 196°C, d: 1.064.

Hazard Analysis

Toxic Hazard Rating:

Acute Local: Irritant 3; Ingestion 3; Inhalation 3.

Acute Systemic: U.

Chronic Local: U.

Chronic Systemic: U.

Disaster Hazard: Dangerous; see chlorides.

Countermeasures

Ventilation Control: Section 2.

Personnel Protection: Section 3.

Storage and Handling: Section 7.

o-XYLYL CHLORIDE. See o-methylbenzyl chloride. See also m-xylyl chloride.

p-XYLYL CHLORIDE

General Information

Description: Fuming liquid, irritating odor.

Formula: C_8H_9Cl.

Constants: Mol wt: 140.6, bp: 200–202°C.

Hazard Analysis and Countermeasures

See m-xylyl chloride.

Y

YELLOW BEESWAX. See beeswax.

YELLOW JASMINE. See gelsemium sempervirens.

YELLOW PLUMBOUS OXIDE. See litharge.

YELLOW PRUSSIATE OF POTASH. See potassium ferrocyanide.

YELLOW PRUSSIATE OF SODA. See sodium ferrocyanide.

YELLOW RESIN. See rosin.

YEW
General Information
Description: A wood.
Hazard Analysis
Toxic Hazard Rating:
 Acute Local: Irritant 1; Allergen 1; Inhalation 1.
 Acute Systemic: U.
 Chronic Local: Allergen 1.
 Chronic Systemic: U.
Fire Hazard: Moderate, when exposed to heat or flame; can react with oxidizing materials (Section 6).
Explosion Hazard: Moderate, in the form of dust when exposed to flame.
Countermeasures
Personal Hygiene: Section 3.
Storage and Handling: Section 7.

YOHIMBENE
General Information
Synonym: Isomer of yohimbine.
Description: Light-sensitive crystals.
Formula: $C_{21}H_{26}N_2O_3$.
Constants: Mol wt: 354.2, mp: 278°C (decomposes).
Hazard Analysis
Toxic Hazard Rating:
 Acute Local: U.
 Acute Systemic: Ingestion 3.
 Chronic Local: U.
 Chronic Systemic: U.
Caution: An alkaloid poison. See yohimbine.
Disaster Hazard: Moderately dangerous; when heated to decomposition it emits toxic fumes.
Countermeasures
Ventilation Control: Section 2.
Storage and Handling: Section 7.

YOHIMBINE
General Information
Synonyms: Corynine; aphrodine; quebrachine.
Description: Colorless needles from water and alcohol.
Formula: $C_{21}H_{26}N_2O_3$.
Constants: Mol wt: 354.2, mp: 235°C.
Hazard Analysis
Toxic Hazard Rating:

Acute Local: U.
Acute Systemic: Ingestion 2.
Chronic Local: U.
Chronic Systemic: U.
Toxicology: This material is a poison. Cases of poisoning have occurred from its use as an aphrodisiac. Upon local application, it produces anesthesia. However, absorption of it can give rise to toxic symptoms, such as salivation, increased respiration, and repeated defecation. With reference to the circulatory system, there may be a fall in blood pressure and sometimes myocardial damage, involving particularly the conduction system of the heart, with a resultant decrease in the efficiency of the heart.
Disaster Hazard: Moderately dangerous; when heated to decomposition it emits toxic fumes.
Countermeasures
Personal Hygiene: Section 3.
Storage and Handling: Section 7.

YOHIMBINE HYDROCHLORIDE. See yohimbine.

YTTERBIUM
General Information
Formula: Yb.
Constants: At wt: 173.04, mp: 1800°C.
Hazard Analysis
Toxicity: Unknown. As a lanthanon it may have an anticoagulant action on blood. See also lanthanum.
Radiation Hazard: Section 5. For permissible levels, see Table 5, p. 150.
 Artificial isotope ^{175}Yb, half life 4.2 d. Decays to stable ^{175}Lu by emitting beta particles of 0.07–0.47 MeV. Also emits gamma rays of 0.11–0.40 MeV.
Fire Hazard: Moderate, in the form of dust when exposed to heat or flame (Section 6).

YTTERBIUM SELENATE
General Information
Description: Hexagonal plates.
Formula: $Yb_2(SeO_4)_3 \cdot 8H_2O$.
Constants: Mol wt: 919.1, d: 3.30.
Hazard Analysis and Countermeasures
See selenium and ytterbium.

YTTERBIUM SELENITE
General Information
Formula: $YB_2(SeO_3)_3$.
Constant: Mol wt: 727.0.
Hazard Analysis and Countermeasures
See selenium.

YTTRIA. See yttrium oxide.

YTTRIUM
General Information
Description: Hexagonal, gray-black metallic element.

TOXIC HAZARD RATING CODE (For detailed discussion, see Section 1.)

0 NONE: (a) No harm under any conditions; (b) Harmful only under unusual conditions or overwhelming dosage.

1 SLIGHT: Causes readily reversible changes which disappear after end of exposure.

2 MODERATE: May involve both irreversible and reversible changes; not severe enough to cause death or permanent injury.

3 HIGH: May cause death or permanent injury after very short exposure to small quantities.

U UNKNOWN: No information on humans considered valid by authors.

Formula: Y.

Constants: At wt: 88.9, mp: 1500°C, bp: 3200°C, d: 5.51.

Hazard Analysis

Toxicity: Unknown. As a lanthanon it may have an anticoagulant effect on the blood. See also lanthanum.

TLV: ACGIH (recommended); 1 milligram per cubic meter of air.

Radiation Hazard: Section 5. For permissible levels, see Table 5, p. 150.

Artificial isotope ^{90}Y, half life 64 h. Decays to stable ^{90}Zr by emitting beta particles of 2.27 MeV. Yttrium exists in radioactive equilibrium with its parent, ^{90}Sr. Permissible levels are given for the equilibrium mixture.

Artificial isotope ^{91m}Y, half life 50 m. Decays to radioactive ^{91}Y by emitting gamma rays of 0.55 MeV.

Artificial isotope ^{91}Y, half life 59 d. Decays to stable ^{91}Zr by emitting beta particles of 1.54 MeV.

Artificial isotope ^{92}Y, half life 3.6 h. Decays to stable ^{92}Zr by emitting beta particles of 1.26 (9%), 3.60 (88%) MeV. Also emits gamma rays of 0.07–2.4 MeV.

Artificial isotope ^{93}Y, half life 10 h. Decays to radioactive ^{93}Zr by emitting beta particles of 2.89 (90%) MeV and others. Also emits gamma rays of 0.26–2.14 MeV.

Fire Hazard: Moderate, in the form of dust when exposed to heat or flame (Section 6).

YTTRIUM BROMATE

General Information

Description: Hexagonal, prismatic crystals.

Formula: $Y(BrO_3)_3 \cdot 9H_2O$.

Constants: Mol wt: 634.8, mp: 74°C, $-6H_2O$ at 100°C.

Hazard Analysis and Countermeasures

See bromates and yttrium.

YTTRIUM BROMIDE

General Information

Description: Deliquescent crystals.

Formula: YBr_3.

Constant: Mol wt: 328.7.

Hazard Analysis and Countermeasures

See bromides and yttrium.

YTTRIUM FLUORIDE

General Information

Description: Gelatinous material.

Formula: $YF_3 \cdot \frac{1}{2}H_2O$.

Constant: Mol wt: 154.9.

Hazard Analysis and Countermeasures

See fluorides and yttrium.

YTTRIUM NITRATE

General Information

Description: Reddish-white prisms.

Formula: $Y(NO_3)_3 \cdot 6H_2O$.

Constants: Mol wt: 383.0, d: 2.682.

Hazard Analysis and Countermeasures

See nitrates and yttrium.

YTTRIUM OXIDE

General Information

Synonym: Yttria.

Description: White powder.

Formula: Y_2O_3.

Constants: Mol wt: 225.8, d: 4.84.

Hazard Analysis

Toxicity: Unknown. See yttrium.

YTTRIUM SULFIDE

General Information

Description: Yellow-gray powder.

Formula: Y_2S_3.

Constant: Mol wt: 274.04.

Hazard Analysis and Countermeasures

See sulfides and yttrium.

Z

ZEOLITE
General Information
Description: A hydrated alkali aluminum silicate, capable of exchanging alkali for calcium and magnesium; used for softening water. There are a number of artificial zeolites now on the market.

Formula: $Na_2O \cdot Al_2O_3(SiO_2)_4 \cdot (H_2O)_x$.

Hazard Analysis
Toxic Hazard Rating:
  Acute Local: Inhalation 1.
  Acute Systemic: 0.
  Chronic Local: Inhalation 1.
  Chronic Systemic: 0.

Toxicity: Can act as a nuisance dust.

ZEPHIRAN CHLORIDE. See benzalkonium chloride.

ZINC
General Information
Description: Hexagonal, crystalline bluish-white metal.

Formula: Zn.

Constants: At wt: 65.38, mp: 419.4°C, bp: 907°C, d: 7.14, vap. press.: 1 mm at 487°C.

Hazard Analysis
Toxicity: See zinc compounds.

Radiation Hazard: Section 5. For permissible levels, see Table 5, p. 150.

    Artificial isotope ^{65}Zn, half life 245 d. Decays to stable ^{65}Cu by electron capture. Emits gamma rays of 1.12 MeV and X-rays.

    Artificial isotope ^{69m}Zn, half life 14 h. Decays to radioactive ^{69}Zn by emitting gamma rays of 0.44 MeV.

    Artificial isotope ^{69}Zn, half life 55 m. Decays to stable ^{69}Ga by emitting beta particles of 0.90 MeV.

Fire Hazard: Moderate in the form of dust when exposed to heat or flame.

Spontaneous Heating: No.

Explosion Hazard: Slight, in the form of dust when exposed to flame. See also powdered metals.

Countermeasures
To Fight Fire: Special mixtures of dry chemical (Section 6).

ZINC ACETATE
General Information
Description: Monoclinic crystals.

Formula: $Zn(C_2H_3O_2)_2$.

Constants: Mol wt: 183.47, mp: 242°C, bp: sublimes in vacuum, d: 1.84.

Hazard Analysis and Countermeasures
See zinc compounds. A trace mineral added to animal feeds (Section 10).

ZINC ALKYL AMINE o-PHENYL PHENATE
Hazard Analysis
Toxicity: Unknown. See phenols and zinc compounds. A fungicide.

Disaster Hazard: Dangerous; when heated to decomposition it emits highly toxic fumes.

Countermeasures
Storage and Handling: Section 7.

ZINC AMIDE
General Information
Description: Amorphous, white powder.

Formula: $Zn(NH_2)_2$.

Constants: Mol wt: 97.43, mp: decomposes at 200°C, d: 2.13 at 25°C.

Hazard Analysis and Countermeasures
See zinc compounds.

ZINC AMMONIUM NITRITE
General Information
Description: Solid.

Formula: $ZnNH_4(NO_2)_3$.

Constant: Mol wt: 221.5.

Hazard Analysis
Toxicity: See nitrites and zinc compounds.

Fire Hazard: Moderate, by spontaneous chemical reaction. A powerful oxidizing agent.

Explosion Hazard: Unknown.

Disaster Hazard: See nitrites.

Countermeasures
Storage and Handling: Section 7.

Shipping Regulations: Section 11.
  I.C.C.: Oxidizing material; yellow label, 100 pounds.
  Coast Guard Classification: Oxidizing material; yellow label.

ZINC ARSENATE
General Information
Description: White, odorless powder.

Composition: Variable; approximately $5ZnO$, $2As_2O_5$, $4H_2O$.

Hazard Analysis and Countermeasures
See arsenic compounds.

Shipping Regulations: Section 11.
  I.C.C.: Poison B; poison label, 200 pounds.
  Coast Guard Classification: Poison B; poison label.
  IATA: Poison B, poison label, 25 kilograms (passenger), 95 kilograms (cargo).

ZINC o-ARSENATE
General Information
Description: Monoclinic crystals.

Formula: $Zn_3(AsO_4)_2 \cdot 8H_2O$.

Constants: Mol wt: 618.09, mp: decomposes at 100°C, d: 3.309 at 15°C.

Hazard Analysis and Countermeasures
See arsenic compounds.

ZINC ARSENIDE
General Information
Description: Cubic crystals.

Formula: Zn_3As_2.
Constant: Mol wt: 345.96.
Hazard Analysis
Toxicity: Highly toxic. See arsenic compounds.
Fire Hazard: Moderate; it can evolve arsine upon contact with moisture or acids. See also arsine.
Explosion Hazard: Moderate; it can evolve arsine upon contact with moisture or acid. See also arsine.
Disaster Hazard: Dangerous; when heated to decomposition or on contact with acid or acid fumes, it emits highly toxic fumes of arsenic.
Countermeasures
Ventilation Control: Section 2.
Storage and Handling: Section 7.

ZINC ARSENITE. See zinc m-arsenite.

ZINC m-ARSENITE
General Information
Synonyms: ZMA; zinc arsenite.
Description: White powder.
Formula: $Zn(AsO_2)_2$.
Constant: Mol wt: 279.2.
Hazard Analysis
Toxicity: Highly toxic. A wood preservative, insecticide. See arsenic compounds.
Disaster Hazard: See arsenic compounds.
Countermeasures
Storage and Handling: Section 7.
Shipping Regulations: Section 11.
  I.C.C.: Poison B; poison label, 200 pounds.
  Coast Guard Classification: Poison B; poison label.
  MCA warning label.
  IATA: Poison B, poison label, 25 kilograms (passenger), 95 kilograms (cargo).

ZINC ASHES
Shipping Regulations: Section 11.
  IATA: Flammable solid, yellow label, 12 kilograms (passenger), 45 kilograms (cargo).

ZINC BACITRACIN
General Information
Description: Creamy white powder; slightly soluble in water.
Hazard Analysis
Toxic Hazard Rating: U. An additive permitted in food for human consumption. Also permitted in the feed and drinking water of animals and/or for the treatment of food producing animals (Section 10).

ZINC BENZOATE
General Information
Description: White powder; soluble in water (40 parts).
Formula: $Zn(C_7H_5O_2)_2$.
Constant: Mol wt: 307.60.
Hazard Analysis
Toxic Hazard Rating:
  Acute Local: Inhalation 1.
  Acute Systemic: U.
  Chronic Local: Inhalation 1.
  Chronic Systemic: U.
Toxicity: A mild respiratory irritant. See also zinc compounds.

ZINC BORATE
General Information
Description: White, amorphous powder or triclinic crystals.
Formula: $3ZnO \cdot 2B_2O_3$.
Constants: Mol wt: 383.42; mp: 980°C; d: (amor.) 3.64, (crystal) 4.22.
Hazard Analysis and Countermeasures
See boron and zinc compounds.

ZINC BROMATE
General Information
Description: White, deliquescent powder.

Formula: $Zn(BrO_3)_2 \cdot 6H_2O$.
Constants: Mol wt: 429.3, mp: 100°C, bp: $-6H_2O$ at 200°C, d: 2.566.
Hazard Analysis and Countermeasures
See bromates and zinc compounds.
Shipping Regulations: Section 11.
  IATA: Oxidizing material, yellow label, 12 kilograms (passenger), 45 kilograms (cargo).

ZINC BROMIDE
General Information
Description: Rhombic, colorless, hygroscopic crystals.
Formula: $ZnBr_2$.
Constants: Mol wt: 225.2, mp: 394°C, bp: 650°C, d: 4.219 at 4°C.
Hazard Analysis and Countermeasures
See bromides and zinc compounds.

ZINC CAPRYLATE
General Information
Description: Lustrous scales; slightly soluble in boiling water; moderately soluble in boiling alcohol.
Formula: $Zn(C_8H_{15}O_2)_2$.
Constants: Mol wt: 351.78, mp: 136°C.
Hazard Analysis
Toxicity: Unknown. On decomposition it releases irritating fumes of caprylic acid. See also zinc compounds.

ZINC CARBONATE
General Information
Description: White crystalline powder.
Formula: $ZnCO_3$.
Constants: Mol wt: 125.4, mp: $-CO_2$ at 300°C, d: 4.42 to 4.45.
Hazard Analysis
Toxicity: See zinc compounds. A trace mineral added to animal feeds (Section 10).

ZINC CHLORATE
General Information
Description: Colorless, very deliquescent crystals.
Formula: $Zn(ClO_3)_2 \cdot 4H_2O$.
Constants: Mol wt: 304.36, mp: decomposes at 60°C, bp: decomposes, d: 2.15.
Hazard Analysis and Countermeasures
See chlorates and zinc compounds.
Shipping Regulations: Section 11.
  I.C.C.: Oxidizing material; yellow label, 100 pounds.
  Coast Guard Classification: Oxidizing material; yellow label.
  IATA: Oxidizing material, yellow label, 12 kilograms (passenger), 45 kilograms (cargo).

ZINC CHLORIDE
General Information
Synonym: Butter of zinc.
Description: Cubic, white, deliquescent crystals.
Formula: $ZnCl_2$.
Constants: Mol wt: 136.30, mp: 262°C, bp: 732°C, d: 2.91 at 25°C, vap. press.: 1 mm at 428°C.
Hazard Analysis and Countermeasures
See zinc compounds and chlorides. Used as a trace mineral added to animal feeds. Also as a nutrient and/or dietary supplement food additive. It is a substance that migrates to food from packaging materials (Section 10).
TLV: ACGIH (tentative); 1 milligram per cubic meter of air.
Shipping Regulations: Section 11.
  MCA warning label.
  IATA (solid): Other restricted Articles, Class B, no label required, 12 kilograms (passenger), 45 kilograms (cargo).
      (solution): Corrosive liquid, white label, 1 liter (passenger and cargo).

ZINC CHLORIDE, CHROMATED
General Information
Description: Solid.
Hazard Analysis and Countermeasures
See chromium compounds and chlorides and zinc compounds.

ZINC 5-CHLORO-2-MERCAPTOBENZOTHIAZOLE
Hazard Analysis and Countermeasures
See zinc compounds and chlorides.

ZINC CHROMATE
General Information
Description: Lemon-yellow prisms.
Formula: $ZnCrO_4$.
Constant: Mol wt: 181.4.
Hazard Analysis and Countermeasures
Toxicity: See chromium compounds and zinc compounds.

ZINC COMPOUNDS
Hazard Analysis
Toxicity: Variable, generally of low toxicity.
Toxicology: Zinc is not inherently a toxic element. However, when heated, it evolves a fume of zinc oxide which, when inhaled fresh, can cause a disease known as "brass founders' ague," or "brass chills." It is possible for poeple to become immune to it. However, this immunity can be broken by cessation of exposure of only a few days. Zinc oxide dust which is not freshly formed is virtually innocuous. There is no cumulative effect to the inhalation of zinc fumes. Fatalities, however, have resulted from lung damage caused by the inhalation of high concentrations of zinc chloride fumes. Soluble salts of zinc have a harsh metallic taste; small doses can cause nausea and vomiting, while larger doses cause violent vomiting and purging. So far as can be determined, the continued administration of zinc salts in small doses has no effect in man except those of disordered digestion and constipation. Exposure to zinc chloride fumes can cause damage to the mucous membrane of the nasopharynx and respiratory tract and give rise to a pale gray cyanosis. Workers in zinc refining have been reported as suffering from a variety of non-specific intestinal, respiratory and nervous symptoms. Ulceration of the nasal septum and eczematous dermatosis are also reported.

It has been stated that zinc oxide dust can block the ducts of the sebaceous glands and give rise to a papular, pustular eczema in men engaged in packing this compound into barrels. Sensitivity to zinc oxide in man is extremely rare. Zinc chloride, because of its caustic action, can cause ulceration of the fingers, hands and forearms of those who use it as a flux in soldering. This condition has even been observed in men who handle railway ties which have been impregnated with this material. It is the opinion of some who work with it that it is carcinogenic.
A common air contaminant (Section 4).
Countermeasures
Treatment and Antidotes: Personnel exposed to zinc chloride fumes should immediately wash the area of contact with copious quantities of warm water and soap. Remove all contaminated clothing at once and if the area of contact is large, subject patient to a deluge-type of shower as quickly as possible. If the eyes are involved in exposure to zinc chloride fumes, they should be irrigated for at least 15 minutes with warm water.

Ventilation Control: Section 2.
Personal Hygiene: Section 3.
Storage and Handling: Section 7.

ZINC CYANIDE
General Information
Description: Rhombic, colorless crystals.
Formula: $Zn(CN)_2$.
Constants: Mol wt: 117.4, mp: decomposes at 800°C.
Hazard Analysis and Countermeasures
See cyanides and zinc compounds.
Shipping Regulations: Section 11.
    I.C.C. Classification: See Section 11, § 73.370.
    Coast Guard Classification: Poison B; poison label.
    IATA: Poison B, poison label, 12 kilograms (passenger), no limit (cargo).

ZINC DIBUTYLDITHIOCARBAMATE
General Information
Description: White powder.
Formula: $Zn[SC(S)N(C_4H_9)_2]_2$.
Constants: Mol wt: 474.2, mp: 104–108°C, d: 1.24 at 20°/20°C.
Hazard Analysis
Toxicity: Details unknown. See zinc ethylene bis(dithiocarbamate) and bis(diethylthiocarbamyl) disulfide.
Disaster Hazard: Dangerous; see sulfides.
Countermeasures
Storage and Handling: Section 7.

ZINC DICHROMATE
General Information
Description: Orange-yellow powder; reddish-brown, hygroscopic crystals.
Formula: $ZnCr_2O_7 \cdot 3H_2O$.
Constant: Mol wt: 335.45.
Hazard Analysis and Countermeasures
See chromium compounds and zinc compounds.

ZINC DIETHYL
General Information
Synonym: Zinc ethyl.
Description: Liquid.
Formula: $Zn(C_2H_5)_2$.
Constants: Mol wt: 123.5, mp: −28°C, bp: 118°C, d: 1.2065 at 20°/4°C.
Hazard Analysis
Toxicity: See zinc compounds.
Fire Hazard: Dangerous, by spontaneous chemical reaction. Spontaneously flammable in air.
Explosion Hazard: Unknown.
Disaster Hazard: Dangerous fire hazard; can react vigorously with oxidizing materials.
Countermeasures
Storage and Handling: Section 7.
To Fight Fire: Do   use water, foam or halogenated extinguishing agents.
Shipping Regulations: Section 11.
    ICC: Flammable liquid; red label (not accepted).
    Coast Guard Classification: Inflammable liquid; red label.

ZINC DIETHYLDITHIOCARBAMATE
General Information
Description: White powder.
Formula: $Zn[SC(S)N(C_2H_5)_2]_2$.
Constants: Mol wt: 361.9, d: 1.47 at 20°/20°C.

TOXIC HAZARD RATING CODE (For detailed discussion, see Section 1.)

0 NONE: (a) No harm under any conditions; (b) Harmful only under unusual conditions or overwhelming dosage.

1 SLIGHT: Causes readily reversible changes which disappear after end of exposure.

2 MODERATE: May involve both irreversible and reversible changes; not severe enough to cause death or permanent injury.

3 HIGH: May cause death or permanent injury after very short exposure to small quantities.

U UNKNOWN: No information on humans considered valid by authors.

Toxicity: Details unknown. See zinc ethylene bis(dithio-carbamate) and bis(diethylthiocarbamyl) disulfide.

Toxicology: This material is very irritating to the eyes, nose and throat. Several hours after the material may have gotten into the eyes, it causes an unbearable pain. A seed disinfectant, fungicide, rubber accelerator.

Disaster Hazard: Dangerous; see sulfides and cyanides.

Countermeasures

Storage and Handling: Section 7.

ZINC DIMETHYL
General Information
Description: Mobile liquid.
Formula: $Zn(CH_3)_2$.
Constants: Mol wt: 95.5, mp: $-40°C$, bp: $46°C$, d: 1.386 at $10.5°/4°C$.
Hazard Analysis
Toxicity: See zinc compounds.
Fire Hazard: Dangerous, by chemical reaction. Spontaneously flammable in air (Section 6).
Explosion Hazard: Unknown.
Disaster Hazard: Dangerous; when heated to decomposition it burns and emits toxic fumes of zinc compounds; can react with oxidizing materials.
Countermeasures
Storage and Handling: Section 7.

ZINC DIMETHYLDITHIOCARBAMATE
General Information
Synonyms: Ziram.
Description: White powder.
Formula: $Zn[SC(S)N(CH_3)_2]_2$.
Constants: Mol wt: 306.0, mp: 248–250°C, d: 1.65 at $20°/20°C$.
Hazard Analysis
Toxicity: Details unknown. See zinc ethylene bis(dithio-carbamate) and bis(diethylthiocarbamyldisulfide).
Toxicology: Zinc dimethyldithiocarbamate is very irritating to the eyes, nose and throat. Several hours after this material gets into the eyes, the pain becomes unbearable. A seed disinfectant and fungicide.
Disaster Hazard: Dangerous; see sulfides and cyanides.
Countermeasures
Storage and Handling: Section 7.

ZINC DIMETHYLDITHIOCARBAMATE-CYCLO-HEXYLAMINE COMPLEX
General Information
Description: White powder.
Hazard Analysis
Toxicity: A fungicide. See individual components.
Fire Hazard: Moderate, when exposed to heat or flame (Section 6).
Explosion Hazard: Unknown.
Disaster Hazard: Dangerous; see sulfides and cyanides; can react with oxidizing materials.
Countermeasures
Storage and Handling: Section 7.

ZINC DITHIOCARBAMATE
Hazard Analysis
Toxicity: See zinc compounds.
Disaster Hazard: Dangerous; see sulfide and cyanides.
Countermeasures
Storage and Handling: Section 7.

ZINC ETHYL. See zinc diethyl.

ZINC ETHYLENEBIS(DITHIOCARBAMATE)
General Information
Synonym: Zineb.
Description: Light colored powder. Water insoluble.
Formula: $Zn(CS_2NHCH_2)_2$.
Constant: Mol wt: 275.8.
Hazard Analysis
Toxic Hazard Rating:

Acute Local: Irritant 2; Ingestion 2; Inhalation 2.
Acute Systemic: Ingestion 1.
Chronic Local: Irritant 2.
Chronic Systemic: U.
Toxicology: A fungicide. Irritating to skin and mucous membranes. Animal experiments suggest low toxicity. See also bis(diethylthiocarbamyl) disulfide.
Disaster Hazard: Dangerous. When heated to decomposition it emits highly toxic fumes of oxides of nitrogen and sulfur.

ZINC FERROCYANIDE
General Information
Description: White powder.
Formula: $Zn_2Fe(CN)_6$.
Constant: Mol wt: 342.72.
Hazard Analysis and Countermeasures
See zinc compounds and ferrocyanides.

ZINC FLUOGALLATE
General Information
Description: Colorless crystals.
Formula: $[Zn(H_2O)_6](GaF_5 5H_2O)$.
Constants: Mol wt: 356.2, mp: $-5H_2O$ at $110°C$, d: 2.33.
Hazard Analysis and Countermeasures
See fluorides and zinc compounds.

ZINC FLUORIDE
General Information
Description: White powder.
Formula: ZnF_2.
Constants: Mol wt: 103.38, mp: $872°C$, bp: $1497°C$, d: 4.84 at $15°C$, vap. press.: 1 mm at $970°C$.
Hazard Analysis and Countermeasures
See fluorides and zinc compounds.

ZINC FLUOSILICATE
General Information
Description: Hexagonal, colorless prisms.
Formula: $ZnSiF_6 \cdot 6H_2O$.
Constants: Mol wt: 315.5, d: 2.104.
Hazard Analysis and Countermeasures
See fluosilicates.

ZINC FORMALDEHYDE SULFOXYLATE
General Information
Description: Rhombic prisms.
Formula: $Zn(HSO_2 \cdot CH_2O)_2$.
Constants: Mol wt: 255.6, mp: decomposes.
Hazard Analysis
Toxicity: See zinc compounds and formaldehyde.
Disaster Hazard: Dangerous; see sulfonates.
Countermeasures
Storage and Handling: Section 7.

ZINC FORMATE
General Information
Description: White crystals.
Formula: $Zn(CHO_2)_2$.
Constants: Mol wt: 155.4, bp: decomposes, d: 2.36.
Hazard Analysis and Countermeasures
See zinc compounds and formic acid.

ZINC GALLATE
General Information
Description: White crystals.
Formula: $ZnGa_2O_4$.
Constants: Mol wt: 268.8, mp: $< 800°C$, d: 6.15 (theor.).
Hazard Analysis and Countermeasures
See zinc compounds and gallium compounds.

ZINC GLUCONATE
Hazard Analysis
Toxic Hazard Rating: U. A dietary supplement food additive (Section 10).

ZINC HYDROSULFITE
General Information
Synonym: Zinc dithionite.
Description: White amorphous solid; soluble in water.
Formula: ZnS_2O_4.
Constant: Mol wt: 193.
Hazard Analysis
Toxic Hazard Rating: U. A substance which migrates to food from packaging materials (Section 10).
Disaster Hazard: Dangerous. See sulfites.

ZINC HYDROXIDE
General Information
Description: White powder.
Formula: $Zn(OH)_2$.
Constants: Mol wt: 99.4, mp: 125°C (decomposes), d: 3.053.
Hazard Analysis and Countermeasures
See zinc compounds.

ZINC HYPOPHOSPHITE
General Information
Description: Colorless, hygroscopic crystalline powder.
Formula: $Zn(H_2PO_2)_2 \cdot H_2O$.
Constant: Mol wt: 213.4.
Hazard Analysis and Countermeasures
See zinc compounds and hypophosphites.

ZINC IODATE
General Information
Description: White crystalline powder.
Formula: $Zn(IO_3)_2$.
Constants: Mol wt: 415.2, mp: decomposes, d: 4.98.
Hazard Analysis and Countermeasures
See iodates and zinc compounds.

ZINC IODIDE
General Information
Description: Cubic, colorless or white, deliquescent powder.
Formula: ZnI_2.
Constants: Mol wt: 319.22, mp: 446°C, bp: 624°C, d: 4.666 at 14.2°C.
Hazard Analysis and Countermeasures
See iodides and zinc compounds.

ZINCITE. See zinc oxide.

ZINC LAURATE
General Information
Description: White powder.
Formula: $Zn(C_{12}H_{23}O_2)_2$.
Constants: Mol wt: 464.0, mp: 128°C.
Hazard Analysis
Toxicity: See zinc compounds.
Fire Hazard: Slight, when exposed to heat or flame; can react with oxidizing materials (Section 6).
Countermeasures
Storage and Handling: Section 7.

ZINC MERCAPTOBENZIMIDAZOLATE
Hazard Analysis
Toxicity: Unknown. Animal experiments suggest low acute toxicity.
Disaster Hazard: Dangerous. See mercaptans.
Countermeasures
Storage and Handling: Section 7.

ZINC MERCAPTOBENZOTHIAZOLE
Hazard Analysis
Toxicity: Details unknown; a fungicide. See zinc compounds.

Disaster Hazard: Dangerous; when heated to decomposition or on contact with acid or acid fumes, it emits highly toxic fumes of oxides of sulfur.
Countermeasures
Storage and Handling: Section 7.

ZINC MERCURY CHROMATE
Hazard Analysis and Countermeasures
See chromium compounds and mercury compounds.

ZINC NAPHTHENATE
General Information
Description: A solid.
Formula: $Zn(C_6H_5COO)_2$.
Constant: Mol wt: 319.7.
Hazard Analysis
Toxicity: A fungicide and mildew preventive. See zinc compounds.
Fire Hazard: Slight, when exposed to heat or flame (Section 6).
Countermeasures
Storage and Handling: Section 7.

ZINC NITRATE
General Information
Description: A: Needles; B: tetragonal, colorless crystals.
Formula: A: $Zn(NO_3)_2 \cdot 3H_2O$; B: $Zn(NO_3)_2 \cdot 6H_2O$.
Constants: Mol wt: A: 243.33; B: 297.49, d: B: 2.065 at 14°C, mp:A: 45.5°C; B: 36.4°C, bp: B: $-6H_2O$ at 105–131°C.
Hazard Analysis and Countermeasures
See nitrates and zinc compounds.
Shipping Regulations: Section 11.
    I.C.C.: Oxidizing material; yellow label, 100 pounds.
    Coast Guard Classification: Oxidizing material.
    IATA: Oxidizing material, yellow label, 12 kilograms (passenger), 45 kilograms (cargo).

ZINC NITRODITHIOACETATE
Hazard Analysis
Toxicity: A fungicide. Details unknown. See zinc compounds and thioacetic acid.
Fire Hazard: Unknown. An oxidizer.
Disaster Hazard: Dangerous; when heated to decomposition it emits highly toxic fumes.
Countermeasures
Storage and Handling: Section 7.

ZINC OXALATE
General Information
Description: White powder.
Formula: $ZnC_2O_4 \cdot 2H_2O$.
Constants: Mol wt: 189.43, mp: sublimes 100°C, d: 2.562 at 24.5°C.
Hazard Analysis and Countermeasures
See oxalates and zinc compounds.

ZINC OXIDE
General Information
Synonyms: Zincite; Chinese white; zinc white; flowers of zinc.
Description: White or yellowish powder.
Formula: ZnO.
Constants: Mol wt: 81.38, mp: > 1800°C, d: 5.47.
Hazard Analysis
Toxicity: A seed disinfectant. See zinc compounds. A fungicide. A trace mineral added to animal feeds. Also a dietary supplement food additive (Section 10).

TOXIC HAZARD RATING CODE (For detailed discussion, see Section 1.)

0 NONE: (a) No harm under any conditions; (b) Harmful only under unusual conditions or overwhelming dosage.

1 SLIGHT: Causes readily reversible changes which disappear after end of exposure.

2 MODERATE: May involve both irreversible and reversible changes; not severe enough to cause death or permanent injury.

3 HIGH: May cause death or permanent injury after very short exposure to small quantities.

U UNKNOWN: No information on humans considered valid by authors.

TLV: ACGIH (recommended); (fume) 5 milligrams per cubic meter of air.

ZINC OXYSULFATE
General Information
Description: White powder.
Formula: $ZnO \cdot ZnSO_4$.
Constant: Mol wt: 242.8.
Hazard Analysis and Countermeasures
See zinc compounds and sulfates. A fungicide.

ZINC PERMANGANATE
General Information
Description: Violet-brown or black, hygroscopic crystals.
Formula: $Zn(MnO_4)_2 \cdot 6H_2O$.
Constants: Mol wt: 411.34, mp: $-5H_2O$ at 100°C, d: 2.47.
Hazard Analysis
Toxicity: See manganese compounds and zinc compounds.
Fire Hazard: Moderate, by chemical reaction with reducing agents. A powerful oxidizing agent (Section 6).
Countermeasures
Storage and Handling: Section 7.
Shipping Regulations: Section 11.
   I.C.C.: Oxidizing material; yellow label, 100 pounds.
   Coast Guard Classification: Oxidizing material; yellow label.
   IATA: Oxidizing material, yellow label, 12 kilograms (passenger), 45 kilograms (cargo).

ZINC PEROXIDE
General Information
Description: Yellow-white powder.
Formula: ZnO_2 (theoretical).
Constants: Mol wt: 97.38, d: 1.571 (theoretical).
Hazard Analysis
Toxicity: Systemic toxicity is similar to zinc oxide. See peroxides and zinc compounds.
Fire Hazard: Moderate, when exposed to heat or by chemical reaction with reducing materials. Finely divided powder is slightly soluble in water, decomposes rapidly at 150°C. It is not dangerous unless mixed with highly combustible materials (Section 6).
Explosion Hazard: Dangerous, when exposed to heat.
Explosive Range: At 212°C for peroxide prepared from $ZnSO_4$, NH_3, H_2O_2; $4ZnO \cdot H_2O \cdot 3H_2O_2$ explodes at 190°C.
Disaster Hazard: Very dangerous; it explodes when heated; it will react with water or steam to produce heat; on contact with reducing material, it can react vigorously.
Countermeasures
Storage and Handling: Section 7.
Shipping Regulations: Section 11.
   I.C.C.: Oxidizing material; yellow label, 100 pounds.
   Coast Guard Classification: Oxidizing material; yellow label.
   IATA: Oxidizing material, yellow label, 12 kilograms (passenger), 45 kilograms (cargo).

ZINC PHOSPHIDE
General Information
Description: Cubic, dark gray crystals or powder.
Formula: Zn_3P_2.
Constants: Mol wt: 258.10, mp: > 420°C, bp: 1100°C, d: 4.55 at 13°C.
Hazard Analysis and Countermeasures
See phosphides and zinc compounds.
Shipping Regulations: Section 11.
   MCA warning label.
   IATA: Poison B, poison label, 25 kilograms (passenger), 95 kilograms (cargo).

ZINC PICRATE
General Information
Description: Yellow, crystalline powder.
Formula: $Zn(C_6H_2N_3O_7)_2 \cdot 8H_2O$.

Constants: Mol wt: 665.7, mp: explodes.
Hazard Analysis
Toxicity: See picric acid and zinc compounds.
Fire Hazard: Dangerous. See nitrates.
Explosion Hazard: Severe. See explosives, high and nitrates.
Disaster Hazard: Dangerous. See nitrates and explosives, high.
Countermeasures
Storage and Handling: Section 7.

ZINC POWDER. See zinc.
Shipping Regulations: Section 11.
   IATA: Flammable solid, yellow label, 12 kilograms (passenger), 45 kilograms (cargo).

ZINC RICINOLEATE
General Information
Description: Fine, white powder.
Formula: $Zn[CO_2(CH_2)_7CH:CHCH_2CHOH(CH_2)_5 \cdot CH_3]_2$.
Constants: Mol wt: 660.4, mp: 92°C, d: 1.10 at 25°/25°C.
Hazard Analysis
Toxicity: See zinc compounds.
Disaster Hazard: Slight; when heated it emits acrid fumes.
Countermeasures
Storage and Handling: Section 7.

ZINC SELENATE
General Information
Description: Triclinic crystals.
Formula: $ZnSeO_4 \cdot 5H_2O$.
Constants: Mol wt: 298.42, mp: decomposes > 50°C, d: 2.591.
Hazard Analysis and Countermeasures
See selenium compounds.

ZINC SELENIDE
General Information
Description: Cubic crystals.
Formula: $ZnSe$.
Constants: Mol wt: 144.34, d: 5.42 at 15°C.
Hazard Analysis
Toxicity: See selenium compounds.
Fire Hazard: Moderate, on contact with moisture or acids it can evolve selenium hydride. See hydrides.
Explosion Hazard: Moderate; it can evolve a hydride. See hydrides.
Disaster Hazard: See selenium.
Countermeasures
Storage and Handling: Section 7.

ZINC SILICOFLUORIDE. See zinc fluosilicate.

ZINC STEARATE
General Information
Description: White powder.
Formula: $Zn(C_{18}H_{35}O_2)_2$.
Constants: Mol wt: 632.30, mp: 130°C, flash p: 530°F (O.C.), autoign. temp.: 790°F.
Hazard Analysis
Toxicity: Inhalation of zinc stearate has been reported as causing pulmonary fibrosis. See zinc compounds. A nutrient and/or dietary supplement food additive (Section 10).
Fire Hazard: Slight, when exposed to heat or flame.
Countermeasures
To Fight Fire: Water, foam, carbon dioxide, dry chemical or carbon tetrachloride (Section 6).
Storage and Handling: Section 7.

ZINC SULFATE
General Information
Synonym: Zinkosite.
Description: Rhombic, colorless crystals.
Formula: $ZnSO_4$.

Constants: Mol wt: 161.44, mp: decomposes at 740°C, d: 3.74 at 15°C.
Hazard Analysis and Countermeasures
See zinc compounds and sulfates. A fungicide. A trace mineral added to animal feeds; a nutrient and/or dietary supplement food additive (Section 10).

ZINC SULFIDE, (α)
General Information
Synonym: Wurtzite.
Description: Hexagonal, colorless crystals.
Formula: ZnS.
Constants: Mol wt: 97.45, mp: 1850°C at 150 atm., bp: sublimes at 1185°C, d: 4.087.
Hazard Analysis and Countermeasures
See sulfides and zinc compounds. A fungicide.

ZINC SULFITE
General Information
Description: White, crystalline powder.
Formula: $ZnSO_3 \cdot 2H_2O$.
Constant: Mol wt: 181.48.
Hazard Analysis and Countermeasures
See sulfites and zinc compounds.

ZIRCONIUM TARTRATE
General Information
Description: White powder; slightly soluble in water.
Formula: $ZnC_4H_4O_6 \cdot H_2O$.
Constants: Mol wt: 231.47.
Hazard Analysis
Toxic Hazard Rating:
  Acute Local: Inhalation 1.
  Acute Systemic: U.
  Chronic Local: Inhalation 1.
  Chronic Systemic: U.
Toxicity: See zinc compounds and tartaric acid.

ZINC TELLURATE
General Information
Description: Heavy, granular, white precipitate.
Formula: Zn_3TeO_6.
Constant: Mol wt: 419.75.
Hazard Analysis and Countermeasures
See tellurium compounds and zinc compounds.

ZINC TELLURIDE
General Information
Description: Cubic, red crystals.
Formula: ZnTe.
Constants: Mol wt: 192.99, mp: 1238.5°C, d: 6.34 at 15°C.
Hazard Analysis
Toxicity: See tellurium compounds.
Fire Hazard: Moderate; upon contact with moisture or acids it evolves tellurium hydride (Section 6).
Explosion Hazard: Slight; it can easily evolve the hydride of tellurium. See hydrides.
Disaster Hazard: Dangerous. See tellurium compounds.
Countermeasures
Storage and Handling: Section 7.

ZINC THIOCYANATE
General Information
Description: White powder.
Formula: $Zn(SCN)_2$.
Constant: Mol wt: 181.55.
Hazard Analysis and Countermeasures
See thiocyanates and zinc compounds.

ZINC 2,4,5-TRICHLOROPHENATE
General Information
Description: Colorless crystals.
Formula: $Zn(OC_6H_2Cl_3)_2$.
Constant: Mol wt: 458.3.
Hazard Analysis
Toxicity: A fungicide and seed protectant. See chlorinated phenols.
Disaster Hazard: Dangerous; see chlorides.
Countermeasures
Ventilation Control: Section 2.
Storage and Handling: Section 7.

ZINC WHITE. See zinc oxide.

ZINEB. See zinc ethylenebis(dithiocarbamate).

ZINGIBER. See ginger.

ZINKOSITE. See zinc sulfate.

ZINO PHENOS
General Information
Synonym: o,o-Diethyl o-(2-pyrazinyl) phosphorothioate.
Hazard Analysis
Toxic Hazard Rating:
  Acute Local: U.
  Acute Systemic: Ingestion 3; Inhalation 3; Skin Absorption 3.
  Chronic Local: U.
  Chronic Systemic: Ingestion 2; Inhalation 2; Skin Absorption 2.
Toxicity: A choleresterase inhibitor. See also parathion.
Disaster Hazard: Dangerous; when heated to decomposition it emits highly toxic fumes.

ZIRAM. See zinc dimethyldithiocarbamate.

ZIRCONIUM
General Information
Description: Hard, lustrous, grayish, crystalline scales or gray powder. An element.
Formula: Zr.
Constants: At wt: 91.22, mp: 1900°C, bp: > 2900°C, d: 6.5, autoign. temp.: 500°F.
Hazard Analysis
Toxicity: See zirconium compounds.
TLV: ACGIH (recommended); 5 milligrams per cubic meter of air.
Radiation Hazard: Section 5. For permissible levels, see Table 5, p. 150.
  Artificial isotope ^{93}Zr, half life 9.5×10^5y. Decays to stable ^{93}Nb by emitting beta particles of 0.03 (25%), 0.06 (75%) MeV. Also emits gamma rays of 0.03 MeV and X-rays.
  Artificial isotope ^{95}Zr, half life 65 d. Decays to radioactive ^{95}Nb by emitting beta particles of 0.36 (43%), 0.40 (55%) MeV. Also emits gamma rays of 0.73, 0.76 MeV.
  Artificial isotope ^{97}Zr, half life 17 h. Decays to radioactive ^{97}Nb by emitting beta particles of 1.91 MeV. Also emits gamma rays of 0.75 MeV.
Fire Hazard: Dangerous, in the form of dust when exposed to heat or flame or by chemical reaction with oxidizers.
Spontaneous Heating: No.
Explosion Hazard: Dangerous in the form of dust by chemical reaction with air, oxidizing agents.
Explosive Range: 0.16 g/l.

TOXIC HAZARD RATING CODE (For detailed discussion, see Section 1.)

0 NONE: (a) No harm under any conditions; (b) Harmful only under unusual conditions or overwhelming dosage.

1 SLIGHT: Causes readily reversible changes which disappear after end of exposure.

2 MODERATE: May involve both irreversible and reversible changes; not severe enough to cause death or permanent injury.

3 HIGH: May cause death or permanent injury after very short exposure to small quantities.

U UNKNOWN: No information on humans considered valid by authors.

Countermeasures
Storage and Handling: Section 7.
To Fight Fire: Special mixtures, dry chemical, salt or dry sand (Section 6).

ZIRCONIUM ACETATE
General Information
Formula: $H_2ZrO_2(C_2H_3O_2)_2$.
Constant: Mol wt: 243.
Hazard Analysis
See zirconium compounds.

ZIRCONIUM AMMONIUM FLUORIDE
General Information
Description: A: Rhombic, white crystals; B: colorless, cubic crystals.
Formula: A: $(NH_4)_2ZrF_6$; B: $(NH_4)_3ZrF_7$.
Constants: Mol wt: A: 243.0; B: 278.34, d: 1.154.
Hazard Analysis and Countermeasures
See fluorides.

ZIRCONIUM CARBIDE
General Information
Description: Hard gray, metallic crystals.
Formula: ZrC.
Constants: Mol wt: 103.23, mp: 3540°C, bp: 5100°C, d: 6.73.
Hazard Analysis and Countermeasures
See carbides and zirconium compounds.

ZIRCONIUM COMPOUNDS
Hazard Analysis
Toxicity: Zirconium is not an important poison, and so far as is known, the inherent toxicity of zirconium compounds is low. Deaths in rabbits have been caused by it through intravenous injection of relatively large doses of the order of 150 mg/kg of body weight. Most of the zirconium compounds in common use are insoluble and considered to be inert. Pulmonary granuloma in zirconium workers has been reported and sodium zirconium lactate has been held responsible for skin granulomas.

ZIRCONIUM DIBROMIDE
General Information
Description: Black powder.
Formula: $ZrBr_2$.
Constant: Mol wt: 251.05, mp: decomposes > 350°C.
Hazard Analysis
Toxicity: See bromides.
Fire Hazard: Dangerous by spontaneous chemical reaction. Ignites spontaneously in air (Section 6).
Explosion Hazard: Unknown.
Disaster Hazard: Dangerous; see bromides; can react vigorously with oxidizing materials.
Countermeasures
Storage and Handling: Section 7.

ZIRCONIUM DICHLORIDE
General Information
Description: Black crystals.
Formula: $ZrCl_2$.
Constants: Mol wt: 162.13, mp: decomposes > 350°C.
Hazard Analysis
Toxicity: See hydrochloric acid.
Disaster Hazard: Dangerous; see chlorides.
Countermeasures
Storage and Handling: Section 7.

ZIRCONIUM DIOXIDE
General Information
Synonym: Baddeleyite.
Description: Colorless, yellow or brown, monoclinic crystals.
Formula: ZrO_2.

Constants: Mol wt: 123.22, mp: 2700°C, bp: 4300°C, d: 5.49.
Hazard Analysis
Toxicity: See zirconium compounds.

ZIRCONIUM FLUORIDE. See zirconium tetrafluoride.

ZIRCONIUM HYDRIDE
General Information
Description: Metallic dark-gray to black powder.
Formula: ZrH_2.
Constants: Mol wt: 93.23, d: 5.6, autoign. temp.: 270°C in air.
Hazard Analysis and Countermeasures
See hydrides and zirconium compounds.
Shipping Regulations: Section 11.
  IATA: Flammable solid, yellow label, not acceptable (passenger), 12 kilograms (cargo).

ZIRCONIUM IODIDE
Hazard Analysis and Countermeasures
See zirconium compounds and iodides.

ZIRCONIUM LACTATE
General Information
Description: White, slightly moist pulp; very slightly soluble in water and the common organic solvents; soluble in aqueous alkali solutions with the formation of salts.
Formula: $H_4ZrO(CH_3CHOCOO)_3$.
Constant: Mol wt: 375.
Hazard Analysis
Toxicity: Details unknown. Animal experiments show no toxicity by oral routes. Prolonged inhalation of dust caused interstitial pneumonia. See also zirconium compounds.

ZIRCONIUM, METALLIC DRY. See zirconium.
Shipping Regulations: Section 11.
  I.C.C.: Flammable solid; yellow label, 75 pounds.
  Coast Guard Classification: Inflammable solid; yellow label; (metallic solutions or mixtures thereof, liquid) inflammable liquid; red label.

ZIRCONIUM METAL (Zr METAL) DRY OR WET WITH LESS THAN 25% WATER, MECHANICALLY PRODUCED PARTICLE SIZE NOT EXCEEDING 53 MICRONS (270 MESH)
  IATA: Flammable solid, yellow label, not acceptable (passenger), 35 kilograms (cargo).

ZIRCONIUM METAL, WET WITH NOT LESS THAN 25% WATER, CHEMICALLY PRODUCED, PARTICLE SIZE NOT EXCEEDING 840 MICRONS (20 MESH).
  IATA: Flammable solid, yellow label, not acceptable (passenger), 70 kilograms (cargo).

ZIRCONIUM METAL, WET WITH NOT LESS THAN 25% WATER, CHEMICALLY PRODUCED, PARTICLE SIZE NOT EXCEEDING 53 MICRONS (270 MESH).
  IATA: Flammable solid, yellow label, not acceptable (passenger), 70 kilograms (cargo).

ZIRCONIUM, METALLIC DRY (OTHER THAN SCRAP) RIBBON, SCRAP OR WIRE LESS THAN 0.25 MILLIMETERS (0.010 INCH) THICK.
  IATA: Flammable solid, yellow label, 12 kilograms (passenger), 45 kilograms (cargo).

ZIRCONIUM, METALLIC, SOLUTIONS OR MIXTURES THEREOF, LIQUID
Shipping Regulations: Section 11.
  IATA: Flammable liquid, red label, not acceptable (passenger), 2½ kilograms (cargo).
  I.C.C.: Flammable liquid, red label, 5 pounds.

ZIRCONIUM METAL WET, CHEMICALLY PRODUCED, FINER THAN 20 MESH PARTICLE SIZE
Shipping Regulations: Section 11.
  I.C.C.: Flammable solid, yellow label, 150 pounds.

ZIRCONIUM NITRATE
General Information
Description: White crystals.
Formula: $Zr(NO_3)_4 \cdot 5H_2O$.
Constants: Mol wt: 429.33, mp: decomposes at 100°C.
Hazard Analysis and Countermeasures
See nitrates.
Shipping Regulations: Section 11.
  Coast Guard Classification: Oxidizing material.
  IATA: Oxidizing material, yellow label, 12 kilograms (passenger), 45 kilograms (cargo).

ZIRCONIUM NITRIDE
General Information
Description: Brassy colored powder. Refractory.
Formula: ZrN.
Constants: Mol wt: 105.22, mp: 2930–2980°C, d: 7.09.
Hazard Analysis
Toxicity: Details unknown; probably low.

ZIRCONIUM PHOSPHIDE
General Information
Description: Gray crystals.
Formula: ZrP_2.
Constants: Mol wt: 153.2, d: 4.77 at 25°/4°C.
Hazard Analysis and Countermeasures
See phosphides.

ZIRCONIUM PICRAMATE, WET WITH 20% OF WATER
Hazard Analysis and Countermeasures
See nitrates.
Shipping Regulations: Section 11.
  I.C.C.: Oxidizing material; yellow label, 25 pounds.
  Coast Guard Classification: Oxidizing material; yellow label.

ZIRCONIUM SCRAP (BORINGS, CLIPPINGS, SHAVINGS, SHEETS OR TURNINGS).
Shipping Regulations: Section 11.
  IATA: Flammable solid, not acceptable (passenger and cargo).
  I.C.C.: Flammable solid, yellow label, 100 pounds.

ZIRCONIUM SELENATE
General Information
Description: Hexagonal, transparent crystals.
Formula: $Zr(SeO_4)_2 \cdot 4H_2O$.
Constants: Mol wt: 449.20, mp: $-3H_2O$ at 100°C; $-4H_2O$ at 130°C.
Hazard Analysis and Countermeasures
See selenium compounds.

ZIRCONIUM SODIUM LACTATE
General Information
Synonym: Sodium zirconium lactate.
Description: Straw colored liquid.
Formula: $NaH_3ZrO(CH_3CHOCOO)_3$.
Constants: Mol wt 397, d: 1.28.
Hazard Analysis
Toxic Hazard Rating:
  Acute Local: Irritant 1.

Acute Systemic: U.
Chronic Local: Irritant 1.
Chronic Systemic: U.
Toxicity: Has caused skin granulomas in humans. Inhalation experiments on rabbits produced bronchiolar abcesses, lobulor pneumonia, and peribronchial granulomas.

ZIRCONIUM SULFATE
General Information
Description: White crystalline powder; insoluble in alcohol.
Formula: $Zr(SO_4)_2 \cdot 4H_2O$.
Constant: Mol wt: 355.42.
Hazard Analysis
See zirconium compounds.

ZIRCONIUM SULFIDE
General Information
Description: Steel-gray crystals.
Formula: ZrS_2.
Constants: Mol wt: 155.35, d: 3.87.
Hazard Analysis and Countermeasures
See sulfides.

ZIRCONIUM TETRACHLORIDE
General Information
Description: White, lustrous crystals.
Formula: $ZrCl_4$.
Constants: Mol wt: 233.05, mp: sublimes at 300°C, bp: 331°C, d: 2.80, vap. press.: 1 mm at 190°C.
Hazard Analysis and Countermeasures
See hydrochloric acid.
Shipping Regulations: Section 11.
  IATA: Other restricted articles, class B, no label required, 12 kilograms (passenger), 45 kilograms (cargo).

ZIRCONIUM TETRAFLUORIDE
General Information
Synonym: Zirconium fluoride.
Description: Refractive crystals. Water soluble.
Formula: ZrF_4.
Constants: Mol wt: 167.2, d: 4.6 at 16°C, sublimes < 600°C.
Hazard Analysis and Countermeasures
See fluorides.

ZIRCONIUM, WET OR SLUDGE. See zirconium.
Shipping Regulations: Section 11.
  I.C.C.: Flammable solid; yellow label.
  Coast Guard Classification: Inflammable solid; yellow label.
  IATA: Flammable solid, yellow label, not acceptable (passenger), 70 kilograms (cargo).

ZIRCONYL BROMIDE
General Information
Description: Brilliant, deliquescent crystals.
Formula: $ZrOBr_2 \cdot xH_2O$.
Constant: Mp: $-H_2O$ at 120°C.
Hazard Analysis and Countermeasures
See bromides.

ZIRCONYL SULFIDE
General Information
Description: Yellow powder.
Formula: ZrOS.
Constants: Mol wt: 139.29, d: 4.87.
Hazard Analysis and Countermeasures
See sulfides.

TOXIC HAZARD RATING CODE (For detailed discussion, see Section 1.)

0 NONE: (a) No harm under any conditions; (b) Harmful only under unusual conditions or overwhelming dosage.

1 SLIGHT: Causes readily reversible changes which disappear after end of exposure.

2 MODERATE: May involve both irreversible and reversible changes; not severe enough to cause death or permanent injury.

3 HIGH: May cause death or permanent injury after very short exposure to small quantities.

U UNKNOWN: No information on humans considered valid by authors.

ZMA. See zinc-m-arsenite.

ZOALENE
General Information
Synonym: 3,5-Dinitro-o-toluamide.
Description: Yellowish solid; very slightly soluble in water; soluble in acetone acetonitrile, and dimethyl formamide.

Formula: $(O_2N)_2C_6H_2(CH_3)CONH_2$.
Constants: Mol wt: 225, mp: 177° C.
Hazard Analysis
Toxic Hazard Rating: U. An additive permitted in the feed and drinking water of animals and/or for the treatment of food producing animals. Also permitted in food for human consumption. (Section 10).
Disaster Hazard: See nitrates.